Lipo Wang   Ke Chen   Yew Soon Ong (Eds.)

Advances in Natural Computation

First International Conference, ICNC 2005
Changsha, China, August 27-29, 2005
Proceedings, Part I

Volume Editors

Lipo Wang
Nanyang Technological University
School of Electrical and Electronic Engineering
Block S1, 50 Nanyang Avenue, Singapore 639798
E-mail: elpwang@ntu.edu.sg

Ke Chen
University of Manchester
School of Informatics
P.O. Box 88, Sackville St., Manchester M6O 1QD, UK
E-mail: k.chen@manchester.ac.uk

Yew Soon Ong
Nanyang Technological University
School of Computer Engineering
Blk N4, 2b-39, Nanyang Avenue, Singapore 639798
E-mail: asysong@ntu.edu.sg

Library of Congress Control Number: Applied for

CR Subject Classification (1998): F.1, F.2, I.2, G.2, I.4, I.5, J.3, J.4

ISSN   0302-9743
ISBN-10   3-540-28323-4 Springer Berlin Heidelberg New York
ISBN-13   978-3-540-28323-2 Springer Berlin Heidelberg New York

This work is subject to copyright. All rights are reserved, whether the whole or part of the material is concerned, specifically the rights of translation, reprinting, re-use of illustrations, recitation, broadcasting, reproduction on microfilms or in any other way, and storage in data banks. Duplication of this publication or parts thereof is permitted only under the provisions of the German Copyright Law of September 9, 1965, in its current version, and permission for use must always be obtained from Springer. Violations are liable to prosecution under the German Copyright Law.

Springer is a part of Springer Science+Business Media

springeronline.com

© Springer-Verlag Berlin Heidelberg 2005
Printed in Germany

Typesetting: Camera-ready by author, data conversion by Scientific Publishing Services, Chennai, India
Printed on acid-free paper   SPIN: 11539087   06/3142   5 4 3 2 1 0

Lecture Notes in Computer Science 3610

Commenced Publication in 1973
Founding and Former Series Editors:
Gerhard Goos, Juris Hartmanis, and Jan van Leeuwen

Editorial Board

David Hutchison
  Lancaster University, UK
Takeo Kanade
  Carnegie Mellon University, Pittsburgh, PA, USA
Josef Kittler
  University of Surrey, Guildford, UK
Jon M. Kleinberg
  Cornell University, Ithaca, NY, USA
Friedemann Mattern
  ETH Zurich, Switzerland
John C. Mitchell
  Stanford University, CA, USA
Moni Naor
  Weizmann Institute of Science, Rehovot, Israel
Oscar Nierstrasz
  University of Bern, Switzerland
C. Pandu Rangan
  Indian Institute of Technology, Madras, India
Bernhard Steffen
  University of Dortmund, Germany
Madhu Sudan
  Massachusetts Institute of Technology, MA, USA
Demetri Terzopoulos
  New York University, NY, USA
Doug Tygar
  University of California, Berkeley, CA, USA
Moshe Y. Vardi
  Rice University, Houston, TX, USA
Gerhard Weikum
  Max-Planck Institute of Computer Science, Saarbruecken, Germany

Preface

This book and its sister volumes, i.e., LNCS vols. 3610, 3611, and 3612, are the proceedings of the 1st International Conference on Natural Computation (ICNC 2005), jointly held with the 2nd International Conference on Fuzzy Systems and Knowledge Discovery (FSKD 2005, LNAI vols. 3613 and 3614) from 27 to 29 August 2005 in Changsha, Hunan, China. In its budding run, ICNC 2005 successfully attracted 1887 submissions from 32 countries/regions (the joint ICNC-FSKD 2005 received 3136 submissions). After rigorous reviews, 502 high-quality papers, i.e., 313 long papers and 189 short papers, were included in the ICNC 2005 proceedings, representing an acceptance rate of 26.6%.

The ICNC-FSKD 2005 featured the most up-to-date research results in computational algorithms inspired from nature, including biological, ecological, and physical systems. It is an exciting and emerging interdisciplinary area in which a wide range of techniques and methods are being studied for dealing with large, complex, and dynamic problems. The joint conferences also promoted cross-fertilization over these exciting and yet closely-related areas, which had a significant impact on the advancement of these important technologies. Specific areas included neural computation, quantum computation, evolutionary computation, DNA computation, chemical computation, information processing in cells and tissues, molecular computation, computation with words, fuzzy computation, granular computation, artificial life, swarm intelligence, ants colonies, artificial immune systems, etc., with innovative applications to knowledge discovery, finance, operations research, and more. In addition to the large number of submitted papers, we were blessed with the presence of four renowned keynote speakers and several distinguished panelists.

On behalf of the Organizing Committee, we thank Xiangtan University for sponsorship, and the IEEE Circuits and Systems Society, the IEEE Computational Intelligence Society, and the IEEE Control Systems Society for technical co-sponsorship. We are grateful for the technical cooperation from the International Neural Network Society, the European Neural Network Society, the Chinese Association for Artificial Intelligence, the Japanese Neural Network Society, the International Fuzzy Systems Association, the Asia-Pacific Neural Network Assembly, the Fuzzy Mathematics and Systems Association of China, and the Hunan Computer Federation. We thank the members of the Organizing Committee, the Advisory Board, and the Program Committee for their hard work in the past 18 months. We wish to express our heartfelt appreciation to the keynote and panel speakers, special session organizers, session chairs, reviewers, and student helpers. Our special thanks go to the publisher, Springer, for publishing the ICNC 2005 proceedings as three volumes of the Lecture Notes in Computer Science series (and the FSKD 2005 proceedings as two volumes of the Lecture Notes in Artificial Intelligence series). Finally, we thank all the authors and participants for their great contributions that made this conference possible and all the hard work worthwhile.

August 2005

Lipo Wang
Ke Chen
Yew Soon Ong

Organization

ICNC 2005 was organized by Xiangtan University and technically co-sponsored by the IEEE Circuits and Systems Society, the IEEE Computational Intelligence Society, and the IEEE Control Systems Society, in cooperation with the International Neural Network Society, the European Neural Network Society, the Chinese Association for Artificial Intelligence, the Japanese Neural Network Society, the International Fuzzy Systems Association, the Asia-Pacific Neural Network Assembly, the Fuzzy Mathematics and Systems Association of China, and the Hunan Computer Federation.

Organizing Committee

Honorary Conference Chairs	Shun-ichi Amari, Japan
	Lotfi A. Zadeh, USA
General Chair	He-An Luo, China
General Co-chairs	Lipo Wang, Singapore
	Yunqing Huang, China
Program Chairs	Ke Chen, UK
	Yew Soon Ong, Singapore
Local Arrangements Chairs	Renren Liu, China
	Xieping Gao, China
Proceedings Chair	Fen Xiao, China
Publicity Chair	Hepu Deng, Australia
Sponsorship/Exhibits Chairs	Shaoping Ling, China
	Geok See Ng, Singapore
Webmasters	Linai Kuang, China
	Yanyu Liu, China

Advisory Board

Toshio Fukuda, Japan
Kunihiko Fukushima, Japan
Tom Gedeon, Australia
Aike Guo, China
Zhenya He, China
Janusz Kacprzyk, Poland
Nikola Kasabov, New Zealand
John A. Keane, UK
Soo-Young Lee, Korea
Erkki Oja, Finland
Nikhil R. Pal, India

Witold Pedrycz, Canada
Jose C. Principe, USA
Harold Szu, USA
Shiro Usui, Japan
Xindong Wu, USA
Lei Xu, Hong Kong
Xin Yao, UK
Syozo Yasui, Japan
Bo Zhang, China
Yixin Zhong, China
Jacek M. Zurada, USA

Program Committee

Shigeo Abe, Japan
Kazuyuki Aihara, Japan
Davide Anguita, Italy
Abdesselam Bouzerdoum, Australia
Gavin Brown, UK
Laiwan Chan, Hong Kong
Sheng Chen, UK
Shu-Heng Chen, Taiwan
YanQiu Chen, China
Vladimir Cherkassky, USA
Sung-Bae Cho, Korea
Sungzoon Cho, Korea
Vic Ciesielski, Australia
Keshav Dahal, UK
Kalyanmoy Deb, India
Emilio Del-Moral-Hernandez, Brazil
Andries Engelbrecht, South Africa
Tomoki Fukai, Japan
Lance Fung, Australia
Takeshi Furuhashi, Japan
Hiroshi Furutani, Japan
John Q. Gan, UK
Wen Gao, China
Peter Geczy, Japan
Fanji Gu, China
Zeng-Guang Hou, Canada
Chenyi Hu, USA
Masumi Ishikawa, Japan
Robert John, UK
Mohamed Kamel, Canada
Yoshiki Kashimori, Japan
Samuel Kaski, Finland
Andy Keane, UK
Graham Kendall, UK
Jong-Hwan Kim, Korea
JungWon Kim, UK
Irwin King, Hong Kong
Natalio Krasnogor, UK
Vincent C.S. Lee, Australia
Stan Z. Li, China
XiaoLi Li, UK
Yangmin Li, Macau
Derong Liu, USA

Jian-Qin Liu, Japan
Bao-Liang Lu, China
Simon Lucas, UK
Frederic Maire, Australia
Jacek Mandziuk, Poland
Satoshi Matsuda, Japan
Masakazu Matsugu, Japan
Bob McKay, Australia
Ali A. Minai, USA
Hiromi Miyajima, Japan
Pedja Neskovic, USA
Richard Neville, UK
Tohru Nitta, Japan
Yusuke Nojima, Japan
Takashi Omori, Japan
M. Palaniswami, Australia
Andrew P. Paplinski, Australia
Asim Roy, USA
Bernhard Sendhoff, Germany
Qiang Shen, UK
Jang-Kyoo Shin, Korea
Leslie Smith, UK
Andy Song, Australia
Lambert Spannenburg, Sweden
Mingui Sun, USA
Johan Suykens, Belgium
Hideyuki Takagi, Japan
Kay Chen Tan, Singapore
Kiyoshi Tanaka, Japan
Seow Kiam Tian, Singapore
Peter Tino, UK
Kar-Ann Toh, Singapore
Yasuhiro Tsujimura, Japan
Ganesh Kumar Venayagamoorthy,
  USA
Brijesh Verma, Australia
Ray Walshe, Ireland
Jun Wang, Hong Kong
Rubin Wang, China
Xizhao Wang, China
Sumio Watanabe, Japan
Stefan Wermter, UK
Kok Wai Wong, Australia

Hong Yan, Hong Kong
Ron Yang, UK
Daniel Yeung, Hong Kong
Ali M.S. Zalzala, UK
Xiaojun Zeng, UK

David Zhang, Hong Kong
Huaguang Zhang, China
Liming Zhang, China
Qiangfu Zhao, Japan

Special Sessions Organizers

Ke Chen, UK
Gary Egan, Australia
Masami Hagiya, Japan
Tai-hoon Kim, Korea
Yangmin Li, Macau
Osamu Ono, Japan
Gwi-Tae Park, Korea
John A. Rose, Japan
Xingming Sun, China

Ying Tan, Hong Kong
Peter Tino, UK
Shiro Usui, Japan
Rubin Wang, China
Keming Xie, China
Xiaolan Zhang, USA
Liang Zhao, Brazil
Henghui Zou, USA
Hengming Zou, China

Reviewers

Ajith Abraham
Wensen An
Yisheng An
Jiancong Bai
Gurvinder Baicher
Xiaojuan Ban
Yukun Bao
Helio Barbosa
Zafer Bingul
Liefeng Bo
Yin Bo
Gavin Brown
Nan Bu
Erhan Butun
Chunhong Cao
Huai-Hu Cao
Qixin Cao
Yijia Cao
Yuan-Da Cao
Yuhui Cao
Yigang Cen
Chunlei Chai

Li Chai
Ping-Teng Chang
Kwokwing Chau
Ailing Chen
Chen-Tung Chen
Enqing Chen
Fangjiong Chen
Houjin Chen
Jiah-Shing Chen
Jing Chen
Jingchun Chen
Junying Chen
Li Chen
Shenglei Chen
Wei Chen
Wenbin Chen
Xi Chen
Xiyuan Chen
Xuhui Chen
Yuehui Chen
Zhen-Cheng Chen
Zhong Chen

Jian Cheng
Il-Ahn Cheong
Yiu-Ming Cheung
Yongwha Chung
Lingli Cui
Jian-Hua Dai
Chuangyin Dang
Xiaolong Deng
Hongkai Ding
Zhan Ding
Chao-Jun Dong
Guangbo Dong
Jie Dong
Sheqin Dong
Shoubin Dong
Wenyong Dong
Feng Du
Hai-Feng Du
Yanping Du
Shukai Duan
Metin Ertunc
Liu Fan

Gang Fang
Hui Fang
Chen Feng
Guiyu Feng
Jian Feng
Peng Fu
Yongfeng Fu
Yuli Fu
Naohiro Fukumura
Haichang Gao
Haihua Gao
Zong Geem
Emin Germen
Ling Gong
Maoguo Gong
Tao Gong
Weiguo Gong
Danying Gu
Qiu Guan
Salyh Günet
Dongwei Guo
Tian-Tai Guo
Xinchen Guo
Xiu Ping Guo
Yi'nan Guo
Mohamed Hamada
Jianchao Han
Lixin Han
Soowhan Han
Xiaozhuo Han
Fei Hao
Jingsong He
Jun He
Liqiang He
Xiaoxian He
Xiping He
Yi He
Zhaoshui He
Xingchen Heng
Chao-Fu Hong
Chi-I Hsu
Chunhua Hu
Hai Hu
Hongying Hu
Hua Hu

Jianming Hu
Li Kun Hu
Tao Hu
Ye Hu
Bingqiang Huang
Gaoming Huang
Min Huang
Yanwen Huang
Yilun Huang
Siu Cheung Hui
Changha Hwang
Jun-Cheol Jeon
Hyuncheol Jeong
Guangrong Ji
Mingxing Jia
Sen Jia
Zhuang Jian
Chunhong Jiang
Dongxiang Jiang
Jijiao Jiang
Minghui Jiang
Mingyan Jiang
Quanyuan Jiang
Li Cheng Jiao
Liu Jie
Wuyin Jin
Xu Jin
Ling Jing
Peng Jing
Xing-Jian Jing
Tao Jun
Hosang Jung
Jo Nam Jung
Venu K Murthy
Jaeho Kang
Kyung-Woo Kang
Ali Karci
Hyun-Sung Kim
Jongmin Kim
Jongweon Kim
Kee-Won Kim
Myung Won Kim
Wonil Kim
Heeyong Kwon
Xiang-Wei Lai

Dongwoo Lee
Kwangeui Lee
Seonghoon Lee
Seunggwan Lee
Kaiyou Lei
Xiongguo Lei
Soo Kar Leow
Anping Li
Boyu Li
Cheng Li
Dahu Li
Guanghui Li
Guoyou Li
Hongyan Li
Huanqin Li
Jianhua Li
Jie Li
Jing Li
Kangshun Li
Qiangwei Li
Qian-Mu Li
Qingyong Li
Ruonan Li
Shouju Li
Xiaobin Li
Xihai Li
Xinchun Li
Xiumei Li
Xuming Li
Ye Li
Ying Li
Yongjie Li
Yuangui Li
Yun Li
Yunfeng Li
Yong Li
Bojian Liang
Jiuzhen Liang
Xiao Liang
Yanchun Liang
Yixiong Liang
Guanglan Liao
Yingxin Liao
Sehun Lim
Tong Ming Lim

Jianning Lin
Ling Lin
Pan Lin
Qiu-Hua Lin
Zhi-Ling Lin
Zhou Ling
Benyong Liu
Bing Liu
Bingjie Liu
Dang-Hui Liu
Feng Liu
Hehui Liu
Huayong Liu
Jianchang Liu
Jing Liu
Jun Liu
Lifang Liu
Linlan Liu
Meiqin Liu
Miao Liu
Qicheng Liu
Ruochen Liu
Tianming Liu
Weidong Liu
Xianghui Liu
Xiaoqun Liu
Yong-Lin Liu
Zheng Liu
Zhi Liu
Jianchang Lu
Jun Lu
Xiaobo Lu
Yinan Lu
Dehan Luo
Guiming Luo
Juan Luo
Qiang Lv
Srinivas M.B.
Changshe Ma
Weimin Ma
Wenping Ma
Xuan Ma
Michiharu Maeda
Bertrand Maillet
Toshihiko Matsuka

Hongling Meng
Kehua Miao
Teijun Miao
Shi Min
Hongwei Mo
Dhinaharan Nagamalai
Atulya Nagar
Mi Young Nam
Rongrong Ni
Rui Nian
Ben Niu
Qun Niu
Sun-Kuk Noh
Linlin Ou
Mayumi Oyama-Higa
Cuneyt Oysu
A. Alper Ozalp
Ping-Feng Pai
Li Pan
Tinglong Pan
Zhiming Pan
Xiaohong Pang
Francesco Pappalardo
Hyun-Soo Park
Yongjin Park
Xiaomei Pei
Jun Peng
Wen Peng
Yan Peng
Yuqing Peng
Zeng Peng
Zhenrui Peng
Zhongbo Peng
Daoying Pi
Fangzhong Qi
Tang Qi
Rong Qian
Xiaoyan Qian
Xueming Qian
Baohua Qiang
Bin Qin
Zhengjun Qiu
Wentai Qu
Yunhua Rao
Sundaram Ravi

Phillkyu Rhee
Lili Rong
Fuhua Shang
Ronghua Shang
Zichang Shangguan
Dayong Shen
Xisheng Shen
Daming Shi
Xiaolong Shi
Zhiping Shi
Noritaka Shigei
Jooyong Shim
Dongkyoo Shin
Yongyi Shou
Yang Shu
Valceres Slva
Daniel Smutek
Haiyan Song
Jiaxing Song
Jingyan Song
Wenbin Song
Xiao-Yu Song
Yan Yan Song
Tieming Su
Xiaohong Su
P.N. Suganthan
Guangzhong Sun
Huali Sun
Shiliang Sun
Wei Sun
Yuqiu Sun
Zhanquan Sun
Jin Tang
Jing Tang
Suqin Tang
Zhiqiang Tang
Zhang Tao
Hissam Tawfik
Hakan Temeltas
Nipon Theera-Umpon
Mei Tian
Chung-Li Tseng
Ibrahim Turkoglu
Juan Velasquez
Bin Wang

Chao-Xue Wang
Chaoyong Wang
Deji Wang
Dingcheng Wang
Gi-Nam Wang
Guojiang Wang
Hong Wang
Hongbo Wang
Hong-Gang Wang
Jigang Wang
Lin Wang
Ling Wang
Min Wang
Qingquan Wang
Shangfei Wang
Shaowei Wang
Teng Wang
Weihong Wang
Xin Wang
Xinyu Wang
Yan Wang
Yanbin Wang
Yaonan Wang
Yen-Nien Wang
Yong-Xian Wang
Zhanshan Wang
Zheng-You Wang
Zhurong Wang
Wang Wei
Xun-Kai Wei
Chunguo Wu
Fei Wu
Ji Wu
Qiongshui Wu
Qiuxuan Wu
Sitao Wu
Wei Wu
Yanwen Wu
Ying Wu
Chen Xi
Shi-Hong Xia
Guangming Xian
Binglei Xie
Li Xie
Tao Xie

Shengwu Xiong
Zhangliang Xiong
Chunlin Xu
Jianhua Xu
Jinhua Xu
Junqin Xu
Li Xu
Lin Xu
Shuxiang Xu
Xianyun Xu
Xin Xu
Xu Xu
Xue-Song Xu
Zhiwei Xu
Yiliang Xu
Jianping Xuan
Yaofeng Xue
Yuncan Xue
Hui Yan
Qiao Yan
Xiaohong Yan
Bo Yang
Chunyan Yang
Feng Yang
Guifang Yang
Guoqqing Yang
Guowei Yang
Huihua Yang
Jianwei Yang
Jing Yang
Li-Ying Yang
Qingyun Yang
Xiaohua Yang
Xiaowei Yang
Xuhua Yang
Yingchun Yang
Zhihui Yang
Jingtao Yao
Her-Terng Yau
Chaoqun Ye
He Yi
Ling-Zhi Yi
Li Yin
Rupo Yin
Liang Ying

Chen Yong
Eun-Jun Yoon
Xinge You
Changjie Yu
Fei Yu
Fusheng Yu
Guoyan Yu
Lean Yu
Mian-Shui Yu
Qingjun Yu
Shiwen Yu
Xinjie Yu
Mingwei Yuan
Shenfang Yuan
Xun Yue
Wu Yun
Yeboon Yun
Jin Zeng
C.H. Zhang
Changjiang Zhang
Chunkai Zhang
Da-Peng Zhang
Defu Zhang
Fan Zhang
Fengyue Zhang
Hong Zhang
Hong-Bin Zhang
Ji Zhang
Jiang Zhang
Li Zhang
Liyan Zhang
Li-Yong Zhang
Min Zhang
Ming-Jie Zhang
Rubo Zhang
Ruo-Ying Zhang
Weidong Zhang
Wei-Guo Zhang
Wen Zhang
Xiufeng Zhang
Yangsen Zhang
Yifei Zhang
Yong-Dong Zhang
Yue-Jie Zhang
Yunkai Zhang

Yuntao Zhang
Zhenya Zhang
Hai Zhao
Jian Zhao
Jianxun Zhao
Jianye Zhao
Lianwei Zhao
Lina Zhao
Wencang Zhao
Xingming Zhao
Xuelong Zhao
Yinliang Zhao
Zhidong Zhao

Tiejun Zhao
Liu Zhen
Guibin Zheng
Shiqin Zheng
Yihui Zheng
Weicai Zhong
Zhou Zhong
Dongming Zhou
Gengui Zhou
Hongjun Zhou
Lifang Zhou
Wengang Zhou
Yuren Zhou

Zhiheng Zhou
Zongtan Zhou
Chengzhi Zhu
En Zhu
Li Zhu
Wen Zhu
Yaoqin Zhu
Xiaobin Zou
Xiaobo Zou
Zhenyu Zou
Wenming Zuo

* The term after a name may represent either a country or a region.

Table of Contents – Part I

Neural Network Learning Algorithms

A Novel Learning Algorithm for Wavelet Neural Networks
  Min Huang, Baotong Cui 1

Using Unscented Kalman Filter for Training the Minimal Resource Allocation Neural Network
  Ye Zhang, Yiqiang Wu, Wenquan Zhang, Yi Zheng 8

The Improved CMAC Model and Learning Result Analysis
  Daqi Zhu, Min Kong, YonQing Yang 15

A New Smooth Support Vector Regression Based on ϵ-Insensitive Logistic Loss Function
  Yang Hui-zhong, Shao Xin-guang, Ding Feng 25

Neural Network Classifier Based on the Features of Multi-lead ECG
  Mozhiwen, Feng Jun, Qiu Yazhu, Shu Lan 33

A New Learning Algorithm for Diagonal Recurrent Neural Network
  Deng Xiaolong, Xie Jianying, Guo Weizhong, Liu Jun 44

Study of On-Line Weighted Least Squares Support Vector Machines
  Xiangjun Wen, Xiaoming Xu, Yunze Cai 51

Globally Exponential Stability Analysis and Estimation of the Exponential Convergence Rate for Neural Networks with Multiple Time Varying Delays
  Huaguang Zhang, Zhanshan Wang 61

Locally Determining the Number of Neighbors in the k-Nearest Neighbor Rule Based on Statistical Confidence
  Jigang Wang, Predrag Neskovic, Leon N. Cooper 71

Fuzzy Self-organizing Map Neural Network Using Kernel PCA and the Application
  Qiang Lv, Jin-shou Yu 81

An Evolved Recurrent Neural Network and Its Application
  Chunkai Zhang, Hong Hu 91

Self-organized Locally Linear Embedding for Nonlinear Dimensionality Reduction
    Jian Xiao, Zongtan Zhou, Dewen Hu, Junsong Yin, Shuang Chen ... 101

Active Learning for Probabilistic Neural Networks
    Bülent Bolat, Tülay Yıldırım 110

Adaptive Training of Radial Basis Function Networks Using Particle Swarm Optimization Algorithm
    Hongkai Ding, Yunshi Xiao, Jiguang Yue 119

A Game-Theoretic Approach to Competitive Learning in Self-Organizing Maps
    Joseph Herbert, JingTao Yao 129

A Novel Intrusions Detection Method Based on HMM Embedded Neural Network
    Weijin Jiang, Yusheng Xu, Yuhui Xu 139

Generate Different Neural Networks by Negative Correlation Learning
    Yong Liu 149

New Training Method and Optimal Structure of Backpropagation Networks
    Songyot Sureerattanan, Nidapan Sureerattanan 157

Learning Outliers to Refine a Corpus for Chinese Webpage Categorization
    Dingsheng Luo, Xinhao Wang, Xihong Wu, Huisheng Chi 167

Bio-kernel Self-organizing Map for HIV Drug Resistance Classification
    Zheng Rong Yang, Natasha Young 179

A New Learning Algorithm Based on Lever Principle
    Xiaoguang He, Jie Tian, Xin Yang 187

An Effective Method to Improve Convergence for Sequential Blind Source Separation
    L. Yuan, Enfang. Sang, W. Wang, J.A. Chambers 199

A Novel LDA Approach for High-Dimensional Data
    Guiyu Feng, Dewen Hu, Ming Li, Zongtan Zhou 209

Research and Design of Distributed Neural Networks with Chip Training Algorithm
    Bo Yang, Ya-dong Wang, Xiao-hong Su 213

Support Vector Regression with Smoothing Property
Zhixia Yang, Nong Wang, Ling Jing 217

A Fast SMO Training Algorithm for Support Vector Regression
*Haoran Zhang, Xiaodong Wang, Changjiang Zhang,
Xiuling Xu* 221

Rival Penalized Fuzzy Competitive Learning Algorithm
Xiyang Yang, Fusheng Yu 225

A New Predictive Vector Quantization Method Using a Smaller Codebook
Min Shi, Shengli Xie 229

Performance Improvement of Fuzzy RBF Networks
Kwang-Baek Kim, Dong-Un Lee, Kwee-Bo Sim 237

Neural Network Architectures

Universal Approach to Study Delayed Dynamical Systems
Tianping Chen 245

Long-Range Connections Based Small-World Network and Its Synchronizability
Liu Jie, Lu Jun-an 254

Double Synaptic Weight Neuron Theory and Its Application
Wang Shou-jue, Chen Xu, Qin Hong, Li Weijun, Bian Yi 264

Comparative Study of Chaotic Neural Networks with Different Models of Chaotic Noise
Huidang Zhang, Yuyao He 273

A Learning Model in Qubit Neuron According to Quantum Circuit
Michiharu Maeda, Masaya Suenaga, Hiromi Miyajima 283

An Algorithm for Pruning Redundant Modules in Min-Max Modular Network with GZC Function
Jing Li, Bao-Liang Lu, Michinori Ichikawa 293

A General Procedure for Combining Binary Classifiers and Its Performance Analysis
Hai Zhao, Bao-Liang Lu 303

A Modular Structure of Auto-encoder for the Integration of Different
Kinds of Information
   Naohiro Fukumura, Keitaro Wakaki, Yoji Uno 313

Adaptive and Competitive Committee Machine Architecture
   Jian Yang, Siwei Luo 322

An ART2/RBF Hybrid Neural Networks Research
   Xuhua Yang, Yunbing Wei, Qiu Guan, Wanliang Wang,
   Shengyong Chen 332

Complex Number Procedure Neural Networks
   Liang Jiuzhen, Han Jianmin 336

Urban Traffic Signal Timing Optimization Based on Multi-layer Chaos
Neural Networks Involving Feedback
   Chaojun Dong, Zhiyong Liu, Zulian Qiu 340

Research on a Direct Adaptive Neural Network Control Method of
Nonlinear Systems
   Weijin Jiang, Yusheng Xu, Yuhui Xu 345

Improving the Resultant Quality of Kohonen's Self Organizing Map
Using Stiffness Factor
   Emin Germen 353

A Novel Orthonormal Wavelet Network for Function Learning
   Xieping Gao, Jun Zhang 358

Fuzzy Back-Propagation Network for PCB Sales Forecasting
   Pei-Chann Chang, Yen-Wen Wang, Chen-Hao Liu 364

An Evolutionary Artificial Neural Networks Approach for BF Hot
Metal Silicon Content Prediction
   Zhao Min, Liu Xiang-guan, Luo Shi-hua 374

Application of Chaotic Neural Model Based on Olfactory System on
Pattern Recognitions
   Guang Li, Zhenguo Lou, Le Wang, Xu Li,
   Walter J. Freeman 378

Double Robustness Analysis for Determining Optimal Feedforward
Neural Network Architecture
   Lean Yu, Kin Keung Lai, Shouyang Wang 382

Stochastic Robust Stability Analysis for Markovian Jump Neural
Networks with Time Delay
  Li Xie 386

Neurodynamics

Observation of Crises and Bifurcations in the Hodgkin-Huxley Neuron
Model
  Wuyin Jin, Qian Lin, Yaobing Wei, Ying Wu 390

An Application of Pattern Recognition Based on Optimized RBF-DDA
Neural Networks
  *Guoyou Li, Huiguang Li, Min Dong, Changping Sun,
  Tihua Wu* 397

Global Exponential Stability of Cellular Neural Networks with
Time-Varying Delays
  Qiang Zhang, Dongsheng Zhou, Haijun Wang, Xiaopeng Wei 405

Effect of Noises on Two-Layer Hodgkin-Huxley Neuronal Network
  Jun Liu, Zhengguo Lou, Guang Li 411

Adaptive Co-ordinate Transformation Based on a Spike Timing-
Dependent Plasticity Learning Paradigm
  *QingXiang Wu, T.M. McGinnity, L.P Maguire, A. Belatreche,
  B. Glackin* 420

Modeling of Short-Term Synaptic Plasticity Using Dynamic Synapses
  Biswa Sengupta 429

A Chaotic Model of Hippocampus-Neocortex
  *Takashi Kuremoto, Tsuyoshi Eto, Kunikazu Kobayashi,
  Masanao Obayashi* 439

Stochastic Neuron Model with Dynamic Synapses and Evolution
Equation of Its Density Function
  Wentao Huang, Licheng Jiao, Yuelei Xu, Maoguo Gong 449

Learning Algorithm for Spiking Neural Networks
  Hesham H. Amin, Robert H. Fujii 456

Exponential Convergence of Delayed Neural Networks
  Xiaoping Xue 466

A Neural Network for Constrained Saddle Point Problems: An
Approximation Approach
  Xisheng Shen, Shiji Song, Lixin Cheng 470

Implementing Fuzzy Reasoning by IAF Neurons
  Zhijie Wang, Hong Fan 476

A Method for Quantifying Temporal and Spatial Patterns of Spike Trains
  Shi-min Wang, Qi-Shao Lu, Ying Du 480

A Stochastic Nonlinear Evolution Model and Dynamic Neural Coding
on Spontaneous Behavior of Large-Scale Neuronal Population
  Rubin Wang, Wei Yu 490

Study on Circle Maps Mechanism of Neural Spikes Sequence
  Zhang Hong, Fang Lu-ping, Tong Qin-ye 499

Synchronous Behaviors of Hindmarsh-Rose Neurons with Chemical
Coupling
  Ying Wu, Jianxue Xu, Mi He 508

Statistical Neural Network Models and Support Vector Machines

A Simple Quantile Regression via Support Vector Machine
  Changha Hwang, Jooyong Shim 512

Doubly Regularized Kernel Regression with Heteroscedastic Censored
Data
  Jooyong Shim, Changha Hwang 521

Support Vector Based Prototype Selection Method for Nearest
Neighbor Rules
  Yuangui Li, Zhonghui Hu, Yunze Cai, Weidong Zhang 528

A Prediction Interval Estimation Method for KMSE
  Changha Hwang, Kyung Ha Seok, Daehyeon Cho 536

An Information-Geometrical Approach to Constructing Kernel in
Support Vector Regression Machines
  Wensen An, Yanguang Sun 546

Training Data Selection for Support Vector Machines
  Jigang Wang, Predrag Neskovic, Leon N. Cooper 554

Model Selection for Regularized Least-Squares Classification
Hui-Hua Yang, Xing-Yu Wang, Yong Wang, Hai-Hua Gao 565

Modelling of Chaotic Systems with Recurrent Least Squares Support
Vector Machines Combined with Reconstructed Embedding Phase Space
Zheng Xiang, Taiyi Zhang, Jiancheng Sun 573

Least-Squares Wavelet Kernel Method for Regression Estimation
Xiangjun Wen, Xiaoming Xu, Yunze Cai 582

Fuzzy Support Vector Machines Based on λ—Cut
Shengwu Xiong, Hongbing Liu, Xiaoxiao Niu 592

Mixtures of Kernels for SVM Modeling
Yan-fei Zhu, Lian-fang Tian, Zong-yuan Mao, Li Wei 601

A Novel Parallel Reduced Support Vector Machine
Fangfang Wu, Yinliang Zhao, Zefei Jiang 608

Recurrent Support Vector Machines in Reliability Prediction
*Wei-Chiang Hong, Ping-Feng Pai, Chen-Tung Chen,
Ping-Teng Chang* 619

A Modified SMO Algorithm for SVM Regression and Its Application in
Quality Prediction of HP-LDPE
Hengping Zhao, Jinshou Yu 630

Gait Recognition via Independent Component Analysis Based on
Support Vector Machine and Neural Network
Erhu Zhang, Jiwen Lu, Ganglong Duan......................... 640

Uncertainty Support Vector Method for Ordinal Regression
Liu Guangli, Sun Ruizhi, Gao Wanlin 650

An Incremental Learning Method Based on SVM for Online Sketchy
Shape Recognition
Zhengxing Sun, Lisha Zhang, Enyi Tang 655

Eigenspectra Versus Eigenfaces: Classification with a Kernel-Based
Nonlinear Representor
Benyong Liu, Jing Zhang 660

Blind Extraction of Singularly Mixed Source Signals
Zhigang Zeng, Chaojin Fu 664

Application of Support Vector Machines in Predicting Employee
Turnover Based on Job Performance
  *Wei-Chiang Hong, Ping-Feng Pai, Yu-Ying Huang,
  Shun-Lin Yang* 668

Palmprint Recognition Based on Unsupervised Subspace Analysis
  Guiyu Feng, Dewen Hu, Ming Li, Zongtan Zhou 675

A New Alpha Seeding Method for Support Vector Machine Training
  Du Feng, Wenkang Shi, Huawei Guo, Liangzhou Chen 679

Multiple Acoustic Sources Location Based on Blind Source Separation
  Gaoming Huang, Luxi Yang, Zhenya He 683

Short-Term Load Forecasting Based on Self-organizing Map and
Support Vector Machine
  Zhejing Bao, Daoying Pi, Youxian Sun 688

A Multi-class Classifying Algorithm Based on Nonlinear Dimensionality
Reduction and Support Vector Machines
  Lukui Shi, Qing Wu, Xueqin Shen, Pilian He 692

A VSC Scheme for Linear MIMO Systems Based on SVM
  Zhang Yibo, Yang Chunjie, Pi Daoying, Sun Youxian 696

Global Convergence of FastICA: Theoretical Analysis and Practical
Considerations
  Gang Wang, Xin Xu, Dewen Hu 700

SVM Based Nonparametric Model Identification and Dynamic Model
Control
  Weimin Zhong, Daoying Pi, Youxian Sun 706

Learning SVM Kernel with Semi-definite Programming
  Shuzhong Yang, Siwei Luo 710

Weighted On-Line SVM Regression Algorithm and Its Application
  Hui Wang, Daoying Pi, Youxian Sun 716

Other Topics in Neural Network Models

Convergence of an Online Gradient Method for BP Neural Networks
with Stochastic Inputs
  Zhengxue Li, Wei Wu, Guorui Feng, Huifang Lu 720

A Constructive Algorithm for Wavelet Neural Networks
Jinhua Xu, Daniel W.C. Ho 730

Stochastic High-Order Hopfield Neural Networks
Yi Shen, Guoying Zhao, Minghui Jiang, Shigeng Hu 740

Predicting with Confidence - An Improved Dynamic Cell Structure
Yan Liu, Bojan Cukic, Michael Jiang, Zhiwei Xu 750

An Efficient Score Function Generation Algorithm with Information Maximization
Woong Myung Kim, Hyon Soo Lee 760

A New Criterion on Exponential Stability of a Class of Discrete Cellular Neural Networks with Time Delay
Fei Hao, Long Wang, Tianguang Chu 769

A Novel Local Connection Neural Network
Shuang Cong, Guodong Li, Yisong Zheng 773

An Unsupervised Cooperative Pattern Recognition Model to Identify Anomalous Massive SNMP Data Sending
Álvaro Herrero, Emilio Corchado, José Manuel Sáiz 778

A Fast Nonseparable Wavelet Neural Network for Function Approximation
Jun Zhang, Xieping Gao, Chunhong Cao, Fen Xiao 783

A Visual Cortex Domain Model for Illusory Contour Figures
Keongho Hong, Eunhwa Jeong 789

Cognitive Science

ANN Ensemble Online Learning Strategy in 3D Object Cognition and Recognition Based on Similarity
Rui Nian, Guangrong Ji, Wencang Zhao, Chen Feng 793

Design and Implementation of the Individualized Intelligent Teachable Agent
Sung-il Kim, Sung-Hyun Yun, Dong-Seong Choi, Mi-sun Yoon, Yeon-hee So, Myung-jin Lee, Won-sik Kim, Sun-young Lee, Su-Young Hwang, Cheon-woo Han, Woo-Gul Lee, Karam Lim 797

Comparison of Complexity and Regularity of ERP Recordings Between Single and Dual Tasks Using Sample Entropy Algorithm
Tao Zhang, Xiaojun Tang, Zhuo Yang 806

Representation of a Physio-psychological Index Through Constellation Graphs
Oyama-Higa Mayumi, Tiejun Miao 811

Neural Network Based Emotion Estimation Using Heart Rate Variability and Skin Resistance
Sun K. Yoo, Chung K. Lee, Youn J. Park, Nam H. Kim, Byung C. Lee, Kee S. Jeong 818

Modeling Belief, Capability and Promise for Cognitive Agents - A Modal Logic Approach
Xinyu Zhao, Zuoquan Lin 825

PENCIL: A Framework for Expressing Free-Hand Sketching in 3D
Zhan Ding, Sanyuan Zhang, Wei Peng, Xiuzi Ye, Huaqiang Hu 835

Blocking Artifacts Measurement Based on the Human Visual System
Zhi-Heng Zhou, Sheng-Li Xie 839

A Computation Model of Korean Lexical Processing
Hyungwook Yim, Heuseok Lim, Kinam Park, Kichun Nam 844

Neuroanatomical Analysis for Onomatopoeia and Phainomime Words: fMRI Study
Jong-Hye Han, Wonil Choi, Yongmin Chang, Ok-Ran Jeong, Kichun Nam 850

Cooperative Aspects of Selective Attention
KangWoo Lee 855

Selective Attention Guided Perceptual Grouping Model
Qi Zou, Siwei Luo, Jianyu Li 867

Visual Search for Object Features
Predrag Neskovic, Leon N. Cooper 877

Agent Based Decision Support System Using Reinforcement Learning Under Emergency Circumstances
Devinder Thapa, In-Sung Jung, Gi-Nam Wang 888

Dynamic Inputs and Attraction Force Analysis for Visual Invariance
and Transformation Estimation
  Tomás Maul, Sapiyan Baba, Azwina Yusof 893

Task-Oriented Sparse Coding Model for Pattern Classification
  Qingyong Li, Dacheng Lin, Zhongzhi Shi 903

Robust Face Recognition from One Training Sample per Person
  Weihong Deng, Jiani Hu, Jun Guo 915

Chinese Word Sense Disambiguation Using HowNet
  Yuntao Zhang, Ling Gong, Yongcheng Wang 925

Modeling Human Learning as Context Dependent Knowledge Utility
Optimization
  Toshihiko Matsuka 933

Automatic Text Summarization Based on Lexical Chains
  Yanmin Chen, Xiaolong Wang, Yi Guan 947

A General fMRI LINEAR Convolution Model Based Dynamic
Characteristic
  Hong Yuan, Hong Li, Zhijie Zhang, Jiang Qiu 952

Neuroscience Informatics, Bioinformatics, and Bio-medical Engineering

A KNN-Based Learning Method for Biology Species Categorization
  Yan Dang, Yulei Zhang, Dongmo Zhang, Liping Zhao 956

Application of Emerging Patterns for Multi-source Bio-Data
Classification and Analysis
  Hye-Sung Yoon, Sang-Ho Lee, Ju Han Kim 965

Nonlinear Kernel MSE Methods for Cancer Classification
  L. Shen, E.C. Tan 975

Fusing Face and Fingerprint for Identity Authentication by SVM
  Chunhong Jiang, Guangda Su 985

A New Algorithm of Multi-modality Medical Image Fusion Based on
Pulse-Coupled Neural Networks
  Wei Li, Xue-feng Zhu 995

Cleavage Site Analysis Using Rule Extraction from Neural Networks
  Yeun-Jin Cho, Hyeoncheol Kim 1002

Prediction Rule Generation of MHC Class I Binding Peptides Using ANN and GA
  Yeon-Jin Cho, Hyeoncheol Kim, Heung-Bum Oh 1009

Combined Kernel Function Approach in SVM for Diagnosis of Cancer
  Ha-Nam Nguyen, Syng-Yup Ohn, Jaehyun Park, Kyu-Sik Park 1017

Automatic Liver Segmentation of Contrast Enhanced CT Images Based on Histogram Processing
  Kyung-Sik Seo, Hyung-Bum Kim, Taesu Park, Pan-Koo Kim, Jong-An Park 1027

An Improved Adaptive RBF Network for Classification of Left and Right Hand Motor Imagery Tasks
  Xiao-mei Pei, Jin Xu, Chong-xun Zheng, Guang-yu Bin 1031

Similarity Analysis of DNA Sequences Based on the Relative Entropy
  Wenlu Yang, Xiongjun Pi, Liqing Zhang 1035

Can Circulating Matrix Metalloproteinases Be Predictors of Breast Cancer? A Neural Network Modeling Study
  H. Hu, S.B. Somiari, J. Copper, R.D. Everly, C. Heckman, R. Jordan, R. Somiari, J. Hooke, C.D. Shriver, M.N. Liebman 1039

Blind Clustering of DNA Fragments Based on Kullback-Leibler Divergence
  Xiongjun Pi, Wenlu Yang, Liqing Zhang 1043

Prediction of Protein Subcellular Locations Using Support Vector Machines
  Na-na Li, Xiao-hui Niu, Feng Shi, Xue-yan Li 1047

Neuroinformatics Research in China- Current Status and Future Research Activities
  Guang Li, Jing Zhang, Faji Gu, Ling Yin, Yiyuan Tang, Xiaowei Tang 1052

Australian Neuroinformatics Research – Grid Computing and e-Research
  G.F. Egan, W. Liu, W-S. Soh, D. Hang 1057

Current Status and Future Research Activities in Clinical
Neuroinformatics: Singaporean Perspective
  Wieslaw L. Nowinski 1065

Japanese Neuroinformatics Research: Current Status and Future
Research Program of J-Node
  Shiro Usui 1074

Neural Network Applications: Communications and Computer Networks

Optimal TDMA Frame Scheduling in Broadcasting Packet Radio
Networks Using a Gradual Noisy Chaotic Neural Network
  Haixiang Shi, Lipo Wang 1080

A Fast Online SVM Algorithm for Variable-Step CDMA Power Control
  Yu Zhao, Hongsheng Xi, Zilei Wang 1090

Fourth-Order Cumulants and Neural Network Approach for Robust
Blind Channel Equalization
  Soowhan Han, Kwangeui Lee, Jongkeuk Lee,
  Fredric M. Ham 1100

Equalization of a Wireless ATM Channel with Simplified Complex
Bilinear Recurrent Neural Network
  Dong-Chul Park, Duc-Hoai Nguyen, Sang Jeen Hong,
  Yunsik Lee 1113

A Novel Remote User Authentication Scheme Using Interacting Neural
Network
  Tieming Chen, Jiamei Cai 1117

Genetic Algorithm Simulated Annealing Based Clustering Strategy in
MANET
  Xu Li 1121

Neural Network Applications: Expert System and Informatics

A Gradual Training Algorithm of Incremental Support Vector Machine
Learning
  Jian-Pei Zhang, Zhong-Wei Li, Jing Yang, Yuan Li 1132

An Improved Method of Feature Selection Based on Concept Attributes
in Text Classification
  Shasha Liao, Minghu Jiang 1140

Research on the Decision Method for Enterprise Information Investment
Based on IA-BP Network
  Xiao-Ke Yan, Hai-Dong Yang, He-Jun Wang, Fei-Qi Deng 1150

Process Control and Management of Etching Process Using Data
Mining with Quality Indexes
  Hyeon Bae, Sungshin Kim, Kwang Bang Woo 1160

Automatic Knowledge Configuration by Reticular Activating
System
  JeongYon Shim 1170

An Improved Information Retrieval Method and Input Device Using
Gloves for Wearable Computers
  Jeong-Hoon Shin, Kwang-Seok Hong 1179

Research on Design and Implementation of the Artificial Intelligence
Agent for Smart Home Based on Support Vector Machine
  Jonghwa Choi, Dongkyoo Shin, Dongil Shin 1185

A Self-organized Network for Data Clustering
  *Liang Zhao, Antonio P.G. Damiance Jr.,
  Andre C.P.L.F. Carvalho* 1189

A General Criterion of Synchronization Stability in Ensembles of
Coupled Systems and Its Application
  Qing-Yun Wang, Qi-Shao Lu, Hai-Xia Wang 1199

Complexity of Linear Cellular Automata over $\mathbb{Z}_m$
  Xiaogang Jin, Weihong Wang 1209

Neural Network Applications: Financial Engineering

Applications of Genetic Algorithm for Artificial Neural Network Model
Discovery and Performance Surface Optimization in Finance
  Serge Hayward 1214

Mining Data by Query-Based Error-Propagation
  Liang-Bin Lai, Ray-I Chang, Jen-Shaing Kouh 1224

The Application of Structured Feedforward Neural Networks to the
Modelling of the Daily Series of Currency in Circulation
  Marek Hlaváček, Josef Čada, František Hakl 1234

Time Delay Neural Networks and Genetic Algorithms for Detecting
Temporal Patterns in Stock Markets
  Hyun-jung Kim, Kyung-shik Shin, Kyungdo Park 1247

The Prediction of the Financial Time Series Based on Correlation
Dimension
  Chen Feng, Guangrong Ji, Wencang Zhao, Rui Nian 1256

Gradient-Based FCM and a Neural Network for Clustering of
Incomplete Data
  Dong-Chul Park 1266

Toward Global Optimization of ANN Supported by Instance Selection
for Financial Forecasting
  Sehun Lim 1270

Other Applications of Natural Computations

FranksTree: A Genetic Programming Approach to Evolve Derived
Bracketed L-Systems
  Danilo Mattos Bonfim, Leandro Nunes de Castro 1275

Data Clustering with a Neuro-immune Network
  *Helder Knidel, Leandro Nunes de Castro,
  Fernando J. Von Zuben* 1279

Author Index 1289

Table of Contents – Part II

Neural Network Applications: Pattern Recognition and Diagnostics

Monitoring of Tool Wear Using Feature Vector Selection and Linear Regression
  Zhong Chen, XianMing Zhang 1

Image Synthesis and Face Recognition Based on 3D Face Model and Illumination Model
  Dang-hui Liu, Lan-sun Shen, Kin-man Lam 7

Head-and-Shoulder Detection in Varying Pose
  Yi Sun, Yan Wang, Yinghao He, Yong Hua 12

Principal Component Neural Networks Based Intrusion Feature Extraction and Detection Using SVM
  Hai-Hua Gao, Hui-Hua Yang, Xing-Yu Wang 21

GA-Driven LDA in KPCA Space for Facial Expression Recognition
  Qijun Zhao, Hongtao Lu 28

A New ART Neural Networks for Remote Sensing Image Classification
  AnFei Liu, BiCheng Li, Gang Chen, Xianfei Zhang 37

Modified Color Co-occurrence Matrix for Image Retrieval
  Min Hyuk Chang, Jae Young Pyun, Muhammad Bilal Ahmad, Jong Hoon Chun, Jong An Park 43

A Novel Data Fusion Scheme for Offline Chinese Signature Verification
  Wen-ming Zuo, Ming Qi 51

A Multiple Eigenspaces Constructing Method and Its Application to Face Recognition
  Wu-Jun Li, Bin Luo, Chong-Jun Wang, Xiang-Ping Zhong, Zhao-Qian Chen 55

Quality Estimation of Fingerprint Image Based on Neural Network
  En Zhu, Jianping Yin, Chunfeng Hu, Guomin Zhang 65

Face Recognition Based on PCA/KPCA Plus CCA
  Yunhui He, Li Zhao, Cairong Zou 71

Texture Segmentation Using Intensified Fuzzy Kohonen Clustering
Network
    Dong Liu, Yinggan Tang, Xinping Guan 75

Application of Support Vector Machines in Reciprocating Compressor
Valve Fault Diagnosis
    Quanmin Ren, Xiaojiang Ma, Gang Miao 81

The Implementation of the Emotion Recognition from Speech and
Facial Expression System
    Chang-Hyun Park, Kwang-Sub Byun, Kwee-Bo Sim 85

Kernel PCA Based Network Intrusion Feature Extraction and Detection
Using SVM
    Hai-Hua Gao, Hui-Hua Yang, Xing-Yu Wang 89

Leak Detection in Transport Pipelines Using Enhanced Independent
Component Analysis and Support Vector Machines
    Zhengwei Zhang, Hao Ye, Guizeng Wang, Jie Yang 95

Line-Based PCA and LDA Approaches for Face Recognition
    Vo Dinh Minh Nhat, Sungyoung Lee 101

Comparative Study on Recognition of Transportation Under Real and
UE Status
    Jingxin Dong, Jianping Wu, Yuanfeng Zhou 105

Adaptive Eye Location Using FuzzyART
    Jo Nam Jung, Mi Young Nam, Phill Kyu Rhee 109

Face Recognition Using Gabor Features and Support Vector
Machines
    Yunfeng Li, Zongying Ou, Guoqiang Wang 119

Wavelet Method Combining BP Networks and Time Series ARMA
Modeling for Data Mining Forecasting
    Weimin Tong, Yijun Li 123

On-line Training of Neural Network for Color Image Segmentation
    Yi Fang, Chen Pan, Li Liu 135

Short-Term Prediction on Parameter-Varying Systems by Multiwavelets
Neural Network
    Fen Xiao, Xieping Gao, Chunhong Cao, Jun Zhang 139

VICARED: A Neural Network Based System for the Detection of
Electrical Disturbances in Real Time
   *Iñigo Monedero, Carlos León, Jorge Ropero, José Manuel Elena,
Juan C. Montaño* 147

Speech Recognition by Integrating Audio, Visual and Contextual
Features Based on Neural Networks
   Myung Won Kim, Joung Woo Ryu, Eun Ju Kim 155

A Novel Pattern Classification Method for Multivariate EMG Signals
Using Neural Network
   Nan Bu, Jun Arita, Toshio Tsuji 165

Data Fusion for Fault Diagnosis Using Dempster-Shafer Theory Based
Multi-class SVMs
   Zhonghui Hu, Yunze Cai, Ye Li, Yuangui Li, Xiaoming Xu 175

Modelling of Rolling and Aging Processes in Copper Alloy by
Levenberg-Marquardt BP Algorithm
   Juanhua Su, Hejun Li, Qiming Dong, Ping Liu 185

Neural Network Applications: Robotics and Intelligent Control

An Adaptive Control for AC Servo System Using Recurrent Fuzzy
Neural Network
   Wei Sun, Yaonan Wang 190

PSO-Based Model Predictive Control for Nonlinear Processes
   Xihuai Wang, Jianmei Xiao 196

Low Cost Implementation of Artificial Neural Network Based Space
Vector Modulation
   Tarık Erfidan, Erhan Butun 204

A Novel Multispectral Imaging Analysis Method for White Blood Cell
Detection
   *Hongbo Zhang, Libo Zeng, Hengyu Ke, Hong Zheng,
Qiongshui Wu* 210

Intelligent Optimal Control in Rare-Earth Countercurrent Extraction
Process *via* Soft-Sensor
   Hui Yang, Chunyan Yang, Chonghui Song, Tianyou Chai 214

Three Dimensional Gesture Recognition Using Modified Matching Algorithm
  Hwan-Seok Yang, Jong-Min Kim, Seoung-Kyu Park 224

Direct Adaptive Control for a Class of Uncertain Nonlinear Systems Using Neural Networks
  Tingliang Hu, Jihong Zhu, Chunhua Hu, Zengqi Sun 234

Neural Network Based Feedback Scheduler for Networked Control System with Flexible Workload
  Feng Xia, Shanbin Li, Youxian Sun 242

Humanoid Walking Gait Optimization Using GA-Based Neural Network
  Zhe Tang, Changjiu Zhou, Zengqi Sun 252

Adaptive Neural Network Internal Model Control for Tilt Rotor Aircraft Platform
  Changjie Yu, Jihong Zhu, Zengqi Sun 262

Novel Leaning Feed-Forward Controller for Accurate Robot Trajectory Tracking
  D. Bi, G.L. Wang, J. Zhang, Q. Xue 266

Adaptive Neural Network Control for Multi-fingered Robot Hand Manipulation in the Constrained Environment
  Gang Chen, Shuqing Wang, Jianming Zhang 270

Control of a Giant Swing Robot Using a Neural Oscillator
  Kiyotoshi Matsuoka, Norifumi Ohyama, Atsushi Watanabe, Masataka Ooshima 274

Neural Network Indirect Adaptive Sliding Mode Tracking Control for a Class of Nonlinear Interconnected Systems
  Yanxin Zhang, Xiaofan Wang 283

Sequential Support Vector Machine Control of Nonlinear Systems via Lyapunov Function Derivative Estimation
  Zonghai Sun, Youxian Sun, Yongqiang Wang 292

An Adaptive Control Using Multiple Neural Networks for the Position Control in Hydraulic Servo System
  Yuan Kang, Ming-Hui Chua, Yuan-Liang Liu, Chuan-Wei Chang, Shu-Yen Chien 296

Neural Network Applications: Signal Processing and Multi-media

Exon Structure Analysis via PCA and ICA of Short-Time Fourier Transform
Changha Hwang, David Chiu, Insuk Sohn 306

Nonlinear Adaptive Blind Source Separation Based on Kernel Function
Feng Liu, Cao Zhexin, Qiang Zhi, Shaoqian Li, Min Liang 316

Hybrid Intelligent Forecasting Model Based on Empirical Mode Decomposition, Support Vector Regression and Adaptive Linear Neural Network
Zhengjia He, Qiao Hu, Yanyang Zi, Zhousuo Zhang, Xuefeng Chen 324

A Real Time Color Gamut Mapping Method Using a Neural Network
Hak-Sung Lee, Dongil Han 328

Adaptive Identification of Chaotic Systems and Its Applications in Chaotic Communications
Jiuchao Feng 332

A Time-Series Decomposed Model of Network Traffic
Cheng Guang, Gong Jian, Ding Wei 338

A Novel Wavelet Watermark Algorithm Based on Neural Network Image Scramble
Jian Zhao, Qin Zhao, Ming-quan Zhou, Jianshou Pan 346

A Hybrid Model for Forecasting Aquatic Products Short-Term Price Integrated Wavelet Neural Network with Genetic Algorithm
Tao Hu, Xiaoshuan Zhang, Yunxian Hou, Weisong Mu, Zetian Fu 352

A Multiple Vector Quantization Approach to Image Compression
Noritaka Shigei, Hiromi Miyajima, Michiharu Maeda 361

Segmentation of SAR Image Using Mixture Multiscale ARMA Network
Haixia Xu, Zheng Tian, Fan Meng 371

Brain Activity Analysis of Rat Based on Electroencephalogram Complexity Under General Anesthesia
Jin Xu, Chongxun Zheng, Xueliang Liu, Xiaomei Pei, Guixia Jing 376

Post-nonlinear Blind Source Separation Using Wavelet Neural Networks and Particle Swarm Optimization
  Ying Gao, Shengli Xie 386

An MRF-ICA Based Algorithm for Image Separation
  Sen Jia, Yuntao Qian 391

Multi-view Face Recognition with Min-Max Modular SVMs
  Zhi-Gang Fan, Bao-Liang Lu 396

Texture Segmentation Using Neural Networks and Multi-scale Wavelet Features
  Tae Hyung Kim, Il Kyu Eom, Yoo Shin Kim 400

An In-depth Comparison on FastICA, CuBICA and IC-FastICA
  Bin Wang, Wenkai Lu 410

Characteristics of Equinumber Principle for Adaptive Vector Quantization
  Michiharu Maeda, Noritaka Shigei, Hiromi Miyajima 415

ANFIS Based Dynamic Model Compensator for Tracking and GPS Navigation Applications
  Dah-Jing Jwo, Zong-Ming Chen 425

Dynamic Background Discrimination with a Recurrent Network
  Jieyu Zhao 432

Gender Recognition Using a Min-Max Modular Support Vector Machine
  Hui-Cheng Lian, Bao-Liang Lu, Erina Takikawa, Satoshi Hosoi ... 438

An Application of Support Vector Regression on Narrow-Band Interference Suppression in Spread Spectrum Systems
  Qing Yang, Shengli Xie 442

A Natural Modification of Autocorrelation Based Video Watermarking Scheme Using ICA for Better Geometric Attack Robustness
  Seong-Whan Kim, Hyun Jin Park, HyunSeong Sung 451

Research of Blind Deconvolution Algorithm Based on High-Order Statistics and Quantum Inspired GA
  Jun-an Yang, Bin Zhao, Zhongfu Ye 461

Differential Demodulation of OFDM Based on SOM
  Xuming Li, Lenan Wu 468

Efficient Time Series Matching Based on HMTS Algorithm
    Min Zhang, Ying Tan 476

3D Polar-Radius Invariant Moments and Structure Moment Invariants
    Zongmin Li, Yuanzhen Zhang, Kunpeng Hou, Hua Li 483

A Fast Searching Algorithm of Symmetrical Period Modulation Pattern
Based on Accumulative Transformation Technique
    FuHua Fan, Ying Tan 493

A Granular Analysis Method in Signal Processing
    Lunwen Wang, Ying Tan, Ling Zhang 501

Other Neural Networks Applications

Adaptive Leakage Suppression Based on Recurrent Wavelet Neural
Network
    Zhangliang Xiong, Xiangquan Shi 508

New Multi-server Password Authentication Scheme Using Neural
Networks
    Eun-Jun Yoon, Kee-Young Yoo 512

Time Domain Substructural Post-earthquake Damage Diagnosis
Methodology with Neural Networks
    Bin Xu 520

Conceptual Modeling with Neural Network for Giftedness Identification
and Education
    Kwang Hyuk Im, Tae Hyun Kim, SungMin Bae, Sang Chan Park ... 530

Online Discovery of Quantitative Model for Web Service Management
    Jing Chen, Xiao-chuan Yin, Shui-ping Zhang 539

Judgment of Static Life and Death in Computer Go Using String Graph
    Hyun-Soo Park, Kyung-Woo Kang, Hang-Joon Kim 543

Research on Artificial Intelligence Character Based Physics Engine in
3D Car Game
    Jonghwa Choi, Dongkyoo Shin, Jinsung Choi, Dongil Shin 552

Document Clustering Based on Nonnegative Sparse Matrix Factorization
    C.F. Yang, Mao Ye, Jing Zhao 557

Prediction Modeling for Ingot Manufacturing Process Utilizing Data
Mining Roadmap Including Dynamic Polynomial Neural Network and
Bootstrap Method
Hyeon Bae, Sungshin Kim, Kwang Bang Woo 564

Implicit Rating – A Case Study
Song Wang, Xiu Li, Wenhuang Liu 574

Application of Grey Majorized Model in Tunnel Surrounding Rock
Displacement Forecasting
Xiaohong Li, Yu Zhao, Xiaoguang Jin, Yiyu Lu, Xinfei Wang 584

NN-Based Damage Detection in Multilayer Composites
Zhi Wei, Xiaomin Hu, Muhui Fan, Jun Zhang, D. Bi 592

Application of Support Vector Machine and Similar Day Method for
Load Forecasting
Xunming Li, Changyin Sun, Dengcai Gong 602

Particle Swarm Optimization Neural Network and Its Application in
Soft-Sensing Modeling
Guochu Chen, Jinshou Yu 610

Solution of the Inverse Electromagnetic Problem of Spontaneous
Potential (SP) by Very Fast Simulated Reannealing (VFSR)
Hüseyin Göksu, Mehmet Ali Kaya, Ali Kökçe 618

Using SOFM to Improve Web Site Text Content
*Sebastián A. Ríos, Juan D. Velásquez, Eduardo S. Vera,
Hiroshi Yasuda, Terumasa Aoki* 622

Online Support Vector Regression for System Identification
Zhenhua Yu, Xiao Fu, Yinglu Li 627

Optimization of PTA Crystallization Process Based on Fuzzy GMDH
Networks and Differential Evolutionary Algorithm
Wenli Du, Feng Qian 631

An Application of Support Vector Machines for Customer Churn
Analysis: Credit Card Case
Sun Kim, Kyung-shik Shin, Kyungdo Park 636

e-NOSE Response Classification of Sewage Odors by Neural Networks
and Fuzzy Clustering
Güleda Önkal-Engin, Ibrahim Demir, Seref N. Engin 648

Using a Random Subspace Predictor to Integrate Spatial and Temporal
Information for Traffic Flow Forecasting
  Shiliang Sun, Changshui Zhang 652

Boosting Input/Output Hidden Markov Models for Sequence
Classification
  Ke Chen 656

Learning Beyond Finite Memory in Recurrent Networks of Spiking
Neurons
  Peter Tiňo, Ashley Mills 666

On Non-Markovian Topographic Organization of Receptive Fields in
Recursive Self-organizing Map
  Peter Tiňo, Igor Farkaš 676

Evolutionary Learning

Quantum Reinforcement Learning
  Daoyi Dong, Chunlin Chen, Zonghai Chen 686

Characterization of Evaluation Metrics in Topical Web Crawling Based
on Genetic Algorithm
  Tao Peng, Wanli Zuo, Yilin Liu 690

A Novel Quantum Swarm Evolutionary Algorithm for Solving 0-1
Knapsack Problem
  *Yan Wang, Xiao-Yue Feng, Yan-Xin Huang, Wen-Gang Zhou,
  Yan-Chun Liang, Chun-Guang Zhou* 698

An Evolutionary System and Its Application to Automatic Image
Segmentation
  Yun Wen Chen, Yan Qiu Chen 705

Incorporating Web Intelligence into Website Evolution
  Jang Hee Lee, Gye Hang Hong 710

Evolution of the CPG with Sensory Feedback for Bipedal Locomotion
  Sooyol Ok, DuckSool Kim 714

Immunity-Based Genetic Algorithm for Classification Rule Discovery
  Ziqiang Wang, Dexian Zhang 727

Dynamical Proportion Portfolio Insurance with Genetic Programming
  Jiah-Shing Chen, Chia-Lan Chang 735

Evolution of Reactive Rules in Multi Player Computer Games Based on Imitation
Steffen Priesterjahn, Oliver Kramer, Alexander Weimer, Andreas Goebels 744

Combining Classifiers with Particle Swarms
Li-ying Yang, Zheng Qin 756

Adaptive Normalization Based Highly Efficient Face Recognition Under Uneven Environments
Phill Kyu Rhee, InJa Jeon, EunSung Jeong 764

Artificial Immune Systems

A New Detector Set Generating Algorithm in the Negative Selection Model
Xinhua Ren, Xiufeng Zhang, Yuanyuan Li 774

Intrusion Detection Based on ART and Artificial Immune Network Clustering
Fang Liu, Lin Bai, Licheng Jiao 780

Nature-Inspired Computations Using an Evolving Multi-set of Agents
E.V. Krishnamurthy, V.K. Murthy 784

Adaptive Immune Algorithm for Solving Job-Shop Scheduling Problem
Xinli Xu, Wanliang Wang, Qiu Guan 795

A Weather Forecast System Based on Artificial Immune System
Chunlin Xu, Tao Li, Xuemei Huang, Yaping Jiang 800

A New Model of Immune-Based Network Surveillance and Dynamic Computer Forensics
Tao Li, Juling Ding, Xiaojie Liu, Pin Yang 804

A Two-Phase Clustering Algorithm Based on Artificial Immune Network
Jiang Zhong, Zhong-Fu Wu, Kai-Gui Wu, Ling Ou, Zheng-Zhou Zhu, Ying Zhou 814

Immune Algorithm for Qos Multicast Routing
Ziqiang Wang, Dexian Zhang 822

IFCPA: Immune Forgetting Clonal Programming Algorithm for Large Parameter Optimization Problems
Maoguo Gong, Licheng Jiao, Haifeng Du, Bin Lu, Wentao Huang .. 826

A New Classification Method for Breast Cancer Diagnosis: Feature
Selection Artificial Immune Recognition System (FS-AIRS)
  Kemal Polat, Seral Sahan, Halife Kodaz, Salih Günes 830

Artificial Immune Strategies Improve the Security of Data Storage
  Lei Wang, Yinling Nie, Weike Nie, Licheng Jiao 839

Artificial Immune System for Associative Classification
  Tien Dung Do, Siu Cheung Hui, Alvis C.M. Fong 849

Artificial Immune Algorithm Based Obstacle Avoiding Path Planning
of Mobile Robots
  Yen-Nien Wang, Hao-Hsuan Hsu, Chun-Cheng Lin 859

An Adaptive Hybrid Immune Genetic Algorithm for Maximum Cut
Problem
  Hong Song, Dan Zhang, Ji Liu 863

Algorithms of Non-self Detector by Negative Selection Principle in
Artificial Immune System
  Ying Tan, Zhenhe Guo 867

An Algorithm Based on Antibody Immunodominance for TSP
  Chong Hou, Haifeng Du, Licheng Jiao 876

Flow Shop Scheduling Problems Under Uncertainty Based on Fuzzy
Cut-Set
  Zhenhao Xu, Xingsheng Gu 880

An Optimization Method Based on Chaotic Immune Evolutionary
Algorithm
  Yong Chen, Xiyue Huang 890

An Improved Immune Algorithm and Its Evaluation of Optimization
Efficiency
  Chengzhi Zhu, Bo Zhao, Bin Ye, Yijia Cao 895

Simultaneous Feature Selection and Parameters Optimization for SVM
by Immune Clonal Algorithm
  Xiangrong Zhang, Licheng Jiao 905

Optimizing the Distributed Network Monitoring Model with Bounded
Bandwidth and Delay Constraints by Genetic Algorithm
  *Xianghui Liu, Jianping Yin, Zhiping Cai, Xueyuan Huang,
  Shiming Chen* 913

Modeling and Optimal for Vacuum Annealing Furnace Based on
Wavelet Neural Networks with Adaptive Immune Genetic Algorithm
  Xiaobin Li, Ding Liu 922

Lamarckian Polyclonal Programming Algorithm for Global Numerical
Optimization
  Wuhong He, Haifeng Du, Licheng Jiao, Jing Li 931

Coevolutionary Genetic Algorithms to Simulate the Immune System's
Gene Libraries Evolution
  Grazziela P. Figueredo, Luis A.V. de Carvalho,
  Helio J.C. Barbosa 941

Clone Mind Evolution Algorithm
  Gang Xie, Xinying Xu, Keming Xie, Zehua Chen 945

The Application of IMEA in Nonlinearity Correction of VCO Frequency
Modulation
  Gaowei Yan, Jun Xie, Keming Xie 951

A Quick Optimizing Multi-variables Method with Complex Target
Function Based on the Principle of Artificial Immunology
  Gang Zhang, Keming Xie, Hongbo Guo, Zhefeng Zhao 957

Evolutionary Theory

Operator Dynamics in Molecular Biology
  Tsuyoshi Kato 963

Analysis of Complete Convergence for Genetic Algorithm with Immune
Memory
  Shiqin Zheng, Kongyu Yang, Xiufeng Wang 978

New Operators for Faster Convergence and Better Solution Quality in
Modified Genetic Algorithm
  Pei-Chann Chang, Yen-Wen Wang, Chen-Hao Liu 983

Fuzzy Programming for Multiobjective Fuzzy Job Shop Scheduling
with Alternative Machines Through Genetic Algorithms
  Fu-ming Li, Yun-long Zhu, Chao-wan Yin, Xiao-yu Song 992

The Study of Special Encoding in Genetic Algorithms and a Sufficient
Convergence Condition of GAs
  Bo Yin, Zhiqiang Wei, Qingchun Meng 1005

The Convergence of a Multi-objective Evolutionary Algorithm Based on Grids
  Yuren Zhou, Jun He 1015

Influence of Finite Population Size – Extinction of Favorable Schemata
  Hiroshi Furutani, Makoto Sakamoto, Susumu Katayama 1025

A Theoretical Model and Convergence Analysis of Memetic Evolutionary Algorithms
  Xin Xu, Han-gen He 1035

New Quality Measures for Multiobjective Programming
  Hong-yun Meng, Xiao-hua Zhang, San-yang Liu 1044

An Orthogonal Dynamic Evolutionary Algorithm with Niches
  Sanyou Zeng, Deyou Tang, Lishan Kang, Shuzhen Yao, Lixin Ding 1049

Fitness Sharing Genetic Algorithm with Self-adaptive Annealing Peaks Radii Control Method
  Xinjie Yu 1064

A Novel Clustering Fitness Sharing Genetic Algorithm
  Xinjie Yu 1072

Cooperative Co-evolutionary Differential Evolution for Function Optimization
  Yan-jun Shi, Hong-fei Teng, Zi-qiang Li 1080

Optimal Design for Urban Mass Transit Network Based on Evolutionary Algorithms
  Jianming Hu, Xi Shi, Jingyan Song, Yangsheng Xu 1089

A Method for Solving Nonlinear Programming Models with All Fuzzy Coefficients Based on Genetic Algorithm
  Yexin Song, Yingchun Chen, Xiaoping Wu 1101

An Evolutionary Algorithm Based on Stochastic Weighted Learning for Constrained Optimization
  Jun Ye, Xiande Liu, Lu Han 1105

A Multi-cluster Grid Enabled Evolution Framework for Aerodynamic Airfoil Design Optimization
  Hee-Khiang Ng, Dudy Lim, Yew-Soon Ong, Bu-Sung Lee, Lars Freund, Shuja Parvez, Bernhard Sendhoff 1112

A Search Algorithm for Global Optimisation
S. Chen, X.X. Wang, C.J. Harris 1122

Selection, Space and Diversity: What Can Biological Speciation Tell Us About the Evolution of Modularity?
Suzanne Sadedin 1131

On Evolutionary Optimization of Large Problems Using Small Populations
Yaochu Jin, Markus Olhofer, Bernhard Sendhoff 1145

Membrane, Molecular, and DNA Computing

Reaction-Driven Membrane Systems
Luca Bianco, Federico Fontana, Vincenzo Manca 1155

A Genetic Algorithm Based Method for Molecular Docking
Chun-lian Li, Yu Sun, Dong-yun Long, Xi-cheng Wang 1159

A New Encoding Scheme to Improve the Performance of Protein Structural Class Prediction
Zhen-Hui Zhang, Zheng-Hua Wang, Yong-Xian Wang 1164

DNA Computing Approach to Construction of Semantic Model
Yusei Tsuboi, Zuwairie Ibrahim, Nobuyuki Kasai, Osamu Ono 1174

DNA Computing for Complex Scheduling Problem
Mohd Saufee Muhammad, Zuwairie Ibrahim, Satomi Ueda, Osamu Ono, Marzuki Khalid 1182

On Designing DNA Databases for the Storage and Retrieval of Digital Signals
Sotirios A. Tsaftaris, Aggelos K. Katsaggelos 1192

Composite Module Analyst: Tool for Prediction of DNA Transcription Regulation. Testing on Simulated Data
Tatiana Konovalova, Tagir Valeev, Evgeny Cheremushkin, Alexander Kel 1202

Simulation and Visualization for DNA Computing in Microreactors
Danny van Noort, Yuan Hong, Joseph Ibershoff, Jerzy W. Jaromczyk 1206

Ants Colony

A Novel Ant Clustering Algorithm with Digraph
  Ling Chen, Li Tu, Hongjian Chen 1218

Ant Colony Search Algorithms for Optimal Packing Problem
  Wen Peng, Ruofeng Tong, Min Tang, Jinxiang Dong 1229

Adaptive Parallel Ant Colony Algorithm
  Ling Chen, Chunfang Zhang 1239

Hierarchical Image Segmentation Using Ant Colony and Chemical Computing Approach
  Pooyan Khajehpour, Caro Lucas, Babak N. Araabi 1250

Optimization of Container Load Sequencing by a Hybrid of Ant Colony Optimization and Tabu Search
  Yong Hwan Lee, Jaeho Kang, Kwang Ryel Ryu, Kap Hwan Kim 1259

A Novel Ant Colony System Based on Minimum 1-Tree and Hybrid Mutation for TSP
  Chao-Xue Wang, Du-Wu Cui, Zhu-Rong Wang, Duo Chen 1269

Author Index 1279

Table of Contents – Part III

Evolutionary Methodology

Multi-focus Image Fusion Based on SOFM Neural Networks and Evolution Strategies
Yan Wu, Chongyang Liu, Guisheng Liao 1

Creative Design by Chance Based Interactive Evolutionary Computation
Chao-Fu Hong, Hsiao-Fang Yang, Mu-Hua Lin 11

Design of the Agent-Based Genetic Algorithm
Honggang Wang, Jianchao Zeng, Yubin Xu 22

Drawing Undirected Graphs with Genetic Algorithms
Qing-Guo Zhang, Hua-Yong Liu, Wei Zhang, Ya-Jun Guo 28

A Novel Type of Niching Methods Based on Steady-State Genetic Algorithm
Minqiang Li, Jisong Kou 37

Simulated Annealing Genetic Algorithm for Surface Intersection
Min Tang, Jin-xiang Dong 48

A Web Personalized Service Based on Dual GAs
Zhengyu Zhu, Qihong Xie, Xinghuan Chen, Qingsheng Zhu 57

A Diversity Metric for Multi-objective Evolutionary Algorithms
Xu-yong Li, Jin-hua Zheng, Juan Xue 68

An Immune Partheno-Genetic Algorithm for Winner Determination in Combinatorial Auctions
JianCong Bai, HuiYou Chang, Yang Yi 74

A Novel Genetic Algorithm Based on Cure Mechanism of Traditional Chinese Medicine
Chao-Xue Wang, Du-Wu Cui, Lei Wang, Zhu-Rong Wang 86

An Adaptive GA Based on Information Entropy
Yu Sun, Chun-lian Li, Ai-guo Wang, Jia Zhu, Xi-cheng Wang 93

A Genetic Algorithm of High-Throughput and Low-Jitter Scheduling
for Input-Queued Switches
    Yaohui Jin, Jingjing Zhang, Weisheng Hu 102

Mutation Matrix in Evolutionary Computation: An Application to
Resource Allocation Problem
    Jian Zhang, Kwok Yip Szeto 112

Dependent-Chance Programming Model for Stochastic Network
Bottleneck Capacity Expansion Based on Neural Network and Genetic
Algorithm
    Yun Wu, Jian Zhou, Jun Yang 120

Gray-Encoded Hybrid Accelerating Genetic Algorithm for Global
Optimization of Water Environmental Model
    Xiaohua Yang, Zhifeng Yang, Zhenyao Shen, Guihua Lu 129

Hybrid Chromosome Genetic Algorithm for Generalized Traveling
Salesman Problems
    *Han Huang, Xiaowei Yang, Zhifeng Hao, Chunguo Wu,
    Yanchun Liang, Xi Zhao* 137

A New Approach Belonging to EDAs: Quantum-Inspired Genetic
Algorithm with Only One Chromosome
    Shude Zhou, Zengqi Sun 141

A Fast Fingerprint Matching Approach in Medicare Identity Verification
Based on GAs
    Qingquan Wang, Lili Rong 151

Using Viruses to Improve GAs
    Francesco Pappalardo 161

A Genetic Algorithm for Solving Fuzzy Resource-Constrained Project
Scheduling
    Hong Wang, Dan Lin, Minqiang Li 171

A Hybrid Genetic Algorithm and Application to the Crosstalk Aware
Track Assignment Problem
    *Yici Cai, Bin Liu, Xiong Yan, Qiang Zhou,
    Xianlong Hong* 181

A Genetic Algorithm for Solving Resource-Constrained Project
Scheduling Problem
    Hong Wang, Dan Lin, Minqiang Li 185

Evolutionary Algorithm Based on Overlapped Gene Expression
Jing Peng, Chang-jie Tang, Jing Zhang, Chang-an Yuan 194

Evolving Case-Based Reasoning with Genetic Algorithm in Wholesaler's Returning Book Forecasting
Pei-Chann Chang, Yen-Wen Wang, Ching-Jung Ting, Chien-Yuan Lai, Chen-Hao Liu 205

A Novel Immune Quantum-Inspired Genetic Algorithm
Ying Li, Yanning Zhang, Yinglei Cheng, Xiaoyue Jiang, Rongchun Zhao 215

A Hierarchical Approach for Incremental Floorplan Based on Genetic Algorithms
Yongpan Liu, Huazhong Yang, Rong Luo, Hui Wang 219

A Task Duplication Based Scheduling Algorithm on GA in Grid Computing Systems
Jianning Lin, Huizhong Wu 225

Analysis of a Genetic Model with Finite Populations
Alberto Bertoni, Paola Campadelli, Roberto Posenato 235

Missing Values Imputation for a Clustering Genetic Algorithm
Eduardo R. Hruschka, Estevam R. Hruschka Jr., Nelson F.F. Ebecken 245

A New Organizational Nonlinear Genetic Algorithm for Numerical Optimization
Zhihua Cui, Jianchao Zeng 255

Hybrid Genetic Algorithm for the Flexible Job-Shop Problem Under Maintenance Constraints
Nozha Zribi, Pierre Borne 259

A Genetic Algorithm with Elite Crossover and Dynastic Change Strategies
Yuanpai Zhou, Ray P.S. Han 269

A Game-Theoretic Approach for Designing Mixed Mutation Strategies
Jun He, Xin Yao 279

FIR Frequency Sampling Filters Design Based on Adaptive Particle Swarm Optimization Algorithm
Wanping Huang, Lifang Zhou, Jixin Qian, Longhua Ma 289

A Hybrid Macroevolutionary Algorithm
  Jihui Zhang, Junqin Xu 299

Evolutionary Granular Computing Model and Applications
  Jiang Zhang, Xuewei Li 309

Application of Genetic Programming for Fine Tuning PID Controller
Parameters Designed Through Ziegler-Nichols Technique
  *Gustavo Maia de Almeida, Valceres Vieira Rocha e Silva,
  Erivelton Geraldo Nepomuceno, Ryuichi Yokoyama* 313

Applying Genetic Programming to Evolve Learned Rules for Network
Anomaly Detection
  Chuanhuan Yin, Shengfeng Tian, Houkuan Huang, Jun He 323

A Pattern Combination Based Approach to Two-Dimensional Cutting
Stock Problem
  Jinming Wan, Yadong Wu, Hongwei Dai 332

Fractal and Dynamical Language Methods to Construct Phylogenetic
Tree Based on Protein Sequences from Complete Genomes
  Zu-Guo Yu, Vo Anh, Li-Quan Zhou 337

Evolutionary Hardware Architecture for Division in Elliptic Curve
Cryptosystems over $GF(2^n)$
  Jun-Cheol Jeon, Kee-Won Kim, Kee-Young Yoo 348

An Evolvable Hardware System Under Varying Illumination
Environment
  In Ja Jeon, Phill Kyu Rhee 356

An Evolvable Hardware Chip for Image Enhancement in Surface
Roughness Estimation
  M. Rajaram Narayanan, S. Gowri, S. Ravi 361

Evolutionary Agents for n-Queen Problems
  Weicai Zhong, Jing Liu, Licheng Jiao 366

Fictitious Play and Price-Deviation-Adjust Learning in Electricity
Market
  Xiaoyang Zhou, Li Feng, Xiuming Dong, Jincheng Shang 374

Automatic Discovery of Subgoals for Sequential Decision Problems
Using Potential Fields
  Huanwen Chen, Changming Yin, Lijuan Xie 384

Improving Multiobjective Evolutionary Algorithm by Adaptive Fitness and Space Division
  Yuping Wang, Chuangyin Dang 392

IFMOA: Immune Forgetting Multiobjective Optimization Algorithm
  Bin Lu, Licheng Jiao, Haifeng Du, Maoguo Gong 399

Genetic Algorithm for Multi-objective Optimization Using GDEA
  Yeboon Yun, Min Yoon, Hirotaka Nakayama 409

Quantum Computing

A Quantum-Inspired Genetic Algorithm for Scheduling Problems
  Ling Wang, Hao Wu, Da-zhong Zheng 417

Consensus Control for Networks of Dynamic Agents via Active Switching Topology
  Guangming Xie, Long Wang 424

Quantum Search in Structured Database
  Yuguo He, Jigui Sun 434

Swarm Intelligence and Intelligent Agents

A Fuzzy Trust Model for Multi-agent System
  Guangzhu Chen, Zhishu Li, Zhihong Cheng, Zijiang Zhao, Haifeng Yan 444

Adaptive Particle Swarm Optimization for Reactive Power and Voltage Control in Power Systems
  Wen Zhang, Yutian Liu 449

A Dynamic Task Scheduling Approach Based on Wasp Algorithm in Grid Environment
  Hui-Xian Li, Chun-Tian Cheng 453

A Novel Ant Colony Based QoS-Aware Routing Algorithm for MANETs
  Lianggui Liu, Guangzeng Feng 457

A Differential Evolutionary Particle Swarm Optimization with Controller
  Jianchao Zeng, Zhihua Cui, Lifang Wang 467

A Mountain Clustering Based on Improved PSO Algorithm
  Hong-yuan Shen, Xiao-qi Peng, Jun-nian Wang, Zhi-kun Hu 477

Multi-agent Pursuit-Evasion Algorithm Based on Contract Net
Interaction Protocol
  Ying-Chun Chen, Huan Qi, Shan-Shan Wang 482

Image Compression Method Using Improved PSO Vector Quantization
  Qian Chen, Jiangang Yang, Jin Gou 490

Swarm Intelligence Clustering Algorithm
Based on Attractor
  Qingyong Li, Zhiping Shi, Jun Shi, Zhongzhi Shi 496

An Agent-Based Soft Computing Society with Application in the
Management of Establishment of Hydraulic Fracture in Oil Field
  Fu hua Shang, Xiao feng Li, Jian Xu 505

Two Sub-swarms Particle Swarm Optimization Algorithm
  Guochu Chen, Jinshou Yu 515

A Mobile Agent-Based P2P Autonomous Security Hole Discovery
System
  Ji Zheng, Xin Wang, Xiangyang Xue, C.K. Toh 525

A Modified Clustering Algorithm Based on Swarm Intelligence
  Lei Zhang, Qixin Cao, Jay Lee 535

Parameter Selection of Quantum-Behaved Particle Swarm Optimization
  Jun Sun, Wenbo Xu, Jing Liu 543

An Emotional Particle Swarm Optimization Algorithm
  Yang Ge, Zhang Rubo 553

Multi-model Function Optimization by a New Hybrid Nonlinear
Simplex Search and Particle Swarm Algorithm
  Fang Wang, Yuhui Qiu, Naiqin Feng 562

Adaptive XCSM for Perceptual Aliasing Problems
  Shumei Liu, Tomoharu Nagao 566

Discrete Particle Swarm Optimization (DPSO) Algorithm for
Permutation Flowshop Scheduling to Minimize Makespan
  K. Rameshkumar, R.K. Suresh, K.M. Mohanasundaram 572

Unified Particle Swarm Optimization for Solving Constrained
Engineering Optimization Problems
    K.E. Parsopoulos, M.N. Vrahatis 582

A Modified Particle Swarm Optimizer for Tracking Dynamic Systems
    Xuanping Zhang, Yuping Du, Zheng Qin, Guoqiang Qin, Jiang Lu .. 592

Particle Swarm Optimization for Bipartite Subgraph Problem: A Case
Study
    Dan Zhang, Zeng-Zhi Li, Hong Song, Tao Zhan 602

On the Role of Risk Preference in Survivability
    Shu-Heng Chen, Ya-Chi Huang 612

An Agent-Based Holonic Architecture for Reconfigurable Manufacturing
Systems
    Fang Wang, Zeng-Guang Hou, De Xu, Min Tan 622

Mobile Robot Navigation Using Particle Swarm Optimization and
Adaptive NN
    Yangmin Li, Xin Chen 628

Collision-Free Path Planning for Mobile Robots Using Chaotic Particle
Swarm Optimization
    Qiang Zhao, Shaoze Yan 632

Natural Computation Applications: Bioinformatics and Bio-medical Engineering

Analysis of Toy Model for Protein Folding Based on Particle Swarm
Optimization Algorithm
    Juan Liu, Longhui Wang, Lianlian He, Feng Shi 636

Selective Two-Channel Linear Descriptors for Studying Dynamic
Interaction of Brain Regions
    Xiao-mei Pei, Jin Xu, Chong-xun Zheng, Guang-yu Bin 646

A Computational Pixelization Model Based on Selective Attention for
Artificial Visual Prosthesis
    Ruonan Li, Xudong Zhang, Guangshu Hu 654

Mosaicing the Retinal Fundus Images: A Robust Registration Technique
Based Approach
    Xinge You, Bin Fang, Yuan Yan Tang 663

Typing Aberrance in Signal Transduction
M. Zhang, G.Q. Li, Y.X. Fu, Z.Z. Zhang, L. He 668

Local Search for the Maximum Parsimony Problem
Adrien Goëffon, Jean-Michel Richer, Jin-Kao Hao 678

Natural Computation Applications: Robotics and Intelligent Control

Optimization of Centralized Power Control by Genetic Algorithm in a DS-CDMA Cellular System
J. Zhou, H. Kikuchi, S. Sasaki, H. Luo 684

Cascade AdaBoost Classifiers with Stage Features Optimization for Cellular Phone Embedded Face Detection System
Xusheng Tang, Zongying Ou, Tieming Su, Pengfei Zhao 688

Proper Output Feedback H_∞ Control for Descriptor Systems: A Convex Optimization Approach
Lei Guo, Keyou Zhao, Chunbo Feng 698

Planning Optimal Trajectories for Mobile Robots Using an Evolutionary Method with Fuzzy Components
Serkan Aydin, Hakan Temeltas 703

Hexagon-Based Q-Learning for Object Search with Multiple Robots
Han-Ul Yoon, Kwee-Bo Sim 713

Adaptive Inverse Control of an Omni-Directional Mobile Robot
Yuming Zhang, Qixin Cao, Shouhong Miao 723

Other Applications of Natural Computation

A Closed Loop Algorithms Based on Chaos Theory for Global Optimization
Xinglong Zhu, Hongguang Wang, Mingyang Zhao, Jiping Zhou 727

Harmony Search for Generalized Orienteering Problem: Best Touring in China
Zong Woo Geem, Chung-Li Tseng, Yongjin Park 741

Harmony Search in Water Pump Switching Problem
Zong Woo Geem 751

A Selfish Non-atomic Routing Algorithm Based on Game
Theory
    Jun Tao, Ye Liu, Qingliang Wu 761

Clone Selection Based Multicast Routing Algorithm
    Cuiqin Hou, Licheng Jiao, Maoguo Gong, Bin Lu 768

A Genetic Algorithm-Based Routing Service for Simulation Grid
    Wei Wu, Hai Huang, Zhong Zhou, Zhongshu Liu 772

Clustering Problem Using Adaptive Genetic Algorithm
    *Qingzhan Chen, Jianghong Han, Yungang Lai, Wenxiu He,
    Keji Mao* 782

FCACO: Fuzzy Classification Rules Mining Algorithm with Ant Colony
Optimization
    Bilal Alatas, Erhan Akin 787

Goal-Directed Portfolio Insurance
    Jiah-Shing Chen, Benjamin Penyang Liao 798

A Genetic Algorithm for Solving Portfolio Optimization Problems with
Transaction Costs and Minimum Transaction Lots
    Dan Lin, Xiaoming Li, Minqiang Li 808

Financial Performance Prediction Using Constraint-Based Evolutionary
Classification Tree (CECT) Approach
    Chi-I Hsu, Yuan Lin Hsu, Pei Lun Hsu 812

A Genetic Algorithm with Chromosome-Repairing Technique for
Polygonal Approximation of Digital Curves
    Bin Wang, Yan Qiu Chen 822

Fault Feature Selection Based on Modified Binary PSO with Mutation
and Its Application in Chemical Process Fault Diagnosis
    Ling Wang, Jinshou Yu 832

Genetic Algorithms for Thyroid Gland Ultrasound Image Feature
Reduction
    Ludvík Tesař, Daniel Smutek, Jan Jiskra 841

Improving Nearest Neighbor Classification with Simulated Gravitational
Collapse
    Chen Wang, Yan Qiu Chen 845

Evolutionary Computation and Rough Set-Based Hybrid Approach to Rule Generation
  Lin Shang, Qiong Wan, Zhi-Hong Zhao, Shi-Fu Chen 855

Assessing the Performance of Several Fitness Functions in a Genetic Algorithm for Nonlinear Separation of Sources
  F. Rojas, C.G. Puntonet, J.M. Górriz, O. Valenzuela 863

A Robust Soft Decision Mixture Model for Image Segmentation
  Pan Lin, Feng Zhang, ChongXun Zheng, Yong Yang, Yimin Hou ... 873

A Comparative Study of Finite Word Length Coefficient Optimization of FIR Digital Filters
  Gurvinder S. Baicher, Meinwen Taylor, Hefin Rowlands 877

A Novel Genetic Algorithm for Variable Partition of Dual Memory Bank DSPs
  Dan Zhang, Zeng-Zhi Li, Hai Wang, Tao Zhan 883

Bi-phase Encoded Waveform Design to Deal with the Range Ambiguities for Sparse Space-Based Radar Systems
  Hai-hong Tao, Tao Su, Gui-sheng Liao 893

Analytic Model for Network Viruses
  Lansheng Han, Hui Liu, Baffour Kojo Asiedu 903

Ant Colony Optimization Algorithms for Scheduling the Mixed Model Assembly Lines
  Xin-yu Sun, Lin-yan Sun 911

Adaptive and Robust Design for PID Controller Based on Ant System Algorithm
  Guanzheng Tan, Qingdong Zeng, Shengjun He, Guangchao Cai 915

Job-Shop Scheduling Based on Multiagent Evolutionary Algorithm
  Weicai Zhong, Jing Liu, Licheng Jiao 925

Texture Surface Inspection: An Artificial Immune Approach
  Hong Zheng, Li Pan 934

Intelligent Mosaics Algorithm of Overlapping Images
  Yan Zhang, Wenhui Li, Yu Meng, Haixu Chen, Tong Wang 938

Adaptive Simulated Annealing for Standard Cell Placement
  Guofang Nan, Minqiang Li, Dan Lin, Jisong Kou 943

Application of Particle Swarm Optimization Algorithm on Robust PID
Controller Tuning
 Jun Zhao, Tianpeng Li, Jixin Qian 948

A Natural Language Watermarking Based on Chinese Syntax
 Yuling Liu, Xingming Sun, Yong Wu 958

A Steganographic Scheme in Digital Images Using Information of
Neighboring Pixels
 Young-Ran Park, Hyun-Ho Kang, Sang-Uk Shin, Ki-Ryong Kwon ... 962

Noun-Verb Based Technique of Text Watermarking Using Recursive
Decent Semantic Net Parsers
 Xingming Sun, Alex Jessey Asiimwe 968

A Novel Watermarking Scheme Based on Independent Component
Analysis
 Haifeng Li, Shuxun Wang, Weiwei Song, Quan Wen 972

On Sequence Synchronization Analysis Against Chaos Based Spread
Spectrum Image Steganography
 Guangjie Liu, Jinwei Wang, Yuewei Dai, Zhiquan Wang 976

Microstructure Evolution of the K4169 Superalloy Blade Based on
Cellular Automaton Simulation
 Xin Yan, Zhilong Zhao, Weidong Yan, Lin Liu 980

Mobile Robot Navigation Based on Multisensory Fusion
 Weimin Ge, Zuoliang Cao 984

Self-surviving IT Systems
 Hengming Zou, Leilei Bao 988

PDE-Based Intrusion Forecast
 Hengming Zou, Henghui Zou 996

A Solution to Ragged Dimension Problem in OLAP
 Lin Yuan, Hengming Zou, Zhanhuai Li 1001

Hardware Implementations of Natural Computation

A Convolutional Neural Network VLSI Architecture Using Sorting
Model for Reducing Multiply-and-Accumulation Operations
 *Osamu Nomura, Takashi Morie, Masakazu Matsugu,
Atsushi Iwata* 1006

A 32-Bit Binary Floating Point Neuro-Chip
  Keerthi Laal Kala, M.B. Srinivas 1015

Improved Blocks for CMOS Analog Neuro-fuzzy Network
  Weizhi Wang, Dongming Jin 1022

A Design on the Vector Processor of 2048point MDCT/IMDCT for
MPEG-2 AAC
  Dae-Sung Ku, Jung-Hyun Yun, Jong-Bin Kim 1032

Neuron Operation Using Controlled Chaotic Instabilities in
Brillouin-Active Fiber Based Neural Network in Smart Structures
  Yong-Kab Kim, Jinsu Kim, Soonja Lim, Dong-Hyun Kim 1044

Parallel Genetic Algorithms on Programmable Graphics Hardware
  Qizhi Yu, Chongcheng Chen, Zhigeng Pan 1051

Fuzzy Neural Systems and Soft Computing

A Neuro-fuzzy Approach to Part Fitup Fault Control During Resistance
Spot Welding Using Servo Gun
  Y.S. Zhang, G.L Chen 1060

Automatic Separate Algorithm of Vein and Artery for
Auto-Segmentation Liver-Vessel from Abdominal MDCT Image
Using Morphological Filtering
  *Chun-Ja Park, Eun-kyung Cho, Young-hee Kwon, Moon-sung Park,
  Jong-won Park* 1069

Run-Time Fuzzy Optimization of IEEE 802.11 Wireless LANs
Performance
  Young-Joong Kim, Myo-Taeg Lim 1079

TLCD Semi-active Control Methodology of Fuzzy Neural Network for
Eccentric Buildings
  Hong-Nan Li, Qiao Jin, Gangbing Song, Guo-Xin Wang 1089

Use of Adaptive Learning Radial Basis Function Network in Real-Time
Motion Tracking of a Robot Manipulator
  Dongwon Kim, Sung-Hoe Huh, Sam-Jun Seo, Gwi-Tae Park 1099

Obstacle Avoidance for Redundant Nonholonomic Mobile Modular
Manipulators via Neural Fuzzy Approaches
  Yangmin Li, Yugang Liu 1109

Invasive Connectionist Evolution
   Paulito P. Palmes, Shiro Usui 1119

Applying Advanced Fuzzy Cellular Neural Network AFCNN to
Segmentation of Serial CT Liver Images
   Shitong Wang, Duan Fu, Min Xu, Dewen Hu 1128

New Algorithms of Neural Fuzzy Relation Systems with Min-implication
Composition
   Yanbin Luo, K. Palaniappan, Yongming Li 1132

Neural Networks Combination by Fuzzy Integral in Clinical
Electromyography
   Hongbo Xie, Hai Huang, Zhizhong Wang 1142

Long-Term Prediction of Discharges in Manwan Hydropower Using
Adaptive-Network-Based Fuzzy Inference Systems Models
   Chun-Tian Cheng, Jian-Yi Lin, Ying-Guang Sun, Kwokwing Chau .. 1152

Vector Controlled Permanent Magnet Synchronous Motor Drive with
Adaptive Fuzzy Neural Network Controller
   Xianqing Cao, Jianguang Zhu, Renyuan Tang 1162

Use of Fuzzy Neural Networks with Grey Relations in Fuzzy Rules
Partition Optimization
   Hui-Chen Chang, Yau-Tarng Juang 1172

A Weighted Fuzzy Min-Max Neural Network and Its Application to
Feature Analysis
   Ho-Joon Kim, Hyun-Seung Yang 1178

A Physiological Fuzzy Neural Network
   Kwang-Baek Kim, Hae-Ryong Bea, Chang-Suk Kim 1182

Cluster-Based Self-organizing Neuro-fuzzy System with Hybrid
Learning Approach for Function Approximation
   *Chunshien Li, Kuo-Hsiang Cheng, Chih-Ming Chen,
   Jin-Long Chen* 1186

Fuzzy Output Support Vector Machines for Classification
   Zongxia Xie, Qinghua Hu, Daren Yu 1190

Credit Rating Analysis with AFS Fuzzy Logic
   Xiaodong Liu, Wanquan Liu 1198

A Neural-fuzzy Based Inferential Sensor for Improving the Control of
Boilers in Space Heating Systems
  Zaiyi Liao 1205

A Hybrid Neuro-fuzzy Approach for Spinal Force Evaluation in Manual
Materials Handling Tasks
  *Yanfeng Hou, Jacek M. Zurada, Waldemar Karwowski,
  William S. Marras* 1216

Medicine Composition Analysis Based on PCA and SVM
  Chaoyong Wang, Chunguo Wu, Yanchun Liang 1226

Swarm Double-Tabu Search
  Wanhui Wen, Guangyuan Liu 1231

A Meta-heuristic Algorithm for the Strip Rectangular Packing Problem
  *Defu Zhang, Yanjuan Liu, Shengda Chen,
  Xiaogang Xie* 1235

Music Composition Using Genetic Algorithms (GA) and Multilayer
Perceptrons (MLP)
  Hüseyin Göksu, Paul Pigg, Vikas Dixit 1242

On the Categorizing of Simply Separable Relations in Partial
Four-Valued Logic
  Renren Liu, Zhiwei Gong, Fen Xu 1251

Equivalence of Classification and Regression Under Support Vector
Machine Theory
  Chunguo Wu, Yanchun Liang, Xiaowei Yang, Zhifeng Hao 1257

Fuzzy Description of Topological Relations I: A Unified Fuzzy
9-Intersection Model
  Shihong Du, Qiming Qin, Qiao Wang, Bin Li 1261

Fuzzy Description of Topological Relations II: Computation Methods
and Examples
  Shihong Du, Qiao Wang, Qiming Qin, Yipeng Yang 1274

Modeling and Cost Analysis of Nested Software Rejuvenation Policy
  Jing You, Jian Xu, Xue-long Zhao, Feng-yu Liu 1280

A Fuzzy Multi-criteria Decision Making Model for the Selection of the
Distribution Center
  Hsuan-Shih Lee 1290

Refinement of Clustering Solutions Using a Multi-label Voting
Algorithm for Neuro-fuzzy Ensembles
  Shuai Zhang, Daniel Neagu, Catalin Balescu 1300

Comparison of Meta-heuristic Hybrid Approaches for Two Dimensional
Non-guillotine Rectangular Cutting Problems
  Alev Soke, Zafer Bingul 1304

A Hybrid Immune Evolutionary Computation Based on Immunity and
Clonal Selection for Concurrent Mapping and Localization
  Meiyi Li, Zixing Cai, Yuexiang Shi, Pingan Gao 1308

Author Index... 1313

A Novel Learning Algorithm for Wavelet Neural Networks

Min Huang and Baotong Cui

Control Science and Engineering Research Center,
Southern Yangtze University,
Wuxi, Jiangsu 214122 P.R.China
huangmzqb@163.com

Abstract. Wavelet neural networks(WNN) are a class of neural networks consisting of wavelets. A novel learning method based on immune genetic algorithm(IGA) for continuous wavelet neural networks is presented in this paper. Through adopting multi-encoding, this algorithm can optimize the structure and the parameters of WNN in the same training process. Simulation results show that WNN with novel algorithm has a comparatively simple structure and enhance the probability for global optimization. The study also indicates that the proposed method has the potential to solve a wide range of neural network construction and training problems in a systematic and robust way.

1 Introduction

In recent years, neural networks have been widely studied because of their outstanding capability of fitting nonlinear models. As wavelet has emerged as a new powerful tool for representing nonlinearity, a class of networks combining wavelets and neural networks has recently been investigated [1,2,3].It has been shown that wavelet neural networks(WNN) provide better function approximation ability than the multi-layer perception (MLP) and radial basis function (RBF) networks. However, the learning algorithm of WNN is focused on in this field. Learning of WNN consists of parameters and structural optimization. The training of the network is still mainly based on the gradient-based algorithm, and the local minimum problem has still not been overcome [1,4].Recently, the Genetic Algorithm(GA) has been used to train the networks [5,6]. Immune genetic algorithm(IGA), which combines the immune and GA [7], operates on the memory cells that guarantees the fast convergence toward the global optimum, has affinity calculation routine to embody the diversity of the real immune system and the self-adjustment of the immune response can be embodied by the suppress of production of antibodies. It can avoid the problems which have been found in genetic algorithm.

In this paper, a novel algorithm based on IGA is proposed for training WNN. This algorithm adopted multi-encoding to optimize the structure and the parameters in the same training process. Simulation results show that WNN with novel algorithm has a comparatively simple structure and enhance the probability for global optimization.

2 Wavelet Neural Networks for Function Approximation

Wavelet is a new powerful tool for representing nonlinearity. A function $f(x)$ can be represented by the superposition of daughters $\psi_{a,b}(x)$ of a mother wavelet $\psi(x)$. Where $\psi_{a,b}(x)$ can be expressed as

$$\psi_{a,b}(x) = \frac{1}{\sqrt{a}} \psi(\frac{x-b}{a}) . \tag{1}$$

where $a \in R^+$ and $b \in R$ are, respectively, called dilation and translation parameters.

The continuous wavelet transform of $f(x)$ is defined as

$$w(a,b) = \int_{-\infty}^{\infty} f(x) \overline{\psi}_{a,b}(x) dx . \tag{2}$$

where $\overline{\psi}_{a,b}(x)$ is conjugate complex of $\psi_{a,b}(x)$, and the function $f(x)$ can be reconstructed by the inverse wavelet transform

$$f(x) = \int_{-\infty}^{\infty} \int_{-\infty}^{\infty} w(a,b) \psi_{a,b}(x) \frac{dadb}{a^2} . \tag{3}$$

The continuous wavelet transform and its inverse transform are not directly implemental on digital computers. When the inverse wavelet transform (3) is discreted, $f(x)$ has the following approximative wavelet-based representation form.

$$\hat{f}(x) \approx \sum_{k=1}^{K} w_k \psi(\frac{x-b_k}{a_k}) + \overline{f} . \tag{4}$$

where the w_k, b_k and a_k are weight coefficients, translations and dilations for each daughter wavelet, and K is the number of network nodes. Introducing the parameter $\overline{f}$ into the network can make the network able to approximate the function with a nonzero mean. This approximation can be expressed as the neural network of Fig.1, which contains wavelet nonlinearities in the artificial neurons rather than the standard sigmoid nonlinearities.

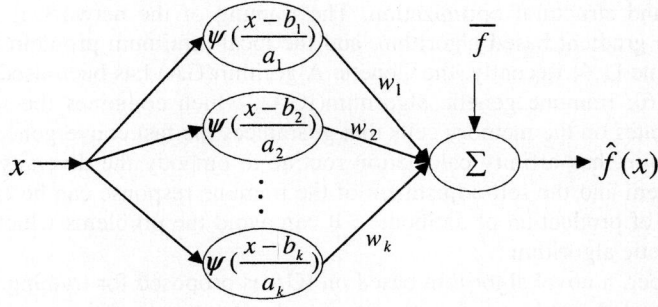

Fig. 1. The structure of wavelet neural networks

The network is learned to determine the minimum value of K and corresponding parameters of network to meet the training error, which is described as follows:

$$E = \frac{1}{2}\sum_{j=1}^{J}(y_j - \hat{f}(x_j))^2 \quad . \tag{5}$$

where $y_j, \hat{f}(x_j)$ are the target output and corresponding network output of the j th sample, respectively, and J is the number of training samples.

3 Immune Genetic Algorithm Based on Multi-encoding for the Training of Wavelet Neural Networks

3.1 Introduction to Immune Genetic Algorithm (IGA)

IGA is an algorithm based on immune principle. It has the same advantages as other stochastic optimization methods possess, but it has the following differences from others for instance GA:

(1) It works on the memory cells, and ensures that it converges on the global optimal solution rapidly.
(2) It uses the computation of affinity to obtain the diversity of the production of antibodies.
(3) It reflects self-adjusting function of the immune system through proliferating and suppressing the emerging of antibodies.

IGA operation composes of recognition of antigens, establishment of coding method, initialization of antibody, computation of affinities and fitness value, proliferation and suppression of antibodies, production of antibodies ,differentiation of memory cells and renovation of group.

3.2 Design of the Algorithm

Step 1. Recognition of antigens: The immune algorithm recognizes the invasion of the antigens which correspond to the input data. In the training problem of WNN, we consider the fitness function $g(x) = E$ as the antigens. E is shown in Eq.(5)

Step 2. Multi-encoding mode: A combined binary system and decimal system is adopted to optimize the structure and the parameters of WNN in the same training process. A chromosome consists of four segments shown in Fig2. Parameters of WNN (b_k, a_k, w_k) are decimal-coded mode and structure of WNN is binary-code mode which describes the validity of corresponding hidden node (1 valid , 0 invalid). M may be selected by experience.

b_1	b_2	...	b_M	a_1	a_2	...	a_M	w_1	w_2	...	w_M	1	0	...	1

Fig. 2. Structure of chromosome multi-encoding

Step 3. Initialization of antibody: In the first iteration, the antibodies are usually produced in the space of solution by random method. The chromosomes corresponding to parameter segment are created from the uniform distribution over the range of (0,1) and the chromosomes corresponding to structure segment are created randomly from a binary string.

Step 4. Computation of affinities: The theory of information entropy is applied to defining affinity here. Suppose there are N antibodies in an immune system, and each antibody has M genes. The information entropy of the j th gene is:

$$H_j(N) = \sum_{i=1}^{N} -p_{ij} \log p_{ij} \quad . \tag{6}$$

where p_{ij} is the probability of the allele of the i th antibody based on the j th gene. For example, if all alleles at j th genes are same, $H_j(N)$ is equal to zero. So, the average information entropy $H(N)$ is given as:

$$H(N) = \frac{1}{M} \sum_{j=1}^{M} H_j(N) \quad . \tag{7}$$

The affinity between antibody v and antibody w is defined as:

$$ay_{v,w} = \frac{1}{1 + H(2)} \quad . \tag{8}$$

When $H(2) = 0$, the genes of the antibodies v and w are identical. And the value of $ay_{v,w}$ is between o and 1.

Similarly, the affinity, ax_v, between the antibody v and the antigen is defined by

$$ax_v = -g \quad . \tag{9}$$

where, g is the value of the fitness function.

Step 5. Proliferation and suppress of antibodies: The antibodies which will perform the next optimization generation are proliferated by crossover and mutation with pre-determined probabilities(Pc, Pm). In this paper, multi-encoding mode is adopted. Crossover and mutation methods in standard GA are applied for binary-code mode. For decimal-coded mode, linear combination crossover method is adopted and mutation operation is defined by

$$x_i^{t+1}(q) = x_i^t(q) + (rand - 0.5) g(x_{best}) \alpha_m \quad . \tag{10}$$

where *rand* is randomly between 0 and 1, $g(x_{best})$ is optimal fitness value until t generation. α_m is mutation operator and q is mutation gene allele. After the proliferation, the size of population is $N + W$, in which W represent the population of newly proliferated antibodies.

For each antibody in the proliferation antibody, if ax_v is less than the threshold T ($T = \min(ax_1, ax_2, \cdots, ax_n)$), the antibody is eliminated. For each antibody in the population, the concentration C_v is calculated by

$$C_v = \frac{\sum_v ay^*_{v,w}}{\text{sum of the antibodies}}. \quad (11)$$

where $\eta \times \max(ay_{v,w}) \leq ay^*_{v,w} \leq \max(ay_{v,w})$, η is a changeable parameter between 0 and 1. The antibody v which has the maximum C_v is eliminated. The procedure will continue unless the population size becomes N.

Step 6. Differentiation of memory cells: The antibodies which have high affinities with the antigen are added to the memory cells.

Step 7. Termination criterion: The termination criterion in this paper is the maximum iteration number I, if the error is less than I, go to step 4, else the optimization procedures stop.

Step 8. Selection of optimal solution: After the iteration stops, the antibody which has the maximum affinity with the antigen in the memory cells is selected as the optimal design parameters.

4 Simulation Results and Analysis

In this section, to investigate the feasibility and effect of the proposed novel algorithm for WNN, one-dimension function approximation is presented. Algorithm is implemented in MATLAB. The selected function is piecewise function defined as follows[1]:

$$f(x) = \begin{cases} -2.186x - 12.864 & -10 \leq x < -2 \\ 4.246x & -2 \leq x < 0 \\ 10e^{-0.05x - 0.5} \times \sin[(0.03x + 0.7)x] & 0 \leq x \leq 10 \end{cases} \quad (12)$$

The wavelet function we have taken is the so-called 'Gaussian-derivative' function $\psi(x) = -xe^{-\frac{1}{2}x^2}$. The maximum number of the hidden nodes is set to 15. The WNN with only one input node and one output node is employed. The parameters are determined as follows: Pop_size=60, Pc=0.85, Pm=0.01, α_m =0.75, $\eta = 0.8$, I =100. 200 of sample are drawn Eq.(12). 150 sets of sample are used to train the network, and the rest are used to test the network. Through evolution by the proposed algorithm, the wavelet network with seven hidden nodes in the hidden layer is obtained. Table 1 shows the results for approximation of the selected piecewise function, and a comparison of the approximation performance with other methods is presented in Table 2.

Table 1. Parameters and structure of wavelet neural networks

M	1	2	3	4	5	6	7	8
b	-4.1472	-3.8215	-4.6872	7.5365	6.4635	-2.1653	5.7420	1.2433
a	6.0528	2.1578	4.8754	3.4237	2.1403	8.7629	2.6135	3.1402
w	-6.3405	6.9365	-3.9365	9.1669	6.7568	5.2816	-10.000	4.2571
structure	1	0	0	1	0	0	1	0
M	9	10	11	12	13	14	15	
b	-1.3405	-6.3472	-3.0614	9.9673	-1.3676	-4.1327	-1.8868	
a	1.4169	0.000	4.1042	5.0197	4.4621	6.1763	2.3473	
w	8.0179	9.9324	-2.8476	-5.1430	1.6230	-6.8178	-4.1979	
structure	0	1	0	1	0	1	1	

Table 2. Comparison of approximation

Models	Number of hidden node	RMS of approximate error
IGA-WNN	7	0.0435
GA-WNN	7	0.0523
WNN(gradient-based)	7	0.0506
BP	9	0.0637

5 Conclusion

This paper adopted the immune genetic algorithm model to solve the learning problems of WNN, which combines the characteristic of both the immune algorithm and the genetic algorithm. Through adopting multi-encoding, this algorithm can optimize the structure and the parameters of WNN in the same training process. The structure of the wavelet neural network can be more reasonable, and the local minimum problem in the training process will be overcome efficiently. Therefore, the wavelet network obtained will give a better approximation and forecasting performance. affinity between antibody v and antibody w is defined as:

References

1. Zhang Q, Benveniste A, Wavelet networks, IEEE Trans. Neural Networks 1992;3(6): 889-898
2. H.H.Szu, B.Telfer, S.Kadambe, Neural network adaptive wavelets for signal representation and classification, Opt. Eng. 31(1992) 1907-1916
3. Y.Fang, T.W.S. Chow, Orthogonal wavelet neural networks applying to identification of Wiener model, IEEE Trans. CAS-I 47(2000) 591-593
4. Zhang J, Gilbert GW, Miao Y. Wavelet neural networks for function learning. IEEE Trans. Signal Processing 1995;43(6): 1485-1496

5. Yao S, Wei CJ, He ZY. Evolving wavelet networks for function approximation. Electronics Letters 1996;32(4): 360-361
6. He Y, Chu F, Zhong B. A Hierarchical Evolutionary Algorithm for Constructing and Training Wavelet Networks. Neural Comput & Applic 2002: 10: 357-366
7. CHU Jang-sung, et al. Study on comparison immune algorithm with other simulated evolution optimization methods[A]. Proceedings of International Conference on Intelligent System Application to Power Systems[C]. Seoul, Korea: 1997.588-592

Using Unscented Kalman Filter for Training the Minimal Resource Allocation Neural Network

Ye Zhang[1], Yiqiang Wu[1], Wenquan Zhang[1], and Yi Zheng[2]

[1] Electronic Information & Engineering Faculty, Nanchang University,
Nanchang 330029, China
zhye901@126.com
[2] Computers information engineering college & Engineering Faculty,
Jiangxi normal university, Nanchang 330027, China

Abstract. The MARN has the same structure as the RBF network and has the ability to grow and prune the hidden neurons to realize a minimal network structure. Several algorithms have been used to training the network. This paper proposes the use of Unscented Kalman Filter (UKF) for training the MRAN parameters i.e. centers, radii and weights of all the hidden neurons. In our simulation, we implemented the MRAN trained with UKF and the MRAN trained with EKF for states estimation. It is shown that the MRAN trained with UKF is superior than the MRAN trained with EKF.

1 Introduction

The radial basis function (RBF) network has been extensively applied to many signal processing, discrete pattern classification, and systems identification problems because of their simple structure and their ability to reveal how learning proceeds in an explicit manner. The MARN is a sequential learning RBF network and has the same structure as a RBF network. The MRAN algorithm uses online learning, and has the ability to grow and prune the hidden neurons to realize a minimal network structure [1]. Fig.1 shows a schematic of a RBF network.

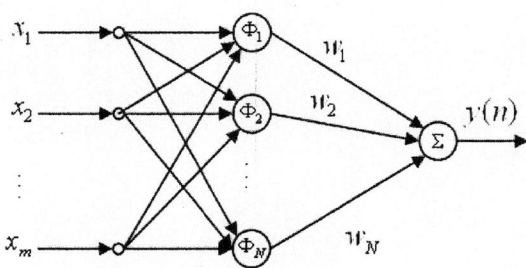

Fig. 1. RAN neural network architecture

The RBF neural network is formed by two layers; hidden layer N local units or basis function, and a linear output layer. The output is given by

$$y(n) = \sum_{i=1}^{N} w_i(n)\Phi_i(\mathbf{x}(n)) \quad (1)$$

where input vector $\mathbf{x}(n) = \begin{bmatrix} x_1 & x_2 & \cdots & x_m \end{bmatrix}^T$, $\Phi_i(\cdot)$ denotes the mapping performed by a local unit, and $w_i(n)$ is the weight associated with that unit. Here n is the time index. The basis function is usually selected as Gaussian function

$$\Phi_i = \exp(-\|\mathbf{x}(n) - \mathbf{c}_i(n)\|^2 / \sigma_i^2(n)) \quad (2)$$

where $\mathbf{c}_i(n)$ and $\sigma_i(n)$ will be referred to as the center and radius, respectively. It can be seen that the design of a RBF requires several decisions, including the centers $\mathbf{c}_i(n)$, the radius $\sigma_i(n)$, the number N, and weight $w_i(n)$. Several training algorithms have been used to train RBF network, including gradient descent [1], back propagation (BP)[5], and extended Kalman filter (EKF) and so on [5]. Major disadvantage of gradient descent and BP methods are slow convergence rates and the long training symbols required. The EKF can be used to determine the centers, radius and weights, but the method provides first-order approximations to optimal nonlinear estimation through the linearization of the nonlinear system. These approximations can include large errors in the true posterior mean and covariance of the transformed (Gaussian) random variable, which may lead to suboptimal performance and sometimes divergence [2]. Using UKF to train the network may have not these problems. In MRAN algorithms, the number of neurons in the hidden layer does not estimate, the network is built based on certain growth criteria. Other network parameters, such as $\mathbf{c}_i(n), \sigma_i(n), w_i(n)$, can be adapted. In section 2, we explain using the EKF for training MRAN network and then present UKF to train the network in section 3. Finally, in section 4, we present simulation results of using the EKF and the UKF for training the MRAN network.

2 Training the MRAN with the EKF

The MRAN network begins with no hidden neuron. As input vector $\mathbf{x}(n)$ are sequentially received, the network builds up based on certain growth and pruning criteria [1]. The following three criteria decide whether a new hidden neuron should be added to the network

$$\|\mathbf{x}(n) - \mathbf{c}_j(n)\| > \varepsilon(n) \quad (3)$$

$$e(n) = d(n) - y(n) > e_{\min} \quad (4)$$

$$e_{rms}(n) = \sqrt{\frac{\sum_{i=n-M+1}^{n}[d(n)-y(n)]^2}{M}} > e'_{min} \qquad (5)$$

where $c_j(n)$ is a centre of the hidden neuron that is nearest to $\mathbf{x}(n)$, the data that was just received, $d(n)$ is the desire output. $\varepsilon(n), e_{min}$ and e'_{min} are threshold to be selected appropriately. M represents the size of a sliding data window that the network has not met the required sum squared error specification. Only when all these criteria are met a new hidden node is added to the network. The parameters associated with it:

$$w_{N+1} = e(n), \quad \mathbf{c}_{N+1} = \mathbf{x}(n), \quad \sigma_{N+1} = \kappa \|\mathbf{x}(n) - \mathbf{c}_j(n)\| \qquad (6)$$

where κ is an overlap factor that determine the overlap of the response of the hidden neuron in the input space. When an input to the network does not meet the criteria for adding a new hidden neuron, EKF will be used to adjust the parameters $\theta = [w_1, \mathbf{c}_1^T, \sigma_1, \cdots, w_N, \mathbf{c}_N^T, \sigma_N]^T$ of the network. The network model to which the EKF can be applied is

$$\theta(n+1) = \theta(n) + \omega(n) \qquad (7)$$

$$y(n) = \sum_{i=1}^{N} w_i(n)\Phi_i(\mathbf{x}(n)) + v(n)$$

$$= g(\theta(n), \mathbf{x}(n)) + v(n)$$

where $\omega(n)$ and $v(n)$ are artificial added noise processes, $\omega(n)$ is the process noise, $v(n)$ is the observation noise. The desired estimate $\hat{\theta}(n)$ can be obtained by the recursion

$$\hat{\theta}(n) = \hat{\theta}(n-1) + \mathbf{k}(n)e(n) \qquad (8)$$

$$\mathbf{k}(n) = \mathbf{P}(n-1)\mathbf{a}(n)\left[\mathbf{R}(n) + \mathbf{a}^T(n)\mathbf{P}(n-1)\mathbf{a}(n)\right]^{-1}$$

$$\mathbf{P}(n) = \left[\mathbf{I} - \mathbf{k}(n)\mathbf{a}^T(n)\right]\mathbf{P}(n-1) + \mathbf{Q}(n)\mathbf{I}$$

where $\mathbf{k}(n)$ is the Kalman gain, $\mathbf{a}(n)$ is the gradient vector and has the following form

$$\mathbf{a}^T(n) = \left.\frac{\partial g(\theta, \mathbf{x}(n))}{\partial \theta}\right|_{\theta=\hat{\theta}(n)} \qquad (9)$$

$\mathbf{P}(n)$ is the error covariance matrix, $\mathbf{R}(n)$ and $\mathbf{Q}(n)$ are the covariance matrices of the artificial noise processes $\omega(n)$ and $v(n)$, respectively. When a new hidden neuron is added the dimensionality of $\mathbf{P}(n)$ is increased by

$$\mathbf{P}(n) = \begin{pmatrix} \mathbf{P}(n-1) & 0 \\ 0 & \mathbf{P}_0 \mathbf{I} \end{pmatrix} \qquad (10)$$

The new rows and columns are initialized by $\mathbf{P}_0$. $\mathbf{P}_0$ is an estimate of the uncertainty in the initial values assigned to the parameters. The dimension of identity matrix $\mathbf{I}$ is equal to the number of new parameters introduced by adding a new hidden neuron.

In order to keep the MRAN in a minimal size and a pruning strategy is employed [1]. According to this, for every observation, each normalized hidden neuron output value $r_k(n)$ is examined to decide whether or not it should be removed.

$$o_k(n) = w_k(n)\exp(-\|\mathbf{x}(n) - \mathbf{c}_k(n)\|^2 / \sigma_k^2(n))$$
$$r_k(n) = \left\| \frac{o_k(n)}{o_{\max}(n)} \right\|, \quad k = 1, \cdots, N \qquad (11)$$

where $o_k(n)$ is the output for kth hidden neuron at time n and $o_{\max}(n)$, the largest absolute hidden neuron output value at n. These normalized values are compared with a threshold δ and if any of them falls below this threshold for M consecutive observation then this particular hidden neuron is removed from the network.

3 Training the MRAN with UKF

The EKF described in the previous section provides first-order approximations to optimal nonlinear estimation through the linearization of the nonlinear system. These approximations can include large errors in the true posterior mean and covariance of the transformed (Gaussian) random variable, which may lead to suboptimal performance and sometimes divergence [2]. The unscented Kalman filter is an alternative to the EKF algorithm. The UKF provides third-order approximation of process and measurement errors for Gaussian distributions and at least second-order approximation for non-Gaussian distributions [5]. Consequently, The UKF may have better performance than the EKF. Foundation to the UKF is the unscented transformation (UT). The UT is a method for calculating the statistic of a random variable that undergoes a nonlinear transformation [2]. Consider propagating a random variable $\mathbf{x}$ (dimension m) through a nonlinear function, $\mathbf{y} = g(\mathbf{x})$. To calculate the statistic of y, a matrix χ of $2m+1$ sigma vectors χ_i is formed as the followings:

$$\begin{aligned}
\chi_0 &= \bar{\mathbf{x}} \\
\chi_i &= \bar{\mathbf{x}} + \left(\sqrt{(m+\lambda)\mathbf{P}_{xx}}\right)_i \quad i = 1, \cdots, m \\
\chi_i &= \bar{\mathbf{x}} - \left(\sqrt{(m+\lambda)\mathbf{P}_{xx}}\right)_{i-L} \quad i = m+1, \cdots, 2m \\
W_0^m &= \lambda/(m+\lambda) \\
W_0^c &= \lambda/(m+\lambda) + (1 - a^2 + \beta) \\
W_i^m &= W_i^c = 1/(2m + 2\lambda) \quad i = 1, \cdots, 2m
\end{aligned} \quad (12)$$

where $\bar{\mathbf{x}}$ and $\mathbf{P}_{xx}$ are the mean and covariance of $\mathbf{x}$, respectively, and $\lambda = a^2(m+\kappa) - m$ is a scaling factor. a determines the spread of the sigma points around $\bar{\mathbf{x}}$ and usually set to a small positive value, typically in the range $0.001 < a < 1$. κ is a secondary scaling parameter which is usually set to 0, and β is used to take account for prior knowledge on the distribution of $\mathbf{x}$, and $\beta = 2$ is the optimal choice for Gaussian distribution[2]. These sigma vectors are propagated through the nonlinear function,

$$y_i = g(\chi_i) \quad i = 0, \cdots, 2m \quad (13)$$

This propagation produces a corresponding vector set that can be used to estimate the mean and covariance matrix of the nonlinear transformed vector $\mathbf{y}$.

$$\begin{aligned}
\bar{y} &\approx \sum_{i=0}^{2m} W_i^m y_i \\
\mathbf{P}_{yy} &\approx \sum_{i=0}^{2m} W_i^c (y_i - \bar{y})(y_i - \bar{y})^T
\end{aligned} \quad (14)$$

From the state-space model of the MRAN given in 7), when an input to the network does not meet the criteria for adding a new hidden neuron, we can use the UKF algorithm to train the network. The algorithms are summarized below.
Initialized with:

$$\begin{aligned}
\hat{\boldsymbol{\theta}}(0) &= E[\boldsymbol{\theta}] \\
\mathbf{P}(0) &= E\left[(\boldsymbol{\theta} - \hat{\boldsymbol{\theta}}(0))(\boldsymbol{\theta} - \hat{\boldsymbol{\theta}}(0))^T\right]
\end{aligned} \quad (15)$$

The sigma-point calculation:

$$\begin{aligned}
\Gamma(n) &= (m + \lambda)(\mathbf{P}(n) + \mathbf{Q}(n)) \\
W(n) &= \left[\hat{\boldsymbol{\theta}}(n), \hat{\boldsymbol{\theta}}(n) + \sqrt{\Gamma(n)}, \hat{\boldsymbol{\theta}}(n) - \sqrt{\Gamma(n)}\right] \\
D(n) &= g(W(n), \mathbf{x}(n)) \\
y(n) &= g(\hat{\boldsymbol{\theta}}(n), \mathbf{x}(n))
\end{aligned} \quad (16)$$

Measurement update equations:

$$\mathbf{P}_{yy}(n) = \sum_{i=0}^{2m} W_i^c (D_i(n) - \overline{\mathbf{y}}(n))(D_i(n) - \overline{\mathbf{y}}(n))^T + \mathbf{R}(n)$$
$$\mathbf{P}_{\theta y}(n) = \sum_{i=0}^{2m} W_i^c (W_i(n) - \hat{\boldsymbol{\theta}}(n))(W_i(n) - \hat{\boldsymbol{\theta}}(n))^T$$
(17)

$$\mathbf{K}(n) = \mathbf{P}_{\theta y}(n) \mathbf{P}_{yy}^{-1}(n) \tag{18}$$

$$\hat{\boldsymbol{\theta}}(n+1) = \hat{\boldsymbol{\theta}}(n) + \mathbf{K}(n) e(n) \tag{19}$$

$$\mathbf{P}(n+1) = \mathbf{P}(n) - \mathbf{K}(n) \mathbf{P}_{yy}(n) \mathbf{K}^T(n) \tag{20}$$

The weight vector of the MRAN is update with the above equations.

4 Experiment Results and Conclusion

In the experiments, the thresholds $e_{\min}, e'_{\min}$, and ε, respectively, set as 0.22, 0.40, and 0.5, the thresholds are chosen largely by trial and error. The other parameters were set as M=10 and δ=0.1. The RAN trained by the UKF and the EKF is used to estimate a time-series corrupted by additive Gaussian white noise (5db SNR). The time-series used is Mackey-Glass chaotic series. In Fig.2 and Fig 3, the dashed is clear time series, "+" is the noise time-series, the solid line is the output of the MRAN trained with EKF or UKF. We can see that the UKF have superior performance compared to that the EKF. After learning, the average number of centers in the hidden layer is 9 nodes.

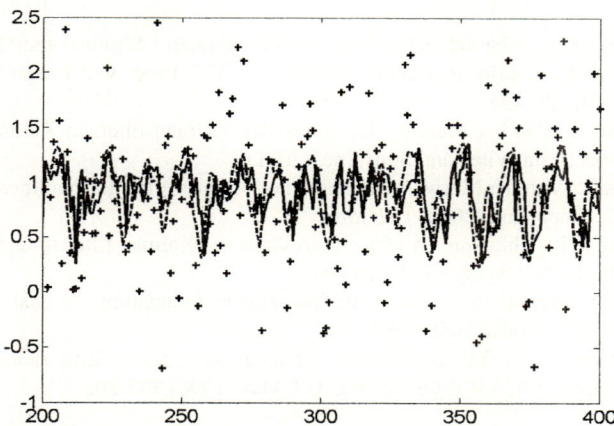

Fig. 2. Estimation of Mackey-Glass time series: EKF

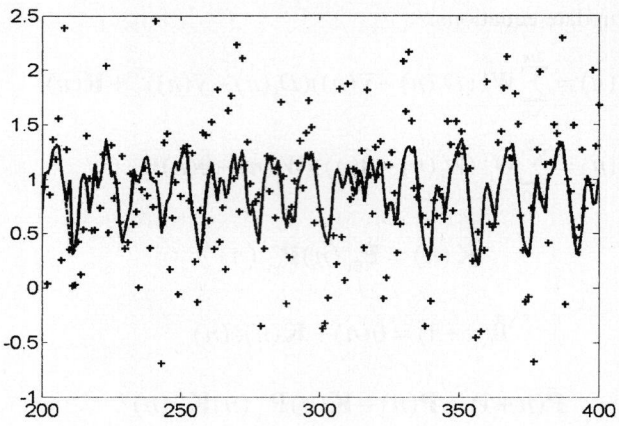

Fig. 3. Estimation of Mackey-Glass time series: UKF

The paper investigated the performance of the MRAN networks. It shows that the MRAN-UKF has better performance than the MRAN-EKF, with much less complexity. In order to reduce the computer load, we can update the parameters of only one hidden neuron instead of all the hidden neurons. This neuron called the winner neuron is chosen as the one closet to the new data received.

Acknowledgments

This work is supported by the foundation of Nanchang University (z02142).

References

1. P. Chandra Kumar, P.Saratchandran and N. Sundararajan.: Minimal radial basis function neural network for nonlinear channel equalization. IEE Proc.-Vis. Image Signal Process, Vol. 147. (2000) 428-435
2. E.A. Wan and R. Van der Merwe.: The unscented Kalman filter, in Kalman filtering and neural networks. John Wiley and Sons, Inc. (2001)
3. E.A. Wan and R. Van der Merwe.; The unscented Kalman filter for nonlinear estimation. In Proc.of IEEE symposium (2000) 152-158
4. S.J. Julier and J. K. Uhlmann.: A new extension of the Kalman filter to nonlinear systems. In Proc. Of the 11th int. symp. on Ascc (1997)
5. Simon Haykin.: Neural networks A comprehensive foundation, Second Edition. Upper Saddle River, NJ: Prentice Hall (1999)
6. J. Lee, C. Beach, and N. Tepedelenlioglu.: Channel equalization using radial basis function network. In Proc. Of ICASSP'96, Atlanta, GA,May (1996) 797-802

The Improved CMAC Model and Learning Result Analysis[*]

Daqi Zhu[1], Min Kong[1], and YonQing Yang[2]

[1] Research Centre of Control Science and Engineering, Southern Yangtze University,
214122 Wu Xi, Jiangshu Province, China
zdq367@yahoo.com.cn
http://www.cc.sytu.edu.cn
[2] Department of mathematics, Southern Yangtze University,
214122 Wu Xi, Jiangshu Province, China
yonqing@sytu.edu.cn

Abstract. An improved neural networks online learning scheme is proposed to speed up the learning process in cerebellar model articulation controllers(CMAC). The improved learning approach is to use the learned times of the addressed hypercubes as the credibility (confidence) of the learned values in the early learning stage, and the updating data for addressed hypercubes is proportional to the inverse of the exponent of learned times, in the later stage the updating data for addressed hypercubes is proportional to the inverse of learned times. With this idea, the learning speed can indeed be improved.

1 Introduction

Speed is very important for the online learning of dynamic nonlinear systems. When learning capability is considered, neural networks are always the first candidates to be taken into account, especially backpropagation (BP) trained multilayer feed forward neural networks. However, owing to the gradient descent nature of BP neural networks, the learning process of BP algorithm may need to iterate many times so as to converge to an acceptable error level, or even cannot converge at all. Another unsuccessful property of BP algorithm is its distributed knowledge representation capability[1-2]. So the BP algorithm can hardly be used for online learning systems. This is because that online learning needs to work within real-time constraints, and training can only be performed for current patterns. As a result, it is hard to find any successful online BP algorithm examples in practical applications.

Another kind of learning approaches termed as cerebellar model articulation controllers(CMAC) was proposed in the literature[3-4], in which several advantages including local generalization and rapid learning convergence have been demonstrated[5-6]. It seems to be a good candidate for online learning. However, when the conventional CMAC approach still needs several cycles(or called epochs) to con-

---

[*] This project is supported by the JiangSu Province Nature Science Foundation (BK 2004021)and the Key Project of Chinese Ministry of Education.(105088).

verge[7-8]. Though the conventional CMAC is much faster than BP algorithm, it still is not good enough for online learning systems. Several approaches have been proposed to improve the learning performance of CMAC[9-10] recently. For instance, the fuzzy concept was introduced into the cell structure of CMAC, it indeed can increase the accuracy of the representation of the stored knowledge. However, the speed of convergence still cannot meet the requirement for real-time applications.

In order to improve the learning speed of CMAC, the learning approach has considered the credibility of the learned values in the literature[11] . In the conventional CMAC learning schemes, the correcting amounts of errors are equally distributed into all addressed hypercubes, regardless of the credibility of those hypercubes. Such an updating algorithm violates the concept of credit assignment, requiring that the updating effects be proportional to the responsibilities of hypercubes. From the literature[11], it is shown that the credit assignment CMAC (CA-CMAC) is faster and more accurate than the conventional CMAC. However, in the literature[11] the times of updating for hypercubes can be viewed as the creditability of those hypercubes, and the updating data for hypercubes is proportional to $\dfrac{1}{f(j)+1}$, $f(j)$ is the learned times of the j th hypercubes. Notice, that the learning times must include the current one to prevent dividing by zero. However in the early learning stage, $f(j)$ is very less, the process of "add one" is unaccepted.

In this paper, A new improved CA-CMAC(ICA-CMAC) learning scheme is presented. The updating data for hypercubes is proportional to $\dfrac{1}{\exp(f(j))}$ when the learned times $f(j)$ =0,1,2 , $f(j)$ >2 the updating data is proportional to $\dfrac{1}{f(j)}$. The example showed that the ICA-CMAC has the best result in learning speed and accuracy.

2 Conventional CMAC and Credit Assigned CMAC

2.1 Conventional CMAC

The basic idea of CMAC is to store learned data into overlapping regions in a way that the data can easily be recalled but use less storage space. Take a two-dimensional(2-D) input vector, or the so-called two-dimensional CMAC(2-D-CMAC),as an example. The input vector is defined by two input variables, x_1 and x_2. The structure of a 2-D-CMAC is shown in Fig .1. In this example, 7 locations, called bits in the literature, are to be distinguished for each variable. For each state variable, three kinds of segmentation, or called floors, are used. For the first floor, the variable x_1 is divided into three blocks, A, B, and C and the variable x_2 is divided into blocks a, b, and c. Then, the areas, Aa, Ab, Ac, Ba, Bb, Bc, Ca, Cb, and Cc are the addresses or the locations that store data, Such areas are often called hypercubes.

Similarly, hypercubes, Dd, De, Df, Ed, Ee, Ef, Fd, Fe, and Ff are defined in the second floor, and Gg, Gh, Gi, Hg, Hh, Hi, Ig, Ih, and Ii are defined in the third floor. Be aware that only the blocks on the same floor can be combined to form a hypercube. Thus, the hypercubes, such as ,Ad and Db, do not exist. In this example, there are 27 hypercubes used to distinguish 49 different states in the 2-D-CMAC.

The basic concept of CMAC is illustrated in Fig.2. There are two phases of operations performed in the CMAC algorithm: the output-producing phase and the learning phase. First, the output-producing phase is discussed. In this phase, CMAC uses a set of indices as an address in accordance with the current input vector(or the so-called state) to extract the stored data. The addressed data are added together to produce the output. Let the number of floors be m, the number of hypercubes be N, and the number of total states be n. Then, the output value y_s for the state s ($s=1,\ldots,n$) is the sum of all addressed data and can be computed as :

$$y_s = \sum_{j=1}^{N} C_s w_j \quad (1)$$

Where w_j is the stored data of the j th hypercube and C_s is the index indicating whether the j th hypercube is addressed by the state s. Since each state addresses exactly m hypercubes, only those addressed C_s are 1, and the others are 0, As shown in Fig1. let the hypercubes Bb, Ee, and Hh be addressed by the state $s(3,3)$, Then only those three C_s, are 1 and the others are 0.

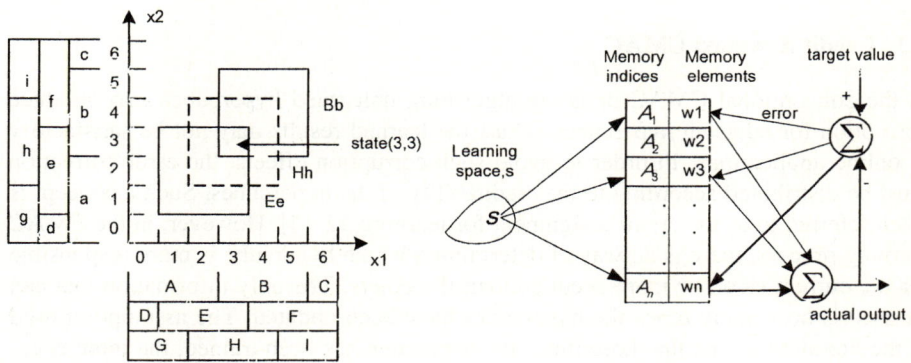

Fig. 1. Structure of a 2-D CMAC   **Fig. 2.** Basic operational concept of CMAC

Whereas the output-producing phase is to generate an output from the CMAC table, the learning phase is to update the data in the CMAC table, according to the error between the desired output and the obtained output. Traditionally, the error is equally distributed to modify the addressed data. Let s be the considered state and

w_j^i be the stored values of the jth hypercube after i iterations. The conventional CMAC updating algorithm for w_j^i is

$$w_j^i = w_j^{i-1} + \frac{\alpha}{m} C_s (\overline{y_s} - \sum_{j=1}^{N} C_s w_j^{i-1}) \qquad (2)$$

Where $\overline{y_s}$ is the desired value for the state s, $\sum_{j=1}^{N} C_s w_j^{i-1}$ is the produced output of the CMAC for the state s, and α is a learning constant. Note that only those addressed hypercubes are updated. It has been proved that if α is not greater than two, then the CMAC learning algorithm will converge[5-6].

In the above learning process, the errors are equally distributed into the hypercubes being addressed. However, after $i-1$ iterations, the original stored data in the CMAC table already contain some knowledge about previous learning. However, not every hypercubes has the same learning history, hence, those hypercubes do not have the same credibility. Disregarding such differences, all addressed hypercubes get equal shares for error correcting in (2) .As a result, previous learned information may be corrupted due to large error caused by an unlearned state. When the training process lasts for several cycles, this situation may actually be "smoothed out". This is evident from successful learning in various CMAC applications, However, when online learning is required, and perhaps only one cycle of training can be performed, there may not have enough time for smoothing out the corrupted data. Thus, the learned results of the updating algorithm may not be acceptable. This can be seen in later simulations.

2.2 Credit Assigned CMAC

In the conventional CMAC updating algorithm, unlearned hypercubes may produce corruption for adjacent hypercubes. Thus, the learned results may not be satisfactory in online applications. In order to avoid such corruption effects, the error correction must be distributed according to the creditability of the hypercubes. Such a concept is often referred to as the credit assignment for learning[12-13]. However, in the CMAC learning process, there is no way of determining which hypercube is more responsible for the current error, or more accurate than the others. The only information that can be used is how many times the hypercubes have been updated. The assumption used in the literature is that the more times the hypercube has been trained, the more accurate the stored value is. Hence, the times of updating for hypercubes can be viewed as the creditability of those hypercubes.

With the above assumption, in the literature[11] formula (2) is rewritten as :

$$w_j^i = w_j^{i-1} + \alpha C_s \{ \frac{(f(j)+1)^{-1}}{\sum_{l=1}^{m}(f(l)+1)^{-1}} \} (\overline{y_s} - \sum_{j=1}^{N} C_s w_j^{i-1}) \qquad (3)$$

Where $f(j)$ is the learned times of the jth hypercube, and m is the number of addressed hypercubes for a state. The idea of the updating algorithm is that the effects of error correcting must be proportional to the inverse of learning times for the addressed hypercubes. Notice, that the learning times must include the current one to prevent dividing by zero. In (3) the equal share of error correcting as 1/m in (2) is replaced by $(f(j)+1)^{-1} \Big/ \sum_{l=1}^{m}(f(l)+1)^{-1}$. With this modification, the learning effects can be appropriately distributed into the addressed hypercubes according to the creditability of hypercubes. However, it is not the best result, because it did not research how to effect learning result by the process of "add one" further.

3 Improved Credit Assigned CMAC(ICA-CMAC)

3.1 Credit Assigned

According to analysis above, in order to prevent dividing by zero, moreover it do not affect the learning speed. in the ICA-CMAC, (3) is rewritten as:

$$w_j^j = \begin{cases} w_j^{j-1} + \alpha C_s \{\dfrac{\exp(f(j))}{\sum_{l=1}^{m}\exp(f(j))}\}(\overline{y}_s - \sum_{j=1}^{N} C_s w_j^{j-1}) & f(j)=0,\ 1,\ 2 \quad j=1,....m \\ \\ w_j^{j-1} + \alpha C_s \{\dfrac{(f(j))^{-1}}{\sum_{l=1}^{m}(f(j))^{-1}}\}(\overline{y}_s - \sum_{j=1}^{N} C_s w_j^{j-1}) & \text{others} \end{cases} \qquad (4)$$

In (4) not only there is the concept of reasonable credit assignment, but also the situation of "dividing by zero" do not existence. From later simulations, it can be seen that the learned results of ICA-CMAC is better than conventional CMAC and CA-CMAC.

To illustrate the learning effects of ICA-CMAC, a simple example is considered. The target function is $d(x_1,x_2) = \sqrt{(x_1-2)^2 + (x_2-2)^2}$. Let the training data be [{(2,0), 2},{(3,0), 2.2361},{(4,1), 2.2361},{(5,2), 3.0000},......]. The CMAC shown in Fig.1 is used. First, the state (2,0) addresses three hypercubes, Aa, Ed, and Hg. Then, y(2,0)=0 and d(2,0)=2. Since those hypercubes are all unlearned, each hypercube gets 1/3 of the error. The weights of Aa, Ed, and Hg all become (2-0)/3=0.6667; Next, (3,0) addresses Ba, Ed, and Hg, d (3,0)=2.2361 and y(3,0)=0.6667+0.6667+0=1.3334. For CMAC, the error is equally distributed into those three hypercubes. Δ=(2.2361-1.3334)/3=0.3009. The weights of Ed and Hg become 0.9676 and the weight of Ba becomes 0.3009. For CA-CMAC, since Ed and Hg are selected the second times, each get 1/4 of the error, and Ba gets 1/2 of the error. The weights of Ed and Hg become 0.6667+(2.2361-1.3334)/4=0.8924 and the weight of Ba becomes (2.2361-1.3334)/2=0.4514. For ICA-CMAC, Ed and Hg are selected the second times also, from (4), each get 1/(e+2) of the error, and Ba gets e/(e+2) of the error. The weights of Ed and Hg become 0.6667+(2.2361-1.3334)/(e+2)=0.8580 and the weight of Ba becomes (2.2361-1.3334)*e/(e+2)=0.5201. Here, it can be found that the error in this step may largely come from the value 0 stored in Ba. For CMAC, all three hypercubes get the same share of the error. For ICA-CMAC, a larger portion of the error goes to the weight of Ba. From the next step, it will be evident that the resultant error of ICA-CMAC will be smaller than others. Now,(4,1) addresses Ba, Fe, and Hg. d(4,1)=2.2361, y(4,1)=0.3009+0+0.9676 =1.2685 for CMAC, and y(4,1)=0.4514+0+0.8924=1.3438 for CA-CMAC, and y(4,1)=0.5201+0+0.8580 =1.3781 for ICA-CMAC, Obviously, the predicted value in ICA-CMAC is more close to the desired value 2.2361 than that in CMAC and CA-CMAC method. Table 1 shows the errors for the first cycles. It can be found that the errors of ICM-CMAC are all lower than others.

Table 1. learning behavior comparison for CMAC, CA-CMAC and ICA-CMAC

State	(2,0)	(3,0)	(4,1)	(5,2)	......
$d(x_1,x_2)$	2.0000	2.2361	2.2361	3.0000	......
CMAC	0	1.3334	1.2685	0.9459	......
CA-CMAC	0	1.3334	1.3438	1.1814	......
ICA-CMAC	0	1.3334	1.3781	1.3009	......

3.2 Adressing Function

In the original CMAC[3-4], a hashing method is used to reduce the storage space. The hashing method is a way of storing data in a more compact manner, but may lead to collisions of data, and then may reduce the accuracy of CMAC. In fact, a paper[14] exists that questions the applicability of the use of hash coding in CMAC. In our approach, an addressing function is used to simultaneously generate the indices to address the required hypercubes[11], This approach is to code all possible hypercubes in an array ,which saves a lot of time and memory when compared to simple addressing approaches, and will not cause any collisions in data retrieval.

Take a three dimensional (3-D) CMAC as an example. Suppose that for each dimension, there are $m*(nb-1)+1$ locations to be distinguished, where m is the number of floors in CMAC and nb is the block number for each floor. In this example, each block covers m states and only $N = m*nb^3$ hypercubes are needed to distinguish $(m*(nb-1)+1)^3$ states. Consider a state s, denoted by (x_1, x_2, x_3) representing the locations of the state for those three dimensions, respectively, Let the m addressed hypercubes by the state s be $s(j)$, for j=1,...,m, The addressing function is to generate $s(j)$, for j=1,...m, The addressing function $s(j) = F(x_1, x_2, x_3, j)$, is

① if j=1,then i=0, else i=m-j+1; ② $ax = \text{int}((x_1 + i)/m)$;
③ $ay = \text{int}((x_2 + i)/m)$; ④ $az = \text{int}((x_3 + i)/m)$;
⑤ $s(j) = F(x_1, x_2, x_3, j) = ax + ay + az * nb^2 + (j-1) * nb^3 + 1$.

When a state is defined, with this addressing function, the addressed hypercubes can directly be obtained, Thus, no matter in the output-producing phase, or in the learning phase, the required data extraction or data updating can be performed with those hypercubes directly.

4 Simulation Results

There are two examples to illustrated the learning effects of ICA-CMAC further, the two examples are conducted to compare the learning speed of conventional CMAC, CA-CMAC, and ICA-CMAC; they are

$$y(x_1, x_2) = (x_1^2 - x_2^2)\sin(x_1)\cos(x_2) \quad -1 \le x_1 \le 1 \text{ and } -1 \le x_2 \le 1 \quad (5)$$

$$y(x_1, x_2) = \sin(x_1) + 2\cos(2x_1) + e^{-x_2} \quad -1 \le x_1 \le 1 \text{ and } -1 \le x_2 \le 1 \quad (6)$$

For both examples, each variable contains 64 locations. For each variable, 9 floors are used, and each floor contains 8 blocks. The total states are 4096= $64 * 64$, and the number of used hypercubes is $9 * 8 * 8$ =576(only 14% of the total states). The learning $\alpha = 1$. The training data is obtained by equally sampling in both variables, and the number of the used training data 4096.

The learning histories for the two examples are illustrated in Fig 3 and Fig 4. The ways of evaluating the errors are considered. The total absolute errors (TAE) from the first cycle to the 6th cycle, and from the 26th to the 30th cycle are tabulated in tables 2 and tables 3.

$$TAE = \sum_{s=1}^{n} |(\overline{y_s} - y_s)| \quad (7)$$

Where n is the number of total states, $\overline{y_s}$ is the desired value for the state s, y_s is the output value for the state s.

Table 2. Total absolute errors (TAE) $y(x_1, x_2) = (x_1^2 - x_2^2)\sin(x_1)\cos(x_2)$

	1	2	3	4	5	6		26	27	28	29	30
CMAC	40.42	33.36	20.57	21.95	15.23	16.28	...	7.08	7.02	6.96	6.89	6.85
CA-CMAC	25.34	15.70	11.27	11.28	9.758	9.868	...	7.34	7.32	7.29	7.27	7.25
ICA-CMAC	20.35	9.001	8.456	8.324	7.925	7.938	...	6.95	6.93	6.92	6.90	6.89

Table 3. Total absolute errors (TAE) $y(x_1, x_2) = \sin(x_1) + 2\cos(2x_1) + e^{-x_2}$

	1	2	3	4	5	6		26	27	28	29	30
CMAC	276.0	252.	237.	225.	209.	184.4	...	7.96	7.50	7.04	6.58	6.14
CA-CMAC	175.0	65.7	51.7	44.6	41.0	37.93	...	8.70	8.39	8.09	7.83	7.56
ICA-CMAC	150.6	22.4	16.2	13.9	12.4	11.01	...	4.09	3.96	3.85	3.75	3.66

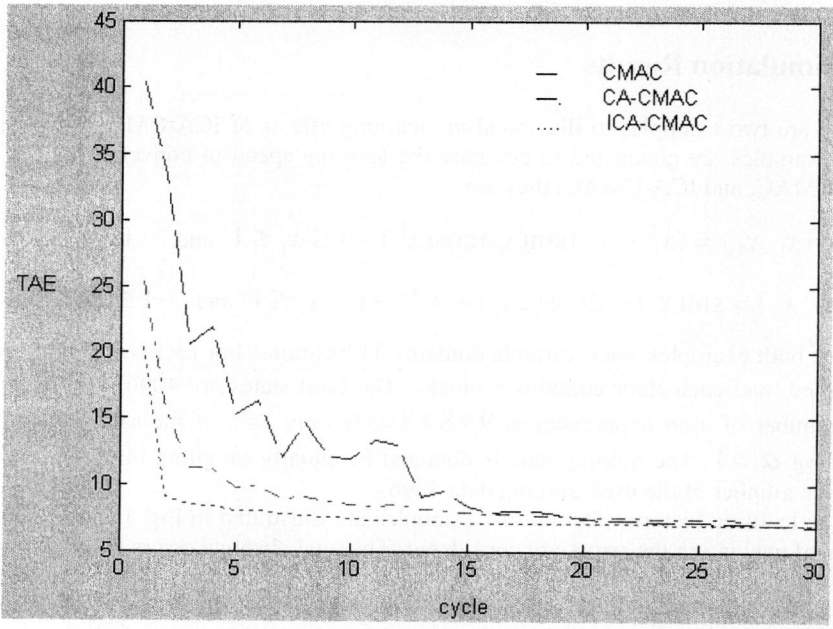

Fig. 3. TAE of learning histories for $y(x_1, x_2) = (x_1^2 - x_2^2)\sin(x_1)\cos(x_2)$

Online learning schemes are typically used for time-varying systems because those schemes can "observe" the changes and then cope with them. When there are changes (time-varying parameters) in the system, errors occur to compensate those changes. Those errors are then distributed into hypercubes according to the used update law. The error correcting ability of ICA-CMAC is not different from conventional CMAC and CA-CMAC for this situation. They may be different only in the distributed amount of the errors. Such a distribution in ICA-CMAC is dependent on the learning times of hypercubes, and the learning times of hypercubes are approximately the same if sufficient learning is conducted. Thus, while facing time-varying systems, there are no differences in different CMAC for long time. From those figures and tables, after 15 cycles, there is a little difference for different neural networks, and all CMAC can learn well.

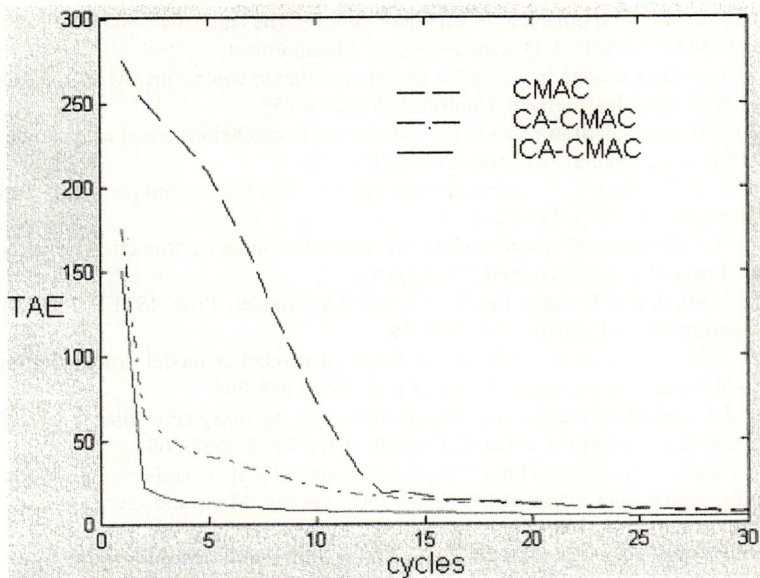

Fig. 4. TAE of learning histories for $y(x_1, x_2) = \sin(x_1) + 2\cos(2x_1) + e^{-x_2}$

But in the early learning stage, the learning results are wholly different. It can be observed that the errors of ICA-CMAC are much smaller than others, such as the conventional CMAC and CA-CMAC. Thus, we can conclude that ICA-CMAC indeed learns faster than conventional CMAC and CA-CMAC in the early learning stage. It compensates inappropriate process of "add one" in the design of credibility.

5 Conclusions

In the paper, the improved CA-CMAC(ICA-CMAC) learning approach is proposed. The updating data for addressed hypercubes is proportional to the inverse of exponent of learned times in the early learning stage (the learned times is one or two), in the later stage the updating data addressed hypercubes is proportional to the inverse of learned times.

With this idea, the learning speed of ICA-CMAC indeed becomes very faster than conventional CMAC and CA-CMAC in the early learning stage. It is very important for successful online learning.

References

1. Hinton G.E., Mcclelland J.L., and Rumelhart D.E.: Distributed representation, Parallel Distributed Processing, Rumelhart et al., Eds: MIP Press, 3(1986).
2. Kosko B.: A dynamical systems approach to machine intelligence, Neural Networks and Fuzzy Systems, Englewood Cliffs, NJ: Prentice-Hall, (1992).

3. Albus J.S.: A new approach to manipulator control: The cerebellar model articulation controller(CMAC), ASME J. Dynamic Systems, Measurement, Control, (1975)220-227.
4. Albus J.S.: Data storage in cerebellar model articulation controller(CMAC), ASME J. Dynamic Systems, Measurement, Control, (1975)228-233.
5. Wong Y.F. and Sideris A.: Learning convergence in cerebellar model articulation controller, IEEE Trans. Neural Networks, 2(1992)115-121.
6. Lin C.S. and Chiang C.T., Learning convergence of CMAC technique, IEEE Trans. Neural Networks, 6(1997)1281-1292.
7. Iiguni Y.: Hierarchical image coding via cerebellar model arithmetic computers, IEEE Trans. Image Processing, 6(1996)1393-1401.
8. Hong C.M., Lin C.H., and Tao T.: Grey-CMAC model, Proc. ISTED Int. Conf. High Technology Power Industry, (1997) 39-44.
9. Nie J. and Linkens D.A.: FCMAC: A fuzzified cerebellar model articulation controller with self-organizing capacity, Automatica, 4 (1994) 655-664.
10. Geng Z.J. and Mccullough C.L.: Missile control using fuzzy cerebellar model arithmetic computer neural networks, J. Guid., Control, Dyn., 3(1997)557-565.
11. Shun-Feng S., Ted T.: and Hung T.H., Credit assigned CMAC and its application to online learning robust controllers, IEEE Trans. On Systems, Man, and Cybernetics----Part B: Cybernetics, 2(2003) 202-213.
12. Smalz R. and Conrad M.: Combining evolution with credit apportionment : A new learning algorithm for neural nets, Neural Networks, 2(1994) 341-351.
13. Plantamura V.L., Soucek B., and Visaggio G.: Holographic fuzzy learning for credit scoring, Proc. Int. Joint Conf. Neural Networks, 6(1993)729-732.
14. Wang Z.Q., Schiano J.L., and Ginsberg M.: Hash-coding in CMAC neural networks, IEEE Int'l Conf. Neural Networks, (1996)1698-1703.

A New Smooth Support Vector Regression Based on ϵ-Insensitive Logistic Loss Function

Yang Hui-zhong*, Shao Xin-guang, and Ding Feng

Research Center of Control Science and Engineering,
Southern Yangtze University, 214122, Wuxi, P. R. China
yhz@sytu.edu.cn

Abstract. A new smooth support vector regression based on ϵ-insensitive logistic loss function, shortly Lϵ-SSVR, was proposed in this paper, which is similar to SSVR, but without adding any heuristic smoothing parameters and with robust absolute loss. Taking advantage of Lϵ-SSVR, one can now consider SVM as linear programming, and efficiently solve large-scale regression problems without any optimization packages. Details of this algorithm and its implementation were presented in this paper. Simulation results for both artificial and real data show remarkable improvement of generalization performance and training time.

1 Introduction

Support Vector Machine (SVM) was first proposed by Vapnik and had been one of the most developed topics in Machine Learning [1,2,3]. The nature of the conventional SVM is solving a standard convex quadratic programming (QP) problem [4], with linear constraints, which depends on the training data set and the selection of a few of SVM parameters. For a small training set (less than few hundreds points), the solution of the QP problem can be obtained straightly by using standard QP packages such as CPLEX and LOQO. However, with the massive datasets, the memory space will increase with the level of $O(m^2)$, where m is the number of the training points. This indicates that the optimization techniques mentioned above may be unsuitable to solve the large-scale problems. Besides the size of training set, the influence of SVM parameters on the performance is also great [5]. It is true that we do not have any analytical method for parameter selection. Hence, designing effective SVM training algorithms for massive datasets with less heuristic parameters will be of momentous practical significance.

At present, a number of SVM algorithms based on iteration or decomposition strategies have been extensively developed to handle large datasets, such as kernel adatron algorithm [6], successive over relaxation algorithm (SOR) [7] and sequential minimal optimization algorithm (SMO) [8]. Although these methods, to a certain extent, can decrease the size and the degree of the difficulty of

---

* This research was supported by the fund of Hi-Tech Research and Development Program of China (863 Program) No. 2002AA412120.

QP problem by partitioning datasets and solving small sub-problems iteratively, most of them still need an optimization package and long CPU time to complete the whole iterative procedure. Another method in which SVM was solved as linear programming without any optimization package was proposed in literature [9] and [10]. In this case, one employed the smoothing techniques to transform the primal QP to a smooth unconstrained minimization problem, and then used a fast Newton-Arjmor algorithm to solve it. Although SSVR yielded a great improvement on training speed, a heuristic smoothing parameter was added during transformation, this would increase the difficulty of model selection, which is very important for obtaining better generalization [5,11,12]. In additional, the squared loss used in SSVR is not the better choice for robust regression either [13].

In order to avoid SSVR's disadvantages, a new smooth support vector regression based on ϵ-insensitive logistic loss function was proposed in this paper.

The paper is organized as follows: Section 2 provides a brief review over support vector regression. A new smooth support vector regression based on ϵ-insensitive logistic loss function is derived in section 3. Section 4 describes the implementation details based on pure Newton method. Section 5 gives the experiments results, and the conclusion of the paper lies in the last section.

2 Support Vector Regression

The basic idea in SVR is to map an input data x into a higher dimensional feature space F via a nonlinear mapping ϕ and then a linear regression problem is obtained and solved in the feature space. Therefore, the regression approximation addresses the problem of estimating a function based on a given data set $G = \{(x_i, y_i)\}_{i=1}^m$ ($x_i \in R^n$ is the input vector, $y_i \in R$ is the desired real-value). In SVM method, the regression function is approximated by

$$f(x) = \langle \omega, \phi(x) \rangle + b \tag{1}$$

where $\{\phi_i(x)\}_{i=1}^m$ are the features of inputs, ω and b are coefficients. The coefficients are estimated by minimizing the regularized risk function:

$$R(\omega) = \frac{1}{2} \|\omega\|^2 + C \sum_{i=1}^m L_\epsilon(f(x_i), y_i) \tag{2}$$

where regularized term $\frac{1}{2} \|\omega\|^2$ is used as a flatness measurement of function (1), C is a fixed constant determining the tradeoff between the training error and the model complexity, and $L_\epsilon(\cdot)$ is the ϵ-insensitive loss function defined by Vapnik [1]:

$$L_\epsilon(f(x), y) = max\{|f(x) - y| - \epsilon, 0\} \tag{3}$$

where ϵ is a prescribed parameter.

There are two common approaches for regression minimization either the sum of the absolute discrepancies over samples ($\sum_i |f(x_i) - y_i|_\epsilon$) or the square of the discrepancies ($\sum_i |f(x_i) - y_i|_\epsilon^2$). It has been proved that the squared loss is sensitive to outliers, hence robust regression algorithms often employ the absolute loss [13].

An introduction of slack variables ξ, ξ^* leads Eq.(2) to the following quadratic programming (QP) problem with $2m$ constraints and $n + 1 + 2m$ variables:

$$\min_{(\omega,b,\xi,\xi^*) \in R^{n+1+2m}} \frac{1}{2}\|\omega\|^2 + C\sum_{i=1}^{m}(\xi_i + \xi_i^*) \quad (4)$$

$$\text{s.t.} \quad \begin{aligned} \langle \omega, \phi(x_i) \rangle + b - y_i &\leq \epsilon + \xi_i \\ y_i - \langle \omega, \phi(x_i) \rangle - b &\leq \epsilon + \xi_i^* \\ \xi_i, \xi_i^* &\geq 0 \quad i = 1, \cdots, m \end{aligned} \quad (5)$$

The classical Lagrange Duality enables above problem to be transformed to its dual problem with $2m$ Lagrange multipliers:

$$\min_{(\alpha,\alpha^*) \in R^{2m}} \frac{1}{2} \sum_{i=1}^{m}\sum_{j=1}^{m}(\alpha_i - \alpha_i^*)(\alpha_j - \alpha_j^*)\langle \phi(x_i), \phi(x_j) \rangle + \sum_{i=1}^{m}\alpha(\epsilon - y_i) + \sum_{i=1}^{m}\alpha_i^*(\epsilon + y_i) \quad (6)$$

$$\text{s.t.} \quad \begin{aligned} \sum_{i=1}^{m}(\alpha_i - \alpha_i^*) &= 0 \\ 0 \leq \alpha_i, \alpha_i^* &\leq C \quad i = 1, \cdots, m \end{aligned} \quad (7)$$

Based on the nature of quadratic programming, only a few of coefficients among α_i, α_i^* will be nonzero, and the data points associated with them refer to support vectors. For computational convenience, the form $\langle \phi(x), \phi(x) \rangle$ in formula (6) is often replaced by a so-called kernel function with the following form,

$$K(x,y) = \langle \phi(x), \phi(x) \rangle \quad (8)$$

And so, all the computations are carried on via kernel function in the input space. Any function that satisfies Mercer's Theorem can be used as a kernel function such as Gaussian kernel $K(x,y) = exp(-\mu\|x-y\|^2)$ and polynomial kernel $K(x,y) = (x^T y + 1)^p$.

3 Smooth Support Vector Regression

The basic idea in smooth support vector machine consists of converting the primal QP (Eq. 4&5) to a non-smooth unconstrained minimization problem, and then using standard smoothing techniques of mathematical programming [14,15,16] to smooth the problem.

Based on Karush-Kuhn-Tucker optimality conditions, the nonzero slacks can occur outlier only, i.e. at the solution of QP, we have

$$\xi^{(*)} = \max\{|\phi(x)\omega + b - y| - \epsilon, 0\} \quad (9)$$

where $\xi^{(*)}$ denotes ξ and ξ^*.

As described in [17] we minimize $\|(\omega,b)\|^2$ at the same time, and transform the primal QP problem to unconstrained minimization problem with $n+1+m$ variables,

$$\min_{(\omega,b,\delta)\in R^{n+1+m}} \frac{1}{2}(\|\omega\|^2+b^2) + C\sum_{i=1}^{m}(|\delta|_\epsilon)_i \tag{10}$$

where $\delta = (\phi(x)\omega + b) - y$.

Based on duality theorem we have $\omega = \phi(x)\alpha, \alpha \in R^m$, redefined (10) as follows

$$\min_{(\alpha,b)\in R^{m+1}} \frac{1}{2}(\|\alpha\|^2+b^2) + C\sum_{i=1}^{m}(|K(x_i,x)\alpha + b - y_i|_\epsilon) \tag{11}$$

Given that the objective function of this unconstrained optimization problem is not smooth as its derivative is discontinuous at $\delta = \pm\epsilon$. The Ref. [10] employed a squared ϵ-insensitive p function to replace the last term in (11), where the ϵ-insensitive p function is defined by

$$p_\epsilon(x,\beta) = p(x-\epsilon,\beta) = (x-\epsilon) + \frac{1}{\beta}\log(1+e^{-\beta(x-\delta)}) \tag{12}$$

Based on the squared ϵ-insensitive p function, redefined (10) as follows

$$\min_{(\alpha,b)\in R^{m+1}} \frac{1}{2}(\|\alpha\|^2+b^2) + C\sum_{i=1}^{m}(p_\epsilon^2(K(x_i,x)\alpha + b - y_i,\beta)) \tag{13}$$

where β is a smoothing parameter and the Eq.(13) was called as SSVR in [10].

The disadvantages of the SSVR (13) include twofold. Firstly the squared loss is always sensitive to outliers. Secondly the selection of the smoothing parameter β is heuristic, which would increase the difficulty of model selection of SVM.

In order to avoid SSVR's disadvantages, we employed another smooth approximation, defined as ϵ-insensitive logistic loss function:

$$L_{log}(|\delta|_\epsilon) = log(1+e^{|\delta|_\epsilon}) \tag{14}$$

We can describe $|\delta|_\epsilon$ as the form of $|\delta|_\epsilon = ((\delta-\epsilon)_+ + (-\delta-\epsilon)_+)$ (see Fig.1). So, based on (14) we have following equation to approximate $|\delta|_\epsilon$

$$L_{log}(\delta,\epsilon) = log(1+e^{\delta-\epsilon}) + log(1+e^{-\delta-\epsilon}) - 2log(1+e^{-\delta}) \tag{15}$$

where the constant term $2log(1+e^{-\delta})$ is set so that $L(0,\epsilon) = 0$ (see Fig.2).

In Fig.2 we can observe that the ϵ-insensitive logistic loss function provides a smooth upper bound on the ϵ-insensitive loss.

Since the additive constants do not change the results of the optimal regression, the constant is omitted in (15). Redefined (10) as follows,

$$\min_{(\alpha,b)\in R^{m+1}} \Psi(\alpha,b) = \frac{1}{2}(\|\alpha\|^2+b^2) + C\sum_{i=1}^{m}(log(1+e^{\delta_i-\epsilon}) + log(1+e^{-\delta_i-\epsilon})) \tag{16}$$

where $\delta_i = K(x_i,x)\alpha + b - y_i$.

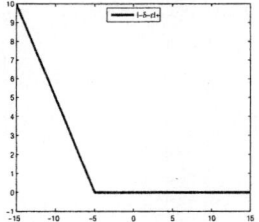 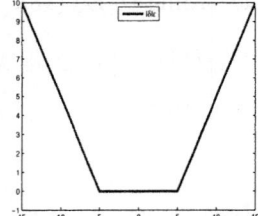 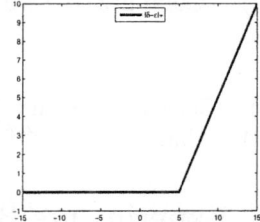

Fig. 1. Constructing ϵ-insensitive loss function $|\delta|_\epsilon$ (second) by $(-\delta - \epsilon)_+$ (first) and $(\delta - \epsilon)_+$ (third) with $\epsilon = 5$

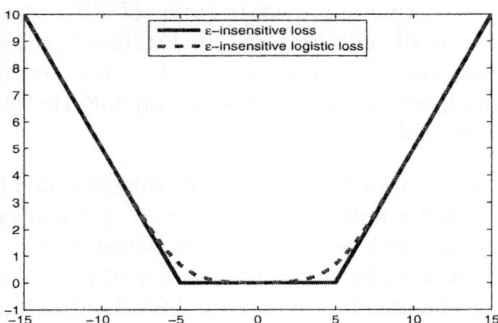

Fig. 2. Approximating ϵ-insensitive loss function (real) by ϵ-insensitive logistic loss function (dot) with $\epsilon = 5$

The major properties of this smooth unconstrained optimization problem (16) are strong convexity and infinitely often differentiability, so we can solve it by Newton method instead of QP packages used in conventional SVM.

4 Implementation of Lϵ-SSVR

By making use of the results of the previous section and taking advantage of the twice differentiability of the objective function of Lϵ-SSVR (16), we prescribed a pure Newton algorithm to implement Lϵ-SSVR.

Algorithm 4.1: Newton Method Algorithm for Lϵ-SSVR
(i) Initialization: Start with any $(\alpha^0, b^0) \in R^{m+1}$, set $\lambda = 1, i = 0$ and $e = 1 \times 10^{-6}$;
(ii) Having (α^i, b^i), go to step (vi) if the gradient of the objective function of (16) is not more than e, i.e. $\nabla \Psi(\alpha^i, b^i) \leq e$; else, go to step (iii);
(iii) Newton Direction: Determine direction $d^i \in R^{m+1}$ according to Eq.(17), in which gives $m + 1$ linear equations in $m + 1$ variables:

$$d^i = -[\nabla^2 \Psi(\alpha^i, b^i)]^{-1} \cdot \nabla \Psi(\alpha^i, b^i) \qquad (17)$$

(iv) Compute (α^{i+1}, b^{i+1}) according to Eq.(18):

$$(\alpha^{i+1}, b^{i+1}) = (\alpha^i, b^i) + \lambda d^i \qquad (18)$$

(v) Set $i = i + 1$ and go to step (ii);
(vi) End

5 Experiments Results

The purpose of the experiments results carried out here is twofold. Firstly it has to be demonstrated that the algorithm proposed here has better generalization capability than SSVR. Secondly it has to be shown that it is really an improvement over the exiting approach in terms of CPU time.

The simulations were done in Matlab 7.0. Joachims' package SVM^{light} with a default working set size of 10 was used to test the decomposition method. The CPU time of all algorithms were measured on 3.0GHz P4 processor running Windows 2000 professional.

Example 1. The training data were generated using the sinc function corrupted by Gaussian noise. Picked x uniformly from $[-3, 3]$, $y = \sin(\pi x)/(\pi x) + \nu$, where ν drawn from Gaussian with zero mean and variance σ^2. We generated 100 samples for train-set and 50 for test-set from this additive noise model.

We approximated the true function by Lϵ-SSVR with Gaussian RBF kernel, $C = 100, \mu = 0.5, \sigma = 0.2$ and $\epsilon = 0.01$. In addition, we also implemented the SSVR with different smoothing parameters. The simulation results are shown in Fig. 3. From Fig. 3 we can observe that the SSVR is quite sensitive to the choice of the smoothing parameter. Table 1 illustrates the number of support vectors, the train-set RMSE, the test-set RMSE and the time consumption for different algorithms on sinc function datasets. From Tab.1 we can conclude that the generalization capability of Lϵ-SSVR is better than SSVR without compromising train error and CPU time.

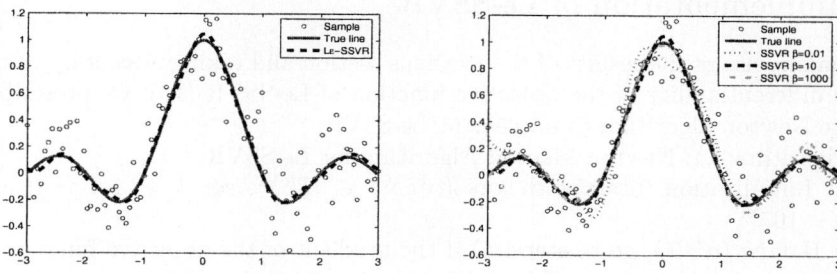

Fig. 3. Approximating *sinc* function by Lϵ-SSVR (left) and SSVR (right) with different smoothing parameters

Table 1. Average results (50 trials) on *sinc* datasets with Root Mean Square Error

Dataset	Method	SV Num.	Train Error	Test Error	CPU Sec.
$y = sic(x) + \nu$	Lϵ-SSVR	75	0.0481	0.0328	0.3
	SSVR	76	0.0481	0.0386	0.3

Table 2. Average results (100 trials) on two real-word datasets with Mean Square Error

Dataset	Method	(C, μ, ϵ)	SV Num.	Test Error	CPU Sec.
Boston Housing	Lϵ-SSVR		173	8.9	0.89
Train size:481	SMO	(500,1.5,2)	173	9.7	2.30
Test size:25	SVM^{light}		178	8.8	4.90
Abalone	Lϵ-SSVR		1315	2.25	6.74
Train size:3000	SMO	(1000,5,3.5)	1316	2.23	12.63
Test size:1177	SVM^{light}		1317	2.65	88.37

Example 2. In this experiment, we chose the Boston Housing and the Abalone datasets from the UCI Repository [18]. The data were rescaled to zero mean and unit variance coordinate-wise. Finally, the gender encoding in Abalone (male/female/infant) was mapped into $\{(1,0,0), (0,1,0), (0,0,1)\}$. We used the same kernel function as Example 1. Table 2 illustrates the training set size, the number of support vectors, the test-set MSE and the time consumption for different algorithms on two real-word datasets.Here we can conclude that Lϵ-SSVR is faster than other algorithms.

6   Conclusion

Based on the absolute loss of ϵ-insensitive logistic loss function, we have proposed a new smooth support vector regression formulation, which is a smooth unconstrained optimization reformulation of the conventional quadratic program associated with an SVR. Taking advantage of this reformulation, we solved SVR as a system of linear equations iteratively with the pure Newton Method. Compared with SSVR, our new method demonstrated better generalization capability without compromising the train error and CPU time. We also got the conclusion that the new method is much faster than any other decomposition methods mentioned in this paper.

References

1. Vapnik, V.: The Nature of Statistical Learning Theory. John Wiley. New York, USA, 1995
2. Kecman, V.: Learning and Soft Computing, Support Vector machines, Neural Networks and Fuzzy Logic Models. The MIT Press, Cambridge, MA, 2001

3. Wang, L.P. (Ed.): Support Vector Machines: Theory and Application. Springer, Berlin Heidelberg New York, 2005
4. Drucker, H., Burges, C.J.C., Kaufman, L., Smola, A. and Vapnik, V.: Support Vector Regression Machines. Advances in Neural Information Processing Systems 9, M.C. Mozer, M.I. Jordan, and T. Petsche, eds., pp. 155-161, Cambridge, Mass.: MIT Press, 1997
5. Chappelle, O., Vapnik, V., Bousquet, O., Mukhcrjee., S.: Choosing Multiple Parameters for Support Vector Machines, Machine Learning, Vol.46, No.1, pp: 131-160, 2002
6. FrieV, T.T., Chistianini, N., Campbell, C.: The kernel adatron algorithm: A fast and simple learning procedure for support vector machines. in: Proceedings of the 15th International Conference of Ma-chine Learning, Morgan Kaufmann, San Fransisco, CA, 1998
7. Mangasarian, O. Musicant, D.: Successive overrelaxation for support vector machines, IEEE Trans. Neural networks 10(1999), 1032-1037
8. Platt J.: Fast training of support vector machines using sequential minimal optimization [A]. Advances in Kernel Method: Support Vector Learning [C]. Cambridge:MIT Press, 1999, 185-208
9. Lee, Y.J., Mangasarian, O. L.: SSVM: A Smooth Support Vector Machine for Classification, Data Mining Institute Technical Report 99-03, September 1999, Computational Optimization and Applications 20(1), October 2001, 5-22.
10. Lee, Y. J., Hsieh, W. F., Huang, C. M.: epsilon-SSVR: A Smooth Support Vector Machine for epsi-lon-Insensitive Regression. IEEE Trans. on Knowledge and Data Engineering, 17(5) (2005) 678-685
11. Chapelle, O., Vapnik, V.: Model selection for support vector machines, in: S.A. Solla, T.K. Leen, K.-R. MVuller (Eds.), Advances in Neural Information Processing Systems, Vol. 12, MIT Press, Cambridge, MA, 2000, 230-236
12. Keerthi. S.S.: Efficient Tuning of SVM Hyperparameters Using Radius/Margin Bound and iterative Algorithm. IEEE Transactions on Neural Networks, Vol.13, pp. 1225-1229, sep. 2002
13. Huber, P.J.: Robust Statistics. John Wiley and Sons, New York, 1981
14. Fukshima, M. and Qi, L.: Reformulation: Nonsmooth, Piecewise Smooth, Semismooth and Smooth-ing Methods. Kluwer Academic Publishers, Dordrecht, The Netherlands, 1999
15. Chen B. and Harker, P.T.: Smooth approximations to nonlinear complementarity problems. SIAM Journal of Optimization, 7:403-420, 1997
16. Ofer, D., Shai, S.S., Yoram S.: Smooth ϵ-insensitive regression by loss symmetrization, in: Learning Theory and Kernel Machines: 16th Annual Conference on Learning Theory and 7th Kernel Workshop, COLT/Kernel 2003, Washington, DC, USA, August 24-27, 2003
17. Musicant, D.R. and Feinberg, A.: Active Set Support Vector Regression, IEEE Trans. Neural Net-works, vol. 15, no. 2, pp.268-275, 2004
18. Blake, C.L. and Merz, C.J.: UCI Repository of machine learning databases, [http://www.ics.uci.edu/ mlearn/MLRepository.html]. Irvine, CA: University of California, Department of Information and Computer Science, 1998

Neural Network Classifier Based on the Features of Multi-lead ECG[*]

Mozhiwen[1,2], Feng Jun[2], Qiu Yazhu[2], and Shu Lan[3]

[1] Department of Applied Mathematics, Southwest Jiaotong University,
Chengdu, Sichuan 610031, P.R.China
mozhiwen@263.net
[2] College of Math. & Software Science, Sichuan Normal University,
Chengdu, Sichuan 610066, P.R.China
[3] School of Applied Mathematics,
University of Electronic Science and Technology of China, Sichuan,
Chengsu, 610054, P.R.China
shul@uestc.edu.cn

Abstract. In this study, two methods for the electrocardiogram (ECG) QRS waves detection were presented and compared. One hand, a modified approach of the linear approximation distance thresholding (LADT) algorithm was studied and the features of the ECG were gained for the later work.. The other hand, Mexican-hat wavelet transform was adopted to detect the character points of ECG. A part of the features of the ECG were used to train the RBF network, and then all of them were used to examine the performance of the network. The algorithms were tested with ECG signals of MIT-BIH, and compared with other tests, the result shows that the detection ability of the Mexican-hat wavelet transform is very good for its quality of time-frequency representation and the ECG character points was represented by the local extremes of the transformed signals and the correct rate of QRS detection rises up to 99.9%. Also, the classification performance with its result is so good that the correct rate with the trained wave is 100%, and untrained wave is 86.6%.

1 Introduction

Classification of the ECG using Neural Networks has being a widely used method in recent years [1, 2, 3, 6]. The far-ranging adopted method has represented its inimitable superiority in the field of signal processing. But the recorded ECG signals with much continues small-amplitude noise of various origins, are weak non-smooth, nonlinear signals. If inputting the ECG signals into the network directly, the redundant information would make the structure of the network much complex. But if only inputting the features of the ECG, the data would be reduced much, and this is also

---

[*] This Work Supported by the Natural Science Foundation of China(No.60074014).

accorded with the processing course of humans that first extracting the features from the stimulators and then transmit it up to the centre neural. In general there are two main aspects to get the features by analyzing the ECG signals, the single lead method and the multi-lead method. Analyzing with multi-lead ECG signals, the information of all leads are integrated and the result is always better than that with single lead.In this paper, multi-lead signals were adopted in detection of the QRS complex. According to the relation that the character points of the ECG were homologous with the local extreme points of the ECG signals preprocessed by multi-scale wavelet transform, the character points of the ECG were determined and more, many features represent the trait of the waves were gained [8]: heart-rate, the QRS complex width, the Q-T. intervals, and the amplitudes of all the waves etc.

In comparison, another ECG detects method: a modified approach of the linear approximation distance thresholding (LADT) [4, 10, 11] algorithm was studied. Also with multi-lead ECG signals, first the modified fast LADT algorithm was adopted to approximate the ECG signals with radials, and thus get the feature vectors representing the signals: the slope of the segment, the length. And then, calculate the vectors to determine the position of the R peak, and more, get the position of the whole QRS complex and its duration and amplitudes.

All ECG features attained from these two methods were putted separately into a RBF network to classification. A part of the features were used to train the network, and then all of them were used to examine the performance of the network. As tested, the two classification methods both performed well, not only in the training speed, but also in the classification result. And as the great feature extraction powers of the wavelet transform, it performed better than the LADT method, and the classification performance with its result with the trained wave is 100%, untrained wave is 86.6%.

2 Detection Algorithm

The detection and the features extraction are the key for ECG analysis. And the detection of the QRS complex is the chief problem in the ECG analysis, only when the position of the R wave is determined can the other details of the ECG be analyzed.

In R wave detection, the fast LADT algorithm and wavelet transformation method were innovated.

In recent time, the methods on the QRS detection had flourished: signal filter, independent component analysis, wavelet transform, and neural network. Especially wavelet transform method has a peculiarity that it has finite-compact support sets in the time-scale domain, it can form an orthogonal basis with the translation in the position and the alternate of the scale and it has alterable time-scale resolving power, thus it has a splendid feature extraction power.

On the study of the wavelet transformation method[5-7, 9], it is found that with the transform using spline wavelet, the zero-crossing points of the modulus maximum pairs should be detected. But the detection of zero-crossing points was always encumbered by the noise of the ECG signals. And the detection of the modulus maximum pairs is not easy yet. However it is found that the Mexican-hat wavelet has many advantages in detecting the R waves and even other waves such as Q and S waves.

In the study on the LADT, it was found that the fast LADT algorithm has some weakness.

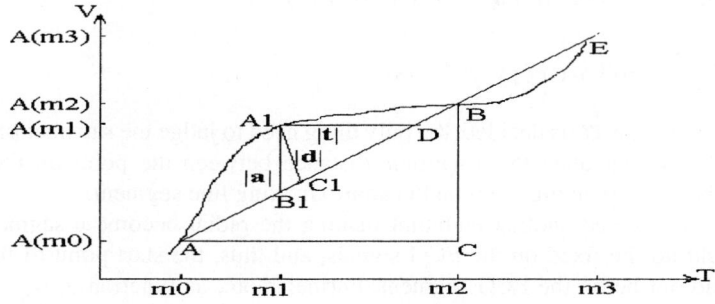

Fig. 1. Theory of LADT

The fast LADT Approximation theory is as follow [10, 11]:

$$|a| = |(A(m_1) - A(m_0)) - (A(m_2) - A(m_0))\frac{m_1 - m_0}{m_2 - m_0}| \quad (1)$$

$$|t| = |\frac{A(m_1) - A(m_0)}{A(m_2) - A(m_0)}(m_2 - m_0) - (m_1 - m_0)| \quad (2)$$

$$|d| = \frac{|a||t|}{(a^2 + t^2)^{1/2}} \quad (3)$$

A (m) is the amplitude of the ECG signal at time m, $\frac{A(m_2) - A(m_0)}{m_2 - m_0}$ is the slope of the approximation line segment. Thus as the relation of similar triangles,

$$\frac{|A(m_2) - A(m_0)|}{|m_2 - m_0|} = \frac{|a|}{|t|} = k = fixedvalue \quad (4)$$

To get the distance the points on the AA_1B from the line segment AB, a precision as the maximum distance σ was decided,

According to (1) (2) (3):

$$|d| = \frac{|a|}{(1+(a/t)^2)^{1/2}},$$

thus $|d| < \sigma$ if and only if $|a| < \sigma(1+(a/t)^2)^{1/2}$,

viz. $d_{max} = \sigma(1+(a/t)^2)^{1/2}$ (5)

When the precision σ is decided, the only thing need to judge the satisfaction of the line segment is to calculate the amplitude distance between the point on the ECG signals and the corresponding point on the approximating line segment.

But when determined another endpoint making the radial become a segment, the endpoint could not be fixed on the ECG signals, and thus, the start point of the next segment could not be on the ECG segment. Further more, as function 9, d_{max} is the distance threshold, σ is the precision determined at first, and k is the slope of the approximation segment. It can be seen that the d_{max} is determined by the k each time.

$$d_{max} = \sigma(1+k^2)^{1/2} \approx \sigma|k|$$ (9)

In the instance when the slope of the segment is very big, such as at the R wave period, the threshold could be very big too. Especially, when the ECG signal changes from the R wave to the base-line, as the slope of the segment approximates the R wave is very big, the segment will cross the ECG signal and only a few points can satisfy the precision, thus the saw-tooth like approximation appears. The reason is that the endpoints of the segment cannot be determined properly.

In order to amend this disadvantage, we fixed the endpoints on the ECG waves and performed the new fast LADT algorithm and got the features of the ECG. As the slope and length of each position of the ECG has their peculiarities, we can detect the positions of the R waves.

3 Detection of Waves

3.1 Detection with Wavelet Transform Method

When the signals being processed by the wavelet transform, the noise of the signal was restrained and the feature information was extruded. As on the scale of $2^3, 2^4$, the high frequency noise was well restrained, these two scales were selected out for the detection work. As the former theory, the position of the R peak was corresponding with the local extreme of the transformed signal on the scale 2^4, so a threshold could judge the R peak, and if any, it could be located in this field as a window.

As figure 1, the transform result of the ECG T103 with a spline wavelet:

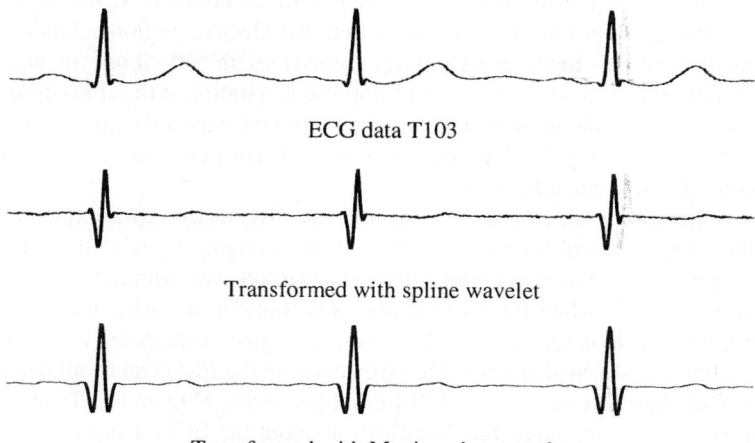

ECG data T103

Transformed with spline wavelet

Transformed with Mexican-hat wavelet

Fig. 2. Results of two wavelet transformations to ECG

The position of the R peak is just the obvious peak thus the local extreme of the signals transformed with the Mexican-hat wavelet, and then, to determine the position of the R peak is to find the local extreme. Then it overcomes the complexity that if transformed with spline wavelet, the modulus maximum pairs should be found first, and then zero-crossing points of in it should be detected. As the Mexican-hat wavelet was adopted, the process of the R detection could be simplified and this also improved the detection performance, the correct rate had achieved at 99.9%.

The process is as bellow:

(1) Read the ECG signal randomly, transform the signals with the Mexican-hat wavelet using the Mallat algorithm and get the signal $d_2^{\,j}(n)$ representing the details of multi-scale.
(2) Select a part of the ECG signals, decided the precisions Rth^j of the detail signals $d_2^{\,j}(n)$ on each scale, and detect the local extremes Mo of $d_2^{\,3}(n)$ with the threshold Rth^3 on the scale of 2^3.
(3) Modify the local extremes Mo according to the refractory period to Mo1.
(4) Detect the local extreme of the original signal to get the position of the R peak in the field of 10ms corresponding to the local extremes Mo1 and calculate the mean time between two R peaks Tm.
(5) Examine whether the interval of the two R peaks is bigger than 1.7Tm, if it is, that means some R peak was failed and then halve the threshold, and detect again as former steps and thus get the R peaks

3.2 Detection of Other Peaks

The ECG signals are constituted by a sequence of component-waves separated by regions of the zero electrical activity, called iso-electric regions. Under normal conditions, the compo- nent-waves repeats themselves in a rhythmic manner with a periodicity determined by the frequency of impulse generation at the sino-atrial node. A single ECG beat is made up of three distinct component-waves designated as, P, QRS and T-waves, respectively. Each compo- nent-wave corresponds to the certain moment of the electro-physiological activity.

The wavelet transform has the quality of time-frequency representation in local period, that it has the peculiarity to analysis the time changing signals. In the analysis of the ECG signals, the binary wavelet transform method was adopted. As multi-scale transform is adopted, when the ECG signal was transformed with multi scales, the character points such as Q, S, P and T waves, were just correspond with some local extremes of the transformed signals. Thus the waves of the ECG can be all detected and the features of the ECG such as the QRS-complex width, P-Q and Q-T intervals, the height of each wave etc. Thus the classification about the ECG using neural network can be done. The transformed signals in various scales as in figure 2, the top signals is the original signal of ECG, and the transformed signals of $2^1 \square 2^6$ are underneath it in turn.

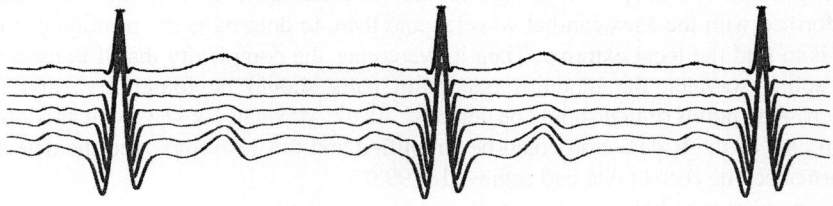

Fig. 3. Multi-scale ECG signal

Wave Q and S is always with high frequency and low amplitude, that their energies are chiefly on these small scales of the transform. Thus they could be detected on these scales. The Q wave is a downward wave before the R wave and the S wave is a downward wave after the R wave. Thus the local extreme in a certain period (about 100ms) left to the R peak is corresponding to the Q wave, the local extreme in the certain period right to the R peak is corresponding to the S wave. If there is no local extremes in these periods, that means the wave Q and wave S is not exist.

The detection of P wave is very important to ECG analysis, but the amplitude of it is small, and the frequency is low too. It is difficult to separate it from the noise. Analysis from the figure 1, it can be found that at the position of the P waves and T waves, on signals of the scales of 2^5, 2^6, there are some distinct waves accordingly. Thus the work to detect the wave P and wave T could be achieved respectively on the signals of the scales 2^5, 2^6.

And more, the features of the wave P and wave T are not so obvious as the QRS complex, and the boundaries of the wave P and T are misty that the study on wave P and T is also not as consummate as it of the QRS complex. But, the P wave, QRS complex, T wave come forth by turns, that when the start and end positions were determined, it could be conclude that the two most distinctive wave between the successive two QRS complex are the wave P and wave T.

The extraction of the features of the ECG, such as the width of the QRS complex, the P-Q and Q-T intervals, need the accurately determining of the start and end point of the waves, is a classical problem in the analysis of the ECG signals. The start and end point are also corresponding to the local extremes according to its frequency as transformed by wavelet. The start point of QRS complex is just the start point of the wave Q; if there is no wave Q, it is the start point of wave R. And the end point of the QRS complex is the end point of wave S; if there is no wave S, it is the end point of wave R. The start point of the QRS complex is corresponding to the sudden slope change point before the wave Q on the signal of scale 2^1, or the sudden slope change point before the wave R if the wave Q is not exist. And the end point of the QRS complex is just corresponding to the sudden slope change point behind the wave S, or the point behind the wave R, if the wave S is not exist. The start and end point of wave P and wave T is just corresponding to some local extremes on the scale 2^5. Thus the start and end point could be determined by detecting the character points of the ECG.

And more, as the start and end points of each waves were determined, the parameters significant to indicate the meaning of the ECG, such as the hear-rate, P-R and Q-T interval, the QRS complex width, the VAT, the time of a beat, and the amplitude. In the analysis of ECG, these parameters are very momentous to estimate the types of the ECG signals.

3.3 Detection with Modified LADT Method

As the ECG signal was approximated, the slope and the length of the approximating line segments can be formed in a vector. This vector contained the information of the ECG, can be used to detect the R peak. As in experimentation, two channel ECG signals was adopted and 40 sects of signals was picked each channel. Each sect is 3 seconds long. The process is as follows:

(1) Approximate the signal sects with the former algorithm, and gain the vectors of the slopes and the lengths of the segment.
(2) Decide the threshold of the slope, for that the R peak is the most sharp part in the whole heart beat periodicity; Decide the threshold of the length, for that the R peak always is the highest peak and the approximating segment is always the longest. And then, get the segments of the R peak. And according the LADT theory, the local extreme point of the endpoint of the segments is probably the R peak.
(3) According to the refractory period, remove the peaks falsely detected such as too close to another one and get the ultimate position of R peak.

When examine with the signals from MIT-BIH, the position of the R peak could be detected correctly with the rate better than 99.5%.

The position of the other waves and their start and end points were be determined with the method similar to the method of the wavelet transform. The position of the character points are just corresponding to some local extremes of the approximating line segments.

To determine the position of the peak, start and end points of each wave, a different method was adopted that with a local coordinate transformation, analysis the potential position of those waves. According to the feature of the ECG, the most sharpest position is corresponding with the biggest slope line segment, and that the absolute value of the first derivative is the biggest. And at the position of the start and end point of each wave, the slope of the line segment changes most acutely, that the second derivative there is the biggest. And thus, the characters of the waves were obtained.

4 Classification Experiment

4.1 Classification Network

With the high capability of classification from Radial-basis network, the features of the ECG were classified in a high dimensions space.

In the experiment, first step, the multi-scale wavelet transform method was adopted separately to detect the R and other waves as well as the modified LADT algorithm. Second step, the features of the ECG were extracted according the positions of all the waves. Third step, some features of each ECG were picked out randomly to train the network. Thus the disease classification knowledge was stored in the conjunctions of the network. Then the trained network could be adopted to classify the whole feature vectors. Last step, all features were calculated by the RBF network; the output of the network is just the result of the classification.

Accordingly, a multi-layer perception was adopted to classify the ECG signals. But as the weight of the conjunction was modified with the negative-grads-descend method, when the network was trained with the BP algorithm, the convergence speed was low and had the short of local extreme. Thus the time for training must be tens or even hundreds times of the training time for the RBF network. Therefore, the ability and the learning speed of the RBF network were some better than the BP network.

Thus this method exerts the excellence of both wavelet transform and neural network, gained the feature vectors well and truly, thus presents a high quality classify network.

4.2 Classification Experiment

To test the classification system present above, some ECG signals from the MIT-BIH database from the MIT-BME USA were classified in MATLAB toolbox.

The RBF network in the system has 20 cells in the input layer, corresponding to the features of the two channels ECG; there are 10 cells in the output layer, corresponding to the 10 types of selected ECG signals. 40 sects of each disease case were selected randomly, and analysis them with the wavelet-transform method and the LADT method, and thus the features of each case were obtained. Then 20 sects of each case were selected randomly to train the network. At last, the trained network to test the efficiency of the system classified all sects.

The classification result was presented in the table 1 and table 2. Table 1 is the result of classification with the features extracted with the wavelet transformation. Table 2 is the result of classification with the LADT method.

All the ten ECG signals were elected according to the article [1], signal T100 is mostly normal, T105, T108 and T219 have several PVC, T106 and T221 have many PVC, T111 and T112 are BBB, T217 has several PVC and FUS, was paced style, T220 has several APC.

Table 1. Results of the classification with wavelet and neural network

Rec. No.	Waves learned	Waves tested	Correct rate (trained)	Correct rate (untrained)
T100	20	40	100%	95.2%
T105	20	40	100%	81.0%
T106	20	40	100%	69.2%
T108	20	40	100%	75.0%
T111	20	40	100%	95.0%
T112	20	40	100%	100%
T217	20	40	100%	82.6%
T219	20	40	100%	87.0%
T220	20	40	100%	100%
T221	20	40	100%	81.0%
Total/average	20	40	**100%**	**86.6%**

Table 2. Results of the classification with LADT and neural network

Rec. No.	Waves learned	Waves tested	Correct rate (trained)	Correct rate (untrained)
T100	20	40	100%	95.2%
T105	20	40	100%	65.4%
T106	20	40	100%	51.9%
T108	20	40	100%	81.0%
T111	20	40	100%	100.0%
T112	20	40	100%	86.4%
T217	20	40	100%	71.4%
T219	20	40	100%	90.5%
T220	20	40	100%	69.6%
T221	20	40	100%	70.8%
Total/average	**20**	**40**	**100%**	**78.2%**

From the tables above it can be conclude that the features of the ECG of two leads were integrated for the classification. It simulates the situation of the real world situation that it classified the ECG according to the relations of the amplitude and width of each waves with a RBF network. This method exerted the splendid character extraction ability and the excellent peculiarity of the network on the classification, which managed the classification work to a good level both in speed and the veracity. In the experiment, the classification system gave a good performance. To the waves been used to train the network, the classification ability is perfect that the correct achieved 100%. To the waves not used to train the network, the performance is also good that the correct rate is 78.2 using the LADT method, with the wavelet-transform method, the correct rate is 86.6%, which are both much better than other classified system, and the wavelet-transform method is better than the LADT method for its accuracy feature extraction ability.

In order to compare, the experiment with BP network and the wavelet-transform method was test accordingly. As the accuracy feature extraction ability of wavelet-transform, the correct rate is also very good, but as the speed of the BP is very slow, the training time of it is hundreds times of the RBF network.

5 Conclusion

In this study, two feature extracting method were compared. First Mexican-hat wavelet transform was adopted to detect the character points of ECG for it has the quality of time-frequency representation and the ECG character points was represented by the local extremes of the transformed signals. In succession, the modified LADT method is adopted to detect the character points.

And with the high capability of classification from Radial-basis network, the features of the ECG were classified in a high dimensions space along the theory of the ECG diagnose and the situation of ECG diagnose in practice.. This method exerts the excellence of both feature extraction methods and neural network, gained the feature vectors well and truly, thus presents a high quality classify network. Thus take a new idea for the ECG automatic analysis.

References

1. Wang JiCheng: A Pattern classification system based on fuzzy neural network, Journal of computer research & development, Jan.1999,36(1): 26-30
2. Chris D Nugent, Jesus A Lopez, Ann E Smith and Norman D Black: Prediction models in the design of neural network based ECG classifiers: A neural network and genetic programming approach, BMC Medical Informatics and Decision Making 2002, 2:1
3. Zumray Dokur, Tamer Olmez: ECG beat classification by a novel hybrid neural network, Computer methods and programs in biomedicine 2001; 66: 167-181
4. Jospehs. Paul, et al: A QRS estimator using linear prediction approach, Signal processing 1999(72): 15-22

5. Li Cuiwei, Zheng Chongxun, Tai Changfeng: Detection of ECG characteristic points using wavelet transform, IEEE Trans BME, 1995;42(1):21-29
6. Zumray Dokur, Tamer Olmez, Ertugrul Yazagan, et al: Detection of ECG waveforms by neural networks [J] Medical Engineering and physics, 1997, 19(8): 738:741
7. Mallat Stephen: Singular Detection and Processing with wavelet, IEEE Trans Information Theory, 1992; 38(2): 617-643
8. I.K.Daskalov, I.I.Christov: Electrocardiogram signal preprocessing for automatic detection of QRS boundaries, Medical Engineering & Physics 21(1999) 37-44
9. Li Changqing, Wang Shuyan: ECG detection method based on adaptive wavelet neural network, Journal of Biomedical Engineering, 2002; 19(3): 452-454 Xing HX, Basic Electrocardiograph, Beijing. PLA Publishing, 1988; 35-42,128-141
10. Li G, Feng J, Lin L, et al: Fast realization of the LADT ECG data compression method, IEEE Eng. Med. Biol. Mag., 1994; 13(2): 255
11. QI J, Mo ZW: Number of classes from ECG and its application to ECG analysis, Journal of Biomedical Engineering, 2002; 19(2): 225

A New Learning Algorithm for Diagonal Recurrent Neural Network[*]

Deng Xiaolong[1], Xie Jianying[1], Guo Weizhong[2], and Liu Jun[3]

[1] Department of Automation, Shanghai Jiaotong University, 200030 Shanghai, China
xL_Deng@sjtu.edu.cn
[2] Department of Mechanical Engineering, Shanghai Jiaotong University, 200030 Shanghai
[3] First Research Institute of Corps of Engineers, General Armaments Department, PLA, 214035 Wuxi, Jiangsu, China

Abstract. A new hybrid learning algorithm combining the extended Kalman filter (EKF) and particle filter is presented. The new algorithm is firstly applied to train diagonal recurrent neural network (DRNN). The EKF is used to train DRNN and particle filter applies the resampling algorithm to optimize the particles, namely DRNNs, with the relative network weights. These methods make the training shorter and DRNN convergent more quickly. Simulation results of the nonlinear dynamical identification verify the validity of the new algorithm.

1 Introduction

Diagonal recurrent neural network (DRNN) was firstly put forward by Chao-Chee Ku, etc [1]. It only has self-feedback connections among the neurons in the hidden layer and it has been becoming one of the hottest research topics for it may obtain the tradeoff between the training cost and accuracy.

Chao-Chee Ku, et al applied the dynamical BP algorithm to train DRNN [1]. But the dynamical BP algorithm needs to adjust the learning rates. The tuning of the learning rates is relatively complex and the convergent speed is also very slow. Williams R.J. introduced the extended Kalman filter (EKF) algorithm for recurrent neural network (RNN) [2]. Although having high convergent speed, the EKF has low accuracy. And he augmented the output variable to the state vector in [2]. Thus, the calculations of the covariance of the state vector and the filtering gain, etc are relatively complex. de Freitas J.F.G., et al combined the EKF and particle filter to train a multilayer perceptron (MLP) [3]. But MLP is a feed-forward neural network. And DRNN is not a static mapping as MLP dose, outputs of DRNN are affected by inputs of both the current and the previous time steps. So it is not suitable to train DRNN by the means in [3] in each training cycle.

In this paper, we firstly combine the EKF and particle filter to train DRNN. We use an effective method to exactly evaluate the weights of particles, and then the resampling step may be just run to optimize particles with respective network weights. Thus, the fast convergent speed of the EKF and the optimization function of particle filter are incorporated into training DRNN. The nonlinear dynamical identification

---

[*] This work is supported by the National Natural Science Foundation of China, No. 50405017.

experiments demonstrate that the new algorithm can effectively be applied to train DRNN.

2 Diagonal Recurrent Neural Network

The model architecture of DRNN is shown as Fig. 1. Suppose DRNN has P input neurons, R recurrent neurons and M output neurons. W^I, W^D or W^O represents input, recurrent or output weight vectors respectively.

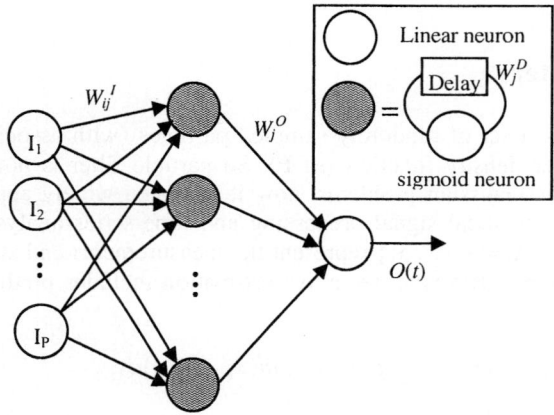

Fig. 1. The model architecture of DRNN

For each discrete time k, $I_i(k)$ is the ith input, $S_j(k)$ is the sum of inputs of the jth recurrent neuron, and $O_m(k)$ is the output of the mth output neuron. The mathematical model of DRNN can be inferred as [1]:

$$S_j(k) = W_j^D X_j(k-1) + \sum_{i=1}^{P} W_{ij}^I I_i(k) \tag{1}$$

$$X_j(k) = f(S_j(k)) \tag{2}$$

$$O_m(k) = \sum_{j=1}^{R} W_{mj}^o X_j(k) \tag{3}$$

where $f(.)$ is sigmoid function which is often $f(x)=1/(1+e^{-x})$. From the negative gradient descent rule, the weight vector of DRNN is updated as follows:

$$W(k+1) = W(k) - \eta(\partial J_m / \partial W) = W(k) + \eta e_m(k)(\partial O(k)/\partial W) \tag{4}$$

where η is the learning rate, J_m represents the function of error, e_m, of output between the plant and DRNN. From the chain rule of deriving the difference, we can have the output gradients with respect to input, recurrent and output weights respectively:

$$\partial O_m(k)/\partial W_{mj}^o(k) = X_j(k) \qquad (5)$$

$$\partial O_m(k)/\partial W_j^D(k) = W_{mj}^o(k)f'(S_j(k))X_j(k-1) \qquad (6)$$

$$\partial O_m(k)/\partial W_{ij}^I(k) = W_{mj}^o(k)f'(S_j(k))I_i(k) \qquad (7)$$

3 Particle Filter

Particle filter uses a set of randomly sampled particles (with associated weights) to approximate poster density function (PDF). So particle filter is not affected by the non-linear and non-Gaussian problems, now it has been widely applied to robotics, computer vision, statistical signal processing and time-series analysis, etc [4]. Suppose $Y_k=\{y_1,\ldots,y_k\}$, $X_k=\{x_1,\ldots,x_k\}$ represent the measurements and state sequences up to time k respectively. Bayesian recursive estimation includes prediction and updating:

$$p(x_k|Y_{k-1}) = \int p(x_k|x_{k-1})p(x_{k-1}|Y_{k-1})dx_{k-1} \qquad (8)$$

$$p(x_k|Y_k) = p(y_k|x_k)p(x_k|Y_{k-1})/p(y_k|Y_{k-1}) \qquad (9)$$

where $p(x_k|x_{k-1})$ is the transition density of the state, $p(y_k|x_k)$ is the likelihood and the denominator $p(y_k|Y_{k-1})$ is the normalized constant.

The analytical solutions to the above integrals are generally hard to be acquired. If we can sample particles from PDF, PDF may be approximately represented by these particles.

$$p(x_k|Y_k) = \frac{1}{N}\sum_{i=1}^{N} w_k^i \delta(x_k - x_k^i) \qquad (10)$$

where x_k^i is the ith particle with the relative weight w_k^i, randomly sampled from the PDF. $\delta(\cdot)$ is Dirac delta function.

It is often not possible to directly sample from PDF, but we can approximate PDF by sampling from a known proposal distribution, $q(.)$, that is easy to sample. From the large number theorem, the randomly sampled discrete particles are convergent to true distribution. The weight w_k is defined as:

$$w_k(x_k) = p(x_k|Y_k)p(Y_k)/q(x_k|Y_k) \qquad (11)$$

where $q(x_k|Y_k)$ is the proposal distribution (function).

As the states follow a first-order Markov process, we can obtain a recursive estimate of the importance weights [3]:

$$w_k = w_{k-1} p(y_k | x_k) p(x_k | x_{k-1}) / q(x_k | X_{k-1}, Y_k) \tag{12}$$

To reduce the effect of the degeneracy in the algorithm, Gordon, et al [5] introduced the resmpling step, which evaluates weights of particles and resamples particles to eliminate particles with small weights and to multiply particles with large weights. Thus, prediction, updating, evaluating and resampling constitute the basic particle filter.

Particle filters require the design of proposal distributions that can approximate PDF as well as possible. The optimal proposal distribution requires it to sample from the integrals [6] and it is often hard to be implemented in practice. Some suboptimal proposals including the prior proposal [5], the EKF proposal [3], etc are presented. The prior proposal, $q(x_k | x_{k-1}, Y_k) = p(x_k | x_{k-1})$, has no considerations of the latest measurements and the evaluation of the weight is simplified as evaluating the likelihood, $w_k = w_{k-1} p(y_k | x_k)$. The EKF proposal uses the EKF to update each particle and is firstly used to train a MLP [3]. For DRNN is very different from MLP, we develop a new hybrid learning algorithm combining the EKF and particle filter to train DRNN.

4 A New Hybrid Learning Algorithm

For DRNN may memorize previous network states, it is not suitable to simply appraise the performance of DRNN and resample particles in every training cycle. In each training cycle, the EKF is used to update network weights of every particle (DRNN). When DRNN has been trained after some training cycles, weights of particles are just exactly evaluated in this certain fixed-length training period. And then, the resampling algorithm is run to multiply good particles and reduce bad ones. Thus, a new algorithm incorporating fast convergent speed and high accuracy is developed.

Now the updating of network weights is represented in the form of state space model:

$$W(k) = W(k-1) + v(k-1) \tag{13}$$

$$y(k) = h(u(k), W(k)) + r(k) \tag{14}$$

where the state vector $W(k) = [W^I(k)\ W^D(k)\ W^O(k)]^T$ contains all network weights. $u(k)$ is the input signal, $y(k)$ is the output of DRNN. $v(k)$, $r(k)$ is the uncorrelated white Gaussian process, measurement noise respectively.

Main steps of the new algorithm are described as follows:

1) Initialize network weights of each particle (DRNN).
2) In the start training cycle of a fixed-length training cycles, update network weights of every particle with the EKF.

$$\hat{W}_k^- = \hat{W}_{k-1} \tag{15}$$

$$P_k^- = P_{k-1} + Q_{k-1} \tag{16}$$

$$K_k = P_k^- H_k^T \left(H_k P_k^- H_k^T + R_k\right)^{-1} \tag{17}$$

$$P_k = P_k^- - K_k H_k P_k^- \tag{18}$$

$$\hat{W}_k = W_k^- + K_k(y_k - h_k(u(k), \hat{W}_k^-)) \tag{19}$$

where P_k is the covariance of the state, Q_k, R_k is the covariance of process, measurement noise respectively. H_k is the local linearized measurement matrix which is calculated as:

$$H_k = \partial h_k(u(k), \hat{W}_k^-)/\partial W = \partial O_k / \partial W \tag{20}$$

3) When DRNN is trained to the end training cycle of a certain fixed number of training cycles, weights of particles, that is, performances of DRNNs, are evaluated. The weight of ith particle is defined as:

$$w_i = \prod_{j=1}^{L} (\sqrt{2\pi R})^{-1} e^{-0.5 R^{-1} \xi_{ij}^2} \tag{21}$$

where $\xi_{ij} = y^j - \hat{y}_i^j$, L denotes the number of training cycles. $\hat{y}_j^i$ is the output of ith particle (DRNN) and y^j is the desired output of the plant in the jth training cycle.

4) The multinomial resampling algorithm [5] is run to produce new discrete particles with optimized network weights. Particles with all relative network weights, which have large weights, are multiplied. Particles with small weights are eliminated. After resampling, all weights of particles are set as being identical.

5) If the training error is decreased into the desired error bounds, the training is ended. Otherwise, move to the next start training cycle. Repeat step 2), 3), 4) and 5) up to the end of training.

5 Simulations

In this paper, we adopt the series-parallel identification model to simulate with two typical plants [7]. In the simulations, we compare the EKF training algorithm and our new hybrid training algorithm (EKF-PF).

The training accuracy is raised with the increased particles and the shorter length of training cycles. But at the same time, the computational cost becomes greatly higher. Considering the tradeoff between the computational cost and the accuracy, we set the values of all parameters empirically by a great deal of simulations. The covariance of process noise, measurement noise is $Q=q\delta_{ij}$, $q=0.0001$, $R=100$ respectively. The initial covariance of the state is set as $P_0=p\delta_{ij}$, $p=1000$. The learning rate is 0.5.

Example 1: A nonlinear plant is described by the first-order difference equation:

$$y(k+1) = \frac{y(k)}{1+y^2(k)} + u^3(k) \tag{22}$$

Series-parallel identification model has two DRNN, $N_f[y(k)]$ and $N_g[u(k)]$, that are to be identified. $N_f[y(k)]$ represents that a DRNN with one network input variable $y(k)$ would approximate the function $f[.]$. Each DRNN has 1 input neuron, 10 recurrent neurons and 1 output neuron. The number of particles is 8 and the fixed length of training cycles is 8.

The training input $u(k)$ is chosen as an i.i.d. random signal uniformly distributed in the interval [-2, 2]. When trained with only 600 random data, the training error of DRNN is convergent into the desired error range. After training, the input test signal is $u(k)=\sin(2\pi k/25)+\sin(2\pi k/10)$, where $k=1,2,\ldots,100$, output of the plant and outputs of DRNN trained by the EKF and our new EKF-PF algorithm respectively are shown as Fig. 2.

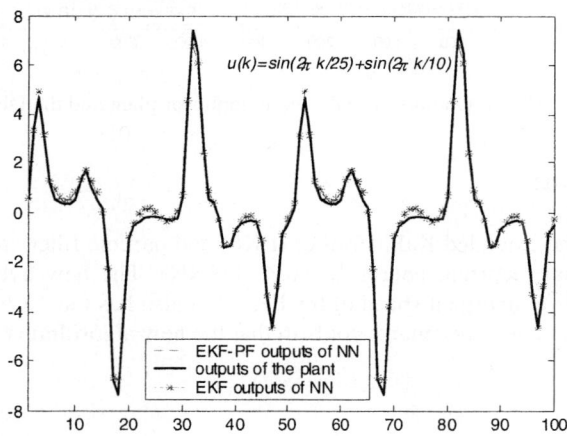

Fig. 2. Outputs of single-input nonlinear plant and the DRNN

Example 2: A multi-input nonlinear dynamical plant is governed by the following form:

$$f[x_1, x_2, x_3, x_4, x_5] = \frac{x_1 x_2 x_3 x_5 (x_3 - 1) + x_4}{1 + x_2^2 + x_3^2} \tag{23}$$

The training input $u(k)$ is chosen as an i.i.d. random signal uniformly distributed in the interval [-1, 1]. The DRNN has 5 neurons, 20 recurrent neurons and 1 output neuron. After training with 600 random data, the input test signal is selected as $u(k)=\sin(2\pi k/250)$ for $k \leq 500$ and $u(k)=0.8\sin(2\pi k/250) + 0.2\sin(2\pi k/25)$ for $k>500$. Outputs of DRNN trained by the EKF and our EKF-PF respectively and output of the plant are shown as Fig. 3.

As seen from the figures, the DRNN trained by our new algorithm can approximate the plant quite accurately.

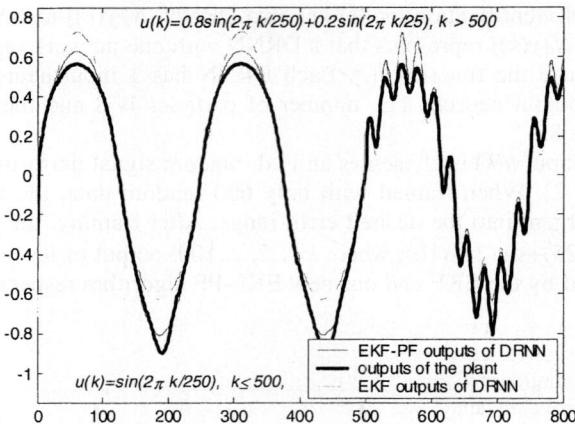

Fig. 3. Outputs of multi-input nonlinear plant and the DRNN

6 Conclusion

In this paper, the extended Kalman filter (EKF) and particle filter are firstly combined to train diagonal recurrent neural network (DRNN). The new hybrid algorithm not only has the fast convergent speed of the EKF, but also has the "survival of the fittest" of particle filter. The experiments confirm that the new algorithm is valid.

References

1. Chao-Chee, Ku, Kwang, Y Lee.: Diagonal Recurrent Neural Networks for Dynamic Systems Control. IEEE Trans on Neural Networks. Vol. 6, No. 1 (1995) 144-155
2. Williams, R.J.: Training Recurrent Networks Using the Extended Kalman Filter. Proc. Intl. Joint Conf. Neural Networks. Vol. 4. Baltimore (1992) 241-246
3. de Freitas, J.F.G., Niranjan, M., Gee, A.H., Doucet, A.: Sequential Monte Carlo Methods to Train Neural Network Models. Neural Computation. Vol. 12, No. 4 (2000) 955-993
4. Doucet, A., de Freitas, J.F.G., Gordon, N.J. (ed.): Sequential Monte Carlo Methods in Practice, Springer-Verlag, New York (2002)
5. Gordon, N., Salmond, D. J., Smith, A.F.M.: Novel Approach to Nonlinear and Non-Gaussian Bayesian State Estimation. IEE Proceedings-F, Vol. 140, No. 2 (1993) 107-113
6. Doucet, A., Godsill, S.J., Andrieu, C.: on Sequential Monte Carlo Sampling Methods for Bayesian Filtering. Statistics and Computing. Vol. 10, No. 3 (2000) 197-208
7. Narendra, K.S., Parthasarathy, K.: Identification and Control of Dynamical Systems Using Neural Networks. IEEE Trans on Neural Networks. Vol. 1, No. 1 (1990) 4-27

Study of On-Line Weighted Least Squares Support Vector Machines

Xiangjun Wen, Xiaoming Xu, and Yunze Cai

Automation Department, Shanghai Jiaotong University, Shanghai, 200030, China
{Wenxiangjun, xmxu, yzcai}@sjtu.edu.cn

Abstract. Based on rolling optimization method and on-line learning strategies, a novel weighted least squares support vector machines (WLS-SVM) are proposed for nonlinear system identification in this paper. The good robust property of the novel approach enhances the generalization ability of LS-SVM method, and a real world nonlinear time-variant system is presented to test the feasibility and the potential utility of the proposed method.

1 Introduction

As a novel breakthrough to neural network, Support Vector Machines (SVM), originally introduced by Vapnik [1] within the frame of the statistical learning theory, has been frequently used in a wide range of fields, including pattern recognition [2], regression [3] and others [4], [5]. In this kernel-based method, one starts formulating the problem in a primal weight space, but maps the input data into a higher dimensional hypothesis space (so-called feature space) and constructs an optimal separating hyper plane by solving a quadratic programming (QP) in the dual space, where kernel functions and regularization parameters are chosen such that a regularized empirical risk instead a conventional empirical risk is minimized. The solution of this convex optimization problem leads to the sparse and robust solutions (or good generalization capability) of the model.

Despite many of these advances, the present SVM methods were basically restricted to static problems. It is known that the use of SVM in a dynamical system and control context becomes quite complicated [8], due to the fact that it is a very stringent requirement to solve online for a large-scale QP problem in standard SVM. As a reformulation of standard SVM, a least squares version of SVM (LS-SVM) that leads to solve linear KKT systems has been extended to dynamical problems of recurrent neural networks [6] and used in optimal control [7]. While comparing with neural network and standard SVM, LS-SVM based control has many advantages such as: no number of hidden units has to be determined for the controller, no centers has to be specified for the Gaussian kernel, fewer parameters have to be prescribed via the training process, and the linear KKT systems can be efficiently solved by iterative methods. It is well known that it is very convenient and straightforward to construct a learning model of static (or time-invariant) problems via LS-SVM, however, noting the learning process is off-line, and the train data is selected as a batch before the whole process, the present LS-SVM methods were basically restricted when extended

L. Wang, K. Chen, and Y.S. Ong (Eds.): ICNC 2005, LNCS 3610, pp. 51–60, 2005.
© Springer-Verlag Berlin Heidelberg 2005

to time-variant dynamic system and on-line learning process. Therefore, a practical on-line learning approach based on weighted LS-SVM (WLS-SVM) method is mainly elaborated for nonlinear system identification in this paper.

This paper is organized as follows. In the next section we first give a brief review on LS-SVM method, then we focus on a practical approach to construct an on-line WLS-SVM method for nonlinear dynamic system modeling. In section 4, a numerical experiment is presented to assess the applicability and the feasibility of the proposed method. Finally, Section 5 concludes the work done.

2 Least Squares Support Vector Machines

Given a training data set D of l samples independent and identically drawn (i.i.d.) from an unknown probability distribution $\mu(X, Y)$ on the product space $Z = X \times Y$:

$$D = \{z_1 = (x_1, y_1), \ldots z_n = (x_l, y_l)\} \tag{1}$$

where the input data X is assumed to be a compact domain in a Euclidean space R^d and the output data Y is assumed to be a closed subset of R.

In the case of Least Squares Support Vector Machines (LS-SVM), function estimation is defined:

$$f(x) = w^T \Phi(x) + b \tag{2}$$

One defines the optimization problem.

$$\min_{w,b,e} J(w, e) = \frac{1}{2} w^T w + \gamma \frac{1}{2} e^T e \tag{3}$$

s.t.

$$y_k = w^T \Phi(x_k) + b + e_k, k = 1, \ldots, l \tag{4}$$

where $e \in R^{l \times 1}$ denotes the error vector, regularization parameter γ denotes an arbitrary positive real constant.

The conditions for optimality lead to a set of linear equations:

$$\begin{bmatrix} 0 & \vec{1}^T \\ \vec{1} & \Omega + \gamma^{-1} I \end{bmatrix} \begin{bmatrix} b \\ \alpha \end{bmatrix} = \begin{bmatrix} 0 \\ y \end{bmatrix} \tag{5}$$

where $y = [y_1, y_2, \ldots, y_l]^T$, $\vec{1} = [1, \ldots, 1]_{1 \times l}^T$, $\alpha = [\alpha_1, \ldots, \alpha_l]^T$, $\Omega_{ij} = \Phi(x_i)^T \Phi(x_j) = K(x_i, x_j)$, $i, j = 1, \ldots, l$.

The resulting LS-SVM model for function estimation becomes:

$$f(x) = \sum_{k=1}^{l} \alpha_k K(x, x_k) + b \qquad (6)$$

where α_k, b are the solution to the linear system (5).

Due to page limitation, more details of standard SVM and LS-SVM please further the reference [1], [8].

3 On-Line Weighted LS-SVM Method

In empirical data-based modeling, learning process of LS-SVM is used to build up some general model off-line based on the input and output data-pairs of the system, from which it is hoped to deduce the prediction responses of the system that have yet to be observed. As we know, the model of the system can be expressed with regard to the basis elements of the hypothesis space, and it will obtain "good" generalization if the hypothesis space can cover most of the target space. However, the observational nature data obtained is frequently finite and sampled non-uniform over the whole domain in practical. The hypothesis space, in which we select some function f based on the empirical (training) data to construct the model of the nonlinear system, is frequently only a subspace of the target space. Hence, the model of the system will obtain "bad" generalization capability while using it to predict the response beyond the hypothesis space. In order to solve this problem, we have to learn on-line with the shifting of the work domain. Inspired by the rolling optimization method in control area, we adapt a sliding window method to solve this problem.

Given a nonlinear system with input and output pairs:

$$\{(x_1, y_1), \ldots, (x_i, y_i), (x_{i+1}, y_{i+1}), \ldots (x_l, y_l), \ldots\} \in R^d \times R \qquad (7)$$

Let assume the response of system at certain work domain is completely illustrated with the past observational data in a sliding window with the length W.

i) Recursive Incremental learning method

When the data points arrive at the system less than W, we propose a recursive incremental algorithm for learning process. The train data set is as follows:

$$\{(x_1, y_1), \ldots, (x_i, y_i), \ldots, (x_m, y_m)\}, m \leq W \qquad (8)$$

where $x_i \in R^d, y_i \in R, i = 1, \ldots, m$.

In order to obtain an on-line robust estimate based on the precious LS-SVM, in a subsequent step, one can weighted the error variable $e_k = \alpha_k / \gamma$ by weighting factors v_k in (3).

A similar derivation as the standard LS-SVM can be made. The conditions for optimality lead to a set of linear equations:

$$\begin{bmatrix} 0 & 1 & \cdots & 1 \\ 1 & K(x_1,x_1)+1/\gamma v_1 & \cdots & K(x_1,x_m) \\ \vdots & \vdots & & \vdots \\ 1 & K(x_m,x_1) & \cdots & K(x_m,x_m)+1/\gamma v_m \end{bmatrix} \begin{bmatrix} b \\ \alpha_1 \\ \vdots \\ \alpha_m \end{bmatrix} = \begin{bmatrix} 0 \\ y_1 \\ \vdots \\ y_m \end{bmatrix} \quad (9)$$

For incremental learning process, the sampled points of the train set increase step by step with the time, hence the Grammar matrix of kernel Ω, the Lagrange multipliers α and bias term b in (9) can be identified as the function of the time m. From (9), we obtain

$$\begin{bmatrix} 0 & e1^T \\ e1 & H(m) \end{bmatrix} \begin{bmatrix} b(m) \\ \alpha(m) \end{bmatrix} = \begin{bmatrix} 0 \\ y(m) \end{bmatrix} \quad (10)$$

where e1 is the column vector with appreciate dimension of elements "1", $\alpha(m) = (\alpha_1,\ldots,\alpha_m)^T$, $b(m) = b_m$, and $H(m) = \Omega_m(x_i, x_j) + diag\{\frac{1}{\gamma v_1},\ldots,\frac{1}{\gamma v_m}\}$, $i, j = 1,\ldots,m$.

Rewritten (10), it is easy to deduce

$$\begin{cases} b(m) = \dfrac{e1^T H(m)^{-1} y(m)}{e1^T H(m)^{-1} e1} \\ \alpha(m) = H(m)^{-1}(y(m) - \dfrac{e1 e1^T H(m)^{-1} y(m)}{e1^T H(m)^{-1} e1}) \end{cases} \quad (11)$$

In order to compute the factors of $\alpha(m), b(m)$ recursively, let define

$$U(m) = H(m)^{-1} \quad (12)$$

The dimension of matrix in (12) is $m \times m$. It is known that we can select direct inverse method when dimension is small or a Hestene-Stiefel conjugate gradient algorithm for solving the inverse of a large-scale matrix [9]. However, we have to calculate (12) at every time when a new sample comes to the sliding window, and it leads to heavy computation burden of the on-line learning algorithms. Here we select a recursive algorithm to solve this problem.

From (10), we obtain

$$H(m) = \begin{bmatrix} K(x_1,x_1)+1/\gamma v_1 & \cdots & K(x_1,x_m) \\ \vdots & \vdots & \vdots \\ K(x_m,x_1) & \cdots & K(x_m,x_m)+1/\gamma v_m \end{bmatrix} \quad (13)$$

For the next moment $m+1$, we get

$$H(m+1) = \begin{bmatrix} K(x_1,x_1)+1/\gamma v_1 & \cdots & K(x_1,x_m) & K(x_1,x_{m+1}) \\ \vdots & \cdots & \vdots & \vdots \\ K(x_m,x_1) & \vdots & K(x_m,x_m)+1/\gamma v_m & K(x_m,x_{m+1}) \\ K(x_{m+1},x_1) & \cdots & K(x_{m+1},x_m) & K(x_{m+1},x_{m+1})+1/\gamma v_{m+1} \end{bmatrix} \quad (14)$$

Substitute (13) in (14), with the symmetric positive definite properties of kernel function, we obtain

$$H(m+1) = \begin{bmatrix} H(m) & V(m+1) \\ V(m+1)^T & h(m+1) \end{bmatrix} \quad (15)$$

where

$$V(m+1) = [K(x_{m+1},x_1),\ldots,K(x_{m+1},x_m)]^T, \quad h(m+1) = K(x_{m+1},x_{m+1}) + \frac{1}{\gamma v_{m+1}}$$

According to the inverse of sub-block matrix computation, it can be deduced that:

$$U(m+1) = H(m+1)^{-1}$$

$$= \begin{bmatrix} U(m) & 0 \\ 0 & 0 \end{bmatrix} + \begin{bmatrix} U(m)V(m+1) \\ -1 \end{bmatrix} B^{-1} \begin{bmatrix} V(m+1)^T U(m) & -1 \end{bmatrix} \quad (16)$$

where $B = h(m+1) - V(m+1)^T U(m) V(m+1)$ is a non-zero scalar factor.

By substitute (16) in (11), it is easy to deduce the factors of $\alpha(m+1), b(m+1)$.

Apparently, if the dimension of matrix in (12) is small enough (for example m=2), we can compute its direct inverse easily via the method as mention above. Hence, we can learn the new samples recursively based on the previous results.

ii) First In First Out (FIFO) strategy for on-line modeling

If the data points arrive at the system beyond the length of sliding window, in order to obtain an on-line robust estimate based on the previous WLS-SVM, we have to throw off some old samples. For simplification, we assume that the new (last) points are more important than the old (first) points, and we adapt First In First Out (FIFO) strategy for selecting the train data. In a subsequent step, one can weighted the error variable $e_k = \alpha_k/\gamma$ by weighting factors v_k, this leads to the optimization problem:

$$\min_{w,b,e} J(w,e) = \frac{1}{2} w^T w + \gamma \frac{1}{2} \sum_{k=i+1}^{i+W} v_k e_k^2 \quad (17)$$

Such that

$$y_k = w^T \Phi(x_k) + b + e_k, k = i+1,\ldots,i+W \quad (18)$$

A similar derivation as previous section can be made. The conditions for optimality leads to a set of linear equations:

$$\begin{bmatrix} 0 & 1 & \cdots & 1 \\ 1 & K(x_{i+1},x_{i+1})+\dfrac{1}{\gamma v_{i+1}} & \cdots & K(x_{i+1},x_{i+W}) \\ \vdots & \vdots & & \vdots \\ 1 & K(x_{i+W},x_{i+1}) & \cdots & K(x_{i+W},x_{i+W})+\dfrac{1}{\gamma v_{i+W}} \end{bmatrix} \begin{bmatrix} b \\ \alpha_{i+1} \\ \vdots \\ \alpha_{i+W} \end{bmatrix} = \begin{bmatrix} 0 \\ y_{i+1} \\ \vdots \\ y_{i+W} \end{bmatrix} \quad (19)$$

That is

$$\begin{bmatrix} 0 & \vec{1}_v^T \\ \vec{1}_v & \Omega^* + V_\gamma \end{bmatrix} \begin{bmatrix} b^* \\ \alpha^* \end{bmatrix} = \begin{bmatrix} 0 \\ y^* \end{bmatrix} \quad (20)$$

where the diagonal matrix V_γ is given by

$$V_\gamma = diag\{\dfrac{1}{\gamma v_{i+1}},\ldots,\dfrac{1}{\gamma v_{i+W}}\} \quad (21)$$

The resulting WLS-SVM model for robust function estimation becomes:

$$f^*(x) = \sum_{k=i+1}^{i+W} \alpha_k^* K(x,x_k) + b^* \quad (22)$$

where α_k^*, b^* are the solutions to the linear system(20).

For simplification, let identify these weighted factors v_{k-i} be a function of time t_n for nonlinear system:

$$v_{k-i} = g(t_n) \quad (23)$$

where t_n, $1 \le n \le W$ is the time that the point arrived in the sliding window of the system. We make the last (new) point be the most important and choose $v_w = \theta$, and make the first (old) point be the least important and choose $v_1 = \theta_0$. If we want to make it be a linear weighted function of the time, we can select

$$v_{k-i} = g(t_n) = at_n + b \quad (24)$$

By application of the boundary conditions, we can get

$$v_{k-i} = \frac{\theta - \theta_0}{t_w - t_1} t_n + \frac{\theta_0 t_w - \theta t_1}{t_w - t_1} \qquad (25)$$

So far, a recursive incremental algorithm based on sliding window for on-line modeling is given as follows:

(1) Select samples point with the length of sliding window W for modeling.
(2) Given initial samples $\{(x_1, y_1), (x_2, y_2)\}$ and initial kernel parameters and regularization parameters, set m=2;
(3) Compute $U(m), b(m), \alpha(m)$, m=m+1;
(4) Sample new data point $\{x_m, y_m\}$ and compute the U (m+1) in (16);
(5) Recursive compute $b(m+1), \alpha(m+1)$;
(6) If $m \leq W$, go to (3); otherwise next
(7) Modeling via (22) and produce prediction output;
(8) Optimization the kernel parameters and regularization parameters in (20);
(9) Add the new data points while discarding the old data with FIFO strategy;
(10) Go to (7); otherwise exit

Note it is straightforward for FIFO strategy when the new data points arrive at the system are more than one point in step (9). For briefness, only the linear weighted function in (25) is considered in this paper. Empirically, we can select the weight parameters $\frac{v_1}{v_w}$ from $\frac{1}{2}$ to $\frac{1}{10}$. When $\frac{v_1}{v_w} = 1$, it leads to standard LS-SVM with sliding windows.

4 Application Study

In this section, we construct nonlinear dynamic model with WLS-SVM from a real world data set sampled from a water plant with interval 10 minutes. Water treatment system is a time-variant nonlinear dynamic system with 40 minutes to 120 minutes delay, the main process can be illustrated simply as follows: the raw water is pumped into the water plant, then dosage for coagulation and flocculation, after clarification and filtration treatments, we can obtain the drinking water in the end. The quality of the output water depends on the quality of the raw water (flow, turbidity, temperature, pH, total organic carbon (TOC), etc), appropriate coagulate dosage and the purification facilities. Since the system involves many complex physical, chemical and biological processes, and it is frequently affected by the natural perturbation or occasional pollution in the whole process. It is well known as a challenge work to construct an accurate prediction model of the water plant. After selecting the primary variables via conventional methods such as principal component analysis (PCA) method, the prediction model of the turbidity of the output water is assumed to be the form:

$$y_p(k+1) = F(u_1(k), u_2(k), u_3(k), y(k)) \qquad (26)$$

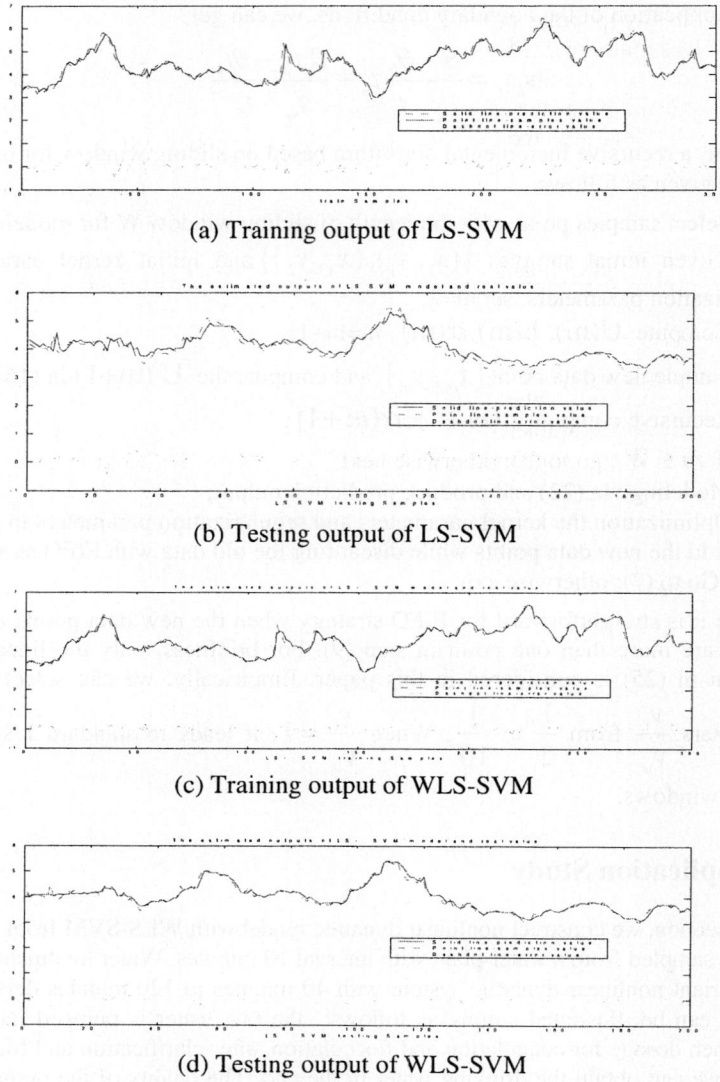

Fig. 1. The identification result of training and testing data via LS-SVM and WLS-SVM

where $u_1(k), u_2(k), u_3(k), y(k)$ denote the flow of raw water, the turbidity of raw water, coagulation dosage, and the turbidity of the output water, respectively. And $y_p(k+1)$ is prediction output of next moment.

We attempted to construct the model in (26) based on WLS-SVM method as mentioned above. The length of sliding window is 30. For comparison, standard LS-SVM with batch learning is also presented here. The training data set consists of 300 samples and another 200 samples in subsequent were used as test data. We compared the results with two popular criteria in control area: the mean square error (MSE) and maximal-absolute-error (MAXE).

The simulation results are illustrated in fig. 1 and table 1. In fig. 1, the solid line represent the output of the identification model and the dashed line show the practical output of the plant, the modeling error is illustrated at the bottom of the figures with dash-dot line.

Table 1. Comparison results of nonlinear system identifaction

Method	MSE (train)	MAXE (train)	MSE (test)	MAXE (test)
LS-SVM	0.0300	1.1436	0.0348	0.6304
WLS-SVM	0.0154	0.48614	0.0109	0.3999

In this simulation, although we adopted cross-validate method for optimization on the regularization parameter and Gaussian kernel parameter of standard LS-SVM, however, the novel approach has greatly outperformed it. Due to on-line learning strategies and optimizing the parameters with the shifting of the work domain, it is not surprising that the WLS-SVM has better performance and generalization ability than the standard LS-SVM based on batching learning.

5 Conclusions

In this paper, we proposed a practical way for nonlinear dynamic system identification based on WLS-SVM, and an on-line algorithm and rolling optimization strategy is discussed. This work provides a novel approach for nonlinear dynamic system modeling, and the experimental results show that the proposed method is feasible. It is worth noting that the length of sliding window is user-prescribed before the learning process while it has a critical influence on the performance of WLS-SVM associated with certain hypothesis space, how to select a suitable sliding window effectively is still remain to be further explored for the future work. In general, this on-line least squares kernel methodology might offer a better opportunity in the area of control.

Acknowledgement

This research is supported in part by the National 973 Key Fundamental Research Project of China (2002CB312200) , the N ational 863 High Technology Projects Foundation of China (2 0 0 2 A A 4 1 2 0 1 0) , and the National Natural Science Foundation of China (60174038).

References

1. Vapnik, V.: The Nature of Statistical Learning Theory (the second edition). New York: Springer-Verlag (1998)
2. Burges, C. J. C.: A tutorial on support vector machines for pattern recognition. Data Mining Knowl. Disc., Vol. 2(2). (1998) 1–47

3. Drucker, H., Burges, C. J. C., Kaufman, L. (ed.): Support vector regression machines. In: Mozer, M., Jordan, M. , Petsche, T.(eds.): Advances in Neural Information Processing Systems,Vol. 9. Cambridge, MA, MIT Press. (1997) 155-161
4. Kecman, V.: Learning and Soft Computing, Support Vector machines, Neural Networks and Fuzzy Logic Models, The MIT Press, Cambridge, MA (2001).
5. Wang, L.P. (Ed.): Support Vector Machines: Theory and Application. Springer, Berlin Heidelberg New York (2005)
6. Suykens, J. A. K. and Vandewalle, J.: Recurrent least squares support vector machines. IEEE Transactions on Circuits and Systems, part I, 47 (7). (2000) 1109-1114.
7. Suykens, J. A. K. and Vandewalle, J.:, Moor, B. De: Optimal control by least squares support vector machines, *Neural Networks*, 14 (1). (2001) 23-35.
8. Suykens, J. A. K.: Support vector machines: a nonlinear modeling and control perspective. European Journal of Control, 7 (2-3). (2001) 311-327.
9. Golub G.H., Van Loan C. F. Matrix Computations, Baltimore MD: Johns Hopkins University Press (1989)

Globally Exponential Stability Analysis and Estimation of the Exponential Convergence Rate for Neural Networks with Multiple Time Varying Delays

Huaguang Zhang[1] and Zhanshan Wang[1,2]

[1] Key Laboratory of Process Industry Automation, Ministry of Education,
Northeastern University, Shenyang 110004, P. R. China
hgzhang@ieee.org
[2] Department of Information Engineering,
Shenyang Ligong University, Shenyang, 110045, P. R. China
zhanshan_wang@163.com

Abstract. Some sufficient conditions for the globally exponential stability of the equilibrium point of neural networks with multiple time varying delays are developed, and the estimation of the exponential convergence rate is presented. The obtained criteria are dependent on time delay, and consist of all the information on the neural networks. The effects of time delay and number of connection matrices of the neural networks on the exponential convergence rate are analyzed, which can give a clear insight into the relation between the exponential convergence rate and the parameters of the neural networks. Two numerical examples are used to demonstrate the effectiveness of the obtained the results.

1 Introduction

In recent years, stability of different classes of neural networks with time delay, such as Hopfield neural networks, cellular networks, bi-directional associative networks, has been extensively studied and various stability conditions have been obtained for these models of neural networks [1-34]. The conditions obtained in those papers establish various types of stability such as complete stability, asymptotic stability, absolute stability and exponential stability, etc. It should be noted that the exponential stability property is particularly important when the exponential convergence rate is used to determine the speed of neural computation and the convergence to the equilibrium in associative memory. Thus, it is important to determine the exponential stability and to estimate the exponential convergence rate for dynamical neural networks.

In general, there are two important notions concerning stability of time-delay systems discussed in the current literatures. One is referred to as delay-independent stability [1,3,9,11,12,13,21,31], the other is delay-dependent stability. As pointed out in [2,10] that the delay independent stability criteria may be overly restrictive when the delays are comparatively small. In many practical applications, the time delays in the neural networks are time varying, or only known to be bounded but nothing else [12,13,25]. Therefore, the study of stability analysis for time varying neural networks has become more important than that with constant delays [2,10,12]. References [1,15] studied the delay-independent stability and [12] studied the delay-dependent stability for single time varying delayed systems, but those papers only concern with stability property, without providing any information on exponential convergence rate of the system's states. References [2,7,8] studied the problem of delay-dependent stability for single time varying delayed systems. References [22,23,24] studied the exponential stability and the estimation of exponential convergence rate for neural networks without time delay. For the case of multiple time varying delays, to the best of our knowledge, few results have been reported.

In this paper, we present some results ensuring the globally exponential stability of delayed neural networks with multiple time varying delays dependent on time delay based on LMI technique, and analyze the effects of time delay and connection matrices on the exponential convergence rate and give an estimate of the exponential convergence rate.

2 Problem Formulation and Preliminaries

Consider the following neural networks with multiple time varying delays

$$\frac{du(t)}{dt} = -Au(t) + W_0 g(u(t)) + \sum_{i=1}^{N} W_i g(u(t - \tau_i(t))) + U, \qquad (1)$$

where $u(t) = [u_1(t), u_2(t), \ldots, u_n(t)]^T$ is the neuron state vector, $A = \text{diag}(a_1, a_2, \ldots, a_n)$ is a positive diagonal matrix with positive entries, $W_0 \in \Re^{n \times n}$ and $W_i \in \Re^{n \times n}$ ($i = 1, 2, \ldots, N$) are the connection weight matrix and delayed connection weight matrices, respectively, $\tau_i(t) \geq 0$ denotes the bounded delay, $\dot{\tau}_i(t) < 1$, $i = 1, 2, \ldots, N$, $U = [U_1, U_2, \ldots, U_n]^T$ denotes the external constant input vector, $g(u(t)) = [g_1(u_1(t)), g_2(u_2(t)), \ldots, g_n(u_n(t))]^T$ denotes the neuron activation function.

Throughout the paper, we need the following notations and preliminaries.

Let B^T, B^{-1}, $\lambda_M(B)$, $\lambda_m(B)$ and $\|B\| = \sqrt{\lambda_M(B^T B)}$ denote the transpose, the inverse, the smallest eigenvalue, the largest eigenvalue, and the Euclidean norm of a square matrix B. Let $B > 0 (B < 0)$ denote the positive (negative) definite symmetric matrix. Let 0 denote a zero matrix or a zero vector with suitable dimension. Let u^* denote a equilibrium point of system (1). Let $\rho_i = \max\{\tau_i(t), t = 0, \cdots, \infty\}$, $\rho = \max\{\rho_i\}$, $0 < \eta_i = 1 - \dot{\tau}_i(t)$, $i = 1, \ldots, N$.

Assumption 1. The activation function, $g_j(u_j)$, satisfies the following condition

$$0 \leq \frac{g_j(\xi) - g_j(\zeta)}{\xi - \zeta} \leq \sigma_j, \qquad (2)$$

for arbitrary $\xi, \zeta \in \Re$, $\xi \neq \zeta$, and for any positive constant $\sigma_j > 0, j = 1, 2, \ldots, n$. Let $\Delta = \text{diag}(\sigma_1, \sigma_2, \ldots, \sigma_n)$. Obviously, Δ is a nonsingular matrix.

Remark 1. Many popular activation functions satisfy the Assumption 1, for example, sigmoid functions, arctan(u), linear piecewise function $0.5(|u+1| - |u-1|)$, and linear function $g(u) = u$, etc. As can be seen from these functions, the function under the Assumption 1 may be bounded or unbounded.

Assumption 2. The equilibrium point set of system (1) is a non-empty set when $\tau_i(t) = 0$, $i = 1, 2, \ldots, N$.

Lemma 1. For any two vectors X and Y, any matrix M, any positive definite matrices Q with same dimensions, and any two positive constants m, n, the following inequality holds,

$$-mX^TQX + 2nX^TMY \leq n^2 Y^T M^T (mQ)^{-1} MY. \tag{3}$$

Lemma 2. Given any real matrices A, B, $Q = Q^T > 0$ with appropriate dimensions, and any scalar $h > 0$, the following inequality holds,

$$A^T B + B^T A \leq h A^T Q A + h^{-1} B^T Q^{-1} B. \tag{4}$$

Lemma 3. For functions $g(u)$ satisfying $0 \leq \frac{g(u)-g(v)}{u-v} \leq p$, the following inequality holds,

$$\int_0^u f(s)ds \leq \frac{1}{2} pu^2, \tag{5}$$

where $u, v \in \Re$, $u \neq v$, $f(x) = g(x+u) - g(u)$.

Definition 1. Consider the system defined by (1), if there exist positive constants $k > 0$ and $\gamma > 0$ such that $\|u(t) - u^\star\| \leq \gamma e^{-kt} \sup_{-\rho \leq \theta \leq 0} \|u(\theta) - u^\star\|$, $\forall t > 0$, then the system (1) is exponential stable, where k is called the exponential convergence rate.

3 Uniqueness of Equilibrium Point

In this section, we will present a sufficient condition to guarantee the uniqueness of the equilibrium point of system (1).

Theorem 1. For a given positive constant k, if the following inequalities

$$\Xi_1 = 2kP - PA - AP + PW_0 Q_0^{-1} W_0^T P + 2k\Delta D$$
$$+ \sum_{i=1}^{N} \frac{1}{\eta_i} e^{2k\tau_i(t)} PW_i Q_i^{-1} W_i^T P < 0, \tag{6}$$

$$\Xi_2 = Q_0 - 2D A \Delta^{-1} + DW_0 + W_0^T D + 2 \sum_{i=1}^{N} Q_i$$
$$+ \sum_{i=1}^{N} \frac{1}{\eta_i} e^{2k\tau_i(t)} DW_i Q_i^{-1} W_i^T D < 0, \tag{7}$$

exist $P > 0$, $Q_j > 0$, $j = 0, 1, \ldots, N$, and positive diagonal matrices $D = \text{diag}(d_1, \ldots, d_n)$, then the system (1) has a unique equilibrium point.

Proof. We take contradiction method. Suppose that there exist two equilibrium points $u^* = [u_1^*, u_2^*, \ldots, u_n^*]^T$ and $v^* = [v_1^*, v_2^*, \ldots, v_n^*]^T$, i.e., $u^* \neq v^*$, satisfying system (1). Then we have

$$-A(u^* - v^*) + \sum_{i=0}^{N} W_i(g(u^*) - g(v^*)) = 0. \tag{8}$$

Let $\tilde{z} = u^* - v^*$, then $\tilde{z} \neq 0$. Let $g(u^*) - g(v^*) = g(\tilde{z} + v^*) - g(v^*) = f(\tilde{z})$, then $f(\tilde{z})$ satisfies Assumption 1 and $f(0) = 0$. Thus, (8) can be converted into the following form

$$-A\tilde{z} + \sum_{i=0}^{N} W_i f(\tilde{z}) = 0. \tag{9}$$

That is to say, $\tilde{z} \neq 0$ is the equilibrium point of the following dynamical system,

$$\frac{dz(t)}{dt} = -Az(t) + \sum_{i=0}^{N} W_i f(z(t)). \tag{10}$$

In the following, we will prove that $\tilde{z} \neq 0$ is not the equilibrium point of system (10).

We choose Lyapunov function as follows,

$$V(z(t)) = e^{2kt} z(t)^T P z(t) + 2e^{2kt} \sum_{i=1}^{n} \int_0^{z_i(t)} d_i f(s) ds. \tag{11}$$

Then the derivative of $V(z(t))$ along the solution of system (10) is

$$\dot{V}(z(t)) \leq e^{2kt} z^T(t)(2kP - PA - AP + 2k\Delta D)z(t)$$
$$+ 2e^{2kt} z^T(t)(PW_0 + PW_1 \cdots + PW_N)f(z(t))$$
$$- 2e^{2kt} f^T(z(t)) DA\Delta^{-1} f(z(t)) + 2e^{2kt} f^T(z(t))(DW_0 + DW_1 \cdots DW_N)f(z(t))$$
$$\leq e^{2kt} z^T(t)(2kP - PA - AP + 2k\Delta D)z(t)$$
$$+ e^{2kt}(z^T(t) \sum_{i=0}^{N} \frac{e^{2k\tau_i(t)}}{\eta_i} PW_i Q_i^{-1} W_i^T P z(t)$$
$$+ \sum_{i=0}^{N} \eta_i e^{-2k\tau_i(t)} f^T(z(t)) Q_i f(z(t)))$$
$$- 2e^{2kt} f^T(z(t)) DA\Delta^{-1} f(z(t)) + 2e^{2kt} f^T(z(t))(DW_0) f(z(t))$$
$$+ e^{2kt}(f^T(z(t))(\sum_{i=1}^{N} \frac{e^{2k\tau_i(t)}}{\eta_i} DW_i Q_i^{-1} W_i^T D)f(z(t)))$$
$$+ \sum_{i=1}^{N} \eta_i e^{-2k\tau_i(t)} f^T(z(t)) Q_i f(z(t))), \tag{12}$$

where $\tau_0(t) = 0$ and $\eta_0 = 1$.

Since $\eta_i e^{-2k\tau_i(t)} \leq 1$, then (12) is equivalent to the following form

$$\dot{V}(z(t)) \leq e^{2kt} z^T(t)(2kP - 2PA + 2k\Delta D + \sum_{i=0}^{N} \frac{e^{2k\tau_i(t)}}{\eta_i} PW_i Q_i^{-1} W_i^T P) z(t)$$

$$+ e^{2kt} f^T(z(t))(Q_0 - 2DA\Delta^{-1} + DW_0 + W_0^T D)$$
$$+ \sum_{i=1}^{N} \frac{e^{2k\tau_i(t)}}{\eta_i} DW_i Q_i^{-1} W_i^T D + 2\sum_{i=1}^{N} Q_i) f(z(t))$$
$$= e^{2kt}(z^T(t)\Xi_1 z(t) + f^T(z(t))\Xi_2 f(z(t))). \tag{13}$$

Thus, $\dot{V}(z(t)) < 0$ if $z(t) \neq 0$ and $f(z(t)) \neq 0$. $\dot{V}(z(t)) = 0$ if and only if $z(t) = 0$ and $f(z(t)) = 0$. By Lyapunov theory, $\tilde{z} = 0$ is the equilibrium point of system (10), which is a contradiction with (9). Therefore, system (1) has a unique equilibrium point if conditions (6) and (7) hold. This completes the proof.

4 Globally Exponential Stability

The transformation $x(\cdot) = u(\cdot) - u^*$ changes system (1) into the following form

$$\frac{dx(t)}{dt} = -Ax(t) + W_0 f(x(t)) + \sum_{i=1}^{N} W_i f(x(t - \tau_i(t))), \tag{14}$$
$$x(t) = \phi(t), \ t \in [-\rho, 0),$$

where $x(t) = [x_1(t), x_2(t), \ldots, x_n(t)]^T$ is the state vector of the system (14), $f_j(x_j(t)) = g_j(x_j(t) + u_j^*) - g_j(u_j^*)$ with $f_j(0) = 0$, $j = 1, 2, \ldots, n$. $\phi(t)$ is a continuous vector-valued function with the maximum norm $\|\phi\|$. Obviously, $f(x(t))$ satisfies the Assumption 1.

Clearly, the equilibrium point u^* is globally exponentially stable for system (1) if and only if the zero solution of system (14) is globally exponentially stable.

Theorem 2. If the conditions in Theorem 1 are satisfied, then the unique equilibrium point u^* of system (1) is globally exponentially stable. Moreover,

$$\|u(t) - u^*\| \leq \sqrt{\frac{Z}{\lambda_m(P)}} \|\phi\| e^{-kt}, \tag{15}$$

where $Z = \lambda_M(P) + \sum_{i=1}^{N} \lambda_M(\Delta^T Q_i \Delta) \frac{1 - e^{-2k\tau_i(0)}}{k} + \lambda_M(D)\lambda_M(\Delta)$.

Proof. Consider the following Lyapunov-Krasovskii functional

$$V(x(t)) = e^{2kt} x^T(t) P x(t) + 2\sum_{i=1}^{N} \int_{t-\tau_i(t)}^{t} e^{2ks} f^T(x(s)) Q_i f(x(s)) ds$$
$$+ 2e^{2kt} \sum_{i=1}^{n} d_i \int_{0}^{x_i(t)} f(s) ds. \tag{16}$$

The time derivative of the functional (16) along the trajectories of system (14) is obtained as follows

$\dot{V}(x(t))$
$= 2ke^{2kt}x^T(t)Px(t) + 2e^{2kt}x^T(t)P\dot{x}(t) + 2\sum_{i=1}^{N} e^{2kt}f^T(x(t))Q_i f(x(t))$
$-2\sum_{i=1}^{N}(1-\dot{\tau}_i(t))e^{2k(t-\tau_i(t))}f^T(x(t-\tau_i(t)))Q_i f(x(t-\tau_i(t)))$
$+4ke^{2kt}\sum_{i=1}^{n} d_i \int_0^{x_i(t)} f(s)ds + 2e^{2kt}f^T(x(t))D\dot{x}(t)$
$\leq 2ke^{2kt}x^T(t)Px(t) + 2e^{2kt}x^T(t)P(-Ax(t) + W_0 f(x(t))$
$+ \sum_{i=1}^{N} W_i f(x(t-\tau_i(t)))) + 2\sum_{i=1}^{N} e^{2kt}f^T(x(t))Q_i f(x(t))$
$-2\sum_{i=1}^{N} \eta_i e^{2k(t-\tau_i(t))}f^T(x(t-\tau_i(t)))Q_i f(x(t-\tau_i(t)))$
$+2ke^{2kt}x^T(t)\Delta Dx(t) + 2e^{2kt}f^T(x(t))D(-Ax(t) + W_0 f(x(t)))$
$+ \sum_{i=1}^{N} W_i f(x(t-\tau_i(t)))) \tag{17}$

where we have applied the inequality $\int_0^{x_i(t)} f(s)ds \leq \frac{1}{2}\sigma_i x_i^2(t)$ obtained from Lemma 3.

By Assumption 1, Lemma 2 and Lemma 3, we have from (17)
$\dot{V}(x(t))$
$\leq e^{2kt}x^T(t)[2kP - PA - AP + PW_0 Q_0^{-1} W_0^T P + 2k\Delta D$
$+ \sum_{i=1}^{N} \frac{1}{\eta_i} e^{2k\tau_i(t)} PW_i Q_i^{-1} W_i^T P]x(t) + e^{2kt}f^T(x(t))[Q_0 - 2D A\Delta^{-1}$
$+ DW_0 + W_0^T D + \sum_{i=1}^{N} \frac{1}{\eta_i} e^{2k\tau_i(t)} DW_i Q_i^{-1} W_i^T D + 2\sum_{i=1}^{N} Q_i]f(x(t))$
$= e^{2kt}\left\{x^T(t)\Xi_1 x(t) + f^T(x(t))\Xi_2 f(x(t))\right\}, \tag{18}$

Thus, if condition (6) and (7) hold, $\dot{V}(x(t)) < 0$ if $x(t) \neq 0$ and $f(x(t)) \neq 0$. Besides, for the case $f(x(t)) = 0$ and $x(t) \neq 0$, or $f(x(t)) = x(t) = 0$, we still have $\dot{V}(x(t)) < 0$. $\dot{V}(x(t)) = 0$ if and only if $f(x(t)) = x(t) = f(x(t-\tau_i(t))) = 0$. Therefore, we have $V(x(t)) \leq V(x(0))$. Furthermore, $V(x(t)) \geq e^{2kt}\lambda_m(P)\|x(t)\|^2$, and

$V(x(0)) = x^T(0)Px(0) + 2\sum_{i=1}^{N} \int_{-\tau_i(0)}^{0} e^{2ks}f^T(x(s))Q_i f(x(s))ds$
$+2\sum_{i=1}^{n} d_i \int_0^{x_i(0)} f(s)ds$
$\leq \lambda_M(P)\|\phi\|^2 + 2\sum_{i=1}^{N} \lambda_M(\Delta^T Q_i \Delta)\|\phi\|^2 \int_{-\tau_i(0)}^{0} e^{2ks}ds$
$+\lambda_M(D)\lambda_M(\Delta)\|\phi\|^2$
$\leq (\lambda_M(P) + \sum_{i=1}^{N} \lambda_M(\Delta^T Q_i \Delta) \cdot \frac{1-e^{-2k\cdot\tau_i(0)}}{k} + \lambda_M(D)\lambda_M(\Delta))\|\phi\|^2, \tag{19}$

then we have
$$\|x(t)\| \le \sqrt{\frac{Z}{\lambda_m(P)}} \|\phi\| e^{-kt}. \tag{20}$$

On the other hand, $V(x(t))$ is radically unbounded, that is $V(x(t)) \to \infty$ as $\|x(t)\| \to \infty$. Thus, by Lyapunov theory and Definition 1, it follows that the origin of (14) is globally exponentially stable, i.e., the unique equilibrium point u^* of system (1) is globally exponentially stable. This completes the proof.

Because of the complexity of time varying delay, it is difficult to solve the inequality (6) and (7) for a given constant k. Therefore, in order to check the applicability of the results conveniently, we have the following corollary.

Corollary 1. For a given positive constant k, if the following inequalities
$$\Xi_3 = 2kP - PA - AP + PW_0 Q_0^{-1} W_0^T P + 2k\Delta D$$
$$+ \sum_{i=1}^{N} \frac{1}{\eta_i} e^{2k\rho_i} PW_i Q_i^{-1} W_i^T P < 0, \tag{21}$$

$$\Xi_4 = Q_0 - 2D\Delta\Delta^{-1} + DW_0 + W_0^T D + 2\sum_{i=1}^{N} Q_i$$
$$+ \sum_{i=1}^{N} \frac{1}{\eta_i} e^{2k\rho_i} DW_i Q_i^{-1} W_i^T D < 0, \tag{22}$$

exist $P > 0$, $Q_j > 0$, $j = 0, 1, \ldots, N$, and positive diagonal matrices $D = \mathrm{diag}(d_1, \ldots, d_n)$, then the system (1) has a unique equilibrium point and it is globally exponentially stable.

In what follows, Theorem 2 will be particularized for the case of constant time delay.

Theorem 3. In the case of $\tau_i(t) = \tau_i = $ constant, $i = 1, \ldots, N$, if for a given positive constant k, there exist $P > 0$, $Q_j > 0$, $j = 0, 1, \ldots, N$, and positive diagonal matrices $D = \mathrm{diag}(d_1, \ldots, d_n)$, such that

$$\Xi_5 = 2kP - PA - AP + PW_0 Q_0^{-1} W_0^T P + \sum_{i=1}^{N} e^{2k\tau_i} PW_i Q_i^{-1} W_i^T P + 2k\Delta D < 0, \tag{23}$$

$$\Xi_6 = Q_0 - 2D\Delta\Delta^{-1} + DW_0 + W_0^T D + \sum_{i=1}^{N} e^{2k\tau_i} DW_i Q_i^{-1} W_i^T D + 2\sum_{i=1}^{N} Q_i < 0, \tag{24}$$

then the system (1) has a unique equilibrium point and it is globally exponentially stable.

In the case of constant delay, we will discuss the relation between A and k. For simplicity, we assume $\Delta = I$ and $P = Q_i = D = \alpha I$ satisfying (23) and (24), $\alpha > 0$ is a constant, and let $\lambda_i = \lambda_M(W_i W_i^T)$, $i = 0, 1, \ldots, N$. Then (23) can be expressed as

$$2kI - 2AI + \alpha \sum_{i=0}^{N} e^{2k\tau_i} \lambda_i I + 2kI < 0, \text{ or } 2k + 0.5\alpha(N+1)e^{2k\rho}\lambda_{\max} < \lambda_m(A), \tag{25}$$

where $\lambda_{\max} = \max\{\lambda_i\}$. (25) restricts k, ρ, N, A and W_i in an inequality, from which we can see that for fixed A and ρ, with the number of delay interconnection term N increasing, the exponential convergence rate k decreases. Similarly, for fixed A and N, the increase of time delay will decrease the exponential convergence rate k.

A more conservative estimate of exponential convergence rate may be obtained from (25), i.e.,

$$\Theta_7 = 2kI - 2A + 2kI < 0, \text{ or } k < 0.5\lambda_m(A). \tag{26}$$

from which we can also conclude that the larger the smallest eigenvalue of A is, the greater the exponential convergence rate k is.

To estimate the exponential convergence rate k, we must know the upper bound of time delay $\tau_i(t)$. In this case, we may solve the following optimization problem

$$\begin{cases} \max(k) \\ \text{s.t. Corollary 1 is satisfied, } \rho_i \text{ is fixed} \end{cases} \tag{27}$$

The solution of (27) determines the maximum exponential convergence rate $k \leq k^*$, which is useful in real-time optimization and neural computation. Note that this is a quasi-convex optimization problem.

5 Numerical Examples

Example 1. Consider the following delayed neural network

$$\dot{x}(t) = -Ax(t) + W_0 g(x(t)) + W_1 g(x(t - \tau_1)) + U, \tag{28}$$

where $g(x(t)) = 0.5(|x(t) + 1| - |x(t) - 1|)$,

$$A = \begin{bmatrix} 5 & 0 \\ 0 & 9 \end{bmatrix}, \quad W_0 = \begin{bmatrix} 2 & -1 \\ -2 & 3 \end{bmatrix}, \quad W_1 = \begin{bmatrix} 3 & 1 \\ 0.5 & 2 \end{bmatrix}, \quad U = \begin{bmatrix} 1 \\ 2 \end{bmatrix}.$$

In this case, the results in [1,29] and Theorem 1-2 in [2] cannot be able to ensure the stability. Take $\tau_1 = 0.2$, the maximum exponential convergence rate is $k \leq 0.83$ from (27). When $k = 0.8$, the parameters in Theorem 3 are

$$P = \begin{bmatrix} 0.4641 & 0.0219 \\ 0.0219 & 0.2518 \end{bmatrix}, \quad D = \begin{bmatrix} 1.3923 & 0 \\ 0 & 0.7576 \end{bmatrix},$$

$$Q_0 = \begin{bmatrix} 1.6975 & -0.2231 \\ -0.2231 & 2.4865 \end{bmatrix}, \quad Q_1 = \begin{bmatrix} 2.6343 & 1.2420 \\ 1.2420 & 2.2424 \end{bmatrix}.$$

The unique equilibrium point is $(1.2000, 0.1250)$.

Example 2. Consider the system (28) except $A = \begin{bmatrix} 9 & 0 \\ 0 & 9 \end{bmatrix}$, $\tau_1(t) = \dfrac{2(1 - e^{-0.5t})}{1 + e^{-0.5t}}$. It is easy to observe that $\rho_1 = 2$, $\eta_1 = 1 - \dot{\tau}_1(t) = 0.5$. In this case, Corollary 1 holds for appropriate exponential convergence rate, and the estimate of exponential convergence rate is $k \leq 0.681$. As comparison, the Theorem 3 in [2] can estimate the maximum exponential convergence rate $k \leq 0.2$. The unique equilibrium point is $(0.2435, 0.4075)$.

6 Conclusions

In this paper, globally exponential stability dependent on time delay and estimation of exponential convergence rate for neural networks with multiple time varying delays are investigated. The obtained criteria are computationally efficient than those based on matrix measure and algebraic inequality techniques. In addition, compared with the results based on M-matrix theory and matrix norm, the stability conditions contain all the information on the connection matrix of neural networks, therefore, the differences between excitatory and inhibitory effects on the neural networks have been eliminated. Moreover, the effects of parameters in the delayed neural networks on the exponential convergence rate are analyzed. Two numerical examples are presented to illustrate the validity of the obtained results.

Acknowledgement. This work was supported by the National Natural Science Foundation of China under grant 60274017 and 60325311.

References

1. Cao J.,Wang J.: Global asymptotic stability of a general class of recurrent neural networks with time-varying delays. IEEE Trans. Circuits Sys. **50** (2003) 34–44
2. Liao X., Chen G., Sanchez E. N.: Delay-dependent exponential stability analysis of delayed neural networks: an LMI approach. Neural Networks. **15** (2002) 855–866
3. Arik S.: Stability analysis of delayed neural networks. IEEE Trans. Circuits Sys. **47** (2000) 1089–1092
4. Arik S.: An improved global stability result for delayed cellular neural networks. IEEE Trans. Circuits Sys. **49** (2002) 1211–1214
5. Ye H., Michel A.N.: Robust stability of nonlinear time delay systems with applications to neural networks. IEEE Trans. Circuits Sys. **43** (1996) 532–543
6. Cao J., Wang L.: Exponential stability and periodic oscillatory solution in BAM networks with delays. IEEE Trans. Neural Networks. **13** (2002) 457–463
7. Arik S.: An analysis of exponential stability of delayed neural networks with time varying delays. Neural Networks. **17** (2004) 1027–1031
8. Yucel E., Arik S.: New exponential stability results for delayed neural networks with time varying delays. Physica D. **191** (2004) 314–322
9. Cao J.: Global stability conditions for delayed CNNs. IEEE Trans. Circuits Sys. **48** (2001) 1330–1333
10. Liao X., Chen G., Sanchez E.N.: LMI-based approach for asymptotic stability analysis of delayed neural networks. IEEE Trans. Circuits Sys. **49** (2002) 1033–1039
11. Lu H.: On stability of nonlinear continuous-time neural networks with delays. Neural Networks. **13** (2000) 1135–1143
12. Joy M.: Results concerning the absolute stability of delayed neural networks. Neural Networks. **13** (2000) 613–616
13. Liao X., Yu J.: Robust stability for interval Hopfield neural networks with time delay. IEEE Trans. Neural Networks. **9** (1998) 1042–1046
14. Zhang J., Jin X.: Global stability analysis in delayed Hopfield neural network models. Neural Networks. **13** (2000) 745–753

15. Huang H., Cao J.: On global asymptotic stability of recurrent neural networks with time- varying delays. Applied Math. Comput. **142** (2003) 143–154
16. Cao J.: Exponential stability and periodic solution of delayed cellular neural networks. Science in China (Series E). **43** (2000) 328–336
17. Zhang Y., Heng P., Vadakkepat P.: Absolute periodicity and absolute stability of delayed neural networks. IEEE Trans. Circuits Sys. **49** (2002) 256–261
18. Zhang J., Yang Y.: Global stability analysis of bi-directional associative memory neural networks with time delay. Int. J. Circ. Theor. Appl. **29** (2001) 185–196
19. Zhang J.: Absolute exponential stability in delayed cellular neural networks. Int. J. Circ. Theor. Appl. **30** (2002) 395–409
20. Takahashi N.: A new sufficient condition for complete stability of cellular neural networks with delay. IEEE Trans. Circuits Sys. **47** (2000) 793–799
21. Lu W., Rong L., Chen T.: Global convergence of delayed neural network systems. Int. J. Neural Sys. **13** (2003) 193–204
22. Michel A. N., Farrell J. A., Porod W.: Qualitative analysis of neural networks. IEEE Trans. Circuits Sys. **36** (1989) 229–243
23. Zhang Y., Heng P. A., Fu Ade W. C.: Estimate of exponential convergence rate and exponential stability for neural networks. IEEE Trans. Neural Networks. **10** (1999) 1487–1493
24. Liang X., Wu L.: Global exponential stability of Hopfield-type neural networks and its applications. Science in China (Series A). **25** (1995) 523–532
25. Zhang Q., Wei X., Xu J.: Global exponential stability of Hopfield neural networks with continuously distributed delays. Physics Letters A, **315** (2003) 431–436
26. Liao T., Wang F.: Global stability for cellular neural networks with time delay. IEEE Trans. Neural Networks. **11** (2000) 1481–1484
27. Peng J., Qiao H., Zu Z.: A new approach to stability of neural networks with time varying delays. Neural Networks. **15** (2002) 95–103
28. Sree Hari Rao V., Phaneendra B.: Global dynamics of bi-directional associative memory neural networks involving transmission delays and dead zones. Neural Networks. **12** (1999) 455–465
29. Chen T.: Global exponential stability of delayed Hopfield neural networks. Neural Networks. **14** (2001) 977–980
30. Feng C., Plamondon R.: On the stability analysis of delayed neural networks systems. Neural Networks. **14** (2001) 1181–1188
31. Van Den Driessche P., Zou X.: Global attractivity in delayed Hopfield neural networks models. SIAM J. Appl. Math. **58** (1998) 1878–1890
32. Liao X., Wong K., Wu Z. et. al.: Novel robust stability criteria for interval delayed Hopfield neural networks. IEEE Trans. Circuits Sys. **48** (2001) 1355–1358
33. Liao X., Wang J., Cao J.: Global and robust stability of interval Hopfield neural networks with time-varying delays. Int. J. Neural Systems. **13** (2003) 171–182
34. Michel A. N., Liu D.: Qualitative analysis and synthesis of recurrent neural networks. New York: Marcel Dekker (2002)

Locally Determining the Number of Neighbors in the k-Nearest Neighbor Rule Based on Statistical Confidence

Jigang Wang, Predrag Neskovic, and Leon N. Cooper*

Institute for Brain and Neural Systems,
Department of Physics, Brown University, Providence RI 02912, USA
{jigang, pedja, Leon_Cooper}@brown.edu
http://physics.brown.edu/physics/researchpages/Ibns/index.html

Abstract. The k-nearest neighbor rule is one of the most attractive pattern classification algorithms. In practice, the value of k is usually determined by the cross-validation method. In this work, we propose a new method that locally determines the number of nearest neighbors based on the concept of statistical confidence. We define the confidence associated with decisions that are made by the majority rule from a finite number of observations and use it as a criterion to determine the number of nearest neighbors needed. The new algorithm is tested on several real-world datasets and yields results comparable to those obtained by the k-nearest neighbor rule. In contrast to the k-nearest neighbor rule that uses a fixed number of nearest neighbors throughout the feature space, our method locally adjusts the number of neighbors until a satisfactory level of confidence is reached. In addition, the statistical confidence provides a natural way to balance the trade-off between the reject rate and the error rate by excluding patterns that have low confidence levels.

1 Introduction

In a typical non-parametric classification problem, one is given a set of n observations $D_n = \{(\boldsymbol{X}_1, Y_1), \ldots, (\boldsymbol{X}_n, Y_n)\}$, where $\boldsymbol{X}_i$ are the feature vectors and Y_i are the corresponding class labels and $(\boldsymbol{X}_i, Y_i)$ are assumed to be i.i.d. from some unknown distribution P of $(\boldsymbol{X}, Y)$ on $R^d \times \{\omega_1, \ldots, \omega_M\}$. The goal is to design a function $\phi_n : R^d \to \{\omega_1, \ldots, \omega_M\}$ that maps a feature vector $\boldsymbol{X}$ to its desired class from $\{\omega_1, \ldots, \omega_M\}$. The performance of a classifier ϕ_n can be measured by the probability of error, defined as

$$L(\phi_n) = P\{(\boldsymbol{X}, Y) : \phi_n(\boldsymbol{X}) \neq Y\} \ . \tag{1}$$

If the underlying distribution is known, the optimal decision rule for minimizing the probability of error is the Bayes decision rule [1]:

$$\phi^*(\boldsymbol{X}) = \arg \max_{Y \in \{\omega_1, \ldots, \omega_M\}} P(Y|\boldsymbol{X}) \ . \tag{2}$$

---

* This work is partially supported by ARO under grant DAAD19-01-1-0754. Jigang Wang is supported by a dissertation fellowship from Brown University.

One of the most attractive classification algorithms is the nearest neighbor rule, first proposed by Fix and Hodges in 1951 [2]. It classifies an unseen pattern X into the class of its nearest neighbor in the training data. Geometrically, each labeled observation in the training dataset serves as a prototype to represent all the points in its Voronoi cell.

It can be shown that at any given point X the probability that its nearest neighbor X' belongs to class ω_i converges to the corresponding a posteriori probability $P(\omega_i|X)$ as the number of reference observations goes to infinity, i.e., $P(\omega_i|X) = \lim_{n\to\infty} P(\omega_i|X')$. Furthermore, it was shown in [3,4] that under certain continuity conditions on the underlying distributions, the asymptotic probability of error L_{NN} of the nearest neighbor rule is bounded by

$$L^* \leq L_{\text{NN}} \leq L^*(2 - \frac{M}{M-1}L^*) \;, \tag{3}$$

where L^* is the optimal Bayes probability of error. Therefore, the nearest neighbor rule, despite its extreme simplicity, is asymptotically optimal when the classes do not overlap. However, when the classes do overlap, the nearest neighbor rule is suboptimal. In these situations, the problem occurs at overlapping regions where $P(\omega_i|X) > 0$ for more than one class ω_i. In those regions, the nearest neighbor rule deviates from the Bayes decision rule by classifying X into class ω_i with probability $P(\omega_i|X)$ instead of assigning X to the majority class with probability one.

In principle, this shortcoming can be overcome by a natural extension, the k-nearest neighbor rule. As the name suggests, this rule classifies X by assigning it to the class that appears most frequently among its k nearest neighbors. Indeed, as shown by Stone and Devroye in [5,6], the k-nearest neighbor rule is universally consistent provided that the speed of k approaching n is properly controlled, i.e., $k \to \infty$ and $k/n \to 0$ as $n \to \infty$. However, choosing an optimal value k in a practical application is always a problem, due to the fact that only a finite amount of training data is available. This problem is known as the bias/variance dilemma in the statistical learning community [7]. In practice, one usually uses methods such as cross-validation to pick the best value for k.

In this work, we address the problem of neighborhood size selection. In the k-nearest neighbor rule, the value of k, once determined by minimizing the estimated probability of error through cross-validation, is the same for all query points in the space. However, there is no a priori reason to believe that the optimal value of k has to be the same for different query points. In general, it might be advantageous to have the value of k determined locally. The question is: what criterion should be used to determine the optimal value of k?

In this paper, we propose an approach to neighborhood size selection based on the concept of statistical confidence. The approach stems from the following observations. When a decision is made from a finite number of observations, there is always a certain non-zero probability that the decision is wrong. Therefore, it is desirable to know what the probability of error is when making a decision and to keep this probability of error under control. For example, in many applications, such as in medical diagnosis, the confidence with which a system makes a decision is of crucial importance. Similarly, in situations where not every class has the same importance, one may require different levels of confidence for decisions regarding different classes. Instead of using a fixed value of k throughout the feature space, in such applications, it is more natural to fix the confidence level. Based on these observations, we propose a method that locally adjusts the number of nearest neighbors until a satisfactory level of confidence is reached.

This paper is organized as follows. In section 2 we define the probability of error for decisions made by the majority rule based on a finite number of observations, and show that the probability of error is bounded by a decreasing function of a confidence measure. We then define the statistical confidence as the complement of the probability of error and use it as a criterion for determining the neighborhood size in the k-nearest neighbor rule. This leads to a new algorithm, which we call the confident-nearest neighbor rule. In section 3 we test the new algorithm on several real-world datasets and compare it with the original k-nearest neighbor rule. Concluding remarks are given in section 4.

2 Probability of Error and Statistical Confidence

One of the main reasons for the success of the k-nearest neighbor rule is the fact that for an arbitrary query point X, the class labels Y' of its k nearest neighbors can be treated as approximately distributed from the desired a posteriori probability $P(Y|X)$. Therefore, the empirical frequency with which each class ω_i appears within the neighborhood provides an estimate of the a posteriori probability $P(\omega_i|X)$. The k-nearest neighbor rule can thus be viewed as an empirical Bayes decision rule based on the estimate of $P(Y|X)$ from the k nearest neighbors. There are two sources of error in this procedure. One results from whether or not the class labels Y' of the neighbors can be approximated as i.i.d. as Y. The other source of error is caused by the fact that, even if Y' can be approximated as i.i.d. as Y, there is still a probability that the empirical majority class differs from the true majority class based on the underlying distribution. In this section, we will address the second issue.

2.1 Probability of Error and Confidence Measure

For simplicity we consider a two-class classification problem. Let $R \in \Omega$ be a neighborhood of X in the feature space, and $p = P(Y = \omega_1|X)$ be the a posteriori probability of the class being ω_1 given the observation X. Let $X_1, \ldots, X_n \in R$ be n i.i.d. random variables and assume that they have the same a posteriori probability as X. The n corresponding labels $Y_1, \ldots, Y_n$ can then be treated as i.i.d. from the Bernoulli distribution Bern(p). According to the binomial law, the probability that n_1 of them belongs to class ω_1 (therefore $n_2 = n - n_1$ belongs to ω_2) is $\binom{n}{n_1} p^{n_1} (1-p)^{n_2}$. Therefore, the probability of observing δ more samples from class ω_2 than from class ω_1 is given by:

$$\sum_{n_1=0}^{[(n-\delta)/2]} \binom{n}{n_1} p^{n_1} (1-p)^{n-n_1} . \qquad (4)$$

We define the probability

$$P_{err}(p;\delta;n) = \sum_{n_1=0}^{[(n-\delta)/2]} \binom{n}{n_1} p^{n_1}(1-p)^{n-n_1} \quad (5)$$

under the condition that $p \in (0.5, 1]$ to be the probability of error for the following reasons: since $p \in (0.5, 1]$, according to the Bayes decision rule, $\boldsymbol{X}$ should be associated with the true majority class ω_1; however, if $n_2 - n_1 = \delta > 0$, $\boldsymbol{X}$ will be classified into class ω_2 by the majority rule, therefore leading to an error. In other words, given $p \in (0.5, 1]$, $P_{err}(p;\delta;n)$ is defined to be the probability of observing δ more samples from class ω_2 than ω_1. Using simple symmetry argument, it is easy to check that $P_{err}(1-p;\delta;n)$ is the probability that one will observe δ more samples from class ω_1 than from ω_2 while $1-p \in (0.5, 1]$. Regardless of whether $p \in (0.5, 1]$ or $1-p \in (0.5, 1]$, if we let $\bar{p} = \max\{p, 1-p\}$, $P_{err}(\bar{p};\delta;n)$ is the probability that one observes δ more samples from the true minority class than from the true majority class.

In practice, p is unknown; hence $\bar{p}$ is also unknown. Fortunately, it is easy to show that $P_{err}(\bar{p};\delta;n)$ is a decreasing function of $\bar{p}$, which means that it is bounded above by

$$P_{err}(\delta;n)_{max} = \frac{1}{2^n}\sum_{n_1=0}^{[(n-\delta)/2]} \binom{n}{n_1} \approx \Phi(-\frac{\delta-1}{\sqrt{n}}) , \quad (6)$$

where Φ is the cumulative distribution function (CDF) of a standard Gaussian random variable. The probability of error $P_{err}(\bar{p};\delta;n)$ can also be bounded by applying concentration of measure inequalities.

Let us consider the relationship between $P_{err}(\delta;n)_{max}$ and $(\delta-1)/\sqrt{n}$. Obviously, $P_{err}(\delta;n)_{max}$ is decreasing in $(\delta-1)/\sqrt{n}$ because as a cumulative distribution function, $\Phi(x)$ is an increasing function of x. Therefore, the larger $(\delta-1)/\sqrt{n}$, the smaller the probability of error. Equation (6) also quantitatively tells us how large $(\delta-1)/\sqrt{n}$ should be in order to keep the probability of error under some preset value. For $n < 200$, we enumerate all possible values of δ and n and calculate $(\delta-1)/\sqrt{n}$ and the corresponding value of $P_{err}(\delta;n)_{max}$. The result is shown in Fig. 1.

Since $P_{err}(\bar{p};\delta;n)$ is the probability that the observation is at odds with the true state of nature, $1-P_{err}(\bar{p};\delta;n)$ is the probability that the observation agrees with the true state of nature. We therefore define

$$CFD(\bar{p};\delta;n) \equiv 1 - P_{err}(\bar{p};\delta;n) \quad (7)$$

to be the confidence (CFD) level. From Eq. (6), it follows that the confidence level is bounded below by

$$CFD(\delta;n) = 1 - P_{err}(\delta;n)_{max} \approx erf(\frac{\delta-1}{\sqrt{n}}) . \quad (8)$$

The larger $(\delta-1)/\sqrt{n}$, the higher the confidence level. For this reason and for convenience, we will call $(\delta-1)/\sqrt{n}$ the confidence measure.

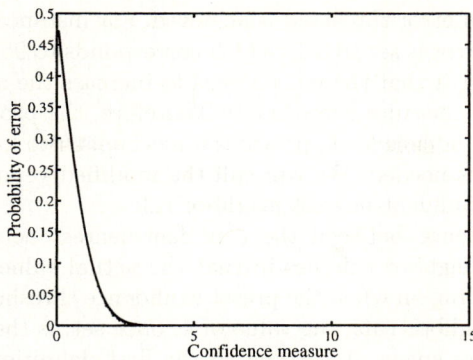

Fig. 1. Probability of error as a function of the confidence measure

An alternative way to define the probability of error for a decision that is made by the majority rule based on a finite number of observations is to use the Beta prior model for the binomial distribution. Using the same argument, the probability of error can be defined as

$$P_{err}(\delta;n) = \frac{\int_{0.5}^{1} p^{\frac{n-\delta}{2}} (1-p)^{\frac{n+\delta}{2}} dp}{\int_{0}^{1} p^{\frac{n-\delta}{2}} (1-p)^{\frac{n+\delta}{2}} dp} \quad , \tag{9}$$

which gives the probability that the actual majority class of the posterior probability distribution differs from the one that is concluded empirically from the majority rule based on n and δ. Likewise, the confidence level can be defined as

$$CFD(\delta;n) = 1 - P_{err}(\delta;n) = \frac{\int_{0}^{0.5} p^{\frac{n-\delta}{2}} (1-p)^{\frac{n+\delta}{2}} dp}{\int_{0}^{1} p^{\frac{n-\delta}{2}} (1-p)^{\frac{n+\delta}{2}} dp} \quad . \tag{10}$$

Numerically, these two different definitions give roughly the same results. More precisely, compared to the second definition, the first definition of the probability of error can be better approximated as a function of the confidence measure, which is easily computable. In addition, for the same values of n and δ, the first definition also gives a higher probability of error value because it is based on the worst case consideration.

2.2 Determining the Number of Neighbors in the k-Nearest Neighbor Rule

In the k-nearest neighbor rule, the only advantage of choosing a large k value is to reduce the variance of the a posteriori probability estimate. Similarly, as we have shown, a large k value can potentially lead to a large confidence measure, and therefore to a small probability of error. Note that at each query point, the probability of error can be easily computed for different numbers of neighbors. Therefore, one can choose to increase the number of nearest neighbors k until a preset probability of error threshold is achieved. For instance, if the threshold of the probability of error is set to 5% (which corresponds to 95% confidence level), one can see from Fig. 1 that there is no need to increase the number of neighbors once the confidence measure exceeds 3.0. Therefore, the probability of error, or equivalently the confidence level, provides a mechanism to locally determine the number of neighbors needed. We will call the modified version of the k-nearest neighbor rule the confident-nearest neighbor rule.

The main difference between the confident-nearest neighbor rule and the original k-nearest neighbor rule lies in that the actual value of k at each query point varies, depending on when the preset confidence threshold is reached, while in the k-nearest neighbor rule, the value of k, once set, is the same for all query points in the feature space. According to the first definition of the confidence level (see Eq. (8)), the confident-nearest neighbor rule reduces to the 1-nearest neighbor rule when the confidence level is set to 50%.

It should be noted that the confident-nearest neighbor rule differs significantly from previous methods that have been developed for adapting neighborhoods in the k-NN rule, such as the flexible metric method by Friedman [8], the discriminant adaptive method by Hastie and Hibshirani [9], and the adaptive metric method by Domeniconi et. al [10]. Although differing in their approaches, the common idea underlying these methods is that they estimate feature relevance locally at each query point and compute a weighted metric for measuring the distance between a query point and the training data. These adaptive metric methods improve the original k-NN rule because they are capable of producing local neighborhoods in which the a posteriori probabilities are approximately constant. However, none of these methods adapts the number of neighbors locally. In fact, these methods fix the number of neighbors in advance, as in the k-nearest neighbor rule, which is in direct contrast with our method that locally determines the number of neighbors. Furthermore, these methods usually need to introduce more model parameters, which are usually optimized along with the value of k through cross-validation, and therefore leading to high computational complexity. It is worth pointing out that our proposed method and previous adaptive metric methods are complementary in that, the adaptive metric methods are able to produce neighborhoods in which the a posteriori probabilities are approximately constant. This constant a posteriori probability property is a key assumption in our probability of error analysis. In this paper, we focus on locally adapting the number of neighbors while using the standard Euclidean metric.

3 Results and Discussion

In this section, we present experimental results of our algorithm on several real-world datasets from the UCI Machine Learning Repository [11]. We used the leave-one-out method to estimate the classification error of the confident-nearest neighbor (confident-NN) rule and the k-nearest neighbor (k-NN) rule. In Table 1, for each dataset, we report the lowest error rate achieved by the k-NN rule, together with the corresponding k value in parentheses, and the lowest error rate obtained by the confident-NN rule, together with the corresponding confidence level and the average nearest neighbor number $\bar{k}$ in parentheses.

Table 1. Comparison of results

Dataset	k-NN (k)	Confident-NN (CFD;$\bar{k}$)
BreastCancer	2.49 (5)	2.78 (75%;2.1)
Ionosphere	13.39 (1)	13.39 (70%;1.0)
Liver	30.43 (9)	31.30 (90%;19.9)
Pima	23.96 (19)	24.61 (90%;20.0)
Sonar	17.31 (1)	16.83 (75%;2.5)

As we can see, in terms of the overall classification accuracy, the confident-NN rule and the k-NN rule are comparable. However, there are several important points we would like to make. First, in many applications, the overall error is not the only important goal. For instance, in medical diagnosis, in addition to the overall error rate, the statistical confidence with which a decision is made is critically important. The confident-NN rule, unlike the k-NN rule, locally adapts the number of nearest neighbors until the desired statistical confidence requirement is met. Second, in many applications, the cost of misclassifying different classes might be significantly different. For example, compared to the consequence of a false alarm, it is more costly to fail to detect a cancer when a patient actually has one. Therefore, the acceptable confidence level for a non-cancer decision should be much higher than for a cancer decision. This consideration can be easily taken into account in the confident-NN rule by setting a higher statistical confidence level for a non-cancer decision, while the k-NN rule, which is based on minimizing the overall error rate, does not address this issue naturally.

Using the data from the Wisconsin Breast Cancer dataset, Figure 2 illustrates how the error rate of the confident-nearest neighbor rule changes as a function of the confidence level. As can be seen, the error rate does not necessarily decrease as the confidence level increases. For example, the lowest error rate of 2.78% is achieved at confidence level 75% − 80%.

The number of nearest neighbors used in the confident-NN rule varies from point to point, as manifested in the non-integer values of the average nearest neighbor number $\bar{k}$ in Table 1. Figure 3 shows the average number $\bar{k}$ at different confidence levels. As one can see, more neighbors are used as one increases the confidence level. However, combining Figs. 2 and 3, it is clear that more neighbors do not necessarily lead to lower error rates. This is because as one increases the confidence level requirement, more and more neighbors are needed in order to satisfy the higher confidence level requirement. However, since the number of training data is limited, it is not always possible to find a sufficient number of training samples in the close neighborhood of the query points. Therefore, in order to satisfy a high confidence requirement, one has to include points that are farther away from the query points and these points may have very different a posteriori probability distributions.

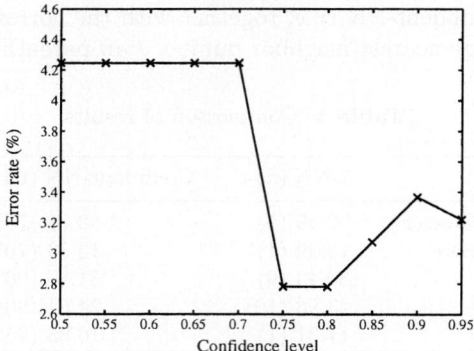

Fig. 2. Leave-one-out estimate of the classification error at different confidence levels

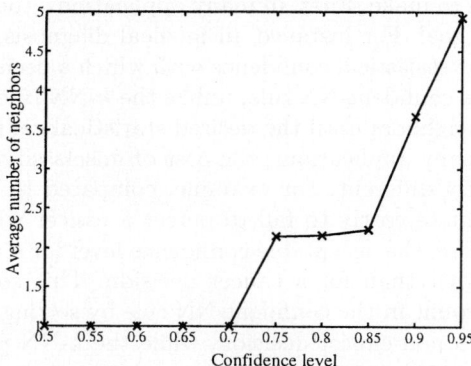

Fig. 3. Average number of neighbors used at different confidence levels

In the most general pattern classification scenario, classes may overlap in some regions of the feature space. Without knowledge of the underlying distribution, it is hard to tell whether a training point is actually misclassified by a classifier, or whether the data point itself is not labeled as the true majority class. We use the k-NN rule to illustrate this point. We fix the value of k to 5 and compute the mean confidence level of the misclassified data and the correctly classified data respectively. The results are reported in Table 2. The numbers in parentheses are the corresponding standard deviations. As one can see, on all datasets that have been tested, the misclassified data have significantly lower confidence levels than the correctly classified data. Since the number of nearest neighbors used is the same, this is a clear indication that the misclassified data are lying in the regions where two different classes overlap and attempts to further reduce the classification error may run into the risk of overfitting the training data.

Table 2. Comparison of mean confidence levels of the misclassified and correctly classified data

Dataset	Misclassified	Correctly Classified
BreastCancer	83.55 (3.44)	97.07 (0.21)
Ionosphere	85.33 (1.76)	95.04 (0.50)
Liver	76.68 (1.21)	81.17 (0.84)
Pima	79.45 (0.91)	86.13 (0.58)
Sonar	82.16 (2.17)	89.83 (0.96)

Table 3. Trade-off between the reject rate and error rate

Confidence Level (%)	Reject Rate	Error Rate
50	0	2.49
60	0	2.49
70	3.37	1.67
80	3.37	1.67
90	9.81	0.81
95	9.81	0.81

In many applications, misclassifications are costly. An important result of this work is the realization that for a given dataset and a given level of confidence, there is always a limit in reducing the error rate. Therefore, instead of making decisions regardless of the confidence level, a better alternative would be to reject patterns with low confidence levels and make a decision only when confidence is high. Since the misclassified data tend to have lower confidence levels than the correctly classified data, rejecting patterns with low confidence levels will lead to a reduction in the error rate on the remaining data. This implies that further reduction of the overall error rate, while keeping the same confidence level in making decisions, can be achieved, but at the expense of reducing the size of the region over which decisions will be made. We illustrate this point on the Breast Cancer dataset, where the lowest error rate, using the k-nearest neighbor rule, is achieved when k is set to 5 (see Table 1). In order to assure that every decision in the k-nearest neighbor rule is made with acceptable confidence, we rejected the patterns whose confidence levels from their 5 nearest neighbors were below a specific confidence level. The reject rate and the error rate on the remaining data for a range of different confidence levels are illustrated in Table 3. As can be easily seen, the reject rate increases monotonically with the confidence level, whereas the error rate on the remaining data is decreasing. At the 95% confidence level, a recognition accuracy of 99% is achieved with a reject rate less than 10%.

4 Conclusion

In this paper, we presented a new method that locally determines the number of nearest neighbors based on the concept of statistical confidence. We introduced two different definitions of the probability of error of decisions made by the majority rule from a finite number of observations, and showed that the probability of error is bounded by a rapidly decreasing function of its confidence measure. The statistical confidence is defined to be the complement of the probability of error, and it is used as a criterion to determine the number of neighbors needed.

We tested the confident-nearest neighbor rule on several real-world datasets and showed that it is comparable to the k-nearest neighbor rule. In contrast to the k-nearest neighbor rule, which uses a fixed number of nearest neighbors over the whole feature space, our method locally adjusts the number of nearest neighbors until a satisfactory level of confidence is reached. In addition, the statistical confidence provides a natural way to balance the trade-off between the reject rate and the error rate by excluding patterns that have low confidence levels. We believe that the statistical confidence can be of great importance in applications where the confidence with which a decision is made is equally or more important than the overall error rate.

References

1. Duda, R.O., Hart, P.E., Stock, D.G.: Pattern Classification. John Wiley & Sons, New York (2000)
2. Fix, E., Hodges, J.: Discriminatory analysis, nonparametric discrimination: consistency properties. Technical Report 4, USAF School of Aviation Medicine, Randolph Field, Texas (1951)
3. Cover, T.M., Hart, P.E.: Nearest neighbor pattern classification. IEEE Transactions on Information Theory **13** (1967) 21–27
4. Devroye, L.: On the inequality of cover and hart. IEEE Transactions on Pattern Analysis and Machine Intelligence **3** (1981) 75–78
5. Stone, C.J.: Consistent nonparametric regression. Annals of Statistics **5** (1977) 595–645
6. Devroye, L., Györfi, L., Krzyżak, A., Lugosi, G.: On the strong universal consistency of nearest neighbor regression function estimates. Annals of Statistics **22** (1994) 1371–1385
7. Geman, S., Bienenstock, E., Doursat, R.: Neural networks and the bias/variance dilemma. Neural Computation **4** (1992) 1–58
8. Friedman, J.: Flexible metric nearest neighbor classification. Technical Report 113, Stanford University Statistics Department (1994)
9. Hastie, T., Tibshirani, R.: Discriminant adaptive nearest neighbor classification. IEEE Transactions on Pattern Analysis and Machine Intelligence **18** (1996) 607–615
10. Domeniconi, C., Peng, J., Gunopulos, D.: Locally adaptive metric nearest-neighbor classification. IEEE Transactions on Pattern Analysis and Machine Intelligence **24** (2002) 1281–1285
11. Blake, C., Merz, C.: UCI repository of machine learning databases (1998) http://www.ics.uci.edu/~mlearn/MLRepository.html

Fuzzy Self-organizing Map Neural Network Using Kernel PCA and the Application

Qiang Lv and Jin-shou Yu

Research Institution of Automation, East China University of Science & Technology
200237, Shanghai, China
luqiang77@yahoo.com.cn

Abstract. The fuzzy self-organizing map neural network using kernel principal component analysis is presented and a hybrid-learning algorithm (KPCA-FSOM) divided into two stages is proposed to train this network. The first stage, the KPCA algorithm is applied to extract the features of nonlinear data. The second stage, combining both the fuzzy theory and locally-weight distortion index to extend SOM basic algorithm, the fuzzy SOM algorithm is presented to train the SOM network with features gained. A real life application of KPCA-FSOM algorithm in classifying data of acrylonitrile reactor is provided. The experimental results show this algorithm can obtain better clustering and network after training can more effectively monitor yields.

1 Introduction

The SOM is an unsupervised learning neural network [1]. It provides a mapping from a high-dimensional input data space into the lower dimensional output map, usually a one- or two-dimensional map [2]. As a result of this process, SOM is widely used for the visualization of high-dimensional data. Moreover, a distinguishing feature of the SOM is that it preserves the topology of the input data from the high-dimensional input space onto the output map in such a way that relative distance between input data are more or less preserved [3]. The input data points, located close to each other in the input space, are mapped to the nearby neuron on the output map [4]. The SOM visualization methods are versatile tools for data exploration. They are widely used in data mining as a tool for exploration and analysis of large amounts of data, to discover meaningful information from the data [5].

There are many research efforts to enhance SOMs for visualization and cluster analysis. Some methods focus on how to visualize neurons clearly and classify data [6]. Others concentrate on better topology preservation. Most of the methods enhancing topology preservation use the squared-norm to measure similarity between weight values and data points [7]. So, they can only be effective in clustering 'spherical' clusters [8]. To cluster more general dataset, Wu and Yang (2002) proposes an algorithm by replacing the squared-norm with other similarity measures. A recent development is to use kernel method to construct the kernel version of the SOM (called KSOM training algorithm) [9]. A common ground of these algorithms is that clustering is performed in the transformed feature space after the nonlinear data transformation. However, kernel method projects data into the feature space of which the dimensions is higher than those of original input space. So, computational complexity is increased. In this paper, kernel principal component analysis (KPCA) instead of kernel method is introduced to deal with the problem.

The basic SOM training algorithm is simply presented as an acceptable heuristic, but one would naturally require more substantial support. So, Kohonen (1995) derives what we term here basic SOM algorithm using the Robbins and Monro (1951) method of stochastic approximation. This general approach involves use of the estimated weight values at each iteration to provide an approximation to true gradient of the distortion index that is defined in Eq. (4)[5]. Such a result is immediately re-assuring, in that the algorithm is no longer based merely on a plausible heuristic, and can be established, albeit as an approximation, according to certain general principles. However, several points still need to be noted. First, the fact is that the basic algorithm is derived only by a method of approximation. Second, the algorithm belongs to hard partition method. For dealing with above two problems, SOM basic algorithm is modified and extended through using fuzzy theory.

The remainder of this paper is organized as follows. Section 2 describes kernel principal component analysis (KPCA) algorithms. In Section 3, the self-organizing map (SOM) using fuzzy theory training (FSOM) algorithm is proposed and one-quality index is defined. To demonstrate the performance of the proposed algorithm, a simulated experiment and one real life application on monitoring the yield of acrylonitrile reactor is conducted and the performance comparison between SOM algorithm and FSOM algorithm is given in Section 4 and Section 5.At last, conclusions and discussions are given in Section 6.

2 Kernel Principal Component Analysis (KPCA)

In the paper, KPCA algorithm [10] is introduced mainly for following two points. First, SOM basic algorithm cannot correctly cluster nonlinear data [11]. Second, although traditional KSOM algorithm can deal with nonlinear data, this method does increase computational complexity for the sample after transforming being equal to the number of samples in dimension. However, KPCA algorithm can make use of both the advantages of kernel function and characteristics of PCA algorithm so that it may not only deal with nonlinear data but also make the dimension of data and complexity of calculation decrease dramatically.

According to the idea of KPCA algorithm, any sample can be transformed using Equation (1) .

$$y_k = \sum_{i=1}^{M} \alpha_i^k K(x_i, x). \tag{1}$$

Where M is the number of samples, a_i^k is the ith value of the kth eigenvector of kernel matrix K, y_k is the kth ($k = p \cdots M$) value of the sample after transforming, x_i is the ith original sample, x is original sample that need to be transformed, p is the sequence number of the first nonzero eigenvalue by ordering eigenvalue in accordance with sort ascending.

The dot product of samples in the feature space is defined as below:

$$K(x_i, x_j) = \Phi(x_i)^T \Phi(x_j). \tag{2}$$

Where x_i, $x_j (i, j = 1, 2\cdots, M)$ are random samples of data set.

We use Gaussian's kernel function, which is defined as:

$$k_{ij} = K(x_i, x_j) = \exp\left(-\|x_i - x_j\|^2 \Big/ 2\sigma^2\right). \tag{3}$$

From above narration, the basic procedures of KPCA algorithm can be summed up as following:

(1) Calculate matrix K according to formulation (3).
(2) Calculate the eigenvalues and the eigenvectors of matrix K, and standardize them.
(3) Calculate the PC (principal component) matrix transformed according to formulation (1).
(4) Calculate the sum of contribution rates from the first PC to the kth PC, and then carry out reduction of data.

3 Fuzzy Self-organizing Map (FSOM)

The traditional SOM basic algorithm belongs to a kind of hard partition methods. The aim of the algorithm is that the sets of objects are strictly grouped into clusters [12]. However, all objects have not strict attributes and both attributes and characters are always fuzzy. For clustering these objects, soft partition is proper [13].

3.1 Theory Foundation of SOM Basic Algorithm

Professor Kohonen presented theory basis that is defined as Eq (4) for his SOM basic algorithm in 1995 and 1999.

$$D = E[p(x)]. \tag{4}$$

$$p(x) = \sum_{j=1}^{M} h_{cj}(t) \|x - w_j\|. \tag{5}$$

Where E denotes the expectation operator, x is input vector, w_j is nerve cell weight at coordinate (k_1, k_2), $h_{cj}(t)$ is neighborhood function, $c(c_1, c_2)$ is the coordinate of the winning neuron, M is the number of neuron. The neighborhood function is defined as below:

$$h_{cj}(t) = \exp\left(-\frac{\|d_c - d_j\|^2}{2\delta(t)^2}\right). \tag{6}$$

Where d_c is the position of the winning neuron, d_j is the position of the jth neuron, $\delta(t)$ is the variance of the neighboring neurons at time t. $\delta(t)$ decreases with time, in order to control the size of the neighboring neurons at time t.

$E[p(x)]$ is called locally-weighted distortion index (LWDI) and we can see that if the neighboring function is erased from the index, the rest of index is the same as similar as mathematic equation of k-means algorithm. Therefore, SOM network can be explained using following idea. Set M numbers of cluster centers (M neurons) and these centers will be organized like SOM lattice array. According to the main principal $\min(E[p(x)])$, these cluster centers are continually updated until a certain condition can be met. Now, M micro-clusters are formed and then we will merge the M clusters to gain the final results. In addition, neighboring function has the most important influence on the visualization of data clustering.

3.2 FSOM Algorithm

Assuming appearance probability of samples to be equal, the LWDI is rewritten to be the following form:

$$D = \frac{1}{N} \sum_{i=1}^{N} \sum_{j=1}^{M} u_{ij}{}^a h_{cj}(t) \|x_i - w_j\|. \tag{7}$$

$$a = m \times h_x(c_1 - k_1, c_2 - k_2). \tag{8}$$

$$m = 2 - t/T. \tag{9}$$

Where (c_1, c_2) is the coordinate of the winning neuron, (k_1, k_2) is the coordinate of the jth neuron, the effect of m is as similar as learning parameter of SOM basic algorithm. The function $h_x(c_1 - k_1, c_2 - k_2)$ takes the following values.

$h_x(0,0)=0$, $h_x(1,1)=0$, $h_x(1,-1)=0$, $h_x(-1,1)=0$, $h_x(-1,-1)=0$, $h_x(1,0)=0$, $h_x(0,1)=0$, $h_x(0,-1)=0$, $h_x(-1,0)=0$

and one for all other values of its arguments. This function is introduced mainly for strengthening visualization and for making the network gain better topology structure.

Where N is number of samples and the constrained condition is $\sum_{j=1}^{M} u_{ij} = 1$. To derive the necessary conditions for the minimization of (7), a lagrangian is constructed and Eq (7) is modified for existence of derivative w_j.

$$L = \frac{1}{N}\sum_{i=1}^{N}\sum_{j=1}^{M} u_{ij}^{a} h_{cj}(t) \|x_i - w_j\|^2 + \lambda\left(\sum_{j=1}^{M} u_{ij} - 1\right). \quad (10)$$

$\frac{\partial L}{\partial w_j}$ and $\frac{\partial L}{\partial u_{ij}}$ is calculated respectively.

At last, weight-updated equation and membership-updated equation is expressed as below.

$$w_j = \frac{\sum_{i=1}^{N} u_{ij}^{a} h_{cj}(t) x_i}{\sum_{i=1}^{N} u_{ij}^{a} h_{cj}(t)} . \quad (11)$$

$$u_{ij} = \frac{\left[\dfrac{1}{h_{cj}(t)\|x_i - w_j\|^2}\right]^{1/(a-1)}}{\sum_{k=1}^{M}\left[\dfrac{1}{h_{ck}(t)\|x_i - w_k\|^2}\right]^{1/(a-1)}} . \quad (12)$$

$$u_{ij} = \frac{\left[\dfrac{1}{h_{cj}(t)\|x_i - w_j\|^2}\right]^{1/(a+1)}}{\sum_{k=1}^{M}\left[\dfrac{1}{h_{ck}(t)\|x_i - w_k\|^2}\right]^{1/(a+1)}} . \quad (13)$$

Because $a-1$ maybe is zero, membership needs to be modified again. Membership equation after modifying is showed in Eq (13).

3.3 The Basic Process of FSOM Algorithm

Considering the idea of FSOM algorithm, its detail procedures are listed as follows.

Step1. Initialize network weights. Select number M to be initial network weight from input vectors.

Step2. Search the winning neuron for input vectors by using Equation (14).

$$\|X_i(t)-W_{win}(t)\| = \min_j \{\|X_i(t)-W_j(t)\|\}. \tag{14}$$

Step3. Membership calculation. Calculate all membership u_{ij} based on Eq (13)。

Step4. Update network weights by using Eq (11).

Step5. If iterative number equal to maximum number, then the algorithm is over. Otherwise, go to step 2.

3.4 Network Quality

For validating effect of algorithm, one criterion, topographic error, is defined in the paper.

Definition 1: Topology Error (TE)

$$TE = \frac{\sum_{j=1}^{N} u(x_k)}{N}. \tag{15}$$

Where $u(x_k)$ is 1, if the neurons of the smallest and second smallest distance between input vector x_k and the weight vector of the neuron are not adjacent. Otherwise, $u(x_k)$ is zero. The topographic error is used to measure the continuity mapping. After the training, the map is evaluated for the topology accuracy, in order to analyze how the map can preserve the topology of the input data. A common measure that calculates the precision of the mapping is the topographic error over all input data.

4 Simulated Experiments

To demonstrate the effectiveness of the proposed clustering algorithm, three data sets are used in our experiments. The first dataset is iris flower one. The iris flower dataset has been widely used as a benchmark dataset for many classification algorithms due to two (iris-versicolor and iris-virginica) of its three (iris-versicolor,iris-virginica and iris-setosa) classes are not linearly separable. Each cluster includes 50 data with four dimensions. The second dataset is wine one. The dataset is thirteen-dimensional with 178 data entries positioned into three clusters. The third dataset is Olitos one. It consists of four clusters having 120 points each in R^{25}.

For clustering above datasets, three steps are performed. First, the features of input data are extracted using KPCA algorithm. Second, the features are normalized such that the value of each feature in each dimension lies in [0,1]. Third, 8×8 neural network is used and using SOM basic algorithm and FSOM algorithm respectively clusters three datasets. Table 1 shows elevating index and error rates on average in ten independent runs of two algorithms.

Table 1. The clustering effect and average results of elevating index in three data set

Elevating index	SOM basic algorithm			FSOM algorithm		
	Iris data set	Wine data set	Olitos data set	Iris data set	Wine data set	Olitos data set
TE	0. 04667	0. 2809	0. 19167	0. 02667	0. 05056	0. 06667
Principal component	3	6	10	3	6	10
Number of clusters	3	3	4	3	3	4
Error rate	6%	8.74%	20.33%	4.667%	7.68%	14.92%

On an average, the topological errors of the FSOM for the three datasets are 0.02667,0.05056,0.06667, which are smaller than those of the SOM. The average error rates of the FSOM are 4.667%, 7.68%, and 14.92% respectively. They are smaller than those of SOM as shown in Table 1. So, we conclude that the FSOM can obtain better topology mappings and the lower error rates.

5 Practical Application

The fluidized-bed reactor shown in Fig.1 is the most important part of acrylonitrile equipment and its yield can directly influence the economic benefit of this equipment. In this section, the data of fluidized-bed reactor will be analyzed by using clustering technique. The prospective number of cluster is two classes that include optimal class and bad class. If the parameters of fluidized-bed reactor are set on the basis of the objects of optimal class, then the yield of reactor will be higher, and the SOM network after training can monitor fluidized-bed reactor yields.

In the following sector, two algorithms, SOM algorithm and FSOM algorithm, are used to cluster reactor data. The number of data is n=344; clustering number is 244 and testing number is 100 with 7dimensions.The first step, KPCA algorithm is used to extract the features of the dataset. The second step, the features of the reactor data are normalized such that the value of each feature in each dimension lies in [0,1]. The final step, SOM algorithm and FSOM algorithm is used to analyze the data. Table 2 shows average elevating indexes and other information in twenty runs of two algorithms respectively. The best clustering structure using SOM algorithm is shown in Fig 2. The best clustering structure using FSOM algorithm is shown in Fig 3.

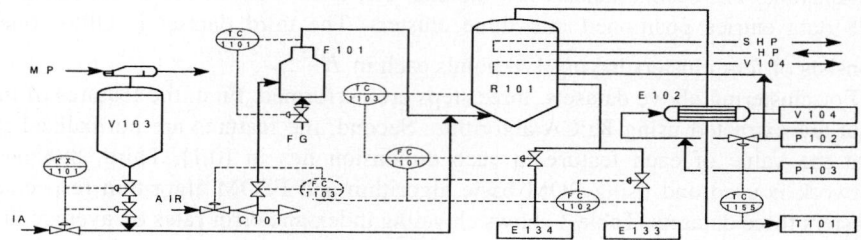

Fig. 1. The fluidized-bed reactor

Table 2. The average results of elevating index in data set of fluidized-bed reactor

Elevating index	SOM basic algorithm	FSOM algorithm
Principal component	4	4
Number of cluster	2	2
TE	0. 22515	0. 14035

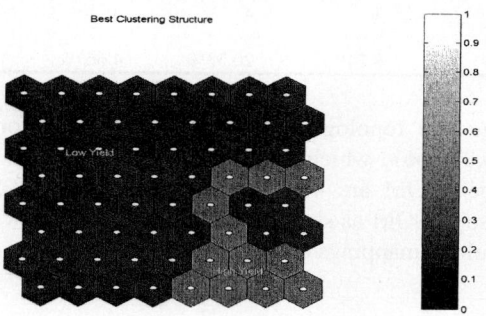

Fig. 2. Best clustering structure using SOM algorithm

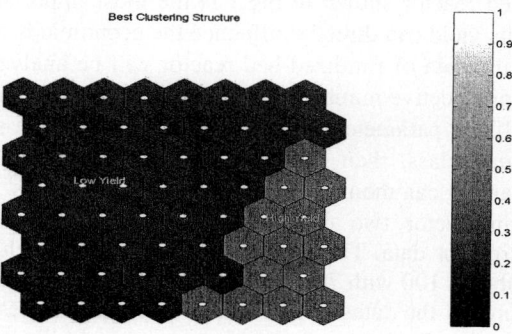

Fig. 3. Best clustering structure using FSOM algorithm

The SOM neural network are divided into high yield filed (77.89-80.95) and low yield field (75.81-77.89) by synthetically considering clustering results and the corresponding yield in the Fig 2 and Fig 3.From the clustering results based on the above division, we will see that the error rate of FSOM algorithm is 5.9% and that of SOM algorithm is 10.56%. Using 100 testing data, the testing results of FSOM algorithm are better. Therefore, The network using FSOM algorithm can monitor reactor yields as shown Fig 3 that show better topological structure. The average cluster centers and the average yields are listed in Table 3.

Table 3. The cluster center and yield of two algorithms

The parameters	FSOM algorithm		SOM basic algorithm	
	Optimal cluster center	Bad cluster center	Optimal cluster center	Bad cluster center
Pressure(Mpa)	0.7479	0.6813	0.7604	0.7366
Temperature ($^{\circ}C$)	434.0819	429.6533	433.4875	430.2254
Propylene(NM^3/H)	2569.1	2245	2544.7	2349.2
Air/Propylene	9.4625	7.785	9.4487	8.8897
Ammonia/Propylene	1.1515	1.1594	1.1687	1.151
Catalyst(KG)	57.795	54.518	56.897	54.4847
Velocity(M/S)	0.6952	0.5267	0.6669	0.6155
Yield (%)	79.0355	77.2379	78.3907	78.0120

The yield corresponding to center of the optimal cluster of FSOM algorithm is higher than that of SOM algorithm, as shown in Table 3. So, the optimal cluster center of FSOM algorithm can guide how to adjust reactor parameters. In other words, if the reactor parameters are set according to this optimal center, then the reactor yield will be higher.

6 Conclusions

In this paper, distortion index is directly extended through fuzzy theory. So, Network quality is enhanced and clustering results is better than that of SOM basic algorithm. However, industry data are always nonlinear. For dealing with nonlinear data, KPCA algorithm is introduced and it is proper for large data, especially industry data. The above experiments demonstrate that better clustering results and topological structure can be obtained by using KPCA-FSOM algorithm. The SOM neural network after training can be used to monitor reactor yields and cluster centers can guide the optimization of parameters.

References

1. Kohonen, T.: Self-Organizing Maps. Springer-Verlag New York (1987)
2. Kohonen, T.: The Self-Organizing Map. Neurocomputing. 21(1998) 1-6

3. Jin, H., Shum, W.H., Leung, K.S.: Expanding Self-Organizing Map for data visulization and Cluster analysis. Information Sciences. 163 (2004) 157-173
4. Guha, S., Rastogi, R., Shim, K.: An efficient clustering for large database. Proceedings of the ACM SIGMOD International Conference on Management of Data. (1998) 73-84
5. Curry, B., Morgan, P. H.: Evaluating Kohonen's learning rule: An approach through genetic algorithms. European Journal of Operational Research. 154 (2004) 191-205
6. Vesanto, J., Alhoniemi, E.: Clustering of the self-organizing map.IEEE Transactions on Neural Networks. 11(2000) 586-600
7. Wu, S., Chow, T.W.S.: Clustering of the self-organizing map using a clustering validity index based on inter-cluster and intra-cluster density. Pattern Recognition. 37 (2004) 175-188
8. Su, M.C., Chang, H.T.: A new model of self-organizing neural networks and its application in data projection. IEEE Transaction on Neural Network. 12 (2001) 153-158
9. Cristianini, N., Taylor, J.S.: An Introduction to Support Vector Machines and Other Kernel-Based Learning Methods. Cambridge Press (2003)
10. Jade, A.M., Srikanth, B., Jayaraman, V.K.: Feature extraction and denoising using kernel PCA. Chemical Engineering Science. 58 (2003) 4441-4448
11. Chen, S., Zhu, Y.L.: Subpattern-based principal component analysis. Pattern Recognition. 37 (2004) 1081-1083
12. Kuo, R.T., Chi, S.C., Teng, P.W.: Generalized part family formation through fuzzy self-organizing feature map neural network. Computers&Industrial Engineering. 40 (2001) 79-100
13. Li, S.T., Shue, L.Y.: Data mining to aid policy making in air pollution management. Expert Systems with Applications. 27 (2004) 331-340

An Evolved Recurrent Neural Network and Its Application

Chunkai Zhang and Hong Hu

Member, IEEE
Department of Mechanical Engineering and Automation, Harbin Institute of Technology,
Shenzhen Graduate School, Shenzhen, China, 518055
ckzhang@hotmail.com

Abstract. An evolved recurrent neural network is proposed which automates the design of the network architecture and the connection weights using a new evolutionary learning algorithm. This new algorithm is based on a cooperative system of evolutionary algorithm (EA) and particle swarm optimisation (PSO), and is thus called REAPSO. In REAPSO, the network architecture is adaptively adjusted by PSO, and then EA is employed to evolve the connection weights with this network architecture, and this process is alternated until the best neural network is accepted or the maximum number of generations has been reached. In addition, the strategy of EAC and ET are proposed to maintain the behavioral link between a parent and its offspring, which improves the efficiency of evolving recurrent neural networks. A recurrent neural network is evolved by REAPSO and applied to the state estimation of the CSTR System. The performance of REAPSO is compared to TDRB, GA, PSO and HGAPSO in these recurrent networks design problems, demonstrating its superiority.

1 Introduction

Modeling complex dynamic relationships are required in many real world applications, such as state estimation, pattern recognition, communication and control etc. One effective approach is the use of recurrent neural network (RNN) [1]. RNN has self-loops and backward connections in their topologies, and these feedback loops are used to memorize past information. Therefore, it can be used to deal with dynamic mapping problems. But the difficulty is that the training algorithm must take into account temporal as well as spatial dependence of the network weights on the mapping error.

Many types of recurrent networks have been proposed, such as back propagation through time (BPTT) [2], real-time recurrent learning (RTRL) [3], and time-dependent recurrent back propagation (TDRB) [4]. But all of them have several limitations: 1) a complex set of gradient equations must be derived and implemented, 2) it is easy to be gets trapped in a local minimum of the error function. One way to overcome the above problems is to adopt genetic algorithm (GA) or evolutionary algorithm (EA) [5, 6, 7, 8], because GA and EA are stochastic search procedures based on the mechanics of natural selection, genetics, and evolution, which make them find the global solution of a given problem. In addition, they use only a simple scalar performance measure that does not require or use derivative information. In order to farther improve the performance of these algorithms, such as avoiding the permutation problem and the structural / functional mapping problem, hybridization of genetic algorithm (GA) and evolutionary algorithm (EA) with particle swarm optimization (PSO), respectively named hybrid PSO+EA and HGAPSO, have been investigated to evolve the fully connected recurrent neural network [9, 10]. But all of them have following limitations: 1) the appropriate network architecture must be determined, and 2) the structure may or may not capable of representing a given dynamic mapping. It means that the above problems depend heavily on the expert experience and a tedious trial-and-error process. There have been many attempts in designing network architectures automatically, such as various constructive and pruning algorithms [11, 12]. However, "Such structural hill climbing methods are susceptible to becoming trapped at structure local optima, and the result depends on initial network architectures." [13]

To overcome all these problems, this paper proposes a new evolutionary learning algorithm (REAPSO) based on a cooperative system of EA and PSO, which combines the architectural evolution with weight learning. In REAPSO, the evolution of architecture and weight learning are alternated, which can avoid a moving target problem resulted from the simultaneous evolution of both architectures and weights [14]. And the network architectures are adaptively evolved by PSO, starting from the parent's weights instead of randomly initialized weights, so this can preferably solve the problem of the noisy fitness evaluation that can mislead the evolution. Since PSO possesses some attractive properties comparing with EA, such as memory, constructive cooperation between individuals, so no selection and crossover operator exist [15], which can avoid the permutation problem in the evolution of architectures. In order to improve the generalization ability, the data sets are partitioned into three sets: training set, validation set, and testing set. The training set is used to evolve the nodes with a given network architecture, and the fitness evaluation is equal to the root mean squared error E of RNN. But in evolving the architecture of network, the fitness evaluation is determined through a validation set which does not overlap with the train set.

The rest of this paper is organised as follows. Section 2 describes the REAPSO algorithm and the motivations on how to evolve the RNN. Section 2 presents experimental results on REAPSO. The paper is concluded in Section 4.

2 REAPSO Algorithm

2.1 Evolutionary Algorithm (EA)

EA refer to a class of population-based stochastic search algorithms that are developed from ideas and principles of natural evolution. One important feature of all these algorithms is their population based search strategy. Individuals in a population compete and exchange information with each other in order to perform certain tasks. A general framework of EA can be described as following:

   1) Initialize the number of individuals in a population, and encode each individual in term of real problems. Each individual represents a point in the search space;

2) Evaluate the fitness of each individual. Each individual is decided by an evaluating mechanism to obtain its fitness value;
3) Select parents for reproduction based on their fitness;
4) Apply search operators, such as crossover and/or mutation, to parents to generate offspring, which form the next generation.

EA are particularly useful for dealing with large complex problems which generate many local optima, such as training artificial neural networks. They are less likely to be trapped in local minima than traditional gradient-based search algorithms. They do not depend on gradient information and thus are quite suitable for problems where such information is unavailable or very costly to obtain or estimate.

2.2 Particle Swarm Optimization (PSO)

PSO is a population based optimization algorithm that is motivated from the simulation of social behaviour. PSO algorithm possesses some attractive properties such as memory and constructive cooperation between individuals, so each individual flies in the search space with a velocity that is dynamically adjusted according to its own flying experience and its companions' flying experience. In this paper we propose an improved PSO algorithm, which is as follows:

1) Initialise positions Pesentx and associated velocity v of all individuals (potential solutions) in the population randomly in the D dimension space.
2) Evaluate the fitness value of all individuals.
3) Compare the PBEST[] of every individual with its current fitness value. If the current fitness value is better, assign the current fitness value to PBEST[] and assign the current coordinates to PBESTx[][d]. Here PBEST[] represents the best fitness value of the nth individual, PBESTx[][d] represents the dth component of an individual.
4) Determine the current best fitness value in the entire population and its coordinates. If the current best fitness value is better than the GBEST, then assign the current best fitness value to GBEST and assign the current coordinates to GBESTx[d].
5) Change velocities and positions using the following rules:

$$v[][d] = W * v[][d] + C1 * rand * (PBESTx[][d] - Pesentx[][d]) + C2 * rand * (GBESTx[d] - Pesentx[][d]) \quad (1)$$

$$Pesentx[][d] = Pesentx[][d] + v[][d]$$

$$W = W_\infty + (W_0 - W_\infty)[1 - t/K]$$

where $C1 = C2 = 2.0$, t and K are the number of current iterations and total generation. The balance between the global and local search is adjusted through the parameter $W \in (W_0, W_\infty)$.

6) Repeat step 2) - 6) until a stop criterion is satisfied or a predefined number of iteration is completed.

Because there is not a selection operator in PSO, each individual in an original population has a corresponding partner in a new population. From the view of the diversity of population, this property is better than EA, so it can avoid the premature convergence and stagnation in GA to some extent.

2.3 REAPSO Algorithm

In REAPSO, the evolution of RNN's architectures and weight learning are alternated. The major steps of PSOEA can be described as follows:

1) Generate an initial population of M networks.
   - The direct encoding scheme is applying to encode the architecture of each network. The architecture of each network is uniformly generated at random within certain ranges. In the direct encoding scheme, a $n \times n$ matrix $C = (c_{ij})_{n \times n}$ can represent a RNN architecture with n nodes, where c_{ij} indicates presence or absence of the connection from ith node to jth node. Here, $c_{ij} = 1$ indicates a connection and $c_{ij} = 0$ indicates no connection. It is shown in Fig. 1.

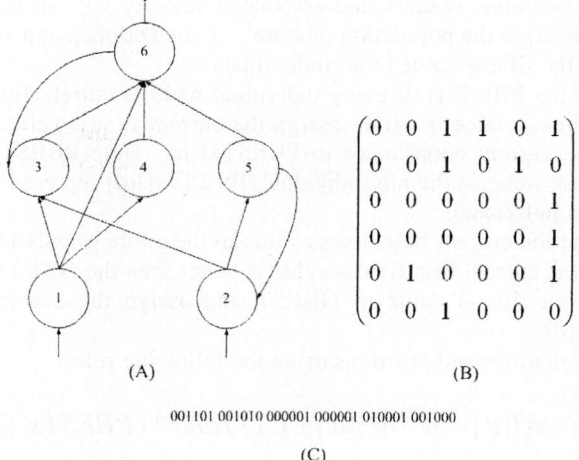

Fig. 1. The direct encoding scheme of a recurrent neural network. (A), (B) and (C) show the architecture, its connectivity matrix, and its binary string representation, respectively

   - The initial weights are uniformly distributed inside a small range.
2) Use the Extended Training (ET) algorithm to train each network in the population on the training set, which is as follows:
   - Choose a network as a parent network, and then randomly generate N-1 initial individuals as a population where each individual's initial weights uniformly generated at random within certain ranges, but their network architectures are the same as the parent network architecture. And then the parent network is added into the population. Here each individual in this population

is to parameterise a whole group of g nodes in a RNN, this means that every component of each individual represents a connection weight.
- Employ EA to evolve this population until the best individual found is accepted or the maximum number of generations has been reached.
- The best individual that survived will join the network architecture evolution.

3) All survived networks form a new population. Evaluate the fitness values of every individual in this population. Here the mean squared error value E of each network on the validation set represents the fitness evaluation of each individual.
4) If the best network found is accepted or the maximum number of generations has been reached, stop and go to step 7). Otherwise continue.
5) Employ the PSO to evolve the network architecture of each individual. Here each individual represents the binary string representation of network architecture.
6) When the network architecture of an individual changes, employ the strategy of Evolving Added Connection (EAC) to decide how to evolve its connection weights with the ET algorithm. There are two choice:
    - If some connections need to be added to this network, under the strategy of EAC, the ET algorithm only evolves the new added connections to explain as much of the remaining output variance as possible. In this case the cost function that is minimised at each step of algorithm is the residual sum squared error that will remain after the addition of the new nodes, and the existing connections are left unchanged during the search for the best new added connections. Compared with the existing connections, the added connections will represent or explain the finer details of this mapping that the entire network is trying to approximate between the inputs and outputs of the training data. This strategy can decrease the computation time for evolving the entire network and prevent destruction of the behaviour already learned by the parent.
    - If some connections need to be deleted from a network, EAC strategy can remove the connections in the reverse order in which they were originally added to the network, then the ET algorithm evolves the connection weights of the entire network, but sometimes a few of jump in fitness from the parent to the offspring is not avoided.

    Then go to Step 3).
7) After the evolutionary process, train the best RNN further with the ET algorithm on the combined training and validation set until it "converges".

In step 7), the generalisation ability of RNN can be further improved by training the best RNN with the ET algorithm on the combined training and validation set. The logic diagram of coevolution between network architecture and weights is shown in Fig. 2.

After evolving the architecture of networks every time, the strategy of EAC and ET algorithm are used to optimise the connection weights of nodes with a given network architecture which has been evolved by PSO. In other words, the purpose of this process is to evaluate the quality of this given network architecture and maintain the behavioural link between a parent and its offspring.

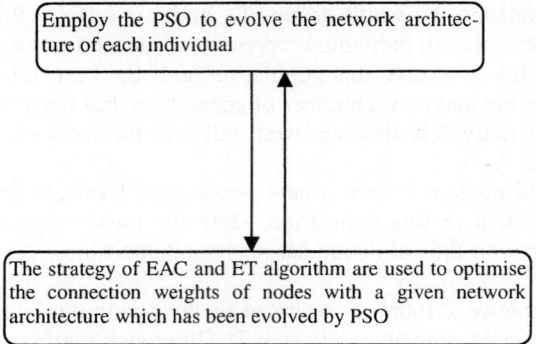

Fig. 2. The logic diagram of coevolution between network architecture and weights

In ET algorithm, each individual of the population in EA is to parameterise a whole group of g nodes in RNN, this means that every component of each individual represents a connection weight. Compared with the encoding scheme that each individual represents a single node, and then the individuals are bundled together in the groups of g individuals, this scheme is simple and easily implemented, and does not need a combinatorial search strategy.

3 Experimental Studies

In order to evaluate the ability of REAPSO in evolving RNN, it was applied to estimate the state of the CSTR system.

3.1 Continuous Stirred Tank Reactor System (CSTR)

Continuous Stirred Tank Reactor System (CSTR) is a chemical reactor system with typical nonlinear dynamic characteristics.

In fig.3, $C_{A,1}$ and $C_{B,1}$ are the concentration of product A and B in tank 1 respectively; $C_{A,2}$ and $C_{B,2}$ are the concentration of product A and B in tank 2 respectively; T_1 and T_2 are the reaction temperature in tank 1 and 2 respectively; F is the flux from tank 1 to tank 2; α is the coefficient of feedback from tank 2 to tank 1. On the basis of the knowledge of thermodynamics and chemical kinetics, mathematical model is obtained:

$$C_{B,2} = f(C_{A,0}, T_1, T_2) \qquad (2)$$

where f is the dynamic nonlinear function, the inputs are $C_{A,0}$, T_1 and T_2, the output is $C_{B,2}$.

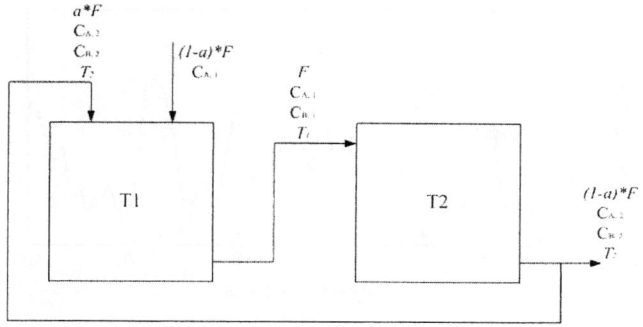

Fig. 3. The continuous stirred tank reactor (CSTR)

In order to forecast $C_{B,2}$ in CSTR system, an evolved RNN is selected. The network inputs are $C_{A,0}$, T_1 and T_2, the output is $C_{B,2}$. The number of hidden nodes is 30. During training, the discrete-time step $\Delta t = 0.2$ is used, the root mean square error (RMSE) in time interval $(t_0, t_1] = (0,100]$ is calculated by

$$RMSE = (\frac{1}{N}\sum_{i=1}^{N}\sum_{k=1}^{500}(y_i(k) - y_{r_i}(k))^2)^{1/2} \qquad (3)$$

where $y_{r_i}(k)$ is the desired target value at kth time step, and $y_i(k)$ is the output of network at the same time, here $N=1$. And the fitness value is defined to be $1/RMSE$.

3.2 Simulation

To demonstrate the superiority of REAPSO, the performance of REAPSO is compared with TDBR, GA, PSO and HGAPSO.

We collected about 500 sets of sample data of $C_{A,0}$, T_1, T_2 and $C_{B,2}$. The sample data from the site often accompany random measurement noise and gross error, and must be processed before they are employed to train the network. For these sets of sample data, the first 250 sets of sample data were used for training set, and the following 150 sets of sample data for the validation set, and the final 100 examples for the testing set.

In REAPSO, the population size is 200, $C1 = C2 = 2.0$, $(W_0, W_\infty) = (0,1)$, and $K = 300$. After 300 epochs off-line learning, the best and averaged RMSEs for the 50 runs for $C_{B,2}$ in the tested 100 date sets are listed in Table 1. Fig. 5 shows the desired target values and estimation values of $C_{B,2}$.

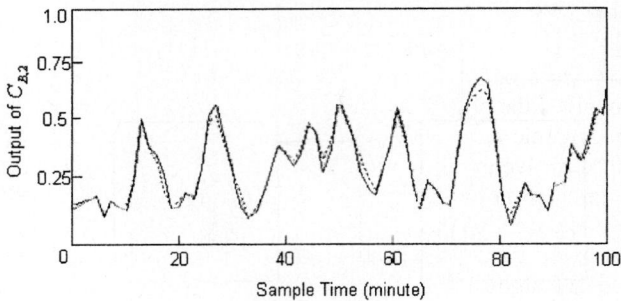

Fig. 4. The desired target values and estimation values of $C_{B,2}$

To show the effectiveness and efficiency of REAPSO, TDRB, GA, PSO, and HGAPSO are applied to the fully connected RNN for the same problem of the state estimation of CSTR system.

In TDRB, the learning constant η is set to 0.3, the iteration is 10000, and the best training result is listed in Table 1.

Table 1. Performance comparisons for different methods of RNN design for the state estimation for CSTR system

	TDRB	GA (Pc=0.4)	PSO	HGAPSO	REAPSO
RMSE(Ave)	-	0.2539	0.1658	0.1083	0.0862
RMSE(Best)	0.0258	0.2240	0.0253	0.0216	0.0127

In GA, the population size is 200, and the parents for crossover are selected from the whole population instead of from only the elites, and the tournament selection is used. The elite strategy is used, where the best individual of each generation is copied into the succeeding generation. The crossover probability P_c is 0.4, the mutation probability P_m is 0.1, and the evolution is processed for 1200 generations. The results after 50 runs are listed in Table 1.

In PSO, the population size is 200, the parameters $C1 = C2 = 2.0$, $(W_0, W_\infty) = (0,1)$, $K = 1200$. The results after 50 runs are listed in Table 1.

In HGAPSO, the population size and initial individuals are the same as those used in GA and PSO. The parameters of Pc, Pm, $C1$, $C2$, W_0 and W_∞ are the same as those used in GA and PSO, and the evolution is processed for 1200 generations. The best and averaged RMSEs for the 50 runs are listed in Table 1.

From the simulation results, it is illustrated that both the averaged and best RMSEs of REAPSO and HGAPSO are obviously smaller than those of GA, PSO and TDRB. Although the result of REAPSO is little better than those of HGAPSO, the evolution generation of REAPSO is smaller than those of HGAPSO, and REAPSO possesses good generalisation ability.

4 Conclusion

This paper describes a cooperative system named REAPSO - a hybrid of EA and PSO, which combines the architectural evolution with weight learning. It means that PSO constructs dynamic architectures without requiring any software redesign, then EA is employed to evolve the network nodes with this architecture, and this process is automatically alternated. It can effectively alleviate the noisy fitness evaluation problem and the moving target problem. And no selection and crossover operator exist in PSO, which can avoid the permutation problem in the evolution of architectures. In addition of these, ET algorithm and EAC strategy, can maintain a closer behavioural link between the parents and their offspring, which improves the efficiency of evolving RNN.

REAPSO has been tested in modeling the state estimation of the CSTR system. To show the effectiveness and efficiency of REAPSO, the algorithms of TDRB, GA, PSO, and HGAPSO applied to the fully connected RNN is used to the same problem. The results show that REAPSO is able to evolve both the architecture and weights of RNN, and the RNN evolved by REAPSO has good accuracy and generalisation ability.

References

1. F. J. Pineda: Generalization of backpropagation to recurrent neural networks. Physical Review Letters, 59(19). (1987) 2229--2232.
2. D. E. Rumelhart, G. E. Hinton, and R. J. Williams: Learning internal representations by error propagation. Parallel Distributed Processing. Cambridge, MA: MIT Press, vol. 1. (1986) 318–362.
3. R. J. Williams and D. Zipser: A learning algorithm for continually running recurrent neural networks. Neural Comput., vol. 1, no. 2. (1989) 270–280.
4. B. A. Pearlmutter: Learning state space trajectories in recurrent neural networks. Neural Comput., vol. 1. (1989) 263–269.
5. Jia Lei, Guangdong He, and Jing Ping Jiang: The State Estimation of the CSTR System Based on a Recurrent Neural Network Trained by HGAs. Neural Networks, International Conference on Vol. 2. (1997) 779 – 782.
6. F. Heimes, G. Zalesski, W. L. Jr., and M. Oshima: Traditional and evolved dynamic neural networks for aircraft simulation. Proc. IEEE Int. Conf. Systems, Man, and Cybernetics, Part 3 (of 5). (1997) 1995–2000.
7. D. Whitley: Genetic algorithms and neural networks. Genetic Algorithms Engineering and Computer Science, G. Winter, J. Periaux, M. Galan, and P. Cuesta, Eds. New York: Wiley, (1995) 191–201.
8. Jaszkiewicz: Comparison of local search-based metaheuristics on the multiple-objective knapsack problem. Found. Comput. Decision Sci., vol. 26. (2001) 99–120.
9. Juang, C.-F.: A hybrid of genetic algorithm and particle swarm optimization for recurrent network design. IEEE Transactions on Systems, Man, and Cubernetics - Part B: Cybernetics, vol. 34, no. 2. (2004) 997-1006.
10. X. Cai, N. Zhang, G. Venayagamoorthy and D. Wunsch: Time Series Prediction with Recurrent Neural Networks Using a Hybrid PSO-EA Algorithm. IJCNN'04, Budapest. (2004).

11. N.Burgess: A constructive algorithm that converges for real-valued input patterns. Int. J. Neural Syst., vol 5, no. 1. (1994) 59-66.
12. R. Reed: Pruning algorithms-A survey. IEEE trans. Neural Networks, vol 4. (1995) 740-747.
13. P.J. Angeline, G.M.Sauders, and J.B. Pollack: An evolutionary algorithm that constructs recurrent neural networks. IEEE Trans. Neural Networks, vol. 5. (1994) 54-65.
14. X. Yao: A review of evolutionary artificial neural networks. Int. J. Intell. Syst., vol. 8, no. 4. (1993) 539-567.
15. Kennedy, J., and Eberhart, R. C.: Particle swarm optimization. Proc. IEEE International Conference on Neural Networks, IEEE Service Center, Piscataway, NJ. (1995) 39-43.

Self-organized Locally Linear Embedding for Nonlinear Dimensionality Reduction

Jian Xiao, Zongtan Zhou, Dewen Hu*, Junsong Yin, and Shuang Chen

Department of Automatic Control, College of Mechatronics and Automation,
National University of Defense Technology, Changsha, Hunan, 410073, P.R.C.
dwhu@nudt.edu.cn

Abstract. Locally Linear Embedding (LLE) is an efficient nonlinear algorithm for mapping high-dimensional data to a low-dimensional observed space. However, the algorithm is sensitive to several parameters that should be set artificially, and the resulting maps may be invalid in case of noises. In this paper, the original LLE algorithm is improved by introducing the self-organizing features of a novel SOM model we proposed recently called DGSOM to overcome these shortages. In the improved algorithm, nearest neighbors are selected automatically according to the topology connections derived from DGSOM. The proposed algorithm can also estimate the intrinsic dimensionality of the manifold and eliminate noises simultaneously. All these advantages are illustrated with abundant experiments and simulations.

1 Introduction

In most machine learning problems, dimensionality reduction is an important and necessary preprocessing step to cope with high-dimensional data set, such as face images with varying pose and expression changes [1]. The purpose of dimensionality reduction is to project high-dimensional data to a lower dimensional space while discovering compact representations of high-dimensional data. Many methods have been presented to cope with high dimensionality of data sets and pattern recognition, including geometric preservation, neural network and genetic algorithms [2], [3], [4], [5]. Two traditional methods of dimensionality reduction are Principal Component Analysis (PCA) and Multidimensional Scaling (MDS). Both of them are linear methods and are widely used, but in the situation of nonlinear input data, they often fail to preserve the structures and relationships in the high-dimensional space when data are mapped into a low-dimensional one [6].

While in these cases, Nonlinear Dimensionality Reduction (NLDR) methods can achieve better results. Locally Linear Embedding (LLE), first proposed by Roweis and Saul in 2000 [7], has attracted more and more attention among such NLDR techniques. LLE reduces the dimension by solving the problem of mapping high-dimensional data (possibly in a nonlinear manifold) into a single global coordinate system of lower dimensionality. The most attractive virtues of LLE are that there are only two parameters to be set, and the computation can avoid converging to a local minimum. However, there have yet been very few reports of application of LLE since it was proposed [8], which, in our opinion while applying it, may be because the mapping results are quite sensitive to parameters, and it may be useless when adequate noises were included in the raw data set.

To tackle the problems, here we introduce features of a novel SOM model proposed by the co-authors of this paper recently. The model, called Diffusing and Growing Self-Organizing Maps (DGSOM) [9], increases units through competition mechanism, generates and updates the topology of network using Competitive Hebbian Learning (CHL) fashion, and uses NO diffusion model with dynamic balance mechanism to define the neighborhoods of unites and the fine-tuning manner. Topological connections among neurons generated by DGSOM, which reflect the dimensionality and structure of input signals, can adapt to the changes of the dynamic distribution [10]. The new algorithm proposed in this paper firstly applies DGSOM to reduce the large amount of high-dimensional input data to a set of data points with connections between neighboring ones to reflect the structure of the original input data set rationally and efficiently. Secondly, the resulting neighboring relationships between data points are adopted directly instead of the neighborhood searching in the original LLE [7], while the following steps are similar. Experiments will show the impressive performance of the combined algorithm.

The rest of this paper is organized as follows: Based on the algorithms of LLE and DGSOM, the unified algorithm is proposed in section 2. Section 3 expatiates on abundant experiments and some theoretical analysis. Conclusions and discussions are propagated in section 4.

2 Algorithms

2.1 Locally Linear Embedding

Supposing that the original data set consists of N vectors $\vec{X_i}(\vec{X_i} \in R^D)$, the purpose of LLE is to find N vectors $\vec{Y_i}(\vec{Y_i} \in R^d)$ in a low-dimensional space while preserving local neighborhood relations of data in both the embedding space and the intrinsic one. The basic algorithm is described as follows [7]:

Step 1: Compute the neighbors of each data point $\vec{X_i}$, by finding K nearest neighbors of each point or choosing points within a hypersphere of fixed radius.

Step 2: Compute the weights W_{ij} that best reconstruct each data point $\vec{X_i}$ from its neighbors. Reconstruction errors are measured by the cost function

$$\varepsilon(W) = \sum_i |\vec{X_i} - \sum_j W_{ij} \vec{X_j}|^2 \quad (1)$$

where W_{ij} summarize the contribution of the j-th data to the i-th reconstruction, and the weight matrix W satisfies two constraints: First, enforcing $W_{ij}=0$ if $\vec{X_j}$

does not belong to the neighbors of $\vec{X_i}$; Second, the rows of W sum to one: $\sum_j W_{ij}=1$. The weights W_{ij} are obtained by finding the minimum of the cost function.

Step 3: Compute the vectors $\vec{Y_i}$ best reconstructed by the weights W_{ij}. Fix the weights W_{ij}, and then compute the d-dimensional coordinates $\vec{Y_i}$ by minimizing the embedding cost function

$$\begin{aligned}\varepsilon(Y) &= \sum_i |\vec{Y_i} - \sum_j W_{ij}\vec{Y_j}|^2 = \|(I-W)Y\|^2 \\ &= Y^T(I-W)^T(I-W)Y\end{aligned} \quad (2)$$

This process can be done by finding the bottom d nonzero eigenvectors of a sparse $N \times N$ matrix $(I-W)^T(I-W)$.

LLE, being a powerful method solving the nonlinear dimensionality reduction problem, however, still has some disadvantages: Quality of manifold characterization is dependent on neighborhood choices and sensitive to noises. Improvements are on demand to solve the problems.

2.2 Diffusion and Growing Self-organizing Maps (DGSOM)

The newly proposed model DGSOM [9] consists of four mechanisms to make it as applicable as possible: Growing mechanism for resource competition, Competitive Hebbian Learning (CHL) method and aging mechanism for topology updating, forgetting mechanism for avoiding data saturation, and diffusion/dynamic balance mechanism for node adaptation. A detailed account of the four mechanisms is given as follows:

i) Mechanism 1: If one unit holds too many resources, a new unit will be generated and compete with it for redistribution of resources rationally.

ii) Mechanism 2: Ensure the formation of topology, with making a relationship between the two nearest nodes to the current input.

iii) Mechanism 3: Make the influence of early input signals be forgotten and prevent the winning times of a particular node from increasing infinitely.

iv) Mechanism 4: Combine the mechanism of topological connection adaptation with the mechanism of NO diffusion to build and keep the balance of the network.

The DGSOM model does not depend on any transcendental knowledge about the input distributions because of the growing nature of nodes and connections. It is not only able to compartmentalize the input space correctly, but also reflect the topology relations and the intrinsic dimensionality. Though there are many mechanisms incorporated in DGSOM model, the description of the model itself is simple and the whole structure of DGSOM algorithm is compact. The detailed steps of DGSOM algorithm are available by consulting [9].

2.3 The Proposed Self-organized LLE Algorithm

In the new algorithm, we unify DGSOM and LLE algorithms in one framework. Firstly, before the operation of LLE, large amount of original data samples are fed as the input of DGSOM. An obviously reduced number of samples with topology connections are achieved as the result of DGSOM, which reflect the intrinsic structure of the manifold adaptively and efficiently.

Secondly, run LLE on the reduced samples. In place of the first step of the original LLE method, the neighborhood relationships of sample points are defined directly according to the connections generated by DGSOM, e.g., if one point has connections with four other points in the DGSOM mapping, the four data points are considered in LLE as the neighborhoods of that point. So the neighborhood number of every sample may be different from each other, and achieved automatically, rather than fixed and being set artificially as in the original LLE.

The third and fourth steps of the unified algorithm are similar to the original LLE, apart from the fact that weight matrix W' is of size $N \times N$ instead of $N \times K$. For the i-th column of W', the number of nonzero entries equals to the number of nearest neighbors of the i-th sample. Finally, the low-dimensional vectors $\overrightarrow{Y_i}$ are computed the same way as in Step 3 of the original LLE algorithm.

In our Self-Organized LLE algorithm, D-dimensional samples u_i are derived from DGSOM while the number of which is remarkably reduced. The deviation between u and the original data X expressed as $E[X-u]$ is achieving zero asymptotically [11] which verifies that the reduced samples are good representations of the input data set.

3 Experiments and Comparisons

In this section, we will discuss several applications of our Self-Organized LLE algorithm to the selection of the nearest neighbors, estimation of the intrinsic dimensionality and the original data added with noises, through abundant experiments on the synthesized manifold S-curve and face database.

3.1 Selection of the Nearest Neighbors

Though LLE has very few parameters to be set, they can impact the result dramatically. One parameter is the number of nearest neighbors K, which is fixed in some range according to the manifold. Fig.1 shows how LLE unfolds the S-curve rationally.

However, if K is set too small, the mapping will cause disjoint sub-manifolds divided from continuous manifold and can not reflect any global properties; while on the other hand, if K is too large, the whole data set is seen as local neighborhood and the mapping will lose nonlinear character [8]. Fig.2 illustrates this problem using the example of S-curve, with $K=4$ and $K=80$.

In the new algorithm, numbers of nearest neighbors K are defined automatically according to the topology connections generated by DGSOM. Since the topology connections can reflect the intrinsic structure of the manifold, this method of defining the neighborhood is adaptive. Fig.3 is the result of S-curve with neighborhood relationship between reduced points automatically generated.

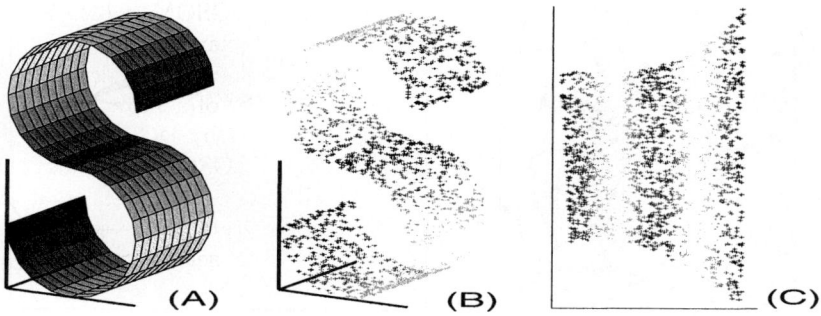

Fig. 1. LLE algorithm maps three-dimensional data into two-dimensional embedding space. Three-dimensional data points (B) are sampled from two-dimensional manifold (A). Neighborhood-preserving mappings are shown (C), with the color coding reveals how the data is embedded in two dimensions. (N=2000, K=12, d=2).

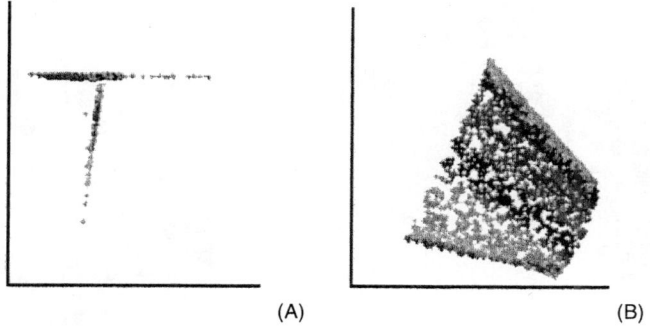

Fig. 2. For S-curve shown in Fig.1, choose K=4 (A) and K=80 (B). There is obviously deformation and incorrect color coding either.

Fig. 4 is another example of our method carried on Frey Face Database[1]. Though the result of Fig. 4 is similar to that in [7], our algorithm avoids the problem of setting the number of nearest neighbors K.

3.2 Estimation of the Intrinsic Dimensionality

Considering n-dimensional input data, if the dimension in the embedding space is m ($m \ll n$), then the intrinsic dimensionality of the input data is m. In the original LLE algorithm, if we don't know the intrinsic dimensionality as a prior, it should be established in advance. This problem can be solved by

---

[1] Available at http://www.cs.toronto.edu/ roweis/data.html

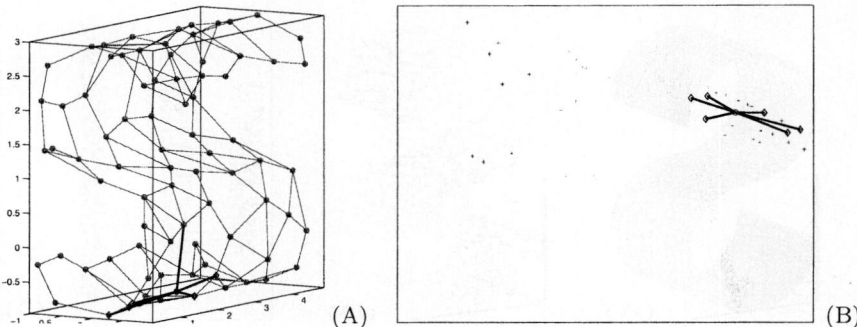

Fig. 3. (A) The reduced samples of S-curve (shown in Fig.1) which number is 82 as the result of DGSOM, with the nodes in red and topology connections in blue. (B) Map the reduced samples into a two-dimensional space using the algorithm we proposed. The black nodes and their connections in (A) and (B) show a single point (represented by circle) and its neighborhood (represented by diamonds).

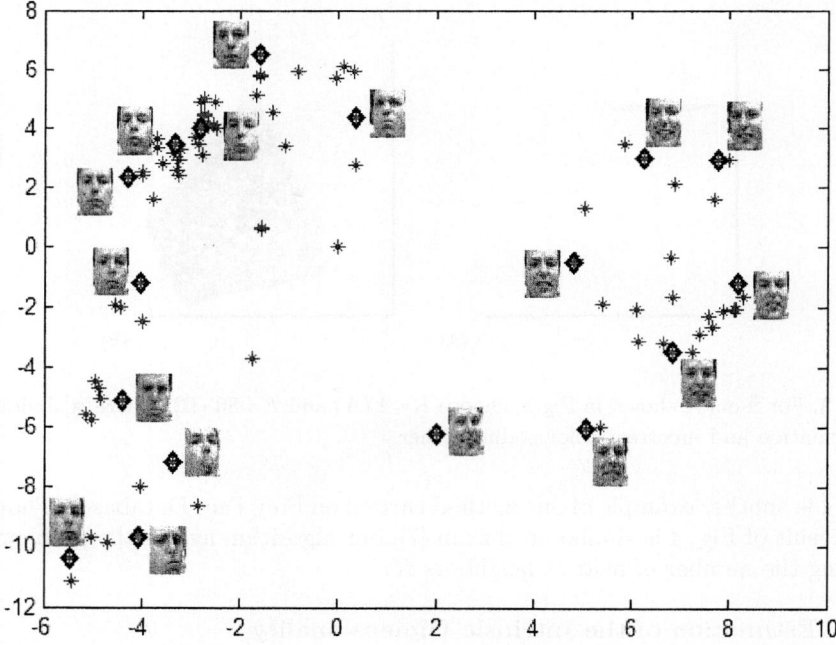

Fig. 4. Images of faces mapped into a two-dimensional embedding space, with the neighborhood defined automatically. In the original data, $N=1965$, $D=560$ (each image has 28¡Á20 pixels). The number of the images is reduced to 92 after running our algorithm. Representative points are marked by diamond with corresponding faces next to them. The variability in expression and pose is illustrated clearly along the two coordinates.

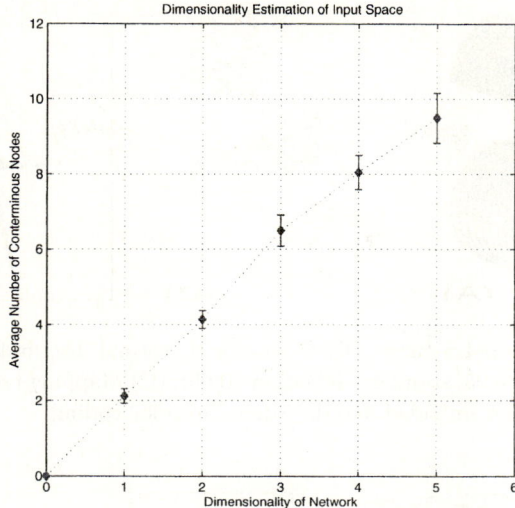

Fig. 5. The statistical graph of correlation between dimensionality of network and average of coterminous neighbor nodes. The mean of average of coterminous neighbor nodes for each node is represented by diamond and the variance by short line.

DGSOM, since the topology of DGSOM will reflect the dimension m rather than n. The correlation between dimensionality of network and average of coterminous neighbor nodes of each node in the network is obtained through Monte-Carlo method [12], shown in Fig.5.

From Fig.5 it can be seen that for one-dimensional network, each node has about two coterminous neighbor nodes averagely, and for two-dimensional network correspondingly, each node has about four coterminous neighbor nodes averagely. Then we can establish the intrinsic dimension by calculating the average coterminous neighbor nodes of the network results from DGSOM. For the S-curve in Fig 3(A), the average number of coterminous neighbor nodes is 4.1205, which is in good agreement with the result derived from Fig.5.

3.3 Eliminating Noises

Though the original LLE algorithm is efficient and robust when the data lie on a smooth and well-sampled single manifold, the embedding result can be affected and destroyed significantly when noises exist. Fig.6 demonstrates the result with S-curve distribution with normal distributed random noises (mean=0, variance=0.05, standard deviation=0.05). The variance and standard deviation has a critical value 0.05, above which the mapping result will distort terribly.

However, in the unified algorithm, samples from DGSOM can reflect the intrinsic distribution of input data in case of noises. To demonstrate this character, same data set as shown in Fig.6 (B) is fed in the Self-Organized LLE algorithm.

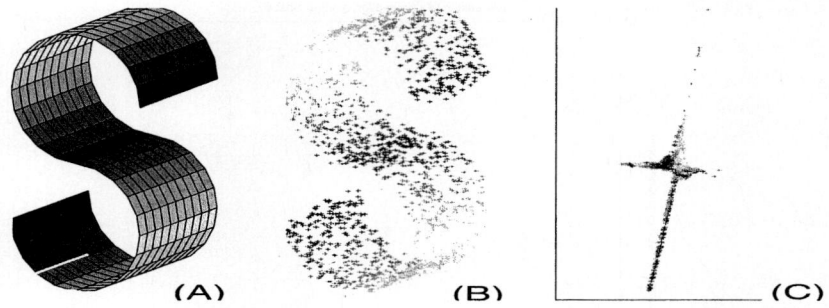

Fig. 6. (A) The original S-curve. (B) S-curve with normal distributed random noises, (mean=0, variance=0.05, standard deviation=0.05). (C) Mapping result using the original LLE, which is not unfolded and disordered in color coding.

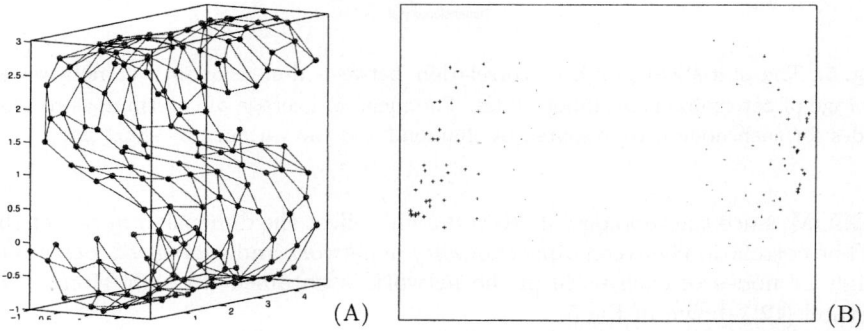

Fig. 7. (A) The network derived from Fig.6 (B) after running DGSOM, reflecting the intrinsic distribution of S-curve, with the reduced number of samples 155. (B) Mapping result in the embedding space after the algorithm we proposed, which is unfolded compared with Fig.6 (C)

The mapping generated by DGSOM and the final result in the embedding space are illustrated in Fig. 7.

From these experiments and comparisons, some conclusions can be drawn.

4 Conclusions

Self-Organized LLE algorithm, as proposed in this paper for nonlinear dimensionality reduction, is the integration of two new but different kinds of approaches. The algorithm first applies DGSOM to high-dimensional input data set, deriving a reduced number of samples with topology connections. Then instead of the first step of LLE, neighborhood relationship is obtained automatically from the topology mapping generated by DGSOM. Experiments and simulations indicate that the integrated algorithm outperforms the original LLE in following aspects: 1) The number of nearest neighbors is achieved automatically and efficiently rather than being set arbitrarily. 2) It provides an efficient way of estimating the intrinsic dimensionality. 3) In case of noises, when the original LLE can not work, our algorithm will still give satisfactory results.

Acknowledgement

This work is partially supported by Specialized Research Fund for the Doctoral Program of Higher Education of China (20049998012), Distinguished Young Scholars Fund of China (60225015), National Natural Science Fund of China (30370416, 60234030), and Ministry of Education of China (TRAPOYT Project).

References

1. Jun, W., Zhang, C.S., Kou, Z.B.: An Analytical Mapping for LLE and Its Applications in Multi-Pose Face Synthesis. The 14th British Machine Vision Conference, University of East Anglia, Norwich, UK (2003)
2. Oh, I.-S., Lee, J.-S., Moon, B.-R.: Hybrid genetic algorithms for feature selection. IEEE Trans. Pattern Analysis and Machine Intelligence 26 (2004) 1424-1437
3. Fu, X.J., Wang, L.P.: Data dimensionality reduction with application to simplifying RBF network structure and improving classification performance. IEEE Trans. System, Man, Cybern, Part B - Cybernetics 33 (2003) 399-409
4. Raymer, M.L., Punch, W.F., Goodman, E.D., Kuhn, L.A., Jain, A.K.: Dimensionality reduction using genetic algorithms. IEEE Trans. Evolutionary Computation 4 (2000) 164-171
5. Fu, X.J., Wang, L.P.: A GA-Based Novel RBF Classifier with Class-Dependent Features. Proc. 2002 IEEE Congress on Evolutionary Computation (CEC 2002), vol.2 (2002) 1890-1894
6. Lawrence, K.S., Roweis, S.T.: An Introduction to Locally Linear Embedding. Technical Report, Gatsby Computational Neuroscience Unit, UCL (2001)
7. Roweis, S.T., Lawrence, K.S.: Nonlinear Dimensionality Reduction by Locally Linear Embedding. Science, vol. 290, no.5500 (2000) 2323-2326
8. Ridder, D.D., Duin, R.P.W.: Locally Linear Embedding for Classification. Technical Report PH-2002-01, Pattern Recognition Group, Dept. of Imaging Science and Technology. Delft University of Technology (2002) 1-15
9. Chen, S., Zhou, Z.T., Hu, D.W.: Diffusion and Growing Self-Organizing Map: A Nitric Oxide Based Neural Model. ISNN (1) (2004) 199-204
10. Kohonen, T.: Self-Organizing Maps. 2nd edition. Springer, Berlin Heidelberg New York (1997)
11. Hirose, A., Nagashima, T.: Predictive Self-Organizing Map for Vector Quantization of Migratory Signals and Its Application to Mobile Communication. IEEE Trans. Neural Networks. Nov. (2003) 1532-1540
12. Woller, J.: The Basics of Monte Carlo Simulations. Available at http://www.chem.unl.edu/zeng/joy/mclab/mcintro.html

Active Learning for Probabilistic Neural Networks

Bülent Bolat and Tülay Yıldırım

Yildiz Technical University, Dept. of Electronics and Communications Eng.,
34349 Besiktas, Istanbul, Turkey
{bbolat, tulay}@yildiz.edu.tr

Abstract. In many neural network applications, the selection of best training set to represent the entire sample space is one of the most important problems. Active learning algorithms in the literature for neural networks are not appropriate for Probabilistic Neural Networks (PNN). In this paper, a new active learning method is proposed for PNN. The method was applied to several benchmark problems.

1 Introduction

In the traditional learning algorithms, the learner learns through observing its environment. The training data is a set of input-output pairs generated by an unknown source. The probability distribution of the source is also unknown. The generalization ability of the learner depends on a number of factors among them the architecture of the learner, the training procedure and the training data [1]. In recent years, most of the researchers focused on the optimization of the learning process with regard to both the learning efficiency and generalization performance. Generally, the training data is selected from the sample space randomly. With growing size of the training set, the learner's knowledge about large regions of the input space becomes increasingly confident so that the additional samples from these regions are redundant. For this reason, the average information per instance decreases as learning proceeds [1, 2, 3].

In the active learning, the learner is not just a passive observer. The learner has the ability of selecting new instances, which are necessary to raise the generalization performance. Similarly, the learner can refuse the redundant instances from the training set [1-5]. By combining these two new abilities, the active learner can collect a better training set which is representing the entire sample space well.

The learning task is a mapping operation between a subset x of the input space X and a subset y of the output space Y. The student realizes a function $X \rightarrow Y: s_w(x)=y$. The subscript w denotes a set of adaptive internal parameters of the student that are adjusted during the learning process [1, 6]. The goal of the training process is minimization of a suitably chosen error function E by adjusting the student's parameters w. The error term is defined as the disagreement of the teacher and the student. The adjustment of w is performed until the student makes decisions close to the teacher's ones. The main goal of the training is not to learn an exact representation of the training data but rather to exact a model of the teacher's function [1]. The student must be able to make good predictions for new samples. This ability is called as generalization. In this paper, non-random selection of the training set is considered. The main goal of the selection of the training set by some rules is to improve the learner's generalization ability. In Section 2, active learning paradigm was discussed. In Sections 3 and 4, Probabilistic Neural Networks (PNN) and a new active learning algorithm for PNN were proposed. Sample applications of the algorithm were given in the Section 5; and the results were discussed in the Section 6.

2 Active Learning

Figure 1 shows a binary classification problem. Class 1 is represented by circles; class 2 is represented by rectangles. The black filled circles and rectangles are the training set. Left to right hatch area is the teacher's decision boundary and right to left hatched area is the learner's approximation. The area, which the learner and teacher decides different represents the generalization error. The learner classifies all of the training data correctly, but there are regions where teacher and learner disagree (Figure 1a). The generalization error of the learner can be reduced if it receives additional new training instances from the error region (Figure 1b). The selection of the new instances might be done randomly or by some rule. An efficient active learning algorithm must, ideally, minimize both its generalization error and amount of training data [7,8]. For classification purposes, it is not necessary to minimize the mean square error, but to estimate the correct boundary between the classes, so called decision boundary [3].

The active learning strategies might be separated into two classes: active sampling and active selection. In active sampling, new training instances are constructed or generated from the existing training set by using some transformation rules. Selecting a concise subset of the entire dataset is called as active selection. There are several active selection approaches in the literature. Most of these approaches are separated into two groups: those that start with a small subset of the training data and sequentially add new instances, and those that start with a large subset of the training data and sequentially remove instances from the training set.

Plutowski and Halbert [9] propose an algorithm that adds new training instances to the training set. A new training instance is added to the training set with the aim to maximize the expected decrement in mean square error that would result from adding the training instance to the training set and training upon the resulting set. Another incremental algorithm was described by [5]. In this algorithm, the network is first trained by a training set. An unused pattern x_n, which has the maximum error, is found and this pattern is added to the training set. Various stopping criterions can be used for this algorithm. The Query-By-Committee [1] approach uses a committee of learners to find a new sample, which has the maximum disagreement among the members of the committee. Once this sample is found, it is added to the training set, members of the committee are retrained and the entire process is repeated. [2] introduces a similar approach for minimization of data collection. [10] represents an active learning scheme for parameter estimation in Bayesian networks. The method tries to find the sample, which has the minimum risk factor based on Kullback-Leibler divergence and adds it to the training set.

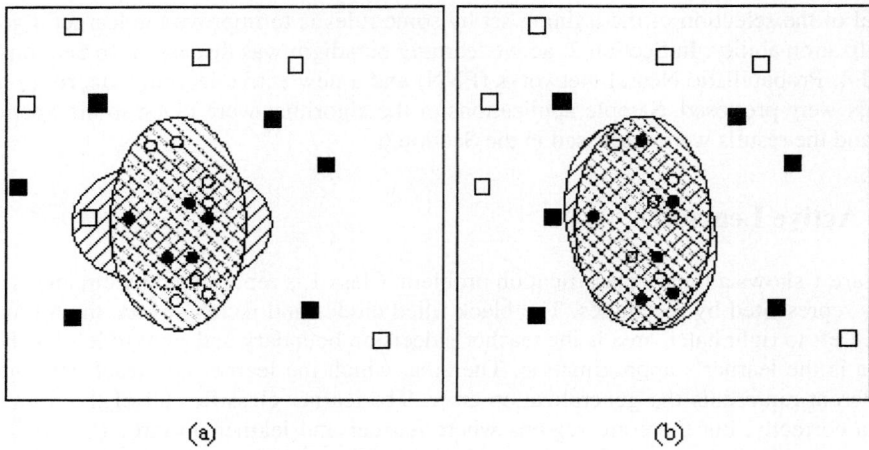

Fig. 1. (a) The passive learning. Learner acts as an observer and there are big dissimilarities between the teacher's and learner's decision boundaries. (b) The active learning. The training set is chosen by the learner.

Pruning of training set can be achieved in a natural way by using Support Vector Machines (SVM) [11]. A SVM tries to find hyper planes, which separate the classes from one to another by maximizing the margin between these classes. A small number of the training instances, those so-called support-vectors, suffice to define the hyper planes. These support vectors are highly informative training instances. Tong and Koller [12] introduced a new algorithm for performing active learning with support vector machines.

Another useful algorithm is Repeat Until Bored (RUB) which is introduced by Munro [13]. In this algorithm, if the current training sample generates a high error (i.e. greater than a fixed criterion value), it is repeated; otherwise, another one is randomly selected. This approach was motivated by casual observations of behavior in small children.

3 Probabilistic Neural Networks (PNN)

Consider a pattern vector x with m dimensions that belongs to one of two categories K_1 and K_2. Let $F_1(x)$ and $F_2(x)$ be the probability density functions (pdf) for the classification categories K_1 and K_2, respectively. From Bayes' decision rule, x belongs to K_1 if (1) is true, or belongs to K_2 if (1) is false;

$$\frac{F_1(x)}{F_2(x)} > \frac{L_1 P_2}{L_2 P_1} \tag{1}$$

where L_1 is the loss or cost function associated with misclassifying the vector as belonging to category K_1 while it belongs to category K_2, L_2 is the loss function associated with misclassifying the vector as belonging to category K_2 while it belongs to category K_1, P_1 is the prior probability of occurrence of category K_1, and P_2 is the prior probability of occurrence of category K_2. In many situations, the loss functions and the prior probabilities can be considered equal. Hence the key to using the decision rule given by (1) is to estimate the probability density functions from the training patterns [14].

In the PNN, a nonparametric estimation technique known as Parzen windows [15] is used to construct the class-dependent probability density functions for each classification category required by Bayes' theory. This allows determination of the chance a given vector pattern lies within a given category. Combining this with the relative frequency of each category, the PNN selects the most likely category for the given pattern vector. Both Bayes' theory and Parzen windows are theoretically well established, have been in use for decades in many engineering applications, and are treated at length in a variety of statistical textbooks. If the jth training pattern for category K_1 is x_j, then the Parzen estimate of the pdf for category K_1 is

$$F_1(x) = \frac{1}{(2\pi)^{m/2} \sigma^m n} \sum_{j=1}^{n} \exp\left[-\frac{(\mathbf{x}-\mathbf{x_j})^T (\mathbf{x}-\mathbf{x_j})}{2\sigma^2} \right] \qquad (2)$$

where n is the number of training patterns, m is the input space dimension, j is the pattern number, and σ is an adjustable smoothing parameter [14].

Figure 2 shows the basic architecture of the PNN. The first layer is the input layer, which represents the m input variables (x_1, x_2, ... x_m). The input neurons merely distribute all of the variables x to all neurons in the second layer. The pattern layer is fully connected to the input layer, with one neuron for each pattern in the training set. The weight values of the neurons in this layer are set equal to the different training patterns. The summation of the exponential term in (2) is carried out by the summation layer neurons. There is one summation layer neuron for each category. The weights on the connections to the summation layer are fixed at unity so that the summation layer simply adds the outputs from the pattern layer neurons. Each neuron in the summation layer sums the output from the pattern layer neurons, which correspond to the category from which the training pattern was selected. The output layer neuron produces a binary output value corresponding to the highest pdf given by (2). This indicates the best classification for that pattern [14].

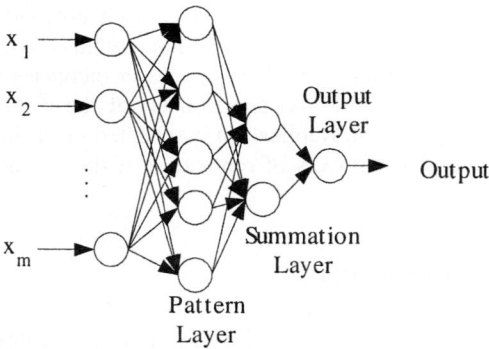

Fig. 2. The basic architecture of the PNN. This case is a binary decision problem. Therefore, the output layer has just one neuron and summation layer has two neurons.

4 Active Learning with Probabilistic Neural Networks

Since the known active learning strategies require an error measure (i.e. mean square, maximum absolute, sum square etc.) or a randomization in the learning phase (i.e. initial weight vector of the Multi Layer Perceptron is random), active learning with PNN is slightly different from other neural networks such as MLP (Multi Layer Perceptron), RBF (Radial Basis Function Networks), etc. The output of PNN is a binary value, not continuous. By using a binary output value, it is hard to develop any useful training data selection criteria such as maximum or mean square error. On the other hand, learning phase of the PNN takes only one sweep and unlike the MLP, the PNN learning is not iterative; if the same training set is used, the learning phase of the PNN always produces the same weights. For these reasons a new and useful approach for PNN is offered below.

The first step of the PNN learning is to find an acceptable spread value. The spread is found by using a trial-by-error process. When the optimum spread value is found, the next step is to find a better training set by using a data exchange algorithm.

The exchange process starts with a random selected training set. After first training process, the test data is applied to the network. A randomly selected true classified instance in the training set (I1) is thrown into the test set; a wrong classified instance in the test set (I2) is put into the training set and the network re-trained. If I2 is false classified, it is marked as a "bad case", I2 is put into the original location, and another false classified test instance is selected and the network retrained. Retraining is repeated until finding a true classified I2. When it is found, I1 is considered. If I2 is true classified and the test accuracy is reduced or not changed (I1 is false classified), I1 is put into the original location and another true classified training instance, say I3, is put into the test set and the process is repeated. If the accuracy is improved, the exchange process is applied to another training and test pairs. When an instance marked as "bad" once, it is left out of the selection process. The process is repeated until reaching the maximum training and test accuracy.

The last step is to find loss functions which give the best accuracy. In general, the loss functions are considered equal. However, if the dataset has rare classes, well selection of the loss functions improves the accuracy of the network. In this work, the MATLAB Neural Network Toolbox was used. The loss functions are adjusted by data replication when needed, because of this toolbox does not permit to adjust the loss functions. The classes of the training set which have relatively less instances are used two or more times in the training set. This operation increases the occurrence probability P_i (or equally, reduces the loss function L_i) of the ith class (see Eq. 1). This operation is called as replication. Replication is repeated until finding better test accuracy. After the considerable numbers of replication if the test accuracy is not raised, the replication is cancelled.

5 Applications and Results

The methods described above were tested by using several datasets, which are Glass data, Lenses data, Lung data, Wisconsin Diagnostic Breast Cancer (WDBC) data, Wisconsin Prognostic Breast Cancer (WPBC) data, Cleveland Heart data, BUPA data and Escheria Coli data. The datasets were taken from UCI Machine Learning Repository Database [16]. The simulations were realized by using MATLAB Neural Network Toolbox.

Glass dataset has 214 instances separated into six classes. Each instance identified by nine continuous chemical measures. The biggest class of this dataset includes 76 instances; the least class has only 9. Escheria coli dataset was taken from Nakai and maintained by Horton in 1996. Dataset has 336 instances divided into eight classes. Each instance is identified by eight continuous attributes. The biggest class of this dataset has 143 instances and the least two classes have only two. Lenses database is a complete database; all possible combinations of attribute value pairs are represented. The dataset has only 24 instances, which are represented by four discrete attributes. Class 1 has 5, class 2 has 4 instances. The remaining 15 instances belong to class 3. BUPA Liver Disorder dataset was originally created by BUPA Medical Research Ltd. This dataset has 345 instances. Each instance has six attributes which are taken from blood tests and a class number which shows if a liver disorder exists or not. All of the attributes are real and continuous. BUPA dataset has only two classes. WDBC dataset has 359 instances separated into two classes. Each instance has 30 continuous features. Features are computed from a digitized image of a fine needle aspiration (FNA) of a breast mass. WPBC dataset has 198 instances separated into two classes. Each instance has 32 continuous features. First 30 features are computed from a digitized image of a FNA of a breast mass. Lung Cancer dataset describes 3 types of pathological lung cancers. The Authors give no information on the individual variables. Dataset has 32 instances (including 5 missing) which are described by 56 nominal features. The last dataset, Speech/Music is not a part of UCI database. It was created by Bolat and Kucuk; and first appeared in [17]. Speech/Music dataset has 150 instances divided into two categories. Speech class has 50 speech samples from four different males. Music class has 100 samples: 50 instrumental music pieces and 50 music pieces with male and/or female singers. Each instance is represented by six continuous values which are means and standard deviations of zero cross rate, RMS energy and spectral centroid. Table 1 shows some past usage of the datasets described above.

Table 1. Recent works based on the datasets used in experiments

Dataset	Network	Accuracy
Glass [18]	Bayesian EM	89.6%
Glass [19]	1-NN	78.8%
E.coli [20]	k-NN	86%
E.coli [21]	PNN	82.97%
Lens [22]	RQuantizer	81.3%
WPBC [23]	9-NN	81.0%
WDBC [23]	5-NN	93.3%
WDBC [24]	XCS	96.67%±2.2
Lung [25]	C4.5	86.65%
BUPA [26]	MLP	74.36%
BUPA [27]	Weighted aggregation	67.83%
BUPA [26]	PNN	42.11%
Speech/Music [17]	PNN	84.45%

Approximately 70% of the datasets were used as training sets; remaining portions were used as test sets. Instances that have missing attributes were deleted from datasets. In the first stage of the learning phase, the optimum spread values were found by a trial-by-error strategy by using randomly selected training sets. In the second stage, the data selection method was applied and the best test accuracies obtained. In the last stage, if needed, a data replication process was applied to achieve better accuracies. Table 2 shows results of data exchange method with no replication and Table 3 shows the results of data exchange with replication.

Table 2. Training and Test Accuracies (Exchange only)

Database	Training	Test
Glass	98.67%	89.06%
E.coli	94.60%	90.35%
Lens	93.33%	75%
WPBC	100%	84.48%
WDBC	100%	98.83%
Lung	100%	87.5%
BUPA	100%	94.78%
Speech/Music	95%	97,78%

Table 3. Training and Test Accuracies (Exchange and Replication)

Database	Training	Test
Glass	100%	95.31%
E.coli	95.5%	90.35%
Lens	100%	87.5%

According to the simulation results, it is seen that the good selection of the training data boosts the accuracy of the network. Data replication also offers an improvement on the classes which have relatively less instance numbers. As an example, test accuracy of the class 3 of the Glass dataset was raised from 50% to 100% by repeating the training portion of this class three times. It is raised from 66% to 100% for class 6 of the Glass dataset with a replication rate of 4. Data replication did not improve the test accuracy of the E.coli dataset, but improved the training accuracy a little. Training part of the class 6 repeated two times for this dataset. Both training and test accuracies of the Lens data were improved by using data replication. The class distributions of the other datasets are not so imbalanced, or accuracies were not improved by adjusting the loss functions. Therefore, the replication was not applied to them.

For large, real-world inductive learning problems, the number of training examples often must be limited due to the costs associated with procuring, preparing, and storing the training examples and/or the computational costs associated with learning from them [28]. Another advantage of this algorithm is that the selection method (data exchange) does not change the amount of the training data. The method presented here is useful for these kinds of difficult learning tasks.

6 Concluding Remarks

Generalization performance of a neural network usually depends on the selection of instances. This also affects the learning efficiency of the network. Traditional learning algorithms generally select the training data randomly from the sample space and the learner is a passive observer here. In the active learning, the learner has the ability of selecting new instances, which are necessary to raise the generalization performance. Similarly, the learner can refuse the redundant instances from the training set. Hence, the active learner can collect a better training set which is representing the entire sample space well.

There is no active learning algorithm for Probabilistic Neural Networks in literature. Active learning with PNN is slightly different from other neural networks such as MLP, RBF, etc. The main reason of this dissimilarity is that the output of the PNN is a binary value, not continuous. By using a binary output value, it is hard to develop any useful training data selection criteria such as maximum error. In the other hand, learning phase of the PNN takes only one sweep and unlike the MLP, the PNN learning is not random; if the same training set is used, the learning phase of the PNN always produces the same network.

In this paper, a new active learning method for PNN is introduced. Firstly, a data exchange method is considered and, secondly, a data replication is applied to increase the performance. A comparative study with benchmark problems is also presented. Concerning the results, it is seen that the good selection of the training data boosts the accuracy of the Probabilistic Neural Network. Data replication also offers an improvement on the classes which have relatively less instance numbers.

References

1. Hasenjager, M., Ritter, H.: Active Learning In Neural Networks. In: Jain L. (ed.): New Learning Techniques in Computational Intelligence Paradigms. CRC Press, Florida, FL (2000)
2. RayChaudhuri, T., Hamey, L. G. C.: Minimization Of Data Collection By Active Learning. In: Proc. of the IEEE Int. Conf. Neural Networks (1995)
3. Takizawa, H., Nakajima, T., Kobayashi, H., Nakamura, T.: An Active Learning Algorithm Based On Existing Training Data. IEICE Trans. Inf. & Sys. E83-D (1) (2000) 90-99
4. Thrun S.: Exploration In Active Learning. In: Arbib M. (ed.): Handbook of Brain Science and Neural Networks. MIT Press, Cambridge, MA (1995)
5. Leisch, F., Jain, L. C., Hornik, K.: Cross-Validation With Active Pattern Selection For Neural Network Classifiers. IEEE Trans. Neural Networks 9 (1) (1998) 35-41
6. Sugiyama, M., Ogawa, H.: Active Learning With Model Selection For Optimal Generalization. In: Proc. of the IBIS 2000 Workshop on Information Based Induction Science. Izu, Japan (2000) 87-92
7. RayChaudhuri, T., Hamey, L. G. C.: Active Learning For Nonlinear System Identification And Control. In: Gertler, J. J., Cruz, J. B., Peshkin, M. (eds): Proc. IFAC World Congress 1996. San Fransisco, USA (1996) 193-197
8. Saar-Tsechansky, M., Provost, F.: Active Learning For Class Probability Estimation And Ranking. In: Proc. of Seventeenth International Joint Conference on Artificial Intelligence (IJCAI-01). Seattle, WA, USA (2001)
9. Plutowski, M., Halbert, W.: Selecting Exemplars For Training Feedforward Networks From Clean Data. IEEE Trans. on Neural Networks 4 (3) (1993) 305-318

10. Tong, S., Koller, D.: Active Learning For Parameter Estimation In Bayesian Networks. In: Proc. of Advances in Neural Information Processing Systems. Denver, Colorado, USA (2000)
11. Cortes, C., Vapnik, V.: Support-Vector Networks. Machine Learning 20 (1995) 273-297
12. Tong, S., Koller, D.: Support Vector Machine Active Learning With Applications To Text Classification. Journal of Machine Learning Research 2 (2001) 45-66
13. Munro, P. W.: Repeat Until Bored: A Pattern Selection Strategy. In: Moody, J., Hanson, S., Lippmann, R. (eds): Proc. Advances in Neural Information Processing Systems (NIPS'91). (1991) 1001-1008
14. Goh, T. C.: Probabilistic Neural Network For Evaluating Seismic Liquefaction Potential. Canadian Geotechnology Journal 39 (2002) 219-232
15. Parzen, E.: On Estimation Of A Probability Density Function And Model. Annals of Mathematical Statistics 36 (1962) 1065-1076
16. Murphy, P. M., Aha, D. W.: UCI Repository of Machine Learning Databases. University of California, Department of Information and Computer Science (1994) Available at url:http://www.ics.uci.edu/~mlearn/MLRepository.html.
17. Bolat, B., Kucuk, U.: Speech/Music Classification By Using Statistical Neural Networks (In Turkish). In: Proc. of IEEE 12th Signal Processing and Communications Applications Conference (SIU 2004). Kusadasi, Turkey (2004) 227-229
18. Holst, A.: The Use Of A Bayesian Neural Network Model For Classification Tasks. Ph.D. Thesis. Stockholm University (1997)
19. Agre, G., Koprinska, I.: Case-Based Refinement Of Knowledge Based Neural Networks. In: Albus, J., Meystel, A., Quintero R. (eds): Proc. of International Conference on Intelligent Systems: A Semiotic Perspective. Gaithersburg, MD, USA (1996) 221-226
20. Horton, P., Nakai, K.: A Probabilistic Classification System For Predicting The Cellular Localization Sites Of Proteins. Intelligent Systems in Molecular Biology 4 (1996) 109-115
21. Avci, M., Yildirim, T.: Classification Of Escheria Coli Bacteria By Artificial Neural Networks. In: Proc. of IEEE International Symposium on Intelligent Systems. Varna, Bulgaria (2002) 16-20
22. Yang, J., Honavar, V.: A Simple Random Quantization Algorithm For Neural Network Pattern Classifiers. In: Proc. of World Congress on Neural Network. San Diego, CA, USA (1996) 223-228
23. Zavrel, J.: An Empirical Re-examination Of Weighted Voting For k-NN. In: Proc. of the 7th Belgian-Dutch Conference on Machine Learning (BENELEARN-97). Tilburg, Netherlands (1997)
24. Bacardit, J., Butz, M. V.: Data Mining In Learning Classifier Systems: Comparing XCS With GAssist. Technical Report No. 2004030. Illinois Genetic Algorithms Laboratory, University of Illinois at Urbana, Urbana, IL, USA (2004)
25. Yu, L., Liu, H.: Efficient Feature Selection Via Analysis Of Relevance And Redundancy. Journal of Machine Learning Research 5 (2004) 1205-1244
26. Yalcin, M., Yildirim, T.: Diagnosis Of Liver Disorders By Artificial Neural Networks (in Turkish). In: Proc. of IX. National Biomedical Engineering Meeting BIYOMUT 2003, Istanbul, Turkey (2003) 293-297
27. Takae, T., Chikamune, M., Arimura, H., Shinohara, A., Inoue, H., Takeya, S., Uezono, K., Kawasaki, T.: Knowledge Discovery From Health Data Using Weighted Aggregation Classifiers. In: Proc. of the 2nd International Conference on Discovery Science. Tokyo, Japan (1999) 359-360
28. Weiss, G. M., Provost, F.: Learning When Training Data Are Costly: the Effect Of Class Distribution on Tree induction. Journal of Artificial Intelligence Research 19 (2003) 315-354

Adaptive Training of Radial Basis Function Networks Using Particle Swarm Optimization Algorithm

Hongkai Ding, Yunshi Xiao, and Jiguang Yue

School of Electric & Information Engineering, Tongji University, ShangHai 200092, China
```
hkding1977@hotmail.com
xys@gs.tongji.edu.cn
yuejiguang@sian.com.cn
```

Abstract. A novel methodology to determine the optimum number of centers and the network parameters simultaneously based on Particle Swarm Optimization (PSO) algorithm with matrix encoding is proposed in this paper. For tackling structure matching problem, a random structure updating rule is employed for determining the current structure at each epoch. The effectiveness of the method is illustrated through the nonlinear system identification problem.

1 Introduction

Radial basis function (RBF) neural networks became very popular due to a number of advantages compared with other types of artificial neural networks, such as better approximation capabilities, simpler network structures and faster learning algorithm [1]. As is well known, the performance of an RBF network critically depends on the choice of the number and centers of hidden units. More specifically, most of the traditional training methods require from the designer to fix the structure of the network and then proceed with the calculation of model parameters. Most natural choice of centers is to let each data point in the training set correspond to a center. However, if data are contaminated by noise, then over-fitting phenomena will occur, which leads to a poor generalization ability of the network. For improving generalization performance, some approaches decompose the training into two stages: the centers of hidden units are determined first in self-organizing manner (structure identification stage), followed by the computation of the weights that connect the hidden layer with output layer (parameters estimation stage) [1],[2]. This is a time consuming procedure as it requires evaluation of many different structures based on trial and error procedure. Another drawback is the centers of hidden units are determined only based on local information. It is desirable combined the structure identification with parameters estimation as a whole optimization problem. However, this results in a rather difficult problem which cannot be solved easily by the standard optimization methods. An interesting alternative for solving this complicated problem can be offered by recently developed swarm intelligent strategies. Genetic algorithms (GA), the typical representative among others, have been successfully utilized for the selection of the optimal structure of RBF network [3],[4]. But GA have some defects such as more predefined parameters, more intensive programming burden etc.

Particle swarm optimization (PSO) algorithm is a recently proposed algorithm by Kennedy and Eberhart, motivated by social behavior of organisms such as bird flocking and fish schooling [5],[6] . PSO as an optimization tool, combines local search methods with global search methods, attempting to balance exploration and exploitation. It is demonstrated that PSO gets better results in a faster, cheaper way compared with other methods. Another reason that PSO is attractive is that there are few parameters to adjust [6].

In this paper, we propose a novel methodology to determine the optimum number of centers and the network parameters simultaneously based on PSO with matrix encoding. The method gives more freedom in the selection of hidden units' centers. The algorithm starts with a random swarm of particles, which are coded as centers of RBF network in the form of matrix. Then, a structure updating operator is employed to determine the structural state of all particles at each epoch. The fitness value of each particle is calculated based on prediction error criterion. In addition, each particle may be grown or pruned a unit for improving diversity. The algorithm is terminated after a predefined number of iterations are completed or prediction error threshold is met. The particle corresponds to the best fitness value throughout the entire training procedure is finally selected as the optimal model.

This paper is organized as follows. In section 2 we formulate the training of RBF network as a whole optimization problem by combined structure identification with parameters estimation. The details of proposed algorithm are described in section 3. Simulation results are shown in section 4. The results are compared with other existing similar algorithm. Finally, the conclusions are summarized in section 5.

2   Formulation of the Whole Optimization Problem

RBF networks form a special neural network architecture which consists of three layers, namely input, hidden, output layer. The input layer is only used to communicate with its environment. The nodes in the hidden layer are associated with centers, which character the structure of network. The response from a hidden unit is activated through a radial basis function, such as Gaussian function. Finally, the output layer is linear and serves as a summation unit.

Assume that we have a training set of N samples $\{\mathbf{x}_i, y_i\}, i=1,2,\cdots,N$ where y_i is the desired output value corresponding to the network input vector $\mathbf{x}_i = [x_1, x_2 \cdots x_d]^\mathrm{T}$ with dimension d. The RBF network training problem can be formulated as an optimization problem, where the sum of squared errors (SSE) between the desired outputs y_i and the network predictions $\hat{y}_i$ must be minimized with respect to both the network structure (the number of units M in the hidden layer) and the network parameters (the hidden unit center locations $\mathbf{c}_j$, width γ_j and the weights w_j, $j=1,2,\cdots,M$):

$$SSE(M, \mathbf{c}_j, \gamma_j, w_j) = \sum_{i=1}^{N}\left(y_i - \hat{y}_i\right)^2. \tag{1}$$

The predicted out $\hat{y}_i$ depends on the input vector and network parameters as follows:

$$\hat{y}_i = \sum_{j=1}^{M} w_j \cdot \phi_j(\mathbf{x}_i), i = 1, 2, \cdots, N, \qquad (2)$$

where $\phi_j = \phi(\|\mathbf{x} - \mathbf{c}_j\| / \gamma_j)$ does non-linear transformation performed by the j th hidden unit and $\|\cdot\|$ denotes Euclidean norm in R^d. The Gaussian function $\phi(r) = \exp(-r^2/2)$ is used in our work.

The whole optimization problem requires minimization of the above error function (1). This is rather difficult using the traditional optimization techniques, especially due to the presence of the number M. PSO algorithm can be used for solving any type of optimization problem, where the objective function can be discontinuous, non-convex or non-differentiable [6].

3 Adaptive Training of RBF Networks Using PSO Algorithm

PSO algorithm is an adaptive method based on a social-psychological metaphor, a population of individuals adapts by backing stochastically toward previously successful regions in the search space, and is influenced by the successes of their topological neighbors [6]. A swarm consists of many particles, where each particle keeps track of its position, velocity, best position thus far, best fitness thus far, current fitness. The velocity keeps track of the speed and direction the particle is currently traveling. The current position is the most important attribute, which corresponds to a potential solution of the function to be minimized.

For RBF networks implementation, a specially designed PSO algorithm with matrix encoding is used to determine the optimum number of hidden units and the network parameters simultaneously. The algorithm starts with an initial swarm of particles, which represent possible networks structure and associated center locations. The centers are determined by current position of particle. The widths are determined using a nearest neighbor heuristic discussed later. After the determination of centers and widths, the weights between the hidden and the output layer are calculated using linear regression. Then the objective function can be computed. New position of particles is produced using PSO algorithm after structural updating operation and growing or pruning operation. The algorithm terminates after a predefined number of iterations are completed or error threshold is met. The particle that has minimum fitness is selected as the optimum RBF network.

The detailed description of the proposed algorithm that follows assumes that N input-output samples $\{\mathbf{x}_i, y_i\}, i = 1, 2, \cdots, N$ are available, which can be grouped into two data sets: the input set $\mathbf{X}$ and the output set $\mathbf{Y}$. The dimension of the input vector is d. While only one output variable is used in our paper, the algorithm can be easily generalized for more than one output variables. Before the application of the algorithm, the training data are processed as follows.

3.1 Data Division

The samples are divided into two subsets $(\mathbf{X}_1, \mathbf{Y}_1)$, $(\mathbf{X}_2, \mathbf{Y}_2)$ of size N_1, N_2, which are the training and validation sets. The first subset $(\mathbf{X}_1, \mathbf{Y}_1)$ is used in the training procedure to calculate the connection weights of the different RBF networks that constitute the whole swarm. The second subset $(\mathbf{X}_2, \mathbf{Y}_2)$ is also used during the training epoch to evaluate the fitness of each particle. This is crucial for the success of the proposed algorithm, since it incorporates a testing procedure into the training process. This strategy can avoid over-fitting effectively.

3.2 Data Scaling

The RBF network to obtain a predicted value at a given input proceeds by doing weighted summation using all centers that are close to the given input. Thus the performance of network depends critically on the metric used to define closeness. This has the consequence that if you have more than one input variable and these input variables have significantly different scales, then closeness depends almost entirely on the variable with the largest scaling.

To circumvent this problem, it is necessary to standardize the scale of the input variables. When all input variables are of the same order of magnitude, the algorithm performs better.

3.3 Particle Swarm Optimization Algorithm for RBF Networks

3.3.1 Particle with Matrix Encoding

The problem of interest to us consists of how to design a particle as the RBF network that performs a desired function. To encode a RBF network, we used the novel matrix encoding to represent a particle. L matrices (particles) of size $M \times d$ are created, each particle corresponds to a set of centers of RBF network, where M is maximum number of hidden units and d is the dimension of input variables. We employ a special label 'N' to indicate invalid center location. Assumed that the ith particle with m^i $(1 \leq m^i \leq M)$ hidden units, then position matrix $\mathbf{C}_i$ and velocity matrix $\mathbf{V}_i$ can be expressed as follows:

$$\mathbf{C}^i = \begin{bmatrix} c^i_{11} & c^i_{12} & \cdots & c^i_{1d} \\ c^i_{21} & c^i_{22} & \cdots & c^i_{2d} \\ \vdots & \vdots & \ddots & \vdots \\ c^i_{m^i 1} & c^i_{m^i 2} & \cdots & c^i_{m^i d} \\ N & N & \cdots & N \\ \vdots & \vdots & \vdots & \vdots \\ N & N & \cdots & N \end{bmatrix}, \quad \mathbf{V}^i = \begin{bmatrix} v^i_{11} & v^i_{12} & \cdots & v^i_{1d} \\ v^i_{21} & v^i_{22} & \cdots & v^i_{2d} \\ \vdots & \vdots & \ddots & \vdots \\ v^i_{m^i 1} & v^i_{m^i 2} & \cdots & v^i_{m^i d} \\ N & N & \cdots & N \\ \vdots & \vdots & \vdots & \vdots \\ N & N & \cdots & N \end{bmatrix}, \quad i = 1, 2 \cdots L \; . \tag{3}$$

The rows labeled by 'N' do not involve any algebraic operation and indicate invalid centers location.

3.3.2 Estimation of Widths and Weights

For each particle, we can identify the valid centers from the position matrix. Assumed that m^i hidden units with centers location of the *ith* particle are identified, then widths γ_j^i are determined using a nearest neighbor heuristic suggested in Moody and Darken [7]. That is

$$\gamma_j^i = \left[\frac{1}{p}\sum_{l=1}^{p}\left\|\mathbf{c}_j^i - \mathbf{c}_l^i\right\|^2\right]^{1/2}, \quad j=1,2,\cdots,m^i, i=1,2,\cdots L, \tag{4}$$

where $\mathbf{c}_l^i \in R^d$ $(l=1,2,\cdots,p)$ are the p nearest neighbors of the center $\mathbf{c}_j^i$ ($p=2$ in our work). Once the locations of centers and widths are defined, the RBF network can be seen as a linear model [8], and the weights $\mathbf{w}^i$ can be calculated either by an algebraic single-shot process or by a gradient descent methods as in [9]. By introducing the notation $\phi_j^i(k) = \exp\left(-\left\|\mathbf{x}_k - \mathbf{c}_j^i\right\|^2 / \left(2(\gamma_j^i)^2\right)\right)$ and $\gamma^i = \left[\gamma_1^i, \gamma_2^i, \cdots \gamma_{m^i}^i\right]^T$, we can express the predicted output $y_i(k)$ given input $\mathbf{x}_k$ of training data $(\mathbf{X}_1, \mathbf{Y}_1)$ as

$$y_i(k) = \sum_{j=1}^{m^i} w_j^i \cdot \phi_j^i(k), \quad k=1,2,\cdots,N_1, i=1,2,\cdots L. \tag{5}$$

By applying all N_1 training sample to equation (5) and employing matrix representation, equation (5) can be rewritten as

$$\mathbf{w}^i = \left(\Phi_i^T \Phi_i\right)^{-1} \cdot \left(\Phi_i^T \mathbf{y}\right), \quad i=1,2\cdots L, \tag{6}$$

where Φ_i is the $N_1 \times m^i$ matrix containing the response of hidden layer identified by the *ith* particle to the N_1 training samples. The calculation of the weights and widths completes the formulation of L RBF networks, which can be represented by the triples $(\mathbf{C}^1, \gamma^1, \mathbf{w}^1), (\mathbf{C}^2, \gamma^2, \mathbf{w}^2), \cdots, (\mathbf{C}^L, \gamma^L, \mathbf{w}^L)$.

3.3.3 Fitness Value Estimation

Fitness value gives an indication how good one particle is relative to the others. For alleviating the occurrence of over-fitting phenomena, fitness value estimation is based on the prediction error criterion by introducing the validation data $(\mathbf{X}_2, \mathbf{Y}_2)$ in the training procedure. The prediction $\hat{\mathbf{Y}}_1, \hat{\mathbf{Y}}_2, \cdots, \hat{\mathbf{Y}}_L$ of the L particles formulated in the previous section and the prediction error SSE_i^{pred} are computed as follows:

$$SSE_{pred}^i = \left\|\mathbf{Y} - \hat{\mathbf{Y}}_i\right\|^2, \quad i=1,2,\cdots L. \tag{7}$$

According to fitness estimation (7), we can determine the best personal position matrix $\mathbf{C}^i_{pbest}$ of each particle and the best global position matrix $\mathbf{C}_{gbest}$.

3.3.4 Particles Updating Operations

During each epoch every particle is accelerated towards its best personal position as well as in the direction of the global best position. This is achieved by calculating a new velocity matrix for each particle based on its current velocity, the distance from its best personal position, as well as the distance from the global best position. An inertia weight ω, reduced linearly by epoch, is multiplied by the current velocity and the other two components are weighted randomly to produce the new velocity matrix for this particle, which in turn affects the next position matrix of the particle during the next epoch. In summary, the L particles interact and move according to the following equations [10]:

$$\mathbf{V}^i(t+1) = \omega \times \mathbf{V}^i(t) + c_1 \times rand(\cdot) \times \left(\mathbf{C}^i_{pbest} - \mathbf{C}^i(t)\right) + c_2 \times rand(\cdot) \times \left(\mathbf{C}_{gbest} - \mathbf{C}^i(t)\right) \tag{8}$$

$$\mathbf{C}^i(t+1) = \mathbf{C}^i(t) + \mathbf{V}^i(t+1) \tag{9}$$

where $rand(\cdot)$ is random number generator between zero and one, $0 \leq c_1, c_2 \leq 2$ and ω is an inertia weight.

For implementation of algorithm, we must tackle the structure matching problem, i.e. operands in equations (8,9) should have identical hidden units number at any epoch. In the spirit of the PSO searching mechanism, a random strategy for determining identical hidden units number is used here. It balances the tradeoff between the approximation ability and the diversity. Assumed that the effective hidden units of the individual particle, best personal and global best are m^i, m^i_{pbest}, m_{gbest} respectively. We can determine the current centers number as the following equation:

$$m = m_1 + rand(\cdot)(m_2 - m_1) \tag{10}$$

where $rand(\cdot)$ is random number generator between zero and one, $m_1 = \min\left(m^i, m^i_{pbest}, m_{gbest}\right)$, $m_2 = \max\left(m^i, m^i_{pbest}, m_{gbest}\right)$.

Once the current structure m is determined, then operands in equation (8) can do some transformation. If the rows number of the matrix is greater than m, it will collapse (i.e. replace additional rows with 'N'). If the rows number of the matrix is less than m, it will expand (i.e. replace additional rows labeled by 'N' with '0'). If the rows number of the matrix is equal to m, it keeps fixed.

Since row labeled by 'N' does not involve any algebraic operation, Equations (8,9) can be calculated after transformation. With the proceeding of optimization process, m will converge to a constant and corresponding centers of RBF network can be identified. It should be noted that updating operation not only communicates the information among the global best position, best personal position and current position, but also converges the optimal structure step by step.

3.3.5 Growing and Pruning

As mentioned in the description of the section 3.3.4, different structures of RBF networks can be determined by updating operation. For faster convergence and additional flexibility, we introduce two more operators: growing and pruning of hidden units. For simplicity, only one hidden unit is grown or pruned depended on growing probability ρ_{grow} or pruning probability ρ_{prune} at one epoch. In order to apply these operators, we generate randomly a binary value and a number r between 0 and 1 for each particle. If the binary value is 0 and $r > \rho_{grow}$, one additional unit is attached to the first row labeled by 'N'. If $r > \rho_{prune}$ and the binary value is 1, the last row unlabeled by 'N' is replaced with 'N'.

4 Simulation Results

The simulation clearly demonstrates the ability of the RBF network trained by PSO to learn the dynamics of the unknown system. The system to be identified is described by the second-order difference equation [11]:

$$y(t+1) = \frac{y(t)y(t-1)\left[y(t)+2.5\right]}{1+y^2(t)+y^2(t-1)} + u(t). \tag{11}$$

The equilibrium states of the unforced system are $(0,0)$ and $(2,2)$ on the state space. If a series-parallel identification model is used for identifying the nonlinear system, the model can be described by the equation

$$\hat{y}(t+1) = f\left(y(t), y(t-1), u(t)\right), \tag{12}$$

where f is an RBF network trained by PSO with three inputs and one output.

For comparison purposes, we developed an additional number of RBF network models using the standard training method, which is based on the k-means clustering algorithm. RBF networks trained with the standard procedure require a fixed number of hidden units, so in order to make a fair basis for comparison, networks with different structures (a.k.a. hidden units number) were developed and evaluated using the validation set. The input and output data have been collected in such a way that for an initial condition i.i.d random input signal uniformly distributed in the region of [-2,2] forces the given system.

The simulation produced 200 data points which were then separated into two sets: 100 data were assigned to the training set, 100 to the testing set. The k-means two stage training algorithm uses validation set to determine the best structure of RBF network within the set $m \in [1, 2, \cdots 50]$. The algorithm was implemented using the parameters: $L = 35$, $M = 50$, $t = 300$, $\rho_{grow} = 0.15$, $\rho_{prune} = 0.15$, $\omega = [0.4, 0.95]$ $c_1 = 2, c_2 = 2$.

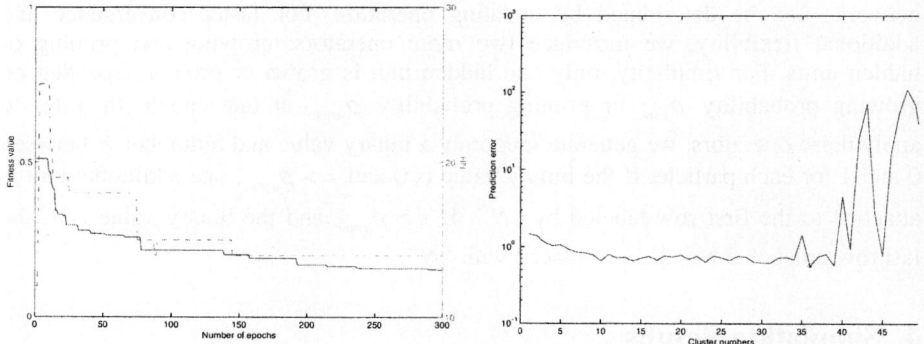

Fig. 1. The evolving process of hidden units number and fitness value corresponding to the best network structure at each epoch. Left is fitness value, right is hidden units number. Solid line: fitness value, dashed line: hidden units number

Fig. 2. The prediction error varies with clusters number using two stage training algorithm, the optimum hidden units number is labeled by star

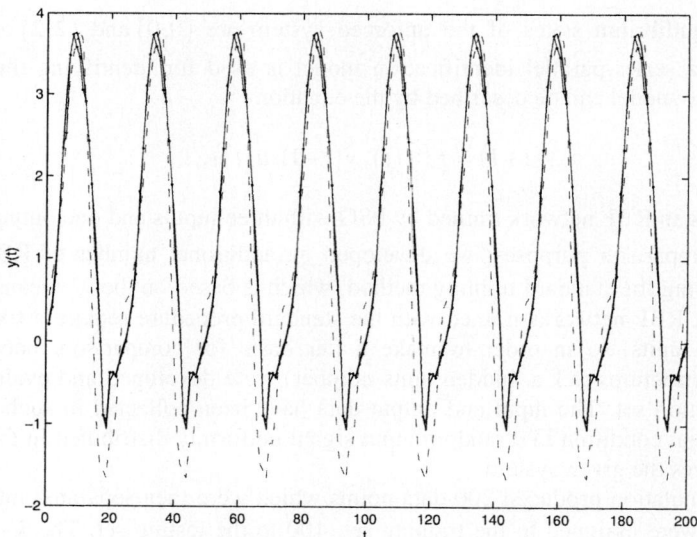

Fig. 3. The results of identification obtained from the RBF network based on PSO algorithm and k-means two stage training algorithm. Solid line: desired output, dotted line: output of RBF network based on PSO training, dashed line: output using k-means two stage training.

Table 1. Comparision of root mean squared errors (RMSE) and hidden units number of RBF network constructed by proposed algorithm and k-means two stage training algorithm

Training algorithm	Hidden units number	RMSE of testing
Proposed algorithm	14	0.11767
k-means two stage training	36	0.72916

Fig. 1 depicts the evolving process of the hidden units number and fitness value corresponding to the best network structure at each epoch. Fig. 2 shows the prediction error varies with clusters number using two stage training algorithm and the optimum hidden units number is labeled by star ($m = 36$). For testing the identified model, the sinusoidal input signal $u(t) = \sin(2\pi t / 25)$ has been applied to both the system and the model and generated 200 data. Fig. 3 shows the results of identification obtained from the RBF network based on PSO algorithm and k-means. Table 1 illustrates the performance of RBF network based on two methods. It can be seen from Table 1 and Fig.3 that the RBF network based on the proposed PSO algorithm has not only generated the most parsimonious structure but also provided the most accurate outcome.

5 Conclusions

In this paper, a novel PSO algorithm with matrix encoding is presented for training RBF network models based on input-output data. After encoding a RBF network with matrix representation, we employ a structure matching rule to update the structure of particles at any epoch. Its main difference with respect to traditional PSO has to do with the update of the position of the particle in each of its effective dimensions.

The superiority of the proposed algorithm over one of the conventional methods for training RBF networks was demonstrated through solving benchmark problems of nonlinear system identification. The results showed that the RBF networks produced by the proposed algorithm possess more parsimonious structure and achieve smaller prediction error compared with those obtained using the k-means two stage training algorithm.

References

1. Bishop, C.: Improving the Generalization Properties of Radial Basis Function Neural Networks. Neural Computation. 3(1991) 579–588
2. Chen, S., Cowan, C. F. N., Grant, P. M.: Orthogonal Least Squares Algorithm for Radial Basis Function Networks. IEEE Trans. on Neural Networks. 2 (1991) 302–309
3. Billings, S. A., Zheng, G. L.: Radial Basis Function Network Configuration using Genetic Algorithms. Neural Networks. 8 (6)(1995)877-890
4. Haralambos, S., Alex, A., Mazarakis, S., Bafas, G.: A New Algorithm for Developing Dynamic Radial Basis Function Neural Network Models based on Genetic Algorithms. Computers and Chemical Engineering. 28 (2004)209-217

5. Kennedy, J., Eberhart, R.C.: Particle Swarm Optimization. International Conference on Neural Networks, IV. (1995)1942–1948
6. Kennedy, J., Eberhart, R.C.: Swarm Intelligence. Morgan Kaufmann Publishers, San Francisco California (2001)
7. Moody, J., Darken, C. J.: Fast-learning in Networks of Locally Tuned Processing Units. Neural Computation. 1 (1989)281–294
8. Orr, M.: Regularization in the Selection of Radial Basis Function Centers. Neural Computation. 3(1995)606–623
9. Whitehead, B. A., Choate, D.: Evolving Space-Filling Curves to Distribute Radial Basis Functions over an Input Space. IEEE Trans. on Neural Networks. 5(1993)15–23
10. Y.H. Shi., Eberhart, R.C.: A Modified Particle Swarm Optimizer. IEEE International Conference on Evolutionary Computation. Anchorage, Alaska (1998)
11. Narendra, K.S., Parthasarathy, K.: Identification and Control of Dynamical Systems using Neural Networks. IEEE Trans. on Neural Networks. 1(1990)4 – 27

A Game-Theoretic Approach to Competitive Learning in Self-Organizing Maps

Joseph Herbert and JingTao Yao

Department of Computer Science,
University of Regina, Regina, Saskatchewan, Canada, S4S 0A2
{herbertj, jtyao}@cs.uregina.ca

Abstract. Self-Organizing Maps (SOM) is a powerful tool for clustering and discovering patterns in data. Competitive learning in the SOM training process focuses on finding a neuron that is most similar to that of an input vector. Since an update of a neuron only benefits part of the feature map, it can be thought of as a local optimization problem. The ability to move away from a local optimization model into a global optimization model requires the use of game theory techniques to analyze overall *quality* of the SOM. A new algorithm GTSOM is introduced to take into account cluster quality measurements and dynamically modify learning rates to ensure improved quality through successive iterations.

1 Introduction

Self-Organizing Maps (SOM), introduced by Kohonen [1], is an approach to clustering similar patterns found within data [2,3]. Used primarily to cluster attribute data for pattern recognition, SOMs offer a robust model with many configurable aspects to suit many different applications.

The training of a SOM does not take into consideration certain advantages that could be obtained if multiple measures were used in deciding which neuron to update. Recent research that makes use of dynamic adaptive and structure-adaptive techniques have been proposed [4,5]. Game theory offers techniques for formulating competition between parties that wish to reach an optimal position. By defining competitive learning in terms of finding a neuron that can perform an action that will improve not only its own position, but also the entire SOM, we may be able to improve the quality of clusters and increase the efficiency of the entire process, moving towards a global optimization process from local optimization found in traditional SOM methods.

This article proposes a new algorithm GTSOM that utilize aspects of game theory. This allows for global optimization of the feature map. This technique could be used to ensure that competitive learning results in the modification of neurons that are truly suitable for improving the training results.

2 A Brief Review of Self-Organizing Maps

At the heart of SOM theory is the concept of creating artificial neurons that are computational duplicates of biological neurons within the human brain [6]. Artificial neural networks follow the model of their biological counterparts. A SOM consists of neurons with weight vectors. Weight vectors are adjusted according to a learning rate α that is decreased over time to allow for fast, vague training in the beginning and specific, accurate training during the remaining runtime.

A SOM model contains three fundamental procedures that are required in order to discover clusters of data. These procedures are similar to that of the knowledge discovery in database process [7,8]. The first procedure consists of all preprocessing tasks that are required to be completed before training can take place. This includes initializing the weights vectors of each neuron either randomly or by some other method [9,10]. Another task to be performed is that of input vector creation. Training data for the SOM must be arranged in input vectors, where each vector represents a tuple in an information system or other similarly organized data set.

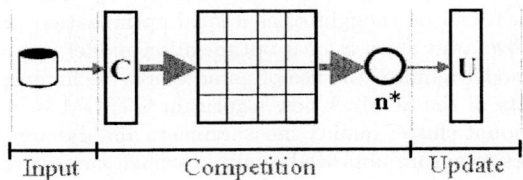

Fig. 1. The layers of a SOM during the training process

2.1 SOM Training

In order for a SOM to cluster data, it must be trained with suitable training data. Training a SOM requires the combination of three layers that work in tandem, where an output of one layer is treated as the input to the next layer, as shown in Figure 1.

The first layer, denoted as the input layer, consists of a data store to be formatted into a set of input vectors P. An input vector represents a *tuple* within the data set. Each input vector $p \in P$ is used as input for the next layer of a SOM. The second layer, denoted as the competition layer, manages the competitive learning methods within the SOM [11]. This layer determines which neuron n_i has a weight vector w_i with minimum distance (maximum similarity) to p. From this layer, a winning neuron n_i^* is marked to be updated in the third layer. The third layer, denoted as the update layer, updates the weight vector associated with the winning neuron that was used as input. After the updating of the neuron, the weight vector is more attuned to that of the input vector.

A data set P contains individual *tuples* of an information system translated into input vectors, $P = \{p_1, \ldots, p_m\}$. A set of artificial neurons, $W = \{n_1, \ldots, n_n\}$, is arranged in a grid-like topology of fixed dimensionality. Each neuron in W has a weight vector w_i of the same dimensionality as the input vectors p_j.

Data: A set of m input vectors $P = \{\boldsymbol{p}_1, \ldots, \boldsymbol{p}_m\}$
Input: A threshold q_m for maximum iterations to be executed.
Output: A feature map A'
1 **for each** neuron $n_i \in W$ **do**
2   initialize $\boldsymbol{w}_i$ randomly ;
3 **end**
4 **while** $(q \leq q_m)$ or $(\forall\ \boldsymbol{p}_k \in P, n_i^*(q) = n_i^*(q-1))$ **do**
5   α_q = adjusted α_{q-1} for iteration q ;
6   d_q = adjusted d_{q-1} for iteration q ;
7   **for each** $\boldsymbol{p}_k \in P$ **do**
8        $n_i^*(q) = \text{Compet}(\boldsymbol{p}_k, W)$;
9   Update_w$(n_i^*(q), \boldsymbol{p}_k, \alpha_q)$;
10   Update_N$(N_{n_i^*(q)}(d_q), \boldsymbol{p}_k, \alpha_q)$;
11   **end**
12 **end**

Algorithm 1: The SOM Training Method

Each neuron $n_i \in W$ has a set of neurons whose proximity is within that defined by d, a scalar whose value is changed according to an iteration q. Therefore, for each neuron n_i, the neighborhood $N_i(d) = \{n_r, \ldots, n_s\}$ consists of all neurons that have connectivity to n_i within distance d. The learning rate is used as a scalar that determines how much a weight vector $\boldsymbol{w}_i$ is changed to become more similar to that of the current input vector.

2.2 Competitive Learning in SOM

To find a neuron $n_i \in W$ that has a weight vector closest to $\boldsymbol{p}_k$, similarity measures [12] are observed between each neuron and the input vector.

Once a winning neuron n_i^* has been identified, the weight vector must be updated according to the learning rate α_q corresponding to iteration q. In addition, the neighborhood of that neuron must be updated so that neurons connected to the winner reflect continued similarity to the new information presented to the network. In Algorithm 1, this process is done with functions Update_w and Update_N, functions that update the winning neuron and its neighborhood respectively. The update of a winning neuron and the update of the winning neuron's neighborhood is shown in Equation 1 and Equation 2 respectively. Equation 1 is known as the Kohonen rule [6].

$$\boldsymbol{w}_i^*(q) = \boldsymbol{w}_i^*(q-1) + \alpha(\boldsymbol{p}_k(q) - \boldsymbol{w}_i^*(q-1)) \ . \tag{1}$$

$$\boldsymbol{w}_{N_{i^*}(d)}(q) = \boldsymbol{w}_{N_{i^*}(d)}(q-1) + \alpha\prime(\boldsymbol{p}_k(q) - \boldsymbol{w}_{N_{i^*}(d)}(q-1)) \ . \tag{2}$$

The modified learning rate $\alpha\prime$ denotes a smaller learning rate that is used on the neurons in $N_{i^*}(d)$. We wish to use a smaller learning rate to signify that although these neurons did not win the competition for the input vector, they do have some connectivity to the neuron that did. The learning rate α in Equation 1 is derived from a decreasing polynomial formula [13].

Algorithm 1 shows the steps taken to train the SOM. The process of updating a neuron and its neighbors can be thought of as a local optimization procedure. For any given input vector, the update layer in Figure 1 only adjusts neurons based on a very small instance of the overall patterns in the full data set.

3 Incorporating Game Theory into SOM Training

Although individual neurons have the ability to improve their situation during each competition, a collective goal for the entire SOM is not considered. The transition between local optimization techniques to those of global optimization must occur in order to solve problems of density mismatch and physical adjacency errors. The concept of overall SOM quality must be defined in order to progress to a state in which properties between overall neuron relationships and input vectors can be measured.

3.1 Measuring SOM Quality

The competitive layer in the traditional SOM model does not have the ability to find a neuron which best represents the current input vector as well as having the ability to improve the quality of neuron placement and density. Improving quality in a SOM could include an increased ability to create and define better clusters. In order to determine the quality of a SOM, definitions on what is considered a high-quality cluster must be discovered. Clusters can be defined in two ways: by the actual input data that was used to adjust the weight vectors or by the neurons associated with that data.

With the two abilities to define clusters, two methods of representing clusters arise. A centroid vector can be used as a representation of the cluster. This vector could be calculated by taking the average of all weight vectors that the cluster includes. Second, a neuron whose weight vector is most similar to that of the average weight vector of all neurons could be given representation status. In addition to the two methods of representing clusters in a SOM, two methods can be used in order to find a neuron required in the latter method:

1. If a centroid input vector for a cluster is known, we can simply discover which neuron that centroid input vector is most similar to.
2. If we wish for the calculations of centroid to be strictly neuron based, we can find groups of neurons and determine which of those neurons have won more competitions.

The above methods allow us to measure the overall quality of a SOM. Using the ability to calculate physical distance between clusters on the feature map as well as the ability to calculate the density of a particular cluster can enable a new algorithm to determine which neuron is best suited to be updated. These quality measures can be used together to see how much a particular neuron, if updated, can improve the overall quality of the feature map.

3.2 Game Theory

In order to facilitate global optimization techniques in competitive learning, a method must be employed that can take into consideration possible improvements of overall SOM quality. Game theory provides a suitable infrastructure to determine which neurons provide the best increase in feature map quality. By manipulating the learning rate applied to both the winning neuron and its neighbors, as well as the size of a neighborhood that should be taken into consideration, a set of strategies with expected payoffs can be calculated.

Game theory, introduced by von Neumann and Morgenstern [14], has been used successfully in many areas, including economics [15,16], networking [17], and cryptography [18,19]. Game theory offers a powerful framework for organizing neurons and to determine which neuron may provide the greatest increase in overall SOM quality.

In a simple game put into formulation, a set of players $O = \{o_1, \ldots, o_n\}$, a set of actions $S = \{a_1, \ldots, a_m\}$ for each player, and the respective payoff functions for each action $F = \{\mu_1, \ldots, \mu_m\}$ are observed from the governing rules of the game. Each player chooses actions from S to be performed according to expected payoff from F, usually some a_i maximizing payoff $\mu_i(a_i)$ while minimizing other player's payoff. A payoff table is created in order to formulate certain payoffs for player strategies, which is shown in Table 1.

3.3 Game-Theoretic Competitive Learning in SOM

With the ability to precisely define neuron clusters within a SOM, measures can be used in order to define overall quality of the network. These measures, such as the size of clusters, the distance between clusters, and the appropriate cluster size to represent input can be combined to give a certain payoff value to a particular neuron, if chosen as a winner. When the competitive phase begins, a ranking can be associated with each neuron according to its distance from the input vector. Using the ranked list of neurons, a new competition layer is constructed in order to determine which neuron and which strategy or action should be taken. This new model architecture is shown in Figure 2.

The first Competition layer is modified so that instead of determining which neuron is most similar to the current input vector, the layer now ranks neurons according to each similarity measure obtained. There is an opportunity here to include a dynamic, user-defined threshold value t_1 that can deter any neurons

Table 1. Payoff table created by second Competition layer

		$n_j^*(q)$		
		$a_{j,1}$	...	$a_{j,r}$
$n_i^*(q)$	$a_{i,1}$	$<\mu_{i,1}, \mu_{j,1}>$	...	$<\mu_{i,1}, \mu_{j,r}>$
	$\vdots$	$\vdots$	...	$\vdots$
	$a_{i,r}$	$<\mu_{i,r}, \mu_{j,1}>$	...	$<\mu_{i,r}, \mu_{j,r}>$

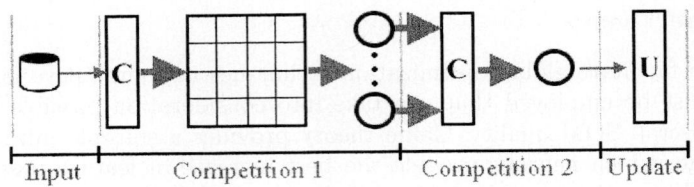

Fig. 2. The layers of GTSOM including the addition of another competition layer used during the training process

that are beyond a certain similarity measure to be included in the ranked set as shown in Equation 3 and Equation 4:

$$W' = \{n_1^*(q), \ldots, n_n^*(q)\} , \qquad (3)$$

where $\forall n_i^*(q) \in W$,

$$|w_i^*(q) - p_i| \leq t_1 , \qquad (4)$$

and $1 \leq i \leq n$. This allows the user to specify a degree of minimum similarity desired when having the first competition layer computing which neurons should enter the second competition layer.

Once a ranked set of neurons has been created, the second competition layer starts to create competition tables of the form shown in Table 1. A neuron n_i^* with possible actions $S = \{a_{i,1}, \ldots, a_{i,r}\}$ and payoffs calculated from corresponding utility functions $U = \{\mu_{i,1}, \ldots, \mu_{i,r}\}$ competes against neuron n_j^* with the same action and utility sets. The neuron whose specific action $a_{i,k}$ results in the greatest overall SOM quality is chosen to be the winner.

With the addition of quality measures, neurons are now ranked in partial order. For example, a particular neuron n_i^* may have a higher ranking than n_j^* in terms of a particular similarity measure between itself and the input vector, but the neuron may not have that same ranking when additional quality measures are taken into account. The second competition layer must take into consideration not only similarity to input, but also how much each neuron can increase or decrease feature map quality. Many different ranking of neurons in W' may occur when more than one measure is used.

There are two possible ways of creating tables to govern the second phase of competition. First, neurons can be initially paired randomly with each other. Victors of each "round" move on to the next round, where new tables are created for the neurons that have been awarded victories. This process proceeds until a total victory is declared for one neuron. Second, for a set $W = \{n_1^*(q), \ldots, n_n^*(q)\}$ of ranked neurons, an *n-dimensional* payoff table can be created. With n neurons ranked and entering competition, each with r possible actions, a total of r^n cells must be observed to determine which neuron gives the best quality or utility value for this iteration.

3.4 A Strategy to Adjust the Learning Rate

Actions performed by a particular neuron could possibly include parameters such as adjustable learning rates or adjust neighborhood size. Such actions can be called *strategies* to describe an action that can be modified in order to create new actions. A strategy of adjust the learning rate α can be modified so that there is an action for an increased adjustment, decreased adjustment, and a nochange scenario. This strategy can improve clusters by forcing subsequent input vectors that are similar to the current input to have a greater possibility to be more similar to a different neuron than it did on a previous iteration in the case of an increased learning rate. That is, the input vector will have an increased likelihood to be closer to a different neuron next iteration. A decreased learning rate will result in a diminished similarity adjustment between the victor and the current input vector, resulting in negligible change from subsequent iterations.

A set of actions detailing neighborhood size for a particular neuron is useful when cluster sizes are desired to either grow or diminish. An increased neighborhood size will modify a larger number of neurons to become more similar to the current input vector. This may result in less dense clusters if desired. In contrast, a decreased neighborhood size could have an exact opposite effect, decreasing the size and increasing the density of clusters. If clusters are too far apart, the density of a particular cluster could be dismissed so that cluster boundaries become closer. Also, if clusters are too compact, the density of some clusters could be increased in order to increase distance between centroids.

4 GTSOM Implementation

The process of ranking neurons according to similarity, creating payoff tables, and determining winning neurons is introduced in Algorithm 2. Training will stop when either of the following three conditions are met on line 4.

1. If a maximum number of specified iterations have been performed.
2. If no neurons have won competitions for new input vectors that were not won before during previous iterations.
3. If the overall quality of the SOM has reached or moved beyond that of a user-defined threshold.

A traditional SOM stops when the first two stopping conditions are met. With the addition of the third condition, training time may be reduced if a certain quality has been reached. For example, if the desired quality of the feature map has been reached before q_m iterations have been performed, training may stop ahead of schedule. This threshold may correlate with the number of iterations that are to be performed or it may represent the desired precision of weight vectors belonging to individual neurons. A lower threshold will most likely result in a lower number of iterations performed. As precision increases with respect to the number of iterations performed (smaller learning rate), a lower number

Data: A set of m input vectors $P = \{\boldsymbol{p}_1,\ldots,\boldsymbol{p}_m\}$
Input: A threshold q_m for maximum iterations to be executed.
Output: A feature map A'

1 **for each** *neuron* $n_i \in W$ **do**
2   Initialize $\boldsymbol{w}_i$ randomly ;
3 **end**
4 **while** $(q \leq q_m)$ *or* $(\forall\ \boldsymbol{p}_i \in P, n_i^*(q) = n_i^*(q-1))$ *or* $(\mu(A) \geq t_2)$ **do**
5     $\alpha_q = $ adjusted α_{q-1} for iteration q ;
6     $d_q = $ adjusted d_{q-1} for iteration q // neighborhood distance ;
7   **for each** $\boldsymbol{p}_k \in P$ **do**
8   Find set $W' = \{n_1^*(q),\ldots,n_n^*(q)\}$;
9   **for each** $< n_i^*(q), n_j^*(q) >$ *pair in* W' **do**
10             $T_{i,j} = (N, S_{i,j}, F_{i,j})$, where
11             $N = \{n_i^*(q), n_j^*(q)\}$,
12             $S_{i,j} = $set of actions for $n_i^*(q)$ and $n_j^*(q)$,
13             $F_{i,j} = $set of utility functions returning quality of A.
14             $\alpha_q = \pm a_i^*$, where $a_i^* = $the action that best improves A. ;
15   **end**
16   Choose $n_q^*(\boldsymbol{p}_i)$ whose utility function μ_i has maximum payoff action a_i ;
17   Update_w$(n_i^*(q), \boldsymbol{p}_k, \alpha_q)$ // update winning neuron ;
18   Update_N$(N_{n_i^*(q)}(d_q), \boldsymbol{p}_k, \alpha_q)$ // update neighborhood of n^* ;
19   **end**
20 **end**

Algorithm 2: The Training Method GTSOM

of iterations will result in the algorithm completing with a learning rate above that of the final desired learning rate.

Lines **7-19** iterate the first and second competition layers for every input vector in P. Line **8**, executing the first competition layer, creates a set of ranked neurons according to their similarity to the input vector. The third embedded repetitive structure ranks neurons according to their similarity to the current input vector. An interesting opportunity arises here when clusters are starting to be defined. There may be an option to include centroid neurons in this set once they have been discovered. This leads to the eventuality that no new clusters will be formed. Another user-defined threshold could be specified if this method is used, comparable to the maximum number of clusters desired. This also decreases the number of distance measures to be calculated between the neuron weight vectors and the current input vector.

The second competition layer is shown in lines **9-15**. Using the set of ranked neurons, tables are created for each neuron pair within W'. This table $T_{i,j} = (N, S_{i,j}, F_{i,j})$, the *payoff table for neurons n_i and n_j*, includes the neurons themselves, a set containing actions S_i and S_j for the neurons, and a set containing utility functions F_i and F_j that returns the quality of the feature map given action $a_i \in S_i$ and $a_j \in S_j$. Once these tables have been created, the neuron with the action that provides the greatest increase in feature map quality through the utility function is chosen as the final winner in the competition process. The action is executed (learning rate modification or neighborhood size) and update procedures are performed.

A large value for t_1 may result in increased computation time as it will result in a larger W'. Since tables are created and observed for each distinct pair of neurons within W', the similarity threshold must be considered carefully. A value too small for t_1 may result in incomplete competition, where neurons that may offer valuable actions could be ignored based on their dissimilarity to the current input vector.

The threshold t_2 found on line **4** gives the option of stopping the training process when a certain overall SOM quality has been reached. Too high of a threshold, although perhaps representing a high quality preference, may result in no computational efficiency improvement. This threshold may never be reached before maximum iterations have occurred. Too low of a threshold could result in too few iterations being performed. Since the learning rate α is adjusted during each iteration, it may not get an opportunity to become sufficiently small for precise weight vector updating.

5 Conclusion

We have proposed a new approach to competitive learning in SOMs. The opportunity to create a model to facilitate global optimization of the feature map requires methods to acquire the overall quality of the feature map. These methods take the form of measuring distance between clusters, cluster density and cluster size.

An additional competitive layer has been added to the traditional SOM model as well as modifying the original competition that results in the proposed GT-SOM algorithm. A similarity ranking within a user-defined threshold between neuron weight vectors and input vectors is used as a basis for the creation of payoff tables between neurons. Payoffs are calculated according to strategy set containing possible actions for each neuron. Each action results in a numeric utility or payoff which may improve or diminish SOM quality. Finding the neuron whose action maximizes the quality of the SOM for that iteration is now possible, enabling neurons to be picked not only on similarity but on strength. Clusters can be increased or decreased in size or density in order to attempt to reach a user-defined threshold for overall desired quality of the SOM. Future research will focus on training result analysis between the traditional SOFM training method and the proposed GTSOFM training algorithm.

References

1. Kohonen, T.: Automatic formation of topological maps of patterns in a self-organizing system. In: Proceedings of the Scandinavian Conference on Image Analysis. (1981) 214–220
2. Huntsberger, T., Ajjimarangsee, P.: Parallel self-organizing feature maps for unsupervised pattern recognition. International Journal of General Systems **16**(4) (1990) 357–372

3. Tsao, E., Lin, W., Chen, C.: Constraint satisfaction neural networks for image recognition. Pattern Recognition **26**(4) (1993) 553–567
4. Cho, S.B.: Ensemble of structure-adaptive self-organizing maps for high performance classification. Inf. Sci. **123**(1-2) (2000) 103–114
5. Hung, C., Wermter, S.: A dynamic adaptive self-organising hybrid model for text clustering. In: Proceedings of the Third IEEE International Conference on Data Mining. (2003) 75–82
6. Hagan, M.T., Demuth, H.B., Beale, M.H. In: Neural Network Design. PWS Publishing Company, Boston (1996)
7. Brachman, R.J., Anand, T.: The process of knowledge discovery in databases: A human-centered approach. In: Advances in knowledge discovery and data mining, American Association for Artificial Intelligence (1996) 37–58
8. Fayyad, U.M., Piatetsky-Shapiro, G., Smyth, P.: From data mining to knowledge discovery: an overview. In: Advances in knowledge discovery and data mining, American Association for Artificial Intelligence (1996) 1–34
9. Chandrasekaran, V., Liu, Z.Q.: Topology constraint free fuzzy gated neural networks for pattern recognition. IEEE Transactions on Neural Networks **9**(3) (1998) 483–502
10. Pal, S., Dasgupta, B., Mitra, P.: Rough self organizing map. Applied Intelligence **21**(3) (2004) 289–299
11. Fritzke, B.: Some competitive learning methods. Technical report, Institute for Neural Computation. Ruhr-Universit at Bochum (1997)
12. Santini, S., Jain, R.: Similarity measures. IEEE Transactions: Pattern Analysis and Machine Intelligence **21**(9) (1999) 871–883
13. Kolen, J.F., Pollack, J.B.: Back propagation is sensitive to initial conditions. In: Advances in Neural Information Processing Systems. Volume 3. (1991) 860–867
14. von Neumann, J., Morgenstern, O.: Theory of Games and Economic Behavior. Princeton University Press, Princeton (1944)
15. Nash, J.: The bargaining problem. Econometrica **18**(2) (1950) 155–162
16. Roth, A.: The evolution of the labor market for medical interns and residents: a case study in game theory. Political Economy **92** (1984) 991–1016
17. Bell, M.: The use of game theory to measure the vulnerability of stochastic networks. IEEE Transactions on Reliability **52**(1) (2003) 63–68
18. Fischer, J., Wright, R.N.: An application of game-theoretic techniques to cryptography. Discrete Mathematics and Theoretical Computer Science **13** (1993) 99–118
19. Gossner, O.: Repeated games played by cryptographically sophisticated players. Technical report, Catholique de Louvain - Center for Operations Research and Economics (1998)

A Novel Intrusions Detection Method Based on HMM Embedded Neural Network

Weijin Jiang[1], Yusheng Xu[2], and Yuhui Xu[1]

[1] Department of Computer, Zhuzhou Institute of Technology, Zhuzhou 412008, P.R.China
jwjnudt@163.com
[2] College of Mechanical Engineering and Applied Electronics,
Beijing University of Technology, Beijing 100022, P.R.China
yshxu520@163.com

Abstract. Due to the excellent performance of the HMM(Hidden Markov Model) in pattern recognition, it has been widely used in voice recognition, text recognition. In recent years, the HMM has also been applied to the intrusion detection. The intrusion detection method based on the HMM is more efficient than other methods. The HMM based intrusion detection method is composed by two processes: one is the HMM process; the other is the hard decision process, which is based on the profile database. Because of the dynamical behavior of system calls, the hard decision process based on the profile database cannot be efficient to detect novel intrusions. On the other hand, the profile database will consume many computer resources. For these reasons, the combined detection method was provided in this paper. The neural network is a kind of artificial intelligence tools and is combined with the HMM to make soft decision. In the implementation, radial basis function model is used, because of its simplicity and its flexibility to adapt pattern changes. With the soft decision based on the neural network, the robustness and accurate rate of detection model network, the robustness and accurate rate of detection model are greatly improved. The efficiency of this method has been evaluated by the data set originated from Hunan Technology University.

1 Introduction

IDS (Intrusion Detection System)is a system that attempts to detect intrusions, which are defined to be unauthorized uses, misuses, or abuses of computer systems by either authorized users or external perpetrators [1]. There are a lot of technologies being used as anomaly detection methods, such as the neural network [2-4],the data mining[5],the support vector machine[]3[6], and the hidden Markov model [7-10], Each kind of these technologies has shown its advantages to detect novel intrusions, but it still has some shortcomings. Therefore, the hybrid architecture is provided.

War render et al. [7] introduced a simple anomaly detection method based on monitoring the system calls used by active, privileged process. It is based on the idea that the normal trace of a privileged process has a different pattern to that of the anomaly process. This pattern can be expressed by a short sequence of system calls. In sequence time delay embedding (STIDE), a profile of normal behavior is built by enumerating all unique, contiguous sequence of a predetermined ,fixed length k that occurs in the training data. The method is efficient for sendmail, lpr, ftpd of Unix. But the normal database typically includes thousands of short sequences. The process of building a normal database is time-consuming. Y.Qiao et al. [9] introduced an anomaly intrusion detection method based on the HMM which has three advantages. First, the profile database of the HMM method is smaller than that of the STIDE method, so the HMM method can detect intrusions more quickly. Second, the HMM method can build a nearly complete database with only small parts of normal data. At the end, the mismatch rate difference between the normal process and the anomaly process of the HMM method is larger than that of the STIDE method. But their method also has insufficiency. It needs very large memory to store the database, though their method has a compressed profile database compared with that of the STIDE method.

Meanwhile, neural network was widely used to detect intrusions and achieved some good results. Debar el al.[11]used a neural network component for an intrusion detection system. Susan C.L. and David V.H. [2] used a neural-network based intrusion detector to recognize intrusion. In their paper, it describes an experiment with and IDS composed of a hierarchy of neural networks(NN) that function as a true anomaly detector. The result is achieved by monitoring selected areas of network behavior such as protocols, that are predictable in advance. The NNs are trained using data that spanned the entire normal space. These detectors are able to recognize attacks that are not specifically presented during training. It shows that using small detectors in a hierarchy gives a better result than using a single large detector.

In this paper, a hybrid architecture, which is composed by the hidden Markov model and neural network, is developed. The hybrid architecture is designed to monitor the system calls used by the active, privileged process. The hybrid architecture can greatly reduce the detection time and simplify the design of the software. The profile database, which is a key component in the Warrender et al. [7] and Y. Qiao et al. [9] method, consumes lots of the system's rare resources. In our method, the profile database will be deleted and the final detection is decided by the neural work. So the detection speed will be greatly increased and many computer resources are saved.

The rest of this paper is organized as follows. In Section II, we give a brief introduction to the hidden Markov model and neural network. The overall design composed of hidden Markov model and neural network is described in Section III. Experimental results are shown in Section IV. Section V is the conclusions.

2 Brief Introduction to the HMM and the Neural Network

2.1 HMM Theory [12]

The HMM is a very powerful statistical method of characterizing the observed data sample arranged in a discrete-time series. It has the ability to process nonlinear and time-variant systems and is viewed as a double-embedded stochastic process.

Given a form of HMM of the previous section, there are three basic problem of interest that must be solved for the model to be useful in real-world applications. These problems are the following:

Problem 1: Given the observation sequence $O=O_1O_2...O_T$, and a model $\lambda = (A, B, \pi)$, how do we efficiently compute $P(O|\lambda)$, the probability of the observation sequence, for the given model?

Problem 2: Given the observation sequence $O=O_1O_2...O_T$, and a model $\lambda = (A, B, \pi)$, how do we choose a corresponding state sequence $Q=q_1q_2...q_T$ which is optimal in some meaningful sense (i.e., best "explains" the observations)?

Problem 3: how do we adjust the model parameters $\lambda = (A, B, \pi)$ to maximize $P(O|\lambda)$.

Problem 1 is the evaluation problem, namely given a model and a sequence of observations, how do we compute the probability of which the observed sequence was produced by the model. We can also view the problem as one of scoring how well a given model matches a given observation sequence. The latter viewpoint is extremely useful.

Problem 2 is the one in which we attempt to uncover the hidden part of the model, i.e., to find the "correct" state sequence. It should be clear that for all but the case of degenerate models, there is no "correct" state sequence to be found. Hence for practical situations, we usually use an optimality criterion to solve this problem as best as possible. Unfortunately, as we still see, there are several reasonable optimality criteria that can be imposed, and hence the choice of criterion is a strong function of the intended use for the uncovered state sequence.

Problem 3 is the one in which we attempt to optimize the model parameters so as to best describe how a given observation sequence comes about. The observation sequence used to adjust the model parameters is called a training sequence since it is used to "train" the HMM. The training problem is the crucial one for most applications of the HMM, since it allows us to optimally adapt model parameters to the observed training data, i.e., to create best model for real phenomena.

To solve these problems, Forward-Backward Algorithm, Baum-Welch Algorithm and Viterbi Algorithm were developed. Forward-Backward Algorithm was developed to solve the first problem: namely, given the observation sequence $O=O_1O_2...O_T$, and a model $\lambda = (A, B, \pi)$, it was used to efficiently compute $P(O|\lambda)$, the probability of the observation sequence. Viterbi Algorithm was developed to solve the second problem: namely, given the observation sequence $O=O_1O_2...O_T$, and a model $\lambda = (A, B, \pi)$, it was used to choose a corresponding state sequence $Q=q_1q_2...q_T$ which is optimal in some meaningful sense. Baum-Welch Algorithm was developed to solve the third problem, namely, it was used to adjust the model parameters $\lambda = (A, B, \pi)$ to maximize $P(O|\lambda)$. In fact, the Baum-Welch Algorithm acted as the HMM training algorithm to maximize $P(O|\lambda)$.

2.2 Forward-Backward Algorithm

Forward-Backward Algorithm consider the forward variable $a_t(i)$ defined as

$$a_t(i) = P(O_1O_2\cdots O_T, q_t = S_i | \lambda)$$

i.e., the probability of the partial observation sequence, $O_1O_2...O_t$, (until time t) and state S_t at time t, given the model λ. We can solve for $\alpha_t(i)$ inductively, as follows:

I Initialization:

$$\alpha_1(i) = \pi_i b_i(O_1), \quad 1 \le i \le N$$

II Induction:

$$\alpha_{t+1}(j) = \left[\sum_{i=1}^{N} \alpha_t(j) a_{ij}\right] b_j(O_{t+1})$$

$$1 \le i \le N, 1 \le t \le T-1$$

III Termination

$$P(O|\lambda) = \sum_{i=1}^{N} \alpha_T(i)$$

Step I initializes the forward probabilities as the joint probability of state S_i and initial observation O_1. The induction step is the heart of the forward calculation. Step III gives the desired calculation of $P(O|\lambda)$ as the sum of the terminal forward variables $\alpha_T(j)$.

2.3 Neural Networks

The neural network consists of a collection of processing elements that are highly interconnected and transform a set of inputs to a set of desired outputs. The result of

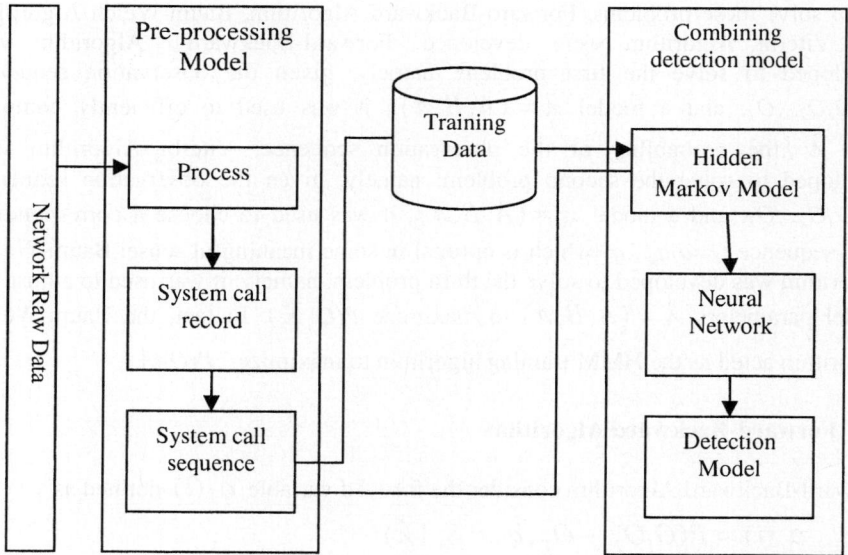

Fig. 1. The Combined detection model

the transformation is determined by the characteristics of the elements and the weights associated with the interconnections among them. By modifying the connections between the nodes, the network is able to adapt to the desired outputs[13].

A neural network conducts an analysis of the information and provides a probability estimate that the data match the characteristics that have been trained to recognize. While the probability of a match determined by a neural network can be 100%, the accuracy of its decisions relies totally on the experience the system gains in analyzing examples of the stated problem.

The neural network gains the experience initially by training the system to correctly identify reselected examples of the problem. The response of the neural network is reviewed and the configuration of the system is refined until the neural network's analysis of the training data reaches a satisfactory level. In addition to the initial training period, the neural network also gains experience over time as it conducts analyses on data related to the problem.

3 The HMM and the Neural Network Based Intrusion Detection Model

3.1 Neural Networks

The model is composed of two components: the first component is hidden Markov model; the other is neural network, as is shown in Fig.1.

For the hidden Markov model, there are two important algorithms that are key to the intrusion detection model, namely the Baum-Welch algorithm and Viterbi algorithm. The Baum- Welch algorithm is used to compute the output. We use a HMM with 21states to perform the experiment. Data for training the prototype is generated by monitoring the system calls used by active, privileged process.

In the neural network, the number of hidden layers, and the number of nodes in the hidden layers, was determined based on the process of trial and error. In our experiment, 5-layer and feed-forward BP neural network are used. The first layer, second layer and third layer apply a Tausig transfer function to the various connection weights. And the fourth layer and fifth layer apply a Sigmoid transfer function ($1/(1 + \exp(-x))$) to the various connection weights. The neural network is designed to provide an output value of 0.0 in the output layer when the analysis indicated no attack and 1.0 in the output layer in the event of an attack. Data for training and testing the prototype is generated by the HMM.

3.2 Intrusion Detection Method

Every program implicitly specified a set of system call sequences that it can produce. These sequences are determined by the ordering of system calls in the set of the possible execution paths. In this model, the system calls are monitoring by a program designed by the Hunan technology University and the system call sequences are recorded.

First, part of the system call sequences are used to training the HMM. To train the HMM, the Baum-Welch Algorithm is used.

Baum-Welch Algorithm[12]

Here we present the Baum-Welch re-estimation formulas:

$$\overline{\pi}_i = r_t(i), 1 \leq i \leq N$$

$$\overline{a}_{ij} = \sum_{t=1}^{T-1} \xi_t(i,j) / \sum_{t=1}^{T-1} r_t(i)$$

$$\overline{b}_j(k) = \sum_{\substack{t=1 \\ O_t=k}}^{T} r_t(j) / \sum_{t=1}^{T} r_t(j)$$

If we denote the initial model by λ and re-estimation model by $\overline{\lambda}$ consisting of the parameters estimated above, then it can be shown that either:

$$P(o \mid \overline{\lambda}) > P(o \mid \lambda)$$

When the HMM has been trained, all the system call sequences are input to the HMM. Through the Viterbi Algorithm, the HMM will output sequences that have n different symbols. N is the number of the HMM states and it is greatly less then the number of symbols in the original sequences. Those sequences with n different symbols can express the intrinsic difference between normal action and intrusion behavior more tersely and stably than the original sequences[9].

Viterbi Algorithm[11]:

Viterbi Algorithm is famous algorithm to find I that will maximize $P(O, I/\lambda)$. It is inductive algorithm in which at each instant you keep the best (i.e. the one giving maximum probability) possible state sequence for each of the N states as the intermediate state for the desired observation sequence $O=O_1O_2...O_T$. In this way you finally have the best path for each of the N states as the last state for the desired observation sequence. Out of these, we select one that has highest probability.

$$P(O, I/\lambda) = P(O/I, \lambda)P(I/\lambda) = \pi_{i_1} b_{i_1}(O_1) a_{i_1 i_2} b_{i_2}(O_2) \cdots a_{i_{T-1} i_T} b_{i_T}(O_T)$$

Now we define

$$U(i_1, i_2, \ldots, i_T) = -\left[In(\pi_{i_1} b_{i_1}(O_1)) + \sum_{t=2}^{T} In(a_{i_{t-1} i_t} b_{i_t}(O_t)) \right]$$

Then it is easily seen that

$$P(O, I/\lambda) = exp(-U(i_1, i_2, \cdots, i_T))$$

Consequently the problem of optimal state estimation, namely,

$$\max_{\{i_t\}_{t=1}^T} p(o, i_1, i_2, \cdots, i_T / \lambda)$$

becomes equivalent to

$$\min_{\{i_t\}_{t=1}^T} U(i_1, i_2, \cdots, i_T)$$

Now the Viterbi Algorithm can be used a dynamic programming approach for minimizing $U(i_1,i_2,...,i_T)$. So the Viterbi Algorithm has four steps:

i. Initialization

For $1 \leq i \leq N$

$$\delta_1(i) = -In(\pi_i) - In(b_i(O_1)), \quad \varphi_1(i) = 0$$

ii. Recursive computation

For $1 \leq i \leq T$ for $1 \leq i \leq N$

$$\delta_t(j) = \min_{1 \leq i \leq N} \lfloor \delta_{t-1}(i) - In(a_{ij}) \rfloor - In(b_j(O_t))$$

$$\varphi_t(j) = \arg \min_{1 \leq i \leq N} \lfloor \delta_{t-1}(i) - In(a_{ij}) \rfloor$$

iii. Termination

$$P^* = \min_{1 \leq i \leq N}[\delta_t(i)]$$

$$q_T^* = \arg \min_{1 \leq i \leq N}[\delta_t(i)]$$

iv. Trace back the optimal state sequence

For $t=T-1, T-2,...,1$

$$q_t^* = \varphi_{t+1}(q_{t+1}^*)$$

So the

$Q^* = \{q_1^*, q_2^*, \cdots q_t^*\}$ is the optimal state sequence.

After that, we must determine whether the sequence is normal or not. Due to the excellent classification ability, the neural network is used to classify the normal sequences and abnormal sequences. The neural network has 51 nodes in the input layer and 1 node in the output layer. The neural network's input is from the HMM's output, but the sequences generated by the HMM are different from each other in length. So we use a sliding window of length n (the HMM's states) with sliding (shift) step of 5 to create the neural network's input sequences. A long sequence is cut into a set of short sequences with fixed length n .Then we can get the output symbols(0 or 1). To determine whether the sequence is normal or not, we must compute the number of 0 symbol and symbol respectively. If the percentage of the 0 symbol exceeds the threshold we put forward in advance, we may draw a conclusion that the sequence is a normal sequence. Otherwise, if the percentage of 1 symbol exceeds the threshold, we consider that sequence is an intrusion sequence.

4 Experiment and Results

4.1 Experiment Setup

Our data set comes from the Hunan technology University and it was considered as the benchmark data set widely used in anomaly detection research. Each trace is the list of system calls issued by a single process from the beginning of its execution to the end .Each file lists pairs of numbers, one pair per line. The first number in a pair is the PID of the executing process, and the second is a number representing the system call. The mapping between system calls and actual system call names is given in a separate file. In our experiment, mail-sending system call is used as the normal data set and syslog-local-1, syslog-local-1, syslog-remote-1,syslocal-remote-2 act as the intrusion data.

The experiment follows these steps:

First, we use 60 percent of all the data(include normal and abnormal data)to train the HMM.

Second, the data generated by the HMM are used to train the neural network.

The third, the other40 percent of the data are used to test the model.

Finally, we must compute the output of the neural network for each sequence. Namely, we must compute the percentage of 0 and 1 for each sequence. If the percentage of 1 exceeds the threshold, we should mark the sequence as an intrusion action. On the other hand, if the percentage of 0 exceeds the threshold, we should consider the sequence as a normal action.

4.2 Experiment Results

Through the experiment, we can see that our intrusion detection method is useful in detecting novel intrusions.

Table 1. Detection results of normal Process

Traces	Percent age of 1	Percent age of 0	Threshold	Nor mal rate
Boun ce	0.2125	0.7875	0.6	0.99
Boun ce1	0.2864	0.7136	0.6	0.95
Boun ce2	0.3001	0.6999	0.6	0.96
Send mail	0.3521	0.6479	0.6	0.95
Queu e	0.2994	0.7006	0.6	0.97
Plus	0.3102	0.6898	0.6	0.94

Table 2. Detection results of abnormal Process

Traces	Percent age of 1	Percent age of 0	Threshold	Abnormal rate
syslog-local-1	0.7107	0.2893	0.6	0.94
syslog-local-2	0.6993	0.3007	0.6	0.91
syslog-remote-1	0.7220	0.278	0.6	0.9
syslog-remote-2	0.7225	0.2775	0.6	0.93

The table 1 is the detection results of normal process and the table 2 is the detection result of the abnormal process. Table1 shows that the combined detection model can make detection very accurately. When the threshold is 60 per cent, the normal rates are more than 90 per cent and the abnormal rates are less than 10 percent. Table 2 shows that our detection method is efficient to detect the anomaly process. When the threshold is 60 per cent, the detection rates of the abnormal process are more than 90 per cent. While the error rates are less than 10 per cent.

Table 3. Comparing detection of anomalies

Traces	Bo Gao el al.	Our results
syslog-local-1	45.8%	71.07%
syslog-local-2	44.6%	69.93%
syslog-remote-1	53.5%	72.20%
syslog-remote-2	54.4%	72.25%

Through the experiment, we find that detection rate will be different when the threshold is different. For the normal process, if the threshold is higher, the positive error(positive error is that the normal process is branded anomaly process) rate will be higher. But, if the threshold is too lower, the negative error (negative error is that the abnormal process is branded normal process) rate will be increased. For the anomaly process, the error rate will be different with the threshold also. If the threshold is higher, the negative error rate will be increased. Otherwise, if the threshold is lower, the positive error rate will be increased. So the choice of the threshold is very important to the intrusion detection. For different processes, the different thresholds can be utilized to intrusion detection and more accurate detection rate can be achieved.

Table 3 shows the comparison of anomaly detection rate among three methods. The results of Forrest and Bo Gao come from reference[14]. From Table 3, we can see that our method greatly increases the anomaly process detection rate[15,16].

5 Conclusions

A combined intrusion detection method based on the HMM and the neural network is proposed in this paper. The experiment results showed that our method is efficiently to classify the anomaly profile from the normal profile. Comparing with other methods based on the HMM only, our method has following advantages. First, it needs less storage without the profile database. With the processes being used by more and more users, the profile database will be greatly enlarged. So the profile database will occupy much storage with the larger and larger and larger profile database. Second, the detection speed will be faster than the other HMM based methods. When the profile database is very large, the detection speed will be slower as the sequence must be compared with all the records in the profile database. In our method, if the HMM and the neural network have been trained, the detection speed only relates with the neural network and it is constant.

References

1. B.Mukherjee, L.T. Heberlein, and K.N. Levitt: "Network intrusion detection," IEEE Network, May/June, 8(3) ,(1994): 26-41
2. C.L. Susan, V.H. DAVID, "Training a neural –network based intrusion detector to recognize novel attacks", IEEE Transactions on systems, man and cybernetics-part a: System and Humans,31(4) , (2001):294-299

3. S. Mukkamala, G .Janoski, A. Sung, "Intrusion detection using neural networks and support vector machines", IEEE IJCNN (2002)
4. H. Debar, M. Becke, D. Siboni: "A neural network component for an intrusion detection system", In Proceedings of the IEEE Computer Society Symposium on Research in Security and Privacy, (1992)
5. Manganaris: "A data mining analysis of RTID alarms[J]",Computer Networks, 34(4) , (2000): 571-577
6. Q. Tran, Q.L. Zhang, X. Li: "SVM classification-based intrusion detection system", Journal of China Institute of Communications,vol23 no.5(2002)
7. C. Warrender, S. Forrest, B. Pealmutt: "Detecting intrusion using system calls: alternative data mode". IEEE Symposium on Security and Privacy, (1999) 133-145
8. S.B.Cho, H.J. Park, "Efficient anomaly detection by modeling privilege flows using hidden Markov model", Computers and Security, 22(1) , (2003):45-55
9. Y. Qiao, X. W. Xin , Y.Bin and S.Ge: "Anomaly intrusion detection method based on HMM", Electronics Letters, 20th,Vol.38,no.13 (2002)
10. X.Q.Zhang, P.Z.Fan, Z.L.Zhu. "A new anomaly detection method based on hierarchical HMM". PDCAT2003,China, (2003), 249-252
11. H.Debar, M. Becker, D.Siboni: "A neural network component for an intrusion detection system", Research in Security and Privacy, 1992,Proceedings,1992 IEEE Computer Society Symposium, (1992), 240-250
12. R. Dugad and U.B. Desai: "A tutorial on hidden Markov models", Technical Report No.:SPANN-96.1(1996)
13. Fox, L.Kevin, Henning, R.Rhonda, and H.Jonathan: "A Neural network approach Towards Intrusion Detection", Proceeding of the 13th National Computer Security Conference(2000)
14. B. Gao, H.Y.Ma, Y.H.Yang: "HMMS based on anomaly intrusion detection method", Proceedings of the First International Conference on Machine Learning and Cybernetics, Beijing, (2002)
15. Weijin Jiang: Hybird Genetic algorithm research and its application in problem optimization. Proceedings of 2004 International Conference on Manchine Learning and Cybernetics, (2004)222-227
16. Weijin Jiang: Research on Optimize Prediction Model and Algorithm about Chaotic Time Series, Wuhan University Journal of Natural Sciences,9(5) ,(2004): 735-740

Generate Different Neural Networks by Negative Correlation Learning

Yong Liu

School of Computer Science, The University of Aizu,
Aizu-Wakamatsu, Fukushima 965-8580, Japan
yliu@u-aizu.ac.jp

Abstract. This paper describes two methods on how to generate different neural networks in an ensemble. One is based on negative correlation learning. The other is based on cross-validation with negative correlation learning, i.e., bagging with negative correlation learning. In negative correlation learning, all individual networks are trained simultaneously on the same training set. In bagging with negative correlation learning, different individual networks are trained on the different sampled data set with replacement from the training set. The performance and correct response sets are compared between two learning methods. The purpose of this paper is to find how to design more effective neural network ensembles.

1 Introduction

The idea of designing an ensemble learning system consisting of many subsystems can be traced back to as early as 1958. Since the early 1990's, algorithms based on similar ideas have been developed in many different but related forms, such as neural network ensembles [1,2], mixtures of experts [3,4,5,6], various boosting and bagging methods [7,8,9], and many others. It is essential to find different neural networks in an ensemble because there is no improvement by combing the same neural networks. There are a number of methods of finding different neural networks including independent training, sequential training, and simultaneous training.

A number of methods have been proposed to train a set of neural networks independently by varying initial random weights, the architectures, the learning algorithm used, and the data [1,10]. Experimental results have showed that networks obtained from a given network architecture for different initial random weights often correctly recognize different subsets of a given test set [1,10]. As argued in [1], because each network makes generalisation errors on different subsets of the input space, the collective decision produced by the ensemble is less likely to be in error than the decision made by any of the individual networks.

Most independent training methods emphasised independence among individual neural networks in an ensemble. One of the disadvantages of such a method is the loss of interaction among the individual networks during learning. There is no consideration of whether what one individual learns has already been learned by other individuals. The errors of independently trained neural networks may still be positively correlated. It has been found that the combining results are weakened if the errors of individual networks are positively correlated [11]. In order to decorrelate the individual neural networks, sequential training methods train a set of networks in a particular order [9,12,13]. Drucker et al. [9] suggested training the neural networks using the boosting algorithm. The boosting algorithm was originally proposed by Schapire [8]. Schapire proved that it is theoretically possible to convert a weak learning algorithm that performs only slightly better than random guessing into one that achieves arbitrary accuracy. The proof presented by Schapire [8] is constructive. The construction uses filtering to modify the distribution of examples in such a way as to force the weak learning algorithm to focus on the harder-to-learn parts of the distribution.

Most of the independent training methods and sequential training methods follow a two-stage design process: first generating individual networks, and then combining them. The possible interactions among the individual networks cannot be exploited until the integration stage. There is no feedback from the integration stage to the individual network design stage. It is possible that some of the independently designed networks do not make much contribution to the integrated system. In order to use the feedback from the integration, simultaneous training methods train a set of networks together. Negative correlation learning [14,15,16] is an example of simultaneous training methods. The idea of negative correlation learning is to encourage different individual networks in the ensemble to learn different parts or aspects of the training data, so that the ensemble can better learn the entire training data. In negative correlation learning, the individual networks are trained simultaneously rather than independently or sequentially. This provides an opportunity for the individual networks to interact with each other and to specialise.

In this paper, two methods are described on how to generate different neural networks in an ensemble. One is based on negative correlation learning. The other is based on cross-validation with negative correlation learning, i.e., bagging with negative correlation learning. In negative correlation learning, all individual networks are trained simultaneously on the same training set. In bagging with negative correlation learning, different individual networks are trained on the different sampled data set with replacement from the training set. The performance and correct response sets are compared between two learning methods. The purpose of this paper is to find how to design more effective neural network ensembles.

The rest of this paper is organised as follows: Section 2 describes negative correlation learning; Section 3 explains how to introduce negative correlation learning into cross-validation so that the bagging predictors would not be independently trained but trained simultaneously; Section 4 discusses how negative correlation learning generates different neural networks on a pattern classification problem; and finally Section 5 concludes with a summary of the paper and a few remarks.

2 Negative Correlation Learning

Given the training data set $D = \{(\mathbf{x}(1), y(1)), \cdots, (\mathbf{x}(N), y(N))\}$, we consider estimating y by forming an neural network ensemble whose output is a simple averaging of outputs F_i of a set of neural networks. All the individual networks in the ensemble are trained on the same training data set D

$$F(n) = \frac{1}{M} \Sigma_{i=1}^{M} F_i(n) \qquad (1)$$

where $F_i(n)$ is the output of individual network i on the nth training pattern $\mathbf{x}(n)$, $F(n)$ is the output of the neural network ensemble on the nth training pattern, and M is the number of individual networks in the neural network ensemble.

The idea of negative correlation learning is to introduce a correlation penalty term into the error function of each individual network so that the individual network can be trained simultaneously and interactively. The error function E_i for individual i on the training data set $D = \{(\mathbf{x}(1), y(1)), \cdots, (\mathbf{x}(N), y(N))\}$ in negative correlation learning is defined by

$$\begin{aligned} E_i &= \frac{1}{N} \Sigma_{n=1}^{N} E_i(n) \\ &= \frac{1}{N} \Sigma_{n=1}^{N} \left[\frac{1}{2}(F_i(n) - y(n))^2 + \lambda p_i(n) \right] \end{aligned} \qquad (2)$$

where N is the number of training patterns, $E_i(n)$ is the value of the error function of network i at presentation of the nth training pattern, and $y(n)$ is the desired output of the nth training pattern. The first term in the right side of Eq.(2) is the mean-squared error of individual network i. The second term p_i is a correlation penalty function. The purpose of minimising p_i is to negatively correlate each individual's error with errors for the rest of the ensemble. The parameter λ is used to adjust the strength of the penalty.

The penalty function p_i has the form

$$pave_i(n) = -\frac{1}{2}(F_i(n) - F(n))^2 \qquad (3)$$

The partial derivative of E_i with respect to the output of individual i on the nth training pattern is

$$\begin{aligned} \frac{\partial E_i(n)}{\partial F_i(n)} &= F_i(n) - y(n) - \lambda(F_i(n) - F(n)) \\ &= (1-\lambda)(F_i(n) - y(n)) + \lambda(F(n) - y(n)) \end{aligned} \qquad (4)$$

where we have made use of the assumption that the output of ensemble $F(n)$ has constant value with respect to $F_i(n)$. The value of parameter λ lies inside the range $0 \leq \lambda \leq 1$ so that both $(1-\lambda)$ and λ have nonnegative values. BP [17] algorithm has been used for weight adjustments in the mode of pattern-by-pattern updating. That is, weight updating of all the individual networks is performed simultaneously using Eq.(4) after the presentation of each training pattern. One complete presentation of the entire training set during the learning process is called an *epoch*. Negative correlation learning from Eq.(4) is a simple extension to the standard BP algorithm. In fact, the only modification that is needed is to calculate an extra term of the form $\lambda(F_i(n) - F(n))$ for the ith neural network.

From Eq. (4), we may make the following observations. During the training process, all the individual networks interact with each other through their penalty terms in the error functions. Each network F_i minimizes not only the difference between $F_i(n)$ and $y(n)$, but also the difference between $F(n)$ and $y(n)$. That is, negative correlation learning considers errors what all other neural networks have learned while training an neural network.

For $\lambda = 1$, from Eq.(4) we get

$$\frac{\partial E_i(n)}{\partial F_i(n)} = F(n) - y(n) \tag{5}$$

Note that the error of the ensemble for the nth training pattern is defined by

$$E_{ensemble} = \frac{1}{2}(\frac{1}{M}\Sigma_{i=1}^{M}F_i(n) - y(n))^2 \tag{6}$$

The partial derivative of $E_{ensemble}$ with respect to F_i on the nth training pattern is

$$\frac{\partial E_{ensemble}}{\partial F_i(n)} = \frac{1}{M}(\frac{1}{M}\Sigma_{i=1}^{M}F_i(n) - y(n))$$
$$= \frac{1}{M}(F(n) - y(n)) \tag{7}$$

In this case, we get

$$\frac{\partial E_i(n)}{\partial F_i(n)} \propto \frac{\partial E_{ensemble}}{\partial F_i(n)} \tag{8}$$

The minimisation of the error function of the ensemble is achieved by minimising the error functions of the individual networks. From this point of view, negative correlation learning provides a novel way to decompose the learning task of the ensemble into a number of subtasks for different individual networks.

3 Cross-Validation with Negative Correlation Learning

Cross-validation is a method of estimating prediction error. Cross-validation can be used to create a set of networks. Split the data into m roughly equal-sized parts, and train each network on the different parts independently. When the data set is small and noisy, such independence will help to reduce the correlation among the m networks more drastically than in the case where each network is trained on the full data.

When a larger set of independent networks are needed, splitting the training data into non-overlapping parts may cause each data part to be too small to train each network if no more data are available. In this case, data reuse methods, such as bootstrap, can help. Bootstrap was introduced as a computer-based method for estimating the standard error of a statistic $s(x)$. B bootstrap samples are generated from the original data set. Each bootstrap sample has n elements, generated by sampling with replacement n times from the original data set. Bootstrap replicates $s(x^{*1}), s(x^{*2}), \ldots, s(x^{*B})$ are obtained by calculating the value of the statistic $s(x)$ on each bootstrap sample. Finally, the standard deviation of the values $s(x^{*1}), s(x^{*2}), \ldots, s(x^{*B})$ is the estimate of the standard error of $s(x)$. The idea of bootstrap has been used in bagging predictors. In bagging predictors, a training set containing N patterns is perturbed by sampling with replacement N times from the training set. The perturbed data set may contain repeats. This procedure can be repeated several times to create a number of different, although overlapping, data sets.

One of the disadvantages of bagging predictors is the loss of interaction among the individual networks during learning. There is no consideration of whether what one individual learns has already been learned by other individuals. The errors of independently trained neural networks may still be positively correlated. It has been found that the combining results are weakened if the errors of individual networks are positively correlated. In order to decorrelate the individual neural networks, each individual neural network can be trained by negative correlation learning in bagging. In the origianl negative correlation learning, each neural network is trained on the same training set. In bagging by negative correlation learning, each neural network is trained on the different sampled data with replacement from the training set.

4 Experimental Studies

This section describes the application of negative correlation learning to the Australian credit card assessment problem. The problem is to assess applications for credit cards based on a number of attributes. There are 690 patterns in total. The output has two classes. The 14 attributes include 6 numeric values and 8 discrete ones, the latter having from 2 to 14 possible values. The Australian credit card assessment problem is a classification problem which is different from the regression type of tasks, such as the chlorophyll-a prediction problem, whose outputs are continuous. The data set was obtained from the UCI machine learning benchmark repository. It is available by anonymous ftp at ics.uci.edu (128.195.1.1) in directory /pub/machine-learning-databases.

Experimental Setup. The data set was partitioned into two sets: a training set and a testing set. The first 518 examples were used for the training set, and the remaining 172 examples for the testing set. The input attributes were rescaled to between 0.0 and 1.0 by a linear function. The output attributes of all the problems were encoded using a 1-of-m output representation for m classes. The output with the highest activation designated the class.

Table 1. Comparison of error rates among negative correlation learning (NCL) and bagging with NCL on the Australian credit card assessment problem. The results were averaged over 25 runs. "Simple Averaging" and "Winner-Takes-All" indicate two different combination methods used in negative correlation learning. *Mean, SD, Min* and *Max* indicate the mean value, standard deviation, minimum and maximum value, respectively.

		Simple Averaging		Winner-Takes-All	
	Error Rate	Training	Test	Training	Test
NCL	Mean	0.0679	0.1323	0.1220	0.1293
	SD	0.0078	0.0072	0.0312	0.0099
	Min	0.0463	0.1163	0.0946	0.1105
	Max	0.0772	0.1454	0.1448	0.1512
Bagging with NCL	Mean	0.0458	0.1346	0.0469	0.1372
	SD	0.0046	0.0111	0.0243	0.0104
	Min	0.0367	0.1163	0.0348	0.1105
	Max	0.0579	0.1570	0.0541	0.1628

The ensemble architecture used in the experiments has four networks. Each individual network is a feedforward network with one hidden layer. All the individual networks have ten hidden nodes.

Experimental Results. Table 1 shows the average results of negative correlation learning and bagging with negative correlation learning over 25 runs. Each run of the experiments was from different initial weights. The simple averaging was first applied to decide the output of the ensemble system. For the simple averaging, the results of bagging with negative correlation learning were slightly worse than those of negative correlation learning.

In simple averaging, all the individual networks have the same combination weights and are treated equally. However, not all the networks are equally important. Because different individual networks created by negative correlation learning were able to specialise to different parts of the testing set, only the outputs of these specialists should be considered to make the final decision of the ensemble for this part of the testing set. In this experiment, a winner-takes-all method was applied to select such networks. For each pattern of the testing set, the output of the ensemble was only decided by the network whose output had the highest activation. Table 1 shows the average results of negative correlation learning and bagging with negative correlation learning over 25 runs using the winner-takes-all combination method. The winner-takes-all combination method improved negative correlation learning because there were good and poor networks for each pattern in the testing set and winner-takes-all selected the best one. However it did not improved bagging with negative correlation learning.

In order to see how different neural networks generated by negative correlation learning are, we compared the outputs of the individual networks trained by negative correlation learning and bagging with negative correlation learning. Two notions were introduced to analyse negative correlation learning. They are

Table 2. The sizes of the correct response sets of individual networks created respectively by negative correlation learning (NCL) and bagging with NCL on the testing set and the sizes of their intersections for the Australian credit card assessment problem. The results were obtained from the first run among the 25 runs.

NCL	$\Omega_1 = 147$	$\Omega_2 = 150$	$\Omega_3 = 138$	$\Omega_4 = 142$	$\Omega_{12} = 142$
	$\Omega_{13} = 126$	$\Omega_{14} = 136$	$\Omega_{23} = 125$	$\Omega_{24} = 136$	$\Omega_{34} = 123$
	$\Omega_{123} = 121$	$\Omega_{124} = 134$	$\Omega_{134} = 118$	$\Omega_{234} = 118$	$\Omega_{1234} = 116$
Bagging with NCL	$\Omega_1 = 150$	$\Omega_2 = 145$	$\Omega_3 = 137$	$\Omega_4 = 143$	$\Omega_{12} = 140$
	$\Omega_{13} = 132$	$\Omega_{14} = 138$	$\Omega_{23} = 127$	$\Omega_{24} = 132$	$\Omega_{34} = 128$
	$\Omega_{123} = 125$	$\Omega_{124} = 128$	$\Omega_{134} = 125$	$\Omega_{234} = 120$	$\Omega_{1234} = 118$

the correct response sets of individual networks and their intersections. The correct response set S_i of individual network i on the testing set consists of all the patterns in the testing set which are classified correctly by the individual network i. Let Ω_i denote the size of set S_i, and $\Omega_{i_1 i_2 \cdots i_k}$ denote the size of set $S_{i_1} \cap S_{i_2} \cap \cdots \cap S_{i_k}$. Table 2 shows the sizes of the correct response sets of individual networks and their intersections on the testing set, where the individual networks were respectively created by negative correlation learning and bagging with negative correlation training. It is evident from Table 2 that different individual networks created by negative correlation learning were able to specialise to different parts of the testing set. For instance, in negative correlation learning with pave, in Table 2 the sizes of both correct response sets S_1 and S_3 were 147 and 138, but the size of their intersection $S_1 \cap S_3$ was 126. The size of $S_1 \cap S_2 \cap S_3 \cap S_4$ was only 116. In comparison, bagging with negative correlation learning can create rather different neural networks as well.

5 Conclusions

This paper describes negative correlation learning and bagging with negative correlation learning for generating different neural networks in an ensemble. Negative correlation learning can be regarded as one way of decomposing a large problem into smaller and specialised ones, so that each subproblem can be dealt with by an individual neural network relatively easily. Bagging with negative correlation learning were proposed to encourage the formation of different neural networks.

The experimental results on a classification task show that both negative correlation learning and bagging with negative correlation learning tend to generate different neural networks. However, bagging with negative correlation learning failed in achieveing the expected good generalisation. More study is needed on how to make bagging more efficient by negative correlation learning.

References

1. L. K. Hansen and P. Salamon. Neural network ensembles. *IEEE Trans. on Pattern Analysis and Machine Intelligence*, 12(10):993–1001, 1990.

2. A. J. C. Sharkey. On combining artificial neural nets. *Connection Science*, 8:299–313, 1996.
3. R. A. Jacobs, M. I. Jordan, S. J. Nowlan, and G. E. Hinton. Adaptive mixtures of local experts. *Neural Computation*, 3:79–87, 1991.
4. R. A. Jacobs and M. I. Jordan. A competitive modular connectionist architecture. In R. P. Lippmann, J. E. Moody, and D. S. Touretzky, editors, *Advances in Neural Information Processing Systems 3*, pages 767–773. Morgan Kaufmann, San Mateo, CA, 1991.
5. R. A. Jacobs, M. I. Jordan, and A. G. Barto. Task decomposition through competition in a modular connectionist architecture: the what and where vision task. *Cognitive Science*, 15:219–250, 1991.
6. R. A. Jacobs. Bias/variance analyses of mixture-of-experts architectures. *Neural Computation*, 9:369–383, 1997.
7. H. Drucker, C. Cortes, L. D. Jackel, Y. LeCun, and V. Vapnik. Boosting and other ensemble methods. *Neural Computation*, 6:1289–1301, 1994.
8. R. E. Schapire. The strength of weak learnability. *Machine Learning*, 5:197–227, 1990.
9. H. Drucker, R. Schapire, and P. Simard. Improving performance in neural networks using a boosting algorithm. In S. J. Hanson, J. D. Cowan, and C. L. Giles, editors, *Advances in Neural Information Processing Systems 5*, pages 42–49. Morgan Kaufmann, San Mateo, CA, 1993.
10. D. Sarkar. Randomness in generalization ability: a source to improve it. *IEEE Trans. on Neural Networks*, 7(3):676–685, 1996.
11. R. T. Clemen and R. .L Winkler. Limits for the precision and value of information from dependent sources. *Operations Research*, 33:427–442, 1985.
12. D. W. Opitz and J. W. Shavlik. Actively searching for an effective neural network ensemble. *Connection Science*, 8:337–353, 1996.
13. B. E. Rosen. Ensemble learning using decorrelated neural networks. *Connection Science*, 8:373–383, 1996.
14. Y. Liu and X. Yao. Negatively correlated neural networks can produce best ensembles. *Australian Journal of Intelligent Information Processing Systems*, 4:176–185, 1998.
15. Y. Liu and X. Yao. A cooperative ensemble learning system. In *Proc. of the 1998 IEEE International Joint Conference on Neural Networks (IJCNN'98)*, pages 2202–2207. IEEE Press, Piscataway, NJ, USA, 1998.
16. Y. Liu and X. Yao. Simultaneous training of negatively correlated neural networks in an ensemble. *IEEE Trans. on Systems, Man, and Cybernetics, Part B: Cybernetics*, 29(6):716–725, 1999.
17. D. E. Rumelhart, G. E. Hinton, and R. J. Williams. Learning internal representations by error propagation. In D. E. Rumelhart and J. L. McClelland, editors, *Parallel Distributed Processing: Explorations in the Microstructures of Cognition, Vol. I*, pages 318–362. MIT Press, Cambridge, MA, 1986.

New Training Method and Optimal Structure of Backpropagation Networks

Songyot Sureerattanan[1] and Nidapan Sureerattanan[2]

[1] Information Technology Department, Bank of Thailand,
273 Samsen Rd., Pranakorn, Bangkok 10200 Thailand
songyots@bot.or.th
[2] Department of Computer Education, Faculty of Technical Education,
King Mongkut's Institute of Technology North Bangkok,
1518 Piboolsongkram Rd., Bangsue, Bangkok 10800 Thailand
nidapan@gmail.com

Abstract. New algorithm was devised to speed up the convergence of backpropagation networks and the Bayesian Information Criterion was presented to obtain the optimal network structure. Nonlinear neural network problem can be partitioned into the nonlinear part in the weights of the hidden layers and the linear part in the weights of the output layer. We proposed the algorithm for speeding up the convergence by employing the conjugate gradient method for the nonlinear part and the Kalman filter algorithm for the linear part. From simulation experiments with daily data on the stock prices in the Thai market, it was found that the algorithm and the Bayesian Information Criterion could perform satisfactorily.

1 Introduction

Backpropagation (BP) method, discovered at different times by Werbose [1], Parker [2], and Rumelhart et al. [3], is *a supervised learning technique* for training multilayer neural networks. The gradient descent (steepest descent) method is used to train BP networks by adjusting the weights in order to minimize the system error between the known output given by user (actual output) and the output from the network (model output). To train a BP network, each input pattern is presented to the network and propagated forward layer by layer, starting from the input layer until the model output is computed. An error is then determined by comparing the actual output with the model output. The error signals are used to readjust the weights in the backward direction starting from the output layer and backtracking layer by layer until the input layer is reached. This process is repeated for all training patterns until the system error converges to a minimum.

Although the BP method is widely and successfully used in many applications [4, 5, 6], there have been several problems encountered. One is its slow convergence, with which many iterations are required to train even a simple network [7]. Another problem is how to determine the appropriate network structure for a particular problem. Generally, *trial-and-error* is used to determine the structure of a network in practice.

Normally, the number of nodes in the input and output layers depend upon the application under consideration. Moreover, one hidden layer suffices for many applications [8, 9]. Therefore, for the appropriate network structure, the remaining problem is how to obtain the number of hidden nodes.

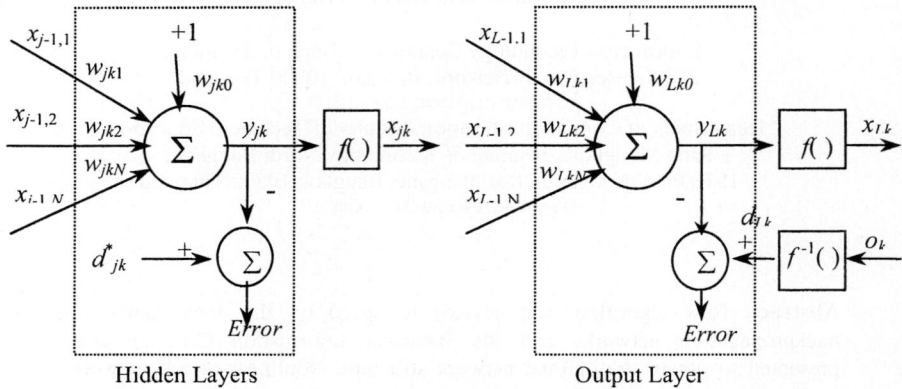

Fig. 1. Linear portions of a neuron in the hidden and output layers (in dotted blocks) [10]

In this paper, we present the algorithm for speeding up the convergence rate of BP networks and the method for choosing the optimal network structure. Experimental results of our simulation studies are given to assess the performance of these methods.

In the following section, we present the algorithm for speeding up the convergence. In Section 3, the method to determine the optimal network structure based upon the Bayesian Information Criterion (BIC) is described. Experimental results are presented in Section 4 for demonstrating the capability of the algorithms and the BIC. Finally, conclusions are given in Section 5.

2 Proposed Algorithm for Speeding Up the Convergence

Even though the original nonlinear problem is reduced to a linear problem that can readily be solved using Kalman filter (KF) (See Fig. 1), the KF algorithm still uses error signals generated by the BP algorithm to estimate the actual pre-image outputs of the hidden layers [10]. Since they are only known at the output layer, the actual pre-image outputs are estimated in the same way as in the BP algorithm at the hidden layers.

By partitioning the nonlinear neural network problem into the nonlinear part in the weights of the hidden layers and the linear part in the weights of the output layer, we propose a new algorithm obtained by combining the conjugate gradient method and the KF algorithm. The conjugate gradient method, which represents a major improvement over steepest descent with only *a marginal increase in computational effort* [11], is employed to solve the nonlinear part and the KF algorithm is employed to solve the linear part.

The system error (overall patterns) between the actual and model pre-image outputs at the output layer is given as:

$$E = \frac{1}{2}\sum_{p=1}^{M}\sum_{k=1}^{N_L}(d_{pk} - y_{pLk})^2 \tag{1}$$

where d_{pk} and y_{pLk} are the actual and model pre-image outputs for the kth node in the output layer L at the pth training pattern, respectively, M is the total number of training patterns (number of data points) and N_L is the number of nodes in the output layer. Substituting the model pre-image output at the output layer, Eq. 1 becomes

$$E = \frac{1}{2}\sum_{p=1}^{M}\sum_{k=1}^{N_L}\left(d_{pk} - \sum_{i=0}^{N_{L-1}} w_{Lki} f\left(\sum_{l=0}^{N_{L-2}} w_{L-1,il} x_{p,L-2,l}\right)\right)^2 \tag{2}$$

where $x_{p,L-2,l}$ is the model output for the lth node in layer $L-2$ at the pth training pattern. Equation 2 is substituted with $x_{p,L-2,l}$ until the input layer. It is noted that the model pre-image output at the output layer is linear in the weights of the output layer, but still nonlinear in the weights of the hidden layers.

The conjugate gradient method is employed to solve the nonlinear problem in the weights of the hidden layers and then the KF algorithm is employed to find the weights of the output layer. Minimizing the system error E with respect to the weights for the kth node in the output layer results in

$$\frac{\partial E}{\partial w_{Lki}} = 0$$

for $i = 0$ through N_{L-1}.

As

$$\frac{\partial E}{\partial w_{Lki}} = \frac{\partial E}{\partial y_{pLk}} \frac{\partial y_{pLk}}{\partial w_{Lki}},$$

we have

$$\frac{\partial E}{\partial w_{Lki}} = -\sum_{p=1}^{M}(d_{pk} - y_{pLk}) x_{p,L-1,i} = 0 \tag{3}$$

Equation 3 can be rewritten as

$$\sum_{p=1}^{M} d_{pk} x_{p,L-1,i} = \sum_{p=1}^{M} y_{pLk} x_{p,L-1,i}$$

Substituting the model pre-image output at the output layer by its expression gives:

$$\sum_{p=1}^{M} d_{pk} x_{p,L-1,i} = \sum_{p=1}^{M}\sum_{r=0}^{N_{L-1}} w_{Lkr} x_{p,L-1,r} x_{p,L-1,i}$$

or

$$\sum_{p=1}^{M} d_{pk} x_{p,L-1,i} = \sum_{p=1}^{M} x_{p,L-1,i} \sum_{r=0}^{N_{L-1}} w_{Lkr} x_{p,L-1,r} \tag{4}$$

Changing the summation on the right-hand side to a vector in Eq. 4, we have

$$\sum_{p=1}^{M} d_{pk} x_{p,L-1,i} = \sum_{p=1}^{M} x_{p,L-1,i} x_{p,L-1}^T w_{Lk} \tag{5}$$

for $i = 0$ through N_{L-1}.

Defining

$$R = \sum_{p=1}^{M} x_{p,L-1} x_{p,L-1}^T$$

and

$$p = \sum_{p=1}^{M} d_{pk} x_{p,L-1}$$

Then Eq. 5 becomes $\quad p = Rw_{Lk}$

or $\quad\quad\quad\quad\quad\quad w_{Lk} = R^{-1}p$

This results in the proposed algorithm, which can be summarized as follows:

1. Randomize all weights and biases as well as set the initial value to the inverse matrix R^{-1}, where R is the correlation matrix of the model outputs in the last hidden layer.
2. For each training pattern pair (x_{po}, o_p) where x_{po} is the input vector and o_p is the actual output vector at the pth training pattern:

(a) Calculate the model pre-image output y_{pjk} and the model output x_{pjk} starting with layer j from 1 and proceeding layer by layer toward the output layer L for every node k. In this case, the sigmoid function is selected as an activation function:

$$y_{pjk} = \sum_{i=0}^{N_{j-1}} x_{p,j-1,i} w_{jki}$$

$$x_{pjk} = f(y_{pjk}) = \frac{1}{1+\exp(-\rho y_{pjk})}$$

where N_j is the number of nodes in the jth layer and ρ is the sigmoid slope.

(b) Calculate the error signals for the weights at the output layer L and backtracking layer by layer from L-1 through 1:

$$e_{pLk} = f'(y_{pLk})(o_{pk} - x_{pLk}) = x_{pLk}(1 - x_{pLk})(o_{pk} - x_{pLk})$$

$$e_{pjk} = f'(y_{pjk})\sum_i e_{p,j+1,i} w_{j+1,i,k} = x_{pjk}(1 - x_{pjk})\sum_i e_{p,j+1,i} w_{j+1,i,k}$$

(c) Calculate the gradient vector for each layer j from 1 through L-1:

$$\nabla E_p(w^t_{jk}) = \frac{\partial E}{\partial w^t_{jk}} = -e_{pjk} x_{p,j-1}$$

where t denotes the present iteration number.

3. Calculate the gradient vector of all training patterns for each layer j from 1 through L-1: $\quad \nabla E(w^t_{jk}) = \sum_{p=1}^{M} \nabla E_p(w^t_{jk})$

where M is the total number of training patterns.

4. Calculate the search direction for each layer j from 1 through L-1:

$$s^t_{jk} = -\nabla E(w^t_{jk})$$

if t is the first iteration or an integral multiple of the dimension of w; otherwise

$$s^t_{jk} = -\nabla E(w^t_{jk}) + \beta^t s^{t-1}_{jk}$$

where β^t is computed using a form of Fletcher-Reeves [12], Polak-Ribiere [13], or Hestenes-Stiefel [14].

5. Calculate the learning rate (step size) λ^t determined by an approximate line search to minimize the error function $E(w^t + \lambda^t s^t)$ along the search direction s^t at the tth iteration.

6. Update the weight vector for each hidden layer j from 1 through L-1:

$$w^{t+1}_{jk} = w^t_{jk} + \lambda^t s^t_{jk}$$

7. For each training pattern:

(a) Calculate the model pre-image output y_{pjk} and the model output x_{pjk} starting with the layer j from 1 through the output layer L.
(b) Calculate the Kalman gain k_{pL} and update the inverse matrix R_{pL}^{-1} for the output layer L:

$$k_{pL} = \frac{R_{pL}^{-1} x_{p,L-1}}{b_L + x_{p,L-1}^T R_{pL}^{-1} x_{p,L-1}}$$

$$R_{pL}^{-1} = \left[R_{pL}^{-1} - k_{pL} x_{p,L-1}^T R_{pL}^{-1} \right] b_L^{-1}$$

where b_L is *the forgetting factor* of the output layer.
(c) Calculate the actual pre-image output at the output layer:

$$d_{pk} = f^{-1}(o_{pk}) = \frac{1}{\rho} \ln\left(\frac{o_{pk}}{1 - o_{pk}} \right)$$

(d) Update the weight vector at the output layer L:

$$w_{Lk}^{t+1} = w_{Lk}^t + k_{pL}(d_{pk} - y_{pLk})\lambda_L$$

where λ_L is the learning rate of the output layer.
8. Repeat steps 2-7 until the system error has reached an acceptable criterion.

3 Proposed Method for Optimal Network Structure

Generally, the number of nodes in the input and output layers depend upon the application under consideration. In most applications, BP network with one hidden layer is used [8, 9]. Thus, the important and difficult problem is how to choose the number of hidden nodes. The optimal number of hidden nodes is usually determined by *trial-and-error*, which starts with choosing an architecture of the network based on experience and tests the network performance after each training phase. This process is continued as long as the network performance increases and stopped whenever the network performance begins to decrease.

Basically, network complexity measures are useful both to assess the relative contributions of different models and to decide when to terminate the network training. The performance measure should balance the complexity of the model with the number of training data and the reduction in the mean squared error (MSE) [15].

Since different numbers of parameters may be involved [16], a straight MSE cannot be used to compare two different models directly. Instead of the MSE, Akaike Information Criterion (AIC) [17] and Bayesian Information Criterion (BIC) [18, 19] can be employed to choose the best among candidate models having different numbers of parameters. While the MSE is expected to progressively improve as more parameters are added to the model, the AIC and BIC penalize the model for having more parameters and therefore tend to result in smaller models. Both criteria can be used to assess the overall network performance, as they balance modelling error against network complexity. The AIC, proposed by Akaike [17], has been extensively used. This criterion incorporates the parsimony criterion suggested by Box and Jenkins [20] to use a model with as few parameters as possible by penalizing the model for having a large number of parameters. The simplified and most commonly used form of the AIC is as follows:

$$\text{AIC} = M \ln(\text{MSE}) + 2P \qquad (6)$$

where M is the number of data points used to train the network, MSE is the mean squared error, and P is the number of parameters involved in the model. In Eq. 6, the first term is a measure of fit and the second term is a penalty term to prevent overfitting. When there are several competing models to choose from, select the one that gives the minimum value of the AIC.

Even if it is commonly used, when viewed as an estimator of the model order, the AIC has been found to be inconsistent [21]. Another model selection criterion, known as the Bayesian Information Criterion (BIC) or the posterior possibility criterion (PPC), was developed independently by Kashyap [18] and Schwarz [19]. The BIC can be expressed as follows:

$$\text{BIC} = M \ln(\text{MSE}) + P \ln(M)$$

The BIC also expresses parsimony but penalizes more heavily than the AIC models having a large number of parameters. As for the AIC, one selects the model that minimizes the BIC. It is known that the BIC gives a consistent decision rule for selecting the true model. As the BIC is more consistent [21], we propose a new method to systematically determine the optimal number of hidden nodes using a procedure that gradually increases the network complexity and employs the BIC for terminating the training phase. The proposed algorithm can be summarized as follows:

1. Create an initial network with one hidden node and randomize the weights.
2. Train the network using with a chosen method e.g. the original BP algorithm, or the proposed algorithm described in Section 2 until the system error has reached an acceptable error criterion. A simple stopping rule is introduced to indicate the convergence of the algorithm. It is based upon the relative error of the sum of squared errors (SE):
$$\left| \frac{\text{SE}(t+1) - \text{SE}(t)}{\text{SE}(t)} \right| \leq \varepsilon_1$$

   where ε_1 is a constant that indicates the acceptable level of the algorithm and $\text{SE}(t)$ denotes the value of SE at iteration t.
3. Check for terminating the training of the network. A termination criterion is suggested based on the relative BIC:
$$\left| \frac{\text{BIC}(k+1) - \text{BIC}(k)}{\text{BIC}(k)} \right| \leq \varepsilon_2$$

   where ε_2 is a constant that indicates the acceptable level for the structure of the network and k denotes the number of hidden nodes. If the relative BIC is less than or equal to ε_2 or the current BIC is greater than the previous, go to step 4; otherwise add a hidden node and randomize the weights then go to step 2.
4. Reject the current network model and replace it by the previous one, then terminate the training phase.

4 Experiment

4.1 Data Employed

The stock market is an important institution serving as a channel that transforms savings into real capital formation. It will stimulate economic growth and also increases the gross national product (GNP). In this study, daily data on the stock prices and volumes in the Thai market from 1993 to 1996 were used. For the gap from Friday to Monday (weekend) and holidays when the stock exchange is closed, the data are treated as being consecutive. Three different types of common stocks; namely, Bangkok Bank Public Company Limited (BBL) in the banking sector, Shin Corporations Public Company Limited (SHIN) in the communication sector, and Land and Houses Public Company Limited (LH) in the property development sector, were selected.

The data were obtained from the Stock Exchange of Thailand (SET). In each case, the data are divided into a calibration part for training and validation part for testing: 1993 to 1994 and 1995 to 1996, respectively. Before being presented to the network, the data are transformed by a linear (affine) transformation to the range [0.05, 0.95]. In this study, the input to the network may consist of the past values of stock price (P) and stock volume (V). The stock price at time $t+1$ is treated as a function of past values of stock price at times t, t-1, and t-2 and stock volume at times t, t-1 and t-2 as follows:

$$P(t+1) = \varphi(P(t), P(t-1), P(t-2), V(t), V(t-1), V(t-2))$$

where φ stands for "function of".

4.2 Experimental Conditions

To compare the performance of the algorithms, the same initial weights were used. During the training process, both the learning rate and temperature learning rate constants were set to 0.01 to avoid oscillation of the search path. The momentum and temperature momentum constants were chosen to be 0.5 to smooth out the descent path. The forgetting factor of 0.99 was found suitable. The temperature of each neuron was set at random to lie within a narrow range of [0.9, 1.1]. The sigmoid slope was set to 1. An architecture of the 6-1-1 network consisting of 6 input nodes, 1 hidden node, and 1 output node was selected as the initial network. We employed the proposed algorithm as described in Section 3 for training the network to demonstrate the determination of the number of hidden nodes. The values adopted for ε_1 and ε_2 were 0.0001 and 0.01, respectively. The conjugate gradient method employed the approximate line search method with backtracking by quadratic and cubic interpolations of Dennis and Schnabel [22] to find the optimal step size. For calculating the search direction, the formula of Fletcher-Reeves [12] was used, based on preliminary experiments.

4.3 Performance Criterion

For measuring the performance of a given model, we employ the efficiency index (EI) defined by Nash and Sutcliffee [23]:

$$EI = \left(ST - \sum_{i=1}^{M} \left(y_i - \hat{y}_i \right)^2 \right) / ST$$

$$ST = \sum_{i=1}^{M} \left(y_i - (1/M) \sum_{i=1}^{M} y_i \right)^2$$

where ST = Total variation,

y_i = Actual output, i.e. observed value at time i,
$\hat{y}_i$ = Model output, i.e. forecast value at time i,
M = Number of data points.

4.4 Results

As mentioned in Section 3, we selected the algorithm described in Section 2 for training the network by using the BIC to obtain the optimal structure. The algorithm is terminated when the relative BIC is less than or equal to ε_2, or the current BIC is greater than the previous one. The algorithm is stopped with the structure 6-4-1 (6 input nodes, 4 hidden nodes, and one output node). Thus the 6-3-1 network is the best as shown in Table 1 for all data sets.

Earlier, Sureerattanan and Phien [24] proposed an algorithm (referred to as Algorithm 1) to speed up the convergence of BP networks by applying the adaptive neural model with the temperature momentum term to the KF algorithm with the momentum term. With the optimal structure obtained from the BIC method, we compare between BP, KF, CG (conjugate gradient), Algorithm 1, and the proposed algorithm (referred to as Algorithm 2). Figures 2-4 show the learning curve between these system error and the iteration numbers for the algorithms during training of BBL, SHIN, and LH, respectively. The calculated results of the efficiency index of each algorithm are provided in Tables 2 and 3 for training and testing phases, respectively. The total computation time of the algorithms for their convergence is given in Table 4. It should be noted that the KF algorithm, the CG method (except in the case of applying to BBL), Algorithms 1 and 2 converge with small value of the system error, but Algorithm 2 required the least computation time when convergence is achieved.

5 Conclusions

New algorithm to speed up the convergence and a method to determine the optimal network structure were presented. The proposed training algorithm can improve its convergence speed since the nonlinear problem in the weights of all layers are reduced to be nonlinear part in the hidden layers and linear part in the output layer. As we know, solving linear problem is less time consume than that of nonlinear problem and Kalman filter technique is employed to solve the linear problem. Moreover, the algorithm still solves the left nonlinear problem by applying the conjugate gradient (CG) method. The potential of CG method can overwhelmingly overcome the gradient descent method, used in the original BP algorithm, with a marginal increase in computation time [11]. From the above reasons, the training algorithm for solving convergence rate is quite effective in the data employed. The experimental results show that Algorithm 1 and the proposed algorithm (Algorithm 2) can greatly speed up the convergence of BP networks. In fact, they are the fastest algorithms among the methods considered, with the proposed algorithm being the best of all. Furthermore, the Bayesian information criterion (BIC) can be employed to determine the optimal network structure, and the best structure network also gives good performance.

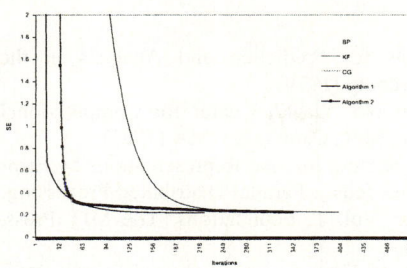

Fig. 2. Learning curve of BBL for BP, KF, CG, Algorithms 1 and 2

Fig. 3. Learning curve of SHIN for BP, KF, CG, Algorithms 1 and 2

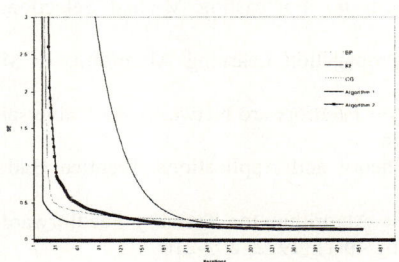

Fig. 4. Learning curve of LH for BP, KF, CG, Algorithms 1 and 2

Table 1. Computed values of BIC

Stock company	6-1-1	6-2-1	**6-3-1**	6-4-1
BBL	-3310.96	-3431.71	**-3593.22**	-3579.97
SHIN	-3057.91	-3227.42	**-3332.04**	-3240.40
LH	-3459.62	-3669.86	**-3863.04**	-3726.92

Table 2. Comparison between BP, KF, CG, Algorithms 1 and 2 for training phase

Stock company	BP			KF			CG			Algorithm 1			**Algorithm 2**		
	SE	EI	epoch	SE	EI	epoch	SE	EI	epoch	SE	EI	epoch	SE	EI	epoch
BBL	0.35	0.98	3253	0.22	0.99	402	5.15	0.72	19	0.21	0.99	224	**0.22**	**0.99**	**341**
SHIN	0.83	0.94	3251	0.60	0.96	459	0.38	0.97	1822	0.56	0.96	278	**0.38**	**0.97**	**293**
LH	0.32	0.98	3254	0.18	0.99	416	0.16	0.99	1406	0.19	0.99	342	**0.13**	**0.99**	**455**

Table 3. Efficiency indices of BP, KF, CG, and Algorithms 1 and 2 for testing phase

Stock company	BP	KF	CG	Algorithm 1	**Algorithm 2**
BBL	0.80	0.93	0.27	0.93	**0.93**
SHIN	0.95	0.94	0.98	0.96	**0.98**
LH	0.94	0.96	0.95	0.96	**0.98**

Table 4. Total computation time (in seconds) for BP, KF, CG, Algorithms 1 and 2

Stock company	BP	KF	CG	Algorithm 1	**Algorithm 2**
BBL	251	173	3*	100	**94**
SHIN	250	217	220	138	**77**
LH	251	188	170	152	**126**

---

* In this case, the value of SE is slightly high and the algorithm gets struck to local minimum.

References

1. Werbos, P.J.: Beyond Regression: New Tools for Prediction and Analysis in the Behavioral Sciences. Ph.D. Thesis. Harvard University (1974)
2. Parker, D.B.: Learning-Logic. Technical Report No. TR-47, Center for Computational Research in Economics and Management Science, MIT, Cambridge, MA (1985)
3. Rumelhart, D.E., Hinton, G.E., Williams, R.J.: Learning Internal Representations by Error Propagation. In: Rumelhert, D.E., McClelland, J.L. (eds.): Parallel Distributed Processing: Exploration in the Microstructure of Cognition: Vol. 1: Foundations. The MIT Press, Cambridge, Massachusetts (1986) 318-362
4. Gershenfield, N.A., Weigend, A.S.: The Future of Time Series. Technical Report, Palo Alto Research Center (1993)
5. Phien, H.N., Siang, J.J.: Forecasting Monthly Flows of the Mekong River using Back Propagation. In: Proc. of the IASTED Int. Conf. (1993) 17-20
6. Chu, C.H., Widjaja, D.: Neural Network System for Forecasting Method Selection. Decision Support Systems **12** (1994) 13-24
7. Sarkar, D.: Methods to speed up Error Back-Propagation Learning Algorithm. ACM Computing Surveys **27**(4) (1995) 519-542
8. Hornik, K., Stinchcombe, M., White, H.: Multilayer Feedforward Networks are Universal Approximators. Neural Networks **2** (1989) 359-366
9. Patterson, D.W.: Artificial Neural Networks: Theory and Applications. Prentice Hall, Singapore (1996)
10. Scalero, R.S., Tepedelenlioglu, N.: A Fast New Algorithm for Training Feedforward Neural Networks. IEEE Trans. on Signal Processing **40**(1) (1992) 202-210
11. Edgar, T.F., Himmelblau, D.H.: Optimization of Chemical Processes. McGraw-Hill, New York (1988)
12. Fletcher, R., Reeves, C.M.: Function Minimization by Conjugate Gradients. Computer J. **7** (1964) 149-154
13. Polak, E., Ribiere, G.: Note sur la Convergence de Methods de Directions Conjugres. Revue Francaise Informat Recherche Operationnelle **16** (1969) 35-43
14. Hestense, M.R., Stierel, E.: Methods of Conjugate Gradients for Solving Linear Systems. J. Res. Nat. Bur. Standards Sec. B **48** (1952) 409-436
15. Pottmann, M., Seborg, D.E.: Identification of Nonlinear Processes using Reciprocal Multiquadratic Functions. J. Proc. Cont. **2**(4) (1992) 189-203
16. Brown, M., Harris, C.: Neurofuzzy Adaptive Modelling and Control. Prentice Hall, UK (1994)
17. Akaike, H.: A New Look at the Statistical Model Identification. IEEE Trans. on Automatic Control AC-**19**(6) (1974) 716-723
18. Kashyap, R.L.: A Bayesian Comparison of Different Classes of Dynamic Models using Empirical Data. IEEE Trans. on Automatic Control AC-**22**(5) (1977) 715-727
19. Schwarz, G.: Estimating the Dimension of a Model. The Annals of Statistics **6**(2) (1978) 461-464
20. Box, G.E.P., Jenkins, G.M.: Time Series Analysis Forecasting and Control, Revised Edition. Holden-Day, San Francisco (1976)
21. Kashyap, R.L.: Inconsistency of the AIC Rule for Estimating the Order of Autoregressive Models. Technical Report, Dep. of Electr. Eng., Purdue University, Lafayette (1980)
22. Dennis, J.E., Schnabel, R.B.: Numerical Methods for Unconstrained Optimization and Nonlinear Equations. Prentice-Hall, New Jersey (1983)
23. Nash, J.E. and Sutcliffee, J.V.: River Flow Forecasting through Conceptual Models. J. Hydrology **10** (1970) 282-290
24. Sureerattanan, S., Phien, H.N.: Speeding up the Convergence of Backpropagation Networks. Proc. IEEE Asia-Pacific Conf. on Circuits and Systems: Microelectronics and Integration Systems, Chiangmai, Thailand (1998) 651-654

Learning Outliers to Refine a Corpus for Chinese Webpage Categorization

Dingsheng Luo[*], Xinhao Wang, Xihong Wu[*], and Huisheng Chi

National Laboratory on Machine Perception, School of Electronics
Engineering & Computer Science, Peking University, Beijing, 100871, China
{DsLuo, Wangxh, Wxh}@cis.pku.edu.cn, Chi@pku.edu.cn

Abstract. Webpage categorization has turned out to be an important topic in recent years. In a webpage, text is usually the main content, so that *auto text categorization* (ATC) becomes the key technique to such a task. For Chinese text categorization as well as Chinese webpage categorization, one of the basic and urgent problems is the construction of a good benchmark corpus. In this study, a machine learning approach is presented to refine a corpus for Chinese webpage categorization, where the AdaBoost algorithm is adopted to identify outliers in the corpus. The standard *k nearest neighbor* (kNN) algorithm under a *vector space model* (VSM) is adopted to construct a webpage categorization system. Simulation results as well as manual investigation of the identified outliers reveal that the presented method works well.

1 Introduction

Webpage categorization, which involves assigning one or more predefined categories to a free webpage according to its content, has turned out to be one of the very important and basic components in web information management, such as web mining, web information retrieval, topic identification, and so on. Most webpages are text oriented. Thus, *auto text categorization* (ATC) becomes the main technique used for webpage categorization, which is also called text-based webpage categorization [1]. ATC has been studied for several years, and a number of efficient machine learning approaches have been proposed, such as Bayesian classifiers [2], nearest neighbor classifiers [3], decision trees [2], rule learning [4], *support vector machines* (SVM)[5], ensemble learning methods [6], neural networks [7], and so on. However, for *Chinese* webpage categorization as well as *Chinese* ATC, even though some studies have been performed [8], because of the unique properties and difficulties of the Chinese language, there still exist a lot of problems. One of the basic and key problems is that a good benchmark corpus is still unavailable. A more refined corpus for the research of Chinese webpage categorization is urgently needed.

One of the main difficulties is the existence of the outliers, which are patterns that are either mislabeled in the training data, or are inherently ambiguous and hard to recognize [9]. It is already known that boosting, a typical ensemble learning method

---

[*] Corresponding authors.

proposed by Schapire [10], is a good method for identifying outliers. In this study, it has been adopted to deal with learning outliers. The basic idea of this study is learning the outliers in the original corpus at the first step and then eliminating those identified outliers to build a refined corpus.

The *k nearest neighbor* (kNN) algorithm is a typical memory-based learning methodology, where past experiences are explicitly stored in a large memory for prediction. Thus, to make an evaluation for this study, a kNN-based webpage categorization system is a desirable selection for comparing the performance between the original corpus and the refined corpus. Since some of the training samples have been eliminated in the refined corpus, the past experience will be reduced, and then the learning model trained based on the kNN algorithm could lead to a worse performance if those past experiences are truly correct prior knowledge. However, the results demonstrate that the learning model trained on the refined corpus outperforms the learning model trained on the original corpus.

The reminder of the paper is organized as follows. In section 2, the boosting-based outlier learning process is presented. In section 3, the system description will be introduced. In section 4, simulations as well as analyses are given. Section 5 is the conclusion.

2 Learning Outliers via Boosting

2.1 Outlier Problem

In machine learning, incomplete data is a big problem. There are many possibilities that can cause the training data to be incomplete, such as mislabeling, biases, omissions, non-sufficiency, imbalance, noise, outliers, etc. This paper mainly tackles the outlier problem. An outlier is a pattern that was either mislabeled in the training data, or inherently ambiguous and hard to recognize. In the course of collecting training data, two circumstances can occur, one is the absence of information that may truly represent the pattern, while the other is the presence of additional information that may not be relevant to the patterns to be recognized. The former addresses the problem of signal collection, feature extraction or feature selection, etc., while the latter deals with noise and outlier problems. Fig. 1 illustrates an outlier x_A in sample space X, where two categories of patterns labeled as "*" and "+" respectively are classified by hypersurface $h(x)$.

2.2 Ensemble Learning Methodologies and Boosting

In recent years, statistical ensemble methodologies, which take advantage of capabilities of individual classifiers via some combining strategy, have turned out to be an effective way to improve the accuracy of a learning system. In general, an ensemble learning system contains two parts: an ensemble of classifiers and a combiner. The key issue is how to build an ensemble based on the original training set. Usually, some re-sampling or re-weighting technique is adopted to produce several new training data sets through which the classifiers are trained to make up an ensemble.

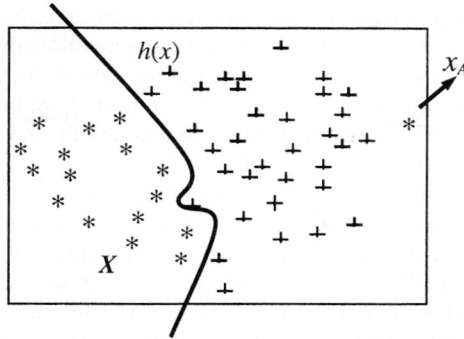

Fig. 1. A demonstration on outlier. Where X is the sample space, $h(x)$ is the hypersurface, x_A is an outlier

One of the standard ensemble learning methods is boosting, proposed by Schipire in 1990, which represents a family of algorithms [10]. The main idea of boosting lies in paying more attention to hard samples in the process of ensemble construction. The AdaBoost algorithm, introduced in 1995 by Freund and Schapire [11], solved many of the practical difficulties of the earlier boosting algorithms and has become a more popular boosting algorithm. In AdaBoost, *hard* samples are paid more attention by receiving larger weights. The basic idea of AdaBoost is as follows. The given training set is used to first learn a classifier, where the weight assigned to each training sample is the same. Suppose a classifier is obtained and then re-weighting is performed, that is, larger weights are assigned to those *hard* samples that are incorrectly predicted and smaller weights to those *easy* samples that are correctly classified. Lastly, by training with the re-weighted samples, a new classifier can be obtained. Repeating the above re-weighting procedure T-1 times produces a T-sized AdaBoost ensemble. Using some combination strategy, one can then build an AdaBoost ensemble learning system.

2.3 Learning Outliers via Boosting

Compared with other methods, the outlier problem is more serious in ensemble learning systems [11][12], since the main idea of most ensemble methods is to pay attention repeatedly to the hard samples in the training phase, while hard is the inherent property of the outliers. Dietterich demonstrated very convincingly that when the number of outliers becomes very large, the emphasis placed on the hard examples could become detrimental to the performance of AdaBoost [12]. To restrain the harmful influence of outliers, Friedman et al suggested a variant of AdaBoost called "Gentle AdaBoost" that puts less emphasis on outliers [13]. Rätsch et al gave a regularization method for AdaBoost to handle outliers and noisy data [14]. A so-called "BrownBoost" algorithm was proposed by Freund that took a more radical approach to de-emphasize outliers when they were "too hard" to be classified correctly [15]. To conquer the limitation of the sample-based weighting strategy, which was adopted in those outlier-solving methods mentioned above, a unit-weighting strategy was proposed in our previous studies [9].

170   D. Luo et al.

To tackle the outlier problem in the incomplete training data, two methods may be used, i.e. "restrain" and "eliminate". Those studies mentioned in the above paragraph mainly focus on a "restrain" strategy. In this corpus refining study, however, our viewpoint in outlier-solving is not to "restrain" but only "eliminate". In spite of what kind of strategy is used for tackling the outlier problem, the first step is to *find* the outliers in the incomplete training samples.

An AdaBoost-based outlier learning procedure is presented as illustrated in Fig. 2., where four subfigures describe a boosting learning process in sequence on a binary classification problem. Fig. 2(a) expresses a sample space where two categories of patterns are labeled as "*" and "+" respectively. After several iterations, the sample space will be divided by a decision boundary of the combined classifier shown as the solid black curve in Fig. 2(b), where those samples that are misclassified or near the decision boundary are comparatively hard samples and hence get a higher weight, such as patterns P_1, P_4 and N_4, etc., where P and N stand for positive and negative respectively. The diameters of the circles that have the corresponding samples as their centers are proportional to the weights of those patterns. In order to simplify the problem, only those relevant samples' weights (which are mainly "*"-labeled class patterns) are illustrated in Fig. 2. As the boosting ensemble-construction process continues, the learning system reaches the state shown in Fig. 2(c), where those samples such as P_1 and P_2 that are misclassified by previous ensemble classifiers have

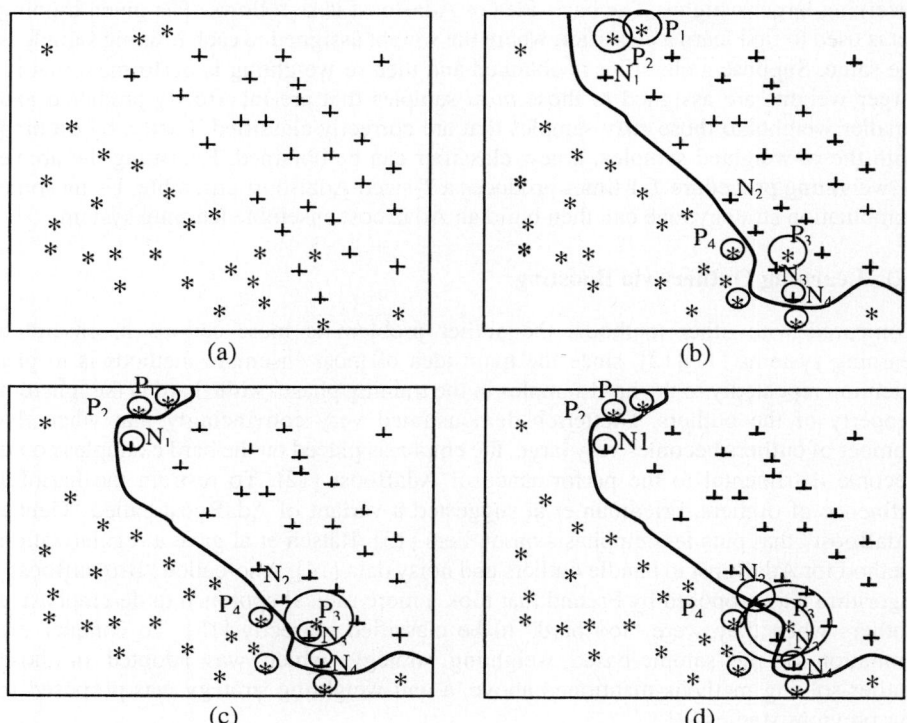

Fig. 2. Illustration of the outlier learning process

been correctly classified and thus, their weights decrease. And those samples such as P$_3$ get higher weights since they are still misclassified by the current component classifiers.

At the same time, the patterns of the other class such as N$_1$ and N$_3$ get higher weights since they are now closer to the decision boundary. When reaching the state illustrated in Figure 3(d), it can be observed that the decision line moves closer to those samples such as P$_3$, but still, they are misclassified. Thus, very large weights are assigned to them, and in fact, they are outliers.

Usually, some criteria will be adopted to stop the learning process as well as fix outliers, such as iteration times, weight value level, error rate, etc. Then, when the learning process is completed, outliers could be identified according to the selected criteria.

In this study, the centers for each category are firstly calculated, and then the nearest neighbor algorithm is used as weak classifier for the AdaBoost learning process. The iteration times is set to 10 for stopping the learning process, and a weight value threshold method is adopted for mark the outliers according to formula (1).

$$\alpha_{ik} = \frac{1}{N_k \cdot w_{ik}} \sum_{j=1}^{N_k} w_{jk} < \alpha_0 , \qquad (1)$$

where N_k is the number of webpages in category k, w_{ik} is the weight value of ith webpage in category k, and α_0 is the predetermined experimental threshold which is set to 0.2 in our simulations. In other words, for the ith webpage in category k, if its calculated value α_{ik} is less than $\alpha_0 = 0.2$, it will be marked as an outlier.

3 System Description

3.1 Webpage Representation

A webpage document written in HTML or XML etc. usually consists of plain text, various text fonts and styles, tags, links to other webpages as well as links to image, audio and video files, etc. In this study, we ignore all other information carriers and only use the text content as the main information for webpage categorization. Thus, a preprocessing step must be performed to remove all those other tags but text. However, the key issue to the Chinese webpage categorization problem is to transform the text document into some kind of representation that is more suitable for learning. We also call this process *webpage representation*, which includes two steps, feature extraction and feature selection.

The most commonly used feature extraction is the *vector space model* (VSM) [16], where a webpage is represented as a vector of terms, each of which may be a single Chinese character, a Chinese word or a *WordsGroup* (WG). Thus, according to the *term frequency* (TF) or other useful information of the text term, we could easily represent the webpage as a feature vector. Those terms used to express the documents form a dictionary. Usually, for a corpus of webpages to be handled, we would have a word-by-document matrix $A = (a_{ik})$, where each element a_{ik} is the weight of term i in webpage k. The number of rows of the matrix corresponds to the number of terms in the dictionary, which is denoted as M, while the number of columns of the matrix corresponds to the number of webpages in the corpus, which is denoted as N. There are several ways to determine the value of each element a_{ik}; the main goal is to search for a more seemly representation for webpages, so that a good learning performance can be reached. Two basic principles are usually considered in determining the value of a_{ik}: (i) the higher the term frequency in a webpage is, the more important it is to the category the webpage belongs to, (ii) the higher the term frequency is in all webpages in the corpus, the more unimportant it is [17].

The matrix A is usually sparse because each webpage only can contain a small number of terms. Meanwhile, the number of rows of A, M, which is also the size of the formed dictionary is typically very large. As a consequence, webpage representations have to face the sparse and high dimensionality problem, which is another key problem in webpage categorization tasks. A *feature selection* procedure therefore becomes very important. There are several methods proposed to perform feature selection, such as DF *Thresholding*, *information gain* (IF), *mutual information* (MI) and χ^2 *statistics* etc [2].

Unlike English and other Indo-European languages, Chinese text does not have a natural delimiter between words, which leads Chinese word segmentation to be another key issue in Chinese text processing tasks [18]. Thus, two common schemes were formed in Chinese web page categorization tasks. One is single Chinese character based mode; another is Chinese words based mode. In the Chinese language, the word is the basic unit of a concept. Frequently, each word will contain more than one Chinese character, although sometimes a single Chinese character will be a word. The former scheme avoids the Chinese word segmentation problem but ignores the utilization of the word meanings. The latter scheme encounters a more serious sparse and high dimensionality problem because there are many more words than individual characters. In order to improve the performance while utilizing knowledge of the Chinese language, some additional knowledge dictionaries were imported in recent studies, such as a thesaurus dictionary, etc. In this paper, a new scheme so-called WordsGroup (WG) is adopted which was introduced in our previous studies [19], where knowledge of Chinese linguistics was imported according to *The Modern Chinese Classification Dictionary* [20]. In the WG scheme, there are about 49,000 words classified to 3,717 WordGroups, which are the selected terms for webpage representation. That means a webpage can be represented as a 3,717 dimensional vector. In contrast, using two traditional schemes, without extra processing, the feature vector dimension will be about 6,000 and 10,000, respectively. Thus, the WG scheme becomes a desirable representation for webpages and we chose it for our studies because of the following two reasons. First, our previous studies have shown that the WG scheme outperforms two traditional schemes in webpage categorization problems [19]. Secondly, in this scheme, the feature vectors have a fixed and comparably lower dimension such that extra processing is not required any more.

3.2 kNN-Based Webpage Categorization System

kNN is one of the top-performing methods for the webpage categorization task. The procedure for building a kNN-based webpage categorization system is very simple. In the training phase, all feature vectors extracted from the training webpages are stored in a large memory. When classifying an input test webpage, the same representation processes are performed on it as during model building. The similarity between the test webpage and every stored feature vector are measured by some distance metric such as the Euclidean distance or the cosine distance, based on which k nearest neighbors can be obtained. By sorting the scores of candidate categories, a ranked list is obtained for the test webpage.

In such a memory-based learning system, more training webpages means more past experiences and more prior knowledge from which the system can learn. Thus, a model based on the refined corpus could potentially perform worse because it contains less training samples. Our purpose is to prove that although the refined corpus has less "past experience", the model trained on it still outperforms the model trained on the original corpus. If this case happens, we could conclude that the refined corpus is truly better than the original corpus by eliminating outliers in the original corpus, and those eliminated outliers could be regarded as error or abnormal "past experience".

4 Simulations

4.1 Original Corpus and the Refined Corpus

The YQ-WPBENCH-V1.0 corpus is a webpage database for the webpage categorization task, which was first collected for an automatic Chinese webpage categorization competition hosted by the Computer Network and Distributed Systems Laboratory at Peking University in 2003. Thereafter, it was freely provided for research on webpage categorization tasks. It contains 12,533 webpages that are to be classified into 11 categories, in which 9,905 webpages were randomly selected as the training data set and the remaining 2,628 webpages were used for the test data.

Table 1 shows information regarding the 11 categories of the original corpus, where for a fixed category label, its category name, number of training webpages as well as number of test webpages are indicated. Apparently the corpus is imbalanced.

Table 1. Category information of the corpus

Category No.	Name of Category	Size of Training Set	Size of Test Set
01	Literature and Art	378	97
02	Journalism	118	17
03	Commerce and Economy	781	201
04	Entertainment	1,417	356
05	Government and Politics	259	76
06	Society and Culture	987	278
07	Education	276	79
08	Natural Science	1,693	443
09	Social Science	1,567	425
10	Computer and Internet	792	210
11	Medical and Health	1,637	446
Total	-----	9,905	2,628

Through elimination of the learned outliers, the refined corpus was obtained. Table 2 shows the comparison between the original corpus and the refined corpus. Here, the number of outliers also refers to the number of webpages in the original corpus that have been eliminated from the corresponding category after outlier learning. From Table 2, one can see that altogether 597 webpages were learned as outliers and were eliminated from the original 12,533 webpages' corpus; about 5% of the original training webpages were removed.

Table 2. The number of webpages in the original corpus and the refined corpus as well as the number of identified outliers

Category No.	Original Corpus		Identified Outliers		Refined Corpus	
	Training Set	Test Set	Training Set	Test set	Training Set	Test Set
01	378	97	14	3	364	94
02	118	17	8	1	110	16
03	781	201	16	4	765	197
04	1,417	356	99	6	1,318	350
05	259	76	4	4	255	72
06	987	278	68	13	919	265
07	276	79	6	0	270	79
08	1,693	443	118	14	1,575	429
09	1,567	425	118	19	1,449	406
10	792	210	29	1	763	209
11	1,637	446	26	26	1,611	420
Total	9,905	2,628	506	91	9,399	2,537

4.2 Performance Comparison

Measure. The evaluation of the performance of a webpage categorization system is based on two aspects: one is the performance on each category, and the other is the overall performance. Three commonly used indexes *Precision*, *Recall* and *F1* have been introduced to measure different aspects of the learning performance on each category [21]. Given a category labeled as i, assume that there are n_i test webpages that belong to the category. Also, assume m_i is the number of test webpages classified to category i by the system, where l_i test webpages are correctly classified. Three indexes are expressed as below.

$$Precision_i = \frac{l_i}{m_i} \times 100\%, \qquad (2)$$

$$Recall_i = \frac{l_i}{n_i} \times 100\%, \qquad (3)$$

$$F1_i = \frac{Recall_i \times Precision_i \times 2}{Recall_i + Precision_i}. \qquad (4)$$

For evaluating overall performance averaged across categories, there are two conventional methods, namely Macro-averaging and Micro-averaging [22]. For the former, three indexes *Macro-Precision*, *Macro-Recall* and *Macro-F1* are determined as the global means of the local measures for each category. For the latter, n_i, m_i and l_i, are first summed up over all i, and then three indexes *Micro-Precision*, *Micro-Recall* and *Micro-F1* can be calculated by substituting these sums into the formulas (2), (3) and (4), where now the subscript i should be ignored. There is an important distinction between the two types of averaging. Macro-averaging gives equal weight to each category, while Micro-averaging gives equal weight to every webpage.

When evaluating the performance of the system, three indexes are calculated for each category based on which overall performance could be obtained. In our experiments, each webpage is classified to only one category and therefore, the three indexes of Micro-averaging are the same.

Performance Comparison. Based on the original corpus and the refined corpus, two models were trained using the kNN-based categorization system. Their performances were compared by predicting webpages in the test set of the original corpus, where all measures discussed above are adopted, as shown in Fig. 3.

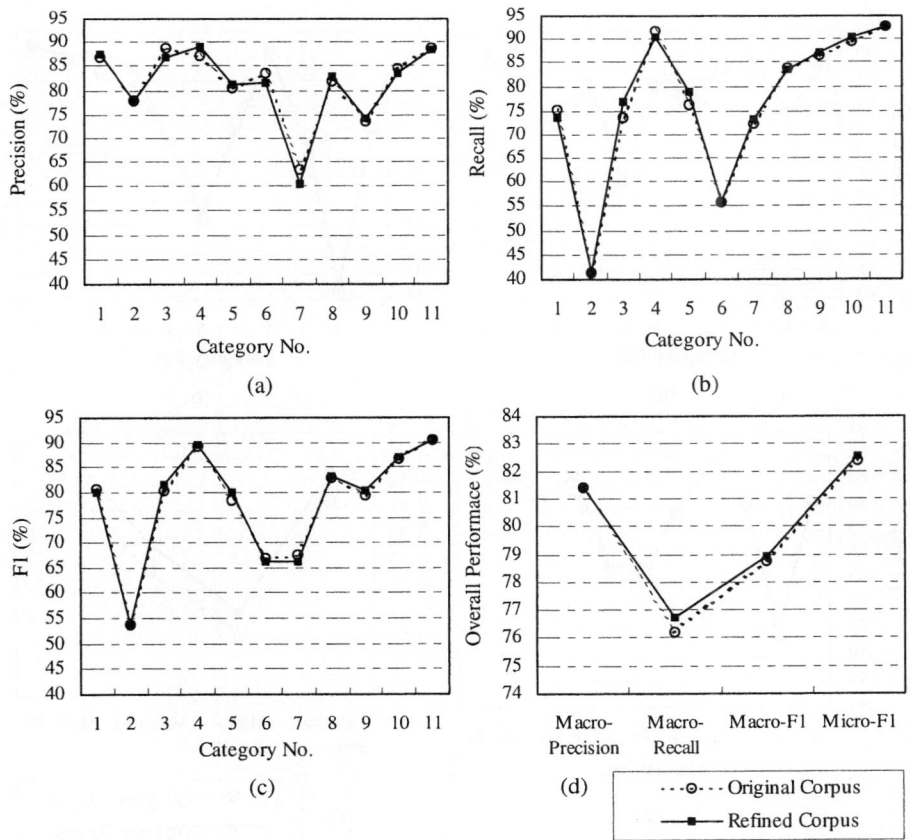

Fig. 3. Performance Comparison between models trained on the original corpus and the refined corpus, where testing webpages are from the test set in the original corpus

There are 4 subfigures in Fig. 3, Precision, Recall and F1 value of each category between two models were compared in Fig. 3 (a), Fig. 3 (b) and Fig. 3 (c), respectively. Fig. 3 (d) shows the overall performance comparison, where the three indexes both under Macro-averaging and Micro-averaging are compared. As mentioned above, under Micro-averaging, the three indexes are equivalent, so we only report comparison results for Micro-F1 of the two models.

From Fig. 3, one can see that the model trained on the refined corpus did not perform worse than the model trained on the original corpus. On the contrary, its overall performance was slightly better. According to the analysis made above, most of those eliminated samples therefore are truly outliers, they are either error or abnormal "past experiences" in the original corpus. In other words, the original corpus for webpage categorization was successfully refined.

To obtain more robust results, the same performance comparisons were performed on the test set of the refined corpus, as shown in Fig. 4., where the same conclusion could be drawn. Moreover, it is observed from Fig. 4 that the performance

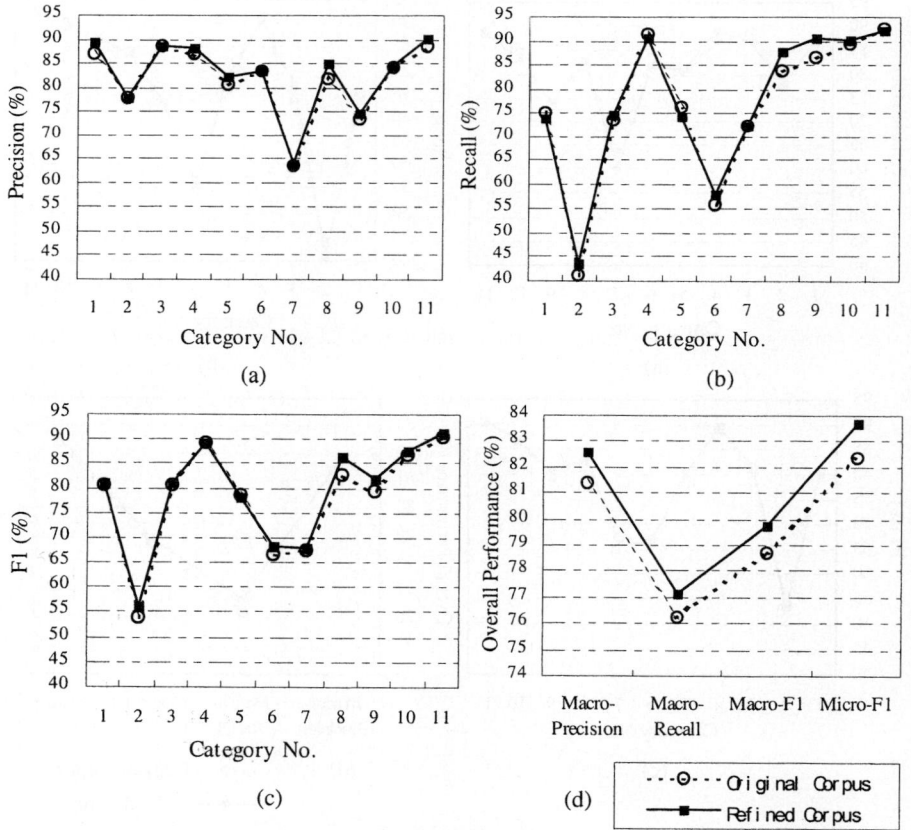

Fig. 4. Performance comparison between models trained on the original corpus and on the refined corpus, where testing webpages are from the test set in the refined corpus

improvement is more distinct than that in Fig. 3. This is because the test set in the original corpus may still have some outliers, which however may have been eliminated in the refined corpus.

Furthermore, a manual investigation has been performed to check those learned outliers. And the investigation results show that eliminated webpages could be divided as four kinds: webpages mislabeled or lying on the border between different categories, webpages that are out of the defined categories, non-sense webpages as well as some regular webpages, where the first three kinds of webpages are truly outliers. Figure 5 shows the manual investigation results, from which one can see there 83.75% eliminated webpages are truly outliers, which further reveals that the presented outlier learning process is effective.

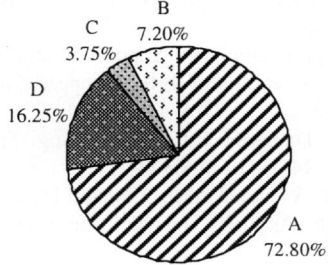

Fig. 5. The distribution of eliminated outliers, where 'A' denotes the webpages mislabeled or lying on the border between different categories, 'B' denotes the webpages that are out of the defined categories, 'C' denotes the non-sense webpages and 'D' denotes the regular webpages

5 Conclusion

In this paper, a machine learning approach was presented to refine a corpus for Chinese webpage categorization, where the Adaboost algorithm was adopted to learn outliers in the corpus. A kNN algorithm based classifier was integrated in building a Chinese webpage categorization system to make an evaluation between the original corpus and the refined corpus. Comparative results demonstrate that the model trained on the refined corpus, where the learned outliers were eliminated, did not perform worse than the model trained on the original corpus. On the contrary, its overall performance was slightly better. One explanation is that there exists some abnormal knowledge manifested as outliers in the original corpus, which could be successfully identified by the presented outlier learning method. Further analysis shows that abnormal knowledge in the original webpage corpus could be roughly divided as: non-sense webpages, webpages lying on the border between different categories, and webpages that are out of the defined categories. This further reveals the effectiveness of the presented outlier learning method for corpus refining. Still, among the learned outliers, there exists a small number of regular webpages that are misreported as outliers. How to deal with them is the topic of future work.

Acknowledgment

The work was supported in part by NSFC 60435010 and NKBRPC 2004CB318000. The authors would like to thank David Paul Wipf for constructive discussions and his patient helps to this work as well as 2 anonymous reviewers for their helpful comments.

References

1. Susan Dumais and Hao Chen: Hierarchical classification of Web content. In Proceedings of the 23rd annual International ACM SIGIR Conference on Research and Development in Information Retrieval (2000) 256-263
2. Lewis D. and Ringuette M.: A Comparison of Two Learning Algorithms for Text Classification. In Third Annual Symposium on Document Analysis and Information Retrieval (1994) 81-93
3. Yang Y., Pedersen J.P.: A Comparative Study on Feature Selection in Text Categorization. In the 14th International Conference on Machine Learning (1997) 412-420
4. Cohen W.J., Singer Y.: Context-sensitive Learning Methods for Text Categorization. In Proceedings of 19th Annual International ACM SIGIR Conference on Research and Development in Information Retrieval (1996) 307-315
5. Joachims T.: Text Categorization with Support Vector Machines: Learning with Many Relevant Features. In Proceedings of 10th European Conference on Machine Learning (ECML), (1998) 137-142
6. Weiss S.M., Apte C., Damerau F.J.: Maximizing Text-mining Performance. IEEE Intelligent Systems, Vol. 14(4) (1999) 63-69
7. Wiener E., Pedersen J.O., Weigend A.S.: A Neural Network Approach to Topic Spotting. In Proceedings of the 4th Annual Symposium on Document Analysis and Information Retrieval (1993) 22-34
8. He J., Tan A.H., Tan C.L.: On Machine Learning Methods for Chinese Document Categorization. Applied Intelligence, Vol. 18(3) (2003) 311-322
9. Luo D.S., Wu X.H., Chi H.S.: On Outlier Problem of Statistical Ensemble Learning. In Proceedings of the IASTED International Conference on Artificial Intelligence and Applications (2004) 281-286
10. Schapire R.E.: The Strength of Weak Learnability. In Machine Learning, Vol. 5 (1990) 197-227
11. Freund, Y., Schapire, R.E.: Experiments with a New Boosting Algorithm. In: Machine Learning: Proceedings of the Thirteenth International Conference (1996) 148-156
12. Friedman J., Hastie T. and Tibshirani R., Additive Logistic Regression: A Statistical View of Boosting. The Annals of Statistics, Vol. 38(2) (2000) 337-374
13. Rätsch G., Onoda T. and Müller K. R.: Soft Margins for AdaBoost. Machine Learning, Vol. 42(3) (2001) 287-320
14. Freund Y.: An Adaptive Version of the Boost by Majority Algorithm. Machine Learning, Vol. 43(3) (2001) 293-318
15. Salton G., McGill M.J.: An Introduction to Modern Information Retrieval. McGraw-Hill (1983)
16. Aas K., Eikvil L.: Text Categorization: A Survey. Technique Report, No. 941, Norwegian Computing Center (1999), http://citeseer.nj.nec.com/aas99text.html
17. Zhang H.P., Liu Q., Cheng X.Q., Zhang H. and Yu H.K.: Chinese Lexical Analysis Using Hierarchical Hidden Markov Model. In Second SIGHAN Workshop on Chinese Language Processing (2003) 63-70
18. Wu X.H., Luo D.S., Wang X.H. and Chi H.S.: WrodsGroup based Scheme for Chinese Text Categorization. Submitted to Journal of Chinese Information Processing
19. Dong D.N.: The Modern Chinese Classification Dictionary. The Publishing House of the Unabridged Chinese Dictionary (1999)
20. Yang Y.M.: An Evaluation of Statistical Approaches to Text Categorization. In Journal of Information Retrieval Vol. 1 (1/2) (1999) 67-88
21. Yang Y.M. and Liu X.: A Re-examination of Text Categorization Methods. In Proceedings of ACM SIGIR Conference on Research and Development in Information Retrieval (1999) 42-49

Bio-kernel Self-organizing Map for HIV Drug Resistance Classification

Zheng Rong Yang and Natasha Young

Department of Computer Science, University of Exeter,
Exeter EX4 4QF, UK
z.r.yang@ex.ac.uk
http://www.dcs.ex.ac.uk/~zryang

Abstract. Kernel self-organizing map has been recently studied by Fyfe and his colleagues [1]. This paper investigates the use of a novel bio-kernel function for the kernel self-organizing map. For verification, the application of the proposed new kernel self-organizing map to HIV drug resistance classification using mutation patterns in protease sequences is presented. The original self-organizing map together with the distributed encoding method was compared. It has been found that the use of the kernel self-organizing map with the novel bio-kernel function leads to better classification and faster convergence rate...

1 Introduction

In analysing molecular sequences, we need to select a proper feature extraction which can convert the non-numerical attributes in sequences to numerical features prior to using a machine learning algorithm. Suppose we denote by **x** a sequence and $\phi(\mathbf{x})$ a feature extraction function, the mapping using a feature extraction function is $F\ (F: S \to \phi) \in R^d$. Finding an appropriate feature extraction approach is a non-trivial task.

It is known that each protein sequence is an ordered list of 20 amino acids while a DNA sequence is an ordered list of four nucleic acids. Both amino acids and nucleic acids are non-numerical attributes. In order to analyze molecular sequences, these non-numerical attributes must be converted to numerical attributes through a feature extraction process for using a machine learning algorithm. The distributed encoding method [2] was proposed in 1988 for extracting features for molecular sequences. The principle is to find orthogonal binary vectors to represent amino (nucleic) acids. With this method, amino acid Alanine is represented by 0000000000 0000000001 while Cystine 0000000000 0000000010, etc. With the introduction of this feature extraction method, the application of machine learning algorithms to bioinformatics has been very successful. For instance, this method has been applied to the prediction of protease cleavage sites [3], signal peptide cleavage sites [4], linkage sites in glycoproteins [5], enzyme active sites [6], phosphorylation sites [7] and water active sites [8].

However, as indicated in the earlier work [9], [10], [11] such a method has its inherent limit in two aspects. First, the dimension of an input space has been enlarged 20 times weakening the significance of a set of training data. Second, the biological content in a molecule sequence may not be efficiently coded. This is because the similarity between any pair of different amino (nucleic) acids varies while the distance between such encoded orthogonal vectors of two different amino (nucleic) acids is fixed.

The second method for extracting features from protein sequences is to calculate the frequency. It has been used for the prediction of membrane protein types [12], the prediction of protein structural classes [13], subcellular location prediction [14] and the prediction of secondary structures [15]. However, the method ignores the coupling effects among the neighbouring residues in sequences leading to potential bias in modelling. Therefore, di-peptides method was proposed where the frequency of each pair of amino acids occurred as neighbouring residues is counted and is regarded as a feature. Dipeptides, gapped (up to two gaps) transitions and the occurrence of some motifs as additive numerical attributes were used for the prediction of subcellular locations [16] and gene identification [17]. Descriptors were also used, for instance, to predict multi-class protein folds [18], to classify proteins [19] and to recognise rRNA-, RNA-, and DNA-binding proteins [20], [21]. Taking into account the high order interaction among the residues, multi-peptides can also be used. It can be seen that there are 400 di-peptides, 8,000 tri-peptides and 16,000 tetra-peptides. Such a feature space can be therefore computational impractical for modelling.

The third class of methods is using profile measurement. A profile of a sequence can be generated by subjecting it to a homology alignment method or Hidden Markov Models (HMMs) [22], [23], [24], [25].

It can be seen that either finding an appropriate approach to define $\phi(\mathbf{x})$ is difficult or the defined approach may lead to a very large dimension, i.e., $d \to \infty$. If an approach which can quantify the distance or similarity between two molecular sequences is available, an alternative learning method can be proposed to avoid the difficulty in searching for a proper and efficient feature extraction method. This means that we can define a reference system to quantify the distance among the molecular sequences. With such a reference system, all the sequences are quantitatively featured by measuring the distance or similarity with the reference sequences.

One of the important issues in using machine learning algorithms for analysing molecular sequences is investigating sequence distribution or visualising sequence space. Self-organizing map [26] has been one of the most important machine learning algorithms for this purpose. For instance, SOM has been employed to identify motifs and families in the context of unsupervised learning [27], [28], [29], [30], [31]. SOM has also been used for partitioning gene data [32]. In these applications, feature extraction methods like the distributed encoding method were used.

In order to enable SOM to deal with complicated applications where feature extraction is difficult, kernel method has been introduced recently by Fyfe and his colleagues [1]. Kernel methods were firstly used in cluster analysis for K-means algorithms [33], where the Euclidean distance between an input vector $\mathbf{x}$ and a mean vector $\mathbf{m}$ is minimized in a feature space spanned by kernels. In the kernel feature space, both $\mathbf{x}$ and $\mathbf{m}$ were the expansion on the training data. Fyfe and his colleagues developed so-called kernel self-organizing maps [34], [35]. This paper aims to introduce a bio-kernel function for kernel SOM. The method is verified on HIV drug resistance classification. A stochastic learning process is used with a regularization term.

2 Methods

A training data set $D = \{s_n\}_{n=1}^h$, where $s_n \in S^D$ (S is a set of possible values and $|S|$ can be either definite or indefinite) and a mapping function which can map a sequence to a numerical feature vector is defined as $F(\phi: S \to F) \in R^d$, $x_n = \varphi(s_n)$. In most situations, $x_n = \varphi(s_n) = (\phi_1(s_n), \phi_2(s_n), \ldots, \phi_d(s_n))^T$ is unknown and possibly, $d \to \infty$. This then causes the difficulty in modelling. In using self-organizing map for unsupervised learning of protein sequences, the error function in the feature space F can be defined as $L = |x_n - w_m|^2$, where $w_m \in R^d$ is the weight vector connecting the mth output neuron. Suppose w_m can be expanded on the training sequences ($w_m = \Phi \alpha_m$). Note that $\alpha_m \in R^h$ is an expansion vector and $\Phi = \{\phi_i(s_j)\}_{1 \leq i \leq d, 1 \leq j \leq h}$. The error function can re-written as $L = K_{nn} - 2k_n \alpha_m + \alpha_m^T K \alpha_m$. Note that $K_{ij} = K(s_i, s_j)$ is the kernel, $k_n = (K_{n1}, K_{n2}, \ldots, K_{nh})$ is a row kernel vector and $K = \{K_{ij}\}_{1 \leq i, j \leq h}$ a kernel matrix. The error function can be as follows if we use L_2 norm regarded as a regularization term

$$L = \frac{1}{2}(K_{nn} - 2k_n \alpha_m + \alpha_m^T K \alpha_m + \lambda \alpha_m^T \alpha_m),$$

where λ is the regularization factor. The update rule is then defined as $\Delta \alpha_m = \eta(t)(k_n - (K + \lambda I)\alpha_m)$. In designing the bio-kernel machine, a key issue is the design of an appropriate kernel function for analysing protein or DNA sequences. Similar as in [9], [10], [11], we use the bio-basis function as the bio-kernel function

$$K(x, b_i) = \exp\left(\frac{M(x, b_i) - M(b_i, b_i)}{M(b_i, b_i)}\right)$$

where x is a training sequence and b_i is a basis sequence, both have D residues. Note that $M(x, b_i) = \sum_{d=1}^{D} M(x_d, b_{id})$ with x_d and b_{id} and the dth residue in sequences. The value of $M(x_d, b_{id})$ can be found in a mutation matrix [36], [37]. The bio-basis function has been successfully used for the prediction of Trypsin cleavage sites [8], HIV cleavage sites [9], signal peptide cleavage site prediction [10], Hepatitis C virus protease cleavage sites [38], disordered protein prediction [39], [40], phosphorylation site prediction [41], the prediction of the O-linkage sites in glycoproteins [42], the prediction of Caspase cleavage sites [43], the prediction of SARS-CoV protease cleavage sites [44] and the prediction of signal peptides [45].

3 Results

Drug resistance modeling is a wide phenomenon and drug resistance modeling is a very important issue in medicine. In computer aided drug design, it is desired to study how the genomic information is related with therapy effect [46]. To predict if HIV drug may fail in therapy using the information contained in viral protease sequences is regarded as genotype-phenotype correlation. In order to discover such relationship, many researchers have done a lot of work in this area. For instance, the original self-organizing map was used on two types of data, i.e., structural information and sequence information [46]. In using sequence information, frequency features were used as the inputs to SOM. The prediction accuracy was between 68% and 85%. Instead of neural networks, statistical methods and decision trees were also used [47], [48], [49].

Data (46 mutation patterns) were obtained from [50]. Based on this data set, bio-kernel SOM was running using different value for the regularization factor. The original SOM was also used for comparison. Both SOMs used the same structure (36 output neurons) and the same learning parameters, i.e. the initial learning rate ($\eta_h = 0.01$). Both algorithms were terminated when the mean square error was less than 0.001 or 1000 learning iterations.

Fig. 1 shows the error curves for two SOMs. It can be seen that the bio-kernel SOM (bkSOM) converged much faster with very small errors.

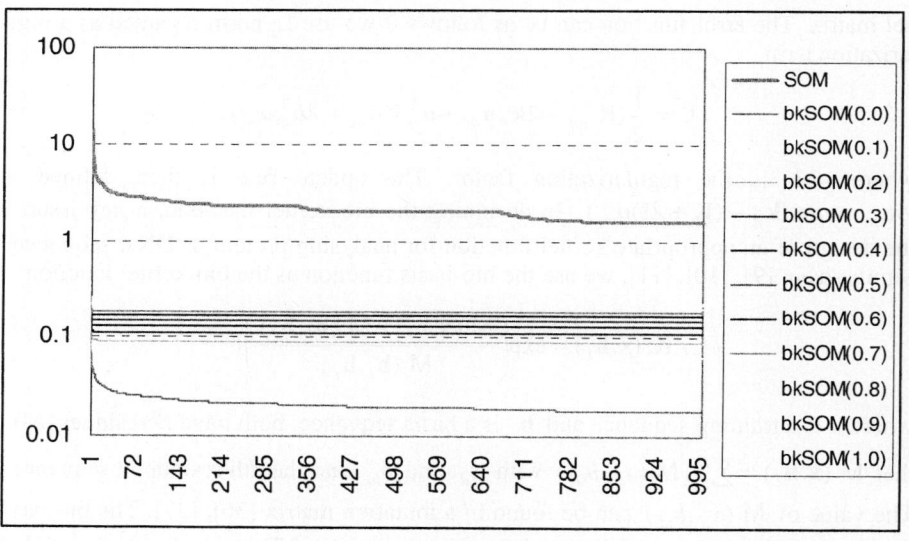

Fig. 1. The error curves for two SOMs. The horizontal axis is the learning iterations and the vertical one (logarithm scale) the errors. The numbers within the brackets of bkSOM mean the regularization factor values.

Fig. 2 shows a map of bkSOM, where "n.a." means that there is no patterns mapped onto the corresponding output neuron, "5:5" means that all the five patterns mapped onto the corresponding neuron are corresponding to the mutation patterns which are resistant to the drug and "0:9" means that all the nine patterns mapped onto the corresponding neuron are corresponding to the mutation patterns which are not resistant to the drug.

```
5:5    n.a.   5:5    1:1    n.a.   0:9
n.a.   1:1    n.a.   n.a.   n.a.   0:1
n.a.   n.a.   n.a.   0:1    0:3    n.a.
2:2    n.a.   n.a.   0:1    n.a.   n.a.
n.a.   0:1    0:1    1:1    n.a.   n.a.
2:2    0:1    0:2    n.a.   0:7    0:2
```

Fig. 2. The feature map of bkSOM.

Table 1 shows the comparison in terms of the classification accuracy, where "NR" means non-resistance and "R" resistance. It can be seen that bkSOM performed better than SOM in terms of classification accuracy. The non-resistance prediction power indicates the likelihood that a predicted non-resistance pattern is a true non-resistance pattern. The resistance prediction power therefore indicates the likelihood that a predicted resistance pattern is a true non-resistance pattern. For instance, the non-resistance prediction power using SOM is 90%. It means that for every 100 predicted non-resistance patterns, 10 would be actually resistance patterns.

Table 1. The classification accuracy of two SOMs

	SOM				bkSOM		
	NR	R	Precision		NR	R	Precision
NR	28	0	100%	NR	28	0	100%
R	3	15	83%	R	0	18	100%
Power	90%	100%	93%	Power	100%	100%	100%

4 Summary

This paper has presented a novel method referred to as bio-kernel self-organizing map (bkSOM) for embedding the bio-kernel function into the kernel self-organizing map for the purpose of modeling protein sequences. The basic principle of the method is using the "kernel trick" to avoid tedious feature extraction work for protein sequences, which has been proven a non-trivial task. The computational simulation on the HIV drug resistance classification task has shown that bkSOM outperformed SOM in two aspects, convergence rate and classification accuracy.

References

1. Corchado, E., Fyfe, C. Relevance and kernel self-organising maps. International Conference on Artificial Neural Networks, (2003)
2. Qian, N., Sejnowski, T.J.: Predicting the secondary structure of globular proteins using neural network models. J. Mol. Biol., 202 (1988) 865-884
3. Thompson, T.B., Chou, K.C., Zhang, C.: Neural network prediction of the HIV-1 protease cleavage sites. Journal of Theoretical Biology, 177 (1995) 369-379.

4. Nielsen, M., Lundegaard, C., Worning, P., Lauemoller, S.L., Lamberth, K., Buss, S., Brukak, S., Lund, O.: Reliable prediction of T-cell epitopes using neural networks with novel sequence representations. Protein Science, 12 (2003) 1007- 1017
5. Hansen, J.E., Lund, O., Engelbrecht, J., Bohr, H., Nielsen, J.O.: Prediction of O-glycosylation of mammalian proteins: specificity patterns of UDP-GalNAc:polypeptide N-acetylgalactosaminyltransferase. Biochem J. 30 (1995) 801-13
6. Gutteridge, A., Bartlett, G.J., Thornton, J.M.: Using a neural network and spatial clustering to predict the location of active sites in enzymes. Journal of Molecular Biology, 330 (2003) 719-734
7. Blom, N., Gammeltoft, S., Brunak, S.: Sequence and structure based prediction of eukaryotic protein phosphorylation sites. J. Mol. Biol. 294 (1999) 1351-1362
8. Ehrlich, L., Reczko, M., Bohr, H., Wade, R.C.: Prediction of protein hydration sites from sequence by modular neural networks. Protein Eng., 11 (1998) 11-19
9. Thomson, R., Hodgman, T. C., Yang, Z. R., Doyle, A. K.: Characterising proteolytic cleavage site activity using bio-basis function neural networks. Bioinformatics, 19 (2003) 1741-1747
10. Yang, Z.R., Thomson, R.: A novel neural network method in mining molecular sequence data. IEEE Trans. on Neural Networks, 16 (2005) 263- 274
11. Yang, Z.R.: Orthogonal kernel machine in prediction of functional sites in preteins. IEEE Trans on Systems, Man and Cybernetics, 35 (2005) 100-106
12. Cai, Y.D., Ricardo, P.W., Jen, C.H., Chou, K.C.:Application of SVMs to predict membrane protein types. Journal of Theoretical Biology, 226 (2004) 373-376
13. Cai, Y.D., Lin, X.J., Xu, X.B., Chou, K.C.:Prediction of protein structural classes by support vector machines. Computers & Chemistry, 26 (2002) 293-296
14. Hua, S., Sun, Z.: Support vector machine approach for protein subcellular localization prediction. Bioinformatics, 17 (2001) 721-728
15. Chu, F., Jin, G., Wang, L.: Cancer diagnosis and protein secondary structure prediction using support vector machines. in Wang, L. (ed) Support Vector Machines, Theory and Applications, Springer-Verlag (2004)
16. Park, K., Kanehisa, M.: Prediction of protein subcellular locations by support vector machines using compositions of amino acids and amino acid pairs. Bioinformatics, 19 (2003) 1656-1663
17. Carter, R.J., Dubchak, I., Holbrook, S.R.: A computational approach to identify genes for functional RNAs in genomic sequences. Nucleic Acids Res., 29 (2001) 3928-3938
18. Ding, C.H.Q, Dubchak, I.: Multi-class protein fold recognition using support vector machines and neural networks. Bioinformatics, 17 (2001) 349-358
19. Cai, C.Z., Wang, W.L., Sun, L.Z., Chen, Y.Z.: Protein function classification via support vector machine approach. Mathematical Biosciences, 185 (2003) 111-122
20. Cai, Y.D., Lin, S.L.: Support vector machines for predicting rRNA-, RNA-, and DNA-binding proteins from amino acid sequence. Biochimica et Biophysica Acta (BBA) - Proteins & Proteomics, 1648 (2003) 127-133
21. Lin, K., Kuang, Y., Joseph, J.S., Kolatkar, P.R.: Conserved codon composition of ribosomal protein coding genes in Escherichia coli, Mycobacterium tuberculosis and Saccharomyces cerevisiae: lessons from supervised machine learning in functional genomics. Nucleic Acids Res., 30 (2002) 2599-2607
22. Jaakkola, T., Diekhans, M., Haussler, D.: Using the Fisher kernel method to detect remote protein homologies. Proceedings of the 7th International Conference on Intelligent Systems for Molecular Biology, (1999) 149-158
23. Jaakkola, T., Diekhans, M., Haussler, D.: A Discriminative Framework for Detecting Remote Protein Homologies. Journal of Computational Biology, 7 (2000) 95–114
24. Karchin, R., Karplus, K., Haussler, D.: Classifying G-protein coupled receptors with support vector machines. Bioinformatics, 18 (2002) 147-159

25. Guermeur, Y., Pollastri, G., Elisseeff, A., Zelus, D., Paugam-Moisy, H., Baldi, P.: Combining protein secondary structure prediction models with ensemble methods of optimal complexity. Neurocomputing, 56 (2004) 305-327
26. Kohonen, T.: Self organization and associative Memory, 3rd Ed (1989) Springer, Berling.
27. Arrigo, P., Giuliano, F., Scalia, F., Rapallo, A., Damiani, G.: Identification of a new motif on nucleic acid sequence data using Kohonen's self-organising map. CABIOS, 7 (1991) 353-357
28. Bengio, Y., Pouliot, Y.: Efficient recognition of immunoglobulin domains from amino acid sequences using a neural network. CABIOS, 6 (1990) 319-324
29. Ferran, E.A., Ferrara, P.: Topological maps of protein sequences. Biological Cybernetics, 65 (1991) 451-458
30. Wang, H. C., Dopazo, J., Carazo, J.M.: Self-organising tree growing network for classifying amino acids. Bioinformatics, 14 (1998) 376-377
31. Ferran, E. A. and Pflugfelder, B. A hybrid method to cluster protein sequences based on statistics and artificial neural networks. CABIOS 1993, 9, 671-680
32. Tamayo, P.; Slonim, D.; Mesirov, J.; Zhu, Q.; Kitareewan, S.; Dmitrovsky, E.; Lander, E. S. and Golub, T. R. Interpreting patterns of gene expression with self-organizing maps: methods and application to hematopoietic differentiation. PNAS 1999, 96, 2907-2912.
33. Scholkopf, B.: The kernel trick for distances, Technical Report. Microsoft Research, May (2000)
34. MacDonald, D., Koetsier, J., Corchado, E., Fyfe, C.: A kernel method for classification LNCS, 2972 (2003)
35. Fyfe, C., MacDonald, D.: Epsilon-insensitive Hebbian learning. Neuralcomputing, 47 (2002) 35-57
36. Dayhoff, M.O., Schwartz, R.M., Orcutt, B.C.: A model of evolutionary change in proteins. matrices for detecting distant relationships. Atlas of protein sequence and structure, 5 (1978) 345-358
37. Johnson, M.S., Overington, J.P.: A structural basis for sequence comparisons-an evaluation of scoring methodologies. J. Molec. Biol., 233 (1993) 716-738
38. Yang, Z. R., Berry, E.: Reduced bio-basis function neural networks for protease cleavage site prediction. Journal of Computational Biology and Bioinformatics, 2 (2004) 511-531
39. Thomson, R., Esnouf, R.: Predict disordered proteins using bio-basis function neural networks. Lecture Notes in Computer Science, 3177 (2004) 19-27
40. Yang Z.R., Thomson R., Esnouf R.: RONN: use of the bio-basis function neural network technique for the detection of natively disordered regions in proteins. Bioinformatics, (accepted)
41. Berry E., Dalby A. and Yang Z.R.: Reduced bio basis function neural network for identification of protein phosphorylation sites: Comparison with pattern recognition algorithms. Computational Biology and Chemistry, 28 (2004) 75-85
42. Yang, Z.R., Chou, K.C.: Bio-basis function neural networks for the prediction of the O-linkage sites in glyco-proteins. Bioinformatics, 20 (2004) 903-908
43. Yang, Z.R.: Prediction of Caspase Cleavage Sites Using Bayesian Bio-Basis Function Neural Networks. Bioinformatics (in press)
44. Yang, Z.R.: Mining SARS-CoV protease cleavage data using decision trees, a novel method for decisive template searching. Bioinformatics (accepted)
45. Sidhu, A., Yang, Z.R.: Predict signal peptides using bio-basis function neural networks. Applied Bioinformatics, (accepted)
46. Draghici, S., Potter, R.B.: Predicting HIV drug resistance with neural networks. Bioinformatics, 19 (2003) 98-107
47. Beerenwinkel, N., Daumer, M., Oette, M., Korn, K., Hoffmann, D., Kaiser, R., Lengauer, T., Selbig, J., Walter, H.: Geno2pheno: estimating phenotypic drug resistance from HIV-1 genotypes. NAR, 31 (2003) 3850-3855

48. Beerenwinkel, N., Schmidt, B., Walter, H., Kaiser, R., Lengauer, T., Hoffmann, D., Korn, K.,Selbig, J.,: Diversity and complexity of HIV-1 drug resistance: a bioinformatics approach to predicting phenotype from genotype. PNAS, 99 (2002) 8271-8276
49. Zazzi, M., Romano, L., Giulietta, V., Shafer, R.W., Reid, C., Bello, F., Parolin, C., Palu, G., Valensin, P.: Comparative evaluation of three computerized algorithms for prediction of antiretroviral susceptibility from HIV type 1 genotype. Journal of Antiimicrobial Chemotherapy, 53 (2004) 356-360
50. Sa-Filho, D. J., Costa, L.J., de Oliceira, C.F., Guimaraes, A.P.C., Accetturi, C.A., Tanuri, A., Diaz, R.S.: Analysis of the protease sequences of HIV-1 infected individuals after Indinavir monotherapy. Journal of Clinical Virology, 28 (2003) 186-202

A New Learning Algorithm Based on Lever Principle[*]

Xiaoguang He, Jie Tian[**], and Xin Yang

Center for Biometrics and Security Research, Key Laboratory of Complex Systems and
Intelligence Science,Institute of Automation, Chinese Academy of Sciences,
Graduate School of the Chinese Academy of Science,P.O.Box 2728, Beijing 100080 China
Tel: 8610-62532105, Fax: 8610-62527995
tian@doctor.com

Abstract. In this paper a new learning algorithm, Lever Training Machine (LTM), is presented for binary classification. LTM is a supervised learning algorithm and its main idea is inspired from a physics principle: Lever Principle. Figuratively, LTM involves rolling a hyper-plane around the convex hull of the target training set, and using the equilibrium position of the hyper-plane to define a decision surfaces. In theory, the optimal goal of LTM is to maximize the correct rejection rate. If the distribution of target set is convex, a set of such decision surfaces can be trained for exact discrimination without false alarm. Two mathematic experiments and the practical application of face detection confirm that LTM is an effective learning algorithm.

1 Introduction

Target detection is an important research field of computer vision especially with the specific subject, e.g. face detection and vehicle detection. Actually, target detection is a binary classification problem, and the goal is to find a binary classifier. Binary classifiers can be sorted into two categories: nonlinear and linear. The nonlinear classifiers, such as neural network [1, 2], and nonlinear SVM [3, 4], are more powerful than linear classifier, but they're computation expensive. On the other hand, linear classifier is the most simple and fast one. An individual linear classifier is weak, but a set of linear classifiers can be constructed to a piecewise linear classifier, which combines the advantage of both linear and nonlinear classifiers, and results in a not only fast but also powerful classifier.

Fisher's LDA [5], SVM [1], and MRC [6], are the examples of linear classifier. They train a linear classifier in some optimal manners. The goal of LDA is to maximize the Mahalanobis distance of the target and non-target classes. And the object of SVM is maximizing the margin between the two classes. In the both algorithms, it's assumed that the two classes are linearly separable and equally important [6]. How-

---

[*] This paper is supported by the Project of National Science Fund for Distinguished Young Scholars of China under Grant No. 60225008, the Key Project of National Natural Science Foundation of China under Grant No. 60332010, the Project for Young Scientists' Fund of National Natural Science Foundation of China under Grant No.60303022, and the Project of Natural Science Foundation of Beijing under Grant No.4052026.
[**] Corresponding author.

ever, the assumptions are invalid in many applications. M. Elad et al. have proposed MRC to overcome those limitations. MRC exploits the property that the probability of target is substantially smaller than non-target; the property is common in many target detection issues. And MRC processes nonlinearly separable classes with the idea of piecewise linear classifier. But as pointed by M. Elad, even if MRC is used to deal with a convex target set, false alarm may exist in practice [6], because its optimal object only considers second moments with neglecting higher ones.

LTM has been developed to pursue a linear classifier in a more direct and novel manner. The idea of LTM is inspired from Lever Principle in physics. Its optimal goal is to maximize the correct reject rate directly. Given the training sets of target and non-target, LTM trains a decision hyper-plane stage by stage to separate the non-target data as many as possible from the target set. Prior the first training stage, LTM generates an initial hyper-plane randomly. In each training stage, LTM aligns the hyper-plane to an advantageous position based on Lever Principle, where the hyper-plane can distinguish more non-target data. When the hyper-plane keeps the balance, the equilibrium hyper-plane is defined as the output linear classifier by LTM.

If the distribution of target set is convex, a sequence of decision surfaces can be found by LTM to exactly discriminate the both training sets without false-alarm. That is confirmed by two mathematic experiments and the practical application of front face detection. Compared with other training algorithms of linear classifier, LTM has direct physical meaning and direct optimal goal. It admits the high probability of non-target and is suitable to deal with nonlinearly separable classes.

In the paper, section 2 describes the theory of LTM in detail. Section 3 gives two mathematic experiments. The application of LTM to face detection is presented in section 4. The last section makes a conclusion with future perspectives.

2 Lever Training Machine

There are two concepts should be reviewed prior to present LTM: linear classifier and Lever Principle.

2.1 Linear Classifier and Lever Principle

Linear classifier is a simple and fast pattern classification approach. It can be defined by the linear-threshold formula:

$$h(x) = \begin{cases} 1, & x \cdot U - d > 0 \\ 0, & otherwise \end{cases} \quad (1)$$

where d is the threshold and U is the unit projection vector. A linear classifier can define a hyper-plane, of which U is the unit normal vector. Likewise a hyper-plane corresponds to a linear classifier.

Lever Principle is a basic physical law and Archimedes stated it vividly: "Give me a place to stand and I will move the earth". The product of a force F by its effort arm L is the moment of F:

$$M = L \times F \quad (2)$$

As demonstrated in Fig. 1, the movement state of the plane can be analyzed according to Lever Principle. Each force Fi that acts on the plane will generate a moment M_i, where $M_i = L_i \times F_i$. If $\sum_{i=1}^{N} M_i = 0$, the plane will keep the balance. Otherwise the plane will rotate, and the direction can be determined with the right-hand rule. The normal vector U of the plane will change in the direction of $\left(\sum_{i=1}^{N} M_i\right) \times U$.

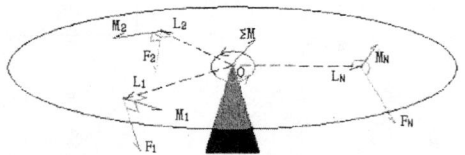

Fig. 1. The forces acted on the plane yield a moment sum $\|\Sigma M\|$ that determines the movement state of the plane

The plane can be replaced by a hyper-plane and naturally Lever Principle is introduced to high dimensional space, where the moment of a force is still defined by Formula 2. With the high dimension extension, Lever Principle can be adapted to optimize a decision hyper-plane by LTM.

2.2 Lever Training Machine

LTM is a supervised learning algorithm. Given the target training sets $X = \{x_i\}_{i=1,2,...M}$ and the non-target training sets $Y = \{y_i\}_{i=1,2,...N}$, the purpose of LTM is to find a decision hyper-plane to separate non-target data as many as possible from target set.

LTM is designed for aligning an initial hyper-plane P_0 to an optimal decision hyper-plane P_{opt} by a serial of training stages. In the training stage n, LTM modifies the hyper-plane P_n to a move advantageous position P_{n+1}, where P_{n+1} can separate more non-target data from target set than P_n. Generally, the dimension of data, e.g. face image data and car image data, is typically high, which means that the hyper-plane has high degree of freedom. Therefore it is quite difficult to determine the advantageous direction to rotate the hyper-plane. Lever principle is adapted to address this issue.

As shown in Fig. 2, P_0 is generated randomly as the initial input, U is the unit normal vector of P_0. Then the fulcrum O of P_0 is located by following:

Step 1. For each $x_i \in X$, calculate the projection value with formula 3:

$$v_{pro}(x_i) = x_i \cdot U \qquad (3)$$

Step 2. Find the target data with the lowest projection values and those data are named as fulcrum data. Denote the fulcrum data set as $X_{fulcrum} = \{x_j^{fulcrum}\}_{i=1,2,...M'}$. It satisfies $X_{fulcrum} \subset X$, and $\forall x_i \in X_{fulcrum}$ and $\forall x_j \in X - X_{fulcrum}$, $v_{pro}(x_i) \leq v_{pro}(x_j)$. r is defined as the proportion factor, and $r = \|X_{fulcrum}\| / \|X\| = M'/M$. It is less than 1.0 and usually set to 0.1 or less;

Step 3. Calculate the mean vector of the fulcrum data: $C = \frac{1}{M'} \sum_{i=1}^{M'} X_i^{fulcrum}$;

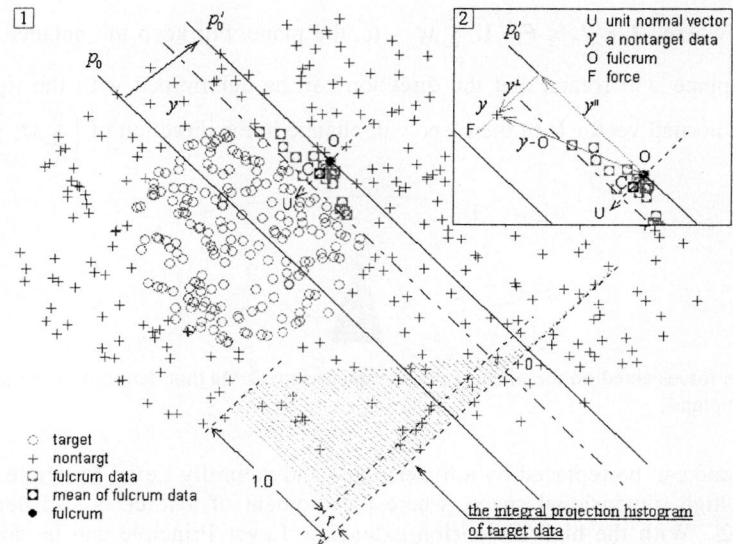

Fig. 2. Image 1 indicates how to locate the fulcrum O of the hyper-plane P_0. The integral projection histogram can be expressed as $His(v) = \|\{x | x \in X, (x-O) \cdot U < v\}\| / \|x\|$. Image 2 shows how to calculate the force that is defined in formula 5.

Step 4. Move C along reverse direction of U until all the target data reach the positive side of the hyper-plane $(x-C) \cdot U = 0$, here we can move C farther to enhance the generalization ability of the classifier. The final position of C is defined as the fulcrum O. Mathematically O is defined by:

$$O = C - \left(v_{pro}(M) - v_{pro}^{min} + g\right)U \quad (4)$$

where $v_{pro}^{min} = \min_{x \in X}\{v_{pro}(x)\}$ and g is the generalization factor. The generalization ability of LTM can be strengthened by augmenting g.

After locating O, P_0 is aligned to P_0': $(X-O) \cdot U = 0$ by moving it onto O. Then the force acted on P_0' is determined by:

$$F(y) = \frac{1}{\sqrt{2\pi}\sigma} exp\left(-\frac{\|y^\perp\|^2}{\sigma^2}\right)U \quad (5)$$

where, $y \in Y$ and $y^\perp$ is the vertical component of y-O to P_0':

$$y^\perp = ((y-O) \cdot U)U \quad (6)$$

In LTM, all forces are generated by non-target data, so only non-target data influence the rotating direction of the hyper-plane. The definition of force satisfies that the longer the $y^\perp$ is, the weaker the force is, as the yellow arrows indicated in Fig. 3. In formula 5, σ is named as distance-insensitivity factor. And the larger it is, the more insensitive the power of force is to $y^\perp$.

The effort arm and the moment of a force can be calculated by formula 7 and formula 8 respectively:

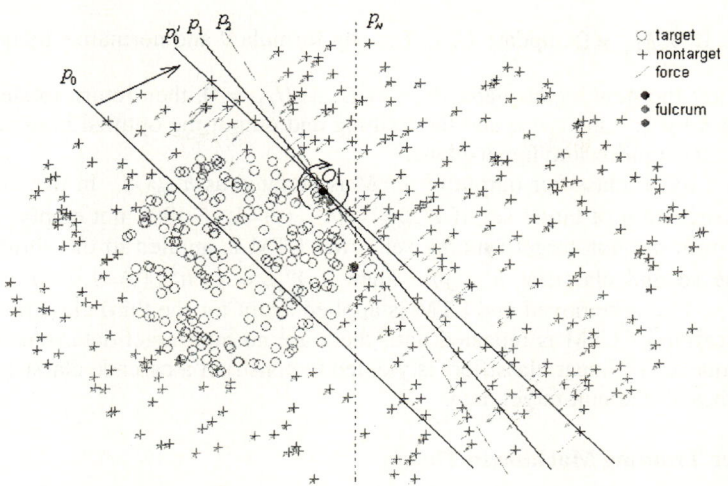

Fig. 3. Training a linear classifier with LTM. P_0 is the initial hyper-plane, and its fulcrum is located at O_1. It is moved onto O_1 as P_0' indicated, and then rotated to P_1 under the effect of forces. The fulcrum of P_1 is located at O_2. P_1 is moved to O_2 and rotated to P_2. The training is repeated until the hyper-plane keeps the balance, see P_N.

$$y^{\|} = (y - O) - y^{\perp} \qquad (7)$$

$$M(y) = y^{\|} \times F(y) \qquad (8)$$

Then according to Lever Principle, the normal vector U will change in the direction of $\left(\sum_{i=1}^{N} M(y_i)\right) \times U$. As shown in Fig. 3, the pose of P_0' is modified to P_1 by updating U:

$$U_{n+1} = U_n - \delta\left(\left(\sum_{i=1}^{N} M(y_i)\right) \times U_n\right) \qquad (9)$$

where δ, named as modification factor, is an empirical value. If δ is too large, U will be over modified; if δ is too small, U will change very slowly. Both cases will slow down the convergence rate of the training.

Fig. 3 indicates the training process. First P_0 is generated randomly. It is aligned to P_1 in the first training stage, and then P_1 is modified to P_2 in the second training stages. The training repeats until the moment sum deceases to zero. As P_N illustrated in Fig. 3, the hyper-plane keeps the balance, P_N is the optimal hyper-plane P_{opt}, it defines the optimal linear classifier. The training process, as above described, can be summarized as below:

Step 1. Generate initial hyper-plane P_0: $(x-O_0) \cdot U_0 = 0$ randomly; Set the stage number n to 0;

Step 2. Find the fulcrum O_{n+1} of P_n, and move P_n onto O_{n+1}, therefore P_n is updated to P_n': $(x-O_{n+1}) \cdot U_n = 0$;

Step 3. Calculate the forces acted on P_n' by formula 5: $F_i = F(y_i)$, where $i=1,2...n$;

Step 4. Calculate the moments of the forces by formula 8: $M_i = M(F_i)$, where $i=1,2...n$;

Step 5. If $\sum_{i=1}^{N} M_i \neq 0$, update U_n to U_{n+1} by formula 9 and normalize its norm to 1, therefore get the new hyper-plane P_{n+1}: $(x-O_{n+1}) \cdot U_{n+1} = 0$, then return to step 2 with increasing n by 1; Otherwise end the training and output the optimal linear classifier defined by the equilibrium hyper-plane.

The first linear classifier output by LTM is denoted as $h^1_{opt}(x)$. In the case of the convex distribution of target set, if $Y^1_{err} = \{y | h^1_{opt}(y)=1, y \in Y\}$ is not empty, $Y-Y^1_{err}$ is removed from the non-target training set. And LTM is applied to the abridged data set for the second classifier $h^2_{opt}(y)$. Then if $Y^2_{err} = \{y | h^2_{opt}(y)=1, y \in Y^1_{err}\}$ is not empty, $Y^1_{err}-Y^2_{err}$ is removed and LTM is applied again for the third classifier $h^3_{opt}(y)$. The application of LTM is repeated until there are no non-target data remained. Finally a sequence of linear classifiers is yielded to construct a cascade classifier, which distinguishes all the non-target data.

2.4 Lever Training Machine in Theory

In this section, it's proved that the update strategy of LTM ensures that R_N is monotonically increasing, therefore the final output classifier is a local optimal solution that maximizes R_N.

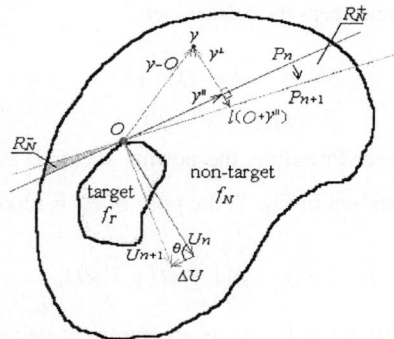

Fig. 4. f_T and f_N are the probability density functions of target and non-target respectively; R_N^+ and R_N^- is the increase and decrease parts of the correct rejection rates R_N respectively with the decision line is rotated form P_n to P_{n+1}; U_n and U_{n+1} are the unit normal vector of P_n and P_{n+1}; ΔU is the differential value where $U_{n+1}=U_n+\Delta U$; $l(O+y^{\|})$ is the line from point $O+y^{\|}$ to P_{n+1} and it is vertical to P_n; y is a non-target sample; $y^{\|}$ and $y^{\perp}$ are the parallel and vertical components of $y-O$ to P_n respectively

As demonstrated in Fig. 4, in the training stage n U_n is updated to U_{n+1} where $U_{n+1}=U_n+\Delta U$. The differential value of R_N is denoted as ΔR_N:

$$\Delta R_N = \int_{y \in l(z), z \in P_n} f_N(y) dy \tag{11}$$

where f_N is the non-target probability density function, $l(z)$ is the line from point z to P_{n+1} and it is vertical to P_n.

First we prove that $\Delta U \perp U_n$ when $\|\Delta U\| \to 0$. The angle between U_n and U_{n+1} is denoted as θ, then according to cosine low, $\|U_{n+1}\|^2 = \|U_n\|^2 + \|\Delta U\|^2 - 2\|U_n\|\|\Delta U\|\cos\theta$.

Because $\|U_{n+1}\|=\|U_n\|=1$, $\cos\theta =\|\Delta U\|/2$. Therefore when $\|\Delta U\|\rightarrow 0$, $\cos\theta \rightarrow 0$ and $\theta=\pi/2$. So when $\|\Delta U\|$ is small enough, it is approximative that $\Delta U \perp U_n$. Then $l(z)$ can be approximatively expressed as $z+\varepsilon U_n$ where $\varepsilon \in [0,(-\Delta U)\cdot(z-O)]$ and $z\in P_n$. And the non-target probability density on $l(z)$ approximate to $f_N(z)$. Therefore:

$$\Delta R_N \approx \int_{z\in Pn}(-\Delta U)\cdot(z-O)f_N(z)dz \quad (12)$$

Then we estimate f_N with the method of Parzen window, and shoose Gaussian distribution as the smoothing function. Therefore:

$$f_N(y)=\sum_{i=1}^{N}\varphi(y-y_i) \quad (13)$$

where $\varphi(y)=\frac{1}{(2\pi\sigma^2)^{\frac{n}{2}}}exp\left(-\frac{y\cdot y}{\sigma^{2n}}\right)$, n is the dimension of data space. Therefore:

$$\begin{aligned}\Delta R_N &\approx \int_{z\in P_N}(-\Delta U)\cdot(z-O)\left(\sum_{i=1}^{N}\varphi(z-y_i)\right)dz \\ &= \sum_{i=1}^{N}\left(\int_{z\in P_N}(-\Delta U)\cdot(z-O)\varphi((z-O-y_i^{\|})-y_i^{\perp})dz\right) \\ &= \sum_{i=1}^{N}\left(\frac{1}{(2\pi\sigma^2)^{\frac{1}{2}}}exp\left(-\frac{\|y^{\perp}\|^2}{\sigma^2}\right)(-\Delta U)\cdot(y_i^{\|})\right) \\ &= (-\Delta U)\cdot\left(\sum_{i=1}^{N}\frac{1}{(2\pi\sigma^2)^{\frac{1}{2}}}exp\left(-\frac{\|y^{\perp}\|^2}{\sigma^2}\right)(y_i^{\|})\right)\end{aligned} \quad (14)$$

The direction of vector $\sum_{i=1}^{N}\frac{1}{(2\pi\sigma^2)^{\frac{1}{2}}}exp\left(-\frac{\|y^{\perp}\|^2}{\sigma^2}\right)y_i^{\|}$ is the optimal direction that ensure $\Delta R_N>0$. Therefore:

$$\Delta U = \delta\sum_{i=1}^{N}\frac{1}{(2\pi\sigma^2)^{\frac{1}{2}}}exp\left(-\frac{\|y^{\perp}\|^2}{\sigma^2}\right)y_i^{\|} \quad (15)$$

where δ is a small positive value to ensure the assumption that $\|\Delta U\|$ is small enough. The update rule defined by formula 15 equals to that defined by formula 9. So the optimal object of LTM is to maximize the correct rejection rate, when Parzen window is used to estimate probability density and Gaussian distribution is chosen as smooth function.

If $\sum_{i=1}^{N}\frac{1}{(2\pi\sigma^2)^{\frac{1}{2}}}exp\left(-\frac{\|y^{\perp}\|^2}{\sigma^2}\right)y_i^{\|}=0$ then $\Delta R_N=0$, training is over and we get an optimal solution that maximizes R_N. The classifier output by LTM is not global optimal, it depends upon the initial hyper-plane. Although the drawback slows the classification speed, it doesn't decrease the correct detection rate. As long as the non-target data is outside the convex hull of the target set, a decision hyper-plane can be found to separate it form the target set.

3 Experiments

The performance of LTM is evaluated in 2D and 3D Euclidean space respectively.

3.1 Experiment in 2D Space

To illuminate the training process of LTM, the experiment in 2D space is conducted. The whole data area is an ellipse centered at point (0, 0) with an x-axis of 300 and a y-axis of 200; the target area is a circle centered at point (0, -100) with a radius of 100. 200 points are sampled randomly at the target area as the target training set; 300 points are sampled randomly at the non-target area as the non-target training set.

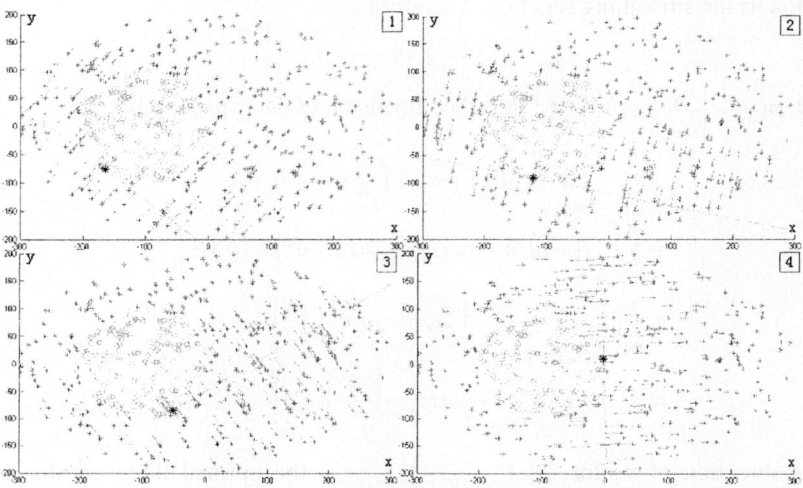

Fig. 5. Training the first classifier with LTM. The blue sparkle is the fulcrum, and yellow arrow stands for force. The length of arrow varies directly proportional to the force.

In the experiment, r, δ, and σ are set as 0.1, 0.003, and 120 respectively. As shown in Fig. 5, the line is aligned to the balanced position stage by stage under the effect of forces. Fig. 5 (1) shows the line after the 1st training stage, and the norm of moment sum $||\Sigma M||$ is 368.1; Fig. 5 (2) indicates the line after the 4th training step, and $||\Sigma M|| = 1124.9$; Fig. 5 (3) shows the line after the 8th training stage, and $||\Sigma M|| = 2114.5$; Fig. 5 (4) presents the line after the 27th training stage, $||\Sigma M||$ is reduced to 0.0011, it is small enough and the training is ended. The first trained classifier distinguishes 62.0% of the non-target data.

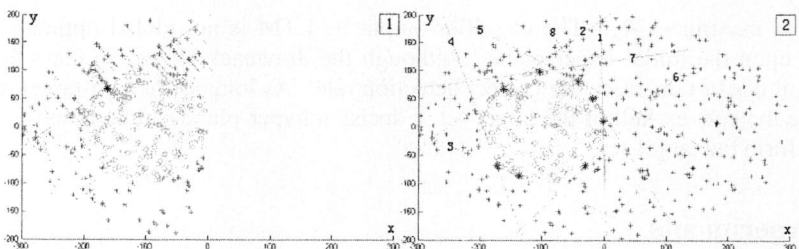

Fig. 6. Training a sequence of linear classifiers. A sequence of linear classifiers is yielded by LTM for exact discrimination.

As shown in Fig. 6 (1), the non-target data rejected by the first classifier is removed. LTM is applied to the abridged data set for the second classifier. The second classifier distinguishes 14.7% of the non-target data. Finally 7 classifiers are trained in turn for accurate discrimination without false alarm as indicated in Fig. 6 (2).

3.2 Experiment in 3D Space

The second experiment is performed in 3D space. The whole data area is an ellipsoid centered at point (0, 0, 0) with an x-axis of 300, a y-axis of 400 and a z-axis of 500; the target area is a sphere centered at (0, -100, 0) with a radius of 100. 200 points are gathered at the target area as the target training set; 400 points are gathered at the non-target area as the non-target training set.

In this experiment, r, δ, and σ are set as same as the first experiment. The first decision plane trained by LTM distinguishes 53.5% of the non-target data; the first three decision plane trained by LTM reject 90.0% of the non-target data, as shown in Fig. 7 (1); the first five decision planes output by LTM cut away 96.0% of the non-target data. Totally LTM has trained 9 planes that separate all non-target data from target set as illustrated in Fig. 7 (2).

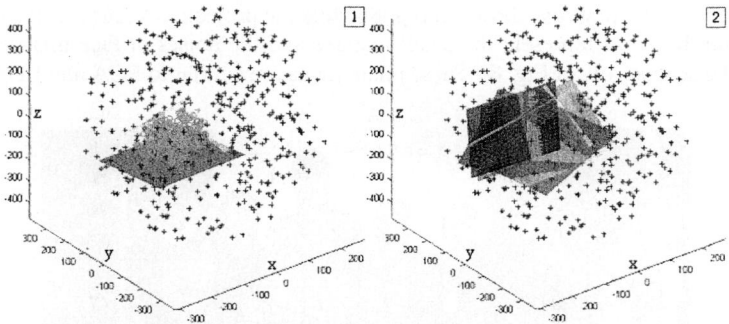

Fig. 7. Training a set of decision planes in 3D space. Finally 9 decision planes are found by LTM to separate target set from non-target set as shown in image 2.

4 Application of LTM to Face Detection

Face detection is an important topic in computer vision and in human computer interaction. The task is locating and extracting face region from all kinds of background. It is an essential technology in face processing in terms of face recognition, tracking, pose estimation, as well as expression recognition [7]. Face detection is a challenging task for the great variance of light condition, 3-D pose, and expression etc [7]. Different methods, such as neural networks [2], Bayesian decision [8], SVM [4], MRC [6], and Adaboost [9], are used to address the issue of face detection.

In the paper, LTM is applied to front face detection for a new attempt. The training sets include a face set with 5412 front face images, and a non-face set with 23573 non-face images. The face set comes from two resources. The first is the face training set of MIT-CBCL [10], which has 2429 images. And the second is the database for Viola's face detector [9], 2983 front images are selected to our face set. The non-face set comes from the non-face test set of MIT-CBCL [10], which contains 23573 non-face images. All of the images are scaled to 19×19 pixels. The size of training set has been doubled by adding the mirror image of each sample, so there are totally 10824 face training data and 47146 non-face training data.

Fig. 8. Some unit projection vectors of the linear classifiers trained by LTM

Finally LTM yielded 209 linear classifiers. Those classifiers separate all the non-face training data from face training set. Fig. 9 and Fig. 10 present some results of the face detection with our classifier. Generally speaking, LTM works well in training a face detector.

Interestingly, LTM has the ability of learning holistic features. The unit projection vector of the classifier trained by LTM presents some kind of holistic feature of face, as indicated in Fig. 8. The projection vectors of the classifiers yielded earlier by LTM look like human faces, as the first row shown in Fig. 8, while the projection vectors of the classifiers yielded later by LTM represent the detail features and the noises of face images, as illustrated in the second row of Fig. 8. These phenomena are similar to PCA and LDA.

Fig. 9. Examples of face detection with LTM
Fig. 10. Examples of face detection with LTM

5 Conclusion and Feature Work

A new learning algorithm, LTM, is introduced in this paper. LTM has direct physical meaning since it is derived from a well-known concept in physics: Lever Principle. And its optimal goal is to maximize the correct rejection rate in theory. If the distribution of target set is convex, a sequence of decision surfaces can be found to exactly discriminate the both sets without false-alarm. That is confirmed by two mathematic experiments and by the application of front face detection. In the application of face detection, it performs well with satisfactory result and it illustrates that LTM has the ability of learning holistic features.

However, LTM should be research further to perfect the theory and to promote the application. One of the future works is on non-linear LTM. LTM can't deal with the case of concave distribution of target set. The issue can be addressed by a nonlinear extension of LTM. A promising way is to develop nonlinear LTM based on kernel method as the same in nonlinear SVM [3, 4], kernel LDA [11] and kernel PCA [12]. Another future work is on the initial algorithm. The linear classifier is not global optimal, this issue can be improved by modifying the generating algorithm of initial hyper-plane. If the initial hyper-plane is near the global optimal position, it is highly likely that the initial hyper-plane will converge to the global optimal solution. Therefore a smarter algorithm should be developed to replace the random initial algorithm. The third future work is about the generalization ability. As described in section 2.2, the generalization ability of LTM can be improved by increasing the generalization factor g. However, if g is too large, false alarm will be caused even when the distribution of target set is convex. Therefore how to optimize g is another important issue.

References

1. S. Haykin, Neural Networks, Second Edition, Published by Prentice-hall, Inc., 1999.
2. H. Rowley, S. Baluja, and T. Kanade, Neural Network-Based Face Detection, IEEE Transactions on Pattern Analysis and Machine Intelligence, Vol. 20 (1): 23-38, January 1998.
3. B. Boser, I. Guyon and V. Vapnik, A training algorithm for optimal margin classifier, *Proceedings of the 5th ACM Workshop on Computational Learning Theory*, pp. 144-152, July 1992.
4. E. Osuna, R. Freund, and F. Girosi, Support Vector Machines: Training and Applications, *A.I. Memo No. 1602*, CBCL paper No. 144, March 1997.
5. R. Duda and P. Hart, *Pattern Classification and Scene Analysis*, 1st Edition, Wiley-Interscience Publication, 1973.
6. M. Elad, Y. Hel-Or, and R. Keshet, Pattern Detection Using a Maximal Rejection Classifier, *Pattern Recognition Letters*, Vol. 23 (12): 1459-1471, 2002.
7. M. Yang, D. Kriegman, and N. Ahuja, Detecting faces in images: a survey, *IEEE Transactions on Pattern Analysis and Machine Intelligence*, Vol. 24 (1): 34 – 58, 2002.
8. C. Liu, A Bayesian Discriminating Features Method for Face Detection, *IEEE Transactions on Pattern Analysis and Machine Intelligence*, vol. 25 (6): 725-740, 2003.
9. P. Viola and M. Jones, Rapid Object Detection Using a Boosted Cascade of Simple Features, *Proceedings of IEEE CS Conference on Computer Vision and Pattern Recognition*, pp. 511–518, December 2001.
10. MIT Center for Biological and Computation Learning, *CBCL face database #1*, http://www.ai.mit.edu/projects.
11. S. Mika, G. Ratsch, J. Weston, B. Scholkopf, and K. Mullers, fisher discriminant analysis with kernels, *Proceedings of the 1999 IEEE Signal Processing Society Workshop*, pp. 23-25, August 1999.
12. B. Scholkopf, A. Smola, and K. Muller, Nonlinear component analysis as a kernel eigenvalue problem, *Neural Computation*, Vol. 10: 1299-1319, 1998.

An Effective Method to Improve Convergence for Sequential Blind Source Separation

L. Yuan[1], Enfang. Sang[1], W. Wang[2], and J.A. Chambers[2]

[1] Lianxi Yuan and Enfang Sang, Harbin Acoustic Engineering College,
Harbin Engineering University, China
yuan_lianxi@hotmail.com
sangef@vip.sina.com

[2] W. Wang and J. A. Chambers, the Centre of Digital Signal Processing,
Cardiff University, United Kingdom,
wangw2@cf.ac.uk, chambersj@cf.ac.uk

Abstract. Based on conventional natural gradient algorithm (NGA) and equivariant adaptive separation via independence algorithm (EASI), a novel sign algorithm for on-line blind separation of independent sources is presented. A sign operator for the adaptation of the separation model is obtained from the derivation of a generalized dynamic separation model. A variable step-size sign algorithm rooted in NGA is also derived to better match the dynamics of the input signals and unmixing matrix. The proposed algorithms are appealing in practice due to their computational simplicity. Experimental results verify the superior convergence performance over conventional NGA and EASI algorithm in both stationary and non-stationary environments.

1 Introduction

Blind signal separation (BSS) is concerned with recovering the original unknown sources from their observed mixtures without. The algorithm operates blindly in the sense that except for statistical independence, no a prior information about either the sources or the transmission medium is available. BSS algorithms separate the sources by forcing the dependent mixed signals to become independent. This method has several applications in communications and signal processing. Suppose n unknown statistically independent zero mean source signals, with at most one having a Gaussian distribution, contained within $s \in \Re^n$ pass through an unknown mixing channel $A \in \Re^{m \times n}$ $(m \geq n)$, such that m mixed signals $x \in \Re^m$ are therefore observed which can be modeled as $x = As + e$, where $e \in \Re^m$ is the possible contaminating noise vector, which is usually ignored for simplicity in this study. The objective of BSS is to recover the original sources given only the observed mixtures, using the separation model $y = Wx$, where $y \notin \Re^n$ is an estimate of s to within the well-known permutation and scaling ambiguities, and $W \in \Re^{n \times m}$ is the separation matrix. The crucial assumption with conventional BSS is that the source signals are statistically independent. In this paper, we further assume that the sources have unit variance and the number of sources matches that of the number of mixtures, i.e. $m = n$, the exactly determined problem. To recover the source signals, it is frequently necessary to estimate an unmixing channel which performs the inverse operation of the mixing process, as subsequently used in the separation model. In this paper, we are particularly concerning with a family of sequential BSS algorithms. Fig.1 shows a block diagram of sequential BSS. The separating coefficients $W(k)$ are updated iteratively according to some estimate of the independence between the estimated signal components in $y(k)$. The sensor signal components in $x(k)$ are fed into the algorithm in order to estimate iteratively the source signal components, i.e. $y(k)$. Compared with block (batch)-based BSS algorithms, sequential approaches have particular practical advantage due to their computational simplicity and potentially improved performance in tracking a nonstationary environment [2]. The focus of this study is therefore the natural gradient algorithm (NGA) [1],[7] and the equivariant adaptive separation via independence algorithm (EASI)[6].

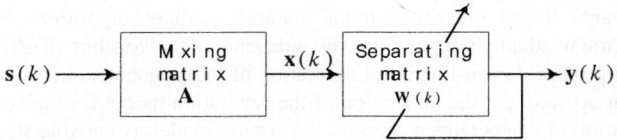

Fig. 1. Diagram of sequential blind source separation

Among important issues affecting the performance of sequential algorithms such as equation (1) are the convergence rate and the misadjustment in steady state [3]. A fixed step-size can restrict the convergence rate and can lead to poor tracking performance [2]. In contrast, an adaptive step-size can exploit the on-line measurements of the state of the separation system, from the outputs and the parameter updates. This means, the step-size can be increased for a higher convergence rate, but can be systematically decreased for reducing any misadjustment of the parameters away from their optimum settings. To improve the convergence rate, we consider using a normalization technique (leading to a sign algorithm) together with gradient-based time-varying step-size (leading to a variable step-size algorithm) in the updating process. Both techniques are shown to increase the convergence speed of the algorithm, and the sign operation can simultaneously reduce the computational complexity of the whole algorithm, additionally introduced by the adaptive step-size. The remainder of this paper is organized as follows. A sign algorithm using a normalization technique based on the standard NGA algorithm is proposed in section 2. Section 3 is dedicated to deriving a variable step-size algorithm for NGA, where the step-size is estimated from the input data and the separation matrix. Following both of the section, S-EASI algorithm was introduced. Then numerical experiments are presented in section 5 to compare the convergence performance of the proposed algorithms with that of the conventional NGA. Finally, section 6 concludes the paper.

2 Sign NGA (S-NGA)

Gradient techniques are established and well known methods for adjusting a set of parameters to minimize or maximize a chosen cost function [4]. However, simple standard gradient descent techniques is usually very slow. On these years, many novel gradient algorithms have been proposed and their better performance properties which can improve convergence speed have been proved. Here, we expect to propose a new sign-algorithm, which is based on NGA. In NGA algorithm, the discrete-time on-line updating equation of the separation matrix is denoted as

$$W(k+1) = W(k) + \mu[I - \psi(k)]W(k) \qquad (1)$$

where k is the discrete-time index, μ is a positive parameter known generally as the step-size, I is an identity matrix, and $\psi(k)$ is given by

$$\psi(k) = f(y(k))y^T(k) \qquad (2)$$

where $f(y(k))$ is an odd nonlinear function which acts element-wise on the output vector $y(k)$, and $(.)^T$ is the vector transpose operator.

In this section, we consider using normalization of the output vector $y(k)$ for the off-diagonal terms of $\psi(k)$. This thereby results in a sign operation on the elements of $Q(k)\psi$ which restricts the norm of the matrix $W(k)$. Our expectation is that, this will lead to faster convergence and better robustness in the adaptation. For mathematical formulation, let us consider a continuous matrix dynamic system

$$\frac{d}{dt}W(k) = -\mu \frac{\partial J(y(k), W(k))}{\partial W(k)} W^T(k)\Pi(y(k))W(k) \qquad (3)$$

where $J(.)$ is a cost function from which NGA is derived, and $\Pi(y)$ is a diagonal matrix with positive elements. Equation (3) can be deemed as an extension of the standard NGA [4], since (1) is a result of $\Pi(y) = I$. By a straightforward differential matrix calculation as in [1], we obtain

$$\frac{d}{dt}W(k) = \mu\Pi(y(k)[I - \Pi^{-1}(y(k)\tilde{f}(y(k))y^T(k)\Pi(y(k))]W(k) \qquad (4)$$

where $f(y(k))$ is a vector of nonlinear activation functions. Defining $\Pi^{-1}(y(k))\tilde{f}(y(k)) = f(y(k))$ and $\mu\Pi(y(k)) = \mu(k)$, we have

$$\frac{d}{dt}W(k) = \mu(k)[I - f(y(k))y^T(k)\Pi(y(k))]W(k) \qquad (5)$$

In parallel with (1), from (5), we have

$$\psi(k) \equiv f(y(k))y^T(k)\Pi(y(k)) \qquad (6)$$

Denote by $f_i(y_i)$ and y_i, $i = 1, \cdots, n$, the entries of $f(y)$, and y, and by π_{ij} the elements of Π, $\psi(k)$ can be re-written element-wise as

$$\psi_{ij}(k) = f_i(y_i) y_j \pi_{ij} \qquad (7)$$

If π_{ij} takes the form of the normalization by y_j, i.e. $\pi_{ij} = |y_j|^{-1}$ then (6) is reduced to

$$\psi(k) \equiv f(y(k))[sign(y(k))]^T \qquad (8)$$

where $sign(y(k)) = [sign(y_1(k)), \cdots, sign(y_n(k))]^T$, and

$$sign(z) = \begin{cases} 1 & z > 0 \\ -1 & z < 0 \\ 0 & z = 0 \end{cases} \qquad (9)$$

Note that, (8) could be deemed as a degenerate form of the median learning rule discussed in [4]. The introduced normalization could potentially lead to faster convergence rate because of the resulting sign activation function of the output data y increasing the magnitude of small values, which could, on the other hand, reduce the accuracy of statistics within the adaptation process, leading to inaccurate separation. To optimize both the convergence rate and separation performance, we suggest to use different normalization schemes for the elements of $\psi(k)$. Particularly, Π does not hold fixed values at its diagonal elements, but these change according to the association between $f(y(k))$ and $y(k)$. That is, (7) is re-written in the discrete-time form as

$$\psi_{ij} = \begin{cases} f_i(y_i(k)) y_i(k) & i = j \\ f_i(y_i(k)) sign(y_j(k)) & i \neq j \end{cases} \qquad (10)$$

Using the Kronecker dot product Θ (element-wise product of matrices), we have the following concise expression

$$\psi(k) \equiv f(y(k)) y^T(k) \Theta \Phi(y(k)) \qquad (11)$$

where $\Phi(y(k))$ is derived from Π and (10), i.e. the entries of Φ are denoted as

$$\varphi_{ij} = \begin{cases} 1 & i = j \\ |y_j|^{-1} & i \neq j \end{cases} \qquad (12)$$

Note that, (11) can also be written as

$$\Phi(k) \equiv diag[f(y(k)) y^T(k)] + off[f(y(k)) sign(y^T(k))] \qquad (13)$$

where $diag[.]$ and $off[.]$ denote the operation of taking the diagonal elements and off-diagonal elements of a matrix respectively.

We call the adaptation procedure of using (11) and (12) the sign natural gradient algorithm (S-NGA). Compared with the NGA using (2), the sign algorithm (SA) has reduced computational complexity, i.e. n(n-1) multiplications in (2) are replaced with simple sign tests which are easily implementable. However, for each k, the off-diagonal elements of $\psi(k)$ are not continuous (see equation (10)), this where makes the analysis of such an algorithm more difficult than that of (1). However, it is straightforward to show the algorithm is Lyapunov stable. Noticing that $W^T \Pi W = (\sqrt{\Pi}W)^T(\sqrt{\Pi}W)$ in (3), where $\sqrt{\Pi}$ represents a diagonal matrix whose diagonal entries are the square root of the corresponding diagonal elements of Π, and denoting by w_{ij}, γ_{ij}, and ψ_{ij} $i,j = 1, \cdots n$, the elements of W, $\sqrt{\Pi}W$, and $\frac{\partial J}{\partial W}(\sqrt{\Pi}W)^T$, we obtain from (3) that

$$\frac{d}{dt}J(y(k),W(k)) = \sum_{i,j}\frac{\partial J}{\partial w_{ij}}\frac{dw_{ij}}{dt}$$
$$= -\sum_{i,j}\frac{\partial J}{\partial w_{ij}}\sum_k \psi_{ik}\gamma_{kj} \qquad (14)$$
$$= -\sum_{i,k}\psi_{ik}^2 \leq 0$$

where zero is obtained if and only if $dW(k)/dk = 0$, which means the solution to W W is an equilibrium of (3).

3 Variable Step-Size Sign NGA (VS-S-NGA)

It has been shown [2] that, as compared with using a fixed step-size which would restrict convergence rate, the algorithm with an adaptive step-size has an improved tracking performance for a non-stationary environment, i.e., the value of which is adjusted according to the time-varying dynamics of the input signals and the separating matrix. As another contribution, we therefore derive a gradient adaptive step-size algorithm for the NGA algorithm, which adapts the step-size in the form of

$$\mu(k) = \mu(k-1) = \rho \nabla_\mu J(k)\big|_{\mu=u(k-1)} \qquad (15)$$

where ρ is a small constant, and $J(k)$ is an instantaneous estimate of the cost function from which the NGA algorithm is derived. To proceed, we use an inner product of matrices defined as [2],

$$\langle C,D \rangle = tr(C^T D) \qquad (16)$$

where $\langle \cdot \rangle$ denotes the inner product, $tr(\cdot)$ is the trace operator, and $C, D \in \Re^{m \times n}$ Therefore, exploiting (16), the gradient term on the right hand side of (5) can be evaluated as

$$\nabla_\mu J(k)|_{\mu=\mu(k-1)} = \langle \partial J(k)/\partial W(k), \partial W(k)/\partial \mu(k-1) \rangle \qquad (17)$$
$$= tr(\partial J(k)/W(k)^T \times \partial W(k)/\partial \mu(k-1))$$
$$\partial J(k)/\partial W(k) = -[I - f(y(k))y^T(k)]W(k) \qquad (18)$$

which is the instantaneous estimate of the natural gradient of the cost function of $J(k)$. From the equation (1), the separating matrix W at time k is obtained as

$$W(k) = W(k-1) + \mu(k-1)[I - f(y(k-1))y^T(k-1)]W(k-1) \quad (19)$$

Following the approach from [2] and [5], from the above equation, we have

$$\partial W(k) / \partial \mu(k-1) = [I - f(y(k-1))y^T(k-1)]W(k-1) \quad (20)$$

Using the notation of (2) for $\psi(k)$ in the standard NGA algorithm and denoting we have

$$\Gamma(k) \equiv [I - \psi(k)]W(k) \quad (21)$$

$$\nabla_\mu J(k)|_{\mu=\mu(k-1)} = -tr(\Gamma^T(k)\Gamma(k-1)) \quad (22)$$

Hence, an adaptive step-size with the form of (15) can be written as

$$\mu(k) = \mu(k-1) + \rho tr(\Gamma^T(k)\Gamma(k-1)) \quad (23)$$

which can be estimated from the input signals and the separation matrix. (21) has a similar form as the equation (7) in [2], which was derived for an equivariant adaptive source separation via independence (EASI) algorithm[6]. The separation procedure using (1), (2), (21) and (23) represents the proposed variable step-size NGA algorithm (VS-NGA). Following a similar procedure as in section 2, see (6) and (11), and as in this section, see (18) and (20), it is straightforward to derive an adaptive step-size algorithm using different normalization for the off-diagonal elements of $\psi(k)$. In this case, $\psi(k)$ takes the form of (11). We represent (1), (11), (21) and (23) the sign version of the variable step-size NGA algorithm, i.e., VS-S-NGA for notational simplicity.

4 Sign-EASI

Cardoso proposed EASI algorithm in 1996. EASI algorithm is a kind of adaptive algorithms for source separation which implements an adaptive version of equivariant estimation. It is based on the idea of serial updating: this specific form of matrix updates systematically yields algorithms with a simple structure, for both real and complex mixtures, and its performance does not depend on the mixing matrix. So convergence rates, stability conditions and interference rejection levels of EASI algorithm only depend on distributions of the source signals. In order to reduce computation complexity of the algorithm and obtain a satisfied stability, sign function is applied to this kind of algorithm. Firstly, the separating matrix update equation for EASI algorithm is given by

$$W(k+1) = W(k) + \mu[I - y(k)y^T(k) - f(y(k)y^T(k) + y(k)f(y(k))^T]W(k) \quad (24)$$

Here we also set a parameter $\psi(k) = f(y(k))y^T(k)$ to substitute $f(y(k))$ and $y(k)$ in the upper equation, then (24) can be rewritten as:

$$W(k+1) = W(k) + \mu[I - y(k)y^T(k) - \psi(k) + \psi(k)^T]W(k) \tag{25}$$

In order to easily understand and keep consistent with the NGA algorithm, all of the parameters in the above equation are defined just as in the section 2, i.e. $Q(k)$ in the (25) takes the same form as in the section 2:

$$\psi(k) \equiv diag[f(y(k))y^T(k)] + off[f(y(k))sign(y^T(k))] \tag{26}$$

Therefore, seeing in the section 2, we can omit some middle procedures and directly derive the final algorithm what we expect. Equation (25) and (26) are all together called the Sign EASI algorithm, namely S-EASI.

5 Numerical Experiments

In the first experiment, we mix a fixed sinusoidal signal with a randomly selected uniform source signal by using a 2-by-2 ($m = n = 2$) matrix $A_0 = randn(m, n)$, i.e. Zero mean, independent white Gaussian noise with standard deviation 0.1 was added to the mixtures. A cubic non-linearity $f(.)$ was used as the activation function. The performance index (PI) [1], as a function of the system matrix $G = WA$, was used to evaluate the proposed algorithm

$$PI(G) = \left[\frac{1}{n}\sum_{i=1}^{n}(\sum_{k=1}^{m}\frac{|g_{ik}|}{\max_k |g_{ik}|}) - 1\right] + \left[\frac{1}{m}\sum_{k=1}^{m}(\sum_{i=1}^{n}\frac{|g_{ik}|}{\max_i |g_{ik}|}) - 1\right] \tag{27}$$

where g_{ik} is the ik-th element of G. The initial value of μ for all the algorithms was set to 0.0045, $\rho = 2\times10^{-5}$, and 200 Monte Carlo trials were run for an averaged performance. The same simulation conditions were used for all the algorithms to allow fair comparison. Fig.2 shows convergence behavior of the various approaches. From Fig.2, it is found that the proposed sign algorithms have much faster convergence speed. For example, for the fixed step size, S-NGA needs approximately 2000 samples to converge, whereas the conventional NGA needs approximately 3250 samples. Note that, we mean the convergence by the PI reduced to 0.02 (corresponding to an approximately successful separation). For the adaptive step-size, VS-S-NGA only requires approximately 1050 samples for convergence, however, VS-NGA requires approximately 1700 samples. It is clear that VS-S-NGA has the fastest convergence rate, which is a very promising property for sequential algorithms. Without any change of parameters, we continued to realize the second group of simulation with S-EASI and EASI algorithms on the same conditions. Fig.3 showed a compared result between them. S-EASI arrived its steady convergence near the approximate 1300 samples, while EASI had to need around 1800 samples to satisfy this requirement. From Fig.3, it clearly proved that the convergence rate of the S-EASI algorithm was faster than EASI. Here, we only provided the simulation results with a fixed step size. For the varying adaptive step-size, we also gained a similar conclusion, but it was not very stable. So we still need further experiments to verify it.

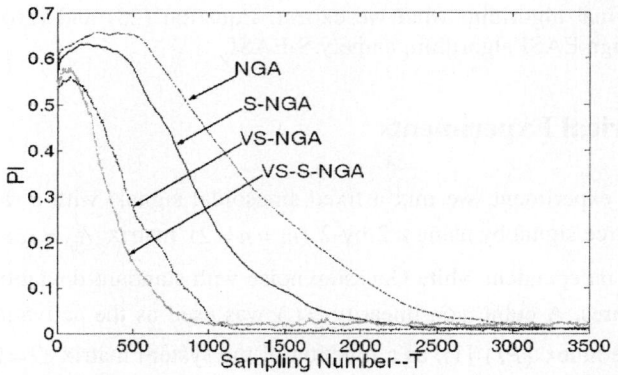

Fig. 2. Comparison of convergence rate by performance index in a stationary environment

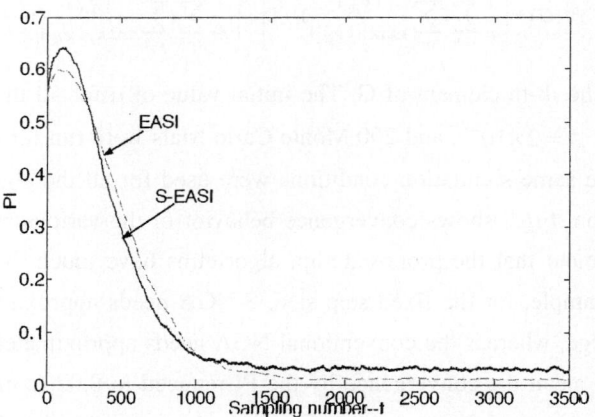

Fig. 3. Comparison of convergence rate between S-EASI and EASI in a stationary environment

In the second experiment, the different approaches were examined for a non-stationary environment. To this end, we use the following time-varying mixing matrix

$$A = A_0 + \Psi \tag{28}$$

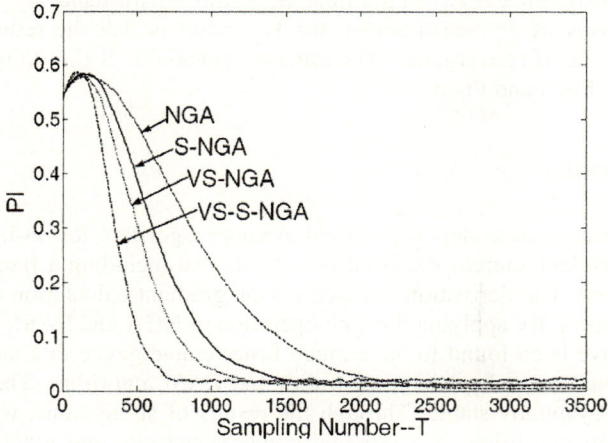

Fig. 4. Comparison of convergence rate by performance index in a non-stationary environment

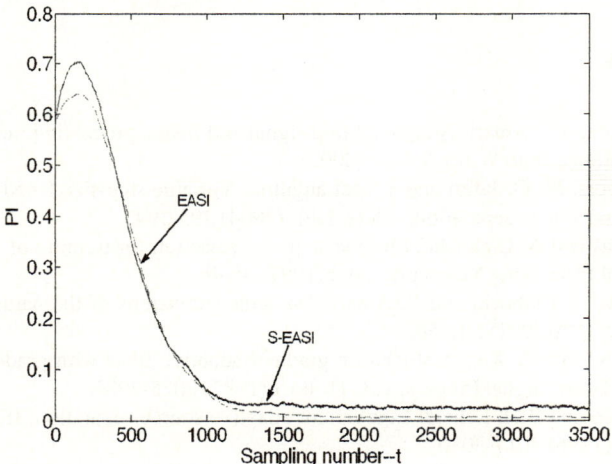

Fig. 5. Comparison of convergence rate between S-EASI and EASI in a non-stationary environment

where $\Psi = \alpha\Psi + \beta.randn(size(A,1))$, $randn(.)$ and $size(.)$ are MATLAB built-in functions, and the initial Ξ is set to a null matrix. A_0 is the same as in (27). Here α is set to 0.95 and β to 0.001. Other parameters are the same as those in the first experiment. Again, their convergence performances are compared in Fig.4 and Fig.5 respectively. For the Fig.4, we observed similar performance improvement gained for the proposed approaches in a non-stationary environment. Note that, lower *PI* generally indicates a better separation performance. In both Fig.2 and Fig.4, although we have not observed much difference between the final separation performance by S-NGA and VS-S-NGA in terms of *PI* measurement, the key point is that the reduced complexity improves the rate of convergence. The same conclusion on S-EASI algorithm also can be made from Fig.3 and Fig.5.

6 Conclusions

A new sign and variable step-size natural gradient algorithm for on-line blind separation of independent sources has been presented, also including a fixed step-size sign EASI algorithm. The derivation is based on the gradient calculation of a generalized dynamic equation. By applying the sign operation to NGA and EASI, these separation algorithms have been found to have much faster convergence rate as compared with the conventional natural gradient algorithm and EASI algorithm. The algorithm was shown to be Lyapunov stable. Through the results of simulations, we prove both of new algorithms can bring us a satisfied convergence rate and reduced computation complexity. Although variable step-size sign EASE algorithm need further testing, we still derived a variable step-size algorithm for the natural gradient learning which was also shown to have faster convergence rate and than using a fixed step-size algorithm.

References

1. A. Cichocki and S. Amari :Adaptive blind signal and image processing: learning algorithms and applications, John Wiley & Sons(2002)
2. J.A. Chambers, M. G. Jafari and S. McLaughlin.: Variable step-size EASI algorithm for sequential blind source separation, Elect. Lett. (2004) 393-394
3. S.C. Douglas and A. Cichocki : On-line step-size selection for training of adaptive systems, IEEE Signal Processing Magazine, no. 6(1997) 45-46
4. P. Georgiev, A. Cichocki and S. Amari :On some extensions of the natural gradient algorithm, Proc. ICA(2001) 581-585
5. V.J. Mathews and Z. Xie: A stochastic gradient adaptive filter with gradient adaptive step size, IEEE Trans. Signal Process, vol. 41, no. 6(1993) 2075-2087
6. J.-F. Cardoso and B.H. Laheld: Equivariant adaptive source separation, IEEE Trans. Signal Process, 44(1996) 3017-3030
7. S.Amari: Natural Gradient Works Efficiently in Learning, Neural Computation 10(1998) 251-276
8. S.Amari and S.C.Douglas: Why Natural Gradient, In Proc. IEEE International Conference Acoustics, Speech, Signal Processing, volume II, Seattle, WA, May (1998) 1213-1216

A Novel LDA Approach for High-Dimensional Data

Guiyu Feng, Dewen Hu*, Ming Li, and Zongtan Zhou

Department of Automatic Control, College of Mechatronics and Automation,
National University of Defense Technology, Changsha, Hunan, 410073, China
dwhu@nudt.edu.cn

Abstract. Linear Discriminant Analysis (LDA) is one of the most popular linear projection techniques for feature extraction. The major drawback of this method is that it may encounter the small sample size problem in practice. In this paper, we present a novel LDA approach for high-dimensional data. Instead of direct dimension reduction using PCA as the first step, the high-dimensional data are mapped into a relatively lower dimensional similarity space, and then the LDA technique is applied. The preliminary experimental results on the ORL face database verify the effectiveness of the proposed approach.

1 Introduction

In pattern recognition applications, how to obtain the most discriminant features is a very significant problem. To this end, Linear Discriminant Analysis (LDA) [2] serves as an important technique for linear feature extraction, the objective of which is to find the set of the most discriminant projection vectors and map high-dimensional samples onto a low-dimensional space. In the projective feature space, all mapped samples will get the maximum between-class scatter and the minimum within-class scatter simultaneously, and then the test samples from different classes should be easily classified.

However, small sample size problem is the possible obstacle for applying LDA whenever the number of samples is smaller than the dimensionality of the samples, which makes the between-class scatter matrix become singular. In recent years, many researchers have noticed this problem and tried to solve it using different methods. In [1], the well-known fisherfaces method was proposed, which is a two step PCA+LDA approach: Principal Component Analysis (PCA)is used as a preprocessing step for dimensionality reduction so as to discard the null space of the within-class scatter matrix of the training data set; the potential problem of this method is that it may result in the loss of some significant discriminatory information in its first step. Contrary to [1], Yu and Yang presented a direct LDA (D-LDA) algorithm [2] for high-dimensional data set, which has been proved to be suboptimal in theory [4]. In 2000, Chen et al. proposed the

---
* Corresponding author.

LDA +PCA method [5], although this method solves the small sample size problem, it is obviously suboptimal because it maximizes the between-class scatter in the null space instead of the original input space.

In this paper, we present a novel LDA approach to deal with the small sample size problem for high-dimensional data. The main idea is described as follows: the high-dimensional data are transformed into a relatively lower dimensional space via similarity analysis, and then the LDA technique is applied. The experimental results on the ORL face database verify the effectiveness of the proposed approach. The advantages of our approach are two-folds: on one hand, the original data may be very high dimensional and computing intractable, after transformed into the similarity space, this problem is avoided; on the other hand, in the relatively lower-dimensional similarity space, the small sample size problem for applying LDA is also alleviated.

The rest of the paper is organized as follows: Section 2 gives the detail of the proposed approach. In Section 3 we describe the database and experiments carried out and analyze the results. Section 4 presents some conclusions and future work.

2 The Proposed LDA Approach for High-Dimensional Data

This novel approach includes two steps. In the first step, all the data in the original space are mapped into the similarity space via similarity analysis. Secondly, traditional LDA is applied for feature extraction. More specifically, suppose training samples $\{x_1, x_2, \ldots, x_M\}$, with class labels $\{X_1, X_2, \ldots, X_c\}$, are given, and each column $x_i (i = 1, 2, \ldots, M)$ vector has n dimensions. The distance (dissimilarity measure) between arbitrary two samples can be expressed as:

$$d(i,j) = \|x_i - x_j\|_2 (i = 1, 2, \ldots, M, j = 1, 2, \ldots, M) \tag{1}$$

which in fact means the Euclidean distance between these two samples in the original space. The similarity between the two samples is then defined as:

$$s(i,j) = (\frac{1}{e})^{\frac{1}{r} d^2(i,j)} (i = 1, 2, \ldots, M, j = 1, 2, \ldots, M) \tag{2}$$

Here r is a positive constant, and it is obvious that $s(i,j) \in (0, 1]$, which can be regarded as the similarity indicator of the two samples in the original data space. By calculating all the similarity indicators $s(i,j)(i = 1, 2, \ldots, M, j = 1, 2, \ldots, M)$ from all the M training samples, a similarity matrix S can be obtained, here,

$$S_{ij} = s(i,j)(i = 1, 2, \ldots, M, j = 1, 2, \ldots, M) \tag{3}$$

the class label of every row of S_i is the same as that of $x_i (i = 1, 2, \ldots, M)$, then there are M mapped samples and the corresponding similarity space is M-dimensional, and then the LDA technique can be applied to maximize the Fisher criterion $J(\Phi)$ in similarity space , and then the projection matrix $A =$

$\{\Phi_1, \Phi_2, \ldots, \Phi_{(c-1)}\}$ are obtained(for details, please refer to Ref.[2]). For a test sample x_t, the corresponding similarity vector $S_t = [S(t,1), S(t,2) \ldots, S(t,M)]$ is projected onto the vectors:

$$f = S_t * A \tag{4}$$

Strictly speaking, $(M+c)$ samples at least are needed to have a nonsingular between-class scatter for the similarity vectors [3]; therefore we should rewrite the S_w matrix:

$$S_w := S_w + \varepsilon I \tag{5}$$

to avoid this problem. Here ε is a small positive constant, I is an M by M identity matrix.

3 Experimental Results and Analysis

We tested this novel LDA approach on the ORL face database which is available at $http://www.cam\_orl.co.uk/facedatabase.html$. This database was built at the Olivetti Research Laboratory in Cambridge, UK. The database consists of 400 different images, 10 for each of 400 distinct subjects. There are 4 female and 36 male subjects. For some subjects, the images were taken at different sessions; varying the lighting, facial expression and facial details.The size of each image is 92*112 pixels with 256 grey levels.

In our experiment, five images from each subject are selected at random to comprise the training set, and the left are the test set (the partition is the same as in [2]). Therefore, there are equally 200 images in both the training and the test set. As far as calculating cost is concerned, the size of all the images is resized to 46*56. We extract the 39 most discriminant vectors by using the proposed approach and the nearest neighbor classifier in L2 norm sense is adopted. The results are shown in Table 1, note that our approach is usually better than the other two methods on other number of training samples. To save space,

Table 1. Face Recognition performance results

Methods	Recognition accuracy rate
Fisherfaces	92.5%
D-LDA	90.8%
Proposed approach	93.5%

we do not show all the results here. From Table 1, it can be seen that using our proposed approach, 93.5% recognition accuracy rate is obtained, and as a comparison, the fisherfaces method, 92.5% and the D-LDA method, 90.8%. It should also be pointed out that whether S_W is revised or not, the performance of the proposed approach changes little, and it can be seen that with the M training samples at hand the small sample size problem is to some extent alleviated .

The results show that the proposed approach achieves better performance for face recognition than the fisherfaces method and D-LDA method, the possible reasons may be as follows: the similarity analysis retains as much information as possible, in addition, the small sample size problem is alleviated, therefore the LDA method can be deployed and discriminative features are obtained, and such features is more suitable for the classification task.

4 Conclusions and Future Work

In this paper, we present a novel LDA approach for high-dimensional data. The proposed approach is verified effective on the ORL face database. It is also found in the experiments that the small sample size problem is to some extent alleviated in the similarity space. In fact, there are several methods to make the within-class scatter matrix entirely nonsingular in the similarity space, for example, divide the training set into two parts, first fix the number of one part as M, and then other training data are also compared with the M samples, hence more than M training vectors can be obtained, which could be $(M + c)$ and larger than the dimensions of the similarity space, however, such experiments are beyond the scope of this short paper. The direction of our future work is to improve this approach and extend its applicable scope to other larger and real applications, and we will also test it in our multi-modal biometrics system[6].

Acknowledgement

This work is partially supported by National Science Foundation of China (30370416 and 60225015), Specialized Research Fund for the Doctoral of Higher Education (20049998012), Ministry of Education of China (TRAPOYT Project).

References

1. Belhumeur P., Hespanha J., and Kriegman D.: Eigenfaces vs. Fisherfaces: Recognition using Class Specific Linear Projection. IEEE Trans. Pattern Anal. Mach. Intell.19 (7) (2001) 711–720.
2. Yu H. and Yang J.: A Direct LDA Algorithm for High-dimensional Data with Application to Face Recognition. Pattern Recognition 34 (2001) 2067–2070.
3. Jin Z., Yang J.Y., Hu Z.S. and Lou Z.: Face Recognition Based on the Uncorrelated Discriminant Transformation. Pattern Recognition 34 (2001) 1405–1416.
4. Yang J., Frangi A. F., Yang J.Y., Zhang D. and Jin Z.: Kernel PCA plus LDA: A Complete Kernel Fisher Discriminant Framework for Feature Extraction and Representation, IEEE Trans. Pattern Anal. Mach. Intell.27 (2) (2005) 230–244.
5. Chen L., Liao H., Ko M., Lin J., and Yu G.: A New LDA-based Face Recognition System Which Can Solve the Small Sample Size Problem. Pattern Recognition 33 (2000) 1713–1726.
6. Feng G., Dong K., Hu D., Zhang D.: When Faces Are Combined with Palmprints: A Novel Biometric Fusion Strategy. In Proc. of the 1st International Conference on Biometric Authentication, Hong Kong, China, LNCS 3072 (2004) 701–707.

Research and Design of Distributed Neural Networks with Chip Training Algorithm[1]

Bo Yang, Ya-dong Wang, and Xiao-hong Su

School of Computer Science and Technology, Harbin Institute of Technology,
150001 Harbin, China
{YangBo03, YDWang, SXH}@hit.edu.cn
http://www.cs.hit.edu.cn/english/index_en.htm

Abstract. To solve the bottleneck of memory in current prediction of protein secondary structure program, a chip training algorithm for a Distributed Neural Networks based on multi-agents is proposed in this paper. This algorithm evolves the global optimum by competition from a group of neural network agents by processing different groups of sample chips. The experimental results demonstrate that this method can effectively improve the convergent speed, has good expansibility, and can be applied to the prediction of protein secondary structure of middle and large size of amino-acid sequence.

1 Introduction

In recent years, more and more distributed problem-solved methods had been proposed to solve a large-scale computing work, such as the multi-agent system (MAS) [1] and parallel virtual machine (PVM) [2].

Those can be concluded in a searching method. The distributed computation aims to improve searching ability. It can be subdivided into two categories:

1. The previous experience knowledge to a problem is known, the key to achieve the answer is to speed-up the convergence to the optimal solution. The often-used method is the hill-climbing algorithm, such as the gradient descent method, simulation anneal method and etc.
2. There is no any previous experience knowledge of a problem, or there are many local optimal solutions in the searching-space. The key to approach the global optimal solution is to avoid the interference from local optimum, and reveal the direction to the global optimum. The evolution algorithm such as genetic algorithm [3] is availability in this situation.

Neural network [4] is a computational model, which consists of many simple units working in parallel with no central control. BP learning algorithm is successfully to train multilayer feed-forward networks, however, there is no guarantee to the global optimum, and its convergence speed is often slow especially when the training set is very large. The distributed neural network aims improve the training algorithm's performance. Recently, the distributed neural networks achieve new developments on image processing and other fields [5].

---

[1] Sponsored by the National Natural Science Foundation of China under Grant No. 60273083.

In this paper, a Distributed Neural Networks (DNN) based on MAS is modeled for learning from a large-scale data set. And a learning algorithm based on chip training is proposed to work on the distributed neural networks, which evolve the global optimum from a group of neural network agents. The experimental results in the prediction of protein secondary structure show that this Distributed Neural Networks with Chip Training Algorithm (DNNCTA) can effectively avoid BP network converging to local optimum. It is found by comparison that the neural network obtained from the DNNCTA can effectively improve the convergent speed.

2 Agent-Based Cooperative Model

In distributed applications, some methods have been proposed to adapt to the distributed environment through changes in the structure of neural network. Jean-Francois Arcand researched on ADN [6], which regard the sub-networks as agents, and establish a whole neural network by combined with those trained sub-networks, but it's just a plan and only some parts has been realized, because it is difficult to combine with the agents with different training target. Another distributed neural network (DNN) is proposed [7], which built a virtual neural network using the communication network. Every computer in the model simulates one or more neural nodes. But the training complexity is increased with the size of the virtual neural network; the rapidly increased communication among the neural nodes will cause the problem of lacking of status consistency and communication disruption [7].

In this paper, a new DNN model based on multi-agent from another point of distributed is proposed. It improves the convergent speed through making use of the current network resources. The model is build based on a Hybrid Model combined the Master-Slave Model [2] with the Node-Only Model [2], every agent in this model is peer to peer that can offer computation service and get help from other agents when training its sample set. So many distributed agents in different location can process the large sample. Those free agents formed a Node-Only Model when there is no computation mission as Fig.1 described. They will change to be a Master-Slave Model when one of them informs a computation mission as shown in Fig.2. Those resources are engaged in computation will be released at the end of the mission, and return to be a free agent as a new Node-Only Model waiting for the next mission.

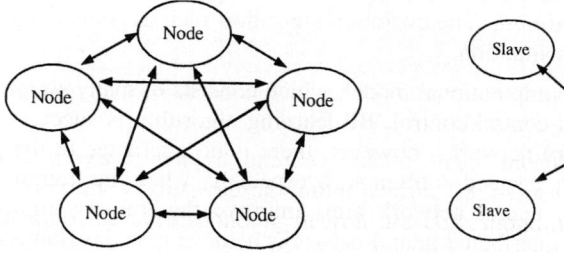

Fig. 1. Node-only model

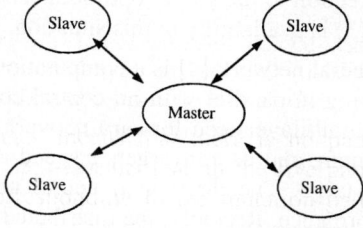

Fig. 2. Master-Slaver Model

3 Chip Training Algorithm for Distributed Neural Network

The Chip Training Algorithm (CTA) is proposed for the above-described distributed neural network structure. CTA comes from the considering of chips, sample chips and training chips. In the distributed environment, every computing resource is charged by an agent that evolved the answer to the question through the competition among the agents. Every agent trains the local neural network with its local sample, and provides the best results to the master. The master collects its cooperators' advices, and chooses one of them as the answer to the question.

Thereby, the more resources, the more advices have been provided, and the probability to the global optimum is increased.

CTA is implemented as follows:

Step1. Set neural network's structure, mutation rate Pm, the training's Termination-Conditions: maximum iteration $Times$, the expected precision

Step2. Collect the information of the current distributed environment, such as the available resource number, the cooperate-agent's state, etc.

Step3. Input training set, initiate a computation mission in the current Node-Only Model and form a Master-Slave computation environment.

Step4. Make partition of training set and distribute it to the cooperators for the new mission.

Step5. Each agent executes the computation mission using the local data set that come from the Master, and returns the result of each phase, a training chip.

Step6. Master evaluates the results come from its cooperators and check the Termination-Conditions:
    If Termination-Conditions = True Then
        Return the best result
    Else
        Evolved the cooperator's weights set through the Select & Mutate operations, which is the starting point in the next phase of cooperate-agents;
    End if

Step7. Repeat the above steps from step5 to step7.

4 Simulation Results

The performance of CTA with the above-described DNN structure is tested on the Protein Secondary Structure Prediction (PSSP) problem in the accuracy of convergence, accuracy of prediction and CPU running times.

The following parameter settings is used in the experiment: sample encoded with orthogonal matrix of 20 nodes, all-zero means the blanks at the N- or C-terminal of a chain, window Length is 13, the sample chip size is 250, the training chip size is 5 and the maximum iteration number is 20. The mutation rate Pm is 0.05.The neural network is three-layer architecture with 260-12-3 and sigmoid function. The training data set contains 2182 samples come from 9 protein sequences, and the test data set contains 2043 samples come from 8 protein sequences.

Table 1. The average results of three different situations

Resources	False Acceptance Rate of Training	Training Time (Sec.)	False Acceptance Rate of Prediction
1	0.6375	1930.9	0.6329
2	0.49	692.3	0.5172
3	0.5159	407.5	0.5089

The results show that DNNCTA algorithm cost-time is less, as the number of resource is more. Because the sample had been proceeded by each cooperate-agent is less than the whole training sample set, which is the key to reduce computing time-cost. And the best result comes from the one with more cooperate-agent in the training, which indicates that the distributed neural network with CTA method has improved the generalization because it has more chances to approach the global optimum.

5 Summary

This article has attempted to describe the DNN how to work on a large-scale dataset with CTA algorithm. First, a DNN environment must be built to be a dynamic model combined the Node-Only Model and Master-Slave Model. Then a training algorithm based on the chips is model, which enable the DNN to learn from a distributed environment. As demonstrated in the experiments about the PSSP problem, the time-cost for training is reduced with the resource's increasing, and the performance of the trained neural network keeps as well as before. This algorithm resolves the memory bottleneck problem, and provides a method to build a Computing Grid for NN.

References

1. Stuart Russell, Peter Norvig, Artificial Intelligence: A Modern Approach. English Reprint Edition by Pearson Education, Inc., publishing as Prentice-Hall, (2002)
2. A.Geist, et al, PVM: A User's Guide and Tutorial for Networked Parallel Computing, MIT Press, (1994)
3. Xiaohong Su, Bo Yang, et al, A genetic algorithm based on evolutionarily stable strategy, Journal of Software, Vol.14, No.11, (2003) 1863-1868
4. Bo Yang, et al, An Algorithm for Fast Convergence of Back Propagation by Enlarging Error, Computer Research and Development, Vol.41, No.5, (2004) 774-779
5. Rosangela Canino-Koning, Dr.Greg Wolffe, Distributed neural network image processing for single image asteroid detection, 7th Annual NASA/Michigan Space Grant Consortium, Ann Arbor, October, (2002)
6. Jean-Francois Arcand, Sophie-Julie Pelletier, Analysis and development of distributed neural networks for intelligent applications, (1994)
7. Constantin Milea, Paul Svasta, Using Distributed neural networks in automated optical inspection, 24th International spring seminar on electronics technology, May 5-9, Calimanesti-Caciulata, Romania, (2001)

Support Vector Regression with Smoothing Property*

Zhixia Yang[1,**], Nong Wang[1], and Ling Jing[1,***]

[1]College of Science, China Agricultural University,
100083, Beijing, China
jingling_aaa@163.com

Abstract. The problem of construction of smoothing curve is actually regression problem. How to use SVR to solve the problem of curve smoothing reconstruction in reverse engineering is discussed in this paper. A modified support vector regression model is proposed. Numerical result shows that the smoothness of curves fitted by modified method is better than by the standard SVR, when there are some bad measure points in the data.

1 Introduction

The freeform curve construction is one of the typical problems in reverse engineering (see [1]-[3]). Essentially speaking, this problem belongs to regression. But there is a particular requirement; the curve produced have to be smoothing.

In this paper, support vector regression (SVR) (see [4]-[5]) is used to deal with the above smoothing curve problem. But the standard SVR must be modified due to the smoothing requirement.

This paper is organized as follows. Section 2 introduces our algorithm: The smoothing SVR. In section 3 some numerical experiments are given. At last, in section 4 we give the conclusion.

2 The Smoothing SVR

Suppose the training set is

$$T = \{(x_1, y_1), (x_2, y_2), \cdots, (x_l, y_l)\} \in (R \times R)^l \qquad (1)$$

with $x_1 \leq x_2 \leq \cdots \leq x_l$.

The key point of our algorithm is to replace the constant C in the standard SVR by C_i which depends on the property of the i-th training point, $i = 1, \cdots, l$. More precisely, our smoothing SVR solves the dual problem:

---

* Supported by the National Natural Science foundation of China (No.10371131).
** PhD student entrusted by Xinjiang University.
*** Corresponding author.

$$\min_{\alpha,\alpha^*} \frac{1}{2} \sum_{i=1}^{l} \sum_{j=1}^{l} (\alpha_i - \alpha_i^*)^{\mathrm{T}} K(x_i, x_j)(\alpha_i - \alpha_i^*) - \sum_{i=1}^{l} y_i(\alpha_i - \alpha_i^*)$$

$$+ \varepsilon \sum_{i=1}^{l} (\alpha_i + \alpha_i^*)$$

$$\text{s.t.} \quad \sum_{j=1}^{l} (\alpha_i - \alpha_i^*) = 0 \quad (2)$$

$$0 \leq \alpha_i, \alpha_i^* \leq C_i, \ i = 1, \cdots, l$$

where $\alpha = (\alpha_1, \alpha_2, \cdots, \alpha_l)^{\mathrm{T}}$, $\alpha^* = (\alpha_1^*, \alpha_2^*, \cdots, \alpha_l^*)^{\mathrm{T}}$, and $K(x_i, x_j) = (\Phi(x_i), \Phi(x_j))$ is the kernel function.

In order to select the penalty factor C_i, $i = 1, \cdots, l$, we first consider the corresponding curvature K_i introduced in [6]. For $i = 2, \cdots, l-1$, the absolute value of K_i is approximated by

$$|K_i| = \frac{2 \sin \frac{\beta_i}{2}}{|P_{i-1} P_{i+1}|}, \ (0 \leq \beta_i \leq \pi) \quad (3)$$

where β_i is the included angle between $\overrightarrow{P_{i-1}P_i}$ and $\overrightarrow{P_i P_{i+1}}$. And the sign of K_i is defined as follows: The sign of K_i is positive when the circular arc $\widehat{P_{i-1}P_i P_{i+1}}$ is inverted hour; otherwise the sign of K_i is negative. As for K_1 and K_l, they are defined by $K_1 = K_2, K_l = K_{l-1}$ respectively. In this way, we get the sequence of curvature $\{K_i\}, i = 1, \cdots, l$.

Now we give the distinguishing criterion of "a bad point" and "a good point". Suppose that both P_1 and P_l are "good point". For $i = 2, \cdots, l-1$, consider sign sequence of the curvature $\{\text{sign}(K_j) | j = 1, \cdots, l\}$. The i-th point P_i is called as "a bad point" if both the signs of K_{i-1} and K_{i+1} are different from the sign of K_i, otherwise P_i is called as "a good point".

So it is reasonable to select C_i by

$$C_i = \begin{cases} C, & \text{If } P_i \text{ is "a good point"}; \\ C(\frac{1-\cos(\pi-\beta_i)}{2})^p, & \text{If } P_i \text{ is "a bad point"}, \end{cases} \quad (4)$$

where β_i is the included angle between $\overrightarrow{P_{i-1}P_i}$ and $\overrightarrow{P_i P_{i+1}}$, and both p and C are a positive parameters.

According to the selection (4), the problem (2) is defined. After getting its solution α and α^*, we obtain the decision function as $f(x) = \sum_{i=1}^{l}(\alpha_i - \alpha_i^*)K(x_i, x) + b$, where b is determined by KKT conditions.

Algorithm: the smoothing SVR

1. Given a training set

$$T = \{(x_1, y_1), (x_2, y_2), \cdots, (x_l, y_l)\} \in (R \times R)^l \quad (5)$$

with $x_1 \leq x_2 \leq \cdots \leq x_l$;

2. Select $\varepsilon > 0$, $C > 0$, and a kernel function $K(x, x')$;
3. According to the above rule, calculate the sign sequence of curvature $\{\text{sign}(K_i) | i = 1, 2, \cdots, l\}$ for training points $\{P_i | i = 1, 2, \cdots, l\}$;
4. For $i = 1, \cdots, l$, decide P_i is "a good point" or "a bad point" by the above distinguishing criterion;
5. For $i = 1, \cdots, l$, select C_i by

$$C_i = \begin{cases} C, & \text{If } P_i \text{ is "a good point"}; \\ C(\frac{1-\cos(\pi-\beta_i)}{2})^p, & \text{If } P_i \text{ is "a bad point"}, \end{cases} \quad (6)$$

where β_i is the included angle between $\overrightarrow{P_{i-1}P_i}$ and $\overrightarrow{P_i P_{i+1}}$, and both p and C are positive parameters;
6. Solve the following optimization problem:

$$\min_{\alpha, \alpha^*} \frac{1}{2} \sum_{i=1}^{l} \sum_{j=1}^{l} (\alpha_i - \alpha_i^*)^{\mathrm{T}} K(x_i, x_j)(\alpha_i - \alpha_i^*) - \sum_{i=1}^{l} y_i(\alpha_i - \alpha_i^*)$$

$$+ \varepsilon \sum_{i=1}^{l} (\alpha_i + \alpha_i^*)$$

$$\text{s.t.} \sum_{j=1}^{l} (\alpha_i - \alpha_i^*) = 0 \quad (7)$$

$$0 \leq \alpha_i, \alpha_i^* \leq C_i, \ i = 1, \cdots, l$$

and get its solution $\alpha^{(*)} = (\alpha_1, \alpha_1^*, \cdots, \alpha_l, \alpha_l^*)$ of problem;
7. Construct the decision function as

$$f(x) = \sum_{i=1}^{l} (\alpha_i - \alpha_i^*) K(x_i, x) + b \quad (8)$$

where b is determined by KKT conditions.

3 Numerical Experiments

Consider the half round curve $y = -\sqrt{1-x^2}$, $x \in [-1, 1]$. The inputs are given by $x_i = -\frac{11}{10} + \frac{1}{10}i$, $i = 1, 2, \cdots, 21$. And the corresponding outputs are given by $y_i = -\sqrt{1-x_i^2} + \xi_i$, where the noise ξ_i obeys normal distribution with $E\xi_i = 0$, $E\xi_i^2 = 0.1$.

Both the standard SVR and our smoothing SVR with Gaussian kernel are executed, while the parameters of $\sigma, C, \varepsilon, p$ are shown in Fig.1. The regression curves obtained by two ways are shown in Fig.1(a) and 1(c). We have also calculated the absolute value of curvature of both regression curves and shown them in Fig.1(b) and 1(d).

It is easy to see that the absolute value of curvature corresponding to smoothing SVR is flatter than the one corresponding to the standard SVR. So the regression curves obtained by the smoothing SVR is smoother than that obtained by the standard SVR.

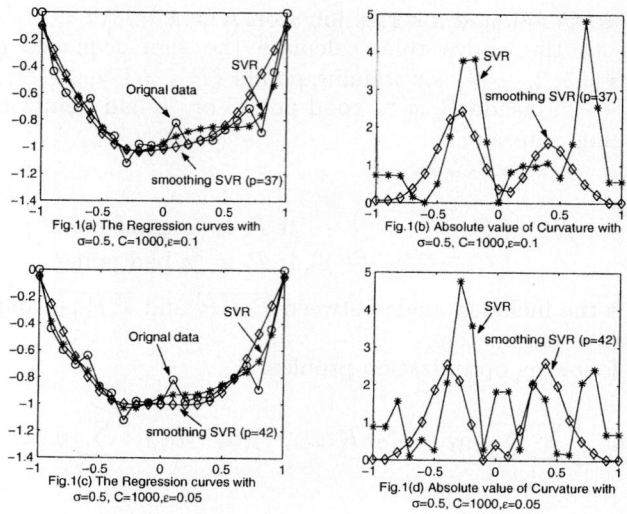

Fig. 1.

4 Conclusion

This paper is concerned with smoothing fitting in one dimensional space in the area of reverse engineering of CAD/CAM. We propose a modified support vector regression by replacing its penalty factor C by C_i depending on the training point. Preliminary numerical experiments show that this approach is promising. An interesting problem is, as an extension of the one dimensional case, to study the corresponding smoothing fit problem in two dimensional space.

Acknowledgement. We would like to thank Professor Naiyang Deng who shares his insights with us in discussions.

References

1. Hermann, G.: Free-form shapes: An integrated CAD/CAM system. Computer Industry[J], 5(1984) 205-210
2. Varady, T., Martin, R.R., Cox, J.: Reverse engineering of geometric models-an introduction. Computer Aided Design[J], 4(1997) 255-268
3. Tang, R.X.: The techniques of modern graph[M]. Shandong Science and Technology Press,China (2001)
4. Smola, A.J., Schölkopf, B.: A tutorial on support vector regression[R]. NeuroCOLT Technical Report Series NC-TR 1998-030,Royal Holloway College University of London, UK (1998)
5. Deng, N.Y., Tian, Y.J.: The new method in data mining—Support Vector Machine. Science Press, Beijing China (2004)
6. Su, B.Q., Liu, D.Y.: Computational geometry[M]. Shanghai Scientific and Technical Publishiers, China (1981) 17-227

A Fast SMO Training Algorithm for Support Vector Regression

Haoran Zhang, Xiaodong Wang, Changjiang Zhang, and Xiuling Xu

College of Information Science and Engineering, Zhejiang Normal University,
Jinhua 321004, China
{hylt, wxd, zcj74922, jkxxl}@zjnu.cn

Abstract. Support vector regression (SVR) is a powerful tool to solve regression problem, this paper proposes a fast Sequential Minimal Optimization (SMO) algorithm for training support vector regression (SVR), firstly gives a analytical solution to the size two quadratic programming (QP) problem, then proposes a new heuristic method to select the working set which leads to algorithm's faster convergence. The simulation results indicate that the proposed SMO algorithm can reduce the training time of SVR, and the performance of proposed SMO algorithm is better than that of original SMO algorithm.

1 Introduction

Support Vector Machine (SVM)[1] is an elegant tool for solving pattern recognition and regression problems, it has attracted a lot of researchers. Training a support vector machine requires the solution of a very large quadratic programming (QP) problem. Conventional QP methods is not impractical due to its high memory requirement and slow computation speed. Platt. J.C.[2] proposed a new algorithm for training classification SVM: Sequential Minimal Optimization, or SMO. SMO dramatically reduce the computational time of training SVM. Recently Smola and Scholkopf[3] proposed an new SMO algorithm for solving the regression problem using SVMs. In this paper, we make some improvement to SMO algorithm for SVR.

2 A Fast SMO Algorithm for SVR

SVR can transform to the following constraint optimization problem [4]:

$$\min \frac{1}{2}\sum_{i=1}^{l}\sum_{j=1}^{l}\overline{\alpha}_i\overline{\alpha}_j k_{ij} + \varepsilon\sum_{i=1}^{l}|\overline{\alpha}_i| - \sum_{i=1}^{l} y_i\overline{\alpha}_i \quad (1)$$

$$s.t. \quad \sum_{i=1}^{l}\overline{\alpha}_i = 0, \quad -C \le \overline{\alpha}_i \le C \quad i=1,...,l$$

SVR output can be written as:

$$f(x) = \sum_{i=1}^{l}\overline{\alpha}_i k(x_i, x) + \overline{\alpha}_0 \quad (2)$$

SMO breaks large QP problem into a series of size two QP problems. The analytical solution to the size two QP problem must be solved. We refer to [4] to give the analytical solution, and correct some mistakes in it.

We use k_{ij} denote $k(x_i, x_j)$, and rewrite the objective function of Formula (1) as a function of two parameters $\overline{\alpha}_a, \overline{\alpha}_b$:

$$L(\overline{\alpha}_a, \overline{\alpha}_b) = \frac{1}{2}\overline{\alpha}_a^2 k_{aa} + \frac{1}{2}\overline{\alpha}_b^2 k_{bb} + \overline{\alpha}_a \overline{\alpha}_b k_{ab} + \varepsilon|\overline{\alpha}_a| + \varepsilon|\overline{\alpha}_b| - \overline{\alpha}_a y_a - \overline{\alpha}_b y_b \\ + \overline{\alpha}_a v_a^* + \overline{\alpha}_b v_b^* + L' \quad (3)$$

Where $v_i^* = \sum_{j=1, j\neq a,b}^{l} \overline{\alpha}_j^* k_{ij} = f_i^* - \overline{\alpha}_a^* k_{ai} - \overline{\alpha}_b^* k_{bj} - \overline{\alpha}_0$, L' is a term that is strictly constant with respect to $\overline{\alpha}_a, \overline{\alpha}_b$, $f_i^* = f(x_i, \overline{\alpha}^*)$, Note that a superscript * is used above to explicitly indicate that values are computed with the old parameters values.

We let: $s^* = \overline{\alpha}_a^* + \overline{\alpha}_b^* = \overline{\alpha}_a + \overline{\alpha}_b$, and substitute it into (3):

$$L(\overline{\alpha}_b) = \frac{1}{2}(s^* - \overline{\alpha}_b)^2 k_{aa} + \frac{1}{2}\overline{\alpha}_b^2 k_{bb} + (s^* - \overline{\alpha}_b)\overline{\alpha}_b k_{ab} + \varepsilon|s^* - \overline{\alpha}_b| + \varepsilon|\overline{\alpha}_b| - (s^* - \overline{\alpha}_b)y_a - \overline{\alpha}_b y_b + (s^* - \overline{\alpha}_b)v_a^* + \overline{\alpha}_b v_b^* + L' \quad (4)$$

To solve the optimization value of Equation (4), we need to compute its partial derivative with respect to $\overline{\alpha}_b$:

$$\partial L(\overline{\alpha}_b)/\partial \overline{\alpha}_b = \varepsilon(\text{sgn}(\overline{\alpha}_b) - \text{sgn}(s^* - \overline{\alpha}_b)) + y_a - y_b + (\overline{\alpha}_b - s^*)k_{aa} + \overline{\alpha}_b k_{bb} + (s^* - 2\overline{\alpha}_b k_{ab}) - v_a^* + v_b^* \quad (5)$$

Now, by $\partial L(\overline{\alpha}_b)/\partial \overline{\alpha}_b = 0$, we get:

$$\overline{\alpha}_b(k_{bb} + k_{aa} - 2k_{ab}) = y_b - y_a + f_a^* - f_b^* + \varepsilon(\text{sgn}(s^* - \overline{\alpha}_b) - \text{sgn}(\overline{\alpha}_b)) + \overline{\alpha}_b^*(k_{aa} + k_{bb} - 2k_{ab}) \quad (6)$$

From formula (6), we can write a recursive update rule for $\overline{\alpha}_b$ in terms of its old value:

$$\overline{\alpha}_b = \overline{\alpha}_b^* + \eta[y_b - y_a + f_a^* - f_b^* + \varepsilon(\text{sgn}(s^* - \overline{\alpha}_b) - \text{sgn}(\overline{\alpha}_b))] \quad (7)$$

where $\eta = 1/(k_{aa} + k_{bb} - 2k_{ab})$. While formula (7) is recursive because of the two sgn() functions, so we must solve it. If the kernel function of the SVM obeys Mercer's condition, then we are guaranteed that $\eta \geq 0$ will always be true. If η is strictly positive, then Equation (5) will always be increasing. Moreover, if s^* is not zero, then it will be piecewise linear with two discrete jumps. Putting these facts together means that we only have to consider five possible solutions for Equation (7). When $(\text{sgn}(\overline{\alpha}_a) - \text{sgn}(\overline{\alpha}_b))$ sets to $-2, 0, 2$ respectively, we can get three possible solutions, the other two candidates correspond to setting $\overline{\alpha}_b$ to one of the transitions: $\overline{\alpha}_b = 0$ or $\overline{\alpha}_b = s^*$. Try Formula (7) with $(\text{sgn}(\overline{\alpha}_a) - \text{sgn}(\overline{\alpha}))$ equal to $-2, 0, 2$. If the new value is a zero to Formula (5), then accept that as the new value. If the above step failed, try

$\bar{\alpha}_b$ equal to 0 or s^*. Accept the value that has the property such that all positive (negative) perturbations yield a positive (negative) value for Formula (5). We can get:

$$\bar{\alpha}_b = \begin{cases} \bar{\alpha}_b^* + \eta[E_a^* - E_b^* + 2\varepsilon], & \bar{\alpha}_a > 0 > \bar{\alpha}_b \\ \bar{\alpha}_b^* + \eta[E_a^* - E_b^* - 2\varepsilon], & \bar{\alpha}_a < 0 < \bar{\alpha}_b \\ \bar{\alpha}_b^* + \eta[E_a^* - E_b^*], & \text{sgn}(\bar{\alpha}_a) = \text{sgn}(\bar{\alpha}_b) \\ 0, & \frac{dL(\bar{\alpha}_b)}{d\bar{\alpha}_b}(\bar{\alpha}_b = 0^+) > 0 \text{ and } \frac{dL(\bar{\alpha}_b)}{d\bar{\alpha}_b}(\bar{\alpha}_b = 0^-) < 0 \\ s^*, & \frac{dL(\bar{\alpha}_b)}{d\bar{\alpha}_b}(\bar{\alpha}_b = s^{*+}) > 0 \text{ and } \frac{dL(\bar{\alpha}_b)}{d\bar{\alpha}_b}(\bar{\alpha}_b = s^{*-}) < 0 \end{cases} \quad (8)$$

Where $E_i^* = f_i^* - y_i$. We also need to consider how the linear and boxed constraints relate to one another. Using: $L = \max(s^* - C, -C), H = \min(C, s^* + C)$, with L and H being the lower and upper bounds, respectively, guarantees that both parameters will obey the boxed constraints. To update the SVR threshold $\bar{\alpha}_0$, according to KKT condition, forces the SVR to have $f_a = y_a$, the second forces $f_b = y_b$, we average the candidate updates.

$$\bar{\alpha}_0^a = \bar{\alpha}_0^{old} + f_a^* - y_a + (\bar{\alpha}_a^{new} - \bar{\alpha}_a^{old})k_{aa} + (\bar{\alpha}_b^{new} - \bar{\alpha}_b^{old})k_{ab}$$

$$\bar{\alpha}_0^b = \bar{\alpha}_0^{old} + f_b^* - y_b + (\bar{\alpha}_a^{new} - \bar{\alpha}_a^{old})k_{ab} + (\bar{\alpha}_b^{new} - \bar{\alpha}_b^{old})k_{bb}$$

$\bar{\alpha}_0$ can be solved by:

$$\bar{\alpha}_0 = 0.5(\bar{\alpha}_0^a + \bar{\alpha}_0^b)$$

In order to accelerating convergence, we propose a new heuristic method to choose two Lagrange multipliers. For optimal problem(1), when we only consider its equality constraint condition, we may get the following Langrage function:

$$L = \frac{1}{2}\sum_{i=1}^{l}\sum_{j=1}^{l}\bar{\alpha}_i\bar{\alpha}_j k_{ij} + \varepsilon\sum_{i=1}^{l}|\bar{\alpha}_i| - \sum_{i=1}^{l}\bar{\alpha}_i y_i + \lambda\sum_{i=1}^{l}\bar{\alpha}_i$$

According to [5], Lagrange multiplier λ equals to the parameter $\bar{\alpha}_0$ in formula (2), namely: $\lambda = \bar{\alpha}_0$, So we can get:

$$L = \frac{1}{2}\sum_{i=1}^{l}\sum_{j=1}^{l}\bar{\alpha}_i\bar{\alpha}_j k_{ij} + \varepsilon\sum_{i=1}^{l}|\bar{\alpha}_i| - \sum_{i=1}^{l}\bar{\alpha}_i y_i + \bar{\alpha}_0\sum_{i=1}^{l}\bar{\alpha}_i \quad (9)$$

For (9), we solve objective function's gradient for $\bar{\alpha}_b$:

$$\partial L/\partial \bar{\alpha}_b = \sum_{i=1}^{l}\bar{\alpha}_i k_{ib} + \bar{\alpha}_0 - y_b + \varepsilon\,\text{sgn}(\bar{\alpha}_b) = E_b + \varepsilon\,\text{sgn}(\bar{\alpha}_b)$$

For an objective function, the bigger its gradient's absolute value for a variable, the bigger its variation when the optimal variable change. So according to this point we choose the first optimal variable. For the other optimal variable's choose, just like

Smola's method, to make the variable's variation the biggest. According to above statement, the strategy to working pair selection is as following: firstly select the first variable $\overline{\alpha}_b$ which makes $Max(|E_b + \varepsilon\,\mathrm{sgn}(\overline{\alpha}_b)|)$ holding, then choose the second variable $\overline{\alpha}_a$ which makes $Max(|E_a - E_b|)$ holding.

3 Simulation and Comparisons

In this section we will take a concrete example to illustrate the proposed method. The fitting function under consideration is as follow:

$$y(x) = 4\sin(2x) - x$$

Then we can train SVM by $\{x(k), y(k)\}$. We take Gaussian function as kernel, the simulating results are as follows:

Table 1. The running time of original SMO and improved SMO

Sample number		100	300	500	800	1000	2000	3000
Improved SMO	Iteration number	3964	5935	6533	9557	9852	8938	9668
	Running time	0.27	0.61	0.81	2.1	2.4	4.81	10.82
Original SMO	Iteration number	19098	21228	24192	36296	34656	30143	31025
	Running time	1.0	2.0	2.8	7.0	8.1	16.1	32.0

Both trained SVM can exactly fit the function, their training and testing error are almost the same, but their running time are very different, The proposed algorithm is several times faster than the original SMO.

References

1. Vapnik, V.N.: The nature of statistical learning theory. Springer, New York(1995).
2. Platt, J.C. : Fast training of support vector machines using sequential minimal optimization. Advances in Kernel Methods: Support vector Machines. MIT Press. Cambridge, MA(1998).
3. Smola, A.J. : Learning with Kernels. PhD Thesis, GMD, Birlinghoven, Germany (1998).
4. Gary William Flake, Steve Lawrence : Efficient SVM Regression Training with SMO. Machine Learning special issue on SVMs (2000).
5. Mario Martin : On-line Support Vector Machines for Function Approximation. Technical Report LSI-02-11-R, Software Department, Universitat Politecnica de Catalunya (2002).

Rival Penalized Fuzzy Competitive Learning Algorithm

Xiyang Yang[1,2] and Fusheng Yu[1]

[1] College of Mathematics, Beijing Normal University,
Beijing 100875, P.R.China
yufusheng@263.net
[2] Math Department, Quanzhou Normal University,
Quanzhou 362000, P.R.China

Abstract. In most of the clustering algorithms the number of clusters must be given in advance. However it's hard to do so without prior knowledge. The RPCL algorithm solves the problem by delearning the rival(the 2nd winner) every step, but its performance is too sensitive to the delearning rate. Moreover, when the clusters are not well separated, RPCL's performance is poor. In this paper We propose a RPFCL algorithm by associating a Fuzzy Inference System to the RPCL algorithm to tune the delearning rate. Experimental results show that RPFCL outperforms RPCL both in clustering speed and in achieving correct number of clusters.

1 Introduction

Given a data set $D = \{x_t\}_{t=1}^{N}$ in a multi-dimension space, the task of clustering D is a classical problem in many fields. The k-means[1] and FCM(Fuzzy C-Means)[2] algorithms are probably the most frequently used algorithms. As an adaptive version of the k-means algorithm, Competitive Learning(CL)[3] and FSCL algorithms[4] have their applications when the data set D is large, and many other algorithms derive from them. In all these algorithms, the number of clusters k should be preselected, and a wrong guess of it will make the algorithms perform poorly. Unfortunately, it is hard to choose it sometimes. Some efforts had been made to tackle this problem in the past decades. A typical example is the RPCL algorithm[5], in which for each input, not only the winner is modified to adapt to the input, but also its rival(the 2nd winner) is delearned by a smaller learning rate. The experiments[5] and other papers[6,7] show that RPCL algorithm works well in determining the number of clusters for well-separated clusters, but how to tune the delearning rate in RPCL is still a problem, to our best knowledge. Xu Lei noted that the delearning rate α_r should be much less than the learning rate α_c, otherwise the RPCL algorithm may fail to work[5]. But if α_r is too small, the speed of RPCL algorithm is slow. To improve the performance of RPCL, some scholars[6] attempted to change α_c, α_r after M steps, but failed to give an appropriate criteria for the selection of M. Some others[7] tried to incorporate full covariance matrices into the original RPCL algorithm, but some parameters are as hard to decide as the number of clusters.

In this paper, we propose a new way to tune the delearning rate in RPCL, to improve both the clustering speed and its performance for data set with over-lapped clusters. In RPCL, the learning rate and delearning rate are set to constants. In fact, the distances of the winner and the rival to the input datum can help much in determining the delearning rate, for there exist some fuzzy rules to determine α_r such that the rival should be penalized more after a close competition, otherwise it should remain almost intact. To utilize these information, we associate a fuzzy inference system to the original RPCL algorithm, then a new improved cluster algorithm, named Rival Penalized Fuzzy Competitive learning(RPFCL), is proposed.

This paper is organized as follows: In Sect. 2, we propose the RPFCL algorithm by associating a Fuzzy Inference System into RPCL. Simulation result of the RPFCL is shown in Sect. 3. Finally, we conclude this paper in Sect. 4.

2 RPFCL Algorithm

In order to explain the RPFCL algorithm, let us consider such a situation: suppose there are k tigers in a mountainous area competing for their food, N goats. When a goat appears, the closest two tigers raise a competition. Obviously, it is a close competition if the two tigers are as close to the goat as each other, and the winner(the tiger that is closer to the goat) gets the food and becomes the host of the domain, while the loser will be driven far away under the threat of the winner. In the contrary cases, if a tiger is the only one who is close to the goat, the second closest tiger will almost definitely quit the hopeless competition and stand still to save its strength.

We imitate the phenomenon above with a fuzzy inference system to adjust the delearning rate in RPCL algorithm. Refer to [8] for the details of fuzzy inference. Here we use language variables "x_0: almost equal; x_1: fairly equal; and x_2: far from equal" to describe the comparability of the distances between competitors and the input datum in the universe $(0, 1]$(the range of $u(m_c, m_r, d_t) = \frac{\|m_c - d_t\|}{\|m_r - d_t\|}, d_t \in D$). We also use other language variables "y_0: serious; y_1: moderate; y_2: slight" to describe the penalty degree in the universe $(0, \alpha_c]$. Here, the larger u is, the more equal the distances($\|m_c - d_t\|, \|m_r - d_t\|$) are. Since $\alpha_r \ll \alpha_c$, we regard $\alpha_r = \alpha_c$ as a very serious punishment. The membership functions of the language are shown in Fig. 1 and Fig. 2.

The rules in the Fuzzy Inference System for choosing the delearning rate α_r are listed below: Rule 1: if u is x_0, then α_r is y_0; Rule 2: if u is x_1, then α_r is y_1; Rule 3: if u is x_2, then α_r is y_2. And the formula to calculate the delearning rate α_r in the Fuzzy Inference System is[8]:

$$\alpha_r = \sum_{i=1}^{3} x_i(u(m_c, m_r, d_t))y_i . \qquad (1)$$

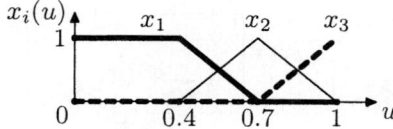
Fig. 1. Membership functions of x_i

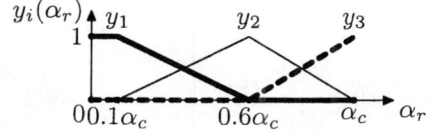
Fig. 2. Membership functions of y_i

After associating this mechanism to RPCL, we modify the RPCL algorithm into our RPFCL algorithm, which is consisted of the following three steps.

Step 1: Preselect an enough large number of clusters k, initiate k centers, set an appropriate α_c. Randomly take a sample d_t from data set D, calculate c and r by formula (2).

$$c = \underset{j}{\operatorname{argmin}} \gamma_j \|d_t - m_j\|, \quad r = \underset{j,j \neq c}{\operatorname{argmin}} \gamma_j \|d_t - m_j\|. \quad (2)$$

Where $\gamma_j = \frac{n_j}{\sum_{l=1}^{k} n_l}$, n_j is the cumulative number of the times when m_j wins a competition. We name m_c the winner, and m_r the rival. Here γ_j, called conscience strategy[4], reduces the winning rate of the frequent winners, and thus solves the problem of dead units. We select the first k input data as initial centers.

Step 2: Update the winner m_c and the rival m_r by the following formulas,

$$m_c^{t+1} = m_c^t + \alpha_c(d_t - m_c^t); \quad m_r^{t+1} = m_r^t - \alpha_r(d_t - m_r^t),$$

where α_r is given by formula (1).

Step 3: The algorithm will stop after p epochs. one epoch means that all the input data are scanned once. Here, p should be preselected. A datum closest to center m_j will belong to the cluster represented by m_j.

Obviously, if α_r is constant, then RPFCL degenerates to RPCL algorithm[5]. So our RPFCL algorithm is an extension of the RPCL algorithm.

3 Simulation Results

Due to the limitation of space, here we present one of the experiments we made. we choose a sample set same to that in [5], a data set of 4 clusters, each cluster having 100 samples from 4 Gaussian distributions centered at $(-1, 0), (1, 0), (0, -1), (0, -1)$. The only difference is that its variance is 0.3 instead of 0.1. We set the initial number of clusters $k = 8$, 2 times that of the real number of clusters. The simulation results of RPCL and RPFCL are shown in Fig. 3-5. Fig. 3 shows that RPCL fails to select the correct number of clusters even after 1000 epochs. As shown in Fig. 4 however, if we set α_r 10 times larger, the correct number of clusters may be achieved, but the positions of the centers are not guaranteed. If the variance of the data is set to 0.4, we find that RPFCL performs as good as usual; but RPCL fails to work on such data set no matter how large α_r is.

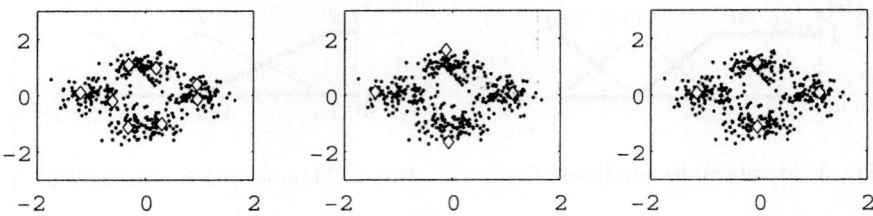

Fig. 3. RPCL: $\alpha_c = 0.05$; $\alpha_r = 0.002$, simulation result at 1000th epoch

Fig. 4. RPCL: $\alpha_c = 0.05$, $\alpha_r = 0.02$, correct cluster number, wrong centers at 200th epoch

Fig. 5. RPFCL: $\alpha_c = 0.05$, the result at 10th epoch, extra centers driven away

4 Conclusion

In this paper, we proposed the RPFCL algorithm by associating to RPCL algorithm a Fuzzy Inference System to tune the delearning rate. Experiments show that RPFCL algorithm not only clusters in a higher speed than RPCL, but also works well on overlapped data set. The Fuzzy Inference System is easy to establish. In our future work, we will further extend the Fuzzy Inference System to tune the learning rate α_c at each step, and create a mechanism to automatically stop the RPFCL algorithm.

References

1. Fan M., Meng XF. (translated): Data Mining Concepts and Techniques. China Machine Press(2001) (in Chinese)
2. Kevin M. Passino, Stephen Yurkovich: Fuzzy Control. Tsinghua University Press (2001) 260-262
3. Martin T.Hagan, Howard B.Demuth, Mark Beale: Neural Network Design. China Machine Press (2002)
4. S.C.Ahalt, A.K.Keishnamurty, P.Chen, and D.E.Melton: Competitive Learning Algorithms for Vector Quantization. Neural Networks, Vol.3. (1990) 227-291
5. Xu Lei, Adam Krzyzak, Erkki Oja.: Rival Penalized Competitive Learning for Clustering Analysis, RBF Net, and Curve Detection. IEEE Transaction on Neural Networks, Vol. 4. (1993) 636-648
6. Xie Weixin, Gao Xinbo: A Method of Extracting Fuzzy Rules Based on Neural Networks with Cluster Validity. Journal of Shenzhen University (Science & Engineering), 4(1). (1997)
7. LI Xin, JIANG Fang-ze, ZHENG Yu: An Improved RPCL Algorithm for Clustering. JOURNAL OF SHANGHAI UNIVERSITY, 5(5). (1999)
8. Li Hongxing: The Interpolate Mechanism of Fuzzy Control. SCIENCE IN CHINA (Series E), 28(3). (1998) 259-267

A New Predictive Vector Quantization Method Using a Smaller Codebook[1]

Min Shi[1,2] and Shengli Xie[1]

[1] School of Electronics & Information Engineering, South China University of Technology, Guangzhou 510640, Guangdong, China
[2] Department of Electronic Engineering, Jinan University, Guangzhou 510632, Guangdong, China

Abstract. For improving coding efficiency, a new predictive vector quantization (VQ) method was proposed in this paper. Two codebooks with different dimensionalities and different size were employed in our algorithm. The defined blocks are first classified based on variance. For smooth areas, the current processing vectors are sampled into even column vectors and odd column vectors. The even column vectors are encoded with the lower-dimensional and smaller size codebook. The odd ones are predicted using the decoded pixels from intra-blocks and inter-blocks at the decoder. For edge areas, the current processing vectors are encoded with traditional codebook to maintain the image quality. An efficient method for codebook design was also presented to improve the quality of the resulted codebook. The experimental comparisons with the other methods show good performance of our algorithm.

1 Introduction

VQ, which has been widely applied in speech and image coding, provides an efficient technique for data compression [1]. VQ is defined as a mapping Q of k-dimensional Euclidean space R^k into a finite subset Y of R^k. Thus $Q: R^k \to Y$ where $Y = \{\mathbf{y}_1, \mathbf{y}_2, \cdots, \mathbf{y}_N \mid \mathbf{y}_i \in R^k\}$ is called a codebook, and N is the codebook size. The distortion between the input vector $\mathbf{x} = (x_1, x_2, \cdots x_k)^T$ and codeword $\mathbf{y}_i = (y_{i1}, y_{i2}, \cdots y_{ik})^T$ is usually defined as squared Euclidean distance:

$$d(\mathbf{x}, \mathbf{y}_i) = \|\mathbf{x} - \mathbf{y}_i\|_2^2 = \sum_{l=1}^{k}(x_l - y_{il})^2 \tag{1}$$

The minimum distortion codeword is just the best-matched codeword for the input vector. Given vector dimensionality, the larger the codebook size is, the smaller the distortion is. However, a large size codebook will obviously result in high bit-rate and

---

[1] The work is supported by the Guang Dong Province Science Foundation for Program of Research Team (grant 04205783), the National Natural Science Foundation of China (Grant 60274006), the Natural Science Key Fund of Guang Dong Province, China (Grant 020826), the National Natural Science Foundation of China for Excellent Youth (Grant 60325310) and the Trans—Century Training Program.

long encoding time. To achieve high encoding efficiency, we always try to make the codebook size as small as possible while still maintaining almost constant distortion. Various VQ schemes had been developed for this aim. The nearest-neighbor search algorithms[2] do not introduce extra distortion. However, the bit-rate of these algorithms is $\log_2 N/k$, which is invariable. To reduce the bit-rate, SMVQ [3], PPM [4] and SB_PVQ[5], were also successfully applied in image encoding.

In the proposed algorithm, these factors, e.g., encoding time, decoded image quality and bit-rate, are sufficiently considered. In contrast to other VQ methods, the defined image blocks are classified based on variance. Our method sufficiently exploits correlation between pixels in intra-blocks and inter-blocks to recover image. Codeword candidate scheme is employed to find the corresponding codeword. Aiming at the drawbacks of LBG[6], a modified codebook design method, which reduces overall distortion, is proposed. Compared with other methods, the new one achieves significant improvement in terms of rate-distortion performance while maintaining comparable computation complexity.

2 Block Classification

Suppose the original image is divided into many sub-blocks, the block size is 4×4 and each block represents a 16-dimensional vector. The relationship of adjacent image blocks is shown as Fig. 1, where $\mathbf{x}$ denotes the current processing block, $\mathbf{u}$, $\mathbf{l}$ and $\mathbf{n}$ are the upper, left and upper left neighboring blocks, respectively. Here the block $\hat{\mathbf{x}}$ confined in dashed square box is defined for every $\mathbf{x}$, because the pixel information around $\mathbf{x}$ will be utilized in the following prediction. If $\mathbf{x}$ is located in the first row and first column of image, $\hat{\mathbf{x}}$ is just $\mathbf{x}$ itself. The mean and variance of $\hat{\mathbf{x}}$ are defined as (2) and (3).

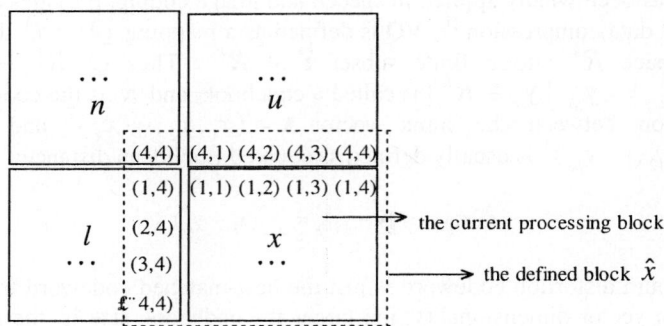

Fig. 1. Block classification

$$m_{\hat{x}} = \frac{n_{4,4} + \sum_{i=1}^{4} u_{4,i} + \sum_{i=1}^{4} l_{i,4} + \sum_{i=1}^{4}\sum_{j=1}^{4} x_{i,j}}{25} \qquad (2)$$

$$v_{\hat{x}} = \frac{(n_{4,4} - m_{\hat{x}})^2 + \sum_{i=1}^{4}(u_{4,i} - m_{\hat{x}})^2 + \sum_{i=1}^{4}(l_{i,4} - m_{\hat{x}})^2 + \sum_{i=1}^{4}\sum_{j=1}^{4}(x_{i,j} - m_{\hat{x}})^2}{25} \qquad (3)$$

If $v_{\hat{x}}$ is relatively large, this means the pixels in $\hat{\mathbf{x}}$ have a high fluctuation of intensities. Then block $\hat{\mathbf{x}}$ is not smooth and probably located in edge areas. If $v_{\hat{x}}$ is relatively small, then block $\hat{\mathbf{x}}$ is smooth. If $v_{\hat{x}}$ is less than threshold value T_x, $\hat{\mathbf{x}}$ can be considered as smooth block. Otherwise $\hat{\mathbf{x}}$ is not smooth.

3 Prediction Method

In this paper, the input vector is divided into two parts of even column vector $\mathbf{x}_{even}$ and odd column vector $\mathbf{x}_{odd}$. $\mathbf{x}_{even}$ is just the vector that needs be encoded. So a 4×4-dimensional vector is sampled as a 4×2-dimensional vector to be quantized, and the vector dimensionality is reduced to the half of original ones.

The correlation between pixels in smooth areas behaves that the pixels can be predicted using neighboring ones. If $v_{\bar{x}}$ calculated is less than the threshold value T_x, the current block $\mathbf{x}$ is even sampled to produce $\mathbf{x}_{even}$. It is quantized with 4×2-dimensional codebook C_s whose size is N_s, and N_s is smaller than the traditional codebook size N. The index of the best-matched codeword for $\mathbf{x}_{even}$ is transmitted to the receiver. In the decoding end, the codeword corresponding to the index is chosen from the same codebook to replace $\mathbf{x}_{even}$. $\mathbf{x}_{odd}$ is predicated using neighboring decoded pixels. Combining decoded $\mathbf{x}_{even}$ with $\mathbf{x}_{odd}$ forms the constructed block, which is called the predication block $\overline{\mathbf{x}}$. The prediction at the receiver is shown as Fig 2, where ● denotes the decoded pixel of even column vector, ○ denotes the decoding pixel of odd column vector and × denotes the decoded pixel of odd column vector. For simplicity, the prediction of $\mathbf{x}_{odd}$ only uses neighboring pixels of decoded even column vectors. The predication equations are described as (4) and (5).

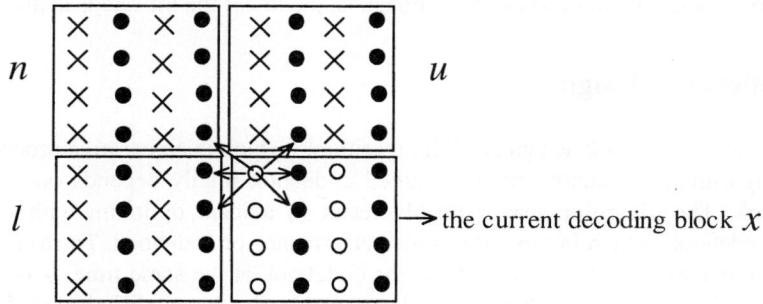

Fig. 2. The prediction of the odd vector

$$x_{i,j} = \left\lfloor \frac{x_{i-1,j+1} + x_{i-1,j} + x_{i-1,j-1} + x_{i+1,j+1} + x_{i+1,j} + x_{i+1,j-1}}{R(x_{i-1,j+1}) + R(x_{i-1,j}) + R(x_{i-1,j-1}) + R(x_{i+1,j+1}) + R(x_{i+1,j}) + R(x_{i+1,j-1})} \right\rfloor \quad (4)$$

$$R(x_{i,j}) = \begin{cases} 1 & \text{if } x_{i,j} \text{ is available} \\ 0 & \text{if } x_{i,j} \text{ is not available} \end{cases} \quad (5)$$

From above, we know that some decimated pixels are obtained by averaging their available neighboring pixels. Though the block variance need be calculated beforehand, this preliminary processing can save much time when a block is not adapt to be encoded with the smaller codebook. To maintain encoding quality, the distortion between $\mathbf{x}$ and $\bar{\mathbf{x}}$ is calculated beforehand. Suppose the distortion is greater than the predetermined threshold value T_y, which means the prediction fails to provide a small distortion result. We switch to using traditional VQ coding.

In the experiment, we find that the best-matched codeword may not reconstruct the best-predicted result for the input block, sometimes next-nearest codeword may result in better predicted vector for reproduction. Considering the computation complexity increased due to more predictions, we select the best and second best codewords from C_s as two candidates for obtaining the best prediction. The prediction blocks are calculated using the two codewords, respectively. The index of the codeword that yields the best prediction is chosen for transmission.

To distinguish the two different coding types, a flag should be applied to inform the decoder as to which codebook of the two is employed for encoding an input block. A simple way is to attach a prefix flag-bit ahead of the index of a codeword selected. Though this will introduce extra bits, the bit-rate is less than that of traditional VQ if N is far more larger than N_s. Assume that the probability of the smooth block is α in an image, the associated bit-rate for every pixel is calculated as:

$$BR_1 = \frac{\alpha(1 + \log_2 N_s) + (1-\alpha)(1 + \log_2 N)}{16} \quad (6)$$

Compared with the bit-rate of the traditional VQ, we can easily see that given the same codebooks, the more smooth the image is, the lower the bit-rate is achieved.

4 Codebook Design

The common codebook design is LBG algorithm[6] which is an iterative procedure. In LBG algorithm, the quality of the resulted codebook highly depends on the initial codebook. The algorithm may probably result in a local optimum with improper initial codebook, which in turn affects the performance of codebook. To overcome the problem, our method does not generate the codebook at the same time. It is efficient for LBG algorithm to generate a small codebook. So the modified algorithm first generates a small codebook by the LBG algorithm. The training vectors whose associated codeword has maximum average distortion will then be split into more codewords by LBG algorithm to reduce overall distortion. In this way, even if the initial small codebook is improper, it can adjust and improve the codebook step by step. The steps to design a codebook with S codewords from a training set $X = \{\mathbf{x}_m \mid m = 1, 2, \cdots, M\}$ of k-dimensional vectors are described as follows:

Step 1: Generate a codebook with L codewords from the training set with LBG algorithm, here L is much smaller than S. For example, L may be 2, 3, or 4.

Step 2: Calculate the average distortion between every codeword and each training vector of the associated cluster. Find the codeword that has the maximum average distortion in the codebook and denote the codeword as $\mathbf{y}_{max}$. The cluster of the training vectors associated with $\mathbf{y}_{max}$ is denoted by T.

Step 3: Employ the LBG algorithm to generate a codebook with L codewords from the training set T.

Step 4: Replace the codeword $\mathbf{y}_{max}$ with above new L codewords. Thus the number of the codewords in the codebook will be expanded by $L-1$ codewords.

Step 5: If the desired size of the codebook is reached, then stop the iteration. Otherwise, go back to Step 2.5

5 The Proposed Algorithm

For encoding an image, encoder partitions it into a set of blocks (or vectors) first. The blocks are processed from top to bottom and left to right. The encoding steps are depicted as follows:

a): For the current processing block $\mathbf{x}$, $v_{\hat{x}}$ of defined block $\hat{\mathbf{x}}$ is calculated. If $v_{\hat{x}}$ is less than the threshold value T_x, go to the next step, otherwise go to step c.

b): The even column vector $\mathbf{x}_{even}$ is produced after sampling from $\mathbf{x}$, and $\mathbf{x}_{even}$ is quantized with codebook C_s. The best and second best codewords for $\mathbf{x}_{even}$ are searched from C_s. Two prediction blocks are produced using the two codewords. Then calculate the distortion between x and two prediction blocks, respectively. The index associated with the minimum distortion d_{min} is marked with p. If $d_{min} < T_y$, p will take place of $\mathbf{x}$ to be transmitted. Otherwise, go to the next step.

c): $\mathbf{x}$ is encoded with traditional codebook C to maintain the image quality. The index q of the best-matched codeword for $\mathbf{x}$ is searched from C and q will take place of $\mathbf{x}$ to be transmitted.

d): To distinguish the different coding types, a one-bit checking flag is appended with "0" for coding using codebook C and "1" for coding using codebook C_s. Encoder transmits the combination of one "0" bit and the index q and the combination of one "1" bit and index p to the decoder.

The decoding procedure is quite straightforward with reference to the encoding one. For each input bit-string, there are two branches to describe the decoding steps:

a): Select the first bit as the check-bit. Suppose the check-bit is "0", this means the traditional VQ is used to process the current block. In this case, the following $\log_2 N$ bits are read as an index from the bit-string. The original block is replaced with the codeword associated with the index in C.

b): Suppose the check-bit is "1", the following $\log_2 N_s$ bits are read as an index from the bit-string. We retrieve the even column vector corresponding to the index from C_s and then predict the odd column vector using (4) and (5) . Then combine them to recover original input block.

After all the blocks are reconstructed from top to bottom and left to right, we can piece the blocks together to obtain the decoded image. SB_PVQ is improved using three amendments in our algorithm. The block classification based on variance reduces encoding time. The number of the encoded pixels is the same as that of SB_PVQ, while our method can sufficiently utilize the neighboring pixels to recover origin block. So it needs less bit-rate than SB_PVQ with the same image quality. We also propose a modified codebook design to improve VQ performance. So our method is superior to SB_PVQ.

6 Experimental Results

Some experiments were conducted to test the efficiency of our proposed method. All images in our experiments are of size 512×512 with 256 gray level. We employ Lena and Pepper as our training images, and apply modified LBG algorithm to generate both the traditional and lower-dimensional codebooks. The splitting number L is fixed to 4 and the size N is 256. We compare our algorithm, the traditional VQ, SMVQ in encoding time, bit-rate and PSNR. Table 1 lists the experimental results for the testing images. In SMVQ algorithm, the size of state codebook N_f is 32. In our algorithm, the threshold value T_x is 2000 and T_y is 1000. The threshold values are

Table 1. The results of the proposed algorithm, the traditional VQ, SMVQ for comparison

Image	Factors	VQ	SMVQ N_f=32	Our algorithm N_s=16	N_s=32	N_s=64
Lena	Time(s)	30.27	120.18	7.18	5.24	6.56
	Bit-rate	0.500	0.315	0.323	0.369	0.415
	PSNR(dB)	31.167	29.332	29.758	30.102	30.502
Pepper	Time(s)	30.68	120.42	8.45	6.23	6.28
	Bit-rate	0.500	0.315	0.324	0.358	0.417
	PSNR(dB)	29.476	27.951	28.256	28.655	28.962
Boat	Time(s)	30.49	119.72	10.56	8.73	9.01
	Bit-rate	0.500	0.315	0.332	0.362	0.419
	PSNR(dB)	29.451	28.011	28.893	29.121	29.378
Airplane	Time(s)	29.78	120.03	9.07	6.51	7.32
	Bit-rate	0.500	0.315	0.363	0.387	0.430
	PSNR(dB)	30.010	28.709	29.031	29.687	29.986

decided by experiments. In fact the bit-rate is variable along with different threshold values. When threshold values are relatively large, the bit-rate will be decreased while the image quality is reduced. PDE[7] is adopted in all searches of codebooks. From the experimental results, we can see that the traditional VQ obtain the best-decoded image quality at the cost of maximum bit-rate and long encoding time. Though SMVQ needs low bit-rate, this algorithm cost so long time to produce state codebook for every input vector and the decoded image quality is not satisfied. The proposed algorithm not only needs least encoding time, but also outperforms SB_PVQ as a prediction method in PSNR and bit-rate. Figure 3 illustrates the rate-PSNR performance curves between our method and SB_PVQ with N is 512. The curves show that our method is superior to SB_PVQ in term of rate-PSNR performance. All the experiments prove that our method outperforms VQ, SMVQ and SB_PVQ in total performance.

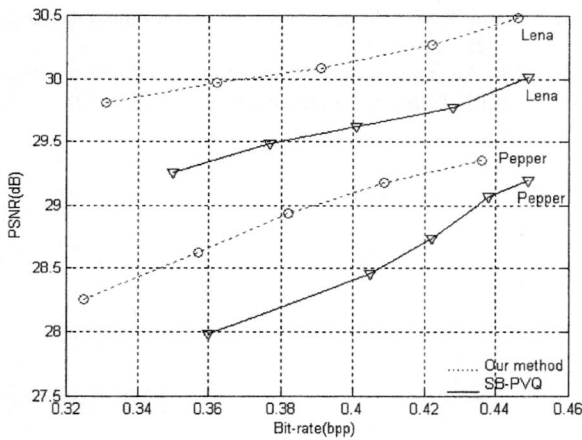

Fig. 3. Rate-PSNR curves for our method and SB_PVQ in encoding Lena and Pepper

7 Conclusion

For improving image coding quality, increasing encoding speed and reducing bit-rate on transmission, a new prediction-based vector quantization method for image coding is presented in this paper. In contrast to other VQ methods, the defined image blocks are classified based on variance, which can speed up encoding time. Correlation between pixels in intra-blocks and inter-blocks is sufficiently utilized to recover image. Codeword candidate scheme is employed to find the corresponding codeword that can generate better-reconstructed images. Aiming at the drawbacks of LBG, a modified codebook design method, which reduces overall distortion and reduces the dependence on initial codebook, is proposed to generate efficient codebook. The experimental result shows that the proposed encoding outperforms VQ, SMVQ and SB_PVQ in total performance.

References

1. Gersho A., Gray R. M.: Vector quantization and signal compression. Boston: Kluwer Academic Publishers, (1992)
2. Chaur H. H., Liu Y. J.: Fast search algorithms for vector quantization of images using multiple triangle inequalities and wavelet transform. IEEE Trans on Image Processing, vol. 9, (2000) 321-328
3. Kim T.: Side match and overlap match vector quantization for image. IEEE Trans on Image Processing, vol. 1, (1992) 170-185
4. Chang C. C.: Chou J. S., Chen T. S.: A predictive image coding scheme using a smaller codebook. Signal Processing: Image Communication, vol. 12, (1998) 23-32
5. Zhu C.: A new subsampling-based predictive vector quantization for image coding. Signal Processing: Image Communication, vol. 17, (2002) 477-484
6. Linde Y., Buzo A., Gray R. M.: An algorithm for vector qauantizer design. IEEE Trans on Communications, vol. 28, (1980) 84-95
7. Bei C. D., Gray R.M.: An improvement of the minimum distortion encoding algorithm for vector quantization. IEEE Trans on Communications, vol. 33, (1985) 1132-1133

Performance Improvement of Fuzzy RBF Networks

Kwang-Baek Kim[1], Dong-Un Lee[2], and Kwee-Bo Sim[3]

[1] Dept. of Computer Engineering, Silla University, Korea
gbkim@silla.ac.kr
[2] School of Architectural Engineering, Pusan National University, Korea
ldu21@hanmail.net
[3] School of Electrical and Electronic Engineering, Chung-Ang Univ., Korea
kbsim@cau.ac.kr

Abstract. In this paper, we propose an improved fuzzy RBF network which dynamically adjusts the rate of learning by applying the Delta-bar-Delta algorithm in order to improve the learning performance of fuzzy RBF networks. The proposed learning algorithm, which combines the fuzzy C-Means algorithm with the generalized delta learning method, improves its learning performance by dynamically adjusting the rate of learning. The adjustment of learning rate is achieved by self-generating middle-layered nodes and applying the Delta-bar-Delta algorithm to the generalized delta learning method for the learning of middle and output layers. To evaluate the learning performance of the proposed RBF network, we used 40 identifiers extracted from a container image as the training data. Our experimental results show that the proposed method consumes less training time and improves the convergence of learning, compared to the conventional ART2-based RBF network and fuzzy RBF network.

1 Introduction

Recently, RBF networks, which have the characteristics of fast training time, generality and simplicity, have been applied to the classification of training data and nonlinear system modeling[1]. RBF networks avoid the problems with algorithms such as error backpropagation learning algorithm. RBF networks reduces training time and prevents training patterns from not being well-classified, which is caused by the weights of multilayer perceptrons falling into local minimum[2]. The middle-layer of RBF Networks is the clustering layer. That is, the purpose of this layer is to classify a given data set into homogeneous clusters. This means that if in the feature vector space of input data, the distance between vectors in a cluster is within the range of the predetermined radius, the cluster is classified as homogeneous. Otherwise, the cluster is classified as heterogeneous[3]. However, clustering within the prescribed radius has the risk of selecting wrong clusters. Thus, the determination of middle-layer has a great effect on the overall efficiency of RBF networks. If learning of a new pattern is processed at the state in which learning is completed, that is, at the state in which the connection weight is fixed, RBF networks have an effect on prescribed weights. This effect leads to the problem of taking a lot of time to retrain a RBF network. It also serves to decrease recognition rate by classifying untrained new patterns as the homogeneous pattern when they are entered into the RBF network[4].

In this paper, we propose a method to improve the learning structure of RBF networks. In the proposed learning structure, the connection structure between input layer and middle layer is the same as the fuzzy C-Means structure. Though the proposed learning structure is a complete connection structure, it compares target vector with output vector in the output layer, and thus avoids the problem of classifying the new patterns as the previously trained pattern since it adjusts connection weight by back-propagating the weight connected with the representative class. And the generalized delta method is applied to the representative class of middle layer and the output layer nodes in terms of supervised learning. In doing this, the rate of learning is dynamically adjusted by the application of Delta-bar-Delta method to reduce training time. This paper comparatively analyzes learning performance between ART2-based RBF networks, fuzzy RBF networks and the proposed learning method in terms of applying them to the identifier extracted from a container image.

2 Related Research

2.1 ART2-Based RBF Networks

In the ART2-based RBF networks, the number of middle-layer nodes is determined according to the boundary parameter setting in the process of generating the middle layer. The boundary parameter is the value of radius that classifies clusters. If the boundary parameter is set with a low value, a small difference between the input pattern and the stored pattern leads to the generation of new clusters in terms of classifying them as different patterns. On the other hand, if the boundary parameter has a high value, the input pattern and the stored pattern are classified as the same in spite of a big difference between them. Thus, this reveals a problem that recognition performance varies depending on boundary parameter setting[5].

2.2 Fuzzy C-Means-Based RBF Networks

Fuzzy C-Means-based RBF networks uses the fuzzy C-Means algorithm to generate the middle layer. It has a disadvantage of consuming too much time when applied to character recognition. In character recognition, a binary pattern is usually used as the input pattern. Thus, when the fuzzy C-Means algorithm is applied to the training pattern composed of 0 and 1, it is not only difficult to precisely classify input patterns but also takes a lot of training time compared to other clustering algorithms[6]. In this paper, we use the Delta-bar-Delta algorithm to improve the learning performance of fuzzy C-Means-based RBF networks. It reduces the training time by dynamically adjust the rate of learning in the process of adjusting the connection weight between middle layer and output layer.

2.3 Delta-Bar-Delta Algorithm

Delta-bar-delta algorithm[7], which improved the quality of backpropagation algorithm, enhances learning quality by arbitrating learning rates dynamically for individual connected weights by means of making delta and delta-bar. The formula of making delta is as follows: In this expression, i, j and k indicate the input layer, the middle layer and the output layer, respectively.

$$\Delta_{ji} = \frac{\partial E}{\partial w_{ji}} = -\delta_j x_i \tag{1}$$

$$\Delta_{kj} = \frac{\partial E}{\partial w_{kj}} = -\delta_k z_j \tag{2}$$

The formula of making delta-bar is as follows:

$$\overline{\Delta}_{ji}(t) = (1-\beta)\cdot\overline{\Delta}_{ji}(t) + \beta\cdot\overline{\Delta}_{ji}(t-1) \tag{3}$$

$$\overline{\Delta}_{kj}(t) = (1-\beta)\cdot\overline{\Delta}_{kj}(t) + \beta\cdot\overline{\Delta}_{kj}(t-1) \tag{4}$$

The value of parameter β in formula (4) is the fixed constant between 0 and 1.0. The variation of learning rate in terms of the change direction of delta and delta-bar is as follows: If the connected weight changes to the same direction in the successive learning process, the learning rate will increase. At this point delta and delta-bar has the same sign. On the other hand, if the signs of delta and delta-bar are different, the learning rate will decrease as much as the ratio of 1-γ of the present value.
The formula of the variable learning rate for each layer is as follows:

$$\begin{aligned}\alpha_{ji}(t+1) &= \alpha_{ji}(t) + \kappa & &\text{if } \overline{\Delta}_{ji}(t-1)\cdot\Delta_{ji}(t) > 0 \\ &= (1-\gamma)\cdot\alpha_{ji}(t) & &\text{if } \overline{\Delta}_{ji}(t-1)\cdot\Delta_{ji}(t) < 0 \\ &= \alpha_{ji}(t) & &\text{if } \overline{\Delta}_{ji}(t-1)\cdot\Delta_{ji}(t) = 0\end{aligned} \tag{5}$$

$$\begin{aligned}\alpha_{kj}(t+1) &= \alpha_{kj}(t) + \kappa & &\text{if } \overline{\Delta}_{kj}(t-1)\cdot\Delta_{kj}(t) > 0 \\ &= (1-\gamma)\cdot\alpha_{kj}(t) & &\text{if } \overline{\Delta}_{kj}(t-1)\cdot\Delta_{kj}(t) < 0 \\ &= \alpha_{kj}(t) & &\text{if } \overline{\Delta}_{kj}(t-1)\cdot\Delta_{kj}(t) = 0\end{aligned} \tag{6}$$

3 Improved Fuzzy RBF Networks

The middle layer of an RBF network is a layer that clusters training patterns. The purpose of this middle layer is to classify the given training patterns into homogeneous clusters. If in the feature vector space of training patterns, the distance between vectors in a cluster is within the range of the prescribed radius, they belong to the same cluster. Otherwise, they belong to the different cluster. Clustering within the range of the prescribed radius can select wrong clusters and take them as the input value of output layer, thus decreasing the learning performance of RBF networks. Since the node of middle layer does not know its target vector in the process of learning, Since the node of middle layer does not know its target node in the process of learning, it takes a lot of training time caused by the paralysis resulting from credit assignment by which the errors of nodes of output layer is inversely assigned to nodes of middle layer.

In this paper, we propose a learning structure that selects the node with the highest membership degree as the winner node in terms of the application of C-Means algorithm, and transmit it to the output layer. We also apply the generalized delta learning method for learning of the middle and output layer. The Delta-bar-Delta algorithm is applied to improve training time. The learning model that we propose is depicted in Fig. 1.

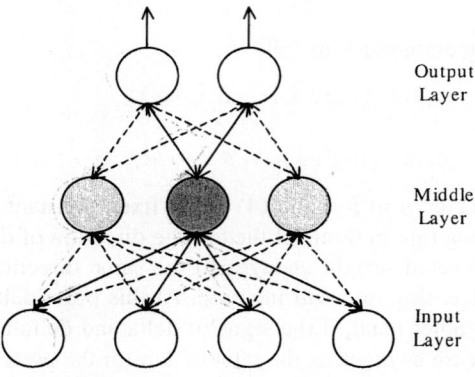

Fig. 1. The proposed learning model

The proposed learning method for fuzzy RBF networks can be summarized as follows.

1. The connection structure of input layer and middle layer is the same as in the fuzzy C-Means algorithm whose output layer is the middle layer of the proposed learning structure.

2. The node of middle layer denotes a class. Thus, though being a complete connection structure as a whole, we adopts the winner node method which back-propagates the weight connected with the representative class in terms of comparing the target vector with the actual output vector.

3. The fuzzy C-Means algorithm selects the node middle layer with the highest membership degree as the winner node.

4. The generalized delta learning method is applied to the learning structure of middle layer and output layer in terms of supervised learning.

5. To improve learning performance, the Delta-bar-Delta algorithm is applied to the general delta learning, dynamically adjusting the rate of learning. If the difference between target vector and output vector is less than 0.1, it is defined as having accuracy. Otherwise, it is defined as having inaccuracy. The Delta-bar-Delta algorithm is applied only in the case that the number of accuracies of the entire patterns is equal to or greater than the number of inaccuracies. The reason for this is

to prevent learning from being stagnant or vibrating, arising from premature saturation due to competitive steps in the learning process, which in turn make the error rate stable. The proposed fuzzy RBF network is given in Fig. 2.

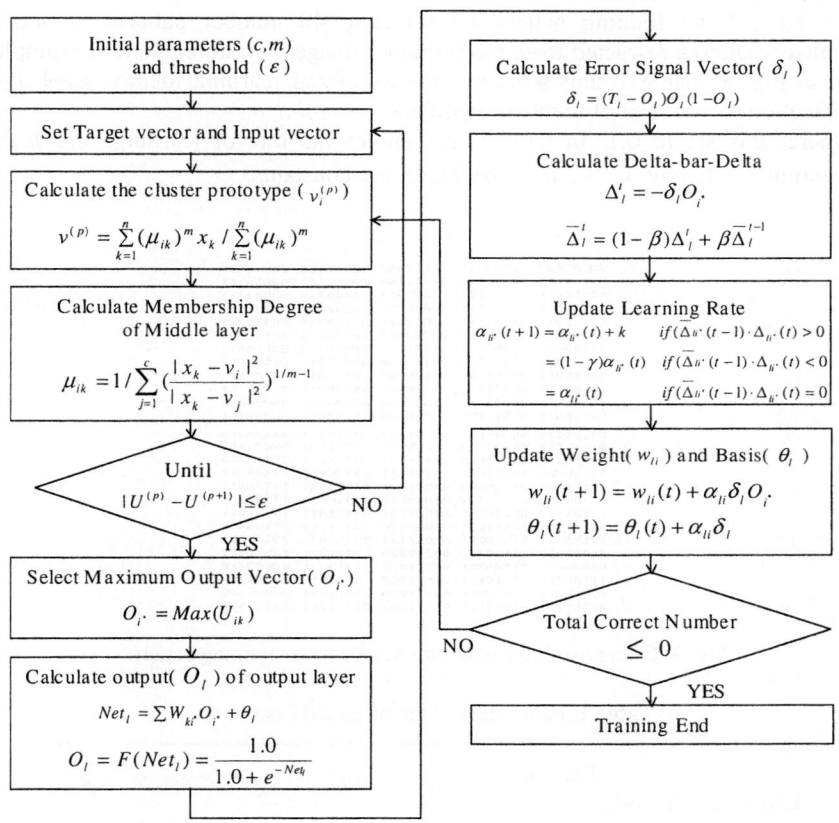

Fig. 2. The proposed fuzzy RBF learning algorithm

4 Experimental Results and Analysis

We have implemented our learning algorithm in Visual C++ 6.0 on an Intel Pentium-IV machine with 2 GHz CPU and 512 MB of RAM. In order to analyze the learning performance of our learning algorithm, we chose the classification problem for 40 number identifiers extracted from a container image, and comparatively analyzed ART2-based RBF networks with fuzzy RBF networks in terms of the number of repetitive learning and the number of recognitions.

To extract container identifiers, the method proposed in [8] was applied and individual number identifiers were extracted. The edge was detected by applying

Canny mask. To eliminate from the detected edge information vertically long noises produced by the outer light source when obtaining the image, we applied fuzzy inference. After removing the noises, we extracted the domain of identifiers and then made them binary-coded. Individual identifiers were extracted by applying the contour-tracking algorithm.

We formed the training pattern by selecting 40 number patterns among the normalized patterns extracted from the container image. A training pattern example is given in Fig. 3. Table 1 shows the parameters of our learning method used in the classification experiment of number identifiers.

Where, ε is set to 0.1. In table 1, α denotes the rate of learning, and μ, the momentum coefficient κ, γ, β are the Delta-bar-constants.

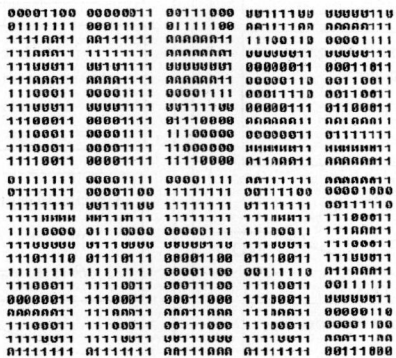

Fig. 3. Example of container number identifier training pattern

Table 1. Parameters of the fuzzy RBF network

Parameter Learning Method	α	μ	κ	γ	β
Fuzzy RBF Network	0.9	0.5	0.05	0.2	0.7

Table 2. Learning results of each learning method

	# of middle layer's nodes	# of Epoch	TSS	# of recognition
ART2-based RBF Network	13	950	0.067065	40
Fuzzy RBF Network	10	822	0.082591	40
Proposed Fuzzy RBF Network	10	526	0.085005	40

In table 2, the Epoch number is the repetitions of learning, and TSS is the sum of the square of the total errors. In the ART2-based network and fuzzy RBF network, the rate of learning and momentum were set to 0.5 and 0.6, respectively. In the proposed fuzzy RBF network, the initial rate of learning was set to 0.5, and we applied the Delta-bar-Delta algorithm if the number of accuracies of the total patterns is equal to or greater than the number of inaccuracies. The momentum was set to 0.6.

Table 2 shows that learning is terminated only in the case where the number of inaccuracies is equal to or less than 0. As can be seen from table 2, the proposed fuzzy RBF network is improved in terms of learning speed, compared to the ART2-based RBF network and the existing fuzzy RBF network. In the ART2-based RBF network, the number of nodes of middle layer was increased or decreased according the value of the boundary parameter, which is considered to be a problem of the network. In table 2, the boundary parameter set to 0.5 proved to be most optimal.

In the conventional fuzzy RBF network and the proposed fuzzy RBF network, both of which apply the fuzzy C-Means algorithm to the middle layer, they generated less middle-layered nodes than the ART2-based RBF network because it generates clusters according to the membership degree of nodes of middle layer. The proposed fuzzy RBF network dynamically adjusts the rate of learning to reduce the premature saturation corresponding to the competitive stage of learning process, thus consuming less training time than the conventional fuzzy RBF network. Fig. 4 shows the curve that is the sum of the square of errors in the conventional methods and the proposed method.

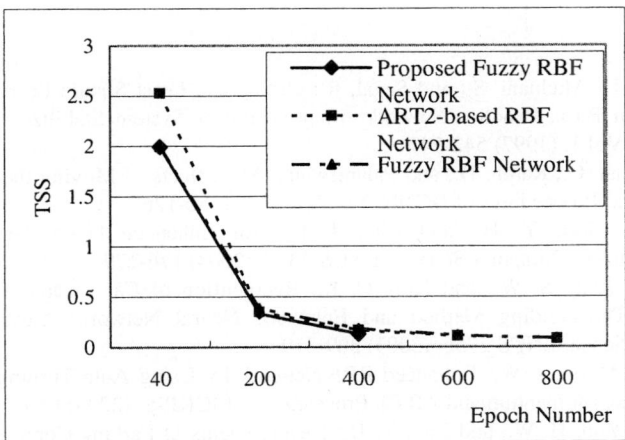

Fig. 4. Variance of TSS according to Learning Methods

As shown in Fig. 4, the proposed method wins the conventional methods in terms of the speed of the initial convergence and training time.

5 Conclusion

In this paper, we proposed an improved fuzzy RBF network which dynamically adjusts rate of learning by applying Delta-bar-Delta algorithm in order to improve learning performance of fuzzy RBF networks. The learning structure of the fuzzy RBF network has also been improved. In the proposed learning structure, the fuzzy C-Means algorithm is applied to the connection structure of input layer and middle layer.

Though the proposed learning structure is a complete connection structure, it compares target vector with output vector in the output layer, and thus avoids the problem of classifying the new patterns as the previously trained pattern since it adjusts the weight by back-propagating the weight connected with the representative class. And the generalized delta method is applied to the representative class of middle layer and the nodes of output layer in terms of supervised learning. In doing this, the rate of learning is dynamically adjusted by the application of Delta-bar-Delta method to reduce training time. The proposed method also avoids the problem of taking too much training time caused by the paralysis resulting from credit assignment by which the errors of nodes of output layer is inversely assigned to the nodes of middle layer. This paper comparatively analyzed learning performance between ART2-based RBF networks, fuzzy RBF networks and the proposed learning method in terms of applying them to the identifier extracted from a container image. The experimental results confirmed that the proposed method improved learning performance of fuzzy RBF networks.

References

1. Kothari, M. L., Madnani, S., and Segal, R.: Orthogonal Least Square Learning Algorithm Based Radial Basis Function Network Adaptive Power System Stabilizer. Proceeding of IEEE SMC. Vol.1. (1997) 542-547
2. Panchapakesan, C., Ralph, D., and Palaniswami, M.: Effects of Moving the Centers in an RBF Network. Proceedings of IJCNN. Vol.2. (1998) 1256-1260
3. Kim, K. B., Joo, Y. H., and Cho, J. H.: An Enhanced Fuzzy Neural Network. Lecture Notes in Computer Science. LNCS 3320. (2004) 176-179
4. Kim, K. B., Jang, S. W., and Kim, C. K.: Recognition of Car License Plate by Using Dynamical Thresholding Method and Enhanced Neural Networks. Lecture Notes in Computer Science. LNCS 2756. (2003) 309-319
5. Kim, K. B., Moon, J. W.: Enhanced RBF Network by Using Auto-Turning Method of Learning Rate, Momentum and ART2. Proceedings of ICKISS. (2003) 85-87
6. Kim, K. B., Yun, H. W., and Lho, Y. U.: Developments of Parking Control System using Color Information and Fuzzy C-Means Algorithm. Journal of Intelligent Information Systems. Vol.8. No.1. (2002) 87-102
7. Jacobs, R. A.: Increased rates of convergence through learning rate adaptation. IEEE Transactions on Neural Networks. Vol.1. No.4. (1988) 295-308
8. Kim, K. B.: The Identifier Recognition from Shipping Container Image by Using Contour Tracking and Self-Generation Supervised Learning Algorithm Based on Enhanced ART1. Journal of Intelligent Information Systems. Vol.9. No.3. (2003) 65-80

Universal Approach to Study Delayed Dynamical Systems

Tianping Chen

Laboratory of Nonlinear Science, Institute of Mathematics,
Fudan University, Shanghai, 200433, P.R. China
tchen@fudan.edu.cn

Abstract. In this paper, we propose a universal approach to study dynamical behaviors of various neural networks with time-varying delays. A universal model is proposed, which includes most of the existing models as special cases. An effective approach, which was first proposed in [1], to investigate global stability is given, too. It is pointed out that the approach proposed in the paper [1] applies to the systems with time-varying delays, too.

1 Introduction

Recurrently connected neural networks have been extensively studied in past years and found many applications in different areas. Such applications heavily depend on the dynamic behavior of the networks. Therefore, the analysis of these dynamic behaviors is a necessary step for practical design of neural networks. Recently, there are dozens papers discussing recurrent neural networks with delays. For example, see [1,2,3] for constant delays; For time-varying delays, see [4,5,6] and [8,9,10,11].

It is natural to raise following question: **Can we propose a unified model and an effective approach to investigate all these models in a universal framework?**

The purpose of this paper is to give an affirmative answer to this question.

We consider the following system

$$\frac{du_i(t)}{dt} = -d_i(t)u_i(t) + \sum_{j=1}^{n} \int_0^{\infty} g_j(u_j(t-s))dJ_{ij}(t,s)$$

$$+ \sum_{k=1}^{m}\sum_{j=1}^{n} \int_0^{\infty} f_j(u_j(t-\tau_{ij}^k(t)-s))dK_{ij}^k(t,s) + I_i(t) \quad (1)$$

where for any $t > 0$, $dJ_{ij}(t,s)$, $dK_{ij}^k(t,s)$ are Lebesgue-Stieljies measures with respect to s for each $i,j = 1, \cdots, n$, $k = 1, \cdots, m$, and satisfy $\int_0^{\infty} |dJ_{ij}(t,s)| < \infty$ and $\int_0^{\infty} |dK_{ij}(t,s)| < \infty$. $\tau = \max_{1 \leq i,j \leq n} \tau_{i,j}^k$, $\tau_{ij}^k = \max_t \tau_{i,j}^k(t)$.

The initial values are

$$u_i(s) = \phi_i(s) \quad for \quad s \in [-\infty, 0],$$

where $\phi_{ij}(t)$ are bounded continuous functions. Some variant of (1) has been proposed in [12,21] to discuss periodic systems.

In particular, we consider following system

$$\frac{du_i(t)}{dt} = -d_i u_i(t) + \sum_{j=1}^{n} \int_0^\infty g_j(u_j(t-s)) dJ_{ij}(s)$$
$$+ \sum_{j=1}^{n} \int_0^\infty f_j(u_j(t-\tau_{ij}(t)-s)) dK_{ij}(s) + I_i(t) \quad (2)$$

where $\lim_{t\to\infty} I_i(t) = I_i$.

We also consider the following system without delays:

$$\dot{v}_i(t) = -d_i v_i(t) + \sum_{j=1}^{n} \int_0^\infty dJ_{ij}(s) g_j(v_j(t))$$
$$+ \sum_{j=1}^{n} \int_0^\infty dK_{ij}(s) f_j(v_j(t)) + I_i \quad (3)$$

For the convenience, we call $g \in Lip(G)$, if $\frac{|g(x+u)-g(x)|}{|u|} \leq G$, where $G > 0$.

2 Main Results

In this section, we will give several theorems and corollaries.

Theorem 1 Suppose that $g_i \in Lip(G_i)$, $f_i \in Lip(F_i)$. If for $i=1,\cdots,n$, there hold

$$-\xi_i d_i + \sum_{j=1}^{n} \xi_j \left\{ \int_0^\infty \left[G_j |dJ_{ij}(s)| + F_j |dK_{ij}(s)| \right] \right\} < 0 \quad (4)$$

Then there is at least a v^* such that for any solution $u(t)$ of (2), there holds

$$\lim_{t\to\infty} u(t) = v^* \quad (5)$$

Remark 1 By transforms $x_i(t) = \xi_i^{-1} u_i(t)$, $J_i(t) = \xi_i^{-1} I_i(t)$, we have

$$\frac{dx_i(t)}{dt} = -d_i x_i(t) + \sum_{j=1}^{n} \int_0^\infty g_j(\xi_j x_j(t-s)) dJ_{ij}(s)$$
$$+ \sum_{j=1}^{n} \int_0^\infty f_j(\xi_j x_j(t-\tau_{ij}(t)-s)) dK_{ij}(s) + J_i(t) \quad (6)$$

Therefore, without loss of generality, in the following proof, we assume all $\xi_i = 1$ for $i = 1, \cdots, n$.

Lemma 1 Under the assumptions in Theorem 1, the dynamical system (3) has at least an equilibrium v^*.

Proof If $g_i \in Lip(G_i)$, $f_i \in Lip(F_i)$, then g_i, f_i as well as all $v_i(t)$, $i = 1, \cdots, n$, are absolutely continuous and differentiable almost everywhere with respect to Lebesgue measure. Therefore, for almost $t \in [0, \infty)$, saying $t \in S$, $i = 1, \cdots, n$, following equalities hold

$$\frac{d\dot{v}_i(t)}{dt} = -d_i \dot{v}_i(t)$$
$$+ \sum_{j=1}^{n} \left\{ \int_0^\infty dJ_{ij}(s) g'_j(v_j(t)) + \int_0^\infty dK_{ij}(s) f'_j(v_j(t)) \right\} \dot{v}_j(t) \quad (7)$$

Pick a small constant $\alpha > 0$, such that

$$-(d_i - \alpha) + \sum_{j=1}^{n} \left\{ \int_0^\infty \left[G_j |dJ_{ij}(s)| + F_j |dK_{ij}(s)| \right] \right\} < 0 \quad (8)$$

Let $z_i(t) = e^{\alpha t} \dot{v}_i(t)$, which is a continuous function. And for every $t \in S$, $i_t = i_t(t)$ is an index such that

$$|z_{i_t}(t)| = ||z(t)|| = \max_{i=1,\cdots,n} |z_i(t)|$$

Then, under (4), it is easy to see that

$$\frac{d|z_i(t)|}{dt} \leq (-d_i + \alpha)|z_i(t)| + \sum_{j=1}^{n} \int_0^\infty \left[G_j |dJ_{ij}(s)| + F_j |dK_{ij}(s)| \right] |z_j(t)|$$
$$\leq \left\{ (-d_i + \alpha) + \sum_{j=1}^{n} \int_0^\infty \left[G_j |dJ_{ij}(s)| + F_j |dK_{ij}(s)| \right] \right\} ||z(t)|| \leq 0$$

Thus, $||z(t)||$ is non-increasing at every $t \in S$. Because S is dense in $[0, \infty)$ and $||z(t)||$ is continuous. Then $||z(t)||$ is bounded and

$$||\dot{v}(t)|| = O(e^{-\alpha t}) \quad (9)$$

By Cauchy convergence principle, there is an equilibrium point $v^* \in R^n$ such that

$$\lim_{t \to \infty} v(t) = v^* \quad (10)$$

and

$$- d_i v_i^* + \sum_{j=1}^{n} \int_0^\infty dJ_{ij}(s) g_j(v_j^*) + \sum_{j=1}^{n} \int_0^\infty dK_{ij}(s) f_j(v_j^*) + I_i = 0 \quad (11)$$

Lemma 1 is proven.

Remark 2 Proof of existence of v^* does not depend on any complicated theories (topology degree theorem, fixed point theorem, Lasalle theorem and so on), which were adopted in many papers. Moreover, we do not assume that the activation functions are bounded or continuous differentiable.

Proof of Theorem 1 Suppose $u(t)$ is a solution of (2), v^* is the equilibrium given in the Lemma 1.

Pick a small number $\eta > 0$ and a sufficient large T_1 such that $||I(t) - I|| < \eta$ for all $t > T_1$.

For $t > T_1$, let $w(t) = u(t) - v^*$, $M(t) = \sup_{-\infty < s \leq t} ||w(t)||$, and $t_0 \in (-\infty, t]$, $i_{t_0} = i_{t_0}(t_0)$ be an index such that $|w_{i_{t_0}}(t_0)| = ||w(t_0)|| = M(t)$. Then, we have

$$D^+ M(t) \leq -d_{i_{t_0}} |w_{i_{t_0}}(t_0)| + \sum_{j=1}^{n} \int_0^{\infty} \Big[G_j |w_j(t_0 - s)| ||dJ_{i_{t_0} j}(s)|$$

$$+ F_j |w_j(t_0 - \tau_{i_{t_0} j}(t) - s)| ||dK_{i_{t_0} j}(s)| \Big] + \eta$$

$$\leq \Big\{ -d_{i_{t_0}} + \sum_{j=1}^{n} \int_0^{\infty} \Big[G_j |dJ_{i_{t_0} j}(s)| + F_j |dK_{i_{t_0} j}(s)| \Big] \Big\} M(t) + \eta \quad (12)$$

which means that if

$$M(t) > \frac{\eta}{d_{i_{t_0}} - \sum_{j=1}^{n} \int_0^{\infty} \Big[G_j |dJ_{i_{t_0} j}(s)| + F_j |dK_{i_{t_0} j}(s)| \Big]} \quad (13)$$

$M(t)$ is non-increasing. Therefore, there is a constant M, such that $||w(t)|| \leq M$.

For any small $\epsilon > 0$, pick a sufficiently large T, such that

$$\eta(t) = ||I(t) - I|| < \frac{\eta \epsilon}{4} \quad if \; t > T \quad (14)$$

and

$$M \sum_{j=1}^{n} \int_T^{\infty} \Big[G_j |dJ_{ij}(s)| + F_j |dK_{ij}(s)| \Big] < \frac{\eta \epsilon}{4} \quad (15)$$

Now, denote $M_1(t) = \sup_{t-T \leq s \leq t} ||y(t)||$. Let $t_1 \in (t-T, t]$, $i_{t_1} = i_{t_1}(t_1)$ be an index such that $|w_{i_{t_1}}(t_1)| = ||w(t_1)|| = M_1(t)$. By same approach, we have

$$D^+ M_1(t) \leq \Big\{ -d_{i_{t_1}} + \sum_{j=1}^{n} \int_0^{T} \Big[G_j |dJ_{i_{t_1} j}(s)| + F_j |dK_{i_{t_1} j}(s)| \Big] \Big\} M_1(t)$$

$$+ M \sum_{j=1}^{n} \int_T^{\infty} \Big[G_j |dJ_{i_{t_1} j}(s)| + F_j |dK_{i_{t_1} j}(s)| \Big] + \frac{\eta \epsilon}{4}$$

$$\leq \Big\{ -d_{i_{t_1}} + \sum_{j=1}^{n} \int_0^{T} \Big[G_j |dJ_{i_{t_1} j}(s)| + F_j |dK_{i_{t_1} j}(s)| \Big] \Big\} M_1(t) + \frac{\eta \epsilon}{2}$$

Thus, if $M(t) \geq \epsilon$, then

$$D^+ M(t) < -\frac{\eta \epsilon}{2} \quad (16)$$

Therefore, there must exists $\bar{t}$ such that $M(\bar{t}) < \epsilon$. It is clear that $M(t) < \epsilon$ for all $t > \bar{t}$. Because ϵ is arbitrary, we conclude

$$\lim_{t \to \infty} u(t) = v^* \tag{17}$$

Theorem 1 is proved.

Theorem 2 Suppose that $g_i \in Lip(G_i)$, $f_i \in Lip(F_i)$, $I_i(t) = I_i$. If for $i = 1, \cdots, n$, there hold

$$\xi_i(-d_i + \alpha) + \sum_{j=1}^{n} \xi_j \int_0^{\infty} e^{\alpha s} \left[G_j |dJ_{ij}(s)| + e^{\alpha \tau_{ij}} F_j |dK_{ij}(s)| \right] \le 0 \tag{18}$$

Then system (2) has a unique equilibrium point v^* such that for any solution $u(t)$ of (2), there holds

$$||u(t) - v^*|| = O(e^{-\alpha t}) \tag{19}$$

Proof Suppose $u(t)$ is any solution of the system (2), and let $y(t) = e^{\alpha t}[u(t) - v^*]$, $M_2(t) = \sup_{-\infty \le s \le t} ||y(t)||_{\{\xi,\infty\}}$.

If for some $t_2 \le t$ and some index i_{t_2} such that $|y_{i_{t_2}}(t_2)| = ||y(t_2)|| = M_2(t)$. Then by the same arguments used in the proof of Theorem 1, we have

$$D^+ M_2(t)$$
$$\le \left\{ (-d_{i_{t_2}} + \alpha) + \sum_{j=1}^{n} \int_0^{\infty} e^{\alpha s} \left[G_j |dJ_{i_{t_2}j}(s)| + F_j e^{\alpha \tau_{ij}} |dK_{i_{t_2}j}(s)| \right] \right\} M_2(t) \le 0$$

Therefore, $M_2(t)$ is bounded and

$$||u(t) - v^*||_{\{\xi,\infty\}} = e^{-\alpha t} ||y(t)||_{\{\xi,\infty\}} = O(e^{-\alpha t}) \tag{20}$$

Corollary 1 Suppose that $g_i \in Lip(G_i)$, $f_i \in Lip(F_i)$. If there are positive constants $\xi_1, \cdots, \xi_n$ and $\alpha > 0$ such that for $i = 1, \cdots, n$,

$$\xi_i(-d_i(t) + \alpha) + \sum_{j=1}^{n} \xi_j \int_0^{\infty} e^{\alpha s} \left[G_j |dJ_{ij}(t,s)| + e^{\alpha \tau_{ij}} F_j |dK_{ij}(t,s)| \right] \le 0 \tag{21}$$

Then the dynamical system (1) is globally exponentially stable. It means that if $u_1(t)$ and $u_2(t)$ are two solutions of (1), then

$$||u_1(t) - u_2(t)|| = O(e^{-\alpha t})$$

In fact, let $u_1(t)$ and $u_2(t)$ are two solutions of (1). Replacing $y(t)$ by $\bar{y}(t) = e^{\alpha t}[u_1(t) - u_2(t)]$, by the same arguments used in the proof of Theorem 2, Corollary 1 can be obtained directly.

3 Comparisons

In this section, we will discuss the relationship among the results given in this paper and those given the references.

Case 1. $dJ_{ij}(s) = a_{ij}\delta(s)$, $dK_{ij}(s) = b_{ij}\delta(s)$, where $\delta(s)$ is the Dirac-delta function. In this case, (2) reduces to the system with time-varying delays

$$\frac{du_i(t)}{dt} = -d_i u_i(t) + \sum_{j=1}^{n} a_{ij} g_j(u_j(t)) + \sum_{j=1}^{n} b_{ij} f_j(u_j(t - \tau_{ij}(t))) + I_i(t) \quad (22)$$

It is clear that the results obtained in [13] are special cases of Theorem 1. Moreover, the model (2) is much more general than that in [13].

If $I_i(t) = I_i$, then (2) reduces to

$$\frac{du_i(t)}{dt} = -d_i u_i(t) + \sum_{j=1}^{n} a_{ij} g_j(u_j(t)) + \sum_{j=1}^{n} b_{ij} f_j(u_j(t - \tau_{ij}(t))) + I_i \quad (23)$$

In this case, conditions in (18) become

$$\xi_i(-d_i + \alpha) + \sum_{j=1}^{n} \xi_j \left\{ G_j |a_{ij}| + e^{\alpha \tau_{ij}} F_j |b_{ij}| \right\} \le 0 \quad (24)$$

Therefore, all stability analysis on the system (22) in [2,3,4,5,6,7,8,9,10,11] and many others are direct consequences of Theorem 2.

On the other hand, if $dJ_{ij}(t,s) = a_{ij}(t)\delta(s)$, $dK_{ij}^k(t,s) = b_{ij}^k(t)\delta(s)$, delayed system (1) reduces to the system with time-varying delays

$$\dot{u}_i(t) = -d_i(t) u_i(t) + \sum_{j=1}^{n} a_{ij}(t) g_j(u_j(t))$$

$$+ \sum_{k=1}^{m} \sum_{j=1}^{n} b_{ij}^k(t) f_j(u_j(t - \tau_{ij}^k(t))) + I_i(t) \quad (25)$$

In this case, by the same method to prove Theorem 1, we can prove that under

$$\xi_i(-d_i(t) + \alpha) + \sum_{j=1}^{n} \xi_j \left\{ G_j |a_{ij}(t)| + \sum_{k=1}^{m} \sum_{j=1}^{n} e^{\alpha \tau_{ij}^k} F_j |b_{ij}^k(t)| \right\} \le 0, \quad (26)$$

for $t > 0$. Delayed system (25) with time-varying coefficients and delays is globally exponentially stable. It is clear that the conditions in (26) is more natural than those given in [18]. Moreover, we do not assume that $\tau_{ij}^k(t)$ are differentiable.

Case 2. $dJ_{ij}(s) = a_{ij}\delta(s)$, $dK_{ij}(s) = b_{ij}k_{ij}(s)ds$, and $\tau_{ij}(t) = 0$. Then system (2) reduces to systems with distributed delays

$$\frac{du_i(t)}{dt} = -d_i u_i(t) + \sum_{j=1}^{n} a_{ij} g_j(u_j(t)) + \sum_{j=1}^{n} b_{ij} \int_{0}^{\infty} f_j(u_j(t-s)) k_{ij}(s) ds + I_i \quad (27)$$

In this case, we have

Proposition 1 Suppose that $g_i \in Lip(G_i)$, $f_i \in Lip(F_i)$. If there are positive constants $\xi_1, \cdots, \xi_n$ and $\alpha > 0$ such that for $i = 1, \cdots, n$,

$$\xi_i(-d_i + \alpha) + \sum_{j=1}^{n} \xi_j \left\{ G_j |a_{ij}| + \int_0^\infty k_{ij}(s) ds F_j |b_{ij}| \right\} < 0 \qquad (28)$$

Then the dynamical system (27) has an equilibrium point v^* and for any solution $u(t)$ of (27), there holds

$$\lim_{t \to \infty} u(t) = v^* \qquad (29)$$

Furthermore, if

$$\xi_i(-d_i + \alpha) + \sum_{j=1}^{n} \xi_j \left\{ G_j |a_{ij}| + \int_0^\infty e^{\alpha s} k_{ij}(s) ds F_j |b_{ij}| \right\} \leq 0 \qquad (30)$$

Then the dynamical system (27) is globally exponentially stable. It means

$$\|u(t) - v^*\| = O(e^{-\alpha t})$$

Therefore, all results in [14,15,16] and many other can be derived from Theorem 2. It is also clear that the results obtained in [17] under more restrictions can be derived directly from Theorem 1 in this paper.

Case 3. $dJ_{ij}(s) = a_{ij}\delta(s) + c_{ij}k_{ij}(s)ds$, $dK_{ij}(s) = b_{ij}\delta(s)$, and $f_j = g_j$. Then system (2) reduces to systems (see [19])

$$\frac{du_i(t)}{dt} = -d_i u_i(t) + \sum_{j=1}^{n} a_{ij} f_j(u_j(t)) + \sum_{j=1}^{n} b_{ij} f_j(u_j(t - \tau_{ij}(t)))$$

$$+ \sum_{j=1}^{n} c_{ij} \int_0^\infty f_j(u_j(t-s)) k_{ij}(s) ds + I_i$$

Thus, the results on stability given in [19] can be derived from Theorem 2.

Remark 3 In proposition 1, we do not assume that

$$\int_0^\infty s |k_{ij}(s)| ds < \infty$$

which was assumed in many papers.

Remark 4 The approach, which was first proposed in [1], used in this paper is very effective. It does not depend on any complicated theory. Derivations are simple. Instead, Conclusions are universal. Moreover, this approach applies to periodic systems with time delays (see [21]).

Remark 5 Theorem 1 and Theorem 2 explore an interesting phenomenon, i.e., concerning stability analysis, there is no difference between the delayed systems with constant delays and time-varying delays. Therefore, theorems for delayed systems with constant delays given in [1] apply to the case with time-varying delays without any difficulty.

Recently, several researchers also investigated stability criteria with $L_p(1 \leq p \leq \infty)$ norm (for example, see [3,7]). Therefore, it is necessary to compare capability of criteria with $L_p(1 < p < \infty)$ norm and with L_1 norm or L_∞. This comparison was given in a recent paper [20]. It was explored in [20] that criteria with L_1 norm or L_∞ are the best. Therefore, the results given with L_p norm can be derived from Theorems in this paper.

4 Conclusions

In this paper, we study dynamical behaviors of delayed systems with time-varying delays. A universal model is proposed, which includes most of the existing models as special cases. An effective approach to investigate global stability is given, too. It is pointed out that the results and approach proposed in [1] also apply to the systems with time-varying delays. We also verify the effectiveness by comparing the results obtained by this approach and those obtained in literature.

Acknowledgements

This work is supported by National Science Foundation of China 69982003 and 60074005.

References

1. Tianping Chen, "Global Exponential Stability of Delayed Hopfield Neural Networks", Neural Networks, 14(8), 2001, p.p. 977-980
2. X. Liao, J. Wang, "Algebraic Criteria for Global Exponential Stability of Cellular Neural Networks With multiple Time Delays," IEEE Tran. on Circuits and Systems-I, 50(2), 2003, 268-275
3. Hongtao Lu, Fu-Lai Chung, zhenya He, "Some sufficient conditions for global exponential stability of delayed Hopfield neural networks," Neural Networks, 17(2004), 537-544
4. Z. Zeng, J. Wang and X. Liao, "Global Exponential Stability of a General Class of Recurrent Neural Networks With Time-Varying Delays," IEEE Tran. on Circuits and Systems-I, 50(10), 2003, 1353-1358
5. J. Zhang, "Globally Exponential Stability of Neural Networks With varying Delays," IEEE Tran. on Circuits and Systems-I, 50(2), 2003, 288-291
6. Z. Yi, "Global exponential convergence of recurrent neural networks with variable delays," Theor. Comput. Sci. 312, 2004, 281–293.
7. He Huang, Daniel W.C.Ho, Jinda Cao, "Analysis of global exponential stability and periodic solutions of neural networks with time-varying delays, Neural Networks, 18(2), 2005 161-170

8. He Huang, Jinde Cao, "On global symptotic stability of recurrent neural networks with time-varying delays" Applied Mathematics and Computation, 142(2003)143-154
9. J. Cao, J. Wang, "Global Asymptotic Stability of a General Class of Recurrent Neural Networks With Time-Varying Delays," IEEE Tran. on Circuits and Systems-I, 50(1), 2003, 34-44
10. Jinde Cao, Jun Wang, "Absolute exponential stability of recurrent neural networks with Lipschitz-continuous activation functions and time delays", Neural Networks, 17(2004), 379-390
11. Jigen Peng, Hong Qiao, Zong-ben Xu, "A new approach to stability of neural networks with time-varying delays", Neural Networks 15(2002)95-103
12. Tianping Chen, Wenlian Lu and Guanrong Chen, "Dynamical Behaviors of a Large Class of General Delayed Neural Networks," Neural Computation, 17(4),2005, 949-968
13. Sanqing Hu, Derong Liu, "On the global output convergence of a class of recurrent neural networks with time-varying inputs, Neural Networks, 18(2), 2005, 171-178
14. Gopalsamy, K. and He, X. "Stability in Asymmetric Hopfield Nets with Transmission Delays," Phys. D. 76, 1994, 344-358.
15. Hongyong Zhao, "Global stability of neural networks with distributed delays" Physical Review E 68, 051909 (2003)
16. Qiang Zhang, Xiaopeng Wei, Jin Xu, "Global exponential stability of Hopfield neural networks with continuous distributed delays" Physics Letters A, 315(2003), 431-436
17. Hongyong Zhao "Global asymptotic stability of Hopfield neural network involving distributed delays' Neural Networks 17(2004) 47-53
18. Haijun Jiang, Zhidong Teng, "Global exponential stability of cellular neural networks with time-varying coefficients and delays", Neural Networks, 17 (2004), 1415-1425
19. Jiye Zhang et.al. "Absolutely exponential stability of a class of neural networks with unbounded delay", Neural Networks, 17(2004) 391-397
20. Yanxu Zheng, Tianping Chen, "Global exponential stability of delayed periodic dynamical systems," Physics Letters A, (2004) 322(5-6), 344-355.
21. W. Lu, T.P. Chen, On Periodic Dynamical Systems, Chinese Annals of Mathematics Series B, 25 B:4 (2004), 455. 51(12),

Long-Range Connections Based Small-World Network and Its Synchronizability

Liu Jie[1,2] and Lu Jun-an[2]

[1] Wuhan University of Science and Engineering,
Wuhan, 430073, P.R. China
[2] School of Mathematics and Statistics,
Wuhan University, Wuhan, 430072, P.R. China
liujie_hch@163.com

Abstract. How crucial is the long-distance connections in small-world networks produced by the semi-random SW strategy? In this paper, we attempted to investigate some related questions by constructing a semi-random small-world network through only randomly adding 'long-range lattice distance connections' to a regular network. The modified network model is compared with the most used NW small-world network. It can be found that, by using the new modified small-worldify algorithm, one can obtain a better clustered small-world network with similar average path length. Further more, we numerically found that, for a dynamical network on typical coupling scheme, the synchronizability of the small-world network formed by our procedure is no better than that of the small-world network formed by NW's algorithm, although the two classes of network constructed at the same constructing prices and having similar average path length. These results further confirmed that, the random coupling in some sense the best candidate for such nonlocal coupling in the semi-random strategy. Main results are confirmed by extensive numerical simulations.

1 Introduction

Small-world network is highly clustered networks with small distances among the nodes. There are many real-world networks that present this kind of connection, such as the WWW, Transportation systems, Biological or Social networks, achieve both a strong local clustering (nodes have many mutual neighbors) and a small diameter (maximum distance between any two nodes). These networks now have been verified and characterized as small-world (SW) networks. In the context of network design, the semi-random SW strategy (typically described as modelling related real networks by the addition of randomness to regular structures) now is shown to be an efficient way of producing synchronically networks when compared with some standard deterministic graphs networks and even to fully random and constructive schemes. A great deal of research interest in the theory and applications of small-world networks has arisen since the pioneering work of D Watts and H Strogatz [1]-[23].

'How crucial is the long-distance connections in such networks produced by the semi-random SW strategy', this question is indeed worth reasoning. Recently, Adilson E. Motter et al., have investigated the range-based attack on connections in scale-free networks, they found that, the small world property of scale free networks is mainly due to short range connections [2]. Further more, Takashi Nishikawa et al., in the same research group, numerically and analytically studied the synchronizability of heterogeneous networks [22].

In this paper, we will try to investigate some related questions by constructing a modified version of small-world network through only randomly adding 'long range connections' (in the sense of 'lattice-space-distance') to a regular network. We will compare the modified model with the most used small-world networks in nowadays research works introduced by M E J Newman and D Watts[13].

The arrangement of the rest of this paper is as follows: in the following section 2, firstly, we provide a brief summary about the most used NW small-world network algorithms as a preliminary. Then the modified version of small-world network based on the NW small-world algorithm is introduced in section 3. Some basic properties of this modified model, such as clustering coefficient, average distance are discussed. In section 4, numerical investigation of the synchronizability of a dynamical network under special coupling schemes on different networks are compared with each other. In section 5, brief conclusion concludes the investigation.

2 Mathematical Model of Main Types of Small-World Networks

In 1998, Watts and Strogatz [5] proposed a single-parameter small-world network model that bridges the gap between a regular network and a random graph. With the WS small-world model, one can link a regular lattice with pure random network by a semirandom network with high clustering coefficient and short average path length. The original WS model is described as follows:

(I) **Initialize:** Start with a nearest-neighbor coupled ring lattice with N nodes, in which each node i is connected to its K neighboring nodes $i \pm 1; i \pm 2; \cdots; i \pm K/2$, where K is an even integer. (Assume that $N \gg K \gg ln(N) \gg 1$, which guarantees that the network is connected but sparse at all times.)

(II) **Randomize:** Randomly rewire each link of the network with probability p such that self-connections and duplicated links are excluded. Rewiring in this sense means transferring one end of the connection to a randomly chosen node. (This process introduces $\frac{pNK}{2}$ long-range links, which connect some nodes that otherwise would not have direct connections. One thus can closely monitor the transition between order ($p = 0$) and randomness ($p = 1$) by adjusting p.)

A small-world network lies along a continuum of network models between the two extreme networks: regular and random ones. Recently, M E J Newman and Watts modified the original WS model. In the NW modelling, instead of rewiring links between nodes, extra links called shortcuts are added between pairs of nodes chosen at random, but no links are removed from the existing network. Clearly, the NW model reduces to the originally nearest-neighbor coupled network if

$p = 0$; while it becomes a globally coupled network if $p = 1$. However, the NW model is equivalent to the WS model for suciently small p and sufficiently large N values. The WS and NW small-world models show a transition with an increasing number of nodes, from a large-world regime in which the average distance between two nodes increases linearly with the system size, to a small-world model in which it increases only logarithmically.

Different from the semi-randomly constructing ways used by D. J. Watts, M. E. J. Newman et al., very recently, F. Comellas and his colleagues show that small-world networks can also be constructed in a deterministic way. Their exact approach permits a direct calculation of relevant network parameters allowing their immediate contrast with real world networks and avoiding complex computer simulations[6]. For example, one of their procedures to create a small-world network is described below. Starting with a regular nearest neighbored coupled networks, they construct a deterministic small-world network by selecting h nodes to be hubs and then using a globally coupled network to interconnect those hubs[7]. These approaches also attract much attention of researchers.

In what follows, We will only consider those small-world network created by semi-random operations (For simplicity, we exclude further comparisons study with the deterministic small-world networks and this will be done elsewhere in the future). We will mainly concern on the NW small-world model for comparison with our new small-world model, since no matter how many nodes the networks has, it keeps to be connected during the randomizing procedure (It is not necessary to assume that $N \gg K \gg ln(N) \gg 1$ as that in WS model described before). This assumption can guarantee the basic condition (connected) required in our research of these networks synchronizability.

3 The Modified Small-World Network Based on Adding Long-Lattice-Distance Connections

The aim of constructing such model is to investigate the impaction of long-range contracts in the NW small-world lattice network. What will happen, if only add some space long range connection between nodes in a regular lattice during the same procedure described in NW small-worldify process? This question attracts our attention much during the research of the synchronization in a circle chain of chaotic oscillators firstly. We then tried to construct such a related model to further investigate its characteristics and the effectiveness of just adding long-range shortcuts to the original lattice. In another words, two main aspects are highly concerned in the whole investigation: One is, whether such a procedure can make a "small-world" network with high cluster coefficient and low average path length or not. And the other is, what effects will it takes on the dynamical behaviors of the original lattice after such re-choosing operations?

Aim to these targets, we construct the modified version of the two semi-random small-world networks by the following two main steps:

(I) **Initialize:** Start with a nearest-neighbor coupled ring lattice with N nodes, in which each node i is connected to its K neighboring nodes: $i\pm 1, i\pm 2, \cdots, i\pm K/2$,

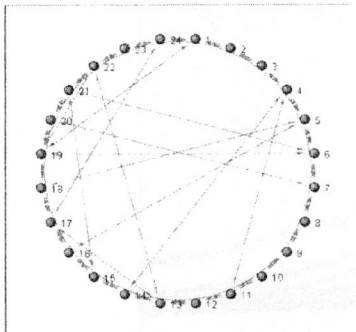 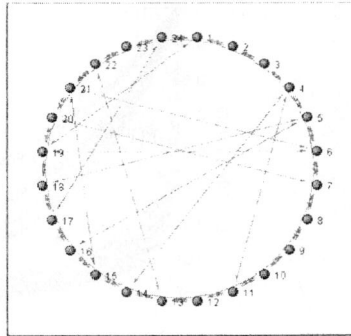

Fig. 1. NW small-world network, where $N = 24, K = 4, p = 0.05435$ for illustration

Fig. 2. Modified SW network, where $D^*(d) = N/4$, links $13-17, 19-22$ of Fig.1 are unchosen in modified process

where K is an even integer. The nodes are numbered sequentially from 1 to N (For simplicity, we suppose that N is a multiple of 4), thus, the "lattice distance" [8]-[9] between two nodes numbered i and j can be calculate by: $d_{i,j} = N/2 - ||i-j| - N/2|$;

(II) **Randomize with re-choosing:** Randomly adding connections between a pair of nodes in the network with probability p, during the whole process, duplicated links are excluded. Then, we re-choose shortcuts through the following procedure: Firstly, defining the space distance of arbitrary two nodes as the "lattice distance" d_{ij} defined in (I). Given a setting value $D^*(d)$, where d is the diameter of the original regular lattice, and $D^*(\bullet)$ represents a function of d. Only reserving those connections added randomly in the above operations, which links two nodes and their "lattice distance" longer than the setting value $D^*(d)$, if d_{ij} is larger than or equal to $D^*(d)$.

Thus, one can obtain a semi-random network after several times of such modified operations.

Remark I: In NW model algorithm, their process will introduce about $\frac{pN(N-1)}{2}$ shortcut links between nodes of the original lattice when N is sufficient large. In our modified algorithm, the number of new links added in the original lattice is obviously much dependent on the parameter value $D^*(d)$, and it will be certainly much fewer than the expectation value $\frac{pN(N-1)}{2}$ in NW small-world network. This fact is caused by the re-choosing strategy used in the new procedure.

Remark II: Obviously, the parameter $D^*(d)$ can be set on the interval $[\frac{K}{2}, \frac{N}{2}]$. (There are two extremal situations in our modified procedure: if setting $D^*(d)$ at $D^*(d) > \frac{N}{2}$, the re-choosing strategy loses its effect on changing the structure of the original lattice. If setting at $\frac{K}{2}$, it reserves all the edges randomly added in, and the operation generates the same network as that generated by NW small-worldify algorithm.)

In the following context, we will further consider some characteristics of the new type of semi-random networks. With regard of practical usage, we only con-

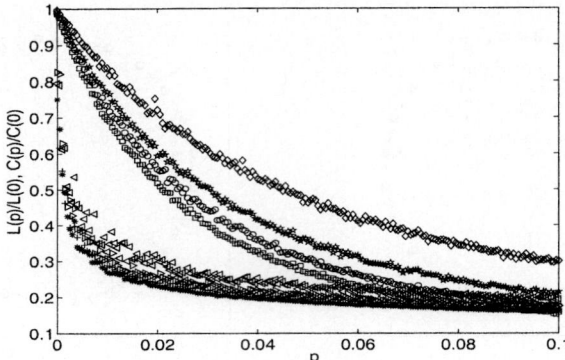

Fig. 3. Comparison of the change of average path lengths and clustering coefficients of M-SW network, where $N = 200$, $'*'$ and $'\square'$ represent the average path lengths and the clustering coefficients of the NW model, $'\triangleleft'$ and $'\diamond'$ represent those of the M-SW model when setting $D(\cdot)$ as $N/3$, $'\triangleright'$ and $'\star'$ represent those of the M-SW model when setting $D(\cdot)$ as $N/4$, $'+'$ and $'\circ'$ represent those of the M-SW model when setting $D(\cdot)$ as $N/5$, respectively

sider adding 10 percent of possible number of edges to the original K neighbored lattices. It should be pointed out that, the situation when choosing $K = 2$ was not appropriate for our discussion, since in this case the clustering coefficient of the original lattice is zero. In what follows, we will set $K \geq 4$ in all simulations. In this section, we are especially interested in the long range connections' effect on the network characteristics, such as the average clustering coefficients, the average networks diameter, and the average shortest path length, etc. We will compare those characteristics with the most used NW models. In all of the numerical experiments shown below, we take the average results of 20 runs at each parameters setting.

The related two semi-randomized network models are illustrated respectively as follows (for the straight intuitive purpose, choosing $N = 24, K = 4, p \approx 0.05, D^*(d) = N/4$; see Figure 1-2).

In Figure 3-4, we give the linear-linear and log-linear scale graphs for the related changes. We compared the basic characteristics of NW model and the modified networks. In these figures, the parameter of re-choosing criterion is set at $D^*(d) = 0$, and $N/4$, respectively. For simplicity, we only give p changes in $[0, 0.1]$. In these figures, the results are obtained by averaging the results of 20 runs and the step change of p is set at 0.0005. In these graphs, $'*'$ and $'\square'$ represent the average path lengths and the clustering coefficients of the NW model, $'\triangleleft'$ and $'\diamond'$ represent those of the M-SW model when setting $D(\cdot)$ as $N/3$, $'\triangleright'$ and $'\star'$ represent those of the M-SW model when setting $D(\cdot)$ as $N/4$, $'+'$ and $'\circ'$ represent those of the M-SW model when setting $D(\cdot)$ as $N/5$, respectively.

Remark III: We have done large amount of numerical experiments about the change of C and L about the modified model besides Figure 5-6. All of the experiments lead to similar results. The results show that the new networks behaves as a typical type of small-world network (Networks those with high average clustering coefficient and short average path length).

Remark IV: It can be seen that: for any given value of p, Cluster Coefficients $C(N,p)$ clearly increase with the increasing of $D^*(d)$; but the average path lengths L increase very slightly with the increasing of $D^*(d)$ ($\leq N/2$). For similar average path length L and L', the cluster coefficients C_{MNW} in the modified model is much larger than that C_{NW} in the NW small-world models. That is to say, by using the modified procedure, one can obtain better clustered networks with similar average path length. (eg., $\frac{\Delta C_{NNW}}{\Delta N_{add}} > \frac{\Delta C_{NW}}{\Delta N_{add}}$, where ΔC is the changed fraction of cluster coefficients, δN_{add} is the changed fraction of added edges.)

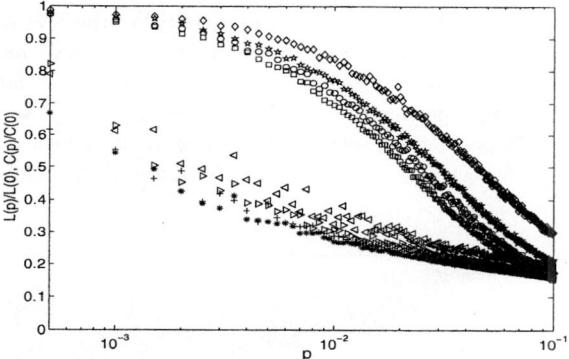

Fig. 4. Figure 3 in log-linear scale

We also studied the "Probability distribution of the connectivity" of the modified model for $K = 4$, $p = 0.01$. The curves are similar with each other, but we can see that, the distribution curves are apparently transported from "right to left", see Fig.5-6. These two figures are obtained by setting "added links fraction" parameter p at $p = 0.01$ and $p = 0.05$, respectively.

4 About Different Networks' Synchronizability

In this section, we begin to discuss the synchronizability of the two types of semi-random networks: the NW model and the modified small-world model. We will only consider a network of N identical nodes, linearly coupled through the first state variable of each node, with each node being an n-dimensional dynamical subsystem. The dynamics of the whole network are

$$\begin{cases} \dot{x}_{i1} = f_1(x_i) + c\sum_{j=1}^{N} a_{ij}x_{j1} \\ \dot{x}_{i2} = f_2(x_i) \\ \vdots \\ \dot{x}_{in} = f_n(x_i) \end{cases} \quad i = 1, 2, \cdots, N. \tag{1}$$

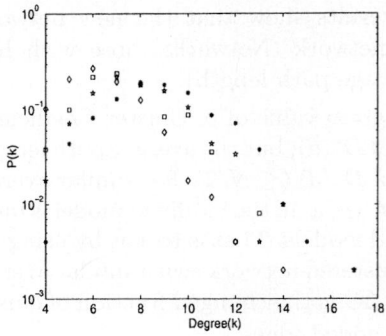 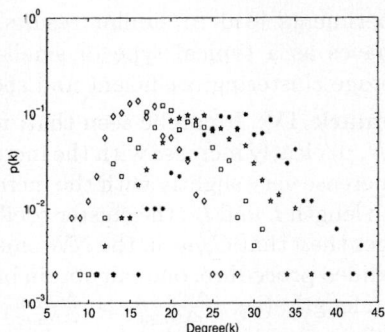

Fig. 5. The "Probability distribution of the connectivity" of the NW small-world model and the modified NW model for $K = 4$, $p = 0.01$, where $N = 500$, $'*'$ represents that of NW's, $'\diamond'$, $'\square'$, $'\bigstar'$ represent that of the M-NW models with parameters setting $N/3$, $N/4$, and $N/5$, respectively

Fig. 6. The "Probability distribution of the connectivity" of the NW small-world model and the modified NW model for $K = 4$, $p = 0.05$, where $N = 500$, $'*'$ represents that of NW's, $'\diamond'$, $'\square'$, $'\bigstar'$ represent that of the M-NW models with parameters setting $N/3$, $N/4$, and $N/5$, respectively

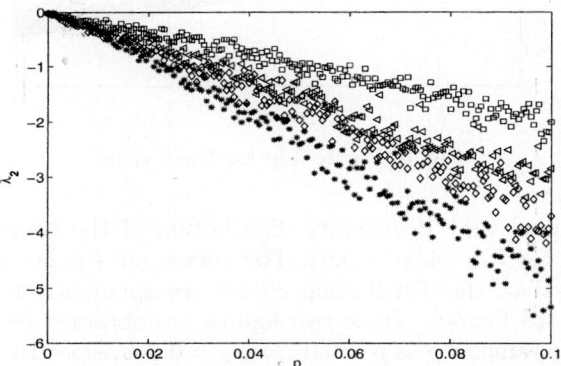

Fig. 7. The change of the largest non-zero eigenvalues of M-SW network and NW network v.s. p, where $D^* = 0$: $'*'$, $D^* = N/3$: $'\square'$, $D^* = N/4$: $'\triangleleft'$, $D^* = N/5$: $'\diamond'$, and the parameter $p = \frac{Number_{added\ edges}}{(N(N-1))/2}$

where $x_i = (x_{i1}, x_{i2}, \cdots, x_{in}) \in \mathbb{R}^n$ are the state variables of node i, $f_k(\mathbf{0}) = 0$, $k = 1, 2, \cdots, n$, $c > 0$ represents the coupling strength, and $A = (a_{ij})_{N \times N}$ is the coupling matrix. If there is a connection between node i and node j, then $a_{ij} = 1$; otherwise, $a_{ij} = 0$ $(i \neq j)$. The diagonal elements of A is defined as

$$a_{ii} = -\sum_{j=1, j\neq i}^{N} a_{ij} = -\sum_{j=1, j\neq i}^{N} a_{ji}, \quad i = 1, 2, \cdots, N \quad (2)$$

Suppose that the network is connected in the sense that there are no isolated clusters. Then the coupling matrix A is a symmetric irreducible matrix. In this case, it can be shown that zero is an eigenvalue of A with multiplicity 1 and all the other eigenvalues of A are strictly negative. Wang X F and Chen G [23] have proved the following result.

Lemma 1 *Consider dynamical network (1). Let λ_2 be the largest nonzero eigenvalue of the coupling matrix A of the network. The synchronization state of network (1) defined by $x_1 = x_2 = \cdots = x_n$ is asymptotically stable, if*

$$\lambda_2 \leq -\frac{T}{c} \tag{3}$$

where $c > 0$ is the coupling strength of (1) and $T > 0$ is a positive constant such that zero is an exponentially stable point of the n-dimensional system

$$\begin{cases} \dot{z}_1 = f_1(z) - Tz_1 \\ \dot{z}_2 = f_2(z) \\ \vdots \\ \dot{z}_n = f_n(z) \end{cases} \tag{4}$$

Note that system (4) is actually a single cell model with self-feedback $-Tz_1$. Condition (3) means that the entire network will synchronize provided that

$$c \geq -T/\lambda_2 \tag{5}$$

We now can compare the changes of the largest nonzero eigenvalue of the coupling matrix of our model and the NW model. Figure 7 shows the results. It can be seen that:

(i) For any given value of N, λ_2 decreases with the increasing of number of added edges;

(ii) Adding the same number of new edges, the value of λ_2 increases with the increasing of $D^*(d)$;

(iii) It is strange that, from Figure 7, we found that the contribution for synchronization of dynamical networks (1) caused by intentionally adding n long-lattice-distance connections is almost no difference when randomly adding n edges to the original regular lattice, when $p \in [0, 0.01]$. If $p \in [0.01, 0.1]$, we found that additional long-lattice-distance connections have not special effects for improving the synchronization for the dynamical networks (1). It is even worse than randomly adding the same number of connections to the original lattice from the viewpoint of considering constructing price. This fact hints us that, in practice, according to both physical and synchronizing mechanism reasons, we should not constructing too many long-lattice-distance connections to obtain better synchronization of networks (1), although it will cause more clustered small-world structure as mentioned before. (eg., $\frac{\Delta \lambda_2^{MNW}}{\Delta N_{add}} < \frac{\Delta \lambda_2^{NW}}{\Delta N_{add}}$, where

$\Delta\lambda_2$ is the changed fraction of second largest eigenvalue, ΔN_{add} is the changed fraction of added connections.)

Remark 3: From the synchronization criterion (3), we can conclude that, although the average path length and clustering coefficients are typical in the range of small world category, the long-lattice-distance connections seem causing lower synchronousness of coupled networks described by (1) without some shorter-lattice-distance connections being added. Thus, the 'short connections' may have similar and equally important effectiveness with the 'long connections' behaved for improving the synchronizability of a dynamical network. In a recent research of Barahona and Pecora [17], they state the small-world property does not guarantee in general that a network will be synchronizable. Further comparison with the results in [17] will proposed later in another paper.

5 Conclusion

In this paper, we proposed a modified small-world lattice network model based on the classical small-world models. Some basic characters are discussed based on numerical experiments with comparison to each other. It can be found that, by using our modified small-worldify algorithm, one can obtain a better clustered small-world network with similar average path length. The result gives us some hints: we can construct a small-world lattice with lower physical price through the proposed modified method. We also discussed the synchronizability of different networks on certain coupling scheme. The numerical results show that, the random coupling in some sense the best candidate for such nonlocal coupling in the semi-random strategy. Main results proposed in this paper are all confirmed by extensive numerical simulations. As we know, sometimes, λ_2 does not guarantee the synchronization and it need more conditions [11], [12]. We will continue our study on the long range for synchronization mechanism in the near future.

Acknowledgement

Author Liu J is partly supported by the Youth Project of Hubei Education Department (Q200517001), the Nature Science Foundation Project of WUSE (20043230). Lu J-A is partly supported the National Key Basic Research and Development 973 Program of China (Grant No. 2003CB415200).

References

1. A, -L, Barabasi.: Linked: The new science of networks. Perseus Publishing, Cambridge. 2002.
2. Adilson, E., Motter., Takashi, Nishikawa., Ying, Cheng, Lai.: Range-based attrack on links in scale-free networks: are long-range links responsible for the small-world phenomenon? Phys. Rev. E. **66**(2002) 065103-4.

3. Chai, Wah, Wu.: Perturbation of coupling matrices and its effect on the synchronizability in array of coupled chaotic systems. Phys. Lett. A. **319** (2003) 495-503.
4. D, J, Watts.: Small Worlds: The Dynamics of Networks between Order and Randomness, Princeton University Press, Princeton, NJ, 1999.
5. D, J., Watts., S, H, Strogatz.: Collective dynamics of small-world. Nature. (1998) 393: 440-442.
6. F, Comellas., Michael, Sampelsb.: Deterministic small-world networks. Phys. A. **309** (2002) 231-235.
7. F, Comellas., J, Ozon., J, G, Peters.: Deterministic small-world communication networks, Inf. Process. Lett. **76** (2000) 83-90.
8. Jon, M, Kleinberg.: The small-world phenomenon: An algorithmic perspective. Proceeding of the 32nd ACM Symposium on theory of computing. May, (2000): 163-170.
9. Jon, M, Kleinberg.: Navigation in a small-world. Nature. **406** (2000) 845.
10. J, Lü., Henry, Leung., G, Chen.: Complex dynamical networks: modeling, synchronization and control,Dynamics of Continuous, Discrete and Impulsive Systems Series B: Applications and Algorithms. **11a** (2004) 70-77.
11. J, Lü., X, Yu., G, Chen., D, Cheng.: Characterizing the synchronizability of small-world dynamical networks. IEEE Trans. Circ. Syst. I. **514** (2004) 787-796.
12. J, Lü., G, Chen.: A time varying complex dynamical network model and its controlled synchronization criteria. IEEE Trans. on Auto. Cont, (2004) accepted.
13. M, E, J, Newman., D, J, Watts.: Renormalization group analysis of the small-world network model. Phys. Lett. A. **263** (1999) 341-346.
14. M, E, J, Newman., Watts, D, J.: Scaling and percolation in the small-world network model. Phys. Rev. E. **60**(1999) 7332-7342.
15. M, E, J, Newman.: Models of the small world: a review, J. Stat. Phys. **101**(2000) 819-841.
16. M, E, J, Newman., C, Moore., D, J, Watts.: Mean-field solution of the small-world network model. Phys. Rev. Lett. **84** 14 (2000) 3201-3204.
17. M, Barahnoa., L, M, Pecora.: Synchronization in small-world systems. Phys. Rev. E. **29** (2002) 054101-4.
18. R, Albert., H, Jeong., A, -L, Barabasi.: Diameter of the world-wide web. Nature, **401** (1999) 130-131.
19. R, Albert., H, Jeong., A, -L, Barabasi.: Error and attack tolerance of complex networks. Nature. **406**(2000) 378-382.
20. R, Albert., A, -L, Barabasi.: Statistical mechanics of complex networks. Rev. Mod. Phys. **74** (2002) 47-97.
21. S, Milgram.: The small-world problem. Psychol. Today. **1** (1967) 60-67.
22. Takashi, Nishikawa., Adilson, E, Motter., Ying, Cheng, Lai., Frank., C, Hoppensteadt.: Hoterogeneity in oscillator networks: are smaller worlds easier to synchronize? Phys. Rev. Lett. **91** (2003) 014101-4.
23. X, F, Wang., G, Chen.: Synchronization in small-world dynamical networks. Int. J. Bifurcation and Chaos. **12** (2002) 187-192.

Double Synaptic Weight Neuron Theory and Its Application*

Wang Shou-jue[1], Chen Xu[1], Qin Hong[1], Li Weijun[1] and Bian Yi[1]

[1] Laboratory of Artificial Neural Networks, Institute of Semiconductors,
Chinese Academy of Sciences
{wsjue, qinhong, wjli}@red.semi.ac.cn
shendacx@163.com
Bianyi@yeah.net

Abstract. In this paper, a novel mathematical model of neuron-Double Synaptic Weight Neuron (DSWN)[1] is presented. The DSWN can simulate many kinds of neuron architectures, including Radial-Basis-Function (RBF), Hyper Sausage and Hyper Ellipsoid models, etc. Moreover, this new model has been implemented in the new CASSANN-II neurocomputer that can be used to form various types of neural networks with multiple mathematical models of neurons. The flexibility of the DSWN has also been described in constructing neural networks. Based on the theory of Biomimetic Pattern Recognition (BPR) and high-dimensional space covering, a recognition system of omni directionally oriented rigid objects on the horizontal surface and a face recognition system had been implemented on CASSANN-II neurocomputer. In these two special cases, the result showed DSWN neural network had great potential in pattern recognition.

1 Introduction

Neural network models consists of a large number of simple processing units (called neuron) densely interconnected to each other through a synaptic interconnection network. In the last decade, the level of interest in Artificial Neural Network (ANNs) has been steadily growing. Although software simulation can be useful, designers have been induced to face hardware solutions in order to meet the required performance of massive computing possibly mission-critical applications. Chip integration, multi-chips system, wafer scale integration (WSI), and even multi - wafers system are the common methods [1], [2]. Despite many models and variations, a common feature for most of them is the basic data progressing unit or artificial neuron. In accordance, the neural networks' performance is primarily decided by the basic computation method and function of neurons.

Neural networks are aimed to mimic biological neural networks often attributed by learning, adaptation, robustness, association and self-organization. In the beginning of 1940's, a classical mathematical model of neuron was presented [3], which was given by the formula (1).

---

* This work was supported by the National Natural Science Foundation of China (No.60135010).

$$Y = f(\sum_{i=0}^{n} W_i X_i - \theta) \qquad (1)$$

where Y is the output vector. f is an activation function (nonlinear function). X is the input vector. W is the weight vector and θ is the activation threshold.

According to formula (1), the neuron's output is decided by two factors: one is the activation function f, and the other is the radix of the function ($\sum_{i=0}^{n} W_i X_i - \theta$), which represents the distance from an input point (in the input space) to a decision hyperplane (one side is positive, and the other is negative). The equation of the decision hyperplane was given by the formula (2).

$$\sum_{i=0}^{n} W_i X_i - \theta = 0 \qquad (2)$$

If the activation function is a step function, the neuron constructs a decision hyperplane in multi- dimensional input spaces. Then the value of output is equal to one when the input point is in one side of this hyperplane. It is zero otherwise. Pattern classifier usually used this kind of neural network [4].

Researchers were at all the time purposing to create closed hypersurface to replace the hyperplane defined by (2) in multi-dimensional space [5]. The RBF neural network is a supervised feed-forward back-propagation neural net with only one hidden layer. While rather than trying to find a boundary between different classes of instances, it forms clusters in the multi-dimensional space with a "center" for each cluster. These clusters are then used to classify different groups of data instances. The number of centers and the nonlinear functions used to determine the distance away from each center dictate the performance of a RBF neural net[6]. It was testified by experiments that in applications of pattern recognition and function fitting, the RBF neural network had the better performance than the neural network described by formula (1).

The mathematical model of RBF neural network is expressed as formula (3):

$$Y = f(\sum_{i=0}^{n} (W_i - X_i)^2 - \theta^2) \qquad (3)$$

According to formula (3), if the activation function of neuron is a step function, the RBF neuron constructs a hypersphere with W_i as the center and θ as the radius. When the input points fall into the inner of this hypersphere, the output is equal to zero. The output is one otherwise. Therefore, RBF neural network can be regarded as one of the simplest high-order neural networks. As its performance is better than the ones described by formula (1), the superiority of high-order hypersurface neural networks is evident.

This paper intends to create a new mathematical model of neuron with better commonality, universal functionality, and easy implementation. The latter text will discuss a novel basic algorithm for high-order hypersurface neural networks-Double

Synaptic Weight Neuron (DSWN) networks, which is applied in the design of CASSANN-II neurocomputer.

2 Early Research Work on General-Purpose Neurocomputer in Author's Lab

A general-purpose neural network hardware should adapt to various neural network connections, diverse activation functions, and flexible algorithm models of neurons. A kind of its implemented method is to use changeable parameters to represent all various factors in a general computing formula. Neurocomputer synchronously calculates the general computing formula repeatedly. And those parameters are adjusted according to practical requirement. Thus, a neural network with complex flexible architecture can be created [7]. For example, CASSANDRA-I neurocomputer, which was created in China, 1995, is readily based on general computing formula under mentioned:

$$O_{mi}(t+1) = F_{k_i}[C_i(\sum_{j=0}^{n-1} S_{ji}W_{ji}I_{mj} + \sum_{g=0}^{n-1} S'_{gi}W'_{gi}O_{mg}(t)) - \theta_i] \qquad (4)$$

where $O_{mi}(t+1)$ is output state value from the i-th neuron at the moment $t+1$ when the m-th example is input. n is the number of input nodes (i.e. the dimension of input space) and the maximum neuron number. F_{k_i} is non-linear function from the i-th neuron, whose subscript k_i is serial number of non-linear function used by the i-th neuron, in function library. I_{mj} is the j-th input value (i.e. the j-th dimension) in the m-th input pattern. $O_{mg}(t)$ is output state value at the moment t from the g-th neuron when the m-th example is input. W_{ji} is the weight from the j-th input node to the i-th neuron. W'_{gi} is the weight from the g-th neuron to the i-th neuron. S_{ji} and S'_{gi} are parameters that determine topological structural model of networks. And if $S_{ji}=0$, there has no connection between the j-th input node and the i-th neuron. If $S'_{gi}=0$, there has no connection between output of the g-th neuron and input of the i-th neuron. On the contrary, there has connection. θ_i is the threshold of the i-th neuron. C_i is scale factor for enlarging dynamic range of calculation.

According to formula (4), CASSANDRA-I neurocomputer can calculate feedforward networks and feedback networks with arbitrary topological architectures. Each neuron can random select various activation functions from non-linear function library. In CASSANDRA-I, there have different thresholds and scale factors. So, it has very flexible and adaptable in topological architecture of networks and parameters of neuron. However, we can easily see that the neuron of formula (4) is based on mathematic model of formula (1). So, it can only calculate "hyperplane" neural networks and the calculation of "hypersurface" neural networks is too hard for it. We can use the basic calculation of formula (1) to calculate included angle between vectors. And then we can construct Direction-basis function (DBF) neural network with the radix of non-linear function being the included angle. This DBF neural network can implement the functions of high-order neurons [8]. But the dimension of closed hypersurface achieved by it was $n-1$ dimension and all modules of input vectors were abandoned during normalization.

In the following sections, a new model-DSWN with various functions and flexibility will be discussed in detail.

3 The Double Synaptic Weight Neuron (DSWN) with Commonality of Hypersurface

The basic mathematical model of hypersurface neuron for the neurocomputer must satisfy the following conditions [9]:

(1)   The model has functions of the traditional hyperplane neuron and RBF neural network.
(2)   The model has the possibility to implement many various hypersurface.
(3)   The model can implement character modification by adjusting minority parameters.
(4)   The model can easily implement high-speed calculation with hardware methods.

According to the conditions, the basic mathematical model of general neuron must be with high flexibility include calculations of both formula (1) and (3). A DSWN model has been proposed. In this model, the signal for each input node has two weights: one is direction weight and the other is core weight. The formula of this model is given as follow:

$$Y = f[\sum_{j=1}^{n} (\frac{W_{ji}(X_j - W'_{ji})}{|W_{ji}(X_j - W'_{ji})|})^S |W_{ji}(X_j - W'_{ji})|^P - \theta] \tag{5}$$

where Y is the output of neuron. f is the activation function. θ is the threshold. W_{ji} and W'_{ji} are two weights from the j-th input node to neuron. X_j is a input value from the j-th input node. n is the dimension of input space. S is a parameter for determining the sign of single entry. If $S=0$, the sign of single entry is always positive and if $S=1$, its sign is the same as the sign of $W_{ji}(X_j - W'_{ji})$. p is a exponent parameter.

Obviously, if all $W'=0$, $S=1$, $p=1$, the formula (5) is the same as the formula (1). If all $W=1$, $S=0$, $p=2$, the formula (5) is the same as the formula (3). Therefore, the formula (5) satisfies the condition (1).

If we assume that $S=0$, this formula defined a closed hypersurface neuron. When the radix of function f is fixed to a definite value, the locus of input points is a closed hypersurface. And its center is decided by W'. Its shape can change according to the value of p on the assumption that all values of W are equal. The case in three-dimensional space can be illustrated. The various shapes of this closed hypersurface according to the value of p being 1/3, 1/2, 1, 2, 3, 4 are showed in figure1, 2, ..., 6, respectively.

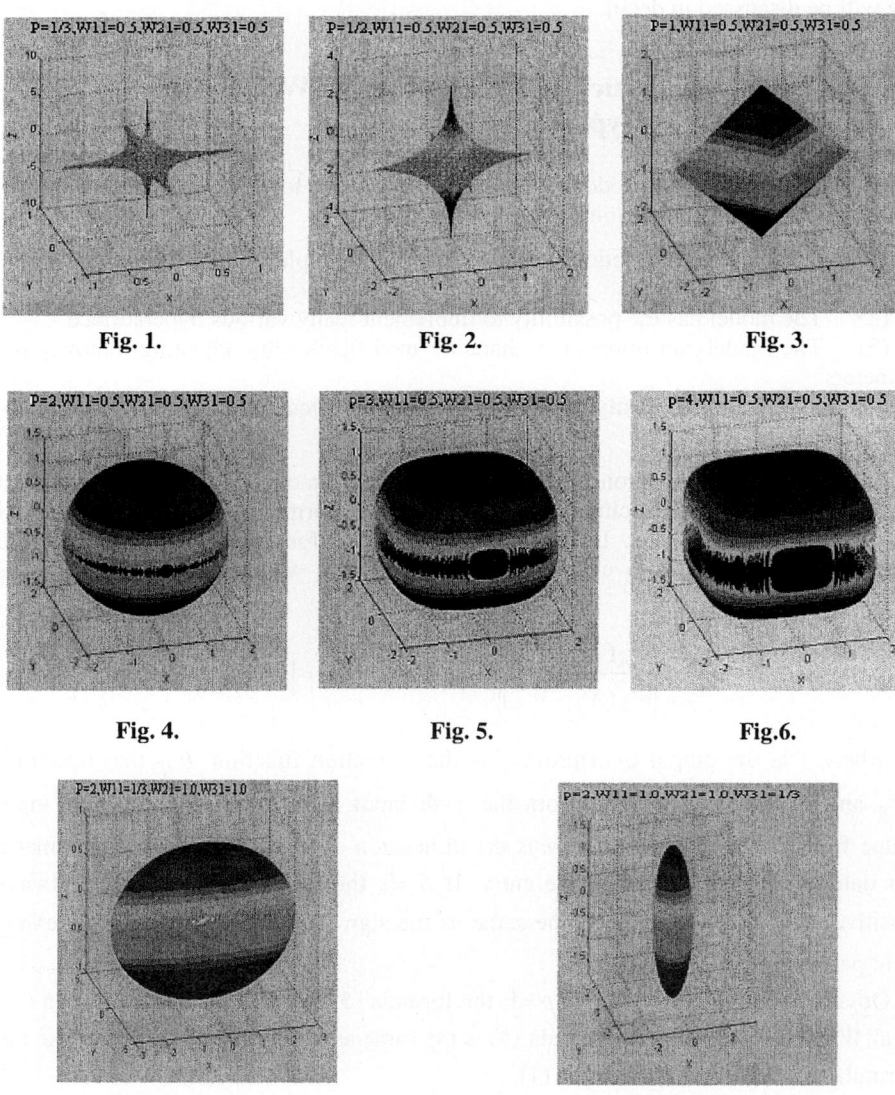

Fig. 1.   Fig. 2.   Fig. 3.

Fig. 4.   Fig. 5.   Fig.6.

Fig. 7.   Fig. 8.

If the weight W has different values, the closed hypersurface will be extended or compressed on various directions. If $p=2$, the hypersphere will be extended or compressed to construct various hyper-sausage or hyper-ellipsoids on different dimensional directions as illustrated in figure 7,8.

Thus, the formula (5) satisfies the condition (2) and (3).

4 The General Formula of Neural Network Hardware Based on DSWN Neurons

The author created general purpose neurocomputer CASSANN–II based on formula (5), which is suitable for traditional BP networks, RBF networks, hyper sausage networks, hyper ellipsoid networks and various high-order hypersurface neural networks. Its general formula is defined as follow:

$$O_{mi}(t+1) = F_{k_i}\{\lambda_i[C_i \cdot (\Re) - \theta_i]\} \tag{6}$$

Where:

$$\Re = [\sum_{j=1}^{n}(\frac{W_{ji}(I_{mj}-W_{ji}^{'})}{|W_{ji}(I_{mj}-W_{ji}^{'})|})^S |W_{ji}(I_{mj}-W_{ji}^{'})|^P + \sum_{g=1}^{n}(\frac{W_{gi}(O_{mg}-W_{gi}^{'})}{|W_{gi}(O_{mg}-W_{gi}^{'})|})^S |W_{gi}(O_{mg}-W_{gi}^{'})|^P]$$

where $O_{mi}(t+1)$ is the output state value from i-th neuron at the moment $t+1$ when the m-th example is input. F_{k_i} is the output non-linear function of the i-th neuron and its subscript k_i is the serial number of non-linear function in function library used by the i-th neuron. I_{mj} is the j-th (i.e. the j-th dimension) input value in the m-th input example. W_{ji} and $W_{ji}^{'}$ are "direction" weight and "core" weight from the j-th input node to the i-th neuron, respectively. W_{gi} and $W_{gi}^{'}$ are "direction" weight and "core" weight from the output of the g-th neuron to the i-th neuron, respectively. p is a exponent parameter (1/3, 1/2, 1, 2, 3, 4). S (0 or 1) is a parameter for determining the sign of single entry. $O_{mg}(t)$ is the output value from the g-th neuron at the moment t when the m-th example is input. θ_i is the threshold of the i-th neuron. C_i is a scale factor. λ_i is a scale factor of coordinate of non-linear function.

According to the formula (6), CASSANN–II neurocomputer can simulate arbitrary neural network architectures with various neuron features (including hyperplane, hypersphere, hyper sausage, various hyper-ellipsoids, hypercube and so on).

5 Applications of Biomimetic Pattern Recognition Implemented on CASSANN-II Neurocomputer

CASSANN-II neurocomputer is composed of DSWN. Based on the theory of Biomimetic Pattern Recognition (BPR) [10] and high-dimensional space covering, many applications of pattern recognition have been implemented on CASSANN-II neurocomputer successfully.

Fig. 9. training set

Fig. 10. test set

The first application of CASSANN-II neurocomputer is a recognition system of omni directionally oriented rigid objects on the horizontal surface [10] based on BPR. Ignoring the disturbance, the distribution region of a certain class is topologically homomorphical to a circle. So Hyper sausage neuron (HSN) networks are used to construct the covering sets of different classes. The SVM method with RBF kernel is also used as control experiment. The samples for training and test are divided in three sample sets. The first one contains 3200 samples of 8 objects (lion, rhinoceros, tiger, dog, tank, bus, car, and pumper,Fig.9), while the second one contains another 3200 samples that are collected later from the same 8 objects. A third one, which comprise 2400 samples of another 6 objects (cat, pug, zebra, little lion, polar bear and elephant,Fig.10), is used for the false acceptance test. All the samples are mapped into a 256-dimensional feature space. The HSN networks are constructed according to the training samples, which are selected from the first sample set. Under the condition that no one sample in the first and second set is misclassified and no one in the third set is accepted falsely, the correct recognition rates of BPR and RBF-SVM with different training set are shown in Table 1.

Table 1. The results of RBF-SVM and BPR

Amount of Training Samples	RBF-SVM		BPR	
	SV	Correct rate	HSN	Correct rate
338	2598	99.72%	338	99.87%
251	1925	99.28%	251	99.87%
216	1646	94.56%	216	99.41%
192	1483	88.38%	192	98.98%
182	1378	80.95%	182	98.22%
169	1307	78.28%	169	98.22%

Another application of CASSANN-II neurocomputer is a face recognition system [11]. If the changes of face appearance are considered as disturbance, the distribution region is topologically homomorphical to an arc when he turns his face horizontally. So the HSN network is very fit to construct the covering set. Ninety-one face pictures of 3 persons are used to construct three HSN networks, and 226 face pictures were used to test the correct recognition rate of the same class, while 728 pictures were used to test the rejection rate of the other classes. The correct recognition rate of the same class reaches 97%, while the rejection rate of the other classes is 99.7%. As the contrast, the correct recognition rate of the same class reaches 89.82%, while the rejection rate of the other classes is 97.94% in K-NN method. [11]

6 Conclusions

This paper proposed a novel general-purpose neuron model- DSWN, which can construct both the hyper sausage and some other more complex shapes. At the same time, this new model is realized using hardware and implemented in the new CASSANN-II neurocomputer.

Based on the theory of BPR and high-dimensional space covering, a recognition system of omni directionally oriented rigid objects on the horizontal surface and a face recognition system had been implemented on CASSANN-II neurocomputer. The result showed DSWN neural network had great potential in pattern recognition.

References

1. Clark S. Lindsey and Thomas Lindblad: Review of Hardware Neural Networks: A User's Perspective [A]. Proceedings of the Neural Networks [C]. From Biology to High Energy Physics, *International Journal of Neural System*, Supp. (1995), 6:215-224.
2. Yuzo Hirai: VLSI Neural Networks Systems[A]. Proceedings of the Neural Networks [C]; From Biology to High Energy Physics, *International Journal of Neural System*, Supp. (1995), 6:203-213.
3. W.S.McCulloch, and W.Pitts : A logic Calculus of the Ideas Imminent in Nerves Activity [J]. *Bulletin of Mathematical Biophysics*, (1943), 5:115-133.

4. M.A.Cohen, and S.Grossberg.:Absolute Stability of Global Pattern Formation and Parallel Memory Storage by Competitive Neural Networks [J]. *IEEE Trans. Syst. Man Cybern.*, (1983), SMC-13:815-826.
5. R.L.Hardy: Multiquadri equations of topography and other irregular surfaces [J].J.Geophys, Re., 76:1905-1915.
6. M.J.D.Powell: Radial basis functions for multivariable interpolation: A review [A]. in J.C. Mason and M.G.Cox, editors, Algorithms for Approximation. Clarendon Press, Oxford, (1987).
7. D.R.Collin and P.A.Penz: Considerations for neural network hardware implementations [J]. *IEEE symp. Circuit and System.* (1989) 834-836.
8. Wang Shoujue, Shi Jingpu, Chen Chuan, Li Yujian: Direction-Basis-Function Neural Networks [A].#58 Session:4.2, *IJCNN'99*.[C](1999).
9. Wang shoujue, Li zhaozhou, Chen xiangdong, Wang bainan: Discussion on the Basic Mathematical Models of Neurons in General Purpose Neurocomputer, (in Chinese), *Acta Electronica Sinica*, (2001).Vol.29 No.5,577-580,
10. Wang shoujue: Biomimetic Pattern Recognition, *Neural Networks Society (INNS, ENNS, JNNS) Newsletters,* (2003).Vol.1, No.1, 3-5
11. Wang zhihai, Zhao zhanqiang and Wang Shoujue: A method of biomimetic pattern recognition for face recognition, (in Chinese), *Pattern Recognize & Artificial Intellegence*, (2003). Vol.16, N0.4

Comparative Study of Chaotic Neural Networks with Different Models of Chaotic Noise

Huidang Zhang and Yuyao He

College of Marine, Northwestern Polytechnical University,
Xi'an, 710072, P.R. China
huidang@haut.edu.cn
heyyao@nwpu.edu.cn

Abstract. In order to explore the search mechanism of chaotic neural network(CNN), this paper first investigates the time evolutions of four chaotic noise models, namely Logistic map, Circle map, Henon map, and a Special Two-Dimension (2-D) Discrete Chaotic System. Second, based on the CNN proposed by Y. He, we obtain three alternate CNN through replacing the chaotic noise source (Logistic map) with Circle map, Henon map, and a Special 2-D Discrete Chaotic System. Third, We apply all of them to TSP with 4-city and TSP with 10-city, respectively. The time evolutions of energy functions and outputs of typical neurons for each model are obtained in terms of TSP with 4-city. The rate of global optimization(GM) for TSP with 10-city are shown in tables by changing chaotic noise scaling parameter γ and decreasing speed parameter β. Finally, the features and effectiveness of four models are discussed and evaluated according to the simulation results. We confirm that the chaotic noise with the symmetry structure property of reverse bifurcation is necessary for chaotic neural network to search efficiently, and the performance of the CNN may depend on the nature of the chaotic noise.

1 Introduction

Recently, many artificial neural networks with chaotic dynamics have been investigated for optimization [2]-[13]. One of the well-known neural networks for combinatorial optimization problems is the Hopfield neural network (HNN) [1]. The HNN model may converge to a stable equilibrium point, but suffers from severe local minima due to its gradient descent dynamics. In order to take advantage of both the Hopfield network's convergent dynamics and chaotic dynamics, some network models composed of chaotic elements have been proposed for information processing [2]-[12]. It may be useful to combine chaotic neurodynamics with heuristic algorithm, high efficiency of which has already been well confirmed [6] [7][13]. For the purpose of harnessing chaos, a kind of chaotic simulated annealing algorithm was derived by extending the original chaotic neural network to a transiently chaotic neural network by introducing the self-feedback connection weight [3]. A more sophisticated adaptive annealing scheme was also considered for practical applications, where the network dynamics is changed from chaotic to convergent by adjusting some parameters [8].

Wang and Smith presented an alternate approach to chaotic simulated annealing by decaying the time step Δt [5]. In order to combine the best of both stochastic wandering and efficient chaotic searching, Wang et al obtained a stochastic chaotic simulated annealing by adding a decaying stochastic noise in the transiently chaotic neural network of Chen and Aihara [3]. The previous approaches are all based on continuous HNN. Based on the discrete-time continuous-output Hopfield neural network (DTCO-HNN) model, Y. He proposed an approach for the TSP by adding chaotic noise to each neuron of the DTCO-HNN and gradually reducing [9][10]. As the chaotic noise approaches zero, the network becomes the DTCO-HNN, thereby stabilizing and minimizing the energy.

In this paper, we harness chaotic behavior for convergence to a stable equilibrium point and attempt to clarify the search mechanism of CNN. We obtain three alternate CNN through replacing the chaotic noise source (Logistic map) in He's CNN with Circle map, Henon map, and a Special 2-D Discrete Chaotic System. According to the computer simulation results of solving the TSP with various approaches, the four CNN all can search global optimal solutions, but the GM is different in terms of different control parameters. Comparisons of solution quality, optimization performance, efficiency of the chaotic search etc. are discussed to try to gain the understanding of chaotic search mechanism.

2 Chaotic Neural Network Models

Based on the discrete chaotic neural network proposed [9][10] by Y. He, the effects of additive chaotic noise are checked. The chaotic neural network based on in this paper is defined as follows:

$$u_i(t+1) = \alpha(\sum w_{ij} v_j + I_i) + \gamma \eta_i(t) \tag{1}$$

$$v_i(t) = f(u_i(t)) = \frac{1}{1+e^{-u_i(t)/\varepsilon}} \tag{2}$$

$$\eta_i(t) = z_i(t) - h \tag{3}$$

where (i=1, 2, ... , n)

$v_i(t)$ Output of neuron i ;

$u_i(t)$ Internal state of neuron i;

w_{ij} Connection weight from neuron j to neuron i , $w_{ij} = w_{ji}$;

I_i Input bias of neuron i;

γ Positive scaling parameter for the chaotic noise;

α Positive scaling parameter for neural inputs;

$\eta_i(t)$ Chaotic noise for neuron i;

ε Gain of the output function, $\varepsilon > 0$;

$z_i(t)$ Chaotic noise source;

h Input bias of chaotic noise

The chaotic noise source $z_i(t)$ can be the Logistic map [10],

$$z_i(t+1) = a(t) z_i(t)(1 - z_i(t)) \tag{4}$$

Let $a(t)$ decay exponentially so that $z_i(t)$ is initially chaotic and eventually settles to a fixed point z^* and $h = z^*$.

$$a(t+1) = (1-\beta)a(t) + \beta \cdot a_0 \quad (5)$$

In order to study the dynamics of the chaotic neural network, we firstly investigate the time evolutions of the previous Logistic map in this section. In terms of the decaying rule (5) ($\beta = 0.005, a_0 = 2.5$) for $a(t)$ and the initial value of $a(0)$ (=3.9), $a(t)$ is decreased by one step after each of iteration. The time evolution of $z(t)$ is shown in Fig. 1(a) with respect to control parameter $a(t)$. The other three type of chaotic noise source take as follows:

(a) Circle map

$$V_{n+1} = f(V_n) = V_n + \Omega - \frac{k_n}{2\pi}\sin(2\pi V_n) \quad (\text{mod} \quad 1) \quad (6)$$

where $V_n \in [-1, +1]$, $\Omega = 0.01$, $V_0 = 0.50$.

$$k_{n+1} = k_n(1-\beta) \quad (7)$$

where β is decreasing rate.

The initial values of k, β are $k = 5.0$, $\beta = 0.003$, respectively, and k is decreased by one step after each of iteration according to the decreasing rule (7). Fig. 1(b) shows the time evolution of V_n according to control parameter k.

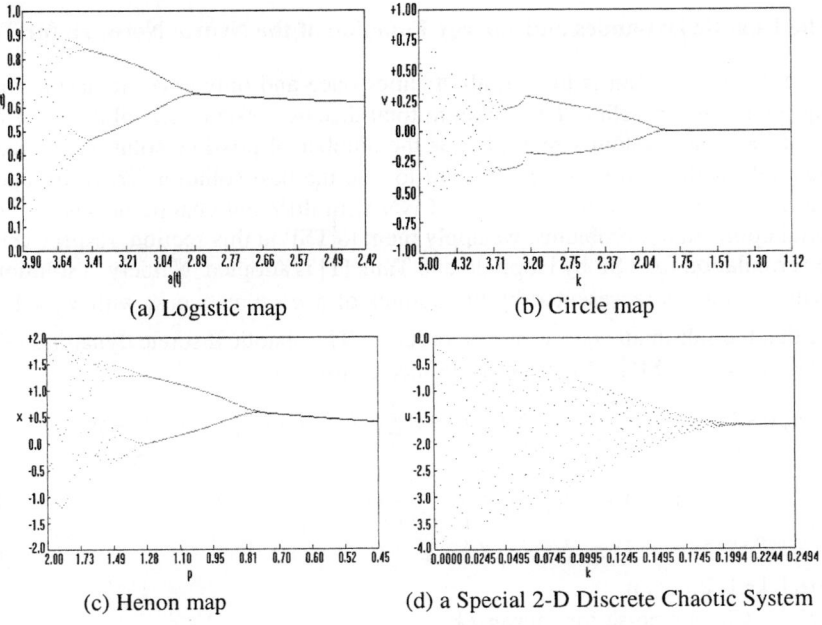

Fig. 1. the time evolutions of chaotic noise within 500 iterations

(b) Henon map

$$x_{n+1} = p_n - x_n^2 + 0.3 y_n \\ y_{n+1} = x_n \} \quad (8)$$

where $x_0 = 0.5, y_0 = 0.5$.

$$p_{n+1} = (1-\beta) p_n \quad (9)$$

The initial values of p, β are $p = 2.0$, $\beta = 0.003$, and p is decreased by one step after each of iteration in terms of the decaying rule (8). Fig. 1(c) shows the time evolution of x_n according to control parameter p.

(c) a Special Two-Dimension (2-D) Discrete Chaotic System [14]

$$u_{n+1} = -c u_n + a v_n + k_n (u_{n-1} - u_n) \\ v_{n+1} = \xi u_n^2 - b / \xi \} \quad (10)$$

where $\xi = 0.25 \quad a = -0.1 \quad b = 1.6 \quad c = 0.1 \quad u_0 = 0.0 \quad v_0 = 0.0$.

$$k_{n+1} = k_n + \beta \quad (11)$$

The initial values of k, β are $k = 5.0$, $\beta = 0.0005$, and k is increased by one step after each of iteration in terms of the rule (11). Fig. 1(d) shows the time evolution of u_n according to control parameter k.

3 Application to the TSP

3.1 The Chaotic Dynamics and Energy Function of the Neural Network for TSP

In the TSP, the salesman is to visit all n cities once and only once, returning to his starting point after traveling the minimum total distance. The exact solution is easily found for the small system size n, but as the number of possible solutions increases exponentially with n, it becomes difficult to find the best solution. To verify and illustrate the features and effectiveness of CNN with different chaotic noises for combinatorial optimization problems, we apply them to TSP in this section, respectively.

The formulation for TSP by Hopfield and Tank [1] is adopted. Namely, a solution of TSP with n cities is represented by the outputs of a $n \times n$ network, with $v_{ik} = 1$ signifying that the salesman visits city i in order k. The chaotic discrete dynamics of the neural network for TSP in this paper is defined as follows:

$$u_{ik}(t+1) = \alpha \left\{ -A(\sum_{l \neq k}^{n} v_{il}(t) + \sum_{j \neq i}^{n} v_{jk}(t)) - B \sum_{j \neq i}^{n} d_{ij}(v_{jk+1}(t) + v_{jk-1}(t)) + A \right\} + \gamma \eta_{ik}(t) \quad (12)$$

$$v_{ik}(t) = f(u_{ik}(t)) = \frac{1}{1 + e^{-u_{ik}(t)/\varepsilon}} \quad (13)$$

$$\eta_{ik}(t) = z_{ik}(t) - h \quad (14)$$

where i, k=1, 2, ... , n
$\eta_{ik}(t)$ Chaotic noise for neuron i,k;
$z_{ik}(t)$ Chaotic noise source;
h Input bias of chaotic noise;

A computational energy function used to minimize the total tour length while simultaneously assuring that all the constraints are satisfied takes the following form:

$$E = \frac{A}{2}\left\{\sum_{i=1}^{n}\left(\sum_{k=1}^{n}v_{ik}-1\right)^2 + \sum_{k=1}^{n}\left(\sum_{i=1}^{n}v_{ik}-1\right)^2\right\} + \frac{B}{2}\sum_{i=1}^{n}\sum_{j=1}^{n}\sum_{k=1}^{n}(v_{jk+1}+v_{jk-1})v_{ik}d_{ij} \quad (15)$$

where $v_{i0} = v_{in}$ and $v_{in+1} = v_{i1}$, A and B are the positive parameters corresponding to the constraints and the tour length, respectively, and d_{ij} is the distance between city i and city j.

Although the dynamics of (12) is discrete, the output v_{ik} from (13) takes a continuous value between zero and one. The corresponding energy function of (15) is also continuous. Since a solution to TSP requires the states of the neurons to be either zero or one, a discrete output is introduced as follows:

$$v_{ik}^{d}(t) = \begin{cases} 1, & \text{iff} \quad v_{ik}(t) > \tilde{v}_{i}(t) \\ 0, & \text{otherwise} \end{cases} \quad (16)$$

where $\tilde{v}_i(t)$ is nth value of $\sum_{k}^{n} v_{ik}(t)$ in order to reduce the number of the iteration. In the simulations, the continuous energy function E^C using (15) and the discrete energy function E^d by replacing v_{ik} with v_{ik}^d in (16) are simultaneously calculated.

In the following studies, a four-city TSP is examined with data originally used by Hopfield and Tank [1], and a ten-city TSP is analyzed with data in [1] with 1000 randomly generated initial conditions of $u_{ik} \in [1,-1]$ and $v_{ik} \in (0,1)$. The constant values of A and B are both one in (12) and (15). The asynchronous cyclic updating of the neural-network model is employed. A iteration means one cyclic updating of all neuron states. The chaotic noise $\eta_{ik}(t)$ in (14) is assigned to each neuron and they are independent of each other. In this paper, the focus of the simulations is on the optimization performance (the rate of global optimization) with different chaotic noise. So, the term $z_{ik}(t)$ in (14) will be replaced with Logistic map (Eq4), circle map (Eq6), Henon map (Eq8) and a Special 2-D Discrete Chaotic System (Eq10), respectively. The term h in (14) is adopted the different fixed point z^* according to Eq4, Eq6, Eq8, Eq10. The other parameters α, β, γ are adjusted with the different chaotic noise model.

3.2 Simulations on TSP with 4-City

The performance of Y. He' method is firstly investigated as a reference for later comparison. The same decaying rule for $a(t)$ ($a_0 = 2.5$) in (5) and the same initial value of $a(0)(= 3.9)$ are used for all neurons, and $a(t)$ is decreased by one step after each of iteration. Fig. 2 shows the time evolutions of (a) the discrete energy function E^d and (b) neuron v_{44} for the four-city TSP with $\alpha = 0.015, \beta = 0.003, \gamma = 0.1, h = 0.6, \varepsilon = 0.004$.

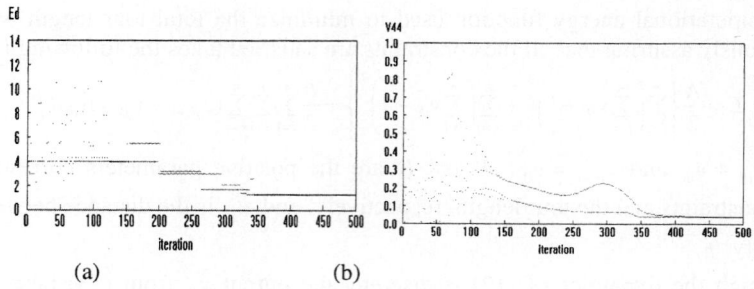

Fig. 2. Time evolutions of (a) E^d and (b) v_{44} with logistic map as chaotic noise

Second, the performance of CNN with the circle map is examined in this subsection. The same decaying rule for k in (7) and the same initial value of k ($=5.0$) are used for all neurons, and k is decreased by one step after each of iteration. Fig. 3 shows the time evolutions of (a) the discrete energy function E^d and (b) neuron v_{34} for the four-city TSP with $\alpha = 0.05, \beta = 0.005, \gamma = 0.5, h = 0.024, \varepsilon = 0.004$.

Furthermore, the performance of another CNN model is explored, which the chaotic noise source is Henon map. The same decaying rule for p in (9) and the same initial value of p ($=2.0$) are used for all neurons, and p is decreased by one step after each of iteration. Fig. 4 shows the time evolutions of (a) the discrete energy function E^d and (b) neuron v_{44} for the four-city TSP with $\alpha = 0.015, \beta = 0.003, \gamma = 0.5, h = 0.4, \varepsilon = 0.004$.

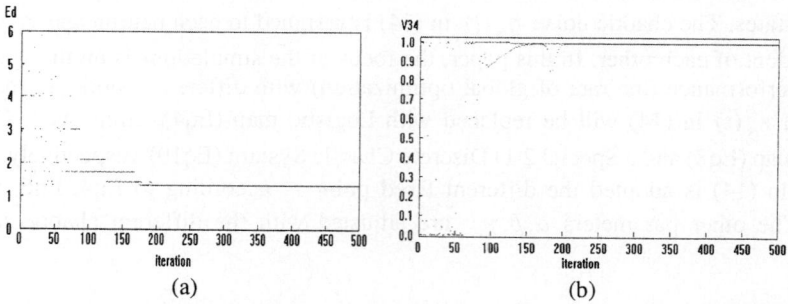

Fig. 3. Time evolutions of (a) E^d and (b) v_{34} with circle map as chaotic noise

Finally, we check the performance of the CNN with a Special 2-D Discrete Chaotic System as chaotic noise. The same increasing rule for k in (11) and the same initial value of k ($=0.0$) are used for all neurons, and k is increased 0.0005 ($\beta = 0.0005$) by one step after each of iteration. Fig. 5 shows the time evolutions of (a) the discrete energy function E^d and (b) neuron v_{12} for the four-city TSP with $\alpha = 0.05, \gamma = 1.5, h = -2.33, \varepsilon = 0.004$.

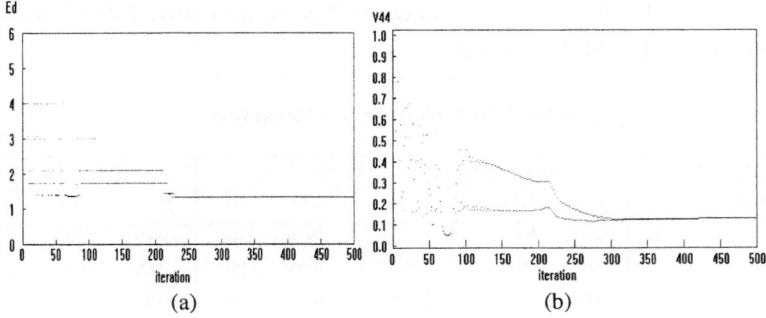

Fig. 4. Time evolutions of (a) E^d and (b) v_{44} with Henon map as chaotic noise

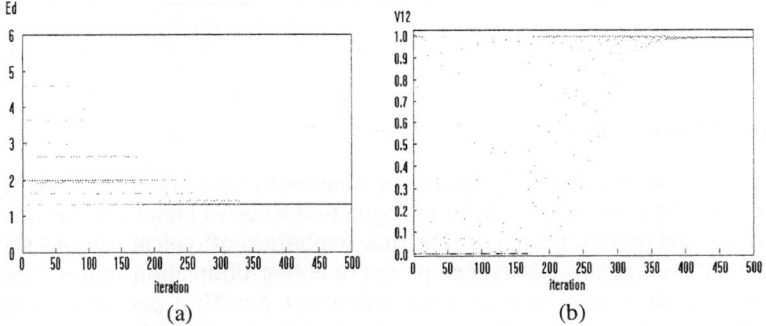

Fig. 5. Time evolutions of (a) E^d and (b) v_{12} with a Special 2-D Discrete Chaotic System as chaotic noise

3.3 Simulations on TSP with 10-City

For the instance, 1000 simulations were performed for each CNN with different initial neuron states. While using logistic map as chaotic noise source, $\alpha = 0.015, h = 0.6$ are adopted. The $\alpha = 0.06, h = 0.008$ are used in CNN when using circle map as chaotic noise source. The coefficients α, h are fixed to 0.015, 0.18 in CNN with Henon map. The coefficients α, h are fixed to 0.02, -2.33 in CNN with 2-D discrete system. The rate of reaching the global minimum (GM) and the number of iterations (NI) required for the network to converge are presented in Table 1 with different coefficient γ while coefficient β is bounded to 0.015, 0.003, 0.005 and 0.001, respectively.

Table 1. Simulation results for each γ

Logistic Map			Circle Map			Henon Map			2-D Map		
γ	GM	NI	γ	GM	NI	γ	GM	NI	γ	GM	NI
0.1	100%	87	0.09	0%		0.005	34%	50	0.015	100%	95
0.3	100%	97	0.1	40%	193	0.007	99%	68	0.10	100%	214
0.6	100%	121	0.15	0%		0.01	100%	186	1.00	100%	261
0.8	100%	144				0.015	100%	233	5.00	100%	285
1.0	100%	165				0.02	100%	254	10.0	100%	297

Bounded γ to 0.1, 0.1, 0.01 and 0.1 in each CNN, respectively, Table 2 shows the GM and NI on 10-city TSP for each β.

Table 2. Simulation results for each β

Logistic Map			Circle Map			Henon Map			2-D Map		
β	GM	NI	β	GM	NI	β	GM	NI	β	GM	NI
0.003	100%	346	0.002	0%		0.002	100%	266	0.0005	100%	368
0.008	100%	143	0.003	40%	193	0.003	100%	186	0.0008	100%	255
0.015	100%	87	0.005	0%		0.005	100%	372	0.001	100%	214
0.05	100%	41				0.01	100%	64	0.003	100%	101
0.1	100%	31				0.03	0%		0.005	100%	75

4 Discussion

4.1 Transient Chaos Scenario

On the basis of the several numerical studies, the time evolution of typical neuron (Fig 2 (b), Fig 3(b), Fig 4 (b) and Fig 5 (b)) is analogue to the one of chaotic noise source (Fig 1 (a), (b), (c) and (d)), respectively. The time evolutions of typical neurons show that the output of each neuron undergoes period-doubling bifurcation routes, which lead to neuronal stable state to be 0 or 1 according to the control parameter. Those figures show that the initial output of neuron is chaotic between 0 and 1. As the control parameter is further decreased, the neuron state switches among smaller scale, and finally merges into a single stable state, which corresponds to a neuron chaotic attractor. The merging process gives rise to the corresponding wandering of the energy among local minima, which can be observed in Fig 2 (a), Fig 3 (a), Fig 4 (a) and Fig 5 (a). We use the control parameter (p) in CNN with Henon map for 4-city TSP to illustrate the phenomenon. At the first stage, $2.0 \geq p > 0.95$, the neuron v_{44} output is chaotic between 1 and 0, and the corresponding value of discrete energy wanders between 4.00 and 1.39379724; at the second stage, $0.95 \geq p > 0.70$, the neuron v_{44} output is period-doubling bifurcation, and the corresponding value of discrete energy switches between 2.110667 and 1.760779; at the third stage, $0.70 \geq P > 0.45$, the neuron v_{44} output is 0, and the corresponding value of discrete energy is the minima (1.341768). The simulations confirm that transient chaos takes a key role for the global optimization of TSP during the chaotic search of CNN.

4.2 Parameter Tuning

Parameter tuning is one of the important issues to improve the performance of such kinds of networks. In fact, total performance for finding the best solution strongly depends on the set of coefficients in E, bias h, and the decaying rate β of chaotic noise. In section 2, we adjust the decaying parameter β to control the chaos to equilibrium via period-doubling bifurcation in terms of Eq4, Eq6, Eq8, and Eq10, respectively.

While incorporating the different chaotic noise into CNN for 10-city TSP, the positive scaling parameter γ and decreasing speed β should be adjusted to gain the global optimization solution. From Table1, Table2, the observations can be made: when β is set to a small value, it uses more steps to converge to a stable state; when γ is set to a large value, it uses more steps to converge to a stable state. The problem to be settled for 'chaotic search' is the difficulty of choosing good parameter values for ($A, B, \alpha, \beta, \gamma$) that may give rise to efficient 'chaotic search'. For 'chaotic search', it may be necessary to adjust the parameters to obtain the symmetric reverse bifurcation structure.

4.3 Efficiency of the Chaotic Search

In this paper, we intend to clarify the search mechanism of CNN designed to solve optimization problems and the role of chaotic noise during the process of chaotic search. It is clear that the searching dynamics of the CNN is made up of two combined dynamics in Eq12. The first term is the input of neuron. The second term is the input of chaotic noise. The second term makes much contribution to the searching dynamics because positive scaling parameter α for neural inputs is small. For example, the Fig5 (b) represents that neuron v_{12} undergoes chaos to equilibrium via reverse bifurcation for the four-city TSP. The other three models show the same effects. From Table1, Table2, the observations can be made: the CNN with circle map has worse effects than the other three methods. The numerical simulation implies the 'better' the transient chaos process is the higher the 'chaotic search' capability. As long as the external noise is appropriate, the energy function (Eq15) of the CNN could sense the force of additive chaotic noise. In this experiment, it is well confirmed that the performance of the CNN may depend on the nature of the chaotic noise.

5 Conclusion

In this paper, possible functional roles of transient chaos are explored. Chaotic effects of four chaotic models are firstly investigated, namely, Logistic map, Circle map, Henon map and a Special 2-D Discrete Chaotic System. The time evolutions of each chaotic noise model are given in Fig1. Second, based on He's CNN model, three alternate approaches are obtained by replacing the chaotic noise source of He's method with Circle map, Henon map and a Special 2-D Discrete Chaotic System, respectively. While applying them to TSP, we obtain that the time evolutions of the discrete energy function E^d and typical neuron for the TSP with four-city. All of them are also applied to TSP with 10-city, respectively. The GM of 1000 different initial conditions are obtained for each of γ with fixed β and for each of β with fixed γ. The simulation results show that the symmetric bifurcation property can improve the efficiency of the chaotic search. Applying the chaotic dynamics to larger scale optimization problems will be studied in forthcoming papers. The systematic way for determining good parameter values for ($A, B, \alpha, \beta, \gamma$) is the subject of future research.

Acknowledgements

This work is supported by A Foundation for the Author of National Excellent Doctoral Dissertation of P.R. China (200250) and Natural Science Foundation of Henan Province (411012400).

References

1. Hopfield, J.J., Tank, D.W.: Neural Computation of Decisions in Optimization Problems. Biological Cybernetics, Vol.52, No.4 (1985) 141–152
2. Nozawa, H.: A Neural Network Model as a Globally Coupled Map and Applications Based on Chaos. Chaos, Vol.2 No.3 (1992) 377–386
3. Chen, L., Aihara, K.: Chaotic Simulated Annealing by a Neural Network Model with Transient Chaos. Neural Networks, Vol. 8, No.6 (1995) 915–930
4. Hayakawa, Y., Marumoto, A., Sawada, Y.: Effects of the chaotic noise on the performance of a neural network model for optimization problems. Physical Review E, Vol.51, No.4 (1995) 2693–2696
5. Wang, L., Smith, K.: On Chaotic Simulated Annealing. IEEE Transaction on Neural Networks, vol. 9, No.4 (1998) 716–718
6. Hasegawa, M., Ikeguchi, T., Aihara, K.: Combination of Chaotic Neurodynamics with the 2-opt Algorithm to Solve Traveling Salesman Problems. Physics Review Letter, Vol.79, No.12 (1997) 2344–2347
7. Hasegawa, M., Ikeguchi, T., Aihara, K.: Exponential and Chaotic Neurodynamical Tabu Searches for Quadratic Assignment Problems. Control and Cybernetics, Vol.29, No.3 (2000) 774–788
8. Tokuda, I., Aihara, K., Nagashima, T.: Adaptive annealing for chaotic optimization. Physic Review E, Vol.58, No.4 (1998) 5157–5160
9. He, Y., Wang, L.: Chaotic Neural Networks and Their Application to Optimization Problems. Control Theory and Applications, Vol.17, No.6 (2000) 847–852
10. He, Y.: Chaotic Simulated Annealing with Decaying Chaotic Noise. IEEE Transaction on Neural Networks, vol.13, No.6 (2002) 1526–1531
11. Lee, G., Farhat, N.H.: The Bifurcating Neuron network 1. Neural Networks, Vol.14, No.1 (2001) 115–131
12. Wang, L., Li, S., Tian, F., Fu, X.: A Noisy Chaotic Neural Networks for Solving Combinatorial Optimization Problem: Stochastic Chaotic Simulated Annealing. IEEE Transaction on System, Man and Cybernetics, Vol.34, No.5 (2004) 2119– 2125
13. Aihara, K.: Chaos Engineering and Its Application to Parallel Distributed Processing With Chaotic Neural Networks. Vol.90, No.5 (2002) 919–930
14. Liu, Bingzheng, Peng, Jianhua: Nonlinear Dynamics. Higher Education Press (in chinese)

A Learning Model in Qubit Neuron According to Quantum Circuit

Michiharu Maeda[1], Masaya Suenaga[1], and Hiromi Miyajima[2]

[1] Kurume National College of Technology,
1-1-1 Komorino, Kurume, Japan
maedami@kurume-nct.ac.jp
[2] Faculty of Engineering, Kagoshima University,
1-21-40 Korimoto, Kagoshima, Japan
miya@eee.kagoshima-u.ac.jp

Abstract. This paper presents a novel learning model in qubit neuron according to quantum circuit and describes the influence to learning with gradient descent by changing the number of neurons. The first approach is to reduce the number of neurons in the output layer for the conventional technique. The second is to present a novel model, which has a 3-qubit neuron including a work qubit in the input layer. For the number of neurons in the output layer, the convergence rate and the average iteration for learning are examined. Experimental results are presented in order to show that the present method is effective in the convergence rate and the average iteration for learning.

1 Introduction

For quantum computation [1],[2] and neural network [3], a number of approaches have been studied. A neural network model dealt with the quantum circuit has been devised for the quantum computation and has been known to exhibit the high capability for learning [4]. However this model has many neurons placed in the output layer so as to correspond to the generall quantum circuit. The neuron is the model which rewrites the computation in quantum mechanics, instead of the real number calculation in neural network, and takes the structure unlike the actual quantum circuit. For the qubit neuron according to the quantum circuit, thus the number of neurons is expected to the less number than in the conventional technique.

In this study, we present a novel learning model in qubit neuron according to quantum circuit and describe the influence to learning with gradient descent by changing the number of neurons. The first approach is to reduce the number of neurons in the output layer for the conventional technique. The second is to present a novel model, which has a 3-qubit neuron including a work qubit in the input layer. For the number of neurons in the output layer, the convergence rate and the average iteration for learning are examined. Experimental results are presented in order to show that the present method is effective in the convergence rate and the average iteration.

2 Quantum Computation

2.1 Quantum Bit

The bit expression in quantum computer is presented with a quantum bit (qubit). For the qubit, the state 0 represents $|0\rangle$, and the state 1, $|1\rangle$. The qubit $|\phi\rangle$ with the superposition of two states is shown as follows.

$$|\phi\rangle = \alpha|0\rangle + \beta|1\rangle, \qquad (1)$$

where α and β are the complex number called the probability amplitude.

In the field of quantum mechanics, the probabilities that $|0\rangle$ and $|1\rangle$ are observed become the square of the absolute value for α and β, respectively. Here α and β satisfy the following relation.

$$|\alpha|^2 + |\beta|^2 = 1 \qquad (2)$$

As $|0\rangle$ and $|1\rangle$ for the qubit are strictly described, these are expressed with the following matrix.

$$|0\rangle = \begin{pmatrix} 1 \\ 0 \end{pmatrix}, \quad |1\rangle = \begin{pmatrix} 0 \\ 1 \end{pmatrix} \qquad (3)$$

2.2 Quantum Gate

The quantum circuit is constituted by the quantum logic gate, such as the rotation gate and the control NOT gate, as shown in Figs. 1 and 2, respectively. For the rotation gate, the state of 1-qubit is rotated. For the control NOT gate, if the qubit a is $|1\rangle$, the output b' becomes the reversal sate of the qubit b. Thus the control NOT gate carries out the XOR operation.

In order to describe the state of qubit, the complex function is used as the quantum state in the following equation, in which the probability amplitude $|0\rangle$ corresponds the real part and the $|1\rangle$, the imaginary part.

$$f(\theta) = e^{i\theta} = \cos\theta + i\sin\theta, \qquad (4)$$

where i is the imaginary unit $\sqrt{-1}$.

Therefore the quantum state is presented as follows.

$$|\phi\rangle = \cos\theta|0\rangle + \sin\theta|1\rangle \qquad (5)$$

Fig. 1. Rotation gate for 1-qubit

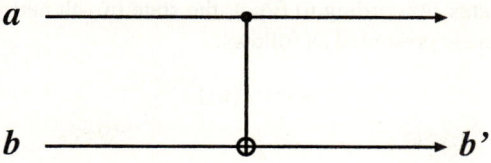

Fig. 2. Control NOT gate for 2-qubit

According to the representation of the quantum state, the rotation gate and the control NOT gate are described. The rotation gate is presented as follows.

$$f(\theta_1 + \theta_2) = f(\theta_1)f(\theta_2) \tag{6}$$

The control NOT gate is expressed as follows.

$$f\left(\frac{\pi}{2}\gamma - \theta\right) = \begin{cases} \sin\theta + i\cos\theta & (\gamma = 1) \\ \cos\theta - i\sin\theta & (\gamma = 0) \end{cases} \tag{7}$$

where γ is the control variable. $\gamma = 1$ implies the reversal state and $\gamma = 0$ means the non-reversal.

2.3 3-Qubit Circuit

Figure 3 shows the 3-qubit circuit. The qubit circuit has the logic state, namely, the work qubit c has the state of logical operation for qubits a and b, by changing the values of θ_1, θ_2, θ_3, and θ_4. For the qubit c, AND state $|a \cdot b\rangle$, OR state $|a + b\rangle$, and XOR state $|a \oplus b\rangle$ can be realized in the quantum circuit.

2.4 Qubit Circuit in Neural Network

As the state $|1\rangle$ is corresponded to an excitatory state and the state $|0\rangle$, an inhibitory state, the quantum state of neurons is considered as the superposition of the excitatory

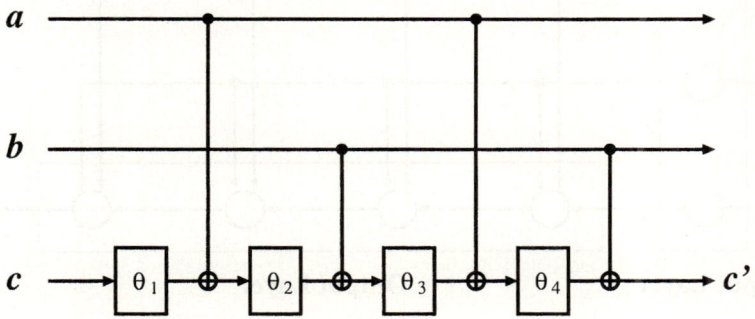

Fig. 3. 3-qubit circuit

and the inhibitory states. According to Eq. 4, the state of i-th neuron s_i which receives inputs from n neurons is presented as follows.

$$s_i = f(u_i). \tag{8}$$

Then

$$u_i = \frac{\pi}{2}g(\delta_i) - \arg(v_i) \tag{9}$$

$$v_i = \sum_{j=1}^{n} f(\theta_{ji})f(u_j), \tag{10}$$

where $g(x)$ is the sigmoid function in the following equation.

$$g(x) = \frac{1}{1+e^{-x}} \tag{11}$$

For the qubit neural network, there exist two variables, the phase variable θ and the reversal rate variable δ. θ is among neurons and δ is in each neuron. θ and δ correspond to the phase of the rotation gate and the reversal rate of the control NOT gate, respectively.

3 Learning Model of Qubit Neuron

In this section, a novel learning model of qubit neuron is presented. Figure 4 shows the network model according to the qubit circuit. x_j ($j = 1, 2, \cdots, m$) and x_0 ($= 0$) in the input layer and y_k ($k = 1, 2, \cdots, n$) in the output layer represent neurons. The input-output properties of neurons in each layer are concretely exhibited as the following description. For the suffix with the top here, a neuron in the input layer represents I, and in the output layer, O.

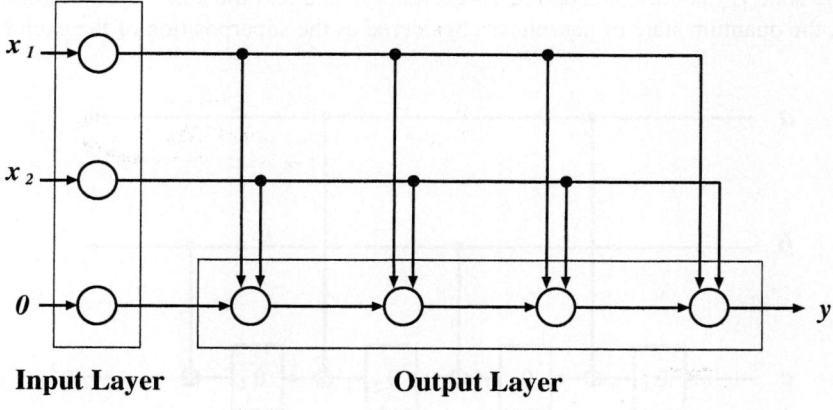

Fig. 4. Model for qubit neuron

(i) Input-output of neurons in the input layer
The output function of neurons in the input layer is written as follows.

$$u_m^I = \frac{\pi}{2} x_m \tag{12}$$

$$s_m^I = f(u_m^I), \tag{13}$$

where x_i has the input $\{0, 1\}$. The output function $f(x)$ corresponds to Eq. 4. For the input 0, the input to the network contains the input of $|0\rangle$, since $u_m^I = 0$ holds and the phase exists on the real axis. For the input 1, the input to the network corresponds to the input $|1\rangle$, because $u_m^I = \pi/2$ holds and the phase exists on the imaginary axis.

(ii) Input-output of neurons in the output layer
According to Eqs. 8, 9, and 10, the output function of neurons in the output layer is presented as follows.

$$v_k^O = e^{i\theta_{k-1,k}} s_{k-1}^O + \sum_{j=1}^{n} e^{i\theta_{j,k}} s_j^I \tag{14}$$

$$u_k^O = \frac{\pi}{2} g(\delta_k) - \arg(v_k^O) \tag{15}$$

$$s_k^O = f(u_k^O) \tag{16}$$

where

$$v_1^O = e^{i\theta_{0,1}} e^{i0} + \sum_{j=1}^{n} e^{i\theta_{j,1}} s_j^I \tag{17}$$

(iii) Final output
The final output is used the probability which is observed the state $|1\rangle$. As the imaginary part represents the probabilistic amplitude of the state $|1\rangle$, the output is the square of the absolute value in the following equation.

$$y = \text{Im}(s_n^O)\text{Im}(s_n^O) \tag{18}$$

For learning in the qubit neuron, the gradient descent is used in this study. The evaluation function is presented as follows.

$$E = \frac{1}{2} \sum_{p=1}^{M} (y_p^t - y_p)^2 \tag{19}$$

where M is the number of sample data, y_p^t is the desired output, and y_p is the final output of neurons.

In order to decrease the value of the evaluation function E, θ and δ are updated as follows.

$$\theta(t+1) = \theta(t) + \Delta\theta(t) \tag{20}$$

$$\delta(t+1) = \delta(t) + \Delta\delta(t) \tag{21}$$

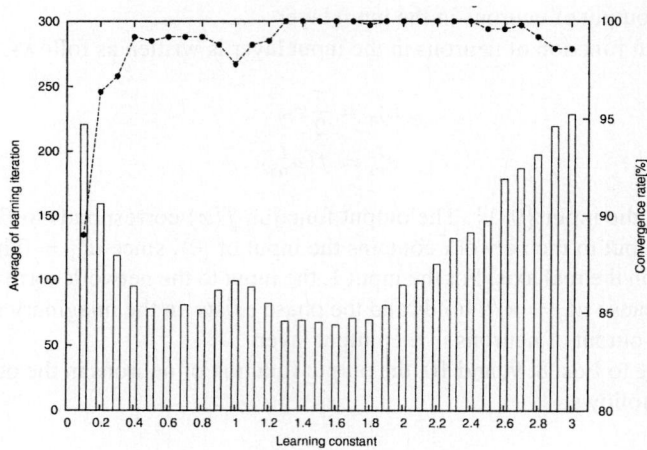

Fig. 5. Dependence of learning constant for conventional model. Number of neurons in output layer is four.

Subsequently $\Delta\theta$ and $\Delta\delta$ are calculated as follows.

$$\Delta\theta(t+1) = -\eta \frac{\partial E}{\partial \theta} \tag{22}$$

$$\Delta\delta(t+1) = -\eta \frac{\partial E}{\partial \delta} \tag{23}$$

where η is the learning constant.

4 Experimental Results

In the numerical experiments, the phase variable θ_{jk} is randomly assigned in $[0, 2\pi)$ and the reversal rate variable δ_k is distributed in $[-1, 1]$ at random for the initial stage. In order to evaluate the performance of the present model, the network learns the basic logical-operation XOR. For XOR operation, four kinds of patterns, $(x_1, x_2 : y^t) = \{(0,0:0), (0,1:1), (1,0:1), (1,1:0)\}$, are given to the network every learning. The results are averages of 500 trials.

Figure 5 shows the dependence of learning constant on the convergence rate and the average of learning iteration for the conventional model (The number of neurons in

Table 1. Learning in conventional approach

Number of neurons	4	3	2	1
Convergence rate [%]	100	100	97.2	0
Average iteration	63.9	66.8	175.2	—

Table 2. Learning in present approach

Number of neurons	4	3	2	1
Convergence rate [%]	100	100	99.0	55.8
Average iteration	45.4	47.3	165.8	445.1

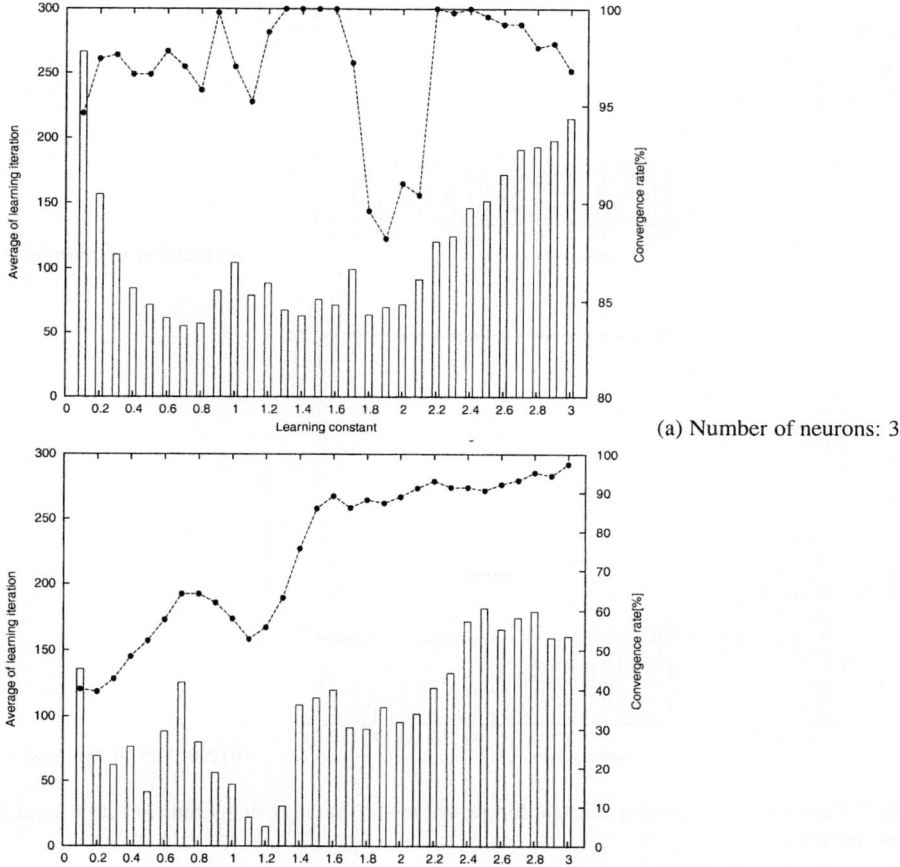

(a) Number of neurons: 3

(b) Number of neurons: 2

Fig. 6. Number of neurons in output layer is three and two for conventional model

the output layer is four). Figure 6 exhibits the number of neurons in the output layer is three and two for the conventional model. Here we confirmed that the network cannot learn for one neuron in the output layer for the conventional model. When the number of neurons increases, the model exhibits high qualities.

Figures 7 and 8 show the dependence of learning constant on the convergence rate and the average of learning iteration for the present model. When the number of neurons increases, the model exhibits high qualities. Especially the present model can learn for one neuron in the output layer.

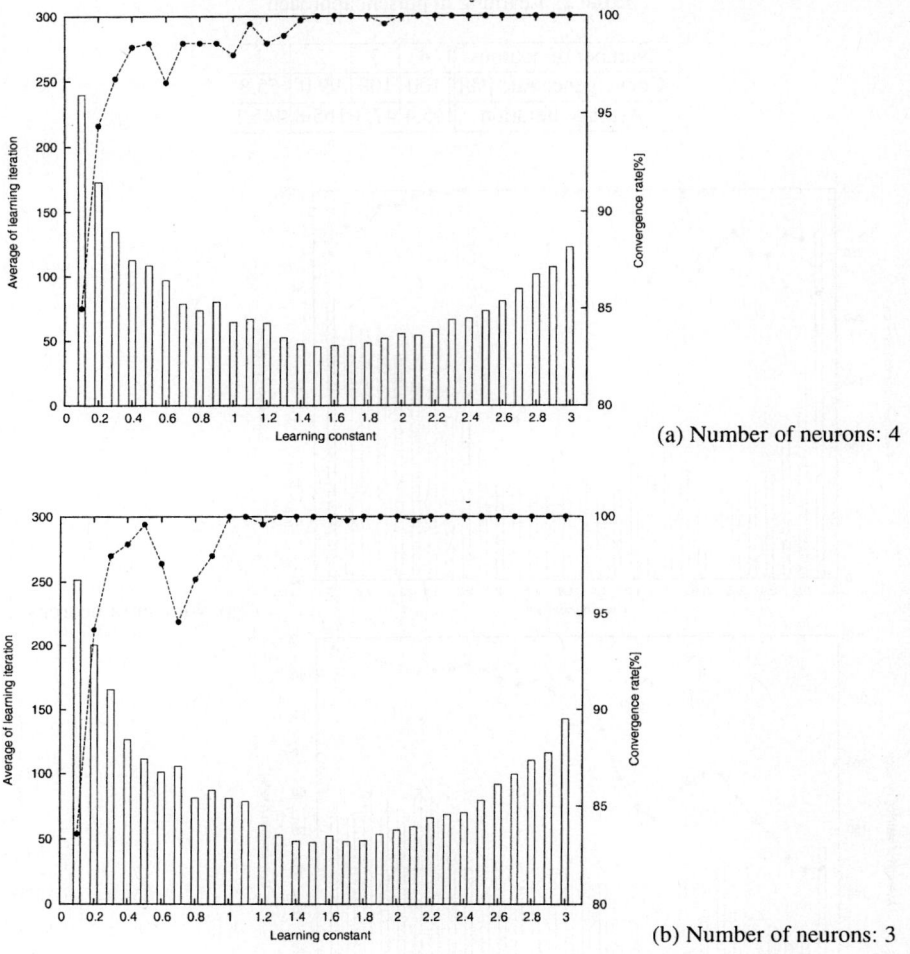

Fig. 7. Dependence of learning constant for present model. Number of neurons in output layer is four and three.

Tables 1 and 2 summarize the convergence rate and the average iteration for XOR problem in the conventional model and the present model, respectively. Here the values are described for the best results in the learning constant. The present method exhibits high qualities.

5 Conclusions

In this paper, we have presented a novel learning model in qubit neuron according to quantum circuit and have described the influence to learning with gradient descent by changing the number of neurons. The first approach was to reduce the number of neurons in the output layer for the conventional technique. The second was to present

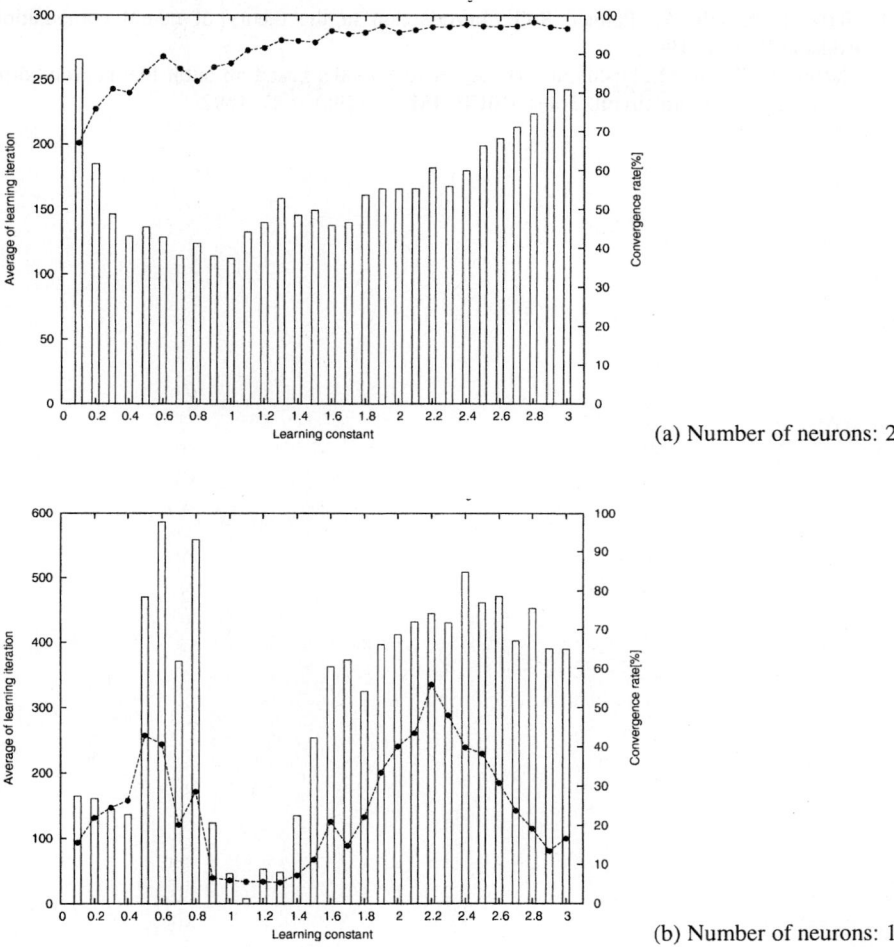

Fig. 8. Dependence of learning constant for present model. Number of neurons in output layer is two and one.

a novel model, which had a 3-qubit neuron including a work qubit in the input layer. For the convergence rate and the average iteration, it has been shown that the present method is more effective than the conventional method. Finally, we will study more effective techniques for the future works.

References

1. Gruska, J.: Quantum computing. McGraw-Hill (1999)
2. Yamada, T., Kinoshita, Y., Kasa, S., Hasegawa, H., Amemiya, Y.: Quantum-dot logic circuits based on the shared binary-decision diagram. Jpn. J. Appl. Phys. **40** (2001) 4485–4488

3. Hertz, J., Krogh, A., Palmer, R.G.: Introduction to the theory of neural computation. Addison-Wesley (1991)
4. Matsui, N., Takai, M., Nishimura, H.: A network model based on qubit-like neuron corresponding to quantum circuit. Trans. IEICE, **J81-A** (1998) 1687–1692

An Algorithm for Pruning Redundant Modules in Min-Max Modular Network with GZC Function

Jing Li[1], Bao-Liang Lu[1,*], and Michinori Ichikawa[2]

[1] Department of Computer Science and Engineering,
Shanghai Jiao Tong University, 1954 Hua Shan Road,
Shanghai 200030, China
jinglee@sjtu.edu.cn , blu@cs.sjtu.edu.cn
[2] Lab. for Brain-Operative Device, RIKEN Brain Science Institue,
2-1 Hirosawa, Wako-shi, Saitama, 351-0198, Japan
ichikawa@brain.riken.go.jp

Abstract. The min-max modular neural network with Gaussian zero-crossing function (M^3-GZC) has locally tuned response characteristic and emergent incremental learning ability, but it suffers from quadratic complexity in storage space and response time. Redundant Sample pruning and redundant structure pruning can be considered to overcome these weaknesses. This paper aims at the latter; it analyzes the properties of receptive field in M^3-GZC network, and then proposes a strategy for pruning redundant modules. Experiments on both structure pruning and integrated with sample pruning are performed. The results show that our algorithm reduces both the size of the network and the response time notably while not changing the decision boundaries.

1 Introduction

The min-max modular (M^3) neural network [1,2] is an alternative modular neural network model for pattern classification. It has been used in real-world problems such as part-of-speech tagging [3] and single-trial EEG signal classification [4]. The fundamental idea of M^3 network is divide-and-conquer strategy: decomposition of a complex problem into easy subproblems; learning all the subproblems by using smaller network modules in parallel; and integration of the trained individual network modules into a M^3 network.

Using linear discriminant function [5] as the base network module, the M^3 network (M^3-Linear) has the same decision boundaries as that of the nearest neighbor classifier (NN) [6]. And M^3-Linear is a specialization of M^3 network with Gaussian zero-crossing function (M^3-GZC) [7], so M^3-GZC can be viewed as a generalization of nearest neighbor classifier. The most attracting attributes of

---

* To whome correspondence should be addressed. This work was supported in part by the National Natural Science Foundation of China via the grants NSFC 60375022 and NSFC 60473040.

M^3-GZC are its locally tuned response characteristic and emergent incremental learning ability. But it suffers from quadratic complexity in space and time, and may be inefficient in large-scale, real-world pattern classification problems.

To decrease the storage space and response time of M^3-GZC network, two ways of redundancy pruning can be considered. One is sample pruning, which is inspired by pruning strategies in NN [8,9,10,11,12,13,14,15]. We have proposed the Enhanced Threshold Incremental Check algorithm [16] for M^3-GZC network in our previous work. The other way is structure pruning, which is correlative with detailed network and can not borrow ideas from NN. In this paper we will analyze the structure of M^3-GZC network and propose a pruning algorithm.

The rest of the paper is organized as follows: In Section 2, M^3-GZC network is introduced briefly. In Sections 3 and 4 we analyze the properties of receptive field and redundant modules in M^3-GZC network. In Section 5 pruning algorithm is described. Experiments are presented in Section 6. Finally, conclusions are presented in Section 7.

2 Min-Max Modular Network with GZC Function

Let $\mathcal{T}$ be the training set for a K-class problem,

$$\mathcal{T} = \{(X_l, D_l)\}_{l=1}^{L}, \tag{1}$$

where $X_l \in R^n$ is the input vector, $D_l \in R^K$ is the desired output, and L is the total number of training data.

According to the min-max modular network [1,2], a K-class problem defined in equation (1) can be divided into $K \times (K-1)/2$ two-class problems that are trained independently, and then integrated according to a module combination rule, namely the minimization principle. Fig.1(a) shows the structure of M^3 network to a K-class problem, where L_i denotes the number of data belonging to class C_i.

A two-class problem can be further decomposed into a number of subproblems and be integrated according to the minimization principle and the maximization principle. These subproblems can be learned by some base network modules, such as SVM[17], back-propagation algorithm[3,4], and so on. Suppose the training set of each subproblem has only two different samples, and the base network module is Gaussian zero-crossing discriminate function as defined in equation (2), the corresponding network is called M^3-GZC network.

$$f_{ij}(x) = exp\left[-\left(\frac{\|x-c_i\|}{\sigma}\right)^2\right] - exp\left[-\left(\frac{\|x-c_j\|}{\sigma}\right)^2\right], \tag{2}$$

where x is the input vector, c_i and c_j are the given training inputs belonging to class C_i and class C_j ($i \neq j$), respectively, $\sigma = \lambda \|c_i - c_j\|$, and λ is a user-defined constant.

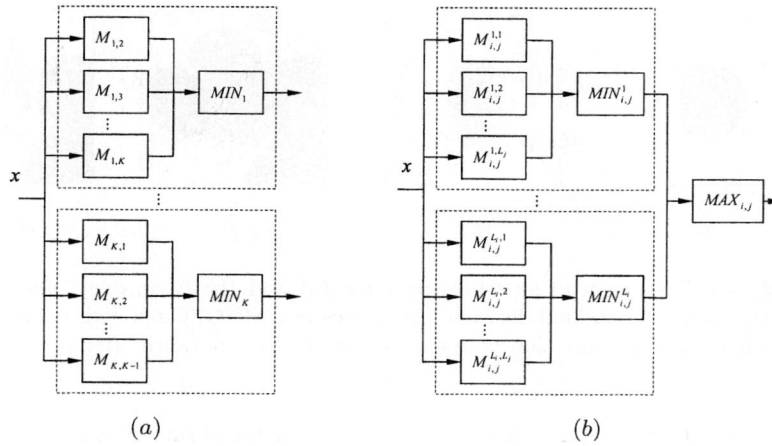

Fig. 1. Structure of M^3-GZC network. (a) A K-class problem; (b) Further decomposition of a two-class problem.

The output of M^3-GZC network is defined as follows.

$$g_i(x) = \begin{cases} 1 & \text{if } y_i(x) > \theta^+ \\ Unknown & \text{if } \theta^- \leq y_i(x) \leq \theta^+ \\ -1 & \text{if } y_i(x) < \theta^- \end{cases} \quad (3)$$

where θ^+ and θ^- are the upper and lower threshold limits, and y_i denotes the transfer function of the M^3 network for class C_i, which discriminates the pattern of the M^3 network for class C_i from those of the rest of the classes.

The structure of M^3-GZC network is shown in Fig.1. It is clear that the total number of modules in a M^3-GZC network is

$$\sum_{i=1}^{K} \sum_{j=1, j \neq i}^{K} L_i \times L_j, \quad (4)$$

which means quadratic complexity in storage space and response time.

3 Properties of Receptive Field in M^3-GZC Network

The receptive field in a M^3-GZC network is defined as the input space that can be classified to one class.

$$RF = \{x | x \epsilon R^n, \exists i, g_i(x) = 1\}. \quad (5)$$

Suppose there are only two samples c_i and c_j, and we only concentrate on the receptive field around c_i. According to the axiom of norm, the following equation is satisfied.

$$\|c_i - c_j\| - \|x - c_i\| \leq \|x - c_j\| \leq \|c_i - c_j\| + \|x - c_i\|. \quad (6)$$

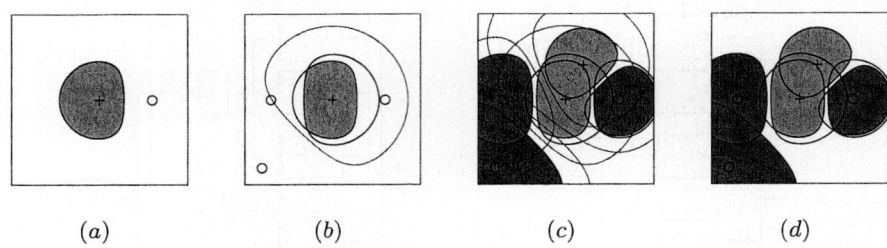

Fig. 2. An illustration of structure pruning. (a) and (b) Receptive fields of a MIN unit; (c) Modules and final decision boundaries in a M^3-GZC network before pruning; (d) Modules and final decision boundaries in a M^3-GZC network after pruning.

So the shortest receptive field radius r_{min} can be obtained when $\|x - c_j\| = \|c_i - c_j\| - \|x - c_i\|$, while the longest receptive field radius r_{max} can be achieved when $\|x - c_j\| = \|c_i - c_j\| + \|x - c_i\|$, as depicted in Fig.2 (a).

From equations (2), (3), (5), and (6), we can prove that the relationship between r_{max} and $\|c_i - c_j\|$ can be expressed as

$$r_{max} = k_1 \|c_i - c_j\|, \qquad (7)$$

where k_1 is only correlated with λ and θ^+.

Proof: Suppose x is on the direction of r_{max} and on the margin of the receptive field, which means $\theta^+ = f_{ij}(x)$. From equations (2) and (7), we get:

$$\begin{aligned}\theta^+ &= exp\left[-\left(\frac{k_1\|c_i - c_j\|}{\lambda\|c_i - c_j\|}\right)^2\right] - exp\left[-\left(\frac{k_1\|c_i - c_j\| + \|c_i - c_j\|}{\lambda\|c_i - c_j\|}\right)^2\right] \\ &= exp\left[-\left(\frac{k_1}{\lambda}\right)^2\right] - exp\left[-\left(\frac{k_1+1}{\lambda}\right)^2\right].\end{aligned} \qquad (8)$$

So k_1 is a function of λ and θ^+.

Also, we can prove that the relationship between r_{min} and $\|c_i - c_j\|$ can be expressed as:

$$r_{min} = k_2 \|c_i - c_j\|. \qquad (9)$$

where k_2 satisfies the following equation.

$$\theta^+ = exp\left[-\left(\frac{k_2}{\lambda}\right)^2\right] - exp\left[-\left(\frac{1-k_2}{\lambda}\right)^2\right]. \qquad (10)$$

4 Redundancy Analysis

When another sample c'_j belonging to class C_j is available, module $M_{i,j'}$ will be established, which determines another receptive field RF_2 around c_i. Then the minimization principle will be used to combine RF_2 and the field RF_1 that was previously determined by c_j and c_i. Since the role of minimization principle is similar to the logical AND [1], only those fields that contained in both RF_1 and RF_2 will be the final receptive field RF, as shown in Fig.2 (b). In other word, if RF_2 includes RF_1, RF will be equal to RF_1. In this case, sample c'_j has no contribution to the final receptive fields around c_i, and module $M_{i,j'}$ is a redundant module.

Now the question of under what circumstances RF_2 will include RF_1 arises. Here we give a sufficient proposition.

Proposition 1: Suppose sample c_j is the nearest sample in class C_j to sample c_i, if sample c'_j in class C_j satisfies equation (11), then module $M_{i,j'}$ is a redundant module.

$$\|c_i - c'_j\| \geq \frac{k_1}{k_2}\|c_i - c_j\| \tag{11}$$

The proof is straightforward. From equation (11) we can get $k_2\|c_i-c_{j'}\| \geq k_1\|c_i-c_j\|$, which means that r_{min} of RF_2 is larger than r_{max} of RF_1, so $RF_1 \subseteq RF_2$, and module $M_{i,j'}$ is a redundant module.

For a k-class classification problem, proposition 1 can be extended to proposition 2 according to the minimization principle in K-class classification problems[1].

Proposition 2: Suppose sample c_j is the nearest sample in class C_j ($1 \leq j \leq K, j \neq i$) to sample c_i, if sample c_k in class C_k ($1 \leq k \leq K, k \neq i$) satisfies equation (12), then module $M_{i,k}$ is a redundant module.

$$\|c_i - c_k\| \geq \frac{k_1}{k_2}\|c_i - c_j\| \tag{12}$$

5 Pruning Algorithm

For a K-class problem defined in equation (1), according to proposition 2, our pruning algorithm works as below.

1. Calculate k_1 and k_2 according to λ and θ^+;
2. For each sample (x, d) in T,
   (a) Find the nearest neighbor (x', d') in T, $d \neq d'$ and $\|x-x'\| = MIN\{\|x''-x\|\}$, $(x'', d'')\epsilon T$, $d'' \neq d$.
   (b) For each sample (x'', d'') in T ($d'' \neq d$), if $\|x'' - x\| \geq \frac{k_1}{k_2}\|x' - x\|$, prune the module based on (x, d) and (x'', d'').

The final structure of pruned M^3-GZC network is composed of L MIN units, as shown in Fig.3. Each MIN unit is composed of a center sample and some neighbors in different classes around it. When a test sample x is presented, if it is in the receptive field of one MIN unit, then the calculation is completed, and the output is the same as the class of the center sample. If x is rejected by all the MIN units, then the output is '$Unknown$'.

Suppose there are N_i neighbors around one center sample, N_i is determined by the distribution of training samples. The total number of modules in the

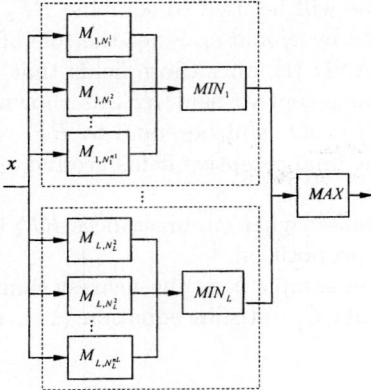

Fig. 3. Structure of pruned M^3-GZC network. N_i^j denotes the jth neighbor around sample i

pruned M^3-GZC network is $\sum_{i=1}^{L} N_i$, which is less than that in the original M^3-GZC network: $\sum_{i=1}^{K} \sum_{j=1, j \neq i}^{K} L_i \times L_j$.

An illustration of our pruning algorithm is depicted in Fig.2 (c) and (d). Each circle line represents a module in M^3-GZC network. The black and grey areas denote the receptive field of each class, while the white area denotes the '$Unknown$' output. Form the results, we can see that the decision boundaries are identical, while 41.7% modules are pruned.

6 Experimental Results

In order to verify our method, we present experiments on three data sets. The first is an artificial problem and the other two are real-world problems. We also do the experiments that integrating our method with sample pruning. All the experiments were performed on a 2.8GHz Pentium 4 PC with 1GB RAM.

6.1 Two-Spiral Problem

We test our structure pruning algorithm on the popular two-spiral benchmark problem firstly. The data include 192 training samples and test samples respectively (non-overlapping). The parameters of the experiment are given as follows: $\lambda = 0.5$; $\theta^+ = 0.01$; $\theta^- = -0.01$. The correspond k_1 and k_2 is 1.073 and 0.497, respectively. Fig.4 (a) shows the original problem, Fig.4 (b) shows the decision boundaries before pruning and Fig.4 (c) shows the decision boundaries after pruning. As we have expected, they are identical, but the number of modules and response time are greatly reduced, as listed in Table 1.

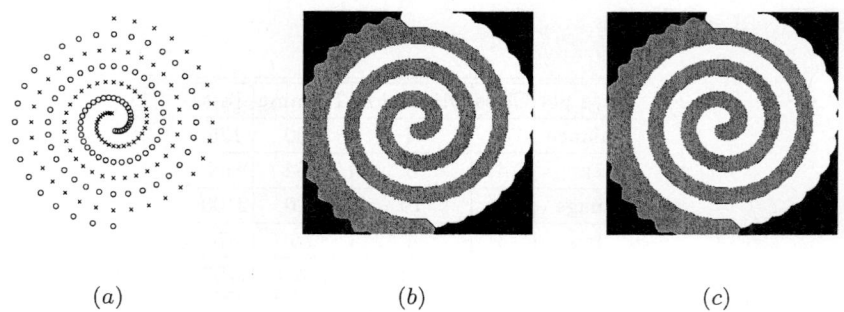

(a) (b) (c)

Fig. 4. Results on two-spiral problem. (a) Training samples; (b) Decision boundaries before pruning; (c) Decision boundaries after pruning. Here black area denotes 'Unknown' output.

Table 1. Experimental results. The upper row in each experiment denotes the pruned net while the lower row denotes the original net. The unit of 'Time' is ms.

Data set	Accuracy	Unknown	False	Size	Time	Size Ratio	Speed Up
two-spirals	100%	0.0%	0.0%	2516	18	13.7%	129
	100%	0.0%	0.0%	18432	2315		
balance	92.0%	0.0%	8.0%	39377	42	31.6%	137
	92.0%	0.0%	8.0%	124800	5767		
car	57.87%	42.13%	0.0%	126079	1805	37.7%	60
	57.87%	42.13%	0.0%	334006	107878		
image	84.0%	7.33%	8.67%	11280	449	33.0%	66
	84.0%	7.33%	8.67%	34200	29730		
Iris	94.67%	1.33%	4.0%	1843	3	49.1%	84
	94.67%	1.33%	4.0%	3750	252		
optdigits	97.22%	1.45%	1.34%	11454592	10784	89.1%	700
	97.22%	1.45%	1.34%	12862520	7548237		
glass image	86.0%	2.0%	12.0%	1167989	18817	43.7%	125
	86.0%	2.0%	12.0%	2673000	2349796		

6.2 UCI Database

In this experiment, our algorithm is tested on five benchmark data sets from the Machine Learning Database Repository[18]: Balance, Car, Image Segmentation, Iris and Optdigits. The detailed information of each problem is described in Table 2. The parameters of each experiments are same as those in the two-spiral problem, and results are listed in Table 1.

Table 2. Number of class, dimension, training samples and test samples in UCI database

Data Set	Class	Dimension	Training	Test
balance	2	4	500	125
car	4	6	864	864
image	5	19	210	2100
Iris	3	4	75	75
optdigits	9	64	3823	1797

Table 3. Experimental results of integrating sample pruning and structure pruning. The upper row in each experiment denotes the net after sample pruning and structure pruning while the lower row denotes the net only with sample pruning. The unit of 'Time' is ms.

Data set	Accuracy	Unknown	False	Size	Time	Size Ratio	Speed Up
two-spirals	100%	0.0%	0.0%	794	11	4.31%	208.3
	100%	0.0%	0.0%	8192	1268	44.4%	1.82
balance	92.0%	0.0%	8.0%	9878	15	7.92%	384.6
	92.0%	0.0%	8.0%	44676	2510	35.8%	2.30
car	62.15%	34.14%	3.70%	31322	645	9.38%	166.7
	62.15%	34.14%	3.70%	111138	36392	33.3%	2.97
image	82.0%	9.24%	8.76%	3280	478	9.59%	625
	82.0%	9.24%	8.76%	11162	12035	32.6%	2.47
Iris	94.67%	1.33%	4.0%	345	3	9.2%	1.19%
	94.67%	1.33%	4.0%	570	125	15.2%	84.0
optdigits	96.05%	2.62%	1.34%	1137798	3714	8.85%	2000
	96.05%	2.62%	1.34%	1378048	840613	10.7%	9.0
glass image	85.55%	2.59%	11.86%	46397	16049	1.74%	147.1
	85.55%	2.59%	11.86%	176928	151796	6.62%	15.5

6.3 Industry Image Classification

Due to its locally tuned response characteristic and incremental learning ability, M^3-GZC has been used in an industry fault detection project. The purpose of this project is to choose out eligible glass-boards in an industrial product line, which is done by trained workers in practice. It is a boring work; workers are easy to be tired and then make wrong decisions. With the help of M^3-GZC network, workers need only judge the glass-boards that are classified to 'Unknown' by the network. In our experiment, each glass-board image is converted into a 4096 dimension vector, 3420 images are used as training data while 1197 images as test data. The parameters are same as those in the two-spiral problem, and results are listed in Table 1.

From Table 1, several observations can be made. Our pruning method has no influence on the classification accuracy, but the size and response time can be decreased notably, by an average of 42.6% and 0.975%, respectively. The response time is saved much further than the size. This is due to that in the pruned net it need not calculate all the modules to get the answer, if there is a MIN unit accepts it, the calculation can be finished. Only those inputs that the correspond result is '$Unknown$' will calculate all the modules. But in most cases, the '$Unknown$' ratio is very low. So the response time can be cut down greatly.

6.4 Integrated with Sample Pruning

Experiments of integrating sample pruning (Enhanced Threshold Incremental Check)[16] and structure pruning are also conducted on the data sets mentioned above. First we use ETIC to prune redundant samples in each training data set; then we use our structure pruning algorithm to prune redundant models. The results are listed in Table 3. We can see that the size and response time are decreased much further, by an average of 7.28% and 0.49%, respectively.

7 Conclusions

M^3-GZC network has the locally tuned response characteristic and emergent incremental learning ability. But it suffers from sample redundancy and module redundancy. In this paper we have presented a novel structure pruning algorithm to reduce the redundant modules based on the properties of receptive field in M^3-GZC network. The decision boundaries of the pruned net are identical with the original network, but the storage and response time requirement decreased significantly. Experiments on structure pruning and integrated with sample pruning verified the effectiveness of our pruning algorithm. We believe that module redundancy reflects sample redundancy, our future work is to investigate the relationship between them and combine them more effectively.

References

1. Lu, B.L. and Ito, M.: Task decomposition based on class relations: a modular neural network architecture for pattern classification. Lecture Notes in Computer Science, Springer vol.1240(1997)330-339
2. Lu, B.L. and Ito, M.: Task Decomposition and Module Combination Based on Class Relations: A Modular Neural Network for Pattern Classification. IEEE Trans. Neural Networks, vol.10 (1999) 1244–1256
3. Lu, B.L., Ma, Q., Ichikawa,M. and Isahara, H. :Efficient Part-of-Speech Tagging with a Min-Max Modular Neural Network Model. Applied Intelligence, vol.19 (2003)65–81

4. Lu, B.L., Shin, J., and Ichikawa, M.: Massively Parallel Classification of Single-Trial EEG Signals Using a Min-Max Modular Neural Network. IEEE Trans. Biomedical Engineering, vol.51, (2004) 551–558
5. Duda, R.O., Hart, P.E. and Stork, D.G.: Pattern Classification, 2nd Ed. John Wiley & Sons, Inc. 2001.
6. Wang, Y.F.: A study of incremental learning method based on Min-Max Module networks (in Chinese). Bachelor Dissertation of SJTU, 2003.
7. Lu, B.L. and Ichikawa, M.: Emergent On-Line Learning with a Gaussian Zero-Crossing Discriminant Function. IJCNN '02, vol.2 (2002) 1263–1268
8. Hart, P.E.: The Condensed Nearest Neighbor Rule. IEEE Trans. Information Theory, vol.14 (1968) 515–516
9. Gates, G.W.: The Reduced Nearest Neighbor Rule. IEEE Trans. Information Theory, vol.18 (1972) 431–433
10. Wilson, D.L.: Asymptotic propoties of nearest neighbor rules using edited data. IEEE trans. System, Man, and Cybernetics, vol.2, No.3 (1972) 431–433
11. Aha, D.W., Dennis, K. and Mack, K.A.: Instance-based learning algorithm. Machine Learning, vol.6 (1991) 37–66
12. Zhang, J.P.: Selecting typical instances in instance-based learning. Proceedings of the Ninth International Conference on Machine Learning, (1992) 470–479
13. Wai, L., Keung, C.K. and Liu, D.Y.: Discovering useful concept prototypes for classification based on filtering and abstraction. IEEE Trans. Pattern Analysis and Machine Intelligence, vol.24, No.8 (2002) 1075–1090
14. Skalak, D.B.: Prototype and feature selection by sampling and random mutation hill climbing algorithm. Proceedings of the Eleventh International Conference on Machine Learning, (1994) 293–301
15. Cameron-Jones, R.M.: Instance selection by encoding length heuristic with random mutation hill climbing. Proceddings of the Eightn Australian Joint Conference on Artificial Intelligence, (1995) 99–106
16. Li, J., Lu, B.L. and Ichikawa, M.: Typical Sample Selection and Redundancy Reduction for Min-Max Modular Network with GZC Function. ISNN'05, Lecture Notes in Computer Science, (2005) 467-472
17. Lu, B.L., Wang, K.A., Utiyama, M. and Isahara, H.: A part-versus-part method for massively parallel training of support vector machines. Proceedings of IJCNN'04, Budapast, July25-29 (2004) 735-740.
18. Murphy, P.M. and Aha, D.W.: UCI Repository of Machine Learning Database. Dept. of Information and Computer Science, Univ. of Calif., Irvine, 1994.

A General Procedure for Combining Binary Classifiers and Its Performance Analysis

Hai Zhao and Bao-Liang Lu*

Departmart of Computer Science and Engineering,
Shanghai Jiao Tong University, 1954 Hua Shan Road,
Shanghai 200030, China
{zhaohai, blu}@cs.sjtu.edu.cn

Abstract. A general procedure for combining binary classifiers for multiclass classification problems with one-against-one decomposition policy is presented in this paper. Two existing schemes, namely the min-max combination and the most-winning combination, may be regarded as its two special cases. We show that the accuracy of the combination procedure will increase and time complexity will decrease as its main parameter increases under a proposed selection algorithm. The experiments verify our main results, and our theoretical analysis gives a valuable criterion for choosing different schemes of combining binary classifiers.

1 Introduction

The construction of a solution to a multiclass classification problem by combining the outputs of binary classifiers is one of fundamental issues in pattern recognition research. For example, many popular pattern classification algorithms such as support vector machine (SVM) and AdaBoosting are originally designed for binary classification problems and strongly depend on the technologies of multiclass task decomposition and binary classifier combination. Basically, there are two methods for decomposing multiclass problems. One is one-against-rest policy, and the other is one-against-one policy. The former is computationally more expensive, the latter is more popular in practical application and will be concerned in this paper.

There are three main combination policies for one-against-one scheme according to reported studies. a) the most-winning combination (round robin rule (R^3) learning [1]); b) the min-max combination that comes from one of two stages in min-max modular (M^3) neural network [2]; and c) decision directed acyclic graph (DDAG) [3]. In comparison with one-against-rest scheme, a shortcoming of one-against-one decomposition procedure is that it will yield too many binary classifier modules, precisely the quantity is $K(K-1)/2$, that is, the quadratic

---

* To whome correspondence should be addressed. This work was supported in part by the National Natural Science Foundation of China via the grants NSFC 60375022 and NSFC 60473040.

function of the number of classes, K. In the recognition phase, however, it is observed that only a part of binary classifiers will be called to produce a solution to the original multiclass problem.

In order to improve the response performance of this kind of classifiers, it is necessary and meaningful to develop an efficient algorithms for selecting necessary binary classifiers in the recognition phase. Therefore, we focus on binary classifier selection problem under a novel general combination procedure of binary classifiers proposed in this paper. Here, we will only care the module based time complexity, which means our work will be independent of the classification algorithms and then it earns more generality. On the contrary, a related work in [4] focuses on an optimized combining policy for margin-based classification, which strongly depends on classification methods used in binary classifiers.

One of our previous work [5] gives a comparison between DDAG combination and the min-max combination and proves that DDAG can be seen as a partial version of the min-max combination. With ulterior study in this paper, we may obtain a more comprehensive understanding of combination of binary classifiers.

The rest of the paper is organized as follows: In Sections 2 we briefly introduce the min-max combination and the most-winning combination for binary classifiers. In Section 3, a generalized combination procedure is presented and two equal relations are proved. A selection algorithm is presented for the general combination procedure is presented in Section 4. The experimental results and comments on theoretical and experimental results are presented in Section 5. Conclusions of our work and the current line of research are outlined in Section 6.

2 Min-Max and Most-Winning Combinations for Binary Classifiers

Suppose a K-class classification problem is divided with one-against-one task decomposition, then $K(K-1)/2$ individual two-class sub-problems will be produced.

We use M_{ij} to denote a binary classifier that learns from training samples of class i and class j, while $0 \leq i, j < K$. The output coding of binary classifier M_{ij} in the min-max combination is defined as 0 and 1, where 1 stands for its output of class i and 0 stands for class j. M_{ij} will be reused as M_{ji} in the min-max combination, and they output contrary results for the same sample. Thus, though $K(K-1)$ binary classifiers will be concerned in the min-max combination, only one half of them need to be trained.

Before combination, we sort all $K(K-1)$ binary classifier M_{ij} into K groups according to the same first subscript i, which is also regarded as the group label. Combination of outputs of all binary classifiers is performed through two steps. Firstly, the minimization combination rule is applied to all binary classifiers of each group to produce the outputs of K groups. Secondly, the maximization combination rule is applied to all groups outputs. If the result of the maximization procedure is 1, then the label of that group which contribute to such result will be the class label of combining output, otherwise, the combining output is unknown. We name the group which leads to the class label of combining output as "winning group", and the others as " failure groups".

A min-max combination procedure is illustrated in Fig 1.

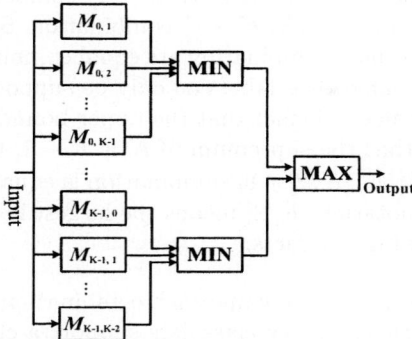

Fig. 1. Illustration of K-class min-max combination of $(K-1) \times K$ binary classifiers with K MIN units and one MAX unit

For the most-winning combination of binary classifiers, a direct output coding is applied. The output of each M_{ij} is just i or j, instead of 0 or 1. And the combination policy is concise, too. The class label supported by the most binary classifiers is the combining output of $K(K-1)/2$ binary classifiers.

3 A General Combining Procedure for Binary Classifiers

For $K(K-1)/2$ binary classifiers produced by one-against-one decomposition procedure, we present a general combination procedure, named N-voting combination, denoted by $V(K, N)$, where N is an additional parameter. A direct class output coding is used in the combination, that is, the output of a binary classifier M_{ij} will just be class i or class j. Combination rule is defined as follows. If there are at least N binary classifiers support a class label, e.g. class i, and no more binary classifiers support any other class label, then the combining output is just class i. Otherwise, the combining output is unknown class.

We will show that N-voting combination $V(K, K-1)$ is equal to the min-max combination. In fact, if there is a class, e.g. class i, with consistent support of $K-1$ binary classifiers under $V(K, K-1)$ combination, then this means that only these binary classifiers, M_{ij}, $0 \leq j \leq K-1$, and $i \neq j$, must all support the same class i. In other words, their output must all be class 1 under coding method of the min-max combination. These $K-1$ binary classifiers just form a group under the min-max combination. Thus, it must be the group with label i that wins the combination, which means the combining output is class i under the min-max combination. On the contrary, if there is one winning group with a label i, under the min-max combination, then these $K-1$ binary classifiers must support the same class i. Notice that since the classifier M_{ij} has output class i, then the symmetrical classifier M_{ji} must output the same result class i, namely only $K-2$ binary classifiers support class j in the group that are supposed to supported class j as the combining output, which leads to a failure and means that no more binary classifiers support any other class except for class i. According to the definition of $V(K, K-1)$ combination, the combining output must be class i under $V(K, K-1)$ combination. So the conclusion that $V(K, K-1)$ and the min-max combination are equal combinations can be drawn. What's more, since the same class label can only be supported by at most $K-1$ binary classifiers, this comes the fact that the upper bound of N must be $K-1$. It is easy to recognize that the supremum of N is $K-1$, too.

We also show that $V(K, [K/2]+1)$ combination is equal to the most-winning combination, where denotation $[K/2]$ means the largest integer below $K/2$. It is induced from the following two facts.

a) For convenient description, we name such combination as $v(K, N)$ combination. If there are just N binary classifiers support a class label, e.g. class i, and no more binary classifiers support any other class label, then the combining output is just class i. Otherwise, the combining output is unknown class. Suppose the set of combining outputs of all defined class labels by $v(K, N)$ combination is denoted by s_N, and the set of combining outputs of all defined class labels by $V(K, N)$ combination is denoted by S_N. For the same test sets and trained binary classifiers, there must be $S_N = s_{K-1} \cup s_{K-2} \cup ... \cup s_N$. Then it is obvious that $S_{N_1} \subseteq S_{N_2}$ when $N_1 > N_2$, for all $0 \le N_1, N_2 < K$. That is, for the larger N, the corresponding $V(K, N)$ combination will give the less outputs of defined class labels. The reason is that the condition to finish a combining output of defined class label is more and more strict as the value of N increases. Turn to the case of the most-winning combination, such result can be obtain according to its definition:

$$S_{mw} = s_{K-1} \cup s_{K-2} \cup ... \cup s_1, or \tag{1}$$
$$S_{mw} = S_1.$$

b) To give a combining output of defined class label under $V(K, N)$ or the most-winning combination, such condition must be satisfied: after N binary classifiers are excluded in $K(K-1)/2$ binary classifiers, the remaining classifiers are divided into $K-1$ groups, in which the numbers of binary classifiers all are less than N, that is, the following inequality should be satisfied.

$$N > \frac{K(K-1)/2 - N}{K-1}. \tag{2}$$

The solution to the above inequality is $N > (K-1)/2$. Consider N must be an integer, we have $N \ge [(K-1)/2]+1$, that is, $N \ge [K/2]+1$. This result suggests

$$s_N = \phi, \forall N, 0 < N < [K/2]+1. \tag{3}$$

According to (1) and (3), we obtain

$$S_{mw} = s_{K-1} \cup s_{K-2} \cup ... \cup s_{[K/2]+1}, or \qquad (4)$$
$$S_{mw} = S_{[K/2]+1},$$

and consider all undefined class labels will be output as unknown classes. Therefore, the equality between $V(K, [K/2]+1)$ and the most-winning combination is obvious.

However, the fact that $[K/2]+1$ is a lower bound of N is not necessary to lead to the fact that $[K/2]+1$ is the infimum of N just like the case of upper bound of N. Actually, many sets s_N are empty for some $N > [K/2]+1$ in practical classification tasks. To find a larger lower bound of N is still remained as an open problem.

4 Selection Algorithm for Combining Binary Classifiers

The original N-voting combination needs $K(K-1)/2$ binary classifiers to be tested for a sample before the mostly supported class label is found. But if we consider the constraint of the value of N, then it is possible to reduce the number of binary classifiers for testing, which give an improvement of response performance.

As mentioned in Section 2, $K-1$ binary classifiers with the same first subscript i are regarded as one group with the group label i. If there exists more than $K-N$ binary classifiers without supporting the group label in a group for a given value of N, then it is meaningless for checking the remained classifiers in the group since this group loses the chance of being a winning one, that is to say, the remained classifiers in the group can be skipped.

The selection algorithm for N-voting combination $V(K, N)$ is described as follows.

1. For a sample, let $i = 0$ and $j = 1$.
2. Set all counters $R[i] = 0$, which stands for the number of binary classifiers rejecting group label i, for $0 \leq i < K$.
3. While $i \leq K$, do
   (a) While $j \leq K$ and $R[i] \leq K - N$, do
      i. Check the binary classifier M_{ij}.
      ii. If M_{ij} rejects class label i, then $R[i] = R[i] + 1$, else $R[j] = R[j] + 1$.
      iii. Let $j = j + 1$, if $j = i$, then let $j = j + 1$ again.
   (b) Let $i = i + 1$ and $j = 1$.
4. Compare each number of binary classifiers rejecting the same class to find the lest-rejected class label as combining output. If all $R[i] > K - N$, for $0 \leq i < K$, then output unknown class as combining classification result.

It is obvious that the chance of a group to be removed by selection algorithm will increase as the value of N increases. This means the efficiency of selection procedure will increase, too. Thus, with the highest value of N, $V(K, K-1)$, or the min-max combination, has the best test performance in the combination series.

Notice that the strictness of voting for a combining output of defined class label will be increase as the value of N increases from $[K/2]+1$ to $K-1$, monotonously. The chance to complete such combination will decrease, simultaneously. This means the accuracy of $V(K,N)$ combination will decrease, monotonously, and the unknown rate will increase, monotonously. Thus, $V(K,[K/2]+1)$ or the most-winning combination is of the highest accuracy in the combination series.

It is hard to directly estimate the performance of N-voting combination selection algorithm. Here we give an experimental estimation. The number of checked binary classifiers under $V(K,K-1)$ or the min-max combination will be

$$n_M = K(\alpha log(K) + \beta), \qquad (5)$$

where α and β are two constants that depend on features of binary classifier, experimentally, $0 < \alpha \leq 1$ and $-0.5 < \beta < 0.5$. And the number of checked binary classifiers under $V(K,[K/2]+1)$ (or the most-winning policy in some cases) combination will be

$$n_R = \gamma K^2, \qquad (6)$$

where γ is a constant that depends on features of binary classifier, experimentally, $0 < \gamma < 0.3$. According to above analysis, performance of $V(K,N)$ combination should be between n_M and n_R.

According to above performance estimation, our selection algorithm can improve the response performance of one-against-one method from quadratical complexity to logarithmal complexity at the number of binary classifiers in the best case, namely the min-max combination or 1.67 times at least in the worst case, namely the most-winning combination policy.

5 Experimental Results

Two data sets shown in Table 1 from UCI Repository[6] are chosen for this study. Two algorithms, k-NN with $k=4$ and SVM with RBF kernel are taken as each binary classifier, respectively. The kernel parameters in SVM training are shown in Table 1, too. The experimental results of N-voting combination with different values of N are shown in Tables 2-5. These tables list the numbers of checked binary classifiers, which show the performance comparison independent of running platform.

It is necessary to access 45 and 325 binary classifiers for two data sets respectively for testing a sample without any module selection. while there is only one half of binary classifiers or less to be checked under presented selection algorithm. This demonstrates an outstanding improvement of test performance. Consider the generality of N-voting combination, the selection algorithm presented has actually included selection procedure of the min-max combination

Table 1. Distributions of data sets and corresponding parameters for SVMs

Data sets	#Class	Number of Samples		Parameters of SVM	
		Train	Test	γ	C
Optdigits	10	3823	1797	0.0008	8
Letter	26	15000	5000	0.0125	8

Table 2. Performance of Optdigits data set on N-voting combination: k-NN algorithm

N	Accuracy	Incorrect rate	Unknown rate	#checked modules
6	98.39	1.61	0.00	25.55
7	98.39	1.61	0.00	26.16
8	98.39	1.61	0.00	24.84
9	98.39	1.61	0.00	20.74

Table 3. Performance of Optdigits data set on N-voting combination: SVM algorithm

N	Accuracy	Incorrect rate	Unknown rate	#checked modules
6	99.00	1.00	0.00	24.91
7	99.00	1.00	0.00	25.53
8	99.00	1.00	0.00	24.61
9	98.94	0.78	0.28	20.69

Table 4. Performance of Letter data set on N-voting combination: k-NN algorithm

N	Accuracy	Incorrect rate	Unknown rate	#checked modules
14	95.78	4.22	0.00	191.15
15	95.78	4.22	0.00	191.05
16	95.78	4.22	0.00	189.47
17	95.78	4.22	0.00	186.34
18	95.78	4.22	0.00	181.51
19	95.78	4.22	0.00	174.85
20	95.78	4.22	0.00	165.73
21	95.78	4.22	0.00	154.19
22	95.78	4.22	0.00	139.74
23	95.78	4.22	0.00	121.98
24	95.78	4.22	0.00	99.49
25	95.74	4.02	0.24	73.41

Table 5. Performance of Letter data set on N-voting combination: SVM algorithm

N	Accuracy	Incorrect rate	Unknown rate	#checked modules
14	97.18	2.82	0.00	188.77
15	97.18	2.82	0.00	189.02
16	97.18	2.82	0.00	187.45
17	97.18	2.82	0.00	184.54
18	97.18	2.82	0.00	180.00
19	97.18	2.82	0.00	173.62
20	97.18	2.82	0.00	165.33
21	97.18	2.82	0.00	155.04
22	97.18	2.82	0.00	141.46
23	97.18	2.80	0.02	124.68
24	97.16	2.80	0.04	103.26
25	96.80	2.34	0.86	76.27

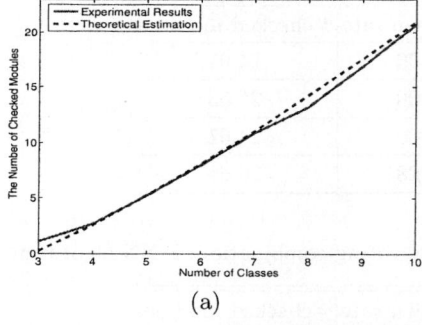
(a)

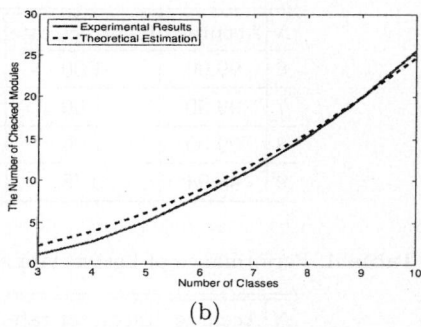
(b)

Fig. 2. Comparison of theoretical estimation and experimental result of N-voting combination on Optdigits data set under k-NN algorithm, where $\alpha = 1.05, \beta = -0.32$ and $\gamma = 0.247$. (a) $V(K, K-1)$ combination and (b) $V(K, \lceil K/2 \rceil + 1)$ combination

and the most-winning combination. If we regard selected $V(K, \lceil K/2 \rceil + 1)$ combination as selected the most-winning combination in the worst case, then there comes nearly 1.7 times improvement at least. If a larger N is taken, then the speeding is much more. In addition, the accuracy and unknown rate do decrease and increase, respectively, while the value of N increases just as expected. However, the decreasing of accuracy or increasing of unknown rate is not outstanding when N is small enough. This suggests that the most-winning combination is equal to $V(K, N)$ combination with a value of N which may be many larger than $\lceil K/2 \rceil + 1$.

By removing samples of the last class continuously from each data set, we obtain a 3-26 data sets for Letter data and 3-10 data sets for Optdigits data.

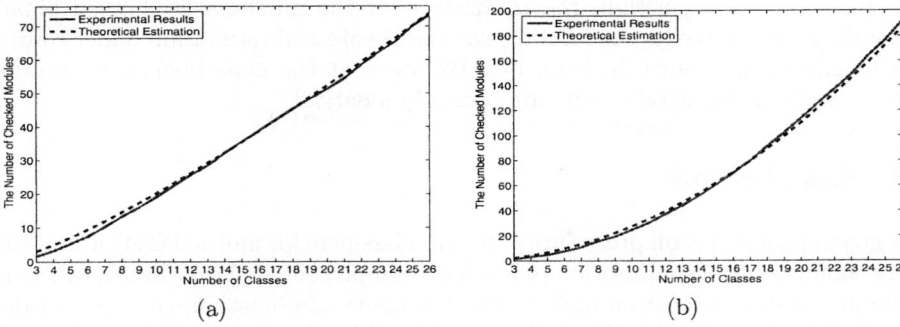

Fig. 3. Comparison of theoretical estimation and experimental result of N-voting combination on Letter data set under k-NN algorithm, where $\alpha = 0.87, \beta = 0.0077$ and $\gamma = 0.275$. (a) $V(K, K-1)$ combination and (b) $V(K, [K/2]+1)$ combination

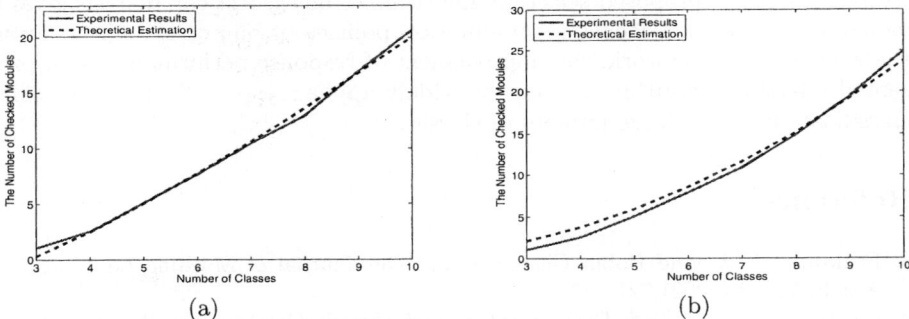

Fig. 4. Comparison of theoretical estimation and experimental result of N-voting combination on Optdigits data set under SVM algorithm, where $\alpha = 1, \beta = -0.3$ and $\gamma = 0.237$. (a) $V(K, K-1)$ combination and (b) $V(K, [K/2]+1)$ combination

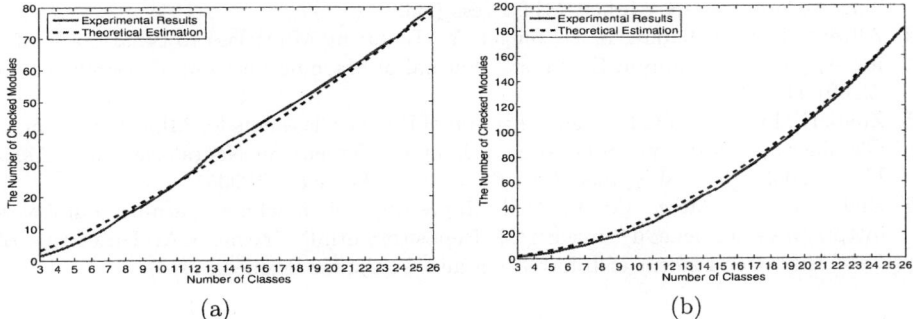

Fig. 5. Comparison of theoretical estimation and experimental result of N-voting combination on Letter data set under SVM algorithm, where $\alpha = 0.92, \beta = -0.0385$ and $\gamma = 0.27$. (a) $V(K, K-1)$ combination and (b) $V(K, [K/2]+1)$ combination

Under selection algorithm, the comparison of the numbers of checked binary classifiers between experimental results and theoretical estimation under continuous classes are shown in Figs. 3-4. We see that the experimental estimation value and experimental results are basically identical.

6 Conclusions

A general combination procedure of binary classifiers for multi-classification with one-against-one decomposition policy has been presented. Two existing schemes, the min-max combination and the most-winning combination, can be regarded as its two special cases. For such general combination procedure, we ulteriorly propose a selection algorithm. An improvement of response performance to the original combining procedure is demonstrated. The experimental performance estimation of selection algorithm is given, too. The experiments verify the effectiveness of the proposed selection algorithm. Our theoretical analysis gives a valuable criterion for choosing combination policies of binary classifiers. From the generality of our work, the improvement of response performance with presented selection algorithm can also be widely applied, especially for multi-class classification with a large number of classes.

References

1. Frnkranz, J.: Round Robin Classification. The Journal of Machine Learning Research, Vol. 2 (2002) 721-747
2. Lu, B. L., Ito, M.: Task Decomposition and Module Combination Based on Class Relations: a Modular Neural Network for Pattern Classification. IEEE Transactions on Neural Networks, Vol. 10 (1999) 1244-1256
3. Platt, J., Cristianini, N., Shawe-Taylor, J.: Large Margin DAGS for Multiclass Classification, Advances in Neural Information Processing Systems, 12 ed. S.A. Solla, T.K. Leen and K.-R. Muller, MIT Press (2000)
4. Allwein, E. L., Schapire, R. E., Singer, Y.: Reducing Multiclass to Binary: a Unifying Approach for Margin Classifiers, Journal of Machine Learning Research, Vol. 1 (2000) 113-141
5. Zhao, H., Lu, B. L.: On Efficient Selection of Binary Classifiers for Min-Max Modular Classifier, Accepted by International Joint Conference on Neural Networks 2005-IJCNN2005, Montral, Qubec, Canada, July 31-August 4, (2005)
6. Blake, C. L., Merz, C. J.: UCI Repository of machine learning databases [http://www.ics.uci.edu/ mlearn/MLRepository.html]. Irvine, CA: University of California, Department of Information and Computer Science (1998)

A Modular Structure of Auto-encoder for the Integration of Different Kinds of Information

Naohiro Fukumura[1,2], Keitaro Wakaki[1], and Yoji Uno[1,2]

[1] Department of Information and Computer Sciences,
Toyohashi University of Technology,
1-1 Hibarigaoka, Tempaku, Toyohashi, Aichi 441-8580, Japan
{fkm, wakaki, uno}@system.tutics.tut.ac.jp
[2] Intelligent Sensing System Research Center,
Toyohashi University of Technology

Abstract. Humans use many different kinds of information from different sensory organs in motion tasks. It is important in human sensing to extract useful information and effectively use the multiple kinds of information. From the viewpoint of a computational theory, we approach the integration mechanism of human sensory and motor information. In this study, the modular structure of auto-encoder is introduced to extract the intrinsic properties about a recognized object that are contained commonly in multiple kind of information. After the learning, the relaxation method using the learned model can solve the transformation between the integrated kinds of information. This model was applied to the problem how a locomotive robot decides a leg's height to climb over an obstacle from the visual information.

1 Introduction

It is supposed that the human recognizes various objects in the real-world by integrating multiple kinds of sensory information. Consider that the human recognizes a cup to drink water. It has been pointed out that not only visual information about the cup but also somatosensory information (e.g., hand configuration when grasping it) concerns the object's shape recognition[1]. We hypothesize that the internal representation of a grasped object is formed in the brain by integrating visual, somatosensory and other sensory information while the human repeats such grasping movements.

When the human recognizes a cup, perceived data consists of the intrinsic property of the cup and the condition of sensing. For example, visual image changes depending on not only the size or the shape of the cup but also the direction or the distance from eyes to the cup. Somatosensory information also changes depending on how human grasps the cup. Consequently, when these different kinds of sensory information are integrated, the relation between them is many-to-many and the recognition process must include the extraction of intrinsic properties of the objects. We focus on the fact that the intrinsic properties are contained constantly and commonly in multiple kinds of sensory information. From this viewpoint, we think that the most important purpose of sensory integration is the extraction of the information commonly contained in different kinds of sensory information, which is called the correlated information[2].

Base on a five-layered auto-encoder[3], we have proposed a neural network model that integrates different kinds of information and shown that this model can solve the many-to-many problem by a relaxation method with a penalty method[4]. In our successive work, the former model has been modified in order to extract the correlated information through the learning process of the integration[5]. However, these models always require both data integrated in the learning phase, which is not natural in the biological system.

In this study, a modular structure of auto-encoder is introduced. Each module corresponds to a kind of information and when the multiple kinds of information about a recognized object are gained, corresponding modules learn the correlated relation. Even if only a kind of information is gained, the corresponding module can learns to achieve the identity map as a simple auto-encoder. Moreover, we show that the proposed model can solve a many-to-many problem without penalty method. This model is applied to the problem how a locomotive robot decides the leg's height to climb over an obstacle from the camera data.

2 Architecture of the Neural Network

2.1 Extraction of the Correlated Information

The proposed model consists of the multiple five-layered auto-encoder models as shown in Fig.1. In this work, we consider the case that two kinds of the sensory information $(\mathbf{x}, \mathbf{y})$ are integrated. Each auto-encoder corresponds to a sensory modality. The numbers of neurons in the third layer in the every auto-encoder are set to the intrinsic dimension of the input data. In the learning phase, each auto-encoder model learns to realize the identity map. Simultaneously, the several neurons in the third layer of each module, which are called the correlated neurons(ζ), must learns to have the same value as the correlated neurons in the other module for the sensory modality that shares the correlated information. The other neurons in the third layer(ξ, η) are called non-correlated neurons. The number of the correlated neurons sets to be the same as the dimension of the correlated information. Consequently, the error functions that must be minimized in the learning phase are as follows:

$$E^x = \sum_i^M (x_i - x_i')^2 + \lambda \sum_i^K (\zeta_i^x - \zeta_i^y)^2 \qquad (1)$$

$$E^y = \sum_i^N (y_i - y_i')^2 + \lambda \sum_i^K (\zeta_i^y - \zeta_i^x)^2 \qquad (2)$$

Here, M and N are dimensions of the input data and K is the number of the correlated neurons. $\mathbf{x}'$ and $\mathbf{y}'$ are outputs of the auto-encoders. When only one kind of the information is gained in the recognition process, the second term in

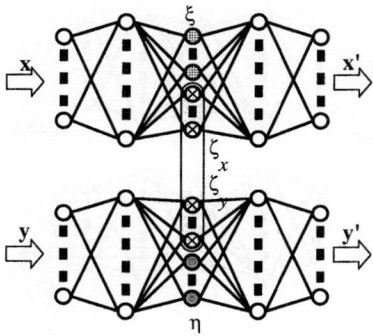

Fig. 1. The structure of a neural network model that extracts correlated information between two different kinds of information

(1) or (2) is omitted. The network learns to minimize the objective functions by the back-propagation method.

Through the learning of the identity map, arbitrary transformation and its inverse transformation will be obtained from the input layer to the third layer and from the third layer to the fifth layer, respectively. When there is the correlated information between the both kinds of the sensory information and its dimension is K, the correlated information should be extracted in the correlation neurons since the correlation neurons in every auto-encoder have the same value for a recognized object after the learning phase.

Generally, it is difficult to determine the number of neurons in the hidden layer of the layered neural network. In this work, it is assumed that the intrinsic dimensions of **x** and **y** and the dimension of the correlated information between **x** and **y** are known; therefore, the number of neurons in the third layer in every auto-encoder can be properly set.

2.2 Relaxation Method for a Many-to-Many Transformation Problem

As described before, the transformation from a pattern of one kind of sensory information to a pattern of another kind of sensory information is a many-to-many transformation problem. A "many-to-many transformation problem" is an ill-posed problem since many corresponding output patterns exist even if an input pattern is specified. In the previous study[4], we have shown that a relaxation computation applying to the learned neural network model can solve such a transformation problem as an optimization with constraints by introducing a criterion of the output pattern in order to determine an optimal solution from many candidates. This relaxation method is equivalent to the penalty function method in the optimization theory and a different solution also can be computed by employing a different criterion. However, this relaxation computation needs a great deal of computation time and it is difficult to choose

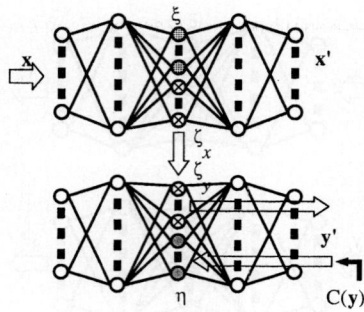

Fig. 2. A schematic of the relaxation method to solve a many-to-many problem

suitable values of some parameters such as a regularization parameter for the criterion and its decreasing rate.

Generally, the condition of constraints means that both kinds of the information are obtained from the same object. In other words, the both sensory data contains the same correlated information that is the object's property. In the case of the proposed model in this study, if $\zeta_x = \zeta_y$, $\mathbf{x}'$ and $\mathbf{y}'$ in the fifth layer that are computed from ζ are always obtained from the same object. That is, the constraint is satisfied.

Consequently, when a kind of the sensory data($\mathbf{x}$) is obtained, the values of the correlated neurons(ζ_x) computed from $\mathbf{x}$ are copied to the correlated neurons of the another module(ζ_y). After that, we may search an output pattern $\mathbf{y}'$ that optimizes the criterion based on the values of ζ_y by adjusting η(Fig.2). In this step, the gradient method can be used to optimize the criterion $C(\mathbf{y})$ as follows:

$$c\frac{d\eta_k}{ds} = -\frac{\partial C(\mathbf{y}')}{\partial \eta_k} \qquad (3)$$

Here, s is relaxation time and c is a positive time constant. After η reaches the equilibrium state, an optimized pattern $\mathbf{y}^*$ that minimizes the criterion is formed in the fifth layer.

In this model, the transformation from the sensory data to the coordinates representing the correlated information is acquired through the learning process. Since the constraints are always satisfied when $\zeta_x = \zeta_y$, the transformation problem can be solved as not the optimization with constraints, but simple optimization problem. Thus, an adjustment of the regularization parameter and its decreasing rate in the penalty method is unnecessary and the computation is expected to be stable. Moreover, since the search space in the third layer become small, less computational time is expected.

3 Experimental Results

3.1 Integration Task

We confirmed the plausibility of the proposed model by a computer simulation and an experiment of a real robot, AIBO. We employed the recognition of obstacle's heights by integrating the visual information when a locomotive robot see it by a camera from different distances and the joint angle of the robot's leg when it touches the top of the obstacle by its front leg in various postures. The visual image changes according to the obstacle's height as the correlated information and the distance from the obstacle. Instead of the raw image data, the width and length of the obstacle in the camera image were used as the visual information in order to reduce the computational cost. Since the AIBO has only one CCD camera, the width of every obstacle was set to be the same such that the height of the obstacle and the distance could be calculated from a camera image. In this experiment, the joint for abduction of the leg was fixed so that the AIBO's leg was considered as a 2-link manipulator. The shoulder and elbow joint angle when the robot touches the top of the obstacle were used as the leg's posture data. These angles change according to the height of the obstacle and the distance from the body to the toe. Thus, it is expected that the height of the obstacle is extracted in the correlated neurons by integration of the visual data and leg's posture of the robot.

3.2 Simulation Experiment

At first, we investigated the proposed model by a computer simulation. Nine obstacles of width 80.0mm were prepared, heights of which were from 8.0mm to 60.0mm.

It was supposed that the image data was obtained by a camera at the same height of the center of the obstacle's side plane. The widths of the side plane in the camera image, which are inversely proportional to the distance, were set to be 30, 50, 70, and 90 pixels. Their lengths in the camera image were calculated from the widths in the camera images of the obstacles and the real proportions of length to width.

About the joint angle data, eleven elbow angles from -25 to 25 degrees (plus value means forward flexion) were prepared. Each shoulder angle, which is orthogonal to the body at zero degree, was calculated from obstacle's height and elbow angle using the kinematics equation.

Thus, 396 data sets, all combination of four image data and eleven posture data for nine obstacles, were used for training. The number of neurons in the first and fifth layer was two in each auto-encoder model. The number of correlated neurons and the non-correlated neurons in the third layer was one, respectively.

Figure3 shows the activities of the correlated neurons in each module when the training data were fed to the model after the sufficient learning. The both values of correlated neurons have almost same for an obstacle and increase monotonically with the height of the obstacle. These results indicate that the correlated neurons extract the height of the obstacles without a supervised signal about the height information.

Using the learned model, the adequate posture of the robot' leg was computed from the image data by the proposed method. In this simulation, (4) was employed as the criterion.

$$C(\boldsymbol{y}) = y_s^2 + y_e^2 \qquad (4)$$

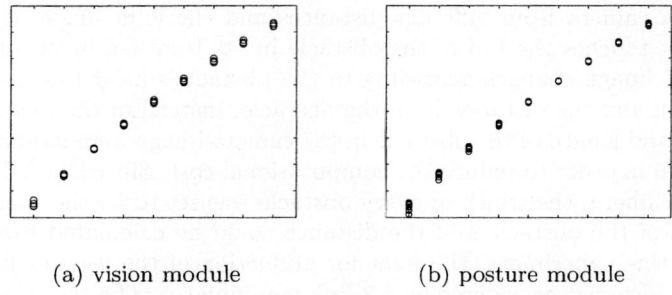

(a) vision module   (b) posture module

Fig. 3. Activities of correlated neurons in the vision and posture module in the simulation

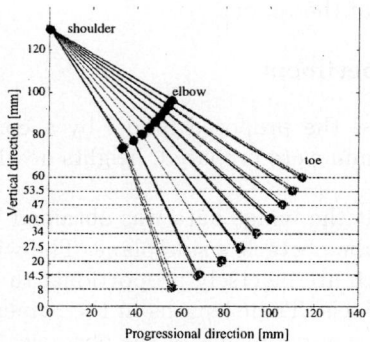

Fig. 4. Estimated postures and optimized postures for each obstacle in the simulation

Here, y_s and y_e means the shoulder and elbow angle. The results of the estimation of leg's posture from the every training data of image module, which was obtained at the different distance, are shown in Fig.4. Dotted lines indicate the optimized leg's postures by Optimization Toolbox of MATLAB. The postures estimated from the different image data for the same obstacle are almost same and very close to the optimized postures, too. These results indicate that the proposed model can extract the correlated information and solve the many-to-many problem with less computational cost.

3.3 Robot Experiment

We tested the proposed model by a real robot. AIBO(RS-210A) made by SONY Corporation was employed in this experiment. Nine rectangular parallelepiped obstacles, width and height of which were the same as those of the obstacles in the simulation were prepared. AIBO has a CCD camera the size of which was 176 by 144pixels. In order to extract the size of the obstacle in the camera image easily, the

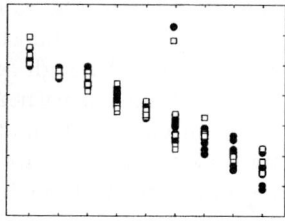

Fig. 5. Activities of correlated neurons in the vision and posture module in the robot experiment

Fig. 6. Estimated postures of AIBO. (a) Obstacle of the height 145mm was estimated from an image data taken at about 100mm away. (b) Obstacle of the height 535mm was estimated from an image data taken at about 200mm away

side planes of these obstacles were painted in pink. At first, AIBO was put at about 100, 200, 300, and 400mm away from the obstacle. When AIBO detected the pink area from the front of the side plane and looks at its center, the robot took each obstacle's image three times. The length and width were computed by counting the number of the pixels of the length and width of the pink region. After that, in order to prepare the leg's posture data for the learning, the sitting robot was put near the obstacle. The elbow joint of the former leg was fixed at seven angle patterns from -24 to 24 degrees by PID controller and when a touch sensor on the leg's toe became on during the shoulder was swung down, the shoulder and elbow angles of the former leg were measured by a potentio-meter for every obstacles. However, some postures for some obstacles were too high for AIBO to touch the top of the obstacles. In such cases, network learning was executed only for the vision module. Since twelve image patterns and seven leg's posture patterns were prepared for nine obstacles, 756 combinations of the image data and leg's posture data were used to train the proposed model. However, 216 sets were trained by only image data because of lack of the posture data. The network configuration was the same as that in the simulation.

After the sufficient network learning, the neuron activation patterns formed in the correlation neurons are shown in Fig.5. The values of ζ for the same obstacle but different visual data or posture data are almost same and monotonically decreased with the height of the obstacle. Although deviation is larger than that in the simulation experiment because of the sensory noise and quantization error, the correlated neuron seems to be acquired the information of the obstacle's height.

After the learning, AIBO took the image data from several distances from the obstacles and the adequate postures to their heights were computed by the proposed relaxation method when (4) was employed as the criterion. AIBO could raise its leg as high as the obstacle(Fig.6).

4 Conclusion

A neural network model that extracts correlated information between different kinds of information has been proposed. The proposed model uses an auto-encoder architecture and, therefore, a supervised signal for internal representation is unnecessary. Although we used the simple back-propagation method to train this network, other learning methods to realize the identity map can be applied.

We have also proposed the new relaxation computation to solve the many-to-many transformation problem using the gradient descend method without the penalty method. The simulation result shows that the proposed method can reduce the iteration number and is not so sensitive to the values of parameters in the gradient method.

Comparing the previously proposed model[5], this model is difficult to learn by back-propagation method. The second terms of the objective functions of learning((1) and (2)) can decrease by depression of the range of the correlated neuron activity. Therefore, the case in which ζ_x and ζ_y always have the constant value for every obstacle and the identity map is not realized is a local minimum in the learning process. Fine adjustment of λ in (1) and (2) and the learning rate by trial-and-error is necessary to escape the local minimum.

The critical problem in our neural network model is how many neurons should be set for the correlated neurons and other neurons in the third layer. To determine an adequate number of neurons for the hidden layer in an auto-encoder model, several methods have been proposed[6,7]. In our model, even if the total number of neurons for the third layer can be decided, the more important problem is how to divide the neurons in the third layer into the correlated and non-correlated neurons. In the present stage, we have no method to decide an adequate number of neurons for each subset in the third layer. This is an important task for the future.

Acknowledgments

This work was partially supported by The 21st Century COE Program "Intelligent Human Sensing." from the Ministry of Education, Culture, Sports, Science and Technology.

References

1. Sakata, H., Taira, M., Kusunoki, M., Murata, A. and Tanaka, Y.: The Parietal Association Cortex in Depth Perception and Visual Control of Hand Action. Trend in Neuroscience Vol.20 8 (1997) 350–357
2. Shibata, K.: A Neural-Network to Get Correlated Information among Multiple Inputs. Proc. of IJCNN'93 NAGOYA Vol.3 (1993) 2532–2535
3. Cottrell, G.W., Munro, P. and Zipser, D.: Image Compression by Back–Propagation: an Example of Extensional Programming. In Sherky, N.E(eds.): Advances in Cognitive Science, Vol.3. Norwood, NJ: Ablex (1988)
4. Uno, Y., Fukumura, N., Suzuki, R. and Kawato, M.: A Computational Model For Recognizing Objects and Planning Hand Shapes in Grasping Movements. Neural Networks Vol.8 6 (1995) 839–851
5. Fukumura, N., Otane, S., Uno, Y. and Suzuki, R.: A Neural Network Model for Extracting Correlated Information in Sensory Integration. Proceedings of ICONIP Vol.2 (1998) 873–876
6. DeMers, D. and Cottrell, G.: Non-Linear Dimensionality Reduction. Advances in Neural Information Processing Systems Vol.5 (1992) 580–587
7. Noda, I.: Acquisition of Internal Representation using Overload Learning. –Case Study of Learning Identity-function–. Japan IEICE Technical Report NC94-34 (1994) 15–22 (in Japanese)

Adaptive and Competitive Committee Machine Architecture

Jian Yang and Siwei Luo

School of Computer and Information Technology, Beijing Jiaotong University,
Beijing, 100044, China
yj_swendy@yahoo.com.cn
swluo@center.bjtu.edu.cn

Abstract. Learning problem has three distinct phases, that is, model representation, learning criterion (target function) and implementation algorithm. This paper focuses on the close relation between the selection of learning criterion for committee machine and network approximation and competitive adaptation. By minimizing the KL deviation between posterior distributions, we give a general posterior modular architecture and the corresponding learning criterion form, which reflects remarkable adaptation and scalability. Besides this, we point out, from the generalized KL deviation defined on finite measure manifold in information geometry theory, that the proposed learning criterion reduces to so-called Mahalanobis deviation of which ordinary mean square error approximation is a special case, when each module is assumed Gaussian.

1 Introduction

Committee machines have been frequently employed to improve results in classification and regression problems [1-10]. Among the key issues are how to design the architecture and scale of the networks; how to make best use of a limited data set; how the results of the various component networks should be combined to give the best estimate; and how to make each component adaptive etc. In this paper we address the last two issues, which are closely related to the learning criterion adopted, through minimization of generalized Kullback-Leibler (KL) divergence.

This paper is organized as follows; the first section discusses influences of learning criterion on approximation and adaptation; the second section introduces the generalized information divergence and KL divergence in information geometry; the necessity of posterior average over all components is in section three; the fourth section describes construction of committee machine using the cost function derived from KL divergence between posterior distributions, and then gives a general form of learning criterion, which not only makes committee machine give best approximation but also have good adaptation; the following section points out that the learning criterion given in section four reduces to so-called Mahalanobis divergence of which ordinary mean square error approximation is a special case, when each module is assumed Gaussian. The last one is conclusion.

2 Influences of Learning Criterion on Approximation and Adaptation

2.1 Learning Criterion and Approximation

It's well known that learning problem is to construct a learning machine using given training data set under a certain learning criterion such that the machine approximates the underlying rule reflected by the data set as best as possible. Learning problem includes three distinct phases [11,12]:model representation, that is available computing resource or model representation ability such as Generalized Linear Model (GLIM) and various Nonlinear Regression Models; the learning criterion used to measure the quality of learning results like Square Sum of Error (SSE) and Likelihood Function (ML); and the implementation algorithm Like the Gradient Descent, etc. As a general nonlinear approximation model neural network has powerful representation ability; and the dynamic committee machine, which is nonlinear combination of nonlinear functions like Mixture of Experts (ME) and Hierarchical Mixture of Experts (HME), can represent almost usual statistical models, although models with same representation power, such as Projection Pursuit Regression Model (PPR) and Multilayer Perceptron (MLP) with single hidden layer, may show different learning effects due to different learning criterion.

The most frequently used learning criterion is to minimize SSE and its variants that are R-error norm when $R = 2$. It has proven that MSSE is to make the network approximate the conditional expectation of the target, whose effect equivalent to taking same variance but mean which is the Gaussian distribution of input function as the input-output conditional probability model. On the other hand, ML of the joint probability can be reduced to MSSE when the target variables are Gaussian; however, ML is no longer effective on some distributions [13]. Of course, we need not assume Gaussian distribution when take SSE as learning criterion, but the results may deviate from the best one if not. Another usual learning criterion is the cross entropy measuring the difference between two distributions, to minimize cross entropy is equivalent to minimizing KL (MKL) divergence. Cross entropy is the function of relative error of network output, but SSE is related to absolute error of network output. Both MSSE and MKL approximate the conditional expectation of network output [1].

2.2 Learning Criterion and Adaptation

If back propagation is used to train a single, multiplayer network to perform different subtasks on different occasions, there will generally be strong interference effects that lead to slow learning and poor generalization. Using committee machine with appropriate learning criterion can efficiently solve this problem. In this system different module inputs correspond to different regions of input space which is realized by gate network; and we expect that, if the output is incorrect, the weight changes are localized to these modules and gating network. So there is no interference with the weights of other modules that specialized in quite different cases. The modules are therefore local in the sense that the weighs in one module are decoupled from the weights in other modules. In addition they will often be local in the sense that each module will be allocated to only a small local region of the possible input vectors. This is determined by the learning criterion adopted. For example, if we use the following learning criterion

$$E^c = \| d^c - \sum_i p_i^c o_i^c \|^2. \tag{1}$$

where E^c is the final error on case c, o_i^c is the output vector of module i on case c, p_i^c is the proportional contribution of module i to the combined output vector, and d^c is the desired output vector in case. So, to minimize E^c, each local module must make its output cancel the residual error that is left by the combined effects of all the other modules. When the weights in one module change, the residual error changes, so the error derivatives for all the other local modules change. This strong coupling between the modules causes them to cooperate nicely, but tends to lead to solutions in which many modules are used for each case. It is possible to encourage competition by adding penalty term to the objective function to encourage solutions in which only one module is active, but a simpler remedy is to redefine the error function so that the local modules are encouraged to compete rather than cooperate. We imagine that the gating network makes a stochastic decision about which single module to use on each occasion. Now the error is the expected value of the squared difference between the desired and actual output vectors.

$$E^c = \langle \| d^c - o_i^c \|^2 \rangle = \sum_i p_i^c \| d^c - o_i^c \|^2. \tag{2}$$

Notice that in this new learning criterion, each module is required to produce the whole of the output vector rather than a residual. As a result, the goal of a local module on a given training case is not directly affected by the weights within other local modules. There is still some indirect coupling because if some other module changes its weights, it may cause the gating network to alter the responsibility that get assigned to the modules. If both the gating network and the local modules are trained by gradient descent in this new learning criterion, the system tends to devote a single module to each training case. Whenever a module gives less error than the weighted average of the errors of all the modules its responsibility for that case will be increased, and whenever it does worse than the weighted average its responsibility will be decreased.

Jacobs in his paper [5] gave the following learning criterion based on the above mentioned learning criterion, which showed better performance in the simulations:

$$E^c = -\ln \sum_i p_i^c \exp(-\frac{1}{2} \| d^c - o_i^c \|^2). \tag{3}$$

To see why this learning criterion works better, it is helpful to compare the derivatives of the two with respect to the output o_i^c of a module. The resultant derivatives of equation (2) and (3) are both R-error norm when $R = 1$ multiplied by a new weighting term, for the former the weighting term is p_i^c, but the weighting term for the latter not only takes into account how well module i does relative to other modules but also adapt the best-fitting module the fastest. This feature is very useful especially in early training.

3 Information Divergence and Information Geometry

Information geometry [14,15] emerged from investigating the natural differential geometric structure possessed by families of parameterized probability distributions, aiming to show information processing capability of systems. A point on the manifold denotes a distribution. Amari introduces α-connection with single parameter, and proved that exponential family corresponds to $\alpha=1$, mixture family to $\alpha=-1$, and they are dually flat. An important divergence on statistical manifold is α-divergence or δ-divergence D_δ ($\delta = \frac{1-\alpha}{2}$). For the set $\tilde{P}$ of all positive finite probability measures, the divergence of any two points q, p is D_δ [16]

$$D_\delta(q,p) = [\delta(1-\delta)]^{-1} \int [\delta q + (1-\delta)p - q^\delta p^{1-\delta}], \delta \in (0,1). \tag{4}$$

Therefore the KL divergence with respect to $\tilde{P}$ is

$$KL(q,p) = \lim_{\delta \to 0} D_\delta(p,q) = \lim_{\delta \to 1} D_\delta(q,p) = \int (q - p + \ln \frac{q}{p}). \tag{5}$$

Obviously when $\int q = 1$ and $\int p = 1$, the above equation reduces to ordinary KL divergence. Like the norm in function space, information divergence enables us consider a set of finite measures as some well-behaved space, not only a set of points.

4 Posterior Averages

The main reason of using committee machine is that when we select only one best model and discard the others, we lost all those knowledge contained in the discarded models, because the selected model only contain a fraction of the whole probability mass. This means that the selected model can explain the observations well, but may not explain future data due to the parameters of the model very sensitive to parameter values. On the other hand, Probability theory tells us that the optimal generalization is the one resulting from a Bayesian approach. A neural network (either deterministic or stochastic) can be regarded as a parameterized model $p(y|x,H)$ denoted as $p_{y|x,H}$ in the following, where x is the input, y is the output and H is the model structure including weights. In the Bayesian framework, knowledge is contained in the conditional probability distributions of the models. Each model can be seen as a hypothesis, or explanation. We can use Bayes theorem to evaluate the conditional probability distributions for the unknown quantities y given the set of observed quantities x. Bayesian rule states that prior becomes posterior after another set of observations and suggests an iterative and "gradient ascend" procedure.

If in some sense over-fitting is caused by selecting only one model, then it's necessary to average over all possible models to gain good generalization. However, the most usual case is to combine models using softmax function of gating network as a priori distribution, it doesn't explicitly reflect the features of each model, and use MSSE or ML as learning criterion; in addition, as we show that the learning criterion should avoid too strong coupling, otherwise lose scalability.

5 Construction of Committee Machine

As we know that KL divergence can be used to measure the misfit between two distributions. The true posterior distribution $p_{y|x}$ is approximated with the $q_{y|x}$ by minimizing the KL divergence

$$D(q_{y|x}, p_{y|x}) = \int q_{y|x} \ln \frac{q_{y|x}}{p_{y|x}} dy = \int q_{y|x} \ln \frac{p_x q_{y|x}}{p_{x,y}} dy \qquad (6)$$

$$= \int q_{y|x} \ln \frac{q_{y|x}}{p_{x,y}} dy + \ln p_x.$$

Since the term p_x is a constant over all the models, we can define a cost function $C_y(x)$, which we are required to minimize to obtain the optimum approximating distribution

$$C_y(x) = D(q_{y|x}, p_{y|x}) - \ln p_x = \int q_{y|x} \ln \frac{q_{y|x}}{p_{y|x}} dy. \qquad (7)$$

It is easy to see that the cost function gives an upper bound for $-\ln p_x$. In the following we denote $q(y \mid x, H)$ by $q_{y|x,H}$ and $p(x \mid H)$ by $p_{x,H}$, use the same notation as with probability distribution, that is, $C_y(x|H)$ means

$$C_y(x \mid H) = D(q_{y|x,H}, p_{y|x,H}) - \ln p_{x|H} = \int q_{y|x,H} \ln \frac{q_{y|x,H}}{p_{x,y|H}}. \qquad (8)$$

where H stands for a model. Obviously since this cost function yields the lower bound for $\ln p_{x|H}$, and $p_{H|x} = \dfrac{p_{x|H} p_H}{p_x} = \dfrac{p_{x|H} p_H}{\sum_H p_{x|H} p_H}$. It is natural that we may use $C_y(x|H)$ to approximate $p_{H|x}$, that is

$$A = \frac{\exp[-C_y(x|H)] \cdot p_H}{\sum_{H'}[-C_y(x|H')] \cdot p_{H'}} \to p_{H|x}. \tag{9}$$

In fact we have the following theorem

Theorem 1. Assume $\sum_H p_{H|x} = 1$. If $p_{H|x} = A$, that is, the posterior about model structure H and the posterior about output y satisfies the relation $p_{H|x} = A$, then $q_{y,H|x}$ is the best approximation for $p_{y,H|x}$, or $C_{y,H(x)}$ is minimized with respect to $p_{H|x}$.

Proof: Without losing any generality, we have

$$q_{y,H|x} = Q_{H|x} \cdot q_{y|x,H}. \tag{10}$$

Now the cost function can be written as

$$C_{y,H(x)} = \sum_H \int q_{y,H|x} \ln \frac{q_{y,H|x}}{p_{x,y,H}} dy = \sum_H Q_{H|x} \int q_{y|x,H} \ln \frac{Q_{H|x} q_{y|x,H}}{p_H p_{x,y|H}} dy \tag{11}$$

$$= \sum_H Q_{H|x} \cdot \left[\ln \frac{Q_{H|x}}{p_H} + C_y(x|H) \right].$$

Minimizing $C_{y,H}(x)$ with respect to $Q_{H,x}$ under the constraint $\sum_H Q_{H|x} = 1$, it is easy to evaluate that when

$$Q_{H,x} = \frac{\exp[-C_y(x|H)] \cdot p_H}{\sum_{H'} \exp[-C_y(x|H')] \cdot p_H}. \tag{12}$$

$C_{y,H}(x)$ with respect to $Q_{H|x}$ arrives at its minimum value which is

$$C_{y,H}(x)|_Q = -\ln \sum_H \exp[-C_y(x|H)] \cdot p_H. \tag{13}$$

This completes the theorem.

ME and HME as typical examples of modular networks, we know that all modules receive the same input. The gating network typically receives the same input as the expert networks, and it has normalized outputs as a prior probability to select the output from each expert to form the final weighted output. The dynamical role of gating network according to input can be considered as a division of input space and is crucial to ME and HME, typically the output (combination proportion or activation function) of gating network is a softmax function of inner product of input and weights of the gating network.

On the other hand, from the above theorem and the poof procedure we can see the posterior about the structure H of a single model can be represented in terms of $C_{y,H}(x)|_Q$, or a group of N models (with different structures, with the same structure but different initial parameter values or be trained with different learning algorithms). So, if we want to use a part of those models that minimize $C_{y,H}(x)|_Q$, obviously each of them may be viewed as a module (or an expert), then these selected modules can be combined with the posterior about structure $Q_{H|x}$ as the mixing proportion. Using a posterior not a prior has some advantages: the division of input space may be more precise; it seems more reasonable that the combination proportion is determined not only by input but also by module, this makes possible to coordinate different tasks according to different features of each module. Here we don't discuss how to do model selection and implementation algorithm. For present task, we assume there are N modules, then the goal of learning or the learning criterion is to minimize equation (13). It's important to see that this learning criterion has the same form as the learning criterion in subsection 1.2. In fact, we will show that the latter is the special case of the former from information geometry point.

6 Committee Machine with Gaussian Regression Models

We suppose there have N trained modules and each module with single output unit, and the corresponding regression model is: $y_k = f_k(x, w_k) + \varepsilon_k$, where subindex k stands for a module in the hybrid networks, ε_k is Gaussian noise with zero mean and variance σ_k; also assume the corresponding true model is Gaussian with mean μ and variance σ; therefore, the input-output relation of each module can be represented as the following conditional probability distribution

$$p_{y_k|x,w_k} = \frac{1}{\sqrt{2\pi}\sigma_k} \exp\left[-\frac{1}{2\sigma_k^2}(y_k - f_k(x, w_k))^2\right]. \tag{14}$$

It's an exponential family. Obviously minimizing equation (13) is equivalent to minimizing $C_y(x|H_k)p_{H_k}$ for each module or minimizing the product of corresponding KL divergence and a prior $D(q_{y|x,H_k}, p_{y|x,H_k})p_{H_k}$. The following theorem [17] is useful for our present task.

Theorem 2. Let $p_k \sim N(\mu_k \mid h_k), k \in \{1,2\}$ be two Gaussian distributions. Denote

$$d_0(h_1,h_2) := \left(\frac{h_1^\delta h_2^{1-\delta}}{\delta h_1 + (1-\delta)h_2} \right)^{1/2}, \quad d_1(\mu \mid h) := \exp\left(-\frac{h}{2}\mu^2\right). \tag{15}$$

Then the δ-divergence is given by

$$D_\delta(p_1,p_2) = \frac{1}{\delta(1-\delta)}[1 - d_0(h_1,h_2)d_1(\mu_1 - \mu_2 \mid V)]. \tag{16}$$

where $\delta \in (0,1)$, $V^{-1} = (\delta h_1)^{-1} + [(1-\delta)h_2]^{-1}$. From section 2, the extreme case $\delta = 1$ corresponds to the ordinary KL divergence. Let $\delta \to 1$, then

$$d_0(h_1,h_2) = \left[\frac{(h_2/h_1)^{1-\delta}}{1-(1-\delta)\frac{h_1-h_2}{h_1}} \right]^{1/2} \to \left[\frac{h_1}{h_2} \right]^{(1-\delta)/2} \exp\left[\frac{1-\delta}{2}\left[\frac{h_1-h_2}{h_1} \right] \right] \tag{17}$$

therefore as $\delta \to 1$, and use $\ln x^{-1} \approx 1-x$, we have

$$-\frac{1}{1-\delta}\ln d_0 \to -\frac{1}{2}\left[\ln \frac{h_2}{h_1} + 1 - \frac{h_2}{h_1} \right], \quad -\frac{1}{1-\delta}\ln d_1 \to \frac{h_2}{2}(\mu_1-\mu_2)^2. \tag{18}$$

Hence we have

$$D_1 \to -\frac{1}{1-\delta}(\ln d_0 + \ln d_1) \to \frac{1}{2}\left[\ln\frac{h_1}{h_2} + \frac{h_2}{h_1} - 1 + h_2(\mu_1-\mu_2)^2 \right]. \tag{19}$$

From equation (14), the mean and variance for module k is $f_k(x,w_k)$ and σ_k respectively, and the assumptions for real model, we have KL divergence for module k

$$D_1^k \approx \frac{1}{2}\left[\ln\frac{\sigma}{\sigma_k} + \frac{\sigma_k}{\sigma} - 1 + \sigma^{-1}(f_k-\mu)^2 \right]. \tag{20}$$

Notice that if $\sigma_k = \sigma$, then the learning criterion is written as

$$C_{y,H}(x)\mid_Q = -\ln \sum_k p_{H_k} \exp\left[-\frac{1}{2}\sigma_k^{-1}(f_k-\mu)^2 \right]. \tag{21}$$

In fact, the content in the square brackets is equivalent to so-called Mahalanobis divergence between two distribution masses [18]. If use the desired output given by data samples as the instantaneous value of μ, now the content in the square brackets becomes equivalent to the Mahalanobis divergence between samples in one mass and another mass. Furthermore, if assume σ_k is constant and p_{H_k}, $k=1...N$ is softmax, then the obtained learning criterion is the same as learning criterion (3). it can be trained by stochastic gradient methods. In this sense, learning criterion (3) is just a special case of our framework.

If models are not Gaussian then it's not easy to get explicit form of KL divergence. However, when model is of general exponential family, Amari in his paper [19] has proven that by introducing suitable hidden variables, combination of multi exponential families is an exponential family. Generally, since the true distribution is unknown, the choice of "target" poses a logical dilemma in itself. It is usually chosen according to some asymptotic properties. Paper [20] explored Bayesian inference using information geometry and suggested using the empirical distribution or the MLE for approximation under $D_1 = KL$ for exponential family manifold. Another implicit implementation algorithm is so-called em algorithm [19], which is equivalent to EM algorithm in most cases.

7 Conclusion

This paper focuses on the close relation between the selection of learning criterion for committee machine and network approximation and competitive adaptation. By minimizing the KL deviation between posterior distributions, we give a general posterior modular structure and the corresponding learning criterion form, which reflects remarkable adaptation and scalability. Besides this, we point out that, from the generalized KL deviation defined on finite measure manifold in information geometry theory, the proposed learning criterion reduces to so-called Mahalanobis deviation of which ordinary mean square error approximation is a special case, when each module is assumed Gaussian. Our future work is to find an appropriate incremental learning implementation algorithm.

Acknowledgements

The research is supported by: National Natural Science Foundations (No. 60373029) and Doctoral Foundations of China(No. 20020004020).

References

[1] Yan pingfan, zhang changshui: Neural network and simulated evolutionary computing (in Chinese), Tsinghua university press,china, 2001.
[2] Ye shiwei and Shi zhongzhi: Foundations of neural network (in Chinese), Mechanics Industry press, china, 2004.

[3] Hashem. S. (1993). Optimal Linear Combinations of Neural Networks. Unpublished PhD thesis, School of Industrial Engineering, Purdue University.
[4] Bast, W. G.(1992). Improving the Accuracy of an Artificial Neural Network Using Multiple Different Trained Networks. Neural Computation, 4(5).
[5] Jacobs, R.A., and Jordan, M.I., G.E.(1991). Adaptive Mixtures of Local Experts. Neural Computation, 3(1), 79-87.
[6] Jacobs, R.A. and Jordan, M.: Hierachical Mixtures of Experts and the EM Alogorithm. Neural Computation, 6, 181-214.
[7] Hansen, L.K. and Salamon, P.: Neural Network Ensembles. IEEE Transactions on Pattern Analysis and Machine Intelligence, 12(10):933-1000.
[8] Michael P.P. and Leon N. C.: When networks disgree ensemble methods for hybrid neural networks,1993
[9] Waterhouse, S.R. and Mackay, D.J.C.: Bayesian Methods for Mixtures of Experts, in Touretzky et al. [226], pp. 351-357.
[10] Jacobs, R.A. and Peng, F.C. and Tanner, M.A: A Bayesian Approach to Model Selection in Hierarchical Mixtures-of-Experts Architectures, Neural Networks, 10(2), 231-241. 243-248.)
[11] T. Poggio and F. Girosi: Networks for Approximation and Learning. Proceedings of the IEEE, Vol.78, No. 9, pp. 1481-1497, 1990.
[12] U. M. Frayyad, Piatetsky-Shapiro, G. and Smyth, P.:From data-mining to knowledge discovery: An overview., in Fayyad, Piatetsky-Shapiro, and Smyth and Uthurusamy [52], chapter 1, pp. 1-37
[13] Vkadimir N. Vapnik : The Essence of Statistical Learning (in Chinese), translated by zhang xuegong, Tsinghua university press, china, 2000, 9。
[14] S. Amari : Differential-Geometrical Methods in Statistics, Lecture Notes in Statistics, Vol. 28, Springer, 1985.
[15] S. Amari and H. Nagaoka : Methods of Information Geometry, AMS, Vol. 191, Oxford University Press, 2000.
[16] Huaiyu Zhu, Richard Rohwer. Information Geometric Measurements of Generalisation. Technical Report NCRG/4350,Aug,1995.
[17] Huaiyu Zhu: Bayesian invariant measurements of generalization for continuous distributions Technical Report NCRG/4352, Aston University. ftp://cs.aston.ac
[18] Fan jincheng and Mei changlin: Data Analysis, Science press, china, 2002.7.
[19] Amari, S.I.: imformation geometry of the EM and em algorithms for neural network, Neural Networks, 8, No. 9, pp.1379-1408, 1995.
[20] Huaiyu Zhu and Rohwer, R.: information geometry, Bayeasian inferance, ideal estimates and error decomposition. Working Paper SFI-98-06-45, Santa Fe Institute, 1998.

An ART2/RBF Hybrid Neural Networks Research*

Xuhua Yang, Yunbing Wei, Qiu Guan, Wanliang Wang, and Shengyong Chen

College of Information Engineering,
Zhejiang University of Technology,
Hangzhou, Zhejiang,
P.R.China, 310032
xhyang@zjut.edu.cn

Abstract. The radial basis function(RBF) neural networks have been widely used for approximation and learning due to its structural simplicity. However, there exist two difficulties in using traditional RBF networks: How to select the optimal number of intermediate layer nodes and centers of these nodes? This paper proposes a novel ART2/RBF hybrid neural networks to solve the two problems. Using the ART2 neural networks to select the optimal number of intermediate layer nodes and centers of these nodes at the same time and further get the RBF network model. Comparing with the traditional RBF networks, the ART2/RBF networks have the optimal number of intermediate layer nodes , optimal centers of these nodes and less error.

1 Introduction

Radial Basis Function (RBF) networks are powerful computational tools that have been used extensively in the areas of systems modeling and pattern recognition. The difficulties of applying RBF networks consist in how to select the optimal number of intermediate layer nodes and centers of these nodes. In general, it is desirable to have less nodes networks that can generalize better and are faster to train. This calls for an optimal number of intermediate layer nodes and optimal positioning of intermediate layer nodes i.e., the location of centers.

This paper proposes a novel ART2/RBF hybrid neural networks. Using the ART2[1] neural networks to select the optimal number of intermediate layer nodes and centers of these nodes at the same time and solves the two difficulties in traditional RBF networks effectively.

The rest of this paper is organized as follows. Section 2 gives a brief introduction to the traditional RBF networks. Section 3 provides a brief introduction to the ART2 networks. Section 4 introduces the principle of the ART2/RBF hybrid neural networks. Simulation results are presented and discussed in Section 5. Finally, conclusion is given in Section 6.

---

* This work was supported by the National Natural Science Foundation of China under Grant 60374056,60405009,50307011.

2 RBF Neural Networks

The RBF feedforward neural networks achieve a nonlinear mapping as following:

$$f_n(x) = W_0 + \sum_{i=1}^{n} W_i \varphi(\|X - c_i\|) \qquad (1)$$

$X \in R^n$ is the input vector. $\varphi(\bullet)$ is the radial basis function which achieve a mapping: $R^+ \to R$. $\|\bullet\|$ is the euclidean norm. W_i is the weight. c_i is the center of the intermediate layer ith node. n is the number of centers. Select the radial basis function as Gaussian function. Then, a RBF networks can be expressed as

$$f_n(X) = \sum_{i=1}^{n} w_i \exp(-\frac{\|X - c_i\|^2}{\sigma_i^2}) \qquad (2)$$

When c_i is known, we can use the following formula to determine σ_i^2

$$\sigma_i^2 = \frac{1}{M_i} \sum_{X \in \theta_i} \|X - c_i\|^2 \qquad (3)$$

M_i is the sample number of θ_i class. Finally, we can use the least square method to solve w_i. We can see that the difficulties of the RBF networks consist in how to select the optimal number of intermediate layer nodes and centers of these nodes.

3 ART2 Neural Networks

ART2 is a class of adaptive resonance architectures which rapidly self-organize pattern recognition categories in response to arbitrary sequences of either analog or binary input patterns. It can not only rapidly recognize the learned pattern but also fleetly adapt a new object which has not been learned formerly. The number of all winning neuron in F_2 layer is the clustered categories number.

The algorithm of the ART2 neural networks was completely described in reference [1]. We can see that the classification precision is determined by the vigilance parameter ρ ($0 < \rho < 1$) and higher ρ corresponds to finer categories. We can determine how fine the categories will be by adjusting the vigilance parameter ρ.

4 ART2/RBF Hybrid Neural Networks

The ART2/RBF hybrid neural networks can select the optimal number of intermediate layer nodes and centers of these nodes at the same time and further get the RBF network model.

Its learning algorithm is as following:

(1)Using the ART2 neural networks to cluster input sample data under a vigilance parameter, the number of all winning neuron in F_2 layer is the clustered categories number and the number is the optimal number of RBF networks' intermediate layer .The top-down weights of these winning neuron are the centers of the nodes of RBF networks' intermediate layer.
(2)Using the least square method to get weights between the RBF networks' intermediate layer and output layer can get the model of RBF networks.

If the RBF networks' model requires higher accuracy, we should increase the vigilance parameter ρ and repeat (1)and (2) up to satisfying the demand.

The ART2/RBF hybrid networks can adaptively get the intermediate layer of RBF networks. When we want to get the different nonlinear model under different input and output sample data, the networks can adaptively get the intermediate layer node's number and centers according to the required precision and input sample.

5 Simulation Research

Suppose input sample data is $\{x_i, i = 1,\cdots, N\}$, output sample data is $\{y_i, i = 1,\cdots, N\}$, the trained neural networks model is $f(\bullet)$, then the approximation error can be gotten with the following formula.

$$E = \frac{1}{2}\sum_{i=1}^{N}(y_i - f(x_i))^2 \quad (4)$$

In traditional RBF networks, people usually adopt trial-and-error method to determine the number of intermediate layer nodes. The traditional method of selecting centers of intermediate layer nodes can be adopted as k-means clustering algorithm[2] ,Konhonen self-organizing map[3] ,Orthogonal least square learning algorithm[4]. This paper compares the three methods' the performances with the ART2/RBF networks' by a nonlinear system identification example.

Example: The nonlinear identification object is as following:

$$y(x) = 1.6x^2 \sin(8x)\cos(x) \qquad x \in [0,2] \quad (5)$$

Training input sample set : $x = 0 : 0.005 : 2$,training output sample set $y(x)$ can be gotten according to the formula (5).The number of training sample is 401.

Traditional RBF networks adopt the trial-and-error method to determine the number of intermediate layer nodes and use above three method to select centers of these nodes.

The ART2/RBF hybrid neural networks adopt ART2 networks to select the optimal number of intermediate layer nodes and centers of these nodes at the same time.

The parameters of ART2 networks are
$M = 2, N = 35, a = 4, b = 4, c = 0.13, d = 0.8, \theta = 0.15, \rho = 0.93$.

All kinds of algorithm's performances are as table 1.

Table 1. Four kinds of algorithm's performances

Algorithm	the Number of Centers	Clustering Measuring and Norm	Error
k-means clustering	16	Euclidean distance 0.01	0.4563
Konhonen self-oganizing Map	16	Euclidean distance 0.01	0.4143
orthogonal least square	16	Euclidean distance 0.01	0.3837
ART2/RBF	26	Euclidean distance 0.01	0.1309

Observing the simulation results in table 1,we can conclude that the ART2/RBF hybrid networks can achieve the optimal number of intermediate layer nodes and less approximation error than the traditional RBF networks.

6 Conclusion

This paper proposes a novel ART2/RBF hybrid networks which can adaptively get the intermediate layer node's number and centers according to the required precision and input sample. This networks structure solves the difficulties of applying RBF networks. Simulation results also show the validity of this networks.

References

1. Carpenter G A,Grossberg S.:ART2:self-organization of stable category recognition codes for analog input patterns.Allpied Optics,(1987),26:4919-4930.
2. Duda,R.O., P.E.Hart.: Pattern Classification and Scene Analysis, New York:Wiley. (1973)
3. Kohonen T.:Self-Organized Formation of Topologically Correct Feature Maps. Biol Cyber,(1982),43:59-69.
4. S.Chen,C.F.N.Cowan,P.M.Grant.:Orthogonal least square learning algorithm for radial basis function networks.IEEE Trans.Neural Networks.(1991).2:302-309.

Complex Number Procedure Neural Networks[1]

Liang Jiuzhen and Han Jianmin

College of Information Science & Engineering, Zhejiang Normal University,
321004 Jinhua, China
{Liangjz, hanjm}@zjnu.cn

Abstract. This paper deals complex number procedure neural networks and its learning algorithm. The conception and mathematic description of complex number procedure neurons are proposed based on traditional complex number neuron and procedure neuron. Feed-forward complex number neural networks models are considered. Grads-descent learning algorithm is deduced according to the supervising learning, and its learning procedure consists of two parallel procedures, the real part and imaginary part. An application example is given which show that the complex procedure neural network is suitable for signal processing problem.

1 Introduction

Generally, traditional neural networks deal with real number data. Neuron state, input/output and weight are all real, which limits its application. Early in 1990s, in the time of neural networks, people began to pay attention to complex number procedure neural networks (CNNN). In 1990, Gordon extended BP algorithm to complex number weights [1]. In 1991, Benvenuto applied CNNN in signal sorting [2]. In 1992, Kechriotis used CNNN in simulating equilibrium as non-linear channels [3]. Later, CNNN were employed in designing FIR digital filters [4] and recognizing non-linear time series model [5], etc.

Procedure neural network is the extension in time domain of traditional neural network, which takes the effect that time factor causes to the system into account, whose inputs are time functions and output is a space vector. Many forms of procedure neural networks and application have been proposed [6, 7].

This paper proposes a model of complex number procedure neural networks (CNPNN) in section 2. The third section deals with CNPNN learning and in the forth part an application example is given for signal processing problem.

2 Complex Number Feed-Forward Procedure Neural Networks

The difference between real procedure neuron and CNPNN is that the inputs, outputs and weights of procedure neuron are all complex number. And its aggregation opera-

---

[1] This paper is supported by Zhejiang Province Nature Science Foundation of China (No.Y104107).

tion includes not only multi-input aggregation on complex number space, but also accumulation on time procedure. The structure of a single complex number procedure neuron can be showed in Fig.1.

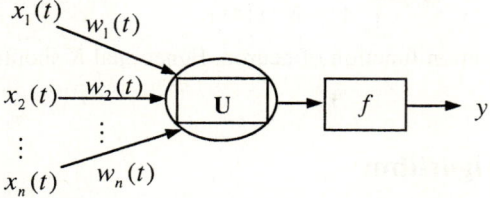

Fig. 1. Model of complex number procedure neuron

Where $x_1(t), x_2(t),\ldots,x_n(t) \in [0, T]$ are vectors of input complex number function of procedure neuron, and

$$x_k(t) = x_{Rk}(t) + ix_{Ik}(t), k=1,2,\ldots,n, \tag{1}$$

where $x_{Rk}(t)$ and $x_{Ik}(t)$ are $x_k(t)$ real part and imaginary part respectively, which are all real functions; $w_1(t), w_2(t),\ldots,w_n(t) \in [0, T]$ are complex number weight functions, also take the form as formula (1).

Complex number procedure neurons can constitute many forms of CNPNN according to their organization and topology structure. In this paper, we consider a CNPNN(Fig. 2) whose weight function can be extended by a set of basis function $B(t)$.

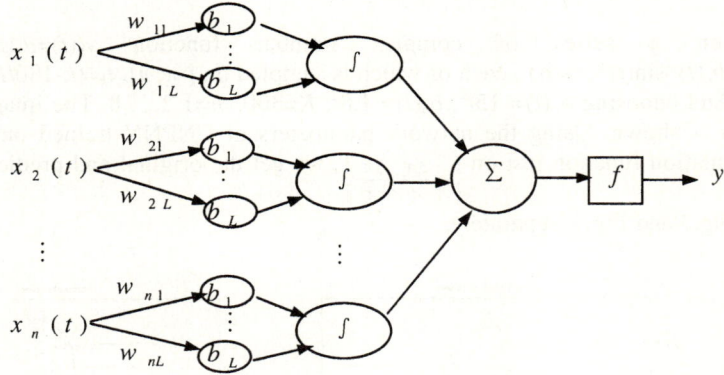

Fig. 2. Procedure neural networks of expansion of basis function

Where L is the number of basis function and

$$w_j(t) = \sum_{l=1}^{L}(w_{Rjl}b_{Rl}(t) - w_{Ijl}b_{Il}(t)) + i\sum_{l=1}^{L}(w_{Rjl}b_{Il}(t) + w_{Ijl}b_{Rl}(t)) \tag{2}$$

$$A(t) = \sum_{j=1}^{n} (w_{Rj}(t)x_{Rk}(t) - w_{Ij}(t)x_{Ik}(t)) + i\sum_{j=1}^{n} (w_{Rj}(t)x_{Ik}(t) + w_{Ij}(t)x_{Rk}(t)) \quad (3)$$

The output of complex number procedure neural networks is as following.

$$y = f(\int_0^T A(t)K(t)dt) \quad (4)$$

Where f is activation function of neuron. Functional K should be defined according to practical need.

3 Learning Algorithm

The learning algorithm of CNPNN can be deduced by supervised algorithm of real number procedure neural network. Suppose

$$\hat{y} = \hat{y}_R + i\hat{y}_I \quad (5)$$

is the desire output of complex procedure neural networks. Define the error function of complex number procedure neural network as

$$E = \frac{1}{2}\|y - \hat{y}\| = \frac{1}{2}(y - \hat{y})\overline{(y - \hat{y})} = \frac{1}{2}(y\overline{y} + \hat{y}\overline{\hat{y}} - y\overline{\hat{y}} - \overline{y}\hat{y}) \quad (6)$$

According to grads-descent learning algorithm, the rules of iterative learning of networks weights can be deduced.

4 Application Example

Given a series of complex number function $y_n(t)=a_n(t)*\cos(t)*\sin(n) +i*b_n(t)*\sin(t)*\cos(n)$, each of which is sampled on $[-\pi, \pi]$, $t_k=(k-180/K)$, $k=0,1,\ldots,K-1$. And choosing $a_n(t)= 15t^2$, $b_n(t)= 1.8t$, $K=360$, $n=1,2,\ldots,8$. The image of the function is shown. Using the network parameters of CNPNN trained on $[-\pi, \pi]$ as the simulation function test on $\left[-\frac{3\pi}{4}, \frac{5\pi}{4}\right]$, we get the original and predictive images as in Fig.3 and Fig.4, separately.

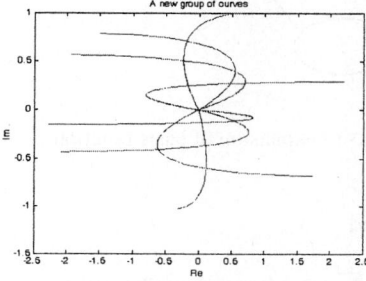

Fig. 3. Original image of $\{y_n\}$

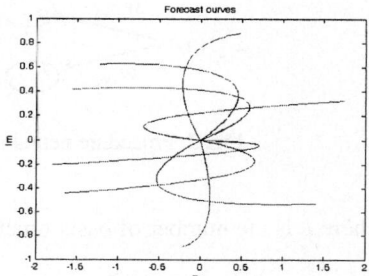

Fig. 4. Predictive image of $\{y_n\}$

5 Conclusions

CNPNN extends the traditional procedure neural network to the complex number field, which many typical problems, such as the signal processing, can be solved directly by single networks. Although the fashion of the CNPNN is a little complicated and the calculation complexity is doubled, the area and ability of solving problems has been highly extended.

References

1. Gordon, R.L., Gustafson, S.C., Senn, R. A.: Generalization of The Back Propagation Neural Network Learning Algorithm to Permit Complex Weights. Applied Optics, Vol.29. (1990) 1591–1592
2. Leung, H., Haykin, S: The Complex Back Propagation Algorithm. IEEE Trans on Signal Processing, Vol.39. (1991) 2101–2104
3. Benvenuto N., Piazza, F: On The Complex Back Propagation Algorithm. IEEE Trans on Signal Processing, Vol.40. (1992) 967–969
4. Wang, Y., Liao, X. F., Yu, Q. B: Complex Neural Networks on Designing FIR Digital Filter. Signal processing, Vol.1. (1999) 193–198
5. Xiong, S.S., Zhou, Z. Y: Non-Linear Time Series Model for Shape Recognition Using Neural Networks. ACTA AUTOMATICA SINICA, Vol.25. (1999) 467–475.
6. He, X. G., Liang, J. Z: Procedure Neural Networks. In: Proceedings of Conference on Intelligent Information Proceeding, 16th World Computer Congress 2000. Publishing House of Electronic Industry of China. (2000) 143–146
7. Liang, J. Z: Research on Approximation Theory of Fuzzy Neural Networks And Learning Algorithm. PhD Dissertation. Beijing University of Aeronautics & Astronautics (2001) 83–99
8. Shen, X. C: Approximation of Complex Number Variable Function. Chinese Science Press (1998) 2–15

Urban Traffic Signal Timing Optimization Based on Multi-layer Chaos Neural Networks Involving Feedback

Chaojun Dong[1,2], Zhiyong Liu[2], and Zulian Qiu[1]

[1] Institute of Electronic & Information Engineering,
Xi'an Jiaotong University,
Xi'an 710049, China
Cjun-dong@163.com
[2] Insititute of Information, Wuyi University, Jiangmen 529020, China
Zliu@mail.wyu.edu.cn

Abstract. Urban traffic system is a complex system in a random way, it is necessary to optimize traffic control signals to cope with so many urban traffic problems. A multi-layer chaotic neural networks involving feedback (ML-CNN) was developed based on Hopfield networks and chaos theory, it was effectively used in dealing with the optimization of urban traffic signal timing. Also an energy function on the network and an equation on the average delay per vehicle for optimal computation were developed. Simulation research was carried out at the intersection in Jiangmen city in China, and which indicates that urban traffic signal timing's optimization by using ML-CNN could reduce 25.1% of the average delay per vehicle at intersection by using the conventional timing methods. The ML-CNN could also be used in other fields.

1 Introduction

With the development of productivity, traffic jam is becoming a tougher and tougher problem in modern cities. It is necessary to develop a kind of high efficiency traffic signal controller with intelligent technologies for better and efficient urban traffic control. It is known that Chaos phenomenon exists in various dynamical systems. Urban traffic system has a typical chaotic characteristic. Chaos theory should be a kind of effective methods to deal with the problem. There is a Hopfield network which is a ripe one and fit for optimization especially. Yet it can't be used for solving complex traffic problems because it is single-layer. Thus here, a ML-CNN using the basic theory of both chaos theory and Hopfield network will be put forward in this paper. It can be used to optimize traffic control signal timing on a single intersection.

In recent years, a lot of research has been carried out on chaotic neural networks (CNN). Zhenya He (2002) developed a Multistage Self-Organizing Algorithm Combined Transiently CNN for Cellular Channel Assignment; Cao Zhitong, Jacob (2003) used the Nagumo-Sato model to construct a chaotic CNN; Ohta, Masaya(2002) proposed a CNN with reinforced self-feedbacks; Lipo Wang(2004) proposed a noise CNN for solving combinatorial optimization problems. And so on.

2 Multi-layer Chaotic Neural Networks Involving Feedbacks

Combining Hopfield network with chaos map, a ML-CNN was put forward here, the ML-CNN has a characteristic of escaping from a local minimum of the energy function, so that it can find a global minimum more easily as compared with the Hopfield's model.

As an example, fig.1 shows a ML-CNN's framework that can be used in an intersection with a standard four signal phases. Compared with Hopfield networks, several major characteristics of the networks consist of: (i) it is a three-layer network including an input layer, an output layer and a hidden layer; (ii) all the outputs in the output layer are returned to the input layer; (iii) the hidden layer consists of many chaos neurons with self-feedback.

As shown in fig.1, $g_1 \sim g_5$ represent respectively the effective green time of the signal phase 1, 2, 3, 4 and the cycle time; $s_1 \sim s_4$ represent respectively the saturation flows of the signal phase 1, 2, 3, and 4; and $s_5 \sim s_{12}$ represent respectively actual average vehicle flows of traffic flow 1, 2 in the signal phase 1, 2, 3, and 4.

The outputs in the output layer are a linear combination of the hidden layer's outputs, which can be expressed as follows,

$$g_j = \tau \sum_{i=1}^{6} \beta_{ji} x_i \qquad j = 1 \sim 5 . \tag{1}$$

Several output models of the hidden layer's can be expressed as follows,

$$x_i(t) = \frac{1}{1 + e^{-y_i(t)/\varepsilon}} , \tag{2}$$

$$y_i(t+1) = k y_i(t) + \alpha \left[\sum_{j=1}^{n} \omega_{ij} u_j(t) + I_i \right] - z_i(t)[x_i(t) - I_0] , \tag{3}$$

$$z_i(t+1) = (1-\beta) z_i(t) \qquad i = 1,2,\ldots, . \tag{4}$$

Some output models of the input layer's can be expressed as follows,

$$u_i = \sum_{j=1}^{5} \rho_{ji} g_j + \lambda s_i \qquad i = 1 \sim 12 . \tag{5}$$

where, t is discrete time step ($t = 0,1,2,\cdots$); x_i and y_i are respectively the output and the internal state variable of the ith chaotic neuron in hidden layer; u_i is the output of the ith neuron in input layer; z_i is the self-feedback's dynamic weight of the ith neuron in hidden layer; β_{ji}, ω_{ij} and ρ_{ji} are respectively the weights between the ith neuron in hidden layer and jth neuron in output layer, the jth neuron in output layer and ith neuron in hidden layer, the jth neuron in output layer and ith neuron in input layer; I_i is the ith neuron's input deviation in hidden layer; ε is a gradient parameter of $x_i(t)$; k is a neuron's attenuation factor in hidden layer; α is a scaling factor; β is the attenuation factor of $z_i(t)$; τ and λ are all dimension uniform factors.

As shown in the equation (4), $z_i(t) \rightarrow 0$, when $t \rightarrow 0$, and the network will become a discrete feedback neural networks without chaotic self-feedback's neurons and converge at a steady balance point, and then we get the optimization.

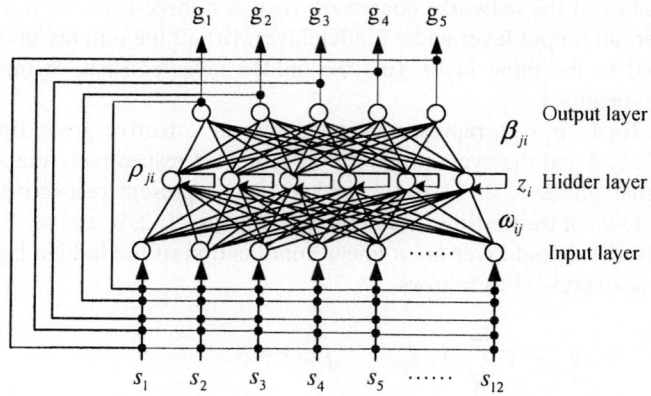

Fig. 1. Multi-layer chaotic neural networks involving feedback

3 Energy Function

It is a key to define energy function in chaotic neural networks involving feedback.

3.1 Delay Model

The Webster equation, shown below (when degree of saturation is smaller than 97%),

$$d = \frac{c(1-\lambda)^2}{2(1-\lambda x)} + \frac{x^2}{2q(1-x)} - 0.65(\frac{c}{q^2})^{\frac{1}{3}} x^{(2+5\lambda)} . \qquad (6)$$

where, d is the average delay per vehicle(s/veh), c the cycle time(s), λ the proportion of a cycle than effectively green for the phase under consideration (i.e., g/c), q the flow(vehicles per h), s the saturation flow (vehicles per second of green), x the degree of saturation (i.e., $x = q/\lambda s$).

Robertson retained Webster's first term for uniform vehicle arrivals but replaced the last two terms using the terms below (when degree of saturation is larger than 97%),

$$D_2 + D_3 = [(\frac{2(1-x)+xz}{4z-z^2})^2 + \frac{x^2}{4z-z^2}]^{\frac{1}{2}} - \frac{2(1-x)+xz}{4z-z^2}. \quad (7)$$

where, $D_2 + D_3$ is the added delay because of vehicle random variety; $z = ((2x/v).(60/T)$, v is the approach volume(veh/h), T the period length.

3.2 Energy Function

The energy function includes the total delay and some restriction conditions of the cycle's and the effective green time. The energy function is in this form,

$$E = \sum_{i=1}^{m}\sum_{j=1}^{n_i} d_{ij}^2 + A_1(c - c_{max}) \mid c - c_{max} \mid + A_2(c_{min} - c) \mid c - c_{min} \mid +$$

$$\sum_{i=1}^{m}[A_{i+2}(g_{min} - g_i) \mid g_i - g_{min} \mid + A_{i+m+2}(g_i - g_{max}) \mid g_i - g_{max} \mid] + \quad (8)$$

$$A_0[\sum_{i=1}^{m}(g_i + I_i) - c]^2.$$

where, m is the numbers of phase, n_i approach number of the ith phase, d_{ij} the average delay per vehicle of the jth lane in the ith phase within a cycle (s/veh), c the cycle length (s), c_{min} and c_{max} the upper limit and the lower limit of cycle length (s), g_i the effective green time of the ith phase (s), g_{min} and g_{max} the upper limit and the lower limit of the effective green time (s), I_i losing time of the ith phase, A_i ($i = 0,1,\cdots,2m+2$) the punishment coefficient.

4 Simulation Research and Conclusion

The object simulated is an intersection in Jiangmen city in China, whose traffic is controlled by four phases, and each phase includes two various traffic flows, turning right isn't controlled. Simulation is done respectively by vehicle actuated control, adaptive control and ML-CNN method under the same traffic condition. Suppose: the upper limit of cycle length is 120s, the lower limit 47s, the upper limit of green time is 60s, the lower limit 10s, the yellow time is 1s. Table1 shows the simulation result, which shows that it can reduce the total delay at intersection by using ML-CNN for timing optimization.

Urban traffic signal timing optimization based on ML-CNN can reduce25.1% of the average delay per vehicle at intersection based on the conventional timing means, and can improve the traffic efficiency. It is a key to fix restriction conditions that have a huge influence on optimization results. The ML-CNN can also been used in other fields. Consider that ML-CNN can effectively be used in a single intersection, we believe that it could also be used in area traffic control.

Table 1. Simulation result. "*"ML-CNN; "**"vehicle induce control; "***" adaptive control; "#"the reduce of average delay per vehicle of ML-CNN compared with vehicle induce control; "##" the reduce of average delay per vehicle of ML-CNN compared with self-optimization control (This table has been abridged)

	7:45 ~8:00	9:55 ~10:10	12:50 ~13:05	14:10 ~14:25	15:10 ~15:25	18:25 ~18:40
Cycle length (s)	89	95	53	90	88	60
Green time in phase1 (s)	11	10	10	10	10	10
Green time in phase2 (s)	23	31	14	27	30	18
Green time in phase3 (s)	17	20	10	20	18	10
Green time in phase4 (s)	31	27	12	26	22	15
Average delay(s/veh) *	67	70	48	76	62	53
Average delay(s/veh) **	86	87	67	95	88	78
Reduce(%) #	22.1	19.5	28.4	20.0	29.5	32.0
Average delay(s/veh) ***	91	96	68	92	81	76
Reduce(%) ##	26.3	27.1	29.4	17.4	23.4	30.2

References

1. S.C.Wong, et al.: Group_ based optimization of a time-dependent TRANSYT traffic model for are traffic control. Transportation Research Part B. 36(2002) 191-312
2. Kazuhiko Takahashi: Remarks on multi layer neural networks involving chaos neurons. International Journal of Applied Electromagnetic and Mechanics. 18(2003) 165-176
3. Lipo Wang, Li Sa, Tian Fuyu, Fu Xiuju: A Noisy Chaotic Neural Network for Solving Combinatorial Optimization Problems: Stochastic Chaotic Simulated Annealing. IEEE Transactions on Systems, Man & Cybernetics: Part B. 5(2004) 2119-2125
4. Michalewicz, Z.: Genetic Algorithms + Data Structures = Evolution Programs. 3rd edn. Springer-Verlag, Berlin Heidelberg New York (1996)
5. Cao Zhitong, Chen Hongping, He Guoguang: Associative Dynamics and Its Control of Chaotic Neural Network. International Journal of Modern Physics B: Condensed Matter Physics. 22-24(2003) 4176-4181
6. Ohta, Masaya: Chaotic neural networks with reinforced self-feedbacks and its application to <f>N</f>Queen problem. Mathematics & Computers in Simulation. 4(2002) 305-317

Research on a Direct Adaptive Neural Network Control Method of Nonlinear Systems

Weijin Jiang[1], Yusheng Xu[2], and Yuhui Xu[1]

[1] Department of Computer, Zhuzhou Institute of Technology, Zhuzhou 412008, P.R.China
jwjnudt@163.com
[2] College of Mechanical Engineering and Applied Electronics, Beijing University of Technology, Beijing 100022, P.R.China
yshxu520@163.com

Abstract. The problem of direct adaptive neural control for a class of nonlinear systems with an unknown gain sign and nonlinear uncertainty is discussed in this paper. Based on the principle of sliding mode control and the approximation capability of multilayer neural networks (MNNs), and using Nussbaum-type function, a novel design scheme of direct adaptive neural control is proposed. By adopting the adaptive compensation term of the upper bound function of the sum of residual and approximation error, the closed-loop control system is shown to be globally stable, with tracking error converging to zero. Simulation results show the effectiveness of the proposed approach.

1 Introduction

In recent years, robust adaptive control of nonlinear systems has received much attention[1-8]. Typically, these methods use neural networks as approximation models for the unknown system nonlinearities[2-7].Using the approximation capability of radial basis function neural networks, which are the linear function of adjustable output weights, a stable adaptive controller was proposed in [2].In order to improve the approximation of radial basis function neural networks, an adaptive neural network control with variable variance parameters was proposed in [3]. But the approximation errors were assumed to be bounded in [2,3].Based on multilayer neural networks, the adaptive controllers proposed by [4-6] ensured tracking error converging to residual set only. A direct adaptive controller was developed based on multilayer neural networks and sliding mode control technique in [7],but the approximation error was assumed to be bounded in stability analysis (See equation(26)).Using the approximation capability of the second–type fuzzy system, which is the nonlinear function of adjustable parameters, the design scheme of a stable adaptive fuzzy controller was proposed in [8].The projection algorithm was adopted for the parameter estimation in [8].However, the control gain signs were assumed to be known in [1-8].By using Nussbaum –type function, two control schemes were presented for a class of strict-feedback nonlinear systems with unknown virtual coefficients signs in [9,13].

In this paper, a new design scheme of adaptive neural controller for a class of nonlinear systems with an unknown gain sign is proposed. The design is based on the principle of sliding mode control and the approximation capability of multilayer neural networks. By utilizing the robust adaptive control and Nussbaum function, an adaptive law is derived to adjust the gain of sliding mode control term to adaptively compensate for the residual and the approximation error of MNNs. By theoretical analysis, the closed-loop neural control system is proved to be stable and the tracking errors asymptotically converge to zero.

2 Problem Statement and Basic Assumptions

Consider the neural adaptive control problem for a class of nonlinear systems in the following form:

$$\begin{cases} \dot{x}_i = x_{i+1}, i=1,\ldots,n-1 \\ \dot{x}_n = f(x) + g(x)u + d(x,t), \\ y = x_1, \end{cases} \quad (1)$$

where $x=(x_1, x_2, \ldots, x_n)^T \in R^n$ is the state vector, u is the control input, f is the unknown continuous function, g is the unknown continuous function control gain, y is the system output, d denotes external disturbance.

The control objective is to force the system output y to follow the specified trajectory y_d. Therefore we should design a neural network control u(t) such that $y-y_d$ converges to zero.

Define x_d, e and a filtered tracking error s as follows:

$$\begin{aligned} x_d &= (y_d, \dot{y}_d, \cdots, y_d^{(n-1)})^T, \\ e &= (e_1, e_2, \cdots, e_n)^T = x - x_d = (x_1 - y_d, x_2 - \dot{y}_d, \ldots, x_n - y_d^{(n-1)})^T, \\ s &= (\frac{d}{dt}+\lambda)^{n-1} e_1 = \sum_{i=1}^{n-1} c_i e_i + e_n, \end{aligned} \quad (2)$$

where $c^i = C_{n-1}^{i-1} \lambda^{n-i}$, $i=1,\ldots,$ λ is a positive constant, specified by the designer.

Lemma 1[6] : Let s be defined by (2), then

1) if s =0, then $\lim_{t \to \infty} e_1 = 0$;
2) if $|s| \leq c$, $e(0) \in \Omega_c$, then $e(t) \in \Omega_c \ \forall \ t \geq 0$;
3) if $|s| \leq c$, $e(0) \notin \Omega_c$, then $\exists T = (n-1)/\lambda$, $\ni \forall \geq T$, $e(t) \in \Omega_c$,

where $c>0$, $\Omega_c = \{ e(t) | |e_j| \leq 2^{j-1} \lambda^{j-n} c, j=1,\ldots,n \}$.

From (1),(2), we have

$$\dot{s} = f(x) + g(x)u + d(x,t) + \gamma, \quad (3)$$

where $\gamma = \sum_{j=1}^{n-1} c_j e_{j+1} - y_d^{(n)}$.

In order to design adaptive neural network control, we make the following assumptions:

1) $0 < g_0 \le |g(x)| \le g_1$, $\forall x \in R^n$;

2) $\dfrac{\partial g(x)}{\partial x_n} = 0, \forall x \in R^n$, and $\dfrac{\partial g(x)}{\partial x}$ is continuous function;

3) $(x_d^T, y_d^{(n)})^T \in \Omega_d \subset R^{n+1}$;

4) $|d(x,t)| \le D(x)$, $\forall t \ge 0$,

where g_0 and g_1 are known positive constants, Ω_d is a known bounded compact set

In order to copy with the unknown control gain sign, the Nussbaum gain technique is employed in this paper. A function $N(\varsigma)$ is called a Nussbaum-type function if it has the following properties:

1) $\lim\limits_{s \to +\infty} \sup \dfrac{1}{s} \int_0^s N(\varsigma) d\varsigma = +\infty$;

2) $\lim\limits_{s \to -\infty} \sup \dfrac{1}{s} \int_0^s N(\varsigma) d\varsigma = -\infty$

Lemma 2[12]: Let $V(\cdot)$ and $\varsigma(\cdot)$ be smooth function defined on $[0,t_f]$ with $V(t) \ge 0$, $\forall t \in [0,t_f)$, and $N(\cdot)$ be an even smooth Nussbaum-type function. If the following inequality holds:

$$V(t) \le c_0 + \int_0^t (gN(\varsigma)+1)\dot{\varsigma} d\tau, \quad \forall t \in [0,t_f)$$

where g is a nonzero constant, c_0 represents some suitable constant, then $V(t)$, $\varsigma(t)$ and $\int_0^t [gN(\varsigma)+1]\dot{\varsigma} d\tau$ must be bounded on $[0,t_f)$.

3 Adaptive Neural Network Controller Design

Let

$$h(z) = \dfrac{f(x)+\gamma}{|g(x)|} - \dfrac{\dot{g}(x)\operatorname{sgn}(g(x))s}{2g^2(x)}, \tag{4}$$

where $z = (x,s,\gamma)^T$, we first define a compact set as

$\Omega_z = \{(x^T, s, \gamma)^T \mid x \in \Omega_\mu, (x_d^T, y_d^{(n)})^T \in \Omega_d\} \subset R^{n+2}$, with compact subst Ω_μ to be specified later. Let h(z,W,V) be the approximation of the three-layer neural networks on the compact Ω_z to h(z),i.e.

$$h(z,W,V)=W^T S(V^T \bar{z}) , \qquad (5)$$

where $z=(z_1, \ldots, z_{n+2})^T$, $\bar{z}=(z^T,1)^T$; $V=(v_1,\ldots,v_l) \in R^{(p+1)\times l}$, $W=(w_1,\ldots,w_l)^T \in R^l$ are the first –to second layer and the second-third layer weights, respectively; p=n+2, l>1 is the NN node number and constant $\gamma>0$; $S(V^T \bar{z})=(s(v_1^T \bar{z}),\cdots,s(v_{l-1}^T \bar{z}),1)^T$ with $s(z_\alpha)=1/(1+e^{-z_\alpha})$. Let

$$(W^*,V^*) = \arg\min_{(W,V)}[\sup_{z\in\Omega z} | h(z,W,V)-h(z) |] , \qquad (6)$$

then we have

$$h(z) = h(z,W^*,V^*) + \varepsilon(z), z \in \Omega_z , \qquad (7)$$

where W^*, V^* are ideal NN weights and $\varepsilon(z)$ is the NN approximation error. Since $h(z), h(z,W^*,V^*)$ are the continuous function on compact region Ω_z, $\exists \varepsilon > 0$ with

$$|\varepsilon(z)| \le \varepsilon, z \in \Omega_z , \qquad (8)$$

Let $\hat{W}(t)$ and $\hat{V}(t)$ be the estimations of W^* and V^* at time t, and the weight estimation errors denoted as $\tilde{W}(t) = \hat{W}(t) - W^*, \tilde{V}(t) = \hat{V}(t) - V^*$. From reference[4], we have

$$h(z,W^*,V^*) - h(z,\hat{W},\hat{V}) = -\tilde{W}^T (\hat{S} - \hat{S}'\hat{V}^T \bar{z}) - \hat{W}^T \hat{S}'\tilde{V}^T \bar{z} + d_u, \qquad (9)$$

$$\forall z \in \Omega_z$$

where $\hat{S}=S(\hat{V}^T \bar{z}); \hat{S}'=\text{diag}(\hat{s}'_1, \hat{s}'_2,\ldots, \hat{s}'_l)$
with $\hat{s}'_k = s'(\hat{v}_k^T \bar{z}) = ds(z_\alpha)/dz_\alpha | z_\alpha = \hat{v}_k^T \bar{z}, k=1,\ldots,l\}$, and the residual term d_u is bounded by

$$|d_u| \le \|V^*\|_F \|\bar{z}\hat{W}^T \hat{S}'\|_F + \|W^*\| \|\hat{S}'\hat{V}^T \bar{z}\| + \|W^*\|_1 \qquad (10)$$

Let $\|\cdot\|_F$ denote the Frobenus norm, $\|\cdot\|$ denote the 2-norm and $\|\cdot\|_1$ denote the 1-norm, i.e.

$A \in R^{m \times n}, \|A\|_F = \sqrt{tr(A^T A)}$; $a = (a_1, \ldots, a_n)^T \in R^n$,

$\|a\| = \sqrt{\sum_{i=1}^{n} a_i^2}, \|a\|_1 = \sum_{i=1}^{n} |a_i|$. From (8) and (10), we have

$$|d_u(z)| + |\varepsilon(z)| \leq \|V^*\|_F \|z\hat{W}^T \hat{S}'\|_F + \|W^*\| \|\hat{S}'\hat{V}^T \bar{z}\| + \|W^*\|_1 + \varepsilon = K^T \phi(z,t), z \in \Omega_z, \quad (11)$$

where $K = (\|V^*\|_F, \|W^*\|, \|W^*\|_1 + \varepsilon)^T$,

$\phi(z,t) = (\|\bar{z}\hat{W}^T \hat{S}'\|_F, \|\hat{S}'\hat{V}^T \bar{z}\|, 1)^T$.

Adopting the following control law:

$$u(t) = N(\varsigma)[k_0 s + \hat{W}^T \hat{S}(\hat{V}^T \bar{z}) + \hat{K}^T \phi(z,t) \operatorname{sgn}(s) + \frac{D(x)}{g_0} \operatorname{sgn}(s)], \quad (12)$$

$$\dot{\varsigma} = k_0 s^2 + \hat{W}^T \hat{S}(\hat{V}^T \bar{z}) s + [\hat{K}^T \phi(z,t) + \frac{D(x)}{g_0}] |s|, \quad (13)$$

where $\hat{K}$ is the estimation of K at time t, $N(\varsigma) = \exp(\varsigma^2) \cos(\frac{\pi}{2}\varsigma)$.

Choose the adaptive law as follows:

$$\dot{\hat{W}} = \Gamma_w (\hat{S} - \hat{S}'\hat{V}^T \bar{z}) s, \quad (14)$$

$$\dot{\hat{V}} = \Gamma_v \bar{z} \hat{W}^T \hat{S}' s, \quad (15)$$

$$\dot{\hat{K}} = \Gamma_k \bar{z} |s| \phi(z,t), \quad (16)$$

where $\Gamma_W > 0, \Gamma_V > 0$ and $\Gamma_K > 0$ are gain matrices which determine the rate of adaptation.

4 Stability Analysis

Define a smooth scalar function $V_s = \dfrac{s^2}{2|g(x)|}$.

Differentiating V_s with respect to t and applying (3),(4) and (7), we obtain

$$\dot{V}_s = \frac{2s\dot{s}|g(x)| - g(x)\operatorname{sgn}(g(x))s^2}{2g^2(x)} \tag{17}$$

$$= s\left[\operatorname{sgn}(g(x))u(t) + \frac{d(x,t)}{|g(x)|} + h(z,W^*,V^*) + \varepsilon(z)\right],$$

Theorem Consider the nonlinear systems (1) with the control law defined by (2),(5),(12) and (13). Let the weights $\hat{W}$, $\hat{V}$ and sliding mode gain $\hat{K}$ be adjusted by the adaptation law determined by (14)-(16) and let the assumptions 1)-5) be true. Then, for any bounded initial conditions, all the signals in the direct adaptive control system will remain bounded; moreover, the tracking error $e_1(t)$ will asymptotically converge to zero, i.e.

(1) The overall closed-loop neural control system is globally stable in the sense that all of the closed –loop signal are bounded, and the state vector

$$x \in \Omega_\mu = \{x(t) \mid |e_j(t)| \le 2^{j-1}\lambda^{j-n}\mu, j = 1,\cdots,n, x_d \in \Omega_d\}, t \ge T;$$

(2) $\lim_{t\to\infty} s = 0$, i.e. $\lim_{t\to\infty} e_1(t) = 0$, where $\mu = 2g_1 \sup_{t\ge 0} V(t)$, $T = (n-1)/\lambda$.

Proof (1) Define the Lyapunov function candidate

$$V(t) = \frac{1}{2|g(x)|}s^2 + \frac{1}{2}[\tilde{W}^T \Gamma_W^{-1}\tilde{W} + tr(\tilde{V}^T \Gamma_V^{-1}\tilde{V}) + \tilde{K}^T \Gamma_K^{-1}\tilde{K}], \tag{18}$$

where $\tilde{K} = \tilde{K} - \hat{K}$. Differentiating V(t) with respect to time t, we have

$$\dot{V}(t) = \dot{V}_s + \tilde{W}^T \Gamma_W^{-1}\dot{\hat{W}} + tr(\tilde{V}^T \Gamma_V^{-1}\dot{\hat{V}}) + \tilde{K}^T \Gamma_K^{-1}\dot{\hat{K}}, \tag{19}$$

Substituting (12) and (14)-(17) into (19), we have

$$\dot{V}(t) = \operatorname{sgn}(g(x))N(\zeta)\dot{\zeta} + \dot{\zeta} - \dot{\zeta} + \frac{d(x,t)}{|g(x)|}s +$$

$$h(z,W^*,V^*)s + \varepsilon(z)s] + \tilde{W}^T \Gamma_W^{-1}\dot{\hat{W}} + tr(\tilde{V}^T \Gamma_V^{-1}\dot{\hat{V}}) + \tag{20}$$

$$\tilde{K}^T \Gamma_K^{-1} \hat{K} \leq \mathrm{sgn}(g(x))N(\zeta)\dot{\zeta} + \dot{\zeta} - k_0 s^2 + [-h(z,\hat{W},\hat{V}) +$$

$$h(z,W^*,V^*) + \varepsilon(z)]s - \hat{K}^T \phi(z,t)|s| +$$

$$\tilde{W}^T \Gamma_W^{-1} \dot{\hat{W}} + tr(\tilde{V}^T \Gamma_V^{-1} \dot{\hat{V}}) + \tilde{K}^T \Gamma_K^{-1} \dot{\hat{K}} \tag{20}$$

$$\leq -k_0 s^2 + \mathrm{sgn}(g(x))N(\zeta)\dot{\zeta} + \dot{\zeta} \tilde{W}^T [\Gamma_W^{-1}, \dot{\hat{W}} -$$

$$s(\hat{S} - \hat{S}'\hat{V}^T \bar{z})] + tr(\tilde{V}^T [\Gamma_V^{-1} \dot{\hat{V}} - s\bar{z}\hat{W}^T \hat{S}']) +$$

$$\tilde{K}^T [\Gamma_K^{-1} \dot{\hat{K}} - |s|\phi(z,t)] \leq -k_0 s^2 + \mathrm{sgn}(g(x))N(\zeta)\dot{\zeta} + \dot{\zeta},$$

Therefore we know that

$$V(t) \leq V(t) + \int_0^t k_0 s^2 \leq V(0) + \int_0^t [\mathrm{sgn}(g(x))N(\zeta) + 1]\dot{\zeta} dt, \forall t \in [0, t_f), \text{According}$$

to Lemma 2, we obtain that $V(t)$, $\zeta(t)$ and $\int_0^t [\mathrm{sgn}(g(x))N(\zeta) + 1]\dot{\zeta} d\tau$ are bounded on $[0, t_f)$. Similar to the discussion in [12], we know that the above conclusion is true for $t_f = \infty$. It is easy to show that $\int_0^\infty s^2 dt$ exists. From (18), we have that

$$\|\hat{W}(t)\| \in L_\infty, \|\hat{K}(t)\| \in L_\infty, \|\hat{V}(t)\|_F \in L_\infty.$$ According to assumption 1) and (18), we have that $s \in L_\infty$. Since a continuous function is always bounded on a compact set, using (3), we have that $\dot{s}(t)$ is bounded and $ds^2(t)/dt = 2s(t)\dot{s}(t)$ is also bounded. Therefore, $s^2(t)$ is uniformly continuous on $[0,\infty)$. According to Barbalat's lemma, it is easy see that $\lim_{t \to \infty}|s(t)|=0$. From (2), we obtain that

$$e_1(t) = \frac{1}{(s+\lambda)^{n-1}} s(t) \text{ with } \lambda > 0. \text{ This means that } \lim_{t \to \infty}|e_1(t)|=0.$$

5 Conclusions

Based on Nussbaum function property and multilayer neural networks, a new direct adaptive control scheme for a class of nonlinear systems with an unknown gain sign and nonlinear uncertainties has been presented in this paper. The adaptive law of the adjustable parameter vector and the matrix of weights in the neural networks and the gain of sliding mode control term are determined by using a Lyapunov method. The developed controller can guarantee the global stability of the resulting closed –loop system in the sense that all signals involved are uniformly bounded and the asymptotic convergence of the tracking error to zero. Since the direct adaptive control technology is used, the controller singularity is avoided in our proposed controller design.

References

1. L. X. Wang : Adaptive Fuzzy Systems and Control –Design and Stability Analysis, Prentice Hall, New Jersey, (1994)
2. R. M. Sanner, and J. J. E. Slotine: "Gaussian networks for direct adaptive control", IEEE Trans . on Neural Networks, Vol. 3, No. 6, 837-863, (1992)
3. S. C. Chen, and W. L Chen: "Adaptive radial basis function neural network control wit variable parameters", Int. J. Systems Science, Vol. 32, No.4, 413-424, (2001)
4. T. Zhang, S. S. Ge, and C. C. Hang: "Design and performance analysis of a direct adaptive control for nonlinear systems", Automatic a, Vol. 35, No. 11, 1809-1817,(1999)
5. T. Zhang, S. S. Ge, and C.C .Hang: " Adaptive neural network control for strict-feedback nonlinear systems using back stepping design", Automatic a, Vol.36, No. 10, pp. 1835-1846, (2000)
6. T.Zhang, S. S. Ge, and C. C. Hang: "Stable adaptive control for a class of nonlinear systems using a modified Lyapunov function", IEEE Trans. on Automatic Control, Vol.45, No. 1, pp.129-132,(2000)
7. S. S. Ge, C. C. Hang, and T Zhang: "A direct approach to adaptive controller design and its application to inverted pendulum tracking", Proceedings of the American Control Conference, Philadelphia,, 1043-1047, (1998)
8. T. P. Zhang, Y. Q. Yang, and H. Y. Zhang: "Direct adaptive sliding mode control with nonlinearly parameterized fuzzy approximates", Proceedings of 4th World Congress on Intelligent Control and Automation, Shanghai, Vol.3, 915-1919,(2002)
9. S. S. Ge, and J. Wang: "Robust adaptive neural control for a class of perturbed strict feedback nonlinear systems", IEEE Trans on Neural Network.., Vol. 13,No. 6, 1409-1419, (2002)
10. S. S. Ge, F. Hong and T. H. Lee: "Adaptive neural control of nonlinear time-delay systems with unknown virtual control coefficients", IEEE Trans on System, Man and Cybernetics-Part B, Vol. 31,No. 1, 499-516, (2004)
11. Weijin Jiang: Hybird Genetic algorithm research and its application in problem optimization. Proceedings of 2004 International Conference on Manchine Learning and Cybernetics, (2004)222-227
12. Weijin Jiang: Research on Extracting Medical Diagnosis Rules Based on Rough Sets Theory. Journal of computer science, 31(11) , (2004): 93-96
13. Weijin Jiang: Research on Optimize Prediction Model and Algorithm about Chaotic Time Series, Wuhan University Journal of Natural Sciences, 9(5) ,(2004): 735-740

Improving the Resultant Quality of Kohonen's Self Organizing Map Using Stiffness Factor

Emin Germen

Anadolu University Electrical & Electronics Engineering Department
egermen@anadolu.edu.tr

Abstract. The performance of Self Organizing Map (SOM) is always influenced by learn methods. The resultant quality of the topological formation of the SOM is also highly dependent onto the learning rate and the neighborhood function. In literature, there are plenty of studies to find a proper method to improve the quality of SOM. However, a new term "stiffness factor" has been proposed and was used in SOM training in this paper. The effect of the stiffness factor has also been tested with a real-world problem and got positive influence.

1 Introduction

Kohonen's Self-Organizing Map (SOM) is a neural network which projects the high dimensional input space to one or two-dimensional array in nonlinear fashion [1, 2]. The codebook vectors (neurons) connected in a lattice structure in a two dimensional plane which forms the resultant topology provide insights about possible relationships among the data items. This idea is inspired from the structure of the cortical map of the brain. Although various disciplines use the SOM model in order to find solutions to broad spectrum of problems, however, there is not so much clue about the how the resultant maps are supposed to look after training or what kind of learning parameters and a neighborhood functions have to be used according to the nature of data itself. In literature, there are plenty of studies to determine the optimum learning rate and neighborhood function [3,4,5,6].

Although a lot of effort has been made to analyze the organization of the topology of SOM, the delineation of data dependent learning rate and neighborhood function is a cumbersome task. The introduction of a hit term in order to improve the topological quality concerning data statistics for two dimensional topographical SOM has been defined by Germen [7] for rectangular lattice. Here in this paper, the same hit term is used to track the density localizations of data points in multi dimensional space, however much more adequate method the, "*stiffness factor*", has been proposed to use it in training. The main insight of this term is, decreasing the fluctuations of the neurons in lattice, if those get much hit ratio than the others. In Newtonian physic, the mass with higher density attracts the others to it. Similarly the "stiffness factor" simulates this phenomenon, and the statistical characteristic of data can be conserved in topology.

In this paper, I address the SOM algorithm and the proposal for the novel term "stiffness factor" and its usage with the learning rate and neighborhood function is given in section 2. The results of the proposal are examined in Section 3. In Section 4 there is a brief conclusion.

2 SOM Algorithm and the Stiffness Factor

In Kohnen's SOM, the learning is an iterative procedure defined as:

$$M_i(k) = M_i(k-1) + \alpha(k) \cdot \beta(c,i,k) \cdot \left(\Lambda(k) - M_i(k-1) \right) \quad (1)$$

Here $M_i(k)$ denotes the modified neuron and $\Lambda(k)$ shows the training data presented in the iteration step k. The subscript i is used to show the Neuron index in the planar lattice. The $\alpha(k)$ and $\beta(c,i,k)$ are used to denote the learning rate and the neighborhood function parameters around the Best Matching Unit (BMU) where the index value is c and found as :

$$c = \arg\min_i \| \Lambda(k) - M_i(k) \| \quad (2)$$

The learning rate usually gets value 1 at he beginning and diminishes gradually during the training phase in order to first find the global localizations and then do local adjustments. Similarly the size of the neighborhood function shrinks with the lapse of time and the training is done in two phases: first with large neighborhood radius, and then fine tuning with small radius.

The automatic formation of the topologically correct mapping is the direct consequence of the localization of the BMU and its direct influence onto the other neurons around the neighborhood of it determined by the neighborhood function. In learning process, if a number of excitations of a particular neuron is more than the others, it is possible to deduce that the weights of that neuron points out the much denser localization in the training data. If those particular neurons' weights are changed as much as the weights of neurons which didn't get so many hits, will cause the loss of this information. However in conventional approaches, this phenomenon doesn't be taken into the consideration.

Here in this paper, a new term has been proposed to effect the change of weights of a neuron according to its past number of hits. The hit term will be used to explain that the neuron is "on" (selected as BMU) at an instant. The main idea of stiffness factor is increasing the inertia of a neuron proportionally with the number of hits during training. Although it seems quite reasonable to count the hits per neuron and use the proportions of the hit rates between BMU and the updated neuron during training, it can easily be prove that, this technique causes twists and butterfly effects which have to be avoided [2]. In order to get rid of this problem, the planar movement of the updated neuron into the direction of BMU, should have to be less than the other neurons' movements which are located around the close vicinity of BMU. The motivation behind the stiffness factor is finding an updating scheme which takes BMU hits into consideration without affecting the twist-free organized topology. The method which is proposed in this work is first finding the closest neurons as "*impact neurons*" $M_n(M_i)$ for the updating neuron M_i in planar lattice "*in the direction*" of BMU. Here either one impact neuron or two impact neurons are selected according to the topological locations of the updated neuron $M_i(k)$ and BMU $M_c(k)$. The Fig. 1 explains this idea. In the figure, possible three different updated neurons M_i, M_j, M_k and their impact neurons $M_{n1}(M_i)$ for M_i, $M_{n1}(M_k)$ and $M_{n2}(M_k)$ for M_k and $M_{n1}(M_j)$ for M_j has been shown.

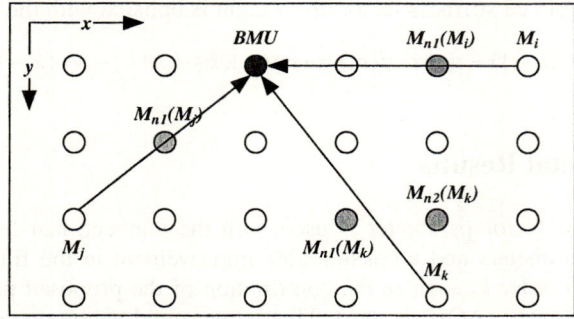

Fig. 1. Different impact neurons for different updated neurons

The Impact Neurons' $M_n(M_i)$ topological indexes are calculated as:

$$x_n(x_{M_i}(k)) = x_{M_i} + \text{sgn}(x_c - x_{M_i})$$
$$y_n(y_{M_i}(k)) = y_{M_i} + \text{sgn}(y_c - y_{M_i}) \quad if\left(\left|x_{M_i} - x_c\right| = \left|y_{M_i} - y_c\right|\right)$$

$$x_n(x_{M_i}(k)) = x_{M_i} + \text{sgn}(x_c - x_{M_i}), y_n = y_c \quad if\left((x_{M_i} \neq x_c) \wedge y_{M_i} = y_c\right)$$
$$y_n(y_{M_i}(k)) = y_{M_i} + \text{sgn}(y_c - y_{M_i}), x_n = x_c \quad if\left((y_{M_i} \neq y_c) \wedge x_{M_i} = x_c\right)$$

$$\begin{cases} x_{n1} = x_{M_i} \\ y_{n1} = y_{M_i} + \text{sgn}(y_c - y_{M_i}) \\ x_{n2} = x_{M_i} + \text{sgn}(x_c - x_{M_i}) \\ y_{n2} = y_{M_i} + \text{sgn}(y_c - y_{M_i}) \end{cases} if\left(\left|x_{M_i} - x_c\right| < \left|y_{M_i} - y_c\right|\right) \quad (3)$$

$$\begin{cases} x_{n1} = x_{M_i} + \text{sgn}(x_c - x_{M_i}) \\ y_{n1} = y_{M_i} \\ x_{n2} = x_{M_i} + \text{sgn}(x_c - x_{M_i}) \\ y_{n2} = y_{M_i} + \text{sgn}(y_c - y_{M_i}) \end{cases} if\left(\left|x_{M_i} - x_c\right| > \left|y_{M_i} - y_c\right|\right)$$

After finding the Impact Neurons, the average impact hit ratio has to be found. If there is only one neuron as an Impact Neuron, the impact hit can be calculated as: $Hit_{impact}(k) = h(M_n(k))$ otherwise $Hit_{impact}(k) = \left(h(M_{n1}(k)) + h(M_{n2}(k))\right)/2$ where $h(M_n)$ represents the number of hits of the neuron M_n.

By using the hit ratio of the Impact Neurons, the Stiffness Factor is defined as:

$$\sigma(c,i,k) = \frac{Hit_{impact}}{Hit_{impact} + h(M_i(k))} \qquad (4)$$

Using the calculated stiffness factor, the weight is updated with the formula:

$$M_i(k) = M_i(k-1) + \alpha(k) \cdot \beta(c,i,k) \cdot \sigma(c,i,k) \cdot \left(\Lambda(k) - M_i(k-1)\right) \qquad (5)$$

3 Experimental Results

The new stiffness factor parameter is used with the conventional learning rate and neighborhood parameters and a considerable improvement in the final topology has been obtained. In order to analyze the contribution of the proposed parameter, Average Quantization Error (AQE) is used. This is measured using average quantization error between data vectors and their corresponding BMU's.

In the experiment, two-dimensional 10x10 neuron map is trained with two dimensional data. The neurons are connected in a rectangular lattice. The training set consisted of 10,000 samples with a normal distribution of Mean = 0, and Standard Deviation = 5. The training set is randomly sampled 10,000 times. Fig. 2 shows the AQE comparisons of training the map when the stiffness factor is applied after training steps of 1000 and 5000 data. The resultant maps and the data have been shown in Fig. 3. Here it is observed that, the Stiffness Factor has considerable positive influence on the final maps.

The effect of the Stiffness Factor also has been tested with a real-world problem in order to classify the power-quality data. At the end of the experiments a considerable improvement on the classification borders of SOM has been observed.

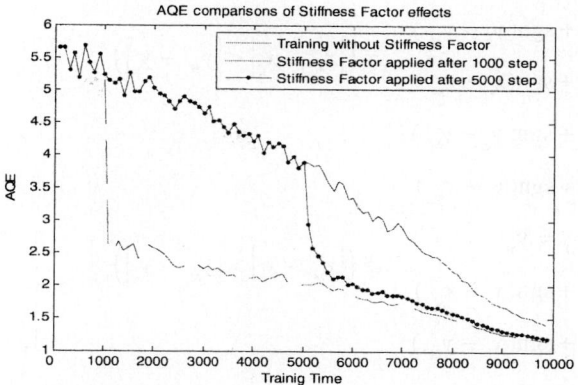

Fig. 2. Average Quantization Error comparisons of Stiffness Factor effects

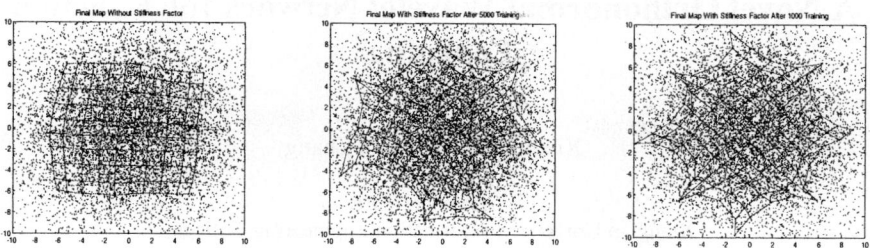

Fig. 3. Effects of Stiffness Factor on the final topologies

4 Conclusion

During the training period of SOM, the stability of the weights of a neuron has been increased directly in proportion with the number of getting hits. While updating a neuron, according to the relative positions of it and the BMU, the impact neurons are found. According to the average hits, the hit ratio $\sigma(i,c,k)$ parameter has been defined and used with different learning rate and neighborhood function parameters. It has been observed that this novel parameter has an improving effect for different kind of SOM parameters from the point of view of the quality of resultant topology.

Another asset of this hit ratio term is, it can easily be applied with conventional parameters, which are used in SOM training. This term enforces the power of self-organization idea and data dependent topological formation of the net.

References

1. Kohonen, T., The Self Organizing Map, Proc.of IEEE, 78, 9, (1990), 1464-1480
2. Kohonen, T., Self-Organizing Maps, Springer-Verlag, (1995)
3. Mulier, F., M., Cherkassky V., Statistical Analyses of Self-organization, Neural Networks, Vol. 8, No. 5, (1995) 717-727
4. Flanagan, J., A., Self-organization in Kohonen's SOM, Neural Networks, Vol. 9, No 7, (1996) 1185-1197
5. Germen, E., Bilgen, S., A Statistical Approach to Determine the Neighborhood Func-tion and Learning Rate in Self-Organizing Maps, Proc. ICONIP'97, Springer, (1997) 334-337.
6. Germen, E., Statistical Self-Organizing Map, Ph. D. Thesis, METU 1999
7. Germen, E., Increasing the Topological Quality of Kohonen's Self Organizing Map by Using a Hit Term, Proc. ICONIP'02, Singapore (2002)

A Novel Orthonormal Wavelet Network for Function Learning[*]

Xieping Gao and Jun Zhang

Member, IEEE,
Xiangtan University, Information Engineering College,
411105, Xiangtan, China
xpgao@xtu.edu.cn, zhangjun7907@hotmail.com

Abstract. This paper proposed a novel self-adaptive wavelet network model for Regression Analysis. The structure of this network is distinguished from those of the present models. It has four layers. This model not only can overcome the structural redundancy which the present wavelet network cannot do, but also can solve the complicated problems respectively. Thus, generalization performance has been greatly improved; moreover, rapid learning can be realized. Some experiments on regression analysis are presented for illustration. Compared with the existing results, the model reaches a hundredfold improvement in speed and its generalization performance has been greatly improved.

1 Introduction

Wavelet networks that has been proposed recently by Zhang, Benveniste, Pati and Krishnaprasad [1]~[3] are a class of neural networks consisting of wavelets. The wavelet network provides a unique and efficient representation of the signal. At the same time, it preserves most of the advantages of the RBF network. The wavelet neural network has shown its excellent performance in many fields and now it has been widely used [1][2][3][6][7]. According to the theory of Multiresolution, Baskshi B R and Stepphanopoulous proposed a novel orthonormal wavelet network model and corresponding learning algorithm [4]. In the network, the hidden layer replaces the sigmoid active function by wavelet function and Scaling function.

Since the present wavelet networks successfully preserve most of the advantages of the RBF network, few researches are focused on the structure of wavelet network. In fact, as to the whole signal, the orthogonal wavelet based network can be constructed and it is not redundant. However, as to some parts of signal, only some of neurons are useful, and the others are redundant. If the structure of the present wavelet network is changed properly, the various advantages of RBF network can be preserved and at the same time, the redundancy can be overcome effectively.

In this paper, a novel self-adaptively wavelet network and algorithm are proposed. Some experiments on Regression Analysis problems have been done to verify the

---

[*] This work was supported by the National Science Foundation of China (Grant No.60375021) and the Science Foundation of Hunan Province (Grant No.00JJY3096) and the Key Project of Hunan Provincial Education Department (Grant No.04A056).

performance (learning speed and generalization performance) of the model. Comparing the experimental results with the ones that are published in references [1] ~ [3] show that the model can reach better generalization performance and can reach a thousandfold improvement in speed.

2 A Novel Wavelet Network

Throughout this paper, let R, Z and N denote the set of real, all integers and natural numbers respectively. As everyone knows, the construction of wavelet is associated with multi-resolution analysis (MRA) developed by Mallat and Meyer.

Suppose function $\varphi(t) \in L^2(R)$ satisfied $\int_{-\infty}^{+\infty} \varphi(t)dt = 1$, $\varphi(x)$ can span multi-resolution analysis (MRA) of $L^2(R)$, which is a nest sequence of closed subspaces V_j in $L^2(R)$. $\{2^{m/2}\varphi(2^m t - n)\}_{n=-\infty}^{n=+\infty}$ or $\{\varphi_{m,n}(t)\}_{n=-\infty}^{n=+\infty}$ are the bases of V_m. $\varphi(t)$ known as the scaling function (the 'father wavelet"), specifically, there exists a function $\psi(t)$ (the "mother wavelet") and $\{\psi_{m,n}(t) = 2^{m/2}\psi(2^m t - n)\}_{n=-\infty}^{n=+\infty}$ which are the bases of space W_m. Space V_m is related to W_m by $V_{m+1} = V_m \oplus W_m$.

It induces a decomposition of $L^2(R)$

$$L^2(R) = V_J \bigoplus_{m \geq J} W_m \tag{1}$$

The above discussions suggest a scheme for decomposing a L^2 function $f(t)$, namely,

$$f(t) = \sum_n <f, \varphi(J,n)> \varphi_{J,n}(t) + \sum_{m>J,n} <f, \psi_{m,n}> \psi_{m,n}(t) \tag{2}$$

Without loss of the generality, for the analyzed signal $f(t)$, a following two-hidden layer wavelet network is set up, which has realized a $R^d \rightarrow R$ mapping. Its structure is as Fig.1.

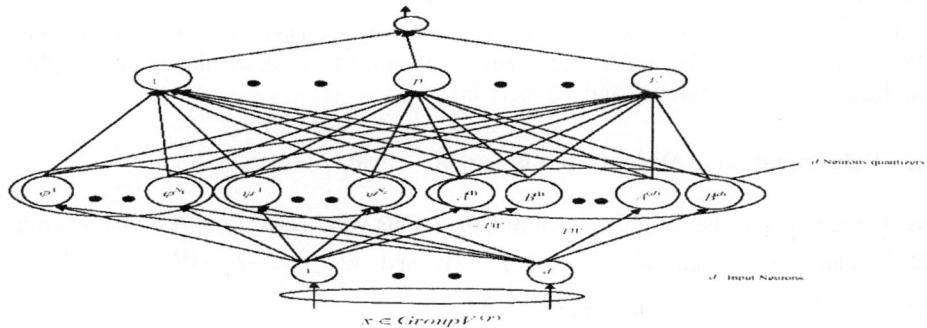

Fig. 1. Four layer wavelet network

As shown in Fig.1, the novel wavelet network has two hidden layers. The first-hidden layer consists of N_1 "φ neurons", N_2 "ψ neurons" and d neurons quantizers. d is the dimension of the input data. Each neuron quantizer consists of two sigmoidal neurons called $A-type$ neuron and $B-type$ neuron respectively. $A-type$ Neuron and $B-type$ neuron of $j-th$ quantizer are denoted as neuron $A^{(j)}$ and neuron $B^{(j)}$ respectively. The output of $j-th$ quantizer is denoted as $O_{A_p^{(j)}} + O_{B_p^{(j)}}$ $1 \leq j \leq d$. All of the neurons in the input layer are linked with all of the "φ neurons" and "ψ neurons". But the $i-th$ input neuron is just linked with the $i-th$ quantizer. $i = 1,...,d$. The second layer has L^d neurons (L is arbitrarily plus integral value). The neurons of the first hidden layer link to all of the neurons of the second hidden layer.

Definition 1: For a $d-$ dimension signal f, suppose $\Gamma_L \overset{\Delta}{=} \{f \mid f = f_1 \oplus f_2 \oplus ... \oplus f_{L^d}\}$, Let f_i denotes the i-th sub-signal of f that is divided continuously into L^d equidistant shares according to the support of each dimension.

As to the distinct samples (x_i, t_i), where $x_i = [x_{i1}, x_{i2},..., x_{id}]^T \in R^d$ and $t_i \in R$, without loss of the generality, suppose the support of signal is $[0, \alpha]^d$, a plus vector W can be chosen randomly. It can separate them into L^d groups according to the support of each dimension.

$$V_j^{(p)} = \{x_{ij} \mid W \cdot \frac{(p-1) \cdot \alpha}{L} \leq W \cdot x_{ij} \leq W \cdot \frac{p \cdot \alpha}{L}\} \quad 1 \leq p \leq L, \quad 1 \leq j \leq d$$

$$GroupV^{(q)} = \{x_i \mid W \cdot \frac{(p_j - 1)\alpha}{L} \leq W \cdot x_{ij} \leq W \cdot \frac{p_j \alpha}{L}, \quad 1 \leq q = \sum_{j=1}^{d} p_j \cdot L^{(j-1)} + 1 \leq L^d$$
$$1 \leq i \leq n, 1 \leq j \leq d, 1 \leq p_j \leq L\}$$

According to the above discuss, $GroupV^{(i)} \subseteq f_i$

The training of novel wavelet network mainly consists of two phases: (1) Determination of weights and biases of neural quantizers. (2) Determination of weights and bias of the "φ neurons" and "ψ neurons"

2.1 Determination of Weights and Biases of Neural Quantizers

As shown in Fig.1, the weights which link the inputs with the neuron $A^{(p)}$ and neuron $B^{(p)}$ can be chosen as $\overline{W}_{A^{(p)}} = T_p \cdot W$ and $\overline{W}_{B^{(p)}} = -T_p \cdot W$. The T_p, $1 \leq p \leq d$ can be set as following:

$Tset = \{T_p \mid T_p = \dfrac{\ln(\dfrac{1-\eta}{\eta})}{\min\limits_{\substack{1\le i\le n \\ 1\le p\le L}}(|W\cdot x_{i,j} - W\cdot \dfrac{p\cdot\alpha}{L}|)}, 1\le j\le d\}$. $\eta < 0.5$ is arbitrarily small positive value. When $T_j \ge T_j^A$, $1\le j\le d$, for $\forall x_i \in GroupV^{(q)}, q = 1,\ldots,L^d$, the corresponding input $Input_m(x_i)$, $m = 1,\ldots L^d$ of m th-neuron in the second-hidden layer satisfies

$$Input_m(x_i)\begin{cases} \le \eta, & \text{if } q = m \\ \ge 1-\eta, & \text{if } q \ne m \end{cases}$$

The biases $\overline{b}_{A^{(p)}}$ and $\overline{b}_{B^{(p)}}$ of neuron $A^{(p)}$ and neuron $B^{(p)}$ are simply analytically calculated as

$$\begin{cases} \overline{b}_{A_p^{(j)}} = -T(W\cdot \dfrac{p\cdot\alpha}{L}) & 1\le j\le d, 1\le p\le L \\ \overline{b}_{B_p^{(j)}} = T(W\cdot \dfrac{(p-1)\cdot\alpha}{L}) & 1\le j\le d, 1\le p\le L \end{cases}$$

For any input x_i within input vector group $GroupV^{(q)}$, $q = 1,\ldots,L^d$, only the q th-neuron's input are almost zero while one of the inputs of other neurons is almost one in the second-hidden layer.

2.2 Determination of Weights and Bias of the "φ neurons" and "ψ neurons"

According to the theory of the wavelet, the weight and bias of the "φ neurons" and "ψ neurons" can be determined as the reference [4].

3 Experimental Results

The scaling function $N^2(x)$ and wavelet $\psi^2(x)$ are selected as the activating function of the N_1 "φ neurons" and N_2 "ψ neurons" of the first hidden layer in the model of the paper respectively.

$$N^2(x) = \begin{cases} x, & [0,1] \\ 2-x, & [1,2] \\ 0, & others \end{cases}$$

$$\psi^2(x) = \begin{cases} \frac{x}{6}, & [0,\frac{1}{2}] \\ \frac{2}{3}-\frac{7}{6}x, & [\frac{1}{2},1] \\ \frac{3}{8}x-\frac{19}{6}, & [1,\frac{3}{2}] \\ \frac{29}{6}-\frac{8}{3}x, & [\frac{3}{2},2] \\ \frac{7}{6}-\frac{17}{6}x, & [2,\frac{5}{2}] \\ \frac{1}{2}-\frac{1}{6}x, & [\frac{5}{2},3] \\ 0 & others \end{cases}$$

Then the same non-liner functions in the reference [2][3] are chosen as the approximate functions. To assess the approximation results, a figure of merit is needed. We select the same figure of merit in the [2].

For the input datum $T_n = \{(x_i, t_i)\}_{i=1}^n$ and the network output $\hat{t}_{M,i}$.

$$error = \sqrt{\frac{\sum_{i=1}^{n}[\hat{t}_{M,i} - t_i]^2}{\sum_{i=1}^{n}(\bar{t} - t_i)^2}}, \quad \bar{t} = \frac{1}{n}\sum_{i}^{n} t_i$$

The computing environment as following: Intel P4 1.7G CPU, 256M RAM, and MATLAB 6.5. At first, the functions that are chosen by the reference [2][3] are chosen to do the experiments and compare the results with those that are shown in the reference [2][3].

Function 1:

$$y = \begin{cases} -43.72x + 8.996, & 0 \leq x < 0.4 \\ 84.92x - 42.46, & 0.4 \leq x \leq 0.5 \\ 10e^{x-1}\sin(12x^2 + 2x - 4) & 0.5 < x \leq 1 \end{cases}$$

Model	Hidden Neurons	Epochs	RMSE Of Testing	Time(s)
Zhang [2]	7	10000	0.05057	1100
Pati [3]	31	800	0.024	101.7
BP	7	10000	0.13286	1150
Our WN	41	1	0.0013	1.8600

Function 2:

$$z = 400(x^2 - y^2 - x + y)\sin(10x - 5) \quad x, y \in [0,1]$$

Model	Hidden Neurons	Epochs	RMSE Of Testing	Time(s)
Zhang [2]	49	40000	0.03395	21300
Pati [3]	187	500	0.023	500
BP	225	40000	0.29381	95640
Our WN	217	1	0.0085	3.2180

From the datum, both the generalization performance and the learning speed of the network in this paper are more ascendant than those of the previous wavelet network models.

4 Summery

In this paper, a novel model and a rapid algorithm of wavelet neural network are described. For the more rational and effective structure is adapted in the model, Comparing with the present wavelet network, this model not only has a hundredfold improvement in speed, but also obtains better generalization performance. For future work, to investigate the model in some real world large-scale applications are of great interest.

References

1. Q.Zhang., A.Benveniste.: Wavelet network [J]. IEEE Trans. Neural Networks, (1992) 3, 889-898.
2. J.Zhang et al.: Wavelet neural networks for function learning. [J]. IEEE Trans. Signal Process, (1995) 43, 1485-1497.
3. Y.C.Pati., P.S.Krishnaprasad.: Analysis and synthesis of feed-forward neural networks using discrete affine wavelet transformations. IEEE Trans. Neural Networks, (1993) 4, 73-85.
4. Baskshi B R., Stepphanopoulous.: Wavelet-net: A Multiresolution, Hierarchical Neural Network with Localization Learning [J]. American Institute Chemical, Engineering Journal, (1993) 39 57-81.
5. X.P.Gao.,B.Zhang.: Interval-wavelets neural networks(1)——theory and implements, Journal of software, (1998) 9, 217-221.
6. X.P.Gao., B.Zhang.: Interval-wavelets Neural Networks(□)——Properties and Experiment, Journal of software, (1998) 9, 246-250.
7. Barrett., R., M., Berry., et al: Templates for the Solution of Linear Systems: Building Blocks for Iterative Methods, SIAM, Philadelphia (1994).

Fuzzy Back-Propagation Network for PCB Sales Forecasting

Pei-Chann Chang, Yen-Wen Wang, and Chen-Hao Liu

Department of Industrial Engineering and Management, Yuan-Ze University,
135 Yuan-Dong Rd., Taoyuan 32026, Taiwan, R.O.C.
iepchang@saturn.yzu.edu.tw

Abstract. Reliable prediction of sales can improve the quality of business strategy. In this research, fuzzy logic and artificial neural network are integrated into the fuzzy back-propagation network (FBPN) for printed circuit board industry. The fuzzy back propagation network is constructed to incorporate production-control expert judgments in enhancing the model's performance. Parameters chosen as inputs to the FBPN are no longer considered as of equal importance, but some sales managers and production control experts are requested to express their opinions about the importance of each input parameter in predicting the sales with linguistic terms, which can be converted into pre-specified fuzzy numbers, aggregated and corresponding input parameters when fed into the FBPN. The proposed system is evaluated through the real life data provided by a printed circuit board company. Model evaluation results for research indicate that the Fuzzy back-propagation outperforms the other three different forecasting models in MAPE.

1 Introduction

Sales forecasting is a very general topic of research. When dealing with the problems of sales forecasting, many researchers have used hybrid artificial intelligent algorithms to forecast, and the most rewarding method is the application integrating artificial neural networks (ANNs) and fuzzy theory. This method is applied by incorporating the experience-based principal and logic-explanation capacity of fuzzy theory and the capacity of memory and error-allowance of ANNs, as well as self learning by numeral data.

This research focuses on the sales forecasting of printed circuit board (PCB) and modifies the fuzzy back-propagation network system (FBPN) proposed by Chen[2003], to select variables with a better and more systematic way from expert experience, with the purpose of improving the forecasting accuracy and using this information to help managers make decisions.

2 Literature Review

Although the traditional sales forecasting methods have been proved effective, they still have certain shortcomings. (Kuo, 2001, Tang, 2003, Luxhøj, 1996) As claimed by Kuo[1998], the new developed Artificial Intelligent (AI) models have more flexibilities and can be used to estimate the non-linear relationship, without the limits of traditional Time Series models. Therefore, more and more researchers tend to use AI forecasting models to deal with problem.

Fuzzy theory has been broadly applied in forecasting. (Chen, 1999, Hwang, 1998, Huarng, 2001) Fuzzy theory is first combined with ANNs by Lin and Lee[1991], who incorporated the traditional fuzzy controller and ANNs to a network structure to proceed appropriate non-linear planning of unplanned control systems based on the relationship of input and output through the learning capacity of ANNs. Following them, many researchers started doing different relative research based on the combination of fuzzy theory and ANNs. Fuzzy theory combining with ANNs is applied in different areas and has positive performance. (Xue, 1994, Dash, 1995, Chen, 2003, Kuo, 1998)

3 Methodology

There are three main stages in this research and the first stage is the variables selection stage. This stage is to select many possible variables, which may influence PCB product sales amount. In order to eliminate the unrelated variables, Stepwise Regression Analysis (SRA) and Fuzzy Delphi Method (FDM) were used to choose the key variables to be considered in the forecasting model. The second stage is the data preprocessing stage and Rescaled Range Analysis (R/S) was used to judge the effects of trend from serial observation values appearing as the time order. If the effect of trend is observed, Winter's method will be applied to remove the trend effect and reduce the forecast error. The third stage is the FBPN forecasting stage, which was developed to forecast the demand of PCB sales amount in this research and will be described in details in the following section. After being compared with other three forecasting models, the superior model will be recommended to the decision makers. The details of each stage will be described as follows:

3.1 Variable Selection Stage

In this stage, fewer factors were considered in order to increase the efficiency of network learning. Many researchers have used several methods to select key factors in their forecast system. (Chang, 2000, Kaufmann, 1988, Lin, 2003 and Hsu, 2003) In this research, the following two methods were used to determine the main factors that would influence the PCB sales amount.

1. SRA (Stepwise Regression Analysis)

Stepwise regression procedure determines the set of independent variables that most closely determine the dependent variable. This is accomplished by the repetition of a variable selection. At each of these steps, a single variable is either entered or removed from the model. For each step, simple regression is performed using the previously included independent variables and one of the excluded variables. Each of these regressions is subjected to an 'F-test'. If the variable small F value, is greater than a user defined threshold (0.05), it is added to the model. When the variable large F value, is smaller than a user defined threshold (0.1), it is removed from the model. This general procedure is easily applied to polynomials by using powers of the independent variable as pseudo-independent variables. The statistical software SPSS for Windows 10.0 was used for stepwise regression analysis in this research.

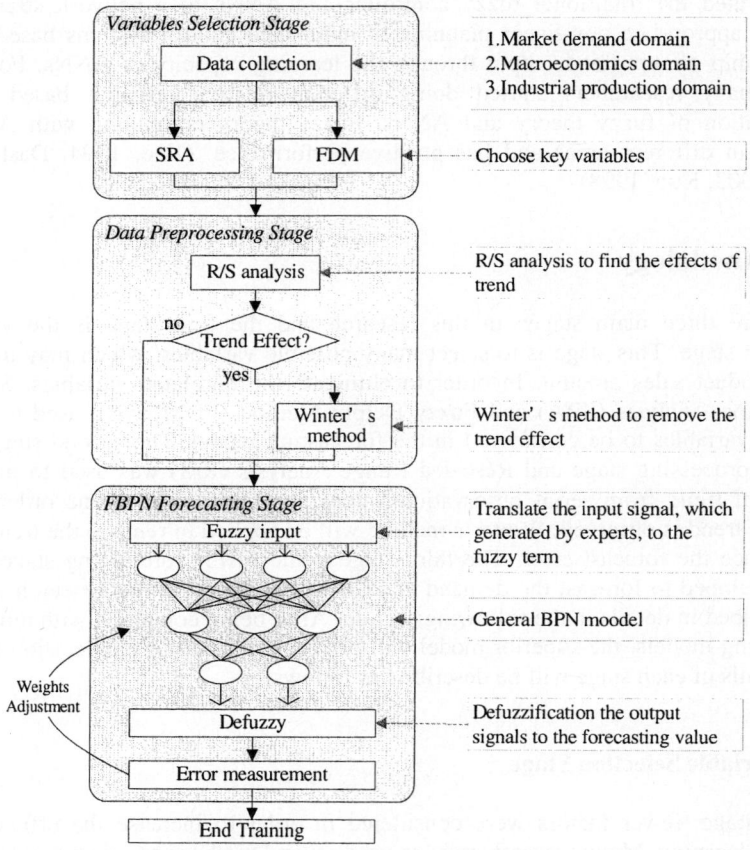

Fig. 1. Architecture of Three Main Stages in the Research

2. FDM (Fuzzy Delphi Method)

The modified procedures of the proposed FDM for the search are listed as follows:

Step 1:
    Collect all the possible factors that may affect the PCB product sales quantity. The domain experts select the important factors and give each a fuzzy number. This is the first questionnaire survey.

Step 2:
    Formulate the questionnaire, which is a set of IF-THEN rules.

Step 3:
Fuzzify the questionnaires that are returned by the domain experts and determine the following indices:

(1). Pessimistic (Minimum) index

$$\ell_A = \frac{\ell_{A1} + \ell_{A2} + \cdots + \ell_{An}}{n} \quad (1)$$

where ℓ_{Ai} means the pessimistic index of the $i-th$ expert and n is the number of the experts.

(2). Optimistic (Maximum) index

$$u_A = \frac{u_{A1} + u_{A2} + \cdots + u_{An}}{n} \quad (2)$$

where u_{Ai} means the pessimistic index of the $i-th$ expert.

(3). Average (Most appropriate) index

For each interval $\ell_{Ai} + u_{Ai}$, calculate the midpoint, $m_{Ai} = (\ell_{Ai} + u_{Ai})/2$, then find $\mu_A = (m_{A1} \times m_{A2} \times \cdots \times m_{An})^{1/n}$.

Step 4:
Therefore, the fuzzy number $A = (\mu, \sigma^R, \sigma^L)$, which represents the mean, right width, and left width, respectively, for an asymmetric bell shaped function that can be determined through the above indices:

$$\sigma^R = \frac{\ell_A - \mu_A}{3} \quad (3)$$

$$\sigma^L = \frac{u_A - \mu_A}{3} \quad (4)$$

Step 5:
Formulate the next questionnaire with the above indices and conduct the survey.

Step 6:
Repeat 3 to 5. Use the following formulas as the stopping criteria to confirm that all experts have the consentaneous importance of each factor.

$$\delta(\overline{A}, \overline{B}) = \int_{\alpha=0}^{1} \delta(\overline{A}[\alpha], \overline{B}[\alpha]) d\alpha \quad (5)$$

$$= \frac{1}{2}(\beta_2 - \beta_1) \int_{\alpha=0}^{1} (|\overline{A}[\alpha]^L - \overline{B}[\alpha]^L|) + (|\overline{A}[\alpha]^U - \overline{B}[\alpha]^u|) d\alpha$$

where $\overline{A}$ and $\overline{B}$ are the fuzzy numbers, $\overline{A}[\ ]$ and $\overline{B}[\ ]$ denote the membership function of fuzzy numbers. The α-cut of the fuzzy number is defined as $\overline{A}[\alpha] = \{x | \overline{A}[x] \geq \alpha, x \in R\}$ for $0 < \alpha \leq 1$. The distance between the two fuzzy numbers is $\delta(\overline{A},\overline{B})$. β_1 and β_2 are any given convenient values in order to surround both $\overline{A}[\alpha] = 0$ and $\overline{B}[\alpha] = 0$.

3.2 Data Preprocessing Stage

When the seasonal and trend variation is present in the time serious data, the accuracy of forecasting will be influenced. This stage will use R/S analysis to detect if there is this kind of effects of serious data. If the effects are observed, Winter's exponential smoothing will be used to take the effects of seasonality and trend into consideration.

1. R/S analysis (Rescaled Range Analysis)

For eliminating possible trend influence, the rescaled range analysis, invented by Hurst (1965), is used to study records in time or a series of observations in different time. Hurst spent his lifetime studying the Nile and the problems related to water storage. The problem is to determine the design of an ideal reservoir on the basis of the given record of observed discharges from the lake. The detail process of R/S analysis will be omitted here.

2. Winters Exponential Smoothing

In order to take the effects of seasonality and trend into consideration, Winter's exponential smoothing is used to preliminarily forecast the quantity of PCB production. According to this method, three components to the model are assumed: a permanent component, a trend, and a seasonal component. Each component is continuously updated using a smoothing constant applied to the most recent observation and the last estimate. Luxh[1996] and Mills[1990] compared Winter's Exponential Smoothing with other forecasting methods, like ARIMA, and all showed that the Winter's method had a superior performance. In this research we assume $\alpha = 0.1$, $\beta = 0.1$ and $\gamma = 0.9$.

3.3 Fuzzy Neural Network Forecasting Stage

There are three main layers, input layer, hidden layer and output layer, and two training stages in our FBPN. In the feedforward stage, FBPN use the data on hand to forecast the PCB sales amount, and the forecasting error will be recalled to adjust the weights between layers in the backprooagation of error stage. The details will be described in the following:

Step0. Initial weights between layers are randomly generated.
Step1. While stopping condition is false, do step 2-11.
Step2. For each training pair, do step 3-8.
Feedforward stage:
Step3. Each input unit I_j, which was generated by many experts, receives input signal $\tilde{s}_{(i)} x_{(i)}$ and broadcasts this signal to all units in the hidden layer.

Where $\tilde{s}_{(i)}$ is the fuzzy membership function, which is supported by the experts, and $x_{(i)}$ is the normalized input signal.

Step4. Sum the weighted input signals of each hidden unit.
Step5. Apply the translation function to compute its output signals.
Step6. Sum the weighted input signals of each output unit.
Step7. Apply the translation function to compute its output signals.
Step8. Defuzzify the output signals to the forecasting value, and compute its MAPE.

Backpropagation of error stage:
Step9. Compare the forecasted output with the actual sales amount and compute the error term between hidden layer and output layer. Next, calculate its weight correction term, (used to update connection weights later). Finally, calculate its bias correction term, and update weights and biases.
Step10. Compute its error information term for hidden nodes. Then, update the information term of each hidden node.
Step11. Calculate its weight correction term between hidden layer and input layer. Then, calculate its bias correction term. Finally, update weights and biases.

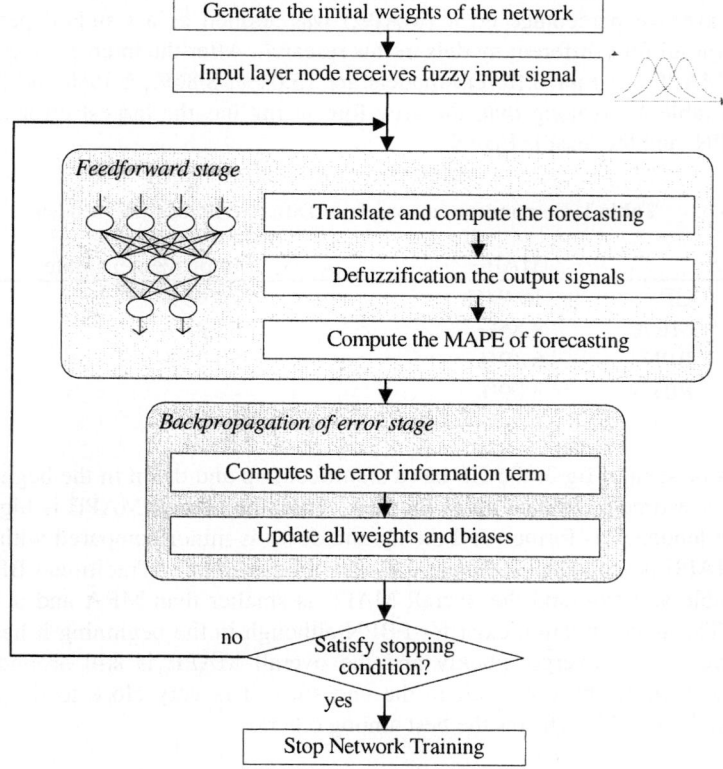

Fig. 2. The detailed flow diagram of FBPN

The configuration of the FBPN is established as follows:

- number of neurons in the input layer: 5
- number of neurons in the output layer: 1
- single hidden layer
- number of neurons in the hidden layer: 5
- network-learning rule: delta rule
- transformation function: sigmoid function
- learning rate: 0.1
- learning times: 30000

4 Experimental Results

The data in this research are from an electronic company in Taiwan from 1999/1 to 2003/12. Monthly sales amount is considered as an objective of the forecasting model. This research develops a FBPN for sales forecasting in PCB industries and we will compare this method with other traditional methods such as Grey Forecasting (GF), Multiple Regression Analysis (MRA) and Back-propagation network (BPN), etc.

Mean average percentage error (MAPE) was applied as a standard performance measure for all four different models in this research. After the intensive experimental test, the MAPEs of four different models are 15.04%, 8.86%, 6.19% and 3.09% (as shown in table 1). Among that, the grey forecasting has the largest errors, and then MRA, BPN, and the least is FBPN.

Table 1. Comparisons among Four Different Forecasting Models

	MAPE	Improvement Rate
GF	15.04%	74.95%
MRA	8.86%	65.21%
BPN	6.19%	50.08%
FBPN	3.09%	-

As can be seen in fig 3, the GF has a significant up and down in the beginning and it also over estimate the data up to the end. Thus the overall MAPE is high. As for MRA, the tendency is formed and the up and down is minor compared with GF. The overall MAPE is around 0~20% and it is also a little higher. Traditional BPN model is in a stable situation and the overall MAPE is smaller than MRA and it is around 0~10%. The same situation exist for FBPN although in the beginning it has a larger error however it converge quickly and the overall MAPE is still around 0~10%. Especially, it performs very well in the end since it is very close to the real data. Therefore, the FBPN performs the best among others.

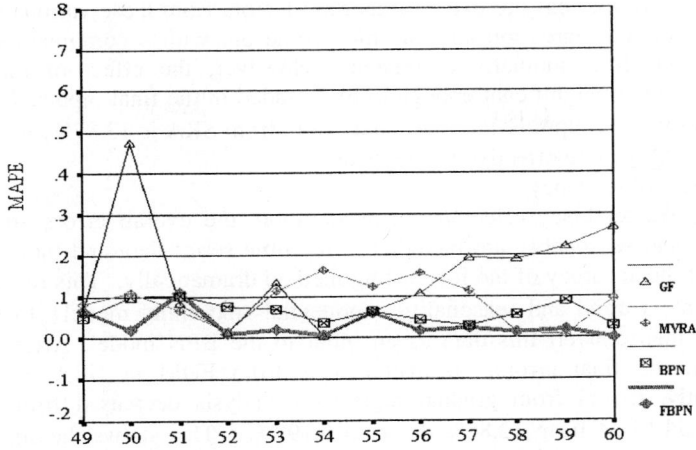

Fig. 3. The MAPE Values for Four Different Forecasting Models

According to the various criteria, i.e., encompassing test, MAPE, and forecasting accuracy, the best model among these four different models is FBPN with a MAPE of 3.09% and accuracy of 97.61%. Therefore, we can claim that by combining the fuzzy theory and BPN the hybrid model can be applied in forecasting the sales of PCB industry and the result is very convincing and deserve further investigation in the future for applications in other areas.

Although, the GF and MRA is very powerful when the data is very scarce and they claim that with only four data points and they can be applied to forecast the future result. However, after intensive experimental test, these two methods did not perform very well especially for those non-linear and highly dynamic data. As for the fuzzy Delphi back-propagation model since it can include the opinion from various experts in PCB sales and production department. It seems the assignment of different weight to the factor really improve the forecasting errors and perform much better than other models.

5 Conclusions

The experimental results in section 4 demonstrated the effectiveness of the FBPN that is superior to other traditional approaches. The FBPN approach also provides another informing tool to the decision maker in PCB industries. In summary, this research has the following important contribution in the sales forecasting area and these contributions might be interested to other academic researchers and industrial practitioners:

1. Feature Selections:
To filter out significant factors from a series of input variables, the FDM is superior to the SRA method. FDM will collect the opinion from various experts and assign different weights to these variables according to their experiences in this field.

Therefore, it is very easy to extract important factors from these various variables. In contrary, gradual regression analysis may come out with a combination of various variables which is mutually correlated. However, the effect of these selected variables may not significant enough to be included in the final inputs. The errors for input from fuzzy Delphi is 12.88%, and errors from SRA is 13.87%. It is obvious to see that FDM is more effective for applications.

2. The effect of tendency:

When take tendency effect into consideration, the overall errors are decreased. Tendency and seasonality are included in the time series data and these two factors will affect the accuracy of the forecasting method dramatically. This research applies the 「Winters trend and seasonality exponential smoothing model」 to forecast the sales and then convert this data as an input to the BPN model. After the training procedure, the final errors, no matter it is from FDM or SRA, are decreased significantly. Errors from gradual regression analysis decreased from 13.84% to 7.15%, and FDM from 12.88% down to 6.19%%. This shows the significance of including 「Winters trend and seasonality exponential smoothing model」 in the model.

3. Comparisons of different forecasting models:

This research applies three different performance measures, i.e., encompassing test, forecasting errors and accuracy of forecasting to compare the FBPN with other three methods, i.e., GF, MRA and BPN. The intensive experimental results show the following: 1. In encompassing test, FBPN and BPN models are superior to GF and MRA. 2. As for MAPE, FBPN has the smallest MAPE and it is only 3.09%. Therefore, FBPN model by combining FDM and BPN model is a very powerful and effective forecasting tool that can be further applied in other field of applications since expert's opinion can be incorporated into the model.

References

1. Chang, P. T., Huang, L. C., Lin, H. J.: The Fuzzy Delphi Method via fuzzy statistics and membership function fitting and an application to the human resources, Fuzzy Sets and Systems, Vol. 112 (2000) 511-520.
2. Chen, T.: A Fuzzy Back Propagation Network for Output Time Prediction in a Wafer Fab, Applied Soft Computing Journal (2003) 211-222.
3. Chen, T., Wang, M. J. J.: Forecasting Methods using Fuzzy Concepts, Fuzzy Sets and Systems, Vol. 105 (1999) 339-352.
4. Dash, P. K., Liew, A. C., Rahman, S.: Peak load forecasting using a fuzzy neural network, Electric Power Systems Research, Vol. 32 (1995) 19-23.
5. Hsu, C. C., Chen, C. Y.: Applications of improved grey prediction model for power demand forecasting, Energy Conversion and Management, Vol. 44 (2003) 2241-2249.
6. Huarng, K.: Heuristic models of fuzzy time series for forecasting, Fuzzy Sets and Systems, Vol. 123 (2001) 369-386.
7. Hwang, J. R., Chen, S. M., Lee, C. H.: Handling forecasting problems using fuzzy time series, Fuzzy Sets and Systems, Vol. 100 (1998) 217-228.
8. Kuo, R. J.: A Sales Forecasting System Based on Fuzzy Neural Network with Initial Weights Generated by Genetic Algorithm, European Journal of Operational Research, Vol. 129 (2001) 496-517.

9. Kuo, R. J., Xue, K. C.: A decision support system for sales forecasting through fuzzy neural networks with asymmetric fuzzy weights, Decisions Support Systems, Vol. 24 (1998) 105-126.
10. Lin, C. T., Lee, C. S. G.: Neural-Network-Based Fuzzy Inference Systems, IEEE Trans. On Computer, Vol. 40, No. 12 (1991) 1320-1336.
11. Lin, C. T., Yang, S. Y.: Forecast of the output value of Taiwan's opto-electronics industry using the Grey forecasting model, Technological Forecasting & Social Change, Vol. 70 (2003) 177-186.
12. Luxh, J. T., Riis, J. O., Stensballe, B.: A hybrid econometric-neural network modeling approach for sales forecasting, The International Journal of Production Economics, Vol. 43 (1996) 175-192.
13. Mills, T. C.: Time series techniques for economists, Cambridge University Press (1990).
14. Tang, J. W.: Application of neural network in cause and effect model of time series data, Chung-Huwa University, Civil Engineering, Unpublished master thesis, Taiwan (2003)
15. Xue, K. Q.: An Intelligent Sales Forecasting System through Artificial Neural Networks and Fuzzy Neural Network, I-Shou University, Department of Management, Unpublished master thesis, Taiwan (1994).

An Evolutionary Artificial Neural Networks Approach for BF Hot Metal Silicon Content Prediction

Zhao Min, Liu Xiang-guan, and Luo Shi-hua

Institute of System Optimum Technique, Zhejiang University,
Hangzhou, 310027, China
Zhaomin_04@yahoo.com.cn

Abstract. This paper presents an evolutionary artificial neural network (EANN) to the prediction of the BF hot metal silicon content. The pareto differential evolution (PDE) algorithm is used to optimize the connection weights and the network's architecture (number of hidden nodes) simultaneously to improve the prediction precision. The application results show that the prediction of hot metal silicon content is successful. Data, used in this paper, were collected from No.1 BF at Laiwu Iron and Steel Group Co..

1 Introduction

In blast furnace (BF) ironmaking process, hot metal silicon content is important both for quality and control purposes [1]. Not only is silicon content an significant quality variable, but also reflects the internal state of the high-temperature lower region of the blast furnace, so its accurate prediction can greatly help to control the thermal state of a BF, which is one of the significant factor ensuing the BF stable operation.

The multi-layer neural network is emerging as an important tool to predict the silicon content of hot metal [2,3], while BP algorithm suffers the disadvantage of being easily trapped in a local minimum and another problem with BP is the choice of a correct architecture. Evolutionary approach is used over traditional learning algorithms to optimize the architecture of neural networks. However, most of the research undertaken in the EANN literatures does not emphasize the trade-off between the architecture and the generalization ability of the network. With the trade-off, the EANN problem is actually a Multi-objective Optimization Problem. The PDE algorithm [4] was designed for vector optimization problems. So the PDE algorithm will be used to evolve the weights and the networks architecture simultaneously here.

2 An Artificial Neural Network Based on the PDE Algorithm

A three-layer feed forward neural network is selected in this paper. Now we have a multi-objective problem with two objectives; one is to minimize the error and the other is to minimize the number of hidden units. Our chromosome is a class that contains one matrix Ω and one vector P. The matrix Ω is of dimension $(I+O) \times (H+O)$, where I, H and O are the number of input, hidden and output units respectively. Each element $\omega_{ij} \in \Omega$ is the weight connecting unit i and j, where $i=0,...,(I-1)$ is the input unit i, $i=I,...,(I+O-1)$ is the output unit I-I, $j=0,...,(H-1)$ is the hidden unit j, $j=H,...,(H+O-1)$ is the output unit j-H. The vector ρ is of dimension H, where P_h is a binary value used to indicate if hidden unit h exists in the network or not. Then we can apply PDE to our neural network as follows:

Step1: Create a random initial population. The elements of the weight matrix Ω are assigned random values according to a Gaussian distribution $N(0, 1)$. The elements of the binary vector P are assigned the value with probability 0.5 based on a randomly generated number according to a uniform distribution between (0, 1); otherwise 0.

Step2: Evaluate the individuals in the population and label those who are non-dominated. If the number of non-dominated individuals is less than 3 repeat the following until the number of non-dominated individuals is greater than or equal to 3:

Find a non-dominated solution among those who are not labeled. Label this solution as a non-dominated.

Step3: Delete all dominated solutions from the population.

Step4: Select at random an individual as the main parent a_1, and two individuals, a_2, a_3, as supporting parents.

Step5: Crossover: With some probability, do

$$\omega_{ih}^{child} \leftarrow \omega_{ih}^{\alpha_1} + N(0,1)(\omega_{ih}^{\alpha_2} - \omega_{ih}^{\alpha_3}) , \qquad (1)$$

$$\rho_h^{child} = \begin{cases} 1 & if \ |\rho_h^{\alpha_1} + N(0,1)(\rho_h^{\alpha_2} - \rho_h^{\alpha_3})| \geq 0.5 \\ 0 & otherwise \end{cases} . \qquad (2)$$

Otherwise

$$\omega_{ih}^{child} = \omega_{ih}^{a_1}, \ \rho_h^{child} = \rho_h^{a_1} . \qquad (3)$$

And with some probability, do

$$\omega_{ho}^{child} = \omega_{ho}^{a_1} + N(0,1)(\omega_{ho}^{a_2} - \omega_{ho}^{a_3}) . \qquad (4)$$

Otherwise

$$\omega_{ho}^{child} = \omega_{ho}^{a_1} . \qquad (5)$$

Each weight in the main parent is perturbed by adding to it a ratio, F $N(0,1)$, of the difference between the two values of this variable in the two supporting parents.

Step6: Mutation: with some probability Uniform (0, 1), do

$$\omega_{ih}^{child} = \omega_{ih}^{child} + N(0, mutation\_rate) , \qquad (6)$$

$$\omega_{ho}^{child} = \omega_{ho}^{child} + N(0, mutation\_rate) ,\qquad(7)$$

$$\rho_h^{child} = \begin{cases} 1 & if\ \rho_h^{child} = 0 \\ 0 & otherwise \end{cases}.\qquad(8)$$

Step7: Apply BP to the child.
Step8: If the child dominates the main parent, place it into the population.
Step9: If the population size is not completed, repeat step 4-8.
Step10: If the termination conditions are not satisfied, repeat step 2-9.

In the following, we will outline the performance of this method on predicting silicon content in hot metal.

3 Practical Applications to Hot Metal Silicon Content Prediction

In this section, firstly, we select six key variables (see Table1) affecting the hot metal silicon content [Si] as the input nodes of our neural network.

Table 1. Input variables

VC(t/h)	PI(m^3/min.kPa)	PC(t/h)	BT($^\circ$C)	[Si]$_{n-1}$(%)	BQ(m^3/min)
Charging mixture velocity	Permeability index	Pulveized coal injection	Blast temperature	Last [Si]	Blast quantity

Secondly, two important criterions used in practice are considered here to evaluate our method: the hit ratio J:

$$J = \frac{1}{N_p}(\sum_{j=1}^{N_p} H_j) \times 100\% ,\ H_j = \begin{cases} 1 & if\ |x'_j - x_j| \le 0.1 \\ 0 & otherwise. \end{cases}.\qquad(9)$$

N_P is the total predicted tap numbers; another criterion which indicate consistency of the method $Perr$:

$$Perr = \sum_{j=1}^{N_p}(x'_j - x_j)^2 / \sum_{j=1}^{N_p} x_j^2 ,\qquad(10)$$

where x_j is the predicted value and x_j the observed value. According to the proposed method, we varied the crossover probability from 0 to 1 with an increment of 0.1. Mutation probability is varied from 0 to1 with an increment of 0.05. The maximum number of hidden units is set to 12, the population size 20, the learning rate for BP 0.03. A total of 1000 patterns were used in optimizing our model. The optimal ANN obtained after 100 generations evolution was tested through another 50 sets of data. When crossover probability is 0.8 and mutation probability is 0.1, we got the optimum solution. The results are shown in Fig. 1. The hit ratio J is calculated to be 88%

and *P*err is in the magnitude of 10^{-2} (0.0286), which is helpful for operator to make right decision to operate blast furnace.

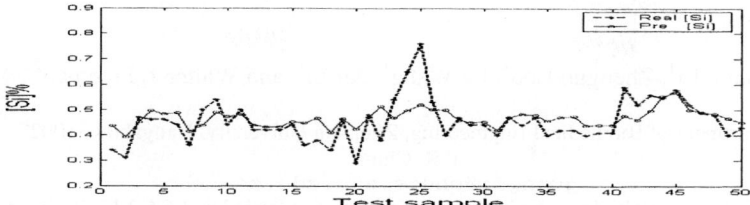

Fig. 1. The silicon content comparison between predicted and real data

Lastly, according to a conventional evolutionary approach [2,3,5], we will need to run the algorithm (e.g. BP algorithm) a number of times with different weights while varying the number of hidden units to select the optimum neural network. This is not an efficient way to solve the problem. The proposed method evolves the weights and the networks architecture simultaneously. Therefore, in terms of the amount of computations, it is much faster than the traditional methods which run for a fixed architecture and could be effectively used for online control of hot metal silicon content.

4 Conclusions

In this paper, we introduced an evolutionary multi-objective approach to artificial neural networks. It not only preserves the advantages of genetic algorithm, but also overcomes some disadvantages of previous approaches by considering the trade-off between the architecture and the generalization ability of the network. However, more work is needed in evaluating the performance of the proposed method and extend the selection of input variables can enhance the quality of prediction further.

References

1. Biswas, A.K.: Principles of Blast Furnace Ironmaking [M], SBA Publication, Calcutta (1984)
2. Singh, H., Sridhar, N.V., Deo, B.: Artificial neural nets for prediction of silicon content of blast furnace hot metal. Steel Research, vol. 521–527 (1996)
3. Juan, J., Javier, M., Jesus, S.A.: Blast furnace hot metal temperature prediction through neural networks-based models. ISIJ International 44 (2004) 573–580
4. Abbass, H.A., Sarker, R., and Newton, C.: A pareto differential evolution approach to vector optimization problems. IEEE Congress on Evolutionary Computation, Seoul, Korea (2001) 971–978
5. Yao, X.: Evolving Artificial Neural Networks. Proceedings of the IEEE (1999) 1423–1447

Application of Chaotic Neural Model Based on Olfactory System on Pattern Recognitions

Guang Li[1], Zhenguo Lou[1], Le Wang[1], Xu Li[1], and Walter J. Freeman[2]

[1] Department of Biomedical Engineering, Zhejiang University, Hangzhou 310027, P.R. China
guangli@cbeis.zju.edu.cn
[2] Division of Neurobiology, University of California at Berkeley, LSA 142, Berkeley, CA, 94720-3200, USA

Abstract. This paper presents a simulation of a biological olfactory neural system with a KIII set, which is a high-dimensional chaotic neural network. The KIII set differs from conventional artificial neural networks by use of chaotic attractors for memory locations that are accessed by, chaotic trajectories. It was designed to simulate the patterns of action potentials and EEG waveforms observed in electrophysioloical experiments, and has proved its utility as a model for biological intelligence in pattern classification. An application on recognition of handwritten numerals is presented here, in which the classification performance of the KIII network under different noise levels was investigated.

1 Introduction

In recent years, the theory of chaos has been used to understand the mesoscopic neural dynamics, which is at the level of self-organization at which neural populations can create novel activity patterns [1]. According to the architecture of the olfactory neural system, to simulate the output waveforms observed in biological experiments with EEG and unit recording, the KIII model, which is a high dimensional chaotic network, in which the interactions of globally connected nodes lead to a global landscape of high-dimensional chaotic attractors, was built.

In this paper we present two application examples of the KIII network for recognitions of image patterns and handwriting numerals [2].

2 Chaotic Neural Model Based on Olfactory System

The central olfactory neural system is composed of olfactory bulb (OB), anterior nucleus (AON) and prepyriform cortex (PC). In accordance with the anatomic architecture, KIII network is a multi-layer neural network model, which is composed of heirarchichal K0, KI and KII units. Fig.1 shows the topology of KIII model, in which M, G represent mitral cells and granule cells in olfactory bulb. E, I, A, B represent excitatory and inhibitory cells in anterior nucleus and prepyriform cortex respectively.

3 Application on Image Pattern and Handwriting Numeral Recognitions

Pattern recognition is an important subject of artificial intelligence, also a primary field for the application of Artificial Neural Network (ANN). KIII network is a more accurate simulation of the biological neural network than conventional ANN.

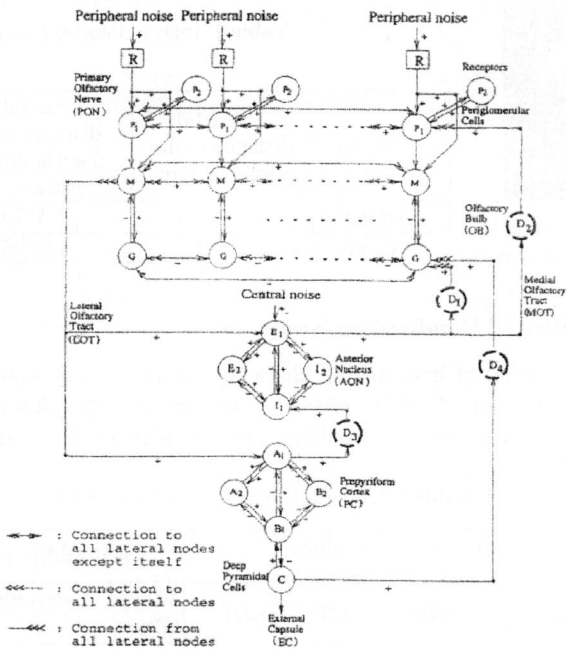

Fig. 1. Topology of the KIII network (Adapted from Chang & Freeman [3].)

Derived from the study of olfactory system, the distributed KIII-set is a high dimensional chaotic network, in which the interactions of globally connected nodes lead to a global landscape of high-dimensional chaotic attractors. After reinforcement learning to discriminate classes of different patterns, the system forms a landscape of low-dimensional local basins, with one basin for each pattern class [4]. The output of the system is controlled by the attractor, which signifies the class to which the stimulus belonged [5].

3.1 Classification of Image Patterns

In this article, we used the KIII model to classify image patterns. The parameters involved in our simulation in this paper were taken from the document [3].

First, the KIII model learned the desired patterns --- the 8*8 binary bitmap image of circle and isosceles triangle. Both patterns were learned for three times in turn. Second, the novel input images need to be preprocessed before classification: image segmentation, image zooming, edge detection, etc. Finally, we input the preprocessed patterns in the R layer of the KIII model and simulate its output, as well as calculate the categories of the input patterns. Only if the difference between the Euclid distances from the novel input pattern to the two kinds of stored patterns reaches the pre-defined threshold, the classification can be viewed as valid and persuasive.

Taking Fig. 2 as an example, Table 1 contains the final result of classification.

Fig. 2. Example image patterns to be classified

Table 1. Image classification result

object	Euclid distance to the circle pattern	Euclid distance to the triangle pattern	Central point of the object
Triangle	0.3559	7.3113	[152,318]
Circle	6.6196	1.9795	[322,111]

3.3 Classification of Handwriting Numerals

Automatic recognition of handwriting characters is a practical problem in the field of pattern recognition, and was here selected to test the classification performance of the KIII network. The test data set contains 200 samples in 20 groups of handwritten

Table 2. Classification Result – Using KIII

Pattern	Correct	Incorrect	Failure	Reliability (%)			
	KIII	KIII	KIII	KIII	Linear filter	Perceptron	Hopfield
0	196	3	1	98.49	74.50	100	59.79
1	185	10	5	94.87	55.85	89.5	78.89
2	192	4	4	97.96	71.0	53.68	78.42
3	177	12	11	93.65	35.5	67.37	79.87
4	179	11	10	94.21	39.44	44.13	41.99
5	181	7	12	96.28	48.73	49.36	21.17
6	191	1	8	99.48	83.5	69.95	89.23
7	189	7	4	96.43	58.59	51.59	64.0
8	174	9	17	95.08	76.53	46.88	87.93
9	186	9	5	95.38	64.06	63.5	64.29
Total	1850	73	77	96.20	60.99	64.84	66.76
Rate (%)	92.5	3.65	3.85	96.20			

numeric characters written by 20 different students. One group included 10 characters from zero to nine. In this application, a 64-channel KIII network was used with system parameters as reference [3]. Every character in the test data was preprocessed to get the 1x64 feature vector and to place a point in a 64-dimensional feature space. Thus the 64 features are given as input to the KIII network as a stimulus pattern in the form of a 1x64 feature vector.

As can be seen in the Table 2, while a high overall reliability of 96.20% was gained using KIII, the reliability of the linear filter, the perceptron and the Hopfield network was merely around 60%. Obviously, the KIII model shows its excellence in practical pattern classification.

4 Discussion

Derived directly from the biological neural system, KIII network gives a more complicated and more accurate model in simulating the biological neural system in comparison with conventional ANN. The KIII model has good capability for pattern recognition as a form of the biological intelligence. It needs much fewer learning trials than ANN when solving problems of pattern recognition. Although when considering the processing speed, the KIII network still could not replace the conventional ANN for solving practical problems, it is surely a promising research for building more intelligent and powerful artificial neural network when the speed is increased by implementing the KIII in analog VLSI [6].

Acknowledgment

This project was supported by the National Basic Research Program of China (973 Program, project No. 2002CCA01800)and the National Natural Science Foundation of China (No. 30470470).

References

1. Freeman, W. J.:Mesoscopic neurodynamics: from neuron to brain. J. Physiology – Paris 94(2000) 303-322
2. Freeman, W. J., Chang H.J., Burke, B.C., Rose, P. A., Badler, J.: Taming chaos: Stabilization of aperiodic attractors by noise. IEEE Transactions on Circuits and Systems 44(1997) 989-996
3. Chang, H. J., Freeman, W. J.: Biologically modeled noise stabilizing neurodynamics for pattern recognition. Int. J. Bifurcation and Chaos 8(1998) 321-345
4. Kozma, R., Freeman, W. J.: Chaotic resonance – methods and applications for robust classification of noisy and variable patterns. Int. J. Bifurcation and Chaos 11(2001) 1607-1629
5. Chang, H.J., Freeman, W. J.: Local homeostasis stabilizes a model of the olfactory system globally in respect to perturbations by input during pattern classification. Int. J. Bifurcation and Chaos 8(1998) 2107-2123
6. Principe, J. C., Tavares, V. G., Harris, J. G., Freeman, W. J.: Design and Implementation of a Biologically Realistic Olfactory Cortex in Analog VLSI. Proceedings of the IEEE 89(2001b) 1030 – 1051

Double Robustness Analysis for Determining Optimal Feedforward Neural Network Architecture

Lean Yu[1], Kin Keung Lai[2,3], and Shouyang Wang[1,2]

[1] Institute of Systems Science, Academy of Mathematics and Systems Sciences,
Chinese Academy of Sciences, Beijing 100080, China
{yulean,sywang}@amss.ac.cn
[2] College of Business Administration, Hunan University, Changsha 410082, China
[3] Department of Management Sciences, City University of Hong Kong,
Tat Chee Avenue, Kowloon, Hong Kong
mskklai@cityu.edu.hk

Abstract. This paper incorporates robustness into neural network modeling and proposes a novel two-phase robustness analysis approach for determining the optimal feedforward neural network (FNN) architecture in terms of Hellinger distance of probability density function (PDF) of error distribution. The proposed approach is illustrated with an example in this paper.

1 Introduction

Generally, the feedforward neural network (FNN) architecture consists of an input layer, an output layer and one or more intervening layers, also referred to as hidden layers. The number of nodes in the input and output layers can be determined by the practical problems. But it is difficult to determine the number of hidden layers and their hidden units per hidden layer. Usually, a three-layer FNN with sufficiently many neurons in a single hidden layer has been proven to be capable of approximating any Borel measurable functions in any accuracy [1]. A focus, thus, is how to determine the hidden neurons in a single hidden layer of FNN modeling and applications.

In the past, many researchers have proposed a variety of methods, such as the upstart algorithm [2] and pruning method [3], to try to determine the number of hidden nodes in a neural network. These methods, however, are not perfect. For example, the algorithms in [2] are likely to disrupt the approximated solution already found. A common problem with the above-mentioned methods is that they do not consider the model robustness – this is one of the important considerations in modeling. A solution that uses a local robustness property is proposed in [4], but such an analytical approach is only suitable for local robustness problems. Here we extend the method and propose a two-phase robustness analysis procedure to determine the optimal FNN architecture. Here "robustness" of the models can be defined in such a way as follows. The set of selected models should be robust in the sense that they are indifferent to radical change of a small portion of the data or a small change in all of the data [4]. Here we use Hellinger distance (HD) between probability density functions (PDF) of error distribution as a model selection criterion. HD can be calculated as

$$HD = \sum\nolimits_{i=0}^{k}(\sqrt{pdf(x_i^{(1)})} - \sqrt{pdf(x_i^{(2)})})^2 \tag{1}$$

where $pdf(x_i^{(1)})$ and $pdf(x_i^{(2)})$ are two PDFs of error distribution [6], the error distribution PDF of FNN approximation for each candidate's architecture is computed using the method described by [5]. In this study, we choose the HD between PDFs as the robustness evaluation criterion to determine the optimal FNN model architecture.

The rest of the study is organized as follows. Section 2 presents the proposed two-phase robustness analysis approach. To demonstrate the efficiency of the proposed approach, a simulated study is given in Section 3. Section 4 concludes the paper.

2 The Proposed Double Robustness Analysis Approach

2.1 Intrapolated Phase

For convenience, the FNN model within the range of error goal is called as "initial model", robust FNN model based on in-sample data set is called as "medial model"; and the robust FNN model based upon out-of-sample data set is called as "final model". This phase contains three steps based upon in-sample data set as follows:

Step 1. Initially, we build a class of network with different hidden nodes, and train the network over the entire training data set (with an increasing number of hidden neurons) in order to learn as many associations as possible. Within the error goal range, some FNN models with different architectures (i.e., initial models) are obtained. Assume that these initial models have the same input and output nodes and different hidden nodes in the single hidden layer.

Step 2. For every "initial model", we change the size of the in-sample data set to check the HD values. If the HD values are unstable, then the corresponding model is discarded, and the models with small fluctuations (i.e., medial models) are retained. Note that we use standard deviation of HD as a measurement of stability.

Step 3. If the HD values of all "medial models" are not stable, then go to Step 1 and select more initial models over again. If we obtain some robust models from Step 2, then go to the next phase.

Since we only check the robustness of the FNN model in terms of the in-sample data set in this phase, we further check the robustness of FNN model using the out-of-sample data set in order to improve the generalization of the FNN model.

2.2 Extrapolated Phase

The main extrapolated phase procedure of this phase includes the following steps.

Step 1. As for every "medial model" from Step 3 in the previous phase, we apply the "medial model" to the out-of-sample data set. Thus, the approximated error series between actual values and approximated values can be obtained.

Step 2. When changing the size of the out-of-sample data set, different HD values of every "medial model" are achieved.

Step 3. If the HD values show little fluctuation, then the models (i.e., final models) are transferred to Step 4, otherwise this model is excluded. If all "medial models" are discarded, then go to Step 1 in the first phase.

Step 4. If the HD values of a certain model are stable, then the model is identified as the "true" model. Accordingly, this model's architecture is the optimal one. If there are several "final models", we select the FNN architecture with minimal standard deviation as the optimal FNN architecture.

To illustrate the efficiency of the approach, an experiment is performed.

3 Simulations

In order to test the efficiency of the proposed approach, a problem of predicting the JPY/USD exchange rate series is considered. The JPY/USD data used are daily and are obtained from Pacific Exchange Rate Service (http://fx.sauder.ubc.ca). The entire data set covers the period from 1 January 2000 until 31 December 2003 with a total of 1121 observations. For convenience, we take daily data from 1 January 2000 to 31 August 2003 as in-sample data sets (999 observations), which are used for the first phase, and meanwhile we take the data from 1 September 2003 to 31 December 2003 as out-of-sample data sets (122 observations), which are used for the second phase. In this experiment, the neural network architecture has the form of "4-x-1".

First of all, according to the predefined error goal (The predefined error NMSE < 0.15), several candidate models (i.e., "initial models") for in-sample data set with different hidden neurons (x) are generated based on NMSE, as shown in Table 1.

Table 1. NMSE of the JPY/USD predictions with different FNN architectures

JPY (x)	JPY(5)	JPY(8)	JPY(9)	JPY(10)	JPY(11)	JPY(13)	JPY(14)	JPY(15)	JPY(16)	JPY(17)
NMSE	0.109	0.118	0.125	0.119	0.135	0.127	0.139	0.133	0.136	0.139

Subsequently, we test the robustness of candidate network architectures by changing the size of the in-sample data set. The results are shown in Table 2.

Table 2. Robustness testing of FNN architecture for JPY predictions with in-sample data

Criterion	Data size	JPY (5)	JPY (8)	JPY (9)	JPY (10)	JPY (11)	JPY (13)	JPY (14)	JPY (15)	JPY (16)	JPY (17)
	999	0.145	0.081	*0.612*	0.228	*0.265*	0.097	0.187	*0.393*	0.158	*0.377*
	989	0.133	0.085	*0.089*	0.219	*0.696*	0.101	0.181	*0.456*	0.166	*0.548*
HD	979	0.147	0.079	*1.258*	0.227	*1.021*	0.093	0.195	*0.558*	0.159	*0.551*
	969	0.151	0.072	*0.556*	0.226	*0.891*	0.114	0.182	*0.987*	0.161	*0.972*
	959	0.144	0.083	*0.157*	0.233	*0.338*	0.087	0.193	*0.118*	0.155	*0.547*
	Stdev.	0.0067	0.0050	*0.4666*	0.0050	*0.3328*	0.0101	0.0063	*0.3162*	0.0041	*0.2214*

From Table 2, we find that JPY (9), JPY (11), JPY (15) and JPY (17) are discarded in view of HD criterion. Therefore, some "medial models" can be obtained.

Finally, we apply "medial models" to out-of-sample data sets in order to check the robustness of "medial models". The results obtained are given in Table 3.

Table 3. Robustness testing of FNN models for JPY predictions with out-of-sample data

Currency	Criterion	Data size	JPY(5)	JPY(8)	JPY(10)	JPY(13)	JPY(14)	JPY(16)
JPY	HD	122	0.074	*0.072*	0.140	0.073	0.123	0.125
		117	0.081	*0.081*	0.109	0.081	0.141	0.058
		112	0.088	*0.077*	0.162	0.093	0.155	0.156
		107	0.078	*0.069*	0.145	0.084	0.102	0.116
		102	0.084	*0.078*	0.158	0.077	0.093	0.163
		Stdev.	0.0054	*0.0048*	0.0209	0.0076	0.0259	0.0417

Table 3 shows that JPY(5), JPY(8) and JPY(13) are robust. According to previous procedure, we select JPY(8) as an optimal model from the smallest standard deviation. In such a way, an optimal FNN architecture is determined using a two-phase robustness analysis approach.

4 Conclusions

In this study, we present a novel and efficient approach for determining the optimal feedforward neural network architecture in terms of model robustness. The proposed approach includes two phases: intrapolated phase and extrapolated phase. Relying on the two-phase robustness analysis approach, an optimal FNN architecture can be obtained. In the meantime, a simulated experiment demonstrates the efficiency and feasibility of the proposed approach.

References

1. Hornik, K., Stinchonbe, M., White, H.: Multilayer feedforward networks are universal approximators. Neural Networks 2 (1989) 359-366
2. Frean, M.: The upstart algorithm: a method for constructing and training feed-forward networks. Neural Computation 2 (1990) 198-209
3. Reed, R.: Pruning algorithms – a survey. IEEE Transactions on Neural Networks 4(5) (1993) 740-747
4. Allende, H., Moraga, C. Salas, R.: Neural model identification using local robustness analysis. In Reusch, B. (ed.): Fuzzy Days, Lecture Notes in Computer Science. Springer-Verlag, Berlin Heidelberg New York 2206 (2001) 162-173
5. Tierney, L., Kass, R., Kadane, J.B.: Approximate marginal densities of nonlinear functions. Biometrika 76 (1989) 425-433
6. Lindsay, B.G.: Efficiency versus robustness: The case for minimum Hellinger distance and related method. The Annals of Statistics 22 (1994) 1081-1114

Stochastic Robust Stability Analysis for Markovian Jump Neural Networks with Time Delay[1]

Li Xie

Department of Information and Electronic Engineering, Zhejiang University,
310027 Hangzhou, P.R.China
Xiehan@zju.edu.cn

Abstract. The problem of stochastic robust stability analysis for Markovian jump neural networks with time delay has been investigated via stochastic stability theory. The neural network under consideration is subject to norm-bounded stochastic nonlinear perturbation. The sufficient conditions for robust stability of Markovian jumping stochastic neural networks with time delay have been developed for all admissible perturbations. All the results are given in terms of linear matrix inequalities.

1 Introduction

The stability analysis for neural networks has received considerable attentions in recent years [1]. When the parameters of neural network are subject to random abrupt changes and stochastic nonlinear perturbations, the neural network can be modeled as stochastic jumping time-delayed systems with the transition jumps described as finite-state Markov chains [2]. These parameters changes may deteriorate the stability as well as the systems performance of the neural networks.

In this paper, we will investigate the problem of stochastic robust stability analysis for Markovian jump neural networks with time delay. The sufficient conditions for the robust stability of the neural networks will be developed. Based on stochastic Lyapunov theory, stable criteria for the neural networks are presented in terms of linear matrix inequalities (LMIs) [3, 4]. In section 2, the system model is described. Some necessary assumptions are given. In section 3, the robust stochastic stable criteria are developed. Finally, conclusions are provided in section 4.

2 Systems Descriptions

Consider the Markovian jump stochastic neural network with time delay, which can be represented in the form of vector state space as follows:

$$dx(t) = \{-A(\theta(t))x(t) - A_1(\theta(t))x(t-\tau) + B(\theta(t))\sigma[x(t)] \\ + B_1(\theta(t))\sigma[x(t-\tau)]\}dt + C(\theta(t))f(x(t), x(t-\tau))dw(t). \quad (1)$$

---

[1] This work was supported by Chinese Nature Science Foundation (60473129).

where $x(t) \in R^n$ is the state vector of the neural network, $x(t-\tau)$ is the delayed state vector of the neural networks with the time delay $\tau \geq 0$. $w(t)$ is standard Wiener process, and $f(x(t), x(t-\tau))$ is stochastic nonlinear perturbation, $\sigma[x(t)]$ is the activation function. $\{\theta(t), t \geq 0\}$ is a time homogeneous Markov process with right continuous trajectories taking values in a finite set $S = \{1,\ldots,N\}$ with stationary transition probabilities:

$$P\{\theta(t+\Delta t) = j \mid \theta(t) = i\} = \begin{cases} \pi_{ij}\Delta t + o(\Delta t) & i \neq j \\ 1 + \pi_{ij}\Delta t + o(\Delta t) & i = j \end{cases} \quad (2)$$

where $\Delta t > 0$, $\lim_{\Delta t \to 0} o(\Delta t)/\Delta t = 0$, and $\pi_{ii} = -\sum_{j=1, j\neq i}^{N} \pi_{ij}$. Here $\pi_{ij} \geq 0$ is the transition rate from mode i at time t to mode $j \neq i$ at time $t + \Delta t$ for $i, j \in S$.

$A(\theta(t))$, $A_1(\theta(t))$, $B(\theta(t))$, $B_1(\theta(t))$, $C(\theta(t))$ are known real constant matrices of appropriate dimensions for all $\theta(t) \in S$. In the sequel, we denote the parameter matrices $A(\theta(t))$, $A_1(\theta(t))$, $B(\theta(t))$, $B_1(\theta(t))$, $C(\theta(t))$ as A_i, A_{1i}, B_i, B_{1i}, C_i when $\theta(t) = i$.

Though out this paper, we assume that the activation function $\sigma[x(t)]$ and the perturbation function $f(x(t), x(t-\tau))$ satisfy the following conditions:

(A.1) If there exist positive constant diagonal matrix K, such that

$$0 < \frac{\sigma(x_1) - \sigma(x_2)}{x_1 - x_2} \leq K, \ \forall x_1, x_2 \in R, \ x_1 \neq x_2. \quad (3)$$

(A.2) There exist positive constant matrices M and M_1, such that

$$f^T(x(t), x(t-\tau)) f(x(t), x(t-\tau)) \leq x^T(t) M^T M x(t) \\ + x^T(t-\tau) M_1^T M_1 x(t-\tau). \quad (4)$$

3 Main Results

In this section, robust stability criteria for Markovian jumping neural networks with time delay and stochastic nonlinear perturbation are given.

Theorem 1. Consider the Markovian jumping stochastic neural networks with time delay (1), if there exist matrices $X_i > 0$, $W > 0$, $S > 0$, $S_1 > 0$ and constants $\rho_j > 0$ ($j = 1, 2$), satisfying the LMIs (a). $\Omega = \begin{bmatrix} \Omega_{11} & X_i A_{1i}^T & X_i K^T & X_i K^T & \rho_i X_i M^T & \rho_i X_i M_1^T & \Omega_{17} \\ A_{1i} X_i & -W & 0 & 0 & 0 & 0 & 0 \\ K X_i & 0 & -S & 0 & 0 & 0 & 0 \\ K X_i & 0 & 0 & -S_1 & 0 & 0 & 0 \\ \rho_i M X_i & 0 & 0 & 0 & -\rho_i I & 0 & 0 \\ \rho_i M_1 X_i & 0 & 0 & 0 & 0 & -\rho_i I & 0 \\ \Omega_{71} & 0 & 0 & 0 & 0 & 0 & \Omega_{77} \end{bmatrix} < 0.$ (5)

where

$$\Omega_{11} = -X_i A_i^T - A_i X_i + W + B_i S B_i^T + B_{1i} S_1 B_{1i}^T$$
$$\Omega_{17} = \Omega_{71}^T = [\pi_{i1} X_i \ \ldots \ \pi_{iN} X_i]$$
$$\Omega_{77} = diag\{-\pi_{i1} X_1, \ldots, -\pi_{iN} X_N\}.$$

(b). $\begin{bmatrix} -\rho_i I & C_i^T \\ C_i & -X_i \end{bmatrix} \leq 0, \ i = 1, \ldots, N.$ (6)

then the neural network (1) is robust stochastic stable for all admissible perturbations.

Proof. Let the mode at time t be i, that is $\theta(t) = i \in S$, and introduce a Lyapunov functional as

$$V(x(t), i) = x^T(t) P_i x(t) + \int_{t-\tau}^{t} x^T(s) R_i x(s) ds.$$ (7)

From (7), it is easy to obtain $0 \leq \varepsilon_{1i} \|x(t)\|^2 \leq V(x(t), i) \leq (\varepsilon_{2i} + \varepsilon_{3i} q\tau) \|x(t)\|^2$, where $q \geq 1$, $\varepsilon_{1i} = \lambda_{\min}(P_i)$, $\varepsilon_{2i} = \lambda_{\max}(P_i)$, $\varepsilon_{3i} = \lambda_{\max}(R_i)$. For simplicity, we denote $x(t)$ and $x(t-\tau)$ as x and x_τ.

By using Ito's formula, the weak infinitesimal operator of the Lyapunov functional along the solution of system (1) is

$$LV(x, i) = \dot{x}^T P_i x + x^T P_i \dot{x} + \sum_{j=1}^{N} \pi_{ij} x^T P_j x + x^T R_i x - x_\tau^T R_i x_\tau$$
$$+ f^T(x, x_\tau) C_i^T P_i C_i f(x, x_\tau)$$
$$\leq -x^T A_i^T P_i x - x^T P_i A_i x + x_\tau^T A_{1i}^T W^{-1} A_{1i} x_\tau + x^T P_i W P_i x + x^T R_i x$$
$$+ \sigma^T(x) S^{-1} \sigma(x) + x^T P_i B_i S B_i^T P_i + \sigma^T(x_\tau) S_1^{-1} \sigma(x_\tau) - x_\tau^T R_i x_\tau$$ (8)
$$+ x^T P_i B_{1i} S_1 B_{1i}^T P_i x + \rho_i f^T(x, x_\tau) f(x, x_\tau) + \sum_{j=1}^{N} \pi_{ij} x^T P_j x.$$

In view of inequality (6), and by using Schur complement, we have

$$C_i^T P_i C_i \leq \rho_i I.$$ (9)

Hence

$$LV(x,i) \le -x^T A_i^T P_i x - x^T P_i A_i x + x_\tau^T A_{1i}^T W^{-1} A_{1i} x_\tau + x^T P_i W P_i x$$
$$+ x^T K^T S^{-1} K x + x^T P_i B_i S B_i^T P_i x + x_\tau^T K^T S_1^{-1} K x_\tau$$
$$+ x^T P_i B_{1i} S_1 B_{1i}^T P_i x + \rho_i x^T M^T M x + \rho_i x_\tau^T M_1^T M_1 x_\tau \qquad (10)$$
$$+ x^T R_i x - x_\tau^T R_i x_\tau + \sum_{j=1}^{N} \pi_{ij} x^T P_j x.$$

Let

$$R_i = A_{1i}^T W^{-1} A_{1i} + K^T S_1^{-1} K + \rho_i M_1^T M_1. \qquad (11)$$

Then, we have

$$LV(x,i) \le x^T \Xi_i x. \qquad (12)$$

where

$$\Xi_i = -A_i^T P_i - P_i A_i + A_{1i}^T W^{-1} A_{1i} + P_i W P_i + K^T S^{-1} K + K^T S_1^{-1} K$$
$$+ P_i B_i S B_i^T P_i + P_i B_{1i} S_1 B_{1i}^T P_i + \rho_i M^T M + \rho_i M_1^T M_1 + \sum_{j=1}^{N} \pi_{ij} P_j. \qquad (13)$$

Pre- and post-multiply (13) with $X_i = P_i^{-1}$. By the Schur complement, $\Xi_i < 0$ holds if and only if inequality (5) holds. It is easy to obtain $LV(x,i) < 0$, that is, the Markovian jumping stochastic neural networks with time delay are robust stable for all the admissible perturbations.

This completes the proof. □

4 Conclusions

In this paper, the problem of robust stability analysis for Markovian jumping neural networks with stochastic nonlinear perturbations and time delay is investigated. The stability criteria are given in terms of linear matrix inequalities for all admissible pertubantions.

References

1. Michel A.N., Farrell J.A., Porod W.: Qualitative analysis of neural networks. IEEE Trans. CAS. 36 (1989) 229-243
2. Feng X., Loparo K.A., Ji Y.: Stochastic stability properties of jump linear system. IEEE Trans. AC. 37 (1992) 38–53
3. Liao, X., Chen, G.R., Sanchez, E.N.: LMI-Based Approach for Asymptotically Stability Analysis of Delayed Neural Networks. IEEE Trans. CAS-I. 49 (2002) 1033-1039
4. Liao, X., Chen, G.R., Sanchez, E.N.: Delay-Dependent Exponential Stability Analysis of Delayed Neural Networks: An LMI Approach. Neural Network, 15 (2002) 855–866

Observation of Crises and Bifurcations in the Hodgkin-Huxley Neuron Model

Wuyin Jin[1], Qian Lin[2], Yaobing Wei[1], and Ying Wu[3]

[1] College of Mechano-Electronic Engineering,
Lanzhou University of Technology, Lanzhou 730050, China
jinwuyin@263.net
[2] College of Petrochemical Technology, Lanzhou University of Technology,
Lanzhou 730050, China
[3] College of Science, Lanzhou University of Technology,
Lanzhou 730050, China

Abstract. With the changing of the stimulus frequency, there are a lot of firing dynamics behaviors of interspike intervals (ISIs), such as quasi-periodic, bursting, period-chaotic, chaotic, periodic and the bifurcations of the chaotic attractor appear alternatively in Hodgkin-Huxley (H-H) neuron model. The chaotic behavior is realized over a wide range of frequency and is visualized by using ISIs, and many kinds of abrupt undergoing changes of the ISIs are observed in deferent frequency regions, such as boundary crisis, interior crisis and merging crisis displaying alternately along with the changes changes of external signal frequency, too. And there are many periodic windows and fractal structures in ISIs dynamics behaviors. The saddle node bifurcation resulted collapses of chaos to period-12 orbit in dynamics of ISIs is identified.

1 Introduction

The bifurcation and crisis of neural system have been an object of major attention since the beginning of the study of chaos theory. As we all known, the neural systems have strong nonlinear characters, and are usually able to display different dynamics according to system parameters or external inputs in ISIs sequences. When these parameters are slightly modified, the system's dynamics usually experience also little modification, except when these changes occur in the vicinity of a critical point, in which case an abrupt qualitative change or transition in the dynamics occurs [1-3]. These transitions, for example, may be from periodic to chaotic, from chaotic to chaotic, and their inverse transitions [4].

And, the numerical evidence and theoretical reasoning has proved that there is a chaos-chaos transition in the neuron, in which the change of the attractor size is sudden but continuous, different from general discontinues chaos-chaos transitions, and which occurs in the Hindmarsh-Rose model of a neuron. This transition corresponds to different neural dynamics, i.e. the chaotic dynamics of bursting and spiking dynamics [3]. The crisis resulted from homoclinic bifurcation and the chaos collapsing to a period-3 orbit in the dynamics of a quadratic Logistic map neuron have also been studied [5,6]. Xie et al introduced periodic orbit theory to characterize the dynamical behavior of aperiodic firing neurons, and considered that bifurcations, crises and sensitive dependence of chaotic motions on control parameters can be the underlying mechanisms [7], and there are many chaotic activities have been observed in experimental studies of electroencephalogram(EEG) signals and neuronal ISIs sequence [8-10].

The transitions between different dynamic behaviors of ISIs sequence of H-H neuron model under external periodic stimulus and the saddle-node bifurcation are studied in this work, which is relevant both to the theory of nonlinear dynamics and to biophysics.

2 The Hodgkin-Huxley (H-H) Neuron Model

The equations that describe the H-H neuron model have been derived from a squid giant axon. These equations can describe the spiking behavior and refractoriness of real neuron very well, so that this kind of model is employed in this work. The H-H model for the action potential of a space clamped squid axon is defined by the four-dimensional vector field [11]

$$\begin{cases} \dot{u} = I_{ext} - [120m^3 h(u+115) + 36n^4(u-12) + 0.3(u+10.6)] \\ \dot{m} = (1-m)\Psi(\frac{u+25}{10}) - m(4exp(\frac{u}{18})) \\ \dot{n} = 0.1(1-n)\Psi(\frac{u+10}{10}) - n(0.125exp(\frac{u}{80})) \\ \dot{h} = 0.07(1-h)\Psi(\frac{u}{20}) - h(\frac{1}{1+exp(\frac{u+30}{10})}) \end{cases}, \quad (1)$$

where

$$\Psi(x) = \frac{x}{exp(x)-1} \quad (2)$$

and variables u, m, n, and h represent membrane potential, activation of a sodium current, activation of a potassium current, and inactivation of the sodium current. There is also a current parameter I_{ext} that is an external periodic signal current into the space-clamped axon in this work, i.e.

$$I_{ext} = I_{shift} + sin(2\pi f_0 t), \quad (3)$$

where $I_{shift} = 10\mu A/cm^2$, being the amplitude of current shift, and $f_0 = 1/3$ Hz being the basic stimulus frequency in this work.

Recalling that the H-H convention for membrane potential reverses the sign from modern conventions, and so the voltage spikes of action potentials are negative in the H-H model. When improved models for the membrane potential of the squid axon have been formulated, the H-H model remains the paradigm for conductance-based models of neural systems. From a mathematical viewpoint, varied properties of the dynamics of the H-H vector field have been studied. Nonetheless, we remain far from a comprehensive understanding of the dynamics displayed by this vector field.

392 W. Jin et al.

In this work, the ordinary differential equations (1) is integrated by using double precision fourth-order Runge-Kutta method, with integration time step 0.01, the rest membrane potential equals to 0 mV.

3 Bifurcations and Crises of ISIs

In this work, the H-H neuron model has been simulated numerically in the absence of noise, using the ISIs as a state variable. The ISIs are registered by the membrane potential crossing a threshold (at 60 mV) with positive derivative (Poincaré surface of section). The controlled frequency of stimulus ranges $f \in [0.01, 10]f_0$. There are a lot of firing dynamic behaviors of ISIs, such as quasi-periodic, bursting, period-chaotic, chaotic and periodic appears alternatively with the changing of the stimulus frequency f (see Fig.1). Associating with our previous works [1,2], we could conclude that the time scale of the external signal (including periodic and chaotic) play an important role in transition of neural information.

One typical detailed bifurcation diagram of ISIs is shown in Fig.2a representing a classical route to chaos through a inverse period doubling cascade located at $f \approx 2.9128f_0, 2.9155f_0$, and $2.935f_0$ respectively. Inside the chaotic regions, we observed several periodic windows located at $2.9071f_0, 2.9102f_0, 2.9104f_0$, and $2.911f_0$, all of which are opened by a saddle-node bifurcation and closed by a global bifurcation, namely being an interior crisis (see section 4). At the same time, several other typical crises occur as the stimulus frequency varies. The first

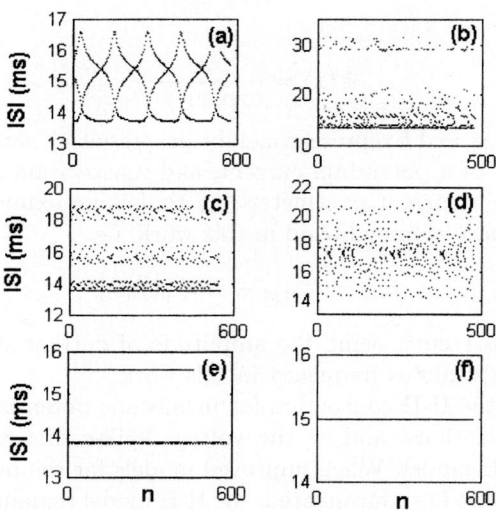

Fig. 1. Scattered ISIs sequences for stimulus frequency $f = 1.35f_0$ represents qusi-periodic firing (a), $2.9f_0$ is bursting (b), $2.9114f_0$ is period-3 chaotic (c), $3.4685f_0$ is chaotic (d), $5.1f_0$ is period-2 (e), and $6.0f_0$ is period-1 (f), respectively

type is boundary crisis, resulted from the attractor colliding with an unstable periodic orbit that was on the basin boundary before the crisis, a chaotic is suddenly destroyed as the parameter passes through its critical value (e.g., $f \approx 2.9104 f_0$). The second type is merging crisis, two or more chaotic attractors, simultaneously colliding with a period orbit (or orbits) on the basin boundary which separates them, and merging to form one chaotic attractor (e.g., $f \approx 2.9080 f_0, 2.9087 f_0$, and $2.9114 f_0$). The last type is interior crisis, i.e., the periodic orbit with which the chaotic attractor collision is in the interior of its basin result in the size of the attractor in phase space suddenly change (e.g., $f \approx 2.9055 f_0$). Certainly, here, we just list a few of cases as examples.

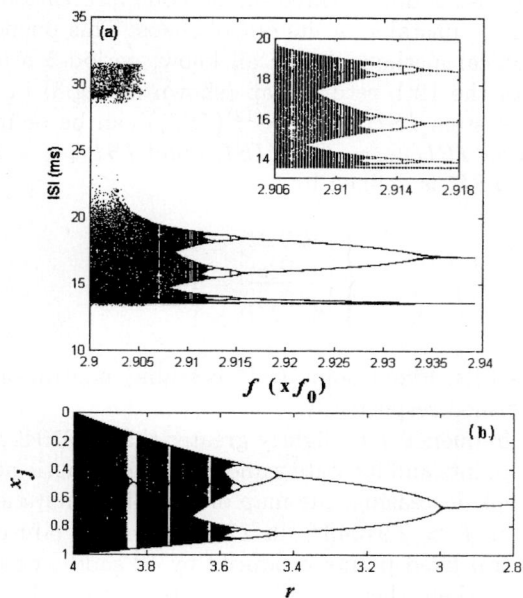

Fig. 2. Detailed bifurcation diagrams of ISIs and its part enlargement, the bifurcation parameter f being increased with step $0.0001 f_0$ (a), and bifurcation diagram of Logistic map $x_{j+1} = r x_j (1 - x_j)$, $r \in [4 \sim 2.8]$ (b)

The pattern of bifurcation diagram of ISI shown in Fig.2a is being very similar to that of the Logistic map $x_{j+1} = r x_j (1 - x_j)$, its bifurcation diagram shown in Fig.2b. Both of them have similar pattern in return map, e.g., the first return map of them with single one-hump structure. In some cases, the single one-hump is replaced by two one-hump pattern corresponding to two chaotic attractors, and so on. For an example, The shape of the 12th return map (shown in Fig.3 b-e) of ISIs is four curves with multiple extremum, each of which is similar to that of the third iterate of Logistic map.

4 Saddle-Node Bifurcation

In this section, we aim at one of the numerous bifurcation processes collapse of chaos to a period-12 orbit in the H-H spiking dynamics, which emerges around $f = 2.9103 f_0$, the bifurcation diagram of ISIs of H-H model shown in Fig.3 suggests that saddle-node, period doubling and other common basic bifurcations which underlie ISIs of H-H neuronal dynamics as Logistic map.

Seen from bifurcation diagram shown in Fig.3a, when stimulus frequency locates within $f = 2.9102 f_0$, four period-3 orbit are embedded in four chaotic attractors, and their shapes of 12th return map $ISI_{n+12} = F^{(12)}(ISI_n)$ are similar to that of the third iterate of Logistic map respectively, shown in Fig.3b-e. Appearance of period-3 is due to three saddle-node bifurcations, giving birth to three stable and three unstable orbits out of chaos. This phenomena can easily be seen with graphical method. As we all know, period-3 orbit correspond to three fixed point of the 12th return map (shown in Fig.3) in this work. Fixed point ISI^* of the system $ISI_{n+12} = F^{(12)}(ISI_n)$ can be defined as a point of intersection of curves $ISI_{n+12} = F^{(12)}(ISI_n)$ and $ISI_{n+12} = ISI_n$ (cf., Fig.3), and its stability of ISI^* is defined by

$$\begin{cases} |\frac{dF^{12}(x)}{x}| < 1 \\ |\frac{dF^{12}(x)}{x}| = 1 \\ |\frac{dF^{12}(x)}{x}| > 1 \end{cases}, \quad (4)$$

where x represents ISIs. Fixed point ISI^* is stable, neutral or unstable if condition Eq.(4) is satisfied respectively.

After stimulus frequency f is slightly greater than $2.91049 f_0$, the map $F^{(12)}$ has no stable fixed points and its state wanders within chaotic attractor as shown in Fig.3b. As f keeps decreasing, the map of $F^{(12)}$ is simultaneously tangent to $ISI_{n+12} = ISI_n$ at $f \approx 2.91049 f_0$ in 12 saddle-node bifurcation points, all of them being neutral fixed points produced by 12 saddle-node bifurcations as shown in Fig.3c, and then, these points split into to six stable and six unstable fixed points as shown in Fig.3d. The stability of stable fixed points keep up to the extremum of the parabolic passed through $ISI_{n+12} = ISI_n$, they loose their stability via period-doubling bifurcations as shown in Fig.3e.

With f decreasing further, the system undergoes period-doubling cascade and around the critical point $f = 2.9101 f_0$ it becomes chaotic. At last, this periodic windows is closed by interior crisis (see Fig.3a). Due to the fractal structure of the bifurcation, there are a larger number of f, and with their chaotic attractors the system lives on collapse, producing stable periodic orbits, such as periodic windows locating at $f = 2.9071 f_0$, $f = 2.90932 f_0$, and $f = 2.9011 f_0$.

5 Conclusions

The study of transitions between different dynamic behaviors in neural systems is an issue of major interest for biophysicist. Chaos-chaos transitions will help

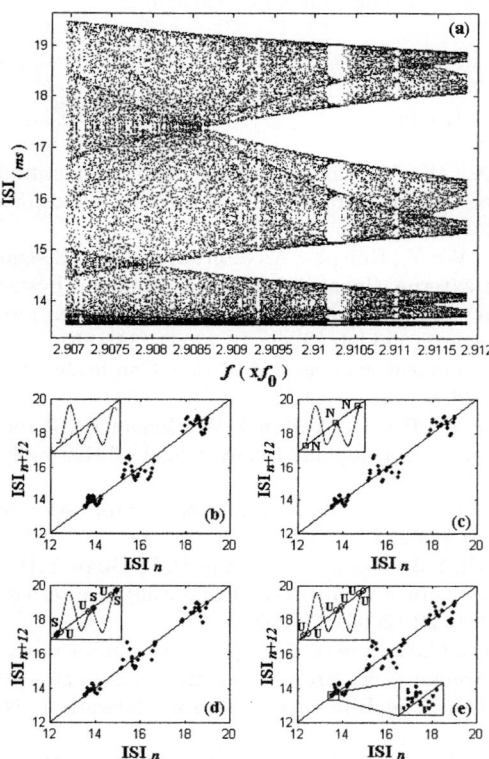

Fig. 3. The detailed bifurcation diagram of ISIs, the bifurcation parameter f is increased with step $0.00001 f_0$ (a), The 12th return map of ISIs for $f = 2.9107 f_0$ (b), $f = 2.91049 f_0$ (c), $f = 2.91043 f_0$ (d), $f = 2.91012 f_0$ (with a partly enlargement)(e); the insets schematically show the emergence and disappearance of fixed points via saddle-node bifurcations with different inputs, rectangle, black and empty circles correspond respectively to neutral, stable and unstable fixed points (marked by 'N', 'S', 'U')

us to understand how the neural system is able to give quick responses to the different external or internal stimulus, and neuronal potential computational and learning properties. The observation of bifurcations and cries in this work is relevant to the theory of nonlinear dynamics and chaos, and to biophysics, particularly to neurobiology.

Acknowledgments

We thank the supporting of the Natural Science Foundation of Gansu Province

(Grant No.Z02200401) and the National Natural Science Foundation of China (Grant No.10432010).

References

1. Jin W.Y., Xu J.X., Wu Y., Hong L.: Rate of afferent stimulus dependent synchronziation andcoding in coupled neurons system. Chaos, Solitons and Fractals **21** (2004) 1221-1229.
2. Jin W.Y., Xu J.X., Wu Y., Hong L.: An alternating periodic-chaotic ISI sequence of H-H neuron under external sinusoidal stimulus. Chinese Physics **13** (2004) 335-340.
3. González-Miranda J.M.: Observation of a continuous interior crisis in the Hindmarsh-Rose neuron model. Chaos **13** (2003) 845-852.
4. Ott E.: Chaos in dynamical systems, 2ed edition, Cambridge: Cambridge University Press (2002) 304-344.
5. Feudel U., Neiman A., Pei X., Wojtenek W., Braun H., Huber M., Moss F.: Homoclinic bifurcation in a Hodgkin-Huxley model of thermally sensitive neurons. Chaos **10** (2000) 231-239.
6. Lysetskiy M., Zurada J.M.: Bifurcating neuron: computation and learning. Neural Networks **17** (2004) 225-232.
7. Xie Y., Xu J.X., Hu S.J., Kang Y.M., Yang H.J., Duan Y.B.: Dynamical mechanisms for sensitive response of aperiodic firing cells to external stimulation Chaos, Soiltons and Fractals **22** (2004) 151-160.
8. Freeman W., Skarda C.A.: Spatial EEG patterns, nonlinear dynamics and perception: neo-Sherringtonian view. Brain Research Reviews **10** (1985) 147-175.
9. Lee G., Farhat N.H.: The Bifurcating Neuron Network 1. Neural Networks **14** (2001) 115-131.
10. Feudel U., Neiman A., Pei X., Wojtenek W., Braun H., Huber M., Moss F.: Homoclinic bifurcation in a Hodgkin-Huxley model of thermally sensitive neurons. Chaos **10** (2000) 231-239.
11. Guckenheimer J., Oliva R.A.: Chaos in the Hodgkin-Huxley Mode. SIAM Journal of Appled Dynamical Systems **1** (2002) 105-114.

An Application of Pattern Recognition Based on Optimized RBF-DDA Neural Networks

Guoyou Li, Huiguang Li, Min Dong, Changping Sun, and Tihua Wu

College of Electrical Engineering, Yanshan University,
Qinhuangdao 066004, China
dongmin@ysu.edu.cn

Abstract. An algorithm of Dynamic Decay Adjustment Radial Basis Function (RBF-DDA) neural networks is presented. It can adaptively get the number of the hidden layer nodes and the center values of data. It resolve the problem of deciding RBF parameters randomly and generalization ability of RBF is improved. When is applied to the system of image pattern recognition, the experimental results show that the recognition rate of the improved RBF neural network still achieves 97.4% even under stronger disturbance. It verifies the good performance of improved algorithm.

1 Introduction

RBF neural networks is one kind of feed forward neural networks. It has the advantage of simple structure, powerful ability of approximation of overall situation and quick simple training method [1,2]. So, it has been broadly applied to prediction, signal process, pattern recognition and so on [3].

At present, generally, the method of the design of RBF neural networks can be grouped into two categories, first, is the random selection of the data centers of the hidden layer nodes, e.g., OLS algorithm, and ROLS algorithm [4]. The advantages of it include easier completing and fixing the number of hidden layer nodes during the weighted value is learning. But it can't design the networks with smallest structure. Second is the positions of data centers are adjusted dynamic. The advantage of it is that it can fix the extended constant of each hidden node according to the distance between cluster centers. The defects are that it also can't fix the number of hidden layer nodes of the networks and the speed of cluster process is slower [5,6]. So, finding the reasonable method that can fix the number of cluster and corresponding data centers is a problem of top concern for the design of RBF neural networks.

In order to tackle the problem above, an improved method of adjusting data centers based on dynamic decay is presented. The method overcomes the defects that mainly depend on prior knowledge to design parameters in the former algorithm. It can adaptively fix the number of hidden layer nodes and the center values of Gaussian function. So it greatly increases the speed and accuracy of the networks. At the same time, the improved networks is applied to the system of image pattern recognition.

2 The Structure and Principle of RBF Neural Networks

RBF neural networks with topological structure is of feedforward neural networks. It is made up of input layers, hidden layers and output layers. The function of hidden layer nodes respond to the input signal only in local area. Only when the input signal is near to the center area of the effective function, the hidden layer nodes will produce higher output. The Gaussian function is selected as the radial basis function in hidden layers

$$G = \exp(-\frac{\|x-c_i\|^2}{2\delta_i^2}) \quad i = 1,2,\cdots,m. \tag{1}$$

In the formula, x is a input vector; c_i is the data center of ith node's function in hidden layers. δ_i means the width of the function nearby the center point. m shows the number of the hidden layer nodes, that is, the number of data center of the effective functions; $\|x-c_i\|$ is the norm of the vector, shows the edclidean distance between x and c_i.

The output layers are linear mapping of the output data of the hidden layers when the sum function is used as the effective function of output layers the output are

$$y_k = \sum_{i=1}^{m} w_{ik} R_i(x) \quad k = 1,2,\cdots,p. \tag{2}$$

In the formula, p - the number of the output nodes, w_{ik} - the output weighted value of the hidden layers of the radial basis networks.

There are three parameters to be processed in the RBF neural networks, that is, the data center of Gaussian function in the hidden layers, the width corresponding to the data center of Gaussian function and the weighted values between the hidden layers and the output layers. The most important parameter of the three is the data center of Gaussian function in the hidden layers. When the data center is fixed, the width of the radial basis function is fixed according to the following expression

$$\delta_i = d_i / \sqrt{2M}. \tag{3}$$

In the formula, d_i - the largest distance between the ith data center and other data centers; M -the number of data centers. After fixing the data center and Gaussian function's width, we can use methods of least square to fix the weighted value, because the relation between the hidden layers and the output layers is linear.

3 The Optimized Design of DDA Algorithm

German scholar–Berthold proposed a changeable structure dynamic RBF networks model [7]. The topological structure of the network is adjusted dynamic in the learning process, which is based on DDA technology. According to the space distribution of the learning samples, the number of the hidden layer nodes, the data center and width of Gaussian function are adjusted dynamic in the learning course [8].

First, to adjust the range of width of RBF neural networks, set two parameters, that is, activated threshold value α and suppressed threshold value β. When the RBF neural networks is training, α and β will make sure that all learning samples fall into the range of width of the Gaussian function, that is, when a sample is inputted, if it is in the coverage of the data center c_j, it will meet the following formula

$$G(\|x_i - c_j\|) \geq \alpha \quad G(\|x_i - c_j\|) \leq \beta. \qquad (4)$$

In the formula: G - the output of RBF neural networks, $k, j \in M$. M - the number of the hidden layers nodes. $k \neq j$.

The former algorithm of dynamic decay adjustment radial basis function networks mainly depends on the prior knowledge to design parameters. In order to conquer the defect of the former algorithm, the improved algorithm of RBF – DDA is proposed in this paper. Because the new algorithm can adaptively fix the number of RBF hidden layer nodes and the center value of the Gaussian function, it can largely increase the algorithm is as follows:

1. Initialize parameters: α and β, set the step length to ρ, and the time of circulation. One parameter varies with step length ρ, the other is fixed.
2. Select one from the input samples randomly as the initial data center c_i, set the width of Gaussian function randomly θ_1, set parameter T, and order $T = 1$;
3. Input the second sample (x_2, y_2), calculate the output of the hidden layers $G(\|x_i - c_j\|)$, if $G(\|x_i - c_j\|) \geq \alpha$, then the sample (x_2, y_2) falls into the coverage of the data center c_j; if $G(\|x_i - c_j\|) \leq \beta$, sample (x_2, y_2) will be the second data center, The width of Gaussian function- θ_2 meets the formula: $G(\|x_i - c_j\|) \leq \beta$.
4. Assume that the number of the fixed data center is M. For inputted training sample (x_i, y_i) randomly, calculate its output at each data center of the Gaussian functions, if $\max (G(\|x_i - C_j\|)) \geq \alpha$, $j = 1, 2, \cdots, M$ then the sample (x_i, y_i) falls into the coverage of data center of the Gaussian whose output is highest. If $\max (G\|x_i - c_j\|) \leq \beta$, then the input sample (x_i, y_i) will become another new data center, C_{M+1}, whose Gaussian function's width meet maximal $\max (G\|x_j - c_{j+1}\|) \leq \beta$;
5. Order the fixed deta centers, adjust the of each data center, adjust the width of each data center, make all adjoined data centers meet the formula: $(G\|x_i - c_j\|) \leq \beta$, then jump(4), When all data centers don't change, jump(6).

6. According to the fixed parameters, construct the RBF neural networks, calculate the error sum of squares of the output of the networks, that is, $E(T) = \sum (y_i - Y_i)^2$, $T = T+1$; The changeable parameter increases one step ,jump(2).When the times of circulation is satisfied, jump (7).
7. Find out the value of the changeable parameter corresponding to the minimum of E, then fix the value. When the minimum of E is less than required error, jump (8); else another parameter varies with step length ρ ,jump (2);
8. Fix α and β, construct the RBF neural networks, output the values, the algorithm ends.

4 Simulating Experiment

Ship recognition has attracted much attention of researchers who study on pattern recognition. For testing the effectiveness of the improved algorithm, we take shipbase as the subject of study, and build the mathematical model of it.

4.1 Build the Mathematical Model

First, recognize the subject, A-type ship that is one of the ship-base with thirty ships. It is shown in Fig. 1. The moment invariants of the template are shown in Table 1.

Fig. 1. The A-type ship template

Table 1. The moment invariants of the A-type ship template

ϕ_1	ϕ_2	ϕ_3	ϕ_4	ϕ_5	ϕ_6	ϕ_7
1.181	0.406	1.256	1.006	0.696	0.624	0.122

As the input of the network, the moment invariants of A - type ship are transformed to a 7×1 column vector. The corresponding expectation output is a 30×1 column vector. The code of the corresponding ship is 1, and the others are 0. The corresponding output is

$$Y_S = [0\,0\,0\,0\,0\,0\,0\,0\,0\,0\,0\,0\,0\,0\,0\,0\,0\,1\,0\,0\,0\,0\,0\,0\,0\,0\,0\,0\,0\,0]^T. \qquad (5)$$

Likewise, adopting this coding rule, we can get the inputs and expectation outputs of the 29 types left. Because the value of normal random noise is between −1 and 1 randomly, we add the noise which is between 0 and 0.4, to the 30 groups of samples. It effectively simulates the actual data with disturbance. When the number of hidden layer nodes of the RBF network and the data center are fixed, a system of linear equations is formed from the input layers to the output layers in the RBF neural networks. Thus we can get the output weighted values by the methods of least square.

In addition, the outputs may not necessarily be a vector solely consisting of 0 and 1.So we adopt the competition rule, the element with the highest output value will win the competition. While the others will fail. To the network in this paper, we set the element with the highest output value to 1,and others to 0.

4.2 The Analysis of the Experimental Results

We respectively test random noise with the disturbance 0.1,0.2,0.3,0.4. As fig.4 shows, the vertical axis shows the error sum of square of the output vector and the horizontal axis shows the number of tested data. As Fig. 2 shows, when the disturbance is lower, for example 0.1,0.2,the output can track the expectation value exactly. As Fig. 2(a) and 2(b) show, when the disturbance is higher, for example 0.3,0.4,the output can't track the expectation value exactly, that is, the fault recognition appears. It is shown in Fig. 2(c) and 2(d).

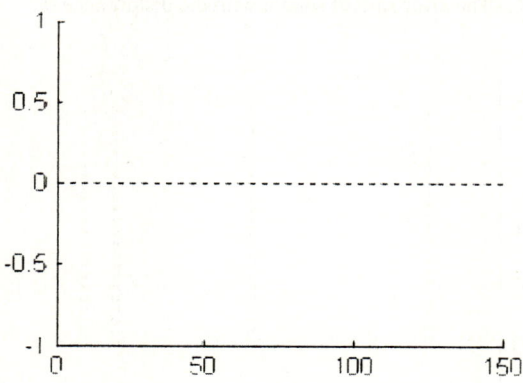

(a) The error sum of square with the disturbance 0.1

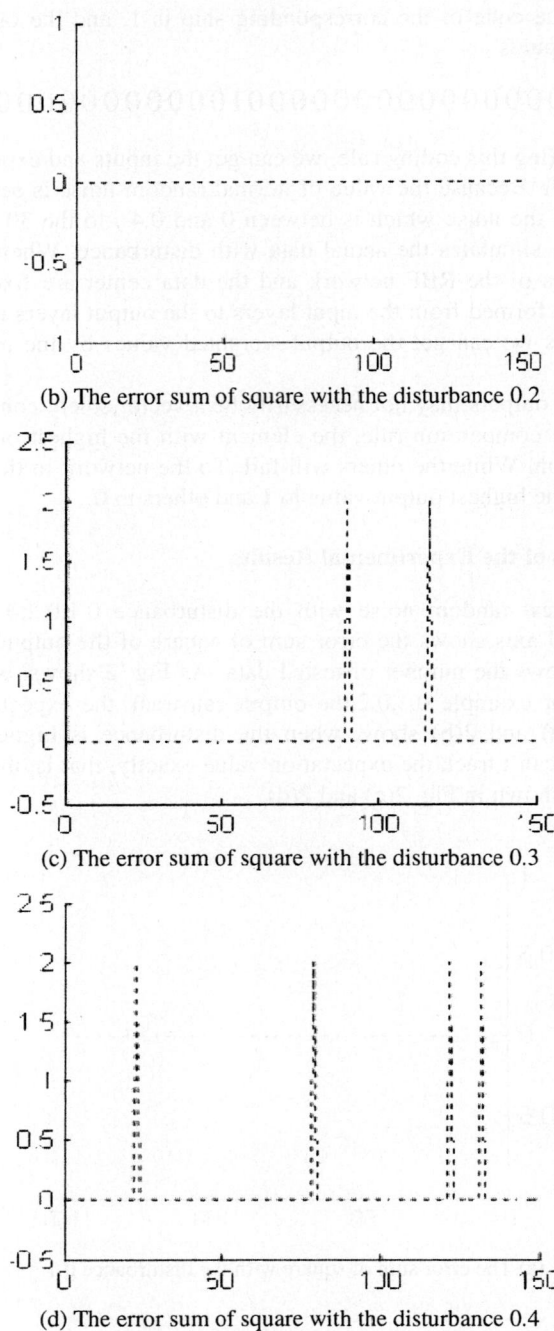

(b) The error sum of square with the disturbance 0.2

(c) The error sum of square with the disturbance 0.3

(d) The error sum of square with the disturbance 0.4

Fig. 2. The error sum of square of the actual output and the expectation value with the different disturbance

To understand the recognition ability of the improved algorithm, we give the curve which shows the recognition rate varies with the disturbance .It is shown in Fig.3.

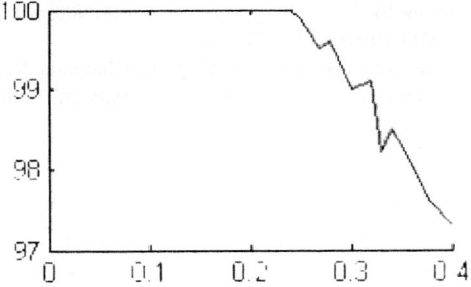

Fig. 3. The recognition rate of the improved algorithm varies with the disturbances

As fig.3 shows, when the disturbance is under 0.24, the system can recognize the input data correctly. With the increasing of the disturbance, the recognition rate begins to decline. The recognition rate is still 97% above although the disturbance is 0.4. Experimental results show the improved RBF neural networks have better performance. Powerful recognition ability even in stronger disturbance is obtained.

5 Conclusions

An optimized RBF-DDA neural networks algorithm is proposed in this paper. The optimized networks can adaptively fix the number of the hidden layer nodes and the data center of Gaussian function. It conquers the defect that parameters to be fixed mainly depend on prior knowledge in the original algorithm. In order to test the performance of optimized algorithm, it has been applied to curve fitting and ship pattern recognition. The experimental results prove that the learning rate and recognition accuracy of the optimized RBF-DDA neural networks is greatly improved.

References

1. A. Jonathan Howell, Hilary Buxton. Learning identity with radial basis function networks. Neuro computing (1998)20:15~34
2. Fu, X. J., Wang L.p.: A GA-Based Novel RBF Classifier with Class-Dependent Features. Proc. 2002 IEEE Congress on Evolutionary Computation (CEC 2002), vol.2(2002) 1890-1894
3. Panchapakeasan, C.; Palaniswami, M; Palph, D.; Manzie, C.:Effects of moving the center's in an RBF network. IEEE Trans. Neural Networks 13(2002) 1299-1307
4. Orr M J L. Regularization in the selection of radial basis function centers[J]. Neural Computation (1995)7:606-623
5. La Poutre, H.; Kok, J.N.; Unsupervised clustering with spiking neurous by sparse temporal coding and multiplayer RBF networks. IEEE Trans. Neural Networks 13 (2002) 426-435

6. Pablo Zegers and Malur K. Sundareshan. Trajectory Generation and Modulation Using Dynamic Neural Networks. IEEE Transactions on Neural Networks, Vol. 14, NO. 3, May 2003.520-533
7. Berthold Michael R.Diamond Jay. Boosting the performance of RBF networks with dynamic decay adjustment In: Proceeding of Advances in Neural Information Processing System .Cambrige MA:MIT Press.1995:521-528
8. Jin Lianwen, Xu Bingzheng. Handwritten Chinese Character Recognition with RBF-DDA Neural Networks[J], Journal of South China University of Technology (Natural Science), 1997,25(9):97-101

Global Exponential Stability of Cellular Neural Networks with Time-Varying Delays*

Qiang Zhang[1,2], Dongsheng Zhou[2], Haijun Wang[2], and Xiaopeng Wei[1]

[1] University Key Lab of Information Science & Engineering,
Dalian University, Dalian, 116622, China
[2] School of Mechanical Engineering,
Dalian University of Technology, Dalian, 116024, China
zhangq26@126.com

Abstract. The problem of global exponential stability of cellular neural networks with time-varying delays is discussed by employing a method of delay differential inequality. A simple sufficient condition is given for global exponential stability of the cellular neural networks with time-varying delays. The result obtained here improves some results in the previous works.

1 Introduction

In recent years, the stability properties of cellular neural networks (CNNs) and delayed cellular neural networks (DCNNs) introduced by Chua et al. [1]-[2] have been extensively studied and many global asymptotic stability and global exponential stability criteria for cellular neural networks with constant or time-varying delays have been proposed[3]-[14]. In this paper, by making use of a delay differential inequality, we present a new sufficient condition which guarantees global exponential stability of the unique equilibrium point of cellular neural networks with time-varying delays. Since it does not assume the delay to be differentiable, this condition is less conservative than some given in the earlier works. An example is illustrated to show the applicability of the result obtained here.

2 Preliminaries

The dynamic behavior of a continuous time cellular neural networks with variable delays can be described by the following state equations:

$$x'_i(t) = -c_i x_i(t) + \sum_{j=1}^{n} a_{ij} f_j(x_j(t)) + \sum_{j=1}^{n} b_{ij} f_j(x_j(t - \tau_j(t))) \\ + I_i, \quad i = 1, 2, \cdots, n. \qquad (1)$$

or equivalently

---

* The project supported by the National Natural Science Foundation of China (grant no. 60403001.) and China Postdoctoral Science Foundation (grant no.200303448).

$$x'(t) = -Cx(t) + Af(x(t)) + Bf(x(t-\tau(t))) + I \qquad (2)$$

where $x(t) = [x_1(t), \cdots, x_n(t)]^T \in R^n$, $f(x(t)) = [f_1(x_1(t)), \cdots, f_n(x_n(t))]^T \in R^n$, $f(x(t-\tau(t))) = [f_1(x_1(t-\tau_1(t))), \cdots, f_n(x_n(t-\tau_n(t)))]^T \in R^n$. $A = \{a_{ij}\}$ is referred to as the feedback matrix, $B = \{b_{ij}\}$ represents the delayed feedback matrix, while $I = [I_1, \cdots, I_n]^T$ is a constant input vector and time delays τ_j are bounded nonnegative functions satisfying $0 \leq \tau_j(t) \leq \tau$ for all $j = 1, 2, \cdots, n$. The activation function $f_i, i = 1, 2, \cdots, n$ satisfy the following condition

(H) Each f_i is bounded continuous and satisfies

$$|f_i(\xi_1) - f_i(\xi_2)| \leq L_i |\xi_1 - \xi_2|$$

for each $\xi_1, \xi_2 \in R$.

This type of activation functions is clearly more general than both the usual sigmoid activation functions in Hopfield networks and the piecewise linear function(PWL): $f_i(x) = \frac{1}{2}(|x+1| - |x-1|)$ in standard cellular networks [1].

Assume that the system (1) is supplemented with initial conditions of the form

$$x_i(s) = \phi_i(s), \qquad s \in [-\tau, 0], \ i = 1, 2, \cdots, n.$$

in which $\phi_i(s)$ is continuous for $s \in [-\tau, 0]$.

Due to the boundedness of the activation function f_i, by employing the well-known Brouwer's fixed point theorem, we can easily obtain that there exists an equilibrium point of Eq.(1). Besides, the uniqueness of the equilibrium point can be derived from the global exponential stability established below.

Suppose that (1) has a unique equilibrium $x^* = (x_1^*, x_2^*, \cdots, x_n^*)$. Denote

$$||\phi - x^*|| = \sup_{-\tau \leq s \leq 0} \left[\sum_{i=1}^n |\phi_i(s) - x_i^*|^2\right]^{1/2}$$

We say that an equilibrium point $x^* = (x_1^*, x_2^*, \cdots, x_n^*)$ is globally exponentially stable if there exist constants $\epsilon > 0$ and $M \geq 1$ such that

$$||x(t) - x^*|| \leq M ||\phi - x^*|| e^{-\epsilon t}, \quad t \geq 0$$

Let $y(t) = x(t) - x^*$, then Eq.(1) can be rewritten as

$$y_i'(t) = -c_i y_i(t) + \sum_{j=1}^n a_{ij} g_j(y_j(t)) + \sum_{j=1}^n b_{ij} g_j(y_j(t-\tau_j(t))) \qquad (3)$$

where $g_j(y_j) = f_j(y_j + x_j^*) - f_j(x_j^*)$, $j = 1, 2, \cdots, n$. It is obvious that the function $g_j(\cdot)$ also satisfies the hypothesis (H).

To prove the stability of the equilibrium point x^* of Eq.(1), it is sufficient to prove the stability of the trivial solution of Eq.(3).

Definition 1. *[15] Let the $n \times n$ matrix $A = (a_{ij})$ have non-positive off-diagonal elements and all principal minors of A are positive, then A is said to be an M-matrix.*

The following lemma will be used to study the global exponential convergence of (1).

Lemma 1. *[16] Let $x(t) = (x_1(t), x_2(t), \cdots, x_n(t))^T$ be a solution of the differential inequality (4).*

$$x'(t) \leq Ax(t) + B\bar{x}(t), \quad t \geq t_0 \tag{4}$$

where
$$\bar{x}(t) = (\sup_{t-\tau \leq s \leq t} \{x_1(s)\}, \sup_{t-\tau \leq s \leq t} \{x_2(s)\}, \cdots, \sup_{t-\tau \leq s \leq t} \{x_n(s)\})^T$$
$A = (a_{ij})_{n \times n}$, $B = (a_{ij})_{n \times n}$. *If :*
(H1) $a_{ij} \geq 0$ $(i \neq j), b_{ij} \geq 0$, $i,j = 1, 2, \cdots, n$, $\sum_{j=1}^{n} \bar{x}_j(t_0) > 0$;
(H2) *The matrix $-(A + B)$ is an M-matrix.*
then there always exist constants $\lambda > 0, r_i > 0$ $(i = 1, 2, \cdots, n)$ such that

$$x_i(t) \leq r_i \sum_{j=1}^{n} \bar{x}_j(t_0) e^{-\lambda(t-t_0)}. \tag{5}$$

3 Stability Analysis

Theorem 1. *If there exist positive constants $\alpha_i > 0$ $(i = 1, 2, \cdots, n)$ such that*

$$\Xi_{ij} = -\left(\left[-2c_i + 2\sum_{j=1}^{n} L_j^2\right]\delta_{ij} + \frac{\alpha_i}{\alpha_j}(a_{ij}^2 + b_{ij}^2)\right)_{n \times n}$$

is an M-matrix, where $\delta_{ij} = \begin{cases} 1, i = j \\ 0, i \neq j \end{cases}$, then the equilibrium point x^* of system (1) is globally exponentially stable.

Proof. Let $z_i(t) = \frac{1}{2}\alpha_i y_i^2(t)$, calculating the $z_i'(t)$ along the solution of (1) as follows:

$$z_i'(t) = \alpha_i y_i(t) y_i'(t)$$
$$= \alpha_i y_i(t) \left\{ -c_i y_i(t) + \sum_{j=1}^{n} a_{ij} g_j(y_j(t)) \right.$$
$$\left. + \sum_{j=1}^{n} b_{ij} g_j(y_j(t - \tau_j(t))) \right\}$$
$$= -c_i \alpha_i y_i^2(t) + \sum_{j=1}^{n} \alpha_i y_i(t) a_{ij} g_j(y_j(t))$$
$$+ \sum_{j=1}^{n} \alpha_i y_i(t) b_{ij} g_j(y_j(t - \tau_j(t)))$$

$$\leq -c_i\alpha_i y_i^2(t) + \sum_{j=1}^{n} \alpha_i L_j |y_i(t)||a_{ij}||y_j(t)|$$

$$+ \sum_{j=1}^{n} \alpha_i L_j |y_i(t)||b_{ij}||\bar{y}_j(t)|$$

$$\leq -c_i\alpha_i y_i^2(t) + \frac{1}{2}\sum_{j=1}^{n} \alpha_i \left(L_j^2 y_i^2(t) + a_{ij}^2 y_j^2(t)\right)$$

$$+ \frac{1}{2}\sum_{j=1}^{n} \alpha_i \left(L_j^2 y_i^2(t) + b_{ij}^2 \bar{y}_j^2(t)\right)$$

$$= \left[-c_i + \sum_{j=1}^{n} L_j^2\right]\alpha_i y_i^2(t) + \frac{1}{2}\sum_{j=1}^{n} \alpha_i a_{ij}^2 y_j^2(t)$$

$$+ \frac{1}{2}\sum_{j=1}^{n} \alpha_i b_{ij}^2 \bar{y}_j^2(t)$$

$$= \sum_{j=1}^{n}\left\{\left[-2c_i + 2\sum_{j=1}^{n} L_j^2\right]\delta_{ij} + a_{ij}^2 \frac{\alpha_i}{\alpha_j}\right\}\frac{1}{2}\alpha_j y_j^2(t)$$

$$+ \sum_{j=1}^{n}\left\{\frac{\alpha_i}{\alpha_j} b_{ij}^2\right\}\frac{1}{2}\alpha_j \bar{y}_j^2(t)$$

$$= \sum_{j=1}^{n}\left\{\left[-2c_i + 2\sum_{j=1}^{n} L_j^2\right]\delta_{ij} + a_{ij}^2 \frac{\alpha_i}{\alpha_j}\right\}z_j(t)$$

$$+ \sum_{j=1}^{n}\left\{\frac{\alpha_i}{\alpha_j} b_{ij}^2\right\}\bar{z}_j(t)$$

Let $\Xi_1 = \left(\left[-2c_i + 2\sum_{j=1}^{n} L_j^2\right]\delta_{ij} + a_{ij}^2 \frac{\alpha_i}{\alpha_j}\right)$, $\Xi_2 = \left(\frac{\alpha_i}{\alpha_j} b_{ij}^2\right)$, then the above inequality can be rewritten as

$$D^+ z(t) \leq \Xi_1 z(t) + \Xi_2 \bar{z}(t)$$

According to Lemma 1, if the matrix $\Xi = -(\Xi_1 + \Xi_2)$ is an M-matrix, then there must exist constants $\lambda > 0, r_i > 0$ $(i = 1, 2, \cdots, n)$ such that

$$\frac{1}{2}\alpha_{\min} y_i^2(t) \leq z_i(t) = \frac{1}{2}\alpha_i y_i^2(t)$$

$$\leq r_i \sum_{j=1}^{n} \bar{z}_j(t_0) e^{-\lambda(t-t_0)}$$

$$= r_i \sum_{j=1}^{n} \frac{1}{2}\alpha_j \bar{y}_j^2(t_0) e^{-\lambda(t-t_0)}$$

$$\leq \frac{1}{2} r_i \alpha_{\max} \sum_{j=1}^{n} \bar{y}_j^2(t_0) e^{-\lambda(t-t_0)}$$

Thus, we have

$$y_i^2(t) \leq \frac{\alpha_{\max}}{\alpha_{\min}} r_i \sum_{j=1}^{n} \bar{y}_j^2(t_0) e^{-\lambda(t-t_0)}$$

that is,

$$||x_i(t) - x_i^*|| \leq \left(\frac{\alpha_{\max}}{\alpha_{\min}} r_i\right)^{1/2} ||\bar{x}_i(t_0) - x_i^*|| e^{-\lambda(t-t_0)/2}$$

This implies that the unique equilibrium point of Eq.(1) is globally exponentially stable.

Remark 1. In [12]-[14], some results on the global asymptotic stability of Eq.(1) are presented by constructing Lyapunov functional. Different from our results, all of their results require that the delay function $\tau(t)$ be differentiable. Thus, compared with the results presented here, their conditions are more restrictive and conservative.

4   An Example

In this section, we will give an example to show the applicability of the condition given here.

Example 1. Consider cellular neural networks with variable delays

$$\begin{aligned}
x_1'(t) &= -c_1 x_1(t) + a_{11} f(x_1(t)) + a_{12} f(x_2(t)) \\
&\quad + b_{11} f(x_1(t - \tau_1(t))) + b_{12} f(x_2(t - \tau_2(t))) + I_1 \\
x_2'(t) &= -c_2 x_2(t) + a_{21} f(x_1(t)) + a_{22} f(x_2(t)) \\
&\quad + b_{21} f(x_1(t - \tau_1(t))) + b_{22} f(x_2(t - \tau_2(t))) + I_2
\end{aligned} \quad (6)$$

where the activation function is described by PWL function: $f_i(x) = \frac{1}{2}(|x+1| - |x-1|)$. Obviously, this function satisfies (H) with $L_1 = L_2 = 1$.

In (6), taking $a_{11} = 0.5, a_{12} = -0.1, a_{21} = 0.3, a_{22} = -0.2; b_{11} = 0.5, b_{12} = 0.1, b_{21} = -0.1, b_{22} = 0.1; c_1 = 2.3, c_2 = 2.04; \tau_1(t) = \tau_2(t) = |t+1| - |t-1|$. i.e., $C = \begin{bmatrix} 2.3 & 0 \\ 0 & 2.04 \end{bmatrix}$, $A = \begin{bmatrix} 0.5 & -0.1 \\ 0.3 & -0.2 \end{bmatrix}$, $B = \begin{bmatrix} 0.5 & 0.1 \\ -0.1 & 0.1 \end{bmatrix}$. Let $\alpha_1 = \alpha_2 = 1$, we can easily check that the matrix in Theorem 1 above

$$\Xi = \begin{bmatrix} 0.1 & -0.02 \\ -0.1 & 0.03 \end{bmatrix}$$

is an M-matrix. Hence, the equilibrium point of Eq.(6) is globally exponentially stable.

5 Conclusions

A new sufficient condition is given ensuring the global exponential stability of cellular neural networks with variable delays by using an approach based on delay differential inequality. The result established here extends some in the previous references.

References

1. Chua, L.O., Yang, L.: Cellular Neural Networks: Theory and Applications. IEEE Trans.Circuits Syst.I **35** (1988) 1257–1290
2. Roska, T., Boros, T., Thiran, P., Chua, L.O.: Detecting Simple Motion Using Cellular Neural Networks. Proc.1990 IEEE Int. Workshop on Cellular Neural Networks and Their Applications (1990) 127–138
3. Zhang, Y., Zhong, S.M., Li, Z.L.: Periodic Solutions and Stabiltiy of Hopfield Neural Networks with Variable Delays. Int. J. Syst. Sci. **27** (1996) 895–901
4. Qiang, Z., Ma, R., Chao, W., Jin, X.: On the Global Stability of Delayed Neural Networks. IEEE Trans.Automatic Control **48** (2003) 794–797
5. Zhang, Q., Wei, X.P. Xu, J.: Global Exponential Convergence Analysis of Delayed Neural Networks with Time-Varying Delays. Phys.Lett.A **318** (2003) 537–544
6. Zhang, Q., Wei, X.P. Xu, J.: On Global Exponential Stability of Delayed Cellular Neural Networks with Time-Varying Delays. Appl.Math.Comput. **162** (2005) 679–686
7. Zhang, Q., Wei, X.P. Xu, J.: Delay-Dependent Exponential Stability of Cellular Neural Networks with Time-Varying Delays. Chaos, Solitons & Fractals **23** (2005) 1363–1369
8. Zhang, J.: Globally Exponential Stability of Neural Networks with Variable Delays. IEEE Trans.Circuits Syst.I **50** (2003) 288–290
9. Lu, H.: On Stability of Nonlinear Continuous-Time Neural Networks with Delays. Neural Networks **13** (2000) 1135–1143
10. Xu, D., Zhao, H., Zhu, H.: Global Dynamics of Hopfield Neural Networks Involving Variable Delays. Comput. Math. Applicat. **42** (2001) 39–45
11. Zhou, D., Cao, J.: Globally Exponential Stability Conditions for Cellular Neural Networks with Time-Varying Delays. Appl.Math.Comput. **131** (2002) 487–496
12. Liao, X., Chen, G., Sanchez, E.N.: LMI-Based Approach for Asymptotically Stability Analysis of Delayed Neural Networks. IEEE Trans.Circuits Syst.I **49** (2002) 1033–1039
13. Joy, M. Results Concerning the Absolute Stability of Delayed Neural Networks. Neural Networks **13** (2000) 613–616
14. Cao, J., Wang, J.: Global Asymptotic Stability of a General Class of Recurrent Neural Networks with Time-Varying Delays. IEEE Trans.Circuits Syst.I **50** (2003) 34–44
15. Berman, A., Plemmons, R.J.: Nonnegative Matrices in the Mathematical Science. Academic Press (1979)
16. Tokumarn, H., Adachi, N., Amemiya, T.: Macroscopic Stability of Interconnected Systems. 6th IFAC Congr., Paper ID44.4 (1975) 1–7

Effect of Noises on Two-Layer Hodgkin-Huxley Neuronal Network

Jun Liu, Zhengguo Lou, and Guang Li

Key Lab of Biomedical Engineering of Ministry of Education of China,
Department of Biomedical Engineering,
Zhejiang University,
Hangzhou 310027, P.R. China
guangli@cbeis.zju.edu.cn

Abstract. Stochastic resonance (SR) effect has been discovered in non-dynamical threshold systems such as sensory systems. This paper presents a network simulating basic structure of a sensory system to study SR. The neuronal network consists of two layers of the Hodgkin-Huxley (HH) neurons. Compared with single HH model, subthreshold stimulating signals do not modulate output signal-noise ratio, thus a fixed level of noise from circumstance can induce SR for the various stimulating signals. Numeric experimental results also show that noises do not always deteriorate the capability of the detection of suprathreshold input signals.

1 Introduction

Stochastic resonance (SR) [1, 2] is a counter intuitive nonlinear phenomenon wherein transmission or detection of a signal can be enhanced by addition of a non-zero level noise. The effect has been discovered in bistable dynamical systems [3] and non-dynamical threshold systems [4].

Neural sensory systems are typical non-dynamical threshold systems [5, 6], thus they have been studied to understand how biological sensory systems utilize SR to improve their sensitivity to external inputs. Although many simulation studies on neuron model and neuronal network model have been carried out to investigate SR [7, 8], and the results indicate that the optimal intensity of noises must be altered with the different stimulating signals. The intensity of background noises depends on the average energy of random noises, so it is approximately constant. This is a limitation for neural systems to utilize external noises to detect changeful signals based SR. Collins et al. [8] investigated the dynamics of the ensemble of FitzHugh-Nagumo (FHN) model and concluded that noises positively affect the FHN model without controlling the intensity of noises, which means that the optimal intensity of noises is not necessary to be adjusted with the change of the stimulating signals. However, the model of Collins et al. adopted a summing network of excitable units, which includes only one-layer so that the effect on the next layer of the network was ignored.

In this paper, a two-layer network, differing from one-layer network in aforementioned papers, has been used to simulate the sensory systems. This neuronal network consists of the Hodgkin-Huxley (HH) model that is physiologically closer to real neuron than the FHN model [9]. Compared with the single HH model, SR in the two-layer network has a wider range of optimal intensity of noises for subthreshold input signals, while the noises do not deteriorate the capability of the detection of the suprathreshold input signals which is consistent with the result Collins obtained [8].

2 Model Description

2.1 Hodgkin-Huxley Model of Single Neuron

The HH neuronal model is a useful paradigm that accounts naturally for both the spiking behavior and refractory properties of real neurons [10], which is described by four nonlinear coupled equations: one for the membrane potential V and the other three for the gating variables: m, n, and h as following:

$$C_m \frac{dV}{dt} = I_{ext} - g_{Na}m^3h(V-V_{Na}) - g_K n^4(V-V_K) - g_L(V-V_L) + I_0 + I_1\sin(2\pi ft) + \xi(t) \tag{1}$$

$$\frac{dm}{dt} = \frac{(m_\infty(V) - m)}{\tau_m(V)} \tag{2}$$

$$\frac{dh}{dt} = \frac{(h_\infty(V) - h)}{\tau_h(V)} \tag{3}$$

$$\frac{dn}{dt} = \frac{(n_\infty(V) - n)}{\tau_n(V)} \tag{4}$$

where the ionic current includes the usual sodium, potassium, and leak currents; the parameters g_{Na}, g_K and g_L are the maximal conductance for the ions, sodium and potassium, and the leakage channels; V_{Na}, V_K and V_L are the corresponding reversal potentials; $m_\infty(V)$, $h_\infty(V)$, $n_\infty(V)$ and $\tau_m(V)$, $\tau_h(V)$, $\tau_n(V)$ represent the saturated values and the relaxation times of the gating variables, respectively. The values of parameters are listed in the appendix of this paper.

$I_1 \sin(2\pi f_s t)$ is a periodic signal with I_1 and f_s being the amplitude and the frequency of the signal respectively. I_0 is a constant stimulus being regarded as the simplest modulation to the neuron. $\xi(t)$ is the Gaussian white noise, satisfying $<\xi(t)> = 0$, $<\xi(t_1)\xi(t_2)> = 2D\delta(t_1-t_2)$, D is intensity of noises.

2.2 Two-Layer HH Neuronal Network Model

Fig.1 shows the structure of the two-layer HH neuronal network model. The first layer network consists of N parallel neurons represented by n_{11}, n_{12}, ...n_{1N}. The second layer has one neuron represented by n_2 and act as the output part of network. The total network has an analogical structure of sensory systems, in which the first layer can be considered as the part of the reception and transmission of external stimulus and converges on the neuron of the second layer.

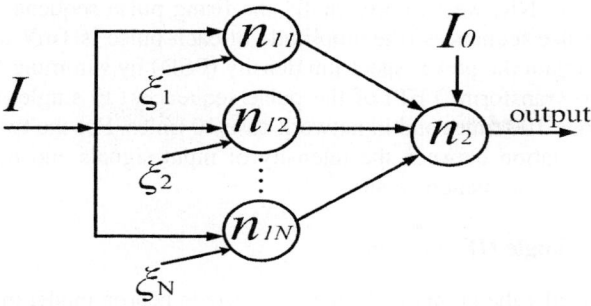

Fig. 1. The structure of the two-layer network

The input signal I of the first layer consists of the periodic stimulating signals $I_1 \sin(2\pi f_s t)$ and constant stimulating signals I_0. The former denotes the external stimulating signals including detection information and the latter is regarded as the average effect of the internal environment in sensory systems. Each neuron in the first layer is subjected to the external input noises represented by $\xi_1, \xi_2, \ldots, \xi_N$, which are assumed as an independent or uncorrelated Gaussian white noise. The neuron of the second layer receives all outputs of the neurons in the first layer and the same internal environment stimulating signals I_0. The neurons in the first layer are parallel connected with the second layer through a synapse. The synaptic current of the neuron n_{1i} is described as [11]:

$$I_{syn}^i(t) = -g_{syn}\alpha(t-t^i)[V(t) - V_{syn}^i] \;, \tag{5}$$

where $\alpha(t) = (t/\tau)e^{-t/\tau}$, g_{syn} is the maximal value of synaptic conductance, V_{syn}^i is the synaptic potential between the neuron n_2 and the neuron n_{1i}. The parameters τ and t^i represent the characteristic time of excitatory postsynaptic potential and the firing time of the neuron n_{1i}, respectively. The corresponding values of the parameters are: g_{syn}= 2mS/cm^2; V_{syn}^i= 0mV (i=1, 2…N), representing the excitatory connection between two layers; τ=2ms; t^i denoting the time when action potentials arrive at the maximal value. The total synaptic currents $I_{syn}(t)$ added on the second layer can be written as:

$$I_{syn}(t) = (1/N)\sum_{i=1}^{N} I_{syn}^i(t) \cdot \tag{6}$$

3 Results and Discussion

In this section, we will discuss the single neuron case and the two-layer HH network case. The relevant equations of two cases are solved by using a second-order algorithm suggested in Reference [12] and the integration step is taken as 0.02ms. The results of two cases are measured through the output signal-noise ratio (SNR). This SNR is defined as 10log10(G/B) with G and B representing the height of the signal peak and the mean amplitude of background noise at the input signal frequency f_s in the power spectrum respectively.

Calculating the SNR, we simplify firstly the firing pulse sequences into the standard rectangle pulse sequences (the amplitude of each pulse is 1mV and the width is 2ms), and then obtain the power spectrum density (PSD) by summing the results from the Fast Fourier Transform (FFT) of the pulse sequences. In single neuron case the summation is done 100 times and in network case 10 times. For the two cases, we pay attention to the relation between the intensity of input signals and noises when stochastic resonance phenomenon occurs.

3.1 Results for Single HH Neuron

Firstly, let us consider the output performance of single neuron model in the presence of the aforementioned input signal and the external Gaussian white noise. When single neuron is subjected to the subthreshold input signal (e.g., the amplitude of the signal I_1 is $1 \mu A/cm^2$ and the threshold of the neuron is about $I_t=1.4 \mu A/cm^2$.) and the noise (the intensity of the noise ranges between 0 and 50), the corresponding characteristics of SR are shown in Fig.2, i.e., the output SNR first rises up to a maximum around $D=2$ and then drops as D increases. On the contrary, if the stimulating signals (e.g., in fig.2 the amplitude of signal I_1 is $1.5 \mu A/cm^2$) is larger than the threshold, then SR disappears. Though SR occurs for the subthreshold stimulating signals, the bell shaped curve of SR is narrow and the optimal intensity of the noise is restricted within small range. Fig. 3 shows that change of stimulating signals exerts influence on the output SNR in present of fixed intensity of noises (D=2). It can be found that the optimal intensity of noises would be adjusted as the nature of the signal to be detected changes, i.e., the optimal detection capacity is modulated by the different stimulating signals.

Based on the central limits theorem, the integration of a variety of noises existing in environment can gain the Gaussian white noise with the steady variance represented by the intensity. Similarly, the noise imposed on neurons can be view as Gaussian white noise with a fixed intensity. Therefore, the simulation results have been thought to add a limitation when SR is used to detect changeful signal.

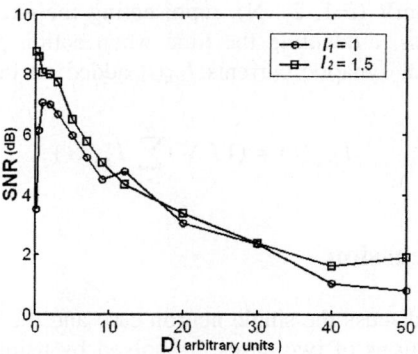

Fig. 2. The output SNR of single neuron varying with the intensity of noises D for $I_1 =1 \mu A/cm^2$ (subthreshold) and $I_1 =1.5 \mu A/cm^2$ (suprathreshold), respectively. The rest parameters: I_0 =1µA/cm², f=50Hz.

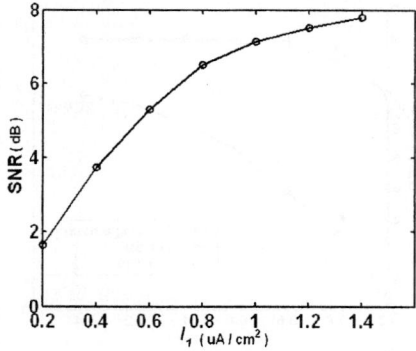

Fig. 3. The output SNR of single neuron varying with the amplitude of the subthreshold signal. I_1 The rest Parameters: $I_0 = 1$ μ A/cm², $f = 50$hz, $D = 2$.

3.2 Results for Two-Layer HH Neuronal Network

Secondly, we investigate the two-layer network described in section 2.2. The first layer of the network has N neurons parallel coupling to a neuron of the second layer. N neurons are subject to the common sinusoidal signal and independent noise.

Fig. 4 shows the output SNR versus the intensity of noises D in the case of $N=1$, $N=50$ and $N=100$. Three curves exhibit the typical characteristic of SR: first a rise and then a drop. Differently, the optimal intensity of noises in the case of $N=50$ and $N=100$ varies from 1 to10 and has much wider range than that in the case of $N=1$. This means that the range and the amplitude of optimal output SNRs (the SNR corresponding to the optimal intensity of noises) increase with the number of neurons in first layer.

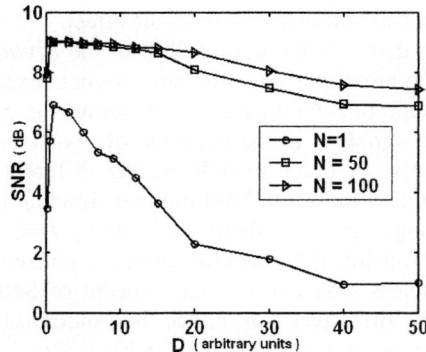

Fig. 4. The output SNR of network varying with the intensity of noises D for $N=1$, 50 and 100, respectively. The rest parameters: $I_1 = 1$μA/cm², $I_0 = 1$μA/cm², $f=50$Hz.

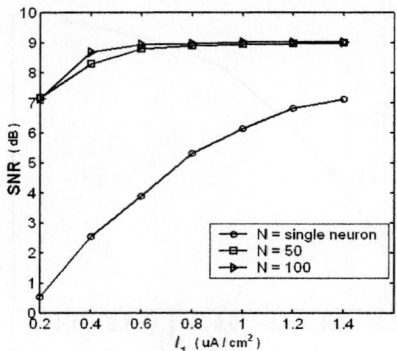

Fig. 5. The output SNR of network varying with the amplitude of periodic signal I_1 for $N=1$, 50 and 100, respectively. The rest parameter: $D=2$, $I_0=1\mu A/cm^2$, $f=50Hz$.

Fig.5 shows output SNR versus amplitude of stimulating signals I_1 in the presence of the fixed intensity of noise $D=2$. The amplitude of signals I_1 varies from 0.2 to 1.5 and is lower than that of the subthrehold signals. The intensity of noises D is during the range of the optimal intensity of noises of fig.4. In contrast with the case of $N=1$, the output SNRs in the case of $N=50$ and $N=100$ has almost constant values when I_1 varies from 0.5 to 1.5. Clearly, the SNR is not necessary to change with the signals when the noise is fixed. It is suggested that the ability of sensory systems to detect a certain range of weak (subthreshold) signals can be optimized by a fixed level of noise, irrespective of the nature of the input signal if such a network is considered as the basic structure of information processing in sensory systems. It is worth noting that the two-layer network based on HH model is more close to the real nature of sensory neuronal systems which exhibits that neurons of previous layer converge at the synapse of a neuron of next layer though the same results as Collins et al. can be obtained. It is convergence of the neurons that decreases the negative effect of noise on the synapse and ensures the rationalization of SR effect.

It is also worth noting that how the output SNR of the network varies with the intensity of noises when the amplitude of stimulating signals exceeds the firing threshold of neuron (i.e., the suprathreshold case). Fig.6 shows the output SNR versus the amplitude of stimulating signals I_1 in the presence of two fixed intensities of noises $D=2$ and $D=0$ respectively. In order to indicate the different output characteristic between the subthreshold and the suprathreshold, we simulate the output SNR in the present of the stimulating signals without noise (i.e., $D=0$). Obviously, the suprathreshold case can be illustrated by occurrence of the nonzero output SNR. It need be emphasized that noiseless case can use the concept of SNR because the output from many neurons of the first layer can induce the randomized input of the second layer. According to noiseless case, a vertical line in Fig.6 illustrates the position of the threshold. For the stimulating signal with the intensity of noise $D=2$, two curves show that the cooperative effect of many neurons can improve the output SNR in the case of the suprathreshold stimulating signals.

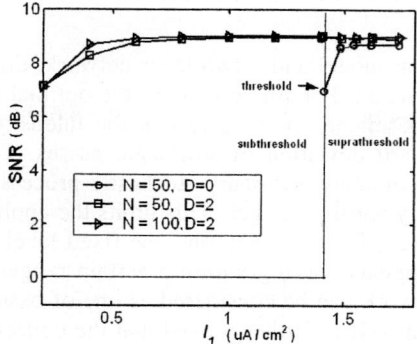

Fig. 6. The output SNR of network varying with the amplitude of periodic signal I_1 for $N=50$ and 100, $D=0$ and 2, respectively. The rest parameter: $I_0=1\mu A/cm^2$, $f=50Hz$.

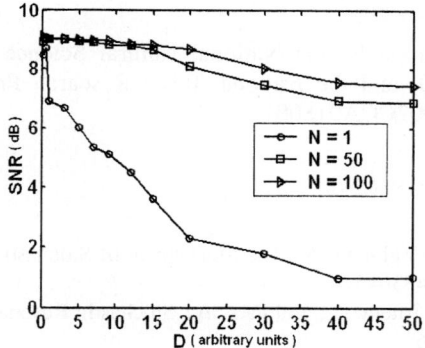

Fig. 7. The output SNR of network varying with the intensity of noises D for $N=1$, 50 and 100, respectively. The rest parameter: $I_1=1.5\mu A/cm^2$, $I_0=1\mu A/cm^2$, $f=50Hz$.

Fig.7 shows the output SNR versus the intensity of noises D in the case of $N=1$, $N=50$ and $N=100$ for the suprathreshold stimulating signals (e.g., a signal amplitude I_1 takes 1.5μA/cm².). Two curves representing $N=50$ and $N=100$ demonstrate that many neurons in the first layer can improve the output performance in contrast with the case of $N=1$. This implies that many neurons employ a certain collective effect on the synapse of one neuron.

Let us to analyze the mechanism that noise can enhances transmission of information. According to the essence of SR, noises and signals have a cooperative and competitive effect on the system. For single neuron in the case of the subthreshold input signals, randomicity of noises exerts great influence on the outputs thus the optimal intensity of SR is prone to change. For the two-layer HH network, many neurons of the first layer can produce the summation effect on the synapse of the neuron of the second layer, which can decrease the randomicity and increase the probability of signal transmission. Consequently, the output SNR of the network can be improved.

4 Conclusion

Based on the HH neuron model and a two-layer network, the effect of noises on the sensory systems is discussed. For single neuron, the optimal intensity of noises must adapt to the stimulating signals. It is noted that the intensity of noises has a linear relation with the standard deviation of stochastic noises. The fluctuation of background noises is approximately stationary stochastic process with constant standard deviation, so its intensity hardly changes. This limits the application of SR. However, for the cooperative effect of a set of neurons, the fixed level of noise can induce SR while the stimulating signals varying within a certain range. According to these results, the two-layer network can be considered as one of basic structure of signal detection in sensory systems. It is further proved that the collective behavior of a set of neurons can restrain the noises by analyzing the suprathreshold cases for the networks with different quantities of neurons.

Acknowledgment

This project was supported by the National Natural Science Foundation of China (projects No. 30470470) and the National Basic Research Program of China (973 Program, project No. 2002CCA01800).

References

1. Benzi, R., Sutera, A., Vulpiani, A.: The Mechanism of Stochastic Resonance. J. Phys. A: Math. Gen. 14(1981) 453~457
2. Benzi, R., Parisi, G., Sutera, A., Vulpiani, A.: Stochastic Resonance in Climatic Change. Tellus 34(1982) 10~16
3. Fauve, S., Heslot, F.: Stochastic Resonance in a Bistable System. Phys. Lett. A 97(1983) 5~7
4. Gingl, Z., Kiss, L. B., Moss, F.: Non-dynamical Stochastic Resonance—Theory and Experiment with White and Arbitrarily Colored Noise. Europhys Lett.. 29(1995) 191-196
5. Douglass, J.K., Wilkens, L., Pantazelou, E., Moss, F.: Noise Enhancement of Information Transfer in Crayfish Mechanoreceptors by Stochastic Resonance. Nature 365(1993) 337~340
6. Gammaitoni, L., Hänggi, P., Jung, P., Marchesoni, F.: Stochastic Resonance. Rev. Mod. Phys. 70(1998) 223~287
7. Shimokawa, T., Rogel, A., Pakdaman, K., Sato, S.: Stochastic Resonance and Spike-timing Precision in an Ensemble of Leaky Integrate and Fire Neuron Models. Phys. Rev. E 59(1999) 3461~3470
8. Collins, J. J., Chow, C. C., Imhoff, T. T.: Stochastic Resonance Without Tuning. Nature 376(1995) 236~238
9. Nagumo, J. S., Arimoto, S., Yoshizawa, S.: An Active Pulse Transmission Line Simulating Nerve Axon. Proc. IRE 50(1962) 2061-2070
10. Hodgkin, A. L., Huxley, A. F.: A Quantitative Description of Membrane Current and Its Application to Conduction and Excitation in Nerve. J. Physiol. 117(1952) 500~544

11. Koch, C., Segev, I.: Methods in Neuronal Modeling: From Ions to Networks (second edition). Cambridge (Massachusetts, USA): The MIT Press (1998) 98~100
12. Ronald, F. F: Second-order algorithm for the numerical integration of colored-noise problems. Phys. Rev. A 43(1991) 2649-2654

Appendix: Detailed Parameters of HH model

Detailed values of parameters are as follows:
V_{Na}=50mV, V_K= -77mV, V_L= -54.4mV ; g_{Na}=120mS/cm^2, g_K=36mS/cm^2, g_L=0.3mS/cm^2 ; C_m=1µF/cm^2 ;
$x_\infty(V)=a_x/(a_x+b_x)$, $\tau_x(V)=1/(a_x+b_x)$ with $x=m,h,n$;
$a_m = 0.1(V+40)/(1-e^{(-V-40)/10})$, $b_m = 4e^{(-V-65)/18}$,
$a_h = 0.07e^{(-V-65)/20}$, $b_h = 1/(1+e^{(-V-35)/10})$,
$a_n = 0.01(V+55)/(1-e^{(-V-55)/10})$, $b_n = 0.125e^{(-V-65)/80}$.

Adaptive Co-ordinate Transformation Based on a Spike Timing-Dependent Plasticity Learning Paradigm

QingXiang Wu, T.M. McGinnity, L.P Maguire,
A. Belatreche, and B. Glackin

School of Computing and Intelligent Systems, University of Ulster, Magee Campus
Derry, BT48 7JL, N.Ireland, UK
Q.Wu@ulster.ac.uk

Abstract. A spiking neural network (SNN) model trained with spiking-timing-dependent-plasticity (STDP) is proposed to perform a 2D co-ordinate transformation of the polar representation of an arm position to a Cartesian representation in order to create a virtual image map of a haptic input. The position of the haptic input is used to train the SNN using STDP such that after learning the SNN can perform the co-ordinate transformation to generate a representation of the haptic input with the same co-ordinates as a visual image. This principle can be applied to complex co-ordinate transformations in artificial intelligent systems to process biological stimuli.

1 Introduction

The brain receives multiple sensory data from environments where the different senses do not operate independently, but there are strong links between modalities [1]. Electrophysiological studies have shown that the somatosensory cortex SI neurons in monkeys respond not only to touch stimulus but also to other modalities. Strong links between vision and touch have been found in behavioural [2] and electrophysiological [3] studies, and at the level of single neurons [4]. For example, neurons in the somatosensory cortex (SI) may respond to visual stimuli [5] and other modalities [6]. Neurons in monkey primary SI may fire both in response to a tactile stimulus and also in response to a visual stimulus [5].

A new interaction between vision and touch in human perception is proposed in [7]. These perceptions may particularly interact during fine manipulation tasks using the fingers under visual and sensory control [8]. Different sensors convey spatial information to the brain with different spatial coordinate frames. In order to plan accurate motor actions, the brain needs to build an integrated spatial representation. Therefore, cross-modal sensory integration and sensory-motor coordinate transformations must occur [9]. Multimodal neurons using non-retinal bodycentred reference frames are found in the posterior parietal and frontal cortices of monkeys [10-12]. Basis function networks with multidimensional attractors [13] are proposed to simulate the cue integration and co-ordinate transformation properties that are observed in several multimodal cortical areas. Adaptive regulation of synaptic strengths within

SI could explain modulation of touch by both vision [14] and attention [15]. Learned associations between visual and tactile stimuli may influence bimodal neurons.

Based on these concepts, a spiking neural network (SNN) model is proposed to perform the co-ordinate transformation required to convert a time-coded haptic input to a space-coded visual image. The SNN model contains STDP synapses from haptic intermediate neurons to the bimodal neurons. In Section 2, the SNN model is presented. The spiking neuron model and STDP implementation is described in Section 3. The training approach is described in Section 4. After training, the strength of synapses between haptic intermediate neurons and bimodal neurons is obtained. A simplified model is provided in this paper to demonstrate that neural networks based on integrate-and-fire neurons with STDP are capable of performing 2D co-ordinate transformation. The implication for a biological system and applications in artificial intelligent systems are discussed in Section 5.

2 Spiking Neural Network Model for Co-ordinate Transformation

In order to simulate location related neurons in the somatosensory cortex (SI), suppose that x' and y' are single layers of bimodal neurons that represent the Cartesian co-ordinates of the output. A point (X, Y) at the touch area can provide both visual and haptic stimuli that reach x' and y' bimodal neuron layers through a visual pathway and a haptic pathway respectively. Fig.1 shows a simplified SNN model for building associations between visual and haptic stimuli. When a finger touches a point in the touch area, visual attention focuses on the point and the retinal neurons corresponding to this point are activated. These neurons provide the training stimulus to x' and y' bimodal neuron layers through the visual pathway. When the finger touches the point, the arms activate the corresponding neurons in θ and Φ neuron layers. These stimuli are fed into haptic pathway. Actually, θ and Φ are based on bodycentred co-ordinates, which are polar co-ordinates. The neurons in θ and Φ layers transfer haptic location signals to the intermediate layer, and then this intermediate layer transfers the bodycentred co-ordinate to the integrated co-ordinate x' and y' neuron layers. In the SNN model, x' and y' bimodal neurons have a receptive field corresponding to the vertical and horizontal lines on the retinal neuron layer respectively, and receive haptic stimuli from all the intermediate neurons through STDP synapses. These STDP synapses make it possible to learn and transform bodycentred co-ordinate (θ, Φ) to co-ordinate (x', y'). The co-ordinate (x', y') can be regarded as integrated co-ordinates in the brain. For simplicity, the synapse strength from retinal neuron layer to (x', y') neurons is fixed. Under this situation, co-ordinate (x', y') is actually the retina-centred co-ordinate. The transformation is equivalent to transformation from a haptic bodycentred co-ordinate to a retina-centred co-ordinate. Each neuron in the θ and Φ layers is connected to an intermediate layer within a vertical field and a horizontal field with fixed synapse strength respectively, as shown in Fig.1.

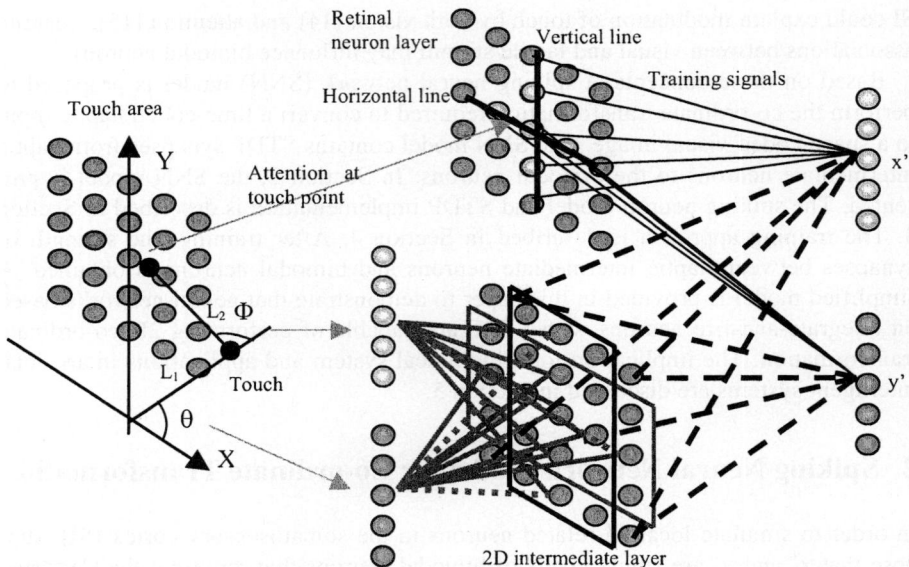

Fig. 1. A SNN model for 2D co-ordinate transformation. (X,Y) is co-ordinate for touch area. (a) Visual pathway: the retinal neuron layer is represented by 2D layer with 40X40 neurons that are connected to x' and y' neuron layer with a fixed weights. (b) Haptic pathway: L1 and L2 are arms. θ and Φ are arm angles represented by a 1D neuron layer respectively. Each θ neuron is connected to the neurons within a corresponding vertical rectangle in the 2D intermediate layer. Each Φ neuron is connected to the neurons within a corresponding horizontal rectangle in the 2D intermediate layer. The neurons in the intermediate layer are fully connected to the x' and y' neuron layers with STDP synapses. These connections are adapted in response to the attention visual stimulus and haptic stimulus under STDP rules.

3 Spiking Neuron Model and STDP Implementation

3.1 Integrate-and-Fire Neuron Model

The integrate-and-fire model is applied to each neuron in the SNN. In a conductance based integrate-and-fire model, the membrane potential $v(t)$ is governed by the following equations [16], [17], [18], [19].

$$c_m \frac{dv(t)}{dt} = g_l(E_l - v(t)) + \sum_j \frac{w^j g_s^j(t)}{A_s}(E_s - v(t)) \qquad (1)$$

where c_m is the specific membrane capacitance, E_l is the membrane reversal potential, E_s is the reversal potential ($s \in \{i,e\}$, i and e indicate inhibitory and excitatory synapses respectively), w^j is a weight for synapse j, and A_s is the membrane surface area connected to a synapse. If the membrane potential v exceeds the threshold voltage, v_{th}, v

is reset to v_{reset} for a time τ_{ref} and an action potential event is generated. Fig. 2 shows that a neuron receives spike trains from three afferent neurons.

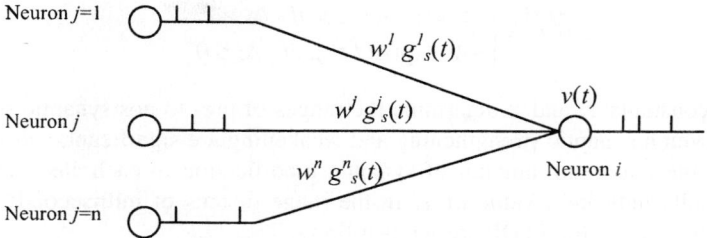

Fig. 2. Conductance based synapses connections in a SNN

The valuable $g^j_s(t)$ is the conductance of synapse j. When an action potential reaches the synapse at t_{ap}, the conductance is increased by the following expression.

$$g^j_s(t_{ap} + t^j_{delay} + dt) = g^j_s(t_{ap} + t^j_{delay}) + q_s \qquad (2)$$

Otherwise, the conductance decays as illustrated in the following equation.

$$\frac{g^j_s(t)}{dt} = -\frac{1}{\tau_s} g^j_s(t) \qquad (3)$$

where q_s is the peak conductance. Neuron i integrates the currents from afferent synapses and increases the membrane potential according to Equation (1). In our simulation, the parameters are set as follows. t^j_{delay}=0. v_{th} =-54 mv. v_{reset} =-70 mv. Ee= 0 mv. Ei=-75 mv. q_{e_max}=0.01 µs. q_{i_max}=0.01 µs. q_e=0.002 µs. q_i=0.002 µs. El=-70 mv. g_l =1.0 µs/mm². c_m=10 nF/mm². τ_e=3 ms. τ_i=10 ms. A_e=0.028953 mm². A_i=0.014103 mm².

3.2 STDP Implementation Approach

In order to perform STDP learning in the SNN, the implementation approach in [20],[21] is applied. Each synapse in an SNN is characterized by a peak conductance q_s (the peak value of the synaptic conductance following a single presynaptic action potential) that is constrained to lie between 0 and a maximum value q_{s_max}. Every pair of pre- and postsynaptic spikes can potentially modify the value of q_s, and the changes due to each spike pair are continually summed to determine how q_s changes over time. The simplifying assumption is that the modifications produced by individual spike pairs combine linearly.

A presynaptic spike occurring at time t_{pre} and a postsynaptic spike at time t_{post} modify the corresponding synaptic conductance by

$$q_s \rightarrow q_s + q_{s\_max} F(\Delta t) \qquad (4)$$

where $\Delta t = t_{post} - t_{pre}$ and

$$F(\Delta t) = \begin{cases} A_+ \exp(\Delta t / \tau_+), & if \; \Delta t > 0 \\ -A_- \exp(\Delta t / \tau_-), & if \; \Delta t \leq 0 \end{cases} \qquad (5)$$

The time constants τ_+ and τ_- determine the ranges of pre- to postsynaptic spike intervals over which synaptic strengthening and weakening are significant, and A_+ and A_- determine the maximum amount of synaptic modification in each case. The experimental results indicate a value of τ_+ in the range of tens of milliseconds (about 20 ms). The parameters for STDP are set as follows.

$q_{s\_max} = 0.01$, $A_+ = 0.01$, $A_- = 0.005$, $\tau_+ = 20$ ms, $\tau_- = 100$ ms.

The function $F(\Delta t)$ for synaptic modification is shown in. Fig. 3.

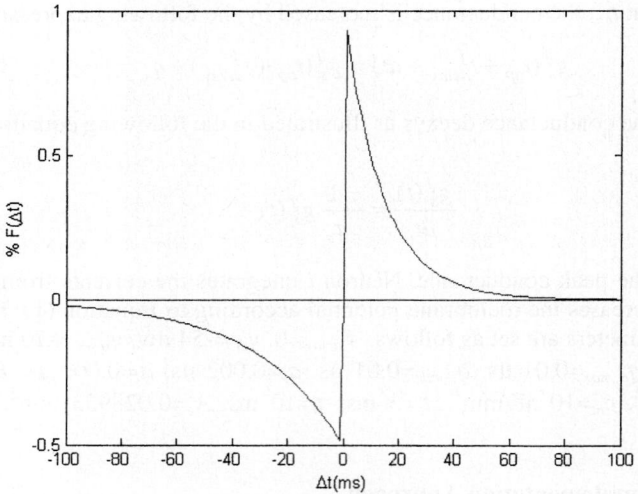

Fig. 3. Synaptic modification

4 Learning and Simulation Results

This network can be trained using an unsupervised approach. When a finger touches a point in the touch area, the haptic stimulus triggers (θ, Φ) stimuli that are fed into the haptic pathway. At the same time, the visual attention focuses on the tip of the finger and this position signal is transferred to (x', y') neuron layer through the visual pathway. The STDP synapses between intermediate layer and (x', y') neuron layer are trained under STDP rules. The finger randomly touches different points for a Poisson distribution period with a mean of 20ms. The STDP synapses from the intermediate layer to (x', y') neurons can adapt synapse strength in response to the stimulus and form a weight distribution for association between haptic and visual training stimuli. By repeating the finger touching within the whole touch area randomly, the weight distribution is adapted in response to the haptic and visual stimuli and reaches a stable state after 800s training time. The weight distribution is shown in Fig. 4. The stimuli are represented by Poissonian spike trains whose firing rate is drawn from a Gaussian distribution. The centre of the stimulus corresponds to the finger position within the touch area.

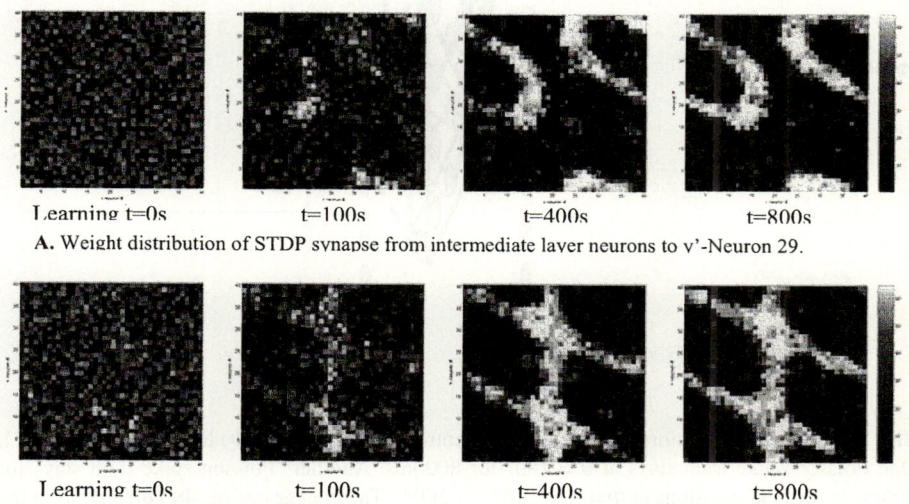

Learning t=0s   t=100s   t=400s   t=800s
A. Weight distribution of STDP synapse from intermediate layer neurons to v'-Neuron 29.

Learning t=0s   t=100s   t=400s   t=800s
B. Weight distribution of STDP synapse from intermediate layer neurons to y'-Neuron 40.

Fig. 4. Change of weight distribution during STDP learning. During the learning process, the weight distribution is recorded each 100s time interval. The distributions at moment 0, 100, 400, and 800s are shown in row A for y'-neuron 29 and row B for y'-neuron 40. Colour yellow indicates maximal weights.

In our experiments, 40 neurons are employed to encode θ and Φ layers respectively. 1600 neurons are applied to the 2D intermediate layer and training layer respectively. 80 neurons are applied to x' and y' layers respectively. After training, (x', y') neurons can respond to both visual and haptic stimuli. When the visual pathway is blocked, (x', y') neurons respond only to haptic stimulus at the correct position, i.e. (θ, Φ) layers and the intermediate layer can perform a co-ordinate transformation from the bodycentred co-ordinate (θ, Φ) to co-ordinate (x', y'). If two Poisson procedure spike trains with bell-shaped distributions are fed into the (θ, Φ) layers respectively, the responses of (x', y') neurons, representing the result of the co-ordinate transformation, are shown in Fig.5.

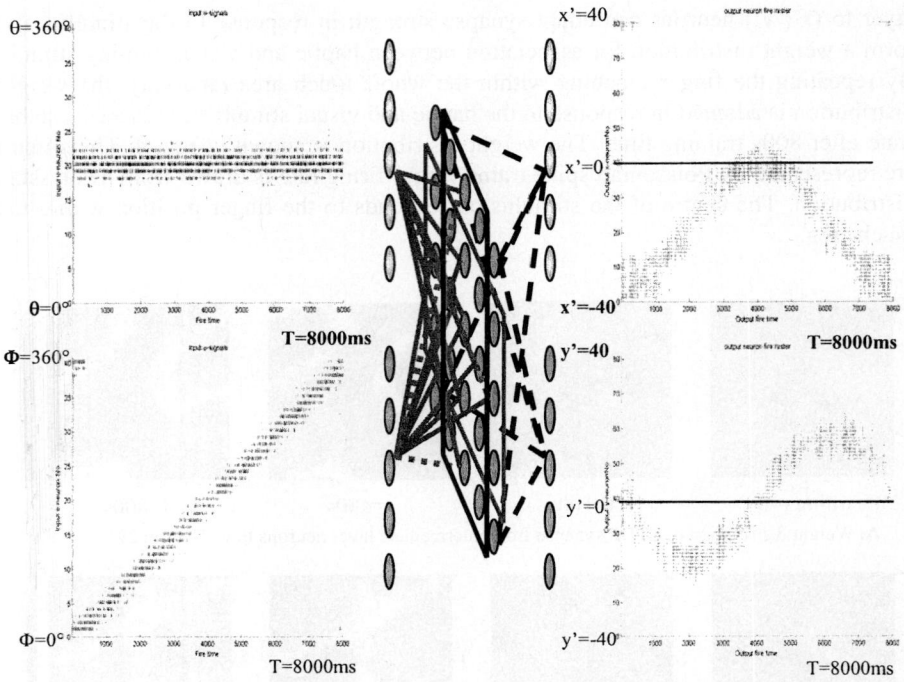

Fig. 5. Co-ordinate transformation from bodycentred co-ordinate (θ, Φ) to co-ordinate (x', y'). One Poisson spike train stays at θ = 180° for 8000ms. Another Poisson spike train stays for 200ms in sequent positions at Φ=0°, 9°, 18°, ...360°. The changes of (θ, Φ) correspond to the finger moving along a circle with radius L. The output x' = L (Sin(θ) − Cos(Φ)), y'=L(Cos(θ) + Sin(Φ)).

5 Conclusion

In the presented SNN model, the network is trained by the arm angles of the haptic stimuli position fed to the input layer, and a position signal, which is regarded as a supervising signal, fed to the output layer via the visual pathway. The strength of the synapses between the intermediate layer and output layer is trained under the STDP learning paradigm. A firing rate encoding scheme is applied in the network. The input stimulus is represented by Poissonian spike trains whose rates are drawn from a two-dimensional Gaussian distribution at the input layer and a one-dimensional Gaussian distribution at the output layer. The conceived network is able to perform a 2D coordinate transformation by learning the Cartesian coordinates (x, y) from the angular positions of the haptic stimulus. The network is more robust and provides better noise immunity than classical neural networks as even if some of the neurons do not work, the network can still perform the transformation function. The model can provide a biologically plausible approach for designing artificial intelligent systems.

Acknowledgement

The authors gratefully acknowledge the valuable discussions on 1D co-ordinate transformation model with Dr. Andrew Davison and Dr. Yves Frégnac from UNIC (Unité de Neurosciences Intégratives et Computationnelles, Centre National de la Recherche Scientifique, France). The authors also acknowledge the financial and technical contribution of the SenseMaker project (IST-2001-34712), which is funded by the EC under the FET life like perception initiative.

References

1. Marisa Taylor-Clarke, Steffan Kennett, Patrick Haggard, Persistence of visual-tactile enhancement in humans, Neuroscience Letters, Vol. 354, No.1, Elsevier Science Ltd, (2004) 22–25.
2. Spence, C., Pavani, F., Driver, J., Crossmodal links between vision and touch in covert endogenous spatial attention, J. Exp. Psychol. Hum. Percept. Perform. 26 (2000) 1298–1319.
3. Eimer M., Driver, J., An event-related brain potential study of crossmodal links in spatial attention between vision and touch, Psychophysiology, 37 (2000) 697–705.
4. Graziano, M.S.A., Gross, C.G., The representation of extrapersonal space: A possible role for bimodal, visual–tactile neurons, in: M.S. Gazzaniga (Ed.), The Cognitive Neurosciences, MIT Press, Cambridge, MA, (1994) 1021–1034.
5. Zhou, Y.D., Fuster, J.M., Visuo-tactile cross-modal associations in cortical somatosensory cells, Proc. Natl. Acad. Sci. USA 97 (2000) 9777–9782.
6. Meftah, E.M., Shenasa, J., Chapman, C.E., Effects of a cross-modal manipulation of attention on somatosensory cortical neuronal responses to tactile stimuli in the monkey, J. Neurophysiol. 88 (2002) 3133–3149.
7. Kennett, S., Taylor-Clarke, M., Haggard, P., Noninformative vision improves the spatial resolution of touch in humans, Curr. Biol. 11 (2001) 1188–1191.
8. Johansson, R.S., Westling, G., Signals in tactile afferents from the fingers eliciting adaptive motor-responses during precision grip, Exp. Brain. Res. 66 (1987) 141–154.
9. Galati, Gaspare-Committeri, Giorgia - Sanes, Jerome N. - Pizzamiglio, Luigi, Spatial coding of visual and somatic sensory information in body-centred coordinates, European Journal of Neuroscience, Vol.14, No.4, Blackwell Publishing, (2001) 737-748.
10. Colby, C.L. & Goldberg, M.E., Space and attention in parietal cortex. Annu. Rev. Neurosci., 22 (1999) 319-349.
11. Gross, C.G. & Graziano, M.S.A. Multiple representations of space in the brain. Neuroscientist, 1 (1995) 43-50.
12. Rizzolatti, G., Fogassi, L.& Gallese, V. Parietal cortex: from sight to action. Curr. Opin. Neurobiol., 7 (1997) 562-567.
13. Deneve S., Latham P. E. and Pouget A., Efficient computation and cue integration with noisy population codes, Nature Neuroscience, 4 (2001) 826-831.
14. Taylor-Clarke M., Kennett S., Haggard P., Vision modulates somatosensory cortical processing, Curr. Biol. 12 (2002) 233–236.
15. Iriki, A., Tanaka, M., Iwamura, Y., Attention-induced neuronal activity in the monkey somatosensory cortex revealed by pupillometrics, Neurosci. Res. 25 (1996) 173–181.
16. Christof Koch, Biophysics of Computation: Information Processing in Single Neurons. Oxford University Press, (1999).

17. Peter Dayan and Abbott, L.F.. Theoretical Neuroscience: Computational and Mathematical Modeling of Neural Systems. The MIT Press, Cambridge, Massachusetts, (2001).
18. Wulfram Gerstner and Werner Kistler. Spiking Neuron Models: Single Neurons, pulations,Plasticity. Cambridge University Press, (2002).
19. Müller, E. Simulation of High-Conductance States in Cortical Neural Networks, Masters thesis, University of Heidelberg, HD-KIP-03-22, (2003).
20. Song, S., Miller, K. D., and Abbott, L. F. Competitive Hebbian learning though spike-timing dependent synaptic plasticity. Nature Neuroscince, 3 (2000) 919-926.
21. Song, S. and Abbott, L.F. Column and Map Development and Cortical Re-Mapping Through Spike-Timing Dependent Plasticity. Neuron 32 (2001) 339-350.
22. Joseph E. Atkins, Robert A. Jacobs, and David C. Knill, Experience-dependent visual cue recalibration based on discrepancies between visual and haptic percepts, Vision Research, Volume 43, Issue 25, (2003) 2603-2613.

Modeling of Short-Term Synaptic Plasticity Using Dynamic Synapses

Biswa Sengupta

Department of Computer Science (Advanced Computer Architectures Group),
Department of Psychology (York Neuroimaging Centre),
University of York, Heslington, York YO10 5DD, England
sengupta@ieee.org

Abstract. This work presents a model of minimal time-continuous target-cell specific use-dependent short-term synaptic plasticity (STP) observed in the pyramidal cells that can account for both short-term depression and facilitation. In general it provides a concise and portable description that is useful for predicting synaptic responses to more complex patterns of simulation, for studies relating to circuit dynamics and for equating dynamic properties across different synaptic pathways between or within preparations. This model allows computation of postsynaptic responses by either facilitation or depression in the synapse thus exhibiting characteristics of dynamic synapses as that found during short-term synaptic plasticity, for any arbitrary pre-synaptic spike train in the presence of realistic background synaptic noise. Thus it allows us to see specific effect of the spike train on a neuronal lattice both small-scale and large-scale, so as to reveal the short-term plastic behavior in neurons.

1 Introduction

Among the various hallmarks in brain science, memory and learning are the most researched because they transmute a brain into a mind. Learning & memory demands the exploration of two levels of modeling computation in neural systems: level of individual synapses & spiking neurons, and the network level i.e., overlap of neurons in ensembles and the dynamics of synaptic connections. The signalling between neurons is central to the functioning of the brain, but we still do not understand how the code used in signalling depends on the properties of synaptic transmission [1]. Generally neurons communicate with each other primarily through fast chemical synapses. Such synapses have action potential (AP) generated near the cell body that propagates down the axon where it opens voltage-gated Ca^{2+} channels. The entering Ca^{2+} ions trigger the rapid release of vesicles containing neurotransmitter, which is ultimately detected by receptors on the postsynaptic cell [2]. Short-term synaptic plasticity refers to this change in the synaptic efficacy on a timescale that is inverse to the mean firing rate and thus of the order of milliseconds *(ms)* [3]. The experimental observation that forms the basis of the short-term plasticity [2] lies in the fact that the transmission of an action potential across a synapse has influence on the postsynaptic potential (PSP) induced by the subsequent spikes [3]. One of the vital features of short-term plasticity is the dependence of the steady-state amplitude on stimulation frequency (Table 1). Also the amplitude of

Table 1. Popular types of short-term plastic behavior

Type	Onset	Decay
Fast Facilitation	1-5 spikes	10-100 ms
Fast Depression	1-10 spikes	100 ms - 1 s
Slow Depression	> 10 spikes	> 1 s
Augmentation	1-10 spikes	1-5 s
Post-tetanic Potentiation (PTP)	> 10 spikes	> 5 s

the postsynaptic response is proportional to the probability that a synapse transmits a given presynaptic spike. Refs. [4, 5, 6] have shown that the synaptic transmission probability for a presynaptic spike train of frequency r is approximately proportional to $\frac{1}{r}$ for $r > 10 - 20\ Hz$.

Typically, the synapses of most artificial neural networks are static, in the sense that the single value characterizing it remains fixed except on short timescales. The model implemented here adds to our understanding of how neural circuits process complex temporal patterns. It is the balance between the facilitation & depression of the synaptic strength in short time scales, that determine the temporal dynamics and the basis for each computation in the synapse.

2 Computational Model

To acquire the coarse grained character of the neuronal dynamics, we compute the postsynaptic current (PSC) using a detailed compartmental model of a hippocampal neuron (based on the data from ModelDB [7]) depicting the phenomenological model of pyramidal neocortical neurons [1]. The NEURON model (basic network structure depicted in Fig. 1) formalism used in this work broadly describes the data on short-term plasticity [4, 1, 5, 6, 2] and, at the same time, is compact enough to be incorporated easily into network models. STP, in particular depression & facilitation strongly influence neuronal activity in cerebral cortical circuits. Facilitation & depression mechanisms in a synapse are quite interconnected as stronger facilitation will lead to higher utilization of synaptic efficacy which subsequently leads to stronger depression. When receiving high frequency input, as during a presynaptic burst, a depressing synapse will only transmit the onset of the signal efficiently. Synapses which are facilitating, on the other hand, will transmit the signal with increasing strength, until a maximum is reached some 30-300 ms after burst onset.

Synaptic short-term plasticity is shown through the proper quantification of features of the action potential activity of the presynaptic neurons and populations transmitted in pyramidal cells and interneurons [1, 5, 6, 4, 2]. Refs. [1, 5] & [6] have also shown the derivation of mean-field dynamics of neocortical networks to understand the dynamic behavior of large neuronal populations. This work facilitates description of neuronal dynamics by calculating the postsynaptic current (PSC) generated by a neuronal lattice with a particular firing rate in response to both deterministic & Poisson spike trains. It is important to note that the coupling strength between two neurons depend on: no.

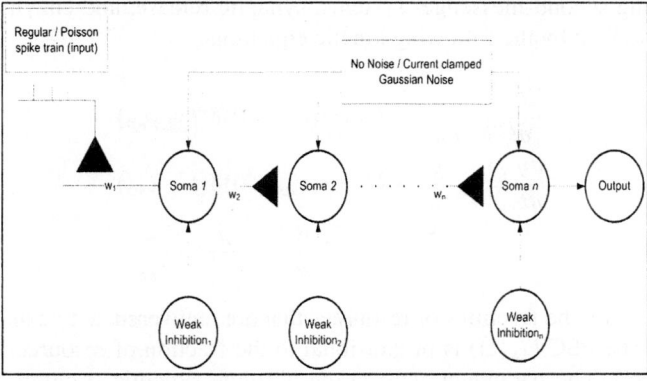

Fig. 1. The neural network in NEURON

of release sites (n), probability of release of a vesicle following a presynaptic action potential (AP) (p) & some measure of postsynaptic response to a single vesicle (q).

We start with a simple implementation to display STP behavior of a dynamic synapse by producing three sub-models with the synapses implemented with conduction changes rather than current sources [5] because real synapses are associated with conductance changes. The models concentrate on:

1. Layer V pyramidal neuron interaction to show the effect of depression
2. Layer V pyramidal neuron to an inhibitory interneuron interaction to show the effects of facilitation
3. Layer V pyramidal neuron to an inhibitory interneuron interaction with different model dynamics to show the effect of facilitation & early depression

It is vital to be aware of the fact that it is experimentally quite difficult to isolate the response of a single synapse, and the data have become available just very recently [8]. The results are quite startling. Those single synapses (synaptic release sites) in the Central Nervous System (CNS) exhibit a binary response to each spike from the presynaptic neuron - either the synapse releases a single neurotransmitter filled vesicle or it doesn't respond at all. In the case when a vesicle is released, its content enters the synaptic cleft and opens ion channels in the postsynaptic membrane, thereby creating an electrical pulse in the postsynaptic neuron. To capture this realism, the model incorporates both deterministic & probabilistic model for the dynamic synapses.

2.1 Model Mechanism

Deterministic Dynamic Synapse Model: The deterministic dynamic synapse model is based on the idea of finite amount of resources available for transmission. Each presynaptic spike, at time t_{sp} activates a fraction of resources (U_{SE}, utilization or synaptic efficacy). This then inactivates within few milliseconds (time constant of τ_{in}) and recovers about 1 second later (time constant of τ_{rec}). The important biochemical reactions, including second-messenger systems, synaptic release, and enzymatic cascades are partly described by the following kinetic equations:

$$\frac{dx}{dt} = \frac{z}{\tau_{rec}} - U_{SE}x(t_{sp} - 0)\delta(t - t_{sp})$$
$$\frac{dy}{dt} = -\frac{y}{\tau_{in}} + U_{SE}x(t_{sp} - 0)\delta(t - t_{sp})$$
$$\frac{dz}{dt} = \frac{y}{\tau_{in}} - \frac{z}{\tau_{rec}}, \qquad (1)$$

where x, y & z are the fractions of resources that are recovered, active and the inactive respectively. The PSC ($I_s(t)$) is proportional to the fraction of resources in the active state ($A_{SE}y(t)$). The parameter A_{SE} is the absolute synaptic strength, which is determined by activating all the resources. U_{SE} determines the dynamics of the synaptic response [5]. The fraction of synaptic resources available for transmission is determined by the following differential equation,

$$\frac{dR}{dt} = \frac{(1-R)}{\tau_{rec}} - U_{SE}.R.\delta(t - t_{sp}). \qquad (2)$$

The amplitude of the postsynaptic response at time t_{sp} is a dynamic variable given by the product PSR (Postsynaptic Response) $= A_{SE} * R(t_{sp})$.

The chemical kinetics represented by (1) doesn't include facilitating mechanism which is only evident in synapses between the pyramidal neurons and inhibitory interneurons. Short-term facilitation is through the introduction of a *facilitation factor*. It increases by the advent of each spike and in the same time decays in between the spikes. Hence for this change, we need to assume that U_{SE} is not fixed but rather increased by a certain amount due to each presynaptic spike. This running value of U_{SE} is referred to as U_{SE}^1. Generally, an increase in U_{SE} would mean an accumulation of calcium ions caused by spikes arriving in the presynaptic zone. If we take an example of a simple kinetic scheme, in which an AP causes a fraction of U_{SE} calcium channel to open, that later closes with a time constant of τ_{facil}, the fraction of opened calcium channel is then determined by the current value of U_{SE}^1. The final kinetic equation then becomes

$$\frac{dU_{SE}^1}{dt} = -\frac{U_{SE}^1}{\tau_{facil}} + U_{SE}(1 - U_{SE}^1)\delta(t - t_{sp}). \qquad (3)$$

U_{SE} determines the increase in the value of U_{SE}^1 due to each spike and coincides with the value of U_{SE}^1 reached upon the arrival of the first spike [5]. The iterative expression for the value of U_{SE}^1 reached upon the arrival of the nth spike, which determines the response according to (1) is:

$$U_{Se}^{1(n+1)} = U_{SE}^{1(n)}(1 - U_{SE})\exp(-\delta t/\tau_{facil}) + U_{SE}$$
$$EPSC_{n+1} = EPSC_n(1 - U_{SE})e^{-\delta t/\tau_{rec}}$$
$$+ A_{SE}.U_{SE}(1 - e^{-\delta t/\tau_{rec}}), \qquad (4)$$

where δt is the time interval between n^{th} and $(n+1)^{th}$ spikes. If the presynaptic neuron releases a regular spike train (as that shown in the first stage of the simulation) at frequency r, then U_{SE} reaches a steady state value, as shown in (5). Hence in this kinetics, U_{SE}^1 becomes a frequency-dependent variable, and U_{SE} is treated as the kinetic parameter characterizing the activity-dependant transmission of this particular synapse. It is evident that U_{SE} is responsible for determining the contribution of facilitation in generating subsequent synaptic responses. Smaller values of U_{SE} display facilitation but this is not observed for higher values of U_{SE}.

$$\frac{U_{SE}}{1 - (1 - U_{SE})\exp(-1/r\tau_{facil})}. \qquad (5)$$

We know that there are an infinite number of ways by which a neuronal population can fire relative to each other. These are usually Poisson processes. The equations for regular spike activities can be adjusted for Poisson processes as under:

$$\frac{d\langle x\rangle}{dt} = \frac{1 - \langle x\rangle}{\tau_{rec}} - \langle U_{SE}^1\rangle \langle x\rangle r(t)$$
$$\frac{d\langle U_{SE}^-\rangle}{dt} = -\frac{\langle U_{SE}^-\rangle}{\tau_{facil}} + U_{SE}(1 - \langle U_{SE}^-\rangle)r(t)$$
$$\langle U_{SE}^1\rangle = \langle U_{SE}^-\rangle(1 - U_{SE}) + U_{SE}. \qquad (6)$$

Here, $r(t)$ denotes the Poisson rate of the spike train for the neuron at time t. $\langle U_{SE}^-\rangle$ is the average value of U_{SE}^1 just before the spike. Depressing synapses are described by (6) with fixed value of U_{SE}^1. To make the model simpler it is assumed that the inactivation time constant τ_{in} is faster than the recovery time τ_{rec}. This assumption is made to adjust the biological data found in pyramidal interneuron synapses. To find the postsynaptic current, we simply use:

$$\frac{d\langle y\rangle}{dt} = -\frac{\langle y\rangle}{\tau_{in}} + \langle U_{SE}^1\rangle \langle x\rangle r(t). \qquad (7)$$

To account for the timescales that are slower than τ_{in}, (7) can be reduced to

$$y = r\tau_{in}U_{SE}^1 \langle x\rangle. \qquad (8)$$

Probabilistic Dynamic Synapse Model: The probabilistic model (accounts for the inter-trial fluctuations) is based on the deterministic model — the probability of a vesicle at release site, P_v is similar to the fraction of resources available, R in the deterministic model. The probability of release of docked vesicle in the former is also analogous to the fraction of available resources in the latter. The kinetics of the probabilistic model is represented by

$$\frac{dP_v}{dt} = \frac{(1-P_v)}{\tau_{rec}} - U_{SE}.P_v.\delta(t-t_{sp})$$

$$P_r(t_{sp}) = U_{SE}.P_v. \qquad (9)$$

$P_r(t_{sp})$ represented the probability of release for every release site [9].

Mean-Field Network Dynamics: Though this study facilitates computation of the postsynaptic responses of facilitating and depressing synapses for any arbitrary presynaptic spike train, it becomes challenging when the need of mean-field equations for describing the neocortical dynamics of large networks arises. The firing rates of a closed population of neurons with two sub-populations of cortical pyramidal excitatory & inhibitory interneurons, where each of the population can be considered as a cortical column having neurons with similar receptive field (RF) properties [5] can be formulated using (10) & (11).

$$\tau_e \frac{dE_r}{dt} = -E_r + g(\sum_{r'} J_{rr'}^{ee} y_{r'}^{ee} - J_{rr'}^{ei} y_{rr'}^{ei} + I_r^e), \qquad (10)$$

$$\tau_i \frac{dI_r}{dt} = -I_r + g(\sum_{r'} J_{rr'}^{ie} y_{r'}^{ie} - J_{rr'}^{ii} y_{rr'}^{ii} + I_r^i), \qquad (11)$$

$E_r(I_r)$ is the firing rate of the excitatory or the inhibitory sub-population located at r,
$g(x)$ is the response function which is assumed to be monotonously increasing,
$J_{rr'}^{ee}$ is the absolute strength of the synaptic connection between excitatory neurons located at r and r' times the average number of such connections per postsynaptic neuron, and
$I_r^e(I_r^i)$ is the external input to the excitatory (inhibitory) population.

2.2 Model Dynamics and Implementation

The NEURON implementation is based on the *synchrony generation model* [5, 6]. The basic scheme is:
1) $x \rightarrow y$ *(Instantaneous, spike triggered)* & the increment here is $u * x$. Here, x is the fraction of *"synaptic resources"* that has *"recovered"* (fraction of transmitter pool that is ready for release, or fraction of postsynaptic channels that are ready to be opened, or some joint function of these two factors) & y is the fraction of *"synaptic resources"* that are in the *"active state"*. This is proportional to the number of channels that are open, or the fraction of maximum synaptic current that is being delivered.

2) $y \xrightarrow{\tau_1} z$, z is the fraction of *"synaptic resources"* that are in the *"inactive state"*.

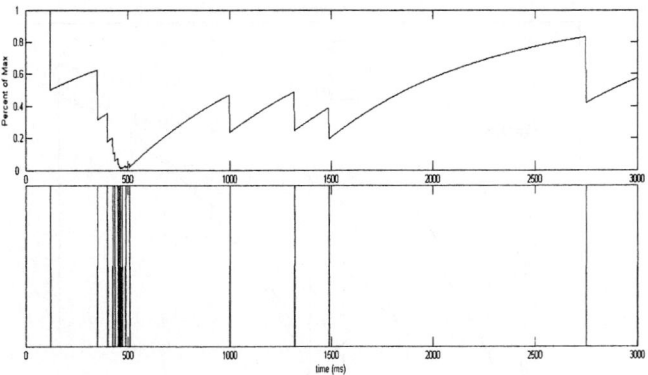

Fig. 2. Relationship of resource pool to that of the spike amplitude

3) $z \stackrel{T_{rec}}{\rightarrow} x$, where $x + y + z = 1$. The synapses represent a conductance change. Active state y is multiplied by a synaptic weight to compute the actual synaptic conductance (or current, in the original form of the model). Additionally, facilitation term u that governs fraction of x, is converted to y on each synaptic activation. It should be noted that u is incremented before x is converted to y. If u is incremented after x is converted to y then the first synaptic activation after a long interval of silence will produce smaller and smaller postsynaptic effect as the length of the silent interval increases, eventually becoming vanishingly small.

4) $\rightarrow u$ (*Instantaneous, spike triggered*). This happens before x is converted to y. Increment is $U * (1 - u)$ where U and u both lie in the range 0 - 1.

5) $u \stackrel{\tau_{facil}}{\rightarrow}$ (*Decay of facilitation*).

This implementation for NEURON offers the user a parameter $u0$ that has a default value of 0 but can be used to specify a nonzero initial value for u. When $\tau_{facil} = 0$, u is supposed to equal U. Note that the synaptic conductance in this mechanism has the same kinetics as y i.e., decays with time constant τ_1. This mechanism can receive multiple streams of synaptic input and each stream keeps track of its own weight and activation history.

3 Result

This model illustrates a viable means to account for the dynamic changes in the post-synaptic response resulting from the timing of pre-synaptic inputs under a constrained synaptic transmission i.e., an amount of neurotransmitter is used each time the pre-synaptic cell is stimulated, it then recovers with a particular rate & probability, binding the amount of neurotransmitter to post-synaptic receptor, resulting in a PSR.

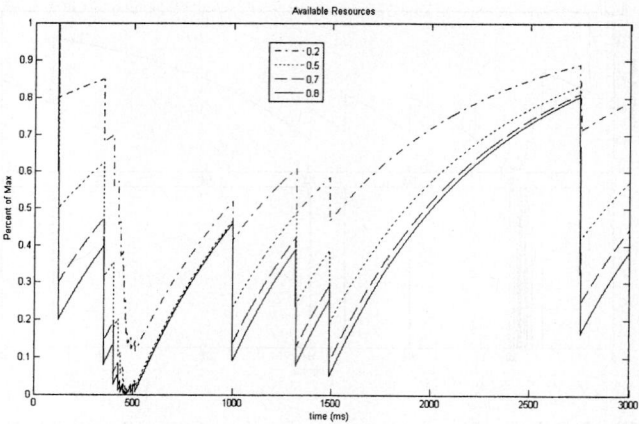

Fig. 3. Different values of U_{SE} for the deterministic model

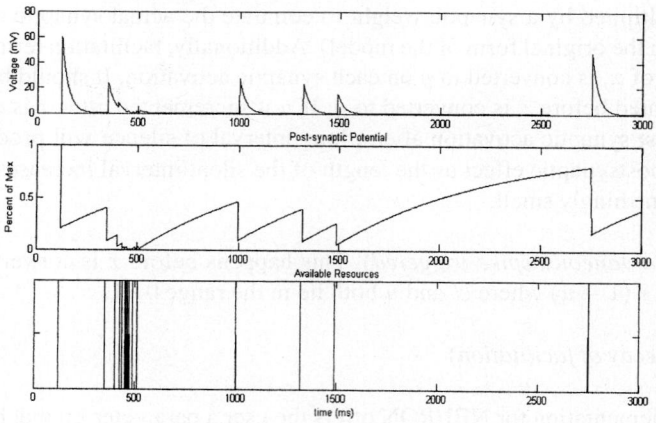

Fig. 4. Post-synaptic response is proportional to available resources

The amplitude of the spike is dictated by the amount of resources available. After a spike has occurred, the resources follow the same recovery equation, increasing the pool available for the next spike as shown in Fig.2. U_{SE} is analogous to the probability of release in the quantal model of synaptic transmission. A comparison of this is made in Fig.3. The PSR on the other hand, is directly proportional to the amount of available resources (Fig.4).

The probabilistic model on the other hand, accounts for the trial-to-trial fluctuations in observed synaptic responses. Each release site has at most one vesicle available for release with a release probability of U_{SE}. A comparison of the probabilistic value to that of the deterministic one can be noted in Fig. 5 for different number of vesicles.

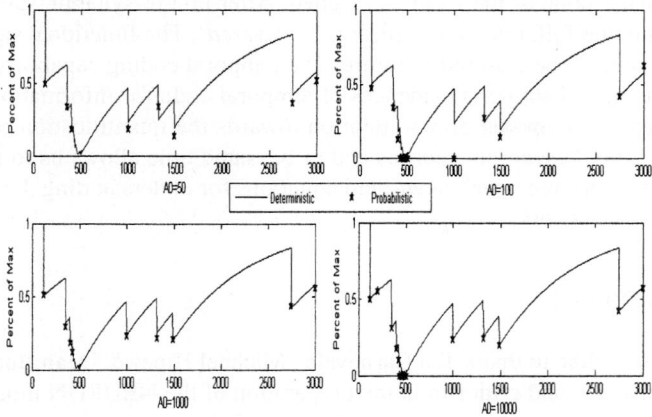

Fig. 5. Comparison of deterministic & probabilistic model. $A0$ is the no. of vesicles

4 Discussion

This theoretical study looks into a simplistic view of single responses from neocortical synapses & their usage in encoding temporal information about the presynaptic spikes. Typically information is potentially represented by a combination of two mechanisms. Rate coding allows the information to be conveyed by the average firing rate of pre-synaptic input. The problem with this type of coding is that it is possible for completely different distributions of spikes to result in the same mean firing rate. On the other hand, in temporal coding the information is conveyed by the timing of the presynaptic input. The PSR is influenced by the interspike interval (ISI). In all these effort, it is seen that the secondary dynamics of the network is quite rightfully portrayed by short-term plasticity. This work also captures the short-term activity-dependent changes in the amplitudes of the PSR that characterize different types of synaptic connections. Probabilistic models *(classical quantal)* were used for studying the behavior of single synapses whereas, the deterministic ones were generally used during the study of big neuron population. In fact, in the big networks having probabilistic transmission does not qualitatively change the behavior of the network. Variability of quantal response amplitudes of single CNS synapse is taken into consideration and hence we assume that the PSR to the release of each vesicle (q) is not a constant value but has a Gaussian distribution with mean μ and variance σ^2.

The model captures the fact that the common induction protocol with repeated pairings of pre- and post- synaptic spikes in a specific temporal relation does not change the scaling factors i.e., the weights or the synaptic efficacy of the amplitudes of the Excitatory Post Synaptic Current (EPSC), but rather the synaptic release probabilities U for the first spike in the spike train. An increase of this parameter U will increase the amplitude of the EPSP for the first spike but it tends to decrease the amplitudes of the following EPSPs. The typical manifestation of this short-term, adaptation mechanism is the rapid decrease in the successive values of EPSPs induced by a fast, regular pre-synaptic train, until a stationary value of the EPSP is reached. After no pre-synaptic spikes occur for about 1 second, the full, maximum EPSP is *'recovered'*. The functional significance of this study remains to be clarified in future. The temporal coding capability of dynamic synapse in the model supports evidences of temporal code for information processing in the neocortex. The model draws attention towards the quantification of features in AP activity of the presynaptic neurons and in the same time allows us to instigate further analysis by deriving novel mean-field equations for understanding the dynamics of large neocortical networks.

Acknowledgment

The author would like to thank Ted Carnevale, Michael Hines & Dean Buonomano for providing assistance and criticism in the preparation of the NEURON model. DeLiang Wang, Misha Tsodyks & Wolfgang Maass were all helpful to the numerous queries with regard to their respective views of the dynamic synapses.

References

1. Tsodyks, M., Markram, H.: The neural code between neocortical pyramidal neurons depends on neurotransmitter release probability. Proc Natl Acad Sci U S A **94** (1997) 719–723
2. Zucker, R., Regehr, W.: Short-term synaptic plasticity. Annu Rev Physiol **64** (2002) 355–405
3. Kistler, W., van Hemmen, J.: Short-term synaptic plasticity and network behavior. Neural Computation **11** (1999) 1579–1594
4. Varela, J., Sen, K., Gibson, J., Fost, J., Abbott, L., Nelson, S.: A quantitative description of short-term plasticity at excitatory synapses in layer 2/3 of rat primary visual cortex. J Neurosci **17** (1997) 7926–7940
5. Tsodyks, M., Pawelzik, K., Markram, H.: Neural networks with dynamic synapses. Neural Computation **10** (1998) 821–835
6. Tsodyks, M., Uziel, A., Markram, H.: Synchrony generation in recurrent networks with frequency-dependent synapses. J Neurosci **20** (2000)
7. Sengupta, B.: Modeling of short-term synaptic plasticity. Undergraduate Thesis (2004)
8. Dobrunz, L., Stevens, C.: Heterogeneity of release probability, facilitation and depletion at central synapses. Neuron, **18** (1997) 995–1008
9. Fuhrmann, G., Segev, I., Markram, H., Tsodyks, M.: Coding of temporal information by activity-dependent synapses. J Neurophysiol **87** (2002) 140–148

A Chaotic Model of Hippocampus-Neocortex

Takashi Kuremoto[1], Tsuyoshi Eto[2], Kunikazu Kobayashi[1], and Masanao Obayashi[1]

[1] Yamaguchi Univ., Tokiwadai 2-16-1, Ube, Japan
{wu, koba, m.obayas}@yamaguchi-u.ac.jp
[2] CEC-OITA Ltd., Kumano Ohira 21-1, Kitsuki, Japan

Abstract. To realize mutual association function, we propose a hippoca- mpus-neocortex model with multi-layered chaotic neural network ($MCNN$). The model is based on Ito et al.'s hippocampus-cortex model (2000), which is able to recall temporal patterns, and form long-term memory. The $MCNN$ consists of plural chaotic neural networks (CNNs), whose each CNN layer is a classical association model proposed by Aihara. $MCNN$ realizes mutual association using incremental and relational learning between layers, and it is introduced into $CA3$ of hippocampus. This chaotic hippocampus-neocortex model intends to retrieve relative multiple time series patterns which are stored (experienced) before when one common pattern is represented. Computer simulations verified the efficiency of proposed model.

1 Introduction

The experimental studies on physiological and anatomical suggest that memory functions of brain are executed in neocortex and hippocampus [1,2,3,4]. Although the mechanism of learning and memory is not understood completely, the process of memorization can be considered roughly as: sensory receptor → sensory memory (in primary cortex) → short-term memory (in neocortex) → intermediate-term memory (in a dialogue between the hippocampus and the neocortex) → long-term memory (in neocortex) [1,3,4,7]. Based on the knowledge of facts in nature, Ito et al. proposed a hippocampus-neocortex model for episodic memory [5,6], and a hippocampus-cortex model for long-term memory [7]. Meanwhile, as chaotic phenomena are observed in neurons activity, there have been many chaotic neural networks were proposed for decades [8,9,10,11,12,13,14,15,16]. For chaotic memory systems, especially, there also exit *chaotic neural networks* (CNN) given by Aihara and his fellows [10,11], *transient-associative network* ($TCAN$) given by Lee [14], adveced Aihara's models and their applications [12,13,16], and so on. These models provide auto-associative function, recall input patterns as short-term memory.

Though all facts of neocortex, hippocampus and the communication between them are understood poorly, recent researches show the important role of hippocampus in the formation of long-term memory in neocortex [3,4]. Here, we assume there is a chaotic circuit in $CA3$ of hippocampus, and improve Ito et al.'s model [7] using a multi-layered chaotic neural network ($MCNN$)[16]. The new chaotic model provides one-to-many retrieval of time-series patterns by its incremental and relational learning between chaotic neural network (CNN) layers. So it is able to realize mutual association which exists in the humans brain but the mechanism is not understood yet.

2 Model

2.1 Model of Ito *et al.* [7]

The original hippocampus-cortex model of Ito *et al.* is presented by Fig. 1 [7]. The signal flow of the system is: input patterns (Input layer) → sensory memory (Cortex 1) → short-term memory (Cortex 2) and intermediate-term memory (DG) → Hebbian learning ($CA3$) → decoding ($CA1$) → long-term memory (Cortex 2). The long-term memory are stored in Cortex 2 at last, and as output of system, the stored temporal patterns are recalled when one of the patterns is represent as input. We repeated computer simulation of this model and obtained the same results as Ref. [7]. When we presented an input pattern which was stored in two different time-series patterns, however, the system failed to retrieve two temporal patterns correctly. The reason could be considered that energy of internal state function dropped at a convergence point corresponding to the input pattern.

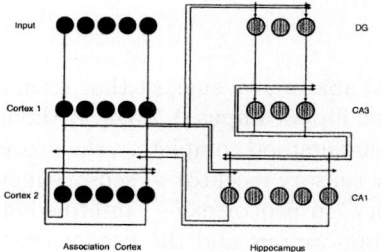

Fig. 1. Structure of hippocampus-cortex model proposed by Ito *et al.* (2000)

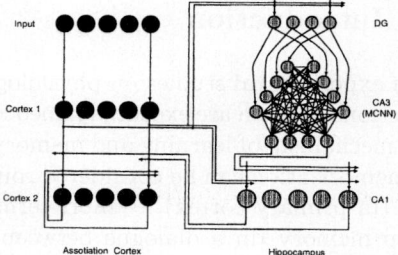

Fig. 2. Structure of a chaotic model of hippocampus-neocortex proposed here

Meanwhile, there are many other remarkable approaches of theoretical studies for associative memory [10,11,14]. Classical chaotic neural models are able to retrieve stored time-series patterns by external stimulus. However, the retrieval is a dynamical short-term memory. Considering the ability of exchanging short-term memory into long-term memory function of hippocampus [1,2,3,4], here we introduce a multi-layered chaotic neural network ($MCNN$) [16] into

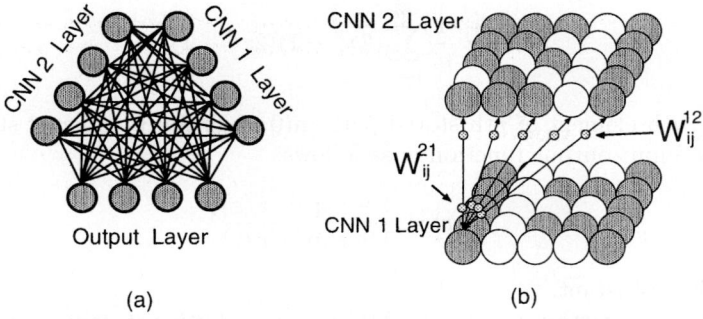

Fig. 3. Proposal structure of $CA3$ layer model: (a) Multi-layered chaotic neural network ($MCNN$); (b) Connections between $MCNN$ layers

conventional hippocampus-cortex model to realize mutual association of different time-series patterns (one-to-many retrieval). The new chaotic hippocampus-neocortex model is expected to form long-term memory in neocortex.

2.2 MCNN [16]

For real neurons active far more complicatedly than artificial neurons constructured with simple threshold elements, chaotic neural models are proposed also [10,11]. To realize mutual association function, for instance, the formation of conditional reflex (Ivan Pavlov), we proposed to combine multiple classical CNN layers as an associative model $MCNN$ (Fig. 3). In $MCNN$, neurons on each CNN layer and between the layers connect each other completely, and the dynamics is as follows:

$$x_i(t+1) = f\left(y_i(t+1) + z_i(t+1) + \gamma \cdot v_i(t+1)\right) \quad (1)$$

$$y_i(t+1) = k_r y_i(t) - \alpha x_i(t) + a_i \quad (2)$$

$$z_i(t+1) = k_f z_i(t) + \sum_{j=1}^{n} w_{ij} x_j(t) \quad (3)$$

$$v_i(t+1) = k_e v_i(t) + \sum_{j=1}^{n} W_{ij}^* x_j^*(t) \quad (4)$$

where $x_i(t)$: output value of ith neuron at time t, n: number of input, w_{ij}: connection weight from jth neuron to ith neuron, $y_i(t)$: internal state of ith neuron as to factory, $z_i(t)$: internal state of ith neuron as to reciprocal action, $v_i(t)$: internal state of ith neuron as to reciprocal action from another layer, α: threshold of ith neuron, k_f, k_r, k_e: damping rate, a_i: item given by the summation of threshold and external input, γ: the rate of effectiveness from another layer, W_{ij}^*: connection weight from jth neuron of another layer to ith neuron, $x_j^*(t)$: output value of jth neuron of another layer at time t. The connection weight w_{ij} is define as:

$$w_{ij} = \frac{1}{m}\sum_{p=1}^{m}(2x_i^p - 1)(2x_j^p - 1) \tag{5}$$

where, x_i^p: ith element of pth stored pattern(0 or 1), m: number of stored patterns. The input-output function is as follows:

$$f(x) = \frac{1 - \exp(-x/\varepsilon)}{1 + \exp(-x/\varepsilon)} \tag{6}$$

where, ε is a constant.

When a new pattern is input to $MCNN$, an additive storage is executed on each CNN layer through a_i ($i = 1, ..., n$). After states of the system store the pattern, Hebb learning, Δw_{ij}, is executed as:

$$\Delta w_{ij} = \frac{1}{m} x_i x_j \tag{7}$$

here, m is a number of the stored patterns.

The connection weights, W_{ij}^{12} and W_{ij}^{21} relate patterns between what stored in different layers of $MCNN$. Using relational Hebbian learning, a 2-layer $MCNN$, for example, stores the time-series patterns as:

$$\Delta W_{ij}^{12} = \beta \cdot x_i^1 x_j^2, \; \Delta W_{ij}^{21} = \beta \cdot x_i^2 x_j^1 \tag{8}$$

, where, β is the learning rate, x_i^1 is output value of ith neuron of $CNN1$, x_i^2 is output value of ith neuron of $CNN2$.

2.3 Chaotic Model of Hippocampus-Neocortex

Hippocampus is considered availably an exchange organ between short-term memory and long-term memory [3,4]. Long term potentiation (LTP), phenomena observed in $CA3$ layer of hippocampus especially, maybe give the key of long-term memory formation. Here, we propose a chaotic model of hippocampus-neocortex by introducing $MCNN$ into $CA3$ of Ito et al. model (Fig. 2). Neurons on each layer of $MCNN$ accept signals from DG, then provide output of sparse representation from its Output layer to $CA1$. The dynamics of this model will be described in this section.

Association Cortex. The dynamics of association cortex (Left of Fig. 1) is described as same as Ito et al. model [7]:

$$I_i(t) = \begin{cases} 1 \cdots \text{excitatory} \\ 0 \cdots \text{inhibitory} \end{cases} \tag{9}$$

$$x_i^{cx1}(t) = I_i(t) \tag{10}$$

$$x_i^{cx2}(t) = f\left(\sum_{j=0}^{N} w_{ij}^{cx2 \cdot cx2} x_j^{cx2}(t-1) \right.$$
$$\left. + w^{cx2 \cdot cx1} x_i^{cx1}(t) + w^{cx2 \cdot c1}(t) x_i^{c1}(t) - \theta^{cx}\right) \tag{11}$$

here, $I_i(t)$: ith input number, $x_i^{cx1}(t)$: output of ith neuron in $CX1$, $x_i^{cx2}(t)$: output of ith neuron in $CX2$, $w_{ij}^{cx2 \cdot cx2}$: weight of connection from jth to ith neuron in $CX2$ (variable), $w^{cx2 \cdot cx1}$: weight of connection from $CX1$ to $CX2$ (fixed), $w^{cx2 \cdot c1}$: weight of connection from $CA1$ to $CX2$ (fixed), $x_i^{c1}(t)$: output of ith neuron in $CA1$ in hippocampus, θ^{cx}: threshold, N: number of neurons in $CX1$ and $CX2$, f: step function.

The learning of connection weights in $CX2$ is according to Hebb rule:

$$\Delta w_{ij}^{cx2 \cdot cx2} = \alpha_{hc} \cdot x_i^{cx2}(t) x_j^{cx2}(t-1) \tag{12}$$

where α_{hc} is a learning rate.

Hippocampus

- DG
  Competition learning is executed in this layer. The input from association cortex is exchanged into internal states (pattern-encoding).

$$k = \arg\max_i \sum_{j=0}^{N} w^{dg \cdot cx2} x_j^{cx2}(t) \tag{13}$$

$$x_i^{dg}(t) = \begin{cases} random & \cdots \text{ initial} \\ f\left(\sum_{j=0}^{N} w_{ij}^{dg \cdot cx1} x_j^{cx1} - \theta^{dg}\right) & \cdots \text{ usual} \end{cases} \tag{14}$$

The learning rule of connection weight from $CX2$ to DG $w_{ij}^{dg \cdot cx2}$ is,

$$\Delta w_{ij}^{dg \cdot cx2} = \beta_{hc} \cdot x_i^{dg}(t) x_j^{cx2}(t) \tag{15}$$

. Here, β_{hc} is a constant, $\alpha_{hc} < \beta_{hc}$.

- CA3
  Feedback connections exist in $CA3$, and they result association function like Hopfield model. Ito et al. noticed in this respect, however just presented the dynamics of $CA3$ by a step function only. We suppose chaotic memorization phenomena exist in $CA3$, and apply $MCNN$ which provides mutual association on $CA3$ layer. By learning of $CA3$ (self-feedback connections), the intermediate patterns are formed.

$$k = \arg\max_i \sum_{j=0}^{n} w_{ij}^{c3out \cdot cnn1}(2x_j^{cnn1}(t) - 1) \tag{16}$$

or

$$k = \arg\max_i \sum_{j=0}^{n} w_{ij}^{c3out \cdot cnn2}(2x_j^{cnn2}(t) - 1) \tag{17}$$

$$x_i^{c3out}(t) = \begin{cases} 1 \cdots (i = k) \\ 0 \cdots (i \neq k) \end{cases} \qquad (18)$$

where,
$w_{ij}^{c3out \cdot cnn1}$: weight of connection from jth neuron of $CNN1$ layer
to ith neuron of Output layer in $CA3$
$w_{ij}^{c3out \cdot cnn2}$: weight of connection from jth neuron of $CNN2$ layer
to ith neuron of Output layer in $CA3$
$x_i^{cnn1}(t)$: output of ith neuron of $CNN1$ layer in $CA3$ (given by 1-4)
$x_i^{cnn2}(t)$: output of ith neuron of $CNN2$ layer in $CA3$ (given by 1-4)
$x_i^{c3out}(t)$: output of ith neuron of Output layer in $CA3$.

For time-series patterns are stored in $MCNN$ alternatively, $CNN1$ layer and $CNN2$ layer are excitative alternately, Eq. 16 and Eq. 17 are adopted alternatively. This structure intends to result mutual association like successful behavior of bidirectional associative memory model (BAM).
The rule of learning of self-feedback connection weights is

$$\Delta w_{ij}^{c3 \cdot c3} = \beta_{hc} \cdot x_i^{c3}(t) x_j^{c3}(t-1). \qquad (19)$$

Here, β_{hc} is as same as DG, n is number of neurons in hippocampus.
– $CA1$
Internal states in hippocampus is decoded into output patterns. The input from association cortex performs as a teacher signal.

$$x_i^{c1}(t) = f\left(\sum_{j=0}^{n} w_{ij}^{c1 \cdot c3} x_j^{c3}(t) + w^{c1 \cdot c1} x_j^{cx1}(t) - \theta^{c1} \right). \qquad (20)$$

From $CA3$ to $CA1$, connections $w_{ij}^{c1 \cdot c3}$ learn according to

$$\Delta w_{ij}^{c1 \cdot c3} = \beta_{hc} \cdot x_i^{c1}(t) x_j^{c3}(t). \qquad (21)$$

3 Computer Simulation

3.1 One-to-Many Time-Series Patterns

We define One-to-many time-series patterns retrieval as : there is a same pattern exists in different time-series patterns, and by representing the pattern to associative memory model (proposed chaotic model of hippocampus-neocortex, at here), all patterns in the different time-series are recalled as output of system. Fig. 4 shows input patterns in two time-series, where the first pattern is common **Pattern A**, are used in our computer simulation. Each time-series ($Time\ Series\ A$ and $Time\ Series\ B$) includes 4 orthogonal patterns. An input pattern is presented temporally to model, and with more than 1 interval, it is classified to be from different time-series (Fig. 5).

3.2 Simulation Process

The purpose and process of computer simulation of proposed model is described as follow:

- Input time-series patterns whether and how to be processed in model: To Cortex 1 layer, external stimulus is time-series patterns described in last section. DG transforms sensory pattern into internal state in hippocampus. $CA3(MCNN)$ compresses the signals of DG, stores internal states in dynamical networks and outputs in simple forms. $CA1$ decodes the signals from Output layer of $MCNN(CA3)$.

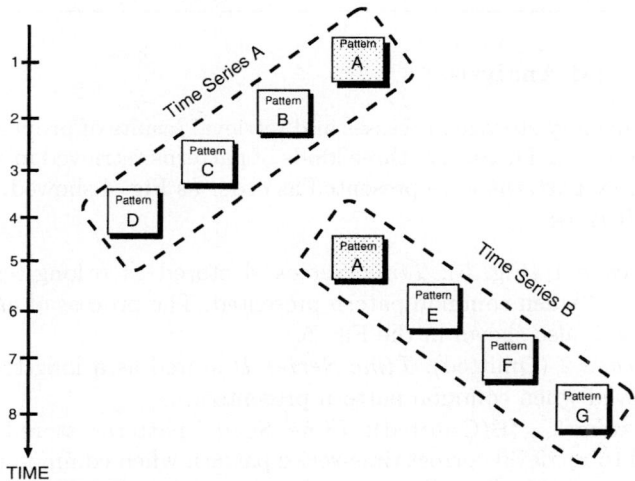

Fig. 4. Time-series patterns for one-to-many retrieval

- Long-term memory whether and how to be formed in model: To repeat to input a holding pattern, which is a common pattern exists in different time-series, to form intermediate term memory in $CA3$ and Cortex 2, and long-term memory becomes to be stored in Cortex 2 at last. The repetition stimulation can be considered as long-term potentiation (LTP) like-hood phenomenon which is observed in brain.
- One-to-many time-series patterns retrieval result: After different time-series patterns are presented, and a common pattern represented, whether proposed chaotic model retrieves all patterns or not.

3.3 Parameters

Parameters of proposed model in simulation is given as follow:

N	$= 50$	: number of neurons in association cortex
n	$= 30$	: number of neurons in hippocampus
$w_{ij}^{cx2 \cdot cx1}$	$= 1.0$	: weight of connection from Cortex 1 to Cortex 2
$w_{ij}^{cx2 \cdot c1}$	$= 1.0$	: weight of connection from hippocampus to Cortex 2
α_{hc}	$= 0.0015$	: learning rate in association cortex
β_{hc}	$= 1.0$	: learning rate in hippocampus
θ^{c3}	$= 0.5$	: threshold of neuron in $CA3$ layer
θ^{c1}	$= 0.5$	: threshold of neuron in $CA1$ layer
S	$= 0.07$	: maximum correlation between random patterns
θ^{dg}	$= 5.5$	: threshold of neuron in DG
ε	$= 0.15$	: slope of sigmoid function
γ	$= 0.3$	: influential rate from other CNN layer in $CA3$

3.4 Result and Analysis

Fig. 5 shows memory storage processes and retrieval results of proposed model in computer simulation. There were three kinds of patterns retrieved in 86 iterations when orthogonal patterns were presented as order as Fig. 4 showed. The rate of different result is as:

- result in *case 1* (Fig. 5): *Time Series A* stored as a long-term memory and retrieved when common patern presented. The process of encoding and decoding was also shown in the Fig. 5.
- result in *case 2* (Omitted): *Time Series B* stored as a long-term memory and retrieved when common pattern presented.
- result in *case 3* : (B(Omitted): *Time Series* patterns stored confusedly, and failed to retrieval correct time-series pattern when common pattern presented. The reason of confusion can be observed on TIME A stage in $CA3$ (Output layer of $MCNN$), where same situation occured for different time-series patterns signals from DG.

The ratio of these different kinds of retrieval is shown in Tab. 1. We also repeated computer simulation of Ito *et al.* model using the same time-series patterns (Tab. 1). *Time Series B* could not be stored as long-term memory for confusion with *Time Series A* which was input before.

Table 1. Simulation result: retrieval rate for one-to-many time-series patterns

Kind of retrieval	case 1 ($TimeSeriesA$)	case 2 ($TimeSeriesB$)	case 3 ($Failed$)
Convetional Model	0.0	0.0	100.0
Proposed Model	60.0	6.0	34.0

unit : (B[%])

A Chaotic Model of Hippocampus-Neocortex 447

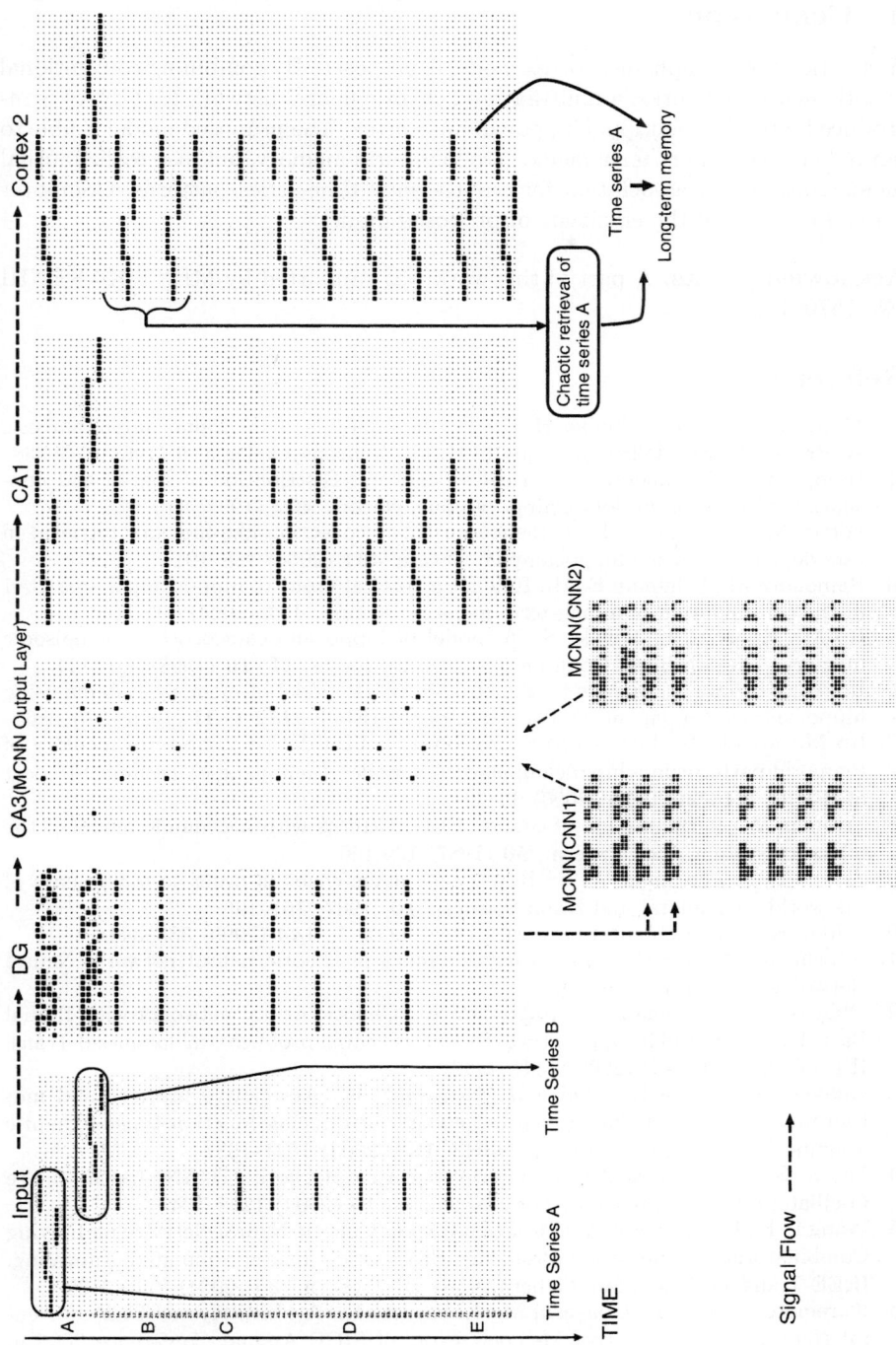

Fig. 5. Simulation result in *case* 1 (*Time Series A* is retrieved)

4 Conclusion

A chaotic hippocampus-neocortex model is proposed. By combining conventional chaotic neural networks, a multi-layered chaotic neural network ($MCNN$) is introduced into a conventional hippocampus-cortex. The proposed model is able to not only convert short-term memory to long-term memory, but also realiz mutual memorization and association for one-to-many time-series patterns. Computer simulation verified the efficiency of proposed model.

Acknowledgments. A part of this work was supported by MEXT-KAKENHI (No.15700161).

References

1. Milner B., Corkin S., Teuber H. L.,: Further analysis of the hippocampal amnesic syndrome: 14-year follow-up study of H. M.. Neuropsychologia. **6** (1968) 317-338
2. Hampson R. E., Simeral J. D., Deadwyler S. A.: Distribution of spatial and non-spatial information in dorsal hippocampus. Nature. **402** (1999) 610-614
3. Fortin N. J., Wright S. P., Eichenbaum H.: Recollection-like memory retrieval in rats dependent on the hippocampus. Nature. **431** (2004) 188-191
4. Remondes M., Schuman E. M.: Role for a cortical input to hippocampal area CA1 in the consolidation of a long-term memory. Nature. **431** (2004) 699-703
5. Ito M., Kuroiwa J., Miyake S.: A model of hippocampus-neocortex for episodic memory. The 5th Inter. Conf. on Neural Inf. Process, **1P-16** (1998) 431-434
6. Ito M., Kuroiwa J., Miyake S.: A neural network model of memory systems using hippocampus. (in Japanese) The Tran. of the IEICE, **J82-D-II** (1999) 276-286
7. Ito M., Miyake S., Inawashiro S., Kuroiwa J., Sawada Y.: Long-term memory of temporal patterns in a hippocampus-cortex model. (in Japanese) Technical Report of IEICE, **NLP2000-18** (2000) 25-32
8. Freeman W. J.: Simulation of chaotic EEG patterns with a dynamic model of the olfactory system. Bio. Cybern., **56** (1987) 139-150
9. Skarda C. A., Freeman W. J.: How brains make chaos in order to make sense of the world. Behavioral and Brain Sciences, **10** (1987) 161-195
10. Aihara K.: Chaotic neural networks. Phys. Lett. A. **144** (1990) 333-340
11. Adachi M., Aihara K.: Associative dynamics in chaotic neural network. Neural Networks. **10** (1997) 83-98
12. Obayashi M., Watanabe K., Kobayashi K.: Chaotic neural networks with Radial Basis Functions and its application to memory search problem. (in Japanese) Trans. IEE of Japan, **120-C** (2000) 1441-1446
13. Obayashi M., Yuda K., Omiya R., Kobayashi K.: Associative memory and mutual information in a chaotic neural network introducing function typed synaptic weights. (in Japanese) IEEJ Trans. EIS, **123** (2003) 1631-1637
14. Lee R. S. T.: A transient-chaotic autoassociative network (TCAN) based on Lee Oscillators. IEEE Trans. on Neural Networks. **15** (2004) 1228-1243
15. Wang L. P., Li S., Tian F. Y., Fu X. J.: A Noisy Chaotic Neural Network for Solving Combinatorial Optimization Problems: Stochastic Chaotic Simulated Annealing. IEEE Trans. on Sys., Man, Cybern., Part B - Cybern., **34** (2004) 2119-2125
16. Kuremoto T., Eto T., Obayashi M., Kobayashi K.: A Multi-layered Chaotic Neural Network for Associative Memory. Proc. of SICE Annual Conference 2005 in Okayama

Stochastic Neuron Model with Dynamic Synapses and Evolution Equation of Its Density Function

Wentao Huang[1,2], Licheng Jiao[1], Yuelei Xu[2], and Maoguo Gong[1]

[1] Institute of Intelligent Information Processing and Key Laboratory of Radar
Signal Processing, Xidian University,
Xi'an 710071, China
wthuang@mail.xidian.edu.cn
[2] Signal and Information Processing Laboratory, Avionic Engineering Department,
College of Engineering, Air Force Engineering University,
Xi'an 710038, China

Abstract. In most neural network models, neurons are viewed as the only computational units, while the synapses are treated as passive scalar parameters (weights). It has, however, long been recognized that biological synapses can exhibit rich temporal dynamics. These dynamics may have important consequences for computing and learning in biological neural systems. This paper proposes a novel stochastic model of single neuron with synaptic dynamics, which is characterized by several stochastic differential equations. From this model, we obtain the evolution equation of their density function. Furthermore, we give an approach to cut the evolution equation of the high dimensional function down to the evolution equation of one dimension function.

1 Introduction

In most neural network models, synapses are treated as static weights that change only with the slow time scales of learning. It is well known, however, that synapses are highly dynamic and show use-dependent plasticity over a wide range of time scales [1]. Moreover, synaptic transmission is an inherently stochastic process: a spike arriving at a pre-synaptic terminal triggers the release of a vesicle of neurotransmitter from a release site with a probability that can be much less than one. The diverse types of synaptic plasticity and the range of timescales over which they operate suggest that synapses have a more active role in information processing. Long-term changes in the transmission properties of synapses provide a physiological substrate for learning and memory, whereas short-term changes support a variety of computations [2]. In this paper, we present a novel stochastic model to descript the single neuron model which considers the synaptic dynamics. This stochastic dynamics model is characterized by several stochastic differential equations, from which we get the evolution equation of density function. Moreover, we reduce the density function of high dimension to one dimension.

2 Models and Methods

2.1 The Integrate-and-Fire Model Neurons and Synaptic Dynamics

The integrate-and-fire (IF) model was introduced long ago by Lapicque (1907). Due to its simplicity, it has become one of the canonical spiking renewal models, since it represents one of the few neuronal models for which analytical calculations can be performed. It describes basic sub-threshold electrical properties of the neuron. It is completely characterized by its membrane potential below threshold. Details of the generation of an action potential above the threshold are ignored. Synaptic and external inputs are summed until it reaches a threshold where a spike is emitted. The general form of the dynamics of the membrane potential $v(t)$ in IF models can be written as

$$\frac{dv}{dt} = -\frac{v(t)}{\tau_v} + \sum_{k=1}^{N} S_k(t) + I_e(t) + I_n(t) ; \quad 0 \leq v \leq 1, \tag{1}$$

where τ_v is the membrane time constant, S_k is the synaptic current, N is the number of synaptic connections, I_e is an external current directly injected in the neuron, I_n is the fluctuating current aroused by noise and assume it is a Gaussian random process

$$I_n(t) = \sigma_v \xi_v(t), \tag{2}$$

where $\xi_v(t)$ is a Gaussian random variable satisfying $<\xi_v(t)>=0$ and $<\xi_v(t)\xi_v(t')>=\delta(t-t')$, and σ_v characterizes the amplitude of the noise. The transmembrane potential, v, has been normalized so that $v = 0$ marks the rest state, and $v = 1$ the threshold for firing. When the latter is achieved v is reset to zero.

The postsynaptic currents have a finite width that can be of the same order of magnitude or even larger than the membrane time constant. An accurate representation of synaptic inputs consists of an instantaneous jump followed by an exponential decay with a time constant τ_s. The realistic models of the synaptic current can be described by the following equation:

$$\frac{dS_k}{dt} = -\frac{S_k(t)}{\tau_s} + J_k(t)\delta(t-t_{sp}), \tag{3}$$

where $J_k(t)$ is the efficacy of synapse k in mV (amplitude of the postsynaptic potential), t_{sp} is occurring time of the firing of a pre-synaptic neuron, the sum over i corresponds to a sum over pre-synaptic spikes of each synapse. In reality, $J_k(t)$ act in accordance with complex dynamics rule. In recent in vitro studies it was found that the short-term synaptic dynamics in the neocortex are specific to the types of neurons involved. For example, pyramidal-to-pyramidal connections typically consist of depressing synapses, whereas pyramidal-to-interneuron connections typically bear facilitating synapses [3], [4], [5]. We use the phenomenological model by Markram *et al.* [4] to simulate short-term synaptic plasticity:

$$\frac{dD_k}{dt} = \frac{(1-D_k(t))}{\tau_r} - F_k(t)D_k(t)\delta(t-t_{sp}), \tag{4}$$

and

$$\frac{dF_k}{dt} = \frac{(U - F_k(t))}{\tau_f} + U(1 - F_k(t))\delta(t - t_{sp}) ,\quad (5)$$

where D_k is a 'depression' variable, $D_k \in [0,1]$, F_k is a 'facilitation' variable, $F_k \in [0,1]$, U is a constant determining the step increase in F_k, τ_r is the recovery time constant, and τ_f is the relaxation time constant of facilitation. The product $D_k F_k$ is the fractional amount of neurotransmitter available at time t. Each firing of a presynaptic neuron, occurring at time t_{sp}, decreases the 'depression' variable F_k by $D_k F_k$, and increases the 'facilitation' variable w by $U(1 - F_k)$. The amplitude of the postsynaptic response (PSR) $J_k(t)$ at time t_{sp} is therefore a dynamic variable given by the product

$$J_k(t) = A F_k(t) D_k(t) ,\quad (6)$$

where A is a constant representing the absolute synaptic efficacy corresponding to the maximal PSR obtained if all the synaptic resources are released at once.

2.2 Diffusion Approximation

Neurons usually have synaptic connections from tens of thousands of other neurons. Thus, even when neurons fire at low rates, a neuron receives a large amount of spikes in an interval corresponding to its integration time constant. If we assume these inputs are Poissonian and uncorrelated and the amplitude of the depolarization due to each input is small, we can use the diffusion approximation [6]. The equations (3), (4) and (5) can be approximated by

$$\begin{cases} \dfrac{dS_k}{dt} = -\dfrac{S_k(t)}{\tau_s} + \eta_{S_k}(t) = -S_k(t) + J_k(t)\lambda_k + J_k(t)\sqrt{\lambda_k}\xi_k(t) \\ \eta_{S_k}(t) = J_k(t)\lambda_k + J_k(t)\sqrt{\lambda_k}\xi_k(t) \end{cases} ,\quad (7)$$

and

$$\begin{cases} \dfrac{dD_k}{dt} \approx \dfrac{(1 - D_k(t))}{\tau_r} - \eta_k^d(t) = \dfrac{(1 - D_k(t))}{\tau_r} - [F_k(t)D_k(t)\lambda_k + F_k(t)D_k(t)\sqrt{\lambda_k}\xi_k(t)] \\ \eta_k^d(t) = F_k(t)D_k(t)\lambda_k + F_k(t)D_k(t)\sqrt{\lambda_k}\xi_k(t) \end{cases} ,\quad (8)$$

and

$$\begin{cases} \dfrac{dF_k}{dt} \approx \dfrac{(U - F_k(t))}{\tau_f} + \eta_k^f(t) = \dfrac{(U - F_k(t))}{\tau_f} + U(1 - F_k(t))\lambda_k + U(1 - F_k(t))\sqrt{\lambda_k}\xi_k(t) \\ \eta_k^f(t) = U(1 - F_k(t))\lambda_k + U(1 - F_k(t))\sqrt{\lambda_k}\xi_k(t) \end{cases} ,\quad (9)$$

where λ_k is the mean intensity of the kth synaptic input, $\xi_k(t)$ is a gaussian random variable satisfying $<\xi_k(t)> = 0$ and $<\xi_k(t)\xi_{k'}(t')> = \delta_{kk'}\delta(t-t')$. We can prove $\eta_k^s(t)$, $\eta_k^d(t)$ and $\eta_k^f(t)$ share identical first and second order statistics with I_s, D_k, and F_k. D_k is taken as an example. Considering a small time interval Δt, since t_{sp} obey Poissonian distribution, $\int_t^{t+\Delta t} F_k(t')D_k(t')\delta(t'-t_{sp})dt'$ equal to $F_k(t)D_k(t)$ with probability $\Delta t \lambda_k$, and equal to zero with probability $(1-\Delta t \lambda_k)$, then we have

$$\lim_{\Delta t \to 0}\frac{<\int_t^{t+\Delta t} F_k(t')D_k(t')\delta(t'-t_{sp})dt'>}{\Delta t} = F_k(t)D_k(t)\lambda_k = \lim_{\Delta t \to 0}\frac{<\int_t^{t+\Delta t} \eta_k^d(t')dt'>}{\Delta t}, \quad (10)$$

and

$$\lim_{\Delta t \to 0}\frac{<(\int_t^{t+\Delta t} F_k(t')D_k(t')\delta(t'-t_{sp})dt')^2> - <\int_t^{t+\Delta t} F_k(t')D_k(t')\delta(t'-t_{sp})dt'>^2}{\Delta t}$$
$$= (F_k(t)D_k(t)\sqrt{\lambda_k})^2 = \lim_{\Delta t \to 0}\frac{<(\int_t^{t+\Delta t} \eta_k^d(t')dt')^2> - <\int_t^{t+\Delta t} \eta_k^d(t')dt'>^2}{\Delta t}. \quad (11)$$

From equation (1), (2), (7), (8) and (9), we can write out their Ito stochastic differential equations:

$$\begin{cases} dv = (-\frac{1}{\tau_v}v(t) + I_e(t) + \sum_{k=1}^{N} S_k(t))dt + \sigma_v \xi_v(t)dt = K_v dt + \sigma_v dW_0(t) \\ K_v = -\frac{1}{\tau_v}v(t) + I_e(t) + \sum_{k=1}^{N} S_k(t) \end{cases}, \quad (12)$$

and

$$\begin{cases} dS_k = (-\frac{S_k(t)}{\tau_s} + AF_k(t)D_k(t)\lambda_k)dt + AF_k(t)D_k(t)\sqrt{\lambda_k}\xi_k(t)dt = K_{S_k} dt + \sigma_{S_k} dW_k(t) \\ K_{S_k} = -\frac{S_k(t)}{\tau_s} + AF_k(t)D_k(t)\lambda_k \\ \sigma_{S_k} = AF_k(t)D_k(t)\sqrt{\lambda_k} \end{cases}, \quad (13)$$

and

$$\begin{cases} dD_k = [\frac{(1-D_k(t))}{\tau_r} - F_k(t)D_k(t)\lambda_k]dt - F_k(t)D_k(t)\sqrt{\lambda_k}\xi_k(t)dt = K_{D_k} dt - \sigma_{D_k} dW_k(t) \\ K_{D_k} = \frac{(1-D_k(t))}{\tau_r} - F_k(t)D_k(t)\lambda_k \\ \sigma_{D_k} = F_k(t)D_k(t)\sqrt{\lambda_k} \end{cases}, \quad (14)$$

and

$$\begin{cases} dF_k = [\frac{(U-F_k(t))}{\tau_f} + U(1-F_k(t))\lambda_k]dt + U(1-F_k(t))\sqrt{\lambda_k}\xi_k(t)dt = K_{F_k} dt + \sigma_{F_k} dW_k(t) \\ K_{F_k} = \frac{(U-F_k(t))}{\tau_f} + U(1-F_k(t))\lambda_k \\ \sigma_{F_k} = U(1-F_k(t))\sqrt{\lambda_k} \end{cases}. \quad (15)$$

The Fokker-Planck equation of equations (12)~(15) is given by[7]

$$\frac{\partial \rho(v,\mathbf{S},\mathbf{D},\mathbf{F})}{\partial t} = -\left\{\frac{\partial}{\partial v}(K_v\rho) + \sum_{k=1}^{N}[\frac{\partial}{\partial S_k}(K_{S_k}\rho) + \frac{\partial}{\partial D_k}(K_{D_k}\rho) + \frac{\partial}{\partial F_k}(K_{F_k}\rho)]\right\}$$
$$+ \frac{1}{2}\left\{\frac{\partial^2}{\partial v^2}(\sigma_v^2\rho) + \sum_{k=1}^{N}[\frac{\partial^2}{\partial S_k^2}(\sigma_{S_k}^2\rho) + \frac{\partial^2}{\partial D_k^2}(\sigma_{D_k}^2\rho) + \frac{\partial^2}{\partial F_k^2}(\sigma_{F_k}^2\rho)]\right\} ,$$
$$+ \left\{\sum_{k=1}^{N}[\frac{\partial^2}{\partial S_k \partial F_k}(\sigma_{S_k}\sigma_{F_k}\rho) - \frac{\partial^2}{\partial S_k \partial D_k}(\sigma_{S_k}\sigma_{D_k}\rho) - \frac{\partial^2}{\partial F_k \partial D_k}(\sigma_{D_k}\sigma_{F_k}\rho)]\right\}$$
(16)

where $\rho(v,\mathbf{S},\mathbf{D},\mathbf{F})$ is the joint distribution density function, $\mathbf{S} = (S_1, S_2, ..., S_N)$, $\mathbf{D} = (D_1, D_2, ..., D_N)$, $\mathbf{F} = (F_1, F_2, ..., F_N)$.

2.3 Reduce to One Dimension

The dimension of joint distribution density $\rho(v,\mathbf{S},\mathbf{D},\mathbf{F})$ is huge, and is discommodiousness for us to analyze its performance. Sometimes, we are more interesting the density evolution of membrane potential v, so, in what following, we discuss how to get the density evolution equation of membrane potential v.

Due to

$$\rho(v,\mathbf{S},\mathbf{D},\mathbf{F}) = \rho_2(\mathbf{S},\mathbf{D},\mathbf{F}|v)\rho_1(v) ,$$
(17)

and

$$\int \rho_2(\mathbf{S},\mathbf{D},\mathbf{F}|v)d\mathbf{S}d\mathbf{D}d\mathbf{F} = 1 ,$$
(18)

substituting (17) into (16) yields:

$$\rho_1\frac{\partial \rho_2}{\partial t} + \rho_2\frac{\partial \rho_1}{\partial t} = -\left\{\frac{\partial}{\partial v}(K_v\rho_1\rho_2) + \sum_{k=1}^{N}\rho_1[\frac{\partial}{\partial S_k}(K_{S_k}\rho_2) + \frac{\partial}{\partial D_k}(K_{D_k}\rho_2) + \frac{\partial}{\partial F_k}(K_{F_k}\rho_2)]\right\}$$
$$+ \frac{1}{2}\left\{\frac{\partial^2}{\partial v^2}(\sigma_v^2\rho_1\rho_2) + \sum_{k=1}^{N}\rho_1[\frac{\partial^2}{\partial S_k^2}(\sigma_{S_k}^2\rho_2) + \frac{\partial^2}{\partial D_k^2}(\sigma_{D_k}^2\rho_2) + \frac{\partial^2}{\partial F_k^2}(\sigma_{F_k}^2\rho_2)]\right\} .$$
$$+ \left\{\sum_{k=1}^{N}\rho_1[\frac{\partial^2}{\partial S_k \partial F_k}(\sigma_{S_k}\sigma_{F_k}\rho_2) - \frac{\partial^2}{\partial S_k \partial D_k}(\sigma_{S_k}\sigma_{D_k}\rho_2) - \frac{\partial^2}{\partial F_k \partial D_k}(\sigma_{D_k}\sigma_{F_k}\rho_2)]\right\}$$
(19)

Integrating with $\mathbf{S},\mathbf{D},\mathbf{F}$ in (19) two side and using equation (18) we can get

$$\frac{\partial \rho_1}{\partial t} = -\frac{\partial}{\partial v}(P_v\rho_1) + \frac{\sigma_v^2}{2}\frac{\partial^2}{\partial v^2}(\rho_1) ,$$
(20)

where

$$P_v = \int K_v \rho_2 d\mathbf{S}d\mathbf{D}d\mathbf{F} .$$
(21)

Because $\xi_k(t)$ ($k=1,...,N$) are uncorrelated, then we have

$$\rho_2(\mathbf{S},\mathbf{D},\mathbf{F}|v) = \prod_{k=1}^{N} \rho_3^k(S_k, D_k, F_k | v) .$$
(22)

Moreover, we can assume

$$\rho_3^k(S_k, D_k, F_k | v) \approx \rho_4^k(S_k, D_k, F_k) .$$
(23)

From (13), (14) and (15), we can get the Fokker-Planck equation of $\rho_4^k(S_k, D_k, F_k)$,

$$\frac{\partial \rho_4^k(S_k, D_k, F_k)}{\partial t} = -\left\{\frac{\partial}{\partial S_k}(K_{S_k}\rho_4^k) + \frac{\partial}{\partial D_k}(K_{D_k}\rho_4^k) + \frac{\partial}{\partial F_k}(K_{F_k}\rho_4^k)\right\} + \frac{1}{2}\left\{\frac{\partial^2}{\partial S_k^2}(\sigma_{S_k}^2\rho_4^k) + \frac{\partial^2}{\partial D_k^2}(\sigma_{D_k}^2\rho_4^k) + \frac{\partial^2}{\partial F_k^2}(\sigma_{F_k}^2\rho_4^k)\right\} - \frac{\partial^2}{\partial S_k \partial D_k}(\sigma_{D_k}\sigma_{S_k}\rho_4^k) - \frac{\partial^2}{\partial F_k \partial D_k}(\sigma_{D_k}\sigma_{F_k}\rho_4^k) \quad (24)$$

Since

$$\rho_4^k(S_k, D_k, F_k) = \rho_6^k(S_k \mid D_k, F_k)\rho_5^k(D_k, F_k), \quad (25)$$

substituting (25) into (24) yields:

$$\rho_5^k \frac{\partial \rho_6^k}{\partial t} + \rho_6^k \frac{\partial \rho_5^k}{\partial t} = -\left\{\rho_5^k \frac{\partial}{\partial S_k}(K_{S_k}\rho_6^k) + \frac{\partial}{\partial D_k}(K_{D_k}\rho_5^k\rho_6^k) + \frac{\partial}{\partial F_k}(K_{F_k}\rho_5^k\rho_6^k)\right\}
+ \frac{1}{2}\left\{\rho_5^k \frac{\partial^2}{\partial S_k^2}(\sigma_{S_k}^2\rho_6^k) + \frac{\partial^2}{\partial D_k^2}(\sigma_{D_k}^2\rho_5^k\rho_6^k) + \frac{\partial^2}{\partial F_k^2}(\sigma_{F_k}^2\rho_5^k\rho_6^k)\right\}
- \left\{\frac{\partial^2}{\partial S_k \partial D_k}(\sigma_{D_k}\sigma_{S_k}\rho_5^k\rho_6^k) + \frac{\partial^2}{\partial F_k \partial D_k}(\sigma_{D_k}\sigma_{F_k}\rho_5^k\rho_6^k)\right\} \quad (26)$$

Integrating with S_k in (26) two side and using normalization condition we can get

$$\frac{\partial \rho_5^k}{\partial t} = -\left\{\frac{\partial}{\partial D_k}(K_{D_k}\rho_5^k) + \frac{\partial}{\partial F_k}(K_{F_k}\rho_5^k)\right\} + \frac{1}{2}\left\{\frac{\partial^2}{\partial D_k^2}(\sigma_{D_k}^2\rho_5^k) + \frac{\partial^2}{\partial F_k^2}(\sigma_{F_k}^2\rho_5^k)\right\} - \frac{\partial^2}{\partial F_k \partial D_k}(\sigma_{D_k}\sigma_{F_k}\rho_5^k) \quad (27)$$

Because of $\tau_s \ll \tau_d$ and $\tau_s \ll \tau_f$, we can assume [8]

$$\begin{cases} \dfrac{\partial \rho_6^k}{\partial D_k} \ll \dfrac{\partial \rho_6^k}{\partial S_k}, & \dfrac{\partial \rho_6^k}{\partial F_k} \ll \dfrac{\partial \rho_6^k}{\partial S_k} \\ \dfrac{\partial^2 \rho_6^k}{\partial D_k^2} \ll \dfrac{\partial^2 \rho_6^k}{\partial S_k^2}, & \dfrac{\partial^2 \rho_6^k}{\partial F_k^2} \ll \dfrac{\partial^2 \rho_6^k}{\partial S_k^2} \end{cases} \quad (28)$$

Then from (26), (27) and (28), and omitting the small terms of high order, we obtain

$$\frac{\partial \rho_6^k}{\partial t} = -\frac{\partial}{\partial S_k}(K_{S_k}\rho_6^k) + \frac{1}{2}\frac{\partial^2}{\partial S_k^2}(\sigma_{S_k}^2\rho_6^k). \quad (29)$$

If we adopt the adiabatic approximation [8], $\dfrac{\partial \rho_6^k}{\partial t} = 0$, we have

$$0 = -\frac{\partial}{\partial S_k}(K_{S_k}\rho_6^k) + \frac{1}{2}\frac{\partial^2}{\partial S_k^2}(\sigma_{S_k}^2\rho_6^k). \quad (30)$$

From (12), (21), (22), (23) and (25), we have

$$P_v = -\frac{v(t)}{\tau_v} + I_e(t) + \sum_{k=1}^{N} \int S_k(t)\rho_6^k(S_k \mid D_k, F_k)\rho_5^k(D_k, F_k)dS_k dD_k dF_k. \quad (31)$$

If we have solved the value of P_v from (27), (30) and (31), substituting it in (20), we can immediately get the probability density evolution equation of the membrane potential v.

3 Conclusion

In this paper we have presented a novel model to descript the single neuron model using the stochastic differential equations. The model has considered the synaptic dynamics. We adopt the diffusion approximation and get the Ito stochastic differential equations from which we can obtain the Fokker-Planck equation to descript the evolution of joint distribution density function. However, the dimension of joint distribution density, $\rho(v,\mathbf{S},\mathbf{D},\mathbf{F})$, is huge and is discommodiousness for us to analyze it. For obtaining the evolution equation of the density function of membrane potential v, we adopted the adiabatic approximation and other approaches to approximation. Tis model is useful for us to analyze the behavior of neural system, such as neural coding, oscillations and synchronization.

References

1. Destexhe1 A., Marder E.: Plasticity in single neuron and circuit computations. Nature, 431 (2004) 789–795
2. Abbott L.F., Regehr W.G.: Synaptic computation. Nature, 431 (2004) 796–803
3. Gupta A., Wang Y., Markram H.: Organizing principles for a diversity of GABAergic interneurons and synapses in the neocortex. Science, 287 (2000) 273–278
4. Markram H., Wang Y., Tsodyks M.: Differential signaling via the same axon from neocortical layer 5 pyramidal neurons. Proc Natl Acad Sci, 95 (1998) 5323–5328
5. Galarreta M., Hestrin S.: Frequency-dependent synaptic depression and the balance of excitation and inhibition in the neocortex. Nature Neuroscience, 1 (1998) 587–594
6. Tuckwell, H. C.: Introduction to theoretical neurobiology. Cambridge University Press, Cambridge, 1988
7. Gardiner C.W.: Handbook of stochastic methods: for physics, chemistry & the natural sciences. Springer-Verlag, Berlin Heidelberg, New York, 1983
8. Haken H.: Advanced Synergetics. Springer-Verlag, Berlin Heidelberg, New York, 1983

Learning Algorithm for Spiking Neural Networks

Hesham H. Amin and Robert H. Fujii

The University of Aizu, Aizu-Wakamatsu, Fukushima, Japan
{d8042201, fujii}@u-aizu.ac.jp

Abstract. Spiking Neural Networks (SNNs) use inter-spike time coding to process input data. In this paper, a new learning algorithm for SNNs that uses the inter-spike times within a spike train is introduced. The learning algorithm utilizes the spatio-temporal pattern produced by the spike train input mapping unit and adjusts synaptic weights during learning. The approach was applied to classification problems.

1 Introduction

Spiking Neural Networks (SNN) can be considered as the third generation of ANNs, after multi-layer perceptron neural networks and neurons which employ activation functions such as sigmoid functions [10]. The latter two types of neural networks use synchronized analog or digital amplitude values as inputs. SNNs do not require a synchronizing system clock (although they may use a local synchronizing signal) and utilize input inter-spike time data to process information. An SNN is composed of spiking neurons as processing units which are connected together with synapses. A spiking neuron receives spikes at its inputs and fires an output spike at a time dependent on the inter-spike times of the input spikes. Thus, SNNs use temporal information in coding and processing input data. Synaptic spike inputs with only one spike per each input synapse during a given time window are called spatio-temporal inputs. A synaptic input which consists of a sequence of spikes with various inter-spike intervals (ISIs) during a given time window is called a spike train. The ISI times within a spike train has a much larger encoding space than the rate code used in traditional neural networks [11]. Accordingly, the processing efficiency of SNNs can be higher than traditional rate code based ANNs for most applications.

Learning how to recognize the temporal information contained in spike trains is the main goal of this research. The literature is scant regarding this area of research. Some SNN learning models have been proposed in the past which make it possible to process spike trains in close to real-time [8], [9], [12], [13]. However, these models used recurrent networks and a large number of synapses which needed a relatively long time to map and process input spike trains. In this paper, a new learning algorithm for spiking neurons which use spike trains inputs is proposed. This learning algorithm utilizes the input spike mapping scheme, described in [1],[2], and input synapses with dynamically changeable weights.

2 Spiking Neural Network

The spiking neuron model employed in this paper is based on the Spike Response Model (SRM) [6] with some modifications. Input spikes come at times $\{t_1...t_n\}$ into the input synapse(s) of a neuron. The neuron outputs a spike when the internal neuron membrane potential $x_j(t)$ crosses the threshold potential ϑ from below at firing time $t_j = min\{t : x_j(t) \geq \vartheta\}$. The threshold potential ϑ is assumed to be constant for the neuron.

The relationship between input spike times and the internal potential of neuron j (or Post Synaptic Potential (PSP)) $x_j(t)$ can be described as follows:

$$x_j(t) = \sum_{i=1}^{n} W_i.\alpha(t - t_i), \quad \alpha(t) = \frac{t}{\tau}e^{1-\frac{t}{\tau}} \tag{1}$$

i represents the ith synapse, W_i is the ith synaptic weight variable which can change the amplitude of the neuron potential $x_j(t)$, t_i is the ith input spike arrival-time, $\alpha(t)$ is the spike response function, and τ represents the membrane potential decay time constant.

In this paper, the $\alpha(t)$ function is approximated as a linear function for $t \ll \tau$. It then follows that the internal neuron potential Equation 1, can be re-written as:

$$x_j(t) = \frac{t}{\tau_1} \sum_{i=1}^{n} W_i.u(t - t_i); \quad t \ll \tau_1 \tag{2}$$

$u(t)$ is the Heaviside function and $\tau_1 = \frac{e}{\tau}$.

3 Mapping-Learning Scheme for Spiking Neural Networks

A one-to-one correspondence between input spike trains and output spike firing times is necessary for the learning algorithm proposed in this paper. By selecting an appropriate set of synaptic weights for a neuron, a particular spike train or a set of spike trains which belong to the same class can be distinguished by the output firing time of the neuron because of the one-to-one correspondence between the input and output. The combined mapping-learning organization is shown in Figure 1.

Learning is performed in two stages: (1) The *mapping stage* is composed of neural mapping units (MUs) as shown in Figure 1. This stage was described in [1],[2] and it is used for mapping the input spike train(s) into unique spatio-temporal output patterns. The one-to-one relationship between the inputs and outputs of the mapping stage was proved in Appendix A of [2]. (2) The *learning stage* consists of several learning units (LUs) as shown in Figure 1. The learning stage receives the spatio-temporal output pattern produced by the mapping stage. Each learning unit is composed of sub-learning units as shown in Figure 2(A). Each sub-learning unit (e.g LUA1) takes inputs from one mapping unit (MU) as shown in Figure 2(B). As shown in Figure 2(B), the outputs t_1 and t_2 from the mapping unit are input into the sub-learning unit ISI blocks. The ISI block performs the same function as the ISI block used in the mapping units used in [1],[2]; the learning unit ISI block input synaptic weights are assigned using $W_i = \beta.t_i$ and $W_i = \frac{\beta}{t_i}$ for the ISI1 and ISI2 blocks respectively. It should be noted that in a

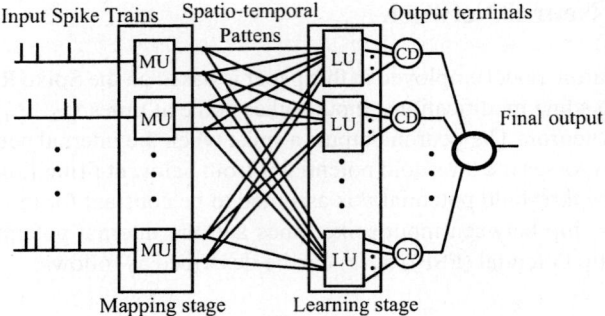

Fig. 1. Combined Mapping-Learning Organization

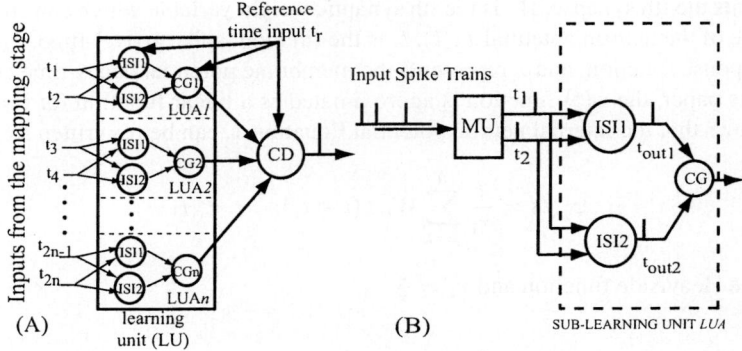

Fig. 2. (A) Details of Learning Unit (LU). (B) MU and sub-learning unit

learning unit there are $2n$ ISI blocks where n is the number of input spike trains. A one-to-one mapping between inputs and outputs is also necessary in the learning unit. The t_r reference time input shown in Figure 2(A) is used as a local reference signal for the combined mapping-learning organization shown in Figure 1. The coincidence generation (CG) neurons in a sub-learning unit perform the function of aligning their output spike times. When all CG neurons in an LU fire simultaneously, the coincidence detection (CD) neuron fires.

Past learning algorithms for spiking neural networks such as back-propagation (SpikeProp) [4], self-organizing map (SOM) [15], and radial basis function (RBF) [14] used synaptic weights and delays as well as multiple sub-synapses as the learning parameters. The learning algorithm proposed in this paper can perform learning in one step and utilizes only synaptic weights for learning. Hence, the proposed algorithm is simpler than past approaches and more practical to implement in hardware.

3.1 The Learning Algorithm

The spatio-temporal patterns generated by the ISI1 and ISI2 blocks in the mapping stage, described in [1],[2], are used as inputs for the learning stage where a supervised learning method is used to classify input patterns. Clustering of input patterns which belong to the same class is achieved by setting the synaptic weights for a learning unit (LU) so that its output fires at approximately the same time for as many input spike trains as possible that belong to the same class.

The supervised learning algorithm works as follows:

1. Choose an input pattern vector (say P_A) at random from the set of $P_l = (P_A, P_B, ...)$ pattern vectors to be used for the learning phase. Each pattern P_l consists of the spatio-temporal outputs generated by the mapping stage. The randomly chosen pattern P_A is used to assign weights to all the ISI blocks in a learning unit. This learning unit will represent the class to which pattern P_A belongs. Once the weights have been assigned, they are temporarily fixed. The weights selected for the initial input pattern works as a center vector which can later be modified slightly to accommodate more than one input pattern; in this manner, similar input patterns can then be clustered together and fewer learning units will be needed.
2. Another input pattern (say P_B) belonging to the same class as pattern P_A chosen in step 1 above is selected. This new pattern is applied to the learning unit for P_A and the output of the ISI blocks times for P_B $\{t_{out1}, t_{out2}, ..., t_{out2n}\}$ are compared against the output times for $P_A \{t^*_{out1}, t^*_{out2}, ...t^*_{out2n}\}$. This new pattern ($P_B$) is assigned to the learning unit (e.g. learning unit for P_A) with which each of the output times differ by less than ϵ.

$$|t^*_{out1} - t_{out1}| \leq \epsilon \quad , |t^*_{out2} - t_{out2}| \leq \epsilon \quad , ... \quad \text{and} \quad |t^*_{out2n} - t_{out2n}| \leq \epsilon \quad (3)$$

ϵ is a small error value determined empirically. If the error is larger than ϵ for any one of the error conditions in Equation 3, a new learning unit is added as is done in incremental learning.
3. Steps 1 and 2 are repeated for all the remaining input patterns in the learning set P_l.

In this learning scheme, all input spike train samples used for learning must be known a priori. However, the total number of learning units (clusters) which will be needed for classification with clustering cannot be known a priori. It may be possible to cluster m input patterns belonging to one class into a single learning unit (cluster) or as many as m learning units may be needed.

This learning scheme is similar to the algorithm proposed in [14] but without the need for synapse delays. This could help to make the model more practical for an IC circuit design implementation. Furthermore, each synapse in the model is not composed of multiple sub-synapses as proposed in [3], [14] and this leads to a reduction in complexity.

The proposed learning algorithm produces locally optimal input clustering because the input patterns for a given class are sequentially chosen at random; the consequence of this is that a larger neural network than necessary may result.

3.2 Learning Unit Output Time Uniqueness

A one-to-one relationship between inputs and outputs for each of the learning units must be achieved in order to guarantee that each learning unit outputs a spike at a time

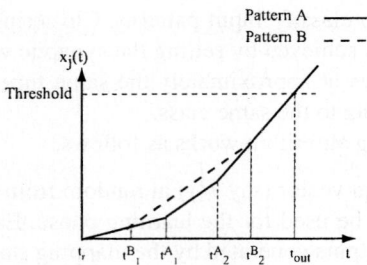

Fig. 3. Two Different Input Patterns Producing an Identical Output Time

which is different from the output times corresponding to other inputs. This one-to-one relationship will be shown using one MU and one sub-learning unit. When a new pattern (e.g. pattern P_B with MU output times t_1^B and t_2^B) is input into a sub-learning unit within an LU which had its synaptic weights fixed during the learning of pattern P_A, the following will result: $(\{t_{out1}^A, t_{out2}^A\} \neq \{t_{out1}^B, t_{out2}^B\}$, where t_{out} is the output firing time of an ISI block. This can be proved by contradiction:

Assume that P_B produces the same t_{out1} or t_{out2} as P_A. For the moment, t_{out1} and t_{out2} will not be distinguished and they will simply be referred to as t_{out}.

Then the internal neuron potentials $x_j(t)$ (Equation 2) for P_A and P_B at time t_{out} can be written as follows:

$$\sum_{i=1}^{2} W_i^A . u(t - t_i^A) = \sum_{i=1}^{2} W_i^A . u(t - t_i^B) \quad (4)$$

W_i^A's are the synaptic weights which have been fixed for the learning unit P_A. Two different input patterns P_A and P_B producing an identical output at time t_{out} can occur only if the neuron internal potential $x_j(t)$ for one of the input patterns becomes equal to the other input pattern's neuron internal potential (at t_2^B) and then both increase at identical rates until crossing the threshold potential ϑ at time t_{out} as shown in Figure 3.

For the ISI1 block, $W_i^A = \beta . t_i^A$; Equation 4 can be rewritten as follows:

$$\frac{t_1^A}{t_2^A} = \frac{u(t - t_2^B) - u(t - t_2^A)}{u(t - t_1^A) - u(t - t_1^B)} \quad (5)$$

For the ISI2 block, $W_i^A = \frac{\beta}{t_i^A}$; thus Equation 4 can be rewritten as follows:

$$\frac{t_1^A}{t_2^A} = \frac{u(t - t_1^A) - u(t - t_1^B)}{u(t - t_2^B) - u(t - t_2^A)} \quad (6)$$

Equations 5 and 6 can have a solution only if $t_1^A = t_2^A$ which cannot happen because 2 distinct spikes output times from the mapping unit are assumed[1]. In other words, if the ISI1 block outputs a spike at the same t_{out} time for both P_A and P_B, the ISI2 output times will be not equal and vice versa.

---

[1] $t_1^A = t_2^A$ can happen only if an input spike train consists of only two spikes at times 0 and 1 when the input spike train time window size is assumed to be equal to 1.

3.3 Firing of Only One Learning Unit

Assume that patten P_A was learned by the learning unit A(LUA) and that patten P_B was learned by the learning unit B (LUB). Assume that the sub-learning units LUA1 and LUB1 get inputs from the same mapping unit (MU). If pattern P_A is input into both LUA1 and LUB1, the neuron internal potentials for LUA's ISI1 or ISI2 and LUB's ISI1 or ISI2 will increase according to equation 2. If $t^A_{out1} = t^B_{out1} = t_{out1}$ and $t^A_{out2} = t^B_{out2} = t_{out2}$ are assumed, the following relationship will be established:

$$\sum_{i=1}^{2} W_i^A \cdot u(t - t_i^A) = \sum_{i=1}^{2} W_i^B \cdot u(t - t_i^A) \quad (7)$$

The only way for LUA1 and LUB1 to produce an output spike at the same t_{out1} (t_{out2}) time is to have the following condition satisfied:

$$\sum_{i=1}^{2} W_i^A = \sum_{i=1}^{2} W_i^B \quad (8)$$

Thus, if the condition specified by Equation 8 is not satisfied by any one of the sub-learning units, only one of the learning units will respond to an input pattern. The learning algorithm has to include a checking phase to guarantee that the condition specified by Equation 8 is not satisfied.

3.4 Coincidence Detection Neuron

In order to have only one learning unit fire for a given input pattern, output times of the CG neurons in the sub-learning units (Figure 2(A)) have to be made coincident by changing the input synaptic weight values of the coincidence generation (CG) neurons. The coincidence detection neuron (CD), shown in Figure 2(A), uses the exponential response function (Equation 1) of a spiking neuron.

The outputs of the ISI1 and ISI2 blocks of each sub-learning unit (Figure 2(A)) fire at certain times according to the assigned synaptic weight centers. The other patterns which have been joined to the same learning unit cause the outputs to fire at times which are close to the ones corresponding to the center pattern. The coincidence detection neuron threshold value ϑ is adjusted so as to allow some fuzziness in the input spike times.

3.5 Local Reference Time

In section 3.2 it was proved that the output combination $\{t_{out1}, t_{out2}\}$ for the ISI1 and ISI2 blocks will be unique for each sub-learning unit; however, the relative time $|t_{out1} - t_{out2}|$ should also be considered for all the sub-learning units of different learning units (LUs). In other words, two different sub-learning units in two different learning units can fire at different output times, t_{out1} and t_{out2}, but the relative time $|t_{out1} - t_{out2}|$ may be the same; this would lead to two (or more) learning units firing outputs for the same input pattern. Thus, a reference time (bias) t_r input is necessary to differentiate these outputs as shown in Figure 2(A). This reference time t_r is the time when the first input spike arrives at one of the mapping stage inputs.

4 Simulations

4.1 Realization of the XOR Function

Due to its nonlinearly separable input characteristics, a two-input exclusive OR (XOR) function has often been used to test the function approximation or classification capability of a neural network [7]. The XOR problem has non-linearly separable classes. One of these classes is represented by x_1x_2 inputs 00 and 11. The other class is represented by x_1x_2 inputs 01 and 10. The logical inputs "0" and "1" are represented by spikes at times 0 and 0.1 respectively in Table 1. The spike time can be defined with any appropriate unit of time (e.g. ms, ns).

For a spiking neural network, the inputs $x_1x_2 = 00$ and $x_1x_2 = 11$ are not distinguishable in the time domain because the inputs are not referenced to a clock. Thus, in order to distinguish the $x_1x_2 = 00$ and $x_1x_2 = 11$ cases, a third reference (bias) input $x_0 = 0$ is used as shown in Figure 4. Thus, the logical input $x1x2 = 00$ and $x1x2 = 11$ can for example be distinguished in the time domain as "$0sec, 0sec, 0sec$" and "$0sec, 0.1sec, 0.1sec$" respectively.

As describes in section 3.4, each learning unit in conjunction with a coincident detection neuron generates a spike when the appropriate spatio-temporal pattern is input.

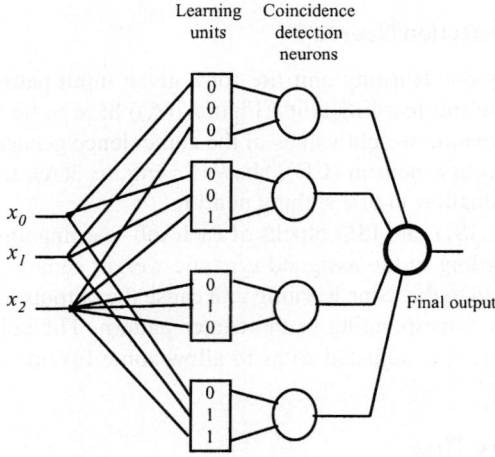

Fig. 4. Spiking neural network for XOR function. Details of learning unit also shown.

Table 1. XOR Input spike times (including the bias) and output times

Input Patterns			Coincident firing time	Final output time
0	0	0	1.464	4
0	0	0.1	1.910	2
0	0.1	0	1.910	2
0	0.1	0.1	3.013	4

The XOR neural network organization is shown in Figure 4. The final output neuron, shown in Figure 4, is used to represent the XOR output value in the time domain (e.g. output time = 2 corresponds to the logical output "1").

4.2 Classification of Spike Trains

The robustness of the learning algorithm was tested using a set of randomly generated spike trains as inputs. These spike trains were generated by adding noise to the original spike trains. Noise consisted of input spike shifts in time or addition/deletion of spikes within a spike train. These types of noise are realistic since a correct spike sequence can be altered by short-lived interferences. Spike time skews were produced by adding Gaussian white noise (GWN) to the spike train, or by time shifting one or two spikes in a spike train randomly. The deletion/addition of spikes was also done randomly.

The spike trains used in the simulations were generated using Poisson distributed inter-spike intervals [5] at a low frequency. By injecting various amounts of GWN into a spike train, noisy time shifted versions of the original spike trains could be generated as shown in Figure 5, where spike train number 1 is the original spike train for each class.

Each of the generated spike trains shown in Figure 5 was used as an input to the mapping stage (shown in Figure 1). The spatio-temporal pattern output from the mapping stage was then used as an input to the learning stage. The mapping stage used multiple mapping units with different β values in the range of $[0.25, 1.0]$ in order to increase the input dimension of the learning stage.

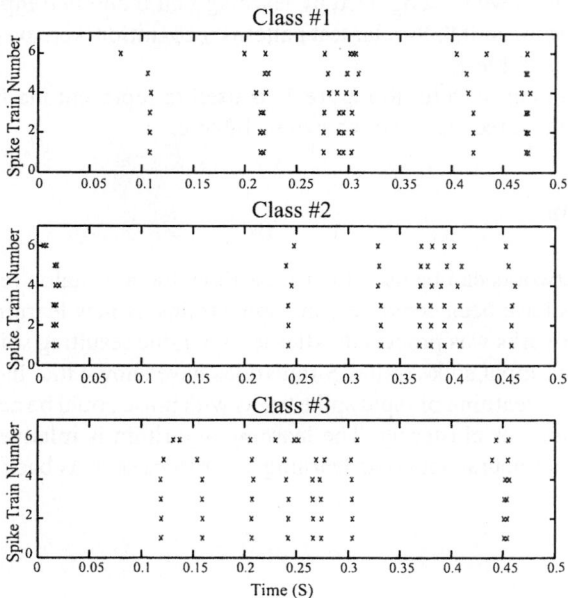

Fig. 5. Three classes of input spike trains. The original spike train for each class is spike train number 1 and the other five trains are noisy versions of it.

Table 2. Input spike train classification, clustering, and final output times

Class No.	Learning unit #	# Learning patterns	# Test patterns	Final output time
1	1	5	5	4.0
	2	1	-	
2	3	4	3	5.0
	4	2	2	
3	5	4	3	6.0
	6	2	2	

After generating the noisy versions of each of the original spike trains, all the patterns including the noisy patterns were used as a learning set. The closer the noisy versions were to the original spike train, the likelihood of being able to use an already assigned learning unit increased.

The learning and input pattern clustering simulation results are shown in Table 2. For example, for the three classes a total of six clusters were needed. For input class 1, learning unit 1 was used for clustering five input patterns and learning unit 2 was used for clustering one input pattern. Similar clusterings were possible for classes 2 and 3 as shown in Table 2.

After the learning phase was completed, additional noisy spike trains for each of the three classes were used to test the neural network. These additional noisy spike trains are called test patterns in Table 2. The testing phase spike trains were generated with the same range of noise used during the learning phase. For example, for input class 3, three input patterns were recognized by learning unit 5 and two input patterns were recognized by learning unit 6. Similar test patterns recognition were possible for classes 1 and 2 as shown in Table 2.

A final output neuron (refer to Figure 1) is used to represent the final output time value for each of the three classes as shown in Table 2.

5 Conclusions

Spiking neural networks can be used to process time domain analog real world signals once these signals have been converted into spike trains. A new learning algorithm for spiking neural networks was proposed. After learning, the resulting spiking neural network could classify input spike trains. Simulations have shown that incremental learning for classification learning of input spike trains with noise could be achieved by either adding learning units or clustering. The learning algorithm is relatively simple when compared with other neural networks learning algorithms such as back-propagation.

References

1. Hesham H. Amin and Robert H. Fujii: Input Arrival-Time-Dependent Decoding Scheme for a Spiking Neural Network. Proceeding of The 12th European Symosium of Artificial Neural Networks (ESANN 2004). (2004) 355–360.

2. Hesham H. Amin and Robert H. Fujii: Spike Train Decoding Scheme for a Spiking Neural Network. Proceedings of the 2004 International Joint Conference on Neural Networks. IEEE. (2004) 477–482.
3. S.M. Bohte, H. La Poutré and J.N. Kok: Unsupervised classification in a Network of spiking neurons. IEEE Transactions on Neural Networks. **13, 2** (2002) 426–435.
4. S.M. Bohte, J.N. Kok and H. La Poutré: Spike-prop: error-backprogation in multi-layer networks of spiking neurons. Proceedings of the European Symposium on Artificial Neural Networks ESANN'2000. (2000) 419–425.
5. Peter Dayan, and L. F. Abbott: Theoretical Neuroscience: Computational and Mathematical Modeling of Neural Systems. MIT Press (2001).
6. W. Gerstner and W. Kistler: Spiking Neuron Models. Single Neurons, Populations, Plasticity. Cambridge University Press (2002).
7. Simon Haykin: Neural Networks, A Comprehensive Foundation. Prentice Hall International Inc. (1999).
8. J. J. Hopfield and C. D. Brody: What is a Moment? Cortical Sensory Integration Over a Brief Interval. Proc. Natl. Acad. Sci. **97(25)** (2000) 13919–13924.
9. J. J. Hopfield and C. D. Brody: What is a Moment? Transient Synchrony as a Collective Mechanism for Spatiotemporal Integration. Proc. Natl. Acad. Sci. **98(3)** (2001) 1282–1287.
10. W. Maass, Networks of spiking neurons: the third generation of neural network models. Neural Networks. **10**. (1997) 1659–1671
11. W. Maass and C. Bishop, editors: Pulsed Neural Networks. MIT press. Cambridge (1999).
12. W. Maass, T. Natschläger, and H. Markram: A Model for Real-Time Computation in Generic Neural Microcircuits, in *Proc. of NIPS 2002*, Advances in Neural Information Processing Systems, volume 15, ed. S. Becker, S. Thrun, and K. Obermayer. MIT Press (2003) 229–236.
13. W. Maass, T. Natschläger, and H. Markram: Computational Models for Generic Cortical Microcircuits, in *Computational Neuroscience: A Comprehensive Approach*, chapter 18, ed. J. Feng. CRC-Press (2003).
14. Berthold Ruf: Computing and Learning with Spiking Neurons - Theory and Simulations, Chapter (8), *Doctoral Thesis*, Technische Universitaet Graz, Austria (1997). Available: ftp://ftp.eccc.uni-trier.de/pub/eccc/theses/ruf.ps.gz
15. B. Ruf and M. Schmitt: Self-organization of spiking neurons using action potential timing. IEEE-Trans. Neural Networks. **9(3)** (1998) 575–578

Exponential Convergence of Delayed Neural Networks*

Xiaoping Xue

Department of Mathematics, Harbin Institute of Technology,
Harbin 150001, P.R. China
xiaopingxue@263.net

Abstract. Several new conditions for exponential convergence of DNN were proposed in this paper. These conditions guarantee the existence and uniqueness of equilibrium of DNN with certain different activation functions. To demonstrate the differences and features of the new criteria, some remarks are presented.

1 Introduction

There are a great of research activities associated with different stability properties of neural NN(see, for example,[1,2,3,4,5,6,7]). However, most of the early work just discussed the asymptotic stability and exponential stability without delayed parameters. During the last few years, a large number of papers addressed the problems of exponential stability with delay parameters([4,5]).

In this paper, we provide two new results for exponential convergence of equilibrium of DNN with different activation functions, and activation functions herein may not be bounded.

2 Preliminaries

We consider the following delayed neural networks(DNN):

$$\frac{dx(t)}{dt} = -Dx(t) + Ah(x(t)) + Bh(x(t-\tau)) + U \qquad (N)$$

where $x(t) = (x_1(t), x_2(t), \cdots, x_n(t))^T \in \mathbb{R}^n$ is the state vector, $D = diag(d_1, d_2, \cdots, d_n)$ is a positive diagonal matrix, $A = (a_{ij})_{n \times n}$ and $B = (b_{ij})_{n \times n}$ are the $n \times n$ matrices, $h(x) = (h_1(x_1), h_2(x_2), \cdots, h_n(x_n))^T : \mathbb{R}^n \longrightarrow \mathbb{R}^n$ denotes the neuron activation vector function, and $U = (u_1, u_2, \cdots, u_n)^T \in \mathbb{R}^n$ is a constant vector, while $\tau > 0$ is the delay parameter.

A continuous function, $h : \mathbb{R} \longrightarrow \mathbb{R}$ is said to be of class $\mathcal{H}(\alpha)$ if (i) h is an increasing function; (ii) there exists a positive constant $\alpha > 0$ such that for any

---

* This work is supported by the Natural Science Foundation of China under grant 10271035 and by the Foundation of HIT under grant 2002.53.

$\rho \in \mathbb{R}$ there exist two numbers $q_\rho > 0$ and $a_\rho > 0$ satisfying $|h(\theta) - h(\rho)| \geq q_\rho |\theta - \rho|^\alpha$ whenever $|\theta - \rho| \leq a_\rho$. For example, $h(\rho) = \rho$, $h(\rho) = \arctan(\rho)$, and $h(\rho) = [1 - \exp(-\lambda\rho)]/[1 + \exp(-\lambda\rho)](\lambda > 0)$ are all in $\mathcal{H}(1)$.

We define $\mathcal{GL}$ as class of globally Lipschitz. Note that $\mathcal{H}(\alpha) \cap \mathcal{GL} \neq \phi$, for example, $h(\rho) = \arctan(\rho) \in \mathcal{H}(1) \cap \mathcal{GL}$.

An equilibrium x^* is a constant solution of (N), i.e., it satisfies the algebraic equation $-Dx^* + Ah(x^*) + Bx^* + U = 0$.

Definition 1. *(N) is said to be exponentially convergent if it has a unique equilibrium x^* and there exist two constants $M > 0$ and $\beta > 0$ such that for any initial continuous function $\varphi(t)(-\tau \leq t \leq 0)$, there exist a solution $x(t, \varphi)$ in $[0, +\infty)$ of (N) and $T(\varphi) > 0$ satisfying $\|x(t, \varphi) - x^*\| \leq Me^{-\beta t}$ $(t \geq T(\varphi))$.*

Lemma 1. *If $h \in \mathcal{H}(\alpha)$, then for any $\rho_0 \in \mathbb{R}$, one has*

$$\lim_{|\rho| \to \infty} \int_{\rho_0}^{\rho} [h(\theta) - h(\rho_0)] d\theta = +\infty.$$

3 Main Results

Theorem 1. *Suppose $h_i \in \mathcal{H}(\alpha)(i = 1, 2, \cdots, n)$ and there exists a positive diagonal matrix $P = diag(p_1, p_2, \cdots, p_n)$ such that $PA + A^T P + (PB)(PB)^T + I < 0$. Then the system (N) is exponentially convergent.*

Proof. We first prove that the system (N) has a unique equilibrium. Let $V(x) = Dx - (A + B)h(x) - U$, then x^* is an equilibrium iff $V(x^*) = 0$. We can rewrite $V(x)$ as the form $V(x) = Dx - (A+B)f(x) + V_0$, where $f(x) = h(x) - h(0) \in \mathbb{R}^n$ and $f_i \in \mathcal{H}(\alpha)$ satisfying $f_i(0) = 0$, $V_0 = -(A+B)h(0) - U \in \mathbb{R}^n$. Construct the open subset $\Omega_r = \{x \in \mathbb{R}^n : \|x\| < r\}$ for some $r > 0$ and the homotopy $H(\lambda, x)$ defined as $H(\lambda, x) = \lambda Dx + (1 - \lambda)V(x)$, $x \in \overline{\Omega_r} = \{x : \|x\| \leq r\}, \lambda \in [0, 1]$. By computing, we have

$$f^T(x) PH(\lambda, x) = f^T(x)PDx - (1-\lambda)f^T(x)P(A+B)f(x) + f^T(x)PV_0$$
$$\geq \sum_{i=1}^n p_i d_i |f_i(x_i)| \left[|x_i| - \frac{|(PV_0)_i|}{p_i d_i} \right]. \quad (1)$$

Since $f_i \in \mathcal{H}(\alpha)$, then there exist $M > 0$ and $b > 0$ such that $|f_i(x_i)| \geq M$ when $|x_i| \geq b$, $i = 1, 2, \cdots, n$. Let $a = \max_{1 \leq i \leq n} \frac{|(PV_0)_i|}{p_i d_i}$. For every index set $Q \subset \{1, 2, \cdots, n\}$, the function $f_Q = \sum_{i \in Q} p_i d_i [|f_i(x_i)| - a]$ is continuous on $\Omega_Q = \{x_i \in \mathbb{R}^{|Q|} : |x_i| \leq a, i \in Q\}$, then it can attain to the minimum. Let $l = \min_{1 \leq i \leq n} p_i d_i M$, μ_Q be the minimum of f_Q on Ω_Q, and $\mu = \min\{\mu_Q : Q \subsetneq \{1, 2, \cdots, n\}\}$. Thus, if $r > max\{n(a + \frac{|\mu|}{l}), nb\}$ and $\|x\| = \left(\sum_{i=1}^n |x_i|^2\right)^{\frac{1}{2}} = r$, then there exist two index sets Q_1 and Q_2 such that

$$|x_i| \leq a \text{ when } i \in Q_1, \ |x_i| > a \text{ when } i \in Q_2, \ Q_1 \cup Q_2 = \{1, 2, \cdots, n\}.$$

On the other hand, we can find an index i_0 in Q_2 such that $|x_{i_0}| \geq \frac{r}{n}$. From (1), we obtain

$$f^T PH \geq \sum_{i \in Q_1} k_i + \sum_{i \in Q_2} k_i \geq p_{i_0} d_{i_0} M \left[|x_{i_0}| - a - \frac{|\mu|}{\rho} \right] > 0,$$

where $k_i = p_i d_i |f_i(x_i)| [\, |x_i| - a]$. Then, we get that for $x \in \partial \Omega_r = \{x \in \mathbb{R}^n : ||x|| = r\}$ and $\lambda \in [0,1]$, $f^T(x) PH(\lambda, x) > 0$, which implies that $H(\lambda, x) \neq 0$. By topological degree theory, it following that $deg(H(0,x), \Omega_r, 0) = deg(H(1,x), \Omega_r, 0)$, i.e., that $deg(V(x), \Omega_r, x) = deg(Dx, \Omega_r, 0) = sgn|D| \neq 0$. Thus, $V(x) = 0$ has at least one solution in Ω_r. We obtain easily that (N) has a unique equilibrium.

Next we will prove that the global existence of solutions of (N). We can easily see that local existence of the solutions of (N) with initial values $\varphi(t)(-\tau \leq t \leq 0)$ Let x^* be the unique equilibrium and $y(t, \widetilde{\varphi}) = x(t, \varphi) - x^*$, where $x(t, \varphi)$ is the local solution of (N), $\widetilde{\varphi}(t) = \varphi(t) - x^*$, then $y(t, \widetilde{\varphi})$ is the local solution of (N_1):

$$\frac{dy(t)}{dt} = -Dy(t) + Ag(y(t)) + Bg(y(t-\tau)), \quad (N_1)$$

where $g(y) = (g_1(y_1), g_2(y_2), \cdots, g_n(y_n))^T$ and $g_i(y_i) = h_i(y_i + x_i^*) - h_i(x_i^*)$. Since $PA + A^T P + (PB)(PB)^T + I < 0$, we can choose a small $\delta > 0$, such that $PA + A^T P + (PB)(PB)^T + e^{\delta \tau} I < 0$ and $\delta < \min\{d_i : 1 \leq i \leq n\}$. Construct the following functional

$$V \equiv V(t, y(t)) = e^{\delta t} \left[2 \sum_{i=1}^n p_i \int_0^{y_i(t)} g_i(\theta) d\theta \right] + \int_{t-\tau}^t g^T(y(\theta)) g(y(\theta)) e^{\delta(\theta+\tau)} d\theta. \quad (2)$$

By the assumption on h_i, there exist $r_0 > 0$ and $M_0 > 0$ satisfying

$$g_i(0) = 0 \text{ and } |g_i(\theta)| \geq M_0 |\theta|^\alpha \text{ if } \theta \in [-r_0, r_0], \ i = 1, 2, \cdots, n. \quad (3)$$

By computing the derivative $\dot{V}(t)$ of V along the solution $y(t, \widetilde{\varphi})$, we obtain $\dot{V}(t) \leq 0$. This implies that $V(t) \leq V(0)$, Hence

$$2 \sum_{i=1}^n p_i \int_0^{y_i(t, \widetilde{\varphi})} g_i(\theta) d\theta \leq V(0) e^{-\delta t}. \quad (4)$$

According to the Lemma 1, it implies that $y_i(t, \widetilde{\varphi})$ are bounded. Therefore, by virtue of the continuation theorem, we can conclude that $y(t, \widetilde{\varphi})$ exists on $[0, +\infty)$, then, $x(t, \varphi)$ is also.

Moreover, by (4), we have $\lim_{t \to \infty} y_i(t, \widetilde{\varphi}) = 0$. Thus, there exists a time constant T, such that $y_i(t, \widetilde{\varphi}) \in [-r_0, r_0]$, $t \geq T$. Let $p = \min_{1 \leq i \leq n} p_i$, then by (2) and (4), we have $\max_{1 \leq i \leq n} |y_i(t, \widetilde{\varphi})| \leq \left[\frac{\alpha+1}{2p M_0} V(0) \right]^{\frac{1}{\alpha+1}} e^{-\frac{\delta}{\alpha+1} t}$ This means that the system (N) is exponentially convergent. □

Remark 1. Forti([2], $B = 0$) and Joy([3]) obtained that the 0-solution of system (N_1) is asymptotically convergent. However, in general case, the system (N) is not equivalent to the system (N_1). In Theorem 1, the functions g_i are stronger than that given in [2,3], but the results are also stronger.

If $h_i \in \mathcal{GL}$, then they satisfy that $|h_i(\rho) - h_i(\rho')| \leq \mu_i |\rho - \rho'|$ ($\forall \rho, \rho' \in R$). Let $\Gamma = diag(\mu_1, \mu_2, \cdots, \mu_n)$.

Theorem 2. *If $h_i \in \mathcal{H}(\alpha) \bigcap \mathcal{GL}$, and assume further that there exists a positive diagonal matrix P and $\beta > 0$ such that*

$$-2PD\Gamma^{-1} + PA + A^T P + \frac{1}{\beta}(PB)(PB)^\Gamma + \beta I < 0 \qquad (*)$$

then the system (N) is exponentially convergent.

Remark 2. The proof is similar to that of Theorem 1 and omitted. The asymptotic convergence of 0-solution of (N_1) is proved in [1,5] based on the stronger conditions of Matrix inequality. In our Theorem 2, the matrix inequality $(*)$ is less restrictive than with [1,5].

4 Conclusion

In this paper, some conditions for existence and uniqueness of equilibrium and its exponential convergence are derived. The results herein impose constraints on the inter connection matrix of the neural networks independently of delay parameter. Our Theorems show that the properties of activation functions play the key role in the convergence of neural networks.

References

1. Arik, S.: An improved global stability result for delayed cellular neural networks. IEEE Trans Circuits Syst. I, **49** (2002) 1211–1214
2. Forti, M.: On global asymptotic stability of a class of nonlinear systems arising in neural network theory. J. Differential Equations. **113** (1994) 246-264
3. Joy, M.: On the global convergence of a class of functional differential equatoins with applications in neural network theory. J. Math. Anal. Appl. **232** (1999) 61-81
4. Liao, X., Wang, J.: Algebraic criteria for global exponential stability of cellular neural networks with multiple time delays. IEEE Trans Circuits Syst. I. **50** (2003) 268-275
5. Liao, T., Wang, F.: Global stability for cellular neural networks with time delay. IEEE Trans Neural Networks. **11** (2000) 1481-1484
6. Haykin, S.: Neural Networks: A Comprehensive Foundation. Prentice Hall, New Jersey, 2nd ed. (1999)
7. Rajapakse, J.C., Wang, L.P. (Eds.): Neural Information Processing: Research and Development. Springer, Berlin (2004)

A Neural Network for Constrained Saddle Point Problems: An Approximation Approach*

Xisheng Shen[1,2], Shiji Song[1], and Lixin Cheng[2]

[1] Department of Automation, Tsinghua University, Beijing 100084, China
[2] School of Mathematical Sciences, Xiamen University, Fujian Xiamen 361005, China

Abstract. This paper proposes a neural network for saddle point problems(SPP) by an approximation approach. It first proves both the existence and the convergence property of approximate solutions, and then shows that the proposed network is globally exponentially stable and the solution of (SPP) is approximated. Simulation results are given to demonstrate further the effectiveness of the proposed network.

1 Introduction

Saddle point problems(SPP) provide a useful reformulation of optimality conditions and also arise in many different areas, such as game theory, automatic control, function approximation, and so on(see e.g., [1]). Recently, many neural networks have been constructed for optimization problems(see e.g., [2,3,4,5]). Among them, Ye [2] proposed a neural network for unconstraint minimax problems, and proved its stability under some convexity assumptions; both Gao [3] and Tao [4] focused on quadratic minimax problems, and established several neural networks in assuming that the matrices in the models are positive definite. All these models solve minimax problems by searching the saddle points of the objective functions. The aim of this paper is to develop a new neural network to solve general constrained saddle point problems by an approximation approach. Without any additional assumptions, the proposed network can exponentially solve (SPP), including those the existing ones can not solve(see e.g., Sect. 4).

Let

$$U = \{x \in \mathbb{R}^n : a_i \leq x_i \leq b_i, i = 1, 2, \cdots, n\} ,$$

$$V = \{y \in \mathbb{R}^m : c_j \leq y_j \leq d_j, j = 1, 2, \cdots, m\} ,$$

for $-a_i, b_i, -c_j, d_j \in \mathbb{R} \cup \{+\infty\}$, and let $f : \mathbb{R}^{n+m} \to \mathbb{R} \cup \{\pm\infty\}$ satisfying f is twice continuously differentiable on some open convex set $D_1 \times D_2 (\supset U \times V)$ and is a saddle function on $U \times V$(i.e., for fixed $(x, y) \in U \times V$, both $f(\cdot, y)$ and $-f(x, \cdot)$ are convex on U and V respectively). Let $g = (g_1, g_2, \cdots, g_{l_1}) : \mathbb{R}^n \to \mathbb{R}^{l_1}, p = (p_1, p_2, \cdots, p_{l_2}) : \mathbb{R}^m \to \mathbb{R}^{l_2}$, with both $-g_i$ and $-p_j$ proper convex on U and V and twice continuously differentiable on D_1 and D_2 respectively. And

---

* Supported by the National Key Basic Research Project (973 Project)(2002cb312205), the Grant of the NSF of China(10471114), and the Grant of the NSF of Fujian Province, China (A04100021).

let
$$\Omega_1 = \{x \in U : g(x) \geq 0, h(x) \equiv A_1 x - b^1 = 0\},$$
$$\Omega_2 = \{y \in V : p(y) \geq 0, q(y) \equiv A_2 y - b^2 = 0\},$$

where $A_1 \in \mathbb{R}^{k_1 \times n}, A_2 \in \mathbb{R}^{k_2 \times m}$ with $\text{rank}(A_1) = k_1 < n$, $\text{rank}(A_2) = k_2 < m$, $b^1 \in \mathbb{R}^{k_1}$ and $b^2 \in \mathbb{R}^{k_2}$. Then we have the following saddle point problem:

$$\text{(SPP)} \quad \begin{cases} \text{Find a point } (x^*, y^*) \in \Omega_1 \times \Omega_2, \text{ such that } (x^*, y^*) \text{ is a} \\ \text{saddle point of } f(x,y) \text{ on } \Omega_1 \times \Omega_2, \text{ that is,} \\ f(x^*, y) \leq f(x^*, y^*) \leq f(x, y^*), \forall (x, y) \in \Omega_1 \times \Omega_2. \end{cases} \quad (1)$$

Throughout this paper, we assume that (SPP) has a solution and satisfies the Slater condition(see e.g., [1, p325]). Let $\|\cdot\|$ be the Euclidean norm, $\mathbb{R}_+^n = \{x \in \mathbb{R}^n : x_i \geq 0, i = 1, 2, \cdots, n\}$, $\nabla_x f(x,y) = (\partial f(x,y)\backslash \partial x_1, \partial f(x,y)\backslash \partial x_2, \cdots, \partial f(x,y)\backslash \partial x_n)^T$, and $[\cdot]^+ = \max\{0, \cdot\}$. A vector $x \in \mathbb{R}^n$ will be the column form, and x^T denotes its transpose.

2 Convergence of Approximate Solutions

In this section, we will show some results about the approximate solutions, which are the theoretical key links in the construction of the network.

For simplicity, let $u = (x^T, \xi^T, \eta^T)^T, v = (y^T, \lambda^T, \mu^T)^T, z = (u^T, v^T)^T, U_1 = U \times \mathbb{R}_+^{l_2} \times \mathbb{R}^{k_2}, U_2 = V \times \mathbb{R}_+^{l_1} \times \mathbb{R}^{k_1}, \Omega = D_1 \times \mathbb{R}^{l_2+k_2} \times D_2 \times \mathbb{R}^{l_1+k_1}$.

Let $\psi(u,v) = \psi_1(u) - \psi_2(v)$, where $\psi_1(u)$ and $\psi_2(v)$ are uniformly convex and twice continuously differentiable on $\mathbb{R}^{n+l_2+k_2}$ and $\mathbb{R}^{m+l_1+k_1}$ respectively. For example, we can take $\psi(u,v) = 1/2 \|u\|^2 - 1/2 \|v\|^2$.

Let L be the "Lagrange function" of (SPP) defined by
$$L(u,v) = f(x,y) - \lambda^T g(x) - \mu^T h(x) + \xi^T p(y) + \eta^T q(y), \forall (u,v) \in \Omega.$$

And for every $k \in \mathbb{N}$, let $L_k(u,v) \equiv L(u,v) + 1/k \psi(u,v)$. Then, for every $k \in \mathbb{N}$, we have the following saddle point problem associated with (SPP):

$$\text{(SPPk)} \quad \begin{cases} \text{Find a point } (u_k^*, v_k^*) \in U_1 \times U_2, \text{ such that} \\ (u_k^*, v_k^*) \text{ is a saddle point of } L_k(u,v) \text{ on } U_1 \times U_2. \end{cases}$$

Lemma 1. *[1] Let $C \subset \mathbb{R}^n$ be closed and convex. Then $u^* \in C$ is equal to the projection $P_C(u)$ of u on C if and only if $[u - u^*]^T[u^* - v] \geq 0, \forall v \in C$.*

Lemma 2. *Suppose that $C \subset \mathbb{R}^n, D \subset \mathbb{R}^m$ are closed and convex. Then*
$$P_{C \times D}(z) = (P_C(x)^T, P_D(y)^T)^T, \forall z = (x^T, y^T)^T \in \mathbb{R}^{n+m}.$$

Theorem 1. *Take $\psi(u,v) = 1/2 \|u\|^2 - 1/2 \|v\|^2$. Then (u_k^*, v_k^*) is a solution of (SPPk) if and only if (u_k^*, v_k^*) satisfying*

$$\begin{cases} x_k^* = P_U[(1 - \alpha/k)x_k^* - \alpha(\nabla_x f(x_k^*, y_k^*) - \nabla g(x_k^*)^T \lambda_k^* - \nabla h(x_k^*)^T \mu_k^*)], \\ y_k^* = P_V[(1 - \alpha/k)y_k^* + \alpha(\nabla_y f(x_k^*, y_k^*) + \nabla p(y_k^*)^T \xi_k^* + \nabla q(y_k^*)^T \eta_k^*)], \\ \xi_k^* = [(1 - \alpha/k)\xi_k^* - \alpha p(y_k^*)]^+, \lambda_k^* = [(1 - \alpha/k)\lambda_k^* - \alpha g(x_k^*)]^+, \\ 1/k \mu_k^* + h(x_k^*) = 0, 1/k \eta_k^* + q(y_k^*) = 0, \end{cases} \quad (2)$$

where $\alpha > 0$ is a constant, $u_k^* = (x_k^{*T}, \xi_k^{*T}, \eta_k^{*T})^T, v_k^* = (y_k^{*T}, \lambda_k^{*T}, \mu_k^{*T})^T$.

Proof. By Lemma 1 and Lemma 2,

$$(2) \iff u_k^* = P_{U_1}[u_k^* - \alpha \nabla_u L_k(u_k^*, v_k^*)], v_k^* = P_{U_2}[v_k^* + \alpha \nabla_v L_k(u_k^*, v_k^*)]$$
$$\iff (u - u_k^*)^T \nabla_u L_k(u_k^*, v_k^*) \geq 0, (v - v_k^*)^T \nabla_v L_k(u_k^*, v_k^*) \leq 0, \forall u \in U_1, v \in U_2$$
$$\iff L_k(u_k^*, v_k^*) \leq L_k(u, v_k^*), L_k(u_k^*, v) \leq L_k(u_k^*, v_k^*), \forall u \in U_1, v \in U_2$$
$$\iff (u_k^*, v_k^*) \text{ is a solution of (SPPk)} .$$
□

Lemma 3. *[6] If $\varphi : \mathbb{R}^n \to \mathbb{R}$ is continuously differentiable. Then*
i) $\varphi(x)$ is uniformly convex if and only if $\nabla \varphi(x)$ is strongly monotone, i.e., there exits a constant $c > 0$, such that

$$(x - y)^T [\nabla \varphi(x) - \nabla \varphi(y)] \geq c \|x - y\|^2, \forall x, y \in \mathbb{R}^n ; \tag{3}$$

ii) if $\varphi(x)$ is uniformly convex, then $\{x \in \mathbb{R}^n : \varphi(x) \leq \gamma\}$ is a closed bounded convex set for every $\gamma \in \mathbb{R}$, and $\varphi(x)$ has a unique global minimum on every nonempty closed convex set $C \subset \mathbb{R}^n$.

Lemma 4. *[7] Suppose that $C \subset \mathbb{R}^n$ is closed and convex, and $T : C \to \mathbb{R}^n$ is continuous and strongly monotone. Then there exists a unique point $x^* \in C$, such that $(x - x^*)^T T(x^*) \geq 0, \forall x \in C$.*

Theorem 2. *For $\psi(u, v) = \psi_1(u) - \psi_2(v)$ given as above, the followings are true*
i) (SPPk) has a unique solution (u_k^*, v_k^*), for every $k \in \mathbb{N}$;
ii) if (SPP) has a solution, then (u_k^*, v_k^*) converges to a point (u^*, v^*), as $k \to \infty$, such that (x^*, y^*) is a solution of (SPP) and $\lambda^*, \mu^*, \xi^*, \eta^*$ are the corresponding Lagrange multipliers, where $u^* = (x^{*T}, \xi^{*T}, \eta^{*T})^T, v^* = (y^{*T}, \lambda^{*T}, \mu^{*T})^T$.

Proof. i) Let $G(u,v) = (\nabla_u L(u,v)^T, -\nabla_v L(u,v)^T)^T, G_k(u,v) = (\nabla_u L_k(u,v)^T, -\nabla_v L_k(u,v)^T)^T$. By Lemma 3, there is $c > 0$, such that (3) holds for both ψ_1 and ψ_2. Noting that $G(u,v)$ is monotone, we obtain $G_k(u,v)$ is strongly monotone on $U_1 \times U_2$. By Lemma 4, there is a unique point $(u_k^*, v_k^*) \in U_1 \times U_2$, such that

$$(u - u_k^*)^T \nabla_u L_k(u_k^*, v_k^*) - (v - v_k^*)^T \nabla_v L_k(u_k^*, v_k^*) \geq 0 , \tag{4}$$

for all $(u, v) \in U_1 \times U_2$. Then

$$(4) \iff (u - u_k^*)^T \nabla_u L_k(u_k^*, v_k^*) \geq 0, (v - v_k^*)^T \nabla_v L_k(u_k^*, v_k^*) \leq 0, \forall u \in U_1, v \in U_2$$
$$\iff L_k(u_k^*, v_k^*) \leq L_k(u, v_k^*), L_k(u_k^*, v) \leq L_k(u_k^*, v_k^*), \forall u \in U_1, v \in U_2$$
$$\iff (u_k^*, v_k^*) \text{ is a saddle point of (SPPk)} .$$

ii) Denote $\Omega^* \equiv \{(u,v) : (u,v) \text{ is a saddle point of } L(u,v) \text{ on } U_1 \times U_2\}$. Then Ω^* is a closed convex set [8]. Suppose (SPP) has a solution $(\overline{x}, \overline{y})$(i.e., (1) satisfies for $x^* = \overline{x}, y^* = \overline{y}$). Then $\overline{x}$ is a solution of the following convex programming:

$$\begin{cases} \min f(x, \overline{y}) \\ s.t. \ g(x) \geq 0, \ h(x) = 0, \ x \in U \end{cases}.$$

By Kuhn-Tucker saddlepoint Theorem [1], there are $\overline{\lambda} \in \mathbb{R}_+^{l_1}, \overline{\mu} \in \mathbb{R}^{k_1}$, such that

$$f(\overline{x}, \overline{y}) - \lambda^T g(\overline{x}) - \mu^T h(\overline{x}) \leq f(\overline{x}, \overline{y}) - \overline{\lambda}^T g(\overline{x}) - \overline{\mu}^T h(\overline{x}) \leq f(x, \overline{y}) - \overline{\lambda}^T g(x) - \overline{\mu}^T h(x),$$

for all $x \in U, \lambda \in \mathbb{R}_+^{l_1}, \mu \in \mathbb{R}^{k_1}$. Similarly, there are $\overline{\xi} \in \mathbb{R}_+^{l_2}, \overline{\eta} \in \mathbb{R}^{k_2}$, such that

$$f(\overline{x},y) + \overline{\xi}^T p(y) + \overline{\eta}^T q(y) \leq f(\overline{x},\overline{y}) + \overline{\xi}^T p(\overline{y}) + \overline{\eta}^T q(\overline{y}) \leq f(\overline{x},\overline{y}) + \xi^T p(\overline{y}) + \eta^T q(\overline{y}),$$

for all $y \in V, \xi \in \mathbb{R}_+^{l_2}, \eta \in \mathbb{R}^{k_2}$. Adding the above two inequalities, we obtain that Ω^* is a nonempty closed convex set.

Since $\psi_1(u)$ and $\psi_2(v)$ are uniformly convex, by Lemma 3 ii), there exists a point $(u^*, v^*) \in \Omega^*$, such that

$$\psi_1(u^*) + \psi_2(v^*) < \psi_1(u) + \psi_2(v), \ \forall\ (u,v) \in \Omega^* \setminus \{(u^*,v^*)\}, \quad (5)$$

and $W \equiv \{(u,v) \in U_1 \times U_2 : \psi_1(u) + \psi_2(v) \leq \psi_1(u^*) + \psi_2(v^*)\}$ is a nonempty bounded closed convex set. Since (u_k^*, v_k^*) is a solution of (SPPk), we have

$$L_k(u_k^*, v) \leq L_k(u_k^*, v_k^*) \leq L_k(u, v_k^*), \ \forall (u,v) \in U_1 \times U_2. \quad (6)$$

Substituting (u,v) by (u^*, v^*) in (6), and noting $L(u^*, v_k^*) \leq L(u^*, v^*) \leq L(u_k^*, v^*)$, we get $\psi_1(u_k^*) + \psi_2(v_k^*) \leq \psi_1(u^*) + \psi_2(v^*)$. That is, $(u_k^*, v_k^*) \in W, \forall k \in \mathbb{N}$.

We claim that (u_k^*, v_k^*) converges to (u^*, v^*). If not, by taking a subsequence, we can assume that $(u_k^*, v_k^*) \to (\overline{u}, \overline{v}) \in W \setminus \{(u^*, v^*)\}$. Letting $k \to \infty$ in (6), we have $(\overline{u}, \overline{v}) \in \Omega^*$, which contradicts (5).

Noting $L(u^*, v) \leq L(u^*, v^*), \forall v \in U_2$, and letting $y = y^*$, we obtain

$$(\lambda^* - \lambda)^T g(x^*) + (\mu^* - \mu)^T h(x^*) \leq 0, \ \forall \lambda \in \mathbb{R}_+^{l_1}, \mu \in \mathbb{R}^{k_1}.$$

Letting $\lambda = \lambda^*$, we get $(\mu - \mu^*)^T h(x^*) \geq 0, \forall \mu \in \mathbb{R}^{k_1}$. Thus $h(x^*) = 0$. Letting $\mu = \mu^*, \lambda = 0$, we get $\lambda^{*T} g(x^*) \leq 0$. For every i, letting $\mu = \mu^*, \lambda_i = \lambda_i^* + 1$, $\lambda_j = \lambda_j^* (j \neq i)$, we get $g_i(x^*) \geq 0, \lambda_i^* g_i(x^*) \geq 0$. Thus, we obtain $h(x^*) = 0$, $g(x^*) \geq 0, \lambda^{*T} g(x^*) = 0$. Similarly, we get $q(y^*) = 0, p(y^*) \geq 0, \xi^{*T} p(y^*) = 0$.

For each $(x,y) \in \Omega_1 \times \Omega_2$, noting $L(u^*, v) \leq L(u^*, v^*) \leq L(u, v^*), \forall u \in U_1, v \in U_2$, and letting $\lambda = 0, \mu = 0$ and $\xi = 0, \eta = 0$ respectively, we get $f(x^*, y) \leq f(x^*, y^*) \leq f(x, y^*)$, which completes the proof. □

3 Neural Network Model with Globally Exponential Stability

In this section, we will construct a neural network model for (SPP) and will show the globally exponential stability for the proposed network. Especially, take $\psi(u,v) = 1/2 \|u\|^2 - 1/2 \|v\|^2$. Then by Theorem 1 and Theorem 2, we have the following dynamic system as a neural network model to solve (SPP):

$$\frac{d}{dt} \begin{pmatrix} x \\ \xi \\ \eta \\ y \\ \lambda \\ \mu \end{pmatrix} = - \begin{pmatrix} x - \widetilde{x} \\ \xi - \widetilde{\xi} \\ \alpha/k\eta + \alpha q(y) \\ y - \widetilde{y} \\ \lambda - \widetilde{\lambda} \\ \alpha/k\mu + \alpha h(x) \end{pmatrix}, \quad (7)$$

where $k \in \mathbb{N}$, $\alpha > 0$ are constants, $\tilde{x} = P_U[(1-\alpha/k)x - \alpha(\nabla_x f(x,y) - \nabla g(x)^T \lambda - \nabla h(x)^T \mu)]$, $\tilde{\xi} = [(1-\alpha/k)\xi - \alpha p(y)]^+$, $\tilde{y} = P_V[(1-\alpha/k)y + \alpha(\nabla_y f(x,y) + \nabla p(y)^T \xi + \nabla q(y)^T \eta)]$, $\tilde{\lambda} = [(1-\alpha/k)\lambda - \alpha g(x)]^+$.

It can be easily seen that model (7) has one-layer structure, and the projection operators $P_U(\cdot), P_V(\cdot)$ and $[\cdot]^+$ can be easily implemented by piecewise-activation functions(see e.g., [9]). Thus, the complexity of (7) depends only on $\nabla_x f(x,y), \nabla_y f(x,y), \nabla g(x), \nabla p(y), g(x)$, and $p(y)$ in the original problem.

Theorem 3. *For any initial point in $U_1 \times U_2$, the solution of (7) will converge to the unique solution (u_k^*, v_k^*) of (SPPk) exponentially. Moreover, for k large enough, (x_k^*, y_k^*) is an approximate solution of (SPP), and $\lambda_k^*, \mu_k^*, \xi_k^*, \eta_k^*$ are the corresponding approximate Lagrange multipliers, where $u_k^* = (x_k^{*T}, \xi_k^{*T}, \eta_k^{*T})^T$, $v_k^* = (y_k^{*T}, \lambda_k^{*T}, \mu_k^{*T})^T$.*

Proof. By Lemma 2, (7) is equivalent to $dz/dt = P_{U_1 \times U_2}[z - \alpha G_k(z)] - z$, where $z = (u^T, v^T)^T$ and $G_k(z) \equiv G_k(u,v)$ is defined in the proof of Theorem 2 i). Since $G_k(u,v)$ is strongly monotone, by Theorem 2 and [5, Thm 2], we completes the proof. □

4 Simulation Examples

In this section, two illustrative examples are given to compare model (7) with the existing one in [3]. The simulations are conducted in MATLAB.

Example 1. Consider (SPP) with $f(x,y) = xy$ on $\mathbb{R} \times \mathbb{R}$, $\Omega_1 = U = \mathbb{R}$, $\Omega_2 = V = \mathbb{R}$. This problem has a unique saddle point $(0,0)$. For initial point $(x_0, y_0) \in \mathbb{R}^2$, the solution of model (7) for it is $x(t) = e^{-t}(x_0 \cos t - y_0 \sin t)$, $y(t) = e^{-t}(y_0 \cos t + x_0 \sin t)$, which converges to $(0,0)$ exponentially. To make a comparison, the solution of model in [3] for this problem is $x(t) = -y_0 \sin t + x_0 \cos t$, $y(t) = x_0 \sin t + y_0 \cos t$, which doesn't converge whenever $(x_0, y_0) \neq (0, 0)$.

Example 2. Consider (SPP) with $f(x,y) = 1/2 x_1^2 - x_1 + x_2(y_2 - 1) + 3x_2 y_3 - 1/2 y_1^2 + 2y_1$, $\Omega_1 = U = \{x \in \mathbb{R}^2 : -5 \leq x_i \leq 5, i = 1, 2\}$, and $\Omega_2 = V = \{y \in \mathbb{R}^3 : -2 \leq y_j \leq 4, j = 1, 2, 3\}$. This problem has infinite saddle points $\{(x,y) \in \mathbb{R}^2 \times \mathbb{R}^3 : x_1 = 1, y_1 = 2, x_2(y_2 + 3y_3 - 1) = 0, -5 \leq x_2 \leq 5, -2 \leq y_2 \leq 4, -2 \leq y_3 \leq 4\}$.

We first use model (7) to solve this problem. All simulation results show that it converges to one saddle point of the problem. As an example, Fig. 1 (a) shows that the trajectories of (7) converge to $(0.9901, -0.0011, 1.9802, -0.1089, -0.3267)$ with the initial point $(4.4, -4.1, -1.5, -1.0, 2.5)$ for $\alpha = 1$, $k = 100$.

Then we solve this problem by the model in [3]. But, Fig. 1 (b) shows that this model with the initial point $(4.4, -4.1, -1.5, -1.0, 2.5)$ is not stable.

Thus, from the above simulation results, we can conclude that the proposed network (7) is feasible and has a good stability performance.

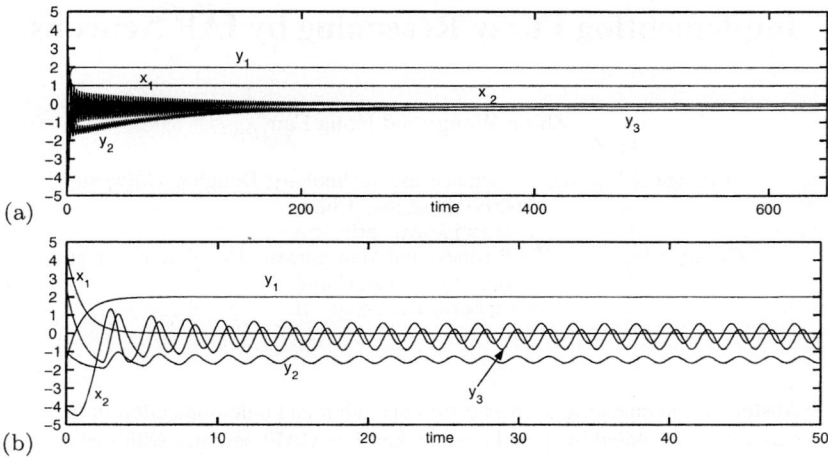

Fig. 1. (a) Transient behavior of (7) for Example 2 with $k = 100, \alpha = 1$. (b) Transient behavior of the model in [3] for Example 2.

5 Conclusion

A neural network model with globally exponential stability is constructed for (SPP) after showing both the existence and the convergence property of approximate solutions. In contrast to the existing ones, the proposed network requires no additional assumptions and has globally exponential stability automatically. The simulation results demonstrate further its effectiveness.

References

1. Bertsekas, D. P.: Nonlinear Programming. 2nd ed. Athena Scientific, (1999)
2. Ye, Z., Zhang, B., Cao, C.: Convergence Analysis on Minimax Neural Networks(in Chinese). Information and Control. **26** (1997) 1–6
3. Gao, X.-B., Liao, L.-Z., Xue, W.: A Neural Network for a Class of Convex Quadratic Minimax Problems with Constraints. IEEE Trans. on Neural Networks. **15** (2004) 622–628
4. Tao, Q., Fang, T.: The Neural Network Model for Solving Minimax Problems with Constraints(in Chinese), Control Theory Applicat. **17** (2000) 82–84
5. Gao, X.-B.: Exponential Stability of Globally Projected Dynamic Systems. IEEE Trans. on Neural Networks. **14** (2003) 426–431
6. Ortega, J. M., Rheinboldt, W. C.: Interative Solution of Nonlinear Equations in Several Variables. Academic, New York, (1970)
7. Browder, F. E.: Nonlinear Monotone Operators and Convex Sets in Banach spaces, Bull. Amer. Math. Soc. **71** (1965) 780–785
8. Ekeland, I., Teman, R.: Convex Analysis and Variational Problems. North-Holland, Amsterdam, (1976)
9. Cichocki, A., Unbehauen, R.: Neural Networks for Optimization and Signal Processing. Wiley, New York, (1993)

Implementing Fuzzy Reasoning by IAF Neurons

Zhijie Wang[1] and Hong Fan[2]

[1] College of Information Science and Technology, Donghua University,
200051 Shanghai, China
wangzj@dhu.edu.cn
[2] Glorious Sun School of Business and Management, Donghua University,
200051 Shanghai, China
hongfan@dhu.edu.cn

Abstract. Implementing of intersection operation and union operation in fuzzy reasoning is explored by three Integrate-And-Fire (IAF) neurons, with two neurons as inputs and the other one as output. We prove that if parameter values of the neurons are set appropriately for intersection operation, firing rate of the output neuron is equal to or is lower than the lower one of two input neurons. We also prove that if parameter values of the neurons are set appropriately for union operation, the firing rate of the output neuron is equal to or is higher than the higher one of the two input neurons. The characteristic of intersection operation and union operation implemented by IAF neurons is discussed.

1 Introduction

Fuzzy logic is considered as one of the information processing mechanisms of the human brain. Fuzzy set theory was proposed to model this mechanism. Numerous successful application systems based on fuzzy set theory are reported. Computation of fuzzy systems is based on mathematical framework of fuzzy set theory, while the computation of fuzzy system in the brain is accomplished by neurons. Though many neuron-fuzzy systems have been proposed, the purpose of these systems is to encode fuzzy rules in artificial neural networks and to tune parameters of fuzzy systems with learning ability of artificial neural networks. Few systems that integrate fuzzy logic and neural network with biological plausible neurons are proposed. We have found a fuzzy-like phenomenon in an autoassociative memory with dynamic neurons [3]. In this work, we explore to implement the reasoning of fuzzy systems with Integrate-And-Fire (IAF) neurons.

2 Fuzzy Reasoning in Fuzzy Systems

Fuzzy systems are usually expressed by fuzzy rules that take a form as follows:
  Rule 1: if X_1 is A_1 and X_2 is B_1 then y is C_1,
  Rule 2: if X_1 is A_2 and X_2 is B_2 then y is C_2, ...
  Where X_1 and X_2 are input linguistic variables, y is an output linguistic variable, A_1, A_2, B_1, B_2, C_1 and C_2 are fuzzy sets that are defined by membership functions.

When crisp inputs $x_1(0)$ and $x_2(0)$ are supplied to the system, firing strengths of Rule1 (α_1) and Rule2 (α_2) are computed by:

$$\alpha_1 = \min(\mu_{A_1}(x_1(0)), \mu_{B_1}(x_2(0))) \tag{1}$$

$$\alpha_2 = \min(\mu_{A_2}(x_1(0)), \mu_{B_2}(x_2(0))) \tag{2}$$

Where $\mu_{A_1}(x_1(0)), \mu_{B_1}(x_2(0)), \mu_{A_2}(x_1(0)), \mu_{B_2}(x_2(0))$ are calculated by the membership functions of the fuzzy sets A_1, B_1, A_2 and B_2 respectively. Based on the firing strengths of rules, membership function of the fuzzy set of the output due to the i th rule is figured out:

$$\mu'_{C_i}(y) = \min(\alpha_i, \mu_{C_i}(y)) \tag{3}$$

The overall membership function of the fuzzy set of the output is given by:

$$\mu_C(y) = \max(\mu'_{C_1}, \mu'_{C_2}) \tag{4}$$

The fuzzy set of output described by the membership function of $\mu_C(y)$ will be defuzzied to obtain a crisp value of the output. Lee [1] gave a diagrammatic representation of the fuzzy reasoning approach discussed in this section.

3 Fuzzy Reasoning Implemented by IAF Neurons

As explained in section 2, min operation is commonly used for fuzzy intersection operation and max operation is commonly used for fuzzy union operation. We discuss how to implement the two operations by IAF neurons in this section. We name fuzzy intersection operation implemented by IAF neurons fuzzy intersection-like operation, and name fuzzy union operation implemented by IAF neurons fuzzy union-like operation.

A simplified version of models of IAF neuron [2] is used in this paper:

$$\frac{dx_i(t)}{dt} = I - \lambda x_i(t) \tag{5}$$

Where x_i is the state variable of i th neuron, I is the external input, λ is the parameter of dissipation. When $x_i = 1$, the i th neuron fires and x_i jumps back to 0. When a given neuron fires, it emits a spike and pulls other neurons by an amount of ε:

$$x_i(t) = 1 \Rightarrow x_j(t^+) = \min(1, x_j(t) + \varepsilon) \tag{6}$$

As shown in Fig. 1, assume that the firing rates of neuron 1 and Neuron 2 correspond to the degrees of membership of two fuzzy sets respectively, the firing rate of neuron 3 is the result of fuzzy intersection-like operation of neuron 1 and neuron 2. Let neuron 1 and neuron 2 receive constant input I_1 and I_2 respectively and let $I_1 > I_2$. Neuron 1 and neuron 2 will fire periodically with period T_1 and T_2 ($T_1 < T_2$).

The spikes of Neuron 1 and Neuron 2 are fed to Neuron 3. If the parameters of IAF neurons are carefully set, the number of spikes generated by neuron 3 is equal to or is smaller than the number of spikes in S2 (see Theorem 1 and Fig. 1).

Theorem 1: If the parameters of neuron 3 in Fig.1 is so set that spike sequence of S1 itself can not make neuron 3 fire, then the number of spikes generated by neuron 3 is equal to or is smaller than the number of spikes in S2.

Proof: Suppose that the number of spikes in S3 is larger than that in S2. As spike sequence S1 itself can not make neuron 3 fire, there are no spikes generated by neuron 3 before the first spike of S2, and there is at most one spike generated by neuron 3 after the last spike of S2. Therefore, there must exist two spikes of S3 that is between two spikes of S2. There are two cases (see Fig. 1). When neuron 3 generates a spike, the state of the neuron resets to zero. Therefore, in these two cases, the spike that is marked by thick line must be generated by neuron 3 under the stimulation of spike sequence of S1 only. This contradicts with the assumption of the theorem. □

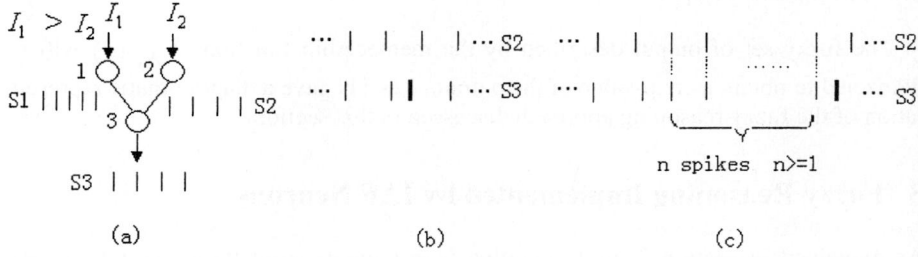

Fig. 1. Figure for the proof of theorem 1. (a) Three neurons (b) Case 1 (c) Case 2.

As we know, firing rate is defined by a temporal average over many spikes of a single neuron. Since degree of membership to a fuzzy set is between 0 and 1, we used a normalized firing rate (FR) in this work:

$$FR = \frac{the \ number \ of \ the \ spikes \ in \ spike \ sequence}{upper \ bound \ of \ the \ number \ of \ spikes \ of \ spike \ sequences} \quad (7)$$

Theorem 1 tells us if the parameter values of neuron 3 are satisfied with the assumption, the firing rate of neuron 3 ($FR3$) is equal to or is lower than that of neuron 2 ($FR2$). Stated in other way, $FR3 \leq \min(FR1, FR2)$.

Theorem 2: The parameters of the neuron 3 in Fig. 2 are so set that it does not fire with none spikes from neuron 1 and neuron 2. If the spike interval in S1 is larger than T in Fig.2, then the firing rate of neuron 3 is equal to or is higher than that of neuron 1.

Proof: As the neuron 3 does not fire with none spikes from neuron1 and neuron 2 and the spike interval in S1 is larger than T in Fig .2, every spike generated by neuron 1 will trigger a spike of neuron 3 if $I_2 = 0$. Therefore the firing rate of neuron 3 is equal to or is higher than that of neuron 1 if we set $I_2 > 0$. □

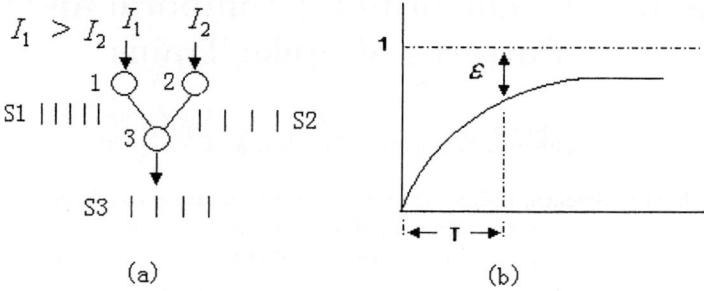

Fig. 2. Figure for the proof of theorem 2. (a) Three neurons. (b) Parameter set of neuron 3.

Theorem 2 tells us if the parameter values of neuron 3 are satisfied with the assumption, $FR3 \geq \max(FR1, FR2)$.

4 Discussion

The analysis shows that the firing rates of neuron 1 ($FR1$), neuron 2 ($FR2$) and neuron 3 ($FR3$) satisfy $FR3 \leq \min(FR1, FR2)$ for fuzzy intersection-like operation, and $FR3 \geq \max(FR1, FR2)$ for fuzzy union-like operation. As for the calculation of the firing strength of Rule 1, min operation is $\alpha_1 = \min(\mu_{A_1}(x_1(0)), \mu_{B_1}(x_2(0)))$. While fuzzy intersection-like operation is $\alpha_1 \leq \min(\mu_{A_1}(x_1(0)), \mu_{B_1}(x_2(0)))$. On the other hand, from the viewpoint of a probabilistic theory, suppose the probabilities for the event A_1 and B_1 are $\mu_{A_1}(x_1(0))$ and $\mu_{B_1}(x_2(0))$ respectively, and suppose that the two events are independent, the firing strength of Rule 1 is $\mu_{A_1}(x_1(0)) * \mu_{B_1}(x_2(0))$. Since $\mu_{A_1}(x_1(0)) * \mu_{B_1}(x_2(0)) < \min(\mu_{A_1}(x_1(0)), \mu_{B_1}(x_2(0)))$, fuzzy intersection-like operation seems to be a compromise between the intersection operation of fuzzy set theory (min operation) and that of a probabilistic theory. Similarly, fuzzy union-like operation seems to be a compromise between the union operation of fuzzy set theory and that of probabilistic theory.

References

1. Lee,C.C.: Fuzzy Logic in Control system: Fuzzy logic controller. IEEE Trans.Systems, Man & Cybernetics, Vol. 20, No. 2. (1990) 404-435
2. Mirollo R.E. , Strogatz S.H.: Synchoronization of pulse-coupled biological oscillators. SIAM J. APPL. MATH., Vol. 50, No. 6. (1990) 1645-1662
3. Wang, Z.J., Aihara, K.: A Fuzzy-Like phenomenon in chaotic autoassociative memory. IEICE TRANS. FUNDAMENTALS, Vol. E85-A, No.3. (2002) 714-722

A Method for Quantifying Temporal and Spatial Patterns of Spike Trains

Shi-min Wang, Qi-Shao Lu, and Ying Du

School of Science, Beijing University of Aeronautics and Astronautics,
Beijing 100083, China
wsmbj@yahoo.com.cn

Abstract. Spike trains are treated as exact time dependent stepwise functions called response functions. Five variables defined at sequential moments with equal interval are introduced to characterize features of response function; and these features can reflect temporal patterns of spike train. These variables have obvious geometric meaning in expressing the response and reasonable coding meaning in describing spike train since the well known 'firing rate' is among them. The dissimilarity or distance between spike trains can be simply defined by means of these variables. The reconstruction of spike train with these variables demonstrates that information carried by spikes is preserved. If spikes of neuron ensemble are taken as a spatial sequence in each time bins, spatial patterns of spikes can also be quantified with a group of variables similar to temporal ones.

1 Introduction

How neurons represent, process and transmit information is of fundamental interest in neuroscience [1]. It is accepted that neural information processing relies on the transmission of a series of stereotyped events called action potentials, or spikes. Temporal recordings of firing events provide inter-spike-interval (ISI) series. It is expected that aspects of the processed information are encoded in the form of structures contained in the ISI series. The basic biophysics that underlie the generation of these action potentials (spike) is well established. However, the features that convey information are not well understood.

An emerging view in neuroscience is that sensory and motor information is processed in a parallel fashion by populations of neurons working in concert [2-4]. Encouraged by this progress many laboratories are investing considerable effort into the development of recording techniques and spike-sorting algorithms that permit simultaneous recording of the activity of multiple neurons [5]. In this context, a fundamental and long-standing question is the type of neural codes used by the population of neurons to represent information in trains of action potentials [1, 6]. The firing rate of spike trains is a candidate for such a neural code [7]; however it is possible that spike timing rather than firing rates plays a significant role in this task [8]. It remains a controversial issue, partly because there are few mathematical methods for directly and quantitatively assessing the temporal structure in spike trains [9]. A key factor in distinguishing among these theories is the temporal precision of individual action potentials. Many existing methods are either qualitative, or limited to examining lower-order structure. Moreover, quantitative techniques used in conjunction with cross-correlations such as the shift predictor can overestimate the number of expected synchronous spikes due to slow rate co-variations which are known to exist [10-12]. In spite of that, and as it was very clearly pointed out by some authors [1, 13, 14], this distinction cannot be pushed too far because both concepts are intrinsically related and the mere introduction of time discretization certainly blurs their differences. Therefore, it is important to measure this precision and to develop new methods to describe population spike trains.

In general£stimuli are time dependent, responses represent dynamic characteristics of stimuli by temporal structures of spike, which are not continuous functions of time. This makes it difficult to relate the time history of stimulus to the temporal patterns of spikes. Therefore, to search how neural responses varying with different stimuli, many researchers turn to measure the statistical signification of temporal structures in spike trains or to determine how much information about stimulus parameter values is contained in neural responses by means of information theory [15]. For example, to investigate the encoding meanings of spike timing and temporal structures or patterns, series expansion approximation method [16], information distortion method [17] and other methods have been used [18-22]. These works show that both the temporal patterns of a spike train and the measurement of dissimilarity between ISI series are important for extracting the information from a neuronal response. However, how to express the varied time histories of temporal patterns and the distance is remained in unsolved.

Taking into account the above considerations, we have developed a way to quantify the temporal and spatial structures of spikes. As we shall explain below, the key feature of this novel method is to express spike trains as stepwise function called response function regarding the dynamic characteristics of spike trains. Several characteristics of the spike train are readily apparent by means of a group of temporal pattern variables deduced from response function, which are defined at sequential moments with equal time interval. Varied temporal patterns of spike train can be uniquely indicated and the time dependent distance between spike trains can also be conveniently defined with these variables. And moreover, these variables may have simple interpretation with neural coding. A tight correlation between dynamic stimuli and cell responses can be expected to set up.

2   Response Function

A recorded spike train can be characterized simply by a stepwise function called response function, the count of spikes fired before time t shown as Fig.1.

It is a monotonic function of time, but has various temporal structures or patterns and exactly reflects the information elicited from stimulus. In geometric viewpoint, a segment of curve can be characterized with its increment, tangent,

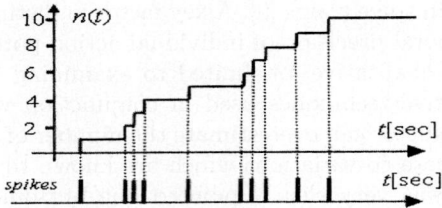

Fig. 1. Spike train expressed as a stepwise function of time

curvature, area surrounded with time axis and other geometric quantities. In neuroscience viewpoint, these aspects may contain coding meanings. Therefore, the temporal pattern variables are introduced basing on geometric quantities of the response function.

3 Variables for Characterizing the Response Function

The first variable is the increment of response function, namely, the spike count of spikes within time interval $T_j = t_j - t_{j-1}, j = 1, 2, \ldots$:

$$q(t_j) = n(t_j) - n(t_{j-1}) \tag{1}$$

The averaged firing rate over can be expressed as:

$$r(t_j) = \frac{q(t_j)}{T_j} \tag{2}$$

It gives the well known rate code or tuning curve of spike train. Evidently, averaged firing rate only reflects part features of the spike train. One cannot tell timing and the order of spikes, namely, the temporal patterns of ISI series. To express the temporal pattern more accurately, we introduced the area surrounded by response function $n(t)$ and time axis in each interval $T_j = t_j - t_{j-1}$:

$$\gamma_j = \int_{T_{j-1}}^{T_j} n(t) dt \tag{3}$$

For illustrating, we divide it into two parts: rectangular and stepwise one shown as Fig.2. The area of rectangular part equals to $n(t_{j-1})T_j$ and another part is:

$$\gamma(t_j) = \sum_{i=1}^{q(t_j)-1} ISI(i) \cdot i + q(t_j)(t_j - t_{q(T_j)}^j) \tag{4}$$

Where represents the order of spikes fired in time interval T_j ; $t_{q(T_j)}^j$ is the firing moment of last spike and $ISI(i)$ is the ith ISI in the same interval.

The response function can be roughly approached by such a sequence of rectangular. Adding stepwise parts, the response function is approximated further

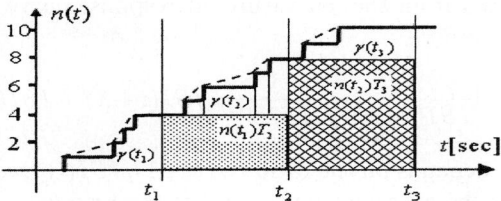

Fig. 2. Area of response function

shown as Fig.2. Variable $\gamma(t_j)$ defined with ISIs and the order of spikes also partly reflects the temporal pattern of spike train. It has statistical meaning, but we prefer to consider it in the view point of functional; it is a functional of response function defined in interval T_j. The coding meaning this variable is that it indicates whether spikes taken placed during gather around moment t_j. For given spike number, the smaller the area $\gamma(t_j)$, the closer to moment the spikes are. Therefore, more information could be extracted. Suppose that is uniformly 16 times of the minimum of ISIs expressed with ISI_{min}, and that each ISI is an integer multiple of ISI_{min}, the total number of possible spike trains is 2^{16}. The count of spikes $q(t_j)$ within T_j can only distinguish 17 kinds of these possible spike trains, and the area $\gamma(t_j)$ can differentiate 137 kinds of them. Using both of these variables, more possible spike trains can be differentiated. The third variable is deduced from the approximate slope of the response function:

$$s(t) = \frac{n(t + \Delta t) - n(t)}{\Delta t} \quad (5)$$

It shows whether a spike fires during sufficiently small Δt around time t. For large Δt it gives an averaged firing rate. Here we take Δt varied as $\Delta t_i = ISI(i)$, and have:

$$s^*(t) = \frac{1}{ISI(i)} \quad (6)$$

Averaging above variable over interval T_j leads:

$$s(t_j) = \sum_{i=1}^{q(t_j)-1} \frac{1}{ISI(i)} \quad (7)$$

This variable shows whether the spikes are close to each other in interval T_j. Since it has the dimension 1/sec, therefore it can be taken as another firing rate. Geometrically, it is the averaged slope of the dashed lines in each time interval T_j shown as Fig.2, where dashed lines connect the 'saw teeth' of the response curve. The coding meaning is that this variable can be used as a 'bursting' index of spikes; the maximum of $s(t_j)$, $s_{max}(t_j) = (q(t_j) - 1)/ISI_{min}$, indicates that all the spikes fire with the smallest ISI_{min}; while the minimum of $s(t_j)$, $s_{min}(t_j) = (q(t_j) - 1)^2/T_j$ means that all spikes are separated uniformly. The next one is originated from the 'curvature' of response curve which is formally defined as:

$$k^* = [\frac{1}{ISI(i+1)} - \frac{1}{ISI(i)}]/[ISI(i+1) + ISI(i)] \qquad (8)$$

Since the sign of k^* depends on the ratio $ISI(i+1)/ISI(i)$, we use the following variable to measure the averaged 'curvature' of the response curve over T_j:

$$k(t_j) = \sum_{i=1}^{q(t_j)-2} \frac{i \cdot ISI(I+1)}{ISI(i)} \qquad (9)$$

It has the code meaning that the ISI increases/decreases. The last one is relevant to the weight center of the area defined by Eq.(4):

$$C(t_j) = \sum_{i=1}^{q(t_j)-1} \frac{1}{2} ISI(i) \cdot i^2 \qquad (10)$$

Its code meaning cab be interpreted as the symmetric degree of ISIs within the interval.

4 The Degree of Reflecting Spike Train

The latter four variables are defined by ISIs and their order; therefore reflect the temporal structures of ISI series in each time interval T_j. Whether these variables characterize the spike train can be examined by reconstructing the spike train with them. In the case of each ISI being integer multiple of the minimum ISI, taking T as 16 times of the minimum ISI, all possible spike trains can be uniquely reconstructed with these variables. For $q(t_j) = 1$ or $q(t_j) = 15$, the possible spike trains are 16, spike trains can be exactly reconstructed only by $\gamma(t_j)$. When $q(t_j) = 8$, there are 12870 possible spike trains, the most variety case, but spike trains can still be exactly reconstructed.

For general cases, ISI varies arbitrarily within a given range. Spike trains cannot be reconstructed exactly. If suppose a spike only takes several possible positions in time bin $\Delta t = ISI_{min}$, for example, each ISI is an integer multiple of one nth of ISI_{min}, reconstruction can still be carried out. The large the number n, the higher the precision is. Here we took n=2 to reconstruct a stochastic spike trains with the following procedures: (A) Finding the minimum ISI $\Delta t = ISI_{min}$ for a give ISI series, and dividing the time span of the spike train into sequential intervals $T = 16ISI_{min}$; (B) Calculating the original values of the group of variables for each intervals T_j with formulae (1), (4), (7), (9) and (10); (C) Placing $q(t_j)$ spikes into T_j by keeping ISIs being integer multiple of $ISI_{min}/2$ and being greater than or equal to ISI_{min}; (D) Calculating the values of the group of variables for all of possible spike trains; and finding the spike train that yields the most approximate values of the group variables comparing to those

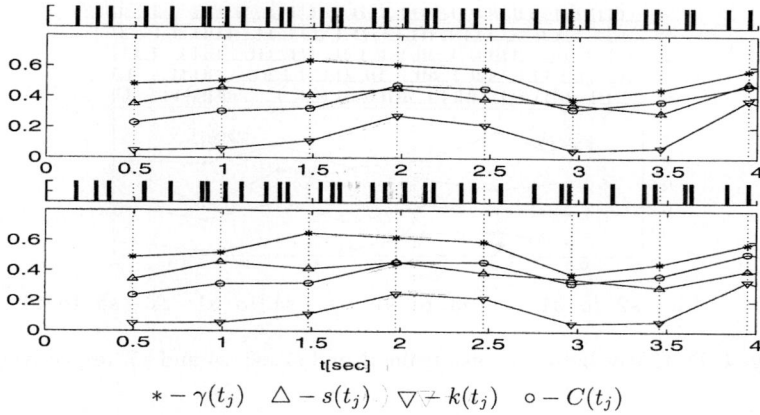

Fig. 3. Reconstruction of a spike train: upper panel is the given spike train and values of the four variables defined at moments shown as dashed lines; lower panel is the reconstructed spike train and values of the four variables

of original spike train. The given and reconstructed stochastic spike trains are shown in Fig. 3.

The validity of reconstruction means that an ISI series is equivalent to these variables defined at sequential moments. Therefore, the analysis on ISI series can be done by dealing with these variables.

5 As a Measurement of Dissimilarity Between Spike Trains

To determine whether a set of ISI series depends systematically on a set of stimuli, it is important to quantify the similarity (or dissimilarity) of two spike trains. Dissimilarity helps characterize neural variability and coding [9]. The distance between spike trains can be defined with this group of variables. At first all the variables except the spike count are scaled to 0-1 by the following equations:

$$\gamma_s(t_j) = \gamma(t_j)/\gamma_{max}^q(t_j) \tag{11}$$

$$s_s(t_j) = s(t_j)/s_{max}^q(t_j) \tag{12}$$

$$k_s(t_j) = k(t_j)/k_{max}^q(t_j) \tag{13}$$

$$C_s(t_j) = C(t_j)/C_{max}^q(t_j) \tag{14}$$

Where $gamma_s(t_j)$, $s_s(t_j)$, $k_s(t_j)$ and $C_s(t_j)$ are scaled variables; $\gamma_{max}^q(t_j)$, $s_{max}^q(t_j)$, $k_{max}^q(t_j)$ and $C_{max}^q(t_j)$ are the maximum values of $\gamma(t_j)$, $s(t_j)$, $k(t_j)$ and $C(t_j)$, respectively, among possible spike trains corresponding to spike count $q(t_j)$. While spike count $q(t_j)$ is scaled as:

$$q_s(t_j) = q(t_j)/(T/ISI_{min}) \tag{15}$$

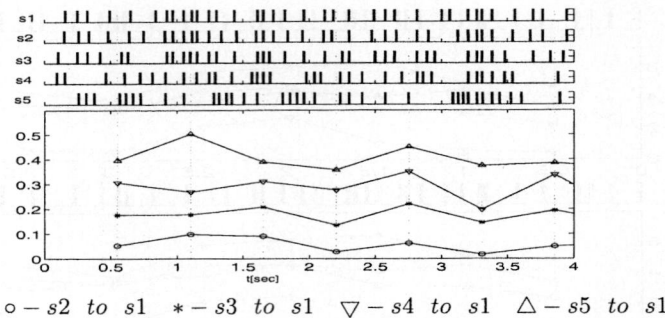

Fig. 4. Distances between spike train s1 and s2, s3, s4 and s5, respectively

Here, we take $T = 16ISI_{min}$ and have: $q_s(t_j) = q(t_j)/16$. Then the Euclidean distance between two groups of scaled variables is defined as the distance of corresponded spike trains. The following example gives distances between one spike train and other four, respectively, shown as Fig.4.

It can be seen that this measurement can conveniently express the varied distance along with time. These variables quantitatively give the temporal patterns of a spike train at discrete moments; it can be expect to relate the varied temporal patterns to the dynamic stimuli.

6 Spatial Patterns

Typically, many neurons respond to a given stimulus, and stimulus features are therefore encoded by the activities of large neural populations. In studying population coding, we must examine not only the firing patterns of individual neurons, but also the relationships of these firing patterns to each other across the population of responding neurons. A group of similar variables are used to quantify the spatial patterns of neurons, along sequential time bin containing only one spike at the most for each neuron.

To quantify the spatial patterns of spikes, neurons are numbered as $i = 1, 2, 3 \ldots N$; here we take N=16. Their spikes can be represented by a spatial-temporal function $n(i,t)$, the count of fired neurons whose number is less than or equal to i, during time bin $\triangle t$ shown as Fig.5.

This spatial pattern varies with time. For given time bin $\triangle t < ISI_{min}$ the spatial pattern can be represented by five variables similar to (1), (4), (7), (9) and (10):
$q(t)$: Count of fired neurons during $\triangle t$;

$$\gamma(t) = \sum_{i=1}^{q(t)-1} ISS(i) \cdot i + q(t)(16 - N_i) \qquad (16)$$

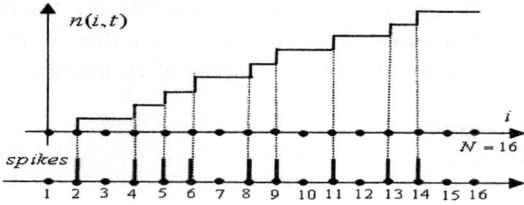

Fig. 5. Response function of neurons, a stepwise function of 'spatial position' (neuron's number) for each time bin

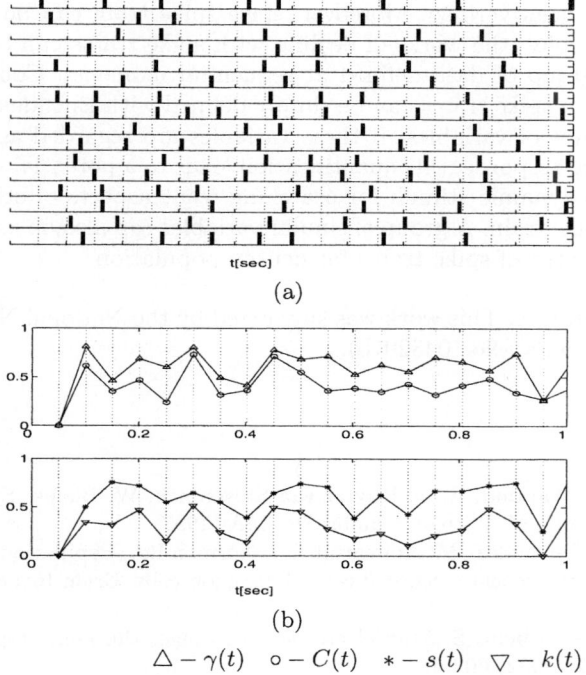

△ − $\gamma(t)$ ○ − $C(t)$ ∗ − $s(t)$ ▽ − $k(t)$

Fig. 6. Time dependent spatial patterns of 16 neurons. Upper panel are 16 stochastic spike trains. Middle and lower panels gives the values of four pattern variable.

$$s(t) = \sum_{i=1}^{q(t)-1} \frac{1}{ISS(i)} \quad (17)$$

$$k(t) = \sum_{i=1}^{q(t)-2} \frac{i \cdot ISS(i+1)}{ISS(i)} \quad (18)$$

$$C(t) = \sum_{i=1}^{q(t)-2} \frac{1}{2} ISS(i) \cdot i^2 \quad (19)$$

Where ISS is taken as the spatial interval of spikes, namely, the number difference between adjacent fired neurons; N_i is the largest number of fired neurons.

In Fig.6 the time history of spatial patterns of 16 neurons is shown.

7 Conclusions

In conclusion we have shown a method to quantify spike trains with a group of variables based on characterizing exact temporal structures of ISI series. The code meaning of these variables can be reflected by one of them, the firing rate that plays equal role as other variables; though the potential use of other variables have not been verified. Moreover, since spike train can be reconstructed with these variables, the works of dealing with spike trains can be transformed to treating a group of data defined at sequential moments. Consequently, the distance between spike trains can be simply defined with Euclidean distance between two groups of variables. If these variables are regarded as discrete time history of neural response, this quantification leads to a potential way of relating spike trains to dynamic stimuli for searching what aspects of stimulus are encoded in the spike train. A group of similar variables can also be used to quantify the spatial patterns of spike trains for neuron population.

Acknowledgement. This work was supported by the National Natural Science foundation of China (No.10432010).

References

1. F. Rieke, D. Warland, R. de Ruyter van Steveninck, W. Bialek, Spike: Exploring the Neural Code MIT Press, Cambridge, MA (1997).
2. Fernndez E, Ferrandez JM, Ammermller J, Norman RA. Population coding in spike trains of simultaneously recorded retinal ganglion cells. Brain Res **887** (2000) 222-9.
3. Nicolelis MA, Ribeiro S. Multielectrode recordings: the next steps. Curr. Opin. Neurobiol **12** (2002) 602-6.
4. Panzeri S, Schultz SR, Trevez A, Rolls ET. Correlation and the encoding of information in the nervous system. Proc R Soc Lond B **266** (1999) 1001-12.
5. Kralik JD, Dimitrov DF, Krupa DJ, Katz DB, Cohen D, Nicolelis MA. Techniques for long-term multisite neuronal ensemble recordings in behaving animals. Methods **25** (2001) 121-51.
6. Meister M, Berry II MJ. The neural code of the retina. Neuron **22** (1999) 435-50.
7. Abbot L, Sejnowsky TJ. Neural code and distributed representations. Cambridge: MIT Press, (1998).
8. Funke E, Worg?tter F. On the significance of temporally structured activity in the dorsal lateral geniculate nucleus (LGN). Prog Neurobiol **53** (1997) 67-119.
9. N. Hatsopoulosa , S. Gemanb , A. Amarasinghamb , E. Bienenstockb At what time scale does the nervous system operate?Neurocomputing **52** (2003) 25 - 29
10. C. Brody, Slow covariations in neuronal resting potentials can lead to artefactually fast cross-correlations in the spike trains, J. Neurophysiol. **80** (1998) 3345-3351.

11. C. Brody, Correlations without synchrony, Neural Comput. **11** (1999) 1553-1577.
12. M.W. Oram, N.G. Hatsopoulos, B.J. Richmond, J.P. Donoghue, Excess synchrony in motor cortical neurons provides direction information that is redundant with the information from coarse temporal response measures, J. Neurophysiol. **86** (2001) 1700-1716.
13. Usrey MW, Reid CR. Synchronous activity in the visual system. Annu Rev Physiol **61** (1999) 435-56.
14. Guillermo J. Ortega a, Markus Bongard b, Enrique Louis c, Eduardo Fernndez Conditioned spikes: a simple and fast method to represent rates and temporal patterns in multielectrode recordings Journal of Neuroscience Methods **133** (2004) 135-141
15. C.M. Gruner, K. Baggerly, D.H. Johnson, C. Seshagiri "Information-theoretic analysis of neural coding." J. of Computational Neuroscience **10** (2001) 47-69
16. S. Panzeri et al., Coding of stimulus location by spike timing in rat somatosensory cortex Neurocomputing **573** (2002) 44-46
17. A. G. Dimitrov et al., Spike pattern-based coding schemes in the cricket cerecal sensory system Neurocomputing **373** (2002) 44-46
18. R. Romero and T. S. Lee, Spike train analysis for single trial data Neurocomputing **597** (2002) 44-46
19. R. Lestienne, Spike timing, synchronization and information processing on the sensory side of the central nervous system. Progress in Neurobiology **65** (2001) 545-591
20. J. P. Segundo, Nonlinear dynamics of point process systems and data I. J. of Bifurcation and Chaos, **13** (2003) 2035
21. M. Christen ea al., Fast spike pattern detection using the correlation integral Physical Review E **70** (2004) 011901
22. D. Aronov and J. D. Victor, Non-Euclidean properties of spike train metric spaces Physical Review E **69** (2004) 061905

A Stochastic Nonlinear Evolution Model and Dynamic Neural Coding on Spontaneous Behavior of Large-Scale Neuronal Population

Rubin Wang and Wei Yu

Institute for Brain Information Processing and Cognitive Neurodynamics, School of
Information Science and Engineering, East China University of Science and Technology, 130
Meilong Road, Shanghai 200237, China
rbwang@163.com or @dhu.edu.cn

Abstract. In this paper we propose a new stochastic nonlinear evolution model that is used to describe activity of neuronal population, we obtain dynamic image of evolution on the average number density in three-dimensioned space along with time, which is used to describe neural synchronization motion. This paper takes into account not only the impact of noise in phase dynamics but also the impact of noise in amplitude dynamics. We analyze how the initial condition and intensity of noise impact on the dynamic evolution of neural coding when the neurons spontaneously interact. The numerical result indicates that the noise acting on the amplitude influences the width of number density distributing around the limit circle of amplitude and the peak value of average number density, but the change of noise intensity cannot make the amplitude to participate in the coding of neural population. The numerical results also indicate that noise acting on the amplitude does not affect phase dynamics.

1 Introduction

In 1960s, Winfree started his famous theoretical investigation of phase resetting with his studies on circadian rhythms [1,2], he showed that an oscillation can be annihilated by a stimulus of a critical intensity and duration administered at a critical initial phase. Haken expanded the theory when he researched the synergetics of the brain [3,4]. P. A. Tass researched Parkinson's disease by the theory of phase resetting [5-8]. Taking into account the noise background within brain, P. A. Tass developed a stochastic approach to phase resetting dynamics. There is abnormal frequency of the action potential in Parkinson's disease. For this reason Tass only considered the curative effect of the frequency in his model. And Tass proposed a stochastic nonlinear model of phase dynamics for the therapy of Parkinson's disease, the amplitude of population of neuronal oscillators is dealt with as a limit circle, namely, amplitude of the action potential is a constant. In this paper, we will apply Tass's model to the research of cognition and propose a new model what base on early our research [9-14]. There are many special structures in nerve nets, for instance, lateral inhibition, presynaptic inhibitory and presynaptic facilitation [15]. Compared with other structures of neurons, the amplitude and the frequency of the action potential of the postsynaptic neuron change stochastically. Furthermore, the background noise in the brain impact on both amplitude and frequency because phase and amplitude are inseparable in the analysis of wave. Though action potential of one neuron does not attenuate when it transmits in the axon, we should consider the impact of the amplitude on the neural coding when we investigate one or more neural functions of the population composed of abundant neurons for the reason of complicated structures and impacts of some uncertain factors. Neural information is transmitted through a couple of cortex areas in the brain information processing, the output of the layer may be the initial condition of the next layer, therefore, it is great important to take into account the impact of the initial condition on the neural coding. In this paper, we numerically analyze the neural coding in the case of different noise intensity and different initial conditions according the stochastic nonlinear dynamic model of the neural population and obtain some important results what can be explained in biological sense.

2 Derivation of Stochastic Model Equation

Setting the amplitude and the phase of N oscillators under the random noise independently are r_j, ψ_j (j=1,2,......, N). The dynamics of phase and the amplitude obey the following evolution equations,

$$\dot{\psi}_j = \Omega + \frac{1}{N}\sum_{k=1}^{N} M(\psi_j - \psi_k, r_j, r_k) + F_{j_2}(t) \quad (1)$$

$$\dot{r}_j = g(r_j) + F_{j_2}(t) \quad (j=1,\cdots,N) \quad (2)$$

We assume that in equation (1) all oscillators have the same eigenfrequency Ω, and the oscillators' mutual interactions are modeled by the term $M(\psi_j - \psi_k, r_j, r_k)$, which model the impact of the kth on the jth oscillator. $\psi_j - \psi_k$ is the difference of their phase. $g(r_j)$ is a nonlinear function of the amplitude. For the sake of simplicity the random force, $F_{j_i}(t)$ (i=1,2) is modeled by Gaussian white noise, which is delta-correlated with zero mean value[1]:

where δ_{jk} denotes the Kroneker delta.

According to Eq. (1) and (2), one obtains the Fokker-Plank equation about probability density f:

$$\frac{\partial f}{\partial t} = \frac{Q_1}{2}\sum_{j=1}^{N}\frac{\partial^2 f}{\partial \Psi_j^2} + \frac{Q_2}{2}\sum_{j=1}^{N}\frac{\partial^2 f}{\partial \Psi_j^2} - \sum_{j=1}^{N}\frac{\partial}{\partial r_j}(g(r_j)f$$

$$-\sum_{j=1}^{N}\frac{\partial}{\partial \Psi_j}[\frac{1}{N}\sum_{k=1}^{N}\Gamma(r_j,r_k,\Psi_j,\Psi_k)f] \quad (3)$$

where $f = f(\Psi_1, \Psi_2, \cdots \Psi_N, r_1, r_2, \cdots r_N, t)$, f is function of the phases and the amplitudes of oscillators, which evolve along with time. By introducing the abbreviation $\Gamma(r_j, r_k, \Psi_j, \Psi_k) = \Omega + M(\psi_j - \psi_k, r_j, r_k)$ and the average number density

$$n(\psi,R,t) = \int_0^{2\pi}\cdots\int_0^{2\pi}d\psi_l\int_0^{\infty}\cdots\int_0^{\infty}\frac{1}{N}\sum_{k=1}^{N}\delta(\psi-\psi_k)\delta(R-r_k)f(\Psi_1,\Psi_2,\cdots\Psi_N,r_1,r_2,\cdots r_N,t)dr_l \quad (4)$$

where R denotes the probability when the amplitude r_j of every oscillator equals R, ψ denotes the probability when the phase ψ_j of every oscillator equals ψ. Inserting (4) into Eq. (3), and then the Fokker-Plank equation with average number density is given by

$$\frac{\partial n}{\partial t} = \frac{Q_1}{2}\frac{\partial^2 n}{\partial \psi^2} + \frac{Q_2}{2}\frac{\partial^2 n}{\partial R^2} - \Omega\frac{\partial n}{\partial \psi} - \frac{\partial(g(R)n)}{\partial R} -$$

$$\frac{\partial}{\partial \psi}(n\int_0^{2\pi}d\psi'\int_0^{\infty}M(\psi-\psi',R,R')n(\psi',R',t)dR') \quad (5)$$

$M(\psi-\psi', R, R')$ is a 2π-period function, it can be expanded as the sum of progression by Fourier transform. We define nonlinear function of amplitude and the mutual interaction term as follows:

$$M(\psi_j-\psi_k,r_j,r_k) = -\sum_{m=1}^{4}2r_j^m r_k^m(K_m\sin m(\psi_j-\psi_k)+C_m\cos m(\psi_j-\psi_k)) \cdot \quad (6)$$

$$g(r_j) = \alpha r_j - \beta r_j^3, \quad \alpha = \beta = 1. \quad (7)$$

As in Tass's reference, we assume $C_m=0$ in function (6), K_m denote the coupling coefficient among neurons. The first term of average number density to be transformed in term Fourier is given by

$$\hat{n}(0,R,t) = B_1 e^{\frac{2}{Q_2}\int_0^R g(x)dx}\int_0^R e^{-\frac{2}{Q_2}\int_0^x g(r)dr}dx. \quad (8)$$

where B_1 fills $\int_0^{R_0}\hat{n}(0,R,t)dR = \frac{1}{2\pi}$, R_0 is the upper limit of the amplitude, $R_0=2$ in this paper.

3 Impact of Noise Intensity on Neural Coding

For the sake of research of noise impact on the dynamics of amplitude, in the case of spontaneous activity, the parameters are chosen as follows:

$$K_1 = 2, \quad K_2 = K_3 = K_4 = 0, \quad Q_1 = 0.4, \Omega = 2\pi,$$

initial condition is given by

$$n(\psi,R,0) = \hat{n}(0,R,0)(1+0.01\sin\psi). \quad (9)$$

We choose two different noise intensity: $Q_2=1$, $Q_2=0.1$. The equation (5) is numerically calculated by difference method and the four groups of figures are obtained as follows:

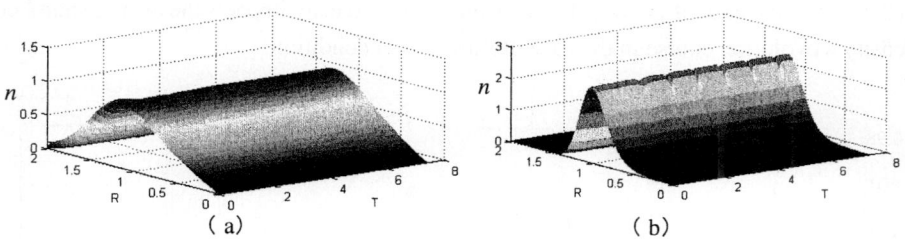

Fig. 1. n about the amplitude change along with time: (a) $Q_2 = 1$, (b) $Q_2 = 0.1$

The limit circle R=1 of the amplitude is confirmed by Eq. (2) and function (7). The distribution of average number density on the amplitude broadens when the intensity Q_2 of Guassian white noise was larger, the larger the intensity Q_2 is, the wider the distribution of average number density is around the limit circle. The distribution will be a narrow band peak when the noise intensity Q_2 is small enough. Though the noise intensity Q_2 are different in Fig.1, probability of the amplitude of action potential centralize around the limit circle, namely, tend to the R=1, in other words, change of noise intensity Q_2 cannot change the configuration of the distribution and the amplitude does not participant in the population coding expect the noise is enough strong to change the original distributing (Fig.1) around the limit circle. This agrees with the conclusion in reference [7].

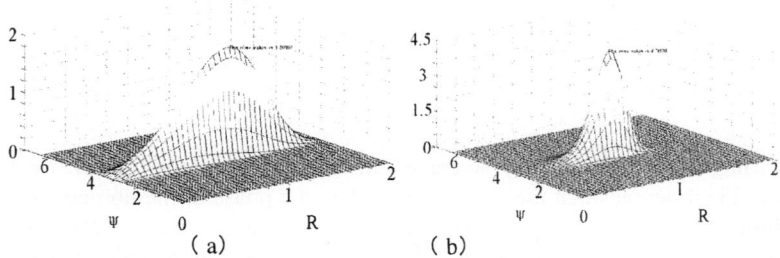

Fig. 2. The figure of $n(\psi, R, t)$ at T=7.351s: (a) $Q_2 = 1$ (b) $Q_2 = 0.1$

Though noise intensities are different in Fig.2.(a) and Fig.2.(b), the wave crests locate at the same phase at the time T=7.351s. But their peak values are different, one is 1.8787 when $Q_2 = 1$, the other is 4.3528 when $Q_2 = 0.1$. It shows that the stronger the noise acting on the amplitude is, the wider the distributing of average number density on the amplitude is, and the smaller the peak value of the wave is.

The stochastic fluctuant range of the amplitude changes on the limit circle R=1, the amplitude fluctuates in 0.82~1 in the case of $Q_2 = 1$ (Fig.3.(a)), but amplitude fluctuates in 2.1~2.5 in the case of $Q_2 = 0.1$. It shows that noise intensity impacts the average number density's evolution course in the case of same initial condition.

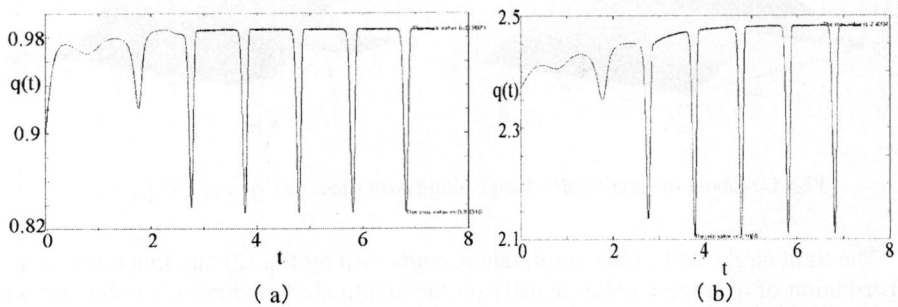

Fig. 3. The evolution figure of average number density on the limit circle: (a) $Q_2 = 1$, (b) $Q_2 = 0.1$, q(t)=n(0,1,t)

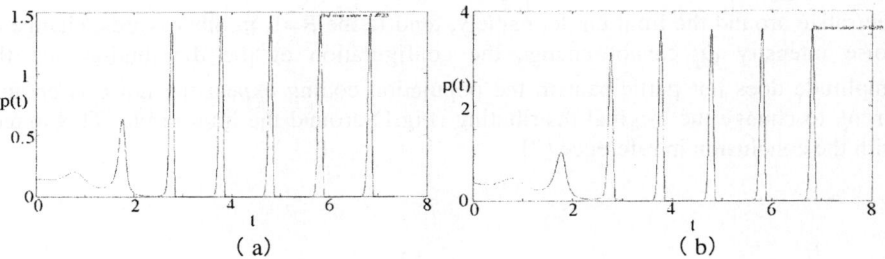

Fig. 4. Firing density on the limit circle. (a) $Q_2 = 1$, (b) $Q_2 = 0.1$, $p(t) = n(0,1,t)$

According to the relation of the average number density and the fire density in Tass's reference, Fig.4 denotes that the neuron fires when the phase of the jth neuron equals 0. According function (4) what is the definition of the average number density, amount of firing neurons at time t is given by $p(t) = n(\rho, r, t)$. $p(t)$ is a macrovariable which corresponds to observable typically measured in experiments. Experimentalists are not able to measure the firing behavior of a large cluster of neurons; they can only assess amount of firing neurons, i.e. how many neurons fire at time t. For this reason the fire density $p(t)$ is introduced. It indicates that researching the phase resetting dynamics of the neural oscillators on the limit circle is feasible, in other words, the noise Q_2 acting on the amplitude does not impact on the neuron's phase dynamics.

According to the numerical results, we can observe that noise in evolution of the amplitude has some affection on the probability distributing of the amplitude. The average number density diminishes on the limit circle when noise intensity acting on the amplitude augments (Fig.3.). A. G. Leventhal [16] pointed out that the energy of the noise within brain augment because the content of GABA diminishes. Our result indicates that the firing density of neurons diminishes on the limit circle of amplitude (Fig.4.).

4 The Impact of Initial Condition on Neural Coding

In order to investigate how the initial condition impacts on the neural coding, we choose the case of one coupling parameter $K_1 \neq 0$ to discuss. The result is also the same with others coupling structures. The parameters are setting as $K_2 = K_3 = K_4 = 0$, $K_1 = 2$, $Q_1 = 0.4$, $\Omega = 2\pi$, $Q_2 = 0.2$. One obtains two groups of figures after computing numerically Eq. (5).

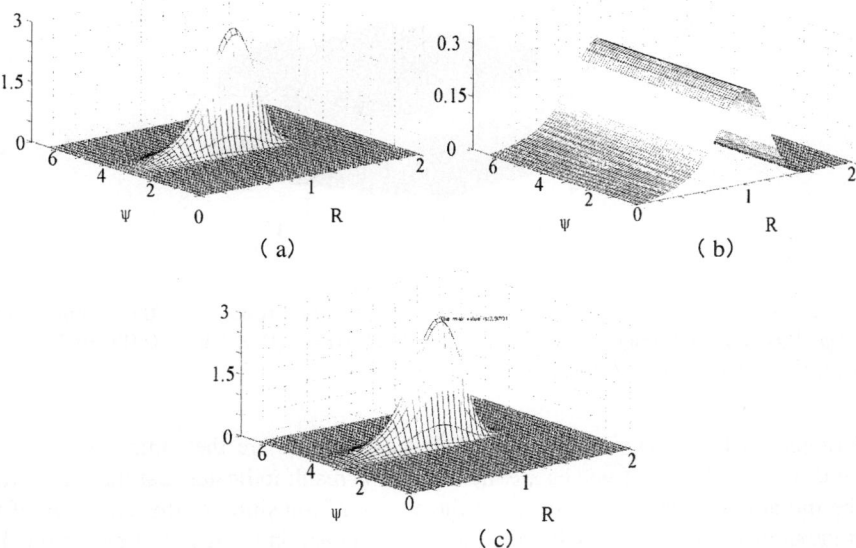

Fig. 5. The figure of $n(\psi, R, t)$ at T=7.351s: the initial condition is
(a) $n(\psi, R, 0) = \hat{n}(0, R, 0)(1 + 0.05 \sin \psi)$ (b) $n(\psi, R, 0) = \hat{n}(0, R, 0)(1 + 0.05 \sin 2\psi)$
(c) $n(\psi, R, 0) = \hat{n}(0, R, 0)(1 + 0.05 \sin \psi + 0.05 \sin 2\psi)$

We compare the three evolution results of the average number density at T=7.351s under case of the stable state. From Fig.5.(a) and Fig.5.(c), one can know that the two figures of the average number density are same though their initial conditions are different. From Fig.5.(b), one can know that the average number densities of the neural oscillators are same at T=7.351s when they are at the same amplitude and different phases, the average number density will keep this figure from now on. The result shows that the initial condition what only contain higher-ordered harmonics terms does not make the average number density change stochastically along with time finally. Fig.6.(b) can approve the conclusion, the coupled structure of the neural population determines its cognitive capability.

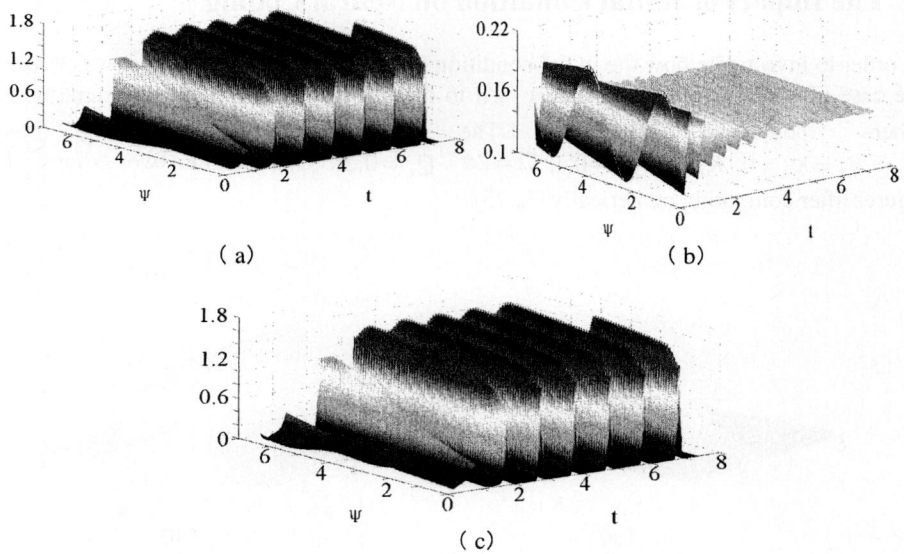

Fig. 6. n about the phase evolutes along with time on the limit circle: (a) $n(\psi, R, 0) = \hat{n}(0, R, 0)(1 + 0.05 \sin \psi)$ (b) $n(\psi, R, 0) = \hat{n}(0, R, 0)(1 + 0.05 \sin 2\psi)$ (c) $n(\psi, R, 0) = \hat{n}(0, R, 0)(1 + 0.05 \sin \psi + 0.05 \sin 2\psi)$

Comparing Fig.6.(a) with Fig.6.(c), their steady states are the same, namely, their periods and the shapes of waves are the same. The result indicates that the higher rank in the initial condition only impacts on the course of transition of the evolution of the average number density, namely that it does not impact on the result of evolution. The different types of synapses increase the complexity when the neurons transmit information. The neural population which has lower-ordered coupling structure can not recognize higher-order initial condition, that is why the neurons can filtrate some information they does not need when they are dealing with information.

5 Conclusion

In this research, we propose a stochastic nonlinear dynamic model which is used to describe the phases and the amplitudes of neurons evolve along with time when neural population actives. We first describe the dynamic evolution course in three-dimensioned space by introducing the average number density $n(\psi, R, t)$.

The result of numerical analysis indicates that the larger the noise intensity Q_2 is, the wider the distributing is. But the configuration of the distributing is not changed, namely, the probability distributing of the amplitude centralize around the limit circle R=1, it is the same with Tass's supposition. The noise on the amplitude has no effect on the neural coding. It indicates change of noise acting on the amplitude does not change the character of the phase dynamics on the limit circle. For the sake of simplicity, we will not take into account the noise's impact on the amplitude in our later research, but it will lose a part of coding information.

The result of numerical analysis also indicates that different initial conditions have distinct impact in the dynamic evolution of the average number density. The evolution course is the coding course. The initial condition and the coupled structure among neurons determine the coding result. It may be explained that the output of the former layer is the initial condition of the later layer. Though the third initial condition contain two-ordered harmonic term, the evolution result only has one peak as result of the first initial condition, namely, it does not impact result of neural coding. It indicates that neural coding is mostly dominated by the structure of the system because the neural system does not have this ability of coding. Though the initial condition has been changed, it does not impact result of coding.

In this paper we research the neural oscillators' mutual interactions and the dynamic evolution when neuronal active spontaneously and obtain the result that amplitude dynamics impacts on the neural coding. We will introduce how the stimulate influences on the neural coding in the other paper.

Project "Phase resetting of neural dynamic system and brain information processing" (30270339) supported by National Natural Science Foundation of China (NSFC)

References

1. Winfree A.T.: An integrated view of the resetting of circadian clock. J. Theor. Biol. 28 (1970) 327-374
2. Winfree, A.T.: The Geometry of Biological Time. Springer, Berlin (1980)
3. Haken, H.: Synopsis and Introduction, In: Synergetics of the Brain. Springer, Berlin (1983)
4. Haken, H.: Principle of Brain Functioning, A Synergetic Approach to Brain Activity, Behavior and Cognition. Springer, Berlin (1996)
5. Tass, P. A.: Phase resetting in Medicine and Biology. Springer, Berlin (1999)
6. Tass, P. A.: Resetting biological oscillators-a stochastic approach. J. Biol. Phys. 22 (1996) 27-64
7. Tass, P. A.: Phase resetting associated with changes of burst shape. J. Biol. Phys. 22 (1996) 122-155
8. Tass, P. A.: Phase and frequency shifts in a population of phase oscillators. Phys. Rev. E. 56 (1997) 2043-2060
9. Wang, R., Zhang, Z.: Nonlinear stochastic models of neurons activities. Neurocomputing. 51C (2003) 401-411
10. Wang, R., Hayashi, H., Zhang, Z.: A stochastic nonlinear evolution model of neuron activity with random amplitude. Proceedings of 9th International Conference on Neural information Processing, Vol. 5. (2002) 2497-2501
11. Wang, R., Zhang, Z.: Analysis of dynamics of the phase resetting on the set of the population of neurons. International Journal of nonlinear science and numerical simulation. 4 (2003) 203-208
12. Wang, R., Zhang, Z.: Nonlinear stochastic models of neurons activities. Proceedings of the 16th International conference on noise in physical systems and 1/f Fluctuations (ICNF2001). World Scientific (2002) 408-411

13. Wang, R.: On the nonlinear stochastic evolution model possessing the populations of neurons of different phase. Proceedings of International conference on neural networks and signal processing (IEEE 2003), Vol. 1. (2003) 139-143
14. Wang, R., et al: Some advance in nonlinear stochastic evolution models for phase resetting dynamics on populations of neuronal oscillators. International Journal of nonlinear sciences and numerical stimulation. 4 (2003) 435-446
15. Sun, J. R. (ed): Introduction of Brain Science. Beijing University Publisher, Beijing (2001) 48-52
16. Leventhal, A. G., Wang, Y. C., Pu, M. L., et al: Ma GABA and Its Agonists Improved Visual Cortical Function in Senescent Monkeys. Science. 300 (2003) 812-815

Study on Circle Maps Mechanism of Neural Spikes Sequence

Zhang Hong[1], Fang Lu-ping[2], and Tong Qin-ye[1]

[1] Institute of Biomedical Engineering, Zhejiang University, Hangzhou,
310027, P.R.China
zhangh@mail.bme.zju.edu.cn
bitong@public.zju.edu.cn
[2] Software College, Zhejiang University of Technology
310032, P.R.China
flp@zjut.edu.cn

Abstract. Till now, the problem of neural coding remains a puzzle. The intrinsic information carried in irregular neural spikes sequence is not known yet. But solution of the problem will have direct influence on the study of neural information mechanism. In this paper, coding mechanism of the neural spike sequence, which is caused by input stimuli of various frequencies, is investigated based on analysis of H-H equation with the method of nonlinear dynamics. The signals of external stimuli -- those continuously varying physical or chemical signals -- are transformed into frequency signals of potential in many sense organs of biological system, and then the frequency signals are transformed into irregular neural coding. This paper analyzes in detail the neuron response of stimuli with various periods and finds the possible rule of coding.

1 Introduction

In a paper published in Science[1], Gilles Laurent remarked: "Studying a neural coding requires asking specific questions, such as the following: What information do the signals carry? What formats are used? Why are such formats used? Although superficially unambiguous, such questions are charged with hidden difficulties and biases."

Till now, it is still not clear what meaning these irregular neural pulse sequences is.

The coding mechanism of the neural discharge spikes sequence caused by the input signals (pulse) of various frequencies is analyzed in this paper using the method of nonlinear dynamics. For a neural system, when sensory organs are sensing the external continuous signals of physical or chemical, the first step taken is to transform these analog signals into the frequency signals of neural response[2], and then the frequency signals are transformed into irregular neural pulse sequence. What is the rule of transmission like? Or how does neuron encode the frequency information? This is the key issue to be discussed in this paper.

2 Circle Maps of Neurons

The main task of neural coding study is to explore how irregular neural pulse sequence varies with the input signal. Once the variation rule is discovered, the rule of signal variation in the neural system can be made clear, and finally the mechanism of neural coding can be found. For this, circle maps can be used as a helpful tool [4].

First, let's consider the response signal of a neuron stimulated by a pulse with constant frequency. In the paper, the classical H-H equation, shown in equation (1), is adopted to describe the potential variation of a neuron. Several decades passed, basic structure of the equation remains unchanged, though many persons contribute various modifications to it [3]. For generality, we take this function as the research target.

$$C\frac{dV}{dt} = -\bar{g}_K n^4(V-E_K) + \bar{g}_{Na} m^3 h(V-E_{Na}) + \bar{g}_l(V-E_l) + I_{ext}$$
$$\frac{dn}{dt} = K_t(A_n(1-n) - B_n n)$$
$$\frac{dm}{dt} = K_t(A_m(1-m) - B_m m)$$
$$\frac{dh}{dt} = K_t(A_h(1-h) - B_h h)$$
(1)

Here, I_{ext} functions as the external input signal.

$$K_t = 3^{(T-6.3)/10}$$
$$V' = V - V_{rest}$$
$$A_n = \frac{0.01 * (10 - V')}{e^{(10-V')/10} - 1}$$
$$B_n = 0.125 * e^{(-V'/80)}$$
$$A_m = \frac{0.1 * (25 - V')}{e^{(25-V')/10} - 1}$$
$$B_m = 4 * e^{(-V'/18)}$$
$$A_h = 0.07 * e^{-V'/20}$$
$$B_h = \frac{1}{e^{(30-V')/10} + 1}$$
(2)

The parameter list used in the emulation is shown in Table 1.

From the emulation, it is found that an irregular pulse sequence (response signal) will come out if a sequence of neural pulses with equal intervals (stimulus signal) is imposed on a neuron:

Input signal takes the form of periodical square wave with amplitude of 20mA/cm² and width of 1ms. The width selection of input pulse signal depends on the general width of potential of neural system. Period of sequence of stimulus input is 5ms.

Runge-Kutta method with variable step size is applied to the numerical solution of H-H equation. In detail, routine ode45 of Matlab is chosen, while the relative error range sets to 1e⁻⁶, and the absolute error range sets to 1e⁻⁹. Meanwhile, the minimal time step period falls into the range from 1e⁻⁸ to 1e⁻⁷, and the maximal time step period falls into the range from 0.01 to 0.1.

Table 1. Parameters list in formula (1)

Parameter	Value	Unit
C	1	uF cm^{-2}
$\overline{g}_K$	36	mS cm^{-2}
$\overline{g}_{Na}$	120	mS cm^{-2}
$\overline{g}_l$	0.3	mS cm^{-2}
E_K	-71.967	mv
E_{Na}	54.98	mv
E_l	-49	mv
T	6.3	℃
V_{rest}	-59.805	mv

Take the time when one action potential reaches its peak as the time point of the potential, the interval between current time point and the nearest preceding one is referred to as period of response pulse. Then the period of response pulse is quite irregular. We extract the circle maps of the signal according to the period of input signal [4].

For a given initial condition, we get a phase sequence:

$$\{\theta_i\} = \theta_1 \theta_2 \theta_3 \ldots \ldots \theta_n . \tag{3}$$

where θ_i takes relative value.

From sequence (3), we obtain the following equation (4)

$$\theta_{n+1} = \Phi(\theta_n) . \tag{4}$$

Since there is no explicit function in it, we are unable to obtain the concrete form of function Φ, however, its curve can be drawn based on equation (3), as shown in Fig.1.

It can be seen from Fig. 1 that the relation between θ_n and θ_{n+1} forms a regular function.

From the point of circle maps theory, this is a monotonic increasing function. If the initial phase θ_1 is given, θ_2 can be determined according to Fig.1. Through similar deduction, a phase sequence is obtained consequently, the phase rule of which coincides with that of actual neural impulses. This rule forms the solid base for us to understand the mechanism of neural coding.

Next, we will analyze the pulse sequence using circle maps method.

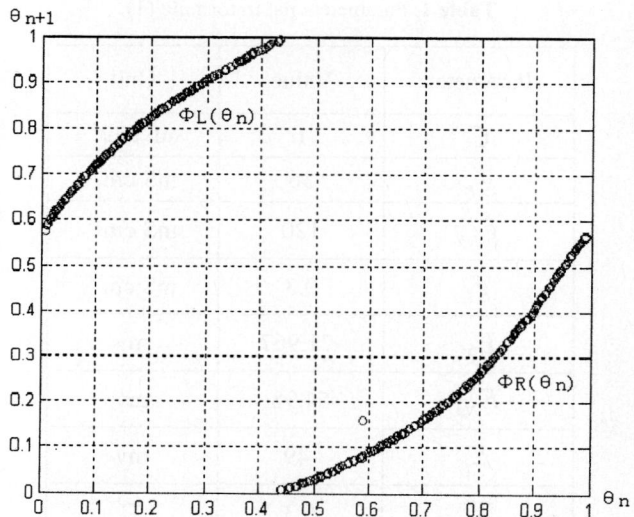

Fig 1. The circle maps of responses of the neuron, when the period of input stimulus is 2.6ms. Relative coordinates $Õ = \theta_i/\tau$ are used, where τ is the period of the input stimulus.

3 Symbolic Dynamics of Circle Map

Symbolic dynamics analysis can be performed with the help of circle map. For it is a monotonic increasing maps function, without descending part, what we should do is to extract the symbol periodically[4]. According to the calculated pulse sequence above, the distance between the ith pulse and the $i+1$th pulse is almost as lengthy as 5 to 6 periods of input pulse. To simplify, pulse within 5 periods is defined as L, and pulse within 6 periods is defined as R. For generality, when input frequency increases, pulse within 5-6 periods may expand to range of 6-7 or 7-8 periods. The approach of extracting symbol remains the same.

$$\{S_i\} = S_1 S_2 S_3 \ldots \ldots S_n \ldots \ldots, \text{ where } S_i = R \text{ or } L. \quad (5)$$

Then, one specific initial phase value θ_1 can determine one pulse sequence and then one phase sequence $\{\theta_i\}$. Similarly, one symbol sequence $\{S_i\}$ can be determined, which can be orderly arranged based on initial phase value θ_1. And its rule is shown as following:

Sequencing method for a symbol sequence is achieved with comparison one by one. Assume there are two sequences:

$$A = a_1 a_2 a_3 \ldots .. a_n$$
$$B = b_1 b_2 b_3 \ldots . b_n . \quad (6)$$

Firstly, the 1st symbols (a_1 & b_1) is compared as follows:
If a_1=R, and b_1=L
   R>L

This indicates that the phase of the 6th period($\theta_1=a_1$) is larger than that of the 5th period ($\theta_1=b_1$).
In the case of more than two symbols, we have
$$(a_1a_2a_3\ldots\ldots\ldots a_ia_{i+1}\ \&\ b_1b_2b_3\text{----------}b_ib_{i+1})$$
$$\Sigma L < \Sigma R . \tag{7}$$

where $\Sigma=a_1a_2a_3\ldots\ldots a_i = b_1b_2b_3\text{-------}b_i$. (8)

It is unnecessary to know the value of every initial phase θ_1. Once symbol sequence corresponding to each of the initial phases is given, the relative value of each initial phase is determined. This is so-called sequencing rule of symbolic dynamics.

4 Principle of Frequency Coding in Neuron

Next, we will study the variation of output phase sequence under the condition of sequence of impulse with various frequencies as input.

Actually the sequencing rule demonstrated in equation (7) reflects the ordering by initial phases. Now the question is: when the frequency of input pulse changes, does the output symbolic sequence order in terms of frequency? From the point of symbolic dynamics, it is an ordering problem by parameters. Therefore, the impact on the curve of function $\Phi(\theta_n)$, shown in Fig.1, which is posed by the variation of input frequency, can be assessed according to equation (1).

The stimulus signals, depicted in the group figures of Fig.2, arranging from top to down and from left to right, have periods ranging from 1.1ms to 2.5ms and pulse width of 0.1ms. From Fig.2, it can be seen that the function curve of circle maps shifts right as the frequency of stimulus descends. In the figure, the function combines two part, $\Phi_L(\theta_n)$ in the left and $\Phi_R(\theta_n)$ in the right. When $\Phi_R(\theta_n)$ moves to the end, $\Phi_L(\theta_n)$ will replace it and at the same time a new $\Phi_L(\theta_n)$ is formed. Refer to the changes between the 1st small figure and the 2nd small figure from the left in the second row, it can be seen that the changes behaves regularly. It seems that the 2nd small figure from the right in the bottom row of Fig.2, which contains only several points, does not match with its surrounding figures. Actually this is due to drawing techniques. The small figure serials of Fig.2 are not depicted point by point according to phase angle calculated one by one originating from the initial phase angle. On the contrary, each point is determined by Poincaré maps after a stimulus sequence with a certain frequency is given. For its property of ergode, we can obtain almost all the points of the curve.

Some figures contain only several points, due to the occurrence of periodical solution under the specific frequency. Thus, Poincaré maps produces limited number of points, while the ordering rule remains unchanged.

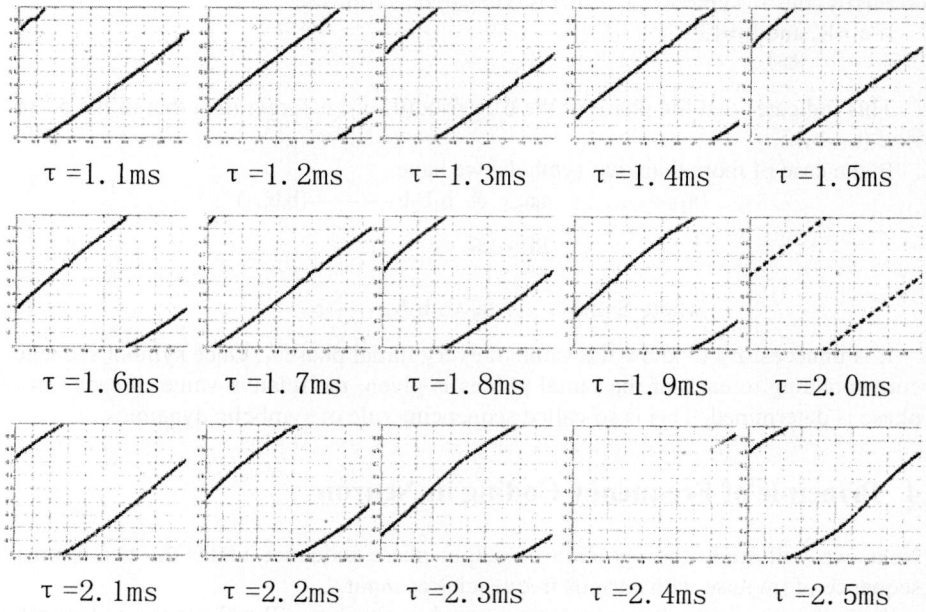

Fig. 2. Variant τ and corresponding frequence map

Because they are all monotonic increasing functions, and the movement and frequency of $\Phi(\theta_n)$ are changed monotonously, it is discernable that under certain initial condition, the yielded phase sequence changes monotonously too.

After pulse sequences with various frequencies are assigned to equation (1), symbols are extracted with the same approach (see Section 3) in the subsequent computer experiment.

Table 2 is obtained using the above method. Within the permission range of stimulus frequencies, the period of system response pulse varies from 5τ to 6τ. Therefore the symbol sequence of the system response can be described simply with two symbols. Symbol 0 (or L) represents that the pulse period is 5τ, whereas symbol 1 (or R) represents that the pulse period is 6τ.

Table 2 exhibits that symbol sequence becomes smaller when the period increases (frequency decreases).

In column 5 and column 6 of table 2, period changes from 2.64 to 2.65. Though it is obvious that the frequency descends, the corresponding 10 symbols are identical, therefore there is no way to distinguish which symbol sequence is bigger. However, the change can be observed in the case of 20 symbol sequence. It reveals that the more lengthy one symbol sequence is, the higher resolution is.

In Fig.2, we notice that system has an approximate periodical solution when the stimulus period is as long as 2.4ms. Meanwhile, the right part of the circle maps function is to be replaced with the left part in the movement process of the circle maps function caused by the variation of stimulus period, i.e. the system evolves to the critical condition, in which the new left branch emerges. As circle maps function changes around the critical zone, the scope of distance between two stimulus periods, where two consecutive system responses occur, changes accordingly, requiring more symbols to implement the symbolization of system response. Fig.3 illustrates the symbol sequence of system response at the time when stimulus period ranges from 2.3ms to 2.5ms; symbol 0 indicates that the distance between two stimulus periods where two consecutive system responses occur is equal to 5; symbol 1 indicates that the distance between two stimulus periods where two consecutive system responses occur is equal to 6; symbol 2 indicates that the distance between two stimulus periods where two consecutive system responses occur is equal to 7. The ordering rule coincides with that of two-symbol sequence. In Table 3, the ordering relation of the symbol sequence is demonstrated, i.e. symbol sequence decreases also once the stimulus frequency decreases.

If symbolization of system response is implemented in a much wider scope of stimulus period variation, more symbols are required. Generally, we can exploit any element in the symbol set to express the symbol sequence of system response.

$$\Delta=\{\delta_1\delta_2\delta_3......\delta_n\}. \tag{9}$$

For example, table 2 is the case of n=2, in other words, the symbol set contains only

Table 2. The symbol sequence of the system response orbit as the period of the stimulus ranges from 2.6ms to 2.7ms, with step 0.1ms. Here, 0 and 1 correspond to L and R in the symbol sequence.

Period of Stimulus (ms)	Corresponding Symbol Sequence (10 symbols)	Corresponding Symbol Sequence (20 symbols)
2.6	1010101101	10101011010101010110
2.61	1010101010	10101010101011010101
2.62	1010101010	10101010101010101010
2.63	1010101010	10101010100101010101
2.64	1010010101	10100101010101010010
2.65	1001010101	10010101010010101010
2.66	1001010100	10010101001010010101
2.67	1001010010	10010010101001010010
2.68	1001001010	10010010100100101001
2.69	1001001001	10010010010010010100
2.7	0100100100	01001001001001001001

two symbols, representing distances between impulses equal to 5 or 6 periods respectively, with $\delta_i= 0$ or $\delta_i= 1$ (L or R) in equation (9).

In Table 3, n equals to 3, indicating the symbol set comprises of 3 symbols, which represents 5,6 or 7 periods of pulse interval respectively, and δ_i is equal to 0,1, or2.

If the scope of stimulus frequency expands further, δ_i varies in the wider range of period, having more change of symbols (see equation (9)).

Table 3. The symbol sequence of the system response orbit as the period of the stimulus ranges from 2.3ms to 2.5ms, with step 0.2ms. Here the frequency scope of the stimulus extends beyond the range, which requires 3 symbols for representation, rather than 2 symbols, with δ_i =0,1 or 2.

Period of Stimulus (ms)	Corresponding Symbol Sequence (10 symbols)	Corresponding Symbol Sequence (20 symbols)
2.30	2111121111	21111211112111211112
2.32	1211111211	12111112111112111112
2.34	1211111111	12111111112111111112
2.36	1121111111	11211111111111111121
2.38	1111211111	11112111111111111111
2.40	1111111111	11111111111111111111
2.42	1111111111	11111111111111110111
2.44	1111111011	11111110111111111011
2.46	1111011111	11110111111011111101
2.48	1110111101	11101111011111011110
2.50	1101111011	11011110111011101111

5 Discussion

According to the above analysis, we have the following points.

1) Though the output response of neuron, stimulated by various frequencies, is quite irregular, it becomes regular under the analysis using method of circle maps and symbolic dynamics. Measures of frequency signal can be determined from symbol sequences. It is also the process of frequency detection for a neural system, various sensory organs change analog signal to frequency signal, and then to chaotic orbits (a sequence of irregular pulses). So, what is discussed in the paper is the general procedure of neural information process.

2) This method can be used to distinguish the seemingly messy neural pulses and then order them. If these orbits which are able to be ordered are assembled together, an orbit space can be constructed. In addition, this orbit space is an orderly one. We think information processing involving neural system can be developed in orderly space. In order space, some operations can be performed, which provide the foundation for further study on neural information processing.

3) If H-H function can reflect the real-world neuron's electric activity in qualitative fashion [3], the above analysis is suitable for the real-world neuron's activity. Even if there are some discrimination between H-H function and the real situation, if only the curve shapes of function $\Phi_L(\theta_n)$ and $\Phi_R(\theta_n)$ in Fig.1 are changed, and if function monotonicity is not changed, then no influence will be imposed on the extraction of symbol. Therefore the above outcomes still take effect. If any change happens on monotonicity, some modification is required for the above analysis

Because of page limitation, we will end our discussion here. Actually this analysis is capable to disclose the information coding mechanism of various sensory organs. In another paper, the information process mechanism of olfactory neural system will be introduced.

Acknowledgement

This program is supported by the program of "the application of chaotic array in the electric nose", which is funded by the dedicated pre-studies of the national basic research program (973 program) (under grant no 2002CCA01800) and national natural science foundation program (under grant no 30170267).

References

1. Gilles Laurent A Systems Perspective on Early Olfactory Coding, SCIENCE VOL 286 22 OCTOBER 1999 725
2. Jean-Pierre R, Petr L, Patricia D V, Andre D. Spiking frequency versus odorant concentration in olfactory receptor neurons. BioSystems, 2000, 58:133-141.
3. Claude Meunier and Idan Sogev , Playing the Dell's advocate: is the Hodgkin-Huxley model useful? Trends in Neuroscience Vol.25 No.11 2002 558-563
4. Zhang Zhong-jin ang Cheng Shi-gang, Symbolic dynamics of the circle map ACTA Physica Sinica Vol.38 No.1 1989 1-8
5. Lestienne R. Spike timing. synchronization and information processing on the sensory side of the central nervous system. Prog. Neurobiol, 2001, 65(6): 545-691
6. Eric R. Kandel and Larry R. Squire. Neuroscience: Breaking Down Scientific Barriers to the Study of Brain and Mind. Science, 2000, Vol 290, Issue 5494, 1113-1120
7. Arun V H. Neural coding by correlation?. Nature, 2004, 428:382-382.
8. Eugene M Izhikevich. Neural Excitability, Spiking and Bursting. Intl J Bif and Chaos, 2000, 10(6), 1171-1266.
9. S. Kim and S.Ostlund. Universal scaling in circle maps, Physica D, 1989, 39, 365-392
10. Averbeck B B, Lee D. Coding and transmission of information by neural ensembles. Trends Neurosci, 2004, 27(4): 225-230.

Synchronous Behaviors of Hindmarsh-Rose Neurons with Chemical Coupling

Ying Wu, Jianxue Xu, and Mi He

School of Architectural Engineering & Mechanics,
Xi'an Jiaotong University, Xi'an 710049, China
Wying36@163.com, jxxu@mail.xjtu.edu.cn

Abstract. We study the synchronization phenomena in a pair of Hindmarsh-Rose(HR) neurons with chemical coupling. We find that excitatory synaptic coupling pushes two neurons towards antisynchrony, and weak or moderate inhibitory synaptic coupling pushes two neurons towards antisynchrony too, but sufficiently strong inhibitory synaptic coupling pushes two neurons towards synchronized periodic oscillations without spikes. And synchronization patterns can't be changed even if the intrinsic frequency of individual cell is changed by modulating external input current. Investigating the effect of synapse on ISIs bifurcation structures shows that whether excitatory synapse or inhibitory synapse, both remarkably influence ISIs structures. That is, the chemical coupling between neurons wholly distorts the neuronal information.

1 Introduction

Synchronization of nonlinear oscillators has been widely study recently [1-5]. Especially, the affection of electrical and chemical coupling on synchrony of coupling neurons has attracted lots of attention.

In Ref. [3], the experimental studies of synchronization phenomena in a pair of biological neurons interacted through electrical coupling were reported. In Ref. [4], the synchronization phenomena in a pair of analog electronic neurons with both direct electrical connections and excitatory and inhibitory chemical connections was studied. Traditionally, it has been assumed that inhibitory synaptic coupling pushes neurons towards antisynchrony. In fact, If the time scale of the synapses is sufficiently slow compared with the intrinsic oscillation period of the individual cells, inhibition can act to synchronize oscillatory activity [5].

In this paper, we investigate dynamics of network of two HR neurons with chemical synapses, the models used were given in Ref. [6]. The results show that excitatory synapses can antisynchronize two neurons and enough inhibition can foster phase synchronization. And the synchronization patterns of two coupled neurons can't be changed with intrinsic frequency of individual cell being changed by modulating external input current,

Investigating the effect of chemical synapse on ISIs bifurcation structure[7] of chemical coupling HR neurons shows that the ISIs bifurcation structures are wholly changed by chemical synapse.

2 Hindmarsh-Rose Models with Electrical and Chemical Synaptic Connections

Consider two identical HR models with reciprocal synaptic connections. The differential equations of the coupled systems are given as[6]

$$\dot{x}_i = y_i + bx_i^2 - ax_i^3 - z_i + I_{dc} + e_s \left(\frac{x_i + V_c}{1 + \exp\frac{x_j - X_0}{Y_0}} \right)$$

$$\dot{y}_i = c - dx_i^2 - y_i \quad (1)$$

$$\dot{z}_i = r[S(x_i - \chi) - z_i]$$

Where $i = 1,2$, $j = 1,2$, $i \neq j$

In the simulation, let $a = 1.0, b = 3.0, c = 1.0, d = 5.0, s = 4.0, r = 0.006, \chi = -1.56$, $I_{dc} = 3.0$, I_{dc} denotes the input constant current. The last term of the first formulation is synaptic current of coupling system, e_s is the strength of the synaptic coupling, and $V_c = 1.4$ is synaptic reverse potential which is selected so that the currents injected into the postsynaptic neuron are always negative for inhibitory synapses and positive for excitatory synapses. Since each neuron must receive an input every time the other neuron produces a spike, we set $Y_0 = 0.01$ and $X_0 = 0.85$ [6]. In numerical simulation, the double precision fourth-order Runge-kutta method with integration time step 0.01 was used, the initial condition is (0.1,1.0,0.2,0.1,0.2,0.3). In each realization, the data for $n < 10^4$ are ignored to avoid transients.

3 Synchronizing Two HR Neurons with Synaptic Connection

The chemical synapse is excitatory for $e_s > 0$ and is inhibitory for $e_s < 0$. The results show that two neurons will be irregular oscillation with small excitatory coupling strength, and will be in full antisynchrony for enough excitatory coupling strength, such as Figs.1(a,b). It is interesting that these results do not agree with those of Ref. [4], and are contrary to traditional view.

Investigating the synchrony course of two neurons with inhibitory synapses shows that two neurons oscillation are irregular for small coupling intensity, and the phase difference between two neurons increase gradually with coupling strength increasing, till $e_s = -0.45$ at which the phase difference between two neurons are biggest, see Fig.1c, which means that two neurons are full antisynchrony. Continuing to increase intensity of inhibitory coupling, the phase difference between two neurons will decrease, till $e_s = -0.9$, when two neurons are full synchronous periodic oscillation without spikes, such as Fig.1d. In Ref. [8], the intrinsic oscillation frequency of the individual cell was increased by increasing external stimulating current, and the systems with inhibitory coupling can evolve to synchronous state. In our paper, numerical results show that the synchronization patterns of membrane potential and synaptic current of two coupling neurons haven't been changed even if the intrinsic oscillation frequency of individual cell has been changed with changing external input current.

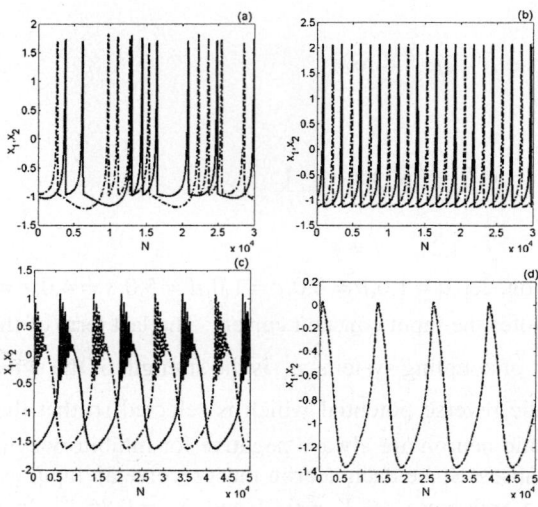

Fig. 1. (a, b) Time courses of membrane potential of two neurons for excitatory synapse, two neurons are irregular activity for $e_s = 0.03$, period 1 antisynchrony for $e_s = 0.3$, respectively; (c,d) Time courses of membrane potential of two neurons with inhibitory synapse., two neurons are full antisynchrony for e_s=-0.45, two neurons are full synchrony periodic oscillation for e_s=-0.9

4 The Effect of Chemical Coupling on ISIs Bifurcation Structure

The neuronal information proceeding and coding are mainly based on ISIs. Figs.2 (b,c) show ISIs bifurcation diagrams of coupled HR neurons with es=0.3 and es=-0.5, respectively. Compared with fig.2a, it is obvious that ISIs structures are remarkably different from those of individual HR neuron without coupling. The difference of ISIs structures between individual HR neuron and coupled HR neuron means that the coupled neurons undergo entirely different firing patterns from those of individual neuron without coupling under the same parameter values; that is, the neuronal information is wholly distorted by the chemical coupling between neurons.

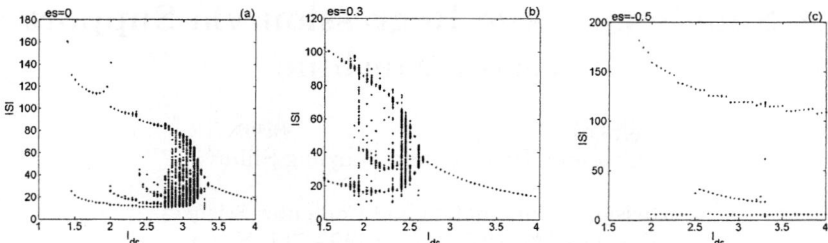

Fig. 2. (a) Bifurcation diagram of ISIs vs the external current I_{dc} in single HR neuron without coupling; (b,c) ISIs bifurcation diagram vs external input current I_{dc} for $e_s=0.3$ and $e_s=-0.5$, respectively

5 Conclusion

In this paper, we investigate synchronization patterns of two HR models with chemical coupling. The results show that excitatory synapses can antisynchronize two neurons, and weak or moderate inhibitory synaptic coupling can antisynchronize two neurons too, but strong inhibitory synapse can foster phase synchrony of two neurons. And the synchronization patterns of membrane potential and synaptic current of two coupling neurons haven't been changed even if the intrinsic oscillation frequency of individual cell has been changed with changing external input current. The ISIs bifurcation structures are wholly changed by chemical synapse. That is, the chemical coupling between two neurons wholly distorts the neuronal information.

References

1. Wu, Y., Xu, J.X., He, D.H., et al: Synchronization in Two Uncoupled Chaotic Neurons. LNCS 3173 (2004) 183-189
2. Wu, Y., Xu, J.X., He, D.H., et al: Generalized synchronization induced by noise and parameter mismatching in Hindmarsh-Rose neurons. Chaos Solitons & Fractals 23 (2005) 1605-1611
3. Elson, R.C., Selverston, A.I., Romon, H., et al: Synchronous Behavior of Two Coupled Biological Neurons. Phys. Rev. lett. 81(1998) 5692-5695
4. Pinto, R.D., Varona, P., Volkovskii, A.R., et al: Synchronous Behavior of Two Coupled electronic neurons. Phys. Rev. E 62(2000) 2644-2656
5. Timothy, J.L., John, R.: Dynamics of Spiking Neurons Connected by Both Inhibitory and Electrical Coupling. Journal of computation neuroscience 14(2003) 283-309
6. Romon, H., Mikhail, I.R.: Spike-train bifurcation in two coupled chaotic neurons. Phys. Rev. E 55(1997) R2108
7. Wu, Y., Xu, J.X., He, D.H., et al: Study on Nonlinear Characteristics of Two Synchronizing Uncoupled Hindmarsh-Rose Neurons. Acta Physica Sinica 54 (2005) accepted

A Simple Quantile Regression via Support Vector Machine

Changha Hwang[1] and Jooyong Shim[2]

[1] Division of Information and Computer Sciences,
Dankook University, Seoul 140 - 714, Korea
chwang@dankook.ac.kr

[2] Corresponding Author, Department of Statistical Information,
Catholic University of Daegu, Kyungbuk 712 - 702, Korea
ds1631@hanmail.net

Abstract. This paper deals with the estimation of the linear and the nonlinear quantile regressions using the idea of support vector machine. Accordingly, the optimization problem is transformed into the Lagrangian dual problem, which is easier to solve. In particular, for the nonlinear quantile regression the idea of kernel function is introduced, which allows us to perform operations in the input space rather than the high dimensional feature space. Experimental results are then presented which illustrate the performance of the proposed method.

1 Introduction

Quantile regression introduced by Koenker and Bassett[6] is gradually envolving into an ensemble of practical statistical methods for estimating and conducting inference about models for conditional quantile functions. Quantile regression is an increasingly popular method for estimating the quantiles of a distribution conditional on the values of covariates. Regression quantiles are robust against the influence of outliers and, taken several at a time, they give a more complete picture of the conditional distribution than a single estimate of the center. Just as classical linear regression methods based on minimizing sum of squared residuals enable one to estimate a wide variety of models for conditional mean functions, quantile regression methods offer a mechanism for estimating models for the conditional median function, and the full range of other conditional quantile functions. By supplementing the estimation of conditional mean functions with techniques for estimating an entire family of conditional quantile functions, quantile regression is capable of providing a more complete statistical analysis of the stochastic relationships among random variables. The introductions and current research areas of the quantile regression can be found in Koenker and Hallock[9], Yu et al.[15].

In this paper we present the estimation methods of linear and nonlinear quantile regression by utilizing support vector machine(SVM). The SVM, firstly developed by Vapnik and his group at AT&T Bell Laboratories, is being used as a new technique for regression and classification problems. SVM is gaining popularity due to many attractive features, and promising empirical performance. SVM was initially developed to solve classification problems but recently it has been extended to the domain of regression problems. SVM is based on the structural risk minimization(SRM) principle, which minimizes an upper bound on the expected risk unlike ERM minimizing the error on the training data. By minimizing this bound, high generalization performance can be achieved. In particular, for the SVM regression case SRM results in the regularized ERM with the ϵ-insensitive loss function. The introductions and overviews of recent developments of SVM regression can be found in Vapnik[12][13], Gunn[4], Smola and Schölkopf[11], Cristianini and Shawe-Taylor[2], Kecman[5], and Wang[14].

The minimization problem associated with linear quantile regression is in essence the linear programming(LP) optimization problem, which is based on simplex algorithm or interior point algorithm. The current state of algorithms for nonlinear quantile regression is far less satisfactory. The widely used algorithm is interior point algorithm. Nonlinear quantile regression poses new algorithmic challenge. Refer to Koenker and Park[8] and Koenker and Hallock[9] for the algorithms. Training an SVM requires the solution of a quadratic programming(QP) optimization problem. Thefore, both the linear and the nonlinear quantile regressions by SVM require solving QP problem to get estimates.

The purpose of this paper is to present the estimation methods of the linear and the nonlinear quantile regressions using SVM. The rest of this paper is organized as follows. In Section 2 we present the estimation methods of quantile regression using SVM. In Section 3 we perform the simulation studies through examples. In Section 4 we give the conclusions.

2 Quantile Regression via SVM

Conditional quantile estimation has long been studied in the literature. Most commonly used approach is quantile regression introduced by Koenker and Basset[6]. In this section we derive the linear and the nonlinear quantile regression methods by implementing the idea of SVM. Consider a random sample $\{\mathbf{x}_i, y_i\}_{i=1}^n$ with input vector $\mathbf{x}_i \in R^d$ and output variable $y_i \in R$. Here the output variable y_i is related to the vector $\mathbf{x}_i$ of covariates, possibly including a constant 1.

2.1 Linear Quantile Regression

In the linear quantile regressin model introduced by Koenker and Bassett[6] the quantile function of the response y_i for a given $\mathbf{x}_i$ is assumed to be linearly related to the input vector $\mathbf{x}_i$ as follows

$$Q(\theta|\mathbf{x}_i) = \boldsymbol{\beta}(\theta)^t \mathbf{x}_i \quad \text{for} \quad \theta \in (0,1), \tag{1}$$

where $\boldsymbol{\beta}(\theta)$ is the θ-th regression quantile and its estimator is defined as any solution to the optimization problem,

$$\min_{\boldsymbol{\beta}} \sum_{i=1}^{n} \rho_\theta(y_i - \boldsymbol{\beta}(\theta)^t \mathbf{x}_i) \quad \text{for } \theta \in (0,1), \tag{2}$$

where $\rho_\theta(\cdot)$ is the check function defined as

$$\rho_\theta(r) = \theta r I(r \geq 0) + (\theta - 1) r I(r < 0).$$

We now describe how to implement the idea of SVM for the linear quantile regression. Since quantile regression is in principle based on absolute deviation loss, to derive quantile regression using the idea of SVM, we should adopt the procedures of the case $\epsilon = 0$ in a standard SVM. In order to follow the basic idea of quantile regressions, we express $\mathbf{x}_i$ as $\mathbf{x}_i = (1, \mathbf{x}_i^t)^t$. We use the same notation for the resulting new vectors to avoid the abuse of notation. Then, we can express the linear quantile regression problem by the formulation for SVM.

$$\text{minimize} \quad \frac{1}{2}\|\mathbf{w}\|^2 + C\sum_{i=1}^{n}(\theta\xi_i + (1-\theta)\xi_i^*) \quad \text{for } \theta \in (0,1), \tag{3}$$

$$\text{subject to} \quad \begin{cases} y_i - \mathbf{w}^t\mathbf{x}_i \leq \xi_i \\ \mathbf{w}^t\mathbf{x}_i - y_i \leq \xi_i^* \\ \xi_i, \xi_i^* \geq 0 \end{cases}.$$

where the θ-th regression quantile $\boldsymbol{\beta}(\theta)$ is expressed in terms of $\mathbf{w}$. The constant $C > 0$ determines the trade off between the flatness of f and the amount up to which deviations larger than 0 are tolerated. We construct a Lagrange function as follows:

$$L = \frac{1}{2}\|\mathbf{w}\|^2 + C\sum_{i=1}^{n}(\theta\xi_i + (1-\theta)\xi_i^*) - \sum_{i=1}^{n}\alpha_i(\xi_i - y_i + \mathbf{w}^t\mathbf{x}_i)$$
$$- \sum_{i=1}^{n}\alpha_i^*(\xi_i^* + y_i - \mathbf{w}^t\mathbf{x}_i) - \sum_{i=1}^{n}(\eta_i\xi_i + \eta_i^*\xi_i^*). \tag{4}$$

We notice that the positivity constraints $\alpha_i, \alpha_i^*, \eta_i, \eta_i^* \geq 0$ should be satisfied. After taking partial derivatives of equation (4) with regard to the primal variables $(\mathbf{w}, \xi_i, \xi_i^*)$ and plugging them into equation (4), we have the optimization problem below.

$$\max_{\boldsymbol{\alpha}, \boldsymbol{\alpha}^*} -\frac{1}{2}\sum_{i,j=1}^{n}(\alpha_i - \alpha_i^*)(\alpha_j - \alpha_j^*)\mathbf{x}_i^t\mathbf{x}_j + \sum_{i=1}^{n}(\alpha_i - \alpha_i^*)y_i \tag{5}$$

with constraints $\alpha_i \in [0, \theta C]$ and $\alpha_i^* \in [0, (1-\theta)C]$.

Solving the above optimization problem with the constraints determines the optimal Lagrange multipliers, $\hat{\alpha}_i$, $\hat{\alpha}_i^*$, the θ-th regression quantile estimators and the θ-th quantile function predictors of the input vector $\mathbf{x}$ are obtained, respectively as follows:

$$\hat{\mathbf{w}} = \sum_{i=1}^{n}(\hat{\alpha}_i - \hat{\alpha}_i^*)\mathbf{x}_i \quad \text{and} \quad \hat{Q}(\theta|\mathbf{x}) = \sum_{i=1}^{n}(\hat{\alpha}_i - \hat{\alpha}_i^*)\mathbf{x}_i^t\mathbf{x}. \tag{6}$$

Here, $\hat{\mathbf{w}}$ and $\hat{Q}(\theta|\mathbf{x})$ depend implicitly on θ through $\hat{\alpha}_i$ and $\hat{\alpha}_i^*$ depending on θ.

2.2 Nonlinear Quantile Regression

In the nonlinear quantile regression model the quantile function of the response y_i for a given $\mathbf{x}_i$ is assumed to be nonlinearly related to the input vector $\mathbf{x}_i \in R^d$. To allow for the nonlinear quantile regression, the input vectors $\mathbf{x}_i$ are nonlinearly transformed into a potentially higher dimensional feature space $\mathcal{F}$ by a nonlinear mapping function $\phi(\cdot)$. Here, similar to SVM for nonlinear regression, the nonlinear regression quantile estimator cannot be given in an explicit form since we use the kernel function of input vectors instead of the dot product of their feature mapping functions. The quantile function of the response y_i for a given $\mathbf{x}_i$ can be given as

$$Q(\theta|\mathbf{x}_i) = \beta(\theta)^t \phi(\mathbf{x}_i) \quad \text{for} \quad \theta \in (0,1), \tag{7}$$

where $\beta(\theta)$ is the θ-th regression quantile. Then, by constructing the Lagrangian function with kernel function $K(\cdot,\cdot)$, we obtain the optimal problem similar to the linear quantile regression case as follows

$$\max_{\boldsymbol{\alpha},\boldsymbol{\alpha}^*} -\frac{1}{2} \sum_{i,j=1}^{n} (\alpha_i - \alpha_i^*)(\alpha_j - \alpha_j^*) K(\mathbf{x}_i, \mathbf{x}_j) + \sum_{i=1}^{n} (\alpha_i - \alpha_i^*) y_i \tag{8}$$

with constraints $\alpha_i \in [0, \theta C]$ and $\alpha_i^* \in [0, (1-\theta)C]$. Solving the above optimization problem with the constraints determines the optimal Lagrange multipliers, $\hat{\alpha}_i$, $\hat{\alpha}_i^*$, then the θ-th quantile function predictor given the input vector $\mathbf{x}$ can be obtained as

$$\hat{Q}(\theta|\mathbf{x}) = \sum_{i=1}^{n} (\hat{\alpha}_i - \hat{\alpha}_i^*) K(\mathbf{x}_i, \mathbf{x}). \tag{9}$$

Likewise the linear case, $\hat{Q}(\theta|\mathbf{x})$ depend implicitly on θ through $\hat{\alpha}_i$ and $\hat{\alpha}_i^*$ depending on θ.

3 Illustrative Examples

We illustrate the performance of the proposed quantile regression methods based on support vector machine(QRSVM) through the numerical studies for $\theta = 0.1, 0.5$, and 0.9. In Example 1, we illustrate the estimation and the prediction performance of regression quantiles and quantile functions for the linear case. In Example 2, we illustrate the prediction performance of quantile functions for the nonlinear case.

Example 1. In this example we illustrate how well QRSVM performs for the linear quantile regression case. We generate 100 training data sets to present the estimation performance of regression quantiles obtained by QRSVM. In addition, we generate a training data set and 100 test data sets to present the prediction performance of these quantile functions. Each data set consists of 200 x's and

200 y's. Here x's are generated from a uniform distribution $U(0,1)$ and y's are generated from a normal distribution $N(1+2x, x^2)$, that is, y's are generated from the heteroscedastic error model.

The θ-th quantile function of y for a given **x** can be modelled as

$$Q(\theta|\mathbf{x}) = \boldsymbol{\beta}(\theta)^t \mathbf{x} \quad \text{for } \theta \in (0,1),$$

where $\mathbf{x} = (1,x)^t$ and $\boldsymbol{\beta}(\theta) = (1, 2+\Phi^{-1}(\theta))^t$ with the θ-th quantile of a standard normal distribution, $\Phi^{-1}(\theta)$. Then true regression quantiles for $\theta = 0.1, 0.5$ and 0.9 are given as $(1, 0.718448)^t, (1,2)^t$, and $(1, 3.281552)^t$, respectively. Since C should be prespecified, we choose $C = 10$ using 10-fold cross validation method for $\theta = 0.5$. Then we use this C for the other θ values. Solving (5) with $C = 10$ we obtain the optimal Lagrange multipliers, $\hat{\alpha}_i, \hat{\alpha}_i^*$, which lead to the regression quantile estimators,

$$\hat{\boldsymbol{\beta}}(\theta) = \sum_{i=1}^{n}(\hat{\alpha}_i - \hat{\alpha}_i^*)\mathbf{x}_i.$$

To illustrate the estimation performance of regression quantiles by QRSVM, we compare QRSVM with the conventional quantile regression(QR) method based on FORTRAN of Koenker and D'Orey[7] by employing the average and the standard error of the regression quantile estimators.

Table 1. Averages and Standard Errors of the Regression Quantile Estimator

		$\theta=0.1$		$\theta=0.5$		$\theta=0.9$	
		Average	SE	Average	SE	Average	SE
$\beta_0(\theta)$	QRSVM	0.99250	0.02867	1.00338	0.01814	1.01745	0.03400
	QR	0.99046	0.03178	1.00190	0.01752	1.01240	0.02965
$\beta_1(\theta)$	QRSVM	0.72260	0.18464	1.98883	0.12134	3.20556	0.18784
	QR	0.73199	0.19076	1.99731	0.11933	3.23370	0.18006

Table 1 shows averages and standard errors of 100 regression quantile estimators obtained by both QRSVM and QR. As seen from Table 1, QRSVM and QR have almost same estimation performance for regression quantiles, and both methods provide reasonable regression quantile estimators of which values are very close to the true values of regression quantiles.

We now illustrate the prediction performance of the quantile functions obtained by both QRSVM and QR. We employ the fraction of variance unexplained(FVU), which is given by

$$FVU = \frac{E(\hat{f}(\mathbf{x}_i) - f(\mathbf{x}_i))^2}{E(f(\mathbf{x}_i) - \bar{f}(\mathbf{x}))^2},$$

where $\hat{f}(\mathbf{x}_i)$ is the predicted value of the function for a given $\mathbf{x}_i$, $f(\mathbf{x}_i)$ is the true value of the function for a given $\mathbf{x}_i$, and $\bar{f}(\mathbf{x})$ is the average of true values of the function for $x_1, \cdots, x_n$. Note that the FVU is the mean squared error for the estimate $\hat{f}(x)$ scaled by the variance of the true function $f(x)$. We evaluate the FVU by replacing the expectations with the average over a set of 200 test set values. With estimates of regression quantiles obtained from the training data set, we obtain the estimates of quantile functions for each of 100 test data sets, and 100 FVUs.

Table 2 shows the averages and the standard errors of the 100 FVUs obtained by QRSVM and QR for the quantile function predictors. As seen from Table 2, QRSVM provides the exact same prediction performance as QR when $\theta = 0.1, 0.5$, and 0.9. This implies that QRSVM performs as well as the conventional QR for these particular data sets.

Table 2. Averages and Standard Errors of 100 FVUs

	Method	$\theta = 0.1$	$\theta = 0.5$	$\theta = 0.9$
Average	QRSVM	0.198387	0.011188	0.009566
	QR	0.204571	0.010958	0.009423
SE	QRSVM	0.025195	0.001247	0.001296
	QR	0.025748	0.001240	0.001359

Example 2. In this example we illustrate how well QRSVM performs for the nonlinear quantile regression case. Similar to the linear case, we generate a training data set and 100 test data sets to present the prediction performance of these quantile functions. Each data set consists of 200 x's and 200 y's. Here x's are generated from a uniform distribution $U(0,1)$, and y's are generated from a normal distribution $N(1+\sin(2\pi x), 0.1)$. The θ-th quantile function of y for a given x can be modelled as

$$Q(\theta|x) = 1 + \sqrt{0.1}\Phi^{-1}(\theta) + \sin(2\pi x), \quad \text{for } \theta \in (0,1),$$

where $x = (1, x)^t$ and $\Phi^{-1}(\theta)$ is the θ-th quantile of a standard normal distribution. True quantile functions are given as $Q(0.1|x) = 0.871845 + \sin(2\pi x)$, $Q(0.5|x) = 1 + \sin(2\pi x)$ and $Q(0.9|x) = 1.128155 + \sin(2\pi x)$ for $\theta = 0.1, 0.5$, and 0.9, respectively. The radial basis function kernel is utilized in this example, which is

$$K(x, z) = \exp(-\frac{1}{\sigma^2}(x-z)^t(x-z)).$$

To apply QRSVM to the nonlinear quantile regression, C and σ should be pre-specified. First, we determine $C = 300$ and $\sigma = 1$ using 10-fold cross validation method for $\theta = 0.5$. Then we use these parameter values for the other θ values. According to our simulation studies, QRSVM is not sensitive to the choice of C. The important parameter that does require careful consideration is the kernel parameter σ. With these values of σ and C, the optimal Lagrange multipliers, $\hat{\alpha}_i, \hat{\alpha}_i^*$, can be obtained by solving (8) based on the training data set, which lead to the θ-th quantile function predictor of the output y for a $\mathbf{x}$ of test data set, defined by

$$\hat{Q}(\theta|\mathbf{x}) = \sum_{i=1}^{n} (\hat{\alpha}_i - \hat{\alpha}_i^*) K(\mathbf{x}_i, \mathbf{x}),$$

where $\mathbf{x} = (1, x)^t$ and $\mathbf{x}_i$'s are the input vectors of the training data set.

With the optimal Lagrange multipliers obtained based on the training data set, we obtain quantile function predictors for 100 test data sets, and 100 FVUs. Table 3 shows averages and standard errors of 100 FVUs obtained by QRSVM for $\theta = 0.1, 0.5$, and 0.9. We can see QRSVM works quite well for prediction.

Figure 1 shows the true quantile function(solid line) for one of 100 test data sets and the quantile function predictor(dotted line) obtained by QRSVM given

Table 3. Averages and Standard Errors of 100 FVUs obtained by QR SVM

	$\theta = 0.1$	$\theta = 0.5$	$\theta = 0.9$
Average	0.04858	0.03738	0.04852
SE	0.02999	0.02788	0.02901

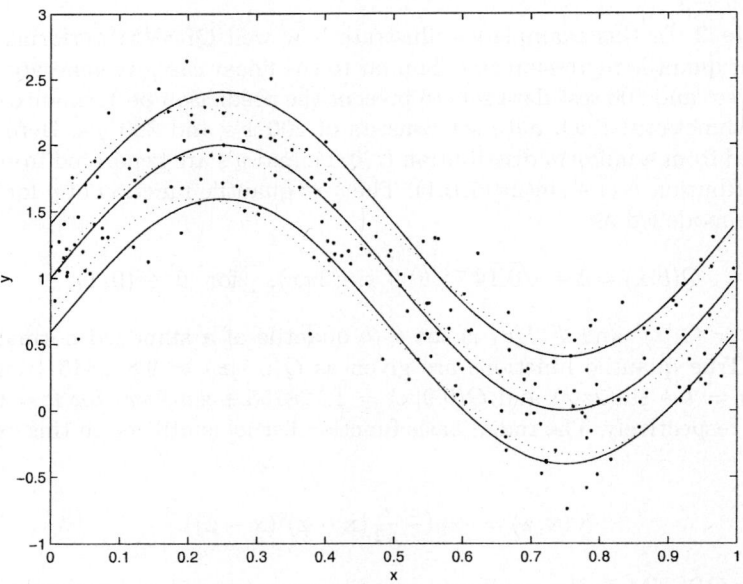

Fig. 1. True Quantile Function and Quantile Function Predictors obtained by QRSVM. The scatter is 200 artificial sample points (x_t, y_t) of testing data set with x_t's generated from a uniform distribution $U(0, 1)$, and y_t's generated from a normal distribution $N(1 + \sin(2\pi x), 0.1)$. True quantile function(solid line) and the quantile function predictor(dotted line) for $\theta = 0.1, 0.5$ and 0.9, respectively, are superimposed on the scatter.

test input data for $\theta = 0.1$, 0.5 and 0.9, respectively. We can see that QRSVM works reasonably well for the nonlinear quantile regression.

4 Conclusions

Quantile regression is an increasingly popular method for estimating the quantiles of a distribution conditional on the values of covariates. In this paper, through two examples - the linear and the nonlinear quantile regressions - we have shown that the proposed QRSVM derives the satisfying solutions and is attractive approaches to modelling the input data and quantile functions of output. In particular, we can apply this method successfully to the case that the linear quantile regression model is inappropriate. Here, we do not need to assume the underlying input structure for the nonlinear quantile regression model. We think this is the first paper presents an algorithm which performs well for both the linear and the nonlinear quantile regressions.

According to Koenker and Hallock[9], the simplex algorithm associated with LP problem for quantile regressions is highly effective on problems with a modest number of observations. But for large data sets the simplex approach eventually becomes considerably slow. For large data sets recent development of interior point methods for LP problems are highly effective. By the way, the proposed quantile regression method should use a time-consuming numerical QP optimization as an inner loop. We can overcome this problem by implementing straightforwardly sequential minimal optimization(SMO) developed by Platt[10], Flake and Lawrence[3] to train SVM regression for particularly large data sets. SMO breaks the large QP problem into a series of smallest possible QP problems. SMO is a fast training algorithm for SVM and is being implemented in some popular training software for SVM, for example LIBSVM developed by Chang and Lin[1].

References

1. Chang, C. and Lin, C. : LIBSVM: A library for support vector machines (2001)
2. Cristianini, N. and Shawe-Taylor, J. : An introduction to support vector machines, Cambridge University Press (2000)
3. Flake, G. W. and Lawrence, S. : Efficient SVM regression training with SMO, Machine Learning **46** (2002) 271–290
4. Gunn, S. : Support vector machines for classification and regression, ISIS Technical Report, University of Southampton (1998)
5. Kecman, V. : Learning and soft computing, support vector machines, neural networks and fuzzy logic moldes. The MIT Press, Cambridge, MA (2001)
6. Koenker, R. and Bassett, G. : Regression quantiles, Econometrica **46** (1978) 33–50
7. Koenker, R. and D'Orey, V. : Computing regression quantiles, Applied Statistics **36** (1987) 383–393
8. Koenker, R. and Park, B. J. : An interior point algorithm for nonlinear quantile regression, Journal of Econometrics **71** (1996) 265–283

9. Koenker, R. and Hallock, K. F. : Quantile regression, Journal of Economic Perspectives **15** (2001) 143–156
10. Platt, J. : Using sparseness and analytic QP to speed training of support vector machines, In: Kearns, M. S., Solla, S. A. and Cohn, D. A. (Eds.), Advances in Neural Information Processing Systems **11** (1999)
11. Smola, A. and Schölkopf, B. : On a kernel-based method for pattern recognition, regression, approximation and operator inversion. Algorithmica **22** (1998) 211–231
12. Vapnik, V. N. : The nature of statistical learning theory. Springer, New York (1995)
13. Vapnik, V. N.: Statistical learning theory. Springer, New York (1998)
14. Wang, L.(Ed.) : Support vector machines: theory and application. Springer, Berlin Heidelberg New York (2005)
15. Yu, K., Lu, Z. and Stander, J. : Quantile regression: applications and current research area, The Statistician **52** (2003) 331–350

Doubly Regularized Kernel Regression with Heteroscedastic Censored Data

Jooyong Shim[1] and Changha Hwang[2]

[1] Department of Statistical Information,
Catholic University of Daegu, Kyungbuk 712 - 702, Korea
ds1631@hanmail.net

[2] Corresponding Author, Division of Information and Computer Sciences,
Dankook University, Seoul 140 - 714, Korea
chwang@dankook.ac.kr

Abstract. A doubly regularized likelihood estimating procedure is introduced for the heteroscedastic censored regression. The proposed procedure provides the estimates of both the conditional mean and the variance of the response variables, which are obtained by two stepwise iterative fashion. The generalized cross validation function and the generalized approximate cross validation function are used alternately to estimate tuning parameters in each step. Experimental results are then presented which indicate the performance of the proposed estimating procedure.

1 Introduction

Minimizing a sum of squares of errors is well known to correspond to the maximum likelihood estimation for the regression model where the errors are assumed to be independently normally distributed with constant variance (homoscedastic). The least squares support vector machine(LS-SVM, Suykens and Vanderwalle[9]) and the kernel ridge regression(Saunders et al.[8]) provide the procedures for the estimation by minimizing a regularized sum of squares of errors which perform the nonlinear regression using a linear model, conducted in a higher dimensional feature space induced by a Mercer kernel(Mercer[6]). The least squares method and the accelerated failure time model to accommodate the censored data seem appealing since they are familiar and well understood. Koul et al.[5] gave a simple least squares type estimation procedure in the censored regression model with the weighted observations and also showed the consistency and asymptotic normality of the estimator. Zhou[12] proposed an M-estimator of the regression parameter based on the censored data using the weights Koul et al.[5] proposed. By introducing the weighting scheme of Zhou[12] into the optimization problem in LS-SVM, Kim et al.[3] obtained the estimate of the conditional mean in the nonlinear censored regression model.

In this paper we introduce a regularized likelihood based approach which can take the heteroscedasticity and the randomly right censoring into account to estimate both the mean function and the variance function simultaneously.

The rest of this paper is organized as follows. In Section 2, we introduce the doubly regularized likelihood estimation for the heteroscedastic kernel regression model. In Section 3, we introduce the regularized likelihood estimation for the censored kernel regression model. In Section 4, we present the estimating procedure for the heteroscedastic censored kernel regression. In Section 5, we perform the numerical studies through examples. In Section 6, we give the conclusions and remarks.

2 Doubly Regularized Kernel Regression

Let the given data set be denoted by $\{\mathbf{x}_i, y_i\}_{i=1}^n$, with $\mathbf{x}_i \in \mathbf{R}^d$ and $y_i \in \mathbf{R}$. For this data set, we can consider the heteroscedastic regression model

$$y_i = \mu(\mathbf{x}_i) + \epsilon_i \quad i = 1, 2, \cdots, n, \tag{1}$$

where $\mathbf{x}_i$ is the covariate vector including a constant 1, ϵ_i is assumed to be independently normally distributed with mean 0 and variance $\sigma^2(\mathbf{x}_i)$ and μ and σ^2 are functions to be estimated. The negative log likelihood of the given data set can be expressed as(constant terms are omitted)

$$L(\mu, \sigma) = \frac{1}{n} \sum_{i=1}^{n} \left\{ \frac{(y_i - \mu(\mathbf{x}_i))^2}{2\sigma^2(\mathbf{x}_i)} + \frac{1}{2} \log \sigma^2(\mathbf{x}_i) \right\}. \tag{2}$$

Due to the positivity of the variance we write the logarithm of $\sigma^2(\mathbf{x}_i)$ as $g(\mathbf{x}_i)$, then the negative log likelihood can reexpressed as

$$L(\mu, g) = \frac{1}{n} \sum_{i=1}^{n} \left\{ (y_i - \mu(\mathbf{x}_i))^2 e^{-g(\mathbf{x}_i)} + g(\mathbf{x}_i) \right\}. \tag{3}$$

The conditional mean is estimated by a linear model, $\mu(\mathbf{x}) = \omega_\mu' \phi_\mu(\mathbf{x})$, conducted in a high dimensional feature space. Here the feature mapping function $\phi_\mu(\cdot) : R^d \to R^{d_f}$ maps the input space to the higher dimensional feature space where the dimension d_f is defined in an implicit way. It is known that $\phi_\mu(\mathbf{x}_i)' \phi_\mu(\mathbf{x}_j) = K_\mu(\mathbf{x}_i, \mathbf{x}_j)$ which are obtained from the application of Mercer[6]'s conditions. Also g is estimated by a linear model, $g(\mathbf{x}) = \omega_\mathbf{g}' \phi_\mathbf{g}(\mathbf{x})$. Then the estimates of $(\mu, g, \omega_\mu, \omega_g)$ are obtained by minimizing the regularized negative log likelihood

$$L(\mu, g, \omega_\mu, \omega_g) = \frac{1}{n} \sum_{i=1}^{n} \left\{ (y_i - \mu(\mathbf{x}_i))^2 e^{-g(\mathbf{x}_i)} + g(\mathbf{x}_i) \right\} + \lambda_\mu \|\omega_\mu\|^2 + \lambda_g \|\omega_g\|^2 \tag{4}$$

where λ_μ and λ_g are nonnegative constants which control the trade-off between the goodness-of-fit on the data and $\|\omega_\mu\|^2$ and $\|\omega_g\|^2$. The representation theorem(Kimeldorf and Wahba[4]) guarantees the minimizer of the regularized negative log likelihood to be

$$\mu(\mathbf{x}) = K_\mu \alpha_\mu \quad \text{and} \quad g(\mathbf{x}) = K_g \alpha_g, \tag{5}$$

for some vectors α_μ and α_g. Now the problem (4) becomes obtaining (α_μ, α_g) to minimize

$$L(\alpha_\mu, \alpha_g) = (\mathbf{y} - K_\mu \alpha_\mu)' D(\mathbf{y} - K_\mu \alpha_\mu) + \mathbf{1}'\mathbf{g} + n\lambda_\mu \alpha_\mu' K_\mu \alpha_\mu$$
$$+ n\lambda_g \alpha_g' K_g \alpha_g, \tag{6}$$

where D is a diagonal matrix with the ith diagonal element $e^{-g(\mathbf{x}_i)}$ and $\mathbf{g}$ is a $(n \times 1)$ vector with the ith element $g(\mathbf{x}_i)$ and $\mathbf{1}$ is a $(n \times 1)$ vector with 1's.

3 Censored Kernel Regression

Consider the censored linear regression model for the response variables t_i's and the covariate vector $\mathbf{x}_i$ including a constant 1,

$$t_i = \mu(\mathbf{x}_i) + \epsilon_i, \quad i = 1, \cdots, n, \tag{7}$$

where ϵ_i is an unobservable error assumed to be independent with mean 0 and constant variance. Here the mean is related to the covariate vector linearly, which can be expressed as $\mu(\mathbf{x}_i) = \mathbf{x}_i'\beta$ with the regression parameter vector of the model, β. Let c_i's be the censoring variables assumed to be independent and identically distributed having a cumulative distribution function $G(y) = P(c_i \leq y)$. The parameter vector of interest is β and t_i is not observed but

$$\delta_i = I_{(t_i < c_i)} \quad \text{and} \quad y_i = \min(t_i, c_i), \tag{8}$$

where $I_{(\cdot)}$ denotes the indicator function. In most practical cases $G(\cdot)$ is not known and needs to be estimated by the Kaplan-Meier estimator(Kaplan and Meier[2]) or its variation, $\hat{G}(\cdot)$. The problem considered here is that of the estimation of β based on $(\delta_1, y_1, \mathbf{x}_1), \cdots, (\delta_n, y_n, \mathbf{x}_n)$. Koul et al.[5] defined a new observable response $\tilde{y}_i$ with weights θ_i as

$$\tilde{y}_i = \theta_i y_i \quad \text{where} \quad \theta_i = \frac{\delta_i}{1 - \hat{G}(y_i)} \tag{9}$$

and showed $\tilde{y}_i$ has the same mean as t_i and thus follows the same linear model as t_i does. And the estimate of β is obtained by minimizing the objective function

$$L(\beta) = \frac{1}{n} \sum_{i=1}^{n} (\tilde{y}_i - \mathbf{x}_i'\beta)^2.$$

By introducing the feature mapping function $\phi_\mu(\cdot)$ for the kernel regression, the estimate of $\mu(\mathbf{x}_i)$ can be obtained as $\sum_{j=1}^{n} K_\mu(\mathbf{x}_i, \mathbf{x}_j) \sqrt{\theta_j} \alpha_j$, where α_j's are obtained by minimizing

$$L(\alpha) = (\mathbf{y} - K_\mu \Theta \alpha)'(\mathbf{y} - K_\mu \Theta \alpha) + n\lambda \alpha' K_\mu \alpha, \tag{10}$$

where $\alpha = (\alpha_1, \cdots, \alpha_n)'$ and Θ is a diagonal matrix with the ith diagonal element $\sqrt{\theta_i}$ defined in (9).

4 Algorithm for Heteroscedastic Censored Kernel Regression

For the censored regression, (9) becomes

$$L(\alpha_\mu, \alpha_g) = (\mathbf{y} - K_\mu \Theta \alpha_\mu)' D(\mathbf{y} - K_\mu \Theta \alpha_\mu) + \mathbf{1}'\mathbf{g} + n\lambda_\mu \alpha_\mu' K_\mu \alpha_\mu$$
$$+ n\lambda_g \alpha_g' K_g \alpha_g, \qquad (11)$$

where μ is the conditional mean of the response t, g is the logarithm of variance of the response t and $\mathbf{y}$ is the observed vector defined in (8). The estimates of μ and $\mathbf{g}$ are obtained via two stepwise iterative procedures.

Fixing the values of $\mathbf{g}$, (11) reduces to

$$L(\alpha_\mu) = (\mathbf{y} - K_\mu \Theta \alpha_\mu)' D(\mathbf{y} - K_\mu \Theta \alpha_\mu) + n\lambda_\mu \alpha_\mu' K_\mu \alpha_\mu, \qquad (12)$$

The solution to (12) is

$$\hat{\mu} = A_\mu \mathbf{y} = (\Theta K_\mu \Theta + n\lambda_\mu D^{-1})^{-1} \Theta \mathbf{y}. \qquad (13)$$

In this step the values of λ_μ and other tuning parameters ν included in the kernel K_μ can be chosen by minimizing the generalized cross validation function (Wahba[10]):

$$GCV(\lambda_\mu, \nu) = \frac{n^{-1} \mathbf{y}' D^{1/2} (I - A_\mu)^2 D^{1/2} \mathbf{y}}{[n^{-1} tr(I - A_\mu)]^2}. \qquad (14)$$

To estimate $g(t_i) = \log \sigma^2(\mathbf{x}_i)$ which is the logarithm of variance of response t_i, we use uncensored data set $\{\mathbf{x}_i^*, y_i^*\}_{i,j=1}^{n^*}$, in this step. Fixing μ, α_{g*} is estimated by minimizing the objective function with uncensored data

$$L(\alpha_{g*}) = \mathbf{1}'(D^*\mathbf{z} + K_{g*}\alpha_{g*}) + n^* \lambda_g \alpha_{g*}' K_{g*} \alpha_{g*}, \qquad (15)$$

where $\mathbf{z}$ is a $(n^* \times 1)$ vector with ith element $(y_i^* - \tilde{\mu}_i^*)^2$, $\tilde{\mu}^*$ is the current estimate of $\mu(\mathbf{x}_i$, and D^* is the diagonal matrix defined in (6) corresponding to uncensored data. α_{g*} is obtained by Newton Raphson method,

$$\alpha_{g*}^{new} = \alpha_{g*}^{old} - H^{-1}G, \qquad (16)$$

where G and H are the gradient vector and Hessian matrix with respect to α_g, respectively. The values of λ_g and other tuning parameters υ included in the kernel $K_{g*} = \{\phi_g(\mathbf{x}^*_j)' \phi_g(\mathbf{x}^*_j)\}_{i=1}^{n^*}$ can be chosen by minimizing the generalized approximate cross validation function(Xiang and Wahba[11]) :

$$GACV(\lambda_g, \upsilon) = \mathbf{1}'(D^*\mathbf{z} + K_{g*}\alpha_{g*}) + \frac{tr(A_{g*})}{1 + tr(D^*A_{g*})}(z - diag(D^*))' D^* z, (17)$$

where $A_{g*} = (D^* + 2n^* \lambda_g K_{g*})^{-1}$ is the influence matrix of (15). Then g corresponding to whole data can be obtained as

$$\mathbf{g} = K_g \alpha_{g*}, \qquad (18)$$

where $K_g = \{\phi_g(\mathbf{x}_i)' \phi_g(\mathbf{x}_j^*)\}$ for $i = 1, \cdots, n, j = 1, \cdots, n^*$.

5 Experimental Results

We illustrate the performance of the proposed estimating procedure through the real data set and the simulated data set.

Stanford heart transplant data set(Miller and Halpern[7]) consists of 152 patient with complete recorded who survived at least 10 days. There were 55 censored observations. Let the response variable be the base 10 logarithm of the survival time of patient and consider the polynomial kernel with degree 2 in this data set. Figure 1 shows the observed values against the ages of patient at transplant. The solid line is the estimate of the conditional mean of response variable, the dashed lines are the estimates of the confidence interval($\mu \pm \sigma$). The estimate of conditional mean behaves similar to the estimate of Buckley and James[1] which are proved to be reliable on this data set(Miller and Halpern[7]). In Figure 1 the heteroscedastic model and the homoscedastic model do not present much difference on estimating of the conditional mean. In Figure 1(a) difference can be seen on younger ages and older ages for the estimate of the confidence interval. The observations are distributed more densely for older ages,

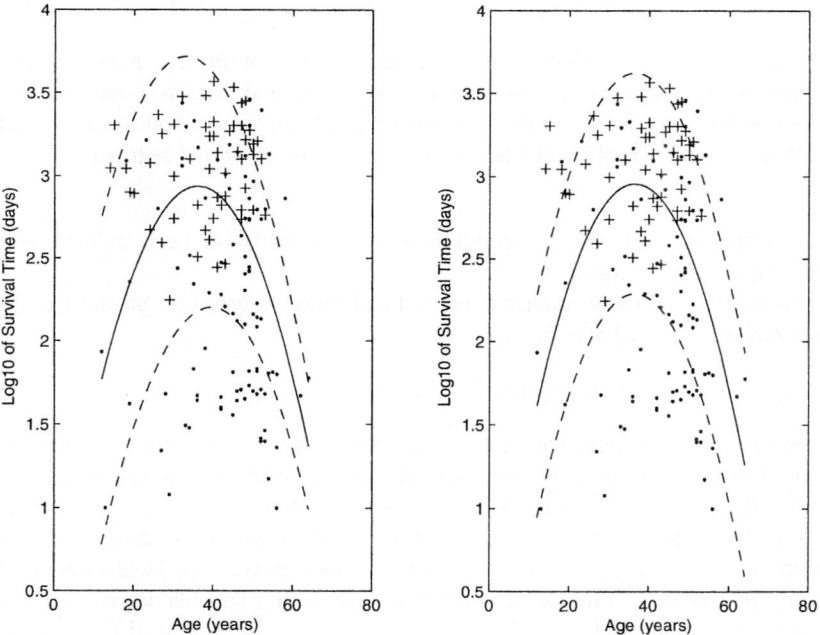

Fig. 1. (a)(left): Scatter plots of $\log_{10}$ survival times(days) versus age at transplant(year) and estimates of the conditional mean(μ) and the confidence interval($\mu \pm \sigma$) by the heteroscedastic censored kernel regression. Patients deceased are denoted by ' · ' and the alive by ' + '. (b)(right): by the homoscedastic censored kernel regression.

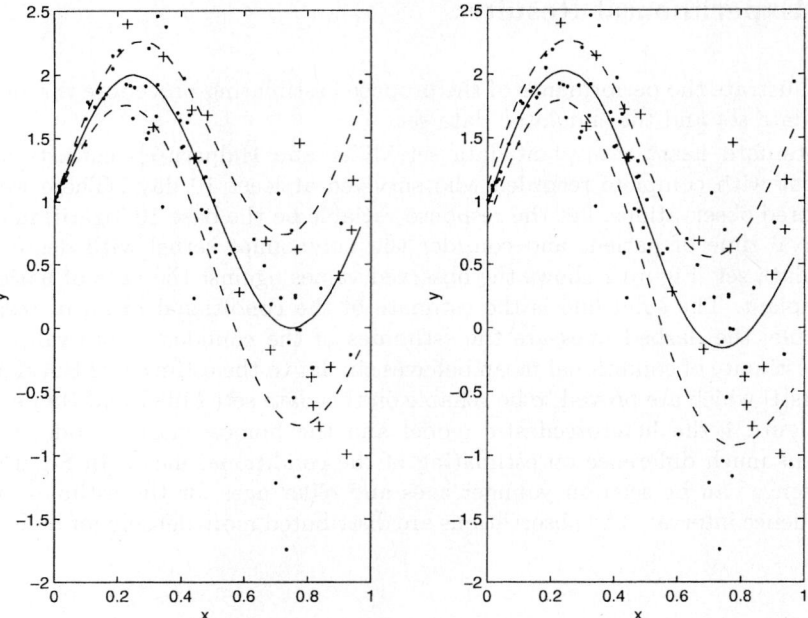

Fig. 2. (a)(left): Scatter plots of y versus x and the conditional mean(μ) and the confidence interval($\mu \pm \sigma$) of true values. The uncensored data are denoted by ' · ' and the censored by ' + '. (b)(right): estimates of the conditional mean(μ) and the confidence interval($\mu \pm \sigma$) by the heteroscedastic censored kernel regression.

this leads the estimate of the confidence interval for older ages to have smaller widths than younger ages.

Consider the heteroscedastic censored nonlinear regression model for the response variables t_i's of the form,

$$t_i = \mu(x_i) + \epsilon_{t_i}, \quad i = 1, \cdots, n.$$

The response variable and the censoring variable given x can be modelled as $t = \mu(x) + \epsilon_t$ and $c = \mu(x) + \epsilon_c$, respectively. We set the true value of the mean of both variables given the covariate x as $\mu(x) = 1 + \sin(2\pi x)$. 100 of x's are generated from a uniform distribution, U(0,1), 100 of (ϵ_t, ϵ_c)'s are generated from normal distributions, $N(0, x^2)$ and $N(0, (x + cc)^2)$, respectively. cc is chosen for 20% censoring proportion. The radial basis function(RBF) kernels are considered for the simulated data set. Figure 2 shows the conditional mean and the confidence interval given x. The confidence interval becomes to have larger width as x increases since the variance of error is set to x^2. The estimate of the conditional mean behave similarly as the true value. For the confidence interval, the estimate has slightly larger width on small values x's but it approaches to the true value as x increases.

6 Conclusions and Remarks

In this paper, we introduce the estimating procedure for the heteroscedastic censored regression model using a doubly regularized kernel regression approach. Through the examples we showed that the proposed method derives the satisfying solutions and is attractive approaches to modelling the data of inputs. By choosing appropriate value of tuning parameters simultaneously, it can used a good tool for explaining not only the heteroscedastic data but also randomly right censored data.

References

1. Buckley, J. and James, I. : Linear regression with censdored dara. Biometrics **30** (1974) 89-99
2. Kaplan, E. L. and Meier, P. : Nonparametric estimation from incomplete observations. Journal of American Statistical Association **53** (1959) 457-481
3. Kim, D., Shim, J. and Oh, K. : Censored Regression by LS -SVM. Journal of of the Korean Statistical Society **33** (2004) 1, 25-34
4. Kimeldorf, G. S. and Wahba, G. : Some results on Tchebycheffian spline functions. Journal of Mathematical Analysis ans its Applications **33** (1971) 82-95
5. Koul, H., Susarla, V. and Van Ryzin J. : Regression analysis with randomly right censored data. The Annal of Statistics **9** (1981) 1276-1288
6. Mercer, J. : Functions of positive and negative type and their connection with the theory of integral equations. Philosophical Transactions of the Royal Society **A** (1909) 415-446
7. Miller, R. G. and Halpern, J. : Regression with censored data. Biometrika **69** (1982) 521-531
8. Saunders, C. , Gammerman, A. and Vovk, V. : Ridege regression learning algorithm in dual variables, In proceeding of 15th International Conference on Machine Learning (1998) 515-521, Madison, WI, July 24-27
9. Suykens, J.A.K. and Vanderwalle, J. : Least Square Support Vector Machine Classifier, Neural Processing Letters **9** (1999) 293-300
10. Wahba, G. : Spline models for observational data. CBMS-NSF Regional Conference Series in Applied Mathematics, **59** (1990) 169 Society for Industrial and Applied Mathematics, Philadelphia, PA
11. Xiang, D. and Wahba, G. : A generalized approximate cross validation for smoothing splines with non-Gaussian data. Statistica Sinica **6** (1996) 675-692
12. Zhou, M. : M-estimation in censored linear models. Biometrika **79** (1992) 4, 837-841

Support Vector Based Prototype Selection Method for Nearest Neighbor Rules

Yuangui Li, Zhonghui Hu, Yunze Cai, and Weidong Zhang

Department of Automation, Shanghai Jiaotong University
1954, Huashan Road, Xuhui, Shanghai, 200030, P. R. China
{li_yuangui, huhzh, yzcai, wdzhang}@sjtu.edu.cn.edu

Abstract. The Support vector machines derive the class decision hyper planes from a few, selected prototypes, the support vectors (SVs) according to the principle of structure risk minimization, so they have good generalization ability. We proposed a new prototype selection method based on support vectors for nearest neighbor rules. It selects prototypes only from support vectors. During classification, for unknown example, it can be classified into the same class as the nearest neighbor in feature space among all the prototypes. Computational results show that our method can obtain higher reduction rate and accuracy than popular condensing or editing instance reduction method.

1 Introduction

For classification problems, complete statistical knowledge regarding the conditional density functions of each class is rarely available, which precludes the application of the optimal Bayes classification methods, while the nearest neighbor(NN) rule and its extension to k neighbors (or k-NN rule) have been in practice one of the most widely used non-parametric classifiers. The advantage of NN rule lies in that it combines its conceptual simplicity with the fact that its asymptotic error rate is conveniently bounded in terms of the optimal Bayes error [1]. However, the main problems of the NN rules lie that it is computationally expensive and the storage requirement is large for large problems because it stores all the training examples in memory and distances between new instance and all the training points is required to be computed to find the nearest neighbor in classifying process; and it is intolerant to noisy instance and irrelevant attributes. Many researches on prototype selection have been done in order to reduce the training set, reduce the effect of noise on accuracy, and obtain the same classification ability as using the whole training set [2-4].

Two different families of prototype selection methods exist in the literature. First, the condensing or reducing algorithm aims at selecting the minimal subset of prototypes that lead to the same performance as using the whole training set. Second, editing algorithm eliminates noisy examples from the original set and "cleans" possible overlapping among classes. The recent condensing algorithm is Minimal Consistent Set(MCS) method proposed by Dasarathy[5] and Dasarathy conjectured MCS was the minimal training-set consistent subset, but the counter-examples to this claim have been found by Kuncheva and Bezdek[6]. The difficulty of condensing algorithm is that the noisy examples are preferred to be selected into prototype set, which harms the accuracy of result classifier. For editing algorithms, it is observed that the asymptotically optimal edited NN-rule, such as well known Multi-edit algorithm, can lead to arbitrarily bad classification result if the number of prototypes is not large enough compared to the intrinsic dimension of feature space[7]. Furthermore the editing algorithm can't reduce the training set effectively. Dasarathy[7] found that the synergy exploitation of condensing and editing algorithm could make the best result on balance of instance reduction with classification accuracy. So an effective prototype selection algorithm should be able to both remove the noise and overlapping out of prototype set and obtain an as small as possible prototype set.

The support vector machine (SVM) is a new kind of learning machine proposed by Vapnik in 1995[8]. It is derived from statistical learning theory and VC-dimension theory [9-12], and has become another research hotspot following neural network. The remarkable advantage of SVM is that it is induced according to the principle of structural risk minimization, so it performs good generalization ability, especially for small sample problems. The decision surface of SVM is parameterized by a set of support vectors and a set of corresponding weights, which indicates that support vectors have the key patterns to define the decision boundaries. So it is possible to develop new prototype selection base on support vectors. Vishwanathan and Murty[15] proposed data reduction method using multi-category proximal SVM, but it simply selected the support vectors with Langrage multipliers larger than 0 and less than the bound. They only indicated that it is feasible to select prototypes for NN with SVM and didn't compare the performance with common instance reduction method.

In order to select prototypes based on support vectors, we should obtain SVM first, why not use SVM to classify new examples? LeeCun et al. [16] found that the classification speed of SVM is substantially slower than that of neural networks, especially for large problems. That is because too many support vectors is required to express the decision boundary and increase the complexity of decision function. To address this problem, Burges[13-14] proposed simplified SVM, which used a new reduced vector set to approximate the decision rule decided by all the support vectors so as to reduce the complexity of SVM and assure the loss in generalization performance is acceptable, and in some cases, the reduced vector set can be computed analytically. But Burges' method is too complex.

This paper is organized as follows. In section 2, the different importance in deciding classification hyper planes between 3 types of support vectors was analyzed. In section 3 we introduced our prototype selection method based on support vectors. Computational results are presented in section 4 to compare performance of our method with that of common instance reduction methods. Section 5 concludes our work.

2 Support Vectors and Decision Hyper Planes of SVM

Suppose that there exists a given training set $\{(\mathbf{x}_i, y_i)\}_{i=1}^{l} \in X^n \times R$, where X^n denotes the space of input vectors. Let ξ be the deviation between $f(x_i)$ and y_i. The optimization problem solved by support vector machine is[17]

$$\min_{w,b} \quad J = \frac{1}{2}\|w\|^2 + C\sum_{i=1}^{n}\xi_i$$
$$s.t. \quad y_i(w\cdot\Phi(x_i)-b) \geq 1-\xi_i \tag{1}$$
$$\xi_i \geq 0, i=1,\cdots,n$$

$\Phi(\cdot)$ is the map from input space into feature space, and it is decided by the kernel function $k(x,\hat{x})$. The Lagrangian for this problem is

$$L = \frac{1}{2}\|w\|^2 + C\sum_{i=1}^{n}\xi_i + \sum_{i=1}^{n}a_i[1-\xi_i-y_i(w\cdot\Phi(x_i)-b)] - \sum_{i=1}^{n}\pi_i\xi_i \tag{2}$$

The Karush-Kuhn-Tucker(KKT) optimal conditions are given by[17]

$$w = \sum_{i=1}^{n} a_i y_i \Phi(x_i) \tag{3}$$

$$\sum_i a_i y_i = 0 \tag{4}$$

$$C - a_i - \pi_i = 0, \forall i \tag{5}$$

$$a_i[1-\xi_i - y_i(w\cdot\Phi(x_i)-b)] = 0, \forall i \tag{6}$$

$$\pi_i \xi_i = 0, \forall i \tag{7}$$

$$a_i \geq 0, \pi_i \geq 0, \forall i \tag{8}$$

According to the above KKT optimal conditions, we can obtain

$$\xi_i = \begin{cases} 0, & \text{if } a_i = 0 \\ 0, & \text{if } 0 < a_i < C \\ \geq 0, & \text{if } a_i = C \end{cases} \tag{9}$$

The separating hyper planes and the distribution of training examples in feature space are similar to that in figure 1.

For a training example $\mathbf{x}_i$, if its corresponding Lagrange multiplier a_i is equal to upper bound of C, such as $\mathbf{x}_5$, this training example must lie between H_1 and H_{-1} or lie among the training examples of other class. Obviously, this example should be dealt with carefully because it seems like noise or 'dangerous' example that may bring on overlapping. If its corresponding Lagrange multiplier a_i is 0, we can see that it can't contribute on the decision of separating hyper plane from (3), so this example should be excluded from prototype set because the class information contained in it is redundant. If its corresponding Lagrange multiplier $0 < a_i < C$, this example must lie on the hyper plane H_1 or H_{-1}, and this example plays most important role in deciding the separating hyper plane, so it is top-priority prototype candidate.

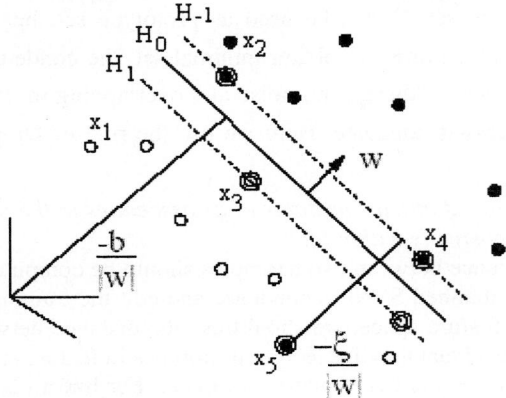

Fig. 1. The separating hyper planes and distribution of training examples in feature space

So, we can find that the support vectors with Lagrange multipliers smaller than C are representative examples and contain most classification information and this throws light on developing new prototype selection method. We also observe that the support vectors with Lagrange multipliers equal to C are 'dangerous' examples and they should be dealt with carefully. In all, by using the classification information contained in support vectors, we can develop new effective prototype selection methods for nearest neighbor rules.

3 Our Prototype Selection Method Based on Support Vectors

In this section, we will introduce our method according to the introduction in section 2. Suppose that there exists a given training set $T = \{(\mathbf{x}_i, y_i)\}_{i=1}^{l} \in X^n \times R$. At first, we will choose proper kernel function $k(x, \hat{x})$ and parameter C, and then training set T is used to learn the SVM. After the SVM is learned, the support vector set noted as S is obtained.

We only select those support vectors on the right side near H_0 as prototype candidates, so we defined prototype candidate set P_c as

$$P_c = \{\mathbf{x}_i \mid y_i \times f(\mathbf{x}_i) > 0, \mathbf{x}_i \in S, f(\mathbf{x}_i) \text{ is output of SVM}\} \tag{10}$$

The process to obtain P_c is both condensing and editing process. Deleting non-support vectors is an condensing process, which can condense prototype set effectively, and excluding support vectors lying among examples of other class can avoid error or overlapping examples' being selected as prototypes. For the SVM performs good generalization ability, and it can select out most representative examples from training set, so prototype selection based on support vectors may obtain better generalization ability than other instance reduction methods.

Prototype candidate set P_c can be used as prototype set, but it may not be the smallest prototype set. In order to obtain minimal set, the condensing process is implemented on P_c. Because there is no noise and overlapping in P_c, simple condensing or deleting process is adequate. Here we use the rule of Drop2 [2] to condense P_c. The rule is

Remove the instance if at least as many of its associates in the original training set would be classified correctly without it.

In Drop2, the distance between two examples should be computed. Because we use the hyper planes of obtained SVM to condense and edit the training set and the hyper planes are linear in feature space, we should use the distance between two examples in feature space. The advantage of selecting prototypes in feature space is obvious.

First, in feature space, the hyper planes are linear. For linear class boundary, fewer points are required to express it than that of nonlinear one, which makes it possible to condense the prototype candidates as small as possible. In another aspect, we can deal with nonlinear and linear SVM with uniform method. The linear SVM can be seen as a special case with kernel function $k(x_i, x_j) = x_i \cdot x_j$.

Let the Euclidean distance between $\mathbf{x}_i$ and $\mathbf{x}_j$ in feature space is $d_{ij}^{(H)}$, if the kernel function is $K(\mathbf{x}_i, \mathbf{x}_j)$, we can obtain:

$$\begin{aligned}(d_{ij}^{(H)})^2 &= \|\Phi(\mathbf{x}_i) - \Phi(\mathbf{x}_j)\|^2 \\ &= \|\Phi(\mathbf{x}_i)\|^2 + \|\Phi(\mathbf{x}_j)\|^2 - 2\|\Phi(\mathbf{x}_i)\|\|\Phi(\mathbf{x}_j)\| \\ &= K(\mathbf{x}_i, \mathbf{x}_i) + K(\mathbf{x}_j, \mathbf{x}_j) - 2K(\mathbf{x}_i, \mathbf{x}_j)\end{aligned} \quad (11)$$

Because the number of prototype candidates is much smaller than the size of training set T, small voting parameter k in Drop2 [2] is adequate and large k may mistake unrelated examples as neighbors. By applying Drop2 on the P_c, we can obtain the result prototype set P_s.

For an unknown example, we compute the distance between it and all the members of P_s in feature space, and classify it into the class of its nearest neighbor in P_s.

4 Computational Results

In this section, experiments are done to illustrate the performance of our method on 3 benchmark data sets from UCI Repository of machine learning databases [19]. They are Johns Hopkins University Ionosphere database, Wisconsin Breast Cancer Database(WBC) and Wisconsin Diagnostic Breast Cancer(WDBC) database. There are 351 instances described by 34 continuous predictor attributes and one binary class attribute in Johns Hopkins University Ionosphere database. For Wisconsin Breast Cancer Database, 463 instances are used and they are described by 9 continuous predictor attributes and one binary class attribute. There are 569 instances described by 30 predictor attributes and one binary class attribute in Wisconsin Diagnostic Breast Cancer database.

Experiment 1 is done to illustrate the performance of our method in prototype selection for nearest neighbor rules. It is compared with other popular prototype selection methods Drop4, Drop5 and MCS. Drop4 and Drop5 are editing algorithms; MCS is training-set-consistent condensing algorithm. These popular algorithms and our method are respectively applied to the same data set in order to compare the performance and 10-fold cross validation method is used to obtain average performance. Keerthi's improved SMO algorithm [18] is used to train SVM with training set. The comparison result is shown in table 1.

Table 1. Comparison result between our method and other condensing and editing algorithms

		Ionosphere	WBC	WDBC
Drop4	Reduction rate	8.04%	4.05%	5.39%
	Accuracy	83.43%	92.61%	94.91%
Drop5	Reduction rate	8.23%	6.15%	5.21%
	Accuracy	76.29%	90.87%	93.51%
MCS	Reduction rate	16.36%	14.1%	10.33%
	Accuracy	85.71%	88.91%	93.68%
Our method	Reduction rate	5.47%	2.13%	1.46%
	Accuracy	87.14%	94.57%	95.61%

The comparison result shows that our method obtains higher reduction rate and higher classification accuracy than those of other popular editing and condensing methods. This indicates that our method is superior to current condensing or editing prototype selection method. It also indicates that SVM can help to improve the reduction rate and accuracy when it is used to develop new prototype selection method.

For the SVMs in experiment 1 on three data sets, we list the average number of support vectors of SVM and the average size of result prototype set based on corresponding SVM in table 2.

Table 2. size of support vectors and prototypes

	Ionosphere	WBC	WDBC
Number of support vectors	132	83.9	103.9
Number of prototypes	17.3	8.9	7.9

As we can see in table 2, a small portion of support vectors are selected as prototypes and used to classify new examples. The number of support vectors is much larger than the size of prototype set, and more support vectors will make the decision function of SVM more complex, as a result, the speed in classification phase will be slow, which is substantial for large problems. So our method supplies new method to simplify the classification of SVM.

5 Conclusion

In this paper, SVM is used to select prototypes in order to obtain higher reduction rate and classification accuracy for nearest neighbor rule. Because all the support vectors can decide the classification boundary, so non-support vectors can be excluded from prototype set. As to the support vectors lying among examples of other class, they may result in overlapping and should be excluded from prototype set in order to improve generalization performance.

The training set is used to train a SVM, and then those support vectors on the right side of H_0 in figure 1 will be selected into prototype candidate set. In order to obtain smaller prototype set, the prototype candidate set is condensed with Wilson's Drop2 instance reduction rule to obtain the resulting prototype set. For an unknown example, the distances in feature space between it and all the member of prototype set are computed and it is classified as the class of its nearest neighbor in the prototype set.

Experiment results show that our method is an effective prototype selection method and it can obtain higher reduction rate and classification accuracy than those of popular editing and condensing algorithms. It combines the condensing process and editing process so as to obtain better performance. The comparison between the number of support vectors and the number of prototypes indicates that our method can simplify support vector decision rule, but it should be improved so as to obtain same generalization ability as that of SVM.

References

1. Cover, T.M., Hart, P.E.: Nearest Neighbor Pattern Classification. IEEE Transaction on Information Theory. Vol. 13, 1(1967) 21-27
2. Wilson, D.R., Martinez, T.R.: Reduction Techniques for Instance-based Learning Algorithm. Machine Learning. Vol. 38, 3(2000) 257-286

3. Toussaint, G.: Proximity Graphs for Nearest Neighbor Decision Rules: Recent Progress. Proceedings of INTERFACE-2002, 34th Symposium on Computing and Statistics, Ritz-Carlton Hotel, Montreal, Canada. (2002) 83-106
4. Aha, D.W., Kibler, D., Albert M.K.: Instance-based Learning Algorithms. Machine Learning. Vol. 6, 19(1991) 37-66
5. Dasarathy, B.V.: Minimal Consistent Set (MCS) Identification for Optimal Nearest Neighbor Decision Systems Design. IEEE transaction on System, Man, and Cybernetics, Vol. 24, 3(1994) 511-517
6. Kuncheva, L.I., Bezdek, J. C.: Nearest Prototype Classification: Clustering, Genetic Algorithms, or Random Search. IEEE transactions on System, Man and Cybernetics, 28(1998) 160-164
7. Dasarathy, B.V., Sánchez, J. S., Townsend S.: Nearest Neighbor Editing and Condensing Tools-Synergy Exploitation. Pattern Analysis & Application. 3(2000) 19-30
8. Vapnik, V. N.: The Nature of Statistical Learning Theory. Springer Verlag, New York. (1995)
9. Vapnik, V. N.: Estimation of Dependences Based on Empirical Data. Springer Verlag, Berlin. (1982)
10. Vapnik, V. N., Chervonenkis A.: Theory of Pattern Recognition. Nauka, Moscow. (1974)
11. Kecman, V.: Learning and Soft Computing, Support Vector machines, Neural Networks and Fuzzy Logic Models. The MIT Press, Cambridge, MA (2001)
12. Wang, L.P. (Ed.): Support Vector Machines: Theory and Application. Springer, Berlin Heidelberg New York (2005)
13. Burges C. J. C.: Simplified Support Vector Decision Rules. the 13th international conference on Machine Learning. (1996) 71-77
14. Burges C. J. C., Schölkopf B.: Improving the Accuracy and Speed of Support Vector Machines. In: M. Mozer, M. Jordan, and T. Petsche (eds.): Neural Information Processing Systems, Vol. 9. MIT Press, Cambridge, MA. (1997)
15. Vishwanathan S.V.N., Murthy M. N.: Use of Multi-category Proximal SVM for Data Set Reduction. In: Ajith A., Mario Köppen(Eds.): Hybrid Information Systems. First International Workshop on Hybrid Intelligent Systems, Adelaide, Australia, December 11-12. (2001) 19-24
16. LeCun Y., Jackel L., Bottou L., Brunot A., Cortes C., Denker J., Drucker H., Guyon I., Müller U., Säckinger E., Simard P., Vapnik V.: Comparison of Learning Algorithms for Handwritten Digit Recognition. In: F. Fogelman, P. Gallinari(Eds): Proc. International Conference on Artificial Neural Networks. (1995) 53-60
17. Burges C. J. C.: A Tutorial on Support Vector Machines for Pattern Recognition. Data Mining and Knowledge Discovery. Vol. 2, 2(1998) 121-167
18. Keerthi S. S., Shevade S. K., Bhattacharyya C., Murthy K. R. K.: Improvements to Platt's SMO Algorithm for SVM Classifier Design. Neural Computation. Vol. 13, (2001) 37-649
19. Blake C., Keogh E., Merz C. J.: UCI Repository of machine learning databases. http://www.ics.uci.edu/~mlearn/MLRepository.html. Irvine, CA. University of California, Department of Information and Computer Science, (1998)

A Prediction Interval Estimation Method for KMSE

Changha Hwang[1], Kyung Ha Seok[2], and Daehyeon Cho[3]

[1] Division of Information and Computer Sciences,
Dankook University, Seoul 140-714, Korea
chwang@dankook.ac.kr
[2] Corresponding Author, Department of Data Science,
Inje University, Kyungnam 621-749, Korea
skh@stat.inje.ac.kr
[3] Department of Data Science,
Inje University, 621-749, Kyungnam, Korea
cho@stat.inje.ac.kr

Abstract. The kernel minimum squared error estimation(KMSE) model can be viewed as a general framework that includes kernel Fisher discriminant analysis(KFDA), least squares support vector machine(LS-SVM), and kernel ridge regression(KRR) as its particular cases. For continuous real output the equivalence of KMSE and LS-SVM is shown in this paper. We apply standard methods for computing prediction intervals in nonlinear regression to KMSE model. The simulation results show that LS-SVM has better performance in terms of the prediction intervals and mean squared error(MSE). The experiment on a real date set indicates that KMSE compares favorably with other method.

1 Introduction

In forecasting tasks the prediction interval gives the range in which you could have a certain level of confidence of finding an individual value of the predicted output for a given input value. Truly reliable prediction systems require the prediction to be qualified by a confidence measure such as prediction or confidence interval. This important issue has received little systematic study. However, this has been paid attention in neural information processing and chemical engineering communities.

Chryssolouris [1] has derived a technique to quantify the confidence intervals for the prediction of neural network models by adopting a variant of the linearisation methodology. Shao et al. [7] have proposed a novel method of computing confidence bounds on predictions from a neural network with determined structure. De Veaux et al. [3] also have proposed a method of computing prediction intervals for neural networks and compared them with prediction intervals based on multivariate adaptive regression splines using generalized additive model (MARS/GAM). Yang et al. [13] have suggested a method of estimating confidence bound for neural networks for the purpose of the prediction of rock porosity values from seismic data for oil reservoir characterization. Seok et al. [6] have presented a Bayesian approach to computing the prediction intervals for support vector machine(SVM) regression and shown SVM regression achieves better performances than the neural networks and MARS in predicting intervals. There are some other literatures related to this issue.

SVM, originally introduced by Vapnik, solves the weak point of neural network such as the existence of local minima in the area of statistical learning theory and structural risk minimization(Vapnik [11]). One of its prominent advantages is the idea of using kernels to realize the nonlinear transforms without knowing the detailed transforms. According to this idea, other authors proposed a class of kernel-based algorithms, such as the kernel Fisher discriminant analysis(KFDA)(Mika et al. [4]), the least squares support vector machine(LS-SVM)(Suykens and Vandewalle [9] , Suykens et al. [10]), and the kernel ridge regression(KRR)(Saunders et al. [5]).

Xu et al. [12] have generalized the conventional minimum squared error method to yield a new type of nonlinear learning machine, by using the kernel idea and adding different regularization terms. They have named the proposed learning machines as KMSE algorithm. KMSE algorithm adopts the idea of kernel function of SVM which is one of the most influential developments in the machine learning.

KMSE model can be viewed as a general framework that includes KFDA, LS-SVM, and KRR as its particular cases. Suykens et al. [8] have proposed a large scale algorithm for LS-SVM by implementing a Hestenes-Stiefel conjugate gradient algorithm for solving the linear equation system. A large scale algorithm for KMSE can be derived without any difficulty by using this idea.

In this paper we discuss a method to compute prediction intervals by applying standard methods for computing prediction intervals in nonlinear regression to the KMSE for regression tasks. The simulation results show that LS-SVM has better performance in terms of the prediction intervals and MSE. The experiment on a real date set indicates that LS-SVM compares favorably with MARS/GAM.

The rest of this paper is organized as follows. Section 2 gives an overview of LS-SVM and KMSE. Section 3 discusses briefly how to compute prediction intervals for KMSE model. Section 4 illustrates the method with a computer generated data and a real data from a polymer process.

2   LS-SVM and KMSE

Let the training data set D be denoted by $\{(\mathbf{x}_k, y_k), k = 1, \ldots, n\}$, with each input $\mathbf{x}_k \in R^d$ and the output $y_k \in R$. It is commonly assumed that

$$y = f(\mathbf{x}, \boldsymbol{\alpha}^*) + \epsilon, \quad (1)$$

where ϵ is independently and identically distributed with zero mean and $\boldsymbol{\alpha}^*$ is the true value of parameters.

Assume a nonlinear function $\boldsymbol{\varphi}(\mathbf{x}) : R^d \to R^h$ maps the input space to a so-called higher dimensional feature space. It is important to note that the dimension h of this space is defined only in an implicit way.

2.1 LS-SVM

The LS-SVM, a modified version of SVM in a least squares sense, has been proposed for the classification and the regression by Suykens and Vanderwalle [9]. The LS-SVM model for function estimation has the following representation in feature space

$$f(\mathbf{x}) = \mathbf{w}^t \varphi(\mathbf{x}) + b \text{ with } \mathbf{w} \in R^h, \ b \in R,$$

where superscript t represents the transpose of a vector. Given a training set D we define now the following optimization problem to get optimal $\mathbf{w}$ and b

$$\min_{\mathbf{w},b,e} T(\mathbf{w},e) = \frac{1}{2}\mathbf{w}^t\mathbf{w} + \gamma\frac{1}{2}\sum_{k=1}^{n} e_k^2 \qquad (2)$$

subject to the equality constraints

$$y_k = \mathbf{w}^t \varphi(\mathbf{x}_k) + b + e_k, \ k = 1,\cdots,n. \qquad (3)$$

The cost function with squared error and regularization corresponds to a form of ridge regression. We construct the Lagrangian

$$L(\mathbf{w},e) = T(\mathbf{w},e) - \sum_{k=1}^{n} \alpha_k \{\mathbf{w}^t \varphi(\mathbf{x}_k) + b + e_k - y_k\}, \qquad (4)$$

where α_k's are Lagrange multipliers. The conditions for optimality $\frac{\partial L}{\partial \mathbf{w}} = \mathbf{0}$, $\frac{\partial L}{\partial b} = 0$, $\frac{\partial L}{\partial e_k} = 0$ and $\frac{\partial L}{\partial \alpha_k} = 0$ yield the linear equation

$$\begin{bmatrix} 0 & \mathbf{1}^t \\ \mathbf{1} & K + \gamma^{-1}I \end{bmatrix} \begin{bmatrix} b \\ \boldsymbol{\alpha} \end{bmatrix} = \begin{bmatrix} 0 \\ \mathbf{y} \end{bmatrix} \qquad (5)$$

with $\mathbf{y} = (y_1, \ldots y_n)^t$, $\mathbf{1} = (1,\ldots,1)^t$, $\boldsymbol{\alpha} = (\alpha_1,\ldots,\alpha_n)^t$ and where

$$\begin{aligned} K_{kl} &= K(\mathbf{x}_k,\mathbf{x}_l), \ k,l = 1,\ldots,n \\ &= \varphi(\mathbf{x}_k)^t \varphi(\mathbf{x}_l) \end{aligned}$$

is a kernel function obtained from the Mercer's condition. Several choices of the kernel function are possible. Here $\mathbf{w}$ and $\varphi(\mathbf{x})$ are not calculated. The resulting LS-SVM model for function estimation becomes

$$f(\mathbf{x}) = \sum_{k=1}^{n} \alpha_k K(\mathbf{x},\mathbf{x}_k) + b, \qquad (6)$$

where α_k's and b are the solution to the linear system.

2.2 KMSE

The KMSE, a new type of nonlinear learning machine based on kernel idea, has been proposed for the classification and the regression by Xu et al. [12]. In the feature space we build a linear estimate whose weight vector and bias are denoted by **w** and b. From the theory of reproducing kernels we can construct an expansion for **w** in the form

$$\mathbf{w} = \sum_{k=1}^{n} \alpha_k \varphi(\mathbf{x}_k), \tag{7}$$

where $\alpha_k \in R$, $k = 1, 2, \ldots, n$. Using the kernel function defined above, we can define the objective function in the feature space as follows.

$$M(\boldsymbol{\alpha}, b) = \frac{1}{2}(\mathbf{y} - K^t\boldsymbol{\alpha} - b\mathbf{1})^t(\mathbf{y} - K^t\boldsymbol{\alpha} - b\mathbf{1}). \tag{8}$$

Note that solution matrix of (8) is always singular and the estimates obtained from (8) tends to be overfitted. In order to avoid this problem, additional regularization term can be added. There exist two usual regularization terms: $\boldsymbol{\alpha}^t\boldsymbol{\alpha}$ in KFD and $\mathbf{w}^t\mathbf{w}$ in SVM, LS-SVM and KRR. We construct different regularized objective functions, that is,

$$M_i(\boldsymbol{\alpha}, b) = \frac{1}{2}\mu_i s + M(\boldsymbol{\alpha}, b), i = 1, 2. \tag{9}$$

where μ_i's are positive regularization parameters and $s = \boldsymbol{\alpha}^t\boldsymbol{\alpha}, i = 1$, $s = \mathbf{w}^t\mathbf{w}, i = 2$.

Minimizing these objective functions, we obtain two sets of linear equation,

$$\begin{bmatrix} KK^t + \mu_1 I & K\mathbf{1} \\ (K\mathbf{1})^t & n \end{bmatrix} \begin{bmatrix} \boldsymbol{\alpha} \\ b \end{bmatrix} = \begin{bmatrix} K\mathbf{y} \\ \mathbf{1}^t\mathbf{y} \end{bmatrix}, i = 1 \tag{10}$$

and

$$\begin{bmatrix} K + \mu_2 I & \mathbf{1} \\ (K\mathbf{1})^t & n \end{bmatrix} \begin{bmatrix} \boldsymbol{\alpha} \\ b \end{bmatrix} = \begin{bmatrix} \mathbf{y} \\ \mathbf{1}^t\mathbf{y} \end{bmatrix}, i = 2. \tag{11}$$

Xu et al. [12] have shown that KMSE is equivalent to KFDA and LS-SVM for classification problem. For continuous real output the equivalence of LS-SVM and KMSE can be shown easily. From (11) we obtain the following two equations,

$$(K + \mu_2 I)\boldsymbol{\alpha} + \mathbf{1}b = \mathbf{y} \tag{12}$$
$$\mathbf{1}^t(K^t\boldsymbol{\alpha} + \mathbf{1}b) = \mathbf{1}^t\mathbf{y}. \tag{13}$$

Now eliminating $\mathbf{1}b$ from (12) and (13), we obtain

$$\mu_2 \mathbf{1}^t \boldsymbol{\alpha} = 0. \tag{14}$$

Comparing (5) to (12) and (13) with $\mu_2 = \gamma^{-1}$, the equivalence follows immediately.

In this paper we denote $\boldsymbol{\alpha} = (\boldsymbol{\alpha}^t, b)^t$ and $K = [K \ \mathbf{1}]$ and the two KMSE estimates with the solutions of (10) and (11) by $f_1(\mathbf{x})$ and $f_2(\mathbf{x})$, respectively.

3 Prediction Interval Estimation

De Veaux et al. [3] have developed a method estimating prediction intervals for neural networks trained by weight decay by implementing standard methods for computing prediction intervals in nonlinear regression. This method can be easily applied to KMSE model for regression tasks. We will briefly state the resulting prediction interval in what follows. The calculation is much simpler than that for neural networks since the estimated regression function is linear with regard to parameters.

The $100(1-\beta)\%$ prediction interval for input $\mathbf{x}_0$ is $f(\mathbf{x}_0) \pm c$, where c is

$$c = t^{n-p}_{1-\frac{\beta}{2}} \, s \, (1 + \mathbf{f}_0^t (\mathbf{F}^t \mathbf{F} + \mu_i \mathbf{I})^{-1} \mathbf{F}^t \mathbf{F} (\mathbf{F}^t \mathbf{F} + \mu_i \mathbf{I})^{-1} \mathbf{f}_0)^{1/2}. \qquad (15)$$

Here $t^{n-p}_{1-\frac{\beta}{2}}$ is $100(1-\frac{\beta}{2})th$ percentile of t-distribution with $n-p$ degrees of freedom, p is the number of parameters and

$$s^2 = \sum_{k=1}^{n}(y_k - f_i(\mathbf{x}_k))^2/(n-p).$$

Given $\mathbf{x}_0$, the vector $\mathbf{f}_0$ is given by

$$\mathbf{f}_0 = \left[\frac{\partial f(\mathbf{x}_0; \boldsymbol{\alpha}^*)}{\partial \alpha_1^*} \; \frac{\partial f(\mathbf{x}_0; \boldsymbol{\alpha}^*)}{\partial \alpha_2^*} \; \cdots \; \frac{\partial f(\mathbf{x}_0; \boldsymbol{\alpha}^*)}{\partial \alpha_{n+1}^*} \right]^t. \qquad (16)$$

$\mathbf{F}$ is Jacobian matrix given by

$$\mathbf{F} = \begin{bmatrix} \frac{\partial f(\mathbf{x}_1; \hat{\boldsymbol{\alpha}})}{\partial \hat{\alpha}_1} & \frac{\partial f(\mathbf{x}_1; \hat{\boldsymbol{\alpha}})}{\partial \hat{\alpha}_2} & \cdots & \frac{\partial f(\mathbf{x}_1; \hat{\boldsymbol{\alpha}})}{\partial \hat{\alpha}_{n+1}} \\ \frac{\partial f(\mathbf{x}_2; \hat{\boldsymbol{\alpha}})}{\partial \hat{\alpha}_1} & \frac{\partial f(\mathbf{x}_2; \hat{\boldsymbol{\alpha}})}{\partial \hat{\alpha}_2} & \cdots & \frac{\partial f(\mathbf{x}_2; \hat{\boldsymbol{\alpha}})}{\partial \hat{\alpha}_{n+1}} \\ \vdots & \vdots & \vdots & \vdots \\ \frac{\partial f(\mathbf{x}_n; \hat{\boldsymbol{\alpha}})}{\partial \hat{\alpha}_1} & \frac{\partial f(\mathbf{x}_n; \hat{\boldsymbol{\alpha}})}{\partial \hat{\alpha}_2} & \cdots & \frac{\partial f(\mathbf{x}_n; \hat{\boldsymbol{\alpha}})}{\partial \hat{\alpha}_{n+1}} \end{bmatrix}^t. \qquad (17)$$

Replacing $\boldsymbol{\alpha}^*$ in (16) by $\hat{\boldsymbol{\alpha}}$, we can get the desired prediction interval. De Veaux et al. [3] have shown that this method is effective on a wide range of problems.

4 Illustrative Examples

In this section we illustrate the behavior of the prediction intervals of KMSE model through a computer generated data and a real data from a polymer process in De Veaux et al. [3]. In Example 1, we evaluate the prediction interval estimation method for f_1 and f_2. In Example 2, we will verify KMSE model by applying the procedure to the polymer process data in De Veaux et al. [3].

Example 1. In this example we illustrate how well the prediction interval estimation method performs for a computer generated data. Similar to the examples in Shao et al. [7] and Yang et al. [13], the sine function $y = \sin(x) + \epsilon$ is

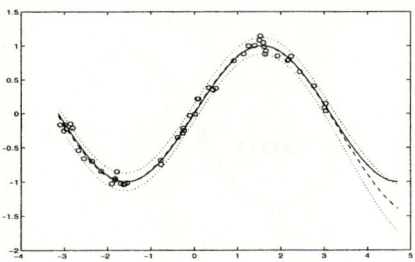 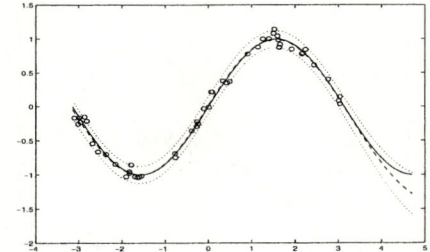

Fig. 1. Sine function estimates and prediction intervals of f_1(left) and f_2(right), for SNR=10% and three inputs case. Circle: training data points. Solid line: target function. Dashed line: function estimates. Dotted line: 90 % prediction intervals.

used as a test function to evaluate the prediction interval method. In order to investigate how the unnecessary inputs affect the confidence intervals and estimations we perform an experiment using the three inputs $x_1 = -x$, $x_2 = x^2$ and $x_3 = 1/(x+4)$ and the output $y = \sin(x) + \epsilon$. The noise ϵ is normally distributed with zero mean and a standard deviation up to 30% of the standard deviation of the data(SNR). The SNR is short for the signal-to-noise ratio.

In this paper, we adopt radial basis function(RBF) kernel

$$K(\mathbf{x}, \mathbf{y}) = \exp(-\frac{\|\mathbf{x} - \mathbf{y}\|}{\sigma^2}).$$

The 10-fold cross validation is used to choose kernel parameter σ and regularization parameters $\mu_i, i = 1, 2$. We use the effective number of parameters(Vapnik(1998)) for p in (15). That is,

$$p = \sum_{k=1}^{n+1} \frac{\lambda_k}{\lambda_k + \mu_i}, \quad i = 1, 2,$$

where λ_k's are eigenvalues of $K^t K$ and μ_i's are regularization parameters in (9).

We first examine how the estimated prediction intervals behave when three input variables are used and the test data extrapolate the training data. The 50 training data points are unevenly distributed between $-\pi$ and π. The 100 test data points are evenly distributed between $-\pi$ and $\pi + \pi/2$. The results are shown in Fig. 1. The prediction intervals tend to be large in the extrapolation area. However, compared with the results of Yang et al.(2000), these intervals are appreciably smaller.

Next we examine when the density of training data varies. The 70% of training points are generated from $[-\pi, 0)$ and 30%'s are from $[0, \pi]$. The results are shown in Fig. 2. We can see from Fig. 2 that the difference of the prediction intervals in the low and high density areas is negligible. In fact, we observe that the sizes of the prediction intervals in the low and high density areas are almost equal. We also observe that the shapes of the prediction intervals of f_1 and f_2 are not significantly different.

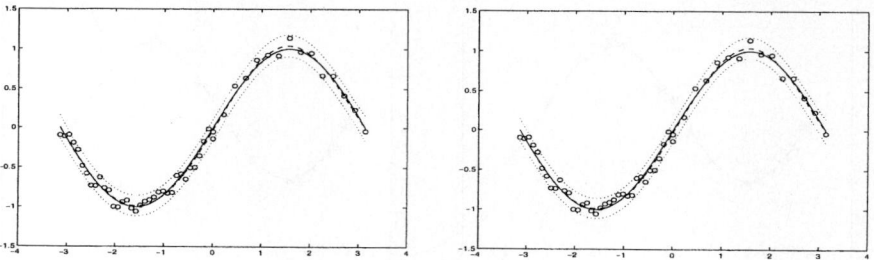

Fig. 2. Sine function estimates and prediction intervals of f_1(left) and f_2(right), for SNR=20% and three inputs case. 35 training data points from $[-\pi, 0)$ and 15 from $[0, \pi]$, Circle: training data points. Solid line: target function. Dashed line: function estimates. Dotted line: 90 % prediction intervals.

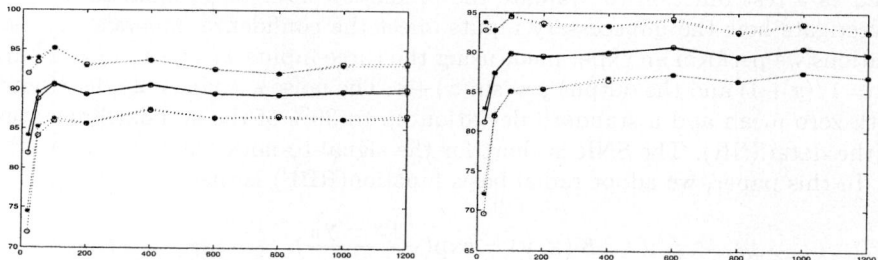

Fig. 3. Average coverage as a function of sample size. Circle: f_1, Star: f_2. Left: SNR=10% and one input case. Right: SNR=30% and three input case. Solid line: Average coverage. Dashed line: Average coverage plus/minus its standard deviation.

To compare the prediction intervals of f_1 and f_2, and to investigate the performance of the prediction intervals we also compute the average of coverage, the average of size and the standard deviation of them. In addition, we compute the average of s^2 to see the performance of f_1 and f_2. Using 100 trials, these statistics are computed for training sample size $n = 20, 50, 100, 200, 400, 600, 800, 1000, 1200$ when SNR= 10%, 20%, 30%, and one or three input variables are used. Since all of the results have similar trend, we will show some of them. The results are shown in Fig. 3, Fig. 4 and Fig. 5.

As seen from Fig. 3, the average coverage becomes close to the desired 90 % coverage if the size of the training set is larger than 50. Moreover, the standard deviation of the coverage is stable.

As seen from Fig. 4, the average size of prediction intervals converges to 0.24 when SNR= 10% and one input variable is used. In this case, the average size for f_2 is appreciably smaller than that for f_1. We also observe that the average size converges to 0.47 regardless of the number of input variables when SNR= 20%. But, this result is not shown in Fig. 4. It tends that the average size depends on the observation noise level rather than the number of input variables.

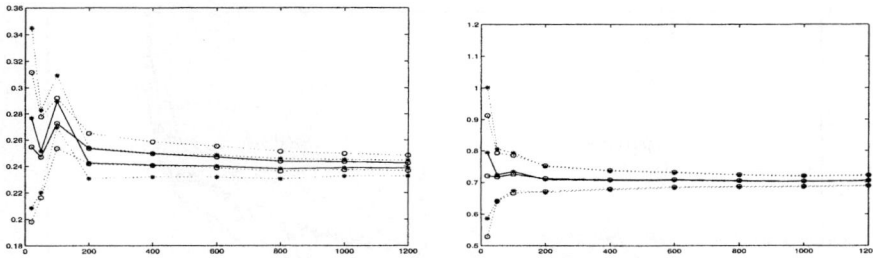

Fig. 4. Average size of prediction intervals as a function of sample size. Circle: f_1, Star: f_2. Left: SNR=10% and one input case. Right: SNR=30% and three inputs case. Solid line: Average size. Dashed line: Average size plus/minus its standard deviation.

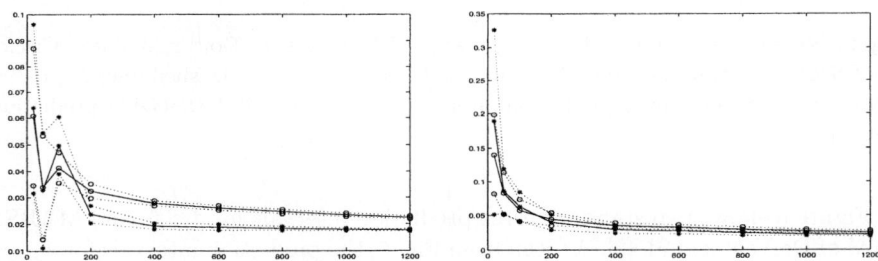

Fig. 5. Average MSE of f_1 and f_2 as a function of sample size. Circle: f_1, Star: f_2. Left: SNR=10% and one input case. Right: SNR=30% and three inputs case. Solid line: Average MSE. Dashed line: Average MSE plus/minus its standard deviation.

We can see from Fig. 5 that the average MSE of f_1 and f_2 converges to 0.018 and 0.022, respectively, for the case of SNR= 10% and one input variable. When SNR= 30% and three input variables are used, each average MSE converges to 0.02 and 0.03, respectively. Therefore we could say that the performance of f_2 is better than that of f_1.

Example 2. In this example we will verify KMSE model by applying the procedure to the polymer process data in De Veaux et al. [3]. We also compare these models with MARS/GAM introduced in De Veaux et al. [2]. This data set is from polymer process with 10 inputs $(x_1, \ldots, x_{10})$ and a single response variable y. Because the data are proprietary, no other information on the variables is available. The data set consists of 61 observations and is available via FTP site. De Veaux et al. [2] have shown that nonlinear regression methods are superior to linear methods for this polymer process data set. Here we choose $\sigma = 1.8$ and $\mu_1 = 6.1 \times 10^{-6}$ for f_1 and $\sigma = 1.6$ and $\mu_2 = 0.0025$ for f_2 based on 6-fold cross validation. Same values of parameters are chosen for MARS/GAM as in De Veaux et al. [3].

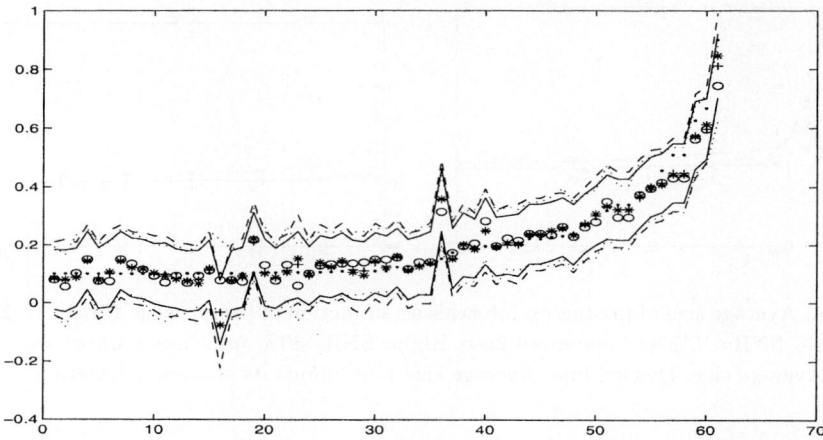

Fig. 6. Prediction intervals for f_1, f_2 and $MARS/GAM$. Dot: real data. Circle: $MARS/GAM$ estimates. Star : f_1 estimates. Plus: f_2 estimates. Dashed line: f_1 prediction intervals. Solid line: f_2 prediction intervals. Dotted line: $MARS/GAM$ prediction interval.

Figure 6 shows two standard error prediction intervals for f_1, f_2 and MARS/GAM of De Veaux et al. [3]. As seen from Fig. 6, the prediction intervals for f_2 are slightly smoother at almost every point. The average size of prediction intervals is 0.2627 for f_1, 0.2098 for f_2 and 0.2341 for MARS/GAM. This experiment indicates that f_2, i.e., LS-SVM compares favorably with f_1 and a sophisticated non-parametric model, MARS/GAM, for regression modelling.

5 Conclusion

The objective of this paper is twofold. The first is to compare two KMSE estimates f_1 and f_2 in terms of the prediction capability through simulation study. The second is to compare these estimates with MARS/GAM based on a real data from a polymer process, which in general works better than neural networks.

The simulation results show that f_2 has better performance in terms of the prediction intervals and MSE. The experiment on a real data set indicates that f_2, i.e., LS-SVM compares favorably with MARS/GAM. To conclude, we recommend LS-SVM as a technique for regression modeling.

Acknowledgement

This work was supported by the 2001 Inje University research grant.

References

1. Chryssolouris, G. Lee and M., Ramsey, A.: Confidence interval prediction for neural network models. IEEE Trans. Neural Networks **7** (1996) 229-232
2. De Veaux, R. D., Psichogios, D. C. and Ungar, L. H.: A comparison of two nonparametric estimation schemes: MARS and neural networks. Computers in Chemical Engineering **17** (1993) 819-837
3. De Veaux, R. D., Schumi, J., Schweinsberg, J. and Ungar, L. H.: Prediction interval for neural networks via nonlinear regression. Technometrics **40** (1998) 273-282
4. Mika, S., Ratsch, G., Weston, J., Scholkopf, B. and Muller, K.-R.: Fisher discriminant analysis with kernels. Neural Networks for Signal Processing IX, 41-48, IEEE press, New York (1999)
5. Saunders, C., Gammerman, A. and Vork, V.: Ridge regression learning algorithm in dual variable. Proceedings of the 15th International Conference on Machine Learning (1998) 515-521
6. Seok, K., Hwang, C. and Cho, D.: Prediction intervals for support vector machine regression. Communications in Statistics : Theory and Methods **31** (2002) 1887-1898
7. Shao, R., Martin, E. B., Zhang, J. and Morris, A. J.: Connfidence bounds for neural network representations. Computers Chem. Engng. **21**(suppl.) (1997): S1173-S1178
8. Suykens J. A. K., Lukas, L., Van Dooren, P., De Moor, B. and Vandewalle, J.: Least squares support vector machine classifiers: a large scale algorithm. European Conference on Circuit Theory and Design, ECCTD'99 (1999) 839-842, Stresa Italy, August
9. Suykens, J. A. K. and Vandewalle, J.: Least squares support vector machine classifiers. Neural Processing Letters **9** (1998) 293-300
10. Suykens J. A. K., Van Gestel, T., De Brabanter, J. D., De Moor, B. and Vandewalle, J.: Least Squares Support Vector Machines. World Scientific, Singapore, 2002
11. Vapnik, V.: Statistical Learning Theory. Springer, New York, 1998
12. Xu, J., Zhang, X. and Li, Y.: Kernel MSE algorithm: A unified framework for KFD, LS-SVM. Proceedings of IJCNN'01 (2001) 2:1486-1491
13. Yang, L., Kavli, T., Carlin, M., Clausen, S. and de Groot, P. F. M.: An evaluation of confidence bound estimation methods for neural networks. Proceeding of ESIT (2000) 322-329

An Information-Geometrical Approach to Constructing Kernel in Support Vector Regression Machines

Wensen An[1,2] and Yanguang Sun[2]

[1] Department of Automation, University of Science and Technology of China,
230027, Hefei, China
anwensen@yahoo.com.cn
[2] Automation Research and Design Institute of Metallurgical Industry,
100071, Beijing, China

Abstract. The type of kernel function has a great important influence on the performance of support vector machines (SVMs); however, there is no theoretical guidance to choose a good kernel. To solve classification problem, Amari presented a method of modifying kernel based on information geometry theory. In the paper, we first review the classical formulation of regression problem, then propose an approach to constructing the kernel function in support vector regression machines from information-geometrical viewpoint, and point out its difference with the method that Amari used in support vector classification machines. Finally some simulation results show the effectiveness of the proposed method.

1 Introduction

The theory of support vector machines (SVMs) is a new promising machine learning technique proposed by Vapnik [1][2][3]. SVMs employ the structural risk minimization (SRM) principle, which has been shown to be superior to the traditional empirical risk minimization (ERM) principle employed in conventional learning algorithms (e.g. neural networks)[4]. SRM minimizes an upper bound on the generalization error as opposed to ERM, which minimizes the error on the training data. It is this difference that equips SVMs with a greater ability to generalize, which is the goal in statistical learning. SVMs were developed to solve the classification problem, but they have been successfully extended to the domain of regression problems [5][6][7].

SVMs are linear learning machines in the parameter space but it is easily extended to nonlinear learning machines by mapping the space S of the input data into a high-dimensional (even infinite-dimensional) feature space F through a nonlinear function ϕ. It is interesting that we need not know the nonlinear function ϕ explicitly; we only need the dot product of input data, which is available from the kernel function K, which generates ϕ. By choosing different kinds of kernels, different kinds of SVMs learning machines can be got such as polynomial, multi-layer preceptor and radial basis function (RBF) learning machines.

It is important to choose a good kernel that is fit for practical problem when training SVMs because the performance of SVMs largely depends on the kernel; however, there is no general theoretical guidance how to choose a kernel function. Chen et al. introduced a construction procedure for sparse kernel modeling based on an approach of directly optimizing model generalization capability [8]; Amari presented a method of modifying a kernel based on the information geometry theory [9][10] and applied the method to solve classification problem. In this paper, under the illumination of the method of Amari, we propose a method of modifying the kernel function, which is applied to solve regression problem.

2 Classical Formulation of Regression Problem

Given the training data set $D=\{(x_k,y_k)|k=1,2,\cdots,l\}, x_k \in R^n, y_k \in R$, of input x_k and associated targets y_k, the goal of regression problem is to fit a flat function $f(x)$ which approximates the relation inherited between the data set points and it can be used later on to infer the output y for a new input data point x.

Suppose the function $f(x)$ is expressed as:

$$f(x)=<\omega,\phi(x)>+b \quad \phi:R^n \to F, \quad \omega \in F, \quad b \in R \tag{1}$$

where $<,>$ is dot product of vector, b is a bias term, and ϕ is a nonlinear map which mapping the input x into a high-dimensional feature space F, thus a linear regression in high-dimensional feature space is corresponding to a nonlinear regression in low-dimensional input space.

According to SRM principle, that function $f(x)$ is flat in the case of Eq. (1) means that one seeks the minimization of the following expression:

$$\tfrac{1}{2}\|\omega\|^2 + C\sum_{k=1}^{l} L(f(x_k), y_k) \tag{2}$$

where $L(\cdot)$ is a loss function, C is a constant.

Many forms for the loss function can be found in the literature: e.g. linear, huber and quadratic loss function, etc. In this paper, Vapnik's loss function is used, which is known as ε-insensitive loss function and defined as:

$$L(y,f(x)) = \begin{cases} 0 & |f(x)-y| < \varepsilon \\ |y-f(x)|-\varepsilon & \text{otherwise} \end{cases} \tag{3}$$

Thus, the regression problem can be written as a convex optimization problem:

$$\text{minimize} \quad \tfrac{1}{2}\|\omega\|^2 + C\sum_{k=1}^{l}(\xi_k + \xi_k^*) \tag{4}$$

$$\text{subject to} \quad \begin{cases} y_k - <\omega,\phi(x_k)> -b \leq \varepsilon + \xi_k \\ <\omega,\phi(x_k)> +b - y_k \leq \varepsilon + \xi_k^* \\ \xi_k, \xi_k^* \geq 0 \end{cases} \tag{5}$$

$\varepsilon>0$ is a predefined constant which controls the noise tolerance, the constant $C>0$ determines the trade-off between the flatness of f and the amount of tolerable deviations, which is larger than ε.

Through introducing a Lagrange function, the optimization problem (4) and (5) can be solved in their dual formulation, which is expressed as follows:

$$\text{maximize} \ -\frac{1}{2}\sum_{k,p=1}^{l}(\alpha_k-\alpha_k^*)(\alpha_p-\alpha_p^*)<\phi(x_k),\phi(x_p)> \\ -\varepsilon\sum_{k=1}^{l}(\alpha_k+\alpha_k^*)+\sum_{k=1}^{l}y_k(\alpha_k-\alpha_k^*) \tag{6}$$

$$\text{subject to} \ \begin{cases}\sum_{k=1}^{l}(\alpha_k-\alpha_k^*)=0 \\ \alpha_k,\alpha_k^*\in[0,C]\end{cases} \tag{7}$$

The optimal value of α_k, α_k^* can be obtained by solving the dual problem (6), (7), accordingly, the ω can be described by:

$$\omega=\sum_{k=1}^{l}(\alpha_k-\alpha_k^*)\phi(x_k) \tag{8}$$

thus, $$f(x)=\sum_{k=1}^{l}(\alpha_k-\alpha_k^*)<\phi(x_k),\phi(x)>+b \tag{9}$$

and the value of b can be computed according to the Karush-Kuhn-Tucker (KKT) conditions. Equation (9) is so-called support vector machines regression expansion.

It can be seen clearly from Eq. (9) that we only need the dot product of input data instead of computing the value of ω and $\phi(x)$. We introduce kernel instead of nonlinear mapping, i.e. $K(x,x')=<\phi(x),\phi(x')>$, then Eq. (9) is rewritten as follows:

$$f(x)=\sum_{k=1}^{l}(\alpha_k-\alpha_k^*)K(x_k,x)+b \tag{10}$$

where kernel $K(x,x')$ are arbitrary symmetric functions, which satisfy the Mercer condition[11].

As stated before, the choice of kernel has a great influence on the performance of the SVMs though there are no theoretical guidance how to choose a kernel function. In the following paper, we will analyze the geometrical structure of kernel from viewpoint of information geometry, then propose a method of constructing the kernel in data-dependent way, which can improve the performance of suppose vector regression machines.

3 Construction of Kernel Based on Information Geometry

3.1 Geometrical Structure of Kernel from Information-Geometrical Viewpoint

Let's give the definition of submanifold and embedding before analyzing the geometrical structure.

Definition 1. M is a submanifold of W, if the following conditions (i), (ii) and (iii) hold, where W and M are manifolds, M is a subset of W, and $[\varsigma^i] = [\varsigma^1, \cdots, \varsigma^n], [\zeta^a] = [\zeta^1, \cdots, \zeta^m]$ are coordinate systems for W and M, respectively, $n = \dim W$, $m = \dim M$.

(i) The restriction $\varsigma^i|_M$ of each ς^i is a C^∞ (infinitely many differentiable) function on M.

(ii) Let $B_a^i = (\frac{\partial \varsigma^i}{\partial \zeta^a})_p$ (more precisely, $(\frac{\partial \varsigma^i|_M}{\partial \zeta^a})_p$) and $B_a = [B_a^1, \cdots, B_a^n] \in R^n$.
Then for each point p in M, $\{B_1, \cdots, B_m\}$ are linearly independent (hence $m \leq n$).

(iii) For any open subset V of M, there exists U, an open subset of W, such that $V = M \cap U$.

These conditions are independent of the choice of coordinate systems $[\varsigma^i], [\zeta^a]$. And, conditions (ii) and (iii) mean that the embedding $\iota: M \to W$ defined by $\iota(p) = p, \forall p \in M$, is a C^∞ mapping.

Now let us look back nonlinear map ϕ, which mapping the input space into a high-dimensional feature space (see Fig. 1.). From the viewpoint of information geometry [12], the mapping ϕ defines an embedding of the space S of the input data

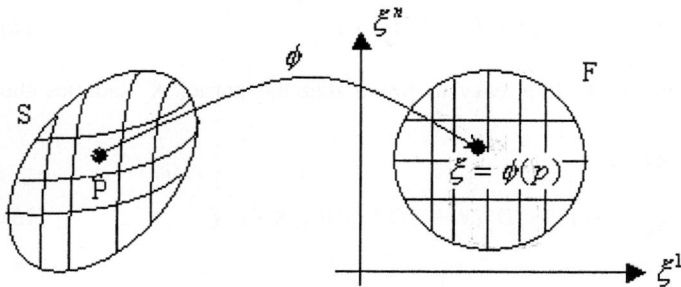

Fig. 1. The nonlinear function ϕ maps the input space into a high-dimensional feature space

into a high-dimensional feature space F as a submanifold. When F is a Euclidean or Hilbert space, a Riemannian metric is thereby induced in the space S.

Denote by g_{ij} the Riemannian metric in space S of input x, we obtain [9][10][12]:

$$g_{ij} = <\frac{\partial}{\partial x_i}\phi(x), \frac{\partial}{\partial x_j}\phi(x)> \qquad (11)$$

and Riemannian distance:

$$ds^2 = g_{ij}dx_i dx_j \qquad (12)$$

Considering $K(x,x')=<\phi(x),\phi(x')>$, Eq. (11) can be rewritten as follows:

$$g_{ij} = \frac{\partial}{\partial x_i}\frac{\partial}{\partial x_j'}K(x,x')\Big|_{x'=x} \qquad (13)$$

Conveniently, the summation sign Σ over indices of i and j is omitted in Eq. (11-13), this notation is known as Einstein's convention.

Let $g(x) = \det|g_{ij}(x)|$, it is clear that $g(x)$ represents how a local area is magnified in F under the mapping ϕ. So we call it the magnification factor. And this is the theoretical basis of our idea constructing the kernel in data-dependent way.

3.2 Constructing the Kernel in Data-Dependent Way

Based on the above analysis, in order to improve the forecasting precision in regression problems, special nonlinear map ϕ (or the related kernel K) can be constructed such that $g_{ij}(x)$ is reduced around the neighboring areas of hyperplane: $|y-f(x)-b|=\varepsilon$, which is contrary to the method of Amari in classification problems. This idea can be implemented by a conformal transformation of the kernel:

$$K^*(x,x') = D(x)D(x')K(x,x') \qquad (14)$$

with a positive function $D(x)$. It is easy to prove that the kernel K^* satisfies the Mercer condition.

From Eq. (13), we obtain

$$g_{ij}^* = D(x)^2 g_{ij}(x) + D_i(x)D_j(x) + 2D_i(x)D(x)K_i(x,x) \qquad (15)$$

where $D_i(x) = \partial D(x)/\partial x_i$ and $K_i(x,x) = \partial K(x,x')/\partial x_i\Big|_{x'=x}$.

However, the positions where the neighboring areas of hyperplane locate are unknown, so we use empirical knowledge that support vectors (SVs) are mostly located around the area and choose $D(x)$ to have the small values at SVs positions.

Taking the above analysis into consideration, we choose $D(x)$ as follows:

$$D(x) = \sum_{i \in SVs} e^{\|x-x_i\|^2/\tau_i^2} \qquad (16)$$

where the parameter τ_i is given by

$$\tau_i = \max_{\alpha} \|x_\alpha - x_i\| \qquad (17)$$

The maximum in Eq. (17) runs over N SVs x_α that are nearest to x_i.

To summarize, the method that training SVMs regression in data-dependent way is: first, choose a kernel, train SVMs and record the information of SVs, then modify kernel according to equations (14), (16) and (17), finally train SVMs with the modified kernel.

4 Simulation Experiment

In this section, some experimental results on the SVMs regression are introduced. In the first experimental, the data set comes from the function: $y = x^2 + noise$, where *noise* comes from a normal distribution with mean zero, variance one and standard deviation one. Fig.2. shows the result of applying the proposed SVMs algorithm for function regression and illustrates the good performance. Here Gaussian RBF is used as the primary kernel function.

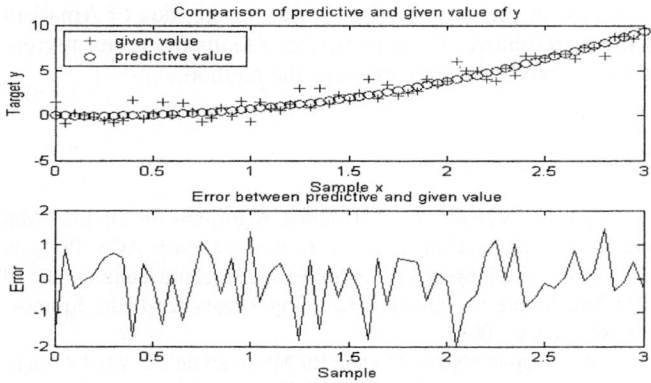

Fig. 2. Function regression using SVMs algorithm

While Fig.3. shows the simulation result of modeling the following data set [13][14]: X=[1.0, 3.0, 4.0, 5.6, 7.8,10.2, 11.0, 11.5, 12.7] and Y=[-1.6, -1.8, -1.0, 1.2, 2.2, 6.8, 10.0, 10.0, 10.0]. The same parameters are used in order to compare the result with [13][14]. The predictive value of Y is got as Y=[-1.4997, -1.6997, -1.1007, 1.3002, 2.0998, 6.9002, 9.6753, 10.0998, 9.9001] with mean squared error 0.02064. It shows the proposed means is effective compared with [13][14].

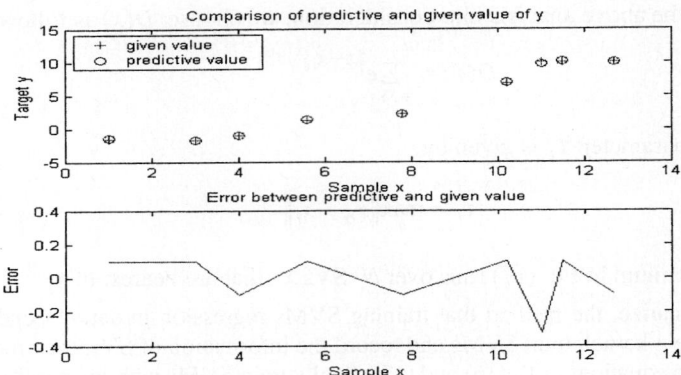

Fig. 3. Predictive result using SVMs regression algorithm

As is indicated in the above two pictures, the proposed method is effective and can improve the performance of the SVMs regression.

5 Conclusions

In the paper we have briefly reviewed the classical formulation of regression problem, and studied a novel approach to constructing kernel function in data-dependent way based on information geometry theory. The kernel function is modified by using the information of SVs, in the way that is contrary to the idea of Amari in classification problems, in order to improve the performance of support vector regression machines, simulation results show the effectiveness of the method.

References

1. Boser, B., Guyon, I., Vapnik, V.: A Training Algorithm for Optimal Margin Classifiers. Fifth Annual Workshop on Computational Learning Theory. ACM Press, New York. 1992
2. Cortes, C., Vapnik, V.: Support Vector Networks. Machine learning. 20 (1995) 273-297
3. Vapnik, V.: The Nature of Statistical Learning Theory. 2nd edn. Springer-Verlag, Berlin Heidelberg New York. 2001
4. Vapnik, V. et al.: Support Vector Method for Multivariate Density Estimation. Advances in Neural Information Processing System. MIT Press, Cambridge, MA. 12 (1999) 659-665
5. Aly, F., Refaat, M.: Regression Using Support Vector Machines: Basic Foundations. Technical Report. Electrical and Computer Engineering Department, University of Louisville. 2004
6. Wang, L.P. (ed.): Support vector machines: Theory and Application. Springer-Verlag, Berlin Heidelberg New York. 2005
7. Kecman, V.: Learning and Soft Computing: Support Vector Machines, Neural Networks, and Fuzzy Logic Models. MIT Press, Cambridge, MA. 2001

8. Chen, S. et al.: Sparse Modeling Using Orthogonal forward Regression with PRESS Statistic and Regularization. IEEE Trans. on Systems, Man and Cybernetics, Part B. 34 (2004) 898-911
9. Amari, S., Wu, S.: Improving Support Vector Machine Classifiers by Modifying Kernel Functions. Neural Networks. 12 (1999) 783-789
10. Wu, S., Amari, S.: Conformal Transformation of Kernel Functions: a Data-dependent Way to Improve Support Vector Machine Classifiers. Neural Processing Letters. 15 (2002) 59-67
11. Colin, C.: Kernel Methods: A Survey of Current Techniques. Neurocomputing. 48 (2002) 63-48
12. Amari, S., Narada, H.: Methods of Information Geometry. American Mathematical Society & Oxford University Press. 2000
13. Wang, D.C.: Support Vector Machines Regression On-line Modelling and Its Application *(In Chinese)*. Control and Decision. 18 (2003) 89-91
14. Kunzhi, H.: A Real-time Model for Forecasting Zinc Output by Support Vector Machines in Imperial Smelting Furnace *(In Chinese)*. Computer Engineering. 30 (2004) 16-18

Training Data Selection for Support Vector Machines

Jigang Wang, Predrag Neskovic, and Leon N. Cooper*

Institute for Brain and Neural Systems,
Physics Department, Brown University, Providence RI 02912, USA
{jigang, pedja, Leon_Cooper}@brown.edu
http://physics.brown.edu/physics/researchpages/Ibns/index.html

Abstract. In recent years, support vector machines (SVMs) have become a popular tool for pattern recognition and machine learning. Training a SVM involves solving a constrained quadratic programming problem, which requires large memory and enormous amounts of training time for large-scale problems. In contrast, the SVM decision function is fully determined by a small subset of the training data, called support vectors. Therefore, it is desirable to remove from the training set the data that is irrelevant to the final decision function. In this paper we propose two new methods that select a subset of data for SVM training. Using real-world datasets, we compare the effectiveness of the proposed data selection strategies in terms of their ability to reduce the training set size while maintaining the generalization performance of the resulting SVM classifiers. Our experimental results show that a significant amount of training data can be removed by our proposed methods without degrading the performance of the resulting SVM classifiers.

1 Introduction

Support vector machines (SVMs), introduced by Vapnik and coworkers in the structural risk minimization (SRM) framework [1,2,3], have gained wide acceptance due to their solid statistical foundation and good generalization performance that has been demonstrated in a wide range of applications.

Training a SVM involves solving a constrained quadratic programming (QP) problem, which requires large memory and takes enormous amounts of training time for large-scale applications [4]. On the other hand, the SVM decision function depends only on a small subset of the training data, called support vectors. Therefore, if one knows in advance which patterns correspond to the support vectors, the same solution can be obtained by solving a much smaller QP problem that involves only the support vectors. The problem is then how to select training examples that are likely to be support vectors. Recently, there has been considerable research on data selection for SVM training. For example,

---

* This work is partially supported by ARO under grant W911NF-04-1-0357. Jigang Wang is supported by a dissertation fellowship from Brown University.

Shin and Cho proposed a method that selects patterns near the decision boundary based on the neighborhood properties [5]. In [6,7,8], k-means clustering is employed to select patterns from the training set. In [9], Zhang and King proposed a β-skeleton algorithm to identify support vectors. In [10], Abe and Inoue used Mahalanobis distance to estimate boundary points. In the reduced SVM (RSVM) setting, Lee and Mangasarian chose a subset of training examples using random sampling [11]. In [12], it was shown that uniform random sampling is the optimal robust selection scheme in terms of several statistical criteria.

In this paper, we introduce two new data selection methods for SVM training. The first method selects training data based on a statistical confidence measure that we will describe later. The second method uses the minimal distance from a training example to the training examples of a different class as a criterion to select patterns near the decision boundary. This method is motivated by the geometrical interpretation of SVMs based on the (reduced) convex hulls. To understand how effective these strategies are in terms of their ability to reduce the training set size while maintaining the generalization performance, we compare the results obtained by the SVM classifiers trained with data selected by these two new methods, by random sampling, and by the data selection method that is based on the distance from a training example to the desired optimal separating hyperplane. Our comparative study shows that a significant amount of training data can be removed from the training set by our methods without degrading the performance of the resulting SVM classifier. We also find that, despite its simplicity, random sampling performs well and often provides results comparable to those obtained by the method based on the desired SVM outputs. Furthermore, in our experiments, we find that incorporating the class distribution information in the training set often improves the efficiency of the data selection methods.

The remainder of the paper is organized as follows. In section 2, we give a brief overview of support vector machines for classification and the corresponding training problem. In section 3, we present the two new methods that select subsets of training examples for training SVMs. In section 4 we report the experimental results on several real-world datasets. Concluding remarks are provided in section 5.

2 Related Background

Given a set of training data $\{(x_1, y_1), \ldots, (x_n, y_n)\}$, where $x_i \in \mathbb{R}^d$ and $y_i \in \{-1, 1\}$, support vector machines seek to construct an optimal separating hyperplane by solving the following quadratic optimization problem:

$$\min_{w,b} \frac{1}{2} \langle w, w \rangle + C \sum_{i=1}^{n} \xi_n \qquad (1)$$

subject to the constraints:

$$y_i(\langle w, x_i \rangle + b) \geq 1 - \xi_i \quad \forall i = 1, \ldots, n \;, \qquad (2)$$

where $\xi_i \geq 0$ for $i = 1, \ldots, n$ are slack variables introduced to handle the non-separable case [2]. The constant $C > 0$ is a parameter that controls the trade-off between the separation margin and the number of training errors. Using the Lagrange multiplier method, one can easily obtain the following Wolfe dual form of the primal quadratic programming problem:

$$\min_{\alpha_i, i=1,\ldots,n} \frac{1}{2} \sum_{i,j=1}^{n} \alpha_i \alpha_j y_i y_j \langle x_i, x_j \rangle - \sum_{i=1}^{n} \alpha_i \tag{3}$$

subject to

$$0 \leq \alpha_i \leq C \quad \forall i = 1, \ldots, n \quad \text{and} \quad \sum_{i=1}^{n} \alpha_i y_i = 0 \ . \tag{4}$$

Solving the dual problem, one obtains the multipliers $\alpha_i, i = 1, \ldots, n$, which give w as an expansion

$$w = \sum_{i=1}^{n} \alpha_i y_i x_i \ . \tag{5}$$

According to the Karush-Kuhn-Tucker (KKT) optimality conditions, we have

$$\alpha_i = 0 \Rightarrow y_i(\langle w, x_i \rangle + b) \geq 1 \quad \text{and} \quad \xi_i = 0$$
$$0 < \alpha_i < C \Rightarrow y_i(\langle w, x_i \rangle + b) = 1 \quad \text{and} \quad \xi_i = 0$$
$$\alpha_i = C \Rightarrow y_i(\langle w, x_i \rangle + b) \leq 1 \quad \text{and} \quad \xi_i \geq 0 \ .$$

Therefore, only α_i that correspond to training examples x_i which lie either on the margin or inside the margin area are non-zero. All the remaining α_i are zero and the corresponding training examples are irrelevant to the final solution.

Knowing the normal vector w, the bias term b can be determined from the KKT conditions $y_i(\langle w, x_i \rangle + b) = 1$ for $0 < \alpha_i < C$. This subsequently leads to the linear decision function $f(x) = \text{sgn}(\sum_{i=1}^{n} \alpha_i y_i \langle x, x_i \rangle + b)$.

In practice, linear decision functions are generally not rich enough for pattern separation. To allow for more general decision surfaces, one can apply the kernel trick by replacing the inner products $\langle x_i, x_j \rangle$ in the dual problem with suitable kernel functions $k(x_i, x_j)$. Effectively, support vector machines implicitly map training vectors x_i in $\mathbb{R}^d$ to feature vectors $\Phi(x_i)$ in some high dimensional feature space $\mathbb{F}$ such that inner products in $\mathbb{F}$ are defined as $\langle \Phi(x_i), \Phi(x_j) \rangle = k(x_i, x_j)$. Consequently, the optimal hyperplane in the feature space $\mathbb{F}$ represents a nonlinear decision functions of the form

$$f(x) = \text{sgn}(\sum_{i=1}^{n} \alpha_i y_i k(x, x_i) + b) \ . \tag{6}$$

To train a SVM classifier, one therefore needs to solve the dual quadratic programming problem (3) under the constraints (4). For a small training set, standard QP solvers, such as CPLEX, LOQO, MINOS and Matlab QP routines, can be readily used to obtain the solution. However, for a large training set, they quickly become intractable because of the large memory requirements and the enormous amounts of training time involved. To alleviate the problem, a number of solutions have been proposed by exploiting the sparsity of the SVM solution and the KKT conditions.

The first such solution, known as chunking [13], uses the fact that only the support vectors are relevant for the final solution. At each step, chunking solves a QP problem that consists of all non-zero Lagrange multipliers α_i from the last step and some of the α_i that violate the KKT conditions. The size of the QP problem varies but finally equals the number of non-zero Lagrange multipliers. At the last step, the entire set of non-zero Lagrange multipliers are identified and the QP problem is solved. Another solution, proposed in [14], solves the large QP problem by breaking it down into a series of smaller QP sub-problems. This decomposition method is justified by the observation that solving a sequence of QP sub-problems that always contain at least one training example that violates the KKT conditions will eventually lead to the optimal solution. Recently, a method called sequential minimal optimization (SMO) was proposed by Platt [15], which approaches the problem by iteratively solving a QP sub-problem of size 2. The key idea is that a QP sub-problem of size 2 can be solved analytically without invoking a quadratic optimizer. This method has been reported to be several orders of magnitude faster than the classical chunking algorithm.

All the above training methods make use of the whole training set. However, according to the KKT optimality conditions, the final separating hyperplane is fully determined by the support vectors. In many real-world applications, the number of support vectors is expected to be much smaller than the total number of training examples. Therefore, the speed of SVM training will be significantly improved if only the set of support vectors is used for training, and the solution will be exactly the same as if the whole training set was used.

In theory, one has to solve the full QP problem in order to identify the support vectors. However, it is easy to see that the support vectors are training examples that are close to decision boundaries. Therefore, if there exists a computationally efficient way to find a small set of training data such that with high probability it contains the desired support vectors, the speed of SVM training will be improved without degrading the generalization performance. The size of the reduced training set can still be larger than the set of desired support vectors. However, as long as its size is much smaller than the size of the total training set, the SVM training speed will be significantly improved because most SVM training algorithms scales quadratically on many problems [4]. In the next section, we propose two new data selection strategies to explore the possibility.

3 Training Data Selection for Support Vector Machines

3.1 Data Selection Based on Confidence Measure

A good heuristic for identifying boundary points is the number of training examples that are contained in the largest sphere centered at a training example without covering an example of a different class.

Centered at each training example x_i, let us draw a sphere that is as large as possible without covering a training example of a different class and count the number of training examples that fall inside the sphere. We denote this number by $N(x_i)$. Obviously, the larger the number $N(x_i)$, the more training examples (of the same class as x_i) will be scattered around x_i, the less likely x_i will be close to the decision boundary, and the less likely x_i will be a support vector. Hence, this number can be used as a criterion to decide which training examples should belong to the reduced training set. For each training example x_i, we compute the number $N(x_i)$ and sort the training data according to the corresponding value of $N(x_i)$ and choose a subset of data with the smallest numbers $N(x_i)$ as the reduced training set. It can be shown that $N(x_i)$ is related to the statistical confidence that can be associated with the class label y_i of the training example x_i. For this reason, we call this data selection scheme the confidence measure-based training set selection.

3.2 Data Selection Based on Hausdorff Distance

Our second data selection strategy is based on the Hausdorff distance. In the separable case, it has been shown that the optimal SVM separating hyperplane is identical to the hyperplane that bisects the line segment which connects the two closest points of the convex hulls of the positive and of the negative training examples [16,17]. The problem of finding the two closest points in the convex hulls can be formulated as

$$\min_{z^+, z^-} \|z^+ - z^-\|^2 \quad (7)$$

subject to

$$z^+ = \sum_{i:y_i=1} \alpha_i x_i \quad \text{and} \quad z^- = \sum_{i:y_i=-1} \alpha_i x_i, \quad (8)$$

where $\alpha_i \geq 0$ satisfies the constraints $\sum_{i:y_i=1} \alpha_i = 1$ and $\sum_{i:y_i=-1} \alpha_i = 1$.

Based on this geometrical interpretation, the support vectors are the training examples that are vertices of the convex hulls that are closest to the convex hull of the training examples from the opposite class. For the non-separable case, a similar result holds by replacing the convex hulls with the reduced convex hulls [16,17]. Therefore, a good heuristic that can be used to determine whether a training example is likely to be a support vector is the distance to the convex hull of the training examples of the opposite class. Computing the distance from a training example x_i to the convex hull of the training examples of the opposite class involves solving a smaller quadratic programming problem. To simplify the computation, the distance from a training example to the closest training examples of the opposite class can be used as an approximation. We denote the minimal distance as

$$d(x_i) = \min_{j:y_j \neq y_i} \|x_i - x_j\|, \quad (9)$$

which is also the Hausdorff distance between the training example x_i and the set of training examples that belong to a different class. To select a subset of training examples, we sort the training set according to $d(x_i)$ and select examples with the smallest Hausdorff distances $d(x_i)$ as the reduced training set. This method will be referred to as the Hausdorff distance-based selection method.

3.3 Data Selection Based on Random Sampling and Desired SVM Outputs

To study the effectiveness of the proposed data selection strategies, we compare them to two other strategies. One is random sampling and the other is a data selection strategy based on the distance from the training examples to the desired separating hyperplane.

The random sampling strategy simply selects a small portion of the training data to form the reduced training set uniformly at random. This method is straightforward to implement and requires no extra computation. The other data selection strategy we compare our methods to is implemented as follows. Given the training set and the parameter setting, we solve the full QP problem to obtained the desired separating hyperplane. Then for each training example x_i, we compute its distance to the desired separating hyperplane as:

$$f(x_i) = y_i(\sum_{j=1}^{n} \alpha_j y_j k(x_i, x_j) + b) \ . \tag{10}$$

Note that Eq. (10) has taken into account the class information and training examples that are misclassified by the desired separating hyperplane will have negative distances. According to the KKT conditions, support vectors are training examples that have relatively small values of distance $f(x_i)$. We sort the training examples according to their distances to the separating hyperplane and select a subset of training examples with the smallest distances as the reduced training set. This strategy, although impractical because one needs to solve the full QP problem first, is ideal for comparison purposes as the distance from a training example to the desired separating hyperplane provides the optimal criterion for selecting the support vectors.

4 Results and Discussion

In this section we report experimental results on several real-world datasets from the UCI Machine Learning repository [18]. The SVM training algorithm was implemented based on the SMO method. For all datasets, Gaussian kernels were used and the generalization error of the SVMs was estimated using the 5-fold cross-validation method. For each training set, according to the data selection method used, a portion of the training set (ranging from 10 to 100 percent) was selected as the reduced training set to train the SVM classifier. The error rate reported is the average error rate of the resulting SVM classifiers on the test sets over the 5 iterations. Due to the space limit, only results on three datasets will be presented.

Note that when the data selection method is based on the desired SVM outputs, the SVM training procedure has to be run twice in each iteration. The

Table 1. Error rates of SVMs on the Breast Cancer dataset when trained with reduced training sets of various sizes

Percent	Confidence	Hausdorff	Random	SVM
10	34.26	5.44	5.44	33.38
20	4.12	7.65	5.15	4.56
30	3.53	5.59	4.71	3.97
40	3.82	5.44	5.00	3.68
50	3.82	5.44	5.00	3.82
60	3.97	5.15	4.41	3.97
70	3.97	4.85	4.12	3.97
80	4.12	4.85	4.26	3.97
90	3.82	4.56	4.41	3.82
100	3.82	3.82	3.82	3.82

first time a SVM classifier is trained with the training set to obtain the desired separating hyperplane. Then a portion of the training examples in the training set is selected to form the reduced training set based on their distances to the desired separating hyperplane (see Eq. (10)). The second time a SVM classifier is trained with the reduced training set.

Given a training set and a particular data selection criterion, there are two ways to form the reduced training set. One can either select training examples regardless of which classes they belong to or select training examples from each class separately while maintaining the class distribution. It was found in our experiments that selecting training examples from each class separately often improves the classification accuracy of the resulting SVM classifiers. Therefore, we only report results in this case.

Table 1 shows the error rates of SVMs on the Wisconsin Breast Cancer dataset when trained with the reduced training sets of various sizes selected by the four different data selection methods. This dataset consists of 683 examples from two classes (excluding the 16 examples with missing attribute values). Each example has 8 attributes. The size of the training set in each iteration is 547 and the size of the test set is 136. The average number of support vectors is 238.6, which is 43.62% of the training set size.

From Table 1 one can see that a significant amount data can be removed from the training set without degrading the performance of the resulting SVM classifier. When more than 10% of the training data is selected, the confidence-based data selection method outperforms the other two methods. Its performance is actually as good as that of the method based on the desired SVM outputs. The method based on the Hausdorff distance gives the worst results. When the data reduction rate is high, e.g., when less than 10 percent of the training data is selected, the results obtained by the Hausdorff distance-based method and random sampling are much better than those based on the confidence measure and the desired SVM outputs.

Table 2 shows the corresponding results obtained on the BUPA Liver dataset, which consists of 345 examples, with each example having 6 attributes. The sizes

Table 2. Results on the BUPA Liver dataset

Percent	Confidence	Hausdorff	Random	SVM
10	42.90	39.71	39.13	63.19
20	44.06	38.55	33.33	62.90
30	41.16	33.62	33.33	51.01
40	40.00	33.62	30.43	45.80
50	40.00	33.62	31.30	42.61
60	35.94	32.75	32.75	42.32
70	33.91	33.33	32.17	37.68
80	31.01	31.88	32.46	32.46
90	31.59	30.72	33.04	31.30
100	31.30	31.30	31.30	31.30

Table 3. Results on the Ionosphere dataset

Percent	Confidence	Hausdorff	Random	SVM
10	26.29	35.71	16.29	33.14
20	21.43	25.71	11.14	22.57
30	18.57	24.00	8.57	6.86
40	11.43	24.00	8.00	6.00
50	7.43	21.43	7.14	5.71
60	6.00	18.86	7.14	5.71
70	5.71	16.00	6.57	6.00
80	5.14	10.29	6.00	6.00
90	6.00	6.57	6.00	5.71
100	5.71	5.71	5.71	5.71

of the training and test sets in each iteration are 276 and 69, respectively. The average number of support vectors is 222.2, which is 80.51% of the size of the training sets. Interestingly, as we can see, the method based on the desired SVM outputs has the worst overall results. When less than 80% of the data is selected for training, the Hausdorff distance-based method and random sampling have similar performance and outperform the methods based on the confidence measure and the desired SVM outputs.

Table 3 provides the results on the Ionosphere dataset, which has a total of 351 examples, with each example having 34 attributes. The sizes of the training and test sets in each iteration are 281 and 70, respectively. The average number of support vectors is 159.8, which is 56.87% of the size of the training sets. From Table 3 we see that the data selection method based on the desired SVM outputs gives the best results when more than 20% of the data is selected. When more than 50% of the data is selected, the results of the confidence-based method are very close to the best achievable results. However, when the reduction rate is high, the performance of random sampling is the best. The Hausdorff distance-based method has the worst overall results.

An interesting finding of the experiments is that the performance of the SVM classifiers deteriorates significantly when the reduction rate is high, e.g., when the size of the reduced training set is much smaller than the number of the desired support vectors. This is especially true for data selection strategies that are based on the desired SVM outputs and the proposed heuristics. On the other hand, the effect is less significant for random sampling, as we have seen that random sampling usually has better relative performance at higher data reduction rates. From a theoretical point of view, this is not surprising because when only a subset of the support vectors is chosen as the reduced training set, there is no guarantee that the solution of the reduced QP problem will still be the same. In fact, if the reduction rate is high and the criterion is based on the desired SVM outputs or the proposed heuristics, the reduced training set is likely to be dominated by 'outliers', therefore leading to worse classification performance. To overcome this problem, we can remove those training examples that lie far inside the margin area since they are likely to be 'outliers'. For the data selection strategy based on the desired SVM outputs, it means that we can discard part of the training data that has extremely small values of the distance to the desired separating hyperplane (see Eq. (10)). For the methods based on the confidence measure and Hausdorff distance, we can similarly discard the part of the training data that has extremely small values of $N(x_i)$ and the Hausdorff distance.

Table 4. Results on the Breast Cancer dataset

Percent	Confidence	Hausdorff	Random	SVM
10	5.74	7.94	5.88	4.56
20	4.26	5.59	4.71	4.71
30	4.12	5.44	4.71	4.71
40	4.12	5.15	4.85	4.56
50	4.26	5.74	5.15	4.26
60	4.12	5.15	4.56	4.41
70	3.97	5.29	4.26	4.26
80	3.82	5.29	4.41	4.26
90	3.82	4.71	4.41	4.12
100	3.82	3.82	3.82	3.82

In Table 4 we show the results of the proposed solution on the Breast Cancer dataset. Comparing Tables 1 and 4, it is easy to see that, when only a very small subset of the training data (compared to the number of the desired support vectors) is selected for SVM training, removing training patterns that are extremely close to the decision boundary according to the confidence measure or according to the underlying SVM outputs significantly improves the performance of the resulting SVM classifiers. The effect is less obvious for the methods based on the Hausdorff measure and random sampling. Similar results have also been observed on other datasets but will not be reported here due to the space limit.

5 Conclusion

In this paper we presented two new data selection methods for SVM training. To analyze their effectiveness in terms of their ability to reduce the training data while maintaining the generalization performance of the resulting SVM classifiers, we conducted a comparative study using several real-world datasets. More specifically, we compared the results obtained by these two new methods with the results of the simple random sampling scheme and the results obtained by the selection method based on the desired SVM outputs. Through our experiments, several important observations have been made: (1) In many applications, significant data reduction can be achieved without degrading the performance of the SVM classifiers. For that purpose, the performance of the confidence measure-based selection method is often comparable to or better than the performance of the method based on the desired SVM outputs. (2) When the reduction rate is high, some of training examples that are 'extremely' close to the decision boundary have to be removed in order to maintain the generalization performance of the resulting SVM classifiers. (3) In spite of its simplicity, random sampling performs consistently well, especially when the reduction rate is high. However, at low reduction rates, random sampling performs noticeably worse compared to the confidence measure-based method. (4) When conducting training data selection, sampling training data from each class separately according to the class distribution often improves the performance of the resulting SVM classifiers.

By directly comparing various data selection schemes with the scheme based on the desired SVM outputs, we are able to conclude that the confidence measure provides a criterion for training data selection that is almost as good as the optimal criterion based on the desired SVM outputs. At high reduction rates, by removing training data that are likely to be outliers, we boost the performance of the resulting SVM classifiers. Random sampling performs consistently well in our experiments, which is consistent with the results obtained by Syed et al. in [19] and the theoretical analysis of Huang and Lee in [12]. The robustness of random sampling at high reduction rates suggests that, although an SVM classifier is fully determined by the support vectors, the generalization performance of an SVM is less reliant on the choice of training data than it appears to be.

References

1. Boser, B. E., Guyon, I. M., Vapnik, V. N.: A training algorithm for optimal margin classifiers. In: Haussler, D. (ed.): Proceedings of the 5th Annual ACM Workshop on Computational Learning Theory (1992) 144–152
2. Cortes, C., Vapnik, V. N.: Support vector networks. Machine Learning. **20** (1995) 273–297
3. Vapnik, V. N.: Statistical Learning Theory. Wiley, New York, NY (1998)
4. Joachims, T.: Making large-scale SVM learning practical. In: Schölkopf, B., Burges, C. J. C., Smola, A. J. (eds.): Advances in Kernel Methods - Support Vector Learning. MIT Press, Cambridge, MA (1999) 169–184

5. Shin, H. J., Cho, S. Z.: Fast pattern selection for support vector classifiers. In: Proceedings of the 7th Pacific-Asia Conference on Knowledge Discovery and Data Mining. Lecture Notes in Artificial Intelligence (LNAI 2637) (2003) 376–387
6. Almeida, M. B., Braga, A. P., Braga, J. P.: SVM-KM: speeding SVMs learning with a priori cluster selection and k-means. In: Proceedings of the 6th Brazilian Symposium on Neural Networks (2000) 162–167
7. Zheng, S. F., Lu, X. F., Zheng, N. N., Xu, W. P.: Unsupervised clustering based reduced support vector machines. In: Proceedings of IEEE International Conference on Acoustics, Speech, and Signal Processing (ICASSP) **2** (2003) 821–824
8. Koggalage, R., Halgamuge, S.: Reducing the number of training samples for fast support vector machine classification. Neural Information Processing - Letters and Reviews **2(3)** (2004) 57–65
9. Zhang, W., King, I.: Locating support vectors via β-skeleton technique. In: Proceedings of the International Conference on Neural Information Processing (ICONIP) (2002) 1423–1427
10. Abe, S., Inoue, T.: Fast training of support vector machines by extracting boundary data. In: Proceedings of the International Conference on Artificial Neural Networks (ICANN) (2001) 308–313
11. Lee, Y. J., Mangasarian, O. L.: RSVM: Reduced support vector machines. In: Proceedings of the First SIAM International Conference on Data Mining (2001)
12. Huang, S. Y., Lee, Y. J.: Reduced support vector machines: a statistical theory. Technical report, Institute of Statistical Science, Academia Sinica, Taiwan. http://www.stat.sinica.edu.tw/syhuang/ (2004)
13. Vapnik, V. N.: Estimation of Dependence Based on Empirical Data. Springer-Verlag, Berlin (1982)
14. Osuna, E., Freund, R., Girosi, R.: Support vector machines: training and applications. A.I. Memo AIM - 1602, MIT A.I. Lab. (1996)
15. Platt, J.: Fast training of support vector machines using sequential minimal optimization. In: Schölkopf, B., Burges, C. J. C., Smola, A. J. (eds.): Advances in Kernel Methods - Support Vector Learning. MIT Press, Cambridge, MA (1999) 185–208
16. Bennett, K. P., Bredensteiner, E. J.: Duality and geometry in SVM classifiers. In: Proceedings of 17th International Conference on Machine Learning. (2000) 57–64
17. Crisp, D. J., Burges, C. J. C.: A geometric interpretation of nu-svm classifiers. Advances in Neural Information Processing Systems. **12** (1999)
18. Blake, C. L., Merz, C. J.: UCI Repository of machine learning databases. http://www.ics.uci.edu/~mlearn/MLRepository.html (1998)
19. Syed, N. A., Liu, H., Sung, K. K.: A study of support vectors on model independent example selection. In: Proceedings of the Workshop on Support Vector Machines at the International Joint Conference on Artificial Intelligence. (1999)

Model Selection for Regularized Least-Squares Classification

Hui-Hua Yang[1,2], Xing-Yu Wang[2,*], Yong Wang[1,2], and Hai-Hua Gao[2]

[1] Department of Computer Science, Guilin University of Electronic Technology, Guilin 541004, China
yang98@gliet.edu.cn, wang@gliet.edu.cn
[2] School of Information Science and Engineering, East China University of Science and Technology, Shanghai 200237, China
xywang@ecust.edu.cn, hhgao@tom.com
+86-21-64253386

Abstract. Regularized Least-Squares Classification (RLSC) can be regarded as a kind of 2 layers neural network using regularized square loss function and kernel trick. Poggio and Smale recently reformulated it in the framework of the mathematical foundations of learning and called it a key algorithm of learning theory. The generalization performance of RLSC depends heavily on the setting of its kernel and hyper parameters. Therefore we presented a novel two-step approach for optimal parameters selection: firstly the optimal kernel parameters are selected by maximizing kernel target alignment, and then the optimal hyper-parameter is determined via minimizing RLSC's leave-one-out bound. Compared with traditional grid search, our method needs no independent validation set. We worked on IDA's benchmark datasets using Gaussian kernel, the results demonstrate that our method is feasible and time efficient.

1 Introduction

It is until recently that Poggio and Smale pointed out that " 'classical' square loss regularization network works also very well for binary classification [1]", and they described a key algorithm(KA) of learning theory in [1]. KA is based on the mathematical foundations of learning [2], which differs greatly from the formulation of support vector machines [3]. KA is originated from Regularized Least-Squares Classification (RLSC, we confuse KA and RLSC in this paper, and use RLSC to stress it is a classification problem) presented by Rifkin and Poggio in [4] and [5], and has the same computational formula with Kernel Ridge Regression and Gaussian Process though from different motivations, and is akin to LS-SVM [6] and PSVM [7] since they all use square loss and solve linear equation. However, RLSC's generalization performance depends heavily on the setting of its kernel and hyper parameters, which is usually tuned by time-consuming grid search on an n-fold cross-validation.

---

* Corresponding Author. This work was supported by Doctorate Foundation of the Education Ministry of China (No.20040251010), National Key Basic Research and Development Program (No. 2002CB312200), and National Science Foundation (No. 69974014), China.

Enlightened by Chapelle and Vapnik et al.'s parameter selection method [8] for SVMs and Cristianini et al.'s idea of kernel target alignment [9], we present in this paper a novel two-step approach for optimal parameters selection for RLSC: firstly the optimal kernel parameters are selected by maximizing kernel target alignment, and then the optimal hyper parameter is determined via minimizing RLSC's leave-one-out bound.

We begin by briefly introducing RLSC to set a base for further discussion, then outline the kernel target alignment method for kernel parameters selection, followed by the leave-one-out (loo) bound of RLSC with application to select hyper-parameter. Next, we formulate our RLSC-AlignLoo model selection method, and apply it to 13 IDA benchmark datasets, and ended with a conclusion.

2 Regularized Least-Squares Classification

Given a training set $(x_i, y_i)_{i=1}^m$ with $x_i \in X, y_i \in Y$, and X is a closed subset of R^n, $Y \subset R$, a key algorithm described in [1] takes the following steps to find a predictive function $f: X \to Y$.

1. Choose a symmetric, positive definite function $K_x(x') = K(x, x')$, which is continuous on $X \times X$. For example, a Gaussian kernel is $K(x, x') = \exp(-\|x - x'\|^2 / (2\sigma^2))$.
2. Let $f: X \to Y$ to be

$$f(x) = \sum_{i=1}^{m} c_i K_x(x_i) \qquad (1)$$

where $c = (c_1, ..., c_m)^T$ (superscript ^{T} denotes transpose), and

$$(m\gamma I + K)c = y \qquad (2)$$

where I is identity matrix, K is an $m \times m$ positive definite matrix with elements $K_{ij} = K(x_i, x_j)$, and y is the vector $(y_1, ..., y_m)^T$, hyper-parameter γ is a positive, real number.

The derivation of KA or RLSC is to minimize the following regularized risk functional,

$$\min_{f \in H_K} R_{reg}(f) = \frac{1}{m} \sum_{i=1}^{m} (y_i - f(x_i))^2 + \gamma \|f\|_K^2 \qquad (3)$$

where $\|f\|_K$ is the norm of H_K - the RKHS defined by the positive kernel K.

When considering classification, that's RLSC, we take $y_i \in \{1, -1\}$ and get the decision function by imposing a sign function to (**1**).

The details of KA or RLSC's formal formulation and theories can be referred from [1, 2, 4, 5].

3 Kernel Target Alignment for Kernel Parameters Selection

The basic idea of kernel target alignment is that the universal kernel gives trivial results in learning task, and the ideal kernel is specialized, or rather, aligns the target labels well [9]. The quantity *alignment* is used to measure the degree of match between two kernels or a kernel and a target.

Definition *Alignment* The alignment of a kernel k_1 and a kernel k_2 is the quantity

$$A(k_1,k_2) = \frac{<k_1,k_2>_P}{\sqrt{<k_1,k_1>_P<k_2,k_2>_P}} \quad (4)$$

where P is the distribution generating the data.

The value of the alignment is in all real problems impossible to estimate, however, it can be empirical estimated once given a sample, $S_m = (x_i, y_i)_{i=1}^m$. We will use the following definition of inner product between gram matrices:

$$<K_1,K_2>_F = \sum_{i,j=1}^m K_1(x_i,x_j)K_2(x_i,x_j) \quad (5)$$

corresponding to the Frobenius inner product.

Definition *Emperical Alignment* The empirical alignment of a kernel k_1 with a kernel k_2 is the quantity

$$\hat{A}(k_1,k_2) = \frac{<K_1,K_2>_F}{\sqrt{<K_1,K_1>_F<K_2,K_2>_F}} \quad (6)$$

As a special case, we consider $K_2 = yy^T$, where y is the vector of labels for the sample, then

$$\hat{A}(K,yy^T) = \frac{<K,yy^T>}{\sqrt{<K,K><yy^T,yy^T>}} \quad (7)$$

A crucial property of the alignment for practical application is that it can be reliably estimated from its empirical estimate $\hat{A}(S)$.

Theorem The sample based estimate of the alignment is concentrated around its expected value. For a kernel with feature vector of norm 1, we have that
$P^m(S : |A(S) - A(y)| \geq \bar{\varepsilon}) \leq \delta$, where $\bar{\varepsilon} = C(S)\sqrt{8\ln(2/\delta)/m}$,
for a non-trivial function $C(S)$ and value $A(y)$.

The alignment has been used for transduction based learning and kernel selection in [9], and we will explore its capability for selecting kernel parameters by maximizing the alignment between the kernel and the labels of y.

4 RLSC's Loo Bound for Hyper-Parameter Selection

Loo bound allows us to estimate the generalization error of a learning machine when there are no independent validation sets. Jaakkola and Haussler introduced an interesting class of simple loo bound [10] for kernel classifiers. Rifkin [4] proved that their bound is valid for RLSC, and that the number of loo errors is bounded by

$$| x_i : y_i \sum_{i \neq j} c_j K_{ij} \leq 0 | \tag{8}$$

This bound can be computed directly given c_i. However, a simple geometric condition underling RLSC leads to a more elegant bound. Here, we present a simpler deduction for the loo bound than the deduction given in [5]. Combining (1) and (2), we get $c_i = (y_i - f(x_i))/m\gamma$. Using this, and considering $y_i y_i = 1$, $m\gamma > 0$, we can eliminate c_i from the bound, the number of loo errors for RLSC is then bounded by

$$\begin{aligned}
& | x_i : y_i (f(x_i) - (\frac{y_i - f(x_i)}{m\gamma}) K_{ii}) \leq 0 | \\
= & | x_i : y_i ((1 + \frac{K_{ii}}{m\gamma}) f(x_i) - \frac{y_i}{m\gamma} K_{ii}) \leq 0 | \\
= & | x_i : y_i (m\gamma + K_{ii}) f(x_i) - K_{ii} \leq 0 | \\
= & | x_i : y_i f(x_i) \leq \frac{K_{ii}}{(m\gamma + K_{ii})} |
\end{aligned} \tag{9}$$

Note that once we have set the optimal kernel parameters, we can use this bound to tune hyper-parameter γ.

5 RLSC-AlignLoo Model Selection Approach

RLSC-AlignLoo Model Selection Approach: Combining the kernel target alignment method and RLSC's loo bound, we put forward our method as follows:

1. Preprocess input data $S = \{x_i, y_i\}_{i=1}^m$, let y_i be 1 or -1, and $X = \{x_i\}$ zero mean and standard derivation 1.
2. Select the optimal kernel parameters which maximize the alignment defined in (7). As an example, for Gaussian kernel, we first compute the corresponding Gram matrix K for each σ^2 in its 1-dimensional search list, and calculate the alignment $A(K, yy^T)$ by (7), then we get the optimal σ^2 which maximizes the alignment A.
3. Select the optimal γ which minimizes the loo error. First calculate the Gram matrix K using the optimal σ^2 got in step 2, and then for each available γ, train the model by (2) and get $f(x)$ by (1), then we get the loo bound for this γ by (9). The optimal γ is the one minimizing the loo bound.

4. Retrain RLSC using the optimal σ^2 and γ, get $c = (K + m\gamma I)^{-1} y$.

5. Test on new input data $x \in X_t$, using $f(x) = sign(\sum_{i=1}^{m} c_i K(x, x_i))$.

Implementation Issues: we find that empirical kernel target alignment can be efficient computed,

$$< K, yy^T > = trace(K^T yy^T) = trace((Ky)y^T) = < y, Ky > = y^T Ky \qquad (10)$$

$$< yy^T, yy^T > = trace((yy^T)^T yy^T) = trace(y(y^T y)y^T) = m \cdot trace(yy^T) = m^2 \qquad (11)$$

Note that for Gaussian kernel, $K(x_i, x_i)$ is always 1, so the loo bound becomes simply $|x_i : (1 + m\gamma) y_i f(x_i) \le 1|$.

6 Experiments

We work our approach on the IDA dataset [11], introduced by Rätsch et al., which is a suit of two-class recognition problems with pre-divided partition of training and testing sets with emphasis on evaluating the model performance by eliminating the discrepancy generated from data division. We use the parameters grid, $\ln \sigma^2 = \{-10:1:10\}$ and $\ln \gamma = \{-15:1:5\}$, for both RLSC grid search and AlignLoo model selection. Like [11], we use the first 5 training/testing datasets for parameters tuning for RLSC's grid search. However, there is no need for us to do so for RLSC-AlignLoo because we can determine the optimal parameters based only on training data. As a result, we select an optimal parameter pair for each training/testing division in the AlignLoo approach.

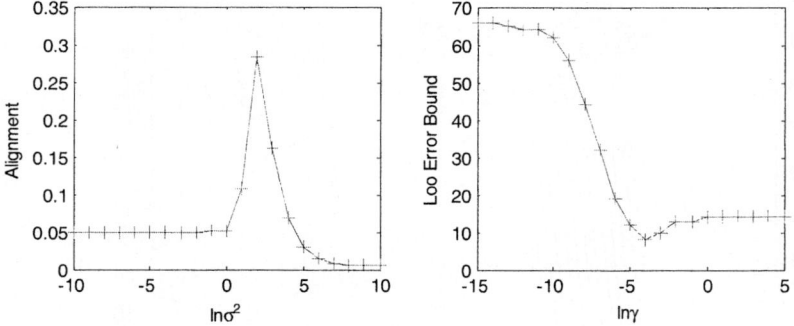

Fig. 1. Left: The alignment of a Gaussian kernel to the target labels of dataset *Twonorm* varies against the choose of different kernel width σ^2. Empirical observations show that the curve is unimodal and peaks at the optimal σ^2. Right: The Loo bound for RLSC built on dataset *Twonorm* changes along with its selection of hyper-parameter γ. Empirical observations demonstrate the curve usually contains a flat segment sharing the same minimal Loo bound, which depicts the range of optimal γ, and we take the middle one in practice.

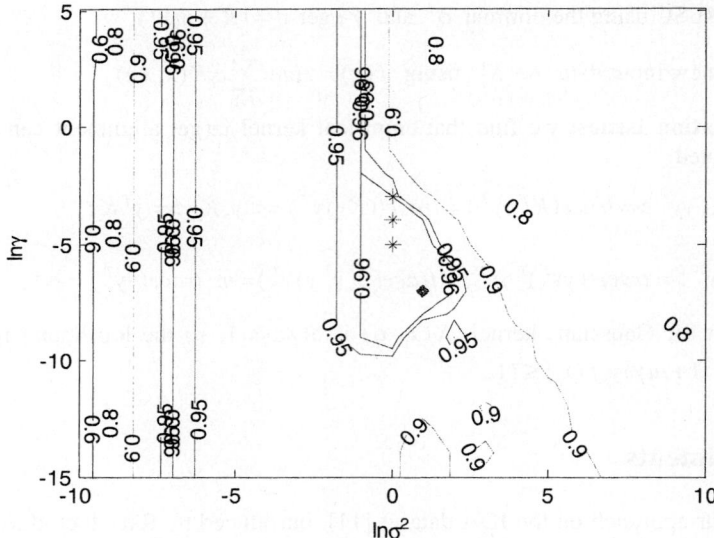

Fig. 2. The contour of the accuracy of the parameter grid on dataset *Thyroid*. All the optimal parameter-pairs of (σ^2, γ) obtained by RLSC AlignLoo algorithm on each division are plotted with blue stars, obviously, they overlap heavily and are located near the optimal point (the red diamond) achieved by grid search on 5-fold CV.

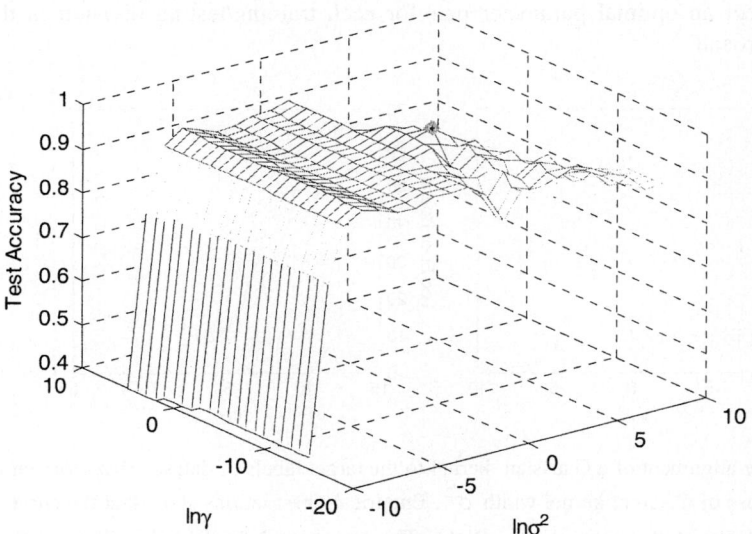

Fig. 3. The mesh plot of the accuracy of the parameter grid on dataset *Thyroid*. The optimal point achieved by grid search on 5-fold CV is indicated specially.

Table 1. A brief description of IDA data set and the time cost on each data set for model selection through RLSC grid search and RLSC-AlignLoo method. Time cost for RLSC Grid is the time consumed on searching on a 21×21 grid using 5-fold CV, and for RLSC-AlignLoo is the time consumed on a single division, which covers time span for searching 21+21 point.

No.	Dataset	Dimension	Train size	Test size	Divisions	RLSC Grid (Minute)	RLSC-AlignLoo (Second)
1	Banana	2	400	4900	100	79.6	1.88
2	Breast Cancer	9	200	77	100	8.7	0.24
3	Diabetes	8	468	300	100	81.6	2.55
4	Flare Sonar	9	666	400	100	149.8	6.68
5	German	20	700	300	100	258.6	7.19
6	Heart	13	170	100	100	6.4	0.14
7	Image	18	1300	1010	20	1410.2	40.30
8	Ringnorm	20	400	7000	100	162.8	1.57
9	Splice	60	1000	2175	20	1009.7	20.21
10	Thyroid	5	140	75	100	3.2	0.11
11	Titanic	3	150	2051	100	10.7	0.19
12	Twonorm	20	400	7000	100	179.1	1.69
13	Waveform	21	400	4600	100	147.3	1.75

Table 2. The mean test cost (with standard derivation) on each data set of RLSC grid search and RLSC AlignLoo with a comparison with methods of RBF Network, SVM and AdaBoost, etc. The mean cost is the average of the costs on the 100 or 20 divisions of each set. The bold-faced text indicates a maximum performance is achieved by this approach.

No.	RBF-Network	AdaBoost RBF	LP_Reg-AdaBoost	QP_Reg-AdaBoost	AdaBoost_Reg	SVM RBF	RLSC Grid	RLSC-AlignLoo
1	10.76±0.42	12.26±0.67	10.73±0.43	10.90±0.46	10.85±0.42	11.53±0.66	**10.41±0.44**	11.01±0.62
2	27.64±4.71	30.36±4.73	26.79±6.08	25.91±4.61	26.51±4.47	26.04±4.74	**25.43±4.03**	25.60±4.26
3	24.29±1.88	26.47±2.29	24.11±1.90	25.39±2.20	23.79±1.80	23.53±1.73	**22.99±1.69**	23.70±1.93
4	34.37±1.95	35.70±1.79	34.74±2.00	36.22±1.80	34.20±2.18	**32.43±1.82**	33.48±1.79	35.86±1.82
5	24.71±2.38	27.45±2.50	24.79±2.22	25.25±2.14	24.34±2.08	**23.61±2.07**	24.61±2.29	26.81±2.91
6	17.55±3.25	20.29±3.44	17.49±3.53	17.17±3.44	16.47±3.51	**15.95±3.26**	16.61±3.95	17.03±3.45
7	3.32±0.65	2.73±0.66	2.76±0.61	2.67±0.63	**2.67±0.61**	2.96±0.60	2.84±0.53	3.15±0.95
8	1.70±0.21	1.93±0.24	2.24±0.46	1.86±0.22	**1.58±0.12**	1.66±0.12	2.44±0.16	2.58±0.24
9	9.95±0.78	10.14±0.51	10.22±1.59	10.11±0.52	**9.50±0.65**	10.88±0.66	10.91±0.81	12.95±0.90
10	4.52±2.12	4.40±2.18	4.59±2.22	4.35±2.18	4.55±2.19	4.80±2.19	**4.21±2.12**	5.24±2.48
11	23.26±1.34	22.58±1.18	23.98±4.38	22.71±1.05	22.64±1.20	**22.42±1.02**	22.55±1.13	23.12±1.46
12	2.85±0.28	3.03±0.28	3.17±0.43	2.97±0.26	2.70±0.24	2.96±0.23	**2.39±0.12**	2.47±0.20
13	10.66±1.08	10.84±0.58	10.53±1.02	10.07±0.51	9.79±0.81	9.88±0.43	**9.54±0.46**	10.43±0.92

The time costs of the two approaches are presented in table 1, the test costs in table 2, and the parameter setting in Fig. 1, Fig. 2, and Fig. 3. We can see RLSC-AlignLoo is much faster than RLSC grid search, although the accuracy is not as good as grid search, yet still comparable with other methods. This discrepancy is generated from their utilization of different empirical criteria for parameters selection. Further study shows it is possible for RLSC-AlignLoo to enhance its accuracy by decreasing search step around the point maximizing the alignment, and the additional time cost is little. For example, we get a mean test cost of 22.38±1.00 for 100 training/testing division of dataset *Titanic*. This is of cause valid for RLSC Grid, but the time cost will be too expensive to pay.

We find that the kernel parameters optimized by kernel target alignment are very stable against different division of the dataset, as showed in Fig. 2, which gives a footnote for the theorem in section 3 that the empirical alignment is a good approximation to its true alignment. We find that the Loo bound is tight in addition, as showed in Fig. 1(right), the expected Loo bound is 8, and 8 divided by sample size 400 corresponds to a 2% Loo error rate bound on this division, and according to table 2, the averaged error rate on 100 divisions is 2.47%.

7 Conclusion

We present a novel 2-step approach for parameter selection for RLSC or KA. Firstly optimal kernel parameters are selected by maximizing the kernel target alignment, and subsequently optimal hyper-parameter is determined by minimizing the loo bound. This 2-step separation greatly reduced the time complexity for model selection. Generally speaking, the AlignLoo method is also applicable for those kernel-based classifiers having been found a generalization bound.

References

1. Poggio, T., Smale, S.: The Mathematics of Learning: Dealing With Data. Notice of American Mathematical Society. 5 (2003) 537-544
2. Cucker, F., and Smale, S.: On the Mathematical Foundations of Learning. Bulletin of American Mathematical Society. 39 (2001) 1-49
3. Vapnik, V.: The Nature of Statistical Learning Theory. (2nd Ed.) Springer, (2000)
4. Rifkin, R.M.: Everything Old Is New Again: A Fresh Look at Historical Approaches to Machine Learning. PhD thesis, Massachusetts Institute of Technology, (2002)
5. Rifkin, R.M., Yeo, G., Poggio, T.: Regularized Least-Squares Classification. In: Advances in Learning Theory: Methods, Model and Applications, NATO Science Series III: Computer and Systems Sciences, l. 190, IOS Press, Amsterdam (2003)
6. Suykens, J.A.K., Gestel, V., De Bra banter, J., et al.: Least Squares Support Vector Machine. Singapore: World Scientific, (2002)
7. Fung, G. and Mangasarian, O.L.: Proximal Support Vector Machine Classifiers. In KDD 2001: Seventh ACM SIGKDD International Conference on Knowledge Discovery and Data Mining, San Francisco, CA, (2001)
8. Chapelle, O., Vapnik, V., Mukjerjee, S.: Choosing Multiple Parameters for Support Vector Machines. Machine Learning. 46 (2002) 131-159
9. Cristianini, N., Kandola, J., Elisseeff, A., Shawe-Taylor, J.: On Kernel Target Alignment. Journal of Machine Learning Research, submitted.
10. Jaakkola, T., Haussler, D.: Probabilistic Kernel Regression Models. In Advances in Neural Information Processing Systems, 11 (1998)
11. Rätsch, G., Onoda, T., Müller, K.-R.: Soft Margin for AdaBoost. Machine Learning, 1 (2000) 1-35

Modelling of Chaotic Systems with Recurrent Least Squares Support Vector Machines Combined with Reconstructed Embedding Phase Space

Zheng Xiang[1,2], Taiyi Zhang[1], and Jiancheng Sun[3]

[1] Dept. of Information and Communication Eng, Xi'an Jiaotong University, Xi'an, Shaanxi, China
tyzhang@mail.xjtu.edu.cn
[2] School of Telecommunication Engineering, Xidian Univ., Xi'an, Shaanxi, China
zhx@mail.xidian.edu.cn
[3] College of Physics and Information Eng., Fuzhou University, Fuzhou, Fujian, China
sunjc@mailst.xjtu.edu.cn

Abstract. A new strategy of modelling of chaotic systems is presented. First, more information is acquired utilizing the reconstructed embedding phase space. Then, based on the Recurrent Least Squares Support Vector Machines (RLS-SVM), modelling of the chaotic system is realized. We use the power spectrum and dynamic invariants involving the Lyapunov exponents and the correlation dimension as criterions, and then apply our method to the Chua`s circuit time series. The simulation of dynamic invariants between the origin and generated time series shows that the proposed method can capture the dynamics of the chaotic time series effectively.

1 Introduction

Building the model of a dynamical system by the time series analysis has been an important issue. Various techniques for modelling and predicting nonlinear time series are developed in past years, there are many traditional methods associated with time series analysis such as linear regression and ARIMA models [1]. Some new methods have been proposed recently as well, such as local linear mode [2], reconstructed embedding phase [3]and wavelets [4]. In recent years, many researcher addressed the nonlinear time series analysis with the artificial neural networks [5,6,7].

In this paper, a new method combining Recurrent Least Squares Support Vector Machines (RLS-SVM) with reconstructed phase space is developed for chaotic time series reconstruction. The strategy is that more information is extracted from high dimension reconstructed phase space, and then modelling of system is realized by RLS-SVM. Support Vector Machines(SVM) have become a subject of intensive study and have been applied successfully to classification tasks as optical character recognition (OCR) [8,9,10]. Least squares (LS) version of SVM can greatly simplify the problem since its solution is characterized by a linear system [11]. To deal with problems where iterative operations are necessary, Recurrent Least Squares Support Vector Machines (RLS-SVM) [12] was developed.

2 Embedding Phase Space of Dynamical System

Deterministic dynamical systems describe the time evolution of a system in some phase space $\Gamma \subset \mathbf{R}^m$. For simplicity it is assumed that the phase space is a finite dimensional vector space. A state is specified by a vector $x \in \mathbf{R}^m$. Then the dynamics can be described by an explicit system of m first-order ordinary differential equations [13]

$$\frac{d}{dt}x(t) = f(t, x(t)), t \in \mathbf{R} \qquad (1)$$

or in discrete time $t = n\Delta t$ by maps of form

$$x_{n+1} = f(x_n) \qquad (2)$$

A time series can be thought of as a sequence of observations $\{s_n\} = h(x_n), n = 1, 2, \ldots, N_T$ performed with some measurement function, where N_T is the number of data points. Since the sequence $\{s_n\}$ in itself does not properly represent the multidimensional phase space of the dynamical system, some techniques are employed to unfold the multidimensional structure using available data. Takens embedding theorem guarantees the reconstruction of a state space representation from a scalar signal alone [14]. A delay coordinate function Φ is defined by

$$\begin{aligned} s_n &= \Phi(x_n) \\ &= [h(x_{n-(m-1)\tau_d}), \ldots, h(x_{n-\tau_d}), h(x_n)] \\ &= [s_{n-(m-1)\tau_d}, \ldots, s_{n-\tau_d}, s_n] \end{aligned} \qquad (3)$$

where s_n are vectors in a new space namely the embedding phase space are formed from time-delayed values of the scalar measurements. The number m of elements is called the embedding dimension, the time τ_d is generally referred to as the time delay or lag. So the trajectory matrix S can be constructed in m-embedding dimensions as

$$S = \begin{bmatrix} s_n^T \\ s_{n+1}^T \\ \vdots \\ s_{n+N-1}^T \end{bmatrix} = \begin{bmatrix} s_{n-(m-1)\tau_d} & s_{n-(m-2)\tau_d} & \cdots & s_n \\ s_{n+1-(m-1)\tau_d} & s_{n+1-(m-2)\tau_d} & \cdots & s_{n+1} \\ \vdots & \vdots & & \vdots \\ s_{n+N-1-(m-1)\tau_d} & s_{n+N-1-(m-2)\tau_d} & \cdots & s_{n+N-1} \end{bmatrix} \qquad (4)$$

where $N = N_T - (m-1)$. Takens state that if the sequence $\{s_n\}$ does consist of scalar measurements of the state of a dynamical system, then under certain genericity assumptions, the time delay embedding provides a one-to-one image of the original set $\{x\}$, provided m is large enough. For almost all τ_d and for some m, Takens embedding theorem ensures that there is a smooth map $f : \mathbf{R}^m - \mathbf{R}$ such that

$$s_{(n+1)\tau_d} = f(s_{n-(m-1)\tau_d}, \ldots, s_{n-\tau_d}, s_n) = f(s_n) \qquad (5)$$

where the number m of elements is called the embedding dimension, the time τ_d is generally referred to as the time delay or lag. There is a large literature on the "optimal" choice of the embedding parameters m and τ_d [15,16,17].

The problem of remodeling becomes equivalent to the problem of estimating the unknown function f in the embedding phase space.

To compare different dynamic modeling methods, rules such as correlation dimension and Lyapunov exponents were proposed because of their ability to describe the global properties of the attractor. In addition, the power spectrum is a conventional method to analyze the time series and it can indicate if the dynamical system is periodic, quasiperiodic or chaotic. Chaotic and stochastic systems are easily distinguishable from periodic or quasiperiodic systems for they have rich broadband power spectra, as well as widely varying phase spectra. It was not possible to distinguish the chaotic system from the stochastic system before the advent of the chaotic theory.

It is proven that the Lyapunov exponent is a useful dynamic invariant to characterize the chaotic dynamic system. The Lyapunov exponent measure the rate at which nearby orbits converge or diverge. It is the time constant that is the expression for the distance between two nearby orbits. If it is negative, then the orbits converge in time, and the dynamical system is insensitive to initial conditions. However, if it is positive, then the distance between nearby orbits grows exponentially in time, and the system exhibits sensitive dependence on initial conditions. Thus very different time series can be produced from the same dynamic system even if the initial condition is different in a slight scale. There are as many Lyapunov exponents as there are dimensions in the state space of the system, but the largest is usually the most important since it indicates the chaotic degree of the dynamic system. Formally, the Lyapunov exponent is defined by Wolf et al: [18] given an n-dimensional phase space, the long-term evolution of an infinitesimal sphere is monitored. As the sphere evolves, it will turn into an n-ellipsoid. The i-th one-dimensional Lyapunov exponent is then defined in terms of the length of the resulting ellipsoid's principal axis.

$$\lambda_i = \lim_{t \to \infty} \frac{1}{t} \log_2 \frac{p_i(t)}{p_i(0)} \qquad (6)$$

The Lyapunov spectrum is then formed by the set $(\lambda_1, \lambda_2, ... \lambda_n)$, where λ_i are arranged in decreasing order. The difference between strange attractors and purely stochastic (random) processes is that the evolution of points in the phase space of a strange attractor has definite structure. The correlation integral provides a measure of the spatial organization of this structure, and is given by

$$C(r) = \lim_{N \to \infty} \frac{1}{N(N-1)} \sum_{i \neq j} \Theta(r - \|s(i) - s(j)\|) \qquad (7)$$

where Θ is the Heaviside function. Grassberger and Privacies found that, for a strange attractor, $C(r) \propto r^v$ for a limited range of r [19]. The power v is called the correlation dimension of the attractor. Thus, we can plot the $\log C(r) - \log r$ graph to identify an attractor.

In addition to above invariants, the Poincar'e map is another parameter to characterize the chaotic system. Since a chaotic system never revisits the same state, it will trace out contours on the Poincar'e map. However, unlike a purely random process, these contours will have definite structure and will graphically indicate the presence of the responsible attractor.

3 Recurrent Least Squares Support Vector Machines and Learnin Algorithm

The foundations of Support Vector Machines (SVM) have been developed since 1990[9]. Suykens and Vandewalle proposed the Recurrent Least Squares Support Vector Machines(RLS-SVM) to deal with problems requiring iterative operations. In following section, we formulated our algorithm based on the RLS-SVM and reconstructed phase space:

1. Generate a time series $x_i, i = 1,2,...,N$, where N is the number of the data points.
2. Choose optimal m, τ_d (scale with discrete sample units) for x_i
3. Constructed the phase trajectory matrix:

$$\mathbf{S} = \begin{bmatrix} \mathbf{s}_n^T \\ \mathbf{s}_{n+1}^T \\ \vdots \\ \mathbf{s}_{n+N-m}^T \end{bmatrix} = \begin{bmatrix} s_{n-(m-1)\tau_d} & s_{n-(m-2)\tau_d} & \cdots & s_n \\ s_{n+1-(m-1)\tau_d} & s_{n+1-(m-2)\tau_d} & \cdots & s_{n+1} \\ \vdots & \vdots & & \vdots \\ s_{n+N-m-(m-1)\tau_d} & s_{n+N-m-(m-2)\tau_d} & \cdots & s_{n+N-m} \end{bmatrix} \quad (8)$$

4. build training data

$$\begin{bmatrix} X_0 \\ X_1 \\ \vdots \\ X_{N-m} \end{bmatrix} = \begin{bmatrix} s_{n-(m-1)\tau_d} & s_{n-(m-2)\tau_d} & \cdots & s_{n-\tau} \\ s_{n+1-(m-1)\tau_d} & s_{n+1-(m-2)\tau_d} & \cdots & s_n \\ \vdots & \vdots & & \vdots \\ s_{n+N-m-(m-1)\tau_d} & s_{n+N-m-(m-2)\tau_d} & \cdots & s_{n+N-m-\tau} \end{bmatrix}, \begin{bmatrix} Y_0 \\ Y_1 \\ \vdots \\ Y_{N-m} \end{bmatrix} = \begin{bmatrix} s_n \\ s_{n+1} \\ \vdots \\ s_{n+N-m} \end{bmatrix} \quad (9)$$

Given initial condition $\hat{s}_i = s_i$ for $i = 1,2,...m$, the prediction problem is given by:

$$\begin{aligned} \hat{s}_{(k+1)\tau_d} &= f\left(\hat{s}_{k-(m-1)\tau_d},...,\hat{s}_{k-\tau_d},\hat{s}_k\right) \\ &= w^T \varphi\left(\left[\hat{s}_{k-(m-1)\tau_d},...,\hat{s}_{k-\tau_d},\hat{s}_k\right]\right) + b \end{aligned} \quad (10)$$

where $\varphi(\cdot): \mathbf{R}^m \to \mathbf{R}^{n_h}$ is a nonlinear mapping in future space, $w \in \mathbf{R}^{n_h}$ is the output weight vector and $b \in \mathbf{R}$ is bias term. Choose the $\varphi(\cdot)$ and estimate the function $f(\cdot)$ by using training data

5. Generate the s_{i+p} by performing the iterative process, where $p = 1, 2, ..., l$, l is the length of reconstruction time series.

In literature [12], Suykens and Vandewalle proposed that the eq.(10) can be convert into optimal problems which can be described as follows.

$$\min_{w,e} J(w,e) = \frac{1}{2} w^T w + \gamma \frac{1}{2} \sum_{k=m+1}^{N+m} e_k^2 \quad (11)$$

$$\text{Subject to } s_{k+1} - e_{k+1} = w^T \varphi(\mathbf{s}_k - \mathbf{e}_k) + b, k = m+1, ..., N+m \quad (12)$$

where $e_k = s_k - \hat{s}_k$, $\mathbf{s}_k = [s_{k-(m-1)\tau_d}; ...; s_{k-\tau_d}, s_k]$, $\mathbf{e}_k = [e_{k-(m-1)\tau_d}; ... e_{k-\tau_d}; e_k]$ and γ is an adjustable constant. The basic idea of mapping function $\varphi(\cdot)$ is to map the data s into a high-dimensional feature space, and to do linear regression in this space.

To resolve the optimal function eq(11) and (12), we define the Lagrangian function

$$L(w, b, e; \alpha) = J(w, e) + \sum_{k=m+1}^{N+m} \alpha_{k-m} \times [h_{k+1} - e_{k+1} - w^T \varphi(\mathbf{s}_k - \mathbf{e}_k) - b] \quad (13)$$

where α_i are Lagrange multipliers.

The resulting recurrent simulation model is described as follows [12]

$$s_{(k+1)\tau_d} = \sum_{l=m+1}^{N+m} \alpha_{l-m} K(\mathbf{z}_l, [\hat{\mathbf{s}}_{k-(m-1)\tau_d}, ..., \hat{\mathbf{s}}_{k-\tau_d}, \hat{\mathbf{s}}_k]) + b \quad (14)$$

where $\mathbf{z}_l = \mathbf{s}_l - \mathbf{e}_l$

The mapping function $\varphi(\cdot)$ can be paraphrased by a kernel function $K(\cdot, \cdot)$ because of the application of Mercer's theorem[9], which means that

$$K(x_i, x_j) = \varphi(x_i)^T \varphi(x_j) \quad (15)$$

with radial basis function (RBF) kernels one employs.

$$K(x_i, x_j) = \exp\left(-\frac{\|x_i - x_j\|}{2\sigma^2}\right) \quad (16)$$

4 Simulation and Results

In the following procedure, we use the data of the Chua's circuit[20] are as follows:

$$\begin{cases} \dot{x} = \alpha[y - h(x)] \\ \dot{y} = x - y + z \\ \dot{z} = -\beta y \end{cases} \quad (17)$$

With piecewise linear characteristic

$$h(x) = m_1 x + \frac{1}{2}(m_0 - m_1)(|x+1| - |x-1|) \tag{18}$$

A double scroll attractor generated by taking $\alpha = 9$, $\beta = 14.286$, $m_0 = -1/7$, $m_1 = 2/7$. A trajectory has been generated for initial condition $[0.1; 0; -0.1]$ by using a Runge–Kutta integration rule.

The reconstructed embedding phase space has been discussed in the Sect.2. The choice of the embedding dimension m and time delay τ_d is the first problem we faced. Cao proposed a method to determine the minimal sufficient embedding dimension that used improved false nearest neighbor method [16]. The time delayed mutual information was suggested by Fraser and Swinney [17] as a effective tool to determine a reasonable τ_d, A certain value of τ correspond to the minimum value of mutual information is a good candidate for a reasonable time delay. For the model structure, we estimated the embedding dimension of the Chua's circuit by the Cao's algorithm. In our simulation, we have not controlled τ_d, the delay parameter of the Takens' embedding, so the sampling gives by default $\tau_d = 1$.

The time series was divided into two subsets referred to the training and test subset: the training subset consists of 1000 time entries and the test subset consists of 1500 entries. In order to solve the constrained nonlinear optimization problem (11), (12), SQP has been applied.

First we resort to estimation of the dynamical invariants of motion from the predicted model. The Poincaré maps for the time series were drawn in Fig.1, where the original attractor (Fig. 1a) is compared with the reconstructed one (Fig. 1b). The similarity of the two graphs suggests that the two time series represent two distinct trajectories on the same attractor.

Figure 2 shows the spectrum of the original and reconstructed time series. The spectrum of the reconstructed time series seems to follow very closely the spectrum of the original time series, only with minor differences in the fine detail of the spectrum. Although these plots lead us to believe that the dynamics have been reasonably captured, we would like to quantify numerically the matching between the dynamics of the two systems. That is the reason we propose to compute the correlation dimension and the largest Lyapunov exponent.

The correlation dimension and the largest Lyapunov exponents were computed using the Grassberger's algorithm[19] and the Wolf's algorithm [18] respectively. In Figure 3, the correlation integral map (CIM) and its slope are depicted both for the original and reconstructed time series. The value of the correlation dimension is defined as the slope of the CIM curves for at least 3 consecutive embeddings. In this case the correlation dimension for the predicted time series is 1.975, and the correlation dimension for the time original time series is 1.894. Figure 4 shows that largest Lyapunov exponent of reconstruction time series is able to follow ones of original time series with a very small error.

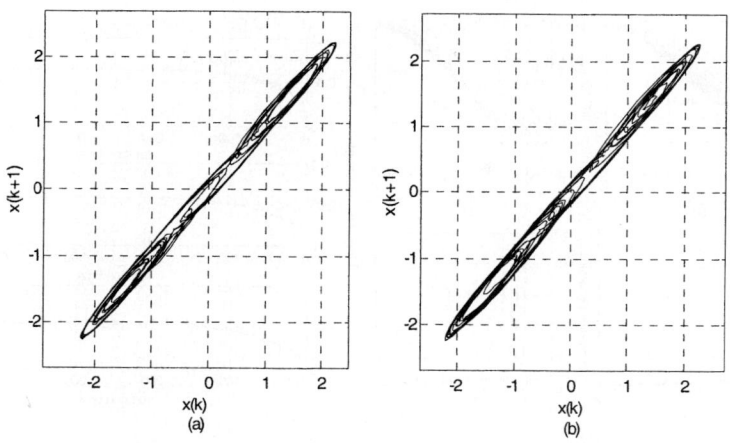

Fig. 1. Poincar'e maps:(a) original system (b) RLS-SVM

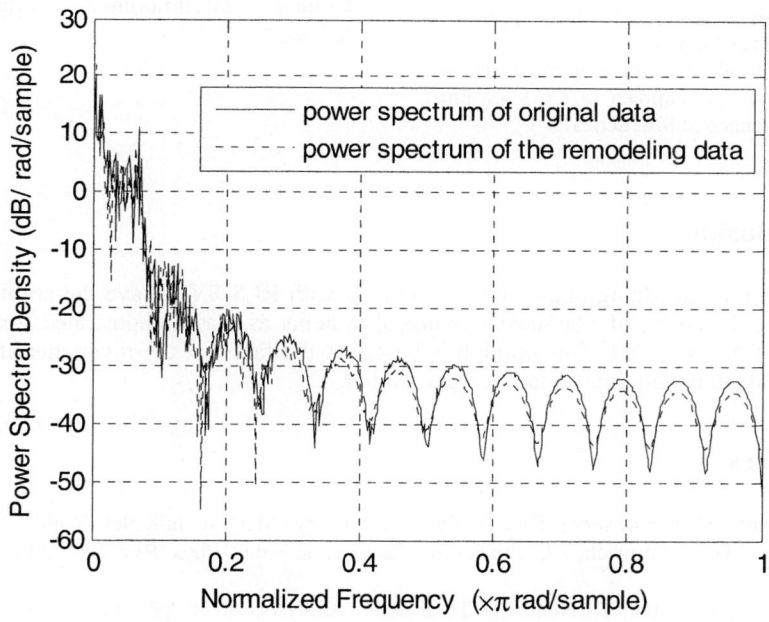

Fig. 2. Power spectrum for the original data and remodeling data of Chua's circuit

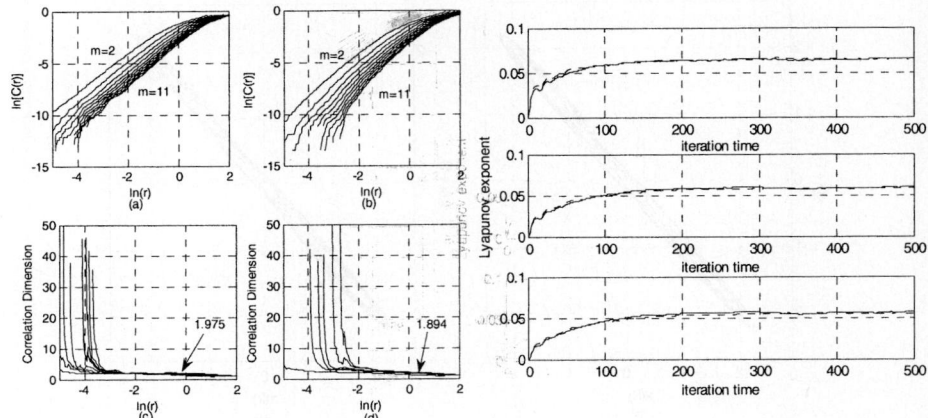

Fig. 3. Correlation dimension of Chua's circuit estimates for embedding dimensions 2~11.(a) Correlation Integral Map (CIM) for original time series. (b) CIM for series generated from proposed method.(c) Slope (correlation dimension) estimate of CIM for original time series.(d) Slope estimate of CIM for proposed generated time series

Fig. 4. Estimation of largest Lyapunov exponent for original (real line) and generated (dotted line) time series of Chua's circuit.(a)m=2, (b)m=4, (c)m=6

5 Conclusion

Reconstructed embedding phase space combined with RLS-SVM have the ability to capture the dynamics of nonlinear dynamical systems as was demonstrated for the system of Chua's circuit. This opinion is based on the fact that the invariants of the original and generated time series are very similar.

References

1. Brillinger, D.R.: Time series, Data Analysis and Theory , McGraw-hill, New York (1981).
2. Farmer, J.D., Sidorowich, J.J.: Predicting chaotic time series, Phys. Rev. Lett. 59 (1987) 845
3. Kantz, H., Schreiber, T.: Nonlinear Time Series Analysis,Cambridge Univ. Press, Cambridge (1997)
4. Meyer, Y., Ryan, R. D.: Wavelets: Algorithms and Applications, Philadelphia: Society for Industrial and Applied Mathematics(SIAM). (1993)
5. Han, M., Xi, J., Xu, S., Yin, F.L.: Prediction of chaotic time series based on the recurrent predictor neural network. IEEE Trans. Signal Processing, 52 (2004) 3409-3416
6. Wang, L.P., Teo, K.K., Lin, Z.: Predicting time series using wavelet packet neural networks. Proc. IJCNN 2001, 1593-1597

7. Castillo, O., Melin, P.: Hybrid intelligent systems for time series prediction using neural networks, fuzzy logic, and fractal theory. IEEE Trans. Neural Networks, 13 (2002) 1395-1408
8. Kecman,V.: Learning and Soft Computing: Support Vector Machines, Neural Networks and Fuzzay Logic Models, The MIT Press, Cambridge, MA(2001)
9. Vapnik, V.: The Nature of Statistical Learning Theory. Springer, New York (1995)
10. Wang, L.P. (Ed.): Support Vector Machines: Theory and Application. Springer, Berli Heidelberg New York (2005)
11. Suykens, J.A.K., Van Gestel, T., De Brabanter, J., De Moor, B., Vandewalle, J.: Least Squares Support Vector Machines, World Scientific, (2002)
12. Suykens , J.A.K., Vandewalle, J.: Recurrent Least Squares Support Vector Machines. IEEE Trans. on Circuits and System-I: Fundamental Theory and Applications, 47 (2000) 1109-1114
13. Kaplan, D., Glass, L.: Understanding nonlinear dynamics. Springer, New York (1995).
14. Takens, F.: Detecting strange attractors in fluid turbulence, In D. Rand and L.S.Young, editors, Dynamical systems and turbulence. Springer-Verlag, Berlin (1981) 366-381
15. Kennel, M.B., Brown, R., Abarbanel, H.D.I.: Determining embedding dimension for phase-space reconstrution using a geometrical construction. Phys. Rev. A , 45 (1992) 3403
16. Cao, L.: Practical method for determining the minimum embedding dimension of a scalar time series. Phys. D, 110 (1997) 43-50
17. Fraser, A.M., Swinney, H.L.: Independent coordinates for strange attractors from mutual information. Phys. Rev. A , 33 (1986) 1134
18. Wolf, A., Swift, J.B., Swinney, H.L., Vastano, J.A.: Determining lyapunov exponents from a time series. Phys. D, 16 (1985) 285–317
19. Grassberger, P., Procaccia, I.: Characterization of strange attractors. Phys. Rev. Letters, 50 (1983) 346–349
20. Chua, L.O., Komuro, M., Matsumoto, T.: The double scroll family. IEEE Trans. Circuits Syst, 33 (1986) 1072–1118

Least-Squares Wavelet Kernel Method for Regression Estimation

Xiangjun Wen, Xiaoming Xu, and Yunze Cai

Automation Department, Shanghai Jiaotong University, Shanghai, 200030, China
{Wenxiangjun, xmxu, yzcai}@sjtu.edu.cn

Abstract. Based on the wavelet decomposition and reproducing kernel Hilbert space (RKHS), a novel notion of least squares wavelet support vector machine (LS-WSVM) with universal reproducing wavelet kernels is proposed for approximating arbitrary nonlinear functions. The good reproducing property of wavelet kernel function enhances the generalization ability of LS-WSVM method and some experimental results are presented to illustrate the feasibility of the proposed method.

1 Introduction

As a new type of network inspired by the neural network and wavelet decomposition, the wavelet networks proposed by Qinghua Z. [1] has been considered as an alternative to the feed forward neural network. It can greatly remedy the weakness of both wavelets and neural networks and has been widely used in the fields of classification and approximation with great success [2]. Despite many of these advances, however, wavelet networks find the solution by minimizing an empirical risk (usually a mean square error) with gradient-based training method such as back-propagation, and they often suffer from the existence of multiple local minima. Presently, Support Vector Machine (SVM) originally introduced by Vapnik [3] has been proved to be a powerful alternative to neural network. In this kernel-based method, one maps the input data into a higher dimensional space (so called feature space) and constructs an optimal separating hyper plane in this feature space. Kernel functions and regularization parameters are chosen such that a regularized empirical risk is minimized.

In fact, kernels (in particular Mercer or reproducing kernels) play a crucial role during the process of solving the convex optimization problem of SVM. Generally, kernels can be considered as choosing an efficient data representation of prior information for a certain classification or approximation problem. How to choose a kernel function with good reproducing properties (generalization ability) is a key issue of data representation, and it is closely related to choose a specific reproducing kernel Hilbert space (RKHS). Noting that kernel methods like wavelet networks also rely on similar basis functions and their behaviors should be closely related. It is a valuable issue whether a better performance could be obtained if we combine the wavelet decompositions with kernel method. Actually it has caused great interest of many researchers in the last few years. In particular, linear splines have been proposed for generating the inner product kernels and solving the function estimation and data compression problems [4]. A reproducing wavelet kernel spanned by Daubechies-2 orthogonal wavelets has also shown promising results on classification problem of some benchmark datasets [5]. Presently, an admissible support vector (SV) kernel based on continuous wavelet function with translation invariant and positive properties is introduced by Li Z. [6], and this approach can give a little better results while comparing with Gaussian kernel both on pattern recognition and regression problem. However, it is very difficult to decompose a translation invariant kernel into the product of two functions and prove it as SV kernel that satisfies the Mercer condition. Our purpose is to take advantage of the wavelet approximation and the kernel method as mentioned above. By this mean, our approach is quite similar since we aim at constructing universal wavelet kernels in RKHS for practical use. Due to the fact that it is a very stringent requirement to solve a large-scale quadratic programming problem, we proposed a least squares version of Support Vector Machine based on reproducing wavelet kernels and develop a framework for regression estimation in this paper.

The rest of the paper is organized as follows. In the next section we first give a brief review on LS-SVM method for function estimation. Then in section 3 we focus on a practical approach to construct the universal wavelet kernels in RKHS. In section 4, numerical experiments are presented to assess the feasibility of the proposed method. Finally, Section 5 concludes the work done.

2 LS-SVM for Function Estimation

Given a training data set D of l samples independent and identically drawn (i.i.d.) from an unknown probability distribution $\mu(X, Y)$ on the product space $Z = X \times Y$:

$$D = \{z_1 = (x_1, y_1), \ldots z_n = (x_l, y_l)\} \tag{1}$$

where the input data X is assumed to be a compact domain in a Euclidean space R^d and the output data Y is assumed to be a closed subset of R. By some nonlinear mapping $\Phi(\cdot)$, input X is mapped into a feature space in which the learning machine (algorithm) selects a certain function f.

In the case of Least Squares Support Vector Machine (LS-SVM), function estimation in RKHS is defined:

$$f(x) = w^T \Phi(x) + b \tag{2}$$

One defines the optimization problem.

$$\min_{w,b,e} J(w,e) = \frac{1}{2} w^T w + \gamma \frac{1}{2} e^T e \tag{3}$$

s.j.

$$y_k = w^T \Phi(x_k) + b + e_k, k = 1, \ldots, l \tag{4}$$

where $e \in R^{l \times 1}$ denotes the error vector, regularization parameter γ denotes an arbitrary positive real constant.

The conditions for optimality lead to a set of linear equations:

$$\begin{bmatrix} 0 & \vec{1}^T \\ \vec{1} & \Omega+\gamma^{-1}I \end{bmatrix} \begin{bmatrix} b \\ \alpha \end{bmatrix} = \begin{bmatrix} 0 \\ y \end{bmatrix} \tag{5}$$

where $y = [y_1, y_2, \ldots, y_l]^T$, $\vec{1} = [1,\ldots,1]_{1\times l}^T$ $\alpha = [\alpha_1,\ldots,\alpha_l]^T$.

The resulting LS-SVM model for function estimation becomes:

$$f(x) = \sum_{k=1}^{l} \alpha_k K(x, x_k) + b \tag{6}$$

where α_k, b are the solution to the linear system (5). $K(\cdot,\cdot)$ is a positive define kernel function which satisfied Mercer condition [7].

$$K(x_i, x_j) = \varphi(x_i)^T \varphi(x_j) \quad k, l = 1, \ldots, l \tag{7}$$

In the novel approach, $K(\cdot,\cdot)$ is a reproducing wavelet kernel constructed in the RKHS, which will be discussed in detail in section 3.

3 Reproducing Wavelet Kernel Method

3.1 Wavelet Frame as Universal Approximants in RKHS

RKHS is a Hilbert space in which all the point evaluations are bounded linear functional. The interest of RKHS arises from its associated kernel functions. According to wavelet theory, the frame establish general conditions under which one can reconstruct a function *f* in RKHS from its inner product with a family of elements function [9], [10]. In this subsection, we assume that the reader is familiar with RKHS theory and the relevant theory of wavelet analysis. Due to page limitation, we have to briefly review some important conclusions on RKHS theory and wavelet frame that we will use in this paper.

A kernel may be characterized as a function from $X \times X$ to R (usually $X \subseteq R^d$). Let H be a $L^2(X,\mu)$ Hilbert space of functions on some domain X, where μ is a measure, that means, for every $x \in X$, there exists $\Gamma_x \in H$, such that

$$f(x) = <\Gamma_x, f>, \forall f \in H \tag{8}$$

where $<\cdot,\cdot>$ denotes the inner product in H. let $x, y \in X$, and set $<\Gamma_x, \Gamma_y> = K(x,y)$. Then the kernel K is called the reproducing kernel for H. Frequently, the kernel K is defined directly as an inner product function, which satisfies the Mercer conditions as follows.

Lemma 1[7]: Supposed any continuous symmetry function $K(x,y) \in L^2 \otimes L^2$ is positive (define) kernel $\Leftrightarrow$

$$\iint_{L^2 \otimes L^2} K(x,y)g(x)g(y)dxdy \geq 0$$
$$\forall g \in L^2, g \neq 0, \int g^2(u)du < \infty \tag{9}$$

The kernel that satisfies this Mercer condition is also called as an admissible support vector (SV) kernel. It belongs to functional space $L^2(X \times X)$ and represents an implicitly nonlinear map from the input space to the feature space.

For further perspective of RKHS, an important kernel operator is given as follows:

Definition 1: (Carleman operator) Let $\Gamma = \{\Gamma_x, x \in X\}$ be a family of $L^2(X, \mu)$ functions. the associated Carleman operator S is

$$S: \begin{array}{l} L^2 \to R^X \\ f \to g(.) = (Sf)(.) = <\Gamma(.), f>_{L^2} = \int_X \Gamma_{(.)} f d\mu \end{array} \tag{10}$$

That is to say $\forall x \in X, g(x) = <\Gamma_x, f>_{L^2}$. this class of integral operators is known as Carleman operators. Note the bijective restriction of operator S is convenient to factorize as follows:

$$S: L^2 \xrightarrow{T} L^2/\ker(S) \xrightarrow{i} \text{Im}(S) \to R^X \tag{11}$$

where $L^2/\ker(S)$ is the quotient set, T is the bijective restriction of S and i the canonical injection. When X is a compact set R or when $\Gamma_x \in L^2(R \times R)$, S is a Hilbert-Schmidt operator [11]. Since the positive kernel is compact, the kernel operator admits a countable spectrum and thus the kernel can be decomposed. A Carleman operator S can map all possible L^2 functions into the set of point-wise valued functions R^X, and it can be built from a family Γ_x of $L^2(X, \mu)$ functions and a linear mapping.

In wavelet theory, a frame is a set of functions $\{\phi_i\}_{i \in I}$ of $L^2(R)$ that satisfied the following condition

$$c_{\min} \|f\|^2 \leq \sum_{i \in I} |<f, \phi_i>|^2 \leq c_{\max} \|f\|^2 \tag{12}$$

with $0 \prec c_{\min} \leq c_{\max} \prec \infty$, for all function f in $L^2(R)$. It is known that when $c_{\min} = c_{\max} = 1$, frame elements consist of an orthonormal basis of $L^2(R)$. In general, wavelet frame can be seen as the extensions of canonical orthonormal basis of $L^2(R)$.

We are interested in some approximation function $\phi(x) \in L^2(R)$ consists of a denumerable family satisfying the frame property in RKHS with the form:

$$\phi(x) = \{\det(A_k^{\frac{1}{2}})\Psi(a_k x - t_k): t_k \in R^d,$$
$$A_k = diag(a_k), a_k \in R_+^d, k \in Z\} \tag{13}$$

where t_k, a_k denote the arbitrary translation vectors and arbitrary dilation vectors specifying the diagonal dilation matrixes A_k. Noting these family functions (13) is proposed as universal approximants from continuous wavelets to orthonormal wavelets [1], let select to construct our reproducing wavelet kernels in this paper.

Based on Carleman operator in (10), it is possible to build reproducing wavelet kernel function from any finite set of wavelet frames in $L^2(R)$. Generally, there exists following proposition.

Proposition 1: Any finite set of wavelet frames of $L^2(R)$ endowed with inner product spans a RKHS, and its reproducing kernel is

$$K(x, y) = <\Gamma_x(.), \Gamma_y(.)>_{L^2} \tag{14}$$

where $\Gamma_x(.) \in L^2(R)$ is a family of functions indexed by $x \in X$ (X being any subset of R), which acts as the evaluation functional in x.

Since $L^2(R)$ is a Hilbert space endowed with inner product $<\cdot, \cdot>_{L^2}$, such that

$$\forall \phi \in L^2(R), \|\phi\|_{L^2} < \infty \tag{15}$$

where ϕ represents any finite set of wavelet frame elements function in $L^2(R)$. Let us define an indexed family of function $\Gamma_x(.) \in L^2(R)$ indexed by $x \in X$ and a linear mapping S:

$$L^2 \to R^X$$
S: $\phi \to g(.)$ so that $\forall x \in X$, \tag{16}
$$g(x) = S\phi(x) \stackrel{\Delta}{=} <\Gamma_x(.), \phi(.)>_{L^2}$$

We can decompose $L^2(R) = Ker(S) \oplus M$ and we obtain

T: $\begin{aligned} M &\to \text{Im}(S) \\ \phi &\to g(.) = T\phi = S\phi \end{aligned}$ \tag{17}

where T is a bijective restriction of Carleman operator S. Let us define $H \stackrel{\Delta}{=} \text{Im}(S)$ and endow with the following inner product:

$$\forall g_1, g_2 \in H \quad <g_1, g_2> = <T\phi_1, T\phi_2>_H = <\phi_1, \phi_2>_{L^2} \tag{18}$$

Hence, H is a RKHS with reproducing kernel K as follows:

$$K(x,y) = <\Gamma_x(.), \Gamma_y(.)>_{L^2} \tag{19}$$

Consider the family $\{e_i\}$ as an orthonormal basis of $L^2(R)$ and ϕ_i be a wavelet basis of $L^2(R)$ (i denote a multi index). One can write:

$$\Gamma_x(.) = \sum_{i,j} \alpha_{i,j} \phi_j(x) e_i(.) \tag{20}$$

where $\alpha_{i,j} = c_j \delta_{i,j}$ is the coefficients combining the orthonormal basis $\{e_i\}$ of $L^2(R)$ with wavelet basis ϕ_j of $L^2(R)$, and c_j is a coefficient depending on the considered wavelet ϕ_j.

So far, we can construct a wavelet kernel in RKHS as follows:

$$K(x,y) = \sum_{i,j,n} \alpha_{i,j} \alpha_{j,n} \phi_j(x) \phi_n(y) \tag{21}$$

For a common multidimensional wavelet kernel function, we can write it as the product of one-dimensional (1-D) wavelet function according to tensor products theory proposed by N. Aronszajn [8].

$$K_d(x,y) = \prod_{i=1}^{d} K(x_i, y_i) \tag{22}$$

Due to page limit, we just sketch the key idea behind these concepts and present here with a brief discussion. A complete study of the properties of reproducing kernel operator and theoretical analysis of the wavelet frame goes beyond the scope of this paper, for a thorough discussion of building RKHS with kernel operator and wavelet frame the reader is further referred to [11], [9], [10].

3.2 Practical Construction of Wavelet Kernel

Let $\varphi(x)$ be a mother wavelet, and let a and t denote the dilation and translation factor, respectively, $a, t \in R$, then according to wavelet theory

$$\phi_i(x) = \varphi_{j,k}(x) = a_0^{-j/2} \varphi(a_0^{-j} x - kt_0) \tag{23}$$

where $a_0, t_0 \in R$, $j, k \in Z$, i denote a multi index. It is know that when the function $\varphi(x)$ satisfies the necessary condition (admissibility of the mother wavelet and suitable parameters with a_0, t_0 such as $a_0=2$, $t_0=1$) will lead to wavelet frames [9].

For practical kernel construction, one has to define a mother wavelet function φ and select suitable parameters according to the problem at hand. In this paper, considering we only use a subset of orthonormal wavelet basis. Moreover, we set coefficients in (20) so that the kernel in RKHS can be written as follows:

$$K(x,y) = \sum_{j=j_{\min}}^{j_{\max}} \sum_{k} \frac{1}{2^j} \varphi_{j,k}(x)\varphi_{j,k}(y) \qquad (24)$$

where j, k are the dilation and translation parameters of a mother wavelet function $\phi(x)$ respectively, $j_{\min}$ and $j_{\max}$ are the minimum and maximum dilatations, respectively.

4 Simulation Results

In this section, we validate the performance of wavelet kernels with two simulation experiments, the approximation of a nonlinear function and the identification of a nonlinear dynamic system.

1) Approximation of a single-variable nonlinear function.
In this experiment, let choose a Mexican hat function

$$f(x) = (1-x^2)\exp(-\frac{x^2}{2}) \qquad (25)$$

over the domain [-10 10].

For comparison, we show the result with our LS-WSVM based on reproducing wavelet kernels spanned by different wavelet frames with *Daubechies*, *Symmlets* and *Coiflets* [10], and with the wavelet network [1]. We have a uniformly sampled example of 300 points, 100 of which are taken as training samples and others for test. We compared the results with two criteria, the normalized root of mean-square-error (NRMSE) and maximal-absolute-error (MAXE). The simulation results are listed in table 1. We can see that our LS-WSVM based different reproducing wavelet kernels gives better approximation results than wavelet network. (Due to space limitation, the figures obtained in this experiment are omitted)

Table 1. Approximation results of Mexican hat function

Method	NRMSE(train)	MAXE(train)	NRMSE(test)	MAXE(test)
Daubechies	0.0152	0.0237	0.0238	0.0408
Symmlets	0.0233	0.0251	0.0233	0.0251
Coiflets	0.0482	0.0426	0.0457	0.0482
Wavenet	0.0949	0.0581	0.0951	0.0581

With parameters: Daubechies, Symmlets and Coiflets with 3 vanishing moment, $j_{\min}$=2, $j_{\max}$=3, gam=571; wavelet network (Wavenet) with 9 wavelons.

2) *Application of nonlinear black-box system identification*
The plant is assumed to be of the form [12]:

$$\hat{y}(k+1) = f[y(k), y(k-1), y(k-2), u(k), u(k-1)] \qquad (26)$$

where the unknown function f has the form

$$f[x_1, x_2, x_3, x_4, x_5] = \frac{x_1 x_2 x_3 x_5 (x_3 - 1) + x_4}{1 + x_3^2 + x_2^2} \quad (27)$$

We generated 200 samples {u (k), y (k)} by using the random input signal uniformly distributed in the interval [-1 1] with the form:

$$u(k) = \sin(2\pi k / 250), \ k \leq 150, \text{ Otherwise} \quad (28)$$
$$u(k) = 0.8\sin(2\pi k / 250) + 0.2\sin(2\pi k / 25).$$

The training data set consists of 100 samples and the other 100 were used as test data. Let compare the results with two popular criteria in control area, MAXE (32) and root of mean-square-error (RMSE). For simplify, only the reproducing kernel spanned by Dubieties wavelet is considered in this experiment. The simulation results are illustrated in table 2. In figure 1, the solid lines represent the approximation and the dashed lines show the function f.

Table 2. Comparison of several algorithms

Method	RMSE(train)	MAXE(train)	RMSE(test)	MAXE(test)
Wavenet	0.0476	0.1948	0.0663	0.3908
LS-SVM	0.00067	0.0972	0.0037	0.1295
LS-WSVM	1.06e-12	4.76e-12	0.0021	0.1754

With parameters: Wavelet network: 16 wavelons. LS-SVM: Gaussian kernel, $\gamma=17.83$, $\delta=0.3$; LS-WSVM: Daubechies wavelet with 4 vanishing moment, $j_{min}=-4, j_{max}=4, \gamma=17.83$

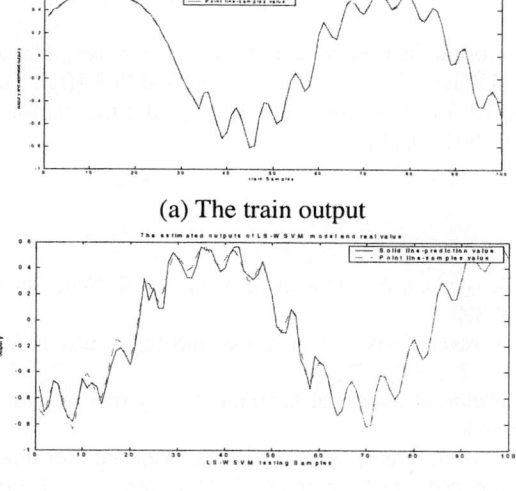

(a) The train output

(b) The test output

Fig. 1. Identification results of nonlinear system via LS-WSVM

So far, we have compared the identification results obtained by wavelet network, LS-SVM with Gaussian kernel and LS-WSVM with reproducing wavelet for nonlinear system, respectively. In this simulation, we adopted cross-validate method for optimization on regularization parameter and kernel parameter of LS-SVM and LS-WSVM. Generally, two kernel methods have greatly outperformed the wavelet networks. Most of all, LS-WSVM based on reproducing wavelet kernel has better performance and generalization ability than LS-SVM based on the Gaussian kernel. It is because our reproducing wavelet kernel based on wavelet decomposition is not only orthonormal (or approximately orthonormal, whereas the Gaussian kernel is correlative or even redundancy), but also suitable for local signal analysis and signal-noise separation for multiresolution analysis, it is not surprising that our LS-WSVM give better approximation results on function estimation and show good generalization ability on dynamic system identification.

5 Conclusions

In this paper, we discussed a practical way to construct wavelet kernel in RKHS and given a brief proof. This work provides a new approach for function estimation and nonlinear system identification, and some numerical experiments show that the proposed method is feasible. In general, the novel methodology inspired by wavelet networks and SVM might offer a new opportunity in the area of automatic control and its application still remain to be further explored for the future.

Acknowledgement

This research is supported in part by the National 973 Key Fundamental Research Project of China (2002CB312200) , t h e N ational 863 High Technology Projects Foundation of China (2 0 0 2 A A 4 1 2 0 1 0) , and the National Natural Science Foundation of China (60174038).

References

1. Qinghua Z. and Benvenisite A.: Wavelet networks. IEEE Tran. On neural Network, Vol 3,No 6. (1992) 889-898
2. Qinghua Z., Using wavelet networks in nonparametric estimation, IEEE Tran. on Neural Network, Vol. 8, No. 2 .(1997)227-236
3. Vapnik, V.: The Nature of Statistical Learning Theory (the second edition). New York: Springer-Verlag (1998)
4. Vapnik, V., Golowich, S., Smola, A.: Support vector method for function approximation, regression estimation, and signal processing. In M. C. Mozer, M. I. Jordan, and T. Petsche, editors, Advances in Neural Information Processing Systems 9, Cambridge, MA, MIT Press. (1997) 281-287
5. Rakotomamonjy A., Mary X., Canu S.: Wavelet Kernel in RKHS. Proc. of Statistical Learning: Theory and Applications, Paris (2002).

6. Li, Z., Weida, Z., Licheng, J.: Wavelet Support Vector Machine. IEEE TRANSACTIONS ON SYSTEMS, MAN, AND CYBERNETICS—PART B: CYBERNETICS, VOL. 34, NO. 1. (2004) 34-39
7. Mercer, J.: Functions of positive and negative type and their connection with the theory of integral equations. Transactions of the London Philosophical Society A, Vol. 209. (1909) 415–446.
8. Aronszajn, N.: Theory of reproducing kernels. Transactions of the American Society. Vol. 68. (1950) 337–404.
9. Daubechies, I.: Ten lectures on wavelets. CBMS-NSF conference series in applied mathematics, Vol.137 (152). SIAM Ed. (1992) 117-119
10. Mallat, S.: A wavelet tour of signal processing. 2nd edn. Academic Press (2003)
11. Canu, S., Mary, X., Rakotomamonjy, A.: Functional learning through kernel. In Suykens, J., Horvath,G., Basu, S., Micchelli, C., Vandewalle, J. (Eds.): Advances in Learning Theory: Methods, Models and Applications. NATO Science Series III: Computer and Systems Sciences, Vol. 190.IOS Press, Amsterdam (2003) 89-110
12. Narendra K., Parthasarathy K.: Identification and control of dynamical systems using neural networks. IEEE tran. On Neural Network. Vol. 1 No 1(1990) 4-27

Fuzzy Support Vector Machines Based on λ—Cut

Shengwu Xiong, Hongbing Liu, and Xiaoxiao Niu

School of Computer Science and Technology, Wuhan University of Technology,
Wuhan 430070, China
xiongsw@mail.whut.edu.cn,
liuhbing@sohu.com, archernxx@hotmail.com

Abstract. A new Fuzzy Support Vector Machines (λ—FSVMs) based on λ—cut is proposed in this paper. The proposed learning machines combine the membership of fuzzy set with support vector machines. The λ—cut set is introduced to distinguish the training samples set in term of the importance of the data. The more important sets are selected as new training sets to construct the fuzzy support vector machines. The benchmark two-class problems and multi-class problems datasets are used to test the effectiveness and validness of λ—FSVMs. The experiment results indicate that λ—FSVMs not only has higher precision but also solves the overfitting problem of the support vector machines more effectively.

1 Introduction

Support vector machines (SVMs) are new machine learning methods, evolving from the statistical learning theory. They embody the principle of the structural risk minimization. Owe to their higher generalization ability and better classification precision, SVMs can solve the overfitting problem effectively and can be applied to a number of issues [1]. Now more and more researches focuses on SVMs as well as the pattern recognition and neural network. SVMs play a more and more important role in classification and regression fields. At present, SVMs have already been applied successively to the problems ranging from hand-written character recognition, face detection, speech recognition to medicine diagnosis [2].

SVMs for pattern classification are based on two-class classification problems. Unclassifiable regions exist when SVMs are extended to multi-class problems. In SVMs, two questions must be paid attention to. One is how to extend two-class problems to multi-class problems. There are many methods to solve this problem, such as one-against-one, ones-against-all and DAGSVMs [3,4]. In order to reduce unclassifiable regions, Inoue and Abe presented Fuzzy Support Vector Machines (FSVMs) [5]. They defined the decision functions according to the membership functions in the directions orthogonal to the hyperplane. The other is how to solve the overfitting problem, which is caused by treating every data point equally during training. Han-Pang Huang and Yi-Hung Liu presented other FSVMs [6]. The performance of SVMs has been enhanced through assigning each training data a membership degree.

SVMs are very sensitive to those training data points, which are far away from their own class center. These points, including the outliers and noises, are sparse [7].

There are two important features about the outliers. Firstly, the outliers are greatly separated from the main body. Secondly, the number of outliers is much less than the number of elements in the main body [8]. We can detect and then discard the outliers by ODM (Outliers Detection Method). The merit of using the ODM is that it can perform the detection in 1-D space so that we can observe the distribution of the training data points along an axis [6]. But ODM also give birth to other problems such as how to deal with the outliers in the case of a large number of them. Because a great deal of useful information will lose simultaneously when all the outliers are discarded.

The aim of this paper is to seek a new method by which the number of training data and the total running time can be reduced and overfitting problem can be avoided. We develop a new FSVMs, λ—FSVMs based on λ—cut, through converse method in which the outliers are regarded as the more important data. The best performance is obtained by selecting suitable parameter λ.

We explain FSVMs in section 2 and λ—FSVMs in section 3 respectively. We compare λ—FSVMs with FSVMs on benchmark data in section 4. Conclusions are drawn finally.

2 Fuzzy Support Vector Machines

The training set S

$$S = \{(x_i, y_i) \mid (x_i, y_i) \in R^d \times R, y_i \in \{-1, 1\}, i = 1, 2, ..., l\}$$

can be linearly separated by a maximum margin classifier named hyperplane (1).

$$w^T x + b = 0 \tag{1}$$

Where w is a vector, b is a scalar, they can be obtained by the constrained optimization problem [1].

FSVMs proposed in [6] solved the overfitting problem by introducing the membership degrees u_i for every data, which are defined as

$$u_i = 1 - \frac{\|x_i - x^*\|}{\max_j \|x_j - x^*\|} + \varepsilon \tag{2}$$

Where x^* is the center of class. The w and b of FSVMs are determined by the constrained optimization problem

$$\min \ W(\alpha) = \frac{1}{2} \sum_{i,j=1}^{l} \alpha_i \alpha_j y_i y_j x_i^T \cdot x_j - \sum_{i=1}^{l} \alpha_i \tag{3a}$$

$$\text{s.t.} \sum_{i=1}^{l} \alpha_i y_i = 0, 0 \leq \alpha_i \leq \mu_i C \ i = 1, 2, ..., l \tag{3b}$$

Some classification problems are non-linearly separable. Namely, in low dimensional space, they are not linearly separable, but they can be classified in higher dimensional space. Such as 0-1 task, these four points can't be separated linearly in two-dimensional space, but they can be separated linearly in three-dimensional space if they are mapped into three-dimensional space. The key to the success of kernel functions lies in the special types of mapping which obeys Mercer's theory and offers an implicit mapping into feature space. That means we needn't to know and calculate the formula of the mapping. The decision function in the higher dimensional space is

$$g(x) = sign\{\sum_{x_i \in SVs} \alpha_i y_i K(x_i, x) + b\} \quad (4)$$

where α_i is determined by the problem

$$\min\ W(\alpha) = \frac{1}{2}\sum_{i,j=1}^{l} \alpha_i \alpha_j y_i y_j K(x_i, x_j) - \sum_{i=1}^{l} \alpha_i \quad (5a)$$

$$s.t. \sum_{i=1}^{l} \alpha_i y_i = 0, 0 \le \alpha_i \le \mu_i C\ \ i=1,2,...,l \quad (5b)$$

3 λ—FSVMs Based on λ—Cut

3.1 Extraction of the More Important Samples

In order to construct λ—FSVMs, the membership functions in this paper are defined as

$$u_i = \frac{\|x_i - x^*\| - \min_j \|x_j - x^*\|}{\max_j \|x_j - x^*\| - \min_j \|x_j - x^*\|} \quad (6)$$

The training set becomes a fuzzy set

$$S_f = \{(x_i, y_i, u_i) | x_i \in R^d, y_i \in \{-1, +1\}, u_i \in [0,1], 1=1,2,...,l\}$$

The membership degrees u_i of fuzzy set S_f are defined through the relative importance of the samples. Each class of the fuzzy set S_f is divided into two subsets by λ—cut. One consists the more important samples, and the other consists the less important ones. The λ—cut is defined as (7)

Definition. Suppose A is a fuzzy subset of domain U. The set (7) is named λ—cut.

$$A_\lambda = \{x | u_A(x) \ge \lambda, x \in U, \lambda \in [0,1]\} \quad (7)$$

Parameter λ is an empirical value. For example, if assigning $\lambda = 0.5$, we regard the samples with membership degree $u_i \ge 0.5$ as the more important one and discard the samples with $u_i < 0.5$.

3.2 Forming λ—FSVMs

The main point of λ—FSVMs is to preserve the more important data and discard the less important ones. The idea comes from the fact that the different training data have different roles in training. The points close to the hyperplane are decisive, such as

SVs, the other points are less important and may bring negative effects to SVMs. In Fig. 1, the points marked '*x*' are regarded as the more important ones, and are considered to have more opportunity to be SVs. Those points marked '*y*' are regarded as the less important ones and can be discarded.

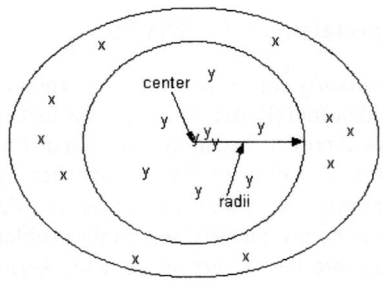

Fig. 1. The λ—cut of one class in the training set. The *center* and *radii* are the center of class and λ respectively.

Algorithm. λ—FSVMs on the training set including class1 and class2.
  Step 1. Searching the centers of two classes

$$x_1^* = \frac{1}{l_1}\sum_{1i=1}^{l_1} x_{1i}, \quad x_2^* = \frac{1}{l_2}\sum_{2i=1}^{l_2} x_{2i}$$

Step 2. Defining the membership degree of each sample in its own class.

$$u_{1i} = \frac{\|x_{1i}-x_1^*\| - \min_{1\leq 1 j\leq l_1}\|x_{1j}-x_1^*\|}{\max_{1\leq 1 j\leq l_1}\|x_{1j}-x_1^*\| - \min_{1\leq 1 j\leq l_1}\|x_{1j}-x_1^*\|}, u_{2i} = \frac{\|x_{2i}-x_2^*\| - \min_{1\leq 2 j\leq l_2}\|x_{2j}-x_2^*\|}{\max_{1\leq 2 j\leq l_2}\|x_{2j}-x_2^*\| - \min_{1\leq 2 j\leq l_2}\|x_{2j}-x_2^*\|}$$

Step 3. Fuzzifying the training set.

$$S_{1f} = \{(x_{1j},y_{1j},u_{1j}) | x_{1j} \in R^d, y_{1j} = +1, u_{1j} \in [0,1], 1j=1,2,...,l_1\}$$

$$S_{2f} = \{(x_{2j},y_{2j},u_{2j}) | x_{2j} \in R^d, y_{2j} = -1, u_{2j} \in [0,1], 2j=1,2,...,l_2\}$$

Step 4. Cutting fuzzy set $S_f = S_{1f} \cup S_{2f}$ with parameter λ, where $\lambda = \{\lambda_1, \lambda_2, ..., \lambda_m\}$, $i=1,2,...,m$

$$S_{1f\lambda_i} = \{(x_{1j}, y_{1j}, u_{1j}) | u_{1j} \geq \lambda_i\}, S_{2f\lambda_i} = \{(x_{2j}, y_{2j}, u_{2j}) | u_{2j} \geq \lambda_i\}$$

Step5. Forming λ_i—FSVMs on λ_i—cut $S_{f\lambda_i} = S_{1f\lambda_i} \cup S_{2f\lambda_i}$

Step6. Verifying the performances of λ_i—FSVMs ($i=1,2,...,m$), and selecting the best λ—FSVMs.

In step 4, the parameter λ_i belongs to the interval [0,1]. The larger λ_i means the smaller training set, and the smaller one means the larger one contrarily. If we cut out too many samples, they will lose a great deal of information. Generally, λ_i focuses the interval [0,0.5]. In step 5, we construct m λ_i—FSVMs on different λ_i—cut.

3.3 Geometrical Interpretation of λ—FSVMs

As mentioned above, we mainly select the more important data set whose elements lie outside of the ball. The idea mainly derives from the fact that the mostly reality data obey normal distributions. Another distributions, such as uniform distributions, can be conversed into normal one. To compare the performance of λ—FSVMs with that of FSVMs clearly, the linearly separable two-class problem is discussed in two-dimensional space. In case of nonlinearly separable problem, λ—FSVMs is formed by mapping the input data into the feature space using kernel methods. In this experiment, two-class training data are generated by two normal distributions, $u_1 = [1,1]$, $u_2 = [5,5]$, $\Sigma_1 = \Sigma_2 = \begin{pmatrix} 1 & -0.3 \\ -0.3 & 1 \end{pmatrix}$. Obviously, the two-class problem is linearly separable. The hyperplane of FSVMs is showed in Fig. 2 (a). In Fig. 2 (b), the two hyperplanes of two kinds FSVMs are identical. The training set using the proposed method includes 43 data points when $\lambda = 0.5$.

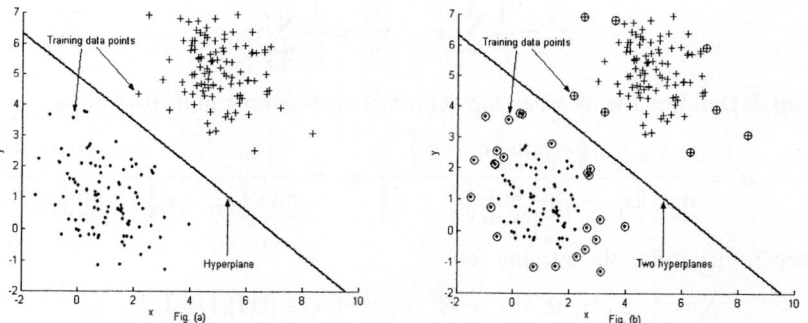

Fig. 2. Fig. (a) indicates the original data and their classification *hyperplane*. Fig. (b) indicates *Two hyperplanes* on the entire training data and the training data extracting from the original data respectively.

4 Comparison of λ—FSVMs and FSVMs

To verify the performance of λ—FSVMs, experiments were performed using some benchmark data in machine learning databases [9]. These data (see table 1) include two-class problem and multi-class problems. In the table, INPUT represents the number of the features, CLASSES represents the number of classes, Ntr and Nts denote the numbers of training and testing data respectively. λ—FSVMs are formed using different λ—cut. FSVMs are formed on the entire training set. For data sets, including WDBC (Wisconsin Diagnostic Breast Cancer), thyroid, iris and wine, two-third data are selected as training data and the rest as testing data at random. The same optimization method [10] is used in all the experiments.

Table 1. Benchmark data specification

DATA	IUPUT	CLASSES	Ntr	Nts
WDBC	30	2	379	190
Thyroid	5	3	143	72
Iris	4	3	99	51
Wine	14	3	118	60
Image	19	7	210	2100

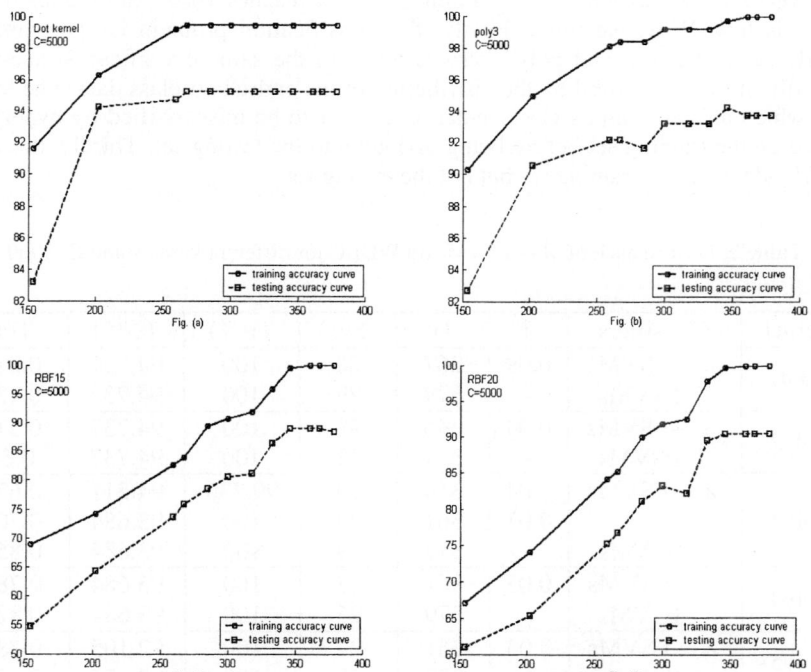

Fig. 3. The relationship between the size of training set and the classification accuracy. The last point in each curve denotes the accuracy of FSVMs. The remaining denote λ —FSVMs when $\lambda =\{0.2, 0.15, 0.1, 0.09, 0.08, 0.07, 0.06, 0.05, 0.04, 0.03, 0.02, 0.01\}$.

As for the selection of parameter, because we mainly discuss the influence that the parameter λ imposes on the proposed FSVMs, we select the constant C using the method in [11] to make the maximal accuracy of FSVMs. In all the tables, poly i denotes the polynomial kernel when order $d = i$, and RBF j denotes the RBF kernel when width $\sigma = j$.

For two-class problems, Fig.3 shows the relationship curves between λ—cut and the classification accuracy on benchmark data WDBC. Horizontal axis denotes the number of training data of the proposed FSVMs though λ—cut when λ ={0.2,0.15, 0.1,0.09,0.08,0.07,0.06,0.05,0.04,0.03,0.02, 0.01}, and the vertical axis denotes the accuracy of different λ—FSVMs. In Fig. 3 (b) and Fig. 3 (c), the overfitting problem emerges. We can obtain better classifier than FSVMs constructed on the entire training set. In table 2, Tr and Svs denote the number of training data and support vectors respectively, Tr(%) and Ts(%) denote the training and testing accuracy respectively, T(s) denotes the total run time including training and testing. The table lists the comparison of λ—FSVMs and FSVMs on WDBC for different kernels with the same C. The best classifier is not FSVMs but λ—FSVMs when $\lambda = 0.09$ for dot product kernel. There are only 267 training data points, including 28 support vectors. For some kernels, such as RBF5, the training accuracy reaches 100%, but the testing accuracy is low. Why, we think: Firstly, the classification problem is approximately linearly separable, dot and poly2 kernels result in the same classification accuracy. Secondly, it may be caused by the distribution of the entire two-class data. The testing data, which are far from its class center, are easier to be misclassified by hyperplane formed on the training set before being divided into the testing set. Thirdly, the width of RBF adapts to the training set but not the testing set.

Table 2. Performance of λ—FSVMs on WDBC for different kernels and C=5000

Kernel	Classifiers	λ	Tr	Svs	Tr(%)	Ts(%)	T(s)
Dot	λ—FSVMs	0.09	267	28	100	94.737	0.219
	FSVMs	-	379	29	100	94.737	0.437
Poly2	λ—FSVMs	0.04	346	32	100	94.737	0.515
	FSVMs	-	379	32	100	94.737	0.61
Poly3	λ—FSVMs	0.04	346	29	99.736	94.211	0.672
		0.03	361	29	100	93.684	0.703
	FSVMs	-	379	29	100	93.684	0.859
Poly4	λ—FSVMs	0.03	361	27	100	93.684	0.781
	FSVMs	-	379	27	100	93.684	0.828
Poly5	λ—FSVMs	0.03	361	27	100	92.105	0.782
	FSVMs	-	379	27	100	92.105	0.875
RBF5	λ—FSVMs	0.01	379	379	100	62.632	29.797
	FSVMs	-	379	379	100	62.632	29.797
RBF10	λ—FSVMs	0.03	361	346	100	88.947	21.875
	FSVMs	-	379	362	100	88.947	26.516
RBF15	λ—FSVMs	0.03	361	306	100	88.947	10.844
	FSVMs	-	379	313	100	88.421	11.828
RBF20	λ—FSVMs	0.03	361	251	100	90.526	6.265
	FSVMs	-	379	256	100	90.526	6.969

Table 3. Performance of λ—FSVMs and FSVMs for benchmark data for different kernels

Data	Parameter Kernel	Classifiers	λ	N_{tr}	Tr (%)	Ts (%)
Thyroid	C=5000 Dot	λ—FSVMs FSVMs	0.04 -	130 143	100 100	94.444 94.444
	C=5000 Poly4	λ—FSVMs FSVMs	0.4 -	23 143	81.818 66.667	77.778 67.13
	C=5000 RBF35	λ—FSVMs FSVMs	0.1 -	116 143	94.406 94.406	93.056 93.056
Iris	C=5000 Dot	λ—FSVMs FSVMs	0.3 -	55 99	100 100	96.078 96.078
	C=5000 Poly4	λ—FSVMs FSVMs	0.3 -	55 99	100 100	96.078 96.078
	C=5000 RBF20	λ—FSVMs FSVMs	0.1 -	82 99	100 100	98.693 98.693
Wine	C=5000 Dot	λ—FSVMs FSVMs	0.04 -	110 118	100 100	100 100
	C=5000 Ploy2	λ—FSVMs FSVMs	0.04 -	110 118	100 100	100 100
	C=5000 RBF20	λ—FSVMs FSVMs	0.04 -	110 118	100 100	83.333 83.333
Image	C=5000 Dot	λ—FSVMs FSVMs	0.01 -	188 210	96.095 97.524	90.667 90.581
	C=5000 Poly 2	λ—FSVMs FSVMs	0.002 -	203 210	97.524 97.524	89.781 89.781
	C=5000 RBF20	λ—FSVMs FSVMs	0.05 -	207 210	96.476 96.476	83.562 83.562

As for the multi-class problems, to reduce the unclassifiable regions for pairwise classification, Decision Directed Acyclic Graph (DDAG) is proposed in [4]. Pontil and Verri [12] proposed to use rules of a tennis tournament to resolve unclassifiable regions. Not knowing their work, Kijsirikul and Ussivakul [13] proposed the same method and called it Adaptive Directed Acyclic Graph (ADAG). There are three different structures of DDAG for the three-class problems. When the number of classes is more than three, the set of ADAGs is included in the set of DDAG. The number of different ADAGs for an n—class problem is given in [14]. In our experiments, we use integration of one against one and DDAG strategies, and randomly select five different structures when the number of classes exceeds three. Namely, for n—class problems, $n(n-1)/2$ FSVMs are constructed, and the class of unknown data is decided by DDAG. Table 3 lists the best results for each kind kernel function, including the number of training data, average training and testing accuracy. In most cases, the accuracy rates of the proposed FSVMs are greater than or equal to those methods by the conventional FSVMs. The number of training data is reduced in the former.

5 Conclusions

In this paper, we present λ—FSVMs based on λ—cut. By computer simulations using five benchmark data sets, we demonstrate the superiority of our method. Firstly, the overfitting problem can be avoided in λ—FSVMs. Secondly, as for the training data points, which obey normal distributions or are compact, λ—FSVMs can find the best results rapidly. Thirdly, in λ—FSVMs, the larger margin of the training data leads to a larger parameter λ and faster running speed. Owe to selecting an array of parameter λ, we must construct many different λ—FSVMs and run the program recurrently, it will cost more running time. The proposed method can be used to many classification problems, such as digit recognition, text classification and face detection, which need a large scale training data.

References

1. Wang, L.P. (Ed.): Support Vector Machines: Theory and Application. Springer, Berlin Heidelberg New York (2005)
2. V. N. Vapnik: Statistical Learning Theory, John Wiley & Sons, New York (1998)
3. C. W. Hsu, C. J. Lin: A Comparison of Methods for Multi-class Support Vector Machines, IEEE Transactions on Neural Networks. Vol. 13, No. 2, (2002) 415-425
4. J. C. Platt, N. Cristianini, and J. Shawe-Taylor: Large Margin DAG's for Multi-class Classification, Advances in Neural Information Processing Systems. Vol. 12, The MIT Press, Cambridge (2000) 547-553
5. T. Inoue, S. Abe: Fuzzy Support Vector Machines for Pattern Classification, Proceeding International Joint Conference on Neural Networks (IJCNN '01), Vol. 2, Washington (2001) 1449–1454
6. Han-Pang Huang and Yi-Hung Liu: Fuzzy Support Vector Machines for Pattern Recognition and Data Mining, International Journal of Fuzzy Systems, Vol. 4, (2002) 826-835
7. I. Guyon, N. Matic, and V. N. Vapnik: Discovering Information Patterns and Data Cleaning, The MIT Press, Cambridge (1996) 181-203
8. V. Barnet, and T. Lewis: Outliers in Statistical Data, Wiley, New York (1994)
9. ftp://ftp.ics.uci.edu/pub/machine-learning-databases
10. Kecman, V.: Learning and Soft Computing, Support Vector machines, Neural Networks and Fuzzy Logic Models, The MIT Press, Cambridge, MA (2001)
11. D. C. Montgomey: Design and Analysis of Experiments, 4th edition, John Wiley and Sons, New York (1997)
12. M. Pontil and A. Verri: Support Vector Machines for 3-d Object Recognition. IEEE Transactions on Pattern Analysis and Machine Intelligence, Vol. 20, No. 6, (1998) 637–646
13. B. Kijsirikul, and N. Ussivakul: Multi-class Support Vector Machines Using Adaptive Directed Acyclic Graph, Proceeding International Joint Conference on Neural Networks (IJCNN2002), Hawaii (2002) 980–985
14. S. Abe: Analysis of Multi-class Support Vector Machines, Proceeding International Conference on Computational Intelligence for Modeling Control and Automation (CIMCA'2003), Vienna (2003) 385-396

Mixtures of Kernels for SVM Modeling[1]

Yan-fei Zhu, Lian-fang Tian, Zong-yuan Mao, and Li Wei

College of Automation Science and Engineering, South China University of
Technology, GuangZhou, China 510640
yanfei_zhu@126.com

Abstract. Kernels are employed in Support Vector Machines (SVM) to map the nonlinear model into a higher dimensional feature space where the linear learning is adopted. The characteristic of kernels has a great impact on learning and predictive results of SVM. Good characteristic for fitting may not represents good characteristic for generalization. After the research on two kinds of typical kernels---global kernel (polynomial kernel) and local kernel (RBF kernel), a new kind of SVM modeling method based on mixtures of kernels is proposed. Through the implementation in Lithopone calcination process, it demonstrates the good performance of the proposed method compared to single kernel.

1 Introduction

Support Vector Machine (SVM) is a kind of machine learning algorithm proposed by V. N. Vapnik, which is based on Statistical Learning Theory (SLT) [1]. It works according to the principle of structural risk minimization (SRM) rather than the principle of empirical risk minimization (ERM) of large samples and has good generalization capability [2].

SVM for nonlinear modeling is to construct a nonlinear mapping to a high dimensional feature space where the linear learning machine is adopted. The nonlinear mapping functions are called Kernels--- $K(x_i, x_j)$, which should satisfy the Mercer's condition.

In the high dimensional feature space, the optimization problem is:

$$\min[\frac{1}{2}\|w\|^2 + C\sum_{i=1}^{n}(\xi_i + \xi_i^{'})] \quad (1)$$

ε - *insensitive* loss function is used to define the distance between points interested ($\varepsilon > 0$):

$$L(f(x), y) = \begin{cases} 0, & |f(x)-y| < \varepsilon \\ |f(x)-y| - \varepsilon, & others \end{cases} \quad (2)$$

When constructing the Lagrangian and transforming this optimization problem into the dual problem, the optimal solution can be presented as follows:

---

[1] Financial supported by Office of Science and Technology of Guangdong province in China (No: C10909) and by Department of Science and Technology of Guangzhou city in China (No: 2003Z3-D0091)

$$\max\left\{-\frac{1}{2}\sum_{i=1}^{n}\sum_{j=1}^{n}(\alpha_i-\alpha_i^{'})(\alpha_j-\alpha_j^{'})\langle x_i,x_j\rangle+\sum_{i=1}^{n}(\alpha_i-\alpha_i^{'})y_i-\sum_{i=1}^{n}(\alpha_i+\alpha_i^{'})\varepsilon\right\} \quad (3)$$

where, $\alpha, \alpha^{'}$ are Lagrange multipliers and subject to:

$$0\leq \alpha_i,\alpha_i^{'}\leq C, i=1,\cdots,n \quad (4)$$

$$\sum_{i=1}^{n}(\alpha_i-\alpha_i^{'})=0 \quad (5)$$

With the kernel $K(x_i,x_j)$, the optimal solution (3) becomes:

$$\max\left\{-\frac{1}{2}\sum_{i=1}^{n}\sum_{j=1}^{n}(\alpha_i-\alpha_i^{'})(\alpha_j-\alpha_j^{'})K(x_i,x_j)+\sum_{i=1}^{n}(\alpha_i-\alpha_i^{'})y_i-\sum_{i=1}^{n}(\alpha_i+\alpha_i^{'})\varepsilon\right\} \quad (6)$$

The nonlinear fitted model is obtained as:

$$f(x)=\sum_{i=1}^{n}(\alpha_i-\alpha_i^{'})K(x,x_i)+b \quad (7)$$

The characteristic of the nonlinear fitting model (7) is mainly determined by the type of the kernels. So it shows that the selection of suitable kernels for different identification systems is very important[4]. To solve the modeling problem of Lithopone calcination process, the modeling performance of SVM with different kernels is analyzed, which is related to the global and local features of the kernels [6]. In the second section, the mapping characteristic of two typical global (polynomial) and local (RBF) kernels is described, then a new kind of SVM modeling method based on mixtures of kernels is proposed in the third section, which not only has a good fitting accuracy, but also can prevent it from the fluctuation of the prediction outputs caused by the local kernel.

Lithopone calcination process is the key part for controlling the product quality. Measures such as adjusting the temperature and the rotational speed and providing suitable reaction conditions can be taken to improve the product quality. However, there still exist some unfavorable factors such as the high temperature of the calcination kiln, the varied inner flow field, limited monitoring points and the closeness of the kiln. So it's very difficult to set up a model for this kind of process [5,8]. The SVM modeling method based on mixtures of kernels provides an effective way to solve such problem with good fitting and prediction ability.

2 Global Kernels and Local Kernels

Kernels used by SVM can be divided into two classes: global and local kernels. In global kernels, points far away from the test point have a great effect on kernel values. While, in local kernels, only those close to the test point have a great effect on kernel values [6]. The polynomial kernel in (8) and the radial basis function (RBF) kernel in (9) are two typical global and local kernels. Figure 1 shows their mapping features.

$$K(x,x_i)=[(x\cdot x_i)+1]^q \quad (8)$$

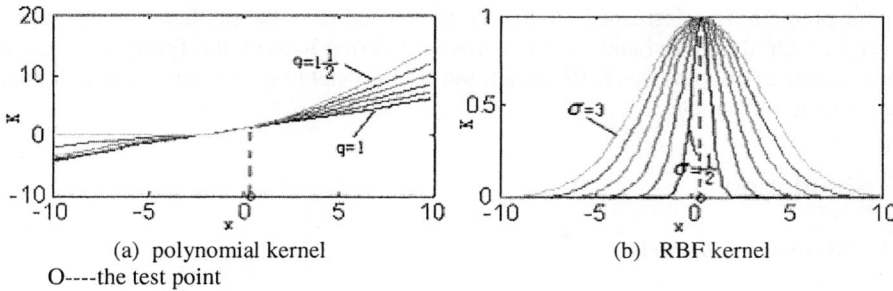

(a) polynomial kernel (b) RBF kernel
O----the test point

Fig. 1. Mapping features of polynomial and RBF kernels

$$K(x, x_i) = \exp(-\frac{|x - x_i|^2}{2\sigma^2}) \tag{9}$$

In polynomial kernel, the kernel parameter q is the operation degree of polynomial to be used. From figure 1 (a), even if various values of q are selected, only the points of the set x whose distances are far enough from the test point have an effective influence on the kernel values, and the further the distance, the greater the influence. In contrast, in the [0,1] range of RBF kernel values, the points adjacent to the test point have a great influence on the kernel values.

Table 1. Errors of SVM fitting and prediction with different kernels

	RBF Kernel					polynomial Kernel			
	Fitting errors		Prediction errors			Fitting errors		Prediction errors	
σ	MAXE	ME	MAXE	ME	q	MAXE	ME	MAXE	ME
15	2.5799	0.2541	16.6614	1.3259	1	2.8729	0.3756	4.4951	0.3280
5	1.2725	0.0268	25.4790	3.9555	1.2	5.6834	1.0732	7.4620	0.7997
1	0.001	0.001	Diverging		1.4	10.4147	1.2555	9.2617	1.4503

This feature is also demonstrated on fitting and prediction performances. For this aim, the data about the temperature and the rotational speed of a lithopone calcination process in Guangzhou of China are collected, which are sampled at every five minutes. After eliminating abnormal data and smooth treatments [7], 400 of them are kept, the former 200 are used for learning study and the latter 200 are for prediction ones.

Parameters of SVM are defined as: $\varepsilon = 0.001, C = 1000$. For different kernels and kernel parameters, the errors of fitting and prediction of SVM are shown in table 1. It can be seen that when using RBF kernel, the good fitting performance of SVM can be achieved. However, with the decrease of σ, both the maximum of absolute errors (MAXE) and the mean of absolute errors (ME) decrease rapidly, which leads to a poor prediction performance. When σ is reduced to 1, the prediction output starts diverging. On the other hand, when polynomial kernel is used, the fitting accuracy of SVM is not as good as the RBF kernel, but it has a stable prediction and the fluctuation is relatively small.

3 Simulation

3.1 Mixtures of Kernels

From the above analysis, it is obvious that RBF kernel (local kernel) can provide a good fitting performance for SVM; while polynomial kernel (global kernel) can restrain the fluctuation and keep a stable prediction. In addition, these two kernels have simple formats and are easy to calculate. On this basis, a new kind of SVM modeling method based on mixtures of kernels is proposed, which is expressed as:

$$K_{mix} = \rho K_{poly} + (1-\rho) K_{rbf} \qquad (10)$$

where, K_{poly} denotes polynomial kernel and K_{rbf} denotes RBF kernel. ρ (constant) is the optimal mixed coefficient to control the effects of these two kernels ($0 \leq \rho \leq 1$).

By using the same data as the former section, the SVM modeling analysis is conducted with the mixtures of kernels. Parameters of SVM are defined as: $\varepsilon = 0.001$, $C = 1000$, $\rho = 0.95$. The fitting and prediction errors of SVM are shown in table 2.

Table 2. Fitting and prediction errors of SVM with mixed kernel

σ, q	Fitting errors		Prediction errors	
	MAXE	ME	MAXE	ME
σ =15, q =1	2.3556	0.1930	4.7776	0.3368
σ =5, q =1	0.5846	0.0071	4.9166	0.3883
σ =1, q =1	0.001	0.0001	4.8381	0.3882
σ =15, q =1.2	3.4343	0.7351	8.0927	0.9530
σ =15, q =1.4	3.9600	0.9909	9.7702	1.4563

Compared with table 1, if decreasing the value of σ from 15 to 1 (q =1), the fitting and prediction errors are both lower than those of SVM with RBF kernel. Especially for prediction, the MAXE and ME decrease greatly. In other words, adding polynomial kernel can effectively restrain the fluctuation of the prediction output caused by RBF kernel. At the same time, adding RBF kernel can perfectly improve the fitting accuracy of SVM, which can't be done by polynomial kernel.

Figure 2 shows the fitting and prediction outputs of the rotational speed with three kinds of kernels---RBF, polynomial and mixed kernels.

In figure 2(a), using RBF kernel, the fitting performance is satisfactory. But the fluctuation of the prediction output is larger. At some sampled points, the rotational speed even reaches to zero value, which is not true in practical applications. Figure 2

(b) gives a relatively stable prediction output with polynomial kernel. In figure 2(c), both fitting and prediction performances are improved. The actual value and fitting value in the fitting curve almost coincide. The prediction error is not more than 5r/min. So, with mixed kernel, SVM has better modeling performance.

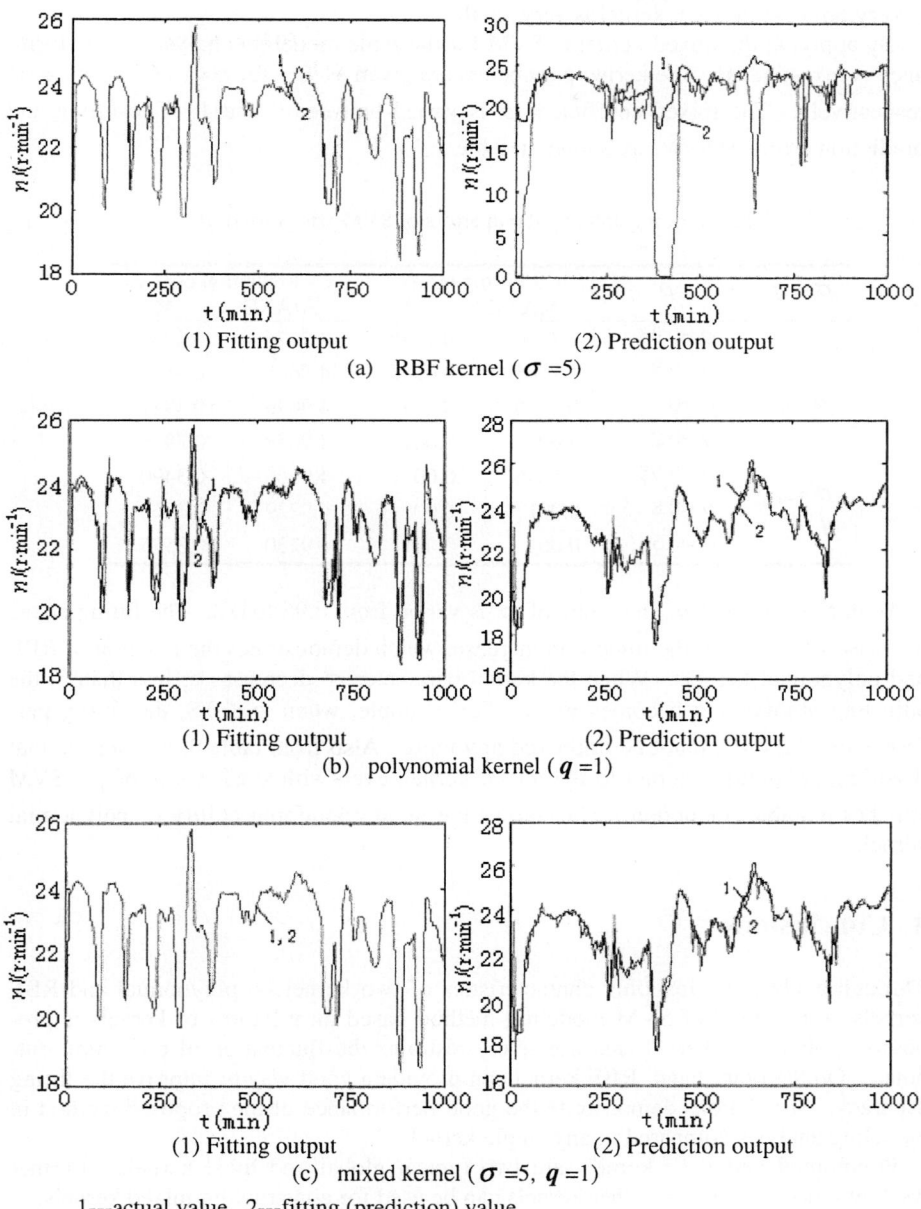

1---actual value   2---fitting (prediction) value

Fig. 2. Fitting and prediction outputs with different kernels

3.2 The Function of ρ

In mixtures of kernels, the contribution of two kernels is adjusted by ρ. For example, when ρ is close to 1, the contribution of polynomial kernel to the mixed kernel is very large while RBF kernel is very small.

By applying the mixed kernel to SVM for the same modeling analysis, its performance is investigated. The kernel parameters are given as $\sigma =15$, $q=1$ and $\sigma =5$, $q=1$ respectively. The mixed coefficient ρ is varied between 0 and 1. The fitting and prediction errors of SVM are shown in table 3.

Table 3. Fitting and prediction errors of SVM with varied ρ

σ, q	ρ	Fitting errors		Prediction errors	
		MAXE	ME	MAXE	ME
$\sigma =15$, $q=1$	$\rho =0.95$	2.3556	0.1930	4.7776	0.3368
	$\rho =0.8$	1.7005	0.0546	4.8605	0.3592
	$\rho =0.5$	0.4166	0.0049	4.9046	0.3921
	$\rho =0.2$	0.001	0.001	4.9058	0.3975
$\sigma =5$, $q=1$	$\rho =0.95$	0.5846	0.0071	4.9166	0.3490
	$\rho =0.8$	0.001	0.001	4.9229	0.3960
	$\rho =0.5$	0.001	0.001	4.9230	0.3575

With fixed σ and q, the value of ρ is varied from 0.95 to 0.2. The fitting errors decrease while the prediction errors increase, which demonstrates the reaction of RBF and polynomial kernels. When the kernel parameter σ decreases to less than 5, the adjusting ability of σ becomes weak. For example, when $\rho \leq 0.8$, the fitting performance of SVM cannot be improved any more. Also from table 3, we can see that if adding polynomial kernel to the mixed kernel, even with small value of ρ, SVM still has a stable prediction. This shows the good restraining ability of polynomial kernel.

4 Conclusions

Through analysis on mapping characteristics of two kernels--- polynomial and RBF kernels, a new kind of SVM modeling method based on mixtures of kernels is proposed. polynomial kernel can effectively restrains the fluctuation of prediction outputs. On the other hand, RBF kernel can provide a good way to improve the fitting accuracy. Simulations demonstrate the good performance of the proposed method in modeling analysis compared to any single kernel.

Polynomial and RBF kernels are two typical global and local kernels. Further work will discuss whether other kernels can be used for constructing mixed kernels.

References

1. Vapnik, V. N.: The Nature of Statistical Learning Theory. Springer-Verlag, New York (1995)
2. Bernhard, S., Alexander, J.S.: Learning with Kernels-Support Vector Machines, Regularization, Optimization and Beyond. The MIT Press, Cambridge, Massachusetts, London (2003)
3. Vapnik, V.N., Golowich, S., Smola, A.J.: Support Vector Method for Function Approximation, Regression Estimation, and Signal Processing. The MIT Press, Cambridge, MA (1997)
4. Smola, A.J.: Learning with Kernels. Ph.D. thesis, TU Belin (1998)
5. Liu, Y.-P.: Development of Process Control System for Lithopone Kiln and Calciner. Master's Degree Thesis, South China University of Technology, GuangZhou, China (2002)
6. Scholkopf, B., Mika, S., Burges, C.J.C., Knirsch, P., Muller, K.R., Ratsch, G., Smola, A.J.: Input Space Versus Feature Space in Kernel-Based Methods. IEEE Trans. on Neural Networks, Vol.10 (1999) 1000-1017
7. Huang, R. T., Liu, Y. P., Di, Z., Mao, Z. Y.: Development of Process Control System for Lithopone Kiln and Calciner (II)---Preprocessing Technology of Sampling Data. Journal of South China University of Technology (Natural Science Edition), Vol. 30. GuangZhou, China (2002) 52-55
8. Huang, R. T., Liu, Y. P., Mao, Z. Y., Di, Z.: Realization of Process Control System for Lithopone Kiln and Calcinator. Journal of South China University of Technology (Natural Science Edition), Vol. 31. GuangZhou, China (2003) 42-45

A Novel Parallel Reduced Support Vector Machine

Fangfang Wu, Yinliang Zhao, and Zefei Jiang

Institute of Neocomputer, Xi'an Jiaotong University,
Xi'an 710049, People's Republic of China
wffbolun@163.com, zhaoy@xjtu.edu.cn, jiangzefei83@163.com

Abstract. Support Vector Machine (SVM) has been applied in many classification systems successfully. However, it is restricted to work well on the small sample sets. This paper presents a novel parallel reduced support vector machine. The proposed algorithm consists of three parts: firstly dividing the training samples into some grids; then training sample subset through density clustering; and finally classifying the samples. After clustering the positive samples and negative samples, this algorithm picks out such samples that locate on the edge of clusters as reduced sample subset. Then, we sum up these reduced sample subsets as reduced sample set. These reduced samples are then used to find the support vectors and the optimal classifying hyperplane by support vector machine. Additionally, it also improves classification precision by reducing the percentage of counterexamples in kernel object ε-area. Experiment results show that not only efficiency but also classification precision are improved, compared with other algorithms.

1 Introduction

Support vector machine is a kind of classifier's studying method on statistic study theory [1,2]. This algorithm derives from linear classifier, and can solve the problem of two kind classifier, later this algorithm applies in non-linear fields, that is to say, we can find the optimal hyperplane (large margin) to classify the samples set. SVM can use the theory of minimizing the structure risk to avoid the problems of excessive study, calamity data, local minimal value and so on. For the small samples set, this algorithm can be generalized well [3].

Support vector machine (SVM) has been successfully used for machine learning with large and high dimensional data sets. This is due to the fact that the generalization property of an SVM does not depend on the complete training data but only a subset thereof, the so-called support vectors. Now, SVM has been applied in many fields as follows: handwriting recognition [4], three-dimension objects recognition, faces recognition [5], text images recognition, voice recognition and so on. However, because of the high cost of kernel function's computation, the training time is long. According to this, the samples set should be small in order to reduce the cost of training time. When the samples set is too large, we need find a new algorithm to reduce the number of this set, at the same time, these samples in the new set can represent the original samples set sufficiently.

For the given samples set $\{(x_1, y_1), ...,(x_m,y_m)\}$, $x_i \in R^n$, $y_i \in \{+1,-1\}$, m is the samples number, n is the number of input dimension. In order to find the optimal hyperplane to classify this data set precisely, SVM use the decision-making function:

$$f(x) = sign\left(\sum_{i=1}^{m} y_i \alpha_i k(x_i, x) - b\right) \quad (1)$$

α_i is the Lagrange factor, and b is the threshold.

The kernel function $k(x_i, x)$ must be satisfied with the condition of Mercer. When we define the kernel function $k(x_i, x)$, we also define the mapping from input to character's space. Training a SVM can be regarded as to solve a problem of protruding quadratic programming:

$$\max_{\alpha} L(\alpha) = \sum_{i=1}^{m} \alpha_i - \frac{1}{2} \sum_{i=1}^{m} \sum_{i=1}^{m} \alpha_i \alpha_j y_i y_j k(x_i, x_j) \quad (2)$$

$k(x_i, x)$ is the kernel function, marking $D_{ij} = y_i y_j k(x_i, x_j)$, D is a $m \times m$ matrix.

According to the above description, for the developing number of samples, the matrix D will be increased at the speed of m^2. If the dimension of x is n, the cost of one time computation of kernel function is $O(n)$, the time complexity of SVM is $O(nm^2)$. When n is big, the cost of one time computation of kernel function is big; therefore, the number of samples limits the training speed of SVM.

In order to accelerate the training speed, the method of random selecting from samples set was proposed at first, the selected samples are usually below 10 percent of original samples, then, we can use these selected samples to solve the classifier problem; however, the random selected samples can not represent the original samples set precisely, the result of hyperplane is bad. Later, the researchers solve this problem with two methods. The first one is that we can reduce the training sample set by clustering before we find the support vectors and the optimal hyperplane, the other one is that we can use parallel algorithm to accelerate the training speed. We will introduce the development of these two methods as follows.

For the first method of clustering, paper [6] first proposed the concept of reduced support vector machine. Later, paper [7] proposed a classifier algorithm of reduced support vector machine on unsupervised clustering (RSVM-UC) which uses the center samples to represent each cluster, and makes these samples become the reduced training samples. But this algorithm has its own disadvantage. It can only get spheric clusters. It is not valid to use the cluster's center to represent this whole cluster. The result of classifier is not good at times.

For the method of parallel dealing with the training procedure, paper [8] proposed a new method to select the support vectors through the analysis of the distance matrix D. It can parallel reduce the matrix D through divide D into several submatrixs. But this method need also compute the matrix D, in fact, the time complexity of computation is also $O(nm^2)$ and not decreased.

In the year of 2002, Zhang Ling and Zhang Bo propose a new theory about SVM [9]. If we select the edge samples of the two kind sample's intersection to find the support vectors, the result of this method is similar with the result of SVM directly. That is to say, we can only use these edge samples to reduce the redundant samples in original training sample set. In the procedure of classifying, we can also find that the most import samples for classifier are the edge samples in a dense samples cluster, the shape of these samples is not normal; according to this, this paper proposes a new classifier algorithm of reduced support vector machine on density clustering (RSVM-DC). It can use density clustering to find the cluster's edge samples to represent all samples in this cluster. At the same time, it can also increase the selected samples through controlling the percent of counterexample in a kernel object's ε-area. This algorithm can solve classifier problem more efficiently and more precisely.

For solving the problem of training the large scale sample set, we also find that we can parallel find particular reduced samples and delete the redundant samples on RSVM-DC. With this method, the training procedure can be run faster. For this reason, we propose a new parallel reduced support vector machine. This algorithm can parallel reduce the sample set and select the edge samples as a reduced sample set after dividing the original training sample set into several subsets. The reduced sample set is much smaller than the original training sample set. Then, the reduced samples can be regarded as new training samples to find the support vectors. In section 2, we propose the classifier algorithm of reduced support vector machine on density clustering (RSVM-DC); in section 3, we propose the parallel reduced support vector machine (PRSVM-DC) and its mended algorithm named mended parallel reduced support vector machine (MPRSVM-DC); in section 4, we do the experiment to compare with other algorithms of SVM; finally, in section 5, we draw the conclusion.

2   The Algorithm of Extracting the Edge Samples on the Basis of Density Clustering

Definition 1: For a given object $x \in R^n$, the field of sphere whose radius is ε with the center of x is named ε-area.

Definition 2: In an object's ε-area, if the number of objects around this object is MinPts θ or more, we name this object *kernel object*.

Definition 3: There are ϕ objects in the ε-area of object m_i. For these objects, m_i belongs to the kind of v, $v \in \{+1, -1\}$; if there are $\phi_v (\phi_v \geq \theta)$ objects belonging to the kind of v and $\phi_v / \phi \geq \eta$, η is a given parameter which means the percent of this kind in m_i ε-area, we name m_i *approximate kernel object*.

Definition 4: A point p is *directly approximate density-reachable* from a object q with respect to ε, MinPts if:1) p belongs to the q ε-area; 2) q should be a approximate kernel object.

Definition 5: A object p is *approximate density-reachable* from a object q with respect to ε and MinPts if there is a chain of objects $p_1,...,p_n$, $p_1=q$, $p_n=p$ such that p_{i+1} is directly approximate density-reachable from p_i.

The theory of density clustering is described as clustering through examining every object ε-area in the samples set. If there are more than θ objects in object p ε-area, and $\phi_v / \phi \geq \eta$, we can found a new cluster around the center p; then, we can continuous find the approximate density-reachable object, so that the shape of the cluster can be arbitrary; when there is no object can be joined this cluster, we can select another object which is not used. If each object is used, this procedure is over.

In order to select the object which has the feature of cluster edge, we can use the above method to do as follows: when an object p is not a approximate density-reachable object, the feature of this object shows that it is the edge object of this cluster, then, we can save this object to the new sample set until the end of procedure of extracting samples. We can describe the algorithm of reduced training samples on Density Clustering as follows:

Algorithm (1):

1) Give the values of ε, MinPts θ and η;
2) Select a non-used object at random from the samples set;
3) Judge the kind of this object, according to the value of ε, MinPts θ and η; we can estimate whether this object is approximate kernel object, if not, jump to (5);
4) Sign the object with used object, select a non-used object in its ε-area, jump to (3);
5) Join this object to the new selected samples set S, Sign it with used object, judge whether there is a non-used object, if there is a non-used object, jump to (2).

For the algorithm (1), if we use the space index to run density clustering, the time complexity is $O(nmlogm)$. As we make the edge samples represent the whole original samples set, algorithm (1) has the characters as follows: (1) it can describe the abnormal shape of cluster; (2) it can increase the selected sample's number through controlling the percent η of counterexamples. In section 4, the experiments describes that the precision can be improved as we increase the value of η. If we compare this algorithm to the reduced algorithm of unsupervised clustering (RSVM-UC), the algorithm (1) is better.

After we get the reduced training sample set, the redundant samples has been deleted. Then we can regard this reduced sample set as new training set to get the support vectors. We give the reduced support vector machine on Density Clustering (RSVM-DC) as follows:

Algorithm (2):

1) Give the training samples set Z, the values of ε, MinPts θ and η;
2) Use the algorithm (1) to get the reduced sample set S;
3) Regard the reduced sample set S as the new training sample set, using SVM, find the optimized classifier hyperplane.
4) For an object x, we can use the equation (1) to tell which kind it is.

For this algorithm, ε, MinPts θ and η are experimental parameters. Generally, firstly we select a small sample set D_r from whole training sample set at random, $D_r=\{x_1,x_2,\ldots,x_r\}$, $r<<m$, then we find the minimal distance d_{ij*} between x_i in D_r and the object x_j in the whole training sample set, $d_{ij*}=\min\{d_{ij}\}(1\leq j\leq m)$, and get the value of d_a, $d_a = 1/r\sum_{i=1}^{r} d_{ij*}$. The value of ε can be given between twice value of d_a and

treble value of d_a. If there are N_i samples in the object x_i's ε-area, $\theta = 1/r \sum_{i=1}^{r} N_i$.
When we want to get the high accuracy of classifier, the value of η can be given between 0.5 and 1. However, when we want to get the high efficiency of classifier, the value of η can be given between 0 and 0.5.

For the algorithm of RSVM-DC, we also use the space index to run density clustering. The time complexity is $O(n(mlogm+l^2))$, $l<<m$, the cost of time is close to RSVM-UC. However, because RSVM-DC make the edge samples of each cluster represent whole cluster, these objects can describe the shape and character of clusters more precisely. Then, we find that the reduced training sample set of RSVM-DC is better than the reduced training sample set of RSVM-UC, and the result of RSVM-DC will be more precisely. More important, if we divide the whole sample set into several sample sets, we can use algorithm (1) to reduce the scale of sample set parallel because of the feature of edge of samples. According to RSVM-DC, the parallel reduced SVM is described in section 3.

3 The Reduced Parallel Support Vector Machine on Density Clustering

3.1 The Reduced Parallel Support Vector Machine on Density Clustering

According to the different attributes of training samples, if we divide the whole training sample set into several subsets by partitioning the values of sample's attributes and the intersection of these subsets is nothing, we can only care about the inner samples and the edge samples within a subset. Then, we can find that the relationship of different subsets focuses on the relationship of edge samples among them. For this reason, after we find the support vectors within a subset, we need also save those edge samples near to the other subsets. We can add these support vectors of subsets and the edge samples to a new reduced training sample set, and use this set to find the support vectors of whole training sample set. Therefore, finding the edge samples within a subset is the key to parallel reduce the training samples. For the above algorithm (1), we can find that the reduced samples of this algorithm can contain the particular samples which reflect the classifier information and the edge samples within a subset, because of this, we can use algorithm (1) to reduce the training samples parallel. For the above introduction, we propose a novel parallel reduced support vector machine on density clustering (PRSVM-DC). We describe the parallel reduced support vector machine on density clustering (PRSVM-DC) as follows:

Algorithm (3):
1. According to the values of training sample set's attributes, divide the set D into subsets $D_1, D_2, \ldots, D_L$;
2. R'=∅;
3. For i=1 to L pardo (deal with sample set D_i parallel)
4. Begin

      a. Use algorithm (1) to select particular edge samples from set D_i as training subset R_i';
      b. Lock(R');

      c. R'=R'+R_i';
      d. Unlock(R');
5. End.
6. According to the new reduced training sample set R', we can get the support vectors and the optimal hyperplane for whole training sample set.

For the algorithm of PRSVM-DC, we also use the space index to run density clustering, E=max{|Di|, 1≤i≤L}, the time complexity of getting reduce sample set R is $O(nElogE)$, $E<<m$, and the time complexity of the algorithm (3) is $O(n(ElogE+b^2))$, $b<<m$, b is the number of reduced samples. Comparing with the time complexity of RSVM-UC and RSVM-DC, the algorithm (3) can accelerate speed of classifier more. At the same time, because the algorithm (3) selects the edge samples, the precision of this algorithm can be warranted in an appropriate range. In section 4, the result of experiment can approve that the speed of classifier can be accelerate deeply.

3.2 The Mended Parallel Reduced Support Vector Machine

If the number of one kind cluster is too many, we find that there are many samples which are not important for classifying, therefore, we propose mended algorithm for algorithm (3).

As the figure 1 shows, the solid round represents the kind of +1, and the solid rectangle represents the kind of -1. After using algorithm (3), we can get the samples as the figure 1 shows. Then, we can get the support vectors within a subset, these support vectors locate at the real line H^+ and H^-, and there are many redundant samples near to these two real lines. We need delete these redundant samples to decrease the scale of reduced training set. We can define the value c, if the distance between a sample and H^+ or H^- is less than c such as the sample between the real line H^+ and 1+c or the real line H^- and 1-c in figure 1, this sample is redundant. At the same time, we need save the samples at edge of this subset. So, after we reduce the training sample set again, we can get the reduced samples as figure 2 shows.

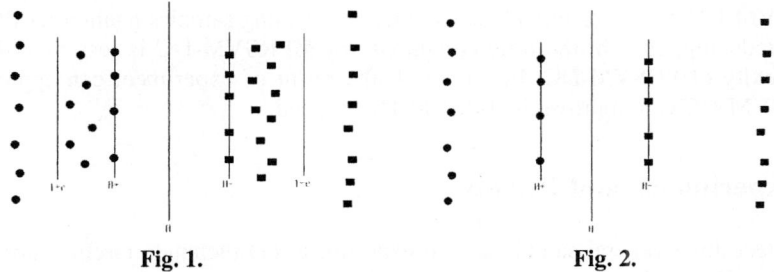

Fig. 1.   Fig. 2.

We describe the mended parallel reduced support vector machine on density clustering (MPRSVM-DC) as follows:

Algorithm (4):
1. According to the values of training sample set's attributes, divide the set D into subsets $D_1, D_2, \ldots, D_L$;
2. R'=∅;

3. for i=1 to L pardo (deal with sample set D_i parallel)
   a. Begin
   b. If $|D_i^+|>2$ and $|D_i^-|>2$ then
      i. Begin
      ii. Use algorithm (1) to select particular edge samples as reduced sample set S_i;
      iii. According to set S_i, find the support vectors and optimal hyperplane H by SVM;
      iv. Give the value of c, save the support vectors, and delete the sample which the distance from H is less than c, finally, we get the new training sample subset R_i';
      v. End
   c. Else Use algorithm (1) to select particular edge samples as training subset R_i' directly ;
   d. Lock(R');
   e. R'=R'+R_i';
   f. Unlock(R');
   g. End.
4. According to the new reduced training sample set R', we can get the support vectors and the optimal hyperplane for whole training sample set.

For this algorithm (4), we use a method to reduce the training sample set again, c is an experimental parameter. Generally, the value of c is less than the half of minimal width of a subset, $c<1/2 width(D_i)$. It can assure that the edge samples of this subset are held. This algorithm is applicable for the subset whose training samples locate near to the classifier's hyperplane mostly.

For the algorithm of MPRSVM-DC, we also use the space index to run density clustering, $E=\max\{|D_i|, 1 \leq i \leq L\}$, the time complexity of getting reduce sample set S is $O(nElogE)$, $E<<m$, and the time complexity of the algorithm (4) is $O(n(ElogE+i^2))$, $i<<m$, i is the number of reduced samples. Comparing with the time complexity of PRRSVM-DC, the algorithm (4) can reduce the training samples again after PRSVM-DC's reducing, so i<b, the time complexity of MPRSVM-DC is less than the time complexity of PRSVM-DC. In section 4, the result of experiment can approve that MPRSVM-DC can improve the PRSVM-DC's speed.

4 Experiments and Results

We select three general sample sets to experiment, (1) rectangle circling sample set, (2) Adult Data [12] whose number is big. We experiment these 2 samples sets as the following introduction.

For set (1), the experiments are done on an Intel P4 PC (with a 2.0GHZ CPU and 512MB memory) running Microsoft Windows 2000 Professional, Matlab6.5.

The samples of set (1) have proportional spacing. This set contains two quadrate circle sample subsets whose kind is -1 and two quadrate circle sample subsets whose kind is +1. The number of samples in this set is 921. As the figure 3 shows, the symbol signed + represents the kind of positive, the symbol signed * represents the kind of negative. The positive samples have 153 samples, and the negative samples have 768 samples.

We use RSVM-UC [7] to classify the set (3), and the clustering radius is 0.1. Then, the centers of +1 kind and -1 kind are the same as the point (1.6, 1.6); if we give the value 0.4 to the incision radius R, we can get the reduced sample set as the figure 4 shows. We can find that this method is not suitable for SVM classifier.

For density clustering, we give the value 0.1 to clustering radius, Minpts θ is 4. We can get the result as the below show in figure 5. There are 464 samples after this experiment. The positive reduced sample set has 80 samples, and the negative reduced sample set has 384 samples. We make these samples become new reduced training samples, then, we can get a hyperplane for the SVM classifier as bold real line in figure 18. The result of this method is available.

For the parallel reduced support vector machine, we divide the training samples into four subsets D_1, D_2, D_3, D_4 as figure 6 shows. If we give 0.1 to ε and 4 to Minpts θ, after we use PRSVM-DC to reduce the training samples, we can get the result as the figure 7 shows. For the reduced result, there are 89 positive samples and 400 negative samples. Obviously, when we use the algorithm of PRSVM-DC, the computing scale of training set is decreased to 1/4 of RSVM-DC, the time complexity is less than RSVM-DC.

For the above dividing method, we use the MPRSVM-DC. If we give 0.4 to c, after we use MPRSVM-DC to reduce the training samples again, we can get the result as the figure 8 shows. In the reduced result, there are 49 positive samples and 200 negative samples. Obviously, when we use the algorithm of MPRSVM-DC, the computing scale of training set is decreased to 1/4 of RSVM-DC, the time complexity is less than PRSVM-DC. The result of this method is available.

After we use the reduced SVM to reduce the training samples, we can get support vectors and optimal hyperplane. As the figure 9 shows, the bold real line represents the optimal hyperplane, and the points near to the bold real line are the support vectors.

For the set (3), the experiment is done on an Intel P4 PC (with a 2.0GHZ CPU and 512MB memory) running Microsoft Windows 2000 Professional, Microsoft VC++6.0 compiled language. The adult data set is the report of census. Every data has 14 attributes. After transforming the value of each attribute between 0 and 1, we have the data set which has six numerical value attributes. The number of this data set is 32561. The set can be trained for forecasting whether man's income is over 50,000$. The kernel function of SVM is the Gauss function that is $k(x_1, x_2) = \exp(-\|x_1 - x_2\|^2 / 2\sigma^2)$, $\sigma^2 = 10$. In this set, there are two kinds of family's income. One kind is over 50,000$, the number of this kind of data is 7841. The other kind is below 50,000$, the number of this kind of data is 24720. Besides training data set, we have a testing data set and the scale of this set is 16281. The testing set examines whether the network structure that has been trained can be generalized. For running PRSVM-DC and MPRSVM-DC, the training sample set is divided into four subsets by the attribute of age. The four subsets are D_1(age<30), D_2(30≤age<40), D_3(40≤age<50) and D_4(age≥50). The subset D_1 has 9711 samples, the subset D_2 has 8631 samples, the subset D_3 has 7175 samples and the subset D_4 has 7062 samples.

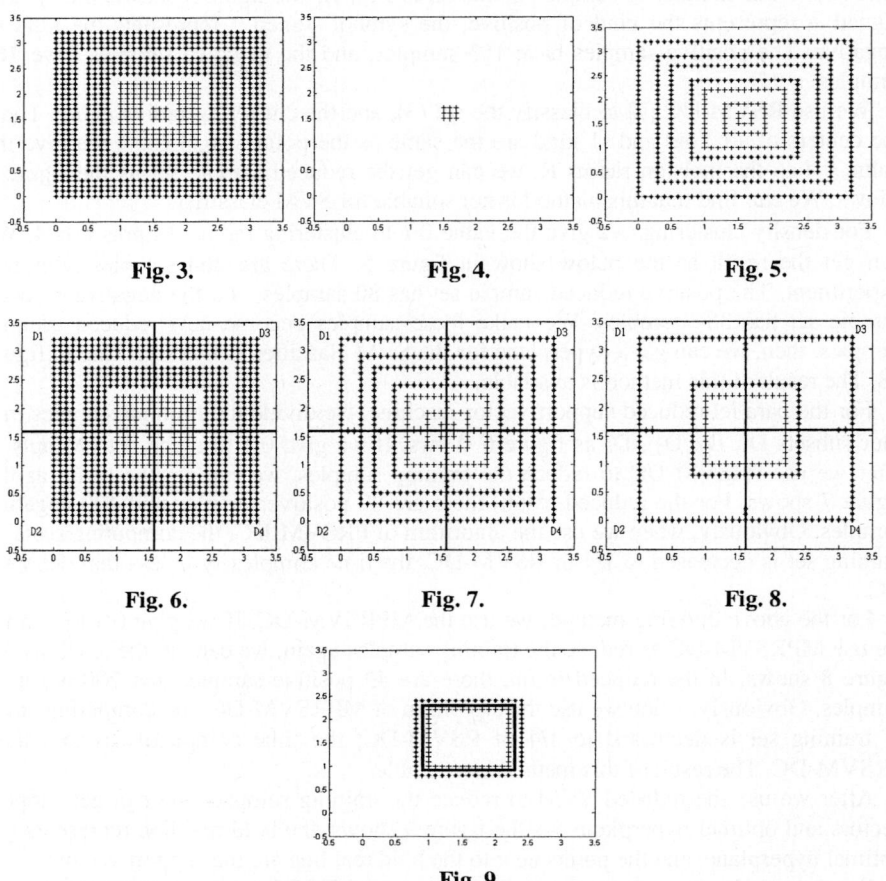

Fig. 3.   Fig. 4.   Fig. 5.

Fig. 6.   Fig. 7.   Fig. 8.

Fig. 9.

Table 1 is the comparison of computation and precision among SVM algorithms. There are 6 kind algorithms as follows: SVM, SVM after selecting 1 percent of samples set at random, RSVM-UC, RSVM-DC, MPRSVM-DC and PRSVM-DC. From the analysis of the data in table 1, the experiment of RSVM-UC and RSVM-DC is done with different parameters. For RSVM-UC, the number of selected samples can be increased by increasing the value of R, and the precision can also be increased; For RSVM-DC, the selected samples set can be increased by improving the value of η, that is to say, when the number of counterexample in a kernel object's ε-area is decreased, we can get more edge vectors, and the precision can be increased. From the result of this experiment, we can see: if we only use SVM to classify, the computation times of kernel are too many. However, if we use other five kind algorithms, the computation times can be decreased from 10^9 to 10^7. When we use the random algorithm, the precision is low; when the number of new selected samples for RSVM-DC is close to RSVM-UC, the precision of RSVM-DC is better than RSVM-UC; when their precisions are similar, the number of new selected samples for RSVM-DC is smaller than RSVM-UC, and the speed of classifier can be accelerated greatly by using RSVM-DC. When we compare RSVM-DC with PRSVM-DC, the time cost of PRSVM-DC is less than RSVM-DC because of parallel running, and the scale of PRSVM-DC's training sample set is smaller than RSVM-DC's. When we compare PRSVM-DC with MPRSVM-DC, because MPRSVM-DC can reduce the result of PRSVM-DC again, the time cost of MPRSVM-DC is less than PRSVM-DC. As the table 1 shows, the result of MPRSVM-DC can not only accelerate the speed of SVM but also assure the precision in a available range.

Table 1. The comparison of computation and precision among Reduced SVM algorithms

	Training samples number	Reduced samples number		The Times of Kernel Function	Accuracy
SVM	32561	32561		1060218721	91.89%
SVM(Random 1%)		326		106276	37.29%
RSVM-UC(r=0.16, R=0.18)		1212		1468944	73.44%
RSVM-UC(r=0.16, R=0.2)		1572		2471184	80.09%
RSVM-DC(ε=0.2, θ=8, η=0)		259		67081	85.37%
RSVM-DC(ε=0.2, θ=8, η=0.8)		884		781456	90.39%
RSVM-DC(ε=0.18, θ=9, η=0)		457		208849	87.55%
RSVM-DC(ε=0.18, θ=9, η=0.8)		1243		1545049	93.43%
PRSVM-DC(ε=0.2, θ=8, η=0)	9711(D_1)	69	372	138384	85.40%
	8613(D_2)	131			
	7175(D_3)	95			
	7062(D_4)	77			
PRSVM-DC(ε=0.18, θ=9, η=0.8)	9711(D_1)	230	1427	2036329	93.24%
	8613(D_2)	462			
	7175(D_3)	375			
	7062(D_4)	360			
MPRSVM-DC(ε=0.2, θ=8, η=0, c=0.2)	9711(D_1)	47	251	63001	84.91%
	8613(D_2)	63			
	7175(D_3)	72			
	7062(D_4)	69			
MPRSVM-DC(ε=0.18, θ=9, η=0.8, c=0.2)	9711(D_1)	53	385	148225	92.98%
	8613(D_2)	77			
	7175(D_3)	95			
	7062(D_4)	160			

5 Conclusion

For the classifier of SVM, this paper proposes an efficient classifier algorithm of parallel reduced support vector machine on density clustering (PRSVM-DC) and it's mended algorithm named MPRSVM-DC. After dividing the whole training sample set into several subsets, these two algorithms can use the edge samples to represent the samples of subset, parallel decreasing the number of training samples, accelerating the speed of training procedure. At the same time, these two algorithms can also assure the highly precision. In high dimension space, the cost for kernel function is high; through parallel decreasing the number of samples, we can decrease the computation times of kernel function, and increase the efficiency of classifier. When we face the large scale training set, the PRSVM-DC can give an available way to solve the classifier problem.

References

1. Vapnik, V.: The Nature of Statistical Learning Theory. Springer-Verlag , New York(1995) 1-175
2. Zhang, X.G.: Introduction to Statistical Learning Therory and Support Vector Machines. Acta Automatica Snica. 26(1), (2000)32-42
3. Burges, C.J.C.: A tutorial on support vector machines for pattern recognition. Data Mining and Knowledge Discovery. 2(2), (1998) 955-974
4. Bernhard S., Sung K.K.: Comparing Support Vector Machines with Gaussian Kernels to Radical Basis Fuction Classifiers. IEEE Transaction on Signal Processing. 45(11), (1997) 2758-2765
5. Edgar O., Robert F., Federico G.: Training Support Vector Machines: An Application to Face Detection. IEEE Conference on Computer Vision and Pattern Recognition. (1997)130-136
6. Y.J. Lee., O. L. Mangasarian.: Reduced Support Vector Machines. First SIAM International Conference on Data Mining, USA, Chicago(2001).
7. Li, X.L., Liu, J.M., Shi, Z.Z.: A Chinese Web Page Classifier Based on Support Vector Machine and Unsupervised Clustering. Journal of Computers. 24(1), China(2002)62-68
8. Jian-xiong, Dong., Adam, Krzyzak., Ching, Y.Suen.: A Fast Parallel Optimization for Training Support Vector Machine. MLDM2003, (2003)96-105.
9. Zhang, Ling., Zhang, Bo.: Relationship between support vector set and kernel functions in SVM. Journal of Computer Science and Technology, vol.17(5), China(2002).
10. Kecman, V.: Learning and Soft Computing, Support Vector machines, Neural Networks and Fuzzy Logic Models, The MIT Press, Cambridge, MA (2001).
11. Wang, L.P.: Support Vector Machines: Theory and Application. Springer, Berlin Heidelberg New York (2005).
12. ftp://ftp.ics.uci.edu.cn/pub/machine-learning-database/adult

Recurrent Support Vector Machines in Reliability Prediction

Wei-Chiang Hong[1], Ping-Feng Pai[2,*], Chen-Tung Chen[3], and Ping-Teng Chang[4]

[1] School of Management, Da-Yeh University
112 Shan-Jiau Rd., Da-Tusen, Chang-hua, 51505, Taiwan
d9230006@mail.dyu.edu.tw
[2] Department of Information Management, National Chi Nan University
1 University Rd. Puli, Nantou, 545, Taiwan
paipf@ncnu.edu.tw
[3] Department of Information Management, Da-Yeh University
112 Shan-Jiau Rd., Da-Tusen, Chang-hua, 51505, Taiwan
chtung@mail.dyu.edu.tw
[4] Department of Industrial Engineering and Enterprise Information, Tunghai University
Box 985, Tunghai University, Taichung, 407, Taiwan
ptchang@ie.thu.edu.tw
Tel. no.: +886-49-2910-960 ext:4141; fax no.: +886-49-2915-205.

Abstract. Support vector machines (SVMs) have been successfully used in solving nonlinear regression and times series problems. However, the application of SVMs for reliability prediction is not widely explored. Traditionally, the recurrent neural networks are trained by the back-propagation algorithms. In the study, SVM learning algorithms are applied to the recurrent neural networks to predict system reliability. In addition, the parameter selection of SVM model is provided by Genetic Algorithms (GAs). A numerical example in an existing literature is used to compare the prediction performance. Empirical results indicate that the proposed model performs better than the other existing approaches.

Keywords: Recurrent neural networks, Support vector machines, Genetic algorithms, Reliability prediction.

1 Introduction

Modeling and forecasting of reliability is a crucial issue in manufacturing systems. In most situations, the reliability of manufacturing systems changes with time. Therefore, the changes can be treated as a time series process. However, it is difficult to predict the variability of reliability with time. The difficulty arises from assumptions of the failure distributions and a lack of suitable reliability models. The forecasting techniques of reliability include lifetime distribution, Markov models, parts count and parts stress, and fault tree analysis. Due to the general nonlinear function mapping capabilities, artificial neural networks have received increasing attentions in time series forecasting. However, the literature on the application of artificial neural networks to reliability forecasting is very limited. Liu et al. [1] showed that feed-forward multilayer perceptron networks are able to identify the failure distribution as well as estimate the distribution parameters. Su

---

* Corresponding author.

et al. [2] proposed an ICBPNN forecasting model (input- combined back-propagation neural networks) which combines time series with neural networks techniques to predict engine reliability. They reported that the ICBPNN outperformed ARIMA models and BPNN models in terms of forecasting accuracy. Amjady and Ehsan [3] proposed a neural-network-based expert system to evaluate the reliability of power systems. The presented systems were able to conquer certain difficulties such as low accuracy, complex modeling and heavy computations. Ho et al. [4] presented a comparative analysis of neural networks and autoregressive-integrated- moving average (ARIMA) techniques in forecasting repairable systems. Their experimental results showed that both recurrent neural networks and multilayer feed-forward neural networks are superior to ARIMA approach in terms of forecasting accuracy. Xu et al. [5] applied feed-forward multilayer perceptron (MLP) neural networks and radial basis function (RBF) neural networks to forecast engine systems reliability. Those researchers compared neural network techniques with the ARIMA approach. Sensitivity analysis of neural networks was performed and appropriate architectures of neural networks were determined.

Originally, SVMs were developed for pattern recognition problems. Recently, with the introduction of Vapnik's ε-insensitive loss function, SVMs have been extended to solve nonlinear regression estimation problems and successfully in dealing with forecasting problems in many fields. Tay and Cao [6] used SVMs in forecasting financial time series. Their numerical results indicated that SVMs are superior to a multi-layer back-propagation neural network in financial time series forecasting. Cao and Gu [7] presented a dynamic SVM model (DSVMs) to deal with non-stationary time series problems. Their experiment results showed that the DSVMs outperform standard SVMs in forecasting nonstationary time series. In the same year, Tay and Cao [8] proposed a C-ascending SVMs to model nonstationary financial time series. Their experimental results showed that C-ascending SVMs with the actual ordered sample data consistently perform better than standard SVMs. Cao [9] used the SVM experts for time series forecasting. A two-stage neural network architecture is contained in the generalized SVM experts. The numerical results indicated that the SVM experts are able to achieve the better generalization in comparison with the single SVM models. Wang et al. [10] applied SVMs to predict air quality. Their experimental results showed that SVMs outperformed the conventional Radial Basis Function networks. Mohandes et al. [11] applied SVMs to the prediction of wind speed. Their experimental results indicated that the SVM model outperformed the multilayer perceptron neural networks in terms of root mean square errors. Pai and Lin [12] used SVMs to forecast production values of machinery industry in Taiwan. They reported that SVMs perform better than the seasonal ARIMA model and general regression neural networks model.

In addition to the feed-forward neural networks, links may be established within the layers of a neural network. These types of networks are called recurrent neural networks (RNNs). The main concept of RNNs is that every unit is considered as the output of the network and provides the adjusted information as input in furthermore training process [13]. RNNs are widely used in time series forecasting. Jordan [14] proposed a recurrent neural network model (Figure 1) for controlling robots. Elman [15] presented a recurrent neural network model (Figure 2) to deal with linguistics problems. A recurrent network model (Figure 3) was proposed by Williams and Zipser [16] to solve nonlinear adaptive filtering and pattern recognition problems. These three models mentioned above all consist of MLP with one hidden layer

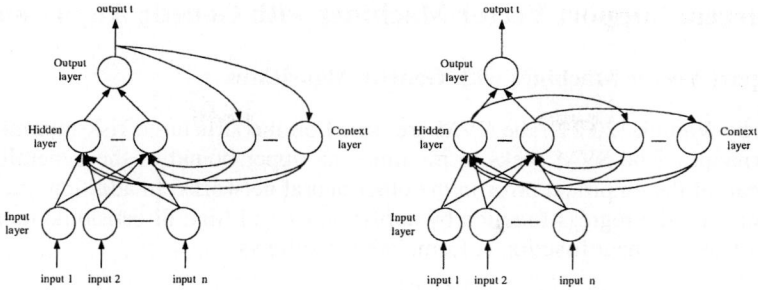

Fig. 1. Jordan networks [14]   **Fig. 2.** Elman networks [15]

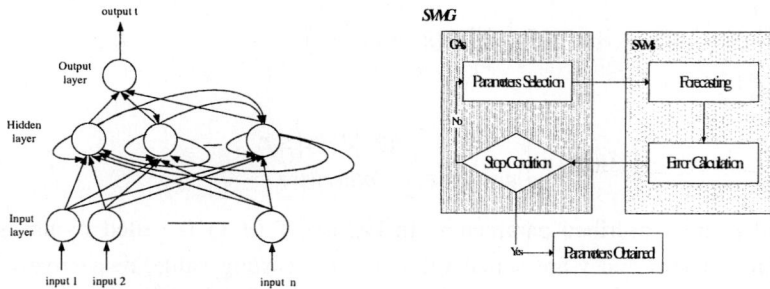

Fig. 3. Williams and Zipser networks [16]   **Fig. 4.** The architecture of SVMG model

In Jordan networks, a feedback loop is from the output layer with past values to an additional input, namely "context layer". Then, output values from the context layer are fed back into the hidden layer. In Elman networks, the feedback loop is from the hidden layer to the context layer. In Williams and Zipser networks, nodes in the hidden layer are full connected with each other. Both Jordan and Elman networks have an additional information source from the output layer or the hidden layer. Hence, these models mainly are based on the past information to capture detailed pattern of information. Williams and Zipser networks have much more information source from the hidden layer and back into themselves. Therefore, Williams and Zipser networks are more sensitive while model implemented [17]. Jordan networks and Elman networks are suitable for time series forecasting [18,19]. In this study, the Jordan network is used as a base to construct the recurrent SVM models.

Traditionally, recurrent neural networks are trained by back-propagation algorithms. In this study, support vector machines with genetic algorithms (SVMG) are used as training algorithms in recurrent neural networks, namely Recurrent Support Vector Machines with Genetic Algorithms (RSVMG), to obtain weights between nodes. Then, the proposed RSVMG model is applied in forecasting system reliability. A numerical example in an existing literature [2] is employed to examine the forecasting performance of the proposed model.

2 Recurrent Support Vector Machines with Genetic Algorithms

2.1 Support Vector Machines with Genetic Algorithms

Proposed by Vapnik [20,21], the SVM are based on the structured risk minimization (SRM) principle. The SVM seeks to minimize an upper bound of the generalization error instead of the empirical error in the other neural networks. In addition, the SVM models generate the regress function by applying a set of high dimensional nonlinear functions. The nonlinear function is formulated as follows.

$$y = f(x) = w\psi(x_i) + b. \quad (1)$$

where $\psi(x_i)$ is called the feature, which is nonlinear mapped from the input space x. The w and b are coefficients estimated by minimizing the regularized risk function Eq. (2):

$$r(C) = C\frac{1}{N}\sum_{i=1}^{N}\Gamma_\varepsilon(d_i, y_i) + \frac{1}{2}\|w\|^2. \quad (2)$$

where

$$\Gamma_\varepsilon(d, y) = \begin{cases} 0, & \text{if } |d-y| \leq \varepsilon \\ |d-y| - \varepsilon, & \text{otherwise} \end{cases} \quad (3)$$

and C and ε are prescribed parameters. In Eq. (2), $\Gamma_\varepsilon(d,y)$ is called ε-insensitive loss function. The d and y are actual value and forecasting value, respectively. The loss is equal to zero if the forecasted value is within the ε-tube (Eq. (3)). The second term, $\|w\|^2/2$, is used as a measure of function flatness. Therefore, C is used as the trade-off between the empirical risk and the model flatness. Both C and ε are parameters determined by users. Two positive slack variables ξ and ξ^*, which represent the distance from actual values to the corresponding boundary values of ε-tube, are introduced. Then, Eq. (2) is transformed into the following constrained form:

Minimize: $r(w,\xi,\xi^*) = \|w\|^2/2 + C\left(\sum_{i=1}^{N}(\xi_i + \xi_i^*)\right).$ (4)

with the constraints:

$$w\psi(x_i) + b_i - d_i \leq \varepsilon + \xi_i^*, \quad d_i - w\psi(x_i) - b_i \leq \varepsilon + \xi_i, \quad \xi_i, \xi_i^* \geq 0, \quad i=1,2,\ldots,N$$

This constrained optimization problem is solved by the following primal Lagrangian form:

$$L(w,b,\xi,\xi^*,\alpha_i,\alpha_i^*,\beta_i,\beta_i^*) = \frac{1}{2}\|w\|^2 + C\left(\sum_{i=1}^{N}(\xi_i + \xi_i^*)\right) - \sum_{i=1}^{N}\beta_i[w\psi(x_i) + b - d_i + \varepsilon + \xi_i]$$
$$- \sum_{i=1}^{N}\beta_i^*[d_i - w\psi(x_i) - b + \varepsilon + \xi_i^*] - \sum_{i=1}^{N}(\alpha_i\xi_i + \alpha_i^*\xi_i^*) \quad (5)$$

Eq. (5) is minimized with respect to primal variables w, b, ξ, and ξ^*, and maximized with respect to nonnegative Lagrangian multipliers α_i, α_i^*, β_i, and β_i^*. Finally, by applying Karush-Kuhn-Tucker conditions for regression, Eq. (4) results in a dual Lagrangian form as Eq. (6).

$$\vartheta(\beta_i,\beta_i^*) = \sum_{i=1}^{N} d_i(\beta_i - \beta_i^*) - \varepsilon \sum_{i=1}^{N}(\beta_i + \beta_i^*) - \frac{1}{2}\sum_{i=1}^{N}\sum_{j=1}^{N}(\beta_i - \beta_i^*)(\beta_j - \beta_j^*)K(x_i,x_j). \quad (6)$$

with the constraints:

$$\sum_{i=1}^{N}(\beta_i - \beta_i^*) = 0, \ 0 \le \beta_i \le C, \ 0 \le \beta_i^* \le C, \ i=1,2,\cdots,N.$$

In Eq. (6), the Lagrange multipliers satisfy the equality $\beta_i * \beta_i^* = 0$. After calculating the Lagrange multipliers β_i and β_i^*, an optimal desired weight vector of the regression hyperplane is

$$w^* = \sum_{i=1}^{N}(\beta_i - \beta_i^*)K(x,x_i). \quad (7)$$

It is shown that minimizing function has the following form

$$f(x,\beta,\beta^*) = \sum_{i=1}^{N}(\beta_i - \beta_i^*)K(x,x_i) + b \quad (8)$$

Here, $K(x,x_i)$ is called the Kernel function. The value of the Kernel is equal to the inner product of two vectors x and x_i in the feature space $\psi(x)$ and $\psi(x_i)$, i.e., $K(x,x_i) = \psi(x)*\psi(x_i)$. Any function that satisfies Mercer's condition [22] can be used as the Kernel function. In this study, the Gaussian function, $\exp(-\|x-x_i\|^2/2\sigma^2)$, is used in the SVMs.

Three free parameters (σ, ε and C) influence the performance of SVM models a lot. Unfortunately, there is lacking of structural approaches to obtain appropriate parameters. Hence, the genetic algorithms (GAs) are employed to determine the parameters in SVM model.

The architecture of the proposed SVMG model is illustrated in Figure 4. The followings are procedures for conducting the SVMG model.

Step 1 Initialization: Construct randomly the initial population of chromosomes. The three free parameters σ, C, and ε should be first encoded into binary format, represented by a "chromosome" composing of "genes".

Step 2 Evaluating fitness: Evaluate the fitness of each chromosome. The random initial chromosomes, σ, C, and ε, first used to forecast, and the forecasting error calculated in a moment. In this study, the negative value of the root mean square error measure (-RMSE) is used as the fitness function in GAs. The fitness function is shown as Eq.(9),

$$Fitness \ function = -\sqrt{\frac{1}{N}\sum_{t=1}^{N}(d_t - y_t)^2}. \quad (9)$$

where, N is the number of forecasting periods; d_t is the actual reliability value at period t; and y_t is the forecasting reliability value.

Step 3 Selection: Select mating pair, #1 parent and #2 parent, for reproduction. Parent selection is a procedure that two mating chromosomes from the parent population based on their fitness function. Chromosomes with higher fitness function have higher probabilities to generate offspring to the next generation. The roulette wheel selection principle [23] is used to select chromosomes for reproduction.

Step 4 Crossover and mutation: Create new offspring by crossover and mutation operations. Then next generation forms a population for the next generation, and number of generation increase one. To perform crossovers, chromosomes are paired randomly. Single-point-crossover principle is employed. Segments of paired chromosomes between two determined break-points are swapped.

Step 5 Next generation: Form a population for the next generation.

Step 6 Stop conditions: If the number of generation is equal to a given scale, then the best chromosomes are presented as a solution, otherwise go back to Step 2.

For simplicity, suppose there are 4 bits in a gene. A chromosome contains 12 bits (Figure 5). Furthermore, supposed the boundaries for σ, C, and ε are 2, 10, and 0.5 respectively. Before the crossover, values of three parameters in #1 parent are 0.625, 1.875, and 0.28125 correspondingly. For #2 parent, the three values are 1.5, 6.25, and 0.21875. After crossover, the three values are 0.5, 1.25, and 0.46875, for #1 offspring. For #2 offspring, the three values are 1.75, 6.875, and 0.03125. Mutations are performed randomly by converting a "1" bit into a "0" bit or a "0" bit in to a "1" bit. The rates of crossover and mutation are determined by probabilities. In this investigation, the probabilities are set to 0.5 and 0.1 for crossover and mutation correspondingly. Figure 6 shows the framework of the SVM model combined with GAs to calculate free parameters in SVM model.

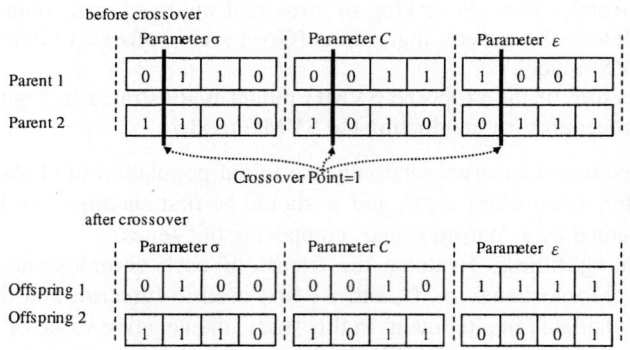

Fig. 5. A simplified example of parameter representation

2.2 Recurrent Support Vector Machines with Genetic Algorithms

In this investigation, the Jordan network is used as a recurrent neural network framework. Figure 1 shows the architecture of a Jordan network [14]. All neurons in one layer are connected with all neurons in the next layer except the context layer. A context layer is special case of a hidden layer. Interactions only happen between neurons of the hidden layer and the context layer. For a Jordan network with p inputs, q hidden and r output neurons, the output of the n th neuron, $f_n(t)$, is [24-28]:

$$f_n(t) = \sum_{i=1}^{q} W_i \varphi_i(t) + b_i(t).\tag{10}$$

where W_i are weights between the hidden and the output layer; $\varphi_i(t)$ is the output function of hidden neurons, which is computed as

$$\varphi_i(t) = g\left(\sum_{j=1}^{P} v_{ij} x_j(t) + \sum_{k=1}^{s}\sum_{v=1}^{r} w_{ikv} f_v(t-k) + b_i(t)\right).\tag{11}$$

where v_{ij} are weights between the input and the hidden layer; w_{ikv} are weights between the context and the hidden layer with delay k periods; s is the total numbers of the context layer of past output data.

Back-propagation is a procedure to obtain gradients for adapting weights of a neural network. Back-propagation algorithm presents as follows. First, the output of the n th neuron in Eq.(11) is rewritten as

$$f_n(t) = h(x^T(t)\phi(t))\tag{12}$$

where $h(\cdot)$ is nonlinearity function of $x^T(t)$ and $f_n(t)$; $x^T(t) = [x_1(t),...,x_P(t)]^T$ is the input vector; $\phi(t) = [\phi_1(t),...\phi_P(t)]^T$ is the weight vector, then, a cost function is proposed to be the instantaneous performance index,

$$J(\phi(t)) = \frac{1}{2}[d(t) - f_n(t)]^2 = \frac{1}{2}[d(t) - h(x^T(t)\phi(t))]^2\tag{13}$$

where $d(t) = [d_1(t),...,d_P(t)]^T$ is the desired output.

Secondly, the instantaneous output error at the output neuron and revised weight vector in the next moment are presented as Eq.(14) and Eq.(15) respectively.

$$e(t) = d(t) - f_n(t)) = d(t) - h(x^T(t)\phi(t))\tag{14}$$

$$\phi(t+1) = \phi(t) - \eta \nabla_\phi J(\phi(t))\tag{15}$$

where η is the learning rate.

Third, the gradient $\nabla_\phi J(\phi(t))$ can be calculated as

$$\nabla_\phi J(\phi(t)) = \frac{\partial J(\phi(t))}{\partial \phi(t)} = e(t) \times \frac{\partial e(t)}{\partial \phi(t)} = -e(t)h'(x^T(t)\phi(t))x(t)\tag{16}$$

where $h'(\cdot)$ is the first derivation of the nonlinearity $h(\cdot)$. Finally, the weight is revised as

$$\phi(t+1) = \phi(t) + \eta e(t)h'(x^T(t)\phi(t))x(t)\tag{17}$$

Figure 6 is the architecture of the proposed RSVMG model. The output of RSVMG ($\tilde{f}_n(t)$) is shown as Eq (18).

$$\tilde{f}_n(t) = \sum_{i=1}^{P} W^T \psi(x^T(t)) + b(t)\tag{18}$$

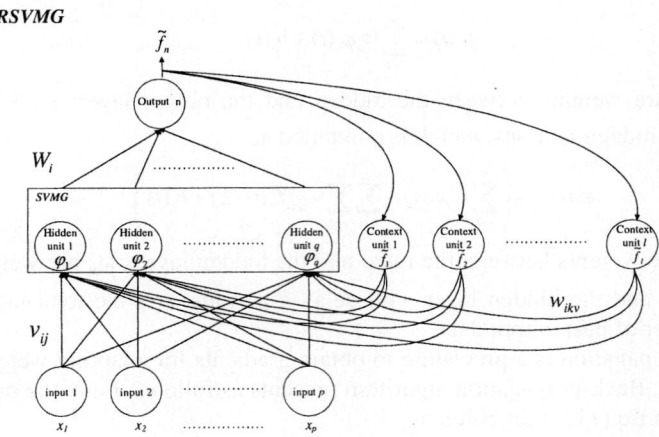

Fig. 6. The architecture of RSVMG model

Then, Eq.(18) replaces the Eq.(1) in the SVMG algorithms to run the loop of SVMG in searching values of three parameters. Finally, the forecasting values $\tilde{f}_n(t)$ are calculated by Eq. (18).

3 A Numerical Example

A numerical data in the work of Su et al. [2] are reprinted in Table 1. The data contains the number of vehicle damages (a_t), the number of damages repaired (b_t), and the period reliability ratio (r_t). In this study, various models are applied to forecast the period reliability ratio. To compare forecasting performance of the proposed models with the models of Su et al. [2], the data division principle is the same as the work of Su et al. [2]. The data are divided into two sets, namely the training data set and the testing data set. Totally, there are 36 numbers of data. The numbers of data are 24 and 12 for the training data set and the testing data set respectively. Then, a rolling-based forecasting procedure is conducted and only one-step-ahead forecasting policy is considered. To achieve better forecasting performance, different numbers of input data are used to forecast output values. In this example, six models with different numbers of input data are employed to forecast output values. The model with the minimum testing RMSE value is selected as the most suitable model for this example.

Table 2 shows the forecasting results of SVMG and RSVMG models. It is indicated that the best forecasting results occur when the numbers of input data are 21 and 22 for SVMG and RSVMG models respectively. The most suitable parameters of SVMG and RSVMG models are obtained to forecast the reliability growth. Table 3 shows the RMSE values of different models. It is observed that SVMG and RSVMG models outperform ARIMA, BPNN and ICBPNN models. In addition, the RSVMG model results in better predictive performance than the SVMG model.

In addition, the sensitivity analysis of three parameters of RSVMG model is performed. In the parameter analysis process, those RSVMG parameters are fixed to examine the change of RMSE values when the third parameter varies. It is indicated that smaller RMSE values cab be obtained only when the parameter is sensitive in this example. For example, the change of parameters C and ε can not improve the solutions. However, changing the values of σ can improve the forecasting accuracies, that RMSE can be reduced from 0.00212 to 0.00175 by moving σ from 10.469 to 18.42.

Table 1. The repair data of a repairable system in automobile industry (Su et al. [2])

Period	Number of vehicle damages (a_t)	Number of damages repaired (b_t)	Period reliability ratio $r_t = b_t/\sum a_t$	Period	Number of vehicle damages (a_t)	Number of damages repaired (b_t)	Period reliability ratio $r_t = b_t/\sum a_t$
1	440	0	0.000000	19	1,444	225	0.007382
2	1,080	0	0.000000	20	1,734	186	0.005774
3	1,002	16	0.006344	21	1,700	354	0.010439
4	1,448	20	0.005038	22	1,355	566	0.016049
5	1,743	15	0.002626	23	1,980	224	0.006014
6	1,201	56	0.008100	24	1,384	519	0.013435
7	2,025	6	0.000671	25	1,276	400	0.010023
8	2,298	70	0.006229	26	2,063	374	0.008911
9	1,665	24	0.001860	27	2,319	508	0.01147
10	2,008	40	0.002683	28	1,772	326	0.007078
11	1,128	78	0.004863	29	1,976	548	0.011408
12	1,372	230	0.013211	30	896	310	0.006335
13	1,696	199	0.010416	31	774	541	0.010884
14	2,106	184	0.008674	32	1,278	545	0.010689
15	1,772	120	0.005221	33	1,766	265	0.005024
16	2,319	70	0.002767	34	865	349	0.006509
17	2,006	199	0.007287	35	665	395	0.007277
18	1,725	94	0.003238	36	386	589	0.010774

Table 2. Forecasting results and parameters of SVMG and RSVMG models

	SVMG				RSVMG			
No. of input	Parameters			Testing RMSE	No. of input	Parameters			Testing RMSE
	σ	C	ε			σ	C	ε	
5	106.48	638.44	0.00361	0.00251	5	30.472	1048.3	0.003354	0.022060
10	46.933	152.78	0.00345	0.00293	10	37.200	1629.7	0.003734	0.002150
15	58.902	294.69	0.00631	0.00252	15	28.001	139.16	0.000300	0.002203
19	45.658	289.73	0.00811	0.00238	19	25.331	1555.2	0.001057	0.002435
20	236.58	682.33	0.00386	0.00253	20	3.1859	1769.2	0.001176	0.002551
21	**208.55**	**183.48**	**0.00444**	**0.00229**	21	0.3257	1016.6	0.000431	0.002159
22	167.91	466.07	0.00517	0.00235	**22**	**10.649**	**2718.2**	**0.000706**	**0.002117**

4 Conclusions

Predicting reliability is one of the most crucial issues in manufacturing systems. The SVMG and RSVMG neural networks are presented in the study to predict system reliability successfully. SVM learning algorithms are embodied in the traditional recurrent neural network structure. It is demonstrated that the proposed SVMG and RSVMG models are able to reach lower prediction errors compared with other forecasting models in an existing literature. The superior performance of the RSVMG model over other approaches is due to the following reasons. First, with the nonlinear mapping capabilities, data patterns of system reliability are more easily captured by the RSVMG model than other approaches. Secondly, the GAs provide a proper selection of free parameters in the proposed RSVMG models to predict system reliability. Secondly, instead of minimizing training errors, a structural risk minimization process is conducted by the RSVMG model. This process seeking to minimize an upper bound of the generalization error results in better generalization performance. Finally, the Jordan networks are capable of capturing the data patterns continually. Therefore, compared with SVMG model, the RSVMG model can obtain more accurate forecasting results than the SVMG model.

This study is a first attempt to combine recurrent neural networks with SVM algorithms to predict system reliability. The promising results obtained in this investigation indicate the proposed model is a valid alternative in predicting system reliability. For future research, some other searching techniques can be used in selecting free parameters in the RSVMG.

Table 3. RMSE values of ARIMA, BPNN, ICBPNN, SVMG and RSVMG models

Periods	Actual values	Forecasting errors				
		ARIMA model	BPNN model	ICBPNN model	SVMG model	RSVMG model
25	0.010023	0.00327	0.00537	0.003925	0.000947	0.001084
26	0.008911	0.00129	0.00139	0.000386	0.000398	0.000194
27	0.011470	0.00195	0.00051	0.00069	0.001928	0.002130
28	0.007078	0.00129	0.00346	0.001958	0.002698	0.002495
29	0.011408	0.00355	0.00146	0.0025	0.001693	0.001602
30	0.006335	0.00202	0.00433	0.00209	0.003560	0.003705
31	0.010884	0.00376	0.00207	0.00316	0.001085	0.000705
32	0.010689	0.00191	0.00401	0.001625	0.001562	0.000377
33	0.005024	0.00442	0.00323	0.00431	0.004121	0.003410
34	0.006509	0.00139	0.00035	0.00195	0.002941	0.002320
35	0.007277	0.00013	0.00078	0.00072	0.002305	0.002650
36	0.010774	0.00341	0.00379	0.002822	0.001073	0.000918
RMSE		**0.002668**	**0.003024**	**0.002484**	**0.002297**	**0.002117**

Acknowledgement

This research was conducted with the support of National Science Council (NSC 92-2213-E-212-001 & NSC 93-2213-E-212-001). Mr. Chih-Shen Lin helped with data analysis.

References

1. Liu, M.C., Sastri, T., Kuo, W.: An exploratory study of a neural network approach for reliability data analysis. Quality and Reliability Engineering International. 11 (1995) 107-112.
2. Su, C.T., Tong, L.I., Leou, C.M.: Combining time series and neural network approaches. Journal of the Chinese Institute of Industrial Engineers. 4 (1997) 419-430.
3. Amjady, N., Ehsan, M.: Evaluation of power systems reliability by artificial neural network. IEEE Transactions on Power Systems. 14 (1999) 287-292.

4. Ho, S.L., Xie, M., Goh, T.N.: A comparative study of neural network and Box-Jenkins ARIMA modeling in time series prediction. Computers & Industrial Engineering. 42 (2002) 371-375.
5. Xu, K., Xie, M., Tang, L.C., Ho, S.L.: Application of neural networks in forecasting engine system reliability. Applied Soft computing. 2 (2003) 255-268.
6. Tay, F.E.H., Cao, L.J.: Application of support vector machines in financial time series forecasting. Omega: The International Journal of Management Science. 29 (2001) 309-317.
7. Cao, L., Gu, Q.: Dynamic support vector machines for non-stationary time series forecasting. Intelligent Data Analysis. 6 (2002) 67-83.
8. Tay, F.E.H., Cao, L.: Modified support vector machines in financial time series forecasting. Neurocomputing. 48 (2002) 847-861.
9. Cao, L.: Support vector machines experts for time series forecasting. Neurocomputting. 51 (2003) 321-339.
10. Wang, W., Xu, Z., Lu, J.W.: Three improved neural network models for air quality forecasting. Engineering Computations. 20 (2003) 192-210.
11. Mohandes, M.A., Halawani, T.O., Rehman, S., Hussain, A.A.: Support vector machines for wind speed prediction. Renewable Energy. 29 (2004) 939-947.
12. Pai, P.F., Lin, C.S.: Using support vector machines in forecasting production values of machinery industry in Taiwan. International Journal of Advanced Manufacturing Technology. (2004) DOI: 10.1007/ s00170-004-2139-y.
13. Kechriotis, G., Zervas, E., Manolakos, E.S.: Using recurrent neural networks for adaptive communication channel equalization. IEEE Transaction on Neural Networks. 5 (1994) 267-278.
14. Jordan, M.I.: Attractor dynamics and parallelism in a connectionist sequential machine. In: Proceeding of 8th Annual Conference of the Cognitive Science Society, Hillsdale (1987) 531-546.
15. Elman, J.L.: Finding structure in time. Cognitive Science. 14 (1990) 179-211.
16. Williams, R., Zipser, D.: A learning algorithm for continually running fully recurrent neural networks. Neural Computation. 1 (1989) 270-280.
17. Tsoi, A.C., Back, A.D.: Locally recurrent globally feedforward networks: Acritical review of architectures. IEEE Transaction on Neural Networks. 5 (1994) 229-239.
18. Jhee, W.C., Lee, J.K.: Performance of Neural Networks in Managerial Forecasting. International Journal of Intelligent Systems in Accounting, Finance, and Management. 2 (1993) 55-71.
19. Suykens, J.A.K., van Gestel, T., De Brabanter, J., De Moor, B., Vandewalle, J., Leu-ven, K.U. (ed.): Least Squares Support Vector Machines. World Scientific Publishing Co., Ltd., Belgium (2002).
20. Kecman, V. (ed.): Learning and Soft Computing, Support Vector machines, Neural Networks and Fuzzy Logic Models, The MIT Press, Cambridge, MA (2001).
21. Wang, L.P. (ed.): Support Vector Machines: Theory and Application. Springer, Berlin Heidelberg New York (2005).
22. Mercer, J.: Function of positive and negative type and their connection with the theory of integral equations. Philosophical Transaction Royal Society London. A209 (1909) 415-446.
23. Holland, J. (ed.): Adaptation in Natural and Artificial System. University of Michigan Press, Ann Arbor (1975).
24. Ayaz, E., Seker, S., Barutcu, B., Türkcan, E.: Comparisons between the various types of neural networks with the data of wide range operational conditions of the Borssele NPP. Progress in Nuclear Energy. 43 (2003) 381-387.
25. Connor, J.T., Martin, R.D., Atlas, L.E.: Recurrent neural networks and robust time series prediction. IEEE Transactions on Neural Networks. 5 (1994) 240-254.
26. Gencay, R., Liu, T.: Nonlinear modeling and prediction with feedforward and recurrent networks. Physica D. 108 (1997) 119-134.
27. Kermanshahi, B.: Recurrent neural network for forecasting next 10 years loads of nine Japanese utilities. Neurocomputing. 23 (1998) 125-133.
28. Mandic, D.P., Chambers, J.A. (ed.): Recurrent Neural Networks for Prediction. John Wiley and Sons, New York (2001).

A Modified SMO Algorithm for SVM Regression and Its Application in Quality Prediction of HP-LDPE

Hengping Zhao and Jinshou Yu

Research Institution of Automation, East China University of Science and Technology
Shanghai, 200237, PRC
zhaohengping@hotmail.com

Abstract. A modified sequential minimal optimization (SMO) algorithm for support vector machine (SVM) regression is proposed based on Shevade's SMO-1 algorithm. The main improvement is that a modified heuristics method is used in this modified SMO algorithm to choose the first Lagrange multiplier when optimizing the Lagrange multipliers corresponding to the non-boundary examples. To illustrate the validity of the proposed modified SMO algorithm, a benchmark dataset and a practical application in predicting the melt index of high-pressure low-density polyethylene (HP-LDPE) are used; the results demonstrate that this modified SMO algorithm is faster in most cases with the same parameters setting and more likely to obtain the better generalization performance than Shevade's SMO-1 algorithm.

1 Introduction

Since the theory of support vector machine (SVM) was proposed by Vapnik [1], it has achieved much more development because it has many advantages such as terse expression, intuitionistic geometry explanation and excellent generalization performances [2]. In fact, the SVM is trained by solving a quadratic programming (QP) problem, which commonly uses traditional optimization algorithms such as interior point algorithm and reduced gradient algorithm. It takes a long time and much memory requirement to train SVM with traditional optimization algorithms when the training set is very large, which prevents its practical application [3]. In order to simplify the training of SVM and reduce its calculation complexity, decomposition algorithm was firstly proposed by Osuna [4], which is an efficient training method. Among various decomposition algorithms, the main differences are in the size of working set and its choosing method. Sequential minimal optimization (SMO) proposed by Platt [5] is a more efficient method for training SVM at present. The basic idea of the SMO algorithm is that a very large QP problem is broken into a series of smallest possible QP problems, which involve two Lagrange multipliers because the Lagrange multipliers must obey a linear equality constraint. These small QP problems are solved analytically, which avoids a time-consuming numerical QP optimization as an inner loop; and the amount of memory required for SMO is linear in the training size, which allows SMO to handle very large training set. For these excellent performances, Smola and Scholkopf [6, 7] proposed a SMO algorithm for training SVM regression, which is an extension of Platt's SMO. The paper [8] pointed that the use of a single threshold value is an important source of inefficiency in Smola and Scholkopf's SMO algorithm for SVM regression and derived modified SMO algorithms by employing two threshold parameters, as SMO-1 and SMO-2, which perform significantly faster for its modified KKT optimality conditions, cache strategy and efficient parameter updating process. But Shevade's SMO-2 algorithm is not always convergent, so only Shevade's SMO-1 algorithm is concerned in this paper.

In general, above-mentioned SMO algorithms use the same heuristic method [5] to choose the first Lagrange multiplier. In this method, the cost of CPU time mainly concentrates on the non-boundary examples [8] and the first Lagrange multiplier is chosen from this subset in turn, which causes such a problem that the change of the first selected Lagrange multiplier may not maximize the change of the objective function in all the possible candidates for the first Lagrange multiplier. In this paper, a new method for the choice of the first Lagrange multiplier is proposed to improve the efficiency of Shevade's SMO-1 algorithm.

The paper is organized as follows. Section II gives a brief overview of SVM regression. Section III describes the modified SMO algorithm for SVM regression. Section IV givens a simulation study by using a benchmark dataset. Section V applies the proposed modified SMO algorithm to build the quality prediction model of high-pressure low-density polyethylene (HP-LDPE). Finally, section VI gives some conclusions.

2 Overview of SVM Regression

In SVM regression, the input data set x is first mapped into a high dimensional feature space by using some fixed (non-linear) mapping function ϕ and then a linear model is constructed in this space [1, 9]. Using mathematical notation, the linear model (in the feature space) $f(x)$ is given by

$$f(x) = \omega^T \phi(x) + b \tag{1}$$

where b is the bias term. In this way, the linear regression in high dimensional feature space is corresponding to the non-linear regression in low dimensional feature space. In order to enhance the robustness of the SVM regression, a new type of loss function called ε-insensitive loss function was proposed by Vapnik [1, 9] and it is characterized by omitting the training errors less than ε and reducing the complexity of functional. The SVM regression is formulated as minimization of the following functional:

$$\text{minimize} \frac{1}{2} \omega^T \omega + C \sum_{i=1}^{l} (\xi_i + \xi_i^*)$$

$$\text{subject to:} \begin{cases} y_i - \omega^T \phi(x_i) - b \leq \varepsilon + \xi_i \\ \omega^T \phi(x_i) + b - y_i \leq \varepsilon + \xi_i^* \\ \xi_i, \xi_i^* \geq 0 \end{cases} \tag{2}$$

where $C > 0$ determines the trade off between the flatness of f and the amount up to which deviations larger than ε are tolerated. And ξ_i and ξ_i^* called slack variables are introduced to cope with the non-strict constraints. In this paper, Eq.(2) is called the primal problem of the SVM and its Lagrangian is represented as

$$L = \frac{1}{2}\omega^T\omega + C\sum_{i=1}^{l}(\xi_i + \xi_i^*) - \sum_{i=1}^{l}\alpha_i(\omega^T\phi(x_i) + b - y_i + \varepsilon + \xi_i)$$
$$- \sum_{i=1}^{l}\alpha_i^*(y_i - \omega^T\phi(x_i) - b + \varepsilon + \xi_i^*) - \sum_{i=1}^{l}(\eta_i\xi_i + \eta_i^*\xi_i^*) \quad (3)$$

where $\alpha \geq 0$ and $\alpha^* \geq 0$ are Lagrange multipliers. Classical Lagrangian duality enables the problem Eq.(3) to be transformed to its dual problem, which is given by:

$$\max_{\alpha,\alpha^*} W(\alpha,\alpha^*) = \max_{\alpha,\alpha^*}(\min_{\omega,b} L(\omega,b,\alpha,\alpha^*)) \quad (4)$$

The minimum with respect to ω and b of the Lagrangian, L, is given by

$$\begin{cases} \frac{\partial L}{\partial \omega} = 0 \Rightarrow \omega = \sum_{i=1}^{l}(\alpha_i - \alpha_i^*)\phi(x_i) \\ \frac{\partial L}{\partial b} = 0 \Rightarrow \sum_{i=1}^{l}(\alpha_i - \alpha_i^*) = 0 \\ \frac{\partial L}{\partial \xi_i} = 0 \Rightarrow C - \alpha_i - \eta_i = 0 \\ \frac{\partial L}{\partial \xi_i^*} = 0 \Rightarrow C - \alpha_i^* - \eta_i^* = 0 \end{cases} \quad (5)$$

Hence from Eq.(3), Eq.(4) and Eq.(5), the dual problem is:

$$\text{maximize} -\frac{1}{2}\sum_{i=1}^{l}\sum_{j=1}^{l}(\alpha_i - \alpha_i^*)(\alpha_j - \alpha_j^*)k(x_i, x_j) - \varepsilon\sum_{i=1}^{l}(\alpha_i + \alpha_i^*) + \sum_{i=1}^{l}y_i(\alpha_i - \alpha_i^*)$$
$$\text{subject to:} \begin{cases} \sum_{i=1}^{l}(\alpha_i - \alpha_i^*) = 0 \\ 0 \leq \alpha_i, \alpha_i^* \leq C \quad i = 1,\cdots,l \end{cases} \quad (6)$$

where $k(x_i, x_j) = \phi(x_i)^T\phi(x_j)$ is called kernel function. And the solution of Eq.(1), or the output of the SVM regression can be expressed as

$$f(x) = \sum_{i=1}^{l}(\alpha_i - \alpha_i^*)k(x_i, x) + b \quad (7)$$

3 SMO for SVM Regression

For easy to do derivation, referring to [10], let $\lambda_i = \alpha_i - \alpha_i^*$ and $|\lambda_i| = \alpha_i + \alpha_i^*$. Then Eq.(6) and Eq.(7) can be rewritten as:

$$\text{maximize} -\frac{1}{2}\sum_{i=1}^{l}\sum_{j=1}^{l}\lambda_i\lambda_j k_{ij} - \varepsilon\sum_{i=1}^{l}|\lambda_i| + \sum_{i=1}^{l}\lambda_i y_i$$
$$\text{subject to} \begin{cases} \sum_{i=1}^{l}\lambda_i = 0 \\ -C \leq \lambda_i \leq C \quad i = 1,\cdots,l \end{cases} \quad (8)$$

$$f(x) = \sum_{i=1}^{l} \lambda_i k(x_i, x) + b \tag{9}$$

where k_{ij} is the abbreviation of kernel function $k(x_i, x_j)$ and satisfies $k_{ij} = k_{ji}$.

3.1 Analytic Solution for the Optimization Sub-problem

The SMO algorithm divides a large QP problem into a series of QP problems with two variables, which can be solved analytically. Let these two variables have indices u and v, so λ_u and λ_v are the two unknowns. From paper [10], we know that λ_v can be updated recursively by

$$\lambda_v = \lambda_v^* + \frac{1}{\eta}(y_v - y_u + f_u^* - f_v^* + \varepsilon(\mathrm{sgn}(\lambda_u) - \mathrm{sgn}(\lambda_v))) \tag{10}$$

where $\eta = (k_{vv} + k_{uu} - 2k_{uv})$, sgn() is signum function, f_u^* and f_v^* are computed by Eq.(9). The superscript * used above indicates that the value is computed with the old parameter values.

3.2 Optimality Conditions for SVM Regression

To improve the efficiency of SMO algorithm, an accurate and quick judgment of whether a Lagrange multiplier violates the optimality conditions is very crucial.

In Smola and Scholkopf's SMO algorithm for SVM regression based on a single threshold parameter b, which needs to be update after each successful optimization step. For using only one single threshold, this inefficient SMO algorithm has two shortcomings: 1) Sometimes the value of b cannot be calculated (for example, all the support vectors lie out of the margin of the ε-tube). In this situation, b is simply chosen as the midpoint of the interval $[b_{low}, b_{up}]$; 2) For the reason that b is updated based on the current two Lagrange multipliers used for joint optimization, while checking whether the remaining examples violate the optimality conditions or not, it is quite possible that a different, shifted choice of b may do a better job. To solve these problems, a modified SMO-1 algorithm employed two threshold parameters for regression is propose in [8]. With the same idea, we derive the optimality conditions based on the parameter λ below.

The Lagrangian for the dual problem Eq.(8) is

$$L_D = \frac{1}{2}\omega^T\omega - \sum_{i=1}^{l}\lambda_i y_i + \varepsilon\sum_{i=1}^{l}|\lambda_i| + \beta\sum_{i=1}^{l}\lambda_i - \sum_{i=1}^{l}\delta_i(C+\lambda_i) - \sum_{i=1}^{l}\tau_i(C-\lambda_i) \tag{11}$$

Let

$$F_i = y_i - \omega^T \phi(x_i) \tag{12}$$

So the KKT conditions for the dual problem are

$$\begin{cases} \dfrac{\partial L_D}{\partial \lambda_i} = -F_i + \varepsilon \dfrac{\partial |\lambda_i|}{\partial \lambda_i} + \beta - \delta_i + \tau_i = 0 \\ \delta_i(C+\lambda_i) = 0, \quad \delta_i \geq 0, \quad \lambda_i \geq -C \\ \tau_i(C-\lambda_i) = 0, \quad \tau_i \geq 0, \quad \lambda_i \leq C \end{cases} \quad (13)$$

These conditions are also referred as optimality conditions, and which can be simplified by considering the following five cases:

$$\begin{cases} F_i - \beta \leq -\varepsilon, & \lambda_i = -C \\ F_i - \beta = -\varepsilon, & -C < \lambda_i < 0 \\ -\varepsilon \leq (F_i - \beta) \leq \varepsilon, & \lambda_i = 0 \\ F_i - \beta = \varepsilon, & 0 < \lambda_i < C \\ F_i - \beta \geq \varepsilon, & \lambda_i = C \end{cases} \quad (14)$$

Based on Eq.(14), we define the following five index sets at a given λ:
$I_{0a} = \{i : 0 < \lambda_i < C\}$;
$I_{0b} = \{i : -C < \lambda_i < 0\}$; $I_1 = \{i : \lambda_i = 0\}$; $I_2 = \{i : \lambda_i = -C\}$; $I_3 = \{i : \lambda_i = C\}$ and let $I_0 = I_{0a} \cup I_{0b}$. We also define $\tilde{F}_i$ and $\overline{F}_i$ as:

$$\tilde{F}_i = \begin{cases} F_i + \varepsilon & \text{if } i \in I_{0b} \cup I_2 \\ F_i - \varepsilon & \text{if } i \in I_{0a} \cup I_1 \end{cases} \quad (15)$$

and

$$\overline{F}_i = \begin{cases} F_i - \varepsilon & \text{if } i \in I_{0a} \cup I_3 \\ F_i + \varepsilon & \text{if } i \in I_{0b} \cup I_1 \end{cases} \quad (16)$$

Using these definitions, the conditions mentioned in Eq.(14) can be rewritten as:

$$\begin{aligned} \beta \geq \tilde{F}_i & \quad \forall i \in I_0 \cup I_1 \cup I_2 \\ \beta \leq \overline{F}_i & \quad \forall i \in I_0 \cup I_1 \cup I_3 \end{aligned} \quad (17)$$

Let us define:

$$\begin{aligned} b_{up} &= \min\{\overline{F}_i : i \in I_0 \cup I_1 \cup I_3\} \\ b_{low} &= \max\{\tilde{F}_i : i \in I_0 \cup I_1 \cup I_2\} \end{aligned} \quad (18)$$

Then it is easily to see that the optimality conditions will hold at some λ if

$$b_{low} \leq b_{up} \quad (19)$$

From paper [8], we know that, at optimality, β and b are identical. So in the rest of this paper, β and b will denote one and the same quantity.

In numerical solution, it is usually not possible to achieve optimality exactly, so the condition Eq.(19) can be replaced by an approximate optimality conditions [8] as

$$b_{low} \leq b_{up} + 2r \quad (20)$$

where r is a positive tolerance parameter and the care is needed in its choosing, see [11] for a related discussion. So an index pair (u,v) defines a violation at λ if one of the following two sets of conditions holds:

$$\begin{aligned} u \in I_0 \cup I_1 \cup I_2, \ v \in I_0 \cup I_1 \cup I_3 \text{ and } \tilde{F}_u > \overline{F}_v + 2r & \quad (a) \\ u \in I_0 \cup I_1 \cup I_3, \ v \in I_0 \cup I_1 \cup I_2 \text{ and } \overline{F}_u < \tilde{F}_v - 2r & \quad (b) \end{aligned} \quad (21)$$

When (u,v) satisfied Eq.(21) then a strict improvement in the dual objective function can be achieved by optimizing only the Lagrange multipliers corresponding to the examples u and v.

3.3 The Choice for the First Lagrange Multiplier

From Platt, Smola and Shevade's heuristic choice method for the first Lagrange multiplier, we found a disadvantage, which is that, in fact, the first Lagrange multiplier is chosen randomly; the change of the first selected Lagrange multiplier may not maximize the change of the objective function in all its possible candidates. In this paper, a new modified heuristic method for the choice of the first Lagrange multiplier is proposed to improve the efficiency of Shevade's SMO-1 algorithm.

The gradient of the dual problem's Lagrangian, Eq.(11), with respect to Lagrange multiplier λ_v is given by:

$$\frac{\partial L}{\partial \lambda_v} = \sum_{i=1}^{l} \lambda_i k_{iv} - y_v + \varepsilon \operatorname{sgn}(\lambda_v) + \beta - \delta_v + \tau_v \qquad (22)$$
$$= -F_v + \varepsilon \operatorname{sgn}(\lambda_v) + \beta - \delta_v + \tau_v$$

Since the larger the absolute gradient value, the more change the objective function with respect to λ_v, we can choose such a Lagrange multiplier in its candidates as the first Lagrange multiplier, which can maximize the absolute value of Eq.(22). When the first Lagrange multiplier is chosen in the outer loop, we determine whether the selected Lagrange multiplier violates the optimality conditions; then if violates, the second Lagrange multiplier can be chosen hand in hand [8]; for example, let i_{low} and i_{up} be indices so that:

$$\tilde{F}_{i_{low}} = b_{low} = \max\{\tilde{F}_i : i \in I_0 \cup I_1 \cup I_2\}$$
$$\overline{F}_{i_{up}} = b_{up} = \min\{\overline{F}_i : i \in I_0 \cup I_1 \cup I_3\} \qquad (23)$$

If $\overline{F}_v < b_{low} - 2r$ then there is a violation and in that case choose $u = i_{low}$; if the first selected Lagrange multiplier doesn't violate any optimality conditions, it should be deleted from the candidates set and go to the next loop.

For that the cost of CPU time mainly concentrates on the non-boundary examples and choosing the first Lagrange multiplier λ_v to maximize $|F_v + \varepsilon \operatorname{sgn}(\lambda_v) + \beta - \delta_v + \tau_v|$ also needs much CPU time. To be efficient, the proposed choice method for the first Lagrange multiplier is used only to alter $\lambda_i, i \in I_0$. From Eq.(13), we know that $\delta_i = 0, \tau_i = 0$ ($i \in I_0$) and because the value of β needs not to be calculated after every optimality step, β in (27) is also omitted to speed up the SMO algorithm. So when optimizing the Lagrange multipliers corresponding to the non-boundary examples, the first Lagrange multiplier λ_v can be chosen to satisfy the following functional:

$$\left|-F_v + \varepsilon \operatorname{sgn}(\lambda_v)\right| = \max_{i \in I_0}\left|-F_i + \varepsilon \operatorname{sgn}(\lambda_i)\right| \qquad (24)$$

4 Simulation Study

In this section, we compare the performance of our modified SMO algorithm against Shevade's SMO-1 algorithm for SVM regression on Boston housing dataset, which is a standard benchmark for regression algorithm and can be available at UCI Repository [12]. These two algorithms are run using Matlab6.5 on a P4 2.4GHz Windows XP Professional machine. In this study, the RBF kernel

$$K(x_i, x_j) = \exp(-\|x_i - x_j\|^2 / 2\sigma^2) \tag{25}$$

is used.

Table 1. Comparison results between the modified SMO and Shevade's SMO-1 when $C = 10$

	r	Modified SMO		Shevade's SMO-1	
		$t(s)$	Generalization error (MSE)	$t(s)$	Generalization error (MSE)
1	0.05	215.39	0.016399	368.844	0.015352
2	0.07	152.406	0.017393	167.937	0.015831
3	0.10	53.015	0.015048	90.266	0.018345
4	0.13	35.5	0.014692	46.282	0.018756
5	0.15	26.265	0.016933	31.453	0.016463
6	0.18	25.422	0.016649	20.11	0.019198
7	0.20	18.719	0.018655	23.375	0.017871

To Boston housing dataset, the dimension of the input is 13 and the size is 506. In the simulation study, the input data as well as the output are all scaled in the intervals [-1,1]. 406 data is used as the training set and the other 100 data is used as the testing set. From simulation, we found that these two algorithms are not insensitive to different choice of C. So the values, $\sigma = 1.3$, $\varepsilon = 0.025$ and $C = 10$ are used.

From Table 1, we can see that our modified SMO algorithm is faster than Shevade's SMO-1 algorithm in most cases. And the minimal generalization mean square error (MSE) is 0.014692, which is obtained by using our modified SMO algorithm.

5 Practical Application in Quality Prediction of HP-LDPE

5.1 Description of Industrial Background

High pressure low-density polyethylene (HP-LDPE) is produced by subjecting ethylene to a large amount of pressure, which is one of the highest-yield, lowest priced and most widely used general-purpose plastics in the world. Each product is produced under specified standard conditions, which means to control melt flow index, density and molecular weight distribution of low-density polyethylene. The melt flow rate measures the viscosity of the polyethylene resin in its molten state. It is a parameter relating to the average molecular weight of the resin chains of polymer extruded through a standard size orifice under specified conditions of pressure and temperature. The greater the lengths of its molecules, the greater the molecular weight and the greater the difficulty in extruding the resin through the standard orifice, the result: resins of greater viscosity are measured by a lower melt flow rate. And the greater the viscosity, the lower the melt index.

So melt index is the prediction of tensile strength, toughness and stress crack resistance and can be used to predict the quality of the end product. According to the analysis of technological mechanisms, the melt index is related with 21 variables, which are: first reference peak in temperature, corresponding tags are $TI212A \sim TI212E$ (°C);); second reference peak in temperature, corresponding tags are $TI217A \sim TI217E$ (°C); front die heads pressure of extruder, corresponding tag is $PI521$ (Mpa); back die heads pressure of extruder, corresponding tag is $PI522$ (Mpa); reactor entrance pressure, corresponding tag is $PI210$ (Mpa); front pressure of pulse valve, corresponding tag is $PI212$ (Mpa); flow rate of fed air, corresponding tag is $FI310A$ (m^3/h); flow rate of catalyzer at the second feeding point, corresponding tag is $FI332$ (kg/h); flow rate of catalyzer at the first feeding point, corresponding tag is $FI334$ (kg/h); flow rate of propane, corresponding tag is $FI322A$ (kg/h); propane content, corresponding tag is $AI141A$ (%); electromotor current of extruder, corresponding tag is $II520$ (A) and rotary speed of electromotor, corresponding tag is $SI520$ (rpm). All 21 variables can be measured and recorded on-line.

5.2 Quality Prediction Model

As mentioned above, the quality of HP-LDPE is determined essentially by the melt index. But the on-line measurement of melt index value is very difficulty, so a model for predicting the melt index on-line would be useful. In this section, the melt index will be studied.

By using method combined correlation analysis and technological mechanisms to analyze the input variables, the melt index is mainly affected by the following nine measurable variables:
$TI212B$、$TI217B$、$PI52$、$PI522$、$PI210$、$FI310A$、$FI334$、$II520$、$SI520$;
So its model structure is represented as:

$$M = g(TI212B, TI217B, PI521, PI522, PI210, FI310A, FI334, II520, SI520) \quad (26)$$

where MI represents the melt index of HP-LDPE, $g()$ is the complex multivariable non-linear function.

Then the proposed algorithm is used to establish the system. 99 data points with nine variables mentioned above are collected in different operating states, 59 data points of which are used as training data set, while the other 40 points are used as testing data set. All data points are scaled in the intervals [-1,1]. For the reason of comparison, we also built soft sensing model based on Shevade's SMO-1 algorithm besides the proposed SMO algorithm. For these two models, RBF kernel, Eq.(25),

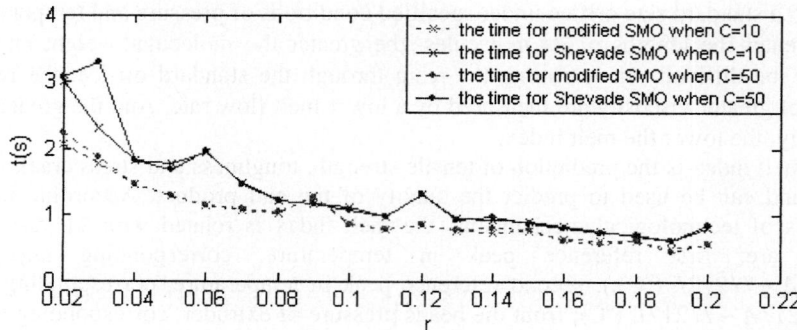

Fig. 1. HP-LDPE data: CPU time (*in second*) as a function of r

and the values $\sigma = 0.9$, $\varepsilon = 0.01$ are used. From simulation, we found that our modified SMO algorithm always has better performance than Shevade's SMO-1 algorithm with different choice of C. Fig.1 shows the performance of these two algorithms with different choice of r. From Fig.1, we can see that these two algorithms are not insensitive to $C = 10$ or 50, in fact, which is true to other choices of C.

Table 2. Comparison results between the modified SMO and Shevade's SMO when $C = 10$

	r	Modified SMO		Sheave's SMO-1	
		$t(s)$	Generalization error (MSE)	$t(s)$	Generalization error (MSE)
1	0.02	2.062	0.028693	3	0.02844
2	0.03	1.766	0.029308	2.297	0.027721
3	0.04	1.468	0.029684	1.828	0.028516
4	0.05	1.281	0.030068	1.765	0.028802
5	0.06	1.156	0.02804	1.937	0.031696
6	0.07	1.078	0.027122	1.484	0.032216
7	0.08	1.032	0.028936	1.219	0.028792
8	0.09	1.281	0.030939	1.172	0.028389
9	0.10	0.891	0.027427	1.109	0.03046
10	0.11	0.781	0.027038	0.969	0.025209
11	0.12	0.844	0.026783	1.312	0.025197
12	0.13	0.765	0.02749	0.953	0.027849
13	0.14	0.734	0.026387	0.938	0.029687
14	0.15	0.766	0.027388	0.907	0.026671
15	0.16	0.641	0.024311	0.828	0.025875
16	0.17	0.61	0.033891	0.765	0.034331
17	0.18	0.609	0.027593	0.687	0.03007
18	0.19	0.484	0.023278	0.609	0.025682
19	0.20	0.578	0.026217	0.797	0.028925

From Table 2, we also can see that in most cases, the modified SMO algorithm proposed in this paper is faster than Shevade's SMO-1 algorithm and the minimal generalization error (MSE) is 0.023278, which is also obtained by using our modified SMO algorithm.

6 Conclusion

In this paper, a new modified SMO algorithm is proposed based on Shevade's SMO-1 algorithm. Against Shevade's SMO-1 algorithm, the proposed SMO algorithm has two differences: the first is that the Lagrange multiplier α is replaced by λ; the second is that a modified heuristics method is used to choose the first Lagrange multiplier when optimizing the Lagrange multipliers corresponding to the non-boundary examples. In section V and section VI, a benchmark data set and a practical application are given to illustrate the validity of the proposed algorithm. The results demonstrate that this new SMO algorithm are faster in most cases with the same parameters setting and more likely to gain the better generalization performance than Shevade's SMO-1 algorithm.

References

1. Vapnik, V.N.: The Nature of Statistical Learning Theory. 2nd edn. Springer-Verlag, New York (2000)
2. Du, S., Wu, T.: Support Vector Machines for Regression. Journal of System Simulation, 15(11) (2003) 1580-1585
3. Li, J., Zhang, B.: An Improvement Algorithm to Sequential Minimal Optimization. Journal of Software, 14(3) (2003) 918-924
4. Osuna, Freund, R., Girosi, F.: An Improved Training Algorithm for Support Vector Machines. In Proc. of IEEE NNSP'97, (1997) 276-285
5. Platt, J.C.: Fast Training of Support Vector Machines Using Sequential Minimal Optimization. In: Schölkopf, B., Burges, C.J.C., and Smola, A.J. (eds.): Advances in Kernel Methods - Support Vector Learning. MIT Press (1998) 185-208
6. Smola, A.J.: Learning with Kernels. Ph.D. dissertation, GMD, Birlinghoven, Germany (1998)
7. Smola, A.J., Schölkopf, B.: A Tutorial on Support Vector Regression. NeuroColt Technical Report, TR-1998-030, Royal Holloway College (1998)
8. Shevade, S.K., Keerthi, S.S., Bhattacharyya, C., and Murthy, K.R.K.: Improvements to SMO Algorithm for SVM Regression. IEEE Transactions on Neural Networks, 11(5) (2000) 1188-1194
9. Vapnik, V.N.: Statistical Learning Theory. Wiley, New York (1998)
10. Flake, G.W., Lawrence, S.: Efficient SVM Regression Training with SMO. Machine Learning, 46 (2002) 271-290
11. Keerthi, S.S., Shevade, S.K., Bhattacharyya, C., and Murthy, K.R.K.: A Fast Iterative Nearest Point Algorithm for Support Vector Machine Classifier Design. IEEE Transaction on Neural Networks, 11(1) (2000) 124–136
12. Blake, C.L. and Merz, C.J.: UCI Repository of Machine Learning Databases. Dept. Inform. Comput. Sci., Univ. California, Irvine, CA, (1998) Online Available: http://www.ics.uci.edu/ ~mlearn/MLRepository.html.

Gait Recognition via Independent Component Analysis Based on Support Vector Machine and Neural Network

Erhu Zhang[1], Jiwen Lu[1], and Ganglong Duan[2]

[1] Department of Information Science, Xi'an University of Technology
Xi'an, Shaanxi, 710048, China
{eh-zhang, lujiwen}@xaut.edu.cn
[2] Department of Information Management, Xi'an University of Technology
Xi'an, Shaanxi, 710048, China
gl-duan@xaut.edu.cn

Abstract. This paper proposes a method of automatic gait recognition using Fourier descriptors and independent component analysis (ICA) for the purpose of human identification at a distance. Firstly, a simple background generation algorithm is introduced to subtract the moving figures accurately and to obtain binary human silhouettes. Secondly, these silhouettes are described with Fourier descriptors and converted into associated one-dimension signals. Then ICA is applied to get the independent components of the signals. For reducing the computational cost, a fast and robust fixed-point algorithm for calculating ICs is adopted and a criterion how to select ICs is put forward. Lastly, the nearest neighbor (NN), support vector machine (SVM) and backpropagation neural network (BPNN) classifiers are chosen for recognition and this method is tested on the small UMD gait database and the NLPR gait database. Experimental results show that our method has encouraging recognition accuracy.

1 Introduction

The demand for automatic human identification systems is strongly increasing in many important applications, especially at a great distance and it has recently gained great interest from the pattern recognition and computer vision researchers for it is widely used in security-sensitive environments, surveillance, access control and smart interfaces such as banks, parks and airports. Biometrics is a new powerful tool for reliable human identification and it makes use of human physiology or behavioral characteristics such as face, iris, fingerprints and hand geometry for identification. As a new behavioral biometric, gait recognition aims at identifying person by the way he or she walk. Compared with the first generational biometrics such as face, fingerprints, speech and iris which are widely applied in some commercial and low applications, gait has some prominent advantages of being non-contact, non-invasive, unobvious, low resolution requirement and it is the only perceivable biometric feature for human identification at a great distance till now though it is also affected by some factors such as drunkenness, pregnancy and injuries involving joints. Unlike face, gait is also difficult to conceal and has great potential applications in many situations especially for human identification at a great distance.

Although gait recognition is a new research field, there have been some studies and researches in recent literatures [1], [2], [3], [4], [5], [6], [7], [8], [9], [10] and [11]. Currently, gait recognition approaches are classified two main classes, namely holistic-based methods [2], [3], [6], [7], [8], [10] and [11] and model-based methods [5] and [9]. Model-based methods aim to model human body by analysis of the parts of body such as hand, torso, thigh, legs, and foot and perform model matching in each frame of a walking sequence to measure these parameters. As the effectiveness of model-based techniques, especially in human body modeling and parameter recovery from a walking sequence, is still limited (e.g. tracking and locating human body accurately in 2D or 3D space has been a long-term challenging and unsolved problem though there are much progresses in the past years even if some researcher have put forward many human tracking approaches), the disadvantages of model-based approaches is typically computational complexity because the movement of human body is non-rigid, therefore most existing gait recognition methods are holistic-based. Hence, like other holistic-based algorithms, we can consider gait being composed of a sequence of body poses and recognize it by the similarity of these body poses and silhouettes with low computational cost. Based on this assumption, this paper proposes an automatic gait recognition method for human identification using Fourier descriptors and independent component analysis, which achieves high recognition accuracy results. The method proposed in the paper can be mainly divided into three procedures including human motion detection, feature representation and gait recognition. The main advantages of our approach in this paper are as bellow: (1) based on Fourier descriptors and ICA, we make a meaningful attempt to human identification through gait information. (2) One-dimensional signals are applied to represent the changing of moving silhouettes which can decrease computational cost effectively. (3) Three classifiers namely nearest neighbor (NN), support vector machine (SVM) and back propagation neural network (BPNN) are applied for recognition and experimental results are compared with the three different classifiers. (4) It is easy to implement and has better recognition accuracy. The remainder sections of this paper are organized as follows: in the next Section, gait feature extraction is proposed while Sections 3 gives the final experimental results and recognition accuracy. At last, Section 4 concludes this paper and put forward future research direction of this field.

2 Gait Feature Extraction

Before training and recognition, each gait sequence involving a walking figure is converted into a sequence of signals which are from Fourier frequency components at this preprocessing stage. This procedure mainly involves background modeling, human motion segmentation and feature representation using Fourier descriptors and feature extraction of one-dimensional signals.

2.1 Segmentation of Human Motion

Human segmentation is the first step of our method and plays a key role in the whole gait recognition system. To extract the silhouettes of walking figures from the background, a simple motion detection approach using the median value is adopted to construct the background image from a small portion of video sequence including moving objects. Let P represents a sequence including N frames. The resulting background $p(x,y)$ can be computed as formulas (1):

$$p(x,y) = median[p_1(x,y), p_2(x,y), \cdots, p_N(x,y)] \qquad (1)$$

The value of $p(x,y)$ is the background brightness to be computed in the location of pixel (x,y) and *median* represents its median value. Here median value is taken rather than mean value of the pixel intensities over N frames, because mean value will be distorted by the large change in pixel intensities when the person moves past that pixel while the median is unaffected by spurious values, and the computational cost of median value is also lower than the least median square value used in literature [2]. The assumption made in this step is that the person does not stand still over the frames which are analyzed as in that case the background extraction will classify the person as a part of background and there is just only one moving person in our scene.

It should be noted that there do not exist a perfect image segmentation algorithm to segment the sequence images effectively up to the present. Here we adopt traditional histogram method to segment the foreground. For each image, the changing pixels can be detected by a suitable threshold T decided by traditional histogram and then we can easily obtain human silhouette through formulas (2):

$$D_{xy} = 1 \cdot (abs(p_i(x,y) - p(x,y)) \geq T) \; i = 1, 2, \cdots, N \qquad (2)$$

It also should be noted that this process is independent for each color component channels (i.e. Red, Green and Blue) in each frame of gait image. For each given pixel, if one of the three components accords with formulas (2), it will be determined as a foreground pixel. After that, there still exist some noises in the foreground, so morphological operators such as erosion and dilation are used to further filter spurious pixels and fill some small holes in human bodies. Two examples of background subtraction can be seen in Fig.1 from (a) to (h). Two separate databases are selected here as the data set for our gait experiments.

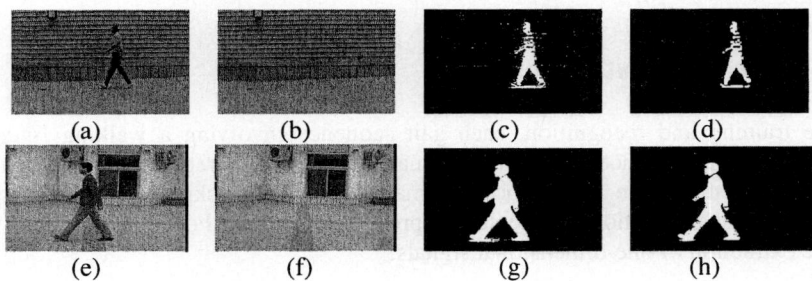

Fig. 1. Examples of background extraction from a sequence. (a), (b) (c) and (d) are from one gait database [4], (e), (f), (g) and (h) are from another database [1]. (a) and (e) are original images, (b) and (f) are the background images extracted through median method, (c) and (g) are the difference images, (d) and (h) are the silhouettes after morphological processing.

2.2 Representation of Human Silhouette Using Fourier Descriptors

An important factor affecting gait recognition is how to represent human silhouettes and extract the feature. To let our method be insensitive to changes of color and texture of clothes, we only use the binary silhouette. There existing some feature extraction methods proposed by early researchers, two typical methods are principal component analysis (PCA) [12] and [13] and Hidden Markov Models (HMM) [14] and [15]. In the above step, we have already obtained each human's boundary. Fourier descriptors have long been established and proved as a good method for representing a two-dimensional shape's boundary and its major advantage is that when representing a shape in Fourier domain, its frequency component can be easily obtained. The general features of the shape are located in the lower frequencies while the details features are located in the higher frequencies. Now, we apply Discrete Fourier descriptors to describe human silhouettes. First, each counter is set in the complex plane and its centroid is set as the origin complex. Each point on the counter can be represented by a complex number $s_i = x_i + j \times y_i$, $(i = 0,1,2,\cdots,N-1)$ where N is the number of counter points. We select the same number of points to represent each counters and unwrap each counter counterclockwise from the top of the counter and convert in into a complex vector $[s_0, s_1, \cdots, s_{N-1}]$. Therefore each gait sequence is transformed into a sequence of complex vectors with the same dimensions and the value of N in our experiment is 512 after re-sampling the boundary points. For a M-length vector, Fourier descriptors can be easily obtained through Fourier transform as formulas (3):

$$a_n = \frac{1}{M}\sum_{i=1}^{M} s_i e^{-2j\pi\frac{ni}{M}}, \text{ for } i = 0,1,\cdots M-1 \quad (3)$$

So Fourier descriptors used for recognition can be represented as formulas (4):

$$F = [\frac{|a_2|}{|a_1|}, \frac{|a_3|}{|a_1|}, \cdots, \frac{|a_{M-1}|}{|a_1|}] \quad (4)$$

Given n classes for training and each class represents a sequence of one subject's gait. Let $F_{i,j}$ be the jth feature signals in class i and N_i is the number of such signals in the ith class. The total numbers of training sequences is $N = N_1 + N_2 + \cdots + N_s$ and the whole training set can be represented into $F = [F_{1,1}, F_{1,2}, \cdots F_{1,N_1}, F_{2,1}, \cdots, F_{i,N_s}]$. Considering there may be different number points in different person's silhouette, we choose a fixed length of points for using discrete Fourier transform (here re-sampling 512 points in each binary silhouette for fast Fourier Transform). After Discrete Fourier transform, there still exists a large number of points in human silhouette. As we know, most energy of human silhouettes is concentrated in low frequency and we can ignore the high frequency as that does not contain much energy. For computational convenience, we only select 50 points of lowest frequency components for the least loss of silhouettes.

2.3 Training and Feature Extraction

At this stage, we will extract and train gait feature using ICA. The concept of ICA can be seen as a generational of principal component analysis (PCA) and its basic idea is to represent a set of random variables using basis functions, where the components are statistically independent or as independent as possible [16].

Let us denote the observed variables x_i as a vector with zero-mean random variable $X = (x_1, x_2, \cdots, x_m)^T$, the component variables s_i as a vector $S = (s_1, s_2, \cdots, s_n)^T$ with the model AS

$$X = AS \qquad (5)$$

Where A is unknown $m \times n$ matrix of full rank, called the mixing or feature matrix. The columns of A represent gait features and s_i signal the amplitude of the ith feature in the observed data x. For reducing computational cost, an algorithm named FastICA [17] using a fix-point iteration algorithm finding the local extrema of the kurtosis of a linear combination of the observed variables is introduced. Applying FatstICA on gait extraction, the random variables will be the training normalized frequency rate of gait images. We select thirty contour images for each class to construct the matrix X and make use of the fixed-point algorithm to calculate matrices A and S. Let $x_i^{'}$ be a distance vector of one contour image, we can construct a training distance set $\{x_1^{'}, x_2^{'}, \cdots, x_m^{'}\}$ with m random variables which are assumed to be linear component of n unknown ICs, denoted by $s_1^{'}, s_2^{'}, \cdots, s_n^{'}$. The relationship between X and S can be modeled as $X = AS$. For this relationship, each vector $x_i^{'}$ can be represented by a linear combination of $s_1, s_2, \cdots, s_n$ with weighting $a_{i1}, a_{i2}, \cdots, a_{in}$. Therefore, the feature matrix A can be considered as the features of all the training images.

According to the ICA theory, the matrix S contains all the independent components, which are calculated from a set of training distances. The matrix AS can reconstruct the original signals X. To reduce the computational cost, we select some ICs from A in the way that the ratio of the within-class scatter and between-class scatter is minimized [18]. The method is proposed as follows.

If the matrix X contains n individual persons and each person has m frames images, a_{ij} represents the entry at the ith row and the jth column. The value SB_j, which is called as the mean of within-class distance in the jth column, is then given by

$$SB_j = \frac{1}{nm(m-1)} \sum_{i=1}^{n} \sum_{u=1}^{m} \sum_{v=1}^{m} (a_{(i-1)m+u,j} - a_{(i-1)m+v,j})^2 \qquad (6)$$

The value SI_j, which is called as the mean of between-class distance in the jth column:

$$SI_j = \frac{1}{n(m-1)} \sum_{s=1}^{n} \sum_{t=1}^{n} \rho(a_{s,j}^{'} - a_{t,j}^{'}) \qquad (7)$$

where

$$a_{i,j}' = \frac{1}{m}\sum_{u=1}^{m} a_{(i-1)m+u,j} \qquad (8)$$

In this paper, we employ the ratio of within-class distance and between-class distance to select stable mixing feature from A. The ratio γ_j is defined as

$$\gamma_j = \frac{SB_j}{SI_j} \qquad (9)$$

From the definition γ_j, the smaller γ_j is, the better the classifier will be. Using (9), we choose the smallest γ_j and select the top k ($k<n$) column features from A and S.

3 Recognition and Experimental Results

Gait recognition is a traditional pattern classification problem which can be solved by measuring similarities between the training database and the test database. The classification process is carried out through three different methods, namely the nearest neighbor (NN), support vector machine (SVM) and classifier derived from the ICs. NN classifier is a very simple classifier and we use the Euclidean distance to evaluate the discriminatory of two gait sequences. Support vector machine classifiers have high generalization capabilities in many tasks especially in the object recognition problem. SVM is based on structural risk minimization, which is the expectation of the test error for the trained machine [19], [20], [21] and [22]. The risk is represented as $R(\alpha)$, α being the parameters of the trained machine. Let n being the number of trained patterns and $0 \leq \eta \leq 1$, with probability $1-\eta$, the following bound on the expected risk holds:

$$R(\alpha) \leq R_{emp}(\alpha) + \sqrt{\frac{h(\log(2l/h)+1) - \log(\eta/4)}{l}} \qquad (10)$$

$R_{emp}(\alpha)$ is the empirical risk and h is the VC dimension. SVM tries to minimize the second term of (10), for a fixed empirical. Here our SVM classifier is a 2-class classifier and there are two options for us: one is using N SVMs (N being the number of classes) while another is separating one class from the rest or using $N(N-1)/2$ SVMs one for each pair of class. We select the first option in our experiments because it is less complex.

Neural network classifier [23], [24] and [25] is a very useful classifier which is widely used in multiple class classification. Generally, BPNN has multiple layers, we can simple it into three layers i.e. input layer, hidden layer and output layer. The output is the recognition result which is the true class of human sequences and input is the feature of the test sequence. After extracting ICs from the training phase, the sequences are stored into the template database. When we input the test sequence, ICs are also need to be extracted. Then we can design a back propagetion Neural Network algorithm which is adopted to train and recognize. Given s classes for training, if the training sequence number is i, the ith output node is expected to be one while others are zeros. When we input the test sequence, the node which has the maximal value will be recognized the corresponding class.

Two famous public gait databases, namely University of Maryland database (UMD) and Chinese National Laboratory of Pattern Recognition (NLPR) database, are chosen to evaluate the capability of the proposed method. Here UMD database adopted is Portion of the whole database, here we use a small database including six persons and one sequence for each person, the walking direction is just perpendicular with our vidicon ocular. (It is just a part of the whole large UMD gait database). NLPR database includes 20 subjects and four sequences for each views angle and have three angles, namely laterally (0°), obliquely (45°) and frontally (90°).Table 1, 2 and 3 give the experiment results separately using different classifiers on the two datasets as follows:

Table 1. The recognition results using the NN classifier

	UMD database (6 persons, 1 view)		NLPR database (20 persons, 3 views)	
	90 ICs selected	Using all ICs	300 ICs selected	Using all ICs
Rank 1	100%	100%	75.0%	75.0%
Rank 5	100%	100%	85.0%	85.0%
Rank 10	100%	100%	90.0%	90.0%

Table 2. The recognition results using the SVM classifier

	UMD database (6 persons, 1 view)		NLPR database (20 persons, 3 views)	
	90 ICs selected	Using all ICs	300 ICs selected	Using all ICs
Rank 1	100%	100%	82.5%	81.9%
Rank 5	100%	100%	87.6%	86.1%
Rank 10	100%	100%	92.1%	89.6%

Table 3. The recognition results using the BPNN classifier

	UMD database (6 persons, 1 view)		NLPR database (20 persons, 3 views)	
	90 ICs selected	Using all ICs	300 ICs selected	Using all ICs
Rank 1	100%	100%	84.6%	84.1%
Rank 5	100%	100%	89.4%	88.6%
Rank 10	100%	100%	95.1%	92.3%

From the above three tables, we can find that BPNN classifier is the best classifier used in gait recognition rather than another two classifiers and NN classifier is the worst, but here we still use NN classifier as it is very simple and it can save a large of computational time for our recognition which is very important for gait recognition system. SVM is a new classifier as it has strong generalization and it is very suit for

2-class classification, so if we apply gait to distinguish the gender of the walkers, SVM classifier is the first choice. Otherwise, BPNN is our best choice as it has the best recognition accuracy though there is some difficult in its recognition speed.

4 Conclusion and Future Work

This paper has proposed a simple gait recognition method based on human silhouettes using Fourier descriptors and independent components analysis. From the analysis, we have found the independent components which are transformed from the frequency components have much better discriminatory capability than other gait feature. Besides these, the median background extraction method is better than the mean method and has less computational cost than the least mean square method. To provide a general approach to automatic human identification based on gait in real environments, much still remains to be done in the future. Although our recognition accuracy is comparatively high and encouraging, we still can not conclude much about gaits. Further evaluation on a much larger and most varied database is still needed. We are building up such a gait database with more subjects, more sequences with more different views and more variation in conditions such as the walkers wear different clothes in different seasons. The lack of general gait database, especially multiview in the gait database, is another limitation to most current gait recognition algorithms. Our proposed method is just recognizing human through one view separately, i.e. perpendicularity, along and oblique with the direction of human walking, in real environment, the angle between the walker's direction and the camera is unpredictable, generally speaking, a useful experiment which can determine the sensitivity of the features from different views should be put forward and that can provide us a more conviction results. Another method of solving this problem is to store more training sequences taken from many multiple views and then classify them. Now we are building such a large gait databases consists of multiviews and work for solving this problem. At last, seeking better maturity measures, designing more sophisticated classifiers, extracting more effect feature, proposing better gait detection and segmentation algorithms and combination of holistic-based and model-based methods deserve more attention in future work.

Acknowledgement

The authors would like to express their thanks to the Institute of Automatic, Chinese Academic of Science (CASIA) for Human ID image database. Portion of the research in this paper uses the CASIA Gait database are collected by Institute of Automatic, Chinese Academic of Science. The author would also like to thank Dr Hyvarinen for providing the fixed-point algorithm for independent component analysis used in our experiments. This work is partly supported by the Science and Technology Innovation Foundation of Xi'an University of Technology (No. 104-210401, 104-210413).

References

1. http://sinobiometrics.com
2. L.Wang, T.Tan, H.Ning, Hu. W: Silhouette Analysis-Based Gait Recognition for Human Identification. IEEE Transactions on Pattern Analysis and Machine Intelligence, 25 (2003) 1505-1518
3. Foster, Jeff P., Mark S. Nixon, Adam Prugel-Bennett.: Automatic Gait Recognition Using Area-based Metrics. Pattern Recognition Letters, 24 (2003) 2489-2497
4. http://www.doc.ic.ac.uk/~mrd98/gait
5. Wagg, D.K., M.S.Nixon: An Automated Model-based Extraction and Analysis of Gait. IEEE Conference on Automatic Face and Gesture Recognition, Seoul, Korea, 5 (2004) 11-16
6. Tian, G., Zhao, R.: Gait Recognition Based on Fourier Descriptors. Journal of Computer Application. 24, (2004)124-125+165 (in Chinese)
7. Stuart. D. Mowbray, Mark S. Nixon. Automatic Gait Recognition via Fourier Descriptors of Deformable Objects. Proceedings of 4th International Conference on Audio and Video-based Biometrics Person Authentication (2003)
8. Yu, S., Wang, L., Hu, W., Tan T.: Gait Analysis for Human Identification in Frequency Domain. Proceedings of the Third International Conference on Image and Graphics. Hong Kong, China (2004)
9. Lee, L.: Gait Analysis for Classification. Doctoral Thesis. Massachusetts Institute of Technology, USA (2002)
10. Han, H.: Study on Gait Extraction and Human Identification Based on Computer Vision. Doctoral thesis. University of Science and Technology Beijing, Beijing (2003) (in Chinese)
11. Li, B.: Study on Gait-based Human Identification. Master Thesis. University of Science and Technology Beijing. Beijing (2004). (in Chinese)
12. Cao, L.J., Chua, K.S., Chong, W.K., Lee H.P., Gu, Q.M.: A Comparison of PCA, KPCA and ICA for Dimensionality Reduction in Support Vector Machine. Neurocomputing 55 (2003) 321-336
13. Karhunen, J., Pajunen, P., E.Oja. The Nonlinear PCA Criterion in Blind Source Separation: Relations with Other Approaches. Neurocomputing 22 (1998) 5-20.
14. Kale, A, Rajagopalan, A. N., Cuntoor, N., Kruger, V.: Gait-based Recognition of Humans Using Continuous HMMs. Proceedings of the Fifth IEEE International Conference on Automatic Face and Gesture Recognition (2002)
15. Kim, M.S., Kim, D. Lee, S. Y.: Face Recognition using the embedded HMM with Second-order Block-specific Observations. Pattern Recognition 36 (2003) 2723-2735
16. Hyvarinen, A., Oja, E.: Independent Component Analysis: Algorithm and Applications. Neural Networks, 13 (2000) 411-430
17. Hyvarinen, A.: A Fast and Robust Fixed-Point Algorithm for Independent Component Analysis. IEEE Transaction on Neural Networks, 3 (1999) 626-634
18. P.C.Yuen, J.H.Lai. Face representation using independent component analysis. Pattern Recognition, Vol.35 (2002) 1247-1257
19. O. Deniz, M.Catrillon and M. Hernandez. Face Recognition Using Independent Component Analysis and Support Vector Machines. Pattern Recognition Letters, 24 (2003) 2153-2157
20. N. Cristianini, J. Shawe-Taylor. An Introduction to Support Vector Machines and Other Kernel-based Learning Methods. Publishing House of Electronics Industry. Beijing (2005) (in Chinese)

21. Kecman, V.: Learning and Soft Computing, Support Vector machines, Neural Networks and Fuzzy Logic Models, The MIT Press, Cambridge, MA (2001)
22. Wang, L.P. (Ed.): Support Vector Machines: Theory and Application. Springer, Berlin Heidelberg New York (2005)
23. Hagan, M.T., Demuth, H. B, Beale, M.: Neural Network Design. China Machine Press and CITIC Publishing House. Beijing (2002)
24. Haykin, S.: Neural Networks: A Comprehensive Foundation. Prentice Hall, New Jersey, 2nd ed. (1999)
25. Rajapakse, J.C., Wang, L.P. (Eds.): Neural Information Processing: Research and Development. Springer, Berlin (2004)

Uncertainty Support Vector Method for Ordinal Regression[1]

Liu Guangli, Sun Ruizhi, and Gao Wanlin

College of Information and Electrical Engineering,
China Agricultural University, Postfach 142,
100083, Beijing, China
{liugl, Sunrz, gaowlin}@cau.edu.cn

Abstract. Ordinal regression is complementary to the standard machine learning tasks of classification and metric regression which goal is to predict variables of ordinal scale. However, every input must be exactly assigned to one of these classes without any uncertainty in standard ordinal regression models. Based on structural risk minimization (SRM) principle, a new support vector learning technique for ordinal regression is proposed, which is able to deal with training data with uncertainty. Firstly, the meaning of the uncertainty is defined. Based on this meaning of uncertainty, two algorithms have been derived. This technique extends the application horizon of ordinal regression greatly. Moreover, the problem about early warning of food security in China is solved by our algorithm.

Keywords: Uncertainty, Ordinal regression, Quadratic Programming, Early-warning.

1 Introduction

Problems of ordinal regression arise in many fields, e.g., in information retrieval and in classical statistics. They can be related to the standard machine learning paradigm as follows [1]. In ordinal regression, we consider a problem which shares properties of both classification and metric regression. A variable of the above type exhibits an ordinal scale and can be thought of as the result of coarse measurement of a continuous variable. The ordinal scale leads to problems in defining an appropriate loss function for our task. Similar to Support Vector methods a learning algorithm can be derived for the task of ordinal regression based on large margin rank boundaries. Maximizing the margin leads to a quadratic programming problem which can be solved from its dual problem [2]-[5].

However, there is some limitation which restricts its applications. For example, it is required that every input x_i must be exactly assigned to one of these classes with full certainty. But sometimes this requirement is too restrictive to be used in practice. A new technique is proposed which is able to deal with the training data with uncertainty

---

[1] This paper is sponsored by China National Science Foundation under grant No. 90412009.

in this paper. Firstly, the meaning of the uncertainty is defined. Based on this meaning of uncertainty, two algorithms have been derived. This technique extends the application horizon of ordinal regression greatly.

2 Ordinal Regression Based on SRM

Let x_i^j be the set of training examples where $j=1,\cdots,k$ denotes the class number, and $i=1,\cdots,i_j$ is the index within each class. Let $l=\sum_i i_j$ be the total number of training examples. The geometric interpretation of this approach is to look for k-1 parallel hyper-planes represented by vector $w \in R^n$ and scalars $b_1 \leq \cdots \leq b_{k-1}$ defining the hyper-planes (w, b_i) such that the data are separated by dividing the space into equally ranked regions by the decision rule f(x)=min{r: $w \cdot x - b_r \leq 0$}.

According to the structural risk minimization principle of 2-classes learning, the margin to be maximized is the one defined by the closest pair of classes. Formally, let (w, b_q) be the hyper-plane separating the two pairs of classes which are the closest among all the neighboring pairs of classes. Let w, b_q be scaled such the distance of the boundary points from the hyper-plane is **1**, i.e., the margin between the classes q,q+1 is 2/||w||.Thus, the fixed margin policy for ranking learning is to find the direction w and the scalars $b_1,\cdots,b_{k-1}$ such that $w \cdot w$ is minimized subject to the separability constraints. So the primal QLP formulation of the OSVR can be obtained based on SRM and the dual problem can also be deduced according to KKT conditions [2].

3 Uncertainty Support Vector Method for Ordinal Regression

It is assumed in the OSVR technique that each input x_i belongs to one class exactly. But in many cases, there exists some uncertainty. The input x_i may not exactly belong to any one, but belongs to one class with a certain probability. The uncertainty can be described by probability. More precisely, it is allowed that x_i belongs to one class with probability z_i^j ($i=1,\cdots,i_j$, $j=1,\cdots,k$). Obviously, the cases z_i^j= +1 or -1 correspond respectively the standard OSVR technique. The cases z_i^j=0 mean that there is no any information at the input x_i and therefore it should be neglected.

Suppose that we are given a triplet set S_1={($x_1, z_1^1,\cdots, z_1^k$) ,$\cdots$, ($x_l, z_l^1,\cdots, z_l^k$)} where $z_i^j \in [0,1]$ are the probability with which the input x_i belongs to j class respectively. First consider the case in which all z_i^j are rational numbers. Let their common denominator be p. Then we have $z_i^j = q_i^j /p$, where q_i^j, p are nonnegative integers and $q_i^j < p$. Therefore, if the input x_i^j is taken repeatedly p times, it is reasonable to expect that the input x_i^j belongs to j class for q_i^j times. Thus, corresponding to the triplet set S_1, we have the training set S_2={(x_1^1, q_1^1) ,$\cdots$, (x_l^1, q_l^1) ,$\cdots$, (x_1^k, q_1^k) ,$\cdots$, (x_l^k, q_l^k) }. For the training set S_2, we have the following quadratic programming problem

$$\min_{w,b,\varepsilon_{im}^j,\varepsilon_{in}^{*j+1}} \tfrac{1}{2}(w\cdot w) + \tfrac{C}{p}\sum_{i,j}(\sum_m \varepsilon_{im}^j + \sum_n \varepsilon_{in}^{*j+1})$$
$$\text{s.t.} \quad w\cdot x_i^j - b_j \leq -1 + \varepsilon_{im}^j, \quad m = 1,\cdots,q_i^j,$$
$$w\cdot x_i^{j+1} - b_j \geq 1 - \varepsilon_{in}^{*j+1}, n = 1,\cdots,q_i^{j+1},$$
$$\varepsilon_{im}^j \geq 0, \varepsilon_{in}^{*j+1} \geq 0, \quad b_j \leq b_{j+1}, j = 1,\cdots,k-2.$$
(1)

Here the constant C in [2] is replaced by the constant C/p since the $\sum_j q_i^j$ examples are considered as one unit. However, this problem is rather complicated. It is easy to know that we need only to solve the following simpler problem

$$\min_{w,b,\varepsilon_i^j,\varepsilon_i^{*j+1}} \tfrac{1}{2}(w\cdot w) + C\sum_{i,j}(q_i^j \varepsilon_i^j + q_i^{j+1}\varepsilon_i^{*j+1})$$
$$\text{s.t.} \quad w\cdot x_i^j - b_j \leq -1 + \varepsilon_i^j, \varepsilon_i^j \geq 0,$$
$$w\cdot x_i^{j+1} - b_j \geq 1 - \varepsilon_i^{*j+1}, \varepsilon_i^{*j+1} \geq 0.$$
$$b_j \leq b_{j+1}, \quad j = 1,\cdots,k-2.$$
(2)

So, for a triplet set S_2, when all z_i^j are rational numbers, we have proposed a reasonable approach. It can be proved that this observation can be obvious to extend to the general case in which z_i^j is real numbers in [0, 1]. For the triplet set S_1, where the input x_i belong to j-class with probability z_i^j, the optimization problem is

$$\min_{w,b,\varepsilon_i^j,\varepsilon_i^{*j+1}} \tfrac{1}{2}(w\cdot w) + C\sum_i\sum_j(z_i^j\varepsilon_i^j + z_i^{j+1}\varepsilon_i^{*j+1})$$
$$\text{s.t.} \quad w\cdot x_i^j - b_j \leq -1 + \varepsilon_i^j, \varepsilon_i^j \geq 0, \text{ for } z_i^j \neq 0,$$
$$w\cdot x_i^{j+1} - b_j \geq 1 - \varepsilon_i^{*j+1}, \varepsilon_i^{*j+1} \geq 0, \text{ for } z_i^{j+1} \neq 0,$$
$$b_j \leq b_{j+1}, \quad j = 1,\cdots,k-2.$$
(3)

For the sake of presenting the dual functional in a compact form, we will introduce some new notations. Let X^j be the $n*i_j$ matrix whose columns are the data points x_i^j ($i=1,\cdots,i_j$). Let λ^j be the vector whose components are the Lagrange multipliers λ_i^j ($i=1,\cdots,i_j$) corresponding to class j. Likewise, let δ^j be the Lagrange multipliers δ_i^j corresponding to class $j+1$. Let $\mu=(\lambda^1,\cdots,\lambda^{k-1},\delta^1,\cdots,\delta^{k-1})$ be the vector holding all λ_i^j and δ_i^j Lagrange multipliers, and let $\mu^1=(\mu_1^1,\cdots,\mu_{k-1}^1)^T=(\lambda^1,\cdots,\lambda^{k-1})^T$ and $\mu^2=(\mu_1^2,\cdots,\mu_{k-1}^2)^T=(\delta^1,\cdots,\delta^{k-1})^T$ the first and second halves of μ. Note that $\mu_j^1=\lambda^j$ is a vector, and likewise so is $\mu_j^2=\delta^j$. Let **1** be the vector of 1's, and finally, let $Q=[-X^1,\cdots,-X^{k-1},X^2,\cdots,X^k]_{n*N}$ be the matrix holding two copies of the training data, where $N=2l-i_1-i_k$.

The dual problem of problem (3) is

$$\max_{\mu} \sum_{i=1}^N \mu_i - \tfrac{1}{2}\mu^T(Q^TQ)\mu$$
$$\text{s.t.} \quad 0 \leq \mu_i \leq z_iC, \quad i = 1,\cdots,(k-2)l,$$
$$\mathbf{1}\cdot\mu_j^1 = \mathbf{1}\cdot\mu_j^2, \quad j = 1,\cdots,k-1,$$
(4)

Suppose we are given a classification problem with uncertainty which is represented by a triplet set S_1. According to the standard OSVR technique and the above two propositions, it is reasonable to establish a classification algorithm. If a nonlinear function $f(x)$ is required, we map the input space X into a high-dimensional feature space $F: \{z = \phi(x)| x \in X\}$. Corresponding to the optimization problem (6), it is easy to get the following dual optimization problem.

$$\max_{\mu} \quad \sum_{i=1}^{N} \mu_i - \mu^T(\bar{Q}^T\bar{Q})\mu$$
$$\text{s.t.} \quad 0 \leq \mu_i \leq z_i C, \quad i = 1,\cdots,(k-2)l, \tag{5}$$
$$1 \cdot \mu_j^1 = 1 \cdot \mu_j^2, \quad j = 1,\cdots,k-1.$$

where $\bar{Q} = \left[-\bar{X}^1,\cdots,-\bar{X}^{k-1},\bar{X}^2,\cdots,\bar{X}^k\right]_{n\times(k-2)l}$ and $\bar{X}^j = (\phi(x_1^j),\cdots,\phi(x_l^j))_{n\times l}$.

When $k=2$, i.e., we have only two classes thus the ranking learning problem is equivalent to the 2-class classification problem with uncertainty, and the dual functional reduces equivalent to the dual form of the SVM with uncertainty. The criteria function involves only inner-products of the training examples, thereby making it possible to work with kernel-based inner-products. From the dual form one can solve for the Lagrange multipliers μ_i and in turn obtain w the direction of the parallel hyper-planes. The scalar b_q can be obtained from the support vectors, but the remaining scalars b_j cannot. Therefore an additional stage is required which amounts to a Linear Programming problem on the original primal functional but this time w is already known.

4 Experiments

For the problem of food security early warning in China, our aim is to predict the degree of warning of grain production by classification. The degree has been classified to two classes, one is warning (+1) and the other is no warning (-1). One is warning (-1) and the other is no warning (+1). According to [6] and [7], the input is selected as a 7-dimensional vector $x=(x^1,\cdots,x^7)$, where x^1 is the government expenditure index for agriculture, x^2 is the expenditure index for capital construction, x^3 is the science & technology promotion funds index, x^4 is the total power index of agriculture machinery, x^5 is the irrigated area index, x^6 is the power index of irrigated machinery, x^7 is the rain price index.

These vectors include the data of 25 years from 1980 to 2004 which are listed in Table 1. From 1980 to 2003, each year has a degree of warning. For some years such as 1982 and 1985, the decision maker assigned them to either the positive class or the negative class. But for some years such as 1980, he could not decide which class to be classified exactly. At last, he concluded that- these years belonged to the positive class with probability z_i^+, and to the negative class with probability z_i^-. These data are also listed in the last two columns of Table 1.

Table 1.

Year	x^1(%)	x^2(%)	x^3(%)	x^4(%)	x^5(%)	x^6(%)	x^7(%)	z_i^-	z_i^+
...	...	...	...	...	...	...	...	...	...
2002	11.25	19.65	-4.366	9.263	0.5648	9.256	1.265	3/7	4/7
2003	15.03	18.33	11.654	8.566	1.2555	9.658	8.654	1	0
2004	14.38	19.25	10.375	8.652	1.9273	9.467	8.083	y=	

Based on the training data from 1980 to 2003 given in Table 1 and using Algorithm UOSVR with k=2 and C=0.1, we obtain the decision function: $f(x)=0.002\ x^1-0.0052\ x^3+0.001\ x^4+0.0003\ x^5-0.0026\ x^7+0.554$.

For 2004, $x_{2004}=(14.38,19.25,10.375,8.654,1.9276,9.467,8.083)^T$, then $f(x_{2004})=0.4834$. So, this year belongs to the positive class, which means we predict its degree of warning is no warning. In fact, this result is reasonable.

5 Conclusion

A new support vector learning technique is proposed, which is able to deal with training data with uncertainty. Moreover, the problem about early warning of food security in China is solved by our algorithm.

References

1. Nello Cristianini and John Shawe-Taylor. An Introduction to Support Vector Machines and Other Kernelbased Learning Methods, Cambridge Univerdity Press, Cambridge, UK, 2000.
2. R.Herbrich, T.Graepel, and K.Obermayer. Large Margin Rank Boundaries for Ordinal Regression. In: Advances in Large Margin Classifiers (2000) 115–132.
3. N.Cristianini, J.Shawe-Taylor. An Introduction to Support Vector Machines and Other Kernel based Learning Methods. Cambridge University Press, Cambridge, UK (2000).
4. Ralf Herbrich, Learning Kernel Classifiers. MIT Press, Cambridge, MA, 2002.
5. Chun-Fu Lin and Sheng-DeWang. Fuzzy Support Vector Machines, IEEE Transactions on Neural Networks, Vol.13, No.2, 2002.
6. Zhiqiang Li, Zhongping Zhao and Yuhua Wu. The Analysis on Early Warning of Food Security in China, China Agriculture Econonmy (in Chinese) Vol.1, 1998.
7. Xiaoning An. Theory, Method and System Design on Early Warning of Food Security, World Agriculture (in Chinese) Vol.231, 1998.

An Incremental Learning Method Based on SVM for Online Sketchy Shape Recognition

Zhengxing Sun, Lisha Zhang, and Enyi Tang

State Key Lab for Novel Software Technology,
Nanjing University, 210093, China
szx@nju.edu.cn

Abstract. This paper presents briefly an incremental learning method based on SVM for online sketchy shape recognition. It can collect all classified results corrected by user and select some important samples as the retraining data according to their distance to the hyper-plane of the SVM-classifier. The classifier can then do incremental learning quickly on the newly added samples, and the retrained classifier can be adaptive to the user's drawing styles. Experiment shows the effectiveness of the proposed method.

1 Introduction

Sketching is a natural and informal interaction mode. But its ambiguity and uncertainty make the deduction of users' intents very difficult. It will be more helpful if the sketchy shape can be online recognized and converted into the user-intended regular shape and user can realize errors or inappropriateness earlier with the online immediate feedback. Though numerous researches have been achieved, the poor efficiency of recognition engines is always frustrating, especially for the newly added users. This is because that the styles of sketching vary with different users, even the same user at different times. Therefore, adaptive sketchy shape recognition should be required [1], where recognition engine should incrementally be trainable and adaptable to a particular user's sketching styles.

In this paper, we will present an incremental learning method based on SVM (Support Vector Machine) for adaptive sketchy shape recognition. In Section 2, our proposed strategy and experiments are described in detail. Conclusions are given in the final Section.

2 Adaptive Sketchy Shape Recognition Based on SVM Classifier

In a broad sense, adaptive sketchy shape recognition means that the recognition engine should be adjustable to fit the variations of user's sketchy shapes dynamically. Accordingly, the classifier should be able to analyze the incremental samples for user's drawing styles and be retrained with the newly added samples obtained.

2.1 Our Strategy for Online Adaptive Sketch Recognition

Our framework for adaptive sketchy shape recognition is shown in **Fig. 1**. The processes can be summarized as three stages as following: raw stroke processing and feature extraction, online sketchy shape recognition based on SVM classifier, newly important samples selection and incremental training of SVM classifier.

The raw strokes pre-processing is firstly used to eliminate the noise caused by input conditions and the inputting sketchy shapes are treated as the composition of some continuous connected strokes. Feature extraction is then applied to obtain the feature vectors of sketchy shapes based on our modified turning function [1].

In succession, online sketchy shape recognition is done by means of the trained or retrained SVM classifier. The SVM classifier is constructed using many binary SVM classifiers, where a Radial Basic Function kernel [2] is used and the training process is the same as in SVMTorch [3]. Additionally, the recognized results must undergo rectification so that the sketchy shapes are easy to be evaluated by user [1].

Although the recognition precisions of SVM classifier could be very high, it still may not be suitable for a specific user's drawing styles, and user would correct some results of recognition. Accordingly, we design a strategy to collect the samples evaluated by user, select some important samples as incremental training data of classifier until the results of classifiers are satisfactory or enough training samples are obtained, and retrain the SVM classifier with the newly added samples obtained, which are named respectively as "sample collection", "sample selection" and "incremental training" as shown in Fig. 1.

For example, it is quite often that a user draws a triangle very quickly such that the angle is not very obvious and the system may recognize it as a quadrangle or ellipse as shown in Fig. 2 (a)(b)(c), and quadrangle and pentagon as ellipse by mistake, as

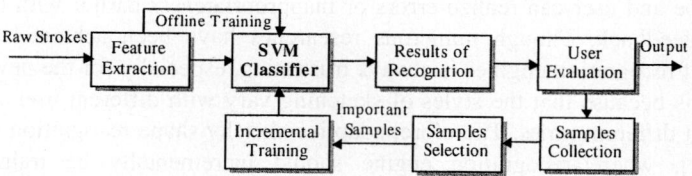

Fig. 1. Framework of adaptive sketch recognition based on SVM incremental learning

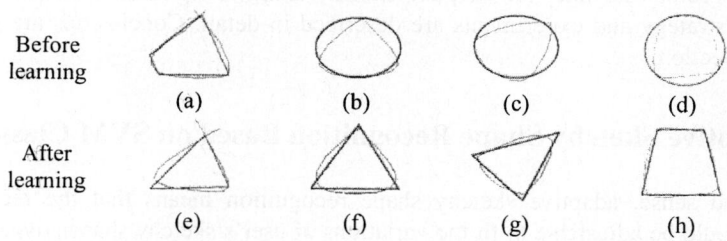

Fig. 2. Example of the adaptive sketch recognition based on SVM incremental learning

shown in Fig. 2 (d). By our strategy, user can correct these errors immediately and the system will collect all these samples. After several samples are corrected, the system will do incremental learning on these newly samples, and the new classifier is adaptive to recognize correctly them, as is shown in Fig. 2 (e)(f)(g)(h).

2.2 Principle of Our SVM Incremental Learning Algorithm

The main idea of SVM is to construct a nonlinear kernel function to map the data from the input space into a possible high-dimensional feature space and then generalize the optimal hyper-plane with maximum margin between the two classes.

Given a set of n training samples: $\{(x_1, y_1), \cdots, (x_n, y_n)\}$, where x_i denotes a vector (samples) in a d-dimensional space and $y_i \in \{+1, -1\}$ represents the class to which x_i belongs. The equation of the optimal hyper-plane separating the two classes can be expressed by $wx + b = 0$, where w denotes the normal of the hyper-plane, b denotes the offset. The normal of the optimal hyper-plane w proved by Vapnik [4] can be expressed by a linear combination of the training samples as in Eq. (1):

$$w = \sum_{i=1}^{n} \alpha_i y_i x_i, \quad 0 < \alpha_i \leq C \qquad (1)$$

The objective of training is to obtain each sample's α value. In most cases, the sample space is not linearly separable. A mapping Φ is usually used to transform non-linearly the input samples into a high dimensional feature space so as to make these samples linearly separable. In other hand, the generalization property of an SVM does not depend on all the training data, but only Support Vectors (in short, SVs). The optimal hyper-plane can then be expressed as:

$$\sum_{x_i \in SV} \alpha_i y_i K(x, x_i) + b = 0, \text{ where, } 0 < \alpha_i \leq C \text{ and } K(x, y) = \Phi(x)\Phi(y) \qquad (2)$$

Accordingly, we develop a new incremental learning algorithm by inserting an evaluating process in the training of SVM incremental learning. The training would only preserve the Support Vectors at each incremental step and then add them instead of all historic data to the training sets collected in the evaluating step. Denoting the training process of SVM as *Train()*, the process of evaluating all newly added samples as *Evaluate()*, the initial and incremental training samples as *IS* and *INS* respectively, the temporary training samples as *NS* and working data as *WS*, the process of our algorithm can be briefly described as: (1). *Γ=Train(IS), WS=IS_{SV}*; (2). *Γ=Train(WS), WS=WS_{SV}*; (3). *NS=Evaluate(INS), WS=WS∪NS*; (4). Retraining - Repeat (2) and (3) until the results of SVM classifiers are satisfactory.

2.3 Selection of Retraining Samples for SVM Incremental Learning

For our strategy for adaptive sketchy shape recognition, the core is how to select some important samples as incremental training data of SVM classifier. According to the principle of our SVM incremental learning described in section 2.2, the candidate training samples are determined by estimating their distance from the hyper-plane as expressed in Eq. (2), and the coefficients α_i can be obtained from solving the following optimization problem:

$$\begin{cases} \text{Minimize}: w(\alpha) = \sum_{i=1}^{l} \alpha_i - \frac{1}{2}\sum_{i,j=1}^{l} \alpha_i \alpha_j y_i y_j \Phi(x_i)\Phi(x_j) \\ \text{Subject to constraints}: \sum_{i=1}^{l} \alpha_i y_i = 0, \alpha_i \geq 0 \end{cases} \quad (3)$$

The distance between the sample x and the hyper-plane can then be defined as [1]:

$$d(x, w) = |w_0 \cdot \Phi(x) + b| = |\sum_{x_i \in SV} \alpha_i y_i K(x, x_i) + b| \quad (4)$$

During sample selection, a threshold of distance must be selected. Intuitively, there is a conflict between the precision and the speed if a constant distance threshold is used to choose the important samples. Hence, a dynamic threshold is considered in our strategy. In the beginning of incremental learning, a larger threshold value is used such that the precision can be a little higher while the time is still acceptable. As the number of added training data increases in later incremental learning steps, the selection process should judge the importance of every newly sample more carefully with a smaller threshold in order to avoid substantial increase of the number samples necessary for retraining of the SVM classifiers. In another aspect, if want to adapt to a new user, a large threshold value can be used to adapt him quickly. Using this dynamic threshold method, we can obtain higher precision with shorter training time.

2.4 Experiment and Performance Evaluation

To validate the performance of our proposed strategy, we have done experiments of four algorithms (the repetitive learning [4], Syed et al's [5], Xiao et al's [6] and ours) based on a 5-class sketchy shape dataset (including triangle, quadrangle, pentagon, hexagon, and ellipse) using one-against-one classifier structure [1]. We collected 1367 sample shapes drawn with pen/tablet and 325 samplers drawn with mouse, and we can have 40 samples for each sample by using a 20-dimension feature vector and after transformation. We select randomly 20210 samples of the pen/tablet to form a test set, TS_1, and use 12000 samples of the mouse style to form another test set, TS_2. Then, we randomly select samples from the remained samples of the pen/tablet style to form 39 incremental training sample sets. The first 6 incremental training sets have 100, 100, 120, 150, 300, 700 samples, respectively, and each of the rest 33 sets has 1000 samples. The remained samples of the mouse style form a sample set with 1000 samples, and it is then added into the incremental learning process as the last training set. All experiments are done on an Intel PC (with a 2.4GHz CPU and 256MB memory) running on Microsoft Windows XP Professional.

Fig. 3 shows the performance comparison among the four algorithms. For each algorithm, we use the same 40 incremental training sample sets to train them incrementally and record the incremental training time in Fig. 3 (a). After each incremental training step, we test it using the two test sets, TS_1 and TS_2, respectively, and record their classification precisions. Fig. 3 (b) shows the curves of the precisions using test set TS_1 (results of the precision on TS_2 are similar with on TS_1 and omitted here). We can see from Fig. 3 that our method can do well in reducing the training time with very little loss of precision. These prove that our strategy for online adaptive sketchy shape recognition is feasible.

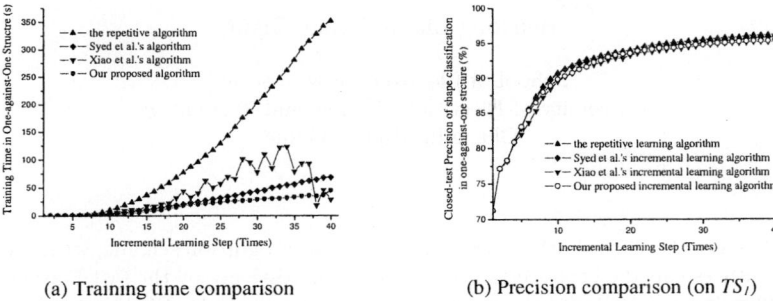

(a) Training time comparison   (b) Precision comparison (on TS_I)

Fig. 3. Performance evaluation among the four algorithms using 1-1 classifier structure

3 Conclusion

By utilizing the advantages of our modified SVM incremental learning algorithm, this paper presents a strategy of adaptive sketchy shape recognition, which can collect the samples evaluated by user during his/her drawing and select some important samples as incremental training data of classifier. This strategy can incrementally adapt the settings of classifier to users' accustomed styles, not only for the patterns they are originally trained for. Experiments show its efficiency for user adaptation. However, main limitation of our method is that it can only be suited for simple shape drawn by some continuous strokes.

Acknowledgement. This paper was supported by grants from National Natural Science Foundation of China (Project No. 69903006 and 60373065) and the Program for New Century Excellent Talents in University of China.

References

1. Sun Z. X., Liu W. Y., Peng B. B., et al: User Adaptation for Online Sketchy Shape Recognition. Lecture Notes in Computer Science, Vol. 3088. Springer-Veralg, Heidelberg New York (2004) 303-314.
2. Ratsaby J.: Incremental Learning with Sample Queries. IEEE Trans. on Pattern Analysis and Machine Intelligence. 20 (1998) 883-888.
3. Collobert R. and Bengio S.: SVMtorch: Support Vector Machines for Large-Scale Regression Problems. Journal of Machine Learning Research. (2001) 143-160.
4. Vapnik V. N.: Statistical Learning Theory. John Wiley and Sons (1998).
5. Syed N., Liu H. and Sung K.: Incremental Learning with Support Vector Machines. In: Procedings of IJCAI-99, Stockholm, Sweden. (1999) 272-276.
6. Xiao R., Wang J. C., Sun Z. X., et al: An Incremental SVM Learning Algorithm Alpha-SVM. Chinese Journal of Software. 12 (2001) 1818-1824.

Eigenspectra Versus Eigenfaces: Classification with a Kernel-Based Nonlinear Representor*

Benyong Liu and Jing Zhang

School of Electronic Engineering,
University of Electronic Science and Technology,
Chengdu 610054, China

Abstract. This short paper proposes a face recognition scheme, wherein features called eigenspectra are extracted successively by the fast Fourier transform (FFT) and the principle component analysis (PCA) and classification results are obtained by a classifier called kernel-based nonlinear representor (KNR). Its effectiveness is shown by experimental results on the Olivetti Research Laboratory (ORL) face database.

1 Introduction

Feature extraction and classification are two key elements of a face recognition system. The PCA and the linear discriminant analysis (LDA) are two popular algorithms for feature extraction [1]. Since an LDA-based method aims at discriminative features while a PCA-based one at representative ones, it is widely believed that the former outperforms the latter, and thus much attention was paid to LDA or its variations [2]. However, some practical applications in face recognition show that this is not always true [3]. In addition, LDA-based algorithms suffer from the so-called "small sample size problem" when the number of samples is far smaller than the dimensionality of samples [1].

Generally face features are not linearly separable and thus we need a suitable nonlinear classifier. In some popular methods, such as the Parzen classifier and a nonlinear support vector machine (SVM) classifier [4],[5], the solution is represented by $f(\boldsymbol{x}) = \sum_{j=1}^{M} a_j k(\boldsymbol{x}, \boldsymbol{x}_j) + b$, where $\{\boldsymbol{x}_j\}_{j=1}^{M}$ is a subset of the training feature vectors, b a constant number, and k an associated kernel function. The coefficient set, $\{\boldsymbol{a}_j\}_{j=1}^{M}$, is determined by the nature of the problem. For example, when Vapnik's ϵ-insensitive cost function is adopted, it is obtained by the SVM approximation scheme, in which a nonlinear programming problem needs to be solved [5]. Recently, we proposed a classifier of the above representation called a kernel-based nonlinear representor (KNR) [6]. Its coefficient set is determined, in a closed form, by the classifier's capability in feature representation.

Practically, the performance of a face recognition system subjects to variations in viewpoint, age, pose, expression, lighting condition, etc., and thus no

---
* Supported by the Key Project of Chinese Ministry of Education (No.105150). Thanks to Prof. H. Ogawa of Tokyo Institute of Technology for helpful discussions.

single algorithm is likely to be the solution to all the above problems. The most successful systems will probably blend several algorithms to achieve satisfactory performance. In this standpoint, we propose a face recognition scheme and demonstrate its effectiveness by experimental results on the ORL face database.

2 Proposed Face Recognition Scheme

The presented face recognition scheme includes three key elements: Preprocessing based on pixel averaging and energy normalization, feature extraction successively conducted by the FFT and the PCA, and classification with a KNR.

2.1 Preprocessing and Feature Extraction

In preprocessing, a face image is down-sampled by pixel averaging; in fact, supposing that there is an $N_0 \times M_0$ image g_0, and the pixel averaged one is an $N_1 \times M_1$ image g, then $g(n,m) = \sum_{(x,y) \in S_{n,m}} g_0(x,y)/D$ for $1 \leq n \leq N_1$, $1 \leq m \leq M_1$, where $S_{n,m}$ is a set of D pixels in a selected block. To reduce the image brightness effect, we further normalize g into a new version represented by $I = g/\|g\|$, where $\|g\|$ is the Frobenius norm of g.

To obtain compact, representative, and robust features, we adopt the discrete Fourier transform (DFT) and the PCA algorithms since they are relatively simple to implement. In DFT realization, the FFT algorithm is adopted. Before FFT, each preprocessed image is row concatenated to form an $N_1 M_1$-dimensional vector, z. Let w denotes the $N_1 M_1$-point FFT of z, then its dimension may be reduced to one half according to symmetry properties of the DFT, and thus we obtain the *spectra vector* $h = [|w(0)| \; |w(1)| \; \cdots \; |w(L)|]^T$, where T denotes the transpose of a vector or a matrix and $L = (N_1 M_1/2 - 1)$ for even $N_1 M_1$ and $(N_1 M_1/2 - 0.5)$ for odd $N_1 M_1$. According to the shift invariant property of the DFT, h is spatially insensitive to face sway within the image plane.

Let C denotes the covariance matrix of the training spectra vectors, then its eigenvectors associated with the $N (N \ll L)$ largest eigenvalues constitute a transform matrix P, and $x = P^T h$ is named an *eigenspectra vector*. For each subject, the training eigenspectra vectors are adopted to train a KNR classifier.

2.2 Classification with a KNR

KNR is a key element in our scheme and thus it should be briefly reviewed. Specifically, assuming that a desirable classifier f_0 is defined on an N-dimensional pattern feature space C^N, and it is an element of a reproducing kernel Hilbert space H which has a kernel function k. Generally f_0 is unknown but its M sampled values are known beforehand and they constitute a teacher vector y, where $y = A f_0$ and A is the sampler. We assume that y is an element of the M-dimensional space C^M. In the viewpoint of an inverse problem, a certain inverse operator X of A is to be found, so that $f = Xy$ becomes an optimal approximation to f_0.

In KNR, the distance between f and f_0 is minimized by setting $X = A^+$, the Moore-Penrose pseudoinverse of A [6]. Since $A^+ = (A^*A)^+A^*$, with A^* the adjoint operator of A, and A relates to k by $A = \sum_{i=1}^{M} e_i \otimes \overline{\Psi_i}$, where $\Psi_i = k(x, x_i)$ denotes a sampling function and $\overline{\Psi_i}$ its complex conjugate, $\{e_i\}_{i=1}^{M}$ the standard basis of C^M, and $\otimes$ the Neuman-Schatten product defined by $(e_i \otimes \overline{\Psi_i})f = <f, \Psi_i>_H e_i$, with $<\cdot,\cdot>_H$ the inner product on H, for a given pattern class, its KNR can be represented by $f(x) = \sum_{i=1}^{M} a_i k(x, x_i)$, where $a = [a_1 \, a_2 \, \ldots \, a_M]^T = K^+y$, K is the kernel matrix determined by k and the M training feature vectors [6].

3 Face Recognition Experiments

We adopt the ORL face database that contains 40 distinct subjects with 10 frontal face images per subject. For a KNR classifier, the Gaussian kernel is adopted and the simple formula of Eq.(9) in Ref.[4] is used to estimate the kernel width. The Euclidean distance classifier is utilized for comparison.

We randomly select five images per subject for training, and the remained ones for test. Ten different runs for different feature dimensions are performed. For each dimension, classification error rates are averaged over the ten runs and the forty subjects. Fig.1 depicts the results. It shows that generally the higher the feature dimension is, the lower the classification error rate becomes. When feature dimension is higher than 20, error rates become lower than 3.8% and 10.2% respectively for a KNR and the Euclidean distance classifiers. For the two classifiers, the best results are 3.0% and 7.8%, respectively, at dimension 70.

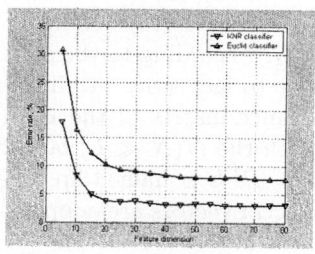

Fig. 1. Error rates of KNR and Euclidean distance classifiers vs. feature dimension

We fix the feature dimension to 70 to study training size effect on recognition rate. Two to eight images per subject are randomly selected for training, and the rest ones for test. The averaged recognition rates, together with those of applying the above two classifiers directly to eigenfaces obtained by PCA from the preprocessed images, are listed in Table 1. It shows that recognition rates increase with training size. Once more, a KNR classifier outperforms the Euclidean distance classifier, and it obtains over 97% recognition rate when more than five images per subject are used for training. Table 1 shows that the proposed

Table 1. Recognition rates (percent) of KNR and Euclidean distance classifiers vs. training size, on 70-dimensional eigenspectra and eigenfaces

Training images per subject		2	3	4	5	6	7	8
KNR classifier	Eigenspectra	87.5	94.0	94.8	97.0	97.9	98.3	97.8
	Eigenfaces	79.7	87.5	90.6	92.4	95.1	96.2	96.5
Euclidean classifier	Eigenspectra	86.7	89.9	90.6	92.2	93.8	94.4	94.8
	Eigenfaces	79.7	84.3	86.3	87.6	90.2	90.5	91.4

scheme are better than some other prospective methods on the same face database, such as the direct LDA (around 94.0%) [2], the Discriminant Waveletface plus Nearest Feature Space classifier (96.1%) [7], and sub-PCA plus simlarity or distance classifier (around 96.5%) [8].

4 Conclusions

A new scheme involving the eigenspectra of a face image and a KNR classifier was proposed for face recognition. The closed-form solution of a KNR avoids the quadratic programming procedure of a nonlinear SVM and the converging problem of some traditional neural networks. Experimental results on the ORL face database showed relatively satisfactory performance of the proposed scheme.

References

1. Belhumeur, P., N., Hespanha, J., P., Kriegman, D., J.: Eigenfaces vs. Fisherfaces: Recognition Using Class Specific Linear Projection. IEEE Trans. PAMI, 19(7) (1997) 711-720
2. Lu, J.,W., Plataniotis, K., N., Venetsanopoulos, A., N.: Face Recognition Using LDA-Based Algorithms. IEEE Trans. NN, 14(1)(2003) 195-200
3. Martinez, A.,M., Kak, A., C.: PCA versus LDA. IEEE Trans. PAMI, 23 (2) (2001): 228-233
4. Kraaijveld, M., A.: A Parzen Classifier with an Improved Robustness against Deviations between Training and Test Data. Pattern Recognition Letters, 17(7) (1996) 679-689
5. Cortes, C., Vapnik, V.: Support-Vector Networks. Machine Learning, 20(3) (1995) 273-297
6. Zhang, J., Liu, B., Y., Tan, H.: A Kernel-based Nonlinear Representor with Application to Eigenface Classification. J. Electronic Science and Technology of China, 2(2) (2004) 19-22
7. Chien, J., T., Wu, C., C.: Discriminant Waveletfaces and Nearest Feature Classifiers for Face Recognition. IEEE Trans. PAMI, 24(12) (2002) 1644-1649
8. Tan, K., R., Chen, S., C.: Adaptively Weighted Sub-pattern PCA for Face Recognition. Neurocomputing, 64 (2005) 505-511

Blind Extraction of Singularly Mixed Source Signals

Zhigang Zeng[1] and Chaojin Fu[2]

[1] School of Automation, Wuhan University of Technology,
Wuhan, Hubei, 430070, China
zhigangzeng@163.com
[2] Department of Mathematics, Hubei Normal University,
Huangshi, Hubei, 435002, China

Abstract. In this paper, a neural network model and its associate learning rule are developed for sequential blind extraction in the case that the number of observable mixed signals is less than the one of sources. This approach is also suitable for the case in which the mixed matrix is non-singular. Using this approach, all separable sources can be extracted one by one. The solvability analysis of the problem is also presented, and the new solvable condition is weaker than existing solvable conditions in some literatures.

1 Introduction

In recent several years, blind separation of independent sources from their mixtures has been an important subject of research in communications [1], medical signal processing [2], speech signal processing [3], and image restoration [4], et al [5-8]. Blind signal separation (BSS) deals with the problem of recovering independent signals using only the observed mixtures. These techniques are referred to as "blind" for the acoustic transfer functions from the sources to the microphones are unknown, and there are no reference signals against which the recovered source signals can be compared. The objective of BSS is to recover sources from their mixtures without the a prior knowledge of the sources and the mixing channels.

In general, depending on the approaches for recovering original sources from instantaneous mixtures, the results can be classified into the simultaneous separation approach [5] and the extraction approach [6]. In the separation approach, all separable sources are separated simultaneously, whereas the sources are extracted one by one in the extraction approach. Simultaneous separation, if possible, is, of course, desirable. In some ill-conditioned cases, simultaneous blind separation cannot be achieved, but sequential blind extraction can be done since sequential blind extraction technique requires weaker solvability conditions than simultaneous blind separation as introduced in [7]. The common presumption is that the signals are emitted from the point sources placed in the far field. In the blind signal separation problem, the array parameters are generally unknown.

Performance is often independent of inaccuracies in the array manifold as well as sensor displacement. A general overview and references on blind equalization can be found in literature [8]. A simplest case of all BSS methods is the instantaneous and linear mixture of signals. Generally, separating linear mixtures of signals is a problem that frequently arises in multiuser communication systems. Typical examples are the extraction of incoming signals from the outputs of an array of sensors or the recovery of transmitted symbols from the outputs of a bank of matched filters in code division multiple access (CDMA) systems. Consider a general linear case of instantaneous mixing of m sources with n observable mixtures

$$x(t) = As(t), \tag{1}$$

where $s(t) = (s_1(t), \cdots, s_m(t))^T$ is an m-dimensional vector of mutually independent unknown sources with zero means; $x(t) = (x_1(t), \cdots, x_n(t))^T$ is an n-dimensional vector of mixed signals; and $A = [a_{ij}]_{n \times m}$ is an unknown constant matrix known as the mixing matrix, $n \leq m$. The task of blind extraction is to recover the sources one by one from the available mixtures $x_1(t), \cdots, x_n(t)$.

Literatures [6,7] were to focus on the case of instantaneous mixtures, where the algorithms of blind extraction have been developed for $n \leq m$. In this paper, a new blind separation method based on neural network model and its associate learning rule are developed for sequential blind extraction in the case with singular matrix A. This approach is also suitable for the case in which A is nonsingular. The solvability analysis of the problem is also discussed.

2 Solvability Analysis

In this section, we discuss the solvability of blind extraction model (1).

Theorem. If there exists constant $\ell \neq 0$ such that

$$\mathrm{rank}(\overline{A}) = \mathrm{rank}(A_\ell), \tag{2}$$

where

$$\overline{A} = \begin{pmatrix} a_{21} & \cdots & a_{n1} \\ \cdots & \cdots & \cdots \\ a_{2m} & \cdots & a_{nm} \end{pmatrix}, \quad A_\ell = \begin{pmatrix} a_{11} - \ell & a_{21} & \cdots & a_{n1} \\ a_{12} & a_{22} & \cdots & a_{n2} \\ \cdots & \cdots & \cdots & \cdots \\ a_{1m} & a_{2m} & \cdots & a_{nm} \end{pmatrix},$$

then one of the source signals can be extracted from the mixed signals.

Proof. Since $\mathrm{rank}(\overline{A}) = \mathrm{rank}(A_\ell)$, there exist $b_{12}, \cdots, b_{1n}$ such that

$$\begin{cases} b_{12}a_{21} + \cdots + b_{1n}a_{n1} = a_{11} - \ell, \\ b_{12}a_{22} + \cdots + b_{1n}a_{n2} = a_{12}, \\ \cdots \\ b_{12}a_{2m} + \cdots + b_{1n}a_{nm} = a_{1m}; \end{cases}$$

i.e., $A^T b_1 = \overline{\ell}$, where $b_1 = (1, -b_{12}, \cdots, -b_{1n})^T$ is an n-dimensional vector, $\overline{\ell} = (\ell, 0, \cdots, 0)^T$ is an n-dimensional vector. Choose n-dimensional vectors

$b_2, \cdots, b_n$. Let $B = (b_1, \cdots, b_n)^T$, $y(t) = Bx(t) = BAs(t)$, then $y_1(t) = \ell s_1(t)$; i.e., a signal corresponding to $y_1(t)$ is extracted.

Corollary 1. If there exist $k_2, k_3, \cdots, k_n$ such that

$$a_{1j} = \sum_{i=2}^{n} k_i a_{ij}, \quad j = 2, 3, \cdots, m, \quad a_{11} \neq \sum_{i=2}^{n} k_i a_{i1},$$

then one of the source signals can be extracted from the mixed signals.

Proof. Choose $\ell = a_{11} - \sum_{i=2}^{n} k_i a_{i1}$, then $\ell \neq 0$ and condition (2) is satisfied. According to Theorem, one of the source signals can be extracted from the mixed signals.

Remark. For example, the following mixed matrices satisfy the above condition (2).

$$\begin{bmatrix} 3 & 2 & 2 \\ 1 & 1 & 1 \\ 2 & 1 & 1 \end{bmatrix}, \quad \begin{bmatrix} 2 & 1 & 1 \\ 1 & 1 & 1 \end{bmatrix}.$$

Corollary 2. If the mixed matrix A is nonsingular, then one of the source signals can be extracted from the mixed signals.

Proof. A is nonsingular implies that $m = n$ and $\text{rank}(\overline{A}) = n-1$, Then in view of $(a_{12}, a_{13}, \cdots, a_{1n})^T$ is an $n-1$ dimensional vector, there exist $\alpha_2, \alpha_2, \cdots, \alpha_n$ such that $a_{1j} = \sum_{i=2}^{n} \alpha_i a_{ij}, j = 2, 3, \cdots, n$. And $a_{11} \neq \sum_{i=2}^{n} \alpha_i a_{i1}$, otherwise $a_{1j} = \sum_{i=2}^{n} \alpha_i a_{ij}, j = 1, 2, \cdots, n$. Then $\text{rank}(A) \leq n - 1$, which contradicts to the conclusion of A being nonsingular. Choose $\ell = a_{11} - \sum_{i=2}^{n} \alpha_i a_{i1}$, then $\ell \neq 0$ and condition (2) is satisfied. According to the above Theorem, one of the source signals can be extracted from the mixed signals.

3 Blind Extraction Algorithm

In the following, the algorithm for a blind extraction from the mixed sources is introduced.

Let $y_1(t), y_2(t), \cdots, y_n(t)$ are the outputs of the neural network, which is defined as follows

$$\begin{cases} y_1(t) = x_1(t) - b_{12}x_2(t) - \cdots - b_{1n}x_n(t) \\ y_2(t) = x_2(t) - b_{21}y_1(t) \\ \cdots \\ y_n(t) = x_n(t) - b_{n1}y_1(t), \end{cases} \quad (3)$$

If the mixed signals satisfy the solvability condition (2), then the task of blind extraction is to look for $b_{12}, \cdots, b_{1n}, b_{21}, \cdots, b_{n1}$ such that y_1 and $y_j, (j = 2, 3, \cdots, n)$ are statistically independent; i.e., a signal corresponding to $y_1(t)$ is extracted. In view of $E(y_1(t)y_j(t)) = E(y_1(t))E(y_j(t)), (j = 2, 3, \cdots, n)$,

$$E(y_1(t)(x_2(t) - b_{21}y_1(t))) = E(y_1(t))E(y_2(t)) = 0,$$
$$\cdots,$$
$$E(y_1(t)(x_n(t) - b_{n1}y_1(t))) = 0.$$

The connection weights $b_{12}, \cdots, b_{1n}, b_{21}, \cdots, b_{n1}$ in (3) are adjusted. If $y_1(t) \neq 0$, then $y_1(t)$ is one estimate of source signal $s_1(t)$. The first time extraction task is finished, and the next extraction should be continued based on the new mixtures of $y_2(t), \cdots, y_n(t)$. If $y_2(t) \neq 0$, then $y_2(t)$ is one estimate of the source signal $s_2(t)$. Thus the second time extraction task is finished. Thus all separable sources can be extracted one by one in the light of this procedure if the separable conditions can be satisfied.

4 Concluding Remarks

In this paper, a general approach was proposed for sequential blind extraction of instantaneously mixed sources in the case that the number of observable mixed signals is less than the one of sources. This method, also referred to as the one based on neural network model and its associate learning rule, is also suitable for the case in which the mixed matrix is nonsingular. Using this approach, all separable sources can be extracted one by one. Some sufficient solvability conditions are also given that are weaker than existing solvable conditions in literature [6].

Acknowledgement

This work was supported by the Natural Science Foundation of China under Grant 60405002, the China Postdoctoral Science Foundation under Grant 2004035579 and the Young Foundation of Hubei Provincial Education Department of China under Grant 2003B001.

References

1. Comon, P.: Independent Component Analysis, A New Concept. Signal Process. **36** (1994) 287-314
2. Hyvarinen, A., Oja, E.: Independent Component Analysis: Algorithms and Applications. Neural Networks, **13** (2000) 411-430
3. Bell, A. J., Sejnowski, T. J.: An information-maximization approach to blind separation and blind deconvolution. Neural Comput., **7** (1995) 1004-1034
4. Kundur, D., Hatzinakos, D.: A Novel Blind Deconvolution Scheme for Image Restoration Using Recursive Filtering. IEEE Trans. Signal Processing, **46** (1998) 375-390
5. Comon, P., Jutten, C., Herault, J.: Blind Separation of Sources, Part II: Problems Statement. Signal Process., **24** (1991) 11-20
6. Li, Y., Wang, J., Zurada, J. M.: Blind Extraction of Singularly Mixed Source Signals. IEEE Trans. Neural Networks, **11** (2000) 1413-1422
7. Li, Y., Wang, J.: Sequential Blind Extraction of Instantaneously Mixed Source. IEEE Trans. Signal Processing, **50** (2000) 997-1006
8. Van Der Veen A. J.: Algebraic Methods for Deterministic Blind Beamforming. Proc. IEEE, **86** (1998) 1987-2008

Application of Support Vector Machines in Predicting Employee Turnover Based on Job Performance

Wei-Chiang Hong[1], Ping-Feng Pai[2,*], Yu-Ying Huang[3], and Shun-Lin Yang[3]

[1] School of Management, Da-Yeh University,
112 Shan-Jiau Rd., Da-Tusen, Chang-hua, 51505, Taiwan
d9230006@mail.dyu.edu.tw
[2] Department of Information Management, National Chi Nan University,
1 University Rd. Puli, Nantou, 545, Taiwan
paipf@ncnu.edu.tw
[3] Department of Industrial Engineering and Technology Management, Da-Yeh University,
112 Shan-Jiau Rd., Da-Tusen, Chang-hua, 51505, Taiwan
{r9315001, r9315018}@mail.dyu.edu.tw

Abstract. Accurate employee turnover prediction plays an important role in providing early information for unanticipated turnover. A novel classification technique, support vector machines (SVMs), has been successfully employed in many fields to deal with classification problems. However, the application of SVMs for employee voluntary turnover prediction has not been widely explored. Therefore, this investigation attempts to examine the feasibility of SVMs in predicting employee turnover. Besides, two other tradition regression models, Logistic and Probability models are used to compare the prediction accuracy with the SVM model. Subsequently, a numerical example of employee voluntary turnover data from a middle motor marketing enterprise in central Taiwan is used to compare the performance of three models. Empirical results reveal that the SVM model outperforms the logit and probit models in predicting the employee turnover based on job performance. Consequently, the SVM model is a promising alternative for predicting employee turnover in human resource management.

1 Introduction

In general, employee turnover can be divided into two types, namely involuntary turnover and voluntary turnover. Involuntary turnover is often defined as the movements across the membership boundary among an organization, over which the employee only conducts slight affections. On the other hand, voluntary turnover is defined as the movements across the membership boundary among an organization, over which the employee conducts heavy affections [1]. Mobley [2] first proposed the structure models regarding employee turnover. Based on the developed voluntary turnover structure, job satisfaction links to the initiate actual voluntary turnover indirectly [3]. In addition, the organizational commitment is treated as intervening variables to explain the stay intentions or employee turnover [4]. Recently, job performance has become one of the most

---

* Corresponding author.

important factors which influence the decision of employee turnover. The relationships between performance and turnover can be divided into four categorizations, a positive relationship, a negative relationship, no relationship and non-linear relationship [5]. Schwab [6] indicated that high performance employees are more likely to leave. Jackofsky [7] showed that low job performance appears high voluntary turnover. Therefore, the turnover displays a non-linear relationship to job performance. Trevor et al. [8] found that high performance employees would be less likely to leave than lower performance ones. Vecchio & Norris [9] concluded the correlations between turnover and job performance are negative. Morrow et al. [1] proved that the negative relationship between the turnover and job performance is statistical significant. Williams & Livingstone [10] showed that the poor performance employees in the marketing department of any organization tend to leave easily.

The voluntary turnover prediction problems can be treated as discrete choice problems. The logistic regression model presented by McFadden [11] is one of the most popular discrete choice models in practical application [12]. Besides, logit models have also been employed in commercial affairs forecasting [13]. However, independence of irrelevant alternatives (IIA) property limits the application of logit models [14]. To relax IIA restriction of the logit model, multinomial probability regression model, namely probit model, allows a free correlation structure among each discrete choice alternative. Dow & Endersby [15] compared the performance of the multinomial logit model and the multinomial probit model in the voting analysis.

Recently, based on statistical learning algorithms, an emerging technique called support vector machines (SVMs) [16] has been widely employed for pattern classification and regression problems. However, the application of SVMs for employee voluntary turnover prediction has not been widely explored.

In this investigation, SVM model, logit model and probit models are employed to compare the prediction performance of employ turnover. The rest of this article is organized as follows. Section 2 briefs three prediction models. Section 3 addresses a numerical example taken from a motor marketing enterprise in Taiwan to compare prediction results of three models. Finally, conclusions are made in section 4.

2 Prediction Models

2.1 Support Vector Machines in Classification

SVMs derive a class decision by determining the separate boundary with maximum distance to the closest points, namely support vectors (SVs), of the training data set. By minimizing structural risk rather than empirical risk, SVMs could efficiently avoid a potential misclassification for testing data. Therefore, SVM classifier has superior generalization performance over that of other conventional classifiers.

Given a training data set $D = \{x_i, y_i\}_{i=1}^N$, where $x_i \in \Re^n$ is the i-th input vector with known binary output label $y_i \in \{-1,+1\}$. Then, the classification function is given by

$$y_i = f(x_i) = w^T \varphi(x_i) + b. \tag{1}$$

where $\varphi: \Re^n \to \Re^m$ is the feature mapping the input space to a high dimensional feature space. The data points become linearly separable by a hyperplane defined by the pair ($w \in \Re^m, b \in \Re$) [16]. The optimal hyperplane that separates the data is formulated as Eq. (2).

$$\text{Minimize} \quad \Phi(w) = \|w\|^2 / 2$$
$$\text{Subject to} \quad y_i[w^T \varphi(x_i) + b] \geq 1 \quad i = 1,...,N \tag{2}$$

where $\|w\|$ is the norm of a normal weights vector of hyperplane. This constrained optimization problem is solved using the following primal Lagrangian form:

$$L(w,b,\alpha) = \frac{1}{2}\|w\|^2 - \sum_{i=1}^{N} \alpha_i [y_i(w^T \varphi(x_i) + b) - 1]. \tag{3}$$

where α_i are the Lagrange multipliers. Applying the Karush-Kuhn-Tucker conditions, the solutions of the dual Lagrangian problem, α_i^0, then determine the parameters w_0 and b_0 of the optimal hyperplane. Then, the decision function is given by Eq. (4):

$$d(x_i) = \text{sgn}(w_0^T \varphi(x_i) + b_0) = \text{sgn}\left(\sum_{i=1}^{N} \alpha_i^0 y_i K(x,x_i) + b_0\right), \quad i = 1,...,N. \tag{4}$$

Here, $K(x,x_i)$ is called the kernel function and should satisfies Mercer's condition [16]. In addition, its value is equal to the inner product of two vectors x and x_i in the feature space $\varphi(x)$ and $\varphi(x_i)$, i.e., $K(x,x_i) = \varphi(x) * \varphi(x_i)$. In this investigation, the Gaussian radical basis function, $\exp(-\|x_i - x_j\|^2 / 2\sigma^2)$, is used in the SVMs classifier model.

To deal with overlapping classes, the concept of a soft margin is applied for the SVM classifier. The width of the soft margin is controlled by a penalty parameter C that determines the trade-off between maximizing the margin and minimizing the training error. Small values of C result in insufficient stress on fitting the training data. On the other hand, too large C leads to the over-fitting of the training data. Therefore, the selection of two positive parameters, σ and C, of a SVM model is important to the classification accuracy. The procedure for selecting two parameters is conducted as follows. **Step 1.** Set a fixed value of the parameter C. Then, adjust the value of σ till a maximum of prediction accuracy is achieved. The finalized σ value is denoted as σ'. **Step 2.** The value of σ is set at σ'. Then, adjust the value of C to achieve a maximum prediction accuracy. The finalized C is defined as C'. Finally, the suitable values of parameters σ and C are determined as σ' and C'.

2.2 Logit and Probit Models

Logit and probit models are used to predict two discrete alternatives, for example, fail or non-fail. Without the preliminary normality assumption of all explanatory variables and with the capability of incorporating nonlinear factors, both logit and probit models are popular in the social science area [17,18]. The following is a brief of logit and probit models.

Assume that the state S_i for each observation appears absolute certainty in the discrete choice models. Thus, the S_i is equal to one when an alternative is selected. On the other hand, the S_i equals to zero if an alternative is not chosen. The decision variable S is a dependent variable in logistic function, represented as Eqs. (5) and (6), respectively,

$$P(S_i = 1, \text{non-fail}) = \exp(\beta x_i)/(1 + \exp(\beta x_i)) = 1/(1 + \exp(-\beta x_i)) \tag{5}$$

$$P(S_i = 0, \text{fail}) = 1/(1 + \exp(\beta x_i)) \qquad (6)$$

where x_i is the explanatory variable for decision makers and β_i is the coefficients of the logit model.

Eqs. (5) and (6) are then estimated by maximum likelihood estimation (MLE). The values of variables coefficients, β_i, are represented as the log odds ratio and obtained by estimated probability. Finally, the prediction values of decision variable S_i is either non-fail (S_i=1) or fail (S_i=0). Instead of using the logit function, a cumulative standard normal distribution functional form is employed in probit model. Therefore, when the alternative is selected, the S_i is equal to one. On the contrary, the S_i equals to zero if the alternative is not chosen. The probit model can be expressed as Eq. (7),

$$P(S_i) = \Phi(\beta x_i) \qquad (7)$$

where $\Phi(\cdot)$ is the cumulative standard normal distribution function; x_i is the explanatory variable, and β_i is the coefficient of the probit model. The remaining procedure for estimating β_i is the same as that of logit model.

3 A Numerical Example

3.1 Data Set and the Measurement of the Prediction Performance

An empirical data regarding employee turnover and job-performance is given by Hui-Lien Motor marketing Co. The Hui-Lien Motor marketing Co. is located in central Taiwan with 300 million NTD annual business volumes and more than 200 marketing specialists since 1992. The data contains totally 132 marketing specialists. The job-performance was evaluated by motor marketing volumes and the voluntary turnover statuses in 2003. For the measurement of prediction accuracy, some indices, such as Cox & Snell R^2 [17], Nagelkerke R^2 [18], McFadden R^2 [11], classification table, and model Chi-square test [19] are often used. However, the core of this study is to predict employees' stay or leave in an organization by the job-performance. Therefore, the measurement of prediction accuracy is the most important. The classification table is employed to compare the total prediction accuracy of three models.

3.2 Experimental Results

The data set is divided into two parts, namely the modeling data set (from 1st employee to 100th employee) and the testing data set (from 101st employee to 132nd employee). The modeling data set is used to train models. The testing data set is applied to estimating model performance for future unseen data. The Maximum likelihood estimation procedure is employed to determine the free parameters for both logit and probit models. Table 1 shows these coefficients β_i for logit and probit models. For both models, the explanatory capability is over 50% level. In addition, the two estimated coefficients β_0 and β_1 are statistical significant.

For the SVM model, set the value of the parameter C at 1. Then, adjust the value of σ till a maximum prediction accuracy is obtained, the finalized σ value equaled to 0.001, therefore, set the σ value as 0.001. Secondly, adjust the value of C. The finalized C value is 2.0 and the maximum prediction accuracy is achieved. Thus, the finalized C value is 2.0. Finally, the predicting accuracy is 84.38% while the values of σ and C are 0.001 and 2.0, respectively.

Table 1. Coefficients of explanatory variables for logit and probit models

	Logit model				Probit model			
Explanatory variables	Estimated coefficients	Standard error	p-value		Explanatory variables	Estimated coefficients	Standard error	p-value
Constant	-0.206	0.041	0.000**		Constant	-0.112	0.020	0.000**
Job-performance	2.343	0.437	0.000**		Job-performance	1.361	0.228	0.000**
Model Chi-square p-value (d.f.=1)		0.000**			Model Chi-square p-value (d.f.=1)		0.000**	
Cox & Snell R^2		0.387			Cox & Snell R^2		0.382	
Nagelkerke R^2		0.529			Nagelkerke R^2		0.521	
McFadden R^2		0.372			McFadden R^2		0.365	

Table 2. The prediction accuracy in training stage for logit, probit and SVM models

	Logit model			Probit model			SVM model		
	$S_i=1$*	$S_i=0$**	Total	$S_i=1$*	$S_i=0$**	Total	$S_i=1$*	$S_i=0$**	Total
$P_i=1$***	48	19	67	49	22	71	49	22	71
$P_i=0$****	7	26	33	6	23	29	6	23	29
Total	55	45	100	55	45	100	55	45	100
Prediction accuracy	87.3%	57.8%	74.0%	89.1%	51.1%	72.0%	89.1%	51.1%	72.0%

Table 3. The prediction accuracy in testing stage for logit, probit and SVM models

	Logit model			Probit model			SVM model		
	$S_i=1$*	$S_i=0$**	Total	$S_i=1$*	$S_i=0$**	Total	$S_i=1$*	$S_i=0$**	Total
$P_i=1$***	19	8	27	19	8	27	16	0	16
$P_i=0$****	2	3	5	2	3	5	5	11	16
Total	21	11	32	21	11	32	21	11	32
Prediction accuracy	90.5%	27.3%	71.9%	90.5%	27.3%	71.9%	76.2%	100.0%	84.38%

*: $S_i=1$ implies actual turnover. **: $S_i=0$ implies actual non-turnover.
: $P_i=1$ implies predicting as turnover. *: $P_i=0$ implies predicting as non-turnover.

Table 2 shows the prediction accuracy of three models in training stages. In the training stage, the total prediction accuracy are 74%, 72%, and 72% for logit model, probit model and SVM model respectively. Table 3 lists the prediction performance of three models in testing stages. It is indicated that the SVM model has higher total prediction accuracy (84.38%) than logit model (71.9%) and probit model (71.9%). Therefore, the SVM model has better generalization ability than the logit and probit models in predicting the employee turn-over.

4 Conclusions

The accurate employee turnover prediction plays an important role in early detection of unanticipated turnover of an organization. Therefore, a suitable model for predicting turnover is vital. In this investigation, the SVM classifier is used to examine the feasibility in predicting the employee turnover. Two other discrete choice models, namely logit model and probit model, are employed to compare the prediction accuracy. A numerical data set of employee turnover is used for the numerical experiment. The simulation results reveal that SVM model outperform the logit model and probit model. Therefore, the SVM model is a valid alternative in dealing with employee turnover prediction problems. In the future, some other factors, such as job-satisfaction, organization commitment, and abnormal absenteeism of employee can be included in the SVM model to predict the employee turnover. In addition, developing a structured way in determining free parameters of SVM model could be another direction for future research.

References

1. Morrow, P.C., McElroy, J.C., Laczniak, K.S.: Using Absenteeism and Performance to Predict Employee Turnover: Early Detection Through Company Records. Journal of Vocational Behavior. 55 (1999) 358-374.
2. Mobley, W.H.: Intermediate Linkages in The Relationship Between Job Satisfaction and Employee Turnover. Journal of Applied Psychology. 63 (1977) 237-240.
3. van Breukelen, W., van der Vlist, R., Steensma, H.: Voluntary Employee Turnover: Combining Variables From the 'Traditional' Turnover Literature with The Theory of Planned Behavior. Journal of Organizational Behavior. 25 (2004) 893-915.
4. Inverson, R.D.: Employee Intent to Stay: An Empirical Test of A Revision of The Price and Muller model. The University of Iowa (unpublished doctoral dissertation), Iowa City, IA (1992).
5. Jackofsky, E.F., Ferris, K.R., Breckenridge, B.G.: Evidence for A Curvilinear Relationship Between Job Performance and Turnover. Journal of Management. 12 (1986) 105-111.
6. Schwab, D.: Contextual Variables in Employee Performance-Turnover Relationships. Academy of Management Journal. 34 (1991) 966-975.
7. Jackofsky, E.: Turnover and Job Performance: An Integrated Process Model. Academy of Management Review. 9 (1984) 74-83.
8. Trevor, C., Gerhart, B., Boundreau, J.: Voluntary Turnover and Job Performance: Curvilinearity and The Moderating Influences of Salary Growth and Promotions. Journal of applied Psychology. 82 (1997) 44-61.
9. Vecchio, R., Norris, W.: Voluntary Turnover and Job Performance, Satisfaction, and Leader-Member Exchange. Journal of Business and Psychology. 11 (1996) 113-125.
10. Williams, C.R., Livingstone, L.P.: Another Look at The Relationship Between Performance and Voluntary Turnover. Academy of Management Journal. 37 (1994) 269-298.
11. McFadden, D.: Conditional Logit Analysis of Qualitative Choice Behavior. In: Zoremmbka, P. (eds.): Frontiers in Econometrics, Academic Press, New York (1973).
12. Tseng, F.M., Yu, C.Y.: Partitioned Fuzzy Integral Multinomial Logit Model for Taiwan's Internet Telephony Market. Omega. 33 (2005) 267-276.

13. Tseng, F.M., Lin, L.: A Quadratic Interval Logit Model for Forecasting Bankruptcy. Omega. 33 (2005) 85-91.
14. Stopher, P.R., Meyburg, A.H., Brg, W.(ed.): New Horizons in Travel Behavior Research. Lexington Books, Lexington, M.A. (1981).
15. Dow, J.K., Endersby, J.W.: Multinomial Probit and Multinomial Logit: A Comparison of Choice Models for Voting Research. Electoral Studies. 23 (2004) 107-122.
16. Vapnik, V. (ed.): The Nature of Statistical Learning Theory. Springer-Verlag, New York (1995).
17. Cox, D.R., Snell, E.J. (ed.): The Analysis of Binary Data. Champman and Hall, London (1989).
18. Nagelkerke, N.J.D.: A Note on A General Definition of The Coefficient of Determination. Biometrica. 7 (1991) 691-692.
19. Greene, W.H. (ed.): Econometric Analysis. Prentice Hall, New Jersey (2003).

Palmprint Recognition Based on Unsupervised Subspace Analysis

Guiyu Feng, Dewen Hu*, Ming Li, and Zongtan Zhou

Department of Automatic Control, College of Mechatronics and Automation,
National University of Defense Technology, Changsha, Hunan, 410073, China
dwhu@nudt.edu.cn

Abstract. As feature extraction techniques, Kernel Principal Component Analysis (KPCA) and Independent Component Analysis (ICA) can both be considered as generalization of Principal Component Analysis (PCA), which has been used for palmprint recognition and gained satisfactory results [3], therefore it is natural to wonder the performances of KPCA and ICA on this issue. In this paper, palmprint recognition using the KPCA and ICA methods is developed and compared with the PCA method. Based on the experimental results, some useful conclusions are drawn, which fits into the scene for a better picture about considering these unsupervised subspace classifiers for palmprint recognition.

1 Introduction

Biometrics has been attracting more and more attentions in recent years. Two possible biometric features, hand geometrical features and palmprint features, can be extracted from hand. Hand geometrical features such as finger width, length, and the thickness are adopted to represent extracted features, but these features frequently vary due to the wearing of rings in fingers, besides, the width of some fingers may vary during pregnancy or illness. Palmprint features have several advantages over such physical characteristics [1]: (1)low-resolution imaging; (2)low intrusiveness; (3)stable line feature and (4)high user acceptance.

Some work on palmprint recognition has been reported in the literature[1-6]. In all this research, the primary focus of attention is on points and lines [2], texture analysis[1][5][6] or second-order statistics features [3][4]. The main contribution of this paper is to consider two novel unsupervised palmprint representation methods using KPCA and ICA, and compare it with PCA method on the same data set. It should be noted that InfoMax algorithm of ICA was deployed for palmprint recognition in [7], however, FastICA algorithm of ICA has been recommended for identity tasks [8], as is adopted in our paper, besides, the topic of this paper is on the use of unsupervised subspace methods for palmprint recognition, therefore the supervised fisherpalms method [4] is not considered. Experimental results show that both ICA and KPCA significantly outperform PCA on the palmprint recognition task, in addition, ICA performs best among the three unsupervised methods. We also provide some possible reasons for this.

---

* Corresponding author.

2 Palmprint Representation Using Unsupervised Subspace Analysis

PCA is a popular feature extraction method, in this paper,the eigenpalm method is used as a benchmark, for details, please refer to Ref.[3].

KPCA is designed for nonlinear dimension reduction method and gained great success in face recognition [10],which is extended to palmprint recognition in this work.

ICA is a technique for extracting statistically independent variables from a mixture of them. Usually, two different architectures of ICA are adopted for feature extraction [8].The FastICA algorithm of both architectures are used for palmprint representation,which are ICA-Arch.1 and ICA-Arch.2 in this paper.

3 Experiment and Results Analysis

3.1 Database

This work is carried on the newly released PolyU Palmprint Database [9]. There are 100 different palms in this database, and six samples from each of these palms were collected in two sessions, where 3 samples were captured in either session. Since the images contain not only the palmprints, but also other parts of the palm and background, a coordinate system is used to align different palmprint images for further processing [5], for examples, see Fig.1(a).

3.2 Experimental Results and Analysis

In the experiment, the images are resized to 60*60 due to computing consideration, and all the palmprint images are divided into two parts, two images of every palm from each session are regarded as the training set, the left comprise the test set. After histogram equalization on the palmprint images, PCA, KPCA and ICA are used to extract the features, and then Nearest Neighbor strategy (L2 norm) and Cosine Angle are adopted to give the final decisions. In our experiment, KPCA is with the radial basis kernel functions.

Due to PCA ahead, the dimension of basis images of ICA-Arch.2 are reduced so much that the basis images can't be provided, and Fig. 1(c)(d) shows the basis images of PCA and ICA-Arch.1. The PCA algorithm is developed in our Lab and the ICA algorithm deployed is the FastICA package($http:/ww.cis.hut.fi/projects/ica/fastica/$). In ICA-Arch.1, considering the computing cost, PCA is firstly applied to the data to obtain an eigenpalm space of dimensions m (here m was set to be 100 considering the representative ability of PCA), which was also adopted in [7]. The FastICA algorithm is then applied to the eigenpalms to get resulting basis images as independent as possible. The goal of ICA-Arch.2 is to find statistically independent coefficients for the input images, thus the source separation is performed on the pixels. In our work, ICA is performed on the PCA coefficients (here m was set to be 100) rather than directly on the raw images to reduce the computing cost as in [8]. Table 1 shows the performance results.

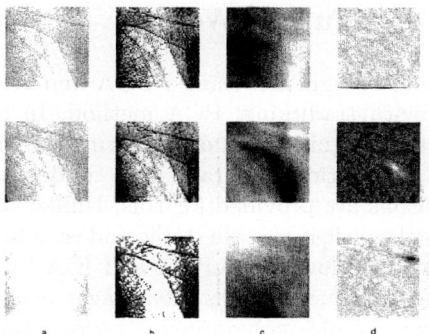

Fig. 1. Three palmprint images(a), palmprint images after histogram equalization(b), three eigenpalms(c), three ICs(Arch.1)(d)

From Table 1, we can see that choice of distance measures for ICA-Arch.1 matters little and it performs better when L2 norm is used, which have not been reported before. When using cosine angle measure, ICA-Arch.2 performs better, as [8] recommended. It should also be pointed out that the two architectures of the FastICA algorithm perform as well as each other with a recognition accuracy rate of 96.5%, and the recognition accuracy using KPCA method is as high as 96.0%, which are obviously more efficient than the accuracy rate of 91.5% using PCA method. In the experimental results, both KPCA and ICA methods outperform the PCA methods. The possible reasons are as follows: for KPCA, through nonlinear mapping using kernel functions, the original space is mapped into an arbitrarily large dimensional space, through which it would be reasonable for us to expect that the distribution of the different persons' palmprint images in he feature space are more sparsely separated, and can be more easily classified; and more importantly, the representation in traditional eigenpalms method in based on second order statistics of palmprint images set, and does not consider high order statistical dependencies such as the relationships among three or more pixels, however, for palmprint recognition, much of the important information may be contained in these relationships among the pixels, and such information is deployed by KPCA and ICA methods, which is robust and more suitable for classification tasks.

Table 1. Recognition accuracy performance results

Methods	L2 norm	Cosine angle
PCA	91.5%	91.0%
KPCA	96.0%	93.0%
ICA-Arch.1	96.5%	96.0%
ICA-Arch.2	94.0%	96.5%

4 Conclusions and Future Work

In this paper, palmprint recognition using KPCA and ICA methods are developed, and compared with traditional PCA method. In this work, the KPCA method and ICA method significantly outperforms the traditional eigenpalms method in term of recognition accuracy rate under the PolyU palmprint database. Possible reasons are provided for this. Unlike structural features and statistical features, algebraic features are stable and easy to extract, PCA, KPCA and ICA are such features, however, KPCA and ICA take higher order correlations into account, which has obvious advantages over the traditional PCA method.Our future work would include investigation other kernel methods for palmprint recognition, and we are also interested in multi-modal biometrics which is our ongoing project [11].

Acknowledgement

This work is partially supported by the Distinguished Young Scholars Fund of China (60225015), Specialized Research Fund for the Doctoral of Higher Education of China (20049998012), Ministry of Education of China (TRAPOYT Project). Portions of the work tested on the PolyU Palmprint Database.

References

1. Kong W.K., Zhang D. and Li W.X.: Palmprint Feature Extraction using 2-D Gabor Filters. Pattern Recognition 36(2003)2339-2347.
2. Zhang D. and Shu W.: Two Novel Characteristics in Palmprint Verification: Datum Point Invariance and Line Feature Matching, Pattern Recognition 32(4)(1999)691-702.
3. Lu G.M., Zhang D., Wang K.Q.: Palmprint Recognition Using Eigenpalms Features. Pattern Recognition Letters 24(2003)1463-1467
4. Wu X.Q., Zhang D. ,Wang K.Q.: Fisherpalms Based Palmprint Recognition. Pattern Recognition Letters 24(2003)2829-2838.
5. Zhang D., Kong W.K., You J., Wong M.: Online Palmprint Identification. IEEE Trans. PAMI 25 (9)(2003)1041-1050
6. Dong K., Feng G. and Hu D.: Digital Curvelet Transform for Palmprint Recognition. LNCS 3338 (2004) 639-645.
7. Connie T., Teoh A., Goh M., Ngo D.: Palmprint Recognition with PCA and ICA. Proc. of Image and Vision Computing New Zealand, New Zealand (2003) 227-232.
8. Draper B., Baek K., Barlett M.S., Beveridge J.: Recognizing Faces with PCA and ICA. Computer Vision and Image Understanding 91(2003)115-137.
9. The PolyU Palmprint Database. http://www.comp.polyu.edu.hk / biometrics/
10. Yang M.H., Ahuja N.and Kriegman D.: Face Recognition using Kernel Eigenfaces, in Proc. of IEEE ICIP, Vancouver, Canada(2000) 37-40.
11. Feng G., Dong K., Hu D., Zhang D.: When Faces Are Combined with Palmprints: A Novel Biometric Fusion Strategy. LNCS 3072(2004)701-707.

A New Alpha Seeding Method for Support Vector Machine Training

Du Feng, Wenkang Shi, Huawei Guo, and Liangzhou Chen

School of Electronics & Electrics Engineering, Shanghai Jiao Tong University,
200030 Shanghai, China
{dfengu, wkshi, hwguo, tiger_chen}@sjtu.edu.cn

Abstract. In order to get good hyperparameters of SVM, user needs to conduct extensive cross-validation such as leave-one-out (**LOO**) cross-validation. Alpha seeding is often used to reduce the cost of SVM training. Compared with the existing schemes of alpha seeding, a new efficient alpha seeding method is proposed. Through some examples, its good performance has been proved. Interpretation from both geometrical and mathematical view is also given.

1 Introduction

Support vector machines (SVMs) [1] have been proven to be very effective methods in the fields of data classification and pattern recognition. In order to get optimal parameter settings, user normally needs to conduct time-consuming cross validation during the SVM training. The hyperparameters which give the minimum estimation of generalization error such as leave-one-out (**LOO**) error are selected as the optimum alternatives. Seeding successive SVM training with the results of previous trainings, called alpha seeding [3], is proved to be an efficient method for computing **LOO** error. What's more, [4] gives a thorough research on reducing the time of computing **LOO** error. In this paper, aiming at exploring good seeding method to expedite the computation of **LOO**, several new alpha seeding schemes are analyzed on detail.

Traditionally, a SVM learns from a set of l N-dimensional example vector x_i, and their associated classed y_i, i.e. $(x_1,y_1),...,(x_l,y_l) \in R^N \times \{-1,1\}$. By using a function ϕ to map x_i into a higher dimensional space, we often solve it from the dual formulation:

$$\min_{\alpha}(0.5\alpha^T Q\alpha - e^T\alpha) \text{ s.t. } y^T\alpha = 0, 0 \leq \alpha_i \leq C, i=1,...,l. \tag{1}$$

where Q is an $l \times l$ positive semi-definite matrix with $Q_{ij}=y_i y_j \varphi(x_i)^T \varphi(x_j)$, and e is the vector with all 1 elements. Usually we call $K(x_i,x_j) = \varphi(x_i)^T \varphi(x_j)$ the kernel function. After (1) is solved, $\omega = \sum_{i=1}^{l}\alpha_i y_i \phi(x_i)$ is obtained. The kth example is classified by

$$O_k = \omega\phi(x_k) - b = \sum_i \alpha_i y_i K(x_i, x_k) - b \ ,$$

where b is the lagrange multiplier according to the equality constraint in (1). The kth example is considered as misclassified in training if $y_k O_k < 0$.

Tuning the hyperparameters such as C and γ in the RBF kernel function is usually done by minimizing an estimate of generalization error such as **LOO** errors [1]. **LOO** error is defined: For a given k, let P_k denote the modified primal problem in which the kth example is omitted, and let D_k be dual problem corresponding to P_k, given by (1) with the kth example is omitted. Let (ω_k', b_k') denote the solution of P_k and define $O_k' = \omega_k' \phi(x_k) - b_k'$ the output of the kth example generated using the solution of (ω_k', b_k'). After obtaining O_k' for every k in a similar way, the **LOO** error can be obtained as $LOO = \text{card}\{k : y_k O_k' < 0\}$, where card denotes cardinality.

LOO error is an unbiased estimator of true generalization error. However, computing it is time-consuming for D_k has to be solved many times.

2 A New Alpha Seeding Method

It is known that conventional quadratic programming algorithm has to be solved by using specially designed iterative decomposition techniques. It has been proved that the update rule of the iterative decomposition techniques is Newton. Alpha seeding is just an effective technique which can give better initialization of alpha for all the algorithm. It is first explored in [3] and can amortize training costs. According to the results of the experiments, Lee [4] proposed a more effective alpha seeding method than that in [3]. The following formula plays an important role in the decomposition method for solving (1)

$$F_i(\alpha) = \sum_j K(x_i, x_j)\alpha_j y_j - y_j \tag{2}$$

Form the view of efficiently computing (2), when new alpha α' is obtained, the following formula should be used to efficiently compute $F_i(\alpha')$:

$$F_i(\alpha') = F_i(\alpha) + \sum_{j:\alpha_j' \neq \alpha_j} K(x_i, x_j)(\alpha_j' - \alpha_j) y_j \tag{3}$$

Avoiding many unnecessary computations in [3], another alpha seeding scheme which only change a few variables is proposed in [4]. Both of the methods redistribute it to in-bound alphas (ie. those greater than 0 and less than C). In order to get a better scheme, three modified schemes are proposed and compared with the one in [4]. The first one is some like the one in [4], however, a maximum possible amount is first allocated to the index that is bottom most in the list. The second one is like the following: compute $K(x_i, x_j)$ for all i such that $i \neq k$, $\alpha_i = 0$ and $y_i = y_k$, and sort these indexes in decreasing order of $K(x_i, x_j)$, then, all the value of α_k is given to the one in the top list. In the third one, which is some like the second one, all the value is given to the one in the bottom list. Obviously, the latter two schemes of redistributing α_k are simpler than the one in [4] and the first one. For our initial experiments to report in this paper we selected Australian (A), diabetes (D) and heart (H) problem, since a lot of related work with these datum have been published [2]. All tests were performed on an AMD PC 1600+ with 256M of RAM. Table 1 contains some initial results of experiments where T denotes the time cost for computing LOO and ITs denote the total iterative number of computing **LOO** with $\gamma=0.5, \tau=0.001$ for RBF kernel.

Table 1. Results of different alpha seeding schemes

Set	C	Scheme in [4]		Scheme 1		Scheme 2		Scheme 3	
		ITs	T(s)	ITs	T(s)	ITs	T(s)	ITs	T(s)
A	100	2064669	98.09	2062743	97.72	2050795	97.29	2055153	97.65
	1	166236	23.20	165906	23.15	165677	22.94	166731	23.12
D	100	9766220	327.1	9660592	324.0	9654127	322.9	9781485	328.3
	1	98514	15.77	98829	15.92	96060	15.41	96561	15.67
H	100	71131	2.94	69829	2.91	69669	2.88	69656	2.87
	1	35492	1.86	35538	1.88	35194	1.82	35077	1.81

From table 1, it is easy to get the following results:

1) More iterations always take more computation time.
2) Alpha seeding scheme in [4] doesn't always give better results than the scheme 1, which means redistributing α_k to the nearest neighbors doesn't guarantee better performance than redistributing it to some other vectors. Then, if the value of α_k is read to be redistributed to the in-bound ones, the sorting process is needless.
3) Another point can be seen from the table 1 is that the scheme 2 always performs better than the others with costing less time and less iterations.

First, they can be interpreted from a simple geometrical view. Suppose point A is a support vector point which is to be omitted, and point B is its nearest non-support vector point whose alpha is zero, and point C is its nearest support vector point whose alpha is $0 < \alpha_k < C$. After A is omitted, B is a nice alternative which will become a new support vector and the possibility of C becoming a new miss-classified point is not great. In fact, the scheme 2 just means to give the value of α_k to its nearest non-support vector like point B which will turn into a new support vector possibly.

Second, from the view of computation, the cost associated with the computation of (3) with the latter two schemes is smaller than the first two when α is redistributed, because they only involve in changing one alpha variable. In Newton algorithm, better initial solution often leads to less iterations. However, it is difficult to prove which scheme will provide better initial solution mathematically. Equation (4) is the object function which is derived from (1) using the decomposing method.

$$\zeta(\alpha_1,\alpha_2) = -\alpha_1 - \alpha_2 + 0.5 K_{11}\alpha_1^2 + 0.5 K_{22}\alpha_2^2 + y_1 y_2 K_{12}\alpha_1\alpha_2 + y_1\alpha_1 v_1 + y_2\alpha_2 v_2 + \zeta_{const} \quad (4)$$

where $v_i = \sum_{j=3}^{l} y_j \alpha_j K_{ij}$, $\zeta_{const} = -\sum_{i=3}^{l} \alpha_i + \frac{1}{2}\sum_{i,j=3}^{l} \alpha_i \alpha_j y_i y_j K_{ij}$, and α_1, α_2 denotes the two selected variables to do iterations in the decomposition algorithm. Let ζ_{min} denote the final value after training all the examples. If a non-support vector is selected (ie. $\alpha_2 = 0$), it can be rewritten as:

$$\zeta_{min} = \zeta(\alpha_1,\alpha_2) = -\alpha_2 + 0.5 K_{22}\alpha_2^2 + y_2\alpha_2 v_2 + \zeta_{const} \quad (5)$$

It's known the vectors whose alpha equals zero give no contribution to ζ_{min}. When α_2 is to be omitted (ie. $\alpha_2'=0$), the object function will become $\zeta(\alpha_1',\alpha_2')=\zeta_{const}$.

If α is redistributed with the first two schemes, it is difficult to judge which object value is smaller because both of them will change a lot of items in the equation (5). And this implies that the process of sorting $K(x_i,x_j)$ may not assure good selection if we want redistribute the value to the in-bound alphas. A simple approximation can be induced according to scheme 2 if a non-support vector is very close to α_2: $\alpha_1' = \alpha_2$, $\phi(x_1) \approx \phi(x_2)$, $K(x_1,x_i) \approx K(x_2,x_i)$ for all i. Then, the object function turns into $\zeta(\alpha_1',\alpha_2') = -\alpha_1' + 0.5 K_{11} \alpha_1'^2 + y_1 \alpha_1 v_1 + \zeta_{const} \approx \zeta_{min}$. In other words, the value of object function after redistributing α will be very possible near to a small value of ζ_{min} if scheme 2 is taken in alpha seeding method, although it may not be the optimum. This means that the possibility of producing a better initial solution with scheme 2 is very great correspondingly.

3 Conclusions

On the base of the existing alpha seeding schemes, a new alpha seeding scheme is proposed in this paper. Through some benchmark examples, performance of this method is proved to be more effective than the existing methods. What's more, our scheme is simpler for only involving in changing one alpha. Trying to understand the nature of the alpha seeding better, some initial mathematic interpretation based on the value of object function is also discussed.

Acknowledgements

The authors thank the anonymous reviewers for their constructive views. This work was supported by National Defence Key Lab grant no.51476040103JW13.

References

1. Vapnik, V.: Statistical Learning Theory. New York: Wiley,(1998)
2. Lee, J.H., Lin, C.J.: Automatic model selection for support vector machines, Dept. Comput. Sci., Inform. Eng,. National Taiwan Univ., Taipei, Tech. Rep., (2000)
3. DeCoste, D., Wagstaff, K.: Alpha seeding for support vector machines, Int. Conf. Knowledge Discovery and Data Mining, (2000).
4. Lee, M.S., Kerrthi, S.S., Ong, C.J.: An efficient method for computing leave-one-out error in support vector machines with Gaussian kernels, IEEE Trancs. Neural Networks, 5 (2004) 750-757

Multiple Acoustic Sources Location Based on Blind Source Separation

Gaoming Huang[1,2], Luxi Yang[2], and Zhenya He[2]

[1] Naval University of Engineering, Wuhan, China, 430033
[2] Department of Radio Engineering, Southeast University, Nanjing, China, 210096
redforce@sohu.com, {lxyang, zyhe}@seu.edu.cn

Abstract. In this paper we study location of multiple acoustic sources by blind source separation (BSS) method, which based on canonical correlation analysis (CCA). The receiving array is a sparse array. This array is composed of three separated subarrays. From the receiving data set, we can obtain the separate components by CCA. After a simple correlation, time difference can be obtained, and then compute the direction of arrival (DOA) of different acoustic sources. The coordinate of different acoustic sources can be obtained at last. The important contribution of this new location method is that it can reduce the effect of inter-sensor spacing and other factors. Simulation result confirms the validity and practicality of the proposed approach. Results of location are more accurate and stable based on this new method.

1 Introduction

Multiple acoustic sources location is of great interest to many applications, such as hearing aids, fault location and target tracking etc. In real environments, multiple acoustic sources are inevitable. The main problem of this topic is how to separate different sources only by the receiving signals and complete the location for different acoustic sources. Naturally, we consider about apply blind source separation (BSS) method to the problem. There are many BSS methods for this problem [1][2]. Consider about the simplicity and practicality, we propose a new location estimation method based on canonical correlation analysis (CCA), which is an important method of multivariate statistical since it was proposed by H.Hotelling [3]. The main character of CCA is it can find the basic vectors from two sets of variables, similarly to our ears. There are detail descriptions of CCA in [4][5], which has been applied in some preliminary work [6][7][8] in recent years. CCA was applied to seek the correlate components of the data from double receiving sensors. After a general correlation of the two canonical components, we can obtain the time difference and then the estimation value of DOA, location will be completed at last.

The organization of this paper is as follows: In Section 2, we formulate the issue of multiple acoustic sources location and describe the problem to be solved. In Section 3, we present the location algorithm. Simulations conducted in Section 4 show the effectiveness of the algorithm, and finally is the conclusion.

2 Problem Formulation

The receiving model is shown as Fig.1, which includes three groups of receiving sensors r_1, r_2, r_3. Each receiving sensor includes two separated receiving units.

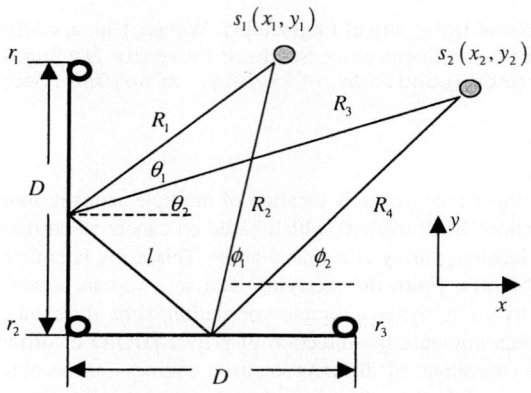

Fig. 1. Signal receiving model

If a source is considered to be far from the array, the source can be said to approximately lie on a line connecting the source to the center of each receiving sensor pair. The model of each receiving sensor pairs can be described as:

$$r(t) = A \cdot s(t) + n(t) \quad (1)$$

where $r(t)$ are the receiving signals. A is a mixing matrix. $s(t)$ are the acoustic source signals. $n(t)$ are noise signals. The key problem of location based time difference is how to eliminate or reduce the affect of noise, interference and mixing signals to time difference estimation by correlation and DOA estimation, which is the main problem that this paper wants to solve.

3 Location Algorithm

The location algorithm includes three steps: First is blind separation by CCA and then obtain the time difference by a correlation. At last we can complete the computation of location by the time difference and corresponding DOA estimation.

The main difference between CCA and the other statistical methods is that CCA is closely related to mutual information [6]. Consider two sets of input data $x_1, x_2, \cdots, x_p$ and $y_1, y_2, \cdots, y_q$, $p \leq q$, The two sets of data can be written as combination of some pairs of variables ξ_i and η_i, which can be described as follows:

$$\begin{cases} \xi_1 = a_{11}x_1' + \cdots + a_{1p}x_p' & \eta_1 = b_{11}y_1' + \cdots + b_{1q}y_q' \\ \quad \cdots \\ \xi_p = a_{p1}x_1' + \cdots + a_{pp}x_p' & \eta_p = b_{p1}y_1' + \cdots + b_{pq}y_q' \end{cases} \quad (2)$$

where x', y' are the standardization value of x, y respectively. Mutual independent variables can be obtained by the method of canonical correlation. Canonical correlation variable and coefficients can be obtained by the follow steps:

Step1: computing the correlation of the two sets of variables as:

$$\Sigma = \begin{bmatrix} \Sigma_{xx} & \Sigma_{xy} \\ \Sigma_{yx} & \Sigma_{yy} \end{bmatrix} \quad (3)$$

Setp2: Computing the canonical correlation coefficients r_i. Firstly, we compute two matrices L and M, where $L = \Sigma_{xx}^{-1}\Sigma_{xy}\Sigma_{yy}^{-1}\Sigma_{yx}, M = \Sigma_{yy}^{-1}\Sigma_{yx}\Sigma_{xx}^{-1}\Sigma_{xy}$, secondly, we compute the eigenvalue λ_i of matrix L and M.

Step3: Computing the canonical variables ξ_i and η_i.

Time difference is based on the correlation of the canonical variables ξ_i and η_i. Then the separate time difference can be obtained by:

$$\tau_i = \arg\left\{\max_{\tau}\left[\hat{R}_i(\tau)\right]\right\}, \quad i = 1...4 \quad (4)$$

Where $\hat{R}_i(\tau)$ is the correlation function of canonical variables. The estimation of DOA is relatively simple. $\Delta D = D\sin\theta$, $\Delta \tau c = \Delta D = D\sin\theta$, where $c = 340.29 m/s$ is the velocity of sound, DOA can be obtained as:

$$\theta = \arccos\left((c \cdot \Delta\tau)/D\right) \quad (5)$$

From Eq.(5) we can see that the precision of DOA is limited by the distance between the phase centers of the subarrays and the time difference estimation precision. As the defined coordinate in the Fig.1, we can obtain R_1, R_2 as:

$$R_1 = l \cdot \sin(180 - \phi_1 - 45)/\sin(\phi_1 - \theta_1), R_2 = l \cdot \sin(45 + \theta_1)/\sin(\phi_1 - \theta_1) \quad (6)$$

R_3, R_4 are similar calculation as R_1, R_2, the position of acoustic sources s_1, s_2 can be written as:

$$\begin{cases} x_1 = R_1\cos\theta_1, y_1 = R_1\sin\theta_1 + D/2 \\ x_2 = R_3\cos\theta_2, y_2 = R_3\sin\theta_2 + + D/2 \end{cases} \quad (7)$$

4 Simulations

The background of the experiments is assumed as: there are three separated receiving sensors as in Fig.1. The distance between the phase centers of the subarrays is $D = 0.3m$, two acoustic sources come from different direction and the sampling frequency is $8000Hz$, time differences are set as $\tau_1 = 3, \tau_2 = 2, \tau_3 = 4, \tau_4 = 3$. Blind separation by CCA was conducted at first, which is the basic of high precision time difference estimation. The separated source details are shown in Fig. 2. Then a cross correlation is conducted. The correlation results are shown as in Fig.3.

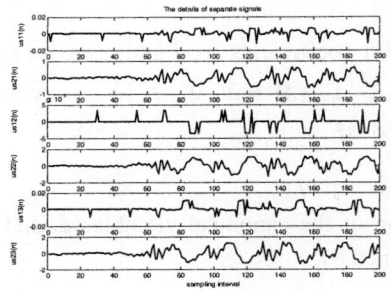 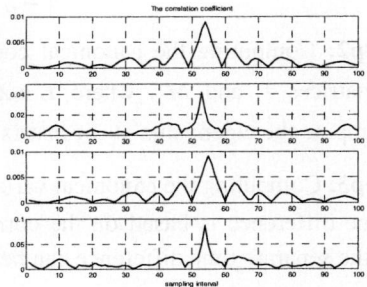

Fig. 2. Blind separation details by CCA   **Fig. 3.** Correlation results

From the correlation results, time difference of different acoustic source can be obtained as: $\tau_1 = 3, \tau_2 = 2, \tau_3 = 4, \tau_4 = 3$, which is corresponding with the experiments setting. Then $\theta_1 = 25.1736°, \theta_2 = 16.4737°, \phi_1 = 34.5517°, \phi_2 = 25.1736°$ can be obtained and $R_1 = 2.5605, R_2 = 2.4493, R_3 = 2.6386, R_4 = 2.4644$. Then the positions of different acoustic sources are: $s_1(\,2.3173, 1.2391)$ and $s_2(2.5303, 0.8982)$ as the coordinate in Fig.1.

5 Conclusion

In this paper, we have investigated the fundamental limitations in multiple acoustic source location. We propose a novel location method by BSS, which based on CCA. This method can effectively overcome the contradictory of phase ambiguity and signal correlation and will play an important role in military and civilian affairs.

References

1. Torkkola,K.: Blind Separation of Delayed Sources Based on Information Maximization. Proc. IEEE ICASSP, (1996) 3509-3512
2. Huang,G.M., Yang,L.X. and He,Z.Y.: Application of Blind Source Separation to a Novel Passive Location. ICA2004, LNCS, Vol.3195, (2004) 1134-1141
3. Hotelling, H.: Relations Between Two Sets of Variates. Biometrika, 1936(28) 321-377
4. Anderson, T. W.: An Introduction to Multivariate Statistical Analysis. John Wiley & Sons, second edition, 1984
5. Zhang, R.T. and Fang ,K.T.: An Introduction to Multivariate Statistical Analysis. Science House, 1982
6. Borga, M.: Learning Multidimensional Signal Processing. PhD thesis, Linköping University, Sweden, SE-581 83 Linköping, Sweden, 1998. Dissertation No 531, ISBN 91-7219-202-X. 1998
7. Bach, F. R. and Jordan,M. I.: Kernel Independent Component Analysis. Journal of Machine Learning Research, 2002(3) 1-48
8. Fyfe, C. and Lai, P. L.: Ica Using Kernel Canonical Correlation Analysis. ICA2000, 2000(8) 279-284

Short-Term Load Forecasting Based on Self-organizing Map and Support Vector Machine

Zhejing Bao, Daoying Pi, and Youxian Sun

National Laboratory of Industrial Control Technology,
Dept. of Control Sci. and Eng.,
Zhejiang University, Hangzhou 310027, P.R. China
{zjbao, dypi, yxsun}@iipc.zju.edu.cn

Abstract. An approach for short-term load forecasting by combining self-organizing map(SOM) and support vector machine(SVM) is proposed in this paper. First, historical load data of same type are clustered using SOM, and then daily 48-point load values are vertically predicted respectively based on SVM. In clustering, factors such as date type, weather conditions and time delay are considered. In addition, influences of kernel function and SVM parameters on load forecasting are discussed and performance of SOM-SVM is compared with pure SVM. It is shown that normal smoothing technique in preprocessing is not suitable to be used in vertical forecasting. Finally, the approach is tested by data from EUNITE network, and results show that the approach runs with high speed and good accuracy.

1 Introduction

Short-term electrical load forecasting is an essential part of Energy Management System(EMS). The key to electrical load forecasting lies in the improvement of accuracy. Recently SVM based short-term load forecasting method is very attractive. During the investigation of SVM forecasting model, it is found that how to choose typical samples is worth further discussing. But clustering methods presented in most papers are too simple. Only date type was considered by totally ignoring weather conditions and time lag [1,2,4]. Moreover, paper [1,2] reported the results of 48-point load values for only a day, and paper [3,4] offered the results of a certain time instant load for consecutive days. All the above approaches can't show good generalization ability.

In this paper, a new algorithm combining SOM and SVM is proposed for load prediction. First, SOM is employed in clustering to find training samples that are similar to the predicted day in date type and weather conditions with the consideration of time delay; second, SVM is applied to vertically predict daily 48-point load values respectively.

This paper consists of four sections. In section 2, the new short-term load predicting method SOM-SVM is presented. Section 3 gives simulation results and discussions. Finally section 4 offers the conclusions.

2 SOM-SVM Based Short-Term Load Forecasting

2.1 Clustering Analysis of Predicted Day Based on SOM

To implement clustering of the predicted day, historical data are selected as follows: 1) 48-point load data of 5 months previous to the predicted day; 2) 48-point load data of a respective month previous to and posterior to the predicted day of last year; 3) 48-point load data of a respective half month previous to and posterior to the predicted day of the year before last.

Date type is classified according to following aspects: i) profile of load curve; ii) magnitude of load; iii) weather information; iv) date type. Of the above four factors, i and ii are unknown for the predicted day, whereas the latter two conditions are already known. So the paper employs iii and iv in clustering. Furthermore, power load system has time delay. As a result, four parameters, composed of temperature of the previous day and this day(normalized), date type of the previous day and this day, are considered in SOM.

2.2 Forecasting 48-Point Load Values Respectively Based on SVM

Extracting the features from samples is of vital importance. Input vectors are chosen to be: 1) load values of predicted time instant k of the previous 7 days obtained by averaging the data of time instant k-1, k and k+1; 2) average load value of predicted time instant k of the previous 7 days; 3) temperature of the predicted day; 4) year type of the predicted day (the year before last: 1; last year: 2; this year: 3). To sum up, the input vectors are 11-dimension, and the output is the load value of time instant k. Daily 48-point load values are predicted respectively. We adopt ε-insensitive loss function with polynomial kernel $K(x_i, x) = ((x \cdot x_i) + 1)^d$ and Gaussian kernel $K(x_i, x) = \exp(-\frac{||x-x_i||^2}{2\sigma^2})$.

3 Simulation Results and Discussions

This approach has been tested in forecasting daily 48-point load values from 981201 to 981215 according to daily temperature and 48-point load data from 970101 to 981130 supplied by EUNITE network [5].

3.1 Selection of Kernel Function and Parameters C and ε

Mean absolute percent error (MAPE) is chosen as evaluation indicator, described as $MAPE = \frac{1}{n}\sum_{i=1}^{n} |\frac{a(i)-f(i)}{f(i)}|$, where $n = 48$, $a(i)$ is the actual load and $f(i)$ is the forecasted load.

Polynomial kernel and Gaussian kernel are taken respectively. We find that Gussian kernel has better results than those of polynomial kernel. Furthermore, the performance is related to the different σ in Gussian kernel case, that is, very low $\sigma(1-10)$ can result in overfittting and very high $\sigma(300-700)$ can lead to underfitting. The suitable value of σ is within 75-150.

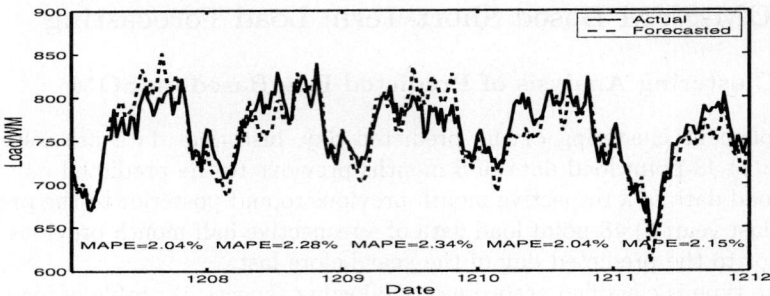

Fig. 1. Actual and forecasted load for 981208-981212 using SOM-SVM

Table 1. Comparison results of DWT(Horizontally)-SOM-SVM,DWT(Vertically)-SOM-SVM,SOM-SVM and Pure SVM(Gussian kernel,$\sigma = 100, C = 10000, \varepsilon = 0.01$)

Date	MAPE/%			
	DWT(H)-SOM-SVM	DWT(V)-SOM-SVM	SOM-SVM	Pure SVM
981203	2.53	2.36	2.48	2.12
981204	2.52	3.67	2.55	2.72
981205	3.11	4.84	2.95	3.61
981206	3.28	5.30	3.28	2.62
981207	2.49	5.36	2.67	2.67
981208	2.23	1.96	2.04	2.96
981209	2.20	2.08	2.28	1.80
981210	2.36	2.64	2.34	3.26
981211	2.13	2.99	2.04	2.30
981212	2.16	2.26	2.15	2.01
Average	2.50	3.35	2.48	2.61

It is also suggested that C and ε can also affect the accuracy when using Gussian kernel($\sigma = 100$). Very low C(100-1000) causes underfitting, while very high C(50000-100000) produces overfitting. In the same way, overfittting can occur when ε is too low(0.0001-0.001) and reversely underfitting arises when ε is too high(0.05-0.1). So the performance is the best when $C = 10000$ and $\varepsilon = 0.01$. Fig.1 gives the predicted curve of 981208-981212(Gussian kernel,$\sigma = 100, C = 10000, \varepsilon = 0.01$).

3.2 Influence of Data Smoothing on Load Prediction Accuracy

In papers [1,3,4], historical data were preprocessed by smoothing as ordinary signal. But as shown in Tab.1, performance with usual horizontal smoothing such as discrete wavelet transform (DWT) before prediction is a little worse than or similar to the one without using smoothing. Discussions are stated as follows:

(1) Due to highly fluctuation of load curve, daily 48-point load values are not estimated as ordinary regression with time t to be one of input vectors, but

vertically predicted by constructing models respectively. Therefore, the ordinary horizontal smoothing doesn't work in this case.

(2) The result of vertically smoothing before SOM-SVM isn't good at all. Smoothing mainly removes the shock component of total load. Shock load usually accounts for 1-2 percent of total load. The result that average MAPE of DWT(V)-SOM-SVM is higher than that of SOM-SVM by almost 1%, shown in Tab.1, exactly explains this point. Forecasting accuracy can certainly be worsened if the objectively existing portion of power load is ignored.

3.3 Role of SOM Clustering to SVM

The training time of pure SVM is 25s for each time instant, yet that of SVM-SOM is composed of SOM clustering and SVM forecasting for each time instant, the former and the latter cost 3s, 500ms respectively (PC: 2.0Ghz, 500M). Apparently, in contrast to pure SOM the speed of SOM-SVM is greatly accelerated. And Tab.1 shows that the accuracy of SOM-SVM is better than pure SVM. In brief, SOM-SVM is better than pure SVM both in speed and accuracy with the interference of different type day eliminated by SOM clustering.

4 Conclusions

The proposed short-term load predicting technique in this paper based on SOM-SVM is tested by data supplied by East-Slovakia Power Distribution Company.

The main conclusions are summarized as follows:

(1) Since shock load certainly exists, ignoring it will affect the forecasting performance for sure. Therefore whether to smooth load data before SVM based forecasting is an open problem.

(2) The proposed SOM-SVM method can incorporate the advantages of SOM and SVM. As a result, the method has high speed and good accuracy.

(3) Choosing different kernel functions and SVM parameters can directly influence forecasting accuracy.

References

1. Yang, J., Xie, H.: Application of SVM and Fourier algorithm to power system short-term load forecast. Relay **32(4)** (2004) 17-19
2. Yang, J., Cheng, H.: Application of SVM to power system short-term load forecast. Electric Power Automation Equipment **24(2)** (2004) 30-32
3. Li, Y., Fang, T., Zheng, G.: Wavelet support vector machines for short-term load forecasting. Journal of University of Science and Technology of China **33(6)** (2003) 726-731
4. Li, Y., Fang, T., Yu, E.: Study of support vector machine for short-term load forecasting. Proceedings of the CSEE **23(6)** (2003) 55-59
5. World-wide Competition Within the EUNITE Network, EUNITE Competition Report. http://neuron.tuke.Sk/competition/index.php

A Multi-class Classifying Algorithm Based on Nonlinear Dimensionality Reduction and Support Vector Machines[1]

Lukui Shi[1,2], Qing Wu[2], Xueqin Shen[2], and Pilian He[1]

[1] School of Electronic and Information Engineering, Tianjin University,
Tianjin 300072, China
[2] School of Computer Science and Engineering, Hebei University of Technology,
Tianjin 300130, China
lkshi@eyou.com, plhe@tju.eud.cn

Abstract. Many problems in pattern classifications involve some form of dimensionality reduction. ISOMAP is a representative nonlinear dimensionality reduction algorithm, which can discover low dimensional manifolds from high dimensional data. To speed ISOMAP and decrease the dependency to the neighborhood size, we propose an improved algorithm. It can automatically select a proper neighborhood size and an appropriate landmark set according to a stress function. A multi-class classifier with high efficiency is obtained through combining the improved ISOMAP with SVM. Experiments show that the classifier presented is effective in fingerprint classifications.

1 Introduction

Dimensionality reduction is an important task in pattern classifications. Its purpose is to discover compact representations of the original data. Two classical methods of dimensionality reduction are PCA and MDS. Both methods are guaranteed to find the true structure of data lying on or near a linear subspace of the high dimensional space. However, these linear algorithms cannot in essence discover complex nonlinear manifold structure [1]. Recently, several algorithms [1], [2], [3] have been developed to perform dimensionality reduction of low dimensional nonlinear manifolds embedded in a high dimensional space. ISOMAP is one of representative techniques, which has been applied to pattern classifications. Nevertheless, ISOMAP not only badly depends on the neighborhood size but also has higher time and space complexity.

In this paper, we put forward an improved ISOMAP (IISOMAP) with respect to flaws of ISOMAP. The algorithm combined with support vector machines (SVM) yields an efficient multi-class classifier. We apply it to fingerprint classifications and obtain better results.

---

[1] This work was supported by Natural Science Foundation of Hebei province of China (No. 603037 and No. E2005000024) and supported by Science-Technology Development Project of Tianjin of China (No. 04310941R).

2 IISOMAP: An Improved ISOMAP

ISOMAP generalizes MDS to nonlinear manifolds. It is based on replacing Euclidean distances with an approximation of the geodesic distances on the manifold. The algorithm is summarized as follows: (i) Determine neighborhoods for each point. (ii) Estimate the geodesic distances between all pairs of points. (iii) Find low dimensional coordinates by applying MDS on the pairwise distances. Details can be referred to [1].

In ISOMAP, the complexity of estimating the geodesic distances is $O(kN^2 \log N)$. The MDS eigenvalue calculation has complexity $O(N^3)$ for involving a full $N \times N$ matrix. LMDS [4] greatly speed up by solving a sparse eigenvalue problem, which only preserves the geodesic distances between each point and some landmark points.

The key is to select a better landmark set because randomly selected landmark sets cannot often represent the true topology of the original data and leads to worse results. To evaluate quantitatively landmark sets, we use the stress function (SF)

$$SF = (\sum_{1 \leq i < j \leq n} |d_M(i,j) - d_m(i,j)|^2) / (\sum_{1 \leq i < j \leq n} d_M^2(i,j)) , \qquad (1)$$

where $d_M(i,j)$ is the geodesic distance in the input space, $d_m(i,j)$ is Euclidean distance in the embedded space and n is the number of landmark points.

We randomly choose 8 landmark sets (LS1-8) with 20 data points each from Swiss roll data set with 1000 samples (Fig. 1(a)). The values of SF are given in Table 1. The results prove that the smaller the value of SF is, the better the selected landmark set is. For example, the result in Fig. 1(b) is much better than that in Fig. 1(c).

Table 1. Values of SF for different landmark sets

Landmark sets	LS1	LS2	LS3	LS4	LS5	LS6	LS7	LS8
Values of SF	0.0181	0.0192	0.0186	0.0173	0.0213	0.0161	0.0297	0.0184

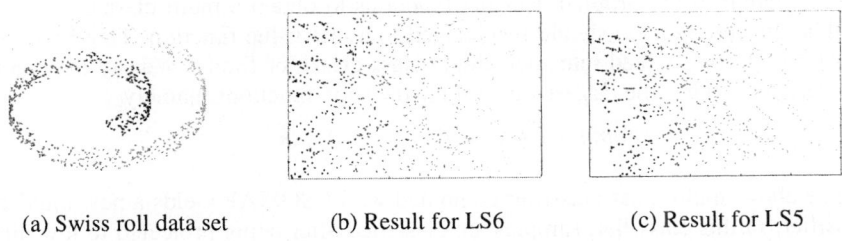

(a) Swiss roll data set   (b) Result for LS6   (c) Result for LS5

Fig. 1. Low dimensional embeddings from ISOMAP for various landmark sets

It is important to select a proper neighborhood size k in ISOMAP because too large or too small neighborhoods cannot reveal the true structure of the manifold. The equation (1) can also be used to determine a proper neighborhood size. The criterion is also that the smaller the value of SF is, the better the low dimensional embedding is.

We acquire an improved ISOMAP by using the stress function to automatically select a neighborhood parameter and a landmark set. The algorithm is as follows.

1) Determine the neighborhood size k. For a given range of k, ISOMAP is executed on a randomly selected landmark set for each k at first. Then we compute the values of SF for each k and chose the k with the smallest SF as the neighborhood parameter.

2) Select the landmark set. Firstly, for m random landmark sets, we calculate the geodesic distances between landmark points and other points. Secondly, each landmark set is mapped into the embedded space with ISOMAP. Lastly, we figure out the values of SF on each set. The set with the smallest SF is taken as the last landmark set.

3) Obtain the low dimensional coordinates of the whole data set with the selected neighborhood size k and landmark set in previous steps.

In IISOMAP, we suppose that each landmark set contains n data points, where $n \ll N$. The time complexity of computing the geodesic distance matrix is $O(knN\log N)$ while MDS runs in $O(n^2 N)$ time. Both are much smaller than those of ISOMAP.

3 Multi-class Classifier Based on IISOMAP and SVM

Single-class classifiers can be constructed using the single-class SVM with hyperplanes, which find an optimal separating hyperplane passing through the origin in the feature space. The decision function is [5]

$$f(x) = \text{sgn}(\sum_i \alpha_i k(\mathbf{x}_i, \mathbf{x}) - \rho) \; , \tag{2}$$

where x_i is a support vector; α_i is Lagrange coefficient of the support vector and ρ is a constant. The kernel function $k(x,x')$ is usually a Gaussian function.

For multi-class classification problems, a multi-class classifier needs to combine all single-class classifiers trained for each class with a suitable way. Here, we directly use the output of the original decision functions to obtain a multi-class classifier. To do this, the equation (2) should be changed to a real value function. To classify samples, testing objects feed into each single-class classifier firstly. We say the object in the class, which has the largest value of the decision functions, namely,

$$output = \arg\max_i (f_i) \qquad i = 1,...,k \; . \tag{3}$$

The above multi-class classifier combined with IISOMAP yields a new multi-class classifier. In this classifier, samples are classified after being projected to low dimensional spaces with IISOMAP.

4 Experiments for Fingerprint Classifications

The classifier was tested on FingerCode [6] preprocessed fingerprints from NIST-4 Database containing 4000 images. We divided it into two subsets with 2000 samples each, one as the training set and another as the testing set. All experiments have been run on a PC with 2.0GHz CPU. IISOMAP was implemented in MATLAB6.5.

The 192-dimensional data was reduced to a 6-dimensional embedding with IISOMAP for k=18 and n=30. Single-class classifiers were trained with LIBSVM [7]. Fingerprint data were classified four classes (W, R, L, AT). Table 2 gives comparisons between with dimensionality reduction and without dimensionality reduction.

An accuracy of 91.5% is achieved for four-class classification by rejecting 1.8% of data. Apparently, the accuracy with dimensionality reduction is close to that without dimensionality reduction. However, the time and space efficiency of the former is greatly superior to that of the latter. Here, the time cost of IISOMAP is ignored.

Table 2. Comparisons of training time, testing time and accuracy for classifications

	Training time (s)	Testing time (s)	Accuracy for classifications
Without dimensionality reduction	16.612	15.421	93.8%
With dimensionality reduction	1.608	1.016	91.5%

5 Conclusions

An efficient multi-class classifier is proposed to handle high dimensional data in pattern classifications. Firstly, the original data are projected to a low dimensional space with IISOMAP. Then objects are classified with the SVM classifier. IISOMAP greatly decreases the dependency to the parameter and the computing complexity by automatically determining a neighborhood size and a landmark set. The application in fingerprint classifications demonstrates that the efficiency of training and testing is improved and memory requirements are reduced under without loss of the accuracy.

References

1. Tenenbaum, J. B., de Silva, V., Langford, J. C.: A Global Geometric Framework for Nonlinear Dimensionality Reduction. Science, Vol. 290, (2000) 2319-2323.
2. Roweis, S. T., Saul, L. K.: Nonlinear Dimensionality Reduction by Locally Linear Embedding. Science, Vol. 290, (2000) 2323-2326.
3. Belkin, M., Niyogi, P.: Laplacian Eigenmaps for Dimensionality Reduction and Data Representation. Neural Computation, Vol. 15, (2003) 1373–1396.
4. de Silva, V., Tenenbaum, J. B.: Global versus Local Methods in Nonlinear Dimensionality Reduction. In NIPS, Vol. 15, (2002) 705-712.
5. Schölkopf, B., Platt, J. C., Shawe-Taylor, J., Smola, A. J., Williamson, R. C.: Estimating the Support of a High-dimensional Distribution. Neural Computation, Vol. 13, (2001) 1443-1471.
6. Jain, A. K., Prabhakar, S., Hong Lin: A Multichannel Approach to Fingerprint Classification. IEEE Transactions on Pattern Analysis and Machine Intelligence, Vol. 21, (1999) 348-359.
7. Chih-Chung Chang, Chih-Jen Lin: LIBSVM: a library for support vector machines. URL: http://www.csie.ntu.edu.tw/~cjlin/libsvm.

A VSC Scheme for Linear MIMO Systems Based on SVM

Zhang Yibo, Yang Chunjie, Pi Daoying, and Sun Youxian

National Laboratory of Industrial Control Technology,
Zhejiang University, Hangzhou 310027, P. R. China
{ybzhang, cjyang, dypi, yxsun}@iipc.zju.edu.cn

Abstract. A variable structure control (VSC) scheme for linear MIMO systems based on support vector machine (SVM) is developed. By analyzing the characters of linear MIMO system, a VSC scheme based on Exponent Reaching Law is adopted to track desired trajectory. Then one input of the system is trained as the output of SVM, while sliding mode function, differences and other inputs of the system are trained as the inputs of SVM. So one VSC input of the black-box system could be obtained directly by trained SVM after other inputs of the system are selected manually, and recognition of system parameters is avoided. A linear MIMO system is used to prove the scheme, and simulation results show that this scheme has high identification precision and quick training speed.

1 Introduction

The design of controller for uncertain systems with extraneous disturbances has been concerned for a long time. One method to resolve the problem is variable structure controller [1-3]. The VSC system is a special kind of nonlinear controller characterized by a series of discontinuous control actions that change the control system structure upon reaching a set of switching surfaces. The most important property of the VSC system is that sliding motion on the switching surface is ensured.

SVM is an elegant tool for solving pattern recognition and function regression problems [4-6]. It has attracted a lot of researchers from the neural network and mathematical programming community, for the main reason is its ability to provide excellent generalization performance. There is only one global minimal point when SVM is training, rather than partial minimal points in neural networks, and its operating speed is much higher than that of the latter. Moreover, SVM can track arbitrary curves with arbitrary precisions, which means it can be easily used in the recognition of linear and nonlinear systems.

A VSC scheme for MIMO systems based on SVM is proposed in this paper. After the exponent reaching law is introduced, the MIMO system is transferred to MISO system, and then SVM is adopted to get the control algorithm. At last, a linear MIMO system is adopted to prove that the scheme is effective.

This paper is organized as follows: In section 2, VSC algorithm and exponent reaching law is introduced; then SVM is proposed, while its regression function is mainly concerned in section 3. In the next section, the algorithm is proposed. In § 4, the scheme is proved in simulation. At last, the conclusions are drawn.

2 Variable Structure Control for MIMO systems

Consider the following discrete black-box MIMO system:
$$y(k+1) = Ay(k) + Bx(k) \qquad (1)$$
where $k \in Z^+$ denotes the sample step, $x = [x_1(k) \; x_2(k) \; ... \; x_m(k)]^T$ and $y = [y_1(k) \; y_2(k) \; ... \; y_n(k)]^T$ denote the input and output matrix respectively, while $A = \{a_{ij} \in R \quad i,j = 1,2,...,n\}$ and $B = \{b_{ij} \in R \quad i = 1,2,...,n \quad j = 1,2,...,m\}$

Suppose
$$e(k) = d(k) - y(k) \qquad (2)$$
$$s(k) = c^T e(k) \qquad (3)$$
where $d(k) = [d_1(k) \; d_2(k) \; ... \; d_n(k)]^T \in R$ denotes the desired output, while $e(k) = [e_1(k) \; e_2(k) \; ... \; e_n(k)]^T \in R$ and $c = [c_1 \; c_2 \; ... \; c_n]^T \in R$ are difference and the parameter of switching surface respectively.

By adopting Exponent Reaching Law
$$s(k+1) = (1 - \delta\tau)s(k) - \varepsilon\tau \operatorname{sgn} s(k) \qquad (4)$$
and consider (1), (2) and (3), the control algorithm can be obtained:
$$A_0 e(k) + B_0 x(k) + D_0 d(k) + cd(k+1) = (1 - \delta\tau)s(k) - \varepsilon\tau \operatorname{sgn} s(k) \qquad (5)$$
where $\tau > 0$, $\varepsilon > 0$ and $\delta > 0$ are sampling period, reaching speed and approaching speed respectively, and what is more, $1 - \delta\tau > 0$, and
$$A_0 = c^T A = [a_{01} \; a_{02} \; ... \; a_{0n}]$$
$$B_0 = c^T B = [b_{01} \; b_{02} \; ... \; b_{0n}]$$
$$D_0 = -c^T A = [d_{01} \; d_{02} \; ... \; d_{0n}]$$

3 Support Vector Machine Regression

Suppose that the training samples are (x_i, y_i), $i = 1,2,....,k$, $\{x_i \in R^n, y_i \in R\}$, the object is to solve the following regress problem:
$$y = f(x) = \langle \omega \cdot x \rangle + b \qquad (6)$$
where $\langle \cdot \rangle$ denotes inner product, and b is bias.

Vapnik [4] suggested the use of ε-insensitive loss function where the error is not penalized if the loss function is less than ε. Using the error function together with a regularizing term, the optimization problem solved by the support vector machine can be formulated as:
$$\min \frac{1}{2}\|\omega\|^2 + C\sum_{i=1}^{}(\xi_i + \xi_i^*) \quad \text{s.t.} \begin{cases} y_i - \langle \omega \cdot x \rangle - b \leq \varepsilon + \xi_i \\ \langle \omega \cdot x_i \rangle + b - y_i \leq \varepsilon + \xi_i^* \\ \xi_i, \xi_i^* \geq 0 \end{cases} \qquad (7)$$

The constant $C > 0$ determines the tradeoff between the smoothness of f and the amount up to which deviations larger than ε are tolerated.

Referring to Lagrange multipliers and KKT conditions, the optimal regression equation can be obtained as following:

$$f(x) = \sum_{i=1}^{nsv} \alpha_i x_i + b \qquad (8)$$

where x_i are the support vectors, and nsv is the number of support vectors, while α_i is the coefficients, b is the threshold value.

4 The Scheme of SVC Based on SVM Regression

There can always find $b_{0j} \neq 0$ $(1 \leq j \leq m)$ that can turns (5) into:

$$x_j(k) = \frac{1}{b_{0j}} s(k+1) - \sum_{i=1}^{n} \frac{a_{0i}}{b_{0j}} e_i(k) - \sum_{i=1, i \neq j}^{m} \frac{b_{0i}}{b_{0j}} x_i(k) - \sum_{i=1}^{n} \frac{d_{0i}}{b_{0j}} d_i(k) - \frac{cd(k+1)}{b_{0j}} \qquad (9)$$

It can be seen that $x_j(k)$ is the linear combination of X. Suppose that:

$$Y = x_j(k)$$
$$X = \begin{bmatrix} s(k+1) & e(k) & x_1(k) & \ldots & x_{j-1}(k) & x_{j+1}(k) & \ldots & x_m(k) & d_n(k) \end{bmatrix}^T$$

Regard (X, Y) as samples, and then the scheme of SVC based on SVM can be obtained. The output Y can be acquired through the model that is trained by SVM.

5 Simulation

Considering the following MIMO system:

$$\begin{bmatrix} y_1(k+1) \\ y_2(k+1) \end{bmatrix} = \begin{bmatrix} 0.2 & 0.1 \\ 0.1 & 0.1 \end{bmatrix} \begin{bmatrix} y_1(k) \\ y_2(k) \end{bmatrix} + \begin{bmatrix} 0.1 & 0.1 \\ 0 & 0.1 \end{bmatrix} \begin{bmatrix} x_1(k) \\ x_2(k) \end{bmatrix}$$

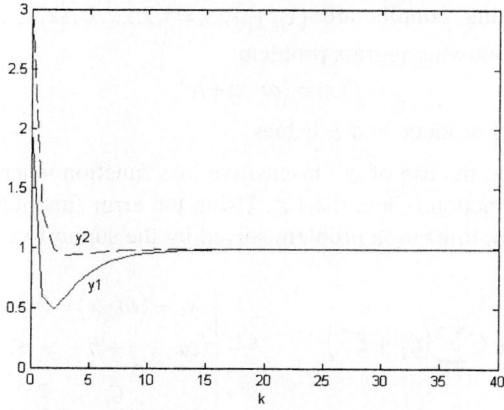

Fig. 1. The figure shows that the outputs track the desired trajectory, and outputs converge in about 15 sample steps. and the stable differences are less than 0.1%

The initial states are $[y_{10} \ y_{20}]^T = [y_1(0) \ y_2(0)]^T = [2 \ 3]^T$, and the sample time $T = 0.1$. In order to ensure the convergence of states, choose $c = [-1 \ 1]^T$ and $\varepsilon = 0.7|s(k)|$ [2]. Suppose that the states track object $d = [1 \ 1]^T$. The simulation result is shown in Fig.1

6 Conclusion

By employing a system input as the output of the linear kernel SVM, this system input turns into the output of SVM. Then the exponent reaching law is adopted to predict the sliding mode function for the next sample time. By combing them, a VSC controller based on a SVM identification algorithm is bring forward. Simulation result shows that the scheme is effective.

References

1. Wei-Bing Gao: Discrete-Time Variable Structure Control Systems, Vol. 42. *IEEE Transactions On Industrial Electronics* (1995)117-121.
2. Chang-Lian Zhai. Zhi-Ming Wu: Variable Structure Control Method for Discrete Time Systems, Vol. 34. *Journal of Shanghai Jiaotong university*(2000) 719-722.
3. Tarek M. M. Nasab: A New Variable Structure Control Design with Stability Analysis for MIMO Systems (2000) 785-788
4. Vapnik V: The nature of statistical learning theory, Springer, NY (1995)
5. Matilde Sanchez-Fernandez, Jeronimo Arenas-Garcia: SVM Multiregression for Nonlinear Channel Estimation in Multiple-Input Multiple-Output Systems, Vol. 52. *IEEE transactions on signal processing*(2004) 2298-2307
6. P.M.L.Drezet, R.F.Harrison: " Support vector machines for system identification," *in UKACC International Conf. on control*, UK(1998)668-692.

Global Convergence of FastICA: Theoretical Analysis and Practical Considerations[*]

Gang Wang[1,2], Xin Xu[1,3], and Dewen Hu[1,*]

[1] College of Mechatronics and Automation,
National University of Defense Technology,
Changsha, Hunan, 410073, P.R.C.
[2] Telecommunication Engineering Institute, Air Force Engineering University,
Xi'an, Shanxi, 710077, P.R.C.
[3] School of Computer, National University of Defense Technology,
Changsha, Hunan, 410073, P.R.C.
dhu@nudt.edu.cn

Abstract. FastICA is now a popular algorithm for independent component analysis (ICA) based on negentropy. However the convergence of FastICA has not been comprehensively studied. This paper provides the global convergence analysis of FastICA and some practical considerations on algorithmic implementations. The exhaustive equilibria are obtained from the iteration first. Then the global convergence property is given on the 2-channel system with cubic nonlinearity function, and the results can also be generalized to the multi-channel system. In addition, two practical considerations, e.g. the convergence threshold for demixing matrix and independence restriction for sources, are evaluated and the influence on the separation solutions is illustrated respectively.

1 Introduction

As a class of data processing methods originated from blind signal separation, independent component analysis (ICA) has been widely applied in blind source separation (BSS) and feature extraction. For various algorithms of ICA, a fundamental problem both in theory and in practice is to ensure an arbitrary initial demixing matrix or vector converge to the stable equilibrium [1,2]. And some researches have been contributed, such as the global and local convergence analysis on the information-theoretic ICA [1], the stability of the general blind source separation methods [2], and the monotonic convergence analysis for fixed-point algorithm [3].

Apart from the cost function for the multi-unit algorithm and stochastic (or natural) gradient optimization involved in the above literatures, FastICA is essentially a one-unit algorithm of maximization nonGaussianity and an approximate Newton's iteration approach [4,5,6]. Hyvärinen et al. elucidated the derivation of FastICA and offered the local stable proof in [6]. Recently Liu et al. discussed the availability of performing ICA in the proof of one-bite-matching

---

[*] Supported by National Natural Science Foundation of China (30370416, 60303012, 60225015), Ministry of Education of China (20049998012, TRAPOYT project).

conjecture in [7] where the cost function involved is negentropy with cubic nonlinearity, and in [8] the global convergence for the kurtosis-based FastICA with circular distributed sources is addressed in ICA's application in CDMA. So far the general global convergence analysis or proof has not been provided. Here we will investigate the important issue and provide the global convergence analysis for equilibrium and stability for separation solution. Also two practical considerations are illustrated to evaluate the influence of convergence threshold and independence restriction on the separation solution respectively.

2 Preliminaries

FastICA is a fast fixed-point algorithm based on approximate negentropy [4]. The iteration for demixing vector w is as follows

$$\mathrm{w} \leftarrow E\{zg(\mathrm{w}^T z)\} - E\{g'(\mathrm{w}^T z)\mathrm{w}\}, \quad \mathrm{w} \leftarrow \mathrm{w}/\|\mathrm{w}\|, \tag{1}$$

where z is the prewhitened form of mixture x, $g(\cdot)$ the differential of a certain nonquadratic function $G(\cdot)$, and "'" the differential symbol [5]. If $g(\cdot)$ is cubic nonlinear, (1) is specialized as

$$\mathrm{w} \leftarrow E\{z(\mathrm{w}^T z)^3\} - 3\mathrm{w}, \quad \mathrm{w} \leftarrow \mathrm{w}/\|\mathrm{w}\|. \tag{2}$$

For the ambiguities of ICA and simplicity, sources are assumed of 0-mean and 1-variance [7]. To investigate the relations between the restoration signals y (y=Wx) and s, relation matrix Q (Q=WA) is introduced. And (1) and (2) can be rewritten as (3) and (4) accordingly

$$\mathrm{q} \leftarrow E\{sg(\mathrm{q}^T s)\} - E\{g'(\mathrm{q}^T s)\mathrm{q}\}, \quad \mathrm{q} \leftarrow \mathrm{q}/\|\mathrm{q}\| \tag{3}$$

in which $\mathrm{q} = (q_1, q_2, \cdots, q_n)^T$ is an element in Q and

$$q_i \leftarrow q_i^3 kurt(s_i), \quad \text{for} \quad i = 1, 2, \cdots, n, \tag{4}$$

where $kurt(s_i)$ denotes the standard kurtosis of s_i. Therefore the performance investigation on w can be transformed to the q-parameter space.

3 Equilibrium and Global Convergence Analysis

3.1 Exhaustive Equilibria

From (1) the following result for the equilibrium is obtained directly.

Theorem 1. The fast fixed-point iteration (1) converges only when it satisfies

$$E\{zg(\mathrm{w}(k_0)^T z)\} - E\{g'(\mathrm{w}(k_0)^T z)\mathrm{w}(k_0)\} = c\mathrm{w}(k_0), \tag{5}$$

where k_0 is a certain positive integer, c nonzero constant, and $\mathrm{w}(k_0)$ the equilibrium.

Proof. Denote $\tilde{w}(k_0) = cw(k_0)$, then (1) can be rewritten as

$$w(k_0+1) \leftarrow \frac{\tilde{w}(k_0)}{\|\tilde{w}(k_0)\|} = \frac{cw(k_0)}{\|cw(k_0+1)\|} = \frac{cw(k_0)}{|c|\|w(k_0)\|}. \tag{6}$$

For the unit-variance constrain on $w(k)$, (6) can be expressed as

$$w(k_0+1) \leftarrow \frac{w(k_0)}{\|c\|} = \begin{cases} w(k), & c > 0 \\ -w(k), & c < 0 \end{cases}. \tag{7}$$

Formula (7) shows that $w(k_0)$ satisfies the convergence condition for demixing vector defined in [4], and thus $w(k_0)$ is the equilibrium of (1). #

For the case of cubic nonlinear function, (6) can be simplified as

$$q_i(k_0+1) = cq_i^3(k_0)kurt(s_i), \quad \text{for } 1 \leq i \leq n. \tag{8}$$

Further researches show that the equilibria in (5) essentially correspond to the extreme points of the constrained cost function $L(q) = J(q) + \lambda(1 - q^T q)$ where λ denotes the Lagrange factor, or directly those of $J(q)$ in the q-parameter space.

3.2 Global Convergence Analysis

First focus on the global convergence on the 2-channel system with cubic nonlinear function. The exhaustive equilibria for q can be obtained from (8) directly as solutions

A_1-A_4: $(0, \pm1)$ and $(\pm1, 0)$,

and when the condition of $kurt(s_1)kurt(s_2) > 0$ is satisfied, additional solutions

B_1-B_4: $\left(\pm\sqrt{\frac{kurt(s_2)}{kurt(s_1)+kurt(s_2)}}, \pm\sqrt{\frac{kurt(s_1)}{kurt(s_1)+kurt(s_2)}}\right)$

and $\left(\pm\sqrt{\frac{kurt(s_2)}{kurt(s_1)+kurt(s_2)}}, \mp\sqrt{\frac{kurt(s_1)}{kurt(s_1)+kurt(s_2)}}\right)$ also exist.

For simplicity, first assume $kurt(s_1)kurt(s_2) > 0$ and then evaluate the global convergence of q in the first quadrant. Denote $\alpha(i) = q_1(i)/q_2(i)$, for $i = 0, 1, \cdots$, and the following equation can be obtained from (8)

$$\alpha(k) = \frac{q_1(k)}{q_2(k)} = \left(\alpha(0) \cdot \left(\frac{kurt(s_1)}{kurt(s_2)}\right)^{1/2}\right)^{3^k-1}. \tag{9}$$

It means that

i) solution B_1 corresponding to $\alpha(0) \cdot \left(\frac{kurt(s_1)}{kurt(s_2)}\right)^{1/2} = 1$ is the critical and unstable point;

ii) when $\alpha(0) \cdot \left(\frac{kurt(s_1)}{kurt(s_2)}\right)^{1/2} > 1$, $\alpha(k) \to \infty$ and $q(k) \to (1, 0)$;

iii) while $0 < \alpha(0) \cdot \left(\frac{kurt(s_1)}{kurt(s_2)}\right)^{1/2} < 1$, $\alpha(k) \to 0$, and $q(k) \to (0, 1)$.

Analogical analysises for the initialization on the unit circle show the stability of solutions A_1-A_4 and B_1-B_4 are unstable equilibria. The relation between initialization and the separation solutions is depicted in Fig.1. Line l_1 and l_2

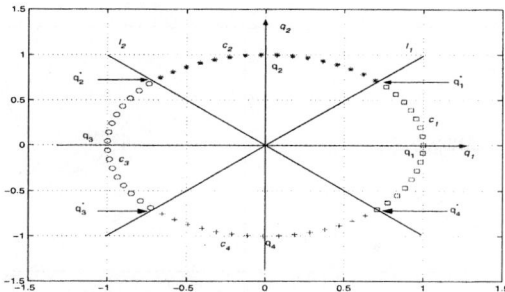

Fig. 1. Relation between initialization and equilibrium for q

separate the unit circle into four arcs as c_1, c_2, c_3 and c_4, and the initial points in each of the four arcs would converge to the stable equilibria (1,0), (0,1), (-1,0) and (0,-1) respectively.

In the case of $kurt(s_1)kurt(s_2) < 0$, solutions B_1-B_4 do not exist. The convergence analysis is simplified and similar results can also be given.

For the m-channel system, the linear representation $y = q_1 s_1 + q_2 s_2 + \cdots + q_m s_m$ can be rewritten as $y = q_1 s_1 + \hat{q}_2 \hat{s}_2$ in which $\hat{q}_2 \hat{s}_2 = q_2 s_2 + \cdots + q_m s_m$. Thus the convergence and stability analysis in the m-dimension can be implemented in a new 2-dimension space since the basic assumptions, such as independence and nonGaussianity, are also satisfied. Therefore the results on the 2-channel system can also be naturally generalized to an m-channel case since s_1 can be arbitrarily substituted by any other element in s.

4 Two Practical Considerations

Despite of the outstanding performance, the classical ICA is really an idea model and in most cases the basic restrictions cannot be strictly satisfied. Here two illustrations are given to evaluate the influence on the separation solution and convergence results when the some of idea assumptions are broken. The first is about the convergence threshold and the second the independence assumption.

Example 1: The idea convergence condition that q($k+1$) equals q(k) or $-$q(k) is implemented by assuming a small positive convergence threshold ε. And when $\|q(k+1) - q(k)\|_2 < \varepsilon$ or $\|q(k+1) + q(k)\|_2 < \varepsilon$ the iteration stops. In a 2-channel system, assume $kurt(s_1) = 1$, $kurt(s_2) = 1.934$, $\varepsilon = 0.01$ and initialize q 2000 times uniformly sampled on the unit circle. Perform the iteration with cubic nonlinear function. Fig.2 (a) shows the spurious equilibria for q, and the detail around the spurious in the first quadrant is provided in Fig.2 (b). The star symbols stand for spurious solutions and square symbols initial points, and all the points are essentially around the theoretical unstable equilibrium B_1 given in section 3.2.

Analysises show that about 2 percents initial points convergence to the spurious equilibria. While for the multi-channel system, the existence of spurious

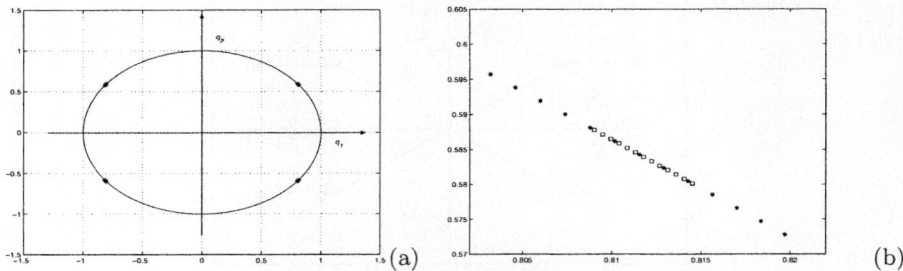

Fig. 2. (a). The influence on separation solution when $\varepsilon = 0.01$; (b). The detail around the spurious in the first quadrant.

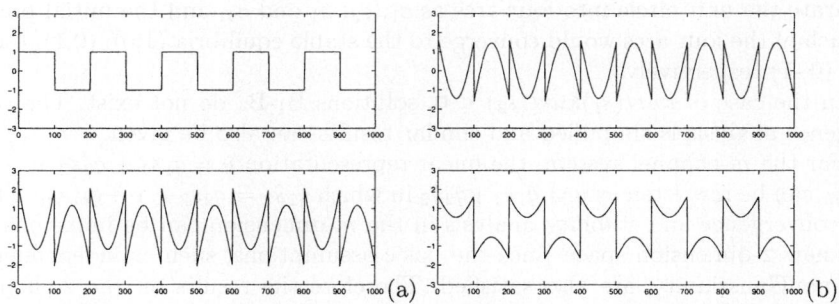

Fig. 3. (a). Original sources s_1 and s_2; (b). Spurious independent components.

equilibrium cannot be neglected when more sources are to be recovered. In BSS for a 20-channel system, the existence of spurious equilibrium is in the probability about $1 - (1 - 0.02)^{20} \approx 0.4$.

Example 2: Two sources estimated by FastICA are introduced as Fig.3 (a), in which s_1 is a unit square signal and s_2 quasi-sawtooth. It shows that they are orthogonal but dependent since the corrcoef between s_1^4 and s_2^4 is 0.2207 but not zero. Perform FastICA with nonlinear activation function $g(x) = \tan(x)$ on x = As where mixing matrix A is randomly selected. The results show that q may convergence to the point of (1,0) or (-1,0) corresponding to the source s_1, or to the spurious solutions of (0.3472,-0.9378) or (-0.3472,0.9378), and the the corresponding spurious independent components are depicted in Fig.3 (b).

Denote $f_1 = \hat{y} = q_1 s_1 + (1 - q_1^2)^{1/2} s_2$ and $f_2 = \tilde{y} = q_1 s_1 - (1 - q_1^2)^{1/2} s_2$. Fig.4 depicts the function of $kurt(\hat{y})$ and $kurt(\tilde{y})$ versus q_1 respectively. Symbols of point stand for the initialized points which would converge to (1,0) or (-1,0) (symbols of circle), while symbols of star to (0.3472,-0.9378) or (-0.3472,0.9378) (symbols of square).

The above two illustrations show the influence on the separation solution when the idea assumptions are broken. By performing the fast fixed-point iteration original sources may be recovered in a certain probability, and spurious

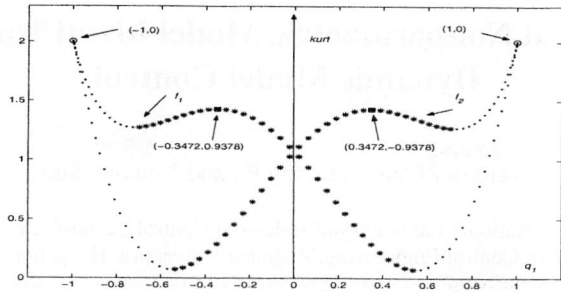

Fig. 4. Function $kurt(\hat{y})$ and $kurt(\tilde{y})$ versus q_1

solutions cannot be excluded at all. Especially in BSS, we cannot ensure the independence even orthogonality, thus whether the recovered signals are the original sources should be doubted when little prior knowledge is available.

5 Conclusions

This paper investigates the global convergence of FastICA, and provides the influence on the separation solution when some idea assumptions are broken. Though ICA has been elucidated well in theory and demonstrated an efficient method in various applications, the practical blind separation or extraction performance is doubted for the severe restriction on basic model. Further researches on the independence restriction are to be carried out.

References

1. Cheung, C.C., Xu, L.: Some Global and Local Convergence Analysis on the Information-Theoretic Independent Component Analysis Approach. Neurocomputing. 30 (2000) 79-102
2. Amari, S., Chen, T.P., Cichocki, A.: Stability Analysis of Adaptive Blind Source Separation. Neural Networks. 10 (1997) 1345-1351
3. Regalia, P.A., Kofidis E.: Monotonic Convergence of Fixed-Point Algorithm for ICA. IEEE Trans. on Neural Networks. 14 (2003) 943-949
4. Hyvärinen, A., Oja, E.: A Fast Fixed-Point Algorithm for Independent Component Analysis. Neural Computation. 9 (1997) 1483-1492
5. Hyvärinen, A.: Fast and Robust Fixed-Point Algorithms for Independent Component Analysis, IEEE Trans. on Neural Networks. 10 (1999) 626-634
6. Hyvärinen, A., Karhunen, J., Oja, E.: Independent Component Analysis. John Wiley, New York (2001)
7. Liu, Z.Y., Chiu, K.C., and Xu, L.: One-Bite-Matching Conjecture for Independent Component Analysis. Neural Computation. 16 (2004) 383-399
8. Ristaniemi, T., Joutsensalo, J.: Advanced ICA-Based Receivers for Block Fading DS-CDMA Channels. Signal Processing. 82 (2002) 417-431

SVM Based Nonparametric Model Identification and Dynamic Model Control

Weimin Zhong, Daoying Pi, and Youxian Sun

National Laboratory of Industrial Control Technology,
Institute of Modern Control Engineering, Zhejiang University, Hangzhou 310027, China
{wmzhong, dypi, yxsun}@iipc.zju.edu.cn

Abstract. In this paper, a support vector machine (SVM) with linear kernel function based nonparametric model identification and dynamic matrix control (SVM_DMC) technique is presented. First, a step response model involving manipulated variables is obtained via system identification by SVM with linear kernel function according to random test data or manufacturing data. Second, an explicit control law of a receding horizon quadric objective is gotten through the predictive control mechanism. Final, the approach is illustrated by a simulation of a system with dead time delay. The results show that SVM_DMC technique has good performance in predictive control with good capability in keeping reference trajectory.

1 Introduction

Model predictive control (MPC) was first developed in the 1970s. And MPC is an optimization-based control algorithm in which a dynamic process model is used to predict and optimize process performance. The most popular MPC techniques are Model Algorithmic Control (MAC) [1] and Dynamic Matrix Control (DMC) [2]. DMC is based a nonparametric model, whose coefficients can be obtained directly from unit step response test without assuming a model order. It is a fact that in practice carrying out a unit step response test is always high-cost and time-consuming. The goal of this paper introduces a new method based on SVM with linear kernel function to obtain system's step response coefficients not by step response test data but manufacturing data or other test data. For more detail about SVM, see reference [3].

2 SVM Based Nonparametric Model with Step Response Coefficients

Assume system can be represented as follows:

$$\hat{y}_m(k+1) = \hat{f}(I_k)) = \hat{f}[\Delta u(k), \Delta u(k-1), \cdots, \Delta u(k-N+1)] \quad (1)$$

Where $\hat{f}$ is a function with SVM architecture, $\hat{y}_m(k+1)$ is the output of SVM model. According to training data pair $\{I_s, y_s\}(s=1,\cdots,d)$, $y_s = y(s+1)$ generated from manufacturing data or other test data, support values α_i and threshold β can be gotten through learning. So the model predictive output at time k is:

$$\hat{y}_m(k+1) = \sum_{i=1}^{nsv} \alpha_i (I_i' \cdot I_k) + \beta \qquad (2)$$

$nsv \leq d$ is the number of support vector, I_i' is the support vector from I_s. Where

$$I_i' \cdot I_k = I_i'(1)\Delta u(k) + I_i'(2)\Delta u(k-1) + \cdots + I_i'(N)\Delta u(k-N+1)$$

So

$$\hat{y}_m(k+1) = w_1 \Delta u(k) + w_2 \Delta u(k-1) + \cdots + w_N \Delta u(k-N+1) + \beta \qquad (3)$$

$$w_i = \sum_{i=1}^{nsv} \alpha_i I_i' \quad i = 1,\cdots,N : \text{SVM based model step response coefficients}$$

3 SVM_DMC

Assume $N \geq P \geq M$, P is the prediction horizon, M is control horizon. When $j \geq M$, assume control value after this time unchangeable. j-step-ahead predictive output is:

$$\hat{y}_m(k+j) = w_1 \Delta u(k+j-1) + w_2 \Delta u(k+j-2) + \cdots + w_N \Delta u(k+j-N) + \beta \qquad (4)$$

Reference trajectory is taken as

$$\begin{cases} y_r(k+j) = a_r^j y(k) + (1-a_r^j) y_{sp} \\ y_r(k) = y(k) \end{cases} \quad j = 1,2\cdots,P \qquad (5)$$

a_r: Tuning factor related with the control system's robustness and convergence
$y_r(k+j)$: Reference value at sample time $k+j$
Introducing feedback correction,

$$\hat{y}_p(k+j) = \hat{y}_m(k+j) + h_j e(k) \quad j = 1,2\cdots,P \qquad (6)$$

$e(k) = y(k) - y_m(k)$: Predictive model's output error at time k
h_j: Error correct coefficients.
Let

$$\hat{Y}_p(k+1) = [\hat{y}_p(k+1), \hat{y}_p(k+2), \cdots, \hat{y}_p(k+P)]^T$$
$$\hat{Y}_m(k+1) = [\hat{y}_m(k+1), \hat{y}_m(k+2), \cdots, \hat{y}_m(k+P)]^T$$
$$Y_r(k+1) = [y_r(k+1), y_r(k+2), \cdots, y_r(k+P)]^T$$

$$\Delta \mathbf{U}(k) = [\Delta u(k), \Delta u(k+1), \cdots, \Delta u(k+M-1)]^T$$
$$\Delta(k-1) = [\Delta u(k-m'+1), \Delta u(k-m'+2), \cdots, \Delta u(k-1)]^T$$
$$\mathbf{h} = [h_1, h_2, \cdots, h_P]^T \text{ In this paper, chose } \mathbf{h} = [1,1,\cdots,1]^T$$
$$\mathbf{B} = [\beta, \cdots, \beta]^T$$

According to (4)- (6), there is

$$\hat{\mathbf{Y}}_p(k+1) = \hat{\mathbf{Y}}_m(k+1) + \mathbf{h}e(k) = W\Delta U(k) + L\Delta(k-1) + \mathbf{h}e(k) + \mathbf{B} \qquad (7)$$

With

$$W = \begin{bmatrix} w_1 & & & 0 \\ w_2 & w_1 & & \\ \vdots & \vdots & & \\ w_P & w_{P-1} & \cdots & w_{P-M+1} \end{bmatrix} \qquad L = \begin{bmatrix} w_N & w_{N-1} & w_{N-2} & \cdots & w_2 \\ w_N & w_N & w_{N-1} & \cdots & w_3 \\ \vdots & \vdots & \ddots & & \vdots \\ w_N & w_N & \cdots w_{N-1} & \cdots w_{P+2} & w_{P+1} \end{bmatrix}$$

Minimize the objective function with receding horizon

$$J(k) = [\hat{\mathbf{Y}}_p(k+1) - Y_r(k+1)]^T Q[\hat{\mathbf{Y}}_p(k+1) - Y_r(k+1)] + \Delta U^T(k) R \Delta U(k) \qquad (8)$$

$$Q = diag[q_1, q_2, \cdots, q_P] \quad R = diag[r_1, r_2, \cdots, r_M]$$

Take derivation of $\Delta \mathbf{U}(k)$, and set $\dfrac{\partial J}{\partial \Delta \mathbf{U}(k)} = 0$, there is

$$\Delta U(k) = (W^T Q W + R)^{-1} W^T Q [Y_r(k+1) - L\Delta(k-1) - \mathbf{h}e(k) - \mathbf{B}] \qquad (9)$$

And only the first computed change in the manipulated variable is implemented.

4 Simulation

Consider a dead time delay system for which it is specified that $N = 24$ and $T = 2s$ described as below [4]:

$$G(s) = \frac{e^{-6s}}{(10s+1)(3s+1)} \qquad (10)$$

Use a series of a series of uniform random numbers between [-1,1] to generate 200 training data pairs. Select $C = 10000$ and $\varepsilon = 0.01$. Set $y_{sp} = 0.9$ $P = 5$, $M = 2$, $Q = diag[1,1,1,1,1]$, $R = 0$, $a_r = 0.9$ and assume system is zero initial state.

Figure 1 gives the results of system's multi-step-ahead SVM_DMC. We can see system output can trace the reference trajectory well and quickly after the dead time.

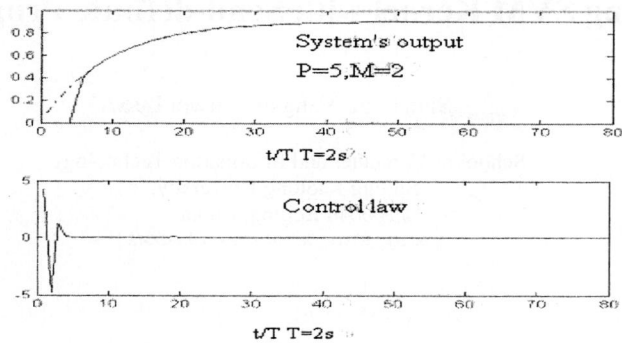

Fig. 1. Result of SVM_DMC with $P = 5$ $M = 2$

5 Conclusions

DMC provides a robust scheme which is directly applicable, and useful, in the field of industrial control. The algorithm introduced in this paper puts forward a new approach to obtain system's impulse response coefficients without special step response test. According to manufacturing data, using SVM to identify system's nonparametric model and do DMC is feasible.

Acknowledgments

This work is supported by China 973 Program under grant No.2002CB312200.

References

1. Rouhani R., Mehra R.K.: Model Algorithmic control (MAC), Basic Theoretical Properties. Automatica, vol.18, no.4 (1982) 401-414
2. Cutler C.R., Ramaker B.L.: Dynamic Matrix Control—A Computer Control Algorithm. Proc. JACC. SanFranciso (1980)
3. Vapnik V.N.: The nature of statistical learning theory. Springer-Verlag, New York, USA (1995)
4. Shu D.Q.: Predictive control system and its application. China Machine Press, Beijing(1996)

Learning SVM Kernel with Semi-definite Programming

Shuzhong Yang and Siwei Luo

School of Computer and Information Technology,
Beijing Jiaotong University,
100044 Beijing, China,
yang_shu_zhong@163.com

Abstract. It is well-known that the major task of the SVM approach lies in the selection of its kernel. The quality of kernel will determine the quality of SVM classifier directly. However, the best choice of a kernel for a given problem is still an open research issue. This paper presents a novel method which learns SVM kernel by transforming it into a standard semi-definite programming (SDP) problem and then solves this SDP problem using various existing methods. Experimental results are presented to prove that SVM with the kernel learned by our proposed method outperforms that with a single common kernel in terms of generalization power.

1 Introduction

In recent years, support vector machines (SVMs) have received considerable attention because of their superior performance in pattern recognition and function regression [1]. The basic principle of SVMs is to find an optimal separating hyperplane so as to separate two classes of patterns with maximal margin [1]. However, it is well-known that the major task of the SVM approach lies in the selection of its kernel. Choosing different kernel functions will produce different SVMs [2] and will result in different performances. The existing kernels include linear kernel, polynomial kernel, radial basis function kernel and many hybrid kernels [6]. Much work has been done on how to learn the SVM kernels, but the best choice of a kernel for a given problem is still an open research issue. In this paper, we will present a novel method to learn the SVM kernels.

This paper is organized as follows: In Section 2, we briefly review the fundamentals of SVM and Semi-Definite Programming. In Section 3, we illustrate how to learn the SVM kernel with Semi-Definite Programming. In Section 4, experimental results are presented to illustrate the proposed method in this paper. Finally, conclusions are given in Section 5.

2 The Fundamentals of SVM and Semi-definite Programming

In this section, we will concisely review the basic principles of SVM and Semi-Definite Programming. For more details, the interested scholars can refer to [1], [4].

Suppose we have some training examples $\{(x_i, y_i), i = 1, \ldots, l\}$. Each training example $x_i \in R^d$, d being the dimension of the input space, belongs to a class labeled by $y_i \in \{-1, 1\}$. The SVM approach can be considered to minimize the following quadratic programming (QP) problem:

$$\min \frac{1}{2}(\mathbf{w}^T\mathbf{w}) + C\sum_{i=1}^{l}\xi_i$$
$$s.t. \quad y_i((w_i \cdot x_i) + b) \geq 1 - \xi_i, \quad i = 1, \cdots, l. \tag{1}$$
$$\xi_i \geq 0, \quad i = 1, \cdots, l$$

The dual problem is as follows:

$$W(\alpha) = \max_{\alpha} \sum_{i=1}^{l}\alpha_i - \frac{1}{2}\sum_{i,j=1}^{l}y_i y_j \alpha_i \alpha_j K(x_i, x_j)$$
$$s.t. \quad 0 \leq \alpha_i \leq C, \quad i = 1, \cdots, l \tag{2}$$
$$\sum_{i=1}^{l}\alpha_i y_i = 0$$

The optimal classification hyperplane can be solved as

$$f(x) = sign[(\sum_{i=1}^{l}\alpha_i y_i K(x, x_i)) + b]. \tag{3}$$

If the kernel function K is determined, we can use various methods to solve the above QP problem and obtain the optimal classification hyperplane.

Semi-definite programming [4] mainly deals with the optimization of convex functions over the convex cone $P = \{X \in R^{p \times p} \mid X = X^T, X \geq 0\}$ of symmetric positive semi-definite matrices, or subsets of this cone. From [5] we know that every positive and symmetric matrix is a kernel matrix and conversely every kernel matrix is symmetric and positive definite. So the above convex cone P can be viewed as a search space for possible positive definite kernel matrices in this paper. We expect to specify a convex cost function that will enable us to learn the optimal SVM kernel matrix from P using semi-definite programming. Because of the convexity, this approach allows us to avoid problems with local minima.

Semi-Definite Programming (SDP) can be defined as

$$\min_{x} c^T x$$
$$s.t. \quad F(x) = F_0 + x_1 F_1 + \cdots + x_n F_n \geq 0. \tag{4}$$
$$Ax = b$$

Where $x \in R^p$ and $F_i = F_i^T \in R^{p \times p}$. $F(x) \geq 0$ (called a linear matrix inequality, LMI) restricts $F(x)$ to be contained in the positive semi-definite cone P. Notice that the objective is linear in the unknowns x, and that both the LMI and the equality constraint are linear in x.

Many tools have been devised to solve SDP problems such as SeDuMi, SDPT3 and so on. Once being transformed into a standard SDP problem, the learning problem of SVM kernel can be solved easily and efficiently. In next section, we will give the detailed transforming procedure.

3 Learning SVM Kernel Matrix with SDP

The most important issue in combining the SDP and the learning of SVM kernel is how to transform the learning of SVM kernel into a standard SDP problem. Then we can use various existing tools such as SeDuMi to solve the SDP problem efficiently. Inspired by hard margin classification problem [5], we will give the detailed derivation of how to transform the learning problem of SVM kernel into a standard SDP problem. The only difference between our derivation and the derivation of hard margin classification problem is that it adds an extra constraint.

In (2) if we consider K as variable too, it can be written as

$$W(K) = \max_{\alpha} 2\alpha^T e - \alpha^T G(K)\alpha : \ 0 \leq \alpha \leq C, \alpha^T y = 0. \tag{5}$$

Where e is the 1-vector of ones, $G(K)$ is defined by $G_{ij}(K) = k(x_i, x_j) y_i y_j$.

Then inspired by hard margin classification problem, the goal of learning the SVM kernel is to solve the following problem.

$$\min_{K \geq 0} W(K) \quad \text{s.t.} \quad trace(K) = c. \tag{6}$$

In order to express (6) as a SDP, we write (6) as

$$\min_{K \geq 0, t} t$$
$$\text{s.t.} \ t \geq \max_{\alpha} 2\alpha^T e - \alpha^T G(K)\alpha, \tag{7}$$
$$0 \leq \alpha \leq C, \alpha^T y = 0, trace(K) = c$$

Now we will transform (7) into a standard SDP problem. Firstly, we define the Lagrangian of the maximization problem (2) by

$$L(\alpha, v, u, \lambda) = 2\alpha^T e - \alpha^T G(K)\alpha + 2v^T \alpha + 2u^T(C-\alpha) + 2\lambda y^T \alpha. \tag{8}$$

At the optimum, we have

$$\frac{\partial L}{\partial \alpha} = 0 \Rightarrow \alpha = G(K)^{-1}(e + v - u + \lambda y). \tag{9}$$

We obtain that for any t>0, the constraint $W(K) \leq t$ is true and only if there exist $v, u \geq 0$ and λ such that

$$(e + v - u + \lambda y)^T G(K)^{-1}(e + v - u + \lambda y) \leq t. \tag{10}$$

or equivalently such that

$$\begin{pmatrix} G(K) & e+v-u+\lambda y \\ (e+v-u+\lambda y)^T & t \end{pmatrix} \geq 0. \tag{11}$$

holds. Stacking all constraints in one single LMI, (7) can be expressed as a standard SDP (4):

$$\min_{K,t,\lambda,v,u} t$$

$$s.t. \ trace(K) = c,$$

$$\begin{pmatrix} K & O & O & O \\ O & G(K) & e+v-u+\lambda y & O \\ O & (e+v-u+\lambda y)^T & t & O \\ O & O & O & diag(v-u) \end{pmatrix} \geq 0 \tag{12}$$

In practice, we commonly consider K as a linear combination $K = \sum_{i=1}^{m} \mu_i K_i$ for a fixed set $\{K_1, K_2, \cdots, K_m\}$, for the consideration can reduce the complexity of space search. Now (12) can be written as

$$\min_{K,t,\lambda,v,u} t$$

$$s.t. \ trace(\sum_{i=1}^{m} \mu_i K_i) = c,$$

$$\begin{pmatrix} \sum_{i=1}^{m} \mu_i K_i & O & O & O \\ O & G(\sum_{i=1}^{m} \mu_i K_i) & e+v-u+\lambda y & O \\ O & (e+v-u+\lambda y)^T & t & O \\ O & O & O & diag(v-u) \end{pmatrix} \geq 0 \tag{13}$$

Thus the learning problem of SVM kernel is transformed into a standard SDP problem. In next section, we will give some experimental results to show the advantages of our method for learning SVM kernel.

4 Experimental Results

To illustrate our proposed method of the paper, we design some experiments to compare the performances (classification margin and test error) of SVM learned by our proposed method and by commonly existing kernels, respectively. The commonly existing kernels include a polynomial kernel $k_1(x_i, x_j) = (1 + x_i^T x_j)^d$ for kernel matrix K_1, a RBF kernel $k_2(x_i, x_j) = \exp(-(x_i - x_j)^T(x_i - x_j)/2\sigma^2)$ for kernel matrix K_2 and a linear kernel $k_3(x_i, x_j) = x_i^T x_j$ for kernel matrix K_3. In order to reduce the complexity of space search, we use $K = \sum_{i=1}^{3} \mu_i K_i$ as initial "bad guesses" of the kernel matrix K^*. The parameters d for K_1 and σ for K_2 are determined in advance.

In experiment, two synthetic datasets and two standard benchmark datasets from the UCI repository are used. The two synthetic datasets are produced by a Gaussian random generator. The first one includes 500 two-dimension data points which can be classified into 2 classes and the second is made up of 500 data points which can be classified into 5 classes. The two benchmark datasets are breast cancer dataset and sonar dataset, respectively. The breast cancer dataset includes 286 instances which can be classified into 2 classes and the sonar dataset includes 208 instances which can be classified into 13 classes. We use the one-against-all method to construct the SVM multi-class classification [7] for the classification of the second and fourth datasets. Each dataset was randomly partitioned into 60% training and 40% test sets. We repeated the random partition 10 times on each dataset. The experimental results are summarized in Table 1 below.

Table 1. Margin and test error rate(TER) for SVMs trained and tested with the initial kernel matrices K_1, K_2, K_3 and with the optimal kernel matrix K^*, learned using semi-definite programming (13). A dash means that no general margin classifier could be found. (N is the number of instances and C is the number of classes).

		K_1	K_2	K_3	K^*
Synthetic dataset 1	d=2 σ=0.5 N=500 C=2				
Margin		0.112	0.189	0.132	0.522
TER		0.0%	0.0%	1.2%	0.0%
Synthetic dataset 2	d=2 σ=0.1 N=500 C=5				
Margin		0.024	0.108	-	0.289
TER		4.5%	7.5%		1.7%
Breast cancer	d=2 σ=0.5 N=286 C=2				
Margin		0.009	0.134	-	0.289
TER		2.8%	4.1%		1.7%
Sonar	d=2 σ=0.1 N=208 C=13				
Margin		0.036	0.190	0.005	0.355
TER		7.5%	9.4%	10.6%	6.7%

From Table 1, we can notice that not every K_i can construct a linearly separable classifier of the training data. The results really show us that the SVM using the kernel K^* has better performance than these using any of the components K_i. The better performance includes a larger margin and a smaller test error rate. We can also learn SVM kernel by using training examples as well as by using test examples, and it will produce better performance.

5 Discussions

In this paper we have proposed a novel method for learning a SVM kernel according to the given dataset, which learns a SVM kernel by transforming it into a standard Semi-Definite Programming (SDP) problem. This method is motivated by two facts.

The first fact is that every symmetric, positive definite matrix can be considered as a kernel matrix and vice versa. The second fact is that SDP mainly deals with the optimization of convex cost functions over the convex cone of positive semi-definite matrices (or convex subsets of this cone). Combining the two facts, a powerful method for learning the SVM kernel with SDP is provided. Experimental results on synthetic datasets and standard benchmark datasets prove the power of our novel approach to SVM kernel learning. In fact, nearly all problems which can be solved by kernel methods can be transformed into a standard SDP problem, so that we can obtain the most suitable kernel according to the corresponding problem by our proposed method.

Acknowledgements

The research is supported by national natural science foundations of china (60373029).

References

1. V.N. Vapnik.: The Nature of Statistical Learning Theory. *New York: Springer Verlag*, 1995.
2. B. Scholkopf, A. Smola, and K.-R. Muler.: Nonlinear Component Analysis as a Kernel Eigenvalue Problem. *Neural Computation*, 1299-1319, 1998.
3. Christopher J.C. Burges.: A Tutorial on Support Vector Machines for Pattern Recognition. *Data Mining and Knowledge Discovery*, 2: 121-167, 1998.
4. Vandenberghe, L. & Boyd S.: Semidefinite Programming. *SIAM Review*, 49-95, 1996.
5. Gert R.G.Lanckriet, Nelo Cristianini, Peter Bartlett, Laurent El Ghaoui, Michael I.Jordan.: Learning the Kernel Matrix with Semi-Definite Programming. *The Journal of Machine Learning Research*, 5: 27-72, 2004.
6. Ying Tan and Jun Wang.: A Support Vector Machine with a Hybrid Kernel and Minimal Vapnik-Chervonenkis Dimension. *IEEE Transactions on Knowledge and Data Engineering*, 385-395, 2004.
7. Chih-Wei Hsu and Chih-Jen Lin.: A Comparison of Methods for Multiclass Support Vector Machines. *IEEE Transactions on Neural Networks*, 415-425, 2002.

Weighted On-Line SVM Regression Algorithm and Its Application

Hui Wang, Daoying Pi, and Youxian Sun

The National Laboratory of Industrial Control Technology,
Institute of Modern Control Engineering,
Zhejiang University, Hangzhou, 310027 P.R. China
plpy@163.com
dypi@iipc.zju.edu.cn
yxsun@iipc.zju.edu.cn

Abstract. Based on KKT condition and Lagrangian multiplier method a weighted SVM regression model and its on-line training algorithm are developed. Standard SVM regression model processes every sample equally with the same error requirement, which is not suitable in the case that different sample has different contribution to the construction of the regression model. In the new weighted model, every training sample is given a weight coefficient to reflect the difference among samples. Moreover, standard online training algorithm couldn't remove redundant samples effectively. A new method is presented to remove the redundant samples. Simulation with a benchmark problem shows that the new algorithm can quickly and accurately approximate nonlinear and time-varying functions with less computer memory needed.

1 Introduction

Support vector machine (SVM) [1] is a new machine learning method and has been used for classification, function regression, and time series prediction, etc [2-3]. Current SVM regression (SVR) training algorithms mostly are off-line, but several on-line SVR algorithms have been researched [4-5]. These on-line algorithms all consider every sample with the same importance to the construction of the SVR model. But in practical application, different sample has different importance. Moreover, current online algorithm can't remove redundant samples effectively. This paper will introduce a weighted SVR online training algorithm. Then a valid algorithm based on weight coefficient is proposed to remove redundant samples. Simulation is used to evaluate the performance of the new training algorithm.

2 Weighted SVM Regression Model

SVM function regression can be expressed as: Given a training sample set:
$T = \{x_i, y_i, i = 1, 2 \cdots, l\}$, $x_i \in R^N$ and $y_i \in R$, a regression function can be construct:

$$f(x) = W^T \Phi(x) + b \qquad (1)$$

on a feature space F. W is a vector in F, and $\Phi(x)$ maps the input x to a vector in F. W and b can be obtained by solving the following optimization problem:

$$\min_{W,b} T = \frac{1}{2} W^T W + C \sum_{i=1}^{l} \rho_i (\xi_i + \xi_i^*), \quad \text{s.t.} \quad \begin{aligned} y_i - (W^T \Phi(x) + b) &\leq \varepsilon + \xi_i \\ (W^T \Phi(x) + b) - y_i &\leq \varepsilon + \xi_i^* \\ \xi_i, \xi_i^* &\geq 0, i = 1, 2, \cdots, l \end{aligned} \quad (2)$$

where ρ_i is weight vector to sample x_i. According to Lagrange Multipliers and Karush-Kuhn-Tucker conditions, training samples can be separated into three sets:
The Error Support Vectors set E: $E = \{i \mid |\theta_i| = \rho_i C, |h(x_i)| \geq \varepsilon\}$;
The Margin Support Vectors set S: $S = \{i \mid 0 < |\theta_i| < \rho_i C, |h(x_i)| = \varepsilon\}$; (3)
The Remaining Support Vectors set R: $R = \{i \mid |\theta_i| = 0, |h(x_i)| \leq \varepsilon\}$

Definitions of coefficient difference θ_i and margin function $h(x_i)$ see reference [5]. The border of set S and E is variable. It varies with the new sample added into training set. In sector 3 a simple method to update ρ_i is introduced. The main idea of the algorithm is that: when a sample x_c is added to sample set, gradually change θ_c and $h(x_c)$ until x_c enter into one of three sets, and during the process some other samples are updated from one of set S,R,E to another because they are influenced by x_c. Formulae used in the algorithm see reference [5].

3 Method of Removing a Redundant Sample

For online SVM algorithm, the number of training sample increases with time, so we must remove redundant samples. Standard online SVM algorithm removes a redundant sample straightly after updating it into set R. Because some samples currently in set R may change into set S in future, it will bring error to SVM model. A new method is proposed here. Assume *num_sp* is max number of samples we wish to keep in training set, $C(x_i) = \rho_i C$, i is sample index according to the order they are added into training set and *cur_i* is the index of the sample added into training set currently. Then $C(x_i)$ can be computed as:

$$C(x_i) = \begin{cases} \dfrac{num\_sp - (cur\_i - i)}{num\_sp} C & if (cur\_i - i) < num\_sp \\ 0 & else \end{cases} \quad (4)$$

Whenever a new sample is added, *cur_i* changes. So $C(x_i)$ will change accordingly. It can change some samples from set S into E if their θ_i s are larger than or equal to $C(x_i)$. Decremental algorithm processes this case. When $C(x_i)$ of sample x_i

changes into 0, the value of θ_i will be zero constantly. This means that sample x_i will not influence the SVM model, so it can be removed from training set safely.

4 Online Training Algorithm

Weighted On-line SVM training algorithm includes two sub-algorithm: incremental and decremental algorithm. The detailed incremental algorithm is as following:

1) Set $\theta_c = 0$; If $|h(x_c)| \leq \varepsilon$, assign x_c to set R, terminate.
2) Increase or decrease θ_c according to the sign of $-h(x_c)$, update b, $\theta_i, i \in S$, and $h(x_i)$, $i \in E \cup R \cup c$, until x_c enters into set S or E:

If $h(x_c) = \varepsilon$, then add x_c into set S, terminate;
If $\theta_c = C$ or $-C$, then add x_c into set E, terminate;
If some sample changes from one of set R, S, E into another, update matrix R.
3) Repeat step 2).

The detailed decremental algorithm is as following:
1) If $x_c \in E$, remove it out of E; If $x_c \in S$, remove it out of S, update matrix R;
2) Increase or decrease θ_c according to the sign of $h(x_c)$, update $b, \theta_i, i \in S$, and $h(x_i)$, $i \in E \cup R \cup c$, until x_c enters into set E:

If $\theta_c = C(x_i)$, change x_c into set E, terminate;
If some sample changes from one of set R,S,E into another, update matrix R;
3) Repeat step 2).

The whole process of online training algorithm is as below:
1) Construct a pair of new data $\{x_c, y_c\}$, use incremental algorithm to add it into training set, and ecompute $C(x_i)$ of all samples according to (4).

2) For samples x_i, if $|\theta_i| \geq C(x_i)$, use decremental algorithm to change it into set E, if $C(x_i) = 0$, remove it out of training set.

3) Update model (1), construct a new data $\{x_c, y_c\}$, and go to step 1).

5 Simulation

A benchmark nonlinear system [6] is described as:

$$y(k+1) = \frac{(1+0.001*k)y(k)}{1+y^2(k)} + u^3(k) \tag{5}$$

Algorithm parameters are: $C = 20$, $\varepsilon = 0.0001$. Kernel function is RBF one. The algorithm runs 200 time step. Weighted vector is computed by (4). Sumulation results are shown in figure 1. We can see that on-line SVM regression algorithm can quickly approximate the system model with high precision. Also when system model varies, on-line SVM regression can modify SVM model rapidly. For 200 training samples, average error between system output y and predicting output y_m is:

$$D = \frac{1}{200} \sum_{i=1}^{200} |y(i) - y_m(i))| = 0.003\ 8.\tag{6}$$

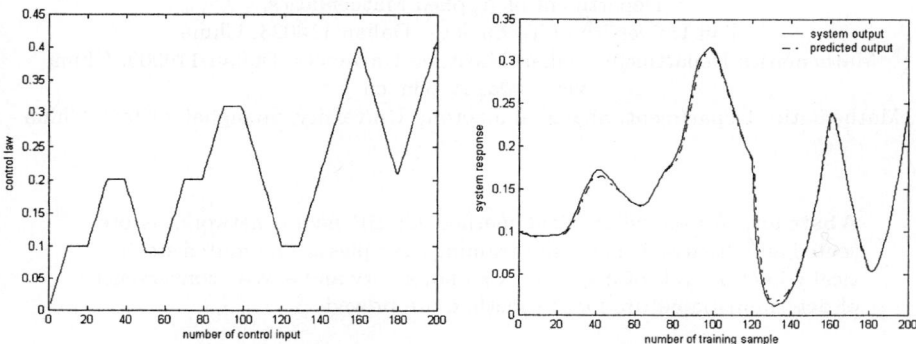

Fig. 1. Input (left) and output response (right) of weighted online algorithm

6 Conclusion

Weighted SVM regression algorithm improves modeling precision with every sample given a weight coefficient. The new algorithm removes any sample with a zero value of weight coefficient, which will construct a more smooth system model. The next work is to analyze generalization ability of the SVM model, and to research the convergence and the rate of convergence of the algorithm.

References

1. Vapnik, V.N.: The nature of statistical learning theory. Berlin: Springer (1995)
2. Smola, A.J., Schölkopf, B.: A tutorial on support vector regression. Neurocolt technical report. Royal Holloway College, Unisversity of London (1998)
3. Müller, K.R., Smola, A.J.: Predicting time series with support vector machines. Proceedings of ICANN'97, Springer LNCS 1327 (1997) 999-1004
4. Cauwenberghs, G., Poggio, T.: Incremental and decremental support vector Machine. Advances in Neural Information Processing Systems, Vol.13 (2001) 409-415
5. Junshui Ma, James Theiler, and Simon Perkins: Accurate On-line Support Vector Regression. Neural Computation, Vol. 15, Issue 11 (2003) 2683-2704
6. Narendra, K.S., Parthasarathy. K. : Identification and control of dynamic systems using neural networks. IEEE Transactions on Neural Networks, I (2000) 4-27.

Convergence of an Online Gradient Method for BP Neural Networks with Stochastic Inputs*

Zhengxue Li[1], Wei Wu[1,2,**], Guorui Feng[3], and Huifang Lu[1]

[1] Department of Applied Mathematics,
Dalian University of Technology, Dalian 116023, China
[2] Mathematics Department, Dalian Maritime University, Dalian 116000, China
wuweiw@dlut.edu.cn
[3] Mathematics Department, Shanghai Jiaotong University, Shanghai 200000, China

Abstract. An online gradient method for BP neural networks is presented and discussed. The input training examples are permuted stochastically in each cycle of iteration. A monotonicity and a weak convergence of deterministic nature for the method are proved.

1 Introduction

BP neural networks have wide applications (see e.g. [6], [9]). Our aim in this note is to investigate the convergence of an online gradient method for the training of the weights. We are concerned with a three layers BP neural network with structure $p - n - 1$. The neural network is supplied with a given set of training examples $\{\xi^j,\ O^j\}_{j=1}^{J} \subset \mathbb{R}^p \times \mathbb{R}$. Assume that $g : \mathbb{R} \to \mathbb{R}$ is a given activation function of the hidden layer, and that $f : \mathbb{R} \to \mathbb{R}$ is a given activation function of the output layer. We denote by $V = (v_{ij})_{n \times p}$ the weight matrix connecting the input and the hidden layers, and write

$$v_i = (v_{i1}, v_{i2}, \cdots, v_{ip})^T, \quad 1 \le i \le n.$$

The weight vector connecting the hidden and the output layers is denoted by

$$w = (w_1, w_2, \cdots, w_n)^T.$$

We also define

$$G(x) = (g(x_1), g(x_2), \cdots, g(x_n))^T, \quad \text{for } x \in \mathbb{R}^n,$$

and

$$\psi_j = G(V\xi^j), \quad j = 1, \cdots, J, \tag{1.1a}$$

---

* Partly supported by the National Natural Science Foundation of China, and the Basic Research Program of the Committee of Science, Technology and Industry of National Defense of China.
** Corresponding author.

$$\psi_j^m = G(V^m \xi^j), \quad j = 1, \cdots, J, \quad m = 0, 1, \cdots. \tag{1.1b}$$

For an input training vector ξ^j, the output of the hidden neurons is ψ_j, and the final output of the network is

$$\zeta^j = f(w \cdot \psi_j), \quad j = 1, \cdots, J. \tag{1.2}$$

Let $E_j(t)$ ($t \in \mathbb{R}$) denote an error function which is a certain kind of measurement of the error $O^j - \zeta^j$. A simple and popular choice is

$$E_j(w \cdot \psi_j) = \frac{1}{2} \left(O^j - \zeta^j \right)^2 = \frac{1}{2} \left[O^j - f(w \cdot \psi_j) \right]^2. \tag{1.3}$$

The total error function usually is specified as

$$E(W) = \sum_{j=1}^{J} E_j(w \cdot \psi_j) = \sum_{j=1}^{J} E_j\left(\sum_{i=1}^{n} w_i g(v_i \cdot \xi^j)\right), \tag{1.4}$$

where the weight matrix $W = (w, V)$.

The purpose of network learning is to obtain W^* such that

$$E(W^*) = \min E(W). \tag{1.5}$$

A often used method solving this kind of problem is the gradient method. The engineering community often prefers using the so-called *online gradient method* (OGM). For simplicity of analysis, we can choose the training examples in a *fixed* order (OGM-F). We can also choose $\{\xi^j\}$ in a *special stochastic* order (OGM-SS) as follows: For each batch $m = 0, 1, \cdots$, let $\{\xi^{m1}, \xi^{m2}, \cdots, \xi^{mJ}\}$ be a stochastic permutation of the set $\{\xi^1, \xi^2, \cdots, \xi^J\}$. Starting from any initial guess W^0, we proceed to refine it iteratively by the formula

$$w^{mJ+j} = w^{mJ+j-1} + \triangle_j^m w^{mJ+j-1}, \tag{1.6a}$$

$$v_i^{mJ+j} = v_i^{mJ+j-1} + \triangle_j^m v_i^{mJ+j-1}, \tag{1.6b}$$

where

$$\triangle_j^m w = -\eta_m E'_{mj}(w \cdot \psi_{mj}) \psi_{mj}, \tag{1.7a}$$

$$\triangle_j^m v_i = -\eta_m E'_{mj}(w \cdot \psi_{mj}) w_i g'(v_i \cdot \xi^{mj}) \xi^{mj}, \tag{1.7b}$$

$E_{mj}(t)$ is the error function of the input example ξ^{mj}, and the learning rate η_m changes its value after each cycle of training.

2 Preliminary Lemmas

We will need the following assumptions in our discussions:

(B1) $|g^{(k)}(t)| \leq C$, $k = 0, 1, 2$, $t \in \mathbb{R}$;

(B2) $E_j(t) \geq 0$, $|E_j^{(k)}(t)| \leq C$, $k = 1, 2$, $1 \leq j \leq J$, $t \in \mathbb{R}$;

(B3) $\|w^k\| \leq C$, $k = 1, 2, \cdots$, where $\|\cdot\|$ is the Euclidean norm on $\mathbb{R}^n$;

(B4) The learning rate $\{\eta_m\}$ is given by

$$\frac{1}{\eta_m} = \frac{1}{\eta_{m-1}} + \beta, \quad m = 1, 2, \cdots, \tag{2.1}$$

where the initial learning rate η_0 and the parameter β are positive constants to be specified later in (3.19) and (3.20) respectively.

For convenient notation, we introduce

$$w_d^m = w^{(m+1)J} - w^{mJ}, \tag{2.2}$$

$$v_{id}^m = v_i^{(m+1)J} - v_i^{mJ}, \quad i = 1, 2, \cdots, n, \tag{2.3}$$

$$\psi_{jd}^m = \psi_j^{(m+1)J} - \psi_j^{mJ}, \quad j = 1, 2, \cdots, J. \tag{2.4}$$

The proofs of Lemma 1-4 below are omitted because they can be proved easily or be found in [5].

Lemma 1. *Let $\{\eta_m\}$ ($m \geq 1$) be given in (2.1), then there holds*

$$\eta_m = \frac{\eta_0}{1 + m\beta\eta_0}; \tag{2.5}$$

$$\frac{\tau}{m} \leq \eta_m < \frac{\lambda}{m}, \quad \tau = \frac{\eta_0}{1 + \beta\eta_0} > 0, \quad \lambda = \frac{1}{\beta} > 0; \tag{2.6}$$

$$0 < \eta_m < \eta_{m-1} \leq 1; \tag{2.7}$$

$$\frac{\eta_{m+1}}{\eta_m} > \frac{1}{2}. \tag{2.8}$$

Remark 1. When (3.20) below holds, we can simply choose τ in (2.6) as $\tau = \eta_0/2$.

Lemma 2. *Assume that $a_n > 0$, that $\sum_{n=1}^{\infty} \frac{a_n^2}{n}$ converges, and that there exists a positive constant μ such that*

$$|a_{n+1} - a_n| < \frac{\mu}{n},$$

then

$$\lim_{n \to \infty} a_n = 0.$$

Lemma 3. *Assume that $a_n \geq 0$, $b_n > 0$ and $\lim_{n \to \infty} \frac{a_n}{b_n} = s$, then for all positive integer n, there exists a constant $C > 0$ such that*

$$a_n < C b_n.$$

Lemma 4. *Assume that Conditions (B1) is valid, then there holds*

$$\left\| \psi_j^m \right\| \leq C; \tag{2.9}$$

$$\left\| \psi_{jd}^m \right\| \leq C \sum_{i=1}^n \left\| v_{id}^m \right\|. \tag{2.10}$$

3 Several Theorems on Error Functions

For $k = 1, 2$, let us write

$$\left\| \sum_{j=1}^J \Delta_j^m w^{mJ} \right\|^k = \sigma_{k,1}^m, \qquad \sum_{i=1}^n \left\| \sum_{j=1}^J \Delta_j^m v_i^{mJ} \right\|^k = \sigma_{k,2}^m,$$

$$\sum_{j=1}^J \left\| \Delta_j^m w^{mJ} \right\|^k = \sigma_{k,3}^m, \qquad \sum_{i=1}^n \sum_{j=1}^J \left\| \Delta_j^m v_i^{mJ} \right\|^k = \sigma_{k,4}^m.$$

The proofs to the following two theorems are straightforward and similar to the corresponding results in [3], and thus are omitted.

Theorem 1. *Assume that Conditions (B1)-(B3) are valid, then there exists a positive constant γ independent of m and η_m such that*

$$E\left(W^{(m+1)J}\right) \leq E\left(W^{mJ}\right) - \frac{1}{\eta_m}\left(\sigma_{2,1}^m + \sigma_{2,2}^m\right) + \gamma\left(\sigma_{2,3}^m + \sigma_{2,4}^m\right). \tag{3.1}$$

From (3.1) we know that for any integer $m \geq 0$, if

$$\frac{1}{\eta_m}\left(\sigma_{2,1}^m + \sigma_{2,2}^m\right) \geq \gamma\left(\sigma_{2,3}^m + \sigma_{2,4}^m\right), \tag{3.2}$$

then

$$E\left(W^{(m+1)J}\right) \leq E\left(W^{mJ}\right). \tag{3.3}$$

Generally, the left-hand-side of (3.2) is greater than zero for $m = 0$. In this case, we can always choose a small enough constant $\delta \in (0, 1)$ and require the initial learning rate η_0 to satisfy

$$0 < \eta_0 < \delta \tag{3.4}$$

such that (3.2) holds. In the sequel, we want to prove (3.2) by an induction on m. After doing that, the important estimate (3.3) will follow from (3.1).

Theorem 2. *Assume that Conditions (B1)-(B3) are valid. If (3.2) holds for an integer $m \geq 0$, then there exists a constant $\gamma_1 > 0$ such that*

$$\max \left\{ \|w_d^m\|^2, \left(\sum_{i=1}^n \|v_{id}^m\|\right)^2 \right\} \leq \gamma_1 \left(\sigma_{2,1}^m + \sigma_{2,2}^m\right). \tag{3.5}$$

Theorem 3. *Assume that Conditions (B1)-(B4) are valid. If (3.2) holds for an integer $m \geq 0$, then we have*

$$\frac{1}{\eta_{m+1}} \left(\sigma_{2,1}^{m+1} + \sigma_{2,2}^{m+1}\right) \geq \gamma \left(\sigma_{2,3}^{m+1} + \sigma_{2,4}^{m+1}\right). \tag{3.6}$$

Proof. Let $\xi^{(m+1)j} = \xi^{mi_j}$ $(1 \leq j \leq J)$, where $\{i_1, i_2, \cdots, i_J\}$ is a stochastic permutation of the subscript set $\{1, 2, \cdots, J\}$. According to (1.7a) and the mean value theorem, we get

$$\triangle_j^{m+1} w^{(m+1)J} = \frac{\eta_{m+1}}{\eta_m} \triangle_{i_j}^m w^{mJ} - \eta_{m+1} E_{mi_j}'' \left(t^{mi_j}\right) \left(w_d^m \cdot \psi_{mi_j}^{(m+1)J}\right) \psi_{mi_j}^{mJ}$$
$$- \eta_{m+1} E_{mi_j}'' \left(t^{mi_j}\right) \left(w^{mJ} \cdot \psi_{mi_jd}^m\right) \psi_{mi_j}^{mJ}$$
$$- \eta_{m+1} E_{mi_j}' \left(w^{(m+1)J} \cdot \psi_{mi_j}^{(m+1)J}\right) \psi_{mi_jd}^m, \tag{3.7}$$

where t^{mi_j} is a value between $w^{mJ} \cdot \psi_{mi_j}^{mJ}$ and $w^{(m+1)J} \cdot \psi_{mi_j}^{(m+1)J}$.

From (3.7), (B2), (B3), (2.9), (2.10) and (2.7), we derive that

$$\left\|\triangle_j^{m+1} w^{(m+1)J}\right\| \leq \frac{\eta_{m+1}}{\eta_m} \left\|\triangle_{i_j}^m w^{mJ}\right\| + C\eta_m \|w_d^m\| + C\eta_m \sum_{i=1}^n \|v_{id}^m\|. \tag{3.8}$$

This together with (2.7) gives

$$\left\|\triangle_j^{m+1} w^{(m+1)J}\right\|^2$$
$$\leq \left(\frac{\eta_{m+1}^2}{\eta_m^2} + C_1\eta_m\right) \left\|\triangle_{i_j}^m w^{mJ}\right\|^2 + C_1\eta_m \left[\|w_d^m\|^2 + \left(\sum_{i=1}^n \|v_{id}^m\|\right)^2\right], \tag{3.9}$$

where $C_1 = \max\{2C, C(1+2C)\}$.

We conclude from (3.9) and (3.5) that

$$\sigma_{2,3}^{m+1} \leq \left(\frac{\eta_{m+1}^2}{\eta_m^2} + C_2\eta_m\right) \sigma_{2,3}^m + C_2\eta_m \sigma_{2,4}^m, \tag{3.10}$$

where γ_1 is the constant that appears in Theorem 2, and $C_2 = (1 + 2\gamma_1 J^2)C_1$.

Combining (3.7), (B2), (B3), (2.9), (2.10) with (2.7) gives

$$\sigma_{1,1}^{m+1} \geq \frac{\eta_{m+1}}{\eta_m}\sigma_{1,1}^m - C_3\eta_m\left(\|w_d^m\| + \sum_{i=1}^n \|v_{id}^m\|\right). \quad (3.11)$$

Note that for any nonnegative numbers x, y and z, if $x \geq y - z$, then

$$x^2 \geq y^2 - 2yz. \quad (3.12)$$

Using (3.11), (3.12) and (3.5), we get

$$\sigma_{2,1}^{m+1} \geq \left(\frac{\eta_{m+1}^2}{\eta_m^2} - C_4\eta_{m+1}\right)\sigma_{2,1}^m - C_4\eta_{m+1}\sigma_{2,2}^m, \quad (3.13)$$

where $C_4 = 2(1+\gamma_1)C_3$.

Similarly as (3.10) and (3.13), we have

$$\sigma_{2,4}^{m+1} \leq \left(\frac{\eta_{m+1}^2}{\eta_m^2} + C_5\eta_m\right)\sigma_{2,4}^m + C_5\eta_m\sigma_{2,3}^m, \quad (3.14)$$

$$\sigma_{2,2}^{m+1} \geq \left(\frac{\eta_{m+1}^2}{\eta_m^2} - C_6\eta_{m+1}\right)\sigma_{2,2}^m - C_6\eta_{m+1}\sigma_{2,1}^m. \quad (3.15)$$

By (3.13) and (3.15)

$$\frac{1}{\eta_{m+1}}\left(\sigma_{2,1}^{m+1} + \sigma_{2,2}^{m+1}\right) \geq \frac{1}{\eta_{m+1}}\left(\frac{\eta_{m+1}^2}{\eta_m^2} - C_7\eta_{m+1}\right)\left(\sigma_{2,1}^m + \sigma_{2,2}^m\right), \quad (3.16)$$

where $C_7 = \max\{2C_4, 2C_6\}$.

It follows from (3.10), (3.14) and (3.2) that

$$\gamma\left(\sigma_{2,3}^{m+1} + \sigma_{2,4}^{m+1}\right) \leq \frac{1}{\eta_m}\left(\frac{\eta_{m+1}^2}{\eta_m^2} + C_8\eta_m\right)\left(\sigma_{2,1}^m + \sigma_{2,2}^m\right), \quad (3.17)$$

where $C_8 = \max\{2C_2, 2C_5\}$.

Write

$$\beta_0 = 4(C_7 + C_8), \quad (3.18)$$

and choose η_0 and β satisfying

$$0 < \eta_0 < \min\{\frac{1}{\beta_0}, \delta\}, \quad (3.19)$$

$$\beta_0 \leq \beta \leq \frac{1}{\eta_0}. \quad (3.20)$$

When $m = 0$, according to (2.5), (3.20), (3.19) and (3.18), we have

$$\beta\left(\frac{\eta_1}{\eta_0}\right)^2 = \frac{\beta}{(1+\beta\eta_0)^2} \geq \frac{\beta_0}{4} = C_7 + C_8.$$

When $m > 0$, by (2.8), (3.20) and (3.18), we obtain

$$\beta \left(\frac{\eta_{m+1}}{\eta_m}\right)^2 \geq C_7 + C_8. \tag{3.21}$$

So, (3.21) is valid for any nonnegative integer m.

By (2.1), we see that inequality (3.21) is equivalent to

$$\frac{1}{\eta_{m+1}} \left(\frac{\eta_{m+1}^2}{\eta_m^2} - C_7 \eta_{m+1}\right) \geq \frac{1}{\eta_m} \left(\frac{\eta_{m+1}^2}{\eta_m^2} + C_8 \eta_m\right). \tag{3.22}$$

Thus, from (3.16), (3.17) and (3.22) we can easily conclude (3.6). This completes the proof. □

Theorem 4. *Assume that Conditions (B1)-(B4) are valid, then*

$$\sum_{m=1}^{\infty} \frac{1}{m} \left\|E_w\left(W^{mJ}\right)\right\|^2 < \infty, \tag{3.23}$$

$$\sum_{m=1}^{\infty} \frac{1}{m} \left\|E_{v_i}\left(W^{mJ}\right)\right\|^2 < \infty, \quad i = 1, 2, \cdots, n. \tag{3.24}$$

Proof. Using an induction argument based on Theorem 3, we get

$$\frac{1}{\eta_m}\left(\sigma_{2,1}^m + \sigma_{2,2}^m\right) - \gamma\left(\sigma_{2,3}^m + \sigma_{2,4}^m\right) \geq 0, \quad m = 1, 2, \cdots. \tag{3.25}$$

For any positive integer M, summing (3.25) over $m = 1, 2, \ldots, M$, and combining (3.1) with Condition (B2), we have

$$\sum_{m=1}^{M} \left[\frac{1}{\eta_m}\left(\sigma_{2,1}^m + \sigma_{2,2}^m\right) - \gamma\left(\sigma_{2,3}^m + \sigma_{2,4}^m\right)\right] \leq E\left(W^J\right). \tag{3.26}$$

Let $M \to \infty$, then

$$\sum_{m=1}^{\infty} \left[\frac{1}{\eta_m}\left(\sigma_{2,1}^m + \sigma_{2,2}^m\right) - \gamma\left(\sigma_{2,3}^m + \sigma_{2,4}^m\right)\right] < \infty. \tag{3.27}$$

According to (1.7a), (1.7b), (B1)-(B3), (2.9) and (2.6),

$$\sum_{m=1}^{\infty} \gamma\left(\sigma_{2,3}^m + \sigma_{2,4}^m\right) < C \sum_{m=1}^{\infty} \eta_m^2 < C_1 \sum_{m=1}^{\infty} \frac{1}{m^2} < \infty.$$

So

$$\sum_{m=1}^{\infty} \frac{1}{\eta_m} \sigma_{2,k}^m < \infty, \quad k = 1, 2. \tag{3.28}$$

According to (1.4), (1.7a) and (1.7b), we obtain

$$E_w(W) = \sum_{j=1}^{J} E'_{mj}(w \cdot \psi_{mj})\psi_{mj} = -\frac{1}{\eta_m}\sum_{j=1}^{J} \triangle_j^m w, \qquad (3.29)$$

$$E_{v_i}(W) = \sum_{j=1}^{J} E'_{mj}(w \cdot \psi_{mj})w_i g'(v_i \cdot \xi^{mj})\xi^{mj} = -\frac{1}{\eta_m}\sum_{j=1}^{J} \triangle_j^m v_i. \qquad (3.30)$$

From (2.6),

$$\frac{1}{m\eta_m} \leq \frac{1}{\tau}. \qquad (3.31)$$

Using (3.29), (3.30), (3.31) and (3.28), we have that for $i = 1, 2, \cdots, n$

$$\sum_{m=1}^{\infty} \frac{1}{m}\left\|E_w\left(W^{mJ}\right)\right\|^2 = \sum_{m=1}^{\infty} \frac{1}{m\eta_m}\left(\frac{1}{\eta_m}\sigma_{2,1}^m\right) \leq \frac{1}{\tau}\sum_{m=1}^{\infty} \frac{1}{\eta_m}\sigma_{2,1}^m < \infty,$$

$$\sum_{m=1}^{\infty} \frac{1}{m}\left\|E_{v_i}\left(W^{mJ}\right)\right\|^2 \leq \sum_{m=1}^{\infty} \frac{1}{m\eta_m}\left(\frac{1}{\eta_m}\sigma_{2,2}^m\right) \leq \frac{1}{\tau}\sum_{m=1}^{\infty} \frac{1}{\eta_m}\sigma_{2,2}^m < \infty.$$

This completes the proof. □

4 The Main Results for OGM-SS

In this section, we present our two main results for OGM-SS.

Theorem 5. *Assume that Conditions (B1)-(B4) are valid, then*

$$E\left(W^{(m+1)J}\right) \leq E\left(W^{mJ}\right), \quad m = 1, 2, \cdots. \qquad (4.1)$$

Proof. Inequality (4.1) is a direct consequence of (3.1) and (3.25).   □

Theorem 6. *Assume that Conditions (B1)-(B4) are valid, then*

$$\lim_{m\to\infty}\left\|E_w\left(W^{mJ+j}\right)\right\| = 0, \quad j = 1, 2, \cdots, J, \qquad (4.2)$$

$$\lim_{m\to\infty}\left\|E_{v_i}\left(W^{mJ+j}\right)\right\| = 0, \quad j = 1, 2, \cdots, J;\ i = 1, 2, \cdots, n. \qquad (4.3)$$

Proof. By (3.29), the expression of the Hessian matrixes $E_{ww}(W)$ and $E_{wv_i}(W)$ are

$$E_{ww}(W) = \sum_{j=1}^{J} E''_{mj}(w \cdot \psi_{mj})\psi_{mj}\left(\psi_{mj}\right)^T, \qquad (4.4)$$

$$E_{wv_i}(W) = \sum_{j=1}^{J} \left[E''_{mj}(w \cdot \psi_{mj})\psi_{mj}\left(w_i g'(v_i \cdot \xi^{mj})\xi^{mj}\right)^T\right.$$

$$+ E'_{mj}(w \cdot \psi_{mj}) \left(0, \cdots, g'(v_i \cdot \xi^{mj}), \cdots, 0\right)^T \left(\xi^{mj}\right)^T\Big]. \quad (4.5)$$

From (4.4), (4.5), (B1)-(B3) and (2.9), we obtain

$$\|E_{ww}(W)\| < C, \quad (4.6)$$

$$\|E_{wv_i}(W)\| < C. \quad (4.7)$$

Using (2.2), (1.6a), (1.7a) and (2.3), (1.6b), (1.7b), we have

$$w_d^m = -\eta_m \sum_{j=1}^J E'_{mj}(w^{mJ+j-1} \cdot \psi_{mj}^{mJ+j-1})\psi_{mj}^{mJ+j-1}, \quad (4.8)$$

$$v_{id}^m = -\eta_m \sum_{j=1}^J E'_{mj}(w^{mJ+j-1} \cdot \psi_{mj}^{mJ+j-1})w_i^{mJ+j-1} g'(v_i^{mJ+j-1} \cdot \xi^{mj})\xi^{mj}. \quad (4.9)$$

Thus, we conclude from (4.8), (4.9), (B1)-(B3), (2.9) and (2.6) that

$$\|w_d^m\| < \frac{C}{m}, \quad (4.10)$$

$$\|v_{id}^m\| < \frac{C}{m}. \quad (4.11)$$

Note that $E_{ww}(W)$ and $E_{wv_i}(W)$ are actually the Fréchet derivatives of the nonlinear mapping $E_w : \mathbb{R}^{n \times (p+1)} \to \mathbb{R}^n$. So by Lemma 3, (4.10) and (4.11), we derive

$$\left\| E_w(W^{(m+1)J}) - E_w(W^{mJ}) - E_{ww}(W^{mJ})w_d^m - \sum_{i=1}^n E_{wv_i}(W^{mJ})v_{id}^m \right\|$$

$$= o\left(\sqrt{\|w_d^m\|^2 + \sum_{i=1}^n \|v_{id}^m\|^2}\right) < C\sqrt{\|w_d^m\|^2 + \sum_{i=1}^n \|v_{id}^m\|^2} < \frac{C_1}{m}. \quad (4.12)$$

Therefore, we have by (4.12), (4.6), (4.7), (4.10) and (4.11) that

$$\left|\left\|E_w(W^{(m+1)J})\right\| - \left\|E_w(W^{mJ})\right\|\right| \leq \left\|E_w(W^{(m+1)J}) - E_w(W^{mJ})\right\|$$

$$\leq \left\|E_w(W^{(m+1)J}) - E_w(W^{mJ}) - E_{ww}(W^{mJ})w_d^m - \sum_{i=1}^n E_{wv_i}(W^{mJ})v_{id}^m\right\|$$

$$+ \left\|E_{ww}(W^{mJ})w_d^m\right\| + \sum_{i=1}^n \left\|E_{wv_i}(W^{mJ})v_{id}^m\right\| < \frac{C_2}{m}. \quad (4.13)$$

According to Lemma 2, Theorem 4 and (4.13), we get

$$\lim_{m \to \infty} \left\|E_w(W^{mJ})\right\| = 0. \quad (4.14)$$

Similarly as (4.13), we have

$$\left|\|E_w(W^{mJ+j})\| - \|E_w(W^{mJ})\|\right| < \frac{C_3}{m}, \quad j = 1, 2, \cdots, J. \quad (4.15)$$

But

$$\|E_w(W^{mJ+j})\| < \|E_w(W^{mJ})\| + \frac{C_3}{m}, \quad (4.16)$$

so

$$\lim_{m \to \infty} \|E_w(W^{mJ+j})\| = 0, \quad j = 1, 2, \cdots, J.$$

We can similarly prove (4.3). This completes the proof. □

References

1. Wu, W., Xu, Y.S.: Deterministic Convergence of an Online Gradient Method for Neural Networks. Journal of Computational and Applied Mathematics. **144** (2002) 335–347
2. Wu, W., Feng, G.R., Li, X.: Training Multilayer Perceptrons via Minimization of Sum of Ridge Functions. Advances in Computational Mathematics. **17** (2002) 331–347
3. Wu, W., Feng, G.R., Li, Z.X., Xu, Y.S.: Convergence of an Online Gradient Method for BP Neural Networks. Accepted by IEEE Trans. Neural Networks
4. Li, Z.X., Wu, W., Zhang H.W.: Convergence of Online Gradient Methods for Two-Layer Feedforward Neural Networks. Journal of Mathematical Research and Exposition. **21** (2001) 219–228
5. Li, Z.X., Wu, W., Tian, Y.L.: Convergence of an Online Gradient Methods for Feedforward Neural Networks with Stochastic Inputs. Journal of Computational and Applied Mathematics. **163** (2004) 165–176
6. Liang, Y.C., et.al.: Proper Orthogonal Decomposition and its Application - Part II: Model Reduction for MEMS Dynamical Analysis. Journal of Sound and Vibration. **256** (2000) 515–532
7. Liang, Y.C., et.al.: Successive Approximation Training Algorithm for Feedforward Neural Networks. Neurocomputing. **42** (2002) 311–322
8. Xu, B.Z., Zhang, B.L, Wei, G.: Theory and Application of Neural Networks. South China University of Technology Press. Guangzhou. 1994
9. Looney, C.G.: Pattern Recognition Using Neural Networks. Oxford University Press. New York. 1997
10. Fine, T.L., Mukherjee, S.: Parameter Convergence and Learning Curves for Neural Networks. Neural Computation. **11** (1999) 747–769
11. Finnoff, W.: Diffusion Approximations for the Constant Learning Rate Backpropagation Algorithm and Resistance to Local Minima. Neural Computation. **6** (1994) 285–295
12. Gori, M., Maggini, M.: Optimal Convergence of Online Backpropagation. IEEE Trans. Neural Networks. **7** (1996) 251–254

A Constructive Algorithm for Wavelet Neural Networks

Jinhua Xu[1] and Daniel W.C. Ho[2]

[1] Institute of System Science, East China Normal University
jhxu@cs.ecnu.edu.cn
[2] Department of Mathematics, City University of Hong Kong
madaniel@cityu.edu.hk

Abstract. In this paper, a new constructive algorithm for wavelet neural networks (WNN) is proposed. Employing the time-frequency localization property of wavelet, the wavelet network is constructed from the low resolution to the high resolution. At each resolution, a new wavelet is initialized as a member of wavelet frames. The input weight freezing technique is used and the Levenberg-Marquardt (LM) algorithm, a quasi-Newton method, is used to train the new wavelet in the WNN. After training, the new wavelet will be added to the wavelet network if the reduction of the residual error between the desired output and WNN output is greater than a threshold. The proposed algorithm is suitable to situations when the wavelet library is very large. The simulations demonstrate the effectiveness of the proposed approach.

1 Introduction

Wavelet transforms have emerged as a means of representing a function in a manner which readily reveals properties of the function in localized regions of the joint time-frequency space. The idea of using wavelets in neural networks has been proposed in [2,3,4,12,13,15,18,22,23,24,25]. The application of wavelet bases and wavelet frames are usually limited to problem of small dimension. The main reason is that they are composed of regularly dilated and translated wavelets. The number of wavelets in a truncated basis and frame drastically increases with the dimension, therefore, constructing and storing wavelet bases or frames of large dimension are of prohibitive cost [25].

Some research have been done on reducing the size of the WNN to handle large dimensional problem. In [3], magnitude based method is used to eliminate of wavelets with small coefficients. In [8], the residual based selection (RBS) algorithm is used for the synthesis of wavelet networks. In [20], an approach for on-line synthesis of wavelet network using recursive least square training is proposed. A new wavelet will be added to the network when the training error becomes stable. The whole network is trained after each new wavelet is added. The optimal number of wavelets is determined by a Bayesian Information Criteria(BIC)[16]. In [25], wavelet network is constructed by some selected wavelets from a wavelet basis (or wavelet frame) by exploring the sparseness of training data and using techniques in regression analysis, so that problems of large dimension can be handled. The orthogonal least square (OLS) algorithm is used to select the wavelets. The computational cost of these algorithms is expensive when there are a number of wavelets in the wavelet library. In [21], an orthogonalized residual based selection (ORBS) algorithm is proposed for wavelet neural networks.

For feedforward neural networks, many research has been done on constructive algorithms, which start with a minimal network and dynamically construct the final network [9]. These algorithms include the dynamic node creation [1], the cascade correlation algorithm [6], projection pursuit regression [7], resource-allocating network [14] and the self-organizing neural network [19].

Wavelet neural network is a special class of feedforward neural network. The most useful property of wavelet transform is the time-frequency localization. It is shown in [5] that the denumerable family

$$\mathcal{W}(\alpha, \beta) := \{\alpha^{\frac{k}{2}} \psi(\alpha^k x - \beta l) : k, l \in \mathbb{Z}\}$$

constitutes a frame of $L^2(\mathbb{R})$ for suitable choices of the parameters (α, β), where α and β are the translation and dilation parameters. The translation simply shifts the wavelet function on the time domain. The dilation means the resolution, that is, the wavelet with larger (smaller) dilation can represent the higher (lower) frequency component of the signal.

In this paper, a new constructive algorithm is proposed for the wavelet neural networks. Employing the time-frequency localization property of wavelet, the wavelet network is constructed from the low resolution to the high resolution. The dilation range can be determined according to the "band-width" of the function and the approximation accuracy. At each resolution, a wavelet with regularly initialized translation parameters is trained until the local minimum is reached. The input weight freezing technique is used and the Levenberg-Marquardt (LM) algorithm is used to train the new wavelet in the WNN. It will be added to the wavelet network if the reduction of the residual error between the desired output and WNN output is greater than a threshold. The proposed algorithm is suitable to situations when the wavelet library is very large.

2 Preliminaries

In [24], wavelet network is first introduced as a class of feedforward networks composed of wavelets for approximating arbitrary nonlinear functions. The wavelet network structure proposed in [24] is shown as follows:

$$g(x) = \sum_{i=1}^{m_l} w_i \psi(D_i R_i (x + t_i)) + g_0, \tag{1}$$

where $x \in \mathbb{R}^n$, the translation parameter $t_i \in \mathbb{R}^n, R_i \in \mathbb{R}^{n \times n}, i = 1, \cdots m_l$, $D_i = diag\{d_i\}, d_i \in \mathbb{R}^n_+, g_0 \in \mathbb{R}$, the wavelet function $\psi : \mathbb{R}^n \to \mathbb{R}$ is multidimensional, which can be chosen as tensor product of one dimensional wavelet or radial wavelet.

In [3,13,25], wavelet network is constructed by some selected wavelets from a wavelet basis (or wavelet frame). The structure is shown as follows:

$$f(x) = \sum_j \sum_k c_{j,k} \psi_{j,k}(x) = \sum_{j=-J_1}^{J_1} \sum_{k_1=-K_1}^{K_1} \cdots \sum_{k_n=-K_n}^{K_n} c_{j,k_1,\cdots,k_n} \psi_{j,k}(x), \quad (2)$$

where $x \in \mathbb{R}^n$, $k = [k_1, \cdots, k_n] \in \mathbb{Z}^n$.

In [23], wavelet network is constructed by orthogonal scaling function according to the theory of multiresolution analysis. The wavelet network structure is shown as follows:

$$f(x) = \sum_k c_k \varphi_{M,k}(x) = \sum_{k_1=-K_1}^{K_1} \cdots \sum_{k_n=-K_n}^{K_n} c_{k_1,\cdots,k_n} \varphi_{M,k}(x) \quad (3)$$

where $x \in \mathbb{R}^n$, $k = [k_1, \cdots, k_n] \in \mathbb{Z}^n$.

In [18,22], multiresolution neural networks composed of the scaling and wavelet functions are constructed based on the multiresolution analysis theory of orthonormal wavelets. The structure is shown as follows:

$$\begin{aligned} f(x) &= \sum_k c_{m,k} \varphi_{m,k}(x) + \sum_j \sum_k d_{j,k} \psi_{j,k}(x) \\ &= \sum_k c_{m,k} \varphi_{m,k}(x) + \sum_{j \geq m} \sum_{k_1} \cdots \sum_{k_n} d_{j,k_1,\cdots,k_n} \psi_{j,k}(x) \end{aligned} \quad (4)$$

where $x \in \mathbb{R}^n$, $k = [k_1, \cdots, k_n] \in \mathbb{Z}^n$.

The WNNs in (2), (3) and (4) have the linear-in-parameters structures. In practice, it is impossible to count infinite frame or basis terms in (2), (3) and (4). However, arbitrary truncations may lead to large errors. In this paper, a constructive algorithm will be proposed for these linear-in-parameter WNN, which can be described in an united form as follows:

$$\hat{f}(x) = \sum_{i=1}^{M} w_i \psi_i(x) \quad (5)$$

or in a compact form

$$\hat{f} = \Phi w \quad (6)$$

where $\Phi = [\psi_1, \cdots, \psi_M]$, ψ_i is the ith wavelet and/or scaling function, $w = [w_1, \cdots, w_M]^T$. M is the number of selected wavelet basis or frame.

In this paper, we assume that the WNN has only one linear output unit. Extension to multiple output units is straightforward. For simplicity, wavelet basis will be used hereafter for wavelet frame(non-orthogonal wavelet) or wavelet basis (orthogonal wavelet).

3 Orthogonalized Residual Based Constructive (ORBC) Algorithm

For large dimensional problem, the number of the basis in the wavelet library may be very large. The heavy computational cost may make the basis selection algorithms in [8,21,25] not feasible in practice. In this section, a new constructive algorithm is proposed for WNN, which starts with no wavelet in the WNN and adds new wavelets with adjustable translation and fixed dilation. The process stops until a satisfactory solution is found.

Now we describe the problems involved in the constructive algorithm.

A. Initialization:
In constructive algorithms for feedforward neural networks, there is no guideline on how to initialize the new neuron, which is therefore randomly initialized. In our constructive algorithm for the WNN, the dilation and translation parameters of the new wavelet can be initialized as the member of a frame. The dilation range can be determined according to the "band-width" of the function and the approximation accuracy.

B. Computation complexity:
There are two training ways for constructive algorithms. A simple-minded approach is to train the whole network completely after the addition of each hidden unit [17,20]. The exact computational requirement depends on the particular nonlinear optimization algorithm used, but most algorithms will require heavy computation when the number of the weights is large. The computational requirement may not be a major concern at the early stage when the network size is small, however, the network will eventually grow to such a size that complete retraining will have serious scale-up problem, especially when more efficient methods like Newton's method are to be used. Another approach is to train only the new hidden units[10,11]. The weights feeding into the hidden units already existing in the network are kept fixed (input weight freezing), and only the weights connected to the new hidden units and the output units will be trained. The number of weights to be optimized, and the time and space requirements for each iteration, can thus be greatly reduced. This input weight freezing technique is also used in our constructive algorithm.

C. Training algorithm:
Instead of using standard backpropagation training algorithm, which is known to have poor convergence rate, Levenberg-Marquardt (LM) algorithm, a quasi-Newton method, is used to train the new wavelet in WNN. The LM algorithm converges much faster than BP algorithm. A potential drawback of the quasi-Newton method is that it requires the heavy computational cost and high memory requirement. This limits the application of quasi-Newton methods to small or medium size problems[17]. However, since we only need to train the translation parameters of the new added neuron with fixed dilation parameter, memory and computation intensive the problems of quasi-Newton methods are not so serious when applied to a construction algorithm for WNN. If

tensor product is used to the multidimensional wavelet, that is, for $x \in \mathbb{R}^n$, $\psi(x) := \psi(x_1, x_2, \cdots, x_n) = \prod_{i=1}^{n} \psi(x_i)$, then for n dimensional problem, there are only n translation parameters to be trained. Note also that with the constructive approach, not one, but a sequence of optimization problem instances need to be solved. It is therefore imperative that a method with fast convergence rate be used. In view of this, the Levenberg-Marquardt (LM) algorithm is chosen for our wavelet network construction algorithm.

A WNN with $i - 1$ wavelets implements the function given by

$$\hat{y}_{i-1}(x) = \sum_{j=1}^{i-1} w_j \psi_j(x) \tag{7}$$

where $\psi_j(x)$ represents the function implemented by the jth wavelet. Moreover, $r_{i-1}(x) = y(x) - \hat{y}_{i-1}(x)$ is the residual error function for the current network with $i - 1$ wavelets. Addition of a new wavelet proceeds in two steps:
1) Input training: Find w_i and ψ_i such that the resultant linear combination of ψ_i with the current network, i.e., $\hat{y}_i = \hat{y}_{i-1} + w_i \psi_i$, gives minimum residual error.
2) Output training: Keeping $\psi_1, \psi_2, \cdots, \psi_i$ fixed, adjust the values of $w_1, w_2, \cdots, w_i$ so as to minimize the residual error. The output training is used to ensure that the residual $r_i = y - \hat{y}_i$ remains orthogonal to the subspace spanned by $\psi_1, \psi_2, \cdots, \psi_i$. This minimization can be performed by computing the pseudo-inverse or orthogonal parameter estimation method described later in this section.

Suppose the dilation is in the range from d^0 to d^1, where d^0 and d^1 are integers which represent the minimum and maximum resolution level respectively. For simplicity, assume that the dilations of all dimension for each wavelet are equal, that is, $d_{i1} = \cdots = d_{in} = d_i$. At first stage, no wavelet is added to the network. Let the dilation of the new wavelet $d = d^0$, and the translation parameters $\Theta = [\theta_1, \theta_2, \cdots, \theta_n]^T$ are regularly initialized as integer numbers. The LM algorithm is then used to train the translation parameter to reduce the residual. When the minimum is reached, check the reduction of the error function. If the reduction of the residual error between the desired output and WNN output is greater than a threshold, then add the wavelet to the network, otherwise, reinitialize and retrain a new wavelet at the current resolution. If no more wavelets can be added to the network at the current resolution, increase the dilation by one. If a wavelet is added to the network successfully, it is orthogonalized to the previously selected wavelet to calculate the optimal weight. Then keep the dilation unchanged and reinitialize and retrain a new wavelet. Since the new wavelet is trained to minimize the residual, and the output training is used to ensure that the residual remains orthogonal to the subspace spanned by the existing wavelets, the proposed method is called orthogonalized residual based constructive (ORBC) algorithm, which can be summarized as follows:
Given N pairs of training sample, $\{(x(1), y(1)\}, \{x(2), y(2)\}, \cdots, \{x(N), y(N)\}$. Set the desired output $y = [y(1), y(2), \cdots, y(N)]^T$.

Step 1: Set the output of the WNN $\hat{y}_0 = 0$, the residual $r_0 = y$; the dilation $d = d^0$(minimum resolution); Set the number of wavelet $i = 1$.

Step 2: Input training. The translation parameters $\Theta = [\theta_1, \theta_2, \cdots, \theta_n]^T$ are regularly initialized as integer numbers, where n is the dimension of the wavelet, which equals to the input dimension of the WNN. Set

$$\phi = [\psi(1), \psi(2), \cdots, \psi(N)]^T$$

where $\psi(t) = \prod_{j=1}^{n} \psi(2^d x_j(t) + \theta_j)$. Let

$$\hat{y}_i = \hat{y}_{i-1} + w\phi$$
$$r_i = y - \hat{y}_i = r_{i-1} - w\phi$$

with $w = (\phi^T \phi)^{-1} \phi^T r_{i-1}$.

The best Θ may be selected to minimize the cost function

$$V_i(\Theta) = r_i^T r_i = (r_{i-1} - w\phi)^T (r_{i-1} - w\phi)$$
$$= r_{i-1}^T r_{i-1} - (\phi^T \phi)^{-1} (\phi^T r_{i-1})^2 \qquad (8)$$

The LM algorithm is used to solve the optimization problem in (8) to find the Θ^* where the minimum is reached.

Step 3: Output training. If $V_{i-1} - V_i(\Theta^*) < \lambda V_{i-1}$, the wavelet is rejected, go to step 4; otherwise, Θ is accepted and set $T_i := [t_{i1}, \cdots, t_{in}]^T = \Theta^*$, $d_i := d$ and $\phi_i := [\psi_i(1), \cdots, \psi_i(N)]$, where $\psi_i(t) = \prod_{j=1}^{n} \psi(2^{d_i} x_j(t) + t_{ij})$. Then ϕ_i is normalized as $v_i = \phi_i / \sqrt{\phi_i^T \phi_i}$. Suppose $i - 1$ wavelets have been obtained and orthonormalized as $q_1, q_2, \cdots q_{i-1}$. The new obtained v_i is orthogonalized to the previous wavelet as follows:

$$p_i = v_i - ((v_i^T q_1)q_1 + \cdots + (v_i^T q_{i-1})q_{i-1}) \qquad (9)$$

$$q_i = p_i / \sqrt{p_i^T p_i} \qquad (10)$$

$$\bar{w}_i = q_i^T y \qquad (11)$$

$$\alpha_{ii} = \sqrt{p_i^T p_i} \qquad (12)$$

$$\alpha_{ki} = v_i^T q_k \qquad (13)$$

and set

$$\hat{y}_i = \hat{y}_{i-1} + \bar{w}_i q_i$$
$$r_i = r_{i-1} - \bar{w}_i q_i$$

If $r_i^T r_i < \epsilon$, the approximation accuracy is reached, go to step 6; Otherwise, set $i := i + 1$.

Step 4: If there is no more wavelet can be added at the current resolution, go to step 5; otherwise, go to step 2 to reinitialize and retrain a new wavelet at the current resolution.

Step 5: Change the dilation parameter $d := d+1$, if $d < d^1$(maximum resolution), go to step 2; otherwise, go to step 6.

Step 6: Set $M := i$(the number of wavelets). Compute

$$[w_1, \cdots, w_M]^T = A^{-1}[\bar{w}_1, \cdots, \bar{w}_M]^T \qquad (14)$$

where A is an upper triangular matrix defined as follows:

$$A = \begin{bmatrix} \alpha_{11} & \alpha_{12} & \cdots & \alpha_{1,M} \\ 0 & \alpha_{22} & \cdots & \alpha_{2,M} \\ 0 & 0 & \ddots & \vdots \\ \vdots & \cdots & \cdots & \alpha_{M-1,M} \\ 0 & \cdots & 0 & \alpha_{MM} \end{bmatrix}, \qquad (15)$$

and its components α_{ij} are given in (12)-(13). Calculate the output of the WNN $\hat{y}_M$ as

$$\hat{y}_M = \sum_{i=1}^{M} w_i v_i . \qquad (16)$$

4 Numerical Examples

Example 1: System identification
The plant to be identified is governed by the difference equation

$$y(i+1) = f(y(i), u(i)) = 1.5y(i)/(1 + y^2(i)) + 0.3\cos(y(i)) + 1.2u(i) \qquad (17)$$

The WNN is used to identify the system. The input and output of the WNN are $\{y(i), u(i)\}$ and $\hat{y}(i+1)$ respectively, that is

$$\hat{y}(i+1) = \hat{f}(y(i), u(i)) \qquad (18)$$

Three different wavelet basis selection algorithms are first used to construct the WNN. Also $N = 150$ random training patterns with $y(i) \in [-2, 2]$ and $u(i) \in [-1.5, 1.5]$ are used to train and construct the WNN. The wavelet function is also taken as the "Gaussian-derivative". The number of candidate wavelet in the wavelet library $\mathcal{W}$ is 611.

The basis selection procedures are stopped when the mean square error (mse) of the training data is less than $\epsilon = 1.0e - 3$. The number of basis selected by OLS, RBS and ORBS are 14, 24, 21 respectively; If $\epsilon = 1.0e - 4$, the number of basis selected by OLS, RBS and ORBS are 26, 160, 45 respectively, as shown in Table 1.

Next, orthogonal residual based constructive algorithm is used to construct the WNN to identify the system. The dilation range is $[-3, 4]$. There are 21 and 31 wavelets required to construct the WNN for approximation accuracy $\epsilon = 1.0e - 3$ and $\epsilon = 1.0e - 4$ respectively.

After the WNNs are constructed, the input u in (19) is used to test the identification performance of WNN. The mse for checking data are also shown

Table 1. Comparison of the different constructive algorithms for Example 1

Algorithms	ϵ	# of wavelets	MSE of training data	MSE of checking data
OLS	1.0e-3	14	9.8892e-4	2.2608e-3
RBS	1.0e-3	24	9.2959e-4	7.0420e-4
ORBS	1.0e-3	21	9.9685e-4	6.0685e-4
ORBC	1.0e-3	21	6.5134e-4	2.9840e-4
OLS	1.0e-4	26	8.6799e-5	7.3171e-4
RBS	1.0e-4	160	1.7136e-4	9.0593e-4
ORBS	1.0e-4	45	9.2801e-5	7.6161e-4
ORBC	1.0e-4	31	9.6805e-5	2.0160e-4

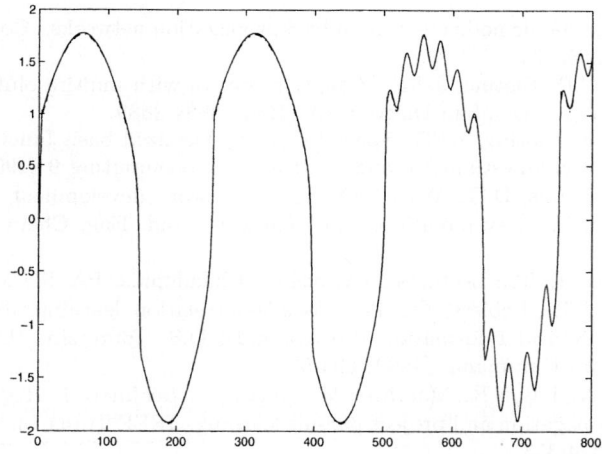

Fig. 1. The simulation result of ORBC algorithm for Example 1

in Table 1. It is obvious that the generalization performance of ORBC is better than other algorithms. The outputs of the plant (solid line) as well as the WNN (dashed line) are shown in Figure 1.

$$u(i) = \begin{cases} \sin(2\pi i/250); & \text{for } i \leq 500 \\ 0.8\sin(2\pi i/250) + 0.2\sin(2\pi i/25); & \text{for } 500 < i \leq 800 \end{cases} \quad (19)$$

Remark: Notice that in this two dimensional example, there are 611 regularly dilated and translated wavelets in the wavelet library. From the basis selection results, it can be seen that many of them are redundant and only a small part of them are required. For large dimensional problems, the RBS, OLS and ORBS algorithms are impractical to be implemented due to the huge number of candidate wavelets in the wavelet library. ORBC is suitable to large dimensional problems since the input dimension has little effect on the new wavelet training algorithm.

5 Conclusion

In this paper, a constructive algorithm (ORBC) is proposed for wavelet neural networks. A wavelet with regularly initialized translation parameters is trained and added to the wavelet network to reduce the residual. Input weight freezing is applied during the training process, and quasi-Newton algorithm is used to train the translation parameter(s) of the new wavelet node. Since the input dimension has little effect on the new wavelet training algorithm, the proposed ORBC algorithm is suitable to large dimensional problems.

References

1. Ash, T.: Dynamic node creation in backpropagation networks. Connection Science **1** (1989) 365-375
2. Bernard, C.P., Slotine, J.J.E.: Adaptive control with multiresolution bases. Proceedings 36th CDC, San Diego, USA (1997) 3884-3889
3. Cannon, M., Slotine, J.J.E.: Space-frequency localized basis function networks for nonlinear system estimation and control. Neurocomputing **9** (1995) 293-342
4. Chen, J., Bruns, D.D.: WaveARX Neural network development for system identification using a systematic design synthesis. Ind. Eng. Chem. Res. **34** (1995) 4420-4435
5. Daubechies, I.: Ten Lectures on Wavelets. Philadephia, PA: SIAM Press, (1992)
6. Fahlman, S.E., Lebiere, C.: The cascade-correlation learning architecture. Advances in Neural Information Processing(2), D.S. Touretzky, (Ed.), San Mateo, CA: Morgan Kaufmann, (1990) 524-532
7. Hwang, J.N., Lay, S.R., Maechler, M., Martin, D., Schimert,J.: Regression modeling in backpropagation and project pursuit learning. IEEE Trans. on Neural Networks **5** (1994) 324-353
8. Kan, K.-C., Wong, K.-W.: Self-construction algorithm for synthesis of wavelet networks. Electronic Letters **34** (1998) 1953-1955
9. Kwok, T.-Y., Yeung, D.-Y.: Constructive algorithms for structuer learning in feedforward neural networks for regression problems. IEEE Trans. on Neural Networks **8** (1997) 630-645
10. Kwok, T.-Y., Yeung, D.-Y.: Objective functions for training new hidden units in constructive neural networks. IEEE Trans. on Neural Networks **8** (1997) 1131-1148
11. Li, C.J., Kim, T.: A new feedforward neural network structural learning algorithm:Augmentation by training with residuals. Journal of Dynamic Systems, Measurement, and Control **117** (1995) 411-415
12. Mukherjee, S., Nayar, S.K.: Automatic generation of RBF networks using wavelets. Pattern Recognition **29** (1996) 1369-1383
13. Pati, Y.C., Krishnaprasad, P.S.: Analysis and synthesis of feedforward neural networks using discrete affine wavelet transformation. IEEE Trans. on Neural Networks **4** (1993) 73-85
14. Platt, J.: A resource-allocating network for function interpolation. Neural Computation **3** (1991) 213-225
15. Sanner, R.M., Slotine, J.J.E.: Structurally dynamic wavelet networks for adaptive control of robotic systems. Int. J. Control **70** (1998) 405-421
16. Schwartz, G: Estimating the dimension of a model. Ann. Statist. **6** (1978) 461-464

17. Setiono, R., Hui, L.C.K.: Use of a Quasi-Newton method in a feedforward neural network construction algorithm. IEEE Trans. on Neural Networks **6** (1995) 273-277
18. Sureshbabu, A. and Farrell, J.A.: Wavelet-based system identification for nonlinear control. IEEE Trans. on Automatic Control **44** (1999) 412-417
19. Tenorio, M.F., Lee, W.T.: Self-organizing network for optimum supervised learning. IEEE Trans. on Neural Networks **1** (1990) 100-110
20. Wong, K.-W., Leung, A. C.-S.: On-line successive synthesis of wavelet networks. Neural Processing Letters **7** (1998) 91-100
21. Xu, J., Ho, D.W.C.: A basis selection algorithm for wavelet neural networks. Neurocomputing **48** (2002) 681-689
22. Yang, Z.J., Sagara, S., Tsuji, T.: System impulse response identification using a mutiresolution neural network. Automatica **33** (1997) 1345-1350
23. Zhang, J., Walter, G.G., Lee, W.N.W.: Wavelet neural networks for function learning. IEEE Trans. on Signal Processing **43** (1995) 1485-1497
24. Zhang, Q., Benveniste, A.: Wavelet networks. IEEE Trans. on Neural Networks **3** (1992) 889-898
25. Zhang, Q.: Using wavelet wetwork in nonparametric estimation. IEEE Trans. on Neural Networks **8** (1997) 227-236

Stochastic High-Order Hopfield Neural Networks

Yi Shen[1], Guoying Zhao[1], Minghui Jiang[1], and Shigeng Hu[2]

[1] Department of Control Science and Engineering,
Huazhong University of Science and Technology,
Wuhan, Hubei, 430074, China
[2] Department of Mathematics,
Huazhong University of Science and Technology,
Wuhan, Hubei, 430074, China

Abstract. In 1984 Hopfield showed that the time evolution of a symmetric Hopfield neural networks are a motion in state space that seeks out minima in the energy function (i.e., equilibrium point set of Hopfield neural networks). Because high-order Hopfield neural networks have more extensive applications than Hopfield neural networks, and have been discussed on the convergence of the networks. In practice, a neural network is often subject to environmental noise. It is therefore useful and interesting to find out whether the high-order neural network system still approacher some limit set under stochastic perturbation. In this paper, we will give a number of useful bounds for the noise intensity under which the stochastic high-order neural network will approach its limit set. Our result cancels the requirement of symmetry of the connection weight matrix and includes the classic result on Hopfield neural networks, which is a special case of stochastic high-order Hopfield neural networks. In the end, A example is given to verify the effective of our results.

1 Introduction

Much of the current interest in artificial networks stems not only from their richness as a theoretical model of collective dynamics but also from the promise they have shown as a practical tool for performing parallel computation [1]. Theoretical understanding of neural networks dynamics has advanced greatly in the past fifteen years [1-11]. The neural networks proposed by Hopfield can be described by an ordinary differential equation of the form

$$C_i \dot{x}_i(t) = -\frac{1}{R_i} x_i(t) + \sum_{j=1}^{n} T_{ij} g_j(x_j(t)) + I_i, i = 1, 2, \cdots, n. \tag{1}$$

on $t \geq 0$. The variable $x_i(t)$ represents the voltage on the input of the ith neuron, and I_i is the external input current to the ith neuron. Each neuron is characterized by an input capacitance C_i and a transfer function g_j. The connection matrix element T_{ij} has a value $+1/R_{ij}$ when the non-inverting output of the jth neuron is connected to the input of the ith neuron through a resistance R_{ij}, and a value $-1/R_{ij}$ when the inverting output of the jth neuron is connected to the input of the ith neuron through a resistance R_{ij}. The parallel resistance at the input of each neuron is defined by $R_i = (\sum_{j=1}^{n}|T_{ij}|)^{-1}$. The nonlinear transfer function g_i is sigmoidal function. By defining

$$b_i = \frac{1}{C_i R_i}, \quad a_{ij} = \frac{T_{ij}}{C_i}, \quad c_i = \frac{I_i}{C_i},$$

network (1) can be rewritten as

$$\dot{x}_i(t) = -b_i x_i(t) + \sum_{j=1}^{n} a_{ij} g_j(x_j(t)) + c_i, \quad i = 1, 2, \cdots, n. \tag{2}$$

We mentioned above that the nonlinear transfer function $g_i(u)$ is sigmoidal, saturation at ± 1 with maximum slope at $u = 0$. To be more precise, let us state the properties of g_i below:

(1) $g_i(u)$ is strictly increasing, $-1 < g_i(u) < 1$ and $g_i(0) = 0$;
(2) $\dot{g}_i(u)$ increases on $u < 0$, reaches its maximum $\beta_i := \dot{g}_i(0)$ at $u = 0$ and then decreases on $u > 0$;
(3) $\ddot{g}_i(u)$ is bounded, $\ddot{g}_i(u) > 0$ for $u < 0$, $\ddot{g}_i(0) = 0$ and $\ddot{g}_i(u) < 0$ for $u > 0$, ;
(4) $g_i(u)$ approaches its asymptotes ± 1 very slowly such that

$$\int_0^{\pm\infty} u \dot{g}_i(u) du = \infty. \tag{3}$$

Let us now define a C^2-function $U : R^n \to R$ by

$$U(x) = \sum_{i=1}^{n} b_i \int_0^{x_i} u \dot{g}_i(u) du - \frac{1}{2} \sum_{i,j=1}^{n} a_{ij} g_i(x_i) g_j(x_j) - \sum_{i=1}^{n} c_i g_i(x_i). \tag{4}$$

Let $x(t)$ be a solution to the symmetric network (2) (i.e., $a_{ij} = a_{ji}, 1 \leq i, j \leq n$). It is easy to get

$$\dot{U}(x(t)) = -\sum_{i=1}^{n} \dot{g}_i(x_i(t))(-b_i x_i(t) + \sum_{j=1}^{n} a_{ij} g_j(x_j(t)) + c_i)^2. \tag{5}$$

Recalling the fact that $\dot{g}_i(x_i) > 0$, we see that

$$\dot{U}(x(t)) < 0, \tag{6}$$

unless $\dot{U}(x(t)) = 0$ iff $-b_i x_i(t) + \sum_{j=1}^{n} a_{ij} g_j(x_j(t)) + c_i = 0$ for all $1 \leq i \leq n$. Owing to the nonpositive property of $\dot{U}(x(t))$, the solution to symmetric network (2) will approach the set M_0 (where M_0 is called the equilibrium point set of network (2)).

$$M_0 := \{x \in R^n : -b_i x_i + \sum_{j=1}^{n} a_{ij} g_j(x_j) + c_i = 0, 1 \leq i \leq n\}. \tag{7}$$

This is the main result in Ref.[1].

If we consider second-order nonlinear interaction term, the network (2) becomes

$$\dot{x}_i(t) = -b_i x_i(t) + \sum_{j=1}^{n} a_{ij} g_j(x_j(t)) + \sum_{j=1}^{n}\sum_{k=1}^{n} d_{ijk} g_j(x_j(t)) g_k(x_k(t)) + c_i,$$
$$i = 1, 2, \cdots, n. \qquad (8)$$

where d_{ijk} are the second-order connection weights, the definition of $g_i(\cdot)$ is seen in the afore assumption. Network (8) is called the high-order Hopfield network which have more extensive applications than the Hopfield network [2-7]. In [12], we have proved that the asymmetric networks (8) converge to the equilibrium set M_1.

$$M_1 := \{x \in R^n : -b_i x_i + \sum_{j=1}^{n} a_{ij} g_j(x_j) +$$
$$\sum_{j=1}^{n}\sum_{k=1}^{n} d_{ijk} g_j(x_j) g_k(x_k) + c_i = 0, 1 \le i \le n\}. \qquad (9)$$

However, network is often subject to environmental noise, then the stochastically perturbed neural network is described by a stochastic differential equation

$$dx_i(t) = [-b_i x_i(t) + \sum_{j=1}^{n} a_{ij} g_j(x_j(t)) + \sum_{j=1}^{n}\sum_{k=1}^{n} d_{ijk} g_j(x_j(t)) g_k(x_k(t))$$
$$+ c_i]dt + \sum_{l=1}^{m} \sigma_{il}(x(t)) dw_l(t), \quad i = 1, 2, \cdots, n. \qquad (10)$$

Here $w(t) = (w_1(t), ..., w_m(t))^T$ is an m-dimensional Brownian motion defined on a complete probability space $(\Omega, F, \{F_t\}_{t\ge 0}, P)$ with a natural filtration $\{F_t\}_{t\ge 0}$ (i.e. $F_t = \sigma\{w(s) : 0 \le s \le t\}$), and $\sigma : R^n \longrightarrow R^{n\times m}$, i.e. $\sigma(x) = (\sigma_{ij}(x))_{n\times m}$ which is called the noise intensity matrix. The question is: does the solution of the network (10) under stochastic perturbation still approach M_1 or a different limit set? The main aim of this paper is to give a positive answer. We will give several bounds for the noise intensity matrix under which the solution of the stochastic network will approach a limit set which is in general different from M_1. Throughout this paper we always assume that $\sigma(x)$ is locally Lipschitz continuous and satisfies the linear growth condition. It is therefore known (Mao [13]) that given any initial value $x_0 \in R^n$, network (10) has a unique global solution on $t \ge 0$ and we denote the solution by $x(t; x_0)$.

For convenience, we always assume that the stochastic high-order networks (10) satisfy the following condition (H).

(H) there exists a positive diagonal matrix $P = \mathrm{diag}(p_1, \cdots, p_n)$ such that

$$p_i a_{ij} = p_j a_{ji}, p_i d_{ijk} = p_j d_{jik} = p_k d_{kji}, c_i \ge 0, \quad i, j, k = 1, 2, \cdots, n,$$

In the sequel, set

$$F_i(x) = -b_i x_i + \sum_{j=1}^n a_{ij} g_j(x_j) + \sum_{j,k=1}^n d_{ijk} g_j(x_j) g_k(x_k) + c_i. \tag{11}$$

Meanwhile, define energy function by

$$V(x) = \sum_{i=1}^n p_i b_i \int_0^{x_i} u \dot{g}_i(u) du - \frac{1}{2} \sum_{i,j=1}^n p_i a_{ij} g_i(x_i) g_j(x_j)$$

$$- \frac{1}{3} \sum_{i,j,k=1}^n p_i d_{ijk} g_i(x_i) g_j(x_j) g_k(x_k) - \sum_{i=1}^n p_i c_i g_i(x_i). \tag{12}$$

2 Stochastic High-Order Neural Networks

The diffusion operator L associated with the network (10) is defined by

$$L = \sum_{i=1}^n F_i(x) \frac{\partial}{\partial x_i} + \frac{1}{2} \sum_{i,j=1}^n (\sigma \sigma^T(x))_{ij} \frac{\partial^2}{\partial x_i \partial x_j},$$

where $(\sigma\sigma^T(x))_{ij} = \sum_{k=1}^n \sigma_{ik}\sigma_{jk}$, $F_i(x)$ is defined as (11).
For the C^2-function V defined by (12) and (H), we compute

$$\frac{\partial V(x)}{\partial x_i} = -p_i F_i(x) \dot{g}_i(x_i),$$

$$\frac{\partial^2 V(x)}{\partial x_i^2} = -p_i F_i(x) \ddot{g}_i(x_i), + p_i[b_i - a_{ii}\dot{g}_i(x_i) - \sum_{k=1}^n d_{iik} \dot{g}_i(x_i) g_k(x_k)$$

$$- \sum_{k=1}^n d_{iki}\dot{g}_i(x_i)g_k(x_k)]\dot{g}_i(x_i),$$

and if $i \neq j$, then

$$\frac{\partial^2 V(x)}{\partial x_i \partial x_j} = -p_i[a_{ij}\dot{g}_j(x_j) + \sum_{k=1}^n d_{ijk}\dot{g}_j(x_j)g_k(x_k) + \sum_{k=1}^n d_{ikj}\dot{g}_j(x_j)g_k(x_k)]\dot{g}_i(x_i).$$

Therefore,

$$LV(x) = -\sum_{i=1}^n p_i \dot{g}_i(x_i) F_i^2(x) - \frac{1}{2} \sum_{i,j=1}^n (\sigma\sigma^T(x))_{ij} p_i[a_{ij}\dot{g}_j(x_j)$$

$$+ \sum_{k=1}^n d_{ijk}\dot{g}_j(x_j)g_k(x_k) + \sum_{k=1}^n d_{ikj}\dot{g}_j(x_j)g_k(x_k)]\dot{g}_i(x_i)$$

$$+ \frac{1}{2} \sum_{i=1}^n (\sigma\sigma^T(x))_{ii} \{-p_i F_i(x)\ddot{g}_i(x_i) + p_i\dot{g}_i(x_i)b_i\}. \tag{13}$$

In the case when there is no stochastic perturbation, i.e. $\sigma = 0$, we have pointed out in section 1 and Ref [12] that $LV \leq 0$ and the solution to the network will approach the set $M_1 = \{x \in R^n : LV(x) = 0\}$. The question is: does the stochastic perturbation change this property? It does, of course, for some type of stochastic perturbation, but it may still preserve the property for a certain class of stochastic perturbation. For example, recalling the property that $x_i \ddot{g}_i(x_i) \leq 0$, $x_i \in R$, $1 \leq i \leq n$. and the boundedness of $\ddot{g}_i, \dot{g}_i$ and g_i, we observe that the sum of the second and third terms on the right-hand side of (13) is bounded by $h|\sigma(x)|^2$ for some constant $h > 0$. Hence

$$LV(x) \leq -\sum_{i=1}^{n} p_i \dot{g}_i(x_i) F_i^2(x) + h|\sigma(x)|^2.$$

If $\sigma(x)$ is sufficiently small, for instance

$$|\sigma(x)|^2 \leq \frac{1}{h} \sum_{i=1}^{n} p_i \dot{g}_i(x_i) F_i^2(x).$$

we should have $LV(x) \leq 0$. In this case, does the solution to the network still approach the set $\{x \in R^n : LV(x) = 0\}$? The following theorem describes the situation.

Theorem 2.1. Assume that the condition (H) is satisfied, if $LV(x) \leq 0$ for all $x \in R^n$, and define

$$M = \{x \in R^n : LV(x) = 0 \quad and \quad H_l(x) = 0, 1 \leq l \leq m\} \quad (14)$$

where

$$H_l(x) = -\sum_{i=1}^{n} p_i \sigma_{il}(x) \dot{g}_i(x_i) F_i(x) \quad (15)$$

then

(A1)

$$M \neq \emptyset. \quad (16)$$

(A2) Define $d(x; M) = min\{|x-y| : y \in M\}$, i.e., the distance between $x \in R^n$ and the set M. Then for any initial value $x_0 \in R^n$, the solution $x(t; x_0)$ of network (10) has the property that

$$\liminf_{t \to \infty} d(x(t; x_0); M) = 0, a.s. \quad (17)$$

that is, almost every sample path of the solution to the network will visit the neighborhood of M infinitely many times.

(A3) If for any $x \in M$, there is a neighborhood Γ_x of x in R^n such that
$$V(y) \neq V(x) \quad \text{for} \quad y \in \Gamma_x \quad y \neq x, \tag{18}$$
then for any initial value $x_0 \in R^n$, the solution $x(t;x_0)$ of network (10) has the property that
$$\lim_{t \to \infty} x(t;x_0) \in M, \quad a.s. \tag{19}$$
that is, almost every sample path of the solution to the network will converge to a point in M.

The proof of theorem 2.1 is omitted due to the restriction of the space. However, the following key lemma 2.2 is required in the proof of the theorem 2.1.

Lemma 2.2. If $LV(x) \leq 0$ holds, then for any initial value $x_0 \in R^n$, the solution of the network (10) has the properties that
$$-\mu \leq \lim_{t \to \infty} V(x(t;x_0)) < \infty, a.s. \tag{20}$$
and
$$\int_0^\infty [-LV(x(t;x_0)) + \sum_{l=1}^m H_l^2(x(t;x_0))]dt < \infty, a.s. \tag{21}$$
where $H_l^2(x)$ have been defined by (15) above and
$$\mu = \frac{1}{2}\sum_{i,j=1}^n p_i|a_{ij}| + \frac{1}{3}\sum_{i,j,k=1}^n p_i|d_{ijk}| + \sum_{i=1}^n p_i c_i.$$

The proof is omitted here.

3 Conditions for $LV \leq 0$

Theorem 2.1 shows that as long as $LV \leq 0$, the nonempty set M exists and the solutions of the neural network under stochastic perturbation will approach this set with probability 1 if the additional condition (18) is satisfied. It is therefore useful to know how large stochastic perturbation the neural network can tolerate without losing the property of $LV(x) \leq 0$. Although we pointed out in the previous section that there is some $h > 0$ such that
$$LV(x) \leq -\sum_{i=1}^n p_i \dot{g}_i(x_i) F_i^2(x) + h|\sigma(x)|^2.$$
we did not estimate the h. If we know more precisely about h, we can estimate the noise intensity, for instance,
$$|\sigma(x)|^2 \leq \frac{1}{h}\sum_{i=1}^n p_i \dot{g}_i(x_i) F_i^2(x).$$
to guarantee $LV \leq 0$.

In section 1 we have listed the properties of functions g_i. Let us now introduce

$$\gamma_i = \max\{|\ddot{g}_i(x_i)| : 0 \wedge (\frac{-\alpha_i + c_i}{b_i}) \leq x_i \leq \frac{\alpha_i + c_i}{b_i}\}, \quad 1 \leq i \leq n. \quad (22)$$

where $\alpha_i = \sum_{j=1}^{n} |a_{ij}| + \sum_{j,k=1}^{n} |d_{ijk}|$. The following lemma explains why γ_i are defined in the way above.

Lemma 3.1. We always have

$$-\ddot{g}_i(x_i)F_i(x) \leq (\alpha_i + c_i)\gamma_i, \quad 1 \leq i \leq n,$$

for all $x_i \in R$, here, $F_i(x)$ is defined as (11). The proof is omitted here.

Theorem 3.2. If

$$\frac{1}{2}|\sigma(x)|^2 \{\max_{1 \leq i \leq n}[p_i(\alpha_i + c_i)\gamma_i + p_i b_i \dot{g}_i(x_i) - \lambda_{\min}(PA)|\dot{g}_i(x_i)|^2] + dn^2\}$$

$$\leq \sum_{i=1}^{n} p_i \dot{g}_i(x_i) F_i^2(x),$$

where $d = \max_{i,j,k} p_i |d_{ijk} + d_{ikj}|\beta_i\beta_j, A = (a_{ij})_{n \times n}$, then $LV(x) \leq 0$. The proof is omitted here.

In the case when $\lambda_{\min}(PA) \geq 0$ we may use the following easier criterion for $LV(x) \leq 0$.

Corollary 3.3. If PA is a symmetric nonnegative-definite matrix and

$$|\sigma(x)|^2 \leq \frac{2}{h} \sum_{i=1}^{n} p_i \dot{g}_i(x_i) F_i^2(x).$$

holds for all $x \in R^n$, where

$$h = \max_{1 \leq i \leq n}[p_i(\alpha_i + c_i)\gamma_i + p_i b_i \beta_i] + dn^2. \quad (23)$$

here, $d = \max_{1 \leq i,j,k \leq n} p_i |d_{ijk} + d_{ikj}|\beta_i\beta_j$, then $LV(x) \leq 0$. (Recall that $\beta_i = \dot{g}_i(0)$ which was defined in section 1.)

In the case when $\lambda_{\min}(PA) < 0$ we may also have the following easier criterion for $LV(x) \leq 0$.

Corollary 3.4. If $\lambda_{\min}(PA) < 0$ and

$$|\sigma(x)|^2 \leq \frac{2}{\bar{h}} \sum_{i=1}^{n} p_i \dot{g}_i(x_i) F_i^2(x).$$

holds for all $x \in R^n$, where

$$\bar{h} = \max_{1 \leq i \leq n}[p_i(\alpha_i + c_i)\gamma_i + p_i b_i \beta_i + |\lambda_{\min}(PA)|\beta_i^2] + dn^2. \quad (24)$$

here, $d = \max_{1 \leq i,j,k \leq n} p_i |d_{ijk} + d_{ikj}|\beta_i\beta_j$, then $LV(x) \leq 0$.

4 Example

In this section, an examples will be given to show the validity of our results. we let the number of neurons be two in order to make the calculations relatively easier but the theory of this paper is illustrated clearly. In what follows we will also let $w(\cdot)$ be a one-dimensional Brownian motion.

Example1. Consider a two-dimensional stochastic neural network

$$\mathrm{d}x_i(t) = [-b_i x_i(t) + \sum_{j=1}^{2} a_{ij} g_j(x_j(t)) + \sum_{j=1}^{2}\sum_{k=1}^{2} d_{ijk} g_j(x_j(t)) g_k(x_k(t))$$

$$+c_i]\mathrm{d}t + \sum_{k=1}^{2} \sigma_i(x)\mathrm{d}w(t), \quad i = 1, 2. \tag{25}$$

where $b_1 = 3, b_2 = 4, a_{11} = a_{12} = 2, a_{21} = a_{22} = 1, c_1 = c_2 = 0, d_{111} = d_{122} = 1, d_{112} = d_{121} = -1, d_{211} = -\frac{1}{2}, d_{222} = 1, d_{212} = d_{221} = \frac{1}{2}, g_i(x_i) = \frac{2}{\pi}\arctan(x_i), i = 1, 2,$ and $\sigma(x)$ is locally Lipschitz continuous and bounded. Compute

$$\dot{g}_i(u) = \frac{2}{\pi(1+u^2)} \quad \text{and} \quad \ddot{g}_i(u) = -\frac{4u}{\pi(1+u^2)^2}, i = 1, 2.$$

Clearly,

$$\beta_i = \frac{2}{\pi} \quad \text{and} \quad \int_0^{\pm\infty} u\dot{g}_i(u) = \infty, i = 1, 2.$$

Moreover, by definition γ_i (see (22)), we have $\gamma_i = \frac{3\sqrt{3}}{4\pi}, i = 1, 2$. Since $|\ddot{g}_i(u)|$ reaches the maximum at $u = \pm\frac{1}{\sqrt{3}}$. Noting that PA is nonnegative-definite, and taking $p_1 = 1, p_2 = 2$, we may apply corollary 3.3 compute by (23)

$$h = \max_{1\le i \le 2}[p_i(\alpha_i + c_i)\gamma_i + b_i\beta_i p_i] + dn^2 = \frac{9\sqrt{3}+16}{\pi} + \frac{64}{\pi^2}.$$

Therefore, if

$$\sigma_1^2(x) + \sigma_2^2(x) \le \frac{2}{h}\sum_{i=1}^{2} p_i \dot{g}_i(x_i) F_i^2(x), \tag{26}$$

where $F_i(x)$ is defined as (11). then $LV(x) \le 0$. The right-hand side of (26) gives a bound for the noise intensity. As long as smaller than the bound, by theorem 2.1, there is a nonempty set M such path of the solution of network (25) will visit the neighborhood of infinitely many times. It is therefore easy to see that the set M defined (14) is contained by the following set: $K_0 = \{x \in R^2 : F_i(x) = 0, \ 1 \le i \le 2\}$.

It is not difficult to show that $K_0 = (0,0)^T$, i.e. K_0 contains only one point in R^2. Since M is nonempty and $M \subseteq K_0$, we must have $M = K_0 = (0,0)^T$. It is not difficult to show that $(0,0)^T$ is the unique minimum point of energy function $V(x)$ in this example (i.e., it is the unique equilibrium of the network (25)). We can therefore conclude by theorem 2.1 that all of the solutions of network (25) will tend to $(0,0)^T$ with probability 1 as long as (26) is satisfied. Note that this conclusion is independent of the form of the noise intensity matrix $\sigma(x)$ but only requires that the norm of $\sigma(x)$ be bounded by the right-hand side of (26). In other words, we obtain a robustness property of the neural network.

5 Concluding Remarks

To close our paper, let us have some further discussions on the way in which noise is introduced into the high-order Hopfield network. It is known that noise has been introduced into the high-order Hopfield network so that the network can avoid getting trapped into a local minima and hence the time evolution of the network is a motion in state space that seeks out its global minima in the system energy function. In such the high-order stochastic Hopfield network, the units are stochastic and the degree is determined by a temperature analogue parameter. The stochastic units are actually introduced to mimic the variable strength with which real neurons fire, delays in synapses and random fluctuations from the release of transmitters in discrete vesicles. By including stochastic units it becomes possible with a simulated annealing technique to try and avoid getting trapped into local minima. By making use of a mean-field approximation the Hopfield network again evolves into a deterministic version, and one can then instead apply mean-field annealing to try and avoid local minima. In the present paper, the introduced high-order Hopfield network is that with continuous-valued transfer functions, but with added terms corresponding to environmental noise. The noise here is not that which is added into the network on purpose to avoid local minima as mentioned above, but it is the environmental noise which the network cannot avoid. Our contribution here is to present some interesting results on the amount of noise that can be tolerated in the high-order Hopfield neural network while still preserving its limit set or experiencing at least another limit set.

Acknowledgments

The work was supported by Natural Science Foundation of Hubei (2004ABA055) and National Natural Science Foundation of China (60274007, 60074008).

References

1. Hopfield J J.: Neurons with Graded Response Have Collective Computational Properties Like those of Two-state Neurons. Porc.Natl Acad.Sci.USA, **81** (1984) 3088-3092
2. Dembo, A., Farotimi O., Kailath T.: High-Order Absolutely Stable Neural Networks. IEEE Trans. Circ. Syst. II, **38** (1991) 57-65

3. Kosmatopoulos E B, Polycarpou M M., Christodoulou M A., et al.: High-Order Neural Networks Structures for Identification of Dynamical Systems. IEEE Trans. Neural Networks , **6** (1995) 422-431
4. Zhang T., Ge S S., Hang C C.: Neural-Based Direct Adaptive Control for a Class of General Nonlinear Systems. International Journal of Systems Science, **28** (1997) 1011-1020
5. Su J., Hu A., He Z.: Solving a Kind of Nonlinear Programming Problems via Analog Neural Networks. Neurocomputing , **18** (1998) 1-9
6. Stringera S.M., Rollsa E.T., TrappenbergbT.P.: Self-organising Continuous Attractor Networks with Multiple Activity Packets, and the Representation of Space. Neural Networks, **17** (2004) 5-27
7. Xu B J., Liu X Z., Liao X X.: Global Asymptotic Stability of High-Order Hopfield Type Neural Networks with Time Delays. Computers and Mathematics with Applications, **45** (2003) 1729-1737
8. Sun C., Feng C. B. Exponential Periodicity of Continuous-Time and Discrete-Time Neural Networks with Delays. Neural Processing Letters, **19** (2004)131-146
9. Sun C., Feng C. B. On Robust Exponential Periodicity of Interval Neural Networks with Delays. Neural Processing Letters, **20** (2004) 53-61
10. Cao,J.: On Exponential Stability and Periodic Solution of CNN's with Delay. Physics Letters A, **267** (2000)312-318
11. Cao J., Wang J.: Global Asymptotic Stability of Recurrent Neural Networks with Lipschitz-continuous Activation Functions and Time-Varying Delays. IEEE Trans. Circuits Syst. I, **50** (2003) 34-44
12. Shen Y., Zong X J., Jiang M H.: High-Order Hopfield Neural Networks. Lecture Notes in Computer Science-ISSN2005, **3496** (2005) 235-240
13. Mao X.: Exponential Stability of Stochastic Differential Equations. New York: Marcel Dekker, 1994

Predicting with Confidence - An Improved Dynamic Cell Structure

Yan Liu[1], Bojan Cukic[1], Michael Jiang[2], and Zhiwei Xu[2]

[1] Lane Department of Computer Science and Electrical Engineering,
West Virginia University, Morgantown, WV 26506, USA
{yanliu, cukic}@csee.wvu.edu
[2] Motorola Labs, Motcrola Inc., Schaumburg, IL 60196, USA
{Michael.Jiang, Zhiwei.Xu}@motorola.com

Abstract. As a special type of Self-Organizing Maps, the Dynamic Cell Structures (DCS) network has topology-preserving adaptive learning capabilities that can, in theory, respond and learn to abstract from a much wider variety of complex data manifolds. However, the highly complex learning algorithm and nonlinearity behind the dynamic learning pattern pose serious challenge to validating the prediction performance of DCS and impede its spread in control applications, safety-critical systems in particular.

In this paper, we improve the performance of DCS networks by providing confidence measures on DCS predictions. We present the validity index, an estimated confidence interval associated with each DCS output, as a reliability-like measure of the network's prediction performance. Our experiments using artificial data and a case study on a flight control application demonstrate an effective validation scheme of DCS networks to achieve better prediction performance with quantified confidence measures.

1 Introduction

Often viewed as black box tools, neural network models have a proven track of record of successful applications in various fields. In safety-critical systems such as flight control, neural networks are adopted as a popular soft-computing paradigm to carry out the adaptive learning. The appeal of including neural networks in these systems is in their ability to cope with a changing environment. Unfortunately, the validation of neural networks is particularly challenging due to their complexity and nonlinearity and thus reliable prediction performance of such models is hard to assure. The uncertainties (low confidence levels) existed in the neural network predictions need to be well analyzed and measured during system operation. In essence, a reliable neural network model should provide not only predictions, but also confidence measures of its predictions.

The Dynamic Cell Structures (DCS) network is derived as a dynamically growing structure in order to achieve better adaptability. DCS is proven to have topology-preserving adaptive learning capabilities that can respond and learn to abstract from a much wider variety of complex data manifolds [1,2]. The structural flexibility of DCS network has gained it a good reputation of adapting faster and better to a new region. A typical application of DCS is the NASA Intelligent Flight Control System (IFCS). DCS is employed in IFCS as online adaptive learner and provides derivative corrections as control adjustments during system operation. Within this application, it has been proven to outperform Radial Basis Function (RBF) and Multi-Layer Perceptron network models [3]. As a crucial component of a safety critical system, DCS network is expected to give robust and reliable prediction performance in operational domains.

Our research focuses on validating and improving the prediction performance of DCS network by investigating the confidence for DCS outputs. We present the Validity Index, as a measure of accuracy imposed on each DCS prediction. Each validity index reflects the confidence level on that particular output. The proposed method is inspired by J. Leonard's paper on the validation of Radial Basis Function (RBF) neural networks [4]. Leonard developed a reliability-like measure called validity index which statistically evaluates each network output. Different from the pre-defined static RBF network structure, the DCS progressively adjusts (grows/prunes) its structure including locations of neurons and connections between them to adapt to the current learning data. Thus, unbiased estimation of confidence interval is impossible to obtain through S-fold cross-validation due to constraints of time and space. Yet, DCS emphasizes topological representation of the data, while RBF does not. By the end of DCS learning, the data domain is divided into Voronoi regions. Every region has a neuron as its centroid. The "locality" of DCS learning is such that the output is determined by only two particular neurons, the best matching unit and the second best matching unit. Intuitively, if the Voronoi region of a neuron does not contain sufficient data, it is expected that the accuracy in that region will be poor. Based on the "local error" computed for each neuron, our approach provides an estimated confidence interval, called the Validity Index for DCS outputs.

The paper is organized as follows. The architecture of DCS network and its learning algorithm are described in Section 2. The concept of validity index and its statistical computation are presented in detail in Section 3. In Section 4, we further illustrate the validity index in DCS networks by presenting the experimental results using an artificial data set. Section 5 describes a case study on a real-world control application, the IFCS, and presents experimental results on the validity index in DCS using flight simulation data. In the end, conclusions are discussed in Section 6.

2 The Dynamic Cell Structures

The Dynamic Cell Structure (DCS) network can be seen as a special case of Self-Organizing Map (SOM) structures. The SOM is introduced by Kohonen [5] and further improved to offer topology-preserving adaptive learning capabilities that can, in theory, respond and learn to abstract from a much wider variety of complex data-manifolds. The DCS network adopts the self-organizing structure and dynamically evolves with respect to the learning data. It approximates the function that maps the input space. At last, the input space is divided into different regions, referred to as the Voronoi regions [1,2,6]. Each Voronoi region is represented by its centroid, a neuron associated with its reference vector known as the "best matching unit (BMU)". Further, a "second best matching unit (SBU)" is defined as the neuron whose reference vector is the second closest to a particular input. Euclidean distance metric is adopted for finding both units. The set of neurons connected to the BMU are considered its neighbors and denoted by NBR.

The training algorithm of the DCS network combines the competitive Hebbian learning rule and the Kohonen learning rule. The competitive Hebbian learning rule is used to adjust the connection strength between two neurons. It induces a Delaunay Triangulation into the network by preserving the neighborhood structure of the feature manifold. Denoted by $C_{ij}(t)$, the connection between neuron i and neuron j at time t is updated as follows:

$$C_{ij}(t+1) = \begin{cases} 1 & (i = BMU) \wedge (j = SBU) \\ 0 & (i = BMU) \wedge (C_{ij} < \theta) \\ & \wedge (j \in NBR \setminus \{SBU\}) \\ \alpha C_{ij}(t) & (i = BMU) \wedge (C_{ij} \geq \theta) \\ & \wedge (j \in NBR \setminus \{SBU\}) \\ C_{ij}(t) & (i, j \neq BMU) \end{cases}$$

where α is a predefined forgetting constant and θ is a threshold preset for dropping connections.

The Kohonen learning rule is used to adjust the weight representations of the neurons which are activated based on the best-matching methods during the learning. Over every training cycle, let $\Delta w_i = w_i(t+1) - w_i(t)$ represent the adjustment of the reference vector needed for neuron i, the Kohonen learning rule followed in DCS computes Δw_i as follows.

$$\Delta w_i = \begin{cases} \varepsilon_{BMU}(m - w_i(t)) & (i = BMU) \\ \varepsilon_{NBR}(m - w_i(t)) & (i \in NBR) \\ 0 & (i \neq BMU) \wedge (i \notin NBR) \end{cases}$$

where m is the desired output, and $0 < \varepsilon_{BMU}, \varepsilon_{NBR} < 1$ are predefined constants known as the learning rates that define the momentum of the update process. For every particular input, the DCS learning algorithm applies the competitive Hebbian rule before any other adjustment to ensure that the SBU is a member of NBR for further structural updates.

The DCS learning algorithm is diplayed in Figure 1. According to the algorithm, N is the number of training examples. Resource values are computed at each epoch as local error measurements associated with each neuron. They are used to determine the sum of squared error of the whole network. Starting initially from two connected neurons randomly selected from the training set, the DCS learning continues adjusting its topologically representative structure until the stopping criterion is met. The adaptation of lateral connections and weights of neurons are updated by the aforementioned Hebbian learning rule and Kohonen learning rule, respectively. The resource values of the neurons are updated using the quantization vector. In the final step of an iteration, the local error is reduced by inserting new neuron(s) in certain area(s) of the input space where the errors are large. The whole neural network is constructed in a dynamic way such that in the end of each learning epoch, the insertion or pruning of a neuron can be triggered if necessary.

```
Initialization;
Repeat until stopping criterion is satisfied
{
    Repeat N times
    {
        Determine the BMU and SBU;
        Update lateral connections;
        Adjust the weights;
        Update resource values;
    }
    If needed, a new neuron is inserted;
    Decrement Resource Values;
}
```

Fig. 1. A brief description of the DCS learning algorithm

It should be noted that while the DCS network is used for prediction, the computation of output is different from that during training. When DCS is in recall, the output is computed based on two neurons for a particular input. One is the BMU of the input; the other is the closest neighbor of the BMU other than the SBU of the input. In the absence of neighboring neurons of the BMU, the output value is calculated using the BMU only.

3 The Validity Index in DCS Networks

As a V&V method, validity check is usually performed through the aide of software tools or manually to to verify the correctness of system functionality and the conformance of system performance to pre-determined standards. The validity index proposed by J. Leonard [4] is a reliability-like measure provided for further validity checking. Validity index is a confidence interval associated with each output predicted by the neural network. Since a poorly fitted region will result in lower accuracy, it should be reflected by poor validity index and later captured through validity checking.

Given a testing input, the validity index in DCS networks is defined as an estimated confidence interval with respect to the DCS output. It can be used to model the accuracy of the DCS network fitting. Based on the primary rules of DCS learning and certain properties of final network structure, we employ the same statistical definition as for confidence intervals and variances for a random variable to calculate the validity index in DCS. The computation of a validity index for a given input x consists of two steps: 1) compute the local error associated with each neuron, and 2) estimate the standard error of the DCS output for x using information obtained from step 1). The detail description of these two steps are as follows.

1. The final form of DCS network structure is represented by neurons as centroids of Voronoi regions. Since the selection of the best matching unit must be unique, only those data points whose BMU are the same will be contained in the same region. Therefore, all Voronoi regions are non-overlapping and cover the entire learned domain. The data points inside each region significantly affect the local fitting accuracy. The local estimate of variance of the network residual in a particular region can be calculated over these data points contained in the region and then be associated with its representative neuron. More specifically, the local estimate of variance s_i^2 associated with neuron i can be computed as:

$$s_i^2 = \frac{1}{(n_i - 1)} \sum_{k=1}^{n_i} E_k,$$

where n_i is the number of data points covered by neuron i and E_k is the residual returned from the DCS recall function for data point k.

In Section 3, we show that the adjustment by competitive Hebbian learning rule concerns connections only between the BMU and its neighbors. The further update of weight values by Kohonen learning rule is performed only on the BMU and its neighbors as well. Consequently, training data points covered by the neighboring neurons of neuron i make proportional contributions to the local error of neuron i. Considering such contributions, we modify the computation of the local estimate of variance, now denoted by $s_i^{'2}$, as follows.

$$s_i^{'2} = \frac{s_i^2 + \sum_{j \in NBR} C_{ij} s_j^2}{1 + \sum_{j \in NBR} C_{ij}}.$$

As a result, the influence of all related data points is taken into account accordingly based on connections, referred to as C_{ij}, between the BMU and its neighbors. It should be noted that since the DCS networks are often adopted for online learning, no cross-validation is allowed. Hence, the residual calculated for each data point is in fact a biased estimate of the expected value of residual due to the fact that each data point itself contributed to its own prediction. Nonetheless, under the assumption that there is no severe multi-collinearity and relatively few outliers exist in the data, the probability that the deviation from the expected value will be significant is very low and thus can be ignored.

2. Recall that the output produced by DCS is determined by the BMU and its closest neighbor (CNB) of the given input. Thus, the local errors associated with these two neurons are the source of inaccuracies of fitting. We use the standard error, a statistic that is often used to place a confidence interval for an estimated statistical value. Provided with the local estimate of variance for every neuron from step 1), we now define the 95% confidence limit for the local prediction error estimate with respect to neuron i as:

$$CL_i = t_{.95} \sqrt{1 + \frac{1}{n_i}} s_i',$$

where $t_{.95}$ is the critical value of the Student's t-distribution with $n_i - 1$ degrees of freedom. The 95% confidence interval for the network output y given a testing input is thus given by:

$$(y - \frac{(CL_i + CL_j)}{2}, y + \frac{(CL_i + CL_j)}{2}),$$

where $i = BMU$ and $j = CNB$ with respect to the input x.

Now we slightly modify the DCS training algorithm in order to calculate the validity index. Note that because all needed information is already saved at the final step of each training cycle, without any additional cost required, we simply calculate $s_i^{'2}$ for each neuron after the learning stops. When the DCS is in recall for prediction, the validity index is computed based on the local errors and then associated with every DCS output. In order to complete the validity check, further examination needs to be done by software tools or system operators. In the case of a control application, a domain specific threshold can be pre-defined to help verify that the accuracy indicated by the validity index is acceptable.

4 An Example with Artificial Data

In order to demonstrate the validity index in DCS network model as an improvement of the network prediction, we present an example using an artificial data set. The DCS is trained on a single-input, single-output function as seen in [4]:

$$f(x) = 0.2\sin(1.5\pi x + 0.5\pi) + 2.0 + \varepsilon,$$

where ε is a Gaussian noise.

We sample x's from the interval $[-1, 1]$ randomly. Therefore, at least initially, there exist regions where the learning data points are not as dense as in the others. We then obtain two different DCS network models by varying the stopping criterion. Figure 2 illustrates the validity index for these two DCS models, one with 13 neurons and the other with 27 neurons, shown as plot (a) and plot (b), respectively. By comparing the prediction performance of these two models using the validity index, which is shown as

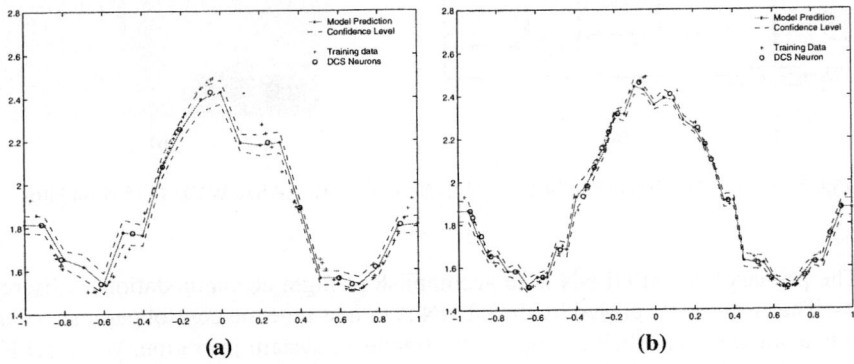

Fig. 2. Examples of validity index for a DCS model. (a): The model with 13 neurons. (b): The model with 27 neurons.

confidence band in both figures, we can conclude that the DCS network model shown in Figure 2 (b) has better prediction performance. Furthermore, we can observe that regions with sparse learning data have low confidence levels.

5 A Case Study

We investigate the prediction performance of DCS networks for the Intelligent Flight Control System (IFCS). The IFCS is an example of adaptive flight control application for NASA F-15 aircraft. As the post-adaptation validation approach, the validity index is a major component of our validation framework for IFCS [7].

5.1 The Intelligent Flight Control System

The Intelligent Flight Control System (IFCS) was developed by NASA with the primary goal to *"flight evaluate control concepts that incorporate emerging soft computing algorithms to provide an extremely robust aircraft capable of handling multiple accident and/or an off-nominal flight scenario"* [8,9].

The diagram in Figure 3 (a) shows the architectural overview of NASA's first generation IFCS implementation using Online Learning Neural Network (OLNN). Figure 3 (b) shows the user interface of an experimental IFCS simulator [10]. The control concept can be briefly described as follows. Notable discrepancies from the outputs of the Baseline Neural Network and the Real-time Parameter Identification (PID), either due to a change in the aircraft dynamics (loss of control surface, aileron, stabilator) or due to sensor noise/failure, are accounted by the Online Learning Neural Network.

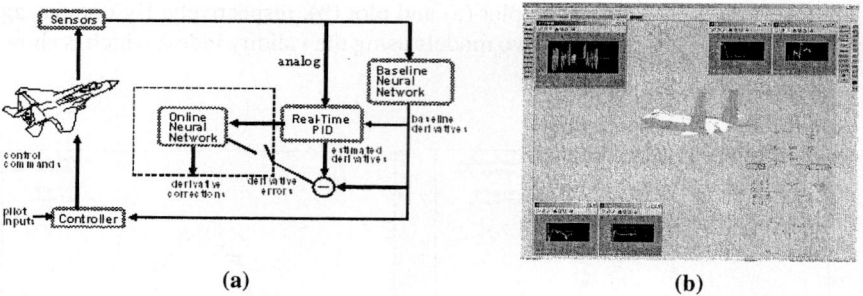

(a) (b)

Fig. 3. (a): The Intelligent Flight Control System and (b): NASA-WVU F-15 Simulator

The primary goal of OLNN is to accomplish in-flight accommodation of discrepancies. The critical role played by the OLNN is to fine-tune the control parameters and provide a smooth and reliable control adjustments to system operation. When OLNN performs adaptation, its behavior has a direct consequence on the performance of the flight control system. In such a safety-critical application, it is necessary to understand and assure the prediction performance of the OLNN.

Our previous research provides a validation framework for validating the OLNN learning. It consists of a novelty detection tool to detect novel (abnormal) conditions entering the OLNN, and online stability monitoring techniques to investigate the NN's stability behavior during adaptation [7,11,12]. Although learning can be closely monitored and analyzed, when the system is in operation, it is probable that the predictions of the OLNN will become unreliable and erroneous due to extrapolation. Therefore, providing a reliability-like measurement with respect to each particular output can further enforce safety of the system in operation. In IFCS, the neural network that implements the OLNN component is the Dynamic Cell Structure (DCS).

5.2 Experimental Results

With the aide of the high-fidelity flight control simulator, we are able to test our approach for adaptive flight control through experimentation in simulated environments. The online neural networks in IFCS learn on the environmental changes and accommodate failures. They generate derivative corrections as compensation to the PTNN output (see Figure 3). We use validity index to evaluate the accommodation performance and validate the predictions of the DCS network. In our experiment, we simulate the online learning of a DCS network on a failure mode condition and compute the validity index in real-time.

We simulate the online learning of the DCS network under two different failure mode conditions. One is the stuck-at-surface type of failure. The simulated failed flight condition in this case is the aircraft¡s stuck left stabilator, which is simulated to stuck at an angle of +3 degree. The other is the loss-of-surface type of failure. This simulated failure has 50% of surface loss at the left stabilator. In each run, the DCS network updates its learning data buffer at every second and learns on the up-to-date data set of size 200 at a frequency of 20Hz. We first start the DCS network under nominal flight conditions with 200 data points. After that, every second, we first set the DCS network in recall mode and calculate the derivative corrections for the freshly generated 20 data points, as well as their validity index. Then we set the DCS network back to the learning mode and update the data buffer. While updating the data buffer, we discard the first incoming 20 data points and add the freshly generated 20 data points to maintain the buffer size, i.e., 200. The DCS network continues learning and repeats the recall-learn procedure.

Figure 4 and Figure 5 show the experimental results of the simulations on these two failures, respectively. Plot (a)'s show the final form of the DCS network structure at the end of the simulation. As a three-dimensional demonstration, the x-axis and y-axis represent two independent variables, α and β, respectively. The z-axis represents one derivative correction, $\Delta Cz\alpha$. The 200 data points in the data buffer at the end of the simulation are shown as crosses in the 3-D space. The network structure is represented by circles (as neurons) connected by lines as a topological mapping to the learning data. Plot (b)'s present the validity index, shown as error bars. The x-axis here represents the time frames. In both simulations, the failure occurs at the 100th data frame. We compute the validity index for the data points that are generated five seconds before and five seconds after the failure occurs. In total, Plot (b) illustrates the validity index for 200 data points.

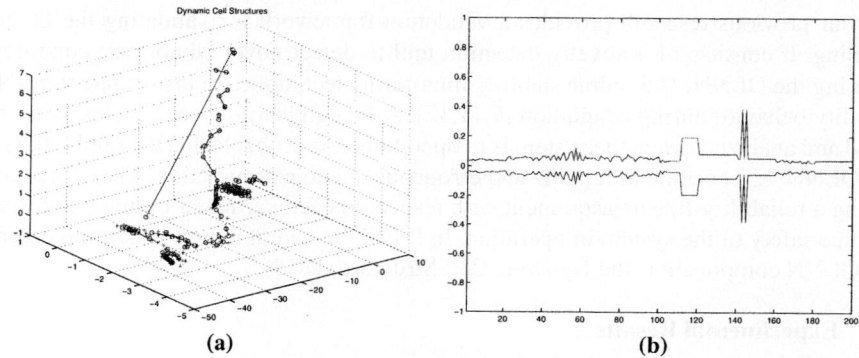

Fig. 4. A stuck-at-surface failure simulation in real-time (20Hz). (a): The final form of DCS network structures. (b): Validity Index shown as error bars for each DCS output.

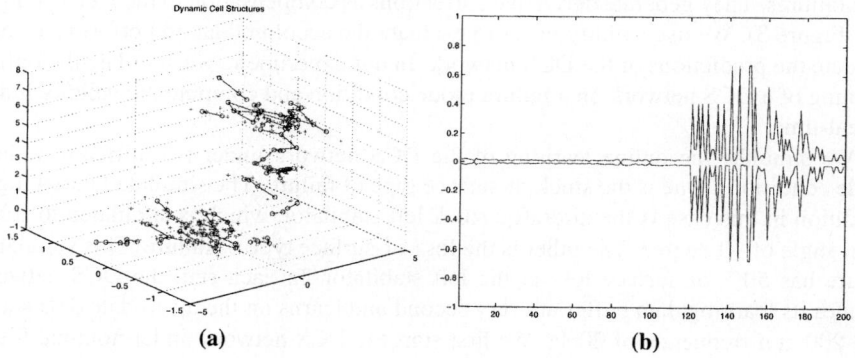

Fig. 5. Testing on loss-of-surface failure simulation data in real-time. (a): The final form of DCS network structures. (b): Validity Index shown as error bars for each DCS output.

A common trend revealed in both figures by the validity index is the increasingly larger error bars after the failure occurs. Then, the error bars start shrinking while the DCS network starts adapting to the new domain and accommodating the failure. After the failure occurs, the change (increase/decrease) of the validity index varies. This depends on the characteristics of the failure as well as the accommodation performance of the DCS network. Nevertheless, the validity index explicitly indicates how well and how fast the DCS network accommodates the failures.

6 Conclusions

Known for its structural flexibility, DCS networks are adopted in safety-critical systems for online learning in order to quickly adapt to a changing environment and provide reliable outputs when needed. However, DCS network predictions cannot be constantly trusted because locally poor fitting will unavoidably occur due to extrapolation. We propose the validity index in DCS for validating its prediction performance as an improvement to DCS network models. The implementation of validity index is straightforward and does not require any additional learning. Experimental results were obtained by running tests on failure flight data collected from the IFCS simulator. The computed validity index effectively indicates poor fitting within regions characterized by sparse data. It demonstrates that the validity index is a feasible improvement to DCS and can be applied to validate the DCS performance by predicting with confidence.

References

1. J. Bruske, and G. Sommer. Dynamic cell structure learns perfectly topology preserving map, *Neural Computation*, Vol. 7, No. 4, 1995, 845-865.
2. T. Martinetz, and K. Schulten. Topology representing networks, *Neural Networks*, Vol. 7, No. 3, 1994, 507-522.
3. Institute of Software Research. Dynamic cell structure neural network report for the intelligent flight control system, Technical report, Document ID: IFC-DCSR-D002-UNCLASS-010401, January 2001.
4. J.A. Leonard, M.A. Kramer, and L.H. Ungar. Using radial basis functions to approximate a function and its error bounds. *IEEE Transactions on Neural Networks*, 3(4):624-627, July 1992.
5. Teuvo Kohonen. The self-organizing map, *Proc. of the IEEE*, Vol. 78, No. 9, September, 1990, 1464-1480.
6. B. Fritzke, Growing cell structures - a self-organizing network for unsupervised and supervised learning, *Neural Networks*, Vol. 7, No. 9, May 1993, 1441-1460.
7. S. Yerramalla, Y. Liu, E. Fuller, B. Cukic and S. Gururajan. An Approach to V&V of Embedded Adaptive Systems, in *Lecture Notes in Computer Science (LNCS) Proceeding of Third NASA-Goddard/IEEE Workshop on Formal Approaches to Agent-Based Systems*, Springer-Verlag, 2004.
8. C.C. Jorgensen. Feedback linearized aircraft control using dynamic cell structures, *World Automation Congress* (ISSCI), Alaska, 1991, 050.1-050.6.
9. The Boeing Company, Intelligent flight control: advanced concept program, Technical report, 1999.
10. M. Napolitano, G. Molinaro, M. Innocenti, and D. Martinelli. A complete hardware package for a fault tolerant flight control system using online learning neural networks. *IEEE Control Systems Technology*, January, 1998.
11. Y. Liu, S. Yerramalla, E. Fuller, B. Cukic and S. Gururajan. Adaptive Control Software: Can We Guarantee Safety? *Proc. of the 28th International Computer Software and Applications Conference, workshop on Software Cybernetics*, Hong Kong, September 2004.
12. S. Yerramalla, E. Fuller, B. Cukic. Lyapunov Analysis of Neural Network Stability in an Adaptive Flight Control System, *6th Symposium on Self Stabilizing Systems (SSS-03)*, San Francisco, CA, June 2003.

An Efficient Score Function Generation Algorithm with Information Maximization

Woong Myung Kim and Hyon Soo Lee

Dept. of Computer Engineering, Kyunghee University,
1, Seocheon-ri, Giheung-eup, Yongin-si, Gyeonggi-do 449-701, Republic of Korea
{wmkim, leehs}@khu.ac.kr

Abstract. In this study, we propose this new algorithm that generates score function in ICA (Independent Component Analysis) using entropy theory. To generate score function, estimation of probability density function about original signals are certainly necessary and density function should be differentiated. Therefore, we used kernel density estimation method in order to derive differential equation of score function by original signals. After changing the formula to convolution form to increase speed of density estimation, we used FFT algorithm which calculates convolution faster. Proposed score function generation method reduces estimation error, it is density difference of recovered signals and original signals. Also, we insert constraint which is able to information maximization using smoothing parameters. In the result of computer simulation, we estimate density function more similar to original signals compared with Extended Infomax algorithm and Fixed Point ICA in blind source separation problem and get improved performance at the SNR (Signal to Noise Ratio) between recovered signals and original signals.

1 Introduction

Independent component analysis is a useful method to separate original signals from linear mixtures. This method is used to solve blind source separation problem, which can find original signals from observed signals. Typically, source separation is to find a minimized contrast function which is related to source densities. These contrast functions are based on maximum likelihood, information maximization and mutual information[1][2][3][4]. Despite many ICA algorithms, we have tried to separate with more accurately estimated signals with inaccurate signals because they do not know the source distribution[8][9][10]. Generally, density of source relates with score function and ICA learning in maximum likelihood estimation which assumes the fixed density function. Therefore, it is important to score function generation based on density estimation so that estimated signals are similar to original signals.

In this study, we focus on two points; one is score function generation and the other is controlling of smoothing parameters in kernel density estimation. In latter, entropy maximization of density estimator is an effective constraint in score function generation.

2 Maximum Likelihood Estimation Setting in ICA

In ICA definition, s is independent sources with size. x is mixture signals by mixing matrix A, where A is mixing and nonsingular matrix with size. Also it is full rank matrix[5].

$$x = As \qquad (1)$$

W matrix is separating matrix that successfully recovers the original sources.

$$u = Wx \qquad (2)$$

And where u is recovered signals. Eq.3 defines as objective (contrast) function using negative log-likelihood function with unknown matrix . If these condition would be, it will be global minimum of objective function[6].

$$L(W) = -\sum_{i=1}^{n} \log p(x) = -n \log |\det W| - \sum_{i=1}^{n}\sum_{j=1}^{m} \log f_j(u_i^{(j)}) \qquad (3)$$

And $p(x) = |detW|f(Wx)$ is defined then $f_j(u_i^{(j)})$ denotes its marginal density. Natural gradient or relative method is adopted to minimize global objective function in updating weights. The updating equation is $L(W)$ in updating weights. The updating equation is

$$W = W - \eta(E[\varphi(u)u^T] - I)W \qquad (4)$$

Where η is constant learning rate, E is expectation operator and I is the $m \times m$ identity matrix. In Eq.6, $\varphi(u)$ is score function which is called nonlinear function with minimized objective function, it is a set of nonlinear function $m \times 1$ size vector

$$\varphi(u) = [\varphi_1(u), ..., \varphi_m(u)]^T \qquad (5)$$

If the density function $f(u)$ can be estimated from finite data set, the score function is defined as Eq.6 through their negative log-derivatives.

$$\varphi(u) = -[\log f(u)]' = -\frac{f'(u)}{f(u)} \qquad (6)$$

3 Design of Score Function Using Kernel Density Estimate

3.1 SFG (Score Function Generation) Algorithm

Kernel density estimation is a method that calculates the probability density function[7][8][10]. This is a nonparametric density estimation which is possible to differential equation[11]. Therefore, we generate a score function using KDE (Kernel Density Estimate) method. KDE is defined as follow:

$$f(u) \cong \tilde{f}(c) = \frac{1}{nh} \sum_{i=1}^{n} K\left(\frac{c - u_i}{h}\right) \qquad (7)$$

$$K(z) = \frac{1}{\sqrt{2\pi}} \exp^{-z^2/2} \tag{8}$$

If we are differentiated about c using Eq.7 and Eq.8, it appears in a form such as Eq.9, because the output dimension of density has c vector size. Also Eq.9 can be omitted on n and π because of constant value.

$$\varphi(u) \cong \tilde{\varphi}(c) = \frac{f'(c)}{f(c)} = \frac{\frac{1}{h^2}\sum_{i=1}^{n}(c-u_i)\exp\left(-\frac{(c-u_i)^2}{2h^2}\right)}{\sum_{i=1}^{n}\exp\left(-\frac{(c-u_i)^2}{2h^2}\right)} \tag{9}$$

In eq 9, n is number of samples and c also is the center of kernels. u_i is recovered signals in ICA learning. h, usually a positive value, and the width in each kernel is called bandwidth or smoothing parameter. We have to change above equation to convolution form because density estimation using kernel, this is which is time consuming. Since convolution is product form in frequency domain, we use Fast Fourier Transform for changing density function to frequency domain. In other words, use FFT for computing convolution to reduce computation time.

We can generate FFT form that use Eq.9, and this expressed in Eq.10 and 11, where G is Gaussian kernel and G' is value that differentiate Gaussian kernel. F corresponds to $f(u)$. If H defines as measured histogram, it is expressed as follows

$$F \equiv ifft(fft(H) * fft(G)) \tag{10}$$

$$F' \equiv ifft(fft(H) * fft(G')) \tag{11}$$

Computational complexity of score function using FFT-convolution is decreased by $O(n \log n)$. Hence, SFG algorithm reduces the time complexity for creating score function. In practical choice, we test with 100 number of c and constant value in set h, described as [7] and [8]. Next section presents a constraint of score function with entropy maximization.

3.2 SFG Algorithm with Entropy Maximization

Bell and Sejnowski proposed a simple learning algorithm for feedforward neural network that blindly separates linear mixtures x of independent sources s using information maximization[2][5]. The joint entropy at the outputs of neural network is

$$H(y_1,...y_m) = H(y_1) + ... + H(y_m) - I(y_1,......,y_m) \tag{12}$$

where $H(y_i)$ are the marginal entropies of the outputs and $I(y_1,....,y_m)$ is their mutual information. Generally, $H(y)$ is joint entropy, sum of marginal entropies. Each of marginal entropy can be written as

$$H(y_i) = -E[\log p(y_i)] \tag{13}$$

The nonlinear mapping between the output density $p(y_i)$ and source estimate density $p(u_i)$ can be described by the absolute value of the derivative with respect to u_i

$$p(y_i) = p(u_i)/|\frac{\partial y_i}{\partial u_i}| \qquad (14)$$

According to derivative of joint entropy is now

$$\partial H(y)/\partial W = \partial(-I(y))/\partial W - \partial \sum_{i=1}^{m} E[\log p(y_i)]/\partial W \qquad (15)$$

In other words, the mutual information $I(y)$ will be minimized when joint entropy $H(y)$ is maximized. In this case the error term vanishes and the maximum of joint entropy $H(y)$ can be found by deriving $H(y)$ with respect to W. Consequently, relation of the gradient of $H(y)$ and W is as follows

$$\partial H(y)/\partial W = \partial(-E[\log|J|])/\partial W \qquad (16)$$

where such entropy can be expressed in terms of $p(y)$ and input density $p(u)$ can be described by the Jacobian matrix[9].

$$p(y) = p(u)/J(u), J = |\partial\varphi(u)/\partial u| \qquad (17)$$

Entropy $H(y)$ that has to be maximized with respect to density function is defined as

$$H(y) = H(u) + E[\ln J] \qquad (18)$$

The main idea is to estimate densities function first using joint entropy maximization then takes their gradient descent equation for estimating score function. Therefore, we need an adaptive estimator of the source densities. That is, finding of the minimized objective function is to compute Eq.6 for performing maximum entropy of density estimation, where k is number of kernel. Constraint that generates adaptive score function has to be considered. Such constraints are expressed as Eq.15 and 18. Generally, kernel method in density estimation is used to transform n dimensional space to k dimensional space. To design kernel density estimator for entropy maximization, we assume that h defines vector as $[h_1, ..., h_k]$. So, gradient descent method can be applied for updating using modified smoothing parameter h. And also c is called to reference vector, it consist of $c = [c_1, ..., c_k]$. Initial values of c are equal to value of δ in $\delta_k = c_1 + (k-1)\delta$, δ is the distance between of center points. As a described Eq.6, we can write a differential equation of to Eq.19.

$$\tilde{\varphi}(u) \cong \frac{f'(c)}{f(c)} = \frac{1}{n}\sum_{i=1}^{n}\frac{1}{h_k}\exp(-\frac{1}{2h^2}(c-u_i)^2), f'(c) = (\partial f(c)/\partial c) \qquad (19)$$

At the ICA learning, a separating matrix W and smoothing parameter h is to maximize joint entropy $H(y)$. The results of the equation are as the following

$$\Delta(W, h) = \eta \frac{\partial H(y)}{\partial(W, h)} = \frac{\partial H(u)}{\partial(W, h)} + \frac{\partial E(\ln J)}{\partial(W, h)} \qquad (20)$$

Consequently, we could find an optimization solution using gradient descent on smoothing parameter h, where α is learning rate of h

$$\Delta h = \alpha \frac{\partial H(y)}{\partial h} = \frac{\partial H(u)}{\partial h} + \frac{\partial E(\ln J)}{\partial h} \qquad (21)$$

If sample points are n and smoothing parameter h is vector that consists of $h = [h_1, ..., h_k]$, we must transform the number of sample and the number of kernel. Therefore, the following equation has to be differentiated by vector c. So we was simplified some of the notation for easier computation.

$$\begin{aligned} K1 &= \exp(-\tfrac{1}{2h_k^2}(c-u_j)^2) \\ K2 &= (c-u_j)\exp(-\tfrac{1}{2h_k^2}(c-u_j)^2) \\ K3 &= (c-u_j)^2 \exp(-\tfrac{1}{2h_k^2}(c-u_j)^2) \\ K4 &= (c-u_j)^3 \exp(-\tfrac{1}{2h_k^2}(c-u_j)^2) \\ K5 &= (c-u_j)^4 \exp(-\tfrac{1}{2h_k^2}(c-u_j)^2) \end{aligned} \qquad (22)$$

$K1, K2, K3, K4$ and $K5$ can be transformed to convolution form. According to the defined Eq.17, Eq.18 and Eq.19, a new kernel density estimator to update suitable value of h as

$$\begin{aligned} &\partial(\varphi(u))/\partial u \cong \partial(\bar{\varphi}(c))/\partial c \\ &= \left(\left[\sum_{i=1}^{n} K1 - \tfrac{1}{h_k^2}K3\right]\left[\sum_{i=1}^{n} K1\right] - \left[\sum_{i=1}^{n} K2\right]\left[\sum_{i=1}^{n} -\tfrac{1}{h_k^2}K2\right]\right) \Big/ \left[\sum_{i=1}^{n} K1\right]^2 \end{aligned} \qquad (23)$$

We can adopt a gradient descent method using the defined above equation to find Δh

$$\begin{aligned} \Delta h &\cong \partial J/\partial h = (\tfrac{\partial(\bar{\varphi}(c))}{\partial c})/\partial h \\ &= \left((P1) \Big/ \left[\sum_{i=1}^{n} K1\right]^2\right) - \left(P2 \Big/ \left[\sum_{i=1}^{n} K1\right]^4\right) - \left(P3 \Big/ \left[\sum_{i=1}^{n} K1\right]^4\right) \end{aligned} \qquad (24)$$

where $P1, P2$ and $P3$ are defined as follows

$$\begin{aligned} P1 &= \left[\sum_{i=1}^{n} \tfrac{1}{h_k^3}K3 - \tfrac{1}{h_k^3}K5\right]\left[\sum_{i=1}^{n} K1\right] - \left[\sum_{i=1}^{n} K1 - \tfrac{1}{h_k^3}K3\right]\left[\sum_{i=1}^{n} \tfrac{1}{h_k^3}K3\right] \\ P2 &= \left(\left[\sum_{i=1}^{n} \tfrac{1}{h_k^3}K4\right]\left[\sum_{i=1}^{n} -\tfrac{1}{h_k^2}K2\right] + \left[\sum_{i=1}^{n} K2\right]\left[\sum_{i=1}^{n} \tfrac{2}{h_k^5}K2 - \tfrac{1}{h_k^3}K4\right]\right)\left[\sum_{i=1}^{n} K1\right]^2 \\ P3 &= \left(\left[\sum_{i=1}^{n} K2\right]\left[\sum_{i=1}^{n} -\tfrac{1}{h_k^2}K2\right]\right) 2\left[K1\right]\left[\sum_{i=1}^{n} \tfrac{1}{h_k^3}K3\right] \end{aligned}$$

In Eq.24, it will be chosen as an updating quantity of smooth parameter h. However, this updating rule leads to unlimited growth of parameter h. We may overcome this problem by saturation or normalization this learning rule. The updating quantity for smoothing parameter and generalized Hebbian Learning rule are similar. Hence, a convenient form that can normalize the smoothing parameter is described by the following equation

$$\Delta h_j^m(t+1) = h_j^m(t) + \alpha(\Delta h_j^m(t)/\sum_{j=1}^{k}\Delta h_j^m(t)) \qquad (25)$$

3.3 Algorithm Summary

In ICA, matrix W after learning has a property that log-likelihood is stabilized with respect to objective function. Therefore, we can derive the following condition, where $L(W)^{(t)}$ is observed at the log-likelihood of current time, and is the observed log-likelihood at $t-1$.

$$||L(W)^{(t)} - L(W)^{(t-1)}|| \leq \varepsilon \qquad (26)$$

If the difference of $L(W)$ is smaller than ε, we interpret it as stable. Thus, SFG algorithms can be switched.

1. Step 1. Initialize W matrix
2. Step 2. Compute score function generation with entropy maximization using Eq.25
3. Step 3. Update $W(t)$ using Eq.4
4. Step 4. If $||L(W)^{(t)} - L(W)^{(t-1)}|| \leq \varepsilon$ then go to Step 5 Else go to Step 2.
5. Step 5. Compute the score function using Eq.9
6. Step 6. Update $W(t+1)$ using Eq.4
7. Step 7. Go to Step 5.

4 Computer Simulations

To estimate performance about proposed SFG algorithm, we tested with 3 source signals which have distribution such as Fig.1. This algorithm is implemented in Matlab language. The first signal is similar to Gaussian distribution, second signal is super-Gaussian distribution having heavily kurtosis and the third signal, skewness is positive tail.

We used matrix A to make mixed signals.

$$A = \begin{bmatrix} -1.5937 & -0.39989 & 0.71191 \\ -1.441 & 0.69 & 1.2902 \\ 0.57115 & 0.81562 & 0.6686 \end{bmatrix} \qquad (27)$$

SFG algorithm performs first updating smoothing parameters to maximize entropy and then log-likelihood judges whether it is stabilized or not. Fig.2

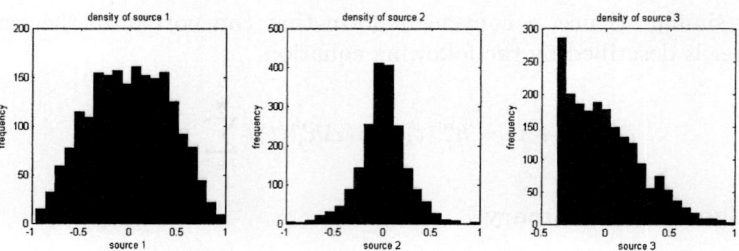

Fig. 1. Distribution of three source signals

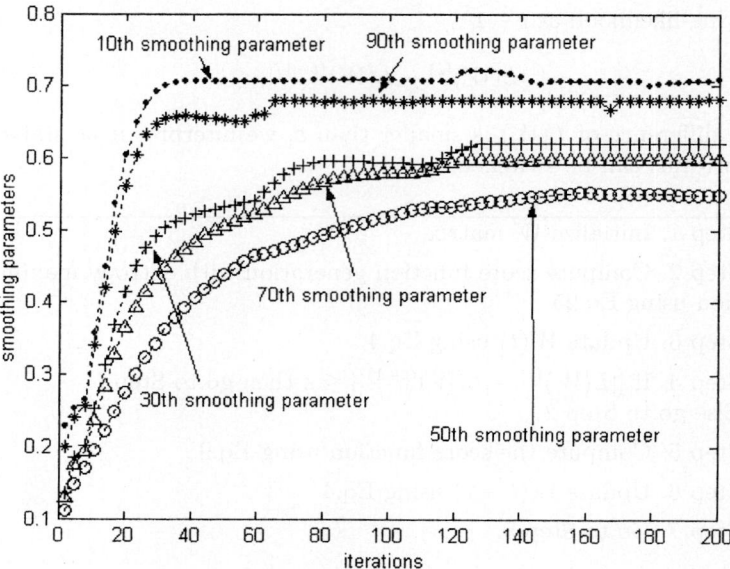

Fig. 2. Smoothing parameters h of source signal 1

displays update of smoothing parameters in SFG algorithm for entropy maximization. The number of kernel is 100. And also learning rate sets $\eta = 0.02$ and $\alpha = 0.05$. We can know that each smoothing parameters that is updated by gradient descent method are converged by constant values.

$$SNR = 10\log_{10}\left(\sum_{j=1}^{m} s_j^2 / \sum_{j=1}^{m}(\hat{s}_j - s_j)^2\right)(dB) \qquad (28)$$

Fig.3 shows on SNR between estimated signals and original signals, where signal with super-Gaussan distribution has good performance in Fixed Point ICA and Extended Infomax algorithm. Such performance appears because use a fixed score function that has super-Gaussian distribution similar to original

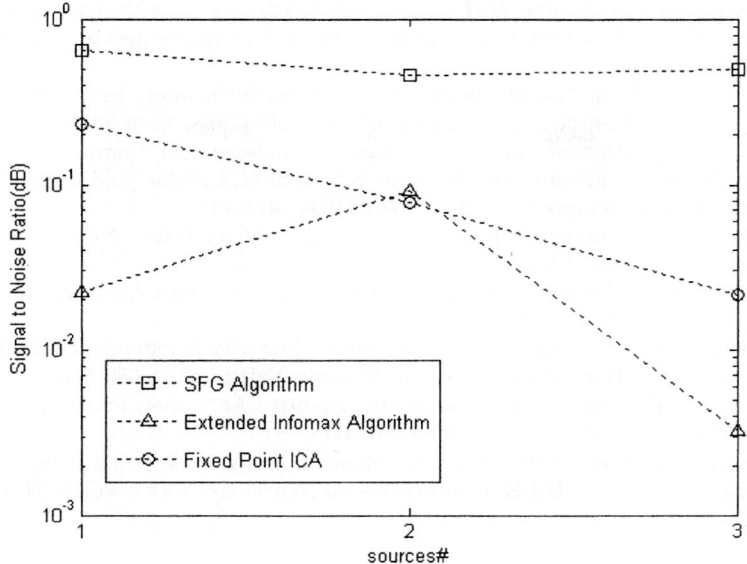

Fig. 3. SNR of SFG, Extended Infomax Algorithm and Fixed Point ICA

source signal at each algorithm. However, SFG algorithm had better performance compare with other methods, even when the third signal was almost perfectly separated.

5 Conclusion

The difference between the source densities and estimated densities, known as errors, will exists because MLE is calculated with score function, which uses fixed densities in ICA. Therefore, new algorithm is required to reduce these errors. We propose SFG algorithm using KDE for solving this problem. To reduce the computation time, we present an equation of score function to convolution form using FFT. Also, we suggest SFG algorithms, which are able to maximize entropy. This Algorithm is new density estimator with constraint and this is focused on controlling smoothing parameters. As a conclusion of computer simulation, smoothing parameters are converged by constant value and SFG algorithm displayed better performance comparing with existent algorithm.

References

1. P. Comon : Independent component analysis, a new concept? : Signal Processing, Vol36 (1994) 287-314
2. A.J.Bell, T.Sejnowski : An Information Maximization Approach to Blind Separation and Blind Deconvolution : Neural computation, Vol7. No6 (1995) 1129-1159

3. S.Amari. A.Cichocki, and H.H.Yang : A new learning algorithm for blind signal separation : in Advances in Neural Information Processing systems, Vol8 (1996) 757-763
4. J-F. Cardoso : Blind Signal Separation , Statistical Principles : Proc. IEEE Special Issue on Blind Identification and Estimation, Vol9 (1998) 2009-2025
5. T-W. Lee, M.Girloami and T. J Sejnowski : Independent Component Analysis using Extended Infomax algorithm for mixed SubGausssian and SuperGaussian sources: Neural Computation, Vol1. No2 (1999) 417-441
6. A. Hyvarinen : Survey on independent component analysis : Neural computing surveys, Vol2 (1999) 94-128
7. B.W.Silverman : Density Estimation for Statistics and Data Analysis : New York, Chapman and Hall (1995)
8. N. Vlassis and Y.Motomura : Efficient source adaptivity in independent component analysis : IEEE transaction on Neural Neworks,Vol12 (2001) 559-566
9. S. Fiori : Blind Signal Processing by the Adaptive Activation Function Neurons : Neural Networks, Vol13. No6 (2000) 597-611
10. R. Boscolo, H. Pan : Independent Component Analysis Based on Nonparametric Density Estimation : IEEE Transactions on Neural Networks, Vol15. No1 (2004) 55-65

A New Criterion on Exponential Stability of a Class of Discrete Cellular Neural Networks with Time Delay

Fei Hao[1], Long Wang[2], and Tianguang Chu[2]

[1] The Seventh Research Division, Beijing University of Aeronautics and Astronautics,
Beijing, 100083, P.R. China
fhao@buaa.edu.cn
[2] Center for Systems and Control, Department of Mechanics and Engineering Science, Peking University,
Beijing 100871, P.R. China

Abstract. A new criterion on exponential stability of the equilibrium point for a class of discrete cellular neural networks (CNNs) with delay is established by Lyapunov-Krasovskii function methods. The obtained result shows a relation between the delayed time and the corresponding parameters of the network. A numerical example is given to illustrate the efficiency of the proposed approach.

1 Introduction

CNNs were first introduced in [1]. CNNs with delay (DCNNs) proposed in [2] have been found applications in many areas, and, recently, have been extensively studied (e.g, see [3]–[11]). For DCNNs, many authors have studied their global exponential stability ([6], [7]). The exponential stability is very important since the exponential convergence rate will determine the speed of neural computations. Also, it is not only theoretically interesting but also practically important to determine the exponential stability and estimate the exponential convergence rate for DCNNs in general. There have been many works devoted to the qualitative analysis of discrete CNNs (see, e.g., [8]–[11] and the references therein). However, to the best of our knowledge, there are few study concerning the global exponential stability for discrete DCNNs.

In this paper, we study the global exponential stability and estimate the exponential convergence rates for discrete CNNs with constant time delays. A new criterion on the global exponential stability of an equilibrium point for discrete DCNNs is established in terms of linear matrix inequality (LMI), which can be solved numerically by efficient LMI toolbox. Finally, a numerical example is given to illustrate the efficiency of the proposed approach.

2 Statement of the Problem and Main Result

The dynamic behavior of a discrete DCNN can be described by

$$y(k+1) = Cy(k) + Ag(y(k)) + Bg(y(k-d)) + u. \qquad (1)$$

where $y(k) = [y_1(k),\ldots,y_n(k)]^T \in \mathbf{R}^n$, $g(y(k)) = [g_1(y_1(k)),\ldots,g_n(y_n(k))]^T \in \mathbf{R}^n$ are the state and the output (or $g(x)$ the activation function of the neurons) vectors respectively. Assume $g(\cdot)$ is bounded and $g(0) = 0$. y_i is the state of neuron i, and n is the number of neurons. d is the transmission delay, A and B are the interconnection matrices, $u = [u_1,\ldots,u_n]^T$ is the constant input vector, and $C = \mathrm{diag}[c_1,\cdots,c_n]$, $|c_i| < 1$. Assume the activation functions satisfying: For each $i \in \{1,\ldots,n\}$ and given $M_i > 0$, $g_i : \mathbf{R} \to \mathbf{R}$ is in a finite sector (see [4] for details), that is, g_i satisfies $0 \le \frac{g_i(u_i)-g_i(v_i)}{u_i-v_i} \le M_i$ for any $u_i, v_i \in \mathbf{R}, u_i \ne v_i$. Let $M = \mathrm{diag}[M_1,\ldots,M_n]$.

As usual, y^* is said to be an equilibrium point of system (1) for a given u if it satisfies $Cy^* + Ag(y^*) + Bg(y^*) + u = y^*$.

By the boundedness of $g(\cdot)$, it is straightforward to show (e.g., by using Brouwer fixed point theorem) that (1) has at least one equilibrium point (see [4]). In the following, we will shift the equilibrium point y^* of system (1) to the origin for the given u. Taking the transformation $x(\cdot) = y(\cdot) - y^*$, we substitute $x(\cdot) + y^*$ for $y(\cdot)$ in system (1) to imply

$$x(k+1) = y(k+1) - y^* = Cx(k) + Af(x(k)) + Bf(x(k-d)). \qquad (2)$$

where $f(x(k)) = g(y(k)) - g(y^*)$ and $f(x(k-d)) = g(y(k-d)) - g(y^*)$.

Remark 1. Obviously, it is easy to obtain $f(0) = 0$, $f_i^2(x_i(t)) \le M_i^2 x_i^2(t)$ and $x_i(t)f_i(x_i(t)) \le M_i x_i^2(t)$. Moreover, this type of functions is more general than the usual activation functions which are continuous, differentiable and monotonically increasing (used in [3], [5]).

Definition 1. If there exist an $\alpha : 0 < \alpha < 1$ and $\gamma(\alpha) > 0$ such that $\|x(k)\| \le \gamma(\alpha)\alpha^k \sup_{-d \le l \le 0} \|x(l)\|$ for $k \in \mathbb{Z}^+$, then system (2) is said to be exponentially stable, where the α called the degree of exponential stability.

We shall establish a new sufficient criterion on the exponential delay-dependent stability of an equilibrium point for DCNNs in the following Theorem.

Theorem 1. Consider (2) with given delay d. If there exist a positive constant $\beta > 1$ and positive definite matrices P, Q satisfying the following inequality

$$\begin{bmatrix} \Phi_{11} & 0 & MA^TP + CP \\ 0 & -\beta^{-2d-2}Q & MB^TP \\ PAM + PC & PBM & -P \end{bmatrix} < 0, \qquad (3)$$

where $\Phi_{11} = -\beta^{-2}P + \beta^{-2}Q + CPAM + MA^TPC$, then system (2) is exponentially stable with the degree β^{-1} of exponential stability. Moreover, $\|x(k)\| \le \alpha^k \sqrt{\frac{\lambda_M(P)+\lambda_M(Q)\frac{\alpha^2-\alpha^{2d+2}}{1-\alpha^2}}{\lambda_m(P)}}\|\phi\|$. where $\alpha = \beta^{-1}$ and $\|\phi\| = \sup_{-d \le l \le 0} \|x(l)\|$.

Proof: By the assumption, in order to show that the origin of (2) is exponentially stable, we can take the following Lyapunov-Krasovskii function:

$$V(x(k)) = \sum_{i=1}^{d} \beta^{2(k-i)} x^T(k-i)Qx(k-i) + \beta^{2k}x^T(k)Px(k), P > 0, Q > 0 \qquad (4)$$

where $\beta > 1$. The difference of this function along the solution of (2) is given by

$$\Delta V = V(x(k+1)) - V(x(k)) \leq \beta^{2(k+1)} \pi^T \Xi \pi$$

where $\pi =: [x^T(k)\ x^T(k-d)]^T$ and $\Xi = \begin{bmatrix} \Xi_{11} & \Xi_{12} \\ \Xi_{21} & \Xi_{22} \end{bmatrix}$ with

$\Xi_{11} =: CPC + MA^T PC + CPAM + MA^T PAM - \beta^{-2} P + \beta^{-2} Q$
$\Xi_{12} =: CPBM + MA^T PBM$
$\Xi_{21} =: MB^T PC + MB^T PAM$
$\Xi_{22} =: MB^T PBM - \beta^{-2d-2} Q$.

Thus, by using Schur complement formula (see [12] for details) with respect to (3), we obtain $\Xi < 0$. This shows that $\Delta V \leq \beta^{2(k+1)} \pi^T \Xi \pi < 0$ and hence $V(x(k)) \leq V(x(0))$ for any $k \in \mathbb{Z}^+$. On the other hand,

$$V(x(0)) \leq \lambda_M(P) \|\phi\|^2 + \lambda_M(Q) \|\phi\|^2 \sum_{i=-d}^{0} \beta^{2i}$$
$$= \left(\lambda_M(P) + \lambda_M(Q) \frac{\beta^{2d} - 1}{\beta^{2d+2} - \beta^{2d}} \right) \|\phi\|^2$$

where $\|\phi\| = \sup_{-d \leq l \leq 0} \|x(l)\|$. Moreover, $V(x(k)) \geq \beta^{2k} \lambda_m(P) \|x(k)\|$. So, $\beta^{2k} \lambda_m(P) \|x(k)\| \leq V(x(k)) \leq V(x(0)) \leq \left(\lambda_M(P) + \lambda_M(Q) \frac{\beta^{2d} - 1}{\beta^{2d+2} - \beta^{2d}} \right) \|\phi\|^2$.

Let $\alpha = \beta^{-1}$, we have $0 < \alpha < 1$ and $\|x(k)\| \leq \alpha^k \sqrt{\frac{\lambda_M(P) + \lambda_M(Q) \frac{\alpha^2 - \alpha^{2d+2}}{1 - \alpha^2}}{\lambda_m(P)}} \|\phi\|$. This guarantees that (2) is exponentially stable by Definition 1. □

Remark 2. For a given $\beta > 1$, (3) is an LMI, which can be solved numerically by LMI Toolbox. Therefore, to solve the matrix inequality, we can first search linearly $\beta > 1$, then we can solve numerically the LMI for this β.

3 A Numerical Example

Consider system (2) with the following parameters: $C = \begin{bmatrix} -0.1 & 0 \\ 0 & 0.1 \end{bmatrix}$, $A = \begin{bmatrix} -0.1 & 1 \\ 0.2 & 0.1 \end{bmatrix}$, $B = \begin{bmatrix} 0.5 & 0.4 \\ 0.15 & 0.5 \end{bmatrix}$. Taking $d = 10$, by searching linearly, the LMI (3) is feasible when $\beta = 1.220$ to obtain $P = \begin{bmatrix} 35.0291 & -6.0597 \\ -6.0597 & 93.4260 \end{bmatrix}$, $Q = \begin{bmatrix} 26.1137 & 7.0860 \\ 7.0860 & 63.9254 \end{bmatrix}$. At this case, $\alpha = 0.8197$ is the exponential convergence rate. In fact, if only $1 < \beta < 1.220$, the LMI (3) is feasible and hence the equilibrium point of (2) is globally exponentially asymptotically stable. Furthermore, $\beta = 1.220$ is corresponding to the fastest speed of exponential convergence with the degree $\alpha = 0.8197$ of exponential stability.

Taking $d = 5$, by searching linearly, the LMI (3) is feasible when $\beta = 1.437$. In this time, $\alpha = 0.6959$ is the exponential convergence rate. Similarly, if only $1 < \beta < 1.437$, the LMI (3) is feasible and hence the equilibrium point of (2) is globally exponentially asymptotically stable.

Remark 3. For a given exponential convergence rate α (corresponding to β), we can obtain the maximum admissible delay through the condition (3) using linear searching. For example, taking $\beta = 1.1$ (i.e., $\alpha = 0.9091$) in the above example, the maximum admissible delay is $d = 22$, that is, the LMI condition is feasible when $d \leq 22$, otherwise it is infeasible. From this example, the exponential convergence rate increases as the delay increases.

4 Conclusions

We have established a sufficient condition on global exponential stability for discrete CNN in terms of LMI. The new result concerning global exponential stability for discrete neural networks presented simultaneously an estimation of the exponential convergence rate. Moreover, a relation between the delayed time and the corresponding parameters of the network was established.

Acknowledgements. This work is supported by the NSFC (No. 60304014, 10372002 and 60274001), and innovation foundation from School of Science, Beihang University.

References

1. Chua, L. O., Yang, L.: Cellular neural networks: Theory. IEEE Trans. Circuits Syst. I. **35** (1988) 1257–1272.
2. Roska, T., Chua, L.O.: Cellular neural networks with nonlinear and delay-type template. Int. J. Circuit Theory Appl. **20** (1992) 469–481.
3. Arik, S., Tavsanoglu, V.: Equilibrium analysis of delayed CNNs. IEEE Trans. Circuits Syst. I. **45** (1998) 168–171.
4. Forti, M., Tesi, A.: New conditions for global stability of neural networks with applocation to linear and quadratic programming problems. IEEE Trans. Circuits Syst. I. **42** (1995) 354–366.
5. Cao, J.: Global stability conditions for delayed CNNs. IEEE Trans. Circuits Syst. I. **48** (2001) 1330–1333.
6. Liao, X., Chen, G., Sanchez, E. N.: LMI-based approach for asymptotically stability analysis of delayed neural networks. IEEE Trans. Circuits Syst. I. **49** (2002) 1033–1039.
7. Chu, T., Wang, Z., Wang, L.: Exponential convergence estimates for neural networks with multiple delays. IEEE Trans. Circuits and Syst. I. **49** (2002) 1829-1832.
8. Liang, J., Gupta, M. M.: Globally asymptotical stability of discrete-time analog neural networks. IEEE Transactions on Neural Networks. **7** (1996) 1024–1031.
9. Michel, A. N., Si, J., Yen, G.: Analysis and synthesis of discrete-time neural networks described on hypercubes. IEEE Transactions on Neural Networks. **2** (1991) 32–46.
10. Chu, T.: Convergence in discrete-time neural networks with specific performance. Phys. Rev. E, **63** (2001) 0519041.
11. Mohamad, S., Gopalsamy, K.: Dynamics of a class of discrete-time neural networks and their continuous-time counterparts. Math. Comput. Simulation. **53** (2000) 1–39.
12. Boyd, S., Ghaoui, L. E., Feron, E., Balakrishnan, V.: Linear matrix inequalities in system and control theory. Philadelphia, PA: SIAM, 1994.

A Novel Local Connection Neural Network*

Shuang Cong, Guodong Li, and Yisong Zheng

Department of Automation, University of Science and Technology of China,
Hefei, Anhui, 230027, P. R. China
scong@ustc.edu.cn

Abstract. A new type of local connection neural network is proposed in this paper. There is a called K-type activation function in its hidden layer so as to have less computation compared with other local connection neural network. First the structure and algorithm of the proposed network are given. Then the function of network and its properties are analyzed theoretically. The proposed network can be used in the function approximation and modeling. Finally, numerical applications are used to verify the advantages of proposed network compared with other local connection neural networks.

1 Introduction

Artificial neural network is a kind of distributed and concurrent network for information processing. Since the eighties of the 20th century, great progress has been made in artificial neural network theory[1]-[3]. Neural networks have been used in the non-linear system modeling and design of controller, pattern classification and recognition, associative memory and computing optimization, etc. If classified from general structure, artificial neural networks can be divided into feed-forward networks and feedback networks. BP network is the typical example of the former, and Hopfield network is that of the latter. However if classified from the point view of connection mode, artificial neural networks can also be divided into the global neural networks and local neural networks. BP network that is used widely in control field is a typical global network. Though it has the advantage of global approximation to the function, this kind of network has relatively slow learning speed. However if only part or even one weight is influenced by each of inputs/outputs, we call the networks 'local neural networks'. These networks have the faster learning speed, which is important to real-time control. Therefore, these networks are extensively used Generally speaking, local neural networks can be divided into two categories according to their fundamental functions. One category consists of Cerebellar Model Articulation Controller (CMAC), B-spline and Radical Basis Function (RBF) networks. Approximation of function and system modeling are their main functions. The other category consists of Adaptive Resonance Theory (ART) and Adaptive Competitive Network. They are used in pattern classification and recognition.

---

* The project supported by National Natural Science Foundation of China (No. 50375148).

This paper proposes a new type of local connection neural network-K-type local connection neural network (shorted by KLCNN). Compared with other common local neural networks, the proposed network has the advantages of less computational cost, higher precision in function approximation and better generalization capability. The paper first analyzes the structure, algorithm and character of KLCNN, then verifies its properties through application of function approximation in system modeling.

2 Structure and Property of KLCNN

The property of a neural network is determined by three parts: network structure, activation function in network and weights update algorithm through network training. Several common local connection networks are similar in the structure and weight training algorithm. They all have common three-layer network structure and use δ algorithm to update the weights. Thus the difference in performance of these networks lies in their different activation functions. In details, RBF networks have better performance in analyticity and generalization capability. CMAC and B-spline networks spend less computational time. In order for the novel network to acquire the advantages of above networks, square-reciprocal formation is chosen here as the activation function in its hidden layer. The input-output relation of the hidden layer is

$$a_i(p_j) = \frac{1}{1 + K*(p_j - \omega_{ij})^2} = f(\sqrt{K}*(p_j - \omega_{ij})) \quad (1)$$

in Equ. (1), p_j is the input vector of network. ω_{ij} is the weight between jth input node and ith node in hidden layer. $a_i(p_j)$ is the output of ith node in hidden layer caused by input p_j. K is parameter that shows the span of the function. We call it K-type local neural network because there is only a parameter K in which it is needed be regulated. Equ. (1) also shows that, when q groups of r-dimension input-vectors p_j^k ($k\square 1,2,\ldots,q; j=1,2$) are put into K-type network, the network calculates the distance $(p_j - \omega_{ij})$ between input vector P_j^k and the weight vector ω_{ij} in the layer, and uses this distance as the input of activation function. Then the input is squared, then added with 1 and finally divided by 1. The result is the output of K-type network.

K-type neural network is built with K-type function and the common structure of local neural network. The network has the three-layer forward structure. The first layer is the input-layer to receive the input vector. The second layer is the hidden layer with K-type function as active function. The third one is the linear output layer. The final output y_i of K-type local network can be obtained as $y_i = \sum_{i=1}^{s} w_i \alpha_i$ .Similar with other local networks, the weights in different layers are modified in their own layers. Through training, the ω_{ij} in hidden layer is equal to input vector or average of vectors. The weights in output layer are modified through δ method: $w_i(k+1) = w_i(k) + \eta*(p_j - \omega_{ij})a_i / a^T a$, In which, η is the learning rate, k is times of learning, a_i is output of hidden layer, a is normalized vector of a_i, p_j and ω_{ij} is input vector of network and weight in hidden layer.

As it is known, the property of local connection is determined by the activation function in hidden layer. Among the common local networks, CMAC and B-spline networks have advantages in less amount of calculation. However, Compared with the B-spline function, Gaussian primary function is smoother, and its derivatives of any higher order exist. Thus, RBF neural network not only has excellent approximation ability, but also has generalization capability. But compared with the CMAC and B-spline neural networks, RBF needs to adjust more connection weights when trained. The K-type network proposed here is hoped to acquire the advantage of above three networks. Fig. 1 shows the curves of first derivatives of K-type function and that of RBF function. It is seen that K-type function has the following property.

(1) Positive-definite character and symmetrical character

In case of K > 0, it is obvious that output of K-type function is above zero constantly. The function is positive and definite. Also the figure of function has the character of bilateral radial symmetry. Thus the K-type function meets the require of non-negative output as to any input.

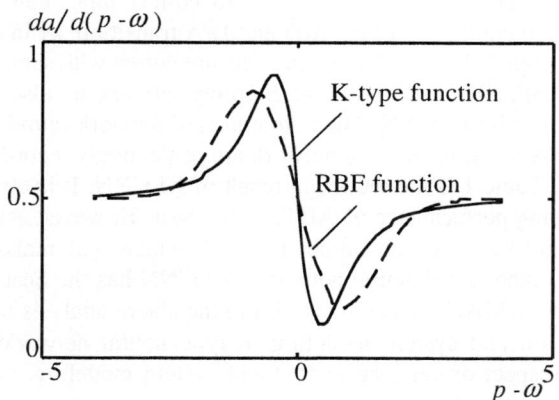

Fig. 1. Comparison of the curves of first derivative of K-type function and that of RBF function

(2) Convergence character

Convergence character ensures the stabilization of function. The integral value of K-type function in real number field can be obtained

$$\int_{-\infty}^{\infty} a_j(x) = \int_{-\infty}^{\infty} \frac{1}{1+K*(X-C_j)^2} dx = tg^{-1}\sqrt{K}(X-C_j)\Big|_{-\infty}^{\infty} \Big/ \sqrt{K}$$

$$= tg^{-1}\sqrt{K}X\Big|_{-\infty}^{\infty} \Big/ \sqrt{K} = \pi/\sqrt{K} < \infty \qquad (2)$$

In Equ. (2), K is positive constant. It is shown that K-type function satisfies the convergence character. As to the limited input, the function will not be divergent. Also by adjusting the lateral extensional coefficient K, the integral value of K-type function can be set optionally in constant field.

(3) Sequence analyticity
K-type network is the first derivative of arctan function. It is proved that derivatives of any order of arctan function are existent and continuous. Thus, derivatives of any order of K-type function exist. And K-type function has excellent ability of generalization as that of RBF function.

(4) Less computational cost
Compared with other local networks, KLCNN avoids the complex exponential calculation in RBF function, doesn't use basic spline to have differential calculation, and also needs not to create the legend space with large dimension as CMAC networks. Therefore, K-type neural network has obvious advantage in calculation amount.

3 Comparison of Numerical Applications and Their Analysis

In this application, KLCNN will be used to model a DC electric machine, which is effected by nonlinear fiction. In order to achieve high precision, the model structure with three-order delay is used. Input data is 4-dimision vector $P = [u\ (k) \Box y\ (k-1) \Box y\ (k-2) \Box y\ (k-3)]$, 4 components of which are the input of system in k, and output of system in k-1,k-2,k-3. The control system used to collect input and output signal includes a Pentium 200 computer, 12 bit A/D and D/A transition set in the computer. Input signal is PRBS signal. Then 500 sampling data are gotten with sampling period of 10ms in 5 seconds. RBF, ART-2, CMAC, and B-spline network are also used to model this system, in contrast with KLCNN. Their structures of network (number of nodes in hidden layer) and corresponding error of network are respectively recorded in Table 2, Table 3 and Table 4. Table 1 is the modeling result of KLCNN. It is obvious in these tables that the modeling performance of ART2 is the best. However, ART-2 network have to be rebuilt and be added an output layer [2], which will make the network complex. Among the other local neural networks, KLCNN has the best performance. RBF ranks the second. CMAC ranks the last. From the above analysis of the result of function approximation and system modeling, K-type neural network has the best performance in the respect of calculation cost and system modeling, compared with several common local networks.

Table 1. The modeling result for electric-machine with KLCNN

Model No.	1	2	3	4	5	6	7
Node No.	**15**	**18**	**24**	**27**	**30**	**35**	**38**
Error	3.16	2.53	2.51	2.49	2.45	1.59	1.23

Table 2. The modeling result for electric-machine with RBF

Model No.	1	2	3	4	5	6	7
Node No.	**19**	**24**	**26**	**29**	**34**	**46**	**52**
Error	3.45	3.11	2.94	2.77	2.69	2.49	2.06

Table 3. The modeling result for electric-machine with ART-2

Model No.	1	2	3	4	5	6	7
Guard value	0.975	0.981	0.983	0.985	0.990	0.994	0.995
Node No.	10	11	14	15	20	31	32
Error	2.57	2.47	2.43	2.36	2.18	2.12	2.03

Table 4. The modeling result for electric-machine with CMAC

Model No.	1	2	3	4	5	6	7
C value	15	20	25	30	40	50	70
Error	2.69	2.68	2.68	2.67	2.67	2.65	2.57

4 Conclusion

A new type of local connection neural network-K-type local connection neural network is proposed in this paper. From theoretical analysis and comparison of applications, KLCNN has several advantages over other common local networks. With little amount of calculation, KLCNN can achieve high degree of preciseness. At the same time, it has good generalization capability. Thus a new way is proposed in this paper for fast design and efficient application of local neural networks.

References

1. Shuang Cong, Yisong Zheng, The Analysis and Comparison of the Structure and Performance of Several Local Neural Networks, Computer Engineering, Vol. 29, No. 22 (2003)11-13
2. Shuang Cong, The Function Analysis and Application Study of Radial Basis Function Network, Computer engineering and application,Vol.38, No. 3 (2002)85-87, 200
3. Zengqi Sun, Intelligent control theory and application, Trsinghua Press, 1997
4. Albus, J. S., Data Storage in the Cerebellar Model Articulation Controller (CMAC), Transaction of ASME, J. Dynam. Syst. Meas. Control, Vol. 97(1975) 228-233
5. Shuang Cong, Yisong Zheng, Yiwen Wang, The Improvement and Modeling Implementation of ART-2 Neural Network, Computer engineering and applications, Vol. 38, No. 14(2002)25-27,4

An Unsupervised Cooperative Pattern Recognition Model to Identify Anomalous Massive SNMP Data Sending

Álvaro Herrero, Emilio Corchado, and José Manuel Sáiz

Department of Civil Engineering, University of Burgos, Spain
escorchado@ubu.es

Abstract. In this paper, we review a visual approach and propose it for analysing computer-network activity, which is based on the use of unsupervised connectionist neural network models and does not rely on any previous knowledge of the data being analysed. The presented Intrusion Detection System (IDS) is used as a method to investigate the traffic which travels along the analysed network, detecting SNMP (Simple Network Management Protocol) anomalous traffic patterns. In this paper we have focused our attention on the study of anomalous situations generated by a MIB (Management Information Base) information transfer.

1 Introduction

IDS are hardware or software systems that monitor the events occurring in a computer system or network, analysing them to automatically identify security problems.

Connectionist models have been identified as a very promising method of addressing the ID problem due to two main features [1]: their generalisation capability and their ability to classify patterns. Up to now, there have been several attempts to apply Artificial Neural Networks (ANN) (such as Self-Organising Maps [2], Elman Network [3]) to the network security field [4, 5].

Our IDS is based on a neural Exploratory Projection Pursuit (EPP) architecture. The aim of EPP [6, 7, 8, 9] is to reveal possible interesting structures hidden in the high-dimensional data so that a human can investigate the projections by eye.

2 A Novel Unsupervised Neural IDS Model

We can classify our IDS as a network-based [1] one because the data for the traffic analysis is obtained from the packets travelling along the whole network. This data can be extracted from the captured packets headers by using a network analyser.

We have developed a system for detecting anomalous traffic patterns; these include proper attacks and dangerous situations without being an attack.

The novel IDS model is structured as follows:

- **1st step.**- Network Traffic Capture: setting up one of the network interfaces as "promiscuous" mode, it can capture all the packets traveling along the network.
- **2nd step.**- Data Pre-processing: the captured data is pre-processed (see Section 3).

- **3rd step**.- Data Classification: once the data has been captured and pre-processed, the connectionist model presented below is used to analyse the data and to identify the anomalous patterns.
- **4th step**.- Result Display: this visualization tool displays data projections highlighting anomalous situations clearly enough to alert the network administrator, taking into account aspects as the traffic density or "anomalous" directions.

The Data Classification and Result Display steps performed by this IDS model are based on the use of a neural EPP architecture called Cooperative Maximum Likelihood Hebbian Learning (CMLHL) [10, 11]. It was initially applied to the field of Artificial Vision to identify local filters in space and time [10, 11]. It is based on the neural architecture called Maximum Likelihood Hebbian Learning (MLHL) [8, 9]. The final neural model (CMLHL) can be described as follows: consider an N-dimensional input vector, x, and an M-dimensional output vector, y, with W_{ij} being the weight linking input j to output i and let η be the learning rate.

$$\text{Feed forward: } y_i = \sum_{j=1}^{N} W_{ij} x_j, \forall i . \tag{1}$$

$$\text{Lateral activation passing: } y_i(t+1) = [y_i(t) + \tau(b - Ay)]^+ . \tag{2}$$

$$\text{Feedback: } e_j = x_j - \sum_{i=1}^{M} W_{ij} y_i . \tag{3}$$

$$\text{Weight change: } \Delta W_{ij} = \eta . y_i . sign(e_j) | e_j |^{p-1} . \tag{4}$$

We use the standard MLHL with lateral connections. These lateral connections [10, 11] have been derived from the Rectified Gaussian Distribution [12] and applied to the negative feedback network [13]. The resultant net [10, 11] can find the independent factors of a data set but do so in a way which captures some type of global ordering in the data set.

3 Real Intrusion Detection Scenario Specific Data Set

Among all the implemented network protocols, there are some of them that can be considered quite dangerous for the network security. Among those, we have focused our effort in the study of SNMP because an attack based on this protocol may severely compromise the network security.

In the short-term, SNMP was oriented to manage nodes in the Internet community [14] and the MIB can be defined as a database which contains information about some elements or devices that can be network-controlled. The data set used in this work contains a transfer of some information contained in a SNMP MIB. This kind of transfer is considered a quite dangerous situation because a person having some free tools, some basic SNMP knowledge and the community password (in SNMP v. 1 and v. 2) can come up with all sorts of interesting and sometimes useful information.

In the Data Pre-processing step the system selects packets based on UDP (User Datagram Protocol). In this step, the system also performs a data selection and only the following 5 variables (extracted from the packet headers) are used: **timestamp** (the time when the packet was sent), **protocol** (we have codified all the protocols contained in the data set), **source port** (the port number of the source host which sent the packet), **destination port** (the port number of the destination host where the packet is sent) and **size** (total packet size in Bytes).

4 Experimental Results, Conclusions and Future Work

Through a simple visual analysis of the figure Fig. 1.a, it is easy to identify several packet groups. Two of them (Groups 1 and 2 in Fig. 1.a) are different from other groups related to normal traffic as it is explained below. The packets belonging to each protocol contained in the data set are identified and visualized in the same group in Fig. 1.a, except in the case of SNMP packets (this case is explained later).

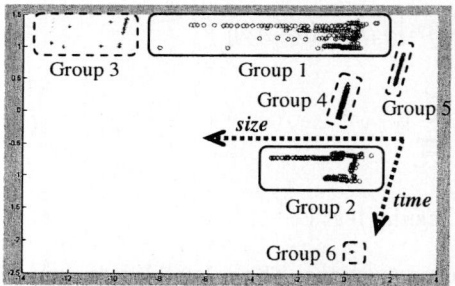 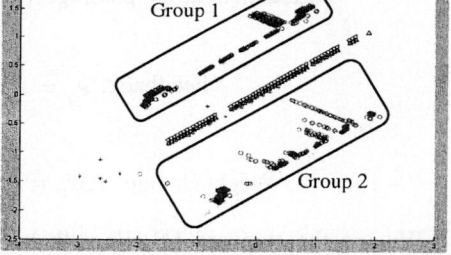

Fig. 1.a. Data projection displayed by the our neural IDS model

Fig. 1.b. Best Maximum Likelihood Hebbian Learning projection

After an analysis (labeling and studying most of the represented points) of the packets belonging to Groups 1 and 2 (Fig. 1.a) we have identified several features:

- These Groups are related to the SNMP MIB transfer mentioned above. They contain packets sent and received during the transfer embedded in the data set. All the packets belonging to SNMP are contained in one of these two groups.
- Group 1 contains all the traffic going from destination to source, while Group 2 contains all the traffic in the other way (from source to destination).
- Each group extends over two main axes: one related to the packet size and the other related to the timestamp.

We have labeled Groups 1 and 2 (Fig. 1.a) as anomalous ones due to two combined issues: the high temporal concentration of packets, and because they are made up of different size packets, situation related to the MIB information transfer.

This IDS model has been previously used to identify a SNMP port sweep [5] and it worked properly, identifying the anomalous situation in a very clear way.

We have applied different ANN such as Principal Component Analysis (PCA) [15, 16] or MLHL (Fig. 1.b) to the same data set. PCA is not able to detect the anomalous situation contained in the data set, because it shows the "anomalous" packets in the same way in which the rest of the traffic is shown. Fig. 1.b shows how MLHL is capable of detecting the anomalous situation (Groups 1 and 2) but it is not detected as clearly as by using CMLHL (Fig. 1.a).

As conclusions, there are some issues to highlight: The visualization tool can show the packets grouped by their protocol and only the network administrator has the authority to decide whether a situation classified as anomalous is dangerous or not.

Future work will have the following work lines: the application of grid computation [17] in both Data Classification and Result Display steps and the use of distributed systems based on agents and multiagents and the exchange of information.

Acknowledgments

This research has been supported by the McyT projects: TIN2004-07033.

References

1. Planquart, J-P.: Application of Neural Networks to Intrusion Detection. Information Security Reading Room - SANS (SysAdmin, Audit, Network, Security) Institute (2002)
2. Hätönen, K., Höglund, A.,Sorvari, A.: A Computer Host-Based User Anomaly Detection System Using the Self-Organizing Map. International Joint Conference of Neural Networks (2000)
3. Ghosh, A. Schwartzbard, A., Schatz, A.: Learning Program Behavior Profiles for Intrusion Detection. Workshop on Intrusion Detection and Network Monitoring (1999)
4. Debar, H., Becker, M., Siboni, D.: A Neural Network Component for an Intrusion Detection System. IEEE Symposium on Research in Computer Security and Privacy (1992)
5. Corchado, E., Herrero, A., Baruque, B., Saiz, J.M.: Intrusion Detection System Based on a Cooperative Topology Preserving Method. International Conference on Adaptive and Natural Computing Algorithms. SpringerComputerScience. SpringerWienNewYork, (2005)
6. Friedman, J., Tukey, J.: A Projection Pursuit Algorithm for Exploratory Data Analysis. IEEE Transaction on Computers (23) (1974) 881-890
7. Hyvärinen, A.: Complexity Pursuit: Separating Interesting Components from Time Series. Neural Computation 13 (2001) 883-898
8. Corchado, E., MacDonald, D., Fyfe, C.: Maximum and Minimum Likelihood Hebbian Learning for Exploratory Projection Pursuit. Data Mining and Knowledge Discovery. Kluwer Academic Publishing 8(3) (2004) 203-225
9. Fyfe, C., Corchado, E.: Maximum Likelihood Hebbian Rules. ESANN(2002)
10. Corchado, E., Han, Y., Fyfe, C.: Structuring Global Responses of Local Filters Using Lateral Connections. JETAI 15 (4) (2003) 473-487
11. Corchado, E., Fyfe, C.: Connectionist Techniques for the Identification and Suppression of Interfering Underlying Factors. International Journal of Pattern Recognition and Artificial Intelligence 17(8) (2003) 1447-1466

12. Seung, H.S., Socci, N.D., Lee, D.: The Rectified Gaussian Distribution. Advances in Neural Information Processing Systems 10 (1998) 350
13. Fyfe, C.: A Neural Network for PCA and Beyond. Neural Processing Letters 6 (1996)
14. Case, J., Fedor, M.S., Schoffstall, M.L., Davin, C.: Simple Network Management (SNMP). RFC-1157 (1990)
15. Postel, J.: IAB Official Protocol Standards. RFC-1100 (1989)
16. Oja, E.: Neural Networks, Principal Components and Subspaces. International Journal of Neural Systems 1 (1989) 61-68
17. Foster, I., Kesselman, C.: The Grid: Blueprint for a New Computing Infrastructure. 1st edition. Morgan Kaufmann Publishers (1998)

A Fast Nonseparable Wavelet Neural Network for Function Approximation[*]

Jun Zhang, Xieping Gao, Chunhong Cao, and Fen Xiao

Member, IEEE
Information Engineering College, Xiangtan University, 411105, China
Zhangjun7907@hotmail.com; xpgao@xtu.edu.cn

Abstract. In this paper, based on the theory of nonseparable wavelet, a novel nonseparable wavelet model has been proposed. The structure of the model is distinguished from that of wavelet network (RBF structure). It is a four-layer structure, which helps overcome the structural redundancy. In the process of the training of the network, in the light of the characteristics of nonseparable wavelet, a novel method of setting the initial value of weight has been proposed. It can overcome the shortcoming of gradient descent methodology that it makes the convergence of the network slow. Some experiments with the novel model for function learning will be shown. Comparing with the present wavelet networks, BP network, the results in this paper show that the speed and generalization performance of the novel model have been greatly improved.

1 Introduction

The idea of using wavelets in neural networks has also been proposed recently by Zhang, Benveniste, Pati, and Krishnaprasad [1][2][3]. The basic idea is to replace the neurons by "wavelons," i.e., computing units obtained by cascading an affine transform and a multidimensional wavelet. For the multidimensional signal, multidimensional "wavelons" have to be constructed in the present wavelet network models. The most commonly used method is the tensor product of univariate wavelets [1][2][3]. This construction leads to a separable wavelet that has a disadvantage of giving a particular importance to the horizontal and vertical directions [8].

At present, the nonseparable wavelet becomes the focus of the wavelet theory [7][8]. In the process of dealing with the multidimensional problems, the characteristics of nonseparable wavelet are better than ones of tensor product. It is reasonable that in the wavelet networks, if better neurons are selected, better results may be obtained.

Since the present wavelet networks successfully preserve most of the advantages of the RBF network, so far, in the field of wavelet network, most researches are focused on various models based on the development of the wavelet theory and learning algorithms. Nevertheless, few researches are focused on the structure of wavelet network. In fact, as to the whole signal, the orthogonal wavelet based on network can be

---

[*] This work was supported by the National Science Foundation of China (Grant No.60375021) and the Science Foundation of Hunan Province (Grant No.00JJY3096) and the Key Project of Hunan Provincial Education Department (Grant No.04A056)

constructed and it is not redundant. However, as to some parts of the signal, only some of neurons are useful, and the others are redundant. If we properly change the structure of the present wavelet network, the various advantages of RBF network can be preserved and at the same time, the redundancy can be overcome effectively.

In this paper, a novel nonseparable wavelet network model and a novel method of setting the initial value of weight have been proposed. Some experiments with the novel model for function learning will be shown. Comparing with the present wavelet networks [1][2][3] and BP network, the results in this paper show that the speed and generalization performance of the novel model have been greatly improved.

2 The Novel Nonseparable Wavelet Network

Throughout this paper, let R, Z and N denote the set of real, all integers and natural numbers respectively. The d-D MRA (multiresolution analysis) with a dilation matrix D is a ladder of closed subspaces $\{V_j\}_{j \in Z}$ which approximates $L^2(R^d)$ and satisfies

$$\{0\} \to \dots V_{-1} \subset V_0 \subset V_1 \dots \to L^2(R^d) \; ; \; f(x) \subset V_{j-1} \Leftrightarrow f(Dx) \subset V_j$$

$\exists \phi \in V_0$ s.t. the set $\{\phi(D^m x - k)\}_{k \in Z^d}$ is an orthonormal basis for V_m

The function $\phi(x)$ is called scaling function.

The above discussions suggest a scheme for decomposing a $L^2(R^d)$ function $f(x)$: For some integer M and any $\varepsilon > 0$, there exists an M sufficiently large such that

$$\left\| f(x) - \sum_n < f, \varphi_{M,n} > \varphi_{M,n}(x) \right\| < \varepsilon \quad \text{Where the norm is the } L^2 \text{ norm.} \tag{1}$$

Without loss of the generality, a following two-hidden layer wavelet networks is set up. It has realized a $R^d \to R$ mapping. Its structure is as Fig.1.

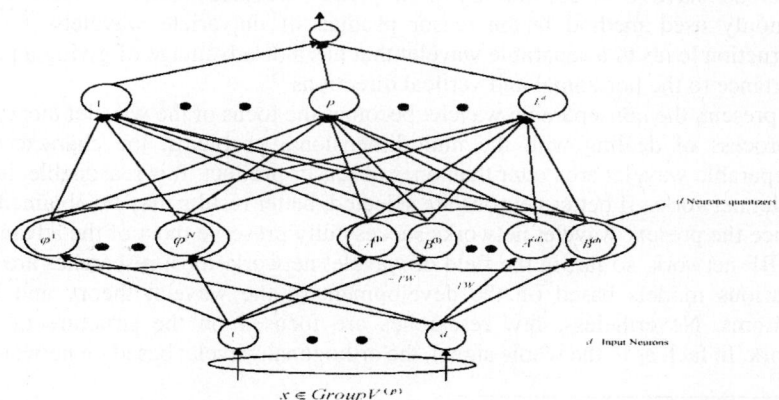

Fig. 1. Four-layer nonseparable wavelet network

As shown in Fig.1, the novel wavelet network has two hidden layers. The first-hidden layer consists of N "φ neurons" and d neurons quantizers; d is the dimension of the input data. Each neuron quantizer consists of two sigmoidal neurons called $A-type$ neuron and $B-type$ neuron respectively. $A-type$ Neuron and $B-type$ neuron of $j-th$ quantizer are denoted as neuron $A^{(j)}$ and neuron $B^{(j)}$ respectively. The outputs of $j-th$ quantizer for each neuron are denoted as $O_{A_p^{(j)}} + O_{B_p^{(j)}}$ $1 \le j \le d$. All of the neurons in the input layer link to all of the "φ neurons", but the $i-th$ input neuron just links to the $i-th$ quantizer; $i = 1,...,d$. The second layer has L^d neurons (L is arbitrarily plus integral value). The neurons of the first hidden layer link to all of the neurons of the second hidden layer.

Definition 1: For a $d-$dimension signal f, suppose $\Gamma_L \stackrel{\Delta}{=} \{f \mid f = f_1 \oplus f_2 \oplus ... \oplus f_{L^d}\}$, Let f_i denote the i-th sub-signal of f that is divided continuously into L^d equidistant shares according to the support of each dimension.

As to the distinct samples (x_i, t_i), where $x_i = [x_{i1}, x_{i2},..., x_{id}]^T \in R^d$ and $t_i \in R$, since all the given x_i are distinctive, without loss of the generality, suppose the support of signal is $[0, \alpha]^d$, a plus vector W can be randomly chosen. They can be separated into L^d groups according to the support of each dimension.

$V_j^{(p)} = \{x_{ij} \mid W \cdot \frac{(p-1) \cdot \alpha}{L} \le W \cdot x_{ij} \le W \cdot \frac{p \cdot \alpha}{L}\}$, Where $1 \le p \le L$, $1 \le j \le d$

$GroupV^{(q)} = \{x_i \mid W \cdot \frac{(p_j - 1)\alpha}{L} \le W \cdot x_{ij} \le W \cdot \frac{p_j \alpha}{L}$, Where $1 \le q = \sum_{j=1}^{d} p_j \cdot L^{(j-1)} + 1 \le L^d$

$1 \le i \le n, 1 \le j \le d, 1 \le p_j \le L\}$

According to the above discuss, we know that $GroupV^{(i)} \subseteq f_i$

The training of novel wavelet networks mainly consists of two phases: (1) determination of weights and bias of the "φ neurons" and (2) determination of weights and biases of neural quantizers.

2.1 Determination of Weights and Bias of the "φ Neurons"

As the theory of the wavelet, the weight (w_i) that joins the inputs and the i-th neuron $i \le N$ in the first hidden layer and the bias b_i of the i-th neuron ($i \le N$) in the first hidden layer can be determined as reference [1]~[3]. The weights that join the "φ neurons" and the second hidden layer can be determined by the Gradient descent methodology that has been widely used in the learning algorithms of the neural networks, wavelet networks, e.g., back propagation (BP) algorithm and its variants. But the parameters of the network are updated gradually in each iteration, the learning is apparently time consuming. In fact, if a reasonable initial value of the weight can be set, we can effectively overcome the shortcoming of gradient descent methodology.

For 2-D nonseparable case, some definitions and notations [4] are made as below:

Definition 2: Let $M = \begin{pmatrix} a & b \\ c & d \end{pmatrix}$, and $\varphi(x)$ satisfies $\varphi(x) = |\det M| \sum_{k \in Z^2} c_k \varphi(Mx - k)$

$\sum_{k \in Z^2} c_k = 1$ and $\int_{R^2} \varphi(x)dx = 1$, then

$c_1 = \frac{1}{Q}[(\det M|d - \det M)\sum k_1 c_k - \det Mb \sum k_2 c_k]$ $c_2 = \frac{1}{Q}[-c \det M \sum k_1 c_k + (\det M|a - \det M)\sum k_2 c_k]$

where $Q = |\det M|^2 - |\det M|(c+d) + \det M \neq 0$, $k = (k_1, k_2)$

For $f \in L^2(R)$, That is, for any $\varepsilon > 0$, there exists an J_0 sufficiently large such that:

$\|f - f_J\|_2 < \varepsilon$ Where the norm is the L^2 norm. $J > J_0$

$$f_J = |\det M|^{-j/2} \sum_{q \in \wedge} f[M^{-j}(q+c)]\varphi_q^j(x,y) \quad j \in Z \tag{2}$$

$$\varphi_q^j(x,y) = |\det M|^{j/2} \varphi(M^j(x,y) - q) \tag{3}$$

$$\wedge = \{q = (q_1, q_2) \in Z^2 \mid \sup p\varphi_q^j(x) \cap \sup pf \neq \phi\} \tag{4}$$

$$f[M^{-j}(q+c)] = 2^j \int_{R^2} f(x,y)\varphi_q^j(x,y)dxdy \quad j \in Z \tag{5}$$

For some samples, the corresponding approximate value of $f[M^{-j}(q+c)]$ can always be obtained, which is regarded as the initial value of weights of the network. [4] proves that for all the $f(x) \in C^N(\overline{\Omega})$, $C^N(\overline{\Omega})$ denotes the space of N-order differentiable function that is finite support:

$$\|f - f_J\|_{L^2(\Omega)} \leq C \|f^{(N)}\|_\infty \sigma_2^{-jN} \tag{6}$$

2.2 Determination of Weights and Biases of Neural Quantizers

As show in Fig.1, the weights of connections linking the inputs to neuron $A^{(p)}$ and neural $B^{(p)}$ can be chosen as $\overline{W}_{A^{(p)}} = T_p \cdot W$ and $\overline{W}_{B^{(p)}} = -T_p \cdot W$. The T_p, $1 \leq p \leq d$ can set as following:

$Tset = \{T_p \mid T_p = \dfrac{\ln(\dfrac{1-\eta}{\eta})}{\min_{\substack{1 \leq i \leq n \\ 1 \leq p \leq L}} (|W \cdot x_{i,j} - W \cdot \dfrac{p \cdot \alpha}{L}|)}, 1 \leq j \leq d\}$. $\eta < 0.5$ is arbitrarily

small positive value. When $T_j \geq T_j^A, 1 \leq j \leq d$, for $\forall x_i \in GroupV^{(q)}, q = 1,...,L^d$, the corresponding input $Input_m(x_i), m = 1,...L^d$ of m th-neuron in the second-hidden layer satisfies

$$Input_m(x_i) \begin{cases} \leq \eta, & \text{if } q = m \\ \geq 1-\eta, & \text{if } q \neq m \end{cases} \tag{8}$$

The biases $\overline{b}_{A^{(p)}}$ and $\overline{b}_{B^{(p)}}$ of neuron $A^{(p)}$ and neuron $B^{(p)}$ are simply analytically calculated as

$$\begin{cases} \overline{b}_{A_p^{(j)}} = -T(W \cdot \dfrac{p \cdot \alpha}{L}) & 1 \le j \le d, 1 \le p \le L \\ \overline{b}_{B_p^{(j)}} = T(W \cdot \dfrac{(p-1) \cdot \alpha}{L}) & 1 \le j \le d, 1 \le p \le L \end{cases} \quad (9)$$

For any input x_i within input vector group $GroupV^{(q)}$, $q = 1,...,L^d$, only the q th-neuron's input are almost zero while one of the inputs of other neurons is almost one in the second-hidden layer.

3 Experimental Results

The nonseparable scaling function is selected from the [8] as the activating function of the former N neurons of the first hidden layer in our model. Then we choose the same non-liner functions in the reference [1]~[3] are chosen as the approximate functions. Additionally, some other non-liner functions are chosen to do the experiments of approximate functions. To assess the approximation results, a figure of merit is needed. We select the same figure of merit in the [2]

For the input datum $T_n = \{(x_i, t_i)\}_{i=1}^{n}$ and the network output $\hat{t}_{M,i}$.

$$error = \sqrt{\dfrac{\sum_{i=1}^{n}[\hat{t}_i - t_i]^2}{\sum_{i=1}^{n}(\bar{t} - t_i)^2}}, \; \bar{t} = \dfrac{1}{n}\sum_{i}^{n} t_i$$

The computing environment as following: Intel P4 1.7G CPU, 256M RAM and MATLAB 6.5.

Function 1[1][2][3] $z = 400 (x^2 - y^2 - x + y)\sin(10x - 5)$ $x, y \in [0,1]$

Function 2
$z = 1/2 \times [((5.12 \times x \times (x+1)/2)^2 - 10 \times \cos(2\pi \times (5.12 \times (x+1)/2)) + 10) + ((5.12 \times y \times (y+1)/2)^2 - 10 \times \cos(2\pi \times (5.12 \times (y+1)/2)) + 10)$ $x, y \in [0,1]$

Function
$z = \left\{\sum_{i=1}^{5} i\cos[(i+1)x + 1]\right\}\left\{\sum_{i=1}^{5} i\cos[(i+1)y + 1]\right\} + 0.5[(x + 1.42513)^2 + (y + 0.80032)^2]$ $x, y \in [0,1]$

Table 1

Model	Number of samples	Hidden Neurons	Epochs	RMSE Of Testing	Time(s)
Zhang [2]	400×400	49	40000	0.03395	21300
Pati [3]	400×400	187	500	0.023	500
BP [3]	400×400	225	40000	0.29381	95640
Our model	256×256	1024	1	0.0198	7.9360

Table 2

Functions	Number of samples	Hidden Neurons	Epochs	RMSE Of Testing	Time(s)
Function 2	512×512	1792	1	0.0117	15.9850
Function 4	512×512	1792	1	0.0485	15.8720

From the datum, our model also suits the function with the high degree of leaping.

4 Conclusions

In this paper, a novel four-layer nonseparable wavelet network for function learning is described. Some of the experiments on function learning had been done. Compared with the result of the reference [1]~[3], the model obtains better generalization performance and has remarkable improvement in speed. At present, multidimensional nonseparable wavelets are far from being well understood. However, the topic that nonseparable wavelets relate with neural network is very attractive. For future work, to investigate nonseparable wavelet networks in some real-world large-scale applications would be of great interest.

References

1. Q.Zhang., A.Benveniste.: Wavelet network [J]. IEEE Trans. On NN. 3 (1992) 889-898
2. J.Zhang. (ed.): Wavelet neural networks for function learning [J]. IEEE Trans.On SP. 6 (1995) 1485-1497
3. Pati Y C., Krishnaprasad P S.: Analysis and synthesis of feedforward neural networks using discrete affine wavelet transformations. IEEE Trans Neural Networks. 4 (1993) 73-85
4. En-Bing Lin., Yi Ling.: 2-D nonseparable scaling function interpolation and approximation. Acta Math.Sci.22 (2002) 19-31
5. X.P.Gao., B.Zhang.: Interval-wavelets neural networks(1)—theory and implements. Journal of software. 9 (1998) 217-221
6. X.P.Gao., B.Zhang.: Interval-wavelets Neural Networks(Ⅱ)—Properties and Experiment. Journal of software. 9 (1998) 246-250
7. E.Belogay,, Y.Wang.: Arbitrarily smooth orthogonal non-separable wavelets in R^2. SIAM J.Math.Anal. 3 (1999) 678-697
8. W. He., M. J. Lai.: Examples of bivariate nonseparable compactly supported orthonormal continuous wavelets. IEEE Trans. Image Processing, 9 (2000) 949--953

A Visual Cortex Domain Model for Illusory Contour Figures

Keongho Hong and Eunhwa Jeong

Information and Communication Division, Cheonan University,
115, Anseodong, Cheonan, Chungnam, 330-794, Republic of Korea
{khhong, ehjeong}@cheonan.ac.kr

Abstract. This study proposes a novel method that can recognize illusory contour figures by using a neural network model referenced on the mechanism of feature extraction found in a visual cortex domain. A common factor in all such illusory contour figures, such as the Kanizsa triangle is the perception of a surface occluding part of a background, i.e. illusory contours are always accompanied by illusory surfaces. In this paper, we propose a neural network model that predicts the shape of illusory surfaces based on features of the visual cortex domain. This model employs an important two-stage process of the Induced Stimuli Extraction System (ISES) and Illusory Surfaces Perception System (ISPS). The former system extracts the induced stimuli for the perception of illusory surfaces, and the latter forms the illusory surfaces from the induced stimuli. The proposed model is demonstrated on a variety of Kanizsa-type illusory contour displays. The results of the experiment shows that the proposed model is successful not only in extracting the induced stimuli for the perception of illusory contours, but also in perceiving the illusory surface figures from the induced stimuli.

1 Introduction

It is common to emphasize the importance of image contours because of their relationship to object boundaries and surface discontinuities in the scene (e.g., Marr, 1982). Often, object boundaries and surface discontinuities exist as luminance changes in the image, but this is not always the case. Although we frequently perceive clear perceptual boundaries between image regions, the physical bases for these percepts might be very slight. Schumann (1904) described the first experiments with stimuli in which contours are perceived without intensity gradients (See Fig.1). This class of figures was virtually forgotten until fifty-five years later when Kanizsa (1955) created a series of new, and more powerful variants of this type of figure which he called "quasi-perceptive margin figures" and which have since been called "illusory contour", "subjective contour", and "anomalous contour" figures. These kinds of illusory contour figures are shown in Fig.1. The Kanizsa square in (a), the distorted triangle in (b), the foot in (c) and the vertical boundary and circle in (d) is all defined by boundaries that are not made explicit. In recent years attention has been increasingly devoted to illusory contours (e.g., Gurnsey & Humphrey & Kapitan, 1992). The Kanizsa-type illusory contours (Fig. 1(a), (b) and (c)) and offset grating contours

(Fig.1 (d) and (e)) are among the most commonly studied illusory contours. Our experimental model deals with the Kanizsa-type illusory contour.

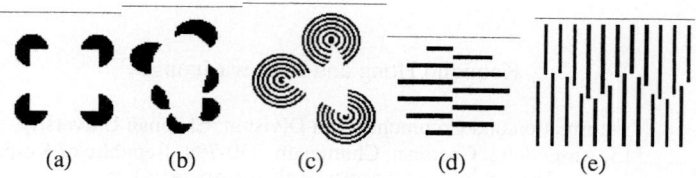

Fig. 1. Some examples of illusory contour Figures. The Kanizsa square in (a), the foot in (b), the distorted triangle in (c), the vertical boundary and circle in (d) and the square boundary and wave in (e) are all defined by boundaries that are not made explicit

Many computational models have been proposed to describe the formation of illusory contours, including Ullman(1976), Heitger & von der Heydt(1993), Grossberg & Mingolla(1985) and Grossberg(1994), and Guy & Medioni(1992), Kellman & Shipley(1995) Williams and Hanson(1994), Williams & Jacobs(1995), Brady & Grimson(1982) and Nitzberg & Mumford(1990). Unlike these computational models, we proposed a neural network model for perceiving illusory surfaces including depth sensations based on the mechanism of feature extraction found in a mammal's visual pathway. This mechanism is proposed by Hubel & Wiesel's paper.

Our model can be divided into two systems which are described as neural networks with multiple layers: (1) the Induced Stimuli Extraction System(ISES) for the perception of illusory contours including depth information and (2) the Illusory Surfaces Perception System(ISPS). The induced stimuli extraction system(ISES) extracts the induced stimuli needed for the perception of occluded surfaces from illusory contours, except for the inducers(background images), which in turn hide parts of the pattern. The illusory surfaces perception system(ISPS) forms the illusory surfaces, which must always be seen to be above the plane of the inducers, from the induced stimuli. Each system composes neural network architecture with multiple layers.

2 The Visual Cortex Domain Model

2.1 Induced Stimuli Extraction System (ISES)

The ISES extracts the induced stimuli needed for the perception of surfaces from illusory contours, except for the inducers, which in turn hide parts of the pattern. This system consists of six layers: image acquisition, contraction extraction by LGN, simple visual features extraction, visual feature restoration, Induced stimuli extraction and image enhancement. In the image acquisition we remove the process of color classification and simply convert a color image to a binary one because the color classification in virtual contour figures is not of great significance. The contrast extraction by LGN detects low-level features such as contrast using spatial filtering. The simple visual features extraction detects the presence of simple visual features, such as lines and edges of a particular orientation. This filter corresponds to a simple cell receptive field found in the mammal's visual system. The visual feature restoration responds to stimuli such as lines and edges of a particular orientation without the exact location of the stimulus. This filter corresponds to a complex cell receptive field found in the mammal's visual system. The Induced stimuli extraction extracts induced stimuli from the illusory contour figures using hypercomplex cells that are light-dark stimuli containing corners, curves and broken lines. The hypothetical arrangement of complex cells can implement an end-stopped hypercomplex receptive field. Image enhancement recovers the weaken or reduced stimuli. A set of two dimentional Gaussian filters is used. For removing unnecessary noises, the system performed image operation with input image.

2.2 Illusory Surfaces Perception System (ISPS)

The ISPS forms the illusory surfaces, which must always be seen to be above the induced stimuli. This system consists of 4 layers: Response Extraction between the induced stimuli, Response Restoration, Illusory surface extraction and Image Smoothing. Response extraction between the induced stimuli detects responses between the induced stimuli of the ISES. A set of asymmetrical two-dimensional three Gaussian filters for eight preferred orientations is used. This filter corresponds to a simple cell receptive field found in mammal's visual cortex domain[4]. Response restoration recovers responses between the induced stimuli using a set of two-dimensional Gaussian filters for eight orientations. This filter corresponds to a complex cell receptive field. Illusory surface extraction forms illusory surface from the extracted stimuli using image operation and feedback process. The output of this layer repeats to input image of ISPS until removing the gap between the induced stimuli and forming surface. For image improvement image smoothing performs.

3 Experimental Results

In order to show the performance of the model, experiments have been carried out using various Kanizsa-type illusory contour figures. The color classification in perceiving occluded surfaces from illusory contours is not of great significance.

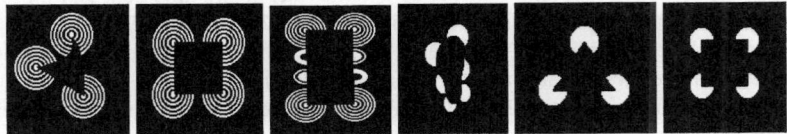

Fig. 2. Original Images

Then, for simplification, we remove the process of color classification, and simply convert a color image to a binary one. Examples of the experiments are shown in Fig.2, Fig.3, and Fig. 4. In most of the cases, illusory surfaces including depth sensation can be extracted. These results show that the performance of the model is sufficiently general.

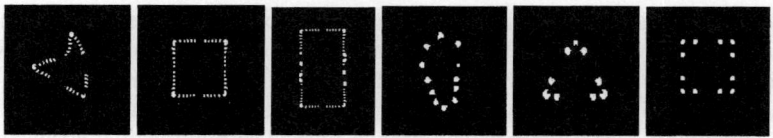

Fig. 3. Images of Induced Stimuli Extraction System

Fig. 4. Output Images of the Illusory Surface Perception System

4 Conclusion

A novel approach for the perception of occluded surface from Kanizsa-type illusory contour figures is proposed using the mechanism of feature extraction in visual cortex domain. Our model consists of two-stage coherent processing systems with multiple layers: the ISES and the ISPS. By the processing steps of each system the former system extracts the induced stimuli for the perception of illusory surfaces and the latter forms the illusory surfaces from the induced stimuli. The proposed model is demonstrated on a variety of Kanizsa-type illusory contour displays. The results of the experiment showed that the proposed neural network model was successful not only in extracting the induced stimuli for the perception of illusory contours, but also in perceiving the illusory surface from the induced stimuli.

References

1. G. Kanizsa..: Organization in Vision, Praeger, New York, (1979)
2. Franco Purghe, Staniey Coren.: Amodal completion, depth straitification, and illusory figures: a test of Kanizsa's explanation, Perception, Volume 21, (1995) 325-335
3. 3. Hubel D. H. and Wiesel T. N.: Receptive fields binocular interaction and functional architecture in the cat's visual cortex, J Physiology, (1962) 106 -154
4. S. Ullman.: Filling in the gaps: The shape of subjective contours and a model for their generation, Biological Cybernetics, (1976) 25:1-6
5. S. Grossberg.: 3-d vision and figure-ground separation by visual cortex. Perception & Psychophysics, 55(1), (1994) 48-120
6. S.Grossberg and E. Mingolla.: Neural dynamics of perceptual grouping: textures, boundaries and emergent segmentations. Perception & Psychophysics, 38(2) (1985) 141-170
7. Heitger F. and R. von der Heydt.: A Computational model of neural contour processing: Figure-ground segregation and illusory contours. Proc. of the IEEE, (1993) 32-40
8. Davi Geiger, Krishnan Kumaran, Leonid Gurvits.: Illusory Surface and Visual Organization, Technical Report, Courant Institute, NYU, (1996)

ANN Ensemble Online Learning Strategy in 3D Object Cognition and Recognition Based on Similarity[1]

Rui Nian, Guangrong Ji, Wencang Zhao, and Chen Feng

College of Information Science and Engineering, Ocean University of China, 266003, China
nianrui_80@163.com, grji@mail.ouc.edu.cn,
wencangzhao@mail.edu.cn, fccjg@sdu.edu.cn

Abstract. In this paper, in aid of ANN ensemble, a supervised online learning strategy continuously achieves omnidirectional information accumulation for 3D object cognition from 2D view sequence. The notion of similarity is introduced to solve the paradox between information simplicity and accuracy. Images are segmented into homogeneous region for training, correspondent to distinct model views characteristic of neighboring generalization. Real-time techniques are adopted to expand knowledge until satisfactory. The insert into joint model views is only needed in case of impartibility. Simulation experiment has achieved encouraging results, and proved the approach effective and feasible.

1 Introduction

Cognitive science is an interdisciplinary study of intelligence and mind, embracing philosophy, psychology, artificial intelligence, neuroscience, brain theory, linguistics, and anthropology. Computer vision provides the most enhancements to intelligent cognition, where pattern recognition plays a main role. Most approaches in object recognition can be categorized as geometry-based and appearance-based. The former explicitly stores volume or surface representation relying on 3D geometry. The latter directly compares and matches 2D images rather than 3D objects by similarity measure based on intensity, geometry, topology or their combination, results in significant reduction in dimensionality [1, 2]. Recently, Artificial Neural Networks (ANN) is also increasingly widely used in object recognition [1, 3].

3D object recognition from 2D view sequence can be treated as human vision simulation within an intelligent machine, which should emphasize most on knowledge cognition one by one, rather than optimal interface classification simultaneously [1,3]. Because of the paradox between information simplicity and accuracy in typical view selection, proper and sufficient sample set is too difficult to fully determine in advance, so the best way is to acquire solution step by step. Based on the notion of similarity, in aid of ANN ensemble with multiple weights, we present a supervised online learning strategy to achieve complete object geometrical coverage gradually and continuously, learning characteristic model views as few as possible, with old knowledge partly replaced and updated by the new one.

---

[1] The National 863 Natural Science Foundation of P. R. China (2001AA635010) fully supported this research.

2 Similarity Measure for Simplicity and Accuracy Tradeoff

Multiple-angle views offer overall object cognition. With high dimensional manifold as topology nature and homologous connectivity law as pre-acquired knowledge, an optimal spatial geometrical coverage is intended to establish for each object [3, 4]. However, simplicity and accuracy is a pair of paradox. Based on similarity measure, strategies such as model view generation and online learning are introduced in order to balance the competing aims. Similarity, or correlation among view points, is the key in separability. Different similarity metrics are employed to measure distance. Due to each metric nature and relative shape weighting, model view generation results in different prototypes or characteristic view. Besides object identity, similarity metric also gives initial pose estimation, which is essential to knowledge expansion on how to effectively combine newly increased samples with already cognized data. The selection of proper similarity metrics still need to pay more attention to in the future.

3 Model View Generation

Let there be N 3D objects $O_1, O_2, \cdots, O_n, \cdots, O_{N-1}, O_N$, each composed of M 2D images sampling the viewing sphere, $I_1^1, \cdots, I_m^n, \cdots, I_M^N$, with I_m^n denoting the mth image of O_n, so the whole image database consists of $N \bullet M$ images. Instead of training the full set of images, a model view learning procedure is introduced to employ few views, each representing a moderate range of possible appearances. Preprocessed images are clustered into groups. Members in each group are then generalized to form a view characteristic of the neighbors. On the assumption of sampling sufficiency, model views are reasonably selected on principle, such as homologous continuity, local monotonicity, cluster distinctiveness and separability, which are imposed to maintain successful recognition [1, 2]. A pair of cluster boundaries could be derived in an iterative scheme. Model view is what minimizes the distance to all others in a cluster. Cluster number depends on object complexity as well as similarity metric sensitivity.

4 Neural Networks Architecture

General neuron models with m weights can be denoted as

$$Y = f[\sum_{i=1}^{n} \Phi(W_{i1}, W_{i2}, \cdots, W_{im}, X_i) - \theta] \tag{1}$$

in multiple weights neural networks (MWNN). Neurons can be considered to be an (n-1)-dimensional hyper plane or curved surface in an n-dimensional space form high dimensional geometry analysis [3, 4]. Neurons like hyper sausages are chosen. Let

$$d(x, \overline{x_1 x_2}) = \min_{\alpha \in [0,1]} d(x, \alpha x_1 + (1-\alpha)x_2) \tag{2}$$

be the distance of x and line segment $\overline{x_1 x_2}$, then the hyper sausage set is

$$S(x_1, x_2; r) = \{x \big| d^2(x, \overline{x_1 x_2}) < r^2\} \quad (3)$$

The input-output transfer function is

$$f(x; x_1, x_2) = \phi(d(x, \overline{x_1 x_2})) \quad (4)$$

$\phi(\cdot)$ is threshold function, $x \in R^n$ input vector, and $x_1, x_2 \in R^n$ two centers. Neural networks consist of an input layer, a single hidden layer and an output layer of linear weights.

5 ANN Ensemble Online Learning

Online and real-time techniques are adopted in order to regrow and develop knowledge. ANN ensemble is first set up, one object an individual neural network. Training set in small size, with transition in explicitly temporal order, is used for object geometrical shape formation. Parallel neural networks organize samples into view categories, whose output converges at object nodes, a clear response to spatial occupation. For test inputs belonging to a learned object, maximal outputs should always come from corresponding object neural networks. So a series of comparison are made to validate whether requirements are met at certain level and to decide whether spatial coverage needs to be improved and adjusted. If tested images could be justified as their counterparts by existing system, there is no need to modify and the images can be released or skipped. Otherwise, they will insert into original training set until satisfactory. In this way, with learning on and on in ensemble, entire information will eventually be constructed in space, and what is acquired previously in other objects could not be affected. Results of multiple images from the same object were input into a working memory for evidence accumulation over time to improve effect, where each occurrence of a view category increases corresponding node's activity and the maximally active node is used to predict or judge the object.

In fact, a $\omega_i / \overline{\omega}_i$ problem is solved here. With $d(x)$ the decision function and N the class number, if $d_i(x) > 0$, $d_j(x) \leq 0$, $j = 1, 2, \cdots, N, j \neq i$, then $x \in \omega_i$. Spatial geometrical coverage V inclusive of embedding set I, is nearly close to the combination of all cases with arbitrary point in multidimensional manifold in set I as the center and a constant k as the radius, i.e., topological product between set I and n-dimensional hyper sphere [3, 4]. In practice, topological set V is defined as below. When indefinite superposition region encounters, views are inserted further in order to subdivide.

$$V = \bigcup_i V_i, \quad V_i = \{x | \rho(x, y) \leq k, y \in J_i, x \in R^n\},$$
$$J_i = \{x | x = \alpha I_i + (1 - \alpha) I_{i+1}, \alpha = [0,1]\} \quad (5)$$

6 Simulation Experiment and Performance Analysis

Image database consists of images covering a sphere surrounding sorts of objects, keeping spatial relationships intact. Based on global similarity, a model view learning procedure is involved. Training sets were formed with different distances between adjoining views. Images from both trained and untrained objects were taken for error `test, trained ones also for correct rate calculation. Some preprocessing was done in advance before formal operation to extract feature. View information was transformed into an invariant presentation under translation, rotation and scale by log-polar. With shift parameters exactly recorded down, an optional inverse transform could be taken to revert into initial state. Object recognition was performed in aid of neural networks ensemble with neurons similar to hyper sausages. Training recognition rates were all 100%. With proper parameters, error rate (mistaken recognition) for images from unlearned object could be 0%, i.e., unknown objects were rejected without incorrect recognition. For test samples from learned objects, with various fraction between images at trained visual angle and all, average correct rate in BP, RBF and MWNN ensemble are shown as Fig. 1. Training time in MWNN ensemble is much faster.

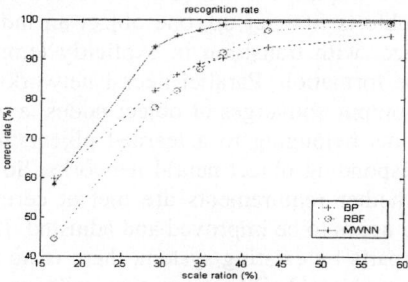

Fig. 1. Recognition rate

7 Conclusions

Information representation depends on similarity metrics and scales. Learning tends to be endless when in pursuit of details. Object spatial shapes are first cognized one by one by model views. If superposition between objects occurs, new views will be inserted to distinguish more subtly. In the recycling course, instead of global learning, only local modification is involved to pursue subtlety until solution is satisfactory. Encouraging results for feasibility test were achieved in simulation experiment.

References

1. Wang, S. J.: Biomimetics pattern recognition. INNS, ENNS, JNNS Newletters Elseviers, (2003)
2. Cyr, C. M., Kimia, B. B.: A Similarity-Based Aspect-Graph Approach to 3D Object Recognition. International Journal of Computer Vision, Volume 57 Issue 1 (2004)
3. Huang, D.S.: Systematic Theory of Neural Networks for Pattern Recognition. Publishing House of Electronic Industry of China, Beijing (1996) 97-99
4. Wang, S. J., Wang, B.N.: Analysis and theory of high-dimension spatial geometry for Artificial Neural Networks. Acta Electronica Sinica, Vol. 30 No.1, (2002) 1-4

Design and Implementation of the Individualized Intelligent Teachable Agent

Sung-il Kim[1], Sung-Hyun Yun[2], Dong-Seong Choi[3], Mi-sun Yoon[4],
Yeon-hee So[1], Myung-jin Lee[1], Won-sik Kim[1], Sun-young Lee[1],
Su-Young Hwang[1], Cheon-woo Han[1], Woo-Gul Lee[1], and Karam Lim[1]

[1] Dept. of Education, Korea University, Seoul, Korea
sungkim@korea.ac.kr
[2] Div. of Information and Communication Engineering, Cheonan University,
Cheonan, Korea
shyoon@cheonan.ac.kr
[3] Div. of Design and Imaging, Cheonan University, Cheonan, Korea
hcilab@cheonan.ac.kr
[4] Dept. of Teacher Education, Jeonju University, Jeonju, Korea
msyoon@jj.ac.kr

Abstract. The traditional ITS have considered the learners as a knowledge receiver. The recent development of teachable agent make it possible to provide the learner with an active role as a knowledge constructor and to take initiatives to persist in learning. In order to make an adaptive teachable agent that responds intelligently for individual learner, it should reflect the individual differences in the level of cognition and motivation, and its ongoing changes. For the purpose of developing individualized teachable agent, it is proposed to a student model based on the correlation among three dimensions: individual differences, learner responses, and learning outcome. A correlation analysis among the log data, questionnaire scores, and learning measurements was conducted. We delineated the relationships among three dimensions, learner responses (mouse-click pattern, duration & frequency at particular task, individual choice etc), individual characteristics (metacognitive awareness, self-efficacy, learning goal, and performance goal), and learning outcomes (interest and comprehension) during interacting with the teachable agent. The results suggest that certain type of learner responses or the combination of the responses would be useful indices to predict the learners' individual characteristics and ongoing learning outcome.

1 Introduction

The researchers in the field of cognitive science and learning science suggest that the teaching activity facilitates not only deeper understanding of the learning material but also enhances motivation to learn [1] [2]. Teaching activity consisted of sub-activities such as memory and comprehension, knowledge reorganization, explanation, demonstration, questioning, answering, and evaluation, and so on.

These sub-activities lead to elaboration, organization, inference, and metacognition, In terms of motivational aspects, students' motivation can be attained by allowing the learner to a tutor role which gives a responsibility, a feeling of engagement, and situational interest to persist in learning [3].

One way of providing the learner with an opportunity for the active engagement in learning is to give them a tutor role. [4] developed the new concept of intelligent agent called Teachable Agent (TA). Teachable agent is the computer program in which students teach the computer agent based on the instructional method of 'learning by teaching'. TA provides student tutors with an active role so that they can have positive attitude toward the subject matter [5].

Although TA is developed to enhance cognitive ability of learners, the effects of the system are not the same for the all learners. Traditional TA did not reflect individual differences in cognitive ability and motivation. The identical interface regardless of the individual differences might be not only less effective in cognitive aspects of learning but also less interesting in terms of motivational aspects of learning. There-fore, individualization is the key concept in developing TA to respond adaptively to individual learner, which reflects the individual differences in the level of cognition and motivation, and its ongoing changes.

The new generation of ITS (Intelligent Tutoring System) would be developed into an adaptive system to maximize the motivation to learn and to optimize the learning by Varying the level of information and affordance to each user. Then, the important questions are what kind of individual differences might play a critical role in learning and motivation, and how to measure those individual differences even if they are identified.

2 User Interface for Measuring Individual Characteristics

The recent development of teachable agent provides the learner with an active role as a knowledge constructor and focuses on the individualization. Individualized agent provides differential interface and responses adaptively depending on the characteristics of user and its behaviors. The aim of the 'individualized agent' is not only to maximize the learner's cognitive functions but also to enhance the interests and motivation to learn. To develop adaptiveness of the agent, it is necessary to assess each user's specific cognitive and motivational characteristics and ongoing response pat-terns during learning.

As the first step of developing individualized teachable agent, the individual characteristics of the learner were measured through the questionnaires. Four variables of individual difference in metacognition and motivation were selected because both the level and type of motivation play a significant role in the persistence and efforts in learning [6]. Among various motivational factors, self-efficacy, learning goal orientation, and performance goal orientation were used in this study. Metacognitive awareness including planning, monitoring, and evaluation was measured since elementary school students may lack of this skill though it is a critical factor for their learning.

Then, various interfaces (see figure 1) were developed to measure the individual characteristics of the user through the previously developed TA, KORI

	Individual Characteristic					
	Self-efficacy	Meta-cognition	Learning goal orientation	Performance goal orientation	Interest	Comprehension
Interface component	KORI's performance prediction	KORI's performance prediction	Learning resource exploration	Duration of teaching activity	Learning resource exploration	Number of correct concepts
	Learning resource exploration	Lesson planning			Response to interruption	

Fig. 1. Summary of Interface Design for Measuring Individual Characteristics

(KORea university Intelligent agent, see details in [5]). It was expected that the user response pattern through the KORI would be correlated with the results of the questionnaires.

In order to measure the level of user self-efficacy, users were required to predict the KORI's future performance score. If the users were highly self-efficacious, they would expect higher level of performance than the low efficacy users since they all knew that KORI's performances would be determined by the users' behavior. Another interface for measuring self-efficacy was the frequency and duration for exploring the learning resources. The icon of learning resource was presented on the right side of the screen so the users can access whenever they want to know the basic and additional knowledge. High self-efficacy users were expected to refer to the learning resources more frequently and longer.

The second individual characteristic is the metacognitive ability. Among the four sub-factors of the metacognition, only the planning and monitoring were focused. In order to measure the user's planning ability, the system asked the user to make a lesson plan before teaching the KORI (see Figure 2). The keywords for basic learning concepts were displayed on the left side of the screen. The user typed the concept to teach, specific teaching activities, and plans in the blank. The quality or duration of lesson plan was expected to reflect the planning ability of the user. The learner's monitoring ability was measured by the prediction of KORI's future performance score. It was also expected that the predicted KORI test score might provide useful information about the learner's monitoring ability if the predicted score and the actual performance score were compared.

The third individual characteristic was learning goal orientation. To get the ongoing measurement of learning goal orientation, the learning resource menu was used. Since the system displays the learning resource menu all the time in

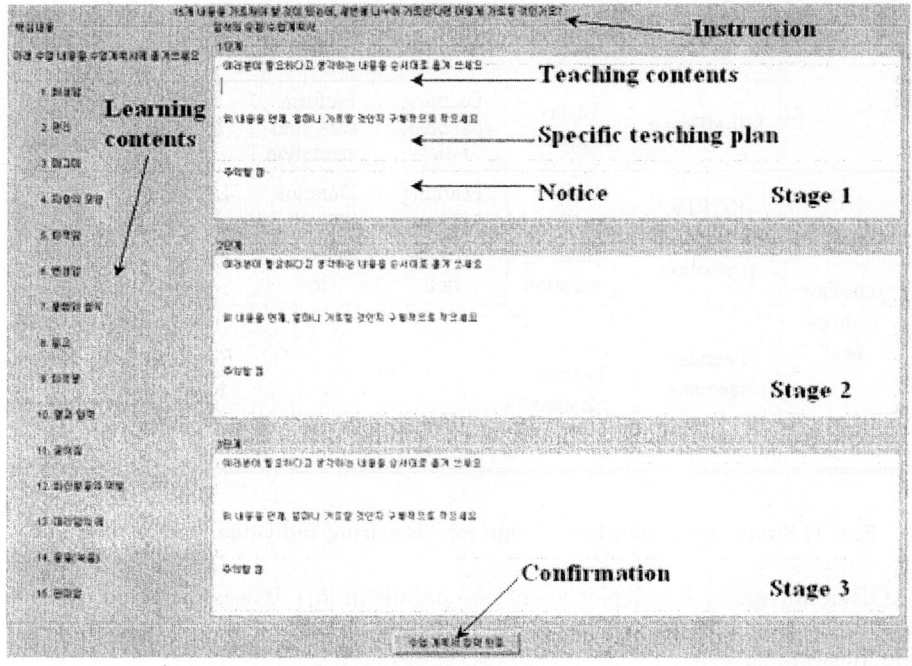

Fig. 2. Lesson Plan to Teach KORI

concept teaching stage (see Figure 3), the frequency and duration for exploring the learning resources might be a good indicator of the learning goal orientation. If the users are highly learning goal oriented, they would want to know more about the learning materials. The last individual characteristic, the performance goal orientation, was measured through the total amount of teaching time for KORI. If the users are highly performance goal orientated, they are expected to spend less time in teaching KORI because they tend to show off their superior intellectual ability to another person by getting the task done quickly.

We also measured the ongoing interests of the user indirectly through the using pat-tern of learning resource and response to the interruption. If the user has high interests in teaching KORI, she may want to click the learning resource more frequently, explore the resources more deeply, and spend longer time. In addition, another interface for measuring user's interests is to make an interruption while the user is teaching KORI. For instance, the KORI sometimes interrupts the user's teaching by displaying her sleepy face and falling asleep suddenly. If the users are highly interested in teach-ing and concentrate on the teaching activity, it might be difficult for them to detect the minor changes in KORI's face and to take longer time to detect the changes than uninterested users.

The teaching activity can be used as a good index for the user's level of comprehension. Thus the ongoing level of comprehension can be measured through

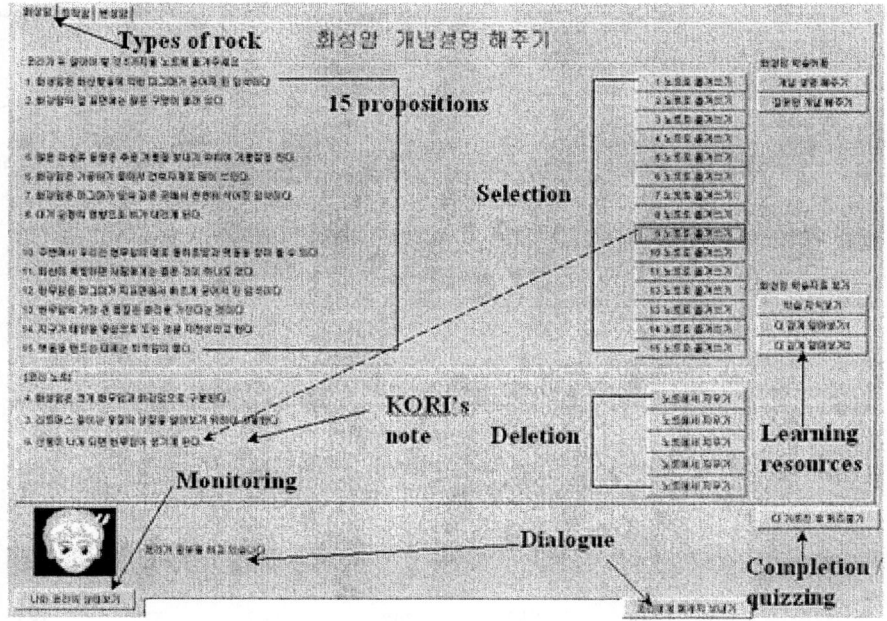

Fig. 3. Concept Teaching

the ratio of selecting the correct propositions and deleting the wrong propositions in the concept teaching stage in Fig. 3.

3 Analysis of User Response Pattern

In order to establish the relationships among user characteristics and response patterns and to extract the algorithm among variables, we measured the individual characteristics, collected log data of interaction with the KORI, and analyzed the relationship among these.

Twelve 5th graders (8 males and 4 females) participated in the student modeling. All of the participants took the 30 minutes lesson on the 'Rock Cycle' which is about the three kinds of rocks and their transformation. Next, participants filled in questionnaires on individual characteristics including self-efficacy, goal orientations, and metacognition. And then, they interacted with the KORI during 30 - 40 minute. During this period, the log data were recorded automatically and participant's behaviors were videotaped. After teaching KORI, participants completed the interest questionnaire and comprehension test.

A correlation analysis among the log data, questionnaire scores, and learning outcomes was conducted (see Figure 4). We delineated the relationship among three dimensions, learner responses (mouse-click pattern, duration & frequency at particular task, individual choice etc), individual characteristics (metacogni-

Learner's response	Individual Characteristic					
	Self-efficacy	Meta-cognition	Learning goal orientation	Performance goal orientation	Interest	Comprehension
Planning (d)	-.056	.003	.151	.015	**.524**	.043
Prediction (s)	-.126	**-.429**	**-.490**	-.158	-.149	-.213
Prediction (d)	-.241	**-.372**	**-.518**	**-.856****	**-.403**	-.022
Difference between prediction and performance	**.766****	**.710****	**.687***	.334	**.529**	**.686***
Learning resources (f)	.060	.193	.197	-.291	**.406**	.125
Learning resources (d)	**.340**	**.382**	**.409**	-.189	**.572**	**.397**
Concept teaching (d)	.111	.288	.177	-.271	.112	**.360**
Concept map teaching (d)	.062	.050	.109	.255	.144	**.404**
Response to interruption	-.156	.190	.104	.136	**-.423**	-.153
Correct concepts put in (n)	**.345**	.166	-.024	-.168	.004	**.724****
Incorrect concepts taken out (n)	**.381**	.107	.061	-.296	-.022	**.371**

$*p < .05$, $**p < .01$
(f = frequency, d = duration, n = number, s = score)

Fig. 4. Correlation Matrix among Individual Characteristics and Learner Responses

tive awareness, self-efficacy, learning goal, and performance goal), and learning outcomes (interest and comprehension) during KORI teaching.

The self-efficacy scores showed a strong correlation with the difference between the predicted and actual performance score ($r = .766$, p ¡ .01), indicating that the higher self-efficacy users are morel likely to predict KORI's performance.

The self-efficacy scores were moderately correlated with the duration of exploring the learning re-sources (r = .340), the ratio of selecting the correct concepts (r = .345), and the ratio of deleting the incorrect concept (r = .381).

Individual learner's metacognitive awareness was highly correlated with both predicted score of KORI performance and exploring of the learning resources. The negative correlation was found between the metacognition scores and the predicted score of the KORI performance (r = -.429) and between the metacognition and the duration of predicting the KORI's performance (r = -.372). However, learner's metacognition was positively correlated with both the difference between the predicted and actual performance score (r = .710, $p < .01$) and the duration of exploring the learning re-sources (r = .382).

Learner's goal orientation scores were also correlated with various log data. The learning goal orientation was negatively correlated with both the predicted score of the KORI performance (r = -.490) and the duration of predicting the KORI's performance (r = -.518). However, learning goal orientation was positively correlated with both the difference between the predicted and actual performance score (r = .687, $p < .05$) and the duration of exploring the learning resources (r = .409).

The performance goal orientation scores showed a strong negative correlation with the duration of predicting the KORI's performance (r = -.856, $p < .01$). However, the correlation between the performance goal orientation and the difference between the predicted and actual performance score (r = .334).

In the motivational learning outcomes, the interest rating scores were negatively correlated with both the duration of predicting the KORI's performance (r = -.403) and the responses to interruption (r = -.423). On the contrary, the interest rating scores showed positive correlation with the duration of teaching planning (r = .524), the difference between the predicted and actual performance score (r = .529), the frequency and duration of exploring the learning resources (r = .406, r = .572 respectively).

On the cognitive aspect of the learning outcome, learner's comprehension test scores were positively correlated with several teaching variables. The comprehension test score were highly correlated with the ratio of selecting the correct concepts (r = .724, $p < .01$) and the ratio of deleting the incorrect concept (r = .371). In addition, the duration of concept teaching including both selection and deletion of propositions was positively correlated with the comprehension test scores (r = .360), and the duration of the concept map teaching was also correlated with the comprehension test scores (r = .404). The difference between the predicted and actual performance score was also significantly correlated with comprehension test scores (r = .686, $p < .05$). And the duration of exploring the learning resources also showed a moderate correlation (r = .397).

The results suggest that certain type of learner responses or the combination of the responses would be useful indices to predict the learners' individual characteristics and ongoing learning outcome. In particular, since the difference between the predicted and actual performance score, the duration of exploring the learning resources, and the duration of predicting the KORI's performance are highly correlated with most of the variables, these user responses can be regarded as the best indices for measuring the individual difference and learning outcomes.

4 Conclusion

Individualization is the key concept in developing computer assisted learning system and intelligent tutoring agent. The ultimate goal of developing the learning agent is to make an adaptive agent respond intelligently for individual learner, which reflects the individual differences in the level of cognition and motivation, and its ongoing changes. Traditional measurements in learning systems include assessing individual differences by standardized test or questionnaires at the beginning or at the end of the learning session. This study proposed a new type of dynamic assessment for individual differences and ongoing cognitive/motivational learning outcomes through the computation of responses without measuring them directly. In near future, various physiological indices such as temperature of fingers, eye-movement, facial expression, and brainwaves combined with the response pattern are likely to be used to measure individual differences or learning outcomes. However, for the time being, it is essential to develop the algorithm of learner response pattern during learning.

Collecting and classifying the indirect log data of the learner that are correlated with the individual differences and learning outcome, and constructing a student model consisted of the structure of nodes may be an useful methodology to understand the learner's dynamic change during the specific learning situation.

Acknowledgments

This research was supported by Brain informatics Research Program sponsored by Korean Ministry of Science and Technology.

References

1. Bargh,J.A.,Schul,Y.: On the cognitive benefits of teaching. Journal of Educational Psychology. **72** (1980) 593–604.
2. Chi,M.T.H.,Siler,S.A.,Jeong,H.,Yamauchi,T.,Hausmann,R.G.: Learning from human tutoring. Cognitive Science. **25(4)** (2001) 471–533.
3. Biswas,G.,Schwartz,D.,Bransford,J.,TAG-V.: Technology support for complex problem solving: From SAD environment to AI. In Forbus and Fel-tovich. (Eds.) Smart machines in education. Menlo Park. CA: AAAI Press. (2001).
4. Kim,S.,Kim,W.,Yoon,M.,So,Y.,Kwon,E.,Choi,J.,Kim,M.,Lee,M., Park,T.: Conceptual understanding and Designing of Teachable Agent. Journal of Korean Cognitive Science. **14(3)** (2003) 13–21.

5. Kim,S.,Yun,S.H.,Yoon,M.,So,Y.,Kim,W.,Lee,M.,Choi,D.,Lee,H.: Design and implementation of the KORI: Intelligent teachable agent and its application to education. Lecture Notes in Computer Science. **3483** (2005) 62–71.
6. Pintrich,P.R.,Schunk,D.H.: Motivation in Education; Theory, Research, and Application. Englewood Cliffs. NJ: Prentice-Hall. (1996).

Comparison of Complexity and Regularity of ERP Recordings between Single and Dual Tasks Using Sample Entropy Algorithm

Tao Zhang[1], Xiaojun Tang[1], and Zhuo Yang[2]

[1] Key Lab of Bioactive Materials of Ministry of Education and College of Life Science, Nankai University, Tianjin, PR China, 300071
zhangtao@nankai.edu.cn
[2] College of Medicine Science, Nankai University, Tianjin, PR China, 300071
zhuoyang@nankai.edu.cn

Abstract. The purpose of this study is to investigate the application of sample entropy (SampEn) measures to electrophysiological studies of single and dual tasking performance. The complexity of short-duration (~s) epochs of EEG data were analysed using SampEn along with the surrogate technique. Individual tasks consisted of an auditory discrimination task and two motor tasks of varying difficulty. Dual task conditions were combinations of one auditory and one motor task. EEG entropies were significantly lower in dual tasks compared to that in the single tasks. The results of this study have demonstrated that entropy measurements can be a useful alternative and nonlinear approach to analyzing short duration EEG signals on a time scale of seconds.

1 Introduction

Research has shown that multiple task performance is of considerably lower quality than when corresponding tasks are performed individually.[1] Whilst traditionally this has been investigated using behavioural methods [2], more recently electrophysiological techniques have been employed as they have been shown to provide additional insights into mechanisms of cognitive processing.[3] However, because of limitations in the analytical techniques, it has been difficult to conclusively distinguish the sometimes-subtle changes in the 125 electroencephalogram (EEG).

There were investigations which not only provided evidence for nonlinearity in EEG time series but indicated that it has a high-dimensional structure.[4] Estimating chaotic characteristics of a high-dimensional dynamic system is difficult. Another approach to measurement of nonlinear trends in EEG is quantification of complexity from the point of view of information theory. To this end, short and noisy EEG data can be analyzed with the help of entropy measurements such as approximate entropy (ApEn) and SampEn.[5] Lower entropy values have been proposed to indicate greater signal regularity corresponding to situations in which communication pathways in a network are poorly developed or system components operate in relative isolation. To date the application of SampEn to analyse ERP data has not been explored.

In the present study, SampEn together with the method of surrogate data was introduced, for estimating complexity or irregularity of short EEG time series collected from participants in both single and dual task conditions. It has been proposed that cognitive control mechanisms are required to orchestrate performance of more than one task at a time [6]. If this is the case then it is expected that the entropy values should be changed in the multiple task condition compared with the single task condition.

2 Method

None of the participants (aged between 18 and 45 years) were taking medication, have a history of head injury, substance abuse, or any significant medical or psychiatric problems. The experimental stimuli and participants behavioural responses were controlled and collected using the *Superlab* (*SL*, Cedrus Corporation, Phoenix, USA) software programme.

2.1 Experimental Tasks

Auditory Single Task (AST): The auditory 'oddball' paradigm consisted of presentation of two tones of 1kHz and 2kHz, with probabilities of 0.8 and 0.2, non-target and target stimuli respectively. Participants responded to the rare stimuli by pressing a pre-designated key, ignoring the frequent stimuli. Motor Task 1 and 2: The tasks consisted of participant executing the motor tasks when the rare vibration occurred. Dual Tasks: Two dual task experiments were designed combining the above single 'oddball' auditory and motor tasks: auditory and motor task 1 (ADT1) and auditory and motor task 2 (ADT2). The stimuli were presented pseudorandomly to ensure that there were no consecutive presentations of target stimuli. Randomisation was carried out using Minitab software.

2.2 Data Acquisition

Brain potentials of 26 electrodes sites positioned according to the International 10-20 system were recorded using a Scan electrophysiological acquisition system (Neuroscan Medical Systems, Virginia, USA). All channels were amplified with a gain of x150 and bandpass filters of 0.01 – 100Hz were employed. The signals were digitised using a 16-bit analogue-to-digital converter and sampled at 500Hz.

2.3 Sample Entropy and Surrogate Data

The improved algorithm of ApEn, SampEn statistics, agree much better than ApEn statistics with the theory of random numbers with known probability characteristics, over a broad range of operating conditions and maintain relative consistency where ApEn statistics do not. The mathematical details of the SampEn are referred to the reference [5]. In the present study, 30 computations of the surrogate algorithm were generated, and SampEn was calculated. Following this the mean, SD and SampEn were calculated, for each 30-member surrogate ensembles, and compared with SampEn for the original time series. The null hypothesis was rejected at a significance level of 0.05.[4]

2.4 Statistics

All the data are expressed as the mean ± SEM. Analysis of variance (ANOVA) was used for statistical analysis of the data, allowing within and between task comparisons to be made and significant differences are presented when P < 0.05.

3 Results

The group values of *SampEn* for the original data from Fp1 position ranged from 0.19 to 0.30. Comparatively, all these data were significantly lower (P < 0.05) than that of the surrogate data group, in which the entropy values are from 0.37 to 0.60.

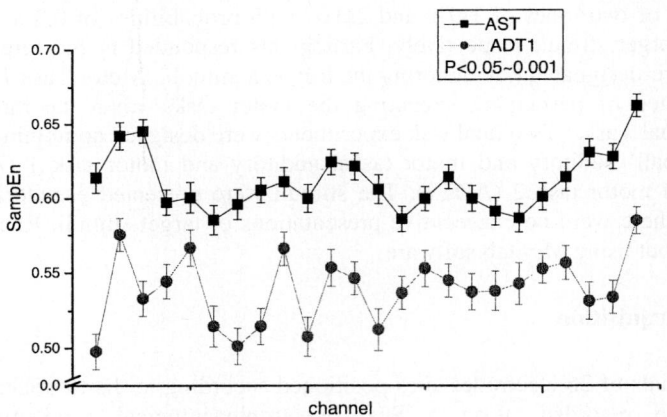

Fig. 1. Comparison of single and dual task rare auditory stimuli. The entropy of ERPs signals in 24 electrodes was found significant differences between AST and ADT1.

Fig 1 gives the *SampEn* for rare tone stimuli in the AST and the ADT1 (n=13) in 24 electrodes, from left to right which are Fp1, Fp2, F7, F3, Fz, F4, F8, FC3, FCz, FC4, T3, C3, C4, T4, CP3, CPz, CP4, T5, P3, P4, T6, O1, Oz and O2. It can be seen that the entropy values in the AST condition were significantly higher than that of the ADT1 condition (P < 0.05). There were only two electrodes, Cz and Pz, in which the *SampEn* didn't detect significant difference between the two tasks.

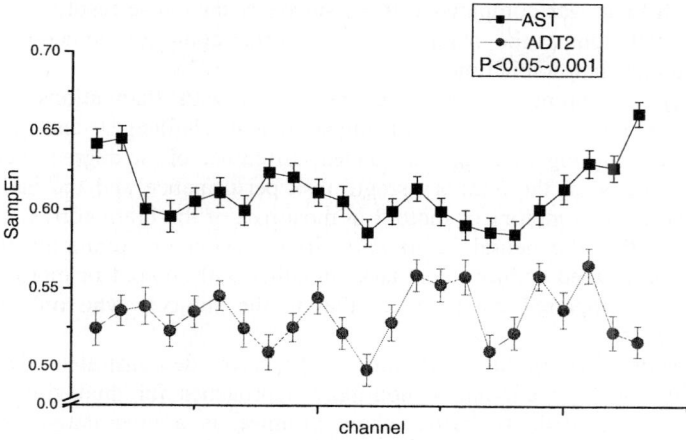

Fig. 2. Comparison of single and dual task rare auditory stimuli. The entropy data of ERPs signals in 23 electrodes were detected significant differences between AST and ADT2.

The data in Fig 2 compares the rare auditory stimuli of the AST condition and the ADT2 (n=14) in 23 electrodes, from left to right which are Fp2, F7, Fz, F8, FC3, FCz, FC4, T3, C3, Cz C4, T4, CP3, CPz, CP4, T5, P3, Pz, P4, T6, O1, Oz and O2. The *SampEn* measurements were significantly high for the single task performance compared with the dual task performance ($P < 0.05 - 0.001$) in all presented electrodes. However, there were still three electrodes, Fp1, F3 and F4, in which the entropy measurement didn't detect significant difference between them.

4 Discussions

The primary aim of the present study was to investigate the potential application of SampEn statistics to analyze short ERP data. Various investigations have shown that applying nonlinear dynamical methods to EEG data provides new information about the complex dynamics of underlying neuronal networks [7]. Within this physical-mathematical framework a variety of measures *e.g* correlation dimension and Largest Lyapunov exponents allow characterization of different static and dynamic properties of a time series However, in a strict sense, well-known problems in extracting nonlinear measures from short, noisy and non-stationary data, generated from potentially high-dimensional systems, would excluded the use of these measures to characterize EEG dynamics.

SampEn does not test for a particular model for ERP dynamics, such as deterministic vs. stochastic; instead, *SampEn* is used to differentiate among data sets on the basis of pattern regularity in the time series. However, the model of surrogate data can be used as a formal test, for individual subjects, for quantifying statistical significance of the rejection of a particular null hypothesis for comparisons between the original time series and dynamic models [4] In the present study, the original time series for each fetus was compared with its surrogate data. The results indicated that there were correlations in the original time series that could not be accounted for by the linear autocorrelation function.

Complexity measurements have been correlated with fluctuations of complex nonlinear variability in a variety of physiological challenges and pathological conditions. The data (Fig.1 and 2) show a clear difference of the degree of complexity or irregularity between the auditory single task performance and the auditory dual task performance with motor1 or motor2 in most part of the brain cortex. The results further suggest that the neural information transmission or communication in the subjects who performed auditory dual tasks in either with motor1 or motor2 could be more isolated or impaired compared to that in the subjects who only performed auditory single task.

In conclusion, the findings of this study have demonstrated that entropy measurements could be alternative nonlinear approaches for analyzing short-term ERP signals. The methods further show promise as a quantitative measure of nonlinear dynamic systems behaviour and its psychological change, such as single or dual tasks challenges, where the validity of traditional nonlinear dynamical approaches such as correlation dimension and Lyapunov exponent measurements have been challenged its validity, recently.

Acknowledgements

This work was partly supported by the National Natural Science Foundation of China (30370386, 30470453), Tianjin Municipal Science and Technology Commission (043611011).

References

1. Pashler, H: Dual-Task Interference in Simple Tasks: Data and Theory. Psychological Bulletin, Vol.116 (2) (1994) 220-244
2. Schumacher, EH., Seymour, TL., Glass, JM., Fencsik, DE., Lauber, EJ., Kieras, DE., Meyer, DE.: Virtually perfect time sharing in dual-task performance: Uncorking the central cognitive bottleneck. Psychological Science, Vol. 12(2) (2001) 101-108
3. Hsieh, S., Yu, Y-T.: Switching between simple response-set: inferences from the lateralized readiness potential. Cognitive Brain Research, Vol. 17 (2003) 228-237
4. Theiler J, Eubank S, Longtin A, Galdrikian B, Farmer JD.: Testing for nonlinearity in time-series - the method of surrogate data. Physica D, Vol. 58 (1992) 77-94
5. Richman J.S Moorman JR.: Physiological time-series analysis using approximate entropy and sample entropy. Am J Physiol, Vol. 278 (2000) H2039-H2049
6. Monsell, S.: Task switching. Trends in Cognitive Science, Vol. 7(3) (2003) 134 -140
7. Pijn JP, van Neerven J., Noest A, Lopes da Silva FH.: Chaos or noise in EEG signals: dependence of state and brain side. Electroen. Clin. Neurophysiol, Vol.79 (1991) 371- 381

Representation of a Physio-psychological Index Through Constellation Graphs

Oyama-Higa Mayumi[1] and Tiejun Miao[2]

[1] Department of Integrated Psychological Science, Kwansei Gakuin University,
1-1-155, Ichibancho, Uegahara, Nishinomiya-City, 662-8501, Japan
oyama@kwansei.ac.jp
[2] Chaos Technical Research Laboratory and CCI Corporation,
3-1-2401, Ryodocho, Nishinomiya-City, 662-0841, Japan
t-miao@tokyo.cci-web.co.jp

Abstract. Chaos theory was applied to analysis of the time series of plethysmograms under various human physio-psychological conditions. It found that the largest Lyapunov exponent could be used to characterize physio-psychological status. A visual representation method based on constellation graphs was developed to indexing temporal changes in the largest Lyapunov exponent. Changes of constellation angles were found to clearly characterizing variations of physio-psychological status in a series of experiments.

1 Introduction

Most biological systems that exist in the natural world are believed to be complex systems with chaotic fluctuations. Chaotic systems appear to be very complex and to behave in a random and unstable manner. But in fact they are systems that change according to simple deterministic rules. The fingertip pulses are easier and less restrictive in measurements in comparing with the other biological signals such as ECG (electroencephalography) and EEG. The generating system of fingertip pulse (plethysmograms) can be described by chaotic dynamics [1]. Recently there were a lot of investigations showing the effectiveness of the chaotic analyzing of plethysmography in relating physio-psychological changes. The effect of work load on fingertip pulsations was studied. Plethysmography was also employed to examine physio-psychological changes of firm employers working on morning, midday and evening. We measured fingertip pulse waves and investigated the changes in chaotic invariants for different age groups. The study of physio-psychological changes caused by decline in communication skill with aging demonstrated the effectiveness of the plethysmography.

In this paper chaos analysis was performed on the measured fingertip pulse waves (plethysmograms). The largest Lyapunov exponents of the time series were computed. Especially we concentrated on the temporal changes in the largest Lyapunov exponent, in relating to indexing physio-psychological status and mental toughness, on the basis of experimental results. In addition, a visually representation of the temporal changes was proposed based on constellation graphic method.

2 Measurement, Analysis and Visual Representation Methods

2.1 Method of Measurement

Fingertip pulses were measured using a photoplethysmography sensor (CCI BC2000) in the following manner. The subjects were allowed to become accustomed to their surroundings for at least 10 minutes in a room maintained at 25°C. They were allowed to sit comfortably in a chair with both hands placed in a relaxed manner on a desk (at a height that was comfortable for writing). The subjects kept their eyes open while measurements were made on the left index finger for a minimum of 60 sec to a maximum of 180 sec. The signals were A/D converted. Digital data sampled at a frequency of 200 Hz with resolution of 12 bits was recorded on a computer.

2.2 Method of Chaos Analysis and Calculation of the Largest Lyapunov Exponent

For the time series data x(i), with i=1,..., N obtained from the fingertip pulses, the phase space was reconstructed using the method of time delays. Assuming that we create a d-dimensional phase space using a constant time delay τ, the vectors in the space are generated as d-tuples from the time series and are given by

$$\mathbf{X}(i) = (x(i),..., x(i-(d-1)\tau)) = \{x_k(i)\} \quad (1)$$

where $x_k(i) = x(i-(k-1)\tau)$, with k=1,...,d. To reconstruct the phase space correctly, the parameters of delay (τ) and embedding dimensions (d) should be chosen optimally [4]. In time series data recorded from human finger photoplethysmograms, we chose the parameters τ=50 ms and d=4, as in references [1] and [2].

In the reconstructed phase space, one of the important measures of complexity is the largest Lyapunov exponent λ_1. If $\mathbf{X}(t)$ is the evolution of some initial orbit $\mathbf{X}(0)$ in the phase space, with time, then

$$\lambda_1 = \lim_{t\to\infty}\lim_{\varepsilon\to 0}\frac{1}{t}\ln\frac{|\delta\mathbf{X}_\varepsilon(t)|}{|\varepsilon|} \quad (2)$$

where $\delta\mathbf{X}_\varepsilon(t) = \mathbf{X}(t) - \mathbf{X}_\varepsilon(t)$ and $\varepsilon = \mathbf{X}(0) - \mathbf{X}_\varepsilon(0)$, for almost all initial difference vectors $\varepsilon = \mathbf{X}(0) - \mathbf{X}_\varepsilon(0)$. We estimate λ_1 using the algorithm of Sano and Sawada method [3], where λ_1 describes the divergence and instability of the orbits in phase space.

Chaotic analysis of finger photoplethysmograms was performed. The largest Lyapunov exponents (λ_1) of were calculated for a basic window of 8,000 points (40 sec). For a longer measured data, temporal changes in λ_1 was obtained by sliding window approach in which the basic window was sequentially shifted in a step of 200 points (1 sec). Accordingly sequentially estimations of λ_1 was determined for each window.

Figure 1 presents a plethysmogram obtained from a 180sec measurement together with its temporal changes in the largest Lyapunov exponent achieved in the sliding window method.

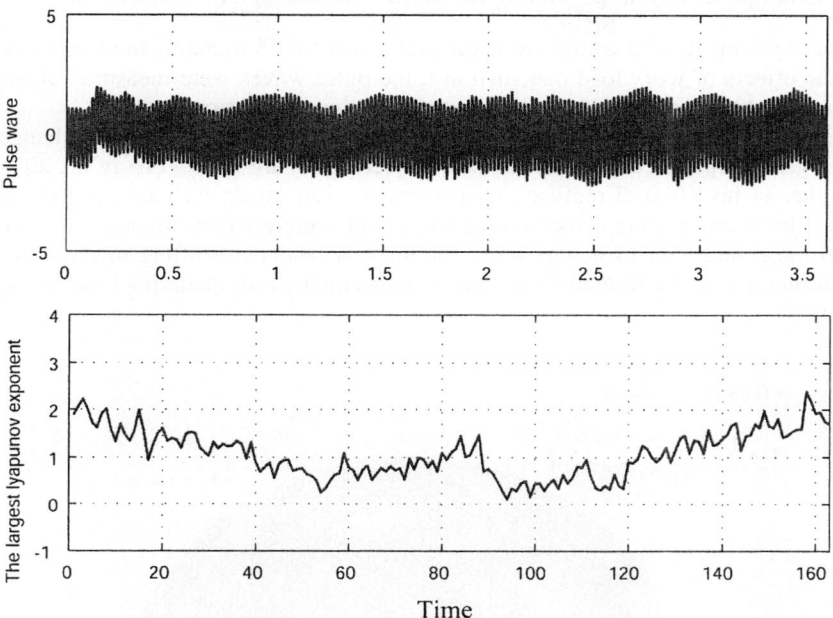

Fig. 1. Plethysmograms (upper) and its temporal changes of λ_1 (bottom)

2.3 Representation Method Through Constellation Graphs

This study developed constellation graphs method to visually describe the temporal variations of the Lyapunov exponents. In these constellation graphs, the numerical data of time variations was converted into variations of angles with minimum=0° and maximum=180°. The vectors of the same length were joined together and depicted on a semicircular graph. The maximum and minimum values could be set automatically or manually from the values of the Lyapunov exponents. Each line on the graph represents the data of one measurement. The smaller the value of the Lyapunov exponent, the closer the vector is to the bottom right of the constellation graph. As the value increases, the line shifts to the left in the graph. The line is straighter when the standard deviation is smaller and kinkier, while bent for a larger one.

3 Experiments and Constellation Graphic Representation Results

We have conducted a series of experiments using measurements of fingertip pulse waves. The experiments were described in details elsewhere. Informed consent was obtained from all the subjects on all experiments. This paper reviews the results by using constellation graphic representation of the temporal changes in the largest Lyapunov exponents which characterize the changes in physio-psychological status.

3.1 Changes in Physio-psychological Status Caused Due to Work Load

In the experiment, a Kraeplin's test was performed for 15 minutes. In order to examine the effects of work load due such test, the pulse waves were measured under the resting condition before and after the test, for 1 minute each time.

Figure 2 shows the constellation graphic representation of the temporal changes in the largest Lyapunov exponents for two subjects (both are males, one in his 20's and the other in his 40's). 2 replicate measurements were made for each subject. There were 4 trials comprising 8 measurements in total. Figure shows the results before (4 green lines) and after (4 orange lines) the tests. A tendency shifting towards the left was found due to the Kraeplin's test that is concerned a task involving brain activities and work.

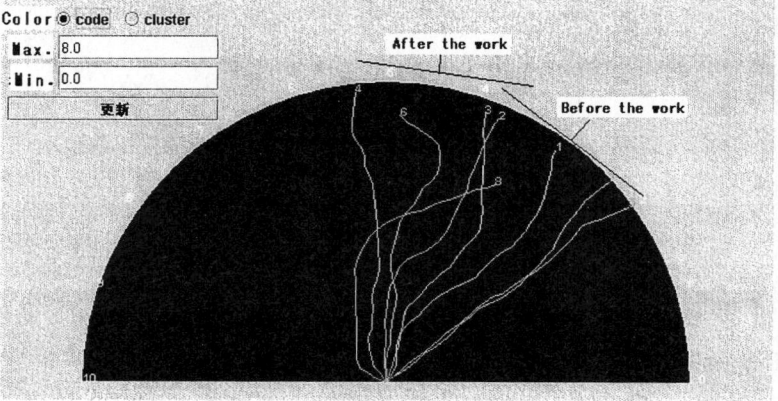

Fig. 2. Constellation graphic representation of Lyapunov exponents before and after brain work. Orange lines: after the work, and green lines: before the work.

It was noted that the subjects commented that their consciousness activities became clearer after the Kraeplin's test.

3.2 Changes in Physio-psychological Index in Firm Employees During a Working Day (Morning, Midday and Evening)

Subjects were 8 employees aged 26 to 34. Measurements were made, for 3 minutes at the time: soon after the employee reached at their office (in the morning), about 1 hour after lunch (midday), and finished work (in evening), respectively.

Fig.3 shows the representation results for "morning" indicated by blue lines, "midday" by red ones, and "evening by green ones, respectively, for 8 subjects corresponding No.1 to No.8. In the constellation graphs, the physio-psychological condition is

Representation of a Physio-psychological Index Through Constellation Graphs 815

better when the line is closer to the left and poorer when it is closer to the right. The average angles of the constellation lines for subject No.1 exhibited the order being "morning" > "midday" > "evening", showing a normal physio-psychological changes from morning to evening. While No. 2 had a reversed order, although the overall tread (angles) was larger. No. 3 had similar tread to No.2. No. 4 had larger angles at midday. No. 5 was relatively better at midday compared to morning or evening. No. 6 was in a good condition in the morning and evening but not at midday. No. 7 showed the same trend as No. 5 but had more subtle changes. No. 8 was fine in the morning, slightly sluggish at midday and not in a good condition in the evening.

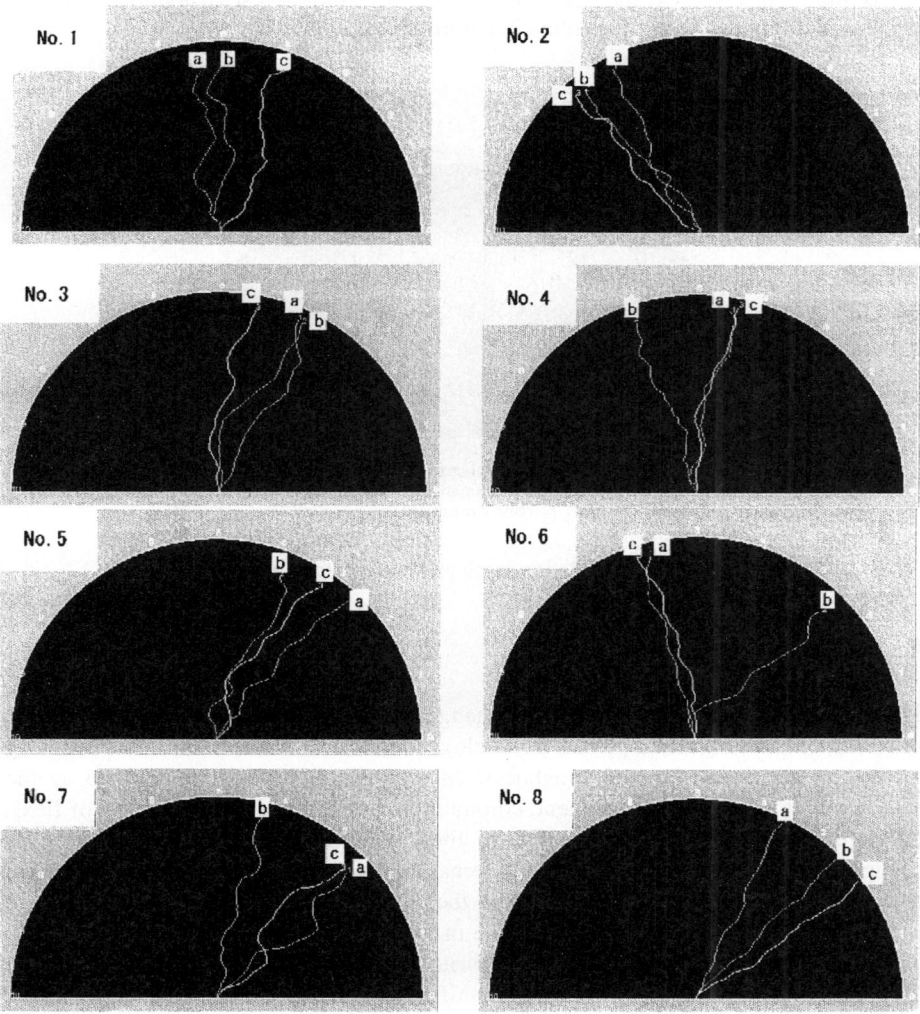

Fig. 3. Constellation graphs during a working day (morning, midday and evening)

3.3 Changes Caused by Decline in Communication Skill with Ageing

Subjects were 179 persons including 139 females and 40 males, aged 65 to 100 with mean of 83.4. Photoplethysmograms were recorded for 3 minutes together with the measurements of body temperature, systolic and diastolic blood pressure and pulse rate. For estimating the communication skill of aged persons, we used Activities of Daily Living (ADL) indices recorded by health care professionals who were looking after the subjects in an old people's house. The communication skill was assigned to one of the three levels. a: Be able to communicate normally; b: Be able to communicate to some extent; and c: Hardly or disable communicate.

As shown in Figure 4, the constellation angles of the largest Lyapunov exponents decreased obviously with the decline in communication skills.

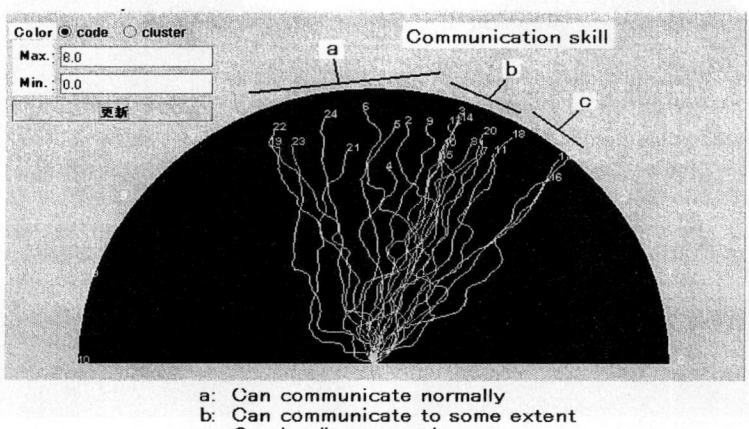

a: Can communicate normally
b: Can communicate to some extent
c: Can hardly communicate

Fig. 4. Constellation graph and communication skills

4 Discussions

This study showed that the temporal changes of Lyapunov exponents of the time series, which was used as a physio-psychological index, could be taken as a promising method to measure of metal toughness. Mental toughness here is defined as the adaptability to external environment, communication skill and a certain level of flexibility (divergence). Human beings have the capability to skillfully avoid various changes, contacts and assaults of the external environment. In some cases, they deal with them, cope with them, and maintain their lives while expressing themselves. In analogs to the phenomena that the decline in biological immunity will results in sick, the mental toughness may well signify "spiritedness," which has been so far described vaguely. "Spiritedness" could not be quantified until now. Our investigations in the studies may give some insight into the problem.

5 Conclusions

Chaos theory was applied to analysis of the time series of plethysmograms under various human physio-psychological conditions in a series of experiments. It found out that the temporal changes in the largest Lyapunov exponent correlated well with changes of physio-psychological status. A visual representation method based on constellation graphs was developed to indexing the temporal variations. Especially the changes of constellation angles were found to clearly characterizing variations in mental/physical status. In concretely, changes in physio-psychological status caused due to work load had a corresponding increase of the constellation angles. Changes in physio-psychological index in firm employees during a working day were explained by the constellation graphs. Changes caused by decline in communication skill with ageing corresponded to a decreased constellation angles.

In our earlier studies, we found that the Lyapunov exponent decreased in aged persons with severe dementia. We also plan to advance the research on changes in mental toughness with growth from birth to early childhood.

References

1. Tsuda I, Tahara T, Iwanaga I: Chaotic pulsation in capillary vessels and its dependence on mental and physical conditions. Int J Bifurcation and Chaos, 2 (1992) 313-324
2. Sumida T, Arimitu Y, Tahara T, Iwanaga H: Mental conditions reflected by the chaos of pulsation in capillary vessels. Int J Bifurcation and Chaos, 10 (2000) 2245-2255
3. Sano M and Sawada Y: Measurement of the Lyapunov spectrum from a chaotic time series. Phys. Rev. Lett., 55 (1985) 1082
4. Abarbanel HDI, Brown R, Sidorowich JJ, Tsimring LS: The analysis of observed chaotic data in physical systems. Rev Mod Phys., 65 (1993) 1331-1392
5. Niwa T, Fujikawa K, Tanaka K, and Oyama M: Visual Data Mining Using a Constellation Graph, ECML/PKDD-2001 Visual Data Mining Working Notes, (2001) 29-44

Neural Network Based Emotion Estimation Using Heart Rate Variability and Skin Resistance

Sun K. Yoo[1], Chung K. Lee[2], Youn J. Park[3], Nam H. Kim[4], Byung C. Lee[5], and Kee S. Jeong[6]

[1] Dept. of Medical Engineering, College of Medicine Yonsei University, Seoul Korea
sunkyoo@yumc.yonsei.ac.kr
[2] Graduate School in Biomedical Engineering, Center for Emergency Medical Informatics, College of Medicine Yonsei University, Seoul Korea
nolegal@yumc.yonsei.ac.kr
[3] Graduate School in Biomedical Engineering, Human Identification Research Center, College of Medicine Yonsei University, Seoul Korea
shydeng@yumc.yonsei.ac.kr
[4] Dept. of Medical Engineering, College of Medicine Yonsei University, Seoul Korea
knh@yumc.yonsei.ac.kr
[5] Dept. of Medical Information System. Yongin Songdam College, Gyeonggi, Korea
bclee@ysc.ac.kr
[6] Dept. of Medical Information System. Yongin Songdam College, Gyeonggi, Korea
ksjeong@ysc.ac.kr

Abstract. In order to build a human-computer interface that is sensitive to a user's expressed emotion, we propose a neural network based emotion estimation algorithm using heart rate variability (HRV) and galvanic skin response (GSR). In this study, a video clip method was used to elicit basic emotions from subjects while electrocardiogram (ECG) and GSR signals were measured. These signals reflect the influence of emotion on the autonomic nervous system (ANS). The extracted features that are emotion-specific characteristics from those signals are applied to an artificial neural network in order to recognize emotions from new signal collections. Results show that the proposed method is able to accurately distinguish a user's emotion.

1 Introduction

Throughout history, humans have made tools to help themselves. Among these tools, none has been greater than the computer. Thanks to the development of computers, humans could be released from complex, difficult work. As computers have evolved at a high speed, human life has changed significantly. However, at present the evolution is facing its limit because users now require a more intelligent system that responds to human emotions. Unlike the existing services dependent on computers, newly required services should be fit to the user's taste by considering the user's emotions. To achieve this purpose, a new computing system must have the ability to detect human emotions. This study focuses on such a computing system. We are especially interested in ANS (autonomic nervous system) signals during emotion changes and in developing a detection algorithm by analyzing ANS changes to establish this computing system.

Since the ANS cannot be controlled artificially, the ANS is used to monitor changes in emotion. In this study, we chose ECG (electrocardiogram) and GSR (galvanic skin resistance) as parameters to measure the ANS. Also, since the number of subjects was limited, it seemed meaningless to generalize the detection algorithm to fit well for every subject. Emotion occurs very differently according to the situation, personality, and growth environment of a person. Furthermore, emotions change from day to day within a specific individual.[9] Therefore, is impossible and meaningless to obtain a statistical output from hundreds of subjects. Instead, we measured two parameters from six subjects and analyzed the results. Then, we established a proper reactance model sent from the ANS, and adapted it to the subjects. Subsequently, using the developed algorithm we may be able to estimate the user's emotion. In conclusion, we can receive individually-tailored services by emotion estimation using only ANS data.

2 Methods

2.1 Classification of Emotions

Most literature references about emotions agree that emotions are complex and are a combination of physical and cognitive factors. The physical aspect is also referred to as bodily or primary emotions, while the cognitive aspect is referred to as mental emotions.[8] Because humans' emotions change continuously, it is difficult to distinguish emotion using a standard value. Also, emotion can be analyzed qualitatively, but quantitative analysis is impossible. In this scheme, emotions are defined in a multi-dimensional space of emotion attributes. A popular concept uses a valence-arousal plane (Fig. 1). Valence defines whether the emotion is positive or negative, and to what degree. Arousal defines the intensity of emotion, ranging from calm (lowest value) to excited (highest value).[8] We defined four emotions shown in the valence-arousal plane (Fig. 1).

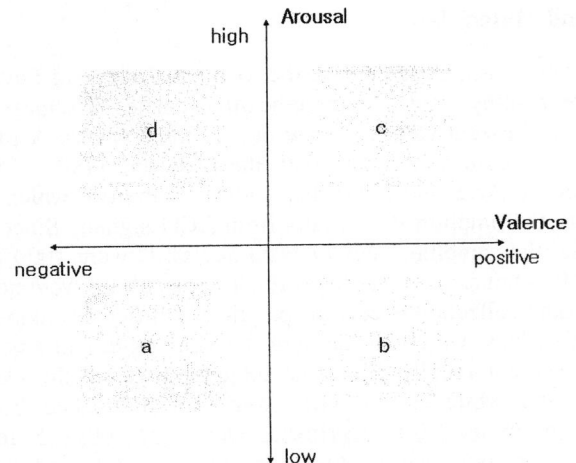

(a): Sad, (b): Calm pleasure, (c): Interesting pleasure, (d): Fear

Fig. 1. Valence-Arousal Plane

2.2 Experimental Sequence

The entire process of the emotion estimation system is shown in Figure 2. Psychological emotion was induced with various video clips. At the end of all experiments, subjects evaluated their degree of valence-arousal. To assess the two dimensions of valence and arousal, we used the Self-Assessment Manikin (SAM), an effective rating system devised by Lang. In this system, a graphic figure depicts reaction values along each dimension[2]. However, SAM estimations are not always in agreement with the physiological change induced by emotions. Pass through upside process and collected data trains the neural network. Using the trained neural network, we could presume emotion.

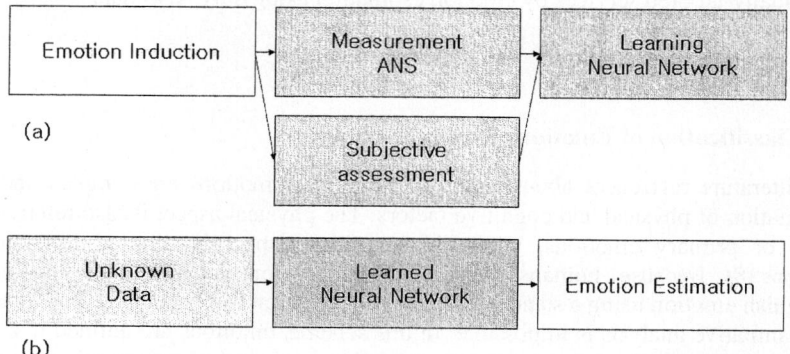

(a): Inducing emotion, physiological analysis, neural network training
(b): Estimating sequence using the trained neural network

Fig. 2. Diagram of the emotion estimation system

2.3 Subjects and Materials

The entire experiment was conducted at the Medical College of Severance. Subjects consisted of three healthy males and three healthy females. We measured three times in a day, for three consecutive days per week, for three weeks. A MP150 (BIOPAC SYSTEM, U.S) was used to measure both the ECG and GSR. ECG signals result from activities in the ANS and HRV (heart rate variability), which is an important factor to estimate the emotion that results from ECG signals. Since HRV is highly sensitive to noise, the sampling rates of ECG and GSR were 1000 Hz and 250 Hz, respectively (GSR is not as sensitive to noise). The electrodes were not removed from the body during data collection except in specific circumstances (eating a meal, using the restroom, etc). Thus, we could measure the vital signals in any state, including normal state, almost continually. Since the subjects stayed in the same environment during stimulus and while resting, they were not influenced by environmental changes. The materials used for experiments were video clips. Some were used to induce a pleasant emotion, while other ones were used to induce an unpleasant emotion. We did not select which movie to show the subjects due to variations in taste. The subjects were shown video clips similar to the television programs they watch at home. The subjects filled out a form after they watched each video clip, which required a full explanation of their emotional state during the video. Figure 3 shows a portion of the form.

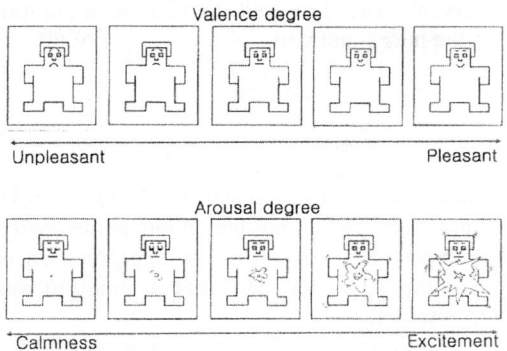

Fig. 3. Emotion assessment using Self-Assessment Manikin (SAM)

2.4 Method of Analysis

The signals emitted from the human body are non-stationary. As a result, ANS activities are affected during emotion changes. Even when a human is in the same emotional state, the non-stationary signals are displayed differently. Thus, we measured three times in a day (morning, afternoon, and evening) in order to ensure the independence to time and subordination to emotion. HRV has been an important factor in the study of both stress and emotions. We also used HRV in the time and frequency domains to analyze the ECG. The HRV was obtained during a given time and we compared the ANS activities in each emotional state, and then estimated the relationship between emotion and the ANS. To detect QRS peaks, we applied the algorithm suggested by Tompkins & Hamilton.[4] Task forces of The European and North American Societies of Cardiology suggest proper times to detect HRV, one of which is a short-term (5 minutes) measurement and the other is a long-term measurement (24 hours).[1][3] Following this recommendation, we measured for 24 hours then divided the results into 5 minute intervals. Using the short-term data, we calculated the SDNN, RMSSD, HR, and LF/HF [1].

ECG and GSR measurements reflect reactions of the ANS, particularly the sympathetic nervous system. When a weak current flows through the two electrodes on the fore finger and the middle finger, changes in skin resistance occurs thus changing the GSR value. This indicated that changes occurred in the ANS. If the emotion of the subject changes and excites the sympathetic nervous system, sweat is secreted from the sweat gland, which increases GSR values.

2.5 Learning and Estimation Using the Neural Network

Numerous advances have been made in the development an intelligent system and some have been inspired by biological neural networks. Researchers from many scientific disciplines are designing neural networks to solve problems in pattern recognition, prediction, optimization, associative memory, and control.[10] The multi-layer perception (MLP) neural network structure was used in this study. The goal of this study was to define the type of physiological data that is appropriate to train the neural network. We distinguished the nuance of emotion, which is analyzed physiologically using the neural network.

3 Results

From the significant amount of data evaluated by SAM, we selected data suitable to distinguish an emotion. If subjects did not feel emotions promptly in SAM, we did not apply the data to the experiment. Physiological interpretation was calculated in the time and frequency domains. We analyzed whether emotion was caused or followed by physiological interpretation.

Figure 4 shows the analyzed value that changed according to the emotion type in the time domain. As indicated, when the emotion is fear (d), SDNN and RMSSD increase more than in other emotions. However, in this case the mean HR decreased, indicating that activity of the sympathetic nervous system increases more than activity

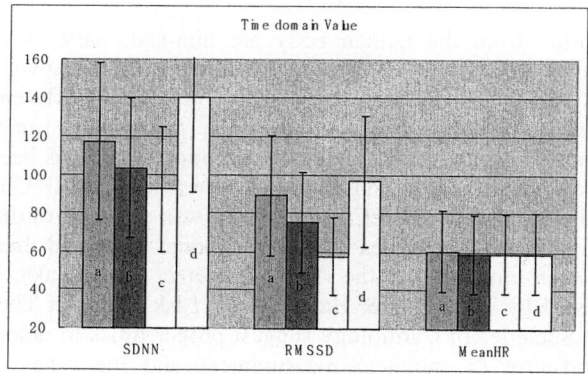

SDNN (ms)   RMSSD (ms)   Heart rate (number/1 min)
a, b, c, d: Reference Fig. 1.

Fig. 4. Statistically-analyzed parameters in the time domain

Table 1. Frequency domain values of HRV and GSR Activity

	A	b	c	d
LF/HF	2.196	2.587	2.743	2.635
GSR	938.35	706.33	643.2	612.45

a, b, c, d: Reference Fig. 1.

the parasympathetic nervous system. Table 1 indicates that the autonomic nervous system was evaluated by HRV in the frequency domain and displays the activity degree of GSR. Fear and interested state (c and d, respectively) demonstrated a LF/HF ratio relatively lower than sad and calm pleasure (a and b, respectively). However, we could not show a statistical significance for emotion using these physiological values. We trained the neural network to estimate four kinds of emotion. Table 2 shows the accuracy of emotion classification using HRV and GSR parameters. Total accuracy was 80.2% and the fear (d) estimation rate was the highest.

Table 2. Accuracy of emotion classification using neural network

Emotion / Test	A	B	c	d
a	84.6	13	2	0.4
b	5.6	77	11.6	5.5
c	0.9	11.3	71.7	16.1
d	3.3	4.1	5.2	87.5
Total accuracy	80.2 %			

a, b, c, d: Reference Fig. 1

4 Discussion and Conclusion

It is very difficult to analyze human emotion. Currently, several research groups are studying emotion and human body response through various methods. We analyzed the relationship between emotion and the autonomic nervous system by HRV and GSR signal. We found an accuracy rate of 80.2%. This study demonstrates the high accuracy in emotion estimation using the neural network. However, it is impossible to analyze emotions that change in real-time. Therefore, a more delicate algorithm is needed to detect real-time emotional changes.

Acknowledgement

This work was supported by the Ministry of Science & Technology(Korea) under Grant M10427030003-04L2703-00300.

References

1. Heart Rate Variability Standard of Measurement physiological interpretation, and clinical use Task Force of The European Society of Cardiology and The North American Society of Pacing and Electrophysiology (Membership of the Task Force listed in the Appendix), European Heart Journal 1996 17, 354-381
2. Lang P J, Bradley, M.M & Cuthbert, B,N NIMH Center for the Study of Emotion and Attention, : international Affective Picture System (IAPS). Technical Manual and Affective Ratings 1995
3. Rollin Mccraty, MA, Mike atkinson, Wiliam A. Tiler, phD : The Effects of Emotions on Short-term power spectrum analysis of heart rate variability. VOL.76 No 14 page 1089~1093 November 15 1995

4. WILLIS J.TOMPKINS : Biomedical Digital Signal processing. prentice Hall international
5. Task Force of the ESC and the NASPE. heart rate variability: Standards of Measurement, Physiological Interpretation, and Clinical Use. Circulation, 1993, 5: 1043-65
6. Moody GB. ECG-based indices of Physical Activity. In: Computers in Cardiology 1992. IEEE Computer Society Press, 1992:403-6
7. Edmund Bourke, James R Sowers : The Autonomic Nervous System and Blood Pressure Regulation in the Elderly. Interdiscipl Top Gerontol. Basel, Karger, 2004, vol 33, pp 45-52
8. Hartwig Holzapfel, Matthias Denecke,Christain Fuegen, Alex Waibel :Integrating Emotional Cues into a Framework for Dialogue Management. Fourth IEEE International Conference on Multimodal Interfaces (ICMI'02) October 14 - 16, 2002
9. John T. Cacioppo, Gary G. Berntson, Jeff T. Larsen, Kirsten M. Poehlmann, Tiffany A. Ito : The psychophysiology of Emotion. The handbook of emotion, 2nd. Edition (pp. 173-191). New York: Guilford Press.
10. Anil K. Jain, Jianchang Mao, K.M. Mohiuddin : Artificial Neural Networks: A tutorial. IEEE Computer Society 1996, March 1996 Vol. 29, No. 3
11. Haykin, S.: Neural Networks : A Comprehensive Foundation. Prentice Hall, New Jersey, 2nd ed. (1999)
12. Rajapakse, J.C., Wang, L.P. (Eds.) : Neural Information Processing: Research and Development. Springer, Berlin (2004)
13. R.W. Picard : Affective Computing. MIT Press, London, England, 1997.

Modeling Belief, Capability and Promise for Cognitive Agents - A Modal Logic Approach*

Xinyu Zhao and Zuoquan Lin

Department of Information Science,
Key Laboratory of Pure and Applied Mathematics,
School of Mathematical Sciences,
Peking University, Beijing 100871, China
{xinyua, lz}@is.pku.edu.cn

Abstract. From the last decade, modeling of cognitive agents have drawn great attention and provide a new paradigm for addressing fundamental questions in cognitive science. In this paper, a logical model for reasoning about cognitive agent's three attitudes *Belief*, *Capability* and *Promise* is proposed. A formalization is provided based on the modal logic to specify and analyze dependencies between the three attitudes. By adopting a set of constraints that describe how the three attitudes are related to each other, we can draw a number of properties of the model. To show the potential applications of the model, we apply the BCP model to a decision-making example in trading agent competition for supply chain management(TAC SCM). The logical model proposed here provides a rigorous semantic basis for modeling cognitive agent and reasoning about multi-agent interactions.

1 Introduction

Cognitive agents and multi-agent interactions play a significant role when building distributed sophisticated systems. There has been much interest in the use of logic for developing formal theories of agents, such as Intention logic[1], *BDI* logics[2], *KARO* framework[3] and *LORA* logic[4]. These logics give agents a substantial base in theory as well as a number of implemented systems that are used for challenging applications such as air-traffic control and manufacturing systems.

When formalizing the properties of cognitive agents, the first fundamental problem is to determine which combination of attitudes is appropriate to modeling the agents' cognitive states and functional components. By considering a type of complex systems, we discuss the necessity of the three attitudes,i.e. belief, promise and capability. These systems commonly take on such characteristics as:

(1) There are a great deal of events from both the outside and the inside of the system when the dynamic environment evolves.

---

* This paper was partially supported by NKBRPC (2004CB318000) and NSFC (60373002, 60496322).

(2) In order to achieve the goal, there must be special components for cognitive agents to perform corresponding actions.

(3) Before cooperation, the participants must make an agreement in some way, for example negotiating a contract, even orally.

We are motivated to investigate how to modeling such agent-based systems. Given characteristic (1), cognitive agents need an informative component to keep the states of the systems and the environments, called *belief*. The agent updates its belief when it perceives the changes of the environment. When taking characteristic (2) into account, it is necessary for agents that there is a functional component, called *capability*, to carry out their plans and bring the plans to success. The agent cannot always do what it intends to because it must have the required capabilities. Considering characteristic (3), a model about *promise* must be built to establish cooperation relationship between participants during the interaction processes. As pointed out in [5], there may not be a unique agent model suitable for all applications, since different domains have different characteristics and thus different requirements regarding rational behavior.

In this paper, a logical model for reasoning about the agent's *Belief, Capability* and *Promise*(abbreviated as BCP) is proposed. We use *modal logic*, which provides an intuitively acceptable, uniform formalization of intensional notions, to model cognitive agent and multi-agent interactions. Following the expression of traditional epistemic logic that $B_i\phi$ means agent i believes ϕ, we will add two modal operators P_{ij} and C_{ij} to the logic. The standard *Kripke*-style semantics for B_i and P_{ij} is combined with almost-standard *neighbourhood* semantics[6] for C_{ij} to interpret the well-formed formulas(*wff*). The intended meaning of $P_{ij}\phi$ is that agent j makes a promise to agent i that j would like to achieve ϕ, while $C_{ij}\phi$ means agent i considers that agent j has the capability to perform action ϕ[1]. In large scale multi-agent systems, the decision-making processes, based on the individual belief, are commonly impacted by whether the cooperating agents would like to make promises to accept the tasks and whether they can fulfill the tasks. Thus, the individual agent is required to be able to reason about other agents' promises and capabilities to decide whether they can accept and accomplish new tasks most effectively in the BCP logic. To tie the three operators up, we place a number of constraints on them. Also, a trading example between the supplier agent and the purchasing agent in the trading agent competition for supply chain management(TAC SCM)[8] is presented to show the potential applications of the logical model.

The remainder of the paper is organized as follows. In section 2, a brief literature review about the three attitudes of agents is firstly given. Then we provide the syntax, semantics and axioms of the BCP logic, with some properties of capability and promise operators. A reasoning and decision-making example in TAC SCM is illustrated in section 3. In section 4, related works are compared and distinguished in detail. Section 5 concludes with a discussion and indicates the future work.

---

[1] Be similar to [7], we will not distinguish between actions and facts, and the occurrence of an action will be represented by the corresponding fact holding.

2 The Logical Model

2.1 Literature Review

Belief expresses the agent's information about the world, which is formalized with possible worlds semantics. The modal logic system $weak - S5_n$ is often chosen as logic of belief. Despite a number of disadvantages, such as the logical omniscience problem, possible worlds are still the semantics of choice for researchers[2][3][4].

Capability, on which we will place strong emphasis, is one of the necessary conditions for agents to interact, cooperate and accomplish tasks successfully. The very beginning of the research on capability can be cast back to Ryle's book[9], in which the author argued the key difference between stupidity, that is, not knowing how, and ignorance, not knowing that. However, there was little considerable work on capability until the late 1980s. *Singh*[10] introduced an abstract concept *Know How* to characterize the agent's capability from the view of external system designers. He suggested that it is not sufficient for an agent to be capable of performing something, moreover the agent must have the knowledge required to form the complete plans before acting. The famous KARO[3] framework tried to deal with the notion of knowledge, ability, result and opportunity. In KARO, dynamic and epistemic logic were combined into one modal system and the ability was considered to be a positive explanatory factor in accounting for the agent's performing an action. *Padgham*[11] extended the well-known BDI architecture by adding *Cap* operator in order to eliminate mismatch between theory and practice for actual systems. A style of commitment was defined to enrich the existing formal models in [11], which allowed a self-aware agent to modify its goals and intentions when its capabilities changed. *Fisher*[12] incorporated more flexible motivational attributes, such as ability and confidence, then introduced ABC model. The main advantage of the ABC modeling is that it provided a simple but flexible foundation for a formal development method.

The study of *promise* is relatively absent in literature. The intuition meaning of promise employed here is that it is a declaration made by the agent assuring that it will be under an obligation to keep the contract persistently. *Liau*[13] proposed a logic of belief, information acquisition and trust (BIT) with some variant axioms. The BIT logic is formulated by using a modal logic approach which is similar to the model we defined here, the main difference is that we additionally introduce the capability and promise operators in order to present capabilities and promises for cognitive agents respectively.

Although such three notions have been separably explored in the literature, there has been very little work on combining them as a whole framework and studying the relationship between them. This is the central issue of what we will do in the BCP logic in the following sections.

2.2 The Logic BCP

The logic BCP is a standard modal logic, extended with operators characterizing agents' capability and promise. Thus, considering a set $A = \{1, ..., n\}$ of agents and a set Φ_0 of atomic propositions, then the set of the well-formed formulas(*wff*) Φ of the *BCP* logic is the least set containing Φ_0 and closed under the following formation rules:

- if φ and ψ are in Φ, then $\varphi \wedge \psi$ is also in Φ
- if φ is in Φ, so are $\neg\varphi, B_i\varphi, C_{ij}\varphi, P_{ij}\varphi, 1 \leq i \neq j \leq n$

The operators $\vee, \supset, \equiv, \top, \bot$ are defined as usual.

We combine standard *Kripke*-style semantics for B_i, P_{ij}, with almost-standard *neighbourhood* semantics[6] for C_{ij} to interpret the wffs in Φ. Formally, a BCP model is a tuple $M = \langle W, \pi, R^{(B_i)_{1 \leq i \leq n}}, R^{(C_{ij})_{1 \leq i \neq j \leq n}}, R^{(P_{ij})_{1 \leq i \neq j \leq n}} \rangle$, where

- W is a none-empty set of possible worlds,
- $\pi : \Phi_0 \to 2^W$ is a truth assignment mapping each atomic proposition to the set of worlds in which it is true,
- $R^{B_i} \subseteq W \times W$ is a serial, transitive and Euclidean relation on W, mapping each an agent's doxastic world to its belief-accessible worlds,
- $R^{C_{ij}} \subseteq W \times 2^W$ is a relation between W and the power set of W,
- $R^{P_{ij}} \subseteq W \times W$ is a binary relation on W.

Informally, $R^{B_i}(w)$ denotes the sets of worlds that are indistinguishable for agent i according to its belief. More specially, in actual world w, agent will consider w' is possible if $w' \in R^{B_i}(w)$. The attributes of relation R^{B_i} ensure the agent's belief to be provided with properties such as consistency, positive introspection and negative introspection. $R^{P_{ij}}(w)$ denotes the sets of worlds that agent i considers possible according to the promises from agent j. It means that, in actual world w, agent i gets a promise from j that w' is possible if $w' \in R^{P_{ij}}(w)$. The idea of $R^{C_{ij}}$ modeling is that each possible world in W has associated it with a collection of subsets of W. It is natural to identify a proposition with a set of possible worlds in W, thus for any $Z \subseteq W$, $Z \in R^{C_{ij}}(w)$ means that agent i considers agent j has the capability according to the proposition corresponding to Z.

The satisfaction relation between wffs and a pair of M, w, consisting of a model M and a world w in M, is defined inductively as follows:

- $M, w \models \varphi$ iff $w \in \pi(\varphi), \varphi \in \Phi_0$,
- $M, w \models \neg\varphi$ iff $M, w \not\models \varphi$,
- $M, w \models \varphi \wedge \psi$ iff $M, w \models \varphi$ and $M, w \models \psi$,
- $M, w \models B_i\varphi$ iff for all $u \in R^{B_i}(w), M, u \models \varphi$,
- $M, w \models C_{ij}\varphi$ iff for some $Z \in R^{C_{ij}}(w)$ for all $u \in Z, M, u \models \varphi$,
- $M, w \models P_{ij}\varphi$ iff for all $u \in R^{P_{ij}}(w), M, u \models \varphi$.

Until now, the relations $R^{B_i}, R^{C_{ij}}$ and $R^{P_{ij}}$ are still independent. In such model, an agent's belief can't be updated with what other agents promise to it and changes of other agents' capabilities. This scenario can't reflect our original motivations, therefore is not what we want. Indeed, agents need to interact, collaborate and negotiate to pursue common-goals and self-interests in large scale multi-agent systems. The decision-making processes, based on the individual belief, are impacted by whether the cooperating agents would like to make promises to accept the tasks and whether they can fulfill the tasks. During the processes of interactions and cooperations, an agent would update its belief, and establish the relationships with those capable agents, whose promises can build on. In such system, the individual agent might be required to be able to reason about other agents' promises and capabilities to decide whether they can accept and accomplish new tasks most effectively. Thus we are more interested in the model satisfying the following conditions:

Con1: $R^{C_{ij}}(w) \neq \varnothing$
Con2: $\varnothing \notin R^{C_{ij}}(w)$
Con3: $R^{C_{ij}}(w) = \bigcap_{u \in R^{B_i}(w)} R^{C_{ij}}(u)$
Con4: for all $Z \in R^{C_{ij}}(w)$, if $R^{B_i} \circ R^{P_{ij}}(w) \subseteq Z$, then $R^{B_i}(w) \subseteq Z$

According to the Con1 and Con2 conditions, we shall say that an agent is capable of doing something at least, as well as performing actions that are not contradictory. The Con3 condition indicates that an agent is self-aware of its attitude towards other agents' capabilities, whereas the Con4 condition ties the three operators up and means that if agent j promises agent i that it will make a goal to be true and agent i also considers that agent j has the capability to achieve the goal, then agent i will believe the goal to be true.

The logic BCP includes the following set of axioms and rules of inference:

P: All tautologies of propositional calculus
B1: $[B_i \varphi \wedge B_i(\varphi \supset \psi)] \supset B_i \psi$
B2: $\neg B_i \bot$
B3: $B_i \varphi \supset B_i B_i \varphi$
B4: $\neg B_i \varphi \supset B_i \neg B_i \varphi$
P1: $[P_{ij} \varphi \wedge P_{ij}(\varphi \supset \psi)] \supset P_{ij} \psi$
P2: $\neg P_{ij} \bot$
C1: $C_{ij} \top$
C2: $\neg C_{ij} \bot$
C3: $C_{ij} \varphi \equiv B_i C_{ij} \varphi$
C4: $B_i P_{ij} \varphi \wedge C_{ij} \varphi \supset B_i \varphi$
R1: From φ and $\varphi \supset \psi$ infer ψ
R2: From φ infer $B_i \varphi$ and $P_{ij} \varphi$
R3: From $\varphi \supset \psi$ infer $C_{ij} \varphi \supset C_{ij} \psi$

In logical terms, the B1-B4 axioms correspond to the $KD45$ modal operator B_i. The B1 axiom formalize that B_i satisfies the K-axiom indicating that agents' beliefs are closed under logical consequence. Moreover B_i satisfies the consistency, positive introspection and negative introspection axioms relating to the serial, transitive and Euclidean properties of the R^{B_i}. The P1 and P2 axioms exhibit the KD dimension of the BCP logical model for the P_{ij} operator. The P1 axiom denotes that if agent j makes a promise of goal φ to agent i, meanwhile it also gives all logical consequence of φ. We use the P2 axiom to eliminate the possibility of agent making contradictory promises. The C1-C4 axioms correspond to Con1-Con4 conditions respectively. We should emphasize here that the C4 axiom ties the three attitudes of cognitive agents up. The agent's belief is affected by what other agents promise to it and changes of other agents' capabilities. More specially, if agent j promises agent i to perform an action, meanwhile agent i considers that agent j has the capability to carry out the action, then agent i will also believe the action could be done. The R2 rule is instance of the rule of necessitation which states that valid wff is believed and promised in advance. The R3 rule indicates that if an agent is considered to be capable of performing a wff, then the consequence of the wff is all the same.

The sound and complete axioms are summarized in the following theorem,

Theorem 1. *The axiomatic system BCP is sound and complete.*

Due to space limitations here we omit the proof of the theorem, which can be gained by the standard technique of canonical model construction in modal logic[6].

2.3 Properties of Capability and Promise

Given the BCP logic, we can draw some useful properties of Capability and Promise operators. For instance, the following formulas are valid in BCP logic:

1. $C_{ij}(\psi \wedge \varphi) \supset C_{ij}\psi \wedge C_{ij}\varphi$
2. $B_i(P_{ij}\psi \wedge P_{ik}\neg\psi) \supset \neg(C_{ij}\psi \wedge C_{ik}\neg\psi)$

Although the first formula is valid, the converse, i.e. $C_{ij}\psi \wedge C_{ij}\varphi \supset C_{ij}(\psi \wedge \varphi)$, is not. For example, let's consider an agent j that is designed for helping a handicapped person i. It is possible for i to consider that j can go either upstairs to serve a cup of milk, or downstairs to fetch an express parcel, depending on the order of the person. But this does not mean that i believes that j has the capability to accomplish the two tasks at the same time.

The second formula implies how agents deal with the inconsistent promises to a certain extent. The intuitive meaning is that if agent i believe that it gets contradictory promises from two agents, i does not believe the two agents simultaneously have the capability of performing the contradictory actions to keep their promises.

Although many valid formulas with intuitive meanings can be deduced from the BCP logic like this, there are still some other none-valid ones worth while further consideration, e.g. $P_{ij}\varphi \supset B_iP_{ij}\varphi$ and $B_iP_{ij}\varphi \supset P_{ij}\varphi$. We exclude the two formulas from the set of axioms in BCP logic on account of the inherent insecurity in Electronic Commerce. It's commonly required that the information must be credible and undeniable during the whole online trading processes. When agent i receives a promise from agent j , if it can not excluded the possibility that someone pretending to be j has made the promise, then it does not necessarily believe that it has received the promise from j. Thus we do not have $P_{ij}\varphi \supset B_i P_{ij}\varphi$. On the other hand, since someone pretending to be j may make a promise to i and cause i wrongly believe that it indeed received the promise from j, $B_i P_{ij}\varphi \supset P_{ij}\varphi$ does not necessarily hold, either. Nowadays, message digesting is combined with digital signatures to provide credible and undeniable transactions, which prevent an agent from claiming that it was really someone else who pretended to do something. Then when i receives promises with j's digital signature, it can believe this is indeed sent by j. When it believes j has made him the promise by recognizing the digital signature of j, it is impossible that it was counterfeit by another. If we want to capture the desired requirement for such domain, we can introduce the following schema,

$$B_i P_{ij}\varphi \equiv P_{ij}\varphi$$

into the BCP logic. The system, composed of the BCP logic and the above axiom, can also be proved to be sound and complete.

In fact, we can introduce diverse axiom schemas into the BCP logic to characterize multi-agent interactions and cooperations in different application domains, meanwhile preserving the soundness and completeness results. For example, agent j is said to be cooperated with agent i, if it would like to promise anything it believes to other agents, i.e. iff the following schema is valid:

$$B_j\varphi \supset P_{ij}\varphi.$$

3 Illustrative Cases

We show the trading agent competition for supply chain management(TAC SCM)[8] as an example of complex system in real life to show the potential applications of the logical model.

Agents that represent their roles in a supply chain perform negotiations with other agents to achieve a common objective. Agents should hold the belief about the information flowing across the supply chain. When making sourcing strategy, the agent has better to affirm that the supplier agent will be provided with sufficient supply capability and will keep its promises to offer materials on time.

TAC SCM provides a competition stage for researchers interested in both artificial intelligence agents and supply chain management. In the game, agents are bidding for supplies and consumer orders as well as planning production and shipment of the end products. Trades with the suppliers, as well as with the customers, are negotiated through a request-for-quotes (RFQ) mechanism. If the supplier can satisfy the order specified in the RFQ in its entirety, the supplier will make promises to the agent meanwhile an offer is sent as a response. One of the more difficult problems in the game is to make sure that the supplies are available at the time they are needed for production. Although the suppliers make promises based on its fixed capacity production per day, their actual production capability is determined by a random walk. Thus, when the supplier produces less than it planned, it delays the deliveries to the agents. It may be necessary for an agent to maintain a model of promise and capability for the supplier in order to compete better.

Let us consider the agent i and the supplier j, whereas φ and ψ denote respectively the facts "delivering the requested quantity specified in the RFQ" and "delivering on the due date". It is one of the possible instances that agent i, when planning production and delivery, will not always believe $\varphi \wedge \psi$ only for the sake of the supplier j makes the promise $\varphi \wedge \psi$ to it. Agent i should bear in mind that it has to stand in competition with the opponents for the limited production capacity of the supplier. Thus it may not believe that ψ is true. But from its historical trading experience, it may consider that the supplier j has the capability to deliver the materials partially on the due data. Then, from $B_i P_{ij}(\varphi \wedge \psi)$, $C_{ij}\varphi$ and $B_i \neg \psi$, agent i may have the following reasoning:

1. $B_i P_{ij}(\varphi \wedge \phi) \supset B_i P_{ij}\varphi \wedge B_i P_{ij}\phi$
2. $B_i P_{ij}\varphi \wedge C_{ij}\varphi \supset B_i \varphi$
3. $B_i P_{ij}(\varphi \wedge \phi) \wedge C_{ij}\varphi \wedge B_i \neg \phi \supset B_i \varphi$

At the same time, this reasoning example shows that agent i can even accept the promise of supplier partially whereas its portion is in contradiction with agent i's belief.

In the preceding discussion, we have mentioned that some formulas are valid whereas their converses are not. Let us dwell on this further with the following example in TAC SCM games. Let φ and ψ denote the facts "delivering the requested quantity specified in the RFQ" and "delivering on the due date", respectively. But the delivered quantity depends upon the different strategies of the supplier(i.e. the *Likelihood* of the agent that sent the RFQ). Thus it is possible for the agent i to consider that the supplier j has the capability to achieve either φ or ψ depending upon j's strategy, but this does not mean that i believes the supplier j can satisfy the order entirely on the due date, i.e. $C_{ij}\varphi \wedge C_{ij}\psi \supset C_{ij}(\varphi \wedge \psi)$ is not valid. On the other hand, if agent i is convinced of both φ and ψ, it will believe that j has the capability to satisfy the order either in quantity or on the due date, i.e. $C_{ij}(\varphi \wedge \psi) \supset C_{ij}\varphi \wedge C_{ij}\psi$ is valid.

4 Related Works

In this section, we briefly compare two related work that are most correlative to our logical model.

Shoham presented the *agent-oriented programming* (AOP) [7] framework emphasizing on an interpreted programming language. He proposed a set of mental state components, such as obligation and capability. Just as Shoham indicated that specific mental properties were requisite for different applications, the selection of mental categories is not objectively right. In this paper, we describe a logical model incorporating agents' belief, capability and promise, by considering a type of multi-agent based complex systems. The addition of the later two modalities and constraints placed on them characterizes the interactions between cooperation agents appropriate, and as a result, enriches the existing agent models. Although McCarthy has debated that all statements including modalities should be viewed in context, both the BCP logic and the AOP framework ignore the topic of context sensitivity, which is worthy of future investigations.

[14] proposed a logic to reason about perceptions and belief. The logic contains three modalities: B stands for belief, P for actual perception, whereas C for the sets of perceptions agent can perceive. Similar to ours, the modalities of [14] use the standard and the neighborhood semantics in modal logic. Various agent types are defined to capture the diverse characters of the application domain. Their formalization deals only with a single agent. Moreover the capability [14] introduced merely refers to perception capability. Capability delivered in the BCP logic is an abstract, high-level ability of agents, and not limited to any kind of special abilities. Also, the axioms proposed here, such as $C4$ axiom, provide a means of formalizing the decision-making processes of cognitive agents.

Previously, we have developed applications based on multi-agent system by using these attitudes of cognitive agents, for instance belief and capability. For full details, the reader is referred to [15][16]. In this paper, we depict a formal framework to model belief, capability an d promise for cognitive agent. In so doing, we seek to provide a bridge between the formality and the practical work (see future work).

5 Conclusions and Future Work

This paper is aiming to present a logical model for reasoning about belief, capability and promise of cognitive agents using a modal logic approach. The possible relations among the three attitudes of the agents are explored with intuitive meanings. A reasoning example is presented to show how the logical model is applied in an actual agent competition game.

Although the logical model provides a kind of rigorous semantic basis for specifying multi-agent interactions, it is still a preliminary work. There are several possible directions for future investigations.

Firstly, the attitudes of agents are studied in a static environment. A promising direction for future work is to extend the model to involve the temporal aspect.

Secondly, in order to characterize multi-agent interactions in different application domains, there are other forms of relations between the three attitudes that one would like to impose on the logic model. This requires additional constraints to be placed on the three operators. These additional constraints may capture diverse assumptions about the application domains and enrich the existing models.

Lastly, a potential application of the logical model is to incorporate with the contract net protocol (CNP) [17]. In essence, Contract Net allows tasks to be distributed among a group of agents. The self-interested agents may delude its manager to get some bids to maximize their profits, even if they cannot accomplish some tasks on time. In this case, the manager may need to reason about the capability of its bidders to avoid an unexpected delay of its task. The underway research direction would attempt towards bridging the gap between fundamental theories and practical systems.

References

1. Cohen, P. R. and Levesque, H. J.: Intentions is Choice with Commitment. Artificial Intelligence **42** (1990) 213–261
2. Rao, A.S., and Georgeff, M.P.: Modeling Rational Agents within a BDI Architecture. In: Proc. of the Second Conference on Knowledge Representation and Reasoning (KR91), Morgan Kaufman, (2001) 473–484
3. van Linder, B., van der Hoek, W., Meyer, J.-J. C.: Formalising Abilities and Opportunities of Agents. Fundamenta Informaticae, **34** (1998) 53–101
4. Wooldridg, M.: Reasoning About Rational Agents. The MIT Press: Cambridge, MA, 2000
5. Rao, A.S., and Georgeff, M.P.: Decision procedures for BDI logics. Journal of Logic and Computation, **8**(3) (1998) 293–342
6. Chellas, B.F.: Modal Logic: An Introduction. Cambridge University Press, Cambridge, 1980
7. Shoham, Y.: Agent-oriented programming. Artificial Intelligence, **60**(1) (1993) 51–92
8. The Supply Chain Management Game for the Trading Agent Competition (TAC SCM) 2004. Available at http://www.sics.se.
9. Ryle, G.: The Concept of Mind. Barnes and Noble, New York, 1949.
10. Singh, M.P.: A Logic of Situated Know-how. In Proceedings of the Ninth National Conference on Artificial Intelligence, (AAAI91), Anaheim, California, (1991) 343–348
11. Padgham, L., Lambrix, P.: Agent Capabilities: Extending BDI Theory. In Proceedings of Seventeenth National Conference on Artificial Intelligence, (AAAI00), Austin, Texas, (2000) 68–73
12. Fisher, M. and Ghidini, C.: The ABC of Rational Agent Modelling. In Proceedings of the First International Joint Conference on Autonomous Agents and MultiAgent Systems, (AAMAS02), Bologna, Italy, (2002) 849-856
13. Churn-Jung L.: Belief, Information Acquisition, and Trust in Multi-Agent Systems- a Modal Logic Formulation. Artificial Intelligence, **149**(1) (2003) 31–60
14. del Val, A., Maynard-Reid, P., Shoham, Y.: Qualitative Reasoning about Perception and Belief. In Proceedings of the Fifteenth International Joint Conference on Artificial Intelligence, (IJCAI97), Nagoya, Japan, (1997) 508–513
15. Xinyu, Z., Zuoquan L.: An Agent-based Supply Chain Modeling (In Chinese). Computer Science, **31** (2004) 16–21
16. Xinyu, Z., Cen W., Runjie Z., et al.: A Multi-Agent System for E-Business Processes Monitoring in A Web-Based Environment. In The Fourth International Conference on Electronic Business, (ICEB2004), Beijing, China, (2004) 470–475
17. Smith, R. G.: The Contract Net Protocol – Highlevel Communication and Control in a Distributed Problem Solver. IEEE Transactions on Computers, **29** (1980) 1104–1113

PENCIL: A Framework for Expressing Free-Hand Sketching in 3D

Zhan Ding[1], Sanyuan Zhang[1], Wei Peng[1], Xiuzi Ye[1,2], and Huaqiang Hu[1]

[1] College of Computer Science/State Key Lab of CAD&CG, Zhejiang University,
Hangzhou 310027, P.R. China
dingzh@hotmail.com, weip@zju.edu.cn
[2] SolidWorks Corporation, 300 Baker Avenue, Concord, MA 01742, USA

Abstract. This paper presents a framework for expressing free-hand sketching in 3D for conceptual design input. In the framework, sketch outlines will be recognized as formal rigid shapes first. Then under a group of gestures and DFAs'(deterministic finite automata) control, the framework can express user's free sketching intents freely. Based on this framework, we implemented a sketch-based 3D prototype system supporting conceptual designs. User can easily and rapidly create 3D objects such as hexahedron, sphere, cone, extrusion, swept body, revolved body, lofted body and their assemblies by sketching and gestures.

1 Introduction

Sketch-based modeling by standard mouse operations became popular in the past decade. Instead of creating precise, large-scale objects, a sketching interface provides an easy way to create a rough model to convey the user's idea quickly. There is a growing research interest on using freehand drawings and sketches as a way to create and edit 3D geometric models.

One of the earliest sketching systems was Viking [4]. In Viking, the user draws line segments, and the system automatically generates a number of constraints which then must be satisfied in order to re-create a 3D shape. Later works include SKETCH [5], Quick-Sketch[1],Teddy [2] and GIDeS[3]. The SKETCH is intended to sketch a scene consisting of simple primitives, such as boxes and cones. Quick-Sketch is a computer tool oriented to mechanical design. It consists of a 2D drawing environment based on constraints. It is also possible to generate 3D models through modeling gestures. The Teddy system is designed to create round objects with spherical topology. The GIDeS permits data input from a single-view projection or from several dihedral views. When creating object from a single-view perspective, the system uses a simple gesture alphabet to identify a basic set of modeling primitives such as prisms, pyramids, extrusion and revolution shapes, among others.

This paper describes a framework to express user's sketching intents using an intuitive and efficient way. Based on this framework, we have implemented a prototype system called PENCIL to support conceptual design.

2 System Framework

For industrial designers, the ability to rapidly create 3D objects by sketching 2D outlines with uncertain types, sizes, shapes, and positions is important to innovative design process. This uncertainty, or ambiguity, encourages the designer to explore more ideas without being burdened by concerns for inappropriate details. Leaving a sketch un-interpreted, or at least in its rough states, is key to preserving this fluency.

It's found that conceptual design can usually be split into four steps: 2D sketching, sketch editing, solid/surface body creating and body editing. We use these steps to construct the framework. Figure 1 shows the workflow of the framework.

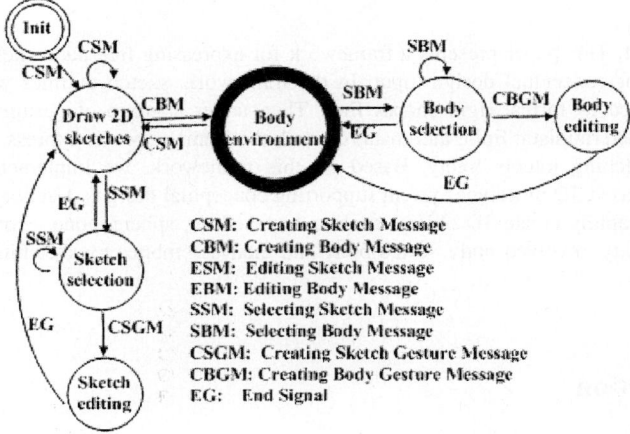

Fig. 1. The workflow of the framework

The workflow is a Deterministic Finite Automata. The initial state is marked as "Init" and the finishing state is marked as "Body environment". The DFA is driven by user's sketching message. When user sketches an object on a plane, the DFA goes to "Draw 2D sketches" state, and the sketched outlines will be recognized as rigid geometric shapes. In this state, user can continuously sketch the profile of a 3D solid/surface body. When switching to "Body environment", the system will receive a Creating Body Message (CBM). Here CBM is not a menu or icon message and it will be explained later in another DFA. "Body environment" can be driven to another two states by message: Creating Sketch Message (CSM) or Selecting Body Message (SBM). After a CSM message, the current state will be driven back to "Draw 2D sketches". After a SBM message, a selection action will be done. Because the framework allows multi-body selection, it means that "Body selection" state can be rolled back to receive other SBM messages. When selection is done, user can sketch specific editing gestures. The result of gesture recognition will drive the DFA to the "Body editing" state. From "Body editing" to "Body environment", the framework needs End Gesture (EG) message. Here it's a right click message.

When current state is "Draw 2D sketches", we can drive it to "Sketch selection" state. It means we can edit the sketches after sketch selection. The editing operations are also controlled by the predefined gestures.

3 Creating Basic 3D Bodies

To rapidly create 3D objects, the framework provides a natural and convenient way to create some basic 3D solid/surface geometries: hexahedron, cone, extruded body, revolved body, swept body, lofted body, and body from silhouettes.

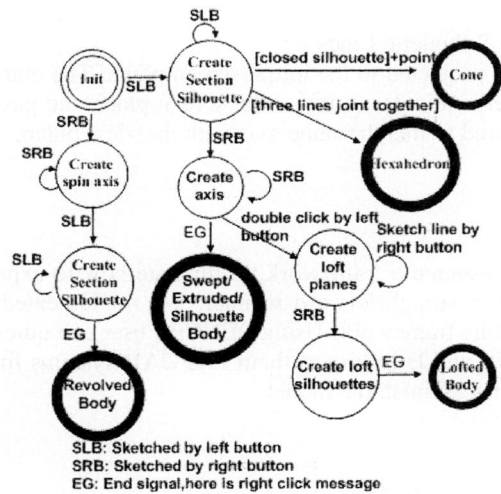

Fig. 2. The DFA of creating basic 3D bodies

The creation of these basic 3D geometric bodies is controlled by another DFA. Figure 2 shows the state converting diagram of this DFA.

- Hexahedron

The first and second lines are projected to the initial screen plane. The third line is projected onto a plane which is perpendicular to the initial screen plane and passes the joint point, and share the same axis with the view plane.

- Cone

The profile is projected to the initial screen plane. The peak is projected to a plane which is perpendicular to the initial screen plane and passes the geometric center of the profile and shares the axis with the view plane.

- Extrusion

The profile is projected to the initial screen plane. The extrude axis is projected to a plane which is perpendicular to the initial screen plane and passes the start point of extrude axis and shares the axis with the view plane.

- Swept Body

The profile is projected to the initial screen plane. The sweep path is projected to the plane which is perpendicular to the initial screen plane and passes the start point of the sweep path and shares the axis with the view plane.

- Revolved Body

Both revolve axis and revolve profile are projected to the initial screen plane.

- Lofted Body

The basic loft profile is projected to the initial screen plane. The loft centerline is projected to a plane that is perpendicular to the initial screen plane, that passes the start point of the centerline and shares the axis with the view plane. Loft planes are generated by projecting the sketched axes using a similar process for loft path generation

- Body from Silhouette Lines

The silhouette is projected to the initial screen plane. The outer line is projected to a plane which is perpendicular to the initial screen plane and passes the start point of the outer axis line, and shares the same axis with the view plane.

4 Conclusion

In this paper, we presented a framework to illustrate how to express user's free-hand sketching intents in a straightforward manner. We implemented a prototype system PENCIL based on this framework. Using PENCIL, user can quickly and easily create complex 3D models, and later import them into CAD systems for detailed constraining and dimensioning to finish the model.

References

1. Eggli L., Hsu C., Brüderlin B.D., Elber G.: Inferring 3D Models from Freehand Sketches and Constraints. Computer-Aided Design, 29(2), 101-112, 1997.
2. Igarashi T, Matsuoka S, Tanaka H.: Teddy: A sketching interface for 3D freeform design. In Computer Graphics Proceedings, Annual Conference Series, ACM SIGGRAPH, Los Angeles, California, 1999. 409-416.
3. Pereira J., Jorge J., Branco V., Nunes F.: Towards calligraphic interfaces: sketching 3D scenes with gestures and context icons. WSCG. Conference Proceedings, Skala V. Ed, 2000.
4. Pugh, D.: Designing Solid Objects Using Interactive Sketch Interpretation. In Computer Graphics (1992 Symposium on Interactive 3D Graphics), 25, 2, (1992) 117~126.
5. Zeleznik R.C, Herndon K.P and Hughesp J.F.: SKETCH: An interface for sketching 3D scenes[A]. In Computer Graphics Proceedings, Annual Conference Series, ACM SIGGRAPH, New Orleans, Louisiana, 1996. 163-170.

Blocking Artifacts Measurement Based on the Human Visual System[1]

Zhi-Heng Zhou and Sheng-Li Xie

College of Electronic & Information Engineering, South China
University of Technology, Guangzhou, 510641, China
crenna@21cn.com

Abstract. The block-based DCT image compression methods usually result in discontinuities called blocking artifacts at the boundaries of blocks due to the coarse quantization. A measurement of blocking artifacts based on Human Vision System (HVS) is proposed. This method separates the blocking effects from the original edges in the image by an adaptive edge detection based on local activity and luminance masking. The blocking artifacts in the non-edge area and on the edge are calculated separately. The weighted sum is regarded as the evaluation result. Simulation results show that the proposed measurement is robust for different kind of images, and has the general performance of image quality evaluation metric.

1 Introduction

The block-based discrete cosine transform (BDCT) scheme is a foundation of many image and video compression standards including JPEG, H.263, MPEG-1, MPEG-2, MPEG-4 and so on. But, usually blocks are coded separately and the correlation among spatially adjacent blocks is not taken into account. It may cause block boundaries being visible in the decoded image. Many numerical metrics for blocking artifacts have been proposed, such as PSNR, MSDS [1][2] and PS-BIM [3]. In this paper, a blocking artifacts metric based on human vision system (HVSBM) is proposed.

2 Measurement of Blocking Artifacts

In order to separate the blocking artifacts from true edges, we first use Sobel operator to detect the edges. Given a pixel $f_{i,j}$ on the border of the current block, we have its local gradient components

$$g_x = f_{i+1,j-1} - f_{i-1,j-1} + 2f_{i+1,j} - 2f_{i-1,j} + f_{i+1,j+1} - f_{i-1,j+1}$$
$$g_y = f_{i-1,j+1} - f_{i-1,j-1} + 2f_{i,j+1} - 2f_{i,j-1} + f_{i+1,j+1} - f_{i+1,j-1} \tag{1}$$

---

[1] This work is supported by the National Natural Science Foundation of China (Grant 60274006), the National Science Foundation of China for Excellent Youth (Grant 60325310), the Natural Science Key Fund of Guangdong Province, PR China (Grant 020826), and the Trans-Century Training Programme Foundation for Talents by the State Education Ministry.

And then the magnitude and angular direction of the gradient at coordinate (i, j) are

$$G_{i,j} = \sqrt{g_x^2 + g_y^2} \quad \theta_{i,j} = \tan^{-1}(g_y/g_x) \tag{2}$$

Blocking artifacts are often mistaken as true edges. But the artificial edges are usually weak edges. If we give a big enough threshold T_0, then only true edges will be extracted. According to human vision system theories, the visibility of edges is due to background activity and luminance [4][5]. So, we construct an alterable threshold

$$T = \left|1 - \frac{\sigma}{m}\right| \cdot T_0 \tag{3}$$

where σ and m are the local variation and mean values of pixel $f_{i,j}$ and its eight neighboring pixels, respectively. σ corresponds to the local activity, and m corresponds to the local luminance. If $G_{i,j} > T$, then $f_{i,j}$ is an edge point, otherwise it is a non-edge point.

We take two horizontal adjacent blocks as example to specify our method. In fact, blocking artifacts will cause discontinuity not only in the non-edge area but also on the edge. So, we use D_1 and D_2 to define the blocking artifacts these two parts respectively

$$D_1 = (f_{i,j} - f_{i,j-1})^2 \tag{4}$$

$$D_2 = (f_{i,j} - f_{i,j}(\theta_{i,j}))^2 \tag{5}$$

where $f_{i,j}(\theta_{i,j})$ is the adjacent pixel of $f_{i,j}$ along the edge direction $\theta_{i,j}$, as shown in Fig.1.

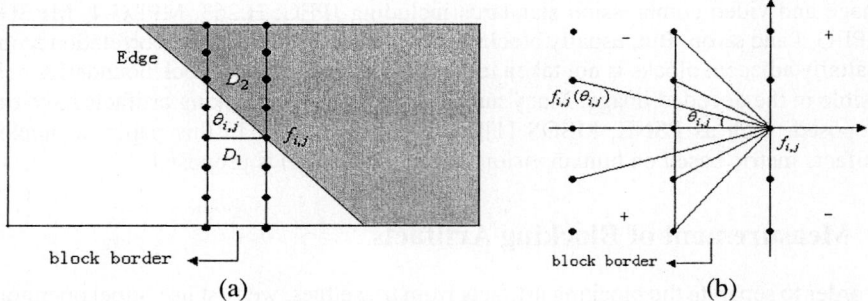

Fig. 1. (a) Definition of D_1 and D_2 (b) Definition of $f_{i,j}(\theta_{i,j})$

In order to select $f_{i,j}(\theta_{i,j})$, the angle of the edge is rounded to the nearest one of 8 directions: $-arctg\,2, -\pi/4, -arctg\,1/2, 0, arctg\,1/2, \pi/4, arctg\,2, \pi/2$. We look $f_{i,j}$ on as the origin of the right-angle coordinate system as shown in Fig.1 (b). If $\theta_{i,j} > 0$, $f_{i,j}(\theta_{i,j})$ locates at the first or third quadrant, otherwise $f_{i,j}(\theta_{i,j})$ locates at the second or fourth quadrant.

And then we can measure the blocking artifacts on the pixel $f_{i,j}$ by D_{HVSBM}

$$D_{HVSBM} = I(G_{i,j} \leq T) \cdot D_1 + \lambda \cdot I(G_{i,j} > T, |\theta_{i,j}| \neq \pi/2) \cdot D_2 + I(G_{i,j} > T, |\theta_{i,j}| = \pi/2) \cdot 0 \quad (6)$$

where λ is an adjusting parameter between two parts. If condition A is satisfied, then $I(A) = 1$. Otherwise, $I(A) = 0$. Similarly, we can measure the blocking artifacts between two vertical adjacent blocks.

The third term of (6) corresponds to the event that a vertical edge happens to exactly locate at the horizontal adjacent blocks border. We cannot take it as blocking artifact, so it will not be included in the calculation.

According to the masking effect theory, human eyes are very sensitive to the blocking artifacts in the non-edge area, rather than that on the edge. We define λ as

$$\lambda = \frac{1}{1 + G_{i,j}/T} \quad (7)$$

where $G_{i,j}$ is the magnitude of the edge, which also reflects the activity of the surrounding background and T is obtained by (3).

3 Simulations and Results

In general, the lower value of PSNR, the greater severity of the blocking effects. But extended MSDS [1][2] and the proposed HVSBM are on the contrary.

Table 1. Comparison of different metrics (o and d stand for original and decoded image)

Metrics	Lena 0.216bpp	Peppers 0.207bpp	Fishingboat 0.244bpp	Barbara 0.232bpp	Test 0.08bpp
PSNR	29.92 (d)	30.22 (d)	27.74 (d)	7.94 (d)	30.28 (d)
Extended MSDS	1639.7 (o) 5693.5 (d)	1881.4 (o) 4666.6 (d)	4760.7 (o) 9262.9 (d)	5269.9 (o) 21666 (d)	4326.5 (o) 3686.9 (d)
PS-BIM	0.6358 (o) 0.4658 (d)	0.5974 (o) 0.4774 (d)	0.7025 (o) 0.5064 (d)	0.8784 (o) 0.5770 (d)	1.8933 (o) 1.2361 (d)
HVSBM	1551.2 (o) 2506.1 (d)	1402.9 (o) 2289.8 (d)	2750.5 (o) 3743.7 (d)	1880.8 (o) 6059.5 (d)	648.8 (o) 652.8 (d)

We want to verify that if the proposed metric has the general performance of the image quality metric. In the simulations, we set $T_0 = 100$. Table 1 shows the comparison of PSNR, extended MSDS [1][2], PS-BIM [3] and the proposed HVSBM using different kinds of images. Because of the low bit rates, the severity of blocking effects in the JPEG decoded image should be greater than that of original image. In table 1, the results of HVSBM have proved this. Another fact found in Table.1 is that PS-BIM does not work as said in its paper. The lower value of PS-BIM, the greater severity of blocking artifacts is taken to be. Fig.2 shows the results of applying HVSBM to different images with different bit rates. It can be shown that in general, the line of HVSBM goes down as bit rates goes up. So, HVSBM achieves consistent results for all these images. In order to verify the performance of different metrics on distinguishing true edges from blocking artifacts, we design a "Test" image as shown in Fig.3. It can be found that the blocking artifacts appear around the edges in "Test" image. The numerical comparison is shown in the last column of Table.1. We find that extended MSDS do not work in this case. And the proposed metric still works well as usual.

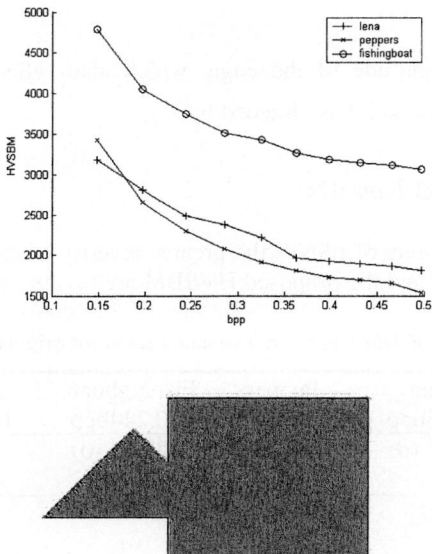

Fig. 2. HVSBM versus bit rates for different images   **Fig. 3.** "Test" image coded at 0.08 bpp

4 Conclusions

This paper presents a measurement for blocking artifacts based on Human Vision System. This method separates the blocking effects from the original edges in the image by an adaptive edge detection based on local activity and luminance masking. Simulation results show that the proposed measurement is robust for different kind of images, and has the general performance of image quality evaluation metric.

References

1. Minami S., Zakhor A.: An optimization approach for removing blocking effects in transform coding. IEEE Trans. Circuits Syst. Video Technol., Vol. 5, (1995) 74-82
2. Triantafyllidis G.A., Tzovaras D., Strintzis M.G.: Blocking artifact detection and reduction in compressed data. IEEE Trans. Circuits Syst. Video Technol., Vol.12, (2002) 877 - 890
3. Suthaharan S.: Perceptual quality metric for digital video coding. IEE Electronics Letters, Vol.39, (2003) 431- 433
4. Ander G.L., Netravali A.N.: Image restoration based on a subjective criterion. IEEE Trans.Syst.,Man, Cybern., SMC-6(12) (1976) 845-853
5. Karunaseker S.A., Kingsbury N.G.: A distortion measure for blocking artifacts in image based on human visual sensitivity. IEEE Trans. Image Processing. (1995) 713-724

A Computation Model of Korean Lexical Processing*

Hyungwook Yim[1], Heuseok Lim[2], Kinam Park[2], and Kichun Nam[1]

[1] Department of Psychology, Korea University, Seoul, Korea
{lapensee, kichun}@korea.ac.kr
[2] Department of Software, Hanshin University, Osan, Korea
limhs@hs.ac.kr, superkn@naver.com

Abstract. This study simulates a lexical decision task in Korean by using a feed forward neural network model with a back propagation learning rule. Reaction time is substituted by a entropy value called 'semantic stress'. The model demonstrates frequency effect, lexical status effect and non-word legality effect, suggesting that lexical decision is made within a structure of orthographic and semantic features. The test implies that the orthographic and semantic features can be automatically applied to lexical information process.

1 Introduction

This study proposes a connectionist model to explain the process of word recognition. To simplify the model, the simulated condition is limited to a task to decide whether a string of letters given is a word or not (lexical decision task: LDT). The results show several lexical effects that have been repeatedly identified in the literature, giving validity to the proposed model. The model is evaluated to have implications on the understanding of the architecture of human word recognition, input and output representations, learning algorithm, and comparison methods of simulation data with human data.

2 The Lexical Effects

Before proposing the model, it is necessary to point out the lexical effects shown in the study. These effects have been repeatedly reported in numerous preceding experiments and are firmly established as characteristics of human lexical processing.

First of all, the recognition speed of a certain word increases along with the frequency of the word in the language. This is called word frequency effect and thought to occur at the lexical access level. The effect implicates that words are directly represented in the mental lexicon in order of the number of times they appear in the practical usage of the language [1].

---

* The research presented in the paper is supported by the BK21 grant(H0041800).

Secondly, lexical status effect refers to a phenomenon where words are recognized faster than pseudo-words are. While words can be identified in the mental lexicon, pseudo-words cannot because they are not registered in it [2]. The effect is also a criterion of the lexical decision task.

Non-word legality effect of the orthography combination occurs when it takes longer to reject legal non-words than illegal non-words. Legality is given when a non-word is produced according to the orthographic combinations rules. This effect implies that orthographic rules exist in combining letters to produce a word and that we are tacitly aware of them[3].

Finally, it takes longer to reject the non-words that look similar to real words in terms of the combination of letters than the non-words that do not. This is word similarity effect and implicates that the lexical decision is carried out based on the orthographic combination rules and the lexical entity itself [4].

3 The Proposed Computational Model

3.1 Network Configuration

A three layer feed-forward neural network is used with 67 orthographical input units, 250 hidden units, and 120 semantic output units which are fully connected. A dot product function is used for the input function and a sigmoid function for the output function.

Input structure The Korean writing system 'HanGeul' is an alphabetic script where several letters make a syllable. Even though, there is not yet a perfect input structure[5,6] if, in HanGeul, the input data is restricted to monosyllables, there is a possible structure that can express every monosyllable without any generalization problem or overlapping problem.

Therefore, the input structure is constructed on the basis of the HanGeul's letter combination property to have a CVC or CV structure and a fixed letter position. As a HanGeul monosyllabic word only contains less than three letters within a syllable and each letter has its own position in a syllable, the input structure can be composed of 67 binary units which is 19 units for the first position, 21 units for the second position, and 27 units for third position. This input structure could not only express every letter combination but also avoid making any generalization problem or overlapping problem.

Output structure Consisting 120 binary units the exact meanings of each word is not represented in the semantic output structure. However, each word meaning has the category common features while having its own characteristic features. In detail, fifty randomly created semantic outputs are used as the category prototypes, where the probability activated in each unit is P_p=0.1. Under these fifty prototypes, ten exemplars are made each, and the probability activated in each dimension under the prototypes is $P_e = 0.05$, producing 500 exemplars. The average number of the activated units for each exemplar word is 16.8(std. 3.56, min 6, max 27).

4 Training

4.1 Training Data

The training data is selected from the Sejong corpus(http://www.sejong.or.kr). Along with usage frequency values, 498 monosyllabic common nouns are selected. These raw frequencies are compressed on the basis of the logarithm function to save computational resources as in (1).

$$P = \text{K} \log{(N+2)} \ . \tag{1}$$

In (1), P is the compressed frequency, K is a constant that is set to 0.2065 (where P of the highest frequency word is 0.93), and N stands for the raw frequency of the word[7]. The training regime is divided into a series of epochs and P is a chance for each word to be presented. For an example, if P is 0.9, the word would be trained 9 times during the total 10 times of training. From 500 randomly generated semantic outputs, two are randomly excluded to match the number of orthographic inputs.

4.2 Training Procedure

The matched orthographical inputs and semantic outputs are trained considering the compressed frequency with a back-propagation algorithm that uses a cross entropy value as an error term. Using LENS as a simulator, the learning rate is 0.1, the momentum 0.9, and Doug's momentum is used [8]. In the case of overtraining, the best trained point is determined[9]. On the base of frequency effect, the highest correlation between the semantic stress and the compressed frequency values is selected as the best trained point(the 3950th trained state, r=0.738 (p<0.01), total error 536.85, error for one input 0.630 in average).

5 Experiment

5.1 Methods

For the experiment, two non-trained data sets are used. One set has 100 non-words where letter combinations existed in the Korean language usage, and the other set has another 100 non-words, where letter combinations do not exist in the Korean language usage.

The output values from the semantic layer are analyzed with semantic stress values [10]. The semantic stress value(Sj) is a kind of entropy value which reflects the error value based on the output value(sj) shown in (2). It is showed that the value could be a good criterion to replace the reaction time reported in human experiments where time factor could not be considered[7,11]. That is, higher error value implicates longer reaction time, and lower error value implicates shorter reaction time.

$$Sj = sj \log_2{(sj)} + (1 - sj) \log_2{(1 - sj)} - \log_2{(0.5)} \ . \tag{2}$$

5.2 Results

To simulate frequency effect, the correlation between the compressed frequency and semantic stress has to be statistically significant. The results show that the correlation coefficient between the compressed frequency and semantic stress values is 0.738 ($p<0.01$) (Fig. 1.). Therefore, it could be indicated that the model simulates frequency effect.

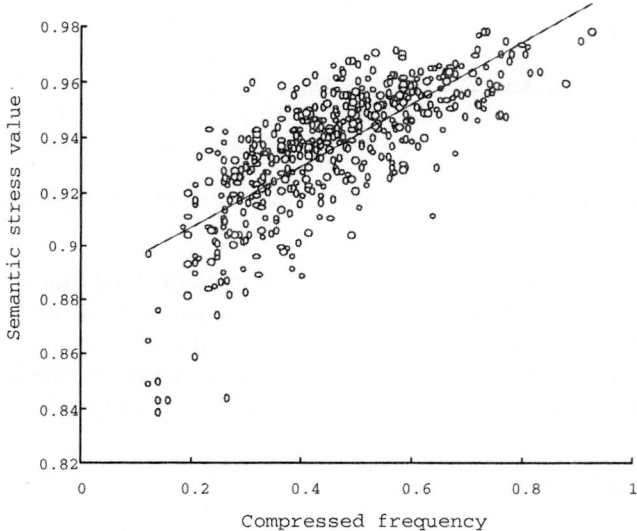

Fig. 1. Semantic stress values and compressed frequency values at the 3950th epoch

As a word set 200 words from the trained data set are randomly selected and as a non-word set 100 words each from the non-trained data set are chosen. If the semantic stress values from the word set show higher values than the nonword set, it can be inferred that the model shows lexical status effect. Moreover, checking the errors between the two conditions can tell the performance degree of the lexical decision task. Results show that the two conditions are significantly different(t[345.65]=30.587, $p<0.01$, where 0.937 for the word condition and 0.847 for the non-word condition in average). What is more, if the decision criterion β is set to 0.905, the error rate in the word condition is 0.95(19/200) and the error rate in the non-word condition is 0.65(13/200), which are not big error rates and resemble the human data. Therefore, it could be said that the proposed model simulates the lexical status effect and performs the lexical decision task naturally (Fig. 2.).

In the present study the non-word legality effect and the word similarity effect are regarded as the same, considering that the matter is the legality in the letter combinations[12]. To know whether the model simulates the effect,

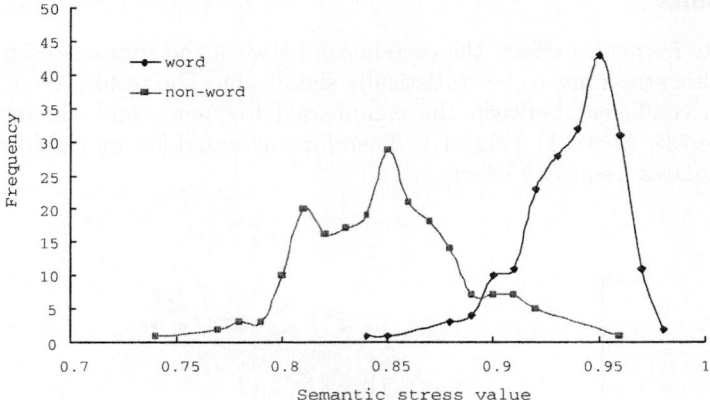

Fig. 2. Distribution of the semantic stress values of word and non-word

semantic stress values of the two non-word data sets should be compared. The results show that the illegal non-words has the semantic stress value of 0.854 and the legal non-words has the semantic stress value of 0.84 in average which is statistically significant(t[198]=3.0, $p<0.01$). Thus, it can be concluded that the model simulates non-word legality effect (or word similarity effect).

6 Conclusions

The frequency effect is explained by the correlation between semantic stress values and compressed frequency values which is caused by the learning procedure. Moreover it reveals that the frequency effect occurs by the difference of the experience on words in the real world.

The lexical status effect is simulated by the difference of the semantic stress values between the word condition and the non-word condition which comes from the difference between the strength of the two conditions. Since the word condition is learned by the network, the strength(the connection values) of the letter combination is strong, whereas the non-word condition is not learned by the network so the strength of the letter combination is weak. This implies that there is a letter combination rule in words and though we do not know the explicit rule of it, we learn it tacitly by learning the word itself.

Non-word legality effect or word similarity effect could be explained by the interruption or competition of the word combinations. Letter combinations that are used in words have strong weight values. However, eventhough non-words with legal combinations resemble the letter combination of the word's letter combination they do not have the exact combination. This leads to unclear output and make low semantic stress values. On the other hand, the non-words that use illegal letter combinations do not resemble the word's letter combination. Consequently, illegal combination non-words are less interrupted by the weight values that were made by the words in the learning procedure leading higher semantic stress values.

The present model also shows that the process of lexical decision occurs with a structure with orthographic and semantic feature unlike several former studies[7]. However, the model only could simulate monosyllabic words which did not reflect the variety of the language usage. Therefore, further study should be taken using novel input structures. Moreover, delicate networks which could deal with time should be proposed to simulate time concerned effects like word context effect.

References

1. Forster, K. I. & Chambers, S. M.: Lexical access and naming time. Journal of Verbal Learning and Verbal Behavior **12** (1973) 627-635
2. Rubenstein, H., Harfield, L., & Millikan, J. A.: Homographic entries in the internal lexicon. Journal of Verbal Learning and Verbal Behavior **9** (1970) 487-494
3. Rubenstein, H., Lewis, S. S., & Rubenstein, M. A.: Evidence for phonemic recoding in visual word recognition. Journal of Verbal Learning and Verbal Behavior **10** (1971) 645-657
4. Chambers, S. M.: Letter and order information in lexical access. Journal of Verbal Learning and Verbal Behavior **18** (1979) 225-241
5. Bullinaria, J. A.: Neural network models of reading: Solving the alignment problem without Wickelfeatures. In: J. P. Levy, D. Bairaktaris, J. A. Bullinaria & P. Cirns (eds.): Connectionist models of memory and language, UCL Press (1995) 161-178
6. Plaut, D. C., McClelland, J. L., Seidenberg, M. S., & Patterson, K.: Understanding normal and impaired word reading: Computational principles in quasi-regular domains. Psychological Review **103** (1996) 56-115
7. Seidenberg, M. S., & McClelland, J. L.: A distributed, developmental model of word recognition and naming. Psychological Review **96**(4) (1989) 523-568
8. Rohde, D. L. T.: Lens: The light, efficient network simulator. (1999) http://tedlab.mit.edu/ dr/Papers/lens.pdf.
9. Mitchell, T. M.: Artificial neural networks. In Mitchell, T. M. (eds.): Machine learning, McGraw-Hill (1997) 81-127
10. Plaut, D. C.: Structure and function in the lexical system: Insights from distributed models of word reading and lexical decision. Language and Cognitive Processes **12**(5/6) (1997) 765-805
11. Ratcliff, R.: A theory of memory retrieval. Psychological Review **85** (1978) 59-108
12. Yim, H., Lim, H., Park, K., & Nam, K.: A Simulation Study Using the Neural Network Model on the Main Effects of the Lexical Decision Task. Proceedings on the 2005 Winter Conference of the Korean Experimental Psychology Society (2005) 101-107

Neuroanatomical Analysis for Onomatopoeia and Phainomime Words: fMRI Study*

Jong-Hye Han[1], Wonil Choi[1],
Yongmin Chang[2], Ok-Ran Jeong[3], and Kichun Nam[1]

[1] Department of Psychology, Korea University, Korea
{jonghyehan, kichun}@korea.ac.kr
[2] Department of Radiology, College of Medicine,
Kyungpook National University, Korea
[3] The Department of Speech-Language Pathology,
College of Rehabilitation Science, Daegu University, Korea

Abstract. The purpose of this study is to examine the Neuroanatomical areas related with onomatopoeia and phainomime word recognition. Using the block-designed fMRI, whole-brain images (N=11) were acquired during lexical decisions. We examined how the lexical information initiatesbrain activation during visual word recognition. The onomatopoeic word recognition activated the bilateral occipital lobes and superior midtemporal-gyrus, whereas the phainomime words recognition activated left SMA and bilateral cerebellum as well as bilateral occipital lobes. Regions more activated for the phainomime word than onomatopoeia included left SMA and bilateral cerebellum. Regions more activated for the onomatopoeia than phainomime word included left superior and midtemporal gyri. The word recognition for onomatopoeia plus phainomime word showed activation on bilateral middle and superior temporal gyrus, right supramarginal gyrus, left middle temporal gyrus, left middle occipital gyrus, and right occipital gyrus. This is the first fMRI research to analyze onomatopoeia and phainomime word.

1 Introduction

Onomatopoeia is a figure of speech in which the sound of word is imitative of the sound of behaviors or appearances of objects. By comparison between Korean and Japanese, Katsuta (2001) concluded that onomatopoeia and phainomime words have special linguistic characteristics. Phonetically they exchange consonant and vowel sounds. In form, many onomatopoeia and or phainomime words were made by duplication, and derived from an adjective or a verb by adding an affix. Korean language has abundant expressions for onomatopoeia and phainomime words [2][4]. Especially Korean has 2196 phainomime words which have more than two syllables [4]. Recent brain imaging researches showed

---

* The research presented in the paper is supported by the KRF grant(2004-074-HM0004).

different activation between motion-related stimulus and sound-related stimulus. Pulvermueller(1999) proposed that some action verbs will activate parts of the motor cortex, whereas animal nouns will activate parts of the visual cortex. Posner and DiGirolamo(1999) said that brain activation depends on the semantic and task context of words. Reading or hearing a word activates linguistic representations as well as associated nonlinguistic information. Relative to non-spatial words, dorsal route of spatial processing (superior occipital and parietal regions) was activated by prepositions [6]. Pictures and verbal descriptions showed bilateral activation of superior occipito-parietal areas that reflect the spatial processing required for the task and activation of the right inferior temporal gyrusfrom complex images [7]. Joe & Nancy. (2000) suggested brain regions like medial temporal/medial superior temporal cortex (MT/MST) were activating during the visual analysis of motion. Kourtzi et al (2000) explained that perceptual analysis involved in the inference of motion from still images involved high level perceptual inferences. Naoyuki et al. (2003) reported an fMRI experiment demonstrating that visualization of onomatopoeia, an emotion-based facial expression word significantly activates both the extrastriate visual cortex near the inferior occipital gyrus and the premotor (PM)/ supplementary motor area (SMA) in the superior frontal gyrus. In this study we will investigate brain activity for onomatopoeia and phainomime word presentation.

2 Method

Subject, Stimulus and Procedure. Eleven right-handed undergraduate students free of medical or neurological problems volunteered to the Experiment. The Experiment consists of three sessions and each session has 9 blocks: 4 activation blocks(30 words, 10 non-words) and 5 control blocks(36 word, 14 non-words). Each block lasts 30 seconds. Usual words were selected as control stimulus. Lexical judgments were required to avoid habituation effects during fMRI scanning. For analysis, we subtract the parts involving the process of lexical decision from whole activation.

Acquisition of Magnetic Resonance Images and Data Analysis. A 1.5T magnetic resonance imaging system (GE) was used. Before acquiring fMRI, anatomical images were acquired (TR/TE 3000/64msec, matrix 256X256, slice thick 5mm, no slice gap, FOV 24 X 24cm). The EPI-BOLD technique was used for acquiring the fMRI of 20 axial slices. A dummy scan of 4 phases (for 12 seconds) was also obtained for correcting any inappropriately high signals before equilibrium state. Parameters for acquiring images are as follows: 3000msec for TR, 64msec for TE, flip angle of 90 degrees, a 64X64 matrix size, slices 5mm thick without separation, and a resolution of 3.75 X 3.75 X 5 mm. The data were analyzed with the SPM. Individual data were conjuncted and ultimate functional imaging was obtained through overlapping brain maps (acquired with the significant level ($p<0.0001$) or ($p<0.00001$)) to standardized T1 image.

3 Results

In our study, following regions were activated by onomatopoeia: Bilateral posterior lobes of cerebellum, bilateral fusiform gyrus (BA19), bilateralinferior frontal gyrus (BA47), Left middle and superior temporal gyrus (BA41, BA42), Right superior occipital gyrus (BA19). Especially onomatopoeia was associated with common cognitive processes like bilateral fusiform gyrus (BA19), right middle temporal occipital gyrus (BA19) and left transverse temporal gyrus. Fusiform gyrus is considered with face-related information [3]. BA 19 activation from our results showed that the onomatopoeia activated sound module which was related to facial expressions. Brain regions activated by phainomime word were Bilateral posterior lobes of cerebellum, Bilateral inferior temporal gyrus (BA20), Bilateral inferior frontal gyrus (BA47), Left middle temporal gyrus, Left precentral gyrus (BA6), left occipital lobe (BA19). Brain regions activated by onomatopoeia and phainomime words were Bilateral middle and superior temporal gyrus (BA41, BA42, BA29), Right supramarginal gyrus, Left middle temporal gyrus (BA21), Left middle occipital gyrus (BA19), Right occipital gyrus (BA18). The results indicated that word properties determine the brain activation in different ways.

4 Discussion

Brain activation for onomatopoeia, phanomime words and onomatopoeia plus phanomime words Kourtzi et al. (2000) found stronger fMRI activation within medial temporal/medial superior temporal cortex (MT/MST) during viewing static photographs with implied motion. Anthony & Carl (2000) explained that "apparent motion" activated BA 9/46 and "Mental rotation" activated BA 8, 19, 39. Our results showed activation on at BA 19 by onomatopoeia. Image formation of onomatopoeia was modulated by BA19. Even by the word presentation of onomatopoeia, subjects associate the onomatopoeia with movement. Our results suggest that superior temporal gyrus (BA41, BA42), inferior frontal gyrus (BA47) and fusiform gyrus (BA19) are involved in high-level recognition of onomatopoeia. In our study, phainomime word presentation activated BA19. In our study, phainomime word showed motion-sensitive fMRI activation. Our study showed that phainomime word activated motor cortex as Pulvermuller (1999) proposed. Subjects associate the phainomime word with movement even by the word presentation. We found that brain activations for onomatopoeia plus phainomime words are not the sum of the activation for onomatopoeia and phainomime words. According to our neuroanatomical analysis, we may categorize Korean words for sound or motion imitated words for three different groups.

Common and Specific Areas. Analysis of word presentations revealed some common brain activations: Bilateral posterior lobes of cerebellum, bilateral inferior frontal gyrus (BA47), Left middle and superior temporal gyrus (BA41, BA42), bilateral superior occipital gyrus (BA19). The same brain regions were

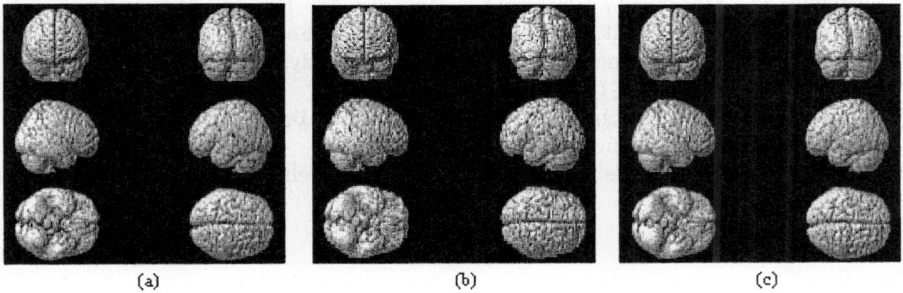

(a) (b) (c)

Fig. 1. Brain regions (a) only activated by onomatopoeia, (b) only activated by phainomime word, and (c) activated by onomatopoeia plus phainomime word

activated by different function of the word presentation. From this result we can analogize those common regions consist of several sub-regions. For Specific cortical areas, we found significantly different brain activations between recognizing onomatopoeia and phainomime words. Sound Specific Areas were Bilateral fusiform gyrus, right middle temporal occipital gyrus (BA19), Left transverse temporal gyrus and Motion Specific Areas were Left precentral gyrus (BA6), Right cingulate, frontal sub-gyral. Regions that were similarly involved in the onomatopoeia words, the phainomime words, and onomatopoeia plus phainomime words, were associated with common cognitive processes. Regions that were especially involved in one of those suggested functional specialization. Further fMRI research about different type of word presentation should be continued to conceptualize the pathway for detailed language mechanism.

References

1. Anthony S. D. & Carl S. :Implicit motion and the brain,Trends in Cognitive Science, Vol. 4, Issue 8, (2000) 293-295
2. Chae, W. :Onomatopoeia in Korea, Seoul National University Press.(2003)
3. Druzgal T. J. and Desposito M. :Activity in fusiform face area modulated as a function of working memory load. Cognitive Brain Research 10,(2001) 355-364.
4. Kim, J. S. :The study of Korean Mimetic words, Ph. D. Dissertation, Kyung Hee University.(1995)
5. Joe K. and Nancy K. :Activation in Human MT/MST by Static images with Implied Motion, Journal of Cognitive Neuroscience **12**:(2000)1, 48-55.
6. Mellet, E., Tzourio, N., Crivello, F., Joliot, M., Denis, M., & Mazoyer, B. :Functional anatomy of spatial mental imagery generated from verbal instruction. The Journal of Neuroscience **16**: (1996) 6504-6512.
7. Mellet, E., Tzourio-Mazoyer, N., Bricogne, S., Mazoyer, B. Kosslyn, S.M., & Denis, M. :Functional anatomy of high-resolution visual mental imagery. Journal of Cognitive Neuroscience **12**: (2000) 98-109.

8. Naoyuki O., Mariko O., Hirohito K., Masanao M., Hidenao F. and Hiroshi S., :An emotion-based facial expression word activates laughter module in the human brain:a functional magnetic resonance imaging study, Neuroscience Letters Vol. 340, Issue 2 ,(2003) 127-130.
9. Posner, M. I., & DiGirolamo. : Flexible neural circuitry in word processing. Behavioral and Brain Sciences **22**:(1999) 299-300.
10. Pulvermueler, F. :Words in the brain's language. Behavioral and Brain Sciences **22**: (1999) 253-270.

Cooperative Aspects of Selective Attention

KangWoo Lee

Human-Robot Interaction Research Center
373-1, KAIST Guseong-dong, Yuseong-gu, Daejeon, Korea
leekw@robot.kaist.ac.kr

Abstract. This paper investigates the cooperative aspects of selective attention in which primary (or bottom-up) information is dynamically integrated by the secondary (top-down or context) information from different channels, and in which the secondary information provides a criterion of what should be many target candidates We present a computational model of selective attention that implements these cooperative behaviors. Simulation results, obtained using still and video images, are presented showing the interesting properties of the model that are not captured by only competitive aspects of selective attention.

1 Limited Capacity and Competition

Due to the intricate and manifold nature of visual attention, it has been investigated by a wide spectrum of approaches that leads controversial issues, on which there is still much debate - early vs. late selection (the issue of 'where the selection process occurs in information processing stages'), spatial vs. object based (the issue of 'whether attention is located in spatial position or objects'), bottom-up vs. top-down driven (the issue of 'the direction of information flow' that constrains the selection process) etc.

Regardless of the controversy surrounding each issue, they are all linked by a common assumption of why attention is needed. The common assumption of the necessity of attention is the limited amount of computational resource that is available for a given task or process. That is, the basic purpose of attention is to avoid a possible information overload in order to protect a mechanism of limited capacity. The assumption was originally conceptualized by Broadbent [1]. In his theory, which is known as filter theory, only a small portion of the incoming information is passed through selective filter and is identified, but other information is shut out from further analysis.

From the limited resource assumption, Desimone and Duncan [2] suggested an influential theory of visual attention on the basis of behavioral and neural studies. According to them, the receptive field (RF) can be viewed as critical visual processing resource for which objects in the visual filed must compete because the information available about any given object will decline as more and more objects are added to RFs. Therefore, a cell's activity is degraded if more than one stimulus falls into an RF in comparison with the activity evoked by a single stimulus presented within the RF. Furthermore, Kastner et al. [3] argued that multiple objects in a restricted RF interact in a mutually suppressive way. That is, the multiple object in an RF compete to get limited processing resource for neural representation, and this competition results in the degraded responses of the cell if more than one object is presented in one RF.

A possible solution to resolve the degraded neural activity is to selectively enhance an object and suppress others through the competition. However, cells in a competition are not equally selectable. Some cells are more likely to win, and others are not. This means that the competition biased toward information that is currently relevant to behaviors.

2 Competition in Computational Models

This section introduces computational models of selective attention in terms of the flow of information processing and selection process. These two dimensions shed light on how competitive mechanism is implemented in computational models.

2.1 Processing Stages

Roughly speaking, any computational model for visual attention has two distinctive processing stages. This distinction is due to the assumption of resource limitation that divides information processing stages into the preattentive and attentive stages.

In preattentive processing, 3 assumptions are commonly made in many computational models: 1) preattentive processing is unlimited in capacity; 2) information is processed in a bottom-up and massively parallel manner; 3) information processing is independent. Therefore, for a given stimulus, different features such as color, intensity, orientation, and movement, are extracted by different processing channels in a parallel way as in, for example, Itti's saliency based model [4, 5].

At this stage, two different mechanisms are widely used for processing a given features. First, a multi-resolution mechanism is used to obtain an image representation from a coarse spatial scale to a finer spatial scale, with the zoom lens metaphor embedded in the mechanism [6, 7]. The information carried by different spatial scales can be used for different purposes. In Deco's model, the coarsest level of spatial resolution is utilized to find the location of an interesting object in a priority map, whereas detailed spatial resolution is used to identify what object is [6]. Second, a center-surround mechanism is used to achieve the contrast within a channel. In computer vision, this mechanism is widely used for detecting local edge in an image. In general, it is assumed that there is homogeneity within an object or a part of an object and discontinuity between objects or parts when detecting a local edge. The homogeneous parts of the image nullify the response of a center-surround filter. Conceptually, the center-surround mechanism for edge-detection is the same as that for bottom-up saliency detection in which attention is directed to a unique object among similar objects.

After the preattentive stage, it is followed by an attentive process that can be characterized by a serial process in which only one item is processed at a time. In this stage, features obtained from different channels are combined to construct a saliency map. Even though saliency can be defined at many different levels from a feature to a semantic level, saliency in most current models is defined at the feature level. The important factor in guiding bottom-up attention is feature contrast rather than absolute feature values.

Once a saliency map has been constructed, a location has to be selected for the deployment of an attentional window. A 'winner-take-all' (WTA) network is commonly used to determine the allocation of attention [4, 5]. In the network that receives input from a saliency map, only one unit is allowed to be active at a given time, and others are suppressed, so that serial processing is accomplished. In other word, biased competition is accomplished through the WTA network. In order to prevent reallocation of attention to this winning location, it is excluded from the saliency map after processing.

2.2 Selection Process

Depending on where selection process occurs, and which level of information is selected, computational model can be discriminated into two classes of models - early and late selection models. First of all, most current computational models are based on early selection. Since, in those models, selection is accomplished by saliency calculated from the center-surround feature contrast, the selected location does not meaningfully correspond to the location of an object. Rather, it simply corresponds to the location where it gives the strongest contrast.

Furthermore, in those models, top-down knowledge is directed to the early stage of information processing, that is, before or soon after the feature extraction process. In contrast, a few models have implemented a late selection. For instance, in Sun and Fisher's model, the feature elements such as color, intensity, and orientation are grouped into more meaningful perceptual units (objects) before attentive process operates [7].

Regarding a competition mechanism, the early vs. late selection has implications for important issues. If we admit that the RF can be viewed as a critical visual processing resource for which objects in the visual field must compete, the logic behind a RF property that shows increased size along visual pathways means: 1) a cell in a higher processing stage that has a relatively larger RF size has a greater chance of having more objects fall inside its RF that a cell in a lower layer; 2) if one moves to higher stage of the visual hierarchy, the competition for processing resource will be stronger; and thus 3) stronger attentional modulation effects will be found at higher stages, compared with lower stages. This also means that attentional effect increases from a lower stage to higher stage, rather than being attenuated from a lower stage to a higher stage. Also, this means that is attention is object-based rather than feature-based.

3 Is the Competition Enough? Necessity of Cooperation

The biased competition hypothesis implies that neurons at a given processing stage take part in an inevitable war to get resources. The relationship among neurons is considered as mutually exclusive and there seem to be little chance for cooperation to solve the limited resource problem, since competition is the main mechanism of the selection process. As noted previously, the concept of competition is embedded in WTA network in which units are mutually interconnected and are inhibited by each other. In those models, only one neuron corresponding to a location or an object in a given visual stimulus is selected at a time.

In spite of the fact that the limited resource assumption provides the logical basis for inevitability of competition, the same logic can be equally applied to the necessity of cooperation. That is, the limited resource assumption may also require the cooperation of different brain areas or neural channels which may help to reduce the burden of processing in various ways. The cooperative information from other brain areas does not simply contribute the enhancement or suppression of neural activities at a given processing stage. It provides general criteria for what or where are selected in a task. Top-down knowledge and contextual information provide critical criteria that allow a system to selectively process current information.

Moreover, neurophysiological evidence is given Rainer et al. [8], who recorded cell activity in the prefrontal cortex of a monkey during a 'delayed-matching-task'. In the task, the monkey was required to find a target object in a stimulus scene containing many objects, and remember its location until a test stimulus was given. They found that the activity of the neurons in the cortex reflected the target location alone and was maintained during the delay. This result suggested that the relevant neural activity corresponding to a target was maintained during the delay and was involved in selection of the location where a matching object would be given. That is, remembering only target location (or cued location) reduces the severe limitation of the capacity of working memory.

4 Integration of Cooperative Information

The argument that the 'cooperative mechanism of visual attention is critical for the selection process on current information' leads to another question - how does the cooperative mechanism work for selection of a location or object? Basically, we argue that the information from other processing channels provides a context or bias to a network that receives incoming neural activities. This context or bias helps the network to interpret the incoming neural signal by setting a criterion for whether the signal is relevant or irrelevant to the current behavior or information processing.

Recently, Spratling [9] investigated differentiated roles of apical and basal dendrites of a pyramidal cell. A typical pyramidal cell has two separate dendritic arbors that receive different information sources. The two set of dendrites of a pyramidal cell may suggest that the dendrites receive information from distinctive sources - feedforward information to the basal dendrite and feedback information to the apical dendrite. Spratling [9] speculated that distal and proximal dendrites of pyramidal cells acts as separate compartments and contribute to different functional roles for information processing. Since apical inputs have weaker effects on the output activity than basal inputs, the apical dendrite is considered to take a role in modulation of responses of the cell. That is, for such neurons, sensory-driven, feedforward information is applied to the basal dendrite while top-down and feedback information arrives at the apical dendrites.

Interestingly, Treue et al. [10] showed that the attentional modulation effect on sensory selectivity is multiplicative. They measured the tuning curve of direction-selective neurons in middle temporal (MT) visual areas of a monkey while the animal was attending to moving random dot patterns guided by a spatial cue. The result showed that attention increased the response to all attended stimuli by the same proportion ('multiplicative modulation') along the different degrees of the orientation, without the width of the tuning curve. Reynold et al. [11] systemically investigated this relationship between the amount of attentional modulation and stimulus strength as they manipulated a range of luminance contrast. They showed that the attentional modulation effect on V4 neurons to a low contrast stimulus is larger than that to a high contrast stimulus. These results are compatible with our models in which the multiplication between two inputs from different information sources, and the amount of attentional modulation effect, varies with the strength of the inputs.

5 Interactive Spiking Neural Network

5.1 Structure of ISNN

Based on this conceptual framework, we have developed an Interactive Spiking Neural Network (ISNN) using a leaky Integrate-and-Fire (IF) neural network [12]. A simple example of the structure of the ISNN is given in Fig. 1. The network consists of bottom-up input units xB, top-down input units xT and output unit o. The output unit o receives two kinds of weighted inputs - bottom-up input and multiplication between both inputs - at time t. The multiplication has an important nonlinear property that correlates the two inputs. The net value for the j-th output unit is given by :

$$net_j(t) = \alpha \sum_{i=1}^{n} w_{ij}(t) x_i^B + \beta \sum_{i=1,r=1}^{n,m} u_{irj}(t)(x_i^B)^2 x_r^T \qquad (1)$$

where B and T stand for the bottom-up and top-down inputs, n and m are the dimensions of the bottom-up and top-down inputs, and w and u are the bottom-up and multiplicative weights, respectively. The constants α and β determine the amount of influence driven by bottom-up and top-down inputs on the net value. If $\alpha = 1$ and $\beta = 0$, the value netj is determined by only bottom-up inputs. If $0 < \alpha < 1$ and $\beta = 1 - \alpha$, the value netj is determined by both to a variable degree.

The second term in Eq. (1) can be considered as a correlation between two input sets because when two inputs are consistent, it produces a certain amount of gain, but when two inputs are inconsistent, it causes a cost to the network.

Another nonlinearity is implemented with a sigmoid function that may correspond to the processing at the level of the soma. The sigmoid function has desirable properties; the output of the function will not be zero or one. So, the amount of activation driven by netj is given by:

$$y_j(t) = \frac{1}{1 + \exp(-net_j(t))} \qquad (2)$$

In the IF model, a postsynaptic spike occurs if the summation of postsynaptic potential produced by the succession of input signals reaches a threshold. Conventionally, the model is described with a circuit that consists of a capacitor C in parallel with a resistance R driven by a current I(t). The trajectory of the membrane potential can be expressed in the following form.

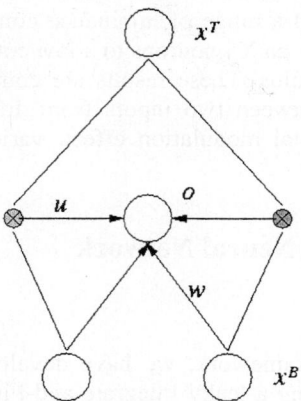

Fig. 1. A simple example of ISNN structure. The model has bottom-up x^B and top-down input units x^T, and output units o. The bottom-up connection w links bottom-up units and output units, and the multiplicative connection u links the two input units and output units. Therefore, an output unit receives two kinds of inputs - one driven by only bottom-up and the other driven by multiplication of both inputs. The output unit produces a spike if the membrane potential of the unit reaches a threshold. The interspike interval is used to measure the response of the unit.

$$V(t+dt) = V(t) + RI(t)\frac{dt}{\tau_m} - V(t)\frac{dt}{\tau_m} \qquad (3)$$

where τ_m is the membrane time constant of a neuron. The equation means the membrane potential V at time $t + dt$ is the sum of the potential V at the previous time t, the amount of ongoing current and the amount of decay.

If we limit our consideration to the special case of a cell firing a train of regularly spaced post synaptic potentials, we may write the voltage trajectory of the membrane potential in the following form by putting the leaky IF model with the sigmoidal activation together :

$$V_j(t) = y_j \frac{1-\exp(-n/k\tau_m)}{1-\exp(-1/k\tau_m)} \qquad (4)$$

where the amplitude yj decays exponentially with the membrane time constant τm and regularly spaced time 1/k. A postsynaptic spike will be generated if the voltage of membrane potential Vj is equal to or larger than a threshold Vth.

$$V_j(t) \le y_j \frac{1-\exp(-n/k\tau_m)}{1-\exp(-1/k\tau_m)} \qquad (5)$$

From the equation above, the interspike interval n/k is determined:

$$T_j = \frac{n}{k} = -\tau_m \ln\left[1 - \frac{V_{th}(1-\exp(-1/k\tau_m))}{y_j}\right] \quad (6)$$

5.2 Learning Equation

In order to derive the learning equation here, we simply define an 'error' as the difference between actual spike interval and desired spike interval. Thus,

$$E = \frac{1}{2}\left(\sum_{j=1}^{l} T_j^d - T_j\right)^2 \quad (7)$$

where T_j^d is the j-th desired spike interval and l is the number of output units. Since we want to find the weight values which minimize the error function, we can differentiate the error function w.r.t. the weight parameters.

$$\frac{\partial E}{\partial w_{ij}} = \eta\alpha(T_j^d - T_j)\left[\frac{\tau_m}{y_j - V_{th}(1-\exp(-1/k\tau_m))}\right]$$
$$\left[\frac{V_{th}(1-\exp(-1/k\tau_m))}{y_j}\right] y_j(1-y_j) x_i^B \quad (8)$$

Similarly, we can apply the learning rule for the secondary connection u_{irj}

$$\frac{\partial E}{\partial u_{irj}} = \eta\beta(T_j^d - T_j)\left[\frac{\tau_m}{y_j - V_{th}(1-\exp(-1/k\tau_m))}\right]$$
$$\left[\frac{V_{th}(1-\exp(-1/k\tau_m))}{y_j}\right] y_j(1-y_j)(x_i^B)^2 x_r^T \quad (9)$$

The preparation of manuscripts which are to be reproduced by photo-offset requires special care. Papers submitted in a technically unsuitable form will be returned for retyping, or canceled if the volume cannot otherwise be finished on time.

5.3 Properties of ISNN

In order to provide an insight to the properties of an ISNN, pilot experiments were carried out. In these experiment, two simple patterns ('x' and '+') that consisted of 3 by 3 pixels with red and blue colors. The pattern 'x' is associated with red color, whereas the pattern '+' is associated with blue color. The network had 9 bottom-up units which corresponded to 3 by 3 pixels, 3 top-down units which corresponds to the rgb values of each pattern, and two output units which corresponds to the patterns.

The desired interspike intervals were set to 10 ms for a corresponding pattern and 100 ms for a non-corresponding pattern.

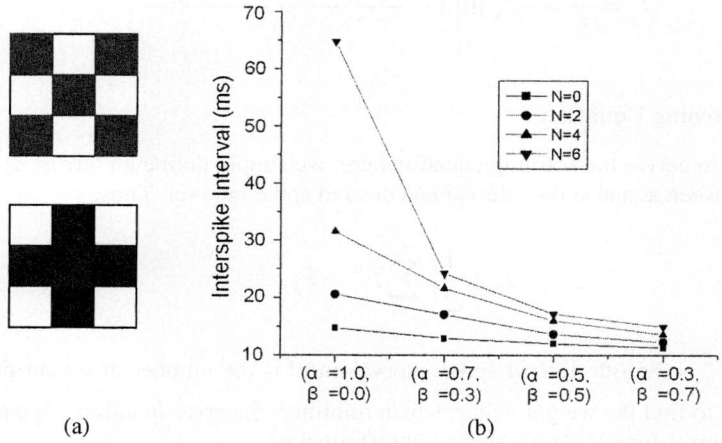

Fig. 2. a) Stimulus b) Interspike intervals with different constant values α and β

After training, the performance of the network was measured by flipping the pixels randomly and varying the constant values. The result is shown in Fig 2. The performance of the network is dramatically changed by introducing additional information (color input). As the β value is larger and lager, it produced shorter interspike intervals if two kinds of inputs are associated. This means additional information (color) is helpful to resolve ambiguities caused by primary input (patterns). However, this is not a whole story. The additional information would interfere with the network's performance if it is not consistent with the shape patterns, and this produces longer interspike intervals.

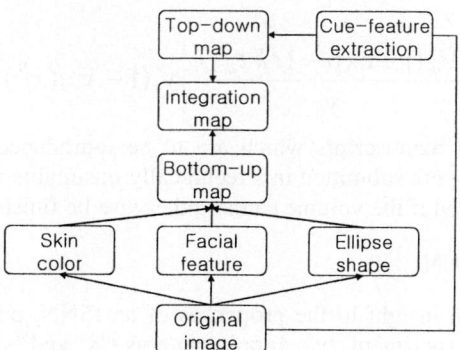

Fig. 3. General model architecture of selective attention. The model consists of 3different submodules that each form a different map -- bottom-up, top-down, and integration map.

6. Model of Selective Attention

We introduce some assumptions that are used as a blueprint for constructing a computational model of selective attention. First, preattentive features including skin color, facial features, and ellipse shape are treated as bottom-up information. Second, top-down knowledge is utilized by extracting a cue from the stimulus itself, representing spatial distance between a cue and a possible target. So, the short distance, the larger top-down input is assigned. Third, both bottom-up and top-down inputs are integrated to construct an integration map via ISNN. Fourth, the attentional allocation is ordered by the consistency of the two inputs.

To illustrate this, consider a situation in which you have to find your friend in a crowded street. If you know where he/she is waiting or what color clothes he/she is wearing, it will be much easier to find this friend. That is, knowledge about a target such as its probable location, physical attributes, context etc. guides us by providing a clue to indicate where a target may appear.

Fig. 4. The attention trajectories varied by different values of α and β. The task for the model is to find the person who is wearing blue color t-shirts. (a) $\alpha = 1.0$, $\beta = 0.0$ (b) $\alpha = 0.6$, $\beta = 0.4$ (c) $\alpha = 0.5$, $\beta = 0.5$ (d) $\alpha = 0.4$, $\beta = 0.6$. Even a small amount of β is enough to change the trajectory of attention. Adding more to β produced a strong tendency that the trajectories of attention are attracted closely to the location of a cue color (blue) region.

Fig 3 shows the general structure of the selective model. The model can be divided into 3 main processing sub-modules - top-down module, bottom-up module and integration module. The original input is given in a form of digitized images. From the original input image some bottom-up features including skin color, facial features (aspect ratio and symmetry), and ellipse shape are extracted in order to construct bottom-up and top-down map. The combination of features into a single bottom-up map is accomplished by a set relation between feature regions defining whether features share a specific region or not. Then, bottom-up input values obtained each feature map are assigned to a corresponding region of target candidate. On the other hand, top-down input values are calculated with Gaussian distance between a cue-color (or motion) segmented region and a region of target candidate. Therefore, a set of bottom-up and top-down inputs for a region of target candidate is obtained, and used to generate output interspikes (more detail see [12, 13]).

The attentional window is allocated in ascending order from the target candidate with the shortest interspike interval to the target to the longest interspike interval calculating from ISNN. In a sense the first location where an attentional window is allocated meets the highest consistency between target properties and indication by a cue.

7 Simulations

The performance of the model was investigated by manipulating amount of bottom-up and top-down influence, and cue conditions. Still and video images obtained from natural environment such as a street, campus and laboratory were used as input images. The first simulation showed that how attentional trajectories are modulated by the interaction between bottom-up and top-down inputs. The second simulation concerned the attentional shift from one face to another and maintenance of attention on a target face guided by a motion cue.

7.1 Top-Down Influence on the Modulation of Attentional Trajectories

The top-down influence on the modulation of attentional trajectories was investigated by assigning different values of α and β. If $\alpha=1.0$ and $\beta=0.0$, then the attentional trajectory is totally dependent on the bottom-up input. However, as the value of β is increased, the influence of top-down input on determination of the trajectory is stronger. Accordingly, the constant values of α and β were changed to (1.0, 0.0) (0.6, 0.4), (0.5, 0.5), (0.4, 0.6).

One of the results was presented in Fig 4. The task for the model is to find the person who is wearing blue color t-shirts. The attentional trajectories were dynamically changed with different values of α and β, since the higher values of constant β force the model to attend to the target candidate at the location of cued color by increasing the correlation gain of the net value, whereas it detains attendance to the location where no cue color indicates by little gain or more interference.

7.2 From One Face to Another

In this simulation video images (5 images per a second) were used to test the performance of the model. The task for this simulation is to find the person who is waving his hand. The motion information obtained from the difference between images at two time frames ($t-1$ and t) was utilized as a cue. The first locations where an attentional window was allocated were marked with the yellow circle. As shown in Fig 5, the attention of the model was maintained at the person's face during he was waving his hand, and then move again to the face of the other person who was waving his hand as well. Interestingly, the locus of attention stays on the location of a particular face by the motion cue, then shifts to the other face indicated by it. The results implied that the knowledge of a cue actively involves in the attentional control to engage attention to a particular location, and to shift to other locations.

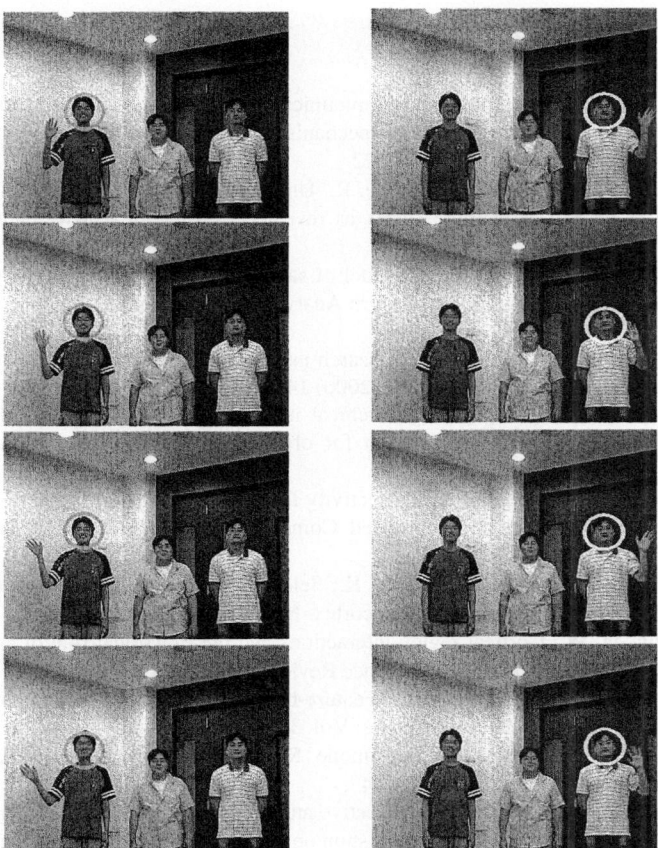

Fig. 5. Sustaining and shifting attention guided by a motion cue. The task for the model is to find the person who is waving his hand. The model's attention is focused on the face of the person waving his hand as shown in each column, and when that person stops waving his hand and a second person starts, the model again shifts its focus to the second person's face.

8 Conclusion

In summary, we demonstrated cooperative aspects of selective attention using a computational model in which two input streams (bottom-up and top-down) cooperate and integrate. The cooperative and integrative aspect of the model not only provides selection criteria for which current incoming information is relevant or not, but also dynamically modulates the information through a multiplicative correlation mechanism by enhancing relevant information or suppressing irrelevant information. In this context, the limited capacity assumption was criticized because of a logical deficiency in supporting the necessity of selective attention as well as in implementing a selection mechanism.

References

1. Broadbent, D. E.: Perception and Communication. Pergamon, London (1958)
2. Desimone, R., Duncan, J.: Neural mechanism of selective attention. Annual Review of Neuroscience, Vol. 18 (1995) 193-222
3. Kastner, S., DeWeerd, P., Desimone, R., Ungerleider, L. G.: Mechanisms of directed attention in ventral extrastriate cortex as revealed by functional MRI. Science, Vol. 282 (1998) 108-111
4. Itti, L., Koch, C., Niebur, E.: A model of saliency-based visual attention for rapid scene analysis. IEEE Transactions on Pattern Analysis and Machine Intelligence, Vol. 20 (1998) 1254-1259
5. Itti, L., Koch, C.: A saliency-based search mechanism for overt and covert shifts of visual attention. Vision Research, Vol. 40 (2000) 1489-1506
6. Deco, G., Schurmann, B.: A hierarchical neural system with attentional top-down enhancement of the spatial resolution for object recognition. Vision Research, Vol. 40 (2000) 2845-2859
7. Sun, Y., Fisher, R.: Hierarchical selectivity for object-based visual attention. In Proc. 2nd Workshop on Biologically Motivated Computer Vision (BMCV'02), Tuebingen, Germany, (2002) 427-438
8. Rainer, G., Assad, W. F., Miller, E. K.: Selective representation of relevant information by neurons in the primate prefrontal cortex. Nature, 393 (1999) 577-579
9. Spratling, M. W.: Cortical region interactions and the functional role of apical dendrites. Behavioral and Cognitive Neuroscience Reviews, Vol. 1 (2002) 219-228
10. Treue. S., Martinez-Trujillo, J. C.: Feature-based attention influences motion processing gain in macaque visual cortex. Nature, Vol. 399 (1999)
11. Reynolds, J. H., Pasternak, T., Desimone, R.: Attention increases sensitivity of v4 neurons. Neuron, Vol. 26 (2000) 703-714
12. Lee, KW., Feng, J., Buxton, H.: Selective attention for cue-guided search using a spiking network. In Proc. International Workshop on Attention and Performance in Computer Vision (WAPCV'03), Graz, Austria (2003)
13. Lee, KW.: Computational Model of Selective Attention: Integrative Approach. PhD thesis, University of Sussex (2004)

Selective Attention Guided Perceptual Grouping Model

Qi Zou[1], Siwei Luo[1], and Jianyu Li[2]

[1] School of Computer and Information Technology,
Beijing Jiaotong University, 100044 Beijing , China
zq_089@163.com
[2] School of Computer Science, Communication University Of China,
100024 Beijing , China
lijianyu163@163.com

Abstract Selective attention works throughout the whole process of vision information processing. Existing attention models concentrate on its role in feature extraction in initial stage, but ignore role of attention in other stages. In this paper, we extend attention to middle stage, especially in guiding perceptual grouping. Selective attention functions in two aspects. One is to select the most salient primitive as grouping seed. The other is to organize groups and decide their pop-out sequence. Compared with traditional attention models, our model judges primitive salience according to global properties rather than local ones. And focus of attention shifts in unit of perceptual object rather than spatial region. These two improvements boost the model's grouping quality and more fit to high stage of vision information processing. Experiments and quantitative analysis testify our model's good performance in certain class of images.

1 Introduction

Attention works throughout all processing stages in vision, starting from feature binding (initial stage)[1], through perceptual grouping (middle stage) [2], to complex tasks like recognition and accessing information in memory (high stage)[3]. Attention mechanism in initial stage has been well studied, but its role in other two stages has by far been mainly researched in psychology and biology domain.

Among attention models in initial stage, the most representative Itti's model [4] built a saliency driven system based on feature binding theory [1]. His model closed to human vision in many ways except that attended location was circular spatial region and focus shifted among fixation points without considering integrity of objects. Rybak's fovea-periphery model [5] simulated retinal imaging theory and organized in concentric circles with increasing radii representing multi-scale analysis. His model measured saliency by contrast and orientation of edges, so it essentially simulated attention in initial vision stage, though applied in recognition. Other space-based models share the same drawback with these two models – focus of attention is organized in unit of circular spatial region. This does not accord with human vision completely. In neurobiology view, after passing middle stage, an image is separated into figure and ground and figure is further organized into objects, so attention is organized in unit of object [6]. In this way, object-based model is more adapted to middle and high vision stages.

While attention works throughout all vision stages, grouping typically performs in middle stage. Input to grouping is initial visual features like edges and patches. Output is perceptual objects waiting for advanced processing. Existing grouping algorithms confront two main obstacles. First, grouping algorithms mostly describe relations with graph structure, but graph cut is generally NP-hard problem which can be solved with finite cost only under specific constraints. Second, graph-based structure inclines to describing local relations, especially relations between a pair of primitives, but global relations are more salient and reliable in vision perception [7,8]. In order to solve the former obstacle, researchers try various approaches to fasten convergence including dynamic programming [9], SA [10]. As to the later, researchers select more reliable local relations by statistic decision strategies like MAP, ML and evidence accumulation [7].

Distinct from previous work, we propose a selective attention guided grouping model. Attention functions in both selecting the most salient primitive from global view and reducing solution space to save computational cost. The two strategies just solve the two obstacles in traditional grouping algorithms. What's more, attention organizing in this way forms an object-based model, which adapts to middle and high vision stage.

In this paper, we'll introduce global relation extraction first, and then detailed grouping algorithm is given together with role of attention. Experiments are carried out to demonstrate efficacy of the model and results are analysed. Direction of future work is discussed in conclusion.

2 Global Relation Extraction

We select closure and parallelism as estimation for global salience. These criterions originate in psychology which found human vision organizes information obeying several laws. Gestalt psychologists summarized these laws including proximity, parallelism, symmetry, good continuation and closure[6]. With these laws, local scattered primitives are organized into global structure reflecting topological relations and having explicit scenery sense.

2.1 Closure

Closure is a global topological property held by object contours. A closed contour means a finite number of edges forming a sequence with head and tail connected. The edges are got from Canny Edge Detector. Outputs of Canny record edges with magnitude of contrast, orientation, position and length. Traditional methods judging closure usually use similarity of region features like color, intensity or texture, but occlusion, shadow or other reasons often parse the region enveloped by a closed contour into sub-regions with dissimilar region features. So taking similarity of region features as measure for closure may lead to untrue result. Better measure is topological relations among edges.

2.1.1 Topological Relations as Measure for Closure
Similar to descriptions adopted by Elder [8], three possible topological relations, namely smoothness, sharp turning and gap caused by occlusion, are considered to

happen to edges belonging to the same closed contour. Thus possibility of any two edges belonging to the same closed contour is computed by

$$p(e_1e_2) = p(e_1e_2 | r_1)p(r_1) + p(e_1e_2 | r_2)p(r_2) + p(e_1e_2 | r_3)p(r_3) \quad (1)$$

where $p(e_1e_2)$ denotes probability of a closed contour passes edge e_1 and e_2 successively without any intermediate edge. r_1, r_2 and r_3 denote smooth connection, sharp turning and gap respectively. As to each situation, (1) can be transformed to posterior probability

$$p(e_1e_2 | r_i) = \frac{p(r_i | e_1e_2)p(e_1e_2)}{p(r_i)}, \quad i = 1, 2, 3 \quad (2)$$

As r_1, r_2 and r_3 are necessary but not adequate condition for e_1 and e_2 belonging to the same closed contour, the complete probability formula of r_i is

$$p(r_i) = p(r_i | e_1e_2)p(e_1e_2) + p(r_i | \overline{e_1e_2})p(\overline{e_1e_2}) \quad (3)$$

where $\overline{e_1e_2}$ denotes e_1 and e_2 not belonging to the same contour. Replace $p(r_i)$ in (2) with (3) we get

$$p(e_1e_2 | r_i) = \frac{1}{1 + \dfrac{p(r_i | \overline{e_1e_2})}{p(r_i | e_1e_2)} \dfrac{p(\overline{e_1e_2})}{p(e_1e_2)}} \quad (4)$$

We set $p(\overline{e_1e_2})/p(e_1e_2)$ to be 5, for only 6 nearest neighbors are considered for each edge. Approximately, among these 6 pairs formed by e_1 and one of its 6 nearest neighbors, the number of edges within the same contour to number of edges not within the same contour is 1:5. Proximity r_i^{prox} and good continuation r_i^{cont} are measures for r_i

$$p(r_i | e_1e_2) = p(r_i^{prox} | e_1e_2) + p(r_i^{cont} | e_1e_2) \quad (5)$$

with $p(r_i^o | e_1e_2) = \dfrac{1}{\sqrt{2\pi}\sigma_i^o} e\left\{-\dfrac{(r_i^o)^2}{2(\sigma_i^o)^2}\right\}$, $\quad i = 1, 2, 3$, $o = prox, cont$

r_i^{prox} is defined as distance between closest ends of e_1 and e_2 divided by shorter length of the two edges. r_i^{cont} is defined as difference of the two edges orientation. And their conditional probabilities are all modeled by Gaussian function but with different variance. When edges are in the same group, under smoothness condition, distance and orientation difference both ought to be small, so $\sigma_1^{prox} = 1/8$, $\sigma_1^{cont} = \pi/18$. Under sharp turning condition, distance is small but orientation difference is big, so $\sigma_2^{prox} = 1/8$, $\sigma_2^{cont} = \pi/2$. Under gap condition, distance is big but orientation difference is small, so $\sigma_3^{prox} = 1/2$, $\sigma_3^{cont} = \pi/18$. When edges are not within the same group, r_i is also measured by proximity and good continuation

$$p(r_i \mid \overline{e_1 e_2}) = p(r_i^{prox} \mid \overline{e_1 e_2}) + p(r_i^{cont} \mid \overline{e_1 e_2}) \qquad (6)$$

At this time distance and orientation difference are stochastic, so they are uniformly distributed, i.e. $p(r_i^{cont} \mid \overline{e_1 e_2}) = 1/\pi$, $p(r_i^{prox} \mid \overline{e_1 e_2}) = r_i^{prox}/R$ where $R = \max\{M, N\}$ of an $M*N$ image. This means distance is proportional to density of edges in an image.

Probability of any pair of edges belonging to the same closed contour $p_{ij} = p(e_i e_j)$ forms element of a square matrix P. We call it relation matrix. It is sparse for only relations between an edge and its 6 nearest neighbors are considered.

2.1.2 Grouping Seed

We define the grouping seed to be the most salient edge, that's the edge most probably lying on closed contours, equivalently the edge with greatest probability of closed contours passing it

$$E = \{e_i \mid i = \arg\max_i \sum_{j=3}^{tol} p(e_i \in CC^j)\} \qquad (7)$$

where E is the most salient edge, CC^j is a closed contour of length j. Contour length means the number of edges constituting the closed contour. *tol* denotes the total number of edges in an image. All closed contours passing e_i can be regarded as forming one contour with infinite length, if closed contour be regarded as recurrent sequence. So sum of probabilities of all closed contours passing e_i is equal to the probability of an infinite length contour passing e_i.

$$E = \{e_i \mid i = \arg\max_i \sum_{j=3}^{tol} p(e_i \in CC^j)\} = \{e_i \mid i = \arg\max_i \lim_{j \to \infty} p(e_i \in CC^j)\} \qquad (8)$$

Denotes p_{ii}^n as the ith diagonal element of n order power of relation matrix P^n, then p_{ii}^n represents probability of n-length closed contours passing e_i [11]. So probability of an infinite length closed contour passing e_i is $\lim_{n \to \infty} p_{ii}^n$. Formula (8) is equivalent to

$$E = \{e_i \mid i = \arg\max_i \lim_{n \to \infty} p_{ii}^n\} \qquad (9)$$

According to theory in matrix analysis [12], for a positive real matrix P

$$\lim_{n \to \infty}(P/\lambda)^n = x \bullet \tilde{x} \qquad (10)$$

where $x, \tilde{x}$ are right and left eigenvectors of P corresponding to the largest eigenvalue λ. For symmetric matrix, $x^T = \tilde{x}$ (x^T denotes transpose of x). As only relative magnitude of salience is to be decided, (9) divided by a constant will not affect selection of the most salient edge.

$$E = \{e_i \mid i = \arg\max_i \lim_{n \to \infty}(P/\lambda)_{ii}^n\} = \{e_i \mid i = \arg\max_i (x_i \bullet \tilde{x}_i)\} = \{e_i \mid i = \arg\max_i (x_i^2)\} \qquad (11)$$

where x_i and $\tilde{x}_i$ are the i th elements of x and $\tilde{x}$. Now the largest element of eigenvectors corresponding to the largest eigenvalue of relation matrix marks the most salient edge in global view.

2.1.3 Closed Contour

Taking the most salient edge as grouping seed, starting from the seed, we search for a closed path on which product of all probabilities between adjacent edges is maximal. Such a path corresponds to a closed contour. This can be expressed as

$$SC = \max \left(\prod_{e_i, e_j, e_s, \ldots e_v \in CC} p_{ij} p_{js} \bullet \bullet \bullet p_{vi} \right)^{1/u} \quad (12)$$

where SC is salience of closed contour CC. $e_i, e_j, e_s, \ldots, e_v$ are sequential adjacent edges of CC. p_{ij} denotes $p(e_i e_j)$ as defined before. u is length of CC. Extracting u order root is to avoid salience decreasing with contour length. For more edges form a contour, more number of p_{ij} joins the product, but notice $0 < p_{ij} < 1$. Above maximizing problem can be transformed into minimizing problem, if $|\log SC|$ is taken

$$|\log SC| = \min \left(-\frac{1}{u} \right) \sum \left(\log p_{ij} + \log p_{js} + \bullet \bullet \bullet + \log p_{vi} \right), \quad e_i, e_j, e_s, \ldots, e_v \in CC \quad (13)$$

Finding a sequence with minimum sum of relations between adjacent edges is equivalent to find the shortest path in relation matrix. Shortest path problem can be easily solved by classic algorithms [8].

Compared with greedy search starting from random edge, our strategy is more saving. Selecting the most salient edge as grouping seed avoids much worthless search. And restricting within 6 nearest neighbors reduces search space. These two strategies are inspired from selective attention which decides resource preferentially provided to important information and satisfies proximity principle.

2.2 Parallelism

Parallelism is defined as

$$SP = \sum_{e_i \in CC_5} len(e_i) \bigg/ \sum_{e_j \in CC} len(e_j) \quad (14)$$

where CC is the set of edges forming a closed contour. CC_5 denotes subset of CC and each edge in CC_5 is within $5°$ orientation difference to at least one edge in CC. $len(e_i)$ is the length of e_i. Obviously, SP is between 0 and 1. When reaches 1, the contour is an absolutely parallel structure like rectangle or parallelogram. Most manmade objects are parallel structure, so parallelism is advantageous to detect artificial objects in natural scenery.

3 Grouping Algorithm

Importance of groups is defined as

$$H = \alpha\ SC + \beta\ SP \tag{15}$$

where α and β are weights of closure and parallelism. They are set by global amplification strategy[4]. After a closed contour emerges, *inhibition of return*[13] rises, which means once an object has been attended, it will never be listed into further search scope. So all relations between edges on the emerged contour are removed. Then global relations are extracted from the remaining elements and new closed contour emerges. Repeat like this until remaining primitives cannot form a contour. Compute importance for every closed contour and rank them decreasingly, then groups pop out sequentially with each corresponding to closed contour of an object.

Algorithm complexity mainly comes from closure computation including relation matrix plus grouping seed plus closed contour, i.e. $o(N) + o(N^3) + o(N)$ in worst situation and $o(N) + o(N^2 \log N) + o(N)$ in optimal situation, where N denotes edge number of the whole image. Parallelism computation complexity is $o(n)$ where n denotes edge number in a contour. Parallelism complexity can be ignored especially when $N \gg n$.

4 Experiment Result

We ran the model on a PC with 2.4G Pentium IV CPU and 256M memory. Some images and grouping results of mini-cut [14], a typical graph based method, used for comparison come from [16]. We experiment on 24 images of 4 classes. 6 natural object in outdoor background, 6 natural object in indoor background, 6 man-made object in outdoor background and 6 man-made object in indoor background. Images are 320*240. It takes about 17 seconds to produce the first group.Fig.1 is a grouping example of natural object in indoor background. We can see grouping results accord with human perception, almost each corresponding to a meaningful object, for global relation of closure plays key role.

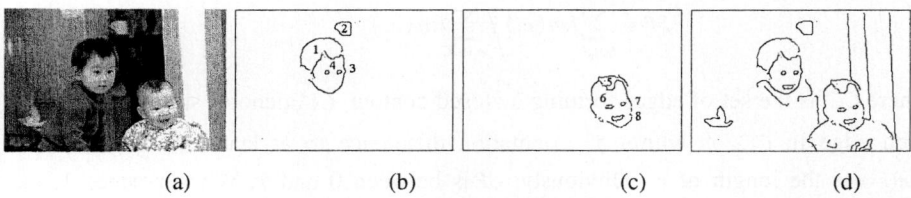

(a) (b) (c) (d)

Fig. 1. a grouping example: (a) source; (b) 1-4 most salient groups with 3 denoting eyes and 4 denoting face; (c) 5-8 most salient groups with 7 denoting eyes and 8 denoting mouth; (d) all salient groups.

We evaluate performance of the model by two indexes: β_{miss} is the ratio of edges belonging to closed contours but missed in grouping, and β_{fal} is the ratio of edges not belonging to the same group but falsely detected as in the same group.

$$\beta_{miss} = \frac{1}{N_{ol}} \sum_{ij} (N_{O_j} - N_{G_i \cap O_j})/N_{O_j}, \quad \beta_{fal} = \frac{1}{N_{ol}} \sum_{ij} (N_{G_i} - N_{G_i \cap O_j})/N_{G_i} \quad (16)$$

where N_{G_i} is the number of edges in detected group G_i, N_{O_j} is the number of edges in true group o_j. An image made up of true groups is called ground truth. $N_{G_i \cap O_j}$ is the number of detected group edges lying in truth group. N_{ol} is the number of overlapping blocks between detected groups and truth groups. Both β should be as small as possible to boost grouping accuracy. Table 1 list comparison of index β among different models. Mini-cut is representative for graph based grouping methods and Itti's model is representative for space-based attention models. We can see in all situations, performance of our model, which uses global relation and is guided by attention, is better than mini-cut using local relation. Large false ratio of Itti's model is due to local feature based spatial attention. Focus is organized in unit of circular region (second column in Fig.2). Circular regions accord with true groups in low probability leading to large β. Miss ratio of Itti's model is even greater and hardly meaningful to measure grouping quality. So they are omitted from the table.

Table 1. comparison of $\beta_{miss} / \beta_{fal}$ among our model, Itti's model and mini-cut model

algorithm	natural obj. outdoor	natural obj. indoor	m-m obj. outdoor	m-m obj. indoor
ours	0.26 / 0.14	0.37 / 0.24	0.39 / 0.32	0.35 / 0.42
Itti's	-- / 0.42	-- / 0.57	-- / 0.39	-- / 0.63
mini-cut	0.37 / 0.46	0.40 / 0.48	0.52 / 0.39	0.41 / 0.55

More examples are shown in Fig. 2. Grouping effect on natural object in outdoor background (first row in Fig.2) is good, especially when the object strongly contrasts

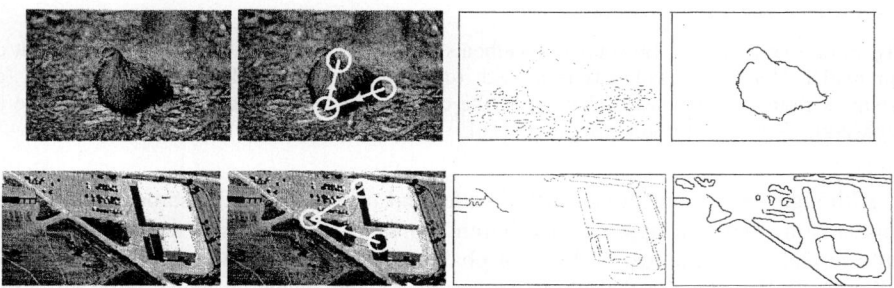

Fig. 2. *left column*: ground truth [16], *second column*: Itti's attention model. circle marks attended region, line with arrow marks focus shift trajectory, *third column*: grouping result of mini-cut [16], *right column*: grouping result of our algorithm

to background. In this situation, integrated closed contour of object is extracted despite cluttered background. Mini-cut cannot segregate background noise from object contour. Itti's model ignores integrality of perceptual object though it does detect the most salient locations. However, grouping effect on some man-made objects in outdoor background, especially those remote images taken from aerial, is relatively bad (second row in Fig.2). Low resolution of this image class may account for the bad effect. But man-made objects with parallel structure outstand clearly due to global relation of parallelism. On the whole, grouping effect of our model is closer to that perceived by human than other two models. It further proves grouping relying on global relations is more suitable for simulating attention in middle vision stage.

Our model may fail when object boundary are weak so that Canny Edge Detector cannot offer enough information to extract global relations. In this situation, either some object contour is missed (first row in Fig.3), or the contour fails to converge to a closed sequence (second row in Fig.3). As the latter occurs, performance of our model nears to or even worse than that of mini-cut.

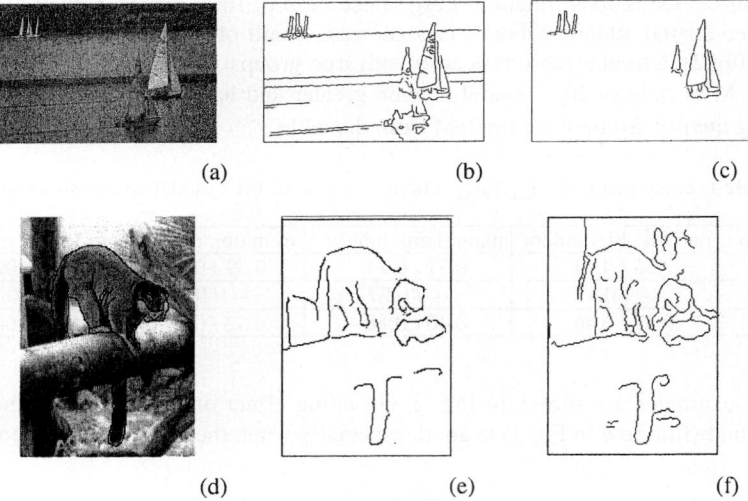

Fig. 3. our model failure (a) source of sailboats image; (b) Canny result; (c) grouping result of our model. One of the sailboats is missed; (d) ground truth of collarlemur image [16]; (e) grouping result of our model. Contour of the collarlemur is not closed and integrated due to occlusion; (f) grouping result of mini-cut [16].

Although we put our model, Itti's model and mini-cut into the same assembly for comparison, they are applied to different domains. Space-based attention, such as Itti's model, is fit for feature binding phase. It is effective when the goal is to find salient locations but not the whole perceptual object. Focus of space-based attention can be used for seed in detection of ROI (region of interest). While results of object-based attention, such as our model, can be directly taken as results of figure-ground segmentation and object detection. Global relations like closure is important for grouping object contours. On the other hand, closure restricts scope which the grouping algorithm applies to. Precondition is that the image should contain objects with explicit closed contours. Mini-cut almost imposes no conditions on images. From this point, mini-cut can be used for grouping on more complex images. Our model exhibits its superiority only within certain class of images.

5 Conclusion

We implement a selective attention guided perceptual grouping model. The global topological relations produce groups of high quality. Attention mechanism decides salience measure in global view and limits solution space to reduce complexity. These are main obstacles of graph-based grouping methods. Experiments on four classes of images testify the model's efficacy on certain class of images. And plausibility of object-based attention in middle and high vision stages is also proved.

Using more precise descriptions of global relation to boost grouping quality is recognized by more and more researchers. In this paper, we exploit statistics, specifically eigenvector to discover global structure embedded in local elements. Recently, manifold and topology are reported to approach essence of global relations more closely [15]. Unifying these tools into grouping framework to design more robust algorithms is our future work. It is also noticed that our model cannot guarantee convergence in strong occlusion. To borrow ideas from other closure extraction methods, such as Mahamud proposed directionality of edges [11], may be attempted. Besides, complexity of our model is under improvement to be applied into larger real images.

Acknowledgements

The research is supported by: National Natural Science Foundations (No. 60373029) and Doctoral Foundations of China(No. 20020004020)

References

1. Tresman, A. M. and Gelade D.: A Feature Integration Theory of Attention. *Cognit. Psychol.* 12(1): 97-136, 1980
2. Shepard R.: Toward a Universal Law of Generalization for Psychological Science. *Science*, 237: 1317-1323, 1987
3. Salinas, E. and Sejnowski, T.J.: Correlated neuronal activity and the flow of neural information. *Nature Review Neuroscience.* 2: 539-550, 2001
4. Itti L.: Models of Bottom-Up Attention and Saliency, In: *Neurobiology of Attention*, Itti L., Rees G. and Tsotsos J.K. Ed. 576-582, San Diego, CA: Elsevier, 2005
5. Rybak I.A., Gusakova V.I. and Golovan A.V., etc.: A model of attention-guided visual perception and recognition. *Vision Research*, 38: 2387–2400, 1998
6. Palmer, S.E.: Modern theories of Gestalt perception. *Understanding Vision.* Humphreys G. W. ed. Blackwell, 1992
7. Amir A. and Lindenbaum M.: A generic grouping algorithm and quantitative analysis. *IEEE Trans. Pattern Analysis and Machine Intelligence*, 20(2): 168-185, 1998

8. Elder J. H. and Zucker S. W.: Computing Contour Closure. *Proc. Fourth European Conf. Computer Vision*, 399-412, 1996
9. Sha'ashua A. and Ullman S.: Grouping Contours by Iterated Pairing Network. *Neural Information Processing Systems (NIPS)* vol. 3, 1990.
10. Herault L. and Horaud R.: Figure-Ground Discrimination: a combinational optimization approach. *IEEE Trans. Pattern Analysis and Machine Intelligence*, 15(9): 899-914, 1993
11. Mahamud S., Williams L. R., Thornber K. K., and Kanglin Xu: Segmentaion of multiple salient closed contour from real images. *IEEE Trans. Pattern Analysis and Machine Intelligence*, 25(4): 433-444, 2003
12. Wang Z. R. and Shi R. C.: *Matrix analysis*. Beijing institute of technology Press, 1989
13. Klein, R. M.: Inhibition of return. *Trends in Cognitive Science*. 4: 138–147, 2000.
14. Soundararajan P. and Sarkar S.: An in-depth study of graph partition measures for perceptual organization *IEEE Trans. Pattern Analysis and Machine Intelligence*, 25(6): 642-660, 2003
15. Seung H.S. and Lee D.D.: The Manifold Way of Perception. *Science*, 290(5500): 268–269, 2000

Visual Search for Object Features

Predrag Neskovic and Leon N. Cooper

Institute for Brain and Neural Systems and Department of Physics,
Brown University, Providence, RI 02912, USA
pedja@@brown.edu, Leon_Cooper@@brown.edu

Abstract. In this work we present the computational algorithm that combines perceptual and cognitive information during the visual search for object features. The algorithm is initially driven purely by the bottom-up information but during the recognition process it becomes more constrained by the top-down information. Furthermore, we propose a concrete model for integrating information from successive saccades and demonstrate the necessity of using two coordinate systems for measuring feature locations. During the search process, across saccades, the network uses an object-based coordinate system, while during a fixation the network uses the retinal coordinate system that is tied to the location of the fixation point. The only information that the network stores during saccadic exploration is the identity of the features on which it has fixated and their locations with respect to the object-centered system.

1 Introduction

When we look at the world around us, we perceive it as highly detailed, full colored and stable. However, our eyes neither process visual information with uniformly high resolution nor are they motionless. The only region of the visual scene that is processed with high resolution is that which is very close to the fixation point. The acuity and color sensitivity of retinal cells rapidly decreases with distance from the fovea, the region of the retina that corresponds to only about the central 2 degrees of the viewed scene. In order to overcome this limitation of the optical structure of the eyes, our visual system uses saccades to reposition the fovea over different locations and thus obtain locally high resolution samples of a visual scene. The question is then what information is retained during saccadic exploration and how detailed is that information? It has been shown in numerous experiments that our visual system is fairly insensitive to visual changes in an image across a saccade, a phenomenon called *change blindness* [1]. As a consequence, it has been proposed [2] that because the "world is its own memory", the visual system does not need to store visual information from fixation to fixation. On the other hand, the visual memory theory of Henderson and Hollingworth [3] posits that a relatively detailed scene representation is built up in memory over time and across successive fixations. The question then becomes: how do we piece together information from different fixations? According to *composite sensory image hypothesis* the sensory images from consecutive fixations are spatially aligned and fused in a system that maps a retinal reference frame onto a spatiotopic frame [4]. However, numerous psychophysical and behavioral data from vision and cognition literature have provided evidence against this hypothesis [5].

Another important question, related to saccadic exploration of the pattern, is what criterion our visual system uses when it select the location on which it is going to fixate - the target location? Is this process driven purely by bottom-up information (by the salient properties of the pattern), or by top-down information (our expectations) or by a combination of the two? It has been known for a long time [6,7] that more informative scene regions receive more fixations and thus informative regions are most likely candidates for being target locations. What is not known is how to define the informative region. Again, one can use only perceptual information (bottom-up), or cognitive information (top-down) or a combination of the two. Experimental evidence suggests that while initial fixations are controlled by bottom-up information, the subsequent fixations are influenced by cognitive expectations [8]. However, how exactly and when (at what stages during the recognition process) these two sources of information interact with each other is still an open question.

In this work we address the previous questions and present the algorithm for searching for object features that combines perceptual and cognitive information. More specifically, the selection of the target feature is initially driven purely by bottom up information but during the recognition process becomes more constrained by the top-down information. Furthermore, we propose a concrete model for integrating information from successive saccades. We show that the only information that is necessary to retain across fixations, in order to segment and recognize an object, is the location and identity of some of the features on which the system has fixated. We demonstrate that our model can also be utilized for building a real-world recognition system. To this end, we constructed a working system and tested it on the difficult task of searching for letters in handwritten words.

The paper is organized as follows: In section 2 we describe the feature-based object representation and the architecture of the network that integrates information from different regions of the pattern. In section 3 we show a detailed algorithm for searching for object features and the mechanism that the system uses to resolve conflicting configurations. We illustrate the results of our algorithm when applied to real-world dataset of cursive script in section 4. Final remarks and summary are given in section 5 .

2 Object Representation

In our model, an object is represented as a collection of features of specific classes arranged at specific locations with respect to one another [9,10]. Detecting an object is then equivalent to detecting constituent features and estimating their locations. The main problem in detecting individual features is that information contained in the local region of an image is often ambiguous and therefore can be interpreted in many different ways. Human visual system, overcomes this ambiguity by incorporating contextual information. During fixation on a particular region of the object, we use contextual information, such as locations and identities of surrounding features, to help determine identity of the fixated region. Similarly, information from the previous fixations is used as contextual information to help determine identity of the region around the current fixation.

Feature Detectors. Let us assume that we have N feature detectors, each selective to a feature of particular class. For example, if an object is a face then features can be the nose, the mouth and the eyes. If an object is a word then features can be the letters from the alphabet, $N = 26$. If we denote with $\boldsymbol{x}_i$ the location of the pattern over which the feature detector is positioned, then the output of the feature detector, $d_i(\boldsymbol{x}_i)$, is proportional to the confidence that the local region around $\boldsymbol{x}_i$ represents the feature of class i, where $i = 1, ..., N$. The closer in appearance the region is to the feature that the detector is selective to, the higher is the output of that feature detector.

Simple Units (SUs). Let us now choose one region of the pattern and assume that it represents a specific feature, say a letter "a" of the word "act". In order to incorporate contextual information, we construct a set of units, called *simple units*, that capture the locations and identities of surrounding features. The sizes and distribution of the receptive fields of the (surrounding) simple units are designed in the following way. The receptive field of the simple unit that is selective to the letter "c" is constructed so that it can capture all possible variations in location of the letter "c" given the location of the letter "a". We will denote this unit as $SU(2|1)$. The simple unit that is selective to the letter "t", $SU(3|1)$, is further away from the central unit and its size is larger than the size of the $SU(2|1)$ since the variations in feature sizes and locations accumulate. In general, both the sizes and the overlapping of the receptive fields of simple units become progressively larger with the distance from the central unit. However, the surrounding SUs that are nearest neighbors to the central SU do not overlap with the central SU and therefore the order is preserved within the local neighborhood around the fixation point.

If we denote with R_{ji} the receptive field of the simple unit that is selective to the jth feature, and $\boldsymbol{x}_j$ is the location of the jth feature, then the simple unit fires if the feature is detected (its value is above some threshold, $d_j(\boldsymbol{x}) > threshold$) and is located within the unit's receptive field ($\boldsymbol{x}_j \in R_{ji}$). In our implementation, we set the threshold to a very small value close to zero. We will denote with symbol y the location of a feature with respect to the location of the fixation point and with symbol x the location of a feature with respect to a coordinate system that is fixed to an object, e.g. a specific object feature. The activation of the simple unit, whose center is at distance $\boldsymbol{y}_j = \boldsymbol{x}_j - \boldsymbol{x}_i$ with respect to the fixation point $\boldsymbol{x}_i$, is calculated as

$$s_{ji}(\boldsymbol{y}_j) = s_{ji}(\boldsymbol{x}_j|\boldsymbol{x}_i) = \max_{\boldsymbol{x}_j \in R_{ji}}[d_j(\boldsymbol{x}_j)], \qquad (1)$$

Complex Units (CUs). Although each simple unit processes only local information, combination of all the simple units associated with a given central feature provides contextual information. Since feature detection and location estimation is much less reliable for features that are further away from the fixation point compared to features that are closer we weigh the contribution of each simple unit differently. The weighing factor in our implementation is set to be inversely proportional to the size of the simple units receptive field, $\omega_{ji} = 1/(size\ of\ R_{ji})$. In this way, the contribution of simple units that are closer to the fixation point (those that have smaller receptive fields) is larger compared to simple units that are further away. The output of a complex unit, associated with the ith object feature is given as

$$c_i(\boldsymbol{x}) = d_i(\boldsymbol{x}_i) \cdot \frac{1}{N-1} \sum_{j=1, j \neq i}^{N} \omega_{ji} \cdot s_{ji}(\boldsymbol{x}_j | \boldsymbol{x}_i), \qquad (2)$$

where $d_i(\boldsymbol{x}_i)$ is the activation of the feature detector positioned over the i-th object feature, and N represents the number of object features. For a given object, there are as many complex units as there are features.

The receptive fields of all the simple units that belong to the same complex unit form complex unit's receptive field. Since the receptive fields of the simple units closer to the central SU are smaller than the receptive fields of those that are further away, the complex unit captures with high accuracy only the locations of the features that are close to the fixation point. As a consequence, a complex unit can determine only whether surrounding features are correctly positioned with respect to the central feature but not whether they are correctly positioned with respect to one another.

Object Units. In order to capture different regions of an object with high resolution and in order to correctly estimate locations of the features with respect to one another, the system has to probe the pattern at different locations. We call these exploratory movements of the system *saccades*. At the top of the processing hierarchy are the object units, one unit representing each object. The outputs of all the complex units are supplied to the object unit and they are combined in the following way

$$o(\boldsymbol{x}) = \sum_{i=1}^{N} c_i(\boldsymbol{x}) \qquad (3)$$

where $\boldsymbol{x} = (\boldsymbol{x}_1, \cdots, \boldsymbol{x}_N)$ is a particular configuration of selected features. An object unit, therefore represents an object regardless of any specific point of view or fixation point. This hierarchical representation consisting of different collections of simple units, complex units and an object unit comprises a neural network-like architecture that we use to represents each object from a library of objects.

In summary, simple units provide local information about presence of specific features within specific regions; complex units integrate information from different regions, given a specific fixation location, whereas object units integrate information from different fixations. In order to accomplish this task of integrating information across fixations, the system has to use two different coordinate systems. One system is tied to the fixation point, the *retinal coordinate system*, while the other system is object-based and can be centered on any object feature. In the following section we will see how the system combines these two coordinate systems during the visual search for object features.

3 The Search Algorithm

In this section we present the algorithm that combines perceptual and cognitive information during the process of searching for object features. In order to simplify description, we will assume that an object is a word and that features are letters. However, the algorithm is general and can be applied to any other object consisting of different features. Instead on operating on the pattern, the network operates on the *detection matrix* that represents sensory input and consists of the outputs of feature detectors (in this case letter detectors) whose receptive fields overlap and completely cover the pattern. Therefore, a row of the detection matrix represents a class of the letter and a column corresponds to the position of the letter within the pattern.

We will call the letter on which the system fixates the central letter, the corresponding location within the matrix the central column, and the (simple) unit that is positioned over the central letter the central unit. All the simple units that are surrounding the central unit are called the surrounding units and they provide contextual information. In the following, we will assume that a particular dictionary word is given and the task of the recognition system (the

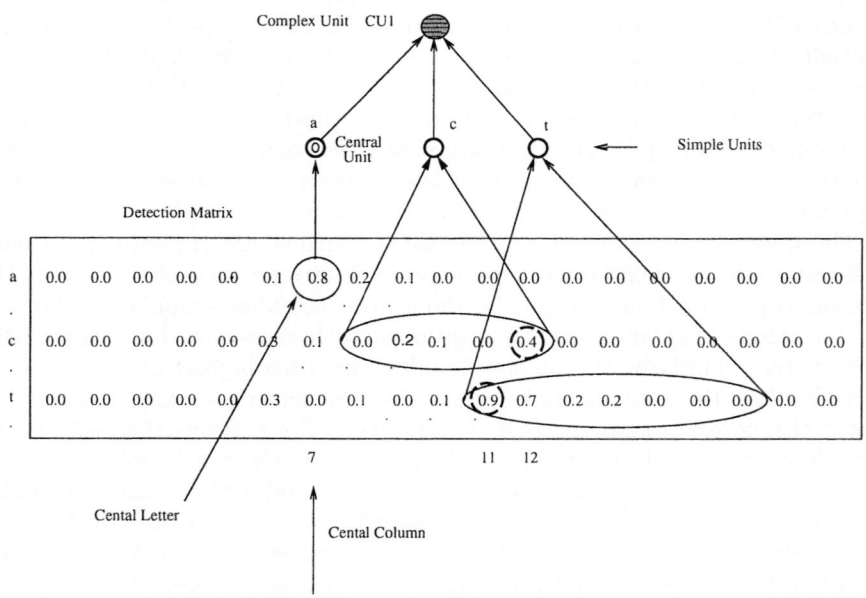

Fig. 1. The complex unit CU1 represents the word "act" from the perspective of the letter "a". The detection matrix corresponds to the pattern representing the word "act". The central letter is the letter "a" from the 7-th column. The simple unit selective to the letter "c" selects the letter "c" from the 12-th column, and the simple unit representing the letter "t" selects the letter "t" from the 11-th column. The segmentation of the pattern determined by the complex unit CU1 is "atc", which is obviously incorrect segmentation.

network) is to select elements from the detection matrix that correspond to the letters of the given word and thus segment a pattern into letters.

The recognition process starts by selecting the most prominent letter from the pattern, the element from the detection matrix that has the highest value. If we think of the detection matrix as the *saliency map* then this procedure is equivalent to *winner-take-all* mechanism proposed by Koch and Ullman [11]. Note that at this stage the feature selection is purely a bottom-up process. Let us now assume that the selected letter, the central letter, is one of the letters of the given dictionary word. All complex units (more specifically their central units) are then positioned over this letter and we say that the system fixates on the central letter. However, not all the complex units will be equally activated and only those complex units whose central unit is selective to the central letter will fire. For example, if the given dictionary word is "again" and the network fixates on the letter "a" then two complex units will have central units that are selective to this letter: the complex unit CU(1) with surrounding simple units selective to letters "-gain", and the complex unit CU(3) with surrounding simple units selective to letters "ag-in". Which of those two complex units would have higher activation would depend on the presence of the letters to which each complex unit (and the corresponding simple units) is selective. If the network finds the letters "-gain" at expected locations then the unit CU(1) becomes activated. On the other hand if the network finds the letters "ag-in" at expected location then the unit CU(3) becomes activated. The complex unit with highest activation then segments the pattern by choosing the letters that activate it the most. We will call this segmentation a *tentative* segmentation. Unfortunately, due to often high overlapping of the simple units' receptive fields, this segmentation can be incorrect in the sense that selected letters are not at correct locations with respect to one another, as illustrated in Figure 1.

The only way to assure that the selected letters are correctly positioned with respect to one another is if the network fixates on each of them since the ordering is preserved only locally, for the nearest neighbor simple units. We will call the letters that are correctly positioned with respect to one another the *active* letters. Similarly, we call the complex unit with highest activation, for a given fixation, the active complex unit. The first letter on which the network fixates therefore becomes the first *active* letter. The network then selects the *target* letter and the location within the pattern on which it is going to fixate. The target letter is selected by one of the simple units of the active complex unit. More specifically, the new fixation point becomes the location of the letter that is selected by the simple unit that has the highest activation. In this way, the network combines top-down information (expectation about the location of the letter) with bottom-up information that is provided by letter detectors.

The question is now what information about the pattern, given the current fixation, is stored in the short term memory? The only information that is retained across fixations is the location of each active letter and its identity. Information about locations of active letters is important for the network so it does not in the future make fixations on the same locations. In effect, in this

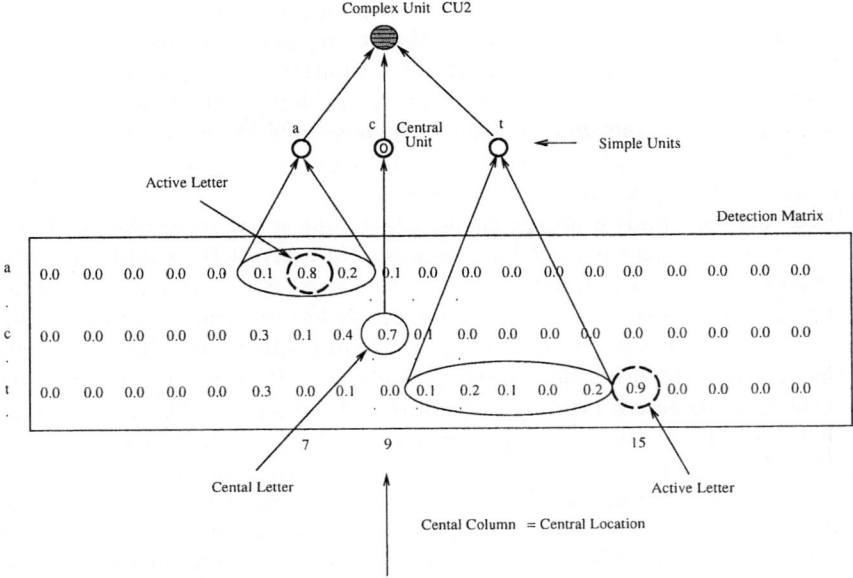

Fig. 2. The complex unit CU2 represents the word "act" from the perspective of the letter "c". The detection matrix corresponds to the pattern representing the word "act". The detection matrix corresponds to the pattern representing the word "act". The central letter is the letter "c" from the 9-th column. The central unit of the complex unit CU2 is selective to the letter "c" and is located over the letter detector that has detected the letter "c" with confidence 0.7. The active letters are the letters "a" and "t". The location of the active letter "a" is within the receptive field of the corresponding simple unit selective to the letter "a", but the location of the active letter "t" is not within the expected region which is the receptive field of the simple unit selective to the letter "t". We say that the central unit is in conflict with active letter "t".

way we implement an *inhibition of return* mechanism which is necessary so that the network doesn't enter an endless loop. The locations of the active letters are measured with respect to some point within the object, for example the location of the first active letter can be used as the center of the coordinate system. It is important to note that the spatial arrangement of the receptive fields of the simple units (that belong to the same complex unit) is fixed (with respect to each other), but the position of the complex unit over the detection matrix is not. With each selection of the new target letter, the network is repositioned so that all the central units (of all the complex units) are placed over the target letter. If the active letter falls within the receptive field of the simple unit that is selective to this letter, then the simple unit does not process information using Eq. 1 but instead immediately chooses the active letter and outputs the confidence with which the active letter is detected.

As we have mentioned earlier, one of the consequences of the "fovea-like" distribution of the receptive fields of the simple units is that feature ordering is preserved only with respect to the fixation point. This means that the central letter is not always at the correct location with respect to one or more active letters, as illustrated in Figure 2. If this situation occurs, then the network probes the local neighborhood around the central letter and tries to find a letter of the same class as the central letter but whose location is not in conflict with active letters. This small repositioning of the network is similar to microsaccades performed by the eyes during the visual search.

If the network finds a new (central) letter that is not in conflict with active letters, then this letter is added to the group of active letters and the network continues to search for remaining letters. Otherwise, if the network cannot find a letter that is not in conflict with existing active letters then it suppresses either the central letter or the active letters that are in conflict with the central letter. This is done by comparing the values of the object unit for those two scenarios: a) when the existing active letters are accepted and the central letter is rejected, and b) when the active letters that are in conflict with the central letter are rejected and the central letter, together with active letters that are not in conflict with the central letter, are accepted. The first configuration we call the old value of the object unit while the second configuration we call the new value of the object unit. In order to calculate the new value (without the conflicting active letters) the network has to fixate on all the active letters that are not in conflict with the central letter since previous fixations on those active letters (previous values of corresponding complex units) did not include the current central letter. This means that for the network the landscape of activations of feature detectors, the pattern, appears differently as a consequence of exploring the pattern and accumulating new information. The central letter is accepted (and conflicting active letters suppressed) only if the new object unit value is strictly greater than the old object unit value. Once the network accepts or rejects the central letter, it continues to search for the new target letter until all letters of a given dictionary word are discovered.

The visual search for object features, when it is not known in advance what object the pattern represents, does not significantly differ from the previously described algorithm for searching the features of the known object. We will again focus on handwriting recognition and assume that a pattern represents an unknown dictionary word. The network first selects the most prominent letter from the detection matrix but this time, instead of using only the complex units of one object unit, the central letters of the complex units of all the object units (all dictionary words) are positioned over this letter. The complex unit that has the highest activation propagates its output to the corresponding object unit that becomes the active object unit. The word that is represented with this active object unit now imposes the structure on the pattern in the sense that the network starts to search for the letters of only the active object unit. In this way, the algorithm reduces to the previously described procedure for searching for the features of a given object. Since the network always searches for the letters of the dictionary word that is associated with object unit that has the highest value, it might happen that during the search process the network switch from searching for the letters of one dictionary word to searching for the letters of some other dictionary word. The visual search is completed when all the letters of the active dictionary word are found and the object unit's value is above some threshold value.

4 Implementation and Results

We tested the search algorithm on a database of online cursive words where the features were letters and objects were dictionary words. The letter detectors were designed using the weight sharing neural network [12] and the receptive fields of simple units were designed using pairwise probabilities of letter locations as described in [9]. In addition to 26 letters from the alphabet, we introduced the features that denote the beginning and end of the word so our alphabet effectively consists of 28 symbols. The beginning and end features are important in order to provide context for one letter words or words that can at the same time be part of longer words such as the word "act" that is also part of the words "actual", "activation", "fact", "exact", and etc.

The only way to verify the accuracy of the search algorithm is to compare the segmentation obtained with our algorithm to groundtruthed data - where the location of every letter of every dictionary word is known. However, since the pre segmented data is not available for this database, another possibility is to compare the recognition rates of our algorithm to recognition rates of some of the best recognition algorithms. We constructed two systems for recognition of online cursive script. One based on Hidden Markov Model (HMM), which is the state-of-the-art model for handwriting recognition, and the other based on the Interactive Parts (IP) model [13]. The objective function for the IP model is exactly the same as the one that we use in this paper except that in the IP model only the first neighbor interactions are considered. This reduced contextual information has important consequences since the model then becomes much more tractable and one can use dynamic programming in order to exactly solve the objective function.

Both the HMM and the IP model give comparable results and they are slightly better compared to our results. The recognition accuracy of our system varies from around 65% to over 90% for different writers, depending how clearly the words are written. On average, the accuracy of our system is about 4% lower compared to HMM and IP models. However, we should emphasize that the recognition rate depends not only on the correct segmentation of the pattern but also on the way the output of the letter detectors are combined - the connections between the units of the network and therefore the recognition accuracy is just one way of testing the performance of the search algorithm.

5 Summary

In this work we presented the computational algorithm that combines both perceptual and cognitive information during the process of searching for object features. Our algorithm, as suggested by numerous experiments [8], is initially driven purely by bottom up information but during the recognition process becomes more constrained by the top-down information. The network that performs the search algorithm utilizes contextual information on two levels. During a fixation, the locations and identities of surrounding features provide context while during the search process contextual information is represented through the locations and identities of visited (active) features.

We showed that in order to capture variations in feature locations, the receptive fields of the units (the simple units) become progressively larger as well as their overlap. Therefore, the network can estimate with high resolution only the locations of features that are close to the fixation point. In order to estimate the locations of all the features with high resolutions, and thus ensure that features are correctly positioned with respect to one another, the network has to make saccadic movements. As a consequence of the foveal distribution of the receptive fields, some of the features that activate simple units may be incorrectly positioned with respect to one another. We described a detailed mechanism for resolving conflicting configurations and showed that in some situations the network benefits from making microsaccades.

We also demonstrated the necessity of using two coordinate systems for measuring feature locations. During the search process, across saccades, the network uses an object-based coordinate system that is centered at any feature/location of the object while during a fixation the network uses the retinal coordinate system that is tied to the location of the fixation point. The only information that the network stores during saccadic exploration is the identity of the active features, on which it has fixated, and their locations with respect to the object-centered system. This information allows the network to effectively implement *inhibition of return* mechanism and therefore enhance processing by withdrawing attention from previously attended locations.

We tested the search algorithm on real world data of online cursive script and achieved very high recognition rates. The performance of the system favorably compares even to the state-of-the-art system for handwriting recognition. We believe that in addition to providing an insight into information processing by the human visual system, one of the major strengths of our algorithm is that it demonstrates that some mechanisms of the human vision can be successfully used in constructing an efficient system for real-world applications.

Acknowledgments. This work is supported in part the ARO under contract W911NF-04-1-0357.

References

1. Simons, D.J., Levin, D.T.: Change blindness. Trends in Cognitive Sciences **1** (1997) 261–267
2. O'Regan, J.K.: Solving the 'real' mysteries of visual perception: The world as an outside memory. Canadian Journal of Psychology **46** (1992) 461–488

3. Hollingworth, A., Henderson, J.M.: Accurate visual memory for previously attended objects in natural scenes. Journal of Experimental Psychology: Human Perception and Performance **28** (2002) 113–136
4. Jonides, J., Irwin, D.E., Yantis, S.: Integrating visual information from succesive fixations. Science **215** (1982) 188
5. Pollatsek, A., Rayner, K.: What is integrated across fixations? In: Eye Movements and Visual Cognition. Springer-Verlag (1992) 166–191
6. Buswell, G.T.: How people look at pictures. Chicago: Univ. Chicago Press (1935)
7. Yarbus, A.L.: Eye movements and vision. New York: Plenum (1967)
8. Henderson, J.M., Hollingworth, A.: High-level scene perception. Annu. Rev. Psychol. **50** (1999) 243–271
9. Neskovic, P., Cooper, L.: Neural network-based context driven recognition of online cursive script. In: 7th International Workshop on Frontiers in Handwriting Recognition. (2000) 352–362
10. Neskovic, P., Schuster, D., Cooper, L.: Biologically inspired recognition system for car detection from real-time video streams. In: Neural Information Processing: Research and Development, J. C. Rajapakse and L. Wang (Eds.), Springer - Verlag (2004) 320–334
11. Koch, C., Ullman, S.: Shifts in selective visual attention: towards the underlying neural circuitry. Hum Neurobiol **4** (1985) 219–227
12. Rumelhart, D.E.: Theory to practice: A case study – recognizing cursive handwriting. In Baum, E.B., ed.: Computational Learning and Cognition: Proceedings of the Third NEC Research Symposium. SIAM, Philadelphia (1993)
13. Neskovic, P., Davis, P., Cooper, L.: Interactive parts model: an application to recognition of on-line cursive script. In: Advances in Neural Information Processing Systems. (2000)

Agent Based Decision Support System Using Reinforcement Learning Under Emergency Circumstances

Devinder Thapa, In-Sung Jung, and Gi-Nam Wang

Department of Industrial and Information Engineering,
Ajou University, South Korea
{debu, gabriel7}@ajou.ac.kr, gnwang@madang.ajou.ac.kr

Abstract. This paper deals with agent based decision support system for patient's right diagnosis and treatment under emergency circumstance. The well known reinforcement learning is utilized for modeling emergency healthcare system. Also designed is a novel interpretation of Markov decision process providing clear mathematical formulation to connect reinforcement learning as well as to express integrated agent system. Computational issues are also discussed with the corresponding solution procedure.

1 Introduction

The objective of this paper is to combine the agent based decision support system with ubiquitous artifacts and make it more intelligent so that it can help the doctors to acquire on time correct diagnosis and select appropriate treatment choices. An attempt is given to supervise the dynamic situation by using agent based ubiquitous artifacts and to find out the appropriate solution for emergency circum-stances providing correct diagnosis and appropriate treatment in time. As per the work done by M. Hauskret, H. Fraser[7], the reason for using the RL (Reinforcement Learning) agent based on MDP (Markov Decision Process) model is that it needs less number of parameters and it also gives approximation method to make trade off between accuracy and speed, in turn, solving the complex number of cases in less time compare to the existing system.

The idea of interface agent has been derived from the concept of [4] although the functional architecture is different but the conceptual idea is similar to our work. The implementation of reinforcement learning agent approach has been utilized in the previous work [7] using the model of partially observable Markov decision process [POMDP]. The concept of ubiquitous healthcare system using agent technology has studied in [2]. All of the existing works have focused on the exploitation of ubiquitous system for the betterment of healthcare system. Our idea is to develop integrated emergency system using agent based approach.

2 Reinforcement Learning Agents

Reinforcement learning (RL) is based on interaction with an environment, from the consequences of action, rather than from explicit teaching [5]. RL could be characterized by a mathematical framework of Markov decision processes (MDPs). Main elements of Reinforcement learning is states s, actions a and rewards r. The reinforcement learning agent (RL-agent) is connected to his environment via sensors. In every step of interaction the agent receives a feedback about the state of the environment (s_{t+1}) and the reward (r_{t+1}) of its latest action at. The agent chooses an action (a_{t+1}) representing the output function, which changes the state (s_{t+2}) of environment. The agent gets a new feedback, through the reinforcement signal (r_{t+2}).

3 Scenario of Reinforcement Learning Agent at Emergency Circumstances

When a high risk patient, far from medical facilities, gets some perilous occurrence in their body the ubiquitous devices attached to their body sends some signals to the hospital knowledge base server.

This signal sends the patient profile to the HIS (Hospital Information Server). Knowledge about the patient will be accumulated by the RL-agent (named as decision maker agent) from the HIS database [1]. The RL-agent compares the patient current status with his existing diagnosis history, RL-agent search for the related physician his scheduling, and sends the patients profile to the related departments. On the bases of this crucial data the decision maker agent, based on reinforcement learning approach, make inference of the data and provide entire data history of the patient with best alternate action(diagnosis and treatment) to the related department with minimal time cost. In this scenario, decision maker agent uses some model based on previous patient's profile, to collect the patient data; however this paper only deals with the processing of decision maker agent based on RL approach.

4 Markov Decision Process

An MDP is defined by a set of states S, and actions A, Reward R, and transition probabilities T.

$$V^*(s) = \max_a (R(s,a) + \Sigma T(s,a,s')V^*(s')), \forall s \in S \tag{1}$$

$$V^*(s) = \max_\pi E(\sum_{t=0}^\infty r_t) \tag{2}$$

$$R(s,a) = \sum_{s' \in S} P(s'|s,a) R(s,a,s') \tag{3}$$

The objective of this model is to find out the optimized action to maximize the reward or cost in a finite horizon (2).

Due to the computation complexities of the pure MDP model we use Bellman's value function recursively; it (1) calculates the total reward value by adding all the suboptimal values (3).

5 Formulations to a Reinforcement Learning Problem

r=Reward, p=Transition probability, a=Action, S=State
  Decision Epochs: [Finite time horizon]
  T={1,2,,N}, $N \leq \infty$
  States: [Patient Condition: Serious, Normal]
  S={S1, S2}
  Actions: [Medication, No action]-
  As1 $\{a_{1,1}, a_{1,2}\}, As2 = \{a_{2,1}\}$
  Rewards: [Cost]
  $rt(S1, a_{1,1}) = r_{1,1}; rt(S1, a_{1,2}) = r_{1,2}; rt(S2, a_{2,1}) = r_{2,1};$r N(S1)=0;rt(S2)=0;
  If $N \leq \infty$
  Transition Probabilities: [Effect of diagnosis and treatment]
  $pt(S1|S1, a_{1,2}) = p_{1,2,1}; pt(S1|S2, a_{2,1}) = p_{2,1,2};$
  $pt(S2|S1, a_{1,1}) = p_{1,1,3}; pt(S2|S1, a_{1,2}) = p_{1,2,4};$
  $pt(S2|S2, a_{2,1}) = p_{2,1,5}$
  Expected Reward/Cost:
  $rt(S1, a_{1,1})= rt(S1, a_{1,1}, S1)$ $pt(S2|S1, a_{1,1})$
  +rt$(S1, a_{1,1}, S2)pt(S2|S1, a_{1,1})$

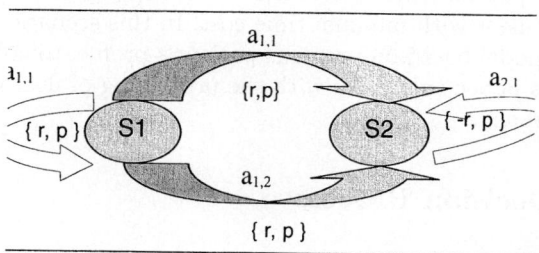

Fig. 1. Symbolic representation of Markov Decision Process

5.1 Finding the Best Policy or the Minimum Cost Function Using DP (Dynamic Programming) Approach

Choose an arbitrary policy loop
  $\Pi := \Pi'$
  compute the value function of policy : solve the linear equations

$$V^\pi(s) = R(s, \pi(s)) + \Sigma_{s' \in S} T(S, \pi(S), S') V^\pi(S') \tag{4}$$

improve the policy at each state:

$$\Pi'(s) := \arg\min_a (R(S, a) + \Sigma_{s' \in s} T(s, a, s') V^\pi(s')) \tag{5}$$

until $\Pi = \Pi'$

Denote a policy as Π, where Π=action selected in current state. Where $V^\pi(s)$ and $\pi'(s)$ are optimal value and control function. We can take Π' as any random policy and $V^\pi(s)$ is reward value starting from current state and following the Π policy. Now we can define another greedy policy in terms of $\Pi'(s)$ and make iteration of the value function $V^\pi(s)$ function until $\Pi = \Pi'$.

6 Conclusions and Future work

This paper presents and describes a Reinforcement Learning agent based model used for information acquiring and real time decision support system at emergency circumstances. Markov decision process is also employed to provide clear mathematical formulation in order to connect reinforcement learning as well as to express integrated agent system. This method will be highly effective for the real time diagnosis and treatment of high risk patient during the emergency circumstances, when they are away from the hospital premises. Further pursuing will be to develop some prototype, and simulate the testing data, planning modules, and find out the actual outcome of this approach.

References

1. Rodriguez,M., Favela,J., Gonzalez,V., and Muoz,M. : Agent Based Mobile Collaboration and Information Access in a Healthcare Environment. *Proceedings of Workshop of E-Health, Applications of Computing Science in Medicine and Health Care.*, ISBN: 970-36-0118-9. Cuernavaca, Mxico, December,(2003)
2. Bardram,J., E. : The Personal Medical Unit – A Ubiquitous Computing Infrastructure for Personal Pervasive Healthcare. *UbiHealth 2004: The 3rd International Workshop on Ubiquitous Computing for Pervasive Healthcare Applications* , (2004)
3. Watrous,R.L. , and Towell,G. : A Patient-adaptive Neural Network ECG Patient Monitoring Algorithm. *In Proceedings Computers in Cardiology, Vienna, Austria* (1995), 229–232
4. Wendelken,S.,M., McGrath,S.,P. and Blike,G.,T. : Medical assessment algorithm for automated remote triage. *International conference of the IEEE EMBS, Mexico*,September (2003)
5. Sutton,R. ,S. , and Barto,A., G.: Reinforcement Learning: An Introduction. *MIT Press, A Bradford Book, Cambridge, MA* , (1998)

6. Kaelbling,L., P. , Littman,M., L. ,Moore, A., W. : Reinforcement Learning: A Survey. *Journal of Artificial Intelligence Research 4*, (1996),237-285
7. Hauskret,M. , Fraser,H. : Planning Treatment of Ischemic Heart Disease with Partially Observable Markov Decision Process. *Artificial Intelligence in Medicine,vol(18)*,(2000), 221-244

Dynamic Inputs and Attraction Force Analysis for Visual Invariance and Transformation Estimation

Tomás Maul[1], Sapiyan Baba[2], and Azwina Yusof[2]

[1] University Malaya, 50603 Kuala Lumpur, Malaysia
tomasmaul@yahoo.co.uk
[2] University Malaya, 50603 Kuala Lumpur, Malaysia
{pian, azwina}@um.edu.my

Abstract. This paper aims to tackle two fundamental problems faced by multiple object recognition systems: invariance and transformation estimation. A neural normalization approach is adopted, which allows for the subsequent incorporation of invariant features. Two new approaches are introduced: dynamic inputs (DI) and attraction force analysis (AFA). The DI concept refers to a cloud of inputs that is allowed to change its configuration in order to latch onto objects thus creating object-based reference frames. AFA is used in order to provide clouds with transformation estimations thus maximizing the efficiency with which they can latch onto objects. AFA analyzes the length and angular properties of the correspondences that are found between stored-patterns and the information conveyed by clouds. The solution provides significant invariance and useful estimations pertaining to translation, scale, rotation and combinations of these. The estimations provided are also considerably resistant to other factors such as deformation, noise, occlusion and clutter.

1 Introduction

One of the fundamental elements of any image understanding system is the ability to recognise multiple cluttered objects under any combination of transformations. Systems that have this ability are called multiple object recognition (MOR) systems. This paper presents a neural approach that addresses arguably the two most fundamental sub-components of any MOR system: invariance and transformation estimation. Invariant object recognition refers to the ability of recognizing objects regardless of variations such as position, orientation and size. Transformation estimation, on the other hand, refers to the ability of making estimates of the very same transformations that the invariance property ignores: e.g. the cat is rotated 45°.

The overall context of this paper lies mainly within biological neural networks (BNN), although the proposed approaches can also be applied outside of this context. According to Wiskott [1], BNN approaches relevant to invariance and transformation estimation can be divided into two general categories: 1) normalization and 2) invariant features. Normalization refers to the application of a particular set of transformations on a reference frame until it is aligned with an image object. Work on normalization by neural systems started as far back as 1947 in a landmark paper by Pitts and McCulloch [2]. Other more recent approaches include: Shifter Circuits [3], Dynamic Routing Circuits [4], the Dynamic Link Architecture [5] and others. The invariant feature approach, in general, pertains to the extraction of certain discriminating characteristics from objects and the utilization of these characteristics for the classification process. These discriminating characteristics are nevertheless invariant to the desired transformations. Some examples from the literature are: the Neocognitron [6], neural traces and temporal sequences [7], slow feature analysis [8], and others.

A synthesis of the normalization and invariant-feature approaches should be fruitful if indeed it combines the advantages of both. Unfortunately, the integration is not as straightforward as expected (see [1] for details). Thus, devising a normalization approach that integrates easily with invariant features forms the underlying motivation for the current work: dynamic inputs (DI) and attraction force analysis (AFA).

2 Dynamic Inputs

The term *dynamic inputs* refers to the inputs of a neural network and to their dynamic/mobility properties: i.e. the inputs of a neural network are allowed to move around an image thus changing what the network "sees" (what a network sees is here onwards referred to as an input-vision). Figure 1 illustrates two input-visions, one originating from a scaled-up and shifted cloud (on the left), and the other one originating from a scaled-down and rotated cloud (on the right).

In the most general concept of dynamic inputs, each input behaves semi-autonomously (conditioned by the image, the neural network's knowledge and other inputs), such that the cloud's global behaviour results in various transformations. This will eventually lead to the cloud latching onto an image object thus

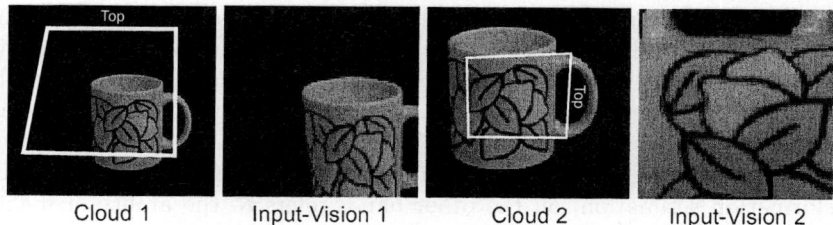

Fig. 1. Two examples of clouds and their corresponding input-visions. Each cloud consists of 100 × 100 inputs and is here represented by its borders. The top of the cloud is indicated by "top". The left-hand example exhibits a scaled-up and shifted cloud, while the right-hand example illustrates a scaled-down and rotated cloud.

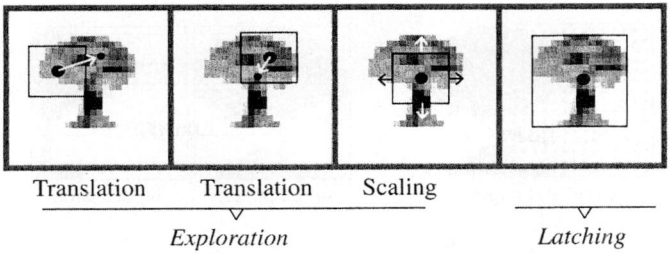

Translation	Translation	Scaling

Exploration   Latching

Fig. 2. This figure illustrates the difference between exploration and latching. Exploration is the process by which a cloud gradually gets closer (in relation to various transformations) to a target-object. Latching corresponds to the final step of the exploration process, when a cloud assumes the exact configuration of the target-object and after which more complex classification processes can be applied.

allowing a subsequent more complex classification process to then take place. In AFA, a specialized version of the above concept is used. Here, the inputs, instead of being semi-autonomous, behave in a global fashion. The process of latching onto an image object can require some exploration on the part of a cloud (see Figure 2). AFA is used in order to provide clouds with transformation estimations, thus minimizing the amount of exploration required before latching. At least two main categories of AFA can be distinguished: one that uses raw brightness values as inputs (AFA-raw) and the other, which can be extended from the first, and that uses neural invariant features as inputs (AFA-NIF). This paper concentrates on AFA-raw.

3 AFA-Raw

AFA-raw functions with a very simple artificial neural network (ANN) consisting of a single subtractive layer where each pattern is stored in a set of connections projecting from all inputs onto the pattern's output node. The two main properties that distinguish this architecture from a simple "template matcher" are: 1) the existence of transparent pixels (these are ignored when matching patterns) and 2) the possibility of performing "single matches". Classification of a whole pattern is performed by selecting the output with the lowest activation (i.e. the output that represents the smallest difference between the connection weights and the input pattern). On the other hand, classification of a single match is performed by considering a single input and activating all the output classes that match the corresponding brightness value at the correct position.

In AFA-raw, each pattern is stored as a fixed-size 2-dimensional matrix. AFA-raw overlaps[1] a stored-pattern and the current input-vision and searches for

---

[1] Recall that the dimensions of a stored-pattern and an input-vision are by definition equivalent. The superposition of both patterns might be easier to visualize if you imagine that the input-cloud (on the left of Fig. 3) is actually the stored-pattern that you are hovering over the test image.

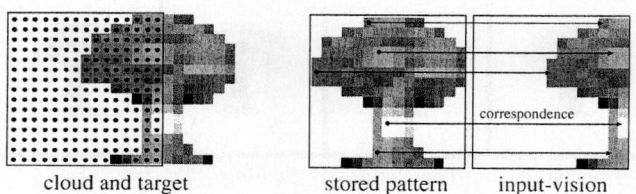

Fig. 3. The detection of correspondences. On the left, a cloud partially overlaps a tree, resulting in the input-vision on the right of the figure. When stored-patterns and input-visions are placed one on top of the other (here they are placed side-by-side in order to facilitate visualization), correspondences can be found, which provide useful information regarding transformations: in this case, translation in a 0° direction, since the correspondences found are all characterised by that angle. Only a small fraction of the total number of possible correspondences has been depicted here.

every pixel-to-pixel correspondence between the two, subsequently analyzing the resulting vectors in terms of angles and lengths in order to perform classification and transformation estimation. A correspondence refers to a vector between a pixel in a stored-pattern and a matching pixel in an input-vision (see Figure 3), where a match, in the strictest sense, means that the brightness values of both pixels are the same. Correspondences, being vectors, are characterized by a length and an angle.

Figure 4 shows the basic angles and lengths that are used as the basis for estimating transformations. Some definitions are necessary at this point:

1. a *cyclop* corresponds to a pixel in a stored-pattern
2. an *attractor* corresponds to a pixel in an input-vision
3. *origin* refers to the geometrical center of a stored-pattern or input-vision
4. a *cyclop-attractor angle* corresponds to the angle of a correspondence relative to the main axes of a stored-pattern
5. an *origin-cyclop-attractor angle* corresponds to the angle between the origin-to-cyclop vector and the origin-to-attractor vector

Cyclop-attractor angles provide information pertaining to translation: if the angle between a cyclop and an attractor is 75° and assuming that there is a "true correspondence"[2] then this means that the stored object is likely to be shifted in a 75° direction within the input-vision (see Figure 4(c)). Origin-cyclop-attractor angles provide information about rotation or whether scaling is required. If an origin-cyclop-attractor angle is for instance 90°, and assuming that it is based on a true correspondence, then this increases one's confidence that the pattern in the input-vision is rotated 90° relative to its stored representation (see Figure 4(d)).

---

[2] A "true correspondence" involves the correct pattern and the correct feature: e.g. "a dog ear corresponding with a dog ear" as opposed to "a dog ear corresponding with a horse tail".

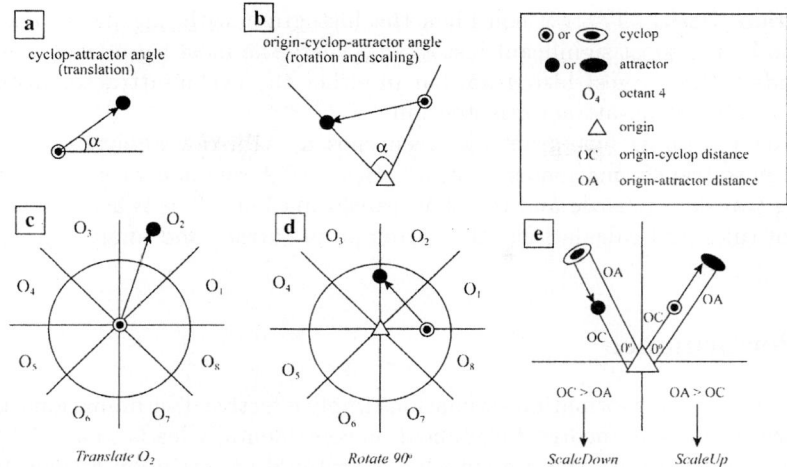

Fig. 4. The angles used by AFA-raw and what they convey. Two fundamental angles are illustrated: 1) the cyclop-attractor angle which provides translation information and 2) the origin-cyclop-attractor angle that provides information on rotation and scaling. The figure also depicts the use of octants.

If the angle is around 0° then this indicates that scaling is required. The relative lengths of near-0° origin-to-cyclop vectors and near-0° origin-to-attractor vectors provide information about the direction of required scaling (i.e. scaling up or down): if the origin-to-attractor vector is longer than the origin-to-cyclop vector then this indicates that the pattern in the input-vision is likely to be scaled up relative to the stored-pattern; the reverse case indicates that the input-vision is likely to be scaled down (see Figure 4(e)).[3]

In order to facilitate the analysis of all the information that can be gathered from the above vectors, AFA-raw is divided into three main stages: 1) histogram generation, 2) histogram selection and 3) analysis of the selected histogram.

The histogram generation phase produces three types of histograms: 1) one for cyclop-attractor angles, 2) another one for origin-cyclop-attractor angles and 3) a final one for the relative lengths of origin-to-cyclop and origin-to-attractor vectors. In order to facilitate histogram generation and subsequent analysis, the angular space of correspondences is divided into octants (i.e. eight slices of 45°: see Figure 4). Each octant count is incremented when a correspondence is found that belongs to that octant.

In the second AFA-raw phase, a winning histogram needs to be selected[4] for subsequent analysis. This selection is done by finding the most significant

---

[3] Note that further information can be extracted for more general cases, and that we are confining ourselves to more constrained cases partly because neural architectures that implement the approach are more easily derived, which serves our demonstration purposes better.

[4] Each stored-pattern can lead to a distinct histogram set.

histogram. Basic AFA-raw considers the histogram with the most significant peak to be the most significant histogram, where the most significant peak corresponds to the largest histogram bar in either the cyclop-attractor histogram or the origin-cyclop-attractor histogram.

Once a winning histogram has been chosen, AFA-raw analyzes it in order to compute transformation estimations. Basic AFA-raw is concerned with estimating translation, scale and rotation transformations. This is accomplished by a set of rules and calculations, that attempt to extract and analyze prominent peaks.

4 Performance

Since effective transformation estimations imply effective discriminations and invariances (unless an incorrect classification coincidentally leads to a valid transformation estimation), the presentation is centered primarily on estimations. It is important to distinguish two main types of estimation: 1) the "direction" of the transformation (e.g. shift 45°, scale up and rotate clockwise) and 2) the "amount" of the transformation (e.g. shift 20 pixels, scale to 60 pixels and rotate 110°). Most of the presentation focuses on the most difficult estimation: i.e. amount. When amount-estimation fails, not all is lost, since direction-estimation is still likely to succeed and the resulting information is sufficient to significantly accelerate the speed of latching. All results are based on AFA-raw using three stored-patterns with dimensionality 15x15 (i.e. cat, dog and tree).

The first performance map (see Figure 5) shows how translation estimation varies for different cloud positions, whilst its scale and rotation are fixed with correct[5] values. The main feature to notice is the broad region of significant positional improvement. The white external region is a result of a "refusal to estimate" when AFA-raw has no information to work on: for exploratory clouds a combination of stochastic, history and goal based factors can be used in contradiction to this static response. The map allows one to conclude that AFA-raw can be considerably resilient to translation.

The next pair of performance graphs (see Figure 6) illustrates how AFA-raw performs when the scale of the cloud changes and the position and orientation are fixed and correct. Various observations can be made: 1) most estimations lead to more than 60% improvement, 2) scaling-up (i.e. when clouds are smaller than a target-object) appears to be somewhat more unstable than scaling-down, 3) the scaling-up graph exhibits an interesting pattern of increasing and decreasing improvements vaguely similar to a bell-shape, 4) a small minority of estimations lead to worse scales and these tend to be close to the target scale and 5) direction-estimation succeeds most of the times when amount-estimation underperforms. It should also be noted that when a cloud is significantly scaled-down, insufficient information is available to form reliable estimates, which can be seen at the leftmost portion of the scaling-up graph where several data points are placed

---

[5] The size and orientation of the cloud are the same as the size and orientation of the target-object.

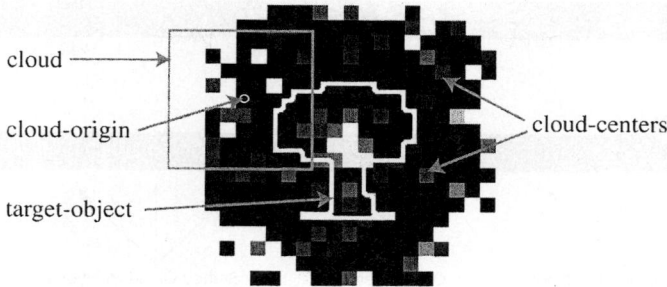

Fig. 5. Illustration of AFA-raw's performance regarding estimations of translation. Each small square represents a different cloud-center. Brightness values represent positional improvement: black represents an improvement of 100%, where a cloud's new position coincides with the target-object's position, white represents no improvement and different gray values represent different degrees of intermediate improvement. The figure has labeled one particular cloud and its respective center, which shows that in spite of possessing limited information (it can "see" only a small part of the left-hand side of the tree) it can still estimate quite accurately.

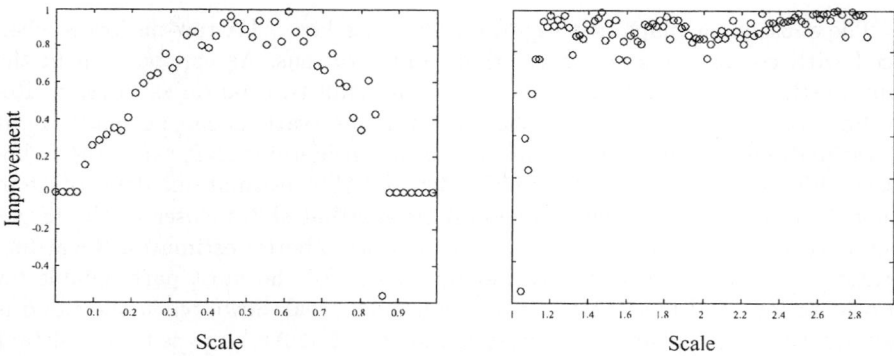

Fig. 6. The effect of a cloud's scale on scale estimation. A tree target-object was used to conduct the tests. The x-axis represents the ratio between the test-cloud's scale and the target-object's scale. When this ratio is smaller than 1 (left-hand graph) this means that the cloud is scaled-down relative to the object and conversely, when this ratio is larger than 1 (right-hand graph) this means that the test-cloud is scaled up relative to the target-object. The y-axis represents a cloud's actual scale improvement relative to its best possible improvement ($actual/best$).

at the 0% improvement level. In spite of the exceptions to perfect estimation encompassed by some of the above observations, the graphs show that AFA-raw, as it is, provides useful scale estimations.

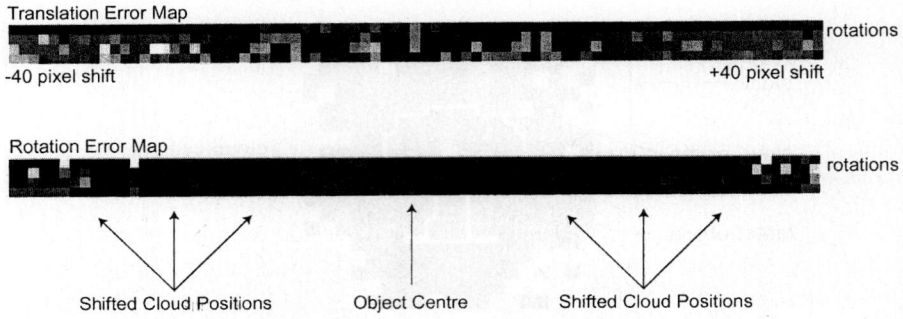

Fig. 7. AFA-raw's performance when faced with combinations of translations and rotations. As before, a tree target-object was used. The top bar represents translation errors while the bottom one represents rotation errors. The x-axis of each bar represents the cloud's horizontal position relative to the target-object (the cloud's vertical position always coincides with the target-object's vertical position). The y-axis of each bar represents cloud orientations: from top to bottom 0°, 90°, 180° and 270°. Pixel brightness is proportional to the error and has been normalized to cover the whole range of grayscale values.

The subsequent maps (see Figure 7) illustrate how AFA-raw performs when faced with combinations of translations and rotations. As can be seen in the figure, estimation of rotation is more resilient than translation estimation. Regarding translation estimations, the following observations can be made: 1) at 0°, estimations are quite robust throughout a considerable shift range (here, between −40 and +40 pixels), 2) at 90°, 180° and 270° estimations deteriorate in general and 3) a darker central region indicates that shifts closer to the target center (regardless of the orientation) tend to lead to better estimates. Regarding rotation estimations, it seems that estimation is for the most part reliable for all orientations and most shifts: it only starts to break down when the cloud is shifted near +/− 40 pixels. These results indicate that AFA-raw is both resistant to rotation by itself and also, albeit less strongly, to combinations of translations and rotations.

The next two maps (see Figure 8) exemplify how AFA-raw performs when faced with combinations of translations and scales. Regarding translation errors, in the left-hand map, the following observations can be made: 1) in general, lower scales allow better estimations and 2) the map exhibits a vague pyramidal shape with larger errors inside and lower errors outside (an equation based on a simple shift/scale ratio should be able to define this shape). In spite of these particularities, the map shows that on average, AFA-raw can estimate translations in spite of scale and translation combinations. In relation to the scale estimation map, the following observations can be made: 1) scale 5 provides insufficient information and thus the error is high across all shifts, 2) errors tend to increase with shift and 3) the map contains a vague circular shape with lower errors inside.

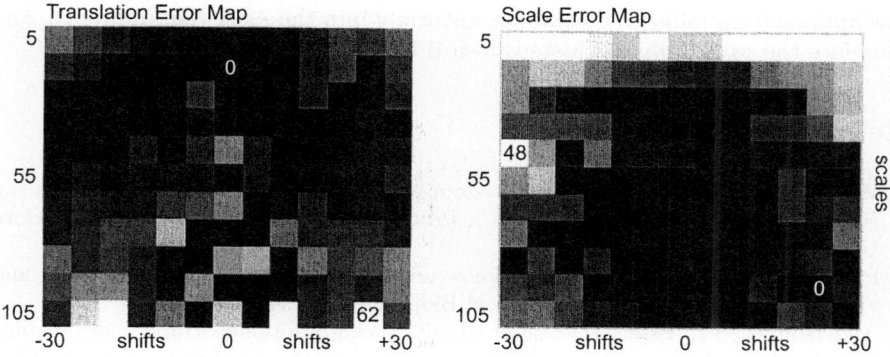

Fig. 8. Illustration of AFA-raw's performance when faced with combinations of translations and scales. The map on the left represents translation errors while the map on the right represents scaling errors. The x-axis represents shifts covering a range from −30 to +30 pixels (with a step of 5) while the y-axis represents different scales starting at 5 at the top and ending at 105 at the bottom (with a step of 10). Note that the target-object's scale is 53 pixels. Brightness values are proportional to the error and have been normalized to the complete range of grayscale values. Each map also depicts its min and max error values (e.g. the max error in the translation map is 62).

Again, on average the map shows that scale estimation is quite resilient in spite of these transformation combinations.

Though less accurately, AFA-raw is still capable of estimating in the case of triplet combinations of translation, scale and rotation.

Though AFA-raw does not estimate jumbling-noise[6], occlusion and clutter variations, it is nevertheless robust to these factors, and it can still perform translation, scale and rotation estimations in spite of them. The same applies to deformation, but less significantly.

5 Conclusion

AFA-raw's main strength lies in its ability to estimate translations, scales and rotations and its robustness to occlusion, clutter and jumbling-noise. A modularity related advantage consists of the possibility of providing AFA-raw with diverse preliminary recognition systems. AFA is also particularly suitable for multiple object tracking (MOT) and image segmentation solutions.

One of AFA-raw's main limitations lies in its weak illumination invariance, which should however be addressable by using illumination invariant features.

It was stated in the introduction that the motivational context for this work lay in the integration of neural normalization and invariant features. The AFA-

---

[6] Jumbling refers to the swapping of nearby pixels.

raw approach explained here extends naturally into the AFA-NIF approach, and therefore the work provides a step toward the above integration.

References

1. Wiskott, L.: How Does Our Visual System Achieve Shift and Size Invariance? in van Hemmen, J.L., Sejnowski, T.J. (eds.): Problems in Systems Neuroscience. Oxford University Press (2004)
2. Pitts, W., McCulloch, W.: How we know universals: the perception of auditory and visual forms. Bulletin of Mathematical Biophysics **9** (1947) 127–147
3. Anderson, C., Van Essen, D.: Shifter circuits: a computational strategy for dynamic aspects of visual processing. Proceedings of the National Academy of Sciences USA **84** (1987) 1148–1167
4. Olshausen, B., Anderson, C., Van Essen, D.: A neurobiological model of visual attention and invariant pattern recognition based on dynamic routing circuits. J. Neuroscience **13-11** (1993) 4700–4719
5. Lades, M., Vorbrüggen, J., Buhmann, J., Lange, J., von der Malsburg, C., Würtz, R., Konen, W.: Distortion invariant object recognition in the dynamic link architecture. IEEE Transactions on Computers **42-3** (1993) 300–311
6. Fukushima, K., Miyake, S., Ito, T.: Neocognitron: A neural network model for a mechanism of visual pattern recognition. IEEE Trans. on Systems, Man, and Cybernetics **13** (2000) 826–834
7. Földiák, P.: Learning invariance from transformation sequences. Neural Computation **3** (1991) 194–200
8. Wiskott, L., Sejnowski, T.: Slow feature analysis: Unsupervised learning of invariances. Neural Computation **14-4** (2002) 715–770

Task-Oriented Sparse Coding Model for Pattern Classification[*]

Qingyong Li[1,2], Dacheng Lin[1,2], and Zhongzhi Shi[1]

[1] Key Laboratory of Intelligent Information Processing, Institute of Computing Technology, Chinese Academy of Sciences, 100080, Beijing, China
{liqy, lindc, shizz}@ics.ict.ac.cn
[2] Graduate School of the Chinese Academy of Sciences, 100039, Beijing, China

Abstract. Although the basic sparse coding model has been quite successful at explaining the receptive fields of simple cells in V1, it ignores an important constrain: perception task. We put forward a novel sparse coding model, called task-oriented sparse coding (**TOSC**) model, combining the discriminability constrain supervised by classification task, besides the sparseness criteria. Simulation experiments are performed using real images including class of scene and class of building. The results show that **TOSC** can organize some significant receptive fields with distinct topological structure which will favor the classification task. Moreover, the coefficients of **TOSC** notably improve the classification accuracy, from the 53.5% of pixel-based model to 86.7%, in the case of none distinct damage on the performance of reconstruction error and sparseness. **TOSC** model, complementing the feedback sparse coding model, is more consistent with biological mechanism, and shows good potential in the feature extraction for pattern classification.

1 Introduction

At any given moment, our visual systems are receiving vast amounts of information about the environment in the form of light intensities. How the brain makes sense of this flood of time-varying information and forms useful internal representations for mediating behavior remains one of the outstanding mysteries in neuroscience. In recent years, a combination of experimental, computational, and theoretical studies have pointed to the existence of a common underlying principle involved in sensory information processing, commonly referred to as 'efficient coding' or 'sparse coding', namely that information is represented by a relatively small number of simultaneously active neurons out of a large population.

Aiming to reveal underlying relationships between the environmental information and internal representation of the visual cortex, Olshausen and Field [1] introduced a learning algorithm for sparse coding. It was shown that, by seeking sparse code for natural images, the network could develop a set of receptive

---

[*] This paper is supported by National Natural Science Foundation of China No. 60435010 and National Basic Research Priorities Programme No. 2003CB317004.

fields similar to those simple cells found in the striate cortex (V1). Though such network successfully captures sparse nature of the input data, it is a feedforward model and it ignores one primary constraint on the visual system: task. The tasks faced by the organism are likely to be an even more important constraint. That is to say, sparse coding states only that information must be represented sparsely, it does not say anything about what information should be represented.

Many biological evidences showed that there are many recurrent horizontal connections from higher areas to the primary visual cortex, i.e, Top-down feedback commonly exists in the visual system. Vinje and Gallant [2,3] have demonstrated that neurons in V1 produce sparse responses when stimulated with natural image sequences. Moreover, when the same neurons are stimulated only within their classical receptive fields, the responses are much more dense or evenly distributed over time. Thus, it would appear that context sparsifies the responses of V1 neurons. Martin [4] argued that cortical responses should be determined mostly by cortical input (feeding back from the higher area or the same level) and not by thalamic inputs (representing the input stimuli). Furthermore, Douglas [5] put forward a recurrent model. According to Barlow's theory [6]: the neurons in the higher area always attempt to form neural representations with higher degrees of specificity, so we can reason that it is necessary to form sparse code with feedback of task.

In this paper, we investigate a new supervised model, called here task-oriented sparse coding (briefly as **TOSC**) model, which evolves the information representation being consistent with the biological finding and valuable in practice. Based on Olshausen's sparse coding scheme [1], **TOSC** model incorporates an additional constraint, named discriminability, so as to form the neural representation which is valuable for pattern classification. The main contribution of **TOSC** is that it probes into what information should be represented under feedback of perception tasks, correspondingly, efficient coding model just states that information must be represented efficiently.

In the following section, we explain the details of our **TOSC** model. Section 3 shows the simulation results obtained by the present model using real natural images. The related findings of **TOSC** model are discussed in section 4. At last, section 5 concludes the present work.

2   Task-Oriented Sparse Coding Model

2.1   Linear Image Synthesis Model

The starting point is from Olshausen [1]. A perceptual system is exposed to a series of small image patches, drawn from one or more large images, just like the classical receptive field(CRF) of neurons. Imagine that each image patch, represented by the vector I (numbered row-wise), has been formed by the linear combination of N basis functions. The basis functions form the columns of a fixed matrix, A. The weighting of this linear combination is given by a vector, s. Each component of this vector has its own associated basis function, and represents a response value of a neuron in vision system. The linear synthesis model is therefore given by:

$$I(x,y) = As = \sum_{i=1}^{\infty} s_i * a_i(x,y) \tag{1}$$

In a cortical interpretation, the s model the responses of (signed) simple cells, and the column of matrix A closely related to their CRF's.

This model can be represented by a simple neural network, where x is an n-dimensional vector denoting the input to the network, s_i denotes the activity of the i-th neuron, and a_i (the i-th column of the A) is an n-dimensional vector composed of the connection weights between the i-th neuron and the input.

The goal of sparse coding is to find a set of a_i that forms a complete code (that is, spans the image space) and results in coefficients being as statistically independent as possible over an ensemble of natural images. The reason for statistical independence have been elaborated else where [9], but it can be summarized briefly as providing a strategy for extracting the intrinsic structure in visual signal.

2.2 Sparse Coding Model

In an influential paper, Olshausen and Field [1] applied two criteria to seek the optimal basis vector and the coefficients. One of the criteria is how well the code describes the input. It can be measured by the squared error between the input and its reconstruction by the network:

$$Error(x,y) = \sum_{x,y}[I(x,y) - \sum_i s_i a_i(x,y)]^2 \tag{2}$$

As an additional criteria for sparse coding, Olshausen and Field proposed the 'sparseness' cost for seeking sparse codes. The sparseness cost function is given by

$$Sparseness(s,A) = \sum_i S(\frac{s_i}{\sigma_i}) \tag{3}$$

where $S(x)$ is a nonlinear function such as $|x|$, $\exp(-x^2)$, and $\log(1+x^2)$. The cost sparseness favors the codes which consist of minimal number of non-zero coefficients. As a result, the network seeks the coefficients which are statistically independent each other over an ensemble of input data. In the case that the data contains some forms of higher-order statistical structure as found in natural images, it can be captured by using this sparseness cost function.

So the search for a sparse code can be formulated as an optimization problem by constructing the following cost function to be minimized:

$$E(s,A) = \sum_{x,y}[I(x,y) - \sum_i s_i a_i(x,y)]^2 + \lambda_s \sum_i S(\frac{s_i}{\sigma_i}) \tag{4}$$

2.3 Discriminability Constrain

In order to code the sensory visual information supervised by the pattern classification task, it is necessary to incorporate a constraint for the classification task. Intuitively, it is very important for the coded coefficients to be good for classification, so the coefficients (or neuron responses) produced by the sparse coding model can be easily utilized by the higher neurons which process such task. Linear discriminant analysis, using within-class scatter and between-class scatter to choose coordinate for transformation, is broadly used for pattern classification. We investigate a somewhat similar approach, and incorporate the 'discriminability' cost function which constrains the neuron activities to be more valuable for classification.

Supposed that $X_1 = \{I_1^1, I_2^1, \ldots, I_{N1}^1\}$ and $X_2 = \{I_1^2, I_2^2, \ldots, I_{N2}^2\}$ represent the pattern sets, here we only consider the two-class classification. And $I_i^j = [s_1, s_2, \ldots, s_n]$, where $s_k(1 \leq k \leq n)$ is the coefficient produced by sparse coding model. N_1 and N_2 are the number of patterns in class X_1 and X_2.

The sparse coding model transforms the input stimuli into code coefficient vectors, I. We define the distance between two coefficient vectors as below:

$$D(I_1, I_2) = \sqrt{\sum_{i=1}^{n}(s_i^1 - s_i^2)^2} \qquad (5)$$

Within-class distance measures the distance between a coefficient vector and the center of class which includes the vector. The formula is

$$D_W = D(I_i^j, \tilde{m}_j) \qquad (6)$$

where $\tilde{m}_j$ is the center of the class j. On the contrary, between-class distance measures the distance between a coefficient vector and the center of class which excludes the vector. The equation is

$$D_B = D(I_i^j, \tilde{m}_{\bar{j}}) \qquad (7)$$

where $\tilde{m}_{\bar{j}}$ represent the center of the excluding class.

In order to make the patterns be correctly classified, we expect: 1) the within-class distance is smaller, so the class is more compact in the N-dimensional coefficient space; 2) the between-class distance is greater, thus the interval between the class 1 and class 2 is bigger. That is to say, we should maximize the between-class distance in the same time minimize the within-class distance. So we make a tradeoff to optimize a ratio. The ratio is given by

$$DR = \frac{D_W}{D_B} \qquad (8)$$

When we look into the Eq.8 we can find that its derivative for coefficient s is too complex to optimize. So we smartly transform the magnitude of the ratio by logarithm. The transformed ratio is as below:

$$Dis(s) = \ln(DR^2) = \ln\Big(\sum_{i=1}^{n}(s_i^j - \tilde{m}_j)^2\Big) - \ln\Big(\sum_{i=1}^{n}(s_i^j - \tilde{m}_{\bar{j}})^2\Big) \quad (9)$$

As a result, the model produces an N-dimensional coefficient space in which the coefficients in the same class tightly locate in a subspace and are apart from the other class.

2.4 Learning

Learning is accomplished by minimizing the total cost function:

$$E(s, A) = Error(s, A) + \lambda_s Sparseness(s, A) + \lambda_d Dis(s) \quad (10)$$

where λ_s and λ_d are positive weights. The function to be minimized, $E(s, A)$, is the sum of three terms: the first term computes the reconstruction error, which forces the basis functions, A, to span the input space; the second term incurs a penalty on the coefficient activities, which encourages sparse representation; and the third term calculates the discriminability which drives the coefficients to be more efficient for pattern classification.

The process for minimizing $E(s, A)$ can be divided into two nested stages. In the inner stage, E is minimized with respect to the s_i for a batch of pattern, holding the A fixed. In the outer stage (i.e, on a long timescale, over many image presentations), E is minimized with respect to the A. The inner stage minimization over the s_i can be performed by conjugate gradient method, so the s_i are determined by the differential equation:

$$\frac{\partial E}{\partial s_i} = -2b_i + \frac{\lambda_s}{\sigma}S'\Big(\frac{s_i}{\sigma}\Big) + \lambda_d\Big(\frac{2(s_i^j - \tilde{m}_j)}{\sum_{i=1}^{n}(s_i^j - \tilde{m}_j)^2} - \frac{2(s_i^j - \tilde{m}_{\bar{j}})}{\sum_{i=1}^{n}(s_i^j - \tilde{m}_{\bar{j}})^2}\Big) \quad (11)$$

where $b_i = \sum_{x,y}\big(I(x,y) - \sum_j s_j a_i(x,y)\big)a_i(x,y)$. According to Eq.11, the s_i are drived by a sum of three terms. The first term takes a spatially weighted sum of the current residual image using the basis function $a_i(x,y)$ as the weights. The second term applies the derivative of sparseness. Especially, the third term incurs a movement which makes the s_i near the center of the included class and apart from the excluded class.

The outer stage minimization over the A may be finished by simple gradient descent method. The learning rule for it is given by

$$\Delta a_i(x,y) = \eta\{s_i[I(x,y) - \sum_j s_j a_j(x,y)]\} \quad (12)$$

where η is the learning rate. In the neural network view, a_i are updated by Hebbian learing between the outputs coefficients, s_j, and the resulting residual image.

Because there is no closed-form solution for the s_i in terms of the input $I(x, y)$, so s_i is calculated by recurrent computation similar to an *analysis/synthesis* loop [7]. An intuitive interpretation for this algorithm is that in the inner stage, the gradient of '*sparseness*' sparsifies the distribution of s by differentially reducing the value of small coefficients more than great coefficient, at the same time, the gradient of '*discriminability*' makes the coefficient near to homogeneous center and apart from the unhomogeneous center. Then, the a_i learn on the error induced by the sparseness criteria and discriminability criteria, resulting in a basis function set which can tolerate sparseness and discriminability in the condition of minimizing mean square reconstruction error.

3 Simulation

In order to confirm that the model is capable of coding the input pattern with good discriminability for classification task, besides sparseness. We tested it on a number of natural images including two classes: scene and building, shown in Fig.1. There are one hundred scene images and one hundred building images, with seventy for training and thirty for testing.

Fig. 1. Samples of the images used in the simulation. (a) scene images; (b) building images.

3.1 Preprocess and Experimental Conditions

The natural images are color images, because here the model just focuses on the gray images, so we first change the color images into gray images.

These data in the raw form will pose potential problems. First, the variance along the low-frequency eigenvectors will be much lager than the variance along the high-frequency eigenvectors. It will be troublesome for gradient descent techniques searching for coefficient in this space. Second, the highest spatial frequencies in most images will typically be corrupted by noise. Furthermore, the energy present in the corners of the $2D$ frequency domain is an artifact of working on a rectangular sampling lattice, because there is a higher sampling density along the diagonals than along the vertical or horizontal. For these reasons, we sphere the data by equalizing the variance in all directions [8], accomplished by filtering with a circularly symmetric 'whitening filter' with frequency response,

$W(f) = f$, thereby attenuating the low frequencies and boosting the high frequencies so as to yield a roughly flat amplitude spectrum across all spatial-frequencies. Then, a lowpass filter is used to cut out the energy at the highest spatial frequencies and also in the corners of the $2D$ Fourier plane. At last, the combined whitening/lowpass used to preprocess the data is given by

$$R(f) = W(f)L(f) = fe^{-(f/f_0)} \tag{13}$$

where $f_0 = 200$ cycles/picture and n is experimented with 4. Atick [9] have shown that such a whitening filter corresponds well to the response properties of retinal ganglion cells.

Training data were obtained by extracting $12 * 12$ image patches at random from the preprocessed images. The network was trained so as to acquire efficient codes for above data by minimizing the cost Eq.10. The parameters s and d were set to 0.01 and 0.5, respectively, and $s(x) = \log(1 + x^2)$ was chosen as the function for Eq.3.

A set of 144 basis functions was initialized to random values and was updated according to Eq.12. The learning rate η was gradually decreased during learning with a initial setting of 3.0. In order to speed up the learning, we first calculated the sparse codes as Eq.4 for the images set of scene and building, respectively, and initialized the center of the scene and building with the mean of the sparse codes of scene and building.

3.2 Results

Basis functions. A stable solution was usually arrived at after about 10,000 updates. The result is shown in Fig.2. The vast majority of basis functions have become well localized, oriented and broken into different spatial-frequency bands, just like the functions of SC model. However, there are quite a few basis functions which have geometric structure, usually appearing in the building images, such as horizontal line, vertical line. The functions labeled with arrow shows some typical examples. This result should not come as a surprise, because the cost function of **TOSC** model includes sparseness criteria, so it can get most sparseness characteristics. Furthermore, the cost function also combines the discriminability criteria, which shifts the responsibility for coding such structure that is most discriminable from this class to the other class. For example, the horizontal structure is one of the most distinct feature in the building image, and it is another case for the scene image, so it is not accidental for the geometric-like basis functions.

It should be noted that the preprocessing steps mentioned above do not affect the overall, qualitative appearance of the basis function (i.e., localized, oriented, and geometric-like functions). It just decreases the time required for learning.

Sparseness and reconstruction error. Sparse representation is a ubiquitous and most important property of primary sensory cortical areas. It produces a more simple flattened representation of the curved manifold structure of data,

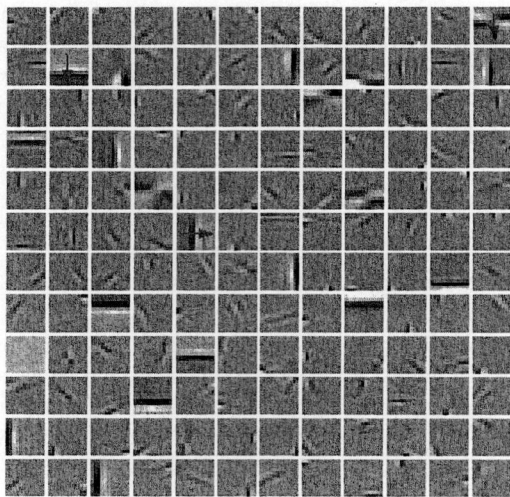

Fig. 2. The set of 144 basis functions learned by the TOSC. All have been normalized to fill the grey scale, but with zero always represented by the same grey level. The functions labeled by arrow are typical geometric-like structure.

furthermore, it is energy efficient [11,12]. In this section we measure the sparseness of the **TOSC** model.

Fig.3 demonstrates the output coefficients computed by efficient models to a given input pattern. Though the sparseness of **TOSC** model is somewhat worse than **SC** model, it is notablely better than the pixel-based model. That is to say, **TOSC** preserves the sparseness characteristic. We can have a deeper look into the sparseness values calculated by the Eq.3 from Table.1. We can readily get that the sparseness cost rate is about 10%, and the maximum and minimum don't have great change. So we can conclude that **TOSC** model doesn't greatly destroy the sparseness of the neural representation.

The reconstruction error is measured by the squared error between the input and its reconstruction by the efficient codes, as the Eq.2. Because we combine the sparse coding model with additional constrain, discriminability, simulating the feedback in the visual neural system, we compare our task-oriented sparse coding model (**TOSC**) with the sparse coding model (**SC**) about reconstruction error.

Table 1. Sparseness value of the sparse coding model (**SC**) and task-oriented sparse coding model (**TOSC**)

	Mean sparseness	Maximum sparseness	Minimum sparseness
TOSC	10.4504	29.1280	0.1003
SC	9.3770	28.2394	0.0704

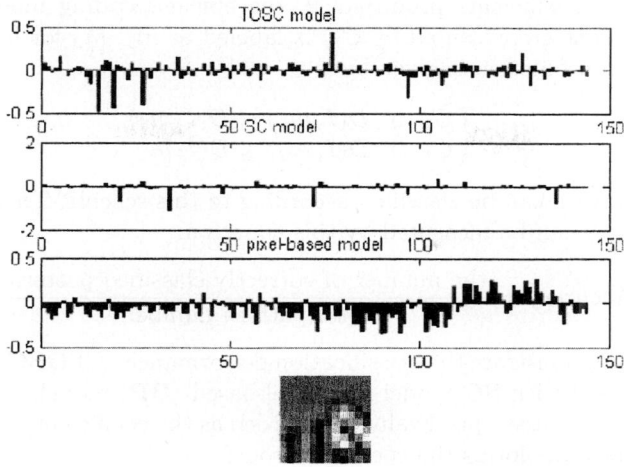

Fig. 3. Outputs computed in response to a given input pattern. Bottom plot is the raw pixel values; middle plot is the coefficients of the SC model; and the top one is the coefficients of TOSC model.

Table 2. Reconstruction errors of the sparse coding model (**SC**) and task-oriented sparse coding model ((**TOSC**)

	Mean error	Maximum error	Minimum error
TOSC	0.1085	0.3413	0.0040
SC	0.1002	0.3209	0.0032

Table.2 shows that **TOSC** has a small increase for the reconstruction error, but the increase extent is very little, for example, the increasing rate for mean error is just 8.3%. The loss of reconstruction error can be well interpreted: in the cost function of Eq.10, the additional term of *'discriminability'* certainly influences the reconstruction error compared with SC without this term. In practice, it is acceptable if the reconstruction is under certain tolerance. After all, the purpose of efficient coding is not for optimal compression.

Classification accuracy. We examine the information representation of the model in term of the discrminability in this section. 100 image patches (12 ∗ 12) were chosen from the test images for scene and building, respectively. Note that because the information in the center of image is always representative and important, the tested image patches are extracted from the center part.

Discriminability is the most characteristic differing **TOSC** model from SC and other efficient coding model. The classification scheme experimented here is similar to Bayesian decision which is extensively used for classification. Supposed that every tested pattern was represented by $I = [s_1, s_2, \ldots, s_n]$, where $s_k (1 \leq$

$k \leq n$) are the coefficients produced by the efficient coding model. And the center of the scene (represented by C_1) is labeled as m_1, m_2 for building (C_2). The classification function is given by

$$I \in \begin{cases} C_1 & \text{if } D(I, m_1) \leq D(I, m_2) \\ C_1 & \text{if } D(I, m_2) \leq D(I, m_1) \end{cases} \quad (14)$$

so every pattern can be classified according to this scheme. Then, the classification accuracy can be measured by this equation:

$$\text{Accuracy} = \frac{\text{the number of correctly classified pattern}}{\text{total pattern number}} \quad (15)$$

In order to demonstrate the classification performance of **TOSC**, we compare the **TOSC** model with **SC** model and pixel-based (**BP**) model. In pixel-based model every preprocessed pixel value is regarded as the coefficient, in other word, every raw pixel value forms the coefficient code.

Table.3 demonstrates that the classification performance of **BP** model is a little better than random selection, which has 50% probability. SC model improves the classification performance compared with **BP** model, because SC model broadly tunes to some stimulus dimensions (e.g., spatial-frequency), or other local feature (e.g., position and orientation), obviously, SC coefficients capture some sparse image structure and have a good discriminability in some extent compared with the raw data. Excitingly, **TOSC** model notably enhances the classification accuracy. It implies that the discriminability term in the cost function guides the codes for the stimulus dimensions which have good discriminability, besides the sparseness. It seems to be qualitatively consistent with the physiological finding [2,3,4].

Table 3. Comparison of the classification accuracy

	TOSE	SC	PB
Accuracy	86.7%	69.8%	53.5%

4 Discussion

A substantial literature exists on the efficient coding of visual sensory information [1,14], which bridges the statistics of the natural images and the neural representation in the primary visual cortex. Most of them are feedforward Hebbian/anti-Hebbian algorithms based on the idea of finding independent components or sparse structures. Interestingly, they can self-organize the visual receptive fields, such as simple cell in V1, and have the same selection for location, orientation and spatial frequency. However, the transformation from retina to V1 is clearly much more complex than the efficient models, and it involves a back-propagation of information from, or within, the output layer or higher layer [13].

The visual cortex, like any other part of the neocortex, is primarily a two-dimensional sheet of neurons and connections. At any location on the cortical sheet, there are many lateral connections from the same sheet or upper sheet [15].

Lateral connections may play a direct role in forming visual representation. Miikkulainen [16] showed that the inhibitory lateral connections encode the correlation of activity in the map and perform redundancy reduction. Kurtosis measures of the activities before and after lateral interaction showed that the settled activity after lateral interaction is sparser than before the lateral interaction. The lateral connections facilitate the feature extraction and binding.

It is natural to try to imagine a mechanism capable of performing such a back-propagation, simulating the lateral connections, so the neural representation can not only reflect the statistics of the natural images which represent the input stimuli, but also the cortical feedback which always represents the higher perception tasks or prior knowledge. However, since it is difficult to identify the parameters of our efficient coding models with 'true' biophysical parameters, we prefer to imagine that potentially real biophysical processes occur in local spatial media where the feedforward and the feedback of information are tightly functionally coupled, and where some microscopic and dynamic analogue of Eq.11 may operate. In **TOSC** model, the reconstruction error and sparseness in Eq.10 are the typical feedforward information, and the term of discriminability is the feedback mechanism which is controlled by the perception task.

5 Conclusion

We have put forward a novel sparse coding model, called task-oriented sparse coding (**TOSC**) model, based on the notion that visual cortex is trying to produce an efficient representation, in terms of extracting the structure with good discriminability, besides the sparse structure in the stimuli. The receptive fields that emerge from **TOSC** have some topological structures representing the most distinct features of certain class, which is consistent and valuable for the classification task. Furthermore, the coefficients of **TOSC** haven't distinctly damaged the performance of reconstruction error and sparseness, but notably improved the classification accuracy, from the 53.5% of pixel-based to 86.7% of our **TOSC**. To our knowledge, **TOSC** is the first efficient coding model to produce the neural representation acted by the classification task. It has a good potential in feature extraction for pattern classification.

References

1. Bruno A. Olshausen and David J. Field: Emergence of simple-cell receptive field properties by learning a sparse code for natural images. Nature. 381 (1996) 131-133.
2. Vinje W.E, Gallant J.L: Sparse coding and decorrelation in primary visual cortex during natural vision. Science. 287 (2000) 1273-1276.

3. Vinje W.E, Gallant J.L: Natural stimulation of the nonclassical receptive field increases information transmission efficiency in V1. Journal Neuronscience. 22 (2002) 2904-2915.
4. Martin Kac: Microciruits in visual cortex. Current Opin Neuronbiol. 12 (2002) 418-425,
5. Douglas R.J, Koch C, Mahowald M, Martin K.A, Suarez H.H: Recurrent excitation in neocortical circuits. Science. 269 (1995) 981-985.
6. Barlow H.B: Single units and sensation: a neuron doctrine for perceptual psychology? Perception, 1 (1972) 371-394.
7. Mumford D: Neuronal architectures for pattern-theoretic problems. In: Large scale neuronal theories of the brain, Koch C, Davis,J.L,eds., MIT Press. (1997) 125-152.
8. Friedman J.H: Exploratory projection pursuit. Journal of the American Statistical Association. 82 (1987) 249-266.
9. Atick J.J, Redlich A.N: Towards a theory of early visual processing. Neural Computation. 2 (1990) 308-320.
10. Olshausen B.A, Field D.J: Sparse coding with an overcomplete basis set: A strategy employed by V1? Vision Research. 37 (1997) 3311-3325.
11. Olshausen B.A, David J.F: Sparse coding of sensory inputs. Current Opinion in Neurobilogy. 14 (2004) 481-487.
12. Levy W.B, Baxter R.A: Energy efficient neural codes. Neural Computation. , 8 (1996) 531-543.
13. Atick, J.J, Redlich, A.N: Convergent algorithm for sensory receptive field development. Neural Computation. 5 (1993) 45-60.
14. Anthony J.Bell, Terrence J. Sejnowski: The "Independent components" of natural scenes are edge filters. Vision Research. 37 (1997) 3327-3338.
15. Gilbert, C.D, and Wiesel, T.N: Morphology and intracortical projections of functionally identified neurons in cat visual cortex. Nature. 280 (1979) 120-125.
16. Miikkulainen, R, and Sirosh,j: Introduction: the emerging understanding of lateral interactions in the cortex. Lateral interaction in the cortex: structure and function. Electronic book, ISBN 0-9647060-0-8.

Robust Face Recognition from One Training Sample per Person

Weihong Deng, Jiani Hu, and Jun Guo

Beijing University of Posts and Telecommunications, 100876, Beijing, China
{cvpr_dwh, cughu}@126.com, junguo@bupt.edu.cn

Abstract. This paper proposes a Gabor-based PCA method using Whiten Cosine Similarity Measure (WCSM) for Face Recognition from One training Sample per Person. Gabor wavelet representation of face images first derives desirable features, which is robust to the variations due to illumination, facial expression changes. PCA is then employed to reduce the dimensionality of the Gabor features. Whiten Cosine Similarity Measure is finally proposed for classification to integrate the virtues of the whiten translation and the cosine similarity measure. The effectiveness and robustness of the proposed method are successfully tested on CAS-PEAL dataset using one training sample per person, which contains 6609 frontal images of 1040 subjects. The performance enhancement power of the Gabor-based PCA feature and WCSM is shown in term of comparative performance against PCA feature, Mahalanobis distance and Euclidean distance. In particular, the proposed method achieves much higher accuracy than the standard Eigenface technique in our large-scale experiment.

1 Introduction

The importance of research on face recognition is fueled by both its scientific challenges and its potential applications. Face recognition is a challenging problem as there are numerous varying factors such as illumination conditions, facial expression, aging, accessory, capture devices, etc., which affect the appearance of an individual's facial features. The typical approach in handling these variations is to use large and representative training sample sets. However, many real-life face recognition applications could only offer one training sample per person. In such situation, the performance of most of the face recognition methods would be degraded dramatically.

In the past, several studies were performed to address the problem of one training sample per class. Wu et al. [1] proposed a $(PC)^2A$ method, which is an extension of the Eigenface Technique. It performs Principal Component Analysis (PCA) on horizontal and vertical projection images. Martinez [2] proposed a local Eigenface-based method for face recognition with a single sample per class. Huang et al. [3] used a component-based Linear Discriminant Analysis (LDA) method to solve the one training sample problem. Chen et al. [4] proposed an Adaptive Principal Component Analysis (APCA) method, which first applies PCA to construct a subspace for face representation; then warps the subspace according to the within-class covariance and between-class covariance of sample to improve class separability.

Instead of focusing on the feature extraction and learning stage as above methods, our works attach importance to the image representation and the similarity measures of the face recognition system. This paper presents a Whiten Cosine Similarity Measure (WCSM) for Gabor-based PCA method to enhanced face recognition performance. To demonstrate the effectiveness and robustness of our proposed method, we have employed a rigid evaluation methodology: training on a natural gallery image and testing on various probe images across illumination conditions, directions of illumination, facial expressions, and accessories. The superiority of the Gabor-Based PCA features and WCSM has been successfully demonstrated through the test on the CAS-PEAL face database [5] with 1040 subjects, by comparing with PCA features, Euclidean distance and Mahalanobis distance.

2   Gabor-Based PCA Using Whiten Cosine Similarity Measure

This section details the Gabor-based PCA method using Whiten Cosine Similarity Measure. First, Gabor wavelet representation of face images derives desirable features, which is robust to variations due to illumination, facial expression changes [10], and high frequency noise. Second, PCA works on the Gabor wavelet representation and derives low-dimension discriminating features using one natural sample per class. Finally, Whiten Cosine Similarity Measure is proposed for the classification stage to integrate the virtues of the whiten translation and the cosine similarity measure.

2.1   Gabor Features for Face Representation

2D Gabor wavelet representation in computer vision was first utilized by Daugman in 1980s [7]. Recently, the Gabor wavelet representation becomes popular in face recognition community for its significantly superior performance over intensity image representation [8][9][10][11].

To extract information about facial appearance, the face image was convolved with a multiple spatial resolution, multiple orientation set of Gabor filters. Gabor filter can capture salient visual properties such as spatial localization, orientation selectivity, and spatial frequency characteristics. Specially, Gabor filters are defined as follow:

$$\psi_{\mu,\upsilon}(z) = \frac{\|k_{\mu,\upsilon}\|^2}{\sigma^2} e^{-\left(\|k_{\mu,\upsilon}\|^2 \|z\|^2 / 2\sigma^2\right)} \left[e^{ik_{\mu,\upsilon} z} - e^{-\sigma^2/2} \right] \tag{1}$$

where $k_{u,v} = k_v e^{i\phi_\mu}$, $k_v = k_{max}/f^v$ gives the frequency, $\phi_\mu = \mu\pi/8$, $\phi_\mu \in [0,\pi)$ gives the orientation, and $z = (x, y)$. Note that, in equation (1), v controls the scale of the Gabor filters, which mainly determines the center of the Gabor filter in the frequency domain; μ controls the orientation of the Gabor filters. This can be observed intuitively from the visualization of the real part of the Gabor filters. The parameters for the Gabor filters are as follows: $\sigma = 2\pi$, $k_{max} = \pi/2$, $f = \sqrt{2}$, five scales $v \in \{0,1,2,3,4\}$ and eight orientations $\mu \in \{0,1,2,3,4,5,6,7\}$. These Gabor kernels form a bank of 40 different filters and exhibit desirable characteristics of spatial frequency, spatial locality, and orientation selectivity.

In our experiments, the Gabor representation for face images is derived as follow:

1) Masked images are derived by first using the centers of two eyes as control points for alignment, and then masking them then to yield 65×75 images. 2) Histogram equalization is then used to smooth the distribution of grey values for the non-masked pixels, making the masked images insensitive to overall level of illumination conditions. Examples of original image and normalized masked image are shown in Fig.1. 3) The normalized masked image is convolved with the 40 Gabor filters and the magnitudes of the complex-value filter responses are sampled at 156 points on settled facial grid and combined into a 6240 dimension Gabor-based feature vector to form the Gabor representation for a face image.

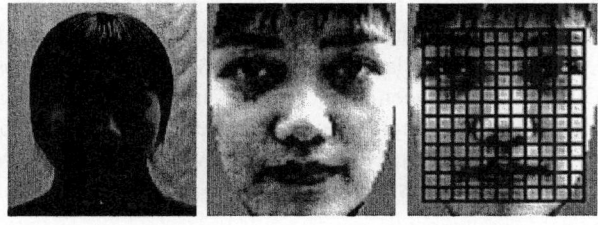

Fig. 1. Example of facial image during the derivation of Gabor representation. Left: the original image. Middle: the normalized masked image. Right: the settled grid for sampling Gabor features.

2.2 PCA for Discriminating Gabor-Based Feature Exaction in "One Training Sample per Class" Scenario

PCA generates a set of orthonormal basis vectors, known as principal components, which maximize the scatter of all the training samples. Let $\mathbf{X} = [x_1, x_2, ..., x_M]$ be the normalized representation set of the original images with unity norm and zero-mean. Each x_i represents a normalized vector with dimensionality N, $x_i = (x_{i_1}, x_{i_2}, ..., x_{i_N})^T$, $(i = 1, 2, ..., M)$. The covariance matrix of the normalized representation set, or total scatter matrix, is defined as

$$\Sigma_X = \frac{1}{M} \sum_{i=1}^{M} x_i x_i^T = \frac{1}{M} \mathbf{X}\mathbf{X}^T \quad (2)$$

and the eigenvector and eigenvalue matrices $\mathbf{\Phi}$, $\mathbf{\Lambda}$ are computed as:

$$\Sigma_X \mathbf{\Phi} = \mathbf{\Phi}\mathbf{\Lambda} \quad (3)$$

where $\Sigma_X \in \Re^{N \times N}$. The first m leading eigenvectors define a matrix $\mathbf{U}_{PCA}$

$$\mathbf{U}_{PCA} = [\Phi_1, \Phi_2, ..., \Phi_m] \quad (4)$$

For a normalized test pattern x, $x \in \Re^N$, the PCA features of x are derived as follow:

$$u = \mathbf{U}_{PCA}^T x \quad (5)$$

The PCA subspace, $\mathbf{U}_{PCA}$, characterizes the distribution of face difference between any two face images in the training set [15]. Therefore, in the usual scenario, in which the training set has multiple samples per class, the projection matrix $\mathbf{U}_{PCA}$ will capture the variation from illumination, expression, and accessory change. Consequently, the points projected in the PCA subspace will not be clustered well. A straightforward way to solve the above problem is to maximize the between-class scatter while minimizing the within-class scatter. Fisherface method [12] is proposed based on this idea.

However, there is only one training sample per class available in our experiments, Fisherface method cannot work. Fortunately, all training face images are captured in the natural expression and illumination under controlled situation, PCA does not suffer from the within-class translation any more. In this scenario, most the variation retained by PCA is the between-class scatter, indicating PCA can extract the discriminating features. Accordingly, we could expect an acceptable performance on PCA-based face recognition in this scenario when using appropriate image representation and similarity measure.

2.3 Whiten Cosine Similarity Measure for Classification

When a face image is presented to the Gabor-based PCA classifier, the augmented Gabor feature vector of the image is first calculated as detailed in Section 2.1, and the low dimensional Gabor-based PCA features, u, is derived using (5). The next step is the whitening transformation and it counteracts the fact that Mean-Square-Error (MSE) principle underlying PCA preferentially weights low frequencies. u is subject to the whitening transformation and yields yet another feature set :

$$w = \Gamma u \quad (6)$$

where $\Gamma = diag\{\lambda_1^{-1/2}, \lambda_2^{-1/2}, ..., \lambda_m^{-1/2}\}$ and $\Gamma \in \Re^{m \times m}$.

The high dimensional components are equalized by the whitening transformation. Therefore, the classifier can make full use of the discriminating power of whole feature space, instead of preferring the leading components with large eigenvalue. However, noise contained in the high frequency components is magnified synchronously, which will deteriorate the recognition results. To compensate for the magnified whiten noise, the cosine similarity measure is employed, due to its rotation and dilation invariance properties. We define the cosine similarity measure combined with the whiten transformation as Whiten Cosine Similarity Measure (WCSM):

$$\delta_{WCSM}(u_1,u_2) = \frac{-w_1^T w_2}{\|w_1\| \cdot \|w_2\|} = \frac{-u_1^T \Gamma^2 u_2}{\|\Gamma u_1\| \cdot \|\Gamma u_2\|} \qquad (7)$$

Let $u_k^0, k=1,2,...,L$, be the prototype for class ω_k after the PCA projection. The classifier applies, then, the nearest neighbor rule for classification using Whiten Cosine Similarity Measure δ_{WCSM}

$$\delta_{WCSM}(u, u_k^0) = \min_j \delta_{WCSM}(u, u_j^0) \rightarrow u \in \omega_k \qquad (8)$$

The image feature vector, u, is classified as belonging to the class of the closest prototype, u_k^0, using δ_{WCSM}.

In addition, The Euclidean distance and Mahalanobis distance measure are employed to evaluate the efficiency of WCSM combined with Gabor-based PCA features. The Euclidean distance is the most popular similarity measure for subspace face recognition, since most researches use it to evaluate the algorithm; Mahalanobis distance measure is also sometime used, and some researchers report that it outperform Euclidean distance [10] [11]. The Mahalanobis distance is measured with respect to a common covariance matrix for all classes in order to treat variations along all axes as equally significant by giving more weight to components corresponding to smaller eigenvalues [12]. Note that the weighting procedure of the Mahalanobis distance is performed by the whitening transform. Therefore, the Mahalanobis distance would suffer from the magnified whitening noise.

3 Experiments and Analysis

We assessed the feasibility and performance enhancement power of the WCSM for Gabor-Based PCA method on face recognition task, using a data set from CAS-PEAL database [5], which contains 99,594 images (30,871 images is publicly available) of 1040 individuals with varying Pose, Expression, Accessory, Lighting (PEAL), and aging. Specially, we used a subset of 6609 frontal images in CAS-PEAL database corresponding to 1040 subjects, which is used to form a gallery set and two probe sets.

3.1 Evaluation Methodology and Dataset

Many face recognition algorithms have been developed and excellent performances are also reported when sufficient number of representative is available. The origin idea of these algorithms is to make the within-class variance in the training set invariant to recognition. However, if the training set cannot cover all the variance in the probe set, the effort to suppress the within-class variance might, on the contrary, deteriorate the performance, which might be always the case in the real-life applications. Consequently, we consider evaluation the algorithm using one training sample per person is meaningful for the face recognition community. To further avoid the sample selection problem, we use the natural image, which is acquired under natural illumination, facial expression, and accessory, to training and as gallery. The rest of the dataset, which varies in illumination, facial expression, and accessory, is used to be the probe. In particular, the CAS-PEAL dataset is divided into three parts: gallery (training) set, Pretty Variation Probe, and Grand Variation Probe.

The gallery set, also the training set, contains 1040 fontal images acquired under natural illumination and facial expression corresponding to 1040 subjects. The gallery set is also used for training, with a single sample per class; The probe set is divided into two subsets: Pretty Variation Probe (PVP) and Grand Variation Probe (GVP). Fig. 2 shows some typical examples used in our experiments.

Pretty Variation Probe contains 2274 frontal images (unseen during training) acquired under variable illumination and facial expressions. We assume PVP cover the similar variations as the FERET database [13]. The Eigenface method achieve about 80% recognition rate.

In order to cover more challenge variations and avoid the possible performance limit (100%) on the PVP, Grand Variation Probe is constructed to cover the variations accessory wears and directions of illumination, consisting of 3295 frontal images. We assume GVP cover the similar variations as the AR database [14], but with much more gallery subjects. The Eigenface method only achieves accuracy around 27%.

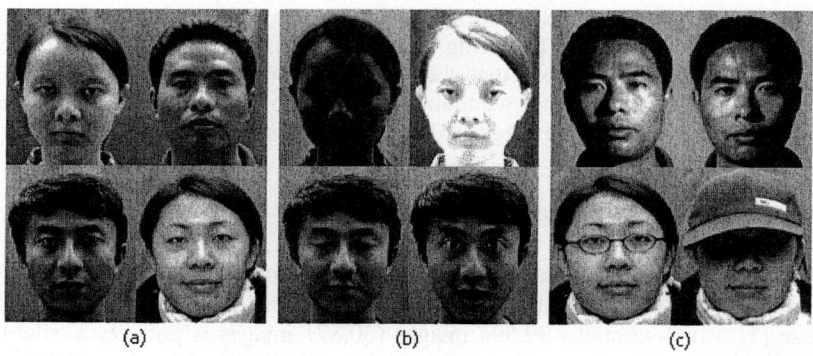

Fig. 2. Some example CAS-PEAL images used in our experiments. (a) The gallery set, also the training set. It contains 1040 fontal images acquired under natural illumination and facial expression corresponding to 1040 subjects. (b) Pretty Variation Probe. It contains 2274 frontal images (unseen during training) acquired under variable illumination and facial expressions. (c) Grand Variation Probe. It contains 3295 frontal images (unseen during training) acquired under variable directions of illumination and accessory wears.

3.2 Experiments on Pretty Variation Probe

For comparison purpose, we first apply PCA method on pretty variation probe and the comparative face recognition performance of the three similarity measures in shown in Fig.3 using three thin curves. We found that 1) the high dimension features make little effect on the performance of Euclidean distance, which testifies the theorem that PCA derives features that preferentially weight low frequencies and 2) by use whiten transformation to counteract the above theorem, Mahalanobis distance measure reaches its peak accuracy (84.39%) with around 140 features. However, its accuracy then drops with further increase of dimensionality, similar to the results reported by

Wang [15]. The reason is that the high dimensional components with small eigenvalues are significantly magnified in whitening. Since these dimensions tend to contain more noise than structural signal, they will deteriorate the recognition results and 3) WCSM also suffer from the magnified noise caused by the whiten transformation, however, the accuracy of the WCSM drops less dramatically than the Mahalanobis distance, indicating cosine similarity measure is less sensitive to the whiten noise.

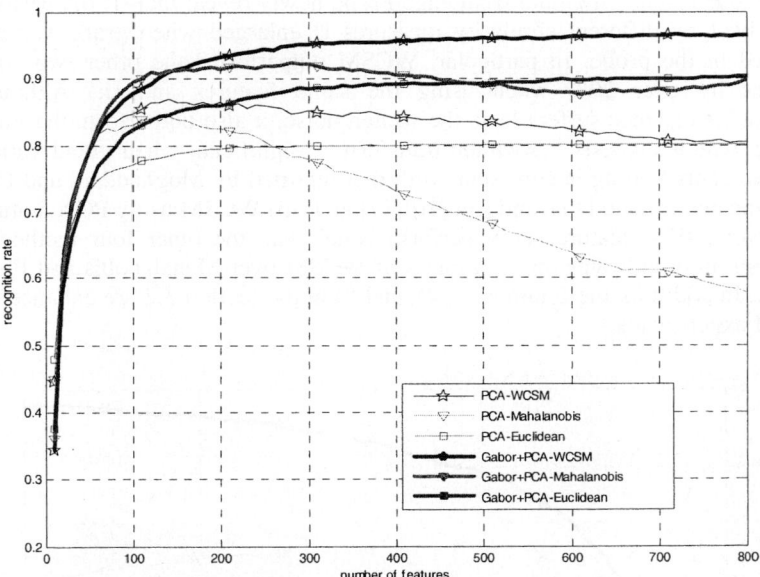

Fig. 3. Face recognition performance of the PCA (thin curves) and Gabor-based PCA (thick curves) method on the Pretty Variation Probe using the three different similarity measures: Whiten Cosine Similarity Measure (star mark), Mahalanobis distance (triangle mark), and Euclidean distance (square mark). The horizontal axis indicates the number of feature used, and the vertical axis represents the correct face recognition rate.

To further improve face recognition performance, we combine the Gabor wavelet representation and the PCA method. Fig.3 shows face recognition performance of the Gabor-based PCA method using the three distance measures using thick curves. Comparing the thick curves with the thin curves, we found that 4) the face recognition performance improves by the large margin (about 10%) for all the three similarity measures, which qualifies the Gabor representation as a discriminating representation method and 5) Mahalanobis distance measure drops less drastically when using Gabor wavelet representation instead of the intensity images. Meanwhile, the accuracy of WCSM ascends slowly with increase of dimensionality. This finding indicates Gabor wavelet has the ability to suppress the high frequency noise, which will be magnified by whitening and then deteriorate the performance. In addition, WCSM using Gabor-based PCA features performs better than the Mahalanobis distance, which show again that the WCSM is more robust to the whiten noise.

3.3 Experiments on Grand Variation Probe

Grand Variation Probe captures more challenge variations than PVP set. Each class has different variations of illumination directions and accessories, as show in Fig. 2(c), which would damage the PCA-based face recognition algorithms [12]. As one would expect, the Eigenface method, which uses Euclidean distance, only achieves 27% accuracy. Fig.4 shows the face recognition performance of PCA and Gabor-based PCA using the three similarity measures. Besides the findings in the same testes on PVP, our results on Grand Challenge Probe newly reveal that (i) the performance gap derived by different similarity measures is enlarged when grand variations is presented in the probe. In particular, WCSM outperforms the other two similarity measures by over 20% when using the same features and (ii) Although the Mahalanobis distance suffers from the whiten noise, it also outperform the Euclidean distance, which suggests the whiten transform is significant when grand variation is presented. This finding is consistent with that reported by Moghaddam and Pentland [16], Sung and Poggio [17], and Liu [10][11] and (iii) WCSM using PCA features and Gabor-based PCA features both perform better than the other four methods. This finding strongly indicates the robustness of WCSM over Mahalanobis and Euclidean distance. In addition, the findings 3), 4), and 5) in the section 3.2 are enhanced in this series of experiments.

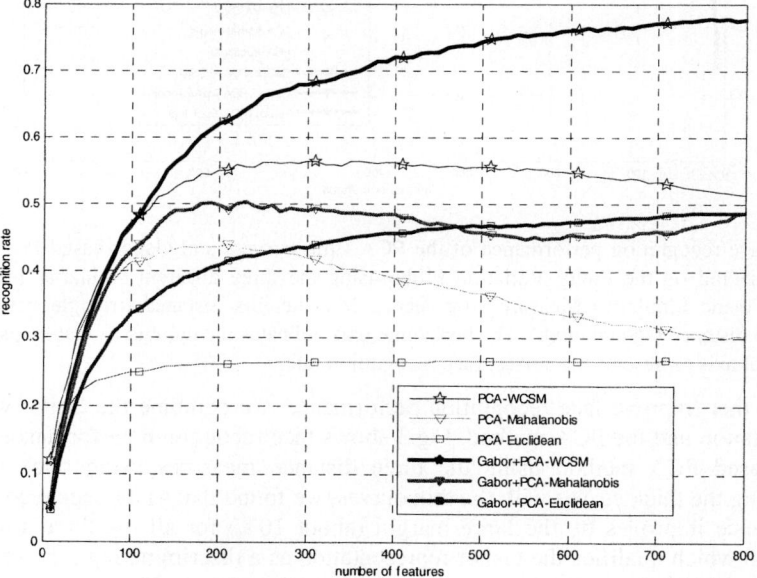

Fig. 4. Face recognition performance of the PCA (thin curves) and Gabor-based PCA (thick curves) method on the Grand Variation Probe using the three different similarity measures: Whiten Cosine Similarity Measure (star mark), Mahalanobis distance (triangle mark), and Euclidean distance (square mark).

By combining the discriminating power of the Gabor representation and WCSM, the face recognition performance is boosted drastically. The Gabor-based PCA method using WCSM achieve 78% accuracy, outperforming the Eigenface method by 50%.

4 Conclusion

We have shown that using WCSM combined with Gabor-based PCA feature can outperform the Eigenface method [6] by a large margin when one single training sample per class is available. It also shows that the more challenge variation is presented in the probe, the larger performance gap is observed, which indicates the robustness of the proposed method. The superiority of the proposed method lies in: 1) Gabor wavelet based face representation extracts desirable facial features characterized by spatial frequency, spatial locality, and orientation selective, which is demonstrated to be able to cope with the variations due to illumination and facial expression changes and suppress the high frequency noise. 2) Whiten transform is performed to counteract the fact the PCA space weight preferentially for low frequencies, making the classifier make full use of all Gabor-based PCA features. 3) Cosine similarity measure is employed to restrain the magnified whiten noise in the high dimension principal component.

Our work strongly indicates the image representation and similarity measurement is essential for face recognition besides the feature exaction method. More research work should focus on image representation and similarity measurement for face recognition.

Acknowledgements

Portions of the research in this paper use the CAS-PEAL face database collected under the sponsor of the Chinese National Hi-Tech Program and ISVISION Tech. Co. Ltd.

References

1. Jianxin Wu and Zhi-Hua Zhou: Face Recognition with One Training Image per Person. Pattern Recognition Letters, vol.23, no.14. (2002) 1711–1719
2. A M Martinez, "Recognizing imprecisely localized partially occluded and expression variant faces from a single sample per class", IEEE Trans. PAMI, Vol. 24, no. 6. (2002) 748–763
3. Jian Huang, Pong C Yuen, Wen-Sheng Chen and J H Lai: Component-based LDA Method for Face Recognition with One Training Sample. Proceedings of the IEEE International Workshop on Analysis and Modeling of Faces and Gestures (AMFG'03)
4. Shaokang Chen and Brian C. Lovell: Illumination and Expression Invariant Face Recognition with One Sample Image. Proceedings of the 17th International Conference on Pattern Recognition (ICPR'04)

5. Wen Gao, Bo Cao, Shiguang Shan, Xiaohua Zhang, Delong Zhou: The CAS-PEAL Large-Scale Chinese Face Database and Baseline Evaluations. technical report of JDL. (2004) http://www.jdl.ac.cn/~peal/peal_tr.pdf
6. M. Turk and A. Pentland: Eigenfaces for recognition. Journal of Cognitive Neuroscience, 3(1). (1991) 71–86.
7. J.G. Daugman, "Uncertainty Relation for Resolution in Space, Spatial Frequency, and Orientation Optimized by Two-Dimensional Visual Cortical Filters. J. Optical Soc. Am. A, vol. 2. (1985) 1,160–1,169
8. Laurenz Wiskott, Jean-Marc Fellous, Norbert Krüger, and Christoph von der Malsburg: Face Recognition by Elastic Bunch Graph Matching. IEEE Trans. PAMI, vol.19, no. 7. (1997) 775–779
9. Michael J. Lyons, Julien Budynek, and Shigeru Akamatsu: Automatic Classification of Single Facial Images", IEEE Trans. PAMI, vol. 21, no. 12. (1999) 1357–1362
10. C. Liu and H. Wechsler: Gabor Feature Based Classification Using the Enhanced Fisher Linear Discriminant Model for Face Recognition. IEEE Trans. Image Processing, 11(4). (2002) 467–476
11. Chengjun Liu: Gabor-Based Kernel PCA with Fractional Power Polynomial Models for Face Recognition", IEEE Trans. PAMI, vol.26, no. 5. (2004) 572-581
12. . N. Belhumeur, J. P. Hespanha, and D. J. Kriegman: Eigenfaces vs. Fisherfaces: Recognition using class specific linear projection," IEEE Trans. PAMI, vol. 19, no. 7. (1997) 711–720
13. P. J. Phillips, H. Wechsler, and P. Rauss: The FERET Database and Evaluation Procedure for Face-Recognition Algorithms. Image and Vision Computing, vol. 16, no. 5. (1998) 295–306
14. A.M. Martinez and R. Benavente: The AR-face database. CVC Technical Report 24. (1998)
15. Xiaogang Wang and Xiaoou Tang: A Unified Framework for Subspace Face Recognition. IEEE Trans. PAMI, vol. 26, no. 9. (2004) 1222–1228
16. B. Moghaddam and A. Pentland: Probabilistic visual learning for object representation. IEEE Trans. PAMI, vol. 19, no. 7. (1997) 696–710
17. K. K. Sung and T. Poggio: Example-based learning for view-based human face detection. IEEE Trans. PAMI, vol. 20, no. 1. (1998) 39–51

Chinese Word Sense Disambiguation Using HowNet

Yuntao Zhang[1,2], Ling Gong[2], and Yongcheng Wang[2]

[1] Network & information Center, Shanghai Jiaotong University,
200030 Shanghai, China
ytzhang@sjtu.edu.cn
[2] School of Electronic & Information Technology, Shanghai Jiaotong University,
200030 Shanghai, China
{lgong, ycwang}@sjtu.edu.cn

Abstract. Word sense disambiguation plays an important role in natural language processing, such as information retrieval, text summarization, machine translation etc. This paper proposes a corpus-based Chinese word sense disambiguation approach using HowNet. The method is based on the co-occurrence frequency between the relatives (such as synonym, antonymy, meronymy) of target word and each word in the context. Further, domains have been used to characterize the senses of polysemous word. To our knowledge, this is the first time a Chinese word sense disambiguation method using domain knowledge is reported. The accuracy is 73.2% at present. The experimental result shows that the method is very promising for Chinese word sense disambiguation.

1 Introduction

Polysemy implies presence of more than one sense of a particular word both in its context-bound and context-free situation[1]. Multiplicity of sense of words is a very general characteristic of natural languages. For example, the polysemous words ratio in Modern Chinese Dictionary is 14.8%. In Chinese corpus, the occurrence frequency of polysemous words is about 42%. It shows that words are highly ambiguous, particularly for frequently occurring words.

Word sense disambiguation (shortly WSD) is the process of assigning a sense to a polysemous word based on the context in which it occurs. Word sense disambiguation plays an important role in natural language processing, such as information retrieval, text summarization, machine translation etc. The word sense ambiguity has a detrimental effect on the precision of automatic text processing. At the same time, word sense disambiguation is a difficult issue in these areas. Indeed, even between expert lexicographers, the inter-annotator agreement can vary hugely[2].

Most WSD research has focused on English word disambiguation. However, the WSD community has no desire to be narrowly monolingual. The paper will research the Chinese word sense disambiguation.

The remainder of this paper is organized as follows. Section 2 describes the method that we adopt to disambiguate the sense of polysemous word. Section 3 presents the experiment result. Section 4 concludes the paper and provides some future research directions.

2 Methodology

The approach is based on the following observation: the different sense of polysemous word tend to appear in cognizably different contexts. In other word, various senses of word can be best reflected by their context. Therefore, we can use the context of the word to be disambiguated together with information about each of its word senses.

It has been observed that WSD is more accurate when multiple knowledge sources are combines. The method described in this paper is based on the co-occurrence frequency between the relatives (such as synonym, antonymy, meronymy) of target word and each word in the context. Further, domains have been used to characterize the senses of polysemous word. The method is similar to the combination of two methods are described in literature [3] and [4]. However, [3] and [4] target at English. For Chinese, the feasibility of this method is needed to confirm. Moreover, [3] and [4] are based on WordNet. WordNet is an online lexical reference system designed for English. In this paper, we use a new knowledge base---HowNet as knowledge resources for Chinese word sense disambiguation.

2.1 HowNet

HowNet is an online common-sense knowledge base unveiling inter-conceptual relations and inter-atttibute relations of concepts as connoting in lexicons of the Chinese and their English equivalents[5]. HowNet is not only like a machine readable dictionary which lists a set of possible senses for each word, but also provides other semantic relations, such as synonymy, antonymy, meronymy.

In HowNet, the basic data of concepts definitions is similar to WordNet; its taxonomies of upper classes are similar to SUMO. Aimportant concept used in HowNet is sememe which is the basic unit of senses.

In HowNet, the format of the lexical entry is as follow[6]:
NO.=053234
W_C=
G_C=N
E_C=
W_E=teacher
G_E=N
E_E=
DEF={human|人 :HostOf={Occupation|职位}, domain={education|教育}, {teach|教:agent={~}}}

Where NO. is the number of this lexical entry; W_C is the Chinese word; G_C is the pos of this Chinese word; E_C contains some example of this Chinese word; W_E is the corresponding English word; G_E is the pos of the corresponding English word; E_E contains some example of this English word; DEF is the definition of this Chinese word. In the example, the definition of the concept of 教师 (teacher) means that: a teacher is (a) a person who teaches, (b)a person who has a special attribute of occupation, (c) a person who belongs to the domain of education[6].

2.2 Without Context Versus. Context

In WSD system, the baseline method ignores context and simply assigns the most common sense in all case. In other words, the most frequent sense of the polysemous word in the corpus will be assigned as the sense of word appearing in the new text. The baseline is estimated as the lower bound of the performance of WSD. However, it is observed that it is sometime quite difficult for WSD system to beat the most frequent sense baseline. The reason is probable that the most frequent sense provides a good default value for polysemous words which do not obviously have another sense[7]. Nevertheless, the baseline perform has a large variation for different polysemous words because some words are relatively easy and others are harder.

The limits of a linguistic context can be defined arbitrarily, but it is common to define it in terms of sentences. In our method, the context is classified into two types: global context and local context.

Although local context is more important, Yarowsky[8] observes that there seems to be only one sense per collocation per collocation and that words tend to keep the same sense during a discourse.

In other hand, it is observed that human seem to be able to disambiguate word sense based on very little context. It is to say, two words co-occur in the same context if they occur in the same sentence, even in the same phrase or clause.

HowNet used in our WSD work contains the corresponding two types of information: domain information and lexical information. It is the reason why we choose it as knowledge resource for Chinese word sense disambiguation.

2.3 Preprocessing

In document, some words have low content discriminating power such as prepositions, conjunctions, articles and pronouns. These words are non-contextual words occurring in the text. They don't directly contribute to the content. Therefore, they are listed at the stop-list. It is clear those words appearing at stop-list will be deleted from set of context words and co-occurrence sets of words before WSD. The step of preprocessing is necessary for Chinese as well as English[9]. It is the first step of preprocessing for WSD.

If we can determine which words co-occur frequently with each sense of polysemous word, we can use the context of target word to disambiguate the word. Hence, we need to realize quantitatively the relation between senses of polysemous word and co-occurrence words in the corpus.

In the Internet era, the availability of large electronic language corpus has strengthened the capability of WSD. We can obtain all possible instances of senses of polysemous word from the corpus (collection of sentences). In corpus, the sentences containing the polysemous word are grouped together according to their occurring in the same sense.

It is the second step of preprocessing for WSD to generate dataset about the co-occurrence frequency. When generating the statistical dataset, the corpus will be segmented into sentences. Here, a sentence is a string of Chinese characters delimited by punctuations. If the polysemous word x and y appear in the same sentence, and i^{th}

sense of x is s_i, the pair $<s_i, y>$ is regarded as co-occurrence in the sentence. For the co-occurrence dataset, each pair of each sense of each polysemous word and each context word co-occurring in the corpus is respectively counted.

2.4 WSD Using Domain Argument in HowNet

The context can be lexical, grammatical, or domain-based. For example, 案 has senses 案件（*law case*) and 文书*(document)*. From lexical angel, the sense is probably *law case* if the word is surrounded by *police, cop or robber* et al; the sense is probably *document* if the word is surrounded by *secretary* or *designer*. From grammatical angel, the sense is probably *document* if the word is modified by *grace*; the sense is probably *law case* if the word is modified by *atrocious* or *barbarous*. From domain angel, the sense is probably *law case* if the text containing the word is about policing; the sense is probably *document* if the text is about advertisement.

The grammatical and lexical knowledge can be exploited from large corpus. The domain knowledge can be obtained from HowNet. In HowNet, word is annotated with one label.

We utilize domain knowledge for Chinese word sense disambiguation. To our knowledge, this is the first time a Chinese word sense disambiguation method using domain knowledge is reported.

The basic idea is to calculate the correlation between word contexts and domain because it is observed that different word senses tend to belong to different conceptual classes, and such classes tend to appear in recognizably different contexts. The polysemous word in same context is assigned the same sense as the sense with high probability.

The relevance between domain and text may be estimated in the corpus. The text can be considered as "bag of words". Therefore, the text can be denoted according to formula:

$$D_i = (T_1, W_1; T_2, W_2; ...; T_n, W_n) \qquad (1)$$

Here, D_i denotes a text; T_i denotes the word appearing in text and W_i denotes the corresponding weight of word T_i.

The high dimensionality of the word space will cause some problems, such as the computational complex and over-fitting. We use the TF-IDF method to have a term selection (or called subject-word selection). The TF-IDF function can be estimated as following:

$$TFIDF(T_i, D_j) = TF(T_i, D_j) * \log \frac{|D|}{|DF(T_i)|} \qquad (2)$$

Here, $TFIDF(T_i, D_j)$ is the weight measure of word T_i within text D_j. $TF(T_i, D_j)$ is the number of times word T_i occurs in text D_j. $|D|$ denotes the total number of all text and $|DF(T_i)|$ is the number of texts in which word T_i occurs at least once. The function encodes the intuitions: (1) the more often a word occurs in a document, the more it is representative of the content of the text; (2) the more text the word occurs in, the less discriminating it is[10].

2.5 WSD Using HowNet Relatives

Corpus is essential resource for WSD. In terms of Zipf's law, a majority of words, no matter how corpus size increases, still do not appear even once at all. Hence, the main problem of distribution-based approach is obtained from the data sparseness problem of statistical methods[11]. For solving the problem, we adopt HowNet to provide the relative of polysemous word.

How Net describes relations between concepts and relations between relations between the attributes of concepts. HowNet provides the semantic relations including synonymy, antonymy, meronymy and hyponymy/hyponymy et al. Relatives, especially those synonymy words, usually have related meanings and tend to share similar contexts[3].

In our Chinese word sense disambiguation method, we consider the context of relative of target word as the context of target word. In other words, the corpus sample relevant with the relatives of target will be augmented as the corpus sample of target word. The relatives can be extracted from HowNet automatically. By this approach, the problem of data sparseness will be solved to a large degree.

2.6 Determining Sense by Combining Knowledge Sources

When disambiguate a polysemous word, we use Bayes theorem to calculate the likelihood of assigning a specific sense. Bayesian statistics provide a theoretically sound method. For the global context (domain knowledge) and local context (lexical knowledge), the calculating procedure is same. The following will illustrate the procedures calculating the relevancy between local context of target word and the sense of target word.

Let $S \in \{S_1, S_2, ... S_n\}$ be the set of possible senses of a polysemous word W, $p(S_i | Local\_Context(W))$ be the probability of sense S_i in the local context of target word W. It is obvious:

$$\sum_{i=1}^{n} p(S_i | Local\_Context(W)) = 1 \qquad (3)$$

Where n is the number of all senses of a target polysemous word.

According to the Bayes' theorem, the probabilities $p(S_i | Local\_Context(W))$ can be evaluated by following formula:

$$p(S_i | Local\_Context(W)) = \frac{p(Local\_Context(W | S_i)) * p(S_i)}{p(Local\_Context(W))} \qquad (4)$$

In the above formula, $p(Local\_Context(W))$ is constant for all sense, only $p(Local\_Context(W | S_i)) * p(S_i)$ need be calculated. $p(S_i)$ is the prior probability.

It is to say that the prior probability is based on the corpus. For each target word W, the prior probability may be estimated by following formula:

$$P(S_i) = \frac{|C_i|}{\sum_i |C_i|} \tag{5}$$

Where C_i is the segment of corpus corresponding to i^{th} sense S_i of target word W, $|C_i|$ is the number of word W with sense S_i in the corpus.

Assuming the local context (the sentence in which the target word) contains words $W_1, W_2, ... W_m$. $p(Local\_Context(W|S_i))$ may be estimated by following formula:

$$p(Local\_Context(W|S_i)) = \prod_{j=1}^{m} p(W_j | S_i) \tag{6}$$

The probabilities $p(W_1|S_i), p(W_2|S_i), ... p(W_m|S_i)$ can be obtained from the co-occurrence frequency dataset described in section 2.3.

The calculation of the relevancy between global context of target word and the sense of target is similar. The distinction is only that the content of context is the subject words corresponding to the domain. Assuming the relevancy between global context of target word is denoted as $p(S_i | Global\_Context(W))$. The uniform probability for all contexts may be estimated by following formula:

$$p(S_i | Context(W)) = p(S_i | Local\_Context(W)) * p(S_i | Global\_Context(W)) \tag{7}$$

For target word W, the sense S_i with the highest posterior probability $p(S_i | Context(W))$, condition on $Local\_Context(W)$ and $Global\_Context(W)$ respectively, will be assigned as sense of target word W.

3 Experimental Result

The experiment is based on real-world data collected from InfoBank, the biggest Chinese information base. Firstly, the corpus is segmented, and then the co-occurrence dataset is generated.

Using a predefined sense inventory and comparing answers against a gold standard is still the most frequent method for evaluating WSD. Therefore, experiment adopts the most commonly used performance measure, namely accuracy. We use the following equation to evaluate the accuracy of WSD:

$$accuracy = \frac{the\ number\ of\ words\ assigned\ correctly\ a\ sense}{the\ number\ of\ words\ assigned\ a\ sense} \tag{8}$$

The accuracy is 73.2% at present. The experimental result shows that the method is very promising for Chinese word sense disambiguation.

4 Conclusions and Future Works

We describe a method which performs Chinese word sense disambiguation by combining lexical co-occurrence knowledge, semantic knowledge and domain knowledge. It is showed that the hybrid method based on large scale corpus and HowNet can be used to disambiguate the Chinese polysemous words.

At the time when the research is conducted, HowNet has not provided all examples (E_C and E_E) for each word. When HowNet completes all examples, the example-based method will be experimented. By comparing the context words of the polysemous word with the examples of each sense of the word in HowNet, the corresponding sense with the highest similarity will be chosen as the assigned sense. Example-based method will be compared with the method presented in this paper.

Now, there are many methods and algorithms have been proposed to automatically assigning a sense to a polysemous word from a given inventory of sense. Because combining multiple knowledge sources is usually beneficial to WSD task, many combination methods are proposed, such as majority voting[12], hierarchy decision lists[13] or Bayesian statistics[14]. However, the optimal method combing these knowledge sources is a challenging task.

For WSD, a quantitative evaluation exercise is required. The experiment results reported in literatures are varied widely in Chinese word sense disambiguation studies. [15] reports that the accuracy of the experiment is 91.89% in open test and 99.4% in close test. [16] reports that the average accuracy is 83.13% for 10 polysemous words in open test by their method. However, these figures are not directly comparable because they use different corpus, different polysemous word, different word sense inventories, and different definitions of accuracy. The dataset of correct answers to evaluate against and a framework for administering the evaluation with the requisite credibility and accountability to the research community are essential[17]. For English, SENSEVAL et al can be use to evaluate the word sense disambiguation. However, the evaluation exercises SENSEVAL and SemCor don't run in Chinese. Hence, a "gold standard" dataset for Chinese word disambiguation is an on-going and challenging task.

References

1. Dash,N.S., Chaudhuri, B.B.: Using Text Corpora for Understanding Polysemy in Bangla. Proceedings of the Language Engineering Conference (2002) 99-109
2. Ahlswede T., Lorand D.: Word Sense Disambiguation by Human Subjects: Computational and Psycholinguistic Implications. In: Proceedings of the Workshop on Acquisitions of Lexical Knowledge from Text. Columbus,Ohio (1993) 1-9
3. Seo, H.C., Chung, H.J., Rim, H.C., Myaeng, S.H., Kim, S.H.: Unsupervised Word Sense Disambiguation Using WordNet Relatives. Computer Speech and Language 18 (2004) 253-273
4. Gliozzo, A., Strapparava, C., Dagan, I.: Unsupervised and Supervised Exploitation of Semantic Domains in Lexical Disambiguation. Computer Speech and Language 18 (2004) 275-299
5. Dong, Z.D., Dong, Q.: HowNet. http://www.keenage.com

6. Dong, Z.D., Dong. Q.: HowNet – A Hybrid Language and Knowledge Resource. In: Proceeding of the International Conference on Natural Language Process and Knowledge Engineering (2003) 820-824
7. Preiss, J.: Probabilistic Word Sense Disambiguation. Computer Speech and Language 18 (2004) 319-337
8. Yarowsky, D.: Unsupervised Word-Sense Disambiguation Rival Supervised Methods. In: Proceeding of the 33rd Annual Meeting of the Association for Computational Languistics (1995) 189-196
9. Zhang, Y.T., Gong, L., Wang, Y.C., Yin. Z.H.: An Effective Concept Extraction Method for Improving Text Classification Performance. Geo-spatial Information Science, 6(4) (2003) 66-72
10. Fabrizio Sebastiani: Machine Learning in Automated Text Categorization. ACM Computing Surveys, Vol.34, No.1, March 2002, 1-47
11. Yi, G., Wang, X.L., Kong, X.Y., Zhao, J.: Quantifying Semantic Similarity of Chinese Words from HowNet. In: Proceedings of the First International Conference on Machine Learning and Cybernetics (2002) 234-239
12. Stevenson M., Wilks Y.: Combining Weak Knowledge Sources for Sense Disambiguation. In: Proceeding of the International Joint Conference for Artificial Intelligence (1999), 884-889
13. Yarowsky D.: Hierarchical Decision Lists for Word Sense Disambiguation. Computers and the Humanities 34(1/2), (2001) 321-349
14. Judita Preiss: Probabilistic Word Sense Disambiguation. Computer Speech and Language 18 (2004) 319-337
15. Liu, Z.M., Liu, T., Zhang, G., Li, S.: Word Sense Disambiguation Based on Dependency Relationship Analysis and Bayes Model (in Chinese). High Technology Letter 5 (2003) 1-7
16. Lu, S., Bai, S., Huang, X.: An Unsuptervised Approach to Word Sense Disambiguation Based on Sense-words in Vector Space Model (in Chinese). Journal of Software, Vol13, No.6 (2002)1082-1089
17. Adam Kilgarriff. Gold Standard Datasets for Evaluating Word Sense Disambiguation Programs. Computer Speech and Language 12 (1998) 453-472

Modeling Human Learning as Context Dependent Knowledge Utility Optimization

Toshihiko Matsuka

Rutgers University, Newark NJ 07102, USA

Abstract. Humans have the ability to flexibly adjust their information processing strategy according to situational characteristics. However, such ability has been largely overlooked in computational modeling research in high-order human cognition, particularly in learning. The present work introduces frameworks of cognitive models of human learning that take contextual factors into account. The framework assumes that human learning processes are not strictly error minimization, but optimization of knowledge. A simulation study was conducted and showed that the present framework successfully replicated observed psychological phenomena.

1 Introduction

Computational high-order cognitive modeling is a field of research trying to understand the nature of real human cognitive processes or to evaluate theories on human cognition by developing descriptive Artificial Intelligence models of ordinary people (henceforth termed AOI for artificial ordinary intelligence) who often can be characterized as having erroneous, irrational, and/or sub-optimal intelligence. The effectiveness of these cognitive models is, therefore, evaluated by comparing their predictions with the results of empirical studies conducted with human subjects. Thus, it is not of great interest and importance to examine how satisfactorily a cognitive model solves complex problems independent of behavioral phenomena of interest (i.e., tendencies in human cognition). This approach is very distinct from some other fields of AI and machine learning research whose objectives are most likely characterized by accurate and efficient identifications of solutions for, in general, numerically represented problems, or by the development of artificial "superior" intelligence (ASI) to solve very complex problems with algorithms that qualitatively have no implication with regard to how humans really do process information. The present research focuses on introducing AOI models of human learning that can be interpreted as apparently sub-optimal processes in terms of absolute performance in a given task, but can result in refined outcomes with regards to given contextual factors.

A conventional AOI approach for modeling human sub-optimal cognitive processes is to incorporate some structural and/or algorithmic constraints on information flow (mostly in forward processes) in order to impose limitations in the computational capability of simulated cognitive processes. Some AOI models of human cognition showed successful results in replicating apparently irrational psychological phenomena (e.g. [1]). However, as Matsuka [2][3][4] pointed out, many AOI models of human learning often incorporate computational processes that appear to have a normative (i.e., how we should think or ASI approach) than descriptive (i.e., what we do really think or AOI approach) orientation. In particular, although, human cognitive competence is characterized by adaptability, which manifests itself prominently in context dependent thinking skills, many computational models of human learning assume that in every instance we have a rather rigid objective, namely error minimization for a given task, throughout learning processes. Certainly, such model approach fails to account qualitatively and probably quantitatively for interesting and important factors comprising the heart of human intelligence as apparently irrational, erroneous, and sub-optimal processes in absolute sense, yet as an efficient, unique (i.e., individually different), and refined system in relative (i.e., context dependent) sense.

The main theme of the present study is to introduce and test new AOI models of human learning that take the contextual factors into account. In particular, ordinary humans are assumed to be adaptive thinkers, being influenced by intra, inter, and/or extra-personal contextual factors, who would attempt to improve their contextually-defined knowledge utility in the course of learning, which does not necessary result in minimization of error. This, however, does not imply that humans are unintelligent animals incapable of minimizing error in learning. Rather, its more sensible interpretation is that humans are intelligent and complex animals who are consciously or unconsciously receptive to contextual factors at a given moment in the course of utility-enhancing learning processes. The main ingredient of the present AOI model is, therefore, its sensitivity to its internal and/or external contextual factors and its capability of revising its knowledge (i.e., learning) to make it suitable for particular circumstances. Consequently, the learning of the same matter can result in different trajectories depending on the person and the circumstances. Perhaps, this is one of the factors making our intelligence extremely adaptive and an apparently heuristic one. Note that although the AOI models would learn to form contextually apt knowledge and concepts, it does not necessarily result in effective outcomes. Rather, some contextual factors could have aversive effects on concept formation (e.g. searching for overly simple concepts from inherently complex concepts).

Since categorization is considered to be a fundamental higher-order cognition that serves as a gateway to many other high-order cognitive processes, as an initial attempt, I choose to apply this theory to AOI models of category learning. The new proposed learning algorithms are general models of learning without any feedforward algorithm, and thus does not restrict one to choose one network architecture over the other (i.e., it is a model of learning or optimization, and thus independent of feedforward algorithm). Without loss of generality, in the present study, the new leaning model is applied to ALCOVE [5].

1.1 ALCOVE

Before describing the new models of learning in detail, ALCOVE [5], one of the most successful models of category learning, will be introduced as the particular example architecture selected in which to embed the new learning models. For detailed descriptions and theoretical foundations of ALCOVE, readers are advised to refer to the original work by Kruschke [5].

Feedforward Algorithm: ALCOVE, for Attention Learning COVEring map, is a computational model of high-order human cognition, namely categorization, based on the exemplar theory on mental representation. It is a type of RBF network that scales each feature dimension independently, and this scaling process is interpreted as a selective attention process. Its basis units are the memorized exemplars introduced during training. Each unit corresponds to a particular exemplar, described by the multidimensional stimulus feature space, and becomes activated when a stimulus enters the network. The strength of activation depends on psychologically biased similarity, based on selective attention processes, between input stimulus and a particular exemplar. Operationally, the activation of exemplar unit j, denoted h_j, is calculated based on its scaled distance to the presented stimulus x:

$$h_j(x) = \exp\left[-c \cdot \sum_i \alpha_i |\psi_{ji} - x_i|\right] \quad (1)$$

where ψ_{ji} is the feature value of exemplar unit j on dimension i, x_i is the input feature value on dimension i, c is a constant called the *specificity* that controls overall sensitivity, and $\alpha_i \geq 0$ is the selective attention strength for dimension i. The activities of the exemplar units are fed forward to the category layer, whose nodes correspond to the categories being learned. The activation of category node k is then computed as the sum of weighted activations of all exemplars, or

$$O_k(x) = \sum_j w_{jk} \cdot h_j(x) \quad (2)$$

where w_{kj} is the strength of association between category node k and exemplar unit j.

The probability that a particular stimulus is classified as category C, denoted as $P(C)$, is assumed equal to the activity of category C relative to the summed activations of all categories, where the activations are first transformed by the exponential function:

$$P(C) = \frac{\exp(\phi \cdot O_c(x))}{\sum_k \exp(\phi \cdot O_k(x))} \quad (3)$$

where ϕ is a real-value mapping constant that controls *decisiveness* of classification responses.

Original ALCOVE Learning Algorithm: The standard version of ALCOVE uses an online version of gradient descent for updating its coefficients. Note that this standard gradient descent approach is introduced only for an illustrative purpose here.

Its objective function is defined as the sum of squared differences between the desired and predicted outputs:

$$E(\theta) = 1/2 \sum_k e_k^2 = 1/2 \sum_k (d_k - O_k)^2 \quad (4)$$

Partial derivatives of the error function with respect to the association weights w_{kj} and the attention strengths α_i are used to compute coefficients updates, or $\Delta w_{kj} = -\lambda_w \frac{\partial E}{\partial w_{kj}}$ and $\Delta \alpha_i = -\lambda_\alpha \frac{\partial E}{\partial \alpha_i}$.

This type of learning algorithm can be qualitatively interpreted as that human tries to learn about the categories by adjusting all possible relevant knowledge or concepts about the categories (i.e., model coefficients) *optimally* in order to minimize misclassification whenever they receive (negative) feedback on their performance in categorization tasks. In other words, locally optimal changes are incrementally and correctly made, given the current stimulus instance, in the model's multi-faceted coefficient space in every learning trial, or $\bigcap_t P(E(\theta + \Delta\theta) < E(\theta + \Delta\vartheta), \forall \Delta\vartheta \neq \Delta\theta | x, \theta, \{|\Delta\theta|, |\Delta\vartheta|\} < \epsilon) = 1$, where t indicates time, and ϵ is a small number, implying locality. However as an AOI model of human cognition, such learning algorithm appears too normatively justified and more ASI oriented. In addition, the standard gradient descent does not account for psychological, physical, and ecological contextual factors, which can be considered as highly influential in high-order human cognition. In response to this type of concerns, two families of learning algorithms that take into account contextual factors are introduced in the following sections.

2 New Learning Algorithms – Learning as Context Dependent Knowledge Utility Optimization

Fundamental Concepts: The present model assumes that people try to optimize utility of knowledge or concepts about category (e.g. establishing simple concepts resulting in sufficiently accurate categorization) in a course of learning, instead of minimizing classification error. Here, the utility of category concepts is assumed situation-dependent. For example, in several real world category learning events, people perhaps implicitly, try to minimize categorization error with smaller amount of effort, or try to have manageably simple, acceptably accurate category concepts, simply because it is not worth the effort, or because having comprehensive knowledge is not critical for some categories. For such categories and/or situations, simpler or more abstract concepts have better utility than more comprehensive concepts, because such concepts require less mental resources (e.g. memory & computation). Conversely, in other cases, some people may put effort in becoming a domain expert who can distinguish one exemplar from others. For domain experts, more comprehensive knowledge or the ability to pay attention to feature dimensions that help distinguish exemplars have better utility than more compact concepts. In short, the learning trajectories of the present model are determined by the utility of category concepts as defined by contextual factors at a given moment, but not accuracy in learning itself.

The present model tries to optimize context dependent category knowledge utility (operationally, it minimizes concept futility, thus smaller values have better utility) defined by the weighted sum of classification error and other contextual factors, or:

$$U(\theta) = E(\theta) + \sum_m^M \gamma_m \cdot Q_m(\theta) \quad (5)$$

where ($w_{kj}, \alpha_i \in \theta$), the first term is as in Eq. 4, and each Q_m function in the second term can be interpreted as a function defining a particular contextual factor m, γ_m is a scalar weighting importance of context factor m, and M is the number of contextual factors. Note that since a particular contextual factor can have rather different effects on different coefficient types, a Q function for w and α for a particular contextual factor may be defined differently. There are virtually infinitely many contextual functions appropriately defined for Eq. 5. For instance, Matsuka [2][3] discussed about incorporating concept abstraction, domain expertise, conservation (i.e., preparing for unforeseen and novel stimuli), and knowledge commonality as few examples of contextual, mainly motivational, factors possibly influencing learning processes and outcomes.

In the following section, two learning algorithms based on utility optimization are introduced. One is based on a gradient decent optimization method with noise perturbations. The other is based on stochastic optimization methods, namely simulated annealing. The fundamental principles behind the two models are the same (i.e., utility optimization), but their qualitative interpretations of implied cognitive processes are fairly distinctive.

2.1 Noisy Gradient Decent with Contextual Regularizers

The partial derivatives with stochastic processes are used for updating the coefficients.

$$\Delta w_{kj} = -\Lambda_w \frac{\partial U}{\partial w_{kj}} + \nu_{kj} = -\lambda_w^E \frac{\partial E}{\partial w_{kj}} - \sum_m \lambda_w^{Q_m} \frac{\partial Q_m(w_{kj})}{\partial w_{kj}} + \nu_{kj} \quad (6)$$

$$\Delta \alpha_i = -\Lambda_\alpha \frac{\partial U}{\partial \alpha_i} + \nu_i = -\lambda_\alpha^E \frac{\partial E}{\partial \alpha_i} - \sum_m \lambda_\alpha^{Q_m} \frac{\partial Q_m(\alpha_i)}{\partial \alpha_i} + \nu_i \quad (7)$$

where, λs are step sizes weighting different contextual factors (including classification accuracy) with superscript indicating particular factors and subscript indicating particular coefficient types, $\nu \sim N(0, \zeta(\cdot))$ is random noise in learning, and ζ is, in general, a decreasing function. ζ needs to be a decreasing function for convergence. Note that γs in Eq. 5 are subsumed in λs in Eq. 6 & 7. Since the present model incorporates a gradient descent optimization method, the contextual factors need to be defined by some differentiable functions or functions for which the first partial derivative can be defined.

There are two main reasons for including stochastic components (i.e., ν); one is to make gradient descent learning algorithm sound more descriptively plausible; another reason is to avoid symmetrical or parallel utilization of redundant feature dimensions.

Interpretation of noisy learning: The added random noise in coefficient updates can be interpreted as follows: People try to revise their relevant knowledge (i.e., coefficient) about categories, but revision of each coefficient succeeds probabilistically due to the somewhat coarse and imperfect nature of human high-order cognitive processes. In other words, although, on average, concepts can be correctly adjusted, the present model assumes that people are capable of adjusting their concepts only probabilistically. Unlike standard gradient descent optimization, the present method does not assume that people can successfully update all relevant category concepts successfully in every learning instance given a particular input stimulus. Rather, it assumes that people do sometimes update some concepts in "improper" directions, particularly in the early stages of learning, because ζ is a decreasing function. In addition some concepts may be learned faster than others, depending on the level and direction (i.e., + or -) of noise.

Asymmetric utilization of perfectly redundant feature dimensions: As Matsuka and Corter [4] pointed out when there are multiple feature dimensions with identical information, ALCOVE trained with standard gradient descent resulted in utilizing the dimensions identically (or in parallel, depending on initial coefficient configurations). However, an empirical study [6] suggested that people tend to pay more attention to one feature dimension than the other and show asymmetric attention learning trajectories. The added noise creates some inertia to break the symmetric balance between the amounts of attention allocated to the two identical dimensions. It is, however, possible to achieve asymmetric attention allocation, when the initial attention allocations to redundant dimensions differ (and if an appropriate contextual factor exists and is incorporated in learning). But, to offer more descriptive model interpretation (see a section above) and because of the fact that empirical studies (e.g., [7]) showed that people tend to allocate attention evenly to feature dimension in early stages of learning, the random error in learning (i.e., ν) is incorporated in the model rather than the random attention initialization.

2.2 Context Dependent Learning via Hypothesis Testing

Matsuka [3] introduced a learning algorithm for human learning based on a stochastic optimization method termed SCODEL (for Stochastic COntext Dependent Learning). The fundamental concepts underlying SCODEL and the previous models are very similar. But, as compared with the gradient descent method, SCODEL has a more heuristic-orientated and hypothesis-testing like interpretations.

In SCODEL, people are assumed to form a hypothesis about category, which is directly related to the network architecture of forward algorithm (i.e., ALCOVE in the present study), in a random fashion. With ALCOVE-type feedforward model in mind, at the beginning of each training epoch, SCODEL randomly produces hypotheses about the relationship between exemplars and category membership (i.e., association weights w) along with hypotheses about diagnosticities of feature dimensions that determine distribution of attention (i.e., attention strengths, α). This is accomplished by randomly updating each coefficient by an independently sampled term from a prespecified, in general, zero-mean symmetric distribution (e.g. the Gaussian distribution). Thus, the coefficients updates are accomplished by the following simple functions.

$$w_{kj}^t = w_{kj}^s + \Delta w_{kj}, \quad \alpha_i^t = \alpha_i^s + \Delta \alpha_i \tag{8}$$

where the superscript s indicates previously accepted coefficient values, Δw_{kj} and $\Delta \alpha_i$ are random numbers generated from prespecified distributions (i.e., $\Delta w_{kj} \sim \Phi^w(T^t)$, $\Delta \alpha_i \sim \Phi^\alpha(T^t)$). The random distributions, Φ^w and Φ^α take a parameter called "temperature" (see [8]) at time t that controls width of the random distributions (it also affects the probability of accepting a new hypotheses set, see Eq. 10). The temperature decreases across training blocks according to the following annealing schedule function:

$$T^t = \delta(\upsilon, t) \tag{9}$$

where δ is the temperature decreasing function that takes temperature decreasing rate, υ, and time t as inputs. A choice of the annealing function (e.g., Eq. 9) may depend on the particular choice of the random number generation functions [9]. Note that the decrease in the temperature causes decreases in width of the distributions Φ^w and Φ^α.

The new sets of coefficients are updated based on the currently accepted coefficients, and thus Eq. 8 indicates that people use the currently accepted hypothesis as a basis for generating new hypothesis. Thus, the present model does not assume learning is carried out by totally stochastic processes, rather it assumes people utilize their experience with stimuli from given categories. The effect of transition in distribution widths controlled by the temperature can be interpreted as follows: In early stages of learning SCODEL is quite likely to produce "radical" hypotheses (i.e., the new set of hypotheses thus coefficients are very different from the currently valid and accepted hypotheses). But, as learning progresses, the widths of the random distribution decrease, so that it increasingly stabilizes its hypotheses and establishes more concrete and stable knowledge about the category.

The hypotheses set (i.e., the set of new coefficient values) is then accepted or rejected, based on the computed relative utility (Eq. 5) of the new hypotheses set. Specifically, if the new coefficient values result in a greater utility (lesser futility) or meet learner's objective better, then they are accepted. If they result in a poorer utility (greater futility), they are accepted with some probability P, defined by the relative utility of the new hypotheses against that of the currently accepted hypotheses. In a general form, the probability of accepting a new set of coefficients is defined as:

$$P(\Delta \mathbf{w}, \Delta \alpha | T^t) = \left[1 + \exp\left(\frac{U(\mathbf{w}^s + \Delta \mathbf{w}, \alpha^s + \Delta \alpha) - U(\mathbf{w}^s, \alpha^s)}{T^t}\right)\right]^{-1} \quad (10)$$

if $U(\mathbf{w}^s + \Delta \mathbf{w}, \alpha^s + \Delta \alpha) > U(\mathbf{w}^s, \alpha^s)$, or 1 otherwise, where $U(\mathbf{w}^s + \Delta \mathbf{w}, \alpha^s + \Delta \alpha)$ is a function defining the utility of the new coefficient set as in Eq. 5, $U(\mathbf{w}^s, \alpha^s)$ is the utility of the previously accepted set, and T^t is the temperature at time t. Note that in SCODEL, a smaller value of U indicates better utility or less futility. When accepted, the accepted coefficients will be replaced by the current coefficients. Equation 10 indicates the acceptance of hypotheses is mainly influenced by the definition of the utility of the coefficient set and thus the same hypotheses can be accepted or rejected depending on the particular context.

In sum, SCODEL does not assume learning involves computation intensive (back) propagations of classification error in the multi-faceted coefficient space, or calculation of partial derivative for each coefficient for the error hypersurface. Rather, in the present learning model framework, a very simple operation (e.g., comparison of two utility values) along with the operation of stochastic processes is assumed to be the key mechanism in category learning. In SCODEL, the knowledge about the category evolves as learning progresses by permitting mainly "good" sets of hypotheses and occasionally "bad" ones to survive and using such enduring hypotheses as bases for generating a new set of hypotheses. The goodness or badness of a set of hypotheses in the SCODEL framework is purposely designed to be situation-specific, resulting in context dependent learning processes. Unlike the gradient descent method, the utility functions for SCODEL do not need to be differentiable or continuous, and thus it allows more flexibility for defining its contextual factors.

Table 1. Schematic representation of stimulus set used in simulations.

Category	Dim1	Dim2	Dim3	Dim4
A	1	1	3	4
A	1	1	4	1
A	1	1	1	2
B	2	2	2	1
B	2	2	3	2
B	2	2	4	3
C	3	3	1	3
C	3	3	2	4
C	3	3	3	1
D	4	4	4	2
D	4	4	2	3
D	4	4	1	4

3 Simulations

Descriptions of Empirical Study: In the present simulation study, predictions of AOI models of category learning via knowledge utility optimization were compared with real human behaviors. To this end, I simulated the results of an empirical study on classification learning, Study 2 of Matsuka [6]. In this study, there were two perfectly redundant feature dimensions, Dimensions 1 & 2 (see Table 1), and those two dimensions are also perfectly correlated with category membership. Thus, information from only one of the two correlated dimensions was necessary and sufficient for perfect categorization performance. Besides classification accuracy, data on the amount of attention allocated to each feature dimension were collected in the empirical study. The measures of attention used were based on feature viewing time, as measured in a MouseLab-type interface [10]. It should be noted that all feature values are treated as nominal values differentiating each element within the dimension, and thus their numeric differences do not have any meaning.

The empirical results that I am trying to simulate indicated that 13 out of 14 subjects were able to categorize the stimuli almost perfectly. Specifically, mean classification accuracy in the last training was 92.3%. The aggregated results suggest that on average subjects paid attention to both of the redundant dimensions approximately equally (Fig 1 top row, left column). However, more interestingly when the attention data were analyzed per individual, it was found that many subjects tended to pay attention primarily to only one of the two correlated dimensions, particularly in the late learning blocks as shown in Fig 1, top row, right column. The scatter plot shows the relative amounts (thus sum of the amount of attention allocated across all dimensions was fixed at 1) of attention allocated to feature Dimension 1 (on x-axis) and Dimension 2 (on y-axis). This suggests that subjects tended to ignore non-diagnostic feature dimensions, Dim3 and Dim4, and tended to have a bias toward one of the two redundant diagnostic feature dimensions, indicating that they have tried to perform the categorization task with smaller amounts of attention allocation or mental effort.

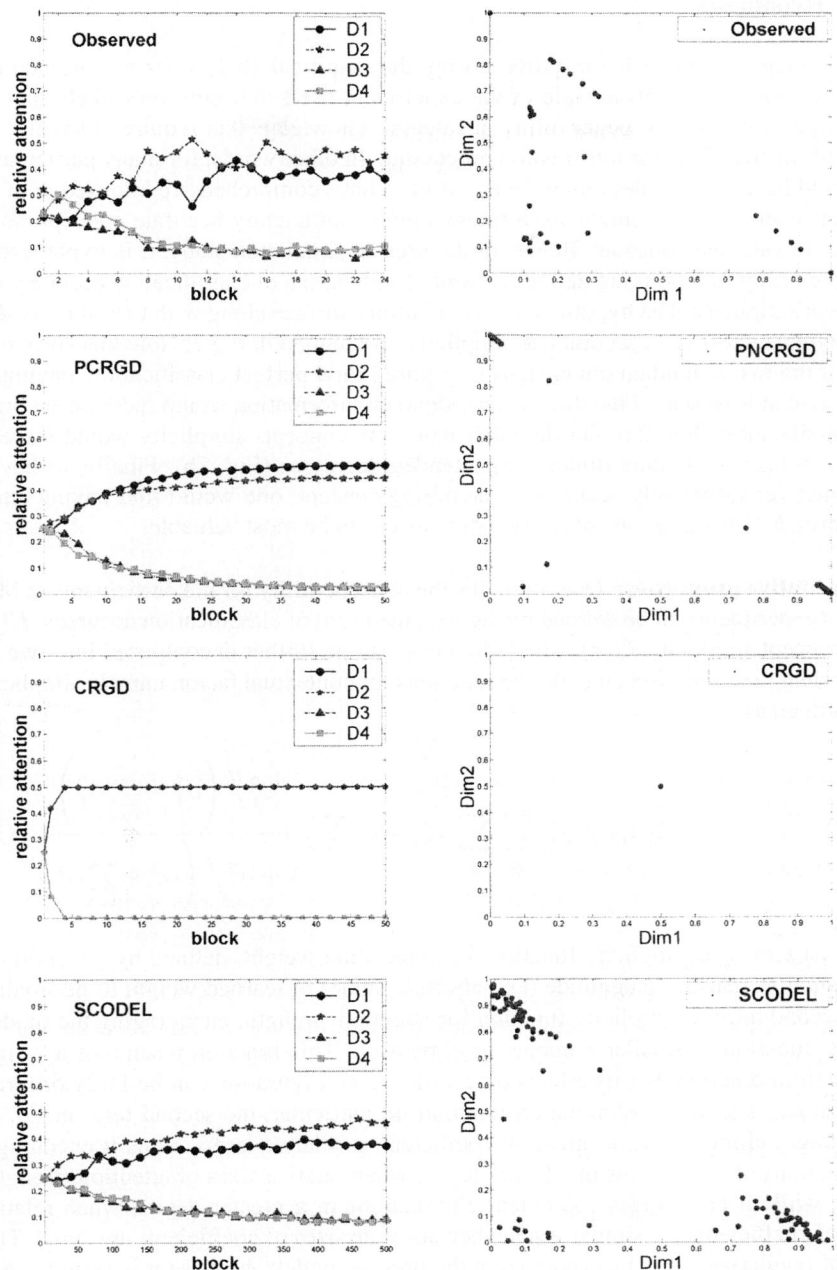

Fig. 1. Empirical finding of Experiment 2 of Matsuka (2002) in top row and the results of simulations in the remaining rows. The graph on left column shows the amounts of relative attention allocated to the four feature dimensions. The scatter plot (right column) compares relative attention allocated to Dimensions 1 and 2 for the last training blocks (for observed data, the last three blocks), where each dot represents an observation.

3.1 Hypotheses

Since category knowledge acquired during the empirical study does not provide any merit to the participants outside of the experiment itself, it seems very likely that the participants would find better utility in category knowledge that requires a smaller degree of mental effort for information processing. In other words, for many participants, it would have been useless to make an effort to have comprehensive knowledge of the stimulus set, and they might have found simple, sufficiently accurate concepts to be the more valuable concepts. Hence, in the present simulation study, it is hypothesized that the utility functions are defined as some combination of classification accuracy and concept simplicity. The hypothetical overall utility surface along with that of individual contextual factors (i.e., accuracy & simplicity) are plotted in Fig 2. Note that since only one of the two redundant dimensions is required for a perfect classification, paying attention to at least one of the diagnostic redundant information would increase accuracy in classification (Fig. 2a). On the other hand, the concepts simplicity would increase as the number of feature dimensions attended decreases (Fig. 2b). Finally, to have a compact yet sufficiently accurately-classifying concept, one would find paying attention primarily to either one of the two dimensions to be most valuable.

Quantitative properties: Operationally, the concept utility for the participants in Matsuka's experiment can be defined by the weighted sum of classification accuracy, $E(\theta)$, and concept simplicity, $Q(\theta)$. The Q function can be further decomposed into two independent functions. Specifically the functions for contextual factor, namely simplicity, are defined as:

$$Q(\theta) = \Omega(w_{kj}) + A(\alpha_i) = \gamma^w \sum_k \sum_j w_{kj}^2 + \gamma^\alpha \sum_i \frac{\alpha_i^2 \Big/ \Big(\alpha_i^2 + \sum_{l \neq i} \alpha_l^2\Big)}{1 + \alpha_i^2 \Big/ \Big(\alpha_i^2 + \sum_{l \neq i} \alpha_l^2\Big)} \quad (11)$$

The first term is a simplicity function for association weight, defined by a weight decay function, causing magnitude (i.e., absolute value) of learned weight to be smaller. The second term is simplicity function for attention strength, encouraging the models to pay attention to smaller numbers of dimension. This function resembles a weight elimination function, but its effects on coefficient configuration can be fairly different from those of a weight elimination function. In particular, the second term in Eq. 11 encourages eliminations of attention coefficient in relative terms. Thus it encourages eliminations of coefficients in a lesser degree when relative sizes of attention strengths differ, while it encourages coefficient eliminations in a greater degree when relative sizes of coefficients are similar, even when absolute sizes of coefficients are small. This type of regularization is incorporated in the present models, because it is assumed that the total amount of attention made available for a particular category learning task dynamically changes (e.g., more available in earlier stages of learning than later stages). To allocate dynamically changing attention resources in the contextually suitable manner, the utility of attention allocation patterns is best defined in relative amounts.

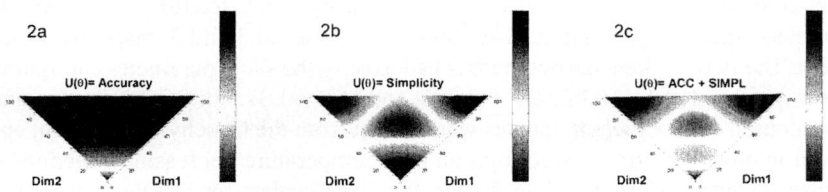

Fig. 2. Hypothetical utility (futility) surfaces relative to the amounts of attention allocated to feature Dimensions 1 and 2 (i.e., redundant dimension). The graphs are, from left, $U(\theta)$ = classification accuracy, $U(\theta)$ = simplicity, $U(\theta)$ = classification accuracy + simplicity.

3.2 Model Implementations

Noisy context regularized gradient decent: The standard gradient descent method along with random error are used to update the model's coefficient. Thus,

$$\Delta w_{kj} = \lambda_w^E \cdot e_k \cdot h_j - 2\lambda_w^{\Omega} \cdot w_{kj} + \nu_{kj} \qquad (12)$$

$$\Delta \alpha_i = -\lambda_\alpha^E \sum_j \sum_k e_k \cdot w_{kj} \cdot h_j \cdot c|\psi_{ji} - x_i| - \lambda_\alpha^A \frac{2\alpha_i \cdot \sum_{l \neq i} \alpha_l^2}{\left(2\alpha_i^2 + \sum_{l \neq i} \alpha_l^2\right)^2} + \nu_i \qquad (13)$$

where $\nu_{kj} \sim N(0, MAX(\Delta w_{kj}^{t-1}))$, $\nu_i \sim N(0, MAX(\Delta \alpha_i^{t-1}))$, MAX is a function that returns the maximum value, and superscript t indicates time.

SCODEL: As in the previous model, SCODEL's utility function consists of two independent functions, namely classification accuracy and concept simplicity. However, since in general, SCODEL requires a stable estimate of classification accuracy of a particular hypotheses set (i.e., model coefficients), it obtains the estimate based on one training block. Thus, for SCODEL, the utility function is defined as:

$$E(\theta) = \sum_n \sum_k (d_k - O_k)^2 + Q(\theta) \qquad (14)$$

where $Q(\theta)$ is as defined in Eq. 11, and N is the number of instances in one training block (i.e., 12).

Methods: Three models with different learning methods were trained in the present simulation study. The models were (a) probabilistic (noise augmented) contextually regularized gradient descent (PCRGD), (b) contextually regularized gradient descent without noise (CRGD), and (c) stochastic context dependent learning (SCODEL). The CRGD model was included for illustrative purposes showing the importance of stochastic noise in AOI modeling. The models were run in a simulated training procedure to learn the correct classification responses for the stimuli of the experiment. Two gradient descent based models were run for 50 blocks of training, where each block consisted of a complete set of the training instances. The SCODEL based model was run for 500 blocks of training. For all models, the final results are based on 100 replications. The model configurations were selected rather arbitrarily. Their feedforward or ALCOVE parameters, specificity, c, and decisiveness, ϕ, were set to 3 and 5, respectively for all models. The two gradient decent models had exactly the same parameter configuration, namely $\lambda_w^E = 0.1, \lambda_\alpha^E = 0.05, \lambda_w^Q = 0.005$, and $\lambda_\alpha^Q = 0.01$. For SCODEL, the random numbers used for coefficient updates were drawn from the Cauchy distribution, and its random number generation algorithm and the temperature decreasing functions (i.e., exponential decay) were based on Ingber [9]. The γ scalars for SCODEL (see Eq. 11) were 0.25 and 0.5 for association weights and attention strength, respectively.

3.3 Results and Discussion

The results of simulations are plotted in Fig. 1, second to fourth rows. All models were successful in learning the stimulus set. The classification accuracies in the last training block were 94%, 93%, 92% for PCRGD, CRGD, and SCODEL, respectively. Note that predicted classification accuracies can be easily adjusted by manipulating ALCOVE's classification decisiveness parameter ϕ and thus are not of the greatest importance other than the fact that all three models reached similar classification accuracies with the same ϕ value. All models qualitatively replicated aggregated observed attention learning curves. However, only PCRGD and SCODEL were able to show the bias in paying attention primarily to one of the two diagnostic redundant dimensions. All 100 simulated CRGD learners paid exactly the same amounts of attention to Dim1 and Dim2, even with a contextual factor, simplicity. As compared with PCRGD, more SCODEL learners resulted in non-minima (e.g. paying little attention to both Dim1 and Dim2 or weaker bias in attending only one diagnostic dimensions, see Fig 2c). Although, such prediction is most likely unacceptable for ASI (artificial superior intelligence) research, as an AOI (artificial ordinal intelligence) research, this SCODEL's prediction appears more preferable than that of PCRGD because of the qualitative resemblance of its prediction to the observed phenomena. One possible explanation for the differences between PCRGD's and SCODEL's predictions, above and beyond the differences in the model parameters (e.g., λs & em γs), is as follows: while PCRGD would continue optimizing concept utility during the entire learning path, thus resulting in better outcomes in terms of utility maximization, SCODEL would stop optimizing or would have trouble finding a better hypotheses set when its temperature decreases "too fast", resulting in annealing at local minima or even at non-minimal points. This is because, while PCRGD updates each coefficient independently, all coefficients are updated interdependently and collectively in SCODEL. That is, SCODEL accepts a hypotheses set only when all coefficients or concepts collectively provide sufficiently worthy utility (however, it probabilistically accepts a hypotheses set with a worse utility value, see Eq. 10). For example, even when a subset of hypothetical concepts is "correct" they may not be accepted if the other subset of hypotheses is "incorrect". Thus, SCODEL may find it progressively difficult or become less motivated to find (marginally) a better set of concepts as learning progresses, if its temperature decreases too fast (i.e., being "too confident" about its learning progress) and/or it is stuck at strong local minima (i.e., being "too confident" about their acquired knowledge). On the other hand, in PCRGD, regardless of "correctness" of updates of the other coefficients or concepts, each coefficient is probabilistically and continuously adjusted in a "correct" manner.

To describe the empirical phenomena observed in Matsuka's experiment with SCODEL's less successful optimization, yet apparently more successful replication of the empirical observations, participants somehow would not mind having less-than-the-optimal concepts or would not bother trying to find better category concepts after they have had "enough" training or experience with the stimuli. Although this is probably true only for learning of mundane and non-critical tasks, yet as a model of ordinary thinker, SCODEL might have captured the essence of the nature of humans' ordinary information processes pattern better than PCRGD.

4 Discussion and Conclusion

Learning Complexity: Both learning mechanisms underlying PCRGD and SCODEL appear equally complex or simple. However, SCODEL's implied cognitive processes can be interpreted as, in general, simpler than that of the PCRGD leaning method. For example, if one could disregard the computational complexity of the random number generation (i.e., random number generation requires very small mental effort), then SCODEL would calculate only $M^{w+\alpha} + M^w + M^\alpha$ numbers of values in learning, where $M^{w+\alpha}, M^w, M^\alpha$ are the number of contextual factors defined for both association weights and attention, association weights alone, and attention alone, respectively. For example, assuming each factor is calculated independently, a SCODEL learner in the simulation study would calculate only three independent values in each step of learning (and this number stays the same, regardless of the numbers of exemplars and the dimensionality of the stimulus set, as long as the model is influenced by the same contextual factors). In contrast, in each learning step, a PCRGD with the same contextual factor requires calculations of $M^{w+\alpha} \times (J \times K + I) + M^w \times J \times K + M^\alpha \times I$ coefficient updates, where, J, K, I are the numbers of exemplars, category types, and feature dimensions, respectively. Thus, for the stimulus set in the simulation study, there would have been 28 + 24 + 4 = 56 values to be calculated.

Even if one cannot disregard the computational complexity of the random number generation, the required mental effort for random number generation appears much less than that for calculation of gradient. That is, a PCRGD based learning model assumes that all coefficient updates would be executed correctly with some complex mathematical operations (e.g. Eq. 6 & 7), while SCODEL's stochastic hypotheses generation does not assume correctness nor any explicit mathematical operations. In these regards, SCODEL models simpler cognitive processes than PCRGD.

There is, however one characteristic of SCODEL that can be considered more complex than PCRGD. That is, SCODEL usually requires multiple instances in training to estimate classification accuracy associated with each hypotheses set. In so doing, SCODEL needs to store information on its performance in multiple instances, whereas in PCRGD, it does not require any memory of past performance. It is, however, uncertain which models are more descriptive in this regard.

Limitations and Extension: Effects of contextual factors in the present models are assumed rather static. However, in reality the effects probably change dynamically. Extending present models to take dynamically changing contexts into account helps them describe more psychological phenomena observed in empirical studies. A related but more important issue is that the present models do not describe how the models came to "realize" different contextual factors. In other words, modelers need to define contexts and supply their functions. Although these seem like very big challenges to be tackled in the field of cognitive modeling research, incorporating such a system is certainly a constructive step towards plausibly and descriptively modeling human category learning processes.

4.1 Conclusions

Although humans are cognitive animals who can adaptively modify their information processes depending on the circumstances, such capability has remained largely underrepresented in previous modeling research in high-order human cognition. The present work is an attempt to outline a framework of cognitive modeling approach that emphasizes context dependent learning strategies and processes. In particular, the present research assumes that the objective in human learning processes is not strictly error minimizing, but the optimization of utility of their knowledge or concepts. Two models with this fundamental concept but with different learning algorithms were introduced and tested. A simulation study qualitatively suggests the descriptive validity of this approach. However, more thorough simulation studies comparing model predictions against empirical phenomena should be conducted in future studies for better understanding the nature of high-order human cognition and for the development of better models of ordinary human intelligence.

References

1. Nosofsky, R. M. (1986). Attention, similarity and the identification –categorization relationship. *Journal of Experimental Psychology: General, 115*, 39-57.
2. Matsuka, T (2004). Biased stochastic learning in computational model of category learning. In *Proc of the 26th Annual Meeting of the Cognitive Science Society.* (pp. 909-914)
3. Matsuka, T (To appear) Simple, Individually-Varying, and Context-Dependent Learning Method for Models of Human Category Learning. *Behavior Research Methods.*
4. Matsuka, T. & Corter, J. E. (2004). Stochastic learning algorithm for modeling human category learning. *International Journal of Computational Intelligence, 1* 40-48.
5. Kruschke, J. E. (1992). ALCOVE: An exemplar-based connectionist model of category learning, *Psychological Review, 99*, 22-44.
6. Matsuka, T. (2002) Attention processes in computational models of category learning. Ph.D. Thesis, Columbia University, New York, NY.
7. Rehder, B. & Hoffman, A. B. (2003). Eyetracking and selective attention in category learning. In *Proc of the 25th Annual Meeting of the Cognitive Science Society.* [CD-ROM].
8. Metropolis, N., Rosenbluth, A. W., Rosenbluth, M. N., Teller, A. H., & Teller, E. (1953). Equation of state calculations by fast computing machines. *Journal of Chemical Physics, 21*, 1087-1092.
9. Ingber, L. (1989). Very fast simulated re-annealing. *Journal of Mathematical Computer Modelling, 12*, 967–973.
10. Bettman, J. R., Johnson, E. J., Luce, M. F.,& Payne, J. W. (1993). Correlation, conflict, and Choice. *Journal of Experimental Psychology: Learning, Memory, and Cognition, 19*, 931-951

Automatic Text Summarization Based on Lexical Chains

Yanmin Chen, Xiaolong Wang, and Yi Guan

Dept. of Computer Science and Engineering, Harbin Institute of Technology,
Harbin 150001, China
{chenyanmin, wangxl, guanyi}@insun.hit.edu.cn

Abstract. The method of lexical chains is the first time introduced to generate summaries from Chinese texts. The algorithm which computes lexical chains based on the HowNet knowledge database is modified to improve the performance and suit Chinese summarization. Moreover, the construction rules of lexical chains are extended, and relationship among more lexical items is used. The algorithm constructs lexical chains first, and then strong chains are identified and significant sentences are extracted from the text to generate the summary. Evaluation results show that the performance of the system has a notable improvement both in precision and recall compared to the original system[1].

1 Introduction

Summarization is a reductive transformation of source text to summary text through content reduction by selection and/or generation on what is important in source text (Jones 1999). It is valuable in the wide landscapes of cognitive science, information systems research, AI, computational linguistics, etc. Since summarization is defined by the related human skills and concepts, cognitive approaches help to establish how a summarization process is organized, which features of the source text influence the resulting summary, how intended uses shape the summary, and so on. One important influence on automatic text summarization has been the psychological study of human summarization in the laboratory (Kintsch and van Dijk 1978). Subjects in these experiments have been found to use conceptual structure for text comprehension. Experiments reveal that humans create a hierarchical discourse organization, which provides retrieval cues for memory; they restore missing information through inference-based reconstruction processes. The other important influence on automatic summarization is the study of professional abstractors. Endres-Niggemeyer et al.(1995, 1998) found that professional abstractors take a top down strategy, exploiting discourse structure. Lexical chains, as a discourse-level approach, are sequences of related words between which lexical cohesion occurs. In fact, cohesion itself is an abstract concept, representing an intuition: some relation help to stick together different parts of the text. This made that some parts of the text are about the same things in a sense. Namely, cohesion is such a method: it admits the appearance of the term directly relevant to the theme as the appearance of the important content. We mainly exploit lexical chains on summarization here. Terms

---

[1] Sponsored by the National Natural Science Key Foundation of China (60435020) and the High Technology Research and Development Programme of China (2002AA117010-09).

are aggregated into lexical chains based on relationships like synonym and hypernym; the indicative summary is built by using these chains to select important sentences.

Some related work is presented in the following section. Section 2 describes our lexical cohesion technology and introduces how to use the HowNet. Section 3 selects the strongest chains from lexical chains constructed above and generates summary. The evaluation of presented system is discussed in section 4. Section 5 gives the conclusion.

2 Algorithm for the Construction of Lexical Chains

The notion of cohesion is introduced by Hasan and Halliday(1995). Cohesion is a device for sticking together different parts of the text by using semantically related terms, co-reference, ellipsis and conjunctions. Lexical cohesion occurs not only between two terms but also among sequences of related words, namely lexical chains [1]. Lexical chains provide a representation of lexical cohesive structure of text. It can be used in information retrieval, topic tracking, summarization, etc[2,3]. Morris and Hirst [1] presented the first computational model for lexical chains based on Roget's Thesaurus. Then Hirst et al. [2,4,5] presented their algorithms for the calculation of lexical chains based on WordNet, respectively. In their algorithms, the relations among words are determined by WordNet. In WordNet, English nouns, verbs, adjectives and adverbs are organized into synonym sets which represent different underlying lexical concept. Different relations, such as synonymy and hyponymy, link the synonym sets. Unlike that in English, calculation of lexical chains uses the HowNet to determine the relation among the Chinese words. HowNet is an on-line common-sense knowledge base unveiling inter-conceptual relations and inter-attribute relations of concepts as connoting in lexicons of the Chinese and their English equivalents. In HowNet, every concept of a word or phrase and its description form one entry. Regardless of the language types, an entry consists of four items: {W_X= word/phrase form, G_X = word/phrase syntactic class, E_X = example of usage, DEF = concept definition}. Just as that in WordNet, there are some relations like hypernym, synonym, antonym, converse, part-whole, attribute-host, location-event, etc., which exist in HowNet and are presented in DEF.

In preprocessing, Hirst et al. chose all words that appear as noun entries in WordNet as candidate words. Barzilay et al. selected simple nouns and noun compounds as candidate words relying on the results of Brill's part-of-speech tagging algorithm to identify, while we experimentally extended the candidate words to nouns, verbs, adjectives, etc. The algorithm we presented for construction lexical chains is as follows:

1. Segment Chinese words in text, and filter stopwords;
2. Select a set of words $w_1 w_2 \cdots w_n$ which exist in HowNet as candidate words;
3. For each candidate word w_j ($j \in [1, n]$), find an appropriate chain L relying on a related criterion among members of the chains;
4. If an appropriate chain L is found, w_j is inserted in the chain together with some elements of its DEF;
5. If no chain is found, construct a new chain for the candidate word w_j.

The algorithm segments Chinese words in text first. Then it filters stopwords with very little semantic contents such as empty words and some high-frequency words, which did not contribute much to information retrieval for frequently appeared in many documents. The words that appear as entries in HowNet are selected as candidate words. For each candidate word, an appropriate chain is found relying on related criterion among members of the chains. The relationship between words is determined by their DEF in the HowNet knowledge dictionary. The related criterion is the sum of relation-weights of w_j and its DEF to w_r ($w_r \in L$) and its DEF is greater than a threshold T which was determined experimentally. Here we assign the relation-weights 1.0 to 0.2 to different relations. For example, *concepts co-relation* has a relation-weight 0.8. If a word has more than one DEF, we will consider all of its DEF in judgement. It is noteworthy that some definitions in DEF are so common that they cannot reflect the relation between two words, such as "attribute", "event", "ProperName". Some even disturb the judgement by ambiguous DEF. They should be filtered from the chains for word DEF disambiguation. Therefore we omit 45 definitions in DEF experimentally. In addition, the algorithm stipulates that the same words belong to the same chains, because same words usually have the same semantic in the same text based on experiment.

3 Summary Generation

When lexical chains are constructed, the strongest chains among the chains must be identified and are selected to generate a summary. The importance of a lexical chain is evaluated by its contribution to the themes of document based on experience and experiment. The score of each lexical chain is calculated by the following formula:

$$S = \sum_{m=1}^{C} \omega_m \times H \qquad (1)$$

$$H = 1 - C \Big/ \sum_{m=1}^{C} \omega_m \qquad (2)$$

Where, S is the score of a lexical chain, m is the frequency of the mth element w_m in chain, H is a homogeneity index, and C is the total occurrences of the difference element in chain. The score of strong chain satisfied with the condition: $S \geq A_S + D_S$. Where, A_S is the average score of lexical chains, D_S is the standard deviation of scores.

Once the score of each chain is calculated, strong chains are ranked in the order of their score. Then the important sentences are extracted from the original text based on the chain score. Considering that all words in a chain reflect the same topic of the chain in different extent, a typical word is selected from each strong chain to represent the topic of it. The frequency of typical word in text satisfied with the follow condition: $\omega \geq \sum_{m=1}^{C} \omega_m / C$. To typical word, given an appropriate measure of strength, we show that picking the concepts represented by above condition gives a better indication of the central topic of a text than simply picking the most frequent words in the lexical chains. The method of building summary is adding sentence that contains the first appearance of a typical word of the strongest chain to the summary, until the summary reaches the specified length or there is no sentence left in strong chains. The specified length of summary is determined by user or a percent of the length of original text, such as 10%.

The sentences extracted from original text are reorganized in its original position in text. The anaphora technology is applied to improve the fluency of summary. The name in text is firstly recognized by fusing the method of natural language modeling and related name rules. 10 rules of substituting pronoun are summarized based on lots of analysis of Chinese sentences, and an anaphora resolution algorithm is provided. When a personal pronoun is appeared in abstract sentence, it is substituted according to related rules. Experiment result shows that our method can process in excess of 80% anaphora phenomena. The method gave almost accurate answers to anaphora in text with simpler character and environment. Thus it satisfied the need of automated summarization.

4 System Evaluation

Generally, there have two kinds of evaluation methods: intrinsic evaluation and extrinsic evaluation. Intrinsic evaluation directly analyzes the summary to judge the quality of summarization. Extrinsic evaluation judges the quality of summarization by its affection to some other task. In this paper, an intrinsic evaluation method is presented. And a series of comparison experiments are carried out to analyze the performance of the system. Evaluation experiments are discussed as follows.

A set of 100 Chinese newswire texts that are various genres is randomly collected from Internet to construct the testing corpus. For each text, three graduate students constructed manually and independently "ideal" summaries in two proportions: 10% and 20%, which are the rates of the summary length to the original text length. The text length is counted in the number of sentences here. Then the summaries generated by lexical chains algorithm are compared with the ideal summaries extracted by human. For each text, the precision and recall are computed to evaluate the quality of the summary. They are defined as follows:

$$\text{Precision} = |S_t \cap S_m|/|S_m| \tag{3}$$

$$\text{Recall} = |S_m \cap S_c|/|S_c| \tag{4}$$

Where, S_m is the set of summary sentences produced by the system, $S_t = |S_1 \cup S_2 \cup S_3|$, $S_c = |S_1 \cap S_2 \cap S_3|$, S_t is the union set of the 3 sets of summary sentences manually extracted by 3 graduate students, and S_c is the intersection set of that. The operator "| |" takes the cardinality of a set. The summaries generated by our system are compared with those obtained from original lexical chains methods. The results are shown in Table 1:

Table 1. Evaluation of summarization system

Summary Rate		Sys0	Sys1	Sys2	Sys3	Sys4
10%	Precision	0.726	0.672	0.725	0.694	0.688
	Recall	0.772	0.729	0.749	0.746	0.751
20%	Precision	0.712	0.654	0.697	0.687	0.669
	Recall	0.75	0.694	0.728	0.743	0.717

Table 1 shows the results of the presented system Sys0 are better on average than that of Sys1, Sys2, Sys3, and Sys4. The lowest average results are obtained with the

Sys1 that use the original method. It indicates that the presented system Sys0 has a notable improvement above the original system both in precision and recall. Lexical chains are good predictors of main topic of a text. By selecting suitable candidate words, omitting some definitions in DEF, selecting right typical words and so forth, the modified algorithm can better reflect the topic of text. Sys2 is similar to Sys0 but not filtering any definitions in DEF. This suggests that some definitions in DEF do not contribute much to the judgement of relation between two words, and even disturb it. In Sys3, typical word is defined as the word that its frequency is the max in the chain. This indicates the word that has the max frequency in the chain is not necessary the most important word in the text. Sys4 choose all words that appear as noun entries in HowNet as candidate words. The result shows that only choosing words that appear as noun entries in HowNet as candidate words may cause some important words to be ignored, such as some useful verbs and adjectives. In addition, the 10%-length summaries are significantly better than the 20%-length summaries. This suggests that difference in summaries is growing as the summary length increasing. In fact, with the summary length increasing, the difference in manually summaries of each expert is growing.

5 Conclusion

The system that produces a summary of Chinese text by exploiting "lexical chains" is presented. The experimental results show that lexical chains are effective for Chinese texts summarization, and the performance of presented system has a notable improvement above the original system both in precision and recall. The approach is highly domain-independent, even though we have illustrated its power mainly for newswire texts. It can be applied to daily web texts. In future, we plan to combine lexical chains with additional knowledge sources to further improve the performance of the system.

References

1. Morris, J., Hirst, G.: Lexical cohesion computed by thesaural relations as an indicator of the structure of the text. Computational Linguistics **17** (1991) 21–48
2. Hirst, G., St-Onge, D.: Lexical chains as representation of context for the detection and correction of malapropisms. In Fellbaum, C., ed.: WordNet: An electronic lexical database and some of its applications. The MIT Press, Cambridge, MA (1998) 305–332
3. Chan, S.W.: Extraction of salient textual patterns: Synergy between lexical cohesion and contextual coherence. IEEE Transactions on Systems, Man, and Cybernetics - Part A: Systems and Humans **34** (2004) 205–218
4. Barzilay, R., Elhadad, M.: Using lexical chains for text summarization. In Mani, I., Maybury, M.T., eds.: Advances in Automatic Text Summarization. The MIT Press, Cambridge, Massachusetts (1999) 111–121
5. Alam, H., Kumar, A., Nakamura, M., Rahman, F., Tarnikova, Y., Wilcox, C.: Structured and unstructured document summarization: Design of a commercial summarizer using lexical chains. In: 7th International Conference on Document Analysis and Recognition. Volume 2., Edinburgh, Scotland, UK (2003) 1147–1150

A General fMRI Linear Convolution Model Based Dynamic Characteristic[*]

Hong Yuan, Hong Li, Zhijie Zhang, and Jiang Qiu

School of psychology, Southwest China normal university, Chongqing, 400715, PR China
yuanyh3927@163.com

Abstract. General linear model (GLM) is a most popularly method of functional magnetic imaging (fMRI) data analysis. The key of this model is how to constitute the design-matrix to model the interesting effects better and separate noises. In this paper, the new general linear convolution model is proposed by introducing dynamic characteristic function as hemodynamic response function for the processing of the fMRI data. The method is implemented by a new dynamic function convolving with stimulus pattern as design-matrix to detect brain active signal. The efficiency of the new method is confirmed by its application into the real-fMRI data. Finally, real- fMRI tests showed that the excited areas evoked by a visual stimuli are mainly in the region of the primary visual cortex.

1 Introduction

Functional magnetic resonance imaging (fMRI) is a new non-hurt measure technique for brain activity, which have been used at the study of brain cognition, locating nerve activity, medicine, psychology and other domains, and has become one of the most important way of the study of brain function[1]. At present, fMRI data processing have two methods: model driven and data driven method[2-9]. The model driven method is mainly the general linear model, which been advanced by Friston , is generally used to identify functionally specialized brain regions and is the most popularly approach to characterize functional anatomy and disease-related changes[6-9]□but its theory is imperfect and in the process of continual development [10-16]. The design matrix is the key of the model analysis. Whether the result of data process is right directly connect with the quality of design matrix. Up to now, it is not a perfect criterion.

In this paper, a new convolution model is firstly presented by a new dynamic function convolving with stimulus pattern as design-matrix to detect brain active signal, then the real-fMRI experiment data is analyzed.

---

[*] Supported by NSFC#90208003 and #30200059, TRAPOYT, the 973 Project No. 2003CB716106.

2 A General Linear Convolution Model

2.1 General Linear Model(GLM) Theory

The general linear model can be represented by

$$y_i = \beta_0 + x_{i1}\beta_1 + \cdots + x_{ik}\beta_k + \cdots + x_{ip}\beta_p + \varepsilon_i, i = 1, 2, \cdots, n \tag{1}$$

Where y_i is the activation value of the voxel in scan i (i.e., at time point i), β_0 denotes the overall mean of the time series, β_k represent the influence of the experimental condition k on scan i, i.e., if at scan i condition k is measured, then x_{ik} equals one and β_k denotes the activation change that is caused by condition k. The terms x_{ik} of conditions k that do not exert an influence on scan i (but may have on other scans) are zero. Nevertheless, it is possible that more than one x_{ik} is one. The measurement error at scan i is denoted by ε_i. ε_i is assumed to be independent and identically normally distributed with the expected value $E(\varepsilon)$ being zero.

The 'least-squares' estimation of parameter is

$$\hat{\beta} = (X^T X)^{-1} X^T y \tag{2}$$

By t-distribution test, Given $\alpha = 0.05$, then $t_\alpha = 1.96$. If $|t| > t_\alpha$, then it is said the voxel is significantly activated by the i-th biological effect. By plotting the absolute value of t in the location of the voxel which is dealing with and repeating the above process with every voxel in the image space, it have obtained the statistic parameter map of the i-th effect[3-6]. However, the design matrix is simply implemented to analyze fMRI data, the result is unsatisfactory.

2.2 Dynamics Convolution Model

In general, the general linear model take Gamma function as hemodynamic response function to get design matrix. But, it is not reasonable in most actually case. The BOLD signal of the cerebral activation is a collective response of an activated region and it can be explained as a mutual interaction process between the neural response to a stimulus and the hemodynamic change due to the activation of a neural cluster. Therefore, the BOLD signal $u(t)$ of a cerebral activation can be expressed by convolution of the neural response cosine function $\cos(t)$ with decay signal $\exp(-t)$ of the CBF hemodynamic response of a neural cluster.

$$u(t) = \cos(t) \otimes \exp(-t) \tag{3}$$

Where ⊗ note the convolution operation. So as to combine biology effects of brain better, instead of box-car stimuli series at the first column, we convolved a new hemodynamic response function $u(t)$ with experiment stimuli patter as the first columns design matrix. The rest of columns model physiological noise which aroused by respiration and palpitation are coincident with Friston model[6-9]. Then we can gain the more realistic design matrix.

3 Application for Real-Human fMRI Data

3.1 Data Description

The fMRI data was collected at Beijing hospital. The stimulus was a red illuminant point presenting at the center of the visual field with frequency 8HZ, light intensity 200cd/cm^2 and visual angle of 2 degrees. Totally six transverse sections were collected in a bottom-up direction. Each section is composed of 128×128 voxels. Each section map was completely collected in 160 seconds resulting in 80 sample images alternating between stimulation and non-stimulation box-car stimuli series. The sample time interval is 2secs.

 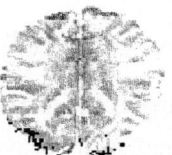

Fig. 1. Design-matrix   Fig. 2. fMRI result image

3.2 Result of Data Analysis

According to 2.2 methods, we get BOLD response of brain activity. The improved design matrix is shown in the Fig. 1. According to the principle of GLM, the cut-off condition T−test >4, gave approximately $P < 0.0001$, and were considered to be the true active voxels. Then the fMRI result image is gained in Fig.2, where the black points denote those activity voxels aroused by stimuli. Result of experiment correspond to the physiology fact that the excited areas evoked by a visual stimuli are mainly in the region of the primary visual cortex which located at outboard of occipital lobe, which consists with the result of independent component analysis (ICA) [2-5].

4 Conclusion

In this paper, we proposed a new General linear Convolution model of fMRI data for processing. The validity of the new method is testified by detecting the brain functional activation from fMRI data.

References

1. Belliveau J. W.,.Kennedy D.N, Mckinstryet R.C. al, Functional mapping of the human visual cortex by magnetic resonance imaging, Science.254(1991)716−719.
2. Chen H, Yao D, Liu Z. Analysis of the fMRI BOLD Response of Spatial Visual. Brain Topography. 17 (2004) 39-46
3. Chen H Yao D. Composite ICA algorithm and the application in localization of brain activities. Neurocomputing, 56 (2004) 429-434.
4. Chen H, Yao D. Discussion on the choice of separation component in fMRI data analysis by Spatial Independent Component analysis. Magnetic Resonance Imaging, 22 (2004) 827-833.
5. Chen H, Yao D, Zhuo Y, Chen L. Analysis of fMRI Data by Blind Separation into Independent temporal Component, Brain topography. 15 (2003) 223-232.
6. Friston K.J., Frith, C.D., Frackowiak R.S.J. et al, Characterizing Dynamic Brain Responses with fMRI: A Multivariate Approach, NeuroImage. 2(2) (1995)166-172.
7. Friston K.J., Holmes A.P., Poline et al J.B., Analysis of fMRI Time-Series Revisited, NeuroImage. 2(1) (1995) 45-53.
8. Friston K.J., Mechelli A., Turner, R et al. Nonlinear Responses in fMRI: The Balloon Model, Volterra Kernels, and Other Hemodynamics, NeuroImage. 12(4) (2000) 466–477.
9. Friston K.J., Fletcher P., Josephs O., et al, Event-Related fMRI: Characterizing Differential Responses, NeuroImage. 7(1) (1998) 30–40.
10. Worsley K.J. and Friston K.J.. Analysis of fMRI Time-Series Revisited—Again. NeuroImage. 2(1) (1995)173-181.
11. Büchel C., Holmes A.P.,.Rees G et al, Characterizing Stimulus– response Functions Using Nonlinear Regressors in Parametric fMRI Experiments, NeuroImage. 8(2) (1998) 140–148.
12. Vazquez A. L. and Noll D. C., Nonlinear Aspects of the BOLD Response in Functional MRI□NeuroImage. 7 (1998) 108–118.
13. Bandettini P. A., Birn R.M., Kelley D. et al ,Dynamic nonlinearities in BOLD contrast: neuronal or hemodynamic? International Congress Series. 1235 (2002) 73–85.
14. Riera J.J., Watanabe J., Kazuki I. et al, A state-space model of the hemodynamic approach : nonlinear filtering of BOLD signals, NeuroImage. 21(2) (2004)547–567.
15. Chen H, Yao, D Liu Z. A comparison of Gamma and Gaussian dynamic convolution models of the fMRI BOLD response. Magnetic Resonance Imaging, 23 (2005) 83–88
16. Chen H, Yao D, Yang L. An extend Gamma dynamic model of functional MRI BOLD response. Neurocomputing, 61 (2004) 395-400
17. Chen H, Yao D, Chen L. A New Method for Detecting Brain Activities from fMRI Dataset, Neurocomputing 48 (2002) 1047-1052

A KNN-Based Learning Method for Biology Species Categorization

Yan Dang[1], Yulei Zhang[2], Dongmo Zhang[1], and Liping Zhao[2]

[1] Computer Science and Engineering Department, Shanghai Jiao Tong University,
Hua Shan Road 1954, Shanghai, P. R. China
dangyan@sjtu.edu.cn, zhang-dm@cs.sjtu.edu.cn
[2] School of Life Science and Biotechnology, Shanghai Jiao Tong University,
Hua Shan Road 1954, Shanghai, P. R. China
zhangyulei@sjtu.edu.cn

Abstract. This paper presents a novel approach toward high precision biology species categorization which is mainly based on KNN algorithm. KNN has been successfully used in natural language processing (NLP). Our work extends the learning method for biological data. We view the DNA or RNA sequences of certain species as special natural language texts. The approach for constructing composition vectors of DNA and RNA sequences is described. A learning method based on KNN algorithm is proposed. An experimental system for biology species categorization is implemented. Forty three different bacteria organisms selected randomly from EMBL are used for evaluation purpose. And the preliminary experiments show promising results on precision.

1 Introduction

Categorization is a very important issue in many fields. In order to identify and thoroughly understand something, people must categorize it first. In natural language processing (NLP), text categorization (TC) is the task of assigning a number of appropriate categories to a text document. This categorization process has many applications such as document routing, document management, or document dissemination [1]. The goal of TC is to learn categorization schemes that can be used to classify texts automatically. There are many categorization schemes addressed in categorization literature. It includes Naïve Bayes (NB) probabilistic classifiers, Decision Tree classifiers, Neural Network, KNN classifiers, Support Vector Machine (SVM), and Rocchio classifiers etc.

Human beings always want to penetrate into and uncover the mystery of the essential of life. The first step is categorization. Traditionally, when we study an unknown life-form, we classify it by its substance, shape and color etc first. And then we can infer its probable character according to its type. With the rapid development of molecular biology technology, nowadays people can easily extract DNA sequence fragments from different kinds of samples. It is faster and more accurate to do researches. But there is also a great problem faced by biologists: Although we have already had the whole genome sequence information of many organisms in Genbank/EMBL which are the most important sources in biology research field, much more unknown DNA sequences are left to be sequenced. When scientists get DNA fragments in their labs, most of them will do BLAST in Genbank. If there is no homogeneous gene sequence stored in the database, they can hardly get any information based on previous records and have no idea of what this unknown sequence might be or come from.

However, there is another theory named Genome Signature to overcome this limitation. Genome signature method focuses on short motifs, such as dinucleodites. It has been proved that dinucleotide relative abundance can be used as a genome signature [2,3,4,5]. Although there are huge amount of DNA sequences submitted and stored in the Genebank, when biologists do BLAST for many of their newly discovered DNA sequence fragments, they can not get any related homogeneous sequence from Genbank. Biologists call these fragments orphan sequences. These orphan sequences are often so important that we have to know where they are from. Using dinucleotide frequency profile vector as a genome signature to portray the DNA sequences, we don't need to have the information of traditional homogeneous gene sequences to predict where the fragments might come from.

So, in this paper, we have developed a KNN-based learning method for biology species categorization. We can use the existed sequence records, no matter whether they are homogeneous or not, to train our model. And then predict what kind the unknown fragment is or is most related to. Figure 1 illustrates the whole learning progress.

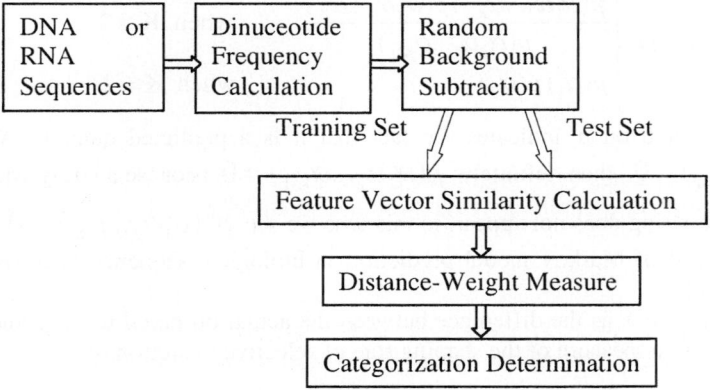

Fig. 1. Learning Progress

We carry out the categorization task by two phases. In the first phase, we map every DNA or RNA sequence into a composition vector. We extract the dinucleotide relative abundance as a genomic signature. We also do subtraction of random background to highlight the selective diversification. Then in the second phase, we build the learning method based on KNN algorithm. The composition vectors we get in the first phase are used as KNN's feature vectors. We use cosine-value between two feature vectors as their similarity.

2 Composition Vector Constitution

Given a DNA or RNA sequence of length L, we count the number of appearances of (overlapping) strings of a fixed length K (for dinucleodite $K=2$) in the sequence. There are N possible types of such strings: $N=4^K$.

For concreteness consider the case of one DNA or RNA sequence of length L. Denote the frequency of appearance of the K-string $\alpha_1\alpha_2...\alpha_K$ by $f(\alpha_1\alpha_2...\alpha_K)$, where each $\alpha_i \in \{A, C, G, T\}$ for DNA sequences or each $\alpha_i \in \{A, C, G, U\}$ for RNA sequences. This frequency divided by the total number $(L - K + 1)$ of K-string in the given sequence may be taken as the probability $p(\alpha_1\alpha_2...\alpha_K)$ of appearance of the string $\alpha_1\alpha_2...\alpha_K$ in the DNA or RNA sequence:

$$p(\alpha_1\alpha_2...\alpha_K) = \frac{f(\alpha_1\alpha_2...\alpha_K)}{(L-K+1)}. \quad (1)$$

Mutation happens in a more or less random manner at the molecular level, while selections shape the direction of evolution. Neutral mutations lead to some randomness in the K-string composition. In order to highlight the selective diversification of sequence composition, we must subtract the random background from the above counting results. This is done as follows:

Suppose we have done the counting for all strings of length $(K - 1)$ and $(K - 2)$. The probability of appearance of K-string is predicted by using a Markov model [6]:

$$p^0(\alpha_1\alpha_2...\alpha_K) = \begin{cases} \dfrac{p(\alpha_1\alpha_2...\alpha_{K-1})p(\alpha_2\alpha_3...\alpha_K)}{p(\alpha_2\alpha_3...\alpha_{K-1})} & \text{when } K > 2 \\ p(\alpha_1)p(\alpha_2) & \text{when } K = 2 \end{cases}. \quad (2)$$

The superscript 0 on p^0 indicates the fact that it is a predicted quantity. When $p(\alpha_2\alpha_3...\alpha_{K-1}) = 0$, then definitely $p(\alpha_1\alpha_2...\alpha_{K-1}) = 0$ because a string will not appear if its sub-string does not appear, in this case we set $p^0(\alpha_1\alpha_2...\alpha_K) = 0$. The usage of this kind of Markov model prediction in biological sequence analysis has been justified [7].

We then calculate X as the difference between the actual observed result p and the predicted value p^0, a measure of the shaping role of selective evolution by:

$$X(\alpha_1\alpha_2...\alpha_K) = \begin{cases} \dfrac{p(\alpha_1\alpha_2...\alpha_K) - p^0(\alpha_1\alpha_2...\alpha_K)}{p^0(\alpha_1\alpha_2...\alpha_K)} & \text{when } p^0 \neq 0 \\ 0 & \text{when } p^0 = 0 \end{cases}. \quad (3)$$

For all possible strings $\alpha_1\alpha_2...\alpha_K$, we use $X(\alpha_1\alpha_2...\alpha_K)$ as components to form a composition vector for a DNA or RNA sequence. To simplify the notations, we use X_i for the i-th component corresponding to the string type i, where $i=1$ to N ($N=4^K$ and the N strings are arranged in a fixed order as the alphabetic order). Hence we construct a composition vector $X=(X_1, X_2,..., X_N)$ for sequence X. As to dinucleotides, the composition vectors are 16-dimensional, in the form of $[X_1(AA), X_2(AC), X_3(AG),$

X_4(AT), X_5(CA), X_6(CC), X_7(CG), X_8(CT), X_9(GA), X_{10}(GC), X_{11}(GG), X_{12}(GT), X_{13}(TA), X_{14}(TC), X_{15}(TG), X_{16}(TT)] for DNA sequences and [X_1(AA), X_2(AC), X_3(AG), X_4(AU), X_5(CA), X_6(CC), X_7(CG), X_8(CU), X_9(GA), X_{10}(GC), X_{11}(GG), X_{12}(GU), X_{13}(UA), X_{14}(UC), X_{15}(UG), X_{16}(UU)] for RNA sequences.

3 KNN-Based Learning Method

KNN is a similarity-based learning algorithm and is known to be very effective for a variety of problem domains. Given a test instance d_t, the KNN algorithm finds its k nearest neighbors among the training instances, which forms a neighborhood of d_t. Majority voting among the instances in the neighborhood is used to decide the category for d_t.

In our novel KNN-based learning approach, we use the composition vector of each sequence as its feature vector. Let D be a collection of n pre-labeled DNA or RNA sequence documents $\{d_1, d_2, \ldots d_n\}$ with m categories $C_1, C_2, \ldots C_m$. Document $d_i \in D$ is represented by a feature vector of the form $<w_{i1}, w_{i2}, \ldots w_{il}>$, where w_{ij} is the numeric weight for the j-th feature and l is the total number of features. Here, we will use cosine similarity as the default similarity metric.

$$Sim(d_i, d_j) = \frac{\sum_{k=1}^{l} w_{ik} \bullet w_{jk}}{\sqrt{(\sum_{k=1}^{l} w_{ik}^2)(\sum_{k=1}^{l} w_{jk}^2)}} \cdot \quad (4)$$

Let $d_1, d_2, \ldots d_k$ denote the k instances from the training set that are nearest to d_q. Then the following formula will return the category of d_q:

$$\hat{f}(d_q) \leftarrow \arg\max_{C_j \in \{C_1, C_2 \ldots C_m\}} \sum_{i=1}^{k} Sim(d_q, d_i) \delta(d_i, C_j) \cdot \quad (5)$$

where $\delta(a,b) = 1$ if $a \in b$ and where $\delta(a,b) = 0$ otherwise.

We show the learning algorithm as follow:

1. Calculate the dinucleotide frequency of each DNA or RNA sequence in both the training set and the test set.
2. Do random background subtraction for every sequence.
3. Create composition vector for each sequence based on the above two steps. Here l, the total number of features, is 16 (2^4).
4. Given a query instance d_q from the test set to be classified, calculate $Sim(d_q, d_i)$ between d_q and each d_i from the training set.
5. For the k nearest neighbors whose similarities rank the top k among the training instances, do the majority voting and return the category of d_t.
6. For each instance in the test set, repeat step 4 and 5 to decide its category.

4 Experiments and Results

Some preliminary experiments have been done to test the validity of our approach.

4.1 Building the DNA Sequence Training Set and Test Set

We select 20 different bacteria randomly from the EMBL [8] database at the genus level. That is to say, the distinct genus of each of the 20 bacteria is selected randomly.

The bacteria we selected are described as follows. Here, the sequence length unit is base pair (bp).

Table 1. The 20 different bacteria selected randomly from the EMBL database at the genus level

Species No.	Accession Number	Total Sequence Length (bp)	Organism
1	AE008923	5175554	Xanthomonas axonopodis pv. citri str. 306
2	AE000512	1860725	Thermotoga maritime
3	AE014291	2107793	Brucella suis 1330
4	AE017125	1799146	Helicobacter hepaticus ATCC 51449
5	BA000016	3031430	Clostridium perfringens str. 13
6	BA000030	9025608	Streptomyces avermitilis
7	AE015924	2343476	Porphyromonas gingivalis W83
8	AL450380	3268203	Mycobacterium leprae
9	BA000039	2593857	Thermosynechococcus elongatus BP-1
10	AE016879	5227293	Bacillus anthracis str. Ames
11	AL592022	3011208	Listeria innocua
12	BA000033	2820462	Staphylococcus aureus subsp. aureus MW2
13	AE014074	1900521	Streptococcus pyogenes MGAS315
14	AE005673	4016947	Caulobacter crescentus CB15
15	BX470250	5339179	Bordetella bronchiseptica
16	AE009952	4600755	Yersinia pestis KIM
17	AE004091	6264403	Pseudomonas aeruginosa PAO1
18	AE016795	3281945	Vibrio vulnificus CMCP6
19	AE000513	2648638	Deinococcus radiodurans
20	AE009951	2174500	Fusobacterium nucleatum subsp. nucleatum ATCC 25586

From each whole genome of the above bacteria, we extract (without overlapping) 150 DNA sequence fragments with the length of 10000 bp. A hundred of them are put into the training set, and the other fifty ones are used as the test instances. The 20 different bacteria are viewed as 20 categories. Therefore, in the training set, there are totally 2000 DNA sequences with the length of 10000 bp and the category of each one is represented. While in the test set, there are totally 1000 DNA sequences with the length of 10000 bp and their categories are unknown. The learning goal is to decide the category of each test instance by training the instances in the training set.

4.2 Experiment

First, for each 10000 bp DNA sequence in both the training set and the test set, we count the number of appearances of (overlapping) the dinucleotides in it. Then for each of the 16 different dinucleotides, we calculate its probability. And we do the random background subtraction in order to get the composition vectors of the 3000 DNA sequences in both the training set and the test set.

Given a sequence d_q in the test set, first we measure $Sim(d_q, d_i)$ between d_q and each d_i from the training set. Then we pick out the sequences whose similarities rank the top k. Finally, we decide the category of d_t by the categories of these k sequences.

For each category C_j, we sum up the $Sim(d_q, d_i)$ where d_i runs from the sequences whose category is C_j in the k nearest neighbors. Then the category with the largest similarity summation is set to be the category of d_q. With the same method, we can decide the category of each DNA sequence in the test set.

43 Test Result

Test results reported in this section are based on precision. The performance of the KNN classifier algorithm also depends on the value of k, the number of nearest neighbors of the test process. Usually the optimal value of k is empirically determined. We set k to 1, 2, 5, 10, 20, 50 and 100 to do the test. The highest mean precision is 93% when k is set to 5 and 10.

Table 2. The precision of each category when k is set to 5 and 10

	Species No. and Precision									
	1	2	3	4	5	6	7	8	9	10
k=5	84%	100%	94%	100%	72%	100%	100%	100%	96%	94%
k=10	84%	100%	94%	100%	72%	100%	100%	100%	96%	92%
	Species No. and Precision									
	11	12	13	14	15	16	17	18	19	20
k=5	82%	94%	82%	96%	96%	94%	94%	86%	100%	92%
k=10	86%	90%	84%	96%	94%	96%	94%	86%	100%	90%

All the mean precisions we get are shown as follows:

Table 3. Mean precisions

	$k=1$	$k=2$	$k=5$	$k=10$	$k=20$	$k=50$	$k=100$
Mean Precision	92%	92%	93%	93%	92%	91%	89%

Admittedly, the experimental precisions will decrease when the lengths of the DNA sequences are shorter or when there are more categories in our classifier. Longer sequences could submerge segments introduced by recent horizontal transfer from another species [9] and highlight the genome signature. Here we use 10000-bp sequences and 20 categories to do our experiment.

4.4 Extended Experiment and Its Result

In the above experiment, we construct the training set and the test set from 20 different bacterial organisms. All the 3000 DNA sequences we used are extracted without overlapping. The test result shows that the mean precision is quite high when k is set to 5 or 10. Although the result is satisfying, there is also a limitation. The limitation is that both the sequences in the training set and the ones in the test set are extracted from the same 20 organisms. That is, although the DNA sequences in the test set are chosen exclusively, they come from the same 20 types of the organisms as in the training set. Suppose there is a new test sequence which has no training sequence of the same type. Of course, our classifier can not return its accurate category. But what result can we get if we add it to the test set? The experiment in this section is built to be in such a case. Our goal is to testify whether the test sequence can find its most closely related species or not. If so, we can extend our classifier to further research and more practical use.

In the experiment, we still use the same training set of 2000 DNA sequences as before. The construction of the test data set needs to be carefully explained.

The first column of Table 4 presents 13 different target bacterial organisms among the 20 training organisms. And the second column shows other 23 different bacterial organisms chosen as test data, making sure that each of them is closely related to, that is in the same genus as, one of the target organisms in the training set. In today's biological field, scientists are still arguing about the real families, orders and classes of one organism. Each side has its own evidence. But in the genus level of organism classification they generally have the same idea. Therefore, we choose our test data in this level to do the experiment. In this way, we extend our KNN-based classifier to identify the genus of biology species, not only their definite categories. The last two columns show the precisions when k is set to 5 and 10.

Most of the precisions the classifier returned are quite satisfying. But there are six low precisions (when both $k=5$ and $k=10$) and three of them are even below 10%. What make this happen? We find out that all the six organisms come from the species that can cause illness to other species. They all have a large number of horizontal transferred genes in their genomes. These horizontally transferred genes cause obvious noise in our experiment. However, as to the relatively conservative organisms, our classifier works well. Especially considering the complexity and irregularity of biological data, the results are satisfying.

Table 4. Target organisms, test organisms and the precision results of the experiment

Target Organisms	Test Organisms	$k=5$	$k=10$
Xanthomonas axonopodis pv. citri str. 306	Xanthomonas campestris pv. campestris str. ATCC 33913	95%	95%
Brucella suis 1330	Brucella melitensis 16M	98%	98%
Helicobacter hepaticus ATCC 51449	Helicobacter pylori 26695	33%	35%
Helicobacter hepaticus ATCC 51449	Helicobacter pylori J99	12%	15%
Clostridium perfringens str. 13	Clostridium acetobutylicum	82%	79%
Clostridium perfringens str. 13	Clostridium tetani E88	95%	96%
Streptomyces avermitilis	Streptomyces coelicolor	100%	100%
Mycobacterium leprae	Mycobacterium bovis AF2122/97	83%	84%
Mycobacterium leprae	Mycobacterium tuberculosis H37Rv	86%	87%
Bacillus anthracis str. Ames	Bacillus halodurans	20%	15%
Bacillus anthracis str. Ames	Bacillus cereus ATCC 14579	92%	93%
Bacillus anthracis str. Ames	Bacillus subtilis subsp. subtilis str. 168	4%	7%
Listeria innocua	Listeria monocytogenes	92%	92%
Staphylococcus aureus subsp. aureus MW2	Staphylococcus aureus subsp. aureus Mu50	83%	82%
Staphylococcus aureus subsp. aureus MW2	Staphylococcus aureus subsp. aureus N315	87%	87%
Staphylococcus aureus subsp. aureus MW2	Staphylococcus epidermidis ATCC 12228	75%	69%
Bordetella bronchiseptica	Bordetella pertussis	91%	90%
Bordetella bronchiseptica	Bordetella parapertussis	94%	92%
Bordetella bronchiseptica	Bordetella bronchiseptica	92%	90%
Yersinia pestis KIM	Yersinia pestis CO92	95%	98%
Pseudomonas aeruginosa PAO1	Pseudomonas putida KT2440	7%	8%
Pseudomonas aeruginosa PAO1	Pseudomonas syringae pv. tomato str. DC3000	7%	9%
Vibrio vulnificus CMCP6	Vibrio vulnificus YJ016	83%	85%

5 Conclusion

In this paper, we have described a KNN-based learning method for biology species categorization. The experiments of the categorization of DNA sequences show that the method is effective and practical.

We can use the KNN-based classifier to give a prediction of what the unknown sequence is, or probably is, or is most related to. This will help biologists a lot. They can get some helpful information to design their wet-lab experiments to predict what the unknown sequences are instead of testing aimlessly, saving both their time and efforts.

References

1. Lam, W., Ho, C.Y.: Using a Generalized Instance Set for Automatic Text Categorization. SIGIR'98, (1998) 81-89
2. Karlin, S., Burge, C.: Dinucleotide Relative Abundance Extremes: a Genomic Signature. Trends Genet, 11 (1995) 283-290
3. Nakashima, H., Nishikawa, K., Ooi, T.: Differences in Dinucleotide Frequencies of Human, Yeast, and Escherichia Coli Genes. DNA Research, 4 (1997) 185-192
4. Nakashima, H., Ota, M., Nishikawa, K., Ooi, T.: Genes from Nine Genomes are Separated into Their Organisms in the Dinucleotide Composition Space. DNA Research, 5 (1998) 251-259
5. Deschavanne, P.J., Giron, A., Vilain, J., Fagot, G., Fertil, B.: Genomic Signature: Characterization and Classification of Species Assessed by Chaos Game Representation of Sequences. Mol. Biol. Evol, 16 (1999) 1391-1399
6. Brendel, V., J.S. Beckman, and E.N. Trifonov. Linguistics of Nucleotide Sequences: Morphology and Comparison of Vocabularies. J. Biomol. Struct, 4 (1986) 11-21
7. Hu, R., Wang, B. Statistically Significant Strings are Related to Regulatory Elements in the Promoter Region of Saccharomyces Cerevisiae. Physica, A 290 (2001) 464-474
8. http://www.ebi.ac.uk/embl/index.html
9. Ochman, H., Lawrence, J.G., Groisman, E.A.: Lateral Gene Transfer and the Nature of Bacterial Innovation. Nature, 405 (2000) 299-304

Application of Emerging Patterns for Multi-source Bio-data Classification and Analysis*

Hye-Sung Yoon[1], Sang-Ho Lee[1], and Ju Han Kim[2]

[1] Ewha Womans University, Department of Computer Science and Engineering,
Seoul 120-750, Korea
comet@ewhain.net, shlee@ewha.ac.kr
[2] Seoul National University Biomedical Informatics (SNUBI),
Seoul National University College of Medicine, Seoul 110-799, Korea
juhan@snu.ac.kr

Abstract. Emerging patterns (EP) represent a class of interaction structures and have recently been proposed as a tool for data mining. Especially, EP have been applied to the production of new types of classifiers during classification in data mining. Traditional clustering and pattern mining algorithms are inadequate for handling the analysis of high dimensional gene expression data or the analysis of multi-source data based on the same variables (e.g. genes), and the experimental results are not easy to understand. In this paper, a simple scheme for using EP to improve the performance of classification procedures in multi-source data is proposed. Also, patterns that make multi-source data easy to understand are obtained as experimental results. A new method for producing EP based on observations (e.g. samples in microarray data) in the search of classification patterns and the use of detected patterns for the classification of variables in multi-source data are presented.

1 Introduction

Microarray experiments have brought innovative technological development to the classification of biological types. But more powerful and efficient analytical strategies need to be developed to carry out complex biological tasks and to classify data sets with various types of information such as mining disease related genes and building genetic networks.

The analytical strategy of bio-data can be classified into two categories according to the form of learning algorithm. First, as an unsupervised learning method such as typical clustering algorithms, the analytical method deals directly with genes while ignoring the biological attributes (labels) when handling DNA data (instance). Supervised learning is a target-driven process in that a suitable induction algorithm is employed to identify the genes that contribute the most toward a specific target, such as the classification of biological types, gene

---

* This work was supported by the Brain Korea 21 Project in 2004.

mining or data-driven gene networking[8]. Among supervised learning methods, rule based approach can be said that this partitions the sample and feature gene space simultaneously and it is especially an efficient method for classification of multi-source data. In statistics, the analysis of gene expression profiles is related to the application of particular supervised learning schemes. The structure of gene expression profiles must be suited for the typical data situation with a small number of patients n (=observations) and a large number of genes p (=variables), the so-called 'small n large p' paradigm in gene expression analysis[2][12].

In this paper, we develop a new rule-based ensemble method using EP for the classification of multi-source data. EP are those whose support changes significantly from one data set to another[19]. EP are among the simplest examples used to understand interaction structures, and are not only highly discriminative in classification problems[19], but can also capture the biologically significant information from the data. However, a very large volume of EP is generated for high dimensional gene expression data[7]. In this paper, we apply concise EP for multi-source data classification based on observations from each individual data set. When dealing with classification methods, microarray data is generally used, but only a few number of approaches are designed to consider explicitly the interaction among the genes being investigated. Interaction is well understood as (co-)expression genes in a cell governing a complicated network of regulatory controls. Hence, the interdependencies of all genes must be taken into consideration in order to achieve optimal classification. We propose a new method that can handle all variables in an appropriate way. It must be noted that the goal of the analyses presented in this paper is not to present correlated interaction genes for multi-source data but rather to illustrate our proposed classification method using EP.

The remainder of this paper is organized as follows. The application of multi-source data and classification methods in bioinformatics, and the analysis of EP are reviewed in section 2. A method for extracting EP from multi-source data sets and their applications are explained in section 3. Furthermore, significant experimental results by applying the proposed method and its details are described in section 4. Finally, concluding remarks and future works are presented in section 5.

2 Related Works and Background

In section 2, multi-source data, classification algorithm applications in bioinformatics and EP for multi-source data classification are reviewed as related works and background.

2.1 Multi-source Data

Bioinformatics not only deals with raw DNA sequences, but also with other various types of data, such as protein sequences, macromolecular structure data, genomes data and gene expression data[19]. The various types of data provide researchers with the opportunity to predict phenoma that were formerly considered unpredictable, and most of these data can be accessed freely on the internet.

We assume that the analysis of combined biological data sets leads to more understandable direction than experimental results derived from a single data set. The purpose for combining and analyzing different types of data is to identify with more accuracy and to provide more correlations using diverse independent attributes in gene classification, clustering and regulatory networks and so on. Among the features of bio-data, one is that the same variables can be used to make various types of multi-source data through a variety of different experiments and under several different experimental conditions. These multi-source data are useful in understanding cellular function at the molecular level and also provide further insight into their biological relatedness by use of information from disparate types of genomic data. In [14], the problem of inferring gene functional classification from a heterogeneous data set consisting of DNA microarray expression measurements and phylogenetic profiles from whole-genome sequence comparisons is considered. As a result, it is proposed that more important information can be extracted by using disparate types of data.

2.2 Classification Problem in Bioinformatics

Classification problems aim at building an efficient and effective model for predicting class membership of data. Initial analysis of multi-source data focused on clustering algorithms, such as hierarchical clustering[9] and self-organizing maps[16]. In these unsupervised learning algorithms, genes that share similar expression patterns form clusters of genes that may show similarities in function[14]. But, because clustering methods ignore biological attributes (labels), they have limitations in the search of attributes or the discovery of rules in observations.

In [10], supervised learning techniques were applied to microarray expression data from yeast genes. It was verified through this application that an algorithm known as support vector machine (SVM)[3][4][13] provides excellent improvement in classification performance compared to a number of other methods, including Parzen windows and Fisher's linear discriminant[10]. Also, the methods used in [10] have been successfully applied to disease genes classification with machine learning approaches such as support vector machines (SVM), artificial neural network(ANN), k-nearest neighbors (kNN), and self-organizing map (SOM). In recent studies on the application of classification methods, supervised learning methods are aiming at showing the existence or nonexistence of disease by searching for disease genes[19].

2.3 Analysis of Emerging Patterns

A wide variety of gene patterns can be found for each data set. In [2] and [19], gene expression profiles were used to individually apply CART algorithm, a supervised learning method, and clustering method, an unsupervised learning

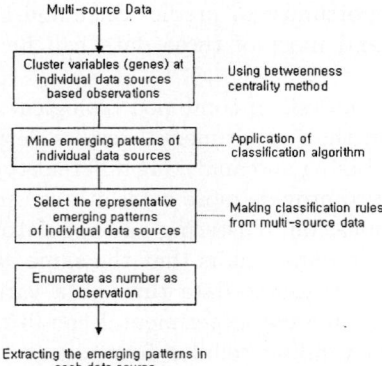

Fig. 1. Flowchart of the experimental method

method, to detect all types of early cancer development. To improve accuracy in classification, EP were applied to express the interaction between cancer-causing genes. Pattern association and clustering are both data mining techniques that are frequently applied in the fields of cancer diagnosis and correlation studies of gene expression[20]. But the results from these methods do not meet our requirements because multi-source data was not considered. EP were first introduced in [5], and they were defined as the item set that significantly increases support in each data sets D_1 and D_2 using the appropriate cut-off value of the growth rate. Unlike frequent patterns in common association analysis, EP are applied to classification problems to provide high discrimination, and are proved to be more useful. Also, EP are easy to understand because they are the collections of attributes in a dataset, and this property is especially important in bioinformatics application problems.

Thus, this paper proposes an efficient classification method using EP that is efficient when using analysis based on the smaller number of observational attributes rather than the very large number of variable attributes.

3 Methods

In this section, the experimental data and experimental methods applied in this paper are explained in detail. The overall framework is illustrated in Figure 1 and it will be explained in order.

3.1 Data

In this paper, two types of genomic data were used as multi-source data for the application of the proposed method. The first data set was derived from a collection of DNA microarray hybridization experiments. Each data point in the microarray data represents the logarithm of the ratio of expression levels of a particular gene under two different experimental conditions. The data consists of a

	The number of 10 samples							
	alpha	elu	cdc	spo1	spo2	spo3	heat	dtt	cold	diau
79 time points	18	14	15	6	3	2	6	4	4	7
2,465 yeast genes										

Fig. 2. Data structure of microarray data

	24 species				
	aero	aful	aquae	......	tpal	worm
2,465 yeast genes						

Fig. 3. Data structure of phylogenetic profile

set of 79-element gene expression vectors across time points for 2,465 yeast genes. These genes were selected by [9] based on accurate functional annotations. The data were collected at various time points during the diauxic shift[6], the mitotic cell division cycle[15], sporulation[17], and temperature and reducing shocks, and are available on the stanford website (http://www-genome.stanford.edu)[14].

In addition to the microarray expression data, we applied data characterized by 24 phylogenetic profiles[11] to each of the 2,465 yeast genes. In this data set, a phylogenetic profile is a bit string, in which the boolean value of each bit reflects whether the gene of interest has a close homolog in the corresponding genome. The profiles employed in this paper contain, at each position, the negative logarithm of the lowest E-value reported by BLAST version 2.0[18] in a search against a complete genome, with negative values truncated to 0. The profiles were constructed using 24 complete genomes, collected from the Institute for Genomic Research website (http://www.tigr.org/tdb) and from the Sanger Centre website (http://www.sanger.au.uk). Prior to learning, the gene expression and phylogenetic profile vectors were adjusted to have a mean of 0 and a variance of 1. The description of each data set, composed of the microarray data and phylogenetic profile data about the 2,465 yeast genes, are as shown in figure 2 and figure 3.

In the experiments of this paper, the betweenness centrality values based on 10 samples (=observations) that have 79-element time points values in the first microarray data set were as shown in figure 2. Thus, gene clusters were formed by extracting the most closely related genes in the order of high betweenness centrality value first.

As a result, a total of 10 clusters were formed (all genes were included in at least one sample in this experiment). And also for the second phylogenetic profile data set, 25 clusters (one additional cluster was formed with the genes that were not included in any of the 24 species) were formed by extracting the most closely related genes in order of high betweenness centrality value first.

3.2 Application of Betweenness Centrality Based on Observation

Bio-data is characterized by having a small number of observations compared to the number of variables. This characteristic found in bio-data can also be observed in microarray data, and this is well reflected in the data where the number of columns corresponding to observations are outnumbered by the number of rows corresponding to variables (=genes). The exclusion of some variables can lead to significant differences in experimental results because a characteristic of bio-data is that interaction among the data is highly dependent. Therefore, in this paper, the characteristic EP are represented by considering all variables in each data set and the results are applied for the classification of multi-source data. Also, since EP are easy to understand and represent, they are useful for judging the features other types of data.

The following explains in order the proposed method of forming EP with a single data set.

1. First, based on the observations, clusters are formed with genes that contribute the most toward these observations, since bio-data sets have a smaller number of observations than variables. Then, the betweenness centrality method used in social network analysis to extract the variables that are closely related to the observations is applied. Social network analysis is a theory in Sociology, and it is the mapping and measuring of relationships and flows between people, groups, organizations, animals, computers or other information/knowledge processing entities. The nodes in the network represent people and groups, and this means that the most active people have the most relationships (=links) with many other people[14]. That is, the entire network is closely related to this node.

   In this experiment, the betweenness centrality value of each observation was computed, then the observation with the highest value was found and genes that were the most closely related were extracted.

2. From the previous experiment, the observation with the highest betweenness centrality value and the genes that were the most closely related to the observation were set aside, and the betweenness centrality value for the remaining observations and genes are computed again. From the resulting values, the observation with the highest betweenness centrality value and the genes that were the most closely related to the observation are clustered.

3. In the same manner, the betweenness centrality value is computed repeatedly as many times as the number of observations, in order to form clusters according to the relations between observations and variables.

4. And finally, one cluster is formed with variables that are not included in any other observation.

The methods mentioned above are shown in figure 4, when applied to a phylogenetic profile data set. In the case of phylogenetic profiles data, 24 species (=observations) and 2,465 genes (=variables) are formed, and the method is repeated to extract the genes that are the most closely related in the order of the observation with the highest betweenness centrality value. And finally, one cluster is formed with genes that are not included in any other species.

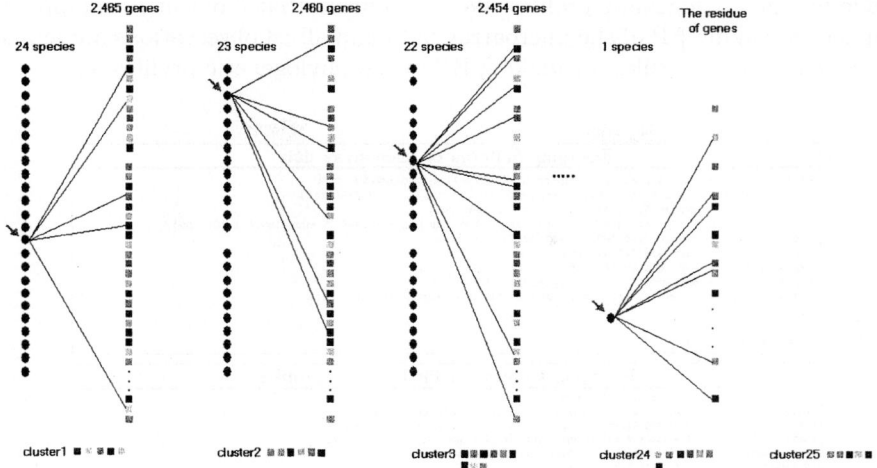

Fig. 4. Application method of betweenness centrality based on observation

3.3 Forming Emerging Patterns Between Individual Datasets

As shown in figure 4, the betweenness centrality value of each observation is computed for the experimental data in section 3.1. Then clusters are formed by extracting variables (=genes) that could explain the observations with the highest betweenness centrality value. As a result, the microarray data of yeast applied to the first experiment are clustered by reducing 79-elements time points to 10 observations (the clusters were formed from 10 samples composed of 79-elements time points, while the experiment handles observations according to sample number and not time point. See figure 2). In the second phylogenetic profiles experimental data, the 24 species corresponding to the columns are regarded as observations, and clusters are formed in as many number as observations. In this paper, EP formed in microarray data sets and phylogenetic profile data are represented in the following way. EP in microarray data sets and phylogenetic profiles are expressed in the form of $exp(X_1) > a_1 \bigwedge exp(X_2) < a_2$ and $phylo(Y_1) > b_1 \bigwedge phylo(Y_2) > b_2$, respectively. In each representation of EP, X_i is the measured expression level of observations in microarray data sets and Y_j is the sequence similarity of observations in phylogenetic profiles data. The a_i and b_j in the representations are boundary constants that can be inferred from each data set, and they represent the threshold value of the expression level in microarray data and the sequence similarity in phylogenetic profiles data.

4 Experimental Results

In this paper, R package was used to compute betweenness centrality and Weka algorithm was applied to make classification rules. The results are shown in figure 5, where 10 rules are made for the microarray data set and 24 rules are made for the phylogenetic profiles. We can confirm that 6 out of 10 samples are applied in making EP of the microarray data and all 24 observations are applied to the classification rules for making EP of the phylogenetic profiles data.

```
                     Emerging Patterns of Microarray data
exp(spo)≥2.24 ∧ exp(spom)≥2.19 ∧ exp(cold)≤0.526 ∧ exp(spom)≥2.69: spom
exp(cold)≥0.712 ∧ exp(heat)≤0.662 ∧ exp(spo)≥2.81 ∧ exp(heat)≤0.236: cold
exp(spo)≥3.33 ∧ exp(spo5)≤-1.35 ∧ exp(elu)≤-0.184: spo5
exp(spo)≤-2.21 ∧ exp(elu)≥0.547 ∧ exp(elu)≤1.08 ∧ exp(heat)≤-0.484 ∧ exp(dieu)≤-1.12: spo5
exp(spo)≥2.42 ∧ exp(cold)≤0.109 ∧ exp(cdc)≥0.818: spo5
exp(spo5)≥1.18 ∧ exp(spo5)≥3.31: spo5
exp(dtt)≥1.12 ∧ exp(cold)≤0.585: dtt
exp(elu)≥1.05 ∧ exp(spom)≤-2.21 ∧ exp(spo)≤-1.77: elu
exp(alpha)≤0.745 ∧ exp(elu)≥0.555 ∧ exp(elu)≥1.24 ∧ exp(spo5)≤0.174: elu
: alpha

Number of Rules : 10

                  Emerging Patterns of Phylogenetic profiles
phylo(cpneu)>1.1 ∧ phylo(tpal)≤1.09: cpneu
phylo(tmar)>1.1 ∧ phylo(tpal)≤1.07 ∧ phylo(ctra)≤1.14 ∧ phylo(mpneu)≤1.11: tmar
phylo(aero)>1.1 ∧ phylo(ctra)≤1.06 ∧ phylo(mpneu)≤0.87 ∧ phylo(tpal)≤1.09: aero
phylo(aful)>1.1 ∧ phylo(tpal)≤0.999 ∧ phylo(ctra)≤1.24 ∧ phylo(mpneu)≤1.75: aful
phylo(mpneu)>1.07 ∧ phylo(ctra)≤1.18 ∧ phylo(tpal)≤0.94: mpneu
phylo(mthe)>1.1 ∧ phylo(tpa)≤ 0.999: mthe
phylo(tpal)≤1.09 ∧ phylo(rpxx)>1.1 ∧ phylo(ctra)≤1.12: rpxx
phylo(tpal)≤1.09 ∧ phylo(ctra)≤1.1 ∧ phylo(mgen)≤1.1 ∧ phylo(atub)>1.1: atub
phylo(tpal)≤1.09 ∧ phylo(ctra)≤1.1 ∧ phylo(mgen)≤1.1 ∧ phylo(hpy199)>1.1: hpy199
phylo(tpal)≤1.09 ∧ phylo(ctra)≤1.1 ∧ phylo(mgen)≤1.1 ∧ phylo(bbur)≤1.1 ∧ phylo(dra)≤1.1 ∧ phylo(synecho)>1.1: synecho
phylo(tpal)≤1.09 ∧ phylo(ctra)≤1.1 ∧ phylo(mgen)≤1.1 ∧ phylo(bbur)>1.1: bbur
phylo(tpal)≤1.09 ∧ phylo(ctra)≤1.1 ∧ phylo(mgen)>1.1: mgen
phylo(tpal)>1.09: tpal
phylo(ctra)≤1.1 ∧ phylo(dra)>1.1: drai
phylo(ctra)>1.1: ctrsu
phylo(aquae)>1.09: aquae
phylo(bsub)>1.1 ∧ phylo(hpy1)≤1.05: bsub
phylo(hpy1)≤1.11 ∧ phylo(ecoli)≤1.08 ∧ phylo(pabyssi)≤1.03 ∧ phylo(njan)≤1.08 ∧ phylo(pyro)≤1.1 ∧ phylo(hinf)≤1.09: worm
phylo(pabyssi)>1.1: pabyssi
phylo(hpy1)≤1.11 ∧ phylo(njan)≤1.09 ∧ phylo(pyro)≤0.981 ∧ phylo(hinf)>1.1: hinf
phylo(hpy1)≤1.06 ∧ phylo(njan)>1.09: njan
phylo(hpy1)≤0.996 ∧ phylo(pyro)≤0.981: ecoli
phylo(hpy1)>0.996: hpy
: pyro

Number of Rules :    24
```

Fig. 5. Emerging patterns of microarray data and phylogenetic profiles

The results in figure 5 can be interpreted as follows: The EP in the 7th line are in the form $exp(dtt) \geq 1.12 \bigwedge exp(cold) \leq 0.585$, and this means that the variables with *dtt* gene expression levels greater than 1.12 and cold values less than 0.585 for *'dtt'* observation in the entire microarray data set can classify the *'dtt'* observations in the entire microarray data set. Also, these EP can be considered as classifiers that can be classified among other observations in the microarray data set. The *alpha* in the last line of the EP of the microarray data shows that the genes that can explain the *'alpha'* observation are those that do not correspond to any of the above rules. The results of the phylogenetic profile can be interpreted in the same way, where the EP of the first line is in the form of $phylo(cpneu) > 1.1 \bigwedge phylo(tpal) \leq 1.09$, and this becomes the classifier that can classify the *'cpneu'* observation in the phylogenetic profile data set.

Validation results of the EP, as to how accurately they can classify the two types of data sets, show that accuracy is 86.76% and 97.79% for microarray data and phylogenetic profile data, respectively. The relatively low accuracy for the microarray data set could be explained by the reduction of 79 time points to 10 observations before the start of the experiment.

5 Conclusions and Future Works

Typical bio-data analysis methods deal directly with genes while ignoring biological attributes, but since the interaction among genes plays an important role in bio-data analysis, new methods must be developed. Also, multi-source data classification and analysis problems are much more complex and have more factors to be considered than single-source data problems. When handling bio-data, disparate types of multi-source data can be made based on the same variables, and we are in need of classifiers that can classify the data sets and methods to easily understand the features of the data sets. Therefore, this paper proposes a new method that considers the characteristics of bio-data, and while existing methods ignore biological attributes and analyze only the genes, the proposed method provides an analysis method based on observations using all variables from each data set. This method makes EP that take into account the relations between genes in the data set and the results are applied to the multi-source data classification. An existing paper introduced a method to map variables to gene function categories by applying the SVM method using the same data set in this paper[14]. But the method introduced in the existing paper differs from the proposed method, which considers both observations and variables, in that the existing method has no regard of the interaction structure between genes in the analysis stage, that it is not easy to interpret and that the analysis is done after variables are removed first by some threshold value in the preprocessing stage.

The experimental methods introduced in this paper suggest several avenues that can be taken for future research. One direction would be to find a better classifier of multi-source data in bio-data. Another direction would be, since only two biological data types were used for multi-source data classification, to include multiple biological data types for discovering EP and for extending the proposed method in multi-source data classification. Also, another important task would be to come up with a theoretically and experimentally justified verification of disparate data.

References

1. Barabasi, A.L.: Link, Penguin,(2003)
2. Boulesteix, A.L., Tutz, G., Strimmer, K.: A CART-based approach to discover emerging patterns in microarray data. Bioinformatics, **19** (2003) 2465–2472
3. Boser, B.E., Guyon, I.M., Vapnik, V.: A training algorithm for optimal margin classifiers. Proceedings of the 5th Annual ACM Workshop on Computational Learning Theory, **5** (1992) 144–152
4. Burges, C.J.C.: A tutorial on support vector machines for pattern recognition. Data Mining and Knowledge Discovery, **2** (1998) 121–167
5. Dong, G., Li, J.: Efficient mining of emerging patterns: discovering trends and differences. Proceedings of the SIGKDD (5th ACM International Conference on Knowledge Discovery and Data Mining), **5** (1999) 43–52
6. DeRisi, J., Iyer, V., Brown, P.: Exploring the metabolic and genetic control of gene expression on a genomic scale. Science, **278** (1997) 680–686

7. Li, J., Wong, L.: Emerging patterns and gene expression data. Proceedings of 12th Workshop on Genome Informatics, **12** (2001) 3–13
8. Xia, L., Shaoqi, R., Yadong, W., Binsheng, G.: Gene mining: a novel and powerful ensemble decision approach to hunting for disease genes using microarray expression profiling. Nucleic Acids Research, **32** (2004) 2685–2694
9. Eisen, M., Spellman, P., Brown, P., Botstein, D.: Cluster analysis and display of genome-wide expression patterns. Proceedings of the National Academy of Sciences of the United States of America, **95** (1998) 14863–14868
10. Brown, M.P.S., Grundy, W.N., Lin, D., Cristianini, N., Sugnet, C., Furey, T., Ares, J.M., Haussler, D.: Knowledge-base analysis of microarray gene expression data using support vector machines. Proceedings of the National Academy of Science of the United States of America, **97** (2000) 262–267
11. Pellegrini, M., Marcotte, E.M., Thompson, M.J., Eisenberg, D., Yeates, T.O.: Assigning protein functions by comparative genome analysis: protein phylogenetic profiles. Proceedings of the National Academy of Sciences of the United States of America, **96** (1999) 4285–4288
12. West, M., Nevins, J.R., Spang, R., Zuzan, H.: Bayesian regression analysis in the 'large p, small n' paradigm with application in DNA microarray studies. Technical Report 15, Institute of Statistics and Decision Sciences, Duke University,(2000)
13. Cristianini, N., Shawe-Taylor, J.: An Introduction to Support Vector Machines. Cambridge UP,(2000)
14. Pavlidis, P., Weston, J., Cai, J., Grundy, W.N.: Learning gene functional classifications from multiple data types. Journal of Computational Biology, **9** (2002) 401–411
15. Spellman, P.T., Sherlock, G., Zhang, M.Q., Iyer, V.R., Anders, K., Eisen, M.B., Brown, P.Q., Botstein, D., Futcher, B.: Comprehensive identification of cell cycle-regulated genes of the yeast Saccharomyces cerevisiae by microarray hybridization. Molecular Biology of the Cell, **9** (1998) 3273–3297
16. Tamayo, P., Slonim, D., Mesirov, J., Zhu, Q., Kitareewan, S., Dmitrovsky, E., Lander, E., Golub, T.: Interpreting patterns of gene expression with self-organizing maps. Proceedings of the National Academy of Sciences of the United States of America, **96** (1999) 2907–2912
17. Chu, S., DeiRisi, J., Eisen, M., Mulholland, J., Botstein, D., Brown, P., Herskowitz, I.: The transcriptional program of sporulation in budding yeast. Science, **282** (1998) 699–705
18. Altschul, S.F., Madden, T.L., Schaffer, A.A., Zhang, Z., Miller, W., Lipman, D.J.: Gapped BLAST and PSI-BLAST: A new generation of protein database search programs. Nucleic Acids Research, **25** (1997) 3389–3402
19. Larray, T. H. Yu., Fu-lai, C., Stephen, C.F.: Using Emerging Pattern Based Projected Clustering and Gene Exptression Data for Cancer Detection. Proceedings of the Asia-Pacific Bioinformatics Conference, **29** (2004) 75–87
20. Yuhang, W., Filla M, M.: Application of Relief-F Feature Filtering Algorithm to Selecting Informative Genes for Cancer Classification Using Microarray Data. International IEEE Computer Society Computational Systems Bioinformatics Conference, **3** (2004) 497–498

Nonlinear Kernel MSE Methods for Cancer Classification

L. Shen and E.C. Tan

School of Computer Engineering, Nanyang Technological University, Nanyang Avenue,
Singapore 639798, Singapore
{PG04480855, asectan}@ntu.edu.sg

Abstract. Combination of kernel PLS (KPLS) and kernel SVD (KSVD) with minimum-squared-error (MSE) criteria has created new machine learning methods for cancer classification and has been successfully applied to seven publicly available cancer datasets. Besides the high accuracy of the new methods, very fast training speed is also obtained because the matrix inversion in the original MSE procedure is avoided. Although the KPLS-MSE and the KSVD-MSE methods have equivalent accuracies, the KPLS achieves the same results using significantly less but more qualitative components.

1 Introduction

Cancer classification based on clinical or histopathological information is subjective and requires highly experienced pathologists for interpretation. Recently, the advent of DNA microarray and protein mass spectra has enabled us to measure thousands of expression levels of genes and mass/charge identities of proteomic patterns simultaneously. These gene expression profiles and proteomic patterns can be used to classify different types of tumors and there have been a lot of activities in this area of cancer classification. Several systematic methods have been applied on these datasets and they have successfully predicted different tumor types [1], [2].

One problem often encountered is that there are a huge number m (thousands) of features but relatively small number n (a few dozens) of samples or arrays due to the high cost of microarray experiment. Therefore, methods which are originally devised for conditions when $m < n$ cannot be directly applied on these datasets. Dimension reduction is one of the solutions for this condition and has been successfully applied on microarray data by other researchers [3], [4].

In past years, a number of kernel-based learning algorithms, e.g. support vector machines (SVM) [5], kernel principal component analysis (KPCA) [6] and KPLS [7] have been proposed. The kernel methods have enabled us to solve nonlinear problems. Successful applications of kernel-based algorithms have been reported in various fields.

MSE linear discriminant function builds a classifier by simple matrix pseudo-inversion and it has received enormous studies in literature [8]. In this paper, we combine the techniques of KSVD and KPLS with MSE. New classifiers called KPLS-MSE and KSVD-MSE are derived and have been validated on seven publicly available cancer datasets. Our baseline method is the SVM which is a very popular kernel method in the area of microarray data analysis and has often achieved superior performance against the other binary classifiers. The results illustrate that our proposed methods are as good as the SVM and require much less training time in the "small sample, large gene" condition.

2 Methods

2.1 General Framework

Given a sample $\mathbf{x}$, the linear discriminant function is defined as $f(\mathbf{x}) = \mathbf{w} \cdot \mathbf{x} + w_0$. Consider only a two-class classification problem. Then $f(\mathbf{x}) \geq 0$ if $\mathbf{x} \in$ class 1 or else $f(\mathbf{x}) < 0$ if $\mathbf{x} \in$ class 2. Suppose that we have a set of training samples, $\mathbf{x}_1, \mathbf{x}_2, \ldots, \mathbf{x}_n \in R^m$, and for convenience, let $\mathbf{X} = [\mathbf{x}_1, \mathbf{x}_2 \ldots, \mathbf{x}_n]^T$ and $\mathbf{Y} = [\mathbf{X}, \mathbf{1}_n]$, where $\mathbf{1}_n$ represents the n-dimensional vector with all elements equal to one. Also, let $\mathbf{a}^T = [\mathbf{w}^T, w_0]$. The weight vector $\mathbf{a}$ can be estimated from the equation

$$\mathbf{Y}\mathbf{a} = \mathbf{b} \tag{1}$$

where $\mathbf{b}$ represents the n-dimensional vector of class labels. The ith element of $\mathbf{b}$ can be set to 1 if $\mathbf{x}_i \in$ class 1 or -1 if $\mathbf{x}_i \in$ class 2. The linear system of (1) is underdetermined if $(m+1) > n$, which is the usual case in microarray data analysis. Principal components analysis (PCA) [9] and partial least squares (PLS) [10] are two popular dimension reduction methods for generating component vectors so that the number of variables can be reduced from $m+1$ to the number of components p. While PCA keeps most variance of the original variable $\mathbf{Y}$, PLS attempts to keep most covariance of both variable $\mathbf{Y}$ and class label $\mathbf{b}$. The original variable $\mathbf{Y}$ can be written as the dot product of the p components and the corresponding loadings vector is

$$\mathbf{Y} = \mathbf{T}_p \mathbf{U}_p^T + \mathbf{R} \tag{2}$$

where the columns of $\mathbf{U}_p$ are the p loading vectors, the columns of $\mathbf{T}_p$ are the p score vectors so that $\mathbf{T}_p = [\mathbf{t}_1, \mathbf{t}_2, \ldots, \mathbf{t}_p]$ and $\mathbf{R}$ are residuals. The score vectors are always assumed to be normalized without influencing the results of classification. Now for the discriminant function (1), we have

$$\mathbf{T}_p \mathbf{U}_p^T \mathbf{a} = \mathbf{b} \tag{3}$$

with residuals $\mathbf{R}$ omitted. $\mathbf{U}_p^T \mathbf{a}$ can then be calculated by the pseudo-inversion of $\mathbf{T}_p$ so that

$$\mathbf{U}_p^T \mathbf{a} = \mathbf{T}_p^+ \mathbf{b} = \mathbf{T}_p^T \mathbf{b} \qquad (4)$$

because of the mutual orthogonality of the component vectors of $\mathbf{T}_p$. For PLS, the algorithm used in this paper produces mutually orthogonal score vectors. Therefore, the matrix inversion which occurs in ordinary MSE process is replaced with transposition. Now the discriminant function can be formulated as

$$f(\mathbf{y}) = \mathbf{d}_p^T (\mathbf{U}_p^T \mathbf{a}) = \mathbf{d}_p^T \mathbf{T}_p^T \mathbf{b} \qquad (5)$$

where $\mathbf{d}_p$ is a length p vector representing the sample vector $\mathbf{y}$ in the reduced dimensional space so that $\mathbf{d}_p = [d_1, d_2, \ldots, d_p]^T$. In the training procedure of cross-validation (CV), because the optimal number of components p is to be sought, the discriminant function is reformulated into the summation of p items as

$$f_p(\mathbf{y}) = \sum_{i=1}^{p} d_i \mathbf{t}_i^T \mathbf{b} \qquad (6)$$

Then the classification results of using different numbers of components can be obtained by adding one item after another to the discriminant function sequentially. In the following sections, we show how to obtain d_i and $\mathbf{t}_i$ in (6) for both KPLS and KPCA.

2.2 Kernel Partial Least Squares

PLS is a technique for modeling a linear relationship between a set of input variables $\{y_i\}_{i=1}^n \in R^m$ and a set of output variables $\{b_i\}_{i=1}^n \in R$. Only one-dimensional output is considered here. Furthermore, we assume centered input and output variables. Let $\mathbf{Y} = [\mathbf{y}_1, \mathbf{y}_2, \ldots, \mathbf{y}_n]^T$ and $\mathbf{b} = [b_1, b_2, \ldots, b_n]^T$. The PLS method finds the weight vectors $\mathbf{w}$ and $\mathbf{c}$ so that

$$\max_{|\mathbf{r}|=|\mathbf{s}|=1} [\text{cov}(\mathbf{Yr}, \mathbf{bs})]^2 = [\text{cov}(\mathbf{Yw}, \mathbf{bc})]^2 = [\text{cov}(\mathbf{t}, \mathbf{u})]^2$$

where $\mathbf{t}$ and $\mathbf{u}$ are score vectors for input vectors $\mathbf{Y}$ and output vector $\mathbf{b}$. After obtaining the pair of $\mathbf{t}$ and $\mathbf{u}$, $\mathbf{Y}$ and $\mathbf{b}$ are deflated by $\mathbf{t}$ and the procedure is repeated to obtain a new pair of $\mathbf{t}$ and $\mathbf{u}$. This can be iterated until the rank of matrix $\mathbf{Y}$. The different forms of deflation correspond to different forms of PLS [11]. The SIMPLS [12] algorithm which provides the same solution as PLS1 in the case of one-dimensional output is used in this paper. The score vectors $\mathbf{t}$ produced by SIMPLS are mutually orthogonal.

Assume a nonlinear transformation of the input variables $\{\mathbf{y}_i\}_{i=1}^n$ into a feature space F; i.e. mapping $\Phi : \mathbf{y}_i \in R^m \to \Phi(\mathbf{y}_i) \in F$. The goal of the kernel PLS method is to construct a linear PLS model in F. Therefore, we effectively obtain a nonlinear KPLS in the space of the original input variables. Denote by $\mathbf{\Phi}$ an $n \times m'$ matrix of input variables whose i th row is the vector $\Phi(\mathbf{y}_i)$. m' is the dimension of $\Phi(\mathbf{y}_i)$

and it could be infinite if the Gaussian kernel is used. The linear SIMPLS method can be written in the feature space F as

$$\mathbf{t} = \mathbf{\Phi}\mathbf{\Phi}^T\mathbf{b}, \quad \|\mathbf{t}\| \to 1 \tag{7}$$

$$\mathbf{u} = \mathbf{b}(\mathbf{b}^T\mathbf{t}) \tag{8}$$

where $\mathbf{t}$ is also called the PLS components. Then the deflation rule is given by

$$\mathbf{\Phi} \leftarrow \mathbf{\Phi} - \mathbf{t}(\mathbf{t}^T\mathbf{\Phi}) \tag{9}$$

$$\mathbf{b} \leftarrow \mathbf{b} - \mathbf{t}(\mathbf{t}^T\mathbf{b}) \tag{10}$$

Instead of explicitly mapping the input data, the "kernel trick" is used resulting in

$$\mathbf{K} = \mathbf{\Phi}\mathbf{\Phi}^T$$

where $\mathbf{K}$ represents the kernel Gram matrix of the dot products between all feature data points $\{\Phi(\mathbf{y}_i)\}_{i=1}^n$; that is

$$\mathbf{K}_{ij} = \Phi(\mathbf{y}_i) \cdot \Phi(\mathbf{y}_j) = K(\mathbf{y}_i, \mathbf{y}_j)$$

where $\mathbf{K}$ is a selected kernel function. $\mathbf{K}$ is now directly used in the deflation instead of $\mathbf{\Phi}$ as follows [7]:

$$\mathbf{K} \leftarrow (\mathbf{I}_n - \mathbf{t}\mathbf{t}^T)\mathbf{K}(\mathbf{I}_n - \mathbf{t}\mathbf{t}^T) \tag{11}$$

where $\mathbf{I}_n$ is an n-dimensional identity matrix. The assumption of zero means of the variables of $\mathbf{Y}$ in linear PLS should also be held in kernel PLS. To centralize the mapped data in a feature space F, the following procedure must be applied [6]:

$$\mathbf{K} \leftarrow (\mathbf{I}_n - \frac{1}{n}\mathbf{1}_n\mathbf{1}_n^T)\mathbf{K}(\mathbf{I}_n - \frac{1}{n}\mathbf{1}_n\mathbf{1}_n^T) \tag{12}$$

where $\mathbf{1}_n$ is a $n \times 1$ vector with all elements equal to one. Let $\mathbf{T} = [\mathbf{t}_1\ \mathbf{t}_2\ \ldots\ \mathbf{t}_p]$, $\mathbf{U} = [\mathbf{u}_1\ \mathbf{u}_2\ \ldots\ \mathbf{u}_p]$ and p is the number of score vectors. Finally, the projection of test samples [7] into the feature space is given by

$$\mathbf{T}_d = \mathbf{K}_d \mathbf{U}(\mathbf{T}^T\mathbf{K}\mathbf{U})^{-1} \tag{13}$$

where $\mathbf{K}_d$ is the test set kernel Gram matrix. $\mathbf{T}^T\mathbf{K}\mathbf{U}$ is an upper triangular matrix and thus invertible. $\mathbf{K}_d$ should also be centered as

$$\mathbf{K}_d \leftarrow (\mathbf{K}_d - \frac{1}{n}\mathbf{1}_{n_t}\mathbf{1}_n^T\mathbf{K})(\mathbf{I}_n - \frac{1}{n}\mathbf{1}_n\mathbf{1}_n^T) \tag{14}$$

2.3 Kernel Principal Components Analysis

Singular value decomposition (SVD) [9] is used to implement PCA in this paper. According to the definition of SVD, the input variables $\mathbf{\Phi}$ in the feature space F can be decomposed as

$$\Phi^T = \mathbf{U}\mathbf{S}\mathbf{T}^T \qquad (15)$$

because $m' \gg n$. $\mathbf{US}$ are loading vectors and $\mathbf{T}$ are score vectors. The classical solution for SVD is to compute $\mathbf{T}$ as the eigenvectors of $\Phi^T\Phi$ because

$$\Phi\Phi^T\mathbf{T} = \mathbf{T}\mathbf{S}^2 \qquad (16)$$

where $\mathbf{S}^2$ is a diagonal matrix with its elements being the eigenvalues. The kernel Gram matrix $\mathbf{K}$ is also used to substitute $\Phi^T\Phi$. The projection of test samples is

$$\mathbf{T}_t = \mathbf{K}_d \mathbf{T}\mathbf{S}^{-2} \qquad (17)$$

where $\mathbf{K}_d$ is again the test set kernel Gram matrix.

2.4 Objective Functions and Training Speed

The KPLS and KSVD based MSE procedure aims to find a weight vector $\mathbf{r}_p = \mathbf{U}_p^T\mathbf{a}$ so that the objective function $J_M = \|\mathbf{T}_p\mathbf{r}_p - \mathbf{b}\|^2$ is minimized. The solution of the weight vectors can easily be obtained even without the pseudoinversion of data matrix by (4). Therefore, the training speeds of KPLS-MSE and KSVD-MSE are very fast and only depend on the size of kernel matrix $\mathbf{K}$. However, the SVM training aims to find a weight vector $\mathbf{w}$ so that the cost function $J_S = \frac{1}{2}\|\mathbf{w}\|^2 + C\sum_i \xi_i$ is minimized with a few other constraints satisfied, where ξ_i are errors caused by the support vectors and C is the regularization parameter. The solution of SVM involves the quadratic programming problem and the training time depends on the number of support vectors. Therefore, SVM is usually slower than the MSE especially when classes are overlapped, originating many support vectors.

3 Results

3.1 Classifiers Accuracy

Seven datasets from [13] were chosen to test the classification accuracies of the KPLS-MSE and the KSVD-MSE. See Table 1 for a brief description of the datasets. As a comparison, SVM of which the implementation is a MATLAB toolbox [14] was also used. A Gaussian kernel $K(\mathbf{x},\mathbf{y}) = \exp(-\|\mathbf{x}-\mathbf{y}\|^2/2\sigma^2)$ was used for our experiments because of its flexibility. Due to the small sizes of the microarray data, the re-sampling technique was used. The original datasets were randomly separated into a training dataset and a testing dataset. The classifier was built on the training dataset and then tested on the testing dataset. This process can be repeated for many times to obtain a stable estimate of the testing errors. Therefore, the evaluation of the three classifiers was designed as follows:

- The numbers of training samples and testing samples were fixed (see Table 1) and then 30 random partitions were carried out.
- For each partition, the optimal parameters for each classifier, including regularization parameter, number of components and kernel parameter σ, were first determined by 10-fold CV. Then the classifiers were built on the training dataset and tested on the testing dataset. The testing errors were recorded.
- Average testing errors with its standard deviations were calculated.
- The performance of the three classifiers on all the datasets was evaluated.

Table 1. Description of the datasets

Dataset	*Genes*	*Partition Setting*
Breast Cancer	24481	40 training vs. 57 testing
Central Nervous System	7129	30 training vs. 30 testing
Colon Tumor	2000	40 training vs. 22 testing
Acute Leukemia	7129	36 training vs. 36 testing
Lung Cancer	12533	50 training vs. 131 testing
Ovarian Cancer	15154	50 training vs. 203 testing
Prostate Cancer	12600	50 training vs. 86 testing

To test whether a classifier is statistically more accurate than another classifier on a particular dataset, the T test between the means of the testing errors of the two classifiers is employed as follows:

$$T = \frac{\mu_1 - \mu_2}{\sqrt{\frac{\sigma_1^2}{n_1} + \frac{\sigma_2^2}{n_2}}} \tag{18}$$

where $\mu_1, \mu_2, \sigma_1, \sigma_2$ are means and standard deviations of the testing errors of the two classifiers, and n_1, n_2 are the numbers of partitions carried out by the two classifiers. The significance level for all tests was set to be $\alpha = 0.05$.

The testing results are listed in Table 2. Very good performance is observed for the KPLS-MSE and the KSVD-MSE classifiers. The minimum average testing errors have been achieved by the KPLS-MSE on two datasets and by the KSVD-MSE on three datasets. The null hypothesis is rejected on the first five datasets between one or two classifiers and the best classifier. Nevertheless, for the last two datasets, the null hypothesis still holds. The classification performance of KPLS-MSE and KSVD-MSE are very similar with the SVM.

Table 2. The number of average testing errors with its standard deviation of KPLS-MSE, KSVD-MSE and SVM on 7 datasets and the number of average components used for KPLS and SVD. The best results are indicated in bold font

Dataset	Testing Errors μ, σ			Components Used	
	PLS	PCA	SVM	PLS	PCA
Breast Cancer	20.9, 3.9	22, 3.5	23.7, 3.7	3.3	9.2
Central Nervous System	12.6, 3.5	10.4, 2.6	11.6, 3.1	3.7	10.8
Colon Tumor	5.3, 2.0	4.5, 1.9	5.2, 2.4	3.3	8.1
Acute Leukemia	3.3, 2.0	3.7, 2.0	2.9, 2.7	2.7	11.6
Lung Cancer	6.8, 5.5	8.9, 5.2	5.0, 4.8	2.8	20.2
Ovarian Cancer	21.4, 10.2	20.1, 12.1	22.9, 6.4	4.6	10.2
Prostate Cancer	28.3, 10.0	30.1, 5.8	28.9, 7.5	6.7	17.1

3.2 Components Selection

The selection of KPLS components and KSVD components is based on minimum validation errors using CV. The number of components used is gradually increased and an optimal number of components is determined to avoid overfitting or underfitting of the training data. KPLS generates components based on the covariance of the predictor variables and the response variables while KSVD generates components only based on the variance of the predictor variables. Therefore, KPLS should produce more qualitative components than KSVD and thus fewer components are required by KPLS to achieve the optimal result. This condition is illustrated in Fig. 1. Ten-fold CV is carried out and the validation errors are obtained by KPLS-MSE and KSVD-MSE on the lung cancer data. KPLS-MSE achieves the optimal result by 4 components while KSVD-MSE needs around 20 components.

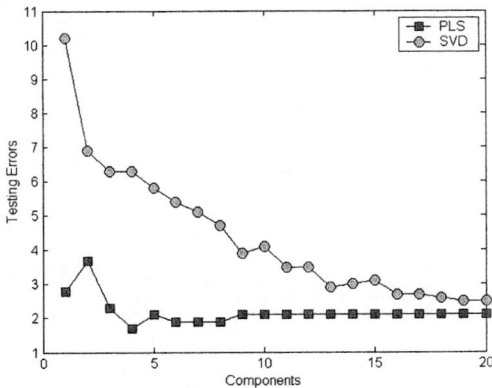

Fig. 1. Average training errors of KPLS-MSE and KSVD-MSE vs. components used on lung cancer data. Ten partitions were carried out.

3.3 Training Speed

In some other authors' research work, feature selection is performed by recursively training a classifier on a series of nested subsets of genes on a particular cancer dataset [15]. Therefore, it is very important to know the training time of a certain classifier on a certain dataset. A five-fold CV to determine the validation errors of a classifier is carried out on the above seven datasets including all genes and samples. For SVM, linear kernel is used and the regularization parameter C is constrained to one. For KPLS-MSE and KSVD-MSE, the validation errors for different numbers of components can be obtained in a single run simultaneously. The results are listed in Table 3. The KPLS-MSE and KSVD-MSE cost significantly less training time than SVM on all the datasets. If the optimal C for SVM is to be chosen by CV, then much more training time is required. KPLS-MSE and KSVD-MSE spend almost the same time for training.

Table 3. The training time of the three classifiers to determine the minimum training errors by five-fold CV on the seven cancer datasets

Dataset	PLS (s)	SVD (s)	SVM (s)
Breast Cancer	2.10	2.09	16.67
Central Nervous System	0.29	0.29	2.24
Colon Tumor	0.12	0.12	2.42
Acute Leukemia	0.37	0.37	3.32
Lung Cancer	2.87	2.99	12.00
Ovarian Cancer	6.30	6.73	50.72
Prostate Cancer	1.77	1.81	15.68

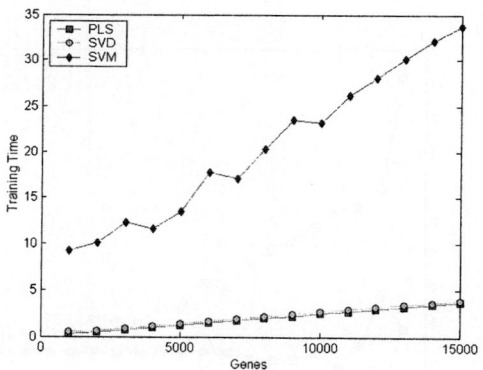

Fig. 2. The training time of KPLS-MSE, KSVD-MSE and SVM on the ovarian cancer dataset versus the number of genes

Another study to determine the impact of the number of genes on the training time of a classifier is also carried out on the ovarian cancer dataset. By increasing the gene number from 1,000 to 15,000, the slope of the training time of a classifier versus the numbers of genes is calculated by least squares regression. The slopes of KPLS-MSE and KSVD-MSE are known as 0.2×10^{-3} but the slope of SVM is as large as 1.8×10^{-3} (see Fig. 2). Thus KPLS-MSE and KSVD-MSE cost much less time (one ninth) than SVM as the number of genes increases.

4 Discussions and Conclusions

The combination of KPLS and KSVD with MSE has created two very effective classifiers for cancer classification using gene expression profiles. The KPLS-MSE and KSVD-MSE have been validated on seven publicly available cancer datasets and very high accuracy has been achieved. Because of the orthogonality of component vectors, no matrix inversion is involved in constructing the discriminant function. Thus, both classifiers have very fast training speed compared with the other pattern classification algorithms. Because KPLS generates more qualitative components than KSVD, KPLS-MSE always uses less component vectors for classification.

One of the main characteristics of the microarray data is $m \gg n$. So high multicollinearity exists among the gene expression profiles, which is illustrated by the relatively small number of components we have extracted for classification. Due to the high cost of microarray experiments, the publicly available microarray samples are relatively few. Testing a classifier on a very small dataset could incur very large bias from the actual accuracy of the classifier. Thus, the random partition method used in this paper is a good approach for evaluation and comparison of the average performance of the classifiers. We have used the KPLS-MSE and the KSVD-MSE to get very similar results using the same training and testing datasets. Comparing the results of the well-established SVM method, our methods are very good and competitive.

In such a very high dimensional datasets with relatively small numbers of samples, all classification algorithms suffer from the "curse of dimensionality" even those have the ability to handle the data when $m > n$ and all of them can benefit from feature selection [15]. Therefore, the fast training speeds of KPLS-MSE and KSVD-MSE should be profitable under this condition.

References

1. Golub,T.R., Slonim,D.K., Tamayo,P., Huard,C., Gaasenbeek,M., Mesirov,J.P., Coller,H., Loh,M.L., Downing,J.R., Caligiuri,M.A., Bloomfeld,C.D., Lander,E.S.: Molecular Classification of Cancer: Class Discovery and Class Prediction by Gene Expression Monitoring. *Science*. Vol 286. (1999) 531-537
2. Khan,J., Wei,J.S., Ringner,M., Saal,L.H., Ladanyi,M., Westermann,F., Berthold,F., Schwab,M., Atonescu,C.R., Peterson,C. and Meltzer,P.S.: Classification and diagnostic prediction of cancers using gene expression profiling and artificial neural networks. *Nature Medicine*, 7. (2001) 673-679
3. Ghosh,D.: Singular value decomposition regression models for classification of tumors from microarray experiments. *PSB*. (2002) 18-29

4. Nguyen,D.V. and Rocke,D.M.: Tumor classification by partial least squares using microarray gene expression data. *Bioinformatics*, Vol. 18. (2002) 39-50
5. Vapnik,V.: *Statistical Learning Theory*. John Wiley and Sons, Inc., New York. (1998)
6. Scholköpf, G., Smola, A. and Müller, K. R.: Nonlinear Component Analysis as a Kernel Eigenvalue Problem. *Neural Computation*, Vol. 10. (1998) 1299-1319
7. Rosipal,R. and Trejo,L.J.: Kernel Partial Least Squares Regression in Reproducing Kernel Hilbert Space. *Journal of Machine Learning Research*, 2. (2001) 97-123
8. Duda,R.O., Hart,P.E. and Stork,D.G.: *Pattern Classification*. A Wiley Interscience Publication. (2000)
9. Golub,G.H. and Van Loan,C.F.: *Matrix Computations*. The Johns Hopkins University Press. (1996)
10. Wold,H.: Soft Modeling by Latent Variables; the Nonlinear Iterative Partial Least Squares Approach. In J. Gani (Ed.), *Perspectives in Probability and Statistics, Papers in Honour of M.S. Bartlett*, Academic Press, London. (1975) 520-540
11. Wegelin,J.A.: *A survey of Partial Least Squares (PLS) methods, with emphasis on the two-block case* (Technical Report). Department of Statistics, University of Washington, Seattle. (2000)
12. de Jong, S.: SIMPLS: an alternative approach to partial least squares regression. *Chemometrics and Intelligent Laboratory Systems*, Vol. 18. (1993) 251-263
13. Li,J., Liu,H.: Kent Ridge Biomedical Data Set Repository. Available: http://sdmc-lit.org.sg/GEDatasets. (2002)
14. Schwaighofer,A.: Available: http://www.cis.tugraz.at/igi/aschwaig/svm_v251.tar.gz. (2001)
15. Guyon,I., Weston J., Barnhill,S., Vapnik,V.: Gene selection for cancer classification using support vector machines. *Maching learning*, 46. (2002) 389-422

Fusing Face and Fingerprint for Identity Authentication by SVM

Chunhong Jiang and Guangda Su

The State Key Laboratory of Intelligent Technology and System,
Electronic Engineering Department, Tsinghua University, Beijing, 100084, China
jiangch@mail.tsinghua.edu.cn, sugd@ee.tsinghua.edu.cn

Abstract. Biometric based person identity authentication is gaining more and more attention. It has been proved that combining multi-biometric modalities enables to achieve better performance than single modality. This paper fused Face and fingerprint (for one identity, face and fingerprint are from the really same person) for person identity authentication, and Support Vector Machine (SVM) is adopted as the fusion strategy. Performances of three SVMs based on three different kernel functions (Polynomial, Radial Based Function and Hyperbolic Tangent) are given out and analyzed in detail. Three different protocols are defined and operated on different data sets. In order to enhance the ability to bear face with bigger pose angle, a client specific SVM classifier is brought forward. Experiment results proved that it can improve the fusion authentication accuracy, and consequently expand the allowable range of face turning degree to some extend in fusion system also.

1 Introduction

A biometric person recognition system can be used for two modes: identification and authentication (or verification). In the identification mode, there is no identity claim from the user. The system should decide who the person is. In the authentication mode, a person claims a certain identity, the system should accept or reject this claim, (the person is really who he claim to be?) [1]. So identification mode involves comparing the acquired biometric information against templates corresponding to all the users in the database (one-to-all), but the authentication mode involves comparison with only those templates corresponding to the claimed identity (one-to-one) [2]. In this paper, we will only focus on the issue of biometric identity authentication. Obviously, the identity authentication problem is a typically binary classification problem, i.e. accept (genuine) or reject (imposter). SVM is well known as a two-class problem classifier with high performance [3,4]. So, in this paper, we adopted SVM as the fusion strategy of the face and fingerprint identity authentication system.

In multimodal biometrics system, the information can be integrated at various levels. A. K. Jain had given an illustration of three levels of fusion when combining two or more biometric systems in reference[5]. The three levels are: the feature extraction level, the matching score level and the decision level. Information fusion based on SVM is operated on the matching score level in this paper, and the experiment results are given out and analyzed in detail in the following sections.

The identity authentication problem can be formulated as a hypothesis testing problem where the two hypotheses are

ω_1 : the person is not from the same identity;

ω_2 : the person is from the same identity.

For an acquired person Uq, the authentication system should decide the person is an impostor or genuine. The decisions are

D_1 : the person is an impostor;

D_2 the person is genuine.

With the hypothesis and decisions above we have the false acceptance rate: $FAR = P(D_2/\omega_1)$, the false rejection rate: $FRR = P(D_1/\omega_2)$, and the genuine acceptance rate: $GAR = 1 - FRR$, the equal error rate (EER) is where FAR=FRR. In this paper, we attach much importance to FAR than to FRR.

If x_1 and x_2 are the outputs of the component classifiers, then Assign $Uq \to \omega_j$ if

$$P(\omega_j/x_1, x_2) = \max_{k=1}^{2} P(\omega_k/x_1, x_2), \quad j = 1, 2 \tag{1}$$

Where the $P(\omega_k/x_1, x_2)$ represents the posteriori of ω_k given x_1 and x_2.

The remainder of this paper is organized as follows. The performance of face and fingerprint authentication system are described in section 2 and 3. The SVM fusion methods and experiment results are presented and analyzed in section 4. Finally we give out the main conclusions and future work in section 5.

2 Face Authentication

Face authentication involves face detection, feature extraction, feature matching process and decision making. In this paper, we use an automatic method for face detection and for eye and chin orientation [6], and adopted multimodal part face recognition method based on principal component analysis (MMP-PCA) to extract feature set [7]. The experiment face images are from the TH (Tsinghua University, China) Database. The TH database contains 270 subjects and 20 face images per subject with every other 5 degree turning from the front face to left (-) or right (+), and 10 fingerprint images from 2 different fingers with 5 images each. In our experiment, 186 subjects were selected for fusion of face and fingerprint authentication. We selected 13 face images and 5 fingerprint images for each subject. For face images, the first one, which is the one with zero turning degree, was selected as template, and the other 12 images as probes. Fig.1 shows the face and fingerprint images in TH database. Table 1 shows the training and testing protocols, the genuine and impostor match numbers. Protocol 1(P1), 2(P2) and 3(P3) are different from the training sets, +5 and +10 degree for P1, -5 and -10 degree for P2, ±5 and ±10 degree for P3. For training face images in P1 and P2, we have 2 genuine match scores per subject, together 2×186 match scores constructing the training genuine distribution, and 2×185×186 impostor match scores constructing the training impostor distribution. Obviously every testing face sets has the same number. From table 1, we can see that each protocol of P1 and P2 has 1 training set and 5 testing sets (Te1~Te5). For P3, Te1 set of P1 or P2 was used for training, so the training set of P3 has 4×186 genuine match scores and 4×185×186 impostor match scores, and the testing sets (Te2~Te5) are same as sets in P1 and P2.

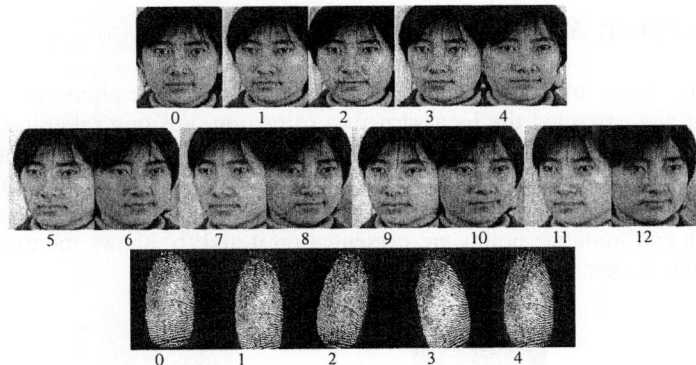

Fig. 1. Samples of Face and Fingerprint Images in the TH Database

Table 1. Authentication Protocols

			Template		Probes							
		SN	0	1	2	3	4	5, 6	7, 8	9, 10	11, 12	
		Degree	0	-5	+5	-10	+10	±15	±20	±25	±30	
Face		Protocol 1 (P1)		Te1	Tr	Te1	Tr	Te2	Te3	Te4	Te5	
		Protocol 2 (P2)		Tr	Te1	Tr	Te1	Te2	Te3	Te4	Te5	
		Protocol 3 (P3)		Tr	Tr	Tr	Tr	Te2	Te3	Te4	Te5	
		Number of genuine match		186	186	186	186	2×186	2×186	2×186	2×186	
		Number of impostor match		185×186	185×186	185×186	185×186	2×185×186	2×185×186	2×185×186	2×185×186	
Finger print	Set	Fusion with		Template	Probes							
	A	Tr, Te1	S N	0	1	2	3	4				
	B	Te2, Te3		1	0	2	3	4				
	C	Te4, Te5		2	0	1	3	4				

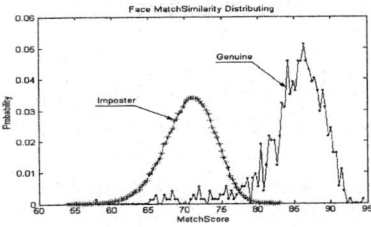

Fig. 2. Face Matching Similarity Distributing

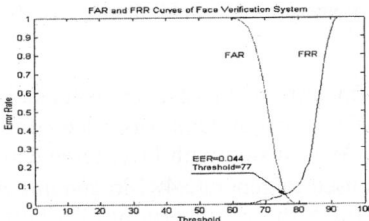

Fig. 3. FAR and FRR Curves of Face

Fig.2 shows face distributing of genuine and impostor match similarity (%) of Tr data set of P3 (Note that the following figures relate to face and fusion systems are all from this set except indicate). It is obvious that the genuine and impostor overlapped each other, and the decision errors are unavoidable. FAR and FRR curves of face authentication system are presented in Fig.3, EER is 0.044 when authentication threshold is 77%.

3 Fingerprint Authentication

In the Fingerprint authentication system, we use an automatic algorithm to locate the core point and extracted the local structure (direction, position relationship with the neighbor minutiaes) and global structure (position in the whole fingerprint) of all the minutiaes [8]. The matching algorithm used local and also global structures of every minutia. Fig.4(a) shows a sample in the TH fingerprint database, the core, the first orientation and minutiae points are presented on it and (b) shows the extracted ridge and minutiae points.

(a) Fingerprint and Its Minutiaes   (b) The Ridge and Minutiaes

Fig. 4. Sample in the TH Fingerprint Database

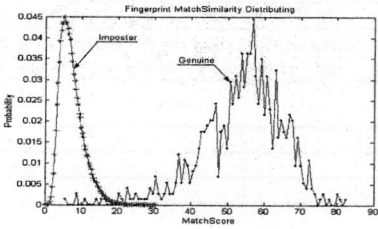
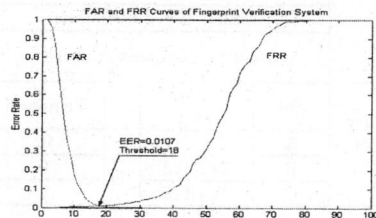

Fig. 5. Fingerprint Match Similarity Distributing   **Fig. 6.** FAR and FRR Curves of Fingerprint

For fingerprint images, we selected 5 images from one finger. Table 1 shows the fingerprint protocol. One was selected to be template and the other four leaved to be probes. As to fusion with face, three data sets are built, i.e. A, B and C. Data in each set was used to generate 4×186 genuine match scores and 4×185×186 impostor match scores. Fig.5 shows fingerprint distributing of genuine and impostor match similarity (%) on data set A. FAR and FRR curves of fingerprint authentication system was presented in Fig.6. EER is 0.0107 when threshold is 18%. See Fig.6, FAR and FRR curves intersect and form a flatter vale, which predicate the range of threshold with respect to smaller FAR and FRR is larger, and the point of intersection is nearer to the x-axis, so the EER is smaller, both compared with face authentication in Fig.3. As a result, the authentication accuracy and robustness of fingerprint outperforms face authentication system obviously. In the next section, we will see fusion systems present a rather better performance than either of face and fingerprint system.

4 Fusion of Face and Fingerprint Authentication

As to fuse face and fingerprint authentication systems, a confidence vector $X(x_1, x_2)$ represents the confidence output of multiple authentication systems was constructed, where x_1 and x_2 correspond to the similarity (score) obtained from the face and fingerprint authentication system respectively. Further more, for multi-biometric modalities more than 2, the problem turns to be N dimensional score vector $X(x_1, x_2, \cdots x_N)$ separated into two classes, genuine or impostor. In other words, the identity authentication problem is always a two-class problem in spite of any number of biometrics.

4.1 Support Vector Machine

Support vector machine (SVM) is based on the principle of structural risk minimization. It aims not only to classifies correctly all the training vectors, but also to maximize the margin from both classes. The optimal hyperplane classifier of a SVM is unique, so the generalization performance of SVM is better than other methods that possible lead to local minimum [3,4]. In reference [9], SVM was compared with other fusion methods, and its performance was the best. And in this paper, we pay attention to the performance of SVM with different protocols and on different data sets. The detailed principle of SVM is not showed in this paper, and it can be seen in reference [3]. And three kernel functions of SVM are used in our study, they are:

$$\text{Polynomials: } K(x,z) = (x^T z + 1)^d, d > 0 \tag{2}$$

$$\text{Radial Basis Functions: } K(x,z) = \exp(-g \|x - z\|^2) \tag{3}$$

$$\text{Hyperbolic Tangent: } K(x,z) = \tanh(\beta x^T z + \gamma) \tag{4}$$

4.2 Experiment Results

4.2.1 Performance of SVMs Based on Different Kernel Functions

In order to test the performances of SVMs based on three kernel functions mentioned above, protocol 3 was selected. And the results can be seen at the following figures.

Fig.7 shows different SVMs with different parameters separating genuine and impostor of *Tr* data set of P3 before normalization. We can see that three SVMs can all separate the two classes correctly. Their performances are similar; however, the number of support vectors and the difficulty to adjust parameters of kernel function are different. In our experiment, the SVM-Pnm is easier to be trained than SVM-RBF and SVM-Tanh; the latter two need more patience during training period.

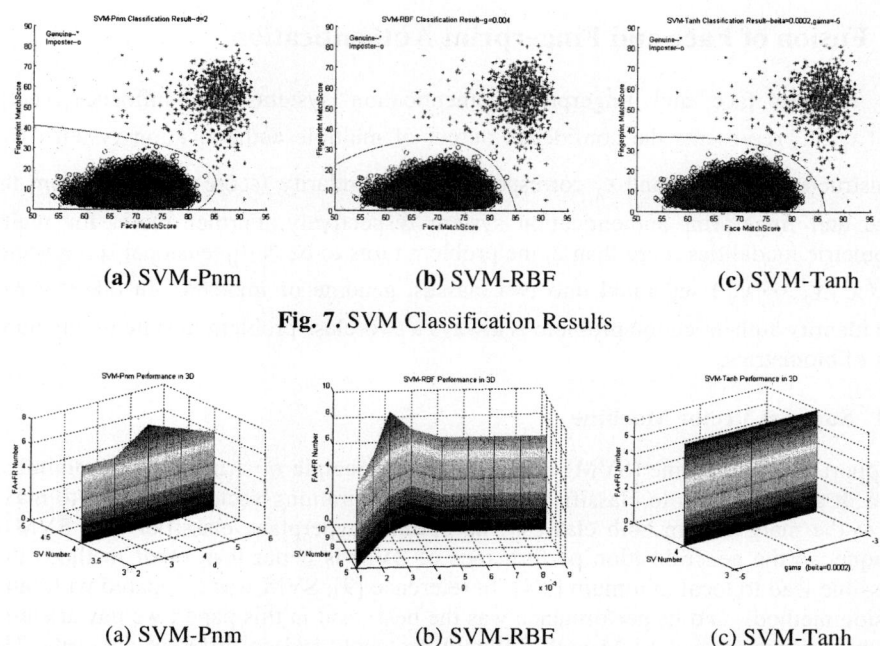

(a) SVM-Pnm   (b) SVM-RBF   (c) SVM-Tanh

Fig. 7. SVM Classification Results

(a) SVM-Pnm   (b) SVM-RBF   (c) SVM-Tanh

Fig. 8. SVM Performance with Different Parameters. (FA+FR—false accepted and false rejected number by classifier (18) on *Tr* data set of P3).

Fig.8 presents different parameters resulting in different number of support vectors and different performances. For Pnm and Tanh kernel, the number of SV is invariable when parameter changed, unlike the former two, the SV number of RBF kernel is fluctuant. The false accepted and rejected number are wavy along with the parameter changing expect the Tanh kernel.

The ROC curves of three SVMs are showed in Fig.9(a). From the ROC curves, we can see that Tanh kernel outperforms the other two, the Pnm is middling and RBF is a shade worse. Notice that the computational complexity of SVM is independent of the dimensionality of the kernel space where the input feature space is mapped. And the high computational burden is required both in training and testing phase. Fortunately, in our experiments, only a relatively small number of training data is enough. Just those points near to the support vectors or near to the hyperplane are inputted into SVM, of course, the hyperplane was not known at beginning, so just estimated it. From Fig.8, we see that the SV numbers of the three kernels are all no more than 10, so the computation quantum is not large during testing phase too.

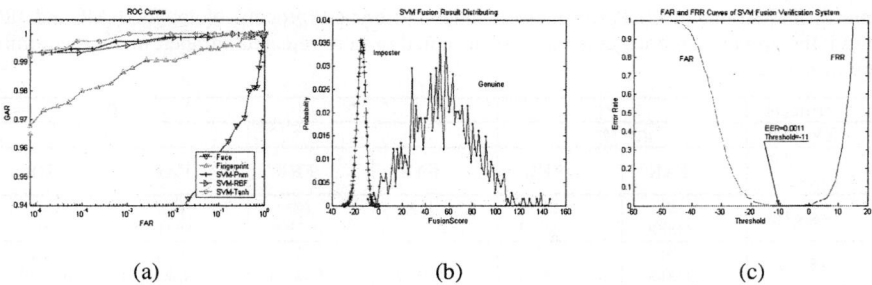

(a) (b) (c)

Fig. 9. (a) is SVM ROC Curves. (b) is SVM-Pnm Result Distributing (c) is FAR and FRR Curves of SVM-Pnm.

Fusion score distributing and FAR and FRR curves of SVM-Pnm (d=2) are showed in Fig.9(b) and (c). See (c), FAR and FRR curves intersect and form a very flat and broad vale (EER is 0.0011), this means that for a large region of threshold value in which the FAR and FRR are both very small. Accordingly, not the accuracy but the robustness of the authentication system are both improved after fusion with SVM.

4.2.2 Performance Under Three Different Protocols

In order to test the performance of SVM under different training data, we defined three protocols, as mentioned in section 2, see table 1. Fig.10 shows the three hyperplanes gained by SVM-Pnm after trained with different data sets. The authentication performances of the three hyperplanes are presented in table 2. From the results, we can see that the three hyperplanes are very similar and their performances are competitive. $P3$ is a shade better than the other two by reason of more training data. This stable characteristic of SVM's hyperplane is very grateful in applications. Furthermore, this result also proved the good generalization ability of SVM. Under $P3$, authentication performances of three SVMs on each data set are presented in table 3.

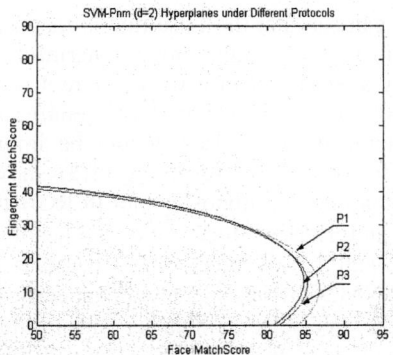

Fig. 10. Different Hyperplanes of SVM-Pnm under Different Protocols

Table 2. SVM-Pnm (d=2) Performances under Different Protocols. Note: for FAR and FRR column, the up value in bracket is the false accepted or false rejected number, under the number is the false rate.

Protocol	P1		P2		P3	
SV num	4		4		4	
FR Degree	FAR	FRR	FAR	FRR	FAR	FRR
+5,+10	(0), 0.0000	(3), 0.0081	(1), 1.4531e-5	(4), 0.0108	(1), 1.4531e-5	(3), 0.0081
-5,-10	(0), 0.0000	(3), 0.0081	(0), 0.000	(2), 0.0054	(0), 0.000	(2), 0.0054
±15	(0), 0.0000	(2), 0.0054	(0), 0.0000	(1), 0.0027	(0), 0.0000	(1), 0.0027
±20	(0), 0.0000	(9), 0.0242	(0), 0.0000	(9), 0.0242	(0), 0.0000	(7), 0.0188
+25	(0), 0.0000	(16), 0.0430	(0), 0.0000	(16), 0.0430	(0), 0.0000	(14), 0.0376
±30	(0), 0.0000	(17), 0.0457	(0), 0.0000	(17), 0.0457	(0), 0.0000	(17), 0.0457

Table 3. Performances of Three SVMs on Each Data Set under P3

FR Set	Pnm		RBF		Tanh	
	FAR	FRR	FAR	FRR	FAR	FRR
Tr	(1), 7.265e-6	(5), 0.0067	(2), 1.453e-5	(5), 0.0067	(1), 7.265e-6	(5), 0.0067
Te2	(0), 0.0000	(1), 0.0027	(0), 0.0000	(1), 0.0027	(0), 0.0000	(1), 0.0027
Te3	(0), 0.0000	(7), 0.0188	(0), 0.0000	(9), 0.0242	(0), 0.0000	(8), 0.0215
Te4	(0), 0.0000	(14), 0.0376	(0), 0.0000	(15), 0.0403	(0), 0.0000	(16), 0.0430
Te5	(0), 0.0000	(17), 0.0457	(0), 0.0000	(16), 0.0430	(0), 0.0000	(16), 0.0430

4.3 Client Specific SVM Classifier

From table 2 and 3, we can see that the false rate is increasing with the face turning degree getting larger. Since face rotating angle is harmful to fusion system, one hyperplane only can hardly adapt to all face with different pose. If we trained a hyperplane for every single identity according to itself genuine and impostor distribution characteristics, the authentication accuracy would be improved. So client specific SVM (CS-SVM) classifier is put forward. Fig.11(a) shows two single identities' genuine and impostor (genuine and impostor plot in RGB three colors, which represent ±5~±10, ±15~±20 and ±25~30 three data sets) scatter plot in 2D place which classified by two SVMs, and that the blue solid hyperplane is the common SVM, the red dash-dot one is CS-SVM. Obviously, the client specific classifier is more reasonable than the common one. We selected data set with turning degree ±25 and ±30 to train CS-SVM, and the others to test in consideration of separating genuine and impostor with larger turning degree. The lowest cyan curve marked with "▷" in Fig.11(b) shows the performance of CS-SVM. We can see that CS-SVM outperforms the other methods. Results proved that CS-SVM can improve the fusion authentication accuracy, and accordingly, expand the range of face turning degree to some extend also. But one disadvantage of CS-SVM is that more memory required for storing the SVM classifier parameters for every identity, off course, the training burden is heavier than common SVM also.

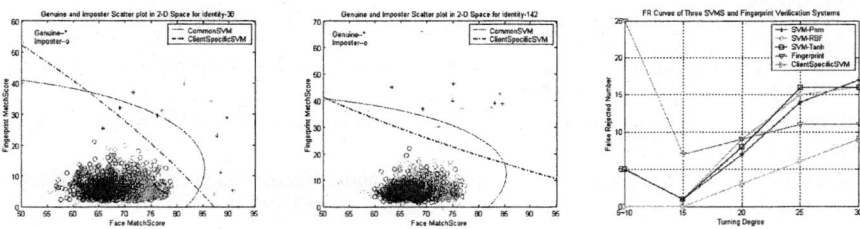

(a) Hyperplanes of Common and CS-SVMs   (b) False Rejected Number Curves

Fig. 11. (a) Hyperplanes of Common and CS-SVMs. (b) False Rejected Number Curves of Three SVMs and Fingerprint Authentication Systems.

5 Conclusion

This paper fuses face and fingerprint for identity authentication by SVM. Differing from some other researches on multi-biometric fusion system, face and fingerprint for one identity are really from the same person other than just let face partnering with fingerprint from other person. Hence, fusion results gained in this paper are closer to practical applications.

Because identity authentication is a two-class problem, so SVM is adopted as the fusion strategy. The performance of SVMs with different kernel functions, different test protocols and different test data sets is given out and analyzed in detail. And a client specific SVM classifier is brought forward to verify genuine and impostor for every identity. From experiment results mentioned above, we can draw the following conclusions.

SVM fusion strategy is effective in face and fingerprint fusion system. And the EER is decreased to 0.0011(face is 0.044 and fingerprint is 0.0107). ROC curve of SVM shows that, not the accuracy but the robustness of the authentication system are both improved after fused with SVM.

Performances of the three SVMs based on different kernel functions are similar, despite different number of support vector. SVM-Pnm is easier to be trained than SVM-RBF and SVM-Tanh; the latter two need more time to adjust the parameters during training period. All-around, SVM-Pnm and SVM-Tanh is a shade of better than SVM-RBF.

We defined three protocols with different training data sets. Results showed that SVM performances under three protocols are very competitive. P3 is a shade better than the other two by reason of more training data. On the other hand, this result also proved the good generalization ability of SVM.

CS-SVM classifier outperforms the common SVM. Result proved that it can improve the fusion authentication accuracy, and also expand the range of face turning degree to some extend.

References

1. Souheil Ben-Yacoub, etc.: Fusion of Face and Speech Data for Person Identity Authentication. IEEE Transactions on Neural Networks, Vol.10(5). (1999) 1065-1074
2. Arun Ross, Anil Jain: Information Fusion in Biometrics. Pattern Recognition Letters, Vol.24. (2003) 2115-2125
3. Kecman, V.: Learning and Soft Computing, Support Vector machines, Neural Networks and Fuzzy Logic Models, The MIT Press, Cambridge, MA (2001)
4. Wang, L.P. (Ed.): Support Vector Machines: Theory and Application. Springer, Berlin Heidelberg New York (2005)
5. Anil K. Jain, Arun Ross etc.: An Introduction to Biometric Recognition. IEEE Transactions on Circuits and Systems for Video Technology, Vol.14(1). (2004) 4-20
6. Gu Hua, Su Guangda, etc.: Automatic Extracting the Key Points of Human Face. Proceeding of the 4th Conference on Biometrics Recognition, Beijing, China (2003)
7. Su Guangda, Zhang Cuiping etc.: MMP-PCA Face Recognition Method. Electronics Letters, Vol.38(25). (2002) 1654-1656
8. Huang Ruke: Research on the Multi-Hierarchical Algorithm for Fast Fingerprint Recognition. Bachelor thesis of Tsinghua University (2002)
9. Jiang Chunhong, Su Guangda: Information Fusion in Face and Fingerprint Identity Authentication System. Proceeding of the 2004 International Conference on Machine Learning and Cybernetics, Vol.1-7. Shanhai, China (2004) 3529-3535

A New Algorithm of Multi-modality Medical Image Fusion Based on Pulse-Coupled Neural Networks

Wei Li and Xue-feng Zhu

College of Automation Science and Engineering, South China University of Technology,
Guangzhou, 510640, China
llewell@163.com

Abstract. In this paper, a new multi-modality medical image fusion algorithm based on pulse-coupled neural networks (PCNN) is presented. Firstly a multi-scale decomposition on each source image is performed, and then the PCNN is used to combine these decomposition coefficients. Finally an inverse multi-scale transform is taken upon the new fused coefficients to reconstruct fusion image. The new algorithm utilizes the global feature of source images because the PCNN has the global couple and pulse synchronization characteristics. Series of experiments are performed about multi-modality medical images fusion such as CT/MRI, CT/SPECT, MRI/PET, etc. The experimental results show that the new algorithm is very effective and provides a good performance in fusing multi-modality medical images.

1 Introduction

Along with the development of medical imaging technology, various imaging modals such as CT, MRI, PET and SPECT are widely applied in clinical diagnosis and therapy. Owing to difference in imaging mechanisms and high complexity of human histology, medical images of different modals provide a variety of non-overlapped and complementary information about human body. These different images have their respective application ranges. For example, functional image (PET, SPECT, etc.) has relative low spatial resolution, but it provides functional information about visceral metabolism and blood circulation; while anatomical image (CT, MRI, B-mode ultrasonic, etc.) provide information about visceral anatomy with relative high spatial resolution. Multi-modality medical image fusion is to combine complementary medical image information of various modals into one image, so as to provides far more comprehensive information and improves reliability of clinical diagnosis and therapy [1].

The pulse-coupled neural networks (PCNN), as a novel artificial neural network model, is different from traditional neural networks. PCNN models have biological background and are based on the experimental observations of synchronous pulse bursts in cat and monkey visual cortex [2]. Due to this, PCNN has been efficiently applied to image processing field, such as image segmentation, image restoration, image recognition, etc. In this paper, we put forward a new algorithm of multi-modality medical image fusion based on PCNN. Firstly we use discrete wavelet frame (DWF) to decompose the original images into a series of frequency channels, and then we use

PCNN to combine the different features and details at multiple decomposition levels and in multi-frequency bands, which is suitable for the human vision system. Finally an inverse DWF transform is taken upon the new fused coefficients to reconstruct fusion image. Compared with some other existing methods for multi-modality medical image fusion purpose, the new algorithm combines the multi-resolution analysis characteristic with the global couple and pulse synchronization characteristics of PCNN, which can extract abundant information from source images. Series of experiments are performed about multi-modality medical images fusion such as CT/MRI, MRI/PET, etc. The experimental results show that the new algorithm is very effective and provides a good performance in merging multi-modality images.

This paper is organized as follows. In section 2, a brief review of the PCNN theory is introduced first, and then the new fusion algorithm based on PCNN is described in detail. In section 3, results of computer simulations of fusion images are shown. Meanwhile, compared with three other existing methods in reference [3~5], the new algorithm has been proved to be very effective and provide a better performance in fusing multi-modality medical images.

2 Image Fusion Algorithm Based on PCNN

2.1 PCNN Basic Model

A PCNN's model neuron consists of three parts: the receptive field, the modulation field, and the pulse generator. The model is shown in Fig.1

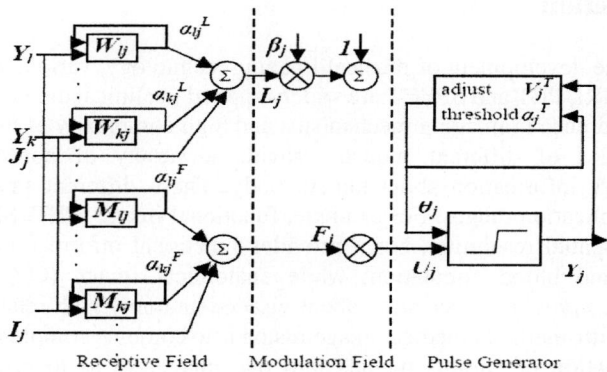

Fig. 1. Internal structure of a single neuron in PCNN

The neuron receives input signals from other neurons and from external sources through the receptive field. Then input signals are divided into two channels. One channel is feeding input F_j, and the other is linking input L_j. I_j and J_j are external inputs; M_{kj} and W_{kj} are the synaptic gain strengths; α_{kj}^F and α_{kj}^L are the decay constants. The total internal activity U_j is the result of modulation. β_j is the linking strength. Y_j is the output of neuron. The threshold θ_j changes with the variation of the neuron's output pulse. V_j^T and a_j^T are the amplitude gain and the time constant of the threshold adjuster. The mathematic models of PCNN are described as:

$$\begin{cases} F_j = \sum_k [M_{kj} \exp(-\alpha_{kj}^F t)] \otimes Y_k(t) + I_j \\ L_j = \sum_k [W_{kj} \exp(-\alpha_{kj}^L t)] \otimes Y_k(t) + J_j \\ U_j = F_j(1 + \beta_j L_j) \\ \dot{\theta}_j = -\alpha_j^T \theta_j + V_j^T Y_j(t) \\ Y_j(t) = step(U_j - \theta_j) \end{cases} \quad (1)$$

Where "$\otimes$" denotes convolution operation. The PCNN used for image processing applications is a single layer two-dimensional array of laterally linked pulse coupled neurons. The number of neurons in the network is equal to the number of pixels in the input image. There exists a one-to-one correspondence between the image pixels and network neurons. Each pixel is connected to a unique neuron and each neuron is connected with the surrounding neurons. Pulse based synchronization is the key characteristic that distinguishes the PCNN from other types of neural networks. The linking connections cause neurons, in close proximity and with related characteristics, to pulse in synchronization.

2.2 Fusion Algorithm Description

The multi-scale decomposition principle of image is the same as that of retina. In this paper, the multi-scale decomposition and the PCNN are combined to give a new fusion algorithm. The fusion scheme is shown in Fig.2.

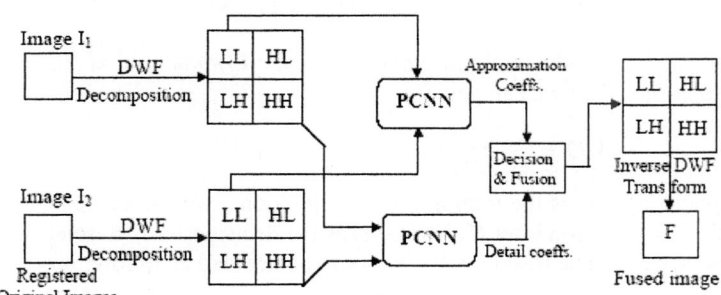

Fig. 2. The image fusion scheme based on PCNN

Supposing that I_1 and I_2 denote the two original images with the same size $M \times N$. They are both accurately registered. F is the fused image. The new fusion algorithm is described as followed:

(1). The DWF decomposition can be continuously applied to the I_1 and I_2 until the desired coarser resolution 2^{-J} ($J>0$). Then we can obtain one low-frequency approximation sub-image and a series of high-frequency detail sub-images:

$$A_{2^{-j}}I_k, D_{2^j}^1 I_k, D_{2^j}^2 I_k, D_{2^j}^3 I_k, \quad (-J \le j \le -1, k = 1,2) \tag{2}$$

In each decomposition stage, due to DWF using dilated analysis filters, each sub-image has the same size with the original images.

(2). Designing a corresponding PCNN with the size $M \times N$ for each sub-image. The pixel's intensity of each sub-image is used as the feeding input F_{ij}^k of the corresponding PCNN neuron. Meanwhile, each neuron which is connected with neurons in the nearest-neighbor 3×3 (or 5×5) field. The sum of responses of the output pulses from surrounding neurons is the linking input.

(3). A simplified discrete model based on equation (1) is used for PCNN fusion network. Each neuron's output Y_{ij}^k is computed by equation (3)~(7):

$$F_{ij}^k(n) = I_{ij}^k \tag{3}$$

$$L_{ij}^k(n) = \exp(-\alpha_L) L_{ij}^k(n-1) + V_L \sum_r W_{r,ij} Y_{r,ij}(n-1) \tag{4}$$

$$U_{ij}^k(n) = F_{ij}^k(n) * (1 + \beta L_{ij}^k(n)) \tag{5}$$

$$Y_{ij}^k(n) = \begin{cases} 1, & if\ : U_{ij}^k(n) > \theta_{ij}^k(n) \\ 0, & otherwise \end{cases} \tag{6}$$

$$\theta_{ij}^k(n) = \exp(-\alpha_\theta) \theta_{ij}^k(n-1) + V_\theta Y_{ij}^k(n) \tag{7}$$

Where k denotes the different low-frequency and high-frequency sub-images. r is the linking radius.

(4). Use N_{ij}^k denotes the total pulse number of the (i,j)th neuron at the kth PCNN of during n iterations, then we can get,

$$N_{ij}^k(n) = N_{ij}^k(n-1) + Y_{ij}^k(n), N_{ij}^k(0) = 0 \tag{8}$$

(5). Suppose that N_p is the total number of iterations, we can utilize the total pulse number of the (i,j) neuron after N_p iterations as the decision basis to select coefficients of fusion sub-image. The fused coefficients can be decided as follow:

$$\begin{cases} I_F^k(i,j) = I_1^k(i,j) & if\ : N_{ij,1}^k(N_p) \ge N_{ij,2}^k(N_p) \\ I_F^k(i,j) = I_2^k(i,j) & if\ : N_{ij,1}^k(N_p) < N_{ij,2}^k(N_p) \end{cases} \tag{9}$$

Then the new fused coefficients are obtained.

(6). Reconstruct image upon new fused coefficients matrix of each sub-image, and the fused image F is obtained by taking an inverse DWF transform.

During image fusion, the most important thing for improving fusion quality is the selection of parameters of PCNN. In simplified PCNN model, the parameters required to be adjusted are α_L, α_θ, W, V_L and β. Since the new algorithm uses the number of output pulses from PCNN's neuron as the basis to select multi-scale decomposition coefficients, the total iteration times N_p can be determined by using image's entropy. The greater the image's entropy is, the more useful information can be got from source images. These can be proved through our experiments.

3 Experimental Results and Performance Evaluation

In order to test and verify the correctness and effectiveness of the proposed new method in this paper, it is compared with three other existing fusion methods in reference [3~5]. Since the useful features in region are usually larger than one pixel, the pixel-by-pixel maximum-based method [3] and the fuzzy-rule-based method [4] are not appropriate though they are very simple. In some fusion themes, the region-feature-based methods [5] are used and obtain the good fusion results, but they have high computation complexity and need huge calculation. The new algorithm utilizes the global coupled property and pulse synchronization characteristic of the PCNN to take an intelligent fusion decision. It can meet the real-time requirement and overcome some shortcomings of the existing fusion methods mentioned above. Series of experiments are performed about multi-modality medical images fusion such as CT/MRI, MRI/PET, etc. A group of the experimental results are shown in Fig.3.

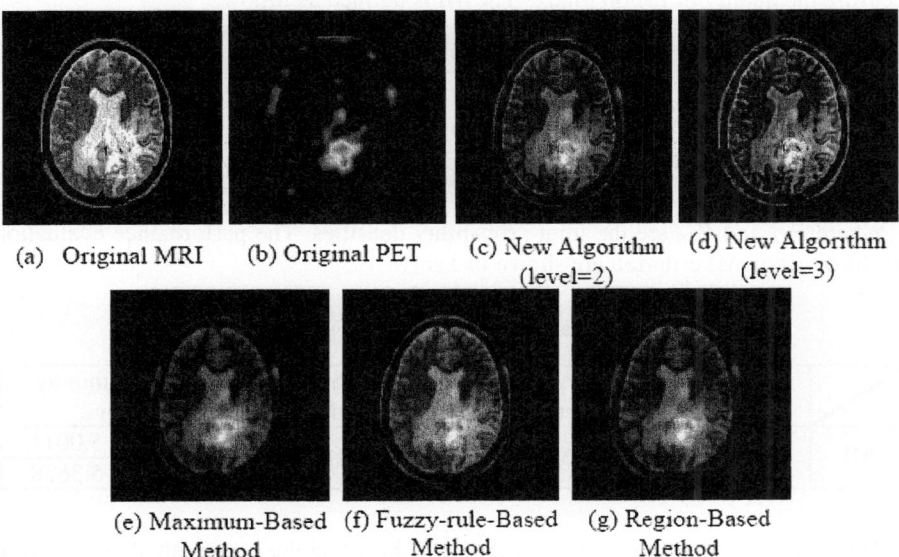

(a) Original MRI (b) Original PET (c) New Algorithm (level=2) (d) New Algorithm (level=3)

(e) Maximum-Based Method (f) Fuzzy-rule-Based Method (g) Region-Based Method

Fig. 3. The experimental results of MRI and PET images fusion

These multi-modality medical images used in our experiments are downloaded from the Whole Brain Atlas of the Harvard University. They have been registered strictly before fusion. In Fig.3, (a), (b) are original MRI and PET images and their sizes both are 256×256. (c) and (d) are the fused images using the new algorithm presented in this paper. (e), (f) and (g) are the fused images respectively by using the Maximum-based, Fuzzy-Rule-based and Region-based fusion methods in Ref. [3~5]. Human visual perception can help judge the effects of fusion result using different fusion methods.

In our new algorithm simulation, we use iteration times $N_p=32$ which is the result that fusion image has the largest entropy. The size of each PCNN is 256×256. The parameter values of PCNN are: $a_L=0.01$, $a_\theta=0.1$, $V_L=1.0$, $V_\theta=50$, and $\beta=0.2$. The linking range is 3×3 and the convolution kernel is a unit matrix with the size of 3×3. The WPF decomposition level is taken from 1 to 5. The experimental results indicate that when decomposition level $m=2$ the fused image has been better than those using three other existing fusion methods. With the increase of the decomposition level number, the fusion results can be improved step by step. But the increase of the decomposition level can cause the computational complexity increase significantly. So we should choice the appropriate decomposition level according to actual requirement. On the other hand, in our experiments, the selection of parameter values in PCNN is a difficult work because there is no efficient approach to select the appropriate parameter values. So the fused images in Fig.3 (c) and (d) are not the optimal results. How to search the optimal parameter values of PCNN will be the next work in our future research.

To evaluate the fusion image objectively, we choose mutual information (MI) to measure the quality of fusion image. The MI is defined as follow:

$$\begin{cases} I_{FA}(f,a) = \sum_{f,a} p_{FA}(f,a) \log \dfrac{p_{FA}(f,a)}{p_F(f)p_A(a)} \\ I_{FB}(f,b) = \sum_{f,b} p_{FB}(f,b) \log \dfrac{p_{FB}(f,b)}{p_F(f)p_B(b)} \end{cases}, M_F^{AB} = I_{FA}(f,a) + I_{FB}(f,b) \qquad (10)$$

Where P_{FA} and P_{FB} are the joint probability densities. The performance evaluation results by using MI criterion are shown in Tab.1.

Table 1. Performance Evaluation of Fusion Results

	Maximum method	Fuzzy-rule-based method	Region-based method	New algorithm by author	
MI	3.3118	3.3989	4.2618	2 level	5.0011
				3 level	5.3628

The experimental results in Fig.3 and Tab.1 show that the new method presented in this paper can improve the fusion effect significantly. The mutual information values of the new algorithm is larger indicating the fused image contains more details and texture features. In PCNN, a neuron is connected and coupled with the neighbor neurons to form a global coupling network. Strength of the coupling relation between neurons is decided by β. Just because of the coupling characteristic of PCNN, information transfers quickly in the whole network. After one neuron fires, it sends pulse signal to the neighbor neurons through β. After N_p iterations, information is transferred, shared and respondent between neurons sufficiently. So the output pulse of a neuron have global characteristic. Using the output of PCNN to make a fusion decision, so the fused results are better.

4 Conclusion

The novel algorithm of multi-modality medical image fusion based on PCNN is presented in this paper. The new algorithm uses DWF to decompose source images into a series of frequency channels. Then the PCNN is used to make intelligent decision on selecting fusion coefficients according to the number of output pulses from the corresponding neuron. Finally the fusion image is obtained by taking an inverse DWF transform. The new method utilizes the global coupling property and the synchronous pulse burst property of PCNN to extract abundant information from different source images. The experimental results show that the new method is very effective and provides good performance in fusing multi-modality medical images.

References

1. Li Yan, Tan ou, Duan Huilong.: Review: Visualization of Three-Dimensional Medical Images. Journal of Image and Graphics, Vol. 6A(2), (2001) 103~110
2. Johnson J L, Padgett M L.: PCNN Models and Applications. IEEE Trans. Neural Networks, Vol. 10 (3), (1999) 480~498
3. Zhong Zhang, Blum R S.: A Categorization of Multiscale-Decomposition-Based Image Fusion Scheme with a Performance Study for a Digital Camera Application. Proceedings of IEEE, Vol. 87(8), (1999) 1315~1326
4. Nejatali A, Ciric I R.: Preserving Internal Contrast Image Fusion Using Fuzzy Sets Theory. IEEE Canadian Conference on Electrical and Computer Engineering, Vol. 2, (1997) 611~613
5. Tao Guanqun, Li Dapeng, Lu Guanghua.: Application of Wavelet Analysis in Medical Image Fusion. Journal of Xidian University, Vol. 31(1), (2004) 82~86

Cleavage Site Analysis Using Rule Extraction from Neural Networks

Yeun-Jin Cho and Hyeoncheol Kim

Department of Computer Science Education,
Korea University, Anam-Dong 5-Ga, Seoul, 136-701, Korea
{jx, hkim}@comedu.korea.ac.kr

Abstract. In this paper, we demonstrate that the machine learning approach of *rule extraction from a trained neural network* can be successfully applied to SARS-coronavirus cleavage site analysis. The extracted rules predict cleavage sites better than consensus patterns. Empirical experiments are also shown.

1 Introduction

The first cases of severe acute respiratory syndrome (SARS) were identified in Guangdong Province, China in November, 2002 and have spread to Hong Kong, Singapore, Vietnam, Canada, the USA and several European countries [20]. An outbreak of a life-threatening disease referred to as SARS has spread to many countries around the world. By late June 2003, the World Health Organization (WHO) has recorded more than 8400 cases of SARS and more than 800 SARS-related deaths, and a global alert for the illness was issued due to the severity of the disease [25]. A growing body of evidence has convincingly shown that SARS is caused by a novel coronavirus, called SARS-coronavirus or SARS-CoV [14,19]. A novel SARS associated with coronavirus (SARS-CoV) has been implicated as the causative agent of a worldwide outbreak of SARS during the first 6 months of 2003 [16,24]. Currently, the complete genome sequences of 11 strains of SARS-CoV isolated from some SARS patients have been sequenced, and more complete genome sequences of SARS-CoV are expected to come [13]. It is also known that the process of cleaving the SARS-CoV polyproteins by a special proteinase, the so-called SARS coronavirus main proteinase (CoV Mpro), is a key step for the replication of SARS-CoV [18]. The importance of the 3CL proteinase cleavage sites not only suggests that this proteinase is a culprit of SARS, but also makes it an attractive target for developing drugs directly against the new disease [3,10,23].

Several machine learning approaches including artificial neural networks have been applied to proteinase cleavage site analysis [1,3,4,7,15]. Even though neural network model has been successfully used for the analysis [1,7], one of the major weakness of the neural network is its lack of explanation capability. It is hidden in a black box and can be used to predict, but not to explain domain knowledge in explicit format. In recent years, there have been studies on rule extraction from feed-forward neural networks [1,5,6,7,8,12,17,21,22]. The extracted rules provide human users with the capability to explain how the patterns are classified and may provide better insights about the domain. Thus, it is used for various data mining applications.

In this paper, we investigate the SARS-CoV cleavage site analysis using feed-forward neural networks. Also we demonstrate how to extract prediction rules for cleavage sites using the approach of *rule extraction from neural networks*. Experimental results compared to other approaches are also shown.

2 Rule Mining for SARS-CoV

Kiemer, *et al.* used feedforward neural networks for SARS-CoV cleavage site analysis [11]. They showed that the neural network outperforms three consensus patterns in terms of classification performance. The three consensus patterns are 'LQ', 'LQ[S/A]' and '[T/S/A]X[L/F]Q[S/A/G]'. However, they used the neural network for just cleavage site prediction, but not for understanding the sites in explicit knowledge. In this paper, we extract If-Then rules from the neural networks and then compare the generated rules to the consensus patterns. Our experiments includes the followings:

- Training of a feedforward neural network with known SASS-CoV cleavage site data.
- Extraction of If-Then rules from the trained neural network.
- Performance comparison of the extracted rules over consensus rules.

In this paper, we use decompositional approach for rule extraction. Decompositional approaches to rule extraction from a trained neural network (i.e., a feed-forward multi-layered neural network) involves the following phases:

1. Intermediate rules are extracted at the level of individual units within the network. At each non-input unit of a trained network, n incoming connection weights and a threshold are given. Rule extraction at the unit searches a set of incoming binary attribute combinations that are valid and maximally-general (i.e., size of each combination is as small as possible).
2. The intermediate rules from each unit are aggregated to form the composite rule base for the neural network. It rewrites rules to eliminate the symbols which refer to hidden units but are not predefined in the domain. In the process, redundancies, subsumptions, and inconsistencies are removed.

There have been many studies for efficient extraction of valid and general rules. One of the issues is time complexity of the rule extraction procedure. The rule extraction is computationally expensive since the rule search space is increased exponentially with the number of input attributes. If a node has n incoming nodes, there are 3^n possible combinations. Kim [12] introduced a computationally efficient algorithm called OAS(Ordered-Attribute Search). In this paper, the OAS is used for extraction of one or two best rules from each node.

3 Experimental Results

3.1 SARS-CoV Cleavage Sites

Twenty-four genomic sequences of coronavirus and the annotation information were downloaded from the GenBank database [2], of which 12 are SARS-CoVs and 12 are other groups of coronaviruses. The former includes SARS-CoV TOR2, Urbani, HKU-39849, CUHK-W1, BJ01, CUHK-Su10, SIN2500, SIN2748, SIN2679, SIN2774, SIN2677 and TW1, whereas the latter comprises IBV, BCoV, bovine coronavirus strain Mebus (BCoVM), bovine coronavirus isolate BCoV-LUN (BCoVL) , bovine coronavirus strain Quebec (BCoVQ), HCoV-229E (NC002645), MHV, murine hepatitis virus strain ML-10 (MHVM), murine hepatitis virus strain 2 (MHV2), murine hepatitis virus strain Penn 97-1 (MHVP), PEDV and TGEV [9].

The data set includes 8 regions (i.e., 8 positions of P4, P3, P2, P1, P1', P2', P3', P4') and one class attribute. Each region represents one of the 20 amino acids and the class attribute tells if the instance with 8 region values belongs to either cleavage (i.e., 1) or non-cleavage (i.e., 0). Each region value that is one of 20 amino acids is converted into 20 binary values in which only one of them is 1 and the rests are 0s. This binary encoding is illustrated in table 1. Thus, we have 160 (i.e., $8*20$) binary input nodes and one output node for our neural network architecture. From each sequence of coronavirus genome, eleven cleavage sites are included and thus, total 264 ($= 24*11$) sites are available. We eliminated duplicated ones out of the total 264 results and identified final seventy cleavage sites. For training a neural network classifier, negative examples (presumed non-cleavage sites) are created by defining all other glutamines in the viral polyproteins as non-cleavable sites [11]. Therefore, the data set include 70 positive (i.e., cleavage) and 281 negative (i.e., non-cleavage) examples. Three-fold cross-validation were used for classifier evaluation. That is, every test set contains 117 examples of which 23 or 24 were positive examples.

3.2 Performance of Extracted Rules

We configured neural networks with 160 input nodes, 2 hidden nodes and 1 output nodes and trained them with training sets. The classification performance of the neural networks is shown in table 2.

We used the OAS algorithm to extract rules from trained neural networks [12] and compared classification performance over consensus rules. The consensus patterns of 'LQ', 'LQ[S/A]' and '[T/S/A]X[L/F]Q[S/A/G]' can be converted into the form of rules. All of the consensus patterns have the 'Q' at position p1, and all of 70 cleavage site examples have the 'Q' at position p1. Thus, we created negative examples with 'Q' at position p1 for fair comparison. Then the consensus patterns we use are 'L.', 'L.[S/A]' and '[T/S/A]X[L/F].[S/A/G]' and they are converted into the form of rules in table 3. The '.' represents the position p1. Their performance on examples are also shown in the table. We extracted five best rules from each of the three trained neural networks. The rules and their performance are shown in table 4. A rule is in the form of *"IF condition,*

Table 1. Each amino acid is assigned to a sequence of 20 bits

amino acid	binary code
a	1 0 0 0 0 0 0 0 0 0 0 0 0 0 0 0 0 0 0 0
c	0 1 0 0 0 0 0 0 0 0 0 0 0 0 0 0 0 0 0 0
d	0 0 1 0 0 0 0 0 0 0 0 0 0 0 0 0 0 0 0 0
e	0 0 1 1 0 0 0 0 0 0 0 0 0 0 0 0 0 0 0 0
f	0 0 0 0 1 0 0 0 0 0 0 0 0 0 0 0 0 0 0 0
g	0 0 0 0 0 1 0 0 0 0 0 0 0 0 0 0 0 0 0 0
h	0 0 0 0 0 0 1 0 0 0 0 0 0 0 0 0 0 0 0 0
i	0 0 0 0 0 0 0 1 0 0 0 0 0 0 0 0 0 0 0 0
k	0 0 0 0 0 0 0 0 1 0 0 0 0 0 0 0 0 0 0 0
l	0 0 0 0 0 0 0 0 0 1 0 0 0 0 0 0 0 0 0 0
m	0 0 0 0 0 0 0 0 0 0 1 0 0 0 0 0 0 0 0 0
n	0 0 0 0 0 0 0 0 0 0 0 1 0 0 0 0 0 0 0 0
p	0 0 0 0 0 0 0 0 0 0 0 0 1 0 0 0 0 0 0 0
q	0 0 0 0 0 0 0 0 0 0 0 0 0 1 0 0 0 0 0 0
r	0 0 0 0 0 0 0 0 0 0 0 0 0 0 1 0 0 0 0 0
s	0 0 0 0 0 0 0 0 0 0 0 0 0 0 0 1 0 0 0 0
t	0 0 0 0 0 0 0 0 0 0 0 0 0 0 0 0 1 0 0 0
v	0 0 0 0 0 0 0 0 0 0 0 0 0 0 0 0 0 1 0 0
w	0 0 0 0 0 0 0 0 0 0 0 0 0 0 0 0 0 0 1 0
y	0 0 0 0 0 0 0 0 0 0 0 0 0 0 0 0 0 0 0 1

Table 2. Performance of three neural networks

data	training accuracy	generalization
dataset1	100%	95.7%
dataset2	100%	97.4%
dataset3	100%	95.7%
average	100%	96.3%

THEN class" where *class* is either of cleavage or non-cleavage. *Coverage* and *accuracy* as defined as follows:

$$Coverage = \frac{\text{Number of examples matched by the } condition \text{ part}}{\text{Total number of examples}}$$

$$Accuracy = \frac{\text{Number of true positive examples}}{\text{Number of examples matched by the } condition \text{ part}}$$

The five rules extracted generally outperforms the consensus rules. Coverage is reasonably high and accuracy is very high. The rule 'L@p2' in consensus patterns actually subsumes 11 other rules in the table 3. While its coverage is high (i.e. 55.6%), its accuracy is low compared to others. The rules that we extracted also contain the 'L' at position p2, but we excluded the rule 'L@p2' by our 90% of rule extraction threshold.

Table 3. Consensus rules and their performance

Consensus Rules	dataset1		dataset2		dataset3	
	coverage	accuracy	coverage	accuracy	coverage	accuracy
$L@p2$	55.6	83.1	55.6	83.1	55.6	83.1
$L@p2 \wedge S@p1'$	23.1	92.6	23.1	92.6	23.1	92.6
$L@p2 \wedge A@p1'$	16.2	94.7	16.2	94.7	16.2	94.7
$T@p4 \wedge L@p2 \wedge S@p1'$	6.0	100.0	6.0	100.0	6.0	100.0
$T@p4 \wedge L@p2 \wedge A@p1'$	2.6	100.0	2.6	100.0	2.6	100.0
$T@p4 \wedge L@p2 \wedge G@p1'$	0.9	100.0	0.9	100.0	0.9	100.0
$T@p4 \wedge F@p2 \wedge S@p1'$	0.0	0.0	0.0	0.0	0.0	0.0
$T@p4 \wedge F@p2 \wedge A@p1'$	0.0	0.0	0.0	0.0	0.0	0.0
$T@p4 \wedge F@p2 \wedge G@p1'$	0.0	0.0	0.0	0.0	0.0	0.0
$S@p4 \wedge L@p2 \wedge S@p1'$	6.0	85.7	6.0	85.7	6.0	85.7
$S@p4 \wedge L@p2 \wedge A@p1'$	1.7	100.0	1.7	100.0	1.7	100.0
$S@p4 \wedge L@p2 \wedge G@p1'$	0.0	0.0	0.0	0.0	0.0	0.0
$S@p4 \wedge F@p2 \wedge S@p1'$	1.7	100.0	1.7	100.0	1.7	100.0
$S@p4 \wedge F@p2 \wedge A@p1'$	0.0	0.0	0.0	0.0	0.0	0.0
$S@p4 \wedge F@p2 \wedge G@p1'$	0.0	0.0	0.0	0.0	0.0	0.0
$A@p4 \wedge L@p2 \wedge S@p1'$	3.4	75.0	3.4	75.0	3.4	75.0
$A@p4 \wedge L@p2 \wedge A@p1'$	1.7	100.0	1.7	100.0	1.7	100.0
$A@p4 \wedge L@p2 \wedge G@p1'$	0.0	0.0	0.0	0.0	0.0	0.0
$A@p4 \wedge F@p2 \wedge S@p1'$	0.9	0.0	0.9	0.0	0.9	0.0
$A@p4 \wedge F@p2 \wedge A@p1'$	0.0	0.0	0.0	0.0	0.0	0.0
$A@p4 \wedge F@p2 \wedge G@p1'$	0.0	0.0	0.0	0.0	0.0	0.0

Table 4. Extracted rules from neural networks trained on three data sets, and their performance

Data	Extracted Rules	Coverage	Accuracy
dataset 1	$L@p2 \wedge S@p1'$	23.1	92.6
	$L@p2 \wedge A@p1'$	16.2	94.7
	$L@p2 \wedge T@p4$	12.0	100.0
	$L@p2 \wedge R@p3$	10.3	100.0
	$L@p2 \wedge S@p4$	7.7	88.9
dataset 2	$L@p2 \wedge E@p2'$	12.8	100.0
	$L@p2 \wedge T@p4$	12.0	100.0
	$L@p2 \wedge V@p4$	12.0	100.0
	$L@p2 \wedge P@p4$	12.0	100.0
	$L@p2 \wedge K@p2'$	3.4	100.0
dataset 3	$L@p2 \wedge V@p4$	12.0	100.0
	$L@p2 \wedge T@p4$	12.0	100.0
	$L@p2 \wedge P@p4$	6.0	100.0
	$L@p2 \wedge T@p3$	4.3	100.0
	$L@p2 \wedge K@p2'$	3.4	100.0

4 Conclusions

For SARS-CoV cleavage site analysis, we used the approach of rule extraction from neural networks. We trained 3-layered feedforward neural networks on genomic sequences of coronaviruses, and then extracted IF-THEN rules from the neural networks. Their performances are compared to consensus patterns. The results are promising. Rule mining using neural network classifier can be a useful tool for cleavage site analysis.

References

1. Andrews, Robert, Diederich, Joachim, Tickle, Alam B.: Survey and critique of techniques for extracting rules from trained artificial neural networks. Knowledge-Based Systems **8(6)** (1995) 373-389
2. Benson DA, Karsch-Mizrachi I, Lipman DJ, Ostell J, Wheeler DL.: GenBank: update. Nucleic Acids Res, 32 Database issue: (2004)D23-26
3. Blom N, Hansen J, Blaas D, Brunak S.: Cleavage site analysis in picornaviral polyproteins: discovering cellular targets by neural networks. Protein Sci (1996) 5:2203-2216.
4. Chen LL, Ou HY, Zhang R, Zhang CT.: ZCURVE-CoV: a new system to recognize protein coding genes in coronavirus genomes, and its applications in analyzing SARS-CoV genomes. SCIENCE DIRECT, BBRC (2003) 382-388
5. Fu, LiMin.: Neural Networks in Computer Intelligence. McGraw Hill, Inc., (1994)
6. Fu, LiMin.: Rule generation from neural networks. IEEE Transactions on Systems, Man, and Cybernetics **24(8)** (1994) 1114-1124
7. Fu, LiMin.: Introduction to knowledge-based neural networks. Knowledge-Based Systems **8(6)** (1995) 299-300
8. Fu, LiMin and Kim, Hyeoncheol.: Abstraction and Representation of Hidden Knowledge in an Adapted Neural Network. unpublished, CISE, University of Florida (1994)
9. Gaoa F, Oua HY, Chena LL, Zhenga WX, Zhanga CT.: Prediction of proteinase cleavage sites in polyproteins of coronaviruses and its applications in analyzing SARS-CoV genomes. FEBS Letters 553 (2003) 451-456
10. Hu LD, Zheng GY, Jiang HS, Xia Y, Zhang Y, Kong XY.: Mutation analysis of 20 SARS virus genome sequences: evidence for negative selection in replicase ORF1b and spike gene. Acta Pharmacol Sin (2003) 741-745
11. Kiemer L, Lund O, Brunak S, Blom N.: Coronavirus 3CL-pro proteinase cleavage sites: Possible relevance to SARS virus pathology. BMC Bioinformatics (2004)
12. Kim, Hyeoncheol.: Computationally Efficient Heuristics for If-Then Rule Extraction from Feed-Forward Neural Networks. Lecture Notes in Artificial Intelligence, Vol. 1967 (2000) 170-182
13. Luo H, Luo J.: Initial SARS Coronavirus Genome Sequence Analysis Using a Bioinformatics Platform. APBC2004, Vol. 29 (2004)

14. Marra MA, Jones SJM, Astell CR, Holt RA, Brooks-Wilson A, Butterfield YSN, Khattra J, Asano JK, Barber SA, Chan SY, CloutierA, Coughlin SM, Freeman D, Girn N, Griffith OL, Leach SR, Mayo M, McDonald H, Montgomery SB, Pandoh PK, Petrescu AS, Robertson AG, Schein JE, Siddiqui A, Smailus DE, Stott JM, Yang GS, Plummer F, Andonov A, Artsob H, Bastien N, Bernard K, Booth TF, Bowness D, Czub M, Drebot M, Fernando L, Flick R, Garbutt M, Gray M, Grolla A, Jones S, Feldmann H, Meyers A, Kabani A, Li Y, Normand S, Stroher U, Tipples GA, Tyler S, Vogrig R, Ward D, Watson B, Brunham RC, Krajden M, Petric M, Skowronski DM, Upton C, Roper RL.: The Genome Sequence of the SARS-Associated Coronavirus. SCIENCE VOL 300 (2003) 1399-1404
15. Narayanan, A., Wu, X., Yang, Z.R.: Mining viral protease data to extract cleavage knowledge. bioinformatics, **18(1)** (2002) s5–s13.
16. Ruan Y, Wei CL, Ee LA, Vega VB, Thoreau H, Yun STS, Chia JM, Ng P, Chiu KP, Lim L, Tao Z, Peng CK, Ean LOL, Lee NM, Sin LY, Ng LFP, Chee RE, Stanton LW, Long PM, Liu ET.: Comparative full-length genome sequence analysis of 14 SARS coronavirus isolates and common mutations associated with putative origins of infection. THE LANCET o Published online (2003)
17. Setino, Rudy, Liu, Huan: Understanding neural networks via rule extraction. Proceedings of the 14th International Conference on Neural Networks. **(1)** Montreal, Canada (1995) 480–485
18. Shi J, Wei Z, Song J.: Dissection Study on the Severe Acute Respiratory Syndrome 3C-like Protease Reveals the Critical Role of the Extra Domain in Dimerization of the Enzyme. THE JOURNAL OF BIOLOGICAL CHEMISTRY Vol. 279, No. 23 (2004) 24765-24773
19. Shi Y, Yi Y, Li P, Kuang T, Li L, Dong M, Ma Q, Cao C.: Diagnosis of Severe Acute Respiratory Syndrome (SARS) by Detection of SARS Coronavirus Nucleocapsid Antibodies in an Antigen-Capturing Enzyme-Linked Immunosorbent Assay. JOURNAL OF CLINICAL MICROBIOLOGY (2003) 5781-5782
20. Stadler K, Masignani V, Eickmann M, Becker S, Abrignani S, Klenk HD, Rappuoli R.: SARS - BEGINNING TO UNDERSTAND A NEW VIRUS. NATURE REVIEWS, MICROBIOLOGY VOLUME 1 (2003) 209-218
21. Taha, Ismali A. and Ghosh, Joydeep: Symbolic interpretation of artificial neural networks. IEEE Transactions on Knowledge and Data Engineering **11(3)** (1999) 443–463
22. Towell, Geoffrey G. and Shavlik, Jude W.: Extracting refined rules from knowledge-based neural networks. Machine Learning **13(1)** (1993)
23. Tsur S.: Data Mining in the Bioinformatics Domain. Proceedings of the 26th VLDB Conference, Cairo, Egypt (2000)
24. Xu D, Zhang Z, Chu F, Li Y, Jin L, Zhang L, Gao GF, Wang FS.: Genetic Variation of SARS Coronavirus in Beijing Hospital. Emerging Infectious Diseases (www.cdc.gov/eid) Vol. 10, No. 5 (2004)
25. Yap YL, Zhang XW, Danchin A.: Relationship of SARS-CoV to other pathogenic RNA viruses explored by tetranucleotide usage profiling. BMC Bioinformatics (2003)

Prediction Rule Generation of MHC Class I Binding Peptides Using ANN and GA

Yeon-Jin Cho[1], Hyeoncheol Kim[1], and Heung-Bum Oh[2]

[1] Dept. of Computer Science Education,
Korea University; Seoul, 136-701, Korea
{jx, hkim}@comedu.korea.ac.kr
[2] Dept. of Laboratory Medicine,
University of Ulsan and Asan Medical Center, Seoul, Korea
hboh@amc.seoul.kr

Abstract. A new method is proposed for generating *if-then* rules to predict peptide binding to class I MHC proteins, from the amino acid sequence of any protein with known binders and non-binders. In this paper, we present an approach based on artificial neural networks (ANN) and knowledge-based genetic algorithm (KBGA) to predict the binding of peptides to MHC class I molecules. Our method includes rule extraction from a trained neural network and then enhancing the extracted rules by genetic evolution. Experimental results show that the method could generate new rules for MHC class I binding peptides prediction.

1 Introduction

T-cell cytotoxicity is initially mediated by recognizing peptides bound to Major Histocompatibility Complex (MHC) class I molecules. It has been generally accepted that only peptides interacting with MHC above a certain affinity threshold are likely to be recognized by T cells. Prediction of the immunodominant peptides, so called T-cell epitopes is a key step toward understanding host-pathogen interactions and hopefully the development of new vaccines against the pathogen as well. Prediction of binding peptide to MHC with high affinity can reduce significantly the number of peptides that have to be tested experimentally [1, 5, 8, 9, 13, 14].

The methods to predict MHC-peptide binding can be divided into two groups: sequence-based and structure-based methods. Profile-based prediction methods, such as SYFPEITHI and HLA_BIND, fall into the first group and the binding prediction based on the fitness of peptide to the binding groove of MHC molecule into the second group [8]. While a structural approach is limited to MHC types with a known structure, the sequence-based approach has advantage of a lot of growing information on peptide binders [2, 8, 21]. In this paper, we used five methods of machine learning based on peptide sequences. Decision tree, artificial neural network and genetic algorithm were used to extract rules for HLA binders.

Decision tree is one of the best-known classification techniques in symbolic machine learning. It uses a heuristic method to select the attribute that separates the samples most effectively into each class using an entropy-based measure [3, 18, 19]. Even though this algorithm has an advantage of extracting rules in simplicity, the performance of predicting HLA binder was not as good as that of neural network. Neural network, even though it can not extract rules in general, has been reported to show a higher specificity in predicting candidate T-cell epitopes than profile methods [5, 8, 9, 14]. However, they used the neural network for just prediction of MHC-peptides, but not for expressing the sites in explicit knowledge such as if-then rules. Genetic algorithm (GA) can be used for searching generalized rules [7, 16, 17]. However performance of GA is very sensitive to the initial population of chromosomes which was generated by random. In this paper, we developed a knowledge-based GA (KBGA) algorithm and applied to predicting HLA binders. Our method includes rule extraction from a trained neural network and then enhancing the extracted rules by genetic evolution to improve its quality. Support vector machine (SVM) is also known to produce better specificity than profile-based methods, because they educate machines with the information related to both of the binder and the non-binder [6, 8, 10]. SVM, however, can not generate explicit rules in general. We compared three algorithms of DT, NN and SVM in terms of the performance and of DT, NN and KBGA in terms of rules extracted.

The methods and results in this study were uploaded at the web site (http://www.koreanhla.net), where sequence logo was implemented to visualize the sequence motif based on DT entropy.

2 Materials and Methods

2.1 Data Set Preparation

The dataset and the size of each data used in our experiment are identical to the ones used in the experiment done by Donnes, et al [8]. (The dataset was kindly given with personal communication). Among the dataset given, we used the dataset generated from the database SYFPEITHI[1]. Briefly, they used 6 alleles of MHC class I molecules in SYFPEITHI. Duplicated data were removed and the number of binder data was maintained to be more than 20 all the time [8]. Protein, randomly extracted from ENSEMBL database, was cut in a fixed size and all the sequences, defined in MHC-peptide database, were removed from it, in order to get non-binders. 20 binders and 40 non-binders are used, maintaining the binder to non-binder ratio 1:2.

Table 1. The six MHC alleles of SYFPEITHI (SYF)

MHC	Length	Size(positive/negative)
HLA-A*0201	9	339(113/226)
HLA-A*0201	10	120(40/80)
HLA-A1	9	63(21/42)
HLA-A3	9	69(23/46)
HLA-B*8	9	75(25/50)
HLA-B*2705	9	87(29/58)

[1] SYFPEITHI is a database comprising more than 4500 peptide sequences known to bind class I and class II MHC molecules.

For neural network training, each region value that is one of 20 amino acids is converted into 20 binary digits as shown in Table 1. Thus, each MHC-peptide position is encoded into 180 bits (9-mer peptide * 20). Class is encoded into either 1 (i.e., binder) or 0 (i.e., non-binder).

Table 2. Each amino acid is assigned to a sequence of 20 bits

Amino Acid	Representation
Alanine (A)	10000000000000000000
Cysteine (C)	01000000000000000000
Aspartate (D)	00100000000000000000
Glutamate (E)	00010000000000000000
Phenylalanine (F)	00001000000000000000
Glycine (G)	00000100000000000000
Histidine (H)	00000010000000000000
Isoleucine (I)	00000001000000000000
Lysine (K)	00000000100000000000
Leucine (L)	00000000010000000000
Methionine (M)	00000000001000000000
Asparagines (N)	00000000000100000000
Proline (P)	00000000000010000000
Glutamine (Q)	00000000000001000000
Arginine (R)	00000000000000100000
Serine (S)	00000000000000010000
Threonine (T)	00000000000000001000
Valine (V)	00000000000000000100
Tryptophan (W)	00000000000000000010
Tyrosine (Y)	00000000000000000001

For SVM training, each amino acid sequence is encoded as follows:

$$d_i\ 1{:}X_{1i}\ 2{:}X_{2i}\ 3{:}X_{3i}\ 4{:}X_{4i}\ 5{:}X_{5i}\ 6{:}X_{6i}\ 7{:}X_{7i}\ 8{:}X_{8i}\ 9{:}X_{9i}$$

where d represents its class (i.e., 1 for binder; 0 for non-binder) and $n{:}X$ represents amino acid code X at position n. The amino acid code at peptide position n is shown in table 3.

2.2 Methods for Rule Extraction

To extract rules of MHC binding, three methods were employed such as decision tree, neural network and genetic algorithm. Quinlan's C5.0 algorithm was used to build a decision tree which was converted into a set of if-then rules [19, 20].

For neural network, a feed-forward neural network was trained, and then we extract *if-then* rules using the OAS (Ordered-Attribute Search) algorithm [12]. The OAS algorithm uses de-compositional approach that involves extracting intermediate rules from each non-input nodes of a trained neural network and aggregating the intermediate rules to form a composite rule base. The details of this algorithm were already published by our group.

Table 3. Encoding methods of SVM (Peptide position: Amino acid at each position)

Amino Acid	Peptide Position								
	P1	P2	P3	P4	P5	P6	P7	P8	P9
Alanine (A)	1:01	2:01	3:01	4:01	5:01	6:01	7:01	8:01	9:01
Cysteine (C)	1:02	2:02	3:02	4:02	5:02	6:02	7:02	8:02	9:02
Aspartate (D)	1:03	2:03	3:03	4:03	5:03	6:03	7:03	8:03	9:03
Glutamate (E)	1:04	2:04	3:04	4:04	5:04	6:04	7:04	8:04	9:04
Phenylalanine (F)	1:05	2:05	3:05	4:05	5:05	6:05	7:05	8:05	9:05
Glycine (G)	1:06	2:06	3:06	4:06	5:06	6:06	7:06	8:06	9:06
Histidine (H)	1:07	2:07	3:07	4:07	5:07	6:07	7:07	8:07	9:07
Isoleucine (I)	1:08	2:08	3:08	4:08	5:08	6:08	7:08	8:08	9:08
Lysine (K)	1:09	2:09	3:09	4:09	5:09	6:09	7:09	8:09	9:09
Leucine (L)	1:10	2:10	3:10	4:10	5:10	6:10	7:10	8:10	9:10
Methionine (M)	1:11	2:11	3:11	4:11	5:11	6:11	7:11	8:11	9:11
Asparagines (N)	1:12	2:12	3:12	4:12	5:12	6:12	7:12	8:12	9:12
Proline (P)	1:13	2:13	3:13	4:13	5:13	6:13	7:13	8:13	9:13
Glutamine (Q)	1:14	2:14	3:14	4:14	5:14	6:14	7:14	8:14	9:14
Arginine (R)	1:15	2:15	3:15	4:15	5:15	6:15	7:15	8:15	9:15
Serine (S)	1:16	2:16	3:16	4:16	5:16	6:16	7:16	8:16	9:16
Threonine (T)	1:17	2:17	3:17	4:17	5:17	6:17	7:17	8:17	9:17
Valine (V)	1:18	2:18	3:18	4:18	5:18	6:18	7:18	8:18	9:18
Tryptophan (W)	1:19	2:19	3:19	4:19	5:19	6:19	7:19	8:19	9:19
Tyrosine (Y)	1:20	2:20	3:20	4:20	5:20	6:20	7:20	8:20	9:20

Genetic Algorithm (GA) was also used to extract rules. Performance of GA, however, is very sensitive to initial population of chromosomes which are randomly generated [4, 7, 11]. It is known that domain knowledge incorporated into the initial population improves the performance of GA [7, 15, 16, 17]. Since domain knowledge is not available, we used the extracted rules from a trained neural network for the domain knowledge. The knowledge-based approach restricts GA's random search space, and the GA-model refines and explores from the initial rules. The Knowledge-based Genetic Algorithm (KBGA) enhances the initial rules from neural network and generates more rules previously unfound.

Individual chromosome in a GA population is a sequence of 9 symbols in which each symbol represents an amino acid or '*' (i.e., don't_care symbol). Therefore the size of rule space is as huge as 21^9. The GA-based model searches for the best fitted set of chromosomes (i.e., rules) among the 21^9 candidates. One-point crossover is used and crossing point is selected randomly. Mutation occurs on each symbol by 1% and changes its symbol to one of other 20 symbols. Fitness function for a chromosome n is defined as follows:

$$f(n) = \frac{nt}{nt + nf + 1} \times 100 + d$$

where *nt (or nf)* is the number of positive (or negative) instances matched by the chromosome rule and *d* is the number of *s (i.e., don't_care symbols) in the chromosome.

3 Results and Discussion

First of all, we evaluated performance of neural network (NN) on the MHC binding dataset, and compared it with the one of support vector machine (SVM). The neural network was configured with 180 input nodes, 4 hidden nodes and 1 output node. 3-fold cross validation was used to evaluate the classification performance of NN and SVM. Table 4 shows the training accuracy and generalization.

Table 4. Classification performance of NN and SVM. 3-fold cross-validation is used to evaluate their performances and its average is shown at the table

MHC alleles	Classifier	Training accuracy	Generalization
HLA-A*0201 (9-mer)	NN	99.7%	86.7%
	SVM	100%	84.96%
HLA-A*0201 (10-mer)	NN	100%	87.50%
	SVM	100%	100%
HLA-A1 (9-mer)	NN	100%	100%
	SVM	100%	96.83%
HLA-A3 (9-mer)	NN	100%	80.63%
	SVM	100%	84.06%
HLA-B*8 (9-mer)	NN	100%	86.77%
	SVM	100%	84.0%
HLA-B*2705 (9-mer)	NN	100%	96.57%
	SVM	100%	99.7%

Rules are generated using DT, NN and KBGA. The rules extracted from each classifier with the accuracy more than 75% were shown in Table 5. The new rules uniquely exist in each algorithm are shaded. DT_Rule was generated by converting decision tree to rules and ANN_Rule was generated from a trained neural network using the OAS algorithm. KBGA_Rule was generated by GA-based rule search with the neural network rules incorporated into the GA initial population. The rules shown in the Table 5 are positive rules in which consequence (i.e. then-part) of if-then rule represents *MHC-binder*. Thus "L@P2 ∧ V@P9" is translated as "if L@P2 ∧ V@P9, then MHC-binder".

Table 5 lists the most general (i.e., most simple) rules with accuracy greater than 75%. For DT-Rule and ANN_Rule, most general rules were generated, eliminating any subsumed rules. For example, rule "if K@p3 ∧ K@p5, then Binder" is subsumed to the rule "if K@p3, then Binder" or "if K@p5, then Binder". For KBGA_Rule, we list all of the rules generated with accuracy greater than 75% so that user can review the rules together.

Our experiment shows that ANN and KBGA can generated more quality rules than DT. The performance of KBGA is equal to or sometimes better than ANN when it comes to rule generation. The *if-then* rule format is very easy to understand to domain experts or molecular biologists.

Table 5. The performance of the rules extracted from each classifier algorithm

- HLA-A*0201 (9-mer) standard accuracy +75%.					
Classifier	Positive Rule		PosNum (113)	NegNum (226)	Accuracy(%)
DT_Rule	L@P2 ^ V@P9	*L******V	36	2	94.74
ANN_Rule	L@P2 ^ V@P9	*L******V	36	2	94.74
KBGA_Rule	L@P2 ^ V@P9	*L******V	36	2	94.74
	L@P2	*L*******	76	24	76.00
	V@P9	********V	53	15	77.94
- HLA-A*0201 (10-mer) standard accuracy +70%.					
Classifier	Positive Rule		PosNum (40)	NegNum (80)	Accuracy(%)
DT_Rule	L@P2	*L********	22	9	70.97
ANN_Rule	L@P2 ^ V@P10	*L*******V	12	1	92.31
KBGA_Rule	L@P2 ^ V@P10	*L*******V	12	1	92.31
	L@P2	*L********	22	9	70.97
- HLA-A1 (9-mer) standard accuracy +75%.					
Classifier	Positive Rule		PosNum (21)	NegNum (42)	Accuracy(%)
DT_Rule	Y@P9	********Y	20	0	100.00
ANN_Rule	E@P3 ^ Y@P9	**E*****Y	7	0	100.00
	L@P7 ^ Y@P9	******L*Y	7	0	100.00
	T@P2 ^ Y@P9	*T******Y	9	0	100.00
	D@P3 ^ Y@P9	**D*****Y	10	0	100.00
KBGA_Rule	T@P2	*T*******	9	3	75.00
	L@P7	******L**	8	2	80.00
	D@P3	**D******	10	2	83.33
	E@P3	**E******	7	1	87.50
	Y@P9	********Y	20	0	100.00
- HLA-A3 (9-mer) standard accuracy +75%.					
Classifier	Positive Rule		PosNum (23)	NegNum (46)	Accuracy(%)
DT_Rule	N/A				
ANN_Rule	L@P2 ^ K@P9	*L******K	8	1	88.89
KBGA_Rule	K@P9	********K	13	3	81.25
	K@P1	K********	9	2	81.82
	L@P2 ^ K@P9	*L******K	8	1	88.89

Table 5. *Continued*

- HLA-B*8 (9-mer) standard accuracy +75%.

Classifier	Positive Rule		PosNum (25)	NegNum (50)	Accuracy(%)
DT_Rule	N/A				
ANN_Rule	K@P3 ^ K@P5	**K*K****	12	0	100.00
	K@P3 ^ L@P9	**K*****L	12	0	100.00
KBGA_Rule	A@P8	*******A*	8	2	80.00
	K@P5	****K****	15	2	88.24
	K@P3	**K******	16	2	88.89
	K@P5 ^ L@P9	****K***L	11	0	100.00
	K@P3 ^ K@P5	**K*K****	12	0	100.00

- HLA-B*2705 (9-mer) standard accuracy +75%.

Classifier	Positive Rule		PosNum (29)	NegNum (58)	Accuracy(%)
DT_Rule	R@P2	*R*******	29	4	87.88
ANN_Rule	R@P1 ^ R@P2	RR******	12	0	100.00
KBGA_Rule	R@P2	*R*******	29	4	87.88
	R@P1 ^ R@P2	RR******	12	0	100.00

4 Conclusion

Prediction (or classification) rules provide us with explanation about the classification and thus better insights about a domain. We presented a new method that generates rules and improves quality of the rules with the subject of MHC class I binding peptides prediction. Rules were extracted from a well-trained neural network and then enhanced by genetic evolution. Our experiment presents the rules generated by three different types of approaches:

- Decision Tree
- Neural networks
- Genetic Algorithm initialized by neural network rules

Neural network could generate the rules of high quality that were not discovered by decision trees. Knowledge-Based Genetic Algorithm (KBGA) model in which the neural network rules were incorporated initially could discover new rules in addition to the neural network rules. The KBGA can be considered as a hybrid model of neural networks and genetic algorithm since knowledge learned by a neural network is enhanced and expanded by GA evolution. The experimental result demonstrates that the hybrid model improves quality of rule generation.

References

1. Adams H.P. and Koziol J.A.: Prediction of binding to MHC class I molecules. J Immunol Methods. (1995) 181-90.

2. Altuvia, Y. Margalit, H.: A structure-based approach for prediction of MHC-binding peptides. Elsevier Inc, Science direct (2004) 454-459.
3. Baldi, P.; Brunak, S. Bioinformatics, the machine learning approach. MIT Press Cambridge Massachusetts, London England. (1998).
4. Bala, J. Huang and H.: Hybrid Learning Using Genetic Algorithms and Decision Trees for Pattern Classification. IJCAI conference, Montreal, August 19-25, (1995).
5. Brusic, V. Rudy, G. Harrison, L.C.: Prediction of MHC binding peptides using artificial neural networks. Complexity International (2), (1995).
6. Cristianini, N. Shawe-Taylor, J.: Support vector machines and other kernel-based learning methods. Cambridge University Press, (2000).
7. De Jong, K.A. and Spears, W.M.: Learning Concept Classification Rules Using Genetic Algorithms. Proceedings of the I Zth. international Conference on Artificial Intelligence. (1991) 651-656.
8. Dönnes, P. Elofsson, A.: Prediction of MHC class I binding peptides, using SVMHC. BMC Bioinformatics. (2002).
9. Honeyman, M. Brusic, V. Stone, N. Harrison, L.: Neural network-based prediction of candidate t-cell epitopes. Nature Biotechnology. (1998) 966–969.
10. Joachims, T.: SVM^{light} 6.01 Online Tutorial, http://svmlight.joachims.org (2004)
11. KA De Jong and WM Spears.: Learning Concept Classification Rules Using Genetic Algorithms. Proceedings of the I Zth. international Conference on Artificial Intelligence, pp. 651-656: (1991).
12. Kim, Hyeoncheol.: Computationally Efficient Heuristics for If-Then Rule Extraction from Feed-Forward Neural Networks. Lecture Notes in Artificial Intelligence, Vol. 1967 (2000) 170-182.
13. Logean, A. Rognan, D.: Recovery of known T-cell epitopes by computational scanning of a viral genome. Journal of Computer-Aided Molecular Design 16(4): (2002) 229-243.
14. Mamitsuka, H.: MHC molecules using supervised learning of hidden Markov models. Proteins: Structure, Function and Genetics. (1998) 460–474.
15. Melanie Mitchell.: An Introduction to Genetic Algorithms. MIT Press, pp92-95.: (1996).
16. Michalewicz, Z.: Genetic Algorithms + Data Structures = Evolution Programs. 3rd edn. Springer-Verlag, Berlin Heidelberg New York (1996).
17. Mitchell, M.: An Introduction to Genetic Algorithms. MIT Press (1996).
18. Narayanan, A. Keedwell, E.C. Olsson, B.: Artificial Intelligence Techniques for Bioinformatics. Bioinformatics: (2002).
19. Quinlan, J.R., Decision trees and decision making, IEEE Trans System, Man and Cybernetics 20(2): (1990) 339–346.
20. Quinlan, J.R. : C5.0 Online Tutorial, http://www.rulequest.com (2003).
21. Schueler-Furman, O. Altuvia, Y. Sette, A. Margalit, H.: Structure-based prediction of binding peptides to MHC class I molecules:application to a broad range of MHC alleles. Protein Sci. (2000) 1838-46.

Combined Kernel Function Approach in SVM for Diagnosis of Cancer

Ha-Nam Nguyen[1], Syng-Yup Ohn[1], Jaehyun Park[2], and Kyu-Sik Park[3]

[1] Department of Computer Engineering
Hankuk Aviation University, Seoul, Korea
{nghanam, syohn}@hau.ac.kr
[2] Department of Electronic Engineering
Myongji University, Seoul, Korea
jhpark@hau.ac.kr
[3] Division of Information and Computer Science
Dankook University, Seoul, Korea
kspark@dankook.ac.kr

Abstract. The problem of determining optimal decision model is a difficult combinatorial task in the fields of pattern classification, machine learning, and especially bioinformatics. Recently, support vector machine (SVM) has shown a higher performance than conventional learning methods in many applications. This paper proposes a new kernel function for support vector machine (SVM) and its learning method that results in fast convergence and good classification performance. The new kernel function is created by combining a set of kernel functions. A new learning method based on evolution algorithm (EA) is proposed to obtain the optimal decision model consisting of an optimal set of features as well as an optimal set of the parameters for combined kernel function. The experiments on clinical datasets such as stomach cancer, colon cancer, and leukemia datasets data sets indicates that the combined kernel function shows higher and more stable classification performance than other kernel functions.

1 Introduction

Support vector machine [1-6] (SVM) is a learning method that uses a hypothesis space of linear functions in a high dimensional feature space. This learning strategy, introduced by Vapnik [2], is a principled and powerful method. In the simplest and linear form, a SVM is the hyperplane that separates a set of positive samples from a set of negative samples with the largest margin. The margin is defined by the distance between the hyperplanes supporting the nearest positive and negative samples. The output formula of a linear case is

$$\mathbf{y} = \mathbf{w} \cdot \mathbf{x} - \mathbf{b}, \tag{1}$$

where $\mathbf{w}$ is a normal vector to the hyperplane and $\mathbf{x}$ is an input vector. The separating hyperplane is the plane $\mathbf{y} = 0$ and two supporting hyperplanes parallel to it with equal distances are

$$H_1: y = w \cdot x - b = +1$$
$$H_2: y = w \cdot x - b = -1 \quad (2)$$

Thus, the margin M is defined as

$$M = \frac{2}{\|w\|} \quad (3)$$

In order to find the optimal separating hyperplane having maximal margin, $\|w\|$ should be minimized subject to inequality constraints. This is a classic nonlinear optimization problem with inequality constraints. The optimization problem can be solved by finding the saddle point of the Lagrange function in the following.

$$L(w, b, \alpha) = \frac{1}{2} w^T w - \sum_{i=1}^{N} \alpha_i y_i ([w^T x + b] - 1) \quad (4)$$

where $\alpha_i \geq 0, i = 0, \ldots, N$, are Lagrange multipliers.

However, the limitation of computational power of linear learning machines was highlighted in the 1960s by Minsky and Papert [7]. It can be easily recognized that real-world applications require more extensive and flexible hypothesis space than linear functions. Such a limitation can be overcome by *multilayer neural networks* proposed by Rumelhart, Hinton and William [3]. Kernel function also offers an alternative solution by projecting the data into high dimensional feature space to increase the computational power of linear learning machines. Non-linear mapping from input space to high dimensional feature space can be implicitly performed by an appropriate kernel function (see Fig. 1). One of the advantages of the kernel method is that a learning algorithm can be exploited to obtain the specifics of application area, which simply can be encoded into the structure of an appropriate kernel function. One of the interesting characteristics on kernel functions is that a new kernel function can be created by combining a set of kernel functions with the operators such as addition or multiplication operators [1].

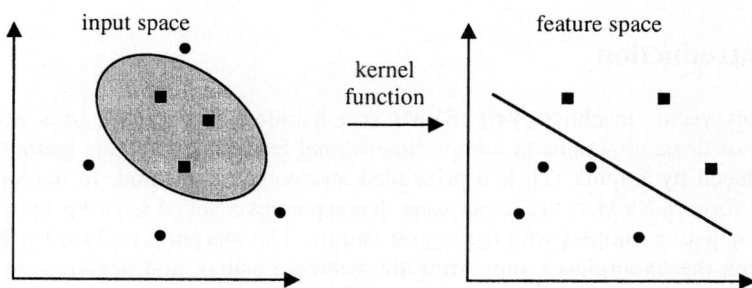

Fig. 1. An input space can be transformed into a linearly separable feature space by an appropriate kernel function

Evolutionary algorithm [8-10] is an optimization algorithms based on the mechanism of natural evolution procedure. Most of evolution algorithms share a common conceptual base of simulating the evolution of individual structures via the processes of selection, mutation, and reproduction. In each generation, a new population is selected based on the fitness values representing the performances of the individuals belonging to the generation, and some individuals of the population are given the chance to undergo alterations by means of crossover and mutation to form new individuals. In this way, EA performs a multi-directional search by maintaining a population of potential solutions and encourages the formation and the exchange of information among different directions. EAs are generally applied to the problems with a large search space. They are different from random algorithms since they combine the elements of directed and stochastic search. Furthermore, EA is also known to be more robust than directed search methods.

Recently, several researches have been working on GA/SVM to improve the performance of classification. Some of them use GA to optimize the number of selected features that are evaluated by classifiers [13, 14] and the best recognition rate of 80% was achieved in case of colon dataset. Other approach used GA to optimize the ensemble of multiple classifiers to improve the performance of classification [15].

In this paper, we propose a new kernel function combining a set of simple kernel functions for SVM and a method to train the combined kernel function. In the new learning method, EA is employed to derive the optimal *decision model* for the classification of patterns, which consists of the optimal set of features and parameters of combined kernel function. The proposed method was applied to the classification of proteome patterns for the identification of cancer, which are extracted from actual clinical samples. The combined kernel function and the learning method showed faster convergence and better classification rate than individual kernel functions.

This paper is organized as follows. In section 2, our new combined kernel and its learning method are presented in detail. In section 3, we compare the performances of combined kernel functions and other individual kernel functions by the experiments with the classification of the datasets of colon cancer dataset, leukemia dataset and proteome pattern samples of stomach cancer. Finally, section 4 is our conclusion.

2 Proposed Learning Method

2.1 Overall Structure

The proposed method is depicted in Fig. 2. Our method consists of preprocessing, learning, and classification phase.

Firstly, in the preprocessing stage, training and testing sets consisting of a number of cancer and normal patterns is produced and passed to the learning phase.

Secondly, we applied a learning method exploiting EA and SVM techniques to figure out optimal *decision model* for the classification of proteome patterns in the learning phase. Here EA generates a set of chromosomes, each of which represents a decision model, by evolutionary procedures. The fitness value of each chromosome is evaluated by measuring the hit ratio from the classification of samples with SVM classifier containing the decision model associated with the chromosome. n-fold validation method is used to evaluate the fitness of a chromosome to reduce overfitting [4]. Then only the chromosomes with a good fitness are selected and given the chance to survive and improve into further generations. This process is repeated for a predefined number of times. At the end of EA procedure, the decision model with the highest hit ratios is chosen as the *optimal decision model*.

Finally, the *optimal decision model* is used to build a SVM for the classification of novel samples and the performance of the model can be evaluated.

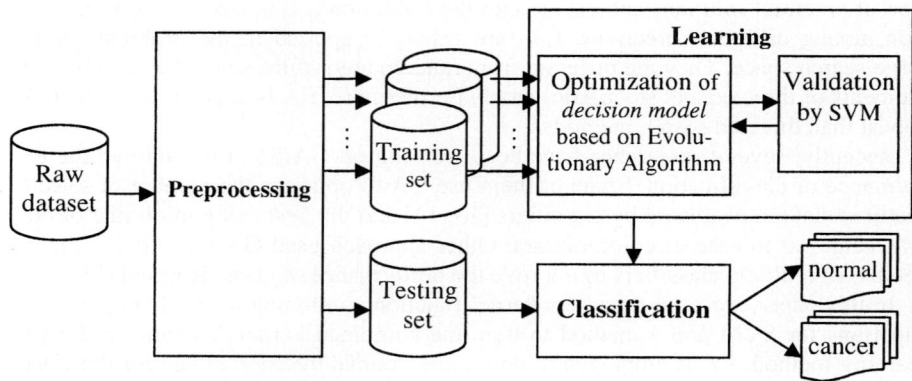

Fig. 2. Overall Framework of Proposed Diagnosis Method

2.2 Training Combined Kernel Function and Feature Selection by Evolution Algorithm

A kernel function provides a flexible and effective learning mechanism in SVM, and the choice of a kernel function should reflect prior knowledge about the problem at hand. However, it is often difficult for us to exploit the prior knowledge on patterns to choose a kernel function, and it is an open question how to choose the best kernel function for a given data set. According to *no free lunch theorem* [4] on machine learning, there is no superior kernel function in general, and the performance of a kernel function rather depends on applications.

Table 1. The types of kernel functions are used to experiments

Kernel function	Formula
Inverse Multi-Quadric	$1/\sqrt{\|x-y\|^2 + c^2}$
Radial	$e^{(-\gamma\|x-y\|^2)}$
Neural	$\tanh(s \cdot \langle x, y \rangle - c)$

In our case, a new kernel function is created by combining the set of kernel functions (see Table 1). The combined kernel function has the form of

$$K_{Combined} = (K_1)^{e_1} \circ \cdots \circ (K_m)^{e_m} \qquad (5)$$

where $\{K_i \mid i = 1, ..., m\}$ is the set of kernel functions to be combined, e_i is the exponent of i-th kernel function, and ∘ denotes an operator between two kernel functions. In our case, three types of the kernel functions listed in Table 1 are combined, and multiplication or addition operators are used to combine kernel functions.

The parameters in a kernel function play the important role of representing the structure of a sample space. The set of the parameters of a combined kernel function consists of three part - i) the exponents of individual kernel functions, ii) the operators between kernel functions, iii) the coefficient in each kernel function. In the learning phase, the structure of a sample space is learned by a kernel function, and the knowledge of a sample space is contained in the set of parameters. Furthermore, the optimal set of features should be chosen in the learning phase. In our case, EA technique is exploited to obtain the optimal set of features as well as the optimal combined kernel function.

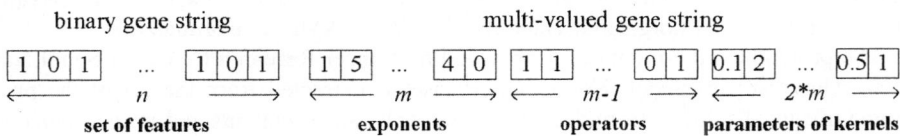

Fig. 3. Structure of a chromosome used in EA procedure

The challenging issue of EA is how to map a real problem into a *chromosome*. In our learning method, we need to map feature space, the set of the parameters for kernels, and the set of operators combining kernels. Firstly, the set of features is encoded into a n-bit binary string to represent an active or non-active state of n features. Then the exponents of m individual kernel functions, the operators between individual kernel functions, and the coefficients in each individual kernel function are encoded into a multi-valued gene string. The combination of the two gene string forms a chromosome in EA procedure which in turn serves as a *decision model* (see Fig. 3). In learning phase, simulating a genetic procedure, EA creates improved decision models containing a combined kernel function and a set of features by the iterative process of reproduction, evaluation, and selection process. At the end of learning stage, the optimal decision model consisting of a combined kernel function and the set of features is obtained, and the optimal decision model is contained in a classifier to be used classify new pattern samples.

3 Experiment Results

In this section, we show the result from the classification based on the model trained by our learning method. Furthermore, the performance of the classification model with combined kernel function is compared to the performances of the models with other kernel functions.

3.1 Datasets

There are several microarray dataset from published cancer gene expression studies. In this paper, we have used two representative datasets (Colon cancer and Leukemia datasets) among them and our own dataset (proteome patterns of Stomach cancer dataset).

The colon cancer dataset [11] contains gene expression information extracted from DNA microarrays. The 62 samples dataset consists of 22 normal and 40 cancer tissue samples and each having 2000 features. 32 samples were chosen randomly as training set and the remaining samples were used as testing set. (Available at: http://sdmc.lit.org.sg/GEDatasets/Data/ColonTumor.zip).

The leukemia dataset [12] consists of 72 samples that have to be discriminated into two classes ALL and AML. There are 47 ALL and 25 AML samples and each sample contains 7129 features. The dataset was divided into a training set with 38 samples (27 ALL and 11 AML) and a test set with 34 samples (20 ALL and 14 AML) (Available at: http://sdmc.lit.org.sg/GEDatasets/ Data/ALL-AML_Leukemia.zip).

The proteome pattern dataset is provided by Cancer Research Center of Seoul National University in Seoul, Korea. The dataset is extracted from the set of the proteome images displayed from the sera of stomach cancer patients and normal persons by 2D PAGE method. Each proteome pattern consists of 119 intensity values, each of which represents the amount of a type of proteome contained in serum. The dataset includes 70 cancer samples and 67 normal samples. From the set of all the samples 72 samples are randomly chosen to form a training set, and the set of the remaining samples serve as a test set.

3.2 Environments for Experiments

All experiments are conducted on a Pentium IV 1.8GHz computer. The experiments are composed preprocessing, learning by EA to obtain optimal decision model, and classification (see Sec. 2). For preprocessing data, we normalize data and randomly build 10 pair of training/testing dataset. For EA, we have used tournament rule for selection method [8]. Some of the selected chromosomes are given the chance to undergo alterations by means of crossover and mutation to form new individuals. One-point crossover is used, and the probabilities for crossover and mutation are 0.8 and 0.015 respectively. The proposed method was done with 100 of generations and 100 of populations. Our combined kernel function and three other kernel functions (Table 1) are trained by EA in learning phase with training set. Three kernel functions are chosen since they were known to have good performances in bioinformatics field [4, 6, 13-15]. Also, 10-fold cross validation is used for the fitness estimating to reduce overfitting problem [4]. The optimal *decision model* obtained after 100 generations of EA is used to classify the set of test samples. The experiments for each kernel function are repeated for 10 times to obtain generalized results. As the result of learning phase, an optimal *decision model* consisting of 15 most important features and optimal combined kernel function is obtained.

3.3 Results of Experiment and Analysis

The classified results of Stomach cancer dataset are shown in Fig. 4. The graph indicated the proposed method with combined kernel function more stable than other cases. The average, highest, and lowest of hit ratios using four kernel functions are shown in Table 2. Here the combined kernel function case shows the best average performance with 79.23% correct classification. The upper and lower bound of it (see Table 2) also are higher and also more stable than other cases.

Table 2. Hit ratioes for the case of stomach cancer dataset in the classification phase based on the decision model obtained after 100 generations of EA

Hit ratio	Combination	Inverse Multi-quadric	Radial	Neural
Average	76.77%	74.62%	69.69%	46.15%
Upper bound	86.15%	86.15%	83.08%	46.15%
Lower bound	73.85%	63.08%	60.00%	46.15%

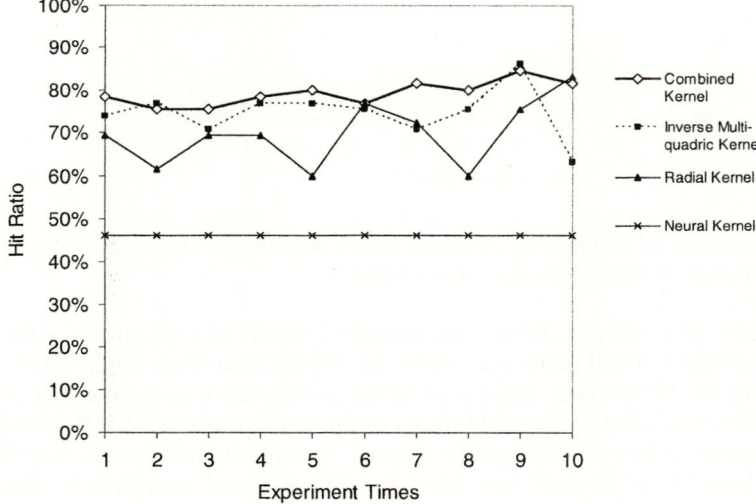

Fig. 4. Comparison of hit ratios by combined kernel function and single kernel functions in classification phase in the case of stomach cancer dataset

In the case of colon dataset, the experiments of proposed method with combined kernel function also show more stable and higher than other cases (see Fig. 5). According to Table 3, the result of combined kernel function case shows the best average performance with 79.00% of recognition rate. The upper and lower bound of it (see Table 3) also are higher than other cases. This indicates that our method is able to discriminate cancer class from normal class.

Table 3. Hit ratios for the case of colon cancer dataset in the classification phase based on the decision model obtained after 100 generations of EA

Hit ratio	Combination	Inverse Multi-quadric	Radial	Neural
Average	75.33%	74%	72.33%	66.67%
Upper bound	96.67%	86.67%	86.67%	66.67%
Lower bound	70%	60%	60%	66.67%

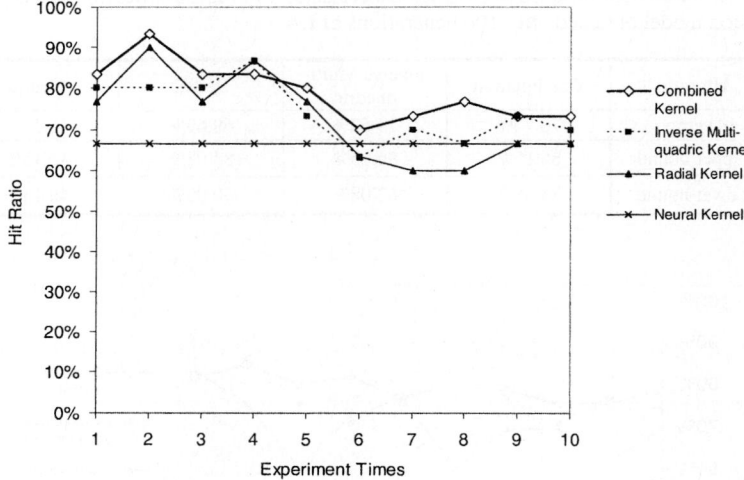

Fig. 5. Comparison of hit ratios by combined kernel function and single kernel functions in classification phase in the case of colon cancer dataset

In the case of Leukemia dataset, the classified results of experiments with combined kernel function still seem more stable and higher than other single kernel function (see Fig. 6). The average, highest, and lowest of hit ratios using four kernel functions are shown in Table 4. The table shows us the best average is 82.35% of recognition rate in case of Inverse Multi-quadric kernel function. In this dataset, even though combined kernel function could not obtain the best average of corrected classification, but the results still stable than other cases.

Table 4. Hit ratioes for the case of leukemia cancer dataset in the classification phase based on the decision model obtained after 100 generations of EA

Hit ratio	Combination	Inverse Multi-quadric	Radial	Neural
Average	77.06%	77.65%	70.57%	58.82%
Upper bound	85.29%	94.12%	79.41%	58.82%
Lower bound	70.59%	67.65%	58.82%	58.82%

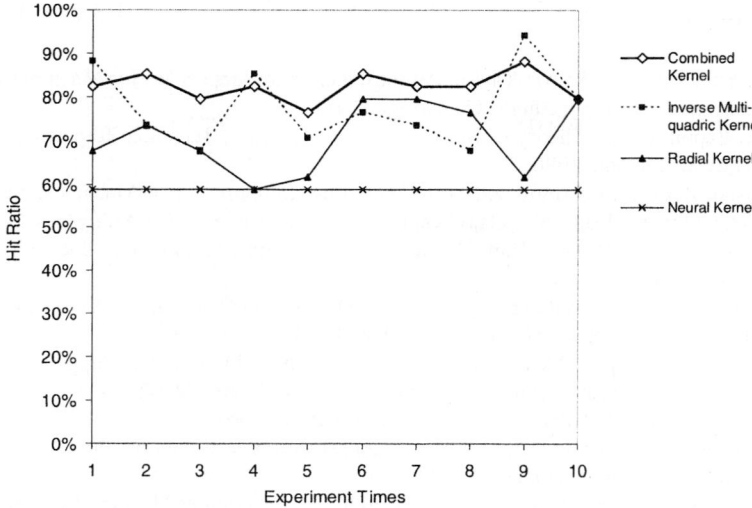

Fig. 6. Comparison of hit ratios by combined kernel function and single kernel functions in classification phase in the case of leukemia dataset

The classification results of three above dataset indicate that the classification ratio of our proposed method is able to get more stable and higher recognition rate than single kernel function cases.

4 Conclusion

In this paper, we proposed a new kernel function combining a set of kernel functions for SVM and its learning method exploiting EA technique to obtain the optimal decision model for classification. A kernel function plays the important role of mapping the problem feature space into a new feature space so that the performance of the SVM classifier is improved. The combined kernel function and the learning method were applied to classify the clinical datasets to identify cancer/normal groups. In the comparison of the classifications by combined kernel and other three kernel functions, the combined kernel function achieved the fastest convergence in learning phase and results in the optimal decision model with the highest hit rate in classification phase. Thus our combined kernel function has greater flexibility in representing a problem space than individual kernel functions.

Acknowledgement

The authors acknowledge that mySVM [16] was used for the implementation of the proposed method in this paper. This research was supported by the Internet Information Retrieval Research Center (IRC) in Hankuk Aviation University. IRC is a Regional Research Center of Kyounggi Province, designated by ITEP and Ministry of Commerce, Industry and Energy.

References

1. N. Cristianini and J. Shawe-Taylor, An introduction to Support Vector Machines and other kernel-based learning methods, Cambridge, 2000.
2. V.N. Vapnik et. al., Theory of Support Vector Machines, Technical Report CSD TR-96-17, Univ. of London, 1996.
3. Vojislav Kecman, Learning and Soft Computing: Support Vector Machines, Neural Networks, and Fuzzy Logic Models (Complex Adaptive Systems), The MIT press, 2001.
4. Richard O. Duda, Peter E. Hart, David G. Stork, Pattern Classification (2nd Edition), John Wiley & Sons Inc., 2001.
5. Joachims, Thorsten, Making large-Scale SVM Learning Practical. In Advances in Kernel Methods - Support Vector Learning, chapter 11. MIT Press, 1999.
6. Bernhard Schökopf, Alexander J. Smola, Learning with Kernels: Support Vector Machines, Regularization, Optimization, and Beyond, MIT press, 2002.
7. M.L. Minsky and S.A.Papert, Perceptrons, MIT Press, 1969.
8. Z.Michalewicz, Genetic Algorithms + Data structures = Evolution Programs, Springer-Verlag, 3 re rev. and extended ed., 1996.
9. D. E. Goldberg, Genetic Algorithms in Search, Optimization & Machine Learning, Adison Wesley, 1989.
10. Melanie Mitchell, Introduction to genetic Algorithms, MIT press, fifth printing, 1999.
11. Alon, U., Barkai, N., Notterman, D., Gish, K., Ybarra, S., Mack, D., and Levine, A.: Broad Patterns of Gene Expression Revealed by Clustering Analysis of Tumor and Normal Colon Tissues Probed by Oligonucleotide Arrays. Proceedings of National Academy of Sciences of the United States of American, vol 96 (1999) Pages: 6745-6750.
12. Golub, T. R., Slonim, D. K., Tamayo, P., Huard, C., Gaasenbeek, M., Mesirov, J. P., Coller, H., Loh, M. L., Downing, J. R., Caligiuri, M. A., Bloomfield, C. D. and Lander, E. S.: Molecular Classification of Cancer: Class Discovery and Class Prediction by Gene Expression Monitoring. Science, vol. 286 (1999) Pages: 531–537.
13. Frohlich, H., Chapelle, O., Scholkopf, B.: Feature selection for support vector machines by means of genetic algorithm, Tools with Artificial Intelligence. Proceedings. 15th. IEEE International Conference (2003) Pages: 142 – 148.
14. Xue-wen Chen: Gene selection for cancer classification using bootstrapped genetic algorithms and support vector machines, The Computational Systems Bioinformatics Conference. Proceedings IEEE International Conference (2003) Pages: 504 – 505.
15. Chanho Park and Sung-Bae Cho: Genetic search for optimal ensemble of feature-classifier pairs in DNA gene expression profiles. Neural Networks, 2003. Proceedings of the International Joint Conference, vol.3 (2003) Pages: 1702 – 1707.
16. Stefan Rüping, mySVM-Manual, University of Dortmund, Lehrstuhl Informatik, 2000. URL: http://www-ai.cs.uni-dortmund.de/SOFTWARE/MYSVM/

Automatic Liver Segmentation of Contrast Enhanced CT Images Based on Histogram Processing

Kyung-Sik Seo[1], Hyung-Bum Kim[1], Taesu Park[1], Pan-Koo Kim[2], and Jong-An Park[1]

[1] Dept. of Information & Communications Engineering,
Chosun University, Gwangju, Korea
japark@chosun.ac.kr
[2] Dept. of Computer Engineering,
Chosun University, Gwangju, Korea
pkkim@chosun.ac.kr

Abstract. Pixel values of contrast enhanced computed tomography (CE-CT) images are randomly changed. Also, the middle liver part has a problem to segregate the liver structure because of similar gray-level values of neighboring organs in the abdomen. In this paper, an automatic liver segmentation method using histogram processing is proposed for overcoming randomness of CE-CT images and removing other abdominal organs. Forty CE-CT slices of ten patients were selected to evaluate the proposed method. As the evaluation measure, the normalized average area and area error rate were used. From the results of experiments, liver segmentation using histogram process has similar performance as the manual method by medical doctor.

1 Introduction

In order to segregate hepatic tumors, the first significant process is to extract the liver structure from other abdominal organs. Liver segmentation using CT images has been dynamically performed because CT is a very conventional and non-invasive technique [1-5]. Generally, in order to improve diagnosis efficiency of the liver, the CT image obtained by contrast media is used. Pixel values of contrast enhanced CT (CE-CT) images acquired from the dose of contrast agent are randomly changed. Also, the middle liver part has a problem to segregate the liver structure because of similar gray-level values of abdominal organs. In this paper, an automatic liver segmentation method using histogram processing is proposed for overcoming pixel variation of CE-CT images and removing adjacent abdominal organs which are contact with liver.

2 Liver Segmentation Using Histogram Processing

Histogram transformation such as convolution and scaling is performed to reduce small noise of a histogram. A convolution method as one dimensional low pass filtering is used to smooth the histogram, even though the histogram's horizontal axis is extended and a vertical axis is very increased [6]. Then, the extended horizontal axis is scaled to gray-level values.

A multi-modal threshold (MMT) method is processed to find a region of interest (ROI) of the liver regardless of histogram variation derived from the contrast enhancement. After removing the background, bones and extremely enhanced organs, several ranges in the multi-modal histogram are found by a piecewise linear interpolation (PLI) method [7]. As a ROI range of the liver is located experimentally in the right side of the histogram, the ROI range is selected easily. Then the ROI of the liver is segmented by using the selected range.

Histogram tail threshold (HTT) is presented to remove the neighboring pancreas, spleen, and left kidney from the ROI. Let $I_{ROI} : Z^2 \rightarrow Z$ be the gray-level ROI. Then, $I_{ROI}(m,n) \in Z$. Let $h_{ROI}(k_1, k_2): Z \rightarrow Z$ be the histogram of I_{ROI} with the range, $[k_1, k_2]$. Let I_{HTT} be the HHT image. Then the HHT algorithm is proposed:

- Find k_{max} where k_{max} is the gray-level value when $h_{ROI}(k)$ is the maximum value.
- Calculate the histogram tail interval $k_{HI} = (k_{max} - k_1)$.
- Find histogram tail threshold $k_{HTT} = (k_{max} - k_{HI} / \gamma)$ where γ is the integer value greater than 0.
- Create the HHT image $I_{HTT} = \{(m,n) \mid k_1 \leq I_{ROI}(m,n) \leq k_{HTT}\}$.

Fig. 1 shows an example of liver segmentation using histogram processing such as histogram transformation, MMT, and HHT.

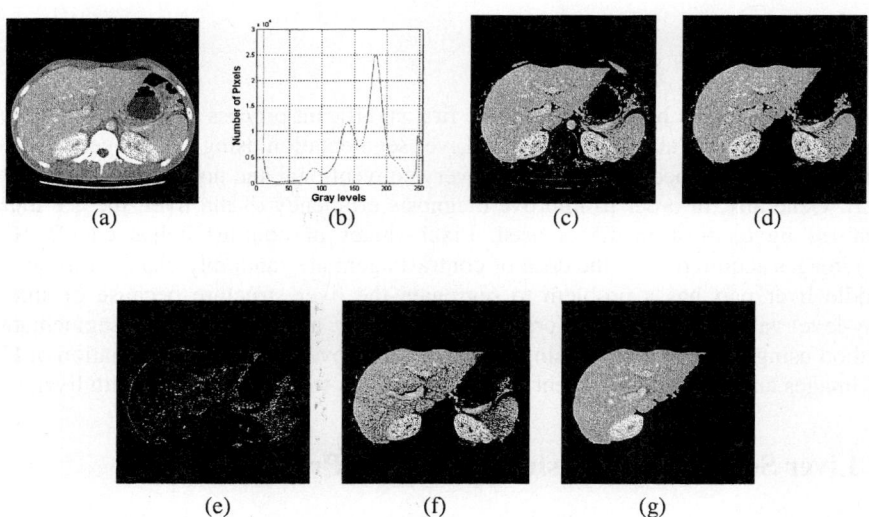

Fig. 1. Liver segmentation using histogram processing: (a) CE-CT image, (b) Transformed histogram, (c) ROI after MMT, (d) ROI after removing small objects, (e) HHT image, (f) Difference image, (g) Segmented liver after removing other objects

3 Experiments and Analysis

Forty CE-CT slices of ten patients were selected for testing the proposed method to segregate a liver structure. Selected each slices were hard to segregate the liver structure in the abdomen. As a criterion, one radiologist took a part in this research to segregate the liver structure by the manual method. In order to evaluate performance of the proposed algorithm, two different methods were compared such as histogram processing and the manual method. Table 1 shows the normalized average area (NAA) segmented by each method. That is, segmented liver area of each patient were averaged and normalized by the image size. From the results of this comparison, we may know histogram processing has almost same area as the manual segmentation. As the average NAA of histogram process and the manual segmentation is 0.1531 and 0.1594, the difference is very small.

Table 1. Comparison of normalized average area between histogram processing and the manual segmentation

	Histogram Processing	Manual Segmentation
Patient 01	0.1198	0.1292
Patient 02	0.1773	0.1791
Patient 03	0.1557	0.1588
Patient 04	0.1922	0.2020
Patient 05	0.1216	0.1304
Patient 06	0.2090	0.2122
Patient 07	0.1856	0.1897
Patient 08	0.1482	0.1547
Patient 09	0.1142	0.1226
Patient 10	0.1080	0.1150
Total Average	0.1531	0.1594

As another comparison method, average area error rate (AER) is used. Average AER is defined as

$$AER = \frac{a_{UR} - a_{IR}}{a_{MSR}} \times 100\% \quad (1)$$

where a_{UR} is the average pixel area of union region, a_{IR} is the average pixel area of intersection, and a_{MSR} is the average pixel area of the manual segmented region. As Fig. 2 shows, the average AER per each patient is 5~13% and the total average AER of all patients is 8.4138 %.

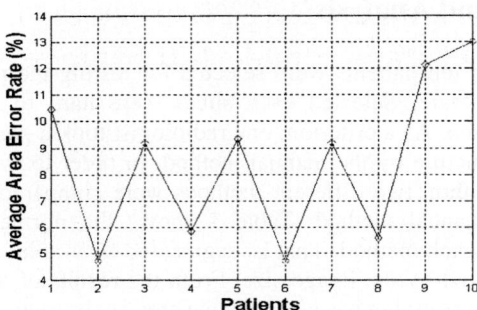

Fig. 2. Average area error rate

4 Conclusions

In this paper, we proposed an automatic liver segmentation method using histogram process. Histogram transformation such as convolution and scaling was first used. Next, multi-modal threshold (MMT) was performed to find the range of the ROI of the liver region. In order to remove other organs which were neighboring with liver, Histogram tail threshold were used. 40 slices from ten patients were selected to evaluate histogram process. As the evaluation measure, the normalized average area and area error rate were used. From the results of experiments, liver segmentation using histogram processing has similar performance as the manual method by medical doctor.

References

1. Bae, K. T., Giger, M. L., Chen, C. T., Kahn, Jr. C. E.: Automatic segmentation of liver structure in CT images. Med. Phys.,Vol. 20. (1993) 71-78
2. Gao, L., Heath, D. G., Kuszyk, B. S., Fishman, E. K.: Automatic liver segmentation technique for three-dimensional visualization of CT data. Radiology, Vol. 201. (1996) 359-364
3. Tsai, D.: Automatic segmentation of liver structure in CT images using a neural network. IEICE Trans. Fundamentals, Vol. E77-A. No. 11. (1994) 1892-1895
4. Husain, S. A., Shigeru, E.: Use of neural networks for feature based recognition of liver region on CT images, Neural Networks for Sig. Proc.-Proceedings of the IEEE Work., Vol. 2. (2000) 831-840
5. Seo, K., Ludeman, L. C., Park S., Park, J.: Efficient liver segmentation based on the spine. LNCS, Vol. 3261. (2004) 400-409
6. Pitas, I.: Digital Image Processing Algorithms and Applications, Wiley & Sons Inc., New York NY (2000)
7. Schilling, R. J., Harris, S. L.: Applied numerical methods for engineers. Brooks/Cole Publishing Com., Pacific Grove CA (2000)

An Improved Adaptive RBF Network for Classification of Left and Right Hand Motor Imagery Tasks

Xiao-mei Pei, Jin Xu, Chong-xun Zheng, and Guang-yu Bin

Institute of Biomedical Engineering of Xi'an Jiaotong University, 710049, Xi'an, China
pei@zy165.com, xujin@mail.xjtu.edu.cn

Abstract. An improved adaptive RBF neural network is proposed to realize the continuous classification of left and right hand motor imagery tasks. Leader-follower clustering is used to initialize the centers and variances of hidden layer neurons, which matches the time-variant input features. Based on the features of multichannel EEG complexity and field power, the time courses of two evaluating indexes i.e. classification accuracy and mutual information (MI) are calculated to obtain the maximum with 87.14% and 0.53bit respectively. The results show that the improved algorithm can provide the flexible initial centers of RBF neural network and could be considered for the continuous classification of mental tasks for BCI (Brain Computer Interface) application.

1 Introduction

BCI technology provides a new non-neuromuscular communication and control channel between brain and external environment [1]. Currently, one important application of BCI is 'thought-control' over functional electrical stimulation (FES) directed by hand motor imagery [2]. Based on event-related desynchronization/ synchronization (ERD/ERS), the left and right hand motor imagery tasks can be easily discriminated, which can be used to implement the control over FES. The classification of mental tasks is the key issue in BCI. The RBF neural network has been widely used in pattern classification, in which K-means clustering is usually used to initialize the centers of network and the number of centers is pre-determined by prior knowledge. In this paper, an improved RBF neural network is proposed for the continuous classification of hand motor imagery tasks, in which Leader-follower clustering is used to initialize the centers of networks, which matches the time-varying input features. The results show that the improved RBF network is effective for continuous classification of hand motor imagery tasks in BCI application.

2 The Experimental Data

The data is available in BCI2003 competition website. The experiment consists of 280 trials. Each trial is 9s length. The first 2s was quite, at t=2s an acoustic stimulus indicates the beginning of the trial and a cross "+" was displayed for 1s; then at t=3s, an arrow (left or right) was displayed. At the same time the subject was asked to move a bar into the direction of the cue. The feedback was based on AAR parameters of channel C3 and C4, the AAR parameters were combined with a discriminant analysis into one output parameter. Three bipolar EEG channels were measured over C3, Cz and C4. EEG was sampled with 128Hz and filtered between 0.5 and 30Hz. The data include train data and test data with equal 140 trials with the details referred to [3].

3 Feature Extraction

Human's EEG frequency is primarily between 0.5~30Hz. Studies by Pfurtscheller and his associates show that when people imagine or prepare for unilateral hand movement, the amplitude of α and β especially μ rhythm (10-12Hz) in contralateral hand area decreases, which is called as ERD; simultaneously the amplitude of the corresponding rhythms in ipsilateral hand area increases, which is called as ERS [4]. The multichannel desynchronized EEG would result in the decrease of local field power and spatial synchrony, and vice versa [5], which can well be characterized by the multichannel EEG linear parameters proposed by Wackermann [5]. Multichannel EEG complexity and field power can reflect the independence between functional processes and field strength of local brain regions [5]. Here, EEG within 10-12Hz from C3, Cz and C4, Cz are selected to calculate two-channel complexity and field power as features, which have good separability for left and right hand motor imagery so as to be considered for classification of two classes of EEG patterns.

4 RBF Neural Network for Classification of Mental Tasks

The Radial basis function usually chooses the Gaussian function.

$$g_k(x) = \exp(-\frac{\|x - u_k\|^2}{2\sigma_k^2}) \qquad (1)$$

where u_k, σ_k is the center and the corresponding variance of the kth cluster ($k = 1, \cdots, K$). The most used K-means clustering for initializing the centers is not suitable for the problem here. Firstly, for the continuous classification of hand motor imagery tasks, the subject's brain state varies with time and then the input EEG features are time-variant, which requires the number of centers of RBF neural network hidden layer changing with time. Secondly, the number of centers is unknown and should change with every new input feature vectors. So it's necessary and reasonable to find another clustering method to initialize the centers u_k and variances σ_k adaptively. For the unknown number of clusters, the simplest clustering method is the Leader-follower clustering [6]. The basic idea of the method is to only change the cluster center nearest to the new input pattern repeatedly and keep others unchanged. So Leader-follower clustering can realize the adaptive initialization of the centers according to the time-varying input patterns.

After initializing the centers of network, the next step is to adjust two parameters u_k, σ_k by the gradient descent in the regularized error function with the regularized coefficient λ to avoid overfitting [7]. Then the optimal RBF neural network output weights $\mathbf{w} = [w_1, \cdots w_K]^T$ are obtained. The output of RBF neural network is computed as a linear combination of K radial basis functions.

$$D(x) = \sum_{k=1}^{K} w_k g_k(x) \qquad (2)$$

where w_k is the weight between the kth hidden layer and output layer. For the two classes of EEG patterns, the classification outputs can be as follows:

$$\begin{cases} if \ D(x) > 0 \ : \ right \\ if \ D(x) < 0 \ : \ left \end{cases} \qquad (3)$$

where $D(x)$ indicates the size of classification margin.

5 Experimental Results

Classification accuracy is an important index for evaluating BCI system performance, which reflects the ability of BCI system identifying the brain consciousness task correctly. Another index MI (Mutual Information) is used to quantify the information transfer performance of BCI, which is proposed by A.Schlogl [8]. Based on the features of two-channel spatial complexity and field power, the improved RBF network is used to classify the left and right hand motor imagery tasks. The classification accuracy and MI time courses for test data are shown in Fig1 (a, b). The parameters η, θ and λ are chosen as 0.1, 0.055 and 10^{-6} respectively by minimizing the error with cross validation.

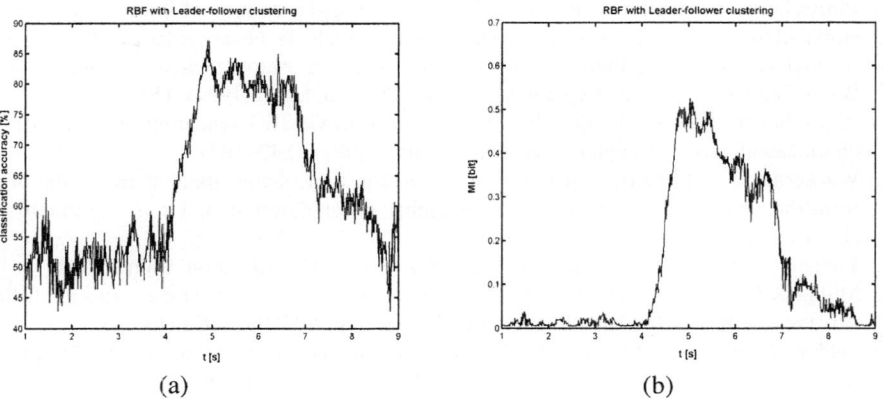

Fig. 1. Classification accuracy time course in (a) and MI time course in (b)

From figure 1 the satisfactory results are obtained with the maximum classification accuracy and MI reaching 87.14% and 0.53bit respectively at about t=5s. MI time course keeps consistent with classification accuracy. The time course of the two indexes by RBF network based on K-means clustering (K=8) are also calculated, which shows a large oscillation. In contrast, the classification accuracy and MI time courses by RBF network with Leader-Follower clustering is more stable and smoother so as to improve the performance of the continuous mental tasks in BCI system.

6 Discussions

In this paper, Leader-follower clustering method provides the flexible time-variant initial parameters of the cluster centers, which could reflect the true clustering centers and meet the requirements of continuous classification of hand motor imagery tasks. Then by the gradient descent, the centers and variances are further tuned adaptively and the weight parameters of network are optimized, which can effectively improve classification accuracy. Combining the advantages of the above two steps, the improved adaptive RBF neural network performs the satisfactory classification so that it could be expected to apply to mental tasks classification in BCI.

Acknowledgements

The work is funded by National Nature Science foundation of China (#30370395, #30400101).

References

1. W olpaw J R, Birbaumer N, McFarland D J, et al. Brain-computer interfaces for communication and control. Clinical Neurophysilolgy. (2002); 767~791.
2. Pfurtscheller G, Muller G R, Pfurtscheller J. 'Thought'-control of functional electrical stimulation to restore hand grasp in a patient with tetraplegia, Neurosci lett, (2003); 22~36
3. Schlögl A, Lugger K, Pfurtscheller G. Using Adaptive Autoregressive Parameters for a Brain-Computer-Interface Experiment, EMBS, Chicago, USA. (1997); 1533~1535
4. Pfurtscheller G, Lopes da Silva F.H. Event-related EEG/MEG synchronization and desynchronization: basic principles. Clin Neurophysiol. (1999);1842~1857
5. Wackermann J. Towards a quantitative characterization of functional states of the brain: from the non-linear methodology to the global linear description. Int. J. Psychophysiol. (1999); 65-80
6. Richard O. Duda, Peter E. Hart, David G. Stork. Pattern Classification. China. 2004.
7. Muller K R, Smola A, Ratsch G. Using support vector machines for time series prediction. Advances in Kernel Methods-Support Vector Learning. MIT Press, Cambridge. 1998.
8. Schlogl A, Neuper C. Pfurtscheller G. Estimating the mutual information of an EEG-based Brain-Computer-Interface], Biomedizinische Technik, (2002); 3~8

Similarity Analysis of DNA Sequences Based on the Relative Entropy

Wenlu Yang[1,2,3], Xiongjun Pi[1,2], and Liqing Zhang[1,*]

[1] Department of Computer Science and Engineering,
1954 Hua Shan Rd., Shanghai Jiaotong University, Shanghai 200030, China
wenluyang@sjtu.edu.cn, zhang-lq@cs.sjtu.edu.cn
[2] Shanghai Institutes for Systems Biology,
Shanghai Institutes for Biological Sciences,
Chinese Academy of Sciences, China
[3] Department of Electronic Engineering,
Shanghai Maritime University, Shanghai 200135, China

Abstract. This paper investigates the similarity of two sequences, one of the main issues for fragments clustering and classification when sequencing the genomes of microbial communities directly sampled from natural environment. In this paper, we use the relative entropy as a criterion of similarity of two sequences and discuss its characteristics in DNA sequences. A method for evaluating the relative entropy is presented and applied to the comparison between two sequences. With combination of the relative entropy and the length of variables defined in this paper, the similarity of sequences is easily obtained. The SOM and PCA are applied to cluster subsequences from different genomes. Computer simulations verify that the method works well.

1 Introduction

In conventional shotgun sequencing projects of microbial isolates, all shotgun fragments are derived from clones of the same genome. When using the shotgun sequencing approach on genomes from an environmental sample, however, the fragments are from different genomes. This makes assembling genomes more complicated [1]. To find whether fragments in a given sample set can be assembled, it is necessary to cluster those fragments likely from the same genome into a group for reduction of computational complexity. The criterion of clustering fragments is no doubt very important. The selection of the criterion is the motivation to investigate the similarity of sequences based on the relative entropy. In this paper, we present a method of evaluating the relative entropy, apply it to the comparison of similarity between two sequences and use it as a measure for clustering fragments sampled from different genomes.

---

[*] To whom correspondence should be addressed. The project was supported by the national natural science foundation of China under grant 60375015.

After simply introducing Kullback-Leibler divergence(KLD), we will apply it to processing the real data from NCBI. We introduce the definition of variables, and present the methods of processing the data, the result and analysis of experiments. Finally, we will discuss the result and applications.

2 Problem Formulation

To understand our method, we will give explanations about the measure of similarity and a method of evaluating KLD of two sequences.

Kullback-Leibler divergence(KLD), also called the relative entropy given by Kullback (1959), $KL(P|Q)$ between two probability distributions P and Q is defined as

$$KL(P \mid Q) = -\sum_{i=1}^{N}\left[p(x_i)log\frac{p(x_i)}{q(x_i)}\right], \quad (1)$$

where, $0 \cdot log0 = 0$. $p(x_i)$ and $q(x_i)$ are probability density function of P and Q, respectively. Therefore $KL(P \mid Q)$ establishes a measure of the distance between the distributions Q and P. However, the Kullback entropy is not symmetric and thus not a distance in the mathematical sense. $KL(P \mid Q) \neq KL(Q \mid P)$. The Kullback entropy $KL(P \mid Q)$ is always greater than or equal to zero and vanishes if and only if the distributions P and Q are identical[2][3].

A sequence X, of length n, is formed as a linear succession of n symbols from a finite alphabet {A, C, G, T}. A substring of K symbols, with $K \leq n$, is defined as a K-length random variable (also defined as K-word or K-tuple [4][5]). For example, K = 2, 2-length random variables are included in the set {AA, AC, AG, AT, CA, ..., TA, TC, TG, TT}, which contains 16 variables in all.

The probability density of a random variable x_i in the sequence X is evaluated by Equation (2)

$$p(x_i) = f(x_i)/(n - K + 1), \quad (2)$$

where, $f(x_i)$ is occurrences of x_i in X. For example, let X be "CGCGT", of length n=5. For K=2, $f(CG) = 2$ and $p(CG) = 2/(5 - 2 + 1) = 0.5$.

Let Y be "CGGT". It is easy to estimate $p_X(GC) = 0.25$, but $p_Y(GC) = 0$. In this case, the standard $KL(X|Y)$ becomes infinite, resulting in no similarity between X and Y according to definition of KLD. But in fact, X is very similar to Y. Therefore, we evaluate $KL(X|Y)$ with the only variables included in X and Y in common.

To let KLD satisfy conditions of symmetry and a distance in the mathematical sense, we evaluate the mean of $KL(X|Y)$ and $KL(Y|X)$ as the relative entropy between X and Y.

3 Experiments and Results

We get some whole genomes from the nucleotide database of GenBank: NC_000117, NC_000853, NC_000854, NC_000868, NC_000907, and NC_001807.

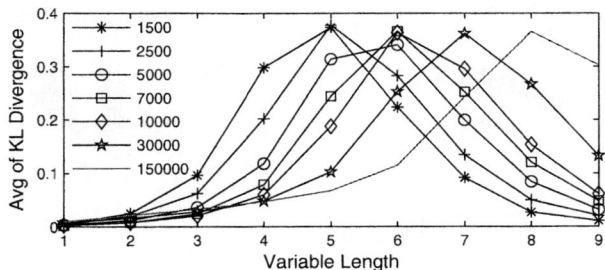

Fig. 1. Average of KLD with different variable length

The whole sequences are cut into subsequences of length ranging from 1.5K to 150K for different sequence analysis, such as selection of variable length and similarity of subsequences from same genomes or different genomes. We will give an application to clustering subsequences.

How long variables are suitable? Experiments indicate that there exists a threshold of variable length that expresses the similarity between two sequences, as shown in Fig. 1. If a variable length is too big, i.e. bigger than the threshold, KLD cannot really reflect the similarity.

Using combination of KLD and variable length, we apply the method to the sequence analysis. Fig. 2 indicates that KLD of all pair-subsequences from a genome are almost same with a variable length. Fig. 3 indicates subsequences(before 10 on X-axis) from a same sequence are more similar to each other than to ones(the rest) from different sequences.

It is noteworthy that in Fig. 3 that when the length is too big, such as 6, KLD of subsequences from different genomes becomes smaller. Obviously, the similarity measure with KLD of such a large length is not suitable. According to the results, we can use the method to cluster the reads to reduce computing complexity of assembling genomes directly sampled from natural environment.

As a similarity-based clustering method for unlabeled data [6], the self-organizing map (SOM) is used to group the subsequences. We use PCA to get three principle components and use them as the input patterns of SOM network

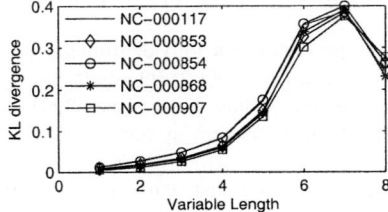

Fig. 2. KLD of 20K-length pair-subsequences from a same genome

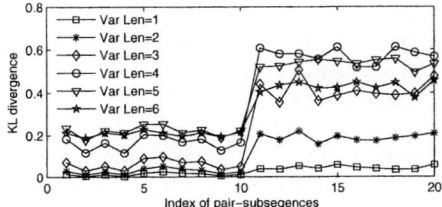

Fig. 3. KLD of two subsequences from different genomes or same genomes

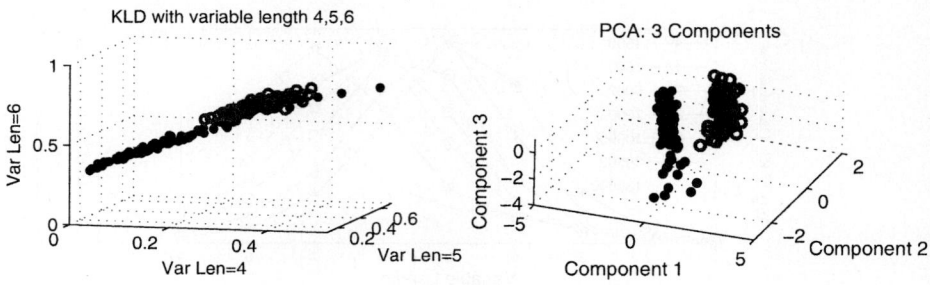

Fig. 4. Grouping by 3 variable length   **Fig. 5.** Grouping by PCA

because, in some cases, we have a little difficulty in grouping the subsequences by the KLD of a single or several variable length, as shown in Fig. 4. Fig.5 indicates that it is easier to classify the subsequences by PCA than by combination of several variable length. Using the method, we can obtain two sets for a subsequence. Subsequences in the first set is more similar to the subsequence than ones in the second. The clustering result is satisfied.

4 Conclusions

In this paper we present and apply KLD with different variable length to the similarity analysis of DNA subsequences from the same genome or different genomes or species. Computer simulations verify that the method works well. The method can be used to cluster the reads from the genomes of microbial communities directly sampled from natural environment to save computation for assembling the whole genomes of varieties of species. The method is also used to search similar sequences from nucleotide or protein databases, to compare homology of sequences to find some similar species, and to cluster genes from gene-expression data.

References

1. Tyson G.W.,Chapman J., et al.:Community structure and metabolism through reconstruction of microbial genomes from the environment. NATURE Vol.428,(2004)37–43
2. Steuer R.,Kurths J., et al.: The mutual information- Detecting and evaluation dependencies between variables. Bioinformatics, Vol.18 Suppl.2(2002) 231–240
3. Thomas M.C.,Joy A.T.:Elements of Information Theory. Wiley, New York (2001)
4. Vinga S., and Almeida J.: Alignment-free sequence comparison - a review. BIoinformatics,Vol.19,(2003) 513-523
5. Basu S., Burma D.P., et al.: Words in DNA sequences- some case studies based on their frequency statistics. Mathematical Biology, 46(6)(2003) 479-503.
6. Strickert M.: Self-Organizing Neural Networks for Sequence Processing. University of Osnabruck, 7(2004) 68

Can Circulating Matrix Metalloproteinases Be Predictors of Breast Cancer? A Neural Network Modeling Study

H. Hu[1], S.B. Somiari[1], J. Copper[2], R.D. Everly[2], C. Heckman[1], R. Jordan[1], R. Somiari[1,*], J. Hooke[3], C.D. Shriver[3], and M.N. Liebman[1]

[1] Windber Research Institute, Windber, PA, USA
[2] NeuralWare, Pittsburgh, PA, USA
[3] Walter Reed Army Medical Center, Washington, DC, USA
*Current address: ITSI Biosciences, Johnstown, PA, USA

Abstract. At Windber Research Institute we have started research programs that use artificial neural networks (ANNs) in the study of breast cancer in order to identify heterogeneous data predictors of patient disease stages. As an initial effort, we have chosen matrix metalloproteinases (MMPs) as potential biomarker predictors. MMPs have been implicated in the early and late stage development of breast cancer. However, it is unclear whether these proteins hold predictive power for breast disease diagnosis, and we are not aware of any exploratory modeling efforts that address the question. Here we report the development of ANN models employing plasma levels of these proteins for breast disease predictions.

1 Introduction

According to estimates by the American Cancer Society, in 2004 breast cancer was the most common cancer among women in the United States and the second leading cause of female deaths. The current screening procedures have a ~20% false negative rate and are effective only after a lesion has developed which biological studies show may take 4-10 years from the onset of the disease. Thus, more sensitive strategies for detection of early stage breast cancers are needed.

We are committed to the research on breast cancer using an integrative approach across clinical, genomic, and proteomic platforms [1-3], and a data warehouse has been developed to harbor the large amount of data produced in the high throughput experiments to facilitate the research [4]. In addition, we have small-scale research programs on identifying biological markers (biomarkers) for breast cancer diagnosis or early detection. One such example is a study on circulatory matrix metalloproteinases (MMP) 2 and 9 [5].

MMPs are involved in extracellular matrix modification. Specifically, MMP2 and MMP9 appear to play critical roles in the early stage of cancer development [6, 7]. We have studied the relative levels of these enzymes, and statistical analysis indicated that MMP2 and MMP9 levels may be used to stratify patients into different disease categories [5]. In this study, we further ask whether Artificial Neural Networks

(ANNs) can help in classifying the disease categories which in the future may be developed for the early prediction of breast cancers.

ANNs comprise a family of powerful machine learning technologies which embody learning mechanisms that are analogous to the ways natural neural networks learn. At their core, ANNs learn by adjusting weights, which are in effect coefficients for non-linear functions in an equation that fits the sample data. A typical ANN consists of an input layer, a so-called "hidden layer" to perform input-to-output mapping, and an output layer which exposes network results. ANNs have been applied to breast cancer studies, and models using mammogram and other image data have been reported [8, 9]. There are also reports of ANN models for breast cancer diagnosis using clinical laboratory blood work data and image features of cells obtained through (fine) needle aspiration [10, 11]. ANN models have also been used for outcome, survival or relapse predictions [12, 13].

All of these studies, however, are relatively specialized, i.e., none of these models used a combination of inputs of multiple categories including biomarkers, patient demographic and medical history, and family medical history data, etc.. Breast cancer is currently considered a heterogeneous disease, and environmental factors appear to play important roles in the development of the disease as well. We believe that an ultimately successful model for breast cancer diagnosis or detection needs to take into account multiple disparate factors. ANNs represent a natural class of candidate models that can use such inputs for breast cancer diagnosis and detection. The MMP study reported in the following reflects our initial ANN modeling efforts in this direction.

2 Methods

Non-control subjects enrolled in this study are fully informed and consenting female patients, 18 years old and above. Subjects were categorized as benign (non-neoplastic) or with breast cancer based on pathological diagnosis. High Risk individuals were diagnosed as disease-free but were determined by the Gail Model to have a 5-year risk ratio of $>= 1.67\%$ of developing breast cancer. The control group is constituted of a group of healthy female volunteers with a 5 year risk ratio of $< 1.67\%$. In the following we denote these groups of subjects as Benign, Cancer, HR, and Control.

A total of 169 subjects were enrolled in this study, which are n=77 for Benign, n=48 for Cancer, n=31 for HR, and n=13 for Control. The blood was drawn from the subjects and processed for plasma following conventional methods, and active and total MMP2 and MMP9 levels were analyzed using a commercial kit [5], denoted as MMP2_Active, MMP2_Total, MMP9_Active, and MMP9_Total respectively, which were used as the initial raw data input to the ANN models.

ANN models were developed using NeuralWorks Predict® from NeuralWare. In developing the input layer, Predict first applies a range of mathematical transformations to the raw data in order to provide a richer pool of potential inputs with which to build a model, then it uses a genetic algorithm optimizer to identify members of the pool which would be most appropriate as model inputs. Next, Predict uses a cascade correlation network construction mechanism [14] to dynamically determine the network architecture by continually monitoring the performance of the model being trained and gradually adding hidden nodes until model performance ceases to improve. The output layer contains four nodes, for the four groups of the subjects. The original dataset was partitioned into training and testing sets, 70% and 30% by default. Scripts were developed to facilitate varying model parameters and allow for fast development of 1000 models in this study.

3 Results and Discussions

Although Predict allows for model parameter adjustment through the interface, we built the first model with all default settings. The model was 61% accurate on the training dataset and 71% accurate on the testing dataset, with a structure of 6-10-4 (6 input nodes, 10 hidden, and 4 outputs) so multiple transformations of one or more of the raw data fields were ultimately chosen by the variable selection operation.

In light of the reasonable, but not exceptional results obtained from the initial model, a script file was then created to guide Predict in building 1000 models by systematically varying model parameters and data partitioning. Of this pool of models, the majority gave a correct classification rate (average of training and test) of over 80% and 43 models yielded a classification rate of 90% or better.

Given the relatively small size of the samples involved in this study we paid special attention to the possibility of over-fitting. We did have seven models producing 100% accurate results with the training dataset but not so with the test dataset, where the accuracies were two at >90% accuracy, 5 others at >80% accuracy, and another at 77% respectively. There was also one model that produced 100% accuracy with the test dataset but showed 85% accuracy with the training set. While the most direct approach to reducing/eliminating issues related to over-fitting is to hold out some of the available data to use in a final validation, limited by our sample size at this stage we were unable to set aside a validation dataset. The validation dataset should be completely inaccessible to the whole training and testing process.

We also compared the ANN model results with logistic models developed using NCSS and SAS (data not shown). After combining some outcomes to form a binary condition (e.g. Cancer and non-Cancer) some logistic models produced an accuracy rate of 80% or higher, whereas when using all the 4 outcomes as in the ANN modeling the best logistic model only produced an accuracy rate of 70.2%.

Overall, the results of the ANN modeling efforts are promising. This screening approach, which is based on the use of circulating blood markers, is attractive because it is minimally invasive. If MMP2 and 9 measured here truly reflect changes due to the cancer or lesion development, then it is possible that they can serve as early predictors before a lesion is detectable. We are now in the process of increasing our sample size, and developing ANN models taking multidisciplinary data types as inputs including other potential biomarkers, demographical, and clinical data etc.. We hope that more comprehensive ANN models can be developed which can potentially find application in clinical practices.

Acknowledgments: We thank Nicholas Jacobs, FACHE, President of WRI for making this study possible. This project was conducted with funds for Clinical Breast

Care Project received from the US Department of Defense (DoD). The comments, opinions, and assertions herein are those of the authors and do not necessarily represent those of the US Army or DoD.

References

1. Ellsworth, R.E., Ellsworth, D.L., Lubert, S.M., Hooke, J., Somiari, R.I. Shriver, C.D.: High-throughput loss of heterozygosity mapping in 26 commonly deleted regions in breast cancer. Cancer Epidemiol Biomarkers Prev. 12 (2003) 915-919
2. Somiari, S.B., Shriver, C.D., He, J., Parikh, K., Jordan, R., Hooke, J., Hu, H., Deyarmin, B., Lubert, S., Malicki, L., et al.: Global search for chromosomal abnormalities in infiltrating ductal carcinoma of the breast using array-comparative genomic hybridization. Cancer Genet Cytogenet. 155 (2004) 108-118
3. Somiari, R.I., Sullivan, A., Russell, S., Somiari, S., Hu, H., Jordan, R., George, A., Katenhusen, R., Buchowiecka, A., Arciero, C., et al.: High-throughput proteomic analysis of human infiltrating ductal carcinoma of the breast. Proteomics. 3 (2003) 1863-1873
4. Hu, H., Brzeski, H., Hutchins, J., Ramaraj, M., Qu, L., Xiong, R., Kalathil, S., Kato, R., Tenkillaya, S., Carney, J., et al.: Biomedical informatics: development of a comprehensive data warehouse for clinical and genomic breast cancer research. Pharmacogenomics. 5 (2004) 933-941
5. Somiari, S., Shriver, C.D., Heckman, C., Olsen, C., Hu, H., Jordan, R., Arciero, C., Russell, S., Garguilo, G., Hooke, J., et al.: Plasma concentration and activity of matrix metalloproteinase 2 & 9 in patients with breast cancer and at risk of developing breast cancer. Cancer Letters (in press). (2005)
6. Ueno, H., Nakamura, H., Inoue, M., Imai, K., Noguchi, M., Sato, H., Seiki, M. Okada, Y.: Expression and tissue localization of membrane-types 1, 2, and 3 matrix metalloproteinases in human invasive breast carcinomas. Cancer Res. 57 (1997) 2055-2060
7. Sheen-Chen, S.M., Chen, H.S., Eng, H.L., Sheen, C.C. Chen, W.J.: Serum levels of matrix metalloproteinase 2 in patients with breast cancer. Cancer Lett. 173 (2001) 79-82
8. Wu, Y., Giger, M.L., Doi, K., Vyborny, C.J., Schmidt, R.A. Metz, C.E.: Artificial neural networks in mammography: application to decision making in the diagnosis of breast cancer. Radiology. 187 (1993) 81-87
9. Goldberg, V., Manduca, A., Ewert, D.L., Gisvold, J.J. Greenleaf, J.F.: Improvement in specificity of ultrasonography for diagnosis of breast tumors by means of artificial intelligence. Med Phys. 19 (1992) 1475-1481
10. Astion, M.L. Wilding, P.: Application of neural networks to the interpretation of laboratory data in cancer diagnosis. Clin Chem. 38 (1992) 34-38
11. Markopoulos, C., Karakitsos, P., Botsoli-Stergiou, E., Pouliakis, A., Ioakim-Liossi, A., Kyrkou, K. Gogas, J.: Application of the learning vector quantizer to the classification of breast lesions. Anal Quant Cytol Histol. 19 (1997) 453-460
12. Ravdin, P.M., Clark, G.M., Hilsenbeck, S.G., Owens, M.A., Vendely, P., Pandian, M.R. McGuire, W.L.: A demonstration that breast cancer recurrence can be predicted by neural network analysis. Breast Cancer Res Treat. 21 (1992) 47-53
13. Ripley, R.M., Harris, A.L. Tarassenko, L.: Non-linear survival analysis using neural networks. Stat Med. 23 (2004) 825-842
14. Fahlman, S.E. Lebiere, C.: The Cascade Correlation Learning Architecture. In: Touretzky, D.S., (ed.): Advances in Neural Information Processing Systems 2. (1990) 524-532

Blind Clustering of DNA Fragments Based on Kullback-Leibler Divergence

Xiongjun Pi[1,2], Wenlu Yang[1], and Liqing Zhang[1,*]

[1] Department of Computer Science and Engineering,
Shanghai Jiao Tong University, 1954 Hua Shan Rd.,
Shanghai 200030, China
pizhou@sjtu.edu.cn, Zhang-lq@cs.sjtu.edu.cn
[2] Shanghai Institute for Systems Biology,
Shanghai Jiao Tong University

Abstract. In whole genome shotgun sequencing when DNA fragments are derived from thousands of microorganisms in the environment sample, traditional alignment methods are impractical to use because of their high computation complexity. In this paper, we take the divergence vector which is consist of Kullback-Leibler divergences of different word lengths as the feature vector. Based on this, we use BP neural network to identify whether two fragments are from the same microorganism and obtain the similarity between fragments. Finally, we develop a new novel method to cluster DNA fragments from different microorganisms into different groups. Experiments show that it performs well.

1 Introduction

The sequences obtained from an environmental sample are from different genomes [1], which makes the tradition assembly method "overlap - layout - consensus" [2] impractical to apply because of large number of DNA fragments and consequently prohibitive computing complexity. It is desirable to cluster fragments which are from the same or similar genomes into a group before assembly. Many rigorous DNA sequence comparison algorithms like BLAST involve sequence alignment at some stage and become computationally prohibitive when comparison against a large number of sequences [6]. The Kullback-Leibler divergence uses word frequencies thus can be computed as fast as Euclidean distance [3]. We take the divergence vector which combines Kullback-Leibler divergence of different word lengths as the feature vector. Then we apply BP(Back-Propagation) neural network to determine whether two sequences are from the same genome and obtain the similarity between sequences. Based on the similarity, we develop a novel method to cluster DNA sequences into several groups and each group is composed of sequences from the same or similar genomes.

The paper is organized as follows: Section 2 introduces the clustering problem we faced in mathematical sense and the concept of Kullback-Leibler divergence.

---

[*] To whom correspondence should be addressed. The project was supported by the national natural science foundation of China under grant 60375015.

Our clustering method is described in Section 3. Section 4 is experiments and results. Finally, Section 5 presents the conclusions.

2 Problem and Definition

2.1 Problem Formulation

In the problem of sequencing fragments from an environmental sample, we have no idea of how many kinds of microorganisms or the coverage rate, and we don't know how many fragments each microorganism contains either [1], so the problem of clustering such fragments can be called blind clustering.

2.2 Words in DNA Sequences and Kullback-Leibler Divergence

A sequence, X, of length n, is defined as a linear succession of n symbols from a finite alphabet $\{A, G, C, T\}$.

L-word is defined as a segment of L symbols, with $L \leq n$. The number of all possible L-words satisfies $K = 4^L$.

The vector of frequencies f_L^X can be obtained by counting occurrences of each L-word in sequence X. The counting is usually performed by sliding a window L-wide that is run through the sequence, from position 1 to $n - L + 1$.

One can then calculate the word probability vector by the equation below.

$$p_L^X = \frac{f_L^X}{n - L + 1} \tag{1}$$

The Kullback-Leibler divergence or relative entropy is a measure of difference between two distributions based on information theory [4]. The Kullback-Leibler divergence between sequence X and sequence Y is defined as

$$KL(X|Y) = \sum_{i=1}^{4^L} p_{L,i}^X \log \frac{p_{L,i}^X}{p_{L,i}^Y} \tag{2}$$

3 Clustering Method of DNA Fragments

3.1 Classification of Kullback-Leibler Divergence Vectors and Measure of Similarity

The Kullback-Leibler divergence vector between two DNA fragments is defined as $(v(i), v(i+1), v(i+2), \ldots, v(j))^T$ where $v(L), L = i, i+1, \ldots, j$ is the divergence of two DNA fragments when the word length is L. BP neural network is a kind of multilayer feedforward network and has been used as an classifier for a long time [7]. We apply BP neural network to determine whether the Kullback-Leibler divergence vector is between two fragments from the same microorganism. In training phase, the output is 1 when two DNA fragments are derived from the same genome and -1 otherwise. In test phase, if the output is not less than 0, it means two fragments come from same or similar microorganisms.

We define the similarity of two fragments as the output of BP neural network when the input is the divergence vector between them. The value of similarity also reflects the possibility whether two fragments are from the same genome.

3.2 Blind Clustering Method

Before introducing our clustering algorithm, we define an important concept first:

veto rate: if in a group there are N such fragments, the similarity between which and the compared fragment is no less than 0. Supposing the size of the group is M. The veto rate defines as N/M.

In our algorithm, there are also three important parameters:

high threshold: if the similarity is equal to or greater than the high threshold, the fragments are from the same genome with high possibility.

low threshold: if the similarity of two fragments is less than the low threshold, the two fragments are from different genomes with high possibility.

veto threshold: If the veto rate between a group and the compared fragment is not less than the veto threshold, the fragment will join the group.

Our clustering method:
1. Compute the divergence of every fragment pair for some word lengths.
2. Input the neural network with divergence vectors to compute similarity.
3. Find out the fragments where the similarity of every two fragments is less than low threshold. We say each of them represents a group.
4. For each group, find out the fragments, the similarity between which and the representation fragment of the group is not less than high threshold.
5. For each fragment, if the veto rate between it and a group is equal or greater than the veto threshold, then it joins the group.
6. For every fragment which does not join any group, join the group, the veto rate between which and the fragment is the greatest.

4 Experiments and Results

According to the taxonomy tree in NCBI [5], we obtain complete genomes of 9 microorganisms from genebank. Their search numbers are NC-000917, NC-003551, NC-002806, NC-002932, NC-005042, NC-006576, NC-001318 and NC-005835 respectively. We cut each genome randomly into sequences of length between $18k$ and $22k$. Then we calculate Kullback-Leibler divergence of word length $2, 3, 4, 5, 6, 7$ respectively. Fig. 1 presents the joint distribution of divergences of word length 5, 6, 7. We do the experiments on a lot of word length combinations. Our conclusion is: when the input is $(v(2), v(3), v(4), v(5), v(6), v(7))^T$, the classification performance of neural network is the best and we obtain the similarity. We adjust the parameters high threshold, low threshold and veto threshold. When their values are 0.95, -0.95 and 0.8 respectively, the final clustering result reaches its best performance 98.55%.

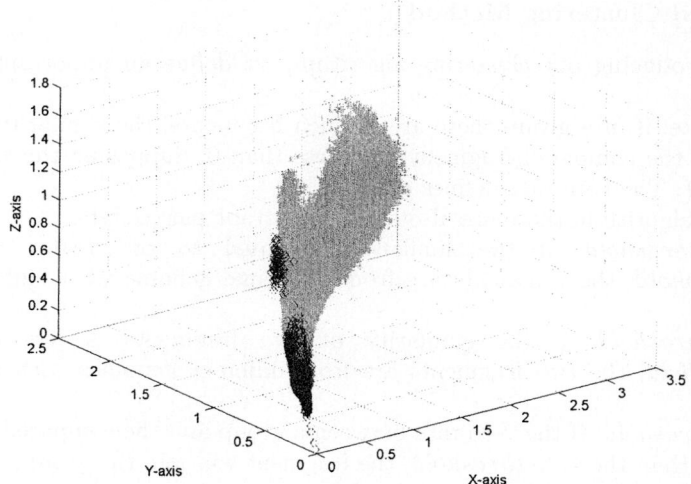

Fig. 1. The joint distribution of Kullback-Leibler divergences. X-axis represents divergence of word length 5, Y-axis represents divergence of word length 6 and Z-axis represents divergence of word length 7. The red diamonds represent divergences between fragments from the same microorganisms and the cyan asterisks represent divergences between fragments from different microorganisms.

5 Conclusions

In this paper, we take the Kullback-Leibler divergence vector as the feature vector which combines divergences of different word lengths. BP neural network performs well to determine whether the vector is between fragments from the same genome. Based on the similarity(the output of BP neural network), we develop a novel blind clustering method. The result of our experiments is satisfying.

References

1. Tyson, G.W., Chapman, J., et al.: Community structure and metabolism through reconstruction of microbial genomes from the environment. NATURE. **428** (2004) 37-43
2. Roach, J., Boysen, C., Wang, K. and Hood., L.: Pariwise end sequencing: a unified appraoch to genomic mapping and sequencing. Genomics. **26** (1995) 345-353
3. Wu, T.J., Hsieh, Y.C. and Li, L.A.: Statistical Measures of DNA Sequence Dissimilarity under Markov Chain Models of Base Composition. BIOMETRICS. **57** (2001) 441-448
4. Thomas, M.C. and Joy, A.T.: Elements of Information Theory. Wiley, New York (2001)
5. http://www.ncbi.nlm.nih.gov/Taxonomy/Browser/wwwtax.cgi?mode=Root
6. Altschul, S.F., Gish, W., Miller, W., Myers, E.W. and Lipman, D.J.: Basic local alignment search tool. J.Mol.Biol. **215** (1990) 403-410
7. Haykin, S.: Neural Network: A Comprehensive Foudation. Prentice-Hall (1999)

Prediction of Protein Subcellular Locations Using Support Vector Machines

Na-na Li[1], Xiao-hui Niu[1], Feng Shi[1], and Xue-yan Li[2]

[1] School of Science, Huazhong Agriculture University
[2] College of Sciences, Civil Aviation University of China

Abstract. In this paper,we constructed a data set of rice proteins with known locations from SWISS-PROT,using the Support Vector Machine to predicte the type of a given rice protein by incorporating sequence information with physics chemistry property of amino acid. Results are assessed through 5-fold cross-validation tests.

1 Introduction

With an enormous amount of raw sequence data accumulating databanks, it raises the challenge of understanding the functions of many genes. Proteins' subcellular locations are valuable to elucidate proteins' functions. Experiments have determined 2744 yeast proteins [1]. However,it is time-consuming and costly. It is highly desirable to predict a protein's subcellular locations automatically from its sequence. There are three basic approaches. One approach is based on amino acid composition, using artificial neural networks (ANN), such as NNPSL [2], or support vector machines (SVM) like SubLoc [3]. A second approach uses the existence of peptide signals, which are short sub-sequences of approximately 3 to 70 amino acids to predict specific cell locations, such as TargetP [4]. A third approach, such as the one used in LOCKey [5], is to do a similarity search on the sequence, extract text from homologs and use a classifier on the text features.

Predictions based only on amino acid composition may lose some sequence-order information. Chou [6] first proposed an augmented covariant discrimination algorithm to incorporate quasi-sequence-order effect, and a remarkable improvement was achieved. Subsequently, Chou[7] further introduced a novel concept, the pseudo-amino acid composition to reflect the sequence-order effect. Recently, Cai et al.[8] used SVM incorporating quasi-sequence-order effect.

We used SVM method here to predict subcellular locations based on amino acid composition, incorporating dipeptide composition and physics chemistry properties of proteins. Dipeptide composition can be considered as another representative of proteins incorporating neighborhood information.

Stability is an important physics chemistry property of protein; it is key to determine the native structure [9]. Guruprasad [10]designed a dipeptide instability weight value (DIWV) according to their influence on the stability of protein native structure. The DIWV is widely used in molecule structure.

Besides it, the occurrence frequency of different amino acid at active sites varied obviously [11], so sequence difference may result in different activity of proteins. Encouraged by the positive impact of these effects, we try to apply Vapnik's Support Vector Machine [12] to approach subcellular location of rice proteins problem.

2 Materials and Methods

2.1 Support Vector Machine

SVM is a type of learning machines based on statistical learning theory. The basic idea of applying SVM to pattern classification can be stated briefly as follows.

Suppose we are given a set of samples, i.e. a series of input vectors $X_i \in R^d$ ($i = 1, 2, \cdots, N$), with corresponding labels $y_i \in \{+1, -1\}$ ($i = 1, 2, \cdots, N$), where -1 and +1 respectively stand for the two classes. Our goal is to construct one binary classifier or derive one decision function from the available samples, which has small probability of misclassifying a future sample. Only the most useful linear non-separable case (for most real problems) are considered here.

SVM performs a nonlinear mapping of the input vector x from the input space into a higher dimensional Hilbert space, where the mapping is determined by the kernel function. It finds the OSH (Optimal Separating Hyperplane, see Cortes and Vapnik[13]) in the space H corresponding to a non-linear boundary in the input space. Two typical kernel functions are listed below:

$$K(x_i, x_j) = (x_i \cdot x_j + 1)^d \qquad (1)$$

$$K(x_i, x_j) = \exp(-r\|x_i - x_j\|^2) \qquad (2)$$

The first one is called the polynomial kernel function of degree d, which will eventually revert to the linear function when $1 = d$, and the latter one is called the RBF (radial basis function) kernel with one parameter λ. Finally, for the selected kernel function, the learning task amounts to solving the following convex Quadratic Programming (QP) problem:

$$\max \sum_{i=1}^{N} \alpha_i - \frac{1}{2} \sum_{i=1}^{N} \sum_{j=1}^{N} \alpha_i \alpha_j \cdot y_i y_j \cdot K(x_i, x_j)$$

subject to:

$$0 \leq \alpha_i \leq C \qquad \sum_{i=1}^{N} \alpha_i y_i = 0$$

where the form of the decision function is

$$f(x) = sgn(\sum_{i=1}^{N} y_i \alpha_i K(x_i, x) + b)$$

For a given data set, only the kernel function and the regularity parameter C must be selected.

In this paper, we apply Vapnik's Support Vector Machine for predicting the types of rice proteins. We have used the SVMlight, which is an implementation (in the C language) of SVM for the problem of pattern recognition.

2.2 Sequence Data

The data were selected from rice proteins with annoted subcellular location in SWISS-PROT (24). 352 rice proteins were classified into the following four groups: [1]chloroplast; [2]cytoplasmic; [3]integral membrane protein; [4]nuclear. There are 79, 54, 55, 154 proteins in four groups, respectively.

2.3 Algorithm

To improve the quality of statistical prediction for protein subcellular location, one of the most important steps is to give an effective representation for a protein. We incorporate sequence information with physics chemistry property of amino acid.

Here, each protein is represented by a 22-D vector $P = [p_1, p_2, \cdots, p_{22}]$. The first 20 components of its vector denote the occurrence frequencies of the 20 amino acids respectively. DIVW can be expressed by a 400-D vector, $D = [d_1 \cdots d_{400}]$, where d_i is a dipeptide instability weight value. Similarly, each protein is expressed by $X = [x_1 \cdots x_{400}]$, where x_i is the occurance frequency of corresponding dipeptide in the sequence. Thus, the 21st component is defined as $p_{21} = \sum_{i=1}^{400} d_i \cdot x_i$, which is the average of all the dipeptides instability values of the protein.

The occurrence frequence of each amino acid at the active position is different, so we can similarly derive the 22nd component, which is the average of the amino acid activity values of the protein. Thus, a protein is expressed by 22-D vector.

Also for the SVM, the width of the Gaussian RBFs is selected to minimize an estimate of the VC-dimension. The parameter C that controls the error-margin tradeoff is set at 1000. After being trained, the hyper-plane output by the SVM was obtained. The SVM method applies to two-class problems.

3 Results and Discussion

The prediction performance was examined by the 5-fold cross-validation test, in which the data set was divided into five subsets of approximately equal size. This means the data was partitioned into training and test data in five different ways. After training the SVMs with a collection of four subsets, the performance of SVMs was tested against the fifth subset. This process is repeated five times so that every subset is once used as the test data.

To assess the accuracy of prediction methods we use the following measures. Let TP is the number of true positive; FP the number of false positive; TN

Table 1. Results for the 352 Rice Proteins Represented by 22-D Vectors Tested by **5-fold cross-validation Test**

	sort1	sort2	sort3	sort4	overall
Sen^+(%)	77.78	85.45	56.96	90.26	79.82
Sen^-(%)	96.18	93.73	92.40	90.96	93.57
Spe^+(%)	79.25	72.31	69.23	89.10	80.53
Spe^-(%)	95.85	97.11	87.73	91.94	93.29
CC	0.754	0.749	0.566	0.820	0.749
accurate rate	0.933	0.924	0.842	0.906	0.901

Table 2. Results for the 352 Rice Proteins Represented by 20-D Vectors (amino acid composition) Tested by **5-fold cross-validation Test**

	sort1	sort2	sort3	sort4	overall
Sen^+(%)	18.52	74.55	21.52	79.22	55.56
Sen^-(%)	95.14	92.33	93.54	88.30	92.69
Spe^+(%)	41.67	65.08	50.00	84.72	71.70
Spe^-(%)	86.16	94.98	79.87	83.84	86.22
CC	0.254	0.652	0.287	0.705	0.564
accurate rate	0.830	0.895	0.769	0.842	0.834

the number of true negative; FN the number of false negatives; N=TP+ TN+ FP+FN. Then we have:

$$Sen^+ = \frac{TP}{TP+FN} \qquad Spe^+ = \frac{TP}{TP+FP}$$

$$Sen^- = \frac{TN}{TN+FP} \qquad Spe^- = \frac{TN}{TN+FN}$$

$$CC = \frac{TP*TN - FP*FN}{\sqrt{(TP+FN)(TP+FP)(TN+FP)(TN+FN)}}$$

The results are given in Table I. Compared with table II, the results obtained by only using single amino acid composition, the overall accurate rate has been improved much from 83.4% to 90.1%. It is tellable that sen^+, spe^+ and CC have been highly enhanced. Only using amino acid composition is unreasonable, for biological molecules is different to linear sequences of discrete units similar to linguistic representations. Sequence-order information and physics chemistry properties of proteins are absolutely necessary factors.

4 Conclusion

From the above results, considering appropriate sequence-order information and physics chemistry properties of proteins is helpful to boost the prediction accurate rate. The current study has further demonstrated that the instability index, the active index and SVM have opened a new and promising approach in dealing with sequence order effect

References

1. Kumar,A. Agarwal,S. Heyman,J.A. Matson,S. Heidtman,M., Piccirillo,S. Umansky,L. Drawid,A. Jansen,R. Liu,Y. et al.(2002) Subcellular localization of the yeast proteome. Genes Dev.,16, 707-719
2. Reinhardt,A. and Hubbard,T. (1998) Using neural networks for prediction of the subcellular location of proteins. Nucleic Acids Res., 26, 2230-2236
3. Hua,S. and Sun,Z. (2001) Support vector machine approach for protein subcellular localization prediction. Bioinformatics, 17, 721-728. http://www.bioinfo.tsinghua.edu.cn/SubLoc/
4. Emanuelsson,O., Nielson,H., Brunak,S. and von Heijne,G. (2000) Predicting subcellular localization of proteins based on their Nterminal amino acid sequence, J. Mol. Biol, 300, 1005-1016
5. Nair,R. and Rost,B. (2002) Inferring subcellular localization through automated lexical analysis. Bioinformatics, 18, S78-S86
6. Chou,K.C. (2000a) Prediction of protein subcellular locations by incorporating quasi-sequence-order effect. Biochem. Biophys. Res. Commun., 278, 477-483
7. Chou,K.C. (2001) Prediction of protein cellular attributes using pseudo-amino acid composition. Proteins Struct. Funct. Genet., 43, 246-255
8. Cai,Y.D., Liu,X.J., Xu,X.B. and Chou,K.C. (2002) Support vector machines for prediction of protein subcellular location by incorporating quasi-sequence-order effect. J. Cell. Biochem., 84, 343-348
9. Wang HC. Essentials of Sequence Analysis . Beijing: Press of Military Medical, 1994 (Ch)
10. Guruprasad K, Reddy BV, Pandit MW. Correlation between stability of a protein and its dipeptide composition: a novel approach for predicting in vivo stability of a protein from its primary sequence. Protein Eng., 1990, 4: 155-161
11. http://www.genome.ad.jp/dbget/AAindex/
12. Vapnik V. Statistical Learning Theory. New York: Wiley-Interscience, 1998.
13. Cortes C, Vapnik V. Support vector networks. Machine Learning, 1995, 20: 273-293

Neuroinformatics Research in China - Current Status and Future Research Activities

Guang Li[1,2,*], Jing Zhang[2], Faji Gu[1,3], Ling Yin[1,4], Yiyuan Tang[1,5], and Xiaowei Tang[1,2]

[1] Neuroinformatics Workgroup of China, 301 Hospital, Beijing, China
  *guangli@cbeis.zju.edu.cn
[2] Center for Neuroinformatics, Zhejiang University, Hangzhou 310027, China
[3] School of Life Science, Fudan University, Shanghai 200433, China
[4] Center of Neuroinformatics, General Hospital of PLA, Beijing 100853, China
[5] Institute of Neuroinformatics, Dalian University of Science and Techology, Dalian 116024, China

Abstract. After the Chinese National Neuroinformatics Working Group was formed in 2001, neuroinformatics research has progressed rapidly in China. This paper reviews the history of neuroinformatics in China, reports current researches and discusses recent trends of neuroinformatics in China.

1 Introduction

The First Conference on Neuroinformatics was held in 2000 in Haikou, China. Neuroinformatics is considered as a frontier and multi-discipline which combines the brain science, the information science and computer science together to investigate the form of neural information carrier, the mechanisms of generation, transmission, processing, coding, storage and retrieving of the neural information, and to construct the data bank system of neuroscience. A lots of Chinese scientists appealed Chinese Government to pay more attention and financial support on neuroinformatics [1, 2]. From then on, a lot of conferences on neuroinformatics have been held in China and a number of research projects have been supported by Chinese government. In 2004, the Chinese government joined the OECD-GSF-NI-WG.

2 Progress of Neuroinformatics in China

In September, 2001, the 168th Xiangshan Science Conference, which focusing on the scientific frontier of basic research and important engineering technology of China, entitled 'Human Brain Project and Neuroinformatics' was held in Beijing [3]. During the conference, Dr Stephen H. Koslow, who was the director of Office on Neuroinformatics Coordinator, NIMH, USA, was invited to introduce the Human Brain Project and discuss how to conduct neuroinformatics research in China with the officers of the Ministry of Science and Technology of China. As a result, the National Neuroinformatics Working Group was formed.

The Sino-Korea-Japan Joint Workshop on Neurobiology and Neurinformatics was held in Hangzhou, China in November 2001.

Authorized by Chinese State Department, the Ministry of Science and Technology of China represented Chinese government joining the Commission of Science and Technology of the Organisation for Economic Co-operation (OECD) in 2001. Prof Xiaowei Tang, Prof Lin Yin and Prof Yiyuan Tang represented China, as an observing country, to attend the meeting of the Organisation for Economic Co-operation and Development - Global Science Forum - Neuroinformatics Working Group (OECD-GSF-NI-WG) within same year. In April 2004, China joined the OECD-GSF-NI-WG [4].

Initially, there were only 6 members and nowadays there are 18 members from 10 different provinces and area. The number of neuroinformatics researchers increase from initial 30 to more than 250. Involved institutes are 25, including Zhejiang University, the PLA General Hospital, Dalian University of Science and Technology, Fudan University, Tsinghua University, Beijing University, Institute of Biophysics Chinese Academy of Science, Shanghai Jiaotong University and so on. The research areas cover neurology, neuropathy, mathematics, physics, chemistry, informatics, computer science etc. al [5].

There are four neuroinformatics research centers in China. They are the Neuroinformatics Center of PLA General Hospital, Center for Neuroinformatics of Zhejiang University, Institute of Neuroinformatics of Dalian University of Science and Technology and Center for Neuroinformatics of Beijing University of Chinese Medicine and Pharmacy.

Neuroinformatics training courses have been held twice. More that 200 people took part in and got basic training for neuroinformatics research.

Since 2001 more than 24 project grants have been funded for research in neuroinformatics by the National Basic Research Program of China (973 Program), the National Natural Science Foundation of China, International Collaboration Scheme and so on. These research programs include the neuroinformatic mechanisms of Chinese Traditional Medicine [6], Chinese recognition, sensory perception, mental diseases, et al. Some of them are listed as following:

- Neuroinformatics Research on fMRI of Acupuncture and Chinese Recognition
- Application of Nonlinear Neuroinformatics on Study of Olfactory Mechanisms
- Construction of Chinese Node of International Neuroinformatics Web-Database and Development of Informatics Tools
- Comparison Research between Structural Functional Atlas of Oriental and Western People's Brain
- Experimental Research on fMRI of Acupuncture at Points of Zhu-San-Li and Nei-Guan
- Study on Mechanism of Depression and Anxiety of Teenage and Strategy for Prevention and Treatment Based on Neuroinformatics Database
- Research on fMRI of Digital Recognition
- Research on Recognition and Nonlinear Analysis of EEG
- Audio Information Coding and Feedback Control and Its Loop Base
- Neural Mechanism of Audio Information Processing

- Signal Measurement and Information Extraction of Neuronal Loop
- Development of Quantitative Sensory Testing System
- Application of Chaotic Array on Artificial Olfaction
- Neural Systematic Function Oriented Computing Model
- ……

3 Data Sharing in Neuroscience in China

In Jan 2004, there was a meeting of OECD committee for scientific and technological policy at ministerial level. A 2002 Report on Neuroinformatics from OECD-GSF-NI-WG was submitted to the ministers attending the meeting. It highly recommended to establish a new global mechanism, the INCF (International Neuroinformatics Coordinating Facility), created an associated funding scheme, the PIN (Program in International Neuroinformatics) and establish national nodes and research programs in Neuroinformatics [5]. According to the plan of OECD-GSF-NI-WG, a lot of works to setup a national web-based neuroinformatics database for data sharing have been carried out in China.

3.1 Data Sharing in China

Data sharing in Science and Technology is a very important policy in China. Management and Sharing System of Scientific Data for Medicine is a key project of the

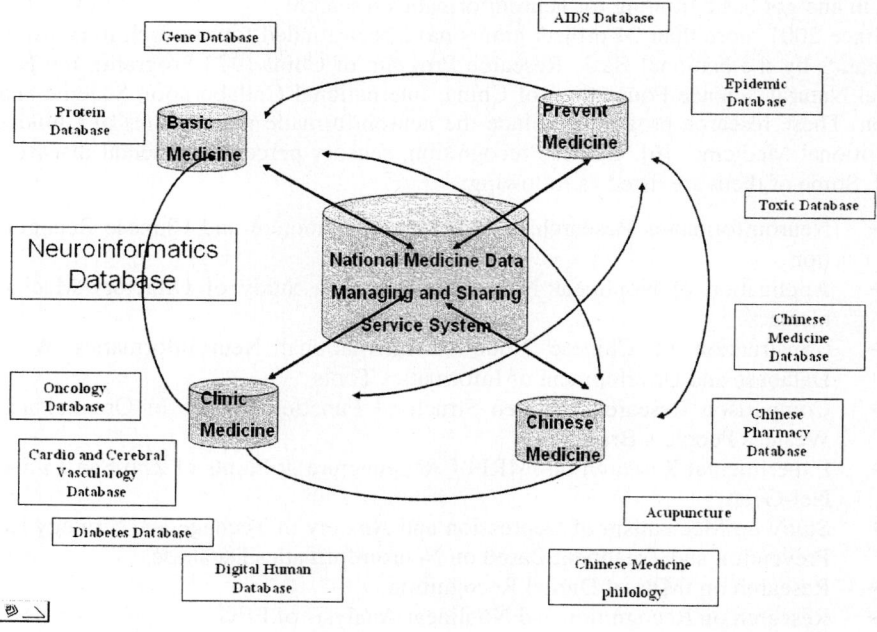

Fig. 1. The structure of the Natioanl Medicine Data Managing and Sharing Service System of China

National Basic Platform for Science and Technology for year 2003 in the Ministry of Science and Technology, China. As the paramount part of the National Engineering of Scientific Data Sharing, the System is undertaken by Chinese Academy of Medical Sciences, Chinese Center for Disease Prevention and Control, Chinese PLA General Hospital, and Chinese Academy of Traditional Chinese Medicine. Scientific data resources in medicine will be integrated together in the way of distribution physically and unification in logic by the system. The system covers most of the fields in medicine, including basic medicine, clinical medicine, public health, traditional Chinese medicine, special medicine, pharmacology and innovated drug etc. As a part of this system, neuroinformatics database is also included (as shown in Fig.1).

3.2 Neuroinfomatics Database in China

Based on previous projects granted by Chinese government, the testing version of neuroinformatics web sites at PLA General Hospital and Dalian University of Science and Technology have been constructed. Some neuroscientific datasets have been stored and shared.

Funded by the National Basic Research Program of China (973 Program), the project, Construction of Chinese National Node of International Neuroinformatics Web-Database and Development of Informatics Tools, has been initialized in January 2005. The aim of this project is to collect and normalize existing neuroinformatics resources all over the country, develop informatics processing platform for neuroscience, setup the national web node of China as a part of international neuroinformatics network and provide neuroinformatics resources with local characteristics.

Following tasks are going to carry out: 1) collect and normalize existing neuroinformatics resources to establish the databases of neural cell biology, neural anatomy-digital brain, neural image, fMRI, neural electrophysiology, neurological function evaluation, clinic neuroinformatics, artificial intelligence, Chinese recognition and sensory perceptions; 2) develop neuroinformatics tool software for Chinese recognition and acupuncture; 3) setup coordinating environment based on internet for neuroinformatics research; 4) establish the national web node of China as a part of international neuroinformatics network sharing the neuroinformatics resources globally; 5) construct databases of Chinese Traditional Medicine and acupuncture to provide local neuroinformatics resources for global network. The project is planned to finish within three years.

4 Future Works

Based on the HH neuron model and a two-layer network, the effect of noises on the sensory systems is discussed. For single neuron, the optimal intensity of noises must adapt to the stimulating signals. It is noted that the intensity of noises has a linear relation with the standard deviation of stochastic noises. The fluctuation of background noises is approximately stationary stochastic process with constant standard deviation, so its intensity hardly changes. This limits the application of SR. However, for the cooperative effect of a set of neurons, the fixed level of noise can induce SR while the stimulating signals varying within a certain range. According to these results, the two-layer network can be considered as one of basic structure of signal detection in sensory systems. It is further proved that the collective behavior of a set of neurons can restrain the noises by analyzing the suprathreshold cases for the networks with different quantities of neurons.

Acknowledgment

This paper was financially supported by the National Basic Research Program of China (No. 2002CCA01800 and 2004CB720302) and the National Natural Science Foundation of China (No. 30470470).

References

1. Shen, J., Human Brain Project and Neuroinformatics, Acta Biohysica Sinica, 17(2001), 607-612
2. Chen, W., Wang, Z., Chen, Z., and Rong, S., Principle and Perspective of Neuroinformatics, Acta Biohysics Sinica, 17(2001), 613-620
3. Han, Z., Human Brain Project and Neuroinformatics—168th Xiangshan Science Conference, China Basic Science, 1 (2001), 47-49
4. Tang, X., Yin, L., Jin X., and Tang Y., Reviews and Perspects for Neuroinformatics of OECD-GSF, China Basic Science, 5 (2004), 11-13
5. OECD-GSF-NI, Report on Neuroinformatics from the Global Science Forum Neuroinformatics Working Group of the Organisation for Economic Co-operation and Development, (2002)
6. Yin, L., Sun, J., Zhu, K., Traditional Chinese Medicine and Human Brain Project, China Basic Science, 2 (2003), 37-38

Australian Neuroinformatics Research – Grid Computing and e-Research

G.F. Egan [1,2,3], W. Liu[1], W-S. Soh[1], and D. Hang[1]

[1] Howard Florey Institute,
[2] Centre for Neuroscience,
[3] National Neuroscience Facility,
University of Melbourne, 3010 Australia
g.egan@hfi.unimelb.edu.au

Abstract. The Australian National Neuroscience Facility (NNF) has been established to provide Australian neuroscientists with access to networks of laboratories offering neuroscience consultancy, technical expertise and state-of-the-art equipment. The facility is fostering neuroscience excellence, combining science, technology, innovation, investment, creativity and the opportunity to advance our understanding and treatment of the brain and mind. Within the NNF a Neuroscience Informatics platform has been established with the objective of enhancing both the national neuroscience research capability, as well as the commercialisation opportunities for the Australian health and biotechnology industries. The Platform has developed a NeuroGrid facility consisting of computational resources and Grid middleware, internet accessible neuroimage databases, and standardised neuroimage analysis tools. A customised NeuroGrid portal is currently under development. It is envisaged that the NeuroGrid facility and software tools will provide the basis for application of Grid computing technologies to other areas of neuroscience research.

1 Introduction

Neuroscience is one of the fastest growing areas of scientific research and of industry development in the biotechnology sector. There is growing interest amongst neuroscientists to equitably share neuroscience data and analytical tools since this sharing affords the opportunity to differently re-analyze previously collected data; secondly, it encourages new neuroscience interpretations; and thirdly, it also fosters otherwise uninitiated collaborations. Sharing of neuroscience data and tools is playing an increasingly important role in human brain research and is leading to innovations in neuroscience, informatics and the diagnosis and treatment of neurological and psychiatric brain disorders [1].

Neuroscience Informatics is the use of information technology to acquire, store, organize, analyse, interpret and computationally model neuroscience data. The term is used to describe both neuroscience databases, as well as computer-based tools for using these databases. In the past decade there has been a rapid growth in the volume and complexity of neuroscience data. The limited degree to which scientific publications can describe the richness of the inter-relationships between these complex data has stimulated the development of internet accessible neuroscience databases. Internet access to these databases permits efficient re-analysis of these data as new concepts are developed, as well as the application of new quantitative modelling approaches to enable more precise investigations of the conceptual understanding that has been derived from the initial research studies.

The Australian National Neuroscience Facility (NNF) has been developed with a Major National Research Facility grant from the Australian Government and with additional supporting funds from the Victorian State Government. A key objective of the National Neuroscience Facility is to develop physical infrastructure to place Australia at the forefront of international neuroscience research, and to provide researchers with efficient access to this infrastructure. The Neuroscience Informatics platform has been developed to address the following issues: the difficulties associated with accessing disparate and limited datasets, the lack of standardised analysis tools and experimental methods, and the lack of integrative analyses across sub-disciplines of neuroscience research. The Neuroscience Informatics platform aims at providing the necessary informatics infrastructure and expertise required by neuroscientists associated with the NNF, or who are working within other NNF platforms, particularly in the neuroimaging and neurogenomics platforms. It is envisaged that the Neuroscience Informatics Platform will become an Australian node of international neuroinformatics projects. The platform is also co-operating with a number of commercial organizations that have developed proprietary databases and are integrating behavioural, psychophysiological, neuroimaging and gene expression datasets.

2 Neuroscience Informatics Platform – Applications to Neuroimaging

The application of neuroimaging methods to the study of human brain structure and function is continuing at an undiminished pace, particularly in the study of psychiatric disorders and higher cognitive functions. Existing MR image datasets from investigations in these fields contain enormously valuable information. The Neuroscience Informatics platform has focussed on the development of neuroimaging databases and specialized neuroimaging software tools. Existing databases within the NNF include a human neonate structural brain image database containing brain images from over 270 human subjects, and with over 1200 structural images stored [2]. Other databases developed by scientists and clinicians in the NNF include a human functional imaging database (including 180 adult subjects containing over 400 structural and over 20,000 functional MR images), and an adolescent and adult human structural MR image database (containing over 1000 structural images) used to investigate brain morphology in schizophrenic patients and individuals at risk of developing schizophrenia [3].

National and indeed international collaboration has been widely recognized [4,5,6,7] as a key requirement to develop effective informatics facilities that will enable researchers to fully reap the potential benefits of neuroimaging research investigations. In order to maximally capture these benefits, analyses of large imaging data sets need to be supported by modern data mining techniques. However, a number of significant research collaboration challenges must firstly be overcome.

The integration of database systems where the data representations are intrinsically different is problematic. Furthermore, the establishment of a standard human neuroanatomical ontology is required in order to, for example, enable unambiguous indexing of images in databases. This would it turn enable retrieval of a specific neuroanatomical segment (or brain structure) from a human brain image. Finally, important enhancements to neuroimaging analyses could be achieved by linking image databases to related subject information such as gene expression and genetic characterisation (using micro-array and/or serial analysis of gene expression) in the same subject. Nevertheless, in spite of these on-going challenges the recent rapid developments in Grid computing and the current high level of interest in e-science and e-research motivate our application of Grid technologies to neuroscience research.

3 Methods

Efficient analyses of large neuroimaging datasets requires layered information technology (IT) systems capable of data management (image database creation and curation), standardized image data processing algorithms (image registration, tissue classification, segmentation, and spatial normalization to name but a few), and standardized access to computational facilities. The NeuroGrid system aims to utilise the computational power and networking technologies of the Grid to perform standardised analyses of large volumes of neuroimaging data that are stored and organized into image databases within the NNF. Eventually it is anticipated that data access and processing operations will be managed through seamless Grid computing facilities that are running multiple operating systems including Mac OS-X, Linux/Solaris, and Windows 2000/XP [8,9].

Importantly, the analysed or processed imaging results will be stored into the originating databases, or to alternative results databases. The NeuroGrid system therefore consists of a cluster of Mac OS-X and Linux desktop computing nodes, a Grid management software (middleware) layer, internet accessible neuroimaging data repositories and neuroimaging analysis algorithms, and a customised Grid graphical user interface (GUI) application (the NeuroGrid portal).

NeuroGrid Computing Infrastructure. The NeuroGrid computing facility currently consists of the following computing hardware: an X-serve G5, an X-serve RAID system, an X-serve G5 Cluster Node, and 12 Power Mac G5 desktop machines (Apple Computer Inc, Cuppertino, CA); and 12 PC desktop machines. The X-serve G5 is configured as the X-serve Raid controller and is operating both as the file server and database server. As demand for the database increases it is envisaged that the database server will be deployed on a separate machine.

The X-serve G5 cluster node is the controller for the internal Grid, accepting and distributing jobs to the other nodes. Ten of the Mac G5 machines are configured on a gigabit switch and are situated in a teaching laboratory. These machines are used as desktop machines and also as remote computing nodes. Two of the Mac G5 machines are used for development and testing of the facilities. The desktop PCs are also configured into he internal Grid and are accessible through the SGE.

Grid Management Software. The requirements for the selection of a suitable Grid middleware application include: ease of setup, maintainence and operation; availability for a heterogeneous computing environment (Mac OS-X and Linux platforms); ability to monitor the available resources at each computing node; load balancing for efficient process distribution across the computing nodes; parallel computation capability; job scheduling and batch tools; and multi-cluster support for future internet based collaboration.

An X-Grid middleware software application (Apple Computer Inc., X-grid beta release 2003) has been tested using the NeuroGrid facilities. The initial testing showed that the X-Grid application has the following disadvantages compared to other middleware applications including the Sun Grid Engine (SGE), Portable Batch System (PBS) and Load Sharing Facility (LSF) middleware applications: inaccurate or non-existent load balancing across computing nodes; remote applications only executable as user "*nobody*" creating file permission problems; applications not executable with options; and applications executed remotely in the *temp* directory, thus creating difficulties with working directories and write permissions in the *temp* directory.

A detailed investigation of alternative Grid management software included: PBS Pro, a commercially available software which is powerful but costly; LSF, a powerful free application, but hard to configure; and SGE, also a powerful but less complex and free software, but with very little support. After evaluation the SGE middleware was selected and has been implemented as an alternative middleware layer, due to its superior performance in the following criteria: ease of installation and initial setup as well as usability; access to the work load list of each active computing node; capability to receive command line jobs; capability to monitor progress of submitted jobs; and the capability to specify both the input and output data directory. The SGE has performed satisfactorily since being configured for use with the NeuroGrid computing hardware.

Neuroimage Database. The key capabilities of the NeuroGrid image database include: the storage of primary and secondary data (raw and processed images); the storage of multiple data types (demographic data, MR images, histology images); and the use of standard neuroimage formats (NIfTI [10], Analyze). The key design features of the image databases include: protection of the confidentiality of subject data through the removal of identifying header information; adoption of secure methods for accessing the system; the ability to remotely access the database for data sharing; adoption of open source database tools (PostgreSQL) and applications; and a remote user data entry and upload capability (via php).

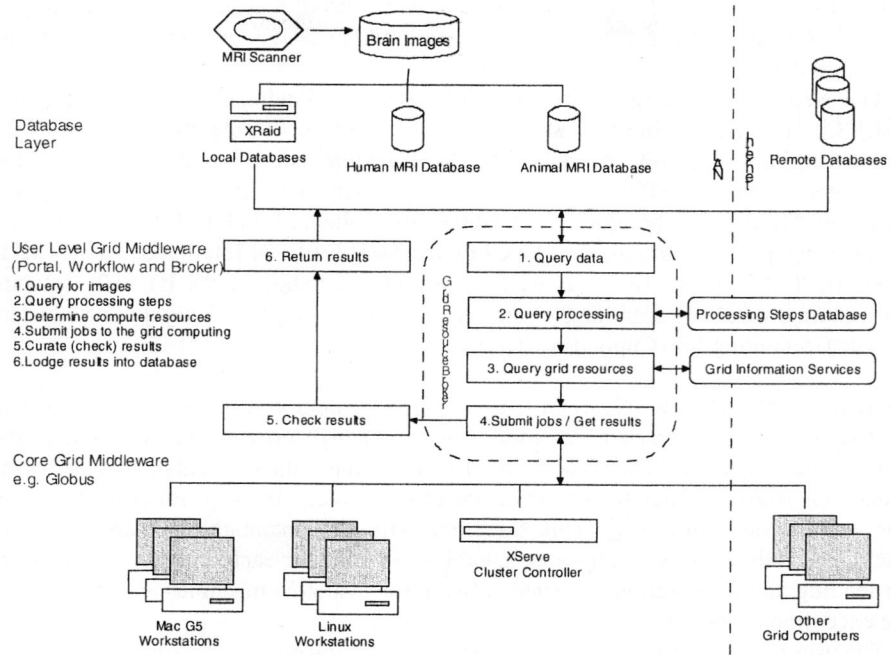

Fig. 1. MRI Grid – Users Schematic View

Further development of the NeuroGrid database includes the creation of tables and tools for data to be re-entered into the database. Specifically, the second phase of the database includes the storage of multiple instances of an image; that is, storage of modified image results from successive processing operations performed on an input image. These image results, together with the details of the processing algorithms used to produce the results, are linked to the input image. Subsequent database queries of the original image will provide an index to processed results of the image. The third phase of the database involves development of a GUI for the semi-automated re-entry of processed image results into the database. A data viewer and data curation GUI will require users to verify the veracity of image analysis results before populating the database with the results.

Neuroimage Analysis Tools. A key objective of the design of NeuroGrid has been to rapidly provide users access to existing well established image analysis tools. This objective has been achieved by re-coding a number of the graphical user interface (GUI) based existing FSL image analysis scripts [11]. The tk/tcl GUIs for the brain extraction tool (BET), linear registration tool (FLIRT) and the automated segmentation tool (FAST) have been re-coded as Java applications. A Java application has been developed to link the output of these applications (after user selection of analysis data sources and processing options) to the SGE middleware. Currently work is aimed at re-coding Java applications of the functional easy analysis tool (FEAT) and FSL diffusion analysis tool (FDT) GUIs for integration with the SGE middleware.

The NeuroGrid software suite also includes the following software applications: FSL 3.2, the native implementation of the tk/tcl tools operating in the X-11 windowing environment; Matlab 6.5 (R13) mathematical modelling tools and associated toolboxes; SPM2, a library of image analysis functions developed by the Functional Imaging Laboratory (FIL, University College, London, UK); Slicer 2.1, a set of image processing tools developed by the MIT Artificial Intelligence Lab, (MIT, Boston); X-Grid Blast (Beta), as well as an implementation of the BLAST software for genomic database searching optimised for use on Apple G5 computing clusters (Apple Computing Inc, Cuppertino, CA).

NeuroGrid Portal. The NeuroGrid user interface, or portal, is being developed to provide a workflow interface for users. The portal will provide direct access to the image database where a user will initially select the data for analysis. A series of processing steps will then be selected from a list of processing algorithms available on the web services directory. This directory will also maintain a database of the machines on the Grid which have executables available for each processing operation. Depending on the operations selected, a list of the available machines is generated for the execution of the job.

The user is also required to select a destination for the output of each step of the job schedule. For intermediate output steps, the user can select to have the job execution halted subject to verification of the intermediate processing result by the user. In these cases the user will typically use an image viewer to view the intermediate result, and choose to continue the job schedule or abort the job. The full NeuroGrid facility will provide an automated means for user to re-enter the processing final results into the originating database, or another customised image database. The selection of the output data destination will require the user to select from a list of possible databases accessible to the portal. The NeuroGrid portal is being developed as a Java application and requires a substantial programming effort.

4 Discussion

Future developments include the parallelizing of codes to gain full advantage of the increased computational power available from the Grid architecture. A number of the FSL processing algorithms are directly parallelizable at the job execution level where multiple datasets are selected with an identical set of processing operations selected for each dataset. These jobs can then be directly shared to multiple machines via the Grid middleware. A Parcellation toolbox of image analysis algorithms has been developed as in-house cortical parcellation codes, based on the Matlab image processing toolboxes. These tools are being coded into c/c++ algorithms for maximally efficient execution speeds, and will be incorporated into the image analysis tools available via NeuroGrid. The analysis of high resolution images (exceeding 250 MB per image) where cortical parcellation image analysis is possible, will require further automation of the image processing protocol developed using the FSL tools BET, FAST, FLIRT and the Parcellation toolbox.

Other future improvements of the computational speeds include the parallelizing of source code for the analysis algorithms, such as a re-code of FSL using the Mac Vector Engine (a native library for the Apple Mac G5 architecture). This could potentially increase the execution speeds by a factor of 5. Furthermore, recompilations of existing tools using the IBM-compiler is also expected to produce significant gains in execution speeds. Future database development plans for the NeuroGrid system include access through the Internet via Java applets and a web portal. Users of the system will be able to retrieve data from external databases and select the desired processing application to apply to the images. These data will use the computing resources on the NeuroGrid to perform the computations.

5 Conclusions

The Australian National Neuroscience Facility (NNF) is providing Australian neuroscientists and commercial organizations with access to networks of laboratories throughout the country. The facility is fostering neuroscience excellence by combining science, technology, innovation, investment, and creativity; and thus jointly providing the opportunity to advance our understanding and treatment of the brain and mind. The Neuroscience Informatics Platform is developing informatics resources for Australian neuroscience researchers, in particular for the neuroimaging research community. A NeuroGrid facility including computational resources and Grid middleware, internet accessible neuroimage databases, and standardised neuroimage analysis tools have been developed. A customised NeuroGrid portal is currently under development that is envisaged will provide the basis for application of Grid computing technologies to other areas of neuroscience research.

Acknowledgements

This research was supported by a Major National Research Facility grant to Neurosciences Victoria, and by the National Neuroscience Facility, Melbourne.

References

1. Amari S-I, et al, "Collaborative Neuroscience: Neuroinformatics for Sharing Data and Tools. A *Communiqué* from the Neuroinformatics Working Group of the Global Science Forum", J of Integrative Neuroscience, 1 (2002) 117-128.

2. Hunt RW, Warfield SK, Wang H, Keane M, Volpe JJ, Inder TE. Assessment of the impact of the removal of cerebrospinal fluid on cerebral tissue volumes by advanced volumetric 3D-MRI in post-hemorrhagic hydrocephalus in a premature infant. J Neurol Neurosurg Psychiatry 74 (2003) 658-660
3. Pantelis C, Velakoulis D, McGorry PD, Wood SJ, Suckling J, Phillips LJ, Yung AR, Bullmore ET, Brewer W, Soulsby B, Desmond P, McGuire P. "Neuroanatomical abnormalities before and after onset of psychosis: a cross-sectional and longitudinal MRI comparison", Lancet 361 (2003) 270-1.
4. www.nbirn.net/Publications
5. Mazziotta JC, et al. Atlases of the human brain. In, Neuroinformatics: An overview of the Human Brain Project, (S.H. Koslow & M.F. Huerta, eds.). Lawrence Erlbaum Associates, (1997) Washington.
6. Koslow, SH. Commentary: Should the neuroscience community make a paradigm shift to sharing primary data? Nature Neuroscience 3 (2000) 863-865.
7. Toga AW. Neuroimage databases: the good, the bad and the ugly. Nature Reviews Neuroscience 3 (2002) 302-309.
8. Buyya R & Venugpoal S. The Gridbus toolkit for service oriented grid and utility computing: an overview and status report. Proceedings of the First IEEE International Workshop on Grid Economics and Business Models (GECON 2004, April 23, 2004, Seoul, Korea), 19-36pp, ISBN 0-7803-8525-X, IEEE Press, New Jersey, USA.
9. Buyya R, Date S, Mizuno-Matsumoto Y, Venugopal S, & Abramson D. Neuroscience instrumentation and distributed analysis of brain activity data: A case for eScience on global Grids, Journal of Concurrency and Computation: Practice and Experience, Wiley Press, (2004) USA *in press*.
10. For information on the NifTI format see http://nifti.nimh.nih.gov
11. For FSL tools see www.fsl.fmrib.oxford.ac.uk

Current Status and Future Research Activities in Clinical Neuroinformatics: Singaporean Perspective

Wieslaw L. Nowinski

Biomedical Imaging Lab; Agency for Science, Technology and Research; Singapore
30 Biopolis Street, #07-01 Matrix, Singapore 138671
`wieslaw@bii.a-star.edu.sg`

Abstract. The Biomedical Imaging Lab in Singapore has been involved in neuroinformatics research for more than a decade. We are focused on clinical neuroinformatics, developing suitable models, tools, and databases. We report here our work on construction of anatomical, vascular, and functional brain atlases as well as development of atlas-assisted neuroscience education, research, and clinical applications. We also present future research activities.

Keywords: Brain atlas, functional neurosurgery, human brain mapping, neuroscience education.

1 Introduction

Singapore is a vibrant research hub, particularly in biomedical sciences. The research efforts in brain sciences and neural networks in Singapore have been summarized recently in an excellent overview by Rajapakse et al [35]. In this paper we are focusing on activities in clinical neuroinformatics being done in the Biomedical Imaging Lab (BIL), Singapore for more than a decade.

The three principal aims of neuroinformatics are models, tools, and databases. BIL activities are aligned with these aims. For instance, we have developed a *Cerefy* brain atlas database which has become a standard in image guided neurosurgery [9,16]. A public domain tool the *Cerefy Neuroradiology Atlas* [13], developed for neuroscience researchers and neuroradiologists, has more than 1,100 users.

BIL major efforts are focused on clinical neuroinformatics. By "clinical neuroinformatics" we mean applying of neuroinformatics methods and tools to address clinical problems related to diagnosis and therapy of the human brain.

This work provides an overview of anatomical, functional, and vascular brain atlases and atlas-assisted education, research and clinical applications developed in BIL.

2 Cerefy Human Brain Atlases

For more than one decade we have developed a family of brain atlases called *Cerefy* atlases. This brain atlas family contains anatomical, vascular, and functional atlases. Recently we have been working on a blood supply territories atlas.

A. *Anatomical atlases*

The anatomical part of the *Cerefy* electronic brain atlas database [9], [18], [28] contains four complementary atlases with gross anatomy, subcortical structures, brain connections, and sulcal patterns. These highly enhanced electronic atlases were derived from the classic print brain atlases edited by Thieme:

1) *Atlas for Stereotaxy of the Human Brain* by Schaltenbrand and Wahren (SW) [37];
2) *Co-Planar Stereotactic Atlas of the Human Brain* by Talairach and Tournoux (TT) [39];
3) *Atlas of the Cerebral Sulci* by Ono, Kubik, and Abernathey [32];
4) *Referentially Oriented Cerebral MRI Anatomy: Atlas of Stereotaxic Anatomical Correlations for Gray and White Matter* by Talairach and Tournoux [40].

These atlases are available in 2D and 3D, and they are mutually co-registered. The anatomical index has about 1,000 structures per hemisphere and more than 400 sulcal patterns. The construction, content, features as well as atlas enhancements and extensions were addressed in detail in our previous papers [9], [18], [28]. The *Cerefy* brain atlas database is suitable for neurosurgery [8], [16], [26], [27], [33]; neuroradiology [7], [28]; human brain mapping [6], [24]; and neuroscience education [23].

To facilitate development of atlas-assisted applications, we have developed two add-on brain atlas libraries, the *Electronic Brain Atlas Library* and *Brain Atlas Geometrical Models*. The *Electronic Brain Atlas Library* [16] comprises the brain atlas database with the SW and TT atlas images, and a browser. The browser provides means for exploring and understanding of the atlas images as well as facilitates building user's own applications. The *Brain Atlas Geometrical Models* [16] is a library with the atlases in contour and polygonal representations. It contains the brain atlas database and a viewer. The database comprises the SW atlas in contour representation and 3D polygonal models of the SW and TT atlases. The detailed specifications of both libraries are at www.cerefy.com .

B. *Vascular atlas*

We have constructed a cerebrovascular atlas from angiography data. A time-of-flight acquisition was done on a 3T MRI scanner. The 3D vascular model was derived from angiographic data in the following steps: 1) vasculature segmentation; 2) extraction of the centerline and radius; 3) centerline editing; 4) centerline smoothing; 5) radius processing; 6) modeling of vascular segments and bifurcations; and 7) labeling of the vascular segments [10]. An application developed allows the user to manipulate the vascular model and get 3D labels. In addition, the vascular atlas is combined with the 3D TT atlas [10].

C. *Functional atlases*

From a functional neurosurgery viewpoint, the main limitations of the *Cerefy* anatomical atlases are sparse image material derived from a few specimens only and 3D inconsistency. More importantly, these atlases are anatomical while the stereotactic target structures are functional. To overcome these limitations, we have developed the probabilistic functional atlas (PFA) generated from pre-, intra-, and post-operative data collected during the surgical treatment of Parkinson's disease patients [15]. So far, we have constructed the PFA for the subthalamic nucleus [11] and ventrointermediate nucleus of the thalamus [12].

3 Brain Atlases in Neuroscience Education

The *Cerefy Atlas of Brain Anatomy* (*CABA*) [23] is a user-friendly application for medical students, residents, and teachers. It is also helpful for neuroinformatics researchers who need some basic neuroanatomical background in their work. The *CABA* contains MRI and atlas images of gross anatomy as well as related textual materials. It also provides testing and scoring capabilities for exam preparation. Its novelty includes:

- atlas-assisted localization of cerebral structures on radiological images;
- atlas-assisted interactive labeling on axial, coronal, and sagittal planes;
- atlas-assisted testing against location and name of cerebral structures;
- saving of the labeled images suitable for preparing teaching materials.

Our recent development includes a Chinese version of the *CABA* [32].

4 Brain Atlases in Research

The usefulness of electronic brain atlases in medical research is growing, particularly in medical image analysis and human brain mapping.

A. Medical image analysis
The most commonly used system for medical image analysis is *ANALYZE* from Mayo [36]: a powerful, comprehensive visualization tool for multi-dimensional display, processing, and analysis of biomedical images from multiple imaging modalities. To facilitate analysis of brain images, the *Cerefy* brain atlas has been integrated with *ANALYZE*.

B. Human brain mapping
Though there are numerous software packages for functional image generation [3], [24], none of them provides atlas-assisted analysis of functional images. The *Brain Atlas for Functional Imaging* (*BAFI*) [24] and *BrainMap* [2] facilitate labeling of functional images (generated by other tools) by means of the TT brain atlas. The *BAFI*, as opposed to *BrainMap*, provides direct access to the TT atlas and the user can display the activation regions superimposed on the atlas, and subsequently place and edit the marks corresponding to the activation regions. Besides labeling, the *BAFI* supports localization analysis of functional images [6]. None of the existing software packages for functional image generation contains the TT atlas, which is the gold standard in human brain mapping research. The *BAFI* also provides numerous functions such as fast data normalization, readout of Talairach coordinates, and data-atlas display. It has several unique features including interactive warping facilitating fine tuning of the data-to-atlas fit, multi-atlas multi-label labeling, navigation on the triplanar formed by the data and the atlas, multiple-images-in-one display with atlas-

anatomy-function blending, loci editing in terms of content and placement, reading and saving of the loci list, and fast locus-controlled generation of results [31].

5 Brain Atlases in Clinical Applications

Electronic brain atlases are prevalent in functional neurosurgery and the interest in them is growing in neuroradiology for computer-aided diagnosis. Pioneering brain atlases in neuroradiological practice is much more difficult than introducing them to functional neurosurgery (as the neurosurgeons were familiar with the print atlases), though the potential they offer and radiology trends will cause atlases to be eventually useful in neuroradiology as well.

A. Computer-aided diagnosis

At present, two major trends are observed in radiology: 1) the number of scans to be read is growing three times faster the number of radiologists, and 2) about 65% of the world is radiologically void while another 30% has only basic technologies. Therefore, new ways for speeding up and facilitating scan interpretation as well as reducing the learning curve of radiologists are needed, and model-enhanced radiology may be one of solutions. We addressed atlas advantages in neuroradiology in [28]; namely, the atlas can potentially:

- reduce time in image interpretation by providing interactive labeling, triplanar display, higher parcellation than the scan itself, multi-modal fusion and displaying the underlying anatomy for functional and molecular images;
- facilitate communication about interpreted scans from the neuroradiologist to other clinicians (neurosurgeons, neurologists, general practitioners) and medical students;
- increase the neuroradiologist's confidence in terms of anatomy and spatial relationships by providing scan labeling on the orthogonal planes;
- reduce the cost by providing information from mutually co-registered multiple atlases which otherwise has to be acquired from other modalities;
- reduce time in learning neuroanatomy and scan interpretation by providing 3D and triplanar displays and labeling of multi-modal scans.

Towards realizing this potential, we have developed the *Cerefy Neuroradiology Atlas (CNA)* [13]. To our best knowledge, at present the *CNA* is the only model-enhanced application for neuroradiology. It is web-enabled, public domain (available from www.cerefy.com) with more than 1,100 users registered. The *CNA* assists in speeding up scan interpretation by rapid labeling of morphological and/or functional scans, displaying underlying anatomy for functional and molecular studies, and facilitating multi-modal fusion. The labeled and annotated (with text and/or graphics) scan can be saved in Dicom and/or XML formats, allowing for storing the atlas-enhanced scan in a PACS and to use it in other web-enabled applications. In this way, the scan interpretation done by the neuroradiologist can easily be communicated to other clinicians and medical students.

While the *CNA* employs interactive atlas-to-data registration, our recent tool [19] provides fully automatic warping of the atlas into morphological images in about 5 seconds. For atlas warping, we use the Talairach transformation [39] which scales the *Cerefy* atlas [25] piecewise linearly based on point landmarks. Their automatic, fast, robust and accurate identification in the data is the core of our approach. We use the modified Talairach landmarks [30]: anterior commissure (AC) and posterior commissure (PC) located on the midsagittal plane (MSP), and 6 cortical landmarks determining the extent of brain. Three component algorithms calculate these landmarks completely automatically based on anatomy and radiologic properties of the scan. The identification of the MSP is detailed in [4], that of AC and PC in [1], and calculation of cortical landmarks in [5].

We have recently extended the above solution to morphological images for atlas-assisted interpretation of stoke images. Magnetic resonance (MR) diffusion and perfusion are key modalities for interpretation of stroke images. They provide information both about the infarcted region and that at risk with high sensitivity and specificity. Their major limitations are the lack of underlying anatomy and blood supply territories as well as low spatial resolution (for instance, a typical slice thickness of diffusion images is 5 mm while that of perfusion is 7.5 mm). We have developed a fast algorithm for overlapping the anatomical atlas as well as an atlas of blood supply territories on MR perfusion and diffusion images [20]. In addition, our solution allows for a simultaneous display of the atlas, diffusion image and one the selected perfusion maps (cerebral blood flow (CBF), cerebral blood volume (CBV), mean transit time (MTT), time to peak (TTP) or peak height (PKHT)).

Another atlas-assisted solution is for analysis of molecular images. The key limitation of PET imaging is the lack of underlying anatomy. A combined PET-CT overcomes this shortcoming, making it one of the fastest growing modality. However, PET-CT scanners are expensive. We have developed a more affordable solution by getting the underlying anatomy from the anatomical atlas warped non-linearly onto a PET scan [20].

B. Stereotactic and functional neurosurgery
The anatomical *Cerefy* brain atlas database [9], [18], [28] has become the standard in stereotactic and functional neurosurgery. It is integrated with major image guided surgery systems including the *StealthStation* (Medtronic/Sofamor-Danek), *Target* (BrainLab), *SurgiPlan* (Elekta) and integration with the *Gamma Knife* is in process, *SNN 3 Image Guided Surgery System* (Surgical Navigation Network), a neurosurgical robot *NeuroMate* (Integrated Surgical Systems/IMMI), and the system of Z-kat.

We have developed several atlas-assisted tools suitable for stereotactic and functional neurosurgery planning. *The Electronic Clinical Brain Atlas* on CD-ROM [17] offers probably the simplest atlas-assisted planning. It generates individualized atlases without loading patient-specific data by applying 2D landmark-based warping; a planning procedure by using this CD-ROM is given in [27].

The *NeuroPlanner* is for preoperative planning and training, intraoperative procedures, and postoperative follow-up [26]. It contains all, mutually co-registered atlases from the anatomical *Cerefy* brain atlas database including their 3D extensions [18]. The *NeuroPlanner* provides four groups of functions: data-related (interpolation, reformatting, image processing); atlas-related (atlas-to-data interactive 3D warping, 2D and 3D interactive multiple labeling); atlas-data exploration-related (interaction in three orthogonal and one 3D views, continuous data-atlas exploration); and neurosurgery-related (targeting, path planning, mensuration, electrode insertion simulation, therapeutic lesioning simulation).

The *BrainBench* [38] is a virtual reality-based neurosurgical planning system with a suite of neurosurgery supporting tools and the 3D TT atlas. It provides 3D, two-hand interaction with the data, atlas, electrode, and stereotactic frame.

The advantages of using the *Cerefy* brain atlases for stereotactic and functional neurosurgery are summarized in [9].

The probabilistic functional atlas (PFA) opens new avenues. A dedicated application is developed for a combined anatomical-functional planning of functional neurosurgery [22]. Moreover, this application provides intraoperative support and also serves as a personal archive. The construction of PFA and development of PFA-based applications are behind two main conceptual breakthroughs. As the PFA is dynamic and can be updated with new cases, knowledge from the currently operated cases can be saved and accumulated continuously. Moreover, the atlas can be built and extended by the community over the internet by using a public domain portal [14], changing in this way the paradigm from the manufacturer-centric to community-centric.

6 Future Directions

We are continuously enhancing our electronic versions of the stereotactic atlases from content, 3D consistency and quality standpoints [33]. We are also in process of interpolating these atlases, which will increase their applicability. An atlas of blood supply territories is important, particularly for atlas-assisted processing of stroke images. We have recently demonstrated its initial version [20]. We are also in process of constructing better brain atlases.

Development of more powerful tools and user friendly applications is another direction of our future research. We have recently developed the *CNA* ver 2.2 with many new features [41], and a new *CNA* ver 3.0, even more powerful, will be ready next year. A web-enabled atlas of cerebrovascular variants is scheduled for this year. Conceptually is it similar to our PFA-based portal for functional neurosurgery. The second version of the *CABA* is under development. It will be extended to 3D cerebral structures, blood supply territories atlas, and will combine the 3D structures with the orthogonal planes to better appreciate tomographic-spatial relationships. Atlas-assisted interpretation of morphological images will be extended from MR to CT images as well; we will also provide a fast non-linear atlas-to-data warping. We will continue enhancing the applications for atlas-assisted interpretation of stroke and molecular images, making them more accurate and even faster. Finally several algorithms developed by us for segmentation of neuroimages will be extended and applied for brain morphometry.

Acknowledgment

The key contributors to the atlases and applications described here include A Thirunavuukarasuu, D Belov, A Fang, A Ananthasubramaniam, G Qian, GL Yang, QM Hu, KN Bhanu Prakash, I Volkau, and L Serra, among many others.

This research is funded by the Biomedical Research Council; Agency for Science, Technology and Research; Singapore.

References

1. Bhanu Prakash, KN., Hu, Q., Volkau, I., Aziz, A., Nowinski, WL.: Rapid and Automatic Localization of the Anterior and Posterior Commissure Point Landmarks in MR Volumetric Neuroimages. Hum Brain Mapp (submitted)
2. Fox, PT., Mikiten, S., Davis, G., Lancaster, JL.: BrainMap: A Database of Human Functional Brain Mapping. In: Thatcher RW., Hallett, M., Zeffiro, T., John, ER., Huerta, M. (eds.): Functional Neuroimaging. Technical Foundations (1994) 95-106
3. Gold, S., Christian, B., Arndt, S., Zeien, G., Cizadlo, T., Johnson, DL., Flaum, M., Andreasen, NC.: Functional MRI Statistical Software Packages: A Comparative Analysis. Hum Brain Mapp 6 (1998) 73-84
4. Hu, Q., Nowinski, WL.: A Rapid Algorithm for Robust and Automatic Extraction of the Midsagittal Plane of the Human Cerebrum from Neuroimages Based on Local Symmetry and Outlier Removal. Neuroimage 20(4) (2003) 2154-2166
5. Hu, Q., Qian, G., Nowinski, WL.: Fast, Accurate and Automatic Extraction of the Modified Talairach Cortical Landmarks from MR Images. Magn Reson Med (2005) 53
6. Nowinski, WL., Thirunavuukarasuu, A.: Atlas-Assisted Localization Analysis of Functional Images. Med Image Anal, Sep 5(3) (2001) 207-220
7. Nowinski, WL., Thirunavuukarasuu, A.: Electronic Atlases Show Value in Brain Studies. Diagn Imaging (Asia Pacific) 8(2) (2001) 35-39
8. Nowinski, WL. : Anatomical Targeting in Functional Neurosurgery by the Simultaneous Use of Multiple Schaltenbrand-Wahren Brain Atlas Microseries. Stereotact Funct Neurosurg 71 (1998) 103-116
9. Nowinski, WL. : Computerized Brain Atlases for Surgery of Movement Disorders. Seminars in Neurosurgery 12(2) (2001) 183-194
10. Nowinski, WL., Thirunavuukarasuu, A., Volkau, I., Baimuratov, R., Hu, Q., Aziz, A., Huang, S.: Three-Dimensional Brain Atlas of Anatomy and Vasculature. Radiographics 25(1) (2005) 263-271
11. Nowinski, WL., Belov, D., Pollak, P., Benabid, AL.: A Probabilistic Functional Atlas of the Human Subthalamic Nucleus. Neuroinformatics 2(4) (2004) 381-98
12. Nowinski, WL., Belov, D., Thirunavuukarasuu, A., Benabid, AL.: A Probabilistic Functional Atlas of the VIM Nucleus. Scientific Program on CD, Poster presentation abstracts, American Association of Neurological Surgeons AANS 2005, 16-21 April 2005, New Orleans, Louisiana, USA (2005)
13. Nowinski, WL., Belov, D.: The Cerefy Neuroradiology Atlas: A Talairach-Tournoux Atlas-Based Tool for Analysis of Neuroimages. Neuroimage 20(1) (2003) 50-57 (available from www.cerefy.com)
14. Nowinski, WL., Belov, D., Benabid, AL.: A Community-Centric Internet Portal for Stereotactic and Functional Neurosurgery with a Probabilistic Functional Atlas. Stereotact Funct Neurosurg 79 (2002) 1-12
15. Nowinski, WL., Belov, D., Benabid, AL.: An Algorithm for Rapid Calculation of a Probabilistic Functional Atlas of Subcortical Structures from Electrophysiological Data Collected During Functional Neurosurgery Procedures. Neuroimage 18(1) (2003) 143-155

16. Nowinski, WL., Benabid, AL.: New Directions in Atlas-Assisted Stereotactic Functional Neurosurgery. In: Germano, IM. (ed.): Advanced Techniques in Image-Guided Brain and Spine Surgery Thieme, New York (2001)
17. Nowinski, WL., Bryan, RN., Raghavan, R.: The Electronic Clinical Brain Atlas. Multiplanar Navigation of the Human Brain. Thieme, New York – Stuttgart (1997)
18. Nowinski, WL., Fang, A., Nguyen, BT., Raphel, JK., Jagannathan, L., Raghavan, R., Bryan, RN., Miller, G.: Multiple Brain Atlas Database and Atlas-Based Neuroimaging System. Comput Aided Surg 2(1) (1997) 42-66
19. Nowinski, WL., Hu, Q., Bhanu Prakash, KN., Qian, G., Thirunavuukarasuu, A., Aziz, A.: Automatic Interpretation of Normal Brain Scans. The Radiological Society of North America, 90th Scientific Assembly & Annual Meeting Program; Chicago, Illinois, USA, (2004) 710
20. Nowinski, WL., Hu, Q., Bhanu Prakash, KN., Volkau, I., Qian, G., Thirunavuukarasuu, A., Liu, J., Aziz, A., Baimouratov, R., Hou, Z., Huang, S., Luo, S., Minoshima, S., Runge, V., Beauchamp, N.: Atlas-Assisted Analysis of Brain Scans. Book of Abstracts of European Congress of Radiology ECR 2005, Eur Radiol, Suppl 1 to Vol 15, (2005) 572
21. Nowinski, WL.: Co-Registration of the Schaltenbrand-Wahren Microseries with the Probabilistic Functional Atlas. Stereotact Funct Neurosurg 82 (2004) 142-146
22. Nowinski, WL., Thirunavuukarasuu, A., Benabid, AL.: The Cerefy Clinical Brain Atlas. Extended Edition with Surgery Planning and Intraoperative Support. Thieme, New York (2005)
23. Nowinski, WL., Thirunavuukarasuu, A., Bryan, RN. : The Cerefy Atlas of Brain Anatomy. An Interactive Reference Tool for Students, Teachers, and Researchers. Thieme, New York – Stuttgart (2002)
24. Nowinski, WL., Thirunavuukarasuu, A., Kennedy, DN.: Brain Atlas for Functional Imaging. Clinical and Research Applications. Thieme, New York – Stuttgart (2000)
25. Nowinski, WL., Thirunavuukarasuu, A.: The Cerefy Clinical Brain Atlas on CD-ROM. Thieme, New York – Stuttgart (2004)
26. Nowinski, WL., Yang, GL., Yeo, TT.: Computer-Aided Stereotactic Functional Neurosurgery Enhanced by the Use of the Multiple Brain Atlas Database. IEEE Trans Med Imaging 19 (1) (2000) 62-69
27. Nowinski, WL., Yeo, TT., Thirunavuukarasuu, A.: Microelectrode-Guided Functional Neurosurgery Assisted by Electronic Clinical Brain Atlas CD-ROM. Comput Aided Surg 3(3) (1998) 115-122
28. Nowinski, WL.: Electronic Brain Atlases: Features And Applications. In: Caramella, D., Bartolozzi, C. (eds.): 3D Image Processing: Techniques and Clinical Applications. Med Radiol series, Springer-Verlag (2002)
29. Nowinski, WL.: Model Enhanced Neuroimaging: Clinical, Research, and Educational Applications. In: Yearbook of Medical Informatics (2002) 132-144
30. Nowinski, WL.: Modified Talairach Landmarks. Acta Neurochir, 143(10) (2001) 1045-1057
31. Nowinski, WL., Thirunavuukarasuu, A.: A Locus-Driven Mechanism for Rapid and Automated Atlas-Assisted Analysis of Functional Images by Using the Brain Atlas for Functional Imaging. Neurosurg Focus Jul. 15(1) Article 3 (2003)
32. Nowinski, WL., Thirunavuukarasuu, A., Fu, Y., Ma, X., Lin, Z., Wang, S.: The Cerefy Clinical Brain Atlas. Chinese Edition. Harbin Institute of Technology, China (2005)
33. Nowinski, WL.: The Cerefy Brain Atlases: Continuous Enhancement of the Electronic Talairach-Tournoux Brain Atlas. Neuroinformatics (2005) (in press)

34. Ono, M., Kubik, S., Abernathey, CD.: Atlas of the Cerebral Sulci. Georg Thieme Verlag/Thieme Medical Publishers, Stuttgart - New York (1990)
35. Rajapakse, JC., Srinivasan, D., Er ,MJ., Huang, GB., Wang, L.: Excerpts of Research in Brain Sciences and Neural Networks in Singapore. Proc. International Joint Conference on Neural Networks IJCNN 2004, Budapest Hungary (2004) 369-375
36. Robb, RA., Hanson, DP.: The ANALYZE Software System for Visualization and Analysis in Surgery Simulation. In: Lavallee, S., Tayor, R., Burdea, G., Mosges, R. (eds): Computer Integrated Surgery. Cambridge MA, MIT Press (1995) 175-190
37. Schaltenbrand, G., Wahren, W.: Atlas for Stereotaxy of the Human Brain. Georg Thieme Verlag, Stuttgart (1977)
38. Serra, L., Nowinski, WL., Poston, T. et al : The Brain Bench: Virtual Tools for Stereotactic Frame Neurosurgery. Med Image Anal 1(4) (1997) 317-329
39. Talairach, J., Tournoux, P.: Co-Planar Stereotactic Atlas of the Human Brain. Georg Thieme Verlag/Thieme Medical Publishers, Stuttgart - New York (1988)
40. Talairach, J., Tournoux, P.: Referentially Oriented Cerebral MRI Anatomy. Atlas of Stereotaxic Anatomical Correlations for Gray and White Matter. Georg Thieme Verlag/Thieme Medical Publishers, Stuttgart - New York (1993)
41. www.cerefy.com

Japanese Neuroinformatics Research: Current Status and Future Research Program of J-Node

Shiro Usui

RIKEN, Brain Science Institute,
2-1 Hirosawa, Wako, Saitama 351-0198 Japan
usuishiro@riken.jp

Abstract. There is a global trend to bring together research resources of the brain in the hope that these collaborations will provide critical information to the understanding of the brain as a system and its functions. Japan, among several countries, is committed to actively participating in this process with the hope that millions of people will greatly benefit from this activity. Currently, we are formulating plans and strategies in order to carry out this objective.. This paper will discuss perspectives of the Japanese Neuroinformatics Node.

1 Introduction

With the advent of the information era, there is a growing trend of global cooperation among communities around the world to tackle issues confronting human society. One of these very important challenges of the 21st century is the study of the human brain. The task of understanding a functional brain system is hindered by the inevitable necessity of tight focus and specialization of researchers in the field. This fragmentation makes the synthesis and integration of disparate lines of evidence exceptionally difficult. In order to address this difficulty, an organized framework is needed that facilitates integration and provides a fertile ground for sharing information. This agenda requires the establishment of a new discipline, aptly named "Neuroinformatics". Neuroinformatics undertakes the challenge of developing the mathematical models, databases, data analyses, and tools necessary for establishing such a framework.

The major emphasis of neuroinformatics is the organization of neuroscience data and knowledge-bases to facilitate the development of computational models and tools. An additional aim is to promote international interdisciplinary cooperation. This becomes especially important with regard to the emerging realization that understanding and developing models of brain processes of one functional area can be significantly facilitated by knowledge of processes in different functional areas. These efforts to integrate the diverse methodologies of neuroscience, if properly carried out, will assist in improving the utility and availability of the vast quantities of high quality data, models, and tools being developed by neuroscience researchers. In turn, this will result in further advancement of scientific research in many disciplines, stimulate promotion of technological and sustainable development, and facilitates the equitable sharing of high quality databases in the brain sciences.

The necessity for the framework to foster international collaboration and sharing of research data was recognized by many countries leading to the establishment of the INCF (International Neuroinformatics Coordinating Facility) under the auspices of the GSF (Global Science Forum) of the OECD (Office for Economic Cooperation and Development) [1]. In the USA, the first 10 years of the Human Brain project has finished in 2004 and the major highlights of their activities were summarized in [2]. They are now starting the next decade of the project building upon their successes and further developments. In Germany, their Federal Ministry of Education has started the initiative of creating a National Network for Computational Neuroscience with the primary aim of understanding cognitive functions through Computational Neuroscience. To carry out this activity, they have established the Bernstein Centers for Computational Neuroscience in four major sites (Berlin, Munich, Gottingen, Freiburg) collaborating and sharing data, computer models, theories, and approaches [3].

2 NRV Project and Visiome Platform

In Japan, Neuroinformatics Research in Vision (**NRV**) is a pioneering project initiated in 1999 under the auspices of Strategic Promotion System for Brain Science of the Special Coordination Funds for Promoting Science and Technology at the Science and Technology Agency (now under the Ministry of Education, Culture, Sports, Science and Technology). The primary aim of NRV is to build the foundation of neuroinformatics research in Japan. NRV's top priority is the promotion of experimental, theoretical, and technical activities related to vision research.

The first goal of the NRV project is the construction of mathematical models for each level of the visual system: single neuron; retinal neural circuit; and higher visual function. The second goal is to build integrated resources for neuroinformatics by utilizing information science technologies within the research support environment that we have named the "**Visiome Platform**" (VP). The third goal is to develop new vision devices based on brain-derived information processing principles. There are five major groups carrying out specific activities summarized below [4]:

G1: Construction of mathematical models of single neurons
G2: Realization of virtual retina based on retinal physiology
G3: Study on the visual function by computational and systems' approaches
G4: Realization of artificial vision devices and utilization of silicon technology
   for recording and stimulation
G5: Fundamental neuroinformatics research and development

The VP (Figure 1) recently became available for public access (http://platform.visiome.org/) in a test mode (May 2004). The VP commencement was one of the major highlights of the NRV project that was recently completed [5]. The platform is designed as a research resource archive that can be accessed from the Internet and provides published references, articles, reusable programs/scripts of mathematical models, experimental data, analytical tools, and many other resources.

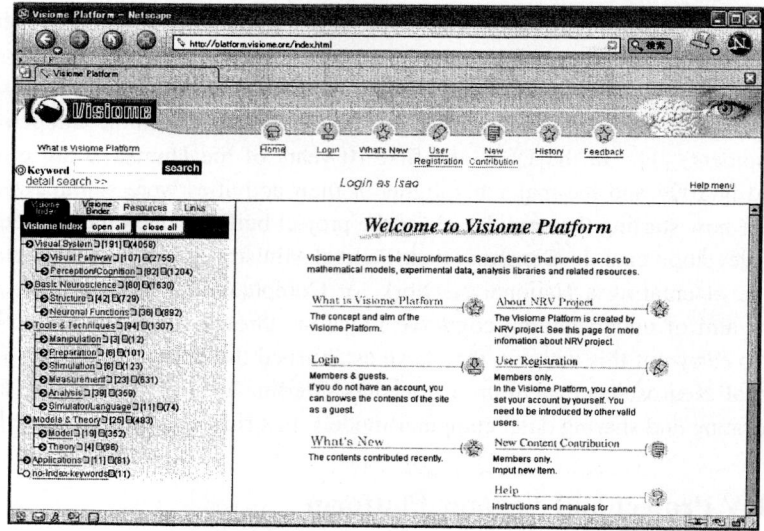

Fig. 1. The top page of the Visiome Platform

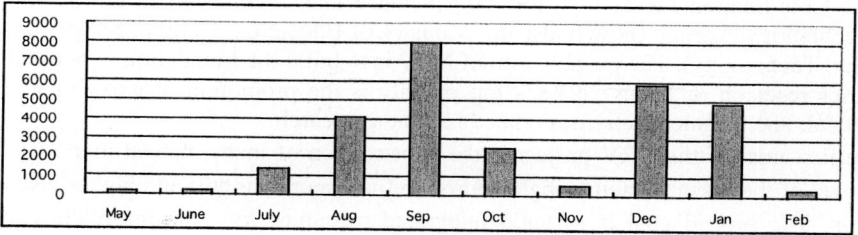

(a) VP Monthly Access Statistics

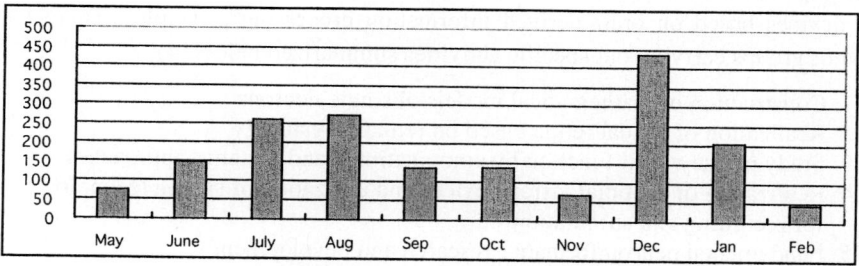

(b) VP Monthly Downloaded Statistics

Fig. 2. Monthly access statistics and downloaded statistics for VP

The platform enables researchers to understand how the published models work or compare their own results with other experimental data. It also allows users to improve existing models by making it easier for them to integrate their new hypotheses into the existing models. Moreover, users can export/import their own models, data, and tools to the database to be shared with other users and colleagues.

An indexed database of literature references, codes or models used in papers registered in VP are easily accessible in a tree structure index. In general, VP helps researchers hasten the process of understanding the visual system from the perspective of visual functions and assists them in the construction of models based on their own hypotheses.

Currently, there are 8 major types of contents in VP, namely: binder, reference, book, model, url, data, tool, and stimulus. As of March 2005, VP contains a total of roughly 3000 registered items. Figures 2(a) and 2(b) show the VP monthly access and downloaded statistics of users during its public test operation. This data excludes statistics containing robots and search agents access to our site. By analyzing this trend, it provides us insights on how to carry out and manage effectively future projects of J-node. Also, more effective policies were formulated based on our experience with this initial phase of VP operation.

3 Perspectives of the Japan Node from NRV Experience

Based on our NRV project experience, we can utilize and extend the basic scheme of the neuroinformatics platform to any other possible research areas. Figure 3 shows a conceptual framework describing what shall constitute the major components of the Japanese Node. We have tentatively identified 10 major platforms, namely:

- Visiome Platform (VP)
- Cerebellar Development Transcriptome Data Base (CDT-DB)
- Neuron/Glia Platform
- Integrative Brain Project
- Invertebrate Brain Platform
- Brain Imaging Platform
- Brain Machine Interface Platform
- Visiome for visual psychophysics
- Clinical Neruroinformatics Platform
- BSI Neuroinformatics Platform

While VP has been accessible to the public for more than a year, CDT-DB will start its operation in late 2005. The Neuron/Glia Platform will be implemented by the National Project for neuron-glia Interactions. The Integrative Brain Project will commence in 2005 while the preliminary implementation for the Invertebrate Brain Platform has already commenced. The implementation details for other platforms are still subject to further discussions.

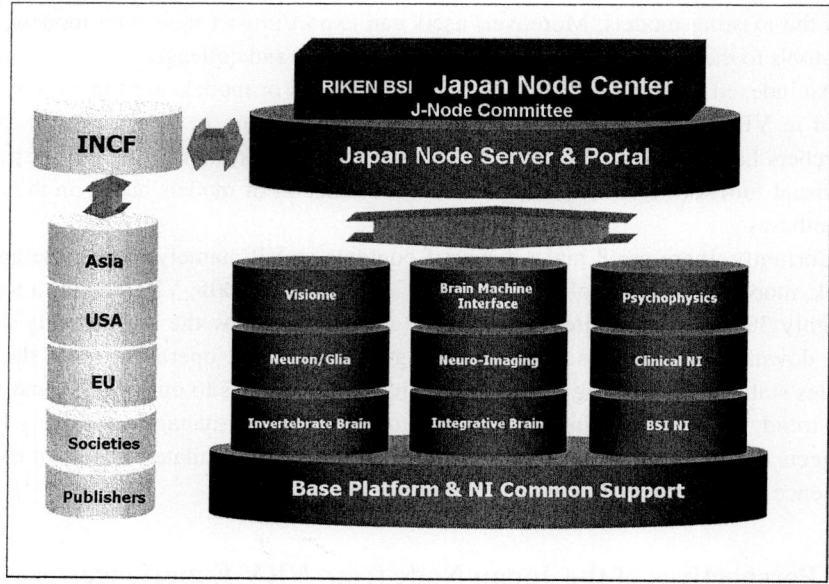

Fig. 3. Conceptual scheme of Japanese node

Our experience in the VP implementation led us to believe that the development of the Japan Node infrastructure requires proper coordination at different organizational levels to ensure standardization of rules for sharing, coherent tools to avoid redundancy and waste of resources, and appropriate guidelines to spell-out a specific target and action plan. One important action plan is the development of core technologies that will serve as the basis of development to ease standardization and integration of data and other resources. Although each platform addresses different areas of specialization, the underlying technology for basic functions such as data storage, retrieval, searching, visualization, statistical analysis, modeling/simulation and other forms of information extraction are similar.

Aside from the basic features such as indexing of research resources, each platform will be more useful if it can provide peer-reviewed sections where authors can directly publish reports on new data. This will pave the way for the standardization of datasets that makes testing of new models more straightforward. More importantly, the future of the Japanese Node lies on the active involvement of its members as well as in the participation of the research communities since the data they provide form the lifeblood of the system. There must be an active promotion of the importance of data sharing and appropriate reward mechanisms to encourage researchers to participate in this endeavor. Relevant government and non-government agencies, journal publication firms, education and research institutes must be encouraged to participate in this project because their influence will enhance the quality of contributions and support to the funding, research, and development of the neuroinformatics resources.

4 Conclusion

Our experiences in NRV and VP provide us with a great opportunity and incentive to discuss issues and to share technical know-how with other groups undertaking neuroinformatics activities to meet the goals mandated by the INCF Committee. Based on our expertise and the results of consultations and surveys we conducted, we have identified the following areas for the initial phase in the establishment of Japanese Node:

- Identify major fields of neuroscience to become development platforms
- Identify potential organizations to work as partners in the building process
- Identify ways to make the development and operation sustainable
- Development of standards for common ontology, data interchange, interoperability, etc.

The fulfillment of knowing that the endeavor is a significant way for the society to continue its quest of understanding the brain and eventually creating the virtual brain [6].

Acknowledgement. The author wishes to thank Drs. Masao Ito and Shun-ichi Amari, the members of Neuroinformatics Laboratory at RIKEN BSI, and NRV project members for their support and collaborations.

References

1. http://www.neuroinformatics.nl/
2. http://www.nimh.nih.gov/neuroinformatics/annmeet2004.cfm
   http://videocast.nih.gov/PastEvents.asp?c=1
3. http://www.bernstein-centers.de/en/index.php
4. Usui S.: Visiome: Neuroinformatics Research in Vision Project. Neural Networks, Special Issue on Neuroinformatics: 16 (9), 1293-1300, 2003.
5. Usui S., et al.: Visiome environment: enterprise solution for neuroinformatics in vision science, Neurocomputing, 58-60, 1097-1101, 2004.
6. International Workshop on Neuroinformatics - for Establishing the Japanese Neuro informatics Node, December 1, 2004, RIKEN Japan

Optimal TDMA Frame Scheduling in Broadcasting Packet Radio Networks Using a Gradual Noisy Chaotic Neural Network

Haixiang Shi[1] and Lipo Wang[1,2]

[1] School of Electrical and Electronic Engineering,
Nanyang Technological University,
Block S1, Nanyang Avenue, Singapore 639798
[2] College of Information Engineering,
Xiangtan University, Xiangtan, China
{pg02782641, elpwang}@ntu.edu.sg

Abstract. In this paper, we propose a novel approach called the gradual noisy chaotic neural network (G-NCNN) to find a collision-free time slot schedule in a time division multiple access (TDMA) frame in packet radio network (PRN). In order to find a minimal average time delay of the network, we aim to find an optimal schedule which has the minimum frame length and provides the maximum channel utilization. The proposed two-phase neural network approach uses two different energy functions, with which the G-NCNN finds the minimal TDMA frame length in the first phase and the NCNN maximizes the node transmissions in the second phase. Numerical examples and comparisons with the previous methods show that the proposed method finds better solutions than previous algorithms. Furthermore, in order to show the difference between the proposed method and the hybrid method of the Hopfield neural network and genetic algorithms, we perform a paired t-test between two of them and show that G-NCNN can make significantly improvements.

1 Introduction

The Packet Radio Network (PRN) gains more attention in recent research and industry as it is a good alternative for the high-speed wireless communication, especially in a broad geographic region [1]. The PRN shares common radio channels as the broadcast medium to interconnect nodes. In order to avoid any collision, a time-division multiple-access (TDMA) protocol has been used to schedule conflict free transmissions. A TDMA cycle is divided into distinct frames consisting of a number of time slots. A time slot has a unit time to transmit one data packet between adjacent nodes. At each time slot, each node can either transmit or receive a packet, but no more than two packets can be received from neighbor nodes. If a node is scheduled to both transmit and receive at the same time slot, a *primary* conflict occurs. If two or more packets reach one node at the same time slot, a *second* conflict occurs.

The BSP has been studied by many researchers [2]-[8]. In [2], Funabiki and Takefuji proposed a parallel algorithm based on an artificial neural network in a TDMA cycle with $n \times m$ neurons. In [3], Wang and Ansari proposed a mean field annealing algorithm to find a TDMA cycle with the minimum delay time. In [4], Chakraborty and Hirano used genetic algorithm with a modified crossover operator to handle large networks with complex connectivity. In [5], Funabiki and Kitamichi proposed a binary neural network with a gradual expansion scheme to find minimum time slots and maximum transmissions through a two-phase process. In [6], Yeo et al proposed a algorithm based on the sequential vertex coloring algorithm. In [7], Salcedo-Sanz et al proposed a hybrid algorithm which combines a Hopfield neural network for constrain satisfaction and a genetic algorithm for achieving a maximal throughput. In [8], Peng et al. used a mixed tabu-greedy algorithm to solve the BSP.

In this paper, we present a novel neural network model for this problem, i.e., gradual noisy chaotic neural network (G-NCNN). Numerical results show that this NCNN method outperforms existing algorithms in both the average delay time and the minimal TDMA length. The organization of this paper is as follows. In section 2, we formulate the broadcast scheduling problem. The noisy chaotic neural network (NCNN) model is proposed in section 3. In section 4, the proposed two-phase neural network is applied to solving the optimal scheduling problem. Numerical results are stated and the performance is evaluated in section 5. In Section 6 we conclude the paper.

2 Broadcast Scheduling Problem

We formulate the packet radio network as a graph, $G = (I,E)$, where I is the set of nodes and E is the set of edges. We follow the assumption in previous research and consider only undirected graphs and the matrix c_{ij} is symmetric. If two nodes are adjacent with $c_{ij} = 1$, then we define two nodes to be one-hop-away, and the two nodes sharing the same neighboring node to be two-hop-away. The compatibility matrix $D = \{d_{ij}\}$ consists of $N \times N$ which represents the network topology by stating the two-hop-away nodes is defined as follows:

$$d_{ij} = \begin{cases} 1 \text{, if node } i \text{ and node } j \text{ are within two-hop-away} \\ 0 \text{, otherwise} \end{cases}$$

We summarize the constraints in the BSP in the following two categories:

1) *No-transmission constraint* [4]: Each node should be scheduled to transmit at least once in a TDMA cycle.

2) *No-conflict constraint*: It excludes the primary conflict (a node cannot have transmission and reception simultaneously) and the secondary conflict (a node is not allowed to receive more than one transmission simultaneously).

The final optimal solution for a N-node network is a conflict-free transmission schedule consisting of M time slots. Additional transmissions can be arranged provided that the transmission does not violate the constrains. We use an $M \times N$ binary matrix $V = (v_{ij})$ to express such a schedule [3], where

$$v_{ij} = \begin{cases} 1\text{ , if node } i \text{ transmits in slot } j \text{ in a frame} \\ 0\text{ , otherwise} \end{cases}$$

The goal of the BSP is to find a transmission schedule with the shortest TDMA frame length (i.e., M should be as small as possible) which satisfies the above constrains, and the total number of node transmissions is maximized in order to maximize the channel utilization.

3 The Proposed Neural Network Model

Since Hopfield and Tank solved the TSP problem using the Hopfield neural network (HNN), many research efforts have been made on solving combinatorial optimizations using the Hopfield-type neural networks. However, since the original Hopfield neural network (HNN) can be easily tramped in local minima, stochastic simulated annealing (SSA) technique has been combined with the HNN [10] [15]. Chen and Aihara [9][10] proposed chaotic simulated annealing (CSA) by starting with a sufficiently large negative self-coupling in the neurons and then gradually reducing the self-coupling to stabilize the network. They called this model the transiently chaotic neural network (TCNN).

In order to improve the searching ability of the TCNN, Wang and Tian [11] proposed a new approach to simulated annealing by adding decaying stochastic noise into the TCNN, i.e., a chaotic neural network with stochastic nature, a noisy chaotic neural network (NCNN). This neural network model has been applied successfully in solving several optimization problems including the traveling salesman problem (TSP) and the channel assignment problem (CAP) [11]-[14]. The NCNN model is described as follows [11]:

$$x_{jk}(t) = \frac{1}{1 + e^{-y_{jk}(t)/\varepsilon}} \qquad (1)$$

$$y_{jk}(t+1) = ky_{jk}(t) + \alpha(\sum_{\substack{i=1\\i\neq j}}^{N}\sum_{\substack{l=1\\l\neq k}}^{M} w_{jkil}x_{jk}(t) + I_{ij})$$
$$-z(t)(x_{jk}(t) - I_0) + n(t) \qquad (2)$$

$$z(t+1) = (1 - \beta_1)z(t) \qquad (3)$$

$$A[n(t+1)] = (1 - \beta_2)A[n(t)] \qquad (4)$$

where
x_{jk} : output of neuron jk ;
y_{jk} : input of neuron jk ;
w_{jkil}: connection weight from neuron jk to neuron il, with $w_{jkil} = w_{iljk}$ and $w_{jkjk} = 0$;

$$\sum_{\substack{i=1\\i\neq j}}^{N}\sum_{\substack{l=1\\l\neq k}}^{M} w_{jkil}x_{jk} + I_{ij} = -\partial E/\partial x_{jk} \text{ , input to neuron } jk. \tag{5}$$

I_{jk} : input bias of neuron jk ;
k : damping factor of nerve membrane ($0 \leq k \leq 1$);
α : positive scaling parameter for inputs ;
β_1 : damping factor for neuronal self-coupling ($0 \leq \beta_1 \leq 1$);
β_2 : damping factor for stochastic noise ($0 \leq \beta_2 \leq 1$);
$z(t)$: self-feedback connection weight or refractory strength ($z(t) \geq 0$) ;
I_0 : positive parameter;
ε : steepness parameter of the output function ($\varepsilon > 0$) ;
E : energy function;
$n(t)$: random noise injected into the neurons, in $[-A, A]$ with a uniform distribution;
$A[n]$: amplitude of noise n.

In this paper, we combined the NCNN with a gradual scheme [5] and propose a new method called the gradual noisy chaotic neural network (G-NCNN). In this method, The number of neurons in the neural networks is not fixed, it starts with a initial number of neurons, and then the additional neurons are gradually added into the existing neural networks until the stop criteria meet. In the next section, we will discuss in detail solving the BSP.

4 The Two-Phase Neural Network for the BSP

4.1 Energy Function in Phase I

The energy function E_1 for phase I is given as following [5]:

$$E_1 = \frac{W_1}{2}\sum_{i=1}^{N}(\sum_{k=1}^{M} v_{ik} - 1)^2 + \frac{W_2}{2}\sum_{i=1}^{N}\sum_{j=1}^{M}\sum_{\substack{k=1\\k\neq i}}^{N} d_{ik}v_{ij}v_{kj} \tag{6}$$

where W_1 and W_2 are weighting coefficients. The W_1 term represents the constraints that each of N nodes must transmit exactly once during each TDMA cycle. The W_2 term indicates the constraint that any pair of nodes which is one-hop away or two-hop away must not transmit simultaneously during each TDMA cycle.

From eqn. (2), eqn. (5), and eqn. (6), we obtain the dynamics of the NCNN for the BSP as below:

$$y_{jk}(t+1) = ky_{jk}(t) + \alpha\{-W_1(\sum_{k=1}^{M} v_{ik} - 1)$$

$$-W_2(\sum_{\substack{k=1\\k\neq i}}^{N} d_{ik}v_{kj})\} - z(t)(x_{jk}(t) - I_0) + n(t) . \tag{7}$$

In order to obtain a minimal frame length which satisfies the constrains, we use a *gradual expansion scheme* in which a initial value of frame length is set with a lower bound value of M. If with current frame length there is no feasible solution which satisfied the constrains, then this value is gradually increased by 1, i.e., $M = M + 1$. The algorithm compute iteratively until every node can transmit at least once in the cycle without conflicts, then the algorithm stopped and the current value of M is the minimal frame length. In this way, the scheduled frame length would be minimized.

4.2 Energy Function in Phase II

In phase II, the objective is to maximize the total number of transmissions based on the minimal TDMA length M obtained in the previous phase. We use the energy function for phase II is defined as follow [5]:

$$E_2 = \frac{W_3}{2} \sum_{i=1}^{N} \sum_{j=1}^{M} \sum_{\substack{k=1 \\ k \neq i}}^{N} d_{ik} v_{ij} v_{kj} + \frac{W_4}{2} \sum_{i=1}^{N} \sum_{j=1}^{M} (1 - v_{ij})^2 \qquad (8)$$

where W_3 and W_4 are coefficients. W_3 represents the constraint term that any pair of nodes which is one-hop away or two-hop away must not transmit simultaneously during each TDMA cycle. W_4 is the optimization term which maximized the total number of output firing neurons.

From eqn. (2), eqn. (5), and eqn. (8), we obtain the dynamics of the NCNN for phase II of the BSP as follow:

$$y_{jk}(t+1) = k y_{jk}(t) + \alpha \{-W_3 \sum_{\substack{k=1 \\ k \neq i}}^{N} d_{ik} v_{kj} + W_4 (1 - v_{ij})\}$$

$$-z(t)(x_{jk}(t) - I_0) + n(t) \qquad (9)$$

In the above models of the BSP, the network with $N \times M$ neurons is updated cyclically and asynchronously. The new state information is immediately available for the other neurons in the next iteration. The iteration is terminated once a feasible transmission schedule is obtained, i.e., the transmission of all nodes are conflict free.

5 Simulation Results

We use three evaluation indices to compare with different algorithms. One is the TDMA cycle length M. The second is the average time delay η defined as [5]:

$$\eta = \frac{1}{N} \sum_{i=1}^{N} \left(\frac{M}{\sum_{j=1}^{M}} v_{ij} \right) = \frac{M}{N} \sum_{i=1}^{N} \left(\frac{1}{\sum_{j=1}^{M} v_{ij}} \right) \qquad (10)$$

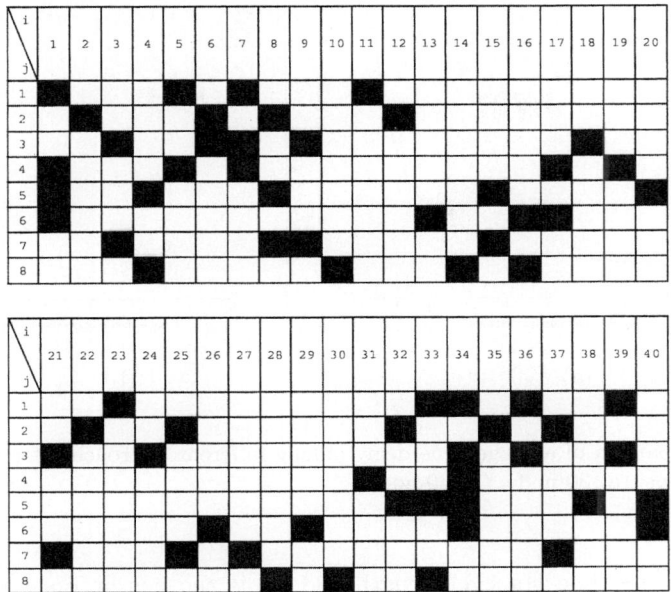

Fig. 1. Broadcasting Schedule for BM #3, the 40-node network

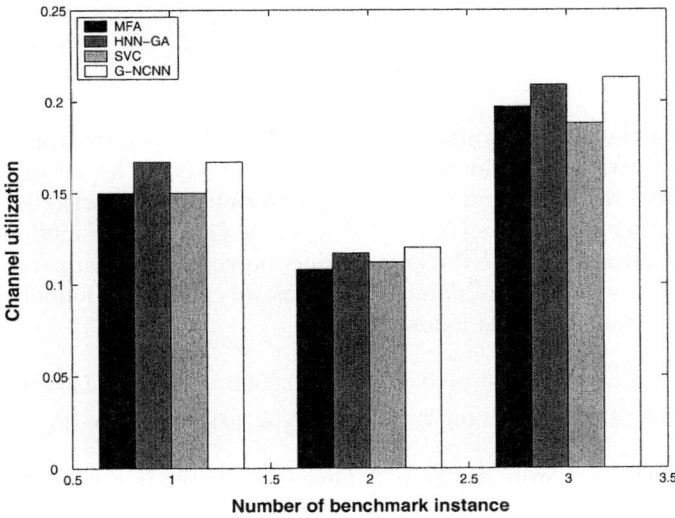

Fig. 2. Comparisons of channel utilization for three benchmark problems. 1, 2, and 3 in the horizontal axis stand for instance with 15, 30, and 40 nodes, respectively.

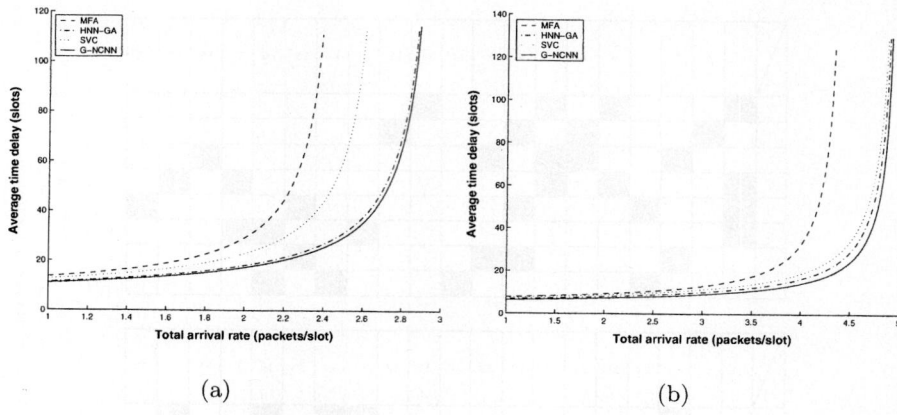

Fig. 3. Comparison of average time delay among different approaches for two benchmark problems: (a) 30-node, (b) 40-node

where M is the time-slot cycle length and v_{ij} is the neuron output. The another definition of average time delay can be found in [3] and [6] which is calculated with the Pollaczek-Khinchin formula [16], which models the network as N $M/D/1$ queues. We will use both definitions in order to compare with other methods. The last index is channel utilization ρ, which is given by [3]:

$$\rho = \frac{1}{NM} \sum_{j=1}^{N} \sum_{i=1}^{M} v_{ij}. \qquad (11)$$

We choose the model parameters in the G-NCNN by investigating the neuron dynamics for various combination of model parameters. The set of parameters which produces the richer and more flexible dynamics will be selected. The selection of weighting coefficients (W_1, W_2, W_3, W_4) in the energy function are based on the rule that all terms in the energy function should be comparable in magnitude, so that none of them dominates. Thus we choose the model parameters and weighting coefficients as follows:

$$k = 0.9, \alpha = 0.015, \beta_1 = 0.001, \beta_2 = 0.0002, \varepsilon = 0.004, I_0 = 0.65$$
$$z_0 = 0.08, A[n(0)] = 0.009, W_1 = 1.0, W_2 = 1.0, W_3 = 1.0, W_4 = 1.0. \qquad (12)$$

Three benchmark problems from [3] have been chosen to compared with other algorithms in [5],[6], and [7]. The three examples are instances with 15-node-29-edge, 30-node-70-edge, and 40-node-66-edge respectively.

Fig. 1 shows the final broadcast schedule for the 40-node network, where the black box represents an assigned time slot. The comparison of channel utilization in eqn. (11) for three benchmark problems is plotted in Fig. 2, which shows that

Table 1. Comparisons of average delay time η and time slot M obtained by the NCNN with other algorithms for the three benchmark problems given by [3]

	NCNN η / M	HNN-GA η / M	SVC η / M	GNN η / M	MFA η / M
#1	6.8 / 8	7.0 / 8	7.2 / 8	7.1 / 8	7.2 / 8
#2	9.0 / 10	9.3 / 10	10.0 / 10	9.5 / 10	10.5 / 12
#3	5.8 / 8	6.3 / 8	6.76 / 8	6.2 / 8	6.9 / 9

Table 2. Paired t-test of average time delay η (second) between the HNN-GA and the G-NCNN

Instances	Node	Edge	HNN-GA	G-NCNN
BM #1	15	29	6.84	6.84
BM #2	30	70	9.17	9.00
BM #3	40	66	6.04	5.81
Case #4	60	277	15.74	13.40
Case #5	80	397	16.33	14.48
Case #6	100	522	17.17	15.16
Case #7	120	647	17.85	16.02
Case #8	150	819	20.47	16.37
Case #9	180	966	20.04	16.38
Case #10	200	1145	20.31	17.22
Case #11	230	1226	20.36	16.58
Case #12	250	1424	20.25	17.17
T-Value = 5.22			
P-Value (one-tail) = 0.0001			
P-Value (two-tail) = 0.0003			

the NCNN can find solutions with the highest channel utilization among all algorithms. The average time delay is plotted in Fig. 3. From this figure, it can be seen that the time delay experienced by the NCNN is much less than that of the MFA algorithm in all three instances. In the 30-node and the 40-node instances, the G-NCNN can find a TDMA schedule with less delay than other methods.

The computational results are summarized in Table 1 in comparison with the hybrid HNN-GA algorithm from [7], the sequential vertex coloring (SVC) from [6], the gradual neural network (GNN) from [5] and the mean field annealing (MFA) from [3]. From this table, we can see that our proposed method can find equal or smaller frame length than other previous methods for all the three examples. In respect of the average time delay, our algorithm outperforms the other algorithms in obtaining the minimal value of η.

In order to show the difference between the HNN-GA and the NCNN, a paired t-test is performed between the two methods, as shown in Table 2. We compared the two methods in 12 cases with node size from 15 to 250, where BM #1 to BM #3 are benchmark examples and case #4 to case #12 are randomly generated instance with edge generation parameter $r = 2/\sqrt{N}$. The results show that the P-value is 0.0001 for one-tail test and 0.0003 for two-tail test. We found that the G-NCNN (mean = 13.7, standard deviation = 4.12) reported having significantly better performance than did the HNN-GA (mean = 15.9, standard deviation = 5.45) did, with T-Value $t(11) = 5.22$, P-Value < 0.05.

6 Conclusion

In this paper, we propose a gradual noisy chaotic neural network for solving the broadcast scheduling problem in packet radio networks. The G-NCNN consists of $N \times M$ noisy chaotic neurons for the N-node-M-slot problem. We evaluate the proposed method in three benchmark examples and several randomly generated instances. We compare our results with previous methods including the mean filed annealing, the HNN-GA, the sequential vertex coloring algorithm, and the gradually neural network. The results of three benchmark instances show that the G-NCNN always finds better solutions with minimal average time delay and maximal channel utilization. We also have performed a paired t-test between the G-NCNN and the HNN-GA in several randomly generated instances, the t-test results show that the G-NCNN is better than the HNN-GA in solving the BSP.

References

1. Leiner, B.M., Nielson, D.L., Tobagi, F.A.: Issues in Packet Radio Network Design. Proc. IEEE, Vol. 75. (1987) 6-20
2. Funabiki, N. , Takefuji, Y.: A Parallel Algorithm for Broadcast Scheduling Problems in Packet Radio Networks. IEEE Trans. on Communications, Vol. 41. (1993) 828-831
3. Wang, G., Ansari, N.: Optimal Broadcast Scheduling in Packet Radio Networks Using Mean Field Annealing. IEEE Journal on selected areas in Communications, Vol. 15. (1997) 250-260
4. Chakraborty, G., Hirano, Y.: Genetic Algorithm for Broadcast Scheduling in Packet Radio Networks. IEEE World Congress on Computational Intelligence. (1998) 183-188
5. Funabiki, N., Kitamichi, J.: A Gradual Neural Network Algorithm for Broadcast Scheduling Problems in Packet Radio Networks. IEICE Trans. Fundamentals, Vol e82-a. (1999) 815-824
6. Yeo, J., Lee, H., Kim, S.: An Efficient Broadcast Scheduling Algorithm for TDMA Ad-hoc Networks. Computer &Operations Research, Vol. 29. (2002) 1793-1806
7. Salcedo-Sanz, S., Bousoño-Calz;on , C., Figueiras-Vidal, A. R.: A Mixed Neural-Genetic Algorithm for the Broadcast Scheduling Problem. IEEE Trans. on wireless communications, Vol. 2. (2003) 277-283
8. Peng, Y.J., Soong, B.H., Wang, L.P.: Broadcast Scheduling in Packet Radio Networks Using a Mixed Tabu-greedy Algorithm. Electronics Letts 40 (2004) 375-376

9. Chen, L., Aihara, K.: Transient Chaotic Neural Networks and Chaotic Simulated Annealing. In: Yamguti, M. (eds.): Towards the Harnessing of chaos. Amsterdam, Elsevier Science Publishers B.V. (1994) 347-352
10. Chen, L., Aihara, K.: Chaotic Simulated Annealing by a Neural Network Model with Transient Chaos. Neural Networks, Vol. 8. (1995) 915-930
11. Wang, L.P., Tian, F.: Noisy Chaotic Neural Networks for Solving Combinatorial Optimization Problems. Proc. International Joint Conference on Neural Networks (2000) 37-40
12. Li, S., Wang, L.P.: Channel Assignment For Mobile Communications Using Stochastic Chaotic Simulated Annealing. The 6th International Work-Conference on Artificial and Natural Neural Networks (2001) 757-764
13. Wang, L.P., Li, S., Wan, C., Soong, B.H.: Minimizing Interference in Cellular Mobile Communications by Optimal Channel Assignment Using Chaotic Simulated Annealing. In : Wang, L. (eds.): Soft Computing in Communications. Springer, Berlin (2003) 131-145
14. Wang, L.P., Li, S., Tian F., Fu, X.: A noisy chaotic neural network for solving combinatorial optimization problems: Stochastic chaotic simulated annealing. IEEE Trans. System, Man, Cybern, Part B - Cybernetics 34 (2004) 2119-2125
15. Wang, L.P., Smith, K.: On Chaotic Simulated Annealing. IEEE Transactions on Neural Networks, Vol. 9. (1998) 716-718
16. Bertsekas, D., Gallager, R.: Data Networks. Englewood Cliffs, NJ: Prentice-Hall (1987)

A Fast Online SVM Algorithm for Variable-Step CDMA Power Control

Yu Zhao, Hongsheng Xi, and Zilei Wang

Network Communication System and Control Laboratory (218),
Department of Automation, University of Science and Technology of China,
230027 Hefei, Anhui, China
xihs@ustc.edu.cn

Abstract. This paper presents a fast online support vector machine (FOSVM) algorithm for variable-step CDMA power control. The FOSVM algorithm distinguishes new added samples and constructs current training sample set using K.K.T. condition in order to reduce the size of training samples. As a result, the training speed is effectively increased. We classify the received signals into two classes with FOSVM algorithm, then according to the output label of FOSVM and the distance from the data points to the SIR decision boundary, variable-step power control command is determined. Simulation results illustrate that the algorithm has a fast training speed and less support vectors. Its convergence performance is better than the fixed-step power control algorithm.

1 Introduction

Power control is one of the most important techniques in CDMA cellular system. Since all the signals in a CDMA system share the same bandwidth, it is critical to use power control to maintain an acceptable signal-to-interference ratio (SIR) for all users, hence maximizing the system capacity. Another critical problem with CDMA is the "near-far effect". Due to propagation characteristics the signals from mobiles closer to the base station could overpower the signals from mobiles located farther away, with power control each mobile adjusts its own transmit power to ensure an desired QoS or SIR at the base station.

Over the past decades, many power control techniques drawn from centralized control [1], distributed control [2], stochastic control [3] etc. have been proposed and applied to cellular radio systems. Most of these available approaches are based on the accurate estimates of signal-to-interference ratio (SIR), bit error rates (BER), or frame error rates (FER).

Support vector machine (SVM) is one of the most effective machine learning methods, which are based on principles of structural risk minimization and statistical learning theory [4]. The standard SVM algorithms are solved using quadratic programming methods, these algorithms are often time consuming and difficult to implement for real-time application, especially for the condition that the size of training data is large. Some modified SVM algorithms [5,6] are proposed to solve the problem of real-time application. Rohwer [7] just applied a least squares support vector machine (LS-SVM) algorithm presented in [6] for fixed-step CDMA power control, but the Lagrangian multipliers for the LS-SVM tend to be all nonzero whereas for the SVM case only support vector are nonzero. On the other hand, the convergence performance of the fixed-step power control is not satisfying when the initial received SIR is far away from the desired SIR.

This paper presents a fast online support vector machine (FOSVM) algorithm for variable-step CDMA power control. The algorithm classify the sets of eigenvalues, from the sample covariance matrices of the received signal, into two SIR sets using FOSVM, then according to the output label of FOSVM and the distance from the data point to the SIR decision boundary, variable-step power control command is determined. For the data points far away from the decision boundary, large step will be considered, otherwise, we will choose small step. In the process of training SIR decision boundary, we choose training samples using K.K.T. condition, hence reducing the size of training data and increasing the training speed effectively. Simulation results illustrate that the FOSVM algorithm has a faster training speed and less support vectors without compromising the generalization capability of the SVM, furthermore its convergence performance is better than fixed-step power control algorithm.

2 Support Vector Machines

SVM can be described firstly considering a binary classification problem. Assume we have a finite set of labeled points in Euclidean n space, they are linearly separable. The goal of SVM is to find the hyperplane, $(\mathbf{w}^T\mathbf{x}) + b = 0$, which maximizes the minimum distance between any point and the hyperplane [4].

Let $(\mathbf{x}_i, y_i)$ $i = 1, \cdots, l$ be a set of l labeled points where $\mathbf{x}_i \in R^N$ and $y_i \in \{+1, -1\}$. Among all hyperplanes separating the data, there exists an optimal one yielding the maximum margin of separation between the classes. Since the margin equals $\frac{2}{\|\mathbf{w}\|}$, maximizing the margin is equivalent to minimizing the magnitude of the weights. If the data is linearly separable, the problem can be described as a 1-norm soft margin SVM [8]

$$\begin{aligned} &\min(\|\mathbf{w}\|^2/2 + C \sum_{i=1}^{l} \xi_i) \\ &s.t. \quad y_i(\langle \mathbf{w}, \mathbf{x}_i \rangle + b) \geq 1 - \xi_i \\ &\quad \xi_i \geq 0, \; i = 1, \cdots, l \;\; C > 0 \end{aligned} \quad (1)$$

where ξ_i denotes slack variables to the quadratic programming problem. Data points are penalized by regularization parameter C if they are misclassified.

For the sets of data points that can not be separated linearly, we need replace inner products $\langle \mathbf{x}_i, y_i \rangle$ with kernel function $K(\mathbf{x}_i, y_i)$ which satisfies Mercers Theorem [4]. The quadratic programming problem can be transformed into a dual problem by introducing Lagrangian multiplier α_i.

$$\min L_D(\alpha) = \frac{1}{2}\sum_{i,j=1}^{l} \alpha_i\alpha_j y_i y_j K(\mathbf{x}_i, \mathbf{x}_j) - \sum_{i=1}^{l}\alpha_i$$

$$s.t. \quad \sum_{i=1}^{l}\alpha_i y_i = 0, i = 1, \cdots, l \qquad 0 \leq \alpha_i \leq C \tag{2}$$

The optimal solution must satisfy the following Karush-Kuhn-Tucker (K.K.T.) conditions

$$\alpha_i[y_i(\langle \mathbf{w}, \mathbf{x}_i\rangle + b) - 1 + \xi_i] = 0, i = 1, \cdots, l \tag{3}$$

The weight $\mathbf{w}$ obtained from (2) is then expressed as

$$\mathbf{w} = \sum_{i=1}^{l}\alpha_i y_i \mathbf{x}_i \tag{4}$$

The decision function can then be written as

$$f(\mathbf{x}) = sgn\left(\sum_{i \in SV} \alpha_i y_i K(\mathbf{x}, \mathbf{x}_i) + b\right) \tag{5}$$

From equation (4), we see that only the inputs associated with nonzero Lagrangian multiplier α_i contribute to the weight vector $\mathbf{w}$. These inputs are called support vectors(SVs) and lie on the margin of the decision region. These support vectors are the critical vectors in determining the optimal margin classifier.

3 FOSVM

In the conventional SVM, training data are supplied and computed in batch by solving the quadratic programming problem, therefore, it is time consuming to classify a large data set and can not satisfy the demands of online application, such as CDMA power control, which needs periodically retraining because of the update of the training data. With this objective in mind, an online training of support vector classifier (OSVC) algorithm is proposed in [5] to overcome the shortcoming of conventional SVM. Although the simulation results show that the training time of the OSVC algorithm is much less than the standard SVM and SMO algorithm [9], it is still inefficient for some kinds of training data. Borrowed the idea from OSVC algorithm, we present a modified fast online SVM (FOSVM) algorithm to improve the performance of training phase. The FOSVM algorithm is summarized as follow:

At the initial stage of online training, obtain the initial optimal hyperplane $f_0(\alpha_0, b_0)$ with the initial training data set S_0, the size of S_0 can be chosen by users according to the actual application. $SV_0 = \{(\mathbf{S}\mathbf{x}_i^0, Sy_i^0)\}$ is the corresponding support vectors set. When a new set of training data S_k is available, we will judge if the training data $(\mathbf{x}_i, y_i) \in S_k$ can be classified by the current optimal hyperplane $f_{k-1}(\alpha_{k-1}, b_{k-1})$ correctly, where the corresponding hyperplane is $f_{k-1}(\mathbf{x}) = sgn(\sum_{i=1}^{|SV_{k-1}|} \alpha_i^{k-1} S y_i^{k-1} K(\mathbf{x}, \mathbf{S}\mathbf{x}_i^{k-1})) + b_{k-1})$. When there is no training data misclassified, set $f_k = f_{k-1}$, $SV_k = SV_{k-1}$, continue for the next step, otherwise we need train the new hyperplane. In the process of training new hyperplane, we should first determine the training data set W_k. If there is a high demand on real-time application and the size of S_k is large, we'd better choose less training data to reduce the training time. Without losing generality, we set $W_k = S_k$. Later, we retrain the current training data set T_k circularly until all the training data $(\mathbf{x}_i, y_i) \in W_k$ satisfy the K.K.T. conditions. The current training data set T_k includes SV_k and the samples violating the K.K.T. conditions corresponding to the current hyperplane f_k.

Compared with OSVC algorithm, the advantage of FOSVM algorithm is that it reduces the training time by decreasing training times. The OSVC algorithm obtain the new sample one by one in a sequence, once obtain a new sample which can not be classified by the current optimal hyperplane correctly, it need carry out a training for a new hyperplane. The training is so frequent that it leads to increase the training time. While in FOSVM algorithm, we obtain a new set of training data and construct current training data set using K.K.T. conditions at every step. With this procedure, the training time can be decreased obviously without compromising the generalization capability of the SVM.

We give the pseudo code of FOSVM algorithm as Algorithm 1:

Algorithm 1 Fast online SVM algorithm

1: Obtain initial training data set S_0.
2: Set $W_0 = S_0$.
3: Minimize (2) with W_0 to obtain an optimal hyperplane $f_0(\alpha_0, b_0)$ and SV_0.
4: **for** $k = 1, 2, 3, \cdots$ **do**
5:    Obtain the kth training data set S_k,
6:    $E_k = \{(\mathbf{x}_i, y_i) \in S_k | (\mathbf{x}_i, y_i) \ are \ misclassified \ by \ f_{k-1}\}$
7:    **if** $|E_k| > 0$ **then**
8:      Determine the training data set W_k.
9:      $V_k = \{(\mathbf{x}_i, y_i) \in W_k | y_i f_k(\mathbf{x}_i) \ violate \ the \ K.K.T. \ conditions\}$
10:     **while** $|V_k| > 0$ **do**
11:       the current training data set $T_k = SV_k \bigcup V_k$
12:       Minimize (2) to obtain a new optimal hyperplane f_k and SV_k with T_k
13:       $V_k = \{(\mathbf{x}_i, y_i) \in W_k/T_k | y_i f_k(\mathbf{x}_i) \ violate \ the \ K.K.T. \ conditions\}$
14:     **end while**
15:    **else**
16:      $f_k = f_{k-1}, SV_k = SV_{k-1}$
17:    **end if**
18: **end for**

4 Convergence of FOSVM

In order to explain the convergence of FOSVM, we rewritten the dual objective function (2) in a matrix as follows:

$$L_D = \frac{1}{2}\alpha^T \mathbf{K}\alpha - \langle \mathbf{c}, \alpha \rangle \tag{6}$$

where $\mathbf{c}$ is an $l \times 1$ vector, $\alpha = (\alpha_1, \cdots, \alpha_l)^T$ and $\mathbf{K} = \{K_{ij}\}$, $K_{ij} = y_i y_j K(\mathbf{x}_i, \mathbf{x}_j)$.

Next we will prove the convergence of the FOSVM algorithm by comparing it with the decomposition algorithm (DA) proposed in [10]. The DA partitioned the training set into two sets B and N. The set B is called working set and N correcting set. Suppose α, $\mathbf{y}$, $\mathbf{c}$ and $\mathbf{K}$ from (6) can be arranged properly as follows:

$$\alpha = \begin{pmatrix} \alpha_B \\ \alpha_N \end{pmatrix}, \mathbf{y} = \begin{pmatrix} \mathbf{y}_B \\ \mathbf{y}_N \end{pmatrix}, \mathbf{c} = \begin{pmatrix} \mathbf{c}_B \\ \mathbf{c}_N \end{pmatrix}, \mathbf{K} = \begin{pmatrix} \mathbf{K}_{BB} & \mathbf{K}_{BN} \\ \mathbf{K}_{NB} & \mathbf{K}_{NN} \end{pmatrix}$$

Then the dual objective function (6) is rewritten involving the working and correcting sets as follows:

$$\min \frac{1}{2}[\alpha_B^T \mathbf{K}_{BB}\alpha_B + \alpha_B^T \mathbf{K}_{BN}\alpha_N + \alpha_N^T \mathbf{K}_{NB}\alpha_B + \alpha_N^T \mathbf{K}_{NN}\alpha_N] - \langle \mathbf{c}_B, \alpha_B \rangle - \langle \mathbf{c}_N, \alpha_N \rangle$$
$$s.t. \quad \langle \mathbf{y}_B, \alpha_B \rangle + \langle \mathbf{y}_N, \alpha_N \rangle = 0, \quad 0 \leq \alpha_B, \alpha_N \leq \mathbf{C}$$

The main idea of DA is that instead of solving the large quadratic programming problem at once, small quadratic programming sub-problem are solved by exchanging elements between the set B and N. Each sub-problem will bring the solution closer to the optimal solution. The process of exchanging elements includes two steps, Build-down and Build-up, which shows in the Fig.1. [11]gives the detailed proof of the convergence of DA.

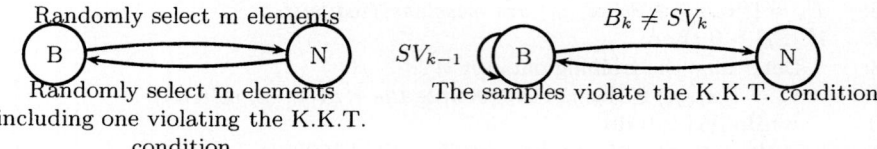

Fig. 1. Decomposition algorithm for SVC   **Fig. 2.** Elements exchange for FOSVM

Compared with DA, the FOSVM keeps the support vector SV_k set in the working set B and removes the other elements of B to the correcting set N. Another difference between FOSVM and DA is that FOSVM takes a new set of elements at each step and put the elements violating the K.K.T. conditions into the working set. Fig.2 shows the state diagram of the FOSVM. Because the elements which are not SVs will farther away from the hyperplane, they have no effect on the optimal solution, which means the parameter $\alpha_N = 0$. Therefore, the optimal solution of FOSVM is unchanged after removing only the elements which are not SVs. In order to show the convergence of the FOSVM, we have the following corollary:

Corollary 1. *Moving elements $\{m\}$ which are not SVs from B to N leaves the cost function unchanged and the solution is feasible in the sub-problem.*

Proof. Let $B' = B - \{m\}$, $N' = N \bigcup \{m\}$, $\{m\} \in B - SV \Rightarrow \alpha_m = 0$ and notice that $\alpha_N = 0$, we have

$$\begin{aligned}
L_D(B', N') &= \frac{1}{2}[\alpha_{B'}^T \mathbf{K}_{B'B'} \alpha_{B'} + 2\alpha_{B'}^T \mathbf{K}_{B'N'} \alpha_{N'} + \alpha_{N'}^T \mathbf{K}_{N'N'} \alpha_{N'}] - \langle \mathbf{c}_{B'}, \alpha_{B'} \rangle - \langle \mathbf{c}_{N'}, \alpha_{N'} \rangle \\
&= \frac{1}{2}[\alpha_{B'}^T \mathbf{K}_{B'B'} \alpha_{B'} + 2\alpha_{B'}^T \mathbf{K}_{B'm} \alpha_m + \alpha_m^T \mathbf{K}_{mm} \alpha_m] - \langle \mathbf{c}_{B'}, \alpha_{B'} \rangle - \langle \mathbf{c}_m, \alpha_m \rangle \\
&= \frac{1}{2} \alpha_B^T \mathbf{K}_{BB} \alpha_B - \langle \mathbf{c}_B, \alpha_B \rangle \\
&= L_D(B, N) \quad (7)
\end{aligned}$$

$$\langle \mathbf{y}_{B'}, \alpha_{B'} \rangle + \langle \mathbf{y}_{N'}, \alpha_{N'} \rangle = \langle \mathbf{y}_{B'}, \alpha_{B'} \rangle + \langle \mathbf{y}_{m, \alpha_m} \rangle = \langle \mathbf{y}_B, \alpha_B \rangle = 0 \quad (8)$$

From (7) (8), we notice that the objection function and constraints conditions both are unchanged, therefore, the sub-problem has the same solution, using the proposed FOSVM algorithm which modifies the build-down and build-up process of the DA.

5 FOSVM for Variable-Step CDMA Power Control

5.1 CDMA Signal Model

The received vector at the output of the ith antenna array detected at the adaptive array processor can be written as [12]

$$x_i(t) = \sum_{j=1}^{M} \sum_{l=1}^{L} \sqrt{P_j G_{ji}} \mathbf{a}_j(\theta_l) g_{ji}^l s_j(t - \tau_j) + \mathbf{n}_i(t) \quad (9)$$

where $s_j(t - \tau_j)$ is the message signal transmitted from the jth user, τ_j is the corresponding time delay, $\mathbf{n}_i(t)$ is the thermal noise vector at the input of antenna array at the ith receiver, and P_j is the power of the jth transmitter. $\mathbf{a}_j(\theta_l)$ is the response of the jth user to the direction θ_l. The attenuation due to shadowing in the lth path is denoted by g_{ji}^l. The link gain between transmitter j and receiver i is denoted by G_{ji}.

The terms relative to the multiple paths are combined as

$$\mathbf{Z}_{ji} = \sum_{l=1}^{L} \mathbf{a}_j(\theta_l) g_{ji}^l \quad (10)$$

Vector Z_{ji} is defined as the spatial signature or array response of the ith antenna array to the jth source. In a spread spectrum system, the message signal is given by

$$s_i(t) = \sum_n b_i(n) c_i(t - nT) \tag{11}$$

where $b_i(n)$ is the ith user information bit stream and $c_i(t)$ is the spreading sequence. The received signal, sampled at the output of the matched filter, is expressed as

$$y_i(n) = \int_{(n-1)T+\tau_i}^{nT+\tau_i} c_i(t - nT - \tau_i) \cdot \left(\sum_j \sqrt{p_j(t) G_{ji}} \sum_m b_j(m) c_j(t - mT - \tau_j) \mathbf{Z}_{ji} + \mathbf{n}_i(t) \right) dt \tag{12}$$

5.2 Variable Step CDMA Power Control

In CDMA system, the signals sampled from the same mobile user at different time are independent, moreover, the power control algorithm have a high demand on the speed of convergence, so we can discard the previous signal samples and only use the current set of signal samples to train the current hyperplane. It means that we can set $W_k = S_k$ in the FOSVM algorithm.

According to the fixed-step power control algorithm mentioned in [7], we can classify the set of eigenvalues from the sample covariance matrices of the received signal into two classes, One represents the SIR greater than the desired SIR, another represents the lower. Variable-step power control command is determined according to the output label of FOSVM and the distance from the data point to the SIR decision boundary. The size of step changes adaptively based on the distance from the data point to the SIR decision boundary. The application of FOSVM for variable-step CDMA power control can be described as follow:

1. Obtain one set of signals sampled from equation (12), suppose the set is S_k, the number of signal sample in S_k is N.
2. Generate the sample covariance matrices with the signals from S_k, each covariance matrix is generated with M samples, $M < N$.
3. Calculate the eigenvalues of sample covariance matrices, obtain $[\frac{N}{M}]$ eigenvalue vectors.
4. Using FOSVM algorithm to classify the eigenvalue vectors into two SIR sets.
5. Generate power control command according to the output label of FOSVM and the distance from the data point to the SIR decision boundary.

6 Simulation and Results

In this section, we compare the performance of various SVMs in terms of classification errors, the number of support vectors and training time. The convergence performance of variable-step power control based on the FOSVM algorithm is also compared with the fixed-step power control.

The signals generated from equation (12) include random noise components and received SIR between 0dB and 15dB. The attenuation of each signal in the different path is randomly generated with a maximum attenuation of 50% of the dominant signal. The simulations include randomly generated link gains for the different paths and the number of paths is 4. The direction of arrival signal (DoA) θ_l is generated randomly from the range 0 to 180 degree and the number of antenna array elements is 8. Simulation results show that the best performance is achieved with the linear kernel, the value of the regularization parameter C is 2. Simulation results with the radial basis function (RBF) includes error rates 30% to 50% higher than error rates from simulations based on the linear kernel.

To evaluate the performance of FOSVM, we compared the FOSVM algorithm with the OSVC and conventional SVM using the same data. We orderly obtain 10 sets of training data based on the signal model, each set includes 100 training samples and 400 test samples. The training time and the number of SVs is shown in Talbe 1 and Talbe 2. Tabel 1 shows that the training speed of

Table 1. The training time of FOSVM, OSVC and conventional SVM (unit: msec)

Training set	1	2	3	4	5	6	7	8	9	10
FOSVM	221	90	0	40	10	10	10	0	10	10
OSVC	221	321	0	234	118	345	187	0	114	110
Conventional SVM	221	445	324	415	287	345	223	218	356	372

Table 2. The number of SVs in FOSVM, OSVC and conventional SVM

Training set	1	2	3	4	5	6	7	8	9	10
FOSVM	34	35	36	32	32	33	31	29	33	32
OSVC	34	36	36	36	35	33	34	29	33	34
Conventional SVM	34	39	42	43	39	38	37	38	42	45

FOSVM is faster than the other two algorithms. Because their initial training set have no difference, the training time of the first training set is the same for these algorithms. When the new training data can be classified by the current hyperplane, FOSVM and OSVC algorithms need not train the new hyperplane, so the training time of FOSVM and OSVC algorithms is zero in column 3 and 8. From Table 2, we find that the number of SVs in FOSVM is less than the other two algorithms.

Table 3 presents the percentage of mean classification error of the three algorithms for 5 different desired SIR. Each simulation includes 100 training samples and 10 set test samples. It can be seen that the percentage of classification error of these SVM algorithms are almost the same, which means their generalization capability are comparative.

Table 3. Percentage of mean classification error of three different SVM algorithms

SIR threshold	FOSVM	OSVC	Conventional SVM
11dB	6.811	6.809	6.813
9 dB	7.606	7.608	7.618
7 dB	8.085	8.091	8.093
5 dB	7.513	7.513	7.515
3 dB	7.130	7.132	7.138

Fig.3(a) shows one received signal with 15dB SIR converges to the desired SIR (7dB) through 100 iterations. There is a little vibration near 7 dB, because at the points which are misclassified by the decision boundary, the outputs of classifier are opposite to the real value, then the power control system will send the wrong commands. Fig.3(b) shows the power control command responding to the output of the FOSVM algorithm. A "-1" represents a received SIR greater than the desired SIR and therefore corresponds to a power down command, a "1" represents a received SIR lower than the desired SIR and therefore corresponds to a power up command.

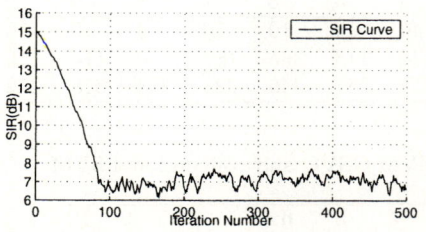
(a) convergence performance of FOSVM variable-step power control algorithm

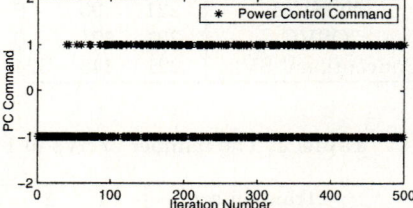
(b) power control command responding to the output of the FOSVM algorithm

Fig. 3. FOSVM algorithm for variable step CDMA power control

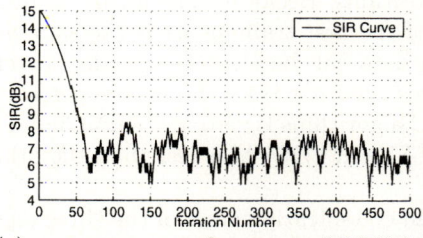
(a) convergence performance of FOSVM power control algorithm with step=0.5

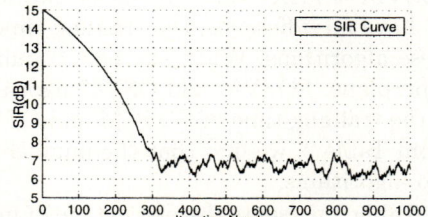
(b) convergence performance of FOSVM power control algorithm with step=0.1

Fig. 4. FOSVM algorithm for fixed step CDMA power control

Fig.4 shows the convergence performance of fixed-step power control with FOSVM algorithm. When the size of fixed step is large, the transmitted signal vibrates sharply near the 7 dB. When the size of step is small, it costs much more time for the transmitted signal converges to 7 dB.

7 Conclusion

In this paper, a FOSVM algorithm was proposed as the modified algorithm of OSVC. The FOSVM algorithm speeds up the training phase by reducing the size of training sample set using K.K.T. conditions. Simulation results show that the FOSVM outperform the OSVC and conventional SVM in term of the number of SVs and training time while keeping the comparable classification errors. According to the distance from the data point to the SIR decision boundary, we present a variable-step CDMA power control method based on the FOSVM algorithm. It offers better convergence performance than fixed-step algorithm.

References

1. Zander, J.: Performance of Optimum Transmitter Power Control in Cellular Radio Systems. IEEE Transaction On Vechicular Technology, vol.41, no.1, 57-62, February 1992.
2. Zander, J.: distributed Cochannel Interference Control in Cellular Radio Systems. IEEE Transaction On Vechicular Technology, vol.41, no.3, 305-311, August 1992.
3. Ulukus, S., Roy D.Yates.: Stochastic Power Control for Cellular Radio Systems. IEEE Transactions On Communications, vol.46, no.6, 784-798, June 1998.
4. Vapnik, V.: Statistical Learning Theory. Wiley-Interscience Publication, 1998.
5. Lau, K.W., Wu, Q.H.: Online training of support vector classifier. Patten Recognition, 36(2003): 1913-1920.
6. Suykens, J., Lukas, L., Vandewalle, J.: Least squares support vector machine classifiers. Neural Processing Letters, Vol.9, 3, 293-300, June 1999.
7. Rohwer, J.A., Abdallah, C.T., Christodoulou, C.G.: Least squares support vector machines for fixed-step and fixed-set CDMA power control. Decision and Control, 2003. Proceedings. 42nd IEEE Conference on, vol. 5 , 5097 - 5102, Dec. 2003.
8. Nello, C., John, S.T.: An Introduction to Support Vector Machines. London: Cambridge University Press, 2000
9. Platt, J.: Fast training of support vector machines using sequential minimal optimization. Scholkopf, B., Burges, C., Smola, A.J.: Advances in Kernel Method-Support Vector Learning. Cambridge, MIT Press, 1999. 185-208
10. Joachims, T.: Making large-scale support vector machine learning practical. Scholkopf, B., Burges, C., Smola, A.J.:Advances in Kernel Methods–Support Vector Learning. Cambridge, MIT Press, 1998.
11. Osuna, E., Freund, R., Girosi, G.: An Improved Training Algorithm for Support Vector Machines. IEEE Workshop on Neural Network for Signal Processing, Amelia Island , 1997. 276-285
12. Farrokh, R.F., Leandros, T., Ray Liu, K.J.: Joint optimum power control and beamforming in wireless network using antenna arrays. IEEE Transaction On Communications, vol.46, no.10, 1313-1324, October 1998.

Fourth-Order Cumulants and Neural Network Approach for Robust Blind Channel Equalization

Soowhan Han[1], Kwangeui Lee[1], Jongkeuk Lee[2], and Fredric M. Ham[3]

[1] Department of Multimedia Engineering,
Dongeui University, Busan, Korea 614-714
{swhan, kelee}@deu.ac.kr
[2] Department of Computer Engineering,
Dongeui University, Busan, Korea 614-714
jklee@deu.ac.kr
[3] Department of Electrical and Computer Engineering,
Florida Institute of Technology, Melbourne, Florida 32901, USA
fmh@ee.fit.edu

Abstract. This study addresses a new blind channel equalization method using fourth-order cumulants of channel inputs and a three-layer neural network equalizer. The proposed algorithm is robust with respect to the existence of heavy Gaussian noise in a channel and does not require the minimum-phase characteristic of the channel. The transmitted signals at the receiver are over-sampled to ensure the channel described by a full-column rank matrix. It changes a single-input/single-output (SISO) finite-impulse response (FIR) channel to a single-input/multi-output (SIMO) channel. Based on the properties of the fourth-order cumulants of the over-sampled channel inputs, the iterative algorithm is derived to estimate the deconvolution matrix which makes the overall transfer matrix transparent, i.e., it can be reduced to the identity matrix by simple reordering and scaling. By using this estimated deconvolution matrix, which is the inverse of the over-sampled unknown channel, a three-layer neural network equalizer is implemented at the receiver. In simulation studies, the stochastic version of the proposed algorithm is tested with three-ray multi-path channels for on-line operation, and its performance is compared with a method based on conventional second-order statistics. Relatively good results, with fast convergence speed, are achieved, even when the transmitted symbols are significantly corrupted with Gaussian noise.

1 Introduction

In digital communication systems, data symbols are transmitted at regular intervals. Time dispersion, which is caused by non-ideal channel frequency response characteristics or multi-path transmission, may create inter-symbol interference (ISI). This has become a limiting factor in many communication environments. Thus, channel equalization is necessary and important with respect to ensuring reliable digital communication links. The conventional approach to channel equalization needs an initial training period with a known data sequence to learn the channel characteristics. In contrast to standard equalization methods, the so-called blind (or self-recovering) channel equalization method does not require a training sequence from the transmitter [1]-[3]. It has two obvious advantages. The first is the bandwidth savings resulting from elimination of training sequences. The second is the self-start capability before the communication link is established or after it experiences an unexpected breakdown. Because of these advantages, blind channel equalization has gained practical interest during the last decade.

Recently, blind channel equalization based on second-order cyclostationary has been receiving increasing interest. The algorithm presented by Tong et al. [4] is one of the first subspace-based methods exploiting only second-order statistics for a system with channel diversity that has a single-input/multi-output (SIMO) discrete-time equivalent model. After their work, a number of different second-order statistical (SOS) methods have been proposed [5]-[10]. However, it should be noted that most SOS methods require a relatively high signal-to-noise ratio (SNR) to achieve reliable performance. In practice, the performance degradation using SOS methods is severe if a received signal is significantly corrupted by noise. In this case, a larger sample size is necessary [4]. To avoid this problem, higher-order statistics (HOS) can be exploited. Several recent works have re-established the robustness of higher-order statistical methods in channel equalization and identification [11]-[13].

In this study, a new iterative algorithm based on the fourth-order cumulants of over-sampled channel inputs is derived to estimate the deconvolution (equalization) matrix which makes the overall transfer matrix transparent, i.e., it can be reduced to the identity matrix by simple reordering and scaling. This solution is chosen so that the fourth-order statistics of the equalized output sequence $\{\hat{s}(k)\}$ is close to the fourth-order statistics of the channel input sequence $\{s(k)\}$. It has a similar formulation with the cumulant-based iterative inversion algorithm which was introduced by Cruces et al. [14] for blind separation of independent source signals, but the iterative solution in our algorithm is extended with an additional constraint (a fourth-order statistical relation between the equalized outputs of over-sampled channels) in order to be applied to the blind channel equalization problem. In the experiments, the proposed iterative solution provides more precise estimates of the deconvolution matrix with fast convergence speeds than a method based on second-order statistics, even when the outputs of a non-minimum phase channel are corrupted by heavy Gaussian noise. However, this deconvolution matrix may yield to an amplification of the noise at the outputs because of noise-corrupted inputs, even though it can be precisely estimated from the noisy channel outputs. To avoid this limitation, a three-layer neural equalizer, instead of the deconvolution matrix itself, is implemented at the receiver by using the over-sampled channel matrix (inverse of estimated deconvolution matrix). It is known that the equalizer made of neural network structure has a better noise-tolerant characteristic [15]-[17].

2 Problem Formulation

In a multi-path digital communication system, a data sequence $\{s(k)\}$, $k=\ldots, -1, 0, 1, 2,\ldots$, is sent over a communication channel with a time interval T. The channel is characterized by a continuous function $h(t)$, and the signals may be corrupted by noise $e(t)$. The received signal $y(t)$ can be expressed as:

$$x(t) = \sum_{-\infty}^{+\infty} s(k)h(t-kT) \tag{1}$$

$$y(t) = x(t) + e(t) \tag{2}$$

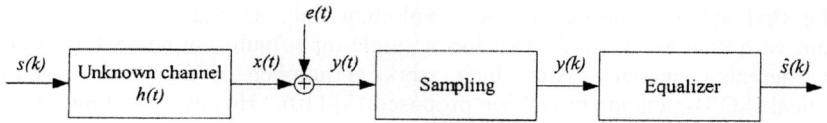

Fig. 1. Blind channel equalization in digital communication

This is shown in Fig. 1. The objective of blind equalization is to recover the transmitted input symbol sequence $\{s(k)\}$ given only the received signal $y(t)$. Instead of choosing the equalizer so that the equalized output sequence $\{\hat{s}(k)\}$ is close to the source symbol sequence $\{s(k)\}$, as in the standard equalization formulation, in blind equalization one chooses the equalizer so that the statistics of the equalized output sequence is close to the statistics of the source symbol sequence. In this study, a robust algorithm with respect to noise is constructed with a higher-order statistical constraint, which makes the fourth-order statistics of $\{\hat{s}(k)\}$ close to the fourth-order statistics of $\{s(k)\}$. For this approach, the following assumption is necessary.

1> The symbol interval T is known and is an integer multiple of the sampling period.
2> The impulse response $h(t)$ has finite support, if the duration of $h(t)$ is L_h, $h(t) = 0$ for $t \prec 0$ or $t \geq L_h$.
3> $\{s(k)\}$ is zero mean, and is driven from a set of i.i.d. random variables, which means the fourth-order zero-lag cumulant or kurtosis of $\{s(k)\}$ can be expressed by

$$C_{s(k),s(l)}^{1,3}(0) = cum(s(k), s(l), s(l), s(l)) = E\{s(k)s^*(l)s(l)s^*(l)\} = \alpha\delta(k-l) \tag{3}$$

where α is non-zero constant and $\delta(t)$ is the discrete time impulse function.
4> $e(t)$ is zero-mean Gaussian noise, and uncorrelated with $\{s(k)\}$.

In the conventional equalizer, the incoming signal, $y(t)$, is spaced by a sampling rate T/N at the receiver, where T is a source symbol interval and N is an positive integer. In this study, the over-sampling technique is applied to change a finite-impulse response (FIR) channel to a SIMO channel, which requires the incoming signal $y(t)$ to be sampled at least as fast as the Nyquist rate ($N \geq 2$). This is illustrated by way of an example shown in Fig. 2, where the channel lasts for 4 adjacent bauds, and the over-sampling rate is $T/4$.

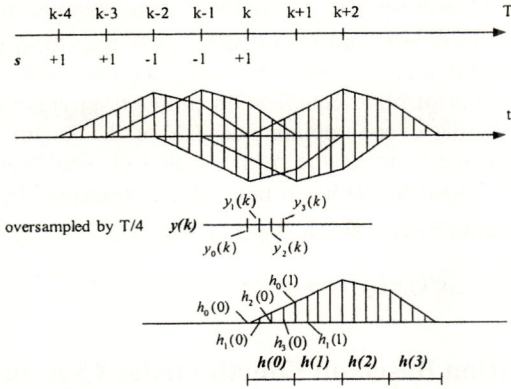

Fig. 2. An over-sampling example of a FIR channel

With over-sampling at rate $T/4$ during observation interval $L=T$ in Fig. 2, a channel output vector at time index k is given by equation (4). If we define a FIR channel $h(t)$ as in equation (5), $y_0(k)$ and $y_1(k)$ can be expressed as in equations (6) and (7), respectively. In the same way, $y_2(k)$ and $y_3(k)$ can be obtained.

$$y(k)=[y_0(k), y_1(k), y_2(k), y_3(k)]^T \qquad (4)$$

$$h(0)=[h_0(0), h_1(0), h_2(0), h_3(0)]^T \qquad (5)$$

$$y_0(k)=h_0(0)s(k)+h_0(1)s(k-1)+h_0(2)s(k-2)+h_0(3)s(k-3)+e_0(k) \qquad (6)$$

$$y_1(k)=h_1(0)s(k)+h_1(1)s(k-1)+h_1(2)s(k-2)+h_1(3)s(k-3)+e_1(k) \qquad (7)$$

Then we have

$$y(k)=Hs(k)+e(k) \qquad (8)$$

where $s(k)=[s(k),s(k-1),s(k-2),s(k-3)]^T$, $e(k)=[e_0(k),e_1(k),e_2(k),e_3(k)]^T$ and

$$H=\begin{bmatrix} h_0(0) & h_0(1) & h_0(2) & h_0(3) \\ h_1(0) & h_1(1) & h_1(2) & h_1(3) \\ h_2(0) & h_2(1) & h_2(2) & h_2(3) \\ h_3(0) & h_3(1) & h_3(2) & h_3(3) \end{bmatrix} = [h(0),h(1),h(2),h(3)] \qquad (9)$$

If the observation interval L is greater than T, for example $L=2T$ in Fig. 2, $y(k)=[y_0(k),y_1(k),y_2(k),y_3(k),y_4(k),y_5(k),y_6(k),y_7(k)]^T$, $s(k)=[s(k+1),s(k),s(k-1),s(k-2),s(k-3)]^T$, $e(k)=[e_0(k),e_1(k),e_2(k),e_3(k),e_4(k),e_5(k),e_6(k),e_7(k)]^T$, and H becomes a 8×5 channel matrix shown in equation (10).

$$H=\begin{bmatrix} 0,h(0),h(1),h(2),h(3) \\ h(0),h(1),h(2),h(3),0 \end{bmatrix} \qquad (10)$$

where $0=[0,0,0,0]^T$.

In our approach to recover the transmitted input symbol sequence $\{s(k)\}$, a deconvolution matrix G in equation (11) is derived to transform the overall transfer function $W=GH$ into the identity matrix by using the observed channel output $y(k)$ only. For the solvability of blind equalization problem, an additional assumption is made throughout, i.e., the over-sampling rate T/N or the length of the observation interval L, qT, is selected to make the over-sampled channel matrix H full column rank. This means if a channel $h(t)$ has p taps, H can described by a $Nq\times(p+q-1)$ matrix, and N or q should be chosen for $Nq \geq (p+q-1)$.

$$\hat{s}(k) = Gy(k) = GHs(k) = Ws(k) \tag{11}$$

3 Iterative Solution Based on Fourth-Order Cumulants

The aim in blind equalization is to select G in equation (11) that recovers the original source sequence $\{s(k)\}$ only from the observations of the sampled channel output $y(k)$. This is obtainable when the overall transfer system W is transparent (or reduced to an identity). Here, for notational simplicity, we consider a special reordering and scaling so that W will always be an identity matrix. If the over-sampled channel H is a $Nq\times(p+q-1)$ matrix and full column rank, its input sequences can be expressed as in equation (12).

$$s = \begin{bmatrix} s_{p+q-2} \\ \vdots \\ s_1 \\ s_0 \end{bmatrix} = \begin{bmatrix} s(p+q-2) & s(p+q-1) & \cdots & s(M-1) \\ \vdots & \vdots & \cdots & \vdots \\ s(1) & s(2) & \cdots & s(M-(p+q-2)) \\ s(0) & s(1) & \cdots & s(M-(p+q-1)) \end{bmatrix} \tag{12}$$

where M is the total number of transmitted sequences and $s_0, s_1, \cdots, s_{p+q-2}$ are the shifted input vectors by time interval T for each of $p+q-1$ over-sampled FIR channels. Then, for the noise-free case, equation (11) can be rewritten as

$$\hat{s} = \begin{bmatrix} \hat{s}_{p+q-2} \\ \vdots \\ \hat{s}_1 \\ \hat{s}_0 \end{bmatrix} = GH \begin{bmatrix} s_{p+q-2} \\ \vdots \\ s_1 \\ s_0 \end{bmatrix} = Ws \tag{13}$$

The identifiability of system W can be guaranteed because the channel H has full column rank and its input vectors, $s_0, s_1, \cdots, s_{p+q-2}$, are mutually independent [18]. Equation (13) can be considered as a blind source separation (BSS) problem. If we properly scale channel input s such that the kurtosis of each of $s_0, s_1, \cdots, s_{p+q-2}$ is equal to +1 or −1 (scaled to $|\alpha|=1$ in equation (3)), its BSS solution by using a preconditioned iteration [19], is given by equation (14) [14].

$$G^{(n+1)} = G^{(n)} - \mu^{(n)}(C^{1,3}_{\hat{s}_k,\hat{s}_l}(0)S_{\hat{s}}^3 - I)G^{(n)} \tag{14}$$

where $C^{1,\,3}_{\hat{s}_k,\hat{s}_l}(0) = cum(\hat{s}_k,\hat{s}_l,\hat{s}_l,\hat{s}_l) = E\{\hat{s}_k\hat{s}_l^*\hat{s}_l\hat{s}_l^*\}$: the fourth-order zero-lag cumulant or kurtosis matrix of $\hat{s}$ ($k,l=0,1,...,p+q-2$), $S^3_{\hat{s}} = diag(sign(diag(C^{1,\,3}_{\hat{s}_k,\hat{s}_l}(0))))$ in the Matlab convention, μ = a step-size of iteration, and I is an identity matrix. The fundamental idea of this solution is based on the fact that the fourth-order statistics of equalizer output $\hat{s}$ should be close enough to the fourth-order statistics of channel input s. However, in order to apply the BSS method in equation (14) to the blind channel equalization problem, an additional constraint must be considered. It is as follows.

The channel input $s = [s_{p+q-2}, \cdots, s_1, s_0]^T$ is constructed by shifting the same sequences with a time interval T, which is shown in equation (12). It means the fourth-order cumulant matrix of s with lag 1 always satisfies the following expression

$$C^{1,3}_{s_k,s_l}(1)J^T S^3_{sJ} = JJ^T \tag{15}$$

where $C^{1,3}_{s_k,s_l}(1) = cum(s_k, s_{l+1}, s_{l+1}, s_{l+1}) = E\{s_k s_{l+1}^* s_{l+1} s_{l+1}^*\}$, J is a shifting matrix denoted by equation (16), and $S^3_{sJ} = diag(sign(diag(C^{1,\,3}_{s_k,s_l}(1)J^T)))$.

$$J = \begin{bmatrix} 0 & 0 & \cdots & 0 & 0 \\ 1 & 0 & \cdots & 0 & 0 \\ 0 & 1 & \cdots & 0 & 0 \\ \vdots & \vdots & \vdots & \vdots & \vdots \\ 0 & 0 & \cdots & 1 & 0 \end{bmatrix} \tag{16}$$

Thus, the fourth-order cumulant matrix of equalizer output $\hat{s}$ with lag 1 should be forced to satisfy equation (15), and its iterative solution can be written as

$$G^{(n+1)} = G^{(n)} - \beta^{(n)}(C^{1,3}_{\hat{s}_k,\hat{s}_l}(1)J^T S^3_{sJ} - JJ^T)G^{(n)} \tag{17}$$

where β = a step-size of iteration. Based on the above analysis, a new iterative solution combining equation (14) with equation (17) is derived for blind channel equalization, which is shown in equation (18).

$$G^{(n+1)} = G^{(n)} - \mu^{(n)}(C^{1,3}_{\hat{s}_k,\hat{s}_l}(0)S^3_{\hat{s}} - I)G^{(n)} - \beta^{(n)}(C^{1,3}_{\hat{s}_k,\hat{s}_l}(1)J^T S^3_{sJ} - JJ^T)G^{(n)} \tag{18}$$

For the stability of equation (18), $G^{(n+1)}$ in equations (14) and (17) should not to be singular [14]. In equation (14), $G^{(n+1)}$ can be rewritten as in equation (19) and $\|\Delta^{(n)}\|$ should be less than 1 to avoid the singularity. Therefore, by taking into account the triangular inequality such as $\|C^{1,3}_{\hat{s}_k,\hat{s}_l}(0)S^3_{\hat{s}} - I\| \leq 1 + \|C^{1,3}_{\hat{s}_k,\hat{s}_l}(0)S^3_{\hat{s}}\| = 1 + \|C^{1,3}_{\hat{s}_k,\hat{s}_l}(0)\|$, the step size $\mu^{(n)}$ is chosen as $\mu^{(n)} < \dfrac{1}{1+\left\|C^{1,3}_{\hat{s}_k,\hat{s}_l}(0)\right\|}$ for the stability. By the same way, $\beta^{(n)}$ is selected as $\beta^{(n)} < \dfrac{1}{1+\left\|C^{1,3}_{\hat{s}_k,\hat{s}_l}(1)J^T\right\|}$ in the experiments.

$$G^{(n+1)} = G^{(n)} - \mu^{(n)}(C^{1,3}_{\hat{s}_k,\hat{s}_l}(0)S_{\hat{s}}^3 - I)G^{(n)} = (I - \Delta^{(n)})G^{(n)} \qquad (19)$$

If the formulation of equation (18) is based on the second-order statistics of equalizer output and the channel input s is scaled to have a unity power, the iterative solution is reduced as

$$G^{(n+1)} = G^{(n)} - \mu^{(n)}(C^{1,1}_{\hat{s}_k,\hat{s}_l}(0) - I)G^{(n)} - \beta^{(n)}(C^{1,1}_{\hat{s}_k,\hat{s}_l}(1)J^T - JJ^T)G^{(n)} \qquad (20)$$

where $C^{1,1}_{\hat{s}_k,\hat{s}_l}(0) = cum(\hat{s}_k,\hat{s}_l) = E\{\hat{s}_k\hat{s}_l^*\}$ and $C^{1,1}_{\hat{s}_k,\hat{s}_l}(1) = cum(\hat{s}_k,\hat{s}_{l+1}) = E\{\hat{s}_k\hat{s}_{l+1}^*\}$: zero-lag and lag 1 correlation function of $\hat{s}$, respectively. These two iterative solutions have been implemented in a batch manner in order to obtain an accurate comparison, and tested with three-ray multi-path channels. In our experiments, their stochastic versions, which are shown in equation (21) for the fourth-order statistics and in equation (22) for the second-order statistics, are evaluated for possible use on-line. These are accomplished by estimating continuously the fourth-order cumulants in equation (18) and the second-order correlations in equation (20) with the over-sampled channel outputs coming in at time interval T. Thus, G gets updated at time interval T. By applying these stochastic versions of algorithm, it is not necessary to wait until a whole block of the sample is received to estimate G. The stochastic version based on second-order statistics in equation (22) is the same as the one used by Fang et al. [5] for their two-layer neural network equalizer. It is compared with our proposed algorithm based on the fourth-order statistics shown in equation (21).

$$G^{(n+1)} = G^{(n)} - \mu^{(n)}(f(\hat{s}_i^{(n-1)})(\hat{s}_i^{(n-1)})^T S_{f\hat{s}}^3 - I)G^{(n)} - \beta^{(n)}(f(\hat{s}_i^{(n)})(\hat{s}_i^{(n-1)})^T J^T S_{f\hat{s}J}^3 - JJ^T)G^{(n)} \qquad (21)$$

$$G^{(n+1)} = G^{(n)} - \mu^{(n)}(\hat{s}_i^{(n-1)}(\hat{s}_i^{(n-1)})^T - I)G^{(n)} - \beta^{(n)}(\hat{s}_i^{(n)}(\hat{s}_i^{(n-1)})^T J^T - JJ^T)G^{(n)} \qquad (22)$$

where $\hat{s}_i^{(n)} = [\hat{s}_{p+q-2}^{(n)}, \cdots, \hat{s}_1^{(n)}, \hat{s}_0^{(n)}]^T$: a $(p+q-1)\times 1$ output vector of $G^{(n)}$, $f(\hat{s}_i^{(n)}) = (\hat{s}_i^{(n)})^3 - 3\hat{s}_i^{(n)}\sigma_{\hat{s}_i}^2$, $\sigma_{\hat{s}_i}^2$: adaptively estimated power of $\hat{s}_i$ at each iteration, $S_{f\hat{s}}^3 = diag(sign(diag(f(\hat{s}_i^{(n-1)})(\hat{s}_i^{(n-1)})^T)))$ and $S_{f\hat{s}J}^3 = diag(sign(diag(f(\hat{s}_i^{(n)})(\hat{s}_i^{(n-1)})^T J^T)))$.

4 Neural Network-Based Equalizer

In the absence of noise, the deconvolution matrix G perfectly recovers the source symbols at the output because of the overall transfer function $W=GH=I$. However, when there is noise, this deconvolution matrix may yield to an amplification of the noise at its outputs, even though it can be precisely estimated from the noisy channel outputs *y* by using our proposed algorithm. To avoid this limitation, a three-layer neural equalizer is employed at the receiver because of its noise robust characteristic [15]-[17]. This is done by using the estimated over-sampled channel as a reference system to train the neural equalizer. It consists of an input layer, a hidden layer, and an output layer of processing elements called neurons [15][16], as shown in Fig. 3.

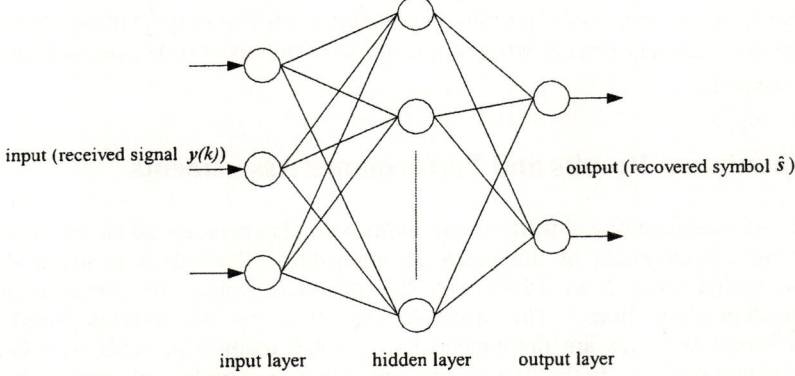

Fig. 3. The structure of three-layer neural equalizer

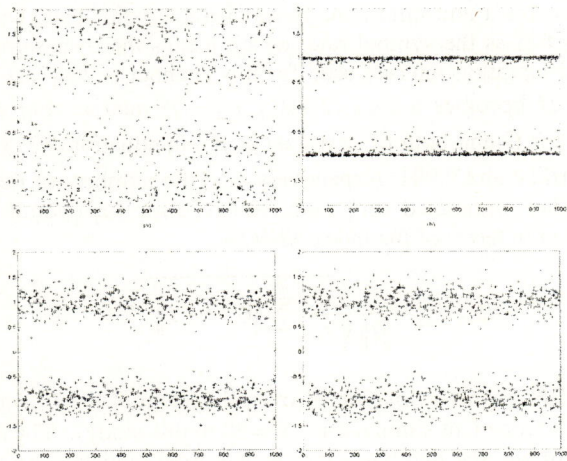

Fig. 4. Samples of received and equalized symbols under 15db SNR: (a) 1000 received symbols, (b) equalization by a neural equalizer, (c) by G itself derived from eq. (21), and (d) by the optimal inverse of H

Once the deconvolution matrix G is estimated, which means the over-sampled channel H is available, the training sequences based on H are generated at the receiver. The three-layer neural equalizer is trained with these sequences by using the back-propagation algorithm. In back-propagation, the output value is compared with the desired output. This results in the generation of an error signal, which is fed *back* through layers in the network, and the weights are adjusted to minimize the error. More details on the back-propagation algorithm can be found in [15][16]. A sample of equalized binary (+1,−1) source symbols under 15dB SNR by this neural equalizer, one by the deconvolution matrix G itself, and one by the optimal inverse of over-sampled channel H are shown in Fig. 4. The deconvolution matrix G used in Fig. 4 is derived from the proposed algorithm in equation (21). The outputs of neural equalizer can be more densely placed onto the transmitted symbols (+1,−1) even in heavy noise environments.

5 Simulation Results and Performance Assessments

The blind equalizations with three-ray multi-path channels are taken into account to show the effectiveness of the proposed algorithm. Performances under different SNRs, varied from 5 to 15dB with 2.5 dB increments, are averaged after 50 independent simulations. The proposed algorithm and the solution based on the second-order statistics are implemented in a batch manner in order to achieve the accurate comparison. In the first experiment, a three-ray multi-path channel truncated up to 2 symbol periods ($p=2$) is tested with 1000 randomly generated binary transmitted symbols (taken from $\{\pm 1\}$). The delays of this channel are *0.5T* and *1.1T*, and its waveform is a raised-cosine pulse with 11% roll-off. It has a *zero* outside unit circle, which indicates a non-minimum phase characteristic. The channel outputs are sampled twice as fast as the symbol rate, which means the over-sampling rate is *T/2* (*N=2*), and the observation interval used for this channel is *T (q=1)*. Thus, the over-sampled channel H becomes a 2×2 ($Nq\times(p+q-1)$) matrix. For each simulation, the initial matrix for G and both of step size (μ, β) in equations (21) and (22) are set to an identity matrix I and 0.001, respectively, and the numbers of iteration is limited to 50 epochs. The normalized root-mean square error for overall transfer system $W=GH$ is measured in terms of the index $NRMSE_w$,

$$NRMSE_w = \frac{1}{\|I\|}\sqrt{\frac{1}{NS}\sum_{j=1}^{NS}\|W^{(j)} - I\|^2} \qquad (23)$$

where $W^{(j)} = G^{(j)}H$ is the estimation of overall system at the jth simulation and NS is the number of independent simulations ($NS=50$ in this study). The $NRMSE_w$ for the proposed algorithm and the one based on second-order statistics are shown in Fig. 5 with different noise levels.

Once G is available, the three-layer neural equalizer is trained with 1000 training sequences which have been generated at the receiver. It has 2 inputs, 4 hidden neurons and 2 outputs, and the learning rate is set to 0.05. The maximum number of iterations for the training process is set to 50 epochs. The output of this neural equalizer is the estimation of transmitted symbols, and its performance measure is defined as follows.

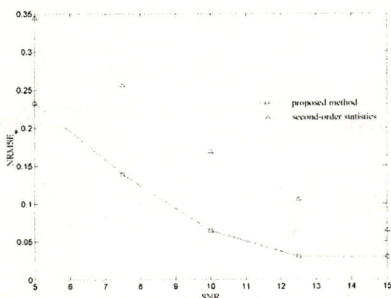

Fig. 5. $NRMSE_w$ with different SNR levels in experiment 1

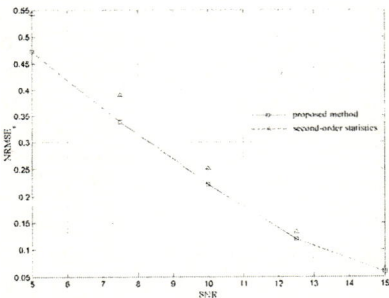

Fig. 6. $NRMSE_s$ with different SNR levels in experiment 1

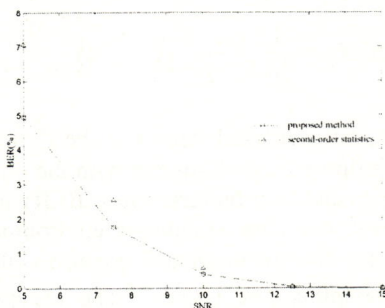

Fig. 7. Averaged BER(%) in experiment 1

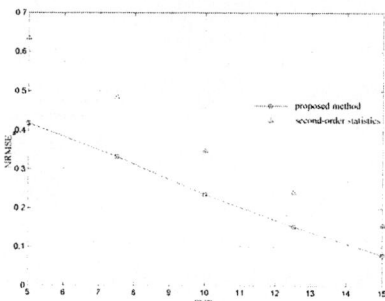

Fig. 8. $NRMSE_w$ with different SNR levels in experiment 2

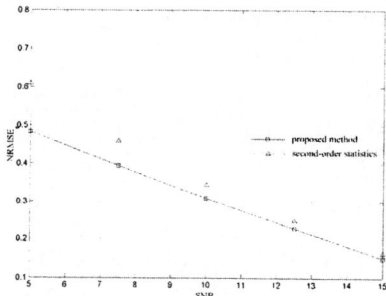

Fig. 9. $NRMSE_s$ with different SNR levels in experiment 2

$$NRMSE_s = \frac{1}{\|s\|}\sqrt{\frac{1}{NS}\sum_{j=1}^{NS}\left\|\hat{s}^{(j)} - s\right\|^2} \qquad (24)$$

where $\hat{s}^{(j)}$ is the estimate of the channel input s at the jth trial. The $NRMSE_s$ by the neural equalizer with the proposed algorithm and with the one based on second-order statistics are shown in Fig. 6, and their bit error rates (BER) are compared in Fig. 7.

In the second experiment, the same simulation environment is used, such as the step size(μ, β), the learning rate for the neural equalizer, the maximum number of iterations, and the over-sampling rate ($N=2$). The exceptions are the length of channel, its delays and the observation interval. The three-ray multi-path channel tested at this time is truncated up to 3 symbol periods ($p=3$), and its delays are T and $1.5T$. It has one *zero* outside unit circle and the other inside. The observation interval used for this non-minimum phase channel is two times longer than one symbol period, $2T$ ($q=2$), and thus, the over-sampled channel H becomes a 4×4 ($Nq\times(p+q-1)$) matrix. The neural equalizer used to recover the transmitted symbols in this experiment has 4 inputs, 8 neurons in the hidden layer, and 4 outputs. The performance measures, $NRMSE_w$, $NRMSE_s$ after 50 independent simulations, and the averaged BER, are presented in Figs. 8-10, respectively.

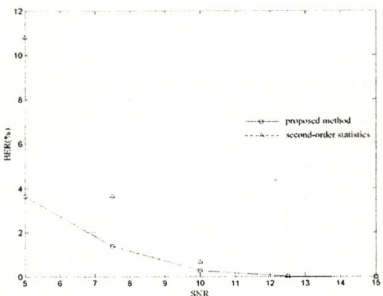

Fig. 10. Averaged BER(%) in experiment 2

From the simulation results for $NRMSE_w$, which are shown in Fig. 5 for experiment 1 and in Fig. 8 for experiment 2, the proposed solution is proved highly effective to estimate G, the inverse of unknown channel H, which makes the overall system $W=GH$ an identity even when the observed symbols are heavily corrupted by noise.

The difference in performance between the proposed solution and the one based on the second-order statistics is not severe if the noise(signal) level is as low(high) as 15dB SNR in our experiments. However, it is observed that, if the noise level is getting higher such as to 10 or 5 db SNR, the proposed algorithm performs relatively well, and the performance difference becomes more serious. It results from the fact that our approach is based on the fourth-order cumulant of the received symbols and it always goes to zero for Gaussian noise. This phenomenon can also be found for $NRMSE_s$ in Figs. 6 and 9, and the averaged BER in Figs. 7 and 10, because the neural equalizer trained with more accurate estimation of H produces the lower symbol estimation error. Therefore, the proposed algorithm in our study can be implemented for on-line operation in a *heavy noise* communication environment.

6 Conclusions

In this paper, a new iterative solution based on the fourth-order cumulants of over-sampled channel inputs is presented for blind channel equalization. It does not require the minimum phase characteristic, and shows relatively high performance results even when the observed symbols are significantly corrupted by Gaussian noise. In addition, it can be implemented for on-line operation for channel estimation without waiting for a whole block of the symbols. The proposed algorithm could possibly be used for heavy noise communication environments. In future work, the proposed iterative solution will be further investigated and applied as a learning algorithm for a neural network so that the transmitted symbols can be directly recovered from the output of a neural-based equalizer without the estimation procedure of the deconvolution matrix, G.

References

1. J.G. Proakis: Digital Communications, Fourth Edition. McGraw-Hill, New York (2001)
2. A. Benveniste, M. Goursat, G. Ruget: Robust identification of a nonminimum phase system: Blind adjustment of a linear equalizer in data communications. IEEE Trans. Automat. Contr., (1980) 385-399
3. Y. Sato: A method of self-recovering equalization for multilevel amplitude modulation. IEEE Trans. Commun., Vol.23. No.6. (1975) 679-682
4. L. Tong, G. Xu, T. Kailath: Blind identification and equalization based on second-order statistics: a time domain approach. IEEE Trans. Inform. Theory, Vol.40. (1994) 340-349
5. Y. Fang, W.S. Chow, K.T. Ng: Linear neural network based blind equalization. Siganl Processing, Vol.76. (1999) 37-42
6. E. Serpedin, G.B. Giannakis: Blind channel identification and equalization with modulation-induced cyclostationarity. IEEE Trans. Siganl Processing, Vol.46 (1998) 1930-1944
7. Y. Hua: Fast maximum likelihood for blind identification of multiple FIR channels. IEEE Trans. Signal Process., Vol.44. (1996) 661-672
8. M. Kristensson, B. Ottersten: Statiscal analysis of a sub-space method for blind channel identification. Proc. IEEE ICASSP, Vol.5. Atlanta, U.S.A. (1996) 2435-2438
9. W. Qiu, Y.Hua: A GCD method for blind channel identification. Digital Signal Process., Vol.7. (1997) 199-205
10. G. Xu, H. Liu, L. Tong, T. Kailath: A least-squares approach to blind channel identification. IEEE Trans. Signal Processing, Vol.43. (1995) 2982-2993
11. Z. Ding, J. Liang: A cumulant matrix subspace algorithm for blind single FIR channel identification. IEEE Trans. Signal Processing, Vol.49. (2001) 325-333
12. D. Boss, K. Kameyer, T. Pertermann: Is blind channel estimation feasible in mobile communication systems? A study based on GSM. IEEE J. Select. Areas Commun., Vol.16. (1998) 1479-1492
13. Z. Ding, G. Li: Single channel blind equalization for GSM cellular systems. IEEE J. Select. Areas Commun., Vol.16. (1998) 1493-1505
14. S. Cruces, L. Castedo, A. Cichocki: Robust blind source separation algorithms using cumulants. Neurocomputing, Vol.49. (2002) 87-118
15. Ham, F.M., Kostanic, I.: Principles of Neurocomputing for Science and Engineering. McGraw-Hill, New York (2001)
16. Fausett, L.: Fundamentals of Neural Networks: Architectures, Algorithm, and Applications. Prentice Hall (1994)
17. Shaomin Mo, Bahram Shafai: Blind equalization using higher-order cumulants and neural network. IEEE Trans. Signal Processing, Vol.42. (1994) 3209-3217
18. X.R. Cao, R.W. Liu: General approach to blind source separation. IEEE Trans. Signal Processing, Vol.44. (1996) 562-571
19. C.T. Kelly: Iterative methods for linear and nonlinear equations. Frontiers in Applied Mathematics, Vol.16. SIAM (1995) 71-78

Equalization of a Wireless ATM Channel with Simplified Complex Bilinear Recurrent Neural Network

Dong-Chul Park[1], Duc-Hoai Nguyen, Sang Jeen Hong, and Yunsik Lee[2]

[1] ICRL, Dept. of Info. Eng., Myong Ji University, Korea
parkd@mju.ac.kr
[2] SoC Research Center, Korea Electronics Tech. Inst.,
Seongnam, Korea

Abstract. A new equalization method for a wireless ATM communication channel using a simplified version of the complex bilinear recurrent neural network (S-CBLRNN) is proposed in this paper. The S-BLRNN is then applied to the equalization of a wireless ATM channel for 8PSK and 16QAM. The results show that the proposed S-CBLRNN converges about 40 % faster than the CBLRNN and gives very favorable results in both of the MSE and SER criteria over the other equalizers.

1 Introduction

One of the major problems in the wireless ATM network is the intersymbol interference that leads to a degradation in performance and capacity of the system. Because of this problem, the received signals tend to get elongated and smeared into each other. Several techniques have been used to build wireless ATM equalizers which aim to correct the received signal. One of the simpler types of equalizers is the Linear Equalizer (LE). With the presence of channel nonlinearities, it appears that techniques based on linear signal processing such as the LE show limited performance. Others types of equalizers such as Decision Feedback Equalizers (DFE)[1], Volterra filter based equalizers, and Viterbi equalizers are based on nonlinear signal processing techniques.

Another promising approach to designing equalizers are neural network (NN) -based approaches[2,3]. Recent researches have shown that a recurrent type NN is more suitable than a feedforward type NN in predicting time series signals[3]. Among recurrent neuron networks, the Complex Bilinear Recurrent Neural network (CBLRNN) -based equalizer gives very favorable results in both the MSE and SER criteria over Volterra filter equalizers, DFEs, and Complex Multiplayered Perceptron type NN (CMLPNN) equalizers[3]. However, the CBLRNN also suffers from slow convergence because of its rather complicated architecture.

In this paper, a simplified version of CBLRNN is proposed and applied to the equalization of wireless ATM channels. The Simplified-CBLRNN (S-CBLRNN) uses only a part of the feedback components for bilinear component calculation. The experiments are performed on 8PSK and 16QAM signals.

2 Simplified Complex Bilinear Recurrent Neural Network

The BLRNN is a recurrent NN which has a robust ability in modelling nonlinear systems and was originally introduced in[4]. CBLRNN is the complex version of BLRNN. CBLRNN has been designed to deal with the problems with complex number operations[3]. Even though the CBLRNN shows very promising results when applied to equalizer problems, it still suffers from slow convergence in practical use. Choi et al. propose a simplified version of the bilinear DFE[5]. In this approach, only a part of feedforward inputs are multiplied to the feedback portion for bilinear components without suffering from performance degradation. By adopting this idea of reduced bilinear components to the CBLRNN, the output of the CBLRNN is derived as follows:

$$s_p[n] = d_p + \sum_{k_2=0}^{N_f-1} a_{pk_2} o_p[n-k_2] \qquad (1)$$

$$+ \sum_{k_1=0}^{N_f-1} \sum_{k_2=S}^{E} b_{pk_1k_2} o_p[n-k_1] x[n-k_2]$$

$$+ \sum_{k_1=0}^{N_i-1} c_{pk_1} x[n-k_1]$$

$$= w_p + A_p{}^T A_p{}^T[n] + Z_p[n] B_p{}^T X[n] + C_p{}^T X[n]$$

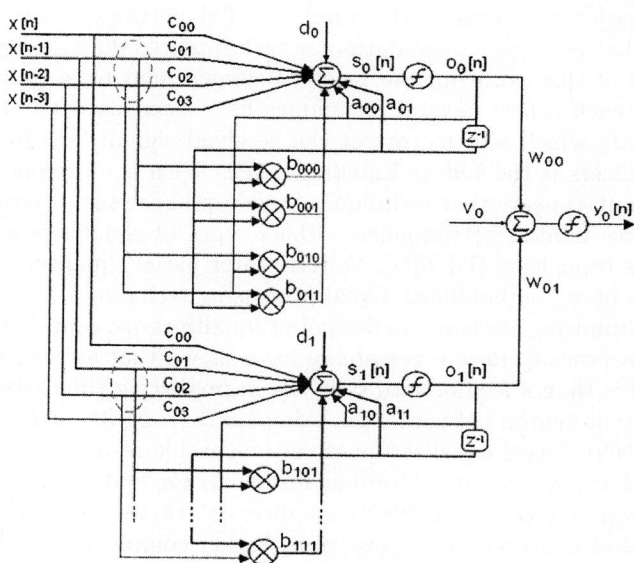

Fig. 1. A simple S-BLRNN with the structure 5-2-1 and 2 feedback lines. Note that only 3 middle inputs are used for bilinear part in this example

where d_p is the weight of bias neuron for the $p-th$ hidden neuron, $p = 1, 2..., N_h$. N_h, N_i, and N_f are the number of hidden neurons, input neurons, and feedback lines, respectively. Note that a part of N_i inputs are used for the bilinear part. A_p is the weight vector for recurrent portion, B_p is the weight matrix for the bilinear recurrent portion, and C_p is the weight vector for the feedforward portion. T represents the transpose of a matrix. More detailed on CBLRNN can be found in[3].

In a CBLRNN architecture (N_i=10, N_h=4, N_f=5), the number of multiplication for CBLRNN and the number of multiplications for S-CBLRNN used in our experiments are 264 and 155, respectively. The number of reduced multiplication is 109 in this case and 41.3% of the multiplications can be reduced.

3 Experiments and Results

The ATM channel for experiments is simulated with 6 propagation paths whose gains and delays are given by: $gain = \{a_i\} = \{$ 1, 0.5, 0.4, 0.32, 0.1, 0.08$\}$, $delay = \{\tau_i\} = \{$ 0, 1/2, 9/8, 13/8, 21/8, 39/8$\}$. The transmitted data symbols are randomly generated 8PSK and 16QAM signals. Transmitted signals over the ATM channel are corrupted with AWGN with various SNRs.

In each experiment, randomly generated 30,000 data symbols are used for training the equalizers and another 100,000 data symbols are used for testing the performance of the equalizers. The proposed equalizer based on a S-CBLRNN is compared with a CDFE, a Volterra equalizer, a CMLPNN-based equalizer, and a CBLRNN-based equalizer.

The selected structures for all equalizers are CDFE(10 inputs, 30 feedbacks), Volterra (10 inputs, $1st$ and $3rd$ orders), CMLPNN(10-25-1), CBLRNN(10-4-1, 5 feedbacks), and S-CBLRNN(10-4-1, 5 feedbacks). We observe that all equalizer

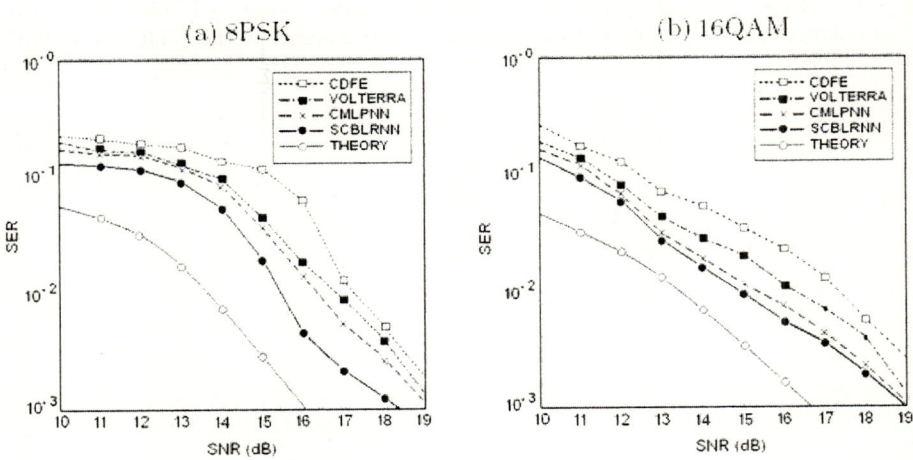

Fig. 2. SER performance of equalizers

models have 10 input tabs and the proposed S-CBLRNN requires much less computational effort when compared with Volterras, CMLPNNs, and CBLRNNs.

In the result shown in Fig. 2, the S-CBLRNN yields much less symbol error rate (SER) than any other equalizer for both cases of 8PSK and 16QAM signals. As can be seen in Fig. 2, the SER of S-CBLRNN gives some improvement over the CMLPNN and other equalizers. This result is very acceptable when compared with other equalizers reported in the literature and implies that the S-CBLRNN-based equalizer is suitable for the wireless ATM channel. Because the performance of the CBLRNN and S-CBLRNN are almost indistinguishable in both the SER and MSE categories throughout the experiments, only the results of the S-CBLRNN are given in this paper. In addition to the SER performance, the performance of mean square error (MSE) of S-CBLRNN is also lower when compared with different equalizer schemes.

Acknowledgement. This research was supported by the Korea Research Foundation (Grant # R05-2003-000-10992-0 (2004)).

References

1. Drewes,C.,Hasholzner,R.,Hammerschmidt,J.S.: Adaptive equalization for wireless ATM, 13th International Conference on Digital Signal Processing, Santorini, Greece (1997) 57-60.
2. You,C.,Hong,C.: Nonlinear blind equalization schemes using complex-valued multilayer feedforward neural networks, IEEE Transaction on Neural Networks **9** (1998) 1442-1445.
3. Park,D.C.,Jeong,T.K: Complex Bilinear Recurrent Neural Network for equalization of a satellite channel, IEEE Transactions on Neural Network **13(3)**(2002) 711-725.
4. Park,D.C.,Zhu,Y.: Bilinear Recurrent Neural Network, Proceedings of IEEE ICNN **3** (1994) 1459-1464.
5. Choi,J.,Ko,K.,Hong,D.: Equalization techniques using a simplified bilinear recursive polynomial perceptron with decision feedback, Proceedings of IEEE IJCNN **4**(2001) 2883-2888.

A Novel Remote User Authentication Scheme Using Interacting Neural Network[*]

Tieming Chen and Jiamei Cai

College of Software, Zhejiang University of Technology,
310014, Hangzhou, China
{tmchen,cjm}@zjut.edu.cn

Abstract. Recently, interacting neural network has been studied out coming a novel result that the two neural networks can synchronize to a stationary weight state with the same initial inputs. In this paper, a simple but novel interacting neural network based authentication scheme is proposed, which can provide a full dynamic and security remote user authentication over a completely insecure communication channel.

1 Introduction

Remote user authentication schemes are used to verify the validity of a login request from a remote user to obtain the access rights on the remote server. Most of schemes make the server and the remote user share a secret as a password. Recently, many password based dynamic authentication schemes using smart cards have been presented[1,2,3,4], but also many security problems discovered[5,6,7]. According to security analysis for remote user identification schemes, latest research paper[8] have discussed two key design rules. One is that at least single unique information must be required to identify an entity, another is that at least single information representing the entity must be static. Anyway, exploring a non-classic cryptography method for authentication scheme can bring some unsurpassable advantages.

The state-of-the-art research on interacting neural network[9,10] has shown a novel phenomenon that both weights of such two networks can be trained to synchronized under some specific Hebbian rules with the identical inputs. Expanding interacting neural network to multi-layer network, such as tree parity machine, a secure authentication model can be designed based on the final synchronization for the truth that only the same input vectors lead to the same weight vectors.

2 Neural Network Based Synchronization

2.1 Synchronization in Interacting Neural Network

A simple interacting system is modeled that two perceptrons receive a common random input vector $\underline{x}$ and modify their weights $\underline{w}$ according to their mutual bit σ. The

---

[*] Supported by Zhejiang Province Natural Science Fund (Y104158).

output bit σ of a single perceptron is given by the equation σ= sign($\underline{w} \cdot \underline{x}$) where $\underline{x}$ is an N-dimensional input vector and $\underline{w}$ is a N-dimensional weight vector.

Let neural network A and B learn from each other, the perceptron learning rules are listed as follows[9]:

$$\underline{w}_A(t+1) = \underline{w}_A(t) - \frac{\eta}{N} \underline{x} \sigma_B \Theta(-\sigma_A \sigma_B) \tag{1}$$

$$\underline{w}_B(t+1) = \underline{w}_B(t) - \frac{\eta}{N} \underline{x} \sigma_A \Theta(-\sigma_A \sigma_B) \tag{2}$$

where $\Theta(x)$ is the step function and η is the learning rate.

Due to the analysis in [9], if η is larger than a critical value the networks can obtain a synchronization parallel weight state, $w_A = w_B$.

2.2 Authentication Using Tree Parity Machine (TPM)

Some modifications and improvements have been done in paper[10] for utilizing tree parity machine to secret key generator. To make system discrete, weight vectors perform random walk with reflecting boundary L: $w_{A(B)i} \in [-L, -L+1, ..., L-1, L]$.

The structure of TPM is illustrated in Fig.1. The two machines A and B receive identical input vectors x_1, x_2, x_3 at each training step. The training algorithm is that only if the two output bits are identical the weights do be modified and only the hidden unit, which is identical to the output bits, do modify its weight[8].

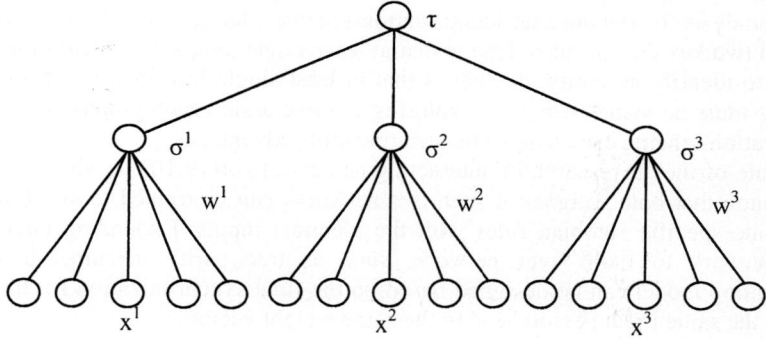

Fig. 1. Tree parity machine with three hidden units

The most important point is that wo partners A and B with their respective initial weights and identical input vectors synchronize to an identical state, $\underline{w}^i_B(t) = \underline{w}^i_B(t)$. But, non-identical inputs lead to non-synchronization.

3 Proposed User Authentication Scheme

3.1 Notations

Although smartcard has been widely used for automating remote user login securely, we shall simplify our authentication scheme without considering smartcard model. Some main notations are shown below.

U	the user
S	the remote server
UP	the user login program
PW_U	the password set by the user
X_S	the secrecy set by the user
W_I	Tree Parity Machine I's weight vector state
IV_I	TPM I's intitial input vector IV_I
$\otimes$	Exclusive Or operation
H	one way hash function
$A \xrightarrow{S} B: M$	A sends M to B over a secure channel
$A \longrightarrow B: M$	A sends M to B over an insecure channel
$A \xleftrightarrow{TPM} B$	Interacting neural network between A and B using TPM

3.2 Authentication Scheme

We separate the remote user authentication into two key phases, User Registration Model and User Authentication Model.

User Registration Model
1. U submits PW_U to S
2. S generates secrecy X_S for U
3. S computes $N_U = H(PW_U) \otimes H(X_S)$
4. $S \xrightarrow{S} U: H, N_U$
5. UP computes $H(X_S) = H(PW_U) \otimes N_U$ using H hash received from S

User Authentication Model
1. UP generates a random R_U
2. UP computes $N_U' = H(PW_U) \otimes H(X_S) \otimes R_U$
3. $U \longrightarrow S: N_U'$
4. S computes $R_U = N_U' \otimes H(PW_U) \otimes H(X_S)$
5. $U \xleftrightarrow{TPM} S$ where $IV_U = IV_S = R_U$, $W_U = H(PW_U)$, $W_S = H(X_S)$
6. TPMs learn from each other for more than 400 steps
7. If $W_U = W_S$ then U logins successfully to S, otherwise fails.

3.3 Analysis

Some concrete analytical results stay in [9,10], from which we know that TPM model with hidden layer is rather secure than simple interacting neural network, since even in case of an opponent breaking into network and obtaining the change of output bits it still can not identify which one of weight vectors is changed. On the other hand, the time for opponents getting synchronization is longer than the partners do, so the agreed secret key is practical secure for private key encryption. In addition, it's true that if the inputs are not identical, the system can never synchronize. So in the proposed scheme, only if the user and server exactly input Ru as initial vector for TPM mutual learning, the authentication can success. During user authentication, two TPMs are running for synchronization, so the exchange information between U and S are completely independent on user identity or password. These full dynamic and ruleless exchanges provide security against any network monitor or adversary.

4 Conclusions

We have explored a novel remote user authentication scheme based on interacting neural network learning. Comparing to other traditional methods, TPM based scheme for secure authentication is full dynamic and secure. However, more security analysis and implementation details for our proposed scheme here should be studied further.

References

1. A. K. Awasthi and S. Lal, "A remote user authentication scheme using smarts cards with forward secrecy", IEEE Trans. Consumer Electronic, vol. 49, no. 4, pp. 1246-1248, Nov 2003.
2. H. Y. Chien, J.K. Jan and Y. M. Tseng, "An efficient and practical solution to remote authentication: smart card," Computer & Security,vol. 21, no. 4, pp. 372-375, 2002.
3. C. C. Lee, L. H. Li and M. S. Hwang, "A remote user authentication scheme using hash functions," ACM Operating Systems Review, vol. 36, no. 4, pp. 23-29, 2002.
4. C. C. Lee, M. S. Hwang and W. P. Yang, "A flexible remote user authentication scheme using smart cards," ACM Operating Systems Review, vol. 36, no. 3, pp. 46-52, 2002.
5. M. S. Hwang and L. H. Li, "A new remote user authentication schemeusing smart cards," IEEE Trans. Consumer Electronic, vol. 46, no. 1, pp.28-30, Feb 2000.
6. S. W. Lee, H. S. Kim and K. Y. Yoo, " Comment on a remote user authentication scheme using smart cards with forward secrecy," IEEE Trans. Consumer Electronic, vol. 50, no. 2, pp. 576-577, May 2004.
7. K. C. Leung, L. M. Cheng, A. S. Fong and C. K. Chen, "Cryptanalysis of a remote user authentication scheme using smart cards", IEEE Trans.Consumer Electronic, vol. 49, no. 3, pp. 1243-1245, Nov 2003.
8. A. K. Awasthi and S. Lal. Security analysis of a dynamic ID-based remote user authentication scheme.
9. R.Metzler, W.Kinzel. Interacting neural networks. Phys. Rev. E. 2000,vol.62, pp.2555-2562.
10. W.Kinzel, I. Kanter. Interacting neural network and cryptography.2002 Advances in Solid State Physics,vol.42, pp.383-391.

Genetic Algorithm Simulated Annealing Based Clustering Strategy in MANET

Xu Li

Dept. of Computer Science, Fujian Normal University, 350007 Fuzhou, China
Xuli@mail.edu.cn

Abstract. MANET (Mobile Ad Hoc Network) is a collection of wireless mobile nodes forming a temporary computer communication network without the aid of any established infrastructure or centralized administration. MANET is characterized by both highly dynamic network topology and limited energy. This makes the efficiency of MANET depending not only on its control protocol, but also on its topology management and energy management. Clustering Strategy can improve the flexibility and scalability in network management. With graph theory model and genetic annealing hybrid optimization algorithm, this paper proposes a new clustering strategy named GASA (Genetic Algorithm Simulated Annealing). Simulation indicates that this strategy can with lower clustering cost and obtain dynamic balance of topology and load inside the whole network, so as to prolong the network lifetime.

1 Introduction

MANET (Mobile ad hoc network) is multi-hop wireless network that are composed of mobile hosts communicating with each other through wireless links [1]. MANET is likely to be use in many practical applications, including personal area networks, home area networking, military environments, and search a rescue operations. The wide range of potential applications has led to a recent rise in research and development activities.

Efficient energy conservation plays an important role in the protocol design of each layer in MANET because mobile host in such networks is usually battery-operated [2]. This paper mainly focuses on the hierarchical topology management. In hierarchical framework, a subset of the network nodes is selected to serve as the network backbone over which essential network control functions are supported. Hierarchical topology management is often called clustering which involves selecting a set of cluster-heads in a way that every node is associated with a cluster-head, and cluster-heads are connected with one another directly by means of gateways. The union of gateways and cluster-heads constitute a connected backbone. Once selected, the cluster-heads and the gateways help to reduce the complexity of maintaining topology information, and can simplify such essential functions as routing, bandwidth allocation, channel access, power control or virtual-circuit support.

---

* Supported Partially by National Natural Science Foundation of China (No.60372107) and Natural Science Foundation of Fujian Province in China (No.A0440001)

The rest of the paper is organized as follows. Section 2 briefly describes previous work and their limitations. Section 3 describes the details of the proposed clustering strategy GASA. Section 4 introduces performance metrics and performance results based on various experiments environment. The paper ends with some conclusions and remarks on future work.

2 Related Work

The key step of hierarchical topology management is cluster-head election. Several algorithms have been proposed to choose cluster-heads in MANET. Let us summarized below these algorithms.

2.1 Max-Degree Algorithm

The node degree is a commonly used algorithm in which nodes with higher degree are more likely to be cluster-heads [3]. The neighbors of cluster-head are members of that cluster and will not participate in the election process. Experiments demonstrate that the system has a low rate of cluster-head changes. However, this approach can result in a high turnover of cluster-heads due to the network topology changes. The high overhead caused by cluster-head's change is undesirable.

2.2 Lowest-ID Algorithm

Several approaches [4-5] utilized the node identifiers to elect the cluster-head with one or multiple hops. This strategy assigns a unique id to each node and chooses the node with the minimum id as a cluster-head. The drawback of this algorithm is its bias towards with smaller id, which leads to the battery drainage of certain nodes. Moreover, it does not attempt to balance the load uniformly across all the nodes.

2.3 Node-Weight Algorithm

Basagni et al. [6-7] introduces two algorithms, named DCA (distributed clustering algorithms) and DMAC (distributed mobility adaptive clustering. A node is chosen to be cluster-head if its node-weight is higher than any of its neighbor's node-weight. The smaller node id is chosen to break a tie. The DCA is suitable for clustering "quasi-static" ad hoc networks. It is easy to implement and the time complexity depends on the topology of the network rather than on its size. The DMAG algorithm is adapted to the changes in the network topology due to the mobility of the nodes. Both algorithms are executed at each node with the sole knowledge of the identity of the one-hop neighbors.

None of the above three kind of algorithms leads to an optimal election of cluster-heads since each deals with only a subset of parameters which can possibly impose constraints on the system. Each of these algorithms is suitable for a specific application rather than for arbitrary wireless mobile networks.

3 System Model and Basic Definition of Our Algorithm

Definition 1. This work assumes that MANET comprises a group of mobile nodes communication through a common broadcast channel using omni-directional antennas with the same transmission range. The topology of MANET is thus presented by an undirected graph G = (V, E), where V is the set of network nodes, and $E \subseteq V \times V$ is the set of links between node. The existence of a link $(u,v) \in E$, and that nodes u and v are within the packet-reception range of each other, in which case u and v are called one-hop neighbors of each other.

Definition 2. Each node has one unique identifier, and all transmissions are omni-directional with the same transmission range. In addition, a reliable neighbor protocol is assumed to enable the quick update neighbor information at each node.

Definition 3. We introduce a function IFClusterHead () for each node. The initial value of this function is –1.If one node is elected as cluster-head, the function value is set to 1. If one node is elected as cluster member, the function value is set to 0.

Definition 4. Average relative mobility M_u is defined as equation (1). Where $d(u,v)$ represents the distance between node u and v, V_v represents the movement velocity of node v.

$$M_u = \frac{1}{|N(u)|} \sum_v \frac{d(u,v)}{V_v} \qquad \forall u \in N(u) \qquad (1)$$

Definition 5. Relative remainder energy $B_u \in [0, 1]$ is defined as equation (2). Where F_u and R_u respectively represents the initial energy and remainder energy at time t of node u.

$$B_u = \frac{R_u(t)}{F_u} \qquad (2)$$

Definition 6. Computes dynamic weighted value DW_u as equation (3). The meaning of Δ_u is the same as in the WCA strategy [6].

$$DW_u = 10^{\lg(\theta_1 M_u)\lg\theta_2(1-B_u)\lg\theta_3(1+\Delta_u)} \qquad (3)$$

The definition 6 decides how well suited a node is for being a cluster-head. The cluster-head election procedure is not periodic and is invoked as rarely as possible. This reduces system updates and hence computation and communication costs. The clustering algorithm is not invoked if the relative distances between the nodes and their cluster-head don not change.

- Mobility is an important factor in deciding the cluster-heads. In order to avoid frequent cluster-head changes, it is desirable to elect a cluster-head that does not move very quickly according its neighbor node.

- Given that cluster-heads provide the backbone for a number of network control function, their energy consumption is more pronounced than that of ordinary hosts. Low-energy nodes must try to avoid serving as cluster-heads to save energy. So the lower remain power of node, the small chance to become cluster-head in our algorithm. This strategy can realize energy consumption balance and avoid the extinction of nodes due to exhaustion of their battery power.
- The more smaller node number implies the more lower management complexity. However, if the number of cluster-head is too lower, the conflict probability is increase and the capacity of network will also decrease. So the Δ_u can resolve this contradiction.

3.2 Steps of Genetic Algorithm Simulated Annealing Based Clustering Strategy

This section provides how the genetic algorithms simulated annealing is applied to DW (Dynamic Weighted) to optimize the total number of cluster-head. As can be seen in Fig.1, we have all the nodes along with the adjacency list as the DW values which are already calculated from the execution of DW definition. This is stored separately in a list where each node is pointing to its neighbor chain. The chain is used to compute the object function.

DW	Node ID	Neighbor node chain			
M1	1 →	4 →	33 →	62 →	...
M2	2 →	21 →	39 →	48 →	...
M3	3 →	9 →	17 →	25 →	...
...	... →	... →	... →	... →	...
M99	99 →	49 →	... →	... →	...
M100	100 →	43 →	68 →	29 →	...

Fig. 1. Adjacency list with DW of network graph

The Genetic Algorithm Simulated Annealing based Cluster procedure consists of four steps as described below:

Step 1. Initial Population: Randomly generates the initial population with the pool size being equal to the number of nodes in the given network. This will produce the same number of chromosomes in the form of integer strings. Suppose the randomly array is:

$$1\ 3\ 33\ 99\ 32\ 19\ 29\ 98\ 68\ 44\ \cdots\ \cdots\ \cdots\ \cdots\ 76\ 39$$

Fig.2 illustrate the procedure of cluster and cluster-head come into being.

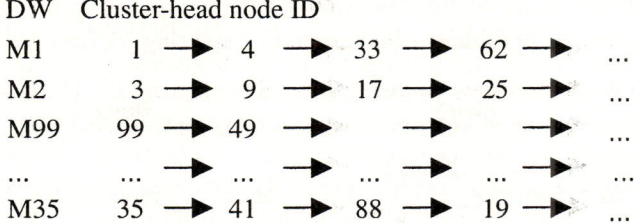

Fig. 2. The procedure of Cluster and Cluster-head come into being

The detailed flow describe as follow.

A. Set object function SDW=0.
B. Select the first node 1 from the random array. Because the IF function ClusterHead () value of this node is –1. So node 1 is elected as cluster-head. Let SDW =SDW +M1, and continuous to visit the whole chain table.
C. If all of the nodes in the randomly array have been visit then go to E, otherwise, visit the next node in the array in turn.
   Case (clusterhead () =0). Skip this node
   Case (clusterhead()=-1). Elect this node (suppose K) as cluster-head and set clusterhead()=1. SDW=SDW+MK. And visit the whole chain table which head node is node K.
D. Complete the clustering procedure. Get the Dominating Set (viz. chromosome), which is made up of all the cluster-head.
E. Repeats from A to D until L times clustering procedures, and get the Initial Population as { DS1, DS2, ...DSL} and fitness value{ SDW1, SDW2, ..., SDWL} .

Step 2. Boltzmann mechanism based selecting strategy and get copulation pool. The conventional select probability defines as

$$P(DS_i) = \frac{SDW_i}{\sum_{i=1}^{L} SDW_i} \qquad i = 1, 2, \cdots L \qquad (4)$$

We introduce boltzmann mechanism into select probability, which defines as

$$P^*(DS_i) = \frac{e^{-SDW_i/T}}{\sum_{i=1}^{L} e^{-SDW_i/T}} \qquad i = 1, 2, \cdots L \qquad (5)$$

Where T>0 is called "temperature" by analogy with physical systems. And temperature is decreased at a constant factor α, $T = \alpha T$. The new select probability strategy can adaptive to the optimize selection procedure.

Step 3. Crossover and Metroplis based elitism

The crossover procedure can be get by replacing $A_i(m)$ and $A_i(n)$ in the randomly array A_i. The new randomly array is defined as A_i^*. The new Population is { DS1*, DS2*, …DSL*} and fitness value is { SDW1*, SDW2*, …, SDWL*}.

If $Random < e^{\frac{|DS_i^* - DS_i|}{T}}$) then replace DS_i with DS_i^* (i=1,2, …L) else discard DS_i^*.

Step 4. Convergence condition

The fitness values of k and k+1 generation is separately SDW_i^k (i=1,2, …L) and SDW_i^{k+1} (i=1,2, …L). if $|\min\{SDW_i^k$ (i=1,2, …L) $\} - \min\{SDW_i^{k+1}$ (i=1,2, …L) $\}| < \varepsilon$ then stops the evolution process.

4 Simulation and Performance Evaluation

4.1 Simulation System and Performance Metrics

The simulation system is build by MATLAB tool and VC++. We compare the GASA-CS clustering strategy with the Max Degree and WCA strategy [8]. There are 30 nodes randomly placed within a rectangle area, whose moving speed ranges from 0 to 10m/s. A random waypoint mobility model [7] is adopted here and pause time set 1 s. the transmission range of node is 30m. We chose the values for constants describe in GASA-CS strategy as shown in Table 1. These parameters are configurable from implementation meaning that the parameter values can be adjusted to serve a particular application.

Table 1. Parameters Used in Simulation

L	Ω	α	T0	ε	δ	θ 1	θ 2	θ 3
10	10	0.9	0.1	0.1	4	1	1	1

We compare the performance of GASA-CS with four performance metrics:

A. The number of clusterhead.
B. CRF (cluster re-affiliation factor), which defines as

$$\text{CRF} = \frac{1}{2}\sum_i |N_{i1} - N_{i2}| \tag{6}$$

where i is the average number of cluster, and N_{i1}, N_{i2} are the degree of node i at different times.

C. We take the same definition of load balancing factor (LBF) [3] as:

$$\text{LBF} = \frac{n_c}{\sum_i (x_i - \mu)^2}, \quad \mu = \frac{(N - n_c)}{n_c} \tag{7}$$

where n_c is the average number of cluster, N is the number of all nodes, and x_i is the practical degree of node i. The larger LBF, the better load balanced among the network.

D. Lifetime of Network. We define this metric as the duration from the beginning of the simulation to the first time a node runs out of energy. The energy consumption of clusterhead is five times as normal node. This set is because the cluster averagely includes about five nodes in our simulations.

4.2 Simulation Result and Evaluation

The first group simulations are set the nodes area as 100m×100m, simulation time is 100 units. As shown in Fig.3, we can see clearly that the GASA-CS has lowest CRF value (mean 0.17). WCA strategy has the highest CRF values (mean 1.56). This indicates by optimization mechanism, GASA-CS decreases the choice of switching from one cluster to another. The Fig.4 illustrates that GASA-CS has the highest LBF (mean 1.1901) and the LBF of Max Degree strategy is lowest (mean 0.0922). This result indicates Max Degree strategy merely considers the number of cluster node. This make some clusters have too many nodes can cause the traffic unbalance. GASA-CS considers not only current cluster node, but also the change. So it can heighten the balance performance.

In the second group simulations, the density of network varied by changing node distributed area. Fig.5 indicates the cluster number of GASA-CS strategy is always fewest. In higher density, the Max Degree strategy is better than WCA. While in lower density, WCA is better than Max Degree. Fig6 illustrates that network lifetime by GASA—CS strategy is longest because the GASA—CS strategy introduces remain energy factor and decrease the chance of lower energy node becoming clusterhead.

5 Conclusion

We proposed an improved clustering algorithm and compared it with original clustering algorithms in terms of average cluster number, CRF, LBF and network lifetime. From the experimental results, we can fairly draw a conclusion that our GASA algorithm performs well compared with other algorithms.. It achieves the best performance on the trial dataset. Further research is required into how to rebuild the simple energy-consuming model proposed in this paper by considering the practical traffic in the application layer and how to control the compute complication in lager scale network which is an opened problem.

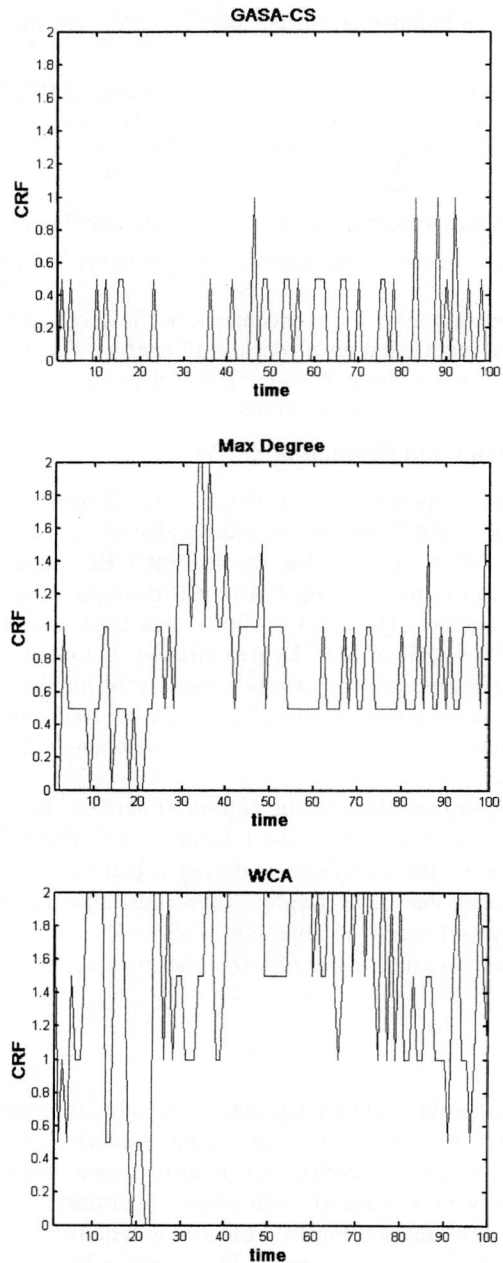

Fig. 3. The cluster re-affiliation factor of GASA-CS, Max Degree and WCA strategy

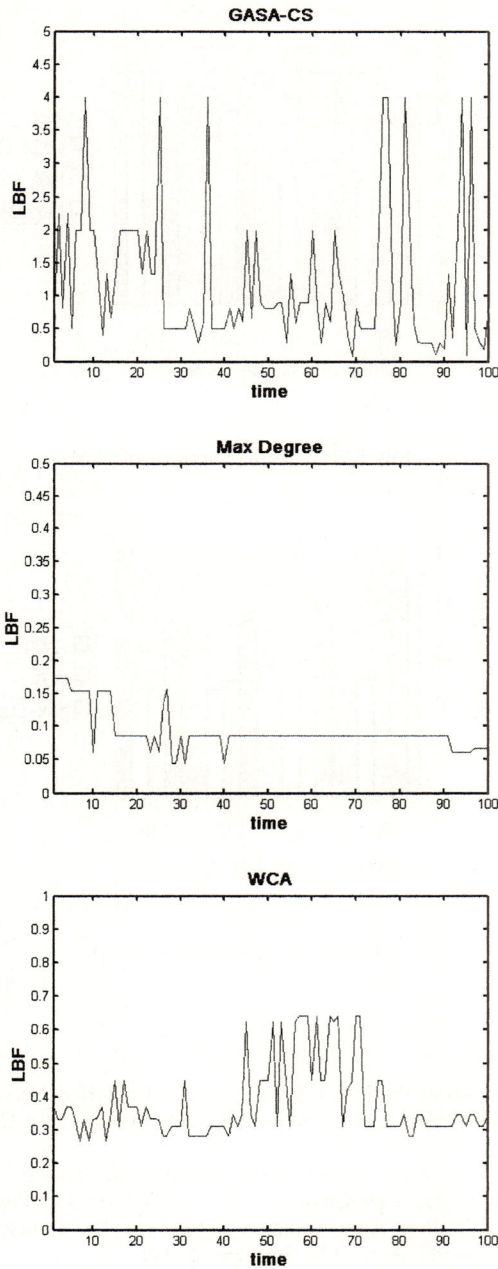

Fig. 4. The load balance factor of GASA-CS, Max Degree and WCA strategy

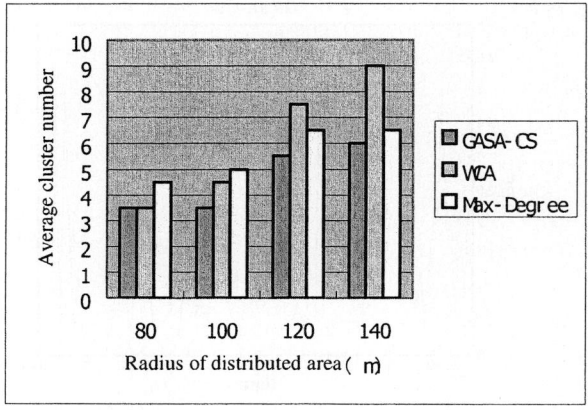

Fig. 5. The average number of cluster

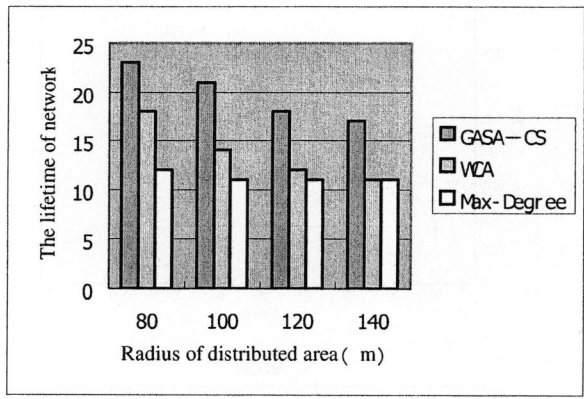

Fig. 6. The lifetime of network

References

1. Ram Ramanathan, Jason Redi. A Brief Overview of Ad Hoc Networks: Challenges and Directions. IEEE Communications Magazine. 50th Anniversary Commemorative Issue, May 2002: 48-53.
2. XU Li, ZHENG Bao-yu. Cross layer coordinated Energy Saving Strategy in MANET, Journal of Electic (in China), 2003,11, 20(6): 455-459
3. S. Guha and S. Khuller. Approximation algorithms for connected dominating sets. Algorithmic, 20, 374-387, Apr. 1998. Springer-Verlag.
4. L. Jia, R. Rajaraman, and T. Suel. An Efficient Distributed Algorithm for Constructing Small Dominating Sets [A]. In Twentieth ACM Symposium on Principles of Distributed Computing PODC'01, Newport, Rhode Island, Aug. 26-29, 2001.

5. C.C. Chiang, H.K. Wu, et al. Routing in clustered multi-hop, mobile wireless networks with fading channel. In IEEE Singapore International Conference on Networks SICON'97, Singapore, Apr, 1997. 197-211.
6. A. Bruce McDonald, Taieb F. Zhati. A Mobility-based Framework for Adaptive Clustering in Wireless Ad Hoc Networks. IEEE Journal on Selected Areas in Communications, VOL.17. NO. 8. August 1999: 1466-1487.
7. Y. Xu. Adaptive Energy Conservation Protocols for Wireless Ad Hoc Routing. PhD thesis, University of Southern Califormia, 2002
8. M.Chatterjee, S.K. Das and D. Turgut. WCA: A Weighted Clustering Algorithm for Mobile Ad Hoc Network. Journal of Clustering Computing. Vol.5, No. 2. April 2002, pp. 193-204.

A Gradual Training Algorithm of Incremental Support Vector Machine Learning[*]

Jian-Pei Zhang, Zhong-Wei Li, Jing Yang, and Yuan Li

College of Computer Science and Technology, Harbin Engineering University,
150001, Harbin, China
davis525@163.com

Abstract. Support Vector Machine(SVM) has become a popular tool for learning with large amounts of high dimensional data, but sometimes we prefer to incremental learning algorithms to handle very vast data for training SVM is very costly in time and memory consumption or because the data available are obtained at different intervals. For its outstanding power to summarize the data space in a concise way, incremental SVM framework is designed to deal with large-scale learning problems. This paper proposes a gradual algorithm for training SVM to incremental learning in a dividable way, taking the possible impact of new training data to history data each other into account. Training data are divided and combined in a crossed way to collect support vectors, and being divided into smaller sets makes it easier to decreases the computation complexity and the gradual process can be trained in a parallel way. The experiment results on test dataset show that the classification accuracy using proposed incremental algorithm is superior to that using batch SVM model, the parallel training method is effective to decrease the training time consumption.

1 Introduction

Support Vector Machine (SVM) is a new approach of pattern recognition based on Structural Risk Minimization which is suitable to deal with magnitude features problem with a given finite amount of training data[1,2]. In recent years, SVM has been given considerable attention because of its superior performance in pattern recognition and been successfully used as a classification tool in a variety of areas, ranging from handwritten digit recognition, face identification, and text categorization[3,4,5].

However, being a very active approach, there still exist some open questions that should be deliberated by further research, such as how to train SVM efficiently to learn from large datasets and stream datasets incrementally. That is to say, when the training datasets are often far too large or always change in practice and new samples data are added in at any moment, incremental learning of SVM should be developed to avoid running time-consuming training process frequently[6,7]. Many researches have been made on incremental learning with SVM.

---

[*] This work is supported by the Natural Science Foundation of Heilongjiang Province under Grant No. F0304.

The basic principle of SVM is to find an optimal separating hyperplane so as to separate two classed of patterns with maximal margin[1]. It tries to find the optimal hyperplane making expected errors minimized to the unknown test data, while the location of the separating hyperplane is specified via only data that lie close to the decision boundary between the two classes, which are support vectors. Obviously, the design of SVM allows the number of support vectors to be small compared with the total number of training data, therefore, SVM seems well suited to be trained according to incremental learning.

The rest of this paper is organized as follows. In section 2, a preliminary review of SVM is given. In section 3, a gradual training algorithm for incremental learning with SVM is proposed after brief review of batch incremental algorithms. In section 4, experiments on real datasets are executed to evaluate the proposed learning algorithm compared with batch learning model. Finally, some concluding remarks are made in section 5.

2 SVM and Incremental Learning Theory

2.1 Brief Review of SVM

The SVM uses the Structural Risk Minimization principle to construct decision rules that generalize well. In doing so, it extracts a small subset of training data. We merely outline its main ideas here. The method of Structural Risk Minimization is based on the fact that the test error rate is bounded by the sum of the training error rate and a term which depends on the so-called VC-dimension of the learning machine[1]. By minimizing the sum of both quantities, high generalization performance can be achieved.

For linear hyperplane decision functions:

$$f(x) = \text{sgn}((w \bullet x) + b) \tag{1}$$

The VC-dimension can be controlled by controlling the norm of the weight vector w. Given training data $(x_1, y_1), \cdots, (x_n, y_n)$, $x_i \in R^N$, $y_i \in \{\pm 1\}$, a separating hyperplane which generalizes well can be found by minimizing

$$\frac{1}{2}\|w\|^2 + \gamma \sum_{i=1}^{n} \xi_i \tag{2}$$

subject to

$$\xi i \geq 0, \ y_i((w \bullet x_i) + b) \geq 1 - \xi_i, \ for \ i = 1, \cdots, n \tag{3}$$

γ is a constant which determines the trade-off between training error and VC-dimension. The solution of this problem can be shown to have an expansion

$$w = \sum_{i=1}^{n} a_i x_i \tag{4}$$

where only those a_i are nonzero which belong to x_i precisely meeting the constraint (3), these x_i lie closest to the decision boundary, called support vectors. The coefficients a_i are found by solving the quadratic programming problem defined by (2) and (3).

Finally, this method can be generalized to nonlinear decision surfaces by first mapping the input nonlinearly into some high-dimension space, and finding the separating hyperplane in that space. This is achieved implicitly by using different type of symmetric functions $K(x, y)$ instead of the ordinary scalar product $(x \cdot y)$. One can get

$$f(x) = \text{sgn}(\sum_{i=1}^{n} a_i K(x \bullet x_i) + b) \tag{5}$$

as a generalization of (1) and (4).

More details of SVM are described in Ref.[1,2].

2.2 Incremental SVM Learning

The development of modern computing and information technologies has enabled that huge amount of information has been produced as digital data format. It is impossible to classify this information one by one by hand in many realistic problems and fields, there is a need to scale up incremental learning algorithms to handle more training data[1]. Therefore, incremental techniques have been developed to facilitate batch SVM learning over very large datasets and stream datasets, and have found widespread use in the SVM community.

Incremental learning algorithm with SVM proposed in Ref. [6,7] consists in learning new data by discarding all past examples except support vectors, it utilizes the property of SVM that only a small fraction of training data end up as support vectors, the SVM is able to summarize the data space in a very concise manner, and assume that the batches of data will be appropriate samples of the data. Although this framework relies on the property that support vectors summarize well the data and it is an approximate solution to learn incrementally with SVM as batch model, the results of experiments proved that it is an effective way to deal with large datasets and stream datasets

It is proved that the key to construct optimal hyperplane is to collect more useful data as support vectors during the incremental learning. Most incremental learning algorithms are based on improving SVM training process by collecting more useful data as support vectors[6,7,8,9]. There aims are also looking for more potential samples as support vectors and take advantage of the outstanding traits of SVM, that is, after learning process, SVM divides data into two classed with a hyperplane. Suppose that x is to be classified, the further x is from the hyperplane, the higher the probability of correct classification is, and vice versa.

2.3 Batch SVM Learning Model

Given that only a small fraction of training data end up as support vectors, the SVM is able to summarize the data space in a very concise manner. This suggests us a feasible method is that we can partition the training samples dataset in batches (subsets) that fit into memory, for each new batch of samples, a SVM is trained on the new samples data and the support vectors from the previous learning step, as the Fig.1 shows[6]. According to important properties of support vectors, we can expect to get an incremental result that is equal to the non-incremental result, if the last training set contains all samples that are support vectors in the non-incremental case.

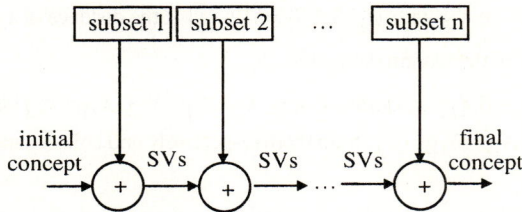

Fig. 1. Batch incremental SVM learning model

The reasoning behind batch learning model is the assumption that the batches of data will be appropriate samples of the data. While the problem is the learning results are subject to numbers of batches and state of data distribution but always the distribution of data is unknown. Another disadvantage of this learning model is the time consuming is not prompted, since all data in any subset should be computed to tell whether it is a support vector or not. Therefore, some researchers proposed their incremental training algorithms to promote learning results[7,8,9,10,11,12,13,14,15].

3 Gradual Training SVM Algorithm

Incremental SVM learning with batch model divides the training dataset into batches (subsets) that may fit into memory suitably. For each new batch of data, a SVM is trained on the new training data and the support vectors from the previous learning step. As mentioned above, separating hyperplane is subject to support vectors of training dataset, batch model learning may not collect all support vectors that incremental learning algorithm need, for the distribution state of batches of training datasets are unknown generally.

According to the distributing characteristic of support vectors, we know that the data located between the face-to-face marges of the two classes of training datasets are not classifiable correctly and easily. Less of these data are, higher of the classification precision is. Therefore, in every step of batch learning model, some actions are taken not only to support vectors but also these data between the marges of the two classes. That is, classify the new coming dataset with current hyperplane defined by training dataset to collect the data that are between the marges of the two classes.

Then divide current training dataset into two or there smaller subsets and combine them with new coming dataset to collect support vectors respectively. Finally, combine the data collected in former two steps to construct new hyperplane, until the classification precision is satisfying.

Given current training data set X_i and new coming data set X_j

Initialize $SV=\Phi$.

Execute

1. Train SVM_i with X_i, construct optimal hyperplane ψ_i,
2. Classify every data x in X_j with ψ_i. Set $SV=SV\cup\{x\}$ if the distance of x to the hyperplane is less than the half of the margin defined by ψ_i,
3. Divide X_i into X_{i1} and X_{i2} randomly, that is, $X_i=X_{i1}\cup X_{i2}$, $X_{i1}\cap X_{i2}=\Phi$,
4. Set $X_{i1}=X_{i1}\cup X_j$, $X_{i2}=X_{i2}\cup X_j$, train them respectively and collect support vectors set SV_{i1} and SV_{i2}.
5. Set $SV=SV\cup SV_{i1}\cup SV_{i2}$, construct ψ_{i+1} with SV and classify the new coming data X_j.
   Go step 2 if the classification precision is less than that in step 2.
6. Stop the algorithm until the precision is satisfying.

Fig. 2. Pseudo-code of proposed gradual increment SVM algorithm

Fig.2 shows the pseudo-code of this gradual increment SVM algorithm. Clearly, computation is almost focused on constructing hyperplanes with new combined datasets twice or thrice, but the classification precision would be improved, as subsets of training dataset may have different distribution status, combining them with new coming dataset may collect different but potential support vectors.

4 Experiments and Results

We conducted experiment using a text database which is pre-labeled by hand and consists of 1830 data points, each having a dimension of 35 to compare the batch incremental SVM learning algorithm with crossed iteration incremental SVM learning algorithm we proposed. All algorithms are programmed with Matlab7.0 in Windows 2000.

The experiment is prepared as following: take 273 data points as initial training set and 427 data points as test set randomly, separate the rest data points into 5 subsets and use the polynomial kernel. Table 1 shows the classification precision results for our experiment.

It can be seen that the classification precision results in incremental steps are improved from the Table 1. The whole classification process needs much more time-cost for computing, but the classification precision is improved in some degree. The numbers of support vectors of two learning model in Table 2 show that our proposed algorithm collects much more support vectors to obtain higher classification precision than batch incremental learning algorithm.

Table 1. The comparison of classification precision and training time of two incremental SVM learning algorithms

	Number	Batch incremental SVM algorithm		Proposed gradual SVM algorithm	
		time /s	precision /%	time /s	precision /%
training dataset	273	92.6	90.21	146.6	92.37
subset 1	367	127.3	91.45	184.2	93.65
subset 2	184	107.8	90.37	175.5	94.87
subset 3	97	76.9	92.64	138.8	94.51
subset 4	219	152.4	92.80	191.7	95.22
subset 5	263	179.2	93.26	233.4	95.93

Table 2. The comparison of numbers of support vectors of two incremental SVM learning algorithms

	Number	SVs	
		Batch incremental SVM algorithm	Proposed gradual SVM algorithm
training dataset	273	55	67
subset 1	367	78	96
subset 2	184	96	127
subset 3	97	108	144
subset 4	219	131	163
subset 5	263	167	212

Table 3. The comparison of classification precision and training time of two learning strategies

	Number	Proposed algorithm with one machine		Proposed algorithm with two parallel machines	
		time /s	precision /%	time /s	precision /%
training dataset	273	146.6	92.37	80.7	92.37
subset 1	367	184.2	93.65	106.6	93.65
subset 2	184	175.5	94.87	88.3	94.87
subset 3	97	138.8	94.51	75.1	94.51
subset 4	219	191.7	95.22	110.5	95.22
subset 5	263	233.4	95.93	133.2	95.93

Another experiment proves it is true that supposing there are two parallel learning machines to train divided datasets respectively, the training time would not increase much more, under the same test condition. Table 3 shows the comparison results of these two strategies with the proposed algorithm.

5 Conclusions

The ability to incremental learning from batches of data is an important feature that makes a learning algorithm more applicable to real-world problems. SVM is adaptable to incremental learning to vast data classification for its outstanding power to summarize the data space. A gradual algorithm for training SVM to incremental learning is proposed based on considering the possibility of new support vectors set works on the history dataset and the incremental dataset. Experiments improved that this approach is efficient to deal with vast data classification problems with higher classification precision.

The algorithm presented in this paper is similar to other heuristic algorithms for their aims are mainly focused on collecting more potential samples as the support vectors. It was experimentally shown that the performance of the new algorithm is comparable with an existing approach in the case of learning without concept changes. More researches will be performed with problems that training dataset contains changing concepts in the future.

References

1. V. Vapnik.: Statistical Learning Theory. Wiley(1998)
2. Christiani, N., Shawe-Taylor, J.: An Introduction to Support Vector Machines and Other Kernel-based Learning Method. Cambridge University Press(2000)
3. Dumais, S., Platt. J., Heckerman. D.eds.: Inductive Learning Algorithms and Representations for Text Categorization. Proceedings of the 7th International Conference on Information and Knowledge Management. Maryland (1998)
4. Osuna. E., Freund. R., Girosi. F.: Training Support Vector Machines: An Application to Face Detection. Proceeding of the IEEE International Conference on Computer Vision and Pattern Recognition. New York(1997)130–136
5. Burges C.: A Tutorial on Support Vector Machines for Pattern Recognition. Knowledge Discovery and Data Mining, Vol.2(1998)1–47
6. N. Syed, H. Liu, and K. Sung.: Incremental Learning with Support Vector Machines. Proceeding of IJCAI Conference. Sweden(1999)
7. P. Mitra, C. A. Murthy, and S. K. Pal.: Data Condensation in Large Databases by Incremental Learning with Support Vector Machines. Proceeding of ICPR Conference. Spain(2000)
8. C. Domeniconi and D. Gunopulos.: Incremental Support Vector Machine Construction. Proceeding of IEEE International Conference on Data Mining, California, USA(2001)
9. G. Cauwenberghs and T. Poggio.: Incremental and Decremental Support Vector Machine Learning. Advances in Neural Information Processing Systems(2000)
10. Rong Xiao, Jicheng Wang, Zhengxing Sun, Fuyan Zhang.: An Apporach to Incremental SVM Learning Algorithm. Journal of NanJing University(Natural Sciences), Vol. 38. NanJing(2002)152–157

11. Yangguang Liu, Qinming He, Qi Chen.: Incremental Batch Learning with Support Vector Machines. Proceedings of the 5th World Congress on Intelligent Control and Automation. China(2004)
12. Kai Li, Houkuan Huang.: Research on Incremental Learning Algorithm of Support Vector Machine. Journal of Northern Jiaotong University, Vol. 27. BeiJing(2003)34–37
13. Fung, G., Mangasarian, O.L..: Incremental Support Vector Machine Classification. Proceedings of the Second SIAM International Conference on Data Mining. SIAM(2002)
14. Tveit, a., Hetland, M.L., Engun,H.: Incremental and Decremental Proximal Support Vector Classification using Decay Coefficients. Proceedings of the 5th International Conference on Data Warehousing and Knowledge Discovery. Czech Repblic(2003)
15. Klinkenberg, R., Joachims, T.: Detecting Concept Drift with Support Vector Machines. Proceedings of the 17th International Conference on Machine Learning, Morgan Kaufmann(2000)

An Improved Method of Feature Selection Based on Concept Attributes in Text Classification

Shasha Liao and Minghu Jiang

Lab of Computational Linguistics, Dept. of Chinese Language,
Tsinghua University, Beijing, 100084, China.
Creative Base of Cognitive Science, Tsinghua Univ., Beijing, 100084, China
jiang.mh@tsinghua.edu.cn

Abstract. The feature selection and weighting are two important parts of automatic text classification. In this paper we give a new method based on concept attributes. We use the *DEF* Terms of the Chinese word to extract concept attributes, and a Concept Tree (C-Tree) to give these attributes proper weighs considering their positions in the C-Tree, as this information describe the expression powers of the attributes. If these attributes are too weak to sustain the main meanings of the words, they will be deserted and the original word will be reserved. Otherwise, the attributes are selected in stead of the original words. Our main research purpose is to make a balance between concept features and word ones by set a shielded level as the threshold of the feature selection after weighting these features. According to the experiment results, we conclude that we can get enough information from the combined feature set for classification and efficiently reduce the useless features and the noises. In our experiment, the feature dimension is reduced to a much smaller space and the category precise is much better than the word selection methods. By choose different shielded levels, we finally select a best one when the average category precise is up to 93.7%. From the results, we find an extra finding that the precise differences between categories are smaller when we use combined features.

1 Introduction

Automatic text classification is a process which classifies the documents to several categories by their content. With the rapid development of the online information, text classification becomes one of the key techniques for handling and organizing the text data. So far, the English text classification is practical and is widely used in E-mail classification, information filer and so on. For example, the E-mail category system of White House, and the Construe system used by Reuter are two successful classification systems.

The Chinese text classification steps much slower than the English one. In 1981, Hou H. discussed the automatic classification and introduced some foreigner research results. After that, the Chinese research begins. Because the Chinese text is different from English, there is much to do before classification, such as particple, speech tagging and so on, which are important techniques of Chinese language process. Therefore, the researchers have developed a lot of techniques which are fit for

Chinese text classification and the text classification begin to boom. For example, the C-ABC system built by Zhongshan Library, and the automatic category system built by Tsinghua University [6].

2   Overview of the Improved Concept Feature Selection

In order to classify the documents, we should reflect them to a vector space. In text classification, the processed data is always called feature, and the original data is sometimes called attribute. In order to get a proper feature set, a process is required which is called feature selection. Feature selection includes two parts, one is called feature reduction, reducing the feature set to a minimum with the essential information of the original attribute, and the other is called feature weighting, giving the features different weight referring to their contribution to the classification[5]. In this paper, we used *HowNet* dictionary to extract concept attributes from the words in the text and weight them by the C-Tree. And then, by setting a shielded level to filter the weak concept attributes, we reduce the feature space with little information lose. After that, we use term frequency and cross entropy (*TF-CE*) as the reduction method, which is based on the difference between categories. With a proper threshold which is called a shielded level, we can reduce the original feature dimension to a much smaller one, which can be easier to calculate, and information noises are mostly filtered out. In this feature selection process, we mainly discuss the expression power of the concept attributes and how to combine the concept features and the original word ones.

3   Principles and Methods

The concept definition in our system comes from *HowNet* [1], which is a knowledge system referring to the concept of Chinese words. When we select feature from the text we will use the *DEF* term of every Chinese word. In this method, we can extract the concept attribute as the feature of the text, which will describe the internal concept information, and get the relationship among the words. There are 24,000 words in the *HowNet* and only 1,500 concept attributes, as a result, the feature will be reduced to a stable dimensionality space with little information lose.

The DEF Term of a certain Chinese word "博士" is "*DEF*=human | research | study | education", so we can extract four attributes from it and add them to the feature set instead of the word "博士",and the attribute "education" is a clear prediction of category "Education". As a result, the concept attributes in *DEF* Terms are sometimes helpful for classification and they are easy for feature weighting because there are some attribute documents in *HowNet* in tree forms, with which we could built a concept Tree (C-tree) with every node for a attribute. Meanwhile, as all the concepts in *HowNet* are definitely positioned, we don't need to worry about that there are redundant concept in a *DEF* term, for example, a *DEF* with both a node and its father node.

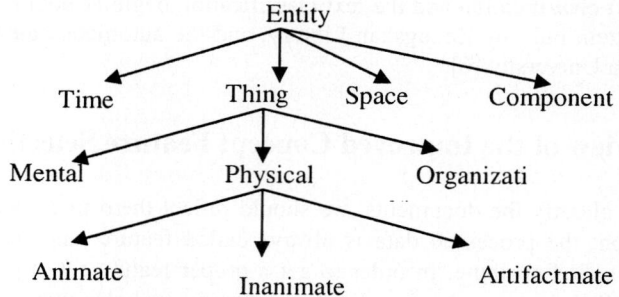

Fig. 1. This is part of the Entity tree of *HowNet* from which we could see the tree form of the concept attributes. As we can see, there are some abstract concept attribute such as Entity, Thing and so on. These attributes exist in the *DEF* terms of many words

Also, there are more than one trees in the *HowNet* and every tree has its special roots and tree high. Below is the Event tree which describes all the event concept in *HowNet*, and it is easy to see that this tree is much higher than the Entity tree above and there should be some strategy to deal with these differences. The strategy will be introduced in the next part of this paper and the experiment shows that the different treatment of different trees is important to make the concept balance.

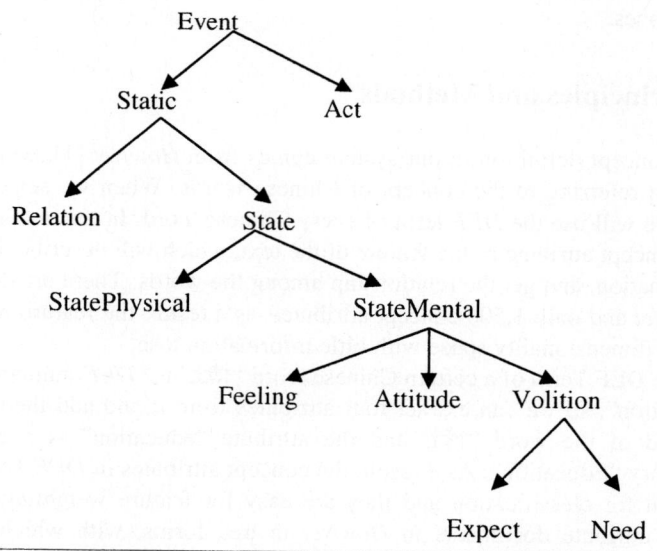

Fig. 2. This is part of the Event tree of *HowNet* from which we could see that it has much more abstract levels than the Entity one. That is why we have to set different weights to different tree roots. In fact, the Event tree is higher than the Entity tree.

3.1 The Weighting of Concept Attributes

The attribute feature table of the *HowNet* lists 1505 concept attributes in tree form, so the level of the node is always considered as the most important reason of the weighting, and the most used weighting formula is as follows[8]:

$$W_{ik} = \log(Droot_{ik} + a) + L \tag{1}$$

In our system, we use a different formula which put much more information into considering. At last, we added two things to the weighting formula, one is the different weighting of different tree roots; the other is the number of the child nodes of the weighting node. Because the weighting of the nodes at the same level in the different trees are not equal, we give a different root weight to treat them differently. For example, in the Entity tree, we give the root "Entity" a weight of 1.0, but in the Event tree, we give the root "Event" a weight of 0.25. Meanwhile, when a node have more child nodes, it means that in the cognize world, this concept is more complex and has more details, so people would use its child nodes more and treat this concept as a more abstract one, so it should have a comparatively lower weight.

At last, we give a weighting formula as follows:

$$W_{ik} = Wtree_i [\log((Droot_{ik} + a)/2) + L + \frac{1}{CNum_k + b}] \tag{2}$$

In this formula, W_{ik} is the weighting of node k in tree i; $Wtree_i$ is the weighting of the tree root i; $Droot_{ik}$ is the distance between node k and the tree root i; $CNum_k$ is the number of the child nodes of node k; L, a and b are the tempering factors, which are used to control the weighting range. According to the experiments, we set $L=0.15$, $a=1$ and $b=5$.

3.2 The Abstract Concept Attributes and the Shielded Level in the C-Tree

Though the *HowNet* is easy to use in text classification, the concept attributes in it exist in a tree form and there are some nodes near the root which are abstract concept attributes. For example, the concept attribute "entity" is an abstract one which does not have much information for category. If the *DEF* term of a certain word contains only abstract concept, it means that the *DEF* term does not describes the word precisely and the information gain is not enough. For example, the DEF term of the Chinese word "平等互利" is "*DEF=ATTRIBUTE VALUE |DIFFERENCE |EQUAL |ENTITY*", but the meaning of this word is that two things(especially two colonies) are equal and should help each other in order to make benefit for themselves. Because none of the attributes in its *DEF* term describe this meaning, they can not express this word efficiently, and because these abstract attributes exist in a lot of Chinese words, they will be calculated a lot and occupy a big scale in feature statistic. As a result, if we extract all concept attributes and ignore the original words, it will not give the correct feature set and the precise of classification will not be perfect. In order to deal with this problem, we have to decide in which case we can extract the concept attributes and in which not. In this paper, we give a strategy to make a balance between the original words and the extracted concept attributes.

We use the tree form of the concept attributes in *HowNet* to make a concept-tree (C-tree), and calculate their expression power according to the level of the node. By a selected shielded, we divided these nodes into two parts, the strong ones and the weak ones. Because we mainly use the level of the node to decide its expression power, we set a level threshold which will ignores the concept attribute nodes above it, otherwise, the attributes will be added to the feature set. This threshold is called shielded level. And for a word, if the levels of the nodes in its *DEF* term are all above the shielded level, we consider that these attributes are weak in expression and give less information than the original word, and it is unsuitable to extract them. As a result, it is much better to add the original word to the feature set, not the concept ones. The formula calculating the concept expression power of a word is as follows:

$$f(c) = \max_{i=0}^{m} k(c_i) \qquad (3)$$

In formula 3, $k(c_i)$ is the weight of attribute i in the *DEF* term of word c; m is the number of attributes in *DEF* term. This formula calculates all the attributes in a *DEF* term and decides whether the attribute or the word should be added to the feature set. If there is at least one attributes whose levels are higher than the shielded level, the expression power of the *DEF* terms are enough and we added them into the feature set. Otherwise, the original word is added. By this strategy, we can reserve the original words which are not fully expressed, and filter the concept which are abstract and scattered in a lot of words.

3.3 The Treatment of the Word not in the HowNet

Because there are so many Chinese words and the scale is getting larger and larger, it is impossible for a dictionary to embody all of them. So there are some words which are not in *HowNet* which we also should deal with. Because these words have no *DEF* terms, we will select them as the feature and give them a weight by a certain strategy. In our system, we give these words weights by their lengths. When a Chinese word is longer, it is much likely to be a proper noun of a certain field, and its occurrence frequency if much lower, as a result, we gives it a higher weight to compensate it[7].

By practical consideration, we use the square root of the word length to avoid depending too much on the word length because this strategy does not have adequate theoretic foundation. The formula is as follows:

$$W_i = 0.5 * \sqrt{length(i)} \qquad (4)$$

3.4 The Improved Feature Selection Method Based on TF-CE

TF-CE(Term Frequency – Cross Entropy) [2] is improved from *TF* method which is widely used in classification and information retrieval. This method considers the information entropy in the document set and the qualification entropy among the words in order to define how much information a certain word provides in classification, which is also called the importance of a word. The formula mainly depends on three factors:

1). *TF* (Term Frequency): *TF* is the occurrence frequency of a certain word.
2). *CE* (Cross Entropy): *CE* is the relationship between the information entropy of the document and that of the certain word:

$$\text{CrossEntropy Txt}(w) = p(w) \sum_l p(c_l | w) \log \frac{p(c_l | w)}{p(c_l)} \quad (5)$$

Here, $p(w)$ is the occurrence rate of word w in the word set; $p(c_l)$ is of category l in document set; $p(c_l|w)$ is of the document which has word w and belongs to category l.
3). *CN* (Cosine Normalization): Cosine Normalization is used to normalize the vectors to an equal measure in order to compare with them. Without it, the comparison is not trusty.

As a result, the full formula of *TF-CE* is as follows:

$$W_{lk} = \frac{t_{lk} * p(w) \sum_l p(c_l | w) \log \frac{p(c_l | w)}{p(c_l)}}{\sqrt{\sum_{j=1}^{t} t_{lk} * p(w) \sum_l p(c_l | w) \log \frac{p(c_l | w)}{p(c_l)}}} \quad (6)$$

The *TF-CE* is used to reduce the original feature set because it is always too large and there are some features which are not essential in classification or are even noise. For example, there are some words which are commonly used in text and have little difference among categories, and some rare words which occur in very little documents. These words always have a low value of *TF-CE*, and will be filtered out of the feature set to make the vector dimension smaller and better[4].

4 Experimental Results

This system is coded in Windows 2000, and the coding tool is *VS.Net*. The corpus comes from the *People Daily* from 1996 to 1998.

Table 1. Below is the corpus of out experiment. These documents are select for training or test in random. And it is easy to see the corpus is a unbalanced one as the number of the documents in different category is different. However, we manage to make the rate of training and test approximately 2:1

	Economy	Polity	Computer	Sport	Education	Law	Total
Train Set	250	175	130	300	150	200	1205
Test Set	121	82	55	282	63	152	755

In order to show the system clearly, we put it into several models, and use the interim files as the interface of two models. In this way, we can see the result of every main step clearly and easily estimate where is incorrect and should be modified.

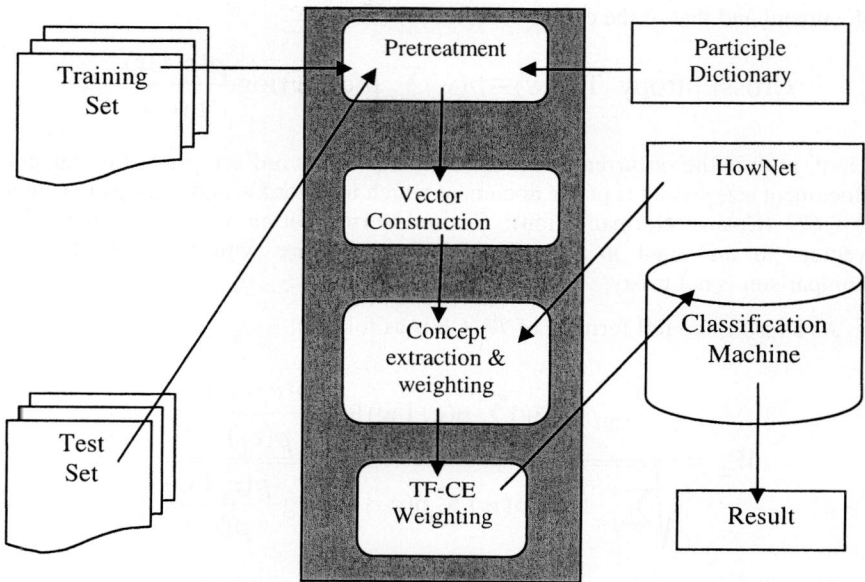

Fig. 3. This is the main flow of the system and the classification machine is based on *SVM* [3] algorithm which is admitted widely as an efficient and quick method of classification

The experiment result shows that, the improved extraction of concept attribute can reduce the feature scale efficiently and with this reduction, the classification precise is much better because it filters the redundant features and the noise in the text which are necessary.

We can see that if we only use original words, the reduction is sharp and many features are filtered while with concept attributes, the reduction is smoother and the difference between different shielded levels are less while the levels are lower. It means that when we reflect the original word to the concept space, we could get a smaller and more stable feature space. For example, when the shielded level is 6, the original feature set is only 40% of the original set. And when we use a threshold (for example the threshold is 3) to reduce the vector dimension, the reduction is small and only 83.6% of the features are cut off, while when we use original words as the feature, we have to cut off 90% of the features with the same threshold. This chart means that using proper concept attribute will reflect the word feature to a smaller and more stable dimension which has a strengthened value in every direction, which is nicer for classification.

4.1 The Reduction of the Improved Feature Selection

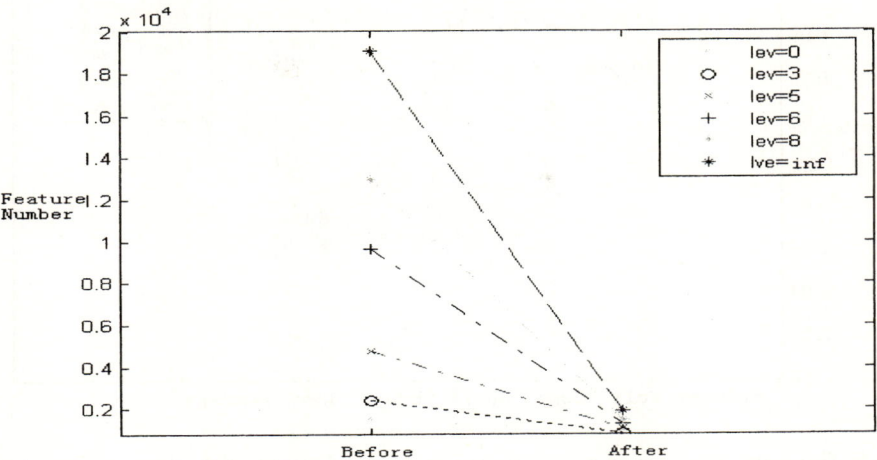

Fig. 4. The *y* axis is the number of the feature set, and it is shown that the different reductions of the features in a threshold of 3.0

4.2 The Classification Precise

We test this method with the *Recall Rate* as the standard of the quality of the classification. The *Recall Rate* is defined as follows:

$$\mathrm{Re}\,call = \frac{CorrectNum}{TotalNum} \tag{7}$$

The *CorrectNum* is the number of the documents which are correctly classified; *TotalNum* is the number of the test set. We used the average value of the six categories as the *Recall Rate* of the system.

From the result we conclude that only uses original words or concept attributes are both not very suitable. In the experiment, if we only use the concept attributes without any shielded levels, the precise is 90.9%, which is the lowest. And when we choose a proper level, for example, level 6, the precise is 93.7%, which is the highest. Also, when the shielded level is ∞, which means we only use original words, the precise is 91.7%, which is higher than the concept one but also not the highest. However, this line is sharper than others, which means that the difference between categories varies largely, from 81.0% to 96.7%. And when we use shielded level, the result is much better and the line is much smoother, that is to say, the precise is much higher and more stable. For example, when the shielded level is 6, the precise varies from 86.2% to 98.2%. This is probably because the feature selection based on original words depends much on the categories because if there are more special words in this field, it is easier to classify it from others. But when we use concept attributes, this difference between categories seems to be smaller and the curve seems to be much smooth.

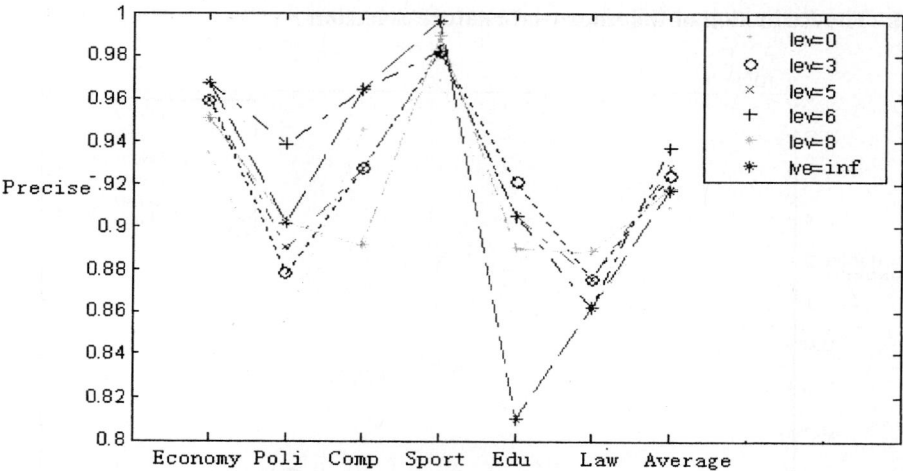

Fig. 5. This is the classification precise of the system with different shielded levels. The x axis is the categories of the classification and we give an average precise as the precise of the system.

5 Conclusions

When we use the concept attributes as the features, we can reduce the feature scale efficiently and reflect the space to a comparatively stable space. However, as there are some weak attributes, it is likely to lose some important information of the words if we put them all into concept. As a result, we set a shielded level to reserve the words with weak concept attributes. Though this would make the feature scale a litter larger, but this addition is reasonable and the classification precise will be improved. Also, the precise difference between different categories will be reduced and the average precise is much more stable. In our experiment, the average precise rises to 93.7% and the difference between categories reduced from 15.7% to 12%.

Acknowledgements

This work was supported by Tsinghua University 985 research fund; the excellent young teacher program, Ministry of Education, China; and State Key Lab of Pattern Recognition open fund, Chinese Academy of Sciences.

References

1. Dong, Z., Dong, Q.: The download of Hownet [EB/OL], http://www.keenage.com
2. Dai. L., Huang, H., The Comparative of Feature Selection in Chinese Text Classification. Journal of Chinese Information Processing, 1(18), (2004) 26-32

3. Cristianini, N., Shawe-Taylor, J.: An Introduction to Support Vector Machines and Other Kernel-based Learning Methods. The Syndicate of the Press of the University of Cambridge, England (2000)
4. Sebastiani, F.: Machine Learning in Automated Text Categorization. ACM Computing Surveys, 34(1), (2002) 1-47
5. Zhou, Y., Zhao M.: The Feature Selection in Text Classification. Journal of Chinese Information Processing, 18(3), (2004) 1-7
6. Lu, M., Li, F., Pang, S.: The Improved Text Classification Method Based on Weighting Adjusting. Tsinghua Science and Technology, 43(4), (2003) 47-52
7. Yang, Y., Pedersen, J. O.: A Comparative Study on Feature Selection in Text Categorization. Proceedings of the 14th International Conference on Machine Learning, (1997) 412-420
8. Salton, G., Buckley, B.: Term-weighting Approaches in Automatic Text Retrieval, Information Processing and Management, 24(5), (1998) 513-523

Research on the Decision Method for Enterprise Information Investment Based on IA-BP Network[*]

Xiao-Ke Yan, Hai-Dong Yang, He-Jun Wang, and Fei-Qi Deng

Institute of Automation,
South China University of Technology,
Guangzhou, 510640, China
yanghd@yeah.net

Abstract: This paper applied the Data Envelopment Analysis (DEA) method to evaluate the input-output efficiency of constructing the enterprise information. Through projecting the inefficacy DEA unit to make the DEA effective in the unit, the data from projection can be used to train the BP network. In the pure BP network model, some flaws exist in BP network model such as slow speed in convergence and easily plunging into the local minima. On the contrary, artificial immune model has a few advantages such as antibody diversity inheritance mechanism and cell-chosen mechanism, which have been applied in this research. In this research, the BP network has been designed, and the IA-BP network model established. By taking the enterprise information application level, enterprise human resources state and information benefit index as the inputs, and the enterprise investing as the output, this model carries out the network training, until to get the satisfied investment decision method. Basing on this model, the enterprise can realize the maximized return on investment. This model not only constructs a new viable method to effectively use the research data, but also overcomes the drawbacks of non-linear description in the traditional information investment decision. The application results show that the model can satisfy the requirements of enterprise information, and provide the best decision method for enterprises as well.

1 Introduction

When implementing the information system for an enterprise, the maximum of input-output benefit on the limited investment could be always deserved. Generally, the maximal benefit is constrained by many factors such as the management, utilization of the information resource, and the quality of the staff. In the traditional process of decision making for establishing the information system for enterprise, the enterprise policy-makers almost always had the highest privilege. In this case, the avoidless personnel state problems always cause the return on investments inefficacy. That is why the scientific decision methods are needed. This kind of scientific decision methods can satisfy the enterprise's requirements, raise the return on investment,

---

[*] Project supported by Guangdong Natural Science Foundation (No. 980150)

improve the enterprise management, promote the application level and strengthen the quality of the staff, in all, it can get the enterprise the maximized input-output benefit[1-3].

Duo to the non-linear function of relationship between the input-output benefit of enterprise information and factors such as utilization level, application level and quality of the staff, the BP network can be well used in the information investment decision[4-5]. This paper firstly employs the DEA method to evaluate the input-output efficiency of constructing the enterprise information. Through projecting the inefficiency DEA unit, the DEA can be made effective in the unit, and the data else can be used to train the BP network. For avoiding the drawbacks of BP network model such as slow speed in convergence and easily plunging into the local minima, the artificial immune system, which has been known having the characteristics such as antibody diversity inheritance mechanism and cell-chosen mechanism, has been taken into account. This paper designs a BP network which is powerful in searching ability in order to avoid the immature constringency. For constructing IA-BP network model, the enterprise information application level, enterprise human resources state and information benefit index are used as input variables and the enterprise's investment is considered as the output variable. The network training should be done repeatedly till the satisfied investment decision is obtained, which means the maximal investment benefit.

2 Data Envelopment Analysis (DEA)

In 1957 Farrell published *The measurement of productive efficiency*, which uses the non-premised production function and the premised production to evaluate the efficiency. It is the first time in the world that the efficiency is evaluated by foreland production. The foreland production is evaluated through the linear programming. In 1978, on the basis of the Farrell theory, Charnes, Cooper and Rhodes established a CCR model, which inverted the efficiency evaluation for more investment and more output into the mathematical ratio, used the linear programming to evaluate the foreland production and calculated relative efficiency in the fixed scale. CCR is used to evaluate the input-output efficiency for more-input and more-output decision units: applying the proportion between input and output to evaluate the efficiency, projecting the input and output to the hyper-plane by the DEA method and acquiring the highest output or the least input efficiency frontier. All *DMUs* that project on the efficiency frontier are called DEA effective and that project within the efficiency frontier is called non-effective DEA[6-8].

Suppose the model CCR has n *DMUs*, each *DMU* has m inputs and p outputs. The vector $X_j = (x_{1j}, x_{2j}, \cdots, x_{mj})^T$ is the input of DMU_j and output is recorded as $Y_j = (y_{1j}, y_{2j}, \cdots, y_{mj})^T$, so the corresponding maximal efficiency value is:

$$\text{Max} \quad V_D = \theta$$

$$\text{Subject to} \quad \sum_{j=1}^{n} X_{ij}\lambda_j + S_i^- = X_o \qquad i = 1,2,\cdots,m$$

$$\sum_{j=1}^{n} Y_{rj}\lambda_j - S_r^+ = \theta Y_{ro} \qquad r = 1,2,\cdots,p \qquad (1)$$

$$\sum_{j=1}^{n} \lambda_j = 1$$

$$\lambda_j \geq 0, j = 1,2,\cdots,n$$

$$S_i^- \geq 0, \ S_r^+ \geq 0$$

θ is the relative efficiency value of the decision-unit DMU_o ($0 \leq \theta \leq 1$), namely the utilization degree of input related with output. θ reflects the reasonable degree to the resource allocation of DMU_o, and the larger value θ, the more reasonable resource allocation is. S_i^- and S_r^+ are slack variables. The former denotes ineffective devotion of the i th resource and the latter is the shortfall of the r th output. The decision unit would be considered under 3 conditions according to the value of θ, S_i^-, S_r^+:

(1) If $\theta^o = 1$ and $S_i^{o-} = S_r^{o+} = 0$, DMU_o is taken for DEA efficiency. In the economic system composed of n decision-units, the resources have be made the full use and the devotion-factors are the best combinations.

(2) If $\theta^o = 1$ and a certain $S_i^{o-} > 0$ or $S_r^{o+} > 0$, DMU_o is considered to be DEA weak efficiency. In the economic system composed of n decision-units, it means that it has not enough data with S_i^{o-} for the i th resource if $S_i^{o-} > 0$ and $S_r^{o+} > 0$, we say that there is less S_r^{o+} between the r output index and the maximal output.

(3) If $\theta^o < 1$, DMU_o is not DEA efficiency. In the system consisting of n $DMUs$, we can make the devotion decrease to the primary θ^o proportion and keep the original output through combination. At this point, the effective units will be connected together to form an efficiency boundary that is the foundation to scale efficiency. It can scale the devotion-redundancy and the scarcity of the output from non-effective DEA units. Through analysis, it can supply the information from every devotion-unit using resource at present. Thus, it may be the benchmark of setting objective, so we can see how much this decision-unit can be improved. For DMU_o with non-effective DEA, its input-output is (X_{io}, Y_{ro}). If the optimal value for formula (1) is $\lambda_j, S_i^{-*}, S_r^{+*}$, and the maximal efficiency value is θ^o, the projection of (X_{io}, Y_{ro}) at the efficiency boundary:

$$\hat{X}_{io} = X_{io}\theta_o^* - S_i^{-*} \quad \hat{Y}_{io} = Y_{io} + S_r^{+*} \quad (i=1,2,\cdots,m), r=(1,2,\cdots,s) \qquad (2)$$

From the above formula, we can figure out the optimal input and output of DMU_o as the improvement objective, so the difference between the primary devotion and the optimal one may be regarded as devotion-redundancy, and the difference between the primary output and optimal one can be considered as the scarcity of the output. DMU_o can decrease the input ΔX_{io} or increase the output ΔY_{io} in order to improve the relative efficiency[8-11].

3 IA-BP Network Model

3.1 BP Network Design

BP network is a kind of popular neural network model. Now there are two major methods to design a BP neural network. One is the incremental or decrement detection method. Using this method, we can get network's structure and weight. On the other hand, it can not avoid the traditional inverse arithmetic's defects, such as network structure hardly confirming, network training slowly, often getting into part convergence and the quality of network by practicing depending on actual detection process. The other is the genetic algorithm (GA),which simulates the survival of the fittest of Darwin and the thinking of random information exchanging to reach the global optimization, so, it is the better choice over the traditional method. But the evolution algorithm still has some limitations. For example, the start population is randomly created so that it is easy to obtain immature convergence and get into the local optimization when the solution candidates distribute unevenly. Therefore, there are still some problems to use the evolution arithmetic to design a BP network. In a word, no satisfied method exists for solving the problem up to now.

Artificial immune model is based on the biological immune system theory and is an optimization model extracting and reflecting the biological immune system. There are several characteristics when using artificial immune model to design BP network. Firstly, the antibody's diversity behaves in cell's splitting and differentiation. The immune system clones a large number of antibodies to resist antigens. If this multiplex inheritance mechanism is used to optimize, it will greatly help to improve capability in global search and avoid the local optimization. Secondly, the self-regulation immunity system has the mechanism to support immunologic balance that can create suitable antibodies by suppressing and stimulating them. Using this functionalities we can improve BP network's capability in local research. Thirdly, immunity memory function can remain in part of cells. If there is the same type of antigen to invade, the corresponding memory cell will be aviated quickly and create lots of antibodies. BP network can use the antigen's identification function to accelerate searching speed and improve BP network's global searching capability.

3.2 An IA-BP Network Model

An idea of using the artificial immunity model to design BP network is considering the network weight and structure as the antibody in biological immunity system and combining with immunity network evolution algorithm to learn. The study is to optimize antibodies continually and to research the best antibody by operating to antibodies variation and adjusting based on antibodies' concentration, which is the same as the vector and network structure that make the error cost function E least.

3.3 The Arithmetic and Coding Based on IA-BP Network Model

During the BP network design, we suppose that the input (including the input node number and input value) and output (including the output node number and output value) are known and the node transition function is given in advance, such as S function.

Antibody coding. Each antibody corresponds to a type of network structure. In this paper, mixed coding is made based on real number and formed of the number of invisible node and network weight. Every antibody's code is formed as following.

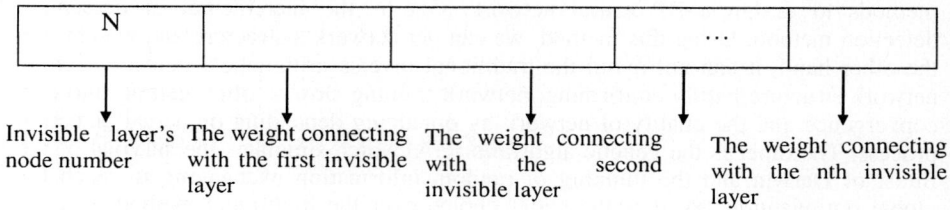

Fig. 1. The coding format of antibody coding

Design of fitness function. Let E_i be the energy function of the network corresponding to antibody P_i, the fitness function $F(i)$ can be simply defined as below.

$$F(i) = 1/(E_i + const) \qquad (3)$$

In the above equation, $const$ is a constant number which is equal to or greater than zero. Introducing $const$ is to avoid the arithmetical overflow when E_i is equal to zero. In this paper the feedback-forward network with a single invisible layer is used.

$$E_i = \sum_p \sum_{out} (T_{p,out} - Y_{p,out})^2 \qquad (4)$$

$T_{p,out}$ and $Y_{p,out}$, respectively, are the expected output and actual output of the out th output node in the P th training sample.

Immune operation. (1) Selection.

Suppose the current filial group is C_{k-1}, the antibody selection for C_{k-1} is based on expected reproduction rate e_i. The expected reproduction rate e_i of the antibody P_i is calculated according to the following steps.

Step1: Calculate the concentration c_i of the antibody P_i.

$$c_i = [\sum_{w=1}^{N} ac_{vw}]/N \qquad (5)$$

Where,

$$ac_{vw} = \begin{cases} 1 & a_{y,w} \geq Tac \\ 0 & a_{y,w} < Tac \end{cases} \qquad (6)$$

$a_{y,w}$ is the affinity between y and w and Tac is a given value.

Step2: Calculate the expected reproduction rate of the antibody P_i

$$e_i = F(i)/c_i \qquad (7)$$

There are n optimal individual C_n determined by e_i. By cloning these C_n, a temporary clone group C is produced. The clone scale is an increasing function in proportion to e_i.

(2) Gaussian mutation. The mutation will be executed only through the component of weight of every antibody among C.

Step1: Each antibody will be decoded into a network structure.

Step2: Mutate all network weights according to the following formula sequentially.

$$W'_{ij} = W_{ij} + a \times \sqrt{F(i)} \times \mu(0,1) \qquad (8)$$

In Equation (8), a is a parameter in [-1,1], $\mu(0,1)$ is a Gaussian operator. Thus C^* will be formed as a mature antibody group, then re-calculate the expected reproduction rate of antibody e_i among C^* and update the antibodies based on e_i to form a new memory group. Some of the group C_{k-1} are replaced by other improved members. d individuals with less e_i will be eliminated to keep variety and the next generation C_k is produced. Then a new antibody is about to be reproduced by the component of invisible node and weight.

4 Information for Enterprises Devotion-Decision-Model Based on IA-BP Network

4.1 The Evaluation of DEA Relative Efficiency

This research analyzed 28 construction corporations including China Road and Bridge Corporation(short for CRBC)(*DMU1*), the First Bureau in Road (*DMU2*), Hebei Road & Bridge Group Co., Ltd(*DMU3*), Beijing Road & Bridge Group Co., Ltd (*DMU4*), the Second Company in Pathway Group(*DMU5*), Fujian Industrial Equipment

Installation Co., Ltd (short for FIEIC)(*DMU6*), South China Engineering of Bridge Group Co., Ltd (*DMU7*) and so on. Their enterprise information scales and levels are unequal.

Every enterprise has the same input index: the application level for enterprises information, human resources state and efficiency index. Every evaluation unit is in the same industry, its character and application field are identical. Therefore, we can use DEA model to evaluate the technology efficiency for information devotion in all enterprises and then determine the effective degree and the level. The non-DEA-efficiency DMU is going to be projected, we can get the Table 1. In this table, there are about 50% enterprises being non-DEA-efficiency, which means that the part information-devotion exists waste and it is necessary to adjust its input.

Table 1. the relevant index of the evaluated unit

Evaluated unit	efficiency index	evaluated unit	efficiency index	evaluated unit	efficiency index	evaluated unit	efficiency index
DMU1	1	DMU8	1	DMU15	1	DMU22	1.21
DMU2	1	DMU9	1.21	DMU16	1	DMU23	1
DMU3	1.31	DMU10	1.32	DMU17	1.15	DMU24	1
DMU4	1	DMU11	1.17	DMU18	1	DMU25	1.31
DMU5	1.22	DMU12	1	DMU19	1.32	DMU26	1.09
DMU6	1.31	DMU13	1.13	DMU20	1	DMU27	1.03
DMU7	1.13	DMU14	1.12	DMU21	1.37	DMU28	1.18

4.2 Establishing IA-BP Network-Decision Model of Information Devotion

(1) Data Processing

The training data and predicted data are normalized in [-1,1] through the function-Premnmx, and all data from 28 entities are divided into two groups: the training data (23 groups) and the predicted data (5 groups).

(2) Network Structure

The neurons in input-layer are respectively the application level of information, the index of human resources, and the information benefit index. The neuron in output-layer is information investment index. The other parameters such as the number of neurons in hidden layer and related weights will be determined by using the artificial immune model.

(3) Network Training

We shall design BP network based on the artificial immune model. Through network training time after time, it is noted that the full match happens when the number of neuron of the hidden layer is eight or more. When there are only 4 neurons, the error will be less and steady for all test data. Hence the number of neurons of the hidden layer is set to be 4. By merging new training data and predicted data, all produced data are used to train the network. After 141 iterations, the training mean absolute error of information-investment index is 4.13, the correlation coefficient is 0.981, and error sum of squares of network is 1.16. Therefore, we can get the decision model of IA-BP network of information-investment with better training accuracy.

5 Practical Application

Now we use the decision model of IA-BP network of information-investment to make information investment decision for CRBC in 2003. The application level of enterprise information and the state of human resources are respectively 0.51 and 0.73. The information efficiency index is in [0.70,0.91], and the sampling interval of the model-decision is set as 0.02. We can get the Fig.2 by setting the information benefit index as the horizontal axis and setting information-devotion index as the vertical axis.

From Fig.2a and Fig.2b, we can know that the information efficiency index is less than 0.81, the characteristic curve of information-investment (the relation curve of the information efficiency index and information-investment) is similar with the curve of function method of information effect. And it shows the demand variation that the increase of the information efficiency index brings information devotion. When the information efficiency index is greater than 0.81, we need more information-devotion as often as the information efficiency index increases 0.02, maybe the scarcity of human resources state in training data creates those.

Take the case of CRBC and FIEIC. The index of human resources condition in CRBC is less than that in FIEIC. The former increases in information-devotion faster than the latter, which means that the characteristic of information-devotion is different in different companies due to the human resources state. Moreover, the demand of information-devotion is not same in identical company.

In Fig 2a, when the information efficiency index of CRBC is less than 0.81, the information-devotion increases slowly, and vice versa.

It reflects the relation between the information efficiency index and the information-devotion in Fig 2b. With the increase of information-devotion, the information efficiency index grows up. Therefore, for FIEIC, the information efficiency index and the competition of enterprises will be improved if the information-devotion increases. Whereas, the information efficiency index of FIEIC is restricted by application level and human resources state so that its value is often less than 0.87.

If the other input restrictions are not in the consideration, the optimal information-devotion of FIEIC is: the hardware investment is 0.677 billions per year and software investment (including personal training) is 0.787 billions per year. In addition, the error is clear between the project of information-devotion and the expect information efficiency index on account of the difference from the application level of enterprise information and the human resources state, and every company has its respective optimal project about the information-devotion.

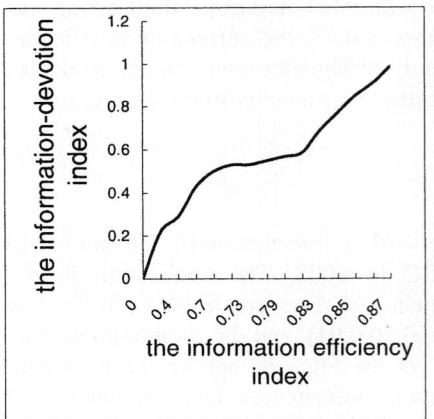 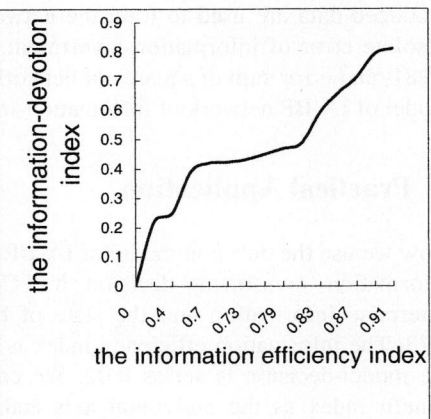

Fig. 2a. The information efficiency index and the information-devotion index of the Primary Company of China Road and Bridge Corporation

Fig. 2b. the information efficiency index and the information-devotion index of Fujian Industrial Equipment Installation Co., Ltd

6 Conclusions

This research established the investment decision-model of IM-BP network information based on statistical network analysis, artificial immune model and BP network. It not only introduced the feasible method for using all data fully, but also overcome the nonlinear deficiency in traditional information methods. Practical application shows that the information efficiency index of CRBC is restricted by application level and human resources state, and commonly the value is less than 0.87. The forecasted result is comparatively reasonable and can reflect the information demand characteristic of CRBC. Many factors, including basic level of information, information security and the recognition for information, affected the output effect of information devotion. Therefore, those factors must not be neglected besides application level and the human resources state, which need further study.

References

1. Jun zhicai.: Data Envelopment Analysis (DEA) Method and Transport Enterprise Technique and Scale Effectiveness [J]. Journal of Highway and Transportation Research and Development. 4 (1994) 44-50.
2. Li Guangjin.: Input-and-output-oriented DEA for assessing relative efficiency[J]. Journal of Management Sciences in China. 2 (2001) 58-62
3. Geng Changsong.: Establishment of Artificial Neural Networks for Optimizing Welding Parameters by Matlab [J]. Jointing Seal. 5 (2001) 14-16
4. Nils J Nilsson.: Artificial Intelligence a New Synthesis[M]. Beijing: China Machine Press (1999)

5. Weymaere, N., Martens, J.P.: On the Initialization and Optimization of Multilayer Perceptrons. IEEE Transactions on Neural Networks. 5 (1994) 738-751
6. Ash. T.: Dynamic Node Creation in Back Propagation Networks. Connection Science. 1 (1989) 365-375
7. Jerne N K.: The immune system. Scientific American. 229 (1973) 51-60
8. Perelson A.: Immune network theory. Immunological Review. 110 (1989) 5-36
9. Hongwei, M.O.: The Principle and Application of Artificial Immune System [M].Haerbin: Harbin University Press of Industry. 11 (2002) 92-136
10. Xing XiaoShuai, Pan jin, Jiao Licheng.: A Novel K-means Clustering Based on the Immune Programming Algorithm. Chinese Journal of Computers. 26 (2003) 50-54
11. Farmer J D., Packard N. H., Perelson A. S.: The Immune System, Adaptation and Machine Learning. Physica D. 22 (1986) 187-204

Process Control and Management of Etching Process Using Data Mining with Quality Indexes

Hyeon Bae[1], Sungshin Kim[1], and Kwang Bang Woo[2]

[1] School of Electrical and Computer Engineering, Pusan National University,
30 Jangjeon-dong, Geumjeong-gu, 609-735 Busan, Korea
{sskim, baehyeon}@pusan.ac.kr
http://icsl.ee.pusan.ac.kr
[2] Automation Technology Research Institute, Yonsei University,
134 Sinchon-dong, Seodaemun-gu, Seoul 120-749, Korea
kbwoo@yonsei.ac.kr

Abstract. As argued in this paper, a decision support system based on data mining and knowledge discovery is an important factor in improving productivity and yield. The proposed decision support system consists of a neural network model and an inference system based on fuzzy logic. First, the product results are predicted by the neural network model constructed by the quality index of the products that represent the quality of the etching process. And the quality indexes are classified according to and expert's knowledge. Finally, the product conditions are estimated by the fuzzy inference system using the rules extracted from the classified patterns. We employed data mining and intelligent techniques to find the best condition for the etching process. The proposed decision support system is efficient and easy to be implemented for process management based on an expert's knowledge.

1 Introduction

Semiconductor is one of the most important elements leading the modern culture. Specially, electronics industries have been developed based on the development of the semiconductor. Fabrication of the semiconductor is achieved by Cleaning, Thermal Treatment, Impurity Doping, Thin Film Deposition, Lithography, and Planarization process. The lithography sub-process sequentially accomplished by photo resistor spread, exposure, development, etching, and photo resistor remove. The fabrication processes are mostly chemical processes, so it is difficult to evaluate the quality of the products on-line. Inspection of the products is completed after fabrication and then the recipe is adjusted by operators. This feature of manufacture can give rise to a serious financial loss in material processing.

In this study, the management system was proposed to solve the mentioned problem. The proposed system was applied in the etching process that is one of the sub-processes of the lithography process (patterning). However, on-line management of the etching process is difficult because inspection of the product quality is taken after fabrication. Therefore, prediction of the product status and inference of the product patterns are necessary to improve the quality and yield in the manufacturing industries. The product quality can be estimated by the prediction model and the prediction result can be information how to adjust the control conditions for quality improvement. In this study, the process management was designed by the prediction models and the inference rules based on data mining techniques such as neural networks, fuzzy clustering, and fuzzy logic. The term *knowledge* can be used separately from or synonymously with *information*. Information is an outcome of data processing and knowledge is accumulative know-how acquired through the systematic use of information [1]. The concept of data mining will be shown to satisfy this need [2]. Data mining is a process that extracts relevant information from a large amount of data utilizing intelligent techniques in decision-making.

Over the past several decades, the clustering method [3], input space partition [4], neuro-fuzzy modeling [5], and the neural network [6] and other methods have been researched in the context of rule extraction and data modeling. Knowledge extraction is the most important aspect of intelligence, and has been much developed. Neuro-fuzzy modeling, an important method to generate rules, was introduced in the early 1990s. In rule generation, clustering [7] and the input space partition method have also been applied. And the extension matrix [8], high-order division [9], decision tree [10] and others have been studied for the purpose of information extraction.

In this study, feed-forward neural networks are used for prediction modeling that is widely applied because of the adaptability. Model inputs consist of control parameters of the process and the quality indexes of the products are predicted by the model. After modeling, the control rules are extracted to adjust the control parameters corresponding to the quality indexes. The rules are generated by fuzzy clustering and inputs of the rules are patterns of the etching products. The main goal of the proposed system is to analyze the process of operation status rapidly and to make a decision easily.

In section 2, the target process such as etching and applied data mining methods are described and section 3 shows the experimental results using the proposed system. Finally, the conclusion is summarized in section 4.

2 Target Process and Methods

The etching process is a very important unit process of semiconductor processes. The etching process is a chemical process that puts slices in a chemical solution and generates a reaction with the solution to shorten the thickness of the slices, or to remove surface defects or surface stress. Etching technologies are broadly classified according to dry and wet etching methods. The process data are not gathered from physical sources, but from a simulation tool. The tool named TCAD is a verified tool by many manufacturing companies of the semiconductor, which use the tool for simulation of processes. We collected dry etching data through simulation in this study and the simulation was designed by Taguchi methods are statistical methods to improve the quality of manufactured goods. Taguchi methods are controversial among many conventional Western statisticians. Taguchi's principle contributions to statistics are

Taguchi loss-function, the philosophy of off-line quality control; and innovations in the design of experiments [11].

Figure 1 shows the proposed management system in this study. The system consists of two parts. One is to predict the quality of the products and the other is to infer the patterns of the products corresponding to the quality. And then the control parameters are handled for quality improvement based on the estimation result. The many goal of the study is to design the management system for etching process. x_1 to x_4 are control parameters, y_1 to y_3 are quality indexes, and p_3 to p_4 are patterns of products. The patterns are classified by quality indexes and the control parameters are adjusted corresponding to the patterns of products. The operating recipe is initially defined by operator's experience with respect to the patterns. Neural networks and fuzzy clustering were applied in modeling and rule generation, respectively.

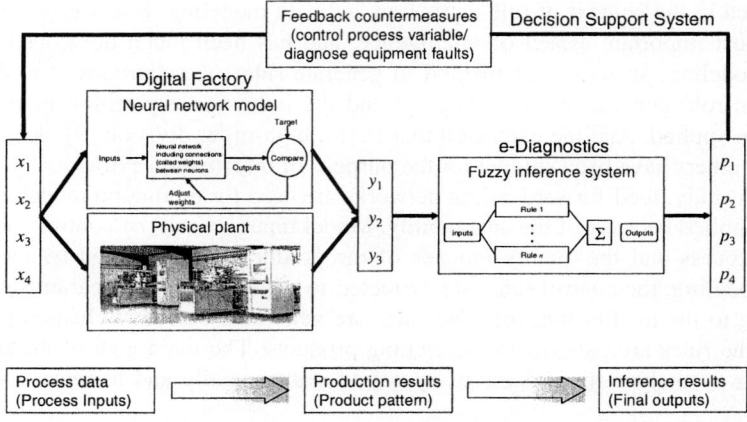

Fig. 1. Schematic concept of the process management system in this research

2.1 Etching Process

Figure 1 shows a schematic diagram of the decision support system that analyzes the process status using data mining and knowledge discovery steps that are called the data model and the inference system, respectively. Sensors or measuring instruments for self-diagnosis are not installed in most equipment assembled in physical processes. Therefore, it is not easy to diagnose processes and equipment through direct detection of the equipment. This paper proposes an indirect diagnosis method to detect the process equipment considering process status and product results.

2.2 Neural Network

The learning property has yielded a new generation of algorithms that can learn from the past to predict the future, extract rules for reasoning in complex environments, and offer solutions when well-defined algorithms and models are unavailable or cumber some [12]. The main advantages of neural networks stem from their ability to recognize patterns in data. This can be achieved without *a priori* knowledge of causal relationships, as would be necessary in knowledge-based systems. The ability of neural networks to generalize relationships from input patterns make them less sensitive to noisy data than other approaches. In the present research, it is difficult to perform a comprehensive inspection in semiconductor assembly lines, so the neural network model was employed to predict the final quality of products with simulations.

2.3 Fuzzy Logic

Fuzzy Logic is a departure from classical two-value sets and logic that uses soft linguistic system variables and a continuous range of true values, rather than strict binary decisions and assignments. Formally, fuzzy logic is a structured, model-free estimator that approximates a function through linguistic input/output associations. Fuzzy rule-based systems apply these methods to solve many types of real-world problems, especially where a system is difficult to model, is controlled by a human operator or expert, or where ambiguity or vagueness is common. A typical fuzzy system consists of a rule base, membership functions, and an inference procedure. Some fuzzy logic applications include control, information systems, and decision support. The key benefits of fuzzy design are ease of implementation, and efficient performance [13].

3 Experimental Results

The final goal of this research was to design a data model for a simulation of the process of manufacture and to build a process management system, as shown in Fig. 2. The algorithm consists of 5 steps that include the input, pre-processing, data modeling, inference system construction and output. The data model and inference system are the main algorithms in the proposed system. Neural networks and fuzzy logic were applied to design the proposed model and inference system. Threshold values are determined by the expert's knowledge according to the data or system.

The proposed system consists of the input, preprocessing to extract features, data modeling, rule extraction and the output stage. The target variables such as quality indexes included the gradient and height of the center barrier of etched area that were applied for modeling as target variables and for rule extraction as input variables. The quality indexes were extracted based on knowledge and the data modeling was achieved to predict the quality indexes of the etched products. After modeling, the rules were generated by the fuzzy clustering to infer the input-output relationship. The rules can indicate the cause of the product quality, so the control parameters can be adjusted based on the inference information. Through the proposed system, the product result is predicted, the cause of the result is inferred, and then the control condition is handled. This sequence of the process is the self-organization and self-learning concept in this study that can improve the flexibility and adaptability.

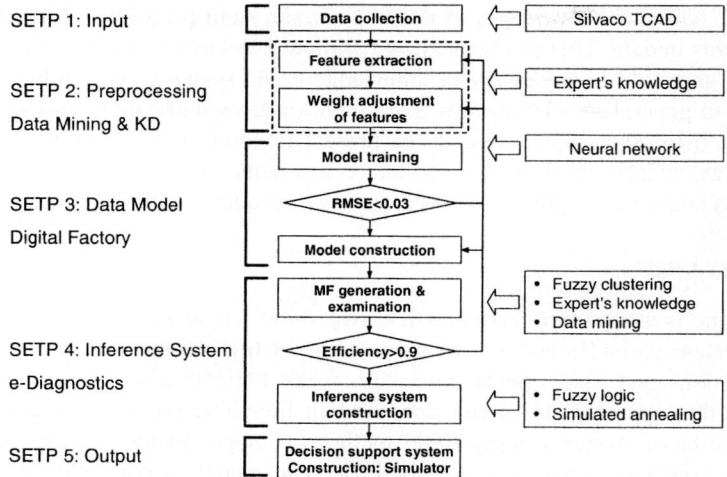

Fig. 2. Whole flow chart of the decision support system for the etching process

3.1 Step 1: Input (Control Parameters)

Test data were generated by the commercial simulator (TCAD) that was used as a substitute for physical plants. The upper layer (Oir32) was the control target of the etching process. The input variables (control parameters) of the etching process consisted of Bake Time, Bake Temp, Circle Sigma, and Projection numerical aperture (na.). The control parameters play an important role in the product quality, so the quality can be handled by adjusting the parameters. The etching status of Oir32 is determined by the four input variables. The ranges of variables are determined by the Taguchi method, which is an optimization method for experimental design. First, the maximum and minimum ranges of each variable are defined, and the range values (discrete ranges) are determined, as shown in Table 1.

Table 1. Input variables of data generation for the etching process using commercial tools

Input variables	Variable name	Range	Values
x_1	Bake Time	2	1(30) 2(60)
x_2	Bake Temp	2	1(125) 2(185)
x_3	Circle Sigma	3	1(0.3) 2(0.5) 3(1.0)
x_4	Projection na	3	1(0.28) 2(0.4) 3(0.52)

3.2 Step 2: Preprocessing (Extraction of Quality Indexes)

Nine feature points are extracted from the approximated pictures, as shown in Fig. 3. The depth of the 4th feature point, the difference between the 4th feature and the 5th feature, and the gradients between the 2nd feature and the 3rd feature are the respective y_1, y_2, and y_3 product variables that express the product quality. Thus, the input variables are four, from x_1 to x_4, and the product variables are three, from y_1 to y_3, as shown in Table 2. The three quality indexes in this study have strong influence on the quality of the etching process.

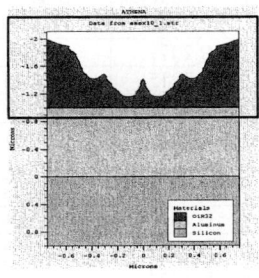
(a) Etching result

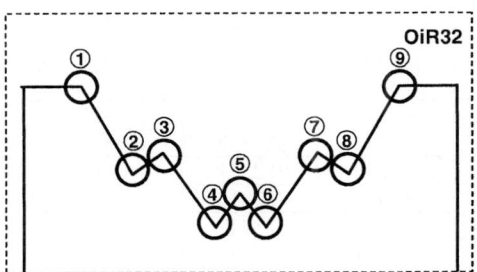
(b) Feature of etching result (quality indexes)

Fig. 3. Feature points of Oir32 in the etching process

Table 2. Product variables generated by the plant or model (quality indexes)

Product variables	Variable description
y_1	Etching depth at the initial point (Depth of 4)
y_2	High of the centre point (Difference between 4 and 5)
y_3	Slope of left side (Gradient between 2 and 3)

3.3 Step 3: Construction of Data Model

The modeling technique was applied to predict the quality index of the etching process with current parameters of the process control. The effective management can be achieved by the prediction model of the product quality.

In this study, the neural network modeling technique was employed to design the data model to match input variables (x_1 to x_4) with quality indexes (y_1 to y_3). Through the traditional mathematical model, it is very difficult to identify the process status in nonlinear and complex systems. But the neural network is a very effective modeling method to express the relationship between the inputs and outputs of complicated systems. The new model can be constructed and the designed model can be implemented by the developed application.

Thirty-six data sets for the input variables and product variables were trained by the neural network model. The number of layers was four and the number of nodes was 7, 9, and 11, respectively. The learning rate was 0.01 and the epoch was 50,000 iterations. The back-propagation algorithm was employed in model learning. The model was trained until the RMSE fell below 0.03. Table 3 shows the prediction results of the product features using the neural network model. The prediction performance is good enough to be applied in the estimation of product qualities.

Table 3. Prediction results of quality indexes using the neural network model

No.	Target values			Prediction results		
	y_1	y_2	y_3	y_1	y_2	y_3
1	0.3594	0.3438	-0.0691	0.33	0.3306	-0.0654
2	0.0057	0.2101	0.0223	0.0246	0.2098	0.0456
3	0.5975	0.2852	0.015	0.5984	0.2748	-0.0009
4	0.3309	0.2926	0.0659	0.3432	0.2979	0.0468
5	0.3639	0.3438	-0.0558	0.3514	0.3215	-0.0845
6	0.013	0.1793	0.0381	0.018	0.1856	0.0131
7	0.6032	0.2738	0.0073	0.5992	0.2673	0.0076
8	0.3444	0.2571	0.0615	0.3347	0.2725	0.0367
9	0.3986	0.2813	-0.076	0.3982	0.2839	-0.0533
10	0.0888	0.1417	-0.0051	0.0913	0.1396	0.0528
11	0.6251	0.2386	0.004	0.6141	0.2357	-0.0265
12	0.3905	0.2029	0.0379	0.463	0.2253	0.0402
13	034064	0.2601	-0.0932	0.4291	0.2841	-0.1098
14	0.1217	0.1313	0.0149	0.1343	0.1307	-0.0222
15	0.63	0.2295	-0.019	0.6597	0.2216	-0.0361
16	0.388	0.1881	0.0387	0.3863	0.1726	0.0427
RMSE (Root Mean Square Error)				0.0252	0.0125	0.025

3.4 Step 4: Construction of Inference System

In this study, the fuzzy inference system was applied to estimate the patterns of the products with respect to quality indexes of the products. The inference rules were extracted according to the results of the model estimation such as quality indexes. This inference system was the primary step in supporting a decision of system operation. If the pattern can be inferred from the quality index, the control parameters can be handled because there is a recipe corresponding to the pattern.

The fuzzy inference system was built by 36 data sets that were employed in the model construction step, as shown in Fig. 4. In addition, 16 test data sets were used to validate the inference performance. Table 4 shows the classification results based on the fuzzy inference system. The classification accuracy was 93.75%. The membership functions were optimized by simulated annealing, which is one of the fastest optimization methods. The optimization was repeated until the performance achieved over 90% accuracy. The threshold was defined by the user according to the performance.

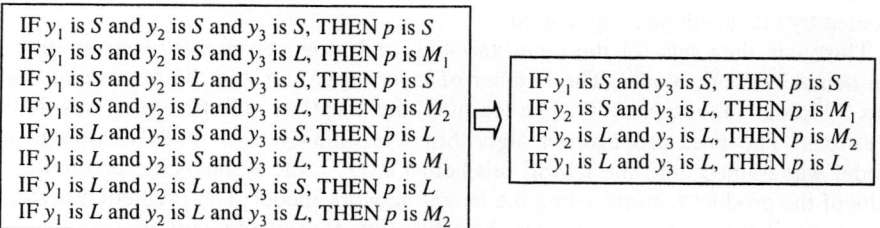

Fig. 4. Extracted rules by the clustering method and reduced rules

Table 4. Classification results of product patterns using the fuzzy inference system

No.	y_1	y_2	y_3	Target pattern	Results	Threshold	Classified results
1	0.3594	0.3438	-0.0691	4	3.526	$3.5 \leq P < 4.5$	4
2	0.0057	0.2101	0.0223	3	2.527	$2.5 \leq P < 3.5$	3
3	0.5975	0.2852	0.015	3	2.96	$2.5 \leq P < 3.5$	3
4	0.3309	0.2926	0.0659	3	2.527	$2.5 \leq P < 3.5$	3
5	0.3639	0.3438	-0.0558	4	3.719	$3.5 \leq P < 4.5$	4
6	0.013	0.1793	0.0381	2	2.475	$1.5 \leq P < 2.5$	2
7	0.6032	0.2738	0.0073	3	2.742	$2.5 \leq P < 3.5$	3
8	0.3444	0.2571	0.0615	3	2.527	$2.5 \leq P < 3.5$	3
9	0.3986	0.2813	-0.076	4	3.512	$3.5 \leq P < 4.5$	4
10	0.0888	0.1417	-0.0051	**2**	**2.527**	Inaccuracy	**3**
11	0.6251	0.2386	0.004	3	3.413	$2.5 \leq P < 3.5$	3
12	0.3805	0.2029	0.0379	3	2.527	$2.5 \leq P < 3.5$	3
13	0.4064	0.2601	-0.0932	4	4	$3.5 \leq P < 4.5$	4
14	0.1217	0.1313	0.0149	2	1.949	$1.5 \leq P < 2.5$	2
15	0.63	0.2295	-0.019	4	3.53	$3.5 \leq P < 4.5$	4
16	0.388	0.1881	0.0387	3	2.527	$2.5 \leq P < 3.5$	3
Accuracy corresponding to the inputs: 93.75%							

3.5 Step 5: Output

The final proposed system can diagnose abnormal situations of the process status and adapt unexpected plant conditions to normal situations. In Step 5, the control commands of the process will be drawn to manage the manufacturing plant effectively and stably. The process simulator and diagnostics were developed to handle the processes and equipment of neural networks and fuzzy logic. This integrated system can make decisions for process and equipment management. Table 5 shows the control command and operating guide based on the results of the fuzzy inference system for faults.

Table 5. Control command and diagnostic action corresponding to output patterns

Output patterns	Control command and diagnosis action
p_1 (pattern 1)	• Decrease x_1 (Bake Time) • Check baking heater or heating controller
p_2 (pattern 2)	• Increase x_3 (Circle Sigma) • Check coater or coating controller
p_3 (pattern 3)	• Good product
p_4 (pattern 4)	• Decrease x_1 (Bake Time) and Increase x_4 (Projection na.) • Check baking heater or heating controller and coater or coating controller

3.6 Simulation Results

By using this simulator, the quality indexes of the products can be estimated and the production pattern can be inferred at the same time. Also, incorrect results of prediction or inference can be adjusted by re-organization and re-training of the neural networks and fuzzy inference system. Figure 5 shows an inferior product case that is inferred to be a Pattern 4 case. In this case, the bake time has to be reduced and the projection numerical aperture must be increased gradually. Figure 5 (a) shows a prediction and inference result that is included in the Pattern 4. There is incorrect operation in the projection numerical aperture. Therefore, the control command is shown in the bottom box for control commands. One command is to increase the projection numerical aperture, so the value is increased shown in Fig. 5 (b). As shown in Fig. 5 (b), the simulation result with handling the control parameter is included in the normal condition. That is, the product quality can be managed by the prediction model and inference rules.

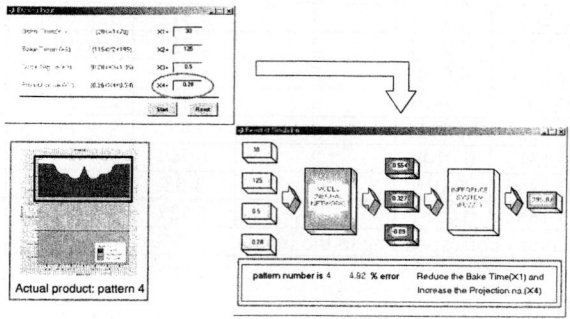

(a) Prediction and inference result about pattern 4

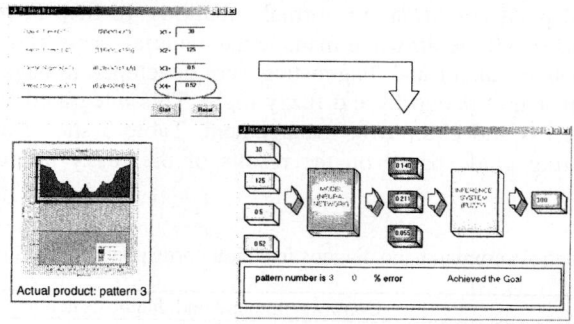

(b) Prediction and inference result about pattern 3

Fig. 5. The simulation result: (a) simulation of Pattern 4 and (b) Pattern 3 with changed x_4

4 Conclusions

The process of production is a complex and dynamic system, and its fabrication line contains other unit processes. Faults or breakdowns of the process can occur under abnormal conditions that influence the product qualities or the production yields. The faults of process usually appear after a specific symptom occurs. Over the past decades, many techniques using artificial intelligence and data mining have been studied to solve these problems. In past research, model-based, case-based, and knowledge-based diagnosis methods have usually been employed. Because each method has strong points and weak points, a suitable method has to be selected according to the features of the target system. A combined technique, furthermore, can be more efficient than a single method. In this study, a prediction model based on neural networks was applied to estimate the special quality of products that are produced under the given input conditions. The final decision support system was built using a fuzzy inference system that contains reasoning functions for the simulation of products.

Acknowledgement

This work was supported by "Intelligent Manufacturing Systems (IMS) Program" hosted by the Ministry of Commerce, Industry and Energy, in Korea.

References

1. Liebowitz, J. (ed.): Knowledge Management Handbook. CRC Press, M.A. (1999)
2. Berry, Michael J. A., Linoff, G. S.: Data Mining Techniques. John Wiley & Sons Inc., Canada (1997)
3. Nadine, T. G.: The neural network model RuleNet and its application to mobile robot navigation. Fuzzy Sets and Systems **85** (1997) 287-303
4. Nozaki, K., Ishibuchi, H., Tanaka, H.: A simple but powerful heuristic method for generating fuzzy rules from numerical data. Fuzzy Sets and Systems **86** (1997) 251-270
5. Lin, C. T.: Neural Fuzzy Control Systems with Structure and Parameter Learning. World Scientific Pub Co., New York (1994)
6. Omatu, S., Khalid, M., Yusof, R.: Neuro-Control and its Applications. Springer (1995)
7. Shi, Y., Mizumoto, M.: An improvement of neuro-fuzzy learning algorithm for tuning fuzzy rules. Fuzzy Sets and Systems **118** (2001) 339-350
8. Wang, X. Z., Wang, Y. D., Xu, X. F., Ling, W. D., Yeung, D. S.: A new approach to fuzzy rule generation-fuzzy extension matrix. Fuzzy Sets and Systems **123** (2001) 291-306
9. Abe, S., Lan, M. S.: A Function Approximator Using Fuzzy Rules Extracted Directly From Numerical Data. Proceedings of 1993 International Joint Conference on Neural Networks **2** (1993) 1887-1892
10. Pal, N. R., Chakraborty, S.: Fuzzy rule extraction from ID3-type decision trees for real data. IEEE Transactions on Systems, Man and Cybernetics, Part B. **31** (2001) 745-754
11. Roy, R. K.: A Primer on the Taguchi Method. Society of Manufacturing Engineers (1990)
12. Tsoukalas, L. H., Uhrig, R. E.: Fuzzy and Neural Approaches in Engineering. John Wiley & Sons Inc., New York (1997)
13. Wang, L. X.: A Course in Fuzzy Systems and Control. Prentice-Hall, NZ (1997)

Automatic Knowledge Configuration by Reticular Activating System

JeongYon Shim

Division of General Studies, Computer Science, Kangnam University,
San 6-2, Kugal-ri, Kihung-up,YongIn Si, KyeongKi Do, Korea
Tel: +82 31 2803 736
mariashim@kangnam.ac.kr

Abstract. Reticular Activating system which has a form of small neural networks in the brain is closely related system with the automatic nervous system. It takes charge of the function that distinguishes between memorizing one and the others, accepts the only selected information and discards the unnecessary things.In this paper, we propose Reticular Activating system which has functions of selective reaction, learning and inference. This system consists of Knowledge acquisition, selection , storing and retrieving part. Reticular Activating layer is connected to Meta knowledge in the high level of this system and takes part in Data Selection. We applied this system to the problem of analyzing the customer's tastes.

1 Introduction

Reticular Activating system which has a form of small neural networks in the brain is closely related system with the automatic nervous system. It takes charge of the function that distinguishes between memorizing one and the others, accepts the only selected information and discards the unnecessary things. In the environment of huge data flood, the requirement for implementing more intelligent system which can select the important information from the huge data pool is getting high. For implementing automatic intelligent smart system, it should be firstly considered to design the efficient structure of component and its intelligent mechanism. Adapting Reticular Activating system is very helpful for making more efficient system. Accordingly, in this paper, we propose Reticular Activating system which has functions of selective reaction, learning and inference. This system consists of Knowledge acquisition, selection , storing and retrieving part. Reticular Activating layer is connected to Meta knowledge in the high level of this system and takes part in Data Selection. We applied this system to the problem of analyzing the customer's tastes.

2 Reticular Activating System

In this section Reticular Activating System which can select and store the information was designed. As shown in Fig.1, this system has a hierarchical structure and it consists of Knowledge acquisition, Selection and Storing to Memory.

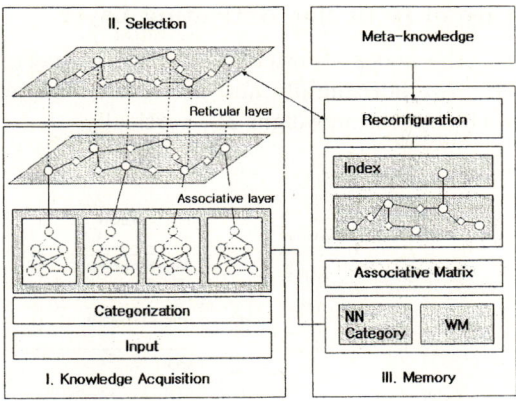

Fig. 1. Associative memory frame

First, Knowledge acquisition part has multi modular NN(neural Network)s and perform the learning process with the training data according to the categories. It uses BP(Back Propagation) algorithm. The output nodes of Modular NN are connected to nodes in Associative layer which has logical network connected by associative relations.

Second, Reticular Activating layer has a knowledge net which consists of nodes and their associative relations. The nodes in knowledge net are connected to the nodes of Associative layer vertically. The importance value is assigned to the connection weight of this vertical relation. Selection module performs selecting process with these values of associative relations and vertical relations using the criteria given by Meta Knowledge.

Third, Storing to Memory consists of two part of Knowledge Reconfiguration and storing the values for NN. In Reconfiguration, the selected nodes and relations are reconfigured and stored in memory. The knowledge net is performed by attaching nodes centering around common node. After reconfiguration the centering node is connected to index which is used in searching process. The another part of memory is storing the values for NN. After finishing the learning process of modular NN, this system stores the values of category, parameters and weight matrix. These stored values are used for perception, inference and knowledge retrieval.

Reticular Activating System performs the functions of Learning, Selection and Knowledge retrieval as these three parts collaborates on a work interactively.

3 Knowledge Retrieval from Memory

In this section ,Selection, Storing and Knowledge Retrieval of Reticular Activating System are described. This system has a same structure of multi modular NN ,learning and functions of Associative layer as explained in the paper [1].

3.1 The Structure of Reticular Activating Layer

As shown in figure 2, the nodes of reticular layer are connected each other with Associative relation, R_{ij} horizontally and are also connected to the nodes in Associative layer with connection weight R_{ij} vertically.

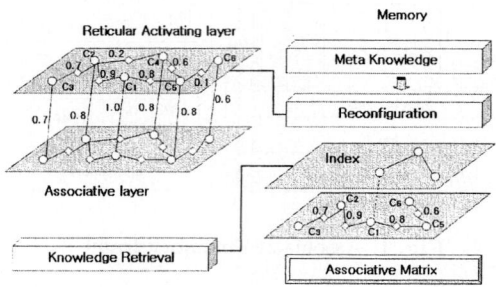

Fig. 2. Reticular Activating Layer

The connection weights R_{ij} and S_{ij} are used for selecting when Meta knowledge gives a criteria.

Storing Criteria:
$V_i \geq 0.5, R_{ij} \geq 0.5$

If Meta knowledge gives a following storing criteria, the nodes and relations satisfying this criteria are selected. The selected nodes and relations are reconfigured by attaching the related nodes centering common node [7]. This centering common node is connected to the index node in Index layer and used for searching as a keyword.

After reconfiguration, new knowledge net is formed and it is transferred to Associative Matrix which is used for knowledge retrieval.

3.2 Associative Matrix

The knowledge net is represented as shown in figure 3 and transferred to Associative Matrix.

Each node has two factors of P(Property) and Vertical Strength,EV, between the node in Reticular Activating layer and node in Associative layer. P has the property of a class and EV retains the empirical value obtained by the mechanism. The empirical value represents the preference or the strength of activation. The relation between the two nodes is represented by both linguistic term and its associative strength.

TRANSFORM-TO : transform in time
AFFECT : partial transform
IS-A : generalization
MADE-OF :component

The linguistic terms are represented as TRANSFER-TO, AFFECT, IS-A, MADE-OF, NOT and SELF. Linguistic term denotes the degree of representing the associative relation. It can be transformed to the associative strength that has a real value of [-1,1]. The positive value denotes the excitatory strength and the negative value represents the inhibitory relation between the nodes. It is used for extracting the related facts. The minus sign of the minus value means the opposite directional relation as -TRANSFORM-TO.

The relational graph is transformed to the forms of AM(Associative Matrix) and vector B in order to process the knowledge retrieval mechanism. AM has the values of associative strengths in the matrix form and vector B denotes the EV of the nodes.

For example, the relational graph in Fig.2. can be transferred to the associative matrix A and vector B as follows.

The associative matrix,A, is :

$$A = \begin{bmatrix} 1.0 & 1.0 & 1.0 & 0.0 & 0.0 \\ -1.0 & 1.0 & 0.9 & 0.0 & 0.0 \\ -1.0 & -0.9 & 1.0 & 0.7 & 0.3 \\ 0.0 & 0.0 & -0.7 & 1.0 & 0.0 \\ 0.0 & 0.0 & -0.3 & 0.0 & 1.0 \end{bmatrix}$$

The matrix, A, has the form of $A = [R_{ij}]$. The associative strength, R_{ij}, between C_i and C_j is calculated by equation (2).

$$R_{ij} = P(a_i|a_j)D \qquad (1)$$

where D is the direction arrow, $D = 1 or -1$, $i = 1, \ldots, n$, $j = 1, \ldots, n$.

The vector B is :

$$B = \begin{bmatrix} -0.9 & 1.0 & 0.3 & 0.8 & 0.2 \end{bmatrix}$$

It consists of the empirical value, V_i.

Using this Associative Matrix,A, and vector B, this system can extract the related facts by following knowledge retrieval algorithm.

Algorithm 1 : knowledge retrieval algorithm from AM

Step 1: Search the corresponding keyword in Index layer
Step 2: Retrieve the corresponding associated nodes and relations in the row of the activated node from AM in memory.

Step 3: IF((not found) AND (found the initial activated node))
Goto Step 4.
ELSE
Output the found fact.
Add the found fact to the list of inference path.
Goto Step 2
Step 2: STOP

When the class,C_i, in the relational graph is assumed to be activated, from the node, C_i, the inferential paths can be extracted using the knowledge retrieval algorithm. The inferential path, I_i has the following form.

$$I_i = \left[C_i\ (V_i)\ (R_{ij})\ C_j\ (V_j) \right]$$

where C_i is i-th class node, V_i is its vertical value, R_{ij} is the associative strength between C_i and C_j. In this step, the vertical values of the classes are also extracted. The following example is the result from the matrix A and vector B using the knowledge retrieval mechanism.

$I_1 = [C_1(-0.9)$ IS-A(1.0) $C_2(1.0)$ TRANSFER-TO(0.9) $C_3(0.3)$ MADE-OF(0.7) $C_4(0.8)]$
$I_2 = [C_1(-0.9)$ IS-A (1.0) $C_2(1.0)$ TRANSFER-TO(0.9)]$
$I_3 = [C_3(0.3)$ MADE-OF(0.7) $C_5(0.2)$]
$I_4 = [C_1(-0.9)$ IS-A(1.0) $C_3(0.3)$ MADE-OF(0.7) $C_4(0.8)]$
$I_5 = [C_1(-0.9)$ IS-A(1.0) $C_3(0.3)$MADE-OF(0.7)$C_5(0.2)]$

From the obtained inferential paths, this system can extract the related facts as much as user wants by masking with the threshold, θ which is given by a criteria of Meta knowledge. In this step, the connected facts that has the value of the associative strength over the threshold are extracted. In the case of I_1, when the threshold is 0.7, the extracted path is $[C_1(-0.9)$ IS-A(1.0) $C_2(1.0)$ TRANSFER-TO(0.9) $C_3(0.3)]$.

The another function of the knowledge retrieval mechanism is to infer the new relations.

From the following extracted inferential path,

$$C_i(V_i)R_{ij}C_j(V_j)R_{jk}C_k(V_k),$$

we can elicit the new inferred path between C_i and C_k. The new associative strength,R_{ik}, is calculated by equation (3).

$$C_i(V_i)R_{ik}C_k(V_k)$$

$$R_{ik} = R_{ij} * R_{jk} \qquad (2)$$

The inferential path, $I_1 : C_1(-0.9)$ IS-A(1.0) C_2 TRANSFER-TO (0.9) C_3 MADE-OF(0.7) C_4, can produce the new relations, C_1 (0.9) C_3, C_1 (0.63) C_4 by its mechanism[4].

The numerical term, -0.9, of $C_1(-0.9)$ denotes the vertical prior knowledge of node C_1. It has the negative value which gives an negative effects on the selecting factor.

3.3 Vertical Selecting Factor

It is generally known that human brain strongly reacts on the familiar facts which are experienced before. In a similar way, this concept can be adopted to the intelligent system and used for developing the more efficient mechanism. To implement this concept, we define the value that represents the vertical prior knowledge as Vertical Selecting factor. Vertical Selecting factor is the value which represents the vertical prior knowledge in the memory. This factor affects the reactive degree of a certain class for the input data in the Selecting layer.

The vertical value(EV) of the node in the Associative layer has an effect on calculating the vertical selecting factor of the Selecting layer.

Vertical Selecting factor, E_i ,is calculated by the equation(3) from the Vertical Values of node i in Associative layer.

$$E_i = \frac{1 - exp^{-V_i}}{1 + exp^{-V_i}} \qquad (3)$$

where E_i is Vertical Selecting factor and V_i is the Vertical Values. The function of E_i is the bipolar sigmoid and its desired range of output values is between -1 and 1.

Positive Vertical Selecting factor denotes the positive vertical prior knowledge and negative Vertical Selecting factor denotes the negative prior knowledge. The value of Vertical Selecting factor is sent to the Reactive layer for deciding the Reactive degree. Reactive degree which represents the degree of activation for the corresponding class, is calculated by Filtering factor and Historical Accessing factor including this Vertical selecting factor.

Filtering factor,F_i is the activated ratio for a certain class and it is already introduced in the equation (1). Historical Accessing factor,H_i, represents the frequency of accessing to a certain class. If a class is not activated for a long time, the value of Historical factor decays.

$$H_i = \frac{1}{1 + e^{-A}}, A = \begin{cases} A + 1 \text{ if } accessed \\ A - 1 \text{ otherwise} \end{cases} \qquad (4)$$

where H_i is Historical Accessing factor and A is the number of being accessed.

$$S_i = \frac{E_i + F_i + H_i}{3} \qquad (5)$$

where S_i is Selecting degree.
This selecting degree is used for filtering the input data in the Selecting layer.

4 Experiments

This system is applied to the area for estimating the purchasing degree from the type of customer's tastes, the pattern of commodities and the evaluation of a company. We tested with three classes. First class consists of ten customer's input term - four types of customer's tastes, second class consists of five input

```
Associative Matrix A

1.000000  0.300000  0.010000
-0.100000  1.000000  0.500000
0.300000  0.700000  1.000000

Vertical Values Vi (vector B)

0.240000  0.3000000  0.830000
```

Fig. 3. Associative Matrix A, vector B

```
Class C2 is fired...

Vi...0.300000
Ei...0.148821
Fi...1.000000
Hi...0.731100
Si...0.626633

Actual Output of NNi:

    T1        T2        T3        T4
0.033432  0.990412  0.017540  0.007431

Output Value : Type2 0.620625
```

Fig. 4. Knowledge retrieval step : output value from the mechanism

```
Retrieved Inferential Path :

C1(0.24) Affected-by(0.30) C2(0.30) Produced-by(0.5) C3(0.83)
C1(0.24) Closely-Related(0.01) C3(0.83)
```

Fig. 5. Knowledge retrieval step : extracted inferential path

Table 1. The output by the selecting degree

V_i	-1.0	-0.5	0.0	0.5	1.0
E_i	-0.4621	-0.2449	0.0000	0.2449	0.4621
S_i	0.4230	0.4954	0.5700	0.6587	0.7311
O_i	0.4199	0.4898	0.5635	0.6512	0.7228

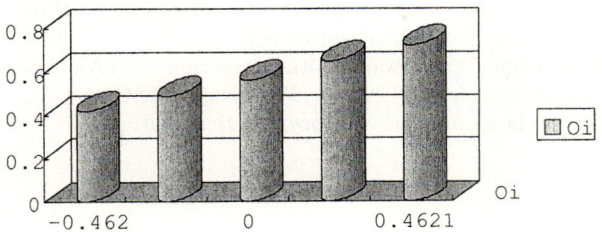

Fig. 6. The variation of output by Vertical selecting factor, $H_i = 0.7311$, $F_i = 1.0$, $O_i = 0.990412$

factors - three patterns of commodities and third class consists of eight evaluating terms - three evaluation degrees of company in the diagnostic area. Fig.3, Fig.4 and Fig.5 represent the results from the data extraction mechanism. Table 1. denotes the variation of output O_i according to the Vertical Value of the node in Associative layer where V_i is Vertical Value, E_i is Vertical Selecting factor, S_i is selecting degree, Historical Accessing factor H_i is 0.7311, Filtering factor F_i is 1.0 and the output value of NNi is 0.990412. Fig.6 shows the variation of output by Vertical Selecting factor. As shown in these figures, this memory is reacted by the Vertical Selecting factor sensitively and produces the different output values according to the Vertical selecting factor.

5 Conclusion

In this paper, we propose Reticular Activating system which has functions of selective reaction, learning and inference. This system consists of Knowledge acquisition, selection , storing and retrieving part. Reticular Activating layer is connected to Meta knowledge in the high level of this system and takes part in Data Selection. We applied this system to the problem of analyzing the customer's tastes.

As a result of testing, we could find that it can extract the related data easily. This system is expected to be applicable to many areas as data mining, pattern recognition and circumspect decision making problem considering associative concepts and prior knowledge.

References

1. Jeong-Yon Shim:Knowledge Retrieval Using Bayesian Associative Relation in the Three Dimensional ModularSystem,Lecture Notes in Computer Science, Vol.3007,Springer-Verlag,(2004)630-635
2. E. Bruce Goldstein,Sensation and Perception,BROOKS/COLE
3. Judea Pearl : Probabilistic reasoning in intelligent systems, networks plausible inference,Morgan kaufman Publishers (1988)
4. Laurene Fausett: Fundamentals of Neural Networks,Prentice Hall
5. Simon Haykin: Neural Networks,Prentice Hall
6. Jeong-Yon Shim, Chong-Sun Hwang,Data Extraction from Associative Matrix based on Selective learning system,IJCNN'99, Washongton D.C
7. John R. Anderson,Learning and Memory,Prentice Hall

An Improved Information Retrieval Method and Input Device Using Gloves for Wearable Computers

Jeong-Hoon Shin and Kwang-Seok Hong

School of Information and Communication Engineering, Sungkyunkwan University,
Suwon, Kyungki-do, 440-746 Korea
only4you@chol.com, kshong@skku.ac.kr
http://only4you.mchol.com, http://hci.skku.ac.kr

Abstract. In this paper, we describe glove-based information retrieval method and input device for wearable computers. We suggest an easy and effective alphanumeric input algorithm using gloves and conduct efficiency test. The key to the development of the proposed device is the use of unique operator-to-key mapping method, key-to-symbol mapping method and simple algorithm. We list and discuss traditional algorithm and method using a glove, then describe an improved newly proposed algorithm using gloves. The efficiency test was conducted and the results were compared with other glove based device and algorithm for wearable computers.

1 Introduction

In this paper, a new gloves-based text input device and improved algorithm are introduced to provide information retrieval method for a wearable computer. Wearable computers are the next generation of portable machine. Worn by people, they provide constant access to various computing and communication resources. Wearable computers are generally composed of small sized PC, display mounted on head, wireless communication hardware and input device. Thus, input to small sized devices is becoming an increasingly crucial factor in development for the ever-more powerful embedded market [1]. The purpose of this paper is to introduce the information retrieval device for the wearable computers using gloves and an improved algorithm, and assess its performance. Because of its device independent characteristic, proposed device could be applied to all kinds of electronic applications. It could be applied to all kinds of wearable computers as well as desktop computers.

Our paper is organized as follows. In section 2, several devices for wearable computers using gloves are introduced. In section 3, we suggest an improved information retrieval method and input device for wearable computer. In section 4 we analyze proposed device and method. And conclusions are given in section 5.

2 Traditional Glove Based Information Retrieval Device and Method

The following subsections explain the main characteristics of traditional glove based alphanumeric input devices. In these sections, we shortly describe the features of each method, and compare between methods.

2.1 Chording Gloves

The Chording Glove employs pressure sensors for each finger of the right hand in a glove to implement chording input device. There is one key for each finger. Multiple keys are pressed simultaneously in various combinations to enter characters. A chord can be made by pressing the fingers against any surface. Almost all possible finger combinations are mapped to symbols, making it potentially hard to type them. Additional buttons, located along the index finger, are used to produce more than the 2^5 distinct characters [2]. Fig 1 shows the external appearance of Chording Glove.

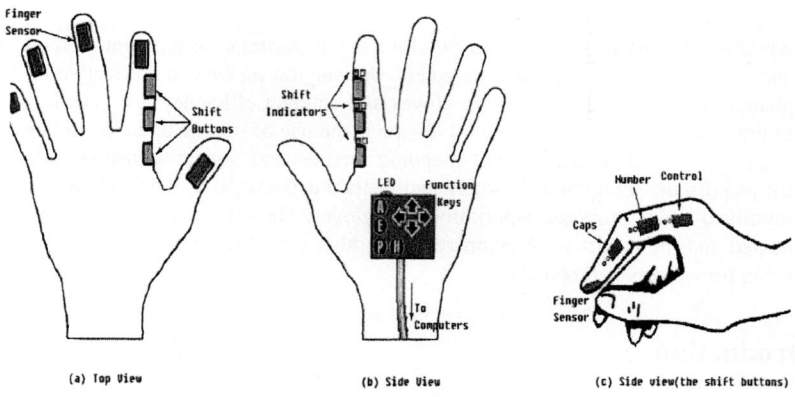

Fig. 1. External appearance of Chording Glove

A weak point of this method is difficult to use. It needs more than 80 minutes to learn the entire chord set. After 11 hours of training, word input speed reached approximately 18 words per minute (wpm) whereas the character error rate amounted to 17%.

2.2 Finger-Joint Gesture Wearable Keypad

The Finger-Joint Gesture Wearable Keypad suggests viewing the phalanges of the fingers (besides the thumb) of one hand as the keys on phone keypad.

By holding the inside of the hand in front of you, and bending the fingers toward you and aligning the fingertips of the four fingers, a 4X3 matrix is similar in shape to the traditional telephone keypad. And FJG keypad employs the same layout as that encountered on any traditional mobile telephone. Nothing else has to be learned. The FJG concept is a generic way of combining the 12 keys of the keypad with 4+1 different functions. It can be used in a variety of different interfaces.

A weak point of this method is the limited number of alphabets can be aligned on the phalanges. To overcome this weak point, if the multiple numbers of alphabets are mapped on the same phalanges (one-to-many characters mapping) in the same mode (EX: ABC, DEF...), the user has to use multiple successive keystrokes on the same phalanx of the fingers.

2.3 Thumbcode

"Thumbcode" method defines the touch of the thumb onto the fingers' phalanges of the same hand as key strokes. Character is signed or thumbed by pressing the tip of the thumb against one of the phalanges. This defines the twelve thumb states of Thumbcode. In combination with the twelve thumb states this gives a total of 96 basic Thumbcode. Fig 2 shows the Thumbcode assignments. Each of the eight 3X4 arrays in Fig 2 should be visualized as being superimposed on the fingers of the right hand.

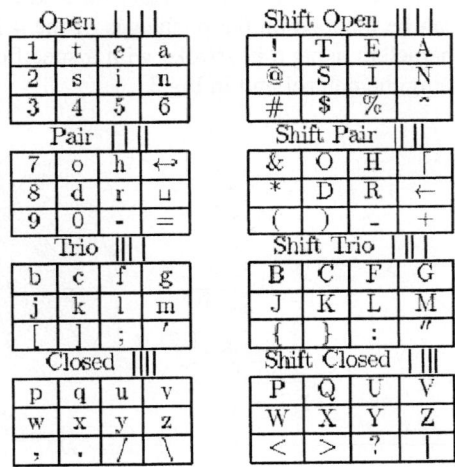

Fig. 2. Thumbcode assignments view of right-hand palm

In Fig 2, the 4 vertical bars mean 4 fingers of right-hand. Narrow space means that the adjacent fingers are closed. And regular space means that the adjacent fingers are opened. The four fingers can touch each other in eight different ways, each basically representing a mode, or modifier key that affects the mapping for the thumb touch [4].

A weak point of this method also can be described as complexity of combining fingers. User has to combine their fingers to generate Thumbcode in complex ways. As a result of this complexity, this method also needs training time to use fluently.

3 An Improved Information Retrieval Method and Input Device

Key-to-symbol mapping methods can be divided into two classes. Exactly one key to one symbol (character) mapping (1 degree of freedom, DOF) method and one-to-many characters mapping (1.5 degree of freedom, DOF) method are typical key-to-symbol mapping methods. In a one-to-many characters mapping method, user has to use multiple successive keystrokes to produce some character. In this letter, we propose an improved one-to-many characters mapping method.

We can produce any character using a keystroke. If the user wants to produce a character "C" in a traditional one-to-many characters mapping method, the user has to use multiple successive keystrokes on the medial phalanx of the index finger. But, in the proposed method, the user can produce a character "C" using a keystroke on the medial phalanx of the index finger with a specific operator (third operator).

First of all, we could decide the number of discrete operators and the layout of the key-to-symbol mapping according to the use of applications. In the proposed text input device, maximum number of the symbols can be mapped on a key depends on the number of using operators (the maximum number of used operators not exceeds 5). If we use 3 operators, we can map 3 characters on a key. Thus, the maximum number of characters can be mapped on the phalanges of the 4 fingers is 36. We can produce 36 different characters using a keystroke with a specific operator. This process could be finished using the control unit in Fig 3.

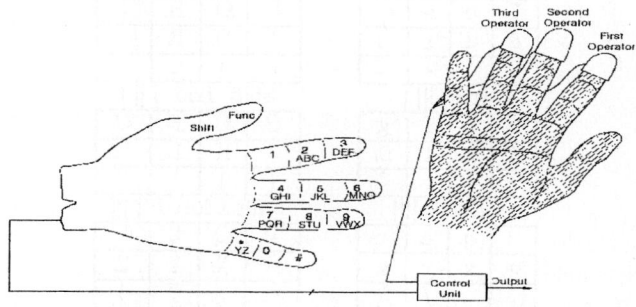

Fig. 3. Key-to-Symbol mapping and the operators

If the user depresses the tip phalanx of the middle finger with a first operator, then the character "M" will be produced. And, if the user depresses the tip phalanx of the middle finger with a second operator, then the character "N" will be produced. And, if the user depress the tip phalanx of the middle finger, then character "O" will be produced, and so on. Key-to-symbol mapping method is very easy and simple. Thus, nothing else has to be learned.

4 Efficiency Test and Results

To verify its efficiency, the proposed glove based text input device was built and assessed. The experiment that we conducted was designed to evaluate the input speed and error rate of the proposed device. 20 subjects were selected from among the respondents to advertisements placed around the university campus. There were 12 males and 8 females, and all were right handed and aged between 24 and 32 years. The initial session consisted of a tutorial which lasted less than a minute, and whose purpose was to teach the subjects the key-map and how to operate the device. Once this session was completed, a sample text was provided to the subjects. Fig 4 shows provided sample text. The complete text to be entered by the subject appeared in the top window of the computer display, while their keyboard input was displayed in the bottom window.

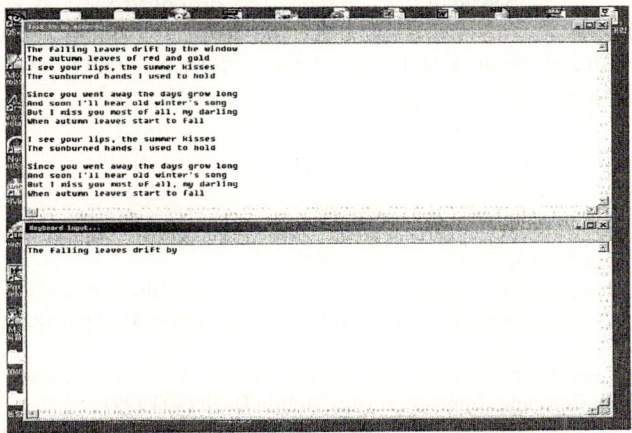

Fig. 4. Sample text for used testing and keyboard input display

We compared the proposed device and its method of utilization with other devices which use other methods, from several points of view, namely the input speed, error rate and the time required to learn the entire key-map. After 1 hour of training, the average input speed for the proposed device was 27.4 words per minute. For comparison, the input speed on a QWERTY keyboard for a previously untrained user after 12 hours of training is 20 words per minute [5], and the input speed for a previously untrained user using a glove after an 80 minutes tutorial is 16.8 words per minute [2]. Therefore, this result means that the proposed device offers a fast and convenient method of inputting text. The error rate was calculated as the ratio of input errors to the total number of characters and was found to be 7.8% after training. Compared with the error rate on a QWERTY keyboard (12.7%) and the traditional method of using a glove (17.4%), the proposed method constitutes an accurate text input method [2 and 6]. Furthermore, the number of keystrokes and the time required to enter the complete sample text were the same as those obtained using a QWERTY keyboard, and were much less than the corresponding values in the case of a traditional glove.

5 Conclusions

Nowadays, many systems adapt multi-modal human computer interfaces. The reason for using multi-modal HCI system is to create a more natural experience for the user by allowing him/her to use other methods of communication than just speech or just mouse, and aid the computer in understanding what the user wants by providing multiple modality streams that can disambiguate each other.

In this paper, we proposed an improved information retrieval method and input device using gloves for the purpose of using as a human computer interface method. Although there are several benefits of using one-handed text input devices, but there are clear-cut lines of input speed and error rate. To overstep these limits, we proposed the method and the device using two hands. The proposed method and experiment gave us possibility of using gloves as a text input device. For the purpose of achieving popular use of the glove as a text input device, more convenient and swift method should be proposed.

References

1. Markus Eisenhauer, Britta Hoffman, Doro Kretschmer.: State of the Art Human-Computer Interaction. Giga Mobile project D2.7.1 (2002)
2. Robert Rosenberg and Mel Slater.: The Chording Glove: A Glove-Based Text Input Device. IEEE Transactions on systems, Man, And Cybernetics-Part C: Applications and reviews (1999)
3. Goldstein, M. and Chincholle, D.: Finger-Joint Gesture Wearable Keypad. Second Workshop on Human Computer Interaction with Mobile Devices (1999)
4. Pratt, V. R.: Thumbcode: A Device-Independent Digital Sign Language. http://boole.stanford.edu/thumbcode (1998)
5. J. Noyes,: The QWERTY keyboard: A review. Int. J. Man-Mach Studies, Vol. 18. No. 3. (1983) 265-281
6. Potosnak K. M. (ed.): Keys and keyboards, Handbook of Human-Computer Interaction. Elsevier, New York (1988) 475-494

Research on Design and Implementation of the Artificial Intelligence Agent for Smart Home Based on Support Vector Machine

Jonghwa Choi, Dongkyoo Shin, and Dongil Shin[*]

Department of Computer Science and Engineering, Sejong University,
98 Kunja-Dong Kwangjin-Gu, Seoul, Korea
com97@gce.sejong.ac.kr, shindk@sejong.ac.kr, dshin@sejong.ac.kr

Abstract. In this paper, we provide information an artificial intelligence agent for a smart home and discuss a context model for implementation in an efficient smart home. An artificial intelligence agent in a smart home learns about the occupants and the smart environment, and predicts the appliance service that they will want. We propose the SVM (Support Vector Machine) for the learning and prediction aspects of the artificial intelligence agent. The experiment was done using three methods. Each of these three methods applies a higher importance to a different set of context data, out of the data related to the occupant, home environment, and the characteristics of the home appliances. Excellent results were seen when the experiment applied a higher importance to the data related to the characteristics of the home appliances.

1 Introduction

A home network integrating sensors, actuators, wireless networks and context-aware middleware will soon become part of our daily life. We define this environment as a smart home [1]. A smart home is a house or living environment that contains the technology to allow devices and systems to be controlled automatically. An artificial intelligence agent in a smart home learns about the occupants and the smart environment, and predicts the appliance services that they will want. We utilize an SVM (Support Vector Machine) for the learning and prediction capabilities of the agent [2]. In order for a smart environment to provide services to its occupants, it must be able to detect its current state or context and determine what actions to take based on the context. Dey and Abowd discuss the requirements for dealing with context in a smart environment and present a software infrastructure solution [3].

In this paper, we discuss the use of a support vector machine for learning and prediction by an artificial intelligence agent in a smart home. Learning human control strategy shows how human control strategy can be represented as a parametric model using a support vector machine [4]. In pattern classification, to improve the limited classification performance of the real support vector machine, it is proposed to use an SVM ensemble with bagging (bootstrap aggregating) or boosting [5].

---

[*] Correspondence Author

2 The System Architecture and Inner Component

The artificial intelligence agent receives six contexts data inputs from the occupant and smart home and one set of context data inputs from occupant commands to recognize the pattern of appliance use by the occupant in the smart home. The artificial intelligence agent gathers information from the occupant and environment and analyzes appliance's choice pattern of the occupant. The artificial intelligence agent acquires status information for the home environment, occupant, and home appliances for the learning and prediction of the agent, and this states information is used to define the context [6]. Figure 1 shows the two-layer context model that presented in this paper. Layer 1 defines 7 context data inputs that are acquired from the occupant, home environment, and home appliance: pulse, body temperature, facial expression value, room temperature, time, and occupant location in the smart home.

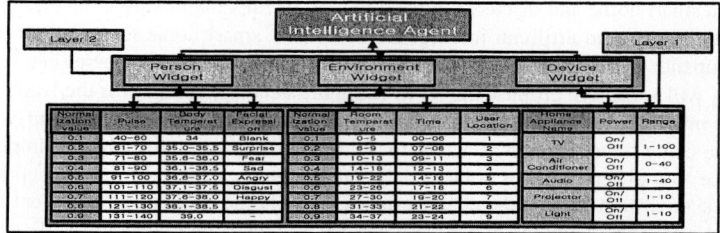

Fig. 1. The two Layer Context model

Six of the seven data inputs in lay 1, excluding only the device context input, are normalized between 0.1 and 0.9. The context related device input is one of the appliances used by the occupant. Pulse rates below 40 and over 180 were eliminated, since they represent abnormal human states. Body temperatures below 34 and over 40 were eliminated for the same reason. Facial expressions are normalized and categorized as seven expressions (by Charles. D). Room temperature is normalized based on the most comfortable temperature, which is between 23 and 24 degrees Celsius. The time is normalized based on a 24-hour clock. The location is the occupant's position in our experimental room.

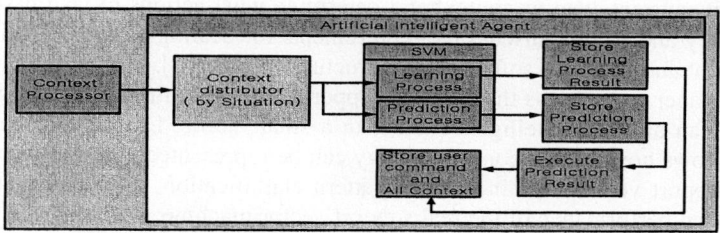

Fig. 2. Structure of artificial intelligence agent

Figure 2 shows the structure of the artificial intelligence agent. The context distributor accepts all the context data from the context model. The context distributor offers context data according to the learning time and prediction time of the SVM. Figure 3 shows the structure of the SVM in the artificial intelligence agent. The structure of the SVM was an applied hierarchical support vector machine classifier for appliance pattern analysis of the occupant.

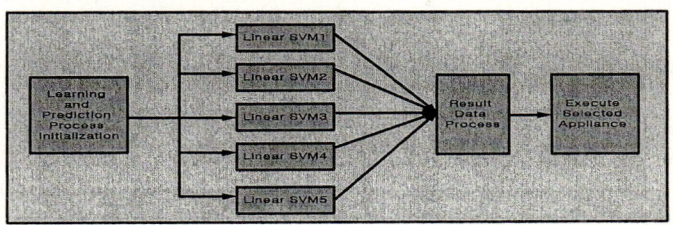

Fig. 3. Structure of SVM in Artificial Intelligence Agent

3 Experiments and Evaluations

The artificial intelligence agent changes the context's importance according to the appliance choice of the occupant. We compared the performance of the artificial intelligence agent based on different levels of context importance. We applied a higher importance to context data related to the occupant in Table 1. In Table 2, we applied higher importance to the context data related to the home environment. In Table 3, we applied a higher importance to the context data related data to the characteristics of the home appliances. Table 1 shows a low pattern recognition rate for the projector, air conditioner, and light. Table 2 shows a low pattern recognition rate for the television, projector, and light. Table 3 shows an excellent pattern recognition rate from all appliances.

Table 1. Higher importance to context data related to the occupant. (sv:number support vector / lv:norm of longest vector / ke:number of kernel evaluations / pt: precision on test set)

	sv	lv	ke	Test Set			Pt
				correct	incorrect	Total	
TV	30	2.25885	12266	981	19	1000	98.10%
Au	51	2.26254	11552	978	22	1000	97.80%
Pr	74	2.11505	12701	343	657	1000	34.30%
Ai	60	2.01333	12377	253	747	1000	25.30%
Li	67	2.14278	12335	275	725	1000	27.50%

Table 2. Higher importance to context data related to the home environment

	sv	lv	ke	Test Set			Pt
				correct	incorrect	Total	
TV	52	2.13000	11912	225	775	1000	22.50%
Au	59	1.95908	12290	909	91	1000	90.90%
Pr	44	1.75454	11342	120	880	1000	12.00%
Ai	55	2.02097	11627	1000	0	1000	100.00%
Li	65	1.96550	12098	208	792	1000	20.80%

Table 3. Higher importance to context data related to the home appliance' characteristics

	sv	lv	ke	Test Set			pt
				correct	incorrect	Total	
TV	95	2.10637	12209	1000	0	1000	100.00%
Au	28	2.02417	10979	1000	0	1000	100.00%
Pr	171	1.75716	14204	886	114	1000	88.60%
Ai	36	2.20606	11213	1000	0	1000	100.00%
Li	135	2.02598	13085	930	70	1000	93.00%

4 Conclusions

In this paper, we provide information on an artificial intelligence agent for a smart home and discuss a context model for implementation in an efficient smart home. An artificial intelligence agent in a smart home learns about the occupants and the smart environment, and predicts the appliance service that they will want. The experiment was done using three methods. Each of these three methods applies a higher importance to a different set of context data, out of the data related occupant, home environment, and the characteristics of the home appliances. It shows that excellent results were obtained from the experiment that applied a higher context's importance to the context data related to the characteristics of the home appliances.

References

[1] Sherif, M. H.: Intelligent homes: a new challenge in telecommunications standardization. Communication Magazine. IEEE, Vol. 40, Issue.1, (2002) 8-8
[2] Vapnik V. N.: The nature of statistical learning theory. New York, Springer-Verlag, (1995)
[3] Dey. A. K, Salber. D, Abowd. G.D.: A context-based infrastructure for smart environments. Proceedings of the 1st International Workshop on Managing Interactions in Smart Environments(MANSE'99), (1999) 114-128
[4] Yongsheng ou, Yangsheng Xu.: Learning human control strategy for dynamically stable robots: support vector machine approach. Robotics and Automation, Proceedings. ICRA'03, IEEE International Conference on. Vol. 3, (2003) 3455-3460
[5] Hyun-Chum Kim, Shaoning Pang, Hong-Mo Je, Daijin Kim.: Pattern classification using support vector machine ensemble. Pattern Recognition, Proceedings. 16th International Conference on, Vol. 2, (2002) 160-163
[6] Schilit.: A system architecture for context-aware computing, Doctoral thesis, Columbia University, (1995)

A Self-organized Network for Data Clustering

Liang Zhao, Antonio P.G. Damiance Jr., and Andre C.P.L.F. Carvalho

Institute of Mathematics and Computer Science, University of São Paulo, Brazil
{zhao, damiance, andre}@icmc.usp.br

Abstract. In this paper, a dynamical model for data clustering is proposed. This approach employs a network consisting of interacting elements with each representing an attribute vector of input data and receiving attractions from other elements within a certain region. Those attractions, determined by a predefined similarity measure, drive the elements to converge to their corresponding cluster center. With this model, neither the number of data clusters nor the initial guessing of cluster centers is required. Computer simulations for clustering of real images and Iris data set are performed. The results obtained so far are very promising.

1 Introduction

A data clustering task can be formulated as follows: partition the N data values into K groups, so that two data points in the attribute space belonging to the same group are more similar than those belonging to different groups [3, 4, 7, 10]. According to [4], data clustering techniques can be classified into two categories: hierarchical techniques and partitional techniques. Hierarchical clustering techniques produce tree type partitions of the data. These techniques do not require the information of the number of clusters beforehand. However, they cannot incorporate a priori knowledge related to the global structure of the clusters, since, in each step, they take into account only local neighbors. Partitional clustering techniques have the advantage of being able to incorporate previous knowledge related to the data set. Nevertheless, they usually require the number of clusters to be previously defined and are sensitive to noise and initialization.

In spite of the success obtained by traditional clustering techniques in several application domains, it must be observed that there has been growing interest in the development of new clustering techniques based on alternative approaches. One of the reasons for such investigation is that traditional techniques are usually based on statistical data analysis and employ serial processing, which suffer from low efficiency and usually need high computational power. Thus, recent techniques exploit parallel architectures and have flexible implementation. Due to elegance and effectiveness of the clustering performed by many biological systems, several of these new techniques are inspired by biological systems [2, 5, 6, 9, 11, 13, 14]. Special interests have been concentrated in developing network of interacting elements for data clustering. This kind of model can not only incorporate global statistical property of the data, but also preserve local geometrical features.

The data-clustering model presented here is composed of a network of interacting elements. Each element in the constructed network corresponds to a data point in the attribute space. When a set of data is supplied to the network, the elements of the network self-organize according to a predefined similarity criterion, such that each group of elements representing a data cluster will be coupled together. Each element receives forces from all other coupled elements, which drive the element toward its corresponding cluster center. With this moving mechanism, the model is designed such that elements representing similar data approximate each other to form a cluster. At the same time, ambiguous elements (those that receive forces from more than one group) will leave other groups and fix themselves in only one group, the group with the strongest attraction to them.

The rest of this paper is organized as follows. Sec. 2 presents the model definition and the clustering strategy. Sec. 3 describes computer simulation results. Finally, Sec. 4 discusses the main conclusions of this work.

2 Model Description

The model presented in this paper can be seen as an improved and quite simplified version of the model described in [14]. It is a network composed of N elements with each corresponding to a data point in the attribute space. Specifically, the model is governed by the following equations:

$$x_{ik}(t+1) = \begin{cases} 0 & \text{if } x_{ik}(t) + \eta_i F_{ik}(t) \leq 0 \\ x_{ik}(t) + \eta_i F_{ik}(t) & \text{if } 0 < x_{ik}(t) + \eta_i F_{ik}(t) < 1 \\ 1 & \text{if } x_{ik}(t) + \eta_i F_{ik}(t) \geq 1 \end{cases} \quad (1)$$

$$\mathbf{F}_i(t) = \frac{\sum_{j \in \Delta_i(t)} \frac{\mathbf{x}_j(t) - \mathbf{x}_i(t)}{\|x_j(t) - x_i(t)\|} e^{-\alpha \|\mathbf{x}_j(t) - \mathbf{x}_i(t)\|}}{M_i(t)} \quad (2)$$

$$j \in \Delta_i(t) \quad \text{if} \quad H\left(e^{-\alpha \|\mathbf{x}_j(t) - \mathbf{x}_i(t)\|} - \theta\right) = 1 \quad (3)$$

$$\eta_i = \frac{\sum_{j \in \Delta_i} d_{ij}}{M_{\Delta_i}} \quad (4)$$

where $i, j = 1, 2, ..., N$ are the element indexes, $\mathbf{x}_i(t) = (x_{i1}(t), x_{i2}(t), ..., x_{ik}(t))^T$ represents the attribute vector of the ith data item at iteration t and k is the number of attributes. For example, for a gray-level image clustering task, $K = 1$ and x_i is a scalar value representing the intensity of the ith pixel; for a color image, $K = 3$ and $\mathbf{x}_i$ is a vector representing the three-color components of the ith pixel. $\mathbf{x}_i(0)$ is the original value of the ith element. $\mathbf{F}_i(t) = (F_{i1}(t), F_{i2}(t), ..., F_{ik}(t))^T$ represents the total force imposed upon the element i from all elements in $\Delta_i(t)$ at iteration t and $M_i(t)$ is the number of elements in $\Delta_i(t)$, where $\Delta_i(t)$ is a pixel region defined by the term $H\left(e^{-\alpha\|\mathbf{x}_j(t)-\mathbf{x}_i(t)\|} - \theta\right)$ in Eq. (3). It means that, all elements in $\Delta_i(t)$ are considered to be similar to i. H is a Heaviside function. It returns a value 1 when the input is larger than zero and returns a value 0, otherwise. The term $e^{-\alpha\|\mathbf{x}_j(t)-\mathbf{x}_i(t)\|}$ is a Gaussian function, which results in a value between 0 and 1, and the parameter α controls its stiffness. $\|\ \|$ is the Euclidean norm. The parameter θ is a threshold, which shifts the Heaviside function. Increasing in the value of θ reduces the chance of the Heaviside function returning the value 1. The term $\dfrac{\mathbf{x}_j(t) - \mathbf{x}_i(t)}{\|\mathbf{x}_j(t) - \mathbf{x}_i(t)\|}$ in Eq. (2) determines the driving direction upon element i from element j, while the term $e^{-\alpha\|\mathbf{x}_j(t)-\mathbf{x}_i(t)\|}$ in the same equation determines the quantity to move. The parameter $0 < \eta_i(t) \leq 1$ controls the moving rate of element i, whose function will be described bellow.

The clustering process can be described as follows: Firstly, the initial data vector $\mathbf{x}_i(0)$ is used by the system as initial condition. As the system runs, elements with similar features are grouped together and finally converge to a unique point. The Heaviside function returns the value 1 if the similarity between $\mathbf{x}_i(t)$ and $\mathbf{x}_j(t)$ is beyond a certain value, which can be adjusted by the parameters α and θ.

It must be observed that, very often, some data are grouped into more than one cluster at the beginning. This ambiguity problem can be solved by the model's adaptive moving process, which is described next.

Initially, $\mathbf{x}_i(0)$ is the original value of ith data item. As the system evolves, each element i receives a force of attraction $\mathbf{F}_i(t)$ from a set of similar elements. This force drives the element to move toward the group with higher similarity. Thus, if any two elements have the same distances to a common element i, and they are exactly on the opposite sides of i, the two forces (positive and negative) imposed on i are cancelled and the element i remains unaffected by them. Otherwise, the element i moves in the direction defined by the sum of the force vectors. Thus, elements at the center of a given group will move slowly, since the majority of forces imposed on them are cancelled. On the other hand, off-center elements will move toward their respective center quickly. Ambiguous elements (those that receive forces from more than one group) will leave other groups and fix themselves in only one group, the group with the strongest attraction to them. Successive iterations will decrease the distance between similar elements (decrease intra-class distance) and increase the distance between very different elements (increase inter-class distance).

At this point, one may perceive that there is still a convergence problem, i.e., as each group of elements get closed, the corresponding force received by each element becomes larger resulting in quickly moving. This is because that the force term $\mathbf{F}_i(t)$ is proportional to similarity measure. What we want is exactly contrary to this situation. Specifically, we want to slow down the movement when the elements of a same group are getting closed and the movement eventually stops when the elements of a same group reach to a unique point. Otherwise, *the moving process may not converge, but oscillate in a certain region*. This problem can be solved by making the moving rate $\eta_i(t)$ time and element dependent, as shown by Eq. (4). Elements move rapidly to their corresponding center with a large value of $\eta_i(t)$, and slowly with a small value. Thus, the model is designed so that $\eta_i(t)$ changes with time and is dependent on the dispersion of the elements in the group $\Delta_i(t)$, characterized by the average Euclidean distance between element i and each element belonging to $\Delta_i(t)$ (see Eq. (4)). When elements in $\Delta_i(t)$ are still far away from one another, the dispersion is large, $\eta_i(t)$ takes a large value. Hence, the element i moves rapidly. On the other hand, at the final stage of the moving process, elements from the same cluster are concentrated, resulting in a low value of $\eta_i(t)$, indicating that the elements move slowly. Finally, when elements of a cluster get closer to one point in the attribute space, $\eta_i(t) \to 0$, i.e., elements do not move at all. Consequently, each group of data is represented by a point whose position in the attributes space is the cluster center.

In comparison to conventional data clustering techniques, this model offers the following interesting characteristics:

- It is not necessary to know the number of clusters in advance;
- The mechanism of adaptive modification of data makes the model robust enough to classify ambiguous elements;
- Due to the model's self-organizing feature, no guessing of initial cluster centers is needed. Thus, combinatorial search can be avoided;
- From the results obtained in the numerical experiments performed by the authors, it is possible to see that the data moving process can be completed in a few iterations. Moreover, due to the model's parallel nature, all elements interact independently. Consequently, the number of iterations needed to form compact groups increases very slowly as the amount of data becomes larger. This feature is especially attractive if the model is implemented in a parallel architecture.

3 Computer Simulations

This section presents the simulation results by using the model to cluster different type of data sets. As previously mentioned, the model has few parameters to be adjusted. The parameter α controls the stiffness of the Gaussian function in Eq. (5) and Eq. (6) and the parameter θ is a hard threshold. Both of them can be adjusted to obtain hierarchical representation of clustering results. A low value for θ or high value for α smoothes the difference between data items and, consequently, results in a small number of clusters. A high value for θ or a low value for α, on the other hand, amplifies the difference and may lead to a large number of clusters. Consider the example of pixel clustering for images, details of the original image can be amplified by setting either a high value to θ or a low value to α. However, if only skeletons of the objects in the image are needed, either a low value can be set to θ or a high value can be set to α. In all simulations to be shown, the hierarchical representation effect is obtained by only changing θ. The parameter α is held constant at $\alpha = 0.8$.

The authors first illustrate the data moving process by using a 2-dimensional artificial data set of various sizes and forms of data groups. From Fig. 1(a) and 2(a), one can perceive that, in a normal situation, the data should be clustered into 3 groups in a gross level and into 5 groups in a fine scale. It is also possible to see that the distances between some elements from the same group are larger than distances between elements belonging to different groups. Without the data moving technique, this feature may result in interconnection among elements from various groups. Due to the moving technique introduced in the previous Section, all elements move correctly to their corresponding group center. Figure 1 and 2 show that 3 and 5 clusters can be correctly extracted by using the present model.

In order to show the complete data clustering process performed by the proposed model, consider Fig. 3, which shows the x-ray of a human head as input data set. Figure 4 shows the pixel clustering results in three different resolutions. As previously observed, when θ is small, a clustering result with few data groups is obtained. Figure 4(a) shows the 2 clusters produced. In this case, one can see that the background and the object are separated. As θ increases, clustering result with more data groups are achieved. Figure 4(b) and 4(c) show cases where 5 and 9 clusters are produced, respectively. In these figures, it is possible to see some details within the object.

Figures 5(a), 5(b) and 5(c) show the evolution of $x_i(t)$ corresponding to the simulation illustrated by Figure 4(a), 4(b) and 4(c), respectively. One can see the pixels initially mixed. As the system evolves, all such elements leave their least attractive group and go to their most attractive group. In this way, overlaps among element groups are eliminated and, consequently, pixel clusters are correctly formed. At the end, all elements move into distinct groups.

Figure 5 also shows that the clustering evolution finishes in a very small number of iterations, demonstrating the high efficiency of the proposed model.

Now we present the clustering results for a data set frequently used in clustering experiments, the Iris data set. The Iris data set has 156 data samples, with 4 attributes each [1]. In the original data set, the correct class associated to each data sample is known. Thus, the clustering results obtained by our model can be compared with the correct results. Table 1 shows the cluster centers found by the model and table 2 shows the clustering results for each class. According to these results, the proposed method obtains an average error rate equal to 4%, which is similar to the best results reported in the literature for the same data set [1].

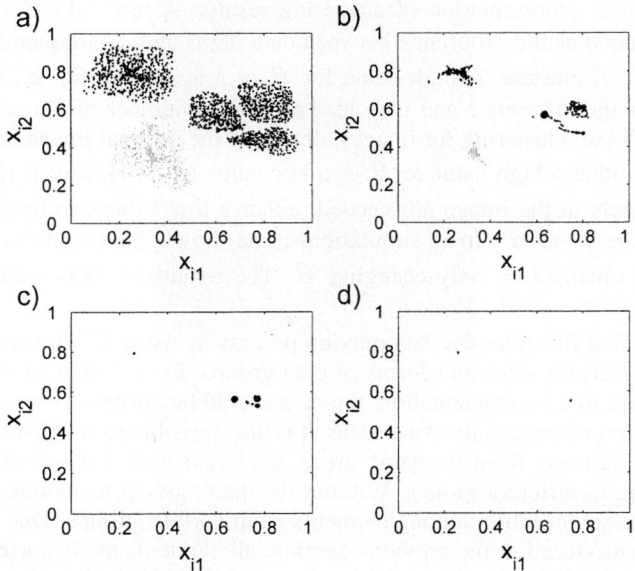

Fig. 1. Data moving process. Three data groups are formed, $\theta = 0.1$

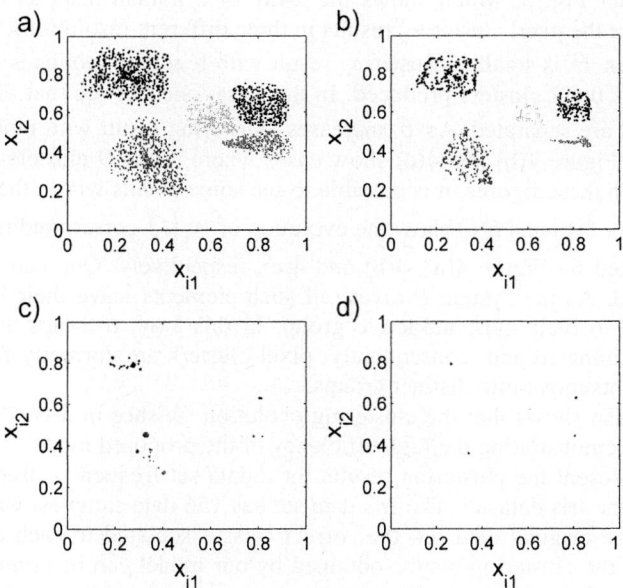

Fig. 2. Data moving process. Five data groups are formed, $\theta = 0.3$

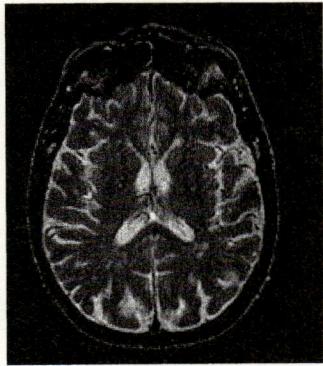

Fig. 3. Original image for pixel clustering

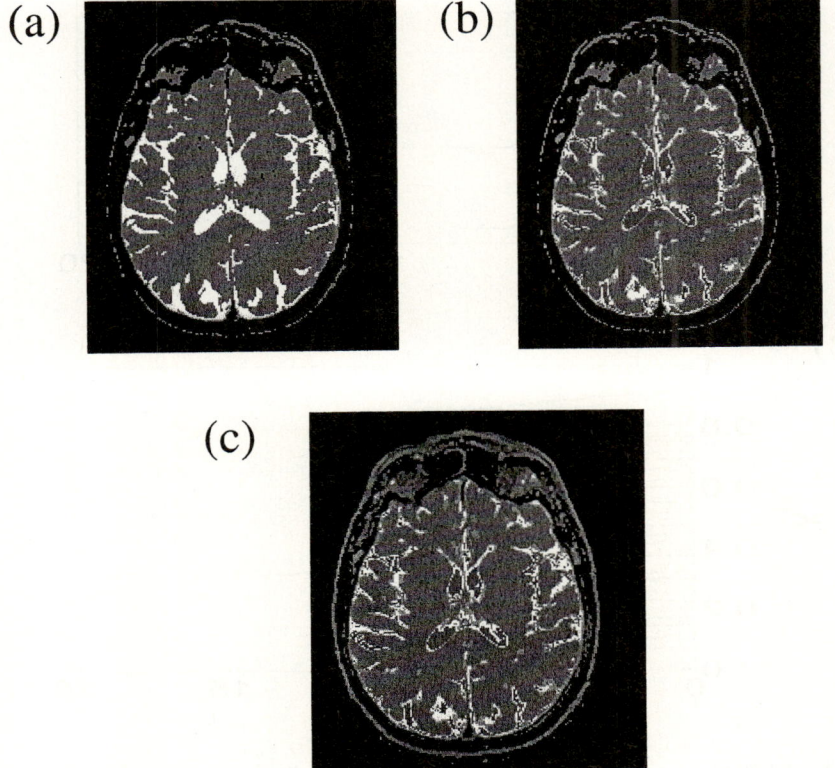

Fig. 4. Clustering results. a) 3 clusters, $\theta = 0.1$; b) 5 clusters, $\theta = 0.3$; c) 7 clusters, $\theta = 0.5$

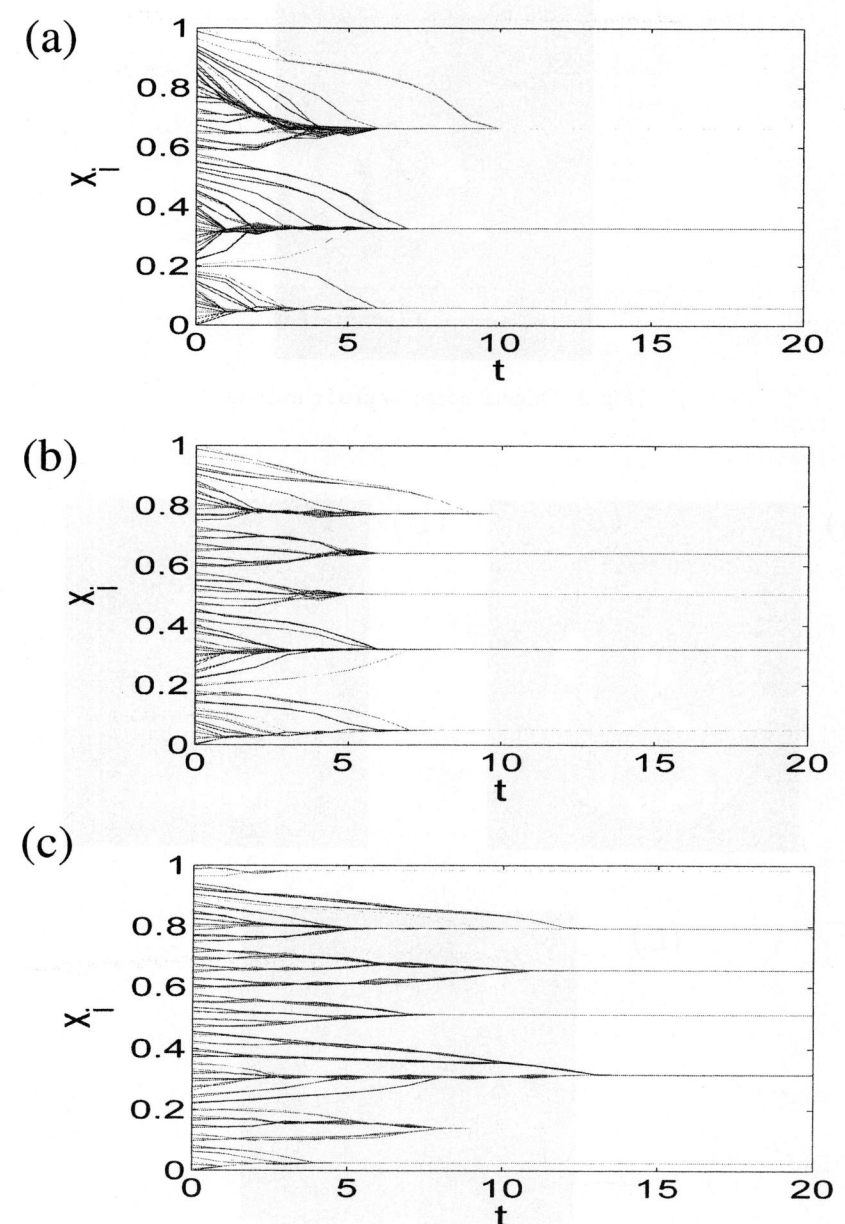

Fig. 5. Evolution of $c_i(t)$ corresponding to the simulation results shown in Fig. 5. a) 3 clusters, $\theta = 0.1$; b) 5 clusters, $\theta = 0.3$; c) 7 clusters, $\theta = 0.5$

Table 1. Cluster centers found (Positions of final fixed points)

Clusters	Attribute 1	Attribute 2	Attribute 3	Attribut3 4
Class 1	0.6305	0.7698	0.2125	0.0938
Class 2	0.7553	0.6312	0.6307	0.5402
Class 3	0.8283	0.6784	0.7866	0.8022

Table 2. Clustering results

	Íris setosa	Íris versicolor	Íris verginica	
Class 1	50	0	0	
Class 2	0	48	4	
Class 3	0	2	46	
R_c	100%	96%	92%	96%

4 Concluding Remarks

This paper presented an adaptive moving mechanism for data clustering. The parallel nature of the model makes it suitable for hardware implementation, providing an efficient alternative approach for solving this general problem. Local interactions among elements in the network have the advantage of simplicity in hardware implementation in contrast to global interactions, where connections between most of the elements should be devised. The results obtained in the set of experiments performed show the potential of this method for two different kinds of datasets.

In general, there is not a unique representation in data clustering problems, i.e., the same data set can be interpreted by different meaningful clustering results. By using this model, one can get hierarchical meaningful results by tuning a very small number of parameters. As described in the text, there are only two parameters (α and θ) to be tuned. Due to the model's self-organizing feature and its free parameters, neither the cluster number, nor the guessing of initial cluster centers is required. Thus, combinatorial search is avoided.

References

1. Blake, C., Keogh, E. and Merz, C.: UCI Irvine Repository of Machine Learning Databases (1998). http://www.ics.uci.edu/~mlearn/MLRepository.html.
2. Chen, K. and Wang, D. L.: A dynamically coupled neural oscillator network for image segmentation. Neural Networks, 15 (2002) 423-439.
3. Duda, R. O., Hart, P. E., and Stork, D. G.: Pattern classification, John Willy & Sons, Inc. (2001).
4. Jain, A. K., Murty, M. N., & Flynn, P. J.: Data clustering: A review. *ACM Computing Surveys, 31(3)* (1999) 264-323.

5. Kaneko, K. Clustering, coding, switching, hierarchical ordering, and coding in a network of chaotic elements. Physica D, 41 (1990) 137-172.
6. König, P. and Schillen, T. B.: Binding by temporal structure in multiple feature domains of an oscillatory neuronal network. Biol. Cybern., 70 (1994) 397-405.
7. Mohan, R. & Nevatia, R.: Perceptual organization for scene segmentation and description. IEEE Trans. PAMI, 14(6) (1992) 616-635.
8. Newman, M. E. J.: The structure and function of complex networks. SIAM Review, 45(2) (2003) 167-256.
9. M. B. H. Rhouma and H. Frigui.: Self-organization of pulse-coupled oscillators with application to clustering, IEEE Trans. Neural Networks, 23(2) (2001) 180-195.
10. Pal, N. R. and Pal, S. K.: A review on image segmentation techniques. Pattern Recognition, 26(9) (1993) 1277-1294.
11. Wang, D. L. and Liu, X. Scene analysis by integrating primitive segmentation and associative memory. IEEE Trans. SMC, Part B: Cybernetics, 32 (2002) 254-268.
12. Zhao, L., Macau, E. E. N., and Omar, N. Scene segmentation of the chaotic oscillator network". Int. J. Bif. Chaos, 10(7) (2000) 1697-1708.
13. Zhao, L. and Macau, E. E. N. A network of globally coupled chaotic maps for scene segmentation. IEEE Trans. Neural Networks, 12(6) (2001) 1375-1385.
14. Zhao, L., Carvalho, A. C. P. L. F., and Li, Z.-H.: Pixel clustering by adaptive moving and chaotic synchronization. IEEE Trans. Neural Networks, 15(5) (2004) 1176-1185.

A General Criterion of Synchronization Stability in Ensembles of Coupled Systems and Its Application

Qing-Yun Wang[1,2], Qi-Shao Lu[1], and Hai-Xia Wang[1]

[1] School of Science, Beijing University of Aeronautics and Astronautics,
Beijing 100083, China
math@ss.buaa.edu.cn, qishaolu@hotmail.com
[2] Inner Mongolia Finance and Economics College,
Huhhot 010051, China

Abstract. Complete synchronization of N coupled systems with symmetric configurations is studied in this paper. The main idea of the synchronization stability criterion is based on stability analysis of zero solution of linearized dynamical systems. By rigorous theoretical analysis, a general synchronization stability criteria is derived for N coupled systems with the first state variable diffusive coupling. This criterion is convenient for us to explore the synchronization of a class of coupled dynamical systems. Finally, the famous Lorenz system and Hindmarsh-Rose(HR) neuron are used to test our theoretical analysis.

1 Introduction

The synchronization of coupled dynamical systems has received much attention since the papers by Pecora and Carrol[1-2] show that chaotic systems can be synchronized. Complete synchronization of coupled identical systems means that motion of coupled elements starting from different initial point in state space is identical with time t going toward infinity, that is to say, when the array is synchronized, the systems are decoupled. Dynamics of coupled systems was extensively studied in many fields of science and technology, such as secure communication[3], chemical systems[4] and neural collective motion[5-7]. Up to now, the study of the synchronization in the coupled systems is still an interesting topic.

However, main problem in the study of the synchronization is how to determine the coupling strength or certain control parameter, where synchronization can occur in the coupled systems. In the past, many authors devoted themselves to the study in this field and some desirable results have been obtained. The master stability function method (MSFM) was introduced by Pecora and Carroll[8-9]. This method is very valid to determine the synchronization threshold of the coupled systems. But The Lyapunov exponents (for chaotic system) or Floquent multipliers (for period system) must be calculated when we resort to MSFM. In general, they can only be detected by numerical method. Based on the Lyapunov function theory, a sufficient condition for synchronization was obtained for symmetric coupling network[10]. In Ref.[11], chaos synchronization of two coupled Lorenz systems was studied by means of asymptotical stability of linearized system and a sufficient condition was given by means of the Routh-Hurwitz rule. Chai Wah Wu and Leon O. Chua gave a synchronization criterion for linearly coupled dynamical systems by mens of construction of the Lyapunov function and asymptotical stability of solution on a single system [12]. In previous works, rigorous calculation is unavoidable.

In this paper, a simple synchronization criterion is developed on basis of the Lyapunov function and matrix theory. This criterion avoids many rigorous calculations for only requiring simple calculation of determinant. It is very valid to detect the synchronization of N coupled systems with regular connection network (See Fig.1). Moreover, it is noted that network we consider in this paper is linked only through the first state variable and the Jacobian matrix of a single system must satisfy the certain conditions as given in the following paper. In short, synchronization stability criterion in this paper is only feasible for a special systems. But most systems studied (including Lorenz system, HR neuron model, Chay neuron model, etc.) satisfy the conditions of developed Theorem in the present paper.

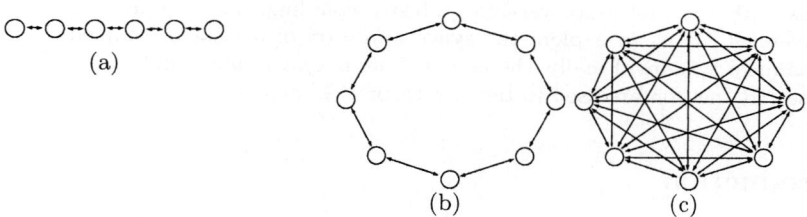

Fig. 1. Sketches of three regular connections (a) chain connection. (b) ring connection. (c) global connection.

2 Synchronization Stability Criterion of N Coupled Systems with Regular Connection

In this section, we develop a criterion, which determines the stability of synchronization manifolds. Here, we consider the regular network structure. Dynamics of N identical linearly and diffusively coupled systems, and with each system being a n-dimensional dynamical system, is governed by the following set of differential equations.

$$\dot{X}_i = F(X_i) + C\sum_{j=1}^{N} a_{ij} X_j \Gamma, \quad i = 1, 2, \ldots, N \qquad (1)$$

where $X_i = (x_{i1}, x_{i2}, \ldots, x_{in}) \in \mathbb{R}^n$ is state variables of the coupled ith cell. C represents the coupling strength and Γ is a $n \times n$ matrix, which has the below form:

$$\varGamma = \begin{pmatrix} 1 & 0 & \cdots & 0 \\ 0 & 0 & \cdots & 0 \\ \cdots & \cdots & \cdots & \cdots \\ 0 & 0 & \cdots & 0 \end{pmatrix}$$

This means that two coupled systems are only linked through their first state variables. The coupling matrix $A = (a_{ij})_{N \times N}$ represents the coupling style of N coupled systems. In the following, we consider three different coupling configurations with chain, ring and global connections. For these cases, the coupling matrix A has the following common properties:

(1) A is a symmetric and irreducible matrix.
(2) The off-diagonal elements, $a_{ij}(i \neq j)$ of A, are either 1 or 0.
(3) The elements of A satisfy

$$a_{ii} = - \sum_{j=1, i \neq j}^{N} a_{ij}, \quad i = 1, 2, 3, ..., N. \tag{2}$$

(4) One eigenvalue $\lambda_1(N)$ of A is zero, with multiplicity 1, and all the other eigenvalues $\lambda_2(N) \geq \lambda_3(N) \geq \ldots \geq \lambda_N(N)$ of A are strictly negative.

Given the dynamics of the single oscillator and the coupling style, the stability of the synchronization state of coupled systems can be characterized by those nonzero eigenvalues of A and the coupling strength C. In the following, a criterion, which determines the synchronization of coupled systems with the matrix A satisfying the above conditions, is analyzed. The coupled system (1) is said to achieve synchronization if the following relation holds:

$$X_1(t) = X_2(t) = \cdots = X_N(t) = s(t) \quad t \to +\infty \tag{3}$$

where $s(t)$ is a solution of an isolate system, namely, the coupling strength C is zero in the system (1).

Lemma 1. *Consider the coupled system (1). Let*

$$0 = \lambda_1 > \lambda_2 \geq \lambda_3 \geq \cdots \geq \lambda_n \tag{4}$$

be the eigenvalues of its coupling matrix A. If the following $N - 1$ of n-dimensional linear time-varying systems are asymptotically stable:

$$\dot{\omega} = (D_X F(s(t)) + C \lambda_k \varGamma) \omega, \quad k = 2, 3, \ldots, N \tag{5}$$

then the synchronization states (3) are asymptotically stable.

Proof of the Lemma 1 can be found in[10].
In what follows, Lemma 2 is given with supposing that Jacobian $DF(s(t))$ satisfies the below conditions.

C1. The matrix $A(s(t)) = (a_{ij}(s(t)))_{n \times n} = DF(s(t)) + (DF(s(t)))^T$ is non-singular, that is, $\det(DF(s(t)) + (DF(s(t)))^T) \neq 0$.

C2. All eigenvalues $\lambda_i(s(t))(i=2,3,\ldots,n)$ of the matrix $A_{22}(s(t))$ are strictly negative for any $s(t)$.

C3. $\frac{1}{2}\frac{\det(A(s(t)))}{\det A_{22}(s(t))}$ is bounded for all $s(t)$.

with

$$A_{22}(s(t)) = \begin{pmatrix} a_{22}(s(t)) & a_{23}(s(t)) & \ldots & a_{2n}(s(t)) \\ a_{32}(s(t)) & a_{33}(s(t)) & \ldots & a_{3n}(s(t)) \\ \vdots & \vdots & \ldots & \vdots \\ a_{n2}(s(t)) & a_{n3}(s(t)) & \ldots & a_{nn}(s(t)) \end{pmatrix}$$

Lemma 2. *If Jacobian* $\mathrm{D}F(s(t))$ *satisfies the conditions C1 and C2, then the matrix* $\mathrm{D}F(s(t))+(\mathrm{D}F(s(t)))^T-2T\Gamma$ *is negative definite when* $T>\frac{1}{2}\frac{\det(A(s(t)))}{\det(A_{22}(s(t)))}$.

Proof. Let

$$A(s(t)) = \begin{pmatrix} a_{11}(s(t)) & \alpha(s(t)) \\ (\alpha(s(t)))^T & A_{22}(s(t)) \end{pmatrix}$$

with $\alpha(s(t))=(a_{12}(s(t)),a_{13}(s(t)),\ldots,a_{1n}(s(t)))$ and $(\alpha(s(t)))^T$ is the transpose of $\alpha(s(t))$.

Because $A_{22}(s(t))$ is symmetric, there exists an orthogonal matrix Q_{22} such that $(Q_{22})^T A_{22}(s(t))Q_{22}=\Lambda_2$, with $\Lambda_2=\mathrm{diag}(\lambda_2(s(t)),\lambda_3(s(t)),\ldots,\lambda_n(s(t)))$.

Let

$$Q = \begin{pmatrix} 1 & \Theta \\ \Theta^T & Q_{22} \end{pmatrix}$$

with $\Theta=(0,0,\ldots,0)$ being a $(n-1)$-dimensional zero vector. Thus, we know that Q is an orthogonal matrix and the following relation holds:

$$Q^T A(s(t))Q = \begin{pmatrix} 1 & \Theta^T \\ \Theta & Q_{22}^T \end{pmatrix} \begin{pmatrix} a_{11}(s(t)) & \alpha(s(t)) \\ (\alpha(s(t)))^T & A_{22}(s(t)) \end{pmatrix} \begin{pmatrix} 1 & \Theta \\ \Theta^T & Q_{22} \end{pmatrix}$$
$$= \begin{pmatrix} a_{11}(s(t)) & \alpha(s(t))Q_{22} \\ Q_{22}^T(\alpha(s(t)))^T & Q_{22}^T A_{22}(s(t))Q_{22} \end{pmatrix} = \begin{pmatrix} a_{11}(s(t)) & \alpha(s(t))Q_{22} \\ Q_{22}^T(\alpha(s(t)))^T & \Lambda_2 \end{pmatrix}$$

Consequently, we have

$$Q^T(A(s(t))-2T\Gamma)Q = \begin{pmatrix} a_{11}(s(t))-2T & \alpha(s(t))Q_{22} \\ Q_{22}^T(\alpha(s(t)))^T & \Lambda_2 \end{pmatrix}$$

Denote $\alpha(s(t))Q_{22}=(q_{12},\ldots,q_{1n})$. Because λ_i $(i=2,3,\ldots,n)$ are nonzero, we have

$$P^T(Q^T(A(s(t))-2T\Gamma)Q)P = \begin{pmatrix} a_{11}(s(t))-2T-(\frac{q_{12}^2}{\lambda_2}+\ldots+\frac{q_{1n}^2}{\lambda_n}) & \Theta \\ \Theta^T & \Lambda_2 \end{pmatrix}$$

by using a series of elementary transformations on the matrix $Q^T(A(s(t))-2T\Gamma)Q$, with P being the product of some elementary matrices.

If the matrix $A(s(t)) - 2T\Gamma$ is negative definite, then $a_{11}(s(t)) - 2T - (\frac{q_{12}^2}{\lambda_2} + \ldots + \frac{q_{1n}^2}{\lambda_n}) < 0$. Hence, $T > \frac{1}{2}(a_{11}(s(t)) - (\frac{q_{12}^2}{\lambda_2} + \ldots + \frac{q_{1n}^2}{\lambda_n}))$. From the above analysis, we know that

$$P^T(Q^T(A(s(t)))Q)P = \begin{pmatrix} a_{11}(s(t)) - (\frac{q_{12}^2}{\lambda_2} + \cdots + \frac{q_{1n}^2}{\lambda_n}) & \Theta \\ \Theta^T & \Lambda_2 \end{pmatrix}$$

In terms of the properties of the matrix P and Q, the following relation is derived:

$$\det(A(s(t)) = \lambda_2 \cdots \lambda_n (a_{11}(s(t)) - (\frac{q_{12}^2}{\lambda_2} + \cdots + \frac{q_{1n}^2}{\lambda_n}))$$

Hence, $T > \frac{1}{2}\frac{\det(A(s(t)))}{\lambda_2 \cdots \lambda_n} = \frac{1}{2}\frac{\det(A(s(t)))}{\det A_{22}(s(t))}$. The proof is complete.

Corollary 1. *If conditions C1, C2 and C3 hold, then when $T \geq \frac{1}{2}\beta$, the matrix $DF(s(t)) + (DF(s(t)))^T - 2T\Gamma$ is negative definite for all $s(t)$, with $\beta \geq \frac{\det(A(s(t)))}{\det A_{22}(s(t))}$ being a constant for all $s(t)$.*

Theorem 1. *Consider the coupled system (1) and suppose that Jacobian $DF(s(t))$ satisfies conditions C1, C2 and C3. The synchronization states of the system (1) defined by (3) are achieved if $C > \frac{T}{|\lambda_2|}$, where λ_2 is the largest nonzero eigenvalue of the coupling matrix A, C is the coupling strength and T is given in Corollary 1.*

Proof. To explore the stability of the synchronization states of the system (2), the perturbed solutions η_i are introduced:

$$X_i = s(t) + \eta_i, \quad i = 1, 2, \ldots, N \tag{6}$$

and linearize system (2) about $s(t)$. This leads to

$$\dot{\eta} = \eta[DF(s(t))] + CA\eta\Gamma \tag{7}$$

where $\eta = (\eta_1, \eta_2, \ldots, \eta_N)^T \in \mathbb{R}^{N \times n}$, $DF(s(t))$ is the Jacobian of $F(X,t)$ at $s(t)$. Since A is a real symmetric matrix, there exists a unitary matrix $\Phi = (\phi_1, \phi_2, \ldots, \phi_N)$ such that:

$$A\phi_i = \lambda_i \phi_i, \quad i = 1, 2, \ldots, N \tag{8}$$

By expending its each column η on the basis Φ, we may derive

$$\eta = \Phi\gamma \tag{9}$$

where $\gamma = (\gamma_1, \gamma_2, \ldots, \gamma_N)^T \in \mathbb{R}^{N \times n}$. So we know that γ satisfies the following differential system

$$\dot{\gamma} = \gamma[DF(s(t))] + C\Lambda\gamma\Gamma \tag{10}$$

where $\Lambda = \text{diag}(\lambda_1, \lambda_2, \ldots, \lambda_N)$. Let γ_k be the kth row of γ. We have

$$\dot{\gamma}_k^T = \gamma_k^T[DF(s(t)) + C\lambda_k\Gamma], \quad k = 1, 2, \ldots, N \tag{11}$$

In terms of the Lemma 1, we have now transformed stability of the synchronization states (3) into the stability problem of the N pieces of n dimensional linear time-varying systems (11). In the following, we will prove that if $C > \frac{T}{|\lambda_2|}$, the N pieces of n dimensional linear time-varying systems (11) are asymptotically stable. Since $\lambda_1 = 0$ corresponds the synchronization state $s(t)$, we only need to prove that the following $N-1$ pieces of n dimensional linear time-varying systems (12) are asymptotically stable when $C > \frac{T}{|\lambda_2|}$.

$$\dot{\gamma}_k^T = \gamma_k^T[DF(s(t)) + C\lambda_k \Gamma], \quad k = 2, \ldots, N \tag{12}$$

According to (4), we have $(C\lambda_k + T) < 0 \ (k = 2, \ldots, N)$ if $C > \frac{T}{|\lambda_2|}$. In order to prove the theorem, let Lyapunov function be

$$V(\gamma_k) = \gamma_k^T \gamma_k, \quad k = 2, \ldots, N \tag{13}$$

and we have

$$\begin{aligned}
\dot{V}(\gamma_k) &= \dot{\gamma}_k^T \gamma_k + \gamma_k^T \dot{\gamma}_k \\
&= \gamma_k^T (DF(s(t)) + (DF(s(t)))^T + 2C\lambda_k \Gamma)\gamma_k \\
&= \gamma_k^T (DF(s(t)) + (DF(s(t)))^T - 2T\Gamma) + 2(C\lambda_k + T)\Gamma)\gamma_k \\
&= \gamma_k^T (DF(s(t)) + (DF(s(t)))^T - 2T\Gamma)\gamma_k + \gamma_k^T (2(C\lambda_k + T)\Gamma)\gamma_k \\
&< \gamma_k^T (2(C\lambda_k + T)\Gamma)\gamma_k \leq 0, (k = 2, \ldots, N)
\end{aligned} \tag{14}$$

So we can conclude that the $N - 1$ pieces of n dimensional linear time-varying systems (13) are asymptotically stable when $C > \frac{T}{|\lambda_2|}$. The proof is complete.

Remark: (1) The above mentioned Theorem is a sufficient condition, not a necessary. Even if the conditions of the Theorem are not satisfied, the synchronization of the coupled systems can occur.

(2) The above conclusion is not only valid for the coupled chaotic systems, but also for coupled limit cycles.

(3) Theorem is also valid to small world network mentioned in Ref.[13], because the corresponding coupling matrix A of this network also satisfies properties (1)-(4).

3 Numerical Simulation

As an illustration, we at first study synchronization of N coupled Lorenz systems with the first state variable coupling. The famous Lorenz system is described by the following differential system:

$$\dot{x} = a(y - x) \tag{15}$$

$$\dot{y} = bx - xz - y \tag{16}$$

$$\dot{z} = xy - cz \tag{17}$$

with the parameters $a = 10, b = 28, c = \frac{8}{3}$, Lorenz system exhibits chaos with double-scroll attractor as shown in Fig.2(a).

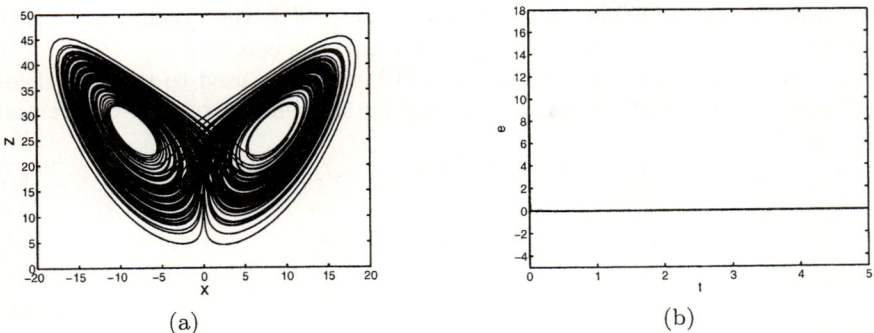

(a)   (b)

Fig. 2. (a) The Lorenz attractor with double-scroll, (b) Temporal evolution of synchronization error $e = x_2 - x_1$ in two coupled Lorenz systems when the coupling strength $C = 190$

By simple calculation, it is derived that the Jacobian

$$DF(s(t)) = \begin{pmatrix} -a & a & 0 \\ b-z & -1 & -x \\ y & x & -c \end{pmatrix}$$

hence,

$$A(s(t)) = DF(s(t)) + (DF(s(t)))^T = \begin{pmatrix} -2a & a+b-z & y \\ a+b-z & -2 & 0 \\ y & 0 & -2c \end{pmatrix}$$

It is seen that all eigenvalues of $A_{22}(s(t))$ are -2 and $-2c$, and $\det A(s(t)) = 4c(-2a + \frac{y^2}{2c} + \frac{(a+b-z)^2}{2})$. Because the solution of the Lorenz system is bounded, there exists a constant β such that $\beta \geq \frac{\det A(s(t))}{\det A_{22}(s(t))} = -2a + \frac{y^2}{2c} + \frac{(a+b-z)^2}{2}$ for all $s(t)$. In Ref.[14], it was proved that there exists a bounded region $\Gamma \subset \mathbb{R}^3$, which includes the whole Lorenz attractor. This region is estimated as follows:

$$\Gamma = \{(x,y,z) \in \mathbb{R}^3 \mid x^2 + y^2 + (z-a-b)^2 = \frac{c^2(a+b)^2}{4(c-1)})\} \quad (18)$$

Hence, $-2a + \frac{y^2}{2c} + \frac{(a+b-z)^2}{2} < -2a + \frac{1}{2}(y^2 + (z-a-b)^2) = -2a + \frac{1}{2}(\frac{c^2(a+b)^2}{4(c-1)} - x^2) < -2a + \frac{1}{2}(\frac{c^2(a+b)^2}{4(c-1)})$. Thus, we can choose $\beta = -2a + \frac{1}{2}(\frac{c^2(a+b)^2}{4(c-1)})$. For two coupled Lorenz systems with the first state variable, it is derived that when the coupling strength $C > \frac{1}{4}(-2a + \frac{1}{2}(\frac{c^2(a+b)^2}{4(c-1)})) = \frac{c^2(a+b)^2}{32(c-1)}) - \frac{a}{2}$, which is in good agreement with result in Ref.[11], the synchronization of two coupled Lorenz systems is achieved(See Fig.2(b)). But it is noted that the proposed method in this paper avoids rigorous calculation of Routh-Hurwitz rule. Here, only the determinant of the matrix $A(s(t))$, $A_{22}(s(t))$ and the largest nonzero eigenvalue of the coupling matrix are considered. It is very convenient to determine the synchronization of a class of coupled systems with the first state variable coupling.

The next is biological application. The HR neuron is used to test our theoretical analysis. The HR neuron is governed by the following set of differential equations[15]:

$$\dot{x} = y - ax^3 + bx^2 - z + I_{ext} \tag{19}$$

$$\dot{y} = c - dx^2 - y \tag{20}$$

$$\dot{z} = r[s(x - X0) - z] \tag{21}$$

where x is the membrane potential, y is associated with the fast current, Na^+, or K^+ and z with the slow current, for example, Ca^{2+}. Here we choose $a = 1.0, b = 3.0, c = 1.0, d = 5.0, s = 4.0, X0 = -1.60$ and external stimulus $I_{ext} = 2.95$. r is used as control parameter. With the parameter r changing, the HR neuron exhibits rich firing behaviour such as periodic and chaotic motions[See Fig.3].

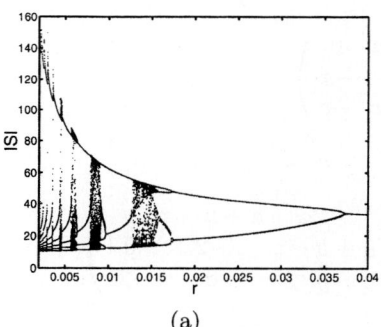
(a)

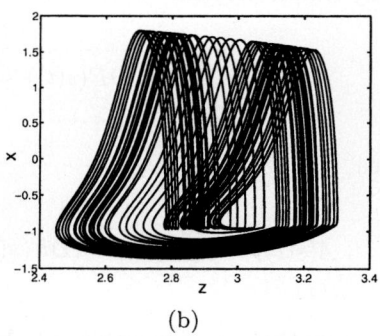
(b)

Fig. 3. (a) Bifurcation diagram of ISI(inter spike interval) of a single HR neuron with respect to the parameter r, (b) Chaotic attractor of a single HR neuron with respect to the parameter $r = 0.015$

The Jacobian of the system (17) is derived as follows:

$$DF(s(t)) = \begin{pmatrix} -3ax^2 + 2bx & 1 & -1 \\ -2dx & -1 & 0 \\ rs & 0 & -r \end{pmatrix}$$

hence,

$$A(s(t)) = DF(s(t)) + (DF(s(t)))^T = \begin{pmatrix} 2(-3ax^2 + 2bx) & -2dx + 1 & rs - 1 \\ -2dx + 1 & -2 & 0 \\ rs - 1 & 0 & -2r \end{pmatrix}$$

By a simple calculation, we know that $A(s(t))$ satisfies all conditions of the Theorem 1 and Corollary. Moreover, $\frac{\det A(s(t))}{\det A_{22}(s(t))} = 2(-3ax^2 + 2bx) + \frac{(rs-1)^2}{2r} + \frac{(-2dx+1)^2}{2}$. The solution of HR neuron is bounded under given parameters in this paper and we know $|x| < 2$. Hence, $\frac{\det A(s(t))}{\det A_{22}(s(t))} = 2(-3ax^2 + 2bx) + \frac{(rs-1)^2}{2r} + \frac{(-2dx+1)^2}{2} \leq 8b + \frac{(1+4d)^2}{2} + \frac{(rs-1)^2}{2r}$ and we can choose $\beta = 8b + \frac{(1+4d)^2}{2} + \frac{(rs-1)^2}{2r}$. According to the Theorem 1, we conclude that when the coupling strength $C > \frac{1}{2}\frac{\beta}{|\lambda_2|}$, the coupled HR neurons can achieve synchronization. As an illustration, we consider four coupled chaotic HR neurons with ring structure. Here, in order to make a single HR neuron chaotic, we take $r = 0.015$. The largest nonzero eigenvalue of the coupling matrix A is $\lambda_2 = -4sin^2\frac{\pi}{4}$. We can infer that when $C > \frac{1}{2}\frac{\beta}{|\lambda_2|} \approx 68.5$, complete synchronization of four coupled chaotic HR neurons is realized (See Fig.4).

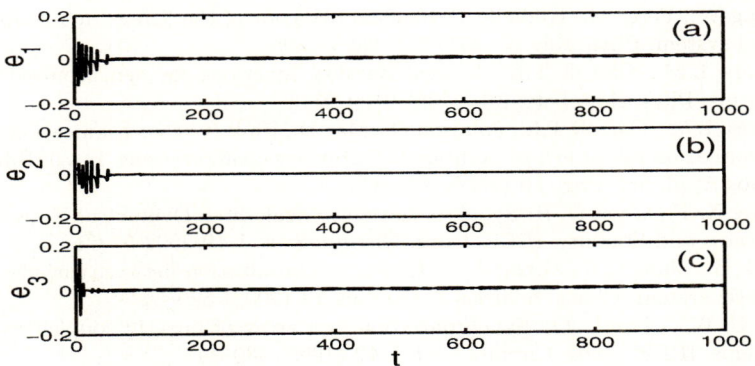

Fig. 4. Temporal evolution of synchronization error $e_i = x_{i+1} - x_1, (i = 1, 2, 3)$ for four coupled chaotic HR neurons when the coupling strength $C = 69$

4 Conclusion

In this paper, a general criterion, which determines synchronization stability of N coupled systems with the first state variable coupling, was developed in terms of the stability theory of dynamical system. This criteria is conveniently implemented with only calculating the determinant of the Jacobian and the largest nonzero eigenvalue of the coupling matrix. It avoid rigorous calculation of the Routh-Hurwitz rule, which determines sign of all eigenvalues of the corresponding Jacobian. Some numerical simulations were conducted to support our theoretical analysis. But it is noted that this is a sufficient condition. In general, the coupling strength, which is derived by means of the Theorem in this paper, is larger than actual critical coupling value. It is our future work to find minimal T in Theorem.

Acknowledgement. This work was supported by the National Natural Science foundation of China (No.10432010).

References

1. Pecora L.M., Carroll T.L.: Driving systems with chaotic signals. Phys. Rev. A. **44** (1991) 2374–2381
2. Pecora L.M., Carroll T.L.: Synchronization in chaotic systems. Phys. Rev. Lett. **64** (1990) 821–824.
3. Strogatz S. H.: Exploring complex networks. Nature. **410** (2001) 268–276
4. Istvn Z. K., Zhai Y. M., Hudson J. L.: Collective dynamics of chaotic chemical oscillators and the law of large numbers. Phys. Rev. Lett. **88** (2002) 238301-1
5. Hansel D., Sompolinsky H.: Synchronization and computation in a chaotic neural network. Phys. Rev. Lett. **68** (1992) 718–721
6. Shuai J. W., Durand D. M.: Phase synchronization in two coupled chaotic neurons. Phys. Lett. A. **264** (1999) 289–297
7. Wang W., Perez G., Hilda A: Dynamical behavior of the firings in a coupled neuronal system. Phys. Rev. E. **47**(1993) 2893–2896
8. Pecora L.M., Carroll T.L.: Master stability functions for synchronized coupled systems. Phys. Rev. Lett. **80** (1998) 2109–2113
9. Pecora L.M., Carroll T.L., Johnson G., Mar D., Fink K.: Synchronization stability in coupled oscillator arrays: solution for arbitrary configurations. Int. J. Bifurcation Chaos Appl. Sci. Eng. **10** (2000) 273–290
10. Wang X. F., Chen G. R: Synchronization in Scale-Free Dynamical Networks: Robustness and Fragility. IEEE Trans. Circuits Syst. I **49** 2002 54-62
11. Lü J. H., Zhou T. S., Zhang S. C.: Chaos synchronization between linearly coupled chaotic system. Chaos, Solitons & Fractals **14** (2002) 529–541
12. Wu C. W., Chua L. O.: Synchronization in a array of linearly coupled dynamical systems. IEEE Trans. Circuits Syst. I **42** (1995) 430–447
13. Newman M. E. J., Watts, D. J.: Renormalization group analysis of the small-world network model. Phys. Lett. A **263** (1999) 341–346
14. Leonov G., Bunin A., Koksch N.: Attractor localization of the Lorenz system, ZAMM. **67** (1987) 649–656
15. Hindmarsh J.L., Rose R. M.: A model of neuronal bursting using three coupled first order differential equations. Proc. R. Soc. Lond. B. **221** (1994) 87–102

Complexity of Linear Cellular Automata over $\mathbb{Z}_m$*

Xiaogang Jin[1] and Weihong Wang[2]

[1] AI Institute, College of Computer Science,
Zhejiang university, Hangzhou 310027, China
xiaogangj@cise.zju.edu.cn
[2] College of Software, Zhejiang University of Technology,
Hangzhou 310032,China

Abstract. Cellular automata(CA) is not only a discrete dynamical system with infinite dimensions, but also an important computational model. How simple can a CA be and yet support interesting and complicated behavior. There are many unsolved problems in the theory of CA, which appeal many researchers to focus their attentions on the field, especially subclass of CA – linear CA. These studies cover the topological properties, chaotical properties, invertibility, attractors and the classification of linear CA etc.. This is a survey of known results and open questions of D-dimensional linear CA over $\mathbb{Z}_m$.

1 Introduction

Cellular automata (CA) was originally proposed by von Neumann. The latter wanted to construct simple mathematical models, capable on the one hand of universal computations, on the other hand of self-reproduction. One of the most remarkable example is the so called "game of life" defined by Conway in 1970.

A decisive contribution to the mathematical study of CA were given by Wolfram at beginning of 1980s[20]. Based on computer simulation, Wolfram proposed an empirical classification of CA. Any way many of the later researchers, however various their means, were aimed at giving a mathematical sense to Wolfram's classification. Unfortunately, some properties of the temporal evolution of general CA are undecidable[3,11].

Linear CA is a subclass of CA. Despite of their apparent simplicity, linear CA may also exhibit complex features and allow a detailed algebraic analysis. Recently, many scientists focus their attentions on linear CA[7,15,16,18,17]. This paper is survey of known results and open questions of D-dimensional linear CA over $\mathbb{Z}_m$. However, we would like to point out that any survey of CA is bound to be incomplete. We only focus on some topological properties, which are closely relative to the chaotical theory.

2 Definitions and Notations

Now we recall the definitions of discrete time dynamical systems(DTDS).

---

* Supported by Natural Science Foundation of China (NSFC) grant 60103015 and The Project-sponsored by SRF for ROCS, SEM.

2.1 Topological Dynamical Systems and Topological Properties

A *topological dynamical system* is a compact metric space X endowed with a continuous self-map $F : X \to X$. If F is surjective, (X, F) is called an endomorphism. It turns out that $(\mathscr{C}_m^D, F)$ endowed with a CA map F is a topological dynamical system.

A dynamical system (X, F) is *sensitive to initial conditions* iff there exists $\delta > 0$ such that $\forall x \in X \ \forall \varepsilon > 0 \ \exists y \in \mathscr{B}(x, \varepsilon) \ \exists n \in \mathbb{Z} : d(F^n(x), F^n(y)) > \delta$. The value δ is called sensitivity constant.

A dynamical system (X, F) is *equicontinuous* at $x \in X$ iff for any $\delta > 0 \ \exists \varepsilon > 0, \ \forall y \in \mathscr{B}(x, \varepsilon) \ \exists n \in \mathbb{N} : d(F^n(x), F^n(y)) < \delta$. A dynamical system (X, F) is equicontinuous iff it is equicontinuous at every $x \in X$. The notions of sensitivity and equicontinuity are related. In fact, F is not sensitive $\Leftrightarrow \exists x : F$ is equicontinuous at x.

A dynamical system (X, F) is *positively expansive* iff there exists $\delta > 0$ such that $\forall x, y \in X, \ x \neq y, \ \exists n \in \mathbb{N} : d(F^n(x), F^n(y)) > \delta$. The value δ is called expansivity constant.

A dynamical system (X, F) is *strongly transitive* iff for all nonempty open set $U \subseteq X$ we have $\cup_{n=0}^{+\infty} F^n(U) = X$.

Let $P(F) = \{x \in X \mid \exists n \in \mathbb{N} : F^n(x) = x\}$ be the set of the periodic points of F. A dynamical system (X, F) is *regular* iff it has dense periodic orbits ($P(F)$ is a dense subset of X), e.g. $\forall x \in X$ and $\varepsilon > o \ \exists y \in P(F)$ such that $d(x, y) < \varepsilon$.

2.2 Linear CA

For $m \geq 2$, let $\mathbb{Z}_m$, denote the ring of integers modulo m. We consider the *space of configurations* $\mathscr{C}_m^D = \{c \mid c : \mathbb{Z}^D \to \mathbb{Z}_m\}$, which consists of all function from $\mathbb{Z}^D$ into $\mathbb{Z}_m$. Each element of $\mathscr{C}_m^D$ can be visualized as an infinite D-dimensional lattice in which each cell contains an element of $\mathbb{Z}_m$.

Let $s \geq 1$. A *neighborhood frame* of size s is an ordered set of distinct vectors $\mathbf{u}_1, \mathbf{u}_2, \cdots, \mathbf{u}_s \in \mathbb{Z}^D$. Given any function $f : \mathbb{Z}_m^s \to \mathbb{Z}_m$, a D-dimensional CA based on *local rule* f is the pair $(\mathscr{C}_m^D, F)$, where $F : \mathscr{C}_m^D \to \mathscr{C}_m^D$, is the *global transition map* defined as follows. For every $c \in \mathscr{C}_m^D$ the configuration $F(c)$ is such that for every $\mathbf{v} \in \mathbb{Z}^D$

$$[F(c)](\mathbf{v}) = f(c(\mathbf{v} + \mathbf{u}_1), \cdots, c(\mathbf{v} + \mathbf{u}_s)). \tag{1}$$

Note that the local rule f and the neighborhood frame completely determine F. The linear CA is CA with a local rule of the form $f(x_1, \cdots, x_s) = \sum_{i=1}^{s} \lambda_i x_i \mod m$, where $\lambda_1, \cdots, \lambda_s \in \mathbb{Z}_m$. Throughout the paper, $\mathscr{B}(x, \varepsilon)$ will denote the set $\{y \in X : d(x, y) \leq \varepsilon\}$ with respect to corresponding distance. $F(c)$ will denote the result of the application of the map F to the configuration c and $F^n(c) = F(F^{(n-1)}(c))$.

3 Classifications of Linear CA

In the beginning of 1980s Stephen Wolfram introduced a heuristic classification of CA based on the qualitative long-term behavior starting from random initial conditions, as observed on computer simulations [20]:

(W1) A spatially homogeneous state.

(W2) A sequence of simple stable or periodic structure.
(W3) Chaotic aperiodic behavior.
(W4) Complicated localized structure, some propagating.

Later work concentrated on formalizing the intuitive classification by Wolfram. From the same line, Čulik and Yu [3], Sutner [19] consider some similar classifications of CA. In recent times, Hurley [9] addressed another classification of CA based on attractor, which has been refined by Kurka [12].

3.1 Classifications According to the Local Behavior

In order to study the topological properties of D-dimensional CA, we need give a distance over the space of configuration, most researches adopted the metric topology induced by the *Tychonoff distance*. Let $\Delta : \mathbb{Z}_m \times \mathbb{Z}_m \to \{0,1\}$ given by $\Delta(i,j) = 0$, if $i = j$ and $\Delta(i,j) = 1$, if $i \neq j$. For any $a,b \in \mathscr{C}_m^D$ the Tychonoff distance $d(a,b)$ is defined by

$$d(a,b) = \sum_{\mathbf{v} \in Z^D} \frac{\Delta(a(\mathbf{v}), b(\mathbf{v}))}{2^{\|\mathbf{v}\|_\infty}} \qquad (2)$$

It is easy to verify that d is a metric on $\mathscr{C}_m^D$. With this topology, $\mathscr{C}_m^D$ is a compact and totally disconnected space and every CA is uniformly continuous map.

For the linear CA over ring $\mathbb{Z}_m$, several important topological properties have been studied during the last few years [1,10,15,18] and in some cases exact characterization have been obtained (see Table.1, $\mathscr{P}$ denotes the set of prime factors of m). According

Table 1. Characterization of topological properties of linear CA over $\mathbb{Z}_m$

Propertry	Characterization
Surjectivity	$gcd(m, \lambda_1, \cdots, \lambda_s) = 1$
Injectivity	$(\forall p \in \mathscr{P}) \exists ! \lambda_i) : p \nmid \lambda_i$
Erodicity	$gcd(m, \lambda_1, \cdots, \lambda_s) = 1$
Transitivity	$gcd(m, \lambda_1, \cdots, \lambda_s) = 1$
Regularity	$gcd(m, a_{-r}, \cdots, a_r) = 1$
Expansitivity	$gcd(m, a_{-r}, !`, a_{-1}, a_1, !`, a_r) = 1$
Sensitivity	$(\exists p \in \mathscr{P}) : \nmid gcd(\lambda_2, !`, \lambda_s)$
Pos. expansivity	$gcd(m, a_1, \cdots, a_r) = 1$
Equicontinuity	$(\forall p \in \mathscr{P}) p \mid gcd(\lambda_2, \cdots, \lambda_s)$
Strong trans.	$(\forall p \in \mathscr{P})(\exists \lambda_i, \lambda_j) : p \nmid \lambda_i \wedge p \nmid \lambda_j)$

to the above results, G. Manzini and L. Margara [15] give a hierarchical classification of linear CA, namely equicontinuous CA, sensitive but not transitive CA, transitive but not strongly transitive CA, strongly transitive but not positively expansive CA, and positive expansive CA, which can be expressed the degrees of chaoticity. The membership of the classification require only gcd computations and it can be checked in polynomial time in $\log m$. Unfortunately, in general case the problem of deciding whether a given CA belongs to one of the above-mentioned class is not even known to be decidable.

3.2 Classifications According to the Attractors

What is an attractor of a CA? Let (X,F) be a dynamical system. A nonempty subset $Z \subseteq X$ is an *attractor* for F iff there exists an open set $U \subseteq X$ such that $F(\overline{U}) \subseteq U$, $Z = \cap_{j \geq 0} F^j(U)$.

Attractors of CA have been studied by Hurley [9], Kukra[12]. For linear CA over $\mathbb{Z}_m$, G. Manzini and L. Margara [16] continue the study on their dynamical properties (see [1,10,15,18]) and furthermore they find that the evolution of a linear nonsurjetive cellular automata F will take place completely within a subspace Y_F of $\mathscr{C}_m^D$ after a transient phase of length at most $\lfloor \log_2 m \rfloor$. Based on the result, they make a further step in the analysis of the long-term behavior of linear CA and prove that for linear CA (Surjective and nonsurjective) it is possible to determine the membership in the Kukra's classification by looking at the coefficients of the associated local rule. The characterization of attractors for linear CA is a basic step towards the computation of their entropy, which is uncomputable for general CA [8].

3.3 Chaos for Linear CA

The notion of chaos is very appealing, and it has intrigued many scientist[5,13]. For DTDS, a universally accepted definition of chaos does not exist. In the popular book by Devaney[5], the author isolates three components as being the essential features of chaos: transitivity, sensitivity to the initial conditions and regularity. For CA, a result has been proved in [2]: topological transitivity implies sensitivity to initial conditions. Kundsen in [13]proposed another definition of chaos which excludes chaos without non-periodicity. According to the Kundsen's definition, the dynamical system is chaotic iff the system has a dense orbit and sensitive to initial conditions.

In [1,6], the authors apply the definition of chaos given by Devaney and by Kundsen to the study of CA. For linear CA, Favati *et al* [6] completely classify one-dimensional additive CA definition over any alphabet of prime cardinality according to the Devaney's definition of chaos. In [1], G. Cattaneo *et al* completely characterize topological transitivity for every D-dimensional linear CA over $\mathbb{Z}_m$ ($m \geq 2$, and $D \geq 1$) which is equivalent to ergodic and regularity for any one-dimensional linear CA over $\mathbb{Z}_m$ ($m \geq 2$). But for D-dimensional linear CA over $\mathbb{Z}_m$ ($D \geq 2$), regularity is equivalent to surjectivity remains open. Furthermore, Finelli *et al* [7] establish a connection between the theory of Lyapunov exponents and the properties of expansivity and sensitivity to initial conditions.

The topological entropy is often interpreted as a measure of the chaotic character of a dynamical system, which measures the uncertainty of the forward evolution of any dynamical system. For general CA, the topological entropy is one of those properties that are known to be undecidable[8]. Nevertheless there a few classes for which it has been computed. In [4], the authors adopt following simpler form:

$$\mathcal{H}(\mathscr{C}_m^D, F) = \lim_{w \to \infty} (\lim_{t \to \infty} \frac{\log R^{(D)}(w,t)}{t}) \qquad (3)$$

where w is the side-length of a D-dimensional region of the lattice, $R^{(D)}(w,t)$ be the number of distinct $D+1$ dimensional hyperrectangles obtained as space-time evolution diagrams of $(\mathscr{C}_m^D, F)$. According to equation(3), the authors prove a closed formula for the topological entropy of D-dimensional linear CA over $\mathbb{Z}_m$.

4 Conclusions

Linear CA have been studied from several different angles other than the ones mentioned here. Computation-theoretic questions in computer science, simulation of natural phenomena in physics and biology are considered important. We focus on topics which are closer to computer science and physics rather than its applications. We hope the survey to be useful to both fresh entrants into this field and to experts working on particular aspects of CA.

Reference

1. G. Cattaneo, E. Formenti, G. Manzini and L.Margara: Erogdicity, transitivity, and regularity for additive cellular automata over $\mathbb{Z}_m$, Theoret. Comput. Sci. 233 (2000) 147-164.
2. B. Codenotti, L. Margara: Transitive cellular automata are sensitive, Amer. Math. Monthly 103 (1996) 58-62.
3. K. Čulik, S. Yu: Undecidability of CA classification achemes, Complex Systems 2 (1988) 177-190.
4. M. D'amico, G. Manzini and L.Margara: On computinmg the entropy of cellular automata, Lecture Notes in Mathematica 1443 (1998) 470-481.
5. R. L. Devaney: An introduction to chaotic dynamical systems, 2nd ed. Addison-Wesley, Reading, MA (1989).
6. P. Favati, G. Lotti and L. Margara: Additive one dimensional cellular automata are chaotic according to Devaney's definition of chaos, Theoret. Comput. Sci. 174(1-2) (1997) 157-170.
7. M. Finelli, G. Manzini and L. Margara: Lyapunov exponents vs expansivity and sensitivity in cellular automata, J.Complexity 14 (1998)210-233.
8. L. Hurd, J. Kari and K. Čulik: The topological entropy of cellular automata is uncomputable, Ergodic Theory and Dynamical Systems 12 (1992) 255-265.
9. M. Hurley: Attractor in cellular automata, Ergodic Theory and Dynamical Systems 10 (1990) 131-140.
10. M. Ito, N. Osato, M. Nasu: Linear cellular automata over $\mathbb{Z}_m$, J. Comput. System Sci. 27 (1983) 125-140.
11. J. Kari: Reversibility of 2D cellular automata is undecidable, Physica D 45 (1990) 379-385.
12. P. Kurka: Languages, equicontinuity and attractors in cellular automata, Ergodic Theory and Dynamical Systems 17 (1997) 417-433.
13. C. Kundsen: Chaos without nonperiodicity, Amer. Math. Monthly 101 (1994) 563-565.
14. G. Manzini: Characterization of sensitive linear cellular automata with respect to the counting distance, Lecture Notes in Computer Science Vol.1450 Springer Verlag (1998) 825-833.
15. G. Manzini, L. Magara: A complete and effciently computable topologyical calssification of D-dimensional linear cellula automata over $\mathbb{Z}_m$, Theoret. Comput. Sci. 221 (1999) 157-177.
16. G. Manzini, L. Margara: Attractor of linear cellular automata, J. Comput. system Sci. 58 (1999) 597-610.
17. T. Sato: Group Structured linear cellular automata over $\mathbb{Z}_m$, J. Comput. System Sci. 27 (1999) 18-23.
18. T. Sato: Erogodicity of linear cellular automata over $\mathbb{Z}_m$, Inform. process. lett. 61(3) (1997) 169-172.
19. K. Sutner: A note on Čulik-Yu classes, Complex Syst. 3 (1989) 107.
20. S. Wolfram: Universility and complexity in cellular automata, Physica D 10 (1984) 1-35.

Applications of Genetic Algorithm for Artificial Neural Network Model Discovery and Performance Surface Optimization in Finance

Serge Hayward

Department of Finance
Ecole Supérieure de Commerce de Dijon, France
shayward@escdijon.com

Abstract. This paper considers a design framework of a computational experiment in finance. The examination of relationships between statistics used for economic forecasts evaluation and profitability of investment decisions reveals that only the 'degree of improvement over efficient prediction' shows robust links with profitability. If profits are not observable, this measure is proposed as an evaluation criterion for an economic prediction. Combined with directional accuracy, it could be used in an estimation technique for economic behavior, as an alternative to conventional least squares. Model discovery and performance surface optimization with genetic algorithm demonstrate profitability improvement with an inconclusive effect on statistical criteria.

1 Introduction

Motivations for this paper come from the ongoing search for the foundation of evolutionary computation (EC) in finance and a claim by [1] that traditional summary statistics are not closely related to a forecast's profit, with the exception of directional accuracy (DA).

Financial prices exhibit non-stationarity, autocovarience and frequent structural brakes, posing problems for their modeling. This paper investigates how data mining benefits from genetic algorithm (GA) model discovery, performance surface optimization and pre/pro-processing, improving predictability or/and profitability.

2 Methodology

For our experiment we build evolutionary / artificial neural network (E/ANN) forecasts and generate a posterior optimal rule. The rule, using future information to determine the best current trading action, returns a buy/sell signal (B/S) today if prices tomorrow have increased/decreased. A posterior optimal rule signal (PORS) is then modeled with ANN forecasts, generating a trading B/S signal. Combining a trading signal with a strategy warrants a position to be taken. We consider a number of market timing strategies, appropriate for different strengths of the B/S signal. If we have a buy (sell) signal on the basis of prices expected to increase (decrease) than we enter a Long (Short) position. Note that our approach is different from standard B/S signal generation by a technical trading rule. In the latter it is only a signal from a technical trading rule that establishes that prices are expected to increase/decrease. In our model we collaborate signal's expectations of price change (given by PORS) with a time-series forecast.

To apply our methodology we develop the dual network structure, presented in Figure 1. The forecasting network feeds into the action network, from which the information set includes the output of the first network and PORS, as well as the inputs used for forecasting, in order to relate the forecast to the data upon which it was based.

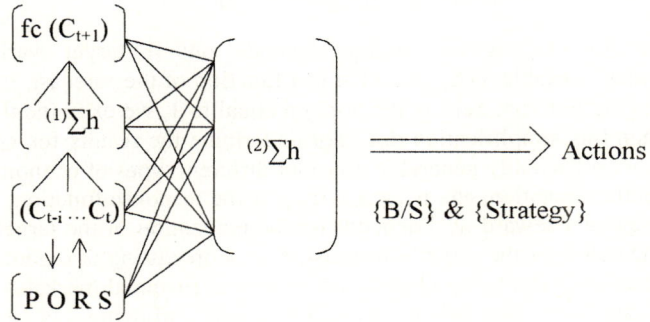

Fig. 1. Dual ANN: (1) forecasting network; (2) acting network

The model is evolutionary in the sense it considers a population of networks (individual agents facing identical problems/instances) that generate different solutions, which are assessed and selected on the basis of their fitness. Backpropagation is used in the forecasting net to learn to approximate the unknown conditional expectation function (without the need to make assumptions about data generating mechanism and beliefs formation). It is also employed in the action net to learn the relationship between forecasts' statistical and actions' economic characteristics. Lastly, agents discover their optimal models with GA; applying it for ANN model discovery makes technical decisions less arbitrary.

2.1 Generating Posterior Optimal Rule Signal

PORS is a function of a trading strategy adopted and based on the amount of minimum profit and the number of samples into the future. Stepping forward one sample at a time, the potential profit is examined. If the profit expected is enough to clear the minimum profit after transaction costs (TC), a PORS is generated. The direction of PORS is governed by the direction of the price movement. Normally, the strength of the signal reflects the size of underlying price changes, although, we also examine signals without this correlation to identify when profit generating conditions begin. Lastly, we consider PORS generated only at the points of highest profit to establish the maximum profit available.

3 Description of the Environment

Let Y be a random variable defined on a probability space $(\Omega, \mathcal{F}, P)$. Ω is a space of outcomes, $\mathcal{F}$ is a σ-field and P is a probability measure. For a space $(\Omega, \mathcal{F}, P)$ a conditional probability $P[A|\mathcal{F}]$ for a set A, defined with respect to a σ-field $\mathcal{F}$, is the conditional probability of the set A, being evaluated in light of the information available in the σ-field $\mathcal{F}$. Suppose economic agents' utility functions are given by a general form:

$$U(W_{t+s}) = g(Y_{t+s}, \delta(fc_{t+s})). \tag{1}$$

According to (1), agents' utility depends on: a target variable Y_{t+s}; a decision/strategy variable, $\delta(fc_{t+s})$, which is a function of the forecast, fc_{t+s}, where $s \geq 1$ is a forecasting horizon. Setting the horizon equal to 1, we examine the next period forecast (when this simplification does not undermine the results for $s \geq 1$). A reward variable W_{t+s} is sufficiently general to consider different types of economic agents and includes wealth, reputation, etc. $w_{t+1}(y_{t+1}, fc_{t+1})$ is the response function, stating that at time $t+1$ an agent's reward w_{t+1} depends on the realization of the target variable y_{t+1} and on the accuracy of the target's forecast, fc_{t+1}. Forecasting is regarded as a major factor of a decision rule, being close to the reality in financial markets. Also, it has a developed statistical foundation in econometrics allowing its application in evolutionary computation.

Let $fct_{+1} = \theta' X_t$ to be a forecast of Y_{t+1} conditional on the information set $\mathcal{F}_t$, where unknown m-vector of parameters, $\theta \chi \Theta$, with Θ to be compact in R^k and observable at time t n-vector of variables, X_t. X_t are $\mathcal{F}_t$-measurable and might include some exogenous variables, indicators, lags of Y_t, etc. An optimal forecast does not exclude model misspecification, which can be due to the form of fc_{t+1} or failure to include all relevant information in X_t. Under imperfect foresight, the response function and, therefore, the utility function are negatively correlated with forecast error, $e_{t+1} \equiv y_{t+1} - fc_{t+1}$; $|e_{t+1}| > 0$. A mapping of the forecast into a strategy rule, $\delta(fc_{t+1})$ (combined with elements of X_t) determines a predictive density g_y, which establishes agents' actions.

In this setting, maximizing expected utility requires us to find an optimal forecast, fc_{t+1} and to establish an optimal decision rule, $\delta(fc_{t+1})$. Note that optimality is with respect to a particular utility function, implemented through a loss function, in the sense that no loss for a correct decision and a positive loss for incorrect one. Given a utility function, expected utility maximization requires minimization of the expected value of a loss function, representing the relationship between the size of the forecast error and the economic loss incurred because of that error. A strategy development (mapping of the forecast into a decision rule) is another way to minimize the expected value of a loss function.

A loss function, $L: R \rightarrow R^+$, related to some economic criteria or a statistical measure of accuracy, takes a general form:

$$L(p, a, e) \equiv [a + (1 - 2a)1(e < 0)]e^p ,\tag{2}$$

where p is a coefficient of risk aversion; e is the forecast error; $\alpha \chi [0,1]$ is the degree of asymmetry in the forecaster's loss function. $L(p, \alpha, e)$ is $\mathcal{F}_t$-measurable. It could also be presented as:

$$L(p, a, \theta) \equiv [a + (1 - 2a)1(Y_{t+1} - fc_{t+1}(\theta) < 0)]|Y_{t+1} - fc_{t+1}(\theta)|^p ,\tag{3}$$

where α and p are shape parameters and a vector of unknown parameters, $\theta \chi \Theta$. For given values of p and α an agent's optimal one-period forecast is

$$\min_{\theta \in \Theta} E[L(\rho, \alpha, \theta)] = E[L(Y_{t+1} - fc_{t+1})] = E[L(e_{t+1})]\tag{4}$$

Training EANN under different criteria allows us to examine relationships between statistical measures and economic characteristics.

4 Experimental Design

4.1 Performance Surface

The performance of ANN learning is monitored by observing how the cost changes over training iterations. The learning curve presents the internal error over each epoch of training, comparing the output of the ANN to the desired output. In price forecasting, the target is the next day closing price, where in signal modeling, the target is the current strategy. Achieving an accurate representation of the mapping between the input and the target might not necessarily lead to a forecast to be exploitable or a strategy using that forecast to be profitable.

Although we train ANN with the goal to minimize internal error function, we test and optimize its generalization ability by comparing its performance with the results of a benchmark, an efficient prediction (EP). In forecasting prices, EP is the last known value. For predicting strategies, it is the buy/hold (B/H) strategy. The degree of improvement over efficient prediction (IEP) is calculated as an error from a denormalized value of the ANN and a desired output, then normalizing the result with the difference between the target and EP value.

Making a prediction using a change or a percentage change, the value of IEP is particularly significant. IEP around 1, implying that the ANN predicted a change or a percentage change of zero, indicates that the network does not have adequate information to make a valid prediction. So, it ends up predicting the mean of all changes, zero. Predicting two samples or more in advance, one can have reduction in value of IEP (in comparison to one sample prediction). This does not mean that there is an improvement, since the change in the desired value is typically larger for a longer prediction. We classify our results using the following scale: IEP<0.8 υ excellent; IEP<0.85 υ very good; IEP<0.9 υ good; IEP<0.95 υ satisfactory; IEPf0.95 υ weak.

4.2 Profitability as Performance Measure

Similar to the performance evaluation criteria of investment managers (total realized returns adjusted for the riskness) the realized total continuously compounded returns or excess returns have been used to review trading rules developed under evolutionary learning. Unlike case-by-case evaluation of actions of portfolio managers, decisions of evolutionary agents are assessed on aggregate, over the entire trading period. Therefore, in computational modeling process/means used by agents need to be explicitly evaluated. Under continuously compounded reinvestment of realized returns, strategies with a higher number of trades and lower returns per trade receive greater fitness. [2] demonstrates that strategies with the lowest mean returns and variances per trade could be evaluated as best.

We examine the following forms of cumulative and individual trades' return measures: non-realized simple aggregate return; profit/loss factor; average, maximum gain/loss. In addition we estimate exit efficiency, measuring whether trades may have been held too long, relative to the maximum amount of profit to be made, as well as the frequency and the length of trades, including out of market position. To assess risk exposure we adopt the Sharpe ratio[1] and the maximum drawdown[2], as well as common 'primitive' statistics. To overcome the Fisher effect we consider trading positions with a one-day delay.

TC is assumed to be paid both when entering and exiting the market, as a percentage of the trade value. TC accounts for broker's fees, taxes, liquidity cost (bid-ask spread), as well as costs of collecting/analysis of information and opportunity costs. According to [3] large institutional investors achieve one-way TC about 0.1-0.2%. Often TC in this range is used in computational models. Since TC (defined above) would differ for heterogeneous agents, we report the break-even TC that offsets trading revenue with costs leading to zero profits.

The classification of the ANN output as different types of B/S signals determines the capability of the model to detect the key turning points of price movement. Evaluating the mapping of a forecast into a strategy, $\delta(fc_{t+1})$, assesses the success in establishing a predictive density, g_y that determines agents' actions.

4.3 Time Horizons and Trading Strategies Styles

Heterogeneous traders use different lengths of past and forward time horizons to build their forecasts/strategies. We have run the experiment on stock indexes from a number of markets and found that 'optimal' length of training/validation period is a function of specific market conditions. In this paper we adopt three memory time horizons, [6; 5; 2½] years. We run the experiment with one year testing horizon, as it seems to be reasonable from the actual trading strategies perspective.

Both long and short trades are allowed in the simulation. Investing total funds for the first trade, subsequent trades (during a year) are made by re-investing all of the money returned from the previous trades. If the account no longer has enough capital to cover TC, trading stops.

---

[1] Given by the average return divided by the standard deviation of that return.
[2] The size of the individual losses occurred while achieving given gains.

5 Genetic Algorithm Optimization

In this research EC is used for ANN model discovery, considering GA optimization for: network's topology; performance surface; learning rules; number of neurons and memory taps; weight update; step size and momentum rate. GA tests the performance of various ANN models. We examine the performance surface optimized with GA for DA, discounting the least recent values and minimizing the number of large errors. For learning rule optimization we consider Steepest Descent; Conjugate Gradient; Quickprop; Delta Bar Delta and Momentum.

With GA optimization we test the integer interval [1, 20] for hidden layers' neurons, expecting that a higher number increases the network's learning ability, although at the expense of harder training and a tendency to overspecialization. GA optimization considers the range [1, 20] for the number of taps, affecting the memory of the net. GA optimization of the weight update for static networks considers whether the weights are updated following all data (batch) or after each piece of data (online) are presented. For dynamic networks GA determines a number of samples to be examined each time ANN updates weights during the training phase.

GA optimizes the step size of the learning rates in the range [0, 1]. The momentum, using the recent weight update, speeds up the learning and helps to avoid local minima. GA searches in the range [0, 1] for the value by which the most recent weight update is multiplied.

In terms of GA parameters, we apply the tournament selection with size 4, {prob=fitness/Σfitness}. Four types of mutation are considered in the experiment: uniform, non-uniform, boundary and Gaussian. Probability of mutation (PM) tested in the range [0, 0.05] and probability of uniform crossover is examined in the range [0.7, 0.95]. We test the effect of the increase in population size in the range [25, 200] on performance and computational time. The training continues until a set of termination criteria is reached, given by maximum generations in the range [100, 500].

When a model lacks information, trading signals' predictions often stay near to the average. If ANN output remains too close to the mean to cross over the thresholds that differentiate entry/exit signals, post-processing is found to be useful (establishing thresholds within the range). Post-processing with GA optimization, examines a predicted signal with simulated trades after each training, searching for the thresholds against the values produced by ANN to generate maximum profit.

GA tests various settings from different initial conditions (in the absence of a priori knowledge and to avoid symmetry that can trap the search algorithm). We use GA optimization with the aim to minimize IEP value and profitability as a measure of overall success.

6 Empirical Application

6.1 Data

We consider daily closing prices for the MTMS (Moscow Times) share index obtained from Yahoo Finance. The time period under investigation is 01/01/97 to 23/01/04. There were altogether 1575 observations in row data sets. Examining the data graphically reveals that the stock prices exhibit a prominent upward, but non-linear trend, with pronounced and persistent fluctuations about it, which increase in variability as the level of the series increases. Asset prices look persistent and close to unit root or non-stationarity. Descriptive statistics confirm that the unit-root hypothesis cannot be rejected at any confidence level. The data also exhibits large and persistent price volatility with significant autocovarience even at high order lags.

Changes in prices increase in amplitude and exhibit clustering volatility. The daily return displays excess kurtosis and the null of no skewness is rejected at 5% critical level. The tests statistics lead to rejection of the Gaussian hypothesis for the distribution of the series. It confirms that high-frequency stock returns follow a leptokurtic and skewed distribution incompatible with normality.

6.2 Experimental Results

ANN with GA optimization was programmed with various topologies[3]. Altogether we have generated and considered 93 forecasting and 143 trading strategies' settings. Effectiveness of search algorithm was examined with multiple trials for each setting. 92% of 10 individual runs produce identical results, confirming the replicability of our models. Efficiency of the search was assessed by the time it takes to find good results. The search with ANN unoptimized genetically took a few minutes, where the search with GA optimization lasted on average 120 minutes on a Pentium 4 processor.

Over a one year testing period 19 trading strategies were able to outperform in economic terms the B/H strategy, with an investment of $10,000 and a TC of 2% of trade value. The average return improvement over B/H strategy was 20%, with the first five outperforming the benchmark by 50% and the last three by 2%. The primary strategy superiority over B/H strategy was 72%.

For the five best performing strategies, the break-even TC was estimated to be 2.75%, increasing to 3.5% for the first three and nearly 5% for the primary strategy. Thus, the break-even TC for at least primary strategy appears to be high enough to exceed actual TC.

The experiment demonstrates that normalization reduces the effect of non-stationarity in the time series. The effect of persistency in prices diminishes with the use of the 'percentage change' in values. Table 1, presenting the average effect of GA post-processing on performance, shows that it has generally improved (positive values) statistical characteristics. Although only accuracy[4] exhibits sizable change, the effects on IEP and correlation[5] were significantly smaller and not always positive.

Table 1. GA Post-Processing Effect

ΔStats./Sets	2000-2004	1998-2004	1997-2004
IEP	0.059	-0.838	0.001
Accuracy (%)	1.3	6.58	0.95
Correlation	0.016	0.011	0.001

---

[3] Programs in Visual C++, v. 6.0 are available upon request. We have run tests on TradingSolutons, v. 2.1, NeuroSolutions v. 4.22 and Matlab v. 6.

[4] Percentage of correct predictions.

[5] Correlation of desired and ANN output.

The experiment with four types of GA mutation did not identify the dominance by a particular type. We have run simulations with different PM to test how the frequency of novel concepts' arrival affects modeling of the environment with structural brakes. The results, presented in Table 2, show that newcomers generally benefit the system. Although we have expected this outcome, its consistency among all (including short time) horizons was not anticipated. In economic terms, runs with a high probability of mutation {PM=0.05} have produced the highest returns. At the same time, this relationship is of non-linear character (e.g. {PM=0.001} consistently outperforms {PM=0.02}).

Some moderate, although consistent relationship between PM and strategies' risk exposure was found. Higher PM resulted in low riskness, given particularly by Sharpe ratio. We have also noticed some positive correlation between PM and annual trades' quantity, although this relationship appears to be of moderate significance and robustness. Trading frequency in simulations without mutation seems to be set at the beginning and stay until the end either at low or high values. The experiments without mutation have produced strong path-dependent dynamics, though not necessarily with sub-optimal outcome. It seems there exist some 'optimal' PM (in our experiment 0.05 and 0.001) and tinkering with this parameter can improve overall profitability.

Table 2. Economic and Statistical Measures under Different Probabilities of Mutation

Meas./PM	0	0.001	0.02	0.05	0	0.001	0.02	0.05	0	0.001	0.02	0.05
Return (%)	76.9	85.7	76.4	99.8	65.6	75.1	62.1	86.8	68.3	74.7	60.8	82
Sharpe R	0.13	0.1	0.15	0.16	0.13	0.13	0.14	0.16	0.13	0.13	0.13	0.14
Trades N°	1	3	3	5	9	1	5	10	7	1	4	3
IEP	1.116	1.126	1.169	1.135	0.949	0.95	0.958	0.936	0.942	1.076	1.077	0.979
Accuracy	51.5	32.9	37.66	54.98	41.2	45.92	40.77	42.06	32.38	32.9	32.9	32.4
Data Sets	2000-2004				1998-2004				1997-2004			

We have not found a robust relationship between the memory length and PM>0. Although, the memory length in simulations without mutation was on average 2.5 times shorter than in experiments with mutation. The relationship between PM and common statistical measures was inconclusive at acceptable significance or robustness.

GA model discovery reveals that MLP and TLRN with Laguarre memory, neurons number in the hidden layer in the range [5, 12] and Conjugate Gradient learning rule generate the best performance in statistical and economic terms for forecasting and acting nets. Generally models discovered with GA have lower trading frequencies, but without reduction in riskness. Annualized returns of those models were improved moderately. The effect of GA discovery on models' statistical performance was not conclusive, with a weak tendency towards accuracy amelioration. An increase in population size for GA optimization didn't lead to improvement in results.

The relationship between statistical measures (accuracy, correlation, IEP) and trading strategies' profitability seems to be of a complicated nature. Among the ten statistically sound price forecasts, there is only one that was used in a trading strategy superior to B/H benchmark. The best five in economic terms strategies are among the worst 50% according to their accuracy. Three of the most accurate strategies are among the worst 25% in terms of their annualized return. Correlation of desired and ANN output characterizes one of the first five strategies with highest return among its best performers, another one among its worst results and the remaining are in the middle. IEP shows some robust relationships with annualized return. All five strategies with highest return have IEP<0.9. Furthermore, one of the first five profitable strategies has one of the three best IEP values. Therefore, if profits are not observable, IEP could be used as an evaluation criterion for an economic prediction.

Regarding the performance surface optimization, two out of the three best strategies included an adjustment to treat directional information as more important than the raw error. We found that training ANN with the performance surface genetically optimized for DA, discounting least recent values or minimizing number of large errors generally improves profitability. Among 25% of the weak (in economic terms) strategies' annualized returns, there is none with learning criteria optimized. Our experiment has shown that among three optimizations of the performance surface considered, strategies trained on learning the sign of the desired output were generally superior to those trained to reduce the number of large errors or focusing learning on recent values. At the same time, the impact of optimization for DA on common statistical measures was insignificant, conforming that DA only weekly relates to conventional statistical criteria.

Our simulation supports a claim that DA relates to forecast profits more than mean squared/absolute errors criteria. The experiment rejects an assertion that all other summary statistics are not related to forecast profit, as was demonstrated by the IEP relationship with profitability. As the results show that DA does not guarantee profitability of trading strategies trained with this criterion, it might be ineffective to base empirical estimates of economic relationships on that measure. If conventional least squares are to be considered inadequate, an alternative estimation technique for economic behavior might use a combination of measures, demonstrated to have certain relationships with profitability; IEP and DA have been identified so far.

7 Conclusion

The performance surface set-up is a crucial factor in search of a profitable prediction with an evolutionary model. Measures of trading strategies' predictive power might significantly differ from criteria leading to its profit maximization.

GA post-processing has generally improved statistical characteristics. Novel concepts' arrival, determined by PM, benefits the system in economic terms, but is inconclusive statistically.

Models discovered with GA have moderately higher profitability, but the impact on their statistical characteristics was inconclusive. GA optimization of performance surface (particularly for DA) has a positive effect on strategies' profitability, though with little impact on their statistical characteristics. Since DA does not guarantee profitability of trading strategies trained with this criterion, it might be ineffective to base empirical estimates of economic relationships only on that measure.

When profits are not observable, IEP is proposed as an evaluation criterion for an economic prediction, due to its robust relationships with returns. If conventional least squares are to be considered inadequate, an alternative estimation technique for economic behavior might use a combination of measures, demonstrated to have certain relationships with profitability; IEP and DA have been identified so far.

References

1. Leitch, G.,Tanner, E.: Economic Forecast Evaluation: Profits Versus the Conventional Error Measures. American Economic Review 81 (1991) 580-590
2. Bhattacharyya, S., Mehta, K.: Evolutionary Induction of Trading Models. In: Chen S.-H. (ed.): Evolutionary Computation in Economics and Finance. Studies in Fuzziness and Soft Computing. Physica-Verlag, (2002) 311-331.
3. Sweeney, R. J.: Some Filter Rule Tests: Methods and Results. J of Financial and Quantitative Analysis. 23 (1988) 285-301

Mining Data by Query-Based Error-Propagation

Liang-Bin Lai, Ray-I Chang[*], and Jen-Shaing Kouh

Department of Engineering Science and Ocean Engineering,
National Taiwan University,
Taipei, Taiwan, ROC
{d93525009, rayichang, kouhjsh}@ntu.edu.tw

Abstract. Neural networks have advantages of the high tolerance to noisy data as well as the ability to classify patterns having not been trained. While being applied in data mining, the time required to induce models from large data sets are one of the most important considerations. In this paper, we introduce a query-based learning scheme to improve neural networks' performance in data mining. Results show that the proposed algorithm can significantly reduce the training set cardinality. Additionally, the quality of training results can be also ensured. Our future work is to apply this concept to other data mining schemes and applications.

1 Introduction

In this paper, we introduce a query-based learning scheme to back-propagation neural networks for data mining. Neural networks have advantages of the high tolerance to noisy data as well as the ability to classify patterns having not been trained [1-2]. They have been applied to a wide variety of problem domains to learn models that are able to perform different interesting tasks [4]. While being applied in data mining, the time required to induce models from large data sets is the most important consideration [5]. Based on our previous paper [3], a query-based back-propagation neural network is introduced for data mining in this paper. Query-based learning is a methodology that requires asking a partially trained neural network to respond to the questions [7]. Usually, only the selective-attention is applied to respond system's goal-directed behavior with self-focus [8]. In some applications, there is no supervisor to verify the self-focus. We need a compromise is then made to environment-focus with self-regulation [3]. Namely Confucius say "To teach students in accordance with their aptitude" [3]. Experiments show that the classification performance can be significantly improved while the quality of training data can be also ensured. The paper is organized as follows. Section 2 presents the data mining problem and introduces the multi-layer perceptron and back-propagation learning. Section 3 introduces query-based learning method that combines into one composite back-propagation neural networks. Section 4 presents the experiments and the results. Section 5 shows the conclusion and future perspectives.

---

[*] This paper is partially supported by NSC under grants NSC93-2213-E-002-086-.

2 Related Works

Data mining is an interdisciplinary field with a general goal of predicting outcomes and uncovering relationships in data. It uses automated tools employing sophisticated algorithms to discover hidden patterns, associations, anomalies and/or structure from large amounts of data stored in data warehouses or other information repositories. Classification of data is one of the most important tasks in data mining. It first builds a model to describe a predetermined set of data classes or concepts [9]. Then, the model is used for classification.

A neural network with multi-layer perceptron (MLP) and back-propagation learning is one of the best known techniques for data classification. MLP is constituted of a set of interconnected neurons organized in layers. The most often employed architecture consists of three layers: an input layer, one hidden layer and an output layer. The output of a neuron as a function of the input signals can thus be written in the following equations.

$$u_i(\ell+1) = \sum_{j=1}^{N_l} W_{ij}(l+1)a_j(l) + \theta_i(l+1) = \sum_{j=0}^{N_l} W_{ij}(l+1)a_j(l) \text{ and } a_i(l+1) = f(u_i(l+1)) \quad (1)$$

θ_i is the bias term and f is the activity function. $a_i(l+1)$ is the output of the generic neuron belonging to layer $(l+1)$ and w_{ij} is the synaptic weight associated with the connection between the generic neurons belonging to layers i and j. The most widely used weight updating algorithm for MLP is the back-propagation learning algorithm [10]. It consists of the repeated application of the rule for computing the influence of each weight in the network with respect to an arbitrary error function E.

$$E = \sum_{i=1}^{N}(t_i - a_i(L))^2 \text{ and } \frac{\partial E}{\partial w_{ij}} = -(t_i - a_i(L))a_j \text{ and } w_{ij} = w_{ij} + \eta(t_i - a_i(L))a_j \quad (2)$$

3 Proposed Method

A learning machine consists of a learning protocol and a deduction procedure [11]. The protocol accumulates information, which is used by the deduction procedure to assimilate the concept. In this paper, the learning protocols allow two kinds of information supply: examples and oracle. When presented with data, oracle tells the learner whether or not the data positively exemplify the concept. In practice, the source of the training information could be modeled as an oracle. This source could be very expensive, such as samples that are simulated by a supercomputer or information that could be gained only by destroying samples. According to Oates [12], when a machine learning algorithm is to learn, what are usually needed are some particular samples for training. With these samples, the algorithm could learn almost completely what it is taught. According to Baum [13], query-based learning is approximate to the way humans learn. It not only employs the samples for training that are presently at hand, but also uses the method of query to produce extra samples. Thus, the oracle should co-work with the query method that initiates training using a small input set, and then produces appropriate training information with minimum cost. Fig. 1 shows the structure of the query-based neural networks. Suppose that the sample for training is $\{x, a(x)\}$, where the input vector is x and $a(x)$ is the output vector. When part of the sample is modeled as an oracle, a query about the input value could modify the ideal value for output. And when the point of query is set as y, the oracle would respond with $a(y)$. Accordingly, $\{y, a(y)\}$ is called the queried sample.

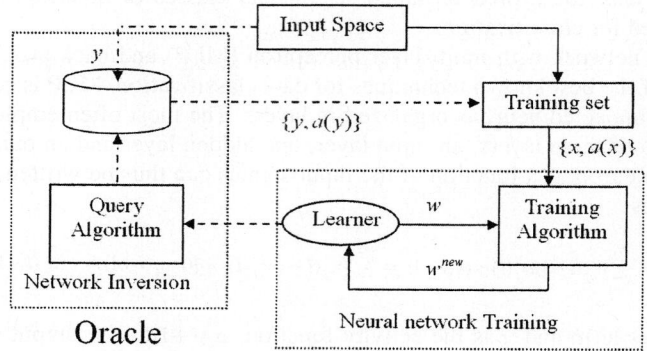

Fig. 1. Query-based Neural Networks Framework

3.1 Region of Maximum Ambiguity

When a neural networks is within a training process of back-propagation, the sore objective is to diminish the difference between the ideal output value and the real output value. In training neural networks, information from the decision boundary has been proved to produce best training results [15]. The so-called "inversion algorithm" would employ the weight in the back-propagation and the desired output value in the decision boundary to gain the output vector $\{a_j(0)\}$.

$$a_j(0) \Leftarrow a_j(0) - \eta \frac{\partial E}{\partial a_j(0)} = a_j(0) - \eta \delta_j(0)$$

Similarly, the inversion algorithm would propagate the error signal backward to the input layer to update the activation value of the input units and eventually to lower the rate at which an error happens to the output value. The activation value of the input segment is updated as follows.

$$u_j(0) \Leftarrow u_j(0) - \eta \frac{\partial E}{\partial u_j(0)} = u_j(0) - \eta \frac{\partial E}{\partial a_j(0)} \frac{\partial a_j(0)}{\partial u_j(0)} = u_j(0) - \eta \delta_j(0) \frac{\partial a_j(0)}{\partial u_j(0)}$$

In each search, gradient descent algorithm would slide along the gradient of the function to search t the smallest error function E. To certain weights w_{ij}, when the lowest point upon the boundary of $E(w_{ij})$ is being searched for, the path of the search is presented as:

$$w_{ij}(l+1) = w_{ij}(l) + \Delta w_{ij}(l) \text{ and } \Delta w_{ij}(l) = -\eta \frac{\partial E(l)}{\partial w_{ij}(l)}$$

This means that the search begins from $w_{ij}(l)$ to where the gradient Δw_{ij} lead, the purpose of which is to make $E(l+1)$ lower than $E(l)$; the search goes on until the value of E does not change anymore.

Suppose that this is applied to classification into more than one group. Each group is represented by a particular output neuron. In the condition that the other output neurons do not influence each other, gradient descent algorithm would search for what could make the activation value of the kth output neuron as 0.5; in other words, it would search among the input pieces of information for the points whose output value is 0.5. The energy function or the error function, which, in the realm of the kth output neuron, is between the target value and the actual value, is represented as:

$$E = \frac{1}{2}(t_k - a_k(L))^2$$

3.1 Query with Selective-Attention

To construct the prediction mode is to draw the boundaries between the groups for the training samples in the field of application; each group would therefore be distinguished from each other. And after the training, the neural network would use the weight value to build up the parametric representation between the input and the output values.

As for each information point in the input spaces, we could calculate the gradient ρ_{kj} of the input neuron that is related to each output neuron. To acquire the modified information at the mixed boundaries, according to Hwang [6], is to employ the inverted boundary points to calculate the gradient and eventually to acquire the conjugate pair. If the input vector is $a_j = (a_{j1}, a_{j2}, a_{j3}, ..., a_{jn})$ and the output vector at the opposite side is $a_k(L)$, then the gradient would be

$$\rho_{kj} = \sqrt{\sum_{j=1}^{n}(\frac{\partial a_k(L)}{\partial a_j})^2}$$

For example, if the input vector is $a_j(0)$, the gradient would be:

$$\rho_{kj}(0) = \frac{\partial a_k(L)}{\partial a_j(0)} \text{ or } |grantient| = \sqrt{\sum_j \rho_{kj}(0)^2}$$

Thus the calculation of the gradient is as follows.

$$\frac{\partial a_k(L)}{\partial a_j} = \frac{\partial a_k(L)}{\partial u_i}\frac{\partial u_i}{\partial a_j} = a_k(L)(1-a_k(L))\sum_{j=1}^{n} w_{oj}\frac{\partial u_i}{\partial a_j}$$

where n means the quantity of neurons at the hidden layer and w_{oj} means the weights between the jth neuron at the hidden level and the oth neurons at the output layer. And

$$u_i = f(\sum_{j=1}^{n} w_{ij}a_j + \theta_i)$$

Where $f(\cdot)$ means the sigmoid function $f(x) = \frac{1}{1+e^{-x}}$. The derivation of u_i with respect to a_j is formulated as

$$\frac{\partial u_i}{\partial a_j} = u_i(1-u_i)w_{ij}$$

Hence, bring (14) into (15).

$$\frac{\partial a_k(L)}{\partial a_j} = a_k(L)(1-a_k(L))\sum_{j=1}^{n}u_i(1-u_i)w_{oj}w_{ij}$$

For each inverted boundary point, a conjugate training data pair { $p+$, $p-$ } based on the magnitude of gradient can be created. The one point is at the one side and the other point is at the other side of the boundary. The two points lying on opposite sides of the line passing through the boundary point and perpendicular to the boundary surface are located with their distance to the corresponding boundary point equal to $1/|gradient|$. The conjugate training data pair is extracted along the reverse boundary: the inverted boundary point p would be used to calculate the reciprocal of the magnitude of gradient, and then along the magnitude of gradient we would pick up two information points in symmetry, that is, p^+ and $p-$. Then we would use the oracle to decide the groups which p^+ and $p-$ belong to. If they belong to the same group, we would ignore the reverse boundary point, and if they belong to two different groups, we would include them into the training information.

3.3 Query with Self-regulation

However, as for the datasets of the problem domain of data mining, the method in which the reverse boundary point is used to obtain the conjugate training data pair is inappropriate. The reason is that the information point obtained via the reverse boundary point does not necessarily exist in the information set of the problem domain and therefore we might fail to obtain information.

Accordingly, in the new method proposed by this research, all of the training information would go through the examination of the oracle, and we would detect the information points that have been put in wrong classes. Further, after the distance between the information point and the boundary is calculated, the information would be saved in the priority queue sorted by min-heap. As Linden [14] and Reed et al. [15] suggest, in the training samples of the neural network, the input data of the noise/jitter or the outlier would not influence the decision boundary produced by the network. Therefore, our research set the quantity of samples in the priority queue as ten percent of the initial training samples. And we employ the sorting method of min-heap to contain only the samples that are spatially the most close to the boundary.

To take Fig. 2 as an example, ■ and ♦ mean respectively class A and class B of the training samples, and the black line means the boundary after the finish training. But in the data sets in the problem domain, different sampling methods would ignore some samples; the actual decision boundary could be wrongly-drawn. In Fig. 2, ○ and ● mean respectively class A and class B of the un-sampled training data, and the pieces of information that are put into wrong class. Therefore, in the new method proposed by this research, all the training samples would be examined by the oracle, in order to detect the samples that are put in wrong class. Further, we would calculate distance between the boundary to each sample—that is, $Dis()$—and then we would pick the samples that are near the boundary, that are spatially close to the boundary.

These samples would be put into the priority queue to become the information to improve the training. That the prediction mode has not learned enough about the sample in question, and that the information should be included in the training information in order to have a better learning. The boundary of the prediction mode would be modified over and over. And when the difference between the square root of mean squared error (RMSE) of the previous prediction and the present calculation is lower than RMSE, it means that the prediction mode has finished the learning process.

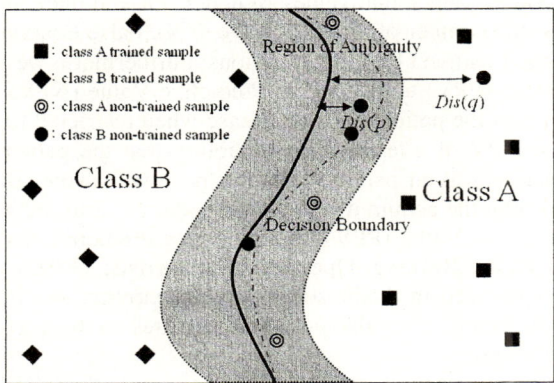

Fig. 2. An example of our query results

3.4 Proposed Query-Based Learning Algorithm

The step-by-step description of the proposed query-based learning algorithm is shown as follows [3].

Step 1: Initialize all weights randomly.
Step 2: Train the neural network using a subset of the training samples $\{a_i\} \in A$,
    where A denotes the training sets, $A = \{a_i \in R^n\}$.
    IF $E <$ RMSE or *Epoch* $>$ Training Threshold
    THEN EXIT.
Step 3: Using Oracle verify A to judge classification whether accuracy.
Step 4: Insert the mistake samples to Priority Queue which sorted by min-heap.
Step 5: Take samples in queue to add training sample.
Step 6: Retaining the network
    IF $E <$ RMSE or *Epoch* $>$ Training Threshold
    THEN EXIT.
    ELSE GOTO Step 2.

4 Experiment Results

The ultimate goal of a data mining process should not be just to produce a model for a problem at hand, but to provide one that is sufficiently credible and accepted and implemented by the decision maker [9]. In order to verify the feasibility and effectiveness of the proposed query-based learning procedure, this research uses database in UCI Machine Learning Repository [16]—Adult Database —to be the test data. We would use the data mining modes constructed by Query-Based Learning Neural Networks (QBL) and Back-propagation Neural Networks (BPN) to examine the correctness and reliability of what the mode predicts.

A common tool for classification analysis is the confusion matrix [17], a matrix of size $L \times L$, where L denotes the number of possible classes. This matrix is created by matching the predicted and actual values. When $L = 2$ and there are four possibilities, as Table I shows: the number of True Negative (TN), False Positive (FP), False Negative (FN) and True Positive (TP) classifications. Furthermore, we applied to the positive predictive value (PPV) and Negative Predictive Value (NPV). The PPV of a test is the probability that the patient has the disease when restricted to those patients who test positive. The NPV of a test is the probability that the patient will not have the disease when restricted to all patients who test negative. Eventually, we employ four criterions to deal with the acquired diagnostic mode; the four are sensitivity, specificity, Positive Predictive Value (PPV), and Negative Predictive Value (NPV).In addition we applied to the Receiver-Operator Characteristic (ROC) analysis that is an evaluation technique used in diagnostic, machine learning, and information-retrieval systems [18]. ROC graphs plot false-positive (FP) rates on the x-axis and true-positive (TP) rates on the y-axis.

In our research, we use the rotation method to sample the training data; that is, the rotation method would train and examine group by group the pieces of training data in the datasets. As for the parameters about the structure of the neural network, the research considers primarily the quantity of the hidden layer, learning rate, and mean squared error. As Rumelhart [10] concluded that lower learning rates tended to give the best network results. The convergence criteria used for training are a root mean squared error (RMSE) less than or equal to 0.01 or a maximum of 10000 iterations. The network topology with the minimum testing RMSE is considered as the optimal network topology. The dataset is extracted from the UCI Adult database that aims to predict whether a person makes over 50k dollars a year. The dataset consists of 48842 instances and 14 attributes. Each record in the dataset contains eight predictor variables, namely, work class, education, marital-status, occupation, relationship, race, sex and native-country. And the response variable is the income status it belongs to, either ">50K" or "<=50K".

The structural design of the network is a multi-layered neural network, as Table II shows. BPN training way to is it train sample is it learn to go on to utilize , and QBL training way in initial to choose 20% do for sample of training at random among training sample when learning, learn , in the course of learning, after inquiring Oracle, and then obtain the new training sample from Priority Queue, strengthen learn.

BPN and QBL are after training the learn cycle of 10000 epoch, the Convergence of network training is as Fig. 3 shows, and Table II states the error rate of carrying out the speed and predicting, among them QBL needs to finish learning promptly in 16344.94 seconds, and verify the results of learning by testing the samples, the error rate of QBL is lower than BPN.

Table 2. The structural design of the network is a multi-layered neural network

		QBL	BPN
Network Model			
	Input-hidden-output	14-28-1	14-28-1
	Learning rate	0.45	0.45
	Momentum	0.9	0.9
	RMSE	0.05	0.05
	Learning epoch	10000	10000
Average Performance			
	Execute time(sec)	16344.94	81090.87
	Accuracy (%)	81.85%	80.70%
	ROC Curve (AUC)	0.756	0.727
Model Estimation			
	Sensitivity	0.872	0.870
	Specificity	0.644	0.61
	PPV	0.888	0.874
	NPV	0.609	0.601

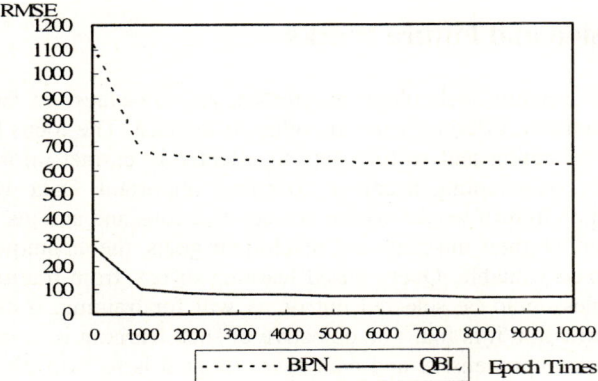

Fig. 3. Convergence behavior for the Census Income

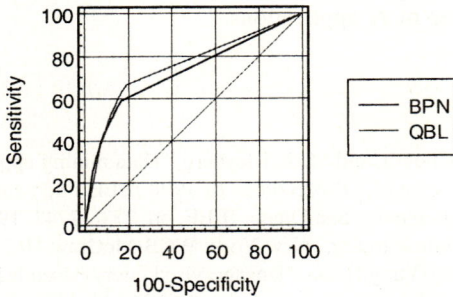

Fig. 4. Comparison of ROC Curves for the Census Income using QBL and BPN

Assessing ROC curve in addition, as Fig. 4 shows, QBL is also better than BPN result. According to confusion matrix show, as Table III , in is it construct Census Income predict model one kind of neural networks, prediction of way while being above-mentioned, under the equally limited number of times of training, QBL is more excellent than BPN.

Table 3. The Census Income confusion matrix

			Census Income	
QBL			Predicted	
	Actual		Positive	Positive
		Positive	10848	1368
		Negative	1587	2478
BPN			Predicted	
	Actual		Positive	Positive
		Positive	10722	1540
		Negative	1601	2418

5 Conclusion and Future Works

Due to the information technology progresses, we nowadays are facing an environment of competition different from any other of the past. The focus has been transfer from merely collecting and sorting data to effectively extract information from the database. Thus, data mining becomes extremely important. Data would become information, which in turn would inspire correct decisions and actions. And when these actions accomplish their missions and reach their goals, the technique of data-mining would be proved valuable. Query-based learning differs from traditional methods in that the samples could be selected out of its will for training, instead of accepting whatever information it is fed. The objective of such learning is to produce a training sample that is comprehensive and educative. Its goal is to help the learner comprehending whatever has appeared to be difficult. When dealing with data-mining, we could combine the techniques of neural network and query-based learning; hence, the neural network could gain the most effective classification with the least training cost. Our future work is to apply this concept for association rule and sequential pattern mining so as to develop more applications.

References

1. D. Ben-Arieh, M. Chopra, and M. Z. Bleyberg, "Data mining application for real-time distributed shop floor control," Proceedings of 1998 IEEE International Conference on Systems, Man, and Cybernetics, San Diego: IEEE, pp. 2738-2743, 1998.
2. T. Mitchell, "Machine learning," New York: WCB/McGraw-Hill, 1997.
3. Ray-I Chang and Pei-Yung Hsiao, "Unsupervised query-based learning of neural networks using selective-attention and self-regulation," IEEE Trans. Neural Networks, vol.8, no.2, pp.205-217, 1997.
4. D. Ridley, and F. Llaugel, "Moving-window spectral neural-network feedforward process control," IEEE Transactions on Engineering Management, 47(3), 393-402, 2000.

5. M.W. Craven, and J. W. Shavlik, "Using neural networks for data mining," Future generation computer systems, 13(2-3) 211-229, 1997.
6. Hongjun Lu, R. Setiono, and Huan Liu, "Effective data mining using neural networks," Knowledge and Data Engineering, IEEE Transactions on, Volume: 8 Issue: 6, pp.957 - 961, 1996.
7. J. Hwang, J. Choi, S. Oh, R. Marks II, "Query-based learning applied to partially trained multilayer perceptrons," IEEE Transactions on Neural Networks, Vol. 2, pp.131-136, 1991.
8. Ray-I Chang, Pei-Yung Hsiao, "VLSI circuit placement with rectilinear modules using three-layer force directed self-organizing maps," IEEE Trans. Neural Networks, vol.8, no.5, pp. 1049-1064, September 1997.
9. J. W. Han and M. Kamber, "Data Mining: Concepts and Techniques," San Francisco: Morgan Kaufmann Publishers, 2001.
10. D. E. Rumelhart, G. E. Hinton, and R. J. Williams, "Learning internal Representations by Error Propagation," In Paralleled Distributed Processing, Vol. 1, pp. 318-362, Cambridge, MA: MIT Press, 1986.
11. L. Valiant, "A theory of the learnable," Communications of ACM, 27, pp.1134-1142, 1984.
12. T. Oates, and D. Jensen, "The Effects of Training Set Size on Decision Tree Complexity," In Proceedings of The Fourteenth International Conference on Machine Learning, pp.254-262, 1997.
13. E. B. Baum, "Neural-net algorithms that learn in polynomial time from examples and queries," IEEE Transactions on Neural Networks, Vol. 1, pp.5-19, 1991.
14. A. Linden, and J. Kindermann, "Inversion of multilayer nets," IJCNN, International Joint Conference on Neural Networks, vol.2, pp.425-430, 1989.
15. R. Reed, R. J. Marks II and S. Oh, "Similarities of error regularization, sigmoid gain scaling, target smoothing, and training with jitter," IEEE Transactions on Neural Networks, Volume: 6 Issue: 3, pp.529-538, 1995.
16. C. Blake, E. Keogh, C. J. Merz, UCI Repository of machine learning databases, http://www.ics.uci.edu/~mlearn/MLRepository.html, Irvine, University of California, Department of Information and Computer Science, 1998.
17. R. Kohavi and F. Provost, "Glossary of Terms. Machine Learning," 30(2/3):271-274, 1998.
18. J. M. DeLeo, and S. J. Rosenfeld, "Essential roles for receiver operating characteristic (ROC) methodology in classifier neural network applications," in Proc. Int. Joint Conf. Neural Networks, Vol.4, pp. 2730-2731, 2001.

The Application of Structured Feedforward Neural Networks to the Modelling of the Daily Series of Currency in Circulation

Marek Hlaváček[1,2], Josef Čada[3], and František Hakl[4]

[1] PPF a.s., Prague, Czech Republic
[2] Czech Technical University, Faculty of Nuclear Science and Physical Engineering, Prague, Czech Republic
[3] Czech National Bank, Prague, Czech Republic
[4] Czech Academy of Science, Institute of Computer Science, Prague, Czech Republic

Abstract. One of the most significant factors influencing the liquidity of financial markets is the amount of currency in circulation. Even the central bank is responsible for the distribution of the currency it could not assess the demand for the currency as it is influenced by the non-banking sector. Therefore the amount of currency in circulation have to be forecasted. This paper introduces feedforward structured neural network model and discusses its applicability to the forecasting of the currency in circulation. The forecasting performance of the new neural network model is compared with an ARIMA model. The results indicates that the performance of the neural network model is slightly better and that both models might be applied at least as supportive tools for the liquidity forecasting.

1 Introduction

Central banks pursue their statutory objectives in maintaining the price stability or foreing exchange rates stability through different sets of monetary policy instruments. Nowadays, central banks mainly control the economic conditions indirectly. They usualy steer money market interest rates through proper liquidity managment. Therefore an accurate estimate of the money market liquidity is essential for the effective monetary policy implementation.

Although only transactions with the central bank have impact on money market liquidity, some of them are out of central bank control. These factors are called autonomous factors. One of the most important autonomous factors is the currency in circulation (CIC). The demand for CIC is influenced by non-banking sector which means it is rather volatile and depends on various seasonal factors. The influence of seasonal factors make the assessment of the demand for CIC very knotty. For that reason central banks employ various mathematical models to deal with this typical seasonal time series.

For example the Federal Reserve System, the European Central Bank and other central banks within the European Monetary Union already use mathematical models of CIC at least as supportive tools. The most of the used models is based on the principles of Box-Jenkins methodology [4] and on its further improvements (e.g [10], [9]). However, recently banks also develop new non-linear models that are supposed to approximate seasonal patterns with higher accuracy then linear models.

This article follows the idea of non-linear model applications and investigates the applicability of a special neural network model called *structured feedforward neural network*. This neural network model is derived from networks with switching units originally developed as a dataclassifiers. However experiments with time series forecasting showed the model might be sufficiently applied in this area too. Moreover the model is based on the combination of linear regression and cluster analysis and hence the analysis of the model is less complicated than the analysis of a multi-layer perceptron or other neural networks models. This fact is also important as the application of neural networks usualy raises many doubts as they are too complicated and dealing with them is like dealing with black-boxes. Therefore structured feedforward neural networks seems to offer an attractive option. This work was partly supported by the Ministry of Education of the Czech Republic under project No. 1M684077004 - Center of Applied Cybernetics.

This work was partly supported by the Ministry of Education of the Czech Republic under project No. 1M684077004 - Center of Applied Cybernetics.

2 Currency in Circulation

For the purposes of this paper currency in circulation is defined as banknotes and coins hold outside the central bank and the series of CIC in Czech Republic from the range between January 1996 and June 2004 is considered (see fig. 1). The selection of Czech data is not very restrictive as the behaviour of CIC in Czech is similar to other countries like the USA, the European Union countries and many others.

The currency is distributed into the system mainly through commercial banks. They tries to follow clients requirments as flexibly as possible not to spend money for needles cash. Hence the demand for cash is mainly influenced

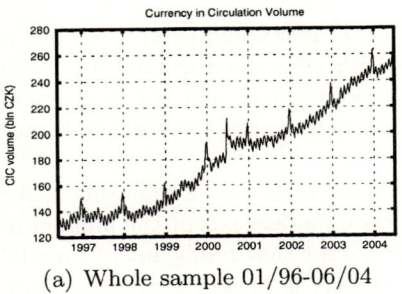
(a) Whole sample 01/96-06/04

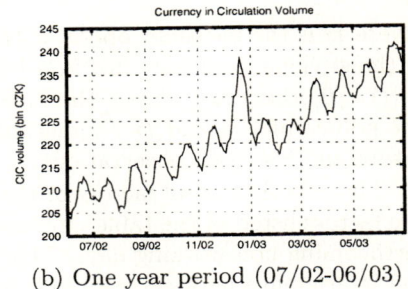
(b) One year period (07/02-06/03)

Fig. 1. Czech currency in circulation - daily volumes

by the non-banking sector including commercial subjects and households as well. It means that the changes of the CIC volume are caused by enourmous number of factors and circumstances that are obviously uncontrollable. Thus the CIC volume is supposed to be a random variable following a compound process with seasonal and stochastic components.

Both models described and compared later on express the series of the CIC volume as a function of historical values, seasonal and shock factors.

The identification of all significant seasonal patterns and shocks and the choice of the correct form for their representation proved to be cruicial for the models forecasting performance. The choice of seasonal factors was predominantly motivated by [6]. However experience of experts who are responsible for the liquidity forecasting in the Czech National Bank was also considered.

The exogenous factors are involved in the models in two different forms following the idea presented in [3]. The first type of seasonal factor is a superposition of goniometric functions and the second is a lagged polynomial of indicator function.

The only factor modeled as a superposition of the goniometric functions is the intramontly effect. The factor has the form:

$$d_t = \sum_{j=1}^{p} \left(a_j \sin\left(\frac{2j\pi m_t}{M_t}\right) + b_j \cos\left(\frac{2j\pi m_t}{M_t}\right) \right), \qquad (1)$$

where d_t is the value of the factor in time t, M_t is the length of the current month and m_t is the position of the day in the current month. Finally, the positive number p sets up the number of different frequences forming the factor. Naturaly, the more frequencies is considered the better the approximation is. However, its also necessary to keep the number of model parameters low hence the number of frequences p have to be chosen carefully. In this paper p is set to eight for both models.

The second group of seasonal factors is mostly applied to model isolated events like national holidays or shocks and is of the form

$$d_{t,i} = \Gamma_i(\boldsymbol{B}) \boldsymbol{B}^{-F_i} \tau_i(t), \qquad (2)$$

where $\boldsymbol{B}$ is the backshift operator ($\boldsymbol{B} y_t = y_{t-1}$), Γ_i is a polynomial in $\boldsymbol{B}$ and τ_i is the seasonal indicator function ($\tau_i(t) = 1$ if ith season occurs at time t and $\tau_i(t) = 0$ otherwise). Finaly the F_i is a positive power of $\boldsymbol{B}$. The combination of the polynomial Γ_i and $\boldsymbol{B}^{-F_i}$ guarantees that particular seasons might influence future and also past observations.

The overview of all factors of the form (2) is summarized in the table 2. All the factors listed in the table were involved in both models except the number of forthcoming non-working days. This additional factor that might help to model the interference between weekends and fixed holiday is included only in neural network model as it was assessed as usignificant factor in case of ARIMA model.

Table 1. Seasonal factors and shocks represented by lagged indicator function

Seasonal factor	Order of Γ	Power F
Monday, Thuesday, Wednesday, Thuersday	3	0
Easter (Good Friday)	10	5
Christmas	15	5
New Year	5	5
Other National Holiday	10	5
Number of forthcoming non-working days	10	5
Bank failure (June 2000)	15	5
Y2K effect (New Year 2000)	10	5

3 Seasonal ARIMA Model

One of the most common classes of seasonal time series models is based on the methodology proposed by Box and Jenkins in [4]. Various generalisations of Box-Jenkins ARMA models are widely approved and applied. The applicability of ARIMA models on CIC forecasting is studied by an ECB research group in [6] where the model is precisely described and all necessary tasks summarized. Therefore the same methodology as that described in [6] was applied for the construction of ARIMA model for Czech data.

The ECB research group follows the idea of Bell and Hillmer ([3]) who suggested to use a linear regression model with ARIMA errors for modeling time series with calendar variations. The model might be witten in the following form:

$$y_t = D_{t,i} + \frac{\Theta(B)}{\Phi(B)\Delta(B)} \epsilon_t. \qquad (3)$$

Here y_t is the modeled series, $D_{t,i}$ is the regression part, B is the backshift operator and Θ, Φ, Δ are polynomials in B. The polynomials Θ and Φ are moving-averages and autoregressive operators, respectively. The polynomial Δ is a difference operator that might also include a seasonal difference operator. The regression part $D_{t,i} = \sum_{i=1}^{s} d_{t,i}$ is the superposition of all seasonal factors $d_{t,i}$ involved in the model as described in the section 2.

The formula (3) defines a quite general model that might still be optimised by least square estimator as shown by Pierce [19]. However, to identify the model it is neccesary to identify the orders of ARIMA process as well as the appropriate seasonal factors and their lags. The model described in this section was identified using the two-step approach proposed by Koreisha and Pukkila in [18].

First, sample autocorrelation and partial autocorrelation functions (SACF, SPACF) were computed and investigated to specify the difference operator Δ.

The correlograms of CIC volumes and CIC daily changes indicated Δ in the form:

$$\Delta = (I - B)\left(I - B^{252}\right)$$

where $\left(I - B^{252}\right)$ is the first order seasonal difference operator that corresponds to the one-year difference.

Appart from the seasonal differencing number of seasonal factors were involved into the regression part of the model to deal with the calendar variations. The included factors were selected from the set of seasonal factors summarized in the section 2. However only factors that were not refused by a significancy test remained in the final model.

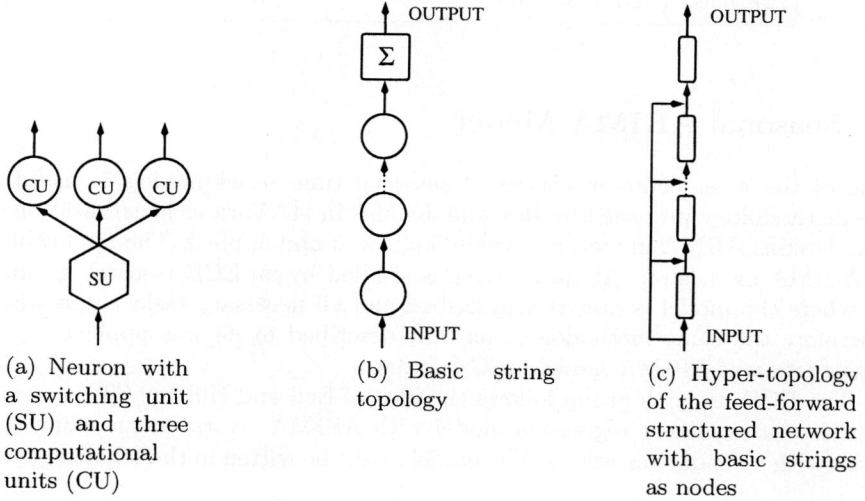

(a) Neuron with a switching unit (SU) and three computational units (CU)

(b) Basic string topology

(c) Hyper-topology of the feed-forward structured network with basic strings as nodes

Fig. 2. Neural Network with Switching Units

The ARMA structure of the stochastic component was investigated simultaneously with the identification of appropriate set of seasonal factors. The lags of MA and AR process was chosen with respect to the SACF and SPACFand results of significancy tests.

The final ARIMA model is described by 71 parameters. The correlogram in the figure 6 shows that a tiny correlation in residuals still remained. Also the Ljung-Box test refer to higher correlation particularly around the lag 60 and 252, however the addition of appropriate lags do not improve the forecasting performance of the model. Jargue-Bera test disproved the normality of standardized residual and verified the difference of $N(0, 1)$ density and the residuals histogram (see figure 3).

4 Neural Network Model

Since the ARIMA model is linear in seasonal effects it might not sufficiently cover the calendar variation. Therefore the application of non-linear models is recently more and more investigated. Unfortunately, functions realized by typical neural networks like MLP or RBF networks are higher-order superpositons of non-linear functions with large number of parameters. Therefore the application of neural networks raises many doubts as they are too complicated and dealing with them is like dealing with black-boxes.

To avoid such doubts a concept of neural networks with switching units was chosen as an alternative approach. Switching units allow to use linear transfer functions and the model at whole is than a combination of two common stochastic methods - linear regression and cluster analysis. Such a model is relatively simple and might be further analysed using common stochastic tools.

The feed-forward structured neural network described in this section is derived from the neural network with switching units introduced in [5]. For the purposes of this paper these original model is refered as *Basic string*. The generalisation is based on the connection of more basic strings into a hyperstructure. In the following subsections the final neural network model is described step by step.

4.1 Neuron with Switching Unit

The main idea of neuron with switching unit is to control the flow of the very input. Any neuron with switching unit consists of one switching unit and several computational units as shown in the figure 2(a). Switching units only choose which computational unit will process the given input while the computational units apply a transfer function to the inputs.

The switching unit splits the input space into several disjoint clusters. The number of clusters is the same as the number of computational units and each cluster is associated with different computational unit. Inputs from a cluster are processed only by the associated computational unit. The clusters are found during the training process using cluster analysis methods. The character of clusters depends on the metric or pseudometric of the input space. This metric or pseudometric together with number of computational units defines the switching unit.

Computational units are the neurons in the common sense. They could be of any type like perceptron or RBF unit, but such general case might lead again to a black-box model. Therefore, only computational units with linear transfer functions are considered. The linear transfer function is defined by the formula:

$$(y_0, y_1, \ldots, y_n) = (\alpha_0, \alpha_1 x_1, \ldots, \alpha_n x_n) \tag{4}$$

where $y = (y_0, \ldots, y_n)$ is an output, $x\,(x_1, \ldots, x_n)$ is an input and parameter $\alpha = (\alpha_0, \ldots, \alpha_n)$ is estimated from the training set using the linear regression equation:

$$Y = X\alpha + \epsilon. \tag{5}$$

where Y is a column vector of a modeled quantity, X is a matrix of appropriate inputs, the column vector α is the estimated parameter and ϵ is the error or residuum. It implies the sum of the components of output y from (4) is the approximation or estimation of the correct value of the modeled quantity. Nevertheless the row vector y instead of the sum of its components is the output of the neuron as it could be further processed rather then a single number.

4.2 Basic String

The original model of neural network with switching units proposed in [5] is quite simple one. Its refered here as a basic string because of its string topology. The topology is defined by a graph which is a path exactly. It means that every neuron except the input and output one has one parent and one child. The input neuron is a neuron with switching unit with Euclidean norm and linear transfer function. The output neuron only sums its inputs and the rest are neurons with switching units with clusters defined by sum of input components and with linear transfer function. The sketch of string architecture is in the figure 2(b). The Euclidean norm is used in the input neuron because it reflects the original structure of network input space. The rest of neurons splits the inputs according to expected output as the sum of input components is the approximation of the modeled quantity.

Even these models are simple they are quite effective. From the theoretical point of view strings are for example capable to approximate any smooth or measurable function ([14]) or they might realize any AR process as well. The model was also applied on several problems with encouraging results ([12]). However, more complicated problems need more complicated models. Therefore more general models called feed-forward structured networks were defined as models with topology described by a hyper-structure with basic strings as its nodes (see 2(c)).

The advantage of the hyper-structure is that the usage of basic strings as nodes helps to define and control the topology. This is particularly important in the context of genetic optimisation (see [17]) but also when the topology is defined according to the results of derived models by an expert.

4.3 Feed-Forward Structured Neural Network for CIC Forecasting

The model used for CIC modeling is a simple feed-forward structured network with the hyper-structure described by a graph that is a path extended by additional connections between network input and each basic string as shown in the figure 2(c). The main role of additional connections is to keep original structure in analysed data because the input space structure might be heavily damaged by non-invertible transfer functions. Even this model could not still realize an MA process the additional connections allows to use more neurons effectively and hence better analysis of the input data is possible.

The number of strings in the very network, their length, and numbers of clusters in neurons might be chosen arbitrary and unfortunately no guideline how to choose the best combination is available. The neural network model described there is a product of iterative process based on analysis and comparison of derived models. First few randomly generated networks were analysed and according to their performances some restrictions were applied. The length of strings was limited up to five neurons and the number of clusters per neuron was restricted to be in the range between two and five. According to these restrictions ten networks were randomly generated and then particular basic strings from best four networks were combined in different order until the final topology described in the table 4.3 was chosen as the best one.

Table 2. Summary of final neural network model topology

string order	string length	number of clusters	string order	string length	number of clusters
1	1	4	6	3	3,3,1
2	4	2,2,2,2	7	2	4,3
3	1	4	8	3	2,2,2
4	4	2,3,2,2	9	3	3,2,2
5	1	3	10	2	2,2

All the inputs summarized in the table 2 together with lagged values of CIC comprise the input of the neural network model. The laggs included into the model were selected according to the SACF and SPACF functions (see fig.7). However, also the incapability of neural network model to realize MA process was considered and hence for example five laggs around one year instead of a single one are used to deal with strong one year correlation. Finally the following laggs were used: 1, 2, 5, 10, 15, 20, 21, 22, 250, 251, 252, 253 and 254.

The model was then applied to the series of daily changes and to its one year seasonal difference. The daily changes were forecasted with significantly higher accuracy then the seasonaly differenced series. The probable explanation is that the seasonally differenced series contains a prominent MA(252) component which might be hardly approximated by the neural network. On the other hand the non-linear neural network model might stabilize the non-stationary series of CIC daily changes as it approximates the seasonal character of the series with higher accuracy then the linear regression model.

Anyway the correlogram of residuals (see fig. 7) do not refer to a strong correlation of the residuals. As in the case of ARIMA model additional laggs do not significantly improve the model performance. Tests for the residuals normality contributed what the histogram (see fig. 4). The normality is also rejected even if the peaks around zero are removed as they are caused by clusters with only few observations that might be classified as outliers.

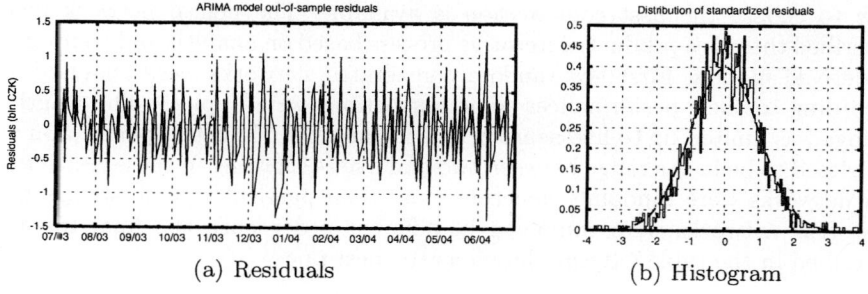

(a) Residuals (b) Histogram

Fig. 3. ARIMA out of sample residuals

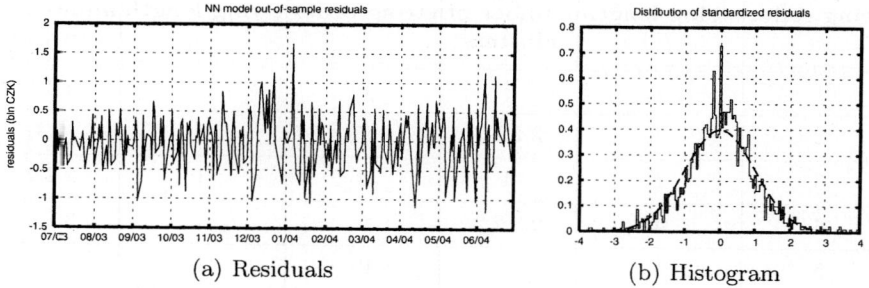

(a) Residuals (b) Histogram

Fig. 4. Neural Network out of sample residuals

5 Comparison

Two models the ARIMA one and the neural netwrok one were described in the previous sections 3 and 4. Appart from the model definitions and description of their functionality the results of the in-sample residuals analysis is also presented there. Moving the analysis along the out-of-sample forecasting performance and the general applicability of both models is discused and compared in this section. First the forecasting performance is compared and then the discussion of the applicability follows.

The sample used for the forecasting performance qualification is the one year period from July 2003 to June 2004 and is the same for both models. The comparison is focused on the one-step-ahead forecasts because the experimetns showed the forecasting horizon do not affect the relative models performance. The initial point for the analysis are the one-step-ahead residual plots (3 and 4 respectively) and the plot of differences between squared residulas of both models (5).

The figures 4 and 5 shows the neural network model miscalculates the forecasts awfully in few cases particularly around Chtristmas. Focusing on these unfitted events it was found they all fell into a cluster with only few observations. It means the application of the neural network is not serious for such

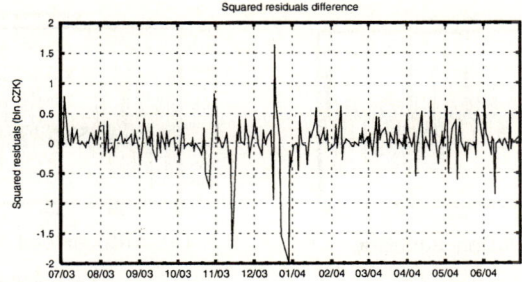

Fig. 5. Differences between the ARIMA and the neural network model squared residuals. The positive values indicates the NN model is more accurate.

Table 3. RMSE and Diebold-Mariano test results

horizon	ARIMA RMSE	NN RMSE	D-M p-value
1	0.491	0.454	0.975
5	0.476	0.442	—
10	0.484	0.448	—

events as the network is strongly overlearned with regard to these sparse events. However the table 5 shows the neural network is more accurate in the average for all considered horizons even the Diebold-Mariano test do not classify the difference to be significant (the test p-value is also reported in the table 5).

The neural network model particularly outperforms the ARIMA model in the beginnig of the testing sample where the neural network RMSE is really low. This might be viewed in the figure 8 that shows RMSE evaluated for the particular months. The figure 8 also indicates that the average error changes during the forecasted period in both models. The growth of the error in the end of the period is probably caused by the obsolescence of the models while the changes in the middle of the period corresponds to the Christmass season which effect might not be well approximated.

The interesting fact is also that the forecast error do not increase with the forecasting horizon. Contrary the RMSE of both models for the five days horizon is lower then that for one day horizon. This apparent paradox means the models could not approximate the intra-weekly effects with sufficient accuracy. The reason is probably that the intra-weekly effect changes a lot during the period.

Moving to the general point of view the interpretability of results is the next important atribute of any model. The regression component of the ARIMA model (3) allows to effectively analyse the approximation of seasonal influence. This could be done for the Neural network model as the influence of a given season might be modeled miscellanously by different computational units. On

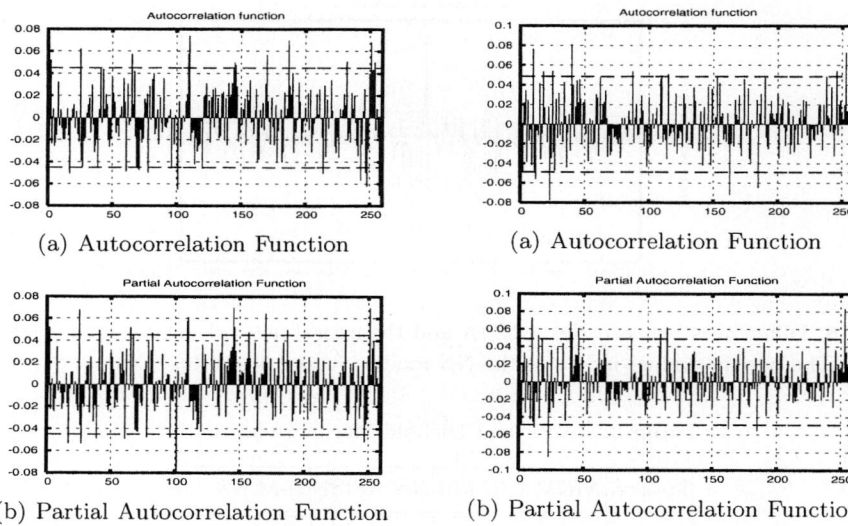

Fig. 6. Correlogram of ARIMA in sample residuals

Fig. 7. Correlogram of NN in sample residuals

the ather hand the proper analysis of neural network model is also possible and it might lead to more complicated findings. The problem is that there is plenty of possible paths for data processing in any neural network model hence the analysis based on the comparison of particular paths is really complicated and it could be hardly done manualy. Unfotunately a serious analysis tool that might analyse the neural network model still missing.

Contrary the advantage of the neural network application is that it might be easily reoptimalised even if the set of exogenous factor changes. In the case of the ARIMA model any change in the set of considered inputs means the model have to be rebuild completly. However the topology of a neural network model might be preserved and it is only necessary to learn the network again using

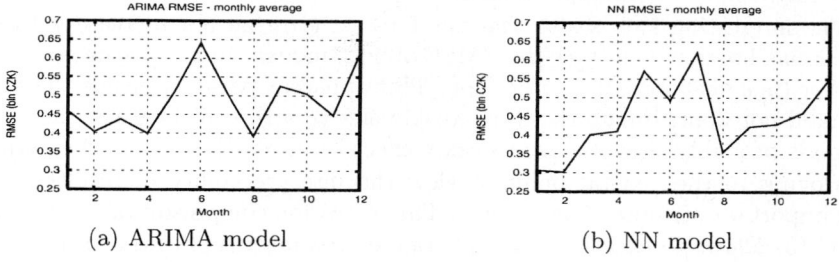

Fig. 8. RMSE in particular months

the new set of inputs. The learning process of a neural network is then a fully automized compact algorithm that might be run without any expert.

6 Conclusion

The paper introduces a new kind of neural network model for currency in circulation forecasting. The feed-forward structured neural network model that might be suitable for the analysis of arbitrary seasonal time series is compared with the conventional Box-Jenkins ARIMA model. Characteristic properties of both models were discussed with the emphasis on the out-of-sample forecasting performance.

The analysis of out-of-sample residuals showed the neural network model outperforms the ARIMA model in the average and particularly in the beginnig of the testing sample. However the Diebold-Mariano test do not classiffy the difference to be significant. On the other hand the neural network model strongly unfit few observations in the testing sample probably due to the model is overlearned in sparse observations. Anyway the neural network model is a competitive alternative to the Box-Jenkins ARIMA models and is worth improving.

Ragarding to the properties of structured neural network model few improvements that might improve the model are obvious. First the feedback would be included to deal with MA processes. Second the selection of relevant inputs might be improved through the application of stochastic tests. Next the more general architecture of network might be considered simultaneously with the usage of genetic algorithms for the architecture optimisation. Finally a tool for analysis of the network, data flows and other model properties might make the model more applicable.

Even these all extension would improve the model performance the application of the neural network model in the current stage or better its combination with the ARIMA model is relevant at least as a supportive tool for the liquidity forecasting.

References

1. Anderson, R.L.: The problems of Autocorrelation in Regression Analysis. Journal of the American Statistical Association, Vol. 49. (1954) 113-129
2. Balkin, S.D.: Using Recurrent Networks for Time Series Forecasting. Working Paper Series, No. 11. Pennsylvania State University (1997)
3. Bell, W.R.,Hillmer, S.C.: Modeling Time Series With Calendar Variation. Journal of the American Statistical Association, Vol. 78, (1983) 526-534
4. Box, G.E.P., Jenkins, G.M.: Time Series Analysis: Forecasting and Control. San Francisco: Holden Day (1976)
5. Bitzan, P., Šmejkalová, J., Kučera, M.: Neural Networks with Switching Units. Neural Network World, Vol.4. (1995) 515-523
6. Cabrero, A., Camba-Mendez, G., Hirsch, A., Nieto F.: Modelling the Daily Series of Banknotes in Circulation in the Context of the Liquidity Management of the Europen Central Bank. ECB Working Paper No. 142. Frankfurt a.M. (2002)

7. Diebold, F.X, Mariano, R. S.: Comparing Predictive Accuracy. Journal of Bussines and Economic Statistics, Vol. 13. (1995) 253-263
8. Dorffner, G.: Neural Networks for Time Series Processing. Neural Network World, Vol. 4. (1996) 447-468.
9. Gourieroux, Ch.: Time Series and Dynamic Models. Cambridge University Press (1997)
10. Hamilton, J.D.: Time Series Analysis. Princeton University Press (1994)
11. Hallas, M., Dorffner, G.: A Comparative Study on Feedforward and Recurrent Neural Networks in Time Series Prediction Using Gradient Descent Learning. Proceedings of 14th European Meeting on Cybernetics and Systems Research. Vienna (1998)
12. Hakl, F., Hlaváček, M., Kalous, R.: Application of Neural Networks to Higgs Boson Search. Nuclear Instruments nad Methods in Physics Research, Sec.A. (2003) 489-491
13. Harvey, A., Koopman, S.J., Riani, M.: The Modeling and Seasonal Adjustment of Weekly Observations. Journal of Bussines & Economic Statistics, Vol. 15, No. 3. (1997) 354-368
14. Hlaváček, M.: Design and Analysis of Neural Network with Switching Units Suitable for the Study of Elementary Particles Decay with the Application of Genetic Algorithms. Master Thesis at Czetch Technical University. Prague (2002)
15. Hornik, K., Stinchcombe, M., and White, H.: Multilayer feedforward networks are universal approximators. Neural Networks, Vol.2. (1989) 359-366
16. Hillmer, S.C., Tiao, G.C.: An ARIMA-Model-Based Approach to Seasonal Adjustment. Journal of the American Statistical Association, Vol. 77, No. 377. (1982) 63-70
17. Kalous, R.: Evolutionary Operators on Neural Networks Architecture. Procceding of PhD. Conference, Institute of Computer Science & MatfyzPress, Prague (2004)
18. Koreisha S.G., Pukkila, T.: A Two-Step Approach for Identifying Seasonal Autoregressive Time Series Forecasting Models. International Journal of Forecasting, Vol. 14. (1989) 483-496
19. Pierce, D.A.: Least Squares Estimation in the Regression Model with Autoregressive-Moving Average Errors. Biometrika, Vol.58, No.2. (1971) 299-312
20. Trapletti, A., Leisch, F., Hornik, K.: On the Ergodicity and Stationarity of the ARMA(1,1) Recurrent Neural Network Process. SFB Working Paper No. 37. Viena (1999)

Time Delay Neural Networks and Genetic Algorithms for Detecting Temporal Patterns in Stock Markets

Hyun-jung Kim, Kyung-shik Shin, and Kyungdo Park

Ewha Womans University, College of Business Administration
11-1 Daehyun-Dong, Seodaemun-Gu, Seoul 120-750, Korea
charitas@empal.com, {ksshin, kyungdo}@ewha.ac.kr

Abstract. This study investigates the effectiveness of a hybrid approach with the time delay neural networks (TDNNs) and the genetic algorithms (GAs) in detecting temporal patterns for stock market prediction tasks. Since TDNN is a multi-layer, feed-forward network whose hidden neurons and output neurons are replicated across time, it has one more estimate of time delays in addition to a number of control variables of the artificial neural network (ANN) design. To estimate these many aspects of the TDNN design, a general method based on trial and error along with various heuristics or statistical techniques is proposed. However, for the reason that determining time delays or network architectural factors in a stand-alone mode doesn't guarantee the illuminating improvement of the performance for building the TDNN models, we apply GAs to support optimization of time delays and network architectural factors simultaneously for the TDNN model. The results show that the accuracy of the integrated approach proposed for this study is higher than that of the standard TDNN and the recurrent neural networks (RNNs).

1 Introduction

Early studies of stock market prediction tended to use statistical techniques. However, studies using only classical statistical techniques for prediction reach their limits in applications with non-linearities in the data set. Compared with statistical methods, an artificial neural network (ANN) has an advantage in handling nonlinear problems by using the hidden layer. Among ANN algorithms, the back-propagation neural network (BPN) is the most popular method in many applications such as classification, forecasting and pattern recognition. A major limitation of BPN, however, is that it can learn only an input-output mapping of static (or spatial) patterns that are independent of time. To overcome this limitation, two methods applying the time property are proposed: the first is use of recurrent links; the second is use of time-delayed links.

This study investigates effectiveness of a hybrid approach with the time delay neural networks (TDNNs) and the genetic algorithms (GAs) to in detecting temporal patterns for stock market prediction tasks. Since TDNN adds a memory to the ANN by use of the time-delayed links for each unit, it has one more estimate of time delays in addition to a number of control variables of the ANN design such as network topologies, the number of hidden nodes, the choice of activation function and so on. To estimate these many aspects of the TDNN design, a general method based on trial and error along with various heuristics or statistical techniques is proposed. However, determining time delays or network architectural factors in a stand-alone mode doesn't guarantee the illuminating improvement of the performance for building the TDNN models. Therefore, the suitable method to select time delays and network architectural factors simultaneously for TDNN is necessary to accurately model the temporal patterns. We apply GAs to support optimization of time delays and network architectural factors at the same time for the TDNN model. Our proposed approach is demonstrated by applications to the stock market's prediction domain.

The rest of the paper is organized as follows: Section 2 proposes ANN for a time series property. Section 3 describes a hybrid approach of TDNN using GA (GA-TDNN), research data and experiments, and in Section 4, empirical results are summarized and analyzed. In the final section, conclusions and the limitations of this study are presented.

2 ANN for a Time Series Property

Regarding previous studies of time-series properties using nonlinear dynamics, most forecasting methods are capable only of picking up general trends, and have difficulty in modeling cycles. Two methods may be applied to add memory to a feed-forward neural network. The first is use of recurrent links [3], while the second is use of time-delayed links [2].

2.1 Recurrent Neural Networks

Unlike BPN, the recurrent neural network (RNN) is permitted to have feedback connections among neurons. Operation of the RNN involves combined use of two network structures that work together, the original recurrent network and the adjoining network. The original network is characterized by the forward propagation equation; for a given input-output, this network computes the error vector. The adjoining network is characterized by the backward propagation equation; the adjoining network takes the error vector from the original network and uses it to compute adjustments to the synaptic weights of the original network.

Due to reliance on feed-forward connections, the standard BPN algorithm suffers from an inability to fill in patterns. Use of the RNN algorithm overcomes the pattern-completion problem by virtue of its inherent feedback connections. Almeida [1] has experimentally confirmed that feedback structures are much better suited to this kind of problem. Feedback connections in the RNN make it less sensitive to noise and lack of synchronization, and also permit it to learn faster than the BPN. However, the BPN is more robust than the RNN with respect to the choice of a high learning rate parameter [5].

2.2 Time Delay Neural Networks

The time delay neural network (TDNN) is much more complex than the BPN, as it is required to explicitly manage activations by storing delays and back-propagated error signals for each unit and for all time delays [7]. The TDNN learns time-based as well as functional relationships that correlate input data with projected neural outputs.

TDNN was first proposed for dealing with speech recognition [6], and is now applied successfully both to speech recognition and phoneme classification. TDNN is described as a multi-layer, feed-forward network whose hidden neurons and output neurons are replicated across time. TDNN is made up by units that get an input at the generic time instant t and an output of the previous level units, in which the input at several time steps $t-1, t-2... t-n$ is summed and fully connected with suitable weights. These delayed inputs let the unit know part of the history of the signal at time t, and enable the solution to more complex decision problems, especially time dependent ones. Training of the TDNN takes place through temporal expansion of time delays over the entire input sequence [2].

We consider a neural network with L levels, containing N_l units in each level l. Let us define the delayed input vector $\mathbf{x}$ for unit i pertaining to level l at discrete time t as:

$$\mathbf{x}_i(t) = [\mathbf{x}_i(t), \mathbf{x}_i(t-1),..., \mathbf{x}_i(t-T_l)]^T, \quad i=1, ..., N_l$$

The vector of variable weights with which each input, $\mathbf{x}_i$, in the previous layer is multiplied and put as an output to unit j, $\mathbf{x}_j$, of level l, is expressed as:

$$\mathbf{w}_{ji} = [\mathbf{w}_{ji}(0), \mathbf{w}_{ji}(1),..., \mathbf{w}_{ji}(T_l)]^T, \quad j=1, ..., N_{l+1}.$$

The contribution, s_j, from unit i to unit j, is given as:

$$s_{ji}(t) = \mathbf{w}_{ji}^T \mathbf{x}_i(t) \tag{1}$$

The output field for unit j is a weighted sum of past-delayed values of the input, expressed as:

$$s_j(t) = \sum_{i=1}^{N_l} s_{ji}(t) \tag{2}$$

Named f for the unit's transfer function, the output of the neuron is given as:

$$\mathbf{x}_j(t) = f(s_j(t)) \tag{3}$$

Assume that at each instant in time, the target vector of the ith element, $\mathbf{d}_i(t)$, is provided to the networks. Accordingly, the instantaneous error is defined as:

$$e_i(t) = \mathbf{d}_i(t) - \mathbf{x}_i(t) \tag{4}$$

The total instantaneous squared error is defined as:

$$e^2(t) = \sum_{i=1}^{N_l} e_i^2(t) \tag{5}$$

The total squared error is defined as:

$$e^2 = \sum_{t=0}^{T} e^2(t). \tag{6}$$

Because training is supervised at each time t, and the net output depends on time delay inputs at times $t, t-1... t-T_l$, the target vector $\mathbf{d}(t)$ changes synchronously with the input at each time t.

Figure 1 illustrates an example of the time delay neural network architecture, where the hidden and output neurons have two and three time delays, respectively (i.e. $L=3$, $T_1=2$, $T_2=3$). For simplicity, only one unit has been shown for each level.

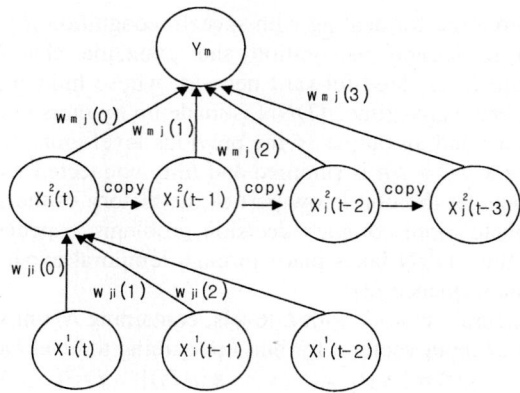

Fig. 1. TDNN algorithm (adapted from Cancelliere, R. and Gemello, R. [2])

3 Hybrid Approach Using GAs

In this study, we propose a hybrid approach of TDNN using GA (GA-TDNN) for optimization of time delays and network architectural factors simultaneously to improve the effectiveness of constructing the TDNN model. Since TDNN is a multilayer, feed-forward network whose hidden neurons and output neurons are replicated across time, it requires one more estimate of time delays to use for each unit in addition to many aspects of the ANN design.

A common method to estimate time delays that will be used to build the TDNN models is to choose the time lag based on trial and error along with various heuristics. On the other hand, it is possible to extrapolate the dynamics of the series based on the autocorrelation function (ACF) or the average mutual information that describe the correlation of the time difference between $x(t)$ and $x(t+T)$, separated by T time lag. The concept of mutual information is similar to that of the ACF, which is a classical statistical measure of linear dependence in a time series, but mutual information is a measure of general dependence in a time series, which can be applied to any time series. However, the process of mutual information is rather cumbersome computationally. So far, the most popular method to choose the time lag for nonlinear time series analysis is to estimate the slope of the curve of the logarithm of the correlation integral.

Considering that to find an appropriate ANN model that can reflect problem characteristics is an art and plays a very important role for the promising performance of an ANN model, a reasonable technique to optimize not only the design of numerous network architectures such as network topologies, the number of hidden nodes, the choice of activation function and so on, but also input selection, learning condition, learning methods, and parameters is also necessary. However, determining time delays or network architectural factors in a stand-alone mode doesn't guarantee the illuminating improvement of the performance for building the neural network models. Therefore, the suitable method to select time delays and network architectural factors simultaneously for TDNN is necessary to accurately model the temporal patterns. This study uses the notion of machine learning, an evolutionary search technique, to find an optimal or near-optimal time lag as well as network architectural factors according to a given task.

GAs have been increasingly applied in conjunction with other AI techniques such as neural networks, rule-based system, fuzzy theory, and CBR. The integration of GAs and neural networks is a rapidly expanding area. Various problems faced by researchers and developers in using neural network can be optimized using GAs [4][8]. Examples include selecting relevant input variables, searching the weight space, determining the optimal number of hidden layers, nodes and connectivity, and tuning the learning parameters.

We apply GAs to learn an optimal and near-optimal set of time delays and network architectural factors at the same time among population searches. By evaluating the fitness of different sets of time delays and network architectural factors, we may find good global solutions for the TDNN model. Notably, the task of defining a fitness function is always application specific. In this study, the objective of the system is to find the more relevant set of time delays and network architectural factors that can lead to the correct solutions. The ability of GA-TDNN to achieve these objectives can be represented by the fitness function that specifies how well the set of time delays and network architectural factors decrease the prediction error. We apply the prediction error rate of the test set to the fitness function for this study.

To calculate the fitness of different sets of time delays and network architectural factors, we use the moving window approach for the train, test and validation set extraction method, which is an efficient way to assess accuracy of the ANN model with sequences of inputs over time. This method starts to train on the defined quantity of training records, tests on the subsequent quantity of test records, and validates on the subsequent quantity of validation records (i.e. the first fold). This process moves all records forward by size of the validation set, retrains, retests and revalidates, and continues through all folds until the end of the data set. Then the average mean square error (AMSE) is computed, yielding the averaged value of the mean square error (MSE) of each fold.

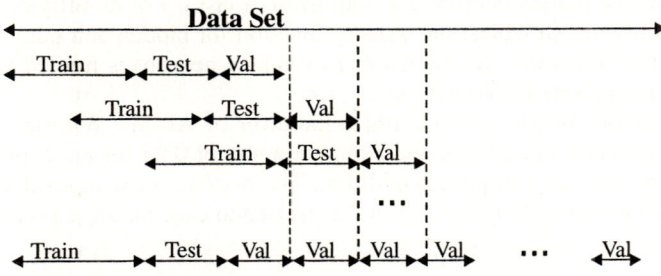

Fig. 2. Moving window approach

Research data in this study comes from the daily Korea Stock Price Index 200 (KOSPI 200) from January 1997 to December 1999. The total number of samples includes 833 trading days. In this study, the training set size is 400 records while the test and validation set size is 15 records each. There are 28 resulting folds through this process, and validation records in the last fold are 13 records. The moving window approach used here is illustrated in Figure 2.

We used the value of the stock price index as an input variable and as output variables. At the generic time instant t, we selected KOSPI 200 at time t as an input variable, and KOSPI at time $t+1$ as an output variable.

To compare with GA-TDNN, we also design the standard TDNN with predefined time delays in the input layer, which is connection of hidden neurons, and the standard RNN with a predefined time lag in input neurons, using the slope of the curve of the logarithm of the correlation integral. In addition, the number of hidden neurons in both algorithms is given as eight, and the other conditions are equally controlled to those of GA-TDNN. Figure 3 represents the RNN architecture in which we have chosen a time lag of two days.

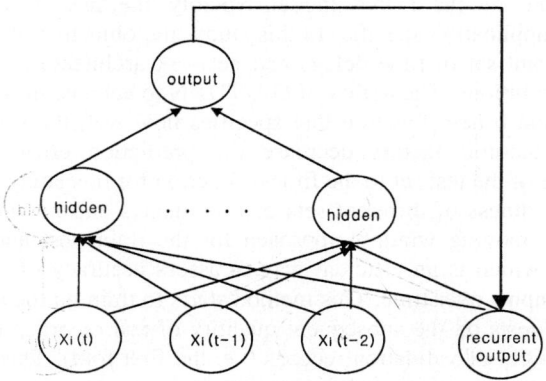

Fig. 3. RNN with time lag

As described above, in this study, GAs search the number of time delays in the input and hidden layer in TDNN, which is connection of hidden and output neurons, and the number of hidden neurons for optimization tasks. For an efficient search, we control the following variables; the transfer function for hidden and output neurons is set to the logistic function, and the number of hidden neurons is limited to eight. The number of hidden layers is given as one.

Fitness function we use is to minimize the error of AMSE. We use a population size of 50, a selection rate of 0.5, and a mutation rate of 0.25 for the experiment. Ten generations are used as a stopping condition. The neuro-genetic algorithms software, NeuroGenetic Optimizer (NGO) version 2.5, is used to execute these processes.

4 Results and Analysis

To investigate the effectiveness of the integrated approach for building neural networks models for a time series property, we set GAs to search the optimal set of time delays and network architectural factors simultaneously in TDNN. The derived results by this genetic search are summarized in Table 1.

Table 1. The optimized architectural factors using GA

Methods	Number of hidden neurons	Connection of hidden neurons	Connection of output neurons	Time delay
GA-TDNN1	5	4	2	4
GA-TDNN2	3	7	1	6
GA-TDNN3	5	1	5	4

To reduce the impact of random variation in GAs search processes, we replicate the experiment several times and suggest the best networks found in each model. As a result, GA-TDNN can be of three types regarding the connection of neurons. GA-TDNN1 is the general TDNN architecture with time delays of both hidden and output neurons. GA-TDNN2 has only have time delays of hidden neurons, while GA-TDNN3 has only time delays of output neurons.

The optimized GA-TDNN models are compared with the standard TDNN and RNN with predefined time delays and the time lag respectively. These predefined values of the standard TDNN and RNN are independently obtained by estimating the slope of the curve of the logarithm of the correlation integral as shown in Figure 4.

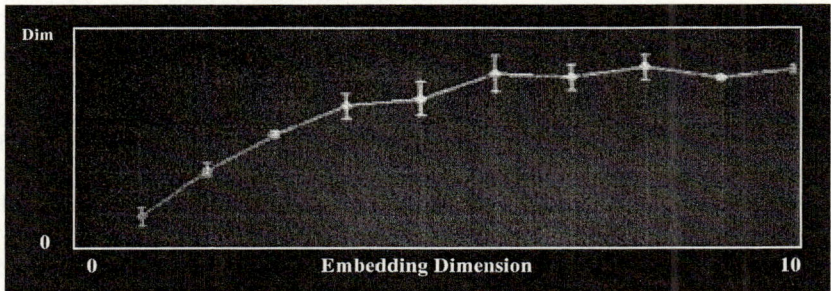

Fig. 4. Function of Dimension

The embedding dimension can be chosen as six, which indicates that we have to train on five successive stock price indices for predicting the sixth value.

In order to compare the enhancement of GA-TDNN with the standard TDNN and RNN, the mean square error (MSE), which is the metric used in this experiment, of each network is calculated between the predicted and actual neural outputs. Comparative methods are represented as the benchmark to verify the application of the proposed model to the domain. Test results are summarized in Table 2.

F-Test results for the comparison of each model are summarized in Table 3. If the F-value is significantly large or small, it is probable that a performance difference exists between the models.

Table 2. The measure of accuracy

Methods	MSE
GA-TDNN1	3.49
GA-TDNN2	3.96
GA-TDNN3	4.23
TDNN	5.30
RNN	7.40

Table 3. F-test result for the comparison of the difference between models

	GA-TDNN2	GA-TDNN3	TDNN	(F-values) RNN
GA-TDNN1	1.13**	1.21***	1.52***	2.12***
GA-TDNN2		1.07	1.34***	1.87***
GA-TDNN3			1.25***	1.75***
TDNN				1.40***

** significant at 5%, *** significant at 1%

On the whole, the MSE of the derived GA-TDNN from the genetic search process is smaller than that of the standard TDNN and RNN. As the F-tests show, predictive ability of all GA-TDNN models is significantly different compared to the standard TDNN and RNN at 1% statistical significance level. Specifically, the MSE of the GA-TDNN1 that has a time delay of both hidden and output neurons is the smallest among other comparative error values. F-test results support that GA-TDNN1 significantly performs better than the other types of GA-TDNN, such as GA-TDNN2 and GA-TDNN3, which are significantly indifferent. It also appears that the standard TDNN outperforms the standard RNN with significant levels.

Based on these empirical results, we conclude that the optimization of time delays and network architectural factors simultaneously plays an important role in improving the performance of the TDNN models as well as the effectiveness of constructing the TDNN models. The overall result shows that the integrated GA-TDNN approach proposed for this study performs better than the standard TDNN and RNN to reflect temporal patterns.

5 Conclusions

This study investigates the effectiveness of a hybrid approach with TDNN and GAs for the time dependent prediction domain. By evaluating the fitness of different sets of time delays and network architectural factors at the same time, we may find good global solutions for the TDNN model. Results show that the accuracy of GA-TDNN is higher than that of the standard TDNN and RNN.

Our study has the following limitations that need further research. Although a hybrid approach with TDNN and GAs intends to optimize time delays and network architectural factors simultaneously and to select feature subsets, we have controlled some variables for search efficiency, such as transfer function and number of hidden layers. However, there are an infinite number of possible combinations with these control variables. In setting up the GAs optimization problem, we have selected several parameters such as stopping condition, population size, crossover rate, mutation rate and so on. Their values of these parameters can greatly influence the performance of the algorithm, and can generate a lot of groups for our general result.

TDNN as applied also has limitations in its inability to learn or adapt time delay values. Time delays are initially fixed, and remain constant throughout training. As a result, TDNN may have poor performance due to inflexibility of time delays and a mismatch between choice of time-delay values and temporal location of important information in the input patterns. In addition, performance may vary depending on the range of the time delay values.

To overcome this limitation, our future work will focus on an adaptive time-delay neural network model. This network adapts its time-delay values as well as its weights during training to better accommodate changing temporal patterns, and provides more flexibility for optimization tasks.

References

1. Almeida, L.: A Learning Rule for Asynchronous Perceptrons with Feedback in a Combinational Environment. Proceedings of 1st IEEE International Conference on Neural Networks, 2 (1987)
2. Cancelliere, R., Gemello, R.: Efficient Training of Time Delay Neural Networks for Sequential Patterns. Neurocomputing. 10 (1996)
3. Elman, J.: Finding structure in time. CRL Technical Report (8801) La Jolla University of California, San Diego (1998)
4. Leigh, W., Purvis, R., Ragusa, J.M.: Forecasting the NYSE Composite Index with Technical Analysis, Pattern Recognizer, Neural Network, and Genetic Algorithm: A Case Study in Romantic Decision Support. Decision Support Systems. 32(4) (2002) 361-377
5. Simard, P., Ottaway, M., Ballard, D.: Fixed Point analysis for Recurrent Networks. Advances in Neural Information Processing Systems, 1. Morgan Kaufmann, San Mateo CA (1989)
6. Waibel, A., Hanazawa, T., Hinton, G., Shikano, K., Lang, K.: Phoneme Recognition Using Time Delay Neural Networks. IEEE Transaction on Acoustics, Speech, Signal Processing. 37(3) (1989)
7. Wan, E.A.: Temporal Backpropagation for FIR Neural Networks. International Joint Conference Neural networks, San Diego CA (1990)
8. Wong, F., Tan, C.: Hybrid Neural, Genetic and Fuzzy Systems. In: Deboeck, G.J.: Trading On the Edge. John Wiley, New York (1994) 245-247

The Prediction of the Financial Time Series Based on Correlation Dimension

Chen Feng[1], Guangrong Ji[1], Wencang Zhao[1,2], and Rui Nian[1]

[1] College of Information Science and Engineering Ocean University of China,
Qingdao, 266003, China
fccjg@sdu.edu.cn, grji@mail.ouc.edu.cn,nianrui_80@163.com
[2] College of Automation and Electronic Engineering, Qingdao University of Science,
&Technology, Qingdao, 266042, China
wencangzhao@mail.edu.cn

Abstract. In this paper we firstly analysis the chaotic characters of three sets of the financial time series (Hang Sheng Index (HIS), Shanghai Stock Index and US gold price) based on the phase space reconstruction. But when we adopt the feedforward neural networks to predict those time series, we found this method run short of a criterion in selecting the training set, so we present a new method: using correlation dimension (CD) as the criterion . By the experiments, the method is proved effective.

1 Introduction

The prediction of the financial time series is a problem which interest the researchers at all time because it has important meaning for macro-economic adjustment and micro-economic management. For predicting the financial time series better researchers made great efforts to find the laws of the time series. In the past the financial time series were considered random walk and the models were built according to this viewpoint, but the predicted results were proved bad by some experiments [1].

In recent years researchers found that some financial time series are chaotic time series rather than the random series in fact. Literature [2] indicated that hourly data of four spot exchange rates (British Pound, Deutschmark, Japanese Yen and Swiss France) are chaotic; literature [3] pointed out American national debt time series has chaotic attractor; literature [4] proved that some metal prices in London market follows a mean process that is dynamic chaotic.

Many methods such as the maximum Lyapunov exponent method [5] and one-rank weighed local method [6] are used to predict the chaotic time series. In maximum Lyapunov exponent method, a teeny error induced by computing the maximum Lyapunov exponent will bring large error in the prediction. The idea of one-rank weighed local method is to use the linear model to resume local chaotic system. But the linear model always has some limits to mirror the nonlinear system. So the predicted effects of the economic time series are not good enough with these methods.

At the same time owing to the strong nonlinear mapping ability of the neural networks, many kinds of neural networks such as BPNN [7], GRNN [8] and RNN [9] etc. were used to predict the financial time series. In this paper we adopt the feedforward neural networks used in the literature [10] as the training networks to predict the financial time series. With this kind of networks introduced in the third section, many classical chaotic systems such as Lorenz system, Henon mapping etc. can be predicted very well.

But in the process of studying the method, we find the training set's choice is hazy and run short of a criterion in this method. So at the forth section, we bring forward a new method to choose the training set. According that the financial time series are chaotic, we choose the correlation dimension -- a kind of fractal dimension that can depict the chaotic characteristics as the criterion to choose the training set. By the experiments the method is proved effective.

If we use the feedforward neural networks to predict the time series, the phase space must be reconstructed firstly, so in the second section we introduce the delay coordinate method adopted to reconstruct the space and compute the financial time series' maximum Lyapunov exponents to prove the three sets of financial time series are chaotic. Then we show the architecture of the neural networks in third section. In the forth section we explain the definition of the correlation dimension simply, and introduce how to choose the training set according to the correlation dimension. At the same time the three sets of economic data are used to prove the effect of the new method in the fifth section. In the last section, we reach the conclusion.

2 Phase Space Reconstruction

2.1 Theory Introduction

For resuming the dynamic characteristics of the original financial systems, the phase space should be reconstructed firstly. Takens' theorem, which opens out some nonlinear systems' dynamic mechanism, is the theoretic base of the phase space reconstruction.

Takens' theorem: M is d dimension manifold, mapping $\varphi: M \to M$ is a smooth differential homeomorphism, mapping $y: M \to R$ has second-order continuous derivative $\phi(\varphi, y): M \to R^{2d+1}$, and

$$\phi(x, y) = (y(x), y(\varphi(x)), y(\varphi^2(x)), \cdots, y(\varphi^{2d}(x))) \tag{1}$$

where the function $\phi(\varphi, y)$ is a embedding from M to R^{2d+1}. The theorem indicates that a suitable embedding dimension can be found to resume the inerratic trajectory [11]. The delay coordinate method is used to reconstruct the phase space in the paper. An embedding dimension m and a delay time τ are determined to create N_m points, and every point Y_i is a m dimension vector,

$$Y_1 = (x_1, x_{1+\tau}, \cdots, x_{1+(m-1)\tau}), \cdots, Y_i = (x_i, x_{i+\tau}, \cdots, x_{i+(m-1)\tau}), \cdots, Y_{N_m} = (x_{N_m}, x_{N_m+1}, \cdots, x_N) \tag{2}$$

where $N_m = N - (m-1)\tau$. The embedding dimension m and the delay time τ are important parameters because they decide the quality of the reconstructed phase space.

In this paper, we use the so-called false nearest-neighbor method [12] to decide the embedding dimension m. The idea of the method is : when the dimension is increased from m to $m+1$, we estimate whether there are false near points in the near points of the point Y_i, if there is none, the geometrical structure of the attractor has been opened. When the dimension is m, supposing that the point $Y_{i'}$ is the nearest point of the point Y_i, the distance between these two points is $\|Y_{i'} - Y_i\|^{(m)}$. When the dimension is increased to $m+1$, their distance is marked $\|Y_{i'} - Y_i\|^{(m+1)}$.

$$\left[\|Y_{i'} - Y_i\|^{(m+1)} - \|Y_{i'} - Y_i\|^{(m)}\right] / \|Y_{i'} - Y_i\|^{(m)} > R_T, 10 \leq R_T \leq 50 \quad (3)$$

The point $Y_{i'}$ is the false neighbor point of the point Y_i where R_T is the threshold. We start at dimension 2 and increase the dimension by one each time. Either the proportion of the nearest neighbor points is smaller than 5% or the number of the nearest neighbor points don't decrease with the increase of the dimension, the dimension m is the optimum.

2.2 Financial Time Series' Phase Space Reconstruction

In the paper, we choose the opening quotation of Hang Sheng Index (HIS) (4067 points from 31 December 1986 to 16 June 2003), Shanghai Stock Index (2729 points from 19 December 1990 to 29 January 2001), and US gold price (7277 points from 2 January 1975 to 8 August 2003) as the experiment data. The three sets of time series are shown in Fig.1.

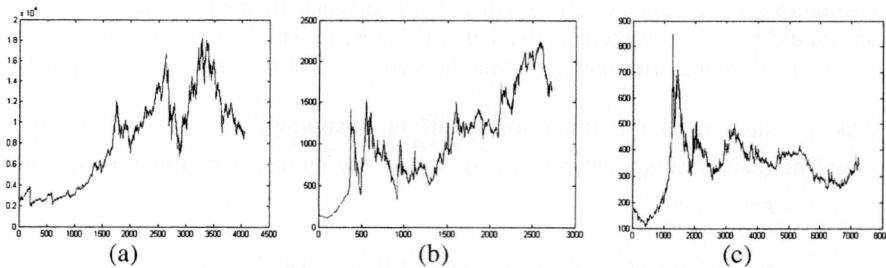

Fig. 1. (a) Opening quotation of Hang Sheng Index (b) Opening quotation of Shanghai Stock Index (c) Opening quotation of US gold price.

From Fig.1 we can observe that in the time series curves some locals have similarity with the whole. For showing the complexity of the three sets economic data, we compute their box dimensions [13]. The box dimension, which always is used to calculate the dimension of the continuous curve, is a kind of fractal dimension. They are shown in Table 1.

According to the theory in the literature [14], if the capital market follows the random walk, the box dimension should be 1.5. The time series whose box dimension is between 1 and 1.5 is called long range correlation fractal time series, which means that the past increment is positive correlative with the future increment. The time series whose box dimension is between 1.5 and 2 is called long range negative correlation fractal time series, which means that the past increment is negative correlative with the future increment. From the Table 1, we can observe that the box dimensions are all between 1 and 1.5, so the financial time series don't follow the random walk entirely, and that there is long range positive correlation in them.

Table 1. The box dimensions of the enconomic time series

	HIS	Shanghai Stock Index	US gold price
Box dimension	1.16016	1.16631	1.18816

We reconstruct the phase space by calculating the embedding dimension m and the delay time τ with the prediction error minimizing method [15].

At the same time, we choose three dimensions data from the every m-dimension reconstructed phase space of the financial time series and plot them which are shown in Fig.2.

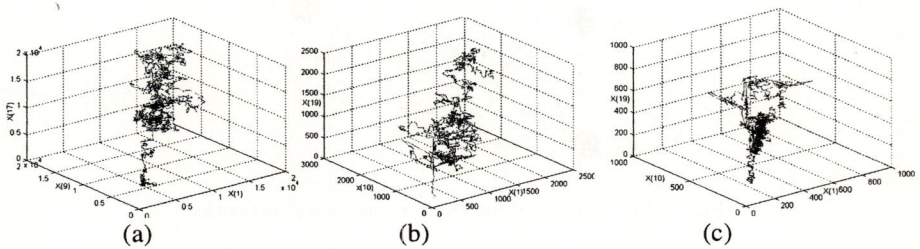

Fig. 2. The 3 dimensions data from the reconstructed phase space of the financial time series (a) the opening quotation of Hang Sheng: 1-dimension, 9-dimension and 17dimension (b) the opening quotation of Shanghai Stock Index:1-dimension, 10-dimension and 19dimension (c) the opening quotation of US gold price:1-dimension, 10-dimension and 19dimension

The maximum Lyapunov exponent λ_{max} is computed with the small data sets method [16] to prove that these financial time series are chaotic. A quantitative measure for the sensitive dependence on the initial conditions is the Lyapunov exponent, which characterizes the average divergence rate of two neighboring trajectories.

It is not necessary to calculate Lyapunove spectrum because a bounded time series with a positive maximum Lyapunove exponent indicates chaos. Moreover, the maximum Lyapunov exponent gives an estimate of the level of chaos in the underlying dynamical system. From Table 2 we can found the maximum Lyapunov exponents are positive, so the financial time series are chaotic.

The chaotic systems are sensitive to the initial values, so the chaotic time series has limited prediction potential. Since the maximum Lyapunove exponent characterizes the average degree of neighboring orbits, its reciprocal $1/\lambda_{max}$ determines the maximum predictable time. The results are all shown in Table 2.

Table 2. The chaotic analyse of the financial time series.

	Embedding dimension	Delay time	maximum Lyapunov exponent	maximum predictable time
HSI	17	6	0.069	14
Shanghai Stock Index	19	4	0.029	30
US gold price	19	7	0.046	20

3 Feedforward Neural Networks

The architecture of the feedforward neural networks used in this lecture is $m:2m:m:1$, where m is the embedding dimension. The topology architecture is shown in Fig.3.

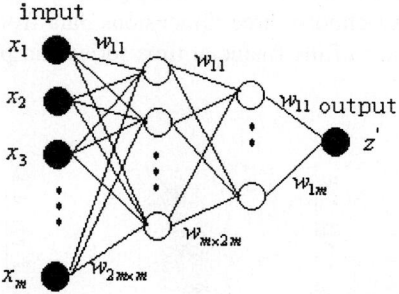

Fig. 3. Architecture of the feedforward neural networks

When the m dimension training set is put into the networks, each hidden unit j in the first hidden layer receives a net input

$$\gamma_j = \sum_i w_{ji} x_i \qquad (4)$$

and produces the output

$$V_j = \tanh(\gamma_j) = \tanh(\sum_i w_{ji} x_i) \qquad (5)$$

where w_{ji} represents the connection weight between the i th input unit and the j th hidden unit in the first layer. Following the same procedure for the other unit in the next layers, the final output is then given by

$$z' = \sum_l w_{sl} \left\{ \tanh \left[\sum_j w_{lj} \tanh \left(\sum_i w_{ij} x_i \right) \right] \right\} \qquad (6)$$

where the hyperbolic tangent activation function is chosen for all hidden unit, and the linear function for the final output unit.

The weights are determined by presenting the networks with the training set and comparing the output of the networks with the real value of the time series. The function of the weights adjusting is

$$w_{qt}^{new} = \alpha w_{qt}^{old} - \eta \Delta w_{qt} \qquad (7)$$

where $\Delta w_{qt} = \partial E(w_{qt})/w_{qt}$, $E(w_{qt})$ is the mean square error function, $0 < \eta \leq 1$ is the learn rate, and $0 < \alpha < 1$ is the inertial term. By setting the delay coordinates of the time series $x(t)$: $(x(t), x(t-\tau), \cdots, x(t-(m-1)\tau)$ as the input pattern and choosing $x(t+\Lambda)$ as the know target, the networks can be trained to predict the future state of the system at a time Λ, which corresponds to a certain number of time steps [10].

4 How to Choose the Training Set

4.1 Method of Choosing Training Set

In the above feedfoward neural networks prediction the input data and the target should be known, but it is impossible to know the target in reality, therefore this prediction is only a systemic simulation. At the same time the literature [10] didn't mention how to choose the training set, thus the choice of the training set has some uncertainty. So if we want to predict the time series authentically, we should choose the training set whose characters are similar with the prediction set's, and use the weights getting from the training set's exercitation to forecast the prediction set. So how to choose the training set became an important problem. We solve this problem in this section.

In the second section we proved the three sets of the financial time series are chaotic time series, so we put forward a new method to choose the training by using the correlation dimension as the criterion. The correlation dimension is a kind of fractal dimension that can depict the chaotic characteristic. The notion of dimension often refers to the degree of complexity of a system expressed by the minimum number of variables that is needed to replicate it.

The steps of how to choose the training set are shown as follows.

1) Reconstructing the phase space.
2) Calculating correlation dimension of the prediction set.
3) Choosing 50 sets which are closest to the prediction set.
4) Computing every set's correlation dimension.
5) Choosing the set whose correlation dimension is nearest to the prediction set's as the training set.

Then use this training set to train the networks, and get the weights. We can adopt these weights to forecast the prediction set.

4.2 Correlation Dimension

The G-P algorithm which was presented by Grassberger and Procaccia is adopted to calculate correlation dimension [17].

For a set of the space points $\{Y_i\}$, defining

$$C_N(r) = \frac{2}{N(N-1)} \sum_{1 \leq i < j \leq N} \theta(r - |Y_i - Y_j|), \tag{8}$$

where $\theta(x) = \begin{cases} 0, & x \leq 0 \\ 1, & x > 0 \end{cases}$ is the Heaviside function.

When we choose different r, we can get different $C_N(r)$. In estimating the correlation dimension from the data, one plots $\log C_N(r)$ against $\log(r)$, where N is the cardinality of the data set. $C_N(r)$ measures the fraction of the total number of pairs (Y_i, Y_j) such that the distance between Y_i and Y_j not longer than r.

5 Experiments

From the embedding dimensions in the Table 2 we can determine the neural networks' architecture, for Shanghai Stock Index the architecture is 19:38:19:1, for HSI the architecture is 17:34:17:1, for US gold price the architecture is 19:38:19:1.

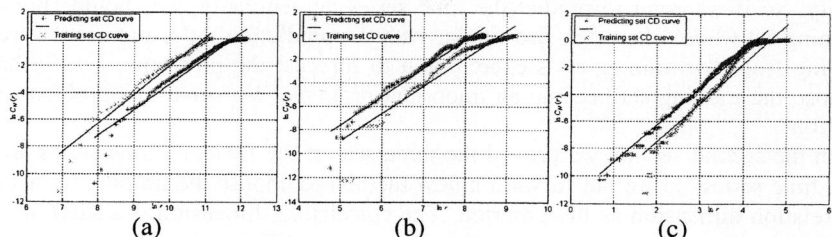

Fig. 4. Fitting curves of prediction set's CD and the Training set's CD (a) the opening quotation of Hang Sheng Index (b) the opening quotation of Shanghai Stock Index (c) the opening quotation of US gold price

Based on the three phase spaces with the financial time series, we choose 100 continuous points in the every phase space as the prediction set. Using the method expatiated in the forth section we determine the training set. The correlation dimensions of the prediction set and training set are listed in Table 3.

Fig.4 shows the fitting curves of prediction set's CD and the Training set's CD. From Table 3 and Fig.4 we can observe for every set of the financial time series that the training sets' correlation dimensions are near to the prediction sets', and their fitting curves are parallel. So in the next step we use these three training sets to train the networks.

Table 3. The correlation dimension comparison between the predicting set data and training data

	HIS	Shanghai Stock Index	US gold price
Predicting set' CD	1.9113	2.3979	2.7269
Training set's CD	2.1078	2.3276	2.7936
CD's difference	0.0965	-0.0703	0.0667

By educating the every training set in the networks, we obtained the weights one by one. Put the prediction set into the networks whose weights have been determined, and the predicted data are calculated. The three sets of the predicted results and the real values are shown in Fig.5.

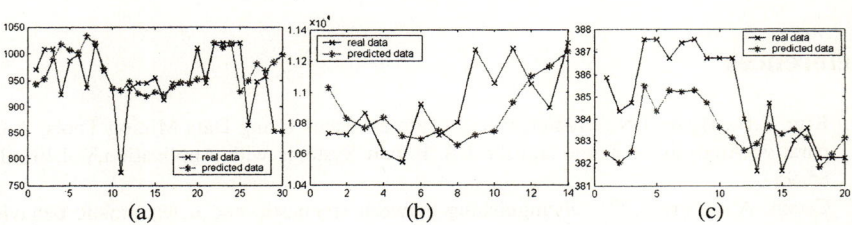

Fig. 5. The prediction of the financial time series (a) the opening quotation of Hang Sheng Index from 2 April 2002 to 22 April 2002 (b) the opening quotation of Shanghai Stock Index from 21 October 1996 to 11 November 1996 (c) the opening quotation of US gold price from 8 September to 5 October 1995

We also calculate the mean absolute percentage error (MAPE) displayed in Table 4 to show the prediction effect.

$$MAPE = \frac{\sum_{i=1}^{n}|x_t - x_t'|/x_t}{n} \tag{9}$$

where x_t is the real data and x_t' is the predicted data.

Table 4. The MAPE between the real data and the predicted data

	HIS	Shanghai Stock Index	US gold price
MAPE	1.9%	3.9%	0.46%

Every MAPE is less than 5%, so the prediction effect is good enough.

6 Conclusions

Though the experiment results, we can find that the predicted data's trend is identical with the real data's on the whole except few exceptional points and the MAPE between the real data and the predicted data are all small. This proved that as the chaotic time series, the financial time series can be predicted by the feedforward neural networks.

On the other hand, we also can prove that the method which is adopted to choose the training set by using the correlation dimension as the criterion is effective from the experiment results. When we predict the chaotic financial time series using this method, the uncertainty of the training set's choice is reduced.

Acknowledgements

The National 863 Natural Science Foundation of P. R. China (2001AA635010) fully supported this research.

References

1. Kim, S.H., Hyun, J.N.: Predictability of Interest Rates Using Data Mining Tools: A Comparative Analysis of Korea and the US. Expert Systems with Application,Vol.13. (1997) 85-95
2. Cecen, A.A., Erkal, C.: Distinguishing between stochastic and deterministic behavior in high frequency foreign exchange rate returns: Can non-linear dynamics help forecasting?. International Journal of Forecasting, Vol.12. (1996) 465-473
3. Harrison, R.G., Yu, D., Oxley, L., Lu, W., George, D.: Non-linear noise reduction and detecting chaos: some evidence from the S&P Composite Price Index. Mathematics and Computers in Simulation, Vol.48. (1999) 407-502
4. Catherine, K., Walter, C.L., Michel, T.: Noisy chaotic dynamics in commodity markets. Empirical Economics, Vol.29. (2004) 489-502
5. Rosenstein, M.T., Collins, J.J, De luca, C.J.: Apractial method for calculating largest-Lyapunov exponents in dynamical systems. Physica D, Vol.65. (1993)117-134
6. Lu, J.H., Zhang, S.C.: Application of adding-weight one-rank local-region method in electric power system short-term load forecast. Control Theory And Application, Vol. 19. (2002) 767-770
7. Oh, K.J., Han, I.: Using change-point detection to support artificial neural networks for interest rates forecasting. Expert Systems with Application, Vol.19. (2000) 105-115
8. Leung, M.T., Chen, A., Daouk, H.: Forecasting exchange rates using general regression neural networks Computer and Operations Research,Vol.27. (2000) 1093-1110
9. Kermanshahi, B.: Recurrent neural network for forecasting next 10 years loads of nine Japanese utilities. Neurocomputing, Vol.23. (1998) 125-133
10. Holger, K., Thomas, S.: Nonlinear Time Series Analysis. Beijing: Qinghua University Press (2000)
11. de Oliveira, Kenya, Andrésia, Vannucci, Álvaro, da Silva, Elton, C.: Using artificial neural networks to forecast chaotic time series. Physica A, Vol.284. (2000) 393-404
12. Kennel, M. B., Abarbanel, H., D., I.: Determining embedding dimension for phase space reconstruction using a geometric construction. Physical Review A, Vol.151. (1990) 225-223
13. Buczkowski, S., Hildgen, P., Cartilier, L.: Measurements of fractal dimension by box-counting: a critical analysis of data scatter. Physica A, Vol.252. (1998) 23–34.

14. Heinz, O.P., Dietmar, S.E.: The science of fractal Image. New York: Springer Verlag New York Inc (1988) 71-94.
15. Wang, H.Y., Zhu, M.: A prediction comparison between univariate and multivariate chaotic time series. Journal of Southeast University (English Edition),Vol.19. (2003) 414-417
16. Zhang, J., Lam, K.C., Yan, W.J., Gao, H., Li, Y.: Time series prediction using Lyapunov exponents in embedding phase space. Computers and Electrical Engineering,Vol.30. (2004)1-15
17. Grassberger, P., Procaccia, I.: Measuring the strangeness of the strange attractors. Physica, 9D (1983) 189-208.

Gradient-Based FCM and a Neural Network for Clustering of Incomplete Data

Dong-Chul Park

Intelligent Computing Research Lab.,
Dept. of Information Engineering,
Myong Ji University, Korea
parkd@mju.ac.kr

Abstract. Clustering of incomplete data using a neural network and the Gradient-Based Fuzzy c-Means (GBFCM) is proposed in this paper. The proposed algorithm is applied to the Iris data to evaluate its performance. When compared with the existing Optimal Completion Strategy FCM (OCSFCM), the proposed algorithm shows 18%-20% improvement of performance over the OCSFCM.

1 Introduction

When a data consists of numerical values, each numerical component of the data usually represents a feature value. The numerical data sometimes contains data that is missing one or more of the feature values. Any data with missing feature values is called an incomplete data. In addition, a data set which consists of at least one incomplete data is called an incomplete data set. Otherwise, it is called a complete data set. Research on clustering and estimating of the missing feature values in the incomplete data was first summarized by Dixon[1]. Hathaway and Bezdek proposed the optimal completion strategy FCM (OCSFCM), a modified fuzzy c-means algorithm for clustering of the incomplete data. They reported how near the centers of the clusters of the original complete data can be to the estimated data by applying the OCSFCM to the incomplete data[2]. The OCSFCM, however, requires the center values of the complete data in advance in its initial stage of the estimation process for the incomplete data and it is not feasible to utilize the OCSFCM when the center values of the complete data are not available as is the case for most practical problems.

This paper proposes a clustering algorithm for incomplete data that does not require the center values of the compete data in advance. The proposed algorithm in this paper first estimates the missing values of the incomplete data using an autoencoder neural network (AENN) and estimates the group centers by using the Gradient-Based FCM (GBFCM).

2 Gradient-Based Fuzzy c-Means Algorithm

One attempt to improve the FCM algorithm was made by minimizing the objective function using one input data at a time instead of the entire input data set. That is, the FCM uses all data to update the center value of the cluster, but the GBFCM that is used in this paper was developed to update the center value of the cluster with a given individual data sequentially.

In GBFCM, the update rule for the center vector v and the membership grade μ_i with a given data x can be summarized as:

$$v(k+1) = v(k) - 2\eta\mu_i^2(v(k) - x) \quad (1)$$

$$\mu_i = \frac{1}{\sum_{j=1}^{c}(\frac{d_i(x)}{d_j(x)})^2} \quad (2)$$

where k denotes the iteration index, η is a learning gain, and $d_i(x)$ is the distance from x to the i-th center.

As can be seen from Eq. (1) and Eq. (2), the GBFCM requires only one data to update the centers and corresponding membership values at a time. More detailed explanation about the GBFCM can be found in [3,4].

3 Optimal Completion Strategy Fuzzy c-Means

The OCSFCM is an effective algorithm for clustering of incomplete data. This algorithm finds a condition that minimizes the objective function with an additional condition from the incomplete data set and estimates the missing portion of the incomplete data. That is, the OCSFCM obtains estimated values of the missing components and the center value of clusters of the optimized incomplete data through adding an additional step given in Eq. (3) over the missing components during iteration [2].

$$x_j = \frac{\sum_{i=1}^{c}(\mu_i)^m v_{ij}}{\sum_{i=1}^{c}(\mu_i)^m} \quad (3)$$

where x_j is the j-th feature value of the incomplete data x and c represents the number of clusters.

4 Optimal Completion Autoencoder Fuzzzy c-Means

Since the OCSFCM utilized the FCM, the OCSFCM requires the center value of clusters of the original complete data set. However, when the original complete data sets are not available, the OCSFCM is not applicable. In this paper, an algorithm that does not require any information on original complete data sets and can estimate the missing components is proposed by using the autoencoder neural network (AENN) [5]. The AENN has a same number of input neurons and output neurons with a smaller number of hidden neurons. In particular, the information of the input pattern after training the AENN is presented by compressing the form of the bottleneck hidden layer.

The OCAEFCM proposed in this paper first estimates missing features by using the trained AENN and clusters the restored incomplete data by using the

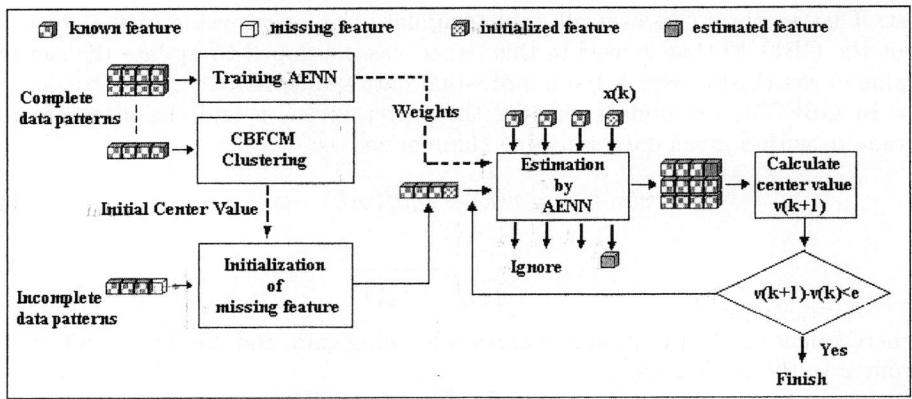

Fig. 1. A schemic block diagram of the OCAEFCM

GBFCM. The estimation of missing features is finished by one incomplete data by one incomplete data and the clustering is performed after estimating all the missing features in the incomplete data set. However, our goal is to find the proper clustering with an incomplete data set and individual missing features that do not affect the clustering results can be ignored. Therefore, the proposed algorithm does not estimate missing feature values by sequential repetition. As shown in Fig. 1, the proposed OCAEFCM estimates and restores the missing feature values only after evaluating the final clustering results by the GBFCM.

5 Experiments and Results

In this paper, Anderson's Iris data [6] was used as the experimental data set to examine the proposed OCAEFCM. The Iris data set has been used extensively to evaluate various clustering and classifier problems. The data set consists of 150 four-dimensional (equivalently, 600 feature) values with 50 vectors for each of three physically labelled classes. Out of the 600 feature values, randomly chosen features are labelled as missing features. The data which contains the missing values is considered an incomplete data. The percentages of incomplete data over complete data are set to vary from 10% to 60% for experiments. Note that an incomplete data must have at least one missing feature value and this experimental method was used in OCSFCM [2]. Experiments are performed 50

Table 1. Results on Iris data completion problem

Algorithm	Data Missing Rate					
	10%	20%	30%	40%	50%	60%
OCSFCM	0.067	0.074	0.085	0.086	0.091	0.109
OCAEFCM	0.043	0.062	0.063	0.067	0.075	0.081

times on each of the complete and incomplete data sets for obtaining unbiased results. In each experiment, the performances of the OCSFCM and OCAEFCM are calculated by the average value of the Mean Prototype Error (MPE). Table 1 shows the MPE of each algorithm compared. As can be seen from these results, the accuracy of prototype estimation by OCAEFCM is always at least as good as the other algorithms. It shows that the OCAEFCM gives 18%-20% less errors over the OCSFCM in MPE. This result shows that the proposed OCAEFCM is a more reliable tool for clustering incomplete data over the OCSFCM.

6 Conclusion

In this paper, an efficient clustering algorithm for incomplete data, called the OCAEFCM (Optimal Completion Autoencoder FCM), is proposed. The proposed OCAEFCM utilizes the AENN for estimating and restoring missing features and the GBFCM for clustering of incomplete data. The proposed OCAEFCM is applied to restoration of artificially deteriorated Iris data and compared with the OCSFCM. The results show that the OCAEFCM gives an improvement in MPE performance of 18% - 20% over the OCSFCM.

Acknowledgement. This research was supported by the Korea Research Foundation (Grant # R05-2003-000-10992-0(2004)).

References

1. Dixon,J.K.: Pattern Recognition with partly Missing Data, IEEE Trans Syst. Man Cybern. **9** (1979) 617-621.
2. Hathaway,R.J.,Bezdek,J.C.: Fuzzy c-Means Clustering of incomplete Data, IEEE Trans. Syst. Man Cybern. **31** (2001) 735-744.
3. Park,D.C.,Dagher,I.: Gradient Based Fuzzy c-Means (GBFCM) Algrithm, Proceedings of IEEE Int. Conf. on Neural Network **3** (1994) 1626-1631.
4. Looney,C.: Pattern Recognition Using Neural Newtorks, New York, Oxford University Press (1997) 252-254.
5. Thomson,B.B.,Marks II,J.J.,Choi,R.J.,El-Sharkawi,M.A.,Yuh,M.A.,Bunje,C.: Implicit Learning in Autoencoder Novelty Assement, Proceedings of IEEE IJCNN **3** (1999) 368-369.
6. Fisher,R.A.: The use of multiple measurement in taxonomic problems, Ann. Eugenics **7** (1936) 179-188.

Toward Global Optimization of ANN Supported by Instance Selection for Financial Forecasting

Sehun Lim

Department of Information System, Chung-Ang University
72-1 Naeri, Deaduck-Myun, Ansung-City, Kyunggi-Province, 456-756, South Korea
slimit@hanmail.net

Abstract. Artificial Neural Network (ANN) is widely used in the business to get on forecasting, but is often low performance for noisy data. Many techniques have been developed to improve ANN outcomes such as adding more algorithms, feature selection and feature weighting in input variables and modification of input case using instance selection. This paper proposes a Euclidean distance matrix approach to instance selection in ANN for financial forecasting. This approach optimizes a selection task for relevant instance. In addition, the technique improves prediction performance. In this research, ANN is applied to solve problems in forecasting a demand for corporate insurance. This research has compared the performance of forecasting a demand for corporate insurance through two types of ANN models; ANN and ISANN (ANN using Instance Selection supported by Euclidean distance metrics). Using ISANN to forecast a demand for corporate insurance is the most outstanding.

1 Introduction

When there is a need of forecasting in business management, ANN (Artificial Neural Network) is an excellent method to use. For example, ANN is widely used in the business to get on forecasting of bankruptcy, churning customers and stock price [10], [8]. Recently, in order to forecast problems more accurately in cooperation management, many techniques have been developed to improve outcomes such as adding more diverse algorithm, feature selection and feature weighting in input variables and modification of input case using instance selection.

The prediction performance using ANN depends on various factors such as the number of hidden layers and nodes as well as learning rate and the number of momentum, etc. Thus, ANN requires control to hidden layer and nodes as well as learning rate, momentum, etc. Also the method shows somewhat lower prediction level when the data set has noise and very large. Thus ANN needs to select relevant case for avoid an over-fitting problem inherent in ANN model.

In developing the most well forecasting model (ISANN : ANN supported by Instance Selection) for a demand for corporate insurance using instance selection supported by Euclidean distance metrics, two sample models (ANN, ISANN) of forecasting performances have been compared.

The result of this research has shown, first, to present more accurate forecasting model to the management of insurance cooperation. Secondly, the guide line will be given to whom a demand for corporate insurance has been used.

2 Research Background

2.1 ANN as a Tool for Forecasting

ANN is used in solving problems of pattern recognition, classification and forecasting after analyzing the linked relation between independent variables and dependent variables [4] [8]. For example, ANN is used in following fields of study; forecasting of stock price, bankruptcy and churning of customer. The advantage of using ANN is that it brings out more outstanding result than statistical technique.

Fletcher, Coss [2] who used ANN in forecasting a bankruptcy of the cooperation proved that ANN has given better performances in forecasting than logit analysis. Tam, Kiang [10] also verified that ANN is more superior in forecasting a bankruptcy of a bank. Recently, many researches are in process using ANN to improve the performance of forecasting with fuzzy membership function, Genetic Algorithm (GA), and Case-based Reasoning (CBR). For example, Yang et al. [10] has proved that Probabilistic Neural Networks (PNN) using neural network in forecasting a bankruptcy is more than existing Back Propagation Neural Networks (BPNN) or MDA.

2.2 Instance Selection

There are feature weighting, instance selection and feature selection techniques to improve the performance of forecasting in data mining. Especially, Instance selection is the technique to improve the performance of predicting in data mining for selecting relevant data. Their methods are proven to be affective in forecasting performance [9].

There are various methods to be applicable in instance selection. Kai et al. [7] has proved the remarkable improvements in forecasting a performance of recommended system based on collaboration with feature weighting and instance selection method. Ahn et al. [6] also verified the innovative progress in predicting of stock price using genetic algorithm and feature selection with instance selecting through case based reasoning. Cardie [1] proposed an instance selection method based on a decision tree approach to improved case-based learning.

3 ANN Supported by Instance Selection

To improve the ANN prediction performance, this research uses ISEDM (Instance Selection technique supported by Euclidean Distance Metrics) as an optimization tool of ANN performance. This research has compared the forecasting performance through ANN models, ISANN as mentioned earlier.

First, the ANN model is a conventional ANN model. This model uses a conventional method. Generally, feature selection and feature weighting of each feature of ANN is not considered and not a used instance selection. Second model is named as ISANN. ISANN is a sequential two step approach. It was carrying out, ISEDM for experimental data set 1 and experimental data set 2. And then, it carried out an ANN. The ISANN model consists of the following stage. For the first stage, we extract experimental data set 1 and experimental data set 2 from a large scale data base for demand forecasting of corporate insurance. For the second stage, we developed the matrix composed of experimental data set 1 and experimental data set 2 the width and length by Euclidean distance to reduce and extract large scale data. And then, we decided the ranking from the small of Euclidean distance to the experimental data set 1 and experimental data set 2 based on the selection principle of minimizing of Euclidean distance. For the third stage, we applied algorism of k-Nearest Neighbor (k-NN) to the instance selection of experimental data set 1 and experimental data set 2. In choosing case, we applied the algorism of k-NN and then chose each 50% in experimental data set 1 and experimental data set 2. Therefore it can improve the forecasting performance by applying to ANN.

4 Research Design and Experiments

4.1 Data Description and Measurement of Variables

Table 1. Descriptive Statistics

Feature Name	Minimum	Maximum	Mean
Firm Size	5558235	64529738387	764826604
Minority Shares	.00	85.70	5.1401
Financial Service Own	.00	78.40	1.0187
Government Ownership	-237.10	302.36	3.2303
Debt-to-Equit Ratio	.00	6.25	.8161
Leverage	-54.56	114.21	.5070
Liquidity	72836.00	14488258048.	144619473.48
Tax shields Depreciation	-143598916.	12290374509.	112897148.88
Tax shields Margin	-.08	4.56	.0122

This paper used KIS-FAS (Korean Investors Services-Financial Analysis System) accounting data of non-financial firms listed in the Korean Stock Exchange from 2000 to 2002. Fortunately the 800 data includes insurance premium expenditures, property casualty losses, ownership structure (shares of major institutional stockholders, the largest individual stockholder, minority stockholders, and foreigners), a number of employee as well as other financial data [3], [4]

In this research, the 9 independent variable is established as based on insurance theory. The original data are scaled into the range of [0, 1]. Linear scaling helps to decrease prediction errors. Table 1 summarizes the definitions of the variables used in this analysis, and also indicates our expectations about the effect of each on insurance demand. Below, we discuss the results of our estimation.

4.2 Research Method

This research has compared the performance of forecasting a demand for corporate insurance through two types of ANN and ISANN models as mentioned earlier. First model is a traditional ANN model which is named as ANN. Second is named as ISANN based on instance selection using Euclidean distance metrics.

ANN, ISANN were given fixed value of 5 for hidden layer for tests. The ratio for experimental data set 1 and experimental data set 2 is 80:20 for the test. Results consisted of 640 of experimental data set 1 and 160 of experimental data set 2. In experimental data set 1, the ratio for training data and holdout data is 80:20 for the test. Results consisted of 512 of training data and 128 of holdout data. Also, in experimental data set 2, the ratio for training data and holdout data is 80:20 for the test. Results consisted of 128 of training data and 32 of holdout data. The rest of conditions were set as given option provided in Clementine 8.1. The rest of default values are used as quick algorithm, Alpha 0.9, Initial Eta 0.3, Eta decay 30, High Eta 0.1, Low Eta 0.01.

5 Results

The prediction performances of ISANN and other alternative model are compared in this section. Table 2 describes the prediction performance of each model. As table 3 shows, in experimental data set 1, ISANN achieves higher prediction accuracy than ANN by 3.190% for the holdout data and by 16.218% for the training data. And in experimental data set 2, ISANN achieves higher prediction accuracy than ANN by 0.656% for the holdout data and by 3.810% for the training data.

According to this research outcome, using ISANN to forecast a demand for corporate insurance is the most outstanding. The order of outstanding performances of forecasting is following; ANN < ISANN. Reflecting on case reduction of Euclidean distance metrics is very important for realizing instance selection.

Table 2. Prediction performance of ANN, ISANN

Model	Experimental data set 1		Experimental data set 2	
	ANN	ISANN	ANN	ISANN
Training Data	53.226	69.444	53.333	57.143
Holdout Data	54.737	57.927	56.944	57.500

6 Conclusion

This research has optimized the forecasting performance a demand for corporate insurance using ISANN. This research result is very significant that forecasting of insurance demand can now have a model in making more accurate decisions.

There are some limits found while carrying out the research. The limitations are followings. First, the generalizability of ISANN should be test further applying them to other area. Second, ANN forecasting performances depend on learning rate, hidden node, hidden layer etc. Therefore, the control method of it remains an interesting further research topic.

References

1. C. Cardie.: Using Decision Trees to Improve Case-Based Learning, Proceedings of the Tenth international Conference on Machine learning, Morgan Kaufmann: San Francisco, CA, 1 (1993) 25-32
2. Flecher, D., and Goss, E.: A Comparative Analysis of Artificial Neural Network using Financial Distress Prediction, Information and Management, 24 (1993) 159-167
3. 3. Garven, J.R., and R.D. MacMinn,.: The Underinvestment Problem, Bond Covenants, and Insurance, The Journal of Risk and Insurance, 60, (1993), 635-646.
4. Hebb, D. O.: The Organization of Behaviors : A Neuropsychological Theory, New York : Wiley (1949)
5. Hornik, K.: Approximation Capability of Multilayer Feedforward Networks, Neural Networks, (1991) 251-257
6. H, Ahn, K, Kim, and I, Han.: Hybrid Genetic Algorithms and Case- Based Reasoning Systems, Lecture Notes in Computer Science 3314 (2004) 922-927
7. Kai et al.: Feature Weighting and Instance Selection for Collaborative Filtering : An Information Theoretic Approach, Knowledge and Information System 5 (2003) 201-224
8. K. Tam, M. Kiang.: Managerial Application of Neural Networks : The Case of Bank Failure Predictions, Management Science, 38 (1992) 926-947
9. Kim, K.: Toward Global Optimization of Case-Based Reasoning Systems for Financial Forecasting, Applied Intelligence, 21 (2004) 239-249.
10. Wilson, R, and Sharda, R.: Bankruptcy Prediction using Neural Networks, Decision Support Systems, 11 (1994) 545-557

FranksTree: A Genetic Programming Approach to Evolve Derived Bracketed L-Systems

Danilo Mattos Bonfim and Leandro Nunes de Castro

Graduate Program in Computer Science, Catholic University of Santos,
R. Carvalho de Mendonça, 144, Vila Mathias, Santos/SP, Brasil
dbonfim@lsin.unisantos.br, lnunes@unisantos.br

Abstract. L-system is a grammar-like formalism introduced to simulate the development of organisms. The L-system grammar can be viewed as a sort of genetic information that will be used to generate a specific structure. However, throughout development, the string (genetic information) that will effectively be used to 'draw' the phenotype of an individual is a result of the derivation of the L-system grammar. This work investigates the effect of applying a genetic programming approach to evolve derived L-systems instead of evolving the L-system grammar. The crossing over of plants from different species results in hybrid plants resembling a 'Frankstree', i.e. plants resultant from phenotypically different parents that present unusual body structures.

1 Introduction

Development at the multicellular level consists of the generation of structures by cell division, enlargement, differentiation, and cell death taking place at determined times and places in the entire life of the organism. It corresponds to the series of changes that animal and vegetable organisms undergo in their passage from the embryonic state to maturity, from a lower to a higher state of organization.

Evolutionary algorithms (Bäck et al., 2000a,b) have been used not only to solve a number of complex problems, but also as a tool for generic evolutionary design. The grammar of an L-system fully describes its development and can thus be considered the genetic information of an L-system. It has been used by many authors as the genetic material to be evolved by an evolutionary algorithm (Runqiang et al., 2002).

This paper introduces a different hybridization of evolutionary algorithms and L-systems. It proposes the use of a genetic programming approach to evolve derived L-systems and discusses its implications on the resultant phenotype. It shows that crossing over derived L-systems is equivalent to performing a graft between the plants.

2 Genetic Programming Design of L-Systems

In his 1968 papers, Lindenmayer (1968) introduced a notation for representing graph-theoretic trees using strings with *brackets*. The *bracketed L-systems* extend an L-system alphabet by the set {[,]}. The motivation was to formally describe branching structures found in many plants, from algae to trees.

The main reason that makes the automatic generation of L-systems a difficult task is the fact that a simple modification on a production rule or produced word, such as the inclusion of an orientation change symbol, may damage the whole plant-like structure. Other changes could affect the plant fatally.

To circumvent these difficulties and due to the tree-like nature of L-systems, Genetic Programming (GP) was used here as a theoretical basis (Koza, 1992; Bahnzaf et al., 1998) for evolving L-systems. Furthermore, instead of evolving the L-system grammars as already performed in the literature, we chose to evolve derived L-systems. In this case, the individuals of the population are the derived L-system words, and the crossover points have to be based on valid tree nodes of the 'adult plants'. No mutation is employed here. If a node is chosen as a cut point during a crossover operation, all nodes from its branch (sub-tree) must also be part of the crossover. In this case, an L-system plant is seen as a genetic programming tree with each bracket representing the interconnection between tree nodes and the initial point of each bracket representing a potential cut point.

2.1 Evolving L-Systems

The approach adopted here to efficiently perform crossover between two derived L-systems (plants) was to identify some *basic units* (branches) of the plants and cross them over. This was implemented through the generation of a pattern table, a bracket table, and a cut table, as follows:

o **Pattern table generation**: For all plants that are going to suffer crossover, find and isolate all the *basic units* that compose the L-system word. The basic units are defined here as the different production rules, which, by default, cannot be divided during a crossover operation. To do this, a *pattern table* is generated for saving the initial and final position of each basic unit within a derived L-system;
o **Bracket table generation**: A new *bracket table* is generated for saving all the initial and final positions of the brackets located outside the basic units found;
o **Cut table generation**: To perform crossover, one *cut point* for each parent plant has to be selected based on the bracket table. Integer numbers between 0 and the word length are randomly generated. All the intervals on the bracket table that contain these numbers are selected and placed on a *cut table* for each plant;
o **Crossover**: With a given probability, the crossover operation has to obey two basic rules. First, the direction of both selected branches (sub-trees) must have a maximum difference of 90°. Second, if the direction of a branch is ±90°, this can only be replaced by another branch pointing in the same direction.

3 Some Computer Experiments

A number of plants are chosen to compose the population for evolution. The user is then responsible for selecting a plant that will serve as a parent and will be crossed over with another plant from the population. Crossover is performed with a probability $pc = 50\%$. Three different types of experiments were performed here. (a) A population of identical individuals (Fig. 1a); (b) A population with two different kinds of plants (Fig. 1b). (c) A population with six different individuals (Fig. 1c).

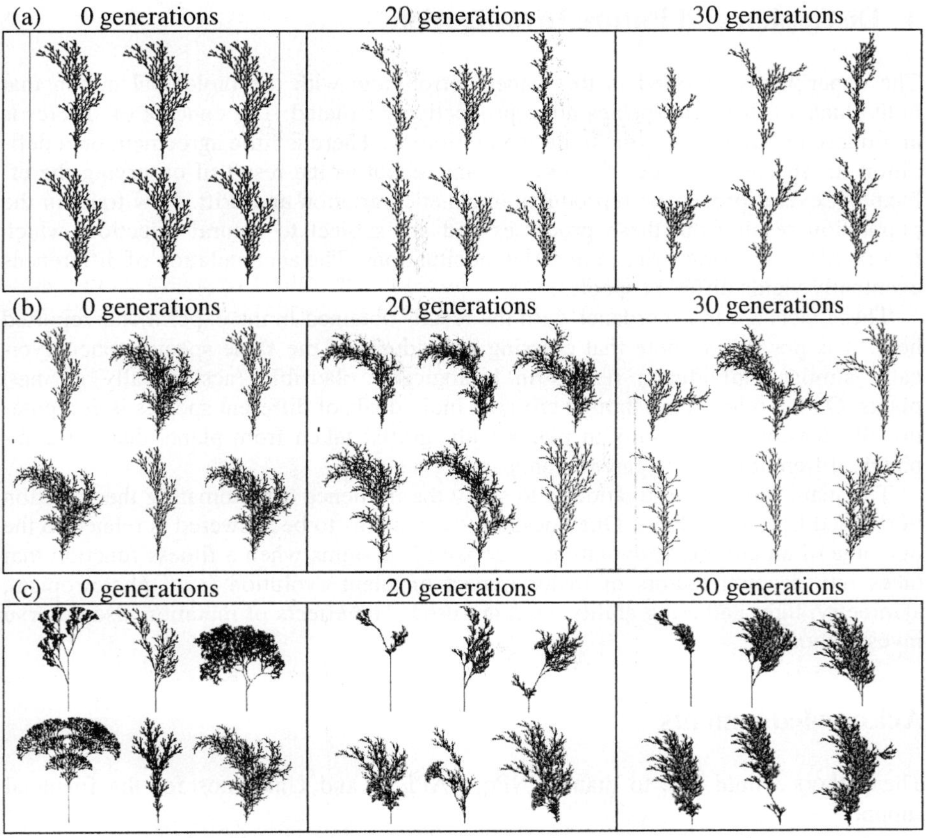

FranksTree: A Genetic Programming Approach to Evolve Derived Bracketed L-Systems 1277

Fig. 1. Aesthetic evolution of (a) six identical plants, (b) six plants of two different species, (c) six different plants

Fig. 2 shows some situations when the crossing of different plants resulted in mongrel offspring that are possibly not biologically plausible. The mix of genetic information from adult parents of different species leads to plants that are actually constructed out of parts from incompatible individuals.

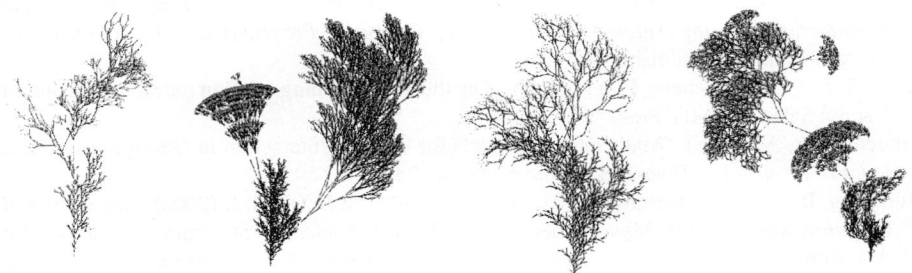

Fig. 2. Hybrid plants (FranksTrees) generated from parents of different species

4 Discussion and Future Investigation

The experiments reported in this paper corroborate with the biological notion that individuals of different species are reproductively isolated. The concept of *species* is an important but difficult one to define in biology. There is little agreement on a definition for species. The idea for species came about as the result of observing the effects of several processes: reproduction, genetic variation and drift. New traits in the population result from these processes and are subject to natural selection, which favors different characteristics in different situations. The accumulation of differences eventually yields different species.

These facts are in accordance with the results obtained in the experiments reported here. It is possible to note that crossing individuals of the same species (phenotypically similar individuals) results in biologically plausible (aesthetically normal) plants. On the other hand, though crossing individuals of different species is computationally feasible; this results in plants with 'grafts' taken from plants that are completely different from the parent plants.

The main future investigation is to study the influence of automating the selection of derived L-system plants. One question that remains to be answered is related to the outcome of an automatic evolution of derived L-systems when a fitness function that takes into account factors of major impact on plant evolution (e.g., phototropism, symmetry, light gathering ability, etc.) are used. The effects of mutation also deserve investigation.

Acknowledgements

The authors would like to thank CNPq, FAPESP and UniSantos for the financial support.

References

Bäck, T., Fogel, D. B. and Michalewicz, Z. (eds.) (2000a), *Evolutionary Computation 1: Basic Algorithms and Operators*, Institut of Physics Publishing.

Bäck, T., Fogel, D. B. and Michalewicz, Z. (eds.) (2000b), *Evolutionary Computation 2: Advanced Algorithms and Operators*, Institute of Physics Publishing.

Banzhaf, W. Nordin, P. Keller, R. E. & Francone, F. D. (1998), *Genetic Programming – An Introduction: On the Automatic Evolution of Computer Programs and Its Applications*, Morgan Kaufmann Publishers.

Koza, J. R. (1992), Genetic Programming: On the Programming of Computers by means of Natural Selection, MIT Press.

Lindenmayer, A. (1968), "Mathematical Models for Cellular Interaction in Development, Parts I and II", *Journal of Theoretical Biology*, **18**, pp. 280-315.

Runqiang, B., Chen, P., Burrage, K., Hanan, J., Room, P. & Belward, J. (2002), "Derivation of L-system Models from Measurements of Biological Branching Structures Using Genetic Algorithms", In T. Hendtlass and M. Ali (Eds.), *Lecture Notes in Artificial Intelligence 2358*, pp. 514-524.

Data Clustering with a Neuro-immune Network

Helder Knidel[1], Leandro Nunes de Castro[2], and Fernando J. Von Zuben[1]

[1] LBiC/DCA/FEEC/UNICAMP PO Box 6101, Campinas/SP, Brazil 13083-970
{knidel, vonzuben}@dca.fee.unicamp.br
[2] Graduate Program in Informatics, Carvalho de Mendonça, 144
Vila Mathias, Santos/SP, Brazil
lnunes@unisantos.edu.br

Abstract. This paper proposes a novel constructive learning algorithm for a competitive neural network. The proposed algorithm is developed by taking ideas from the immune system and demonstrates robustness for data clustering in the initial experiments reported here for three benchmark problems. Comparisons with results from the literature are also provided. To automatically segment the resultant neurons at the output, a tool from graph theory was used with promising results. A brief sensitivity analysis of the algorithm was performed in order to investigate the influence of the main user-defined parameters on the learning speed and accuracy of the results presented. General discussions and avenues for future works are also provided.

1 Introduction

Most living organisms exhibit extremely sophisticated learning and processing abilities that allow them to survive and proliferate, generation after generation, in dynamic and competitive environments. These are some of the reasons why nature has inspired several scientific and technological developments [3]. One such natural system is the human immune system. It can be seen as a parallel and distributed adaptive system with a tremendous potential as a source of inspiration for the development of robust problem solving techniques. This results from the fact that the immune system exhibits features like self-maintenance, pattern recognition and classification, feature extraction, diversity, adaptability, memory to past encounters, distributed detection, and self-regulation [2],[6].

This paper extends the work proposed in [4] on the use of features from the human immune system, and also artificial immune systems [6], to design novel artificial neural network learning algorithms. In particular, the network to be developed here is modeled as a competitive and constructive (i.e., with network growing and pruning phases) artificial neural network. Differently from the network proposed in [4], which was characterized as a Boolean neural network, called ABNET (Antibody Network), this new version uses real-valued vectors to represent the weight connections to the neurons and is, thus, termed RABNET (Real-valued Antibody Network).

The resultant hybrid system has a typical competitive neural network architecture similar to a one-dimensional self-organizing map [11]. In order to adapt to the input patterns, the network makes use of several features of an immune response, such as the clonal expansion of the most stimulated cells, death of the non-stimulated cells and the affinity maturation of the repertoire. The network does not have a predefined number of neurons, which will be dynamically determined based on immune principles. Finally, a minimal spanning tree [14] is used to automatically specify the number of clusters in the neural network after learning, and thus in the input data.

2 RABNET: A Neuro-immune Network

Inspired by ideas from immunology, the RABNET development assumes an antigen population (**Ag**) that should be recognized by an antibody repertoire (**Ab**) modeled as a one-dimensional competitive neural network with real-valued weights. At the beginning of the adaptation process, RABNET contains a single antibody (neuron) in the network and grows when required.

Similar to the work in [4], RABNET presents the following main features: competitive network with unsupervised learning and dynamic network structure, with growing and pruning phases governed by an implementation of the clonal selection principle. As a distinctive aspect, RABNET makes use of real-valued connection strengths in an Euclidean shape-space [12], instead of binary connections [4].

Furthermore, a method based on graph theory is presented as a means to automatically determine the number of groups in the network after training. It is important to remark here that most, if not all, self-organizing neural networks requires an additional technique after training for segmenting the output grid of the network in order to automatically identify the clusters found. An example of an approach typically used in the literature is the U-matrix [13].

2.1 Competitive Phase

The competitive phase of the algorithm involves finding the most similar antibody $\mathbf{Ab}_K$ to a given antigen **Ag**; that is, to find the winner neuron to a given input pattern. This antibody is said to have the highest affinity with the antigen.

$$K = \arg \min_k \|\mathbf{Ag} - \mathbf{Ab}_k\|, \forall k \qquad (1)$$

In the present implementation, there is one variable (v_k) that stores the index of the antibody that has the highest affinity for each antigen. For instance, if the antibody with highest affinity with a given antigen $\mathbf{Ag}_5$ is antibody 9 ($\mathbf{Ab}_9$), then $v_5 = 9$. This is aimed at calculating the *concentration level* of each antibody, what corresponds to the number of antigens recognized by each antibody.

2.2 Network Growing

Network growing is inspired by the clonal selection principle, where the most stimulated cell is selected for cloning. The choice of the most stimulated cell is based on the affinity to the antigen, determined during the competitive phase, and also on the concentration of antigens recognized by an antibody. The antibody with the highest antigen concentration will generate a single offspring (clone) and the two antibodies might turn into memory antibodies after their maturation phase.

Two parameters control the network dynamics and metadynamics: one related to the concentration of antigens (τ) recognized by a given antibody, and the other related

to the affinity threshold (ε) between an antigen and an antibody. All antibodies j stimulated by more than one antigen ($\tau_j > 1$) are potential candidates to be cloned. The network growing process can be described by the following sequence of steps, executed after a sequence of β iterations has been performed.
If the current iteration is a multiple of β,

    a. **Then**, antibody $\mathbf{Ab}_s$ with highest concentration level τ is selected for cloning. If there is more than one antibody with the same concentration level, one of then is chosen randomly.
    b. From all antigens recognized by antibody $\mathbf{Ab}_s$, antigen $\mathbf{Ag}_w$ with lowest affinity to $\mathbf{Ab}_s$ is selected. **If** the distance $(\mathbf{Ab}_s - \mathbf{Ag}_w) > \varepsilon$, **then** antibody $\mathbf{Ab}_s$ is cloned; otherwise, the network structure does not change.

The weight vector of the newly created antibody receives the attributes of the antigen (input pattern) with the smallest affinity to $\mathbf{Ab}_s$, that is, the one with highest Euclidean distance to the antibody selected for cloning.

2.3 Network Pruning

The strategy adopted here to perform network pruning is based on the concentration level of each antibody. If the concentration level of a given antibody $\mathbf{Ab}_i$ is zero ($\tau_i = 0$), it means that antibody i was not stimulated by any of the antigens. In this case, antibody i is pruned. Pruning is performed every iteration.

2.4 Weight Updating

Updating the weights in RABNET is similar to the procedure used in winner-takes-all competitive neural networks. Equation (2) shows the weight updating rule used, where α is the learning rate and $\mathbf{Ab}_k$ is the antibody that recognizes antigen $\mathbf{Ag}$. Thus, antibodies are constantly being moved in the direction of the recognized antigens. After γ iterations, the learning rate α is exponentially decreased by a factor σ.

$$\mathbf{Ab}_k = \mathbf{Ab}_k + \alpha(\mathbf{Ag} - \mathbf{Ab}_k). \tag{2}$$

2.5 Convergence Criterion

Two features are important to define a convergence criterion: the stability of the network topology and of the network weights. To contemplate both, it is proposed a convergence criterion that checks the stability of the number of neurons in the network, and the variation in the weight vectors.

It is assumed that the network topology has reached stability if during the last 10.β iterations there is no variation in the number of neurons. Concerning the weights, they are assumed to have stabilized if the sum of their modules does not vary by more than 10^{-4} from the current iteration to the past 10.β iterations.

2.6 Defining the Number of Clusters

One of the main objectives of RABNET is to cluster unlabelled data. The adaptation procedure builds a network that associates each element of the input data set with a single neuron at the output by means of the inner product between the current input and the weight vector of each neuron. This way, the spatial distribution of the input patterns is represented by the spatial distribution of the network weight vectors.

However, after network learning it is still necessary to extract the information obtained, i.e., to automatically determine the number of clusters detected by the network. In order to accomplish this task, we propose the use of a minimal spanning tree (MST) [14], which is a powerful tool from graph theory for data clustering.

This clustering process can be seen as a two level approach: first the dataset is clustered and quantized using the RABNET, and then the clusters found are segmented by the MST. The MST thus defines a neighborhood relationship among antibodies (the learning algorithm does not account for neighborhood) and determines the optimal number of clusters found by the learning algorithm.

The application of the MST to automatically segment the neurons at the output layer of RABNET is performed as follows. First, link all network antibodies using a minimal spanning tree, what can be implemented, for instance, by Prim's algorithm [10]. Second, determine the inconsistent edges and remove them, thus identifying those edges that are linked together as belonging to the same cluster.

An inconsistent edge may be determined as follows: for each edge of the MST built, its two end points are analyzed; the average and the standard deviation of the length of all edges which are within p steps from each end point are calculated; if the length of an edge is greater than the average plus two standard deviations, then this edge is considered inconsistent [14].

2.7 RABNET Pseudocode

```
1. Initialization and parameter definition
   1.1. Initialize randomly a single antibody in the network
   1.2. Define parameters: α, β, γ and σ
2. While the convergence criterion is not reached do:
   2.1. For each input pattern do:
      2.1.1. Present a random antigen to the network;
      2.1.2. Calculate the Euclidean distance between the anti-
             gen presented and the antibodies in the network;
      2.1.3. Find the winner antibody (Eq. (1));
      2.1.4. Increase the concentration level of the winner;
      2.1.5. Update the weights of the winner antibody: Eq.(2);
   2.2. If iteration > γ
      2.2.1. α = σ * α;
   2.3. If iteration is multiple of β, then
      2.3.1. Grow if necessary (Section 2.2)
   2.4. If the concentration level of a given antibody is zero,
        then prune it from the network.
3. Use the MST criterion proposed to automatically segment
   the neurons at the output layer of the network.
```

3 Performance Evaluation

RABNET was applied to three benchmark problems and the results compared with the literature (when available and appropriate). In particular, RABNET was applied to the Animals data set [11], to the Two-Spirals data set [13], and to the Chain Link data set [13]. The parameters used to run RABNET in all experiments to be reported here were: $\alpha = 0.2$, $\beta = 2$, $\gamma = 100$ and $\sigma = 0.95$, unless otherwise specified.

3.1 Animals Data Set

The Animals data set [11] consists of the description of 16 animals by binary property. By varying the affinity threshold, ε, it is possible to obtain networks with different sizes, as depicted in Figs. 1(a). The right hand side scale represents the weight value of each connection plotted with the corresponding gray level in the network picture; darker lines correspond to connections with larger positive values. Note that higher values of ε result in more generalist networks, i.e., networks with a smaller number of neurons. For instance, for $\varepsilon = 2$ it could be observed that only two neurons resulted from learning and these were responsible for mapping the groups of birds and mammals, respectively, the two major classes of animals in the data set.

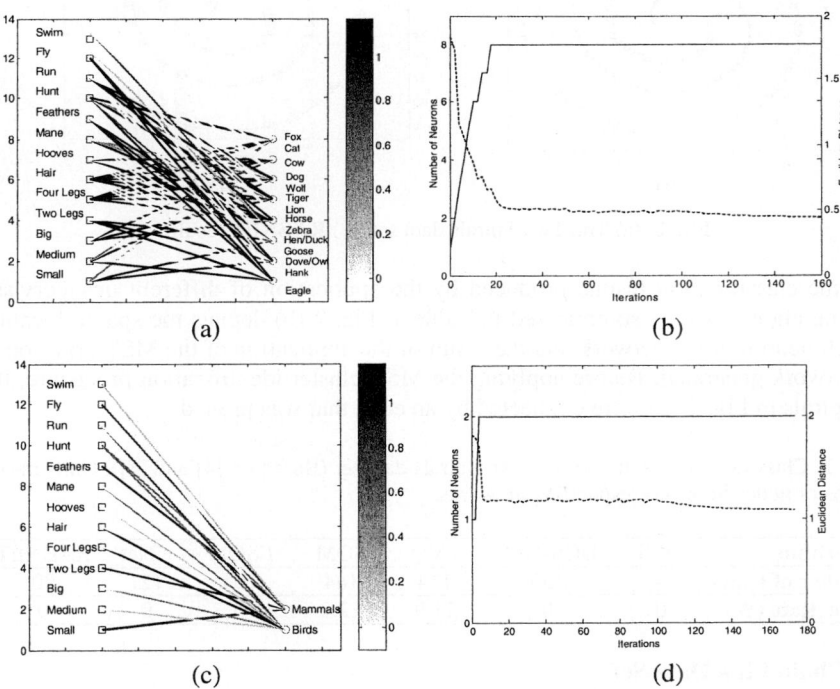

Fig. 1. RABNET applied to the Animals data set. Figs (a) and (b) $\varepsilon = 1$ and Figs (c) and (d) $\varepsilon = 2$. Figs (a) and (c) show the networks obtained with their respective output units. In Figs (b) and (d) the dashed lines correspond to the average error and the solid lines indicate the network size evolution.

The use of an affinity threshold value close to zero, for instance $\varepsilon = 10^{-6}$, should produce a network with the same number of neurons as input patterns. However, when using this value for the Animals data set, this was not verified. After looking for a possible cause, it could be observed that this happened because the Owl and the Hawk present exactly the same attributes, thus being mapped into the same neuron. The same has happened with the Zebra and the Horse.

3.2 Two Spirals Data Set

The two spirals data set has already been used to assess the performance of unsupervised growing neural networks [13],[4]. The data set used here has 95 input patterns in each spiral, as presented in Fig. 2 (a). The input patterns were adjusted within the [0,1] interval, and an affinity threshold $\varepsilon = 0.025$ was empirically obtained.

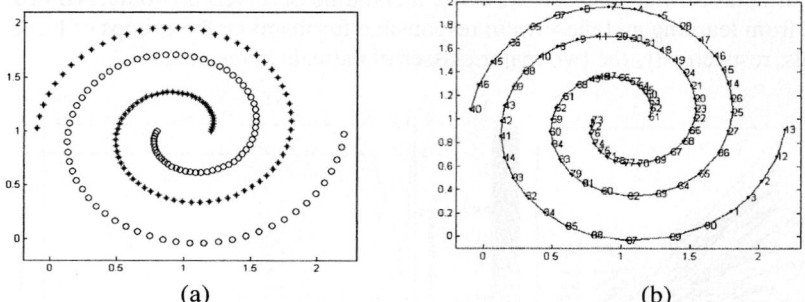

Fig. 2. (a) The Two Spirals data set. (b) RABNET result.

Some classification results produced by the application of different unsupervised learning algorithms are summarized in Table 1. Fig. 2 (b) depicts the spatial location of each neuron in the network and the result of the application of the MST criterion to the network generated. Before applying the MST cluster identification procedure, the two spirals in Fig. 2 (b) were connected by an edge that was pruned.

Table 1. Classification results for the Two Spirals data set (Based on [4] and [8]). Units means neurons or nodes in the corresponding networks.

Algorithm	GCS	DCS-GCS	LVQ	SOM	ESOM	aiNet	RABNET
Number of Units	145	135	114	144	105	121	90
Error Rate (%)	0	0	11.9	22.2	0	0	0

3.3 Chain Link Data Set

The Chain Link data set [13], sometimes called Donuts data set, consists of 1,000 input patterns forming two intertwined 3D rings, as depicted in Fig. 3 (a). The results presented by RABNET, when applied to the Chain Link data set, were also encouraging (see Fig. 3 (b)). This result was obtained using an affinity threshold $\varepsilon = 0.3$ and the other parameters as presented earlier. It was observed that the algorithm required only 14 prototypes to represent the input data with 100% accuracy.

One of the best results available in the literature for the Chain Link problem was obtained by an algorithm named aiNet [4]. In its minimum configuration, aiNet required 55 prototypes (cells) to represent this data set with 100% accuracy. Fig. 3 (b) presents the spatial distribution of the 2 clusters (rings) generated by the network. It is important to remark that aiNet and RABNET are different types of network models. Work in [4] summarizes the differences and similarities between aiNet and competitive neural networks. Similar comparison can be made between RABNET and aiNet.

4 Sensitivity Analysis

The application of RABNET to the solution of any problem requires the definition of parameters α, β, γ, σ and the affinity threshold ε. To study the influence of these parameters in the behavior and final result of RABNET, a sensitivity analysis of the algorithm will be performed by applying it to the Animals and Chain Link data sets.

As discussed, parameter ε influences the specificity of the network cells, and thus the final number of neurons in the network: the lower the value of ε, the larger the number of neurons in the network, and vice-versa. This behavior can be observed in the networks of Figs. 1(a) and 1(c). Table 2 summarizes the trade-off between the number of cells in the network and ε for the Animals data set. In practice, this corresponds to having more generalist or more specific networks. For instance, in the experiments performed with the Animals data set, it can be seen that a network with only two neurons can group together mammals and birds, while a network with eight neurons (Fig. 1(a)) can group together passive birds, mammals, etc.

To study the influence of σ and γ in the final network size and in the number of iterations for convergence, 30 independent runs of RABNET were performed for the Animals data set. Three different values of each of these parameters were adopted: $\gamma \in \{20,100,200\}$; $\sigma \in \{0.1,0.5,0.95\}$. The results are summarized in Figs. 4 to 6.

It can be observed from Figs. 4 and 5 that parameters σ and γ have almost no influence on the final network size, but strongly influence the number of iterations for convergence. Higher values of σ and γ result in longer convergence times.

To evaluate the influence of the initial learning rate and the affinity threshold on the final network size, 10 experiments were run with the Chain Link data set. The other parameters used in this analysis were $\beta = 2$, $\gamma = 100$ and $\sigma = 0.95$.

Fig. 6 shows the results obtained for the Chain Link data set when the initial learning rate and the affinity threshold are varied. In Fig. 6 it can be observed that, in addition to the affinity threshold, the initial learning rate (α) also influences the final number of neurons in the network, but in a much lesser degree. For a constant affinity threshold, it can be observed that the smaller the initial learning rate, the smaller the final number of neurons in the network. In the experiments performed, it could also be observed that when the initial value α is very small, the network may not be able to reach an adequate structure.

Table 2. Tradeoff between the number of cells in the network and the affinity threshold ε

$\varepsilon = 0$	$\varepsilon = 1$	$\varepsilon = 2$
14 Cells	8 Cells	2 Cells

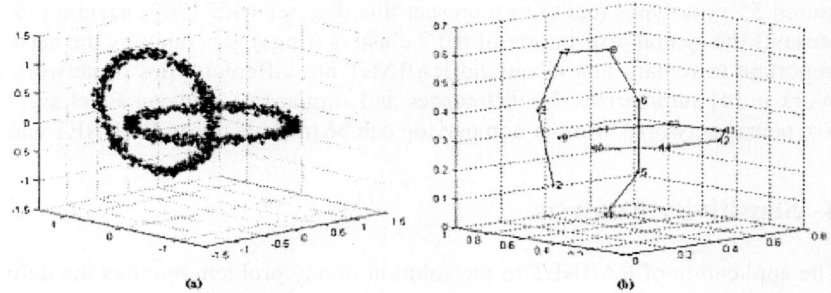

Fig. 3. (a) Chain Link data set. (b) Clusters obtained by the application of RABNET to the Chain Link data set.

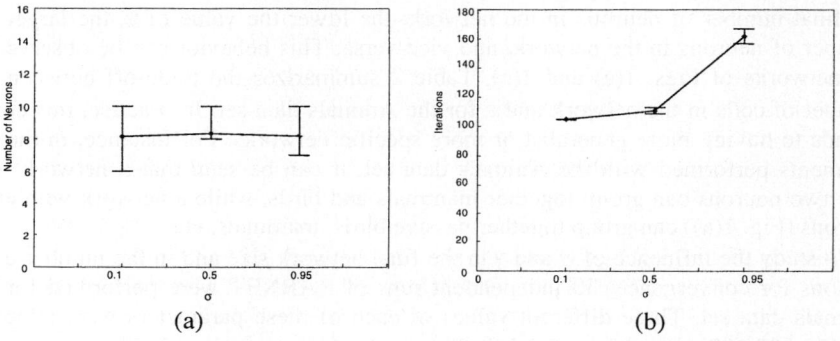

Fig. 4. Final number of neurons in the network (a), and iterations for convergence (b) for three different values of σ: σ = 0.1, σ = 0.5, and σ = 0.95

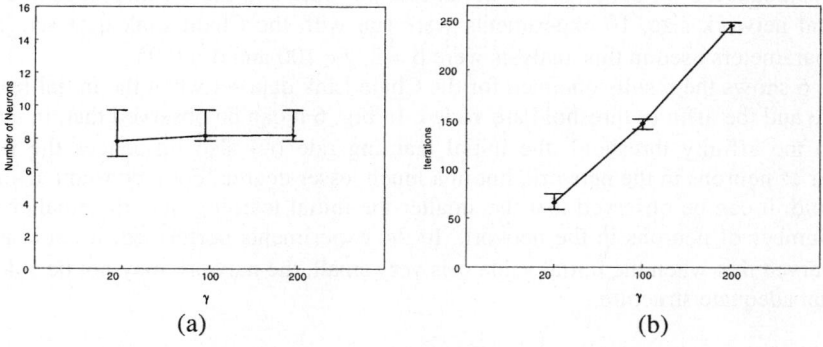

Fig. 5. Final number of neurons in the network (a), and iterations for convergence (b) for three different values of γ: γ = 20, γ = 100, and γ = 200

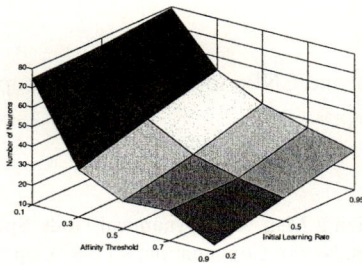

Fig. 6. Number of neurons in the output layer as a function of the initial learning rate and the affinity threshold for the Chain Link data set

5 Discussion and Future Trends

This paper introduced RABNET, a real-valued antibody network, developed by mixing ideas from the immune system with concepts from artificial neural networks. A simple competitive neural network constitutes the basic model, and a learning algorithm developed by taking ideas from the immune system is proposed to determine the network weights and structure according to the input data.

The main concepts from the immune system used in the development of the algorithm are clonal selection and affinity maturation. The resultant network has the following main properties: competitive learning, constructive architecture with growing and pruning phases, real-valued connection strengths, and automatic determination of the number of clusters.

After learning, the number of clusters (groups) identified by the network is determined using a criterion based on a minimum spanning tree (MST) criterion. The MST is built connecting the resultant weight vectors and those edges considered inconsistent are removed from the network. So, the neighborhood definition and the cluster discrimination are both performed a posteriori.

To assess the performance of RABNET, it was applied to three benchmark problems and the obtained results are compared with those from the literature. The network demonstrated to be capable of appropriately grouping the input data with the production of parsimonious networks. A brief sensitivity analysis was also carried out to study the sensitivity of the algorithm to its user-defined parameters. It was observed that most parameters only influence the convergence speed of the algorithm, and not the final quality of the results. The exception is the affinity threshold, which is responsible for the construction of more or less parsimonious solutions.

The proposed algorithm opens up some avenues for future investigation. First, it is necessary to perform exhaustive tests with the system, including the use of more challenging data sets, and broad comparisons with similar networks from the literature, such as the works presented in [13], [1] and [9]. Fundamental differences between these works and our proposal are the absence of the concept of a neighborhood during the process of weight adjustment in RABNET, and the way growing is performed. As already mentioned, the neighborhood is defined a posteriori by an MST algorithm.

Acknowledgements

The authors thank FAPESP and CNPq for their financial support.

References

1. Cho, S. -B.: Self-Organizing Map with Dynamical Node Splitting: Application to Handwritten Digit Recognition, Neural Computation, 9 (1997), pp. 1345-1355
2. Dasgupta, D.: Artificial Neural Networks and Artificial Immune Systems: Similarities and Differences, Proc. of the IEEE SMC, 1 (1997), pp. 873-878
3. de Castro, L. N., Von Zuben, F. J.: Recent Developments in Biologically Inspired Computing, Idea Group Publishing (2004)
4. de Castro, L. N., Von Zuben, F. J., Júnior, G. A. de D.: The Construction of a Boolean Competitive Neural Network Using Ideas From Immunology, Neurocomputing, 50C (2003), pp. 51-85
5. de Castro, L. N., Von Zuben, F. J.: Immune and Neural Network Models: Theoretical and Empirical Comparisons, International Journal of Computational Intelligence and Applications (IJCIA), 1(3), (2001), pp. 239-257
6. de Castro, L. N., Timmis, J.: Artificial Immune Systems: A New Computational Intelligence Approach. Springer-Verlag (2002)
7. Haykin, S.: Neural Networks – A Comprehensive Foundation, Prentice Hall, 2nd (1999)
8. Kasabov, N.: Evolving Connectionist Systems: Methods and Applications in Bioinformatics, Brain Study and Intelligent Machines, Springer Verlag (2002)
9. Marsland, J., Shapiro, S., Nehmzaw, U.: A Self-Organising Network that Grows When Required, Neural Networks, 15 (2002), pp. 1041-1058
10. Prim, R. C.: Shortest Connection Networks and Some Generalizations, Bell Sys. Tech. Journal, 36 (1957), pp. 1389-1401
11. Ritter, J., Kohonen, T.,: Self-Organizing Semantic Maps, Biolog. Cybern., vol. 61 (1989), pp. 241–254
12. Segel, L., Perelson, A. S.: Computations in Shape Space: A New Approach to Immune Network Theory, in: ed. A. S. Perelson, Theoretical Immunology 2 (1988), 321-343
13. Ultsch, A.: Knowledge Extraction from Artificial Neural Networks and Applications, Information and Classification, O. Optiz et al. (Eds.), Springer, pp. 307-313, (1993)
14. Zahn, T.: Graph-Theoretical Methods for Detecting and Describing Gestalt Clusters. IEEE Trans. on Comp. C-20 (1971), pp.68-86

Author Index

Ahmad, Muhammad Bilal II-43
Akin, Erhan III-787
Alatas, Bilal III-787
Almeida, Gustavo Maia de III-313
Amin, Hesham H. I-456
An, Wensen I-546
Anh, Vo III-337
Aoki, Terumasa II-622
Araabi, Babak N. II-1250
Arita, Jun II-165
Asiedu, Baffour Kojo III-903
Asiimwe, Alex Jessey III-968
Aydin, Serkan III-703

Baba, Sapiyan I-893
Bae, Hyeon I-1160, II-564
Bae, SungMin II-530
Bai, JianCong III-74
Bai, Lin II-780
Baicher, Gurvinder S. III-877
Balescu, Catalin III-1300
Bao, Leilei III-988
Bao, Zhejing I-688
Barbosa, Helio J.C. II-941
Bea, Hae-Ryong III-1182
Belatreche, A. I-420
Bertoni, Alberto III-235
Bi, D. II-266, II-592
Bianco, Luca II-1155
Bin, Guang-yu I-1031, III-646
Bingul, Zafer III-1304
Bolat, Bülent I-110
Bonfim, Danilo Mattos I-1275
Borne, Pierre III-259
Bu, Nan II-165
Butun, Erhan II-204
Byun, Kwang-Sub II-85

Čada, Josef I-1234
Cai, Guangchao III-915
Cai, Jiamei I-1117
Cai, Yici III-181
Cai, Yunze I-51, I-528, I-582, II-175
Cai, Zhiping II-913

Cai, Zixing III-1308
Campadelli, Paola III-235
Cao, Chunhong I-783, II-139
Cao, Qixin III-535, III-723
Cao, Xianqing III-1162
Cao, Yijia II-895
Cao, Zuoliang III-984
Carvalho, Andre C.P.L.F. I-1189
Chai, Tianyou II-214
Chambers, J.A. I-199
Chang, Chia-Lan II-735
Chang, Chuan-Wei II-296
Chang, Hui-Chen III-1172
Chang, HuiYou III-74
Chang, Min Hyuk II-43
Chang, Pei-Chann I-364, II-983, III-205
Chang, Ping-Teng I-619
Chang, Ray-I I-1224
Chang, Yongmin I-850
Chau, Kwokwing III-1152
Chen, Chen-Tung I-619
Chen, Chih-Ming III-1186
Chen, Chongcheng III-1051
Chen, Chunlin II-686
Chen, Duo II-1269
Chen, G.L III-1060
Chen, Gang II-37, II-270
Chen, Guangzhu III-444
Chen, Guochu II-610, III-515
Chen, Haixu III-938
Chen, Hongjian II-1218
Chen, Huanwen III-384
Chen, Jiah-Shing II-735, III-798
Chen, Jin-Long III-1186
Chen, Jing II-539
Chen, Ke II-656
Chen, Liangzhou I-679
Chen, Ling II-1218, II-1239
Chen, Qian III-490
Chen, Qingzhan III-782
Chen, S. II-1122
Chen, Shengda III-1235
Chen, Shengyong I-332
Chen, Shi-Fu III-855

Chen, Shiming II-913
Chen, Shu-Heng III-612
Chen, Shuang I-101
Chen, Tianping I-245
Chen, Tieming I-1117
Chen, Xin III-628
Chen, Xinghuan III-57
Chen, Xuefeng II-324
Chen, Yan Qiu II-705, III-822, III-845
Chen, Yanmin I-947
Chen, Ying-Chun III-482
Chen, Yingchun II-1101
Chen, Yong II-890
Chen, Yun Wen II-705
Chen, Zehua II-945
Chen, Zhao-Qian II-55
Chen, Zhong II-1
Chen, Zong-Ming II-425
Chen, Zonghai II-686
Cheng, Chun-Tian III-453, III-1152
Cheng, Kuo-Hsiang III-1186
Cheng, Lixin I-470
Cheng, Yinglei III-215
Cheng, Zhihong III-444
Cheremushkin, Evgeny II-1202
Chi, Huisheng I-167
Chien, Shu-Yen II-296
Chiu, David II-306
Cho, Daehyeon I-536
Cho, Eun-kyung III-1069
Cho, Yeon-Jin I-1009
Cho, Yeun-Jin I-1002
Choi, Dong-Seong I-797
Choi, Jinsung II-552
Choi, Jonghwa I-1185, II-552
Choi, Wonil I-850
Chu, Ming-Hui II-296
Chu, Tianguang I-769
Chua, Ming-Hui II-296
Chun, Jong Hoon II-43
Chunjie, Yang I-696
Cong, Shuang I-773
Cooper, Leon N. I-71, I-554, I-877
Copper, J. I-1039
Corchado, Emilio I-778
Cui, Baotong I-1
Cui, Du-Wu II-1269, III-86
Cui, Zhihua III-255, III-467
Cukic, Bojan I-750

Dai, Hongwei III-332
Dai, Yuewei III-976
Damiance, Antonio P.G., Jr. I-1189
Dang, Chuangyin III-392
Dang, Yan I-956
Daoying, Pi I-696
de Carvalho, Luis A.V. II-941
de Castro, Leandro Nunes I-1275, I-1279
Demir, Ibrahim II-648
Deng, Fei-Qi I-1150
Deng, Weihong I-915
Ding, Hongkai I-119
Ding, Juling II-804
Ding, Lixin II-1049
Ding, Zhan I-835
Dixit, Vikas III-1242
Do, Tien Dung II-849
Dong, Chaojun I-340
Dong, Daoyi II-686
Dong, Jin-xiang III-48
Dong, Jingxin II-105
Dong, Jinxiang II-1229
Dong, Min I-397
Dong, Qiming II-185
Dong, Xiuming III-374
Du, Haifeng II-826, II-876, II-931, III-399
Du, Shihong III-1261, III-1274
Du, Wenli II-631
Du, Ying I-480
Du, Yuping III-592
Duan, Ganglong I-640

Ebecken, Nelson F.F. III-245
Egan, G.F. I-1057
Elena, José Manuel II-147
Engin, Seref N. II-648
Eom, Il Kyu II-400
Erfidan, Tarık II-204
Estevam, R. Hruschka Jr. III-245
Eto, Tsuyoshi I-439
Everly, R.D. I-1039

Fan, FuHua II-493
Fan, Hong I-476
Fan, Muhui II-592
Fan, Zhi-Gang II-396
Fang, Bin III-663
Fang, Yi II-135
Farkaš, Igor II-676

Feng, Chen I-793, I-1256
Feng, Chunbo III-698
Feng, Ding I-25
Feng, Du I-679
Feng, Guangzeng III-457
Feng, Guiyu I-209, I-675
Feng, Guorui I-720
Feng, Jiuchao II-332
Feng, Li III-374
Feng, Naiqin III-562
Feng, Xiao-Yue II-698
Figueredo, Grazziela P. II-941
Fong, Alvis C.M. II-849
Fontana, Federico II-1155
Freeman, Walter J. I-378
Freund, Lars II-1112
Fu, Chaojin I-664
Fu, Duan III-1128
Fu, Xiao II-627
Fu, Y.X. III-668
Fu, Zetian II-352
Fujii, Robert H. I-456
Fukumura, Naohiro I-313
Furutani, Hiroshi II-1025

Gao, Hai-Hua I-565, II-21, II-89
Gao, Pingan III-1308
Gao, Xieping I-358, I-783, II-139
Gao, Ying II-386
Ge, Weimin III-984
Ge, Yang III-553
Geem, Zong Woo III-741, III-751
Germen, Emin I-353
Glackin, B. I-420
Goebels, Andreas II-744
Goëffon, Adrien III-678
Göksu, Hüseyin II-618, III-1242
Gong, Dengcai II-602
Gong, Ling I-925
Gong, Maoguo I-449, II-826, III-399, III-768
Gong, Zhiwei III-1251
Górriz, J.M. III-863
Gou, Jin III-490
Gowri, S. III-361
Gu, Faji I-1052
Gu, Xingsheng II-880
Guan, Qiu I-332, II-795
Guan, Xinping II-75
Guan, Yi I-947

Guang, Cheng II-338
Guangli, Liu I-650
Günes, Salih II-830
Guo, Hongbo II-957
Guo, Huawei I-679
Guo, Jun I-915
Guo, Lei III-698
Guo, Ya-Jun III-28
Guo, Zhenhe II-867

Hakl, František I-1234
Ham, Fredric M. I-1100
Han, Cheon-woo I-797
Han, Dongil II-328
Han, Jianghong III-782
Han, Jong-Hye I-850
Han, Lansheng III-903
Han, Lu II-1105
Han, Ray P.S. III-269
Han, Soowhan I-1100
Hang, D. I-1057
Hao, Fei I-769
Hao, Jin-Kao III-678
Hao, Zhifeng III-137, III-1257
Harris, C.J. II-1122
Hayward, Serge I-1214
He, Han-gen II-1035
He, Jun II-1015, III-279, III-323
He, Lianlian III-636, III-668
He, Mi I-508
He, Pilian I-692
He, Shengjun III-915
He, Wenxiu III-782
He, Wuhong II-931
He, Xiaoguang I-187
He, Yinghao II-12
He, Yuguo III-434
He, Yunhui II-71
He, Yuyao I-273
He, Zhengjia II-324
He, Zhenya I-683
Heckman, C. I-1039
Herbert, Joseph I-129
Herrero, Álvaro I-778
Hlaváček, Marek I-1234
Ho, Daniel W.C. I-730
Hong, Chao-Fu III-11
Hong, Gye Hang II-710
Hong, Keongho I-789
Hong, Kwang-Seok I-1179

Hong, Qin I-264
Hong, Sang Jeen I-1113
Hong, Wei-Chiang I-619, I-668
Hong, Xianlong III-181
Hong, Yuan II-1206
Hong, Zhang I-499
Hooke, J. I-1039
Hosoi, Satoshi II-438
Hou, Chong II-876
Hou, Cuiqin III-768
Hou, Kunpeng II-483
Hou, Yanfeng III-1216
Hou, Yimin III-873
Hou, Yunxian II-352
Hou, Zeng-Guang III-622
Hruschka, Eduardo R. III-245
Hsu, Chi-I III-812
Hsu, Hao-Hsuan II-859
Hsu, Pei Lun III-812
Hsu, Yuan Lin III-812
Hu, Chunfeng II-65
Hu, Chunhua II-234
Hu, Dewen I-101, I-209, I-675, I-700, III-1128
Hu, Guangshu III-654
Hu, Hong I-91, I-1039
Hu, Huaqiang I-835
Hu, Jiani I-915
Hu, Jianming II-1089
Hu, Qiao II-324
Hu, Qinghua III-1190
Hu, Shigeng I-740
Hu, Tao II-352
Hu, Tingliang II-234
Hu, Weisheng III-102
Hu, Xiaomin II-592
Hu, Zhi-kun III-477
Hu, Zhonghui I-528, II-175
Hua, Yong II-12
Huang, Gaoming I-683
Huang, Hai III-772, III-1142
Huang, Han III-137
Huang, Houkuan III-323
Huang, Min I-1
Huang, Wanping III-289
Huang, Wentao I-449, II-826
Huang, Xiyue II-890
Huang, Xuemei II-800
Huang, Xueyuan II-913
Huang, Ya-Chi III-612

Huang, Yan-Xin II-698
Huang, Yu-Ying I-668
Huh, Sung-Hoe III-1099
Hui, Siu Cheung II-849
Hui-zhong, Yang I-25
Hwang, Changha I-512, I-521, I-536, II-306
Hwang, Su-Young I-797

Ibershoff, Joseph II-1206
Ibrahim, Zuwairie II-1174, II-1182
Ichikawa, Michinori I-293
Im, Kwang Hyuk II-530
Iwata, Atsushi III-1006

Jaromczyk, Jerzy W. II-1206
Jeon, In Ja II-764, III-356
Jeon, Jun-Cheol III-348
Jeong, Eunhwa I-789
Jeong, EunSung II-764
Jeong, Kee S. I-818
Jeong, Ok-Ran I-850
Ji, Guangrong I-793, I-1256
Jia, Sen II-391
Jian, Gong II-338
Jiang, Chunhong I-985
Jiang, Michael I-750
Jiang, Minghu I-1140
Jiang, Minghui I-740
Jiang, Weijin I-139, I-345
Jiang, Xiaoyue III-215
Jiang, Yaping II-800
Jiang, Zefei I-608
Jianmin, Han I-336
Jianying, Xie I-44
Jiao, Licheng I-449, II-780, II-826, II-839, II-876, II-905, II-931, III-366, III-399, III-768, III-925
Jie, Liu I-254
Jin, Dongming III-1022
Jin, Qiao III-1089
Jin, Wuyin I-390
Jin, Xiaogang I-1209
Jin, Xiaoguang II-584
Jin, Yaochu II-1145
Jin, Yaohui III-102
Jing, Guixia II-376
Jing, Ling I-217
Jiskra, Jan III-841
Jiuzhen, Liang I-336

Jordan, R. I-1039
Juang, Yau-Tarng III-1172
Jun, Feng I-33
Jun, Liu I-44
Jun-an, Lu I-254
Jung, In-Sung I-888
Jung, Jo Nam II-109
Jwo, Dah-Jing II-425

Kala, Keerthi Laal III-1015
Kang, Hyun-Ho III-962
Kang, Jaeho II-1259
Kang, Kyung-Woo II-543
Kang, Lishan II-1049
Kang, Yuan II-296
Karwowski, Waldemar III-1216
Kasai, Nobuyuki II-1174
Katayama, Susumu II-1025
Kato, Tsuyoshi II-963
Katsaggelos, Aggelos K. II-1192
Kaya, Mehmet Ali II-618
Ke, Hengyu II-210
Kel, Alexander II-1202
Khajehpour, Pooyan II-1250
Khalid, Marzuki II-1182
Kikuchi, H. III-684
Kim, Chang-Suk III-1182
Kim, Dong-Hyun III-1044
Kim, Dongwon III-1099
Kim, DuckSool II-714
Kim, Eun Ju II-155
Kim, Hang-Joon II-543
Kim, Ho-Joon III-1178
Kim, Hyeoncheol I-1002, I-1009
Kim, Hyun-jung I-1247
Kim, Hyung-Bum I-1027
Kim, Jinsu III-1044
Kim, Jong-Bin III-1032
Kim, Jong-Min II-224
Kim, Ju Han I-965
Kim, Kap Hwan II-1259
Kim, Kee-Won III-348
Kim, Kwang-Baek I-237, III-1182
Kim, Myung Won II-155
Kim, Nam H. I-818
Kim, Pan-Koo I-1027
Kim, Seong-Whan II-451
Kim, Sun II-636
Kim, Sung-il I-797
Kim, Sungshin I-1160, II-564

Kim, Tae Hyun II-530
Kim, Tae Hyung II-400
Kim, Won-sik I-797
Kim, Woong Myung I-760
Kim, Yong-Kab III-1044
Kim, Yoo Shin II-400
Kim, Young-Joong III-1079
Knidel, Helder I-1279
Kobayashi, Kunikazu I-439
Kodaz, Halife II-830
Kökçe, Ali II-618
Kong, Min I-15
Konovalova, Tatiana II-1202
Kou, Jisong III-37, III-943
Kouh, Jen-Shaing I-1224
Kramer, Oliver II-744
Krishnamurthy, E.V. II-784
Ku, Dae-Sung III-1032
Kuremoto, Takashi I-439
Kwon, Ki-Ryong III-962
Kwon, Young-hee III-1069

Lai, Chien-Yuan III-205
Lai, Kin Keung I-382
Lai, Liang-Bin I-1224
Lai, Yungang III-782
Lam, Kin-man II-7
Lan, Shu I-33
Lee, Bu-Sung II-1112
Lee, Byung C. I-818
Lee, Chung K. I-818
Lee, Dong-Un I-237
Lee, Hak-Sung II-328
Lee, Hsuan-Shih III-1290
Lee, Hyon Soo I-760
Lee, Jang Hee II-710
Lee, Jay III-535
Lee, Jongkeuk I-1100
Lee, KangWoo I-855
Lee, Kwangeui I-1100
Lee, Myung-jin I-797
Lee, Sang-Ho I-965
Lee, Sun-young I-797
Lee, Sungyoung II-101
Lee, Woo-Gul I-797
Lee, Yong Hwan II-1259
Lee, Yunsik I-1113
León, Carlos II-147
Li, BiCheng II-37
Li, Bin III-1261

Li, Chun-lian II-1159, III-93
Li, Chunshien III-1186
Li, Fu-ming II-992
Li, G.Q. III-668
Li, Guang I-378, I-411, I-1052
Li, Guodong I-773
Li, Guoyou I-397
Li, Haifeng III-972
Li, Hejun II-185
Li, Hong I-952
Li, Hong-Nan III-1089
Li, Hua II-483
Li, Huiguang I-397
Li, Hui-Xian III-453
Li, Jianyu I-867
Li, Jing I-293, II-931
Li, Meiyi III-1308
Li, Ming I-209, I-675
Li, Minqiang III-37, III-171, III-185, III-808, III-943
Li, Na-na I-1047
Li, Qingyong I-903, III-496
Li, Ruonan III-654
Li, Shanbin II-242
Li, Shaoqian II-316
Li, Tao II-800, II-804
Li, Tianpeng III-948
Li, Wei I-995
Li, Wenhui III-938
Li, Wu-Jun II-55
Li, Xiao feng III-505
Li, Xiaobin II-922
Li, Xiaohong II-584
Li, Xiaoming III-808
Li, Xiu II-574
Li, Xu I-378, I-1121
Li, Xu-yong III-68
Li, Xue-yan I-1047
Li, Xuewei III-309
Li, Xuming II-468
Li, Xunming II-602
Li, Yangmin III-628, III-1109
Li, Ye II-175
Li, Yijun II-123
Li, Ying III-215
Li, Yinglu II-627
Li, Yongming III-1132
Li, Yuan I-1132
Li, Yuangui I-528, II-175
Li, Yuanyuan II-774

Li, Yunfeng II-119
Li, Zeng-Zhi III-602, III-883
Li, Zhanhuai III-1001
Li, Zhengxue I-720
Li, Zhishu III-444
Li, Zhong-Wei I-1132
Li, Zi-qiang II-1080
Li, Zongmin II-483
Lian, Hui-Cheng II-438
Liang, Min II-316
Liang, Yan-Chun II-698
Liang, Yanchun III-137, III-1226, III-1257
Liao, Benjamin Penyang III-798
Liao, Guisheng III-1, III-893
Liao, Shasha I-1140
Liao, Zaiyi III-1205
Liebman, M.N. I-1039
Lim, Dudy II-1112
Lim, Heuseok I-844
Lim, Karam I-797
Lim, Myo-Taeg III-1079
Lim, Sehun I-1270
Lim, Soonja III-1044
Lin, Chun-Cheng II-859
Lin, Dacheng I-903
Lin, Dan III-171, III-185, III-808, III-943
Lin, Jian-Yi III-1152
Lin, Jianning III-225
Lin, Mu-Hua III-11
Lin, Pan III-873
Lin, Qian I-390
Lin, Zuoquan I-825
Liu, AnFei II-37
Liu, Benyong I-660
Liu, Bin III-181
Liu, Chen-Hao I-364, II-983, III-205
Liu, Chongyang III-1
Liu, Dang-hui II-7
Liu, Ding II-922
Liu, Dong II-75
Liu, Fang II-780
Liu, Feng II-316
Liu, Feng-yu III-1280
Liu, Guangjie III-976
Liu, Guangyuan III-1231
Liu, Hongbing I-592
Liu, Hua-Yong III-28
Liu, Hui III-903

Liu, Ji   II-863
Liu, Jing   III-366, III-543, III-925
Liu, Juan   III-636
Liu, Jun   I-411
Liu, Li   II-135
Liu, Lianggui   III-457
Liu, Lin   III-980
Liu, Ping   II-185
Liu, Renren   III-1251
Liu, San-yang   II-1044
Liu, Shumei   III-566
Liu, Wanquan   I-1057, III-1198
Liu, Wenhuang   II-574
Liu, Xiande   II-1105
Liu, Xianghui   II-913
Liu, Xiaodong   III-1198
Liu, Xiaojie   II-804
Liu, Xueliang   II-376
Liu, Yan   I-750
Liu, Yanjuan   III-1235
Liu, Ye   III-761
Liu, Yilin   II-690
Liu, Yong   I-149
Liu, Yongpan   III-219
Liu, Yuan-Liang   II-296
Liu, Yugang   III-1109
Liu, Yuling   III-958
Liu, Yutian   III-449
Liu, Zhiyong   I-340
Liu, Zhongshu   III-772
Long, Dong-yun   II-1159
Lou, Zhengguo   I-411
Lou, Zhenguo   I-378
Lu, Bao-Liang   I-293, I-303, II-396, II-438
Lu, Bin   II-826, III-399, III-768
Lu, Guihua   III-129
Lu, Hongtao   II-28
Lu, Huifang   I-720
Lu, Jiang   III-592
Lu, Jiwen   I-640
Lu, Qi-Shao   I-480, I-1199
Lu, Wenkai   II-410
Lu, Yiyu   II-584
Lucas, Caro   II-1250
Luo, Bin   II-55
Luo, H.   III-684
Luo, Rong   III-219
Luo, Siwei   I-322, I-710, I-867
Luo, Yanbin   III-1132

Lu-ping, Fang   I-499
Lv, Qiang   I-81

Ma, Longhua   III-289
Ma, Xiaojiang   II-81
Maeda, Michiharu   I-283, II-361, II-415
Maguire, L.P   I-420
Manca, Vincenzo   II-1155
Mao, Keji   III-782
Mao, Zong-yuan   I-601
Marras, William S.   III-1216
Matsugu, Masakazu   III-1006
Matsuka, Toshihiko   I-933
Matsuoka, Kiyotoshi   II-274
Maul, Tomás   I-893
Mayumi, Oyama-Higa   I-811
McGinnity, T.M.   I-420
Meng, Fan   II-371
Meng, Hong-yun   II-1044
Meng, Qingchun   II-1005
Meng, Yu   III-938
Miao, Gang   II-81
Miao, Shouhong   III-723
Miao, Tiejun   I-811
Mills, Ashley   II-666
Min, Zhao   I-374
Miyajima, Hiromi   I-283, II-361, II-415
Mohanasundaram, K.M.   III-572
Monedero, Iñigo   II-147
Montaño, Juan C.   II-147
Morie, Takashi   III-1006
Mozhiwen   I-33
Mu, Weisong   II-352
Muhammad, Mohd Saufee   II-1182
Murthy, V.K.   II-784

Nagao, Tomoharu   III-566
Nakayama, Hirotaka   III-409
Nam, Kichun   I-844, I-850
Nam, Mi Young   II-109
Nan, Guofang   III-943
Narayanan, M. Rajaram   III-361
Neagu, Daniel   III-1300
Nepomuceno, Erivelton Geraldo   III-313
Neskovic, Predrag   I-71, I-554, I-877
Ng, Hee-Khiang   II-1112
Nguyen, Duc-Hoai   I-1113
Nguyen, Ha-Nam   I-1017
Nhat, Vo Dinh Minh   II-101
Nian, Rui   I-793, I-1256

Nie, Weike II-839
Nie, Yinling II-839
Niu, Xiao-hui I-1047
Niu, Xiaoxiao I-592
Nomura, Osamu III-1006
Nowinski, Wieslaw L. I-1065

Obayashi, Masanao I-439
Oh, Heung-Bum I-1009
Ohn, Syng-Yup I-1017
Ohyama, Norifumi II-274
Ok, Sooyol II-714
Olhofer, Markus II-1145
Ong, Yew-Soon II-1112
Önkal-Engin, Güleda II-648
Ono, Osamu II-1174, II-1182
Ooshima, Masataka II-274
Ou, Ling II-814
Ou, Zongying II-119, III-688

Pai, Ping-Feng I-619, I-668
Palaniappan, K. III-1132
Palmes, Paulito P. III-1119
Pan, Chen II-135
Pan, Li III-934
Pan, Zhigeng III-1051
Pappalardo, Francesco III-161
Park, Chang-Hyun II-85
Park, Chun-Ja III-1069
Park, Dong-Chul I-1113, I-1266
Park, Gwi-Tae III-1099
Park, Hyun Jin II-451
Park, Hyun-Soo II-543
Park, Jaehyun I-1017
Park, Jong-An I-1027, II-43
Park, Jong-won III-1069
Park, Kinam I-844
Park, Kyu-Sik I-1017
Park, Kyungdo I-1247, II-636
Park, Moon-sung III-1069
Park, Sang Chan II-530
Park, Seoung-Kyu II-224
Park, Taesu I-1027
Park, Yongjin III-741
Park, Youn J. I-818
Park, Young-Ran III-962
Parsopoulos, K.E. III-582
Parvez, Shuja II-1112
Pei, Xiao-mei I-1031, III-646
Pei, Xiaomei II-376

Peng, Jing III-194
Peng, Tao II-690
Peng, Wei I-835
Peng, Wen II-1229
Peng, Xiao-qi III-477
Pi, Daoying I-688, I-706, I-716
Pi, Xiongjun I-1035, I-1043
Pigg, Paul III-1242
Polat, Kemal II-830
Posenato, Roberto III-235
Priesterjahn, Steffen II-744
Puntonet, C.G. III-863
Pyun, Jae Young II-43

Qi, Huan III-482
Qi, Ming II-51
Qian, Feng II-631
Qian, Jixin III-289, III-948
Qian, Yuntao II-391
Qin, Guoqiang III-592
Qin, Qiming III-1261, III-1274
Qin, Zheng II-756, III-592
Qin-ye, Tong I-499
Qiu, Jiang I-952
Qiu, Yuhui III-562
Qiu, Zulian I-340

Rameshkumar, K. III-572
Ravi, S. III-361
Ren, Quanmin II-81
Ren, Xinhua II-774
Rhee, Phill Kyu II-109, II-764, III-356
Richer, Jean-Michel III-678
Ríos, Sebastían A. II-622
Rocha e Silva, Valceres Vieira III-313
Rojas, F. III-863
Rong, Lili III-151
Ropero, Jorge II-147
Rowlands, Hefin III-877
Rubo, Zhang III-553
Ruizhi, Sun I-650
Ryu, Joung Woo II-155
Ryu, Kwang Ryel II-1259

Sadedin, Suzanne II-1131
Sahan, Seral II-830
Sáiz, José Manuel I-778
Sakamoto, Makoto II-1025
Sang, Enfang. I-199
Sasaki, S. III-684

Sendhoff, Bernhard   II-1112, II-1145
Sengupta, Biswa   I-429
Seo, Kyung-Sik   I-1027
Seo, Sam-Jun   III-1099
Seok, Kyung Ha   I-536
Shang, Fu hua   III-505
Shang, Jincheng   III-374
Shang, Lin   III-855
Shen, Hong-yuan   III-477
Shen, Lan-sun   I-975, II-7
Shen, Xisheng   I-470
Shen, Xueqin   I-692
Shen, Yi   I-740
Shen, Zhenyao   III-129
Shi, Feng   I-1047, III-636
Shi, Haixiang   I-1080
Shi, Jun   III-496
Shi, Lukui   I-692
Shi, Min   I-229
Shi, Wenkang   I-679
Shi, Xi   II-1089
Shi, Xiangquan   II-508
Shi, Yan-jun   II-1080
Shi, Yuexiang   III-1308
Shi, Zhiping   III-496
Shi, Zhongzhi   I-903, III-496
Shigei, Noritaka   II-361, II-415
Shi-hua, Luo   I-374
Shim, JeongYon   I-1170
Shim, Jooyong   I-512, I-521
Shin, Dongil   I-1185, II-552
Shin, Dongkyoo   I-1185, II-552
Shin, Jeong-Hoon   I-1179
Shin, Kyung-shik   I-1247, II-636
Shin, Sang-Uk   III-962
Shou-jue, Wang   I-264
Shriver, C.D.   I-1039
Sim, Kwee-Bo   I-237, II-85, III-713
Smutek, Daniel   III-841
So, Yeon-hee   I-797
Soh, W-S.   I-1057
Sohn, Insuk   II-306
Soke, Alev   III-1304
Somiari, R.   I-1039
Somiari, S.B.   I-1039
Song   III-1089
Song, Chonghui   II-214
Song, Gangbing   III-1089
Song, Hong   II-863, III-602
Song, Jingyan   II-1089

Song, Shiji   I-470
Song, Weiwei   III-972
Song, Xiao-yu   II-992
Song, Yexin   II-1101
Srinivas, M.B.   III-1015
Su, Guangda   I-985
Su, Juanhua   II-185
Su, Tao   III-893
Su, Tieming   III-688
Su, Xiao-hong   I-213
Suenaga, Masaya   I-283
Sun, Changping   I-397
Sun, Changyin   II-602
Sun, Jiancheng   I-573
Sun, Jigui   III-434
Sun, Jun   III-543
Sun, Lin-yan   III-911
Sun, Shiliang   II-652
Sun, Wei   II-190
Sun, Xin-yu   III-911
Sun, Xingming   III-958, III-968
Sun, Yanguang   I-546
Sun, Yi   II-12
Sun, Ying-Guang   III-1152
Sun, Youxian   I-688, I-706, I-716,
   II-242, II-292
Sun, Yu   II-1159, III-93
Sun, Zengqi   II-234, II-252,
   II-262, III-141
Sun, Zhengxing   I-655
Sun, Zonghai   II-292
Sung, HyunSeong   II-451
Sureerattanan, Nidapan   I-157
Sureerattanan, Songyot   I-157
Suresh, R.K.   III-572
Szeto, Kwok Yip   III-112

Takikawa, Erina   II-438
Tan, E.C.   I-975
Tan, Guanzheng   III-915
Tan, Min   III-622
Tan, Ying   II-476, II-493, II-501, II-867
Tang, Chang-jie   III-194
Tang, Deyou   II-1049
Tang, Enyi   I-655
Tang, Min   II-1229, III-48
Tang, Renyuan   III-1162
Tang, Xiaojun   I-806
Tang, Xiaowei   I-1052
Tang, Xusheng   III-688

Tang, Yinggan   II-75
Tang, Yiyuan   I-1052
Tang, Yuan Yan   III-663
Tang, Zhe   II-252
Tao, Hai-hong   III-893
Tao, Jun   III-761
Taylor, Meinwen   III-877
Temeltas, Hakan   III-703
Teng, Hong-fei   II-1080
Tesař, Ludvík   III-841
Thapa, Devinder   I-888
Tiňo, Peter   II-666, II-676
Tian, Jie   I-187
Tian, Lian-fang   I-601
Tian, Shengfeng   III-323
Tian, Zheng   II-371
Ting, Ching-Jung   III-205
Toh, C.K.   III-525
Tong, Ruofeng   II-1229
Tong, Weimin   II-123
Tsaftaris, Sotirios A.   II-1192
Tseng, Chung-Li   III-741
Tsuboi, Yusei   II-1174
Tsuji, Toshio   II-165
Tu, Li   II-1218

Ueda, Satomi   II-1182
Uno, Yoji   I-313
Usui, Shiro   I-1074, III-1119

Valeev, Tagir   II-1202
Valenzuela, O.   III-863
van Noort, Danny   II-1206
Velásquez, Juan D.   II-622
Vera, Eduardo S.   II-622
Von Zuben, Fernando J.   I-1279
Vrahatis, M.N.   III-582

Wakaki, Keitaro   I-313
Wan, Jinming   III-332
Wan, Qiong   III-855
Wang, Ai-guo   III-93
Wang, Bin   II-410, III-822
Wang, Chao-Xue   II-1269, III-86
Wang, Chaoyong   III-1226
Wang, Chen   III-845
Wang, Chong-Jun   II-55
Wang, Dingsheng Luo Xinhao   I-167
Wang, Fang   III-562, III-622
Wang, G.L.   II-266

Wang, Gang   I-700
Wang, Gi-Nam   I-888
Wang, Guizeng   II-95
Wang, Guo-Xin   III-1089
Wang, Guoqiang   II-119
Wang, Hai   III-883
Wang, Hai-Xia   I-1199
Wang, Haijun   I-405
Wang, He-Jun   I-1150
Wang, Hong   III-171, III-185
Wang, Honggang   III-22
Wang, Hongguang   III-727
Wang, Hui   I-716, III-219
Wang, Jigang   I-71, I-554
Wang, Jinwei   III-976
Wang, Jun-nian   III-477
Wang, Le   I-378
Wang, Lei   II-839, III-86
Wang, Lifang   III-467
Wang, Ling   III-417, III-832
Wang, Lipo   I-1080
Wang, Long   I-769, III-424
Wang, Longhui   III-636
Wang, Lunwen   II-501
Wang, Nong   I-217
Wang, Qiao   III-1261, III-1274
Wang, Qing-Yun   I-1199
Wang, Qingquan   III-151
Wang, Rubin   I-490
Wang, Shan-Shan   III-482
Wang, Shi-min   I-480
Wang, Shitong   III-1128
Wang, Shouyang   I-382
Wang, Shuqing   II-270
Wang, Shuxun   III-972
Wang, Song   II-574
Wang, Tong   III-938
Wang, W.   I-199
Wang, Wanliang   I-332, II-795
Wang, Weihong   I-1209
Wang, Weizhi   III-1022
Wang, X.X.   II-1122
Wang, Xi-cheng   II-1159, III-93
Wang, Xiaodong   I-221
Wang, Xiaofan   II-283
Wang, Xiaolong   I-947
Wang, Xihuai   II-196
Wang, Xin   III-525
Wang, Xinfei   II-584
Wang, Xing-Yu   I-565, II-21, II-89

Wang, Xiufeng II-978
Wang, Ya-dong I-213
Wang, Yan II-12, II-698
Wang, Yaonan II-190
Wang, Yen-Nien II-859
Wang, Yen-Wen I-364, II-983, III-205
Wang, Yong I-565
Wang, Yong-Xian II-1164
Wang, Yongcheng I-925
Wang, Yongqiang II-292
Wang, Yuping III-392
Wang, Zhanshan I-61
Wang, Zheng-Hua II-1164
Wang, Zhijie I-476
Wang, Zhiquan III-976
Wang, Zhizhong III-1142
Wang, Zhu-Rong II-1269, III-86
Wang, Zilei I-1090
Wang, Ziqiang II-727, II-822
Wanlin, Gao I-650
Watanabe, Atsushi II-274
Wei, Ding II-338
Wei, Li I-601
Wei, Xiaopeng I-405
Wei, Yaobing I-390
Wei, Yunbing I-332
Wei, Zhi II-592
Wei, Zhiqiang II-1005
Weijun, Li I-264
Weimer, Alexander II-744
Weizhong, Guo I-44
Wen, Quan III-972
Wen, Wanhui III-1231
Wen, Xiangjun I-51, I-582
Woo, Kwang Bang I-1160, II-564
Wu, Chunguo III-137, III-1226, III-1257
Wu, Fangfang I-608
Wu, Hao III-417
Wu, Huizhong III-225
Wu, Jianping II-105
Wu, Kai-Gui II-814
Wu, Lenan II-468
Wu, Qing I-692
Wu, Qingliang III-761
Wu, QingXiang I-420
Wu, Qiongshui II-210
Wu, Tihua I-397
Wu, Wei I-720, III-772
Wu, Xiaoping II-1101
Wu, Xihong I-167

Wu, Yadong III-332
Wu, Yan III-1
Wu, Ying I-390, I-508
Wu, Yiqiang I-8
Wu, Yong III-958
Wu, Yun III-120
Wu, Zhong-Fu II-814

Xi, Hongsheng I-1090
Xia, Feng II-242
Xiang, Zheng I-573
Xiang-guan, Liu I-374
Xiao, Fen I-783, II-139
Xiao, Jian I-101
Xiao, Jianmei II-196
Xiao, Yunshi I-119
Xiaolong, Deng I-44
Xie, Gang II-945
Xie, Guangming III-424
Xie, Hongbo III-1142
Xie, Jun II-951
Xie, Keming II-945, II-951, II-957
Xie, Li I-386
Xie, Lijuan III-384
Xie, Qihong III-57
Xie, Sheng-Li I-839
Xie, Shengli I-229, II-386, II-442
Xie, Xiaogang III-1235
Xie, Zongxia III-1190
Xin-guang, Shao I-25
Xiong, Shengwu I-592
Xiong, Zhangliang II-508
Xu, Bin II-520
Xu, Chen I-264
Xu, Chunlin II-800
Xu, De III-622
Xu, Fen III-1251
Xu, Haixia II-371
Xu, Jian III-505, III-1280
Xu, Jianxue I-508
Xu, Jin I-1031, II-376, III-646
Xu, Jinhua I-730
Xu, Junqin III-299
Xu, Min III-1128
Xu, Wenbo III-543
Xu, Xiaoming I-51, I-582, II-175
Xu, Xin I-700, II-1035
Xu, Xinli II-795
Xu, Xinying II-945
Xu, Xiuling I-221

Xu, Yangsheng II-1089
Xu, Yubin III-22
Xu, Yuelei I-449
Xu, Yuhui I-139, I-345
Xu, Yusheng I-139, I-345
Xu, Zhenhao II-880
Xu, Zhiwei I-750
Xue, Juan III-68
Xue, Q. II-266
Xue, Xiangyang III-525
Xue, Xiaoping I-466

Yan, Gaowei II-951
Yan, Haifeng III-444
Yan, Shaoze III-632
Yan, Weidong III-980
Yan, Xiao-Ke I-1150
Yan, Xin III-980
Yan, Xiong III-181
Yang, Bo I-213
Yang, C.F. II-557
Yang, Chunyan II-214
Yang, Hai-Dong I-1150
Yang, Hsiao-Fang III-11
Yang, Huazhong III-219
Yang, Hui II-214
Yang, Hui-Hua I-565, II-21, II-89
Yang, Hwan-Seok II-224
Yang, Hyun-Seung III-1178
Yang, Jian I-322
Yang, Jiangang III-490
Yang, Jie II-95
Yang, Jing I-1132
Yang, Jun III-120
Yang, Jun-an II-461
Yang, Kongyu II-978
Yang, Li-ying II-756
Yang, Luxi I-683
Yang, Pin II-804
Yang, Qing II-442
Yang, Shun-Lin I-668
Yang, Shuzhong I-710
Yang, Wenlu I-1043
Yang, Xiaohua III-129
Yang, Xiaowei III-137, III-1257
Yang, Xin I-187
Yang, Xiyang I-225
Yang, Xuhua I-332
Yang, Yipeng III-1274
Yang, Yong III-873

Yang, YonQing I-15
Yang, Zheng Rong I-179
Yang, Zhifeng III-129
Yang, Zhixia I-217
Yang, Zhuo I-806
Yao, JingTao I-129
Yao, Shuzhen II-1049
Yao, Xin III-279
Yasuda, Hiroshi II-622
Yazhu, Qiu I-33
Ye, Bin II-895
Ye, Hao II-95
Ye, Jun II-1105
Ye, Mao II-557
Ye, Xiuzi I-835
Ye, Zhongfu II-461
Yi, Bian I-264
Yi, Yang III-74
Yibo, Zhang I-696
Yim, Hyungwook I-844
Yin, Bo II-1005
Yin, Changming III-384
Yin, Chao-wan II-992
Yin, Chuanhuan III-323
Yin, Jianping II-65, II-913
Yin, Junsong I-101
Yin, Ling I-1052
Yin, Xiao-chuan II-539
Yokoyama, Ryuichi III-313
Yoo, Kee-Young II-512, III-348
Yoo, Sun K. I-818
Yoon, Eun-Jun II-512
Yoon, Han-Ul III-713
Yoon, Hye-Sung I-965
Yoon, Mi-sun I-797
Yoon, Min III-409
You, Jing III-1280
You, Xinge III-663
Young, Natasha I-179
Youxian, Sun I-696
Yu, Changjie II-262
Yu, Daren III-1190
Yu, Fusheng I-225
Yu, Jin-shou I-81
Yu, Jinshou I-630, II-610,
  III-515, III-832
Yu, Lean I-382
Yu, Qizhi III-1051
Yu, Wei I-490
Yu, Xinjie II-1064, II-1072

Yu, Zhenhua II-627
Yu, Zu-Guo III-337
Yuan, Chang-an III-194
Yuan, Hong I-952
Yuan, Lin I-199, III-1001
Yue, Jiguang I-119
Yun, Jung-Hyun III-1032
Yun, Sung-Hyun I-797
Yun, Yeboon III-409
Yusof, Azwina I-893
Yıldırım, Tülay I-110

Zeng, Jianchao III-22, III-255, III-467
Zeng, Libo II-210
Zeng, Qingdong III-915
Zeng, Sanyou II-1049
Zeng, Zhigang I-664
Zhan, Tao III-602, III-883
Zhang, Changjiang I-221
Zhang, Changshui II-652
Zhang, Chunfang II-1239
Zhang, Chunkai I-91
Zhang, Dan II-863, III-602, III-883
Zhang, Defu III-1235
Zhang, Dexian II-727, II-822
Zhang, Dongmo I-956
Zhang, Erhu I-640
Zhang, Feng III-873
Zhang, Gang II-957
Zhang, Guomin II-65
Zhang, Haoran I-221
Zhang, Hongbo II-210
Zhang, Huaguang I-61
Zhang, Huidang I-273
Zhang, Jian II-266, III-112
Zhang, Jiang III-309
Zhang, Jianming II-270
Zhang, Jian-Pei I-1132
Zhang, Jihui III-299
Zhang, Jing I-660, I-1052, III-194
Zhang, Jingjing III-102
Zhang, Jun I-358, I-783, II-139, II-592
Zhang, Lei III-535
Zhang, Ling II-501
Zhang, Liqing I-1043
Zhang, Lisha I-655
Zhang, Min II-476, III-668
Zhang, Qiang I-405
Zhang, Qing-Guo III-28
Zhang, Sanyuan I-835

Zhang, Shuai III-1300
Zhang, Shui-ping II-539
Zhang, Taiyi I-573
Zhang, Tao I-806
Zhang, Wei III-28
Zhang, Weidong I-528
Zhang, Wen III-449
Zhang, Wenquan I-8
Zhang, Xianfei II-37
Zhang, Xiangrong II-905
Zhang, XianMing II-1
Zhang, Xiao-hua II-1044
Zhang, Xiaoshuan II-352
Zhang, Xiufeng II-774
Zhang, Xuanping III-592
Zhang, Xudong III-654
Zhang, Y.S. III-1060
Zhang, Yan III-938
Zhang, Yanning III-215
Zhang, Yanxin II-283
Zhang, Ye I-8
Zhang, Yuanzhen II-483
Zhang, Yulei I-956
Zhang, Yuming III-723
Zhang, Yuntao I-925
Zhang, Z.Z. III-668
Zhang, Zhen-Hui II-1164
Zhang, Zhengwei II-95
Zhang, Zhijie I-952
Zhang, Zhousuo II-324
Zhao, Bin II-461
Zhao, Bo II-895
Zhao, Guoying I-740
Zhao, Hai I-303
Zhao, Hengping I-630
Zhao, Jian II-346
Zhao, Jieyu II-432
Zhao, Jin-cheng II-1159
Zhao, Jing II-557
Zhao, Jun III-948
Zhao, Keyou III-698
Zhao, Li II-71
Zhao, Liang I-1189
Zhao, Liping I-956
Zhao, Mingyang III-727
Zhao, Pengfei III-688
Zhao, Qiang III-632
Zhao, Qijun II-28
Zhao, Qin II-346
Zhao, Rongchun III-215

Zhao, Wencang   I-793, I-1256
Zhao, Xi   III-137
Zhao, Xinyu   I-825
Zhao, Xue-long   III-1280
Zhao, Yinliang   I-608
Zhao, Yu   I-1090, II-584
Zhao, Zhefeng   II-957
Zhao, Zhi-Hong   III-855
Zhao, Zhilong   III-980
Zhao, Zijiang   III-444
Zheng, ChongXun   I-1031, II-376, III-646, III-873
Zheng, Da-zhong   III-417
Zheng, Hong   II-210, III-934
Zheng, Ji   III-525
Zheng, Jin-hua   III-68
Zheng, Shiqin   II-978
Zheng, Yi   I-8
Zheng, Yisong   I-773
Zhexin, Cao   II-316
Zhi, Qiang   II-316
Zhong, Jiang   II-814
Zhong, Weicai   III-366, III-925
Zhong, Weimin   I-706
Zhong, Xiang-Ping   II-55
Zhou, Changjiu   II-252
Zhou, Chun-Guang   II-698
Zhou, Dongsheng   I-405
Zhou, Jian   III-120, III-684
Zhou, Jiping   III-727
Zhou, Li-Quan   III-337
Zhou, Lifang   III-289
Zhou, Ming-quan   II-346

Zhou, Qiang   III-181
Zhou, Shude   III-141
Zhou, Wen-Gang   II-698
Zhou, Xiaoyang   III-374
Zhou, Ying   II-814
Zhou, Yuanfeng   II-105
Zhou, Yuanpai   III-269
Zhou, Yuren   II-1015
Zhou, Zhi-Heng   I-839
Zhou, Zhong   III-772
Zhou, Zongtan   I-101, I-209, I-675
Zhu, Chengzhi   II-895
Zhu, Daqi   I-15
Zhu, En   II-65
Zhu, Jia   III-93
Zhu, Jianguang   III-1162
Zhu, Jihong   II-234, II-262
Zhu, Qingsheng   III-57
Zhu, Xinglong   III-727
Zhu, Xue-feng   I-995
Zhu, Yan-fei   I-601
Zhu, Yun-long   II-992
Zhu, Zheng-Zhou   II-814
Zhu, Zhengyu   III-57
Zi, Yanyang   II-324
Zou, Cairong   II-71
Zou, Henghui   III-996
Zou, Hengming   III-988, III-996, III-1001
Zou, Qi   I-867
Zribi, Nozha   III-259
Zuo, Wanli   II-690
Zuo, Wen-ming   II-51
Zurada, Jacek M.   III-1216

Lecture Notes in Computer Science

For information about Vols. 1–3536

please contact your bookseller or Springer

Vol. 3659: J.R. Rao, B. Sunar (Eds.), Cryptographic Hardware and Embedded Systems – CHES 2005. XIV, 458 pages. 2005.

Vol. 3654: S. Jajodia, D. Wijesekera (Eds.), Data and Applications Security XIX. X, 353 pages. 2005.

Vol. 3653: M. Abadi, L.d. Alfaro (Eds.), CONCUR 2005 – Concurrency Theory. XIV, 578 pages. 2005.

Vol. 3649: W.M.P. van der Aalst, B. Benatallah, F. Casati, F. Curbera (Eds.), Business Process Management. XII, 472 pages. 2005.

Vol. 3639: P. Godefroid (Ed.), Model Checking Software. XI, 289 pages. 2005.

Vol. 3638: A. Butz, B. Fisher, A. Krüger, P. Olivier (Eds.), Smart Graphics. XI, 269 pages. 2005.

Vol. 3636: M.J. Blesa, C. Blum, A. Roli, M. Sampels (Eds.), Hybrid Metaheuristics. XII, 155 pages. 2005.

Vol. 3634: L. Ong (Ed.), Computer Science Logic. XI, 567 pages. 2005.

Vol. 3633: C. Bauzer Medeiros, M. Egenhofer, E. Bertino (Eds.), Advances in Spatial and Temporal Databases. XIII, 433 pages. 2005.

Vol. 3632: R. Nieuwenhuis (Ed.), Automated Deduction – CADE-20. XIII, 459 pages. 2005. (Subseries LNAI).

Vol. 3627: C. Jacob, M.L. Pilat, P.J. Bentley, J. Timmis (Eds.), Artificial Immune Systems. XII, 500 pages. 2005.

Vol. 3626: B. Ganter, G. Stumme, R. Wille (Eds.), Formal Concept Analysis. X, 349 pages. 2005. (Subseries LNAI).

Vol. 3625: S. Kramer, B. Pfahringer (Eds.), Inductive Logic Programming. XIII, 427 pages. 2005. (Subseries LNAI).

Vol. 3624: C. Chekuri, K. Jansen, J.D.P. Rolim, L. Trevisan (Eds.), Approximation, Randomization and Combinatorial Optimization. XI, 495 pages. 2005.

Vol. 3623: M. Liśkiewicz, R. Reischuk (Eds.), Fundamentals of Computation Theory. XV, 576 pages. 2005.

Vol. 3621: V. Shoup (Ed.), Advances in Cryptology – CRYPTO 2005. XI, 568 pages. 2005.

Vol. 3620: H. Muñoz-Avila, F. Ricci (Eds.), Case-Based Reasoning Research and Development. XV, 654 pages. 2005. (Subseries LNAI).

Vol. 3619: X. Lu, W. Zhao (Eds.), Networking and Mobile Computing. XXIV, 1299 pages. 2005.

Vol. 3615: B. Ludäscher, L. Raschid (Eds.), Data Integration in the Life Sciences. XII, 344 pages. 2005. (Subseries LNBI).

Vol. 3614: L. Wang, Y. Jin (Eds.), Fuzzy Systems and Knowledge Discovery, Part II. XLI, 1314 pages. 2005. (Subseries LNAI).

Vol. 3613: L. Wang, Y. Jin (Eds.), Fuzzy Systems and Knowledge Discovery, Part I. XLI, 1334 pages. 2005. (Subseries LNAI).

Vol. 3612: L. Wang, K. Chen, Y. S. Ong (Eds.), Advances in Natural Computation, Part III. LXI, 1326 pages. 2005.

Vol. 3611: L. Wang, K. Chen, Y. S. Ong (Eds.), Advances in Natural Computation, Part II. LXI, 1292 pages. 2005.

Vol. 3610: L. Wang, K. Chen, Y. S. Ong (Eds.), Advances in Natural Computation, Part I. LXI, 1302 pages. 2005.

Vol. 3608: F. Dehne, A. López-Ortiz, J.-R. Sack (Eds.), Algorithms and Data Structures. XIV, 446 pages. 2005.

Vol. 3607: J.-D. Zucker, L. Saitta (Eds.), Abstraction, Reformulation and Approximation. XII, 376 pages. 2005. (Subseries LNAI).

Vol. 3606: V. Malyshkin (Ed.), Parallel Computing Technologies. XII, 470 pages. 2005.

Vol. 3603: J. Hurd, T. Melham (Eds.), Theorem Proving in Higher Order Logics. IX, 409 pages. 2005.

Vol. 3602: R. Eigenmann, Z. Li, S.P. Midkiff (Eds.), Languages and Compilers for High Performance Computing. IX, 486 pages. 2005.

Vol. 3599: U. Aßmann, M. Aksit, A. Rensink (Eds.), Model Driven Architecture. X, 235 pages. 2005.

Vol. 3598: H. Murakami, H. Nakashima, H. Tokuda, M. Yasumura, Ubiquitous Computing Systems. XIII, 275 pages. 2005.

Vol. 3597: S. Shimojo, S. Ichii, T.W. Ling, K.-H. Song (Eds.), Web and Communication Technologies and Internet-Related Social Issues - HSI 2005. XIX, 368 pages. 2005.

Vol. 3596: F. Dau, M.-L. Mugnier, G. Stumme (Eds.), Conceptual Structures: Common Semantics for Sharing Knowledge. XI, 467 pages. 2005. (Subseries LNAI).

Vol. 3595: L. Wang (Ed.), Computing and Combinatorics. XVI, 995 pages. 2005.

Vol. 3594: J.C. Setubal, S. Verjovski-Almeida (Eds.), Advances in Bioinformatics and Computational Biology. XIV, 258 pages. 2005. (Subseries LNBI).

Vol. 3593: V. Mařík, R. W. Brennan, M. Pěchouček (Eds.), Holonic and Multi-Agent Systems for Manufacturing. XI, 269 pages. 2005. (Subseries LNAI).

Vol. 3592: S. Katsikas, J. Lopez, G. Pernul (Eds.), Trust, Privacy and Security in Digital Business. XII, 332 pages. 2005.

Vol. 3591: M.A. Wimmer, R. Traunmüller, Å. Grönlund, K.V. Andersen (Eds.), Electronic Government. XIII, 317 pages. 2005.

Vol. 3590: K. Bauknecht, B. Pröll, H. Werthner (Eds.), E-Commerce and Web Technologies. XIV, 380 pages. 2005. XVII, 631 pages. 2005.

Vol. 3584: X. Li, S. Wang, Z.Y. Dong (Eds.), Advanced Data Mining and Applications. XIX, 835 pages. 2005. (Subseries LNAI).

Vol. 3583: R.W. H. Lau, Q. Li, R. Cheung, W. Liu (Eds.), Advances in Web-Based Learning – ICWL 2005. XIV, 420 pages. 2005.

Vol. 3582: J. Fitzgerald, I.J. Hayes, A. Tarlecki (Eds.), FM 2005: Formal Methods. XIV, 558 pages. 2005.

Vol. 3581: S. Miksch, J. Hunter, E. Keravnou (Eds.), Artificial Intelligence in Medicine. XVII, 547 pages. 2005. (Subseries LNAI).

Vol. 3580: L. Caires, G.F. Italiano, L. Monteiro, C. Palamidessi, M. Yung (Eds.), Automata, Languages and Programming. XXV, 1477 pages. 2005.

Vol. 3579: D. Lowe, M. Gaedke (Eds.), Web Engineering. XXII, 633 pages. 2005.

Vol. 3578: M. Gallagher, J. Hogan, F. Maire (Eds.), Intelligent Data Engineering and Automated Learning - IDEAL 2005. XVI, 599 pages. 2005.

Vol. 3577: R. Falcone, S. Barber, J. Sabater-Mir, M.P. Singh (Eds.), Trusting Agents for Trusting Electronic Societies. VIII, 235 pages. 2005. (Subseries LNAI).

Vol. 3576: K. Etessami, S.K. Rajamani (Eds.), Computer Aided Verification. XV, 564 pages. 2005.

Vol. 3575: S. Wermter, G. Palm, M. Elshaw (Eds.), Biomimetic Neural Learning for Intelligent Robots. IX, 383 pages. 2005. (Subseries LNAI).

Vol. 3574: C. Boyd, J.M. González Nieto (Eds.), Information Security and Privacy. XIII, 586 pages. 2005.

Vol. 3573: S. Etalle (Ed.), Logic Based Program Synthesis and Transformation. VIII, 279 pages. 2005.

Vol. 3572: C. De Felice, A. Restivo (Eds.), Developments in Language Theory. XI, 409 pages. 2005.

Vol. 3571: L. Godo (Ed.), Symbolic and Quantitative Approaches to Reasoning with Uncertainty. XVI, 1028 pages. 2005. (Subseries LNAI).

Vol. 3570: A. S. Patrick, M. Yung (Eds.), Financial Cryptography and Data Security. XII, 376 pages. 2005.

Vol. 3569: F. Bacchus, T. Walsh (Eds.), Theory and Applications of Satisfiability Testing. XII, 492 pages. 2005.

Vol. 3568: W.-K. Leow, M.S. Lew, T.-S. Chua, W.-Y. Ma, L. Chaisorn, E.M. Bakker (Eds.), Image and Video Retrieval. XVII, 672 pages. 2005.

Vol. 3567: M. Jackson, D. Nelson, S. Stirk (Eds.), Database: Enterprise, Skills and Innovation. XII, 185 pages. 2005.

Vol. 3566: J.-P. Banâtre, P. Fradet, J.-L. Giavitto, O. Michel (Eds.), Unconventional Programming Paradigms. XI, 367 pages. 2005.

Vol. 3565: G.E. Christensen, M. Sonka (Eds.), Information Processing in Medical Imaging. XXI, 777 pages. 2005.

Vol. 3564: N. Eisinger, J. Małuszyński (Eds.), Reasoning Web. IX, 319 pages. 2005.

Vol. 3562: J. Mira, J.R. Álvarez (Eds.), Artificial Intelligence and Knowledge Engineering Applications: A Bioinspired Approach, Part II. XXIV, 636 pages. 2005.

Vol. 3561: J. Mira, J.R. Álvarez (Eds.), Mechanisms, Symbols, and Models Underlying Cognition, Part I. XXIV, 532 pages. 2005.

Vol. 3560: V.K. Prasanna, S. Iyengar, P.G. Spirakis, M. Welsh (Eds.), Distributed Computing in Sensor Systems. XV, 423 pages. 2005.

Vol. 3559: P. Auer, R. Meir (Eds.), Learning Theory. XI, 692 pages. 2005. (Subseries LNAI).

Vol. 3558: V. Torra, Y. Narukawa, S. Miyamoto (Eds.), Modeling Decisions for Artificial Intelligence. XII, 470 pages. 2005. (Subseries LNAI).

Vol. 3557: H. Gilbert, H. Handschuh (Eds.), Fast Software Encryption. XI, 443 pages. 2005.

Vol. 3556: H. Baumeister, M. Marchesi, M. Holcombe (Eds.), Extreme Programming and Agile Processes in Software Engineering. XIV, 332 pages. 2005.

Vol. 3555: T. Vardanega, A.J. Wellings (Eds.), Reliable Software Technology – Ada-Europe 2005. XV, 273 pages. 2005.

Vol. 3554: A. Dey, B. Kokinov, D. Leake, R. Turner (Eds.), Modeling and Using Context. XIV, 572 pages. 2005. (Subseries LNAI).

Vol. 3553: T.D. Hämäläinen, A.D. Pimentel, J. Takala, S. Vassiliadis (Eds.), Embedded Computer Systems: Architectures, Modeling, and Simulation. XV, 476 pages. 2005.

Vol. 3552: H. de Meer, N. Bhatti (Eds.), Quality of Service – IWQoS 2005. XVIII, 400 pages. 2005.

Vol. 3551: T. Härder, W. Lehner (Eds.), Data Management in a Connected World. XIX, 371 pages. 2005.

Vol. 3548: K. Julisch, C. Kruegel (Eds.), Intrusion and Malware Detection and Vulnerability Assessment. X, 241 pages. 2005.

Vol. 3547: F. Bomarius, S. Komi-Sirviö (Eds.), Product Focused Software Process Improvement. XIII, 588 pages. 2005.

Vol. 3546: T. Kanade, A. Jain, N.K. Ratha (Eds.), Audio- and Video-Based Biometric Person Authentication. XX, 1134 pages. 2005.

Vol. 3544: T. Higashino (Ed.), Principles of Distributed Systems. XII, 460 pages. 2005.

Vol. 3543: L. Kutvonen, N. Alonistioti (Eds.), Distributed Applications and Interoperable Systems. XI, 235 pages. 2005.

Vol. 3542: H.H. Hoos, D.G. Mitchell (Eds.), Theory and Applications of Satisfiability Testing. XIII, 393 pages. 2005.

Vol. 3541: N.C. Oza, R. Polikar, J. Kittler, F. Roli (Eds.), Multiple Classifier Systems. XII, 430 pages. 2005.

Vol. 3540: H. Kalviainen, J. Parkkinen, A. Kaarna (Eds.), Image Analysis. XXII, 1270 pages. 2005.

Vol. 3539: K. Morik, J.-F. Boulicaut, A. Siebes (Eds.), Local Pattern Detection. XI, 233 pages. 2005. (Subseries LNAI).

Vol. 3538: L. Ardissono, P. Brna, A. Mitrovic (Eds.), User Modeling 2005. XVI, 533 pages. 2005. (Subseries LNAI).

Vol. 3537: A. Apostolico, M. Crochemore, K. Park (Eds.), Combinatorial Pattern Matching. XI, 444 pages. 2005.